General data and fundamental consta

Quantity	Symbol	Value	Power of ten	Units
Speed of light	c	2.997 924 58*	10^{8}	$m\ s^{-1}$
Elementary charge	e	1.602 176	10^{-19}	C
Faraday's constant	$F = N_A e$	9.648 53	10^{4}	$C\ mol^{-1}$
Boltzmann's constant	k	1.380 65	10^{-23}	$J\ K^{-1}$
Gas constant	$R = N_A k$	8.314 47		$J\ K^{-1}\ mol^{-1}$
		8.314 47	10^{-2}	$dm^3\ bar\ K^{-1}\ mol^{-1}$
		8.205 74	10^{-2}	$dm^3\ atm\ K^{-1}\ mol^{-1}$
		6.236 37	10	$dm^3\ Torr\ K^{-1}\ mol^{-1}$
Planck's constant	h	6.626 08	10^{-34}	J s
	$\hbar = h/2\pi$	1.054 57	10^{-34}	J s
Avogadro's constant	N_A	6.022 14	10^{23}	mol^{-1}
Atomic mass constant	m_u	1.660 54	10^{-27}	kg
Mass				
electron	m_e	9.109 38	10^{-31}	kg
proton	m_p	1.672 62	10^{-27}	kg
neutron	m_n	1.674 93	10^{-27}	kg
Vacuum permittivity	$\varepsilon_0 = 1/c^2\mu_0$	8.854 19	10^{-12}	$J^{-1}\ C^2\ m^{-1}$
	$4\pi\varepsilon_0$	1.112 65	10^{-10}	$J^{-1}\ C^2\ m^{-1}$
Vacuum permeability	μ_0	4π	10^{-7}	$J\ s^2\ C^{-2}\ m^{-1}$ $(= T^2\ J^{-1}\ m^3)$
Magneton				
Bohr	$\mu_B = e\hbar/2m_e$	9.274 01	10^{-24}	$J\ T^{-1}$
nuclear	$\mu_N = e\hbar/2m_p$	5.050 78	10^{-27}	$J\ T^{-1}$
g value	g_e	2.002 32		
Bohr radius	$a_0 = 4\pi\varepsilon_0\hbar^2/m_e e^2$	5.291 77	10^{-11}	m
Fine-structure constant	$\alpha = \mu_0 e^2 c/2h$	7.297 35	10^{-3}	
	α^{-1}	1.370 36	10^{2}	
Second radiation constant	$c_2 = hc/k$	1.438 78	10^{-2}	m K
Stefan–Boltzmann constant	$\sigma = 2\pi^5 k^4/15h^3c^2$	5.670 51	10^{-8}	$W\ m^{-2}\ K^{-4}$
Rydberg constant	$R = m_e e^4/8h^3 c\varepsilon_0^2$	1.097 37	10^{5}	cm^{-1}
Standard acceleration of free fall	g	9.806 65*		$m\ s^{-2}$
Gravitational constant	G	6.673	10^{-11}	$N\ m^2\ kg^{-2}$

**Exact value*

The Greek alphabet

A, α	alpha	H, η	eta	N, ν	nu	Y, υ	upsilon
B, β	beta	Θ, θ	theta	Ξ, ξ	xi	Φ, ϕ	phi
Γ, γ	gamma	I, ι	iota	Π, π	pi	X, χ	chi
Δ, δ	delta	K, κ	kappa	P, ρ	rho	Ψ, ψ	psi
E, ε	epsilon	Λ, λ	lambda	Σ, σ	sigma	Ω, ω	omega
Z, ζ	zeta	M, μ	mu	T, τ	tau		

ATKINS'
PHYSICAL CHEMISTRY

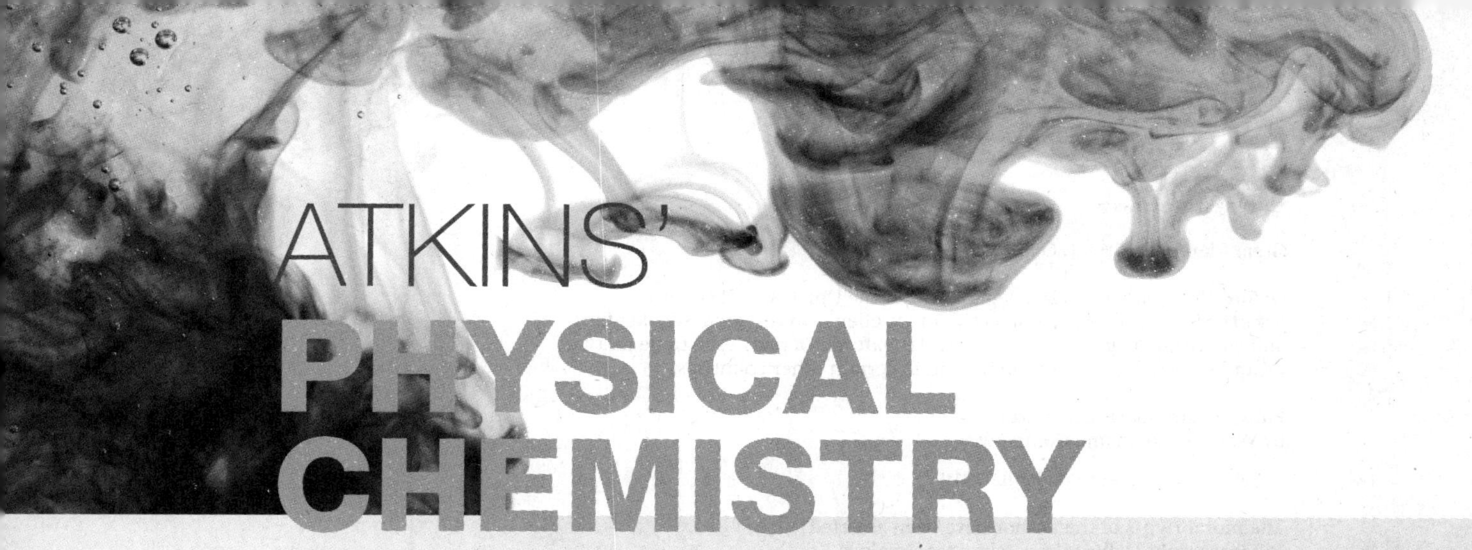

ATKINS' PHYSICAL CHEMISTRY

Ninth Edition

Peter Atkins

Fellow of Lincoln College,
University of Oxford,
Oxford, UK

Julio de Paula

Professor of Chemistry,
Lewis and Clark College,
Portland, Oregon, USA

Great Clarendon Street, Oxford, OX2 6DP,

Oxford University Press is a department of the University of Oxford. It furthers the University's objective of excellence in research, scholarship, and education by publishing worldwide. Oxford is a registered trade mark of Oxford University Press in the UK and in certain other countries.

Published in the United States
by W. H. Freeman and Company

First edition published 1978
Second edition published 1982
Third edition published 1986
Fourth edition published 1990
Fifth edition published 1994
Sixth edition published 1998
Seventh edition published 2002
Eighth edition published 2006
Ninth edition published 2010
First Indian Edition 2011

British Library Cataloguing in Publication Data
Data available

Library of Congress Cataloguing in Publication Data
Data available

Typeset by Graphicraft Limited, Hong Kong

ISBN-13: 978-0-19-959959-2

Reprinted 2011, 2012 (Thrice), 2013

Printed in India by Nutech Print Services

Preface

We have followed our usual tradition in that this new edition of the text is yet another thorough update of the content and its presentation. Our goal is to keep the book flexible to use, accessible to students, broad in scope, and authoritative, without adding bulk. However, it should always be borne in mind that much of the bulk arises from the numerous pedagogical features that we include (such as *Worked examples*, *Checklists of key equations*, and the *Resource section*), not necessarily from density of information.

The text is still divided into three parts, but material has been moved between chapters and the chapters themselves have been reorganized. We continue to respond to the cautious shift in emphasis away from classical thermodynamics by combining several chapters in Part 1 (Equilibrium), bearing in mind that some of the material will already have been covered in earlier courses. For example, material on phase diagrams no longer has its own chapter but is now distributed between Chapters 4 (*Physical transformation of pure substances*) and 5 (*Simple mixtures*). New *Impact* sections highlight the application of principles of thermodynamics to materials science, an area of growing interest to chemists.

In Part 2 (Structure) the chapters have been updated with a discussion of contemporary techniques of materials science—including nanoscience—and spectroscopy. We have also paid more attention to computational chemistry, and have revised the coverage of this topic in Chapter 10.

Part 3 has lost chapters dedicated to kinetics of complex reactions and surface processes, but not the material, which we regard as highly important in a contemporary context. To make the material more readily accessible within the context of courses, descriptions of polymerization, photochemistry, and enzyme- and surface-catalysed reactions are now part of Chapters 21 (*The rates of chemical reactions*) and 22 (*Reaction dynamics*)—already familiar to readers of the text—and a new chapter, Chapter 23, on *Catalysis*.

We have discarded the Appendices of earlier editions. Material on mathematics covered in the appendices is now dispersed through the text in the form of *Mathematical background* sections, which review and expand knowledge of mathematical techniques where they are needed in the text. The review of introductory chemistry and physics, done in earlier editions in appendices, will now be found in a new *Fundamentals* chapter that opens the text, and particular points are developed as *Brief comments* or as part of *Further information* sections throughout the text. By liberating these topics from their appendices and relaxing the style of presentation we believe they are more likely to be used and read.

The vigorous discussion in the physical chemistry community about the choice of a 'quantum first' or a 'thermodynamics first' approach continues. In response we have paid particular attention to making the organization flexible. The strategic aim of this revision is to make it possible to work through the text in a variety of orders and at the end of this Preface we once again include two suggested paths through the text. For those who require a more thorough-going 'quantum first' approach we draw attention to our *Quanta, matter, and change* (with Ron Friedman) which covers similar material to this text in a similar style but, because of the different approach, adopts a different philosophy.

The concern expressed in previous editions about the level of mathematical ability has not evaporated, of course, and we have developed further our strategies for showing the absolute centrality of mathematics to physical chemistry and to make it accessible. In addition to associating *Mathematical background* sections with appropriate chapters, we continue to give more help with the development of equations, motivate them, justify them, and comment on the steps. We have kept in mind the struggling student, and have tried to provide help at every turn.

We are, of course, alert to the developments in electronic resources and have made a special effort in this edition to encourage the use of the resources on our website (at www.oxfordtextbooks.co.uk/orc/pchem9e). In particular, we think it important to encourage students to use the *Living graphs* on the website (and their considerable extension in the electronic book and *Explorations* CD). To do so, wherever we call out a *Living graph* (by an icon attached to a graph in the text), we include an *interActivity* in the figure legend, suggesting how to explore the consequences of changing parameters.

Many other revisions have been designed to make the text more efficient and helpful and the subject more enjoyable. For instance, we have redrawn nearly every one of the 1000 pieces of art in a consistent style. The *Checklists of key equations* at the end of each chapter are a useful distillation of the most important equations from the large number that necessarily appear in the exposition. Another innovation is the collection of *Road maps* in the *Resource section*, which suggest how to select an appropriate expression and trace it back to its roots.

Overall, we have taken this opportunity to refresh the text thoroughly, to integrate applications, to encourage the use of electronic resources, and to make the text even more flexible and up-to-date.

Oxford P.W.A.
Portland J.de P.

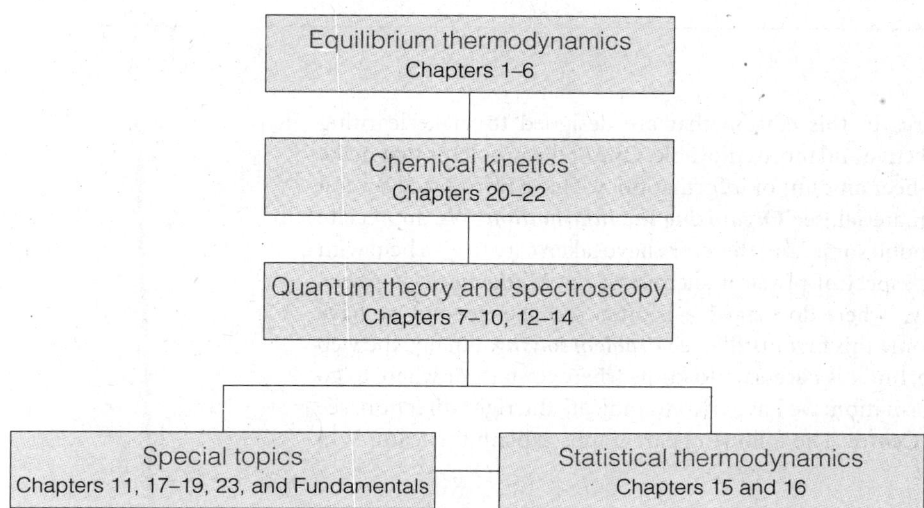

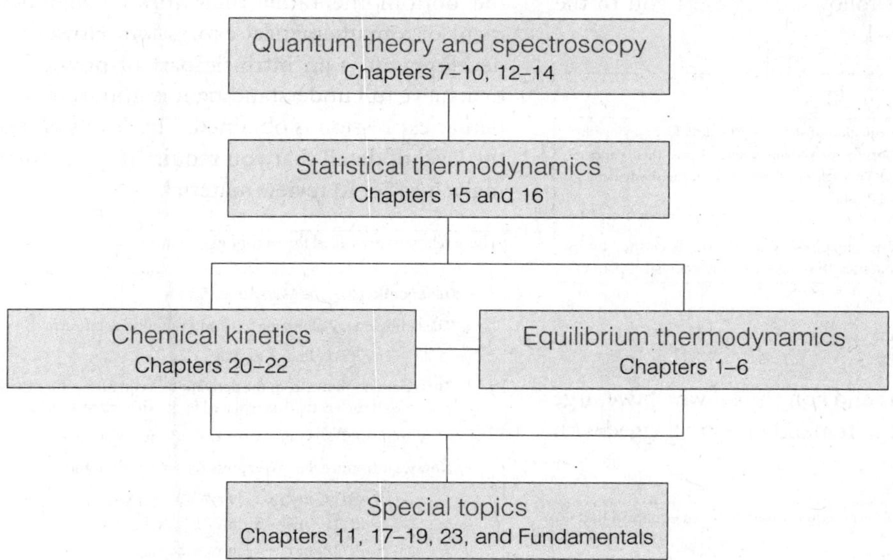

This text is available as a customizable ebook. This text can also be purchased in two volumes. For more information on these options please see pages xiv and xvi.

About the book

There are numerous features in this edition that are designed to make learning physical chemistry more effective and more enjoyable. One of the problems that make the subject daunting is the sheer amount of information: we have introduced several devices for organizing the material: see ***Organizing the information***. We appreciate that mathematics is often troublesome, and therefore have taken care to give help with this enormously important aspect of physical chemistry: see ***Mathematics support***. Problem solving—especially, 'where do I start?'—is often a challenge, and we have done our best to help overcome this first hurdle: see ***Problem solving***. Finally, the web is an extraordinary resource, but it is necessary to know where to start, or where to go for a particular piece of information; we have tried to indicate the right direction: see ***About the Online Resource Centre***. The following paragraphs explain the features in more detail.

Organizing the information

Key points

The *Key points* act as a summary of the main take-home message(s) of the section that follows. They alert you to the principal ideas being introduced.

> 1.1 **The states of gases**
>
> ***Key points*** Each substance is described by an equation of state. (a) Pressure, force divided by area, provides a criterion of mechanical equilibrium for systems free to change their volume. (b) Pressure is measured with a barometer. (c) Through the Zeroth Law of thermodynamics, temperature provides a criterion of thermal equilibrium.
>
> The **physical state** of a sample of a substance, its physical condition, is defined by its physical properties. Two samples of a substance that have the same physical proper-

Equation and concept tags

The most significant equations and concepts—which we urge you to make a particular effort to remember—are flagged with an annotation, as shown here.

> mental fact that each substance is described by an **equation of state**, an equation that interrelates these four variables.
>
> The general form of an equation of state is
>
> $$p = f(T,V,n) \qquad \text{General form of an equation of state} \qquad (1.1)$$

Justifications

On first reading it might be sufficient simply to appreciate the 'bottom line' rather than work through detailed development of a mathematical expression. However, mathematical development is an intrinsic part of physical chemistry, and to achieve full understanding it is important to see how a particular expression is obtained. The *Justifications* let you adjust the level of detail that you require to your current needs, and make it easier to review material.

> These relations are called the **Margules equations.**
>
> **Justification 5.5** *The Margules equations*
>
> The Gibbs energy of mixing to form a nonideal solution is
>
> $$\Delta_{mix}G = nRT\{x_A \ln a_A + x_B \ln a_B\}$$
>
> This relation follows from the derivation of eqn 5.16 with activities in place of mole fractions. If each activity is replaced by γx, this expression becomes
>
> $$\Delta_{mix}G = nRT\{x_A \ln x_A + x_B \ln x_B + x_A \ln \gamma_A + x_B \ln \gamma_B\}$$
>
> Now we introduce the two expressions in eqn 5.64, and use $x_A + x_B = 1$, which gives
>
> $$\begin{aligned}\Delta_{mix}G &= nRT\{x_A \ln x_A + x_B \ln x_B + \xi x_A x_B^2 + \xi x_B x_A^2\}\\ &= nRT\{x_A \ln x_A + x_B \ln x_B + \xi x_A x_B(x_A + x_B)\}\\ &= nRT\{x_A \ln x_A + x_B \ln x_B + \xi x_A x_B\}\end{aligned}$$
>
> as required by eqn 5.29. Note, moreover, that the activity coefficients behave correctly for dilute solutions: $\gamma_A \to 1$ as $x_B \to 0$ and $\gamma_B \to 1$ as $x_A \to 0$.
>
> At this point we can use the Margules equations to write the activity of A as

Checklists of key equations

We have summarized the most important equations introduced in each chapter as a checklist. Where appropriate, we describe the conditions under which an equation applies.

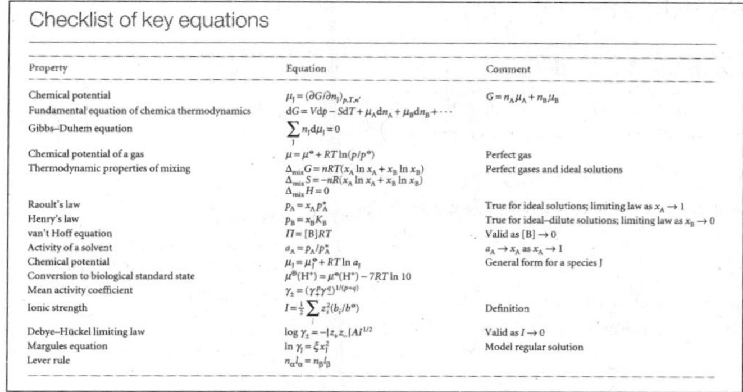

Checklist of key equations

Property	Equation	Comment
Chemical potential	$\mu_J = (\partial G/\partial n_J)_{p,T,n'}$	$G = n_A\mu_A + n_B\mu_B$
Fundamental equation of chemica thermodynamics	$dG = Vdp - SdT + \mu_A dn_A + \mu_B dn_B + \cdots$	
Gibbs–Duhem equation	$\sum_J n_J d\mu_J = 0$	
Chemical potential of a gas	$\mu = \mu^\ominus + RT\ln(p/p^\ominus)$	Perfect gas
Thermodynamic properties of mixing	$\Delta_{mix}G = nRT(x_A \ln x_A + x_B \ln x_B)$ $\Delta_{mix}S = -nR(x_A \ln x_A + x_B \ln x_B)$ $\Delta_{mix}H = 0$	Perfect gases and ideal solutions
Raoult's law	$p_A = x_A p_A^*$	True for ideal solutions; limiting law as $x_A \to 1$
Henry's law	$p_B = x_B K_B$	True for ideal–dilute solutions; limiting law as $x_B \to 0$
van't Hoff equation	$\Pi = [B]RT$	Valid as $[B] \to 0$
Activity of a solvent	$a_A = p_A/p_A^*$	$a_A \to x_A$ as $x_A \to 1$
Chemical potential	$\mu_J = \mu_J^\ominus + RT\ln a_J$	General form for a species J
Conversion to biological standard state	$\mu^\oplus(H^+) = \mu^\ominus(H^+) - 7RT\ln 10$	
Mean activity coefficient	$\gamma_\pm = (\gamma_+^p\gamma_-^q)^{1/(p+q)}$	
Ionic strength	$I = \frac{1}{2}\sum_i z_i^2(b_i/b^\ominus)$	Definition
Debye–Hückel limiting law	$\log \gamma_\pm = -\lvert z_+z_-\rvert AI^{1/2}$	Valid as $I \to 0$
Margules equation	$\ln \gamma_J = \xi x_J^2$	Model regular solution
Lever rule	$n_\alpha l_\alpha = n_\beta l_\beta$	

Road maps

In many cases it is helpful to see the relations between equations. The suite of 'Road maps' summarizing these relations are found in the *Resource section* at the end of the text.

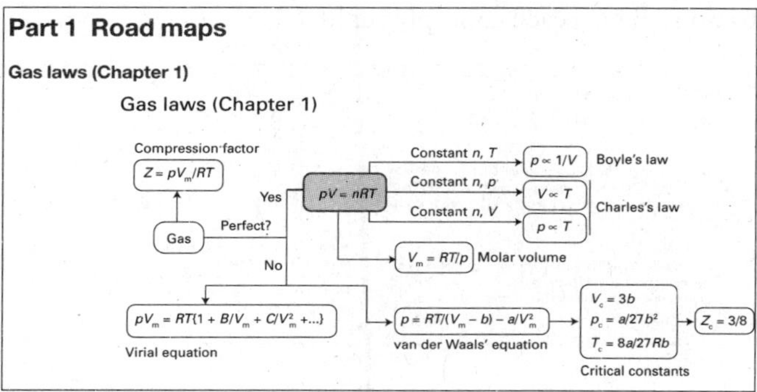

Impact sections

Where appropriate, we have separated the principles from their applications: the principles are constant and straightforward; the applications come and go as the subject progresses. The *Impact* sections show how the principles developed in the chapter are currently being applied in a variety of modern contexts.

IMPACT ON NANOSCIENCE

I8.1 Quantum dots

Nanoscience is the study of atomic and molecular assemblies with dimensions ranging from 1 nm to about 100 nm and nanotechnology is concerned with the incorporation of such assemblies into devices. The future economic impact of nanotechnology could be very significant. For example, increased demand for very small digital electronic devices has driven the design of ever smaller and more powerful microprocessors. However, there is an upper limit on the density of electronic circuits that can be incorporated into silicon-based chips with current fabrication technologies. As the ability to process data increases with the number of components in a chip, it follows that soon chips and the devices that use them will have to become bigger if processing

Notes on good practice

Science is a precise activity and its language should be used accurately. We have used this feature to help encourage the use of the language and procedures of science in conformity to international practice (as specified by IUPAC, the International Union of Pure and Applied Chemistry) and to help avoid common mistakes.

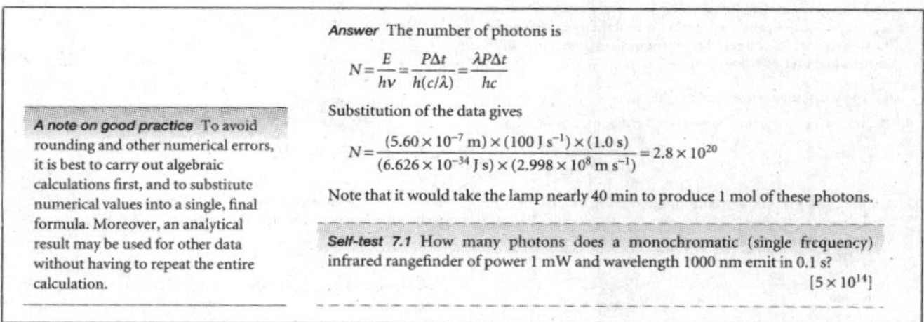

Answer The number of photons is

$$N = \frac{E}{h\nu} = \frac{P\Delta t}{h(c/\lambda)} = \frac{\lambda P\Delta t}{hc}$$

Substitution of the data gives

$$N = \frac{(5.60\times10^{-7}\ \mathrm{m})\times(100\ \mathrm{J\ s^{-1}})\times(1.0\ \mathrm{s})}{(6.626\times10^{-34}\ \mathrm{J\ s})\times(2.998\times10^{8}\ \mathrm{m\ s^{-1}})} = 2.8\times10^{20}$$

Note that it would take the lamp nearly 40 min to produce 1 mol of these photons.

A note on good practice To avoid rounding and other numerical errors, it is best to carry out algebraic calculations first, and to substitute numerical values into a single, final formula. Moreover, an analytical result may be used for other data without having to repeat the entire calculation.

Self-test 7.1 How many photons does a monochromatic (single frequency) infrared rangefinder of power 1 mW and wavelength 1000 nm emit in 0.1 s? [5×10^{14}]

interActivities

You will find that many of the graphs in the text have an interActivity attached: this is a suggestion about how you can explore the consequences of changing various parameters or of carrying out a more elaborate investigation related to the material in the illustration. In many cases, the activities can be completed by using the online resources of the book's website.

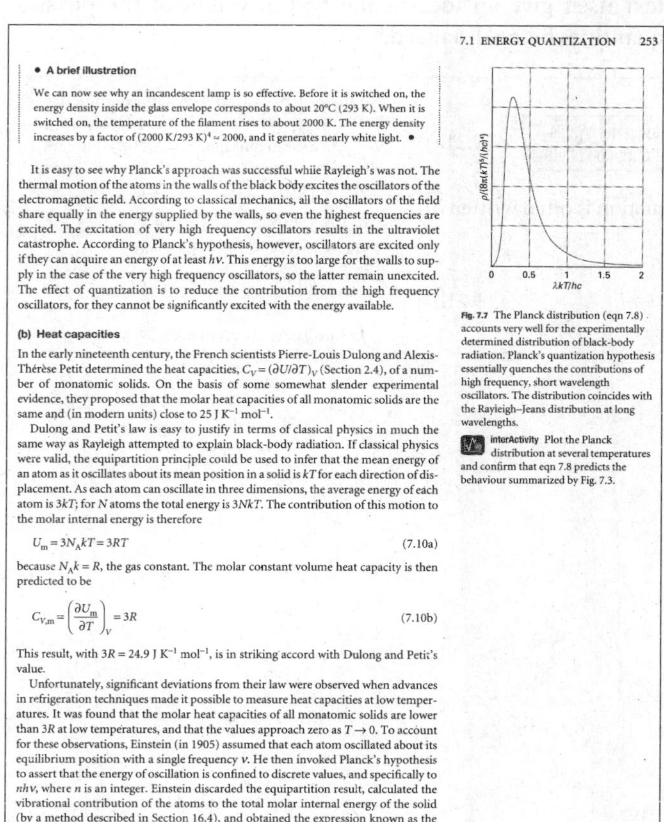

7.1 ENERGY QUANTIZATION 253

• **A brief illustration**

We can now see why an incandescent lamp is so effective. Before it is switched on, the energy density inside the glass envelope corresponds to about 20°C (293 K). When it is switched on, the temperature of the filament rises to about 2000 K. The energy density increases by a factor of $(2000\ \mathrm{K}/293\ \mathrm{K})^4 \approx 2000$, and it generates nearly white light. •

It is easy to see why Planck's approach was successful whiie Rayleigh's was not. The thermal motion of the atoms in the walls of the black body excites the oscillators of the electromagnetic field. According to classical mechanics, all the oscillators of the field share equally in the energy supplied by the walls, so even the highest frequencies are excited. The excitation of very high frequency oscillators results in the ultraviolet catastrophe. According to Planck's hypothesis, however, oscillators are excited only if they can acquire an energy of at least $h\nu$. This energy is too large for the walls to supply in the case of the very high frequency oscillators, so the latter remain unexcited. The effect of quantization is to reduce the contribution from the high frequency oscillators, for they cannot be significantly excited with the energy available.

(b) Heat capacities

In the early nineteenth century, the French scientists Pierre-Louis Dulong and Alexis-Thérèse Petit determined the heat capacities, $C_V = (\partial U/\partial T)_V$ (Section 2.4), of a number of monatomic solids. On the basis of some somewhat slender experimental evidence, they proposed that the molar heat capacities of all monatomic solids are the same and (in modern units) close to $25\ \mathrm{J\ K^{-1}\ mol^{-1}}$.

Dulong and Petit's law is easy to justify in terms of classical physics in much the same way as Rayleigh attempted to explain black-body radiation. If classical physics were valid, the equipartition principle could be used to infer that the mean energy of an atom as it oscillates about its mean position in a solid is kT for each direction of displacement. As each atom can oscillate in three dimensions, the average energy of each atom is $3kT$; for N atoms the total energy is $3NkT$. The contribution of this motion to the molar internal energy is therefore

$$U_m = 3N_A kT = 3RT \qquad (7.10a)$$

because $N_A k = R$, the gas constant. The molar constant volume heat capacity is then predicted to be

$$C_{V,m} = \left(\frac{\partial U_m}{\partial T}\right)_V = 3R \qquad (7.10b)$$

This result, with $3R = 24.9\ \mathrm{J\ K^{-1}\ mol^{-1}}$, is in striking accord with Dulong and Petit's value.

Unfortunately, significant deviations from their law were observed when advances in refrigeration techniques made it possible to measure heat capacities at low temperatures. It was found that the molar heat capacities of all monatomic solids are lower than $3R$ at low temperatures, and that the values approach zero as $T \to 0$. To account for these observations, Einstein (in 1905) assumed that each atom oscillated about its equilibrium position with a single frequency ν. He then invoked Planck's hypothesis to assert that the energy of oscillation is confined to discrete values, and specifically to $nh\nu$, where n is an integer. Einstein discarded the equipartition result, calculated the vibrational contribution of the atoms to the total molar internal energy of the solid (by a method described in Section 16.4), and obtained the expression known as the **Einstein formula:**

Fig. 7.7 The Planck distribution (eqn 7.8) accounts very well for the experimentally determined distribution of black-body radiation. Planck's quantization hypothesis essentially quenches the contributions of high frequency, short wavelength oscillators. The distribution coincides with the Rayleigh–Jeans distribution at long wavelengths.

interActivity Plot the Planck distribution at several temperatures and confirm that eqn 7.8 predicts the behaviour summarized by Fig. 7.3.

Further information

In some cases, we have judged that a derivation is too long, too detailed, or too different in level for it to be included in the text. In these cases, the derivations will be found less obtrusively at the end of the chapter.

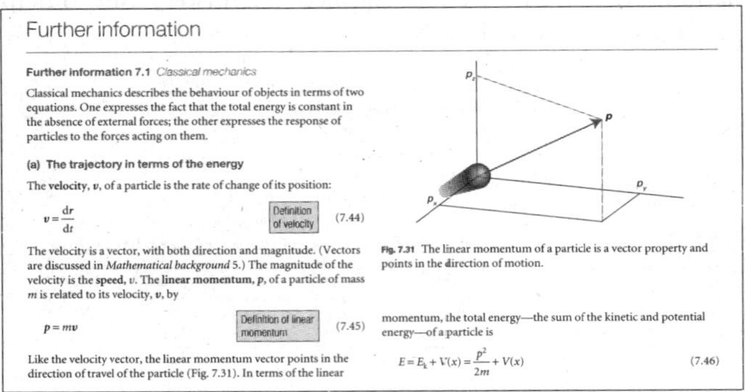
Further information

Further information 7.1 *Classical mechanics*

Classical mechanics describes the behaviour of objects in terms of two equations. One expresses the fact that the total energy is constant in the absence of external forces; the other expresses the response of particles to the forces acting on them.

(a) The trajectory in terms of the energy

The **velocity**, $\boldsymbol{v}$, of a particle is the rate of change of its position:

$$\boldsymbol{v} = \frac{\mathrm{d}\boldsymbol{r}}{\mathrm{d}t} \quad \text{Definition of velocity} \quad (7.44)$$

The velocity is a vector, with both direction and magnitude. (Vectors are discussed in *Mathematical background* 5.) The magnitude of the velocity is the **speed**, v. The **linear momentum**, $\boldsymbol{p}$, of a particle of mass m is related to its velocity, $\boldsymbol{v}$, by

$$\boldsymbol{p} = m\boldsymbol{v} \quad \text{Definition of linear momentum} \quad (7.45)$$

Like the velocity vector, the linear momentum vector points in the direction of travel of the particle (Fig. 7.31). In terms of the linear momentum, the total energy—the sum of the kinetic and potential energy—of a particle is

$$E = E_k + V(x) = \frac{p^2}{2m} + V(x) \quad (7.46)$$

Fig. 7.31 The linear momentum of a particle is a vector property and points in the direction of motion.

Resource section

Long tables of data are helpful for assembling and solving exercises and problems, but can break up the flow of the text. The *Resource section* at the end of the text consists of the *Road maps*, a *Data section* with a lot of useful numerical information, and *Character tables.* Short extracts of the tables in the text itself give an idea of the typical values of the physical quantities being discussed.

van der Waals equation of state (1.21a)

uation is often written in

(1.21b)

Table 1.6* van der Waals coefficients

	a/(atm dm^6 mol^{-2})	b/(10^{-2} dm^3 mol^{-1})
Ar	1.337	3.20
CO_2	3.610	4.29
He	0.0341	2.38
Xe	4.137	5.16

* More values are given in the *Data section.*

Mathematics support

A brief comment

A topic often needs to draw on a mathematical procedure or a concept of physics; a brief comment is a quick reminder of the procedure or concept.

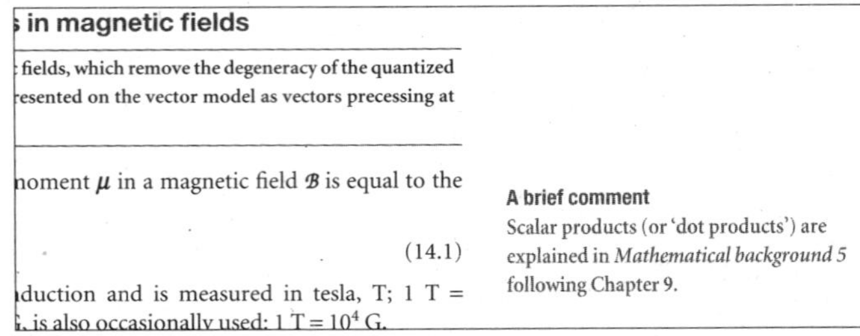
s in magnetic fields

fields, which remove the degeneracy of the quantized
esented on the vector model as vectors precessing at

noment μ in a magnetic field $\mathcal{B}$ is equal to the

(14.1)

duction and is measured in tesla, T; 1 T =
. is also occasionally used: 1 T = 10^4 G.

A brief comment

Scalar products (or 'dot products') are explained in *Mathematical background 5* following Chapter 9.

Mathematical background

It is often the case that you need a more full-bodied account of a mathematical concept, either because it is important to understand the procedure more fully or because you need to use a series of tools to develop an equation. The *Mathematical background* sections are located between some chapters, primarily where they are first needed, and include many illustrations of how each concept is used.

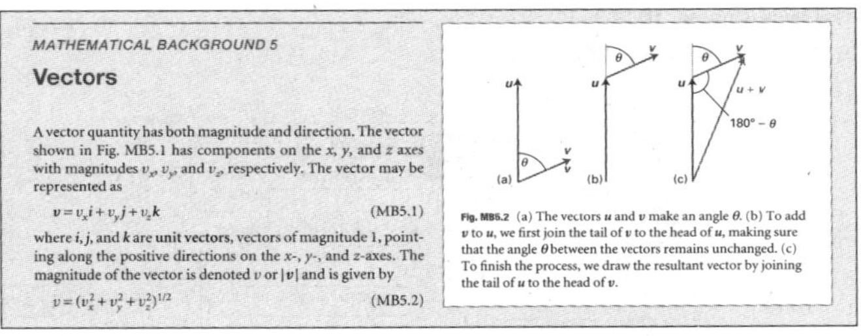
MATHEMATICAL BACKGROUND 5

Vectors

A vector quantity has both magnitude and direction. The vector shown in Fig. MB5.1 has components on the x, y, and z axes with magnitudes v_x, v_y, and v_z, respectively. The vector may be represented as

$$\boldsymbol{v} = v_x\boldsymbol{i} + v_y\boldsymbol{j} + v_z\boldsymbol{k} \quad (MB5.1)$$

where $\boldsymbol{i}$, $\boldsymbol{j}$, and $\boldsymbol{k}$ are **unit vectors**, vectors of magnitude 1, pointing along the positive directions on the x-, y-, and z-axes. The magnitude of the vector is denoted v or $|\boldsymbol{v}|$ and is given by

$$v = (v_x^2 + v_y^2 + v_z^2)^{1/2} \quad (MB5.2)$$

Fig. MB5.2 (a) The vectors $\boldsymbol{u}$ and $\boldsymbol{v}$ make an angle θ. (b) To add $\boldsymbol{v}$ to $\boldsymbol{u}$, we first join the tail of $\boldsymbol{v}$ to the head of $\boldsymbol{u}$, making sure that the angle θ between the vectors remains unchanged. (c) To finish the process, we draw the resultant vector by joining the tail of $\boldsymbol{u}$ to the head of $\boldsymbol{v}$.

Problem solving

A brief illustration

A brief illustration is a short example of how to use an equation that has just been introduced in the text. In particular, we show how to use data and how to manipulate units correctly.

• **A brief illustration**

The unpaired electron in the ground state of an alkali metal atom has $l = 0$, so $j = \frac{1}{2}$. Because the orbital angular momentum is zero in this state, the spin–orbit coupling energy is zero (as is confirmed by setting $j = s$ and $l = 0$ in eqn 9.42). When the electron is excited to an orbital with $l = 1$, it has orbital angular momentum and can give rise to a magnetic field that interacts with its spin. In this configuration the electron can have $j = \frac{3}{2}$ or $j = \frac{1}{2}$, and the energies of these levels are

$$E_{3/2} = \tfrac{1}{2}hc\tilde{A}\{\tfrac{3}{2}\times\tfrac{5}{2} - 1\times 2 - \tfrac{1}{2}\times\tfrac{3}{2}\} = \tfrac{1}{2}hc\tilde{A}$$

$$E_{1/2} = \tfrac{1}{2}hc\tilde{A}\{\tfrac{1}{2}\times\tfrac{3}{2} - 1\times 2 - \tfrac{1}{2}\times\tfrac{3}{2}\} = -hc\tilde{A}$$

The corresponding energies are shown in Fig. 9.30. Note that the baricentre (the 'centre of gravity') of the levels is unchanged, because there are four states of energy $\frac{1}{2}hc\tilde{A}$ and two of energy $-hc\tilde{A}$. •

Examples

We present many worked examples throughout the text to show how concepts are used, sometimes in combination with material from elsewhere in the text. Each worked example has a *Method* section suggesting an approach as well as a fully worked out answer.

Example 9.2 *Calculating the mean radius of an orbital*

Use hydrogenic orbitals to calculate the mean radius of a 1s orbital.

Method The mean radius is the expectation value

$$\langle r\rangle = \int \psi^* r\psi\, d\tau = \int r|\psi|^2 d\tau$$

We therefore need to evaluate the integral using the wavefunctions given in Table 9.1 and $d\tau = r^2 dr \sin\theta\, d\theta\, d\phi$. The angular parts of the wavefunction (Table 8.2) are normalized in the sense that

$$\int_0^{\pi}\int_0^{2\pi} |Y_{l,m_l}|^2 \sin\theta\, d\theta\, d\phi = 1$$

The integral over r required is given in Example 7.4.

Answer With the wavefunction written in the form $\psi = RY$, the integration is

$$\langle r\rangle = \int_0^{\infty}\int_0^{\pi}\int_0^{2\pi} rR_{n,l}^2 |Y_{l,m_l}|^2 r^2\, dr \sin\theta\, d\theta\, d\phi = \int_0^{\infty} r^3 R_{n,l}^2 dr$$

For a 1s orbital

$$R_{1,0} = 2\left(\frac{Z}{a_0}\right)^{3/2} e^{-Zr/a_0}$$

Hence

$$\langle r\rangle = \frac{4Z^3}{a_0^3}\int_0^{\infty} r^3 e^{-2Zr/a_0} dr = \frac{3a_0}{2Z}$$

Self-tests

Each *Example* has a *Self-test* with the answer provided as a check that the procedure has been mastered. There are also a number of free-standing *Self-tests* that are located where we thought it a good idea to provide a question to check your understanding. Think of *Self-tests* as in-chapter exercises designed to help you monitor your progress.

Self-test 9.4 Evaluate the mean radius of a 3s orbital by integration. [$27a_0/2Z$]

Discussion questions

The end-of-chapter material starts with a short set of questions that are intended to encourage reflection on the material and to view it in a broader context than is obtained by solving numerical problems.

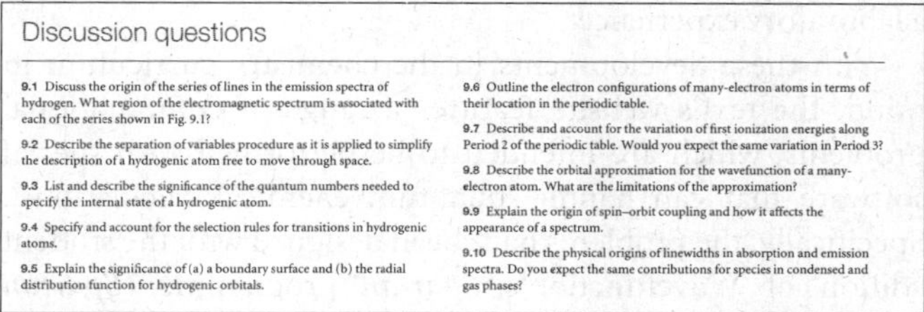

Discussion questions

9.1 Discuss the origin of the series of lines in the emission spectra of hydrogen. What region of the electromagnetic spectrum is associated with each of the series shown in Fig. 9.1?

9.2 Describe the separation of variables procedure as it is applied to simplify the description of a hydrogenic atom free to move through space.

9.3 List and describe the significance of the quantum numbers needed to specify the internal state of a hydrogenic atom.

9.4 Specify and account for the selection rules for transitions in hydrogenic atoms.

9.5 Explain the significance of (a) a boundary surface and (b) the radial distribution function for hydrogenic orbitals.

9.6 Outline the electron configurations of many-electron atoms in terms of their location in the periodic table.

9.7 Describe and account for the variation of first ionization energies along Period 2 of the periodic table. Would you expect the same variation in Period 3?

9.8 Describe the orbital approximation for the wavefunction of a many-electron atom. What are the limitations of the approximation?

9.9 Explain the origin of spin–orbit coupling and how it affects the appearance of a spectrum.

9.10 Describe the physical origins of linewidths in absorption and emission spectra. Do you expect the same contributions for species in condensed and gas phases?

Exercises and Problems

The core of testing understanding is the collection of end-of-chapter *Exercises* and *Problems*. The *Exercises* are straightforward numerical tests that give practice with manipulating numerical data. The *Problems* are more searching. They are divided into 'numerical', where the emphasis is on the manipulation of data, and 'theoretical', where the emphasis is on the manipulation of equations before (in some cases) using numerical data. At the end of the *Problems* are collections of problems that focus on practical applications of various kinds, including the material covered in the *Impact* sections.

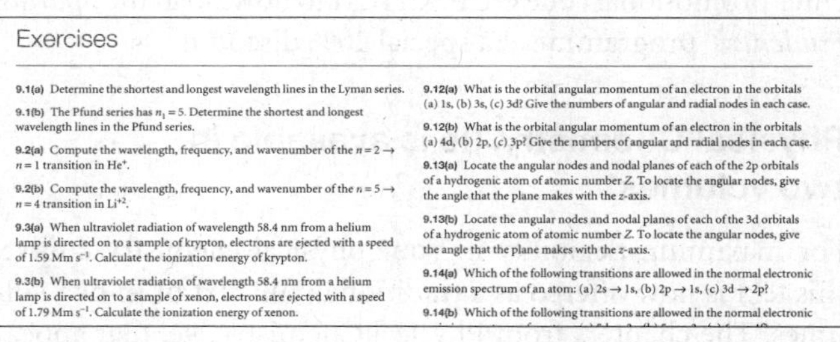

Exercises

9.1(a) Determine the shortest and longest wavelength lines in the Lyman series.

9.1(b) The Pfund series has $n_1 = 5$. Determine the shortest and longest wavelength lines in the Pfund series.

9.2(a) Compute the wavelength, frequency, and wavenumber of the $n = 2 \rightarrow n = 1$ transition in He^+.

9.2(b) Compute the wavelength, frequency, and wavenumber of the $n = 5 \rightarrow n = 4$ transition in Li^{+2}.

9.3(a) When ultraviolet radiation of wavelength 58.4 nm from a helium lamp is directed on to a sample of krypton, electrons are ejected with a speed of 1.59 Mm s^{-1}. Calculate the ionization energy of krypton.

9.3(b) When ultraviolet radiation of wavelength 58.4 nm from a helium lamp is directed on to a sample of xenon, electrons are ejected with a speed of 1.79 Mm s^{-1}. Calculate the ionization energy of xenon.

9.12(a) What is the orbital angular momentum of an electron in the orbitals (a) 1s, (b) 3s, (c) 3d? Give the numbers of angular and radial nodes in each case.

9.12(b) What is the orbital angular momentum of an electron in the orbitals (a) 4d, (b) 2p, (c) 3p? Give the numbers of angular and radial nodes in each case.

9.13(a) Locate the angular nodes and nodal planes of each of the 2p orbitals of a hydrogenic atom of atomic number Z. To locate the angular nodes, give the angle that the plane makes with the z-axis.

9.13(b) Locate the angular nodes and nodal planes of each of the 3d orbitals of a hydrogenic atom of atomic number Z. To locate the angular nodes, give the angle that the plane makes with the z-axis.

9.14(a) Which of the following transitions are allowed in the normal electronic emission spectrum of an atom: (a) 2s → 1s, (b) 2p → 1s, (c) 3d → 2p?

9.14(b) Which of the following transitions are allowed in the normal electronic

Problems*

Numerical problems

9.1 The *Humphreys series* is a group of lines in the spectrum of atomic hydrogen. It begins at 12 368 nm and has been traced to 3281.4 nm. What are the transitions involved? What are the wavelengths of the intermediate transitions?

9.2 A series of lines in the spectrum of atomic hydrogen lies at 656.46 nm, 486.27 nm, 434.17 nm, and 410.29 nm. What is the wavelength of the next line in the series? What is the ionization energy of the atom when it is in the lower state of the transitions?

9.3 The Li^{2+} ion is hydrogenic and has a Lyman series at 740 747 cm^{-1}, 877 924 cm^{-1}, 925 933 cm^{-1}, and beyond. Show that the energy levels are of the form $-hcR/n^2$ and find the value of R for this ion. Go on to predict the wavenumbers of the two longest-wavelength transitions of the Balmer series of the ion and find the ionization energy of the ion.

the spectrum are therefore expected to be hydrogen-like, the differences arising largely from the mass differences. Predict the wavenumbers of the first three lines of the Balmer series of positronium. What is the binding energy of the ground state of positronium?

9.9 The *Zeeman effect* is the modification of an atomic spectrum by the application of a strong magnetic field. It arises from the interaction between applied magnetic fields and the magnetic moments due to orbital and spin angular momenta (recall the evidence provided for electron spin by the Stern–Gerlach experiment, Section 8.8). To gain some appreciation for the so-called *normal Zeeman effect*, which is observed in transitions involving singlet states, consider a p electron, with $l = 1$ and $m_l = 0, \pm 1$. In the absence of a magnetic field, these three states are degenerate. When a field of magnitude $\mathcal{B}$ is present, the degeneracy is removed and it is observed that the state with $m_l = +1$ moves up in energy by $\mu_B\mathcal{B}$, the state with $m_l = 0$ is unchanged, and the state with $m_l = -1$ moves down in energy by $\mu_B\mathcal{B}$, where $\mu_B = e\hbar/2m_e = 9.274 \times 10^{-24}$ J T^{-1} is the Bohr magneton (see Section 13.1). Therefore, a

Molecular modelling and computational chemistry

Over the past two decades computational chemistry has evolved from a highly specialized tool, available to relatively few researchers, into a powerful and practical alternative to experimentation, accessible to all chemists. The driving force behind this evolution is the remarkable progress in computer technology. Calculations that previously required hours or days on giant mainframe computers may now be completed in a fraction of time on a personal computer. It is natural and necessary that computational chemistry finds its way into the undergraduate chemistry curriculum as a hands-on experience, just as teaching experimental chemistry requires a laboratory experience.

With these developments in the chemistry curriculum in mind, the text's website features a range of computational problems, which are intended to be performed with special software that can handle 'quantum chemical calculations'. Specifically, the problems have been designed with the student edition of Wavefunction's ***Spartan*** programme (***Spartan Student***™) in mind, although they could be completed with any electronic structure programme that allows Hartree–Fock, density functional, and MP2 calculations.

It is necessary for students to recognize that calculations are not the same as experiments, and that each 'chemical model' built from calculations has its own strengths and shortcomings. With this caveat in mind, it is important that some of the problems yield results that can be compared directly with experimental data. However, most problems are intended to stand on their own, allowing computational chemistry to serve as an exploratory tool.

Students can visit www.wavefun.com/cart/spartaned.html and enter promotional code OUPPCHEM to download the *Spartan Student*™ programme at a special 20% discount.

Physical Chemistry, 9e is available in two volumes!

For maximum flexibility in your physical chemistry course, this text is now offered as a traditional, full text or in two volumes. The chapters from Physical Chemistry, 9e, that appear each volume are as follows:

Volume 1: Thermodynamics and Kinetics

Chapter 0: Fundamentals
Chapter 1: The properties of gases
Chapter 2: The First Law
Chapter 3: The Second Law
Chapter 4: Physical transformations of pure substances
Chapter 5: Simple mixtures
Chapter 6: Chemical equilibrium
Chapter 20: Molecules in motion
Chapter 21: The rates of chemical reactions
Chapter 22: Reaction dynamics
Chapter 23: Catalysis

Volume 2: Quantum Chemistry, Spectroscopy, and Statistical Thermodynamics

Chapter 7: Quantum theory: introduction and principles
Chapter 8: Quantum theory: techniques and applications
Chapter 9: Atomic structure and spectra
Chapter 10: Molecular structure
Chapter 11: Molecular symmetry
Chapter 12: Molecular spectroscopy 1: rotational and vibrational spectra
Chapter 13: Molecular spectroscopy 2: electronic transitions
Chapter 14: Molecular spectroscopy 3: magnetic resonance
Chapter 15: Statistical thermodynamics 1: the concepts
Chapter 16: Statistical thermodynamics 2: applications

Chapters 17, 18, and 19 are not contained in the two volumes, but can be made available online on request.

Solutions manuals

As with previous editions, Charles Trapp, Carmen Giunta, and Marshall Cady have produced the solutions manuals to accompany this book. A *Student's Solutions Manual* (978–019–958397–3) provides full solutions to the 'a' exercises and the odd-numbered problems. An *Instructor's Solutions Manual* (978–019–958396–6) provides full solutions to the 'b' exercises and the even-numbered problems.

About the Online Resource Centre

The Online Resource Centre to accompany *Physical Chemistry 9/e* provides teaching and learning resources to augment the printed book. It is free of charge, and provides additional material for download, much of which can be incorporated into a virtual learning environment.

The Online Resource Centre can be accessed by visiting

www.oxfordtextbooks.co.uk/orc/pchem9e/

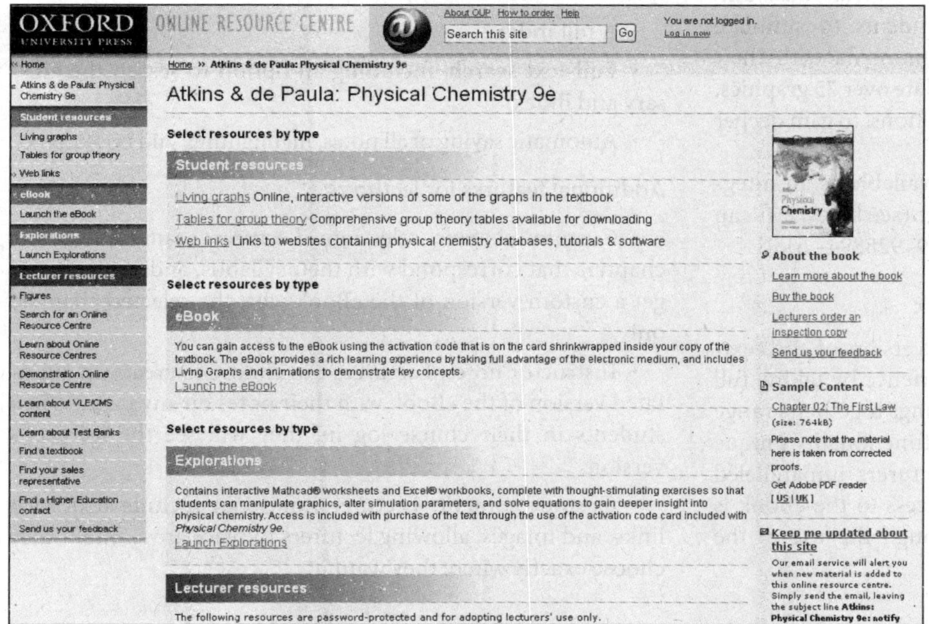

Note that lecturer resources are available only to registered adopters of the textbook. To register, simply visit www.oxfordtextbooks.co.uk/orc/pchem9e/ and follow the appropriate links. You will be given the opportunity to select your own username and password, which will be activated once your adoption has been verified.

Student resources are openly available to all, without registration.

For students

Living graphs

A *Living graph* can be used to explore how a property changes as a variety of parameters are changed. To encourage the use of this resource (and the more extensive *Explorations in physical chemistry*; see below), we have included a suggested *interActivity* with many of the illustrations in the text.

Web links

There is a huge network of information available about physical chemistry which can be bewildering to find your way around. The web links guide you to a collection of websites and web resources related to physical chemistry, organised by chapter to aid navigation.

Group theory tables

Comprehensive group theory tables are available for downloading.

For lecturers

Artwork

An instructor may wish to use the figures from this text in a lecture. Almost all the figures are available in PowerPoint®

format and can be used for lectures without charge (but not for commercial purposes without specific permission).

Tables of data

All the tables of data that appear in the chapter text are available and may be used under the same conditions as the figures.

Other resources

Explorations in Physical Chemistry by Valerie Walters, Julio de Paula, and Peter Atkins

Explorations in Physical Chemistry consists of interactive Mathcad® worksheets, interactive Excel® workbooks, and stimulating exercises. They motivate students to simulate physical, chemical, and biochemical phenomena with their personal computers. Students can manipulate over 75 graphics, alter simulation parameters, and solve equations, to gain deeper insight into physical chemistry.

Explorations in Physical Chemistry is available as an integrated part of the eBook version of the text (see below). It can also be purchased as a CD-ROM (978–019–928894–6).

Physical Chemistry, Ninth Edition eBook

The eBook, which is a complete online version of the textbook itself, provides a rich learning experience by taking full advantage of the electronic medium. It brings together a range of student resources alongside additional functionality unique to the eBook. The eBook also offers lecturers unparalleled flexibility and customization options. Access to the eBook is included with purchase of the text through the use of the activation code included with the book.

Key features of the eBook include:

- Easy access from any Internet-connected computer via a standard Web browser.
- Quick, intuitive navigation to any section or subsection, as well as any printed book page number.
- Living Graph animations.
- Integration of *Explorations in Physical Chemistry*.
- Text **highlighting**, down to the level of individual phrases.
- A **book marking** feature that allows for quick reference to any page.
- A powerful **Notes** feature that allows students or instructors to add notes to any page.
- A full **index**.
- **Full-text search**, including an option to search the glossary and index.
- Automatic saving of all notes, highlighting, and bookmarks.

Additional features for lecturers:

- Custom chapter selection: Lecturers can choose the chapters that correspond with their syllabus, and students will get a custom version of the eBook with the selected chapters only.
- Instructor notes: Lecturers can choose to create an annotated version of the eBook with their notes on any page. When students in their course log in, they will see the lecturer's version.
- Custom content: Lecturer notes can include text, web links, and images, allowing lecturers to place any content they choose exactly where they want it.

About the authors

Professor Peter Atkins is a fellow of Lincoln College, University of Oxford, and the author of more than sixty books for students and a general audience. His texts are market leaders around the globe. A frequent lecturer in the United States and throughout the world, he has held visiting professorships in France, Israel, Japan, China, and New Zealand. He was the founding chairman of the Committee on Chemistry Education of the International Union of Pure and Applied Chemistry and a member of IUPAC's Physical and Biophysical Chemistry Division.

Julio de Paula is Professor of Chemistry at Lewis and Clark College. A native of Brazil, Professor de Paula received a B.A. degree in chemistry from Rutgers, The State University of New Jersey, and a Ph.D. in biophysical chemistry from Yale University. His research activities encompass the areas of molecular spectroscopy, biophysical chemistry, and nanoscience. He has taught courses in general chemistry, physical chemistry, biophysical chemistry, instrumental analysis, and writing.

Acknowledgements

A book as extensive as this could not have been written without significant input from many individuals. We would like to reiterate our thanks to the hundreds of people who contributed to the first eight editions.

Many people gave their advice based on the eighth edition of the text, and others reviewed the draft chapters for the ninth edition as they emerged. We would like to thank the following colleagues:

Adedoyin Adeyiga, Cheyney University of Pennsylvania
David Andrews, University of East Anglia
Richard Ansell, University of Leeds
Colin Bain, University of Durham
Godfrey Beddard, University of Leeds
Magnus Bergstrom, Royal Institute of Technology, Stockholm, Sweden
Mark Bier, Carnegie Mellon University
Robert Bohn, University of Connecticut
Stefan Bon, University of Warwick
Fernando Bresme, Imperial College, London
Melanie Britton, University of Birmingham
Ten Brinke, Groningen, Netherlands
Ria Broer, Groningen, Netherlands
Alexander Burin, Tulane University
Philip J. Camp, University of Edinburgh
David Cedeno, Illinois State University
Alan Chadwick, University of Kent
Li-Heng Chen, Aquinas College
Aurora Clark, Washington State University
Nigel Clarke, University of Durham
Ron Clarke, University of Sydney
David Cooper, University of Liverpool
Garry Crosson, University of Dayton
John Cullen, University of Manitoba
Rajeev Dabke, Columbus State University
Keith Davidson, University of Lancaster
Guy Dennault, University of Southampton
Caroline Dessent, University of York
Thomas DeVore, James Madison University
Michael Doescher, Benedictine University
Randy Dumont, McMaster University
Karen Edler, University of Bath
Timothy Ehler, Buena Vista University
Andrew Ellis, University of Leicester
Cherice Evans, The City University of New York
Ashleigh Fletcher, University of Newcastle
Jiali Gao, University of Minnesota
Sophya Garashchuk, University of South Carolina in Columbia
Benjamin Gherman, California State University
Peter Griffiths, Cardiff, University of Wales
Nick Greeves, University of Liverpool
Gerard Grobner, University of Umeä, Sweden
Anton Guliaev, San Francisco State University
Arun Gupta, University of Alabama
Leonid Gurevich, Aalborg, Denmark
Georg Harhner, St Andrews University
Ian Hamley, University of Reading
Chris Hardacre, Queens University Belfast
Anthony Harriman, University of Newcastle
Torsten Hegmann, University of Manitoba
Richard Henchman, University of Manchester
Ulf Henriksson, Royal Institute of Technology, Stockholm, Sweden
Harald Høiland, Bergen, Norway
Paul Hodgkinson, University of Durham
Phillip John, Heriot-Watt University
Robert Hillman, University of Leicester
Pat Holt, Bellarmine University
Andrew Horn, University of Manchester
Ben Horrocks, University of Newcastle
Rob A. Jackson, University of Keele
Seogjoo Jang, The City University of New York
Don Jenkins, University of Warwick
Matthew Johnson, Copenhagen, Denmark
Mats Johnsson, Royal Institute of Technology, Stockholm, Sweden
Milton Johnston, University of South Florida
Peter Karadakov, University of York
Dale Keefe, Cape Breton University
Jonathan Kenny, Tufts University
Peter Knowles, Cardiff, University of Wales
Ranjit Koodali, University Of South Dakota
Evguenii Kozliak, University of North Dakota
Krish Krishnan, California State University
Peter Kroll, University of Texas at Arlington
Kari Laasonen, University of Oulu, Finland
Ian Lane, Queens University Belfast
Stanley Latesky, University of the Virgin Islands
Daniel Lawson, University of Michigan
Adam Lee, University of York
Donál Leech, Galway, Ireland
Graham Leggett, University of Sheffield
Dewi Lewis, University College London
Goran Lindblom, University of Umeä, Sweden
Lesley Lloyd, University of Birmingham
John Lombardi, City College of New York
Zan Luthey-Schulten, University of Illinois at Urbana-Champaign
Michael Lyons, Trinity College Dublin
Alexander Lyubartsev, University of Stockholm
Jeffrey Mack, California State University
Paul Madden, University of Edinburgh
Arnold Maliniak, University of Stockholm
Herve Marand, Virginia Tech

Louis Massa, Hunter College
Andrew Masters, University of Manchester
Joe McDouall, University of Manchester
Gordon S. McDougall, University of Edinburgh
David McGarvey, University of Keele
Anthony Meijer, University of Sheffield
Robert Metzger, University of Alabama
Sergey Mikhalovsky, University of Brighton
Marcelo de Miranda, University of Leeds
Gerald Morine, Bemidji State University
Damien Murphy, Cardiff, University of Wales
David Newman, Bowling Green State University
Gareth Parkes, University of Huddersfield
Ruben Parra, DePaul University
Enrique Peacock-Lopez, Williams College
Nils-Ola Persson, Linköping University
Barry Pickup, University of Sheffield
Ivan Powis, University of Nottingham
Will Price, University of Wollongong, New South Wales, Australia
Robert Quandt, Illinois State University
Chris Rego, University of Leicester
Scott Reid, Marquette University
Gavin Reid, University of Leeds
Steve Roser, University of Bath
David Rowley, University College London
Alan Ryder, Galway, Ireland
Karl Ryder, University of Leicester
Stephen Saeur, Copenhagen, Denmark
Sven Schroeder, University of Manchester
Jeffrey Shepherd, Laurentian University
Paul Siders, University of Minnesota Duluth
Richard Singer, University of Kingston
Carl Soennischsen, The Johannes Gutenberg University of Mainz
Jie Song, University of Michigan
David Steytler, University of East Anglia
Michael Stockenhuber, Nottingham-Trent University
Sven Stolen, University of Oslo
Emile Charles Sykes, Tufts University
Greg Szulczewski, University of Alabama
Annette Taylor, University of Leeds
Peter Taylor, University of Warwick
Jeremy Titman, University of Nottingham
Jeroen Van-Duijneveldt, University of Bristol
Joop van Lenthe, University of Utrecht
Peter Varnai, University of Sussex
Jay Wadhawan, University of Hull
Palle Waage Jensen, University of Southern Denmark
Darren Walsh, University of Nottingham
Kjell Waltersson, Malarden University, Sweden
Terence E. Warner, University of Southern Denmark
Richard Wells, University of Aberdeen
Ben Whitaker, University of Leeds
Kurt Winkelmann, Florida Institute of Technology
Timothy Wright, University of Nottingham
Yuanzheng Yue, Aalborg, Denmark
David Zax, Cornell University

We would like to thank two colleagues for their special contribution. Kerry Karukstis (Harvey Mudd College) provided many useful suggestions that focused on applications of the material presented in the text. David Smith (University of Bristol) made detailed comments on many of the chapters.

We also thank Claire Eisenhandler and Valerie Walters, who read through the proofs with meticulous attention to detail and caught in private what might have been a public grief. Our warm thanks also go to Charles Trapp, Carmen Giunta, and Marshall Cady who have produced the *Solutions manuals* that accompany this book.

Last, but by no means least, we would also like to thank our two publishers, Oxford University Press and W.H. Freeman & Co., for their constant encouragement, advice, and assistance, and in particular our editors Jonathan Crowe and Jessica Fiorillo. Authors could not wish for a more congenial publishing environment.

Summary of contents

Contents

List of impact sections

Impact on astrophysics

Impact on biochemistry

Impact on biology

Impact on engineering

Impact on environmental science

Impact on materials science

Impact on medicine

Impact on nanoscience

Impact on technology

Fundamentals

Chemistry is the science of matter and the changes it can undergo. **Physical chemistry** is the branch of chemistry that establishes and develops the principles of the subject in terms of the underlying concepts of physics and the language of mathematics. It provides the basis for developing new spectroscopic techniques and their interpretation, for understanding the structures of molecules and the details of their electron distributions, and for relating the bulk properties of matter to their constituent atoms. Physical chemistry also provides a window on to the world of chemical reactions and allows us to understand in detail how they take place. In fact, the subject underpins the whole of chemistry, providing the principles in terms we use to understand structure and change and providing the basis of all techniques of investigation.

Throughout the text we shall draw on a number of concepts, most of which should already be familiar from introductory chemistry. This section reviews them. In almost every case the following chapters will provide a deeper discussion, but we are presuming that we can refer to these concepts at any stage of the presentation. Because physical chemistry lies at the interface between physics and chemistry, we also need to review some of the concepts from elementary physics that we need to draw on in the text.

F.1 Atoms

Key points **(a) The nuclear model is the basis for discussion of atomic structure: negatively charged electrons occupy atomic orbitals, which are arranged in shells around a positively charged nucleus. (b) The periodic table highlights similarities in electronic configurations of atoms, which in turn lead to similarities in their physical and chemical properties. (c) Monatomic ions are electrically charged atoms and are characterized by their oxidation numbers.**

Matter consists of atoms. The atom of an element is characterized by its **atomic number**, Z, which is the number of protons in its nucleus. The number of neutrons in a nucleus is variable to a small extent, and the **nucleon number** (which is also commonly called the *mass number*), A, is the total number of protons and neutrons, which are collectively called **nucleons**, in the nucleus. Atoms of the same atomic number but different nucleon number are the **isotopes** of the element.

According to the **nuclear model**, an atom of atomic number Z consists of a nucleus of charge $+Ze$ surrounded by Z electrons each of charge $-e$ (e is the fundamental charge: see inside the front cover for its value and the values of the other fundamental constants). These electrons occupy **atomic orbitals**, which are regions of space where they are most likely to be found, with no more than two electrons in any one orbital. The atomic orbitals are arranged in **shells** around the nucleus, each shell being characterized by the **principal quantum number**, $n = 1, 2, \ldots$. A shell consists of n^2

individual orbitals, which are grouped together into *n* **subshells**; these subshells, and the orbitals they contain, are denoted s, p, d, and f. For all neutral atoms other than hydrogen, the subshells of a given shell have slightly different energies.

The sequential occupation of the orbitals in successive shells results in periodic similarities in the **electronic configurations**, the specification of the occupied orbitals, of atoms when they are arranged in order of their atomic number, which leads to the formulation of the **periodic table** (a version is shown inside the back cover). The vertical columns of the periodic table are called **groups** and (in the modern convention) numbered from 1 to 18. Successive rows of the periodic table are called **periods**, the number of the period being equal to the principal quantum number of the **valence shell**, the outermost shell of the atom. The periodic table is divided into s, p, d, and f **blocks**, according to the subshell that is last to be occupied in the formulation of the electronic configuration of the atom. The members of the d block (specifically the members of Groups 3–11 in the d block) are also known as the **transition metals**; those of the f block (which is not divided into numbered groups) are sometimes called the **inner transition metals**. The upper row of the f block (Period 6) consists of the **lanthanoids** (still commonly the 'lanthanides') and the lower row (Period 7) consists of the **actinoids** (still commonly the 'actinides'). Some of the groups also have familiar names: Group 1 consists of the **alkali metals**, Group 2 (more specifically, calcium, strontium, and barium) of the **alkaline earth metals**, Group 17 of the **halogens**, and Group 18 of the **noble gases**. Broadly speaking, the elements towards the left of the periodic table are **metals** and those towards the right are **nonmetals**; the two classes of substance meet at a diagonal line running from boron to polonium, which constitute the **metalloids**, with properties intermediate between those of metals and nonmetals.

A monatomic **ion** is an electrically charged atom. When an atom gains one or more electrons it becomes a negatively charged **anion**; when it loses one or more electrons it becomes a positively charged **cation**. The charge number of an ion is called the **oxidation number** of the element in that state (thus, the oxidation number of magnesium in Mg^{2+} is +2 and that of oxygen in O^{2-} is −2). It is appropriate, but not always done, to distinguish between the oxidation number and the **oxidation state**, the latter being the physical state of the atom with a specified oxidation number. Thus, the oxidation number *of* magnesium is +2 when it is present as Mg^{2+}, and it is present *in* the oxidation state Mg^{2+}. The elements form ions that are characteristic of their location in the periodic table: metallic elements typically form cations by losing the electrons of their outermost shell and acquiring the electronic configuration of the preceding noble gas. Nonmetals typically form anions by gaining electrons and attaining the electronic configuration of the following noble gas.

F.2 Molecules

Key points **(a) Covalent compounds consist of discrete molecules in which atoms are linked by covalent bonds. (b) Ionic compounds consist of cations and anions in a crystalline array. (c) Lewis structures are useful models of the pattern of bonding in molecules. (d) The valence-shell electron pair repulsion theory (VSEPR theory) is used to predict the three-dimensional structures of molecules from their Lewis structures. (e) The electrons in polar covalent bonds are shared unevenly between the bonded nuclei.**

A **chemical bond** is the link between atoms. Compounds that contain a metallic element typically, but far from universally, form **ionic compounds** that consist of cations and anions in a crystalline array. The 'chemical bonds' in an ionic compound are due to the Coulombic interactions (Section F.4) between all the ions in the crystal, and it is inappropriate to refer to a bond between a specific pair of neighbouring ions. The smallest unit of an ionic compound is called a **formula unit.** Thus $NaNO_3$, consisting of a Na^+ cation and a NO_3^- anion, is the formula unit of sodium nitrate. Compounds that do not contain a metallic element typically form **covalent compounds** consisting of discrete molecules. In this case, the bonds between the atoms of a molecule are **covalent**, meaning that they consist of shared pairs of electrons.

The pattern of bonds between neighbouring atoms is displayed by drawing a **Lewis structure**, in which bonds are shown as lines and **lone pairs** of electrons, pairs of valence electrons that are not used in bonding, are shown as dots. Lewis structures are constructed by allowing each atom to share electrons until it has acquired an **octet** of eight electrons (for hydrogen, a *duplet* of two electrons). A shared pair of electrons is a **single bond**, two shared pairs constitute a **double bond**, and three shared pairs constitute a **triple bond.** Atoms of elements of Period 3 and later can accommodate more than eight electrons in their valence shell and 'expand their octet' to become **hypervalent**, that is, form more bonds than the octet rule would allow (for example, SF_6), or form more bonds to a small number of atoms (for example, a Lewis structure of SO_4^{2-} with one or more double bonds). When more than one Lewis structure can be written for a given arrangement of atoms, it is supposed that **resonance**, a blending of the structures, may occur and distribute multiple-bond character over the molecule (for example, the two Kekulé structures of benzene). Examples of these aspects of Lewis structures are shown in Fig. F.1.

Except in the simplest cases, a Lewis structure does not express the three-dimensional structure of a molecule. The simplest approach to the prediction of molecular shape is **valence-shell electron pair repulsion theory** (VSEPR theory). In this approach, the regions of high electron density, as represented by bonds—whether single or multiple—and lone pairs, take up orientations around the central atom that maximize their separations. Then the position of the attached atoms (not the lone pairs) is noted and used to classify the shape of the molecule. Thus, four regions of electron density adopt a tetrahedral arrangement; if an atom is at each of these locations (as in CH_4), then the molecule is tetrahedral; if there is an atom at only three of these locations (as in NH_3), then the molecule is trigonal pyramidal; and so on. The names of the various shapes that are commonly found are shown in Fig. F.2. In a refinement of the theory, lone pairs are assumed to repel bonding pairs more strongly than bonding pairs repel each other. The shape a molecule then adopts, if it is not

A note on good practice Some chemists use the term 'molecule' to denote the smallest unit of a compound with the composition of the bulk material regardless of whether it is an ionic or covalent compound and thus speak of 'a molecule of NaCl'. We use the term 'molecule' to denote a discrete covalently bonded entity (as in H_2O); for an ionic compound we use 'formula unit'.

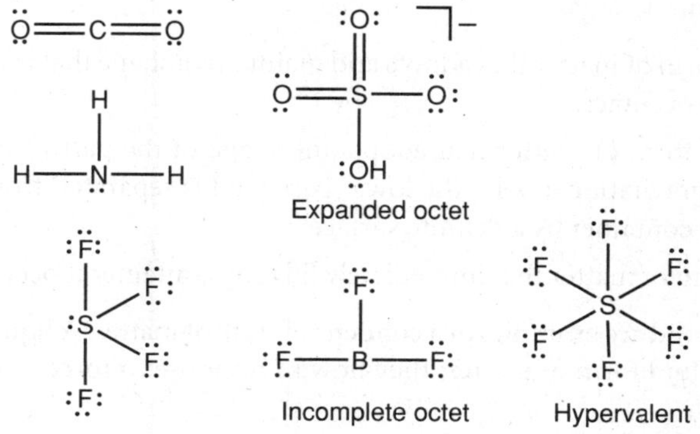

Expanded octet

Incomplete octet

Hypervalent

Fig. F.1 A collection of typical Lewis structures for simple molecules and ions. The structures show the bonding patterns and lone pairs and, except in simple cases, do not express the shape of the species.

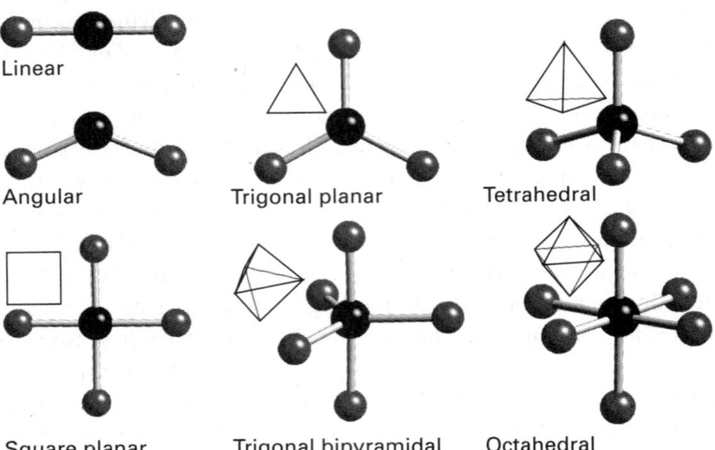

Fig. F.2 The names of the shapes of the geometrical figures used to describe symmetrical polyatomic molecules and ions.

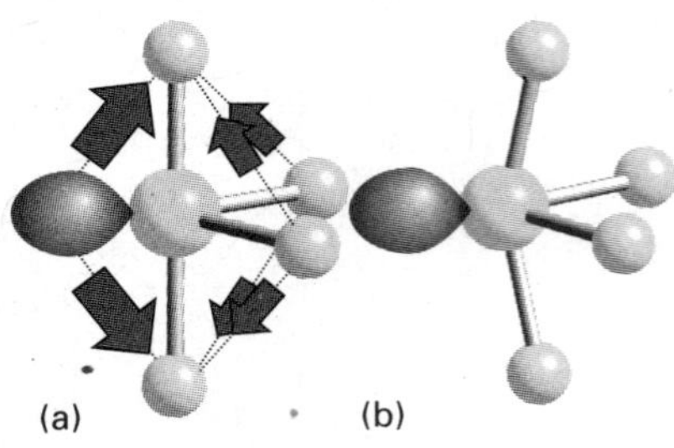

Fig. F.3 (a) The influences on the shape of the SF_4 molecule according to the VSEPR model. (b) As a result the molecule adopts a bent see-saw shape.

determined fully by symmetry, is such as to minimize repulsions from lone pairs. Thus, in SF_4 the lone pair adopts an equatorial position and the two axial S–F bonds bend away from it slightly, to give a bent see-saw shaped molecule (Fig. F.3).

Covalent bonds may be **polar**, or correspond to an unequal sharing of the electron pair, with the result that one atom has a partial positive charge (denoted $\delta+$) and the other a partial negative charge ($\delta-$). The ability of an atom to attract electrons to itself when part of a molecule is measured by the **electronegativity**, χ(chi), of the element. The juxtaposition of equal and opposite partial charges constitutes an **electric dipole**. If those charges are $+Q$ and $-Q$ and they are separated by a distance d, the magnitude of the **electric dipole moment** is $\mu = Qd$. Whether or not a molecule as a whole is polar depends on the arrangement of its bonds, for in highly symmetrical molecules there may be no net dipole. Thus, although the linear CO_2 molecule (which is structurally OCO) has polar CO bonds, their effects cancel and the molecule as a whole is nonpolar.

F.3 Bulk matter

Key points **(a) The physical states of bulk matter are solid, liquid, or gas. (b) The state of a sample of bulk matter is defined by specifying its properties, such as mass, volume, amount, pressure, and temperature. (c) The perfect gas law is a relation between the pressure, volume, amount, and temperature of an idealized gas.**

Bulk matter consists of large numbers of atoms, molecules, or ions. Its physical state may be solid, liquid, or gas:

A **solid** is a form of matter that adopts and maintains a shape that is independent of the container it occupies.

A **liquid** is a form of matter that adopts the shape of the part of the container it occupies (in a gravitational field, the lower part) and is separated from the unoccupied part of the container by a definite surface.

A **gas** is a form of matter that immediately fills any container it occupies.

A liquid and a solid are examples of a **condensed state** of matter. A liquid and a gas are examples of a **fluid** form of matter: they flow in response to forces (such as gravity) that are applied.

The state of a bulk sample of matter is defined by specifying the values of various properties. Among them are:

The **mass**, m, a measure of the quantity of matter present (unit: kilogram, kg).

The **volume**, V, a measure of the quantity of space the sample occupies (unit: cubic metre, m^3).

The **amount of substance**, n, a measure of the number of specified entities (atoms, molecules, or formula units) present (unit: mole, mol).

An **extensive property** of bulk matter is a property that depends on the amount of substance present in the sample; an **intensive property** is a property that is independent of the amount of substance. The volume is extensive; the mass density, ρ (rho), the mass of a sample divided by its volume, $\rho = m/V$, is intensive.

The amount of substance, n (colloquially, 'the number of moles'), is a measure of the number of specified entities present in the sample. 'Amount of substance' is the official name of the quantity; it is commonly simplified to 'chemical amount' or simply 'amount'. The unit 1 mol is defined as the number of carbon atoms in exactly 12 g of carbon-12. The number of entities per mole is called **Avogadro's constant,** N_A; the currently accepted value is 6.022×10^{23} mol^{-1} (note that N_A is a constant with units, not a pure number). The **molar mass of a substance,** M (units: formally kilograms per mole but commonly grams per mole, g mol^{-1}) is the mass per mole of its atoms, its molecules, or its formula units. The amount of substance of specified entities in a sample can readily be calculated from its mass, by noting that

$$n = \frac{m}{M} \tag{F.1}$$

A note on good practice Be careful to distinguish atomic or molecular mass (the mass of a single atom or molecule; units kg) from molar mass (the mass per mole of atoms or molecules; units kg mol^{-1}). *Relative* molecular masses of atoms and molecules, $M_r = m/m_u$, where m is the mass of the atom or molecule and m_u is the atomic mass constant, are still widely called 'atomic weights' and 'molecular weights' even though they are dimensionless quantities and not weights (the gravitational force exerted on an object). Even IUPAC continues to use the terms 'for historical reasons'.

A sample of matter may be subjected to a **pressure**, p (unit: pascal, Pa; 1 Pa = 1 kg m^{-1} s^{-2}), which is defined as the force, F, it is subjected to, divided by the area, A, to which that force is applied. A sample of gas exerts a pressure on the walls of its container because the molecules of gas are in ceaseless, random motion and exert a force when they strike the walls. The frequency of the collisions is normally so great that the force, and therefore the pressure, is perceived as being steady. Although pascal is the SI unit of pressure (Section F.6), it is also common to express pressure in bar (1 bar = 10^5 Pa) or atmospheres (1 atm = 101 325 Pa exactly), both of which correspond to typical atmospheric pressure. We shall see that, because many physical properties depend on the pressure acting on a sample, it is appropriate to select a certain value of the pressure to report their values. The **standard pressure** for reporting physical quantities is currently defined as $p^\ominus = 1$ bar exactly. We shall see the role of the standard pressure starting in Chapter 2.

To specify the state of a sample fully it is also necessary to give its **temperature**, T. The temperature is formally a property that determines in which direction energy will flow as heat when two samples are placed in contact through thermally conducting walls: energy flows from the sample with the higher temperature to the sample with the lower temperature. The symbol T is used to denote the **thermodynamic temperature**, which is an absolute scale with $T = 0$ as the lowest point. Temperatures above $T = 0$ are then most commonly expressed by using the **Kelvin scale**, in which the gradations of temperature are called **kelvin** (K). The Kelvin scale is defined by setting the triple point of water (the temperature at which ice, liquid water, and water vapour are in mutual equilibrium) at exactly 273.16 K. The freezing point of water (the melting point of ice) at 1 atm is then found experimentally to lie 0.01 K below the triple point, so the freezing point of water is 273.15 K. The Kelvin scale is unsuitable for everyday

A note on good practice Note that we write $T = 0$, not $T = 0$ K. General statements in science should be expressed without reference to a specific set of units. Moreover, because T (unlike θ) is absolute, the lowest point is 0 regardless of the scale used to express higher temperatures (such as the Kelvin scale or the Rankine scale). Similarly, we write $m = 0$, not $m = 0$ kg and $l = 0$, not $l = 0$ m.

measurements of temperature, and it is common to use the **Celsius scale**, which is defined in terms of the Kelvin scale as

$$\theta/°\mathrm{C} = T/\mathrm{K} - 273.15$$ Definition of Celsius scale (F.2)

Thus, the freezing point of water is 0°C and its boiling point (at 1 atm) is found to be 100°C (more precisely 99.974°C). Note that in this text T invariably denotes the thermodynamic (absolute) temperature and that temperatures on the Celsius scale are denoted θ (theta).

The properties that define the state of a system are not in general independent of one another. The most important example of a relation between them is provided by the idealized fluid known as a **perfect gas** (also, commonly, an 'ideal gas')

$$pV = nRT$$ Perfect gas equation (F.3)

A note on good practice Although the term 'ideal gas' is almost universally used in place of 'perfect gas', there are reasons for preferring the latter term. In an ideal system (as will be explained in Chapter 5) the interactions between molecules in a mixture are all the same. In a perfect gas not only are the interactions all the same but they are in fact zero. Few, though, make this useful distinction.

Here R is the **gas constant**, a universal constant (in the sense of being independent of the chemical identity of the gas) with the value 8.314 J K^{-1} mol^{-1}. Equation F.3 is central to the development of the description of gases in Chapter 1.

F.4 Energy

Key points **(a) Energy is the capacity to do work. (b) The total energy of a particle is the sum of its kinetic and potential energies. The kinetic energy of a particle is the energy it possesses as a result of its motion. The potential energy of a particle is the energy it possesses as a result of its position. (c) The Coulomb potential energy between two charges separated by a distance *r* varies as 1/*r*.**

Much of chemistry is concerned with transfers and transformations of energy, and it is appropriate to define this familiar quantity precisely: **energy** is the capacity to do work. In turn, work is defined as motion against an opposing force. The SI unit of energy is the joule (J), with

$$1\ \mathrm{J} = 1\ \mathrm{kg\ m^2\ s^{-2}}$$

(see Section F.7).

A body may possess two kinds of energy, kinetic energy and potential energy. The **kinetic energy**, E_k, of a body is the energy the body possesses as a result of its motion. For a body of mass m travelling at a speed v

$$E_\mathrm{k} = \tfrac{1}{2}mv^2$$ Kinetic energy (F.4)

The **potential energy**, E_p or more commonly V, of a body is the energy it possesses as a result of its position. No universal expression for the potential energy can be given because it depends on the type of force that the body experiences. For a particle of mass m at an altitude h close to the surface of the Earth, the gravitational potential energy is

$$V(h) = V(0) + mgh$$ Gravitational potential energy (F.5)

where g is the **acceleration of free fall** ($g = 9.81\ \mathrm{m\ s^{-2}}$). The zero of potential energy is arbitrary, and in this case it is common to set $V(0) = 0$.

One of the most important forms of potential energy in chemistry is the **Coulomb potential energy**, the potential energy of the electrostatic interaction between two point electric charges. For a point charge Q_1 at a distance r in a vacuum from another point charge Q_2

$$V(r) = \frac{Q_1 Q_2}{4\pi\varepsilon_0 r} \quad \text{Coulomb potential energy} \qquad \text{(F.6)}$$

It is conventional (as here) to set the potential energy equal to zero at infinite separation of charges. Then two opposite charges have a negative potential energy at finite separations, whereas two like charges have a positive potential energy. Charge is expressed in coulombs (C), often as a multiple of the fundamental charge, e. Thus, the charge of an electron is $-e$ and that of a proton is $+e$; the charge of an ion is ze, with z the **charge number** (positive for cations, negative for anions). The constant ε_0 (epsilon zero) is the **vacuum permittivity**, a fundamental constant with the value 8.854×10^{-12} C^2 J^{-1} m^{-1}. In a medium other than a vacuum, the potential energy of interaction between two charges is reduced, and the vacuum permittivity is replaced by the **permittivity**, ε, of the medium. The permittivity is commonly expressed as a multiple of the vacuum permittivity

$$\varepsilon = \varepsilon_r \varepsilon_0 \qquad \text{(F.7)}$$

with ε_r the dimensionless **relative permittivity** (formerly, the *dielectric constant*).

The **total energy** of a particle is the sum of its kinetic and potential energies

$$E = E_k + E_p \qquad \text{(F.8)}$$

We make frequent use of the apparently universal law of nature that *energy is conserved*; that is, energy can neither be created nor destroyed. Although energy can be transferred from one location to another and transformed from one form to another, the total energy is constant.

F.5 The relation between molecular and bulk properties

Key points (a) The energy levels of confined particles are quantized. (b) The Boltzmann distribution is a formula for calculating the relative populations of states of various energies. (c) The equipartition theorem provides a way to calculate the energy of some systems.

The energy of a molecule, atom, or subatomic particle that is confined to a region of space is **quantized**, or restricted to certain discrete values. These permitted energies are called **energy levels**. The values of the permitted energies depend on the characteristics of the particle (for instance, its mass) and the extent of the region to which it is confined. The quantization of energy is most important—in the sense that the allowed energies are widest apart—for particles of small mass confined to small regions of space. Consequently, quantization is very important for electrons in atoms and molecules, but usually unimportant for macroscopic bodies. For particles in containers of macroscopic dimensions the separation of energy levels is so small that for all practical purposes the motion of the particles through space—their translational motion—is unquantized and can be varied virtually continuously. As we shall see in detail in Chapter 7, quantization becomes increasingly important as we change focus from rotational to vibrational and then to electronic motion. The separation of rotational energy levels (in small molecules, about 10^{-23} J or 0.01 zJ, corresponding to about 0.01 kJ mol^{-1}) is smaller than that of vibrational energy levels (about 10 kJ mol^{-1}), which itself is smaller than that of electronic energy levels (about 10^{-18} J or 1 aJ, corresponding to about 10^3 kJ mol^{-1}). Figure F.4 depicts these typical energy level separations.

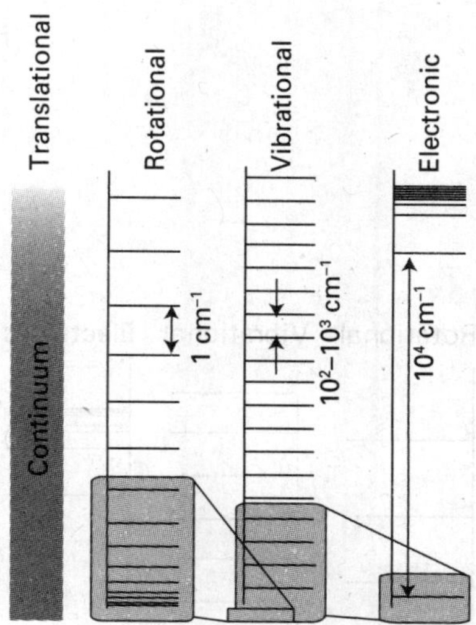

Fig. F.4 The energy level separations (expressed as wavenumbers) typical of four types of system.

A brief comment

The uncommon but useful prefixes z (for zepto) and a (for atto) are explained in Section F.7 on the use of units.

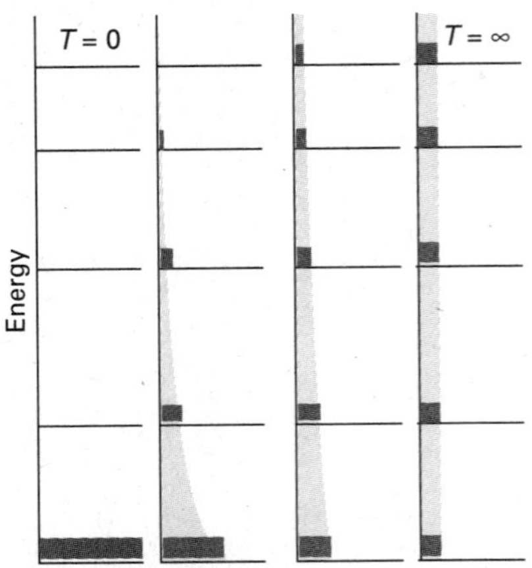

Fig. F.5 The Boltzmann distribution of populations for a system of five energy levels as the temperature is raised from zero to infinity.

(a) The Boltzmann distribution

The continuous thermal agitation that the molecules experience in a sample when $T > 0$ ensures that they are distributed over the available energy levels. One particular molecule may be in a state corresponding to a low energy level at one instant, and then be excited into a high energy state a moment later. Although we cannot keep track of the state of a single molecule, we can speak of the *average* numbers of molecules in each state. Even though individual molecules may be changing their states as a result of collisions, the average number in each state is constant (provided the temperature remains the same).

The average number of molecules in a state is called the **population** of the state. Only the lowest energy state is occupied at $T = 0$. Raising the temperature excites some molecules into higher energy states, and more and more states become accessible as the temperature is raised further (Fig. F.5). The formula for calculating the relative populations of states of various energies is called the **Boltzmann distribution** and was derived by the Austrian scientist Ludwig Boltzmann towards the end of the nineteenth century. Although we shall derive and discuss this distribution in more detail in Chapter 15, at this point it is important to know that it gives the ratio of the numbers of particles in states with energies E_i and E_j as

$$\frac{N_i}{N_j} = e^{-(E_i - E_j)/kT} \qquad \text{Boltzmann distribution} \qquad (F.9)$$

where k is **Boltzmann's constant**, a fundamental constant with the value $k = 1.381 \times 10^{-23}$ J K^{-1}. This constant occurs throughout physical chemistry, often in a disguised (molar) form as the gas constant, for

$$R = N_A k \qquad (F.10)$$

where N_A is Avogadro's constant. We shall see in Chapter 15 that the Boltzmann distribution provides the crucial link for expressing the macroscopic properties of bulk matter in terms of the behaviour of its constituent atoms.

The important features of the Boltzmann distribution to bear in mind are:

- The higher the energy of a state, the lower its population.
- The higher the temperature, the more likely it is that a state of high energy is populated.
- More levels are significantly populated if they are close together in comparison with kT (like rotational and translational states), than if they are far apart (like vibrational and electronic states).

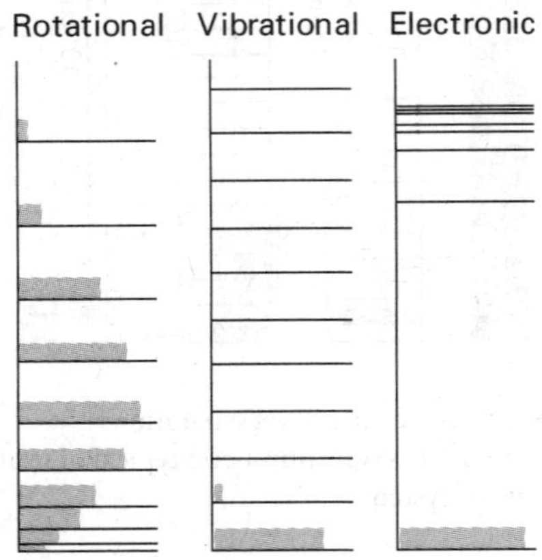

Rotational Vibrational Electronic

Fig. F.6 The Boltzmann distribution of populations for rotation, vibration, and electronic energy levels at room temperature.

Figure F.6 summarizes the form of the Boltzmann distribution for some typical sets of energy levels. The peculiar shape of the population of rotational levels stems from the fact that eqn F.9 applies to *individual states*, and for molecular rotation the number of rotational states corresponding to a given energy increases with energy. Broadly speaking, the number of planes of rotation increases with energy. Therefore, although the population of each *state* decreases with energy, the population of the *levels* goes through a maximum.

One of the simplest examples of the relation between microscopic and bulk properties is provided by **kinetic molecular theory**, a model of a perfect gas. In this model, it is assumed that the molecules, imagined as particles of negligible size, are in ceaseless, random motion and do not interact except during their brief collisions. Different speeds correspond to different kinetic energies, so the Boltzmann formula can be used to predict the proportions of molecules having a specific speed at a particular temperature. The expression giving the fraction of molecules that have a particular speed is called the **Maxwell distribution**, and has the features summarized in Fig. F.7. The Maxwell distribution, which is derived, specified, and discussed more fully in Chapter 20, can be used to show that the average speed, v_{mean}, of the molecules depends on the temperature and their molar mass as

$$v_{\text{mean}} \propto \left(\frac{T}{M}\right)^{1/2} \tag{F.11}$$

That is, the average speed increases as the square-root of the temperature and decreases as the square-root of the molar mass. Thus, the average speed is high for light molecules at high temperatures. The distribution itself gives more information than the average value. For instance, the tail towards high speeds is longer at high temperatures than at low, which indicates that at high temperatures more molecules in a sample have speeds much higher than average.

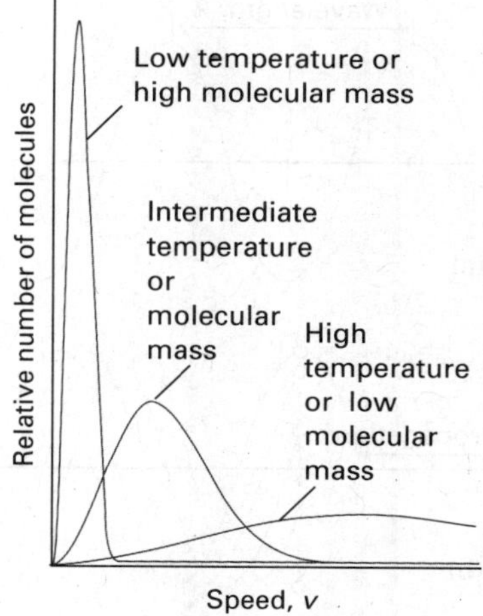

Fig. F.7 The distribution of molecular speeds with temperature and molar mass. Note that the most probable speed (corresponding to the peak of the distribution) increases with temperature and with decreasing molar mass, and simultaneously the distribution becomes broader.

interActivity (a) Plot different distributions by keeping the molar mass constant at 100 g mol^{-1} and varying the temperature of the sample between 200 K and 2000 K. (b) Use mathematical software or the *Living graph* applet from the text's web site to evaluate numerically the fraction of molecules with speeds in the range 100 m s^{-1} to 200 m s^{-1} at 300 K and 1000 K. (c) Based on your observations, provide a molecular interpretation of temperature.

(b) Equipartition

The Boltzmann distribution can be used to calculate the average energy associated with each mode of motion of a molecule (as we shall see in detail in Chapters 15 and 16). However, for certain modes of motion (which in practice means translation of any molecule and the rotation of all except the lightest molecules) there is a short cut, called the **equipartition theorem**. This theorem (which is derived from the Boltzmann distribution) states:

> In a sample at a temperature T, all quadratic contributions to the total energy have the same mean value, namely $\frac{1}{2}kT$.

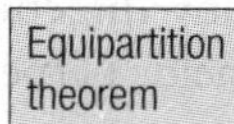

A 'quadratic contribution' simply means a contribution that depends on the square of the position or the velocity (or momentum). For example, because the kinetic energy of a body of mass m free to undergo translation in three dimensions is $E_{\text{k}} = \frac{1}{2}mv_x^2 + \frac{1}{2}mv_y^2 + \frac{1}{2}mv_z^2$, there are three quadratic terms. The theorem implies that the average kinetic energy of motion parallel to the x-axis is the same as the average kinetic energy of motion parallel to the y-axis and to the z-axis. That is, in a normal sample (one at thermal equilibrium throughout), the total energy is equally 'partitioned' over all the available modes of motion. One mode of motion is not especially rich in energy at the expense of another. Because the average contribution of each mode is $\frac{1}{2}kT$, the average kinetic energy of a molecule free to move in three dimensions is $\frac{3}{2}kT$, as there are three quadratic contributions to the kinetic energy.

We shall often use the equipartition theorem to make quick assessments of molecular properties and to judge the outcome of the competition of the ordering effects of intermolecular interactions and the disordering effects of thermal motion.

F.6 The electromagnetic field

Key point **Electromagnetic radiation is characterized by its direction of propagation, its wavelength, frequency, and wavenumber, and its state of polarization.**

Light is a form of electromagnetic radiation. In classical physics, electromagnetic radiation is understood in terms of the **electromagnetic field**, an oscillating electric and magnetic disturbance that spreads as a harmonic wave through empty space, the vacuum. The wave travels at a constant speed called the *speed of light*, *c*, which is about 3×10^8 m s^{-1}. As its name suggests, an electromagnetic field has two components, an **electric field** that acts on charged particles (whether stationary or moving) and a **magnetic field** that acts only on moving charged particles. The electromagnetic field,

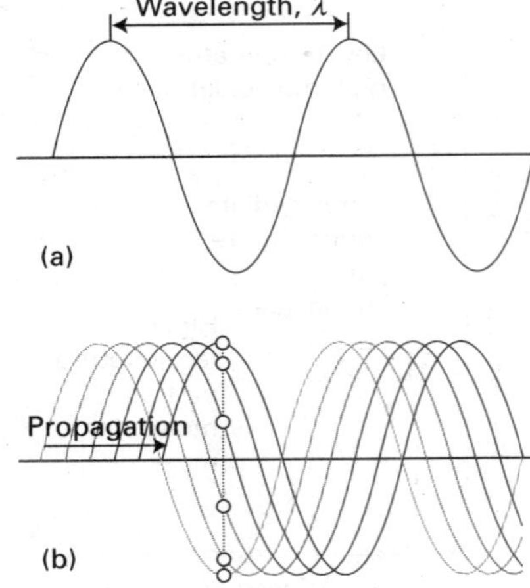

Fig. F.8 (a) The wavelength, λ, of a wave is the peak-to-peak distance. (b) The wave is shown travelling to the right at a speed c. At a given location, the instantaneous amplitude of the wave changes through a complete cycle (the six dots show half a cycle) as it passes a given point. The frequency, ν, is the number of cycles per second that occur at a given point. Wavelength and frequency are related by $\lambda\nu = c$.

A note on good practice You will hear people speaking of 'a frequency of so many wavenumbers'. That is doubly wrong. First, wavenumber and frequency are two different physical observables. Second, wavenumber is a physical quantity, not a unit. The dimensions of wavenumber are 1/length and it is commonly reported in reciprocal centimetres, cm^{-1}.

like any periodic wave, is characterized by a **wavelength**, λ (lambda), the distance between the neighbouring peaks of the wave, and its **frequency**, ν (nu), the number of times in a given time interval at which its displacement at a fixed point returns to its original value divided by the length of the time interval, normally in seconds (Fig. F.8). The frequency is measured in *hertz*, where 1 Hz = 1 s^{-1}. The wavelength and frequency of an electromagnetic wave are related by

$$\lambda\nu = c \tag{F.12}$$

Therefore, the shorter the wavelength, the higher the frequency. The characteristics of a wave are also reported by giving the **wavenumber**, $\tilde{\nu}$ (nu tilde), of the radiation, where

$$\tilde{\nu} = \frac{\nu}{c} = \frac{1}{\lambda} \tag{F.13}$$

A wavenumber can be interpreted as the number of complete wavelengths in a given length. Wavenumbers are normally reported in reciprocal centimetres (cm^{-1}), so a wavenumber of 5 cm^{-1} indicates that there are 5 complete wavelengths in 1 cm. A typical wavenumber of visible light is about 15 000 cm^{-1}, corresponding to 15 000 complete wavelengths in each centimetre. The classification of the electromagnetic field according to its frequency and wavelength is summarized in Fig. F.9.

Electromagnetic radiation is **plane-polarized** if the electric and magnetic fields each oscillate in a single plane (Fig. F.10). The plane of polarization may be orientated in any direction around the direction of propagation with the electric and magnetic fields perpendicular to that direction (and perpendicular to each other). An alternative mode of polarization is **circular polarization**, in which the electric and magnetic fields rotate around the direction of propagation in either a clockwise or a counter-clockwise sense but remain perpendicular to it and each other.

According to classical electromagnetic theory, the intensity of electromagnetic radiation is proportional to the square of the amplitude of the wave. For example, the radiation detectors used in spectroscopy are based on the interaction between the electric field of the incident radiation and the detecting element, so light intensities are proportional to the square of the amplitude of the waves.

F.7 Units

Key points **(a) The measurement of a physical property is expressed as the product of a numerical value and a unit. (b) In the International System of units (SI), the units are formed from seven base units, and all other physical quantities may be expressed as combinations of these physical quantities and reported in terms of derived units.**

The measurement of a physical property is expressed as

Physical property = numerical value × unit

For example, a length (l) may be reported as l = 5.1 m, if it is found to be 5.1 times as great as a defined unit of length, namely, 1 metre (1 m). Units are treated as algebraic quantities, and may be multiplied and divided. Thus, the same length could be reported as l/m = 5.1. The symbols for physical properties are always italic (sloping; thus V for volume, not V), including Greek symbols (thus, μ for electric dipole moment, not μ), but available typefaces are not always so obliging.

In the **International System** of units (SI, from the French *Système International d'Unités*), the units are formed from seven **base units** listed in Table F.1. All other physical quantities may be expressed as combinations of these physical quantities and

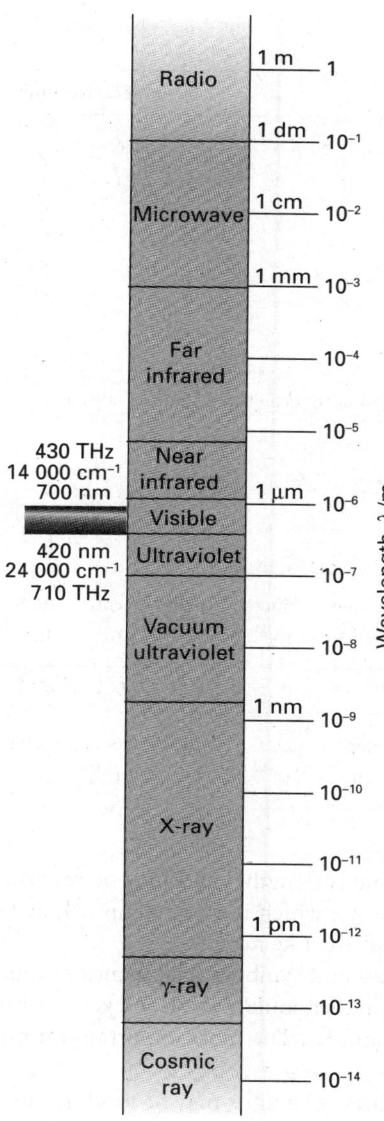

Fig. F.9 The regions of the electromagnetic spectrum. The boundaries are only approximate.

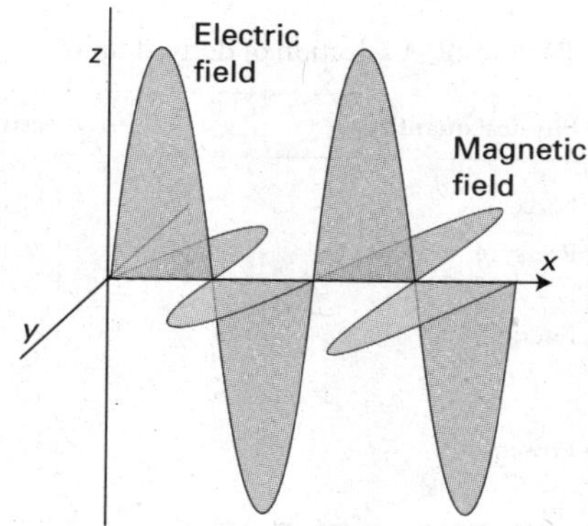

Fig. F.10 Electromagnetic radiation consists of a wave of electric and magnetic fields perpendicular to the direction of propagation (in this case the *x*-direction), and mutually perpendicular to each other. This illustration shows a plane-polarized wave, with the electric and magnetic fields oscillating in the *xz*- and *xy*-planes, respectively.

Table F.1 The SI base units

Physical quantity	Symbol for quantity	Base unit
Length	l	metre, m
Mass	m	kilogram, kg
Time	t	second, s
Electric current	I	ampere, A
Thermodynamic temperature	T	kelvin, K
Amount of substance	n	mole, mol
Luminous intensity	I_v	candela, cd

Table F.2 A selection of derived units

Physical quantity	Derived unit*	Name of derived unit
Force	1 kg m s^{-2}	newton, N
Pressure	1 kg m^{-1} s^{-2} 1 N m^{-2}	pascal, Pa
Energy	1 kg m^2 s^{-2} 1 N m 1 Pa m^3	joule, J
Power	1 kg m^2 s^{-3} 1 J s^{-1}	watt, W

* Equivalent definitions in terms of derived units are given following the definition in terms of base units.

Table F.3 Common SI prefixes

Prefix	y	z	a	f	p	n	μ	m	c	d
Name	yocto	zepto	atto	femto	pico	nano	micro	milli	centi	deci
Factor	10^{-24}	10^{-21}	10^{-18}	10^{-15}	10^{-12}	10^{-9}	10^{-6}	10^{-3}	10^{-2}	10^{-1}
Prefix	da	h	k	M	G	T	P	E	Z	Y
Name	deca	hecto	kilo	mega	giga	tera	peta	exa	zeta	yotta
Factor	10	10^2	10^3	10^6	10^9	10^{12}	10^{15}	10^{18}	10^{21}	10^{24}

reported in terms of **derived units.** Thus, volume is (length)^3 and may be reported as a multiple of 1 metre cubed (1 m^3), and density, which is mass/volume, may be reported as a multiple of 1 kilogram per metre cubed (1 kg m^{-3}).

A number of derived units have special names and symbols. The names of units derived from names of people are lower case (as in torr, joule, pascal, and kelvin), but their symbols are upper case (as in Torr, J, Pa, and K). The most important for our purposes are listed in Table F.2.

In all cases (both for base and derived quantities), the units may be modified by a prefix that denotes a factor of a power of 10. The Greek prefixes of units are upright (as in μm, not μm). Among the most common prefixes are those listed in Table F.3. Examples of the use of these prefixes are

$$1\ \text{nm} = 10^{-9}\ \text{m} \qquad 1\ \text{ps} = 10^{-12}\ \text{s} \qquad 1\ \mu\text{mol} = 10^{-6}\ \text{mol}$$

The kilogram (kg) is anomalous: although it is a base unit, it is interpreted as 10^3 g, and prefixes are attached to the gram (as in 1 mg = 10^{-3} g). Powers of units apply to the prefix as well as the unit they modify

$$1\ \text{cm}^3 = 1\ (\text{cm})^3 = 1\ (10^{-2}\ \text{m})^3 = 10^{-6}\ \text{m}^3$$

Note that 1 cm^3 does not mean 1 c(m^3). When carrying out numerical calculations, it is usually safest to write out the numerical value of an observable as a power of 10.

There are a number of units that are in wide use but are not a part of the International System. Some are exactly equal to multiples of SI units. These include the *litre* (L), which is exactly 10^3 cm^3 (or 1 dm^3) and the *atmosphere* (atm), which is exactly 101.325 kPa. Others rely on the values of fundamental constants, and hence are liable to change when the values of the fundamental constants are modified by more accurate

Table F.4 Some common units

Physical quantity	Name of unit	Symbol for unit	Value*
Time	minute	min	60 s
	hour	h	3600 s
	day	d	86 400 s
	year	a	31 556 952 s
Length	ångström	Å	10^{-10} m
Volume	litre	L, l	1 dm^3
Mass	tonne	t	10^3 kg
Pressure	bar	bar	10^5 Pa
	atmosphere	atm	101.325 kPa
Energy	electronvolt	eV	$1.602\ 176\ 53 \times 10^{-19}$ J
			96.485 31 kJ mol^{-1}

* All values in the final column are exact, except for the definition of 1 eV, which depends on the measured value of e, and the year, which is not a constant and depends on a variety of astronomical assumptions.

or more precise measurements. Thus, the size of the energy unit *electronvolt* (eV), the energy acquired by an electron that is accelerated through a potential difference of exactly 1 V, depends on the value of the charge of the electron, and the present (2008) conversion factor is 1 eV = $1.602\ 176\ 53 \times 10^{-19}$ J. Table F.4 gives the conversion factors for a number of these convenient units.

Exercises

F.1 Atoms

F1.1(a) Summarize the nuclear model of the atom.

F1.1(b) Define the terms atomic number, nucleon number, mass number.

F1.2(a) Express the typical ground-state electron configuration of an atom of an element in (a) Group 2, (b) Group 7, (c) Group 15 of the periodic table.

F1.2(b) Express the typical ground-state electron configuration of an atom of an element in (a) Group 3, (b) Group 5, (c) Group 13 of the periodic table.

F1.3(a) Identify the oxidation numbers of the elements in (a) $MgCl_2$, (b) FeO, (c) Hg_2Cl_2.

F1.3(b) Identify the oxidation numbers of the elements in (a) CaH_2, (b) CaC_2, (c) LiN_3.

F1.4(a) Where in the periodic table are metals and nonmetals found?

F1.4(b) Where in the periodic table are transition metals, lanthanoids, and actinoids found?

F.2 Molecules

F2.1(a) Summarize what is meant by a single and multiple bond.

F2.1(b) Identify a molecule with (a) one, (b) two, (c) three lone pairs on the central atom.

F2.2(a) Draw the Lewis (electron dot) structures of (a) SO_3^{2-}, (b) XeF_4, (c) P_4.

F2.2(b) Draw the Lewis (electron dot) structures of (a) O_3, (b) ClF_3^+, (c) N_3^-.

F2.3(a) Summarize the principal concepts of the VSEPR theory of molecular shape.

F2.3(b) Identify four hypervalent compounds.

F2.4(a) Use VSEPR theory to predict the structures of (a) PCl_3, (b) PCl_5, (c) XeF_2, (d) XeF_4.

F2.4(b) Use VSEPR theory to predict the structures of (a) H_2O_2, (b) FSO_3^-, (c) KrF_2, (d) PCl_4^+.

F2.5(a) Identify the polarities (by attaching partial charges $\delta+$ and $\delta-$) of the bonds (a) C–Cl, (b) P–H, (c) N–O.

F2.5(b) Identify the polarities (by attaching partial charges $\delta+$ and $\delta-$) of the bonds (a) C–H, (b) P–S, (c) N–Cl.

F2.6(a) State whether you expect the following molecules to be polar or nonpolar: (a) CO_2, (b) SO_2, (c) N_2O, (d) SF_4.

F2.6(b) State whether you expect the following molecules to be polar or nonpolar: (a) O_3, (b) XeF_2, (c) NO_2, (d) C_6H_{14}.

F2.7(a) Arrange the molecules in Exercise F2.6a by increasing dipole moment.

F2.7(b) Arrange the molecules in Exercise F2.6b by increasing dipole moment.

F.3 Bulk matter

F3.1(a) Compare and contrast the properties of the solid, liquid, and gas states of matter.

F3.1(b) Compare and contrast the properties of the condensed and gaseous states of matter.

F3.2(a) Classify the following properties as extensive or intensive: (a) mass, (b) mass density, (c) temperature, (d) number density.

F3.2(b) Classify the following properties as extensive or intensive: (a) pressure, (b) specific heat capacity, (c) weight, (d) molality.

F3.3(a) Calculate (a) the amount of C_2H_5OH (in moles) and (b) the number of molecules present in 25.0 g of ethanol.

F3.3(b) Calculate (a) the amount of $C_6H_{12}O_6$ (in moles) and (b) the number of molecules present in 5.0 g of glucose.

F3.4(a) Express a pressure of 1.45 atm in (a) pascal, (b) bar.

F3.4(b) Express a pressure of 222 atm in (a) pascal, (b) bar.

F3.5(a) Convert blood temperature, 37.0°C, to the Kelvin scale.

F3.5(b) Convert the boiling point of oxygen, 90.18 K, to the Celsius scale.

F3.6(a) Equation F.2 is a relation between the Kelvin and Celsius scales. Devise the corresponding equation relating the Fahrenheit and Celsius scales and use it to express the boiling point of ethanol (78.5°C) in degrees Fahrenheit.

F3.6(b) The Rankine scale is a version of the thermodynamic temperature scale in which the degrees (°R) are the same size as degrees Fahrenheit. Derive an expression relating the Rankine and Kelvin scales and express the freezing point of water in degrees Rankine.

F3.7(a) A sample of hydrogen gas was found to have a pressure of 110 kPa when the temperature was 20.0°C. What is its pressure expected to be when the temperature is 7.0°C?

F3.7(b) A sample of 325 mg of neon occupies 2.00 dm^3 at 20.0°C. Use the perfect gas law to calculate the pressure of the gas.

F.4 Energy

F4.1(a) Define energy and work.

F4.1(b) Distinguish between kinetic and potential energy.

F4.2(a) Consider a region of the atmosphere of volume 25 dm^3 that at 20°C contains about 1.0 mol of molecules. Take the average molar mass of the molecules as 29 g mol^{-1} and their average speed as about 400 m s^{-1}. Estimate the energy stored as molecular kinetic energy in this volume of air.

F4.2(b) Calculate the minimum energy that a bird of mass 25 g must expend in order to reach a height of 50 m.

F4.3(a) The potential energy of a charge Q_1 in the presence of another charge Q_2 can be expressed in terms of the *Coulomb potential*, ϕ (phi):

$$V = Q_1\phi \qquad \phi = \frac{Q_2}{4\pi\varepsilon_0 r}$$

The units of potential are joules per coulomb, J C^{-1} so, when ϕ is multiplied by a charge in coulombs, the result is in joules. The combination joules per coulomb occurs widely and is called a volt (V), with 1 V = 1 J C^{-1}. Calculate the Coulomb potential due to the nuclei at a point in a LiH molecule located at 200 pm from the Li nucleus and 150 pm from the H nucleus.

F4.3(b) Plot the Coulomb potential due to the nuclei at a point in a Na^+Cl^- ion pair located on a line halfway between the nuclei (the internuclear separation is 283 pm) as the point approaches from infinity and ends at the midpoint between the nuclei.

F.5 The relation between molecular and bulk properties

F5.1(a) What is meant by quantization of energy?

F5.1(b) In what circumstances are the effects of quantization most important for microscopic systems?

F5.2(a) The unit 1 electronvolt (1 eV) is defined as the energy acquired by an electron as it moves through a potential difference of 1 V. Suppose two states differ in energy by 1.0 eV. What is the ratio of their populations at (a) 300 K, (b) 3000 K?

F5.2(b) Suppose two states differ in energy by 1.0 eV, what can be said about their populations when $T = 0$ and when the temperature is infinite?

F5.3(a) What are the assumptions of the kinetic molecular theory?

F5.3(b) What are the main features of the Maxwell distribution of speeds?

F5.4(a) Suggest a reason why most molecules survive for long periods at room temperature.

F5.4(b) Suggest a reason why the rates of chemical reactions typically increase with increasing temperature.

F5.5(a) Calculate the relative mean speeds of N_2 molecules in air at 0°C and 40°C.

F5.5(b) Calculate the relative mean speeds of CO_2 molecules in air at 20°C and 30°C.

F5.6(a) Use the equipartition theorem to calculate the contribution of translational motion to the total energy of 5.0 g of argon at 25°C.

F5.6(b) Use the equipartition theorem to calculate the contribution of translational motion to the total energy of 10.0 g of helium at 30°C.

F5.7(a) Use the equipartition theorem to calculate the contribution to the total energy of a sample of 10.0 g of (a) carbon dioxide, (b) methane at 20°C; take into account translation and rotation but not vibration.

F5.7(b) Use the equipartition theorem to calculate the contribution to the total internal energy of a sample of 10.0 g of lead at 20°C, taking into account the vibrations of the atoms.

F.6 The electromagnetic field

F6.1(a) Express a wavelength of 230 nm as a frequency.

F6.1(b) Express a wavelength of 720 nm as a frequency.

F6.2(a) Express a frequency of 560 THz as a wavenumber.

F6.2(b) Express a frequency of 160 MHz as a wavenumber.

F6.3(a) A radio station broadcasts at a frequency of 91.7 MHz. What is (a) the wavelength, (b) the wavenumber of the radiation?

F6.3(b) A spectroscopic technique uses microwave radiation of wavelength 3.0 cm. What is (a) the wavenumber, (b) the frequency of the radiation?

F.7 Units

F7.1(a) Express a volume of 1.45 cm^3 in cubic metres.

F7.1(b) Express a volume of 1.45 dm^3 in cubic centimetres.

F7.2(a) Express a mass density of 11.2 g cm^{-3} in kilograms per cubic metre.

F7.2(b) Express a mass density of 1.12 g dm^{-3} in kilograms per cubic metre.

F7.3(a) Express pascal per joule in base units.

F7.3(b) Express $(joule)^2$ per $(newton)^3$ in base units.

F7.4(a) The expression kT/hc sometimes appears in physical chemistry. Evaluate this expression at 298 K in reciprocal centimetres (cm^{-1}).

F7.4(b) The expression kT/e sometimes appears in physical chemistry. Evaluate this expression at 298 K in millielectronvolts (meV).

F7.5(a) Given that $R = 8.3144$ J K^{-1} mol^{-1}, express R in decimetre cubed atmospheres per kelvin per mole.

F7.5(b) Given that $R = 8.3144$ J K^{-1} mol^{-1}, express R in pascal centimetre cubed per kelvin per molecule.

F7.6(a) Convert 1 dm^3 atm into joules.

F7.6(b) Convert 1 J into litre-atmospheres.

F7.7(a) Determine the SI units of $e^2/\varepsilon_0 r^2$. Express them in (a) base units, (b) units containing newtons.

F7.7(b) Determine the SI units of $\mu_B^2/\mu_0 r^3$, where μ_B is the Bohr magneton ($\mu_B = e\hbar/2m_e$) and μ_0 is the vacuum permeability (see inside front cover). Express them in (a) base units, (b) units containing joules.

PART 1 Equilibrium

Part 1 of the text develops the concepts that are needed for the discussion of equilibria in chemistry. Equilibria include physical change, such as fusion and vaporization, and chemical change, including electrochemistry. The discussion is in terms of thermodynamics, and particularly in terms of enthalpy and entropy. We see that we can obtain a unified view of equilibrium and the direction of spontaneous change in terms of the chemical potentials of substances. The chapters in Part 1 deal with the bulk properties of matter; those of Part 2 will show how these properties stem from the behaviour of individual atoms.

The properties of gases

1

This chapter establishes the properties of gases that will be used throughout the text. It begins with an account of an idealized version of a gas, a perfect gas, and shows how its equation of state may be assembled experimentally. We then see how the properties of real gases differ from those of a perfect gas, and construct an approximate equation of state that describes their properties.

The simplest state of matter is a **gas**, a form of matter that fills any container it occupies. Initially we consider only pure gases, but later in the chapter we see that the same ideas and equations apply to mixtures of gases too.

The perfect gas

We shall find it helpful to picture a gas as a collection of molecules (or atoms) in continuous random motion, with average speeds that increase as the temperature is raised. A gas differs from a liquid in that, except during collisions, the molecules of a gas are widely separated from one another and move in paths that are largely unaffected by intermolecular forces.

1.1 The states of gases

Key points **Each substance is described by an equation of state. (a) Pressure, force divided by area, provides a criterion of mechanical equilibrium for systems free to change their volume. (b) Pressure is measured with a barometer. (c) Through the Zeroth Law of thermodynamics, temperature provides a criterion of thermal equilibrium.**

The **physical state** of a sample of a substance, its physical condition, is defined by its physical properties. Two samples of a substance that have the same physical properties are in the same state. The state of a pure gas, for example, is specified by giving its volume, V, amount of substance (number of moles), n, pressure, p, and temperature, T. However, it has been established experimentally that it is sufficient to specify only three of these variables, for then the fourth variable is fixed. That is, it is an experimental fact that each substance is described by an **equation of state**, an equation that interrelates these four variables.

The general form of an equation of state is

$$p = f(T,V,n) \qquad \text{General form of an equation of state} \qquad (1.1)$$

This equation tells us that, if we know the values of n, T, and V for a particular substance, then the pressure has a fixed value. Each substance is described by its own equation of state, but we know the explicit form of the equation in only a few special cases. One very important example is the equation of state of a 'perfect gas', which has the form $p = nRT/V$, where R is a constant (Section F.3). Much of the rest of this chapter will examine the origin of this equation of state and its applications.

(a) Pressure

Pressure, p, is defined as force, F, divided by the area, A, to which the force is applied:

$$p=\frac{F}{A} \qquad \text{Definition of pressure} \qquad [1.2]$$

That is, the greater the force acting on a given area, the greater the pressure. The origin of the force exerted by a gas is the incessant battering of the molecules on the walls of its container. The collisions are so numerous that they exert an effectively steady force, which is experienced as a steady pressure. The SI unit of pressure, the *pascal* (Pa, 1 Pa = 1 N m^{-2}) was introduced in Section F.7. As we saw there, several other units are still widely used (Table 1.1). A pressure of 1 bar is the **standard pressure** for reporting data; we denote it $p^{\ominus}$.

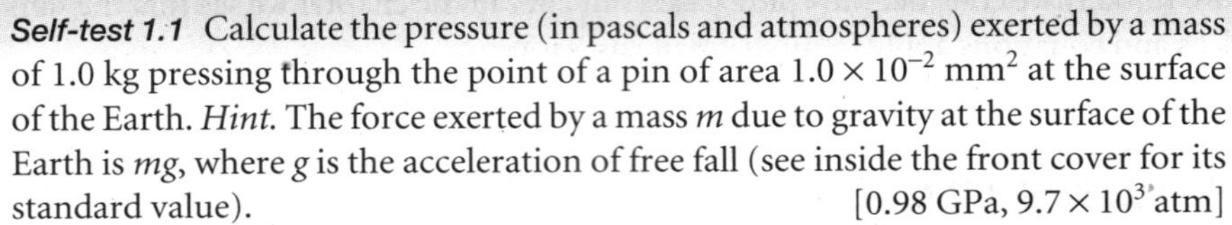
Self-test 1.1 Calculate the pressure (in pascals and atmospheres) exerted by a mass of 1.0 kg pressing through the point of a pin of area 1.0×10^{-2} mm^2 at the surface of the Earth. *Hint.* The force exerted by a mass m due to gravity at the surface of the Earth is mg, where g is the acceleration of free fall (see inside the front cover for its standard value). [0.98 GPa, 9.7×10^3 atm]

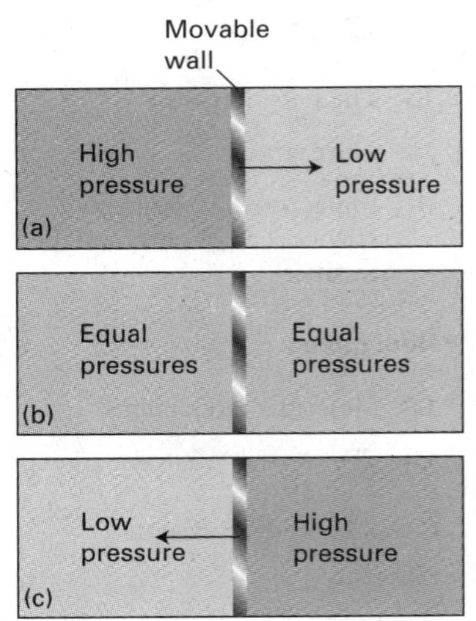

Fig. 1.1 When a region of high pressure is separated from a region of low pressure by a movable wall, the wall will be pushed into one region or the other, as in (a) and (c). However, if the two pressures are identical, the wall will not move (b). The latter condition is one of mechanical equilibrium between the two regions.

If two gases are in separate containers that share a common movable wall (a 'piston', Fig. 1.1), the gas that has the higher pressure will tend to compress (reduce the volume of) the gas that has lower pressure. The pressure of the high-pressure gas will fall as it expands and that of the low-pressure gas will rise as it is compressed. There will come a stage when the two pressures are equal and the wall has no further tendency to move. This condition of equality of pressure on either side of a movable wall is a state of **mechanical equilibrium** between the two gases. The pressure of a gas is therefore an indication of whether a container that contains the gas will be in mechanical equilibrium with another gas with which it shares a movable wall.

Table 1.1 Pressure units

Name	Symbol	Value
pascal	1 Pa	1 N m^{-2}, 1 kg m^{-1} s^{-2}
bar	1 bar	10^5 Pa
atmosphere	1 atm	101.325 kPa
torr	1 Torr	(101 325/760) Pa = 133.32 . . . Pa
millimetres of mercury	1 mmHg	133.322 . . . Pa
pound per square inch	1 psi	6.894 757 . . . kPa

(b) The measurement of pressure

The pressure exerted by the atmosphere is measured with a **barometer**. The original version of a barometer (which was invented by Torricelli, a student of Galileo) was an inverted tube of mercury sealed at the upper end. When the column of mercury is in mechanical equilibrium with the atmosphere, the pressure at its base is equal to that exerted by the atmosphere. It follows that the height of the mercury column is proportional to the external pressure.

Example 1.1 *Calculating the pressure exerted by a column of liquid*

Derive an equation for the pressure at the base of a column of liquid of mass density ρ (rho) and height h at the surface of the Earth. The pressure exerted by a column of liquid is commonly called the 'hydrostatic pressure'.

Method Use the definition of pressure in eqn 1.2 with $F = mg$. To calculate F we need to know the mass m of the column of liquid, which is its mass density, ρ, multiplied by its volume, V: $m = \rho V$. The first step, therefore, is to calculate the volume of a cylindrical column of liquid.

Answer Let the column have cross-sectional area A; then its volume is Ah and its mass is $m = \rho Ah$. The force the column of this mass exerts at its base is

$$F = mg = \rho Ahg$$

The pressure at the base of the column is therefore

$$p = \frac{F}{A} = \frac{\rho Agh}{A} = \rho gh \quad \text{Hydrostatic pressure} \quad (1.3)$$

Note that the hydrostatic pressure is independent of the shape and cross-sectional area of the column. The mass of the column of a given height increases as the area, but so does the area on which the force acts, so the two cancel.

Self-test 1.2 Derive an expression for the pressure at the base of a column of liquid of length l held at an angle θ (theta) to the vertical (1). $[p = \rho gl \cos \theta]$

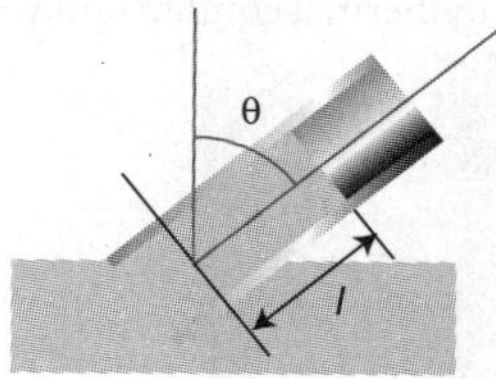

1

The pressure of a sample of gas inside a container is measured by using a pressure gauge, which is a device with electrical properties that depend on the pressure. For instance, a *Bayard–Alpert pressure gauge* is based on the ionization of the molecules present in the gas and the resulting current of ions is interpreted in terms of the pressure. In a *capacitance manometer*, the deflection of a diaphragm relative to a fixed electrode is monitored through its effect on the capacitance of the arrangement. Certain semiconductors also respond to pressure and are used as transducers in solid-state pressure gauges.

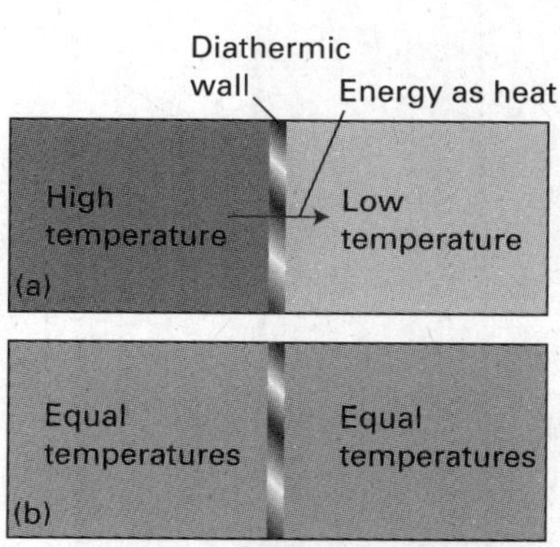

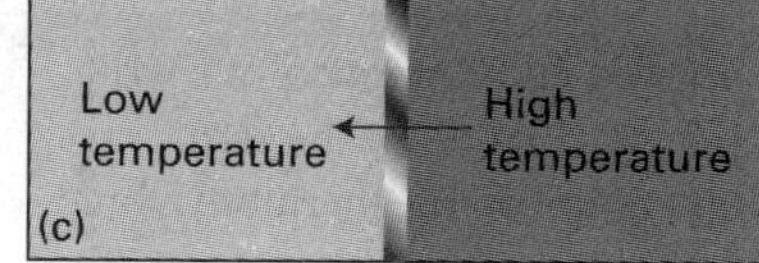

Fig. 1.2 Energy flows as heat from a region at a higher temperature to one at a lower temperature if the two are in contact through a diathermic wall, as in (a) and (c). However, if the two regions have identical temperatures, there is no net transfer of energy as heat even though the two regions are separated by a diathermic wall (b). The latter condition corresponds to the two regions being at thermal equilibrium.

(c) Temperature

The concept of temperature springs from the observation that a change in physical state (for example, a change of volume) can occur when two objects are in contact with one another, as when a red-hot metal is plunged into water. Later (Section 2.1) we shall see that the change in state can be interpreted as arising from a flow of energy as heat from one object to another. The **temperature**, T, is the property that indicates the direction of the flow of energy through a thermally conducting, rigid wall. If energy flows from A to B when they are in contact, then we say that A has a higher temperature than B (Fig. 1.2).

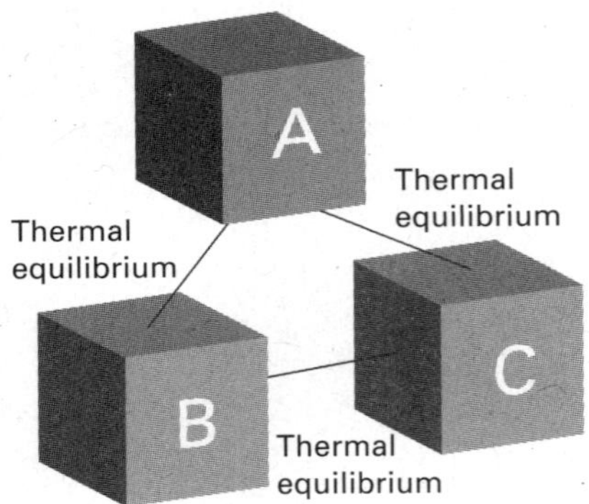

Fig. 1.3 The experience summarized by the Zeroth Law of thermodynamics is that, if an object A is in thermal equilibrium with B and B is in thermal equilibrium with C, then C is in thermal equilibrium with A.

It will prove useful to distinguish between two types of boundary that can separate the objects. A boundary is **diathermic** (thermally conducting; 'dia' is from the Greek word for 'through') if a change of state is observed when two objects at different temperatures are brought into contact. A metal container has diathermic walls. A boundary is **adiabatic** (thermally insulating) if no change occurs even though the two objects have different temperatures. A vacuum flask is an approximation to an adiabatic container.

The temperature is a property that indicates whether two objects would be in 'thermal equilibrium' if they were in contact through a diathermic boundary. **Thermal equilibrium** is established if no change of state occurs when two objects A to B are in contact through a diathermic boundary. Suppose an object A (which we can think of as a block of iron) is in thermal equilibrium with an object B (a block of copper), and that B is also in thermal equilibrium with another object C (a flask of water). Then it has been found experimentally that A and C will also be in thermal equilibrium when they are put in contact (Fig. 1.3). This observation is summarized by the **Zeroth Law of thermodynamics**:

> If A is in thermal equilibrium with B, and B is in thermal equilibrium with C, then C is also in thermal equilibrium with A. [Zeroth Law of thermodynamics]

The Zeroth Law justifies the concept of temperature and the use of a **thermometer**, a device for measuring the temperature. Thus, suppose that B is a glass capillary containing a liquid, such as mercury, that expands significantly as the temperature increases. Then, when A is in contact with B, the mercury column in the latter has a certain length. According to the Zeroth Law, if the mercury column in B has the same length when it is placed in thermal contact with another object C, then we can predict that no change of state of A and C will occur when they are in thermal contact. Moreover, we can use the length of the mercury column as a measure of the temperatures of A and C.

In the early days of thermometry (and still in laboratory practice today), temperatures were related to the length of a column of liquid, and the difference in lengths shown when the thermometer was first in contact with melting ice and then with boiling water was divided into 100 steps called 'degrees', the lower point being labelled 0. This procedure led to the **Celsius scale** of temperature. In this text, temperatures on the Celsius scale are denoted θ (theta) and expressed in *degrees Celsius* (°C). However, because different liquids expand to different extents, and do not always expand uniformly over a given range, thermometers constructed from different materials showed different numerical values of the temperature between their fixed points. The pressure of a gas, however, can be used to construct a **perfect-gas temperature scale** that is independent of the identity of the gas. The perfect-gas scale turns out to be identical to the **thermodynamic temperature scale** to be introduced in Section 3.2d, so we shall use the latter term from now on to avoid a proliferation of names. On the thermodynamic temperature scale, temperatures are denoted T and are normally reported in *kelvins* (K; not °K). Thermodynamic and Celsius temperatures are related by the exact expression

$$T/\mathrm{K} = \theta/{}^\circ\mathrm{C} + 273.15 \qquad \text{Definition of Celsius scale} \qquad (1.4)$$

This relation is the current definition of the Celsius scale in terms of the more fundamental Kelvin scale. It implies that a difference in temperature of 1°C is equivalent to a difference of 1 K.

A note on good practice We write $T = 0$, not $T = 0$ K for the zero temperature on the thermodynamic temperature scale. This scale is absolute, and the lowest temperature is 0 regardless of the size of the divisions on the scale (just as we write $p = 0$ for zero pressure, regardless of the size of the units we adopt, such as bar or pascal). However, we write 0°C because the Celsius scale is not absolute.

• A brief illustration

To express 25.00°C as a temperature in kelvins, we use eqn 1.4 to write

$$T/\text{K} = (25.00°\text{C})/°\text{C} + 273.15 = 25.00 + 273.15 = 298.15$$

Note how the units (in this case, °C) are cancelled like numbers. This is the procedure called 'quantity calculus' in which a physical quantity (such as the temperature) is the product of a numerical value (25.00) and a unit (1°C); see Section F.7. Multiplication of both sides by the unit K then gives $T = 298.15$ K. •

A note on good practice When the units need to be specified in an equation, the approved procedure, which avoids any ambiguity, is to write (physical quantity)/units, which is a dimensionless number, just as (25.00°C)/°C = 25.00 in this *brief illustration*. Units may be multiplied and cancelled just like numbers.

1.2 The gas laws

Key points **(a) The perfect gas law, a limiting law valid in the limit of zero pressure, summarizes Boyle's and Charles's laws and Avogadro's principle. (b) The kinetic theory of gases, in which molecules are in ceaseless random motion, provides a model that accounts for the gas laws and a relation between average speed and temperature. (c) A mixture of perfect gases behaves like a single perfect gas; its components each contribute their partial pressure to the total pressure.**

The equation of state of a gas at low pressure was established by combining a series of empirical laws.

A brief comment

Avogadro's principle is a principle rather than a law (a summary of experience) because it depends on the validity of a model, in this case the existence of molecules. Despite there now being no doubt about the existence of molecules, it is still a model-based principle rather than a law.

(a) The perfect gas law

We assume that the following individual gas laws are familiar:

Boyle's law: $pV = \text{constant}$, at constant n, T (1.5)°

Charles's law: $V = \text{constant} \times T$, at constant n, p (1.6a)°

$p = \text{constant} \times T$, at constant n, V (1.6b)°

Avogadro's principle: $V = \text{constant} \times n$ at constant p, T (1.7)°

Boyle's and Charles's laws are examples of a **limiting law**, a law that is strictly true only in a certain limit, in this case $p \to 0$. Equations valid in this limiting sense will be signalled by a ° on the equation number, as in these expressions. Avogadro's principle is commonly expressed in the form 'equal volumes of gases at the same temperature and pressure contain the same numbers of molecules'. In this form, it is increasingly true as $p \to 0$. Although these relations are strictly true only at $p = 0$, they are reasonably reliable at normal pressures ($p \approx 1$ bar) and are used widely throughout chemistry.

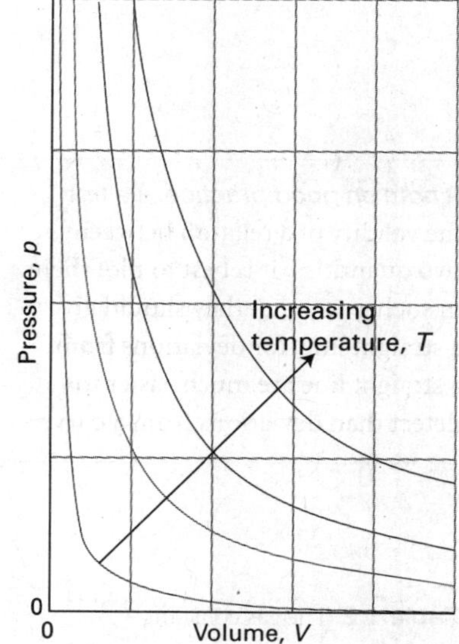

Fig. 1.4 The pressure–volume dependence of a fixed amount of perfect gas at different temperatures. Each curve is a hyperbola ($pV = \text{constant}$) and is called an isotherm.

interActivity Explore how the pressure of 1.5 mol CO_2(g) varies with volume as it is compressed at (a) 273 K, (b) 373 K from 30 dm^3 to 15 dm^3.

Figure 1.4 depicts the variation of the pressure of a sample of gas as the volume is changed. Each of the curves in the graph corresponds to a single temperature and hence is called an **isotherm**. According to Boyle's law, the isotherms of gases are hyperbolas (a curve obtained by plotting y against x with $xy = \text{constant}$). An alternative depiction, a plot of pressure against 1/volume, is shown in Fig. 1.5. The linear variation of volume with temperature summarized by Charles's law is illustrated in Fig. 1.6. The lines in this illustration are examples of **isobars**, or lines showing the variation of properties at constant pressure. Figure 1.7 illustrates the linear variation of pressure with temperature. The lines in this diagram are **isochores**, or lines showing the variation of properties at constant volume.

The empirical observations summarized by eqns 1.5–7 can be combined into a single expression

$$pV = \text{constant} \times nT$$

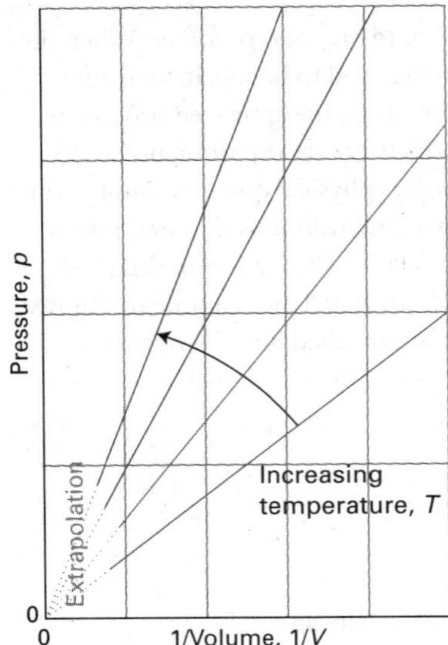

Fig. 1.5 Straight lines are obtained when the pressure is plotted against $1/V$ at constant temperature.

interActivity Repeat *interActivity 1.4*, but plot the data as p against $1/V$.

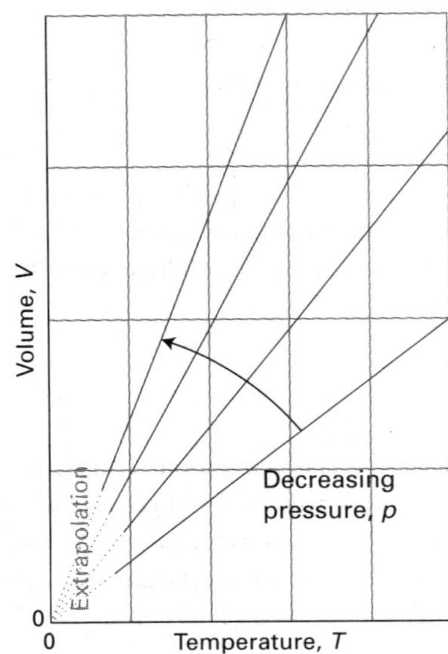

Fig. 1.6 The variation of the volume of a fixed amount of gas with the temperature at constant pressure. Note that in each case the isobars extrapolate to zero volume at $T = 0$ or $\theta = -273°C$.

interActivity Explore how the volume of 1.5 mol CO_2(g) in a container maintained at (a) 1.00 bar, (b) 0.50 bar varies with temperature as it is cooled from 373 K to 273 K.

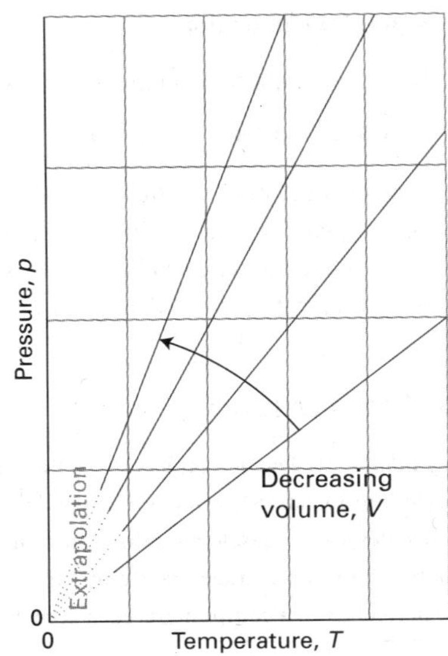

Fig. 1.7 The pressure also varies linearly with the temperature at constant volume, and extrapolates to zero at $T = 0$ (−273°C).

interActivity Explore how the pressure of 1.5 mol CO_2(g) in a container of volume (a) 30 dm^3, (b) 15 dm^3 varies with temperature as it is cooled from 373 K to 273 K.

A note on good practice To test the validity of a relation between two quantities, it is best to plot them in such a way that they should give a straight line, for deviations from a straight line are much easier to detect than deviations from a curve.

This expression is consistent with Boyle's law (pV = constant) when n and T are constant, with both forms of Charles's law ($p \propto T$, $V \propto T$) when n and either V or p are held constant, and with Avogadro's principle ($V \propto n$) when p and T are constant. The constant of proportionality, which is found experimentally to be the same for all gases, is denoted R and called the **gas constant**. The resulting expression

$$pV = nRT \quad \text{Perfect gas law} \qquad (1.8)°$$

is the **perfect gas law** (or *perfect gas equation of state*). It is the approximate equation of state of any gas, and becomes increasingly exact as the pressure of the gas approaches zero. A gas that obeys eqn 1.8 exactly under all conditions is called a **perfect gas** (or *ideal gas*). A **real gas**, an actual gas, behaves more like a perfect gas the lower the pressure, and is described exactly by eqn 1.8 in the limit of $p \rightarrow 0$. The gas constant R can be determined by evaluating $R = pV/nT$ for a gas in the limit of zero pressure (to guarantee that it is behaving perfectly). However, a more accurate value can be obtained by measuring the speed of sound in a low-pressure gas (argon is used in practice) and extrapolating its value to zero pressure. Table 1.2 lists the values of R in a variety of units.

The surface in Fig. 1.8 is a plot of the pressure of a fixed amount of perfect gas against its volume and thermodynamic temperature as given by eqn 1.8. The surface depicts the only possible states of a perfect gas: the gas cannot exist in states that do not correspond to points on the surface. The graphs in Figs. 1.4, 1.6, and 1.7 correspond to the sections through the surface (Fig. 1.9).

Table 1.2 The gas constant

R	
8.314 47	J K^{-1} mol^{-1}
8.205 74 × 10^{-2}	dm^3 atm K^{-1} mol^{-1}
8.314 47 × 10^{-2}	dm^3 bar K^{-1} mol^{-1}
8.314 47	Pa m^3 K^{-1} mol^{-1}
62.364	dm^3 Torr K^{-1} mol^{-1}
1.987 21	cal K^{-1} mol^{-1}

Example 1.2 *Using the perfect gas law*

In an industrial process, nitrogen is heated to 500 K in a vessel of constant volume. If it enters the vessel at 100 atm and 300 K, what pressure would it exert at the working temperature if it behaved as a perfect gas?

Method We expect the pressure to be greater on account of the increase in temperature. The perfect gas law in the form $pV/nT = R$ implies that, if the conditions are changed from one set of values to another, then, because pV/nT is equal to a constant, the two sets of values are related by the 'combined gas law'

$$\frac{p_1V_1}{n_1T_1} = \frac{p_2V_2}{n_2T_2} \qquad \text{Combined gas law} \qquad (1.9)^\circ$$

This expression is easily rearranged to give the unknown quantity (in this case p_2) in terms of the known. The known and unknown data are summarized in (2).

Answer Cancellation of the volumes (because $V_1 = V_2$) and amounts (because $n_1 = n_2$) on each side of the combined gas law results in

$$\frac{p_1}{T_1} = \frac{p_2}{T_2}$$

which can be rearranged into

$$p_2 = \frac{T_2}{T_1} \times p_1$$

Substitution of the data then gives

$$p_2 = \frac{500\ \text{K}}{300\ \text{K}} \times (100\ \text{atm}) = 167\ \text{atm}$$

Experiment shows that the pressure is actually 183 atm under these conditions, so the assumption that the gas is perfect leads to a 10 per cent error.

Self-test 1.3 What temperature would result in the same sample exerting a pressure of 300 atm? [900 K]

The perfect gas law is of the greatest importance in physical chemistry because it is used to derive a wide range of relations that are used throughout thermodynamics. However, it is also of considerable practical utility for calculating the properties of a gas under a variety of conditions. For instance, the molar volume, $V_m = V/n$, of a perfect gas under the conditions called **standard ambient temperature and pressure** (SATP), which means 298.15 K and 1 bar (that is, exactly 10^5 Pa), is easily calculated from $V_m = RT/p$ to be 24.789 $\text{dm}^3\ \text{mol}^{-1}$. An earlier definition, **standard temperature and pressure** (STP), was 0°C and 1 atm; at STP, the molar volume of a perfect gas is 22.414 $\text{dm}^3\ \text{mol}^{-1}$.

(b) The kinetic model of gases

The molecular explanation of Boyle's law is that, if a sample of gas is compressed to half its volume, then twice as many molecules strike the walls in a given period of time than before it was compressed. As a result, the average force exerted on the walls is

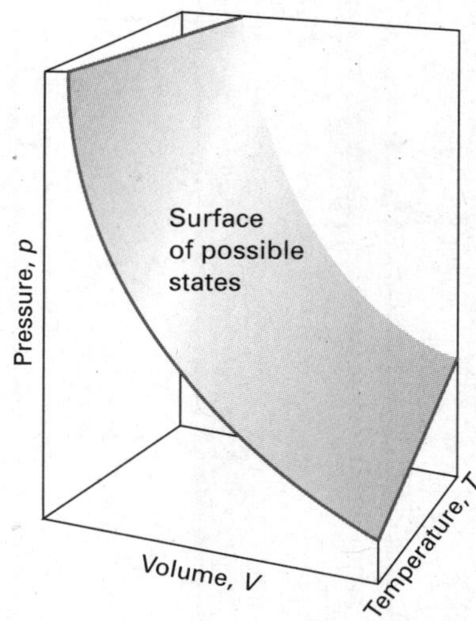

Fig. 1.8 A region of the p,V,T surface of a fixed amount of perfect gas. The points forming the surface represent the only states of the gas that can exist.

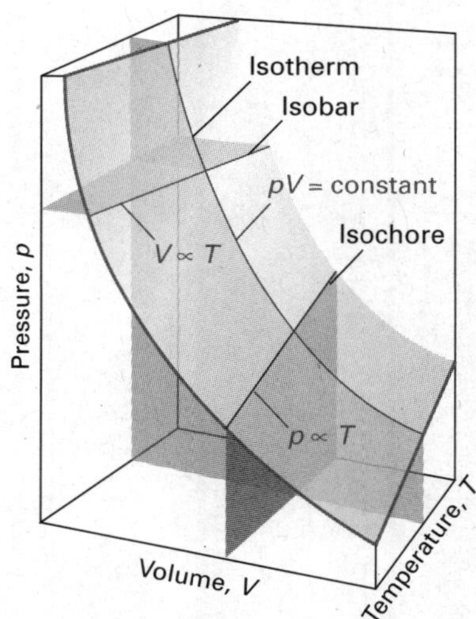

Fig. 1.9 Sections through the surface shown in Fig. 1.8 at constant temperature give the isotherms shown in Fig. 1.4 and the isobars shown in Fig. 1.6.

	n	*p*	*V*	*T*
Initial	Same	100	Same	300
Final	Same	?	Same	500

2

doubled. Hence, when the volume is halved the pressure of the gas is doubled, and $p \times V$ is a constant. Boyle's law applies to all gases regardless of their chemical identity (provided the pressure is low) because at low pressures the average separation of molecules is so great that they exert no influence on one another and hence travel independently. The molecular explanation of Charles's law lies in the fact that raising the temperature of a gas increases the average speed of its molecules. The molecules collide with the walls more frequently and with greater impact. Therefore they exert a greater pressure on the walls of the container.

These qualitative concepts are expressed quantitatively in terms of the kinetic model of gases, which is described more fully in Chapter 20. Briefly, the kinetic model is based on three assumptions:

1. The gas consists of molecules of mass m in ceaseless random motion.
2. The size of the molecules is negligible, in the sense that their diameters are much smaller than the average distance travelled between collisions.
3. The molecules interact only through brief, infrequent, and elastic collisions.

An *elastic collision* is a collision in which the total translational kinetic energy of the molecules is conserved. From the very economical assumptions of the kinetic model, it can be deduced (as we show in detail in Chapter 20) that the pressure and volume of the gas are related by

$$pV = \tfrac{1}{3}nMc^2 \qquad (1.10)°$$

where $M = mN_A$, the molar mass of the molecules, and c is the *root mean square speed* of the molecules, the square root of the mean of the squares of the speeds, v, of the molecules:

$$c = \langle v^2 \rangle^{1/2} \qquad (1.11)$$

We see that, if the root mean square speed of the molecules depends only on the temperature, then at constant temperature pV = constant, which is the content of Boyle's law. Moreover, for eqn 1.10 to be the equation of state of a perfect gas, its right-hand side must be equal to nRT. It follows that the root mean square speed of the molecules in a gas at a temperature T must be

$$c = \left(\frac{3RT}{M}\right)^{1/2} \qquad \text{Relation between molecular speed and temperature} \qquad (1.12)°$$

We can conclude that *the root mean square speed of the molecules of a gas is proportional to the square root of the temperature and inversely proportional to the square root of the molar mass*. That is, the higher the temperature, the higher the root mean square speed of the molecules, and, at a given temperature, heavy molecules travel more slowly than light molecules. The root mean square speed of N_2 molecules, for instance, is found from eqn 1.12 to be 515 m s^{-1} at 298 K.

(c) Mixtures of gases

When dealing with gaseous mixtures, we often need to know the contribution that each component makes to the total pressure of the sample. The **partial pressure**, p_J, of a gas J in a mixture (any gas, not just a perfect gas), is defined as

$$p_J = x_J p \qquad \text{Definition of partial pressure} \qquad [1.13]$$

where x_J is the **mole fraction** of the component J, the amount of J expressed as a fraction of the total amount of molecules, n, in the sample:

$$x_J = \frac{n_J}{n} \qquad n = n_A + n_B + \cdots \qquad \text{Definition of mole fraction} \qquad [1.14]$$

When no J molecules are present, $x_J = 0$; when only J molecules are present, $x_J = 1$. It follows from the definition of x_J that, whatever the composition of the mixture, $x_A + x_B + \cdots = 1$ and therefore that the sum of the partial pressures is equal to the total pressure

$$p_A + p_B + \cdots = (x_A + x_B + \cdots)p = p \qquad (1.15)$$

This relation is true for both real and perfect gases.

When all the gases are perfect, the partial pressure as defined in eqn 1.13 is also the pressure that each gas would exert if it occupied the same container alone at the same temperature. The latter is the original meaning of 'partial pressure'. That identification was the basis of the original formulation of **Dalton's law**:

> The pressure exerted by a mixture of gases is the sum of the pressures that each one would exert if it occupied the container alone.

Now, however, the relation between partial pressure (as defined in eqn 1.13) and total pressure (as given by eqn 1.15) is true for all gases and the identification of partial pressure with the pressure that the gas would exert on its own is valid only for a perfect gas.

Example 1.3 *Calculating partial pressures*

The mass percentage composition of dry air at sea level is approximately N_2: 75.5; O_2: 23.2; Ar: 1.3. What is the partial pressure of each component when the total pressure is 1.20 atm?

Method We expect species with a high mole fraction to have a proportionally high partial pressure. Partial pressures are defined by eqn 1.13. To use the equation, we need the mole fractions of the components. To calculate mole fractions, which are defined by eqn 1.14, we use the fact that the amount of molecules J of molar mass M_J in a sample of mass m_J is $n_J = m_J/M_J$. The mole fractions are independent of the total mass of the sample, so we can choose the latter to be exactly 100 g (which makes the conversion from mass percentages very easy). Thus, the mass of N_2 present is 75.5 per cent of 100 g, which is 75.5 g.

Answer The amounts of each type of molecule present in 100 g of air, in which the masses of N_2, O_2, and Ar are 75.5 g, 23.2 g, and 1.3 g, respectively, are

$$n(N_2) = \frac{75.5\text{ g}}{28.02\text{ g mol}^{-1}} = \frac{75.5}{28.02}\text{ mol}$$

$$n(O_2) = \frac{23.2\text{ g}}{32.00\text{ g mol}^{-1}} = \frac{23.2}{32.00}\text{ mol}$$

$$n(\text{Ar}) = \frac{1.3\text{ g}}{39.95\text{ g mol}^{-1}} = \frac{1.3}{39.95}\text{ mol}$$

These three amounts work out as 2.69 mol, 0.725 mol, and 0.033 mol, respectively, for a total of 3.45 mol. The mole fractions are obtained by dividing each of the above amounts by 3.45 mol and the partial pressures are then obtained by multiplying the mole fraction by the total pressure (1.20 atm):

	N_2	O_2	Ar
Mole fraction:	0.780	0.210	0.0096
Partial pressure/atm:	0.936	0.252	0.012

We have not had to assume that the gases are perfect: partial pressures are defined as $p_J = x_J p$ for any kind of gas.

Self-test 1.4 When carbon dioxide is taken into account, the mass percentages are 75.52 (N_2), 23.15 (O_2), 1.28 (Ar), and 0.046 (CO_2). What are the partial pressures when the total pressure is 0.900 atm? [0.703, 0.189, 0.0084, 0.00027 atm]

IMPACT ON ENVIRONMENTAL SCIENCE
I1.1 The gas laws and the weather

The biggest sample of gas readily accessible to us is the atmosphere, a mixture of gases with the composition summarized in Table 1.3. The composition is maintained moderately constant by diffusion and convection (winds, particularly the local turbulence called *eddies*) but the pressure and temperature vary with altitude and with the local conditions, particularly in the troposphere (the 'sphere of change'), the layer extending up to about 11 km.

In the troposphere the average temperature is 15°C at sea level, falling to −57°C at the bottom of the tropopause at 11 km. This variation is much less pronounced when expressed on the Kelvin scale, ranging from 288 K to 216 K, an average of 268 K. If we suppose that the temperature has its average value all the way up to the tropopause, then the pressure varies with altitude, h, according to the *barometric formula*

Table 1.3 The composition of dry air at sea level

	Percentage	
Component	**By volume**	**By mass**
Nitrogen, N_2	78.08	75.53
Oxygen, O_2	20.95	23.14
Argon, Ar	0.93	1.28
Carbon dioxide, CO_2	0.031	0.047
Hydrogen, H_2	5.0×10^{-3}	2.0×10^{-4}
Neon, Ne	1.8×10^{-3}	1.3×10^{-3}
Helium, He	5.2×10^{-4}	7.2×10^{-5}
Methane, CH_4	2.0×10^{-4}	1.1×10^{-4}
Krypton, Kr	1.1×10^{-4}	3.2×10^{-4}
Nitric oxide, NO	5.0×10^{-5}	1.7×10^{-6}
Xenon, Xe	8.7×10^{-6}	1.2×10^{-5}
Ozone, O_3: summer	7.0×10^{-6}	1.2×10^{-5}
winter	2.0×10^{-6}	3.3×10^{-6}

$$p = p_0 e^{-h/H} \tag{1.16}$$

where p_0 is the pressure at sea level and H is a constant approximately equal to 8 km. More specifically, $H = RT/Mg$, where M is the average molar mass of air and T is the temperature. This formula represents the outcome of the competition between the potential energy of the molecules in the gravitational field of the Earth and the stirring effects of thermal motion; it is derived on the basis of the Boltzmann distribution (Section F.5a). The barometric formula fits the observed pressure distribution quite well even for regions well above the troposphere (Fig. 1.10). It implies that the pressure of the air falls to half its sea-level value at $h = H \ln 2$, or 6 km.

Local variations of pressure, temperature, and composition in the troposphere are manifest as 'weather'. A small region of air is termed a *parcel*. First, we note that a parcel of warm air is less dense than the same parcel of cool air. As a parcel rises, it expands adiabatically (that is, without transfer of heat from its surroundings), so it cools. Cool air can absorb lower concentrations of water vapour than warm air, so the moisture forms clouds. Cloudy skies can therefore be associated with rising air and clear skies are often associated with descending air.

The motion of air in the upper altitudes may lead to an accumulation in some regions and a loss of molecules from other regions. The former result in the formation of regions of high pressure ('highs' or anticyclones) and the latter result in regions of low pressure ('lows', depressions, or cyclones). On a weather map, such as that shown in Fig. 1.11, the lines of constant pressure marked on it are called *isobars*. Elongated regions of high and low pressure are known, respectively, as *ridges* and *troughs*.

Horizontal pressure differentials result in the flow of air that we call *wind* (Fig. 1.12). Winds coming from the north in the Northern hemisphere and from the south in the Southern hemisphere are deflected towards the west as they migrate from a region where the Earth is rotating slowly (at the poles) to where it is rotating most rapidly (at the equator). Winds travel nearly parallel to the isobars, with low pressure to their left in the Northern hemisphere and to the right in the Southern hemisphere. At the surface, where wind speeds are lower, the winds tend to travel perpendicular to the isobars from high to low pressure. This differential motion results in a spiral outward flow of air clockwise in the Northern hemisphere around a high and an inward counterclockwise flow around a low.

The air lost from regions of high pressure is restored as an influx of air converges into the region and descends. As we have seen, descending air is associated with clear skies. It also becomes warmer by compression as it descends, so regions of high pressure are associated with high surface temperatures. In winter, the cold surface air may prevent the complete fall of air, and result in a temperature *inversion*, with a layer of warm air over a layer of cold air. Geographical conditions may also trap cool air, as in Los Angeles, and the photochemical pollutants we know as *smog* may be trapped under the warm layer.

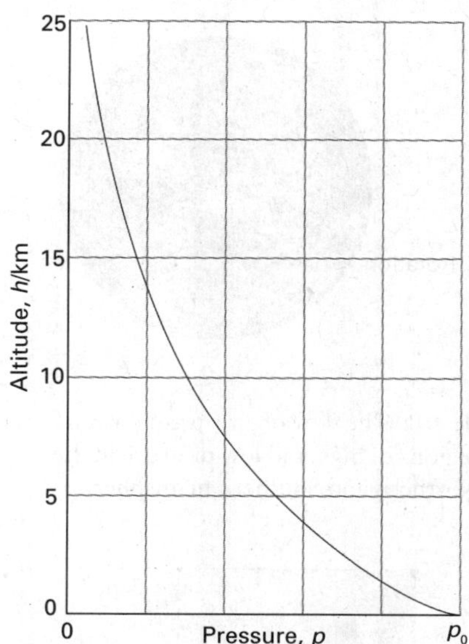

Fig. 1.10 The variation of atmospheric pressure with altitude, as predicted by the barometric formula and as suggested by the 'US Standard Atmosphere', which takes into account the variation of temperature with altitude.

interActivity How would the graph shown in the illustration change if the temperature variation with altitude were taken into account? Construct a graph allowing for a linear decrease in temperature with altitude.

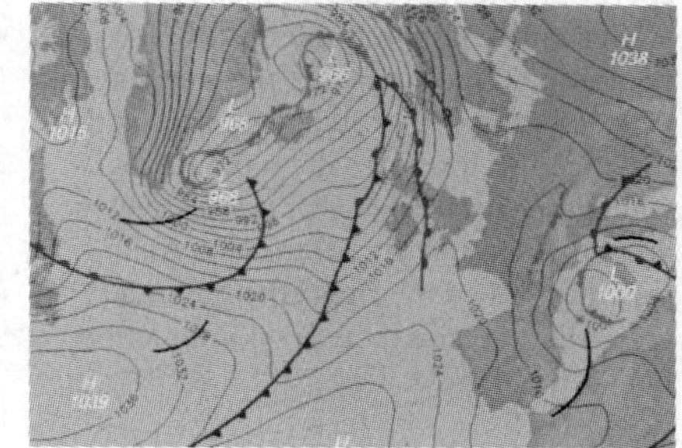

Fig. 1.11 A typical weather map; in this case, for the North Atlantic and neighbouring regions on 16 December 2008.

Real gases

Real gases do not obey the perfect gas law exactly except in the limit of $p \to 0$. Deviations from the law are particularly important at high pressures and low temperatures, especially when a gas is on the point of condensing to liquid.

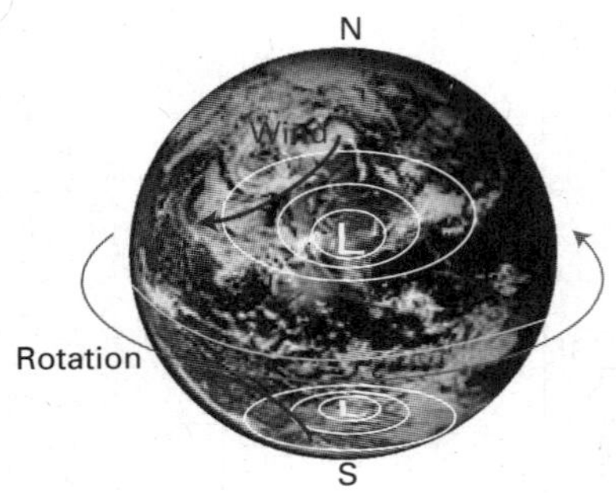

Fig. 1.12 The flow of air ('wind') around regions of high and low pressure in the Northern and Southern hemispheres.

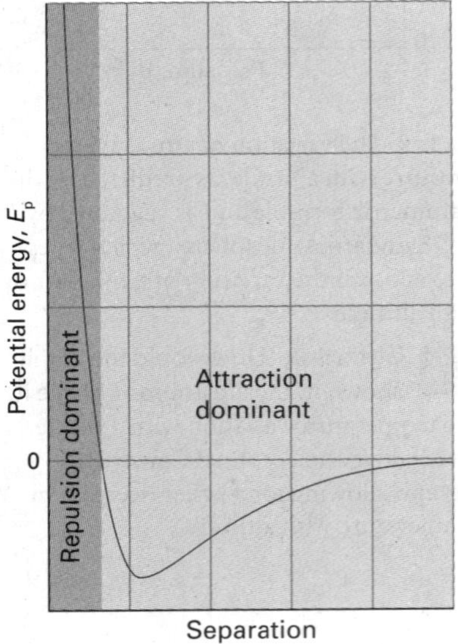

Fig. 1.13 The variation of the potential energy of two molecules on their separation. High positive potential energy (at very small separations) indicates that the interactions between them are strongly repulsive at these distances. At intermediate separations, where the potential energy is negative, the attractive interactions dominate. At large separations (on the right) the potential energy is zero and there is no interaction between the molecules.

1.3 Molecular interactions

Key points (a) The extent of deviations from perfect behaviour is summarized by introducing the compression factor. (b) The virial equation is an empirical extension of the perfect gas equation that summarizes the behaviour of real gases over a range of conditions. (c) The isotherms of a real gas introduce the concept of vapour pressure and critical behaviour. (d) A gas can be liquefied by pressure alone only if its temperature is at or below its critical temperature.

Real gases show deviations from the perfect gas law because molecules interact with one another. A point to keep in mind is that repulsive forces between molecules assist expansion and attractive forces assist compression.

Repulsive forces are significant only when molecules are almost in contact: they are short-range interactions, even on a scale measured in molecular diameters (Fig. 1.13). Because they are short-range interactions, repulsions can be expected to be important only when the average separation of the molecules is small. This is the case at high pressure, when many molecules occupy a small volume. On the other hand, attractive intermolecular forces have a relatively long range and are effective over several molecular diameters. They are important when the molecules are fairly close together but not necessarily touching (at the intermediate separations in Fig. 1.13). Attractive forces are ineffective when the molecules are far apart (well to the right in Fig. 1.13). Intermolecular forces are also important when the temperature is so low that the molecules travel with such low mean speeds that they can be captured by one another.

At low pressures, when the sample occupies a large volume, the molecules are so far apart for most of the time that the intermolecular forces play no significant role, and the gas behaves virtually perfectly. At moderate pressures, when the average separation of the molecules is only a few molecular diameters, the attractive forces dominate the repulsive forces. In this case, the gas can be expected to be more compressible than a perfect gas because the forces help to draw the molecules together. At high pressures, when the average separation of the molecules is small, the repulsive forces dominate and the gas can be expected to be less compressible because now the forces help to drive the molecules apart.

(a) The compression factor

The **compression factor**, Z, of a gas is the ratio of its measured molar volume, $V_m = V/n$, to the molar volume of a perfect gas, V_m^o, at the same pressure and temperature:

$$Z = \frac{V_m}{V_m^o} \qquad \text{Definition of compression factor} \qquad [1.17]$$

Because the molar volume of a perfect gas is equal to RT/p, an equivalent expression is $Z = pV_m/RT$, which we can write as

$$pV_m = RTZ \qquad (1.18)$$

Because, for a perfect gas $Z = 1$ under all conditions, deviation of Z from 1 is a measure of departure from perfect behaviour.

Some experimental values of Z are plotted in Fig. 1.14. At very low pressures, all the gases shown have $Z \approx 1$ and behave nearly perfectly. At high pressures, all the gases have $Z > 1$, signifying that they have a larger molar volume than a perfect gas. Repulsive forces are now dominant. At intermediate pressures, most gases have $Z < 1$, indicating that the attractive forces are reducing the molar volume relative to that of a perfect gas.

(b) Virial coefficients

Figure 1.15 shows the experimental isotherms for carbon dioxide. At large molar volumes and high temperatures the real-gas isotherms do not differ greatly from perfect-gas isotherms. The small differences suggest that the perfect gas law is in fact the first term in an expression of the form

$$pV_m = RT(1 + B'p + C'p^2 + \cdots) \qquad (1.19a)$$

This expression is an example of a common procedure in physical chemistry, in which a simple law that is known to be a good first approximation (in this case $pV = nRT$) is treated as the first term in a series in powers of a variable (in this case p). A more convenient expansion for many applications is

$$pV_m = RT\left(1 + \frac{B}{V_m} + \frac{C}{V_m^2} + \cdots\right) \qquad \text{Virial equation of state} \qquad (1.19b)$$

These two expressions are two versions of the **virial equation of state.**[1] By comparing the expression with eqn 1.18 we see that the term in parentheses in eqn 1.19b is just the compression factor, Z.

The coefficients B, C, . . . , which depend on the temperature, are the second, third, . . . **virial coefficients** (Table 1.4); the first virial coefficient is 1. The third virial coefficient, C, is usually less important than the second coefficient, B, in the sense that at typical molar volumes $C/V_m^2 \ll B/V_m$. The values of the virial coefficients of a gas are determined from measurements of its compression factor.

An important point is that, although the equation of state of a real gas may coincide with the perfect gas law as $p \to 0$, not all its properties necessarily coincide with those of a perfect gas in that limit. Consider, for example, the value of dZ/dp, the slope of the graph of compression factor against pressure. For a perfect gas $dZ/dp = 0$ (because $Z = 1$ at all pressures), but for a real gas from eqn 1.19a we obtain

$$\frac{dZ}{dp} = B' + 2pC' + \cdots \to B' \quad \text{as} \quad p \to 0 \qquad (1.20a)$$

However, B' is not necessarily zero, so the slope of Z with respect to p does not necessarily approach 0 (the perfect gas value), as we can see in Fig. 1.14. Because several physical properties of gases depend on derivatives, the properties of real gases do not always coincide with the perfect gas values at low pressures. By a similar argument

$$\frac{dZ}{d\left(\dfrac{1}{V_m}\right)} \to B \quad \text{as} \quad V_m \to \infty \qquad (1.20b)$$

Because the virial coefficients depend on the temperature, there may be a temperature at which $Z \to 1$ with zero slope at low pressure or high molar volume (Fig. 1.16). At this temperature, which is called the **Boyle temperature**, T_B, the properties of the real gas do coincide with those of a perfect gas as $p \to 0$. According to eqn 1.20a, Z has zero slope as $p \to 0$ if $B = 0$, so we can conclude that $B = 0$ at the Boyle temperature. It then follows from eqn 1.18 that $pV_m \approx RT_B$ over a more extended range of pressures than at other temperatures because the first term after 1 (that is, B/V_m) in the virial equation is zero and C/V_m^2 and higher terms are negligibly small. For helium $T_B = 22.64$ K; for air $T_B = 346.8$ K; more values are given in Table 1.5.

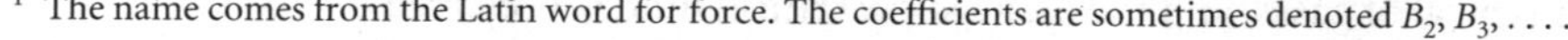

[1] The name comes from the Latin word for force. The coefficients are sometimes denoted B_2, B_3, ...

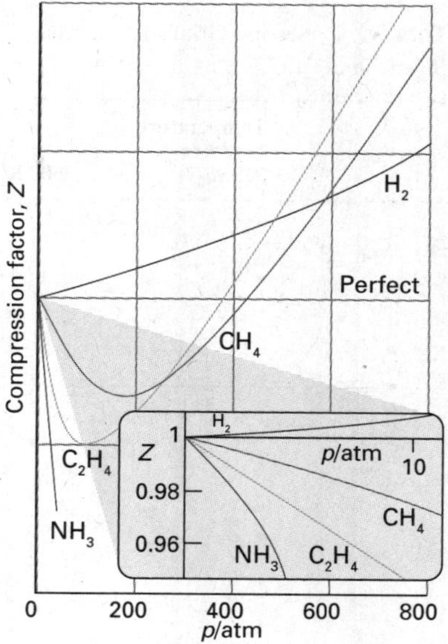

Fig. 1.14 The variation of the compression factor, Z, with pressure for several gases at 0°C. A perfect gas has $Z = 1$ at all pressures. Notice that, although the curves approach 1 as $p \to 0$, they do so with different slopes.

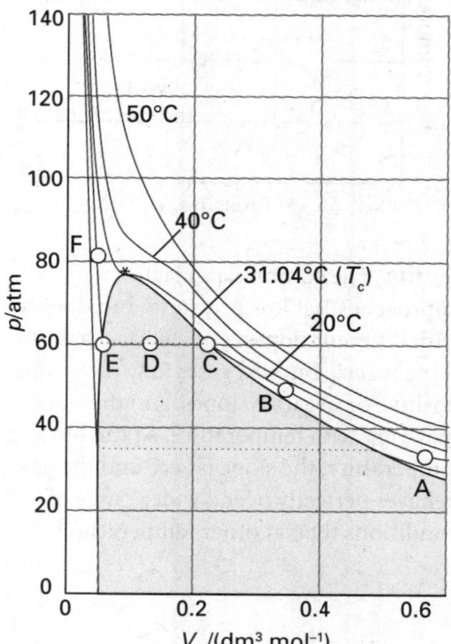

Fig. 1.15 Experimental isotherms of carbon dioxide at several temperatures. The 'critical isotherm', the isotherm at the critical temperature, is at 31.04°C. The critical point is marked with a star.

Table 1.4* Second virial coefficients, $B/(cm^3\ mol^{-1})$

	Temperature	
	273 K	600 K
Ar	−21.7	11.9
CO_2	−142	−12.4
N_2	−10.5	21.7
Xe	−153.7	−19.6

* More values are given in the *Data section*.

Table 1.5* Critical constants of gases

	p_c/atm	$V_c/(cm^3\ mol^{-1})$	T_c/K	Z_c	T_B/K
Ar	48.0	75.3	150.7	0.292	411.5
CO_2	72.9	94.0	304.2	0.274	714.8
He	2.26	57.8	5.2	0.305	22.64
O_2	50.14	78.0	154.8	0.308	405.9

* More values are given in the *Data section*.

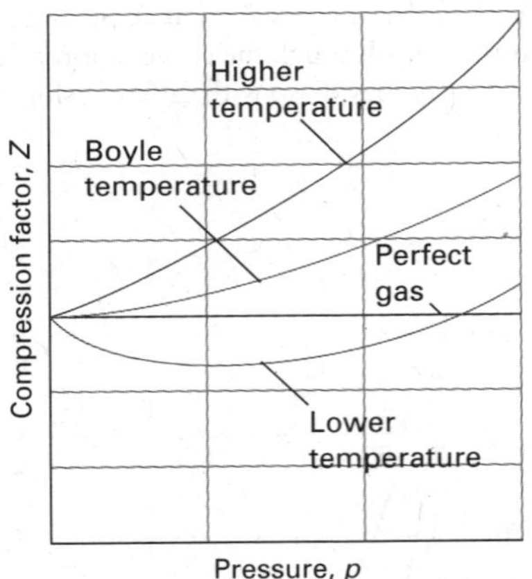

Fig. 1.16 The compression factor, Z, approaches 1 at low pressures, but does so with different slopes. For a perfect gas, the slope is zero, but real gases may have either positive or negative slopes, and the slope may vary with temperature. At the Boyle temperature, the slope is zero and the gas behaves perfectly over a wider range of conditions than at other temperatures.

(c) Condensation

Now consider what happens when we compress (reduce the volume of) a sample of gas initially in the state marked A in Fig. 1.15 at constant temperature by pushing in a piston. Near A, the pressure of the gas rises in approximate agreement with Boyle's law. Serious deviations from that law begin to appear when the volume has been reduced to B.

At C (which corresponds to about 60 atm for carbon dioxide), all similarity to perfect behaviour is lost, for suddenly the piston slides in without any further rise in pressure: this stage is represented by the horizontal line CDE. Examination of the contents of the vessel shows that just to the left of C a liquid appears, and there are two phases separated by a sharply defined surface. As the volume is decreased from C through D to E, the amount of liquid increases. There is no additional resistance to the piston because the gas can respond by condensing. The pressure corresponding to the line CDE, when both liquid and vapour are present in equilibrium, is called the **vapour pressure** of the liquid at the temperature of the experiment.

At E, the sample is entirely liquid and the piston rests on its surface. Any further reduction of volume requires the exertion of considerable pressure, as is indicated by the sharply rising line to the left of E. Even a small reduction of volume from E to F requires a great increase in pressure.

(d) Critical constants

The isotherm at the temperature T_c (304.19 K, or 31.04°C for CO_2) plays a special role in the theory of the states of matter. An isotherm slightly below T_c behaves as we have already described: at a certain pressure, a liquid condenses from the gas and is distinguishable from it by the presence of a visible surface. If, however, the compression takes place at T_c itself, then a surface separating two phases does not appear and the volumes at each end of the horizontal part of the isotherm have merged to a single point, the **critical point** of the gas. The temperature, pressure, and molar volume at the critical point are called, respectively, the **critical temperature,** T_c, **critical pressure,** p_c, and **critical molar volume,** V_c, of the substance. Collectively, p_c, V_c, and T_c are the **critical constants** of a substance (Table 1.5).

At and above T_c, the sample has a single phase that occupies the entire volume of the container. Such a phase is, by definition, a gas. Hence, the liquid phase of a substance does not form above the critical temperature. The critical temperature of oxygen, for instance, signifies that it is impossible to produce liquid oxygen by compression alone if its temperature is greater than 155 K: to liquefy oxygen—to obtain a fluid phase that does not occupy the entire volume—the temperature must first be lowered to below 155 K, and then the gas compressed isothermally. The single phase that fills the entire volume when $T > T_c$ may be much denser that we normally consider typical of gases, and the name **supercritical fluid** is preferred.

1.4 The van der Waals equation

Key points **(a) The van der Waals equation is a model equation of state for a real gas expressed in terms of two parameters, one corresponding to molecular attractions and the other to molecular repulsions. (b) The van der Waals equation captures the general features of the behaviour of real gases, including their critical behaviour. (c) The properties of real gases are coordinated by expressing their equations of state in terms of reduced variables.**

We can draw conclusions from the virial equations of state only by inserting specific values of the coefficients. It is often useful to have a broader, if less precise, view of all gases. Therefore, we introduce the approximate equation of state suggested by J.D. van der Waals in 1873. This equation is an excellent example of an expression that can be obtained by thinking scientifically about a mathematically complicated but physically simple problem; that is, it is a good example of 'model building'.

(a) Formulation of the equation

The van der Waals equation is

$$p=\frac{nRT}{V-nb}-a\frac{n^2}{V^2} \qquad \text{van der Waals equation of state} \qquad (1.21a)$$

and a derivation is given in the following *Justification*. The equation is often written in terms of the molar volume $V_m = V/n$ as

$$p=\frac{RT}{V_m-b}-\frac{a}{V_m^2} \qquad (1.21b)$$

The constants a and b are called the **van der Waals coefficients.** As can be understood from the following *Justification*, a represents the strength of attractive interactions and b that of the repulsive interactions between the molecules. They are characteristic of each gas but independent of the temperature (Table 1.6). Although a and b are not precisely defined molecular properties, they correlate with physical properties such as critical temperature, vapor pressure, and enthalpy of vaporization that reflect the strength of intermolecular interactions. Correlations have also been sought where intermolecular forces might play a role. For example, the potencies of certain general anaesthetics show a correlation in the sense that a higher activity is observed with lower values of a (Fig. 1.17).

Table 1.6* van der Waals coefficients

	a/(atm dm^6 mol^{-2})	b/(10^{-2} dm^3 mol^{-1})
Ar	1.337	3.20
CO_2	3.610	4.29
He	0.0341	2.38
Xe	4.137	5.16

* More values are given in the *Data section*.

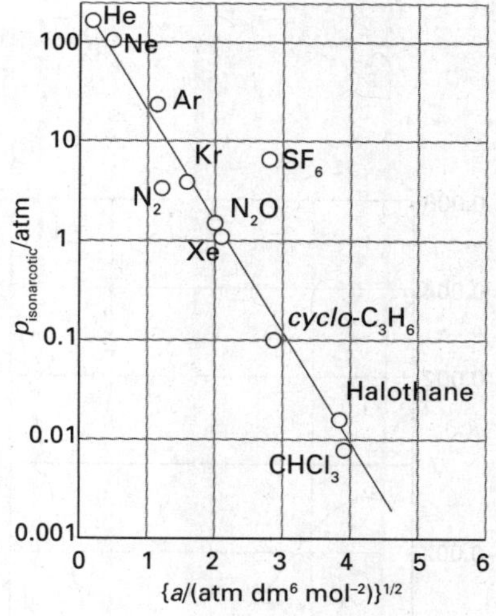

Fig. 1.17 The correlation of the effectiveness of a gas as an anaesthetic and the van der Waals parameter a. (Based on R.J. Wulf and R.M. Featherstone, *Anesthesiology*, 18, 97 (1957).) The isonarcotic pressure is the pressure required to bring about the same degree of anaesthesia.

Justification 1.1 *The van der Waals equation of state*

The repulsive interactions between molecules are taken into account by supposing that they cause the molecules to behave as small but impenetrable spheres. The non-zero volume of the molecules implies that instead of moving in a volume V they are restricted to a smaller volume $V - nb$, where nb is approximately the total volume taken up by the molecules themselves. This argument suggests that the perfect gas law $p = nRT/V$ should be replaced by

$$p=\frac{nRT}{V-nb}$$

when repulsions are significant. To calculate the excluded volume we note that the closest distance of two hard-sphere molecules of radius r, and volume $V_{molecule} = \frac{4}{3}\pi r^3$, is $2r$, so the volume excluded is $\frac{4}{3}\pi(2r)^3$, or $8V_{molecule}$. The volume excluded per molecule is one-half this volume, or $4V_{molecule}$, so $b \approx 4V_{molecule}N_A$.

The pressure depends on both the frequency of collisions with the walls and the force of each collision. Both the frequency of the collisions and their force are reduced by the attractive interactions, which act with a strength proportional to the molar concentration, n/V, of molecules in the sample. Therefore, because both the frequency and the force of the collisions are reduced by the attractive interactions, the pressure is reduced in proportion to the square of this concentration. If the reduction of pressure is written as $-a(n/V)^2$, where a is a positive constant characteristic of each gas, the combined effect of the repulsive and attractive forces is the van der Waals equation of state as expressed in eqn 1.21.

In this *Justification* we have built the van der Waals equation using vague arguments about the volumes of molecules and the effects of forces. The equation can be derived in other ways, but the present method has the advantage that it shows how to derive the form of an equation from general ideas. The derivation also has the advantage of keeping imprecise the significance of the coefficients a and b: they are much better regarded as empirical parameters that represent attractions and repulsions, respectively, rather than as precisely defined molecular properties.

Example 1.4 *Using the van der Waals equation to estimate a molar volume*

Estimate the molar volume of CO_2 at 500 K and 100 atm by treating it as a van der Waals gas.

Method We need to find an expression for the molar volume by solving the van der Waals equation, eqn 1.21b. To do so, we multiply both sides of the equation by $(V_m - b)V_m^2$, to obtain

$$(V_m - b)V_m^2 p = RTV_m^2 - (V_m - b)a$$

Then, after division by p, collect powers of V_m to obtain

$$V_m^3 - \left(b + \frac{RT}{p}\right)V_m^2 + \left(\frac{a}{p}\right)V_m - \frac{ab}{p} = 0$$

Although closed expressions for the roots of a cubic equation can be given, they are very complicated. Unless analytical solutions are essential, it is usually more expedient to solve such equations with commercial software; graphing calculators can also be used to help identify the acceptable root.

Answer According to Table 1.6, $a = 3.610$ dm^6 atm mol^{-2} and $b = 4.29 \times 10^{-2}$ dm^3 mol^{-1}. Under the stated conditions, $RT/p = 0.410$ dm^3 mol^{-1}. The coefficients in the equation for V_m are therefore

$$b + RT/p = 0.453 \text{ dm}^3 \text{ mol}^{-1}$$
$$a/p = 3.61 \times 10^{-2} \text{ (dm}^3 \text{ mol}^{-1})^2$$
$$ab/p = 1.55 \times 10^{-3} \text{ (dm}^3 \text{ mol}^{-1})^3$$

Therefore, on writing $x = V_m/(\text{dm}^3 \text{ mol}^{-1})$, the equation to solve is

$$x^3 - 0.453x^2 + (3.61 \times 10^{-2})x - (1.55 \times 10^{-3}) = 0$$

The acceptable root is $x = 0.366$ (Fig. 1.18), which implies that $V_m = 0.366$ dm^3 mol^{-1}. For a perfect gas under these conditions, the molar volume is 0.410 dm^3 mol^{-1}.

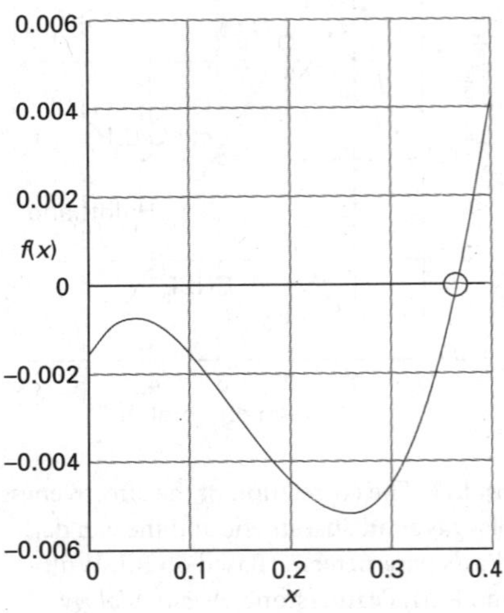

Fig. 1.18 The graphical solution of the cubic equation for V in Example 1.4.

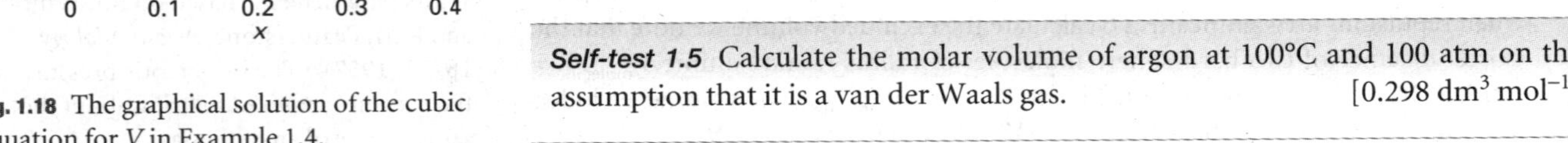

Self-test 1.5 Calculate the molar volume of argon at 100°C and 100 atm on the assumption that it is a van der Waals gas. [0.298 dm^3 mol^{-1}]

Table 1.7 Selected equations of state

	Equation	Reduced form*	Critical constants		
			p_c	V_c	T_c
Perfect gas	$p=\frac{RT}{V_m}$				
van der Waals	$p=\frac{RT}{V_m-b}-\frac{a}{V_m^2}$	$p=\frac{8T_r}{3V_r-1}-\frac{3}{V_r^2}$	$\frac{a}{27b^2}$	$3b$	$\frac{8a}{27bR}$
Berthelot	$p=\frac{RT}{V_m-b}-\frac{a}{TV_m^2}$	$p=\frac{8T_r}{3V_r-1}-\frac{3}{T_rV_r^2}$	$\frac{1}{12}\left(\frac{2aR}{3b^3}\right)^{1/2}$	$3b$	$\frac{2}{3}\left(\frac{2a}{3bR}\right)^{1/2}$
Dieterici	$p=\frac{RTe^{-a/RTV_m}}{V_m-b}$	$p=\frac{e^2T_re^{-2/T_rV_r}}{2V_r-1}$	$\frac{a}{4e^2b^2}$	$2b$	$\frac{a}{4bR}$
Virial	$p=\frac{RT}{V_m}\left\{1+\frac{B(T)}{V_m}+\frac{C(T)}{V_m^2}+\cdots\right\}$				

* Reduced variables are defined in Section 1.4c.

(b) The features of the equation

We now examine to what extent the van der Waals equation predicts the behaviour of real gases. It is too optimistic to expect a single, simple expression to be the true equation of state of all substances, and accurate work on gases must resort to the virial equation, use tabulated values of the coefficients at various temperatures, and analyse the systems numerically. The advantage of the van der Waals equation, however, is that it is analytical (that is, expressed symbolically) and allows us to draw some general conclusions about real gases. When the equation fails we must use one of the other equations of state that have been proposed (some are listed in Table 1.7), invent a new one, or go back to the virial equation.

That having been said, we can begin to judge the reliability of the equation by comparing the isotherms it predicts with the experimental isotherms in Fig. 1.15. Some calculated isotherms are shown in Fig. 1.19 and Fig. 1.20. Apart from the oscillations below the critical temperature, they do resemble experimental isotherms quite well. The oscillations, the **van der Waals loops**, are unrealistic because they suggest that under some conditions an increase of pressure results in an increase of volume. Therefore they are replaced by horizontal lines drawn so the loops define equal areas above and below the lines: this procedure is called the **Maxwell construction** (3). The van der Waals coefficients, such as those in Table 1.6, are found by fitting the calculated curves to the experimental curves.

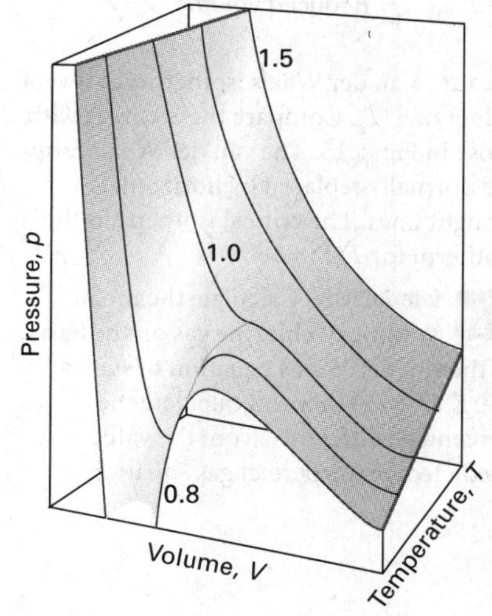

Fig. 1.19 The surface of possible states allowed by the van der Waals equation. Compare this surface with that shown in Fig. 1.8.

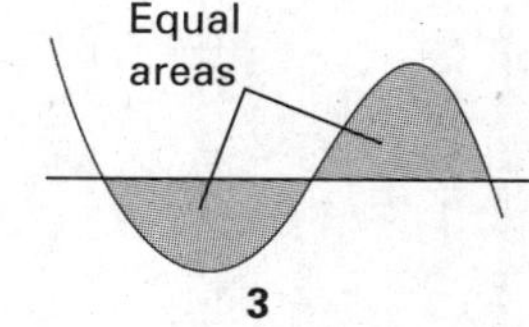

The principal features of the van der Waals equation can be summarized as follows.

(1) Perfect gas isotherms are obtained at high temperatures and large molar volumes.

When the temperature is high, RT may be so large that the first term in eqn 1.21b greatly exceeds the second. Furthermore, if the molar volume is large in the sense $V_m \gg b$, then the denominator $V_m - b \approx V_m$. Under these conditions, the equation reduces to $p = RT/V_m$, the perfect gas equation.

(2) Liquids and gases coexist when cohesive and dispersing effects are in balance.

The van der Waals loops occur when both terms in eqn 1.21b have similar magnitudes. The first term arises from the kinetic energy of the molecules and their repulsive interactions; the second represents the effect of the attractive interactions.

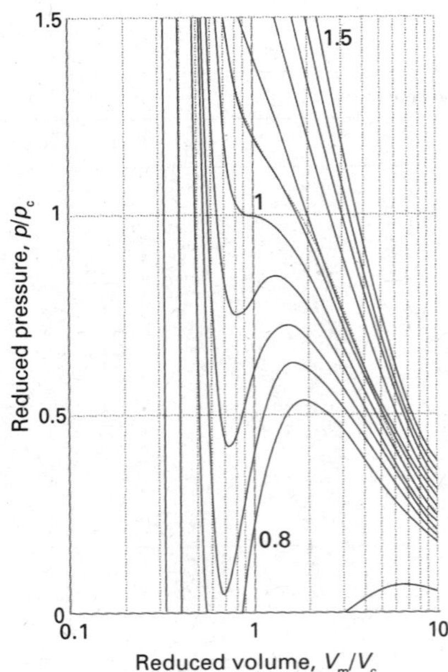

Fig. 1.20 Van der Waals isotherms at several values of T/T_c. Compare these curves with those in Fig. 1.15. The van der Waals loops are normally replaced by horizontal straight lines. The critical isotherm is the isotherm for $T/T_c = 1$.

interActivity Calculate the molar volume of chlorine gas on the basis of the van der Waals equation of state at 250 K and 150 kPa and calculate the percentage difference from the value predicted by the perfect gas equation.

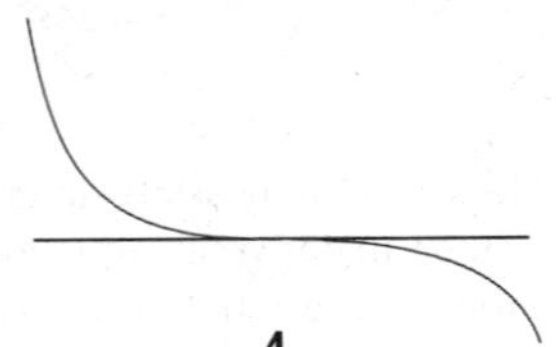

4

(3) The critical constants are related to the van der Waals coefficients.

For $T < T_c$, the calculated isotherms oscillate, and each one passes through a minimum followed by a maximum. These extrema converge as $T \to T_c$ and coincide at $T = T_c$; at the critical point the curve has a flat inflexion (**4**). From the properties of curves, we know that an inflexion of this type occurs when both the first and second derivatives are zero. Hence, we can find the critical constants by calculating these derivatives and setting them equal to zero

$$\frac{dp}{dV_m} = -\frac{RT}{(V_m - b)^2} + \frac{2a}{V_m^3} = 0$$

$$\frac{d^2p}{dV_m^2} = \frac{2RT}{(V_m - b)^3} - \frac{6a}{V_m^4} = 0$$

at the critical point. The solutions of these two equations (and using eqn 1.21b to calculate p_c from V_c and T_c) are

$$V_c = 3b \qquad p_c = \frac{a}{27b^2} \qquad T_c = \frac{8a}{27Rb} \tag{1.22}$$

These relations provide an alternative route to the determination of a and b from the values of the critical constants. They can be tested by noting that the **critical compression factor**, Z_c, is predicted to be equal to

$$Z_c = \frac{p_c V_c}{RT_c} = \frac{3}{8} \tag{1.23}$$

for all gases that are described by the van der Waals equation near the critical point. We see from Table 1.5 that, although $Z_c < \frac{3}{8} = 0.375$, it is approximately constant (at 0.3) and the discrepancy is reasonably small.

(c) The principle of corresponding states

An important general technique in science for comparing the properties of objects is to choose a related fundamental property of the same kind and to set up a relative scale on that basis. We have seen that the critical constants are characteristic properties of gases, so it may be that a scale can be set up by using them as yardsticks. We therefore introduce the dimensionless **reduced variables** of a gas by dividing the actual variable by the corresponding critical constant:

$$V_r = \frac{V_m}{V_c} \qquad p_r = \frac{p}{p_c} \qquad T_r = \frac{T}{T_c} \qquad \text{Definition of reduced variables} \tag{1.24}$$

If the reduced pressure of a gas is given, we can easily calculate its actual pressure by using $p = p_r p_c$, and likewise for the volume and temperature. van der Waals, who first tried this procedure, hoped that gases confined to the same reduced volume, V_r, at the same reduced temperature, T_r, would exert the same reduced pressure, p_r. The hope was largely fulfilled (Fig. 1.21). The illustration shows the dependence of the compression factor on the reduced pressure for a variety of gases at various reduced temperatures. The success of the procedure is strikingly clear: compare this graph with Fig. 1.14, where similar data are plotted without using reduced variables. The observation that real gases at the same reduced volume and reduced temperature exert the same reduced pressure is called the **principle of corresponding states**. The principle is only an approximation. It works best for gases composed of spherical molecules; it fails, sometimes badly, when the molecules are non-spherical or polar.

The van der Waals equation sheds some light on the principle. First, we express eqn 1.21b in terms of the reduced variables, which gives

$$p_r p_c = \frac{RT_rT_c}{V_rV_c - b} - \frac{a}{V_r^2 V_c^2}$$

Then we express the critical constants in terms of a and b by using eqn 1.22:

$$\frac{ap_r}{27b^2} = \frac{8aT_r}{27b(3bV_r - b)} - \frac{a}{9b^2V_r^2}$$

which can be reorganized into

$$p_r = \frac{8T_r}{3V_r - 1} - \frac{3}{V_r^2} \qquad (1.25)$$

This equation has the same form as the original, but the coefficients a and b, which differ from gas to gas, have disappeared. It follows that, if the isotherms are plotted in terms of the reduced variables (as we did in fact in Fig. 1.20 without drawing attention to the fact), then the same curves are obtained whatever the gas. This is precisely the content of the principle of corresponding states, so the van der Waals equation is compatible with it.

Looking for too much significance in this apparent triumph is mistaken, because other equations of state also accommodate the principle (Table 1.7). In fact, all we need are two parameters playing the roles of a and b, for then the equation can always be manipulated into reduced form. The observation that real gases obey the principle approximately amounts to saying that the effects of the attractive and repulsive interactions can each be approximated in terms of a single parameter. The importance of the principle is then not so much its theoretical interpretation but the way in which it enables the properties of a range of gases to be coordinated on to a single diagram (for example, Fig. 1.21 instead of Fig. 1.14).

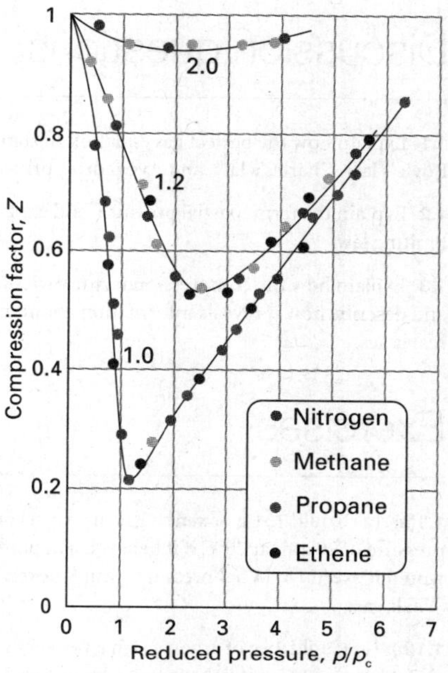

Fig. 1.21 The compression factors of four gases plotted using reduced variables. The curves are labelled with the reduced temperature $T_r = T/T_c$. The use of reduced variables organizes the data on to single curves.

interActivity Is there a set of conditions at which the compression factor of a van der Waals gas passes through a minimum? If so, how do the location and value of the minimum value of Z depend on the coefficients a and b?

Checklist of key equations

Property	Equation	Comment
Equation of state	$p = f(n,V,T)$	
Perfect gas law	$pV = nRT$	Valid for real gases in the limit $p \to 0$
Relation between temperature scales	$T/\text{K} = \theta/°\text{C} + 273.15$	273.15 is exact temperature scales
Partial pressure	$p_J = x_J p$	Valid for all gases
Virial equation of state	$pV_m = RT(1 + B/V_m + C/V_m^2 + \cdots)$	B, C depend on temperature
van der Waals equation of state	$p = nRT/(V - nb) - a(n/V)^2$	a parametrizes attractions; b parametrizes repulsions

➔ For a chart of the relations between principal equations, see the Road map section of the Resource section.

Discussion questions

1.1 Explain how the perfect gas equation of state arises by combination of Boyle's law, Charles's law, and Avogadro's principle.

1.2 Explain the term 'partial pressure' and explain why Dalton's law is a limiting law.

1.3 Explain how the compression factor varies with pressure and temperature and describe how it reveals information about intermolecular interactions in real gases.

1.4 What is the significance of the critical constants?

1.5 Describe the formulation of the van der Waals equation and suggest a rationale for one other equation of state in Table 1.7.

1.6 Explain how the van der Waals equation accounts for critical behaviour.

Exercises

1.1(a) (a) Could 131 g of xenon gas in a vessel of volume 1.0 dm^3 exert a pressure of 20 atm at 25°C if it behaved as a perfect gas? If not, what pressure would it exert? (b) What pressure would it exert if it behaved as a van der Waals gas?

1.1(b) (a) Could 25 g of argon gas in a vessel of volume 1.5 dm^3 exert a pressure of 2.0 bar at 30°C if it behaved as a perfect gas? If not, what pressure would it exert? (b) What pressure would it exert if it behaved as a van der Waals gas?

1.2(a) A perfect gas undergoes isothermal compression, which reduces its volume by 2.20 dm^3. The final pressure and volume of the gas are 5.04 bar and 4.65 dm^3, respectively. Calculate the original pressure of the gas in (a) bar, (b) atm.

1.2(b) A perfect gas undergoes isothermal compression, which reduces its volume by 1.80 dm^3. The final pressure and volume of the gas are 1.97 bar and 2.14 dm^3, respectively. Calculate the original pressure of the gas in (a) bar, (b) Torr.

1.3(a) A car tyre (i.e. an automobile tire) was inflated to a pressure of 24 lb in^{-2} (1.00 atm = 14.7 lb in^{-2}) on a winter's day when the temperature was −5°C. What pressure will be found, assuming no leaks have occurred and that the volume is constant, on a subsequent summer's day when the temperature is 35°C? What complications should be taken into account in practice?

1.3(b) A sample of hydrogen gas was found to have a pressure of 125 kPa when the temperature was 23°C. What can its pressure be expected to be when the temperature is 11°C?

1.4(a) A sample of 255 mg of neon occupies 3.00 dm^3 at 122 K. Use the perfect gas law to calculate the pressure of the gas.

1.4(b) A homeowner uses 4.00×10^3 m^3 of natural gas in a year to heat a home. Assume that natural gas is all methane, CH_4, and that methane is a perfect gas for the conditions of this problem, which are 1.00 atm and 20°C. What is the mass of gas used?

1.5(a) A diving bell has an air space of 3.0 m^3 when on the deck of a boat. What is the volume of the air space when the bell has been lowered to a depth of 50 m? Take the mean density of sea water to be 1.025 g cm^{-3} and assume that the temperature is the same as on the surface.

1.5(b) What pressure difference must be generated across the length of a 15 cm vertical drinking straw in order to drink a water-like liquid of density 1.0 g cm^{-3}?

1.6(a) A manometer consists of a U-shaped tube containing a liquid. One side is connected to the apparatus and the other is open to the atmosphere. The pressure inside the apparatus is then determined from the difference in heights of the liquid. Suppose the liquid is water, the external pressure is 770 Torr, and the open side is 10.0 cm lower than the side connected to the apparatus. What is the pressure in the apparatus? (The density of water at 25°C is 0.997 07 g cm^{-3}.)

1.6(b) A manometer like that described in Exercise 1.6a contained mercury in place of water. Suppose the external pressure is 760 Torr, and the open side is 10.0 cm higher than the side connected to the apparatus. What is the pressure in the apparatus? (The density of mercury at 25°C is 13.55 g cm^{-3}.)

1.7(a) In an attempt to determine an accurate value of the gas constant, R, a student heated a container of volume 20.000 dm^3 filled with 0.251 32 g of helium gas to 500°C and measured the pressure as 206.402 cm of water in a manometer at 25°C. Calculate the value of R from these data. (The density of water at 25°C is 0.997 07 g cm^{-3}; the construction of a manometer is described in Exercise 1.6a.)

1.7(b) The following data have been obtained for oxygen gas at 273.15 K. Calculate the best value of the gas constant R from them and the best value of the molar mass of O_2.

p/atm	0.750 000	0.500 000	0.250 000
$V_m/(dm^3\ mol^{-1})$	29.8649	44.8090	89.6384

1.8(a) At 500°C and 93.2 kPa, the mass density of sulfur vapour is 3.710 kg m^{-3}. What is the molecular formula of sulfur under these conditions?

1.8(b) At 100°C and 16.0 kPa, the mass density of phosphorus vapour is 0.6388 kg m^{-3}. What is the molecular formula of phosphorus under these conditions?

1.9(a) Calculate the mass of water vapour present in a room of volume 400 m^3 that contains air at 27°C on a day when the relative humidity is 60 per cent.

1.9(b) Calculate the mass of water vapour present in a room of volume 250 m^3 that contains air at 23°C on a day when the relative humidity is 53 per cent.

1.10(a) Given that the density of air at 0.987 bar and 27°C is 1.146 kg m^{-3}, calculate the mole fraction and partial pressure of nitrogen and oxygen assuming that (a) air consists only of these two gases, (b) air also contains 1.0 mole per cent Ar.

1.10(b) A gas mixture consists of 320 mg of methane, 175 mg of argon, and 225 mg of neon. The partial pressure of neon at 300 K is 8.87 kPa. Calculate (a) the volume and (b) the total pressure of the mixture.

1.11(a) The density of a gaseous compound was found to be 1.23 kg m^{-3} at 330 K and 20 kPa. What is the molar mass of the compound?

1.11(b) In an experiment to measure the molar mass of a gas, 250 cm^3 of the gas was confined in a glass vessel. The pressure was 152 Torr at 298 K and, after correcting for buoyancy effects, the mass of the gas was 33.5 mg. What is the molar mass of the gas?

1.12(a) The densities of air at −85°C, 0°C, and 100°C are 1.877 g dm^{-3}, 1.294 g dm^{-3}, and 0.946 g dm^{-3}, respectively. From these data, and assuming that air obeys Charles's law, determine a value for the absolute zero of temperature in degrees Celsius.

1.12(b) A certain sample of a gas has a volume of 20.00 dm^3 at 0°C and 1.000 atm. A plot of the experimental data of its volume against the Celsius temperature, θ, at constant p, gives a straight line of slope 0.0741 dm^3 $(°C)^{-1}$. From these data alone (without making use of the perfect gas law), determine the absolute zero of temperature in degrees Celsius.

1.13(a) Calculate the pressure exerted by 1.0 mol C_2H_6 behaving as (a) a perfect gas, (b) a van der Waals gas when it is confined under the following conditions: (i) at 273.15 K in 22.414 dm^3, (ii) at 1000 K in 100 cm^3. Use the data in Table 1.6.

1.13(b) Calculate the pressure exerted by 1.0 mol H_2S behaving as (a) a perfect gas, (b) a van der Waals gas when it is confined under the following conditions: (i) at 273.15 K in 22.414 dm^3, (ii) at 500 K in 150 cm^3. Use the data in Table 1.6.

1.14(a) Express the van der Waals parameters $a = 0.751$ atm dm^6 mol^{-2} and $b = 0.0226$ dm^3 mol^{-1} in SI base units.

1.14(b) Express the van der Waals parameters $a = 1.32$ atm dm^6 mol^{-2} and $b = 0.0436$ dm^3 mol^{-1} in SI base units.

1.15(a) A gas at 250 K and 15 atm has a molar volume 12 per cent smaller than that calculated from the perfect gas law. Calculate (a) the compression factor under these conditions and (b) the molar volume of the gas. Which are dominating in the sample, the attractive or the repulsive forces?

1.15(b) A gas at 350 K and 12 atm has a molar volume 12 per cent larger than that calculated from the perfect gas law. Calculate (a) the compression factor under these conditions and (b) the molar volume of the gas. Which are dominating in the sample, the attractive or the repulsive forces?

1.16(a) In an industrial process, nitrogen is heated to 500 K at a constant volume of 1.000 m^3. The gas enters the container at 300 K and 100 atm. The mass of the gas is 92.4 kg. Use the van der Waals equation to determine the approximate pressure of the gas at its working temperature of 500 K. For nitrogen, $a = 1.352$ dm^6 atm mol^{-2}, $b = 0.0387$ dm^3 mol^{-1}.

1.16(b) Cylinders of compressed gas are typically filled to a pressure of 200 bar. For oxygen, what would be the molar volume at this pressure and 25°C based on (a) the perfect gas equation, (b) the van der Waals equation. For oxygen, $a = 1.364$ dm^6 atm mol^{-2}, $b = 3.19 \times 10^{-2}$ dm^3 mol^{-1}.

1.17(a) Suppose that 10.0 mol C_2H_6(g) is confined to 4.860 dm^3 at 27°C. Predict the pressure exerted by the ethane from (a) the perfect gas and (b) the van der Waals equations of state. Calculate the compression factor based on these calculations. For ethane, $a = 5.507$ dm^6 atm mol^{-2}, $b = 0.0651$ dm^3 mol^{-1}.

1.17(b) At 300 K and 20 atm, the compression factor of a gas is 0.86. Calculate (a) the volume occupied by 8.2 mmol of the gas under these conditions and (b) an approximate value of the second virial coefficient B at 300 K.

1.18(a) A vessel of volume 22.4 dm^3 contains 2.0 mol H_2 and 1.0 mol N_2 at 273.15 K. Calculate (a) the mole fractions of each component, (b) their partial pressures, and (c) their total pressure.

1.18(b) A vessel of volume 22.4 dm^3 contains 1.5 mol H_2 and 2.5 mol N_2 at 273.15 K. Calculate (a) the mole fractions of each component, (b) their partial pressures, and (c) their total pressure.

1.19(a) The critical constants of methane are $p_c = 45.6$ atm, $V_c = 98.7$ cm^3 mol^{-1}, and $T_c = 190.6$ K. Calculate the van der Waals parameters of the gas and estimate the radius of the molecules.

1.19(b) The critical constants of ethane are $p_c = 48.20$ atm, $V_c = 148$ cm^3 mol^{-1}, and $T_c = 305.4$ K. Calculate the van der Waals parameters of the gas and estimate the radius of the molecules.

1.20(a) Use the van der Waals parameters for chlorine to calculate approximate values of (a) the Boyle temperature of chlorine and (b) the radius of a Cl_2 molecule regarded as a sphere.

1.20(b) Use the van der Waals parameters for hydrogen sulfide (Table 1.6 in the *Data section*) to calculate approximate values of (a) the Boyle temperature of the gas and (b) the radius of a H_2S molecule regarded as a sphere.

1.21(a) Suggest the pressure and temperature at which 1.0 mol of (a) NH_3, (b) Xe, (c) He will be in states that correspond to 1.0 mol H_2 at 1.0 atm and 25°C.

1.21(b) Suggest the pressure and temperature at which 1.0 mol of (a) H_2S, (b) CO_2, (c) Ar will be in states that correspond to 1.0 mol N_2 at 1.0 atm and 25°C.

1.22(a) A certain gas obeys the van der Waals equation with $a = 0.50$ m^6 Pa mol^{-2}. Its volume is found to be 5.00×10^{-4} m^3 mol^{-1} at 273 K and 3.0 MPa. From this information calculate the van der Waals constant b. What is the compression factor for this gas at the prevailing temperature and pressure?

1.22(b) A certain gas obeys the van der Waals equation with $a = 0.76$ m^6 Pa mol^{-2}. Its volume is found to be 4.00×10^{-4} m^3 mol^{-1} at 288 K and 4.0 MPa. From this information calculate the van der Waals constant b. What is the compression factor for this gas at the prevailing temperature and pressure?

Problems*

Numerical problems

1.1 Recent communication with the inhabitants of Neptune has revealed that they have a Celsius-type temperature scale, but based on the melting point (0°N) and boiling point (100°N) of their most common substance, hydrogen. Further communications have revealed that the Neptunians know about perfect gas behaviour and they find that, in the limit of zero pressure, the value of pV is 28 dm^3 atm at 0°N and 40 dm^3 atm at 100°N. What is the value of the absolute zero of temperature on their temperature scale?

1.2 Deduce the relation between the pressure and mass density, ρ, of a perfect gas of molar mass M. Confirm graphically, using the following data on dimethyl ether at 25°C, that perfect behaviour is reached at low pressures and find the molar mass of the gas.

p/kPa	12.223	25.20	36.97	60.37	85.23	101.3
ρ/(kg m^{-3})	0.225	0.456	0.664	1.062	1.468	1.734

1.3 Charles's law is sometimes expressed in the form $V = V_0(1 + \alpha\theta)$, where θ is the Celsius temperature, α is a constant, and V_0 is the volume of the sample at 0°C. The following values for α have been reported for nitrogen at 0°C:

p/Torr	749.7	599.6	333.1	98.6
$10^3\alpha/(°C)^{-1}$	3.6717	3.6697	3.6665	3.6643

* Problems denoted with the symbol * were supplied by Charles Trapp, Carmen Giunta, and Marshall Cady.

For these data calculate the best value for the absolute zero of temperature on the Celsius scale.

1.4 The molar mass of a newly synthesized fluorocarbon was measured in a gas microbalance. This device consists of a glass bulb forming one end of a beam, the whole surrounded by a closed container. The beam is pivoted, and the balance point is attained by raising the pressure of gas in the container, so increasing the buoyancy of the enclosed bulb. In one experiment, the balance point was reached when the fluorocarbon pressure was 327.10 Torr; for the same setting of the pivot, a balance was reached when CHF_3 ($M = 70.014$ g mol^{-1}) was introduced at 423.22 Torr. A repeat of the experiment with a different setting of the pivot required a pressure of 293.22 Torr of the fluorocarbon and 427.22 Torr of the CHF_3. What is the molar mass of the fluorocarbon? Suggest a molecular formula.

1.5 A constant-volume perfect gas thermometer indicates a pressure of 6.69 kPa at the triple point temperature of water (273.16 K). (a) What change of pressure indicates a change of 1.00 K at this temperature? (b) What pressure indicates a temperature of 100.00°C? (c) What change of pressure indicates a change of 1.00 K at the latter temperature?

1.6 A vessel of volume 22.4 dm^3 contains 2.0 mol H_2 and 1.0 mol N_2 at 273.15 K initially. All the H_2 reacted with sufficient N_2 to form NH_3. Calculate the partial pressures and the total pressure of the final mixture.

1.7 Calculate the molar volume of chlorine gas at 350 K and 2.30 atm using (a) the perfect gas law and (b) the van der Waals equation. Use the answer to (a) to calculate a first approximation to the correction term for attraction and then use successive approximations to obtain a numerical answer for part (b).

1.8 At 273 K measurements on argon gave $B = -21.7$ cm^3 mol^{-1} and $C = 1200$ cm^6 mol^{-2}, where B and C are the second and third virial coefficients in the expansion of Z in powers of $1/V_m$. Assuming that the perfect gas law holds sufficiently well for the estimation of the second and third terms of the expansion, calculate the compression factor of argon at 100 atm and 273 K. From your result, estimate the molar volume of argon under these conditions.

1.9 Calculate the volume occupied by 1.00 mol N_2 using the van der Waals equation in the form of a virial expansion at (a) its critical temperature, (b) its Boyle temperature, and (c) its inversion temperature. Assume that the pressure is 10 atm throughout. At what temperature is the gas most perfect? Use the following data: $T_c = 126.3$ K, $a = 1.390$ dm^6 atm mol^{-2}, $b = 0.0391$ dm^3 mol^{-1}.

1.10‡ The second virial coefficient of methane can be approximated by the empirical equation $B'(T) = a + be^{-c/T^2}$, where $a = -0.1993$ bar^{-1}, $b = 0.2002$ bar^{-1}, and $c = 1131$ K^2 with 300 K $< T <$ 600 K. What is the Boyle temperature of methane?

1.11 The mass density of water vapour at 327.6 atm and 776.4 K is 133.2 kg m^{-3}. Given that for water $T_c = 647.4$ K, $p_c = 218.3$ atm, $a = 5.464$ dm^6 atm mol^{-2}, $b = 0.03049$ dm^3 mol^{-1}, and $M = 18.02$ g mol^{-1}, calculate (a) the molar volume. Then calculate the compression factor (b) from the data, (c) from the virial expansion of the van der Waals equation.

1.12 The critical volume and critical pressure of a certain gas are 160 cm^3 mol^{-1} and 40 atm, respectively. Estimate the critical temperature by assuming that the gas obeys the Berthelot equation of state. Estimate the radii of the gas molecules on the assumption that they are spheres.

1.13 Estimate the coefficients a and b in the Dieterici equation of state from the critical constants of xenon. Calculate the pressure exerted by 1.0 mol Xe when it is confined to 1.0 dm^3 at 25°C.

Theoretical problems

1.14 Show that the van der Waals equation leads to values of $Z < 1$ and $Z > 1$, and identify the conditions for which these values are obtained.

1.15 Express the van der Waals equation of state as a virial expansion in powers of $1/V_m$ and obtain expressions for B and C in terms of the parameters a and b. The expansion you will need is $(1-x)^{-1} = 1 + x + x^2 + \cdots$. Measurements on argon gave $B = -21.7$ cm^3 mol^{-1} and $C = 1200$ cm^6 mol^{-2} for the virial coefficients at 273 K. What are the values of a and b in the corresponding van der Waals equation of state?

1.16‡ Derive the relation between the critical constants and the Dieterici equation parameters. Show that $Z_c = 2e^{-2}$ and derive the reduced form of the Dieterici equation of state. Compare the van der Waals and Dieterici predictions of the critical compression factor. Which is closer to typical experimental values?

1.17 A scientist proposed the following equation of state:

$$p = \frac{RT}{V_m} - \frac{B}{V_m^2} + \frac{C}{V_m^3}$$

Show that the equation leads to critical behaviour. Find the critical constants of the gas in terms of B and C and an expression for the critical compression factor.

1.18 Equations 1.19a and 1.19b are expansions in p and $1/V_m$, respectively. Find the relation between B, C and B', C'.

1.19 The second virial coefficient B' can be obtained from measurements of the density ρ of a gas at a series of pressures. Show that the graph of p/ρ against p should be a straight line with slope proportional to B'. Use the data on dimethyl ether in Problem 1.2 to find the values of B' and B at 25°C.

1.20 The equation of state of a certain gas is given by $p = RT/V_m + (a + bT)/V_m^2$, where a and b are constants. Find $(\partial V/\partial T)_p$.

1.21 The following equations of state are occasionally used for approximate calculations on gases: (gas A) $pV_m = RT(1 + b/V_m)$, (gas B) $p(V_m - b) = RT$. Assuming that there were gases that actually obeyed these equations of state, would it be possible to liquefy either gas A or B? Would they have a critical temperature? Explain your answer.

1.22 Derive an expression for the compression factor of a gas that obeys the equation of state $p(V - nb) = nRT$, where b and R are constants. If the pressure and temperature are such that $V_m = 10b$, what is the numerical value of the compression factor?

1.23‡ The discovery of the element argon by Lord Rayleigh and Sir William Ramsay had its origins in Rayleigh's measurements of the density of nitrogen with an eye toward accurate determination of its molar mass. Rayleigh prepared some samples of nitrogen by chemical reaction of nitrogen-containing compounds; under his standard conditions, a glass globe filled with this 'chemical nitrogen' had a mass of 2.2990 g. He prepared other samples by removing oxygen, carbon dioxide, and water vapour from atmospheric air; under the same conditions, this 'atmospheric nitrogen' had a mass of 2.3102 g (Lord Rayleigh, *Royal Institution Proceedings* **14**, 524 (1895)). With the hindsight of knowing accurate values for the molar masses of nitrogen and argon, compute the mole fraction of argon in the latter sample on the assumption that the former was pure nitrogen and the latter a mixture of nitrogen and argon.

1.24‡ A substance as elementary and well known as argon still receives research attention. Stewart and Jacobsen have published a review of thermodynamic properties of argon (R.B. Stewart and R.T. Jacobsen, *J. Phys. Chem. Ref. Data* **18**, 639 (1989)) which included the following 300 K isotherm.

p/MPa	0.4000	0.5000	0.6000	0.8000	1.000
V_m/(dm^3 mol^{-1})	6.2208	4.9736	4.1423	3.1031	2.4795
p/MPa	1.500	2.000	2.500	3.000	4.000
V_m/(dm^3 mol^{-1})	1.6483	1.2328	0.98357	0.81746	0.60998

(a) Compute the second virial coefficient, B, at this temperature. (b) Use nonlinear curve-fitting software to compute the third virial coefficient, C, at this temperature.

Applications: to atmospheric science

1.25 Atmospheric pollution is a problem that has received much attention. Not all pollution, however, is from industrial sources. Volcanic eruptions can be a significant source of air pollution. The Kilauea volcano in Hawaii emits 200–300 t of SO_2 per day. If this gas is emitted at 800°C and 1.0 atm, what volume of gas is emitted?

1.26 Ozone is a trace atmospheric gas that plays an important role in screening the Earth from harmful ultraviolet radiation. The abundance of ozone is commonly reported in *Dobson units*. One Dobson unit is the thickness, in thousandths of a centimetre, of a column of gas if it were collected as a pure gas at 1.00 atm and 0°C. What amount of O_3 (in moles) is found in a column of atmosphere with a cross-sectional area of 1.00 dm^2 if the abundance is 250 Dobson units (a typical mid-latitude value)? In the seasonal Antarctic ozone hole, the column abundance drops below 100 Dobson units; how many moles of ozone are found in such a column of air above a 1.00 dm^2 area? Most atmospheric ozone is found between 10 and 50 km above the surface of the Earth. If that ozone is spread uniformly through this portion of the atmosphere, what is the average molar concentration corresponding to (a) 250 Dobson units, (b) 100 Dobson units?

1.27 The barometric formula relates the pressure of a gas of molar mass M at an altitude h to its pressure p_0 at sea level. Derive this relation by showing that the change in pressure dp for an infinitesimal change in altitude dh where the density is ρ is $dp = -\rho g\, dh$. Remember that ρ depends on the pressure. Evaluate (a) the pressure difference between the top and bottom of a laboratory vessel of height 15 cm, and (b) the external atmospheric pressure at a typical cruising altitude of an aircraft (11 km) when the pressure at ground level is 1.0 atm.

1.28 Balloons are still used to deploy sensors that monitor meteorological phenomena and the chemistry of the atmosphere. It is possible to investigate some of the technicalities of ballooning by using the perfect gas law. Suppose your balloon has a radius of 3.0 m and that it is spherical. (a) What amount of H_2 (in moles) is needed to inflate it to 1.0 atm in an ambient temperature of 25°C at sea level? (b) What mass can the balloon lift at sea level, where the density of air is 1.22 kg m^{-3}? (c) What would be the payload if He were used instead of H_2?

1.29‡ The preceding problem is most readily solved (see the *Solutions manual*) with the use of Archimedes' principle, which states that the lifting force is equal to the difference between the weight of the displaced air and the weight of the balloon. Prove Archimedes' principle for the atmosphere from the barometric formula. *Hint.* Assume a simple shape for the balloon, perhaps a right circular cylinder of cross-sectional area A and height h.

1.30‡ Chlorofluorocarbons such as CCl_3F and CCl_2F_2 have been linked to ozone depletion in Antarctica. As of 1994, these gases were found in quantities of 261 and 509 parts per trillion (10^{12}) by volume (World Resources Institute, *World resources* 1996–97). Compute the molar concentration of these gases under conditions typical of (a) the mid-latitude troposphere (10°C and 1.0 atm) and (b) the Antarctic stratosphere (200 K and 0.050 atm).

1.31‡ The composition of the atmosphere is approximately 80 per cent nitrogen and 20 per cent oxygen by mass. At what height above the surface of the Earth would the atmosphere become 90 per cent nitrogen and 10 per cent oxygen by mass? Assume that the temperature of the atmosphere is constant at 25°C. What is the pressure of the atmosphere at that height?

MATHEMATICAL BACKGROUND 1

Differentiation and integration

Rates of change of functions—slopes of their graphs—are best discussed in terms of infinitesimal calculus. The slope of a function, like the slope of a hill, is obtained by dividing the rise of the hill by the horizontal distance (Fig. MB1.1). However, because the slope may vary from point to point, we should make the horizontal distance between the points as small as possible. In fact, we let it become infinitesimally small—hence the name *infinitesimal* calculus. The values of a function f at two locations x and $x + \delta x$ are $f(x)$ and $f(x + \delta x)$, respectively. Therefore, the slope of the function f at x is the vertical distance, which we write δf, divided by the horizontal distance, which we write δx:

$$\text{Slope} = \frac{\text{rise in value}}{\text{horizontal distance}} = \frac{\delta f}{\delta x} = \frac{f(x+\delta x) - f(x)}{\delta x} \qquad \text{(MB1.1)}$$

The slope at x itself is obtained by letting the horizontal distance become zero, which we write $\lim \delta x \to 0$. In this limit, the δ is replaced by a d, and we write

$$\text{Slope at } x = \frac{\mathrm{d}f}{\mathrm{d}x} = \lim_{\delta x \to 0}\left(\frac{f(x+\delta x) - f(x)}{\delta x}\right) \qquad \text{(MB1.2)}$$

To work out the slope of any function, we work out the expression on the right: this process is called **differentiation** and the expression for $\mathrm{d}f/\mathrm{d}x$ is the **derivative** of the function f with respect to the variable x. Some important derivatives are given inside the front cover of the text. Most of the functions encountered in chemistry can be differentiated by using the following rules (noting that in these expressions, derivatives $\mathrm{d}f/\mathrm{d}x$ are written as $\mathrm{d}f$).

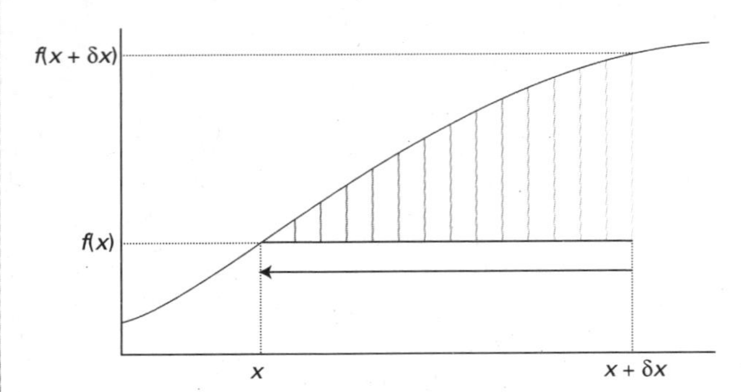

Fig. MB1.1 The slope of $f(x)$ at x, $\mathrm{d}f/\mathrm{d}x$, is obtained by making a series of approximations to the value of $f(x + \delta x) - f(x)$ divided by the change in x, denoted δx, and allowing δx to approach 0 (as indicated by the vertical lines getting closer to x).

Rule 1 For two functions f and g:

$$\mathrm{d}(f+g) = \mathrm{d}f + \mathrm{d}g \qquad \text{[MB1.3]}$$

Rule 2 (the product rule) For two functions f and g:

$$\mathrm{d}(fg) = f\mathrm{d}g + g\mathrm{d}f \qquad \text{[MB1.4]}$$

Rule 3 (the quotient rule) For two functions f and g:

$$\mathrm{d}\frac{f}{g} = \frac{1}{g}\mathrm{d}f - \frac{f}{g^2}\mathrm{d}g \qquad \text{[MB1.5]}$$

Rule no. 4 (the chain rule) For a function $f = f(g)$, where $g = g(t)$:

$$\frac{\mathrm{d}f}{\mathrm{d}t} = \frac{\mathrm{d}f}{\mathrm{d}g}\frac{\mathrm{d}g}{\mathrm{d}t} \qquad \text{[MB1.6]}$$

The area under a graph of any function f is found by the techniques of **integration**. For instance, the area under the graph of the function f drawn in Fig. MB1.2 can be written as the value of f evaluated at a point multiplied by the width of the region, δx, and then all those products $f(x)\delta x$ summed over all the regions:

$$\text{Area between } a \text{ and } b = \sum f(x)\delta x$$

When we allow δx to become infinitesimally small, written $\mathrm{d}x$, and sum an infinite number of strips, we write

$$\text{Area between } a \text{ and } b = \int_a^b f(x)\mathrm{d}x \qquad \text{[MB1.7]}$$

The elongated S symbol on the right is called the **integral** of the function f. When written as $\int$ alone, it is the **indefinite integral** of the function. When written with limits (as in eqn MB1.7), it is the **definite integral** of the function. The definite integral is the indefinite integral evaluated at the upper limit (b) minus the indefinite integral evaluated at the lower limit (a). The **average value** (or *mean value*) of a function $f(x)$ in the range $x = a$ to $x = b$ is

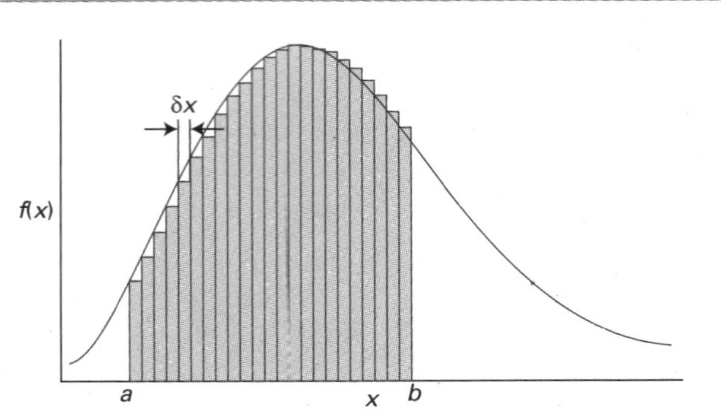

Fig. MB1.2 The shaded area is equal to the definite integral of $f(x)$ between the limits a and b.

$$\text{Average value of } f(x) \text{ from } a \text{ to } b = \frac{1}{b-a}\int_a^b f(x)\mathrm{d}x \quad \text{[MB1.8]}$$

The **mean value theorem** states that a continuous function has its mean value at least once in the range.

Integration is the inverse of differentiation. That is, if we integrate a function and then differentiate the result, we get back the original function. Some important integrals are given inside the front cover of the text. Many other standard forms are found in tables and it is also possible to calculate definite and indefinite integrals with mathematical software. Two integration techniques are useful.

Technique 1 (integration by parts) For two functions f and g:

$$\int f\frac{\mathrm{d}g}{\mathrm{d}x}\mathrm{d}x = fg - \int g\frac{\mathrm{d}f}{\mathrm{d}x}\mathrm{d}x \qquad \text{[MB1.9]}$$

Technique 2 (method of partial fractions) To solve an integral of the form

$$\int \frac{1}{(a-x)(b-x)}\mathrm{d}x$$

where a and b are constants, we write

$$\frac{1}{(a-x)(b-x)} = \frac{1}{b-a}\left(\frac{1}{a-x} - \frac{1}{b-x}\right)$$

and integrate the expression on the right. It follows that

$$\int \frac{\mathrm{d}x}{(a-x)(b-x)} = \frac{1}{b-a}\left[\int \frac{\mathrm{d}x}{a-x} - \int \frac{\mathrm{d}x}{b-x}\right]$$

$$= \frac{1}{b-a}\left(\ln\frac{1}{a-x} - \ln\frac{1}{b-x}\right) + \text{constant} \qquad \text{[MB1.10]}$$

2 The First Law

This chapter introduces some of the basic concepts of thermodynamics. It concentrates on the conservation of energy—the experimental observation that energy can be neither created nor destroyed—and shows how the principle of the conservation of energy can be used to assess the energy changes that accompany physical and chemical processes. Much of this chapter examines the means by which a system can exchange energy with its surroundings in terms of the work it may do or have done on it or the heat that it may produce or absorb. The target concept of the chapter is enthalpy, which is a very useful bookkeeping property for keeping track of the heat output (or requirements) of physical processes and chemical reactions at constant pressure. We also begin to unfold some of the power of thermodynamics by showing how to establish relations between different properties of a system. We shall see that one very useful aspect of thermodynamics is that a property can be measured indirectly by measuring others and then combining their values. The relations we derive also enable us to discuss the liquefaction of gases and to establish the relation between the heat capacities of a substance under different conditions.

The release of energy can be used to provide heat when a fuel burns in a furnace, to produce mechanical work when a fuel burns in an engine, and to generate electrical work when a chemical reaction pumps electrons through a circuit. In chemistry, we encounter reactions that can be harnessed to provide heat and work, reactions that liberate energy that is released unused but which give products we require, and reactions that constitute the processes of life. **Thermodynamics**, the study of the transformations of energy, enables us to discuss all these matters quantitatively and to make useful predictions.

The basic concepts

For the purposes of thermodynamics, the universe is divided into two parts, the system and its surroundings. The **system** is the part of the world in which we have a special interest. It may be a reaction vessel, an engine, an electrochemical cell, a biological cell, and so on. The **surroundings** comprise the region outside the system and are where we make our measurements. The type of system depends on the characteristics of the boundary that divides it from the surroundings (Fig. 2.1). If matter can be transferred through the boundary between the system and its surroundings the system is classified as **open**. If matter cannot pass through the boundary the system is classified as **closed**. Both open and closed systems can exchange energy with their surroundings. For example, a closed system can expand and thereby raise a weight in the surroundings; a closed system may also transfer energy to the surroundings if they are at a lower temperature. An **isolated system** is a closed system that has neither mechanical nor thermal contact with its surroundings.

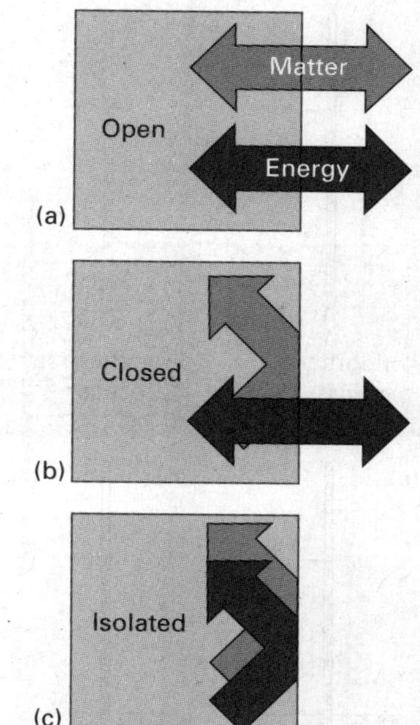

Fig. 2.1 (a) An open system can exchange matter and energy with its surroundings. (b) A closed system can exchange energy with its surroundings, but it cannot exchange matter. (c) An isolated system can exchange neither energy nor matter with its surroundings.

2.1 Work, heat, and energy

Key points **(a) Work is done to achieve motion against an opposing force; energy is the capacity to do work. (b) Heating is the transfer of energy that makes use of disorderly molecular motion; work is the transfer of energy that makes use of organized motion.**

Although thermodynamics deals with observations on bulk systems, it is immeasurably enriched by understanding the molecular origins of these observations. In each case we shall set out the bulk observations on which thermodynamics is based and then describe their molecular interpretations.

(a) Operational definitions

The fundamental physical property in thermodynamics is work: **work** is done to achieve motion against an opposing force. A simple example is the process of raising a weight against the pull of gravity. A process does work if, in principle, it can be harnessed to raise a weight somewhere in the surroundings. An example of doing work is the expansion of a gas that pushes out a piston: the motion of the piston can in principle be used to raise a weight. A chemical reaction that drives an electric current through a resistance also does work, because the same current could be passed through a motor and used to raise a weight.

The **energy** of a system is its capacity to do work. When work is done on an otherwise isolated system (for instance, by compressing a gas or winding a spring), the capacity of the system to do work is increased; in other words, the energy of the system is increased. When the system does work (when the piston moves out or the spring unwinds), the energy of the system is reduced and it can do less work than before.

Experiments have shown that the energy of a system may be changed by means other than work itself. When the energy of a system changes as a result of a temperature difference between the system and its surroundings we say that energy has been transferred as **heat**. When a heater is immersed in a beaker of water (the system), the capacity of the system to do work increases because hot water can be used to do more work than the same amount of cold water. Not all boundaries permit the transfer of energy even though there is a temperature difference between the system and its surroundings. Boundaries that do permit the transfer of energy as heat are called **diathermic**; those that do not are called **adiabatic**.

An **exothermic process** is a process that releases energy as heat into its surroundings. All combustion reactions are exothermic. An **endothermic process** is a process in which energy is acquired from its surroundings as heat. An example of an endothermic process is the vaporization of water. To avoid a lot of awkward language, we say that in an exothermic process energy is transferred 'as heat' to the surroundings and in an endothermic process energy is transferred 'as heat' from the surroundings into the system. However, it must never be forgotten that heat is a process (the transfer of energy as a result of a temperature difference), not an entity. An endothermic process in a diathermic container results in energy flowing into the system as heat to restore the temperature to that of the surroundings. An exothermic process in a similar diathermic container results in a release of energy as heat into the surroundings. When an endothermic process takes place in an adiabatic container, it results in a lowering of temperature of the system; an exothermic process results in a rise of temperature. These features are summarized in Fig. 2.2.

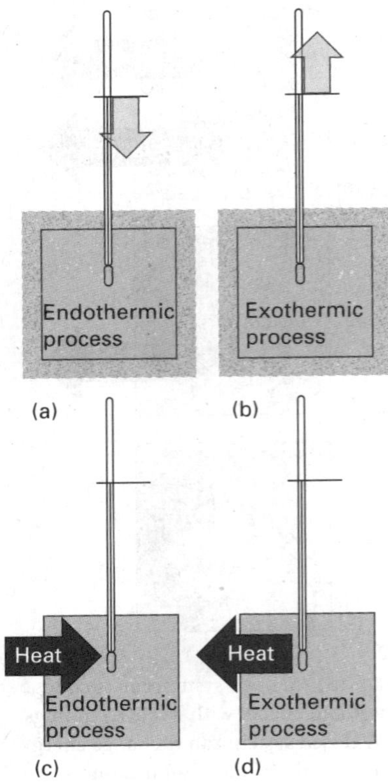

Fig. 2.2 (a) When an endothermic process occurs in an adiabatic system, the temperature falls; (b) if the process is exothermic, the temperature rises. (c) When an endothermic process occurs in a diathermic container, energy enters as heat from the surroundings, and the system remains at the same temperature. (d) If the process is exothermic, energy leaves as heat, and the process is isothermal.

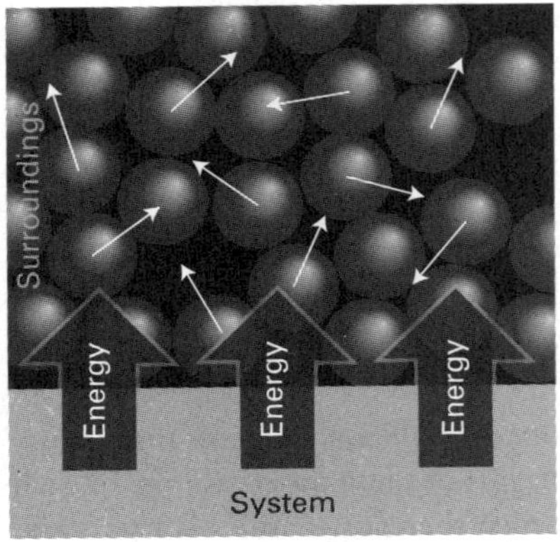

Fig. 2.3 When energy is transferred to the surroundings as heat, the transfer stimulates random motion of the atoms in the surroundings. Transfer of energy from the surroundings to the system makes use of random motion (thermal motion) in the surroundings.

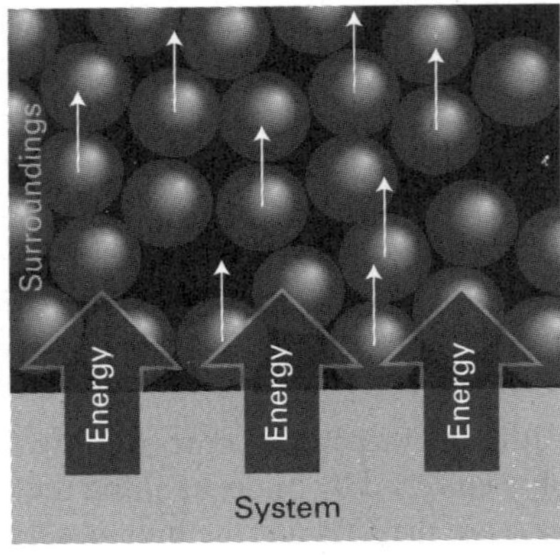

Fig. 2.4 When a system does work, it stimulates orderly motion in the surroundings. For instance, the atoms shown here may be part of a weight that is being raised. The ordered motion of the atoms in a falling weight does work on the system.

(b) The molecular interpretation of heat and work

In molecular terms, heating is the transfer of energy that makes use of disorderly molecular motion in the surroundings. The disorderly motion of molecules is called **thermal motion**. The thermal motion of the molecules in the hot surroundings stimulates the molecules in the cooler system to move more vigorously and, as a result, the energy of the system is increased. When a system heats its surroundings, molecules of the system stimulate the thermal motion of the molecules in the surroundings (Fig. 2.3).

In contrast, work is the transfer of energy that makes use of organized motion in the surroundings (Fig. 2.4). When a weight is raised or lowered, its atoms move in an organized way (up or down). The atoms in a spring move in an orderly way when it is wound; the electrons in an electric current move in an orderly direction. When a system does work it causes atoms or electrons in its surroundings to move in an organized way. Likewise, when work is done on a system, molecules in the surroundings are used to transfer energy to it in an organized way, as the atoms in a weight are lowered or a current of electrons is passed.

The distinction between work and heat is made in the surroundings. The fact that a falling weight may stimulate thermal motion in the system is irrelevant to the distinction between heat and work: work is identified as energy transfer making use of the organized motion of atoms in the surroundings, and heat is identified as energy transfer making use of thermal motion in the surroundings. In the adiabatic compression of a gas, for instance, work is done on the system as the atoms of the compressing weight descend in an orderly way, but the effect of the incoming piston is to accelerate the gas molecules to higher average speeds. Because collisions between molecules quickly randomize their directions, the orderly motion of the atoms of the weight is in effect stimulating thermal motion in the gas. We observe the falling weight, the orderly descent of its atoms, and report that work is being done even though it is stimulating thermal motion.

2.2 The internal energy

Key points Internal energy, the total energy of a system, is a state function. (a) The equipartition theorem can be used to estimate the contribution to the internal energy of classical modes of motion. (b) The First Law states that the internal energy of an isolated system is constant.

In thermodynamics, the total energy of a system is called its **internal energy**, U. The internal energy is the total kinetic and potential energy of the molecules in the system. We denote by ΔU the change in internal energy when a system changes from an initial state i with internal energy U_i to a final state f of internal energy U_f:

$$\Delta U = U_f - U_i \qquad [2.1]$$

Throughout thermodynamics, we use the convention that $\Delta X = X_f - X_i$, where X is a property (a 'state function') of the system.

The internal energy is a **state function** in the sense that its value depends only on the current state of the system and is independent of how that state has been prepared. In other words, internal energy is a function of the properties that determine the current state of the system. Changing any one of the state variables, such as the pressure, results in a change in internal energy. That the internal energy is a state function has consequences of the greatest importance, as we shall start to unfold in Section 2.10.

The internal energy is an extensive property of a system (Section F.3) and is measured in joules ($1\ \text{J} = 1\ \text{kg m}^2\ \text{s}^{-2}$, Section F.4). The molar internal energy, U_m, is the internal energy divided by the amount of substance in a system, $U_m = U/n$; it is an intensive property and commonly reported in kilojoules per mole (kJ mol^{-1}).

A brief comment

The internal energy does not include the kinetic energy arising from the motion of the system as a whole, such as its kinetic energy as it accompanies the Earth on its orbit round the Sun. That is, the internal energy is the energy 'internal' to the system.

(a) Molecular interpretation of internal energy

A molecule has a certain number of motional degrees of freedom, such as the ability to translate (the motion of its centre of mass through space), rotate around its centre of mass, or vibrate (as its bond lengths and angles change, leaving its centre of mass unmoved). Many physical and chemical properties depend on the energy associated with each of these modes of motion. For example, a chemical bond might break if a lot of energy becomes concentrated in it, for instance as vigorous vibration.

The 'equipartition theorem' of classical mechanics was introduced in Section F.5. According to it, the average energy of each quadratic contribution to the energy is $\frac{1}{2}kT$. As we saw in Section F.5, the mean energy of the atoms free to move in three dimensions is $\frac{3}{2}kT$ and the total energy of a monatomic perfect gas is $\frac{3}{2}NkT$, or $\frac{3}{2}nRT$ (because $N = nN_A$ and $R = N_Ak$). We can therefore write

$$U_m(T) = U_m(0) + \tfrac{3}{2}RT \quad \text{(monatomic gas; translation only)} \qquad (2.2a)$$

where $U_m(0)$ is the molar internal energy at $T = 0$, when all translational motion has ceased and the sole contribution to the internal energy arises from the internal structure of the atoms. This equation shows that the internal energy of a perfect gas increases linearly with temperature. At 25°C, $\frac{3}{2}RT = 3.7$ kJ mol^{-1}, so translational motion contributes about 4 kJ mol^{-1} to the molar internal energy of a gaseous sample of atoms or molecules.

When the gas consists of molecules, we need to take into account the effect of rotation and vibration. A linear molecule, such as N_2 and CO_2, can rotate around two axes perpendicular to the line of the atoms (Fig. 2.5), so it has two rotational modes of motion, each contributing a term $\frac{1}{2}kT$ to the internal energy. Therefore, the mean rotational energy is kT and the rotational contribution to the molar internal energy is RT. By adding the translational and rotational contributions, we obtain

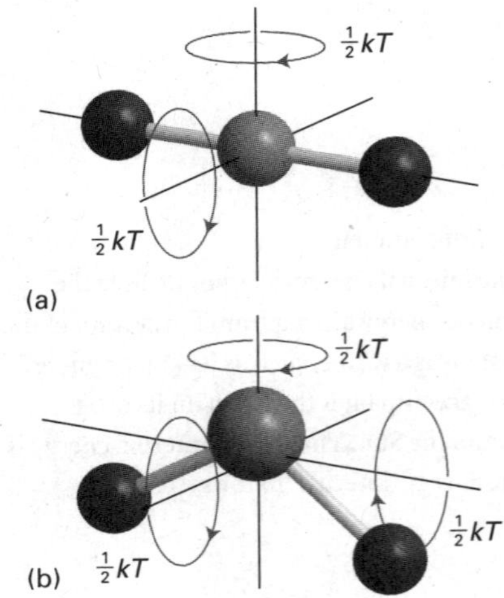

Fig. 2.5 The rotational modes of molecules and the corresponding average energies at a temperature T. (a) A linear molecule can rotate about two axes perpendicular to the line of the atoms. (b) A nonlinear molecule can rotate about three perpendicular axes.

$$U_m(T) = U_m(0) + \tfrac{5}{2}RT \quad \text{(linear molecule; translation and rotation only)} \tag{2.2b}$$

A nonlinear molecule, such as CH_4 or H_2O, can rotate around three axes and, again, each mode of motion contributes a term $\frac{1}{2}kT$ to the internal energy. Therefore, the mean rotational energy is $\frac{3}{2}kT$ and there is a rotational contribution of $\frac{3}{2}RT$ to the molar internal energy. That is,

$$U_m(T) = U_m(0) + 3RT \quad \text{(nonlinear molecule; translation and rotation only)} \tag{2.2c}$$

The internal energy now increases twice as rapidly with temperature compared with the monatomic gas. Put another way: for a gas consisting of 1 mol of nonlinear molecules to undergo the same rise in temperature as 1 mol of monatomic gas, twice as much energy must be supplied. Molecules do not vibrate significantly at room temperature and, as a first approximation, the contribution of molecular vibrations to the internal energy is negligible except for very large molecules such as polymers and biological macromolecules.

None of the expressions we have derived depends on the volume occupied by the molecules: there are no intermolecular interactions in a perfect gas, so the distance between the molecules has no effect on the energy. That is, *the internal energy of a perfect gas is independent of the volume it occupies.* The internal energy of interacting molecules in condensed phases also has a contribution from the potential energy of their interaction. However, no simple expressions can be written down in general. Nevertheless, the crucial molecular point is that, as the temperature of a system is raised, the internal energy increases as the various modes of motion become more highly excited.

(b) The formulation of the First Law

It has been found experimentally that the internal energy of a system may be changed either by doing work on the system or by heating it. Whereas we may know how the energy transfer has occurred (because we can see if a weight has been raised or lowered in the surroundings, indicating transfer of energy by doing work, or if ice has melted in the surroundings, indicating transfer of energy as heat), the system is blind to the mode employed. *Heat and work are equivalent ways of changing a system's internal energy.* A system is like a bank: it accepts deposits in either currency, but stores its reserves as internal energy. It is also found experimentally that, if a system is isolated from its surroundings, then no change in internal energy takes place. This summary of observations is now known as the **First Law of thermodynamics** and is expressed as follows:

The internal energy of an isolated system is constant. [First Law of thermodynamics]

We cannot use a system to do work, leave it isolated, and then come back expecting to find it restored to its original state with the same capacity for doing work. The experimental evidence for this observation is that no 'perpetual motion machine', a machine that does work without consuming fuel or using some other source of energy, has ever been built.

These remarks may be summarized as follows. If we write w for the work done on a system, q for the energy transferred as heat to a system, and ΔU for the resulting change in internal energy, then it follows that

$$\Delta U = q + w \tag{2.3}$$

[Mathematical statement of the First Law]

Equation 2.3 summarizes the equivalence of heat and work and the fact that the internal energy is constant in an isolated system (for which $q=0$ and $w=0$). The equation states that the change in internal energy of a closed system is equal to the energy that passes through its boundary as heat or work. It employs the 'acquisitive convention', in which w and q are positive if energy is transferred to the system as work or heat and are negative if energy is lost from the system. In other words, we view the flow of energy as work or heat from the system's perspective.

• A brief illustration

If an electric motor produced 15 kJ of energy each second as mechanical work and lost 2 kJ as heat to the surroundings, then the change in the internal energy of the motor each second is

$$\Delta U = -2\text{ kJ} - 15\text{ kJ} = -17\text{ kJ}$$

Suppose that, when a spring was wound, 100 J of work was done on it but 15 J escaped to the surroundings as heat. The change in internal energy of the spring is

$$\Delta U = 100\text{ J} - 15\text{ J} = +85\text{ J}$$ •

A note on good practice Always include the sign of ΔU (and of ΔX in general), even if it is positive.

2.3 Expansion work

Key points **(a) Expansion work is proportional to the external pressure. (b) Free expansion (against zero pressure) does no work. (c) The work of expansion against constant pressure is proportional to that pressure and to the change in volume. (d) To achieve reversible expansion, the external pressure is matched at every stage to the pressure of the system. (e) The work of reversible, isothermal expansion of a perfect gas is a logarithmic function of the volume.**

The way is opened to powerful methods of calculation by switching attention to infinitesimal changes of state (such as infinitesimal change in temperature) and infinitesimal changes in the internal energy dU. Then, if the work done on a system is dw and the energy supplied to it as heat is dq, in place of eqn 2.3 we have

$$dU = dq + dw \tag{2.4}$$

To use this expression we must be able to relate dq and dw to events taking place in the surroundings.

We begin by discussing **expansion work**, the work arising from a change in volume. This type of work includes the work done by a gas as it expands and drives back the atmosphere. Many chemical reactions result in the generation of gases (for instance, the thermal decomposition of calcium carbonate or the combustion of octane), and the thermodynamic characteristics of the reaction depend on the work that must be done to make room for the gas it has produced. The term 'expansion work' also includes work associated with negative changes of volume, that is, compression.

(a) The general expression for work

The calculation of expansion work starts from the definition used in physics, which states that the work required to move an object a distance dz against an opposing force of magnitude F is

$$dw = -F dz \qquad \text{General definition of work done} \qquad [2.5]$$

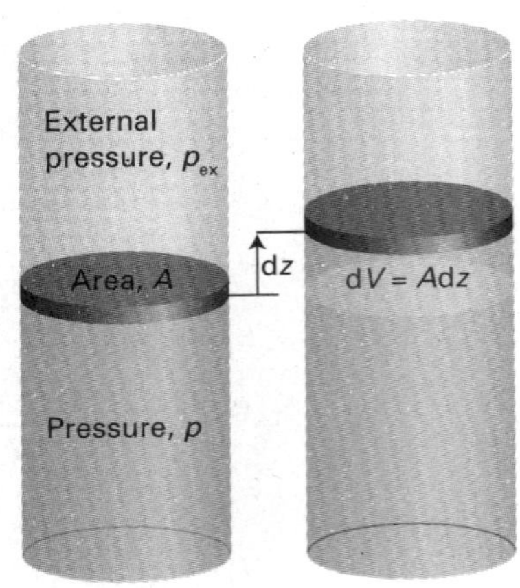

Fig. 2.6 When a piston of area A moves out through a distance dz, it sweeps out a volume dV = Adz. The external pressure p_{ex} is equivalent to a weight pressing on the piston, and the force opposing expansion is $F = p_{ex}A$.

The negative sign tells us that, when the system moves an object against an opposing force of magnitude F, and there are no other changes, then the internal energy of the system doing the work will decrease. That is, if dz is positive (motion to positive z), dw is negative, and the internal energy decreases (dU in eqn 2.4 is negative provided that dq = 0).

Now consider the arrangement shown in Fig. 2.6, in which one wall of a system is a massless, frictionless, rigid, perfectly fitting piston of area A. If the external pressure is p_{ex}, the magnitude of the force acting on the outer face of the piston is $F = p_{ex}A$. When the system expands through a distance dz against an external pressure p_{ex}, it follows that the work done is $\mathrm{d}w = -p_{ex}A\mathrm{d}z$. The quantity Adz is the change in volume, dV, in the course of the expansion. Therefore, the work done when the system expands by dV against a pressure p_{ex} is

$$\mathrm{d}w = -p_{ex}\mathrm{d}V \quad \text{Expansion work} \quad (2.6a)$$

To obtain the total work done when the volume changes from an initial value V_i to a final value V_f we integrate this expression between the initial and final volumes:

$$w = -\int_{V_i}^{V_f} p_{ex}\mathrm{d}V \quad (2.6b)$$

The force acting on the piston, $p_{ex}A$, is equivalent to the force arising from a weight that is raised as the system expands. If the system is compressed instead, then the same weight is lowered in the surroundings and eqn 2.6 can still be used, but now $V_f < V_i$. It is important to note that it is still the external pressure that determines the magnitude of the work. This somewhat perplexing conclusion seems to be inconsistent with the fact that the gas *inside* the container is opposing the compression. However, when a gas is compressed, the ability of the *surroundings* to do work is diminished by an amount determined by the weight that is lowered, and it is this energy that is transferred into the system.

Other types of work (for example, electrical work), which we shall call either **non-expansion work** or **additional work**, have analogous expressions, with each one the product of an intensive factor (the pressure, for instance) and an extensive factor (the change in volume). Some are collected in Table 2.1. For the present we continue with the work associated with changing the volume, the expansion work, and see what we can extract from eqn 2.6.

Table 2.1 Varieties of work*

Type of work	dw	Comments	Units†
Expansion	$-p_{ex}\mathrm{d}V$	p_{ex} is the external pressure dV is the change in volume	Pa m^3
Surface expansion	$\gamma\mathrm{d}\sigma$	γ is the surface tension dσ is the change in area	N m^{-1} m^2
Extension	$f\mathrm{d}l$	f is the tension dl is the change in length	N m
Electrical	$\phi\mathrm{d}Q$	ϕ is the electric potential dQ is the change in charge	V C

* In general, the work done on a system can be expressed in the form $\mathrm{d}w = -F\mathrm{d}z$, where F is a 'generalized force' and dz is a 'generalized displacement'.

† For work in joules (J). Note that 1 N m = 1 J and 1 V C = 1 J.

(b) Free expansion

Free expansion is expansion against zero opposing force. It occurs when $p_{ex} = 0$. According to eqn 2.6a, $dw = 0$ for each stage of the expansion. Hence, overall:

$$w = 0 \qquad \text{Work of free expansion} \qquad (2.7)$$

That is, no work is done when a system expands freely. Expansion of this kind occurs when a gas expands into a vacuum.

(c) Expansion against constant pressure

Now suppose that the external pressure is constant throughout the expansion. For example, the piston may be pressed on by the atmosphere, which exerts the same pressure throughout the expansion. A chemical example of this condition is the expansion of a gas formed in a chemical reaction in a container that can expand. We can evaluate eqn 2.6b by taking the constant p_{ex} outside the integral:

$$w = -p_{ex}\int_{V_i}^{V_f} dV = -p_{ex}(V_f - V_i)$$

Therefore, if we write the change in volume as $\Delta V = V_f - V_i$,

$$w = -p_{ex}\Delta V \qquad \text{Expansion work against constant external pressure} \qquad (2.8)$$

This result is illustrated graphically in Fig. 2.7, which makes use of the fact that an integral can be interpreted as an area. The magnitude of w, denoted $|w|$, is equal to the area beneath the horizontal line at $p = p_{ex}$ lying between the initial and final volumes. A p,V-graph used to illustrate expansion work is called an **indicator diagram**; James Watt first used one to indicate aspects of the operation of his steam engine.

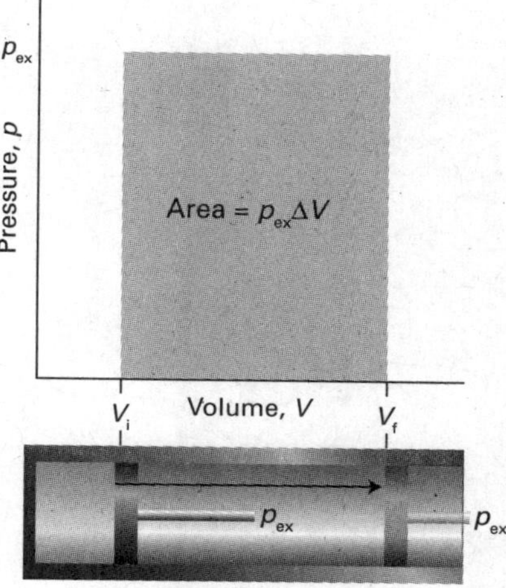

Fig. 2.7 The work done by a gas when it expands against a constant external pressure, p_{ex}, is equal to the shaded area in this example of an indicator diagram.

A brief comment

The value of the integral $\int_a^b f(x)dx$ is equal to the area under the graph of $f(x)$ between $x = a$ and $x = b$. For instance, the area under the curve $f(x) = x^2$ shown in the illustration that lies between $x = 1$ and 3 is

$$\int_1^3 x^2 dx = (\tfrac{1}{3}x^3 + \text{constant})\Big|_1^3 = \tfrac{1}{3}(3^3 - 1^3) = \tfrac{26}{3} \approx 8.67$$

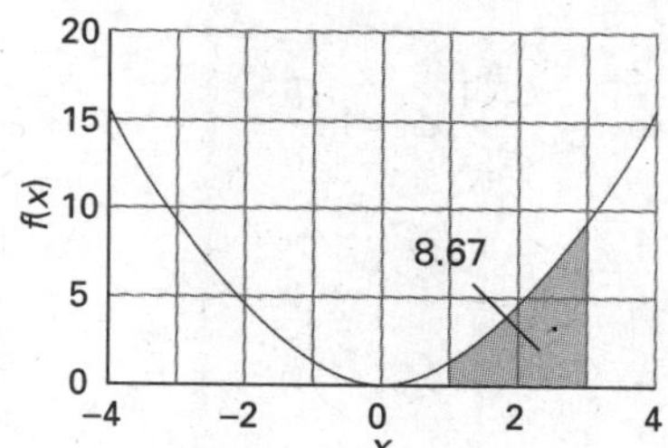

(d) Reversible expansion

A **reversible change** in thermodynamics is a change that can be reversed by an infinitesimal modification of a variable. The key word 'infinitesimal' sharpens the everyday meaning of the word 'reversible' as something that can change direction. One example of reversibility that we have encountered already is the thermal equilibrium of two systems with the same temperature. The transfer of energy as heat between the two is reversible because, if the temperature of either system is lowered infinitesimally, then energy flows into the system with the lower temperature. If the temperature of either system at thermal equilibrium is raised infinitesimally, then energy flows out of the hotter system. There is obviously a very close relationship between reversibility and equilibrium: systems at equilibrium are poised to undergo reversible change.

Suppose a gas is confined by a piston and that the external pressure, p_{ex}, is set equal to the pressure, p, of the confined gas. Such a system is in mechanical equilibrium with its surroundings because an infinitesimal change in the external pressure in either direction causes changes in volume in opposite directions. If the external pressure is reduced infinitesimally, the gas expands slightly. If the external pressure is increased infinitesimally, the gas contracts slightly. In either case the change is reversible in the thermodynamic sense. If, on the other hand, the external pressure differs measurably from the internal pressure, then changing p_{ex} infinitesimally will not decrease it below the pressure of the gas, so will not change the direction of the process. Such a system is not in mechanical equilibrium with its surroundings and the expansion is thermodynamically irreversible.

To achieve reversible expansion we set p_{ex} equal to p at each stage of the expansion. In practice, this equalization could be achieved by gradually removing weights from the piston so that the downward force due to the weights always matches the changing upward force due to the pressure of the gas. When we set $p_{ex}=p$, eqn 2.6a becomes

$$dw=-p_{ex}dV=-pdV$$ Reversible expansion work (2.9a)$_{rev}$

(Equations valid only for reversible processes are labelled with a subscript rev.) Although the pressure inside the system appears in this expression for the work, it does so only because p_{ex} has been set equal to p to ensure reversibility. The total work of reversible expansion from an initial volume V_i to a final volume V_f is therefore

$$w=-\int_{V_i}^{V_f} pdV$$ (2.9b)$_{rev}$

We can evaluate the integral once we know how the pressure of the confined gas depends on its volume. Equation 2.9 is the link with the material covered in Chapter 1 for, if we know the equation of state of the gas, then we can express p in terms of V and evaluate the integral.

(e) Isothermal reversible expansion

Consider the isothermal, reversible expansion of a perfect gas. The expansion is made isothermal by keeping the system in thermal contact with its surroundings (which may be a constant-temperature bath). Because the equation of state is $pV=nRT$, we know that at each stage $p=nRT/V$, with V the volume at that stage of the expansion. The temperature T is constant in an isothermal expansion, so (together with n and R) it may be taken outside the integral. It follows that the work of reversible isothermal expansion of a perfect gas from V_i to V_f at a temperature T is

$$w=-nRT\int_{V_i}^{V_f}\frac{dV}{V}=-nRT\ln\frac{V_f}{V_i}$$ Reversible, isothermal expansion work of a perfect gas (2.10)$^{\circ}_{rev}$

A brief comment

An integral that occurs throughout thermodynamics is

$$\int\frac{1}{x}dx=\ln x+\text{constant},$$

so $$\int_a^b\frac{1}{x}dx=\ln\frac{b}{a}$$

When the final volume is greater than the initial volume, as in an expansion, the logarithm in eqn 2.10 is positive and hence $w<0$. In this case, the system has done work on the surroundings and there is a corresponding reduction in its internal energy. (Note the cautious language: we shall see later that there is a compensating influx of energy as heat, so overall the internal energy is constant for the isothermal expansion of a perfect gas.) The equations also show that more work is done for a given change of volume when the temperature is increased: at a higher temperature the greater pressure of the confined gas needs a higher opposing pressure to ensure reversibility and the work done is correspondingly greater.

We can express the result of the calculation as an indicator diagram, for the magnitude of the work done is equal to the area under the isotherm $p=nRT/V$ (Fig. 2.8). Superimposed on the diagram is the rectangular area obtained for irreversible expansion against constant external pressure fixed at the same final value as that reached in the reversible expansion. More work is obtained when the expansion is reversible (the area is greater) because matching the external pressure to the internal pressure at each stage of the process ensures that none of the system's pushing power is wasted. We cannot obtain more work than for the reversible process because increasing the external pressure even infinitesimally at any stage results in compression. We may infer from this discussion that, because some pushing power is wasted when $p>p_{ex}$, the maximum work available from a system operating between specified initial and final states and passing along a specified path is obtained when the change takes place reversibly.

We have introduced the connection between reversibility and maximum work for the special case of a perfect gas undergoing expansion. Later (in Section 3.5) we shall see that it applies to all substances and to all kinds of work.

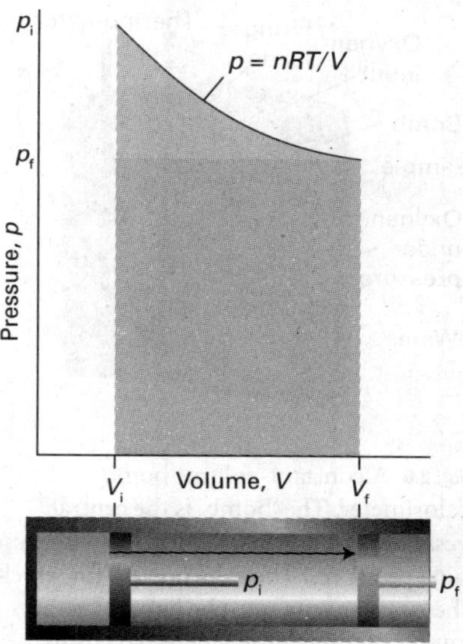

Fig. 2.8 The work done by a perfect gas when it expands reversibly and isothermally is equal to the area under the isotherm $p = nRT/V$. The work done during the irreversible expansion against the same final pressure is equal to the rectangular area shown slightly darker. Note that the reversible work is greater than the irreversible work.

interActivity Calculate the work of isothermal reversible expansion of 1.0 mol $CO_2(g)$ at 298 K from 1.0 m^3 to 3.0 m^3 on the basis that it obeys the van der Waals equation of state.

Example 2.1 *Calculating the work of gas production*

Calculate the work done when 50 g of iron reacts with hydrochloric acid to produce $FeCl_2(aq)$ and hydrogen in (a) a closed vessel of fixed volume, (b) an open beaker at 25°C.

Method We need to judge the magnitude of the volume change and then to decide how the process occurs. If there is no change in volume, there is no expansion work however the process takes place. If the system expands against a constant external pressure, the work can be calculated from eqn 2.8. A general feature of processes in which a condensed phase changes into a gas is that the volume of the former may usually be neglected relative to that of the gas it forms.

Answer In (a) the volume cannot change, so no expansion work is done and $w = 0$. In (b) the gas drives back the atmosphere and therefore $w = -p_{ex}\Delta V$. We can neglect the initial volume because the final volume (after the production of gas) is so much larger and $\Delta V = V_f - V_i \approx V_f = nRT/p_{ex}$, where n is the amount of H_2 produced. Therefore,

$$w = -p_{ex}\Delta V \approx -p_{ex} \times \frac{nRT}{p_{ex}} = -nRT$$

Because the reaction is $Fe(s) + 2\,HCl(aq) \rightarrow FeCl_2(aq) + H_2(g)$, we know that 1 mol H_2 is generated when 1 mol Fe is consumed, and n can be taken as the amount of Fe atoms that react. Because the molar mass of Fe is 55.85 g mol^{-1}, it follows that

$$w \approx -\frac{50\ \text{g}}{55.85\ \text{g mol}^{-1}} \times (8.3145\ \text{J K}^{-1}\ \text{mol}^{-1}) \times (298\ \text{K})$$
$$\approx -2.2\ \text{kJ}$$

The system (the reaction mixture) does 2.2 kJ of work driving back the atmosphere. Note that (for this perfect gas system) the magnitude of the external pressure does not affect the final result: the lower the pressure, the larger the volume occupied by the gas, so the effects cancel.

Self-test 2.1 Calculate the expansion work done when 50 g of water is electrolysed under constant pressure at 25°C. [−10 kJ]

2.4 Heat transactions

Key points **The energy transferred as heat at constant volume is equal to the change in internal energy of the system. (a) Calorimetry is the measurement of heat transactions. (b) The heat capacity at constant volume is the slope of the internal energy with respect to temperature.**

In general, the change in internal energy of a system is

$$dU = dq + dw_{exp} + dw_e \qquad (2.11)$$

where dw_e is work in addition (e for 'extra') to the expansion work, dw_{exp}. For instance, dw_e might be the electrical work of driving a current through a circuit. A

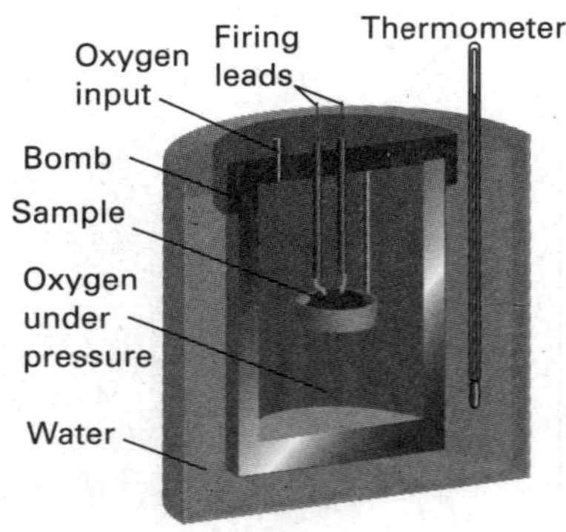

Fig. 2.9 A constant-volume bomb calorimeter. The 'bomb' is the central vessel, which is strong enough to withstand high pressures. The calorimeter (for which the heat capacity must be known) is the entire assembly shown here. To ensure adiabaticity, the calorimeter is immersed in a water bath with a temperature continuously readjusted to that of the calorimeter at each stage of the combustion.

system kept at constant volume can do no expansion work, so $dw_{exp} = 0$. If the system is also incapable of doing any other kind of work (if it is not, for instance, an electrochemical cell connected to an electric motor), then $dw_e = 0$ too. Under these circumstances:

$$dU = dq \qquad \text{Heat transferred at constant volume} \tag{2.12a}$$

We express this relation by writing $dU = dq_V$, where the subscript implies a change at constant volume. For a measurable change,

$$\Delta U = q_V \tag{2.12b}$$

It follows that, by measuring the energy supplied to a constant-volume system as heat ($q_V > 0$) or released from it as heat ($q_V < 0$) when it undergoes a change of state, we are in fact measuring the change in its internal energy.

(a) Calorimetry

Calorimetry is the study of heat transfer during physical and chemical processes. A **calorimeter** is a device for measuring energy transferred as heat. The most common device for measuring ΔU is an **adiabatic bomb calorimeter** (Fig. 2.9). The process we wish to study—which may be a chemical reaction—is initiated inside a constant-volume container, the 'bomb'. The bomb is immersed in a stirred water bath, and the whole device is the calorimeter. The calorimeter is also immersed in an outer water bath. The water in the calorimeter and of the outer bath are both monitored and adjusted to the same temperature. This arrangement ensures that there is no net loss of heat from the calorimeter to the surroundings (the bath) and hence that the calorimeter is adiabatic.

The change in temperature, ΔT, of the calorimeter is proportional to the energy that the reaction releases or absorbs as heat. Therefore, by measuring ΔT we can determine q_V and hence find ΔU. The conversion of ΔT to q_V is best achieved by calibrating the calorimeter using a process of known energy output and determining the **calorimeter constant**, the constant C in the relation

$$q = C\Delta T \tag{2.13}$$

The calorimeter constant may be measured electrically by passing a constant current, I, from a source of known potential difference, $\Delta\phi$, through a heater for a known period of time, t, for then

$$q = It\Delta\phi \tag{2.14}$$

A brief comment

Electrical charge is measured in *coulombs*, C. The motion of charge gives rise to an electric current, I, measured in coulombs per second, or *amperes*, A, where $1\ A = 1\ C\ s^{-1}$. If a constant current I flows through a potential difference $\Delta\phi$ (measured in volts, V), the total energy supplied in an interval t is $It\Delta\phi$. Because $1\ A\ V\ s = 1\ (C\ s^{-1})\ V\ s = 1\ C\ V = 1\ J$, the energy is obtained in joules with the current in amperes, the potential difference in volts, and the time in seconds.

• **A brief illustration**

If we pass a current of 10.0 A from a 12 V supply for 300 s, then from eqn 2.14 the energy supplied as heat is

$$q = (10.0\ A) \times (12\ V) \times (300\ s) = 3.6 \times 10^4\ A\ V\ s = 36\ kJ$$

because $1\ A\ V\ s = 1\ J$. If the observed rise in temperature is 5.5 K, then the calorimeter constant is $C = (36\ kJ)/(5.5\ K) = 6.5\ kJ\ K^{-1}$. •

Alternatively, C may be determined by burning a known mass of substance (benzoic acid is often used) that has a known heat output. With C known, it is simple to interpret an observed temperature rise as a release of heat.

(b) Heat capacity

The internal energy of a system increases when its temperature is raised. The increase depends on the conditions under which the heating takes place and for the present we suppose that the system has a constant volume. For example, it may be a gas in a container of fixed volume. If the internal energy is plotted against temperature, then a curve like that in Fig. 2.10 may be obtained. The slope of the tangent to the curve at any temperature is called the **heat capacity** of the system at that temperature. The **heat capacity at constant volume** is denoted C_V and is defined formally as

$$C_V = \left(\frac{\partial U}{\partial T}\right)_V \qquad \text{Definition of heat capacity at constant volume} \qquad [2.15]$$

In this case, the internal energy varies with the temperature and the volume of the sample, but we are interested only in its variation with the temperature, the volume being held constant (Fig. 2.11).

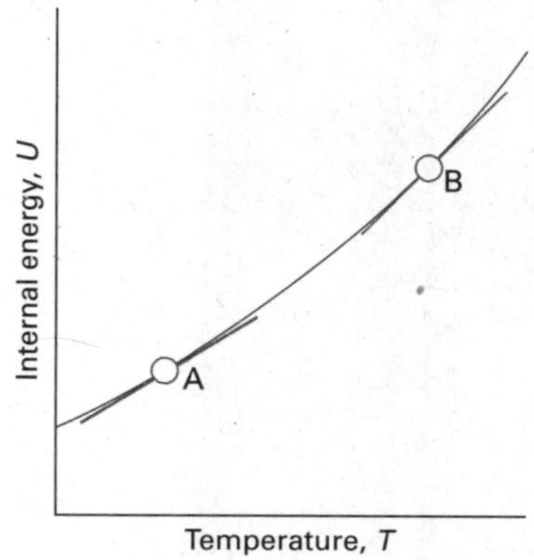

Fig. 2.10 The internal energy of a system increases as the temperature is raised; this graph shows its variation as the system is heated at constant volume. The slope of the tangent to the curve at any temperature is the heat capacity at constant volume at that temperature. Note that, for the system illustrated, the heat capacity is greater at B than at A.

A brief comment

Partial derivatives are reviewed in *Mathematical background* 2 following this chapter.

• A brief illustration

The heat capacity of a monatomic perfect gas can be calculated by inserting the expression for the internal energy derived in Section 2.2a. There we saw that

$$U_m = U_m(0) + \tfrac{3}{2}RT$$

so from eqn 2.15

$$C_{V,m} = \frac{\partial}{\partial T}(U_m(0) + \tfrac{3}{2}RT) = \tfrac{3}{2}R$$

The numerical value is 12.47 J K^{-1} mol^{-1}. •

Heat capacities are extensive properties: 100 g of water, for instance, has 100 times the heat capacity of 1 g of water (and therefore requires 100 times the energy as heat to bring about the same rise in temperature). The **molar heat capacity at constant volume**, $C_{V,m} = C_V/n$, is the heat capacity per mole of substance, and is an intensive property (all molar quantities are intensive). Typical values of $C_{V,m}$ for polyatomic gases are close to 25 J K^{-1} mol^{-1}. For certain applications it is useful to know the **specific heat capacity** (more informally, the 'specific heat') of a substance, which is the heat capacity of the sample divided by the mass, usually in grams: $C_{V,s} = C_V/m$. The specific heat capacity of water at room temperature is close to 4.2 J K^{-1} g^{-1}. In general, heat capacities depend on the temperature and decrease at low temperatures. However, over small ranges of temperature at and above room temperature, the variation is quite small and for approximate calculations heat capacities can be treated as almost independent of temperature.

The heat capacity is used to relate a change in internal energy to a change in temperature of a constant-volume system. It follows from eqn 2.15 that

$$dU = C_V dT \quad \text{(at constant volume)} \qquad (2.16a)$$

That is, at constant volume, an infinitesimal change in temperature brings about an infinitesimal change in internal energy, and the constant of proportionality is C_V. If the heat capacity is independent of temperature over the range of temperatures of interest, a measurable change of temperature, ΔT, brings about a measurable increase in internal energy, ΔU, where

$$\Delta U = C_V \Delta T \quad \text{(at constant volume)} \qquad (2.16b)$$

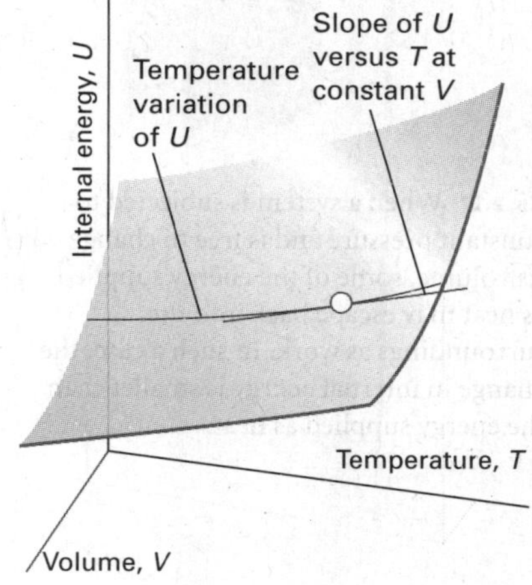

Fig. 2.11 The internal energy of a system varies with volume and temperature, perhaps as shown here by the surface. The variation of the internal energy with temperature at one particular constant volume is illustrated by the curve drawn parallel to T. The slope of this curve at any point is the partial derivative $(\partial U/\partial T)_V$.

Because a change in internal energy can be identified with the heat supplied at constant volume (eqn 2.12b), the last equation can also be written

$$q_V = C_V \Delta T \tag{2.17}$$

This relation provides a simple way of measuring the heat capacity of a sample: a measured quantity of energy is transferred as heat to the sample (electrically, for example), and the resulting increase in temperature is monitored. The ratio of the energy transferred as heat to the temperature rise it causes ($q_V/\Delta T$) is the constant-volume heat capacity of the sample.

A large heat capacity implies that, for a given quantity of energy transferred as heat, there will be only a small increase in temperature (the sample has a large capacity for heat). An infinite heat capacity implies that there will be no increase in temperature however much energy is supplied as heat. At a phase transition, such as at the boiling point of water, the temperature of a substance does not rise as energy is supplied as heat: the energy is used to drive the endothermic transition, in this case to vaporize the water, rather than to increase its temperature. Therefore, at the temperature of a phase transition, the heat capacity of a sample is infinite. The properties of heat capacities close to phase transitions are treated more fully in Section 4.6.

2.5 Enthalpy

Key points **(a) Energy transferred as heat at constant pressure is equal to the change in enthalpy of a system. (b) Enthalpy changes are measured in a constant-pressure calorimeter. (c) The heat capacity at constant pressure is equal to the slope of enthalpy with temperature.**

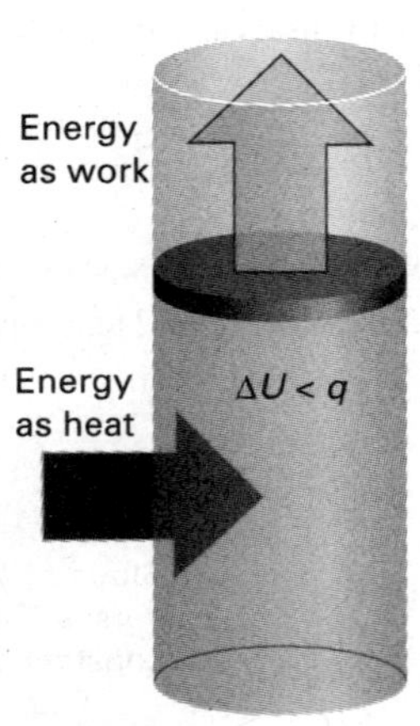

Fig. 2.12 When a system is subjected to constant pressure and is free to change its volume, some of the energy supplied as heat may escape back into the surroundings as work. In such a case, the change in internal energy is smaller than the energy supplied as heat.

The change in internal energy is not equal to the energy transferred as heat when the system is free to change its volume. Under these circumstances some of the energy supplied as heat to the system is returned to the surroundings as expansion work (Fig. 2.12), so $\mathrm{d}U$ is less than $\mathrm{d}q$. However, we shall now show that in this case the energy supplied as heat at constant pressure is equal to the change in another thermodynamic property of the system, the enthalpy.

(a) The definition of enthalpy

The **enthalpy**, H, is defined as

$$H = U + pV \quad \text{Definition of enthalpy} \tag{2.18}$$

where p is the pressure of the system and V is its volume. Because U, p, and V are all state functions, the enthalpy is a state function too. As is true of any state function, the change in enthalpy, ΔH, between any pair of initial and final states is independent of the path between them.

Although the definition of enthalpy may appear arbitrary, it has important implications for thermochemisty. For instance, we show in the following *Justification* that eqn 2.18 implies that *the change in enthalpy is equal to the energy supplied as heat at constant pressure* (provided the system does no additional work):

$$\mathrm{d}H = \mathrm{d}q \quad \text{Heat transferred at constant pressure} \tag{2.19a}$$

For a measurable change

$$\Delta H = q_p \tag{2.19b}$$

Justification 2.1 *The relation $\Delta H = q_p$*

For a general infinitesimal change in the state of the system, U changes to $U + dU$, p changes to $p + dp$, and V changes to $V + dV$, so from the definition in eqn 2.18, H changes from $U + pV$ to

$$\begin{aligned} H + dH &= (U + dU) + (p + dp)(V + dV) \\ &= U + dU + pV + pdV + Vdp + dpdV \end{aligned}$$

The last term is the product of two infinitesimally small quantities and can therefore be neglected. As a result, after recognizing $U + pV = H$ on the right, we find that H changes to

$$H + dH = H + dU + pdV + Vdp$$

and hence that

$$dH = dU + pdV + Vdp$$

If we now substitute $dU = dq + dw$ into this expression, we get

$$dH = dq + dw + pdV + Vdp$$

If the system is in mechanical equilibrium with its surroundings at a pressure p and does only expansion work, we can write $dw = -pdV$ and obtain

$$dH = dq + Vdp$$

Now we impose the condition that the heating occurs at constant pressure by writing $dp = 0$. Then

$$dH = dq \qquad \text{(at constant pressure, no additional work)}$$

as in eqn 2.19a.

The result expressed in eqn 2.19 states that, when a system is subjected to constant pressure and only expansion work can occur, the change in enthalpy is equal to the energy supplied as heat. For example, if we supply 36 kJ of energy through an electric heater immersed in an open beaker of water, then the enthalpy of the water increases by 36 kJ and we write $\Delta H = +36$ kJ.

(b) The measurement of an enthalpy change

An enthalpy change can be measured calorimetrically by monitoring the temperature change that accompanies a physical or chemical change occurring at constant pressure. A calorimeter for studying processes at constant pressure is called an **isobaric calorimeter**. A simple example is a thermally insulated vessel open to the atmosphere: the heat released in the reaction is monitored by measuring the change in temperature of the contents. For a combustion reaction an **adiabatic flame calorimeter** may be used to measure ΔT when a given amount of substance burns in a supply of oxygen (Fig. 2.13). Another route to ΔH is to measure the internal energy change by using a bomb calorimeter, and then to convert ΔU to ΔH. Because solids and liquids have small molar volumes, for them pV_m is so small that the molar enthalpy and molar internal energy are almost identical ($H_m = U_m + pV_m \approx U_m$). Consequently, if a process involves only solids or liquids, the values of ΔH and ΔU are almost identical. Physically, such processes are accompanied by a very small change in volume; the system does negligible work on the surroundings when the process occurs, so the energy supplied as heat stays entirely within the system. The most sophisticated way to measure enthalpy changes, however, is to use a **differential scanning calorimeter**

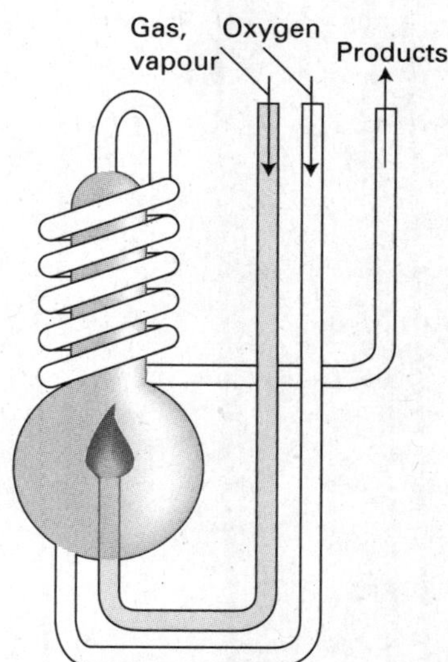

Fig. 2.13 A constant-pressure flame calorimeter consists of this component immersed in a stirred water bath. Combustion occurs as a known amount of reactant is passed through to fuel the flame, and the rise of temperature is monitored.

(DSC). Changes in enthalpy and internal energy may also be measured by noncalorimetric methods (see Chapter 6).

Example 2.2 *Relating ΔH and ΔU*

The change in molar internal energy when $CaCO_3$(s) as calcite converts to another form, aragonite, is +0.21 kJ mol^{-1}. Calculate the difference between the molar enthalpy and internal energy changes when the pressure is 1.0 bar given that the densities of the polymorphs are 2.71 g cm^{-3} and 2.93 g cm^{-3}, respectively.

Method The starting point for the calculation is the relation between the enthalpy of a substance and its internal energy (eqn 2.18). The difference between the two quantities can be expressed in terms of the pressure and the difference of their molar volumes, and the latter can be calculated from their molar masses, M, and their mass densities, ρ, by using $\rho = M/V_m$.

Answer The change in enthalpy when the transition occurs is

$$\begin{aligned}\Delta H_m &= H_m(\text{aragonite}) - H_m(\text{calcite}) \\ &= \{U_m(\text{a}) + pV_m(\text{a})\} - \{U_m(\text{c}) + pV_m(\text{c})\} \\ &= \Delta U_m + p\{V_m(\text{a}) - V_m(\text{c})\}\end{aligned}$$

where a denotes aragonite and c calcite. It follows by substituting $V_m = M/\rho$ that

$$\Delta H_m - \Delta U_m = pM\left(\frac{1}{\rho(\text{a})} - \frac{1}{\rho(\text{c})}\right)$$

Substitution of the data, using $M = 100$ g mol^{-1}, gives

$$\begin{aligned}\Delta H_m - \Delta U_m &= (1.0\times10^5\ \text{Pa})\times(100\ \text{g mol}^{-1})\times\left(\frac{1}{2.93\ \text{g cm}^{-3}} - \frac{1}{2.71\ \text{g cm}^{-3}}\right) \\ &= -2.8\times10^5\ \text{Pa cm}^3\ \text{mol}^{-1} = -0.28\ \text{Pa m}^3\ \text{mol}^{-1}\end{aligned}$$

Hence (because 1 Pa m^3 = 1 J), $\Delta H_m - \Delta U_m = -0.28$ J mol^{-1}, which is only 0.1 per cent of the value of ΔU_m. We see that it is usually justifiable to ignore the difference between the molar enthalpy and internal energy of condensed phases, except at very high pressures, when $p\Delta V_m$ is no longer negligible.

Self-test 2.2 Calculate the difference between ΔH and ΔU when 1.0 mol Sn(s, grey) of density 5.75 g cm^{-3} changes to Sn(s, white) of density 7.31 g cm^{-3} at 10.0 bar. At 298 K, $\Delta H = +2.1$ kJ. [$\Delta H - \Delta U = -4.4$ J]

The enthalpy of a perfect gas is related to its internal energy by using $pV = nRT$ in the definition of H:

$$H = U + pV = U + nRT \qquad (2.20)^\circ$$

This relation implies that the change of enthalpy in a reaction that produces or consumes gas is

$$\Delta H = \Delta U + \Delta n_g RT \qquad (2.21)^\circ$$

where Δn_g is the change in the amount of gas molecules in the reaction.

• A brief illustration

In the reaction $2 H_2(g) + O_2(g) \rightarrow 2 H_2O(l)$, 3 mol of gas-phase molecules is replaced by 2 mol of liquid-phase molecules, so $\Delta n_g = -3$ mol. Therefore, at 298 K, when $RT = 2.48$ kJ mol^{-1}, the enthalpy and internal energy changes taking place in the system are related by

$$\Delta H_m - \Delta U_m = (-3 \text{ mol}) \times RT \approx -7.4 \text{ kJ mol}^{-1}$$

Note that the difference is expressed in kilojoules, not joules as in Example 2.2. The enthalpy change is smaller (in this case, less negative) than the change in internal energy because, although heat escapes from the system when the reaction occurs, the system contracts when the liquid is formed, so energy is restored to it from the surroundings. •

Example 2.3 *Calculating a change in enthalpy*

Water is heated to boiling under a pressure of 1.0 atm. When an electric current of 0.50 A from a 12 V supply is passed for 300 s through a resistance in thermal contact with it, it is found that 0.798 g of water is vaporized. Calculate the molar internal energy and enthalpy changes at the boiling point (373.15 K).

Method Because the vaporization occurs at constant pressure, the enthalpy change is equal to the heat supplied by the heater. Therefore, the strategy is to calculate the energy supplied as heat (from $q = It\Delta\phi$), express that as an enthalpy change, and then convert the result to a molar enthalpy change by division by the amount of H_2O molecules vaporized. To convert from enthalpy change to internal energy change, we assume that the vapour is a perfect gas and use eqn 2.21.

Answer The enthalpy change is

$$\Delta H = q_p = (0.50\text{A}) \times (12 \text{ V}) \times (300 \text{ s}) = 0.50 \times 12 \times 300 \text{ J}$$

Here we have used 1 A V s = 1 J. Because 0.798 g of water is (0.798 g)/(18.02 g mol^{-1}) = (0.798/18.02) mol H_2O, the enthalpy of vaporization per mole of H_2O is

$$\Delta H_m = +\frac{0.50 \times 12 \times 300 \text{ J}}{(0.798/18.02) \text{ mol}} = +41 \text{ kJ mol}^{-1}$$

In the process $H_2O(l) \rightarrow H_2O(g)$ the change in the amount of gas molecules is $\Delta n_g = +1$ mol, so

$$\Delta U_m = \Delta H_m - RT = +38 \text{ kJ mol}^{-1}$$

Notice that the internal energy change is smaller than the enthalpy change because energy has been used to drive back the surrounding atmosphere to make room for the vapour.

Self-test 2.3 The molar enthalpy of vaporization of benzene at its boiling point (353.25 K) is 30.8 kJ mol^{-1}. What is the molar internal energy change? For how long would the same 12 V source need to supply a 0.50 A current in order to vaporize a 10 g sample? [+27.9 kJ mol^{-1}, 6.6×10^2 s]

(c) The variation of enthalpy with temperature

The enthalpy of a substance increases as its temperature is raised. The relation between the increase in enthalpy and the increase in temperature depends on the conditions (for example, constant pressure or constant volume). The most important

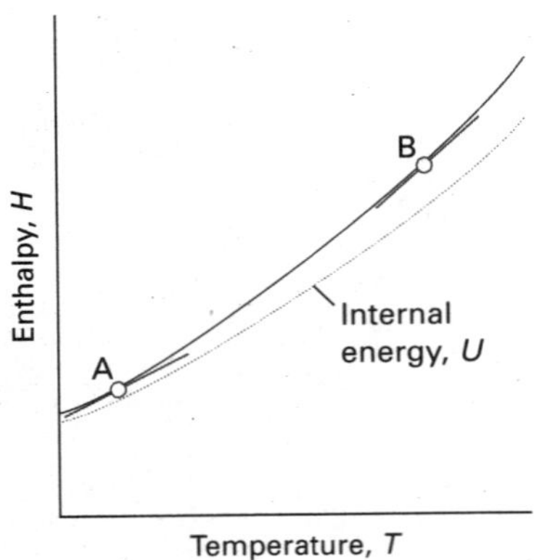

Fig. 2.14 The constant-pressure heat capacity at a particular temperature is the slope of the tangent to a curve of the enthalpy of a system plotted against temperature (at constant pressure). For gases, at a given temperature the slope of enthalpy versus temperature is steeper than that of internal energy versus temperature, and $C_{p,m}$ is larger than $C_{V,m}$.

condition is constant pressure, and the slope of the tangent to a plot of enthalpy against temperature at constant pressure is called the **heat capacity at constant pressure**, C_p, at a given temperature (Fig. 2.14). More formally:

$$C_p = \left(\frac{\partial H}{\partial T}\right)_p \quad \text{Definition of heat capacity at constant pressure} \quad [2.22]$$

The heat capacity at constant pressure is the analogue of the heat capacity at constant volume and is an extensive property. The **molar heat capacity at constant pressure**, $C_{p,m}$, is the heat capacity per mole of material; it is an intensive property.

The heat capacity at constant pressure is used to relate the change in enthalpy to a change in temperature. For infinitesimal changes of temperature

$$dH = C_p dT \quad \text{(at constant pressure)} \tag{2.23a}$$

If the heat capacity is constant over the range of temperatures of interest, then for a measurable increase in temperature

$$\Delta H = C_p \Delta T \quad \text{(at constant pressure)} \tag{2.23b}$$

Because an increase in enthalpy can be equated with the energy supplied as heat at constant pressure, the practical form of the latter equation is

$$q_p = C_p \Delta T \tag{2.24}$$

This expression shows us how to measure the heat capacity of a sample: a measured quantity of energy is supplied as heat under conditions of constant pressure (as in a sample exposed to the atmosphere and free to expand) and the temperature rise is monitored.

The variation of heat capacity with temperature can sometimes be ignored if the temperature range is small; this approximation is highly accurate for a monatomic perfect gas (for instance, one of the noble gases at low pressure). However, when it is necessary to take the variation into account, a convenient approximate empirical expression is

$$C_{p,m} = a + bT + \frac{c}{T^2} \tag{2.25}$$

The empirical parameters a, b, and c are independent of temperature (Table 2.2) and are found by fitting this expression to experimental data.

Table 2.2* Temperature variation of molar heat capacities, $C_{p,m}/(\text{J K}^{-1}\text{ mol}^{-1}) = a + bT + c/T^2$

	a	$b/(10^{-3}\text{ K})$	$c/(10^5\text{ K}^2)$
C(s, graphite)	16.86	4.77	−8.54
CO_2(g)	44.22	8.79	−8.62
H_2O(l)	75.29	0	0
N_2(g)	28.58	3.77	−0.50

* More values are given in the *Data section*.

Example 2.4 *Evaluating an increase in enthalpy with temperature*

What is the change in molar enthalpy of N_2 when it is heated from 25°C to 100°C? Use the heat capacity information in Table 2.2.

Method The heat capacity of N_2 changes with temperature, so we cannot use eqn 2.23b (which assumes that the heat capacity of the substance is constant). Therefore, we must use eqn 2.23a, substitute eqn 2.25 for the temperature dependence of the heat capacity, and integrate the resulting expression from 25°C to 100°C.

Answer For convenience, we denote the two temperatures T_1 (298 K) and T_2 (373 K). The relation we require is

$$\int_{H(T_1)}^{H(T_2)} \mathrm{d}H = \int_{T_1}^{T_2} \left(a + bT + \frac{c}{T^2} \right) \mathrm{d}T$$

and the relevant integrals are

$$\int \mathrm{d}x = x + \text{constant} \qquad \int x\,\mathrm{d}x = \tfrac{1}{2}x^2 + \text{constant} \qquad \int \frac{\mathrm{d}x}{x^2} = -\frac{1}{x} + \text{constant}$$

It follows that

$$H(T_2) - H(T_1) = a(T_2 - T_1) + \tfrac{1}{2}b(T_2^2 - T_1^2) - c\left(\frac{1}{T_2} - \frac{1}{T_1}\right)$$

Substitution of the numerical data results in

$$H(373\ \mathrm{K}) = H(298\ \mathrm{K}) + 2.20\ \mathrm{kJ\ mol^{-1}}$$

If we had assumed a constant heat capacity of 29.14 J K^{-1} mol^{-1} (the value given by eqn 2.25 at 25°C), we would have found that the two enthalpies differed by 2.19 kJ mol^{-1}.

Self-test 2.4 At very low temperatures the heat capacity of a solid is proportional to T^3, and we can write $C_p = aT^3$. What is the change in enthalpy of such a substance when it is heated from 0 to a temperature T (with T close to 0)? [$\Delta H = \tfrac{1}{4}aT^4$]

Most systems expand when heated at constant pressure. Such systems do work on the surroundings and therefore some of the energy supplied to them as heat escapes back to the surroundings. As a result, the temperature of the system rises less than when the heating occurs at constant volume. A smaller increase in temperature implies a larger heat capacity, so we conclude that in most cases the heat capacity at constant pressure of a system is larger than its heat capacity at constant volume. We show later (Section 2.11) that there is a simple relation between the two heat capacities of a perfect gas:

$$C_p - C_V = nR \qquad \text{Relation between heat capacities of a perfect gas} \qquad (2.26)^\circ$$

It follows that the molar heat capacity of a perfect gas is about 8 J K^{-1} mol^{-1} larger at constant pressure than at constant volume. Because the heat capacity at constant volume of a monatomic gas is about 12 J K^{-1} mol^{-1}, the difference is highly significant and must be taken into account.

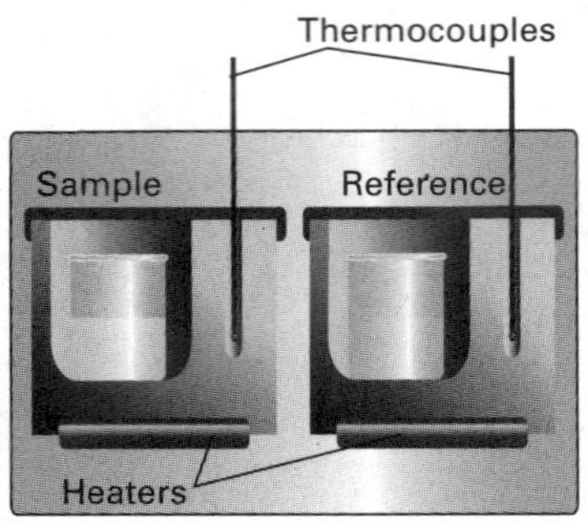

Fig. 2.15 A differential scanning calorimeter. The sample and a reference material are heated in separate but identical metal heat sinks. The output is the difference in power needed to maintain the heat sinks at equal temperatures as the temperature rises.

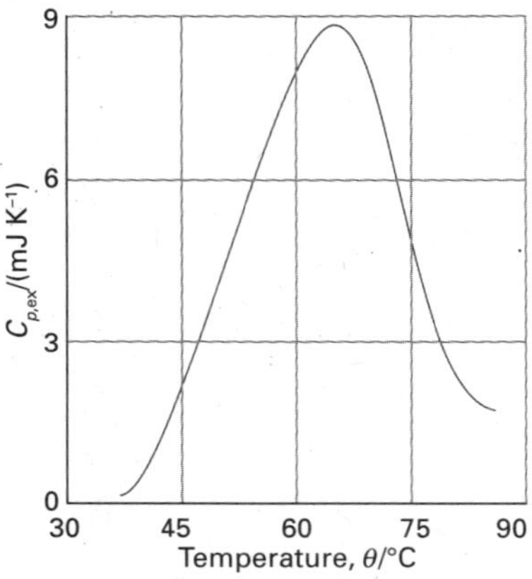

Fig. 2.16 A thermogram for the protein ubiquitin at pH = 2.45. The protein retains its native structure up to about 45°C and then undergoes an endothermic conformational change. (Adapted from B. Chowdhry and S. LeHarne, *J. Chem. Educ.* 74, 236 (1997).)

IMPACT ON BIOCHEMISTRY AND MATERIALS SCIENCE

I2.1 Differential scanning calorimetry

A *differential scanning calorimeter* (DSC) measures the energy transferred as heat to or from a sample at constant pressure during a physical or chemical change. The term 'differential' refers to the fact that the behaviour of the sample is compared to that of a reference material that does not undergo a physical or chemical change during the analysis. The term 'scanning' refers to the fact that the temperatures of the sample and reference material are increased, or scanned, during the analysis.

A DSC consists of two small compartments that are heated electrically at a constant rate. The temperature, T, at time t during a linear scan is $T = T_0 + \alpha t$, where T_0 is the initial temperature and α is the temperature scan rate. A computer controls the electrical power supply that maintains the same temperature in the sample and reference compartments throughout the analysis (Fig. 2.15).

If no physical or chemical change occurs in the sample at temperature T, we write the heat transferred to the sample as $q_p = C_p\Delta T$, where $\Delta T = T - T_0$ and we have assumed that C_p is independent of temperature. Because $T = T_0 + \alpha t$, $\Delta T = \alpha t$. The chemical or physical process requires the transfer of $q_p + q_{p,\text{ex}}$, where $q_{p,\text{ex}}$ is the excess energy transferred as heat needed to attain the same change in temperature of the sample as the control. The quantity $q_{p,\text{ex}}$ is interpreted in terms of an apparent change in the heat capacity at constant pressure of the sample, C_p, during the temperature scan:

$$C_{p,\text{ex}} = \frac{q_{p,\text{ex}}}{\Delta T} = \frac{q_{p,\text{ex}}}{\alpha t} = \frac{P_{\text{ex}}}{\alpha}$$

where $P_{\text{ex}} = q_{p,\text{ex}}/t$ is the excess electrical power necessary to equalize the temperature of the sample and reference compartments. A DSC trace, also called a *thermogram*, consists of a plot of $C_{p,\text{ex}}$ against T (Fig. 2.16). From eqn 2.23a, the enthalpy change associated with the process is

$$\Delta H = \int_{T_1}^{T_2} C_{p,\text{ex}}\text{d}T$$

where T_1 and T_2 are, respectively, the temperatures at which the process begins and ends. This relation shows that the enthalpy change is equal to the area under the plot of $C_{p,\text{ex}}$ against T.

With a DSC, enthalpy changes may be determined in samples of masses as low as 0.5 mg, which is a significant advantage over conventional calorimeters, which require several grams of material. The technique is used in the chemical industry to characterize polymers in terms of their structural integrity, stability, and nanoscale organization. For example, it is possible to detect the ability of certain polymers such as ethylene oxide (EO) and propylene oxide (PO) to self-aggregate as their temperature is raised. These copolymers are widely used as surfactants and detergents with the amphiphilic (both water- and hydrocarbon-attracting) character provided by the hydrophobic central PO block and the more hydrophilic EO blocks attached on either side. They aggregate to form micelles (clusters) as the temperature is raised because the more hydrophobic central PO block becomes less soluble at higher temperature but the terminal EO blocks retain their strong interaction with water. This enhanced amphiphilic character of the molecules at higher temperature drives the copolymers to form micelles that are spherical in shape. The micellization process is strongly endothermic, reflecting the initial destruction of the hydrogen bonds of the PO block with water, and is readily detected by DSC. Further increases in temperature affect the shape of the micelle, changing from spherical to rod-like. A new but weaker DSC signal at higher temperature reflects a small change in enthalpy as micelles aggregate to form the rod-like structure. The marked decrease in the heat capacity accompanying the sphere-to-rod transition presumably reflects an extensive decrease in the degree of hydration of the polymer.

The technique is also used to assess the stability of proteins, nucleic acids, and membranes. For example, the thermogram shown in Fig. 2.16 indicates that the protein ubiquitin undergoes an endothermic conformational change in which a large number of non-covalent interactions (such as hydrogen bonds) are broken simultaneously and result in denaturation, the loss of the protein's three-dimensional structure. The area under the curve represents the heat absorbed in this process and can be identified with the enthalpy change. The thermogram also reveals the formation of new intermolecular interactions in the denatured form. The increase in heat capacity accompanying the native → denatured transition reflects the change from a more compact native conformation to one in which the more exposed amino acid side chains in the denatured form have more extensive interactions with the surrounding water molecules.

2.6 **Adiabatic changes**

Key point **For the reversible adiabatic expansion of a perfect gas, pressure and volume are related by an expression that depends on the ratio of heat capacities.**

We are now equipped to deal with the changes that occur when a perfect gas expands adiabatically. A decrease in temperature should be expected: because work is done but no heat enters the system, the internal energy falls, and therefore the temperature of the working gas also falls. In molecular terms, the kinetic energy of the molecules falls as work is done, so their average speed decreases, and hence the temperature falls.

The change in internal energy of a perfect gas when the temperature is changed from T_i to T_f and the volume is changed from V_i to V_f can be expressed as the sum of two steps (Fig. 2.17). In the first step, only the volume changes and the temperature is held constant at its initial value. However, because the internal energy of a perfect gas is independent of the volume the molecules occupy, the overall change in internal energy arises solely from the second step, the change in temperature at constant volume. Provided the heat capacity is independent of temperature, this change is

$$\Delta U = C_V(T_f - T_i) = C_V\Delta T$$

Because the expansion is adiabatic, we know that $q = 0$; because $\Delta U = q + w$, it then follows that $\Delta U = w_{ad}$. The subscript 'ad' denotes an adiabatic process. Therefore, by equating the two expressions we have obtained for ΔU, we obtain

$$w_{ad} = C_V\Delta T \tag{2.27}$$

That is, the work done during an adiabatic expansion of a perfect gas is proportional to the temperature difference between the initial and final states. That is exactly what we expect on molecular grounds, because the mean kinetic energy is proportional to T, so a change in internal energy arising from temperature alone is also expected to be proportional to ΔT. In *Further information 2.1* we show that the initial and final temperatures of a perfect gas that undergoes reversible adiabatic expansion (reversible expansion in a thermally insulated container) can be calculated from

$$T_f = T_i\left(\frac{V_i}{V_f}\right)^{1/c} \qquad (2.28a)^{\circ}_{rev}$$

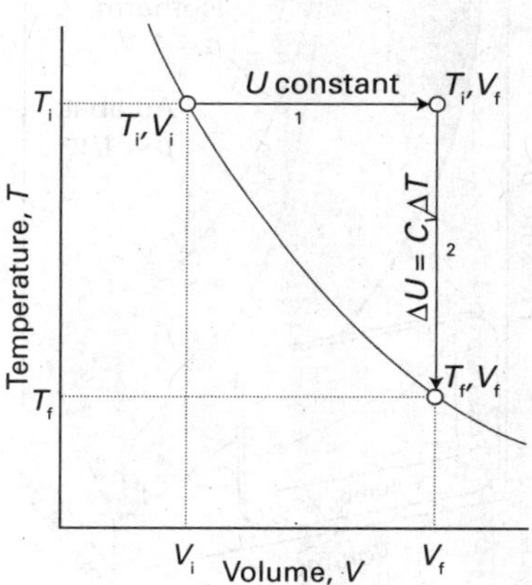

Fig. 2.17 To achieve a change of state from one temperature and volume to another temperature and volume, we may consider the overall change as composed of two steps. In the first step, the system expands at constant temperature; there is no change in internal energy if the system consists of a perfect gas. In the second step, the temperature of the system is reduced at constant volume. The overall change in internal energy is the sum of the changes for the two steps.

where $c = C_{V,\text{m}}/R$. By raising each side of this expression to the power c, an equivalent expression is

$$V_\text{i}T_\text{i}^c = V_\text{f}T_\text{f}^c \qquad (2.28\text{b})^\circ_\text{rev}$$

This result is often summarized in the form VT^c = constant.

• A brief illustration

Consider the adiabatic, reversible expansion of 0.020 mol Ar, initially at 25°C, from 0.50 dm^3 to 1.00 dm^3. The molar heat capacity of argon at constant volume is 12.48 J K^{-1} mol^{-1}, so c = 1.501. Therefore, from eqn 2.28a,

$$T_\text{f} = (298\text{ K}) \times \left(\frac{0.50\text{ dm}^3}{1.00\text{ dm}^3}\right)^{1/1.501} = 188\text{ K}$$

It follows that ΔT = −110 K and, therefore, from eqn 2.27, that

$$w = \{(0.020\text{ mol}) \times (12.48\text{ J K}^{-1}\text{ mol}^{-1})\} \times (-110\text{ K}) = -27\text{ J}$$

Note that temperature change is independent of the amount of gas but the work is not. •

Self-test 2.5 Calculate the final temperature, the work done, and the change of internal energy when ammonia is used in a reversible adiabatic expansion from 0.50 dm^3 to 2.00 dm^3, the other initial conditions being the same.

[195 K, −56 J, −56 J]

We also show in *Further information 2.1* that the pressure of a perfect gas that undergoes reversible adiabatic expansion from a volume V_i to a volume V_f is related to its initial pressure by

$$p_\text{f}V_\text{f}^\gamma = p_\text{i}V_\text{i}^\gamma \qquad \text{Reversible adiabatic expansion of a perfect gas} \qquad (2.29)^\circ_\text{rev}$$

where $\gamma = C_{p,\text{m}}/C_{V,\text{m}}$. This result is commonly summarized in the form pV^γ = constant. For a monatomic perfect gas (Section 2.2a), and from eqn 2.26 $C_{p,\text{m}} = \frac{5}{2}R$, so $\gamma = \frac{5}{3}$. For a gas of nonlinear polyatomic molecules (which can rotate as well as translate), $C_{V,\text{m}} = 3R$, so $\gamma = \frac{4}{3}$. The curves of pressure versus volume for adiabatic change are known as **adiabats**, and one for a reversible path is illustrated in Fig. 2.18. Because $\gamma > 1$, an adiabat falls more steeply ($p \propto 1/V^\gamma$) than the corresponding isotherm ($p \propto 1/V$). The physical reason for the difference is that, in an isothermal expansion, energy flows into the system as heat and maintains the temperature; as a result, the pressure does not fall as much as in an adiabatic expansion.

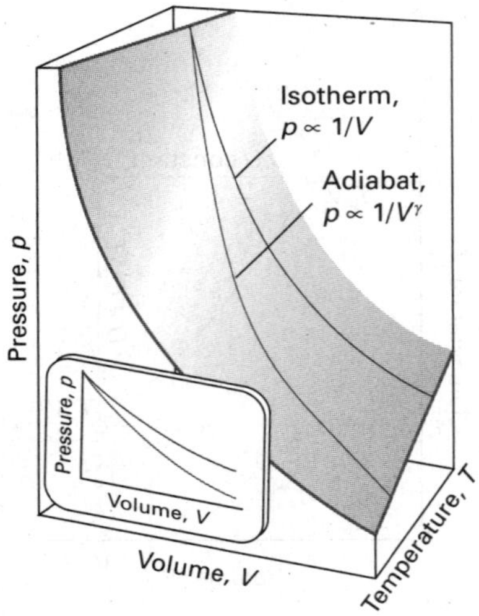

Fig. 2.18 An adiabat depicts the variation of pressure with volume when a gas expands adiabatically. Note that the pressure declines more steeply for an adiabat than it does for an isotherm because the temperature decreases in the former.

interActivity Explore how the parameter γ affects the dependence of the pressure on the volume. Does the pressure–volume dependence become stronger or weaker with increasing volume?

• A brief illustration

When a sample of argon (for which $\gamma = \frac{5}{3}$) at 100 kPa expands reversibly and adiabatically to twice its initial volume the final pressure will be

$$p_\text{f} = \left(\frac{V_\text{i}}{V_\text{f}}\right)^\gamma p_\text{i} = \left(\frac{1}{2}\right)^{5/3} \times (100\text{ kPa}) = 31.5\text{ kPa}$$

For an isothermal doubling of volume, the final pressure would be 50 kPa. •

Thermochemistry

The study of the energy transferred as heat during the course of chemical reactions is called **thermochemistry**. Thermochemistry is a branch of thermodynamics because a reaction vessel and its contents form a system, and chemical reactions result in the exchange of energy between the system and the surroundings. Thus we can use calorimetry to measure the energy supplied or discarded as heat by a reaction, and can identify q with a change in internal energy if the reaction occurs at constant volume or with a change in enthalpy if the reaction occurs at constant pressure. Conversely, if we know ΔU or ΔH for a reaction, we can predict the heat the reaction can produce.

We have already remarked that a process that releases energy as heat into the surroundings is classified as exothermic and one that absorbs energy as heat from the surroundings is classified as endothermic. Because the release of heat signifies a decrease in the enthalpy of a system, we can now see that an exothermic process is one for which $\Delta H < 0$. Conversely, because the absorption of heat results in an increase in enthalpy, an endothermic process has $\Delta H > 0$:

$$\text{exothermic process: } \Delta H < 0 \qquad \text{endothermic process: } \Delta H > 0$$

2.7 Standard enthalpy changes

Key points **(a) The standard enthalpy of transition is equal to the energy transferred as heat at constant pressure in the transition. (b) A thermochemical equation is a chemical equation and its associated change in enthalpy. (c) Hess's law states that the standard enthalpy of an overall reaction is the sum of the standard enthalpies of the individual reactions into which a reaction may be divided.**

Changes in enthalpy are normally reported for processes taking place under a set of standard conditions. In most of our discussions we shall consider the **standard enthalpy change**, $\Delta H^{\ominus}$, the change in enthalpy for a process in which the initial and final substances are in their standard states:

> The **standard state** of a substance at a specified temperature is its pure form at 1 bar. [Specification of standard state]

A brief comment
The definition of standard state is more sophisticated for a real gas (*Further information 3.2*) and for solutions (Sections 5.10 and 5.11).

For example, the standard state of liquid ethanol at 298 K is pure liquid ethanol at 298 K and 1 bar; the standard state of solid iron at 500 K is pure iron at 500 K and 1 bar. The standard enthalpy change for a reaction or a physical process is the difference between the products in their standard states and the reactants in their standard states, all at the same specified temperature.

As an example of a standard enthalpy change, the *standard enthalpy of vaporization*, $\Delta_{vap}H^{\ominus}$, is the enthalpy change per mole when a pure liquid at 1 bar vaporizes to a gas at 1 bar, as in

$$H_2O(l) \rightarrow H_2O(g) \qquad \Delta_{vap}H^{\ominus}(373\text{ K}) = +40.66\text{ kJ mol}^{-1}$$

A note on good practice The attachment of the name of the transition to the symbol Δ, as in $\Delta_{vap}H$, is the modern convention. However, the older convention, ΔH_{vap}, is still widely used. The new convention is more logical because the subscript identifies the type of change, not the physical observable related to the change.

As implied by the examples, standard enthalpies may be reported for any temperature. However, the conventional temperature for reporting thermodynamic data is 298.15 K (corresponding to 25.00°C). Unless otherwise mentioned, all thermodynamic data in this text will refer to this conventional temperature.

(a) Enthalpies of physical change

The standard enthalpy change that accompanies a change of physical state is called the **standard enthalpy of transition** and is denoted $\Delta_{trs}H^{\ominus}$ (Table 2.3). The **standard**

Table 2.3* Standard enthalpies of fusion and vaporization at the transition temperature, $\Delta_{trs}H^{\ominus}/(\text{kJ mol}^{-1})$

	T_f/K	Fusion	T_b/K	Vaporization
Ar	83.81	1.188	87.29	6.506
C_6H_6	278.61	10.59	353.2	30.8
H_2O	273.15	6.008	373.15	40.656 (44.016 at 298 K)
He	3.5	0.021	4.22	0.084

* More values are given in the *Data section*.

Table 2.4 Enthalpies of transition

Transition	Process	Symbol*
Transition	Phase $\alpha \rightarrow$ phase β	$\Delta_{trs}H$
Fusion	$s \rightarrow l$	$\Delta_{fus}H$
Vaporization	$l \rightarrow g$	$\Delta_{vap}H$
Sublimation	$s \rightarrow g$	$\Delta_{sub}H$
Mixing	Pure $\rightarrow$ mixture	$\Delta_{mix}H$
Solution	Solute $\rightarrow$ solution	$\Delta_{sol}H$
Hydration	$X^{\pm}(g) \rightarrow X^{\pm}(aq)$	$\Delta_{hyd}H$
Atomization	Species(s, l, g) $\rightarrow$ atoms(g)	$\Delta_{at}H$
Ionization	$X(g) \rightarrow X^{+}(g) + e^{-}(g)$	$\Delta_{ion}H$
Electron gain	$X(g) + e^{-}(g) \rightarrow X^{-}(g)$	$\Delta_{eg}H$
Reaction	Reactants $\rightarrow$ products	$\Delta_r H$
Combustion	Compound(s, l, g) + $O_2(g) \rightarrow CO_2(g)$, $H_2O(l, g)$	$\Delta_c H$
Formation	Elements $\rightarrow$ compound	$\Delta_f H$
Activation	Reactants $\rightarrow$ activated complex	$\Delta^{\ddagger}H$

* IUPAC recommendations. In common usage, the transition subscript is often attached to ΔH, as in ΔH_{trs}.

enthalpy of vaporization, $\Delta_{vap}H^{\ominus}$, is one example. Another is the **standard enthalpy of fusion**, $\Delta_{fus}H^{\ominus}$, the standard enthalpy change accompanying the conversion of a solid to a liquid, as in

$$H_2O(s) \rightarrow H_2O(l) \qquad \Delta_{fus}H^{\ominus}(273\text{ K}) = +6.01\text{ kJ mol}^{-1}$$

As in this case, it is sometimes convenient to know the standard enthalpy change at the transition temperature as well as at the conventional temperature of 298 K. The different types of enthalpies encountered in thermochemistry are summarized in Table 2.4. We shall meet them again in various locations throughout the text.

Because enthalpy is a state function, a change in enthalpy is independent of the path between the two states. This feature is of great importance in thermochemistry, for it implies that the same value of $\Delta H^{\ominus}$ will be obtained however the change is brought about between the same initial and final states. For example, we can picture the conversion of a solid to a vapour either as occurring by sublimation (the direct conversion from solid to vapour)

$$H_2O(s) \rightarrow H_2O(g) \qquad \Delta_{sub}H^{\ominus}$$

or as occurring in two steps, first fusion (melting) and then vaporization of the resulting liquid:

$$H_2O(s) \rightarrow H_2O(l) \qquad \Delta_{fus}H^{\ominus}$$
$$H_2O(l) \rightarrow H_2O(g) \qquad \Delta_{vap}H^{\ominus}$$
$$\text{Overall: } H_2O(s) \rightarrow H_2O(g) \qquad \Delta_{fus}H^{\ominus} + \Delta_{vap}H^{\ominus}$$

Because the overall result of the indirect path is the same as that of the direct path, the overall enthalpy change is the same in each case (**1**), and we can conclude that (for processes occurring at the same temperature)

$$\Delta_{sub}H^{\ominus} = \Delta_{fus}H^{\ominus} + \Delta_{vap}H^{\ominus} \qquad (2.30)$$

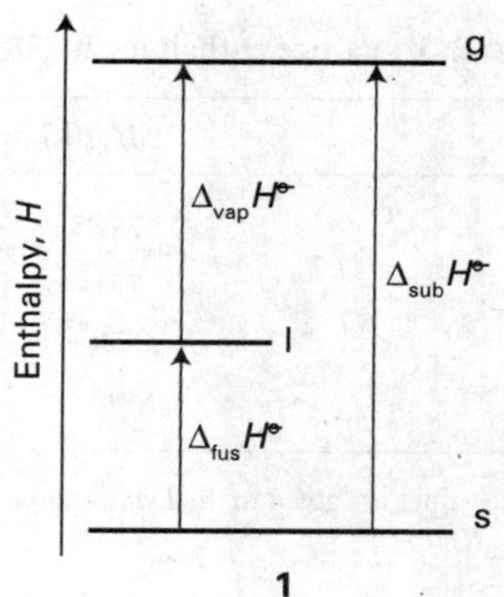

1

An immediate conclusion is that, because all enthalpies of fusion are positive, the enthalpy of sublimation of a substance is greater than its enthalpy of vaporization (at a given temperature).

Another consequence of H being a state function is that the standard enthalpy changes of a forward process and its reverse differ in sign (**2**):

$$\Delta H^{\ominus}(A \rightarrow B) = -\Delta H^{\ominus}(B \rightarrow A) \qquad (2.31)$$

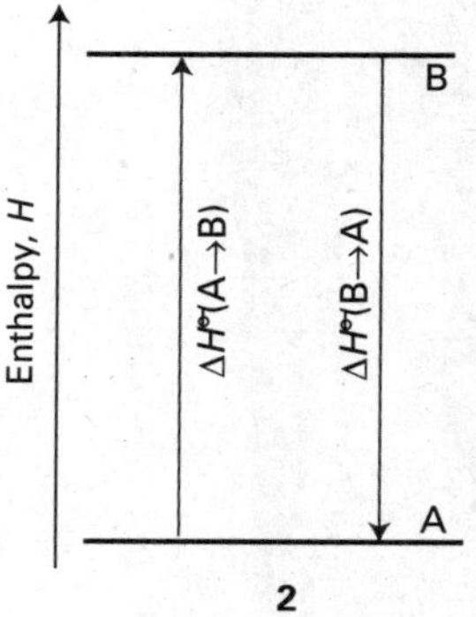

2

For instance, because the enthalpy of vaporization of water is +44 kJ mol^{-1} at 298 K, its enthalpy of condensation at that temperature is −44 kJ mol^{-1}.

The vaporization of a solid often involves a large increase in energy, especially when the solid is ionic and the strong Coulombic interaction of the ions must be overcome in a process such as

$$MX(s) \rightarrow M^+(g) + X^-(g)$$

The **lattice enthalpy**, ΔH_L, is the change in standard molar enthalpy for this process. The lattice enthalpy is equal to the lattice internal energy at $T = 0$; at normal temperatures they differ by only a few kilojoules per mole, and the difference is normally neglected.

Experimental values of the lattice enthalpy are obtained by using a **Born–Haber cycle**, a closed path of transformations starting and ending at the same point, one step of which is the formation of the solid compound from a gas of widely separated ions.

• A brief illustration

A typical Born–Haber cycle, for potassium chloride, is shown in Fig. 2.19. It consists of the following steps (for convenience, starting at the elements):

	$\Delta H^{\ominus}/(\text{kJ mol}^{-1})$	
1. Sublimation of K(s)	+89	[dissociation enthalpy of K(s)]
2. Dissociation of $\frac{1}{2}$ Cl_2(g)	+122	[$\frac{1}{2}\times$ dissociation enthalpy of Cl_2(g)]
3. Ionization of K(g)	+418	[ionization enthalpy of K(g)]
4. Electron attachment to Cl(g)	−349	[electron gain enthalpy of Cl(g)]
5. Formation of solid from gas	$-\Delta H_L/(\text{kJ mol}^{-1})$	
6. Decomposition of compound	+437	[negative of enthalpy of formation of KCl(s)]

Because the sum of these enthalpy changes is equal to zero, we can infer from

$$89 + 122 + 418 - 349 - \Delta H_L/(\text{kJ mol}^{-1}) + 437 = 0$$

that $\Delta H_L = +717$ kJ mol^{-1}. •

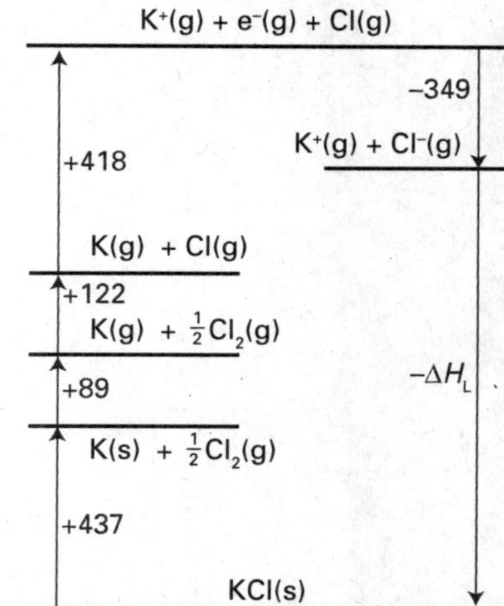

Fig. 2.19 The Born–Haber cycle for KCl at 298 K. Enthalpies changes are in kilojoules per mole.

Some lattice enthalpies obtained in the same way as in the *brief illustration* are listed in Table 2.5. They are large when the ions are highly charged and small, for then they are close together and attract each other strongly. We examine the quantitative relation between lattice enthalpy and structure in Section 19.6.

Table 2.5* Lattice enthalpies at 298 K

	$\Delta H_L/(\text{kJ mol}^{-1})$
NaF	787
NaBr	751
MgO	3850
MgS	3406

* More values are given in the *Data section*.

(b) Enthalpies of chemical change

Now we consider enthalpy changes that accompany chemical reactions. There are two ways of reporting the change in enthalpy that accompanies a chemical reaction. One is to write the **thermochemical equation**, a combination of a chemical equation and the corresponding change in standard enthalpy:

$$CH_4(g) + 2\,O_2(g) \rightarrow CO_2(g) + 2\,H_2O(l) \qquad \Delta H^{\ominus} = -890 \text{ kJ}$$

$\Delta H^{\ominus}$ is the change in enthalpy when reactants in their standard states change to products in their standard states:

Pure, separate reactants in their standard states
→ pure, separate products in their standard states

Except in the case of ionic reactions in solution, the enthalpy changes accompanying mixing and separation are insignificant in comparison with the contribution from the reaction itself. For the combustion of methane, the standard value refers to the reaction in which 1 mol CH_4 in the form of pure methane gas at 1 bar reacts completely with 2 mol O_2 in the form of pure oxygen gas at 1 bar to produce 1 mol CO_2 as pure carbon dioxide gas at 1 bar and 2 mol H_2O as pure liquid water at 1 bar; the numerical value is for the reaction at 298.15 K.

Alternatively, we write the chemical equation and then report the **standard reaction enthalpy**, $\Delta_r H^{\ominus}$ (or 'standard enthalpy of reaction'). Thus, for the combustion of methane, we write

$$CH_4(g) + 2\,O_2(g) \rightarrow CO_2(g) + 2\,H_2O(l) \qquad \Delta_r H^{\ominus} = -890 \text{ kJ mol}^{-1}$$

For a reaction of the form 2 A + B → 3 C + D the standard reaction enthalpy would be

$$\Delta_r H^{\ominus} = \{3H_m^{\ominus}(C) + H_m^{\ominus}(D)\} - \{2H_m^{\ominus}(A) + H_m^{\ominus}(B)\}$$

where $H_m^{\ominus}(J)$ is the standard molar enthalpy of species J at the temperature of interest. Note how the 'per mole' of $\Delta_r H^{\ominus}$ comes directly from the fact that molar enthalpies appear in this expression. We interpret the 'per mole' by noting the stoichiometric coefficients in the chemical equation. In this case 'per mole' in $\Delta_r H^{\ominus}$ means 'per 2 mol A', 'per mole B', 'per 3 mol C', or 'per mol D'. In general,

$$\Delta_r H^{\ominus} = \sum_{\text{Products}} \nu H_m^{\ominus} - \sum_{\text{Reactants}} \nu H_m^{\ominus} \qquad [2.32]$$

Definition of standard reaction enthalpy

where in each case the molar enthalpies of the species are multiplied by their (dimensionless and positive) stoichiometric coefficients, ν.

Some standard reaction enthalpies have special names and a particular significance. For instance, the **standard enthalpy of combustion**, $\Delta_c H^{\ominus}$, is the standard reaction enthalpy for the complete oxidation of an organic compound to CO_2 gas and liquid H_2O if the compound contains C, H, and O, and to N_2 gas if N is also present. An example is the combustion of glucose:

$$C_6H_{12}O_6(s) + 6\,O_2(g) \rightarrow 6\,CO_2(g) + 6\,H_2O(l) \qquad \Delta_c H^{\ominus} = -2808 \text{ kJ mol}^{-1}$$

The value quoted shows that 2808 kJ of heat is released when 1 mol $C_6H_{12}O_6$ burns under standard conditions (at 298 K). More values are given in Table 2.6.

(c) Hess's law

Standard enthalpies of individual reactions can be combined to obtain the enthalpy of another reaction. This application of the First Law is called **Hess's law**:

Table 2.6* Standard enthalpies of formation and combustion of organic compounds at 298 K

	$\Delta_f H^\ominus/(\text{kJ mol}^{-1})$	$\Delta_c H^\ominus/(\text{kJ mol}^{-1})$
Benzene, $C_6H_6(l)$	+49.0	−3268
Ethane, $C_2H_6(g)$	−84.7	−1560
Glucose, $C_6H_{12}O_6(s)$	−1274	−2808
Methane, $CH_4(g)$	−74.8	−890
Methanol, $CH_3OH(l)$	−238.7	−726

* More values are given in the *Data section*.

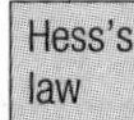

The standard enthalpy of an overall reaction is the sum of the standard enthalpies of the individual reactions into which a reaction may be divided.

The individual steps need not be realizable in practice: they may be hypothetical reactions, the only requirement being that their chemical equations should balance. The thermodynamic basis of the law is the path-independence of the value of $\Delta_r H^\ominus$ and the implication that we may take the specified reactants, pass through any (possibly hypothetical) set of reactions to the specified products, and overall obtain the same change of enthalpy. The importance of Hess's law is that information about a reaction of interest, which may be difficult to determine directly, can be assembled from information on other reactions.

Example 2.5 *Using Hess's law*

The standard reaction enthalpy for the hydrogenation of propene

$$CH_2{=}CHCH_3(g) + H_2(g) \rightarrow CH_3CH_2CH_3(g)$$

is −124 kJ mol^{-1}. The standard reaction enthalpy for the combustion of propane

$$CH_3CH_2CH_3(g) + 5\,O_2(g) \rightarrow 3\,CO_2(g) + 4\,H_2O(l)$$

is −2220 kJ mol^{-1}. Calculate the standard enthalpy of combustion of propene.

Method The skill to develop is the ability to assemble a given thermochemical equation from others. Add or subtract the reactions given, together with any others needed, so as to reproduce the reaction required. Then add or subtract the reaction enthalpies in the same way. Additional data are in Table 2.6.

Answer The combustion reaction we require is

$$C_3H_6(g) + \tfrac{9}{2}O_2(g) \rightarrow 3\,CO_2(g) + 3\,H_2O(l)$$

This reaction can be recreated from the following sum:

	$\Delta_r H^\ominus/(\text{kJ mol}^{-1})$
$C_3H_6(g) + H_2(g) \rightarrow C_3H_8(g)$	−124
$C_3H_8(g) + 5\,O_2(g) \rightarrow 3\,CO_2(g) + 4\,H_2O(l)$	−2220
$H_2O(l) \rightarrow H_2(g) + \tfrac{1}{2}O_2(g)$	+286
$C_3H_6(g) + \tfrac{9}{2}O_2(g) \rightarrow 3\,CO_2(g) + 3\,H_2O(l)$	−2058

Self-test 2.6 Calculate the enthalpy of hydrogenation of benzene from its enthalpy of combustion and the enthalpy of combustion of cyclohexane. [−205 kJ mol^{-1}]

Table 2.7 Thermochemical properties of some fuels

Fuel	Combustion equation	$\Delta_c H^{\ominus}$/(kJ mol^{-1})	Specific enthalpy/(kJ g^{-1})	Enthalpy density/(kJ dm^{-3})
Hydrogen	$H_2(g) + \frac{1}{2}O_2(g) \rightarrow H_2O(l)$	−286	142	13
Methane	$CH_4(g) + 2\,O_2(g) \rightarrow CO_2(g) + 2\,H_2O(l)$	−890	55	40
Octane	$C_8H_{18}(l) + \frac{25}{2}O_2(g) \rightarrow 8\,CO_2(g) + 9\,H_2O(l)$	−5471	48	3.8×10^4
Methanol	$CH_3OH(l) + \frac{3}{2}O_2(g) \rightarrow CO_2(g) + 2\,H_2O(l)$	−726	23	1.8×10^4

IMPACT ON BIOLOGY

I2.2 Food and energy reserves

The thermochemical properties of fuels and foods are commonly discussed in terms of their *specific enthalpy*, the enthalpy of combustion per gram of material. Thus, if the standard enthalpy of combustion is $\Delta_c H^{\ominus}$ and the molar mass of the compound is M, then the specific enthalpy is $\Delta_c H^{\ominus}/M$. Table 2.7 lists the specific enthalpies of several fuels.

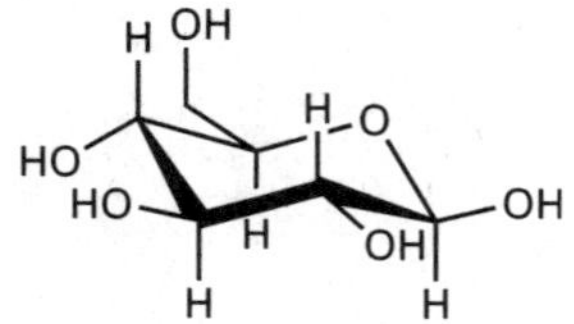

3 α-D-Glucose (α-D-Glucopyranose)

A typical 18–20 year old man requires a daily input of about 12 MJ; a woman of the same age needs about 9 MJ. If the entire consumption were in the form of glucose (3; which has a specific enthalpy of 16 kJ g^{-1}), that would require the consumption of 750 g of glucose for a man and 560 g for a woman. In fact, digestible carbohydrates have a slightly higher specific enthalpy (17 kJ g^{-1}) than glucose itself, so a carbohydrate diet is slightly less daunting than a pure glucose diet, as well as being more appropriate in the form of fibre, the indigestible cellulose that helps move digestion products through the intestine.

Fats are long-chain esters like tristearin (beef fat). The enthalpy of combustion of a fat at around 38 kJ g^{-1} is much greater than that of carbohydrates and only slightly less than that of the hydrocarbon oils used as fuel (48 kJ g^{-1}). Fats are commonly used as an energy store, to be used only when the more readily accessible carbohydrates have fallen into short supply. In Arctic species, the stored fat also acts as a layer of insulation; in desert species (such as the camel), the fat is also a source of water, one of its oxidation products.

Proteins are also used as a source of energy, but their components, the amino acids, are often too valuable to squander in this way, and are used to construct other proteins instead. When proteins are oxidized (to urea, $CO(NH_2)_2$), the equivalent enthalpy density is comparable to that of carbohydrates.

The heat released by the oxidation of foods needs to be discarded in order to maintain body temperature within its typical range of 35.6–37.8°C. A variety of mechanisms contribute to this aspect of homeostasis, the ability of an organism to counteract environmental changes with physiological responses. The general uniformity of temperature throughout the body is maintained largely by the flow of blood. When heat needs to be dissipated rapidly, warm blood is allowed to flow through the capillaries of the skin, so producing flushing. Radiation is one means of discarding heat; another is evaporation and the energy demands of the enthalpy of vaporization of water. Evaporation removes about 2.4 kJ per gram of water perspired. When vigorous exercise promotes sweating (through the influence of heat selectors on the hypothalamus), 1–2 dm^3 of perspired water can be produced per hour, corresponding to a heat loss of 2.4–5.0 MJ h^{-1}.

2.8 Standard enthalpies of formation

Key points **Standard enthalpies of formation are defined in terms of the reference states of elements. (a) The standard reaction enthalpy is expressed as the difference of the standard enthalpies of formation of products and reactants. (b) Computer modelling is used to estimate standard enthalpies of formation.**

The **standard enthalpy of formation**, $\Delta_f H^\ominus$, of a substance is the standard reaction enthalpy for the formation of the compound from its elements in their reference states:

The **reference state** of an element is its most stable state at the specified temperature and 1 bar.

Specification of reference state

For example, at 298 K the reference state of nitrogen is a gas of N_2 molecules, that of mercury is liquid mercury, that of carbon is graphite, and that of tin is the white (metallic) form. There is one exception to this general prescription of reference states: the reference state of phosphorus is taken to be white phosphorus despite this allotrope not being the most stable form but simply the more reproducible form of the element. Standard enthalpies of formation are expressed as enthalpies per mole of molecules or (for ionic substances) formula units of the compound. The standard enthalpy of formation of liquid benzene at 298 K, for example, refers to the reaction

$$6\,\mathrm{C(s, graphite)} + \mathrm{H_2(g)} \rightarrow \mathrm{C_6H_6(l)}$$

and is +49.0 kJ mol^{-1}. The standard enthalpies of formation of elements in their reference states are zero at all temperatures because they are the enthalpies of such 'null' reactions as $N_2(g) \rightarrow N_2(g)$. Some enthalpies of formation are listed in Tables 2.6 and 2.8.

The standard enthalpy of formation of ions in solution poses a special problem because it is impossible to prepare a solution of cations alone or of anions alone. This problem is solved by defining one ion, conventionally the hydrogen ion, to have zero standard enthalpy of formation at all temperatures:

$$\Delta_f H^\ominus(\mathrm{H^+, aq}) = 0 \qquad [2.33]$$

Convention for ions in solution

Thus, if the enthalpy of formation of HBr(aq) is found to be −122 kJ mol^{-1}, then the whole of that value is ascribed to the formation of Br^-(aq), and we write $\Delta_f H^\ominus(\mathrm{Br^-, aq}) = -122$ kJ mol^{-1}. That value may then be combined with, for instance, the enthalpy formation of AgBr(aq) to determine the value of $\Delta_f H^\ominus(\mathrm{Ag^+, aq})$, and so on. In essence, this definition adjusts the actual values of the enthalpies of formation of ions by a fixed amount, which is chosen so that the standard value for one of them, H^+(aq), has the value zero.

Table 2.8* Standard enthalpies of formation of inorganic compounds at 298 K

	$\Delta_f H^\ominus$/(kJ mol^{-1})
H_2O(l)	−285.83
H_2O(g)	−241.82
NH_3(g)	−46.11
N_2H_4(l)	+50.63
NO_2(g)	+33.18
N_2O_4(g)	+9.16
NaCl(s)	−411.15
KCl(s)	−436.75

* More values are given in the *Data section*.

(a) The reaction enthalpy in terms of enthalpies of formation

Conceptually, we can regard a reaction as proceeding by decomposing the reactants into their elements and then forming those elements into the products. The value of $\Delta_r H^\ominus$ for the overall reaction is the sum of these 'unforming' and forming enthalpies. Because 'unforming' is the reverse of forming, the enthalpy of an unforming step is the negative of the enthalpy of formation (4). Hence, in the enthalpies of formation of substances, we have enough information to calculate the enthalpy of any reaction by using

$$\Delta_r H^\ominus = \sum_{\text{Products}} \nu \Delta_f H^\ominus - \sum_{\text{Reactants}} \nu \Delta_f H^\ominus \qquad (2.34a)$$

Procedure for calculating standard reaction enthalpy

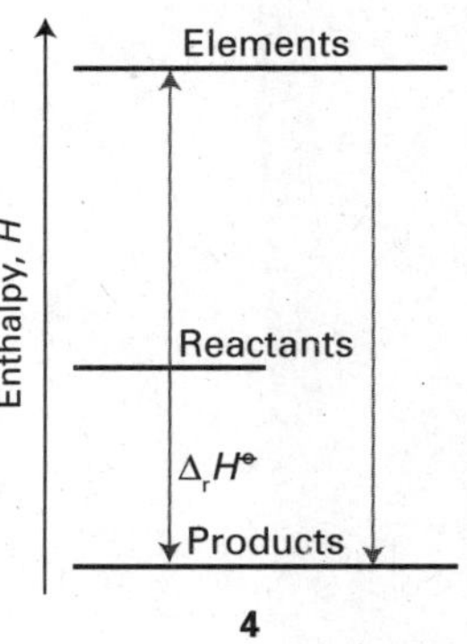

4

A brief comment

Stoichiometric *numbers*, which have a sign, are denoted ν_J or $\nu(J)$. Stoichiometric *coefficients*, which are all positive, are denoted simply ν (with no subscript).

where in each case the enthalpies of formation of the species that occur are multiplied by their stoichiometric coefficients. A more sophisticated way of expressing the same result is to introduce the **stoichiometric numbers** ν_J (as distinct from the stoichiometric coefficients) which are positive for products and negative for reactants. Then we can write

$$\Delta_r H^\ominus = \sum_J \nu_J \Delta_f H^\ominus(J) \qquad (2.34b)$$

• A brief illustration

According to eqn 2.34a, the standard enthalpy of the reaction $2\,HN_3(l) + 2\,NO(g) \rightarrow H_2O_2(l) + 4\,N_2(g)$ is calculated as follows:

$$\begin{aligned}\Delta_r H^\ominus &= \{\Delta_f H^\ominus(H_2O_2,l) + 4\Delta_f H^\ominus(N_2,g)\} - \{2\Delta_f H^\ominus(HN_3,l) + 2\Delta_f H^\ominus(NO,g)\}\\ &= \{-187.78 + 4(0)\}\ \text{kJ mol}^{-1} - \{2(264.0) + 2(90.25)\}\ \text{kJ mol}^{-1}\\ &= -896.3\ \text{kJ mol}^{-1}\end{aligned}$$

To use eqn 2.34b we identify $\nu(HN_3) = -2$, $\nu(NO) = -2$, $\nu(H_2O_2) = +1$, and $\nu(N_2) = +4$, and then write

$$\Delta_r H^\ominus = -2\Delta_f H^\ominus(HN_3,l) - 2\Delta_f H^\ominus(NO,g) + \Delta_f H^\ominus(H_2O_2,l) + 4\Delta_f H^\ominus(N_2,g)$$

which gives the same result. •

(b) Enthalpies of formation and molecular modelling

We have seen how to construct standard reaction enthalpies by combining standard enthalpies of formation. The question that now arises is whether we can construct standard enthalpies of formation from a knowledge of the chemical constitution of the species. The short answer is that there is no thermodynamically exact way of expressing enthalpies of formation in terms of contributions from individual atoms and bonds. In the past, approximate procedures based on **mean bond enthalpies**, $\Delta H(A\text{–}B)$, the average enthalpy change associated with the breaking of a specific A–B bond,

$$A\text{–}B(g) \rightarrow A(g) + B(g) \qquad \Delta H(A\text{–}B)$$

have been used. However, this procedure is notoriously unreliable, in part because the $\Delta H(A\text{–}B)$ are average values for a series of related compounds. Nor does the approach distinguish between geometrical isomers, where the same atoms and bonds may be present but experimentally the enthalpies of formation might be significantly different.

Computer-aided molecular modelling has largely displaced this more primitive approach. Commercial software packages use the principles developed in Chapter 10 to calculate the standard enthalpy of formation of a molecule drawn on the computer screen. These techniques can be applied to different conformations of the same molecule. In the case of methylcyclohexane, for instance, the calculated conformational energy difference ranges from 5.9 to 7.9 kJ mol^{-1}, with the equatorial conformer having the lower standard enthalpy of formation. These estimates compare favourably with the experimental value of 7.5 kJ mol^{-1}. However, good agreement between calculated and experimental values is relatively rare. Computational methods almost always predict correctly which conformer is more stable but do not always predict the correct magnitude of the conformational energy difference. The most reliable technique for the determination of enthalpies of formation remains calorimetry, typically by using enthalpies of combustion.

2.9 The temperature dependence of reaction enthalpies

Key point **The temperature dependence of a reaction enthalpy is expressed by Kirchhoff's law.**

The standard enthalpies of many important reactions have been measured at different temperatures. However, in the absence of this information, standard reaction enthalpies at different temperatures may be calculated from heat capacities and the reaction enthalpy at some other temperature (Fig. 2.20). In many cases heat capacity data are more accurate than reaction enthalpies. Therefore, providing the information is available, the procedure we are about to describe is more accurate than the direct measurement of a reaction enthalpy at an elevated temperature.

It follows from eqn 2.23a that, when a substance is heated from T_1 to T_2, its enthalpy changes from $H(T_1)$ to

$$H(T_2) = H(T_1) + \int_{T_1}^{T_2} C_p \mathrm{d}T \qquad (2.35)$$

(We have assumed that no phase transition takes place in the temperature range of interest.) Because this equation applies to each substance in the reaction, the standard reaction enthalpy changes from $\Delta_r H^\ominus(T_1)$ to

$$\Delta_r H^\ominus(T_2) = \Delta_r H^\ominus(T_1) + \int_{T_1}^{T_2} \Delta_r C_p^\ominus \mathrm{d}T \qquad (2.36a)$$

Kirchhoff's law

where $\Delta_r C_p^\ominus$ is the difference of the molar heat capacities of products and reactants under standard conditions weighted by the stoichiometric coefficients that appear in the chemical equation:

$$\Delta_r C_p^\ominus = \sum_{\text{Products}} \nu C_{p,\mathrm{m}}^\ominus - \sum_{\text{Reactants}} \nu C_{p,\mathrm{m}}^\ominus \qquad (2.36b)$$

Equation 2.36a is known as **Kirchhoff's law**. It is normally a good approximation to assume that $\Delta_r C_p^\ominus$ is independent of the temperature, at least over reasonably limited ranges. Although the individual heat capacities may vary, their difference varies less significantly. In some cases the temperature dependence of heat capacities is taken into account by using eqn 2.25.

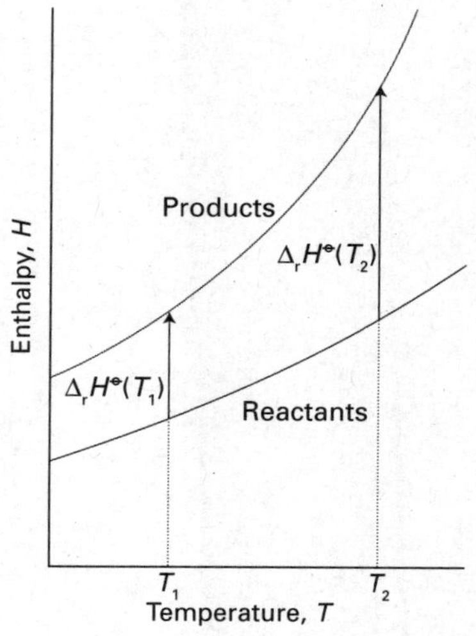

Fig. 2.20 An illustration of the content of Kirchhoff's law. When the temperature is increased, the enthalpy of the products and the reactants both increase, but may do so to different extents. In each case, the change in enthalpy depends on the heat capacities of the substances. The change in reaction enthalpy reflects the difference in the changes of the enthalpies.

Example 2.6 *Using Kirchhoff's law*

The standard enthalpy of formation of H_2O(g) at 298 K is −241.82 kJ mol^{-1}. Estimate its value at 100°C given the following values of the molar heat capacities at constant pressure: H_2O(g): 33.58 J K^{-1} mol^{-1}; H_2(g): 28.82 J K^{-1} mol^{-1}; O_2(g): 29.36 J K^{-1} mol^{-1}. Assume that the heat capacities are independent of temperature.

Method When $\Delta C_p^\ominus$ is independent of temperature in the range T_1 to T_2, the integral in eqn 2.36a evaluates to $(T_2 - T_1)\Delta_r C_p^\ominus$. Therefore,

$$\Delta_r H^\ominus(T_2) = \Delta_r H^\ominus(T_1) + (T_2 - T_1)\Delta_r C_p^\ominus$$

To proceed, write the chemical equation, identify the stoichiometric coefficients, and calculate $\Delta_r C_p^\ominus$ from the data.

Answer The reaction is $H_2(g) + \frac{1}{2}O_2(g) \rightarrow H_2O(g)$, so

$$\Delta_r C_p^\ominus = C_{p,\mathrm{m}}^\ominus(\mathrm{H_2O,g}) - \{C_{p,\mathrm{m}}^\ominus(\mathrm{H_2,g}) + \tfrac{1}{2}C_{p,\mathrm{m}}^\ominus(\mathrm{O_2,g})\} = -9.92\ \mathrm{J\,K^{-1}\,mol^{-1}}$$

It then follows that

$$\Delta_r H^{\ominus}(373\ \text{K}) = -241.82\ \text{kJ mol}^{-1} + (75\ \text{K}) \times (-9.92\ \text{J K}^{-1}\ \text{mol}^{-1}) = -242.6\ \text{kJ mol}^{-1}$$

Self-test 2.7 Estimate the standard enthalpy of formation of cyclohexane, $C_6H_{12}(l)$, at 400 K from the data in Table 2.6. [-163 kJ mol^{-1}]

State functions and exact differentials

We saw in Section 2.2 that a state function is a property that depends only on the current state of a system and is independent of its history. The internal energy and enthalpy are two examples of state functions. Physical quantities that do depend on the path between two states are called **path functions**. Examples of path functions are the work and the heating that are done when preparing a state. We do not speak of a system in a particular state as possessing work or heat. In each case, the energy transferred as work or heat relates to the path being taken between states, not the current state itself.

A part of the richness of thermodynamics is that it uses the mathematical properties of state functions to draw far-reaching conclusions about the relations between physical properties and thereby establish connections that may be completely unexpected. The practical importance of this ability is that we can combine measurements of different properties to obtain the value of a property we require.

2.10 Exact and inexact differentials

Key points **The quantity dU is an exact differential; dw and dq are not.**

Consider a system undergoing the changes depicted in Fig. 2.21. The initial state of the system is i and in this state the internal energy is U_i. Work is done by the system as it expands adiabatically to a state f. In this state the system has an internal energy U_f and the work done on the system as it changes along Path 1 from i to f is w. Notice our use of language: U is a property of the state; w is a property of the path. Now consider another process, Path 2, in which the initial and final states are the same as those in Path 1 but in which the expansion is not adiabatic. The internal energy of both the initial and the final states are the same as before (because U is a state function). However, in the second path an energy q' enters the system as heat and the work w' is not the same as w. The work and the heat are path functions.

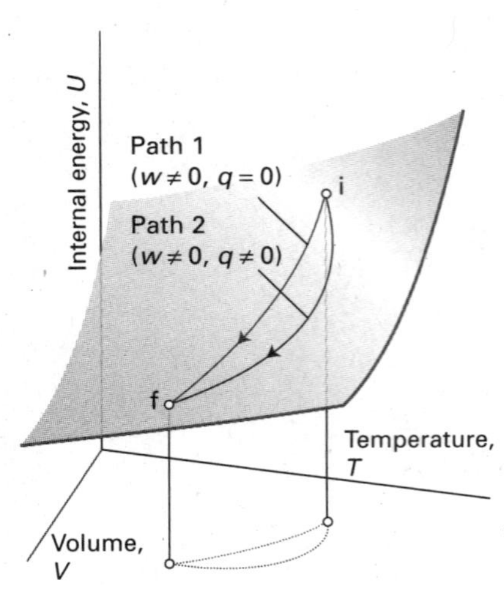

Fig. 2.21 As the volume and temperature of a system are changed, the internal energy changes. An adiabatic and a non-adiabatic path are shown as Path 1 and Path 2, respectively: they correspond to different values of q and w but to the same value of ΔU.

If a system is taken along a path (for example, by heating it), U changes from U_i to U_f, and the overall change is the sum (integral) of all the infinitesimal changes along the path:

$$\Delta U = \int_i^f dU \tag{2.37}$$

The value of ΔU depends on the initial and final states of the system but is independent of the path between them. This path independence of the integral is expressed by saying that dU is an 'exact differential'. In general, an **exact differential** is an infinitesimal quantity that, when integrated, gives a result that is independent of the path between the initial and final states.

When a system is heated, the total energy transferred as heat is the sum of all individual contributions at each point of the path:

$$q = \int_{i,\,\text{path}}^{f} dq \qquad (2.38)$$

Notice the differences between this equation and eqn 2.37. First, we do not write Δq, because q is not a state function and the energy supplied as heat cannot be expressed as $q_f - q_i$. Secondly, we must specify the path of integration because q depends on the path selected (for example, an adiabatic path has $q = 0$, whereas a non-adiabatic path between the same two states would have $q \neq 0$). This path-dependence is expressed by saying that dq is an 'inexact differential'. In general, an **inexact differential** is an infinitesimal quantity that, when integrated, gives a result that depends on the path between the initial and final states. Often dq is written $đq$ to emphasize that it is inexact and requires the specification of a path.

The work done on a system to change it from one state to another depends on the path taken between the two specified states; for example, in general the work is different if the change takes place adiabatically and non-adiabatically. It follows that dw is an inexact differential. It is often written $đw$.

Example 2.7 *Calculating work, heat, and change in internal energy*

Consider a perfect gas inside a cylinder fitted with a piston. Let the initial state be T,V_i and the final state be T,V_f. The change of state can be brought about in many ways, of which the two simplest are the following: Path 1, in which there is free expansion against zero external pressure; Path 2, in which there is reversible, isothermal expansion. Calculate w, q, and ΔU for each process.

Method To find a starting point for a calculation in thermodynamics, it is often a good idea to go back to first principles and to look for a way of expressing the quantity we are asked to calculate in terms of other quantities that are easier to calculate. We saw in Section 2.2a that the internal energy of a perfect gas depends only on the temperature and is independent of the volume those molecules occupy, so for any isothermal change, $\Delta U = 0$. We also know that in general $\Delta U = q + w$. The question depends on being able to combine the two expressions. We have already derived a number of expressions for the work done in a variety of processes, and here we need to select the appropriate ones.

Answer Because $\Delta U = 0$ for both paths and $\Delta U = q + w$, in each case $q = -w$. The work of free expansion is zero (Section 2.3b), so in Path 1, $w = 0$ and therefore $q = 0$ too. For Path 2, the work is given by eqn 2.10, so $w = -nRT\ln(V_f/V_i)$ and consequently $q = nRT\ln(V_f/V_i)$.

Self-test 2.8 Calculate the values of q, w, and ΔU for an irreversible isothermal expansion of a perfect gas against a constant non-zero external pressure.

$[q = p_{ex}\Delta V,\ w = -p_{ex}\Delta V,\ \Delta U = 0]$

2.11 Changes in internal energy

Key points **(a) The change in internal energy may be expressed in terms of changes in temperature and volume. The internal pressure is the variation of internal energy with volume at constant temperature. (b) Joule's experiment showed that the internal pressure of a perfect gas is zero. (c) The change in internal energy with volume and temperature is expressed in terms of the internal pressure and the heat capacity and leads to a general expression for the relation between heat capacities.**

We begin to unfold the consequences of dU being an exact differential by exploring a closed system of constant composition (the only type of system considered in the rest of this chapter). The internal energy U can be regarded as a function of V, T, and p, but, because there is an equation of state, stating the values of two of the variables fixes the value of the third. Therefore, it is possible to write U in terms of just two independent variables: V and T, p and T, or p and V. Expressing U as a function of volume and temperature fits the purpose of our discussion.

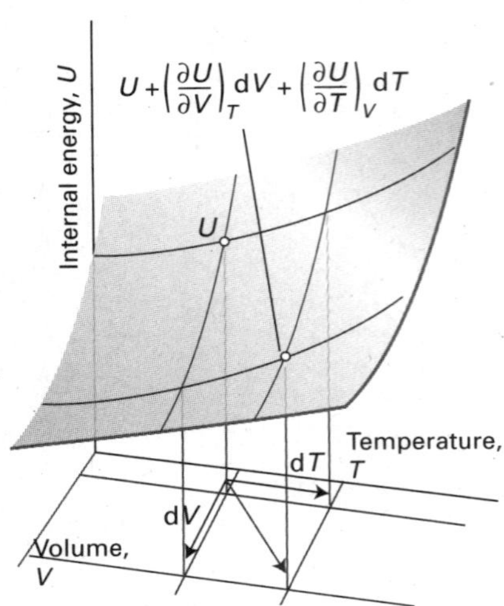

Fig. 2.22 An overall change in U, which is denoted dU, arises when both V and T are allowed to change. If second-order infinitesimals are ignored, the overall change is the sum of changes for each variable separately.

(a) General considerations

Because the internal energy is a function of the volume and the temperature, when these two quantities change, the internal energy changes by

$$dU = \left(\frac{\partial U}{\partial V}\right)_T dV + \left(\frac{\partial U}{\partial T}\right)_V dT \qquad (2.39)$$

General expression for a change in U with T and V

The interpretation of this equation is that, in a closed system of constant composition, any infinitesimal change in the internal energy is proportional to the infinitesimal changes of volume and temperature, the coefficients of proportionality being the two partial derivatives (Fig. 2.22).

In many cases partial derivatives have a straightforward physical interpretation, and thermodynamics gets shapeless and difficult only when that interpretation is not kept in sight. In the present case, we have already met $(\partial U/\partial T)_V$ in eqn 2.15, where we saw that it is the constant-volume heat capacity, C_V. The other coefficient, $(\partial U/\partial V)_T$, plays a major role in thermodynamics because it is a measure of the variation of the internal energy of a substance as its volume is changed at constant temperature (Fig. 2.23). We shall denote it π_T and, because it has the same dimensions as pressure but arises from the interactions between the molecules within the sample, call it the **internal pressure**:

$$\pi_T = \left(\frac{\partial U}{\partial V}\right)_T \qquad [2.40]$$

Definition of internal pressure

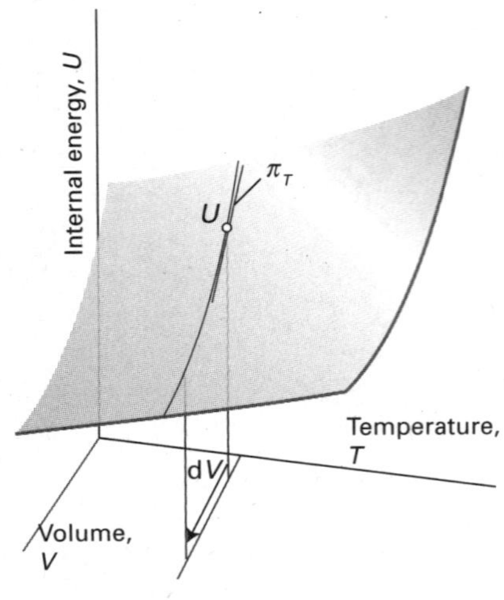

Fig. 2.23 The internal pressure, π_T, is the slope of U with respect to V with the temperature T held constant.

In terms of the notation C_V and π_T, eqn 2.39 can now be written

$$dU = \pi_T dV + C_V dT \qquad (2.41)$$

(b) The Joule experiment

When there are no interactions between the molecules, the internal energy is independent of their separation and hence independent of the volume of the sample (Section 2.2a). Therefore, for a perfect gas we can write $\pi_T = 0$. The statement $\pi_T = 0$ (that is, the internal energy is independent of the volume occupied by the sample) can be taken to be the definition of a perfect gas, for later we shall see that it implies the equation of state $pV \propto T$. If the attractive forces between the particles dominate the repulsive forces, then the internal energy increases ($dU > 0$) as the volume of the sample increases ($dV > 0$) and the molecules attract each other less strongly; in this case a plot of internal energy against volume slopes upwards and $\pi_T > 0$ (Fig. 2.24).

James Joule thought that he could measure π_T by observing the change in temperature of a gas when it is allowed to expand into a vacuum. He used two metal vessels immersed in a water bath (Fig. 2.25). One was filled with air at about 22 atm and the other was evacuated. He then tried to measure the change in temperature of the water of the bath when a stopcock was opened and the air expanded into a vacuum. He observed no change in temperature.

The thermodynamic implications of the experiment are as follows. No work was done in the expansion into a vacuum, so $w = 0$. No energy entered or left the system (the gas) as heat because the temperature of the bath did not change, so $q = 0$. Consequently, within the accuracy of the experiment, $\Delta U = 0$. Joule concluded that U does not change when a gas expands isothermally and therefore that $\pi_T = 0$. His experiment, however, was crude. In particular, the heat capacity of the apparatus was so large that the temperature change that gases do in fact cause was too small to measure. Nevertheless, from his experiment Joule had extracted an essential limiting property of a gas, a property of a perfect gas, without detecting the small deviations characteristic of real gases.

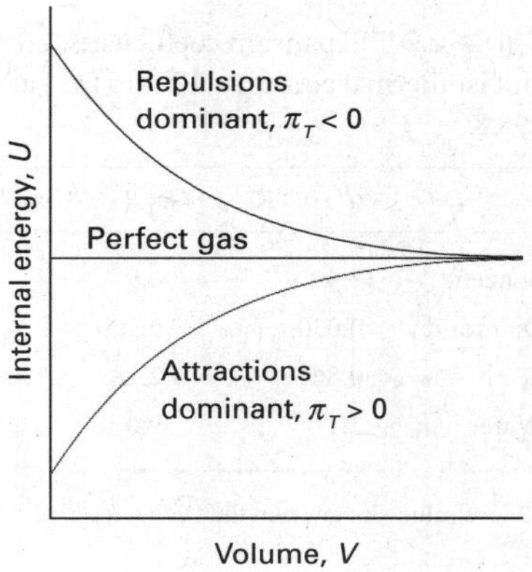

Fig. 2.24 For a perfect gas, the internal energy is independent of the volume (at constant temperature). If attractions are dominant in a real gas, the internal energy increases with volume because the molecules become farther apart on average. If repulsions are dominant, the internal energy decreases as the gas expands.

(c) Changes in internal energy at constant pressure

Partial derivatives have many useful properties and some that we shall draw on frequently are reviewed in *Mathematical background 2*. Skilful use of them can often turn some unfamiliar quantity into a quantity that can be recognized, interpreted, or measured.

As an example, suppose we want to find out how the internal energy varies with temperature when the pressure rather than the volume of the system is kept constant. If we divide both sides of eqn 2.41 ($\mathrm{d}U = \pi_T \mathrm{d}V + C_V \mathrm{d}T$) by $\mathrm{d}T$ and impose the condition of constant pressure on the resulting differentials, so that $\mathrm{d}U/\mathrm{d}T$ on the left becomes $(\partial U/\partial T)_p$, we obtain

$$\left(\frac{\partial U}{\partial T}\right)_p = \pi_T \left(\frac{\partial V}{\partial T}\right)_p + C_V$$

It is usually sensible in thermodynamics to inspect the output of a manipulation like this to see if it contains any recognizable physical quantity. The partial derivative on the right in this expression is the slope of the plot of volume against temperature (at constant pressure). This property is normally tabulated as the **expansion coefficient**, α, of a substance, which is defined as

$$\alpha = \frac{1}{V}\left(\frac{\partial V}{\partial T}\right)_p \qquad \text{Definition of the expansion coefficient} \qquad [2.42]$$

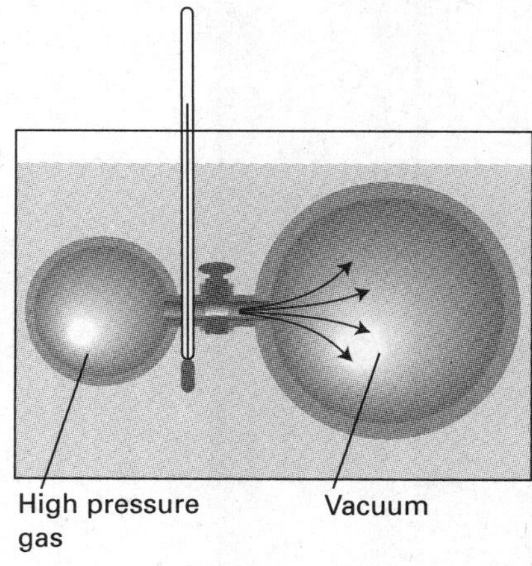

Fig. 2.25 A schematic diagram of the apparatus used by Joule in an attempt to measure the change in internal energy when a gas expands isothermally. The heat absorbed by the gas is proportional to the change in temperature of the bath.

and physically is the fractional change in volume that accompanies a rise in temperature. A large value of α means that the volume of the sample responds strongly to changes in temperature. Table 2.9 lists some experimental values of α. For future reference, it also lists the **isothermal compressibility**, κ_T (kappa), which is defined as

$$\kappa_T = -\frac{1}{V}\left(\frac{\partial V}{\partial p}\right)_T \qquad \text{Definition of the isothermal compressibility} \qquad [2.43]$$

The isothermal compressibility is a measure of the fractional change in volume when the pressure is increased by a small amount; the negative sign in the definition ensures that the compressibility is a positive quantity, because an increase of pressure, implying a positive $\mathrm{d}p$, brings about a reduction of volume, a negative $\mathrm{d}V$.

Table 2.9* Expansion coefficients (α) and isothermal compressibilities (κ_T) at 298 K

	$\alpha/(10^{-4}\ \mathrm{K}^{-1})$	$\kappa_T/(10^{-6}\ \mathrm{bar}^{-1})$
Benzene	12.4	90.9
Diamond	0.030	0.185
Lead	0.861	2.18
Water	2.1	49.0

* More values are given in the *Data section*.

Example 2.8 *Calculating the expansion coefficient of a gas*

Derive an expression for the expansion coefficient of a perfect gas.

Method The expansion coefficient is defined in eqn 2.42. To use this expression, substitute the expression for V in terms of T obtained from the equation of state for the gas. As implied by the subscript in eqn 2.42, the pressure, p, is treated as a constant.

Answer Because $pV = nRT$, we can write

$$\alpha = \frac{1}{V}\left(\frac{\partial(nRT/p)}{\partial T}\right)_p = \frac{1}{V}\times\frac{nR}{p}\frac{\mathrm{d}T}{\mathrm{d}T} = \frac{nR}{pV} = \frac{1}{T}$$

The higher the temperature, the less responsive is the volume of a perfect gas to a change in temperature.

Self-test 2.9 Derive an expression for the isothermal compressibility of a perfect gas. $[\kappa_T = 1/p]$

When we introduce the definition of α into the equation for $(\partial U/\partial T)_p$, we obtain

$$\left(\frac{\partial U}{\partial T}\right)_p = \alpha\pi_T V + C_V \tag{2.44}$$

This equation is entirely general (provided the system is closed and its composition is constant). It expresses the dependence of the internal energy on the temperature at constant pressure in terms of C_V, which can be measured in one experiment, in terms of α, which can be measured in another, and in terms of the quantity π_T. For a perfect gas, $\pi_T = 0$, so then

$$\left(\frac{\partial U}{\partial T}\right)_p = C_V \tag{2.45}^\circ$$

That is, although the constant-volume heat capacity of a perfect gas is defined as the slope of a plot of internal energy against temperature at constant volume, for a perfect gas C_V is also the slope at constant pressure.

Equation 2.45 provides an easy way to derive the relation between C_p and C_V for a perfect gas. Thus, we can use it to express both heat capacities in terms of derivatives at constant pressure:

$$C_p - C_V = \left(\frac{\partial H}{\partial T}\right)_p - \left(\frac{\partial U}{\partial T}\right)_V = \left(\frac{\partial H}{\partial T}\right)_p - \left(\frac{\partial U}{\partial T}\right)_p \tag{2.46}^\circ$$

Then we introduce $H = U + pV = U + nRT$ into the first term, which results in

$$C_p - C_V = \left(\frac{\partial U}{\partial T}\right)_p + nR - \left(\frac{\partial U}{\partial T}\right)_p = nR \tag{2.47}^\circ$$

which is eqn 2.26. We show in *Further information 2.2* that in general

$$C_p - C_V = \frac{\alpha^2 TV}{\kappa_T} \tag{2.48}$$

Equation 2.48 applies to any substance (that is, it is 'universally true'). It reduces to eqn 2.47 for a perfect gas when we set $\alpha = 1/T$ and $\kappa_T = 1/p$. Because expansion coefficients α of liquids and solids are small, it is tempting to deduce from eqn 2.48 that for them $C_p \approx C_V$. But this is not always so, because the compressibility κ_T might also be small, so α^2/κ_T might be large. That is, although only a little work need be done to push back the atmosphere, a great deal of work may have to be done to pull atoms apart from one another as the solid expands. As an illustration, for water at 25°C, eqn 2.48 gives $C_{p,\text{m}} = 75.3\ \text{J K}^{-1}\ \text{mol}^{-1}$ compared with $C_{V,\text{m}} = 74.8\ \text{J K}^{-1}\ \text{mol}^{-1}$. In some cases, the two heat capacities differ by as much as 30 per cent.

2.12 The Joule–Thomson effect

Key point **The Joule–Thomson effect is the change in temperature of a gas when it undergoes isenthalpic expansion.**

We can carry out a similar set of operations on the enthalpy, $H = U + pV$. The quantities U, p, and V are all state functions; therefore H is also a state function and $\text{d}H$ is an exact differential. It turns out that H is a useful thermodynamic function when the pressure is under our control: we saw a sign of that in the relation $\Delta H = q_p$ (eqn 2.19b). We shall therefore regard H as a function of p and T, and adapt the argument in Section 2.11 to find an expression for the variation of H with temperature at constant volume. As explained in the following *Justification*, we find that for a closed system of constant composition

$$\text{d}H = -\mu C_p \text{d}p + C_p \text{d}T \qquad (2.49)$$

where the **Joule–Thomson coefficient**, μ (mu), is defined as

$$\mu = \left(\frac{\partial T}{\partial p}\right)_H \qquad [2.50]$$

Definition of the Joule–Thomson coefficient

This relation will prove useful for relating the heat capacities at constant pressure and volume and for a discussion of the liquefaction of gases.

Justification 2.2 *The variation of enthalpy with pressure and temperature*

Because H is a function of p and T we can write, when these two quantities change by an infinitesimal amount, that the enthalpy changes by

$$\text{d}H = \left(\frac{\partial H}{\partial p}\right)_T \text{d}p + \left(\frac{\partial H}{\partial T}\right)_p \text{d}T \qquad (2.51)$$

The second partial derivative is C_p; our task here is to express $(\partial H/\partial p)_T$ in terms of recognizable quantities. If the enthalpy is constant, $\text{d}H = 0$ and this expression then requires that

$$\left(\frac{\partial H}{\partial p}\right)_T \text{d}p = -C_p \text{d}T \qquad \text{at constant } H$$

Division of both sides by $\text{d}p$ then gives

$$\left(\frac{\partial H}{\partial p}\right)_T = -C_p \left(\frac{\partial T}{\partial p}\right)_H = -C_p \mu$$

Equation 2.49 now follows directly.

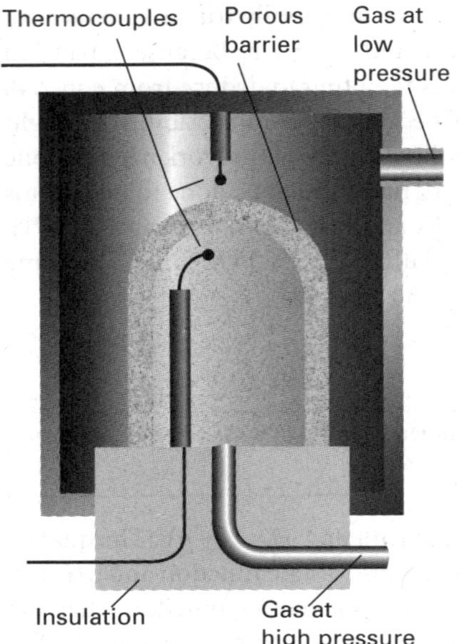

Fig. 2.26 The apparatus used for measuring the Joule–Thomson effect. The gas expands through the porous barrier, which acts as a throttle, and the whole apparatus is thermally insulated. As explained in the text, this arrangement corresponds to an isenthalpic expansion (expansion at constant enthalpy). Whether the expansion results in a heating or a cooling of the gas depends on the conditions.

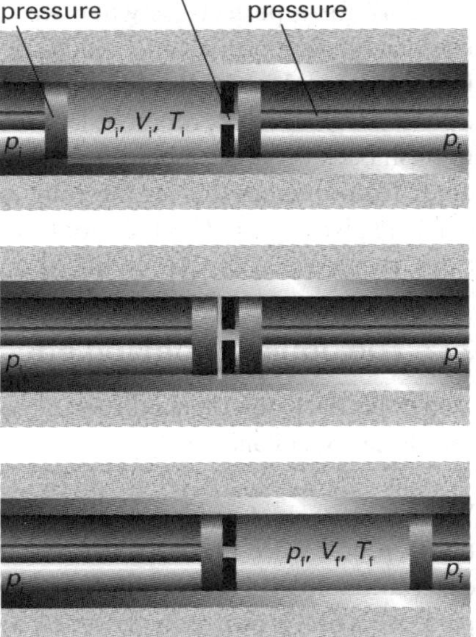

Fig. 2.27 The thermodynamic basis of Joule–Thomson expansion. The pistons represent the upstream and downstream gases, which maintain constant pressures either side of the throttle. The transition from the top diagram to the bottom diagram, which represents the passage of a given amount of gas through the throttle, occurs without change of enthalpy.

(a) Observation of the Joule–Thomson effect

The analysis of the Joule–Thomson coefficient is central to the technological problems associated with the liquefaction of gases. We need to be able to interpret it physically and to measure it. As shown in the following *Justification*, the cunning required to impose the constraint of constant enthalpy, so that the process is **isenthalpic**, was supplied by Joule and William Thomson (later Lord Kelvin). They let a gas expand through a porous barrier from one constant pressure to another and monitored the difference of temperature that arose from the expansion (Fig. 2.26). The whole apparatus was insulated so that the process was adiabatic. They observed a lower temperature on the low pressure side, the difference in temperature being proportional to the pressure difference they maintained. This cooling by isenthalpic expansion is now called the **Joule–Thomson effect**.

Justification 2.3 *The Joule–Thomson effect*

Here we show that the experimental arrangement results in expansion at constant enthalpy. Because all changes to the gas occur adiabatically, $q=0$ implies that $\Delta U=w$. Next, consider the work done as the gas passes through the barrier. We focus on the passage of a fixed amount of gas from the high pressure side, where the pressure is p_i, the temperature T_i, and the gas occupies a volume V_i (Fig. 2.27). The gas emerges on the low pressure side, where the same amount of gas has a pressure p_f, a temperature T_f, and occupies a volume V_f. The gas on the left is compressed isothermally by the upstream gas acting as a piston. The relevant pressure is p_i and the volume changes from V_i to 0; therefore, the work done on the gas is

$$w_1 = -p_i(0 - V_i) = p_iV_i$$

The gas expands isothermally on the right of the barrier (but possibly at a different constant temperature) against the pressure p_f provided by the downstream gas acting as a piston to be driven out. The volume changes from 0 to V_f, so the work done on the gas in this stage is

$$w_2 = -p_f(V_f - 0) = -p_fV_f$$

The total work done on the gas is the sum of these two quantities, or

$$w = w_1 + w_2 = p_iV_i - p_fV_f$$

It follows that the change of internal energy of the gas as it moves adiabatically from one side of the barrier to the other is

$$U_f - U_i = w = p_iV_i - p_fV_f$$

Reorganization of this expression gives

$$U_f + p_fV_f = U_i + p_iV_i \quad \text{or} \quad H_f = H_i$$

Therefore, the expansion occurs without change of enthalpy.

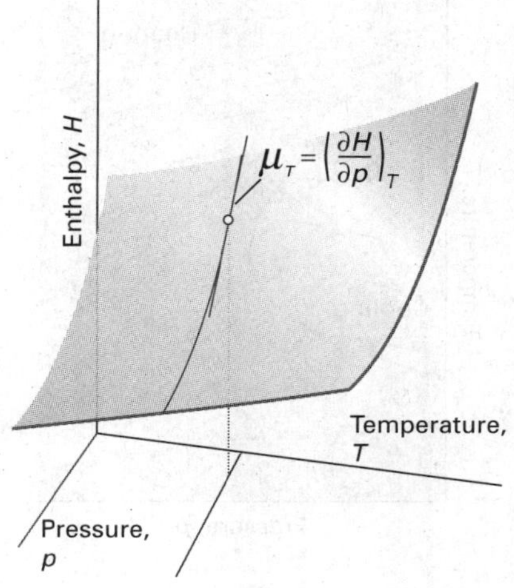

Fig. 2.28 The isothermal Joule–Thomson coefficient is the slope of the enthalpy with respect to changing pressure, the temperature being held constant.

The property measured in the experiment is the ratio of the temperature change to the change of pressure, $\Delta T/\Delta p$. Adding the constraint of constant enthalpy and taking the limit of small Δp implies that the thermodynamic quantity measured is $(\partial T/\partial p)_H$, which is the Joule–Thomson coefficient, μ. In other words, the physical interpretation of μ is that it is the ratio of the change in temperature to the change in pressure when a gas expands under conditions that ensure there is no change in enthalpy.

The modern method of measuring μ is indirect, and involves measuring the **isothermal Joule–Thomson coefficient**, the quantity

$$\mu_T = \left(\frac{\partial H}{\partial p}\right)_T \qquad \text{Definition of the isothermal Joule–Thomson coefficient} \qquad [2.52]$$

which is the slope of a plot of enthalpy against pressure at constant temperature (Fig. 2.28). Comparing eqns 2.51 and 2.52, we see that the two coefficients are related by

$$\mu_T = -C_p\mu \qquad (2.53)$$

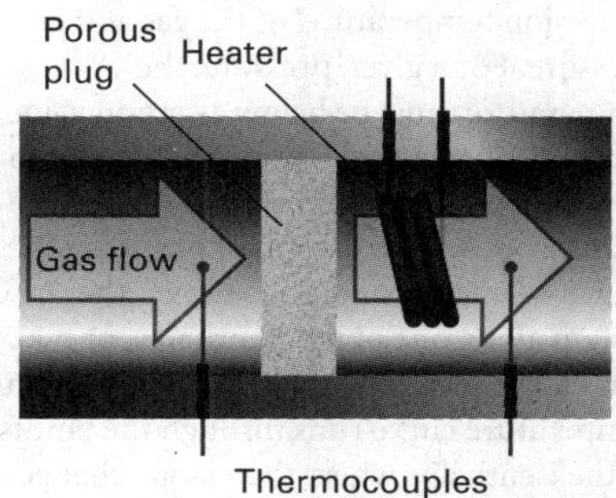

Fig. 2.29 A schematic diagram of the apparatus used for measuring the isothermal Joule–Thomson coefficient. The electrical heating required to offset the cooling arising from expansion is interpreted as ΔH and used to calculate $(\partial H/\partial p)_T$, which is then converted to μ as explained in the text.

To measure μ_T, the gas is pumped continuously at a steady pressure through a heat exchanger, which brings it to the required temperature, and then through a porous plug inside a thermally insulated container. The steep pressure drop is measured and the cooling effect is exactly offset by an electric heater placed immediately after the plug (Fig. 2.29). The energy provided by the heater is monitored. Because $\Delta H = q_p$, the energy transferred as heat can be identified with the value of ΔH. The pressure change Δp is known, so we can find μ_T from the limiting value of $\Delta H/\Delta p$ as $\Delta p \to 0$ and then convert it to μ. Table 2.10 lists some values obtained in this way.

Real gases have nonzero Joule–Thomson coefficients. Depending on the identity of the gas, the pressure, the relative magnitudes of the attractive and repulsive intermolecular forces, and the temperature, the sign of the coefficient may be either positive or negative (Fig. 2.30). A positive sign implies that dT is negative when dp is negative, in which case the gas cools on expansion. Gases that show a heating effect ($\mu < 0$) at one temperature show a cooling effect ($\mu > 0$) when the temperature is below their upper **inversion temperature**, T_I (Table 2.10, Fig. 2.31). As indicated in

Table 2.10* Inversion temperatures (T_I), normal freezing (T_f) and boiling (T_b) points, and Joule–Thomson coefficient (μ) at 1 atm and 298 K

	T_I/K	T_f/K	T_b/K	μ/(K bar^{-1})
Ar	723	83.8	87.3	
CO_2	1500	194.7		+1.10
He	40		4.2	−0.060
N_2	621	63.3	77.4	+0.25

* More values are given in the *Data section*.

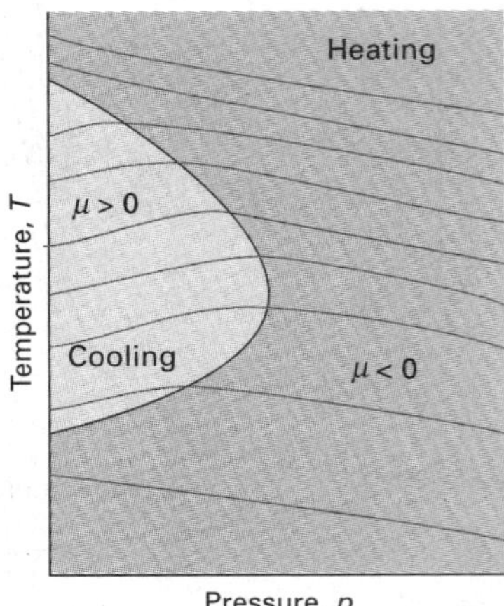

Fig. 2.30 The sign of the Joule–Thomson coefficient, μ, depends on the conditions. Inside the boundary, the blue area, it is positive and outside it is negative. The temperature corresponding to the boundary at a given pressure is the 'inversion temperature' of the gas at that pressure. For a given pressure, the temperature must be below a certain value if cooling is required but, if it becomes too low, the boundary is crossed again and heating occurs. Reduction of pressure under adiabatic conditions moves the system along one of the isenthalps, or curves of constant enthalpy. The inversion temperature curve runs through the points of the isenthalps where their slope changes from negative to positive.

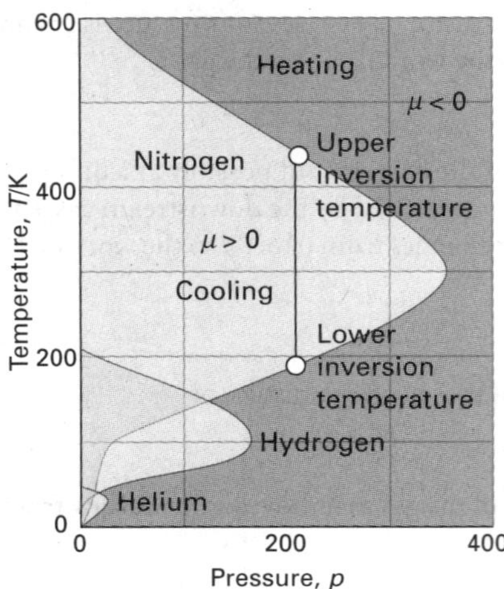

Fig. 2.31 The inversion temperatures for three real gases, nitrogen, hydrogen, and helium.

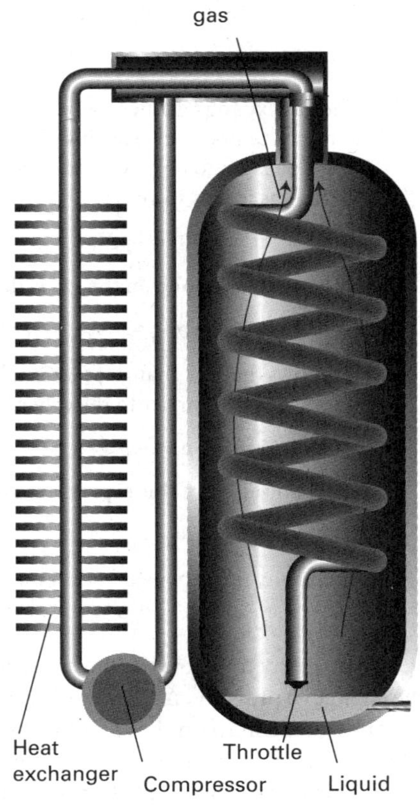

Fig. 2.32 The principle of the Linde refrigerator is shown in this diagram. The gas is recirculated, and, so long as it is beneath its inversion temperature, it cools on expansion through the throttle. The cooled gas cools the high-pressure gas, which cools still further as it expands. Eventually liquefied gas drips from the throttle.

Fig. 2.31, a gas typically has two inversion temperatures, one at high temperature and the other at low.

The 'Linde refrigerator' makes use of Joule–Thompson expansion to liquefy gases (Fig. 2.32). The gas at high pressure is allowed to expand through a throttle; it cools and is circulated past the incoming gas. That gas is cooled, and its subsequent expansion cools it still further. There comes a stage when the circulating gas becomes so cold that it condenses to a liquid.

For a perfect gas, $\mu = 0$; hence, the temperature of a perfect gas is unchanged by Joule–Thomson expansion. (Simple adiabatic expansion does cool a perfect gas, because the gas does work, Section 2.6.) This characteristic points clearly to the involvement of intermolecular forces in determining the size of the effect. However, the Joule–Thomson coefficient of a real gas does not necessarily approach zero as the pressure is reduced even though the equation of state of the gas approaches that of a perfect gas. The coefficient behaves like the properties discussed in Section 1.3b in the sense that it depends on derivatives and not on p, V, and T themselves.

(b) The molecular interpretation of the Joule–Thomson effect

The kinetic model of gases (Section 1.2b) and the equipartition theorem (Section F.5) jointly imply that the mean kinetic energy of molecules in a gas is proportional to the temperature. It follows that reducing the average speed of the molecules is equivalent to cooling the gas. If the speed of the molecules can be reduced to the point that neighbours can capture each other by their intermolecular attractions, then the cooled gas will condense to a liquid.

To slow the gas molecules, we make use of an effect similar to that seen when a ball is thrown into the air: as it rises it slows in response to the gravitational attraction of the Earth and its kinetic energy is converted into potential energy. We saw in Section 1.3 that molecules in a real gas attract each other (the attraction is not gravitational, but the effect is the same). It follows that, if we can cause the molecules to move apart from each other, like a ball rising from a planet, then they should slow. It is very easy to move molecules apart from each other: we simply allow the gas to expand, which increases the average separation of the molecules. To cool a gas, therefore, we allow it to expand without allowing any energy to enter from outside as heat. As the gas expands, the molecules move apart to fill the available volume, struggling as they do so against the attraction of their neighbours. Because some kinetic energy must be converted into potential energy to reach greater separations, the molecules travel more slowly as their separation increases. This sequence of molecular events explains the Joule–Thomson effect: the cooling of a real gas by adiabatic expansion. The cooling effect, which corresponds to $\mu > 0$, is observed under conditions when attractive interactions are dominant ($Z < 1$, eqn 1.17), because the molecules have to climb apart against the attractive force in order for them to travel more slowly. For molecules under conditions when repulsions are dominant ($Z > 1$), the Joule–Thomson effect results in the gas becoming warmer, or $\mu < 0$.

Checklist of key equations

Property	Equation	Comment
First Law of thermodynamics	$\Delta U = q + w$	Acquisitive convention
Work of expansion	$\mathrm{d}w = -p_{\mathrm{ex}}\mathrm{d}V$	
Work of expansion against a constant external pressure	$w = -p_{\mathrm{ex}}\Delta V$	$p_{\mathrm{ex}} = 0$ corresponds to free expansion
Work of isothermal reversible expansion of a perfect gas	$w = -nRT\ln(V_f/V_i)$	Isothermal, reversible, perfect gas
Heat capacity at constant volume	$C_V = (\partial U/\partial T)_V$	Definition
Heat capacity at constant pressure	$C_p = (\partial H/\partial T)_p$	Definition
Relation between heat capacities	$C_p - C_V = nR$	Perfect gas
Enthalpy	$H = U + pV$	Definition
The standard reaction enthalpy	$\Delta_r H^{\ominus} = \sum_{\text{Products}} \nu H_m^{\ominus} - \sum_{\text{Reactants}} \nu H_m^{\ominus}$	
Kirchhoff's law	$\Delta_r H^{\ominus}(T_2) = \Delta_r H^{\ominus}(T_1) + \int_{T_1}^{T_2} \Delta_r C_p^{\ominus}\mathrm{d}T$	
Internal pressure	$\pi_T = (\partial U/\partial V)_T$	For a perfect gas, $\pi_T = 0$
Joule–Thomson coefficient	$\mu = (\partial T/\partial p)_H$	For a perfect gas, $\mu = 0$

→ For a chart of the relations between principal equations, see the Road map section of the Resource section.

Further information

Further information 2.1 *Adiabatic processes*

Consider a stage in a reversible adiabatic expansion when the pressure inside and out is p. The work done when the gas expands by dV is $dw=-pdV$; however, for a perfect gas, $dU=C_V dT$. Therefore, because for an adiabatic change ($dq=0$) $dU=dw+dq=dw$, we can equate these two expressions for dU and write

$$C_V dT=-pdV$$

We are dealing with a perfect gas, so we can replace p by nRT/V and obtain

$$\frac{C_V dT}{T}=-\frac{nRdV}{V}$$

To integrate this expression we note that T is equal to T_i when V is equal to V_i, and is equal to T_f when V is equal to V_f at the end of the expansion. Therefore,

$$C_V\int_{T_i}^{T_f}\frac{dT}{T}=-nR\int_{V_i}^{V_f}\frac{dV}{V}$$

(We are taking C_V to be independent of temperature.) Then, because $\int dx/x=\ln x+\text{constant}$, we obtain

$$C_V\ln\frac{T_f}{T_i}=-nR\ln\frac{V_f}{V_i}$$

Because $\ln(x/y)=-\ln(y/x)$, this expression rearranges to

$$\frac{C_V}{nR}\ln\frac{T_f}{T_i}=\ln\frac{V_i}{V_f}$$

With $c=C_V/nR$ we obtain (because $\ln x^a=a\ln x$)

$$\ln\left(\frac{T_f}{T_i}\right)^c=\ln\left(\frac{V_i}{V_f}\right)$$

which implies that $(T_f/T_i)^c=(V_i/V_f)$ and, upon rearrangement, eqn 2.28.

The initial and final states of a perfect gas satisfy the perfect gas law regardless of how the change of state takes place, so we can use $pV=nRT$ to write

$$\frac{p_iV_i}{p_fV_f}=\frac{T_i}{T_f}$$

However, we have just shown that

$$\frac{T_i}{T_f}=\left(\frac{V_f}{V_i}\right)^{1/c}=\left(\frac{V_f}{V_i}\right)^{\gamma-1}$$

where we use the definition of the heat capacity ratio where $\gamma=C_{p,m}/C_{V,m}$ and the fact that, for a perfect gas, $C_{p,m}-C_{V,m}=R$ (the molar version of eqn 2.26). Then we combine the two expressions, to obtain

$$\frac{p_i}{p_f}=\frac{V_f}{V_i}\times\left(\frac{V_f}{V_i}\right)^{\gamma-1}=\left(\frac{V_f}{V_i}\right)^{\gamma}$$

which rearranges to $p_iV_i^{\gamma}=p_fV_f^{\gamma}$, which is eqn 2.29.

Further information 2.2 *The relation between heat capacities*

A useful rule when doing a problem in thermodynamics is to go back to first principles. In the present problem we do this twice, first by expressing C_p and C_V in terms of their definitions and then by inserting the definition $H=U+pV$:

$$C_p-C_V=\left(\frac{\partial H}{\partial T}\right)_p-\left(\frac{\partial U}{\partial T}\right)_V$$
$$=\left(\frac{\partial U}{\partial T}\right)_p+\left(\frac{\partial (pV)}{\partial T}\right)_p-\left(\frac{\partial U}{\partial T}\right)_V$$

We have already calculated the difference of the first and third terms on the right, and eqn 2.44 lets us write this difference as $\alpha\pi_T V$. The factor αV gives the change in volume when the temperature is raised, and $\pi_T=(\partial U/\partial V)_T$ converts this change in volume into a change in internal energy. We can simplify the remaining term by noting that, because p is constant,

$$\left(\frac{\partial (pV)}{\partial T}\right)_p=p\left(\frac{\partial V}{\partial T}\right)_p=\alpha pV$$

The middle term of this expression identifies it as the contribution to the work of pushing back the atmosphere: $(\partial V/\partial T)_p$ is the change of volume caused by a change of temperature, and multiplication by p converts this expansion into work.

Collecting the two contributions gives

$$C_p-C_V=\alpha(p+\pi_T)V \qquad (2.54)$$

As just remarked, the first term on the right, αpV, is a measure of the work needed to push back the atmosphere; the second term on the right, $\alpha\pi_T V$, is the work required to separate the molecules composing the system.

At this point we can go further by using the result we prove in Section 3.8 that

$$\pi_T=T\left(\frac{\partial p}{\partial T}\right)_V-p$$

When this expression is inserted in the last equation we obtain

$$C_p-C_V=\alpha TV\left(\frac{\partial p}{\partial T}\right)_V \qquad (2.55)$$

We now transform the remaining partial derivative. With V regarded as a function of p and T, when these two quantities change the resulting change in V is

$$dV=\left(\frac{\partial V}{\partial T}\right)_p dT+\left(\frac{\partial V}{\partial p}\right)_T dp \tag{2.56}$$

If (as in eqn 2.56) we require the volume to be constant, $dV=0$ implies that

$$\left(\frac{\partial V}{\partial T}\right)_p dT=-\left(\frac{\partial V}{\partial p}\right)_T dp \quad \text{at constant volume} \tag{2.57}$$

On division by dT, this relation becomes

$$\left(\frac{\partial V}{\partial T}\right)_p=-\left(\frac{\partial V}{\partial p}\right)_T\left(\frac{\partial p}{\partial T}\right)_V \tag{2.58}$$

and therefore

$$\left(\frac{\partial p}{\partial T}\right)_V=-\frac{(\partial V/\partial T)_p}{(\partial V/\partial p)_T}=\frac{\alpha}{\kappa_T} \tag{2.59}$$

Insertion of this relation into eqn 2.55 produces eqn 2.48.

Discussion questions

2.1 Provide mechanical and molecular definitions of work and heat.

2.2 Consider the reversible expansion of a perfect gas. Provide a physical interpretation for the fact that $pV^\gamma=$ constant for an adiabatic change, whereas $pV=$ constant for an isothermal change.

2.3 Explain the difference between the change in internal energy and the change in enthalpy accompanying a chemical or physical process.

2.4 Explain the significance of a physical observable being a state function and compile a list of as many state functions as you can identify.

2.5 Explain the significance of the Joule and Joule–Thomson experiments. What would Joule observe in a more sensitive apparatus?

2.6 Suggest (with explanation) how the internal energy of a van der Waals gas should vary with volume at constant temperature.

Exercises

Assume all gases are perfect unless stated otherwise. Unless otherwise stated, thermodynamic data are for 298.15 K.

2.1(a) Calculate the work needed for a 65 kg person to climb through 4.0 m on the surface of (a) the Earth and (b) the Moon ($g=1.60$ m s^{-2}).

2.1(b) Calculate the work needed for a bird of mass 120 g to fly to a height of 50 m from the surface of the Earth.

2.2(a) A chemical reaction takes place in a container of cross-sectional area 100 cm^2. As a result of the reaction, a piston is pushed out through 10 cm against an external pressure of 1.0 atm. Calculate the work done by the system.

2.2(b) A chemical reaction takes place in a container of cross-sectional area 50.0 cm^2. As a result of the reaction, a piston is pushed out through 15 cm against an external pressure of 121 kPa. Calculate the work done by the system.

2.3(a) A sample consisting of 1.00 mol Ar is expanded isothermally at 0°C from 22.4 dm^3 to 44.8 dm^3 (a) reversibly, (b) against a constant external pressure equal to the final pressure of the gas, and (c) freely (against zero external pressure). For the three processes calculate q, w, ΔU, and ΔH.

2.3(b) A sample consisting of 2.00 mol He is expanded isothermally at 22°C from 22.8 dm^3 to 31.7 dm^3 (a) reversibly, (b) against a constant external pressure equal to the final pressure of the gas, and (c) freely (against zero external pressure). For the three processes calculate q, w, ΔU, and ΔH.

2.4(a) A sample consisting of 1.00 mol of perfect gas atoms, for which $C_{V,m}=\frac{3}{2}R$, initially at $p_1=1.00$ atm and $T_1=300$ K, is heated reversibly to 400 K at constant volume. Calculate the final pressure, ΔU, q, and w.

2.4(b) A sample consisting of 2.00 mol of perfect gas molecules, for which $C_{V,m}=\frac{5}{2}R$, initially at $p_1=111$ kPa and $T_1=277$ K, is heated reversibly to 356 K at constant volume. Calculate the final pressure, ΔU, q, and w.

2.5(a) A sample of 4.50 g of methane occupies 12.7 dm^3 at 310 K. (a) Calculate the work done when the gas expands isothermally against a constant external pressure of 200 Torr until its volume has increased by 3.3 dm^3. (b) Calculate the work that would be done if the same expansion occurred reversibly.

2.5(b) A sample of argon of mass 6.56 g occupies 18.5 dm^3 at 305 K. (a) Calculate the work done when the gas expands isothermally against a constant external pressure of 7.7 kPa until its volume has increased by 2.5 dm^3. (b) Calculate the work that would be done if the same expansion occurred reversibly.

2.6(a) A sample of 1.00 mol H_2O(g) is condensed isothermally and reversibly to liquid water at 100°C. The standard enthalpy of vaporization of water at 100°C is 40.656 kJ mol^{-1}. Find w, q, ΔU, and ΔH for this process.

2.6(b) A sample of 2.00 mol CH_3OH(g) is condensed isothermally and reversibly to liquid at 64°C. The standard enthalpy of vaporization of methanol at 64°C is 35.3 kJ mol^{-1}. Find w, q, ΔU, and ΔH for this process.

2.7(a) A strip of magnesium of mass 15 g is placed in a beaker of dilute hydrochloric acid. Calculate the work done by the system as a result of the reaction. The atmospheric pressure is 1.0 atm and the temperature 25°C.

2.7(b) A piece of zinc of mass 5.0 g is placed in a beaker of dilute hydrochloric acid. Calculate the work done by the system as a result of the reaction. The atmospheric pressure is 1.1 atm and the temperature 23°C.

2.8(a) The constant-pressure heat capacity of a sample of a perfect gas was found to vary with temperature according to the expression $C_p/(\text{J K}^{-1}) = 20.17 + 0.3665(T/\text{K})$. Calculate q, w, ΔU, and ΔH when the temperature is raised from 25°C to 200°C (a) at constant pressure, (b) at constant volume.

2.8(b) The constant-pressure heat capacity of a sample of a perfect gas was found to vary with temperature according to the expression $C_p/(\text{J K}^{-1}) = 20.17 + 0.4001(T/\text{K})$. Calculate q, w, ΔU, and ΔH when the temperature is raised from 0°C to 100°C (a) at constant pressure, (b) at constant volume.

2.9(a) Calculate the final temperature of a sample of argon of mass 12.0 g that is expanded reversibly and adiabatically from 1.0 dm^3 at 273.15 K to 3.0 dm^3.

2.9(b) Calculate the final temperature of a sample of carbon dioxide of mass 16.0 g that is expanded reversibly and adiabatically from 500 cm^3 at 298.15 K to 2.00 dm^3.

2.10(a) A sample of carbon dioxide of mass 2.45 g at 27.0°C is allowed to expand reversibly and adiabatically from 500 cm^3 to 3.00 dm^3. What is the work done by the gas?

2.10(b) A sample of nitrogen of mass 3.12 g at 23.0°C is allowed to expand reversibly and adiabatically from 400 cm^3 to 2.00 dm^3. What is the work done by the gas?

2.11(a) Calculate the final pressure of a sample of carbon dioxide that expands reversibly and adiabatically from 57.4 kPa and 1.0 dm^3 to a final volume of 2.0 dm^3. Take $\gamma = 1.4$.

2.11(b) Calculate the final pressure of a sample of water vapour that expands reversibly and adiabatically from 87.3 Torr and 500 cm^3 to a final volume of 3.0 dm^3. Take $\gamma = 1.3$.

2.12(a) When 229 J of energy is supplied as heat to 3.0 mol Ar(g) at constant pressure, the temperature of the sample increases by 2.55 K. Calculate the molar heat capacities at constant volume and constant pressure of the gas.

2.12(b) When 178 J of energy is supplied as heat to 1.9 mol of gas molecules at constant pressure, the temperature of the sample increases by 1.78 K. Calculate the molar heat capacities at constant volume and constant pressure of the gas.

2.13(a) When 3.0 mol O_2 is heated at a constant pressure of 3.25 atm, its temperature increases from 260 K to 285 K. Given that the molar heat capacity of O_2(g) at constant pressure is 29.4 $\text{J K}^{-1}\text{ mol}^{-1}$, calculate q, ΔH, and ΔU.

2.13(b) When 2.0 mol CO_2 is heated at a constant pressure of 1.25 atm, its temperature increases from 250 K to 277 K. Given that the molar heat capacity of CO_2(g) at constant pressure is 37.11 $\text{J K}^{-1}\text{ mol}^{-1}$, calculate q, ΔH, and ΔU.

2.14(a) A sample of 4.0 mol O_2(g) is originally confined in 20 dm^3 at 270 K and then undergoes adiabatic expansion against a constant pressure of 600 Torr until the volume has increased by a factor of 3.0. Calculate q, w, ΔT, ΔU, and ΔH. (The final pressure of the gas is not necessarily 600 Torr.)

2.14(b) A sample of 5.0 mol CO_2(g) is originally confined in 15 dm^3 at 280 K and then undergoes adiabatic expansion against a constant pressure of 78.5 kPa until the volume has increased by a factor of 4.0. Calculate q, w, ΔT, ΔU, and ΔH. (The final pressure of the gas is not necessarily 78.5 kPa.)

2.15(a) A sample consisting of 1.0 mol of perfect gas molecules with $C_V = 20.8\text{ J K}^{-1}$ is initially at 3.25 atm and 310 K. It undergoes reversible adiabatic expansion until its pressure reaches 2.50 atm. Calculate the final volume and temperature and the work done.

2.15(b) A sample consisting of 1.5 mol of perfect gas molecules with $C_{p,\text{m}} = 20.8\text{ J K}^{-1}\text{ mol}^{-1}$ is initially at 230 kPa and 315 K. It undergoes reversible adiabatic expansion until its pressure reaches 170 kPa. Calculate the final volume and temperature and the work done.

2.16(a) A certain liquid has $\Delta_{\text{vap}}H^\ominus = 26.0\text{ kJ mol}^{-1}$. Calculate q, w, ΔH, and ΔU when 0.50 mol is vaporized at 250 K and 750 Torr.

2.16(b) A certain liquid has $\Delta_{\text{vap}}H^\ominus = 32.0\text{ kJ mol}^{-1}$. Calculate q, w, ΔH, and ΔU when 0.75 mol is vaporized at 260 K and 765 Torr.

2.17(a) Calculate the lattice enthalpy of SrI_2 from the following data:

	$\Delta H/(\text{kJ mol}^{-1})$
Sublimation of Sr(s)	+164
Ionization of Sr(g) to Sr^{2+}(g)	+1626
Sublimation of I_2(s)	+62
Dissociation of I_2(g)	+151
Electron attachment to I(g)	−304
Formation of SrI_2(s) from Sr(s) and I_2(s)	−558

2.17(b) Calculate the lattice enthalpy of $MgBr_2$ from the following data:

	$\Delta H/(\text{kJ mol}^{-1})$
Sublimation of Mg(s)	+148
Ionization of Mg(g) to Mg^{2+}(g)	+2187
Vaporization of Br_2(l)	+31
Dissociation of Br_2(g)	+193
Electron attachment to Br(g)	−331
Formation of $MgBr_2$(s) from Mg(s) and Br_2(l)	−524

2.18(a) The standard enthalpy of formation of ethylbenzene is −12.5 kJ mol^{-1}. Calculate its standard enthalpy of combustion.

2.18(b) The standard enthalpy of formation of phenol is −165.0 kJ mol^{-1}. Calculate its standard enthalpy of combustion.

2.19(a) The standard enthalpy of combustion of cyclopropane is −2091 kJ mol^{-1} at 25°C. From this information and enthalpy of formation data for CO_2(g) and H_2O(g), calculate the enthalpy of formation of cyclopropane. The enthalpy of formation of propene is +20.42 kJ mol^{-1}. Calculate the enthalpy of isomerization of cyclopropane to propene.

2.19(b) From the following data, determine $\Delta_f H^\ominus$ for diborane, B_2H_6(g), at 298 K:

(1) $B_2H_6(g) + 3\,O_2(g) \rightarrow B_2O_3(s) + 3\,H_2O(g)$ $\quad \Delta_r H^\ominus = -2036\text{ kJ mol}^{-1}$

(2) $2\,B(s) + \tfrac{3}{2}O_2(g) \rightarrow B_2O_3(s)$ $\quad \Delta_r H^\ominus = -1274\text{ kJ mol}^{-1}$

(3) $H_2(g) + \tfrac{1}{2}O_2(g) \rightarrow H_2O(g)$ $\quad \Delta_r H^\ominus = -241.8\text{ kJ mol}^{-1}$

2.20(a) When 120 mg of naphthalene, $C_{10}H_8$(s), was burned in a bomb calorimeter the temperature rose by 3.05 K. Calculate the calorimeter constant. By how much will the temperature rise when 10 mg of phenol, C_6H_5OH(s), is burned in the calorimeter under the same conditions?

2.20(b) When 2.25 mg of anthracene, $C_{14}H_{10}$(s), was burned in a bomb calorimeter the temperature rose by 1.35 K. Calculate the calorimeter constant. By how much will the temperature rise when 135 mg of phenol, C_6H_5OH(s), is burned in the calorimeter under the same conditions? ($\Delta_c H^\ominus(C_{14}H_{10},s) = -7061\text{ kJ mol}^{-1}$.)

2.21(a) Calculate the standard enthalpy of solution of AgCl(s) in water from the enthalpies of formation of the solid and the aqueous ions.

2.21(b) Calculate the standard enthalpy of solution of AgBr(s) in water from the enthalpies of formation of the solid and the aqueous ions.

2.22(a) The standard enthalpy of decomposition of the yellow complex H_3NSO_2 into NH_3 and SO_2 is +40 kJ mol^{-1}. Calculate the standard enthalpy of formation of H_3NSO_2.

2.22(b) Given that the standard enthalpy of combustion of graphite is -393.51 kJ mol^{-1} and that of diamond is -395.41 kJ mol^{-1}, calculate the enthalpy of the graphite-to-diamond transition.

2.23(a) Given the reactions (1) and (2) below, determine (a) $\Delta_r H^\ominus$ and $\Delta_r U^\ominus$ for reaction (3), (b) $\Delta_f H^\ominus$ for both HCl(g) and H_2O(g) all at 298 K.

(1) $H_2(g) + Cl_2(g) \rightarrow 2\,HCl(g)$ $\quad \Delta_r H^\ominus = -184.62$ kJ mol^{-1}

(2) $2\,H_2(g) + O_2(g) \rightarrow 2\,H_2O(g)$ $\quad \Delta_r H^\ominus = -483.64$ kJ mol^{-1}

(3) $4\,HCl(g) + O_2(g) \rightarrow 2\,Cl_2(g) + 2\,H_2O(g)$

2.23(b) Given the reactions (1) and (2) below, determine (a) $\Delta_r H^\ominus$ and $\Delta_r U^\ominus$ for reaction (3), (b) $\Delta_f H^\ominus$ for both HI(g) and H_2O(g) all at 298 K.

(1) $H_2(g) + I_2(s) \rightarrow 2\,HI(g)$ $\quad \Delta_r H^\ominus = +52.96$ kJ mol^{-1}

(2) $2\,H_2(g) + O_2(g) \rightarrow 2\,H_2O(g)$ $\quad \Delta_r H^\ominus = -483.64$ kJ mol^{-1}

(3) $4\,HI(g) + O_2(g) \rightarrow 2\,I_2(s) + 2\,H_2O(g)$

2.24(a) For the reaction $C_2H_5OH(l) + 3\,O_2(g) \rightarrow 2\,CO_2(g) + 3\,H_2O(g)$, $\Delta_r U^\ominus = -1373$ kJ mol^{-1} at 298 K, calculate $\Delta_r H^\ominus$.

2.24(b) For the reaction $2\,C_6H_5COOH(s) + 13\,O_2(g) \rightarrow 12\,CO_2(g) + 6\,H_2O(g)$, $\Delta_r U^\ominus = -772.7$ kJ mol^{-1} at 298 K, calculate $\Delta_r H^\ominus$.

2.25(a) Calculate the standard enthalpies of formation of (a) $KClO_3$(s) from the enthalpy of formation of KCl, (b) $NaHCO_3$(s) from the enthalpies of formation of CO_2 and NaOH together with the following information:

$2\,KClO_3(s) \rightarrow 2\,KCl(s) + 3\,O_2(g)$ $\quad \Delta_r H^\ominus = -89.4$ kJ mol^{-1}

$NaOH(s) + CO_2(g) \rightarrow NaHCO_3(s)$ $\quad \Delta_r H^\ominus = -127.5$ kJ mol^{-1}

2.25(b) Calculate the standard enthalpy of formation of NOCl(g) from the enthalpy of formation of NO given in Table 2.8, together with the following information:

$2\,NOCl(g) \rightarrow 2\,NO(g) + Cl_2(g)$ $\quad \Delta_r H^\ominus = +75.5$ kJ mol^{-1}

2.26(a) Use the information in Table 2.8 to predict the standard reaction enthalpy of $2\,NO_2(g) \rightarrow N_2O_4(g)$ at 100°C from its value at 25°C.

2.26(b) Use the information in Table 2.8 to predict the standard reaction enthalpy of $2\,H_2(g) + O_2(g) \rightarrow 2\,H_2O(l)$ at 100°C from its value at 25°C.

2.27(a) From the data in Table 2.8, calculate $\Delta_r H^\ominus$ and $\Delta_r U^\ominus$ at (a) 298 K, (b) 378 K for the reaction C(graphite) + $H_2O(g) \rightarrow CO(g) + H_2(g)$. Assume all heat capacities to be constant over the temperature range of interest.

2.27(b) Calculate $\Delta_r H^\ominus$ and $\Delta_r U^\ominus$ at 298 K and $\Delta_r H^\ominus$ at 348 K for the hydrogenation of ethyne (acetylene) to ethene (ethylene) from the enthalpy of combustion and heat capacity data in Tables 2.6 and 2.8. Assume the heat capacities to be constant over the temperature range involved.

2.28(a) Calculate $\Delta_r H^\ominus$ for the reaction $Zn(s) + CuSO_4(aq) \rightarrow ZnSO_4(aq) + Cu(s)$ from the information in Table 2.8 in the *Data section*.

2.28(b) Calculate $\Delta_r H^\ominus$ for the reaction $NaCl(aq) + AgNO_3(aq) \rightarrow AgCl(s) + NaNO_3(aq)$ from the information in Table 2.8 in the *Data section*.

2.29(a) Set up a thermodynamic cycle for determining the enthalpy of hydration of Mg^{2+} ions using the following data: enthalpy of sublimation of Mg(s), +167.2 kJ mol^{-1}; first and second ionization enthalpies of Mg(g), 7.646 eV and 15.035 eV; dissociation enthalpy of Cl_2(g), +241.6 kJ mol^{-1}; electron gain enthalpy of Cl(g), -3.78 eV; enthalpy of solution of $MgCl_2$(s), -150.5 kJ mol^{-1}; enthalpy of hydration of Cl^-(g), -383.7 kJ mol^{-1}.

2.29(b) Set up a thermodynamic cycle for determining the enthalpy of hydration of Ca^{2+} ions using the following data: enthalpy of sublimation of Ca(s), +178.2 kJ mol^{-1}; first and second ionization enthalpies of Ca(g), 589.7 kJ mol^{-1} and 1145 kJ mol^{-1}; enthalpy of vaporization of bromine, 30.91 kJ mol^{-1}; dissociation enthalpy of Br_2(g), +192.9 kJ mol^{-1}; electron gain enthalpy of Br(g), -331.0 kJ mol^{-1}; enthalpy of solution of $CaBr_2$(s), -103.1 kJ mol^{-1}; enthalpy of hydration of Br^-(g), -97.5 kJ mol^{-1}.

2.30(a) When a certain freon used in refrigeration was expanded adiabatically from an initial pressure of 32 atm and 0°C to a final pressure of 1.00 atm, the temperature fell by 22 K. Calculate the Joule–Thomson coefficient, μ, at 0°C, assuming it remains constant over this temperature range.

2.30(b) A vapour at 22 atm and 5°C was allowed to expand adiabatically to a final pressure of 1.00 atm; the temperature fell by 10 K. Calculate the Joule–Thomson coefficient, μ, at 5°C, assuming it remains constant over this temperature range.

2.31(a) For a van der Waals gas, $\pi_T = a/V_m^2$. Calculate ΔU_m for the isothermal expansion of nitrogen gas from an initial volume of 1.00 dm^3 to 24.8 dm^3 at 298 K. What are the values of q and w?

2.31(b) Repeat Exercise 2.31(a) for argon, from an initial volume of 1.00 dm^3 to 22.1 dm^3 at 298 K.

2.32(a) The volume of a certain liquid varies with temperature as

$$V = V'\{0.75 + 3.9 \times 10^{-4}(T/K) + 1.48 \times 10^{-6}(T/K)^2\}$$

where V' is its volume at 300 K. Calculate its expansion coefficient, α, at 320 K.

2.32(b) The volume of a certain liquid varies with temperature as

$$V = V'\{0.77 + 3.7 \times 10^{-4}(T/K) + 1.52 \times 10^{-6}(T/K)^2\}$$

where V' is its volume at 298 K. Calculate its expansion coefficient, α, at 310 K.

2.33(a) The isothermal compressibility of copper at 293 K is 7.35×10^{-7} atm^{-1}. Calculate the pressure that must be applied in order to increase its density by 0.08 per cent.

2.33(b) The isothermal compressibility of lead at 293 K is 2.21×10^{-6} atm^{-1}. Calculate the pressure that must be applied in order to increase its density by 0.08 per cent.

2.34(a) Given that $\mu = 0.25$ K atm^{-1} for nitrogen, calculate the value of its isothermal Joule–Thomson coefficient. Calculate the energy that must be supplied as heat to maintain constant temperature when 15.0 mol N_2 flows through a throttle in an isothermal Joule–Thomson experiment and the pressure drop is 75 atm.

2.34(b) Given that $\mu = 1.11$ K atm^{-1} for carbon dioxide, calculate the value of its isothermal Joule–Thomson coefficient. Calculate the energy that must be supplied as heat to maintain constant temperature when 12.0 mol CO_2 flows through a throttle in an isothermal Joule–Thomson experiment and the pressure drop is 55 atm.

Problems*

Assume all gases are perfect unless stated otherwise. Note that 1 atm = 1.013 25 bar. Unless otherwise stated, thermochemical data are for 298.15 K.

Numerical problems

2.1 A sample consisting of 1 mol of perfect gas atoms (for which $C_{V,m} = \frac{3}{2}R$) is taken through the cycle shown in Fig. 2.33. (a) Determine the temperature at the points 1, 2, and 3. (b) Calculate q, w, ΔU, and ΔH for each step and for the overall cycle. If a numerical answer cannot be obtained from the information given, then write +, –, 0, or ? as appropriate.

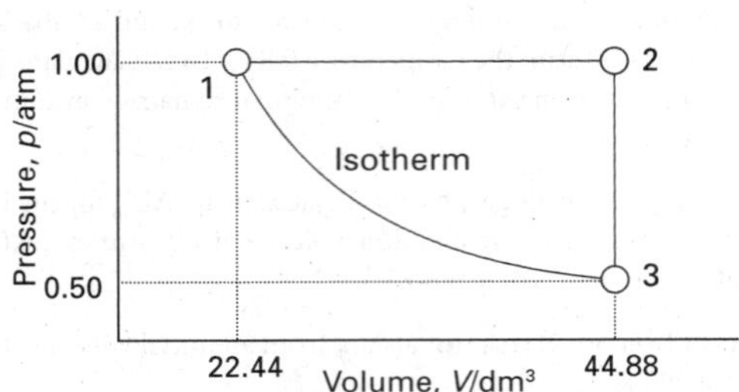

Fig. 2.33

2.2 A sample consisting of 1.0 mol $CaCO_3$(s) was heated to 800°C, when it decomposed. The heating was carried out in a container fitted with a piston that was initially resting on the solid. Calculate the work done during complete decomposition at 1.0 atm. What work would be done if instead of having a piston the container was open to the atmosphere?

2.3 A sample consisting of 2.0 mol CO_2 occupies a fixed volume of 15.0 dm^3 at 300 K. When it is supplied with 2.35 kJ of energy as heat its temperature increases to 341 K. Assume that CO_2 is described by the van der Waals equation of state, and calculate w, ΔU, and ΔH.

2.4 A sample of 70 mmol Kr(g) expands reversibly and isothermally at 373 K from 5.25 cm^3 to 6.29 cm^3, and the internal energy of the sample is known to increase by 83.5 J. Use the virial equation of state up to the second coefficient $B = -28.7$ cm^3 mol^{-1} to calculate w, q, and ΔH for this change of state.

2.5 A sample of 1.00 mol perfect gas molecules with $C_{p,m} = \frac{7}{2}R$ is put through the following cycle: (a) constant-volume heating to twice its initial pressure, (b) reversible, adiabatic expansion back to its initial temperature, (c) reversible isothermal compression back to 1.00 atm. Calculate q, w, ΔU, and ΔH for each step and overall.

2.6 Calculate the work done during the isothermal reversible expansion of a van der Waals gas. Account physically for the way in which the coefficients a and b appear in the final expression. Plot on the same graph the indicator diagrams for the isothermal reversible expansion of (a) a perfect gas, (b) a van der Waals gas in which $a = 0$ and $b = 5.11 \times 10^{-2}$ dm^3 mol^{-1}, and (c) $a = 4.2$ dm^6 atm mol^{-2} and $b = 0$. The values selected exaggerate the imperfections but give rise to significant effects on the indicator diagrams. Take $V_i = 1.0$ dm^3, $n = 1.0$ mol, and $T = 298$ K.

2.7 The molar heat capacity of ethane is represented in the temperature range 298 K to 400 K by the empirical expression $C_{p,m}/(\text{J K}^{-1}\text{ mol}^{-1}) = 14.73 + 0.1272(T/\text{K})$. The corresponding expressions for C(s) and H_2(g) are given in Table 2.2. Calculate the standard enthalpy of formation of ethane at 350 K from its value at 298 K.

2.8 A sample of the sugar D-ribose ($C_5H_{10}O_5$) of mass 0.727 g was placed in a constant-volume calorimeter and then ignited in the presence of excess oxygen. The temperature rose by 0.910 K. In a separate experiment in the same calorimeter, the combustion of 0.825 g of benzoic acid, for which the internal energy of combustion is −3251 kJ mol^{-1}, gave a temperature rise of 1.940 K. Calculate the internal energy of combustion of D-ribose and its enthalpy of formation.

2.9 The standard enthalpy of formation of the metallocene bis(benzene)chromium was measured in a calorimeter. It was found for the reaction $Cr(C_6H_6)_2(s) \rightarrow Cr(s) + 2\,C_6H_6(g)$ that $\Delta_r U^\ominus(583\text{ K}) = +8.0$ kJ mol^{-1}. Find the corresponding reaction enthalpy and estimate the standard enthalpy of formation of the compound at 583 K. The constant-pressure molar heat capacity of benzene is 136.1 J K^{-1} mol^{-1} in its liquid range and 81.67 J K^{-1} mol^{-1} as a gas.

2.10‡ From the enthalpy of combustion data in Table 2.6 for the alkanes methane through octane, test the extent to which the relation $\Delta_c H^\ominus = k\{M/(\text{g mol}^{-1})\}^n$ holds and find the numerical values for k and n. Predict $\Delta_c H^\ominus$ for decane and compare to the known value.

2.11 An average human produces about 10 MJ of heat each day through metabolic activity. If a human body were an isolated system of mass 65 kg with the heat capacity of water, what temperature rise would the body experience? Human bodies are actually open systems, and the main mechanism of heat loss is through the evaporation of water. What mass of water should be evaporated each day to maintain constant temperature?

2.12 Glucose and fructose are simple sugars with the molecular formula $C_6H_{12}O_6$. Sucrose, or table sugar, is a complex sugar with molecular formula $C_{12}H_{22}O_{11}$ that consists of a glucose unit covalently bound to a fructose unit (a water molecule is given off as a result of the reaction between glucose and fructose to form sucrose). (a) Calculate the energy released as heat when a typical table sugar cube of mass 1.5 g is burned in air. (b) To what height could you climb on the energy a table sugar cube provides assuming 25 per cent of the energy is available for work? (c) The mass of a typical glucose tablet is 2.5 g. Calculate the energy released as heat when a glucose tablet is burned in air. (d) To what height could you climb on the energy a cube provides assuming 25 per cent of the energy is available for work?

2.13 It is possible to investigate the thermochemical properties of hydrocarbons with molecular modelling methods. (a) Use electronic structure software to predict $\Delta_c H^\ominus$ values for the alkanes methane through pentane. To calculate $\Delta_c H^\ominus$ values, estimate the standard enthalpy of formation of $C_nH_{2(n+1)}$(g) by performing semi-empirical calculations (for example, AM1 or PM3 methods) and use experimental standard enthalpy of formation values for CO_2(g) and H_2O(l). (b) Compare your estimated values with the experimental values of $\Delta_c H^\ominus$ (Table 2.6) and comment on the reliability of the molecular modelling method. (c) Test the extent to which the relation $\Delta_c H^\ominus = k\{M/(\text{g mol}^{-1})\}^n$ holds and find the numerical values for k and n.

2.14‡ When 1.3584 g of sodium acetate trihydrate was mixed into 100.0 cm^3 of 0.2000 M HCl(aq) at 25°C in a solution calorimeter, its temperature fell by 0.397°C on account of the reaction:

$$H_3O^+(aq) + NaCH_3CO_2 \cdot 3\,H_2O(s) \rightarrow Na^+(aq) + CH_3COOH(aq) + 4\,H_2O(l)$$

The heat capacity of the calorimeter is 91.0 J K^{-1} and the heat capacity density of the acid solution is 4.144 J K^{-1} cm^{-3}. Determine the standard enthalpy of

* Problems denoted with the symbol ‡ were supplied by Charles Trapp, Carmen Giunta, and Marshall Cady.

formation of the aqueous sodium cation. The standard enthalpy of formation of sodium acetate trihydrate is -1604 kJ mol^{-1}.

2.15‡ Since their discovery in 1985, fullerenes have received the attention of many chemical researchers. Kolesov *et al.* (*J. Chem. Thermodynamics* **28**, 1121 (1996)) reported the standard enthalpy of combustion and of formation of crystalline C_{60} based on calorimetric measurements. In one of their runs, they found the standard specific internal energy of combustion to be -36.0334 kJ g^{-1} at 298.15 K Compute $\Delta_c H^{\ominus}$ and $\Delta_f H^{\ominus}$ of C_{60}.

2.16‡ A thermodynamic study of $DyCl_3$ by Cordfunke *et al.* (*J. Chem. Thermodynamics* **28**, 1387 (1996)) determined its standard enthalpy of formation from the following information

(1) $DyCl_3(s) \rightarrow DyCl_3(aq, \text{in } 4.9\ \text{M HCl})$ $\qquad \Delta_r H^{\ominus} = -180.06$ kJ mol^{-1}

(2) $Dy(s) + 3\,HCl(aq, 4.0\ \text{M}) \rightarrow DyCl_3(aq, \text{in } 4.0\ \text{M HCl(aq)}) + \tfrac{3}{2}H_2(g)$ $\qquad \Delta_r H^{\ominus} = -699.43$ kJ mol^{-1}

(3) $\tfrac{1}{2}H_2(g) + \tfrac{1}{2}Cl_2(g) \rightarrow HCl(aq, 4.0\ \text{M})$ $\qquad \Delta_r H^{\ominus} = -158.31$ kJ mol^{-1}

Determine $\Delta_f H^{\ominus}(DyCl_3,s)$ from these data.

2.17‡ Silylene (SiH_2) is a key intermediate in the thermal decomposition of silicon hydrides such as silane (SiH_4) and disilane (Si_2H_6). Moffat *et al.* (*J. Phys. Chem.* **95**, 145 (1991)) report $\Delta_f H^{\ominus}(SiH_2) = +274$ kJ mol^{-1}. If $\Delta_f H^{\ominus}(SiH_4) = +34.3$ kJ mol^{-1} and $\Delta_f H^{\ominus}(Si_2H_6) = +80.3$ kJ mol^{-1} (*CRC Handbook* (2008)), compute the standard enthalpies of the following reactions:

(a) $SiH_4(g) \rightarrow SiH_2(g) + H_2(g)$

(b) $Si_2H_6(g) \rightarrow SiH_2(g) + SiH_4(g)$

2.18‡ Silanone (SiH_2O) and silanol (SiH_3OH) are species believed to be important in the oxidation of silane (SiH_4). These species are much more elusive than their carbon counterparts. C.L. Darling and H.B. Schlegel (*J. Phys. Chem.* **97**, 8207 (1993)) report the following values (converted from calories) from a computational study: $\Delta_f H^{\ominus}(SiH_2O) = -98.3$ kJ mol^{-1} and $\Delta_f H^{\ominus}(SiH_3OH) = -282$ kJ mol^{-1} Compute the standard enthalpies of the following reactions:

(a) $SiH_4(g) + \tfrac{1}{2}O_2(g) \rightarrow SiH_3OH(g)$

(b) $SiH_4(g) + O_2(g) \rightarrow SiH_2O(g) + H_2O(l)$

(c) $SiH_3OH(g) \rightarrow SiH_2O(g) + H_2(g)$

Note that $\Delta_f H^{\ominus}(SiH_4,g) = +34.3$ kJ mol^{-1} (*CRC Handbook* (2008)).

2.19 The constant-volume heat capacity of a gas can be measured by observing the decrease in temperature when it expands adiabatically and reversibly. If the decrease in pressure is also measured, we can use it to infer the value of $\gamma = C_p/C_V$ and hence, by combining the two values, deduce the constant-pressure heat capacity. A fluorocarbon gas was allowed to expand reversibly and adiabatically to twice its volume; as a result, the temperature fell from 298.15 K to 248.44 K and its pressure fell from 202.94 kPa to 81.840 kPa. Evaluate C_p.

2.20 A sample consisting of 1.00 mol of a van der Waals gas is compressed from 20.0 dm^3 to 10.0 dm^3 at 300 K. In the process, 20.2 kJ of work is done on the gas. Given that $\mu = \{(2a/RT) - b\}/C_{p,m}$, with $C_{p,m} = 38.4$ J K^{-1} mol^{-1}, $a = 3.60$ dm^6 atm mol^{-2}, and $b = 0.044$ dm^3 mol^{-1}, calculate ΔH for the process.

2.21 Take nitrogen to be a van der Waals gas with $a = 1.352$ dm^6 atm mol^{-2} and $b = 0.0387$ dm^3 mol^{-1}, and calculate ΔH_m when the pressure on the gas is decreased from 500 atm to 1.00 atm at 300 K. For a van der Waals gas, $\mu = \{(2a/RT) - b\}/C_{p,m}$. Assume $C_{p,m} = \tfrac{7}{2}R$.

Theoretical problems

2.22 Show that the following functions have exact differentials: (a) $x^2y + 3y^2$, (b) $x\cos xy$, (c) x^3y^2, (d) $t(t + e^s) + s$.

2.23 (a) What is the total differential of $z = x^2 + 2y^2 - 2xy + 2x - 4y - 8$? (b) Show that $\partial^2 z/\partial y\partial x = \partial^2 z/\partial x\partial y$ for this function. (c) Let $z = xy - y\ln x + 2$. Find dz and show that it is exact.

2.24 (a) Express $(\partial C_V/\partial V)_T$ as a second derivative of U and find its relation to $(\partial U/\partial V)_T$ and $(\partial C_p/\partial p)_T$ as a second derivative of H and find its relation to $(\partial H/\partial p)_T$. (b) From these relations show that $(\partial C_V/\partial V)_T = 0$ and $(\partial C_p/\partial p)_T = 0$ for a perfect gas.

2.25 (a) Derive the relation $C_V = -(\partial U/\partial V)_T(\partial V/\partial T)_U$ from the expression for the total differential of $U(T,V)$ and (b) starting from the expression for the total differential of $H(T,p)$, express $(\partial H/\partial p)_T$ in terms of C_p and the Joule–Thomson coefficient, μ.

2.26 Starting from the expression $C_p - C_V = T(\partial p/\partial T)_V(\partial V/\partial T)_p$, use the appropriate relations between partial derivatives to show that

$$C_p - C_V = -\frac{T(\partial V/\partial T)_p^2}{(\partial V/\partial p)_T}$$

Evaluate $C_p - C_V$ for a perfect gas.

2.27 (a) By direct differentiation of $H = U + pV$, obtain a relation between $(\partial H/\partial U)_p$ and $(\partial U/\partial V)_p$. (b) Confirm that $(\partial H/\partial U)_p = 1 + p(\partial V/\partial U)_p$ by expressing $(\partial H/\partial U)_p$ as the ratio of two derivatives with respect to volume and then using the definition of enthalpy.

2.28 Use the chain relation and the reciprocal identity of partial derivatives (*Mathematical background 2*) to derive the relation $(\partial H/\partial p)_T = -\mu C_p$.

2.29 Use the chain relation and the reciprocal identity of partial derivatives (*Mathematical background 2*) to derive the relation $(\partial p/\partial T)_V = \alpha/\kappa_T$. Confirm this relation by evaluating all three terms for (a) a perfect gas, (b) a van der Waals gas.

2.30 (a) Write expressions for dV and dp given that V is a function of p and T and p is a function of V and T. (b) Deduce expressions for d ln V and d ln p in terms of the expansion coefficient and the isothermal compressibility.

2.31 Calculate the work done during the isothermal reversible expansion of a gas that satisfies the virial equation of state, eqn 1.19. Evaluate (a) the work for 1.0 mol Ar at 273 K (for data, see Table 1.4) and (b) the same amount of a perfect gas. Let the expansion be from 500 cm^3 to 1000 cm^3 in each case.

2.32 Express the work of isothermal reversible expansion of a van der Waals gas in reduced variables and find a definition of reduced work that makes the overall expression independent of the identity of the gas. Calculate the work of isothermal reversible expansion along the critical isotherm from V_c to xV_c.

2.33‡ A gas obeying the equation of state $p(V - nb) = nRT$ is subjected to a Joule–Thomson expansion. Will the temperature increase, decrease, or remain the same?

2.34 Use the fact that $(\partial U/\partial V)_T = a/V_m^2$ for a van der Waals gas to show that $\mu C_{p,m} \approx (2a/RT) - b$ by using the definition of μ and appropriate relations between partial derivatives. (*Hint.* Use the approximation $pV_m \approx RT$ when it is justifiable to do so.)

2.35 Rearrange the van der Waals equation of state to give an expression for T as a function of p and V (with n constant). Calculate $(\partial T/\partial p)_V$ and confirm that $(\partial T/\partial p)_V = 1/(\partial p/\partial T)_V$. Go on to confirm Euler's chain relation.

2.36 Calculate the isothermal compressibility and the expansion coefficient of a van der Waals gas. Show, using Euler's chain relation, that $\kappa_T R = \alpha(V_m - b)$.

2.37 Given that $\mu C_p = T(\partial V/\partial T)_p - V$, derive an expression for μ in terms of the van der Waals parameters a and b, and express it in terms of reduced variables. Evaluate μ at 25°C and 1.0 atm, when the molar volume of the gas is 24.6 dm^3 mol^{-1}. Use the expression obtained to derive a formula for the inversion temperature of a van der Waals gas in terms of reduced variables, and evaluate it for the xenon sample.

2.38 The thermodynamic equation of state $(\partial U/\partial V)_T = T(\partial p/\partial T)_V - p$ was quoted in the chapter. Derive its partner

$$\left(\frac{\partial H}{\partial p}\right)_T = -T\left(\frac{\partial V}{\partial T}\right)_p + V$$

from it and the general relations between partial differentials.

2.39 Show that for a van der Waals gas,

$$C_{p,\mathrm{m}} - C_{V,\mathrm{m}} = \lambda R \qquad \frac{1}{\lambda} = 1 - \frac{(3V_\mathrm{r} - 1)^2}{4V_\mathrm{r}^3 T_\mathrm{r}}$$

and evaluate the difference for xenon at 25°C and 10.0 atm.

2.40 The speed of sound, c_s, in a gas of molar mass M is related to the ratio of heat capacities γ by $c_s = (\gamma RT/M)^{1/2}$. Show that $c_s = (\gamma p/\rho)^{1/2}$, where ρ is the mass density of the gas. Calculate the speed of sound in argon at 25°C.

2.41‡ A gas obeys the equation of state $V_\mathrm{m} = RT/p + aT^2$ and its constant-pressure heat capacity is given by $C_{p,\mathrm{m}} = A + BT + Cp$, where a, A, B, and C are constants independent of T and p. Obtain expressions for (a) the Joule–Thomson coefficient and (b) its constant-volume heat capacity.

Applications: to biology and the environment

2.42 In biological cells that have a plentiful supply of O_2, glucose is oxidized completely to CO_2 and H_2O by a process called *aerobic oxidation*. Muscle cells may be deprived of O_2 during vigorous exercise and, in that case, one molecule of glucose is converted to two molecules of lactic acid ($CH_3CH(OH)COOH$) by a process called *anaerobic glycolysis* (see *Impact I6.1*). (a) When 0.3212 g of glucose was burned in a bomb calorimeter of calorimeter constant 641 J K^{-1} the temperature rose by 7.793 K. Calculate (i) the standard molar enthalpy of combustion, (ii) the standard internal energy of combustion, and (iii) the standard enthalpy of formation of glucose. (b) What is the biological advantage (in kilojoules per mole of energy released as heat) of complete aerobic oxidation compared with anaerobic glycolysis to lactic acid?

2.43‡ Alkyl radicals are important intermediates in the combustion and atmospheric chemistry of hydrocarbons. Seakins *et al.* (*J. Phys. Chem.* **96**, 9847 (1992)) report $\Delta_\mathrm{f}H^\ominus$ for a variety of alkyl radicals in the gas phase, information that is applicable to studies of pyrolysis and oxidation reactions of hydrocarbons. This information can be combined with thermodynamic data on alkenes to determine the reaction enthalpy for possible fragmentation of a large alkyl radical into smaller radicals and alkenes. Use the following data to compute the standard reaction enthalpies for three possible fates of the *tert*-butyl radical, namely, (a) *tert*-$C_4H_9 \rightarrow$ *sec*-C_4H_9, (b) *tert*-$C_4H_9 \rightarrow C_3H_6 + CH_3$, (c) *tert*-$C_4H_9 \rightarrow C_2H_4 + C_2H_5$.

Species:	C_2H_5	*sec*-C_4H_9	*tert*-C_4H_9
$\Delta_\mathrm{f}H^\ominus$/(kJ mol^{-1})	+121.0	+67.5	+51.3

2.44‡ In 2007, the Intergovernmental Panel on Climate Change (IPCC) considered a global average temperature rise of 1.0–3.5°C likely by the year 2100 with 2.0°C its best estimate. Predict the average rise in sea level due to thermal expansion of sea water based on temperature rises of 1.0°C, 2.0°C, and 3.5°C given that the volume of the Earth's oceans is 1.37×10^9 km^3 and their surface area is 361×10^6 km^2, and state the approximations that go into the estimates.

2.45‡ Concerns over the harmful effects of chlorofluorocarbons on stratospheric ozone have motivated a search for new refrigerants. One such alternative is 2,2-dichloro-1,1,1-trifluoroethane (refrigerant 123). Younglove and McLinden published a compendium of thermophysical properties of this substance (*J. Phys. Chem. Ref. Data* **23**, 7 (1994)), from which properties such as the Joule–Thomson coefficient μ can be computed. (a) Compute μ at 1.00 bar and 50°C given that $(\partial H/\partial p)_T = -3.29 \times 10^3$ J MPa^{-1} mol^{-1} and $C_{p,\mathrm{m}} = 110.0$ J K^{-1} mol^{-1}. (b) Compute the temperature change that would accompany adiabatic expansion of 2.0 mol of this refrigerant from 1.5 bar to 0.5 bar at 50°C.

2.46‡ Another alternative refrigerant (see preceding problem) is 1,1,1,2-tetrafluoroethane (refrigerant HFC-134a). Tillner-Roth and Baehr published a compendium of thermophysical properties of this substance (*J. Phys. Chem. Ref. Data* **23**, 657 (1994)), from which properties such as the Joule–Thomson coefficient μ can be computed. (a) Compute μ at 0.100 MPa and 300 K from the following data (all referring to 300 K):

p/MPa	0.080	0.100	0.12
Specific enthalpy/(kJ kg^{-1})	426.48	426.12	425.76

(The specific constant-pressure heat capacity is 0.7649 kJ K^{-1} kg^{-1}.)
(b) Compute μ at 1.00 MPa and 350 K from the following data (all referring to 350 K):

p/MPa	0.80	1.00	1.2
Specific enthalpy/(kJ kg^{-1})	461.93	459.12	456.15

(The specific constant-pressure heat capacity is 1.0392 kJ K^{-1} kg^{-1}.)

2.47 Differential scanning calorimetry is used to examine the role of solvent–protein interactions in the denaturation process. Figure 2.34 shows the thermogram for ubiquitin in water with the signal observed for ubiquitin in methanol/water mixtures. Suggest an interpretation of the thermograms.

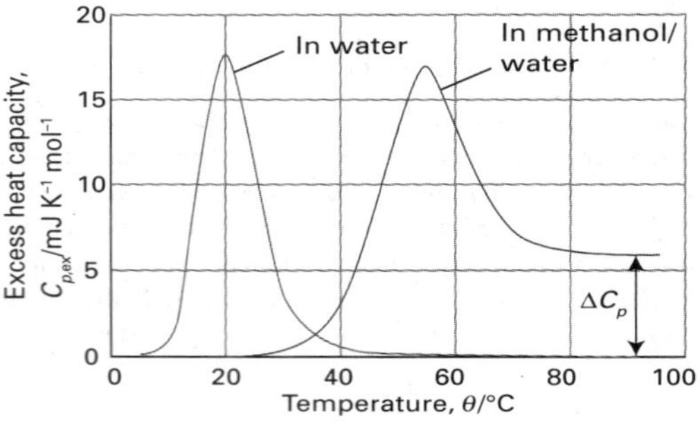

Fig. 2.34

MATHEMATICAL BACKGROUND 2

Multivariate calculus

A thermodynamic property of a system typically depends on a number of variables, such as the internal energy depending on the amount, volume, and temperature. To understand how these properties vary with the conditions we need to understand how to manipulate their derivatives. This is the field of **multivariate calculus**, the calculus of several variables.

MB2.1 Partial derivatives

A **partial derivative** of a function of more than one variable, such as $f(x,y)$, is the slope of the function with respect to one of the variables, all the other variables being held constant (Fig. MB2.1). Although a partial derivative shows how a function changes when one variable changes, it may be used to determine how the function changes when more than one variable changes by an infinitesimal amount. Thus, if f is a function of x and y, then when x and y change by dx and dy, respectively, f changes by

$$\mathrm{d}f = \left(\frac{\partial f}{\partial x}\right)_y \mathrm{d}x + \left(\frac{\partial f}{\partial y}\right)_x \mathrm{d}y \qquad \text{(MB2.1)}$$

where the symbol ∂ is used (instead of d) to denote a partial derivative and the subscript on the parentheses indicates which variable is being held constant. The quantity df is also called the **differential** of f. Successive partial derivatives may be taken in any order:

$$\left(\frac{\partial}{\partial y}\left(\frac{\partial f}{\partial x}\right)_y\right)_x = \left(\frac{\partial}{\partial x}\left(\frac{\partial f}{\partial y}\right)_x\right)_y \qquad \text{(MB2.2)}$$

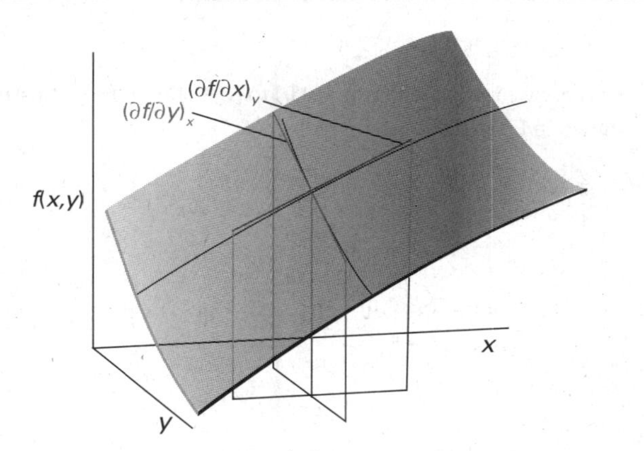

Fig. MB2.1 A function of two variables, $f(x,y)$, as depicted by the coloured surface, and the two partial derivatives, $(\partial f/\partial x)_y$ and $(\partial f/\partial y)_x$, the slope of the function parallel to the x- and y-axes, respectively. The function plotted here is $f(x,y) = ax^3y + by^2$ with $a = 1$ and $b = -2$.

- **A brief illustration**

Suppose that $f(x,y) = ax^3y + by^2$ (the function plotted in Fig. MB2.1) then

$$\left(\frac{\partial f}{\partial x}\right)_y = 3ax^2y \qquad \left(\frac{\partial f}{\partial y}\right)_x = ax^3 + 2by$$

Then, when x and y undergo infinitesimal changes, f changes by

$$\mathrm{d}f = 3ax^2y\mathrm{d}x + (ax^3 + 2by)\mathrm{d}y$$

To verify that the order of taking the second partial derivative is irrelevant, we form

$$\left(\frac{\partial}{\partial y}\left(\frac{\partial f}{\partial x}\right)_y\right)_x = \left(\frac{\partial(3ax^2y)}{\partial y}\right)_x = 3ax^2$$

$$\left(\frac{\partial}{\partial x}\left(\frac{\partial f}{\partial y}\right)_x\right)_y = \left(\frac{\partial(ax^3 + 2by)}{\partial x}\right)_y = 3ax^2 \; \bullet$$

Self test MB2.1 Evaluate df for $f(x,y) = 2x^2 \sin 3y$ and verify that the order of taking the second derivative is irrelevant.
[d$f = 4x \sin 3y \,\mathrm{d}x + 6x^2 \cos 3y \,\mathrm{d}y$]

In the following, z is a variable on which x and y depend (for example, x, y, and z might correspond to p, V, and T).

Relation 1 When x is changed at constant z:

$$\left(\frac{\partial f}{\partial x}\right)_z = \left(\frac{\partial f}{\partial x}\right)_y + \left(\frac{\partial f}{\partial y}\right)_x\left(\frac{\partial y}{\partial x}\right)_z \qquad \text{(MB2.3a)}$$

Relation 2

$$\left(\frac{\partial y}{\partial x}\right)_z = \frac{1}{(\partial x/\partial y)_z} \qquad \text{(MB2.3b)}$$

Relation 3

$$\left(\frac{\partial x}{\partial y}\right)_z = -\left(\frac{\partial x}{\partial z}\right)_y\left(\frac{\partial z}{\partial y}\right)_x \qquad \text{(MB2.3c)}$$

By combining this relation and Relation 2 we obtain the **Euler chain relation**:

$$\left(\frac{\partial y}{\partial x}\right)_z\left(\frac{\partial x}{\partial z}\right)_y\left(\frac{\partial z}{\partial y}\right)_x = -1 \qquad \text{(MB2.4)}$$

Euler chain relation

MB2.2 Exact differentials

The relation in eqn MB2.2 is the basis of a test for an **exact differential**, that is, the test of whether

$$df = g(x,y)dx + h(x,y)dy \qquad \text{(MB2.5)}$$

has the form in eqn MB2.1. If it has that form, then g can be identified with $(\partial f/\partial x)_y$ and h can be identified with $(\partial f/\partial y)_x$. Then eqn MB2.2 becomes

$$\left(\frac{\partial g}{\partial y}\right)_x = \left(\frac{\partial h}{\partial x}\right)_y \qquad \text{(MB2.6)}$$

Test for exact differential

• **A brief illustration**

Suppose, instead of the form $df = 3ax^2ydx + (ax^3 + 2by)dy$ in the previous *brief illustration* we were presented with the expression

$$df = \overbrace{3ax^2y}^{g(x,y)}dx + \overbrace{(ax^2 + 2by)}^{h(x,y)}dy$$

with ax^2 in place of ax^3 inside the second parentheses. To test whether this is an exact differential, we form

$$\left(\frac{\partial g}{\partial y}\right)_x = \left(\frac{\partial(3ax^2y)}{\partial y}\right)_x = 3ax^2$$

$$\left(\frac{\partial h}{\partial x}\right)_y = \left(\frac{\partial(ax^2 + 2by)}{\partial x}\right)_y = 2ax$$

These two expressions are not equal, so this form of df is not an exact differential and there is not a corresponding integrated function of the form $f(x,y)$. •

Self-test MB2.2 Determine whether the expression $df = (2y - x^3)dx + xdy$ is an exact differential. [No]

If df is exact, then we can do two things: (1) from a knowledge of the functions g and h we can reconstruct the function f; (2) we can be confident that the integral of df between specified limits is independent of the path between those limits. The first conclusion is best demonstrated with a specific example.

• **A brief illustration**

We consider the differential $df = 3ax^2ydx + (ax^3 + 2by)dy$, which we know to be exact. Because $(\partial f/\partial x)_y = 3ax^2y$, we can integrate with respect to x with y held constant, to obtain

$$f = \int df = \int 3ax^2ydx = 3ay\int x^2\,dx = ax^{3y} + k$$

where the 'constant' of integration k may depend on y (which has been treated as a constant in the integration), but not on x. To find $k(y)$, we note that $(\partial f/\partial y)_x = ax^3 + 2by$, and therefore

$$\left(\frac{\partial f}{\partial y}\right)_x = \left(\frac{\partial(ax^3y + k)}{\partial y}\right)_x = ax^3 + \frac{dk}{dy} = ax^3 + 2by$$

Therefore

$$\frac{dk}{dy} = 2by$$

from which it follows that $k = by^2 + \text{constant}$. We have found, therefore, that

$$f(x,y) = ax^3y + by^2 + \text{constant}$$

which, apart from the constant, is the original function in the first *brief illustration*. The value of the constant is pinned down by stating the boundary conditions; thus, if it is known that $f(0,0) = 0$, then the constant is zero. •

Self-test MB2.3 Confirm that $df = 3x^2 \cos y\, dx - x^3 \sin y\, dy$ is exact and find the function $f(x,y)$. [$f = x^3 \cos y$]

To demonstrate that the integral of df is independent of the path is now straightforward. Because df is a differential, its integral between the limits a and b is

$$\int_a^b df = f(b) - f(a)$$

The value of the integral depends only on the values at the end points and is independent of the path between them. If df is not an exact differential, the function f does not exist, and this argument no longer holds. In such cases, the integral of df does depend on the path.

• **A brief illustration**

Consider the inexact differential (the expression with ax^2 in place of ax^3 inside the second parentheses):

$$df = 3ax^2ydx + (ax^2 + 2by)dy$$

Suppose we integrate df from (0,0) to (2,2) along the two paths shown in Fig. MB2.2. Along Path 1,

$$\int_{\text{Path 1}} df = \int_{0,0}^{2,0} 3ax^2ydx + \int_{2,0}^{2,2} (ax^2 + 2by)dy$$

$$= 0 + 4a\int_0^2 dy + 2b\int_0^2 ydy = 8a + 4b$$

whereas along Path 2,

$$\int_{\text{Path 2}} df = \int_{0,2}^{2,2} 3ax^2ydx + \int_{0,0}^{0,2} (ax^2 + 2by)dy$$

$$= 6a\int_0^2 x^2dx + 0 + 2b\int_0^2 ydy = 16a + 4b$$

The two integrals are not the same. •

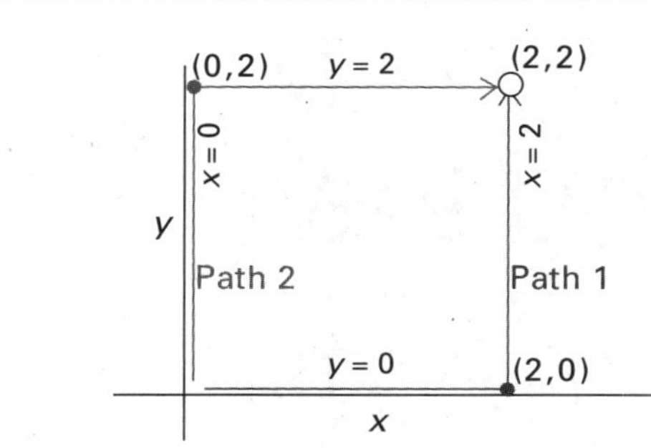

Fig. MB2.2 The two integration paths referred to in the *brief illustration*.

Self-test MB2.4 Confirm that the two paths do give the same value for the exact differential in the first *brief illustration*.

[Both paths: $16a+4b$]

An inexact differential may sometimes be converted into an exact differential by multiplication by a factor known as an *integrating factor*. A physical example is the integrating factor $1/T$ that converts the inexact differential dq_{rev} into the exact differential dS in thermodynamics (see Chapter 3).

- **A brief illustration**

We have seen that the differential $df = 3ax^2y dx + (ax^2 + 2by)dy$ is inexact; the same is true when we set $b = 0$ and consider $df = 3ax^2y dx + ax^2\, dy$ instead. Suppose we multiply this df by $x^m y^n$ and write $x^m y^n df = df'$, then we obtain

$$df' = \overbrace{3ax^{m+2}y^{n+1}}^{g(x,y)}dx + \overbrace{ax^{m+2}y^n}^{h(x,y)}dy$$

We evaluate the following two partial derivatives:

$$\left(\frac{\partial g}{\partial y}\right)_x = \left(\frac{\partial(3ax^{m+2}y^{n+1})}{\partial y}\right)_x = 3a(n+1)x^{m+2}y^n$$

$$\left(\frac{\partial h}{\partial x}\right)_y = \left(\frac{\partial(ax^{m+2}y^n)}{\partial x}\right)_y = a(m+2)x^{m+1}y^n$$

For the new differential to be exact, these two partial derivatives must be equal, so we write

$$3a(n+1)x^{m+2}y^n = a(m+2)x^{m+1}y^n$$

which simplifies to

$$3(n+1)x = m+2$$

The only solution that is independent of x is $n=-1$ and $m=-2$. It follows that

$$df' = 3a dx + (a/y)dy$$

is an exact differential. By the procedure already illustrated, its integrated form is $f'(x,y) = 3ax + a \ln y + \text{constant}$. ●

Self-test MB2.5 Find an integrating factor of the form $x^m y^n$ for the inexact differential $df = (2y - x^3)dx + xdy$ and the integrated form of f'.

[$df' = x df$, $f' = yx^2 - \frac{1}{5}x^5 + \text{constant}$]

3 The Second Law

The purpose of this chapter is to explain the origin of the spontaneity of physical and chemical change. We examine two simple processes and show how to define, measure, and use a property, the entropy, to discuss spontaneous changes quantitatively. The chapter also introduces a major subsidiary thermodynamic property, the Gibbs energy, which lets us express the spontaneity of a process in terms of the properties of a system. The Gibbs energy also enables us to predict the maximum non-expansion work that a process can do. As we began to see in Chapter 2, one application of thermodynamics is to find relations between properties that might not be thought to be related. Several relations of this kind can be established by making use of the fact that the Gibbs energy is a state function. We also see how to derive expressions for the variation of the Gibbs energy with temperature and pressure and how to formulate expressions that are valid for real gases. These expressions will prove useful later when we discuss the effect of temperature and pressure on equilibrium constants.

Some things happen naturally; some things don't. A gas expands to fill the available volume, a hot body cools to the temperature of its surroundings, and a chemical reaction runs in one direction rather than another. Some aspect of the world determines the **spontaneous** direction of change, the direction of change that does not require work to bring it about. A gas can be confined to a smaller volume, an object can be cooled by using a refrigerator, and some reactions can be driven in reverse (as in the electrolysis of water). However, none of these processes is spontaneous; each one must be brought about by doing work. An important point, though, is that throughout this text 'spontaneous' must be interpreted as a natural *tendency* that may or may not be realized in practice. Thermodynamics is silent on the rate at which a spontaneous change in fact occurs, and some spontaneous processes (such as the conversion of diamond to graphite) may be so slow that the tendency is never realized in practice whereas others (such as the expansion of a gas into a vacuum) are almost instantaneous.

The recognition of two classes of process, spontaneous and non-spontaneous, is summarized by the **Second Law of thermodynamics**. This law may be expressed in a variety of equivalent ways. One statement was formulated by Kelvin:

> No process is possible in which the sole result is the absorption of heat from a reservoir and its complete conversion into work.

For example, it has proved impossible to construct an engine like that shown in Fig. 3.1, in which heat is drawn from a hot reservoir and completely converted into work. All real heat engines have both a hot source and a cold sink; some energy is always discarded into the cold sink as heat and not converted into work. The Kelvin statement is a generalization of another everyday observation, that a ball at rest on a surface has never been observed to leap spontaneously upwards. An upward leap of the ball would be equivalent to the conversion of heat from the surface into work.

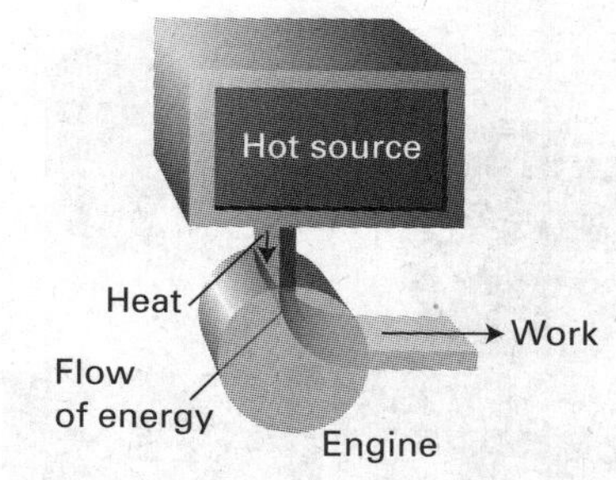

Fig. 3.1 The Kelvin statement of the Second Law denies the possibility of the process illustrated here, in which heat is changed completely into work, there being no other change. The process is not in conflict with the First Law because energy is conserved.

The direction of spontaneous change

What determines the direction of spontaneous change? It is not the total energy of the isolated system. The First Law of thermodynamics states that energy is conserved in any process, and we cannot disregard that law now and say that everything tends towards a state of lower energy: the total energy of an isolated system is constant.

Is it perhaps the energy of the *system* that tends towards a minimum? Two arguments show that this cannot be so. First, a perfect gas expands spontaneously into a vacuum, yet its internal energy remains constant as it does so. Secondly, if the energy of a system does happen to decrease during a spontaneous change, the energy of its surroundings must increase by the same amount (by the First Law). The increase in energy of the surroundings is just as spontaneous a process as the decrease in energy of the system.

When a change occurs, the total energy of an isolated system remains constant but it is parcelled out in different ways. Can it be, therefore, that the direction of change is related to the *distribution* of energy? We shall see that this idea is the key, and that spontaneous changes are always accompanied by a dispersal of energy.

3.1 The dispersal of energy

Key point **During a spontaneous change in an isolated system the total energy is dispersed into random thermal motion of the particles in the system.**

We can begin to understand the role of the distribution of energy by thinking about a ball (the system) bouncing on a floor (the surroundings). The ball does not rise as high after each bounce because there are inelastic losses in the materials of the ball and floor. The kinetic energy of the ball's overall motion is spread out into the energy of thermal motion of its particles and those of the floor that it hits. The direction of spontaneous change is towards a state in which the ball is at rest with all its energy dispersed into disorderly thermal motion of molecules in the air and of the atoms of the virtually infinite floor (Fig. 3.2).

A ball resting on a warm floor has never been observed to start bouncing. For bouncing to begin, something rather special would need to happen. In the first place, some of the thermal motion of the atoms in the floor would have to accumulate in a single, small object, the ball. This accumulation requires a spontaneous localization of energy from the myriad vibrations of the atoms of the floor into the much smaller number of atoms that constitute the ball (Fig. 3.3). Furthermore, whereas the thermal motion is random, for the ball to move upwards its atoms must all move in the same direction. The localization of random, disorderly motion as concerted, ordered motion is so unlikely that we can dismiss it as virtually impossible.[1]

We appear to have found the signpost of spontaneous change: *we look for the direction of change that leads to dispersal of the total energy of the isolated system.* This principle accounts for the direction of change of the bouncing ball, because its energy

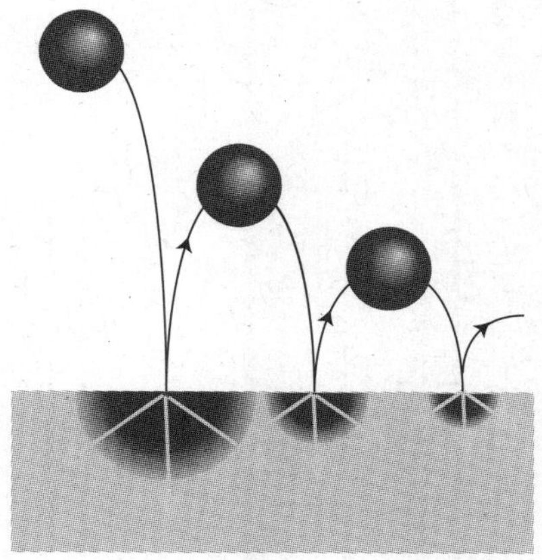

Fig. 3.2 The direction of spontaneous change for a ball bouncing on a floor. On each bounce some of its energy is degraded into the thermal motion of the atoms of the floor, and that energy disperses. The reverse has never been observed to take place on a macroscopic scale.

[1] Concerted motion, but on a much smaller scale, is observed as *Brownian motion*, the jittering motion of small particles suspended in a liquid or gas.

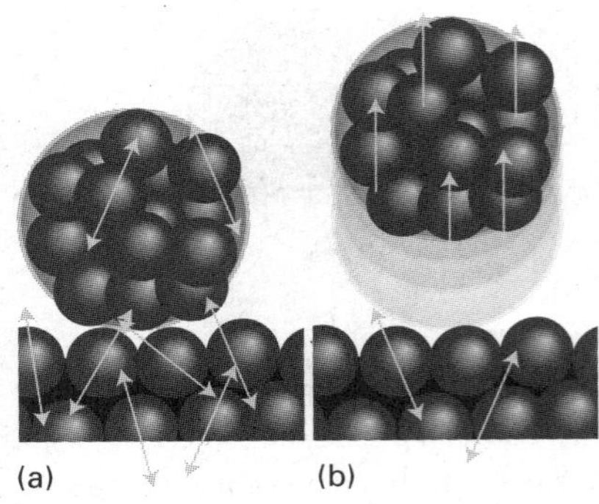

Fig. 3.3 The molecular interpretation of the irreversibility expressed by the Second Law. (a) A ball resting on a warm surface; the atoms are undergoing thermal motion (vibration, in this instance), as indicated by the arrows. (b) For the ball to fly upwards, some of the random vibrational motion would have to change into coordinated, directed motion. Such a conversion is highly improbable.

is spread out as thermal motion of the atoms of the floor. The reverse process is not spontaneous because it is highly improbable that energy will become localized, leading to uniform motion of the ball's atoms. A gas does not contract spontaneously because to do so the random motion of its molecules, which spreads out the distribution of kinetic energy throughout the container, would have to take them all into the same region of the container, thereby localizing the energy. The opposite change, spontaneous expansion, is a natural consequence of energy becoming more dispersed as the gas molecules occupy a larger volume. An object does not spontaneously become warmer than its surroundings because it is highly improbable that the jostling of randomly vibrating atoms in the surroundings will lead to the localization of thermal motion in the object. The opposite change, the spreading of the object's energy into the surroundings as thermal motion, is natural.

It may seem very puzzling that the spreading out of energy and matter can lead to the formation of such ordered structures as crystals or proteins. Nevertheless, in due course, we shall see that dispersal of energy and matter accounts for change in all its forms.

3.2 Entropy

Key points **The entropy acts as a signpost of spontaneous change. (a) Entropy change is defined in terms of heat transactions (the Clausius definition). (b) Absolute entropies are defined in terms of the number of ways of achieving a configuration (the Boltzmann formula). (c) The Carnot cycle is used to prove that entropy is a state function. (d) The efficiency of a heat engine is the basis of the definition of the thermodynamic temperature scale and one realization, the Kelvin scale. (e) The Clausius inequality is used to show that the entropy increases in a spontaneous change and therefore that the Clausius definition is consistent with the Second Law.**

The First Law of thermodynamics led to the introduction of the internal energy, U. The internal energy is a state function that lets us assess whether a change is permissible: only those changes may occur for which the internal energy of an isolated system remains constant. The law that is used to identify the signpost of spontaneous change, the Second Law of thermodynamics, may also be expressed in terms of another state function, the **entropy**, S. We shall see that the entropy (which we shall define shortly, but is a measure of the energy dispersed in a process) lets us assess whether one state is accessible from another by a spontaneous change. The First Law uses the internal energy to identify *permissible* changes; the Second Law uses the entropy to identify the *spontaneous changes* among those permissible changes.

The Second Law of thermodynamics can be expressed in terms of the entropy:

The entropy of an isolated system increases in the course of a spontaneous change: $\Delta S_{tot} > 0$

where S_{tot} is the total entropy of the system and its surroundings. Thermodynamically irreversible processes (like cooling to the temperature of the surroundings and the free expansion of gases) are spontaneous processes, and hence must be accompanied by an increase in total entropy.

(a) The thermodynamic definition of entropy

The thermodynamic definition of entropy concentrates on the change in entropy, dS, that occurs as a result of a physical or chemical change (in general, as a result of a 'process'). The definition is motivated by the idea that a change in the extent to which energy is dispersed depends on how much energy is transferred as heat. As we have remarked, heat stimulates random motion in the surroundings. On the other hand, work stimulates uniform motion of atoms in the surroundings and so does not change their entropy.

The thermodynamic definition of entropy is based on the expression

$$dS = \frac{dq_{rev}}{T} \quad \text{Definition of entropy change} \quad [3.1]$$

where q_{rev} is the heat supplied reversibly. For a measurable change between two states i and f this expression integrates to

$$\Delta S = \int_i^f \frac{dq_{rev}}{T} \quad (3.2)$$

That is, to calculate the difference in entropy between any two states of a system, we find a *reversible* path between them, and integrate the energy supplied as heat at each stage of the path divided by the temperature at which heating occurs.

A note on good practice According to eqn 3.2, when the energy transferred as heat is expressed in joules and the temperature is in kelvins, the units of entropy are joules per kelvin ($J K^{-1}$). Entropy is an extensive property. Molar entropy, the entropy divided by the amount of substance, is expressed in joules per kelvin per mole ($J K^{-1} mol^{-1}$). The units of entropy are the same as those of the gas constant, R, and molar heat capacities. Molar entropy is an intensive property.

Example 3.1 *Calculating the entropy change for the isothermal expansion of a perfect gas*

Calculate the entropy change of a sample of perfect gas when it expands isothermally from a volume V_i to a volume V_f.

Method The definition of entropy instructs us to find the energy supplied as heat for a reversible path between the stated initial and final states regardless of the actual manner in which the process takes place. A simplification is that the expansion is isothermal, so the temperature is a constant and may be taken outside the integral in eqn 3.2. The energy absorbed as heat during a reversible isothermal expansion of a perfect gas can be calculated from $\Delta U = q + w$ and $\Delta U = 0$, which implies that $q = -w$ in general and therefore that $q_{rev} = -w_{rev}$ for a reversible change. The work of reversible isothermal expansion was calculated in Section 2.3.

Answer Because the temperature is constant, eqn 3.2 becomes

$$\Delta S = \frac{1}{T}\int_i^f dq_{rev} = \frac{q_{rev}}{T}$$

From eqn 2.10, we know that

$$q_{rev} = -w_{rev} = nRT \ln \frac{V_f}{V_i}$$

It follows that

$$\Delta S = nR \ln \frac{V_f}{V_i}$$

• **A brief illustration**

When the volume occupied by 1.00 mol of any perfect gas molecules is doubled at any constant temperature, $V_f/V_i = 2$ and

$$\Delta S = (1.00 \text{ mol}) \times (8.3145 \text{ J K}^{-1} \text{ mol}^{-1}) \times \ln 2 = +5.76 \text{ J K}^{-1}$$ •

Self-test 3.1 Calculate the change in entropy when the pressure of a fixed amount of perfect gas is changed isothermally from p_i to p_f. What is this change due to?

[$\Delta S = nR \ln(p_i/p_f)$; the change in volume when the gas is compressed]

We can use the definition in eqn 3.1 to formulate an expression for the change in entropy of the surroundings, ΔS_{sur}. Consider an infinitesimal transfer of heat dq_{sur} to the surroundings. The surroundings consist of a reservoir of constant volume, so the energy supplied to them by heating can be identified with the change in the internal energy of the surroundings, dU_{sur}.[2] The internal energy is a state function, and dU_{sur} is an exact differential. As we have seen, these properties imply that dU_{sur} is independent of how the change is brought about and in particular is independent of whether the process is reversible or irreversible. The same remarks therefore apply to dq_{sur}, to which dU_{sur} is equal. Therefore, we can adapt the definition in eqn 3.1, delete the constraint 'reversible', and write

$$dS_{sur} = \frac{dq_{sur,rev}}{T_{sur}} = \frac{dq_{sur}}{T_{sur}} \qquad \text{Entropy change of the surroundings} \qquad (3.3a)$$

Furthermore, because the temperature of the surroundings is constant whatever the change, for a measurable change

$$\Delta S_{sur} = \frac{q_{sur}}{T_{sur}} \qquad (3.3b)$$

That is, regardless of how the change is brought about in the system, reversibly or irreversibly, we can calculate the change of entropy of the surroundings by dividing the heat transferred by the temperature at which the transfer takes place.

Equation 3.3 makes it very simple to calculate the changes in entropy of the surroundings that accompany any process. For instance, for any adiabatic change, $q_{sur} = 0$, so

For an adiabatic change: $\quad \Delta S_{sur} = 0 \qquad (3.4)$

This expression is true however the change takes place, reversibly or irreversibly, provided no local hot spots are formed in the surroundings. That is, it is true so long as the surroundings remain in internal equilibrium. If hot spots do form, then the localized energy may subsequently disperse spontaneously and hence generate more entropy.

• **A brief illustration**

To calculate the entropy change in the surroundings when 1.00 mol $H_2O(l)$ is formed from its elements under standard conditions at 298 K, we use $\Delta H^{\ominus} = -286$ kJ from Table 2.8. The energy released as heat is supplied to the surroundings, now regarded as being at constant pressure, so $q_{sur} = +286$ kJ. Therefore,

$$\Delta S_{sur} = \frac{2.86 \times 10^5 \text{ J}}{298 \text{ K}} = +960 \text{ J K}^{-1}$$

This strongly exothermic reaction results in an increase in the entropy of the surroundings as energy is released as heat into them. •

Self-test 3.2 Calculate the entropy change in the surroundings when 1.00 mol $N_2O_4(g)$ is formed from 2.00 mol $NO_2(g)$ under standard conditions at 298 K. [-192 J K^{-1}]

[2] Alternatively, the surroundings can be regarded as being at constant pressure, in which case we could equate dq_{sur} to dH_{sur}.

(b) The statistical view of entropy

The entry point into the molecular interpretation of the Second Law of thermodynamics is Boltzmann's insight, first explored in Section F.5a, that an atom or molecule can possess only certain values of the energy, called its 'energy levels'. The continuous thermal agitation that molecules experience in a sample at $T > 0$ ensures that they are distributed over the available energy levels. Boltzmann also made the link between the distribution of molecules over energy levels and the entropy. He proposed that the entropy of a system is given by

$$S = k \ln W \qquad \text{Boltzmann formula for the entropy} \qquad (3.5)$$

where $k = 1.381 \times 10^{-23}$ J K^{-1} and W is the number of *microstates*, the ways in which the molecules of a system can be arranged while keeping the total energy constant. Each microstate lasts only for an instant and corresponds to a certain distribution of molecules over the available energy levels. When we measure the properties of a system, we are measuring an average taken over the many microstates the system can occupy under the conditions of the experiment. The concept of the number of microstates makes quantitative the ill-defined qualitative concepts of 'disorder' and 'the dispersal of matter and energy' that are used widely to introduce the concept of entropy: a more 'disorderly' distribution of energy and matter corresponds to a greater number of microstates associated with the same total energy.

Equation 3.5 is known as the **Boltzmann formula** and the entropy calculated from it is sometimes called the **statistical entropy**. We see that, if $W = 1$, which corresponds to one microstate (only one way of achieving a given energy, all molecules in exactly the same state), then $S = 0$ because $\ln 1 = 0$. However, if the system can exist in more than one microstate, then $W > 1$ and $S > 0$. If the molecules in the system have access to a greater number of energy levels, then there may be more ways of achieving a given total energy, that is, there are more microstates for a given total energy, W is greater, and the entropy is greater than when fewer states are accessible. Therefore, the statistical view of entropy summarized by the Boltzmann formula is consistent with our previous statement that the entropy is related to the dispersal of energy. In particular, for a gas of particles in a container, the energy levels become closer together as the container expands (Fig. 3.4; this is a conclusion from quantum theory that we shall verify in Chapter 8). As a result, more microstates become possible, W increases, and the entropy increases, exactly as we inferred from the thermodynamic definition of entropy.

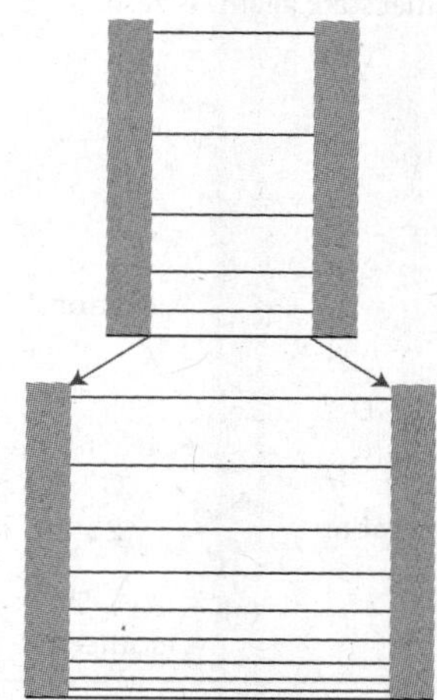

Fig. 3.4 When a box expands, the energy levels move closer together and more become accessible to the molecules. As a result the partition function increases and so does the entropy.

The molecular interpretation of entropy advanced by Boltzmann also suggests the thermodynamic definition given by eqn 3.1. To appreciate this point, consider that molecules in a system at high temperature can occupy a large number of the available energy levels, so a small additional transfer of energy as heat will lead to a relatively small change in the number of accessible energy levels. Consequently, the number of microstates does not increase appreciably and neither does the entropy of the system. In contrast, the molecules in a system at low temperature have access to far fewer energy levels (at $T = 0$, only the lowest level is accessible), and the transfer of the same quantity of energy by heating will increase the number of accessible energy levels and the number of microstates significantly. Hence, the change in entropy upon heating will be greater when the energy is transferred to a cold body than when it is transferred to a hot body. This argument suggests that the change in entropy should be inversely proportional to the temperature at which the transfer takes place, as in eqn 3.1.

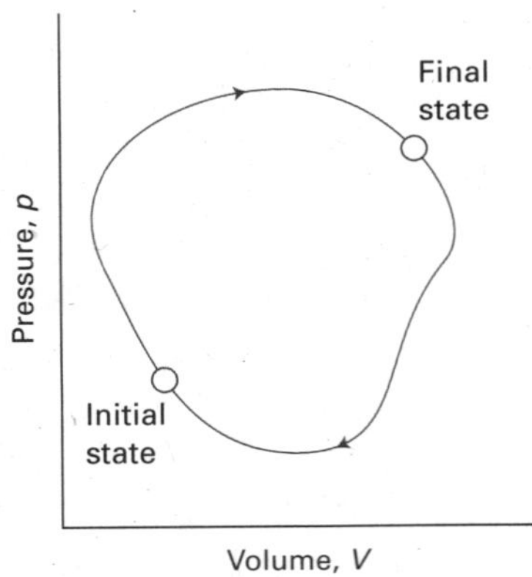

Fig. 3.5 In a thermodynamic cycle, the overall change in a state function (from the initial state to the final state and then back to the initial state again) is zero.

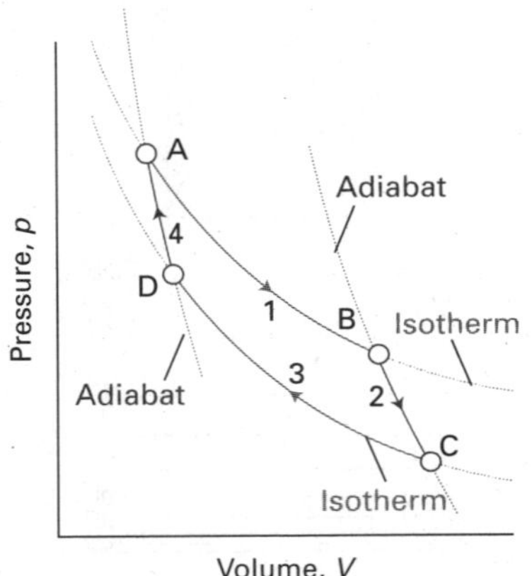

Fig. 3.6 The basic structure of a Carnot cycle. In Step 1, there is an isothermal reversible expansion at the temperature T_h. Step 2 is a reversible adiabatic expansion in which the temperature falls from T_h to T_c. In Step 3 there is an isothermal reversible compression at T_c, and that isothermal step is followed by an adiabatic reversible compression, which restores the system to its initial state.

(c) The entropy as a state function

Entropy is a state function. To prove this assertion, we need to show that the integral of dS is independent of path. To do so, it is sufficient to prove that the integral of eqn 3.1 around an arbitrary cycle is zero, for that guarantees that the entropy is the same at the initial and final states of the system regardless of the path taken between them (Fig. 3.5). That is, we need to show that

$$\oint \frac{\mathrm{d}q_{\mathrm{rev}}}{T_{\mathrm{sur}}} = 0 \tag{3.6}$$

where the symbol $\oint$ denotes integration around a closed path. There are three steps in the argument:

1. First, to show that eqn 3.6 is true for a special cycle (a 'Carnot cycle') involving a perfect gas.
2. Then to show that the result is true whatever the working substance.
3. Finally, to show that the result is true for any cycle.

A **Carnot cycle**, which is named after the French engineer Sadi Carnot, consists of four reversible stages (Fig. 3.6):

1. Reversible isothermal expansion from A to B at T_h; the entropy change is q_h/T_h, where q_h is the energy supplied to the system as heat from the hot source.
2. Reversible adiabatic expansion from B to C. No energy leaves the system as heat, so the change in entropy is zero. In the course of this expansion, the temperature falls from T_h to T_c, the temperature of the cold sink.
3. Reversible isothermal compression from C to D at T_c. Energy is released as heat to the cold sink; the change in entropy of the system is q_c/T_c; in this expression q_c is negative.
4. Reversible adiabatic compression from D to A. No energy enters the system as heat, so the change in entropy is zero. The temperature rises from T_c to T_h.

The total change in entropy around the cycle is the sum of the changes in each of these four steps:

$$\oint \mathrm{d}S = \frac{q_h}{T_h} + \frac{q_c}{T_c}$$

However, we show in the following *Justification* that for a perfect gas

$$\frac{q_h}{q_c} = -\frac{T_h}{T_c} \tag{3.7}$$

Substitution of this relation into the preceding equation gives zero on the right, which is what we wanted to prove.

Justification 3.1 *Heating accompanying reversible adiabatic expansion*

This *Justification* is based on two features of the cycle. One feature is that the two temperatures T_h and T_c in eqn 3.7 lie on the same adiabat in Fig. 3.6. The second feature is that the energies transferred as heat during the two isothermal stages are

$$q_h = nRT_h \ln \frac{V_B}{V_A} \qquad q_c = nRT_c \ln \frac{V_D}{V_C}$$

We now show that the two volume ratios are related in a very simple way. From the relation between temperature and volume for reversible adiabatic processes (VT^c = constant, eqn 2.28):

$$V_A T_h^c = V_D T_c^c \qquad V_C T_c^c = V_B T_h^c$$

Multiplication of the first of these expressions by the second gives

$$V_A V_C T_h^c T_c^c = V_D V_B T_h^c T_c^c$$

which, on cancellation of the temperatures, simplifies to

$$\frac{V_A}{V_B} = \frac{V_D}{V_C}$$

With this relation established, we can write

$$q_c = nRT_c \ln\frac{V_D}{V_C} = nRT_c \ln\frac{V_A}{V_B} = -nRT_c \ln\frac{V_B}{V_A}$$

and therefore

$$\frac{q_h}{q_c} = \frac{nRT_h \ln(V_B/V_A)}{-nRT_c \ln(V_B/V_A)} = -\frac{T_h}{T_c}$$

as in eqn 3.7.

In the second step we need to show that eqn 3.6 applies to any material, not just a perfect gas (which is why, in anticipation, we have not labelled it with a °). We begin this step of the argument by introducing the **efficiency**, η (eta), of a heat engine:

$$\eta = \frac{\text{work performed}}{\text{heat absorbed from hot source}} = \frac{|w|}{|q_h|} \qquad \text{Definition of efficiency} \quad [3.8]$$

We are using modulus signs to avoid complications with signs: all efficiencies are positive numbers. The definition implies that, the greater the work output for a given supply of heat from the hot reservoir, the greater is the efficiency of the engine. We can express the definition in terms of the heat transactions alone, because (as shown in Fig. 3.7), the energy supplied as work by the engine is the difference between the energy supplied as heat by the hot reservoir and returned to the cold reservoir:

$$\eta = \frac{|q_h| - |q_c|}{|q_h|} = 1 - \frac{|q_c|}{|q_h|} \qquad (3.9)$$

It then follows from eqn 3.7 (noting that the modulus signs remove the minus sign) that

$$\eta = 1 - \frac{T_c}{T_h} \qquad \text{Carnot efficiency} \quad (3.10)_{rev}$$

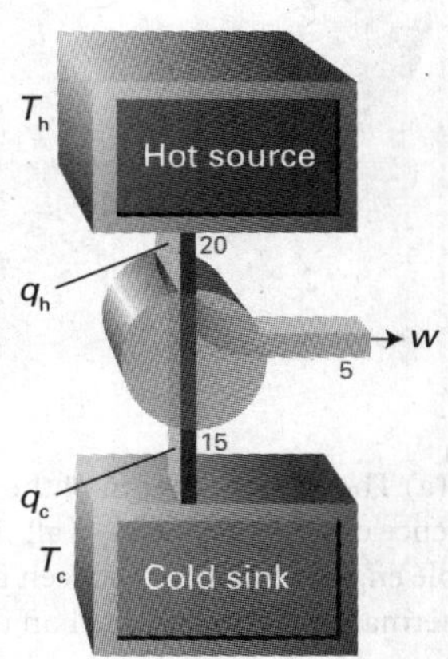

Fig. 3.7 Suppose an energy q_h (for example, 20 kJ) is supplied to the engine and q_c is lost from the engine (for example, $q_c = -15$ kJ) and discarded into the cold reservoir. The work done by the engine is equal to $q_h + q_c$ (for example, 20 kJ + (−15 kJ) = 5 kJ). The efficiency is the work done divided by the energy supplied as heat from the hot source.

Now we are ready to generalize this conclusion. The Second Law of thermodynamics implies that *all reversible engines have the same efficiency regardless of their construction.* To see the truth of this statement, suppose two reversible engines are coupled together and run between the same two reservoirs (Fig. 3.8). The working substances and details of construction of the two engines are entirely arbitrary. Initially, suppose that engine A is more efficient than engine B, and that we choose a setting of the controls that causes engine B to acquire energy as heat q_c from the cold reservoir and to release a certain quantity of energy as heat into the hot reservoir. However, because engine A is more efficient than engine B, not all the work that A produces is needed for

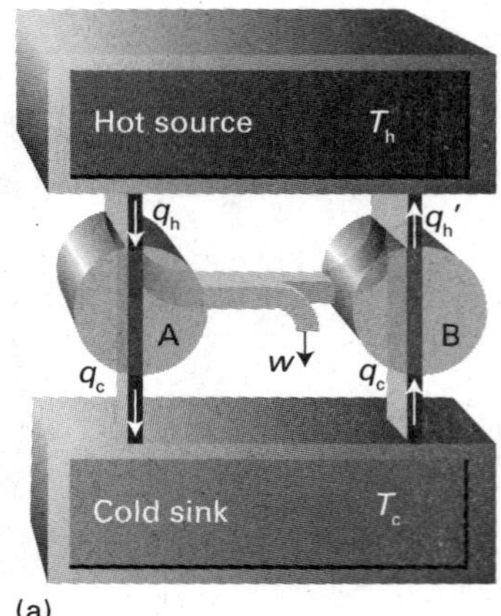

(a)

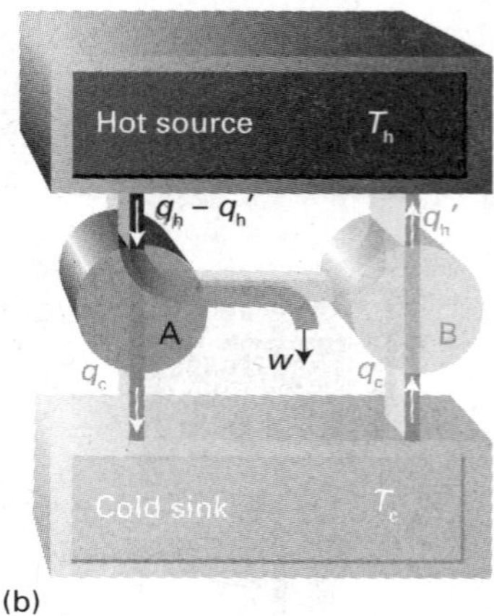

(b)

Fig. 3.8 (a) The demonstration of the equivalence of the efficiencies of all reversible engines working between the same thermal reservoirs is based on the flow of energy represented in this diagram. (b) The net effect of the processes is the conversion of heat into work without there being a need for a cold sink: this is contrary to the Kelvin statement of the Second Law.

this process, and the difference can be used to do work. The net result is that the cold reservoir is unchanged, work has been done, and the hot reservoir has lost a certain amount of energy. This outcome is contrary to the Kelvin statement of the Second Law, because some heat has been converted directly into work. In molecular terms, the random thermal motion of the hot reservoir has been converted into ordered motion characteristic of work. Because the conclusion is contrary to experience, the initial assumption that engines A and B can have different efficiencies must be false. It follows that the relation between the heat transfers and the temperatures must also be independent of the working material, and therefore that eqn 3.10 is always true for any substance involved in a Carnot cycle.

For the final step in the argument, we note that any reversible cycle can be approximated as a collection of Carnot cycles and the integral around an arbitrary path is the sum of the integrals around each of the Carnot cycles (Fig. 3.9). This approximation becomes exact as the individual cycles are allowed to become infinitesimal. The entropy change around each individual cycle is zero (as demonstrated above), so the sum of entropy changes for all the cycles is zero. However, in the sum, the entropy change along any individual path is cancelled by the entropy change along the path it shares with the neighbouring cycle. Therefore, all the entropy changes cancel except for those along the perimeter of the overall cycle. That is,

$$\sum_{\text{all}} \frac{q_{\text{rev}}}{T} = \sum_{\text{perimeter}} \frac{q_{\text{rev}}}{T} = 0$$

In the limit of infinitesimal cycles, the non-cancelling edges of the Carnot cycles match the overall cycle exactly, and the sum becomes an integral. Equation 3.6 then follows immediately. This result implies that $\mathrm{d}S$ is an exact differential and therefore that S is a state function.

(d) The thermodynamic temperature

Suppose we have an engine that is working reversibly between a hot source at a temperature T_h and a cold sink at a temperature T, then we know from eqn 3.10 that

$$T = (1 - \eta)T_\mathrm{h} \tag{3.11}$$

This expression enabled Kelvin to define the **thermodynamic temperature scale** in terms of the efficiency of a heat engine: we construct an engine in which the hot source is at a known temperature and the cold sink is the object of interest. The temperature of the latter can then be inferred from the measured efficiency of the engine. The **Kelvin scale** (which is a special case of the thermodynamic temperature scale) is defined by using water at its triple point as the notional hot source and defining that temperature as 273.16 K exactly. For instance, if it is found that the efficiency of such an engine is 0.20, then the temperature of the cold sink is $0.80 \times 273.16\ \mathrm{K} = 220\ \mathrm{K}$. This result is independent of the working substance of the engine.

(e) The Clausius inequality

We now show that the definition of entropy is consistent with the Second Law. To begin, we recall that more work is done when a change is reversible than when it is irreversible. That is, $|\mathrm{d}w_{\text{rev}}| \geq |\mathrm{d}w|$. Because $\mathrm{d}w$ and $\mathrm{d}w_{\text{rev}}$ are negative when energy leaves the system as work, this expression is the same as $-\mathrm{d}w_{\text{rev}} \geq -\mathrm{d}w$, and hence $\mathrm{d}w - \mathrm{d}w_{\text{rev}} \geq 0$. Because the internal energy is a state function, its change is the same for irreversible and reversible paths between the same two states, so we can also write:

$$\mathrm{d}U = \mathrm{d}q + \mathrm{d}w = \mathrm{d}q_{\text{rev}} + \mathrm{d}w_{\text{rev}}$$

It follows that $dq_{rev} - dq = dw - dw_{rev} \geq 0$, or $dq_{rev} \geq dq$, and therefore that $dq_{rev}/T \geq dq/T$. Now we use the thermodynamic definition of the entropy (eqn 3.1; $dS = dq_{rev}/T$) to write

$$dS \geq \frac{dq}{T} \qquad \text{Clausius inequality} \qquad (3.12)$$

This expression is the **Clausius inequality**. It will prove to be of great importance for the discussion of the spontaneity of chemical reactions, as we shall see in Section 3.5.

• A brief illustration

Consider the transfer of energy as heat from one system—the hot source—at a temperature T_h to another system—the cold sink—at a temperature T_c (Fig. 3.10). When $|dq|$ leaves the hot source (so $dq_h < 0$), the Clausius inequality implies that $dS \geq dq_h/T_h$. When $|dq|$ enters the cold sink the Clausius inequality implies that $dS \geq dq_c/T_c$ (with $dq_c > 0$). Overall, therefore,

$$dS \geq \frac{dq_h}{T_h} + \frac{dq_c}{T_c}$$

However, $dq_h = -dq_c$, so

$$dS \geq -\frac{dq_c}{T_h} + \frac{dq_c}{T_c} = \left(\frac{1}{T_c} - \frac{1}{T_h}\right)dq_c$$

which is positive (because $dq_c > 0$ and $T_h > T_c$). Hence, cooling (the transfer of heat from hot to cold) is spontaneous, as we know from experience. •

We now suppose that the system is isolated from its surroundings, so that $dq = 0$. The Clausius inequality implies that

$$dS \geq 0 \qquad (3.13)$$

and we conclude that *in an isolated system the entropy cannot decrease when a spontaneous change occurs*. This statement captures the content of the Second Law.

IMPACT ON ENGINEERING

I3.1 Refrigeration

The same argument that we have used to discuss the efficiency of a heat engine can be used to discuss the efficiency of a refrigerator, a device for transferring energy as heat from a cold object (the contents of the refrigerator) to a warm sink (typically, the room in which the refrigerator stands). The less work we have to do to bring this transfer about, the more efficient is the refrigerator.

When an energy $|q_c|$ migrates from a cool source at a temperature T_c into a warmer sink at a temperature T_h, the change in entropy is

$$\Delta S = -\frac{|q_c|}{T_c} + \frac{|q_c|}{T_h} < 0 \qquad (3.14)$$

The process is not spontaneous because not enough entropy is generated in the warm sink to overcome the entropy loss from the cold source (Fig. 3.11). To generate more entropy, energy must be added to the stream that enters the warm sink. Our task is to find the minimum energy that needs to be supplied. The outcome is expressed as the **coefficient of performance**, *c*:

$$c = \frac{\text{energy transferred as heat}}{\text{energy transferred as work}} = \frac{|q_c|}{|w|} \qquad \text{Definition of coefficient of performance} \qquad [3.15]$$

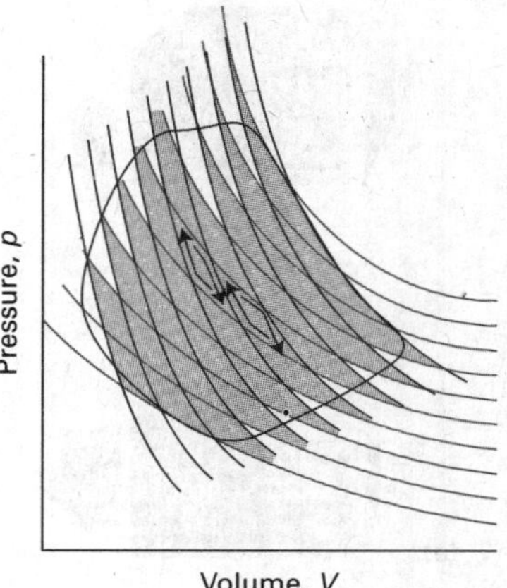

Fig. 3.9 A general cycle can be divided into small Carnot cycles. The match is exact in the limit of infinitesimally small cycles. Paths cancel in the interior of the collection, and only the perimeter, an increasingly good approximation to the true cycle as the number of cycles increases, survives. Because the entropy change around every individual cycle is zero, the integral of the entropy around the perimeter is zero too.

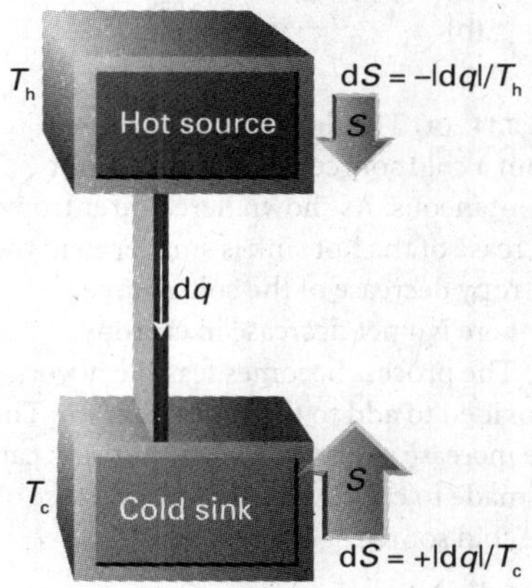

Fig. 3.10 When energy leaves a hot reservoir as heat, the entropy of the reservoir decreases. When the same quantity of energy enters a cooler reservoir, the entropy increases by a larger amount. Hence, overall there is an increase in entropy and the process is spontaneous. Relative changes in entropy are indicated by the sizes of the arrows.

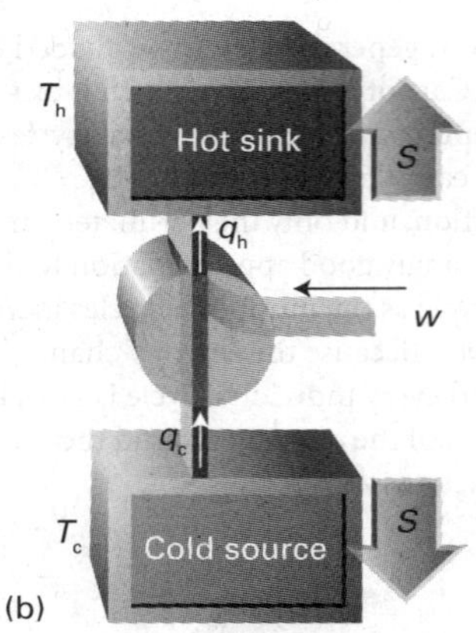

Fig. 3.11 (a) The flow of energy as heat from a cold source to a hot sink is not spontaneous. As shown here, the entropy increase of the hot sink is smaller than the entropy decrease of the cold source, so there is a net decrease in entropy. (b) The process becomes feasible if work is provided to add to the energy stream. Then the increase in entropy of the hot sink can be made to cancel the entropy decrease of the cold source.

The less the work that is required to achieve a given transfer, the greater the coefficient of performance and the more efficient is the refrigerator. For some of this development it will prove best to work with $1/c$.

Because $|q_c|$ is removed from the cold source, and the work $|w|$ is added to the energy stream, the energy deposited as heat in the hot sink is $|q_h| = |q_c| + |w|$. Therefore,

$$\frac{1}{c} = \frac{|w|}{|q_c|} = \frac{|q_h| - |q_c|}{|q_c|} = \frac{|q_h|}{|q_c|} - 1$$

We can now use eqn 3.7 to express this result in terms of the temperatures alone, which is possible if the transfer is performed reversibly. This substitution leads to

$$\frac{1}{c} = \frac{T_h}{T_c} - 1 = \frac{T_h - T_c}{T_c}$$

and therefore

$$c = \frac{T_c}{T_h - T_c} \qquad (3.16)_{rev}$$

for the thermodynamically optimum coefficient of performance.

• A brief illustration

For a refrigerator withdrawing heat from ice-cold water ($T_c = 273$ K) in a typical environment ($T_h = 293$ K), $c = 14$, so, to remove 10 kJ (enough to freeze 30 g of water), requires transfer of at least 0.71 kJ as work. Practical refrigerators, of course, have a lower coefficient of performance. •

3.3 Entropy changes accompanying specific processes

Key points **(a) The entropy of a perfect gas increases when it expands isothermally. (b) The change in entropy of a substance accompanying a change of state at its transition temperature is calculated from its enthalpy of transition. (c) The increase in entropy when a substance is heated is expressed in terms of its heat capacity. (d) The entropy of a substance at a given temperature is determined from measurements of its heat capacity from $T = 0$ up to the temperature of interest, allowing for phase transitions in that range.**

We now see how to calculate the entropy changes that accompany a variety of basic processes.

(a) Expansion

We established in Example 3.1 that the change in entropy of a perfect gas that expands isothermally from V_i to V_f is

$$\Delta S = nR \ln \frac{V_f}{V_i} \qquad (3.17)°$$

Entropy change for the isothermal expansion of a perfect gas

Because S is a state function, the value of ΔS *of the system* is independent of the path between the initial and final states, so this expression applies whether the change of state occurs reversibly or irreversibly. The logarithmic dependence of entropy on volume is illustrated in Fig. 3.12.

The *total* change in entropy, however, does depend on how the expansion takes place. For any process the energy lost as heat from the system is acquired by the surroundings, so $dq_{sur} = -dq$. For a reversible change we use the expression in Example 3.1 ($q_{rev} = nRT\ln(V_f/V_i)$); consequently, from eqn 3.3b

$$\Delta S_{sur} = \frac{q_{sur}}{T} = -\frac{q_{rev}}{T} = -nR\ln\frac{V_f}{V_i} \qquad (3.18)^{\circ}_{rev}$$

This change is the negative of the change in the system, so we can conclude that $\Delta S_{tot} = 0$, which is what we should expect for a reversible process. If, on the other hand, the isothermal expansion occurs freely ($w = 0$), then $q = 0$ (because $\Delta U = 0$). Consequently, $\Delta S_{sur} = 0$, and the total entropy change is given by eqn 3.17 itself:

$$\Delta S_{tot} = nR\ln\frac{V_f}{V_i} \qquad (3.19)^{\circ}$$

In this case, $\Delta S_{tot} > 0$, as we expect for an irreversible process.

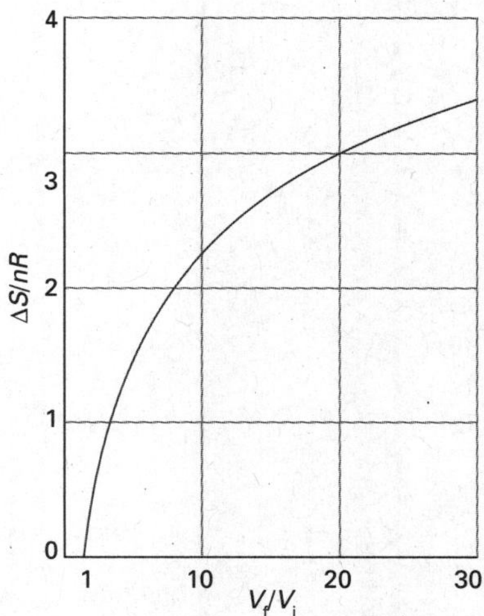

Fig. 3.12 The logarithmic increase in entropy of a perfect gas as it expands isothermally.

interActivity Evaluate the change in expansion of 1.00 mol CO_2 (g) from 0.001 m^3 to 0.010 m^3 at 298 K, treated as a van der Waals gas.

(b) Phase transition

The degree of dispersal of matter and energy changes when a substance freezes or boils as a result of changes in the order with which the molecules pack together and the extent to which the energy is localized or dispersed. Therefore, we should expect the transition to be accompanied by a change in entropy. For example, when a substance vaporizes, a compact condensed phase changes into a widely dispersed gas and we can expect the entropy of the substance to increase considerably. The entropy of a solid also increases when it melts to a liquid and when that liquid turns into a gas.

Consider a system and its surroundings at the **normal transition temperature**, T_{trs}, the temperature at which two phases are in equilibrium at 1 atm. This temperature is 0°C (273 K) for ice in equilibrium with liquid water at 1 atm, and 100°C (373 K) for liquid water in equilibrium with its vapour at 1 atm. At the transition temperature, any transfer of energy as heat between the system and its surroundings is reversible because the two phases in the system are in equilibrium. Because at constant pressure $q = \Delta_{trs}H$, the change in molar entropy *of the system* is[3]

$$\Delta_{trs}S = \frac{\Delta_{trs}H}{T_{trs}} \qquad \text{Entropy of phase transition} \qquad (3.20)$$

If the phase transition is exothermic ($\Delta_{trs}H < 0$, as in freezing or condensing), then the entropy change of the system is negative. This decrease in entropy is consistent with the increased order of a solid compared with a liquid and with the increased order of a liquid compared with a gas. The change in entropy of the surroundings, however, is positive because energy is released as heat into them, and at the transition temperature the total change in entropy is zero. If the transition is endothermic ($\Delta_{trs}H > 0$, as in melting and vaporization), then the entropy change of the system is positive, which is consistent with dispersal of matter in the system. The entropy of the surroundings decreases by the same amount, and overall the total change in entropy is zero.

Table 3.1 lists some experimental entropies of transition. Table 3.2 lists in more detail the standard entropies of vaporization of several liquids at their boiling points. An interesting feature of the data is that a wide range of liquids give approximately the same standard entropy of vaporization (about 85 J K^{-1} mol^{-1}): this empirical observation is called **Trouton's rule**. The explanation of Trouton's rule is that a comparable change in volume occurs when any liquid evaporates and becomes a gas. Hence, all

[3] Recall from Section 2.6 that $\Delta_{trs}H$ is an enthalpy change per mole of substance; so $\Delta_{trs}S$ is also a molar quantity.

Table 3.1* Standard entropies (and temperatures) of phase transitions, $\Delta_{trs}S^{\ominus}/(J\,K^{-1}\,mol^{-1})$

	Fusion (at T_f)	Vaporization (at T_b)
Argon, Ar	14.17 (at 83.8 K)	74.53 (at 87.3 K)
Benzene, C_6H_6	38.00 (at 279 K)	87.19 (at 353 K)
Water, H_2O	22.00 (at 273.15 K)	109.0 (at 373.15 K)
Helium, He	4.8 (at 1.8 K and 30 bar)	19.9 (at 4.22 K)

* More values are given in the *Data section*.

Table 3.2* The standard entropies of vaporization of liquids

	$\Delta_{vap}H^{\ominus}/(kJ\,mol^{-1})$	$\theta_b/°C$	$\Delta_{vap}S^{\ominus}/(J\,K^{-1}\,mol^{-1})$
Benzene	30.8	80.1	87.2
Carbon tetrachloride	30	76.7	85.8
Cyclohexane	30.1	80.7	85.1
Hydrogen sulfide	18.7	−60.4	87.9
Methane	8.18	−161.5	73.2
Water	40.7	100.0	109.1

* More values are given in the *Data section*.

liquids can be expected to have similar standard entropies of vaporization. Liquids that show significant deviations from Trouton's rule do so on account of strong molecular interactions that result in a partial ordering of their molecules. As a result, there is a greater change in disorder when the liquid turns into a vapour than for a fully disordered liquid. An example is water, where the large entropy of vaporization reflects the presence of structure arising from hydrogen-bonding in the liquid. Hydrogen bonds tend to organize the molecules in the liquid so that they are less random than, for example, the molecules in liquid hydrogen sulfide (in which there is no hydrogen bonding). Methane has an unusually low entropy of vaporization. A part of the reason is that the entropy of the gas itself is slightly low (186 J K^{-1} mol^{-1} at 298 K); the entropy of N_2 under the same conditions is 192 J K^{-1} mol^{-1}. As we shall see in Chapter 12, fewer rotational states are accessible at room temperature for light molecules than for heavy molecules.

• A brief illustration

There is no hydrogen bonding in liquid bromine and Br_2 is a heavy molecule that is unlikely to display unusual behaviour in the gas phase, so it is safe to use Trouton's rule. To predict the standard molar enthalpy of vaporization of bromine given that it boils at 59.2°C, we use the rule in the form

$$\Delta_{vap}H^{\ominus} = T_b \times (85\ \mathrm{J\,K^{-1}\,mol^{-1}})$$

Substitution of the data then gives

$$\Delta_{vap}H^{\ominus} = (332.4\ \mathrm{K}) \times (85\ \mathrm{J\,K^{-1}\,mol^{-1}}) = +2.8\times10^{3}\ \mathrm{J\,mol^{-1}} = +28\ \mathrm{kJ\,mol^{-1}}$$

The experimental value is +29.45 kJ mol^{-1}. •

Self-test 3.3 Predict the enthalpy of vaporization of ethane from its boiling point, −88.6°C. [16 kJ mol^{-1}]

(c) Heating

We can use eqn 3.2 to calculate the entropy of a system at a temperature T_f from a knowledge of its entropy at another temperature T_i and the heat supplied to change its temperature from one value to the other:

$$S(T_f) = S(T_i) + \int_{T_i}^{T_f} \frac{dq_{rev}}{T} \tag{3.21}$$

We shall be particularly interested in the entropy change when the system is subjected to constant pressure (such as from the atmosphere) during the heating. Then, from the definition of constant-pressure heat capacity (eqn 2.22, written as $dq_{rev} = C_p dT$). Consequently, at constant pressure:

$$S(T_f) = S(T_i) + \int_{T_i}^{T_f} \frac{C_p dT}{T} \tag{3.22}$$

Entropy variation with temperature

The same expression applies at constant volume, but with C_p replaced by C_V. When C_p is independent of temperature in the temperature range of interest, it can be taken outside the integral and we obtain

$$S(T_f) = S(T_i) + C_p \int_{T_i}^{T_f} \frac{dT}{T} = S(T_i) + C_p \ln \frac{T_f}{T_i} \tag{3.23}$$

with a similar expression for heating at constant volume. The logarithmic dependence of entropy on temperature is illustrated in Fig. 3.13.

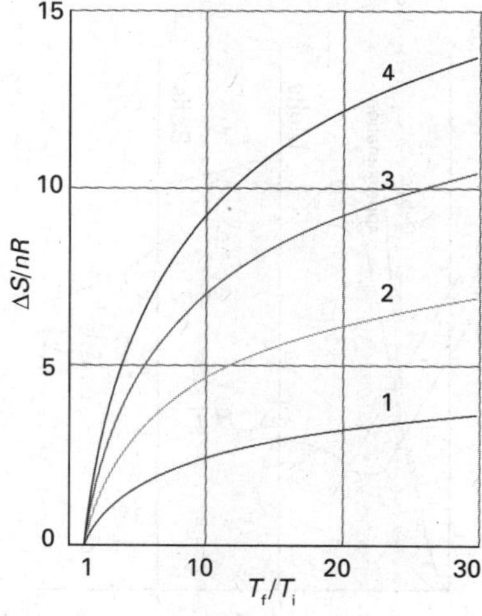

Fig. 3.13 The logarithmic increase in entropy of a substance as it is heated at constant volume. Different curves correspond to different values of the constant-volume heat capacity (which is assumed constant over the temperature range) expressed as $C_{V,m}/R$.

interActivity Plot the change in entropy of a perfect gas of (a) atoms, (b) linear rotors, (c) nonlinear rotors as the sample is heated over the same range under conditions of (i) constant volume, (ii) constant pressure.

Example 3.2 *Calculating the entropy change*

Calculate the entropy change when argon at 25°C and 1.00 bar in a container of volume 0.500 dm^3 is allowed to expand to 1.000 dm^3 and is simultaneously heated to 100°C.

Method Because S is a state function, we are free to choose the most convenient path from the initial state. One such path is reversible isothermal expansion to the final volume, followed by reversible heating at constant volume to the final temperature. The entropy change in the first step is given by eqn 3.17 and that of the second step, provided C_V is independent of temperature, by eqn 3.23 (with C_V in place of C_p). In each case we need to know n, the amount of gas molecules, and can calculate it from the perfect gas equation and the data for the initial state from $n = p_i V_i / RT_i$. The molar heat capacity at constant volume is given by the equipartition theorem as $\frac{3}{2}R$. (The equipartition theorem is reliable for monatomic gases: for others and, in general, use experimental data like those in Table 2.8, converting to the value at constant volume by using the relation $C_{p,m} - C_{V,m} = R$.)

Answer From eqn 3.17 the entropy change of the isothermal expansion from V_i to V_f is

$$\Delta S(\text{Step 1}) = nR \ln \frac{V_f}{V_i}$$

From eqn 3.23, the entropy change in the second step, from T_i to T_f at constant volume, is

$$\Delta S(\text{Step 2}) = nC_{V,m}\ln\frac{T_f}{T_i} = \frac{3}{2}nR\ln\frac{T_f}{T_i} = nR\ln\left(\frac{T_f}{T_i}\right)^{3/2}$$

The overall entropy change of the system, the sum of these two changes, is

$$\Delta S = nR\ln\frac{V_f}{V_i} + nR\ln\left(\frac{T_f}{T_i}\right)^{3/2} = nR\ln\left\{\frac{V_f}{V_i}\left(\frac{T_f}{T_i}\right)^{3/2}\right\}$$

(We have used $\ln x + \ln y = \ln xy$.) Now we substitute $n = p_iV_i/RT_i$ and obtain

$$\Delta S = \frac{p_iV_i}{T_i}\ln\left\{\frac{V_f}{V_i}\left(\frac{T_f}{T_i}\right)^{3/2}\right\}$$

At this point we substitute the data:

$$\Delta S = \frac{(1.00\times10^5\,\text{Pa})\times(0.500\times10^{-3}\,\text{m}^3)}{298\,\text{K}}\times\ln\left\{\frac{1.000}{0.500}\left(\frac{373}{298}\right)^{3/2}\right\}$$

$$= +0.173\,\text{J K}^{-1}$$

A note on good practice It is sensible to proceed as generally as possible before inserting numerical data so that, if required, the formula can be used for other data and to avoid rounding errors.

Self-test 3.4 Calculate the entropy change when the same initial sample is compressed to $0.0500\,\text{dm}^3$ and cooled to $-25°\text{C}$. $[-0.43\,\text{J K}^{-1}]$

(d) The measurement of entropy

The entropy of a system at a temperature T is related to its entropy at $T = 0$ by measuring its heat capacity C_p at different temperatures and evaluating the integral in eqn 3.22, taking care to add the entropy of transition ($\Delta_{trs}H/T_{trs}$) for each phase transition between $T = 0$ and the temperature of interest. For example, if a substance melts at T_f and boils at T_b, then its molar entropy above its boiling temperature is given by

$$S_m(T) = S_m(0) + \int_0^{T_f}\frac{C_{p,m}(s,T)}{T}dT + \frac{\Delta_{fus}H}{T_f} + \int_{T_f}^{T_b}\frac{C_{p,m}(l,T)}{T}dT + \frac{\Delta_{vap}H}{T_b} + \int_{T_b}^{T}\frac{C_{p,m}(g,T)}{T}dT \quad (3.24)$$

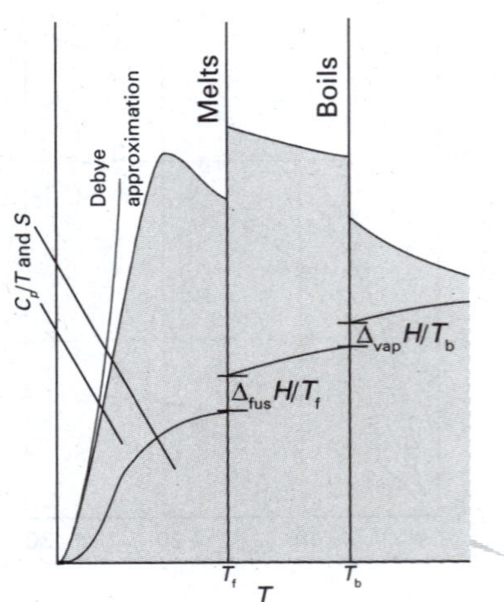

Fig. 3.14 The variation of C_p/T with the temperature for a sample is used to evaluate the entropy, which is equal to the area beneath the upper curve up to the corresponding temperature, plus the entropy of each phase transition passed.

interActivity Allow for the temperature dependence of the heat capacity by writing $C = a + bT + c/T^2$, and plot the change in entropy for different values of the three coefficients (including negative values of c).

All the properties required, except $S_m(0)$, can be measured calorimetrically, and the integrals can be evaluated either graphically or, as is now more usual, by fitting a polynomial to the data and integrating the polynomial analytically. The former procedure is illustrated in Fig. 3.14: the area under the curve of $C_{p,m}/T$ against T is the integral required. Because $dT/T = d\ln T$, an alternative procedure is to evaluate the area under a plot of $C_{p,m}$ against $\ln T$.

One problem with the determination of entropy is the difficulty of measuring heat capacities near $T = 0$. There are good theoretical grounds for assuming that the heat capacity is proportional to T^3 when T is low (see Section 7.1), and this dependence is the basis of the **Debye extrapolation**. In this method, C_p is measured down to as low a temperature as possible, and a curve of the form aT^3 is fitted to the data. That fit determines the value of a, and the expression $C_{p,m} = aT^3$ is assumed valid down to $T = 0$.

• A brief illustration

The standard molar entropy of nitrogen gas at 25°C has been calculated from the following data:

	$S_m^{\ominus}/(\text{J K}^{-1}\text{ mol}^{-1})$
Debye extrapolation	1.92
Integration, from 10 K to 35.61 K	25.25
Phase transition at 35.61 K	6.43
Integration, from 35.61 K to 63.14 K	23.38
Fusion at 63.14 K	11.42
Integration, from 63.14 K to 77.32 K	11.41
Vaporization at 77.32 K	72.13
Integration, from 77.32 K to 298.15 K	39.20
Correction for gas imperfection	0.92
Total	192.06

Therefore

$$S_m^{\ominus}(298.15\text{ K}) = S_m(0) + 192.1\text{ J K}^{-1}\text{ mol}^{-1} \quad \bullet$$

Example 3.3 *Calculating the entropy at low temperatures*

The molar constant-pressure heat capacity of a certain solid at 4.2 K is 0.43 J K^{-1} mol^{-1}. What is its molar entropy at that temperature?

Method Because the temperature is so low, we can assume that the heat capacity varies with temperature as aT^3, in which case we can use eqn 3.22 to calculate the entropy at a temperature T in terms of the entropy at $T = 0$ and the constant a. When the integration is carried out, it turns out that the result can be expressed in terms of the heat capacity at the temperature T, so the data can be used directly to calculate the entropy.

Answer The integration required is

$$S_m(T) = S_m(0) + \int_0^T \frac{aT^3}{T}\,dT = S_m(0) + a\int_0^T T^2 dT$$
$$= S_m(0) + \tfrac{1}{3}aT^3 = S_m(0) + \tfrac{1}{3}C_{p,m}(T)$$

from which it follows that

$$S_m(4.2\text{ K}) = S_m(0) + 0.14\text{ J K}^{-1}\text{ mol}^{-1}$$

Self-test 3.5 For metals, there is also a contribution to the heat capacity from the electrons that is linearly proportional to T when the temperature is low. Find its contribution to the entropy at low temperatures. $[S(T) = S(0) + C_p(T)]$

3.4 The Third Law of thermodynamics

Key points **(a) The Nernst heat theorem implies the Third Law of thermodynamics. (b) The Third Law allows us to define absolute entropies of substances and to define the standard entropy of a reaction.**

At $T = 0$, all energy of thermal motion has been quenched, and in a perfect crystal all the atoms or ions are in a regular, uniform array. The localization of matter and the absence of thermal motion suggest that such materials also have zero entropy. This conclusion is consistent with the molecular interpretation of entropy, because $S = 0$ if there is only one way of arranging the molecules and only one microstate is accessible (all molecules occupy the ground state).

(a) The Nernst heat theorem

The experimental observation that turns out to be consistent with the view that the entropy of a regular array of molecules is zero at $T = 0$ is summarized by the **Nernst heat theorem:**

> The entropy change accompanying any physical or chemical transformation approaches zero as the temperature approaches zero: $\Delta S \to 0$ as $T \to 0$ provided all the substances involved are perfectly ordered.

Nernst heat theorem

• A brief illustration

Consider the entropy of the transition between orthorhombic sulfur, S(α), and monoclinic sulfur, S(β), which can be calculated from the transition enthalpy (-402 J mol^{-1}) at the transition temperature (369 K):

$$\Delta_{trs}S = S_m(\beta) - S_m(\alpha) = \frac{(-402 \text{ J mol}^{-1})}{369 \text{ K}} = -1.09 \text{ J K}^{-1} \text{ mol}^{-1}$$

The two individual entropies can also be determined by measuring the heat capacities from $T = 0$ up to $T = 369$ K. It is found that $S_m(\alpha) = S_m(\alpha, 0) + 37 \text{ J K}^{-1} \text{ mol}^{-1}$ and $S_m(\beta) = S_m(\beta, 0) + 38 \text{ J K}^{-1} \text{ mol}^{-1}$. These two values imply that at the transition temperature

$$\Delta_{trs}S = S_m(\alpha, 0) - S_m(\beta, 0) = -1 \text{ J K}^{-1} \text{ mol}^{-1}$$

On comparing this value with the one above, we conclude that $S_m(\alpha, 0) - S_m(\beta, 0) \approx 0$, in accord with the theorem. •

It follows from the Nernst theorem that, if we arbitrarily ascribe the value zero to the entropies of elements in their perfect crystalline form at $T = 0$, then all perfect crystalline compounds also have zero entropy at $T = 0$ (because the change in entropy that accompanies the formation of the compounds, like the entropy of all transformations at that temperature, is zero). This conclusion is summarized by the **Third Law of thermodynamics:**

> The entropy of all perfect crystalline substances is zero at $T = 0$.

Third Law of thermodynamics

As far as thermodynamics is concerned, choosing this common value as zero is a matter of convenience. The molecular interpretation of entropy, however, justifies the value $S = 0$ at $T = 0$. We saw in Section 3.2b that, according to the Boltzmann formula, the entropy is zero if there is only one accessible microstate ($W = 1$). In most cases, $W = 1$ at $T = 0$ because there is only one way of achieving the lowest total energy: put all the molecules into the same, lowest state. Therefore, $S = 0$ at $T = 0$, in accord with the Third Law of thermodynamics. In certain cases, though, W may differ from 1 at $T = 0$. This is the case if there is no energy advantage in adopting a particular orientation even at absolute zero. For instance, for a diatomic molecule AB there may be almost no energy difference between the arrangements . . . AB AB AB . . . and . . . BA AB BA . . . , so $W > 1$ even at $T = 0$. If $S > 0$ at $T = 0$ we say that the substance has a **residual entropy**. Ice has a residual entropy of 3.4 J K^{-1} mol^{-1}. It stems from the arrangement of the hydrogen bonds between neighbouring water molecules: a given O atom has two short O–H bonds and two long O···H bonds to its neighbours, but there is a degree of randomness in which two bonds are short and which two are long.

(b) Third-Law entropies

Entropies reported on the basis that $S(0) = 0$ are called **Third-Law entropies** (and often just 'entropies'). When the substance is in its standard state at the temperature T, the **standard (Third-Law) entropy** is denoted $S^{\ominus}(T)$. A list of values at 298 K is given in Table 3.3.

The **standard reaction entropy**, $S^{\ominus}(T)$, is defined, like the standard reaction enthalpy, as the difference between the molar entropies of the pure, separated products and the pure, separated reactants, all substances being in their standard states at the specified temperature:

$$\Delta_r S^{\ominus} = \sum_{\text{Products}} \nu S_m^{\ominus} - \sum_{\text{Reactants}} \nu S_m^{\ominus} \qquad (3.25a)$$

Definition of standard reaction entropy

In this expression, each term is weighted by the appropriate stoichiometric coefficient. A more sophisticated approach is to adopt the notation introduced in Section 2.8 and to write

$$\Delta_r S^{\ominus} = \sum_J \nu_J S_m^{\ominus}(J) \qquad (3.25b)$$

Standard reaction entropies are likely to be positive if there is a net formation of gas in a reaction, and are likely to be negative if there is a net consumption of gas.

Table 3.3* Standard Third-Law entropies at 298 K

	$S_m^{\ominus}$/(J K^{-1} mol^{-1})
Solids	
Graphite, C(s)	5.7
Diamond, C(s)	2.4
Sucrose, $C_{12}H_{22}O_{11}$(s)	360.2
Iodine, I_2(s)	116.1
Liquids	
Benzene, C_6H_6(l)	173.3
Water, H_2O(l)	69.9
Mercury, Hg(l)	76.0
Gases	
Methane, CH_4(g)	186.3
Carbon dioxide, CO_2(g)	213.7
Hydrogen, H_2(g)	130.7
Helium, He	126.2
Ammonia, NH_3(g)	192.4

* More values are given in the *Data section*.

• **A brief illustration**

To calculate the standard reaction entropy of $H_2(g) + \frac{1}{2}O_2(g) \rightarrow H_2O(l)$ at 25°C, we use the data in Table 2.8 of the *Data section* to write

$$\begin{aligned}\Delta_r S^{\ominus} &= S_m^{\ominus}(H_2O,l) - \{S_m^{\ominus}(H_2,g) + \tfrac{1}{2}S_m^{\ominus}(O_2,g)\}\\ &= 69.9\ \text{J K}^{-1}\ \text{mol}^{-1} - \{130.7 + \tfrac{1}{2}(205.0)\}\text{J K}^{-1}\ \text{mol}^{-1}\\ &= -163.4\ \text{J K}^{-1}\ \text{mol}^{-1}\end{aligned}$$

The negative value is consistent with the conversion of two gases to a compact liquid. •

A note on good practice Do not make the mistake of setting the standard molar entropies of elements equal to zero: they have non-zero values (provided $T > 0$), as we have already discussed.

Self-test 3.6 Calculate the standard reaction entropy for the combustion of methane to carbon dioxide and liquid water at 25°C. [−243 J K^{-1} mol^{-1}]

Just as in the discussion of enthalpies in Section 2.8, where we acknowledged that solutions of cations cannot be prepared in the absence of anions, the standard molar entropies of ions in solution are reported on a scale in which the standard entropy of the H^+ ions in water is taken zero at all temperatures:

$$S^{\ominus}(H^+, aq) = 0 \qquad [3.26]$$

Convention for ions in solution

The values based on this choice are listed in Table 2.8 in the *Data section*.[4] Because the entropies of ions in water are values relative to the hydrogen ion in water, they may be either positive or negative. A positive entropy means that an ion has a higher molar entropy than H^+ in water and a negative entropy means that the ion has a lower molar entropy than H^+ in water. For instance, the standard molar entropy of $Cl^-(aq)$ is $+57$ J K^{-1} mol^{-1} and that of $Mg^{2+}(aq)$ is -128 J K^{-1} mol^{-1}. Ion entropies vary as expected on the basis that they are related to the degree to which the ions order the water molecules around them in the solution. Small, highly charged ions induce local structure in the surrounding water, and the disorder of the solution is decreased more than in the case of large, singly charged ions. The absolute, Third-Law standard molar entropy of the proton in water can be estimated by proposing a model of the structure it induces, and there is some agreement on the value -21 J K^{-1} mol^{-1}. The negative value indicates that the proton induces order in the solvent.

IMPACT ON MATERIALS CHEMISTRY
I3.2 Crystal defects

The Third Law implies that at $T = 0$ the entropies of perfect crystalline substances are characterized by long-range, regularly repeating arrangements of atoms, ions, or molecules. This regularity, and the accompanying inter- and intramolecular interactions between the subunits of the crystal, govern the physical, optical, and electronic properties of the solid. In reality, however, all crystalline solids possess one or more defects that affect the physical and chemical properties of the substance. In fact, impurities are often introduced to achieve particular desirable properties, such as the colour of a gemstone or enhanced strength of a metal.

One of the main types of crystalline imperfection is a **point defect**, a location where an atom is missing or irregularly placed in the lattice structure. Other terms used to describe point defects include *voids*, or *lattice vacancies, substitutional impurity atoms, dopant sites*, and *interstitial impurity atoms*. Many gemstones feature substitutional solids, such as in rubies and blue sapphires where the Al^{3+} ions in the corundum structure of alumina are replaced with Cr^{3+} and Fe^{3+} ions, respectively. Interstitial solids can result from the random diffusion of dopants in interstices (voids) or from self-diffusion, as in ionic crystals, where a lattice ion can migrate into an interstitial position and leave behind a vacancy known as a Frenkel defect.

Figure 3.15 illustrates the impact of impurities on the heat capacity and thus entropy of a pure crystal. Niobium has become the dominant metal in low-temperature superconductor alloys because it can be manufactured economically in a ductile form that is needed for the high critical current of a superconductor. The purity of the metal, however, is essential to yield superconducting properties. Close to 1 K the heat capacity of pure niobium follows the Debye T^3 law. However, when niobium is treated by allowing H_2 or D_2 to diffuse over the sample at 700°C impurities are introduced and the heat capacity diverges from that of the pure metal. To identify the role of the defects the values of C_p for the pure metal are subtracted from those of the doped samples, divided by T, and plotted against temperature. The area under the resulting curves then represents the contributions to the entropy from the presence of the impurities.

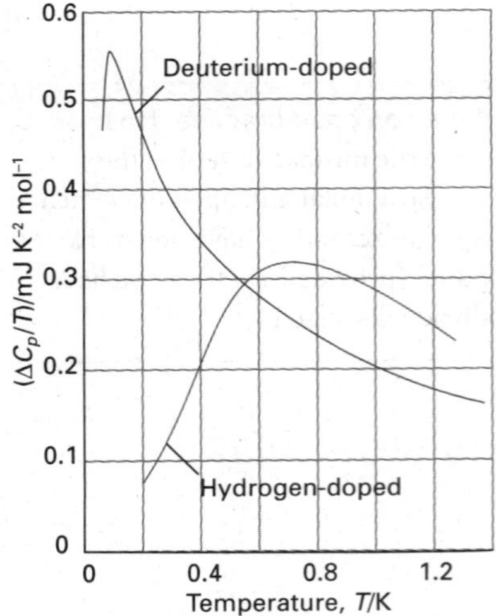

Fig. 3.15 Molar heat capacity contributions of the defects in hydrogen- and deuterium doped niobium. The area under each curve is used to calculate the entropy due to the presence of the defects. (Based on G.J. Sellers and A.C. Anderson, *Phys. Rev.* B. **10**, 2771 (1974).)

[4] In terms of the language to be introduced in Section 5.1, the entropies of ions in solution are actually *partial molar entropies*, for their values include the consequences of their presence on the organization of the solvent molecules around them.

Concentrating on the system

Entropy is the basic concept for discussing the direction of natural change, but to use it we have to analyse changes in both the system and its surroundings. We have seen that it is always very simple to calculate the entropy change in the surroundings, and we shall now see that it is possible to devise a simple method for taking that contribution into account automatically. This approach focuses our attention on the system and simplifies discussions. Moreover, it is the foundation of all the applications of chemical thermodynamics that follow.

3.5 The Helmholtz and Gibbs energies

Key points **(a) The Clausius inequality implies a number of criteria for spontaneous change under a variety of conditions that may be expressed in terms of the properties of the system alone; they are summarized by introducing the Helmholtz and Gibbs energies. (b) A spontaneous process at constant temperature and volume is accompanied by a decrease in the Helmholtz energy. (c) The change in the Helmholtz energy is equal to the maximum work accompanying a process at constant temperature. (d) A spontaneous process at constant temperature and pressure is accompanied by a decrease in the Gibbs energy. (e) The change in the Gibbs energy is equal to the maximum non-expansion work accompanying a process at constant temperature and pressure.**

Consider a system in thermal equilibrium with its surroundings at a temperature T. When a change in the system occurs and there is a transfer of energy as heat between the system and the surroundings, the Clausius inequality ($dS \geq dq/T$, eqn 3.12) reads

$$dS - \frac{dq}{T} \geq 0 \qquad (3.27)$$

We can develop this inequality in two ways according to the conditions (of constant volume or constant pressure) under which the process occurs.

(a) Criteria for spontaneity

First, consider heating at constant volume. Then, in the absence of non-expansion work, we can write $dq_V = dU$; consequently

$$dS - \frac{dU}{T} \geq 0 \qquad (3.28)$$

The importance of the inequality in this form is that it expresses the criterion for spontaneous change solely in terms of the state functions of the system. The inequality is easily rearranged into

$$TdS \geq dU \quad \text{(constant } V\text{, no additional work)}^5 \qquad (3.29)$$

At either constant internal energy ($dU = 0$) or constant entropy ($dS = 0$), this expression becomes, respectively,

$$dS_{U,V} \geq 0 \qquad dU_{S,V} \leq 0 \qquad (3.30)$$

where the subscripts indicate the constant conditions.

Equation 3.30 expresses the criteria for spontaneous change in terms of properties relating to the system. The first inequality states that, in a system at constant volume

[5] Recall that 'additional work' is work other than expansion work.

and constant internal energy (such as an isolated system), the entropy increases in a spontaneous change. That statement is essentially the content of the Second Law. The second inequality is less obvious, for it says that, if the entropy and volume of the system are constant, then the internal energy must decrease in a spontaneous change. Do not interpret this criterion as a tendency of the system to sink to lower energy. It is a disguised statement about entropy and should be interpreted as implying that, if the entropy of the system is unchanged, then there must be an increase in entropy of the surroundings, which can be achieved only if the energy of the system decreases as energy flows out as heat.

When energy is transferred as heat at constant pressure, and there is no work other than expansion work, we can write $dq_p = dH$ and obtain

$$TdS \geq dH \quad \text{(constant } p\text{, no additional work)} \tag{3.31}$$

At either constant enthalpy or constant entropy this inequality becomes, respectively,

$$dS_{H,p} \geq 0 \qquad dH_{S,p} \leq 0 \tag{3.32}$$

The interpretations of these inequalities are similar to those of eqn 3.30. The entropy of the system at constant pressure must increase if its enthalpy remains constant (for there can then be no change in entropy of the surroundings). Alternatively, the enthalpy must decrease if the entropy of the system is constant, for then it is essential to have an increase in entropy of the surroundings.

Because eqns 3.29 and 3.31 have the forms $dU - TdS \leq 0$ and $dH - TdS \leq 0$, respectively, they can be expressed more simply by introducing two more thermodynamic quantities. One is the **Helmholtz energy**, A, which is defined as

$$A = U - TS \qquad \text{Definition of Helmholtz energy} \qquad [3.33]$$

The other is the **Gibbs energy**, G:

$$G = H - TS \qquad \text{Definition of Gibbs energy} \qquad [3.34]$$

All the symbols in these two definitions refer to the system.

When the state of the system changes at constant temperature, the two properties change as follows:

$$\text{(a) } dA = dU - TdS \qquad \text{(b) } dG = dH - TdS \tag{3.35}$$

When we introduce eqns 3.29 and 3.31, respectively, we obtain the criteria of spontaneous change as

$$\text{(a) } dA_{T,V} \leq 0 \qquad \text{(b) } dG_{T,p} \leq 0 \tag{3.36}$$

These inequalities are the most important conclusions from thermodynamics for chemistry. They are developed in subsequent sections and chapters.

(b) Some remarks on the Helmholtz energy

A change in a system at constant temperature and volume is spontaneous if $dA_{T,V} \leq 0$. That is, a change under these conditions is spontaneous if it corresponds to a decrease in the Helmholtz energy. Such systems move spontaneously towards states of lower A if a path is available. The criterion of equilibrium, when neither the forward nor reverse process has a tendency to occur, is

$$\mathrm{d}A_{T,V}=0 \tag{3.37}$$

The expressions $\mathrm{d}A=\mathrm{d}U-T\mathrm{d}S$ and $\mathrm{d}A<0$ are sometimes interpreted as follows. A negative value of $\mathrm{d}A$ is favoured by a negative value of $\mathrm{d}U$ and a positive value of $T\mathrm{d}S$. This observation suggests that the tendency of a system to move to lower A is due to its tendency to move towards states of lower internal energy and higher entropy. However, this interpretation is false (even though it is a good rule of thumb for remembering the expression for $\mathrm{d}A$) because the tendency to lower A is solely a tendency towards states of greater overall entropy. Systems change spontaneously if in doing so the total entropy of the system and its surroundings increases, not because they tend to lower internal energy. The form of $\mathrm{d}A$ may give the impression that systems favour lower energy, but that is misleading: $\mathrm{d}S$ is the entropy change of the system, $-\mathrm{d}U/T$ is the entropy change of the surroundings (when the volume of the system is constant), and their total tends to a maximum.

(c) Maximum work

It turns out, as we show in the following *Justification*, that A carries a greater significance than being simply a signpost of spontaneous change: *the change in the Helmholtz function is equal to the maximum work accompanying a process at constant temperature*:

$$\mathrm{d}w_{\max}=\mathrm{d}A \tag{3.38}$$

As a result, A is sometimes called the 'maximum work function', or the 'work function'.[6]

Justification 3.2 *Maximum work*

To demonstrate that maximum work can be expressed in terms of the changes in Helmholtz energy, we combine the Clausius inequality $\mathrm{d}S \geq \mathrm{d}q/T$ in the form $T\mathrm{d}S \geq \mathrm{d}q$ with the First Law, $\mathrm{d}U=\mathrm{d}q+\mathrm{d}w$, and obtain

$$\mathrm{d}U \leq T\mathrm{d}S+\mathrm{d}w$$

($\mathrm{d}U$ is smaller than the term of the right because we are replacing $\mathrm{d}q$ by $T\mathrm{d}S$, which in general is larger.) This expression rearranges to

$$\mathrm{d}w \geq \mathrm{d}U-T\mathrm{d}S$$

It follows that the most negative value of $\mathrm{d}w$, and therefore the maximum energy that can be obtained from the system as work, is given by

$$\mathrm{d}w_{\max}=\mathrm{d}U-T\mathrm{d}S$$

and that this work is done only when the path is traversed reversibly (because then the equality applies). Because at constant temperature $\mathrm{d}A=\mathrm{d}U-T\mathrm{d}S$, we conclude that $\mathrm{d}w_{\max}=\mathrm{d}A$.

When a macroscopic isothermal change takes place in the system, eqn 3.38 becomes

$$w_{\max}=\Delta A \tag{3.39}$$

Relation between A and maximum work

with

$$\Delta A=\Delta U-T\Delta S \tag{3.40}$$

[6] *Arbeit* is the German word for work; hence the symbol A.

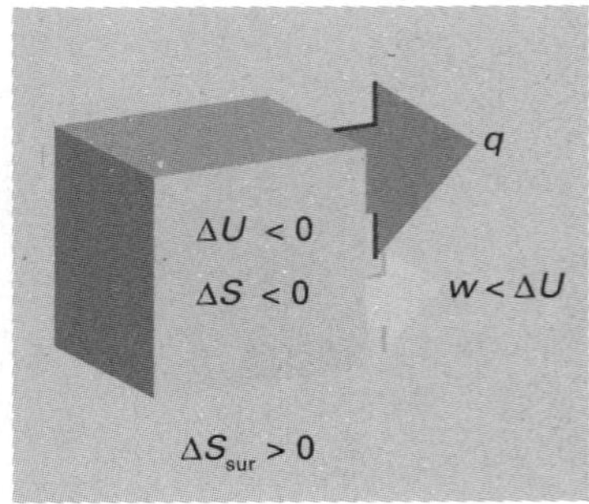

Fig. 3.16 In a system not isolated from its surroundings, the work done may be different from the change in internal energy. Moreover, the process is spontaneous if overall the entropy of the system and its surroundings increases. In the process depicted here, the entropy of the system decreases, so that of the surroundings must increase in order for the process to be spontaneous, which means that energy must pass from the system to the surroundings as heat. Therefore, less work than ΔU can be obtained.

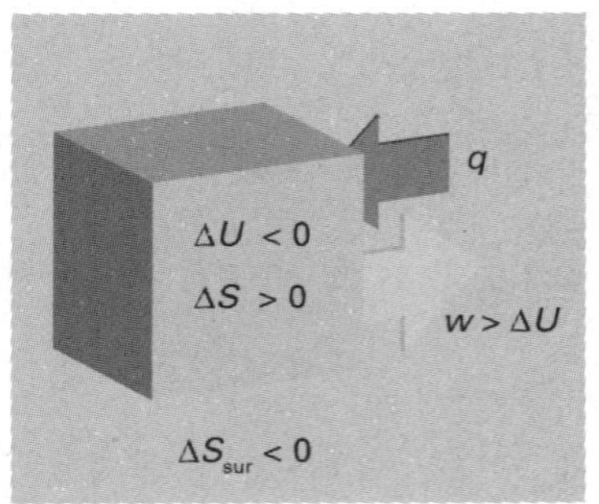

Fig. 3.17 In this process, the entropy of the system increases; hence we can afford to lose some entropy of the surroundings. That is, some of their energy may be lost as heat to the system. This energy can be returned to them as work. Hence the work done can exceed ΔU.

This expression shows that in some cases, depending on the sign of $T\Delta S$, not all the change in internal energy may be available for doing work. If the change occurs with a decrease in entropy (of the system), in which case $T\Delta S < 0$, then the right-hand side of this equation is not as negative as ΔU itself, and consequently the maximum work is less than ΔU. For the change to be spontaneous, some of the energy must escape as heat in order to generate enough entropy in the surroundings to overcome the reduction in entropy in the system (Fig. 3.16). In this case, Nature is demanding a tax on the internal energy as it is converted into work. This is the origin of the alternative name 'Helmholtz free energy' for A, because ΔA is that part of the change in internal energy that we are free to use to do work.

Further insight into the relation between the work that a system can do and the Helmholtz energy is to recall that work is energy transferred to the surroundings as the uniform motion of atoms. We can interpret the expression $A = U - TS$ as showing that A is the total internal energy of the system, U, less a contribution that is stored as energy of thermal motion (the quantity TS). Because energy stored in random thermal motion cannot be used to achieve uniform motion in the surroundings, only the part of U that is not stored in that way, the quantity $U - TS$, is available for conversion into work.

If the change occurs with an increase of entropy of the system (in which case $T\Delta S > 0$), the right-hand side of the equation is more negative than ΔU. In this case, the maximum work that can be obtained from the system is greater than ΔU. The explanation of this apparent paradox is that the system is not isolated and energy may flow in as heat as work is done. Because the entropy of the system increases, we can afford a reduction of the entropy of the surroundings yet still have, overall, a spontaneous process. Therefore, some energy (no more than the value of $T\Delta S$) may leave the surroundings as heat and contribute to the work the change is generating (Fig. 3.17). Nature is now providing a tax refund.

Example 3.4 *Calculating the maximum available work*

When 1.000 mol $C_6H_{12}O_6$ (glucose) is oxidized to carbon dioxide and water at 25°C according to the equation $C_6H_{12}O_6(s) + 6\ O_2(g) \rightarrow 6\ CO_2(g) + 6\ H_2O(l)$, calorimetric measurements give $\Delta_r U^\ominus = -2808$ kJ mol^{-1} and $\Delta_r S^\ominus = +259.1$ J K^{-1} mol^{-1} at 25°C. How much of this energy change can be extracted as (a) heat at constant pressure, (b) work?

Method We know that the heat released at constant pressure is equal to the value of ΔH, so we need to relate $\Delta_r H^\ominus$ to $\Delta_r U^\ominus$, which is given. To do so, we suppose that all the gases involved are perfect, and use eqn 2.21 in the form $\Delta_r H = \Delta_r U + \Delta \nu_g RT$. For the maximum work available from the process we use eqn 3.39.

Answer (a) Because $\Delta \nu_g = 0$, we know that $\Delta_r H^\ominus = \Delta_r U^\ominus = -2808$ kJ mol^{-1}. Therefore, at constant pressure, the energy available as heat is 2808 kJ mol^{-1}. (b) Because $T = 298$ K, the value of $\Delta_r A^\ominus$ is

$$\Delta_r A^\ominus = \Delta_r U^\ominus - T\Delta_r S^\ominus = -2885 \text{ kJ mol}^{-1}$$

Therefore, the combustion of 1.000 mol $C_6H_{12}O_6$ can be used to produce up to 2885 kJ of work. The maximum work available is greater than the change in internal energy on account of the positive entropy of reaction (which is partly due to the generation of a large number of small molecules from one big one). The system can therefore draw in energy from the surroundings (so reducing their entropy) and make it available for doing work.

Self-test 3.7 Repeat the calculation for the combustion of 1.000 mol $CH_4(g)$ under the same conditions, using data from Tables 2.6 and 2.8.
[$|q_p| = 890$ kJ, $|w_{max}| = 818$ kJ]

(d) Some remarks on the Gibbs energy

The Gibbs energy (the 'free energy') is more common in chemistry than the Helmholtz energy because, at least in laboratory chemistry, we are usually more interested in changes occurring at constant pressure than at constant volume. The criterion $dG_{T,p} \le 0$ carries over into chemistry as the observation that, *at constant temperature and pressure, chemical reactions are spontaneous in the direction of decreasing Gibbs energy.* Therefore, if we want to know whether a reaction is spontaneous, the pressure and temperature being constant, we assess the change in the Gibbs energy. If G decreases as the reaction proceeds, then the reaction has a spontaneous tendency to convert the reactants into products. If G increases, then the reverse reaction is spontaneous.

The existence of spontaneous endothermic reactions provides an illustration of the role of G. In such reactions, H increases, the system rises spontaneously to states of higher enthalpy, and $dH > 0$. Because the reaction is spontaneous we know that $dG < 0$ despite $dH > 0$; it follows that the entropy of the system increases so much that TdS outweighs dH in $dG = dH - TdS$. Endothermic reactions are therefore driven by the increase of entropy of the system, and this entropy change overcomes the reduction of entropy brought about in the surroundings by the inflow of heat into the system ($dS_{sur} = -dH/T$ at constant pressure).

(e) Maximum non-expansion work

The analogue of the maximum work interpretation of ΔA, and the origin of the name 'free energy', can be found for ΔG. In the following *Justification*, we show that at constant temperature and pressure, the maximum additional (non-expansion) work, $w_{add,max}$, is given by the change in Gibbs energy:

$$dw_{add,max} = dG \tag{3.41a}$$

The corresponding expression for a measurable change is

$$w_{add,max} = \Delta G \tag{3.41b}$$

Relation between G and maximum non-expansion work

This expression is particularly useful for assessing the electrical work that may be produced by fuel cells and electrochemical cells, and we shall see many applications of it.

Justification 3.3 *Maximum non-expansion work*

Because $H = U + pV$, for a general change in conditions, the change in enthalpy is

$$dH = dq + dw + d(pV)$$

The corresponding change in Gibbs energy ($G = H - TS$) is

$$dG = dH - TdS - SdT = dq + dw + d(pV) - TdS - SdT$$

When the change is isothermal we can set $dT = 0$; then

$$dG = dq + dw + d(pV) - TdS$$

When the change is reversible, $dw = dw_{rev}$ and $dq = dq_{rev} = TdS$, so for a reversible, isothermal process

$$dG = TdS + dw_{rev} + d(pV) - TdS = dw_{rev} + d(pV)$$

The work consists of expansion work, which for a reversible change is given by $-p\mathrm{d}V$, and possibly some other kind of work (for instance, the electrical work of pushing electrons through a circuit or of raising a column of liquid); this additional work we denote $\mathrm{d}w_{add}$. Therefore, with $\mathrm{d}(pV) = p\mathrm{d}V + V\mathrm{d}p$,

$$\mathrm{d}G = (-p\mathrm{d}V + \mathrm{d}w_{add,rev}) + p\mathrm{d}V + V\mathrm{d}p = \mathrm{d}w_{add,rev} + V\mathrm{d}p$$

If the change occurs at constant pressure (as well as constant temperature), we can set $\mathrm{d}p = 0$ and obtain $\mathrm{d}G = \mathrm{d}w_{add,rev}$. Therefore, at constant temperature and pressure, $\mathrm{d}w_{add,rev} = \mathrm{d}G$. However, because the process is reversible, the work done must now have its maximum value, so eqn 3.41 follows.

Example 3.5 *Calculating the maximum non-expansion work of a reaction*

How much energy is available for sustaining muscular and nervous activity from the combustion of 1.00 mol of glucose molecules under standard conditions at 37°C (blood temperature)? The standard entropy of reaction is $+259.1\ \mathrm{J\ K^{-1}\ mol^{-1}}$.

Method The non-expansion work available from the reaction is equal to the change in standard Gibbs energy for the reaction ($\Delta_r G^\ominus$, a quantity defined more fully below). To calculate this quantity, it is legitimate to ignore the temperature-dependence of the reaction enthalpy, to obtain $\Delta_r H^\ominus$ from Tables 2.6 and 2.8, and to substitute the data into $\Delta_r G^\ominus = \Delta_r H^\ominus - T\Delta_r S^\ominus$.

Answer Because the standard reaction enthalpy is $-2808\ \mathrm{kJ\ mol^{-1}}$, it follows that the standard reaction Gibbs energy is

$$\Delta_r G^\ominus = -2808\ \mathrm{kJ\ mol^{-1}} - (310\ \mathrm{K}) \times (259.1\ \mathrm{J\ K^{-1}\ mol^{-1}}) = -2888\ \mathrm{kJ\ mol^{-1}}$$

Therefore, $w_{add,max} = -2888\ \mathrm{kJ}$ for the combustion of 1 mol glucose molecules, and the reaction can be used to do up to 2888 kJ of non-expansion work. To place this result in perspective, consider that a person of mass 70 kg needs to do 2.1 kJ of work to climb vertically through 3.0 m; therefore, at least 0.13 g of glucose is needed to complete the task (and in practice significantly more).

Self-test 3.8 How much non-expansion work can be obtained from the combustion of 1.00 mol $CH_4(g)$ under standard conditions at 298 K? Use $\Delta_r S^\ominus = -243\ \mathrm{J\ K^{-1}\ mol^{-1}}$. [818 kJ]

3.6 Standard molar Gibbs energies

Key points **Standard Gibbs energies of formation are used to calculate the standard Gibbs energies of reactions. The Gibbs energies of formation of ions may be estimated from a thermodynamic cycle and the Born equation.**

Standard entropies and enthalpies of reaction can be combined to obtain the **standard Gibbs energy of reaction** (or 'standard reaction Gibbs energy'), $\Delta_r G^\ominus$:

$$\Delta_r G^\ominus = \Delta_r H^\ominus - T\Delta_r S^\ominus \qquad \text{Definition of standard Gibbs energy of reaction} \qquad [3.42]$$

The standard Gibbs energy of reaction is the difference in standard molar Gibbs energies of the products and reactants in their standard states at the temperature specified for the reaction as written. As in the case of standard reaction enthalpies, it is convenient to define the **standard Gibbs energies of formation**, $\Delta_f G^\ominus$, the standard reaction Gibbs energy for the formation of a compound from its elements in their reference states.[7] Standard Gibbs energies of formation of the elements in their reference states are zero, because their formation is a 'null' reaction. A selection of values for compounds is given in Table 3.4. From the values there, it is a simple matter to obtain the standard Gibbs energy of reaction by taking the appropriate combination:

Table 3.4* Standard Gibbs energies of formation (at 298 K)

	$\Delta_f G^{\ominus}/(\text{kJ mol}^{-1})$
Diamond, C(s)	+2.9
Benzene, C_6H_6(l)	+124.3
Methane, CH_4(g)	−50.7
Carbon dioxide, CO_2(g)	−394.4
Water, H_2O(l)	−237.1
Ammonia, NH_3(g)	−16.5
Sodium chloride, NaCl(s)	−384.1

* More values are given in the *Data section*.

$$\Delta_r G^{\ominus} = \sum_{\text{Products}} \nu \Delta_f G^{\ominus} - \sum_{\text{Reactants}} \nu \Delta_f G^{\ominus} \qquad (3.43a)$$

Procedure for calculating the standard Gibbs energy of reaction

In the notation introduced in Section 2.8,

$$\Delta_r G^{\ominus} = \sum_J \nu_J \Delta_f G^{\ominus}(J) \qquad (3.43b)$$

• A brief illustration

To calculate the standard Gibbs energy of the reaction $CO(g) + \frac{1}{2}O_2(g) \rightarrow CO_2(g)$ at 25°C, we write

$$\begin{aligned}\Delta_r G^{\ominus} &= \Delta_f G^{\ominus}(CO_2,g) - \{\Delta_f G^{\ominus}(CO,g) + \tfrac{1}{2}\Delta_f G^{\ominus}(O_2,g)\}\\ &= -394.4\ \text{kJ mol}^{-1} - \{(-137.2) + \tfrac{1}{2}(0)\}\text{kJ mol}^{-1}\\ &= -257.2\ \text{kJ mol}^{-1}\ \bullet\end{aligned}$$

Self-test 3.9 Calculate the standard reaction Gibbs energy for the combustion of CH_4(g) at 298 K. [$-818\ \text{kJ mol}^{-1}$]

Just as we did in Section 2.8, where we acknowledged that solutions of cations cannot be prepared without their accompanying anions, we define one ion, conventionally the hydrogen ion, to have zero standard Gibbs energy of formation at all temperatures:

$$\Delta_f G^{\ominus}(H^+,aq) = 0 \qquad [3.44]$$

Convention for ions in solution

In essence, this definition adjusts the actual values of the Gibbs energies of formation of ions by a fixed amount that is chosen so that the standard value for one of them, H^+(aq), has the value zero.

• A brief illustration

For the reaction

$$\tfrac{1}{2}H_2(g) + \tfrac{1}{2}Cl_2(g) \rightarrow H^+(aq) + Cl^-(aq) \qquad \Delta_r G^{\ominus} = -131.23\ \text{kJ mol}^{-1}$$

we can write

$$\Delta_r G^{\ominus} = \Delta_f G^{\ominus}(H^+,aq) + \Delta_f G^{\ominus}(Cl^-,aq) = \Delta_f G^{\ominus}(Cl^-,aq)$$

and hence identify $\Delta_f G^{\ominus}(Cl^-,aq)$ as $-131.23\ \text{kJ mol}^{-1}$. With the value of $\Delta_f G^{\ominus}(Cl^-,aq)$ established, we can find the value of $\Delta_f G^{\ominus}(Ag^+,aq)$ from

$$Ag(s) + \tfrac{1}{2}Cl_2(g) \rightarrow Ag^+(aq) + Cl^-(aq) \qquad \Delta_r G^{\ominus} = -54.12\ \text{kJ mol}^{-1}$$

which leads to $\Delta_f G^{\ominus}(Ag^+,aq) = +77.11\ \text{kJ mol}^{-1}$. All the Gibbs energies of formation of ions tabulated in the *Data section* were calculated in the same way. •

[7] The reference state of an element was defined in Section 2.8.

A brief comment

The standard Gibbs energies of formation of the gas-phase ions are unknown. We have therefore used ionization energies and electron affinities and have assumed that any differences from the Gibbs energies arising from conversion to enthalpy and the inclusion of entropies to obtain Gibbs energies in the formation of H^+ are cancelled by the corresponding terms in the electron gain of X. The conclusions from the cycles are therefore only approximate.

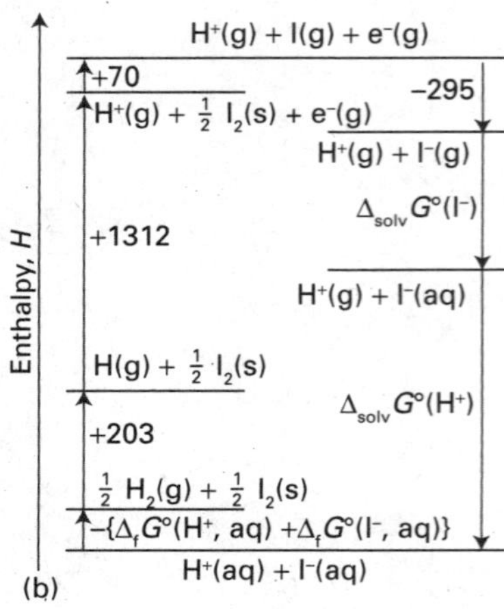

Fig. 3.18 The thermodynamic cycles for the discussion of the Gibbs energies of solvation (hydration) and formation of (a) chloride ions, (b) iodide ions in aqueous solution. The sum of the changes in Gibbs energies around the cycle sum to zero because *G* is a state function.

The factors responsible for the magnitude of the Gibbs energy of formation of an ion in solution can be identified by analysing it in terms of a thermodynamic cycle. As an illustration, we consider the standard Gibbs energy of formation of Cl^- in water, which is -131 kJ mol^{-1}. We do so by treating the formation reaction

$$\tfrac{1}{2}H_2(g) + \tfrac{1}{2}X_2(g) \rightarrow H^+(aq) + X^-(aq)$$

as the outcome of the sequence of steps shown in Fig. 3.18 (with values taken from the *Data section*). The sum of the Gibbs energies for all the steps around a closed cycle is zero, so

$$\Delta_f G^\ominus(Cl^-,aq) = 1272\ \text{kJ mol}^{-1} + \Delta_{solv}G^\ominus(H^+) + \Delta_{solv}G^\ominus(Cl^-)$$

An important point to note is that the value of $\Delta_f G^\ominus$ of an ion X is not determined by the properties of X alone but includes contributions from the dissociation, ionization, and hydration of hydrogen.

Gibbs energies of solvation of individual ions may be estimated from an equation derived by Max Born, who identified $\Delta_{solv}G^\ominus$ with the electrical work of transferring an ion from a vacuum into the solvent treated as a continuous dielectric of relative permittivity ε_r. The resulting **Born equation**, which is derived in *Further information* 3.1, is

$$\Delta_{solv}G^\ominus = -\frac{z_i^2 e^2 N_A}{8\pi\varepsilon_0 r_i}\left(1 - \frac{1}{\varepsilon_r}\right) \qquad \text{Born equation} \qquad (3.45a)$$

where z_i is the charge number of the ion and r_i its radius (N_A is Avogadro's constant). Note that $\Delta_{solv}G^\ominus < 0$, and that $\Delta_{solv}G^\ominus$ is strongly negative for small, highly charged ions in media of high relative permittivity. For water for which $\varepsilon_r = 78.54$ at 25°C,

$$\Delta_{solv}G^\ominus = -\frac{z_i^2}{(r_i/\text{pm})} \times (6.86 \times 10^4\ \text{kJ mol}^{-1}) \qquad (3.45b)$$

• **A brief illustration**

To see how closely the Born equation reproduces the experimental data, we calculate the difference in the values of $\Delta_f G^\ominus$ for Cl^- and I^- in water at 25°C, given their radii as 181 pm and 220 pm (Table 19.3), respectively, is

$$\Delta_{solv}G^\ominus(Cl^-) - \Delta_{solv}G^\ominus(I^-) = -\left(\frac{1}{181} - \frac{1}{220}\right) \times (6.86 \times 10^4\ \text{kJ mol}^{-1})$$
$$= -67\ \text{kJ mol}^{-1}$$

This estimated difference is in good agreement with the experimental difference, which is -61 kJ mol^{-1}. •

Self-test 3.10 Estimate the value of $\Delta_{solv}G^\ominus(Cl^-) - \Delta_{solv}G^\ominus(Br^-)$ in water from experimental data and from the Born equation.

[-26 kJ mol^{-1} experimental; -29 kJ mol^{-1} calculated]

Calorimetry (for ΔH directly, and for S via heat capacities) is only one of the ways of determining Gibbs energies. They may also be obtained from equilibrium constants and electrochemical measurements (Chapter 6), and for gases they may be calculated using data from spectroscopic observations (Chapter 16).

Combining the First and Second Laws

The First and Second Laws of thermodynamics are both relevant to the behaviour of matter, and we can bring the whole force of thermodynamics to bear on a problem by setting up a formulation that combines them.

3.7 The fundamental equation

Key point **The fundamental equation, a combination of the First and Second Laws, is an expression for the change in internal energy that accompanies changes in the volume and entropy of a system.**

We have seen that the First Law of thermodynamics may be written $dU = dq + dw$. For a reversible change in a closed system of constant composition, and in the absence of any additional (non-expansion) work, we may set $dw_{rev} = -pdV$ and (from the definition of entropy) $dq_{rev} = TdS$, where p is the pressure of the system and T its temperature. Therefore, for a reversible change in a closed system,

$$dU = TdS - pdV \qquad \text{The fundamental equation} \qquad (3.46)$$

However, because dU is an exact differential, its value is independent of path. Therefore, the same value of dU is obtained whether the change is brought about irreversibly or reversibly. Consequently, *eqn 3.46 applies to any change—reversible or irreversible—of a closed system that does no additional (non-expansion) work*. We shall call this combination of the First and Second Laws the **fundamental equation.**

The fact that the fundamental equation applies to both reversible and irreversible changes may be puzzling at first sight. The reason is that only in the case of a reversible change may TdS be identified with dq and $-pdV$ with dw. When the change is irreversible, $TdS > dq$ (the Clausius inequality) and $-pdV > dw$. The sum of dw and dq remains equal to the sum of TdS and $-pdV$, provided the composition is constant.

3.8 Properties of the internal energy

Key points **Relations between thermodynamic properties are generated by combining thermodynamic and mathematical expressions for changes in their values. (a) The Maxwell relations are a series of relations between derivatives of thermodynamic properties based on criteria for changes in the properties being exact differentials. (b) The Maxwell relations are used to derive the thermodynamic equation of state and to determine how the internal energy of a substance varies with volume.**

Equation 3.46 shows that the internal energy of a closed system changes in a simple way when either S or V is changed ($dU \propto dS$ and $dU \propto dV$). These simple proportionalities suggest that U is best regarded as a function of S and V. We could regard U as a function of other variables, such as S and p or T and V, because they are all interrelated; but the simplicity of the fundamental equation suggests that $U(S,V)$ is the best choice.

The *mathematical* consequence of U being a function of S and V is that we can express an infinitesimal change dU in terms of changes dS and dV by

$$dU = \left(\frac{\partial U}{\partial S}\right)_V dS + \left(\frac{\partial U}{\partial V}\right)_S dV \qquad (3.47)$$

A brief comment

Partial derivatives were introduced in *Mathematical background 2*. The type of result in eqn 3.47 was first obtained in Section 2.11, where we treated U as a function of T and V.

The two partial derivatives are the slopes of the plots of U against S and V, respectively. When this expression is compared to the *thermodynamic* relation, eqn 3.46, we see that, for systems of constant composition,

$$\left(\frac{\partial U}{\partial S}\right)_V = T \qquad \left(\frac{\partial U}{\partial V}\right)_S = -p \tag{3.48}$$

The first of these two equations is a purely thermodynamic definition of temperature (a Zeroth-Law concept) as the ratio of the changes in the internal energy (a First-Law concept) and entropy (a Second-Law concept) of a constant-volume, closed, constant-composition system. We are beginning to generate relations between the properties of a system and to discover the power of thermodynamics for establishing unexpected relations.

(a) The Maxwell relations

An infinitesimal change in a function $f(x,y)$ can be written $df = g dx + h dy$ where g and h are functions of x and y. The mathematical criterion for df being an exact differential (in the sense that its integral is independent of path) is that

$$\left(\frac{\partial g}{\partial y}\right)_x = \left(\frac{\partial h}{\partial x}\right)_y \tag{3.49}$$

This criterion is discussed in *Mathematical background 2*. Because the fundamental equation, eqn 3.46, is an expression for an exact differential, the functions multiplying dS and dV (namely T and $-p$) must pass this test. Therefore, it must be the case that

$$\left(\frac{\partial T}{\partial V}\right)_S = -\left(\frac{\partial p}{\partial S}\right)_V \qquad \text{A Maxwell relation} \tag{3.50}$$

We have generated a relation between quantities that, at first sight, would not seem to be related.

Equation 3.50 is an example of a **Maxwell relation**. However, apart from being unexpected, it does not look particularly interesting. Nevertheless, it does suggest that there may be other similar relations that are more useful. Indeed, we can use the fact that H, G, and A are all state functions to derive three more Maxwell relations. The argument to obtain them runs in the same way in each case: because H, G, and A are state functions, the expressions for dH, dG, and dA satisfy relations like eqn 3.49. All four relations are listed in Table 3.5 and we put them to work later in the chapter.

Table 3.5 The Maxwell relations

From U:	$\left(\frac{\partial T}{\partial V}\right)_S = -\left(\frac{\partial p}{\partial S}\right)_V$
From H:	$\left(\frac{\partial T}{\partial p}\right)_S = \left(\frac{\partial V}{\partial S}\right)_p$
From A:	$\left(\frac{\partial p}{\partial T}\right)_V = \left(\frac{\partial S}{\partial V}\right)_T$
From G:	$\left(\frac{\partial V}{\partial T}\right)_p = -\left(\frac{\partial S}{\partial p}\right)_T$

(b) The variation of internal energy with volume

The quantity $\pi_T = (\partial U/\partial V)_T$, which represents how the internal energy changes as the volume of a system is changed isothermally, played a central role in the manipulation of the First Law, and in *Further information 2.2* we used the relation

$$\pi_T = T\left(\frac{\partial p}{\partial T}\right)_V - p \qquad \text{A thermodynamic equation of state} \tag{3.51}$$

This relation is called a **thermodynamic equation of state** because it is an expression for pressure in terms of a variety of thermodynamic properties of the system. We are now ready to derive it by using a Maxwell relation.

Justification 3.4 *The thermodynamic equation of state*

We obtain an expression for the coefficient π_T by dividing both sides of eqn 3.47 by dV, imposing the constraint of constant temperature, which gives

$$\left(\frac{\partial U}{\partial V}\right)_T = \left(\frac{\partial U}{\partial S}\right)_V \left(\frac{\partial S}{\partial V}\right)_T + \left(\frac{\partial U}{\partial V}\right)_S$$

Next, we introduce the two relations in eqn 3.48 and the definition of π_T to obtain

$$\pi_T = T\left(\frac{\partial S}{\partial V}\right)_T - p$$

The third Maxwell relation in Table 3.5 turns $(\partial S/\partial V)_T$ into $(\partial p/\partial T)_V$, which completes the proof of eqn 3.51.

Example 3.6 *Deriving a thermodynamic relation*

Show thermodynamically that $\pi_T = 0$ for a perfect gas, and compute its value for a van der Waals gas.

Method Proving a result 'thermodynamically' means basing it entirely on general thermodynamic relations and equations of state, without drawing on molecular arguments (such as the existence of intermolecular forces). We know that for a perfect gas, $p = nRT/V$, so this relation should be used in eqn 3.51. Similarly, the van der Waals equation is given in Table 1.7, and for the second part of the question it should be used in eqn 3.51.

Answer For a perfect gas we write

$$\left(\frac{\partial p}{\partial T}\right)_V = \left(\frac{\partial(nRT/V)}{\partial T}\right)_V = \frac{nR}{V}$$

Then, eqn 3.51 becomes

$$\pi_T = \frac{nRT}{V} - p = 0$$

The equation of state of a van der Waals gas is

$$p = \frac{nRT}{V - nb} - a\frac{n^2}{V^2}$$

Because a and b are independent of temperature,

$$\left(\frac{\partial p}{\partial T}\right)_V = \left(\frac{\partial(nRT/(V-nb))}{\partial T}\right)_V = \frac{nR}{V-nb}$$

Therefore, from eqn 3.51,

$$\pi_T = \frac{nRT}{V-nb} - p = \frac{nRT}{V-nb} - \left(\frac{nRT}{V-nb} - a\frac{n^2}{V^2}\right) = a\frac{n^2}{V^2}$$

This result for π_T implies that the internal energy of a van der Waals gas increases when it expands isothermally (that is, $(\partial U/\partial V)_T > 0$), and that the increase is related to the parameter a, which models the attractive interactions between the particles. A larger molar volume, corresponding to a greater average separation between molecules, implies weaker mean intermolecular attractions, so the total energy is greater.

Self-test 3.11 Calculate π_T for a gas that obeys the virial equation of state (Table 1.7). $[\pi_T = RT^2(\partial B/\partial T)_V/V_m^2 + \cdots]$

3.9 Properties of the Gibbs energy

Key points **(a) The variation of the Gibbs energy of a system suggests that it is best regarded as a function of pressure and temperature. The Gibbs energy of a substance decreases with temperature and increases with pressure. (b) The variation of Gibbs energy with temperature is related to the enthalpy by the Gibbs–Helmholtz equation. (c) The Gibbs energies of solids and liquids are almost independent of pressure; those of gases vary linearly with the logarithm of the pressure.**

The same arguments that we have used for U can be used for the Gibbs energy $G = H - TS$. They lead to expressions showing how G varies with pressure and temperature that are important for discussing phase transitions and chemical reactions.

(a) General considerations

When the system undergoes a change of state, G may change because H, T, and S all change. As in *Justification 2.1*, we write for infinitesimal changes in each property

$$dG = dH - d(TS) = dH - TdS - SdT$$

Because $H = U + pV$, we know that

$$dH = dU + d(pV) = dU + pdV + Vdp$$

and therefore

$$dG = dU + pdV + Vdp - TdS - SdT$$

For a closed system doing no non-expansion work, we can replace dU by the fundamental equation $dU = TdS - pdV$ and obtain

$$dG = TdS - pdV + pdV + Vdp - TdS - SdT$$

Four terms now cancel on the right, and we conclude that for a closed system in the absence of non-expansion work and at constant composition

$$dG = Vdp - SdT \qquad \text{The fundamental equation of chemical thermodynamics} \qquad (3.52)$$

This expression, which shows that a change in G is proportional to a change in p or T, suggests that G may be best regarded as a function of p and T. It may be regarded as the **fundamental equation of chemical thermodynamics** as it is so central to the application of thermodynamics to chemistry: it suggests that G is an important quantity in chemistry because the pressure and temperature are usually the variables under our control. In other words, G carries around the combined consequences of the First and Second Laws in a way that makes it particularly suitable for chemical applications.

The same argument that led to eqn 3.48, when applied to the exact differential $dG = Vdp - SdT$, now gives

$$\left(\frac{\partial G}{\partial T}\right)_p = -S \qquad \left(\frac{\partial G}{\partial p}\right)_T = V \qquad \text{The variation of } G \text{ with } T \text{ and } p \qquad (3.53)$$

These relations show how the Gibbs energy varies with temperature and pressure (Fig. 3.19). The first implies that:

• Because $S > 0$ for all substances, G always *decreases* when the temperature is raised (at constant pressure and composition).

• Because $(\partial G/\partial T)_p$ becomes more negative as S increases, G decreases most sharply when the entropy of the system is large.

Therefore, the Gibbs energy of the gaseous phase of a substance, which has a high molar entropy, is more sensitive to temperature than its liquid and solid phases (Fig. 3.20). Similarly, the second relation implies that:

• Because $V > 0$ for all substances, G always *increases* when the pressure of the system is increased (at constant temperature and composition).

• Because $(\partial G/\partial p)_T$ increases with V, G is more sensitive to pressure when the volume of the system is large.

Because the molar volume of the gaseous phase of a substance is greater than that of its condensed phases, the molar Gibbs energy of a gas is more sensitive to pressure than its liquid and solid phases (Fig. 3.21).

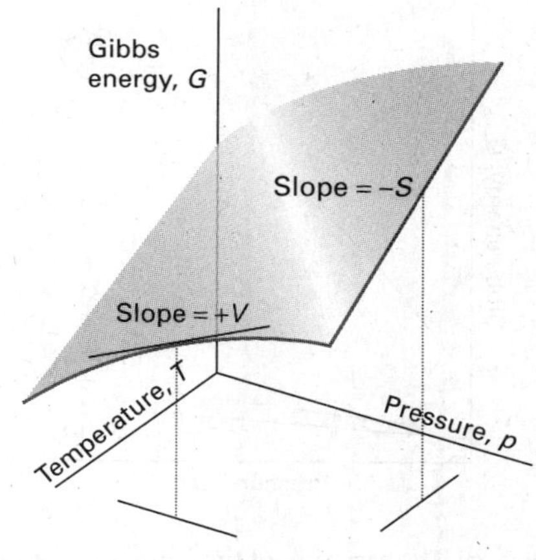

Fig. 3.19 The variation of the Gibbs energy of a system with (a) temperature at constant pressure and (b) pressure at constant temperature. The slope of the former is equal to the negative of the entropy of the system and that of the latter is equal to the volume.

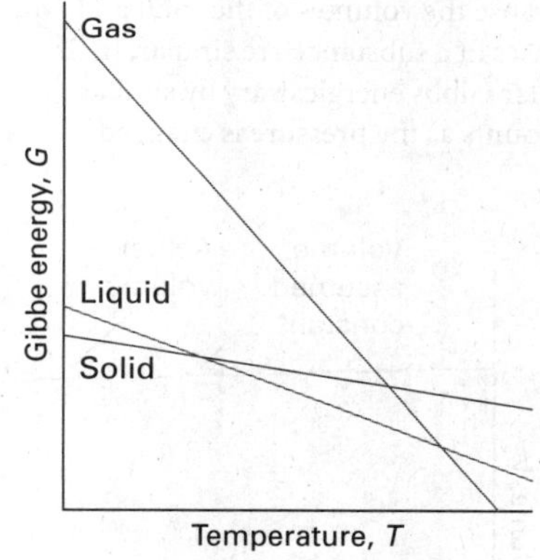

Fig. 3.20 The variation of the Gibbs energy with the temperature is determined by the entropy. Because the entropy of the gaseous phase of a substance is greater than that of the liquid phase, and the entropy of the solid phase is smallest, the Gibbs energy changes most steeply for the gas phase, followed by the liquid phase, and then the solid phase of the substance.

(b) The variation of the Gibbs energy with temperature

As we remarked in the introduction, because the equilibrium composition of a system depends on the Gibbs energy, to discuss the response of the composition to temperature we need to know how G varies with temperature.

The first relation in eqn 3.53, $(\partial G/\partial T)_p = -S$, is our starting point for this discussion. Although it expresses the variation of G in terms of the entropy, we can express it in terms of the enthalpy by using the definition of G to write $S = (H - G)/T$. Then

$$\left(\frac{\partial G}{\partial T}\right)_p = \frac{G-H}{T} \tag{3.54}$$

We shall see later that the equilibrium constant of a reaction is related to G/T rather than to G itself,[8] and it is easy to deduce from the last equation (see the following *Justification*) that

$$\left(\frac{\partial (G/T)}{\partial T}\right)_p = -\frac{H}{T^2} \tag{3.55}$$

Gibbs–Helmholtz equation

This expression is called the **Gibbs–Helmholtz equation.** It shows that, if we know the enthalpy of the system, then we know how G/T varies with temperature.

Justification 3.5 *The Gibbs–Helmholtz equation*

First, we note that

$$\left(\frac{\partial (G/T)}{\partial T}\right)_p = \frac{1}{T}\left(\frac{\partial G}{\partial T}\right)_p + G\frac{\mathrm{d}(1/T)}{\mathrm{d}T} = \frac{1}{T}\left(\frac{\partial G}{\partial T}\right)_p - \frac{G}{T^2} = \frac{1}{T}\left\{\left(\frac{\partial G}{\partial T}\right)_p - \frac{G}{T}\right\}$$

Then we use eqn 3.54 to write

$$\left(\frac{\partial G}{\partial T}\right)_p - \frac{G}{T} = \frac{G-H}{T} - \frac{G}{T} = -\frac{H}{T}$$

When this expression is substituted in the preceding one, we obtain eqn 3.55.

[8] In Section 6.2b we derive the result that the equilibrium constant for a reaction is related to its standard reaction Gibbs energy by $\Delta_r G^{\ominus}/T = -R \ln K$.

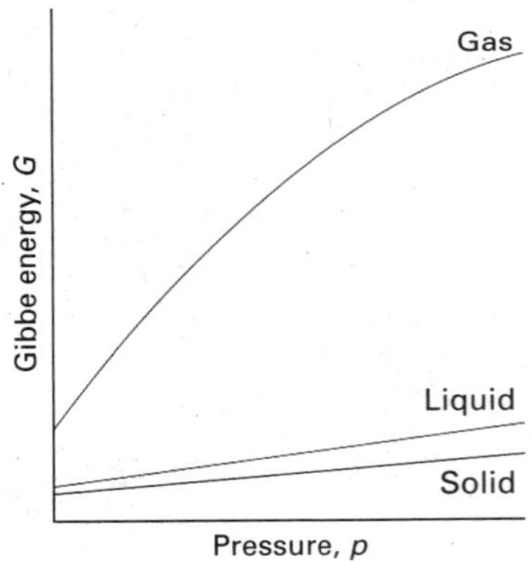

Fig. 3.21 The variation of the Gibbs energy with the pressure is determined by the volume of the sample. Because the volume of the gaseous phase of a substance is greater than that of the same amount of liquid phase, and the entropy of the solid phase is smallest (for most substances), the Gibbs energy changes most steeply for the gas phase, followed by the liquid phase, and then the solid phase of the substance. Because the volumes of the solid and liquid phases of a substance are similar, their molar Gibbs energies vary by similar amounts as the pressure is changed.

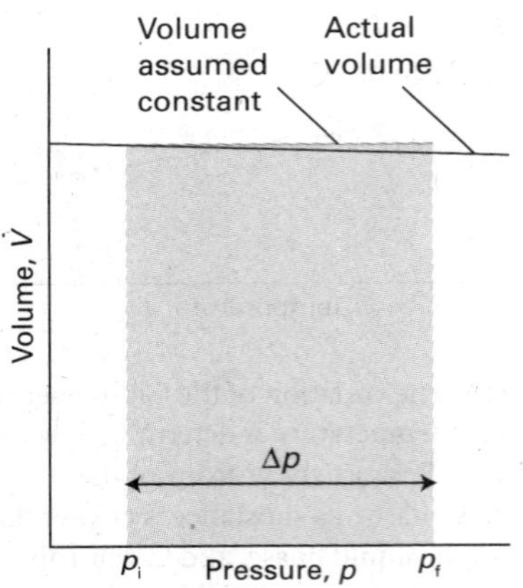

Fig. 3.22 The difference in Gibbs energy of a solid or liquid at two pressures is equal to the rectangular area shown. We have assumed that the variation of volume with pressure is negligible.

The Gibbs–Helmholtz equation is most useful when it is applied to changes, including changes of physical state and chemical reactions at constant pressure. Then, because $\Delta G = G_f - G_i$ for the change of Gibbs energy between the final and initial states and because the equation applies to both G_f and G_i, we can write

$$\left(\frac{\partial(\Delta G/T)}{\partial T}\right)_p = -\frac{\Delta H}{T^2} \tag{3.56}$$

This equation shows that, if we know the change in enthalpy of a system that is undergoing some kind of transformation (such as vaporization or reaction), then we know how the corresponding change in Gibbs energy varies with temperature. As we shall see, this is a crucial piece of information in chemistry.

(c) The variation of the Gibbs energy with pressure

To find the Gibbs energy at one pressure in terms of its value at another pressure, the temperature being constant, we set $dT = 0$ in eqn 3.52, which gives $dG = Vdp$, and integrate:

$$G(p_f) = G(p_i) + \int_{p_i}^{p_f} V dp \tag{3.57a}$$

For molar quantities,

$$G_m(p_f) = G_m(p_i) + \int_{p_i}^{p_f} V_m\, dp \tag{3.57b}$$

This expression is applicable to any phase of matter, but to evaluate it we need to know how the molar volume, V_m, depends on the pressure.

The molar volume of a condensed phase changes only slightly as the pressure changes (Fig. 3.22), so we can treat V_m as a constant and take it outside the integral:

$$G_m(p_f) = G_m(p_i) + V_m \int_{p_i}^{p_f} dp = G_m(p_i) + (p_f - p_i)V_m \tag{3.58}$$

Self-test 3.12 Calculate the change in G_m for ice at −10°C, with density 917 kg m^{-3}, when the pressure is increased from 1.0 bar to 2.0 bar. [+2.0 J mol^{-1}]

Under normal laboratory conditions $(p_f - p_i)V_m$ is very small and may be neglected. Hence, we may usually suppose that the Gibbs energies of solids and liquids are independent of pressure. However, if we are interested in geophysical problems, then, because pressures in the Earth's interior are huge, their effect on the Gibbs energy cannot be ignored. If the pressures are so great that there are substantial volume changes over the range of integration, then we must use the complete expression, eqn 3.57.

• A brief illustration

Suppose that for a certain phase transition of a solid $\Delta_{trs}V = +1.0$ cm^3 mol^{-1} independent of pressure. Then for an increase in pressure to 3.0 Mbar (3.0 × 10^{11} Pa) from 1.0 bar (1.0 × 10^5 Pa), the Gibbs energy of the transition changes from $\Delta_{trs}G$(1 bar) to

$$\begin{aligned}\Delta_{trs}G(3\text{ Mbar}) &= \Delta_{trs}G(1\text{ bar}) + (1.0\times10^{-6}\text{ m}^3\text{ mol}^{-1})\times(3.0\times10^{11}\text{ Pa} - 1.0\times10^5\text{ Pa})\\ &= \Delta_{trs}G(1\text{ bar}) + 3.0\times10^2\text{ kJ mol}^{-1}\end{aligned}$$

where we have used 1 Pa m^3 = 1 J. •

The molar volumes of gases are large, so the Gibbs energy of a gas depends strongly on the pressure. Furthermore, because the volume also varies markedly with the pressure, we cannot treat it as a constant in the integral in eqn 3.57b (Fig. 3.23). For a perfect gas we substitute $V_m = RT/p$ into the integral, treat RT as a constant, and find

$$G_m(p_f) = G_m(p_i) + RT\int_{p_i}^{p_f} \frac{1}{p}\,dp = G_m(p_i) + RT\ln\frac{p_f}{p_i} \qquad (3.59)^\circ$$

This expression shows that, when the pressure is increased tenfold at room temperature, the molar Gibbs energy increases by $RT\ln 10 \approx 6\ \text{kJ mol}^{-1}$. It also follows from this equation that, if we set $p_i = p^\ominus$ (the standard pressure of 1 bar), then the molar Gibbs energy of a perfect gas at a pressure p (set $p_f = p$) is related to its standard value by

$$G_m(p) = G_m^\ominus + RT\ln\frac{p}{p^\ominus} \qquad \text{The molar Gibbs energy of a perfect gas} \qquad (3.60)^\circ$$

Self-test 3.13 Calculate the change in the molar Gibbs energy of water vapour (treated as a perfect gas) when the pressure is increased isothermally from 1.0 bar to 2.0 bar at 298 K. Note that, whereas the change in molar Gibbs energy for a condensed phase (Self-test 3.12) is a few joules per mole, the answer you should get for a gas is of the order of kilojoules per mole [$+1.7\ \text{kJ mol}^{-1}$]

The logarithmic dependence of the molar Gibbs energy on the pressure predicted by eqn 3.60 is illustrated in Fig. 3.24. This very important expression, the consequences of which we unfold in the following chapters, applies to perfect gases (which is usually a good enough approximation). *Further information 3.2* describes how to take into account gas imperfections.

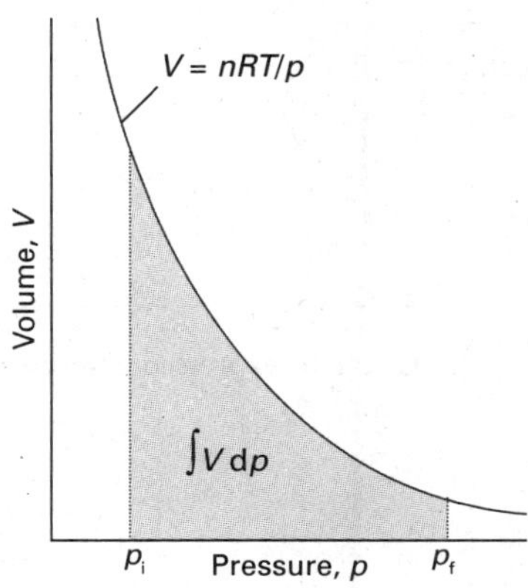

Fig. 3.23 The difference in Gibbs energy for a perfect gas at two pressures is equal to the area shown below the perfect-gas isotherm.

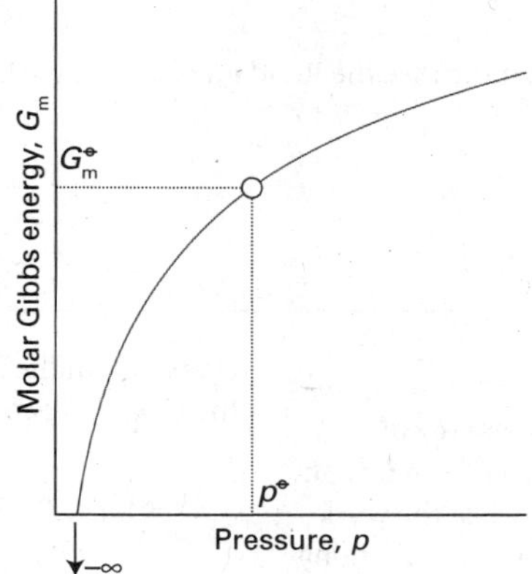

Fig. 3.24 The molar Gibbs energy of a perfect gas is proportional to ln p, and the standard state is reached at $p^\ominus$. Note that, as $p \to 0$, the molar Gibbs energy becomes negatively infinite.

interActivity Show how the first derivative of G, $(\partial G/\partial p)_T$, varies with pressure, and plot the resulting expression over a pressure range. What is the physical significance of $(\partial G/\partial p)_T$?

Checklist of key equations

Property	Equation	Comment
Thermodynamic entropy	$dS = dq_{rev}/T$	Definition
Boltzmann formula	$S = k \ln W$	Definition
Clausius inequality	$dS \geq dq/T$	
Entropy of isothermal expansion	$\Delta S = nR \ln(V_f/V_i)$	Perfect gas
Entropy of transition	$\Delta_{trs}S = \Delta_{trs}H/T_{trs}$	At the transition temperature
Variation of the entropy with temperature	$S(T_f) = S(T_i) + C \ln(T_f/T_i)$	The heat capacity, C, is independent of temperature and no phase transitions occur
Reaction entropy	$\Delta_r S^{\ominus} = \sum_{\text{Products}} \nu S_m^{\ominus} - \sum_{\text{Reactants}} \nu S_m^{\ominus}$	
Helmholtz energy	$A = U - TS$	Definition
Gibbs energy	$G = H - TS$	Definition
Maximum work	$w_{max} = \Delta A$	
Maximum non-expansion work	$w_{add,max} = \Delta G$	Constant p and T
Criteria of spontaneity	(a) $dS_{U,V} \geq 0$ and $dU_{S,V} \leq 0$, or (b) $dA_{T,V} \leq 0$ and $dG_{T,p} \leq 0$	
Reaction Gibbs energy	$\Delta_r G^{\ominus} = \sum_{\text{Products}} \nu \Delta_f G^{\ominus} - \sum_{\text{Reactants}} \nu \Delta_f G^{\ominus}$	
Fundamental equation	$dU = TdS - pdV$	
Fundamental equation of chemical thermodynamics	$dG = Vdp - SdT$ $(\partial G/\partial p)_T = V$ and $(\partial G/\partial T)_p = -S$	
Gibbs–Helmholtz equation	$(\partial(G/T)/\partial T)_p = -H/T^2$	
	$G_m(p_f) = G_m(p_i) + V_m \Delta p$	Incompressible substance
	$G(p_f) = G(p_i) + nRT \ln(p_f/p_i)$	Perfect gas

→ For a chart of the relations between principal equations, see the Road map section of the Resource section.

Further information

Further information 3.1 *The Born equation*

The strategy of the calculation is to identify the Gibbs energy of solvation with the work of transferring an ion from a vacuum into the solvent. That work is calculated by taking the difference of the work of charging an ion when it is in the solution and the work of charging the same ion when it is in a vacuum.

The Coulomb interaction between two charges Q_1 and Q_2 separated by a distance r is described by the *Coulombic potential energy*:

$$V = \frac{Q_1 Q_2}{4\pi\varepsilon r}$$

where ε is the medium's permittivity. The permittivity of vacuum is $\varepsilon_0 = 8.854 \times 10^{-12}\ \text{J}^{-1}\ \text{C}^2\ \text{m}^{-1}$. The relative permittivity (formerly called the 'dielectric constant') of a substance is defined as $\varepsilon_r = \varepsilon/\varepsilon_0$. Ions do not interact as strongly in a solvent of high relative permittivity (such as water, with $\varepsilon_r = 80$ at 293 K) as they do in a solvent of lower relative permittivity (such as ethanol, with $\varepsilon_r = 25$ at 293 K). See Chapter 17 for more details. The potential energy of a charge Q_1 in the presence of a charge Q_2 can be expressed in terms of the *Coulomb potential*, ϕ:

$$V = Q_1\phi \qquad \phi = \frac{Q_2}{4\pi\varepsilon r}$$

We model an ion as a sphere of radius r_i immersed in a medium of permittivity ε. It turns out that, when the charge of the sphere is Q, the electric potential, ϕ, at its surface is the same as the potential due to a point charge at its centre, so we can use the last expression and write

$$\phi = \frac{Q}{4\pi\varepsilon r_i}$$

The work of bringing up a charge dQ to the sphere is ϕdQ. Therefore, the total work of charging the sphere from 0 to $z_i e$ is

$$w = \int_0^{z_i e} \phi dQ = \frac{1}{4\pi\varepsilon r_i}\int_0^{z_i e} Q dQ = \frac{z_i^2 e^2}{8\pi\varepsilon r_i}$$

This electrical work of charging, when multiplied by Avogadro's constant, is the molar Gibbs energy for charging the ions.

The work of charging an ion in a vacuum is obtained by setting $\varepsilon=\varepsilon_0$, the vacuum permittivity. The corresponding value for charging the ion in a medium is obtained by setting $\varepsilon=\varepsilon_r\varepsilon_0$, where ε_r is the relative permittivity of the medium. It follows that the change in molar Gibbs energy that accompanies the transfer of ions from a vacuum to a solvent is the difference of these two quantities:

$$\Delta_{\text{solv}}G^{\ominus}=\frac{z_i^2e^2N_A}{8\pi\varepsilon r_i}-\frac{z_i^2e^2N_A}{8\pi\varepsilon_0 r_i}=\frac{z_i^2e^2N_A}{8\pi\varepsilon_r\varepsilon_0 r_i}-\frac{z_i^2e^2N_A}{8\pi\varepsilon_0 r_i}=-\frac{z_i^2e^2N_A}{8\pi\varepsilon_0 r_i}\left(1-\frac{1}{\varepsilon_r}\right)$$

which is eqn 3.45.

Further information 3.2 *The fugacity*

At various stages in the development of physical chemistry it is necessary to switch from a consideration of idealized systems to real systems. In many cases it is desirable to preserve the form of the expressions that have been derived for an idealized system. Then deviations from the idealized behaviour can be expressed most simply. For instance, the pressure dependence of the molar Gibbs energy of a real gas might resemble that shown in Fig. 3.25. To adapt eqn 3.60 to this case, we replace the true pressure, p, by an effective pressure, called the **fugacity**,[9] f, and write

$$G_m=G_m^{\ominus}+RT\ln\frac{f}{p^{\ominus}} \qquad [3.61]$$

The fugacity, a function of the pressure and temperature, is defined so that this relation is exactly true. Although thermodynamic expressions in terms of fugacities derived from this expression are exact, they are useful only if we know how to interpret fugacities in terms of actual pressures. To develop this relation we write the fugacity as

$$f=\phi p \qquad [3.62]$$

where ϕ is the dimensionless **fugacity coefficient**, which in general depends on the temperature, the pressure, and the identity of the gas.

Equation 3.57b is true for all gases whether real or perfect. Expressing it in terms of the fugacity by using eqn 3.61 turns it into

$$\int_{p'}^{p}V_m\mathrm{d}p=G_m(p)-G_m(p')=\left\{G_m^{\ominus}+RT\ln\frac{f}{p^{\ominus}}\right\}-\left\{G_m^{\ominus}+RT\ln\frac{f'}{p^{\ominus}}\right\}$$

In this expression, f is the fugacity when the pressure is p and f' is the fugacity when the pressure is p'. If the gas were perfect, we would write

$$\int_{p'}^{p}V_{\text{perfect,m}}\mathrm{d}p=RT\int_{p'}^{p}\frac{1}{p}\mathrm{d}p=RT\ln\frac{p}{p'}$$

The difference between the two equations is

$$\int_{p'}^{p}(V_m-V_{\text{perfect,m}})\mathrm{d}p=RT\left(\ln\frac{f}{f'}-\ln\frac{p}{p'}\right)=RT\ln\left(\frac{(f/f')}{(p/p')}\right)$$

which can be rearranged into

$$\ln\left(\frac{f}{p}\times\frac{p'}{f'}\right)=\frac{1}{RT}\int_{p'}^{p}(V_m-V_{\text{perfect,m}})\mathrm{d}p$$

When $p'\to 0$, the gas behaves perfectly and f' becomes equal to the pressure, p'. Therefore, $f'/p'\to 1$ as $p'\to 0$. If we take this limit, which means setting $f'/p'=1$ on the left and $p'=0$ on the right, the last equation becomes

$$\ln\frac{f}{p}=\frac{1}{RT}\int_0^p(V_m-V_{\text{perfect,m}})\mathrm{d}p$$

Then, with $\phi=f/p$,

$$\ln\phi=\frac{1}{RT}\int_0^p(V_m-V_{\text{perfect,m}})\mathrm{d}p$$

For a perfect gas, $V_{\text{perfect,m}}=RT/p$. For a real gas, $V_m=RTZ/p$, where Z is the compression factor of the gas (Section 1.3a). With these two substitutions, we obtain

$$\ln\phi=\int_0^p\frac{Z-1}{p}\mathrm{d}p \qquad (3.63)$$

Provided we know how Z varies with pressure up to the pressure of interest, this expression enables us to determine the fugacity coefficient and hence, through eqn 3.62, to relate the fugacity to the pressure of the gas.

We see from Fig. 1.14 that for most gases $Z<1$ up to moderate pressures, but that $Z>1$ at higher pressures. If $Z<1$ throughout the range of integration, then the integrand in eqn 3.63 is negative and $\phi<1$. This value implies that $f<p$ (the molecules tend to stick

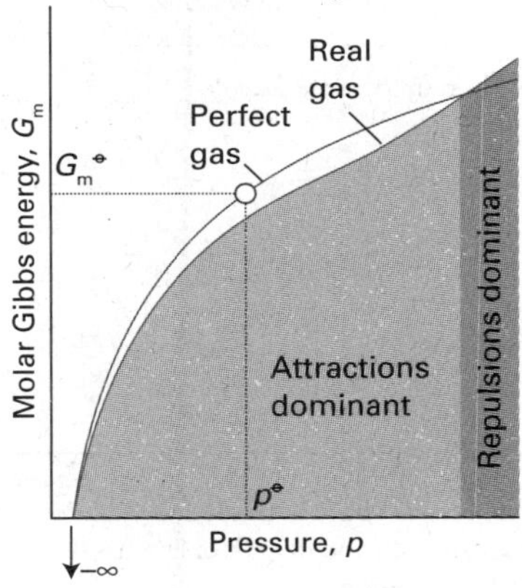

Fig. 3.25 The molar Gibbs energy of a real gas. As $p\to 0$, the molar Gibbs energy coincides with the value for a perfect gas (shown by the black line). When attractive forces are dominant (at intermediate pressures), the molar Gibbs energy is less than that of a perfect gas and the molecules have a lower 'escaping tendency'. At high pressures, when repulsive forces are dominant, the molar Gibbs energy of a real gas is greater than that of a perfect gas. Then the 'escaping tendency' is increased.

9 The name 'fugacity' comes from the Latin for 'fleetness' in the sense of 'escaping tendency'; fugacity has the same dimensions as pressure.

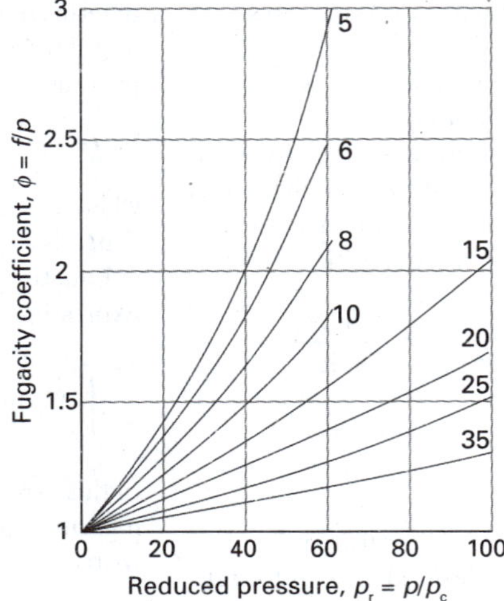

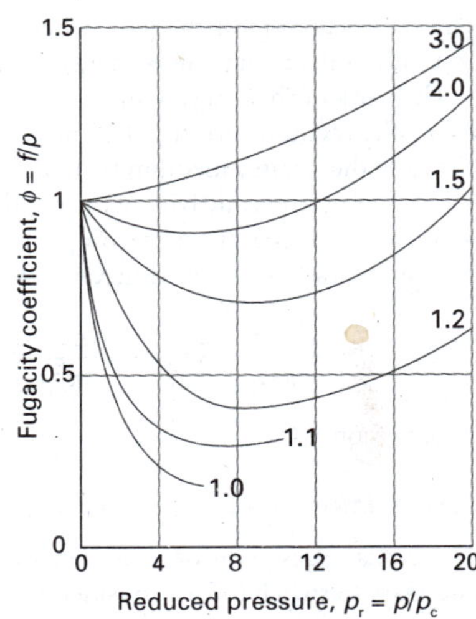

Fig. 3.26 The fugacity coefficient of a van der Waals gas plotted using the reduced variables of the gas. The curves are labelled with the reduced temperature $T_r = T/T_c$.

interActivity Evaluate the fugacity coefficient as a function of the reduced volume of a van der Waals gas and plot the outcome for a selection of reduced temperatures over the range $0.8 \le V_r \le 3$.

together) and that the molar Gibbs energy of the gas is less than that of a perfect gas. At higher pressures, the range over which $Z > 1$ may dominate the range over which $Z < 1$. The integral is then positive, $\phi > 1$, and $f > p$ (the repulsive interactions are dominant and tend to drive the particles apart). Now the molar Gibbs energy of the gas is greater than that of the perfect gas at the same pressure.

Figure 3.26, which has been calculated using the full van der Waals equation of state, shows how the fugacity coefficient depends on the pressure in terms of the reduced variables (Section 1.4). Because critical constants are available in Table 1.5, the graphs can be used for quick estimates of the fugacities of a wide range of gases. Table 3.6 gives some explicit values for nitrogen.

Table 3.6* The fugacity of nitrogen at 273 K

p/atm	f/atm
1	0.999 55
10	9.9560
100	97.03
1000	1839

* More values are given in the *Data section*.

Discussion questions

3.1 The evolution of life requires the organization of a very large number of molecules into biological cells. Does the formation of living organisms violate the Second Law of thermodynamics? State your conclusion clearly and present detailed arguments to support it.

3.2 You received an unsolicited proposal from a self-declared inventor who is seeking investors for the development of his latest idea: a device that uses heat extracted from the ground by a heat pump to boil water into steam that is used to heat a home and to power a steam engine that drives the heat pump. This procedure is potentially very lucrative because, after an initial extraction of energy from the ground, no fossil fuels would be required to keep the device running indefinitely. Would you invest in this idea? State your conclusion clearly and present detailed arguments to support it.

3.3 The following expressions have been used to establish criteria for spontaneous change: $\Delta S_{tot} > 0$, $dS_{U,V} \ge 0$ and $dU_{S,V} \le 0$, $dA_{T,V} \le 0$, and $dG_{T,p} \le 0$. Discuss the origin, significance, and applicability of each criterion.

3.4 The following expressions have been used to establish criteria for spontaneous change: $dA_{T,V} < 0$ and $dG_{T,p} < 0$. Discuss the origin, significance, and applicability of each criterion.

3.5 Discuss the physical interpretation of any one Maxwell relation.

3.6 Account for the dependence of π_T of a van der Waals gas in terms of the significance of the parameters a and b.

3.7 Suggest a physical interpretation of the dependence of the Gibbs energy on the pressure.

3.8 Suggest a physical interpretation of the dependence of the Gibbs energy on the temperature.

Exercises

Assume that all gases are perfect and that data refer to 298.15 K unless otherwise stated.

3.1(a) Calculate the change in entropy when 25 kJ of energy is transferred reversibly and isothermally as heat to a large block of iron at (a) 0°C, (b) 100°C.

3.1(b) Calculate the change in entropy when 50 kJ of energy is transferred reversibly and isothermally as heat to a large block of copper at (a) 0°C, (b) 70°C.

3.2(a) Calculate the molar entropy of a constant-volume sample of neon at 500 K given that it is 146.22 J K^{-1} mol^{-1} at 298 K.

3.2(b) Calculate the molar entropy of a constant-volume sample of argon at 250 K given that it is 154.84 J K^{-1} mol^{-1} at 298 K.

3.3(a) Calculate ΔS (for the system) when the state of 3.00 mol of perfect gas atoms, for which $C_{p,m} = \frac{5}{2}R$, is changed from 25°C and 1.00 atm to 125°C and 5.00 atm. How do you rationalize the sign of ΔS?

3.3(b) Calculate ΔS (for the system) when the state of 2.00 mol diatomic perfect gas molecules, for which $C_{p,m} = \frac{7}{2}R$, is changed from 25°C and 1.50 atm to 135°C and 7.00 atm. How do you rationalize the sign of ΔS?

3.4(a) A sample consisting of 3.00 mol of diatomic perfect gas molecules at 200 K is compressed reversibly and adiabatically until its temperature reaches 250 K. Given that $C_{V,m} = 27.5$ J K^{-1} mol^{-1}, calculate q, w, ΔU, ΔH, and ΔS.

3.4(b) A sample consisting of 2.00 mol of diatomic perfect gas molecules at 250 K is compressed reversibly and adiabatically until its temperature reaches 300 K. Given that $C_{V,m} = 27.5$ J K^{-1} mol^{-1}, calculate q, w, ΔU, ΔH, and ΔS.

3.5(a) Calculate ΔH and ΔS_{tot} when two copper blocks, each of mass 10.0 kg, one at 100°C and the other at 0°C, are placed in contact in an isolated container. The specific heat capacity of copper is 0.385 J K^{-1} g^{-1} and may be assumed constant over the temperature range involved.

3.5(b) Calculate ΔH and ΔS_{tot} when two iron blocks, each of mass 1.00 kg, one at 200°C and the other at 25°C, are placed in contact in an isolated container. The specific heat capacity of iron is 0.449 J K^{-1} g^{-1} and may be assumed constant over the temperature range involved.

3.6(a) Consider a system consisting of 2.0 mol CO_2(g), initially at 25°C and 10 atm and confined to a cylinder of cross-section 10.0 cm^2. It is allowed to expand adiabatically against an external pressure of 1.0 atm until the piston has moved outwards through 20 cm. Assume that carbon dioxide may be considered a perfect gas with $C_{V,m} = 28.8$ J K^{-1} mol^{-1} and calculate (a) q, (b) w, (c) ΔU, (d) ΔT, (e) ΔS.

3.6(b) Consider a system consisting of 1.5 mol CO_2(g), initially at 15°C and 9.0 atm and confined to a cylinder of cross-section 100.0 cm^2. The sample is allowed to expand adiabatically against an external pressure of 1.5 atm until the piston has moved outwards through 15 cm. Assume that carbon dioxide may be considered a perfect gas with $C_{V,m} = 28.8$ J K^{-1} mol^{-1}, and calculate (a) q, (b) w, (c) ΔU, (d) ΔT, (e) ΔS.

3.7(a) The enthalpy of vaporization of chloroform ($CHCl_3$) is 29.4 kJ mol^{-1} at its normal boiling point of 334.88 K. Calculate (a) the entropy of vaporization of chloroform at this temperature and (b) the entropy change of the surroundings.

3.7(b) The enthalpy of vaporization of methanol is 35.27 kJ mol^{-1} at its normal boiling point of 64.1°C. Calculate (a) the entropy of vaporization of methanol at this temperature and (b) the entropy change of the surroundings.

3.8(a) Calculate the standard reaction entropy at 298 K of

(a) $2\,CH_3CHO(g) + O_2(g) \rightarrow 2\,CH_3COOH(l)$
(b) $2\,AgCl(s) + Br_2(l) \rightarrow 2\,AgBr(s) + Cl_2(g)$
(c) $Hg(l) + Cl_2(g) \rightarrow HgCl_2(s)$

3.8(b) Calculate the standard reaction entropy at 298 K of

(a) $Zn(s) + Cu^{2+}(aq) \rightarrow Zn^{2+}(aq) + Cu(s)$
(b) $C_{12}H_{22}O_{11}(s) + 12\,O_2(g) \rightarrow 12\,CO_2(g) + 11\,H_2O(l)$

3.9(a) Combine the reaction entropies calculated in Exercise 3.8a with the reaction enthalpies, and calculate the standard reaction Gibbs energies at 298 K.

3.9(b) Combine the reaction entropies calculated in Exercise 3.8b with the reaction enthalpies, and calculate the standard reaction Gibbs energies at 298 K.

3.10(a) Use standard Gibbs energies of formation to calculate the standard reaction Gibbs energies at 298 K of the reactions in Exercise 3.8a.

3.10(b) Use standard Gibbs energies of formation to calculate the standard reaction Gibbs energies at 298 K of the reactions in Exercise 3.8b.

3.11(a) Calculate the standard Gibbs energy of the reaction $4\,HCl(g) + O_2(g) \rightarrow 2\,Cl_2(g) + 2\,H_2O(l)$ at 298 K, from the standard entropies and enthalpies of formation given in the *Data section*.

3.11(b) Calculate the standard Gibbs energy of the reaction $CO(g) + CH_3OH(l) \rightarrow CH_3COOH(l)$ at 298 K, from the standard entropies and enthalpies of formation given in the *Data section*.

3.12(a) The standard enthalpy of combustion of solid phenol (C_6H_5OH) is −3054 kJ mol^{-1} at 298 K and its standard molar entropy is 144.0 J K^{-1} mol^{-1}. Calculate the standard Gibbs energy of formation of phenol at 298 K.

3.12(b) The standard enthalpy of combustion of solid urea ($CO(NH_2)_2$) is −632 kJ mol^{-1} at 298 K and its standard molar entropy is 104.60 J K^{-1} mol^{-1}. Calculate the standard Gibbs energy of formation of urea at 298 K.

3.13(a) Calculate the change in the entropies of the system and the surroundings, and the total change in entropy, when a sample of nitrogen gas of mass 14 g at 298 K and 1.00 bar doubles its volume in (a) an isothermal reversible expansion, (b) an isothermal irreversible expansion against $p_{ex} = 0$, and (c) an adiabatic reversible expansion.

3.13(b) Calculate the change in the entropies of the system and the surroundings, and the total change in entropy, when the volume of a sample of argon gas of mass 21 g at 298 K and 1.50 bar increases from 1.20 dm^3 to 4.60 dm^3 in (a) an isothermal reversible expansion, (b) an isothermal irreversible expansion against $p_{ex} = 0$, and (c) an adiabatic reversible expansion.

3.14(a) Calculate the maximum non-expansion work per mole that may be obtained from a fuel cell in which the chemical reaction is the combustion of methane at 298 K.

3.14(b) Calculate the maximum non-expansion work per mole that may be obtained from a fuel cell in which the chemical reaction is the combustion of propane at 298 K.

3.15(a) (a) Calculate the Carnot efficiency of a primitive steam engine operating on steam at 100°C and discharging at 60°C. (b) Repeat the calculation for a modern steam turbine that operates with steam at 300°C and discharges at 80°C.

3.15(b) A certain heat engine operates between 1000 K and 500 K. (a) What is the maximum efficiency of the engine? (b) Calculate the maximum work that can be done by for each 1.0 kJ of heat supplied by the hot source. (c) How much heat is discharged into the cold sink in a reversible process for each 1.0 kJ supplied by the hot source?

3.16(a) Suppose that 3.0 mmol N_2(g) occupies 36 cm^3 at 300 K and expands to 60 cm^3. Calculate ΔG for the process.

3.16(b) Suppose that 2.5 mmol Ar(g) occupies 72 dm^3 at 298 K and expands to 100 dm^3. Calculate ΔG for the process.

3.17(a) The change in the Gibbs energy of a certain constant-pressure process was found to fit the expression $\Delta G/J = -85.40 + 36.5(T/K)$. Calculate the value of ΔS for the process.

3.17(b) The change in the Gibbs energy of a certain constant-pressure process was found to fit the expression $\Delta G/J = -73.1 + 42.8(T/K)$. Calculate the value of ΔS for the process.

3.18(a) Calculate the change in Gibbs energy of 35 g of ethanol (mass density 0.789 g cm^{-3}) when the pressure is increased isothermally from 1 atm to 3000 atm.

3.18(b) Calculate the change in Gibbs energy of 25 g of methanol (mass density 0.791 g cm^{-3}) when the pressure is increased isothermally from 100 kPa to 100 MPa. Take $k_T = 1.26 \times 10^{-9}$ Pa^{-1}.

3.19(a) Calculate the change in chemical potential of a perfect gas when its pressure is increased isothermally from 1.8 atm to 29.5 atm at 40°C.

3.19(b) Calculate the change in chemical potential of a perfect gas that its pressure is increased isothermally from 92.0 kPa to 252.0 kPa at 50°C.

3.20(a) The fugacity coefficient of a certain gas at 200 K and 50 bar is 0.72. Calculate the difference of its molar Gibbs energy from that of a perfect gas in the same state.

3.20(b) The fugacity coefficient of a certain gas at 290 K and 2.1 MPa is 0.68. Calculate the difference of its molar Gibbs energy from that of a perfect gas in the same state.

3.21(a) Estimate the change in the Gibbs energy of 1.0 dm^3 of benzene when the pressure acting on it is increased from 1.0 atm to 100 atm.

3.21(b) Estimate the change in the Gibbs energy of 1.0 dm^3 of water when the pressure acting on it is increased from 100 kPa to 300 kPa.

3.22(a) Calculate the change in the molar Gibbs energy of hydrogen gas when its pressure is increased isothermally from 1.0 atm to 100.0 atm at 298 K.

3.22(b) Calculate the change in the molar Gibbs energy of oxygen when its pressure is increased isothermally from 50.0 kPa to 100.0 kPa at 500 K.

Problems*

Assume that all gases are perfect and that data refer to 298 K unless otherwise stated.

Numerical problems

3.1 Calculate the difference in molar entropy (a) between liquid water and ice at −5°C, (b) between liquid water and its vapour at 95°C and 1.00 atm. The differences in heat capacities on melting and on vaporization are 37.3 J K^{-1} mol^{-1} and −41.9 J K^{-1} mol^{-1}, respectively. Distinguish between the entropy changes of the sample, the surroundings, and the total system, and discuss the spontaneity of the transitions at the two temperatures.

3.2 The heat capacity of chloroform (trichloromethane, $CHCl_3$) in the range 240 K to 330 K is given by $C_{p,m}/(J\ K^{-1}\ mol^{-1}) = 91.47 + 7.5 \times 10^{-2}\ (T/K)$. In a particular experiment, 1.00 mol $CHCl_3$ is heated from 273 K to 300 K. Calculate the change in molar entropy of the sample.

3.3 A block of copper of mass 2.00 kg ($C_{p,m} = 24.44$ J K^{-1} mol^{-1}) and temperature 0°C is introduced into an insulated container in which there is 1.00 mol H_2O(g) at 100°C and 1.00 atm. (a) Assuming all the steam is condensed to water, what will be the final temperature of the system, the heat transferred from water to copper, and the entropy change of the water, copper, and the total system? (b) In fact, some water vapour is present at equilibrium. From the vapour pressure of water at the temperature calculated in (a), and assuming that the heat capacities of both gaseous and liquid water are constant and given by their values at that temperature, obtain an improved value of the final temperature, the heat transferred, and the various entropies. (*Hint.* You will need to make plausible approximations.)

3.4 Consider a perfect gas contained in a cylinder and separated by a frictionless adiabatic piston into two sections A and B. All changes in B are isothermal, that is, a thermostat surrounds B to keep its temperature constant. There is 2.00 mol of the gas in each section. Initially $T_A = T_B = 300$ K, $V_A = V_B = 2.00$ dm^3. Energy is supplied as heat to Section A and the piston moves to the right reversibly until the final volume of Section B is 1.00 dm^3. Calculate (a) ΔS_A and ΔS_B, (b) ΔA_A and ΔA_B, (c) ΔG_A and ΔG_B, (d) ΔS of the total system and its surroundings. If numerical values cannot be obtained, indicate whether the values should be positive, negative, or zero or are indeterminate from the information given. (Assume $C_{V,m} = 20$ J K^{-1} mol^{-1}.)

3.5 A Carnot cycle uses 1.00 mol of a monatomic perfect gas as the working substance from an initial state of 10.0 atm and 600 K. It expands isothermally to a pressure of 1.00 atm (Step 1), and then adiabatically to a temperature of 300 K (Step 2). This expansion is followed by an isothermal compression (Step 3), and then an adiabatic compression (Step 4) back to the initial state. Determine the values of q, w, ΔU, ΔH, ΔS, ΔS_{tot}, and ΔG for each stage of the cycle and for the cycle as a whole. Express your answer as a table of values.

3.6 1.00 mol of perfect gas molecules at 27°C is expanded isothermally from an initial pressure of 3.00 atm to a final pressure of 1.00 atm in two ways: (a) reversibly, and (b) against a constant external pressure of 1.00 atm. Determine the values of q, w, ΔU, ΔH, ΔS, ΔS_{sur}, ΔS_{tot} for each path.

3.7 The standard molar entropy of NH_3(g) is 192.45 J K^{-1} mol^{-1} at 298 K, and its heat capacity is given by eqn 2.25 with the coefficients given in Table 2.2. Calculate the standard molar entropy at (a) 100°C and (b) 500°C.

3.8 A block of copper of mass 500 g and initially at 293 K is in thermal contact with an electric heater of resistance 1.00 kΩ and negligible mass. A current of 1.00 A is passed for 15.0 s. Calculate the change in entropy of the copper, taking $C_{p,m} = 24.4$ J K^{-1} mol^{-1}. The experiment is then repeated with the copper immersed in a stream of water that maintains its temperature at 293 K. Calculate the change in entropy of the copper and the water in this case.

3.9 Find an expression for the change in entropy when two blocks of the same substance and of equal mass, one at the temperature T_h and the other at T_c, are brought into thermal contact and allowed to reach equilibrium. Evaluate the

* Problems denoted with the symbol ‡ were supplied by Charles Trapp, Carmen Giunta, and Marshall Cady.

change for two blocks of copper, each of mass 500 g, with $C_{p,m} = 24.4$ J K^{-1} mol^{-1}, taking $T_h = 500$ K and $T_c = 250$ K.

3.10 A gaseous sample consisting of 1.00 mol molecules is described by the equation of state $pV_m = RT(1 + Bp)$. Initially at 373 K, it undergoes Joule–Thomson expansion from 100 atm to 1.00 atm. Given that $C_{p,m} = \frac{5}{2}R$, $\mu = 0.21$ K atm^{-1}, $B = -0.525(\mathrm{K}/T)$ atm^{-1}, and that these are constant over the temperature range involved, calculate ΔT and ΔS for the gas.

3.11 The molar heat capacity of lead varies with temperature as follows:

T/K	10	15	20	25	30	50
$C_{p,m}$/(J K^{-1} mol^{-1})	2.8	7.0	10.8	14.1	16.5	21.4
T/K	70	100	150	200	250	298
$C_{p,m}$/(J K^{-1} mol^{-1})	23.3	24.5	25.3	25.8	26.2	26.6

Calculate the standard Third-Law entropy of lead at (a) 0°C and (b) 25°C.

3.12 From standard enthalpies of formation, standard entropies, and standard heat capacities available from tables in the *Data section*, calculate the standard enthalpies and entropies at 298 K and 398 K for the reaction $CO_2(g) + H_2(g) \rightarrow CO(g) + H_2O(g)$. Assume that the heat capacities are constant over the temperature range involved.

3.13 The heat capacity of anhydrous potassium hexacyanoferrate(II) varies with temperature as follows:

T/K	$C_{p,m}$/(J K^{-1} mol^{-1})	T/K	$C_{p,m}$/(J K^{-1} mol^{-1})
10	2.09	100	179.6
20	14.43	110	192.8
30	36.44	150	237.6
40	62.55	160	247.3
50	87.03	170	256.5
60	111.0	180	265.1
70	131.4	190	273.0
80	149.4	200	280.3
90	165.3		

Calculate the molar enthalpy relative to its value at $T = 0$ and the Third-Law entropy at each of these temperatures.

3.14 The compound 1,3,5-trichloro-2,4,6-trifluorobenzene is an intermediate in the conversion of hexachlorobenzene to hexafluorobenzene, and its thermodynamic properties have been examined by measuring its heat capacity over a wide temperature range (R.L. Andon and J.F. Martin, *J. Chem. Soc. Faraday Trans. I.* 871 (1973)). Some of the data are as follows:

T/K	14.14	16.33	20.03	31.15	44.08	64.81
$C_{p,m}$/(J K^{-1} mol^{-1})	9.492	12.70	18.18	32.54	46.86	66.36
T/K	100.90	140.86	183.59	225.10	262.99	298.06
$C_{p,m}$/(J K^{-1} mol^{-1})	95.05	121.3	144.4	163.7	180.2	196.4

Calculate the molar enthalpy relative to its value at $T = 0$ and the Third-Law molar entropy of the compound at these temperatures.

3.15‡ Given that $S_m^{\ominus} = 29.79$ JK^{-1} mol^{-1} for bismuth at 100 K and the following tabulated heat capacities data (D.G. Archer, *J. Chem. Eng. Data* **40**, 1015 (1995)), compute the standard molar entropy of bismuth at 200 K.

T/K	100	120	140	150	160	180	200
$C_{p,m}$/(J K^{-1} mol^{-1})	23.00	23.74	24.25	24.44	24.61	24.89	25.11

Compare the value to the value that would be obtained by taking the heat capacity to be constant at 24.44 J K^{-1} mol^{-1} over this range.

3.16 Calculate $\Delta_r G^{\ominus}(375\ \mathrm{K})$ for the reaction $2\,CO(g) + O_2(g) \rightarrow 2\,CO_2(g)$ from the value of $\Delta_r G^{\ominus}(298\ \mathrm{K})$, $\Delta_r H^{\ominus}(298\ \mathrm{K})$, and the Gibbs–Helmholtz equation.

3.17 Estimate the standard reaction Gibbs energy of $N_2(g) + 3\,H_2(g) \rightarrow 2\,NH_3(g)$ at (a) 500 K, (b) 1000 K from their values at 298 K.

3.18 At 200 K, the compression factor of oxygen varies with pressure as shown below. Evaluate the fugacity of oxygen at this temperature and 100 atm.

p/atm	1.0000	4.00000	7.00000	10.0000	40.00	70.00	100.0
Z	0.9971	0.98796	0.97880	0.96956	0.8734	0.7764	0.6871

Theoretical problems

3.19 Represent the Carnot cycle on a temperature–entropy diagram and show that the area enclosed by the cycle is equal to the work done.

3.20 Prove that two reversible adiabatic paths can never cross. Assume that the energy of the system under consideration is a function of temperature only. (*Hint.* Suppose that two such paths can intersect, and complete a cycle with the two paths plus one isothermal path. Consider the changes accompanying each stage of the cycle and show that they conflict with the Kelvin statement of the Second Law.)

3.21 Prove that the perfect gas temperature scale and the thermodynamic temperature scale based on the Second Law of thermodynamics differ from each other by at most a constant numerical factor.

3.22 The molar Gibbs energy of a certain gas is given by $G_m = RT \ln p + A + Bp + \frac{1}{2}Cp^2 + \frac{1}{3}Dp^3$, where A, B, C, and D are constants. Obtain the equation of state of the gas.

3.23 Evaluate $(\partial S/\partial V)_T$ for (a) a van der Waals gas, (b) a Dieterici gas (Table 1.7). For an isothermal expansion, for which kind of gas (and a perfect gas) will ΔS be greatest? Explain your conclusion.

3.24 Show that, for a perfect gas, $(\partial U/\partial S)_V = T$ and $(\partial U/\partial V)_S = -p$.

3.25 Two of the four Maxwell relations were derived in the text, but two were not. Complete their derivation by showing that $(\partial S/\partial V)_T = (\partial p/\partial T)_V$ and $(\partial T/\partial p)_S = (\partial V/\partial S)_p$.

3.26 Use the Maxwell relations to express the derivatives (a) $(\partial S/\partial V)_T$ and $(\partial V/\partial S)_p$ and (b) $(\partial p/\partial S)_V$ and $(\partial V/\partial S)_p$ in terms of the heat capacities, the expansion coefficient α, and the isothermal compressibility, κ_T.

3.27 Use the Maxwell relations to show that the entropy of a perfect gas depends on the volume as $S \propto R \ln V$.

3.28 Derive the thermodynamic equation of state

$$\left(\frac{\partial H}{\partial p}\right)_T = V - T\left(\frac{\partial V}{\partial T}\right)_p$$

Derive an expression for $(\partial H/\partial p)_T$ for (a) a perfect gas and (b) a van der Waals gas. In the latter case, estimate its value for 1.0 mol Ar(g) at 298 K and 10 atm. By how much does the enthalpy of the argon change when the pressure is increased isothermally to 11 atm?

3.29 Show that, if $B(T)$ is the second virial coefficient of a gas, and $\Delta B = B(T'') - B(T')$, $\Delta T = T'' - T'$, and T is the mean of T'' and T', then $\pi_T \approx RT^2\Delta B/V_m^2\Delta T$. Estimate π_T for argon given that $B(250\ \mathrm{K}) = -28.0$ cm^3 mol^{-1} and $B(300\ \mathrm{K}) = -15.6$ cm^3 mol^{-1} at 275 K at (a) 1.0 atm, (b) 10.0 atm.

3.30 The Joule coefficient, μ_J, is defined as $\mu_J = (\partial T/\partial V)_U$. Show that $\mu_J C_V = p - \alpha T/\kappa_T$.

3.31 Evaluate π_T for a Dieterici gas (Table 1.7). Justify physically the form of the expression obtained.

3.32 The adiabatic compressibility, κ_S, is defined like κ_T (eqn 2.43) but at constant entropy. Show that for a perfect gas $p\gamma\kappa_S = 1$ (where γ is the ratio of heat capacities).

3.33 Suppose that S is regarded as a function of p and T. Show that $TdS = C_p dT - \alpha TVdp$. Hence, show that the energy transferred as heat when the pressure on an incompressible liquid or solid is increased by Δp is equal to $-\alpha TV\Delta p$. Evaluate q when the pressure acting on 100 cm^3 of mercury at 0°C is increased by 1.0 kbar. ($\alpha = 1.82 \times 10^{-4}$ K^{-1}.)

3.34 Suppose that (a) the attractive interactions between gas particles can be neglected, (b) the attractive interaction is dominant in a van der Waals gas, and the pressure is low enough to make the approximation $4ap/(RT)^2 \ll 1$. Find expressions for the fugacity of a van der Waals gas in terms of the pressure and estimate its value for ammonia at 10.00 atm and 298.15 K in each case.

3.35 Find an expression for the fugacity coefficient of a gas that obeys the equation of state $pV_m = RT(1 + B/V_m + C/V_m^2)$. Use the resulting expression to estimate the fugacity of argon at 1.00 atm and 100 K using $B = -21.13$ cm^3 mol^{-1} and $C = 1054$ cm^6 mol^{-2}.

Applications: to biology, environmental science, polymer science, and engineering

3.36 The protein lysozyme unfolds at a transition temperature of 75.5°C and the standard enthalpy of transition is 509 kJ mol^{-1}. Calculate the entropy of unfolding of lysozyme at 25.0°C, given that the difference in the constant-pressure heat capacities upon unfolding is 6.28 kJ K^{-1} mol^{-1} and can be assumed to be independent of temperature. *Hint.* Imagine that the transition at 25.0°C occurs in three steps: (i) heating of the folded protein from 25.0°C to the transition temperature, (ii) unfolding at the transition temperature, and (iii) cooling of the unfolded protein to 25.0°C. Because the entropy is a state function, the entropy change at 25.0°C is equal to the sum of the entropy changes of the steps.

3.37 At 298 K the standard enthalpy of combustion of sucrose is −5797 kJ mol^{-1} and the standard Gibbs energy of the reaction is −6333 kJ mol^{-1}. Estimate the additional non-expansion work that may be obtained by raising the temperature to blood temperature, 37°C.

3.38 In biological cells, the energy released by the oxidation of foods (*Impact I2.2*) is stored in adenosine triphosphate (ATP or ATP^{4-}). The essence of ATP's action is its ability to lose its terminal phosphate group by hydrolysis and to form adenosine diphosphate (ADP or ADP^{3-}):

$$ATP^{4-}(aq) + H_2O(l) \rightarrow ADP^{3-}(aq) + HPO_4^{2-}(aq) + H_3O^+(aq)$$

At pH = 7.0 and 37°C (310 K, blood temperature) the enthalpy and Gibbs energy of hydrolysis are $\Delta_r H = -20$ kJ mol^{-1} and $\Delta_r G = -31$ kJ mol^{-1}, respectively. Under these conditions, the hydrolysis of 1 mol ATP^{4-}(aq) results in the extraction of up to 31 kJ of energy that can be used to do non-expansion work, such as the synthesis of proteins from amino acids, muscular contraction, and the activation of neuronal circuits in our brains. (a) Calculate and account for the sign of the entropy of hydrolysis of ATP at pH = 7.0 and 310 K. (b) Suppose that the radius of a typical biological cell is 10 μm and that inside it 10^6 ATP molecules are hydrolysed each second. What is the power density of the cell in watts per cubic metre (1 W = 1 J s^{-1})? A computer battery delivers about 15 W and has a volume of 100 cm^3. Which has the greater power density, the cell or the battery? (c) The formation of glutamine from glutamate and ammonium ions requires 14.2 kJ mol^{-1} of energy input. It is driven by the hydrolysis of ATP to ADP mediated by the enzyme glutamine synthetase. How many moles of ATP must be hydrolysed to form 1 mol glutamine?

3.39‡ In 1995, the Intergovernmental Panel on Climate Change (IPCC) considered a global average temperature rise of 1.0–3.5°C likely by the year 2100, with 2.0°C its best estimate. Because water vapour is itself a greenhouse gas, the increase in water vapour content of the atmosphere is of some concern to climate change experts. Predict the relative increase in water vapour in the atmosphere based on a temperature rises of 2.0 K, assuming that the relative humidity remains constant. (The present global mean temperature is 290 K, and the equilibrium vapour pressure of water at that temperature is 0.0189 bar.)

3.40‡ Nitric acid hydrates have received much attention as possible catalysts for heterogeneous reactions that bring about the Antarctic ozone hole. Worsnop *et al.* investigated the thermodynamic stability of these hydrates under conditions typical of the polar winter stratosphere (*Science* **259**, 71 (1993).). They report thermodynamic data for the sublimation of mono-, di-, and trihydrates to nitric acid and water vapours, $HNO_3 \cdot nH_2O(s) \rightarrow HNO_3(g) + nH_2O(g)$, for $n = 1$, 2, and 3. Given $\Delta_r G^\ominus$ and $\Delta_r H^\ominus$ for these reactions at 220 K, use the Gibbs–Helmholtz equation to compute $\Delta_r G^\ominus$ at 190 K.

n	1	2	3
$\Delta_r G^\ominus$/(kJ mol^{-1})	46.2	69.4	93.2
$\Delta_r H^\ominus$/(kJ mol^{-1})	127	188	237

3.41‡ J. Gao and J. H. Weiner in their study of the origin of stress on the atomic level in dense polymer systems (*Science* **266**, 748 (1994)), observe that the tensile force required to maintain the length, l, of a long linear chain of N freely jointed links each of length a, can be interpreted as arising from an entropic spring. For such a chain, $S(l) = -3kl^2/2Na^2 + C$, where k is the Boltzmann constant and C is a constant. Using thermodynamic relations of this and previous chapters, show that the tensile force obeys Hooke's law, $f = -k_f l$, if we assume that the energy U is independent of l.

3.42 Suppose that an internal combustion engine runs on octane, for which the enthalpy of combustion is −5512 kJ mol^{-1} and take the mass of 1 gallon of fuel as 3 kg. What is the maximum height, neglecting all forms of friction, to which a car of mass 1000 kg can be driven on 1.00 gallon of fuel given that the engine cylinder temperature is 2000°C and the exit temperature is 800°C?

3.43 The cycle involved in the operation of an internal combustion engine is called the *Otto cycle*. Air can be considered to be the working substance and can be assumed to be a perfect gas. The cycle consists of the following steps: (1) reversible adiabatic compression from A to B, (2) reversible constant-volume pressure increase from B to C due to the combustion of a small amount of fuel, (3) reversible adiabatic expansion from C to D, and (4) reversible and constant-volume pressure decrease back to state A. Determine the change in entropy (of the system and of the surroundings) for each step of the cycle and determine an expression for the efficiency of the cycle, assuming that the heat is supplied in Step 2. Evaluate the efficiency for a compression ratio of 10:1. Assume that in state A, $V = 4.00$ dm^3, $p = 1.00$ atm, and $T = 300$ K, that $V_A = 10V_B$, $p_C/p_B = 5$, and that $C_{p,m} = \frac{7}{2}R$.

3.44 To calculate the work required to lower the temperature of an object, we need to consider how the coefficient of performance changes with the temperature of the object. (a) Find an expression for the work of cooling an object from T_i to T_f when the refrigerator is in a room at a temperature T_h. *Hint.* Write $dw = dq/c(T)$, relate dq to dT through the heat capacity C_p, and integrate the resulting expression. Assume that the heat capacity is independent of temperature in the range of interest. (b) Use the result in part (a) to calculate the work needed to freeze 250 g of water in a refrigerator at 293 K. How long will it take when the refrigerator operates at 100 W?

3.45 The expressions that apply to the treatment of refrigerators also describe the behaviour of heat pumps, where warmth is obtained from the back of a refrigerator while its front is being used to cool the outside world. Heat pumps are popular home heating devices because they are very efficient. Compare heating of a room at 295 K by each of two methods: (a) direct conversion of 1.00 kJ of electrical energy in an electrical heater, and (b) use of 1.00 kJ of electrical energy to run a reversible heat pump with the outside at 260 K. Discuss the origin of the difference in the energy delivered to the interior of the house by the two methods.

Physical transformations of pure substances

4

The discussion of the phase transitions of pure substances is among the simplest applications of thermodynamics to chemistry. We shall see that one type of phase diagram is a map of the pressures and temperatures at which each phase of a substance is the most stable. The thermodynamic criterion of phase stability enables us to deduce a very general result, the phase rule, that summarizes the constraints on the equilibria between phases. In preparation for later chapters, we express the rule in a general way that can be applied to systems of more than one component. Then, we describe the interpretation of empirically determined phase diagrams for a selection of substances. We then consider the factors that determine the positions and shapes of the boundaries between the regions on a phase diagram. The practical importance of the expressions we derive is that they show how the vapour pressure of a substance varies with temperature and how the melting point varies with pressure. Transitions between phases are classified by noting how various thermodynamic functions change when the transition occurs. This chapter also introduces the chemical potential, a property that will be at the centre of our discussions of mixtures and chemical reactions.

Vaporization, melting (fusion), and the conversion of graphite to diamond are all examples of changes of phase without change of chemical composition. In this chapter we describe such processes thermodynamically, using as the guiding principle the tendency of systems at constant temperature and pressure to minimize their Gibbs energy.

Phase diagrams

One of the most succinct ways of presenting the physical changes of state that a substance can undergo is in terms of its 'phase diagram'. This material is also the basis of the discussion of mixtures in Chapter 5.

4.1 The stabilities of phases

Key points **(a) A phase is a form of matter that is uniform throughout in chemical composition and physical state. (b) A phase transition is the spontaneous conversion of one phase into another and may be studied by techniques that include thermal analysis. (c) The thermodynamic analysis of phases is based on the fact that, at equilibrium, the chemical potential of a substance is the same throughout a sample.**

Thermodynamics provides a powerful language for describing and understanding the stabilities and transformations of phases, but to apply it we need to employ definitions carefully.

(a) The number of phases

A **phase** is a form of matter that is uniform throughout in chemical composition and physical state. Thus, we speak of solid, liquid, and gas phases of a substance, and of its various solid phases, such as the white and black allotropes of phosphorus or the aragonite and calcite polymorphs of calcium carbonate.

A note on good practice An *allotrope* is a particular form of an element (such as O_2 and O_3) and may be solid, liquid, or gas. A *polymorph* is one of a number of solid phases of an element or compound.

The number of phases in a system is denoted P. A gas, or a gaseous mixture, is a single phase ($P = 1$), a crystal of a substance is a single phase, and two fully miscible liquids form a single phase. A solution of sodium chloride in water is a single phase. Ice is a single phase even though it might be chipped into small fragments. A slurry of ice and water is a two-phase system ($P = 2$) even though it is difficult to map the physical boundaries between the phases. A system in which calcium carbonate undergoes the thermal decomposition

$$CaCO_3(s) \rightarrow CaO(s) + CO_2(g)$$

consists of two solid phases (one consisting of calcium carbonate and the other of calcium oxide) and one gaseous phase (consisting of carbon dioxide).

Two metals form a two-phase system ($P = 2$) if they are immiscible, but a single-phase system ($P = 1$), an alloy, if they are miscible. This example shows that it is not always easy to decide whether a system consists of one phase or of two. A solution of solid B in solid A—a homogeneous mixture of the two substances—is uniform on a molecular scale. In a solution, atoms of A are surrounded by atoms of A and B, and any sample cut from the sample, even microscopically small, is representative of the composition of the whole.

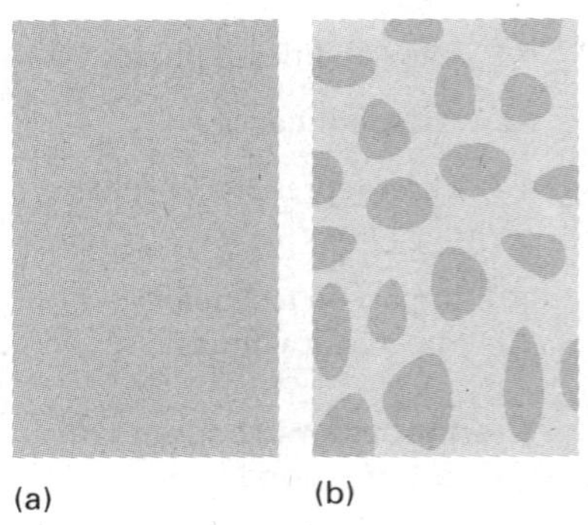

Fig. 4.1 The difference between (a) a single-phase solution, in which the composition is uniform on a microscopic scale, and (b) a dispersion, in which regions of one component are embedded in a matrix of a second component.

A dispersion is uniform on a macroscopic scale but not on a microscopic scale, for it consists of grains or droplets of one substance in a matrix of the other. A small sample could come entirely from one of the minute grains of pure A and would not be representative of the whole (Fig. 4.1). Dispersions are important because, in many advanced materials (including steels), heat treatment cycles are used to achieve the precipitation of a fine dispersion of particles of one phase (such as a carbide phase) within a matrix formed by a saturated solid solution phase. The ability to control this microstructure resulting from phase equilibria makes it possible to tailor the mechanical properties of the materials to a particular application.

(b) Phase transitions

A **phase transition**, the spontaneous conversion of one phase into another phase, occurs at a characteristic temperature for a given pressure. Thus, at 1 atm, ice is the stable phase of water below 0°C, but above 0°C liquid water is more stable. This difference indicates that below 0°C the Gibbs energy decreases as liquid water changes into ice and that above 0°C the Gibbs energy decreases as ice changes into liquid water. The **transition temperature**, T_{trs}, is the temperature at which the two phases are in equilibrium and the Gibbs energy of the system is minimized at the prevailing pressure.

Detecting a phase transition is not always as simple as seeing water boil in a kettle, so special techniques have been developed. One technique is **thermal analysis**, which takes advantage of the heat that is evolved or absorbed during any transition. The transition is detected by noting that the temperature does not change even though heat is being supplied or removed from the sample (Fig. 4.2). Differential scanning calorimetry is also used (see *Impact I2.1*). Thermal techniques are useful for solid–solid transitions, where simple visual inspection of the sample may be inadequate. X-ray diffraction (Section 19.3) also reveals the occurrence of a phase transition in a solid, for different structures are found on either side of the transition temperature.

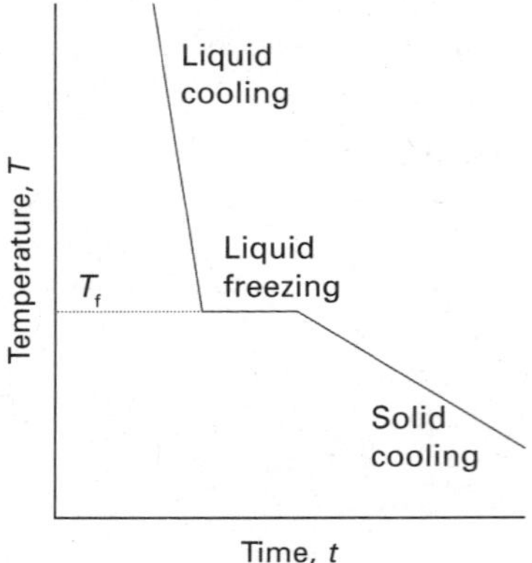

Fig. 4.2 A cooling curve at constant pressure. The halt corresponds to the pause in the fall of temperature while the first-order exothermic transition (freezing) occurs. This pause enables T_f to be located even if the transition cannot be observed visually.

As always, it is important to distinguish between the thermodynamic description of a process and the rate at which the process occurs. A phase transition that is predicted from thermodynamics to be spontaneous may occur too slowly to be significant in practice. For instance, at normal temperatures and pressures the molar Gibbs energy of graphite is lower than that of diamond, so there is a thermodynamic tendency for diamond to change into graphite. However, for this transition to take place, the C atoms must change their locations, which is an immeasurably slow process in a solid except at high temperatures. The discussion of the rate of attainment of equilibrium is a kinetic problem and is outside the range of thermodynamics. In gases and liquids the mobilities of the molecules allow phase transitions to occur rapidly, but in solids thermodynamic instability may be frozen in. Thermodynamically unstable phases that persist because the transition is kinetically hindered are called **metastable phases.** Diamond is a metastable phase of carbon under normal conditions.

(c) Thermodynamic criteria of phase stability

All our considerations will be based on the Gibbs energy of a substance, and in particular on its molar Gibbs energy, G_m. In fact, this quantity will play such an important role in this chapter and the rest of the text that we give it a special name and symbol, the **chemical potential,** μ (mu). For a one-component system, 'molar Gibbs energy' and 'chemical potential' are synonyms, so $\mu = G_m$, but in Chapter 5 we shall see that chemical potential has a broader significance and a more general definition. The name 'chemical potential' is also instructive: as we develop the concept, we shall see that μ is a measure of the potential that a substance has for undergoing change in a system. In this chapter, it reflects the potential of a substance to undergo physical change. In Chapter 6 we shall see that μ is the potential of a substance to undergo chemical change.

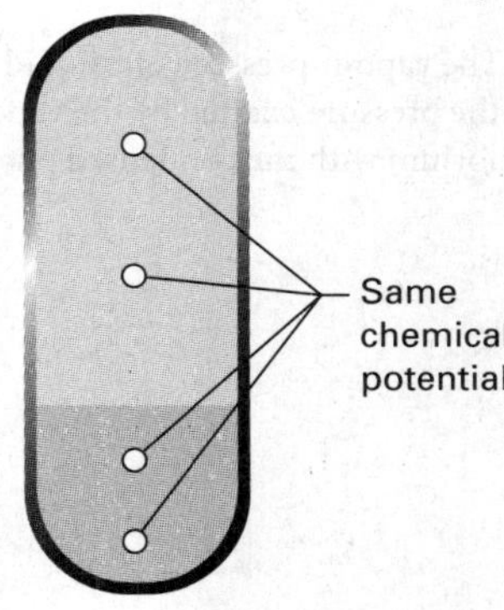

Fig. 4.3 When two or more phases are in equilibrium, the chemical potential of a substance (and, in a mixture, a component) is the same in each phase and is the same at all points in each phase.

We base the entire discussion on the following consequence of the Second Law (Fig. 4.3):

> At equilibrium, the chemical potential of a substance is the same throughout a sample, regardless of how many phases are present.

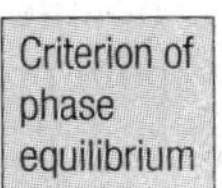

To see the validity of this remark, consider a system in which the chemical potential of a substance is μ_1 at one location and μ_2 at another location. The locations may be in the same or in different phases. When an infinitesimal amount dn of the substance is transferred from one location to the other, the Gibbs energy of the system changes by $-\mu_1 dn$ when material is removed from location 1, and it changes by $+\mu_2 dn$ when that material is added to location 2. The overall change is therefore $dG = (\mu_2 - \mu_1)dn$. If the chemical potential at location 1 is higher than that at location 2, the transfer is accompanied by a decrease in G, and so has a spontaneous tendency to occur. Only if $\mu_1 = \mu_2$ is there no change in G, and only then is the system at equilibrium.

4.2 Phase boundaries

Key points (a) **A substance is characterized by a variety of parameters that can be identified on its phase diagram. (b) The phase rule relates the number of variables that may be changed while the phases of a system remain in mutual equilibrium.**

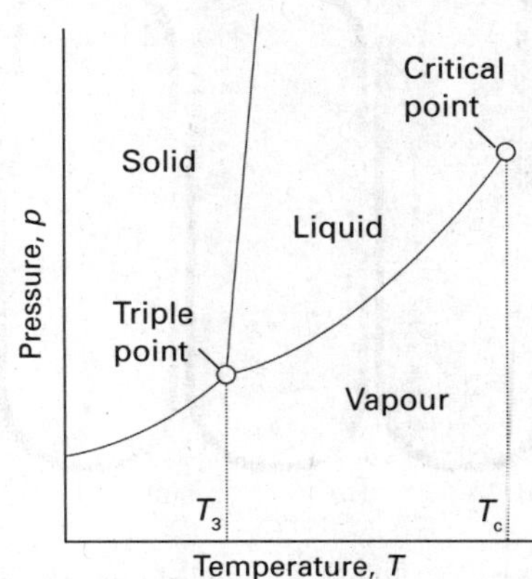

Fig. 4.4 The general regions of pressure and temperature where solid, liquid, or gas is stable (that is, has minimum molar Gibbs energy) are shown on this phase diagram. For example, the solid phase is the most stable phase at low temperatures and high pressures. In the following paragraphs we locate the precise boundaries between the regions.

The **phase diagram** of a pure substance shows the regions of pressure and temperature at which its various phases are thermodynamically stable (Fig. 4.4). In fact, any two intensive variables may be used (such as temperature and magnetic field; in Chapter 5 mole fraction is another variable), but in this chapter we concentrate on pressure and

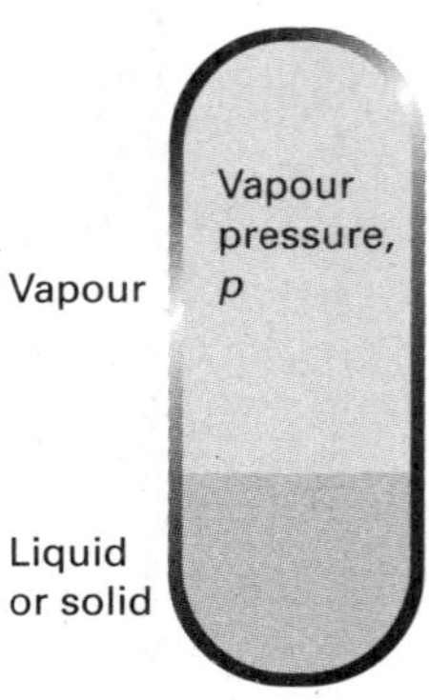

Fig. 4.5 The vapour pressure of a liquid or solid is the pressure exerted by the vapour in equilibrium with the condensed phase.

temperature. The lines separating the regions, which are called **phase boundaries** (or *coexistence curves*), show the values of p and T at which two phases coexist in equilibrium and their chemical potentials are equal.

(a) Characteristic properties related to phase transitions

Consider a liquid sample of a pure substance in a closed vessel. The pressure of a vapour in equilibrium with the liquid is called the **vapour pressure** of the substance (Fig. 4.5). Therefore, the liquid–vapour phase boundary in a phase diagram shows how the vapour pressure of the liquid varies with temperature. Similarly, the solid–vapour phase boundary shows the temperature variation of the **sublimation vapour pressure**, the vapour pressure of the solid phase. The vapour pressure of a substance increases with temperature because at higher temperatures more molecules have sufficient energy to escape from their neighbours.

When a liquid is heated in an open vessel, the liquid vaporizes from its surface. When the vapour pressure is equal to the external pressure, vaporization can occur throughout the bulk of the liquid and the vapour can expand freely into the surroundings. The condition of free vaporization throughout the liquid is called **boiling**. The temperature at which the vapour pressure of a liquid is equal to the external pressure is called the **boiling temperature** at that pressure. For the special case of an external pressure of 1 atm, the boiling temperature is called the **normal boiling point**, T_b. With the replacement of 1 atm by 1 bar as standard pressure, there is some advantage in using the **standard boiling point** instead: this is the temperature at which the vapour pressure reaches 1 bar. Because 1 bar is slightly less than 1 atm (1.00 bar = 0.987 atm), the standard boiling point of a liquid is slightly lower than its normal boiling point. The normal boiling point of water is 100.0°C; its standard boiling point is 99.6°C. We need to distinguish normal and standard properties only for precise work in thermodynamics because any thermodynamic properties that we intend to add together must refer to the same conditions.

Boiling does not occur when a liquid is heated in a rigid, closed vessel. Instead, the vapour pressure, and hence the density of the vapour, rise as the temperature is raised (Fig. 4.6). At the same time, the density of the liquid decreases slightly as a result of its expansion. There comes a stage when the density of the vapour is equal to that of the remaining liquid and the surface between the two phases disappears. The temperature at which the surface disappears is the **critical temperature**, T_c, of the substance. We first encountered this property in Section 1.3d. The vapour pressure at the critical temperature is called the **critical pressure**, p_c. At and above the critical temperature, a single uniform phase called a **supercritical fluid** fills the container and an interface no longer exists. That is, above the critical temperature, the liquid phase of the substance does not exist.

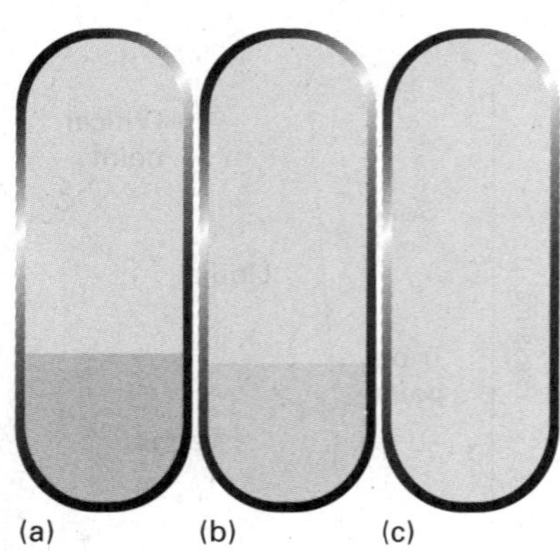

Fig. 4.6 (a) A liquid in equilibrium with its vapour. (b) When a liquid is heated in a sealed container, the density of the vapour phase increases and that of the liquid decreases slightly. There comes a stage, (c), at which the two densities are equal and the interface between the fluids disappears. This disappearance occurs at the critical temperature. The container needs to be strong: the critical temperature of water is 374°C and the vapour pressure is then 218 atm.

The temperature at which, under a specified pressure, the liquid and solid phases of a substance coexist in equilibrium is called the **melting temperature**. Because a substance melts at exactly the same temperature as it freezes, the melting temperature of a substance is the same as its **freezing temperature**. The freezing temperature when the pressure is 1 atm is called the **normal freezing point**, T_f, and its freezing point when the pressure is 1 bar is called the **standard freezing point**. The normal and standard freezing points are negligibly different for most purposes. The normal freezing point is also called the **normal melting point**.

There is a set of conditions under which three different phases of a substance (typically solid, liquid, and vapour) all simultaneously coexist in equilibrium. These conditions are represented by the **triple point**, a point at which the three phase boundaries meet. The temperature at the triple point is denoted T_3. The triple point of a pure substance is outside our control: it occurs at a single definite pressure and temperature characteristic of the substance. The triple point of water lies at 273.16 K and 611 Pa (6.11 mbar, 4.58 Torr), and the three phases of water (ice, liquid water, and water vapour) coexist in equilibrium at no other combination of pressure and temperature. This invariance of the triple point is the basis of its use in the definition of the thermodynamic temperature scale (Section 3.2d).

As we can see from Fig. 4.4, the triple point marks the lowest pressure at which a liquid phase of a substance can exist. If (as is common) the slope of the solid–liquid phase boundary is as shown in the diagram, then the triple point also marks the lowest temperature at which the liquid can exist; the critical temperature is the upper limit.

(b) The phase rule

In one of the most elegant arguments of the whole of chemical thermodynamics, which is presented in the following *Justification*, J.W. Gibbs deduced the **phase rule**, which gives the number of parameters that can be varied independently (at least to a small extent) while the number of phases in equilibrium is preserved. The phase rule is a general relation between the variance, F, the number of components, C, and the number of phases at equilibrium, P, for a system of any composition:

$$F = C - P + 2 \qquad \text{The phase rule} \qquad (4.1)$$

A **component** is a *chemically independent* constituent of a system. The number of components, C, in a system is the minimum number of types of independent species (ions or molecules) necessary to define the composition of all the phases present in the system. In this chapter we deal only with one-component systems ($C = 1$). By a **constituent** of a system we mean a chemical species that is present. Thus, a mixture of ethanol and water has two constituents. A solution of sodium chloride has three constituents—water, Na^+ ions, and Cl^- ions—but only two components because the numbers of Na^+ and Cl^- ions are constrained to be equal by the requirement of charge neutrality. The **variance** (or *number of degrees of freedom*), F, of a system is the number of intensive variables that can be changed independently without disturbing the number of phases in equilibrium.

In a single-component, single-phase system ($C = 1$, $P = 1$), the pressure and temperature may be changed independently without changing the number of phases, so $F = 2$. We say that such a system is **bivariant**, or that it has two **degrees of freedom**. On the other hand, if two phases are in equilibrium (a liquid and its vapour, for instance) in a single-component system ($C = 1$, $P = 2$), the temperature (or the pressure) can be changed at will, but the change in temperature (or pressure) demands an accompanying change in pressure (or temperature) to preserve the number of phases in equilibrium. That is, the variance of the system has fallen to 1.

Justification 4.1 *The phase rule*

Consider first the special case of a one-component system for which the phase rule is $F = 3 - P$. For two phases α and β in equilibrium ($P = 2$, $F = 1$) at a given pressure and temperature, we can write

$$\mu(\alpha; p,T) = \mu(\beta; p,T)$$

(For instance, when ice and water are in equilibrium, we have $\mu(\mathrm{s}; p,T) = \mu(\mathrm{l}; p,T)$ for H_2O.) This is an equation relating p and T, so only one of these variables is independent (just as the equation $x + y = xy$ is a relation for y in terms of x: $y = x/(x-1)$). That conclusion is consistent with $F = 1$. For three phases of a one-component system in mutual equilibrium ($P = 3$, $F = 0$),

$$\mu(\alpha; p,T) = \mu(\beta; p,T) = \mu(\gamma; p,T)$$

This relation is actually two equations for two unknowns, $\mu(\alpha; p,T) = \mu(\beta; p,T)$ and $\mu(\beta; p,T) = \mu(\gamma; p,T)$, and therefore has a solution only for a single value of p and T (just as the pair of equations $x + y = xy$ and $3x - y = xy$ has the single solution $x = 2$ and $y = 2$). That conclusion is consistent with $F = 0$. Four phases cannot be in mutual equilibrium in a one-component system because the three equalities

$$\mu(\alpha; p,T) = \mu(\beta; p,T) \qquad \mu(\beta; p,T) = \mu(\gamma; p,T) \qquad \mu(\gamma; p,T) = \mu(\delta; p,T)$$

are three equations for two unknowns (p and T) and are not consistent (just as $x + y = xy$, $3x - y = xy$, and $x + y = 2xy^2$ have no solution).

Now consider the general case. We begin by counting the total number of intensive variables. The pressure, p, and temperature, T, count as 2. We can specify the composition of a phase by giving the mole fractions of $C - 1$ components. We need specify only $C - 1$ and not all C mole fractions because $x_1 + x_2 + \cdots + x_C = 1$, and all mole fractions are known if all except one are specified. Because there are P phases, the total number of composition variables is $P(C - 1)$. At this stage, the total number of intensive variables is $P(C - 1) + 2$.

At equilibrium, the chemical potential of a component J must be the same in every phase (Section 4.4):

$$\mu(\alpha; p,T) = \mu(\beta; p,T) = \ldots \text{ for } P \text{ phases}$$

That is, there are $P - 1$ equations of this kind to be satisfied for each component J. As there are C components, the total number of equations is $C(P - 1)$. Each equation reduces our freedom to vary one of the $P(C - 1) + 2$ intensive variables. It follows that the total number of degrees of freedom is

$$F = P(C - 1) + 2 - C(P - 1) = C - P + 2$$

which is eqn 4.1.

4.3 Three representative phase diagrams

Key points **(a) Carbon dioxide is a typical substance but shows features that can be traced to its weak intermolecular forces. (b) Water shows anomalies that can be traced to its extensive hydrogen bonding. (c) Helium shows anomalies, including superfluidity, that can be traced to its low mass and weak interactions.**

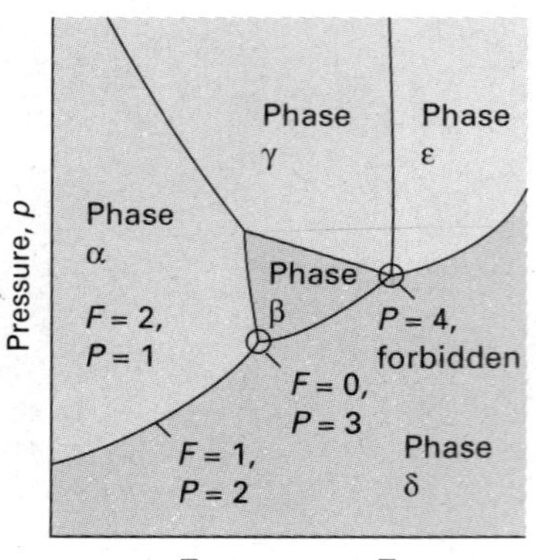

Fig. 4.7 The typical regions of a one-component phase diagram. The lines represent conditions under which the two adjoining phases are in equilibrium. A point represents the unique set of conditions under which three phases coexist in equilibrium. Four phases cannot mutually coexist in equilibrium.

For a one-component system, such as pure water, $F = 3 - P$. When only one phase is present, $F = 2$ and both p and T can be varied independently (at least over a small range) without changing the number of phases. In other words, a single phase is represented by an *area* on a phase diagram. When two phases are in equilibrium $F = 1$, which implies that pressure is not freely variable if the temperature is set; indeed, at a given temperature, a liquid has a characteristic vapour pressure. It follows that the equilibrium of two phases is represented by a *line* in the phase diagram. Instead of selecting the temperature, we could select the pressure, but having done so the two phases would be in equilibrium at a single definite temperature. Therefore, freezing (or any other phase transition) occurs at a definite temperature at a given pressure.

When three phases are in equilibrium, $F = 0$ and the system is invariant. This special condition can be established only at a definite temperature and pressure that is characteristic of the substance and outside our control. The equilibrium of three phases is therefore represented by a *point*, the triple point, on a phase diagram. Four phases cannot be in equilibrium in a one-component system because F cannot be negative. These features are summarized in Fig. 4.7 and should be kept in mind when considering the form of the phase diagrams of the three pure substances treated here.

(a) Carbon dioxide

The phase diagram for carbon dioxide is shown in Fig. 4.8. The features to notice include the positive slope (up from left to right) of the solid–liquid boundary; the direction of this line is characteristic of most substances. This slope indicates that the melting temperature of solid carbon dioxide rises as the pressure is increased. Notice also that, as the triple point lies above 1 atm, the liquid cannot exist at normal atmospheric pressures whatever the temperature. As a result, the solid sublimes when left in the open (hence the name 'dry ice'). To obtain the liquid, it is necessary to exert a pressure of at least 5.11 atm. Cylinders of carbon dioxide generally contain the liquid or compressed gas; at 25°C that implies a vapour pressure of 67 atm if both gas and liquid are present in equilibrium. When the gas squirts through the throttle it cools by the Joule–Thomson effect, so, when it emerges into a region where the pressure is only 1 atm, it condenses into a finely divided snow-like solid. That carbon dioxide gas cannot be liquefied except by applying high pressure reflects the weakness of the intermolecular forces between the nonpolar carbon dioxide molecules (Section 17.5).

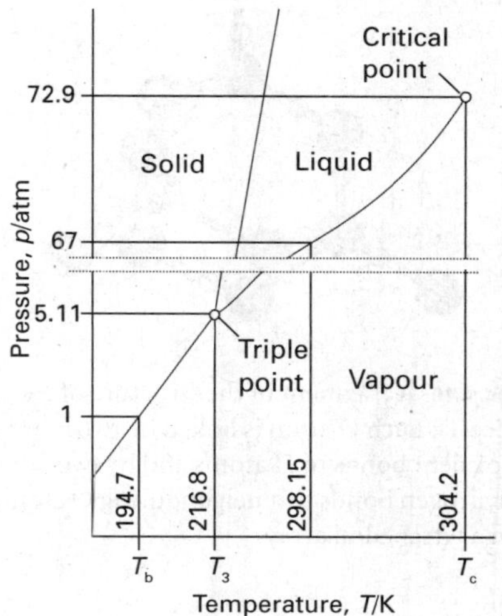

Fig. 4.8 The experimental phase diagram for carbon dioxide. Note that, as the triple point lies at pressures well above atmospheric, liquid carbon dioxide does not exist under normal conditions (a pressure of at least 5.11 atm must be applied).

(b) Water

Figure 4.9 is the phase diagram for water. The liquid–vapour boundary in the phase diagram summarizes how the vapour pressure of liquid water varies with temperature. It also summarizes how the boiling temperature varies with pressure: we simply read off the temperature at which the vapour pressure is equal to the prevailing atmospheric pressure. The solid–liquid boundary shows how the melting temperature varies with the pressure. Its very steep slope indicates that enormous pressures are needed to bring about significant changes. The line has a steep negative slope (down from left to right) up to 2 kbar, which means that the melting temperature falls as the pressure is raised. The reason for this almost unique behaviour can be traced to the decrease in volume that occurs on melting: it is more favourable for the solid to transform into the liquid as the pressure is raised. The decrease in volume is a result of the very open structure of ice: as shown in Fig. 4.10, the water molecules are held apart, as well as together, by the hydrogen bonds between them but the hydrogen-bonded structure partially collapses on melting and the liquid is denser than the solid. Other consequences of its extensive hydrogen bonding are the anomalously high boiling point of water for a molecule of its molar mass and its high critical temperature and pressure.

Figure 4.9 shows that water has one liquid phase but many different solid phases other than ordinary ice ('ice I'). Some of these phases melt at high temperatures. Ice VII, for instance, melts at 100°C but exists only above 25 kbar. Two further phases, Ice XIII and XIV, were identified in 2006 at −160°C but have not yet been allocated regions in the phase diagram. Note that many more triple points occur in the diagram other than the one where vapour, liquid, and ice I coexist. Each one occurs at a definite pressure and temperature that cannot be changed. The solid phases of ice differ in the arrangement of the water molecules: under the influence of very high pressures, hydrogen bonds buckle and the H_2O molecules adopt different arrangements. These polymorphs of ice may contribute to the advance of glaciers, for ice at the bottom of glaciers experiences very high pressures where it rests on jagged rocks.

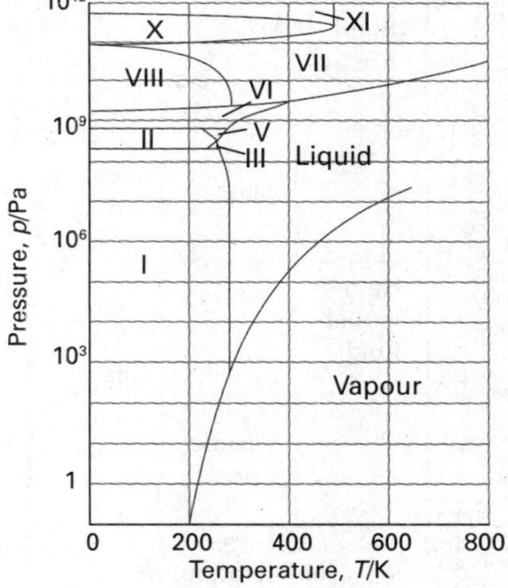

Fig. 4.9 The experimental phase diagram for water showing the different solid phases.

(c) Helium

When considering helium at low temperatures it is necessary to distinguish between the isotopes ^3He and ^4He. Figure 4.11 shows the phase diagram of helium-4. Helium behaves unusually at low temperatures because the mass of its atoms is so low and their small number of electrons results in them interacting only very weakly with their

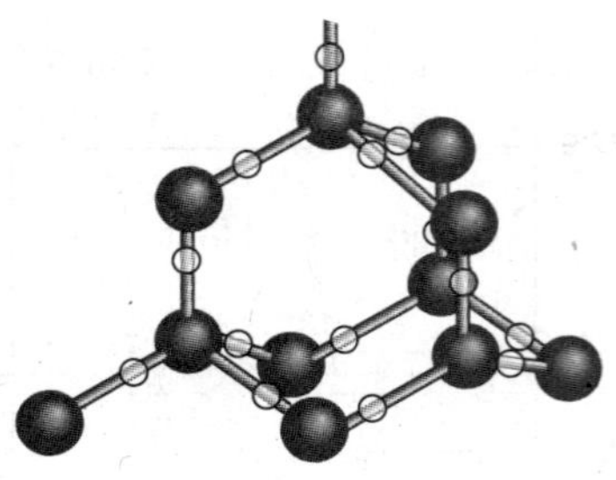

Fig. 4.10 A fragment of the structure of ice (ice-I). Each O atom is linked by two covalent bonds to H atoms and by two hydrogen bonds to a neighbouring O atom, in a tetrahedral array.

neighbours. For instance, the solid and gas phases of helium are never in equilibrium however low the temperature: the atoms are so light that they vibrate with a large-amplitude motion even at very low temperatures and the solid simply shakes itself apart. Solid helium can be obtained, but only by holding the atoms together by applying pressure. The isotopes of helium behave differently for quantum mechanical reasons that will become clear in Part 2.

A brief comment

The difference stems from the different nuclear spins of the isotopes and the role of the Pauli exclusion principle: helium-4 has $I=0$ and is a boson; helium-3 has $I=\frac{1}{2}$ and is a fermion.

Pure helium-4 has two liquid phases. The phase marked He-I in the diagram behaves like a normal liquid; the other phase, He-II, is a **superfluid**; it is so called because it flows without viscosity.[1] Provided we discount the liquid crystalline substances discussed in *Impact I5.2*, helium is the only known substance with a liquid–liquid boundary, shown as the **λ-line** (lambda line) in Fig. 4.11.

The phase diagram of helium-3 differs from the phase diagram of helium-4, but it also possesses a superfluid phase. Helium-3 is unusual in that melting is exothermic ($\Delta_{fus}H < 0$) and therefore (from $\Delta_{fus}S = \Delta_{fus}H/T_f$) at the melting point the entropy of the liquid is lower than that of the solid.

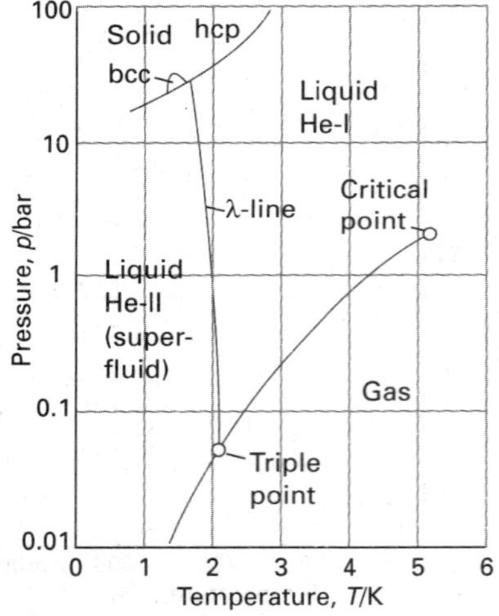

Fig. 4.11 The phase diagram for helium (^{4}He). The λ-line marks the conditions under which the two liquid phases are in equilibrium. Helium-II is the superfluid phase. Note that a pressure of over 20 bar must be exerted before solid helium can be obtained. The labels hcp and bcc denote different solid phases in which the atoms pack together differently: hcp denotes hexagonal closed packing and bcc denotes body-centred cubic (see Section 19.5 for a description of these structures).

IMPACT ON TECHNOLOGY

I4.1 Supercritical fluids

Supercritical carbon dioxide, $scCO_2$, is the centre of attention for an increasing number of solvent-based processes. The critical temperature of CO_2, 304.2 K (31.0°C), and its critical pressure, 72.9 atm, are readily accessible, it is cheap, and it can readily be recycled. The density of $scCO_2$ at its critical point is 0.45 g cm^{-3}. However, the transport properties of any supercritical fluid (its diffusion behaviour, viscosity, and thermal conductivity) depend strongly on its density, which in turn is sensitive to the pressure and temperature. For instance, densities may be adjusted from a gas-like 0.1 g cm^{-3} to a liquid-like 1.2 g cm^{-3}. A useful rule of thumb is that the solubility of a solute is an exponential function of the density of the supercritical fluid, so small increases in pressure, particularly close to the critical point, can have very large effects on solubility. Because the relative permittivity (dielectric constant) of a supercritical fluid is highly sensitive to the pressure and temperature, it is possible to run a reaction in polar and nonpolar conditions without changing the solvent, so solvent effects can be studied.

A great advantage of $scCO_2$ is that there are no noxious residues once the solvent has been allowed to evaporate, so, coupled with its low critical temperature, $scCO_2$ is ideally suited to food processing and the production of pharmaceuticals. It is used, for instance, to remove caffeine from coffee or fats from milk. The supercritical fluid is also increasingly being used for dry cleaning, which avoids the use of carcinogenic and environmentally damaging chlorinated hydrocarbons.

Supercritical CO_2 has been used since the 1960s as a mobile phase in *supercritical fluid chromatography* (SFC), but it fell out of favour when the more convenient technique of high-performance liquid chromatography (HPLC) was introduced. However, interest in SFC has returned, and there are separations possible in SFC that cannot easily be achieved by HPLC, such as the separation of lipids and of phospholipids. Samples as small as 1 pg can be analysed. The essential advantage of SFC is that diffusion coefficients in supercritical fluids are an order of magnitude greater than in liquids. As a result, there is less resistance to the transfer of solutes through the column and separations may be effected rapidly or with high resolution.

The principal problem with $scCO_2$ is that it is not a very good solvent and surfactants are needed to induce many potentially interesting solutes to dissolve. Indeed,

[1] Recent work has suggested that water may also have a superfluid liquid phase.

$scCO_2$-based dry cleaning depends on the availability of cheap surfactants; so too does the use of $scCO_2$ as a solvent for homogeneous catalysts, such as d-metal complexes. There appear to be two principal approaches to solving the solubilization problem. One solution is to use fluorinated and siloxane-based polymeric stabilizers, which allow polymerization reactions to proceed in $scCO_2$. The disadvantage of these stabilizers for commercial use is their great expense. An alternative and much cheaper approach is poly(ether-carbonate) copolymers. The copolymers can be made more soluble in $scCO_2$ by adjusting the ratio of ether and carbonate groups.

The critical temperature of water is 374°C and its pressure is 218 atm. The conditions for using scH_2O are therefore much more demanding than for $scCO_2$ and the properties of the fluid are highly sensitive to pressure. Thus, as the density of scH_2O decreases, the characteristics of a solution change from those of an aqueous solution through those of a non-aqueous solution and eventually to those of a gaseous solution. One consequence is that reaction mechanisms may change from those involving ions to those involving radicals.

Thermodynamic aspects of phase transitions

As we have seen, the thermodynamic criterion of phase equilibrium is the equality of the chemical potentials of each phase. For a one-component system, the chemical potential is the same as the molar Gibbs energy of the phase. As we already know how the Gibbs energy varies with temperature and pressure (Section 3.9); we can expect to be able to deduce how phase equilibria vary as the conditions are changed.

4.4 The dependence of stability on the conditions

Key points **(a) The chemical potential of a substance decreases with increasing temperature at a rate determined by its molar entropy. (b) The chemical potential of a substance increases with increasing pressure at a rate determined by its molar volume. (c) When pressure is applied to a condensed phase, its vapour pressure rises.**

At very low temperatures and provided the pressure is not too low, the solid phase of a substance has the lowest chemical potential and is therefore the most stable phase. However, the chemical potentials of different phases change with temperature in different ways, and above a certain temperature the chemical potential of another phase (perhaps another solid phase, a liquid, or a gas) may turn out to be the lowest. When that happens, a transition to the second phase is spontaneous and occurs if it is kinetically feasible to do so.

(a) The temperature dependence of phase stability

The temperature dependence of the Gibbs energy is expressed in terms of the entropy of the system by eqn 3.53 ($(\partial G/\partial T)_p = -S$). Because the chemical potential of a pure substance is just another name for its molar Gibbs energy, it follows that

$$\left(\frac{\partial \mu}{\partial T}\right)_p = -S_m \qquad (4.2)$$

Variation of chemical potential with T

This relation shows that, as the temperature is raised, the chemical potential of a pure substance decreases: $S_m > 0$ for all substances, so the slope of a plot of μ against T is negative.

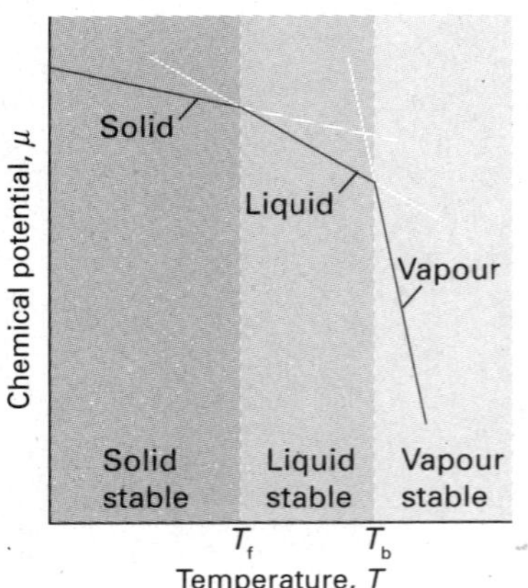

Fig. 4.12 The schematic temperature dependence of the chemical potential of the solid, liquid, and gas phases of a substance (in practice, the lines are curved). The phase with the lowest chemical potential at a specified temperature is the most stable one at that temperature. The transition temperatures, the melting and boiling temperatures (T_f and T_b, respectively), are the temperatures at which the chemical potentials of the two phases are equal.

Equation 4.2 implies that, because $S_m(g) > S_m(l)$, the slope of a plot of μ against temperature is steeper for gases than for liquids. Because $S_m(l) > S_m(s)$ almost always, the slope is also steeper for a liquid than the corresponding solid. These features are illustrated in Fig. 4.12. The steep negative slope of $\mu(l)$ results in it falling below $\mu(s)$ when the temperature is high enough, and then the liquid becomes the stable phase: the solid melts. The chemical potential of the gas phase plunges steeply downwards as the temperature is raised (because the molar entropy of the vapour is so high), and there comes a temperature at which it lies lowest. Then the gas is the stable phase and vaporization is spontaneous.

(b) The response of melting to applied pressure

Most substances melt at a higher temperature when subjected to pressure. It is as though the pressure is preventing the formation of the less dense liquid phase. Exceptions to this behaviour include water, for which the liquid is denser than the solid. Application of pressure to water encourages the formation of the liquid phase. That is, water freezes and ice melts at a lower temperature when it is under pressure.

We can rationalize the response of melting temperatures to pressure as follows. The variation of the chemical potential with pressure is expressed (from the second of eqns 3.53) by

$$\left(\frac{\partial \mu}{\partial p}\right)_T = V_m \qquad \text{Variation of chemical potential with } p \qquad (4.3)$$

This equation shows that the slope of a plot of chemical potential against pressure is equal to the molar volume of the substance. An increase in pressure raises the chemical potential of any pure substance (because $V_m > 0$). In most cases, $V_m(l) > V_m(s)$ and the equation predicts that an increase in pressure increases the chemical potential of the liquid more than that of the solid. As shown in Fig. 4.13a, the effect of pressure in such a case is to raise the melting temperature slightly. For water, however, $V_m(l) < V_m(s)$, and an increase in pressure increases the chemical potential of the solid more than that of the liquid. In this case, the melting temperature is lowered slightly (Fig. 4.13b).

Example 4.1 *Assessing the effect of pressure on the chemical potential*

Calculate the effect on the chemical potentials of ice and water of increasing the pressure from 1.00 bar to 2.00 bar at 0°C. The density of ice is 0.917 g cm^{-3} and that of liquid water is 0.999 g cm^{-3} under these conditions.

Method From eqn 4.3, we know that the change in chemical potential of an incompressible substance when the pressure is changed by Δp is $\Delta\mu = V_m\Delta p$. Therefore, to answer the question, we need to know the molar volumes of the two phases of water. These values are obtained from the mass density, ρ, and the molar mass, M, by using $V_m = M/\rho$. We therefore use the expression $\Delta\mu = M\Delta p/\rho$.

Answer The molar mass of water is 18.02 g mol^{-1} (1.802×10^{-2} kg mol^{-1}); therefore,

$$\Delta\mu(\text{ice}) = \frac{(1.802\times10^{-2}\ \text{kg mol}^{-1})\times(1.00\times10^{5}\ \text{Pa})}{917\ \text{kg m}^{-3}} = +1.97\ \text{J mol}^{-1}$$

$$\Delta\mu(\text{water}) = \frac{(1.802\times10^{-2}\ \text{kg mol}^{-1})\times(1.00\times10^{5}\ \text{Pa})}{999\ \text{kg m}^{-3}} = +1.80\ \text{J mol}^{-1}$$

We interpret the numerical results as follows: the chemical potential of ice rises more sharply than that of water so, if they are initially in equilibrium at 1 bar, then there will be a tendency for the ice to melt at 2 bar.

Self-test 4.1 Calculate the effect of an increase in pressure of 1.00 bar on the liquid and solid phases of carbon dioxide (of molar mass 44.0 g mol^{-1}) in equilibrium with densities 2.35 g cm^{-3} and 2.50 g cm^{-3}, respectively.
[$\Delta\mu(\text{l}) = +1.87$ J mol^{-1}, $\Delta\mu(\text{s}) = +1.76$ J mol^{-1}; solid forms]

(c) The vapour pressure of a liquid subjected to pressure

When pressure is applied to a condensed phase, its vapour pressure rises: in effect, molecules are squeezed out of the phase and escape as a gas. Pressure can be exerted on the condensed phase mechanically or by subjecting it to the applied pressure of an inert gas (Fig. 4.14). In the latter case, the vapour pressure is the partial pressure of the vapour in equilibrium with the condensed phase. We then speak of the **partial vapour pressure** of the substance. One complication (which we ignore here) is that, if the condensed phase is a liquid, then the pressurizing gas might dissolve and change the properties of the liquid. Another complication is that the gas phase molecules might attract molecules out of the liquid by the process of **gas solvation**, the attachment of molecules to gas-phase species.

As shown in the following *Justification*, the quantitative relation between the vapour pressure, p, when a pressure ΔP is applied and the vapour pressure, p^*, of the liquid in the absence of an additional pressure is

$$p = p^* e^{V_m(\text{l})\Delta P/RT} \quad \text{Effect of applied pressure } \Delta P \text{ on vapour pressure } p \qquad (4.4)$$

This equation shows how the vapour pressure increases when the pressure acting on the condensed phase is increased.

Justification 4.2 *The vapour pressure of a pressurized liquid*

We calculate the vapour pressure of a pressurized liquid by using the fact that at equilibrium the chemical potentials of the liquid and its vapour are equal: $\mu(\text{l}) = \mu(\text{g})$. It follows that, for any change that preserves equilibrium, the resulting change in $\mu(\text{l})$ must be equal to the change in $\mu(\text{g})$; therefore, we can write $d\mu(\text{g}) = d\mu(\text{l})$. When the pressure P on the liquid is increased by dP, the chemical potential of the liquid changes by $d\mu(\text{l}) = V_m(\text{l})dP$. The chemical potential of the vapour changes by $d\mu(\text{g}) = V_m(\text{g})dp$ where dp is the change in the vapour pressure we are trying to find. If we treat the vapour as a perfect gas, the molar volume can be replaced by $V_m(\text{g}) = RT/p$, and we obtain $d\mu(\text{g}) = RTdp/p$. Next, we equate the changes in chemical potentials of the vapour and the liquid:

$$\frac{RTdp}{p} = V_m(\text{l})dP$$

We can integrate this expression once we know the limits of integration.

When there is no additional pressure acting on the liquid, P (the pressure experienced by the liquid) is equal to the normal vapour pressure p^*, so when $P = p^*$, $p = p^*$ too. When there is an additional pressure ΔP on the liquid, with the result that $P = p + \Delta P$, the vapour pressure is p (the value we want to find). Provided the effect of pressure on the vapour pressure is small (as will turn out to be the case) a good approximation is to replace the p in $p + \Delta P$ by p^* itself, and to set the upper limit of the integral to $p^* + \Delta P$. The integrations required are therefore as follows:

$$RT\int_{p^*}^{p} \frac{dp}{p} = \int_{p^*}^{p^*+\Delta P} V_m(\text{l})dP$$

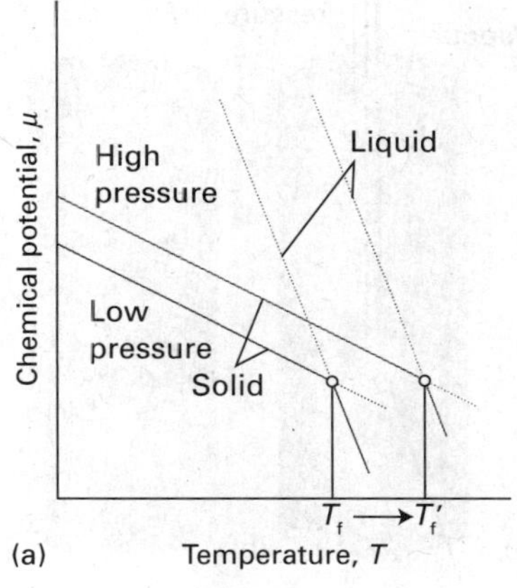

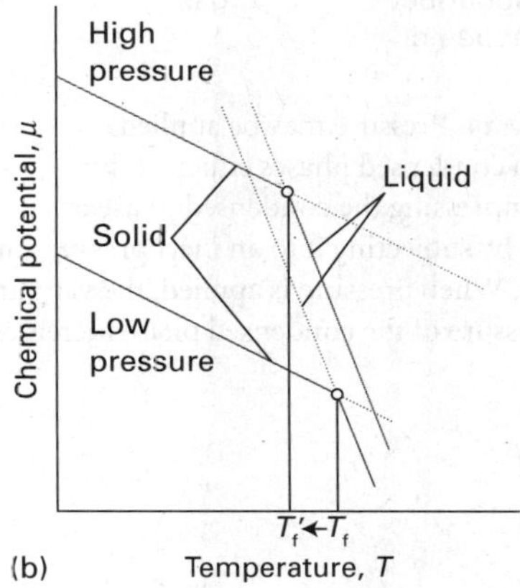

Fig. 4.13 The pressure dependence of the chemical potential of a substance depends on the molar volume of the phase. The lines show schematically the effect of increasing pressure on the chemical potential of the solid and liquid phases (in practice, the lines are curved), and the corresponding effects on the freezing temperatures. (a) In this case the molar volume of the solid is smaller than that of the liquid and $\mu(\text{s})$ increases less than $\mu(\text{l})$. As a result, the freezing temperature rises. (b) Here the molar volume is greater for the solid than the liquid (as for water), $\mu(\text{s})$ increases more strongly than $\mu(\text{l})$, and the freezing temperature is lowered.

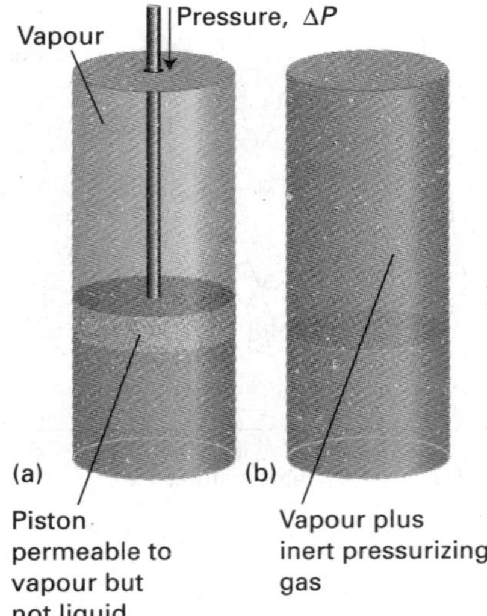

Fig. 4.14 Pressure may be applied to a condensed phases either (a) by compressing the condensed phase or (b) by subjecting it to an inert pressurizing gas. When pressure is applied, the vapour pressure of the condensed phase increases.

We now divide both sides by RT and assume that the molar volume of the liquid is the same throughout the small range of pressures involved:

$$\int_{p^*}^{p} \frac{dp}{p} = \frac{V_m(l)}{RT} \int_{p^*}^{p^*+\Delta P} dP$$

Then both integrations are straightforward, and lead to

$$\ln \frac{p}{p^*} = \frac{V_m(l)}{RT} \Delta P$$

which rearranges to eqn 4.4 because $e^{\ln x} = x$.

- **A brief illustration**

For water, which has density 0.997 g cm^{-3} at 25°C and therefore molar volume 18.1 cm^3 mol^{-1}, when the pressure is increased by 10 bar (that is, $\Delta P = 1.0 \times 10^6$ Pa)

$$\frac{V_m(l)\Delta P}{RT} = \frac{(1.81 \times 10^{-5}\ \text{m}^3\ \text{mol}^{-1}) \times (1.0 \times 10^6\ \text{Pa})}{(8.3145\ \text{J K}^{-1}\ \text{mol}^{-1}) \times (298\ \text{K})} = \frac{1.81 \times 1.0 \times 10}{8.3145 \times 298}$$

where we have used 1 J = 1 Pa m^3. It follows that $p = 1.0073p^*$, an increase of 0.73 per cent. ●

Self-test 4.2 Calculate the effect of an increase in pressure of 100 bar on the vapour pressure of benzene at 25°C, which has density 0.879 g cm^{-3}. [43 per cent]

4.5 The location of phase boundaries

Key points **(a) The Clapeyron equation is an expression for the slope of a phase boundary. (b) The Clapeyron equation gives an expression for the slope of the solid–liquid phase boundary in terms of the enthalpy of fusion. (c) The Clausius–Clapeyron equation is an approximation that relates the slope of the liquid–vapour boundary to the enthalpy of vaporization. (d) The slope of the solid–vapour boundary is similarly related to the enthalpy of sublimation.**

We can find the precise locations of the phase boundaries—the pressures and temperatures at which two phases can coexist—by making use of the fact that, when two phases are in equilibrium, their chemical potentials must be equal. Therefore, where the phases α and β are in equilibrium,

$$\mu(\alpha; p,T) = \mu(\beta; p,T) \tag{4.5}$$

By solving this equation for p in terms of T, we get an equation for the phase boundary.

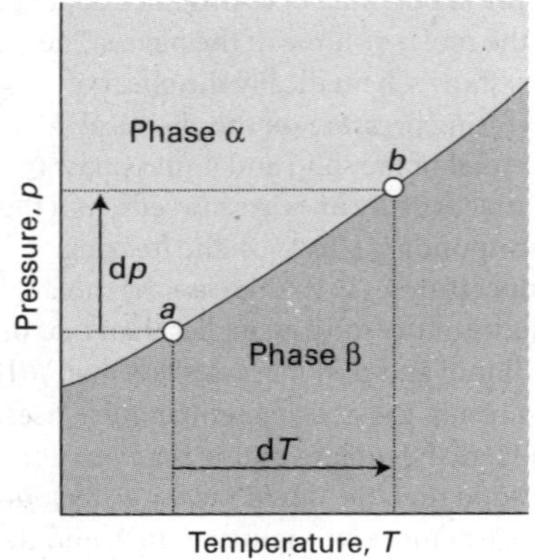

Fig. 4.15 When pressure is applied to a system in which two phases are in equilibrium (at a), the equilibrium is disturbed. It can be restored by changing the temperature, so moving the state of the system to b. It follows that there is a relation between dp and dT that ensures that the system remains in equilibrium as either variable is changed.

(a) The slopes of the phase boundaries

It turns out to be simplest to discuss the phase boundaries in terms of their slopes, dp/dT. Let p and T be changed infinitesimally, but in such a way that the two phases α and β remain in equilibrium. The chemical potentials of the phases are initially equal (the two phases are in equilibrium). They remain equal when the conditions are changed to another point on the phase boundary, where the two phases continue to be in equilibrium (Fig. 4.15). Therefore, the changes in the chemical potentials of the two phases must be equal and we can write $d\mu(\alpha) = d\mu(\beta)$. Because, from eqn 3.52 ($dG = Vdp - SdT$), we know that $d\mu = -S_m dT + V_m dp$ for each phase, it follows that

$$-S_m(\alpha)dT + V_m(\alpha)dp = -S_m(\beta)dT + V_m(\beta)dp$$

where $S_m(\alpha)$ and $S_m(\beta)$ are the molar entropies of the phases and $V_m(\alpha)$ and $V_m(\beta)$ are their molar volumes. Hence

$$\{V_m(\beta) - V_m(\alpha)\}dp = \{S_m(\beta) - S_m(\alpha)\}dT$$

which rearranges into the **Clapeyron equation:**

$$\frac{dp}{dT} = \frac{\Delta_{trs}S}{\Delta_{trs}V} \qquad \text{Clapeyron equation} \qquad (4.6)$$

In this expression $\Delta_{trs}S = S_m(\beta) - S_m(\alpha)$ and $\Delta_{trs}V = V_m(\beta) - V_m(\alpha)$ are the entropy and volume of transition, respectively. The Clapeyron equation is an exact expression for the slope of the tangent to the boundary at any point and applies to any phase equilibrium of any pure substance. It implies that we can use thermodynamic data to predict the appearance of phase diagrams and to understand their form. A more practical application is to the prediction of the response of freezing and boiling points to the application of pressure.

(b) The solid–liquid boundary

Melting (fusion) is accompanied by a molar enthalpy change $\Delta_{fus}H$ and occurs at a temperature T. The molar entropy of melting at T is therefore $\Delta_{fus}H/T$ (Section 3.3), and the Clapeyron equation becomes

$$\frac{dp}{dT} = \frac{\Delta_{fus}H}{T\Delta_{fus}V} \qquad \text{Slope of solid–liquid boundary} \qquad (4.7)$$

where $\Delta_{fus}V$ is the change in molar volume that occurs on melting. The enthalpy of melting is positive (the only exception is helium-3) and the volume change is usually positive and always small. Consequently, the slope dp/dT is steep and usually positive (Fig. 4.16).

We can obtain the formula for the phase boundary by integrating dp/dT, assuming that $\Delta_{fus}H$ and $\Delta_{fus}V$ change so little with temperature and pressure that they can be treated as constant. If the melting temperature is T^* when the pressure is p^*, and T when the pressure is p, the integration required is

$$\int_{p^*}^{p} dp = \frac{\Delta_{fus}H}{\Delta_{fus}V}\int_{T^*}^{T}\frac{dT}{T}$$

Therefore, the approximate equation of the solid–liquid boundary is

$$p = p^* + \frac{\Delta_{fus}H}{\Delta_{fus}V}\ln\frac{T}{T^*} \qquad (4.8)$$

This equation was originally obtained by yet another Thomson—James, the brother of William, Lord Kelvin. When T is close to T^*, the logarithm can be approximated by using

$$\ln\frac{T}{T^*} = \ln\left(1 + \frac{T - T^*}{T^*}\right) \approx \frac{T - T^*}{T^*}$$

Therefore,

$$p = p^* + \frac{\Delta_{fus}H}{T^*\Delta_{fus}V}(T - T^*) \qquad (4.9)$$

This expression is the equation of a steep straight line when p is plotted against T (as in Fig. 4.16).

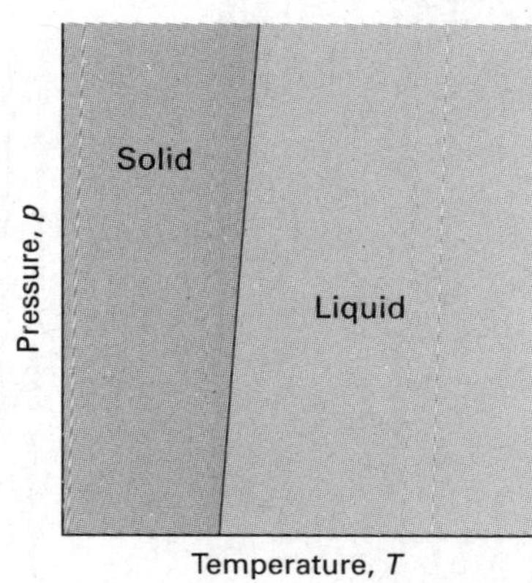

Fig. 4.16 A typical solid–liquid phase boundary slopes steeply upwards. This slope implies that, as the pressure is raised, the melting temperature rises. Most substances behave in this way.

A brief comment

Calculations involving natural logarithms often become simpler if we note that, provided $-1 < x < 1$, $\ln(1+x) = x - \frac{1}{2}x^2 + \frac{1}{3}x^3 \cdots$. If $x \ll 1$, a good approximation is $\ln(1+x) \approx x$.

(c) The liquid–vapour boundary

The entropy of vaporization at a temperature T is equal to $\Delta_{vap}H/T$; the Clapeyron equation for the liquid–vapour boundary is therefore

$$\frac{dp}{dT}=\frac{\Delta_{vap}H}{T\Delta_{vap}V} \qquad (4.10)$$

Slope of liquid–vapour boundary

The enthalpy of vaporization is positive; $\Delta_{vap}V$ is large and positive. Therefore, dp/dT is positive, but it is much smaller than for the solid–liquid boundary. It follows that dT/dp is large, and hence that the boiling temperature is more responsive to pressure than the freezing temperature.

Example 4.2 *Estimating the effect of pressure on the boiling temperature*

Estimate the typical size of the effect of increasing pressure on the boiling point of a liquid.

Method To use eqn 4.10 we need to estimate the right-hand side. At the boiling point, the term $\Delta_{vap}H/T$ is Trouton's constant (Section 3.3b). Because the molar volume of a gas is so much greater than the molar volume of a liquid, we can write $\Delta_{vap}V = V_m(g) - V_m(l) \approx V_m(g)$ and take for $V_m(g)$ the molar volume of a perfect gas (at low pressures, at least).

Answer Trouton's constant has the value 85 J K^{-1} mol^{-1}. The molar volume of a perfect gas is about 25 dm^3 mol^{-1} at 1 atm and near but above room temperature. Therefore,

$$\frac{dp}{dT}\approx\frac{85\ \text{J K}^{-1}\ \text{mol}^{-1}}{2.5\times10^{-2}\ \text{m}^3\ \text{mol}^{-1}}=3.4\times10^3\ \text{Pa K}^{-1}$$

We have used 1 J = 1 Pa m^3. This value corresponds to 0.034 atm K^{-1} and hence to $dT/dp = 29$ K atm^{-1}. Therefore, a change of pressure of +0.1 atm can be expected to change a boiling temperature by about +3 K.

Self-test 4.3 Estimate dT/dp for water at its normal boiling point using the information in Table 3.2 and $V_m(g) = RT/p$. [28 K atm^{-1}]

Because the molar volume of a gas is so much greater than the molar volume of a liquid, we can write $\Delta_{vap}V \approx V_m(g)$ (as in *Example 4.2*). Moreover, if the gas behaves perfectly, $V_m(g) = RT/p$. These two approximations turn the exact Clapeyron equation into

$$\frac{dp}{dT}=\frac{\Delta_{vap}H}{T(RT/p)}$$

which rearranges into the **Clausius–Clapeyron equation** for the variation of vapour pressure with temperature:

$$\frac{d\ln p}{dT}=\frac{\Delta_{vap}H}{RT^2} \qquad (4.11)$$

Clausius–Clapeyron equation

(We have used $dx/x = d\ln x$.) Like the Clapeyron equation (which is exact), the Clausius–Clapeyron equation (which is an approximation) is important for understanding the appearance of phase diagrams, particularly the location and shape of the liquid–vapour and solid–vapour phase boundaries. It lets us predict how the vapour pressure varies with temperature and how the boiling temperature varies with pressure. For instance, if we also assume that the enthalpy of vaporization is independent of temperature, this equation can be integrated as follows:

$$\int_{\ln p^*}^{\ln p} \mathrm{d}\ln p = \frac{\Delta_{\mathrm{vap}}H}{R}\int_{T^*}^{T}\frac{\mathrm{d}T}{T^2} = -\frac{\Delta_{\mathrm{vap}}H}{R}\left(\frac{1}{T}-\frac{1}{T^*}\right)$$

where p^* is the vapour pressure when the temperature is T^* and p the vapour pressure when the temperature is T. Therefore, because the integral on the left evaluates to $\ln(p/p^*)$, the two vapour pressures are related by

$$p = p^* \mathrm{e}^{-\chi} \qquad \chi = \frac{\Delta_{\mathrm{vap}}H}{R}\left(\frac{1}{T}-\frac{1}{T^*}\right) \tag{4.12}$$

Equation 4.12 is plotted as the liquid–vapour boundary in Fig. 4.17. The line does not extend beyond the critical temperature T_c, because above this temperature the liquid does not exist.

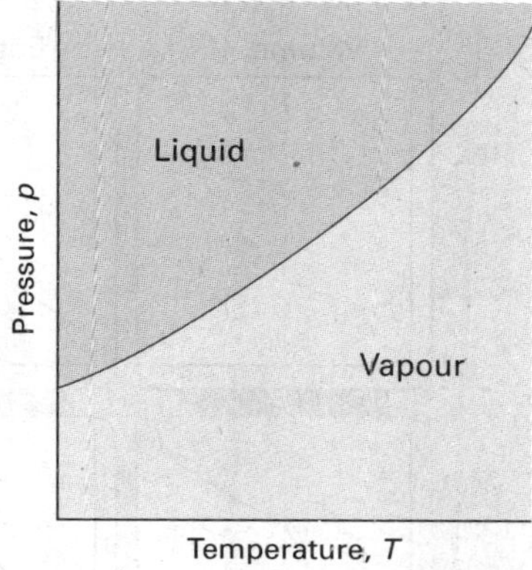

Fig. 4.17 A typical liquid–vapour phase boundary. The boundary can be regarded as a plot of the vapour pressure against the temperature. Note that, in some depictions of phase diagrams in which a logarithmic pressure scale is used, the phase boundary has the opposite curvature (see Fig. 4.11). This phase boundary terminates at the critical point (not shown).

- **A brief illustration**

Equation 4.12 can be used to estimate the vapour pressure of a liquid at any temperature from its normal boiling point, the temperature at which the vapour pressure is 1.00 atm (101 kPa). The normal boiling point of benzene is 80°C (353 K) and (from Table 2.3) $\Delta_{\mathrm{vap}}H^{\ominus} = 30.8$ kJ mol^{-1}. Therefore, to calculate the vapour pressure at 20°C (293 K), we write

$$\chi = \frac{3.08\times10^4\ \mathrm{J\ mol^{-1}}}{8.3145\ \mathrm{J\ K^{-1}\ mol^{-1}}}\left(\frac{1}{293\ \mathrm{K}}-\frac{1}{353\ \mathrm{K}}\right) = \frac{3.08\times10^4}{8.3145}\left(\frac{1}{293}-\frac{1}{353}\right)$$

and substitute this value into eqn 4.12 with $p^* = 101$ kPa. The result is 12 kPa. The experimental value is 10 kPa. ●

A note on good practice Because exponential functions are so sensitive, it is good practice to carry out numerical calculations like this without evaluating the intermediate steps and using rounded values.

(d) The solid–vapour boundary

The only difference between this case and the last is the replacement of the enthalpy of vaporization by the enthalpy of sublimation, $\Delta_{\mathrm{sub}}H$. Because the enthalpy of sublimation is greater than the enthalpy of vaporization (recall that $\Delta_{\mathrm{sub}}H = \Delta_{\mathrm{fus}}H + \Delta_{\mathrm{vap}}H$), the equation predicts a steeper slope for the sublimation curve than for the vaporization curve at similar temperatures, which is near where they meet at the triple point (Fig. 4.18).

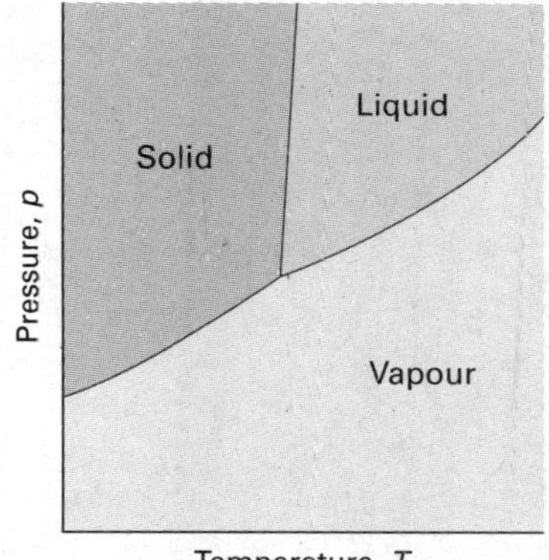

Fig. 4.18 Near the point where they coincide (at the triple point), the solid–gas boundary has a steeper slope than the liquid–gas boundary because the enthalpy of sublimation is greater than the enthalpy of vaporization and the temperatures that occur in the Clausius–Clapeyron equation for the slope have similar values.

4.6 The Ehrenfest classification of phase transitions

Key points **(a) Different types of phase transition are identified by the behaviour of thermodynamic properties at the transition temperature. (b) The classification reveals the type of molecular process occurring at the phase transition.**

There are many different types of phase transition, including the familiar examples of fusion and vaporization and the less familiar examples of solid–solid, conducting–superconducting, and fluid–superfluid transitions. We shall now see that it is possible to use thermodynamic properties of substances, and in particular the behaviour of the chemical potential, to classify phase transitions into different types. The classification scheme was originally proposed by Paul Ehrenfest, and is known as the **Ehrenfest classification**.

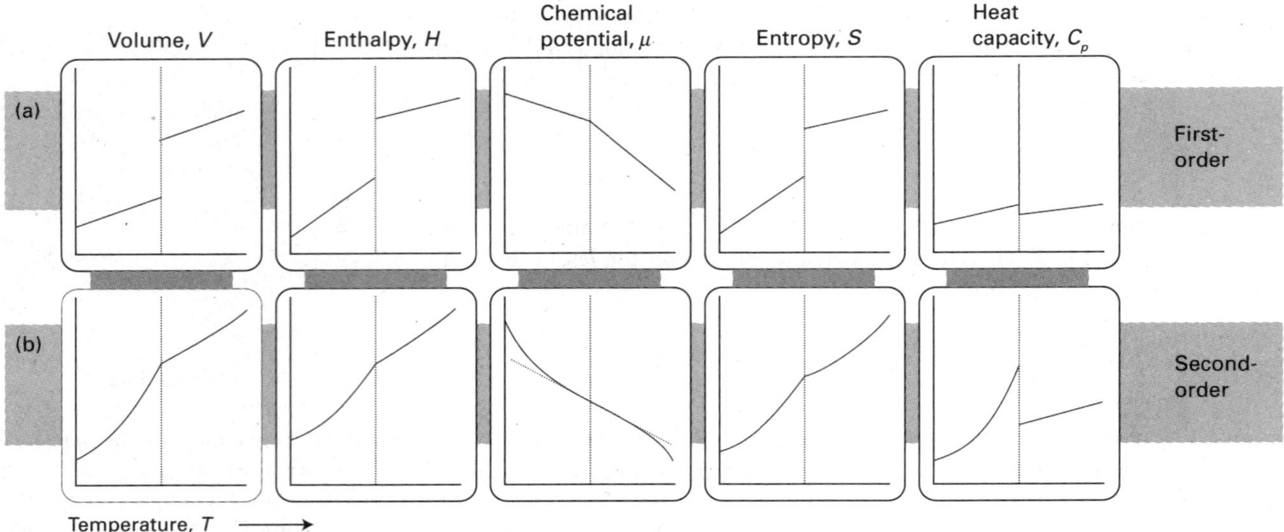

Fig. 4.19 The changes in thermodynamic properties accompanying (a) first-order and (b) second-order phase transitions.

(a) The thermodynamic basis

Many familiar phase transitions, like fusion and vaporization, are accompanied by changes of enthalpy and volume. These changes have implications for the slopes of the chemical potentials of the phases at either side of the phase transition. Thus, at the transition from a phase α to another phase β,

$$\begin{aligned}\left(\frac{\partial\mu(\beta)}{\partial p}\right)_T-\left(\frac{\partial\mu(\alpha)}{\partial p}\right)_T&=V_m(\beta)-V_m(\alpha)=\Delta_{trs}V\\ \left(\frac{\partial\mu(\beta)}{\partial T}\right)_p-\left(\frac{\partial\mu(\alpha)}{\partial T}\right)_p&=-S_m(\beta)+S_m(\alpha)=-\Delta_{trs}S=-\frac{\Delta_{trs}H}{T_{trs}}\end{aligned}\qquad(4.13)$$

Because $\Delta_{trs}V$ and $\Delta_{trs}H$ are non-zero for melting and vaporization, it follows that for such transitions the slopes of the chemical potential plotted against either pressure or temperature are different on either side of the transition (Fig. 4.19a). In other words, the first derivatives of the chemical potentials with respect to pressure and temperature are discontinuous at the transition.

A transition for which the first derivative of the chemical potential with respect to temperature is discontinuous is classified as a **first-order phase transition.** The constant-pressure heat capacity, C_p, of a substance is the slope of a plot of the enthalpy with respect to temperature. At a first-order phase transition, H changes by a finite amount for an infinitesimal change of temperature. Therefore, at the transition the heat capacity is infinite. The physical reason is that heating drives the transition rather than raising the temperature. For example, boiling water stays at the same temperature even though heat is being supplied.

A **second-order phase transition** in the Ehrenfest sense is one in which the first derivative of μ with respect to temperature is continuous but its second derivative is discontinuous. A continuous slope of μ (a graph with the same slope on either side of the transition) implies that the volume and entropy (and hence the enthalpy) do not change at the transition (Fig. 4.19b). The heat capacity is discontinuous at the transition but does not become infinite there. An example of a second-order transition is the conducting–superconducting transition in metals at low temperatures.[2]

The term **λ-transition** is applied to a phase transition that is not first-order yet the heat capacity becomes infinite at the transition temperature. Typically, the heat capacity of a system that shows such a transition begins to increase well before the transition (Fig. 4.20), and the shape of the heat capacity curve resembles the Greek letter lambda. This type of transition includes order–disorder transitions in alloys, the onset of ferromagnetism, and the fluid–superfluid transition of liquid helium.

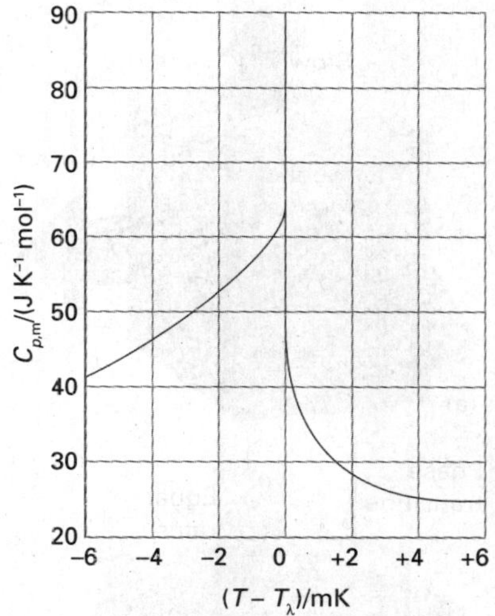

Fig. 4.20 The λ-curve for helium, where the heat capacity rises to infinity. The shape of this curve is the origin of the name λ-transition.

(b) Molecular interpretation

First-order transitions typically involve the relocation of atoms, molecules, or ions with a consequent change in the energies of their interactions. Thus, vaporization eliminates the attractions between molecules and a first-order phase transition from one ionic polymorph to another (as in the conversion of calcite to aragonite) involves the adjustment of the relative positions of ions.

One type of second-order transition is associated with a change in symmetry of the crystal structure of a solid. Thus, suppose the arrangement of atoms in a solid is like that represented in Fig. 4.21a, with one dimension (technically, of the unit cell) longer than the other two, which are equal. Such a crystal structure is classified as tetragonal (see Section 19.1). Moreover, suppose the two shorter dimensions increase more than the long dimension when the temperature is raised. There may come a stage when the three dimensions become equal. At that point the crystal has cubic symmetry (Fig. 4.21b), and at higher temperatures it will expand equally in all three directions (because there is no longer any distinction between them). The tetragonal → cubic phase transition has occurred but, as it has not involved a discontinuity in the interaction energy between the atoms or the volume they occupy, the transition is not first-order.

The order–disorder transition in β-brass (CuZn) is an example of a λ-transition. The low-temperature phase is an orderly array of alternating Cu and Zn atoms. The high-temperature phase is a random array of the atoms (Fig. 4.22). At $T = 0$ the order is perfect, but islands of disorder appear as the temperature is raised. The islands form because the transition is cooperative in the sense that, once two atoms have exchanged locations, it is easier for their neighbours to exchange their locations. The islands grow in extent and merge throughout the crystal at the transition temperature (742 K). The heat capacity increases as the transition temperature is approached because the cooperative nature of the transition means that it is increasingly easy for the heat supplied to drive the phase transition rather than to be stored as thermal motion.

[2] A metallic conductor is a substance with an electrical conductivity that decreases as the temperature increases. A superconductor is a solid that conducts electricity without resistance. See Chapter 19 for more details.

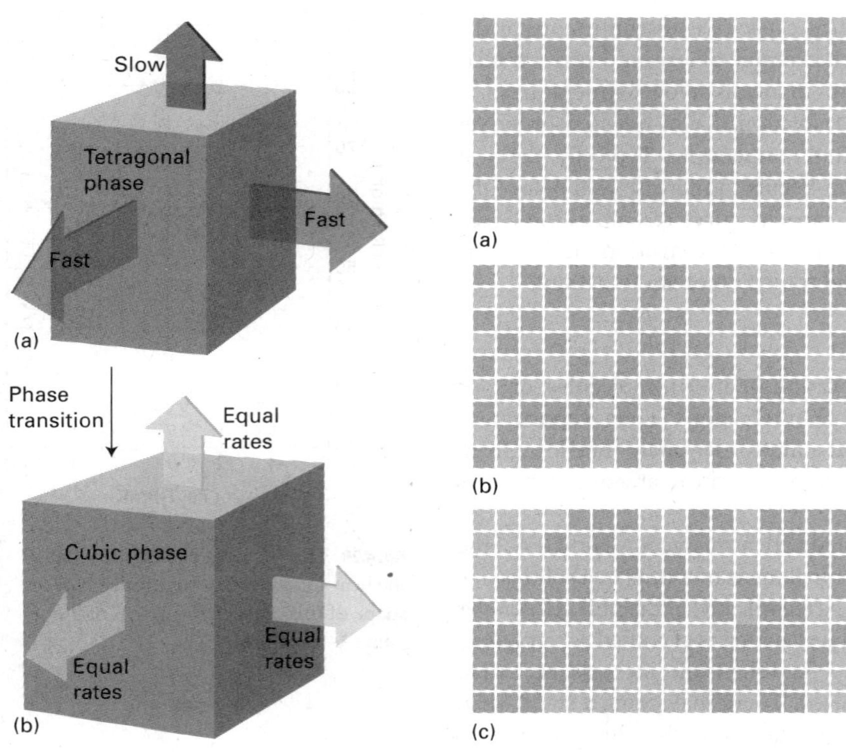

Fig. 4.21 (Left) One version of a second-order phase transition in which (a) a tetragonal phase expands more rapidly in two directions than a third, and hence becomes a cubic phase, which (b) expands uniformly in three directions as the temperature is raised. There is no rearrangement of atoms at the transition temperature, and hence no enthalpy of transition.

Fig. 4.22 (Right) An order–disorder transition. (a) At $T=0$, there is perfect order, with different kinds of atoms occupying alternate sites. (b) As the temperature is increased, atoms exchange locations and islands of each kind of atom form in regions of the solid. Some of the original order survives. (c) At and above the transition temperature, the islands occur at random throughout the sample.

Checklist of key equations

Property	Equation	Comment
Chemical potential	$\mu = G_m$	For a pure substance
Phase rule	$F = C - P + 2$	
Variation of μ with temperature	$(\partial\mu/\partial T)_p = -S_m$	
Variation of μ with pressure	$(\partial\mu/\partial p)_T = V_m$	
Vapour pressure in the presence of applied pressure	$p = p^* e^{V_m \Delta P/RT}$	$\Delta P = P_{applied} - p^*$
Clapeyron equation	$dp/dT = \Delta_{trs}S/\Delta_{trs}V$	
Clausius–Clapeyron equation	$d\ln p/dT = \Delta_{vap}H/RT^2$	Assumes $V_m(g) \gg V_m(l)$ and vapour is a perfect gas

➔ For a chart of the relations between principal equations, see the Road map section of the Resource section.

Discussion questions

4.1 Describe how the concept of chemical potential unifies the discussion of phase equilibria.

4.2 Why does the chemical potential change with pressure even if the system is incompressible (that is, remains at the same volume when pressure is applied)?

4.3 How may DSC be used to identify phase transitions?

4.4 Discuss what would be observed as a sample of water is taken along a path that encircles and is close to its critical point.

4.5 Consult library and internet resources and prepare a discussion of the principles, advantages, disadvantages, and current uses of supercritical fluids.

4.6 Distinguish between a first-order phase transition, a second-order phase transition, and a λ-transition at both molecular and macroscopic levels.

Exercises

4.1(a) How many phases are present at each of the points marked in Fig. 4.23a?

4.1(b) How many phases are present at each of the points marked in Fig. 4.23b?

4.2(a) The difference in chemical potential between two regions of a system is +7.1 kJ mol^{-1}. By how much does the Gibbs energy change when 0.10 mmol of a substance is transferred from one region to the other?

4.2(b) The difference in chemical potential between two regions of a system is −8.3 kJ mol^{-1}. By how much does the Gibbs energy change when 0.15 mmol of a substance is transferred from one region to the other?

4.3(a) Estimate the difference between the normal and standard melting points of ice.

4.3(b) Estimate the difference between the normal and standard boiling points of water.

4.4(a) What is the maximum number of phases that can be in mutual equilibrium in a two-component system?

4.4(b) What is the maximum number of phases that can be in mutual equilibrium in a four-component system?

4.5(a) Water is heated from 25°C to 100°C. By how much does its chemical potential change?

4.5(b) Iron is heated from 100°C to 1000°C. By how much does its chemical potential change? Take $S_m^{\ominus} = 53$ J K^{-1} mol^{-1} for the entire range (its average value).

4.6(a) By how much does the chemical potential of copper change when the pressure exerted on a sample is increased from 100 kPa to 10 MPa?

4.6(b) By how much does the chemical potential of benzene change when the pressure exerted on a sample is increased from 100 kPa to 10 MPa?

4.7(a) Pressure was exerted with a piston on water at 20°C. The vapour pressure of water under 1.0 bar is 2.34 kPa. What is its vapour pressure when the pressure on the liquid is 20 MPa?

4.7(b) Pressure was exerted with a piston on molten naphthalene at 95°C. The vapour pressure of naphthalene under 1.0 bar is 2.0 kPa and its density is 0.962 g cm^{-3}. What is its vapour pressure when the pressure on the liquid is 15 MPa?

4.8(a) The molar volume of a certain solid is 161.0 cm^3 mol^{-1} at 1.00 atm and 350.75 K, its melting temperature. The molar volume of the liquid at this temperature and pressure is 163.3 cm^3 mol^{-1}. At 100 atm the melting temperature changes to 351.26 K. Calculate the enthalpy and entropy of fusion of the solid.

4.8(b) The molar volume of a certain solid is 142.0 cm^3 mol^{-1} at 1.00 atm and 427.15 K, its melting temperature. The molar volume of the liquid at this temperature and pressure is 152.6 cm^3 mol^{-1}. At 1.2 MPa the melting temperature changes to 429.26 K. Calculate the enthalpy and entropy of fusion of the solid.

4.9(a) The vapour pressure of dichloromethane at 24.1°C is 53.3 kPa and its enthalpy of vaporization is 28.7 kJ mol^{-1}. Estimate the temperature at which its vapour pressure is 70.0 kPa.

4.9(b) The vapour pressure of a substance at 20.0°C is 58.0 kPa and its enthalpy of vaporization is 32.7 kJ mol^{-1}. Estimate the temperature at which its vapour pressure is 66.0 kPa.

4.10(a) The vapour pressure of a liquid in the temperature range 200 K to 260 K was found to fit the expression $\ln(p/\text{Torr}) = 16.255 - 2501.8/(T/\text{K})$. What is the enthalpy of vaporization of the liquid?

4.10(b) The vapour pressure of a liquid in the temperature range 200 K to 260 K was found to fit the expression $\ln(p/\text{Torr}) = 18.361 - 3036.8/(T/\text{K})$. What is the enthalpy of vaporization of the liquid?

4.11(a) The vapour pressure of benzene between 10°C and 30°C fits the expression $\log(p/\text{Torr}) = 7.960 - 1780/(T/\text{K})$. Calculate (a) the enthalpy of vaporization and (b) the normal boiling point of benzene.

4.11(b) The vapour pressure of a liquid between 15°C and 35°C fits the expression $\log(p/\text{Torr}) = 8.750 - 1625/(T/\text{K})$. Calculate (a) the enthalpy of vaporization and (b) the normal boiling point of the liquid.

4.12(a) When benzene freezes at 5.5°C its density changes from 0.879 g cm^{-3} to 0.891 g cm^{-3}. Its enthalpy of fusion is 10.59 kJ mol^{-1}. Estimate the freezing point of benzene at 1000 atm.

4.12(b) When a certain liquid of molar mass 46.1 g mol^{-1} freezes at −3.65°C its density changes from 0.789 g cm^{-3} to 0.801 g cm^{-3}. Its enthalpy of fusion is 8.68 kJ mol^{-1}. Estimate the freezing point of the liquid at 100 MPa.

4.13(a) In July in Los Angeles, the incident sunlight at ground level has a power density of 1.2 kW m^{-2} at noon. A swimming pool of area 50 m^2 is directly exposed to the sun. What is the maximum rate of loss of water? Assume that all the radiant energy is absorbed.

4.13(b) Suppose the incident sunlight at ground level has a power density of 0.87 kW m^{-2} at noon. What is the maximum rate of loss of water from a lake of area 1.0 ha? (1 ha = 10^4 m^2.) Assume that all the radiant energy is absorbed.

4.14(a) An open vessel containing (a) water, (b) benzene, (c) mercury stands in a laboratory measuring 5.0 m × 5.0 m × 3.0 m at 25°C. What mass of each substance will be found in the air if there is no ventilation? (The vapour pressures are (a) 3.2 kPa, (b) 13.1 kPa, (c) 0.23 Pa.)

4.14(b) On a cold, dry morning after a frost, the temperature was −5°C and the partial pressure of water in the atmosphere fell to 0.30 kPa. Will the frost sublime? What partial pressure of water would ensure that the frost remained?

4.15(a) Naphthalene, $C_{10}H_8$, melts at 80.2°C. If the vapour pressure of the liquid is 1.3 kPa at 85.8°C and 5.3 kPa at 119.3°C, use the Clausius–Clapeyron

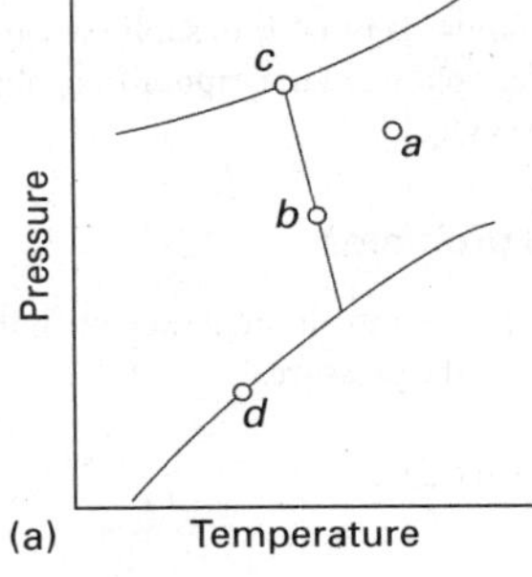

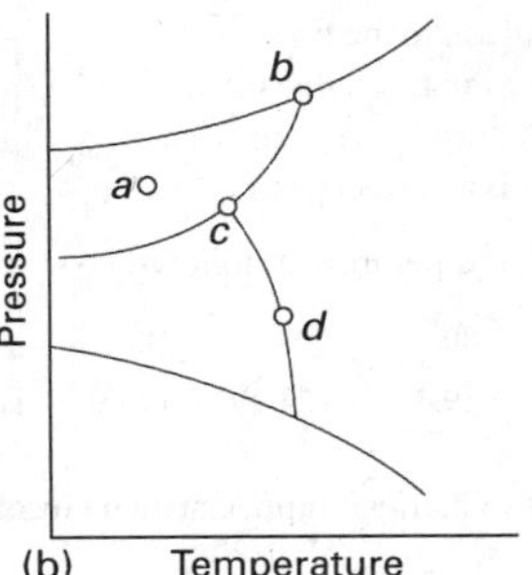

Fig. 4.23

equation to calculate (a) the enthalpy of vaporization, (b) the normal boiling point, and (c) the enthalpy of vaporization at the boiling point.

4.15(b) The normal boiling point of hexane is 69.0°C. Estimate (a) its enthalpy of vaporization and (b) its vapour pressure at 25°C and 60°C.

4.16(a) Calculate the melting point of ice under a pressure of 50 bar. Assume that the density of ice under these conditions is approximately 0.92 g cm^{-3} and that of liquid water is 1.00 g cm^{-3}.

4.16(b) Calculate the melting point of ice under a pressure of 10 MPa. Assume that the density of ice under these conditions is approximately 0.915 g cm^{-3} and that of liquid water is 0.998 g cm^{-3}.

4.17(a) What fraction of the enthalpy of vaporization of water is spent on expanding the water vapour?

4.17(b) What fraction of the enthalpy of vaporization of ethanol is spent on expanding its vapour?

Problems*

Numerical problems

4.1 The temperature dependence of the vapour pressure of solid sulfur dioxide can be approximately represented by the relation $\log(p/\text{Torr}) = 10.5916 - 1871.2/(T/\text{K})$ and that of liquid sulfur dioxide by $\log(p/\text{Torr}) = 8.3186 - 1425.7/(T/\text{K})$. Estimate the temperature and pressure of the triple point of sulfur dioxide.

4.2 Prior to the discovery that freon-12 (CF_2Cl_2) was harmful to the Earth's ozone layer, it was frequently used as the dispersing agent in spray cans for hair spray, etc. Its enthalpy of vaporization at its normal boiling point of −29.2°C is 20.25 kJ mol^{-1}. Estimate the pressure that a can of hair spray using freon-12 had to withstand at 40°C, the temperature of a can that has been standing in sunlight. Assume that $\Delta_{\text{vap}}H$ is a constant over the temperature range involved and equal to its value at −29.2°C.

4.3 The enthalpy of vaporization of a certain liquid is found to be 14.4 kJ mol^{-1} at 180 K, its normal boiling point. The molar volumes of the liquid and the vapour at the boiling point are 115 cm^3 mol^{-1} and 14.5 dm^3 mol^{-1}, respectively. (a) Estimate dp/dT from the Clapeyron equation and (b) the percentage error in its value if the Clausius–Clapeyron equation is used instead.

4.4 Calculate the difference in slope of the chemical potential against temperature on either side of (a) the normal freezing point of water and (b) the normal boiling point of water. (c) By how much does the chemical potential of water supercooled to −5.0°C exceed that of ice at that temperature?

4.5 Calculate the difference in slope of the chemical potential against pressure on either side of (a) the normal freezing point of water and (b) the normal boiling point of water. The densities of ice and water at 0°C are 0.917 g cm^{-3} and 1.000 g cm^{-3}, and those of water and water vapour at 100°C are 0.958 g cm^{-3} and 0.598 g dm^{-3}, respectively. By how much does the chemical potential of water vapour exceed that of liquid water at 1.2 atm and 100°C?

4.6 The enthalpy of fusion of mercury is 2.292 kJ mol^{-1}, and its normal freezing point is 234.3 K with a change in molar volume of +0.517 cm^3 mol^{-1} on melting. At what temperature will the bottom of a column of mercury (density 13.6 g cm^{-3}) of height 10.0 m be expected to freeze?

4.7 50.0 dm^3 of dry air was slowly bubbled through a thermally insulated beaker containing 250 g of water initially at 25°C. Calculate the final temperature. (The vapour pressure of water is approximately constant at 3.17 kPa throughout, and its heat capacity is 75.5 J K^{-1} mol^{-1}. Assume that the air is not heated or cooled and that water vapour is a perfect gas.)

4.8 The vapour pressure, p, of nitric acid varies with temperature as follows:

θ/°C	0	20	40	50	70	80	90	100
p/kPa	1.92	6.38	17.7	27.7	62.3	89.3	124.9	170.9

What are (a) the normal boiling point and (b) the enthalpy of vaporization of nitric acid?

4.9 The vapour pressure of the ketone carvone (M = 150.2 g mol^{-1}), a component of oil of spearmint, is as follows:

θ/°C	57.4	100.4	133.0	157.3	203.5	227.5
p/Torr	1.00	10.0	40.0	100	400	760

What are (a) the normal boiling point and (b) the enthalpy of vaporization of carvone?

4.10 Construct the phase diagram for benzene near its triple point at 36 Torr and 5.50°C using the following data: $\Delta_{\text{fus}}H$ = 10.6 kJ mol^{-1}, $\Delta_{\text{vap}}H$ = 30.8 kJ mol^{-1}, ρ(s) = 0.891 g cm^{-3}, ρ(l) = 0.879 g cm^{-3}.

4.11‡ In an investigation of thermophysical properties of toluene, R.D. Goodwin (*J. Phys. Chem. Ref. Data* **18**, 1565 (1989)) presented expressions for two coexistence curves (phase boundaries). The solid–liquid coexistence curve is given by

$$p/\text{bar} = p_3/\text{bar} + 1000 \times (5.60 + 11.727x)x$$

where $x = T/T_3 - 1$ and the triple point pressure and temperature are p_3 = 0.4362 μbar and T_3 = 178.15 K. The liquid–vapour curve is given by:

$$\ln(p/\text{bar}) = -10.418/y + 21.157 - 15.996y + 14.015y^2 - 5.0120y^3 + 4.7224(1-y)^{1.70}$$

where $y = T/T_c = T/(593.95\text{ K})$. (a) Plot the solid–liquid and liquid–vapour phase boundaries. (b) Estimate the standard melting point of toluene. (c) Estimate the standard boiling point of toluene. (d) Compute the standard enthalpy of vaporization of toluene, given that the molar volumes of the liquid and vapour at the normal boiling point are 0.12 dm^3 mol^{-1} and 30.3 dm^3 mol^{-1}, respectively.

4.12‡ In a study of the vapour pressure of chloromethane, A. Bah and N. Dupont-Pavlovsky (*J. Chem. Eng. Data* **40**, 869 (1995)) presented data for the vapour pressure over solid chloromethane at low temperatures. Some of that data is shown below:

T/K	145.94	147.96	149.93	151.94	153.97	154.94
p/Pa	13.07	18.49	25.99	36.76	50.86	59.56

Estimate the standard enthalpy of sublimation of chloromethane at 150 K. (Take the molar volume of the vapour to be that of a perfect gas, and that of the solid to be negligible.)

Theoretical problems

4.13 Show that, for a transition between two incompressible solid phases, ΔG is independent of the pressure.

* Problems denoted by the symbol ‡ were supplied by Charles Trapp, Carmen Giunta, and Marshall Cady.

4.14 The change in enthalpy is given by $\mathrm{d}H = C_p\mathrm{d}T + V\mathrm{d}p$. The Clapeyron equation relates $\mathrm{d}p$ and $\mathrm{d}T$ at equilibrium, and so in combination the two equations can be used to find how the enthalpy changes along a phase boundary as the temperature changes and the two phases remain in equilibrium. Show that $\mathrm{d}(\Delta H/T) = \Delta C_p \mathrm{d}\ln T$.

4.15 In the 'gas saturation method' for the measurement of vapour pressure, a volume V of gas (as measured at a temperature T and a pressure P) is bubbled slowly through the liquid that is maintained at the temperature T, and a mass loss m is measured. Show that the vapour pressure, p, of the liquid is related to its molar mass, M, by $p = AmP/(1 + Am)$, where $A = RT/MPV$. The vapour pressure of geraniol ($M = 154.2\ \mathrm{g\ mol^{-1}}$), which is a component of oil of roses, was measured at 110°C. It was found that, when 5.00 $\mathrm{dm^3}$ of nitrogen at 760 Torr was passed slowly through the heated liquid, the loss of mass was 0.32 g. Calculate the vapour pressure of geraniol.

4.16 The vapour pressure of a liquid in a gravitational field varies with the depth below the surface on account of the hydrostatic pressure exerted by the overlying liquid. Adapt eqn. 4.4 to predict how the vapour pressure of a liquid of molar mass M varies with depth. Estimate the effect on the vapour pressure of water at 25°C in a column 10 m high.

4.17 Combine the barometric formula (stated in *Impact I1.1*) for the dependence of the pressure on altitude with the Clausius–Clapeyron equation, and predict how the boiling temperature of a liquid depends on the altitude and the ambient temperature. Take the mean ambient temperature as 20°C and predict the boiling temperature of water at 3000 m.

4.18 Figure 4.12 gives a schematic representation of how the chemical potentials of the solid, liquid, and gaseous phases of a substance vary with temperature. All have a negative slope, but it is unlikely that they are truly straight lines as indicated in the illustration. Derive an expression for the curvatures (specifically, the second derivatives with respect to temperature) of these lines. Is there a restriction on the curvature of these lines? Which state of matter shows the greatest curvature?

4.19 The Clapeyron equation does not apply to second-order phase transitions, but there are two analogous equations, the *Ehrenfest equations*, that do. They are:

$$\frac{\mathrm{d}p}{\mathrm{d}T} = \frac{\alpha_2 - \alpha_1}{\kappa_{T,2} - \kappa_{T,1}} \qquad \frac{\mathrm{d}p}{\mathrm{d}T} = \frac{C_{p,\mathrm{m}2} - C_{p,\mathrm{m}1}}{TV_\mathrm{m}(\alpha_2 - \alpha_1)}$$

where α is the expansion coefficient, κ_T the isothermal compressibility, and the subscripts 1 and 2 refer to two different phases. Derive these two equations. Why does the Clapeyron equation not apply to second-order transitions?

4.20 For a first-order phase transition, to which the Clapeyron equation does apply, prove the relation

$$C_S = C_p - \frac{\alpha V \Delta_{\mathrm{trs}} H}{\Delta_{\mathrm{trs}} V}$$

where $C_S = (\partial q/\partial T)_S$ is the heat capacity along the coexistence curve of two phases.

Applications: to biology and engineering

4.21 Proteins are polypeptides, polymers of amino acids, that can exist in ordered structures stabilized by a variety of molecular interactions. However, when certain conditions are changed, the compact structure of a polypeptide chain may collapse into a random coil. This structural change may be regarded as a phase transition occurring at a characteristic transition temperature, the *melting temperature*, T_m, which increases with the strength and number of intermolecular interactions in the chain. A thermodynamic treatment allows predictions to be made of the temperature T_m for the unfolding of a helical polypeptide held together by hydrogen bonds into a random coil. If a polypeptide has n amino acids, $n - 4$ hydrogen bonds are formed to form an α-helix, the most common type of helix in naturally occurring proteins (see Chapter 18). Because the first and last residues in the chain are free to move, $n - 2$ residues form the compact helix and have restricted motion. Based on these ideas, the molar Gibbs energy of unfolding of a polypeptide with $n \geq 5$ may be written as

$$\Delta G_\mathrm{m} = (n-4)\Delta_{\mathrm{hb}}H_\mathrm{m} - (n-2)T\Delta_{\mathrm{hb}}S_\mathrm{m}$$

where $\Delta_{\mathrm{hb}}H_\mathrm{m}$ and $\Delta_{\mathrm{hb}}S_\mathrm{m}$ are, respectively, the molar enthalpy and entropy of dissociation of hydrogen bonds in the polypeptide. (a) Justify the form of the equation for the Gibbs energy of unfolding. That is, why are the enthalpy and entropy terms written as $(n-4)\Delta_{\mathrm{hb}}H_\mathrm{m}$ and $(n-2)\Delta_{\mathrm{hb}}S_\mathrm{m}$, respectively? (b) Show that T_m may be written as

$$T_\mathrm{m} = \frac{(n-4)\Delta_{\mathrm{hb}}H_\mathrm{m}}{(n-2)\Delta_{\mathrm{hb}}S_\mathrm{m}}$$

(c) Plot $T_\mathrm{m}/(\Delta_{\mathrm{hb}}H_\mathrm{m}/\Delta_{\mathrm{hb}}S_\mathrm{m})$ for $5 \leq n \leq 20$. At what value of n does T_m change by less than 1 per cent when n increases by one?

4.22‡ The use of supercritical fluids as mobile phases in SFC depends on their properties as nonpolar solvents. The solubility parameter, δ, is defined as $(\Delta U_{\mathrm{cohesive}}/V_\mathrm{m})^{1/2}$, where $\Delta U_{\mathrm{cohesive}}$ is the cohesive energy of the solvent, the energy per mole needed to increase the volume isothermally to an infinite value. Diethyl ether, carbon tetrachloride, and dioxane have solubility parameter ranges of 7–8, 8–9, and 10–11, respectively. (a) Derive a practical equation for the computation of the isotherms for the reduced internal energy change, $\Delta U_\mathrm{r}(T_\mathrm{r},V_\mathrm{r})$ defined as

$$\Delta U_\mathrm{r}(T_\mathrm{r},V_\mathrm{r}) = \frac{U_\mathrm{r}(T_\mathrm{r},V_\mathrm{r}) - U_\mathrm{r}(T_\mathrm{r},\infty)}{p_\mathrm{c}V_\mathrm{c}}$$

(b) Draw a graph of ΔU_r against p_r for the isotherms $T_\mathrm{r} = 1, 1.2$, and 1.5 in the reduced pressure range for which $0.7 \leq V_\mathrm{r} \leq 2$. (c) Draw a graph of δ against p_r for the carbon dioxide isotherms $T_\mathrm{r} = 1$ and 1.5 in the reduced pressure range for which $1 \leq V_\mathrm{r} \leq 3$. In what pressure range at $T_\mathrm{f} = 1$ will carbon dioxide have solvent properties similar to those of liquid carbon tetrachloride? *Hint.* Use mathematical software or a spreadsheet.

4.23‡ A substance as well known as methane still receives research attention because it is an important component of natural gas, a commonly used fossil fuel . Friend *et al.* have published a review of thermophysical properties of methane (D.G. Friend, J.F. Ely, and H. Ingham, *J. Phys. Chem. Ref. Data* **18**, 583 (1989)), which included the following data describing the liquid–vapour phase boundary.

T/K	100	108	110	112	114	120	130	140	150	160	170	190
p/MPa	0.034	0.074	0.088	0.104	0.122	0.192	0.368	0.642	1.041	1.593	2.329	4.521

(a) Plot the liquid–vapour phase boundary. (b) Estimate the standard boiling point of methane. (c) Compute the standard enthalpy of vaporization of methane, given that the molar volumes of the liquid and vapour at the standard boiling point are 3.80×10^{-2} and 8.89 $\mathrm{dm^3\ mol^{-1}}$, respectively.

4.24‡ Diamond is the hardest substance and the best conductor of heat yet characterized. For these reasons, it is used widely in industrial applications that require a strong abrasive. Unfortunately, it is difficult to synthesize diamond from the more readily available allotropes of carbon, such as graphite. To illustrate this point, calculate the pressure required to convert graphite into diamond at 25°C. The following data apply to 25°C and 100 kPa. Assume the specific volume, V_s, and κ_T are constant with respect to pressure changes.

	Graphite	Diamond
$\Delta_\mathrm{f}G^\ominus/(\mathrm{kJ\ mol^{-1}})$	0	+2.8678
$V_\mathrm{s}/(\mathrm{cm^3\ g^{-1}})$	0.444	0.284
κ_T/kPa	3.04×10^{-8}	0.187×10^{-8}

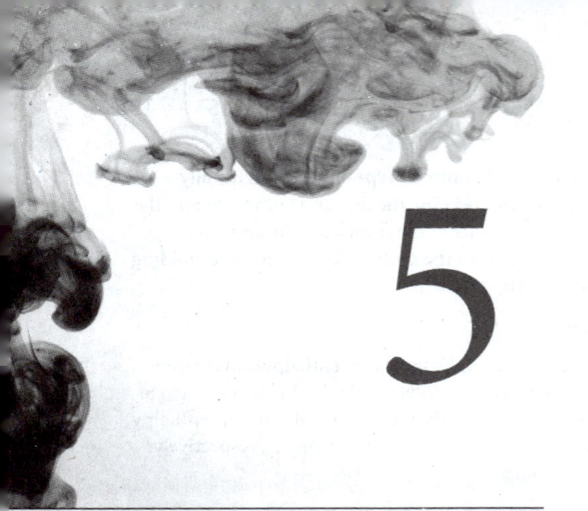

5 Simple mixtures

This chapter begins by developing the concept of chemical potential to show that it is a particular case of a class of properties called partial molar quantities. Then it explores how to use the chemical potential of a substance to describe the physical properties of mixtures. The underlying principle to keep in mind is that at equilibrium the chemical potential of a species is the same in every phase. We see, by making use of the experimental observations known as Raoult's and Henry's laws, how to express the chemical potential of a substance in terms of its mole fraction in a mixture. With this result established, we can calculate the effect of a solute on certain thermodynamic properties of a solution. These properties include the lowering of vapour pressure of the solvent, the elevation of its boiling point, the depression of its freezing point, and the origin of osmotic pressure. We then see how to construct and interpret phase diagrams that summarize the properties of binary mixtures over a wide range of compositions. The chapter introduces systems of gradually increasing complexity. In each case we shall see how the phase diagram for the system summarizes empirical observations on the conditions under which the various phases of the system are stable. Finally, we see how to express the chemical potential of a substance in a real mixture in terms of a property known as the activity. We see how the activity may be measured and conclude with a discussion of how the standard states of solutes and solvents are defined and ion–ion interactions are taken into account in electrolyte solutions.

Chemistry deals with mixtures, including mixtures of substances that can react together. Therefore, we need to generalize the concepts introduced so far to deal with substances that are mingled together. As a first step towards dealing with chemical reactions (which are treated in the next chapter), here we consider mixtures of substances that do not react together. At this stage we deal mainly with **binary mixtures**, which are mixtures of two components, A and B. We shall therefore often be able to simplify equations by making use of the relation $x_A + x_B = 1$.

The thermodynamic description of mixtures

We have already seen that the partial pressure, which is the contribution of one component to the total pressure, is used to discuss the properties of mixtures of gases. For a more general description of the thermodynamics of mixtures we need to introduce other analogous 'partial' properties.

5.1 Partial molar quantities

Key points **(a) The partial molar volume of a substance is the contribution to the volume that a substance makes when it is part of a mixture. (b) The chemical potential is the partial molar Gibbs energy and enables us to express the dependence of the Gibbs energy on the composition of a mixture. (c) The chemical potential also shows how, under a variety of different conditions, the thermodynamic functions vary with composition. (d) The Gibbs–Duhem equation shows how the changes in chemical potential of the components of a mixture are related.**

The easiest partial molar property to visualize is the 'partial molar volume', the contribution that a component of a mixture makes to the total volume of a sample.

(a) Partial molar volume

Imagine a huge volume of pure water at 25°C. When a further 1 mol H_2O is added, the volume increases by 18 cm^3 and we can report that 18 $cm^3\ mol^{-1}$ is the molar volume of pure water. However, when we add 1 mol H_2O to a huge volume of pure ethanol, the volume increases by only 14 cm^3. The reason for the different increase in volume is that the volume occupied by a given number of water molecules depends on the identity of the molecules that surround them. In the latter case there is so much ethanol present that each H_2O molecule is surrounded by ethanol molecules. The network of hydrogen bonds that normally hold H_2O molecules at certain distances from each other in pure water does not form. The packing of the molecules in the mixture results in the H_2O molecules increasing the volume by only 14 cm^3. The quantity 14 $cm^3\ mol^{-1}$ is the partial molar volume of water in pure ethanol. In general, the **partial molar volume** of a substance A in a mixture is the change in volume per mole of A added to a large volume of the mixture.

The partial molar volumes of the components of a mixture vary with composition because the environment of each type of molecule changes as the composition changes from pure A to pure B. It is this changing molecular environment, and the consequential modification of the forces acting between molecules, that results in the variation of the thermodynamic properties of a mixture as its composition is changed. The partial molar volumes of water and ethanol across the full composition range at 25°C are shown in Fig. 5.1.

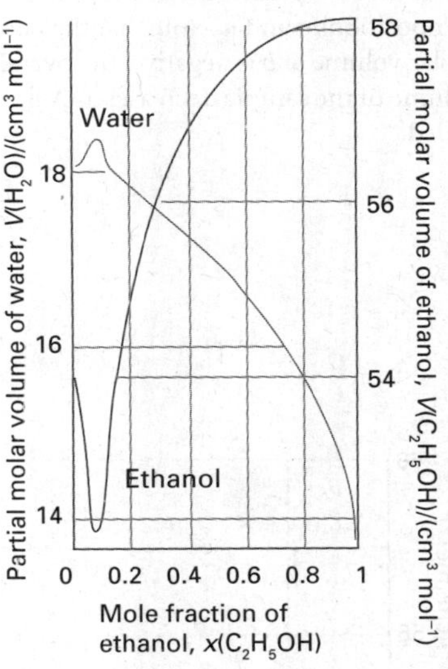

Fig. 5.1 The partial molar volumes of water and ethanol at 25°C. Note the different scales (water on the left, ethanol on the right).

The partial molar volume, V_J, of a substance J at some general composition is defined formally as follows:

$$V_J = \left(\frac{\partial V}{\partial n_J}\right)_{p,T,n'} \qquad \text{Definition of partial molar volume} \qquad (5.1)$$

where the subscript n' signifies that the amounts of all other substances present are constant. The partial molar volume is the slope of the plot of the total volume as the amount of J is changed, the pressure, temperature, and amount of the other components being constant (Fig. 5.2). Its value depends on the composition, as we saw for water and ethanol.

The definition in eqn 5.1 implies that, when the composition of the mixture is changed by the addition of dn_A of A and dn_B of B, then the total volume of the mixture changes by

$$dV = \left(\frac{\partial V}{\partial n_A}\right)_{p,T,n_B} dn_A + \left(\frac{\partial V}{\partial n_B}\right)_{p,T,n_A} dn_B = V_A dn_A + V_B dn_B \qquad (5.2)$$

Provided the relative composition is held constant as the amounts of A and B are increased, we can obtain the final volume by integration:

A note on good practice The IUPAC recommendation is to denote a partial molar quantity by $\bar{X}$, but only when there is the possibility of confusion with the quantity X. For instance, the partial molar volume of NaCl in water could be written $V(\text{NaCl, aq})$ to distinguish it from the volume of the solution, V.

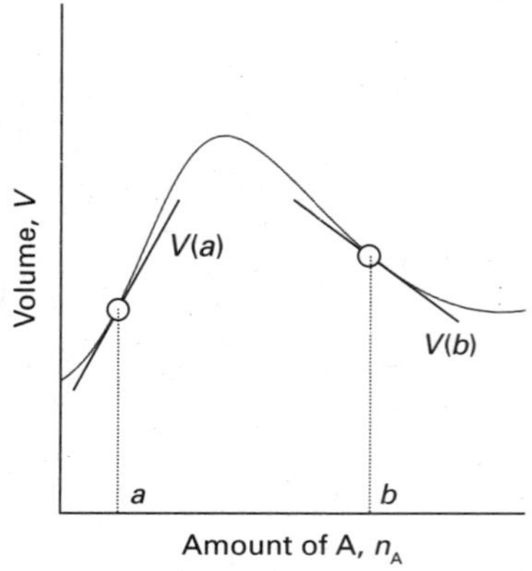

Fig. 5.2 The partial molar volume of a substance is the slope of the variation of the total volume of the sample plotted against the composition. In general, partial molar quantities vary with the composition, as shown by the different slopes at the compositions *a* and *b*. Note that the partial molar volume at *b* is negative: the overall volume of the sample decreases as A is added.

$$V=\int_0^{n_A} V_A \mathrm{d}n_A+\int_0^{n_B} V_B \mathrm{d}n_B=V_A\int_0^{n_A} \mathrm{d}n_A+V_B\int_0^{n_B} \mathrm{d}n_B$$
$$=V_An_A+V_Bn_B \qquad (5.3)$$

Although we have envisaged the two integrations as being linked (in order to preserve constant relative composition), because V is a state function the final result in eqn 5.3 is valid however the solution is in fact prepared.

Partial molar volumes can be measured in several ways. One method is to measure the dependence of the volume on the composition and to fit the observed volume to a function of the amount of the substance. Once the function has been found, its slope can be determined at any composition of interest by differentiation.

- **A brief illustration**

A polynomial fit to measurements of the total volume of a water/ethanol mixture at 25°C that contains 1.000 kg of water is

$$v=1002.93+54.6664x-0.363\,94x^2+0.028\,256x^3$$

where $v=V/\mathrm{cm}^3$, $x=n_E/\mathrm{mol}$, and n_E is the amount of CH_3CH_2OH present. The partial molar volume of ethanol, V_E, is therefore

$$V_E=\left(\frac{\partial V}{\partial n_E}\right)_{p,T,n_W}=\left(\frac{\partial (V/\mathrm{cm}^3)}{\partial (n_E/\mathrm{mol})}\right)_{p,T,n_W}\frac{\mathrm{cm}^3}{\mathrm{mol}}=\left(\frac{\partial v}{\partial x}\right)_{p,T,n_W}\mathrm{cm}^3\,\mathrm{mol}^{-1}$$

Then, because

$$\frac{\mathrm{d}v}{\mathrm{d}x}=54.6664-2(0.363\,94)x+3(0.028\,256)x^2$$

we can conclude that

$$V_E/(\mathrm{cm}^3\,\mathrm{mol}^{-1})=54.6664-0.72788x+0.084768x^2$$

Figure 5.3 is a graph of this function. ●

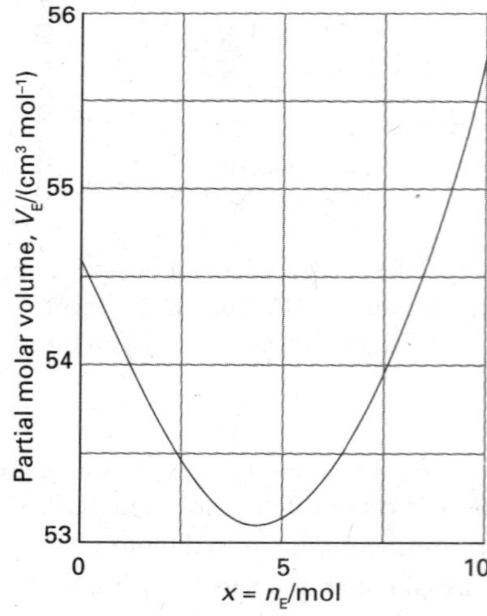

Fig. 5.3 The partial molar volume of ethanol as expressed by the polynomial in the *brief illustration*.

interActivity Using the data from the *brief illustration*, determine the value of *b* at which V_E has a minimum value.

Self-test 5.1 At 25°C, the density of a 50 per cent by mass ethanol/water solution is 0.914 g cm^{-3}. Given that the partial molar volume of water in the solution is 17.4 cm^3 mol^{-1}, what is the partial molar volume of the ethanol? [56.4 cm^3 mol^{-1}]

Molar volumes are always positive, but partial molar quantities need not be. For example, the limiting partial molar volume of $MgSO_4$ in water (its partial molar volume in the limit of zero concentration) is −1.4 cm^3 mol^{-1}, which means that the addition of 1 mol $MgSO_4$ to a large volume of water results in a decrease in volume of 1.4 cm^3. The mixture contracts because the salt breaks up the open structure of water as the Mg^{2+} and SO_4^{2-} ions become hydrated, and it collapses slightly.

(b) Partial molar Gibbs energies

The concept of a partial molar quantity can be extended to any extensive state function. For a substance in a mixture, the chemical potential is *defined* as the partial molar Gibbs energy:

$$\mu_J=\left(\frac{\partial G}{\partial n_J}\right)_{p,T,n'} \qquad \text{Definition of chemical potential} \qquad [5.4]$$

That is, the chemical potential is the slope of a plot of Gibbs energy against the amount of the component J, with the pressure and temperature (and the amounts of the other substances) held constant (Fig. 5.4). For a pure substance we can write $G = n_J G_{J,m}$, and from eqn 5.4 obtain $\mu_J = G_{J,m}$: in this case, the chemical potential is simply the molar Gibbs energy of the substance, as we saw in Chapter 4.

By the same argument that led to eqn 5.3, it follows that the total Gibbs energy of a binary mixture is

$$G = n_A\mu_A + n_B\mu_B \tag{5.5}$$

where μ_A and μ_B are the chemical potentials at the composition of the mixture. That is, the chemical potential of a substance in a mixture is the contribution of that substance to the total Gibbs energy of the mixture. Because the chemical potentials depend on composition (and the pressure and temperature), the Gibbs energy of a mixture may change when these variables change, and, for a system of components A, B, etc., the equation $dG = Vdp - SdT$ becomes

$$dG = Vdp - SdT + \mu_A dn_A + \mu_B dn_B + \cdots \tag{5.6}$$

Fundamental equation of chemical thermodynamics

This expression is the **fundamental equation of chemical thermodynamics**. Its implications and consequences are explored and developed in this and the next two chapters.

At constant pressure and temperature, eqn 5.6 simplifies to

$$dG = \mu_A dn_A + \mu_B dn_B + \cdots \tag{5.7}$$

We saw in Section 3.5e that under the same conditions $dG = dw_{add,max}$. Therefore, at constant temperature and pressure,

$$dw_{add,max} = \mu_A dn_A + \mu_B dn_B + \cdots \tag{5.8}$$

That is, additional (non-expansion) work can arise from the changing composition of a system. For instance, in an electrochemical cell, the chemical reaction is arranged to take place in two distinct sites (at the two electrodes). The electrical work the cell performs can be traced to its changing composition as products are formed from reactants.

(c) The wider significance of the chemical potential

The chemical potential does more than show how G varies with composition. Because $G = U + pV - TS$, and therefore $U = -pV + TS + G$, we can write a general infinitesimal change in U for a system of variable composition as

$$\begin{aligned} dU &= -pdV - Vdp + SdT + TdS + dG \\ &= -pdV - Vdp + SdT + TdS + (Vdp - SdT + \mu_A dn_A + \mu_B dn_B + \cdots) \\ &= -pdV + TdS + \mu_A dn_A + \mu_B dn_B + \cdots \end{aligned}$$

This expression is the generalization of eqn 3.46 (that $dU = TdS - pdV$) to systems in which the composition may change. It follows that, at constant volume and entropy,

$$dU = \mu_A dn_A + \mu_B dn_B + \cdots \tag{5.9}$$

and hence that

$$\mu_J = \left(\frac{\partial U}{\partial n_J}\right)_{S,V,n'} \tag{5.10}$$

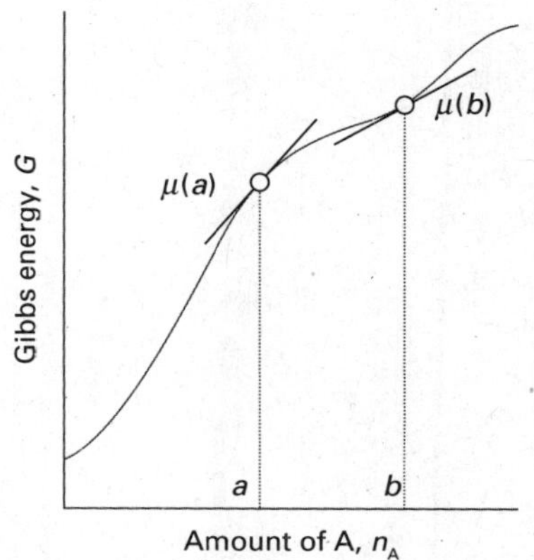

Fig. 5.4 The chemical potential of a substance is the slope of the total Gibbs energy of a mixture with respect to the amount of substance of interest. In general, the chemical potential varies with composition, as shown for the two values at a and b. In this case, both chemical potentials are positive.

Therefore, not only does the chemical potential show how G changes when the composition changes, it also shows how the internal energy changes too (but under a different set of conditions). In the same way it is easy to deduce that

$$\text{(a) } \mu_J = \left(\frac{\partial H}{\partial n_J}\right)_{S,p,n'} \qquad \text{(b) } \mu_J = \left(\frac{\partial A}{\partial n_J}\right)_{T,V,n'} \tag{5.11}$$

Thus we see that the μ_J shows how all the extensive thermodynamic properties U, H, A, and G depend on the composition. This is why the chemical potential is so central to chemistry.

(d) The Gibbs–Duhem equation

Because the total Gibbs energy of a binary mixture is given by eqn 5.5 and the chemical potentials depend on the composition, when the compositions are changed infinitesimally we might expect G of a binary system to change by

$$dG = \mu_A dn_A + \mu_B dn_B + n_A d\mu_A + n_B d\mu_B$$

However, we have seen that at constant pressure and temperature a change in Gibbs energy is given by eqn 5.7. Because G is a state function, these two equations must be equal, which implies that at constant temperature and pressure

$$n_A d\mu_A + n_B d\mu_B = 0 \tag{5.12a}$$

This equation is a special case of the **Gibbs–Duhem equation**:

$$\sum_J n_J d\mu_J = 0 \qquad \text{Gibbs–Duhem equation} \tag{5.12b}$$

The significance of the Gibbs–Duhem equation is that the chemical potential of one component of a mixture cannot change independently of the chemical potentials of the other components. In a binary mixture, if one partial molar quantity increases, then the other must decrease, with the two changes related by

$$d\mu_B = -\frac{n_A}{n_B} d\mu_A \tag{5.13}$$

The same line of reasoning applies to all partial molar quantities. We can see in Fig. 5.1, for example, that where the partial molar volume of water increases, that of ethanol decreases. Moreover, as eqn 5.13 shows, and as we can see from Fig. 5.1, a small change in the partial molar volume of A corresponds to a large change in the partial molar volume of B if n_A/n_B is large, but the opposite is true when this ratio is small. In practice, the Gibbs–Duhem equation is used to determine the partial molar volume of one component of a binary mixture from measurements of the partial molar volume of the second component.

A brief comment

The *molar concentration* (colloquially, the 'molarity', [J] or c_J) is the amount of solute divided by the volume of the solution and is usually expressed in moles per cubic decimetre (mol dm^{-3}). We write $c^{\ominus} = 1$ mol dm^{-3}. The term *molality*, b, is the amount of solute divided by the mass of solvent and is usually expressed in moles per kilogram of solvent (mol kg^{-1}). We write $b^{\ominus} = 1$ mol kg^{-1}.

Example 5.1 *Using the Gibbs–Duhem equation*

The experimental values of the partial molar volume of K_2SO_4(aq) at 298 K are found to fit the expression

$$v_B = 32.280 + 18.216x^{1/2}$$

where $v_B = V_{K_2SO_4}/(\text{cm}^3\ \text{mol}^{-1})$ and x is the numerical value of the molality of K_2SO_4 ($x = b/b^{\ominus}$; see the *brief comment* in the margin). Use the Gibbs–Duhem equation to derive an equation for the molar volume of water in the solution. The molar volume of pure water at 298 K is 18.079 cm^3 mol^{-1}.

$V_B + V_W = V$

Method Let A denote H_2O, the solvent, and B denote K_2SO_4, the solute. The Gibbs–Duhem equation for the partial molar volumes of two components is $n_A dV_A + n_B dV_B = 0$. This relation implies that $dv_A = -(n_B/n_A)dv_B$, and therefore that v_A can be found by integration:

$$v_A = v_A^* - \int_0^{v_B} \frac{n_B}{n_A} dv_B$$

where $v_A^* = V_A/(cm^3 \, mol^{-1})$ is the numerical value of the molar volume of pure A. The first step is to change the variable v_B to $x = b/b^\ominus$ and then to integrate the right-hand side between $x = 0$ (pure B) and the molality of interest.

Answer It follows from the information in the question that, with B = K_2SO_4, $dv_B/dx = 9.108x^{-1/2}$. Therefore, the integration required is

$$v_A = v_A^* - 9.108 \int_0^{b/b^\ominus} \frac{n_B}{n_A} x^{-1/2} dx$$

However, the ratio of amounts of A (H_2O) and B (K_2SO_4) is related to the molality of B, $b = n_B/(1 \text{ kg water})$ and $n_A = (1 \text{ kg water})/M_A$ where M_A is the molar mass of water, by

$$\frac{n_B}{n_A} = \frac{n_B}{(1 \text{ kg})/M_A} = \frac{n_B M_A}{1 \text{ kg}} = bM_A = xb^\ominus M_A$$

and hence

$$v_A = v_A^* - 9.108 M_A b^\ominus \int_0^{b/b^\ominus} x^{1/2} dx = v_A^* - \frac{2}{3}(9.108 M_A b^\ominus)(b/b^\ominus)^{3/2}$$

It then follows, by substituting the data (including $M_A = 1.802 \times 10^{-2}$ kg mol^{-1}, the molar mass of water), that

$$V_A/(cm^3 \, mol^{-1}) = 18.079 - 0.1094(b/b^\ominus)^{3/2}$$

The partial molar volumes are plotted in Fig. 5.5.

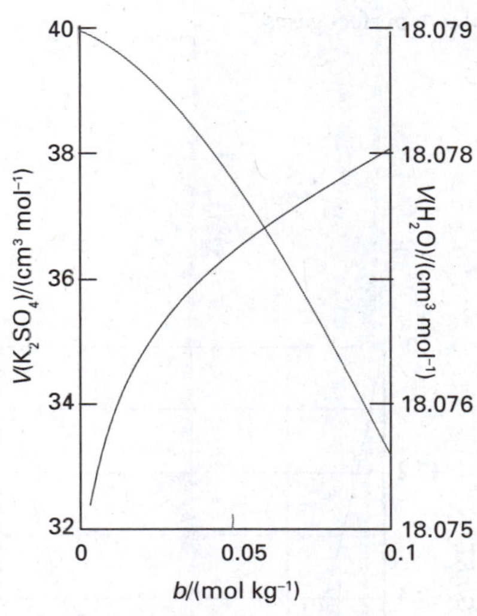

Fig. 5.5 The partial molar volumes of the components of an aqueous solution of potassium sulfate. The blue curve corresponds to water and the purple curve to potassium sulfate.

Self-test 5.2 Repeat the calculation for a salt B for which $V_B/(cm^3 \, mol^{-1}) = 6.218 + 5.146b - 7.147b^2$. $[V_A/(cm^3 \, mol^{-1}) = 18.079 - 0.0464b^2 + 0.0859b^3]$

5.2 The thermodynamics of mixing

Key points **(a) The Gibbs energy of mixing is calculated by forming the difference of the Gibbs energies before and after mixing: the quantity is negative for perfect gases at the same pressure. (b) The entropy of mixing of perfect gases initially at the same pressure is positive and the enthalpy of mixing is zero.**

The dependence of the Gibbs energy of a mixture on its composition is given by eqn 5.5, and we know that at constant temperature and pressure systems tend towards lower Gibbs energy. This is the link we need in order to apply thermodynamics to the discussion of spontaneous changes of composition, as in the mixing of two substances. One simple example of a spontaneous mixing process is that of two gases introduced into the same container. The mixing is spontaneous, so it must correspond to a decrease in G. We shall now see how to express this idea quantitatively.

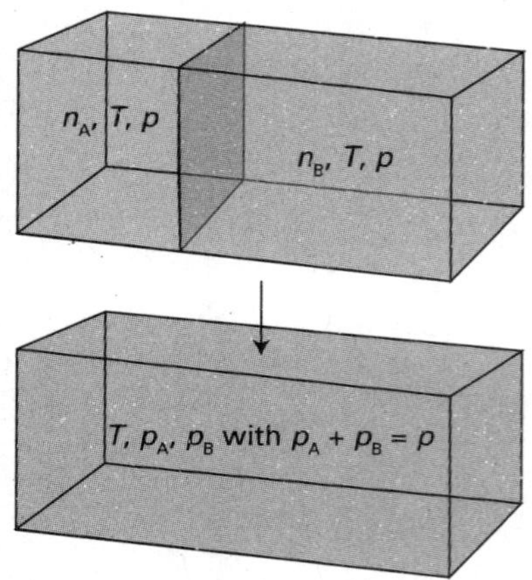

Fig. 5.6 The arrangement for calculating the thermodynamic functions of mixing of two perfect gases.

(a) The Gibbs energy of mixing of perfect gases

Let the amounts of two perfect gases in the two containers be n_A and n_B; both are at a temperature T and a pressure p (Fig. 5.6). At this stage, the chemical potentials of the two gases have their 'pure' values, which are obtained by applying the definition $\mu = G_m$ to eqn 3.60:

$$\mu = \mu^{\ominus} + RT \ln \frac{p}{p^{\ominus}} \qquad (5.14a)^{\circ}$$

Variation of chemical potential of a perfect gas with pressure

where $\mu^{\ominus}$ is the **standard chemical potential**, the chemical potential of the pure gas at 1 bar. It will be much simpler notationally if we agree to let p denote the pressure relative to $p^{\ominus}$; that is, to replace $p/p^{\ominus}$ by p, for then we can write

$$\mu = \mu^{\ominus} + RT \ln p \qquad \{5.14b\}^{\circ}$$

Equations for which this convention is used will be labelled {1}, {2}, . . . ; to use the equations, we have to remember to replace p by $p/p^{\ominus}$ again. In practice, that simply means using the numerical value of p in bars. The Gibbs energy of the total system is then given by eqn 5.5 as

$$G_i = n_A\mu_A + n_B\mu_B = n_A(\mu_A^{\ominus} + RT \ln p) + n_B(\mu_B^{\ominus} + RT \ln p) \qquad \{5.15a\}^{\circ}$$

After mixing, the partial pressures of the gases are p_A and p_B, with $p_A + p_B = p$. The total Gibbs energy changes to

$$G_f = n_A(\mu_A^{\ominus} + RT \ln p_A) + n_B(\mu_B^{\ominus} + RT \ln p_B) \qquad \{5.15b\}^{\circ}$$

The difference $G_f - G_i$, the **Gibbs energy of mixing,** $\Delta_{mix}G$, is therefore

$$\Delta_{mix}G = n_A RT \ln \frac{p_A}{p} + n_B RT \ln \frac{p_B}{p} \qquad (5.15c)^{\circ}$$

At this point we may replace n_J by $x_J n$, where n is the total amount of A and B, and use the relation between partial pressure and mole fraction (Section 1.2c) to write $p_J/p = x_J$ for each component, which gives

$$\Delta_{mix}G = nRT(x_A \ln x_A + x_B \ln x_B) \qquad (5.16)^{\circ}$$

Gibbs energy of mixing of perfect gases

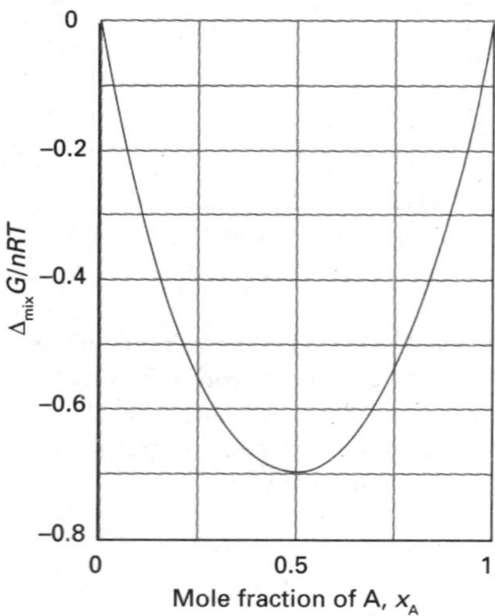

Fig. 5.7 The Gibbs energy of mixing of two perfect gases and (as discussed later) of two liquids that form an ideal solution. The Gibbs energy of mixing is negative for all compositions and temperatures, so perfect gases mix spontaneously in all proportions.

interActivity Draw graphs of $\Delta_{mix}G$ against x_A at different temperatures in the range 298 K to 500 K. For what value of x_A does $\Delta_{mix}G$ depend on temperature most strongly?

Because mole fractions are never greater than 1, the logarithms in this equation are negative, and $\Delta_{mix}G < 0$ (Fig. 5.7). The conclusion that $\Delta_{mix}G$ is negative for all compositions confirms that perfect gases mix spontaneously in all proportions. However, the equation extends common sense by allowing us to discuss the process quantitatively.

Example 5.2 *Calculating a Gibbs energy of mixing*

A container is divided into two equal compartments (Fig. 5.8). One contains 3.0 mol $H_2(g)$ at 25°C; the other contains 1.0 mol $N_2(g)$ at 25°C. Calculate the Gibbs energy of mixing when the partition is removed. Assume perfect behaviour.

Method Equation 5.16 cannot be used directly because the two gases are initially at different pressures. We proceed by calculating the initial Gibbs energy from the chemical potentials. To do so, we need the pressure of each gas. Write the pressure of nitrogen as p; then the pressure of hydrogen as a multiple of p can be found from the gas laws. Next, calculate the Gibbs energy for the system when the partition is removed. The volume occupied by each gas doubles, so its initial partial pressure is halved.

Answer Given that the pressure of nitrogen is p, the pressure of hydrogen is $3p$; therefore, the initial Gibbs energy is

$$G_i = (3.0\text{ mol})\{\mu^\ominus(H_2) + RT\ln 3p\} + (1.0\text{ mol})\{\mu^\ominus(N_2) + RT\ln p\}$$

When the partition is removed and each gas occupies twice the original volume, the partial pressure of nitrogen falls to $\frac{1}{2}p$ and that of hydrogen falls to $\frac{3}{2}p$. Therefore, the Gibbs energy changes to

$$G_f = (3.0\text{ mol})\{\mu^\ominus(H_2) + RT\ln \tfrac{3}{2}p\} + (1.0\text{ mol})\{\mu^\ominus(N_2) + RT\ln \tfrac{1}{2}p\}$$

The Gibbs energy of mixing is the difference of these two quantities:

$$\begin{aligned}\Delta_{mix}G &= (3.0\text{ mol})RT\ln\left(\frac{\frac{3}{2}p}{3p}\right) + (1.0\text{ mol})RT\ln\left(\frac{\frac{1}{2}p}{p}\right)\\ &= -(3.0\text{ mol})RT\ln 2 - (1.0\text{ mol})RT\ln 2\\ &= -(4.0\text{ mol})RT\ln 2 = -6.9\text{ kJ}\end{aligned}$$

In this example, the value of $\Delta_{mix}G$ is the sum of two contributions: the mixing itself, and the changes in pressure of the two gases to their final total pressure, $2p$. When 3.0 mol H_2 mixes with 1.0 mol N_2 at the same pressure, with the volumes of the vessels adjusted accordingly, the change of Gibbs energy is –5.6 kJ. However, do not be misled into interpreting this negative change in Gibbs energy as a sign of spontaneity: in this case, the pressure changes, and $\Delta G < 0$ is a signpost of spontaneous change only at constant temperature and pressure.

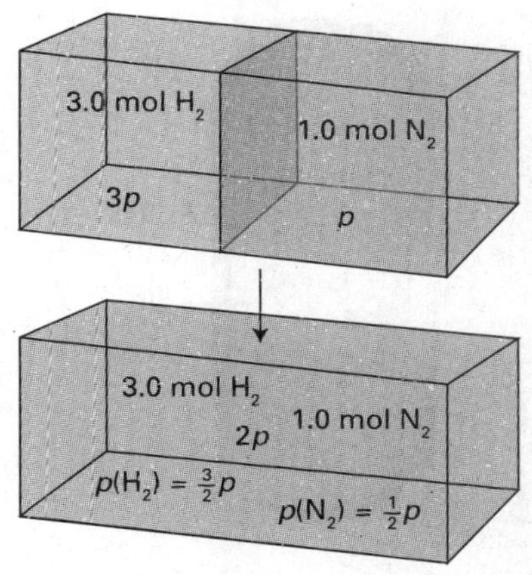

Fig. 5.8 The initial and final states considered in the calculation of the Gibbs energy of mixing of gases at different initial pressures.

Self-test 5.3 Suppose that 2.0 mol H_2 at 2.0 atm and 25°C and 4.0 mol N_2 at 3.0 atm and 25°C are mixed at constant volume. Calculate $\Delta_{mix}G$. What would be the value of $\Delta_{mix}G$ had the pressures been identical initially? [−9.7 kJ, −9.5 kJ]

(b) Other thermodynamic mixing functions

Because $(\partial G/\partial T)_{p,n} = -S$, it follows immediately from eqn 5.16 that, for a mixture of perfect gases initially at the same pressure, the entropy of mixing, $\Delta_{mix}S$, is

$$\Delta_{mix}S = -\left(\frac{\partial \Delta_{mix}G}{\partial T}\right)_{p,n_A,n_B} = -nR(x_A\ln x_A + x_B\ln x_B) \quad \text{Entropy of mixing of perfect gases} \quad (5.17)°$$

Because $\ln x < 0$, it follows that $\Delta_{mix}S > 0$ for all compositions (Fig. 5.9). For equal amounts of gas, for instance, we set $x_A = x_B = \frac{1}{2}$ and obtain $\Delta_{mix}S = nR\ln 2$, with n the total amount of gas molecules. This increase in entropy is what we expect when one gas disperses into the other and the disorder increases.

We can calculate the isothermal, isobaric (constant pressure) **enthalpy of mixing**, $\Delta_{mix}H$, the enthalpy change accompanying mixing, of two perfect gases from $\Delta G = \Delta H - T\Delta S$. It follows from eqns 5.16 and 5.17 that

$$\Delta_{mix}H = 0 \quad \text{Enthalpy of mixing of perfect gases} \quad (5.18)°$$

The enthalpy of mixing is zero, as we should expect for a system in which there are no interactions between the molecules forming the gaseous mixture. It follows that the whole of the driving force for mixing comes from the increase in entropy of the system because the entropy of the surroundings is unchanged.

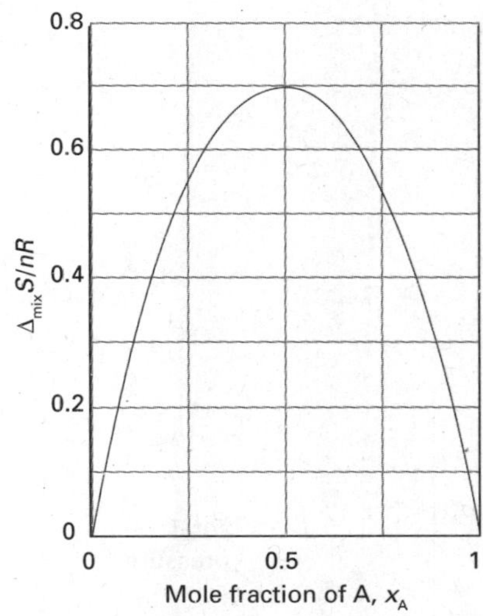

Fig. 5.9 The entropy of mixing of two perfect gases and (as discussed later) of two liquids that form an ideal solution. The entropy increases for all compositions and temperatures, so perfect gases mix spontaneously in all proportions. Because there is no transfer of heat to the surroundings when perfect gases mix, the entropy of the surroundings is unchanged. Hence, the graph also shows the total entropy of the system plus the surroundings when perfect gases mix.

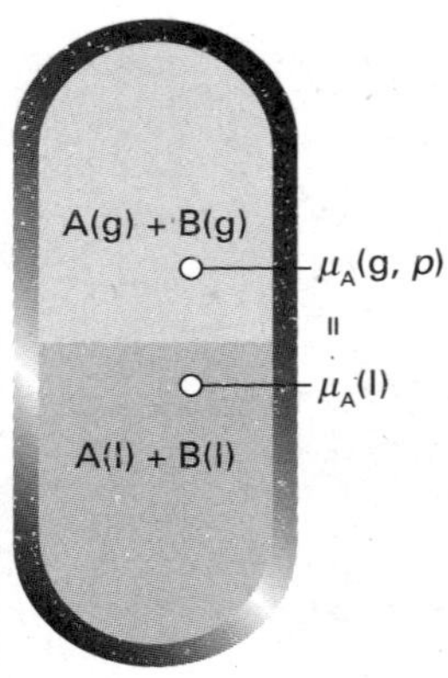

Fig. 5.10 At equilibrium, the chemical potential of the gaseous form of a substance A is equal to the chemical potential of its condensed phase. The equality is preserved if a solute is also present. Because the chemical potential of A in the vapour depends on its partial vapour pressure, it follows that the chemical potential of liquid A can be related to its partial vapour pressure.

5.3 The chemical potentials of liquids

Key points (a) Raoult's law provides a relation between the vapour pressure of a substance and its mole fraction in a mixture; it is the basis of the definition of an ideal solution. (b) Henry's law provides a relation between the vapour pressure of a solute and its mole fraction in a mixture; it is the basis of the definition of an ideal-dilute solution.

To discuss the equilibrium properties of liquid mixtures we need to know how the Gibbs energy of a liquid varies with composition. To calculate its value, we use the fact that, at equilibrium, the chemical potential of a substance present as a vapour must be equal to its chemical potential in the liquid.

(a) Ideal solutions

We shall denote quantities relating to pure substances by a superscript *, so the chemical potential of pure A is written μ_A^* and as $\mu_A^*(\mathrm{l})$ when we need to emphasize that A is a liquid. Because the vapour pressure of the pure liquid is p_A^* it follows from eqn 5.14 that the chemical potential of A in the vapour (treated as a perfect gas) is $\mu_A^\ominus + RT\ln p_A^*$ (with p_A to be interpreted as the relative pressure $p_A/p^\ominus$). These two chemical potentials are equal at equilibrium (Fig. 5.10), so we can write

$$\mu_A^* = \mu_A^\ominus + RT\ln p_A^* \qquad \{5.19a\}^\circ$$

If another substance, a solute, is also present in the liquid, the chemical potential of A in the liquid is changed to μ_A and its vapour pressure is changed to p_A. The vapour and solvent are still in equilibrium, so we can write

$$\mu_A = \mu_A^\ominus + RT\ln p_A \qquad \{5.19b\}^\circ$$

Next, we combine these two equations to eliminate the standard chemical potential of the gas. To do so, we write eqn 5.19a as $\mu_A^\ominus = \mu_A^* - RT\ln p_A^*$ and substitute this expression into eqn 5.19b to obtain

$$\mu_A = \mu_A^* - RT\ln p_A^* + RT\ln p_A = \mu_A^* + RT\ln\frac{p_A}{p_A^*} \qquad (5.20)^\circ$$

In the final step we draw on additional experimental information about the relation between the ratio of vapour pressures and the composition of the liquid. In a series of experiments on mixtures of closely related liquids (such as benzene and methylbenzene), the French chemist François Raoult found that the ratio of the partial vapour pressure of each component to its vapour pressure as a pure liquid, p_A/p_A^*, is approximately equal to the mole fraction of A in the liquid mixture. That is, he established what we now call **Raoult's law**:

$$p_A = x_A p_A^* \qquad \text{Raoult's law} \qquad (5.21)^\circ$$

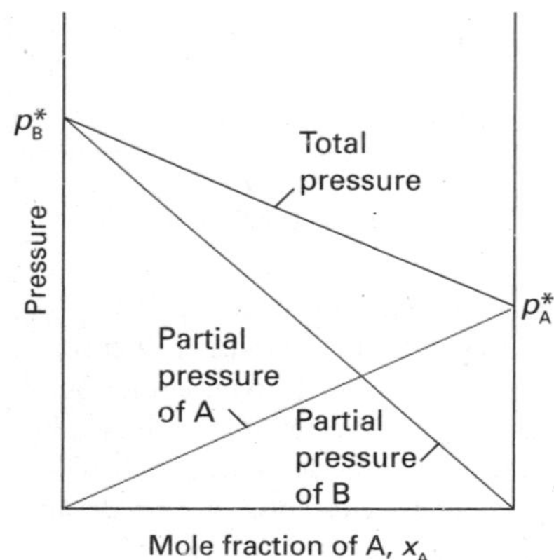

Fig. 5.11 The total vapour pressure and the two partial vapour pressures of an ideal binary mixture are proportional to the mole fractions of the components.

This law is illustrated in Fig. 5.11. Some mixtures obey Raoult's law very well, especially when the components are structurally similar (Fig. 5.12). Mixtures that obey the law throughout the composition range from pure A to pure B are called **ideal solutions**. When we write equations that are valid only for ideal solutions, we shall label them with a superscript $^\circ$, as in eqn 5.21.

For an ideal solution, it follows from eqns 5.20 and 5.21 that

$$\mu_A = \mu_A^* + RT\ln x_A \qquad \text{Chemical potential of component of an ideal solution} \qquad (5.22)^\circ$$

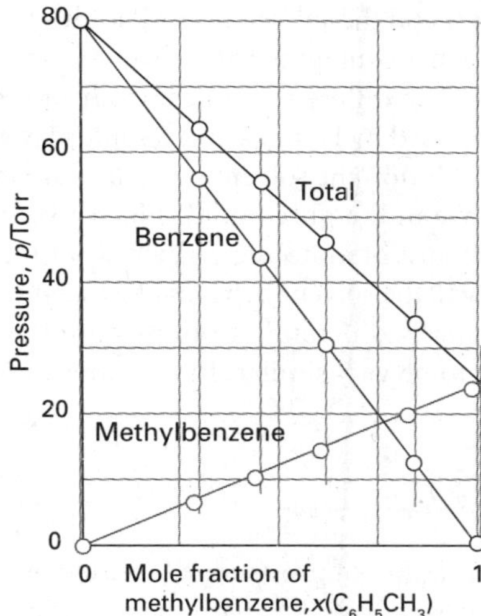

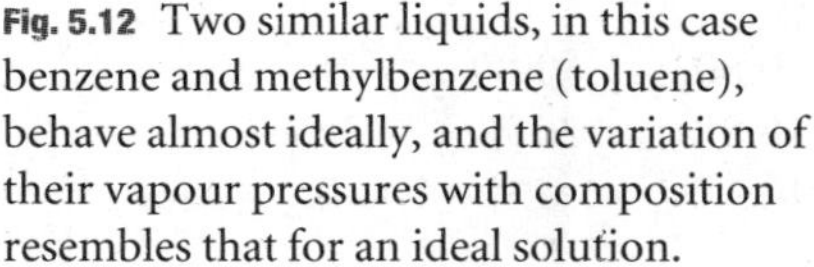

Fig. 5.12 Two similar liquids, in this case benzene and methylbenzene (toluene), behave almost ideally, and the variation of their vapour pressures with composition resembles that for an ideal solution.

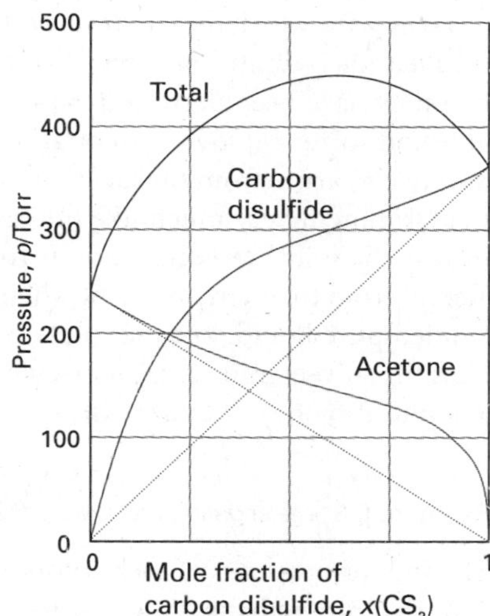

Fig. 5.13 Strong deviations from ideality are shown by dissimilar liquids (in this case carbon disulfide and acetone, propanone).

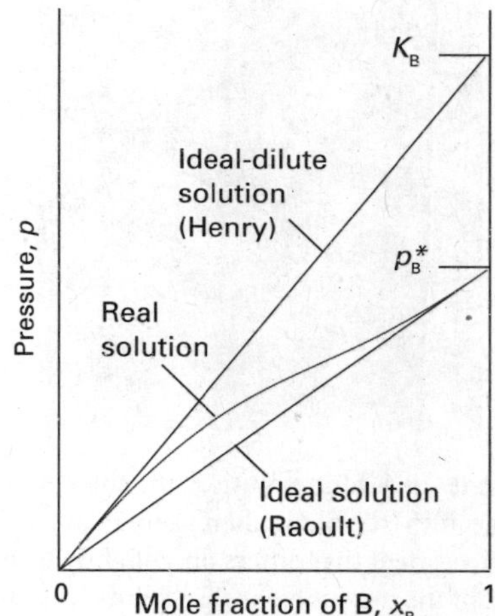

Fig. 5.14 When a component (the solvent) is nearly pure, it has a vapour pressure that is proportional to its mole fraction with a slope p_B^* (Raoult's law). When it is the minor component (the solute) its vapour pressure is still proportional to the mole fraction, but the constant of proportionality is now K_B (Henry's law).

This important equation can be used as the *definition* of an ideal solution (so that it implies Raoult's law rather than stemming from it). It is in fact a better definition than eqn 5.21 because it does not assume that the vapour is a perfect gas.

The molecular origin of Raoult's law is the effect of the solute on the entropy of the solution. In the pure solvent, the molecules have a certain disorder and a corresponding entropy; the vapour pressure then represents the tendency of the system and its surroundings to reach a higher entropy. When a solute is present, the solution has a greater disorder than the pure solvent because we cannot be sure that a molecule chosen at random will be a solvent molecule. Because the entropy of the solution is higher than that of the pure solvent, the solution has a lower tendency to acquire an even higher entropy by the solvent vaporizing. In other words, the vapour pressure of the solvent in the solution is lower than that of the pure solvent.

Some solutions depart significantly from Raoult's law (Fig. 5.13). Nevertheless, even in these cases the law is obeyed increasingly closely for the component in excess (the solvent) as it approaches purity. The law is therefore a good approximation for the properties of the solvent if the solution is dilute.

(b) Ideal-dilute solutions

In ideal solutions the solute, as well as the solvent, obeys Raoult's law. However, the English chemist William Henry found experimentally that, for real solutions at low concentrations, although the vapour pressure of the solute is proportional to its mole fraction, the constant of proportionality is not the vapour pressure of the pure substance (Fig. 5.14). **Henry's law** is:

$$p_B = x_B K_B \qquad \text{Henry's law} \qquad (5.23)^{\circ}$$

In this expression x_B is the mole fraction of the solute and K_B is an empirical constant (with the dimensions of pressure) chosen so that the plot of the vapour pressure of B against its mole fraction is tangent to the experimental curve at $x_B = 0$.

Fig. 5.15 In a dilute solution, the solvent molecules (the blue spheres) are in an environment that differs only slightly from that of the pure solvent. The solute particles, however, are in an environment totally unlike that of the pure solute.

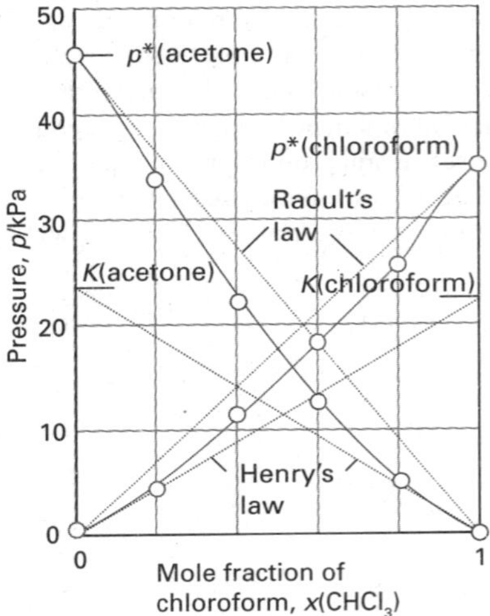

Fig. 5.16 The experimental partial vapour pressures of a mixture of chloroform (trichloromethane) and acetone (propanone) based on the data in *Example 5.3*. The values of K are obtained by extrapolating the dilute solution vapour pressures as explained in the *Example*.

Table 5.1* Henry's law constants for gases in water at 298 K

	K/(kPa kg mol^{-1})
CO_2	3.01×10^3
H_2	1.28×10^5
N_2	1.56×10^5
O_2	7.92×10^4

* More values are given in the *Data section*.

Mixtures for which the solute obeys Henry's law and the solvent obeys Raoult's law are called **ideal-dilute solutions**. We shall also label equations with a superscript ° when they have been derived from Henry's law. The difference in behaviour of the solute and solvent at low concentrations (as expressed by Henry's and Raoult's laws, respectively) arises from the fact that in a dilute solution the solvent molecules are in an environment very much like the one they have in the pure liquid (Fig. 5.15). In contrast, the solute molecules are surrounded by solvent molecules, which is entirely different from their environment when pure. Thus, the solvent behaves like a slightly modified pure liquid, but the solute behaves entirely differently from its pure state unless the solvent and solute molecules happen to be very similar. In the latter case, the solute also obeys Raoult's law.

Example 5.3 *Investigating the validity of Raoult's and Henry's laws*

The vapour pressures of each component in a mixture of propanone (acetone, A) and trichloromethane (chloroform, C) were measured at 35°C with the following results:

x_C	0	0.20	0.40	0.60	0.80	1
p_C/kPa	0	4.7	11	18.9	26.7	36.4
p_A/kPa	46.3	33.3	23.3	12.3	4.9	0

Confirm that the mixture conforms to Raoult's law for the component in large excess and to Henry's law for the minor component. Find the Henry's law constants.

Method Both Raoult's and Henry's laws are statements about the form of the graph of partial vapour pressure against mole fraction. Therefore, plot the partial vapour pressures against mole fraction. Raoult's law is tested by comparing the data with the straight line $p_J = x_J p_J^*$ for each component in the region in which it is in excess (and acting as the solvent). Henry's law is tested by finding a straight line $p_J = x_J K_J^*$ that is tangent to each partial vapour pressure at low x, where the component can be treated as the solute.

Answer The data are plotted in Fig. 5.16 together with the Raoult's law lines. Henry's law requires K = 23.3 kPa for propanone and K = 22.0 kPa for trichloromethane. Notice how the system deviates from both Raoult's and Henry's laws even for quite small departures from $x = 1$ and $x = 0$, respectively. We deal with these deviations in Sections 5.10 and 5.11.

Self-test 5.4 The vapour pressure of chloromethane at various mole fractions in a mixture at 25°C was found to be as follows:

x	0.005	0.009	0.019	0.024
p/kPa	27.3	48.4	101	126

Estimate Henry's law constant. [5 MPa]

For practical applications, Henry's law is expressed in terms of the molality, b, of the solute, $p_B = b_B K_B$. Some Henry's law data for this convention are listed in Table 5.1. As well as providing a link between the mole fraction of solute and its partial pressure, the data in the table may also be used to calculate gas solubilities. A knowledge of Henry's law constants for gases in blood and fats is important for the discussion of respiration, especially when the partial pressure of oxygen is abnormal, as in diving and mountaineering, and for the discussion of the action of gaseous anaesthetics.

• **A brief illustration**

To estimate the molar solubility of oxygen in water at 25°C and a partial pressure of 21 kPa, its partial pressure in the atmosphere at sea level, we write

$$b_{O_2} = \frac{p_{O_2}}{K_{O_2}} = \frac{21\text{ kPa}}{7.9\times 10^4\text{ kPa mol}^{-1}} = 2.7\times 10^{-4}\text{ mol kg}^{-1}$$

The molality of the saturated solution is therefore 0.27 mmol kg^{-1}. To convert this quantity to a molar concentration, we assume that the mass density of this dilute solution is essentially that of pure water at 25°C, or $\rho_{H_2O} = 0.99709\text{ kg dm}^{-3}$. It follows that the molar concentration of oxygen is

$$[O_2] = b_{O_2}\times\rho_{H_2O} = 0.27\text{ mmol kg}^{-1}\times 0.99709\text{ kg dm}^{-3} = 0.27\text{ mmol dm}^{-3}$$ •

Self-test 5.5 Calculate the molar solubility of nitrogen in water exposed to air at 25°C; partial pressures were calculated in *Example 1.3*. [0.51 mmol dm^{-3}]

The properties of solutions

In this section we consider the thermodynamics of mixing of liquids. First, we consider the simple case of mixtures of liquids that mix to form an ideal solution. In this way, we identify the thermodynamic consequences of molecules of one species mingling randomly with molecules of the second species. The calculation provides a background for discussing the deviations from ideal behaviour exhibited by real solutions.

5.4 Liquid mixtures

Key points **(a) The Gibbs energy of mixing of two liquids to form an ideal solution is calculated in the same way as for two perfect gases. The enthalpy of mixing is zero and the Gibbs energy is due entirely to the entropy of mixing. (b) A regular solution is one in which the entropy of mixing is the same as for an ideal solution but the enthalpy of mixing is non-zero.**

Thermodynamics can provide insight into the properties of liquid mixtures, and a few simple ideas can bring the whole field of study together.

(a) Ideal solutions

The Gibbs energy of mixing of two liquids to form an ideal solution is calculated in exactly the same way as for two gases (Section 5.2). The total Gibbs energy before liquids are mixed is

$$G_i = n_A\mu_A^* + n_B\mu_B^* \qquad (5.24a)$$

When they are mixed, the individual chemical potentials are given by eqn 5.22 and the total Gibbs energy is

$$G_f = n_A\{\mu_A^* + RT\ln x_A\} + n_B\{\mu_B^* + RT\ln x_B\} \qquad (5.24b)°$$

Consequently, the Gibbs energy of mixing, the difference of these two quantities, is

$$\Delta_{mix}G = nRT\{x_A\ln x_A + x_B\ln x_B\} \qquad (5.25)°$$

Gibbs energy of mixing to form an ideal solution

A note on good practice It is on the basis of this distinction (in the second paragraph) that the term 'perfect gas' is preferable to the more common 'ideal gas'. In an ideal solution there are interactions, but they are effectively the same between the various species. In a perfect gas, not only are the interactions the same, but they are also zero. Few people, however, trouble to make this valuable distinction.

where $n = n_A + n_B$. As for gases, it follows that the ideal entropy of mixing of two liquids is

$$\Delta_{\text{mix}}S = -nR\{x_A \ln x_A + x_B \ln x_B\}$$ Entropy of mixing to form an ideal solution (5.26)°

Because $\Delta_{\text{mix}}H = \Delta_{\text{mix}}G + T\Delta_{\text{mix}}S = 0$, the ideal enthalpy of mixing is zero. The ideal volume of mixing, the change in volume on mixing, is also zero because it follows from eqn 3.53 ($(\partial G/\partial p)_T = V$) that $\Delta_{\text{mix}}V = (\partial\Delta_{\text{mix}}G/\partial p)_T$, but $\Delta_{\text{mix}}G$ in eqn 5.25 is independent of pressure, so the derivative with respect to pressure is zero.

Equation 5.26 is the same as that for two perfect gases and all the conclusions drawn there are valid here: the driving force for mixing is the increasing entropy of the system as the molecules mingle and the enthalpy of mixing is zero. It should be noted, however, that solution ideality means something different from gas perfection. In a perfect gas there are no forces acting between molecules. In ideal solutions there are interactions, but the average energy of A–B interactions in the mixture is the same as the average energy of A–A and B–B interactions in the pure liquids. The variation of the Gibbs energy of mixing with composition is the same as that already depicted for gases in Fig. 5.7; the same is true of the entropy of mixing, Fig. 5.9.

Real solutions are composed of particles for which A–A, A–B, and B–B interactions are all different. Not only may there be enthalpy and volume changes when liquids mix, but there may also be an additional contribution to the entropy arising from the way in which the molecules of one type might cluster together instead of mingling freely with the others. If the enthalpy change is large and positive or if the entropy change is adverse (because of a reorganization of the molecules that results in an orderly mixture), then the Gibbs energy might be positive for mixing. In that case, separation is spontaneous and the liquids may be immiscible. Alternatively, the liquids might be **partially miscible**, which means that they are miscible only over a certain range of compositions.

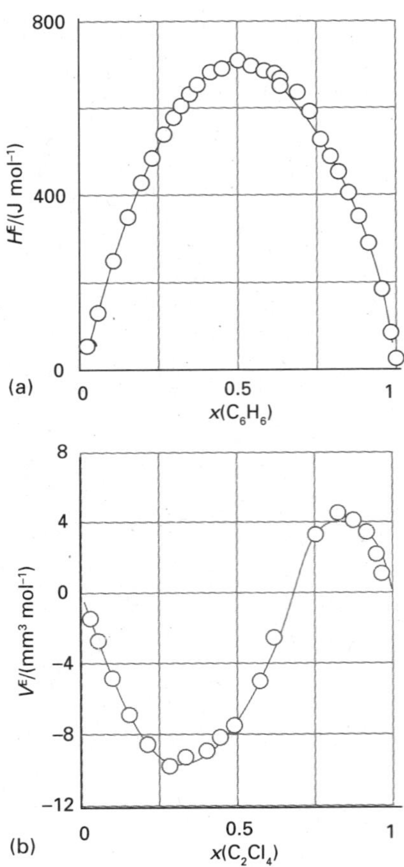

Fig. 5.17 Experimental excess functions at 25°C. (a) H^E for benzene/cyclohexane; this graph shows that the mixing is endothermic (because $\Delta_{\text{mix}}H = 0$ for an ideal solution). (b) The excess volume, V^E, for tetrachloroethene/cyclopentane; this graph shows that there is a contraction at low tetrachloroethene mole fractions, but an expansion at high mole fractions (because $\Delta_{\text{mix}}V = 0$ for an ideal mixture).

(b) Excess functions and regular solutions

The thermodynamic properties of real solutions are expressed in terms of the **excess functions**, X^E, the difference between the observed thermodynamic function of mixing and the function for an ideal solution. The **excess entropy**, S^E, for example, is defined as

$$S^E = \Delta_{\text{mix}}S - \Delta_{\text{mix}}S^{\text{ideal}}$$ Definition of excess entropy [5.27]

where $\Delta_{\text{mix}}S^{\text{ideal}}$ is given by eqn 5.26. The excess enthalpy and volume are both equal to the observed enthalpy and volume of mixing, because the ideal values are zero in each case. Figure 5.17 shows two examples of the composition dependence of molar excess functions. In Fig. 5.17(a), the positive values of H^E indicate that the A–B interactions in the mixture are weaker than the A–A and B–B interactions in the pure liquids (which are benzene and pure cyclohexane). The symmetrical shape of the curve reflects the similar strengths of the A–A and B–B interactions. Figure 5.17(b) shows the composition dependence of the excess volume, V^E, of a mixture of tetrachloroethene and cyclopentane. At high mole fractions of cyclopentane, the solution contracts as tetrachloroethene is added because the ring structure of cyclopentane results in inefficient packing of the molecules but, as tetrachloroethene is added, the molecules in the mixture pack together more tightly. Similarly, at high mole fractions of tetrachloroethene, the solution expands as cyclopentane is added because tetrachloroethene molecules are nearly flat and pack efficiently in the pure liquid but become disrupted as bulky ring cyclopentane is added.

Deviations of the excess energies from zero indicate the extent to which the solutions are nonideal. In this connection a useful model system is the **regular solution**, a solution for which $H^E \neq 0$ but $S^E = 0$. We can think of a regular solution as one in which the two kinds of molecules are distributed randomly (as in an ideal solution) but have different energies of interactions with each other. To express this concept more quantitatively we can suppose that the excess enthalpy depends on composition as

$$H^E = n\xi RTx_Ax_B \qquad (5.28)$$

where ξ (xi) is a dimensionless parameter that is a measure of the energy of AB interactions relative to that of the AA and BB interactions. The function given by eqn 5.28 is plotted in Fig. 5.18, and we see it resembles the experimental curve in Fig. 5.17. If $\xi < 0$, mixing is exothermic and the solute–solvent interactions are more favourable than the solvent–solvent and solute–solute interactions. If $\xi > 0$, then the mixing is endothermic. Because the entropy of mixing has its ideal value for a regular solution, the excess Gibbs energy is equal to the excess enthalpy, and the Gibbs energy of mixing is

$$\Delta_{\text{mix}}G = nRT\{x_A \ln x_A + x_B \ln x_B + \xi x_Ax_B\} \qquad (5.29)$$

Figure 5.19 shows how $\Delta_{\text{mix}}G$ varies with composition for different values of ξ. The important feature is that for $\xi > 2$ the graph shows two minima separated by a maximum. The implication of this observation is that, provided $\xi > 2$, the system will separate spontaneously into two phases with compositions corresponding to the two minima, for that separation corresponds to a reduction in Gibbs energy. We develop this point in Sections 5.6 and 5.10.

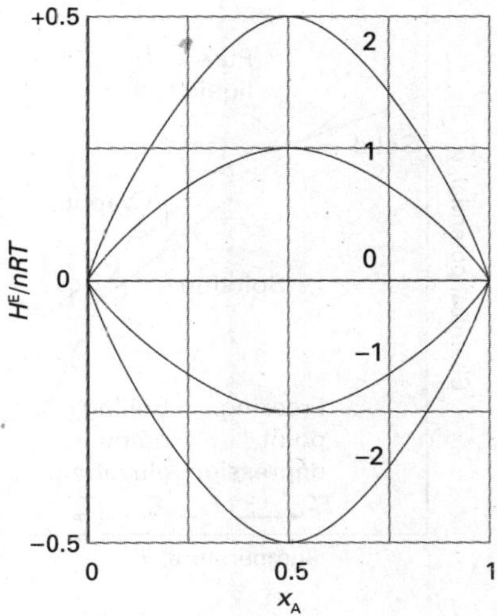

Fig. 5.18 The excess enthalpy according to a model in which it is proportional to ξx_Ax_B, for different values of the parameter ξ.

interActivity Using the graph above, fix ξ and vary the temperature. For what value of x_A does the excess enthalpy depend on temperature most strongly?

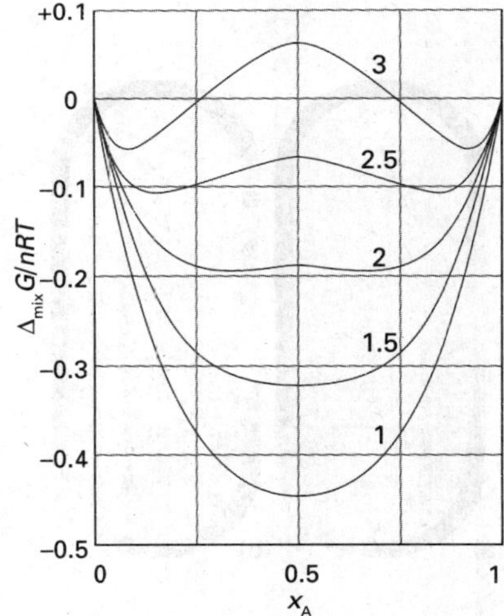

Fig. 5.19 The Gibbs energy of mixing for different values of the parameter ξ.

5.5 Colligative properties

Key points **A colligative property depends only on the number of solute particles present, not their identity. (a) All the colligative properties stem from the reduction of the chemical potential of the liquid solvent as a result of the presence of solute. (b) The elevation of boiling point is proportional to the molality of the solute. (c) The depression of freezing point is also proportional to the molality of the solute. (d) Solutes with high melting points and large enthalpies of melting have low solubilities at normal temperatures. (e) The relation of the osmotic pressure to the molar concentration of the solute is given by the van't Hoff equation and is a sensitive way of determining molar mass.**

The properties we now consider are the lowering of vapour pressure, the elevation of boiling point, the depression of freezing point, and the osmotic pressure arising from the presence of a solute. In dilute solutions these properties depend only on the number of solute particles present, not their identity. For this reason, they are called **colligative properties** (denoting 'depending on the collection').

We assume throughout the following that the solute is not volatile, so it does not contribute to the vapour. We also assume that the solute does not dissolve in the solid solvent: that is, the pure solid solvent separates when the solution is frozen. The latter assumption is quite drastic, although it is true of many mixtures; it can be avoided at the expense of more algebra, but that introduces no new principles.

(a) The common features of colligative properties

All the colligative properties stem from the reduction of the chemical potential of the liquid solvent as a result of the presence of solute. For an ideal-dilute solution, the reduction is from $\mu_A^\star$ for the pure solvent to $\mu_A^\star + RT \ln x_A$ when a solute is present ($\ln x_A$ is negative because $x_A < 1$). There is no direct influence of the solute on the

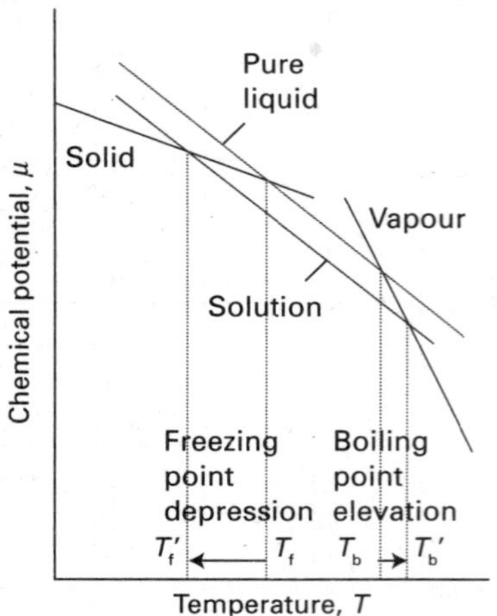

Fig. 5.20 The chemical potential of a solvent in the presence of a solute. The lowering of the liquid's chemical potential has a greater effect on the freezing point than on the boiling point because of the angles at which the lines intersect.

chemical potential of the solvent vapour and the solid solvent because the solute appears in neither the vapour nor the solid. As can be seen from Fig. 5.20, the reduction in chemical potential of the solvent implies that the liquid–vapour equilibrium occurs at a higher temperature (the boiling point is raised) and the solid–liquid equilibrium occurs at a lower temperature (the freezing point is lowered).

The molecular origin of the lowering of the chemical potential is not the energy of interaction of the solute and solvent particles, because the lowering occurs even in an ideal solution (for which the enthalpy of mixing is zero). If it is not an enthalpy effect, it must be an entropy effect. The vapour pressure of the pure liquid reflects the tendency of the solution towards greater entropy, which can be achieved if the liquid vaporizes to form a gas. When a solute is present, there is an additional contribution to the entropy of the liquid, even in an ideal solution. Because the entropy of the liquid is already higher than that of the pure liquid, there is a weaker tendency to form the gas (Fig. 5.21). The effect of the solute appears as a lowered vapour pressure, and hence a higher boiling point. Similarly, the enhanced molecular randomness of the solution opposes the tendency to freeze. Consequently, a lower temperature must be reached before equilibrium between solid and solution is achieved. Hence, the freezing point is lowered.

The strategy for the quantitative discussion of the elevation of boiling point and the depression of freezing point is to look for the temperature at which, at 1 atm, one phase (the pure solvent vapour or the pure solid solvent) has the same chemical potential as the solvent in the solution. This is the new equilibrium temperature for the phase transition at 1 atm, and hence corresponds to the new boiling point or the new freezing point of the solvent.

(b) The elevation of boiling point

The heterogeneous equilibrium of interest when considering boiling is between the solvent vapour and the solvent in solution at 1 atm (Fig. 5.22). We denote the solvent by A and the solute by B. The equilibrium is established at a temperature for which

$$\mu_A^*(g) = \mu_A^*(l) + RT \ln x_A \qquad (5.30)°$$

(The pressure of 1 atm is the same throughout, and will not be written explicitly.) We show in the following *Justification* that this equation implies that the presence of a solute at a mole fraction x_B causes an increase in normal boiling point from T^* to $T^* + \Delta T$, where

$$\Delta T = Kx_B \qquad K = \frac{RT^{*2}}{\Delta_{vap}H} \qquad (5.31)°$$

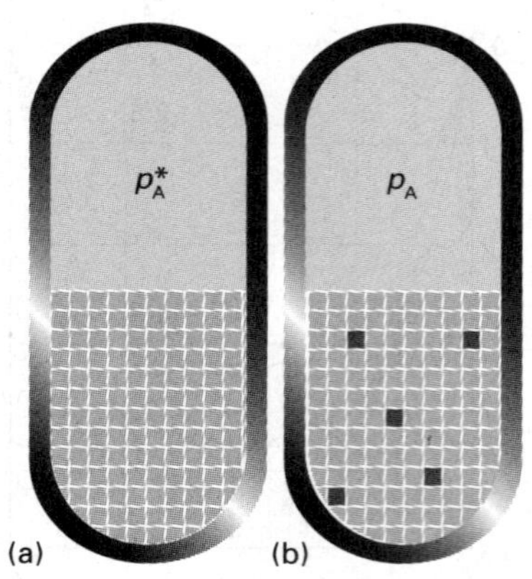

Fig. 5.21 The vapour pressure of a pure liquid represents a balance between the increase in disorder arising from vaporization and the decrease in disorder of the surroundings. (a) Here the structure of the liquid is represented highly schematically by the grid of squares. (b) When solute (the dark squares) is present, the disorder of the condensed phase is higher than that of the pure liquid, and there is a decreased tendency to acquire the disorder characteristic of the vapour.

Justification 5.1 *The elevation of the boiling point of a solvent*

Equation 5.30 can be rearranged into

$$\ln x_A = \frac{\mu_A^*(g) - \mu_A^*(l)}{RT} = \frac{\Delta_{vap}G}{RT}$$

where $\Delta_{vap}G$ is the Gibbs energy of vaporization of the pure solvent (A). First, to find the relation between a change in composition and the resulting change in boiling temperature, we differentiate both sides with respect to temperature and use the Gibbs–Helmholtz equation (eqn 3.55, $(\partial(G/T)/\partial T)_p = -H/T^2$) to express the term on the right:

$$\frac{d \ln x_A}{dT} = \frac{1}{R}\frac{d(\Delta_{vap}G/T)}{dT} = -\frac{\Delta_{vap}H}{RT^2}$$

Now multiply both sides by dT and integrate from $x_A = 1$, corresponding to $\ln x_A = 0$ (and when $T = T^*$, the boiling point of pure A) to x_A (when the boiling point is T):

$$\int_0^{\ln x_A} d\ln x_A = -\frac{1}{R}\int_{T^*}^{T} \frac{\Delta_{vap}H}{T^2}dT$$

The left-hand side integrates to $\ln x_A$, which is equal to $\ln(1 - x_B)$. The right-hand side can be integrated if we assume that the enthalpy of vaporization is a constant over the small range of temperatures involved and can be taken outside the integral. Thus, we obtain

$$\ln(1 - x_B) = -\frac{\Delta_{vap}H}{R}\int_{T^*}^{T} \frac{1}{T^2}dT$$

and therefore

$$\ln(1 - x_B) = \frac{\Delta_{vap}H}{R}\left(\frac{1}{T} - \frac{1}{T^*}\right)$$

We now suppose that the amount of solute present is so small that $x_B \ll 1$. We can then write $\ln(1 - x_B) = -x_B$ and hence obtain

$$x_B = \frac{\Delta_{vap}H}{R}\left(\frac{1}{T^*} - \frac{1}{T}\right)$$

Finally, because $T \approx T^*$, it also follows that

$$\frac{1}{T^*} - \frac{1}{T} = \frac{T - T^*}{TT^*} \approx \frac{\Delta T}{T^{*2}}$$

with $\Delta T = T - T^*$. The previous equation then rearranges into eqn 5.31.

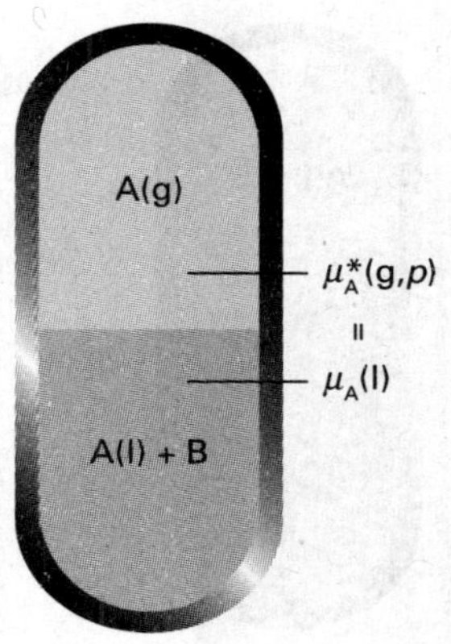

Fig. 5.22 The heterogeneous equilibrium involved in the calculation of the elevation of boiling point is between A in the pure vapour and A in the mixture, A being the solvent and B an involatile solute.

A brief comment

The series expansion of a natural logarithm is

$$\ln(1 - x) = -x - \tfrac{1}{2}x^2 - \tfrac{1}{3}x^3 \cdots$$

provided that $-1 < x < 1$. If $x \ll 1$, then the terms involving x raised to a power greater than 1 are much smaller than x, so $\ln(1 - x) \approx -x$.

Because eqn 5.31 makes no reference to the identity of the solute, only to its mole fraction, we conclude that the elevation of boiling point is a colligative property. The value of ΔT does depend on the properties of the solvent, and the biggest changes occur for solvents with high boiling points.[1] For practical applications of eqn 5.31, we note that the mole fraction of B is proportional to its molality, b, in the solution, and write

$$\Delta T = K_b b \qquad \text{Boiling point elevation} \qquad (5.32)$$

where K_b is the empirical **boiling-point constant** of the solvent (Table 5.2).

Table 5.2* Freezing-point and boiling-point constants

	$K_f/(\text{K kg mol}^{-1})$	$K_b/(\text{K kg mol}^{-1})$
Benzene	5.12	2.53
Camphor	40	
Phenol	7.27	3.04
Water	1.86	0.51

* More values are given in the *Data section*.

[1] By Trouton's rule (Section 3.3b), $\Delta_{vap}H/T^*$ is a constant; therefore eqn 5.31 has the form $\Delta T \propto T^*$ and is independent of $\Delta_{vap}H$ itself.

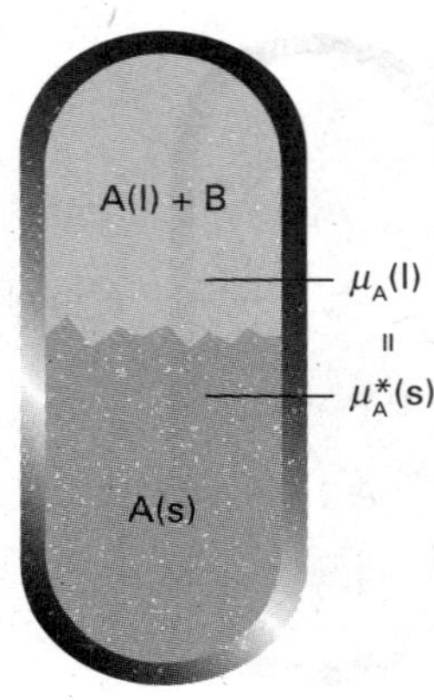

Fig. 5.23 The heterogeneous equilibrium involved in the calculation of the lowering of freezing point is between A in the pure solid and A in the mixture, A being the solvent and B a solute that is insoluble in solid A.

(c) The depression of freezing point

The heterogeneous equilibrium now of interest is between pure solid solvent A and the solution with solute present at a mole fraction x_B (Fig. 5.23). At the freezing point, the chemical potentials of A in the two phases are equal:

$$\mu_A^*(s) = \mu_A^*(l) + RT \ln x_A \qquad (5.33)^\circ$$

The only difference between this calculation and the last is the appearance of the solid's chemical potential in place of the vapour's. Therefore we can write the result directly from eqn 5.31:

$$\Delta T = K' x_B \qquad K' = \frac{RT^{*2}}{\Delta_{fus} H} \qquad (5.34)^\circ$$

where ΔT is the freezing point depression, $T^* - T$, and $\Delta_{fus}H$ is the enthalpy of fusion of the solvent. Larger depressions are observed in solvents with low enthalpies of fusion and high melting points. When the solution is dilute, the mole fraction is proportional to the molality of the solute, b, and it is common to write the last equation as

$$\Delta T = K_f b \qquad \text{Freezing point depression} \qquad (5.35)$$

where K_f is the empirical **freezing-point constant** (Table 5.2). Once the freezing-point constant of a solvent is known, the depression of freezing point may be used to measure the molar mass of a solute in the method known as **cryoscopy**; however, the technique is of little more than historical interest.

(d) Solubility

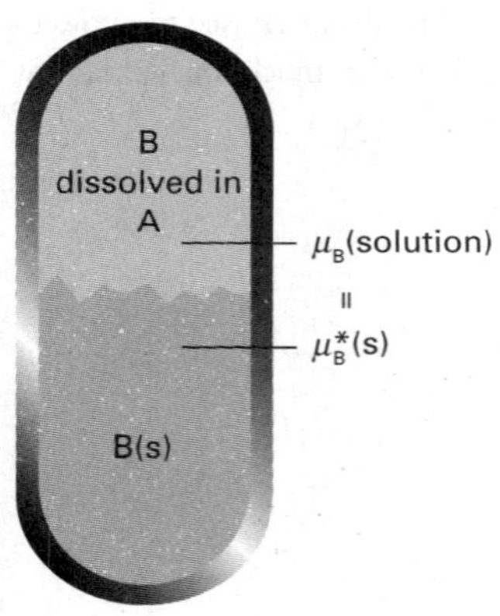

Fig. 5.24 The heterogeneous equilibrium involved in the calculation of the solubility is between pure solid B and B in the mixture.

Although solubility is not a colligative property (because solubility varies with the identity of the solute), it may be estimated by the same techniques as we have been using. When a solid solute is left in contact with a solvent, it dissolves until the solution is saturated. Saturation is a state of equilibrium, with the undissolved solute in equilibrium with the dissolved solute. Therefore, in a saturated solution the chemical potential of the pure solid solute, $\mu_B^*(s)$, and the chemical potential of B in solution, μ_B, are equal (Fig. 5.24). Because the latter is $\mu_B = \mu_B^*(l) + RT \ln x_B$, we can write

$$\mu_B^*(s) = \mu_B^*(l) + RT \ln x_B \qquad (5.36)^\circ$$

This expression is the same as the starting equation of the last section, except that the quantities refer to the solute B, not the solvent A. We now show in the following *Justification* that

$$\ln x_B = \frac{\Delta_{fus} H}{R}\left(\frac{1}{T_f} - \frac{1}{T}\right) \qquad \text{Ideal solubility} \qquad (5.37)^\circ$$

Justification 5.2 *The solubility of an ideal solute*

The starting point is the same as in *Justification 5.1* but the aim is different. In the present case, we want to find the mole fraction of B in solution at equilibrium when the temperature is T. Therefore, we start by rearranging eqn 5.36 into

$$\ln x_B = \frac{\mu_B^*(s) - \mu_B^*(l)}{RT} = -\frac{\Delta_{fus} G}{RT}$$

As in *Justification 5.1*, we relate the change in composition d ln x_B to the change in temperature by differentiation and use of the Gibbs–Helmholtz equation. Then we integrate from the melting temperature of B (when $x_B = 1$ and $\ln x_B = 0$) to the *lower* temperature of interest (when x_B has a value between 0 and 1):

$$\int_0^{\ln x_B} \mathrm{d}\ln x_B = \frac{1}{R}\int_{T_f}^{T} \frac{\Delta_{\text{fus}}H}{T^2}\,\mathrm{d}T$$

If we suppose that the enthalpy of fusion of B is constant over the range of temperatures of interest, it can be taken outside the integral, and we obtain eqn 5.37.

Equation 5.37 is plotted in Fig. 5.25. It shows that the solubility of B decreases exponentially as the temperature is lowered from its melting point. The illustration also shows that solutes with low melting points and large enthalpies of melting have low solubilities at normal temperatures. However, the detailed content of eqn 5.37 should not be treated too seriously because it is based on highly questionable approximations, such as the ideality of the solution. One aspect of its approximate character is that it fails to predict that solutes will have different solubilities in different solvents, for no solvent properties appear in the expression.

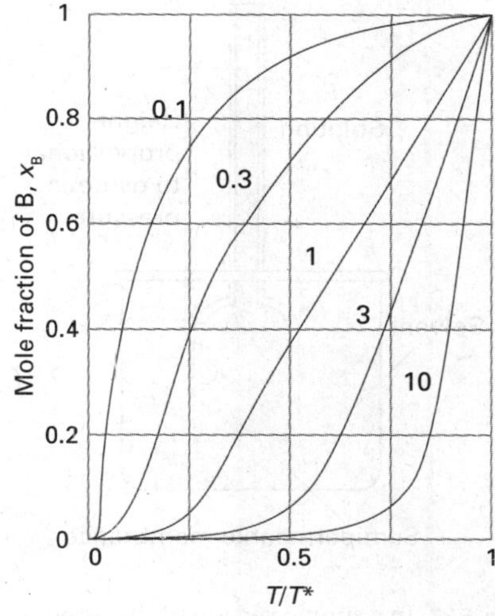

Fig. 5.25 The variation of solubility (the mole fraction of solute in a saturated solution) with temperature (T^* is the freezing temperature of the solute). Individual curves are labelled with the value of $\Delta_{\text{fus}}H/RT^*$.

interActivity Derive an expression for the temperature coefficient of the solubility, $\mathrm{d}x_B/\mathrm{d}T$, and plot it as a function of temperature for several values of the enthalpy of fusion.

(e) Osmosis

The phenomenon of **osmosis** (from the Greek word for 'push') is the spontaneous passage of a pure solvent into a solution separated from it by a **semipermeable membrane**, a membrane permeable to the solvent but not to the solute (Fig. 5.26). The **osmotic pressure**, Π, is the pressure that must be applied to the solution to stop the influx of solvent. Important examples of osmosis include transport of fluids through cell membranes, dialysis, and **osmometry**, the determination of molar mass by the measurement of osmotic pressure. Osmometry is widely used to determine the molar masses of macromolecules.

In the simple arrangement shown in Fig. 5.27, the opposing pressure arises from the head of solution that the osmosis itself produces. Equilibrium is reached when the hydrostatic pressure of the column of solution matches the osmotic pressure. The complicating feature of this arrangement is that the entry of solvent into the solution results in its dilution, and so it is more difficult to treat than the arrangement in Fig. 5.26, in which there is no flow and the concentrations remain unchanged.

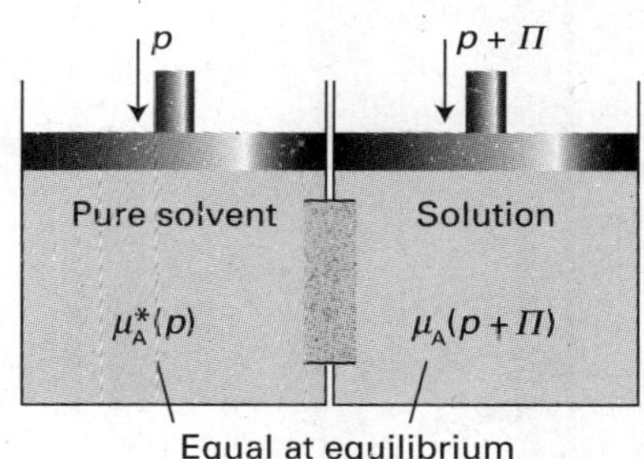

Fig. 5.26 The equilibrium involved in the calculation of osmotic pressure, Π, is between pure solvent A at a pressure p on one side of the semipermeable membrane and A as a component of the mixture on the other side of the membrane, where the pressure is $p + \Pi$.

The thermodynamic treatment of osmosis depends on noting that, at equilibrium, the chemical potential of the solvent must be the same on each side of the membrane. The chemical potential of the solvent is lowered by the solute, but is restored to its 'pure' value by the application of pressure. As shown in the following *Justification*, this equality implies that for dilute solutions the osmotic pressure is given by the **van't Hoff equation**:

$$\Pi = [\mathrm{B}]RT \qquad \text{van't Hoff equation} \qquad (5.38)^\circ$$

where $[\mathrm{B}] = n_B/V$ is the molar concentration of the solute.

Justification 5.3 *The van't Hoff equation*

On the pure solvent side the chemical potential of the solvent, which is at a pressure p, is $\mu_A^*(p)$. On the solution side, the chemical potential is lowered by the presence of the solute, which reduces the mole fraction of the solvent from 1 to x_A. However, the chemical potential of A is raised on account of the greater pressure, $p + \Pi$, that the solution experiences. At equilibrium the chemical potential of A is the same in both compartments, and we can write

$$\mu_A^*(p) = \mu_A(x_A, p + \Pi)$$

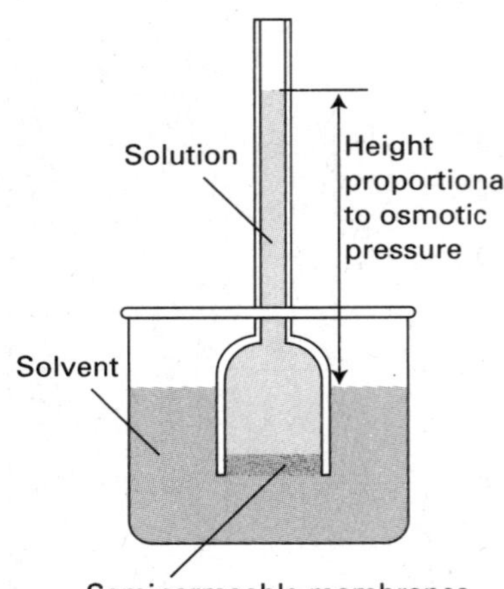

Fig. 5.27 In a simple version of the osmotic pressure experiment, A is at equilibrium on each side of the membrane when enough has passed into the solution to cause a hydrostatic pressure difference.

The presence of solute is taken into account in the normal way:

$$\mu_A(x_A, p+\Pi) = \mu_A^*(p+\Pi) + RT\ln x_A$$

We saw in Section 3.9 (eqn 3.57) how to take the effect of pressure into account:

$$\mu_A^*(p+\Pi) = \mu_A^*(p) + \int_p^{p+\Pi} V_m \mathrm{d}p$$

where V_m is the molar volume of the pure solvent A. When these three equations are combined we get

$$-RT\ln x_A = \int_p^{p+\Pi} V_m \mathrm{d}p \qquad (5.39)°$$

This expression enables us to calculate the additional pressure Π that must be applied to the solution to restore the chemical potential of the solvent to its 'pure' value and thus to restore equilibrium across the semipermeable membrane. For dilute solutions, $\ln x_A$ may be replaced by $\ln(1-x_B) \approx -x_B$. We may also assume that the pressure range in the integration is so small that the molar volume of the solvent is a constant. That being so, V_m may be taken outside the integral, giving

$$RTx_B = \Pi V_m$$

When the solution is dilute, $x_B \approx n_B/n_A$. Moreover, because $n_A V_m = V$, the total volume of the solvent, the equation simplifies to eqn 5.38.

Because the effect of osmotic pressure is so readily measurable and large, one of the most common applications of osmometry is to the measurement of molar masses of macromolecules, such as proteins and synthetic polymers. As these huge molecules dissolve to produce solutions that are far from ideal, it is assumed that the van't Hoff equation is only the first term of a virial-like expansion:

$$\Pi = [\mathrm{J}]RT\{1 + B[\mathrm{J}] + \cdots\} \qquad (5.40)$$

(We have denoted the solute J to avoid too many different Bs in this expression). The additional terms take the nonideality into account; the empirical constant B is called the **osmotic virial coefficient**.

Example 5.4 *Using osmometry to determine the molar mass of a macromolecule*

The osmotic pressures of solutions of poly(vinyl chloride), PVC, in cyclohexanone at 298 K are given below. The pressures are expressed in terms of the heights of solution (of mass density $\rho = 0.980$ g cm^{-3}) in balance with the osmotic pressure. Determine the molar mass of the polymer.

$c/(\mathrm{g\,dm^{-3}})$	1.00	2.00	4.00	7.00	9.00
h/cm	0.28	0.71	2.01	5.10	8.00

Method The osmotic pressure is measured at a series of mass concentrations, c, and a plot of Π/c against c is used to determine the molar mass of the polymer. We use eqn 5.40 with $[\mathrm{J}] = c/M$ where c is the mass concentration of the polymer and M is its molar mass. The osmotic pressure is related to the hydrostatic pressure by $\Pi = \rho gh$ (Example 1.1) with $g = 9.81$ m s^{-2}. With these substitutions, eqn 5.40 becomes

$$\frac{h}{c} = \frac{RT}{\rho gM}\left(1 + \frac{Bc}{M} + \cdots\right) = \frac{RT}{\rho gM} + \left(\frac{RTB}{\rho gM^2}\right)c + \cdots$$

Therefore, to find M, plot h/c against c, and expect a straight line with intercept $RT/\rho gM$ at $c = 0$.

Answer The data give the following values for the quantities to plot:

$c/(\mathrm{g\,dm^{-3}})$	1.00	2.00	4.00	7.00	9.00
$(h/c)/(\mathrm{cm\,g^{-1}\,dm^{3}})$	0.28	0.36	0.503	0.729	0.889

The points are plotted in Fig. 5.28. The intercept is at 0.21. Therefore,

$$M = \frac{RT}{\rho g} \times \frac{1}{0.21\ \mathrm{cm\,g^{-1}\,dm^{3}}}$$

$$= \frac{(8.3145\ \mathrm{J\,K^{-1}\,mol^{-1}}) \times (298\ \mathrm{K})}{(980\ \mathrm{kg\,m^{-1}}) \times (9.81\ \mathrm{m\,s^{-2}})} \times \frac{1}{2.1 \times 10^{-3}\ \mathrm{m^4\,kg^{-1}}}$$

$$= 1.2 \times 10^{2}\ \mathrm{kg\,mol^{-1}}$$

where we have used $1\ \mathrm{kg\,m^2\,s^{-2}} = 1\ \mathrm{J}$. Molecular masses of macromolecules are often reported in daltons (Da), with $1\ \mathrm{Da} = m_u$. The macromolecule in this example has a molecular mass of about 120 kDa. Modern osmometers give readings of osmotic pressure in pascals, so the analysis of the data is more straightforward and eqn 5.40 can be used directly. As we shall see in Chapter 19, the value obtained from osmometry is the 'number average molar mass'.

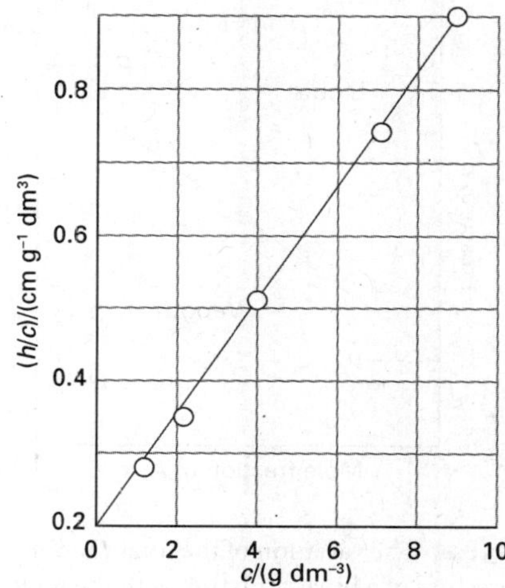

Fig. 5.28 The plot involved in the determination of molar mass by osmometry. The molar mass is calculated from the intercept at $c = 0$.

interActivity Calculate the osmotic virial coefficient B from these data.

Self-test 5.6 Estimate the depression of freezing point of the most concentrated of these solutions, taking K_f as about $10\ \mathrm{K/(mol\,kg^{-1})}$. [0.8 mK]

IMPACT ON BIOLOGY

I5.1 Osmosis in physiology and biochemistry

Osmosis helps biological cells maintain their structure. Cell membranes are semipermeable and allow water, small molecules, and hydrated ions to pass, while blocking the passage of biopolymers synthesized inside the cell. The difference in concentrations of solutes inside and outside the cell gives rise to an osmotic pressure, and water passes into the more concentrated solution in the interior of the cell, carrying small nutrient molecules. The influx of water also keeps the cell swollen, whereas dehydration causes the cell to shrink. These effects are important in everyday medical practice. To maintain the integrity of blood cells, solutions that are injected into the bloodstream for blood transfusions and intravenous feeding must be *isotonic* with the blood, meaning that they must have the same osmotic pressure as blood. If the injected solution is too dilute, or *hypotonic*, the flow of solvent into the cells, required to equalize the osmotic pressure, causes the cells to burst and die by a process called *haemolysis*. If the solution is too concentrated, or *hypertonic*, equalization of the osmotic pressure requires flow of solvent out of the cells, which shrink and die.

Osmosis also forms the basis of *dialysis*, a common technique for the removal of impurities from solutions of biological macromolecules and for the study of binding of small molecules to macromolecules, such as an inhibitor to an enzyme, an antibiotic to DNA, and any other instance of cooperation or inhibition by small molecules attaching to large ones. In a purification experiment, a solution of macromolecules containing impurities, such as ions or small molecules (including small proteins or nucleic acids), is placed in a bag made of a material that acts as a semipermeable membrane and the filled bag is immersed in a solvent. The membrane permits the passage of the small ions and molecules but not the larger macromolecules, so the former

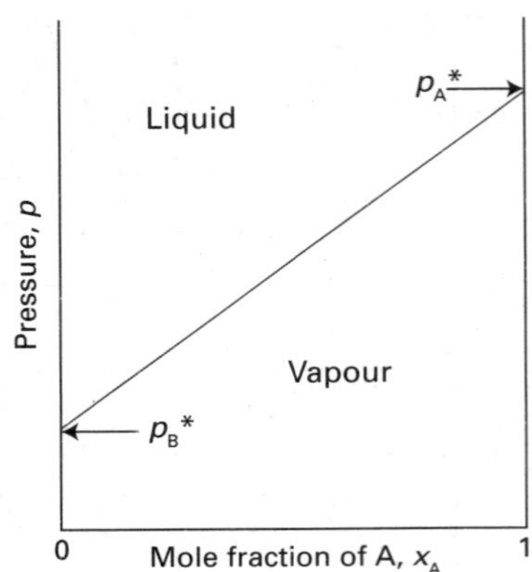

Fig. 5.29 The variation of the total vapour pressure of a binary mixture with the mole fraction of A in the liquid when Raoult's law is obeyed.

migrate through the membrane, leaving the macromolecules behind. In practice, purification of the sample requires several changes of solvent to coax most of the impurities out of the dialysis bag.

Phase diagrams of binary systems

We encountered one-component phase diagrams in Chapter 4. The phase equilibria of binary systems are more complex because composition is an additional variable. However, they provide very useful summaries of phase equilibria for both ideal and empirically established real systems.

5.6 Vapour pressure diagrams

Key points **Raoult's law is used to calculate the total vapour pressure of a binary system of two volatile liquids. (a) The composition of the vapour in equilibrium with a binary mixture is calculated by using Dalton's law. (b) The compositions of the vapour and the liquid phase in equilibrium are located at each end of a tie line. (c) The lever rule is used to deduce the relative abundances of each phase in equilibrium.**

The partial vapour pressures of the components of an ideal solution of two volatile liquids are related to the composition of the liquid mixture by Raoult's law (Section 5.3)

$$p_A = x_A p_A^* \qquad p_B = x_B p_B^* \tag{5.41}^\circ$$

where p_A^* is the vapour pressure of pure A and p_B^* that of pure B. The total vapour pressure p of the mixture is therefore

$$p = p_A + p_B = x_A p_A^* + x_B p_B^* = p_B^* + (p_A^* - p_B^*)x_A \tag{5.42}^\circ$$

This expression shows that the total vapour pressure (at some fixed temperature) changes linearly with the composition from p_B^* to p_A^* as x_A changes from 0 to 1 (Fig. 5.29).

(a) The composition of the vapour

The compositions of the liquid and vapour that are in mutual equilibrium are not necessarily the same. Common sense suggests that the vapour should be richer in the more volatile component. This expectation can be confirmed as follows. The partial pressures of the components are given by eqn 1.13. It follows from Dalton's law that the mole fractions in the gas, y_A and y_B, are

$$y_A = \frac{p_A}{p} \qquad y_B = \frac{p_B}{p} \tag{5.43}$$

Provided the mixture is ideal, the partial pressures and the total pressure may be expressed in terms of the mole fractions in the liquid by using eqn 5.41 for p_J and eqn 5.42 for the total vapour pressure p, which gives

$$y_A = \frac{x_A p_A^*}{p_B^* + (p_A^* - p_B^*)x_A} \qquad y_B = 1 - y_A \tag{5.44}^\circ$$

Figure 5.30 shows the composition of the vapour plotted against the composition of the liquid for various values of $p_A^*/p_B^* > 1$. We see that in all cases $y_A > x_A$, that is, the vapour is richer than the liquid in the more volatile component. Note that if B is non-volatile, so that $p_B^* = 0$ at the temperature of interest, then it makes no contribution to the vapour ($y_B = 0$).

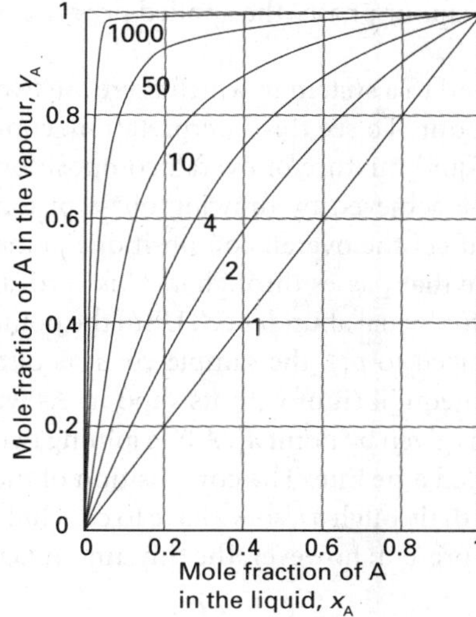

Fig. 5.30 The mole fraction of A in the vapour of a binary ideal solution expressed in terms of its mole fraction in the liquid, calculated using eqn 5.44 for various values of p_A^*/p_B^* (the label on each curve) with A more volatile than B. In all cases the vapour is richer than the liquid in A.

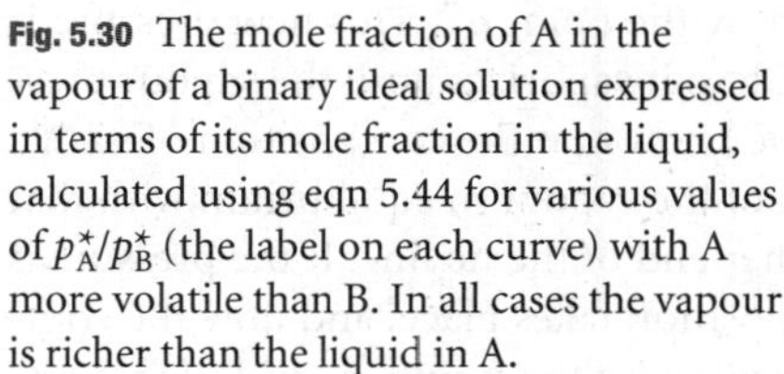

interActivity To reproduce the results of Fig. 5.30, first rearrange eqn 5.44 so that y_A is expressed as a function of x_A and the ratio p_A^*/p_B^*. Then plot y_A against x_A for several values of $p_A^*/p_B^* > 1$.

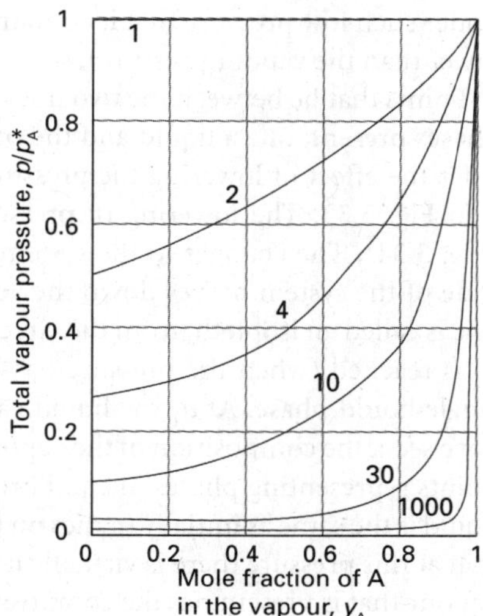

Fig. 5.31 The dependence of the vapour pressure of the same system as in Fig. 5.30, but expressed in terms of the mole fraction of A in the vapour by using eqn 5.45. Individual curves are labelled with the value of p_A^*/p_B^*.

interActivity To reproduce the results of Fig. 5.31, first rearrange eqn 5.45 so that the ratio p_A/p_A^* is expressed as a function of y_A and the ratio p_A^*/p_B^*. Then plot p_A/p_A^* against y_A for several values of $p_A^*/p_B^* > 1$.

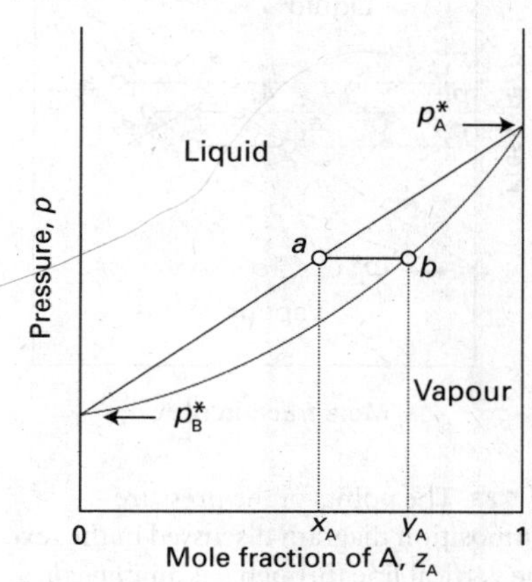

Fig. 5.32 The dependence of the total vapour pressure of an ideal solution on the mole fraction of A in the entire system. A point between the two lines corresponds to both liquid and vapour being present; outside that region there is only one phase present. The mole fraction of A is denoted z_A, as explained below.

Equation 5.42 shows how the total vapour pressure of the mixture varies with the composition of the liquid. Because we can relate the composition of the liquid to the composition of the vapour through eqn 5.44, we can now also relate the total vapour pressure to the composition of the vapour:

$$p = \frac{p_A^* p_B^*}{p_A^* + (p_B^* - p_A^*)y_A} \qquad (5.45)^\circ$$

This expression is plotted in Fig. 5.31.

(b) The interpretation of the diagrams

If we are interested in distillation, both the vapour and the liquid compositions are of equal interest. It is therefore sensible to combine Figs. 5.29 and 5.31 into one (Fig. 5.32). The point *a* indicates the vapour pressure of a mixture of composition x_A, and the point *b* indicates the composition of the vapour that is in equilibrium with the liquid at that pressure. A richer interpretation of the phase diagram is obtained, however, if we interpret the horizontal axis as showing the *overall* composition, z_A, of the system. If the horizontal axis of the vapour pressure diagram is labelled with z_A, then all the points down to the solid diagonal line in the graph correspond to a system that is under such high pressure that it contains only a liquid phase (the applied pressure is higher than the vapour pressure), so $z_A = x_A$, the composition of the liquid. On the other hand, all points below the lower curve correspond to a system that is

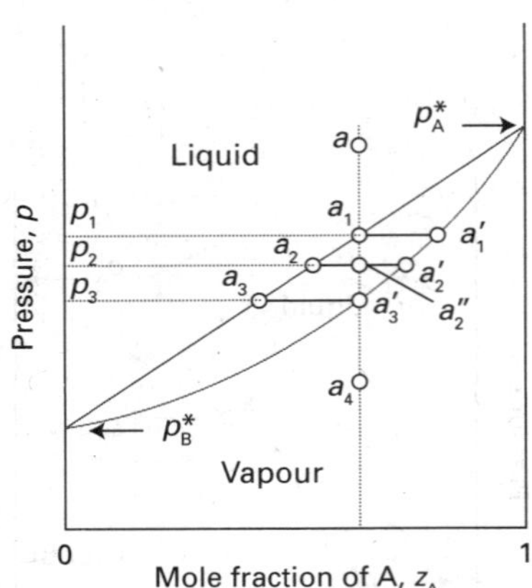

Fig. 5.33 The points of the pressure–composition diagram discussed in the text. The vertical line through a is an *isopleth*, a line of constant composition of the entire system.

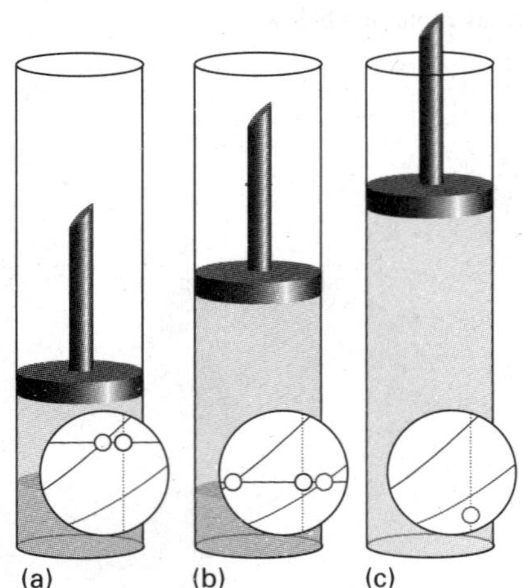

Fig. 5.34 (a) A liquid in a container exists in equilibrium with its vapour. The superimposed fragment of the phase diagram shows the compositions of the two phases and their abundances (by the lever rule). (b) When the pressure is changed by drawing out a piston, the compositions of the phases adjust as shown by the tie line in the phase diagram. (c) When the piston is pulled so far out that all the liquid has vaporized and only the vapour is present, the pressure falls as the piston is withdrawn and the point on the phase diagram moves into the one-phase region.

under such low pressure that it contains only a vapour phase (the applied pressure is lower than the vapour pressure), so $z_A = y_A$.

Points that lie between the two lines correspond to a system in which there are two phases present, one a liquid and the other a vapour. To see this interpretation, consider the effect of lowering the pressure on a liquid mixture of overall composition a in Fig. 5.33. The lowering of pressure can be achieved by drawing out a piston (Fig. 5.34). The changes to the system do not affect the overall composition, so the state of the system moves down the vertical line that passes through a. This vertical line is called an **isopleth**, from the Greek words for 'equal abundance'. Until the point a_1 is reached (when the pressure has been reduced to p_1), the sample consists of a single liquid phase. At a_1 the liquid can exist in equilibrium with its vapour. As we have seen, the composition of the vapour phase is given by point a_1'. A line joining two points representing phases in equilibrium is called a **tie line**. The composition of the liquid is the same as initially (a_1 lies on the isopleth through a), so we have to conclude that at this pressure there is virtually no vapour present; however, the tiny amount of vapour that is present has the composition a_1'.

Now consider the effect of lowering the pressure to p_2, so taking the system to a pressure and overall composition represented by the point a_2''. This new pressure is below the vapour pressure of the original liquid, so it vaporizes until the vapour pressure of the remaining liquid falls to p_2. Now we know that the composition of such a liquid must be a_2. Moreover, the composition of the vapour in equilibrium with that liquid must be given by the point a_2' at the other end of the tie line. If the pressure is reduced to p_3, a similar readjustment in composition takes place, and now the compositions of the liquid and vapour are represented by the points a_3 and a_3', respectively. The latter point corresponds to a system in which the composition of the vapour is the same as the overall composition, so we have to conclude that the amount of liquid present is now virtually zero, but the tiny amount of liquid present has the composition a_3. A further decrease in pressure takes the system to the point a_4; at this stage, only vapour is present and its composition is the same as the initial overall composition of the system (the composition of the original liquid).

(c) The lever rule

A point in the two-phase region of a phase diagram indicates not only qualitatively that both liquid and vapour are present, but represents quantitatively the relative amounts of each. To find the relative amounts of two phases α and β that are in equilibrium, we measure the distances l_α and l_β along the horizontal tie line, and then use the **lever rule** (Fig. 5.35):

$$n_\alpha l_\alpha = n_\beta l_\beta \qquad \text{Lever rule} \qquad (5.46)$$

Here n_α is the amount of phase α and n_β the amount of phase β. In the case illustrated in Fig. 5.35, because $l_\beta \approx 2l_\alpha$, the amount of phase α is about twice the amount of phase β.

Justification 5.4 *The lever rule*

To prove the lever rule we write $n = n_\alpha + n_\beta$ and the overall amount of A as nz_A. The overall amount of A is also the sum of its amounts in the two phases:

$$nz_A = n_\alpha x_A + n_\beta y_A$$

Since also

$$nz_A = n_\alpha z_A + n_\beta z_A$$

by equating these two expressions it follows that

$$n_\alpha(x_A - z_A) = n_\beta(z_A - y_A)$$

which corresponds to eqn 5.46.

• A brief illustration

At p_1 in Fig. 5.33, the ratio l_{vap}/l_{liq} is almost infinite for this tie line, so n_{liq}/n_{vap} is also almost infinite, and there is only a trace of vapour present. When the pressure is reduced to p_2, the value of l_{vap}/l_{liq} is about 0.5, so $n_{liq}/n_{vap} \approx 0.5$ and the amount of liquid is about 0.5 times the amount of vapour. When the pressure has been reduced to p_3, the sample is almost completely gaseous and because $l_{vap}/l_{liq} \approx 0$ we conclude that there is only a trace of liquid present. •

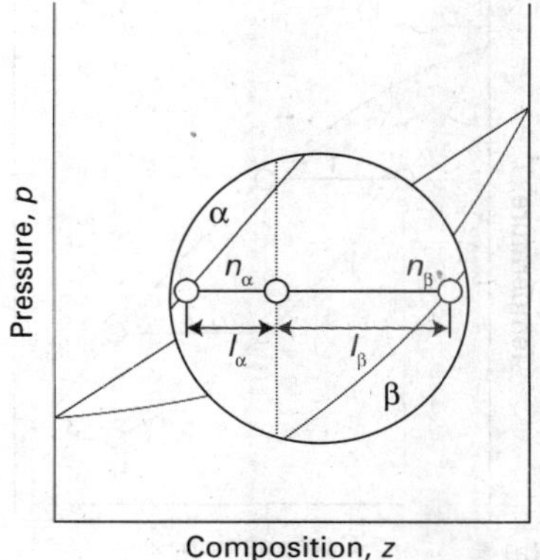

Fig. 5.35 The lever rule. The distances l_α and l_β are used to find the proportions of the amounts of phases α (such as liquid) and β (for example, vapour) present at equilibrium. The lever rule is so called because a similar rule relates the masses at two ends of a lever to their distances from a pivot ($m_\alpha l_\alpha = m_\beta l_\beta$ for balance).

5.7 Temperature–composition diagrams

Key points **(a) A phase diagram can be used to discuss the process of fractional distillation. (b) Depending on the relative strengths of the intermolecular forces, high- or low-boiling azeotropes may be formed. (c) The vapour pressure of a system composed of immiscible liquids is the sum of the vapour pressures of the pure liquids. (d) A phase diagram may be used to discuss the distillation of partially miscible liquids.**

To discuss distillation we need a **temperature–composition diagram**, a phase diagram in which the boundaries show the composition of the phases that are in equilibrium at various temperatures (and a given pressure, typically 1 atm). An example is shown in Fig. 5.36. Note that the liquid phase now lies in the lower part of the diagram.

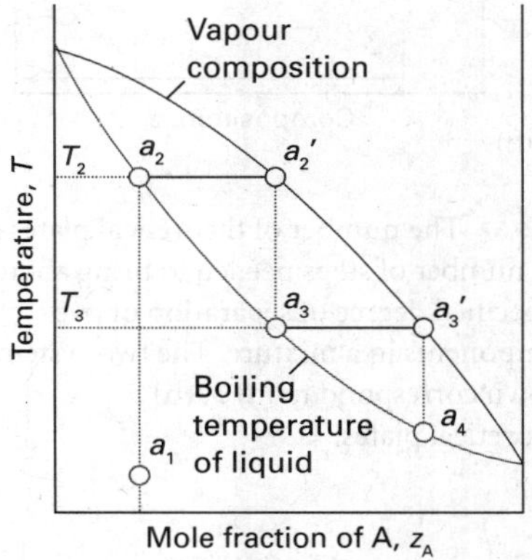

Fig. 5.36 The temperature–composition diagram corresponding to an ideal mixture with the component A more volatile than component B. Successive boilings and condensations of a liquid originally of composition a_1 lead to a condensate that is pure A. The separation technique is called fractional distillation.

(a) The distillation of mixtures

Consider what happens when a liquid of composition a_1 in Fig. 5.36 is heated. It boils when the temperature reaches T_2. Then the liquid has composition a_2 (the same as a_1) and the vapour (which is present only as a trace) has composition a_2'. The vapour is richer in the more volatile component A (the component with the lower boiling point). From the location of a_2, we can state the vapour's composition at the boiling point, and from the location of the tie line joining a_2 and a_2' we can read off the boiling temperature (T_2) of the original liquid mixture.

In a **simple distillation**, the vapour is withdrawn and condensed. This technique is used to separate a volatile liquid from a non-volatile solute or solid. In **fractional distillation**, the boiling and condensation cycle is repeated successively. This technique is used to separate volatile liquids. We can follow the changes that occur by seeing what happens when the first condensate of composition a_3 is reheated. The phase diagram shows that this mixture boils at T_3 and yields a vapour of composition a_3', which is even richer in the more volatile component. That vapour is drawn off, and the first drop condenses to a liquid of composition a_4. The cycle can then be repeated until in due course almost pure A is obtained in the vapour and pure B remains in the liquid.

The efficiency of a fractionating column is expressed in terms of the number of **theoretical plates**, the number of effective vaporization and condensation steps that are required to achieve a condensate of given composition from a given distillate. Thus, to achieve the degree of separation shown in Fig. 5.37a, the fractionating column must correspond to three theoretical plates. To achieve the same separation for the system shown in Fig. 5.37b, in which the components have more similar partial pressures, the fractionating column must be designed to correspond to five theoretical plates.

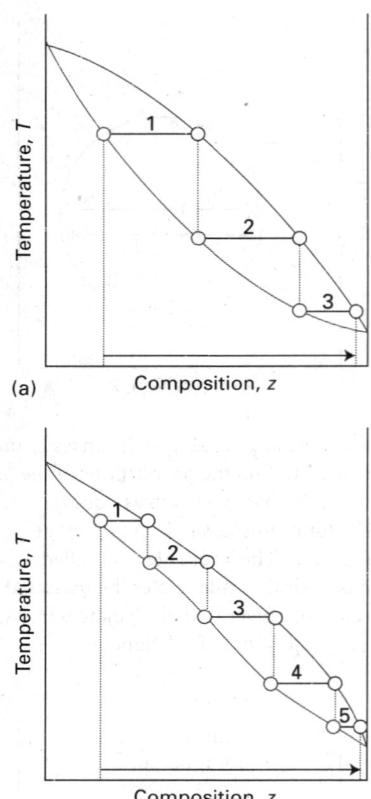

Fig. 5.37 The number of theoretical plates is the number of steps needed to bring about a specified degree of separation of two components in a mixture. The two systems shown correspond to (a) 3, (b) 5 theoretical plates.

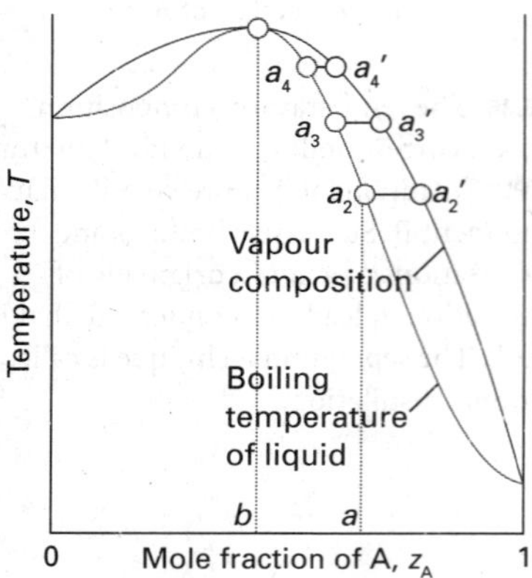

Fig. 5.38 A high-boiling azeotrope. When the liquid of composition a is distilled, the composition of the remaining liquid changes towards b but no further.

(b) Azeotropes

Although many liquids have temperature–composition phase diagrams resembling the ideal version in Fig. 5.36, in a number of important cases there are marked deviations. A maximum in the phase diagram (Fig. 5.38) may occur when the favourable interactions between A and B molecules reduce the vapour pressure of the mixture below the ideal value: in effect, the A–B interactions stabilize the liquid. In such cases the excess Gibbs energy, G^E (Section 5.4), is negative (more favourable to mixing than ideal). Examples of this behaviour include trichloromethane/propanone and nitric acid/water mixtures. Phase diagrams showing a minimum (Fig. 5.39) indicate that the mixture is destabilized relative to the ideal solution, the A–B interactions then being unfavourable. For such mixtures G^E is positive (less favourable to mixing than ideal), and there may be contributions from both enthalpy and entropy effects. Examples include dioxane/water and ethanol/water mixtures.

Deviations from ideality are not always so strong as to lead to a maximum or minimum in the phase diagram, but when they do there are important consequences for distillation. Consider a liquid of composition a on the right of the maximum in Fig. 5.38. The vapour (at a_2') of the boiling mixture (at a_2) is richer in A. If that vapour is removed (and condensed elsewhere), then the remaining liquid will move to a composition that is richer in B, such as that represented by a_3, and the vapour in equilibrium with this mixture will have composition a_3'. As that vapour is removed, the composition of the boiling liquid shifts to a point such as a_4, and the composition of the vapour shifts to a_4'. Hence, as evaporation proceeds, the composition of the remaining liquid shifts towards B as A is drawn off. The boiling point of the liquid rises, and the vapour becomes richer in B. When so much A has been evaporated that the liquid has reached the composition b, the vapour has the same composition as the liquid. Evaporation then occurs without change of composition. The mixture is said to form an **azeotrope**.[2] When the azeotropic composition has been reached, distillation cannot separate the two liquids because the condensate has the same composition as the azeotropic liquid. One example of azeotrope formation is hydrochloric acid/water, which is azeotropic at 80 per cent by mass of water and boils unchanged at 108.6°C.

The system shown in Fig. 5.39 is also azeotropic, but shows its azeotropy in a different way. Suppose we start with a mixture of composition a_1, and follow the changes in the composition of the vapour that rises through a fractionating column (essentially a vertical glass tube packed with glass rings to give a large surface area). The mixture boils at a_2 to give a vapour of composition a_2'. This vapour condenses in the column to a liquid of the same composition (now marked a_3). That liquid reaches equilibrium with its vapour at a_3', which condenses higher up the tube to give a liquid of the same composition, which we now call a_4. The fractionation therefore shifts the vapour towards the azeotropic composition at b, but not beyond, and the azeotropic vapour emerges from the top of the column. An example is ethanol/water, which boils unchanged when the water content is 4 per cent by mass and the temperature is 78°C.

(c) Immiscible liquids

Finally we consider the distillation of two immiscible liquids, such as octane and water. At equilibrium, there is a tiny amount of A dissolved in B, and similarly a tiny amount of B dissolved in A: both liquids are saturated with the other component

[2] The name comes from the Greek words for 'boiling without changing'.

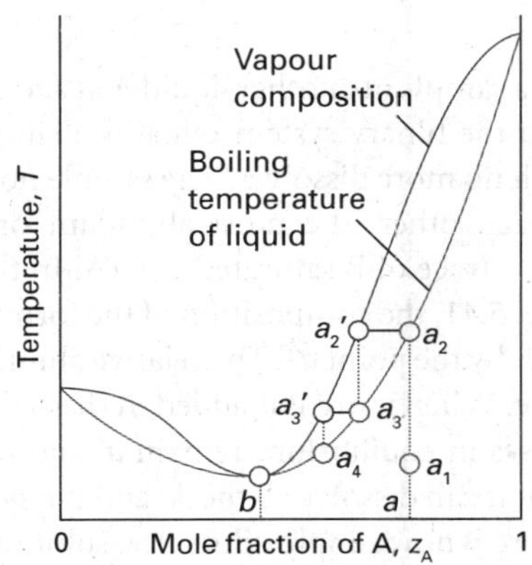

Fig. 5.39 A low-boiling azeotrope. When the mixture at *a* is fractionally distilled, the vapour in equilibrium in the fractionating column moves towards *b* and then remains unchanged.

Fig. 5.40 The distillation of (a) two immiscible liquids can be regarded as (b) the joint distillation of the separated components, and boiling occurs when the sum of the partial pressures equals the external pressure.

(Fig. 5.40a). As a result, the total vapour pressure of the mixture is close to $p = p_A^* + p_B^*$. If the temperature is raised to the value at which this total vapour pressure is equal to the atmospheric pressure, boiling commences and the dissolved substances are purged from their solution. However, this boiling results in a vigorous agitation of the mixture, so each component is kept saturated in the other component, and the purging continues as the very dilute solutions are replenished. This intimate contact is essential: two immiscible liquids heated in a container like that shown in Fig. 5.40b would not boil at the same temperature. The presence of the saturated solutions means that the 'mixture' boils at a lower temperature than either component would alone because boiling begins when the total vapour pressure reaches 1 atm, not when either vapour pressure reaches 1 atm. This distinction is the basis of **steam distillation**, which enables some heat-sensitive, water-insoluble organic compounds to be distilled at a lower temperature than their normal boiling point. The only snag is that the composition of the condensate is in proportion to the vapour pressures of the components, so oils of low volatility distil in low abundance.

5.8 Liquid–liquid phase diagrams

Key points **(a) Phase separation of partially miscible liquids may occur when the temperature is below the upper critical solution temperature or above the lower critical solution temperature; the process may be discussed in terms of the model of a regular solution. (b) The upper critical solution temperature is the highest temperature at which phase separation occurs. The lower critical solution temperature is the temperature below which components mix in all proportions and above which they form two phases. (c) The outcome of a distillation of a low-boiling azeotrope depends on whether the liquids become fully miscible before they boil or boiling occurs before mixing is complete.**

Now we consider temperature–composition diagrams for systems that consist of pairs of **partially miscible** liquids, which are liquids that do not mix in all proportions at all temperatures. An example is hexane and nitrobenzene. The same principles of interpretation apply as to liquid–vapour diagrams.

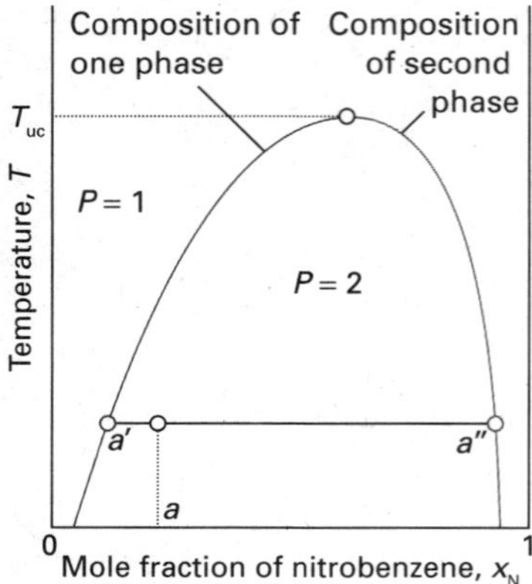

Fig. 5.41 The temperature–composition diagram for hexane and nitrobenzene at 1 atm. The region below the curve corresponds to the compositions and temperatures at which the liquids are partially miscible. The upper critical temperature, T_{uc}, is the temperature above which the two liquids are miscible in all proportions.

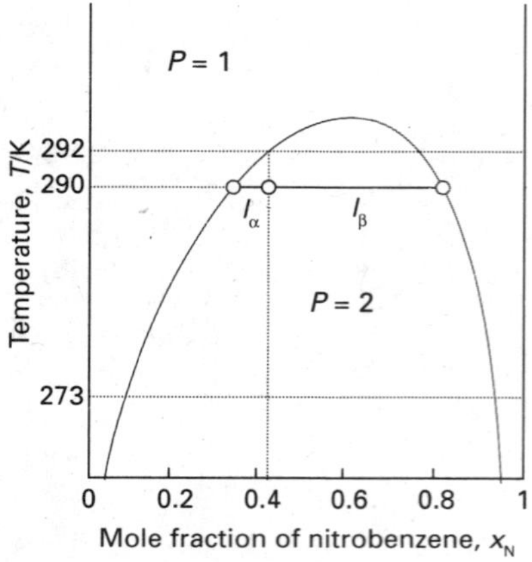

Fig. 5.42 The temperature–composition diagram for hexane and nitrobenzene at 1 atm again, with the points and lengths discussed in the text.

(a) Phase separation

Suppose a small amount of a liquid B is added to a sample of another liquid A at a temperature T'. Liquid B dissolves completely, and the binary system remains a single phase. As more B is added, a stage comes at which no more dissolves. The sample now consists of two phases in equilibrium with each other, the most abundant one consisting of A saturated with B, the minor one a trace of B saturated with A. In the temperature–composition diagram drawn in Fig. 5.41, the composition of the former is represented by the point a' and that of the latter by the point a''. The relative abundances of the two phases are given by the lever rule. When more B is added, A dissolves in it slightly. The compositions of the two phases in equilibrium remain a' and a''. A stage is reached when so much B is present that it can dissolve all the A, and the system reverts to a single phase. The addition of more B now simply dilutes the solution, and from then on a single phase remains.

The composition of the two phases at equilibrium varies with the temperature. For hexane and nitrobenzene, raising the temperature increases their miscibility. The two-phase region therefore covers a narrower range of composition because each phase in equilibrium is richer in its minor component: the A-rich phase is richer in B and the B-rich phase is richer in A. We can construct the entire phase diagram by repeating the observations at different temperatures and drawing the envelope of the two-phase region.

Example 5.5 *Interpreting a liquid–liquid phase diagram*

A mixture of 50 g of hexane (0.58 mol C_6H_{14}) and 50 g of nitrobenzene (0.41 mol $C_6H_5NO_2$) was prepared at 290 K. What are the compositions of the phases, and in what proportions do they occur? To what temperature must the sample be heated in order to obtain a single phase?

Method The compositions of phases in equilibrium are given by the points where the tie line representing the temperature intersects the phase boundary. Their proportions are given by the lever rule (eqn 5.46). The temperature at which the components are completely miscible is found by following the isopleth upwards and noting the temperature at which it enters the one-phase region of the phase diagram.

Answer We denote hexane by H and nitrobenzene by N; refer to Fig. 5.42, which is a simplified version of Fig. 5.41. The point $x_N = 0.41$, $T = 290$ K occurs in the two-phase region of the phase diagram. The horizontal tie line cuts the phase boundary at $x_N = 0.35$ and $x_N = 0.83$, so those are the compositions of the two phases. According to the lever rule, the ratio of amounts of each phase is equal to the ratio of the distances l_α and l_β:

$$\frac{n_\alpha}{n_\beta} = \frac{l_\beta}{l_\alpha} = \frac{0.83 - 0.41}{0.41 - 0.35} = \frac{0.42}{0.06} = 7$$

That is, there is about 7 times more hexane-rich phase than nitrobenzene-rich phase. Heating the sample to 292 K takes it into the single-phase region. Because the phase diagram has been constructed experimentally, these conclusions are not based on any assumptions about ideality. They would be modified if the system were subjected to a different pressure.

Self-test 5.7 Repeat the problem for 50 g of hexane and 100 g of nitrobenzene at 273 K. [$x_N = 0.09$ and 0.95 in ratio 1:1.3; 294 K]

(b) Critical solution temperatures

The **upper critical solution temperature**, T_{uc} (or *upper consolute temperature*), is the highest temperature at which phase separation occurs. Above the upper critical temperature the two components are fully miscible. This temperature exists because the greater thermal motion overcomes any potential energy advantage in molecules of one type being close together. One example is the nitrobenzene/hexane system shown in Fig. 5.41. An example of a solid solution is the palladium/hydrogen system, which shows two phases, one a solid solution of hydrogen in palladium and the other a palladium hydride, up to 300°C but forms a single phase at higher temperatures (Fig. 5.43).

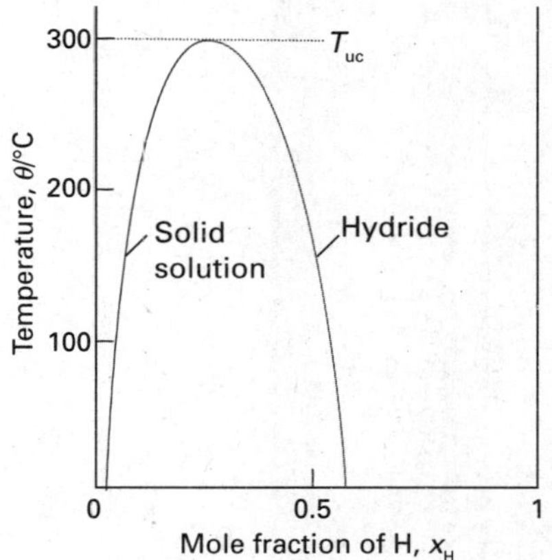

Fig. 5.43 The phase diagram for palladium and palladium hydride, which has an upper critical temperature at 300°C.

The thermodynamic interpretation of the upper critical solution temperature focuses on the Gibbs energy of mixing and its variation with temperature. We saw in Section 5.4 that a simple model of a real solution results in a Gibbs energy of mixing that behaves as shown in Fig. 5.44. Provided the parameter ξ that was introduced in eqn 5.28 is greater than 2, the Gibbs energy of mixing has a double minimum. As a result, for $\xi > 2$ we can expect phase separation to occur. The same model shows that the compositions corresponding to the minima are obtained by looking for the conditions at which $\partial\Delta_{mix}G/\partial x = 0$, and a simple manipulation of eqn 5.29 shows that we have to solve

$$\ln\frac{x}{1-x} + \xi(1-2x) = 0 \qquad (5.47)$$

The solutions are plotted in Fig. 5.45. We see that, as ξ decreases, which can be interpreted as an increase in temperature provided the intermolecular forces remain constant, the two minima move together and merge when $\xi = 2$.

Some systems show a **lower critical solution temperature**, T_{lc} (or *lower consolute temperature*), below which they mix in all proportions and above which they form two phases. An example is water and triethylamine (Fig. 5.46). In this case, at low temperatures the two components are more miscible because they form a weak complex; at higher temperatures the complexes break up and the two components are less miscible.

Some systems have both upper and lower critical solution temperatures. They occur because, after the weak complexes have been disrupted, leading to partial miscibility, the thermal motion at higher temperatures homogenizes the mixture again, just as in the case of ordinary partially miscible liquids. The most famous example is nicotine and water, which are partially miscible between 61°C and 210°C (Fig. 5.47).

A brief comment

Equation 5.47 is an example of a *transcendental equation*, an equation that does not have a solution that can be expressed in a closed form. The solutions can be found numerically by using mathematical software or by plotting the first term against the second and identifying the points of intersection as ξ is changed.

(c) The distillation of partially miscible liquids

Consider a pair of liquids that are partially miscible and form a low-boiling azeotrope. This combination is quite common because both properties reflect the tendency of the two kinds of molecule to avoid each other. There are two possibilities: one in which the liquids become fully miscible before they boil; the other in which boiling occurs before mixing is complete.

Figure 5.48 shows the phase diagram for two components that become fully miscible before they boil. Distillation of a mixture of composition a_1 leads to a vapour of composition b_1, which condenses to the completely miscible single-phase solution at b_2. Phase separation occurs only when this distillate is cooled to a point in the two-phase liquid region, such as b_3. This description applies only to the first drop of distillate. If distillation continues, the composition of the remaining liquid changes. In the end, when the whole sample has evaporated and condensed, the composition is back to a_1.

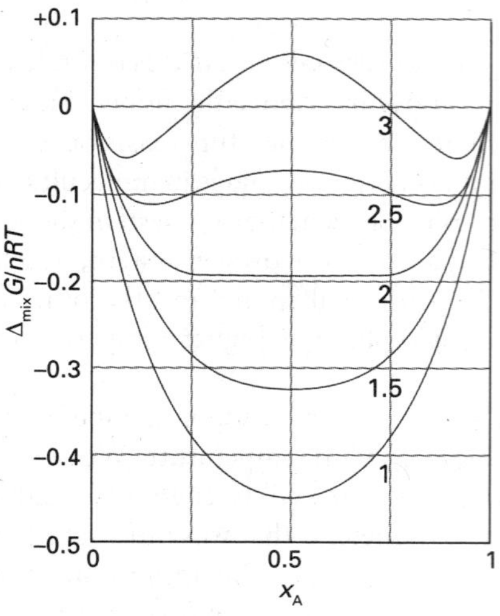

Fig. 5.44 The temperature variation of the Gibbs energy of mixing of a system that is partially miscible at low temperatures. A system of composition in the region $P = 2$ forms two phases with compositions corresponding to the two local minima of the curve. This illustration is a duplicate of Fig. 5.19.

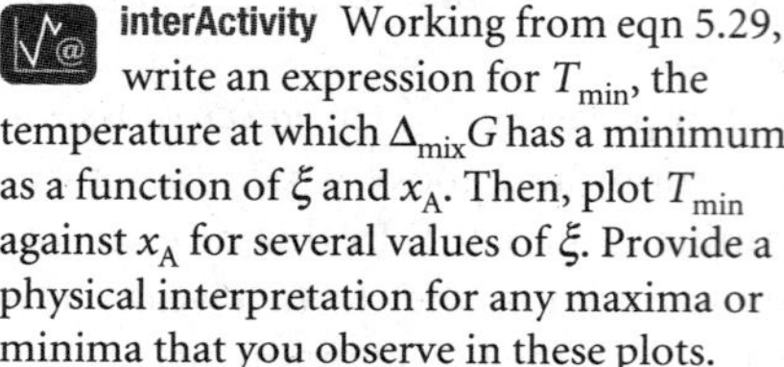

interActivity Working from eqn 5.29, write an expression for T_{min}, the temperature at which $\Delta_{mix}G$ has a minimum, as a function of ξ and x_A. Then, plot T_{min} against x_A for several values of ξ. Provide a physical interpretation for any maxima or minima that you observe in these plots.

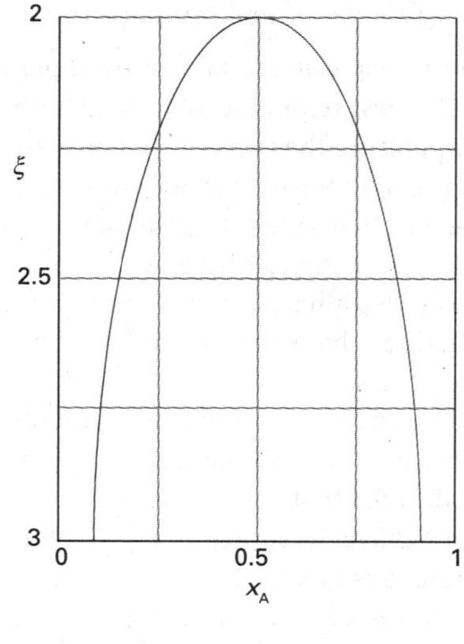

Fig. 5.45 The location of the phase boundary as computed on the basis of the ξ-parameter model introduced in Section 5.4a.

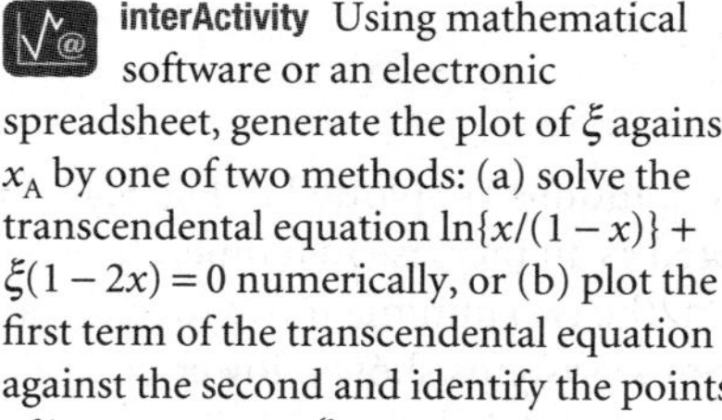

interActivity Using mathematical software or an electronic spreadsheet, generate the plot of ξ against x_A by one of two methods: (a) solve the transcendental equation $\ln\{x/(1-x)\} + \xi(1-2x) = 0$ numerically, or (b) plot the first term of the transcendental equation against the second and identify the points of intersection as ξ is changed.

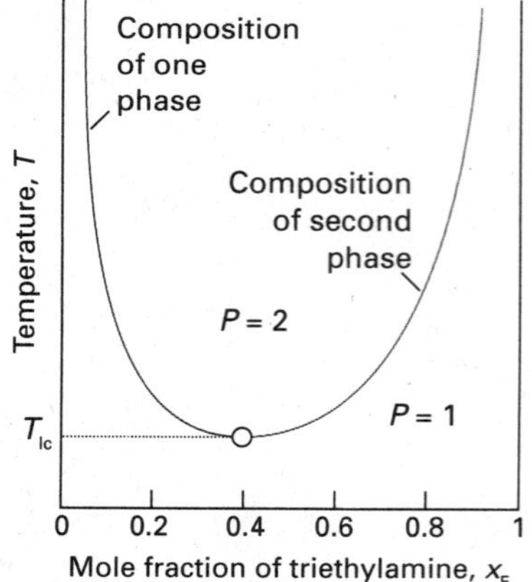

Fig. 5.46 The temperature–composition diagram for water and triethylamine. This system shows a lower critical temperature at 292 K. The labels indicate the interpretation of the boundaries.

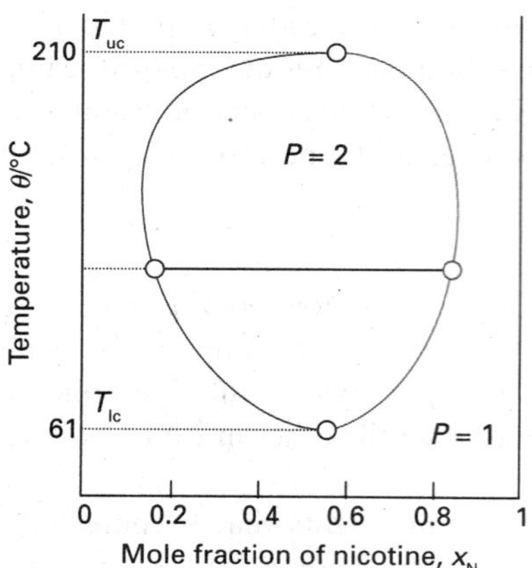

Fig. 5.47 The temperature–composition diagram for water and nicotine, which has both upper and lower critical temperatures. Note the high temperatures for the liquid (especially the water): the diagram corresponds to a sample under pressure.

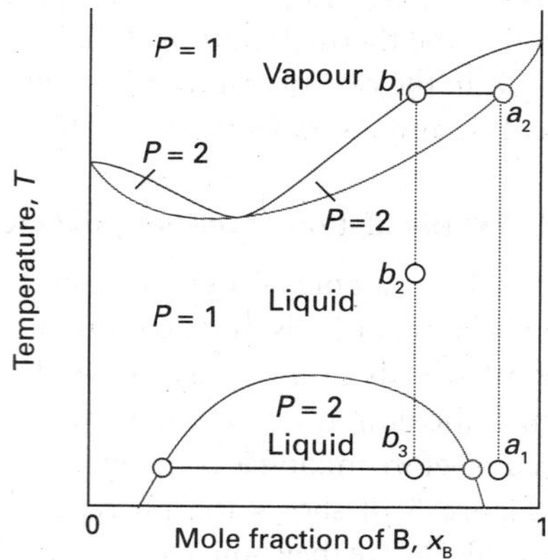

Fig. 5.48 The temperature–composition diagram for a binary system in which the upper critical temperature is less than the boiling point at all compositions. The mixture forms a low-boiling azeotrope.

Figure 5.49 shows the second possibility, in which there is no upper critical solution temperature. The distillate obtained from a liquid initially of composition a_1 has composition b_3 and is a two-phase mixture. One phase has composition b_3' and the other has composition b_3''.

The behaviour of a system of composition represented by the isopleth *e* in Fig. 5.49 is interesting. A system at e_1 forms two phases, which persist (but with changing proportions) up to the boiling point at e_2. The vapour of this mixture has the same composition as the liquid (the liquid is an azeotrope). Similarly, condensing a vapour of composition e_3 gives a two-phase liquid of the same overall composition. At a fixed temperature, the mixture vaporizes and condenses like a single substance.

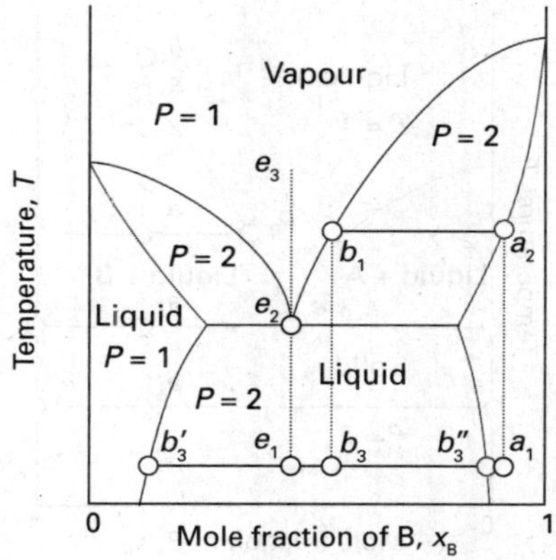

Fig. 5.49 The temperature–composition diagram for a binary system in which boiling occurs before the two liquids are fully miscible.

Example 5.6 *Interpreting a phase diagram*

State the changes that occur when a mixture of composition $x_B = 0.95$ (a_1) in Fig. 5.50 is boiled and the vapour condensed.

Method The area in which the point lies gives the number of phases; the compositions of the phases are given by the points at the intersections of the horizontal tie line with the phase boundaries; the relative abundances are given by the lever rule .

Answer The initial point is in the one-phase region. When heated it boils at 350 K (a_2) giving a vapour of composition $x_B = 0.56$ (b_1). The liquid gets richer in B, and the last drop (of pure B) evaporates at 390 K. The boiling range of the liquid is therefore 350 to 390 K. If the initial vapour is drawn off, it has a composition $x_B = 0.56$. This composition would be maintained if the sample were very large, but for a finite sample it shifts to higher values and ultimately to $x_B = 0.95$. Cooling the distillate corresponds to moving down the $x_B = 0.56$ isopleth. At 330 K, for instance, the liquid phase has composition $x_B = 0.87$, the vapour $x_B = 0.49$; their relative proportions are 1:4·4. At 320 K the sample consists of three phases: the vapour and two liquids. One liquid phase has composition $x_B = 0.30$; the other has composition $x_B = 0.80$ in the ratio 0.92:1. Further cooling moves the system into the two-phase region, and at 298 K the compositions are 0.20 and 0.90 in the ratio 0.94:1. As further distillate boils over, the overall composition of the distillate becomes richer in B. When the last drop has been condensed the phase composition is the same as at the beginning.

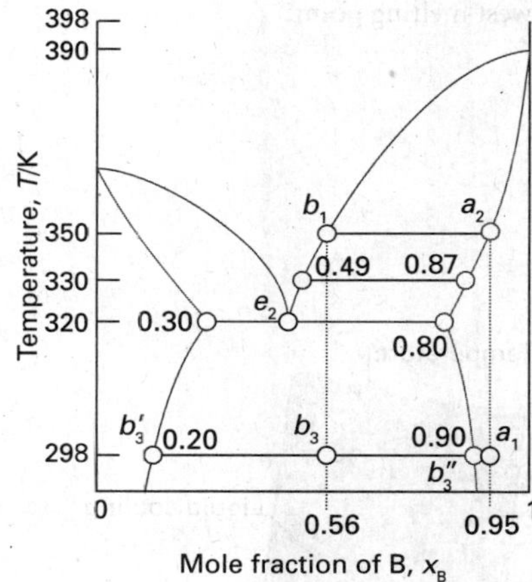

Fig. 5.50 The points of the phase diagram in Fig. 5.49 that are discussed in Example 5.6.

Self-test 5.8 Repeat the discussion, beginning at the point $x_B = 0.4$, $T = 298$ K.

5.9 Liquid–solid phase diagrams

Key points **(a) A phase diagram summarizes the temperature–composition properties of a binary system with solid and liquid phases; at the eutectic composition the liquid phase solidifies without change of composition. (b) The phase equilibria of binary systems in which the components react may also be summarized by a phase diagram. (c) In some cases, a solid compound does not survive melting.**

Knowledge of the temperature–composition diagrams for solid mixtures guides the design of important industrial processes, such as the manufacture of liquid crystal displays and semiconductors. In this section, we shall consider systems where solid and liquid phases may both be present at temperatures below the boiling point.

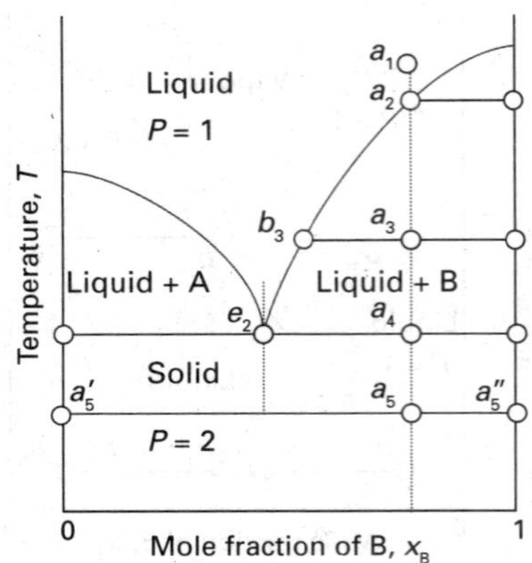

Fig. 5.51 The temperature–composition phase diagram for two almost immiscible solids and their completely miscible liquids. Note the similarity to Fig. 5.49. The isopleth through *e* corresponds to the eutectic composition, the mixture with lowest melting point.

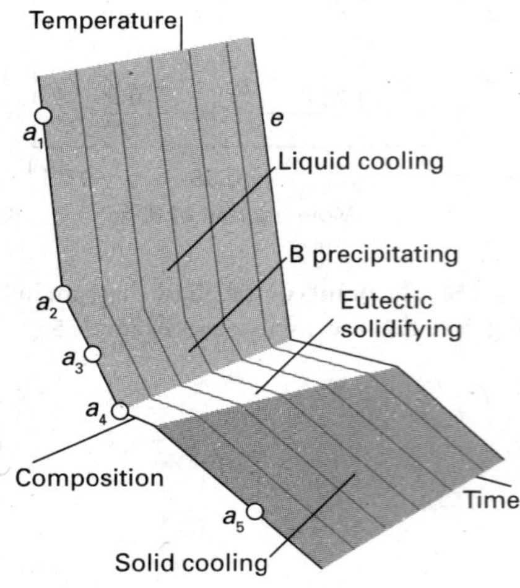

Fig. 5.52 The cooling curves for the system shown in Fig. 5.51. For isopleth *a*, the rate of cooling slows at a_2 because solid B deposits from solution. There is a complete halt at a_4 while the eutectic solidifies. This halt is longest for the eutectic isopleth, *e*. The eutectic halt shortens again for compositions beyond *e* (richer in A). Cooling curves are used to construct the phase diagram.

(a) Eutectics

Consider the two-component liquid of composition a_1 in Fig. 5.51. The changes that occur as the system is cooled may be expressed as follows.

1. $a_1 \rightarrow a_2$. The system enters the two-phase region labelled 'Liquid + B'. Pure solid B begins to come out of solution and the remaining liquid becomes richer in A.
2. $a_2 \rightarrow a_3$. More of the solid B forms, and the relative amounts of the solid and liquid (which are in equilibrium) are given by the lever rule. At this stage there are roughly equal amounts of each. The liquid phase is richer in A than before (its composition is given by b_3) because some B has been deposited.
3. $a_3 \rightarrow a_4$. At the end of this step, there is less liquid than at a_3, and its composition is given by e_2. This liquid now freezes to give a two-phase system of pure B and pure A.

The isopleth at e_2 in Fig. 5.51 corresponds to the **eutectic** composition, the mixture with the lowest melting point.[3] A liquid with the eutectic composition freezes at a single temperature, without previously depositing solid A or B. A solid with the eutectic composition melts, without change of composition, at the lowest temperature of any mixture. Solutions of composition to the right of e_2 deposit B as they cool, and solutions to the left deposit A: only the eutectic mixture (apart from pure A or pure B) solidifies at a single definite temperature without gradually unloading one or other of the components from the liquid.

One technologically important eutectic is solder, which in one form has mass composition of about 67 per cent tin and 33 per cent lead and melts at 183°C. The eutectic formed by 23 per cent NaCl and 77 per cent H_2O by mass melts at −21.1°C. When salt is added to ice under isothermal conditions (for example, when spread on an icy road) the mixture melts if the temperature is above −21.1°C (and the eutectic composition has been achieved). When salt is added to ice under adiabatic conditions (for example, when added to ice in a vacuum flask) the ice melts, but in doing so it absorbs heat from the rest of the mixture. The temperature of the system falls and, if enough salt is added, cooling continues down to the eutectic temperature. Eutectic formation occurs in the great majority of binary alloy systems, and is of great importance for the microstructure of solid materials. Although a eutectic solid is a two-phase system, it crystallizes out in a nearly homogeneous mixture of microcrystals. The two microcrystalline phases can be distinguished by microscopy and structural techniques such as X-ray diffraction (Chapter 19).

Thermal analysis is a very useful practical way of detecting eutectics. We can see how it is used by considering the rate of cooling down the isopleth through a_1 in Fig. 5.51. The liquid cools steadily until it reaches a_2, when B begins to be deposited (Fig. 5.52). Cooling is now slower because the solidification of B is exothermic and retards the cooling. When the remaining liquid reaches the eutectic composition, the temperature remains constant until the whole sample has solidified: this region of constant temperature is the eutectic halt. If the liquid has the eutectic composition *e* initially, the liquid cools steadily down to the freezing temperature of the eutectic, when there is a long **eutectic halt** as the entire sample solidifies (like the freezing of a pure liquid).

Monitoring the cooling curves at different overall compositions gives a clear indication of the structure of the phase diagram. The solid–liquid boundary is given by the points at which the rate of cooling changes. The longest eutectic halt gives the location of the eutectic composition and its melting temperature.

[3] The name comes from the Greek words for 'easily melted'.

(b) Reacting systems

Many binary mixtures react to produce compounds, and technologically important examples of this behaviour include the Group 13/15 (III/V) semiconductors, such as the gallium arsenide system, which forms the compound GaAs. Although three constituents are present, there are only two components because GaAs is formed from the reaction Ga + As ⇌ GaAs. We shall illustrate some of the principles involved with a system that forms a compound C that also forms eutectic mixtures with the species A and B (Fig. 5.53).

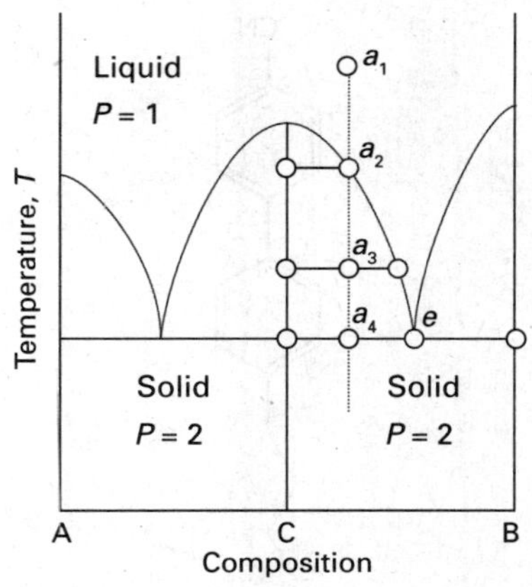

Fig. 5.53 The phase diagram for a system in which A and B react to form a compound C = AB. This resembles two versions of Fig. 5.51 in each half of the diagram. The constituent C is a true compound, not just an equimolar mixture.

A system prepared by mixing an excess of B with A consists of C and unreacted B. This is a binary B, C system, which we suppose forms a eutectic. The principal change from the eutectic phase diagram in Fig. 5.51 is that the whole of the phase diagram is squeezed into the range of compositions lying between equal amounts of A and B ($x_B = 0.5$, marked C in Fig. 5.53) and pure B. The interpretation of the information in the diagram is obtained in the same way as for Fig. 5.51. The solid deposited on cooling along the isopleth a is the compound C. At temperatures below a_4 there are two solid phases, one consisting of C and the other of B. The pure compound C melts **congruently**, that is, the composition of the liquid it forms is the same as that of the solid compound.

(c) Incongruent melting

In some cases the compound C is not stable as a liquid. An example is the alloy Na_2K, which survives only as a solid (Fig. 5.54). Consider what happens as a liquid at a_1 is cooled:

1. $a_1 \rightarrow a_2$. A solid solution rich in Na is deposited, and the remaining liquid is richer in K.

2. $a_2 \rightarrow$ just below a_3. The sample is now entirely solid and consists of a solid solution rich in Na and solid Na_2K.

Now consider the isopleth through b_1:

1. $b_1 \rightarrow b_2$. No obvious change occurs until the phase boundary is reached at b_2 when a solid solution rich in Na begins to deposit.

2. $b_2 \rightarrow b_3$. A solid solution rich in Na deposits, but at b_3 a reaction occurs to form Na_2K: this compound is formed by the K atoms diffusing into the solid Na.

3. b_3. At b_3, three phases are in mutual equilibrium: the liquid, the compound Na_2K, and a solid solution rich in Na. The horizontal line representing this three-phase equilibrium is called a **peritectic line**.

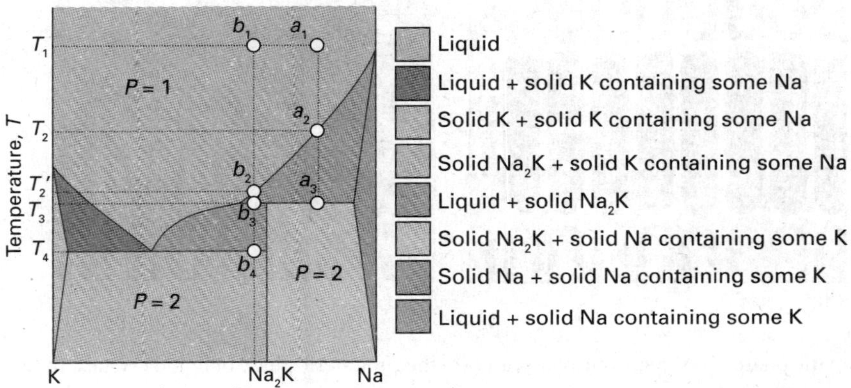

Fig. 5.54 The phase diagram for an actual system (sodium and potassium) like that shown in Fig. 5.53, but with two differences. One is that the compound is Na_2K, corresponding to A_2B and not AB as in that illustration. The second is that the compound exists only as the solid, not as the liquid. The transformation of the compound at its melting point is an example of incongruent melting.

At this stage the liquid Na/K mixture is in equilibrium with a little solid Na_2K, but there is still no liquid compound.

4. $b_3 \rightarrow b_4$. As cooling continues, the amount of solid compound increases until at b_4 the liquid reaches its eutectic composition. It then solidifies to give a two-phase solid consisting of a solid solution rich in K and solid Na_2K.

If the solid is reheated, the sequence of events is reversed. No liquid Na_2K forms at any stage because it is too unstable to exist as a liquid. This behaviour is an example of **incongruent melting**, in which a compound melts into its components and does not itself form a liquid phase.

CN

1

2

IMPACT ON MATERIALS SCIENCE
I5.2 Liquid crystals

A *mesophase* is a phase intermediate between solid and liquid. Mesophases are of great importance in biology, for they occur as lipid bilayers and in vesicular systems. A mesophase may arise when molecules have highly non-spherical shapes, such as being long and thin (**1**), or disc-like (**2**). When the solid melts, some aspects of the long-range order characteristic of the solid may be retained, and the new phase may be a *liquid crystal*, a substance having liquid-like imperfect long-range order in at least one direction in space but positional or orientational order in at least one other direction. *Calamitic liquid crystals* (from the Greek word for reed) are made from long and thin molecules, whereas *discotic liquid crystals* are made from disc-like molecules. A *thermotropic* liquid crystal displays a transition to the liquid crystalline phase as the temperature is changed. A *lyotropic* liquid crystal is a solution that undergoes a transition to the liquid crystalline phase as the composition is changed.

One type of retained long-range order gives rise to a *smectic phase* (from the Greek word for soapy), in which the molecules align themselves in layers (Fig. 5.55). Other materials, and some smectic liquid crystals at higher temperatures, lack the layered structure but retain a parallel alignment; this mesophase is called a *nematic phase* (from the Greek for thread, which refers to the observed defect structure of the phase). In the *cholesteric phase* (from the Greek for bile solid) the molecules lie in sheets at angles that change slightly between each sheet. That is, they form helical structures

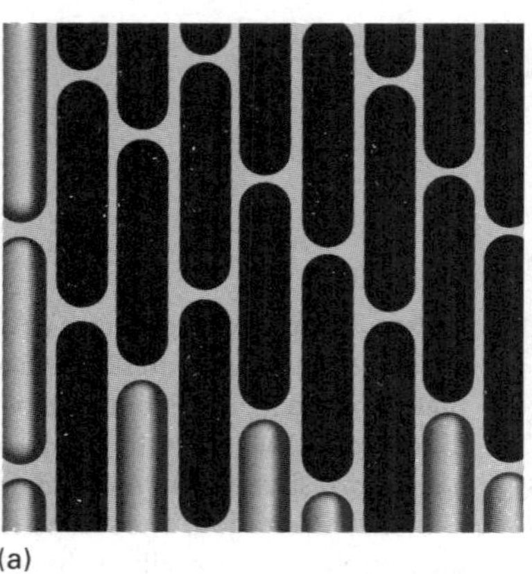
(a)

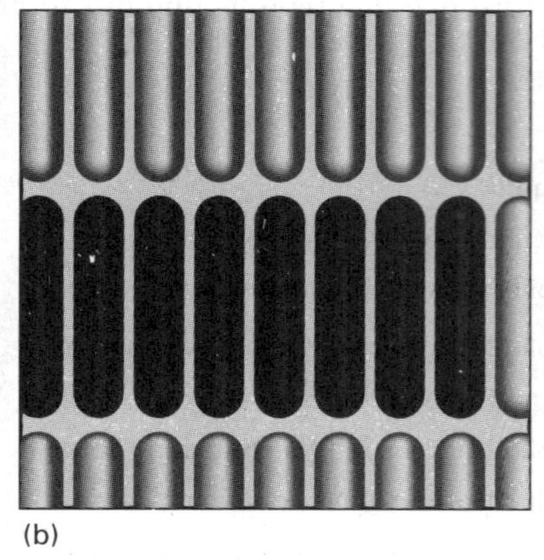
(b)

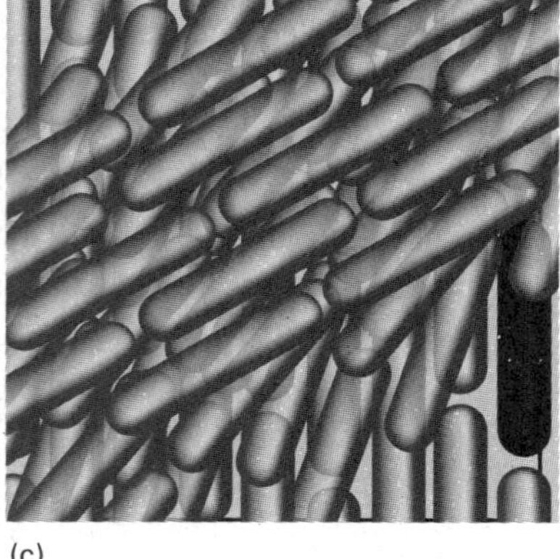
(c)

Fig. 5.55 The arrangement of molecules in (a) the nematic phase, (b) the smectic phase, and (c) the cholesteric phase of liquid crystals. In the cholesteric phase, the stacking of layers continues to give a helical arrangement of molecules.

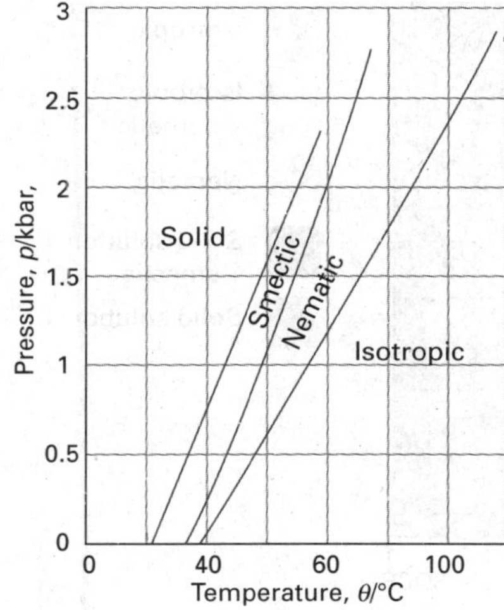

Fig. 5.56 The pressure–temperature diagram of octylcyanobiphenyl (8CB). (Based on R. Shashidhar and G. Venkatesh, *J. de Physique Colloque*, 40, C3 (1979).)

with a pitch that depends on the temperature. As a result, cholesteric liquid crystals diffract light and have colours that depend on the temperature. Disc-like molecules such as (2) can form nematic and *columnar* mesophases. In the latter, the aromatic rings stack one on top of the other and are separated by very small distances (less than 0.5 nm). Figure 5.56 shows the pressure–temperature phase diagram of octylcyanobiphenyl, which is widely used in liquid crystal displays.

The optical properties of nematic liquid crystals are anisotropic, meaning that they depend on the relative orientation of the molecular assemblies with respect to the polarization of the incident beam of light. Nematic liquid crystals also respond in special ways to electric fields. Together, these unique optical and electrical properties form the basis of operation of liquid crystal displays (LCDs). In a 'twisted nematic' LCD, the liquid crystal is held between two flat plates about 10 mm apart. The inner surface of each plate is coated with a transparent conducting material, such as indium–tin oxide. The plates also have a surface that causes the liquid crystal to adopt a particular orientation at its interface and are typically set at 90° to each other but 270° in a 'supertwist' arrangement. The entire assembly is set between two polarizers, optical filters that allow light of only one specific plane of polarization to pass. The incident light passes through the outer polarizer, then its plane of polarization is rotated as it passes through the twisted nematic, and, depending on the setting of the second polarizer, will pass through (if that is how the second polarizer is arranged). When a potential difference is applied across the cell, the helical arrangement is lost and the plane of the light is no longer rotated and will be blocked by the second polarizer.

Although there are many liquid crystalline materials, some difficulty is often experienced in achieving a technologically useful temperature range for the existence of the mesophase. To overcome this difficulty, mixtures can be used. An example of the type of phase diagram that is then obtained is shown in Fig. 5.57. As can be seen, the mesophase exists over a wider range of temperatures than either liquid crystalline material alone.

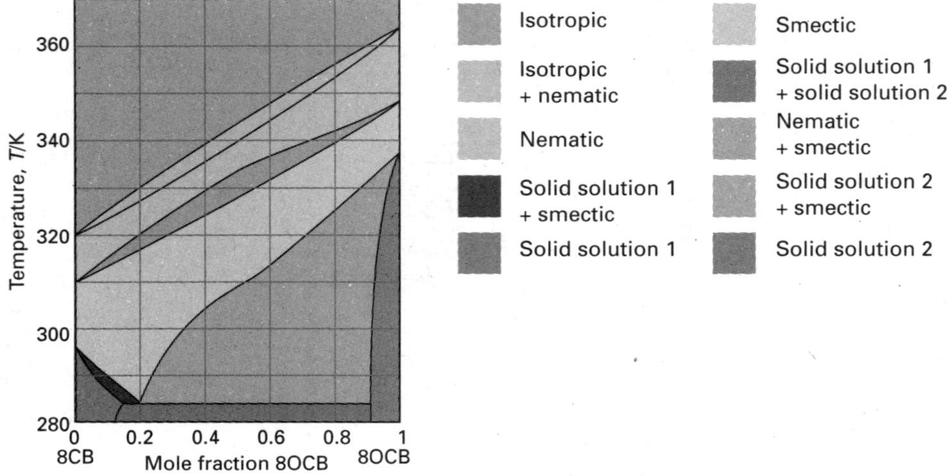

Fig. 5.57 The phase diagram at 1 atm for a binary system of two liquid crystalline materials, octylcyanobiphenyl (8CB) and octyloxycyanobiphenyl (8OCB). (Based on P. Rushikesh, A. Matkar, and T. Kyua, *J. Chem. Phys.*, **124**, 224902 (2006).)

Activities

Now we see how to adjust the expressions developed earlier in the chapter to take into account deviations from ideal behavior that we have encountered during the discussion of phase diagrams. In Chapter 3 (specifically, *Further information 3.2*) we remarked that a quantity called 'fugacity' takes into account the effects of gas imperfections in a manner that resulted in the least upset of the form of equations. Here we see how the expressions encountered in the treatment of ideal solutions can also be preserved almost intact by introducing the concept of 'activity'. It is important to be aware of the different definitions of standard states and activities, and they are summarized in Table 5.3. We shall put them to work in the next few chapters, when we shall see that using them is much easier than defining them.

5.10 The solvent activity

Key point **The activity is an effective concentration that preserves the form of the expression for the chemical potential.**

The general form of the chemical potential of a real or ideal solvent is given by a straightforward modification of eqn 5.20 (that $\mu_A = \mu_A^* + RT\ln(p_A/p_A^*)$, where p_A^* is the vapour pressure of pure A and p_A is the vapour pressure of A when it is a

Table 5.3 Standard states

Component	Basis	Standard state	Activity	Limits
Solid or liquid		Pure	$a = 1$	
Solvent	Raoult	Pure solvent	$a = p/p^*$, $a = \gamma x$	$\gamma \to 1$ as $x \to 1$ (pure solvent)
Solute	Henry	(1) A hypothetical state of the pure solute	$a = p/K$, $a = \gamma x$	$\gamma \to 1$ as $x \to 0$
		(2) A hypothetical state of the solute at molality $b^⦵$	$a = \gamma b/b^⦵$	$\gamma \to 1$ as $b \to 0$

In each case, $\mu = \mu^⦵ + RT\ln a$.

component of a solution. For an ideal solution, as we have seen, the solvent obeys Raoult's law at all concentrations and we can express this relation as eqn 5.22 (that is, as $\mu_A = \mu_A^\star + RT\ln x_A$). The form of this relation can be preserved when the solution does not obey Raoult's law by writing

$$\mu_A = \mu_A^\star + RT\ln a_A \quad \text{Definition of activity of solvent} \quad (5.48)$$

The quantity a_A is the **activity** of A, a kind of 'effective' mole fraction, just as the fugacity is an effective pressure.

Because eqn 5.20 is true for both real and ideal solutions (the only approximation being the use of pressures rather than fugacities), we can conclude by comparing it with eqn 5.48 that

$$a_A = \frac{p_A}{p_A^\star} \quad \text{Procedure for determining activity of solvent} \quad (5.49)$$

We see that there is nothing mysterious about the activity of a solvent: it can be determined experimentally simply by measuring the vapour pressure and then using eqn 5.49.

- **A brief illustration**

The vapour pressure of 0.500 M $KNO_3(aq)$ at 100°C is 99.95 kPa, so the activity of water in the solution at this temperature is

$$a_A = \frac{99.95\text{ kPa}}{101.325\text{ kPa}} = 0.9864 \quad \bullet$$

Because all solvents obey Raoult's law (that $p_A/p_A^\star = x_A$) more closely as the concentration of solute approaches zero, the activity of the solvent approaches the mole fraction as $x_A \to 1$:

$$a_A \to x_A \quad \text{as} \quad x_A \to 1 \quad (5.50)$$

A convenient way of expressing this convergence is to introduce the **activity coefficient**, γ (gamma), by the definition

$$a_A = \gamma_A x_A \quad \gamma_A \to 1 \quad \text{as} \quad x_A \to 1 \quad \text{Definition of activity coefficient of solvent} \quad [5.51]$$

at all temperatures and pressures. The chemical potential of the solvent is then

$$\mu_A = \mu_A^\star + RT\ln x_A + RT\ln \gamma_A \quad (5.52)$$

The standard state of the solvent, the pure liquid solvent at 1 bar, is established when $x_A = 1$.

5.11 The solute activity

Key points **(a) The chemical potential of a solute in an ideal-dilute solution is defined on the basis of Henry's law. (b) The activity of a solute takes into account departures from Henry's law behavior. (c) An alternative approach to the definition of the solute activity is based on the molality of the solute. (d) The biological standard state of a species in solution is defined as pH = 7 (and 1 bar).**

The problem with defining activity coefficients and standard states for solutes is that they approach ideal-dilute (Henry's law) behaviour as $x_B \to 0$, not as $x_B \to 1$ (corresponding to pure solute). We shall show how to set up the definitions for a solute that obeys Henry's law exactly, and then show how to allow for deviations.

(a) Ideal-dilute solutions

A solute B that satisfies Henry's law has a vapour pressure given by $p_B = K_B x_B$, where K_B is an empirical constant. In this case, the chemical potential of B is

$$\mu_B = \mu_B^* + RT \ln \frac{p_B}{p_B^*} = \mu_B^* + RT \ln \frac{K_B}{p_B^*} + RT \ln x_B \qquad (5.53)^\circ$$

Both K_B and p_B^* are characteristics of the solute, so the second term may be combined with the first to give a new standard chemical potential:

$$\mu_B^\ominus = \mu_B^* + RT \ln \frac{K_B}{p_B^*} \qquad [5.54]^\circ$$

It then follows that the chemical potential of a solute in an ideal-dilute solution is related to its mole fraction by

$$\mu_B = \mu_B^\ominus + RT \ln x_B \qquad (5.55)^\circ$$

If the solution is ideal, $K_B = p_B^*$ and eqn 5.54 reduces to $\mu_B^\ominus = \mu_B^*$, as we should expect.

(b) Real solutes

We now permit deviations from ideal-dilute, Henry's law behaviour. For the solute, we introduce a_B in place of x_B in eqn 5.55, and obtain

$$\mu_B = \mu_B^\ominus + RT \ln a_B \qquad [5.56]$$

Definition of activity of solute

The standard state remains unchanged in this last stage, and all the deviations from ideality are captured in the activity a_B. The value of the activity at any concentration can be obtained in the same way as for the solvent, but in place of eqn 5.49 we use

$$a_B = \frac{p_B}{K_B} \qquad (5.57)$$

Procedure for determining activity of solute

As we did for for the solvent, it is sensible to introduce an activity coefficient through

$$a_B = \gamma_B x_B \qquad [5.58]$$

Definition of activity coefficient of solute

Now all the deviations from ideality are captured in the activity coefficient γ_B. Because the solute obeys Henry's law as its concentration goes to zero, it follows that

$$a_B \to x_B \quad \text{and} \quad \gamma_B \to 1 \quad \text{as} \quad x_B \to 0 \qquad (5.59)$$

at all temperatures and pressures. Deviations of the solute from ideality disappear as zero concentration is approached.

Example 5.7 *Measuring activity*

Use the information in Example 5.3 to calculate the activity and activity coefficient of chloroform in acetone at 25°C, treating it first as a solvent and then as a solute. For convenience, the data are repeated here:

x_C	0	0.20	0.40	0.60	0.80	1
p_C/kPa	0	4.7	11	18.9	26.7	36.4
p_A/kPa	46.3	33.3	23.3	12.3	4.9	0

Method For the activity of chloroform as a solvent (the Raoult's law activity), form $a_C = p_C/p_C^*$ and $\gamma_C = a_C/x_C$. For its activity as a solute (the Henry's law activity), form $a_C = p_C/K_C$ and $\gamma_C = a_C/x_C$.

Answer Because $p_C^* = 36.4$ kPa and $K_C = 22.0$ kPa, we can construct the following tables. For instance, at $x_C = 0.20$, in the Raoult's law case we find $a_C = (4.7\ \text{kPa})/(36.4\ \text{kPa}) = 0.13$ and $\gamma_C = 0.13/0.20 = 0.65$; likewise, in the Henry's law case, $a_C = (4.7\ \text{kPa})/(22.0\ \text{kPa}) = 0.21$ and $\gamma_C = 0.21/0.20 = 1.05$.

From Raoult's law (chloroform regarded as the solvent):

a_C	0	0.13	0.30	0.52	0.73	1.00
γ_C		0.65	0.75	0.87	0.91	1.00

From Henry's law (chloroform regarded as the solute):

a_C	0	0.21	0.50	0.86	1.21	1.65
γ_C	1	1.05	1.25	1.43	1.51	1.65

These values are plotted in Fig. 5.58. Notice that $\gamma_C \to 1$ as $x_C \to 1$ in the Raoult's law case, but that $\gamma_C \to 1$ as $x_C \to 0$ in the Henry's law case.

Self-test 5.9 Calculate the activities and activity coefficients for acetone according to the two conventions.

[At $x_A = 0.60$, for instance $a_R = 0.50$; $\gamma_R = 0.83$; $a_H = 1.00$, $\gamma_H = 1.67$]

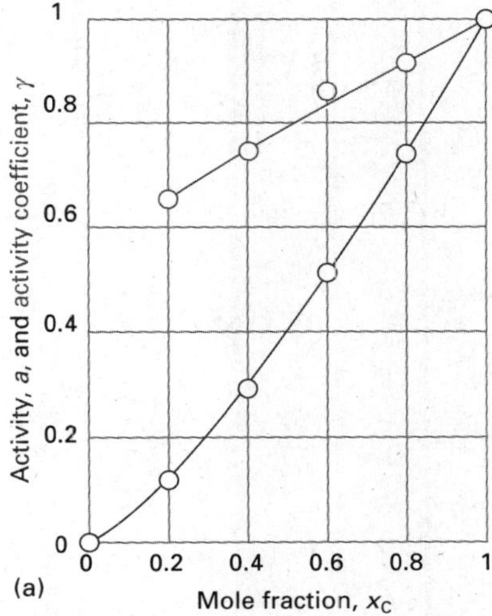

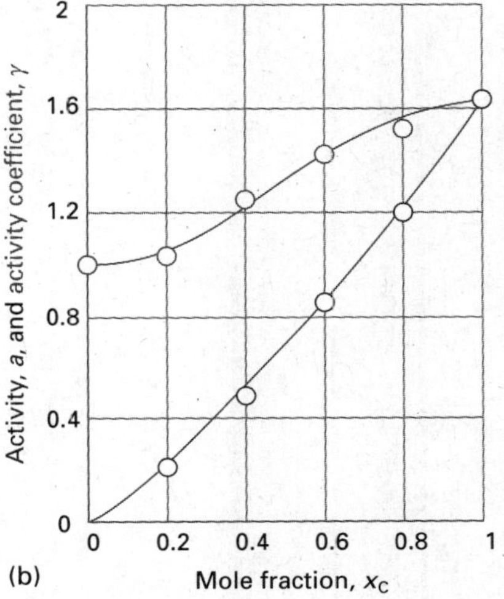

Fig. 5.58 The variation of activity and activity coefficient of chloroform (trichloromethane) with composition according to (a) Raoult's law, (b) Henry's law.

(c) Activities in terms of molalities

The selection of a standard state is entirely arbitrary, so we are free to choose one that best suits our purpose and the description of the composition of the system. In chemistry, compositions are often expressed as molalities, b, in place of mole fractions. It therefore proves convenient to write

$$\mu_B = \mu_B^\ominus + RT \ln b_B \tag{5.60}$$

Where $\mu^\ominus$ has a different value from the standard values introduced earlier. According to this definition, the chemical potential of the solute has its standard value $\mu^\ominus$ when the molality of B is equal to $b^\ominus$ (that is, at 1 mol kg^{-1}). Note that as $b_B \to 0$, $\mu_B \to \infty$; that is, as the solution becomes diluted, so the solute becomes increasingly stabilized. The practical consequence of this result is that it is very difficult to remove the last traces of a solute from a solution.

Now, as before, we incorporate deviations from ideality by introducing a dimensionless activity a_B, a dimensionless activity coefficient γ_B, and writing

$$a_B = \gamma_B \frac{b_B}{b^\ominus} \quad \text{where} \quad \gamma_B \to 1 \quad \text{as} \quad b_B \to 0 \tag{5.61}$$

at all temperatures and pressures. The standard state remains unchanged in this last stage and, as before, all the deviations from ideality are captured in the activity coefficient γ_B. We then arrive at the following succinct expression for the chemical potential of a real solute at any molality:

$$\mu = \mu^\ominus + RT \ln a \tag{5.62}$$

(d) The biological standard state

One important illustration of the ability to choose a standard state to suit the circumstances arises in biological applications. The conventional standard state of hydrogen ions (unit activity, corresponding to pH = 0)[4] is not appropriate to normal biological conditions. Therefore, in biochemistry it is common to adopt the **biological standard state,** in which pH = 7 (an activity of 10^{-7}, neutral solution) and to label the corresponding standard thermodynamic functions as $G^{\oplus}$, $H^{\oplus}$, $\mu^{\oplus}$, and $S^{\oplus}$ (some texts use $X^{o\prime}$).

To find the relation between the thermodynamic and biological standard values of the chemical potential of hydrogen ions we need to note from eqn 5.62 that

$$\mu(H^+) = \mu^{\ominus}(H^+) + RT \ln a(H^+) = \mu^{\ominus}(H^+) - (RT \ln 10) \times pH$$

It follows that

$$\mu^{\oplus}(H^+) = \mu^{\ominus}(H^+) - 7RT \ln 10 \quad \text{Relation between standard state and biological standard state} \quad (5.63)$$

At 298 K, $7RT \ln 10 = 39.96$ kJ mol^{-1}, so the two standard values differ by about 40 kJ mol^{-1}.

5.12 The activities of regular solutions

Key point **The Margules equations relate the activities of the components of a model regular solution to its composition. They lead to expressions for the vapour pressures of the components of a regular solution.**

The material on regular solutions presented in Section 5.4 gives further insight into the origin of deviations from Raoult's law and its relation to activity coefficients. The starting point is the expression for the Gibbs energy of mixing for a regular solution (eqn 5.29). We show in the following *Justification* that eqn 5.29 implies that the activity coefficients are given by expressions of the form

$$\ln \gamma_A = \xi x_B^2 \qquad \ln \gamma_B = \xi x_A^2 \quad \text{Margules equations} \quad (5.64)$$

These relations are called the **Margules equations.**

Justification 5.5 *The Margules equations*

The Gibbs energy of mixing to form a nonideal solution is

$$\Delta_{mix}G = nRT\{x_A \ln a_A + x_B \ln a_B\}$$

This relation follows from the derivation of eqn 5.16 with activities in place of mole fractions. If each activity is replaced by γx, this expression becomes

$$\Delta_{mix}G = nRT\{x_A \ln x_A + x_B \ln x_B + x_A \ln \gamma_A + x_B \ln \gamma_B\}$$

Now we introduce the two expressions in eqn 5.64, and use $x_A + x_B = 1$, which gives

$$\begin{aligned}\Delta_{mix}G &= nRT\{x_A \ln x_A + x_B \ln x_B + \xi x_A x_B^2 + \xi x_B x_A^2\} \\ &= nRT\{x_A \ln x_A + x_B \ln x_B + \xi x_A x_B(x_A + x_B)\} \\ &= nRT\{x_A \ln x_A + x_B \ln x_B + \xi x_A x_B\}\end{aligned}$$

as required by eqn 5.29. Note, moreover, that the activity coefficients behave correctly for dilute solutions: $\gamma_A \to 1$ as $x_B \to 0$ and $\gamma_B \to 1$ as $x_A \to 0$.

At this point we can use the Margules equations to write the activity of A as

$$a_A = \gamma_A x_A = x_A e^{\xi x_B^2} = x_A e^{\xi(1-x_A)^2} \quad (5.65)$$

[4] Recall from introductory chemistry courses that pH = $-\log a(H_3O^+)$.

with a similar expression for a_B. The activity of A, though, is just the ratio of the vapour pressure of A in the solution to the vapour pressure of pure A (eqn 5.49), so we can write

$$p_A = \{x_A e^{\xi(1-x_A)^2}\} p_A^* \qquad (5.66)$$

This function is plotted in Fig. 5.59. We see that $\xi = 0$, corresponding to an ideal solution, gives a straight line, in accord with Raoult's law (indeed, when $\xi = 0$, eqn 5.66 becomes $p_A = x_A p_A^*$, which is Raoult's law). Positive values of ξ (endothermic mixing, unfavourable solute–solvent interactions) give vapour pressures higher than ideal. Negative values of ξ (exothermic mixing, favourable solute–solvent interactions) give a lower vapour pressure. All the curves approach linearity and coincide with the Raoult's law line as $x_A \to 1$ and the exponential function in eqn 5.66 approaches 1. When $x_A \ll 1$, eqn 5.66 approaches

$$p_A = x_A e^{\xi} p_A^* \qquad (5.67)$$

This expression has the form of Henry's law once we identify K with $e^{\xi} p_A^*$, which is different for each solute–solvent system.

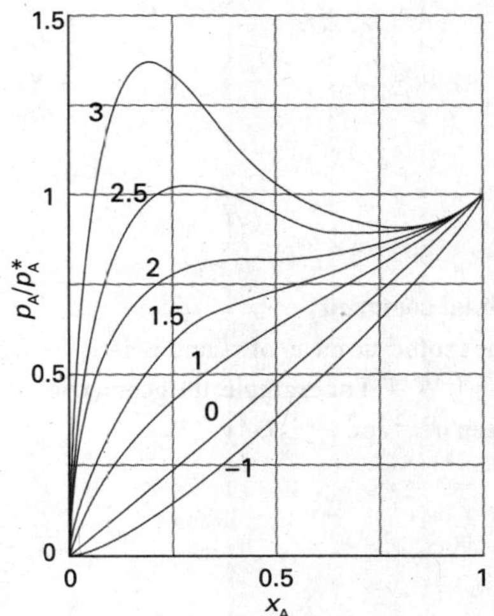

Fig. 5.59 The vapour pressure of a mixture based on a model in which the excess enthalpy is proportional to $\xi RTx_A x_B$. An ideal solution corresponds to $\xi = 0$ and gives a straight line, in accord with Raoult's law. Positive values of ξ give vapour pressures higher than ideal. Negative values of ξ give a lower vapour pressure.

interActivity Plot p_A/p_A^* against x_A with $\xi = 2.5$ by using eqn 5.66 and then eqn 5.67. Above what value of x_A do the values of p_A/p_A^* given by these equations differ by more than 10 per cent?

5.13 The activities of ions in solution

Key points **(a) Mean activity coefficients apportion deviations from ideality equally to the cations and anions in an ionic solution. (b) The Debye–Hückel theory ascribes deviations from ideality to the Coulombic interaction of an ion with the ionic atmosphere that assembles around it. (c) The Debye–Hückel limiting law is extended by including two further empirical constants.**

Interactions between ions are so strong that the approximation of replacing activities by molalities is valid only in very dilute solutions (less than 1 mmol kg^{-1} in total ion concentration) and in precise work activities themselves must be used. We need, therefore, to pay special attention to the activities of ions in solution, especially in preparation for the discussion of electrochemical phenomena.

(a) Mean activity coefficients

If the chemical potential of the cation M^+ is denoted μ_+ and that of the anion X^- is denoted μ_-, the total molar Gibbs energy of the ions in the electrically neutral solution is the sum of these partial molar quantities. The molar Gibbs energy of an *ideal* solution of such ions is

$$G_m^{ideal} = \mu_+^{ideal} + \mu_-^{ideal} \qquad (5.68)^{\circ}$$

However, for a *real* solution of M^+ and X^- of the same molality,

$$G_m = \mu_+ + \mu_- = \mu_+^{ideal} + \mu_-^{ideal} + RT\ln\gamma_+ + RT\ln\gamma_- = G_m^{ideal} + RT\ln\gamma_+\gamma_- \qquad (5.69)$$

All the deviations from ideality are contained in the last term.

There is no experimental way of separating the product $\gamma_+\gamma_-$ into contributions from the cations and the anions. The best we can do experimentally is to assign responsibility for the nonideality equally to both kinds of ion. Therefore, for a 1,1-electrolyte, we introduce the **mean activity coefficient** as the geometric mean of the individual coefficients:

$$\gamma_\pm = (\gamma_+\gamma_-)^{1/2} \qquad [5.70]$$

and express the individual chemical potentials of the ions as

$$\mu_+ = \mu_+^{ideal} + RT\ln\gamma_\pm \qquad \mu_- = \mu_-^{ideal} + RT\ln\gamma_\pm \qquad (5.71)$$

The sum of these two chemical potentials is the same as before, eqn 5.69, but now the nonideality is shared equally.

We can generalize this approach to the case of a compound M_pX_q that dissolves to give a solution of p cations and q anions from each formula unit. The molar Gibbs energy of the ions is the sum of their partial molar Gibbs energies:

$$G_m = p\mu_+ + q\mu_- = G_m^{ideal} + pRT \ln \gamma_+ + qRT \ln \gamma_- \qquad (5.72)$$

A brief comment

The geometric mean of x^p and y^q is $(x^p y^q)^{1/(p+q)}$. For example, the geometric mean of x^2 and y^{-3} is $(x^2 y^{-3})^{-1}$.

If we introduce the mean activity coefficient now defined in a more general way as

$$\gamma_\pm = (\gamma_+^p \gamma_-^q)^{1/s} \qquad s = p + q \qquad \text{Mean activity coefficient} \qquad [5.73]$$

and write the chemical potential of each ion as

$$\mu_i = \mu_i^{ideal} + RT \ln \gamma_\pm \qquad (5.74)$$

we get the same expression as in eqn 5.72 for G_m when we write $G = p\mu_+ + q\mu_-$. However, both types of ion now share equal responsibility for the nonideality.

(b) The Debye–Hückel limiting law

The long range and strength of the Coulombic interaction between ions means that it is likely to be primarily responsible for the departures from ideality in ionic solutions and to dominate all the other contributions to nonideality. This domination is the basis of the **Debye–Hückel theory** of ionic solutions, which was devised by Peter Debye and Erich Hückel in 1923. We give here a qualitative account of the theory and its principal conclusions. The calculation itself, which is a profound example of how a seemingly intractable problem can be formulated and then resolved by drawing on physical insight, is described in *Further information 5.1*.

Oppositely charged ions attract one another. As a result, anions are more likely to be found near cations in solution, and vice versa (Fig. 5.60). Overall, the solution is electrically neutral, but near any given ion there is an excess of counter ions (ions of opposite charge). Averaged over time, counter ions are more likely to be found near any given ion. This time-averaged, spherical haze around the central ion, in which counter ions outnumber ions of the same charge as the central ion, has a net charge equal in magnitude but opposite in sign to that on the central ion, and is called its **ionic atmosphere**. The energy, and therefore the chemical potential, of any given central ion is lowered as a result of its electrostatic interaction with its ionic atmosphere. This lowering of energy appears as the difference between the molar Gibbs energy G_m and the ideal value G_m^{ideal} of the solute, and hence can be identified with $RT \ln \gamma_\pm$. The stabilization of ions by their interaction with their ionic atmospheres is part of the explanation why chemists commonly use dilute solutions, in which the stabilization is less important, to achieve precipitation of ions from electrolyte solutions.

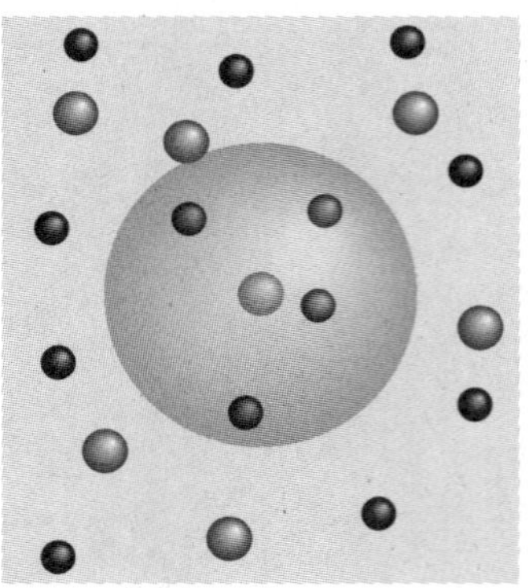

Fig. 5.60 The picture underlying the Debye–Hückel theory is of a tendency for anions to be found around cations, and of cations to be found around anions (one such local clustering region is shown by the circle). The ions are in ceaseless motion, and the diagram represents a snapshot of their motion. The solutions to which the theory applies are far less concentrated than shown here.

The model leads to the result that at very low concentrations the activity coefficient can be calculated from the **Debye–Hückel limiting law**

$$\log \gamma_\pm = -|z_+ z_-| A I^{1/2} \qquad \text{Debye–Hückel limiting law} \qquad (5.75)$$

where $A = 0.509$ for an aqueous solution at 25°C and I is the dimensionless **ionic strength** of the solution:

$$I = \tfrac{1}{2} \sum_i z_i^2 (b_i / b^\ominus) \qquad \text{Definition of ionic strength} \qquad [5.76]$$

In this expression z_i is the charge number of an ion i (positive for cations and negative for anions) and b_i is its molality. The ionic strength occurs widely wherever ionic solutions are discussed, as we shall see. The sum extends over all the ions present in the solution. For solutions consisting of two types of ion at molalities b_+ and b_-,

$$I = \tfrac{1}{2}(b_+z_+^2 + b_-z_-^2)/b^{\ominus} \qquad (5.77)$$

The ionic strength emphasizes the charges of the ions because the charge numbers occur as their squares. Table 5.4 summarizes the relation of ionic strength and molality in an easily usable form.

• A brief illustration

The mean activity coefficient of 5.0 mmol kg^{-1} KCl(aq) at 25°C is calculated by writing

$$I = \tfrac{1}{2}(b_+ + b_-)/b^{\ominus} = b/b^{\ominus}$$

where b is the molality of the solution (and $b_+ = b_- = b$). Then, from eqn 5.75,

$$\log \gamma_{\pm} = -0.509 \times (5.0 \times 10^{-3})^{1/2} = -0.036$$

Hence, $\gamma_{\pm} = 0.92$. The experimental value is 0.927. •

Self-test 5.10 Calculate the ionic strength and the mean activity coefficient of 1.00 mmol kg^{-1} $CaCl_2$(aq) at 25°C. [3.00 mmol kg^{-1}, 0.880]

The name 'limiting law' is applied to eqn 5.75 because ionic solutions of moderate molalities may have activity coefficients that differ from the values given by this expression, yet all solutions are expected to conform as $b \rightarrow 0$. Table 5.5 lists some experimental values of activity coefficients for salts of various valence types. Figure 5.61 shows some

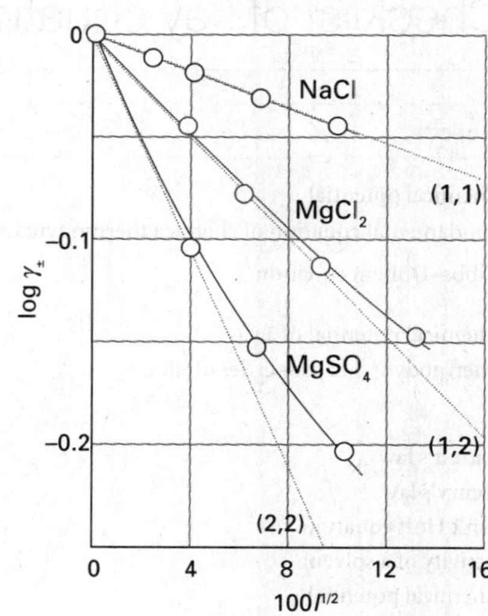

Fig. 5.61 An experimental test of the Debye–Hückel limiting law. Although there are marked deviations for moderate ionic strengths, the limiting slopes as $I \rightarrow 0$ are in good agreement with the theory, so the law can be used for extrapolating data to very low molalities.

Table 5.4 Ionic strength and molality, $I = kb/b^{\ominus}$

k	X^-	X^{2-}	X^{3-}	X^{4-}
M^+	1	3	6	10
M^{2+}	3	4	15	12
M^{3+}	6	15	9	42
M^{4+}	10	12	42	16

For example, the ionic strength of an M_2X_3 solution of molality b, which is understood to give M^{3+} and X^{2-} ions in solution is $15b/b^{\ominus}$.

Table 5.5* Mean activity coefficients in water at 298 K

$b/b^{\ominus}$	KCl	$CaCl_2$
0.001	0.966	0.888
0.01	0.902	0.732
0.1	0.770	0.524
1.0	0.607	0.725

* More values are given in the *Data section*.

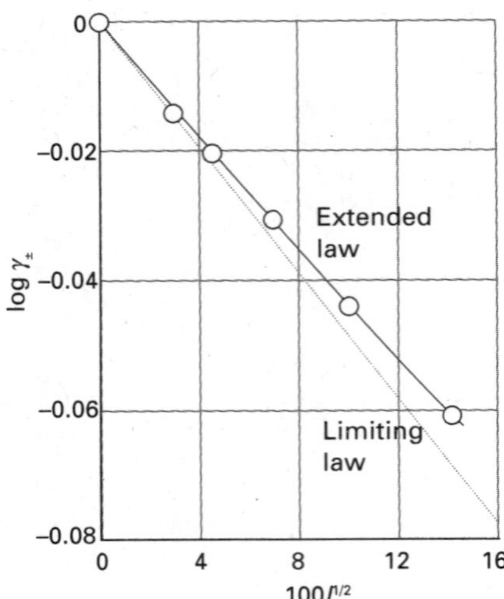

Fig. 5.62 The extended Debye–Hückel law gives agreement with experiment over a wider range of molalities (as shown here for a 1,1-electrolyte), but it fails at higher molalities.

interActivity Consider the plot of log $\gamma_\pm$ against $I^{1/2}$ with $B = 1.50$ and $C = 0$ as a representation of experimental data for a certain 1,1-electrolyte. Over what range of ionic strengths does the application of the limiting law lead to an error in the value of the activity coefficient of less than 10 per cent of the value predicted by the extended law?

of these values plotted against $I^{1/2}$, and compares them with the theoretical straight lines calculated from eqn 5.75. The agreement at very low molalities (less than about 1 mmol kg^{-1}, depending on charge type) is impressive, and convincing evidence in support of the model. Nevertheless, the departures from the theoretical curves above these molalities are large, and show that the approximations are valid only at very low concentrations.

(c) The extended Debye–Hückel law

When the ionic strength of the solution is too high for the limiting law to be valid, the activity coefficient may be estimated from the **extended Debye–Hückel law**:

$$\log \gamma_\pm = -\frac{A|z_+z_-|I^{1/2}}{1+BI^{1/2}} + CI \qquad \text{Extended Debye–Hückel law} \qquad (5.78)$$

where B and C are dimensionless constants. Although B can be interpreted as a measure of the closest approach of the ions, it (like C) is best regarded as an adjustable empirical parameter. A curve drawn in this way is shown in Fig. 5.62. It is clear that eqn 5.78 accounts for some activity coefficients over a moderate range of dilute solutions (up to about 0.1 mol kg^{-1}); nevertheless it remains very poor near 1 mol kg^{-1}.

Current theories of activity coefficients for ionic solutes take an indirect route. They set up a theory for the dependence of the activity coefficient of the solvent on the concentration of the solute, and then use the Gibbs–Duhem equation (eqn 5.12) to estimate the activity coefficient of the solute. The results are reasonably reliable for solutions with molalities greater than about 0.1 mol kg^{-1} and are valuable for the discussion of mixed salt solutions, such as sea water.

Checklist of key equations

Property	Equation	Comment
Chemical potential	$\mu_J = (\partial G/\partial n_J)_{p,T,n'}$	$G = n_A\mu_A + n_B\mu_B$
Fundamental equation of chemica thermodynamics	$dG = Vdp - SdT + \mu_A dn_A + \mu_B dn_B + \cdots$	
Gibbs–Duhem equation	$\sum_J n_J d\mu_J = 0$	
Chemical potential of a gas	$\mu = \mu^{\ominus} + RT\ln(p/p^{\ominus})$	Perfect gas
Thermodynamic properties of mixing	$\Delta_{mix}G = nRT(x_A \ln x_A + x_B \ln x_B)$ $\Delta_{mix}S = -nR(x_A \ln x_A + x_B \ln x_B)$ $\Delta_{mix}H = 0$	Perfect gases and ideal solutions
Raoult's law	$p_A = x_A p_A^*$	True for ideal solutions; limiting law as $x_A \to 1$
Henry's law	$p_B = x_B K_B$	True for ideal–dilute solutions; limiting law as $x_B \to 0$
van't Hoff equation	$\Pi = [B]RT$	Valid as $[B] \to 0$
Activity of a solvent	$a_A = p_A/p_A^*$	$a_A \to x_A$ as $x_A \to 1$
Chemical potential	$\mu_J = \mu_J^{\ominus} + RT\ln a_J$	General form for a species J
Conversion to biological standard state	$\mu^{\oplus}(H^+) = \mu^{\ominus}(H^+) - 7RT\ln 10$	
Mean activity coefficient	$\gamma_\pm = (\gamma_+^p\gamma_-^q)^{1/(p+q)}$	
Ionic strength	$I = \frac{1}{2}\sum_i z_i^2(b_i/b^{\ominus})$	Definition
Debye–Hückel limiting law	$\log \gamma_\pm = -\lvert z_+z_-\rvert AI^{1/2}$	Valid as $I \to 0$
Margules equation	$\ln \gamma_J = \xi x_J^2$	Model regular solution
Lever rule	$n_\alpha l_\alpha = n_\beta l_\beta$	

➔ For a chart of the relations between principal equations, see the Road map section of the Resource section.

Further information

Further information 5.1 *The Debye–Hückel theory of ionic solution*

Imagine a solution in which all the ions have their actual positions, but in which their Coulombic interactions have been turned off. The difference in molar Gibbs energy between the ideal and real solutions is equal to w_e, the electrical work of charging the system in this arrangement. For a salt M_pX_q, we write

$$\begin{aligned} w_e &= \overbrace{(p\mu_+ + q\mu_-)}^{G_m} - \overbrace{(p\mu_+^{\text{ideal}} + q\mu_-^{\text{ideal}})}^{G_m^{\text{ideal}}} \\ &= p(\mu_+ - \mu_+^{\text{ideal}}) + q(\mu_- - \mu_-^{\text{ideal}}) \end{aligned} \tag{5.79}$$

From eqn 5.71 we write

$$\mu_+ - \mu_+^{\text{ideal}} = \mu_- - \mu_-^{\text{ideal}} = RT\ln\gamma_\pm$$

So it follows that

$$\ln\gamma_\pm = \frac{w_e}{sRT} \qquad s = p + q \tag{5.80}$$

This equation tells us that we must first find the final distribution of the ions and then the work of charging them in that distribution.

The Coulomb potential at a distance r from an isolated ion of charge z_ie in a medium of permittivity ε is

$$\phi_i = \frac{Z_i}{r} \qquad Z_i = \frac{z_ie}{4\pi\varepsilon} \tag{5.81}$$

The ionic atmosphere causes the potential to decay with distance more sharply than this expression implies. Such shielding is a familiar problem in electrostatics, and its effect is taken into account by replacing the Coulomb potential by the **shielded Coulomb potential**, an expression of the form

$$\phi_i = \frac{Z_i}{r}e^{-r/r_D} \qquad \text{Shielded Coulomb potential} \tag{5.82}$$

where r_D is called the **Debye length**. When r_D is large, the shielded potential is virtually the same as the unshielded potential. When it is small, the shielded potential is much smaller than the unshielded potential, even for short distances (Fig. 5.63).

To calculate r_D, we need to know how the **charge density**, ρ_i, of the ionic atmosphere, the charge in a small region divided by the volume of the region, varies with distance from the ion. This step draws on another standard result of electrostatics, in which charge density and potential are related by **Poisson's equation**:

$$\nabla^2\phi = -\frac{\rho}{\varepsilon} \qquad \text{Poisson's equation} \tag{5.83a}$$

where $\nabla^2 = (\partial^2/\partial x^2 + \partial^2/\partial y^2 + \partial^2/\partial z^2)$ is called the *laplacian*. Because we are considering only a spherical ionic atmosphere, we can use a simplified form of this equation in which the charge density varies only with distance from the central ion:

$$\frac{1}{r^2}\frac{d}{dr}\left(r^2\frac{d\phi_i}{dr}\right) = -\frac{\rho_i}{\varepsilon} \tag{5.83b}$$

Substitution of the expression for the shielded potential, eqn 5.82, results in

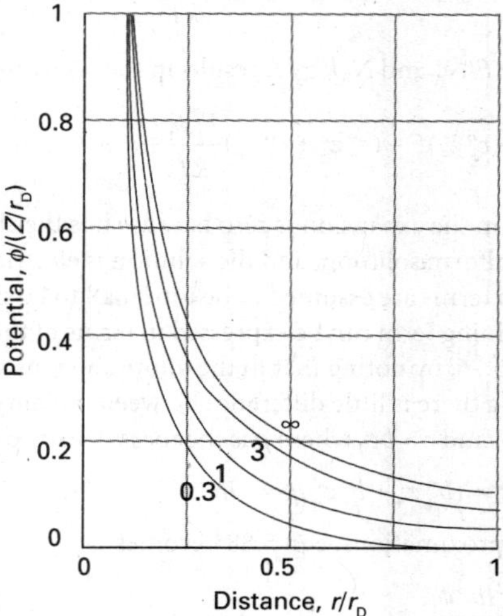

Fig. 5.63 The variation of the shielded Coulomb potential with distance for different values of the Debye length, r_D/a. The smaller the Debye length, the more sharply the potential decays to zero. In each case, a is an arbitrary unit of length.

interActivity Write an expression for the difference between the unshielded and shielded Coulomb potentials evaluated at r_D. Then plot this expression against r_D and provide a physical interpretation for the shape of the plot.

$$r_D^2 = -\frac{\varepsilon\phi_i}{\rho_i} \tag{5.84}$$

To solve this equation we need to relate ρ_i and ϕ_i.

For the next step we draw on the fact that the energy of an ion depends on its closeness to the central ion, and then use the Boltzmann distribution (see *Fundamentals F.5a*) to work out the probability that an ion will be found at each distance. The energy of an ion of charge z_je at a distance where it experiences the potential ϕ_i of the central ion i relative to its energy when it is far away in the bulk solution is its charge times the potential:

$$E = z_je\phi_i \tag{5.85}$$

Therefore, according to the Boltzmann distribution, the ratio of the molar concentration, c_j, of ions at a distance r and the molar concentration in the bulk, c_j°, where the energy is zero, is

$$\frac{c_j}{c_j^\circ} = e^{-E/kT} \tag{5.86}$$

The charge density, ρ_i, at a distance r from the ion i is the molar concentration of each type of ion multiplied by the charge per mole of ions, z_ieN_A. The quantity eN_A, the magnitude of the charge per mole of electrons, is Faraday's constant, $F = 96.48$ kC mol^{-1}. It follows that

$$\rho_i = c_+z_+F + c_-z_-F = c_+^\circ z_+Fe^{-z_+e\phi_i/kT} + c_-^\circ z_-Fe^{-z_-e\phi_i/kT} \tag{5.87}$$

At this stage we need to simplify the expression to avoid the awkward exponential terms. Because the average electrostatic interaction energy is small compared with kT we may use $e^{-x} = 1 - x + \frac{1}{2}x^2 - \cdots$ and ignore all terms higher than x to write eqn 5.87 as

$$\rho_i = c_+^\circ z_+ F\left(1 - \frac{z_+ e\phi_i}{kT} + \cdots\right) + c_-^\circ z_- F\left(1 - \frac{z_- e\phi_i}{kT} + \cdots\right)$$

$$= (c_+^\circ z_+ + c_-^\circ z_-)F - (c_+^\circ z_+^2 + c_-^\circ z_-^2)F\frac{e\phi_i}{kT} + \cdots$$

Replacing e by F/N_A and $N_A k$ by R results in the following expression:

$$\rho_i = (c_+^\circ z_+ + c_-^\circ z_-)F - (c_+^\circ z_+^2 + c_-^\circ z_-^2)\frac{F^2\phi_i}{RT} + \cdots \quad (5.88)$$

The first term in the expansion is zero because it is the charge density in the bulk, uniform solution, and the solution is electrically neutral. The unwritten terms are assumed to be too small to be significant. The one remaining term can be expressed in terms of the ionic strength, eqn 5.76, by noting that in the dilute aqueous solutions we are considering there is little difference between molality and molar concentration, and $c \approx b\rho$, where ρ is the mass density of the solvent

$$c_+^\circ z_+^2 + c_-^\circ z_-^2 \approx (b_+^\circ z_+^2 + b_-^\circ z_-^2)\rho = 2Ib^\ominus\rho \quad (5.89)$$

With these approximations, eqn 5.88 becomes

$$\rho_i = -\frac{2\rho F^2 I b^\ominus \phi_i}{RT} \quad (5.90)$$

We can now solve eqn 5.84 for r_D:

$$r_D = \left(\frac{\varepsilon RT}{2\rho F^2 I b^\ominus}\right)^{1/2} \quad \text{Debye length} \quad (5.91)$$

To calculate the activity coefficient we need to find the electrical work of charging the central ion when it is surrounded by its atmosphere. To do so, we need to know the potential at the ion due to its atmosphere, ϕ_{atmos}. This potential is the difference between the total potential, given by eqn 5.82, and the potential due to the central ion itself:

$$\phi_{\text{atmos}} = \phi - \phi_{\text{central ion}} = Z_i\left(\frac{e^{-r/r_D}}{r} - \frac{1}{r}\right) \quad (5.92a)$$

The potential at the central ion (at $r = 0$) is obtained by taking the limit of this expression as $r \to 0$ and is

$$\phi_{\text{atmos}}(0) = -\frac{Z_i}{r_D} \quad (5.92b)$$

This expression shows us that the potential of the ionic atmosphere is equivalent to the potential arising from a single charge of equal magnitude but opposite sign to that of the central ion and located at a distance r_D from the ion. If the charge of the central ion were Q and not $z_i e$, then the potential due to its atmosphere would be

$$\phi_{\text{atmos}}(0) = -\frac{Q}{4\pi\varepsilon r_D} \quad (5.92c)$$

The work of adding a charge dQ to a region where the electrical potential is $\phi_{\text{atmos}}(0)$ is

$$dw_e = \phi_{\text{atmos}}(0)dQ \quad (5.93)$$

Therefore, the total molar work of fully charging the ions is

$$w_e = N_A\int_0^{z_i e}\phi_{\text{atmos}}(0)\,dQ = -\frac{N_A}{4\pi\varepsilon r_D}\int_0^{z_i e} Q\,dQ$$
$$= -\frac{N_A z_i^2 e^2}{8\pi\varepsilon r_D} = -\frac{z_i^2 F^2}{8\pi\varepsilon N_A r_D} \quad (5.94)$$

where in the last step we have used $F = N_A e$. It follows from eqn 5.80 that the mean activity coefficient of the ions is

$$\ln \gamma_\pm = \frac{pw_{e,+} + qw_{e,-}}{sRT} = -\frac{(pz_+^2 + qz_-^2)F^2}{8\pi\varepsilon s N_A RT r_D} \quad (5.95a)$$

However, for neutrality $pz_+ + qz_- = 0$; therefore

$$\ln \gamma_\pm = -\frac{|z_+z_-|F^2}{8\pi\varepsilon N_A RT r_D} \quad (5.95b)$$

(For this step, multiply $pz_+ + qz_- = 0$ by p and also, separately, by q; add the two expressions and rearrange the result by using $p + q = s$ and $z_+z_- = -|z_+z_-|$.) Replacing r_D with the expression in eqn 5.91 gives

$$\ln \gamma_\pm = -\frac{|z_+z_-|F^2}{8\pi\varepsilon N_A RT}\left(\frac{2\rho F^2 I b^\ominus}{\varepsilon RT}\right)^{1/2}$$
$$= -|z_+z_-|\left\{\frac{F^3}{4\pi N_A}\left(\frac{\rho b^\ominus}{2\varepsilon^3 R^3 T^3}\right)^{1/2}\right\} I^{1/2} \quad (5.96a)$$

where we have grouped terms in such a way as to show that this expression is beginning to take the form of eqn 5.75. Indeed, conversion to common logarithms (by using $\ln x = \ln 10 \times \log x$) gives

$$\log \gamma_\pm = -|z_+z_-|\left\{\frac{F^3}{4\pi N_A \ln 10}\left(\frac{\rho b^\ominus}{2\varepsilon^3 R^3 T^3}\right)^{1/2}\right\} I^{1/2} \quad (5.96b)$$

which is eqn 5.75 ($\log \gamma_\pm = -|z_+z_-|AI^{1/2}$) with

$$A = \frac{F^3}{4\pi N_A \ln 10}\left(\frac{\rho b^\ominus}{2\varepsilon^3 R^3 T^3}\right)^{1/2} \quad (5.97)$$

Discussion questions

5.1 State and justify the thermodynamic criterion for solution–vapour equilibrium.

5.2 How is Raoult's law modified so as to describe the vapour pressure of real solutions?

5.3 Explain the origin of colligative properties.

5.4 Explain what is meant by a regular solution.

5.5 Describe the general features of the Debye–Hückel theory of electrolyte solutions.

5.6 What factors determine the number of theoretical plates required to achieve a desired degree of separation in fractional distillation?

5.7 Draw phase diagrams for the following types of systems. Label the regions and intersections of the diagrams, stating what materials (possibly compounds or azeotropes) are present and whether they are solid, liquid, or gas. (a) Two-component, temperature–composition, solid–liquid diagram, one compound AB formed that melts congruently, negligible solid–solid solubility; (b) two-component, temperature–composition, solid–liquid diagram, one compound of formula AB_2 that melts incongruently, negligible solid–solid solubility; (c) two-component, constant temperature–composition, liquid–vapour diagram, formation of an azeotrope at $x_B = 0.333$, complete miscibility.

Exercises

5.1(a) The partial molar volumes of acetone (propanone) and chloroform (trichloromethane) in a mixture in which the mole fraction of $CHCl_3$ is 0.4693 are 74.166 $cm^3\ mol^{-1}$ and 80.235 $cm^3\ mol^{-1}$, respectively. What is the volume of a solution of mass 1.000 kg?

5.1(b) The partial molar volumes of two liquids A and B in a mixture in which the mole fraction of A is 0.3713 are 188.2 $cm^3\ mol^{-1}$ and 176.14 $cm^3\ mol^{-1}$, respectively. The molar masses of the A and B are 241.1 g mol^{-1} and 198.2 g mol^{-1}. What is the volume of a solution of mass 1.000 kg?

5.2(a) At 25°C, the density of a 50 per cent by mass ethanol–water solution is 0.914 g cm^{-3}. Given that the partial molar volume of water in the solution is 17.4 $cm^3\ mol^{-1}$, calculate the partial molar volume of the ethanol.

5.2(b) At 20°C, the density of a 20 per cent by mass ethanol/water solution is 968.7 kg m^{-3}. Given that the partial molar volume of ethanol in the solution is 52.2 $cm^3\ mol^{-1}$, calculate the partial molar volume of the water.

5.3(a) At 300 K, the partial vapour pressures of HCl (that is, the partial pressure of the HCl vapour) in liquid $GeCl_4$ are as follows:

x_{HCl}	0.005	0.012	0.019
p_{HCl}/kPa	32.0	76.9	121.8

Show that the solution obeys Henry's law in this range of mole fractions, and calculate Henry's law constant at 300 K.

5.3(b) At 310 K, the partial vapour pressures of a substance B dissolved in a liquid A are as follows:

x_B	0.010	0.015	0.020
p_B/kPa	82.0	122.0	166.1

Show that the solution obeys Henry's law in this range of mole fractions, and calculate Henry's law constant at 310 K.

5.4(a) Predict the partial vapour pressure of HCl above its solution in liquid germanium tetrachloride of molality 0.10 mol kg^{-1}. For data, see Exercise 5.3a.

5.4(b) Predict the partial vapour pressure of the component B above its solution in A in Exercise 5.3b when the molality of B is 0.25 mol kg^{-1}. The molar mass of A is 74.1 g mol^{-1}.

5.5(a) The vapour pressure of benzene is 53.3 kPa at 60.6°C, but it fell to 51.5 kPa when 19.0 g of an involatile organic compound was dissolved in 500 g of benzene. Calculate the molar mass of the compound.

5.5(b) The vapour pressure of 2-propanol is 50.00 kPa at 338.8°C, but it fell to 49.62 kPa when 8.69 g of an involatile organic compound was dissolved in 250 g of 2-propanol. Calculate the molar mass of the compound.

5.6(a) The addition of 100 g of a compound to 750 g of CCl_4 lowered the freezing point of the solvent by 10.5 K. Calculate the molar mass of the compound.

5.6(b) The addition of 5.00 g of a compound to 250 g of naphthalene lowered the freezing point of the solvent by 0.780 K. Calculate the molar mass of the compound.

5.7(a) The osmotic pressure of an aqueous solution at 300 K is 120 kPa. Calculate the freezing point of the solution.

5.7(b) The osmotic pressure of an aqueous solution at 288 K is 99.0 kPa. Calculate the freezing point of the solution.

5.8(a) Consider a container of volume 5.0 dm^3 that is divided into two compartments of equal size. In the left compartment there is nitrogen at 1.0 atm and 25°C; in the right compartment there is hydrogen at the same temperature and pressure. Calculate the entropy and Gibbs energy of mixing when the partition is removed. Assume that the gases are perfect.

5.8(b) Consider a container of volume 250 cm^3 that is divided into two compartments of equal size. In the left compartment there is argon at 100 kPa and 0°C; in the right compartment there is neon at the same temperature and pressure. Calculate the entropy and Gibbs energy of mixing when the partition is removed. Assume that the gases are perfect.

5.9(a) Air is a mixture with a composition given in *Example 1.3*. Calculate the entropy of mixing when it is prepared from the pure (and perfect) gases.

5.9(b) Calculate the Gibbs energy, entropy, and enthalpy of mixing when 1.00 mol C_6H_{14} (hexane) is mixed with 1.00 mol C_7H_{16} (heptane) at 298 K; treat the solution as ideal.

5.10(a) What proportions of hexane and heptane should be mixed (a) by mole fraction, (b) by mass in order to achieve the greatest entropy of mixing?

5.10(b) What proportions of benzene and ethylbenzene should be mixed (a) by mole fraction, (b) by mass in order to achieve the greatest entropy of mixing?

5.11(a) Use Henry's law and the data in Table 5.1 to calculate the solubility (as a molality) of CO_2 in water at 25°C when its partial pressure is (a) 0.10 atm, (b) 1.00 atm.

5.11(b) The mole fractions of N_2 and O_2 in air at sea level are approximately 0.78 and 0.21. Calculate the molalities of the solution formed in an open flask of water at 25°C.

5.12(a) A water carbonating plant is available for use in the home and operates by providing carbon dioxide at 5.0 atm. Estimate the molar concentration of the soda water it produces.

5.12(b) After some weeks of use, the pressure in the water carbonating plant mentioned in the previous exercise has fallen to 2.0 atm. Estimate the molar concentration of the soda water it produces at this stage.

5.13(a) The enthalpy of fusion of anthracene is 28.8 kJ mol^{-1} and its melting point is 217°C. Calculate its ideal solubility in benzene at 25°C.

5.13(b) Predict the ideal solubility of lead in bismuth at 280°C given that its melting point is 327°C and its enthalpy of fusion is 5.2 kJ mol^{-1}.

5.14(a) The osmotic pressure of solutions of polystyrene in toluene were measured at 25°C and the pressure was expressed in terms of the height of the solvent of density 1.004 g cm^{-3}:

c/(g dm^{-3})	2.042	6.613	9.521	12.602
h/cm	0.592	1.910	2.750	3.600

Calculate the molar mass of the polymer.

5.14(b) The molar mass of an enzyme was determined by dissolving it in water, measuring the osmotic pressure at 20°C, and extrapolating the data to zero concentration. The following data were obtained:

$c/(\mathrm{mg\,cm^{-3}})$	3.221	4.618	5.112	6.722
h/cm	5.746	8.238	9.119	11.990

Calculate the molar mass of the enzyme.

5.15(a) Substances A and B are both volatile liquids with $p_A^* = 300$ Torr, $p_B^* = 250$ Torr, and $K_B = 200$ Torr (concentration expressed in mole fraction). When $x_A = 0.9$, $b_B = 2.22$ mol kg^{-1}, $p_A = 250$ Torr, and $p_B = 25$ Torr. Calculate the activities and activity coefficients of A and B. Use the mole fraction, Raoult's law basis system for A and the Henry's law basis system (both mole fractions and molalities) for B.

5.15(b) Given that $p^*(H_2O) = 0.02308$ atm and $p(H_2O) = 0.02239$ atm in a solution in which 0.122 kg of a non-volatile solute ($M = 241$ g mol^{-1}) is dissolved in 0.920 kg water at 293 K, calculate the activity and activity coefficient of water in the solution.

5.16(a) A dilute solution of bromine in carbon tetrachloride behaves as an ideal-dilute solution. The vapour pressure of pure CCl_4 is 33.85 Torr at 298 K. The Henry's law constant when the concentration of Br_2 is expressed as a mole fraction is 122.36 Torr. Calculate the vapour pressure of each component, the total pressure, and the composition of the vapour phase when the mole fraction of Br_2 is 0.050, on the assumption that the conditions of the ideal-dilute solution are satisfied at this concentration.

5.16(b) Benzene and toluene form nearly ideal solutions. The boiling point of pure benzene is 80.1°C. Calculate the chemical potential of benzene relative to that of pure benzene when $x_{\text{benzene}} = 0.30$ at its boiling point. If the activity coefficient of benzene in this solution were actually 0.93 rather than 1.00, what would be its vapour pressure?

5.17(a) By measuring the equilibrium between liquid and vapour phases of an acetone(A)/methanol(M) solution at 57.2°C at 1.00 atm, it was found that $x_A = 0.400$ when $y_A = 0.516$. Calculate the activities and activity coefficients of both components in this solution on the Raoult's law basis. The vapour pressures of the pure components at this temperature are: $p_A^* = 105$ kPa and $p_M^* = 73.5$ kPa. (x_A is the mole fraction in the liquid and y_A the mole fraction in the vapour.)

5.17(b) By measuring the equilibrium between liquid and vapour phases of a solution at 30°C at 1.00 atm, it was found that $x_A = 0.220$ when $y_A = 0.314$. Calculate the activities and activity coefficients of both components in this solution on the Raoult's law basis. The vapour pressures of the pure components at this temperature are: $p_A^* = 73.0$ kPa and $p_B^* = 92.1$ kPa. (x_A is the mole fraction in the liquid and y_A the mole fraction in the vapour.)

5.18(a) Calculate the ionic strength of a solution that is 0.10 mol kg^{-1} in KCl(aq) and 0.20 mol kg^{-1} in $CuSO_4$(aq).

5.18(b) Calculate the ionic strength of a solution that is 0.040 mol kg^{-1} in $K_3[Fe(CN)_6]$(aq), 0.030 mol kg^{-1} in KCl(aq), and 0.050 mol kg^{-1} in NaBr(aq).

5.19(a) Calculate the masses of (a) $Ca(NO_3)_2$ and, separately, (b) NaCl to add to a 0.150 mol kg^{-1} solution of KNO_3(aq) containing 500 g of solvent to raise its ionic strength to 0.250.

5.19(b) Calculate the masses of (a) KNO_3 and, separately, (b) $Ba(NO_3)_2$ to add to a 0.110 mol kg^{-1} solution of KNO_3(aq) containing 500 g of solvent to raise its ionic strength to 1.00.

5.20(a) Estimate the mean ionic activity coefficient and activity of $CaCl_2$ in a solution that is 0.010 mol kg^{-1} $CaCl_2$(aq) and 0.030 mol kg^{-1} NaF(aq).

5.20(b) Estimate the mean ionic activity coefficient and activity of NaCl in a solution that is 0.020 mol kg^{-1} NaCl(aq) and 0.035 mol kg^{-1} $Ca(NO_3)_2$(aq).

5.21(a) The mean activity coefficients of HBr in three dilute aqueous solutions at 25°C are 0.930 (at 5.0 mmol kg^{-1}), 0.907 (at 10.0 mmol kg^{-1}), and 0.879 (at 20.0 mmol kg^{-1}). Estimate the value of B in the extended Debye–Hückel law. Set $C = 0$.

5.21(b) The mean activity coefficients of KCl in three dilute aqueous solutions at 25°C are 0.927 (at 5.0 mmol kg^{-1}), 0.902 (at 10.0 mmol kg^{-1}), and 0.816 (at 50.0 mmol kg^{-1}). Estimate the value of B in the extended Debye–Hückel law. Set $C = 0$.

5.22(a) At 90°C, the vapour pressure of methylbenzene is 53.3 kPa and that of 1,2-dimethylbenzene is 20.0 kPa. What is the composition of a liquid mixture that boils at 90°C when the pressure is 0.50 atm? What is the composition of the vapour produced?

5.22(b) At 90°C, the vapour pressure of 1,2-dimethylbenzene is 20 kPa and that of 1,3-dimethylbenzene is 18 kPa. What is the composition of a liquid mixture that boils at 90°C when the pressure is 19 kPa? What is the composition of the vapour produced?

5.23(a) The vapour pressure of pure liquid A at 300 K is 76.7 kPa and that of pure liquid B is 52.0 kPa. These two compounds form ideal liquid and gaseous mixtures. Consider the equilibrium composition of a mixture in which the mole fraction of A in the vapour is 0.350. Calculate the total pressure of the vapour and the composition of the liquid mixture.

5.23(b) The vapour pressure of pure liquid A at 293 K is 68.8 kPa and that of pure liquid B is 82.1 kPa. These two compounds form ideal liquid and gaseous mixtures. Consider the equilibrium composition of a mixture in which the mole fraction of A in the vapour is 0.612. Calculate the total pressure of the vapour and the composition of the liquid mixture.

5.24(a) It is found that the boiling point of a binary solution of A and B with $x_A = 0.6589$ is 88°C. At this temperature the vapour pressures of pure A and B are 127.6 kPa and 50.60 kPa, respectively. (a) Is this solution ideal? (b) What is the initial composition of the vapour above the solution?

5.24(b) It is found that the boiling point of a binary solution of A and B with $x_A = 0.4217$ is 96°C. At this temperature the vapour pressures of pure A and B are 110.1 kPa and 76.5 kPa, respectively. (a) Is this solution ideal? (b) What is the initial composition of the vapour above the solution?

5.25(a) Dibromoethene (DE, $p_{DE}^* = 22.9$ kPa at 358 K) and dibromopropene (DP, $p_{DP}^* = 17.1$ kPa at 358 K) form a nearly ideal solution. If $z_{DE} = 0.60$, what is (a) p_{total} when the system is all liquid, (b) the composition of the vapour when the system is still almost all liquid?

5.25(b) Benzene and toluene form nearly ideal solutions. Consider an equimolar solution of benzene and toluene. At 20°C the vapour pressures of pure benzene and toluene are 9.9 kPa and 2.9 kPa, respectively. The solution is boiled by reducing the external pressure below the vapour pressure. Calculate (a) the pressure when boiling begins, (b) the composition of each component in the vapour, and (c) the vapour pressure when only a few drops of liquid remain. Assume that the rate of vaporization is low enough for the temperature to remain constant at 20°C.

5.26(a) The following temperature/composition data were obtained for a mixture of octane (O) and methylbenzene (M) at 1.00 atm, where x is the mole fraction in the liquid and y the mole fraction in the vapour at equilibrium.

θ/°C	110.9	112.0	114.0	115.8	117.3	119.0	121.1	123.0
x_M	0.908	0.795	0.615	0.527	0.408	0.300	0.203	0.097
y_M	0.923	0.836	0.698	0.624	0.527	0.410	0.297	0.164

The boiling points are 110.6°C and 125.6°C, for M and O, respectively. Plot the temperature/composition diagram for the mixture. What is the composition of the vapour in equilibrium with the liquid of composition (a) $x_M = 0.250$ and (b) $x_O = 0.250$?

5.26(b) The following temperature/composition data were obtained for a mixture of two liquids A and B at 1.00 atm, where x is the mole fraction in the liquid and y the mole fraction in the vapour at equilibrium.

θ/°C	125	130	135	140	145	150
x_A	0.91	0.65	0.45	0.30	0.18	0.098
y_A	0.99	0.91	0.77	0.61	0.45	0.25

The boiling points are 124°C for A and 155°C for B. Plot the temperature/composition diagram for the mixture. What is the composition of the vapour in equilibrium with the liquid of composition (a) $x_A = 0.50$ and (b) $x_B = 0.33$?

5.27(a) Methylethyl ether (A) and diborane, B_2H_6 (B), form a compound which melts congruently at 133 K. The system exhibits two eutectics, one at 25 mol per cent B and 123 K and a second at 90 mol per cent B and 104 K. The melting points of pure A and B are 131 K and 110 K, respectively. Sketch the phase diagram for this system. Assume negligible solid–solid solubility.

5.27(b) Sketch the phase diagram of the system NH_3/N_2H_4 given that the two substances do not form a compound with each other, that NH_3 freezes at −78°C and N_2H_4 freezes at +2°C, and that a eutectic is formed when the mole fraction of N_2H_4 is 0.07 and that the eutectic melts at −80°C.

5.28(a) Figure 5.64 shows the phase diagram for two partially miscible liquids, which can be taken to be that for water (A) and 2-methyl-1-propanol (B). Describe what will be observed when a mixture of composition $x_B = 0.8$ is heated, at each stage giving the number, composition, and relative amounts of the phases present.

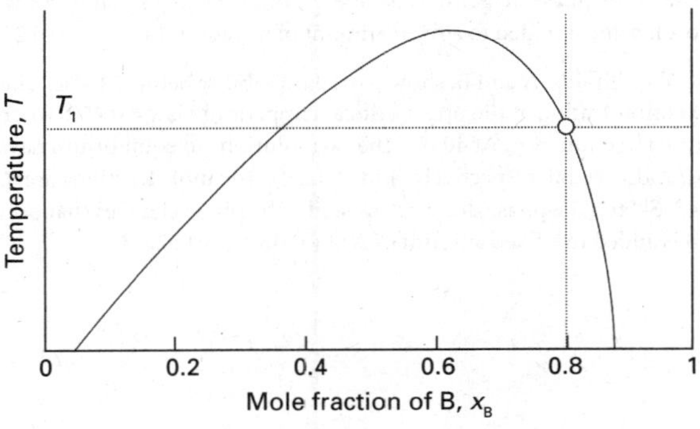

Fig. 5.64

5.28(b) Figure 5.65 is the phase diagram for silver and tin. Label the regions, and describe what will be observed when liquids of compositions a and b are cooled to 200 K.

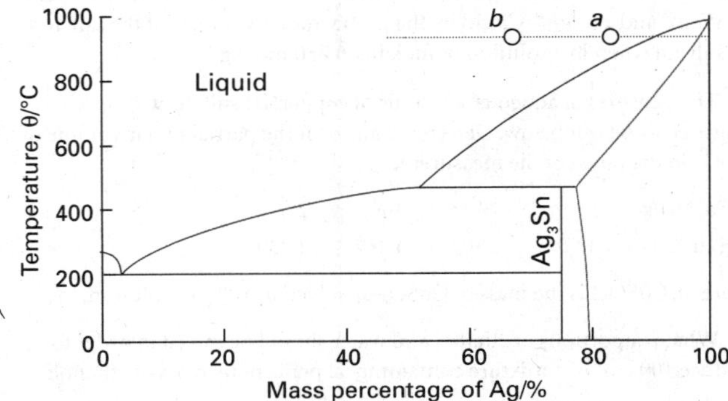

Fig. 5.65

5.29(a) Indicate on the phase diagram in Fig. 5.66 the feature that denotes incongruent melting. What is the composition of the eutectic mixture and at what temperature does it melt?

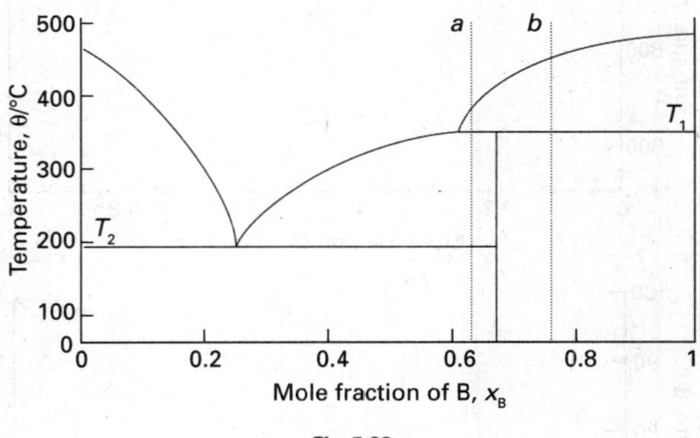

Fig. 5.66

5.29(b) Indicate on the phase diagram in Fig. 5.67 the feature that denotes incongruent melting. What is the composition of the eutectic mixture and at what temperature does it melt?

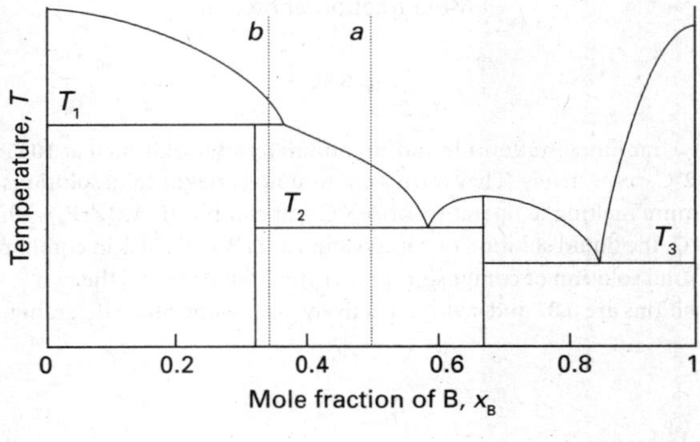

Fig. 5.67

5.30(a) Sketch the cooling curves for the isopleths a and b in Fig. 5.66.

5.30(b) Sketch the cooling curves for the isopleths a and b in Fig. 5.67.

5.31(a) Use the phase diagram in Fig. 5.65 to state (a) the solubility of Ag in Sn at 800°C and (b) the solubility of Ag_3Sn in Ag at 460°C, (c) the solubility of Ag_3Sn in Ag at 300°C.

5.31(b) Use the phase diagram in Fig. 5.66 to state (a) the solubility of B in A at 390°C and (b) the solubility of AB_2 in B at 300°C.

5.32(a) Figure 5.68 shows the experimentally determined phase diagrams for the nearly ideal solution of hexane and heptane. (a) Label the regions of the diagrams as to which phases are present. (b) For a solution containing 1 mol each of hexane and heptane molecules, estimate the vapour pressure at 70°C when vaporization on reduction of the external pressure just begins. (c) What is the vapour pressure of the solution at 70°C when just one drop of liquid remains. (d) Estimate from the figures the mole fraction of hexane in the liquid and vapour phases for the conditions of part b. (e) What are the mole fractions for the conditions of part c? (f) At 85°C and 760 Torr, what are the amounts of substance in the liquid and vapour phases when $z_{\text{heptane}} = 0.40$?

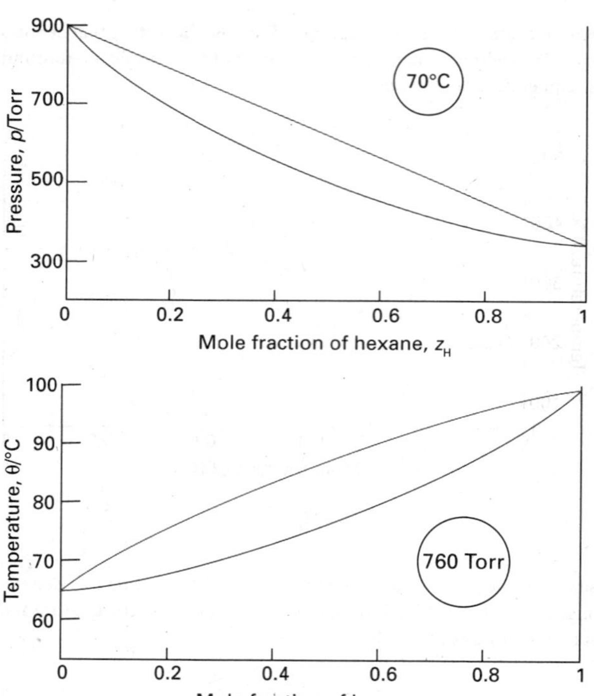

Fig. 5.68

5.32(b) Uranium tetrafluoride and zirconium tetrafluoride melt at 1035°C and 912°C, respectively. They form a continuous series of solid solutions with a minimum melting temperature of 765°C and composition $x(ZrF_4) = 0.77$. At 900°C, the liquid solution of composition $x(ZrF_4) = 0.28$ is in equilibrium with a solid solution of composition $x(ZrF_4) = 0.14$. At 850°C the two compositions are 0.87 and 0.90, respectively. Sketch the phase diagram for this system and state what is observed when a liquid of composition $x(ZrF_4) = 0.40$ is cooled slowly from 900°C to 500°C.

5.33(a) Methane (melting point 91 K) and tetrafluoromethane (melting point 89 K) do not form solid solutions with each other, and as liquids they are only partially miscible. The upper critical temperature of the liquid mixture is 94 K at $x(CF_4) = 0.43$ and the eutectic temperature is 84 K at $x(CF_4) = 0.88$. At 86 K, the phase in equilibrium with the tetrafluoromethane-rich solution changes from solid methane to a methane-rich liquid. At that temperature, the two liquid solutions that are in mutual equilibrium have the compositions $x(CF_4) = 0.10$ and $x(CF_4) = 0.80$. Sketch the phase diagram.

5.33(b) Describe the phase changes that take place when a liquid mixture of 4.0 mol B_2H_6 (melting point 131 K) and 1.0 mol CH_3OCH_3 (melting point 135 K) is cooled from 140 K to 90 K. These substances form a compound $(CH_3)_2OB_2H_6$ that melts congruently at 133 K. The system exhibits one eutectic at $x(B_2H_6) = 0.25$ and 123 K and another at $x(B_2H_6) = 0.90$ and 104 K.

5.34(a) Refer to the information in Exercise 5.33a and sketch the cooling curves for liquid mixtures in which $x(CF_4)$ is (a) 0.10, (b) 0.30, (c) 0.50, (d) 0.80, and (e) 0.95.

5.34(b) Refer to the information in Exercise 5.33b and sketch the cooling curves for liquid mixtures in which $x(B_2H_6)$ is (a) 0.10, (b) 0.30, (c) 0.50, (d) 0.80, and (e) 0.95.

5.35(a) Hexane and perfluorohexane show partial miscibility below 22.70°C. The critical concentration at the upper critical temperature is $x = 0.355$, where x is the mole fraction of C_6F_{14}. At 22.0°C the two solutions in equilibrium have $x = 0.24$ and $x = 0.48$, respectively, and at 21.5°C the mole fractions are 0.22 and 0.51. Sketch the phase diagram. Describe the phase changes that occur when perfluorohexane is added to a fixed amount of hexane at (a) 23°C, (b) 22°C.

5.35(b) Two liquids, A and B, show partial miscibility below 52.4°C. The critical concentration at the upper critical temperature is $x = 0.459$, where x is the mole fraction of A. At 40.0°C the two solutions in equilibrium have $x = 0.22$ and $x = 0.60$, respectively, and at 42.5°C the mole fractions are 0.24 and 0.48. Sketch the phase diagram. Describe the phase changes that occur when B is added to a fixed amount of A at (a) 48°C, (b) 52.4°C.

Problems*

Numerical problems

5.1 The following table gives the mole fraction of methylbenzene (A) in liquid and gaseous mixtures with butanone at equilibrium at 303.15 K and the total pressure p. Take the vapour to be perfect and calculate the partial pressures of the two components. Plot them against their respective mole fractions in the liquid mixture and find the Henry's law constants for the two components.

x_A	0	0.0898	0.2476	0.3577	0.5194	0.6036
y_A	0	0.0410	0.1154	0.1762	0.2772	0.3393
p/kPa	36.066	34.121	30.900	28.626	25.239	23.402

x_A	0.7188	0.8019	0.9105	1
y_A	0.4450	0.5435	0.7284	1
p/kPa	20.6984	18.592	15.496	12.295

5.2 The volume of an aqueous solution of NaCl at 25°C was measured at a series of molalities b, and it was found that the volume fitted the expression $v = 1003 + 16.62x + 1.77x^{3/2} + 0.12x^2$ where $v = V/\text{cm}^3$, V is the volume of a solution formed from 1.000 kg of water, and $x = b/b^{\ominus}$. Calculate the partial molar volume of the components in a solution of molality 0.100 mol kg^{-1}.

5.3 At 18°C the total volume V of a solution formed from $MgSO_4$ and 1.000 kg of water fits the expression $v = 1001.21 + 34.69(x - 0.070)^2$, where $v = V/\text{cm}^3$ and $x = b/b^{\ominus}$. Calculate the partial molar volumes of the salt and the solvent when in a solution of molality 0.050 mol kg^{-1}.

5.4 The densities of aqueous solutions of copper(II) sulfate at 20°C were measured as set out below. Determine and plot the partial molar volume of $CuSO_4$ in the range of the measurements.

$m(CuSO_4)$/g	5	10	15	20
ρ/(g cm^{-3})	1.051	1.107	1.167	1.230

where $m(CuSO_4)$ is the mass of $CuSO_4$ dissolved in 100 g of solution.

5.5 What proportions of ethanol and water should be mixed in order to produce 100 cm^3 of a mixture containing 50 per cent by mass of ethanol?

* Problems denoted with the symbol ‡ were supplied by Charles Trapp, Carmen Giunta, and Marshall Cady.

What change in volume is brought about by adding 1.00 cm^3 of ethanol to the mixture? (Use data from Fig. 5.1.)

5.6 Potassium fluoride is very soluble in glacial acetic acid and the solutions have a number of unusual properties. In an attempt to understand them, freezing point depression data were obtained by taking a solution of known molality and then diluting it several times (J. Emsley, *J. Chem. Soc. A*, 2702 (1971)). The following data were obtained:

b/(mol kg^{-1})	0.015	0.037	0.077	0.295	0.602
ΔT/K	0.115	0.295	0.470	1.381	2.67

Calculate the apparent molar mass of the solute and suggest an interpretation. Use $\Delta_{fus}H = 11.4$ kJ mol^{-1} and $T_f^* = 290$ K.

5.7 In a study of the properties of an aqueous solution of $Th(NO_3)_4$ (by A. Apelblat, D. Azoulay, and A. Sahar, *J. Chem. Soc. Faraday Trans., I*, 1618, (1973)), a freezing point depression of 0.0703 K was observed for an aqueous solution of molality 9.6 mmol kg^{-1}. What is the apparent number of ions per formula unit?

5.8 The table below lists the vapour pressures of mixtures of iodoethane (I) and ethyl acetate (A) at 50°C. Find the activity coefficients of both components on (a) the Raoult's law basis, (b) the Henry's law basis with iodoethane as solute.

x_I	0	0.0579	0.1095	0.1918	0.2353	0.3718
p_I/kPa	0	3.73	7.03	11.7	14.05	20.72
p_A/kPa	37.38	35.48	33.64	30.85	29.44	25.05
x_I	0.5478	0.6349	0.8253	0.9093	1.0000	
p_I/kPa	28.44	31.88	39.58	43.00	47.12	
p_A/kPa	19.23	16.39	8.88	5.09	0	

5.9 Plot the vapour pressure data for a mixture of benzene (B) and acetic acid (A) given below and plot the vapour pressure/composition curve for the mixture at 50°C. Then confirm that Raoult's and Henry's laws are obeyed in the appropriate regions. Deduce the activities and activity coefficients of the components on the Raoult's law basis and then, taking B as the solute, its activity and activity coefficient on a Henry's law basis. Finally, evaluate the excess Gibbs energy of the mixture over the composition range spanned by the data.

x_A	0.0160	0.0439	0.0835	0.1138	0.1714	
p_A/kPa	0.484	0.967	1.535	1.89	2.45	
p_B/kPa	35.05	34.29	33.28	32.64	30.90	
x_A	0.2973	0.3696	0.5834	0.6604	0.8437	0.9931
p_A/kPa	3.31	3.83	4.84	5.36	6.76	7.29
p_B/kPa	28.16	26.08	20.42	18.01	10.0	0.47

5.10‡ Aminabhavi *et al.* examined mixtures of cyclohexane with various long-chain alkanes (T.M. Aminabhavi, *et al.*, *J. Chem. Eng. Data* **41**, 526 (1996)). Among their data are the following measurements of the density of a mixture of cyclohexane and pentadecane as a function of mole fraction of cyclohexane (x_c) at 298.15 K:

x_c	0.6965	0.7988	0.9004
ρ/(g cm^{-3})	0.7661	0.7674	0.7697

Compute the partial molar volume for each component in a mixture which has a mole fraction cyclohexane of 0.7988.

5.11‡ Comelli and Francesconi examined mixtures of propionic acid with various other organic liquids at 313.15 K (F. Comelli and R. Francesconi, *J. Chem. Eng. Data* **41**, 101 (1996)). They report the excess volume of mixing propionic acid with oxane as $V^E = x_1x_2\{a_0 + a_1(x_1 - x_2)\}$, where x_1 is the mole fraction of propionic acid, x_2 that of oxane, $a_0 = -2.4697$ cm^3 mol^{-1}, and $a_1 = 0.0608$ cm^3 mol^{-1}. The density of propionic acid at this temperature is 0.97174 g cm^{-3}; that of oxane is 0.86398 g cm^{-3}. (a) Derive an expression for the partial molar volume of each component at this temperature. (b) Compute the partial molar volume for each component in an equimolar mixture.

5.12‡ Francesconi, Lunelli, and Comelli studied the liquid–vapour equilibria of trichloromethane and 1,2-epoxybutane at several temperatures (Francesconi, B. *et al.*, *J. Chem. Eng. Data* **41**, 310 (1996)). Among their data are the following measurements of the mole fractions of trichloromethane in the liquid phase (x_T) and the vapour phase (y_T) at 298.15 K as a function of pressure.

p/kPa	23.40	21.75	20.25	18.75	18.15	20.25	22.50	26.30
x	0	0.129	0.228	0.353	0.511	0.700	0.810	1
y	0	0.065	0.145	0.285	0.535	0.805	0.915	1

Compute the activity coefficients of both components on the basis of Raoult's law.

5.13‡ Chen and Lee studied the liquid–vapour equilibria of cyclohexanol with several gases at elevated pressures (J.-T. Chen and M.-J. Lee, *J. Chem. Eng. Data* **41**, 339 (1996)). Among their data are the following measurements of the mole fractions of cyclohexanol in the vapour phase (y) and the liquid phase (x) at 393.15 K as a function of pressure.

p/bar	10.0	20.0	30.0	40.0	60.0	80.0
y_{cyc}	0.0267	0.0149	0.0112	0.00947	0.00835	0.00921
x_{cyc}	0.9741	0.9464	0.9204	0.892	0.836	0.773

Determine the Henry's law constant of CO_2 in cyclohexanol, and compute the activity coefficient of CO_2.

5.14‡ Equation 5.37 indicates that solubility is an exponential function of temperature. The data in the table below gives the solubility, S, of calcium acetate in water as a function of temperature.

θ/°C	0	20	40	60	80
S/(mol dm^{-3})	36.4	34.9	33.7	32.7	31.7

Determine the extent to which the data fit the exponential $S = S_0 e^{\tau/T}$ and obtain values for S_0 and τ. Express these constants in terms of properties of the solute.

5.15 The excess Gibbs energy of solutions of methylcyclohexane (MCH) and tetrahydrofuran (THF) at 303.15 K was found to fit the expression

$$G^E = RTx(1-x)\{0.4857 - 0.1077(2x-1) + 0.0191(2x-1)^2\}$$

where x is the mole fraction of the methylcyclohexane. Calculate the Gibbs energy of mixing when a mixture of 1.00 mol of MCH and 3.00 mol of THF is prepared.

5.16 The mean activity coefficients for aqueous solutions of NaCl at 25°C are given below. Confirm that they support the Debye–Hückel limiting law and that an improved fit is obtained with the extended law.

b/(mmol kg^{-1})	1.0	2.0	5.0	10.0	20.0
$\gamma_\pm$	0.9649	0.9519	0.9275	0.9024	0.8712

5.17‡ 1-Butanol and chlorobenzene form a minimum-boiling azeotropic system. The mole fraction of 1-butanol in the liquid (x) and vapour (y) phases at 1.000 atm is given below for a variety of boiling temperatures (H. Artigas, *et al.*, *J. Chem. Eng. Data* **42**, 132 (1997)).

T/K	396.57	393.94	391.60	390.15	389.03	388.66	388.57
x	0.1065	0.1700	0.2646	0.3687	0.5017	0.6091	0.7171
y	0.2859	0.3691	0.4505	0.5138	0.5840	0.6409	0.7070

Pure chlorobenzene boils at 404.86 K. (a) Construct the chlorobenzene-rich portion of the phase diagram from the data. (b) Estimate the temperature at which a solution whose mole fraction of 1-butanol is 0.300 begins to boil.

(c) State the compositions and relative proportions of the two phases present after a solution initially 0.300 1-butanol is heated to 393.94 K.

5.18‡ An, Zhao, Jiang, and Shen investigated the liquid–liquid coexistence curve of *N,N*-dimethylacetamide and heptane (X. An, *et al., J. Chem. Thermodynamics* **28**, 1221 (1996)). Mole fractions of *N,N*-dimethylacetamide in the upper (x_1) and lower (x_2) phases of a two-phase region are given below as a function of temperature:

T/K	309.820	309.422	309.031	308.006	306.686
x_1	0.473	0.400	0.371	0.326	0.293
x_2	0.529	0.601	0.625	0.657	0.690
T/K	304.553	301.803	299.097	296.000	294.534
x_1	0.255	0.218	0.193	0.168	0.157
x_2	0.724	0.758	0.783	0.804	0.814

(a) Plot the phase diagram. (b) State the proportions and compositions of the two phases that form from mixing 0.750 mol of *N,N*-dimethylacetamide with 0.250 mol of heptane at 296.0 K. To what temperature must the mixture be heated to form a single-phase mixture?

5.19‡ The following data have been obtained for the liquid–vapour equilibrium compositions of mixtures of nitrogen and oxygen at 100 kPa.

T/K	77.3	78	80	82	84	86	88	90.2
$x(O_2)$	0	10	34	54	70	82	92	100
$y(O_2)$	0	2	11	22	35	52	73	100
$p^*(O_2)$/Torr	154	171	225	294	377	479	601	760

Plot the data on a temperature–composition diagram and determine the extent to which it fits the predictions for an ideal solution by calculating the activity coefficients of O_2 at each composition.

5.20 Phosphorus and sulfur form a series of binary compounds. The best characterized are P_4S_3, P_4S_7, and P_4S_{10}, all of which melt congruently. Assuming that only these three binary compounds of the two elements exist, (a) draw schematically only the P/S phase diagram. Label each region of the diagram with the substance that exists in that region and indicate its phase. Label the horizontal axis as x_S and give the numerical values of x_S that correspond to the compounds. The melting point of pure phosphorus is 44°C and that of pure sulfur is 119°C. (b) Draw, schematically, the cooling curve for a mixture of composition $x_S = 0.28$. Assume that a eutectic occurs at $x_S = 0.2$ and negligible solid–solid solubility.

5.21 The table below gives the break and halt temperatures found in the cooling curves of two metals A and B. Construct a phase diagram consistent with the data of these curves. Label the regions of the diagram, stating what phases and substances are present. Give the probable formulas of any compounds that form.

$100x_B$	θ_{break}/°C	$\theta_{halt,1}$/°C	$\theta_{halt,2}$/°C
0		1100	
10.0	1060	700	
20.0	1000	700	
30.0	940	700	400
40.0	850	700	400
50.0	750	700	400
60.0	670	400	
70.0	550	400	
80.0		400	
90.0	450	400	
100.0		500	

5.22 Consider the phase diagram in Fig. 5.69, which represents a solid–liquid equilibrium. Label all regions of the diagram according to the chemical species that exist in that region and their phases. Indicate the number of species and phases present at the points labelled *b, d, e, f, g,* and *k*. Sketch cooling curves for compositions $x_B = 0.16, 0.23, 0.57, 0.67$, and 0.84.

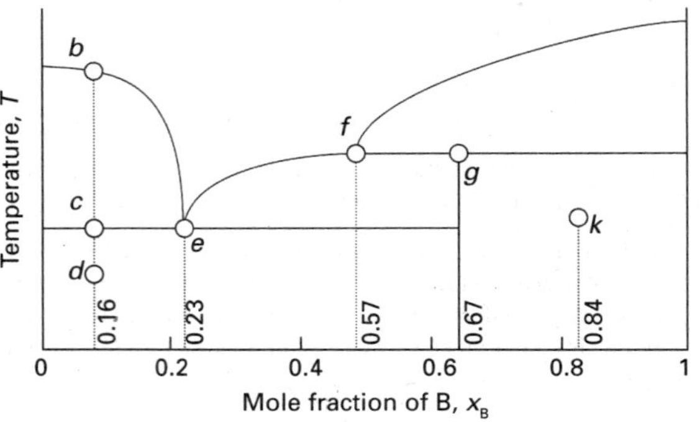

Fig. 5.69

5.23 Sketch the phase diagram for the Mg/Cu system using the following information: $\theta_f(Mg) = 648$°C, $\theta_f(Cu) = 1085$°C; two intermetallic compounds are formed with $\theta_f(MgCu_2) = 800$°C and $\theta_f(Mg_2Cu) = 580$°C; eutectics of mass percentage Mg composition and melting points 10 per cent (690°C), 33 per cent (560°C), and 65 per cent (380°C). A sample of Mg/Cu alloy containing 25 per cent Mg by mass was prepared in a crucible heated to 800°C in an inert atmosphere. Describe what will be observed if the melt is cooled slowly to room temperature. Specify the composition and relative abundances of the phases and sketch the cooling curve.

5.24‡ Figure 5.70 shows $\Delta_{mix}G(x_{Pb}, T)$ for a mixture of copper and lead. (a) What does the graph reveal about the miscibility of copper and lead and the spontaneity of solution formation? What is the variance (F) at (i) 1500 K, (ii) 1100 K? (b) Suppose that at 1500 K a mixture of composition (i) $x_{Pb} = 0.1$, (ii) $x_{Pb} = 0.7$, is slowly cooled to 1100 K. What is the equilibrium composition of the final mixture? Include an estimate of the relative amounts of each phase. (c) What is the solubility of (i) lead in copper, (ii) copper in lead at 1100 K?

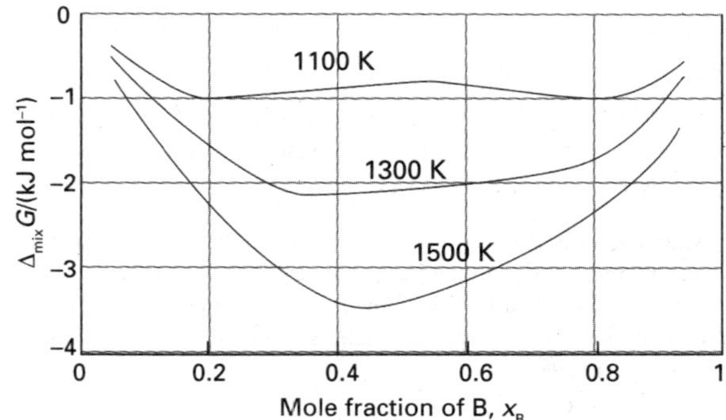

Fig. 5.70

5.25‡ The temperature/composition diagram for the Ca/Si binary system is shown in Fig. 5.71. (a) Identify eutectics, congruent melting compounds, and incongruent melting compounds. (b) If a 20 per cent by atom composition melt of silicon at 1500°C is cooled to 1000°C, what phases (and phase composition) would be at equilibrium? Estimate the relative amounts of each phase. (c) Describe the equilibrium phases observed when an 80 per cent

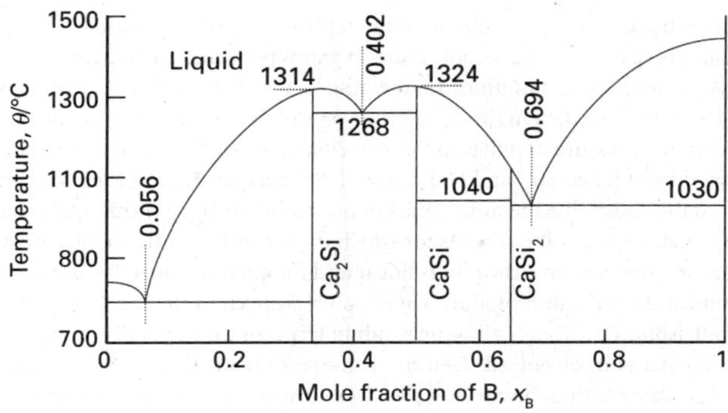

Fig. 5.71

by atom composition Si melt is cooled to 1030°C. What phases, and relative amounts, would be at equilibrium at a temperature (i) slightly higher than 1030°C, (ii) slightly lower than 1030°C? Draw a graph of the mole percentages of both Si(s) and $CaSi_2$(s) as a function of mole percentage of melt that is freezing at 1030°C.

5.26 Iron(II) chloride (melting point 677°C) and potassium chloride (melting point 776°C) form the compounds $KFeCl_3$ and K_2FeCl_4 at elevated temperatures. $KFeCl_3$ melts congruently at 399°C and K_2FeCl_4 melts incongruently at 380°C. Eutectics are formed with compositions $x = 0.38$ (melting point 351°C) and $x = 0.54$ (melting point 393°C), where x is the mole fraction of $FeCl_2$. The KCl solubility curve intersects the K_2FeCl_4 curve at $x = 0.34$. Sketch the phase diagram. State the phases that are in equilibrium when a mixture of composition $x = 0.36$ is cooled from 400°C to 300°C.

Theoretical problems

5.27 The excess Gibbs energy of a certain binary mixture is equal to $gRTx(1-x)$ where g is a constant and x is the mole fraction of a solute A. Find an expression for the chemical potential of A in the mixture and sketch its dependence on the composition.

5.28 Use the Gibbs–Duhem equation to derive the Gibbs–Duhem–Margules equation

$$\left(\frac{\partial \ln f_A}{\partial \ln x_A}\right)_{p,T} = \left(\frac{\partial \ln f_B}{\partial \ln x_B}\right)_{p,T}$$

where f is the fugacity. Use the relation to show that, when the fugacities are replaced by pressures, if Raoult's law applies to one component in a mixture then Henry's law must apply to the other.

5.29 Use the Gibbs–Duhem equation to show that the partial molar volume (or any partial molar property) of a component B can be obtained if the partial molar volume (or other property) of A is known for all compositions up to the one of interest. Do this by proving that

$$V_B = V_B^* - \int_{V_A^*}^{V_A} \frac{x_A}{1-x_A}\,dV_A$$

Use the following data (which are for 298 K) to evaluate the integral graphically to find the partial molar volume of acetone at $x = 0.500$.

$x(CHCl_3)$	0	0.194	0.385	0.559	0.788	0.889	1.000
$V_m/(cm^3\,mol^{-1})$	73.99	75.29	76.50	77.55	79.08	79.82	80.67

5.30 Use the Gibbs–Helmholtz equation to find an expression for $d\ln x_A$ in terms of dT. Integrate $d\ln x_A$ from $x_A = 0$ to the value of interest, and integrate the right-hand side from the transition temperature for the pure liquid A to the value in the solution. Show that, if the enthalpy of transition is constant, then eqns 5.31 and 5.34 are obtained.

5.31 The *osmotic coefficient* ϕ is defined as $\phi = -(x_A/x_B)\ln a_A$. By writing $r = x_B/x_A$, and using the Gibbs–Duhem equation, show that we can calculate the activity of B from the activities of A over a composition range by using the formula

$$\ln\left(\frac{a_B}{r}\right) = \phi - \phi(0) + \int_0^r \left(\frac{\phi - 1}{r}\right)dr$$

5.32 Show that the osmotic pressure of a real solution is given by $\Pi V = -RT\ln a_A$. Go on to show that, provided the concentration of the solution is low, this expression takes the form $\Pi V = \phi RT[B]$ and hence that the osmotic coefficient ϕ (which is defined in Problem 5.31) may be determined from osmometry.

5.33 Show that the freezing-point depression of a real solution in which the solvent of molar mass M has activity a_A obeys

$$\frac{d\ln a_A}{d(\Delta T)} = -\frac{M}{K_f}$$

and use the Gibbs–Duhem equation to show that

$$\frac{d\ln a_B}{d(\Delta T)} = -\frac{1}{b_B K_f}$$

where a_B is the solute activity and b_B is its molality. Use the Debye–Hückel limiting law to show that the osmotic coefficient (ϕ, Problem 5.31) is given by $\phi = 1 - \frac{1}{3}A'I$ with $A' = 2.303A$ and $I = b/b^{\ominus}$.

Applications: to biology and materials science

5.34 Haemoglobin, the red blood protein responsible for oxygen transport, binds about 1.34 cm^3 of oxygen per gram. Normal blood has a haemoglobin concentration of 150 g dm^{-3}. Haemoglobin in the lungs is about 97 per cent saturated with oxygen, but in the capillary is only about 75 per cent saturated. What volume of oxygen is given up by 100 cm^3 of blood flowing from the lungs in the capillary?

5.35 For the calculation of the solubility c of a gas in a solvent, it is often convenient to use the expression $c = Kp$, where K is the Henry's law constant. Breathing air at high pressures, such as in scuba diving, results in an increased concentration of dissolved nitrogen. The Henry's law constant for the solubility of nitrogen is 0.18 μg/(g H_2O atm). What mass of nitrogen is dissolved in 100 g of water saturated with air at 4.0 atm and 20°C? compare your answer to that for 100 g of water saturated with air at 1.0 atm. (Air is 78.08 mole per cent N_2.) If nitrogen is four times as soluble in fatty tissues as in water, what is the increase in nitrogen concentration in fatty tissue in going from 1 atm to 4 atm?

5.36 We saw in *Impact I5.1* that dialysis may be used to study the binding of small molecules to macromolecules, such as an inhibitor to an enzyme, an antibiotic to DNA, and any other instance of cooperation or inhibition by small molecules attaching to large ones. To see how this is possible, suppose that inside the dialysis bag the molar concentration of the macromolecule M is [M] and the total concentration of small molecule A is $[A]_{in}$. This total concentration is the sum of the concentrations of free A and bound A, which we write $[A]_{free}$ and $[A]_{bound}$, respectively. At equilibrium, $\mu_{A,free} = \mu_{A,out}$, which implies that $[A]_{free} = [A]_{out}$, provided the activity coefficient of A is the same in both solutions. Therefore, by measuring the concentration of A in the solution outside the bag, we can find the concentration of unbound A in the macromolecule solution and, from the difference $[A]_{in} - [A]_{free} = [A]_{in} - [A]_{out}$, the concentration of bound A. Now we explore the quantitative consequences of the experimental arrangement just described. (a) The average number of A molecules bound to M molecules, ν, is

$$v = \frac{[A]_{bound}}{[M]} = \frac{[A]_{in} - [A]_{out}}{[M]}$$

The bound and unbound A molecules are in equilibrium, $M + A \rightleftharpoons MA$. Recall from introductory chemistry that we may write the equilibrium constant for binding, K, as

$$K = \frac{[MA]}{[M]_{free}[A]_{free}}$$

Now show that

$$K = \frac{v}{(1-v)[A]_{out}}$$

(b) If there are N *identical* and *independent* binding sites on each macromolecule, each macromolecule behaves like N separate smaller macromolecules, with the same value of K for each site. It follows that the average number of A molecules per site is v/N. Show that, in this case, we may write the *Scatchard equation*:

$$\frac{v}{[A]_{out}} = KN - Kv$$

(c) To apply the Scatchard equation, consider the binding of ethidium bromide (EB) to a short piece of DNA by a process called *intercalation*, in which the aromatic ethidium cation fits between two adjacent DNA base pairs. An equilibrium dialysis experiment was used to study the binding of ethidium bromide (EB) to a short piece of DNA. A 1.00 μmol dm^{-3} aqueous solution of the DNA sample was dialysed against an excess of EB. The following data were obtained for the total concentration of EB:

[EB]/(μmol dm^{-3})					
Side without DNA	0.042	0.092	0.204	0.526	1.150
Side with DNA	0.292	0.590	1.204	2.531	4.150

From these data, make a Scatchard plot and evaluate the intrinsic equilibrium constant, K, and total number of sites per DNA molecule. Is the identical and independent sites model for binding applicable?

5.37 The form of the Scatchard equation given in Problem 5.36 applies only when the macromolecule has identical and independent binding sites. For non-identical independent binding sites, the Scatchard equation is

$$\frac{v}{[A]_{out}} = \sum_i \frac{N_i K_i}{1 + K_i [A]_{out}}$$

Plot $v/[A]$ for the following cases. (a) There are four independent sites on an enzyme molecule and the intrinsic binding constant is $K = 1.0 \times 10^7$. (b) There are a total of six sites per polymer. Four of the sties are identical and have an intrinsic binding constant of 1×10^5. The binding constants for the other two sites are 2×10^6.

5.38 The addition of a small amount of a salt, such as $(NH_4)_2SO_4$, to a solution containing a charged protein increases the solubility of the protein in water. This observation is called the *salting-in effect*. However, the addition of large amounts of salt can decrease the solubility of the protein to such an extent that the protein precipitates from solution. This observation is called the *salting-out effect* and is used widely by biochemists to isolate and purify proteins. Consider the equilibrium $PX_v(s) \rightleftharpoons P^{v+}(aq) + v\,X^-(aq)$, where P^{v+} is a polycationic protein of charge $+v$ and X^- is its counter ion. Use Le Chatelier's principle and the physical principles behind the Debye–Hückel theory to provide a molecular interpretation for the salting-in and salting-out effects.

5.39‡ Polymer scientists often report their data in rather strange units. For example, in the determination of molar masses of polymers in solution by osmometry, osmotic pressures are often reported in grams per square centimetre (g cm^{-2}) and concentrations in grams per cubic centimetre (g cm^{-3}). (a) With these choices of units, what would be the units of R in the van't Hoff equation? (b) The data in the table below on the concentration dependence of the osmotic pressure of polyisobutene in chlorobenzene at 25°C have been adapted from J. Leonard and H. Daoust (*J. Polymer Sci.* 57, 53 (1962)). From these data, determine the molar mass of polyisobutene by plotting Π/c against c. (c) Theta solvents are solvents for which the second osmotic coefficient is zero; for 'poor' solvents the plot is linear and for good solvents the plot is nonlinear. From your plot, how would you classify chlorobenzene as a solvent for polyisobutene? Rationalize the result in terms of the molecular structure of the polymer and solvent. (d) Determine the second and third osmotic virial coefficients by fitting the curve to the virial form of the osmotic pressure equation. (e) Experimentally, it is often found that the virial expansion can be represented as

$$\Pi/c = RT/M\,(1 + B'c + gB'^2c^2 + \cdots)$$

and, in good solvents, the parameter g is often about 0.25. With terms beyond the second power ignored, obtain an equation for $(\Pi/c)^{1/2}$ and plot this quantity against c. Determine the second and third virial coefficients from the plot and compare to the values from the first plot. Does this plot confirm the assumed value of g?

$10^{-2}(\Pi/c)/(\text{g cm}^{-2}/\text{g cm}^{-3})$	2.6	2.9	3.6	4.3	6.0	12.0
$c/(\text{g cm}^{-3})$	0.0050	0.010	0.020	0.033	0.057	0.10
$10^{-2}(\Pi/c)/(\text{g cm}^{-2}/\text{g cm}^{-3})$	19.0	31.0	38.0	52	63	
$c/(\text{g cm}^{-3})$	0.145	0.195	0.245	0.27	0.29	

5.40‡ K. Sato, F.R. Eirich, and J.E. Mark (*J. Polymer Sci., Polym. Phys.* 14, 619 (1976)) have reported the data in the table below for the osmotic pressures of polychloroprene ($\rho = 1.25$ g cm^{-3}) in toluene ($\rho = 0.858$ g cm^{-3}) at 30°C. Determine the molar mass of polychloroprene and its second osmotic virial coefficient.

$c/(\text{mg cm}^{-3})$	1.33	2.10	4.52	7.18	9.87
$\Pi/(\text{N m}^{-2})$	30	51	132	246	390

5.41 The compound *p*-azoxyanisole forms a liquid crystal. 5.0 g of the solid was placed in a tube, which was then evacuated and sealed. Use the phase rule to prove that the solid will melt at a definite temperature and that the liquid crystal phase will make a transition to a normal liquid phase at a definite temperature.

5.42 Some polymers can form liquid crystal mesophases with unusual physical properties. For example, liquid crystalline Kevlar (**3**) is strong enough to be the material of choice for bulletproof vests and is stable at temperatures up to 600 K. What molecular interactions contribute to the formation, thermal stability, and mechanical strength of liquid crystal mesophases in Kevlar?

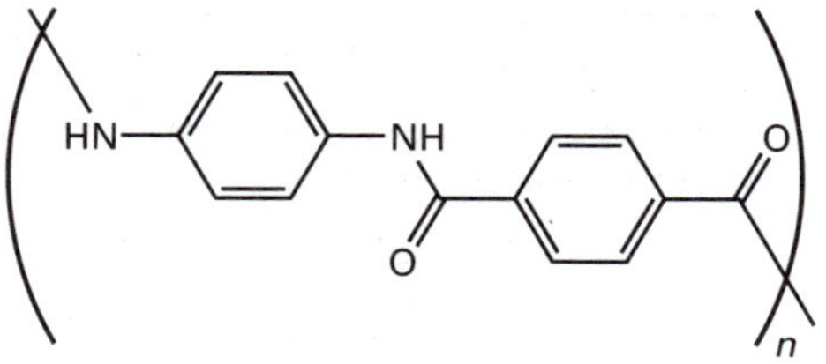

HN NH O O n

3

Chemical equilibrium

6

This chapter develops the concept of chemical potential and shows how it is used to account for the equilibrium composition of chemical reactions. The equilibrium composition corresponds to a minimum in the Gibbs energy plotted against the extent of reaction. By locating this minimum we establish the relation between the equilibrium constant and the standard Gibbs energy of reaction. The thermodynamic formulation of equilibrium enables us to establish the quantitative effects of changes in the conditions. The same principles can be applied to the description of the thermodynamic properties of reactions that take place in electrochemical cells, in which the reaction drives electrons through an external circuit. Thermodynamic arguments can be used to derive an expression for the electric potential of such cells and this potential can be related to their composition. There are two major topics developed in this connection. One is the definition and tabulation of standard potentials; the second is the use of these standard potentials to determine the equilibrium constants and other thermodynamic properties of chemical reactions.

Chemical reactions tend to move towards a dynamic equilibrium in which both reactants and products are present but have no further tendency to undergo net change. In some cases, the concentration of products in the equilibrium mixture is so much greater than that of the unchanged reactants that for all practical purposes the reaction is 'complete'. However, in many important cases the equilibrium mixture has significant concentrations of both reactants and products. In this chapter we see how to use thermodynamics to predict the equilibrium composition under any reaction conditions and understand the underlying molecular processes.

Because many reactions involve the transfer of electrons, they can be studied (and utilized) by allowing them to take place in an electrochemical cell. Electrochemistry is in part a major application of thermodynamic concepts to chemical equilibria as well as being of great technological importance. The final sections of the chapter show how to apply ideas relating to chemical equilibria to this vitally important field.

Spontaneous chemical reactions

We have seen that the direction of spontaneous change at constant temperature and pressure is towards lower values of the Gibbs energy, G. The idea is entirely general, and in this chapter we apply it to the discussion of chemical reactions.

6.1 The Gibbs energy minimum

Key points **(a) The reaction Gibbs energy is the slope of the plot of Gibbs energy against extent of reaction. (b) Reactions are either exergonic or endergonic.**

We locate the equilibrium composition of a reaction mixture by calculating the Gibbs energy of the reaction mixture and identifying the composition that corresponds to minimum *G*. Here we proceed in two steps: first, we consider a very simple equilibrium, and then we generalize it.

(a) The reaction Gibbs energy

Consider the equilibrium $A \rightleftharpoons B$. Even though this reaction looks trivial, there are many examples of it, such as the isomerization of pentane to 2-methylbutane and the conversion of L-alanine to D-alanine. Suppose an infinitesimal amount $d\xi$ of A turns into B; then the change in the amount of A present is $dn_A = -d\xi$ and the change in the amount of B present is $dn_B = +d\xi$. The quantity ξ (xi) is called the **extent of reaction**; it has the dimensions of amount of substance and is reported in moles. When the extent of reaction changes by a finite amount $\Delta\xi$, the amount of A present changes from $n_{A,0}$ to $n_{A,0} - \Delta\xi$ and the amount of B changes from $n_{B,0}$ to $n_{B,0} + \Delta\xi$.

• A brief illustration

If initially 2.0 mol A is present and we wait until $\Delta\xi = +1.5$ mol, then the amount of A remaining will be 0.5 mol. The amount of B formed will be 1.5 mol. •

The **reaction Gibbs energy**, $\Delta_r G$, is defined as the slope of the graph of the Gibbs energy plotted against the extent of reaction:

$$\Delta_r G = \left(\frac{\partial G}{\partial \xi}\right)_{p,T} \qquad \text{Definition of reaction Gibbs energy} \qquad (6.1)$$

Although Δ normally signifies a *difference* in values, here it signifies a *derivative*, the slope of *G* with respect to ξ. However, to see that there is a close relationship with the normal usage, suppose the reaction advances by $d\xi$. The corresponding change in Gibbs energy is

$$dG = \mu_A dn_A + \mu_B dn_B = -\mu_A d\xi + \mu_B d\xi = (\mu_B - \mu_A)d\xi$$

This equation can be reorganized into

$$\left(\frac{\partial G}{\partial \xi}\right)_{p,T} = \mu_B - \mu_A$$

That is,

$$\Delta_r G = \mu_B - \mu_A \qquad (6.2)$$

We see that $\Delta_r G$ can also be interpreted as the difference between the chemical potentials (the partial molar Gibbs energies) of the reactants and products *at the composition of the reaction mixture.*

Because chemical potentials vary with composition, the slope of the plot of Gibbs energy against extent of reaction, and therefore the reaction Gibbs energy, changes as the reaction proceeds. The spontaneous direction of reaction lies in the direction of decreasing *G* (that is, down the slope of *G* plotted against ξ). Thus we see from eqn 6.2 that the reaction $A \rightarrow B$ is spontaneous when $\mu_A > \mu_B$, whereas the reverse reaction is spontaneous when $\mu_B > \mu_A$. The slope is zero, and the reaction is at equilibrium and spontaneous in neither direction, when

$$\Delta_r G = 0 \qquad \text{Condition of equilibrium} \qquad (6.3)$$

This condition occurs when $\mu_B = \mu_A$ (Fig. 6.1). It follows that, if we can find the composition of the reaction mixture that ensures $\mu_B = \mu_A$, then we can identify the composition of the reaction mixture at equilibrium. Note that the chemical potential is now fulfilling the role its name suggests: it represents the potential for chemical change, and equilibrium is attained when these potentials are in balance.

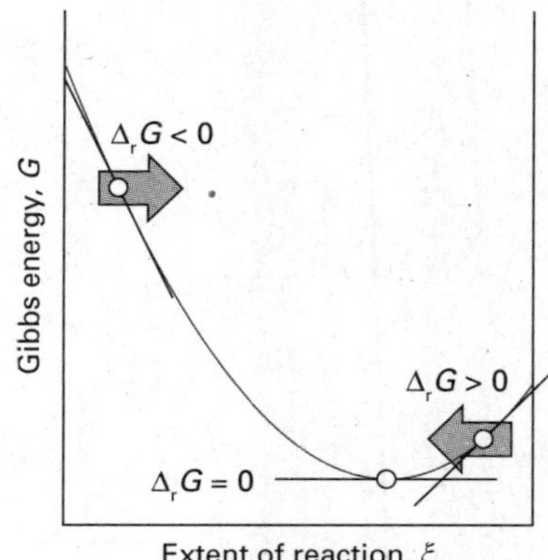

Fig. 6.1 As the reaction advances (represented by motion from left to right along the horizontal axis) the slope of the Gibbs energy changes. Equilibrium corresponds to zero slope, at the foot of the valley.

(b) Exergonic and endergonic reactions

We can express the spontaneity of a reaction at constant temperature and pressure in terms of the reaction Gibbs energy:

If $\Delta_r G < 0$, the forward reaction is spontaneous.
If $\Delta_r G > 0$, the reverse reaction is spontaneous.
If $\Delta_r G = 0$, the reaction is at equilibrium.

A reaction for which $\Delta_r G < 0$ is called **exergonic** (from the Greek words for work-producing). The name signifies that, because the process is spontaneous, it can be used to drive another process, such as another reaction, or used to do non-expansion work. A simple mechanical analogy is a pair of weights joined by a string (Fig. 6.2): the lighter of the pair of weights will be pulled up as the heavier weight falls down. Although the lighter weight has a natural tendency to move downward, its coupling to the heavier weight results in it being raised. In biological cells, the oxidation of carbohydrates act as the heavy weight that drives other reactions forward and results in the formation of proteins from amino acids, muscle contraction, and brain activity. A reaction for which $\Delta_r G > 0$ is called **endergonic** (signifying work-consuming). The reaction can be made to occur only by doing work on it, such as electrolysing water to reverse its spontaneous formation reaction.

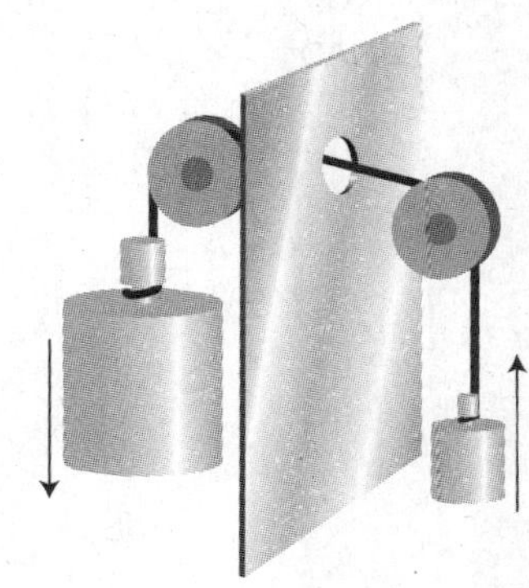

Fig. 6.2 If two weights are coupled as shown here, then the heavier weight will move the lighter weight in its non-spontaneous direction: overall, the process is still spontaneous. The weights are the analogues of two chemical reactions: a reaction with a large negative ΔG can force another reaction with a less negative ΔG to run in its non-spontaneous direction.

IMPACT ON BIOCHEMISTRY

I6.1 Energy conversion in biological cells

The whole of life's activities depends on the coupling of exergonic and endergonic reactions, for the oxidation of food drives other reactions forward. In biological cells, the energy released by the oxidation of foods is stored in adenosine triphosphate (ATP, 1). The essence of the action of ATP is its ability to lose its terminal phosphate group by hydrolysis and to form adenosine diphosphate (ADP):

$$ATP(aq) + H_2O(l) \rightarrow ADP(aq) + P_i^-(aq) + H_3O^+(aq)$$

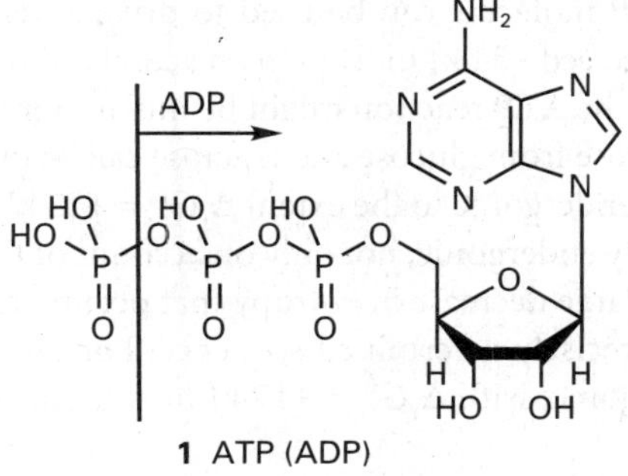

1 ATP (ADP)

where P_i^- denotes an inorganic phosphate group, such as $H_2PO_4^-$. The biological standard values (Section 5.11d) for ATP hydrolysis at 37°C (310 K, blood temperature) are $\Delta_r G^{\oplus} = -31$ kJ mol^{-1}, $\Delta_r H^{\oplus} = -20$ kJ mol^{-1}, and $\Delta_r S^{\oplus} = +34$ J K^{-1} mol^{-1}. The hydrolysis is therefore exergonic ($\Delta_r G^{\oplus} < 0$) under these conditions and 31 kJ mol^{-1} is available for driving other reactions. Moreover, because the reaction entropy is large, the reaction Gibbs energy is sensitive to temperature. In view of its exergonicity the ADP-phosphate bond has been called a 'high-energy phosphate bond'. The name is intended to signify a high tendency to undergo reaction, and should not be confused with 'strong' bond. In fact, even in the biological sense it is not of very 'high energy'. The action of ATP depends on it being intermediate in activity. Thus ATP acts as a phosphate donor to a number of acceptors (for example, glucose), but is recharged by more powerful phosphate donors in a number of biochemical processes.

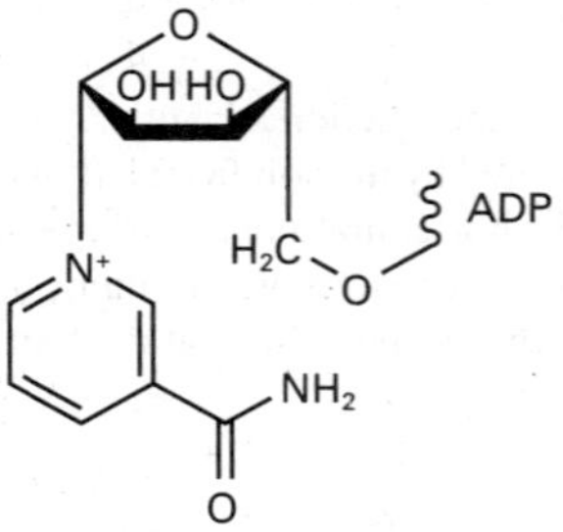

The oxidation of glucose to CO_2 and H_2O by O_2 is an example of how the breakdown of foods is coupled to the formation of ATP in the cell. The process begins with *glycolysis*, a partial oxidation of glucose by nicotinamide adenine dinucleotide (NAD^+, **2**) to pyruvate ion, $CH_3COCO_2^-$, continues with the *citric acid cycle*, which oxidizes pyruvate to CO_2, and ends with *oxidative phosphorylation*, which reduces O_2 to H_2O. Glycolysis is the main source of energy during *anaerobic metabolism*, a form of metabolism in which inhaled O_2 does not play a role. The citric acid cycle and oxidative phosphorylation are the main mechanisms for the extraction of energy from carbohydrates during *aerobic metabolism*, a form of metabolism in which inhaled O_2 does play a role.

2 NAD+

At blood temperature, $\Delta_r G^{\oplus} = -147$ kJ mol^{-1} for the oxidation of glucose by NAD^+ to pyruvate ions. The oxidation of one glucose molecule is coupled to the conversion of two ADP molecules to two ATP molecules, so the net reaction of glycolysis is

$$\begin{aligned}&C_6H_{12}O_6(aq) + 2\,NAD^+(aq) + 2\,ADP(aq) + 2\,P_i^-(aq) + 2\,H_2O(l)\\&\quad \rightarrow 2\,CH_3COCO_2^-(aq) + 2\,NADH(aq) + 2\,ATP(aq) + 2\,H_3O^+(aq)\end{aligned}$$

The standard reaction Gibbs energy is $(-147) - 2(-31)$ kJ mol^{-1} $= -85$ kJ mol^{-1}: the reaction is exergonic and can be used to drive other reactions.

The standard Gibbs energy of combustion of glucose is -2880 kJ mol^{-1}, so terminating its oxidation at pyruvate is a poor use of resources. In the presence of O_2, pyruvate is oxidized further during the citric acid cycle:

$$\begin{aligned}&2\,CH_3COCO_2^-(aq) + 8\,NAD^+(aq) + 2\,FAD(aq) + 2\,ADP(aq) + 2\,P_i(aq) + 8\,H_2O(l)\\&\quad \rightarrow 6\,CO_2(g) + 8\,NADH(aq) + 4\,H_3O^+(aq) + 2\,FADH_2(aq) + 2\,ATP(aq)\end{aligned}$$

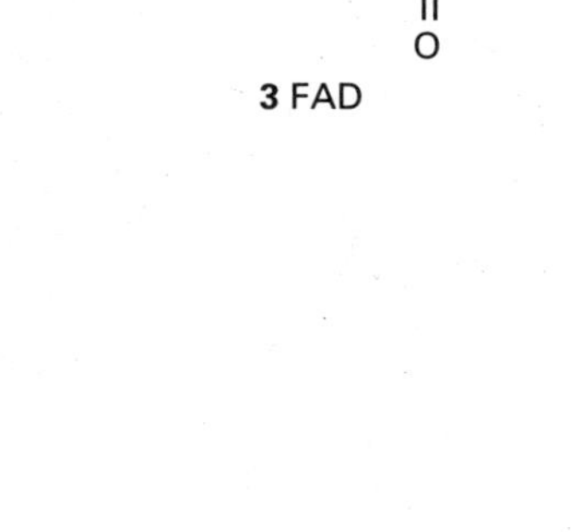

where FAD is flavin adenine dinucleotide (**3**). The NADH and $FADH_2$ go on to reduce O_2 during oxidative phosphorylation, which also produces ATP. The citric acid cycle and oxidative phosphorylation generate as many as 38 ATP molecules for each glucose molecule consumed. Each mole of ATP molecules extracts 31 kJ from the 2880 kJ supplied by 1 mol $C_6H_{12}O_6$ (180 g of glucose), so 1178 kJ is stored for later use. Therefore, aerobic oxidation of glucose is much more effcient than glycolysis.

3 FAD

In the cell, each ATP molecule can be used to drive an endergonic reaction for which $\Delta_r G^{\oplus}$ does not exceed +31 kJ mol^{-1}. (In an actual cell the composition may be far from standard, and the ATP reaction might be much more potent.) For example, the biosynthesis of sucrose from glucose and fructose can be driven by plant enzymes because the reaction is endergonic to the extent $\Delta_r G^{\oplus} = +23$ kJ mol^{-1}. The biosynthesis of proteins is strongly endergonic, not only on account of the enthalpy change but also on account of the large decrease in entropy that occurs when many amino acids are assembled into a precisely determined sequence. For instance, the formation of a peptide link is endergonic, with $\Delta_r G^{\oplus} = +17$ kJ mol^{-1}, but the biosynthesis occurs indirectly and is equivalent to the consumption of three ATP molecules for each link. In a moderately small protein like myoglobin, with about 150 peptide links, the construction alone requires 450 ATP molecules, and therefore about 12 mol of glucose molecules for 1 mol of protein molecules.

6.2 The description of equilibrium

Key points **(a) The reaction Gibbs energy depends logarithmically on the reaction quotient. When the reaction Gibbs energy is zero the reaction quotient has a value called the equilibrium constant. (b) The results are readily extended to a general reaction. (c) Under ideal conditions, the thermodynamic equilibrium constant may be approximated by expressing it in terms of concentrations and partial pressures. (d) The presence of the enthalpy and entropy contributions to *K* are related to the role of the Boltzmann distribution of molecules over the available states. (e) The biological standard state is defined at pH = 7.**

With the background established, we are now ready to see how to apply thermodynamics to the description of chemical equilibrium.

(a) Perfect gas equilibria

When A and B are perfect gases we can use eqn 5.14 ($\mu = \mu^{\ominus} + RT\ln p$, with p interpreted as $p/p^{\ominus}$) to write

$$\Delta_r G = \mu_B - \mu_A = (\mu_B^{\ominus} + RT\ln p_B) - (\mu_A^{\ominus} + RT\ln p_A)$$

$$= \Delta_r G^{\ominus} + RT\ln\frac{p_B}{p_A} \qquad (6.4)°$$

If we denote the ratio of partial pressures by Q, we obtain

$$\Delta_r G = \Delta_r G^{\ominus} + RT\ln Q \qquad Q = \frac{p_B}{p_A} \qquad (6.5)°$$

The ratio Q is an example of a **reaction quotient**. It ranges from 0 when $p_B = 0$ (corresponding to pure A) to infinity when $p_A = 0$ (corresponding to pure B). The **standard reaction Gibbs energy**, $\Delta_r G^{\ominus}$, is defined (like the standard reaction enthalpy) as the difference in the standard molar Gibbs energies of the reactants and products. For our reaction

$$\Delta_r G^{\ominus} = G_m^{\ominus}(B) - G_m^{\ominus}(A) = \mu_B^{\ominus} - \mu_A^{\ominus} \qquad (6.6)$$

Note that in the definition of $\Delta_r G^{\ominus}$, the Δ_r has its normal meaning as the difference 'products – reactants'. In Section 3.6 we saw that the difference in standard molar Gibbs energies of the products and reactants is equal to the difference in their standard Gibbs energies of formation, so in practice we calculate $\Delta_r G^{\ominus}$ from

$$\Delta_r G^{\ominus} = \Delta_f G^{\ominus}(B) - \Delta_f G^{\ominus}(A) \qquad (6.7)$$

At equilibrium $\Delta_r G = 0$. The ratio of partial pressures at equilibrium is denoted K, and eqn 6.5 becomes

$$0 = \Delta_r G^{\ominus} + RT\ln K$$

which rearranges to

$$RT\ln K = -\Delta_r G^{\ominus} \qquad K = \left(\frac{p_B}{p_A}\right)_{\text{equilibrium}} \qquad (6.8)°$$

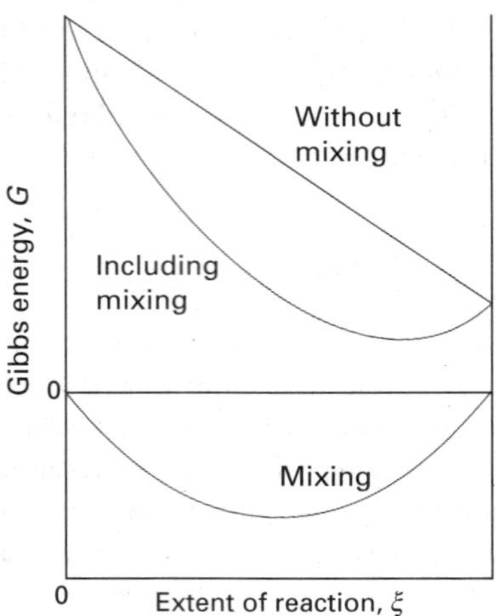

Fig. 6.3 If the mixing of reactants and products is ignored, then the Gibbs energy changes linearly from its initial value (pure reactants) to its final value (pure products) and the slope of the line is $\Delta_r G^\ominus$. However, as products are produced, there is a further contribution to the Gibbs energy arising from their mixing (lowest curve). The sum of the two contributions has a minimum. That minimum corresponds to the equilibrium composition of the system.

This relation is a special case of one of the most important equations in chemical thermodynamics: it is the link between tables of thermodynamic data, such as those in the *Data section* and the chemically important **equilibrium constant**, K.

In molecular terms, the minimum in the Gibbs energy, which corresponds to $\Delta_r G = 0$, stems from the Gibbs energy of mixing of the two gases. To see the role of mixing, consider the reaction A → B. If only the enthalpy were important, then H and therefore G would change linearly from its value for pure reactants to its value for pure products. The slope of this straight line is a constant and equal to $\Delta_r G^\ominus$ at all stages of the reaction and there is no intermediate minimum in the graph (Fig. 6.3). However, when we take entropy into account, there is an additional contribution to the Gibbs energy that is given by eqn 5.25 ($\Delta_{mix}G = nRT(x_A \ln x_A + x_B \ln x_B)$). This expression makes a U-shaped contribution to the total change in Gibbs energy. As can be seen from Fig. 6.3, when it is included there is an intermediate minimum in the total Gibbs energy, and its position corresponds to the equilibrium composition of the reaction mixture.

We see from eqn 6.8 that, when $\Delta_r G^\ominus > 0$, $K < 1$. Therefore, at equilibrium the partial pressure of A exceeds that of B, which means that the reactant A is favoured in the equilibrium. When $\Delta_r G^\ominus < 0$, $K > 1$, so at equilibrium the partial pressure of B exceeds that of A. Now the product B is favoured in the equilibrium.

A note on good practice A common remark is that 'a reaction is spontaneous if $\Delta_r G^\ominus < 0$'. However, whether or not a reaction is spontaneous at a particular composition depends on the value of $\Delta_r G$ at that composition, not $\Delta_r G^\ominus$. It is far better to interpret the sign of $\Delta_r G^\ominus$ as indicating whether K is greater or smaller than 1. The forward reaction is spontaneous ($\Delta_r G < 0$) when $Q < K$ and the reverse reaction is spontaneous when $Q > K$.

(b) The general case of a reaction

We can now extend the argument that led to eqn 6.8 to a general reaction. We saw in Section 2.8a that a chemical reaction may be expressed symbolically in terms of stoichiometric numbers as

$$0 = \sum_J \nu_J J \qquad \text{Symbolic form of a chemical equation} \qquad (6.9)$$

where J denotes the substances and the ν_J are the corresponding stoichiometric numbers in the chemical equation. In the reaction 2 A + B → 3 C + D, for instance, these numbers have the values $\nu_A = -2$, $\nu_B = -1$, $\nu_C = +3$, and $\nu_D = +1$. A stoichiometric number is positive for products and negative for reactants. We define the extent of reaction ξ so that, if it changes by $\Delta\xi$, then the change in the amount of any species J is $\nu_J \Delta\xi$.

With these points in mind and with the reaction Gibbs energy, $\Delta_r G$, defined in the same way as before (eqn 6.1) we show in the following *Justification* that the Gibbs energy of reaction can always be written

$$\Delta_r G = \Delta_r G^{\ominus} + RT \ln Q \qquad \text{Reaction Gibbs energy at an arbitrary stage} \qquad (6.10)$$

with the standard reaction Gibbs energy calculated from

$$\Delta_r G^{\ominus} = \sum_{\text{Products}} \nu \Delta_f G^{\ominus} - \sum_{\text{Reactants}} \nu \Delta_f G^{\ominus} \qquad \text{Procedure for calculating standard reaction Gibbs energy} \qquad (6.11a)$$

where the ν are the (positive) stoichiometric coefficients. More formally,

$$\Delta_r G^{\ominus} = \sum_J \nu_J \Delta_f G^{\ominus}(J) \qquad \text{Formal expression} \qquad (6.11b)$$

where the ν_J are the (signed) stoichiometric numbers. The reaction quotient, Q, has the form

$$Q = \frac{\text{activities of products}}{\text{activities of reactants}} \qquad \text{General form of reaction quotient} \qquad (6.12a)$$

with each species raised to the power given by its stoichiometric coefficient. More formally, to write the general expression for Q we introduce the symbol Π to denote the product of what follows it (just as Σ denotes the sum), and define Q as

$$Q = \prod_J a_J^{\nu_J} \qquad \text{Definition of reaction quotient} \qquad (6.12b)$$

Because reactants have negative stoichiometric numbers, they automatically appear as the denominator when the product is written out explicitly. Recall from Table 5.3 that, for pure solids and liquids, the activity is 1, so such substances make no contribution to Q even though they may appear in the chemical equation.

- **A brief illustration**

Consider the reaction 2 A + 3 B → C + 2 D, in which case $\nu_A = -2$, $\nu_B = -3$, $\nu_C = +1$, and $\nu_D = +2$. The reaction quotient is then

$$Q = a_A^{-2} a_B^{-3} a_C a_D^2 = \frac{a_C a_D^2}{a_A^2 a_B^3} \bullet$$

Justification 6.1 *The dependence of the reaction Gibbs energy on the reaction quotient*

Consider a reaction with stoichiometric numbers ν_J. When the reaction advances by $d\xi$, the amounts of reactants and products change by $dn_J = \nu_J d\xi$. The resulting infinitesimal change in the Gibbs energy at constant temperature and pressure is

$$dG = \sum_J \mu_J dn_J = \sum_J \mu_J \nu_J d\xi = \left(\sum_J \nu_J \mu_J\right) d\xi$$

It follows that

$$\Delta_r G = \left(\frac{\partial G}{\partial \xi}\right)_{p,T} = \sum_J \nu_J \mu_J$$

To make further progress, we note that the chemical potential of a species J is related to its activity by eqn 5.56 ($\mu_J = \mu_J^\ominus + RT \ln a_J$). When this expression is substituted into the expression above for $\Delta_r G$ we obtain

$$\Delta_r G = \overbrace{\sum_J \nu_J \mu_J^\ominus}^{\Delta_r G^\ominus} + RT \sum_J \nu_J \ln a_J$$

$$= \Delta_r G^\ominus + RT \sum_J \ln a_J^{\nu_J} = \Delta_r G^\ominus + RT \ln \overbrace{\prod_J a_J^{\nu_J}}^{Q}$$

$$= \Delta_r G^\ominus + RT \ln Q$$

with Q given by eqn 6.12b.

A brief comment

In the second line we use first $a \ln x = \ln x^a$ and then $\ln x + \ln y + \ldots = \ln xy \ldots$, so $\sum_i \ln x_i = \ln \left(\prod_i x_i \right)$.

Now we conclude the argument, starting from eqn 6.10. At equilibrium, the slope of G is zero: $\Delta_r G = 0$. The activities then have their equilibrium values and we can write

$$K = \left(\prod_J a_J^{\nu_J} \right)_{\text{equilibrium}} \qquad [6.13]$$

Definition of equilibrium constant

This expression has the same form as Q but is evaluated using equilibrium activities. From now on, we shall not write the 'equilibrium' subscript explicitly, and will rely on the context to make it clear that for K we use equilibrium values and for Q we use the values at the specified stage of the reaction. An equilibrium constant K expressed in terms of activities (or fugacities) is called a **thermodynamic equilibrium constant**. Note that, because activities are dimensionless numbers, the thermodynamic equilibrium constant is also dimensionless. In elementary applications, the activities that occur in eqn 6.13 are often replaced by:

- molalities, by replacing a_J by $b_J/b^\ominus$, where $b^\ominus = 1$ mol kg^{-1}
- molar concentrations, by replacing a_J by $[J]/c^\ominus$, where $c^\ominus = 1$ mol dm^{-3}
- partial pressures, by replacing a_J by $p_J/p^\ominus$, where $p^\ominus = 1$ bar

In such cases, the resulting expressions are only approximations. The approximation is particularly severe for electrolyte solutions, for in them activity coefficients differ from 1 even in very dilute solutions (Section 5.13).

• **A brief illustration**

The equilibrium constant for the heterogeneous equilibrium $CaCO_3(s) \rightleftharpoons CaO(s) + CO_2(g)$ is

$$K = a_{CaCO_3(s)}^{-1} a_{CaO(s)} a_{CO_2(g)} = \frac{\overbrace{a_{CaO(s)}}^{1} a_{CO_2(g)}}{\underbrace{a_{CaCO_3(s)}}_{1}} = a_{CO_2}$$

(Table 5.3). Provided the carbon dioxide can be treated as a perfect gas, we can go on to write

$$K \approx p_{CO_2}/p^\ominus$$

and conclude that in this case the equilibrium constant is the numerical value of the decomposition vapour pressure of calcium carbonate. •

At this point we set $\Delta_r G = 0$ in eqn 6.10 and replace Q by K. We immediately obtain

$$RT \ln K = -\Delta_r G^{\ominus} \quad \text{Thermodynamic equilibrium constant} \quad (6.14)$$

This is an exact and highly important thermodynamic relation, for it enables us to calculate the equilibrium constant of any reaction from tables of thermodynamic data, and hence to predict the equilibrium composition of the reaction mixture.

A brief comment

In Chapter 16 we shall see that the right-hand side of eqn 6.14 may be expressed in terms of spectroscopic data for gas-phase species; so this expression also provides a link between spectroscopy and equilibrium composition.

Example 6.1 *Calculating an equilibrium constant*

Calculate the equilibrium constant for the ammonia synthesis reaction, $N_2(g) + 3\,H_2(g) \rightleftharpoons 2\,NH_3(g)$, at 298 K and show how K is related to the partial pressures of the species at equilibrium when the overall pressure is low enough for the gases to be treated as perfect.

Method Calculate the standard reaction Gibbs energy from eqn 6.11 and convert it to the value of the equilibrium constant by using eqn 6.14. The expression for the equilibrium constant is obtained from eqn 6.13, and because the gases are taken to be perfect, we replace each activity by the ratio $p_J/p^{\ominus}$, where p_J is the partial pressure of species J.

Answer The standard Gibbs energy of the reaction is

$$\begin{aligned}\Delta_r G^{\ominus} &= 2\Delta_f G^{\ominus}(NH_3,g) - \{\Delta_f G^{\ominus}(N_2,g) + 3\Delta_f G^{\ominus}(H_2,g)\}\\ &= 2\Delta_f G^{\ominus}(NH_3,g) = 2 \times (-16.5\ \text{kJ mol}^{-1})\end{aligned}$$

Then,

$$\ln K = -\frac{2 \times (-16.5 \times 10^3\ \text{J mol}^{-1})}{(8.3145\ \text{J K}^{-1}\ \text{mol}^{-1}) \times (298\ \text{K})} = \frac{2 \times 16.5 \times 10^3}{8.3145 \times 298}$$

Hence, $K = 6.1 \times 10^5$. This result is thermodynamically exact. The thermodynamic equilibrium constant for the reaction is

$$K = \frac{a_{NH_3}^2}{a_{N_2} a_{H_2}^3}$$

and this ratio has the value we have just calculated. At low overall pressures, the activities can be replaced by the ratios $p_J/p^{\ominus}$ and an approximate form of the equilibrium constant is

$$K = \frac{(p_{NH_3}/p^{\ominus})^2}{(p_{N_2}/p^{\ominus})(p_{H_2}/p^{\ominus})^3} = \frac{p_{NH_3}^2/p^{\ominus 2}}{p_{N_2} p_{H_2}^3}$$

Self-test 6.1 Evaluate the equilibrium constant for $N_2O_4(g) \rightleftharpoons 2\,NO_2(g)$ at 298 K. [$K = 0.15$]

Example 6.2 *Estimating the degree of dissociation at equilibrium*

The *degree of dissociation* (or *extent of dissociation*, α) is defined as the fraction of reactant that has decomposed; if the initial amount of reactant is n and the amount at equilibrium is n_{eq}, then $\alpha = (n - n_{eq})/n$. The standard reaction Gibbs energy for the decomposition $H_2O(g) \rightarrow H_2(g) + \frac{1}{2}O_2(g)$ is $+118.08$ kJ mol^{-1} at 2300 K. What is the degree of dissociation of H_2O at 2300 K and 1.00 bar?

Method The equilibrium constant is obtained from the standard Gibbs energy of reaction by using eqn 6.14, so the task is to relate the degree of dissociation, α, to K and then to find its numerical value. Proceed by expressing the equilibrium compositions in terms of α, and solve for α in terms of K. Because the standard reaction Gibbs energy is large and positive, we can anticipate that K will be small, and hence that $\alpha \ll 1$, which opens the way to making approximations to obtain its numerical value.

Answer The equilibrium constant is obtained from eqn 6.14 in the form

$$\ln K = -\frac{\Delta_r G^{\ominus}}{RT} = -\frac{(+118.08 \times 10^3\ \mathrm{J\ mol^{-1}})}{(8.3145\ \mathrm{J\ K^{-1}\ mol^{-1}}) \times (2300\ \mathrm{K})}$$
$$= -\frac{118.08 \times 10^3}{8.3145 \times 2300}$$

It follows that $K = 2.08 \times 10^{-3}$. The equilibrium composition can be expressed in terms of α by drawing up the following table:

	H_2O	H_2	O_2	
Initial amount	n	0	0	
Change to reach equilibrium	$-\alpha n$	$+\alpha n$	$+\frac{1}{2}\alpha n$	
Amount at equilibrium	$(1-\alpha)n$	αn	$\frac{1}{2}\alpha n$	Total: $(1+\frac{1}{2}\alpha)n$
Mole fraction, x_J	$\dfrac{1-\alpha}{1+\frac{1}{2}\alpha}$	$\dfrac{\alpha}{1+\frac{1}{2}\alpha}$	$\dfrac{\frac{1}{2}\alpha}{1+\frac{1}{2}\alpha}$	
Partial pressure, p_J	$\dfrac{(1-\alpha)p}{1+\frac{1}{2}\alpha}$	$\dfrac{\alpha p}{1+\frac{1}{2}\alpha}$	$\dfrac{\frac{1}{2}\alpha p}{1+\frac{1}{2}\alpha}$	

where, for the entries in the last row, we have used $p_J = x_J p$ (eqn 1.13). The equilibrium constant is therefore

$$K = \frac{p_{H_2} p_{O_2}^{1/2}}{p_{H_2O}} = \frac{\alpha^{3/2} p^{1/2}}{(1-\alpha)(2+\alpha)^{1/2}}$$

In this expression, we have written p in place of $p/p^{\ominus}$, to simplify its appearance. Now make the approximation that $\alpha \ll 1$, and hence obtain

$$K \approx \frac{\alpha^{3/2} p^{1/2}}{2^{1/2}}$$

Under the stated condition, $p = 1.00$ bar (that is, $p/p^{\ominus} = 1.00$), so $\alpha \approx (2^{1/2}K)^{2/3} = 0.0205$. That is, about 2 per cent of the water has decomposed.

A note on good practice Always check that the approximation is consistent with the final answer. In this case $\alpha \ll 1$ in accord with the original assumption.

Self-test 6.2 Given that the standard Gibbs energy of reaction at 2000 K is +135.2 kJ mol^{-1} for the same reaction, suppose that steam at 200 kPa is passed through a furnace tube at that temperature. Calculate the mole fraction of O_2 present in the output gas stream. [0.00221]

(c) The relation between equilibrium constants

Equilibrium constants in terms of activities are exact, but it is often necessary to relate them to concentrations. Formally, we need to know the activity coefficients, and then to use $a_J = \gamma_J x_J$, $a_J = \gamma_J b_J/b^{\ominus}$, or $a_J = [J]/c^{\ominus}$, where x_J is a mole fraction, b_J is a molality, and [J] is a molar concentration. For example, if we were interested in the composition in terms of molality for an equilibrium of the form $A+B \rightleftharpoons C+D$, where all four species are solutes, we would write

$$K=\frac{a_C a_D}{a_A a_B}=\frac{\gamma_C\gamma_D}{\gamma_A\gamma_B}\times\frac{b_C b_D}{b_A b_B}=K_\gamma K_b \quad (6.15)$$

The activity coefficients must be evaluated at the equilibrium composition of the mixture (for instance, by using one of the Debye–Hückel expressions, Section 5.13b), which may involve a complicated calculation, because the activity coefficients are known only if the equilibrium composition is already known. In elementary applications, and to begin the iterative calculation of the concentrations in a real example, the assumption is often made that the activity coefficients are all so close to unity that $K_\gamma=1$. Then we obtain the result widely used in elementary chemistry that $K\approx K_b$, and equilibria are discussed in terms of molalities (or molar concentrations) themselves.

A special case arises when we need to express the equilibrium constant of a gas-phase reaction in terms of molar concentrations instead of the partial pressures that appear in the thermodynamic equilibrium constant. Provided we can treat the gases as perfect, the p_J that appear in K can be replaced by $[J]RT$, and

$$K=\prod_J (a_J)^{\nu_J}=\prod_J\left(\frac{p_J}{p^\ominus}\right)^{\nu_J}=\prod_J [J]^{\nu_J}\left(\frac{RT}{p^\ominus}\right)^{\nu_J}=\prod_J [J]^{\nu_J}\times\prod_J\left(\frac{RT}{p^\ominus}\right)^{\nu_J}$$

The (dimensionless) equilibrium constant K_c is defined as

$$K_c=\prod_J\left(\frac{[J]}{c^\ominus}\right)^{\nu_J} \quad [6.16]$$

Definition of K_c for gas-phase reactions

It follows that

$$K=K_c\times\prod_J\left(\frac{c^\ominus RT}{p^\ominus}\right)^{\nu_J} \quad (6.17a)$$

If now we write $\Delta\nu=\sum_J \nu_J$, which is easier to think of as $\nu(\text{products})-\nu(\text{reactants})$, then the relation between K and K_c for a gas-phase reaction is

$$K=K_c\times\left(\frac{c^\ominus RT}{p^\ominus}\right)^{\Delta\nu} \quad (6.17b)$$

Relation between K and K_c for gas-phase reactions

The term in parentheses works out as $T/(12.03\ \text{K})$.

- **A brief illustration**

For the reaction $N_2(g)+3\,H_2(g)\rightarrow 2\,NH_3(g)$, $\Delta\nu=2-4=-2$, so

$$K=K_c\times\left(\frac{T}{12.03\ \text{K}}\right)^{-2}=K_c\times\left(\frac{12.03\ \text{K}}{T}\right)^{2}$$

At 298.15 K the relation is

$$K=K_c\times\left(\frac{12.03\ \text{K}}{298.15\ \text{K}}\right)^{2}=\frac{K_c}{614.2}$$

so $K_c=614.2K$. Note that both K and K_c are dimensionless. •

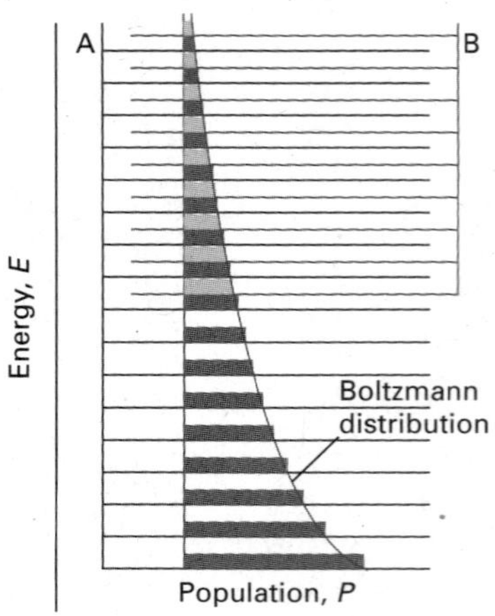

Fig. 6.4 The Boltzmann distribution of populations over the energy levels of two species A and B with similar densities of energy levels; the reaction A → B is endothermic in this example. The bulk of the population is associated with the species A, so that species is dominant at equilibrium.

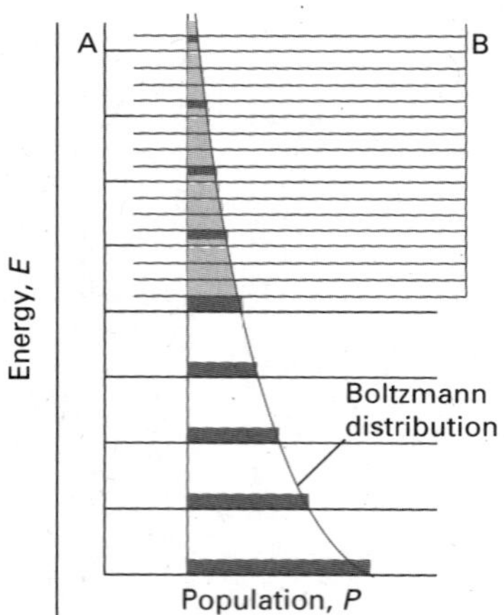

Fig. 6.5 Even though the reaction A → B is endothermic, the density of energy levels in B is so much greater than that in A that the population associated with B is greater than that associated with A, so B is dominant at equilibrium.

(d) Molecular interpretation of the equilibrium constant

We can obtain a deeper insight into the origin and significance of the equilibrium constant by considering the Boltzmann distribution of molecules over the available states of a system composed of reactants and products (*Fundamentals F.5a*). When atoms can exchange partners, as in a reaction, the available states of the system include arrangements in which the atoms are present in the form of reactants and in the form of products: these arrangements have their characteristic sets of energy levels, but the Boltzmann distribution does not distinguish between their identities, only their energies. The atoms distribute themselves over both sets of energy levels in accord with the Boltzmann distribution (Fig. 6.4). At a given temperature, there will be a specific distribution of populations, and hence a specific composition of the reaction mixture.

It can be appreciated from the illustration that, if the reactants and products both have similar arrays of molecular energy levels, then the dominant species in a reaction mixture at equilibrium will be the species with the lower set of energy levels. However, the fact that the Gibbs energy occurs in the expression is a signal that entropy plays a role as well as energy. Its role can be appreciated by referring to Fig. 6.5. We see that, although the B energy levels lie higher than the A energy levels, in this instance they are much more closely spaced. As a result, their total population may be considerable and B could even dominate in the reaction mixture at equilibrium. Closely spaced energy levels correlate with a high entropy (Section 3.2b), so in this case we see that entropy effects dominate adverse energy effects. This competition is mirrored in eqn 6.14, as can be seen most clearly by using $\Delta_r G^\ominus = \Delta_r H^\ominus - T\Delta_r S^\ominus$ and writing it in the form

$$K = e^{-\Delta_r H^\ominus/RT} e^{\Delta_r S^\ominus/R} \tag{6.18}$$

Note that a positive reaction enthalpy results in a lowering of the equilibrium constant (that is, an endothermic reaction can be expected to have an equilibrium composition that favours the reactants). However, if there is positive reaction entropy, then the equilibrium composition may favour products, despite the endothermic character of the reaction.

(e) Equilibria in biological systems

We saw in Section 5.11d that for biological systems it is appropriate to adopt the biological standard state, in which $a_{H^+} = 10^{-7}$ and $pH = -\log a_{H^+} = 7$. The relation between the thermodynamic and biological standard Gibbs energies of reaction for a reaction of the form

$$R + \nu H^+(aq) \rightarrow P \tag{6.19a}$$

can be found by using eqn. 5.63. First, the general expression for the reaction Gibbs energy of this reaction is

$$\Delta_r G = \Delta_r G^\ominus + RT \ln \frac{a_P}{a_R a_{H^+}^{\nu}} = \Delta_r G^\ominus + RT \ln \frac{a_P}{a_R} - \nu RT \ln a_{H^+}$$

In the biological standard state, both P and R are at unit activity. Therefore, by using $\ln x = \ln 10 \log x$, this expression becomes

$$\Delta_r G = \Delta_r G^\ominus - \nu RT \ln 10 \log a_{H^+} = \Delta_r G^\ominus + \nu RT \ln 10 \,\text{pH}$$

For the full specification of the biological state, we set pH = 7, and hence obtain

$$\Delta_r G^{\oplus} = \Delta_r G^\ominus + 7\nu RT \ln 10 \tag{6.19b}$$

Conversion to biological standard value

Note that there is no difference between the two standard values if hydrogen ions are not involved in the reaction ($\nu = 0$).

• A brief illustration

Consider the reaction NADH(aq) + H^+(aq) → NAD^+(aq) + H_2(g) at 37°C, for which $\Delta_r G^{\ominus} = -21.8$ kJ mol^{-1}. It follows that, because $\nu = 1$ and $7 \ln 10 = 16.1$,

$$\begin{aligned}\Delta_r G^{\oplus} &= -21.8 \text{ kJ mol}^{-1} + 16.1 \times (8.3145 \times 10^{-3} \text{ kJ K}^{-1} \text{ mol}^{-1}) \times (310 \text{ K}) \\ &= +19.7 \text{ kJ mol}^{-1}\end{aligned}$$

Note that the biological standard value is opposite in sign (in this example) to the thermodynamic standard value: the much lower concentration of hydronium ions (by seven orders of magnitude) at pH = 7 in place of pH = 0, has resulted in the reverse reaction becoming spontaneous under the new standard conditions. •

Self-test 6.3 For a particular reaction of the form A → B + 2 H^+ in aqueous solution, it was found that $\Delta_r G^{\ominus} = +20$ kJ mol^{-1} at 28°C. Estimate the value of $\Delta_r G^{\oplus}$.

[−61 kJ mol^{-1}]

The response of equilibria to the conditions

Equilibria respond to changes in pressure, temperature, and concentrations of reactants and products. The equilibrium constant for a reaction is not affected by the presence of a catalyst or an enzyme (a biological catalyst). As we shall see in detail in Chapter 22, catalysts increase the rate at which equilibrium is attained but do not affect its position. However, it is important to note that in industry reactions rarely reach equilibrium, partly on account of the rates at which reactants mix.

6.3 How equilibria respond to changes of pressure

Key point **The thermodynamic equilibrium constant is independent of pressure. The response of composition to changes in the conditions is summarized by Le Chatelier's principle.**

The equilibrium constant depends on the value of $\Delta_r G^{\ominus}$, which is defined at a single, standard pressure. The value of $\Delta_r G^{\ominus}$, and hence of K, is therefore independent of the pressure at which the equilibrium is actually established. In other words, at a given temperature K is a constant.

The conclusion that K is independent of pressure does not necessarily mean that the equilibrium composition is independent of the pressure, and its effect depends on how the pressure is applied. The pressure within a reaction vessel can be increased by injecting an inert gas into it. However, so long as the gases are perfect, this addition of gas leaves all the partial pressures of the reacting gases unchanged: the partial pressure of a perfect gas is the pressure it would exert if it were alone in the container, so the presence of another gas has no effect. It follows that pressurization by the addition of an inert gas has no effect on the equilibrium composition of the system (provided the gases are perfect). Alternatively, the pressure of the system may be increased by confining the gases to a smaller volume (that is, by compression). Now the individual partial pressures are changed but their ratio (as it appears in the equilibrium constant) remains the same. Consider, for instance, the perfect gas equilibrium A $\rightleftharpoons$ 2 B, for which the equilibrium constant is

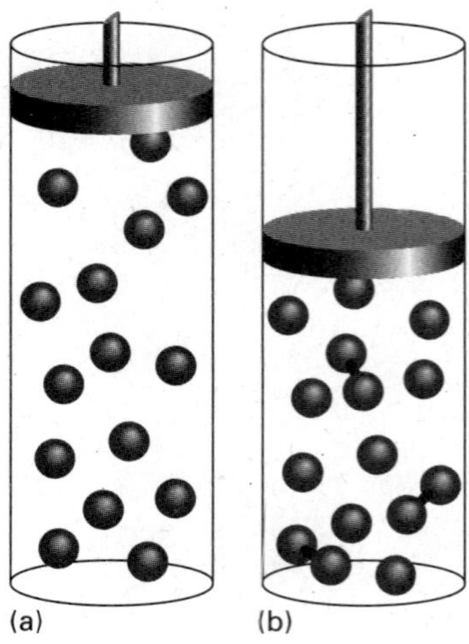

Fig. 6.6 When a reaction at equilibrium is compressed (from *a* to *b*), the reaction responds by reducing the number of molecules in the gas phase (in this case by producing the dimers represented by the linked spheres).

$$K=\frac{p_B^2}{p_A p^{\ominus}}$$

The right-hand side of this expression remains constant only if an increase in p_A cancels an increase in the *square* of p_B. This relatively steep increase of p_A compared to p_B will occur if the equilibrium composition shifts in favour of A at the expense of B. Then the number of A molecules will increase as the volume of the container is decreased and its partial pressure will rise more rapidly than can be ascribed to a simple change in volume alone (Fig. 6.6).

The increase in the number of A molecules and the corresponding decrease in the number of B molecules in the equilibrium A $\rightleftharpoons$ 2 B is a special case of a principle proposed by the French chemist Henri Le Chatelier, which states that:

A system at equilibrium, when subjected to a disturbance, responds in a way that tends to minimize the effect of the disturbance.

Le Chatelier's principle

The principle implies that, if a system at equilibrium is compressed, then the reaction will adjust so as to minimize the increase in pressure. This it can do by reducing the number of particles in the gas phase, which implies a shift A $\leftarrow$ 2 B.

To treat the effect of compression quantitatively, we suppose that there is an amount n of A present initially (and no B). At equilibrium the amount of A is $(1-\alpha)n$ and the amount of B is $2\alpha n$, where α is the degree of dissociation of A into 2B. It follows that the mole fractions present at equilibrium are

$$x_A=\frac{(1-\alpha)n}{(1-\alpha)n+2\alpha n}=\frac{1-\alpha}{1+\alpha} \qquad x_B=\frac{2\alpha}{1+\alpha}$$

The equilibrium constant for the reaction is

$$K=\frac{p_B^2}{p_A p^{\ominus}}=\frac{x_B^2 p^2}{x_A p p^{\ominus}}=\frac{4\alpha^2(p/p^{\ominus})}{1-\alpha^2}$$

which rearranges to

$$\alpha=\left(\frac{1}{1+4p/Kp^{\ominus}}\right)^{1/2} \tag{6.20}$$

This formula shows that, even though K is independent of pressure, the amounts of A and B do depend on pressure (Fig. 6.7). It also shows that, as p is increased, α decreases, in accord with Le Chatelier's principle.

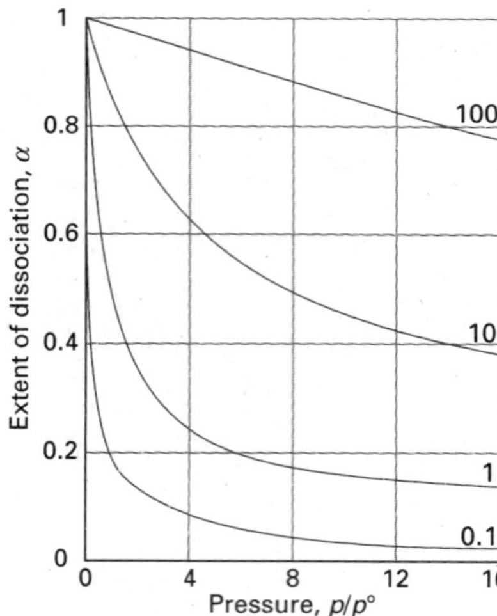

Fig. 6.7 The pressure dependence of the degree of dissociation, α, at equilibrium for an A(g) $\rightleftharpoons$ 2 B(g) reaction for different values of the equilibrium constant K. The value $\alpha=0$ corresponds to pure A; $\alpha=1$ corresponds to pure B.

- **A brief illustration**

To predict the effect of an increase in pressure on the composition of the ammonia synthesis at equilibrium (see Example 6.1), we note that the number of gas molecules decreases (from 4 to 2). So, Le Chatelier's principle predicts that an increase in pressure will favour the product. The equilibrium constant is

$$K=\frac{p_{NH_3}^2 p^{\ominus 2}}{p_{N_2}p_{H_2}^3}=\frac{x_{NH_3}^2 p^2 p^{\ominus 2}}{x_{N_2}x_{H_2}^3 p^4}=\frac{K_x p^{\ominus 2}}{p^2}$$

where K_x is the part of the equilibrium constant expression that contains the equilibrium mole fractions of reactants and products (note that, unlike K itself, K_x is not an equilibrium constant). Therefore, doubling the pressure must increase K_x by a factor of 4 to preserve the value of K. ●

Self-test 6.4 Predict the effect of a tenfold pressure increase on the equilibrium composition of the reaction $3\,N_2(g) + H_2(g) \rightarrow 2\,N_3H(g)$.

[100-fold increase in K_x]

6.4 The response of equilibria to changes of temperature

Key points **(a) The dependence of the equilibrium constant on the temperature is expressed by the van't Hoff equation and can be explained in terms of the distribution of molecules over the available states. (b) Integration of the van't Hoff equation gives an expression that relates the equilibrium constant to temperature.**

Le Chatelier's principle predicts that a system at equilibrium will tend to shift in the endothermic direction if the temperature is raised, for then energy is absorbed as heat and the rise in temperature is opposed. Conversely, an equilibrium can be expected to shift in the exothermic direction if the temperature is lowered, for then energy is released and the reduction in temperature is opposed. These conclusions can be summarized as follows:

Exothermic reactions: increased temperature favours the reactants.
Endothermic reactions: increased temperature favours the products.

We shall now justify these remarks and see how to express the changes quantitatively.

(a) The van 't Hoff equation

The **van't Hoff equation**, which is derived in the *Justification* below, is an expression for the slope of a plot of the equilibrium constant (specifically, ln K) as a function of temperature. It may be expressed in either of two ways:

$$\text{(a)}\quad \frac{\mathrm{d}\ln K}{\mathrm{d}T} = \frac{\Delta_r H^{\ominus}}{RT^2} \qquad \text{(b)}\quad \frac{\mathrm{d}\ln K}{\mathrm{d}(1/T)} = -\frac{\Delta_r H^{\ominus}}{R} \qquad \boxed{\text{van't Hoff equation}} \qquad (6.21)$$

Justification 6.2 *The van't Hoff equation*

From eqn 6.14, we know that

$$\ln K = -\frac{\Delta_r G^{\ominus}}{RT}$$

Differentiation of ln K with respect to temperature then gives

$$\frac{\mathrm{d}\ln K}{\mathrm{d}T} = -\frac{1}{R}\frac{\mathrm{d}(\Delta_r G^{\ominus}/T)}{\mathrm{d}T}$$

The differentials are complete (that is, they are not partial derivatives) because K and $\Delta_r G^{\ominus}$ depend only on temperature, not on pressure. To develop this equation we use the Gibbs–Helmholtz equation (eqn 3.56) in the form

$$\frac{\mathrm{d}(\Delta_r G^{\ominus}/T)}{\mathrm{d}T} = -\frac{\Delta_r H^{\ominus}}{T^2}$$

where $\Delta_r H^{\ominus}$ is the standard reaction enthalpy at the temperature T. Combining the two equations gives the van't Hoff equation, eqn 6.21a. The second form of the equation is obtained by noting that

$$\frac{d(1/T)}{dT} = -\frac{1}{T^2}, \quad \text{so} \quad dT = -T^2\, d(1/T)$$

It follows that eqn 6.21a can be rewritten as

$$-\frac{d \ln K}{T^2 d(1/T)} = \frac{\Delta_r H^\ominus}{RT^2}$$

which simplifies into eqn 6.21b.

Equation 6.21a shows that $d \ln K/dT < 0$ (and therefore that $dK/dT < 0$) for a reaction that is exothermic under standard conditions ($\Delta_r H^\ominus < 0$). A negative slope means that $\ln K$, and therefore K itself, decreases as the temperature rises. Therefore, as asserted above, in the case of an exothermic reaction the equilibrium shifts away from products. The opposite occurs in the case of endothermic reactions.

Insight into the thermodynamic basis of this behaviour comes from the expression $\Delta_r G^\ominus = \Delta_r H^\ominus - T\Delta_r S^\ominus$ written in the form $-\Delta_r G^\ominus/T = -\Delta_r H^\ominus/T + \Delta_r S^\ominus$. When the reaction is exothermic, $-\Delta_r H^\ominus/T$ corresponds to a positive change of entropy of the surroundings and favours the formation of products. When the temperature is raised, $-\Delta_r H^\ominus/T$ decreases, and the increasing entropy of the surroundings has a less important role. As a result, the equilibrium lies less to the right. When the reaction is endothermic, the principal factor is the increasing entropy of the reaction system. The importance of the unfavourable change of entropy of the surroundings is reduced if the temperature is raised (because then $-\Delta_r H^\ominus/T$ is smaller), and the reaction is able to shift towards products.

These remarks have a molecular basis that stems from the Boltzmann distribution of molecules over the available energy levels (*Fundamentals F.5a*). The typical arrangement of energy levels for an endothermic reaction is shown in Fig. 6.8a. When the temperature is increased, the Boltzmann distribution adjusts and the populations change as shown. The change corresponds to an increased population of the higher energy states at the expense of the population of the lower energy states. We see that the states that arise from the B molecules become more populated at the expense of the A molecules. Therefore, the total population of B states increases, and B becomes more abundant in the equilibrium mixture. Conversely, if the reaction is exothermic (Fig. 6.8b), then an increase in temperature increases the population of the A states (which start at higher energy) at the expense of the B states, so the reactants become more abundant.

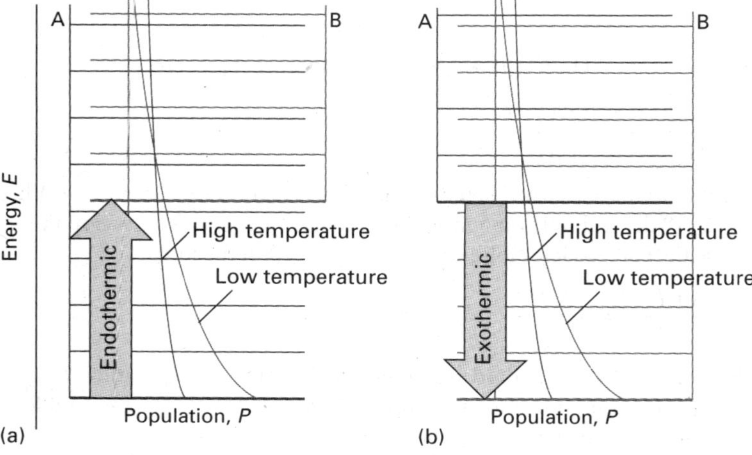

Fig. 6.8 The effect of temperature on a chemical equilibrium can be interpreted in terms of the change in the Boltzmann distribution with temperature and the effect of that change in the population of the species. (a) In an endothermic reaction, the population of B increases at the expense of A as the temperature is raised. (b) In an exothermic reaction, the opposite happens.

Example 6.3 *Measuring a reaction enthalpy*

The data below show the temperature variation of the equilibrium constant of the reaction $Ag_2CO_3(s) \rightleftharpoons Ag_2O(s) + CO_2(g)$. Calculate the standard reaction enthalpy of the decomposition.

T/K	350	400	450	500
K	3.98×10^{-4}	1.41×10^{-2}	1.86×10^{-1}	1.48

Method It follows from eqn 6.21b that, provided the reaction enthalpy can be assumed to be independent of temperature, a plot of $-\ln K$ against $1/T$ should be a straight line of slope $\Delta_r H^\ominus/R$.

Answer We draw up the following table:

T/K	350	400	450	500
$(10^3\ \text{K})/T$	2.86	2.50	2.22	2.00
$-\ln K$	7.83	4.26	1.68	-0.39

These points are plotted in Fig. 6.9. The slope of the graph is $+9.6\times10^3$, so

$$\Delta_r H^\ominus = (+9.6\times10^3\ \text{K})\times R = +80\ \text{kJ mol}^{-1}$$

Self-test 6.5 The equilibrium constant of the reaction $2\,SO_2(g) + O_2(g) \rightleftharpoons 2\,SO_3(g)$ is 4.0×10^{24} at 300 K, 2.5×10^{10} at 500 K, and 3.0×10^4 at 700 K. Estimate the reaction enthalpy at 500 K. [-200 kJ mol^{-1}]

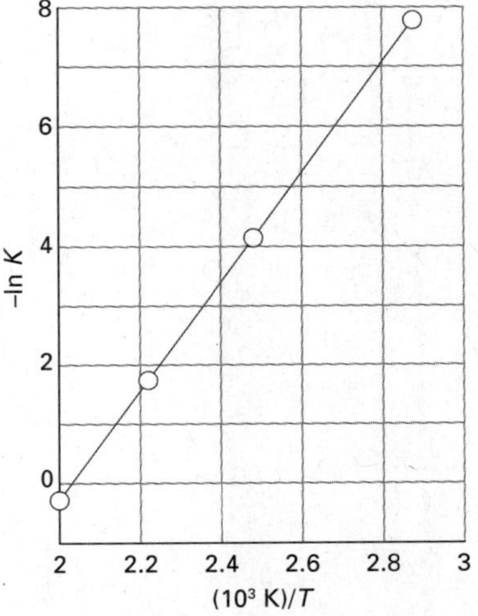

Fig. 6.9 When $-\ln K$ is plotted against $1/T$, a straight line is expected with slope equal to $\Delta_r H^\ominus/R$ if the standard reaction enthalpy does not vary appreciably with temperature. This is a non-calorimetric method for the measurement of reaction enthalpies.

The temperature dependence of the equilibrium constant provides a non-calorimetric method of determining $\Delta_r H^\ominus$. A drawback is that the reaction enthalpy is actually temperature-dependent, so the plot is not expected to be perfectly linear. However, the temperature dependence is weak in many cases, so the plot is reasonably straight. In practice, the method is not very accurate, but it is often the only method available.

(b) The value of *K* at different temperatures

To find the value of the equilibrium constant at a temperature T_2 in terms of its value K_1 at another temperature T_1, we integrate eqn 6.21b between these two temperatures:

$$\ln K_2 - \ln K_1 = -\frac{1}{R}\int_{1/T_1}^{1/T_2} \Delta_r H^\ominus \mathrm{d}(1/T) \qquad (6.22)$$

If we suppose that $\Delta_r H^\ominus$ varies only slightly with temperature over the temperature range of interest, then we may take it outside the integral. It follows that

$$\ln K_2 - \ln K_1 = -\frac{\Delta_r H^\ominus}{R}\left(\frac{1}{T_2} - \frac{1}{T_1}\right) \qquad \text{Temperature dependence of } K \qquad (6.23)$$

- **A brief illustration**

To estimate the equilibrium constant for the synthesis of ammonia at 500 K from its value at 298 K (6.1×10^5 for the reaction as written in Example 6.1) we use the standard reaction enthalpy, which can be obtained from Table 2.8 in the *Data section* by using $\Delta_r H^\ominus = 2\Delta_f H^\ominus(NH_3,g)$, and assume that its value is constant over the range of temperatures. Then, with $\Delta_r H^\ominus = -92.2$ kJ mol^{-1}, from eqn 6.23 we find

$$\ln K_2 = \ln(6.1\times10^5) - \frac{(-92.2\times10^3\ \text{J mol}^{-1})}{8.3145\ \text{J K}^{-1}\ \text{mol}^{-1}}\left(\frac{1}{500\ \text{K}} - \frac{1}{298\ \text{K}}\right)$$
$$= -1.71$$

It follows that $K_2 = 0.18$, a lower value than at 298 K, as expected for this exothermic reaction. ●

Self-test 6.6 The equilibrium constant for $N_2O_4(g) \rightleftharpoons 2\ NO_2(g)$ was calculated in Self-test 6.1. Estimate its value at 100°C. [15]

IMPACT ON TECHNOLOGY

I6.2 Supramolecular chemistry

There is currently considerable interest in assemblies of molecules that are too small to be regarded as bulk matter yet too large to be regarded as individual molecules: this is the domain of supramolecular chemistry. Supramolecular surfactant assemblies and macromolecules are common host systems employed to solubilize guest molecules by taking advantage of intermolecular forces. Numerous applications use these organized media to encapsulate small molecules to create a host–guest system in which the new microenvironment for the guest substantially modifies its properties. Cyclodextrins, for example, are ring-like oligomers composed of glucopyranose units. A cyclodextrin molecule has a hydrophilic exterior and a hydrophobic interior that readily forms inclusion complexes with nonpolar guest molecules. Solubilization of the guest in the cyclodextrin core is governed by a temperature-dependent equilibrium constant that can be studied by making a van't Hoff plot to determine the thermodynamic properties of the complex formation process.

The guest molecule often possesses spectroscopic properties, such as its absorption and fluorescence wavelengths and intensities (Chapter 13), that enable the extent of encapsulation to be measured. For example, for reasons explained in Section 13.4b, the fluorescence wavelength and intensity of the chalcone pigment DMATP (4) are highly sensitive to the polarity of the pigment's environment. The emission spectrum in water is centred on 559 nm, but as the pigment is incorporated in the hydrophobic interior of β-cyclodextrin (the macromolecule formed with seven glucopyranone units) the emission shifts to 543 nm (Fig. 6.10). Incorporation of the pigment in the macromolecule also significantly enhances the DMATP fluorescence emission intensity. The equilibrium constant for the formation of the 1:1 inclusion complex consisting of one DMATP molecule in one β-cyclodextrin cavity can be calculated from a plot of $1/(I_f - I_f^\circ)$ against $1/[\text{CD}]$:

$$\frac{1}{I_f - I_f^\circ} = \frac{1}{I_f^\infty - I_f^\circ} + \frac{1}{(I_f^\infty - I_f^\circ)K_{eq}[\text{CD}]}$$

Benesi–Hildebrand equation

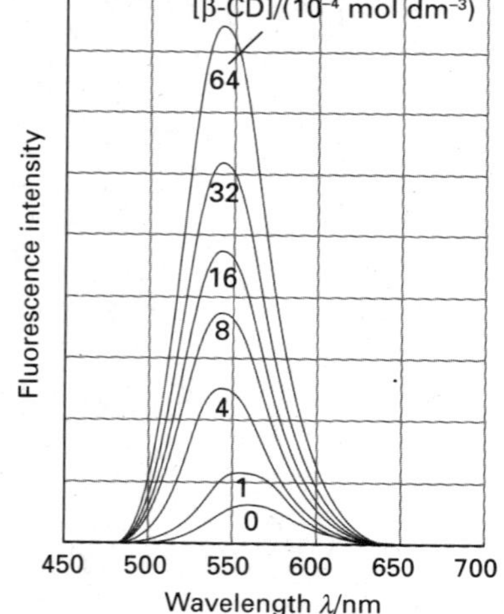

Fig. 6.10 The variation of fluorescence intensity (arbitrary units) of DMATP at 10 μmol dm^{-3} with concentration of β-cyclodextrin. (M. Gaber, T. A. Fayed, S. A. El-Daly, and Y. S. El-Sayed, *Photochem. Photobiol. Sci.*, 2008, 7, 257.)

4 DMATP

I_f is the fluorescence intensity at 543 nm at a given cyclodextrin concentration [CD], I_f^o is the fluorescence intensity at this wavelength in the absence of host macromolecule, and I_f^∞ is the fluorescence emission intensity when all DMATP molecules at a fixed concentration are complexed within hosts.

For DMATP/β-CD (Fig. 6.11) the equilibrium constant falls as temperature is raised, with $K_{eq} = 682, 326, 170$, and 59 at 25, 35, 45, and 55°C, respectively. The corresponding van't Hoff plot of $\ln K_{eq}$ against $1/T$ yields a straight line (Fig. 6.12) from which it can be inferred that the standard enthalpy and entropy of formation of the complex are -64.7 kJ mol^{-1} and -162.3 J K^{-1} mol^{-1}, respectively. The highly exothermic complexation process is consistent with the affinity of the hydrophobic DMATP molecule for the cyclodextrin cavity. The overall negative entropy change upon encapsulation of the guest molecule is expected as a consequence of the restricted motion of the guest within the host cavity. The expulsion of water molecules from the cyclodextrin cavity as DMATP is entrapped gives rise to a positive contribution to the entropy of the water molecules, but the magnitude of this change is significantly less than that of the decrease in entropy of the DMATP guest. Nevertheless, the overall entropy change for the formation of the 1:1 inclusion complex is more negative than often observed for cyclodextrin systems, suggesting that the CD host also experiences restricted motion upon complex formation. Thus, the van't Hoff analysis of the complex formation not only yields the typical thermodynamics parameters for the process but also provides insights into the process on a molecular level.

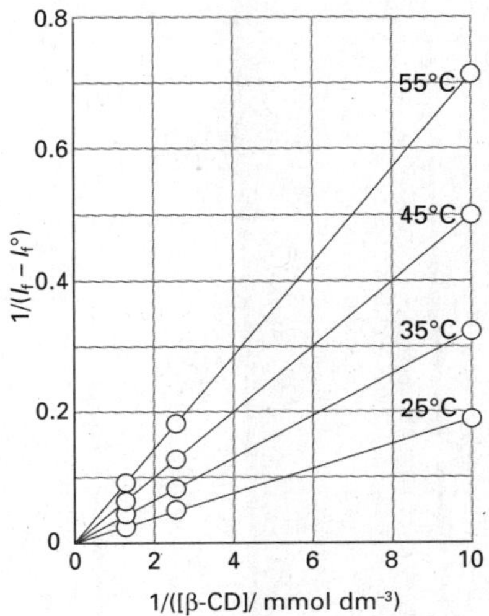

Fig. 6.11 Benesi–Hildebrand plots of the fluorescence intensity from the DMATP/β-CD inclusion complex at 543 nm at various temperatures as a function of β-cyclodextrin concentration (based on the reference for Fig. 6.10).

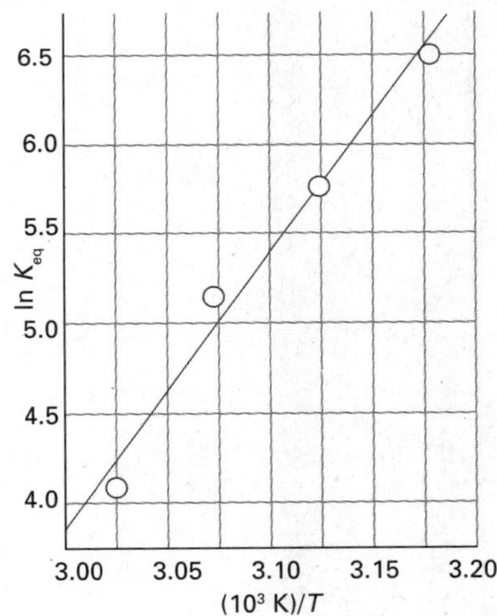

Fig. 6.12 van't Hoff plots for the DMATP/β-CD equilibrium (based on the reference for Fig. 6.10).

Equilibrium electrochemistry

The discussion has been general and applies to all reactions. One very special case that has enormous fundamental, technological, and economic significance concerns reactions that take place in electrochemical cells. Moreover, the ability to make very precise measurements of potential differences ('voltages') means that electrochemical methods can be used to determine thermodynamic properties of reactions that may be inaccessible by other methods.

An **electrochemical cell** consists of two **electrodes**, or metallic conductors, in contact with an **electrolyte**, an ionic conductor (which may be a solution, a liquid, or a solid). An electrode and its electrolyte comprise **an electrode compartment**. The two electrodes may share the same compartment. The various kinds of electrode are summarized in Table 6.1. Any 'inert metal' shown as part of the specification is present to act as a source or sink of electrons, but takes no other part in the reaction other than acting as a catalyst for it. If the electrolytes are different, the two compartments may be joined by a **salt bridge**, which is a tube containing a concentrated electrolyte solution (for instance, potassium chloride in agar jelly) that completes the electrical

Table 6.1 Varieties of electrode

Electrode type	Designation	Redox couple	Half-reaction
Metal/metal ion	$M(s)\vert M^+(aq)$	M^+/M	$M^+(aq) + e^- \rightarrow M(s)$
Gas	$Pt(s)\vert X_2(g)\vert X^+(aq)$	X^+/X_2	$X^+(aq) + e^- \rightarrow \frac{1}{2}X_2(g)$
	$Pt(s)\vert X_2(g)\vert X^-(aq)$	X_2/X^-	$\frac{1}{2}X_2(g) + e^- \rightarrow X^-(aq)$
Metal/insoluble salt	$M(s)\vert MX(s)\vert X^-(aq)$	$MX/M,X^-$	$MX(s) + e^- \rightarrow M(s) + X^-(aq)$
Redox	$Pt(s)\vert M^+(aq),M^{2+}(aq)$	M^{2+}/M^+	$M^{2+}(aq) + e^- \rightarrow M^+(aq)$

circuit and enables the cell to function. A **galvanic cell** is an electrochemical cell that produces electricity as a result of the spontaneous reaction occurring inside it. An **electrolytic cell** is an electrochemical cell in which a non-spontaneous reaction is driven by an external source of current.

6.5 Half-reactions and electrodes

Key point **A redox reaction is expressed as the difference of two reduction half-reactions; each one defines a redox couple.**

It will be familiar from introductory chemistry courses that **oxidation** is the removal of electrons from a species, a **reduction** is the addition of electrons to a species, and a **redox reaction** is a reaction in which there is a transfer of electrons from one species to another. The electron transfer may be accompanied by other events, such as atom or ion transfer, but the net effect is electron transfer and hence a change in oxidation number of an element. The **reducing agent** (or *reductant*) is the electron donor; the **oxidizing agent** (or *oxidant*) is the electron acceptor. It should also be familiar that any redox reaction may be expressed as the difference of two reduction **half-reactions**, which are conceptual reactions showing the gain of electrons. Even reactions that are not redox reactions may often be expressed as the difference of two reduction half-reactions. The reduced and oxidized species in a half-reaction form a **redox couple**. In general we write a couple as Ox/Red and the corresponding reduction half-reaction as

$$\mathrm{Ox} + \nu\,\mathrm{e}^- \rightarrow \mathrm{Red} \tag{6.24}$$

• A brief illustration

The dissolution of silver chloride in water $AgCl(s) \rightarrow Ag^+(aq) + Cl^-(aq)$, which is not a redox reaction, can be expressed as the difference of the following two reduction half-reactions:

$$AgCl(s) + e^- \rightarrow Ag(s) + Cl^-(aq)$$

$$Ag^+(aq) + e^- \rightarrow Ag(s)$$

The redox couples are AgCl/Ag, Cl^- and Ag^+/Ag, respectively. •

Self-test 6.7 Express the formation of H_2O from H_2 and O_2 in acidic solution (a redox reaction) as the difference of two reduction half-reactions.

[$4\,H^+(aq) + 4\,e^- \rightarrow 2\,H_2(g)$, $O_2(g) + 4\,H^+(aq) + 4\,e^- \rightarrow 2\,H_2O(l)$]

We shall often find it useful to express the composition of an electrode compartment in terms of the reaction quotient, Q, for the half-reaction. This quotient is defined like the reaction quotient for the overall reaction, but the electrons are ignored because they are stateless.

• A brief illustration

The reaction quotient for the reduction of O_2 to H_2O in acid solution, $O_2(g) + 4\,H^+(aq) + 4\,e^- \rightarrow 2\,H_2O(l)$, is

$$Q = \frac{a_{H_2O}^2}{a_{H^+}^4 a_{O_2}} \approx \frac{p^\ominus}{a_{H^+}^4 p_{O_2}}$$

The approximations used in the second step are that the activity of water is 1 (because the solution is dilute) and the oxygen behaves as a perfect gas, so $a_{O_2} \approx p_{O_2}/p^\ominus$. •

Self-test 6.8 Write the half-reaction and the reaction quotient for a chlorine gas electrode. $[Cl_2(g) + 2\,e^- \rightarrow 2\,Cl^-(aq),\ Q \approx a_{Cl^-}^2 p^{\ominus}/p_{Cl_2}]$

The reduction and oxidation processes responsible for the overall reaction in a cell are separated in space: oxidation takes place at one electrode and reduction takes place at the other. As the reaction proceeds, the electrons released in the oxidation $Red_1 \rightarrow Ox_1 + \nu\,e^-$ at one electrode travel through the external circuit and re-enter the cell through the other electrode. There they bring about reduction $Ox_2 + \nu\,e^- \rightarrow Red_2$. The electrode at which oxidation occurs is called the **anode**; the electrode at which reduction occurs is called the **cathode**. In a galvanic cell, the cathode has a higher potential than the anode: the species undergoing reduction, Ox_2, withdraws electrons from its electrode (the cathode, Fig. 6.13), so leaving a relative positive charge on it (corresponding to a high potential). At the anode, oxidation results in the transfer of electrons to the electrode, so giving it a relative negative charge (corresponding to a low potential).

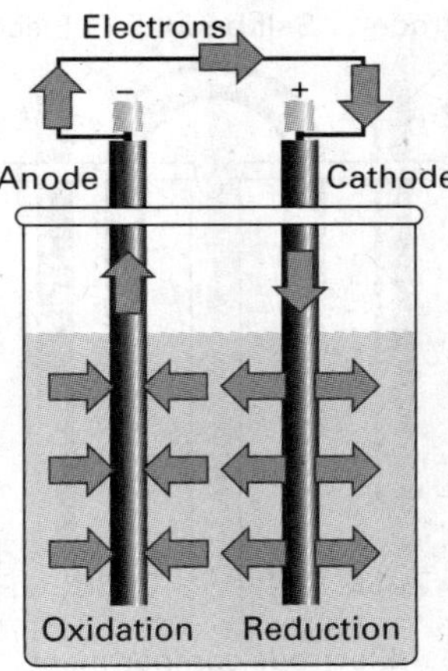

Fig. 6.13 When a spontaneous reaction takes place in a galvanic cell, electrons are deposited in one electrode (the site of oxidation, the anode) and collected from another (the site of reduction, the cathode), and so there is a net flow of current, which can be used to do work. Note that the + sign of the cathode can be interpreted as indicating the electrode at which electrons enter the cell, and the − sign of the anode as where the electrons leave the cell.

6.6 Varieties of cells

Key points Galvanic cells are classified as electrolyte concentration and electrode concentration cells. (a) A liquid junction potential arises at the junction of two electrolyte solutions. (b) The cell notation specifies the structure of a cell.

The simplest type of cell has a single electrolyte common to both electrodes (as in Fig. 6.13). In some cases it is necessary to immerse the electrodes in different electrolytes, as in the 'Daniell cell' in which the redox couple at one electrode is Cu^{2+}/Cu and at the other is Zn^{2+}/Zn (Fig. 6.14). In an **electrolyte concentration cell**, the electrode compartments are identical except for the concentrations of the electrolytes. In an **electrode concentration cell** the electrodes themselves have different concentrations, either because they are gas electrodes operating at different pressures or because they are amalgams (solutions in mercury) with different concentrations.

(a) Liquid junction potentials

In a cell with two different electrolyte solutions in contact, as in the Daniell cell, there is an additional source of potential difference across the interface of the two electrolytes. This potential is called the **liquid junction potential**, E_{lj}. Another example of a junction potential is that between different concentrations of hydrochloric acid. At the junction, the mobile H^+ ions diffuse into the more dilute solution. The bulkier Cl^- ions follow, but initially do so more slowly, which results in a potential difference at the junction. The potential then settles down to a value such that, after that brief initial period, the ions diffuse at the same rates. Electrolyte concentration cells always have a liquid junction; electrode concentration cells do not.

The contribution of the liquid junction to the potential can be reduced (to about 1 to 2 mV) by joining the electrolyte compartments through a salt bridge (Fig. 6.15). The reason for the success of the salt bridge is that, provided the ions dissolved in the jelly have similar mobilities, then the liquid junction potentials at either end are largely independent of the concentrations of the two dilute solutions, and so nearly cancel.

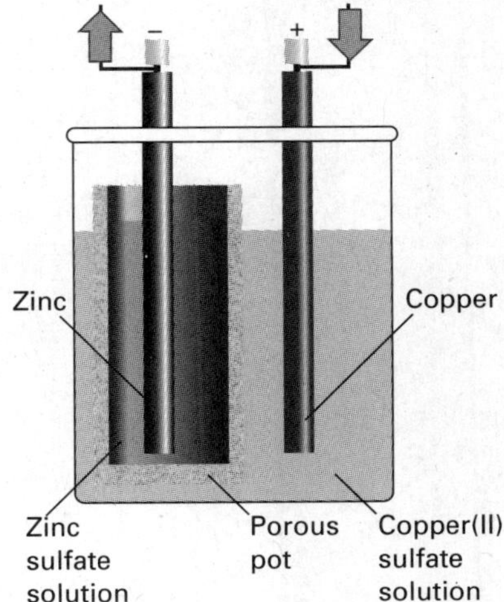

Fig. 6.14 One version of the Daniell cell. The copper electrode is the cathode and the zinc electrode is the anode. Electrons leave the cell from the zinc electrode and enter it again through the copper electrode.

(b) Notation

In the notation for cells, phase boundaries are denoted by a vertical bar. For example,

$$Pt(s)|H_2(g)|HCl(aq)|AgCl(s)|Ag(s)$$

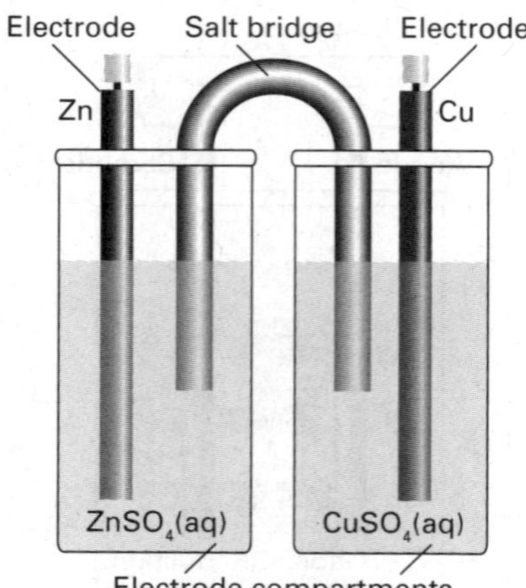

Fig. 6.15 The salt bridge, essentially an inverted U-tube full of concentrated salt solution in a jelly, has two opposing liquid junction potentials that almost cancel.

A liquid junction is denoted by ⋮, so the cell in Fig. 6.14 is denoted

$$Zn(s)|ZnSO_4(aq)\vdots CuSO_4(aq)|Cu(s)$$

A double vertical line, ||, denotes an interface for which it is assumed that the junction potential has been eliminated. Thus the cell in Fig. 6.15 is denoted

$$Zn(s)|ZnSO_4(aq)||CuSO_4(aq)|Cu(s)$$

An example of an electrolyte concentration cell in which the liquid junction potential is assumed to be eliminated is

$$Pt(s)|H_2(g)|HCl(aq,b_1)||HCl(aq,b_2)|H_2(g)|Pt(s)$$

6.7 The cell potential

Key points **(a) The Nernst equation relates the cell potential to the composition of the reaction mixture. (b) The standard cell potential may be used to calculate the equilibrium constant of the cell reaction.**

The current produced by a galvanic cell arises from the spontaneous chemical reaction taking place inside it. The **cell reaction** is the reaction in the cell written on the assumption that the right-hand electrode is the cathode, and hence that the spontaneous reaction is one in which reduction is taking place in the right-hand compartment. Later we see how to predict if the right-hand electrode is in fact the cathode; if it is, then the cell reaction is spontaneous as written. If the left-hand electrode turns out to be the cathode, then the reverse of the corresponding cell reaction is spontaneous.

To write the cell reaction corresponding to a cell diagram, we first write the right-hand half-reaction as a reduction (because we have assumed that to be spontaneous). Then we subtract from it the left-hand reduction half-reaction (for, by implication, that electrode is the site of oxidation). Thus, in the cell $Zn(s)|ZnSO_4(aq)||CuSO_4(aq)|Cu(s)$ the two electrodes and their reduction half-reactions are

Right-hand electrode: $Cu^{2+}(aq) + 2\,e^- \rightarrow Cu(s)$

Left-hand electrode: $Zn^{2+}(aq) + 2\,e^- \rightarrow Zn(s)$

Hence, the overall cell reaction is the difference:

$$Cu^{2+}(aq) + Zn(s) \rightarrow Cu(s) + Zn^{2+}(aq)$$

(a) The Nernst equation

A cell in which the overall cell reaction has not reached chemical equilibrium can do electrical work as the reaction drives electrons through an external circuit. The work that a given transfer of electrons can accomplish depends on the potential difference between the two electrodes. When the potential difference is large, a given number of electrons travelling between the electrodes can do a large amount of electrical work. When the potential difference is small, the same number of electrons can do only a small amount of work. A cell in which the overall reaction is at equilibrium can do no work, and then the potential difference is zero.

According to the discussion in Section 3.5e, we know that the maximum non-expansion work a system can do is given by eqn 3.41b ($w_{add,max} = \Delta G$). In electrochemistry, the non-expansion work is identified with electrical work, the system is the cell, and ΔG is the Gibbs energy of the cell reaction, $\Delta_r G$. Maximum work is produced when a change occurs reversibly. It follows that, to draw thermodynamic conclusions from measurements of the work that a cell can do, we must ensure that the cell is operating reversibly. Moreover, we saw in Section 6.1a that the reaction Gibbs energy is actually a property relating to a specified composition of the reaction mixture. Therefore, to make use of $\Delta_r G$ we must ensure that the cell is operating reversibly at a specific, constant composition. Both these conditions are achieved by measuring the cell potential when it is balanced by an exactly opposing source of potential so that the cell reaction occurs reversibly, the composition is constant, and no current flows: in effect, the cell reaction is poised for change, but not actually changing. The resulting potential difference is called the **cell potential,** E_{cell}, of the cell.

A note on good practice The cell potential was formerly, and is still widely, called the *electromotive force* (emf) of the cell. IUPAC prefers the term 'cell potential' because a potential difference is not a force.

As we show in the *Justification* below, the relation between the reaction Gibbs energy and the cell potential is

$$-\nu F E_{cell} = \Delta_r G \qquad \text{The cell potential} \qquad (6.25)$$

where F is Faraday's constant, $F = eN_A$, and ν is the stoichiometric coefficient of the electrons in the half-reactions into which the cell reaction can be divided. This equation is the key connection between electrical measurements on the one hand and thermodynamic properties on the other. It will be the basis of all that follows.

Justification 6.3 *The relation between the cell potential and the reaction Gibbs energy*

We consider the change in G when the cell reaction advances by an infinitesimal amount $d\xi$ at some composition. From *Justification 6.1* we can write (at constant temperature and pressure)

$$dG = \Delta_r G d\xi$$

The maximum non-expansion (electrical) work that the reaction can do as it advances by $d\xi$ at constant temperature and pressure is therefore

$$dw_e = \Delta_r G d\xi$$

This work is infinitesimal, and the composition of the system is virtually constant when it occurs.

Suppose that the reaction advances by $d\xi$; then $\nu d\xi$ electrons must travel from the anode to the cathode. The total charge transported between the electrodes when this change occurs is $-\nu e N_A d\xi$ (because $\nu d\xi$ is the amount of electrons and the charge per mole of electrons is $-eN_A$). Hence, the total charge transported is $-\nu F d\xi$ because $eN_A = F$. The work done when an infinitesimal charge $-\nu F d\xi$ travels from the anode to the cathode is equal to the product of the charge and the potential difference E_{cell} (see Table 2.1):

$$dw_e = -\nu F E_{cell} d\xi$$

When we equate this relation to the one above ($dw_e = \Delta_r G d\xi$), the advancement $d\xi$ cancels, and we obtain eqn 6.25.

It follows from eqn 6.25 that, by knowing the reaction Gibbs energy at a specified composition, we can state the cell potential at that composition. Note that a negative reaction Gibbs energy, corresponding to a spontaneous cell reaction, corresponds to a positive cell potential. Another way of looking at the content of eqn 6.25 is that it shows that the driving power of a cell (that is, its potential) is proportional to the slope of the Gibbs energy with respect to the extent of reaction. It is plausible that a reaction that is far from equilibrium (when the slope is steep) has a strong tendency to drive electrons through an external circuit (Fig. 6.16). When the slope is close to zero (when the cell reaction is close to equilibrium), the cell potential is small.

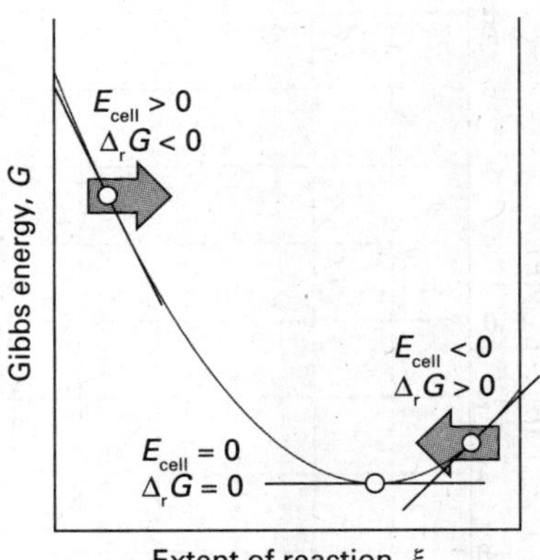

Fig. 6.16 A spontaneous reaction occurs in the direction of decreasing Gibbs energy and can be expressed in terms of the cell potential, E_{cell}. The reaction is spontaneous as written (from left to right on the illustration) when $E_{cell} > 0$. The reverse reaction is spontaneous when $E_{cell} < 0$. When the cell reaction is at equilibrium, the cell potential is zero.

- **A brief illustration**

Equation 6.25 provides an electrical method for measuring a reaction Gibbs energy at any composition of the reaction mixture: we simply measure the cell potential and convert it to $\Delta_r G$. Conversely, if we know the value of $\Delta_r G$ at a particular composition, then we can predict the cell potential. For example, if $\Delta_r G = -1 \times 10^2$ kJ mol^{-1} and $\nu = 1$, then

$$E_{\text{cell}} = -\frac{\Delta_r G}{\nu F} = -\frac{(-1 \times 10^5 \text{ J mol})}{1 \times (9.6485 \times 10^4 \text{ C mol}^{-1})} = 1 \text{ V}$$

where we have used 1 J = 1 C V. •

We can go on to relate the cell potential to the activities of the participants in the cell reaction. We know that the reaction Gibbs energy is related to the composition of the reaction mixture by eqn 6.10 (($\Delta_r G = \Delta_r G^{\ominus} + RT \ln Q$)); it follows, on division of both sides by $-\nu F$, that

$$E_{\text{cell}} = -\frac{\Delta_r G^{\ominus}}{\nu F} - \frac{RT}{\nu F} \ln Q$$

The first term on the right is written

$$E^{\ominus}_{\text{cell}} = -\frac{\Delta_r G^{\ominus}}{\nu F} \qquad \text{Definition of standard cell potential} \qquad [6.26]$$

and called the **standard cell potential**. That is, the standard cell potential is the standard reaction Gibbs energy expressed as a potential difference (in volts). It follows that

$$E_{\text{cell}} = E^{\ominus}_{\text{cell}} - \frac{RT}{\nu F} \ln Q \qquad \text{Nernst equation} \qquad (6.27)$$

This equation for the cell potential in terms of the composition is called the **Nernst equation**; the dependence that it predicts is summarized in Fig. 6.17. One important application of the Nernst equation is to the determination of the pH of a solution and, with a suitable choice of electrodes, of the concentration of other ions (*Impact I6.3*).

We see from eqn 6.27 that the standard cell potential (which will shortly move to centre stage of the exposition) can be interpreted as the cell potential when all the reactants and products in the cell reaction are in their standard states, for then all activities are 1, so $Q = 1$ and $\ln Q = 0$. However, the fact that the standard cell potential is merely a disguised form of the standard reaction Gibbs energy (eqn 6.26) should always be kept in mind and underlies all its applications.

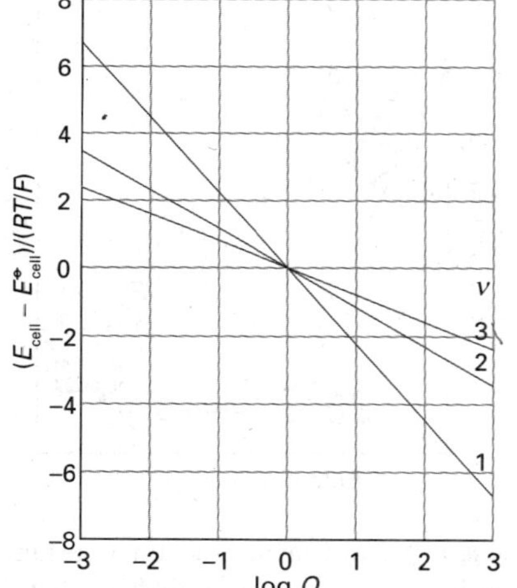

Fig. 6.17 The variation of cell potential with the value of the reaction quotient for the cell reaction for different values of ν (the number of electrons transferred). At 298 K, $RT/F = 25.69$ mV, so the vertical scale refers to multiples of this value.

- **A brief illustration**

Because $RT/F = 25.7$ mV at 25°C, a practical form of the Nernst equation is

$$E_{\text{cell}} = E^{\ominus}_{\text{cell}} - \frac{25.7 \text{ mV}}{\nu} \ln Q$$

It then follows that, for a reaction in which $\nu = 1$, if Q is increased by a factor of 10, then the cell potential decreases by 59.2 mV. •

(b) Cells at equilibrium

A special case of the Nernst equation has great importance in electrochemistry and provides a link to the earlier part of the chapter. Suppose the reaction has reached equilibrium; then $Q = K$, where K is the equilibrium constant of the cell reaction. However, a chemical reaction at equilibrium cannot do work, and hence it generates zero potential difference between the electrodes of a galvanic cell. Therefore, setting $E_{cell} = 0$ and $Q = K$ in the Nernst equation gives

$$\ln K = \frac{\nu F E^{\ominus}_{cell}}{RT} \qquad (6.28)$$

Equilibrium constant and standard cell potential

This very important equation (which could also have been obtained more directly by substituting eqn 6.26 into eqn 6.14) lets us predict equilibrium constants from measured standard cell potentials. However, before we use it extensively, we need to establish a further result.

• A brief illustration

Because the standard potential of the Daniell cell is +1.10 V, the equilibrium constant for the cell reaction $Cu^{2+}(aq) + Zn(s) \rightarrow Cu(s) + Zn^{2+}(aq)$, for which $\nu = 2$, is $K = 1.5 \times 10^{37}$ at 298 K. We conclude that the displacement of copper by zinc goes virtually to completion. Note that a cell potential of about 1 V is easily measurable but corresponds to an equilibrium constant that would be impossible to measure by direct chemical analysis. •

6.8 Standard electrode potentials

Key point **The standard potential of a couple is the cell potential in which it forms the right-hand electrode and the left-hand electrode is a standard hydrogen electrode.**

A galvanic cell is a combination of two electrodes each of which can be considered to make a characteristic contribution to the overall cell potential. Although it is not possible to measure the contribution of a single electrode, we can define the potential of one of the electrodes as zero and then assign values to others on that basis. The specially selected electrode is the **standard hydrogen electrode** (SHE):

$$Pt(s)|H_2(g)|H^+(aq) \qquad E^{\ominus} = 0 \qquad [6.29]$$

Convention for standard potentials

at all temperatures. To achieve the standard conditions, the activity of the hydrogen ions must be 1 (that is, pH = 0) and the pressure (more precisely, the fugacity) of the hydrogen gas must be 1 bar. The **standard potential,** $E^{\ominus}$, of another couple is then assigned by constructing a cell in which it is the right-hand electrode and the standard hydrogen electrode is the left-hand electrode.

The procedure for measuring a standard potential can be illustrated by considering a specific case, the silver chloride electrode. The measurement is made on the 'Harned cell':

$$Pt(s)|H_2(g)|HCl(aq)|AgCl(s)|Ag(s) \qquad \tfrac{1}{2}H_2(g) + AgCl(s) \rightarrow HCl(aq) + Ag(s)$$

$$E^{\ominus}_{cell} = E^{\ominus}(AgCl/Ag,Cl^-) - E^{\ominus}(SHE) = E^{\ominus}(AgCl/Ag,Cl^-)$$

for which the Nernst equation is

$$E_{cell} = E^{\ominus}(AgCl/Ag,Cl^-) - \frac{RT}{F}\ln\frac{a_{H^+}a_{Cl^-}}{a_{H_2}^{1/2}}$$

We shall set $a_{H_2} = 1$ from now on, and for simplicity write the standard potential of the $AgCl/Ag,Cl^-$ electrode as $E^{\ominus}$; then

$$E_{cell} = E^{\ominus} - \frac{RT}{F}\ln a_{H^+}a_{Cl^-}$$

The activities can be expressed in terms of the molality b of HCl(aq) through $a_{H^+} = \gamma_\pm b/b^\ominus$ and $a_{Cl^-} = \gamma_\pm b/b^\ominus$ as we saw in Section 5.13, so

$$E_{cell} = E^\ominus - \frac{RT}{F}\ln b^2 - \frac{RT}{F}\ln\gamma_\pm^2$$

where for simplicity we have replaced $b/b^\ominus$ by b. This expression rearranges to

$$E_{cell} + \frac{2RT}{F}\ln b = E^\ominus - \frac{2RT}{F}\ln \gamma_\pm \qquad (6.30)$$

From the Debye–Hückel limiting law for a 1,1-electrolyte (eqn 5.75; a 1,1-electrolyte is a solution of singly charged M^+ and X^- ions), we know that $\ln \gamma_\pm \propto -b^{1/2}$. The natural logarithm used here is proportional to the common logarithm that appears in eqn 5.75 (because $\ln x = \ln 10 \log x = 2.303 \log x$). Therefore, with the constant of proportionality in this relation written as $(F/2RT)C$, eqn 6.30 becomes

$$E_{cell} + \frac{2RT}{F}\ln b = E^\ominus + Cb^{1/2} \qquad (6.31)$$

The expression on the left is evaluated at a range of molalities, plotted against $b^{1/2}$, and extrapolated to $b = 0$. The intercept at $b^{1/2} = 0$ is the value of $E^\ominus$ for the silver/silver-chloride electrode. In precise work, the $b^{1/2}$ term is brought to the left, and a higher-order correction term from the extended Debye–Hückel law is used on the right.

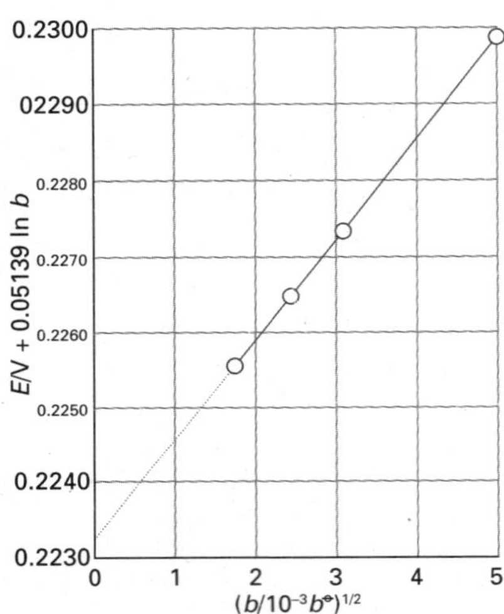

Fig. 6.18 The plot and the extrapolation used for the experimental measurement of a standard cell potential. The intercept at $b^{1/2} = 0$ is $E^\ominus_{cell}$.

• A brief illustration

The cell potential of $Pt(s)|H_2(g,p^\ominus)|HCl(aq,b)|AgCl(s)|Ag(s)$ at 25°C has the following values:

$b/(10^{-3}b^\ominus)$	3.215	5.619	9.138	25.63
E_{cell}/V	0.520 53	0.492 57	0.468 60	0.418 24

To determine the standard potential of the cell we draw up the following table, using $2RT/F = 0.051\ 39$ V:

$b/(10^{-3}b^\ominus)$	3.215	5.619	9.138	25.63
$\{b/(10^{-3}b^\ominus)\}^{1/2}$	1.793	2.370	3.023	5.063
E_{cell}/V	0.520 53	0.492 57	0.468 60	0.418 24
$E_{cell}/V + 0.051\ 39 \ln b$	0.2256	0.2263	0.2273	0.2299

The data are plotted in Fig. 6.18; as can be seen, they extrapolate to $E^\ominus = 0.2232$ V. •

Self-test 6.9 The data below are for the cell $Pt(s)|H_2(g,p^\ominus)|HBr(aq,b)|AgBr(s)|Ag(s)$ at 25°C. Determine the standard cell potential.

$b/(10^{-4}b^\ominus)$	4.042	8.444	37.19
E_{cell}/V	0.047 381	0.043 636	0.036 173

[0.076 V]

Table 6.2* Standard potentials at 298 K

Couple	$E^\ominus/V$
$Ce^{4+}(aq) + e^- \rightarrow Ce^{3+}(aq)$	+1.61
$Cu^{2+}(aq) + 2e^- \rightarrow Cu(s)$	+0.34
$H(aq) + e^- \rightarrow \frac{1}{2}H_2(g)$	0
$AgCl(s) + e^- \rightarrow Ag(s) + Cl^-(aq)$	+0.22
$Zn^{2+}(aq) + 2e^- \rightarrow Zn(s)$	−0.76
$Na^+(aq) + e^- \rightarrow Na(s)$	−2.71

* More values are given in the *Data section*.

Table 6.2 lists standard potentials at 298 K. An important feature of the standard potential of cells and standard potentials of electrodes is that they are unchanged if the chemical equation for the cell reaction or a half-reaction is multiplied by a numerical factor. A numerical factor increases the value of the standard Gibbs energy for the reaction. However, it also increases the number of electrons transferred by the same factor, and by eqn 6.26 the value of $E^\ominus_{cell}$ remains unchanged. A practical consequence is that a cell potential is independent of the physical size of the cell. In other words, the cell potential is an intensive property.

The standard potentials in Table 6.2 may be combined to give values for couples that are not listed there. However, to do so, we must take into account the fact that different couples may correspond to the transfer of different numbers of electrons. The procedure is illustrated in the following *Example.*

Example 6.4 *Evaluating a standard potential from two others*

Given that the standard potentials of the Cu^{2+}/Cu and Cu^{+}/Cu couples are +0.340 V and +0.522 V, respectively, evaluate $E^{\ominus}(Cu^{2+},Cu^{+})$.

Method First, we note that reaction Gibbs energies may be added (as in a Hess's law analysis of reaction enthalpies). Therefore, we should convert the $E^{\ominus}$ values to $\Delta G^{\ominus}$ values by using eqn 6.26, add them appropriately, and then convert the overall $\Delta G^{\ominus}$ to the required $E^{\ominus}$ by using eqn 6.26 again. This roundabout procedure is necessary because, as we shall see, although the factor F cancels, the factor ν in general does not.

Answer The electrode reactions are as follows:

(a) $Cu^{2+}(aq) + 2\,e^- \rightarrow Cu(s)$ $\quad E^{\ominus} = +0.340\text{ V},$ so $\quad \Delta_r G^{\ominus} = -2(0.340\text{ V})F$

(b) $Cu^{+}(aq) + e^- \rightarrow Cu(s)$ $\quad E^{\ominus} = +0.522\text{ V},$ so $\quad \Delta_r G^{\ominus} = -(0.522\text{ V})F$

The required reaction is

(c) $Cu^{2+}(aq) + e^- \rightarrow Cu^{+}(aq)$ $\quad E^{\ominus} = -\Delta_r G^{\ominus}/F$

Because (c) = (a) − (b), the standard Gibbs energy of reaction (c) is

$$\Delta_r G^{\ominus} = \Delta_r G^{\ominus}(a) - \Delta_r G^{\ominus}(b) = (-0.158\text{ V}) \times F$$

Therefore, $E^{\ominus} = +0.158$ V. Note that the generalization of the calculation we just performed is

$$\nu_c E^{\ominus}(c) = \nu_a E^{\ominus}(a) - \nu_b E^{\ominus}(b) \qquad \text{Combination of standard potentials} \qquad (6.32)$$

with the ν_r the stoichiometric coefficients of the electrons in each half-reaction.

6.9 Applications of standard potentials

Key points **(a) The electrochemical series lists the metallic elements in the order of their reducing power as measured by their standard potentials in aqueous solution: low reduces high. (b) The cell potential is used to measure the activity coefficient of electroactive ions. (c) The standard cell potential is used to infer the equilibrium constant of the cell reaction. (d) Species-selective electrodes contribute a potential that is characteristic of certain ions in solution. (e) The temperature coefficient of the cell potential is used to determine the standard entropy and enthalpy of reaction.**

Cell potentials are a convenient source of data on equilibrium constants and the Gibbs energies, enthalpies, and entropies of reactions. In practice the standard values of these quantities are the ones normally determined.

(a) The electrochemical series

We have seen that for two redox couples, Ox_1/Red_1 and Ox_2/Red_2, and the cell

$$Red_1,Ox_1||Red_2,Ox_2 \qquad E^{\ominus}_{cell} = E^{\ominus}_2 - E^{\ominus}_1 \qquad \text{Cell convention} \qquad (6.33a)$$

Table 6.3 The electrochemical series of the metals*

Least strongly reducing
Gold
Platinum
Silver
Mercury
Copper
(Hydrogen)
Lead
Tin
Nickel
Iron
Zinc
Chromium
Aluminium
Magnesium
Sodium
Calcium
Potassium
Most strongly reducing

* The complete series can be inferred from Table 6.2.

that the cell reaction

$$\text{Red}_1 + \text{Ox}_2 \rightarrow \text{Ox}_1 + \text{Red}_2 \tag{6.33b}$$

has $K > 1$ as written if $E^{\ominus}_{\text{cell}} > 0$, and therefore if $E^{\ominus}_2 > E^{\ominus}_1$. Because in the cell reaction Red_1 reduces Ox_2, we can conclude that

Red_1 has a thermodynamic tendency (in the sense $K > 1$) to reduce Ox_2 if $E^{\ominus}_1 < E^{\ominus}_2$

More briefly: low reduces high.

• A brief illustration

Because $E^{\ominus}(\text{Zn}^{2+},\text{Zn}) = -0.76\text{ V} < E^{\ominus}(\text{Cu}^{2+},\text{Cu}) = +0.34\text{ V}$, the reduction of Cu^{2+} by Zn is a reaction with $K > 1$, so zinc has a thermodynamic tendency to reduce Cu^{2+} ions in aqueous solution under standard conditions. •

Table 6.3 shows a part of the **electrochemical series**, the metallic elements (and hydrogen) arranged in the order of their reducing power as measured by their standard potentials in aqueous solution. A metal low in the series (with a lower standard potential) can reduce the ions of metals with higher standard potentials. This conclusion is qualitative. The quantitative value of K is obtained by doing the calculations we have described previously. For example, to determine whether zinc can displace magnesium from aqueous solutions at 298 K, we note that zinc lies above magnesium in the electrochemical series, so zinc cannot reduce magnesium ions in aqueous solution. Zinc can reduce hydrogen ions, because hydrogen lies higher in the series. However, even for reactions that are thermodynamically favourable, there may be kinetic factors that result in very slow rates of reaction.

(b) The determination of activity coefficients

Once the standard potential of an electrode in a cell is known, we can use it to determine mean activity coefficients by measuring the cell potential with the ions at the concentration of interest. For example, the mean activity coefficient of the ions in hydrochloric acid of molality b is obtained from eqn 6.30 in the form

$$\ln \gamma_{\pm} = \frac{E^{\ominus} - E_{\text{cell}}}{2RT/F} - \ln b \qquad \{6.34\}$$

once E_{cell} has been measured.

(c) The determination of equilibrium constants

The principal use for standard potentials is to calculate the standard potential of a cell formed from any two electrodes. To do so, we subtract the standard potential of the left-hand electrode from the standard potential of the right-hand electrode:

$$E^{\ominus}_{\text{cell}} = E^{\ominus}(\text{right}) - E^{\ominus}(\text{left}) \qquad \text{Cell convention} \qquad (6.35)$$

Because $\Delta_r G^{\ominus} = -\nu F E^{\ominus}_{\text{cell}}$, it then follows that, if the result gives $E^{\ominus}_{\text{cell}} > 0$, then the corresponding cell reaction has $K > 1$.

• A brief illustration

A disproportionation is a reaction in which a species is both oxidized and reduced. To study the disproportionation $2\,\text{Cu}^+(\text{aq}) \rightarrow \text{Cu(s)} + \text{Cu}^{2+}(\text{aq})$ we combine the following electrodes:

Right-hand electrode:

$$\mathrm{Cu(s)|Cu^+(aq)} \qquad \mathrm{Cu^+(aq) + e^- \rightarrow Cu(aq)} \qquad E^\ominus = +0.52\ \mathrm{V}$$

Left-hand electrode:

$$\mathrm{Pt(s)|Cu^{2+}(aq), Cu^+(aq)} \qquad \mathrm{Cu^{2+}(aq) + e^- \rightarrow Cu^+(s)} \qquad E^\ominus = +0.16\ \mathrm{V}$$

where the standard potentials are measured at 298 K. The standard potential of the cell is therefore

$$E^\ominus_{\mathrm{cell}} = +0.52\ \mathrm{V} - 0.16\ \mathrm{V} = +0.36\ \mathrm{V}$$

We can now calculate the equilibrium constant of the cell reaction. Because $\nu = 1$, from eqn 6.28,

$$\ln K = \frac{0.36\ \mathrm{V}}{0.025\,693\ \mathrm{V}} = \frac{0.36}{0.025\,693}$$

Hence, $K = 1.2 \times 10^6$. ●

(d) The determination of thermodynamic functions

The standard potential of a cell is related to the standard reaction Gibbs energy through eqn 6.25 ($\Delta_r G^\ominus = -\nu F E^\ominus_{\mathrm{cell}}$). Therefore, by measuring $E^\ominus_{\mathrm{cell}}$ we can obtain this important thermodynamic quantity. Its value can then be used to calculate the Gibbs energy of formation of ions by using the convention explained in Section 3.6.

● A brief illustration

The cell reaction taking place in

$$\mathrm{Pt(s)|H_2|H^+(aq)||Ag^+(aq)|Ag(s)} \qquad E^\ominus_{\mathrm{cell}} = +0.7996\ \mathrm{V}$$

is

$$\mathrm{Ag^+(aq)} + \tfrac{1}{2}\mathrm{H_2(g)} \rightarrow \mathrm{H^+(aq) + Ag(s)} \qquad \Delta_r G^\ominus = -\Delta_f G^\ominus(\mathrm{Ag^+,aq})$$

Therefore, with $\nu = 1$, we find

$$\Delta_f G^\ominus(\mathrm{Ag^+,aq}) = -(-FE^\ominus_{\mathrm{cell}}) = +77.15\ \mathrm{kJ\ mol^{-1}}$$

which is in close agreement with the value in Table 2.8 of the *Data section*. ●

The temperature coefficient of the standard cell potential, $\mathrm{d}E^\ominus_{\mathrm{cell}}/\mathrm{d}T$, gives the standard entropy of the cell reaction. This conclusion follows from the thermodynamic relation $(\partial G/\partial T)_p = -S$ and eqn 6.26, which combine to give

$$\frac{\mathrm{d}E^\ominus_{\mathrm{cell}}}{\mathrm{d}T} = \frac{\Delta_r S^\ominus}{\nu F} \qquad \text{Temperature coefficient of standard cell potential} \qquad (6.36)$$

The derivative is complete (not partial) because $E^\ominus$, like $\Delta_r G^\ominus$, is independent of the pressure. Hence we have an electrochemical technique for obtaining standard reaction entropies and through them the entropies of ions in solution.

Finally, we can combine the results obtained so far and use them to obtain the standard reaction enthalpy:

$$\Delta_r H^\ominus = \Delta_r G^\ominus + T\Delta_r S^\ominus = -\nu F\left(E^\ominus_{\mathrm{cell}} - T\frac{\mathrm{d}E^\ominus_{\mathrm{cell}}}{\mathrm{d}T}\right) \qquad (6.37)$$

This expression provides a non-calorimetric method for measuring $\Delta_r H^\ominus$ and, through the convention $\Delta_f H^\ominus(H^+,aq) = 0$, the standard enthalpies of formation of ions in solution (Section 2.8). Thus, electrical measurements can be used to calculate all the thermodynamic properties with which this chapter began.

Example 6.5 *Using the temperature coefficient of the cell potential*

The standard potential of the cell $Pt(s)|H_2(g)|HBr(aq)|AgBr(s)|Ag(s)$ was measured over a range of temperatures, and the data were found to fit the following polynomial:

$$E^\ominus_{cell}/V = 0.07131 - 4.99\times10^{-4}(T/K - 298) - 3.45\times10^{-6}(T/K - 298)^2$$

The cell reaction is $AgBr(s) + \frac{1}{2}H_2(g) \rightarrow Ag(s) + HBr(aq)$. Evaluate the standard reaction Gibbs energy, enthalpy, and entropy at 298 K.

Method The standard Gibbs energy of reaction is obtained by using eqn 6.26 after evaluating $E^\ominus_{cell}$ at 298 K and by using 1 V C = 1 J. The standard entropy of reaction is obtained by using eqn 6.36, which involves differentiating the polynomial with respect to T and then setting $T = 298$ K. The reaction enthalpy is obtained by combining the values of the standard Gibbs energy and entropy.

Answer At $T = 298$ K, $E^\ominus_{cell} = +0.07131$ V, so

$$\begin{aligned}\Delta_r G^\ominus &= -\nu F E^\ominus_{cell} = -(1)\times(9.6485\times10^4\ \text{C mol}^{-1})\times(+0.07131\ \text{V})\\ &= -6.880\times10^3\ \text{V C mol}^{-1} = -6.880\ \text{kJ mol}^{-1}\end{aligned}$$

The temperature coefficient of the cell potential is

$$\frac{dE^\ominus_{cell}}{dT} = -4.99\times10^{-4}\ \text{V K}^{-1} - 2(3.45\times10^{-6})(T/K - 298)\ \text{V K}^{-1}$$

At $T = 298$ K this expression evaluates to

$$\frac{dE^\ominus_{cell}}{dT} = -4.99\times10^{-4}\ \text{V K}^{-1}$$

So, from eqn 6.36, the reaction entropy is

$$\begin{aligned}\Delta_r S^\ominus &= 1\times(9.6485\times10^4\ \text{C mol}^{-1})\times(-4.99\times10^{-4}\ \text{V K}^{-1})\\ &= -48.1\ \text{J K}^{-1}\ \text{mol}^{-1}\end{aligned}$$

The negative value stems in part from the elimination of gas in the cell reaction. It then follows that

$$\begin{aligned}\Delta_r H^\ominus &= \Delta_r G^\ominus + T\Delta_r S^\ominus = -6.880\ \text{kJ mol}^{-1} + (298\ \text{K})\times(-0.0482\ \text{kJ K}^{-1}\ \text{mol}^{-1})\\ &= -21.2\ \text{kJ mol}^{-1}\end{aligned}$$

One difficulty with this procedure lies in the accurate measurement of small temperature coefficients of cell potential. Nevertheless, it is another example of the striking ability of thermodynamics to relate the apparently unrelated, in this case to relate electrical measurements to thermal properties.

Self-test 6.10 Predict the standard potential of the Harned cell at 303 K from tables of thermodynamic data. [+0.219 V]

IMPACT ON TECHNOLOGY

I6.3 Species-selective electrodes

An **ion-selective electrode** is an electrode that generates a potential in response to the presence of a solution of specific ions. An example is the **glass electrode** (Fig. 6.19), which is sensitive to hydrogen ion activity, and has a potential proportional to pH. It is filled with a phosphate buffer containing Cl^- ions, and conveniently has $E = 0$ when the external medium is at pH = 6. It is necessary to calibrate the glass electrode before use with solutions of known pH.

The responsiveness of a glass electrode to the hydrogen ion activity is a result of complex processes at the interface between the glass membrane and the solutions on either side of it. The membrane itself is permeable to Na^+ and Li^+ ions but not to H^+ ions. Therefore, the potential difference across the glass membrane must arise by a mechanism different from that responsible for biological transmembrane potentials. A clue to the mechanism comes from a detailed inspection of the glass membrane, for each face is coated with a thin layer of hydrated silica (Fig. 6.20). The hydrogen ions in the test solution modify this layer to an extent that depends on their activity in the solution, and the charge modification of the outside layer is transmitted to the inner layer by the Na^+ and Li^+ ions in the glass. The hydrogen ion activity gives rise to a membrane potential by this indirect mechanism.

Electrodes sensitive to hydrogen ions, and hence to pH, are typically glasses based on lithium silicate doped with heavy-metal oxides. The glass can also be made responsive to Na^+, K^+, and NH_4^+ ions by being doped with Al_2O_3 and B_2O_3.

A suitably adapted glass electrode can be used to detect the presence of certain gases. A simple form of a **gas-sensing electrode** consists of a glass electrode contained in an outer sleeve filled with an aqueous solution and separated from the test solution by a membrane that is permeable to gas. When a gas such as sulfur dioxide or ammonia diffuses into the aqueous solution, it modifies its pH, which in turn affects the potential of the glass electrode. The presence of an enzyme that converts a compound, such as urea or an amino acid, into ammonia, which then affects the pH, can be used to detect these organic compounds.

Somewhat more sophisticated devices are used as ion-selective electrodes that give potentials according to the presence of specific ions present in a test solution. In one arrangement, a porous lipophilic (hydrocarbon-attracting) membrane is attached to a small reservoir of a hydrophobic (water-repelling) liquid, such as dioctylphenylphosphonate, that saturates it (Fig. 6.21). The liquid contains an agent, such as $(RO)_2PO_2^-$ with R a C_8 to C_{18} chain, that acts as a kind of solubilizing agent for the ions with which it can form a complex. The complex's ions are able to migrate through the lipophilic membrane, and hence give rise to a transmembrane potential, which is detected by a silver/silver chloride electrode in the interior of the assembly. Electrodes of this construction can be designed to be sensitive to a variety of ionic species, including calcium, zinc, iron, lead, and copper ions.

In theory, the transmembrane potential should be determined entirely by differences in the activity of the species that the electrode was designed to detect. In practice, a small potential difference, called the **asymmetry potential**, is observed even when the activity of the test species is the same on both sides of the membrane. The asymmetry potential is due to the fact that it is not possible to manufacture a membrane material that has the same structure and the same chemical properties throughout. Furthermore, all species-selective electrodes are sensitive to more than one species. For example, a Na^+ selective electrode also responds, albeit less effectively, to the activity of K^+ ions in the test solution. As a result of these effects, the potential

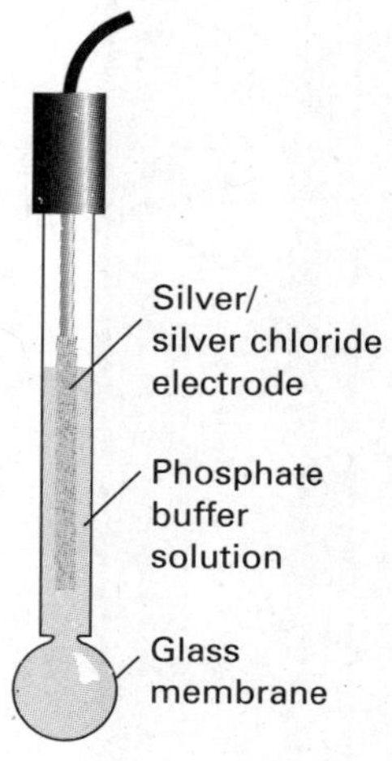

Fig. 6.19 The glass electrode. It is commonly used in conjunction with a calomel electrode that makes contact with the test solution through a salt bridge.

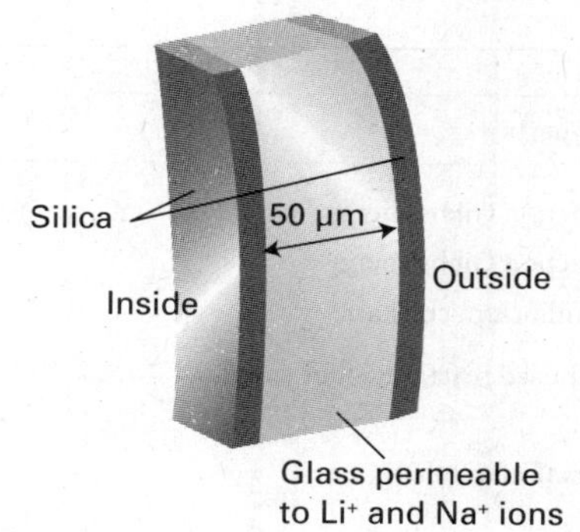

Fig. 6.20 A section through the wall of a glass electrode.

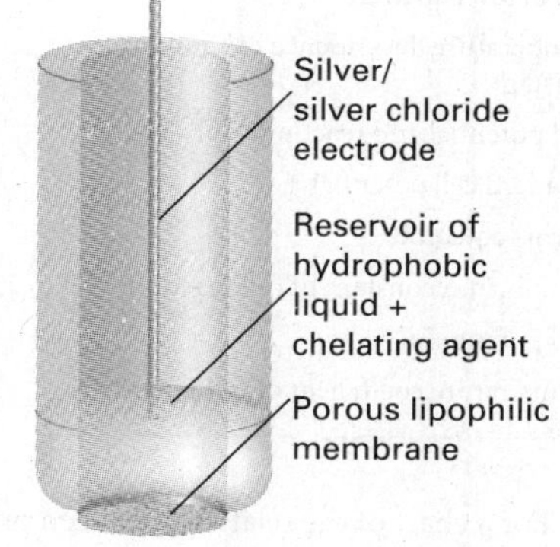

Fig. 6.21 The structure of an ion-selective electrode. Chelated ions are able to migrate through the lipophilic membrane.

of an electrode sensitive to species X^+ that is also susceptible to interference by species Y^+ is given by a modified form of the Nernst equation:

$$E_{cell} = E_{ap} + \beta\frac{RT}{F}\ln(a_{X^+} + k_{X,Y}a_{Y^+}) \qquad (6.38)$$

where E_{ap} is the asymmetry potential, β is an experimental parameter that captures deviations from the Nernst equation, and $k_{X,Y}$ is the **selectivity coefficient** of the electrode and is related to the response of the electrode to the interfering species Y^+. A value of $\beta = 1$ indicates that the electrode responds to the activity of ions in solution in a way that is consistent with the Nernst equation and, in practice, most species-selective electrodes of high quality have $\beta \approx 1$. The selectivity coefficient, and hence interference effects, can be minimized when designing and manufacturing a species-selective electrode. For precise work, it is necessary to calibrate the response of the electrode by measuring E_{ap}, β, and $k_{X,Y}$ before performing experiments on solutions of unknown concentration of X^+.

Checklist of key equations

Property	Equation	Comment
Reaction Gibbs energy	$\Delta_r G = (\partial G/\partial\xi)_{p,T}$	
Reaction Gibbs energy	$\Delta_r G = \Delta_r G^⦵ + RT\ln Q$	
Equilibrium constant	$\Delta_r G^⦵ = -RT\ln K$	
Standard reaction Gibbs energy	$\Delta_r G^⦵ = \sum_{\text{Products}} \nu\Delta_f G^⦵ - \sum_{\text{Reactants}} \nu\Delta_f G^⦵ = \sum_J \nu_J \Delta_f G^⦵(J)$	ν are positive; ν_J are signed
Reaction quotient	$Q = \prod_J a_J^{\nu_J}$	Evaluated at arbitrary stage of reaction
Thermodynamic equilibrium constant	$K = \left(\prod_J a_J^{\nu_J}\right)_{\text{equilibrium}}$	
Relation between K and K_c	$K = K_c(c^⦵RT/p^⦵)^{\Delta\nu}$	Gas-phase reactions
van't Hoff equation	$d\ln K/dT = \Delta_r H^⦵/RT^2$	
Temperature dependence of equilibrium constant	$\ln K_2 - \ln K_1 = -(\Delta_r H^⦵/R)(1/T_2 - 1/T_1)$	Derived from the van't Hoff equation with $\Delta_r H^⦵$ assumed constant
Cell potential and reaction Gibbs energy	$-\nu FE_{cell} = \Delta_r G$	
Standard cell potential	$E^⦵_{cell} = -\Delta_r G^⦵/\nu F$	Definition
Nernst equation	$E_{cell} = E^⦵_{cell} - (RT/\nu F)\ln Q$	
Equilibrium constant of cell reaction	$\ln K = \nu FE^⦵_{cell}/RT$	
Cell potential	$E^⦵_{cell} = E^⦵(\text{right}) - E^⦵(\text{left})$	
Temperature coefficient of cell potential	$dE^⦵_{cell}/dT = \Delta_r S^⦵/\nu F$	

→ For a chart of the relations between principal equations, see the Road map section of the Resource section.

Discussion questions

6.1 Explain how the mixing of reactants and products affects the position of chemical equilibrium.

6.2 What is the justification for not including a pure liquid or solid in the expression for an equilibrium constant?

6.3 Suggest how the thermodynamic equilibrium constant may respond differently to changes in pressure and temperature from the equilibrium constant expressed in terms of partial pressures.

6.4 Account for Le Chatelier's principle in terms of thermodynamic quantities.

6.5 Explain the molecular basis of the van't Hoff equation for the temperature dependence of K.

6.6 Explain why reactions that are not redox reactions may be used to generate an electric current.

6.7 Describe a method for the determination of the standard potential of a redox couple.

6.8 Devise a method for the determination of the pH of an aqueous solution.

Exercises

6.1(a) Consider the reaction $A \rightarrow 2\,B$. Initially 1.50 mol A is present and no B. What are the amounts of A and B when the extent of reaction is 0.60 mol?

6.1(b) Consider the reaction $2\,A \rightarrow B$. Initially 1.75 mol A and 0.12 mol B are present. What are the amounts of A and B when the extent of reaction is 0.30 mol?

6.2(a) When the reaction $A \rightarrow 2\,B$ advances by 0.10 mol (that is, $\Delta\xi = +0.10$ mol) the Gibbs energy of the system changes by -6.4 kJ mol^{-1}. What is the Gibbs energy of reaction at this stage of the reaction?

6.2(b) When the reaction $2\,A \rightarrow B$ advances by 0.051 mol (that is, $\Delta\xi = +0.051$ mol) the Gibbs energy of the system changes by -2.41 kJ mol^{-1}. What is the Gibbs energy of reaction at this stage of the reaction?

6.3(a) The standard Gibbs energy of the reaction $N_2(g) + 3\,H_2(g) \rightarrow 2\,NH_3(g)$ is -32.9 kJ mol^{-1} at 298 K. What is the value of Δ_rG when Q = (a) 0.010, (b) 1.0, (c) 10.0, (d) 100 000, (e) 1 000 000? Estimate (by interpolation) the value of K from the values you calculate. What is the actual value of K?

6.3(b) The standard Gibbs energy of the reaction $2\,NO_2(g) \rightarrow N_2O_4(g)$ is -4.73 kJ mol^{-1} at 298 K. What is the value of Δ_rG when Q = (a) 0.10, (b) 1.0, (c) 10, (d) 100? Estimate (by interpolation) the value of K from the values you calculate. What is the actual value of K?

6.4(a) At 2257 K and 1.00 bar total pressure, water is 1.77 per cent dissociated at equilibrium by way of the reaction $2\,H_2O(g) \rightleftharpoons 2\,H_2(g) + O_2(g)$. Calculate K.

6.4(b) For the equilibrium, $N_2O_4(g) \rightleftharpoons 2\,NO_2(g)$, the degree of dissociation, α, at 298 K is 0.201 at 1.00 bar total pressure. Calculate K.

6.5(a) Dinitrogen tetroxide is 18.46 per cent dissociated at 25°C and 1.00 bar in the equilibrium $N_2O_4(g) \rightleftharpoons 2\,NO_2(g)$. Calculate K at (a) 25°C, (b) 100°C given that $\Delta_rH^{\ominus} = +56.2$ kJ mol^{-1} over the temperature range.

6.5(b) Molecular bromine is 24 per cent dissociated at 1600 K and 1.00 bar in the equilibrium $Br_2(g) \rightleftharpoons 2\,Br(g)$. Calculate K at (a) 1600 K, (b) 2000 K given that $\Delta_rH^{\ominus} = +112$ kJ mol^{-1} over the temperature range.

6.6(a) From information in the *Data section*, calculate the standard Gibbs energy and the equilibrium constant at (a) 298 K and (b) 400 K for the reaction $PbO(s) + CO(g) \rightleftharpoons Pb(s) + CO_2(g)$. Assume that the reaction enthalpy is independent of temperature.

6.6(b) From information in the *Data section*, calculate the standard Gibbs energy and the equilibrium constant at (a) 25°C and (b) 50°C for the reaction $CH_4(g) + 3\,Cl_2(g) \rightleftharpoons CHCl_3(l) + 3\,HCl(g)$. Assume that the reaction enthalpy is independent of temperature.

6.7(a) Establish the relation between K and K_c for the reaction $H_2CO(g) \rightleftharpoons CO(g) + H_2(g)$.

6.7(b) Establish the relation between K and K_c for the reaction $3\,N_2(g) + H_2(g) \rightleftharpoons 2\,HN_3(g)$.

6.8(a) In the gas-phase reaction $2\,A + B \rightleftharpoons 3\,C + 2\,D$, it was found that, when 1.00 mol A, 2.00 mol B, and 1.00 mol D were mixed and allowed to come to equilibrium at 25°C, the resulting mixture contained 0.90 mol C at a total pressure of 1.00 bar. Calculate (a) the mole fractions of each species at equilibrium, (b) K_x, (c) K, and (d) $\Delta_rG^{\ominus}$.

6.8(b) In the gas-phase reaction $A + B \rightleftharpoons C + 2\,D$, it was found that, when 2.00 mol A, 1.00 mol B, and 3.00 mol D were mixed and allowed to come to equilibrium at 25°C, the resulting mixture contained 0.79 mol C at a total pressure of 1.00 bar. Calculate (a) the mole fractions of each species at equilibrium, (b) K_x, (c) K, and (d) $\Delta_rG^{\ominus}$.

6.9(a) The standard reaction enthalpy of $Zn(s) + H_2O(g) \rightarrow ZnO(s) + H_2(g)$ is approximately constant at +224 kJ mol^{-1} from 920 K up to 1280 K. The standard reaction Gibbs energy is +33 kJ mol^{-1} at 1280 K. Estimate the temperature at which the equilibrium constant becomes greater than 1.

6.9(b) The standard enthalpy of a certain reaction is approximately constant at +125 kJ mol^{-1} from 800 K up to 1500 K. The standard reaction Gibbs energy is +22 kJ mol^{-1} at 1120 K. Estimate the temperature at which the equilibrium constant becomes greater than 1.

6.10(a) The equilibrium constant of the reaction $2\,C_3H_6(g) \rightleftharpoons C_2H_4(g) + C_4H_8(g)$ is found to fit the expression $\ln K = A + B/T + C/T^2$ between 300 K and 600 K, with $A = -1.04$, $B = -1088$ K, and $C = 1.51 \times 10^5$ K^2. Calculate the standard reaction enthalpy and standard reaction entropy at 400 K.

6.10(b) The equilibrium constant of a reaction is found to fit the expression $\ln K = A + B/T + C/T^3$ between 400 K and 500 K with $A = -2.04$, $B = -1176$ K, and $C = 2.1 \times 10^7$ K^3. Calculate the standard reaction enthalpy and standard reaction entropy at 450 K.

6.11(a) Establish the relation between K and K_c for the reaction $H_2CO(g) \rightleftharpoons CO(g) + H_2(g)$.

6.11(b) Establish the relation between K and K_c for the reaction $3\,N_2(g) + H_2(g) \rightleftharpoons 2\,HN_3(g)$.

6.12(a) Calculate the values of K and K_c for the reaction $H_2CO(g) \rightleftharpoons CO(g) + H_2(g)$ at (a) 25°C, (b) 100°C.

6.12(b) Calculate the values of K and K_c for the reaction $3\,N_2(g) + H_2(g) \rightleftharpoons 2\,HN_3(g)$ at (a) 25°C, (b) 100°C.

6.13(a) The standard reaction Gibbs energy of the isomerization of borneol ($C_{10}H_{17}OH$) to isoborneol in the gas phase at 503 K is +9.4 kJ mol^{-1}. Calculate the reaction Gibbs energy in a mixture consisting of 0.15 mol of borneol and 0.30 mol of isoborneol when the total pressure is 600 Torr.

6.13(b) The equilibrium pressure of H_2 over solid uranium and uranium hydride, UH_3, at 500 K is 139 Pa. Calculate the standard Gibbs energy of formation of $UH_3(s)$ at 500 K.

6.14(a) Calculate the percentage change in K_x for the reaction $H_2CO(g) \rightleftharpoons CO(g) + H_2(g)$ when the total pressure is increased from 1.0 bar to 2.0 bar at constant temperature.

6.14(b) Calculate the percentage change in K_x for the reaction $CH_3OH(g) + NOCl(g) \rightleftharpoons HCl(g) + CH_3NO_2(g)$ when the total pressure is increased from 1.0 bar to 2.0 bar at constant temperature.

6.15(a) The equilibrium constant for the gas-phase isomerization of borneol ($C_{10}H_{17}OH$) to isoborneol at 503 K is 0.106. A mixture consisting of 7.50 g of borneol and 14.0 g of isoborneol in a container of volume 5.0 dm^3 is heated to 503 K and allowed to come to equilibrium. Calculate the mole fractions of the two substances at equilibrium.

6.15(b) The equilibrium constant for the reaction $N_2(g) + O_2(g) \rightleftharpoons 2\,NO(g)$ is 1.69×10^{-3} at 2300 K. A mixture consisting of 5.0 g of nitrogen and 2.0 g of oxygen in a container of volume 1.0 dm^3 is heated to 2300 K and allowed to come to equilibrium. Calculate the mole fraction of NO at equilibrium.

6.16(a) What is the standard enthalpy of a reaction for which the equilibrium constant is (a) doubled, (b) halved when the temperature is increased by 10 K at 298 K?

6.16(b) What is the standard enthalpy of a reaction for which the equilibrium constant is (a) doubled, (b) halved when the temperature is increased by 15 K at 310 K?

6.17(a) The standard Gibbs energy of formation of $NH_3(g)$ is −16.5 kJ mol^{-1} at 298 K. What is the reaction Gibbs energy when the partial pressures of the N_2, H_2, and NH_3 (treated as perfect gases) are 3.0 bar, 1.0 bar, and 4.0 bar, respectively? What is the spontaneous direction of the reaction in this case?

6.17(b) The dissociation vapour pressure of NH_4Cl at 427°C is 608 kPa but at 459°C it has risen to 1115 kPa. Calculate (a) the equilibrium constant, (b) the standard reaction Gibbs energy, (c) the standard enthalpy, (d) the standard entropy of dissociation, all at 427°C. Assume that the vapour behaves as a perfect gas and that $\Delta H^{\ominus}$ and $\Delta S^{\ominus}$ are independent of temperature in the range given.

6.18(a) Estimate the temperature at which $CaCO_3$(calcite) decomposes.

6.18(b) Estimate the temperature at which $CuSO_4 \cdot 5H_2O$ undergoes dehydration.

6.19(a) For $CaF_2(s) \rightleftharpoons Ca^{2+}(aq) + 2\,F^-(aq)$, $K = 3.9 \times 10^{-11}$ at 25°C and the standard Gibbs energy of formation of $CaF_2(s)$ is −1167 kJ mol^{-1}. Calculate the standard Gibbs energy of formation of $CaF_2(aq)$.

6.19(b) For $PbI_2(s) \rightleftharpoons Pb^{2+}(aq) + 2\,I^-(aq)$, $K = 1.4 \times 10^{-8}$ at 25°C and the standard Gibbs energy of formation of $PbI_2(s)$ is −173.64 kJ mol^{-1}. Calculate the standard Gibbs energy of formation of $PbI_2(aq)$.

6.20(a) Write the cell reaction and electrode half-reactions and calculate the standard potential of each of the following cells:

(a) $Zn|ZnSO_4(aq)||AgNO_3(aq)|Ag$
(b) $Cd|CdCl_2(aq)||HNO_3(aq)|H_2(g)|Pt$
(c) $Pt|K_3[Fe(CN)_6](aq),K_4[Fe(CN)_6](aq)||CrCl_3(aq)|Cr$

6.20(b) Write the cell reaction and electrode half-reactions and calculate the standard potential of each of the following cells:

(a) $Pt|Cl_2(g)|HCl(aq)||K_2CrO_4(aq)|Ag_2CrO_4(s)|Ag$
(b) $Pt|Fe^{3+}(aq),Fe^{2+}(aq)||Sn^{4+}(aq),Sn^{2+}(aq)|Pt$
(c) $Cu|Cu^{2+}(aq)||Mn^{2+}(aq),H^+(aq)|MnO_2(s)|Pt$

6.21(a) Devise cells in which the following are the reactions and calculate the standard cell potential in each case:

(a) $Zn(s) + CuSO_4(aq) \rightarrow ZnSO_4(aq) + Cu(s)$
(b) $2\,AgCl(s) + H_2(g) \rightarrow 2\,HCl(aq) + 2\,Ag(s)$
(c) $2\,H_2(g) + O_2(g) \rightarrow 2\,H_2O(l)$

6.21(b) Devise cells in which the following are the reactions and calculate the standard cell potential in each case:

(a) $2\,Na(s) + 2\,H_2O(l) \rightarrow 2\,NaOH(aq) + H_2(g)$
(b) $H_2(g) + I_2(s) \rightarrow 2\,HI(aq)$
(c) $H_3O^+(aq) + OH^-(aq) \rightarrow 2\,H_2O(l)$

6.22(a) Use the Debye–Hückel limiting law and the Nernst equation to estimate the potential of the cell $Ag|AgBr(s)|KBr(aq, 0.050\ mol\ kg^{-1})||Cd(NO_3)_2(aq, 0.010\ mol\ kg^{-1})|Cd$ at 25°C.

6.22(b) Consider the cell $Pt|H_2(g,p^{\ominus})|HCl(aq)|AgCl(s)|Ag$, for which the cell reaction is $2\,AgCl(s) + H_2(g) \rightarrow 2\,Ag(s) + 2\,HCl(aq)$. At 25°C and a molality of HCl of 0.010 mol kg^{-1}, $E_{cell} = +0.4658$ V. (a) Write the Nernst equation for the cell reaction. (b) Calculate Δ_rG for the cell reaction. (c) Assuming that the Debye–Hückel limiting law holds at this concentration, calculate $E^{\ominus}(Cl^-,AgCl,Ag)$.

6.23(a) Calculate the equilibrium constants of the following reactions at 25°C from standard potential data:

(a) $Sn(s) + Sn^{4+}(aq) \rightleftharpoons 2\,Sn^{2+}(aq)$
(b) $Sn(s) + 2\,AgCl(s) \rightleftharpoons SnCl_2(aq) + 2\,Ag(s)$

6.23(b) Calculate the equilibrium constants of the following reactions at 25°C from standard potential data:

(a) $Sn(s) + CuSO_4(aq) \rightleftharpoons Cu(s) + SnSO_4(aq)$
(b) $Cu^{2+}(aq) + Cu(s) \rightleftharpoons 2\,Cu^+(aq)$

6.24(a) The potential of the cell $Ag|AgI(s)|AgI(aq)|Ag$ is +0.9509 V at 25°C. Calculate (a) the solubility product of AgI and (b) its solubility.

6.24(b) The potential of the cell $Bi|Bi_2S_3(s)|Bi_2S_3(aq)|Bi$ is 0.96 V at 25°C. Calculate (a) the solubility product of Bi_2S_3 and (b) its solubility.

Problems*

Numerical problems

6.1 The equilibrium constant for the reaction, $I_2(s) + Br_2(g) \rightleftharpoons 2\,IBr(g)$ is 0.164 at 25°C. (a) Calculate $\Delta_r G^\ominus$ for this reaction. (b) Bromine gas is introduced into a container with excess solid iodine. The pressure and temperature are held at 0.164 atm and 25°C, respectively. Find the partial pressure of IBr(g) at equilibrium. Assume that all the bromine is in the liquid form and that the vapour pressure of iodine is negligible. (c) In fact, solid iodine has a measurable vapour pressure at 25°C. In this case, how would the calculation have to be modified?

6.2 Consider the dissociation of methane, $CH_4(g)$, into the elements $H_2(g)$ and C(s, graphite). (a) Given that $\Delta_f H^\ominus(CH_4,g) = -74.85\ kJ\ mol^{-1}$ and that $\Delta_f S^\ominus(CH_4,g) = -80.67\ J\ K^{-1}\ mol^{-1}$ at 298 K, calculate the value of the equilibrium constant at 298 K. (b) Assuming that $\Delta_f H^\ominus$ is independent of temperature, calculate K at 50°C. (c) Calculate the degree of dissociation, α, of methane at 25°C and a total pressure of 0.010 bar. (d) Without doing any numerical calculations, explain how the degree of dissociation for this reaction will change as the pressure and temperature are varied.

6.3 The equilibrium pressure of H_2 over U(s) and $UH_3(s)$ between 450 K and 715 K fits the expression $\ln(p/Pa) = A + B/T + C\ln(T/K)$, with $A = 69.32$, $B = -1.464 \times 10^4$ K, and $C = -5.65$. Find an expression for the standard enthalpy of formation of $UH_3(s)$ and from it calculate $\Delta_r C_p^\ominus$.

6.4 The degree of dissociation, α, of $CO_2(g)$ into CO(g) and $O_2(g)$ at high temperatures was found to vary with temperature as follows:

T/K	1395	1443	1498
$\alpha/10^{-4}$	1.44	2.50	4.71

Assuming $\Delta_r H^\ominus$ to be constant over this temperature range, calculate K, $\Delta_r G^\ominus$, $\Delta_r H^\ominus$, and $\Delta_r S^\ominus$. Make any justifiable approximations.

6.5 The standard reaction enthalpy for the decomposition of $CaCl_2{\cdot}NH_3(s)$ into $CaCl_2(s)$ and $NH_3(g)$ is nearly constant at +78 kJ mol^{-1} between 350 K and 470 K. The equilibrium pressure of NH_3 in the presence of $CaCl_2{\cdot}NH_3$ is 1.71 kPa at 400 K. Find an expression for the temperature dependence of $\Delta_r G^\ominus$ in the same range.

6.6 Calculate the equilibrium constant of the reaction $CO(g) + H_2(g) \rightleftharpoons H_2CO(g)$ given that, for the production of liquid formaldehyde, $\Delta_r G^\ominus = +28.95\ kJ\ mol^{-1}$ at 298 K and that the vapour pressure of formaldehyde is 1500 Torr at that temperature.

6.7 Acetic acid was evaporated in a container of volume 21.45 cm^3 at 437 K and at an external pressure of 101.9 kPa, and the container was then sealed. The combined mass of acid monomer and dimer in the sealed container was 0.0463 g. The experiment was repeated with the same container but at 471 K, and the combined mass of acid monomer and dimer was found to be 0.0380 g. Calculate the equilibrium constant for the dimerization of the acid in the vapour and the enthalpy of dimerization.

6.8 A sealed container was filled with 0.300 mol $H_2(g)$, 0.400 mol $I_2(g)$, and 0.200 mol HI(g) at 870 K and total pressure 1.00 bar. Calculate the amounts of the components in the mixture at equilibrium given that $K = 870$ for the reaction $H_2(g) + I_2(g) \rightleftharpoons 2\,HI(g)$.

6.9 The dissociation of I_2 can be monitored by measuring the total pressure, and three sets of results are as follows:

T/K	973	1073	1173
$100p$/atm	6.244	7.500	9.181
$10^4 n_I$	2.4709	2.4555	2.4366

where n_I is the amount of I atoms per mole of I_2 molecules in the mixture, which occupied 342.68 cm^3. Calculate the equilibrium constants of the dissociation and the standard enthalpy of dissociation at the mean temperature.

6.10‡ Thorn *et al.* (*J. Phys. Chem.* **100**, 14178 (1996)) carried out a study of $Cl_2O(g)$ by photoelectron ionization. From their measurements, they report $\Delta_f H^\ominus(Cl_2O) = +77.2\ kJ\ mol^{-1}$. They combined this measurement with literature data on the reaction $Cl_2O(g) + H_2O(g) \rightarrow 2\,HOCl(g)$, for which $K = 8.2 \times 10^{-2}$ and $\Delta_r S^\ominus = +16.38\ J\ K^{-1}\ mol^{-1}$, and with readily available thermodynamic data on water vapour to report a value for $\Delta_f H^\ominus(HOCl)$. Calculate that value. All quantities refer to 298 K.

6.11‡ The 1980s saw reports of $\Delta_f H^\ominus(SiH_2)$ ranging from 243 to 289 kJ mol^{-1}. If the standard enthalpy of formation is uncertain by this amount, by what factor is the equilibrium constant for the formation of SiH_2 from its elements uncertain at (a) 298 K, (b) 700 K?

6.12 Fuel cells provide electrical power for spacecraft (as in the NASA space shuttles) and also show promise as power sources for automobiles. Hydrogen and carbon monoxide have been investigated for use in fuel cells, so their solubilities in molten salts are of interest. Their solubilities in a molten $NaNO_3/KNO_3$ mixture were found to fit the following expressions:

$$\log s_{H_2} = -5.39 - \frac{768}{T/K} \qquad \log s_{CO} = -5.98 - \frac{980}{T/K}$$

where s is the solubility in mol cm^{-3} bar^{-1}. Calculate the standard molar enthalpies of solution of the two gases at 570 K.

6.13 Given that $\Delta_r G^\ominus = -212.7\ kJ\ mol^{-1}$ for the reaction in the Daniell cell at 25°C, and $b(CuSO_4) = 1.0 \times 10^{-3}\ mol\ kg^{-1}$ and $b(ZnSO_4) = 3.0 \times 10^{-3}\ mol\ kg^{-1}$, calculate (a) the ionic strengths of the solutions, (b) the mean ionic activity coefficients in the compartments, (c) the reaction quotient, (d) the standard cell potential, and (e) the cell potential. (Take $\gamma_+ = \gamma_- = \gamma_\pm$ in the respective compartments.)

6.14 A fuel cell develops an electric potential from the chemical reaction between reagents supplied from an outside source. What is the cell potential of a cell fuelled by (a) hydrogen and oxygen, (b) the combustion of butane at 1.0 bar and 298 K?

6.15 Although the hydrogen electrode may be conceptually the simplest electrode and is the basis for our reference state of electrical potential in electrochemical systems, it is cumbersome to use. Therefore, several substitutes for it have been devised. One of these alternatives is the quinhydrone electrode (quinhydrone, $Q{\cdot}QH_2$, is a complex of quinone, $C_6H_4O_2 = Q$, and hydroquinone, $C_6H_4O_2H_2 = QH_2$). The electrode half-reaction is $Q(aq) + 2\,H^+(aq) + 2\,e^- \rightarrow QH_2(aq)$, $E^\ominus = +0.6994$ V. If the cell $Hg|Hg_2Cl_2(s)|HCl(aq)|Q{\cdot}QH_2|Au$ is prepared, and the measured cell potential is +0.190 V, what is the pH of the HCl solution? Assume that the Debye–Hückel limiting law is applicable.

6.16 Consider the cell, $Zn(s)|ZnCl_2\ (0.0050\ mol\ kg^{-1})|Hg_2Cl_2(s)|Hg(l)$, for which the cell reaction is $Hg_2Cl_2(s) + Zn(s) \rightarrow 2\,Hg(l) + 2\,Cl^-(aq) + Zn^{2+}(aq)$. Given that $E^\ominus(Zn^{2+},Zn) = -0.7628$ V, $E^\ominus(Hg_2Cl_2,Hg) = +0.2676$ V, and that the cell potential is +1.2272 V, (a) write the Nernst equation for the cell. Determine (b) the standard cell potential, (c) $\Delta_r G$, $\Delta_r G^\ominus$, and K for the cell reaction, (d) the mean ionic activity and activity coefficient of $ZnCl_2$ from the measured cell potential, and (e) the mean ionic activity coefficient of $ZnCl_2$ from the Debye–Hückel limiting law. (f) Given that $(\partial E_{cell}/\partial T)_p = -4.52 \times 10^{-4}\ V\ K^{-1}$, calculate $\Delta_r S$ and $\Delta_r H$.

* Problems denoted with the symbol ‡ were supplied by Charles Trapp, Carmen Giunta, and Marshall Cady.

6.17 The potential of the cell $Pt|H_2(g,p^{\ominus})|HCl(aq,b)|Hg_2Cl_2(s)|Hg(l)$ has been measured with high precision with the following results at 25°C:

$b/(\text{mmol kg}^{-1})$	1.6077	3.0769	5.0403	7.6938	10.9474
E/V	0.60080	0.56825	0.54366	0.52267	0.50532

Determine the standard cell potential and the mean activity coefficient of HCl at these molalities. (Make a least-squares fit of the data to the best straight line.)

6.18 Careful measurements of the potential of the cell $Pt|H_2(g,p^{\ominus})|NaOH(aq, 0.0100\ \text{mol kg}^{-1}), NaCl(aq, 0.01125\ \text{mol kg}^{-1})|AgCl(s)|Ag$ have been reported. Among the data is the following information:

$\theta/°\text{C}$	20.0	25.0	30.0
E_{cell}/V	1.04774	1.04864	1.04942

Calculate pK_w at these temperatures and the standard enthalpy and entropy of the autoprotolysis of water at 25.0°C.

6.19 Measurements of the potential of cells of the type $Ag|AgX(s)|MX(b_1)|M_xHg|MX(b_2)|AgX(s)|Ag$, where M_xHg denotes an amalgam and the electrolyte is LiCl in ethylene glycol, are given below. Estimate the activity coefficient at the concentration marked * and then use this value to calculate activity coefficients from the measured cell potential at the other concentrations. Base your answer on the following version of the extended Debye–Hückel law:

$$\log \gamma_{\pm} = -\frac{AI^{1/2}}{1+BI^{1/2}} + CI$$

with $A = 1.461$, $B = 1.70$, $C = 0.20$, and $I = b/b^{\ominus}$. For $b_2 = 0.09141\ \text{mol kg}^{-1}$:

$b_1/(\text{mol kg}^{-1})$	0.0555	0.09141*	0.1652	0.2171	1.040	1.350
E/V	−0.0220	0.0000	0.0263	0.0379	0.1156	0.1336

6.20 The standard potential of the $AgCl/Ag,Cl^-$ couple fits the expression

$$E^{\ominus}/\text{V} = 0.23659 - 4.8564 \times 10^{-4}(\theta/°\text{C}) - 3.4205 \times 10^{-6}(\theta/°\text{C})^2 + 5.869 \times 10^{-9}(\theta/°\text{C})^3$$

Calculate the standard Gibbs energy and enthalpy of formation of $Cl^-(aq)$ and its entropy at 298 K.

6.21‡ The table below summarizes the potential of the cell $Pd|H_2(g, 1\ \text{bar})|BH(aq, b), B(aq, b)|AgCl(s)|Ag$. Each measurement is made at equimolar concentrations of 2-aminopyridinium chloride (BH) and 2-aminopyridine (B). The data are for 25°C and it is found that $E^{\ominus}_{\text{cell}} = 0.22251$ V. Use the data to determine pK_a for the acid at 25°C and the mean activity coefficient ($\gamma_{\pm}$) of BH as a function of molality (b) and ionic strength (I). Use the extended Debye–Hückel equation for the mean activity coefficient in the form

$$\log \gamma_{\pm} = -\frac{AI^{1/2}}{1+BI^{1/2}} + Cb$$

where $A = 0.5091$ and B and C are parameters that depend upon the ions. Draw a graph of the mean activity coefficient with $b = 0.04\ \text{mol kg}^{-1}$ and $0 \le I \le 0.1$.

$b/(\text{mol kg}^{-1})$	0.01	0.02	0.03	0.04	0.05
$E_{\text{cell}}(25°\text{C})/\text{V}$	0.74452	0.72853	0.71928	0.71314	0.70809
$b/(\text{mol kg}^{-1})$	0.06	0.07	0.08	0.09	0.10
$E_{\text{cell}}(25°\text{C})/\text{V}$	0.70380	0.70059	0.69790	0.69571	0.69338

Hint. Use mathematical software or a spreadsheet.

Theoretical problems

6.22 Express the equilibrium constant of a gas-phase reaction $A + 3\,B \rightleftharpoons 2\,C$ in terms of the equilibrium value of the extent of reaction, ξ, given that initially A and B were present in stoichiometric proportions. Find an expression for ξ as a function of the total pressure, p, of the reaction mixture and sketch a graph of the expression obtained.

6.23 Find an expression for the standard reaction Gibbs energy at a temperature T' in terms of its value at another temperature T and the coefficients a, b, and c in the expression for the molar heat capacity listed in Table 2.2. Evaluate the standard Gibbs energy of formation of $H_2O(l)$ at 372 K from its value at 298 K.

6.24 Derive an expression for the temperature dependence of K_c for a gas-phase reaction.

Applications: to biology, environmental science, and chemical engineering

6.25 Here we investigate the molecular basis for the observation that the hydrolysis of ATP is exergonic at pH = 7.0 and 310 K. (a) It is thought that the exergonicity of ATP hydrolysis is due in part to the fact that the standard entropies of hydrolysis of polyphosphates are positive. Why would an increase in entropy accompany the hydrolysis of a triphosphate group into a diphosphate and a phosphate group? (b) Under identical conditions, the Gibbs energies of hydrolysis of H_4ATP and $MgATP^{2-}$, a complex between the Mg^{2+} ion and ATP^{4-}, are less negative than the Gibbs energy of hydrolysis of ATP^{4-}. This observation has been used to support the hypothesis that electrostatic repulsion between adjacent phosphate groups is a factor that controls the exergonicity of ATP hydrolysis. Provide a rationale for the hypothesis and discuss how the experimental evidence supports it. Do these electrostatic effects contribute to the $\Delta_r H$ or $\Delta_r S$ terms that determine the exergonicity of the reaction? *Hint.* In the $MgATP^{2-}$ complex, the Mg^{2+} ion and ATP^{4-} anion form two bonds: one that involves a negatively charged oxygen belonging to the terminal phosphate group of ATP^{4-} and another that involves a negatively charged oxygen belonging to the phosphate group adjacent to the terminal phosphate group of ATP^{4-}.

6.26 To get a sense of the effect of cellular conditions on the ability of ATP to drive biochemical processes, compare the standard Gibbs energy of hydrolysis of ATP to ADP with the reaction Gibbs energy in an environment at 37°C in which pH = 7.0 and the ATP, ADP, and P_i^- concentrations are all $1.0\ \mu\text{mol dm}^{-3}$.

6.27 Under biochemical standard conditions, aerobic respiration produces approximately 38 molecules of ATP per molecule of glucose that is completely oxidized. (a) What is the percentage efficiency of aerobic respiration under biochemical standard conditions? (b) The following conditions are more likely to be observed in a living cell: $p_{CO_2} = 5.3 \times 10^{-2}$ atm, $p_{O_2} = 0.132$ atm, $[\text{glucose}] = 5.6 \times 10^{-2}\ \text{mol dm}^{-3}$, $[\text{ATP}] = [\text{ADP}] = [\text{P}_i] = 1.0 \times 10^{-4}\ \text{mol dm}^{-3}$, pH = 7.4, $T = 310$ K. Assuming that activities can be replaced by the numerical values of molar concentrations, calculate the efficiency of aerobic respiration under these physiological conditions. (c) A typical diesel engine operates between $T_c = 873$ K and $T_h = 1923$ K with an efficiency that is approximately 75 per cent of the theoretical limit of $(1 - T_c/T_h)$ (see Section 3.2). Compare the efficiency of a typical diesel engine with that of aerobic respiration under typical physiological conditions (see part b). Why is biological energy conversion more or less efficient than energy conversion in a diesel engine?

6.28 In anaerobic bacteria, the source of carbon may be a molecule other than glucose and the final electron acceptor is some molecule other than O_2. Could a bacterium evolve to use the ethanol/nitrate pair instead of the glucose/O_2 pair as a source of metabolic energy?

6.29 The standard potentials of proteins are not commonly measured by the methods described in this chapter because proteins often lose their native structure and function when they react on the surfaces of electrodes. In an alternative method, the oxidized protein is allowed to react with an appropriate electron donor in solution. The standard potential of the protein is then determined from the Nernst equation, the equilibrium concentrations of all species in solution, and the known standard potential of the electron donor. We illustrate this method with the protein cytochrome *c*. The one-electron reaction between cytochrome *c*, cyt, and 2,6-dichloroindophenol, D, can be followed spectrophotometrically because each of the four species in solution has a distinct absorption spectrum. We write the reaction as $\text{cyt}_{\text{ox}} + \text{D}_{\text{red}} \rightleftharpoons \text{cyt}_{\text{red}} + \text{D}_{\text{ox}}$, where the subscripts 'ox' and 'red' refer to oxidized and reduced states, respectively. (a) Consider $E^{\ominus}_{\text{cyt}}$ and $E^{\ominus}_{\text{D}}$ to be the standard potentials of cytochrome *c* and D, respectively. Show that, at equilibrium, a plot of $\ln([\text{D}_{\text{ox}}]_{\text{eq}}/[\text{D}_{\text{red}}]_{\text{eq}})$ versus $\ln([\text{cyt}_{\text{ox}}]_{\text{eq}}/[\text{cyt}_{\text{red}}]_{\text{eq}})$ is linear with slope of 1 and *y*-intercept $F(E^{\ominus}_{\text{cyt}} - E^{\ominus}_{\text{D}})/RT$, where equilibrium activities are replaced by the numerical values of equilibrium molar concentrations. (b) The following data were obtained for the reaction between oxidized cytochrome *c* and reduced D in a pH 6.5 buffer at 298 K. The ratios $[\text{D}_{\text{ox}}]_{\text{eq}}/[\text{D}_{\text{red}}]_{\text{eq}}$ and $[\text{cyt}_{\text{ox}}]_{\text{eq}}/[\text{cyt}_{\text{red}}]_{\text{eq}}$ were adjusted by titrating a solution containing oxidized cytochrome *c* and reduced D with a solution of sodium ascorbate, which is a strong reductant. From the data and the standard potential of D of 0.237 V, determine the standard potential cytochrome *c* at pH 6.5 and 298K.

$[\text{D}_{\text{ox}}]_{\text{eq}}/[\text{D}_{\text{red}}]_{\text{eq}}$	0.00279	0.00843	0.0257	0.0497	0.0748	0.238	0.534
$[\text{cyt}_{\text{ox}}]_{\text{eq}}/[\text{cyt}_{\text{red}}]_{\text{eq}}$	0.0106	0.0230	0.0894	0.197	0.335	0.809	1.39

6.30‡ The dimerization of ClO in the Antarctic winter stratosphere is believed to play an important part in that region's severe seasonal depletion of ozone. The following equilibrium constants are based on measurements on the reaction $2\,\text{ClO (g)} \rightarrow (\text{ClO})_2\,\text{(g)}$.

T/K	233	248	258	268	273	280	288	295	303
K	4.13×10^8	5.00×10^7	1.45×10^7	5.37×10^6	3.20×10^6	9.62×10^5	4.28×10^5	1.67×10^5	6.02×10^4

(a) Derive the values of $\Delta_r H^{\ominus}$ and $\Delta_r S^{\ominus}$ for this reaction. (b) Compute the standard enthalpy of formation and the standard molar entropy of $(\text{ClO})_2$ given $\Delta_f H^{\ominus}(\text{ClO}) = +101.8\ \text{kJ mol}^{-1}$ and $S^{\ominus}_{\text{m}}(\text{ClO}) = 266.6\ \text{J K}^{-1}\ \text{mol}^{-1}$.

6.31‡ Nitric acid hydrates have received much attention as possible catalysts for heterogeneous reactions that bring about the Antarctic ozone hole. Standard reaction Gibbs energies are as follows:

(i) $\text{H}_2\text{O (g)} \rightarrow \text{H}_2\text{O (s)}$ $\quad \Delta_r G^{\ominus} = -23.6\ \text{kJ mol}^{-1}$

(ii) $\text{H}_2\text{O (g)} + \text{HNO}_3\text{ (g)} \rightarrow \text{HNO}_3\cdot\text{H}_2\text{O (s)}$ $\quad \Delta_r G^{\ominus} = -57.2\ \text{kJ mol}^{-1}$

(iii) $2\,\text{H}_2\text{O (g)} + \text{HNO}_3\text{ (g)} \rightarrow \text{HNO}_3\cdot 2\text{H}_2\text{O (s)}$ $\quad \Delta_r G^{\ominus} = -85.6\ \text{kJ mol}^{-1}$

(iv) $3\,\text{H}_2\text{O (g)} + \text{HNO}_3\text{ (g)} \rightarrow \text{HNO}_3\cdot 3\text{H}_2\text{O (s)}$ $\quad \Delta_r G^{\ominus} = -112.8\ \text{kJ mol}^{-1}$

Which solid is thermodynamically most stable at 190 K if $p_{\text{H}_2\text{O}} = 1.3 \times 10^{-7}$ bar and $p_{\text{HNO}_3} = 4.1 \times 10^{-10}$ bar? *Hint.* Try computing $\Delta_r G$ for each reaction under the prevailing conditions; if more than one solid forms spontaneously, examine $\Delta_r G$ for the conversion of one solid to another.

6.32‡ Suppose that an iron catalyst at a particular manufacturing plant produces ammonia in the most cost-effective manner at 450°C when the pressure is such that $\Delta_r G$ for the reaction $\tfrac{1}{2}\text{N}_2\text{(g)} + \tfrac{3}{2}\text{H}_2\text{(g)} \rightarrow \text{NH}_3\text{(g)}$ is equal to $-500\ \text{J mol}^{-1}$. (a) What pressure is needed? (b) Now suppose that a new catalyst is developed that is most cost-effective at 400°C when the pressure gives the same value of $\Delta_r G$. What pressure is needed when the new catalyst is used? What are the advantages of the new catalyst? Assume that (i) all gases are perfect gases or that (ii) all gases are van der Waals gases. Isotherms of $\Delta_r G(T, p)$ in the pressure range $100\ \text{atm} \le p \le 400\ \text{atm}$ are needed to derive the answer. (c) Do the isotherms you plotted confirm Le Chatelier's principle concerning the response of equilibrium changes in temperature and pressure?

PART 2 Structure

In Part 1 we examined the properties of bulk matter from the viewpoint of thermodynamics. In Part 2 we examine the structures and properties of individual atoms and molecules from the viewpoint of quantum mechanics. The two viewpoints merge in Chapter 15.

Quantum theory: introduction and principles

7

This chapter introduces some of the basic principles of quantum mechanics. First, it reviews the experimental results that overthrew the concepts of classical physics. These experiments led to the conclusion that particles may not have an arbitrary energy and that the classical concepts of 'particle' and 'wave' blend together. The overthrow of classical mechanics inspired the formulation of a new set of concepts and led to the formulation of quantum mechanics. In quantum mechanics, all the properties of a system are expressed in terms of a wavefunction that is obtained by solving the Schrödinger equation. We see how to interpret wavefunctions. Finally, we introduce some of the techniques of quantum mechanics in terms of operators, and see that they lead to the uncertainty principle, one of the most profound departures from classical mechanics.

It was once thought that the motion of atoms and subatomic particles could be expressed using **classical mechanics**, the laws of motion introduced in the seventeenth century by Isaac Newton, for these laws were very successful at explaining the motion of everyday objects and planets. However, towards the end of the nineteenth century, experimental evidence accumulated showing that classical mechanics failed when it was applied to particles as small as electrons, and it took until the 1920s to discover the appropriate concepts and equations for describing them. We describe the concepts of this new mechanics, which is called **quantum mechanics**, in this chapter, and apply them throughout the remainder of the text.

The origins of quantum mechanics

The basic principles of classical mechanics are reviewed in *Further information 7.1*. In brief, they show that classical physics (1) predicts a precise trajectory for particles, with precisely specified locations and momenta at each instant, and (2) allows the translational, rotational, and vibrational modes of motion to be excited to any energy simply by controlling the forces that are applied. These conclusions agree with everyday experience. Everyday experience, however, does not extend to individual atoms, and careful experiments of the type described below have shown that classical mechanics fails when applied to the transfers of very small energies and to objects of very small mass.

We shall also investigate the properties of light. In classical physics, light is described as electromagnetic radiation, which is understood in terms of the **electromagnetic field**, an oscillating electric and magnetic disturbance that spreads as a harmonic wave, wave displacements that can be expressed as sine or cosine functions (see *Fundamentals* F.6), through empty space, the vacuum. Such waves are generated by

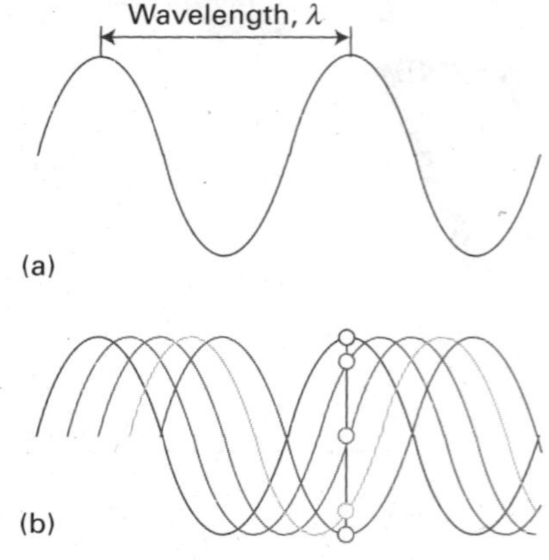

Fig. 7.1 The wavelength, λ, of a wave is the peak-to-peak distance. (b) The wave is shown travelling to the right at a speed c. At a given location, the instantaneous amplitude of the wave changes through a complete cycle (the five dots show half a cycle). The frequency, ν, is the number of cycles per second that occur at a given point.

the acceleration of electric charge, as in the oscillating motion of electrons in the antenna of a radio transmitter. The wave travels at a constant speed called the *speed of light*, c, which is about 3×10^8 m s^{-1}. As its name suggests, an electromagnetic field has two components, an **electric field** that acts on charged particles (whether stationary of moving) and a **magnetic field** that acts only on moving charged particles. The electromagnetic field is characterized by a **wavelength**, λ (lambda), the distance between the neighbouring peaks of the wave, and its **frequency**, ν (nu), the number of times per second at which its displacement at a fixed point returns to its original value (Fig. 7.1). The frequency is measured in *hertz*, where 1 Hz = 1 s^{-1}. The wavelength and frequency of an electromagnetic wave are related by

$$\lambda\nu = c \tag{7.1}$$

Therefore, the shorter the wavelength, the higher the frequency. The characteristics of the wave are also reported by giving the **wavenumber**, $\tilde{\nu}$ (nu tilde), of the radiation, where

$$\tilde{\nu} = \frac{\nu}{c} = \frac{1}{\lambda} \qquad [7.2]$$

Wavenumbers are normally reported in reciprocal centimetres (cm^{-1}).

Figure 7.2 summarizes the **electromagnetic spectrum**, the description and classification of the electromagnetic field according to its frequency and wavelength. 'Light' is electromagnetic radiation that falls in the visible region of the spectrum. White light is a mixture of electromagnetic radiation with wavelengths ranging from about 400 nm to about 700 nm (1 nm = 10^{-9} m). Our eyes perceive different wavelengths of radiation in this range as different colours, so it can be said that white light is a mixture of light of all different colours.

7.1 Energy quantization

Key points **(a) The classical approach to the description of black-body radiation results in the ultraviolet catastrophe. (b) To avoid this catastrophe, Planck proposed that the electromagnetic field could take up energy only in discrete amounts. (c) The thermal properties of solids, specifically their heat capacities, also provide evidence that the vibrations of atoms can take up energy only in discrete amounts. (d) Atomic and molecular spectra show that atoms and molecules can take up energy only in discrete amounts.**

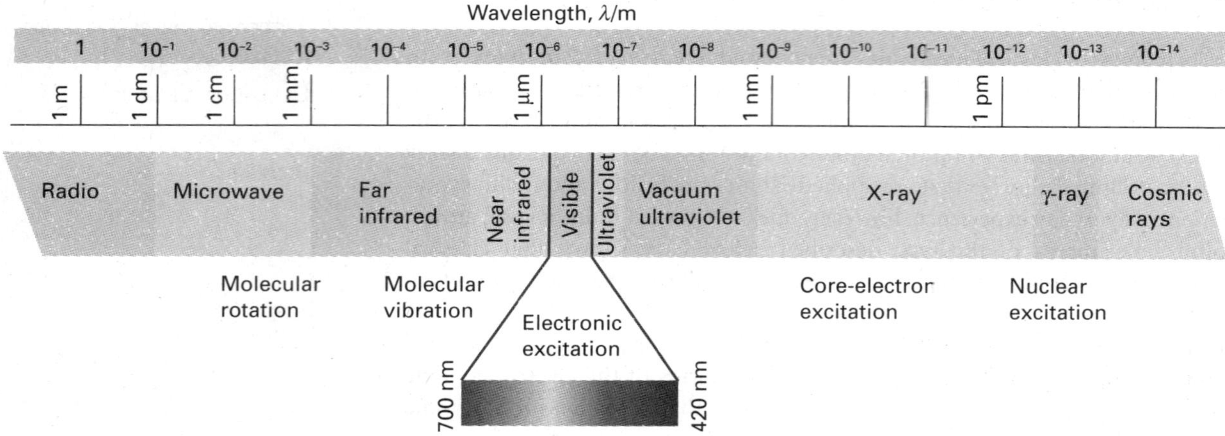

Fig. 7.2 The electromagnetic spectrum and the classification of the spectral regions.

The overthrow of classical mechanics and its replacement by quantum mechanics was driven, as always in science, by noticing that experimental observations conflicted with the predictions of accepted theory. Here we outline three examples of experiment overthrowing current theory, which came to light at the end of the nineteenth century and which drove scientists to the view that energy can be transferred only in discrete amounts.

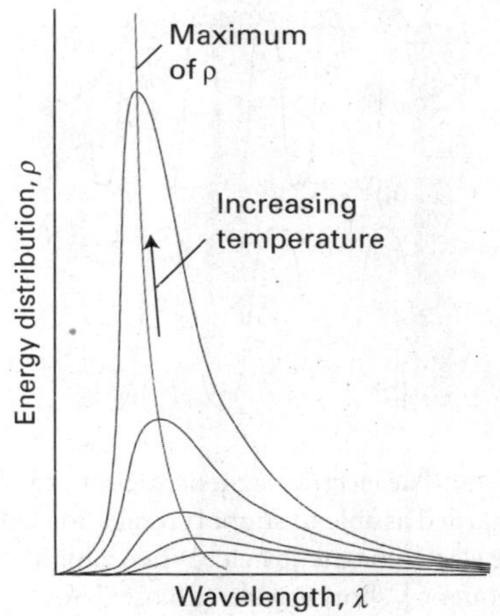

Fig. 7.3 The energy distribution in a black-body cavity at several temperatures. Note how the energy density increases in the region of shorter wavelengths as the temperature is raised, and how the peak shifts to shorter wavelengths. The total energy density (the area under the curve) increases as the temperature is increased (as T^4).

(a) Black-body radiation

A hot object emits electromagnetic radiation. At high temperatures, an appreciable proportion of the radiation is in the visible region of the spectrum, and a higher proportion of short-wavelength blue light is generated as the temperature is raised. This behaviour is seen when a heated metal bar glowing red hot becomes white hot when heated further. The dependence is illustrated in Fig. 7.3, which shows how the energy output varies with wavelength at several temperatures. The curves are those of an ideal emitter called a **black body**, which is an object capable of emitting and absorbing all wavelengths of radiation uniformly. A good approximation to a black body is a pinhole in an empty container maintained at a constant temperature, because any radiation leaking out of the hole has been absorbed and re-emitted inside so many times as it reflected around inside the container that it has come to thermal equilibrium with the walls (Fig. 7.4).

The approach adopted by nineteenth-century scientists to explain black-body radiation was to calculate the **energy density**, $d\mathcal{E}$, the total energy in a region of the electromagnetic field divided by the volume of the region (units: joules per metre-cubed, $J\,m^{-3}$), due to all the oscillators corresponding to wavelengths between λ and $\lambda + d\lambda$. This energy density is proportional to the width, $d\lambda$, of this range, and is written

$$d\mathcal{E}(\lambda,T) = \rho(\lambda,T)d\lambda \tag{7.3}$$

where ρ (rho), the constant of proportionality between $d\mathcal{E}$ and $d\lambda$, is called the **density of states** (units: joules per metre^4, $J\,m^{-4}$). A high density of states at the wavelength λ and temperature T simply means that there is a lot of energy associated with wavelengths lying between λ and $\lambda + d\lambda$ at that temperature. The total energy density in a region is the integral over all wavelengths:

$$\mathcal{E}(T) = \int_0^\infty \rho(\lambda,T)d\lambda \tag{7.4}$$

and depends on the temperature: the higher the temperature, the greater the energy density. Just as the mass of an object is its mass density multiplied by its volume, the total energy within a region of volume V is this energy density multiplied by the volume:

$$E(T) = V\mathcal{E}(T) \tag{7.5}$$

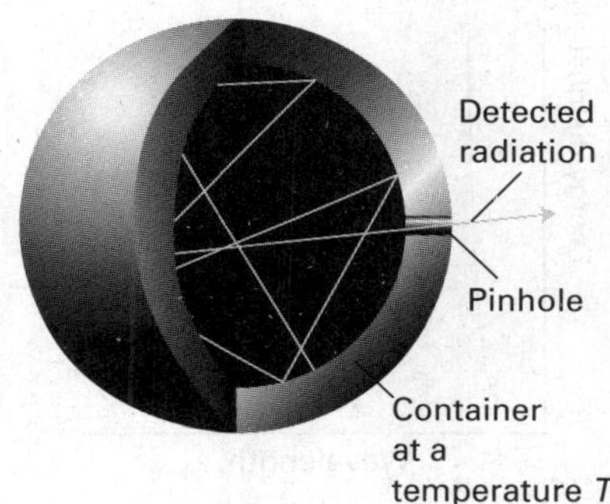

Fig. 7.4 An experimental representation of a black body is a pinhole in an otherwise closed container. The radiation is reflected many times within the container and comes to thermal equilibrium with the walls at a temperature T. Radiation leaking out through the pinhole is characteristic of the radiation within the container.

The physicist Lord Rayleigh thought of the electromagnetic field as a collection of oscillators of all possible frequencies. He regarded the presence of radiation of frequency ν (and therefore of wavelength $\lambda = c/\nu$) as signifying that the electromagnetic oscillator of that frequency had been excited (Fig. 7.5). Rayleigh knew that according to the classical equipartition principle (*Fundamentals* F.5b), the average energy of each oscillator, regardless of its frequency, is kT. On that basis, with minor help from James Jeans, he arrived at the **Rayleigh–Jeans law** for the density of states:

$$\rho(\lambda,T) = \frac{8\pi kT}{\lambda^4} \qquad \text{Rayleigh–Jeans law} \tag{7.6}$$

where k is Boltzmann's constant ($k = 1.381 \times 10^{-23}\,J\,K^{-1}$).

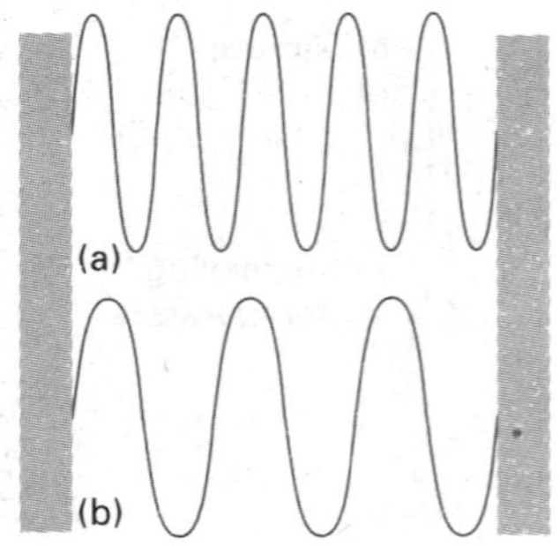

Fig. 7.5 The electromagnetic vacuum can be regarded as able to support oscillations of the electromagnetic field. When a high frequency, short wavelength oscillator (a) is excited, that frequency of radiation is present. The presence of low frequency, long wavelength radiation (b) signifies that an oscillator of the corresponding frequency has been excited.

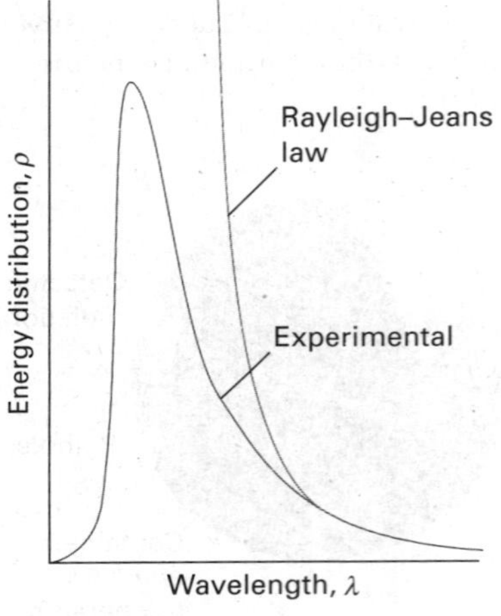

Fig. 7.6 The Rayleigh–Jeans law (eqn 7.6) predicts an infinite energy density at short wavelengths. This approach to infinity is called the *ultraviolet catastrophe.*

A brief comment

The series expansion of an exponential function is $e^x = 1 + x + \frac{1}{2}x^2 + \cdots$. If $x \ll 1$, a good approximation is $e^x \approx 1 + x$. For example, $e^{0.01} = 1.010\,050\ldots \approx 1 + 0.01$.

Although the Rayleigh–Jeans law is quite successful at long wavelengths (low frequencies), it fails badly at short wavelengths (high frequencies). Thus, as λ decreases, ρ increases without going through a maximum (Fig. 7.6). The equation therefore predicts that oscillators of very short wavelength (corresponding to ultraviolet radiation, X-rays, and even γ-rays) are strongly excited even at room temperature. The total energy density in a region, the integral in eqn 7.4, is also predicted to be infinite at all temperatures above zero. This absurd result, which implies that a large amount of energy is radiated in the high-frequency region of the electromagnetic spectrum, is called the **ultraviolet catastrophe.** According to classical physics, even cool objects should radiate in the visible and ultraviolet regions, so objects should glow in the dark; there should in fact be no darkness.

In 1900, the German physicist Max Planck found that he could account for the experimental observations by proposing that the energy of each electromagnetic oscillator is limited to discrete values and cannot be varied arbitrarily. This proposal is contrary to the viewpoint of classical physics in which all possible energies are allowed and every oscillator has a mean energy kT. The limitation of energies to discrete values is called the **quantization of energy.** In particular, Planck found that he could account for the observed distribution of energy if he supposed that the permitted energies of an electromagnetic oscillator of frequency ν are integer multiples of $h\nu$:

$$E = nh\nu \qquad n = 0, 1, 2, \ldots \tag{7.7}$$

where h is a fundamental constant now known as **Planck's constant.** On the basis of this assumption, Planck was able to derive the **Planck distribution:**

$$\rho(\lambda,T) = \frac{8\pi hc}{\lambda^5(e^{hc/\lambda kT} - 1)} \tag{7.8}$$

Planck distribution

This expression fits the experimental curve very well at all wavelengths (Fig. 7.7), and the value of h, which is an undetermined parameter in the theory, may be obtained by varying its value until a best fit is obtained. The currently accepted value for h is 6.626×10^{-34} J s.

As usual, it is a good idea to 'read' the content of an equation:

1. The Planck distribution resembles the Rayleigh–Jeans law (eqn 7.6) apart from the all-important exponential factor in the denominator. For short wavelengths, $hc/\nu kT \gg 1$ and $e^{hc/\lambda kT} \to \infty$ faster than $\lambda^5 \to 0$; therefore $\rho \to 0$ as $\lambda \to 0$ or $\nu \to \infty$. Hence, the energy density approaches zero at high frequencies, in agreement with observation.

2. For long wavelengths, $hc/\lambda kT \ll 1$, and the denominator in the Planck distribution can be replaced by

$$e^{hc/\lambda kT} - 1 = \left(1 + \frac{hc}{\lambda kT} + \cdots\right) - 1 \approx \frac{hc}{\lambda kT}$$

When this approximation is substituted into eqn 7.8, we find that the Planck distribution reduces to the Rayleigh–Jeans law.

3. As we should infer from the graph in Fig. 7.7, the total energy density (the integral in eqn 7.4 and therefore the area under the curve) is no longer infinite, and in fact

$$\mathcal{E}(T) = \int_0^\infty \frac{8\pi hc}{\lambda^5(e^{hc/\lambda kT} - 1)} d\lambda = aT^4 \qquad \text{with} \qquad a = \frac{8\pi^5 k^4}{15(hc)^3} \tag{7.9}$$

That is, the energy density increases as the fourth power of the temperature.

• **A brief illustration**

We can now see why an incandescent lamp is so effective. Before it is switched on, the energy density inside the glass envelope corresponds to about 20°C (293 K). When it is switched on, the temperature of the filament rises to about 2000 K. The energy density increases by a factor of $(2000\ \text{K}/293\ \text{K})^4 \approx 2000$, and it generates nearly white light. •

It is easy to see why Planck's approach was successful while Rayleigh's was not. The thermal motion of the atoms in the walls of the black body excites the oscillators of the electromagnetic field. According to classical mechanics, all the oscillators of the field share equally in the energy supplied by the walls, so even the highest frequencies are excited. The excitation of very high frequency oscillators results in the ultraviolet catastrophe. According to Planck's hypothesis, however, oscillators are excited only if they can acquire an energy of at least $h\nu$. This energy is too large for the walls to supply in the case of the very high frequency oscillators, so the latter remain unexcited. The effect of quantization is to reduce the contribution from the high frequency oscillators, for they cannot be significantly excited with the energy available.

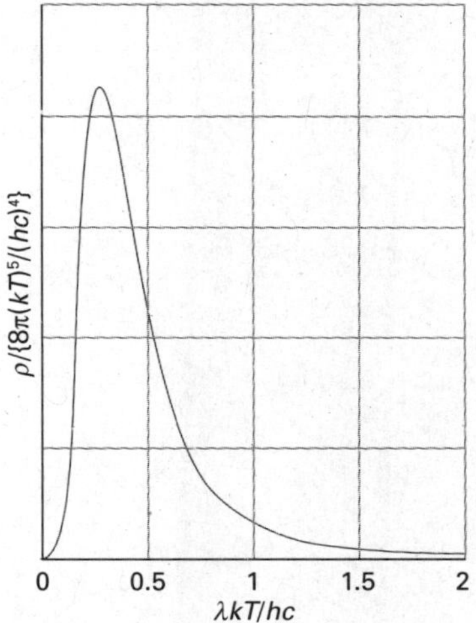

Fig. 7.7 The Planck distribution (eqn 7.8) accounts very well for the experimentally determined distribution of black-body radiation. Planck's quantization hypothesis essentially quenches the contributions of high frequency, short wavelength oscillators. The distribution coincides with the Rayleigh–Jeans distribution at long wavelengths.

interActivity Plot the Planck distribution at several temperatures and confirm that eqn 7.8 predicts the behaviour summarized by Fig. 7.3.

(b) Heat capacities

In the early nineteenth century, the French scientists Pierre-Louis Dulong and Alexis-Thérèse Petit determined the heat capacities, $C_V = (\partial U/\partial T)_V$ (Section 2.4), of a number of monatomic solids. On the basis of some somewhat slender experimental evidence, they proposed that the molar heat capacities of all monatomic solids are the same and (in modern units) close to 25 J K^{-1} mol^{-1}.

Dulong and Petit's law is easy to justify in terms of classical physics in much the same way as Rayleigh attempted to explain black-body radiation. If classical physics were valid, the equipartition principle could be used to infer that the mean energy of an atom as it oscillates about its mean position in a solid is kT for each direction of displacement. As each atom can oscillate in three dimensions, the average energy of each atom is $3kT$; for N atoms the total energy is $3NkT$. The contribution of this motion to the molar internal energy is therefore

$$U_m = 3N_AkT = 3RT \qquad (7.10a)$$

because $N_Ak = R$, the gas constant. The molar constant volume heat capacity is then predicted to be

$$C_{V,m} = \left(\frac{\partial U_m}{\partial T}\right)_V = 3R \qquad (7.10b)$$

This result, with $3R = 24.9$ J K^{-1} mol^{-1}, is in striking accord with Dulong and Petit's value.

Unfortunately, significant deviations from their law were observed when advances in refrigeration techniques made it possible to measure heat capacities at low temperatures. It was found that the molar heat capacities of all monatomic solids are lower than $3R$ at low temperatures, and that the values approach zero as $T \to 0$. To account for these observations, Einstein (in 1905) assumed that each atom oscillated about its equilibrium position with a single frequency ν. He then invoked Planck's hypothesis to assert that the energy of oscillation is confined to discrete values, and specifically to $nh\nu$, where n is an integer. Einstein discarded the equipartition result, calculated the vibrational contribution of the atoms to the total molar internal energy of the solid (by a method described in Section 16.4), and obtained the expression known as the **Einstein formula**:

$$C_{V,\mathrm{m}}(T) = 3Rf_{\mathrm{E}}(T) \qquad f_{\mathrm{E}}(T) = \left(\frac{\theta_{\mathrm{E}}}{T}\right)^2 \left(\frac{\mathrm{e}^{\theta_{\mathrm{E}}/2T}}{\mathrm{e}^{\theta_{\mathrm{E}}/T} - 1}\right)^2 \quad \text{Einstein formula} \tag{7.11}$$

The **Einstein temperature,** $\theta_{\mathrm{E}} = h\nu/k$, is a way of expressing the frequency of oscillation of the atoms as a temperature: a high frequency corresponds to a high Einstein temperature.

As before, we now 'read' this expression:

1. At high temperatures (when $T \gg \theta_{\mathrm{E}}$) the exponentials in f_{E} can be expanded as $1 + \theta_{\mathrm{E}}/T + \cdots$ and higher terms ignored. The result is

$$f_{\mathrm{E}}(T) = \left(\frac{\theta_{\mathrm{E}}}{T}\right)^2 \left\{\frac{1 + \theta_{\mathrm{E}}/2T + \cdots}{(1 + \theta_{\mathrm{E}}/T + \cdots) - 1}\right\}^2 \approx 1 \tag{7.12a}$$

Consequently, the classical result ($C_{V,\mathrm{m}} = 3R$) is obtained at high temperatures.

2. At low temperatures, when $T \ll \theta_{\mathrm{E}}$,

$$f_{\mathrm{E}}(T) \approx \left(\frac{\theta_{\mathrm{E}}}{T}\right)^2 \left(\frac{\mathrm{e}^{\theta_{\mathrm{E}}/2T}}{\mathrm{e}^{\theta_{\mathrm{E}}/T}}\right)^2 = \left(\frac{\theta_{\mathrm{E}}}{T}\right)^2 \mathrm{e}^{-\theta_{\mathrm{E}}/T} \tag{7.12b}$$

The strongly decaying exponential function goes to zero more rapidly than $1/T$ goes to infinity; so $f_{\mathrm{E}} \to 0$ as $T \to 0$, and the heat capacity therefore approaches zero too.

We see that Einstein's formula accounts for the decrease of heat capacity at low temperatures. The physical reason for this success is that at low temperatures only a few oscillators possess enough energy to oscillate significantly so the solid behaves as though it contains far fewer atoms than is actually the case. At higher temperatures, there is enough energy available for all the oscillators to become active: all $3N$ oscillators contribute, many of their energy levels are accessible, and the heat capacity approaches its classical value.

Figure 7.8 shows the temperature dependence of the heat capacity predicted by the Einstein formula. The general shape of the curve is satisfactory, but the numerical agreement is in fact quite poor. The poor fit arises from Einstein's assumption that all the atoms oscillate with the same frequency, whereas in fact they oscillate over a range of frequencies from zero up to a maximum value, ν_{D}. This complication is taken into account by averaging over all the frequencies present, the final result being the **Debye formula:**

$$C_{V,\mathrm{m}} = 3Rf_{\mathrm{D}}(T) \qquad f_{\mathrm{D}}(T) = 3\left(\frac{T}{\theta_{\mathrm{D}}}\right)^3 \int_0^{\theta_{\mathrm{D}}/T} \frac{x^4\mathrm{e}^x}{(\mathrm{e}^x - 1)^2}\mathrm{d}x \quad \text{Debye formula} \tag{7.13}$$

where $\theta_{\mathrm{D}} = h\nu_{\mathrm{D}}/k$ is the Debye temperature. The integral in eqn 7.13 has to be evaluated numerically, but that is simple with mathematical software. The details of this modification, which, as Fig. 7.9 shows, gives improved agreement with experiment, need not distract us at this stage from the main conclusion, which is that quantization must be introduced in order to explain the thermal properties of solids.

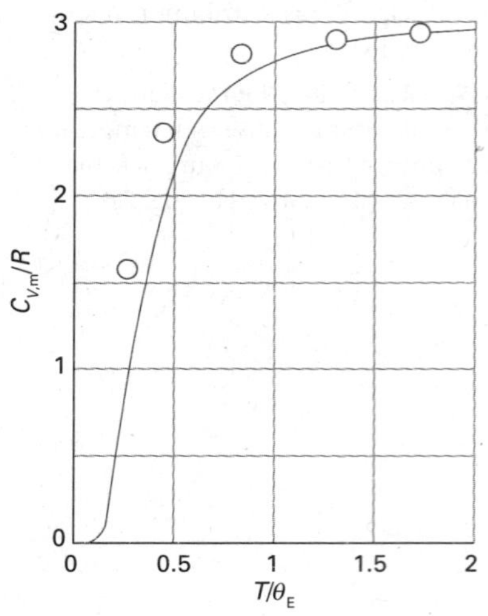

Fig. 7.8 Experimental low-temperature molar heat capacities and the temperature dependence predicted on the basis of Einstein's theory. His equation (eqn 7.11) accounts for the dependence fairly well, but is everywhere too low.

interActivity Using eqn 7.11, plot $C_{V,\mathrm{m}}$ against T for several values of the Einstein temperature θ_{E}. At low temperature, does an increase in θ_{E} result in an increase or decrease of $C_{V,\mathrm{m}}$? Estimate the temperature at which the value of $C_{V,\mathrm{m}}$ reaches the classical value given by eqn 7.10.

• **A brief illustration**

The Debye temperature for lead is 105 K, corresponding to a vibrational frequency of 2.2×10^{12} Hz, whereas that for diamond and its much lighter, more rigidly bonded atoms, is 2230 K, corresponding to 4.6×10^{13} Hz. As we see from Fig. 7.9, $f \approx 1$ for $T > \theta_{\mathrm{D}}$ and the heat capacity is almost classical. For lead at 25°C, corresponding to $T/\theta_{\mathrm{D}} = 2.8$, $f = 0.99$ and the heat capacity has almost its classical value. For diamond at the same temperature, $T/\theta_{\mathrm{D}} = 0.13$, corresponding to $f = 0.15$, and the heat capacity is only 15 per cent of its classical value. •

(c) Atomic and molecular spectra

The most compelling and direct evidence for the quantization of energy comes from **spectroscopy**, the detection and analysis of the electromagnetic radiation absorbed, emitted, or scattered by a substance. The record of light intensity transmitted or scattered by a molecule as a function of frequency (ν), wavelength (λ), or wavenumber ($\tilde{\nu} = \nu/c$) is called its **spectrum** (from the Latin word for appearance).

A typical atomic spectrum is shown in Fig. 7.10, and a typical molecular spectrum is shown in Fig. 7.11. The obvious feature of both is that radiation is emitted or absorbed at a series of discrete frequencies. This observation can be understood if the energy of the atoms or molecules is also confined to discrete values, for then energy can be discarded or absorbed only in discrete amounts (Fig. 7.12). Then, if the energy of an atom decreases by ΔE, the energy is carried away as radiation of frequency ν, and an emission 'line', a sharply defined peak, appears in the spectrum. We say that a molecule undergoes a **spectroscopic transition**, a change of state, when the **Bohr frequency condition**

$$\Delta E = h\nu \qquad \text{Bohr frequency condition} \qquad (7.14)$$

is fulfilled. We develop the principles and applications of atomic spectroscopy in Chapter 9 and of molecular spectroscopy in Chapters 12–14.

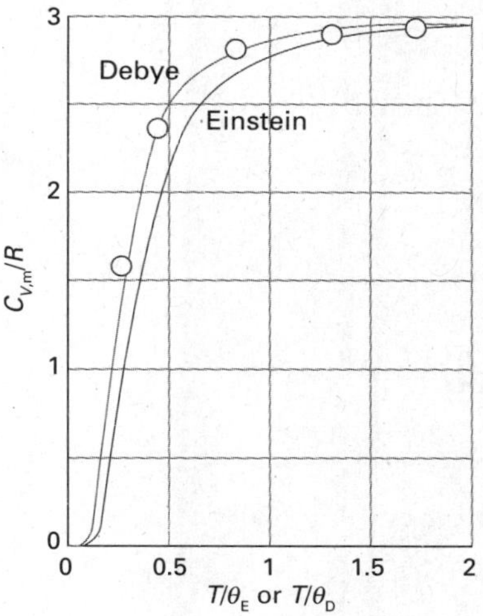

Fig. 7.9 Debye's modification of Einstein's calculation (eqn 7.13) gives very good agreement with experiment. For copper, $T/\theta_D = 2$ corresponds to about 170 K, so the detection of deviations from Dulong and Petit's law had to await advances in low-temperature physics.

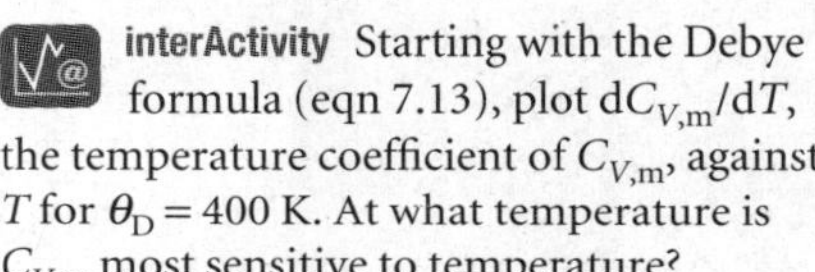

interActivity Starting with the Debye formula (eqn 7.13), plot $dC_{V,m}/dT$, the temperature coefficient of $C_{V,m}$, against T for $\theta_D = 400$ K. At what temperature is $C_{V,m}$ most sensitive to temperature?

7.2 Wave–particle duality

Key points (a) The photoelectric effect establishes the view that electromagnetic radiation, regarded in classical physics as wave-like, consists of particles (photons). (b) The diffraction of electrons establishes the view that electrons, regarded in classical physics as particles, are wave-like with a wavelength given by the de Broglie relation.

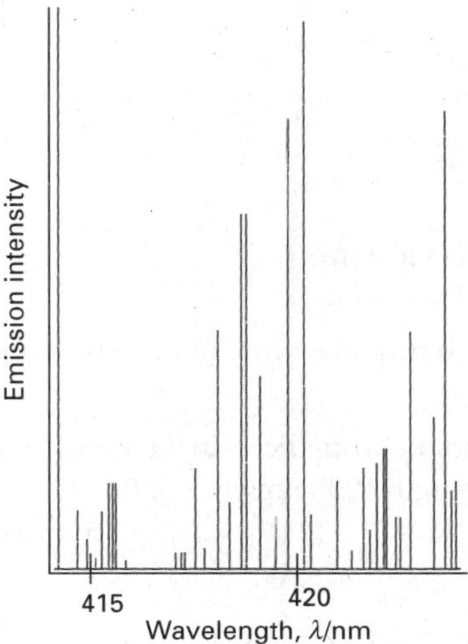

Fig. 7.10 A region of the spectrum of radiation emitted by excited iron atoms consists of radiation at a series of discrete wavelengths (or frequencies).

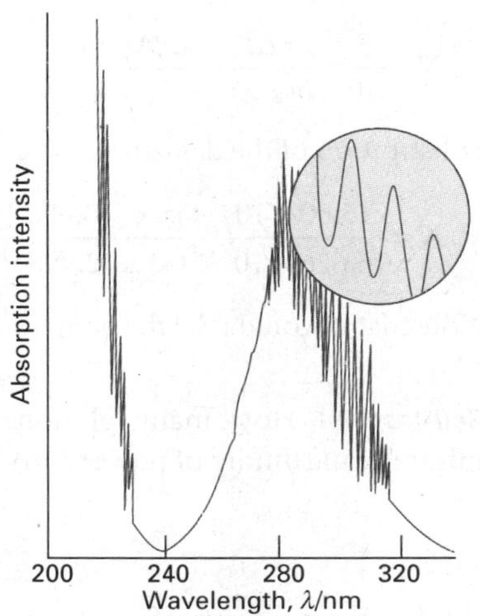

Fig. 7.11 When a molecule changes its state, it does so by absorbing radiation at definite frequencies. This spectrum is part of that due to the electronic, vibrational, and rotational excitation of sulfur dioxide (SO_2) molecules. This observation suggests that molecules can possess only discrete energies, not an arbitrary energy.

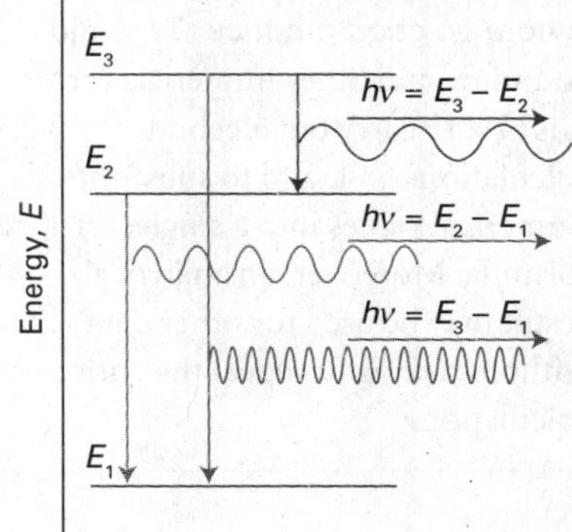

Fig. 7.12 Spectroscopic transitions, such as those shown above, can be accounted for if we assume that a molecule emits a photon as it changes between discrete energy levels. Note that high-frequency radiation is emitted when the energy change is large.

At this stage we have established that the energies of the electromagnetic field and of oscillating atoms are quantized. In this section we shall see the experimental evidence that led to the revision of two other basic concepts concerning natural phenomena. One experiment shows that electromagnetic radiation—which classical physics treats as wave-like—actually also displays the characteristics of particles. Another experiment shows that electrons—which classical physics treats as particles—also display the characteristics of waves.

(a) The particle character of electromagnetic radiation

The observation that electromagnetic radiation of frequency ν can possess only the energies 0, $h\nu$, $2h\nu$, . . . suggests (and at this stage it is only a suggestion) that it can be thought of as consisting of 0, 1, 2, . . . particles, each particle having an energy $h\nu$. Then, if one of these particles is present, the energy is $h\nu$, if two are present the energy is $2h\nu$, and so on. These particles of electromagnetic radiation are now called **photons**. The observation of discrete spectra from atoms and molecules can be pictured as the atom or molecule generating a photon of energy $h\nu$ when it discards an energy of magnitude ΔE, with $\Delta E = h\nu$.

Example 7.1 *Calculating the number of photons*

Calculate the number of photons emitted by a 100 W yellow lamp in 1.0 s. Take the wavelength of yellow light as 560 nm and assume 100 per cent efficiency.

Method Each photon has an energy $h\nu$, so the total number of photons needed to produce an energy E is $E/h\nu$. To use this equation, we need to know the frequency of the radiation (from $\nu = c/\lambda$) and the total energy emitted by the lamp. The latter is given by the product of the power (P, in watts) and the time interval for which the lamp is turned on ($E = P\Delta t$).

Answer The number of photons is

$$N = \frac{E}{h\nu} = \frac{P\Delta t}{h(c/\lambda)} = \frac{\lambda P\Delta t}{hc}$$

Substitution of the data gives

$$N = \frac{(5.60\times10^{-7}\ \text{m})\times(100\ \text{J s}^{-1})\times(1.0\ \text{s})}{(6.626\times10^{-34}\ \text{J s})\times(2.998\times10^{8}\ \text{m s}^{-1})} = 2.8\times10^{20}$$

Note that it would take the lamp nearly 40 min to produce 1 mol of these photons.

A note on good practice To avoid rounding and other numerical errors, it is best to carry out algebraic calculations first, and to substitute numerical values into a single, final formula. Moreover, an analytical result may be used for other data without having to repeat the entire calculation.

Self-test 7.1 How many photons does a monochromatic (single frequency) infrared rangefinder of power 1 mW and wavelength 1000 nm emit in 0.1 s? [5×10^{14}]

So far, the existence of photons is only a suggestion. Experimental evidence for their existence comes from the measurement of the energies of electrons produced in the **photoelectric effect**. This effect is the ejection of electrons from metals when they are exposed to ultraviolet radiation. The experimental characteristics of the photoelectric effect are as follows.

1. No electrons are ejected, regardless of the intensity of the radiation, unless its frequency exceeds a threshold value characteristic of the metal.

2. The kinetic energy of the ejected electrons increases linearly with the frequency of the incident radiation but is independent of the intensity of the radiation.

3. Even at low light intensities, electrons are ejected immediately if the frequency is above the threshold.

Figure 7.13 illustrates the first and second characteristics.

These observations strongly suggest that the photoelectric effect depends on the ejection of an electron when it is involved in a collision with a particle-like projectile that carries enough energy to eject the electron from the metal. If we suppose that the projectile is a photon of energy $h\nu$, where ν is the frequency of the radiation, then the conservation of energy requires that the kinetic energy of the ejected electron ($\frac{1}{2}m_ev^2$) should obey

$$\tfrac{1}{2}m_ev^2 = h\nu - \Phi \qquad (7.15)$$

In this expression Φ (upper-case phi) is a characteristic of the metal called its **work function**, the energy required to remove an electron from the metal to infinity (Fig. 7.14), the analogue of the ionization energy of an individual atom or molecule. We can now see that the existence of photons accounts for the three observations we have summarized:

1. Photoejection cannot occur if $h\nu < \Phi$ because the photon brings insufficient energy.

2. Equation 7.15 predicts that the kinetic energy of an ejected electron should increase linearly with frequency.

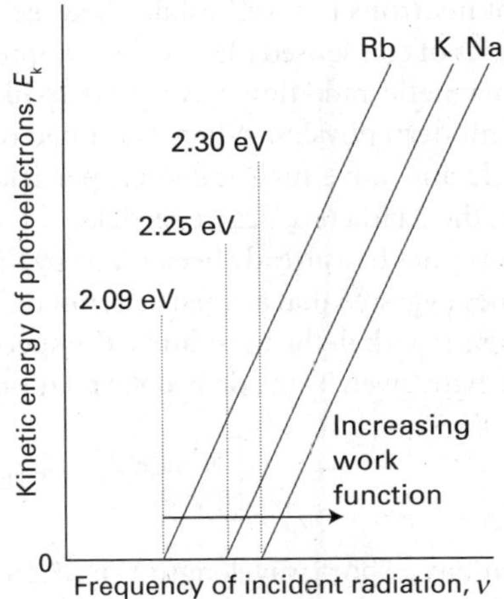

Fig. 7.13 In the photoelectric effect, it is found that no electrons are ejected when the incident radiation has a frequency below a value characteristic of the metal and, above that value, the kinetic energy of the photoelectrons varies linearly with the frequency of the incident radiation.

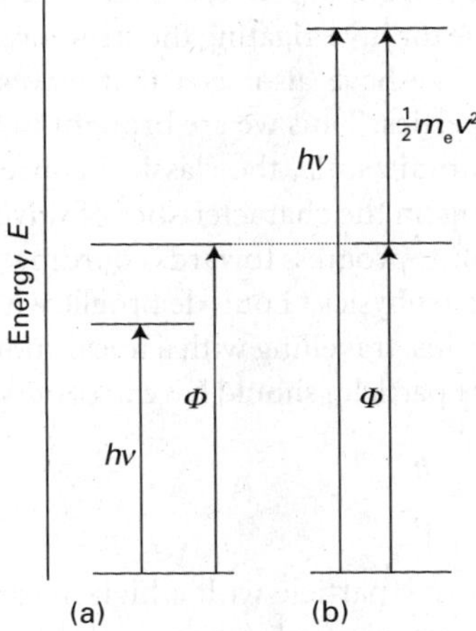

Fig. 7.14 The photoelectric effect can be explained if it is supposed that the incident radiation is composed of photons that have energy proportional to the frequency of the radiation. (a) The energy of the photon is insufficient to drive an electron out of the metal. (b) The energy of the photon is more than enough to eject an electron, and the excess energy is carried away as the kinetic energy of the photoelectron (the ejected electron).

interActivity Calculate the value of Planck's constant given that the following kinetic energies were observed for photoejected electrons irradiated by radiation of the wavelengths noted.

λ_i/nm	320	330	345	360	385
E_k/eV	1.17	1.05	0.885	0.735	0.511

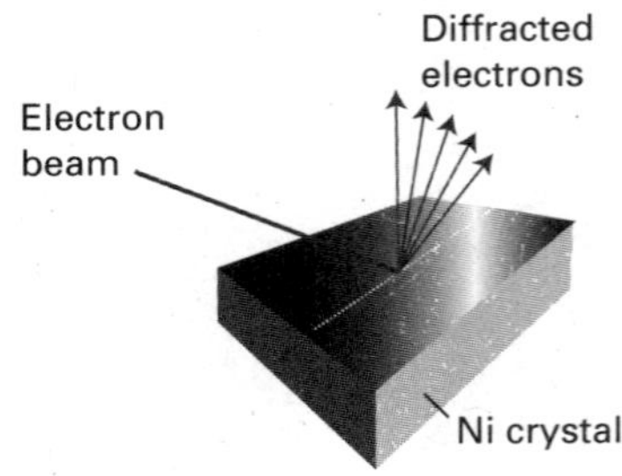

Fig. 7.15 The Davisson–Germer experiment. The scattering of an electron beam from a nickel crystal shows a variation of intensity characteristic of a diffraction experiment in which waves interfere constructively and destructively in different directions.

3. When a photon collides with an electron, it gives up all its energy, so we should expect electrons to appear as soon as the collisions begin, provided the photons have sufficient energy.

A practical application of eqn 7.15 is that it provides a technique for the determination of Planck's constant, for the slopes of the lines in Fig. 7.13 are all equal to h.

(b) The wave character of particles

Although contrary to the long-established wave theory of light, the view that light consists of particles had been held before, but discarded. No significant scientist, however, had taken the view that matter is wave-like. Nevertheless, experiments carried out in 1925 forced people to consider that possibility. The crucial experiment was performed by the American physicists Clinton Davisson and Lester Germer, who observed the diffraction of electrons by a crystal (Fig. 7.15). Diffraction is the interference caused by an object in the path of waves. Depending on whether the interference is constructive or destructive, the result is a region of enhanced or diminished intensity of the wave. Davisson and Germer's success was a lucky accident, because a chance rise of temperature caused their polycrystalline sample to anneal, and the ordered planes of atoms then acted as a diffraction grating. At almost the same time, G.P. Thomson, working in Scotland, showed that a beam of electrons was diffracted when passed through a thin gold foil.

The Davisson–Germer experiment, which has since been repeated with other particles (including α particles and molecular hydrogen), shows clearly that particles have wave-like properties, and the diffraction of neutrons is a well-established technique for investigating the structures and dynamics of condensed phases (see Chapter 19). We have also seen that waves of electromagnetic radiation have particle-like properties. Thus we are brought to the heart of modern physics. When examined on an atomic scale, the classical concepts of particle and wave melt together, particles taking on the characteristics of waves, and waves the characteristics of particles.

Some progress towards coordinating these properties had already been made by the French physicist Louis de Broglie when, in 1924, he suggested that any particle, not only photons, travelling with a linear momentum $p = mv$ (with m the mass and v the speed of the particle) should have in some sense a wavelength given by the **de Broglie relation**:

$$\lambda = \frac{h}{p} \qquad \text{de Broglie relation} \qquad (7.16)$$

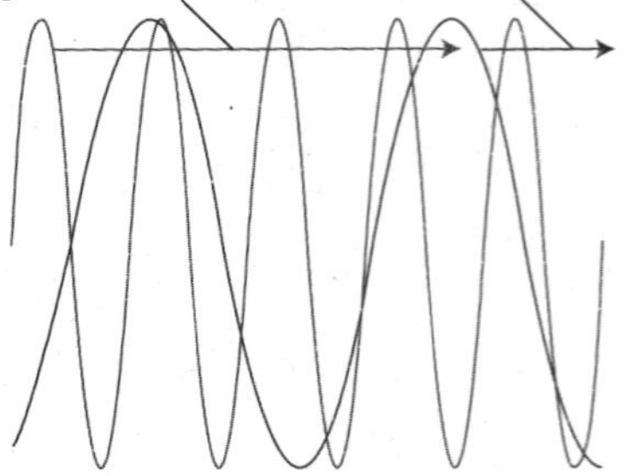

Short wavelength, high momentum
Long wavelength, low momentum

Fig. 7.16 An illustration of the de Broglie relation between momentum and wavelength. The wave is associated with a particle (shortly this wave will be seen to be the wavefunction of the particle). A particle with high momentum has a wavefunction with a short wavelength, and vice versa.

That is, a particle with a high linear momentum has a short wavelength (Fig. 7.16). Macroscopic bodies have such high momenta (because their mass is so great), even when they are moving slowly, that their wavelengths are undetectably small, and the wave-like properties cannot be observed. This undetectability is why, in spite of its deficiencies, classical mechanics can be used to explain the behaviour of macroscopic bodies. It is necessary to invoke quantum mechanics only for microscopic systems, such as atoms and molecules, in which masses are small.

Example 7.2 *Estimating the de Broglie wavelength*

Estimate the wavelength of electrons that have been accelerated from rest through a potential difference of 40 kV.

Method To use the de Broglie relation, we need to know the linear momentum, p, of the electrons. To calculate the linear momentum, we note that the energy acquired by an electron accelerated through a potential difference $\Delta\phi$ is $e\Delta\phi$, where

e is the magnitude of its charge. At the end of the period of acceleration, all the acquired energy is in the form of kinetic energy, $E_k = \frac{1}{2}m_e v^2 = p^2/2m_e$, so we can determine p by setting $p^2/2m_e$ equal to $e\Delta\phi$. As before, carry through the calculation algebraically before substituting the data.

Answer The expression $p^2/2m_e = e\Delta\phi$ solves to $p = (2m_e e\Delta\phi)^{1/2}$; then, from the de Broglie relation $\lambda = h/p$,

$$\lambda = \frac{h}{(2m_e e\Delta\phi)^{1/2}}$$

Substitution of the data and the fundamental constants (from inside the front cover) gives

$$\lambda = \frac{6.626 \times 10^{-34}\ \text{J s}}{\{2 \times (9.109 \times 10^{-31}\ \text{kg}) \times (1.602 \times 10^{-19}\ \text{C}) \times (4.0 \times 10^{4}\ \text{V})\}^{1/2}}$$
$$= 6.1 \times 10^{-12}\ \text{m}$$

where we have used 1 V C = 1 J and 1 J = 1 kg m^2 s^{-2}. The wavelength of 6.1 pm is shorter than typical bond lengths in molecules (about 100 pm). Electrons accelerated in this way are used in the technique of electron diffraction for the determination of the structures of solid surfaces (Section 23.3).

Self-test 7.2 Calculate the wavelength of (a) a neutron with a translational kinetic energy equal to kT at 300 K, (b) a tennis ball of mass 57 g travelling at 80 km h^{-1}.

[(a) 178 pm, (b) 5.2×10^{-34} m]

We now have to conclude that, not only has electromagnetic radiation the character classically ascribed to particles, but electrons (and all other particles) have the characteristics classically ascribed to waves. This joint particle and wave character of matter and radiation is called **wave–particle duality**. Duality strikes at the heart of classical physics, where particles and waves are treated as entirely distinct entities. We have also seen that the energies of electromagnetic radiation and of matter cannot be varied continuously, and that for small objects the discreteness of energy is highly significant. In classical mechanics, in contrast, energies could be varied continuously. Such total failure of classical physics for small objects implied that its basic concepts were false. A new mechanics had to be devised to take its place.

IMPACT ON BIOLOGY

I7.1 Electron microscopy

The basic approach of illuminating a small area of a sample and collecting light with a microscope has been used for many years to image small specimens. However, the *resolution* of a microscope, the minimum distance between two objects that leads to two distinct images, is on the order of the wavelength of light used as a probe. Therefore, conventional microscopes employing visible light have resolutions in the micrometre range and are blind to features on a scale of nanometres.

There is great interest in the development of new experimental probes of very small specimens that cannot be studied by traditional light microscopy. For example, our understanding of biochemical processes, such as enzymatic catalysis, protein folding, and the insertion of DNA into the cell's nucleus, will be enhanced if it becomes possible to image individual biopolymers—with dimensions much smaller than visible wavelengths—at work. One technique that is often used to image nanometre-sized

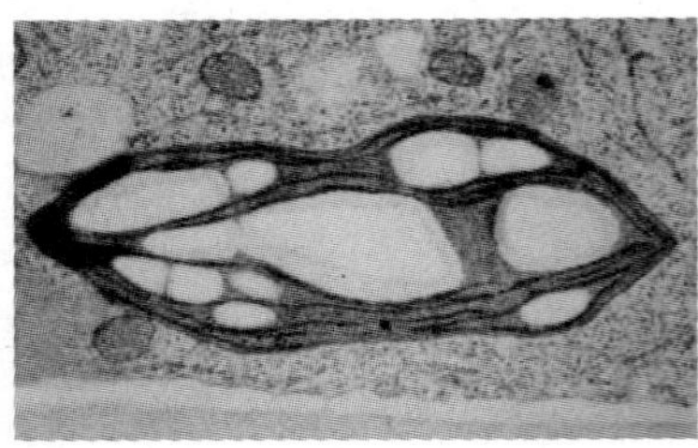

Fig. 7.17 A TEM image of a cross-section of a plant cell showing chloroplasts, organelles responsible for the reactions of photosynthesis (Chapter 21). Chloroplasts are typically 5 μm long. (Image supplied by Brian Bowes.)

objects is *electron microscopy*, in which a beam of electrons with a well-defined de Broglie wavelength replaces the lamp found in traditional light microscopes. Instead of glass or quartz lenses, magnetic fields are used to focus the beam. In *transmission electron microscopy* (TEM), the electron beam passes through the specimen and the image is collected on a screen. In *scanning electron microscopy* (SEM), electrons scattered back from a small irradiated area of the sample are detected and the electrical signal is sent to a video screen. An image of the surface is then obtained by scanning the electron beam across the sample.

As in traditional light microscopy, the wavelength of and the ability to focus the incident beam—in this case a beam of electrons—govern the resolution. Electron wavelengths in typical electron microscopes can be as short as 10 pm, but it is not possible to focus electrons well with magnetic lenses so, in the end, typical resolutions of TEM and SEM instruments are about 2 nm and 50 nm, respectively. It follows that electron microscopes cannot resolve individual atoms (which have diameters of about 0.2 nm). Furthermore, only certain samples can be observed under certain conditions. The measurements must be conducted under high vacuum. For TEM observations, the samples must be very thin cross-sections of a specimen and SEM observations must be made on dry samples. A consequence of these requirements is that neither technique can be used to study living cells. In spite of these limitations, electron microscopy is very useful in studies of the internal structure of cells (Fig. 7.17).

The dynamics of microscopic systems

At this point we have to construct a new mechanics from the ashes of classical physics. **Quantum mechanics** acknowledges the wave–particle duality of matter and the existence of quantization by supposing that, rather than travelling along a definite path, a particle is distributed through space like a wave. This remark may seem mysterious: it will be interpreted more fully shortly. The mathematical representation of the wave that in quantum mechanics replaces the classical concept of trajectory is called a **wavefunction**, ψ (psi).

7.3 The Schrödinger equation

Key point **The Schrödinger equation is a second-order differential equation used to calculate the wavefunction of a system.**

In 1926, the Austrian physicist Erwin Schrödinger proposed an equation for finding the wavefunction of any system. The **time-independent Schrödinger equation** for a particle of mass m moving in one dimension with energy E in a system that does not change with time (for instance, its volume remains constant) is

$$-\frac{\hbar^2}{2m}\frac{d^2\psi}{dx^2}+V(x)\psi=E\psi \quad \text{Time-independent Schrödinger equation} \tag{7.17}$$

The factor $V(x)$ is the potential energy of the particle at the point x; because the total energy E is the sum of potential and kinetic energies, the first term must be related (in a manner we explore later) to the kinetic energy of the particle; $\hbar = h/2\pi$ (which is read h-cross or h-bar) is a convenient modification of Planck's constant with the value 1.055×10^{-34} J s.

The following *Justification* shows that the Schrödinger equation is plausible and the discussions later in the chapter will help to overcome its apparent arbitrariness. For

Table 7.1 The Schrödinger equation

For one-dimensional systems

$$-\frac{\hbar^2}{2m}\frac{d^2\psi}{dx^2}+V(x)\psi=E\psi$$

Where $V(x)$ is the potential energy of the particle and E is its total energy. For three-dimensional systems

$$-\frac{\hbar^2}{2m}\nabla^2\psi+V\psi=E\psi$$

where V may depend on position and ∇^2 ('del squared') is

$$\nabla^2=\frac{\partial^2}{\partial x^2}+\frac{\partial^2}{\partial y^2}+\frac{\partial^2}{\partial z^2}$$

In systems with spherical symmetry three equivalent forms are

$$\nabla^2=\frac{1}{r}\frac{\partial^2}{\partial r^2}r+\frac{1}{r^2}\Lambda^2$$

$$=\frac{1}{r^2}\frac{\partial}{\partial r}r^2\frac{\partial}{\partial r}+\frac{1}{r^2}\Lambda^2$$

$$=\frac{\partial^2}{\partial r^2}+\frac{2}{r}\frac{\partial}{\partial r}+\frac{1}{r^2}\Lambda^2$$

where

$$\Lambda^2=\frac{1}{\sin^2\theta}\frac{\partial^2}{\partial\phi^2}+\frac{1}{\sin\theta}\frac{\partial}{\partial\theta}\sin\theta\frac{\partial}{\partial\theta}$$

In the general case the Schrodinger equation is written

$$\hat{H}\psi=E\psi$$

where $\hat{H}$ is the hamiltonian operator for the system:

$$\hat{H}=-\frac{\hbar^2}{2m}\nabla^2+V$$

For the evolution of a system with time, it is necessary to solve the time-dependent Schrödinger equation:

$$\hat{H}\Psi=i\hbar\frac{\partial\Psi}{\partial t}$$

the present, we shall treat the equation simply as a quantum-mechanical postulate that replaces Newton's postulate of his apparently equally arbitrary equation of motion (that force = mass × acceleration). Various ways of expressing the Schrödinger equation, of incorporating the time dependence of the wavefunction, and of extending it to more dimensions are collected in Table 7.1. In Chapter 8 we shall solve the equation for a number of important cases; in this chapter we are mainly concerned with its significance, the interpretation of its solutions, and seeing how it implies that energy is quantized.

Justification 7.1 *Using the Schrödinger equation to develop the de Broglie relation*

The Schrödinger equation can be seen to be plausible by noting that it implies the de Broglie relation for a freely moving particle in a region where its potential energy V is constant. After writing $V(x) = V$, we can rearrange eqn 7.17 into

$$\frac{d^2\psi}{dx^2}=-\frac{2m}{\hbar^2}(E-V)\psi$$

General strategies for solving differential equations of this and other types that occur frequently in physical chemistry are treated in *Mathematical background 4* following Chapter 8. In this case a solution is

$$\psi = \cos kx \qquad k = \left\{\frac{2m(E-V)}{\hbar^2}\right\}^{1/2}$$

We now recognize that cos kx is a wave of wavelength $\lambda = 2\pi/k$, as can be seen by comparing cos kx with the standard form of a harmonic wave, $\cos(2\pi x/\lambda)$. The quantity $E - V$ is equal to the kinetic energy of the particle, E_k, so $k = (2mE_k/\hbar^2)^{1/2}$, which implies that $E_k = k^2\hbar^2/2m$. Because $E_k = p^2/2m$, it follows that $p = k\hbar$. Therefore, the linear momentum is related to the wavelength of the wavefunction by

$$p = \frac{2\pi}{\lambda} \times \frac{h}{2\pi} = \frac{h}{\lambda}$$

which is the de Broglie relation.

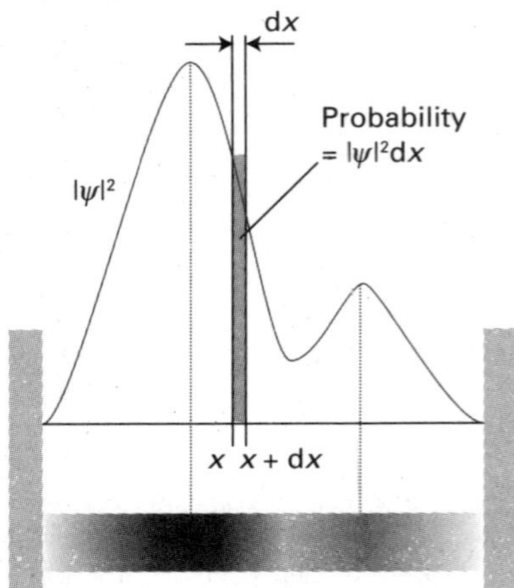

Fig. 7.18 The wavefunction ψ is a probability amplitude in the sense that its square modulus ($\psi^*\psi$ or $|\psi|^2$) is a probability density. The probability of finding a particle in the region dx located at x is proportional to $|\psi|^2$dx. We represent the probability density by the density of shading in the superimposed band.

7.4 The Born interpretation of the wavefunction

Key points **According to the Born interpretation, the probability density is proportional to the square of the wavefunction. (a) A wavefunction is normalized if the integral of its square is equal to 1. (b) The quantization of energy stems from the constraints that an acceptable wavefunction must satisfy.**

A central principle of quantum mechanics is that *the wavefunction contains all the dynamical information about the system it describes*. Here we concentrate on the information it carries about the location of the particle.

The interpretation of the wavefunction in terms of the location of the particle is based on a suggestion made by Max Born. He made use of an analogy with the wave theory of light, in which the square of the amplitude of an electromagnetic wave in a region is interpreted as its intensity and therefore (in quantum terms) as a measure of the probability of finding a photon present in the region. The **Born interpretation** of the wavefunction focuses on the square of the wavefunction (or the square modulus, $|\psi|^2 = \psi^*\psi$, if ψ is complex; see *Mathematical background 3*). For a one-dimensional system (Fig. 7.18):

> If the wavefunction of a particle has the value ψ at some point x, then the probability of finding the particle between x and $x + dx$ is proportional to $|\psi|^2$dx.

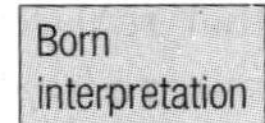

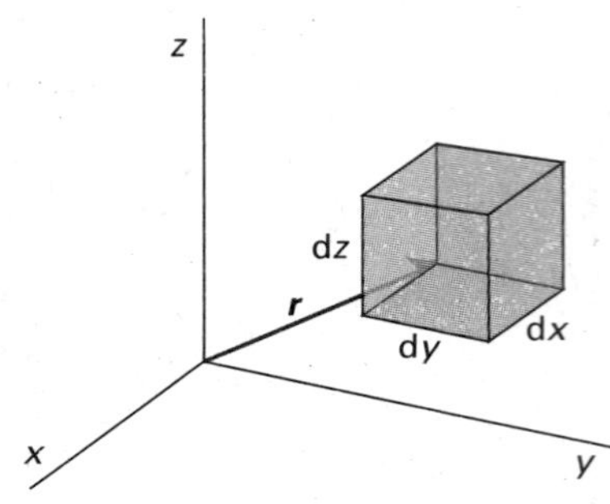

Fig. 7.19 The Born interpretation of the wavefunction in three-dimensional space implies that the probability of finding the particle in the volume element $d\tau = dxdydz$ at some location $\boldsymbol{r}$ is proportional to the product of $d\tau$ and the value of $|\psi|^2$ at that location.

Thus, $|\psi|^2$ is the **probability density**, and to obtain the probability it must be multiplied by the length of the infinitesimal region dx. The wavefunction ψ itself is called the **probability amplitude**. For a particle free to move in three dimensions (for example, an electron near a nucleus in an atom), the wavefunction depends on the point $\boldsymbol{r}$ with coordinates x, y, and z, and the interpretation of $\psi(\boldsymbol{r})$ is as follows (Fig. 7.19):

> If the wavefunction of a particle has the value ψ at some point $\boldsymbol{r}$, then the probability of finding the particle in an infinitesimal volume $d\tau = dxdydz$ at that point is proportional to $|\psi|^2 d\tau$.

The Born interpretation does away with any worry about the significance of a negative (and, in general, complex) value of ψ because $|\psi|^2$ is real and never negative. There is no *direct* significance in the negative (or complex) value of a wavefunction:

only the square modulus, a positive quantity, is directly physically significant, and both negative and positive regions of a wavefunction may correspond to a high probability of finding a particle in a region (Fig. 7.20). However, later we shall see that the presence of positive and negative regions of a wavefunction is of great *indirect* significance, because it gives rise to the possibility of constructive and destructive interference between different wavefunctions.

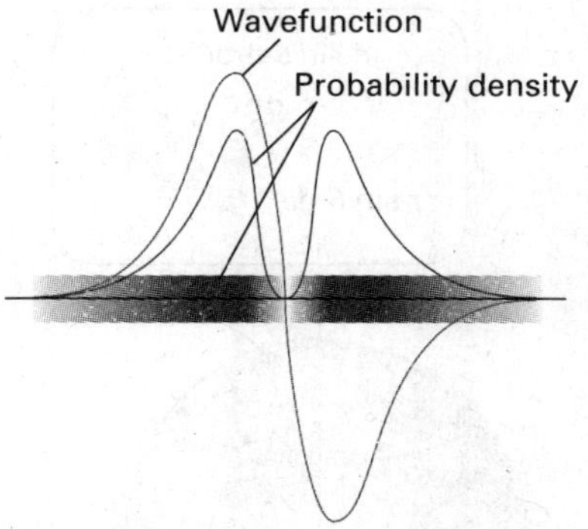

Fig. 7.20 The sign of a wavefunction has no direct physical significance: the positive and negative regions of this wavefunction both correspond to the same probability distribution (as given by the square modulus of ψ and depicted by the density of shading).

Example 7.3 *Interpreting a wavefunction*

We shall see in Chapter 9 that the wavefunction of an electron in the lowest energy state of a hydrogen atom is proportional to e^{-r/a_0}, with a_0 a constant and r the distance from the nucleus. Calculate the relative probabilities of finding the electron inside a region of volume $\delta V = 1.0\ \text{pm}^3$, which is small even on the scale of the atom, located at (a) the nucleus, (b) a distance a_0 from the nucleus.

Method The region of interest is so small on the scale of the atom that we can ignore the variation of ψ within it and write the probability, P, as proportional to the probability density (ψ^2; note that ψ is real) evaluated at the point of interest multiplied by the volume of interest, δV. That is, $P \propto \psi^2 \delta V$, with $\psi^2 \propto e^{-2r/a_0}$.

Answer In each case $\delta V = 1.0\ \text{pm}^3$. (a) At the nucleus, $r = 0$, so

$$P \propto e^0 \times (1.0\ \text{pm}^3) = (1.0) \times (1.0\ \text{pm}^3)$$

(b) At a distance $r = a_0$ in an arbitrary direction,

$$P \propto e^{-2} \times (1.0\ \text{pm}^3) = (0.14) \times (1.0\ \text{pm}^3)$$

Therefore, the ratio of probabilities is $1.0/0.14 = 7.1$. Note that it is more probable (by a factor of 7) that the electron will be found at the nucleus than in a volume element of the same size located at a distance a_0 from the nucleus. The negatively charged electron is attracted to the positively charged nucleus, and is likely to be found close to it.

Self-test 7.3 The wavefunction for the electron in its lowest energy state in the ion He^+ is proportional to e^{-2r/a_0}. Repeat the calculation for this ion. Any comment?

[55; more compact wavefunction]

A note on good practice The square of a wavefunction is a probability density, and (in three dimensions) has the dimensions of 1/length^3. It becomes a (unitless) probability when multiplied by a volume. In general, we have to take into account the variation of the amplitude of the wavefunction over the volume of interest, but here we are supposing that the volume is so small that the variation of ψ in the region can be ignored.

(a) Normalization

A mathematical feature of the Schrödinger equation is that, if ψ is a solution, then so is $N\psi$, where N is any constant. This feature is confirmed by noting that ψ occurs in every term in eqn 7.17, so any constant factor can be cancelled. This freedom to vary the wavefunction by a constant factor means that it is always possible to find a **normalization constant**, N, such that the proportionality of the Born interpretation becomes an equality.

We find the normalization constant by noting that, for a normalized wavefunction $N\psi$, the probability that a particle is in the region dx is equal to $(N\psi^*)(N\psi)\text{d}x$ (we are taking N to be real). Furthermore, the sum over all space of these individual probabilities must be 1 (the probability of the particle being somewhere is 1). Expressed mathematically, the latter requirement is

$$N^2 \int_{-\infty}^{\infty} \psi^* \psi \, \text{d}x = 1 \qquad (7.18)$$

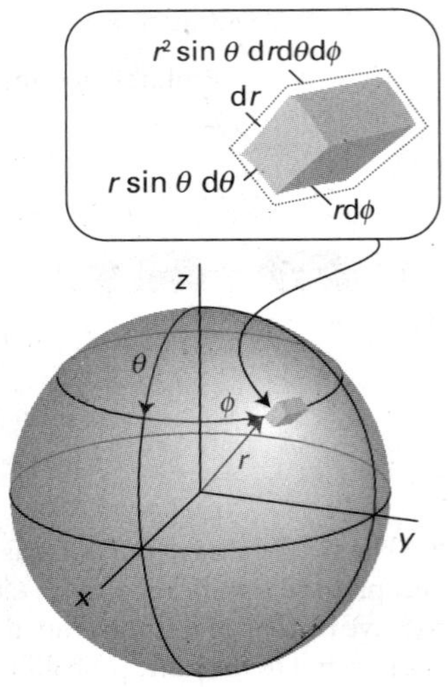

Fig. 7.21 The spherical polar coordinates used for discussing systems with spherical symmetry.

Wavefunctions for which the integral in eqn 7.18 exists (in the sense of having a finite value) are said to be 'square-integrable'. It follows that

$$N = \frac{1}{\left(\int_{-\infty}^{\infty} \psi^* \psi \mathrm{d}x\right)^{1/2}} \tag{7.19}$$

Therefore, by evaluating the integral, we can find the value of N and hence 'normalize' the wavefunction. From now on, unless we state otherwise, we always use wavefunctions that have been normalized to 1; that is, from now on we assume that ψ already includes a factor that ensures that (in one dimension)

$$\int_{-\infty}^{\infty} \psi^* \psi \, \mathrm{d}x = 1 \tag{7.20a}$$

In three dimensions, the wavefunction is normalized if

$$\int_{-\infty}^{\infty}\int_{-\infty}^{\infty}\int_{-\infty}^{\infty} \psi^* \psi \, \mathrm{d}x\mathrm{d}y\mathrm{d}z = 1 \tag{7.20b}$$

or, more succinctly, if

Normalization integral

$$\int \psi^* \psi \, \mathrm{d}\tau = 1 \tag{7.20c}$$

where $\mathrm{d}\tau = \mathrm{d}x\mathrm{d}y\mathrm{d}z$ and the limits of this definite integral are not written explicitly: in all such integrals, the integration is over all the space accessible to the particle. For systems with spherical symmetry it is best to work in **spherical polar coordinates** r, θ, and ϕ (Fig. 7.21):

Spherical polar coordinates

$$x = r \sin\theta \cos\phi, \; y = r \sin\theta \sin\phi, \; z = r\cos\theta$$

r, the radius, ranges from 0 to ∞

θ, the colatitude, ranges from 0 to π

ϕ, the azimuth, ranges from 0 to 2π

That these ranges cover space is illustrated in Fig. 7.22. Standard manipulations then yield

$$\mathrm{d}\tau = r^2 \sin\theta \, \mathrm{d}r\mathrm{d}\theta\mathrm{d}\phi$$

In these coordinates, the explicit form of eqn 7.20c is

$$\int_0^{\infty}\int_0^{\pi}\int_0^{2\pi} \psi^* \psi r^2 \, \mathrm{d}r \sin\theta \, \mathrm{d}\theta \, \mathrm{d}\phi = 1 \tag{7.20d}$$

The limits on the first integral sign refer to r, those on the second to θ, and those on the third to ϕ.

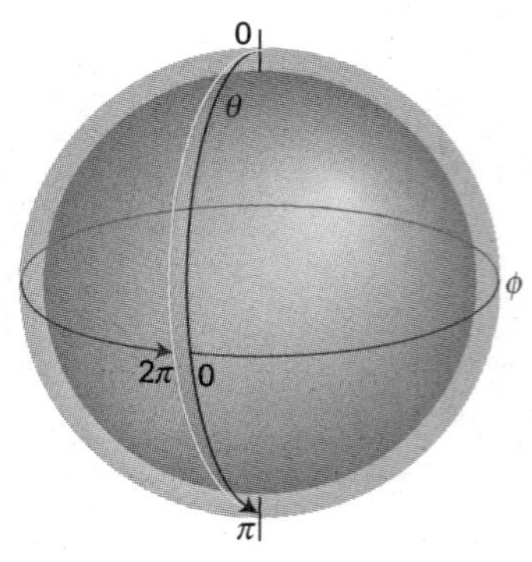

Fig. 7.22 The surface of a sphere is covered by allowing θ to range from 0 to π, and then sweeping that arc around a complete circle by allowing ϕ to range from 0 to 2π.

Example 7.4 *Normalizing a wavefunction*

Normalize the wavefunction used for the hydrogen atom in Example 7.3.

Method We need to find the factor N that guarantees that the integral in eqn 7.20c is equal to 1. Because the system is spherical, it is most convenient to use spherical coordinates and to carry out the integrations specified in eqn 7.20d. A useful integral for calculations on atomic wavefunctions is

$$\int_0^\infty x^n \mathrm{e}^{-ax}\mathrm{d}x = \frac{n!}{a^{n+1}}$$

where $n!$ denotes a factorial: $n! = n(n-1)(n-2)\ldots 1$, and $0! = 1$ by definition.

Answer The integration required is the product of three factors:

$$\int \psi^*\psi\mathrm{d}\tau = N^2 \overbrace{\int_0^\infty r^2\mathrm{e}^{-2r/a_0}\mathrm{d}r}^{\frac{1}{4}a_0^3} \overbrace{\int_0^\pi \sin\theta\,\mathrm{d}\theta}^{2} \overbrace{\int_0^{2\pi}\mathrm{d}\phi}^{2\pi} = \pi a_0^3 N^2$$

Therefore, for this integral to equal 1, we must set

$$N = \left(\frac{1}{\pi a_0^3}\right)^{1/2}$$

and the normalized wavefunction is

$$\psi = \left(\frac{1}{\pi a_0^3}\right)^{1/2} \mathrm{e}^{-r/a_0}$$

Note that, because a_0 is a length, the dimensions of ψ are $1/\text{length}^{3/2}$ and therefore those of ψ^2 are $1/\text{length}^3$ (for instance, $1/\text{m}^3$) as is appropriate for a probability density (in the sense that a probability density times a volume is a probability).

If Example 7.3 is now repeated, we can obtain the actual probabilities of finding the electron in the volume element at each location, not just their relative values. Given (from inside the front cover) that $a_0 = 52.9$ pm, the results are (a) 2.2×10^{-6}, corresponding to 1 chance in about 500 000 inspections of finding the electron in the test volume, and (b) 2.9×10^{-7}, corresponding to 1 chance in 3.4 million.

Self-test 7.4 Normalize the wavefunction given in Self-test 7.3. $[N = (8/\pi a_0^3)^{1/2}]$

(b) Quantization

The Born interpretation puts severe restrictions on the acceptability of wavefunctions. The principal constraint is that ψ must not be infinite anywhere. If it were, the integral in eqn 7.20 would be infinite (in other words, ψ would not be square-integrable) and the normalization constant would be zero. The normalized function would then be zero everywhere, except where it is infinite, which would be unacceptable. The requirement that ψ is finite everywhere rules out many possible solutions of the Schrödinger equation, because many mathematically acceptable solutions rise to infinity and are therefore physically unacceptable. We shall meet several examples shortly.

A brief comment

Infinitely sharp spikes are acceptable provided they have zero width, so it is more appropriate to state that the wavefunction must not be infinite over any finite region. In elementary quantum mechanics the simpler restriction, to finite ψ, is sufficient.

The requirement that ψ is finite everywhere is not the only restriction implied by the Born interpretation. We could imagine (and in Section 8.6a will meet) a solution of the Schrödinger equation that gives rise to more than one value of $|\psi|^2$ at a single point. The Born interpretation implies that such solutions are unacceptable, because it would be absurd to have more than one probability that a particle is at the same point. This restriction is expressed by saying that the wavefunction must be *single-valued*; that is, have only one value at each point of space.

The Schrödinger equation itself also implies some mathematical restrictions on the type of functions that will occur. Because it is a second-order differential equation, the second derivative of ψ must be well-defined if the equation is to be applicable everywhere. We can take the second derivative of a function only if it is continuous

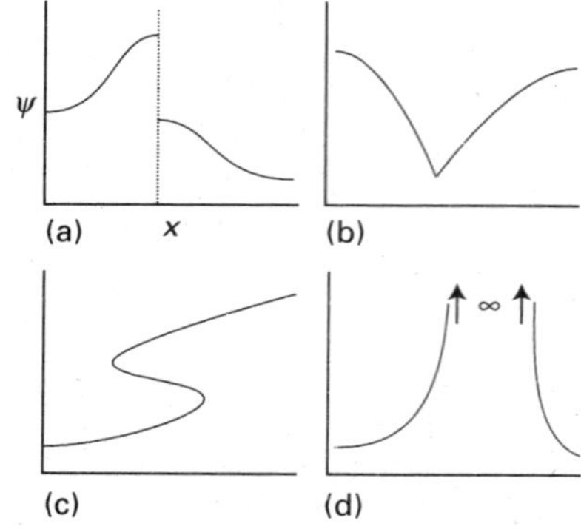

Fig. 7.23 The wavefunction must satisfy stringent conditions for it to be acceptable. (a) Unacceptable because it is not continuous; (b) unacceptable because its slope is discontinuous; (c) unacceptable because it is not single-valued; (d) unacceptable because it is infinite over a finite region.

(so there are no sharp steps in it, Fig. 7.23) and if its first derivative, its slope, is continuous (so there are no kinks).

At this stage we see that ψ must be:

- continuous
- have a continuous slope
- be single-valued
- be square-integrable

Conditions on the wavefunction

An acceptable wavefunction cannot be zero everywhere, because the particle it describes must be somewhere. These are such severe restrictions that acceptable solutions of the Schrödinger equation do not in general exist for arbitrary values of the energy E. In other words, a particle may possess only certain energies, for otherwise its wavefunction would be physically unacceptable. That is, *as a consequence of the restriction on its wavefunction, the energy of a particle is quantized.* We can find the acceptable energies by solving the Schrödinger equation for motion of various kinds, and selecting the solutions that conform to the restrictions listed above. That is the task of the next chapter.

A brief comment

There are cases, and we shall meet them, where acceptable wavefunctions have kinks. These cases arise when the potential energy has peculiar properties, such as rising abruptly to infinity. When the potential energy is smoothly well-behaved and finite, the slope of the wavefunction must be continuous; if the potential energy becomes infinite, then the slope of the wavefunction need not be continuous. There are only two cases of this behaviour in elementary quantum mechanics, and the peculiarity will be mentioned when we meet them.

Quantum mechanical principles

We have claimed that a wavefunction contains all the information it is possible to obtain about the dynamical properties of the particle (for example, its location and momentum). We have seen that the Born interpretation tells us as much as we can know about location, but how do we find any additional dynamical information?

7.5 The information in a wavefunction

Key points (a) The wavefunction of a free particle with a specific linear momentum corresponds to a uniform probability density. (b) The Schrödinger equation is an eigenvalue equation in which the wavefunction is an eigenfunction of the Hamiltonian operator. (c) Observables are represented by operators; the value of an observable is an eigenvalue of the corresponding operator constructed from the operators for position and linear momentum. (d) All operators that correspond to observables are hermitian; their eigenvalues are real and their eigenfunctions are mutually orthogonal. Sets of functions that are normalized and mutually orthogonal are called orthonormal. (e) When the system is not described by an eigenfunction of an operator, it may be expressed as a superposition of such eigenfunctions. The mean value of a series of observations is given by the expectation value of the corresponding operator.

The Schrödinger equation for a particle of mass m free to move parallel to the x-axis with zero potential energy is obtained from eqn 7.17 by setting $V=0$, and is

$$-\frac{\hbar^2}{2m}\frac{d^2\psi}{dx^2}=E\psi \tag{7.21}$$

The solutions of this equation have the form

$$\psi=Ae^{ikx}+Be^{-ikx} \qquad E=\frac{k^2\hbar^2}{2m} \tag{7.22}$$

where A and B are constants. (See *Mathematical background 3* following this chapter for more on complex numbers.) To verify that ψ is a solution of eqn 7.21, we simply substitute it into the left-hand side of the equation and confirm that we obtain $E\psi$:

$$-\frac{\hbar^2}{2m}\frac{d^2\psi}{dx^2} = -\frac{\hbar^2}{2m}\frac{d^2}{dx^2}(Ae^{ikx} + Be^{-ikx})$$
$$= -\frac{\hbar^2}{2m}\{A(ik)^2e^{ikx} + B(-ik)^2e^{-ikx}\}$$
$$= \frac{\hbar^2k^2}{2m}(Ae^{ikx} + Be^{-ikx}) = E\psi$$

(a) The probability density

We shall see later what determines the values of A and B; for the time being we can treat them as arbitrary constants that we can vary at will. Suppose that $B = 0$ in eqn 7.22, then the wavefunction is simply

$$\psi = Ae^{ikx} \tag{7.23}$$

Where is the particle? To find out, we calculate the probability density:

$$|\psi|^2 = (Ae^{ikx})^*(Ae^{ikx}) = (A^*e^{-ikx})(Ae^{ikx}) = |A|^2 \tag{7.24}$$

This probability density is independent of x so, wherever we look along the x-axis, there is an equal probability of finding the particle (Fig. 7.24a). In other words, if the wavefunction of the particle is given by eqn 7.23, then we cannot predict where we will find it. The same would be true if the wavefunction in eqn 7.22 had $A = 0$; then the probability density would be $|B|^2$, a constant.

Now suppose that in the wavefunction $A = B$. Then eqn 7.22 becomes

$$\psi = A(e^{ikx} + e^{-ikx}) = 2A\cos kx \tag{7.25}$$

The probability density now has the form

$$|\psi|^2 = (2A\cos kx)^*(2A\cos kx) = 4|A|^2\cos^2 kx \tag{7.26}$$

This function is illustrated in Fig. 7.24b. As we see, the probability density periodically varies between 0 and $4|A|^2$. The locations where the probability density is zero correspond to *nodes* in the wavefunction. Specifically, a **node** is a point where a wavefunction passes *through* zero. The location where a wavefunction approaches zero without actually passing through zero is not a node.

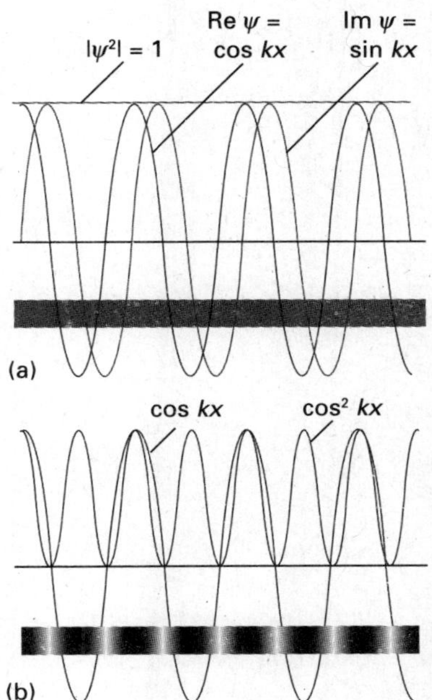

Fig. 7.24 (a) The square modulus of a wavefunction corresponding to a definite state of linear momentum is a constant; so it corresponds to a uniform probability of finding the particle anywhere. (b) The probability distribution corresponding to the superposition of states of equal magnitude of linear momentum but opposite direction of travel.

A brief comment

If the probability density of a particle is a constant, then it follows that, with x ranging from $-\infty$ to $+\infty$, the normalization constants, A or B, are 0. To avoid this embarrassing problem, x is allowed to range from $-L$ to $+L$, and L is allowed to go to infinity at the end of any subsequent calculation. We ignore this complication here.

(b) Operators, eigenvalues, and eigenfunctions

To formulate a systematic way of extracting information from the wavefunction, we first note that any Schrödinger equation (such as those in eqns 7.17 and 7.21) may be written in the succinct form

$$\hat{H}\psi = E\psi \tag{7.27a}$$

Operator form of Schrödinger equation

with (in one dimension)

$$\hat{H} = -\frac{\hbar^2}{2m}\frac{d^2}{dx^2} + V(x) \tag{7.27b}$$

Hamiltonian operator

The quantity $\hat{H}$ (commonly read aitch-hat) is an **operator**, something that carries out a mathematical operation on the function ψ. In this case, the operation is to take the second derivative of ψ and (after multiplication by $-\hbar^2/2m$) to add the result to the outcome of multiplying ψ by V. The operator $\hat{H}$ plays a special role in quantum mechanics, and is called the **hamiltonian operator** after the nineteenth century mathematician William Hamilton, who developed a form of classical mechanics that, it subsequently turned out, is well suited to the formulation of quantum mechanics. The hamiltonian operator is the operator corresponding to the total energy of the system, the sum of the kinetic and potential energies. Consequently, we can infer that the first term in eqn 7.27b (the term proportional to the second derivative) must be the operator for the kinetic energy.

When the Schrödinger equation is written as in eqn 7.27a, it is seen to be an **eigenvalue equation**, an equation of the form

$$\text{(Operator)(function)} = \text{(constant factor)} \times \text{(same function)} \tag{7.28a}$$

If we denote a general operator by $\hat{\Omega}$ (where Ω is upper-case omega) and a constant factor by ω (lower-case omega), then an eigenvalue equation has the form

$$\hat{\Omega}\psi = \omega\psi \qquad \text{Eigenvalue equation} \tag{7.28b}$$

The factor ω is called the **eigenvalue** of the operator. The eigenvalue in eqn 7.27a is the energy. The function ψ in an equation of this kind is called an **eigenfunction** of the operator $\hat{\Omega}$ and is different for each eigenvalue. So, in this technical language, we would write eqn 7.28a as

$$\text{(Operator)(eigenfunction)} = \text{(eigenvalue)} \times \text{(eigenfunction)} \tag{7.28c}$$

The eigenfunction in eqn 7.27a is the wavefunction corresponding to the energy E. It follows that another way of saying 'solve the Schrödinger equation' is to say 'find the eigenvalues and eigenfunctions of the hamiltonian operator for the system'.

Example 7.5 *Identifying an eigenfunction*

Show that e^{ax} is an eigenfunction of the operator d/dx, and find the corresponding eigenvalue. Show that e^{ax^2} is not an eigenfunction of d/dx.

Method We need to operate on the function with the operator and check whether the result is a constant factor times the original function.

Answer For $\hat{\Omega} = d/dx$ (the operation 'differentiate with respect to x') and $\psi = e^{ax}$:

$$\hat{\Omega}\psi = \frac{d}{dx}e^{ax} = ae^{ax} = a\psi$$

Therefore e^{ax} is indeed an eigenfunction of d/dx, and its eigenvalue is a. For $\psi = e^{ax^2}$,

$$\hat{\Omega}\psi = \frac{d}{dx}e^{ax^2} = 2axe^{ax^2} = 2ax \times \psi$$

which is not an eigenvalue equation of $\hat{\Omega}$ even though the same function ψ occurs on the right, because ψ is now multiplied by a variable factor ($2ax$), not a constant factor. Alternatively, if the right-hand side is written $2a(xe^{a^2})$, we see that it is a constant ($2a$) times a *different* function.

Self-test 7.5 Is the function $\cos ax$ an eigenfunction of (a) d/dx, (b) d^2/dx^2?

[(a) No, (b) yes]

(c) The construction of operators

The importance of eigenvalue equations is that the pattern

$$\text{(Energy operator)}\psi = \text{(energy)} \times \psi$$

exemplified by the Schrödinger equation is repeated for other **observables**, or measurable properties of a system, such as the momentum or the electric dipole moment. Thus, it is often the case that we can write

(Operator corresponding to an observable) ψ = (value of observable) $\times\ \psi$

The symbol $\hat{\Omega}$ in eqn 7.28b is then interpreted as an operator (for example, the hamiltonian operator) corresponding to an observable (for example, the energy), and the eigenvalue ω is the value of that observable (for example, the value of the energy, E). Therefore, if we know both the wavefunction ψ and the operator $\hat{\Omega}$ corresponding to the observable Ω of interest, and the wavefunction is an eigenfunction of the operator $\hat{\Omega}$, then we can predict the outcome of an observation of the property Ω (for example, an atom's energy) by picking out the factor ω in the eigenvalue equation, eqn 7.28b.

A basic postulate of quantum mechanics tells us how to set up the operator corresponding to a given observable:

Observables, Ω, are represented by operators, $\hat{\Omega}$, built from the following position and momentum operators:

$$\hat{x} = x\times \qquad \hat{p}_x = \frac{\hbar}{\mathrm{i}}\frac{\mathrm{d}}{\mathrm{d}x} \qquad [7.29]$$

A brief comment

The rules summarized by eqn 7.29 apply to observables that depend on spatial variables; intrinsic properties such as spin (Section 8.8) are treated differently.

That is, the operator for location along the x-axis is multiplication (of the wavefunction) by x and the operator for linear momentum parallel to the x-axis is proportional to taking the derivative (of the wavefunction) with respect to x.

Example 7.6 *Determining the value of an observable*

What is the linear momentum of a particle described by the wavefunction in eqn 7.22 with (a) $B = 0$, (b) $A = 0$?

Method We operate on ψ with the operator corresponding to linear momentum (eqn 7.29), and inspect the result. If the outcome is the original wavefunction multiplied by a constant (that is, we generate an eigenvalue equation), then the constant is identified with the value of the observable.

Answer (a) With the wavefunction given in eqn 7.22 with $B = 0$

$$\hat{p}_x\psi = \frac{\hbar}{\mathrm{i}}\frac{\mathrm{d}\psi}{\mathrm{d}x} = \frac{\hbar}{\mathrm{i}}A\frac{\mathrm{d}\mathrm{e}^{ikx}}{\mathrm{d}x} = \frac{\hbar}{\mathrm{i}}A \times \mathrm{i}k\mathrm{e}^{ikx} = k\hbar A\mathrm{e}^{ikx} = k\hbar\psi$$

This is an eigenvalue equation, and by comparing it with eqn 7.28b we find that $p_x = +k\hbar$. (b) For the wavefunction with $A = 0$

$$\hat{p}_x\psi = \frac{\hbar}{\mathrm{i}}\frac{\mathrm{d}\psi}{\mathrm{d}x} = \frac{\hbar}{\mathrm{i}}B\frac{\mathrm{d}\mathrm{e}^{-ikx}}{\mathrm{d}x} = \frac{\hbar}{\mathrm{i}}B \times (-\mathrm{i}k)\mathrm{e}^{-ikx} = -k\hbar\psi$$

The magnitude of the linear momentum is the same in each case ($k\hbar$), but the signs are different: in (a) the particle is travelling to the right (positive x) but in (b) it is travelling to the left (negative x).

Self-test 7.6 The operator for the angular momentum of a particle travelling in a circle in the xy-plane is $\hat{l}_z = (\hbar/\mathrm{i})\mathrm{d}/\mathrm{d}\phi$, where ϕ is its angular position. What is the angular momentum of a particle described by the wavefunction $\mathrm{e}^{-2\mathrm{i}\phi}$? $[l_z = -2\hbar]$

We use the definitions in eqn 7.29 to construct operators for other spatial observables. For example, suppose we wanted the operator for a potential energy of the form

$V(x) = \frac{1}{2}kx^2$, with k a constant (later, we shall see that this potential energy describes the vibrations of atoms in molecules). Then it follows from eqn 7.29 that the operator corresponding to $V(x)$ is multiplication by x^2:

$$\hat{V} = \frac{1}{2}kx^2 \times \qquad (7.30)$$

In normal practice, the multiplication sign is omitted. To construct the operator for kinetic energy, we make use of the classical relation between kinetic energy and linear momentum, which in one dimension is $E_k = p_x^2/2m$. Then, by using the operator for p_x in eqn 7.29 we find:

$$\hat{E}_k = \frac{1}{2m}\left(\frac{\hbar}{i}\frac{d}{dx}\right)\left(\frac{\hbar}{i}\frac{d}{dx}\right) = -\frac{\hbar^2}{2m}\frac{d^2}{dx^2} \qquad (7.31)$$

It follows that the operator for the total energy, the hamiltonian operator, is

$$\hat{H} = \hat{E}_k + \hat{V} = -\frac{\hbar^2}{2m}\frac{d^2}{dx^2} + \hat{V} \qquad \text{Hamiltonian operator} \qquad (7.32)$$

with $\hat{V}(x)$ the multiplicative operator in eqn 7.30 (or some other appropriate expression for the potential energy).

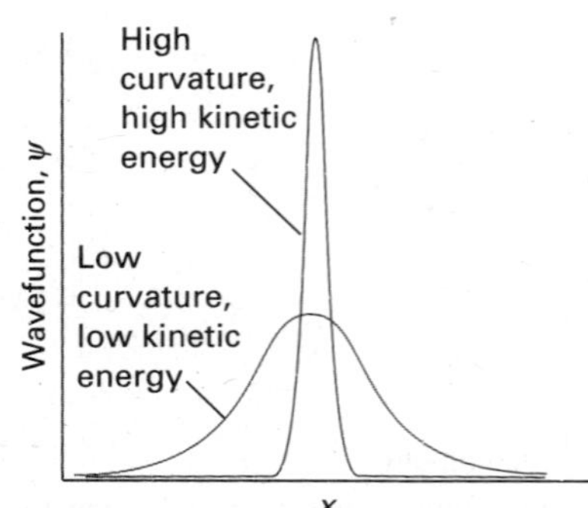

Fig. 7.25 Even if a wavefunction does not have the form of a periodic wave, it is still possible to infer from it the average kinetic energy of a particle by noting its average curvature. This illustration shows two wavefunctions: the sharply curved function corresponds to a higher kinetic energy than the less sharply curved function.

The expression for the kinetic energy operator, eqn 7.31, enables us to develop the point made earlier concerning the interpretation of the Schrödinger equation. In mathematics, the second derivative of a function is a measure of its curvature: a large second derivative indicates a sharply curved function (Fig. 7.25). It follows that a sharply curved wavefunction is associated with a high kinetic energy, and one with a low curvature is associated with a low kinetic energy. This interpretation is consistent with the de Broglie relation, which predicts a short wavelength (a sharply curved wavefunction) when the linear momentum (and hence the kinetic energy) is high. However, it extends the interpretation to wavefunctions that do not spread through space and resemble those shown in Fig. 7.25. The curvature of a wavefunction in general varies from place to place. Wherever a wavefunction is sharply curved, its contribution to the total kinetic energy is large (Fig. 7.26). Wherever the wavefunction is not sharply curved, its contribution to the overall kinetic energy is low. As we shall shortly see, the observed kinetic energy of the particle is an integral of all the contributions of the kinetic energy from each region. Hence, we can expect a particle to have a high kinetic energy if the average curvature of its wavefunction is high. Locally there can be both positive and negative contributions to the kinetic energy (because the curvature can be either positive, ∪, or negative, ∩), but the average is always positive (see Problem 7.26).

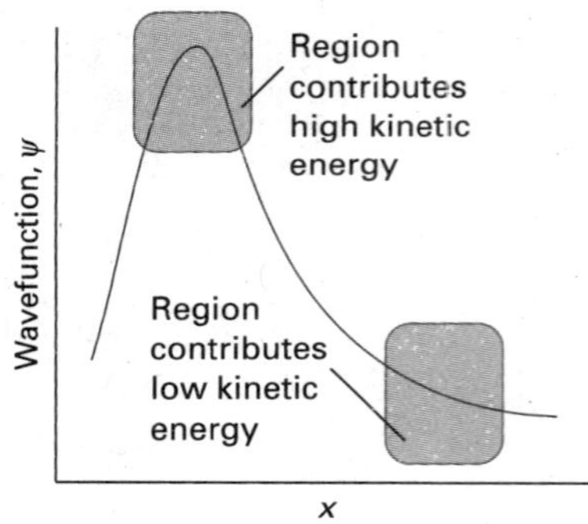

Fig. 7.26 The observed kinetic energy of a particle is an average of contributions from the entire space covered by the wavefunction. Sharply curved regions contribute a high kinetic energy to the average; slightly curved regions contribute only a small kinetic energy.

The association of high curvature with high kinetic energy will turn out to be a valuable guide to the interpretation of wavefunctions and the prediction of their shapes. For example, suppose we need to know the wavefunction of a particle with a given total energy and a potential energy that decreases with increasing x (Fig. 7.27). Because the difference $E - V = E_k$ increases from left to right, the wavefunction must become more sharply curved as x increases: its wavelength decreases as the local contributions to its kinetic energy increase. We can therefore guess that the wavefunction will look like the function sketched in the illustration, and more detailed calculation confirms this to be so.

A brief comment

We are using the term 'curvature' informally: the precise technical definition of the curvature of a function f is $(d^2f/dx^2)/\{1 + (df/dx)^2\}^{3/2}$.

(d) Hermitian operators

All the quantum mechanical operators that correspond to observables have a very special mathematical property: they are 'hermitian'. A **hermitian operator** is one for which the following relation is true:

Hermiticity: $$\int \psi_i^* \hat{\Omega} \psi_j \mathrm{d}\tau = \left\{ \int \psi_j^* \hat{\Omega} \psi_i \mathrm{d}\tau \right\}^*$$ Definition of hermiticity [7.33]

That is, the same result is obtained by letting the operator act on ψ_j and then integrating or by letting it act on ψ_i instead, integrating, and then taking the complex conjugate of the result. One trivial consequence of hermiticity is that it reduces the number of integrals we need to evaluate. However, as we shall see, hermiticity has much more profound implications.

It is easy to confirm that the position operator ($x\times$) is hermitian because we are free to change the order of the factors in the integrand:

$$\int_{-\infty}^{\infty} \psi_i^* x \psi_j \, \mathrm{d}\tau = \int_{-\infty}^{\infty} \psi_j x \psi_i^* \mathrm{d}\tau = \left\{ \int_{-\infty}^{\infty} \psi_j^* x \psi_i \, \mathrm{d}\tau \right\}^*$$

The demonstration that the linear momentum operator is hermitian is more involved because we cannot just alter the order of functions we differentiate; but it is hermitian, as we show in the following *Justification*.

Justification 7.2 *The hermiticity of the linear momentum operator*

Our task is to show that

$$\int_{-\infty}^{\infty} \psi_i^* \hat{p}_x \psi_j \mathrm{d}x = \left\{ \int_{-\infty}^{\infty} \psi_j^* \hat{p}_x \psi_i \mathrm{d}x \right\}^*$$

with $\hat{p}_x$ given in eqn 7.29. To do so, we use 'integration by parts' (see *Mathematical background* 1), the relation

$$\int f \frac{\mathrm{d}g}{\mathrm{d}x} \mathrm{d}x = fg - \int g \frac{\mathrm{d}f}{\mathrm{d}x} \mathrm{d}x$$

In the present case we write

$$\int_{-\infty}^{\infty} \psi_i^* \hat{p}_x \psi_j \mathrm{d}x = \frac{\hbar}{\mathrm{i}} \int_{-\infty}^{\infty} \overbrace{\psi_i^*}^{f} \overbrace{\frac{\mathrm{d}\psi_j}{\mathrm{d}x}}^{\mathrm{d}g/\mathrm{d}x} \mathrm{d}x$$
$$= \frac{\hbar}{\mathrm{i}} \psi_i^* \psi_j \bigg|_{-\infty}^{\infty} - \frac{\hbar}{\mathrm{i}} \int_{-\infty}^{\infty} \psi_j \frac{\mathrm{d}\psi_i^*}{\mathrm{d}x} \mathrm{d}x$$

The first term on the right of the second equality is zero, because all wavefunctions are zero at infinity in either direction, so we are left with

$$\int_{-\infty}^{\infty} \psi_i^* \hat{p}_x \psi_j \mathrm{d}x = -\frac{\hbar}{\mathrm{i}} \int_{-\infty}^{\infty} \psi_j \frac{\mathrm{d}\psi_i^*}{\mathrm{d}x} \mathrm{d}x = \left\{ \frac{\hbar}{\mathrm{i}} \int_{-\infty}^{\infty} \psi_j^* \frac{\mathrm{d}\psi_i}{\mathrm{d}x} \mathrm{d}x \right\}^*$$
$$= \left\{ \int_{-\infty}^{\infty} \psi_j^* \hat{p}_x \psi_i \mathrm{d}x \right\}^*$$

as we set out to prove. In the final line we have used $(\psi^*)^* = \psi$.

Self-test 7.7 Confirm that the operator $\mathrm{d}^2/\mathrm{d}x^2$ is hermitian.

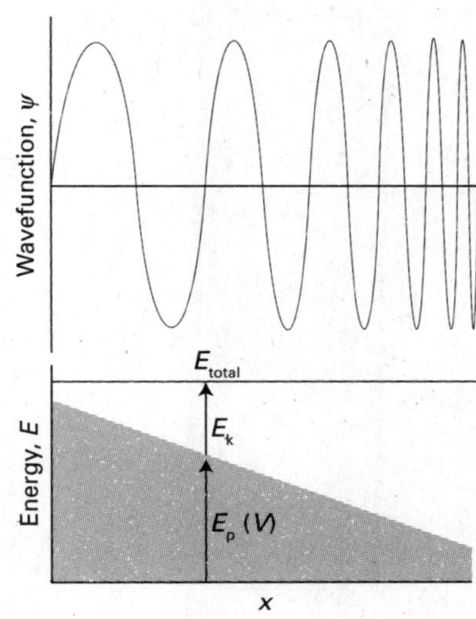

Fig. 7.27 The wavefunction of a particle in a potential decreasing towards the right and hence subjected to a constant force to the right. Only the real part of the wavefunction is shown, the imaginary part is similar, but displaced to the right.

Hermitian operators are enormously important by virtue of two properties: their eigenvalues are real (as we prove in the following *Justification*), and their eigenfunctions are 'orthogonal'. All observables have real values (in the mathematical sense, such as $x = 2$ m and $E = 10$ J), so all observables are represented by hermitian operators.

Justification 7.3 *The reality of eigenvalues*

For a wavefunction ψ that is normalized and is an eigenfunction of a hermitian operator $\hat{\Omega}$ with eigenvalue ω, we can write

$$\int \psi^* \hat{\Omega} \psi \, d\tau = \int \psi^* \omega \psi \, d\tau = \omega \int \psi^* \psi \, d\tau = \omega$$

However, by taking the complex conjugate we can write

$$\omega^* = \left\{ \int \psi^* \hat{\Omega} \psi \, d\tau \right\}^* \overset{\text{hermiticity}}{=} \int \psi^* \hat{\Omega} \psi \, d\tau = \omega$$

The conclusion that $\omega^* = \omega$ confirms that ω is real.

To say that two different functions ψ_i and ψ_j are **orthogonal** means that the integral (over all space) of their product is zero:

$$\int \psi_i^* \psi_j \, d\tau = 0 \qquad \text{for} \qquad i \neq j \qquad \text{Definition of orthogonality} \qquad (7.34)$$

A general feature of quantum mechanics, which we prove in the following *Justification*, is that *wavefunctions corresponding to different eigenvalues of an hermitian operator are orthogonal*. For example, the hamiltonian operator is hermitian (it corresponds to an observable, the energy). Therefore, if ψ_1 corresponds to one energy, and ψ_2 corresponds to a different energy, then we know at once that the two functions are orthogonal and that the integral of their product is zero.

Justification 7.4 *The orthogonality of wavefunctions*

Suppose we have two wavefunctions ψ_n and ψ_m corresponding to two different energies E_n and E_m, respectively. Then we can write

$$\hat{H}\psi_n = E_n \psi_n \qquad \hat{H}\psi_m = E_m \psi_m$$

Now multiply the first of these two Schrödinger equations by ψ_m^* and the second by ψ_n^* and integrate over all space:

$$\int \psi_m^* \hat{H} \psi_n \, d\tau = E_n \int \psi_m^* \psi_n \, d\tau \qquad \int \psi_n^* \hat{H} \psi_m \, d\tau = E_m \int \psi_n^* \psi_m \, d\tau$$

Next, noting that the energies themselves are real, form the complex conjugate of the second expression (for the state m) and subtract it from the first expression (for the state n):

$$\int \psi_m^* \hat{H} \psi_n \, d\tau - \left(\int \psi_n^* \hat{H} \psi_m \, d\tau \right)^* = E_n \int \psi_m^* \psi_n \, d\tau - E_m \int \psi_n \psi_m^* \, d\tau$$

By the hermiticity of the hamiltonian, the two terms on the left are equal, so they cancel and we are left with

$$0=(E_n-E_m)\int \psi_m^*\psi_n\,\mathrm{d}\tau$$

However, the two energies are different; therefore the integral on the right must be zero, which confirms that two wavefunctions belonging to different energies are orthogonal. The same argument applies to eigenfunctions of any Hermitian operator.

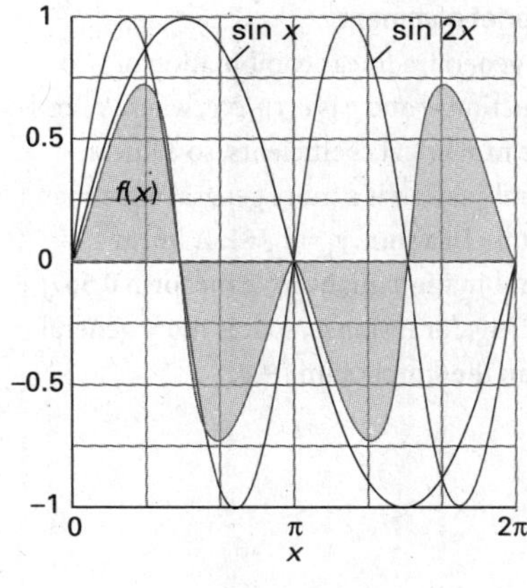

Fig. 7.28 The integral of the function $f(x)=\sin x \sin 2x$ is equal to the area (tinted) below the green curve, and is zero, as can be inferred by symmetry. The function—and the value of the integral—repeats itself for all replications of the section between 0 and 2π, so the integral from $-\infty$ to ∞ is zero.

The property of orthogonality is of great importance in quantum mechanics because it enables us to eliminate a large number of integrals from calculations. Orthogonality plays a central role in the theory of chemical bonding (Chapter 10) and spectroscopy (Chapters 12–14). Sets of functions that are normalized and mutually orthogonal are called **orthonormal**.

• A brief illustration

The wavefunctions $\sin x$ and $\sin 2x$ are eigenfunctions of the hermitian operator $\mathrm{d}^2/\mathrm{d}x^2$, with eigenvalues -1 and -4, respectively. To verify that the two wavefunctions are mutually orthogonal, we integrate the product $(\sin x)(\sin 2x)$ over all space, which we may take to span from $x=0$ to $x=2\pi$, because both functions repeat themselves outside that range. Hence proving that the integral of their product is zero within that range implies that the integral over the whole of space is also zero (Fig. 7.28). A useful integral for this calculation when $a^2\neq b^2$ is

$$\int \sin ax \sin bx\,\mathrm{d}x=\frac{\sin(a-b)x}{2(a-b)}-\frac{\sin(a+b)x}{2(a+b)}+\text{constant}$$

It follows that, for $a=1$ and $b=2$, and the fact that $\sin 0=0$, $\sin 2\pi=0$, and $\sin 6\pi=0$, that

$$\int_0^{2\pi}\sin x\sin 2x\,\mathrm{d}x=0$$

and the two functions are mutually orthogonal. •

Self-test 7.8 Confirm that the functions $\sin x$ and $\sin 3x$ are mutually orthogonal.

$$\left[\int_0^{2\pi}\sin x\sin 3x\,\mathrm{d}x=0\right]$$

(e) Superpositions and expectation values

Suppose now that the wavefunction is the one given in eqn 7.22 (with $A=B$). What is the linear momentum of the particle it describes? We quickly run into trouble if we use the operator technique. When we operate with $\hat{p}_x$, we find

$$\hat{p}_x\psi=\frac{\hbar}{\mathrm{i}}\frac{\mathrm{d}\psi}{\mathrm{d}x}=\frac{2\hbar}{\mathrm{i}}A\frac{\mathrm{d}\cos kx}{\mathrm{d}x}=-\frac{2k\hbar}{\mathrm{i}}A\sin kx \qquad (7.35)$$

This expression is not an eigenvalue equation, because the function on the right ($\sin kx$) is different from that on the left ($\cos kx$).

When the wavefunction of a particle is not an eigenfunction of an operator, the property to which the operator corresponds does not have a definite value. However, in the current example the momentum is not completely indefinite because the cosine wavefunction is a **linear combination**, or sum, of $\mathrm{e}^{\mathrm{i}kx}$ and $\mathrm{e}^{-\mathrm{i}kx}$, and these two functions, as we have seen, individually correspond to definite momentum states. We say that

A brief comment

In general, a linear combination of two functions f and g is $c_1f + c_2g$, where c_1 and c_2 are numerical coefficients, so a linear combination is a more general term than 'sum'. In a sum, $c_1 = c_2 = 1$. A linear combination might have the form $0.567f + 1.234g$, for instance, so it is more general than the simple sum $f + g$.

the total wavefunction is a **superposition** of more than one wavefunction. Symbolically we can write the superposition as

$$\psi = \underbrace{\psi_{\rightarrow}}_{\substack{\text{Particle with}\\\text{linear}\\\text{momentum}\\+k\hbar}} + \underbrace{\psi_{\leftarrow}}_{\substack{\text{Particle with}\\\text{linear}\\\text{momentum}\\-k\hbar}}$$

The interpretation of this composite wavefunction is that, if the momentum of the particle is repeatedly measured in a long series of observations, then its magnitude will found to be $k\hbar$ in all the measurements (because that is the value for each component of the wavefunction). However, because the two component wavefunctions occur equally in the superposition, half the measurements will show that the particle is moving to the right ($p_x = +k\hbar$), and half the measurements will show that it is moving to the left ($p_x = -k\hbar$). According to quantum mechanics, we cannot predict in which direction the particle will in fact be found to be travelling; all we can say is that, in a long series of observations, if the particle is described by this wavefunction, then there are equal probabilities of finding the particle travelling to the right and to the left.

The same interpretation applies to any wavefunction written as a linear combination of eigenfunctions of an operator. Thus, suppose the wavefunction is known to be a superposition of many different linear momentum eigenfunctions and written as the linear combination

$$\psi = c_1\psi_1 + c_2\psi_2 + \cdots = \sum_k c_k\psi_k \qquad \text{Linear combination of basis functions} \qquad (7.36)$$

where the c_k are numerical (possibly complex) coefficients and the ψ_k correspond to different momentum states. The functions ψ_k are said to form a **complete set** in the sense that any arbitrary function can be expressed as a linear combination of them. Then according to quantum mechanics:

1. When the momentum is measured, in a single observation one of the eigenvalues corresponding to the ψ_k that contribute to the superposition will be found.
2. The probability of measuring a particular eigenvalue in a series of observations is proportional to the square modulus ($|c_k|^2$) of the corresponding coefficient in the linear combination.
3. The average value of a large number of observations is given by the expectation value, $\langle\Omega\rangle$, of the operator corresponding to the observable of interest.

The **expectation value** of an operator $\hat{\Omega}$ is defined as

$$\langle\Omega\rangle = \int \psi^*\hat{\Omega}\psi\,\mathrm{d}\tau \qquad \text{Definition of expectation value} \qquad [7.37]$$

This formula is valid only for normalized wavefunctions. As we see in the following *Justification*, an expectation value is the weighted average of a large number of observations of a property.

Justification 7.5 *The expectation value of an operator*

If ψ is an eigenfunction of $\hat{\Omega}$ with eigenvalue ω, the expectation value of $\hat{\Omega}$ is

$$\langle\Omega\rangle = \int \psi^*\overbrace{\hat{\Omega}\psi}^{\omega\psi}\mathrm{d}\tau = \int \psi^*\omega\psi\mathrm{d}\tau = \omega\int \psi^*\psi\mathrm{d}\tau = \omega$$

because ω is a constant and may be taken outside the integral, and the resulting integral is equal to 1 for a normalized wavefunction. The interpretation of this expression is that, because every observation of the property Ω results in the value ω (because the wavefunction is an eigenfunction of $\hat{\Omega}$), the mean value of all the observations is also ω.

A wavefunction that is not an eigenfunction of the operator of interest can be written as a linear combination of eigenfunctions. For simplicity, suppose the wavefunction is the sum of two eigenfunctions (the general case, eqn 7.36, can easily be developed). Then

$$\begin{aligned}\langle\Omega\rangle &= \int(c_1\psi_1+c_2\psi_2)^*\hat{\Omega}(c_1\psi_1+c_2\psi_2)\mathrm{d}\tau\\ &= \int(c_1\psi_1+c_2\psi_2)^*(c_1\hat{\Omega}\psi_1+c_2\hat{\Omega}\psi_2)\mathrm{d}\tau\\ &= \int(c_1\psi_1+c_2\psi_2)^*(c_1\omega_1\psi_1+c_2\omega_2\psi_2)\mathrm{d}\tau\\ &= c_1^*c_1\omega_1\overbrace{\int\psi_1^*\psi_1\mathrm{d}\tau}^{1}+c_2^*c_2\omega_2\overbrace{\int\psi_2^*\psi_2\mathrm{d}\tau}^{1}\\ &\quad+c_2^*c_1\omega_1\overbrace{\int\psi_2^*\psi_1\mathrm{d}\tau}^{0}+c_1^*c_2\omega_2\overbrace{\int\psi_1^*\psi_2\mathrm{d}\tau}^{0}\end{aligned}$$

The first two integrals on the right are both equal to 1 because the wavefunctions are individually normalized. Because ψ_1 and ψ_2 correspond to different eigenvalues of an hermitian operator, they are orthogonal, so the third and fourth integrals on the right are zero. We can conclude that

$$\langle\Omega\rangle=|c_1|^2\omega_1+|c_2|^2\omega_2$$

This expression shows that the expectation value is the sum of the two eigenvalues weighted by the probabilities that each one will be found in a series of measurements. Hence, the expectation value is the weighted mean of a series of observations.

Example 7.7 *Calculating an expectation value*

Calculate the average value of the distance of an electron from the nucleus in the hydrogen atom in its state of lowest energy.

Method The average radius is the expectation value of the operator corresponding to the distance from the nucleus, which is multiplication by r. To evaluate $\langle r\rangle$, we need to know the normalized wavefunction (from Example 7.4) and then evaluate the integral in eqn 7.37.

Answer The average value is given by the expectation value

$$\langle r\rangle=\int\psi^*r\psi\,\mathrm{d}\tau=\int r|\psi|^2\mathrm{d}\tau$$

which we evaluate by using spherical polar coordinates and the appropriate expression for the volume element, $\mathrm{d}\tau=r^2\mathrm{d}r\sin\theta\,\mathrm{d}\theta\,\mathrm{d}\phi$. Using the normalized function in Example 7.4, gives

$$\langle r\rangle = \frac{1}{\pi a_0^3}\overbrace{\int_0^\infty r^3 e^{-2r/a_0}\mathrm{d}r}^{3a_0^4/2^3}\overbrace{\int_0^\pi \sin\theta\,\mathrm{d}\theta}^{2}\overbrace{\int_0^{2\pi}\mathrm{d}\phi}^{2\pi} = \tfrac{3}{2}a_0$$

Because $a_0 = 52.9$ pm (see inside the front cover), $\langle r\rangle = 79.4$ pm. This result means that, if a very large number of measurements of the distance of the electron from the nucleus are made, then their mean value will be 79.4 pm. However, each different observation will give a different and unpredictable individual result because the wavefunction is not an eigenfunction of the operator corresponding to r.

Self-test 7.9 Evaluate the root mean square distance, $\langle r^2\rangle^{1/2}$, of the electron from the nucleus in the hydrogen atom. [$3^{1/2}a_0 = 91.6$ pm]

The mean kinetic energy of a particle in one dimension is the expectation value of the operator given in eqn 7.31. Therefore, we can write

$$\langle E_\mathrm{k}\rangle = \int \psi^* \hat{E}_\mathrm{k}\psi\,\mathrm{d}x = -\frac{\hbar^2}{2m}\int \psi^*\frac{\mathrm{d}^2\psi}{\mathrm{d}x^2}\mathrm{d}x \qquad (7.38)$$

This conclusion confirms the previous assertion that the kinetic energy is a kind of average over the curvature of the wavefunction: we get a large contribution to the observed value from regions where the wavefunction is sharply curved (so $\mathrm{d}^2\psi/\mathrm{d}x^2$ is large) and the wavefunction itself is large (so that ψ^* is large too).

7.6 The uncertainty principle

Key points **The uncertainty principle restricts the precision with which complementary observables may be specified and measured. Complementary observables are observables for which the corresponding operators do not commute.**

We have seen that, if the wavefunction is Ae^{ikx}, then the particle it describes has a definite state of linear momentum, namely travelling to the right with momentum $p_x = +k\hbar$. However, we have also seen that the position of the particle described by this wavefunction is completely unpredictable. In other words, if the momentum is specified precisely, it is impossible to predict the location of the particle. This statement is one-half of a special case of the **Heisenberg uncertainty principle**, one of the most celebrated results of quantum mechanics:

It is impossible to specify simultaneously, with arbitrary precision, both the momentum and the position of a particle. [Heisenberg uncertainty principle]

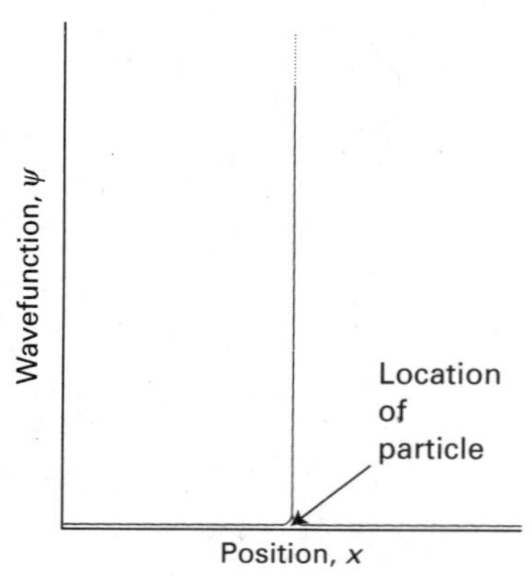

Fig. 7.29 The wavefunction for a particle at a well-defined location is a sharply spiked function that has zero amplitude everywhere except at the particle's position.

Before discussing the principle further, we must establish its other half: that, if we know the position of a particle exactly, then we can say nothing about its momentum. The argument draws on the idea of regarding a wavefunction as a superposition of eigenfunctions, and runs as follows.

If we know that the particle is at a definite location, its wavefunction must be large there and zero everywhere else (Fig. 7.29). Such a wavefunction can be created by superimposing a large number of harmonic (sine and cosine) functions, or, equivalently, a number of e^{ikx} functions. In other words, we can create a sharply localized wavefunction, called a **wave packet**, by forming a linear combination of wavefunctions that correspond to many different linear momenta. The superposition of a few harmonic functions gives a wavefunction that spreads over a range of locations

(Fig. 7.30). However, as the number of wavefunctions in the superposition increases, the wave packet becomes sharper on account of the more complete interference between the positive and negative regions of the individual waves. When an infinite number of components are used, the wave packet is a sharp, infinitely narrow spike, which corresponds to perfect localization of the particle. Now the particle is perfectly localized. However, we have lost all information about its momentum because, as we saw above, a measurement of the momentum will give a result corresponding to any one of the infinite number of waves in the superposition, and which one it will give is unpredictable. Hence, if we know the location of the particle precisely (implying that its wavefunction is a superposition of an infinite number of momentum eigenfunctions), then its momentum is completely unpredictable.

A quantitative version of this result is

$$\Delta p \Delta q \geq \tfrac{1}{2}\hbar \qquad (7.39a)$$

Heisenberg uncertainty principle

In this expression Δp is the 'uncertainty' in the linear momentum parallel to the axis q, and Δq is the uncertainty in position along that axis. These 'uncertainties' are precisely defined, for they are the root mean square deviations of the properties from their mean values:

$$\Delta p = \{\langle p^2\rangle - \langle p\rangle^2\}^{1/2} \qquad \Delta q = \{\langle q^2\rangle - \langle q\rangle^2\}^{1/2} \qquad (7.39b)$$

If there is complete certainty about the position of the particle ($\Delta q = 0$), then the only way that eqn 7.39a can be satisfied is for $\Delta p = \infty$, which implies complete uncertainty about the momentum. Conversely, if the momentum parallel to an axis is known exactly ($\Delta p = 0$), then the position along that axis must be completely uncertain ($\Delta q = \infty$).

The p and q that appear in eqn 7.39 refer to the same direction in space. Therefore, whereas simultaneous specifications of the position on the x-axis and momentum parallel to the x-axis are restricted by the uncertainty relation, simultaneous locations of position on x and motion parallel to y or z are not restricted. The restrictions that the uncertainty principle implies are summarized in Table 7.2.

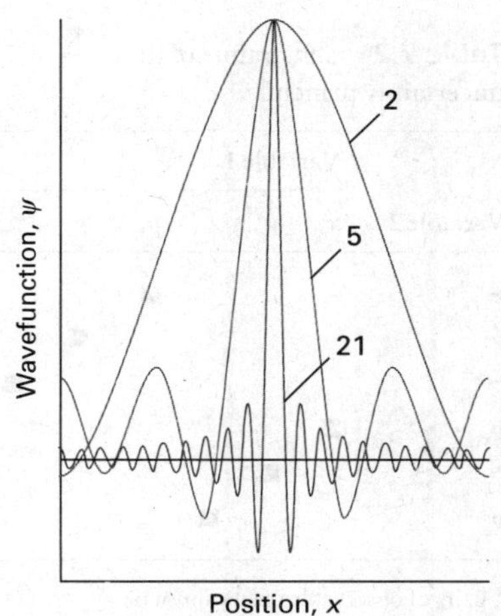

Fig. 7.30 The wavefunction for a particle with an ill-defined location can be regarded as the superposition of several wavefunctions of definite wavelength that interfere constructively in one place but destructively elsewhere. As more waves are used in the superposition (as given by the numbers attached to the curves), the location becomes more precise at the expense of uncertainty in the particle's momentum. An infinite number of waves is needed to construct the wavefunction of a perfectly localized particle.

interActivity Use mathematical software or an electronic spreadsheet to construct superpositions of cosine functions as $\psi(x) = \sum_{k=1}^{N}(1/N)\cos(k\pi x)$, where the constant $1/N$ is introduced to keep the superpositions with the same overall magnitude. Explore how the probability density $\psi^2(x)$ changes with the value of N.

Example 7.8 *Using the uncertainty principle*

Suppose the speed of a projectile of mass 1.0 g is known to within 1 μm s^{-1}. Calculate the minimum uncertainty in its position.

Method Estimate Δp from $m\Delta v$, where Δv is the uncertainty in the speed; then use eqn 7.39a to estimate the minimum uncertainty in position, Δq.

Answer The minimum uncertainty in position is

$$\Delta q = \frac{\hbar}{2m\Delta v}$$

$$= \frac{1.055\times10^{-34}\ \text{J s}}{2\times(1.0\times10^{-3}\ \text{kg})\times(1\times10^{-6}\ \text{m s}^{-1})} = 5\times10^{-26}\ \text{m}$$

where we have used 1 J = 1 kg m^2 s^{-2}. The uncertainty is completely negligible for all practical purposes concerning macroscopic objects. However, if the mass is that of an electron, then the same uncertainty in speed implies an uncertainty in position far larger than the diameter of an atom (the analogous calculation gives $\Delta q = 60$ m); so the concept of a trajectory, the simultaneous possession of a precise position and momentum, is untenable.

Table 7.2* Constraints of the uncertainty principle

	Variable 1					
Variable 2	x	y	z	p_x	p_y	p_z
x				■		
y					■	
z						■
p_x	■					
p_y		■				
p_z			■			

* Pairs of observables that cannot be determined simultaneously with arbitrary precision are marked with a black rectangle; all others are unrestricted.

Self-test 7.10 Estimate the minimum uncertainty in the speed of an electron in a one-dimensional region of length $2a_0$. [547 km s^{-1}]

The Heisenberg uncertainty principle is more general than eqn 7.39 suggests. It applies to any pair of observables called **complementary observables**, which are defined in terms of the properties of their operators. Specifically, two observables Ω_1 and Ω_2 are complementary if

$$\hat{\Omega}_1(\hat{\Omega}_2\psi) \neq \hat{\Omega}_2(\hat{\Omega}_1\psi) \tag{7.40}$$

where the term on the left implies that $\hat{\Omega}_2$ acts first, then $\hat{\Omega}_1$ acts on the result, and the term on the right implies that the operations are performed in the opposite order. When the effect of two operators applied in succession depends on their order (as this equation implies), we say that they do not **commute**. The different outcomes of the effect of applying $\hat{\Omega}_1$ and $\hat{\Omega}_2$ in a different order are expressed by introducing the **commutator** of the two operators, which is defined as

$$[\hat{\Omega}_1,\hat{\Omega}_2] = \hat{\Omega}_1\hat{\Omega}_2 - \hat{\Omega}_2\hat{\Omega}_1 \tag{7.41}$$

Definition of commutator

We show in the following *Justification* that the commutator of the operators for position and linear momentum is

$$[\hat{x},\hat{p}_x] = \mathrm{i}\hbar \tag{7.42}$$

Justification 7.6 *The commutator of position and momentum*

To show that the operators for position and momentum do not commute (and hence are complementary observables) we consider the effect of $\hat{x}\hat{p}_x$ (that is, the effect of $\hat{p}_x$ followed by the effect on the outcome of multiplication by x) on a wavefunction ψ:

$$\hat{x}\hat{p}_x\psi = x\times\frac{\hbar}{\mathrm{i}}\frac{\mathrm{d}\psi}{\mathrm{d}x}$$

Next, we consider the effect of $\hat{p}_x\hat{x}$ on the same function (that is, the effect of multiplication by x followed by the effect of $\hat{p}_x$ on the outcome):

$$\hat{p}_x\hat{x}\psi = \frac{\hbar}{\mathrm{i}}\frac{\mathrm{d}(x\psi)}{\mathrm{d}x} = \frac{\hbar}{\mathrm{i}}\left(\psi + x\frac{\mathrm{d}\psi}{\mathrm{d}x}\right)$$

For this step we have used the standard rule about differentiating a product of functions ($\mathrm{d}(fg)/\mathrm{d}x = f\mathrm{d}g/\mathrm{d}x + g\mathrm{d}f/\mathrm{d}x$). The second expression is clearly different from the first, so the two operators do not commute. Their commutator can be inferred from the difference of the two expressions:

$$\hat{x}\hat{p}_x\psi - \hat{p}_x\hat{x}\psi = -\frac{\hbar}{\mathrm{i}}\psi = \mathrm{i}\hbar\psi$$

This relation is true for any wavefunction ψ, so the operator relation in eqn 7.42 follows immediately.

The commutator in eqn 7.42 is of such vital significance in quantum mechanics that it is taken as a fundamental distinction between classical mechanics and quantum mechanics. In fact, this commutator may be taken as a postulate of quantum mechanics, and is used to justify the choice of the operators for position and linear momentum given in eqn 7.29.

With the concept of commutator established, the Heisenberg uncertainty principle can be given its most general form. For *any* two pairs of observables, Ω_1 and Ω_2, the uncertainties (to be precise, the root mean square deviations of their values from the mean) in simultaneous determinations are related by

$$\Delta\Omega_1\Delta\Omega_2 \geq \tfrac{1}{2}|\langle[\hat{\Omega}_1,\hat{\Omega}_2]\rangle| \qquad (7.43)$$

We obtain the special case of eqn 7.39 when we identify the observables with x and p_x and use eqn 7.42 for their commutator. (See *Mathematical background* 3 for the meaning of the $|\ldots|$ notation.)

Complementary observables are observables with non-commuting operators. With the discovery that some pairs of observables are complementary (we meet more examples in the next chapter), we are at the heart of the difference between classical and quantum mechanics. Classical mechanics supposed, falsely as we now know, that the position and momentum of a particle could be specified simultaneously with arbitrary precision. However, quantum mechanics shows that position and momentum are complementary, and that we have to make a choice: we can specify position at the expense of momentum, or momentum at the expense of position.

The realization that some observables are complementary allows us to make considerable progress with the calculation of atomic and molecular properties; but it does away with some of the most cherished concepts of classical physics.

7.7 The postulates of quantum mechanics

For convenience, we collect here the postulates on which quantum mechanics is based and which have been introduced in the course of this chapter.

The wavefunction. All dynamical information is contained in the wavefunction ψ for the system, which is a mathematical function found by solving the Schrödinger equation for the system. In one dimension:

$$-\frac{\hbar^2}{2m}\frac{d^2\psi}{dx^2} + V(x)\psi = E\psi$$

The Born interpretation. If the wavefunction of a particle has the value ψ at some point r, then the probability of finding the particle in an infinitesimal volume $d\tau = dxdydz$ at that point is proportional to $|\psi|^2 d\tau$.

Acceptable wavefunctions. An acceptable wavefunction must be continuous, have a continuous first derivative, be single-valued, and be square-integrable.

Observables. Observables, Ω, are represented by operators, $\hat{\Omega}$, built from the following position and momentum operators:

$$\hat{x} = x\times \qquad \hat{p}_x = \frac{\hbar}{i}\frac{d}{dx}$$

or, more generally, from operators that satisfy the commutation relation $[\hat{x},\hat{p}_x] = i\hbar$.

The Heisenberg uncertainty relation. It is impossible to specify simultaneously, with arbitrary precision, both the momentum and the position of a particle and, more generally, any pair of observables with operators that do not commute.

Checklist of key equations

Property	Equation	Comment
Bohr frequency condition	$\Delta E = h\nu$	Conservation of energy
Photoelectric effect	$\frac{1}{2}m_e v^2 = h\nu - \Phi$	Φ is the work function
de Broglie relation	$\lambda = h/p$	λ is the wavelength of a particle of linear momentum p
The time-independent Schrödinger equation in one dimension	$-(\hbar^2/2m)(\mathrm{d}^2\psi/\mathrm{d}x^2) + V(x)\psi = E\psi$, or $\hat{H}\psi = E\psi$	
Operators corresponding to observables	$\hat{x} = x\times \quad \hat{p}_x = \frac{\hbar}{\mathrm{i}}\frac{\mathrm{d}}{\mathrm{d}x}$	Position and linear momentum
Expectation value of an operator	$\langle \Omega \rangle = \int \psi^* \hat{\Omega} \psi \, \mathrm{d}\tau$	Mean value of the observable
Normalization	$\int \psi^* \psi \, \mathrm{d}\tau = 1$	
Orthogonality	$\int \psi_i^* \psi_j \, \mathrm{d}\tau = 0$	
Hermiticity	$\int \psi_i^* \hat{\Omega} \psi_j \, \mathrm{d}\tau = \left\{ \int \psi_j^* \hat{\Omega} \psi_i \, \mathrm{d}\tau \right\}^*$	Real eigenvalues, orthogonal eigenfunctions
Heisenberg uncertainty relation	$\Delta\Omega_1 \Delta\Omega_2 \geq \frac{1}{2}\lvert\langle[\hat{\Omega}_1, \hat{\Omega}_2]\rangle\rvert$ Special case: $\Delta p \Delta q \geq \frac{1}{2}\hbar$	
Commutator of two operators	$[\hat{\Omega}_1, \hat{\Omega}_2] = \hat{\Omega}_1\hat{\Omega}_2 - \hat{\Omega}_2\hat{\Omega}_1$ Special case: $[\hat{x}, \hat{p}_x] = \mathrm{i}\hbar$	The observables are complementary if this commutator is zero.

Further information

Further information 7.1 *Classical mechanics*

Classical mechanics describes the behaviour of objects in terms of two equations. One expresses the fact that the total energy is constant in the absence of external forces; the other expresses the response of particles to the forces acting on them.

(a) The trajectory in terms of the energy

The **velocity**, $\boldsymbol{v}$, of a particle is the rate of change of its position:

$$\boldsymbol{v} = \frac{\mathrm{d}\boldsymbol{r}}{\mathrm{d}t} \qquad (7.44)$$

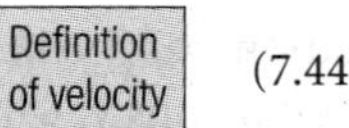

The velocity is a vector, with both direction and magnitude. (Vectors are discussed in *Mathematical background* 5.) The magnitude of the velocity is the **speed**, v. The **linear momentum**, $\boldsymbol{p}$, of a particle of mass m is related to its velocity, $\boldsymbol{v}$, by

$$\boldsymbol{p} = m\boldsymbol{v} \qquad (7.45)$$

Definition of linear momentum

Like the velocity vector, the linear momentum vector points in the direction of travel of the particle (Fig. 7.31). In terms of the linear momentum, the total energy—the sum of the kinetic and potential energy—of a particle is

$$E = E_k + V(x) = \frac{p^2}{2m} + V(x) \qquad (7.46)$$

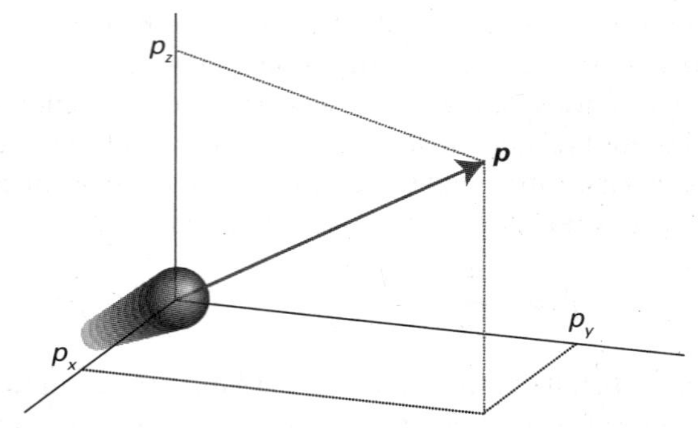

Fig. 7.31 The linear momentum of a particle is a vector property and points in the direction of motion.

This equation can be used to show that a particle will have a definite **trajectory**, or definite position and momentum at each instant. For example, consider a particle free to move in one direction (along the x-axis) in a region where $V=0$ (so the energy is independent of position). From the definition of the kinetic energy, $E_k = \frac{1}{2}mv^2$, and $v = dx/dt$, it follows from eqns 7.45 and 7.46 that

$$\frac{dx}{dt} = \left(\frac{2E_k}{m}\right)^{1/2} \qquad (7.47)$$

A solution of this differential equation is

$$x(t) = x(0) + \left(\frac{2E_k}{m}\right)^{1/2} t \qquad (7.48)$$

The linear momentum is a constant:

$$p(t) = mv(t) = m\frac{dx}{dt} = (2mE_k)^{1/2} \qquad (7.49)$$

Hence, if we know the initial position and momentum, we can predict all later positions and momenta exactly.

(b) Newton's second law

The **force**, F, experienced by a particle free to move in one dimension is related to its potential energy, V, by

$$F = -\frac{dV}{dx} \qquad (7.50a)$$

This relation implies that the direction of the force is towards decreasing potential energy (Fig. 7.32). In three dimensions

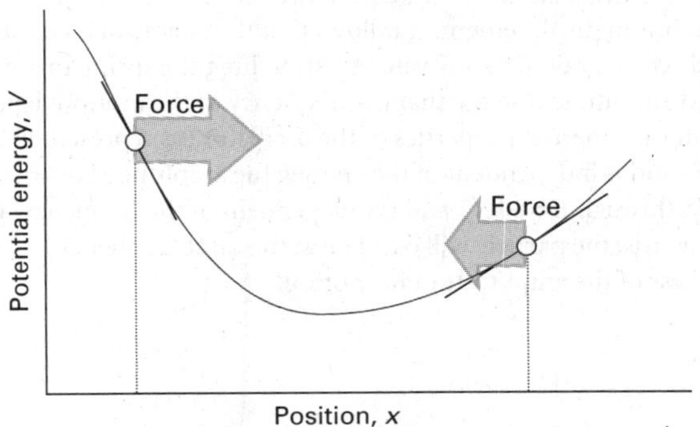

Fig. 7.32 The force acting on a particle is determined by the slope of the potential energy at each point. The force points in the direction of lower potential energy.

$$F = -\nabla V \qquad \nabla = i\frac{\partial}{\partial x} + j\frac{\partial}{\partial y} + k\frac{\partial}{\partial z} \qquad (7.50b)$$

Newton's second law of motion states that *the rate of change of momentum is equal to the force acting on the particle*. In one dimension:

$$\frac{dp}{dt} = F \qquad (7.51a)$$

Newton's second law of motion

Because $p = m(dx/dt)$ in one dimension, it is sometimes more convenient to write this equation as

$$m\frac{d^2x}{dt^2} = F \qquad (7.51b)$$

The second derivative, d^2x/dt^2, is the **acceleration** of the particle, its rate of change of velocity (in this instance, along the x-axis). It follows that, if we know the force acting everywhere and at all times, then solving eqn 7.51 will also give the trajectory. This calculation is equivalent to the one based on E, but is more suitable in some applications. For example, it can be used to show that, if a particle of mass m is initially stationary and is subjected to a constant force F for a time τ, then its kinetic energy increases from zero to

$$E_k = \frac{F^2\tau^2}{2m} \qquad (7.52)$$

and then remains at that energy after the force ceases to act. Because the applied force, F, and the time, τ, for which it acts may be varied at will, the solution implies that the energy of the particle may be increased to any value.

(c) Rotational motion

The rotational motion of a particle about a central point is described by its **angular momentum, J**. The angular momentum is a vector: its magnitude gives the rate at which a particle circulates and its direction indicates the axis of rotation (Fig. 7.33). The magnitude of the angular momentum, J, is given by the expression

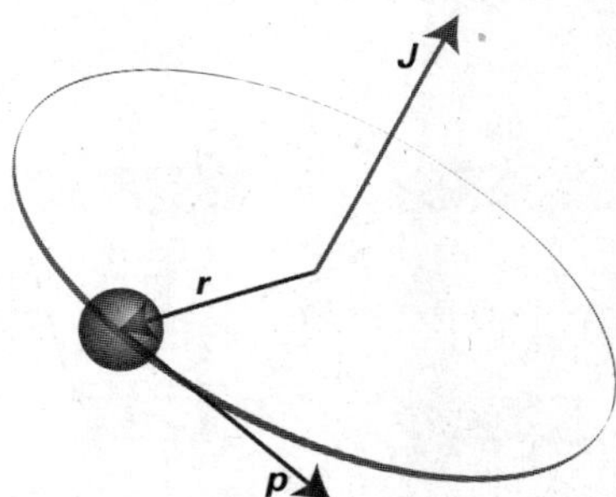

Fig. 7.33 The angular momentum of a particle is represented by a vector along the axis of rotation and perpendicular to the plane of rotation. The length of the vector denotes the magnitude of the angular momentum. The direction of motion is clockwise to an observer looking in the direction of the vector.

$$J = I\omega \qquad (7.53)$$

Magnitude of the angular momentum

where ω is the **angular velocity** of the body, its rate of change of angular position (in radians per second), and I is the **moment of inertia**. The analogous roles of m and I, of v and ω, and of p and J in the translational and rotational cases, respectively, should be remembered, because they provide a ready way of constructing and recalling equations. For a point particle of mass m moving in a circle of radius r, the moment of inertia about the axis of rotation is given by the expression

$$I = mr^2 \quad \text{Moment of inertia of a point particle moving in a circle} \quad (7.54)$$

To accelerate a rotation it is necessary to apply a **torque**, T, a twisting force. Newton's equation is then

$$\frac{dJ}{dt} = T \quad \text{Definition of torque} \quad (7.55)$$

If a constant torque is applied for a time τ, the rotational energy of an initially stationary body is increased to

$$E_k = \frac{T^2\tau^2}{2I} \quad (7.56)$$

The implication of this equation is that an appropriate torque and period for which it is applied can excite the rotation to an arbitrary energy.

(d) The harmonic oscillator

A **harmonic oscillator** consists of a particle that experiences a restoring force proportional to its displacement from its equilibrium position:

$$F = -kx \quad \text{Restoring force} \quad (7.57)$$

An example is a particle joined to a rigid support by a spring. The constant of proportionality k is called the **force constant**, and the stiffer the spring the greater the force constant. The negative sign in F signifies that the direction of the force is opposite to that of the displacement (Fig. 7.34).

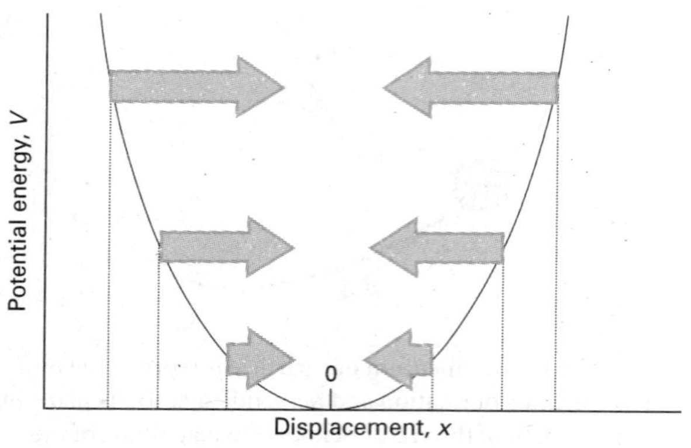

Fig. 7.34 The force acting on a particle that undergoes harmonic motion. The force is directed toward zero displacement and is proportional to the displacement. The corresponding potential energy is parabolic (proportional to x^2).

The motion of a particle that undergoes harmonic motion is found by substituting the expression for the force, eqn 7.57, into Newton's equation, eqn 7.51b. The resulting equation is

$$m\frac{d^2x}{dt^2} = -kx$$

A solution is

$$x(t) = A \sin \omega t \qquad p(t) = m\omega A \cos \omega t \qquad \omega = (k/m)^{1/2} \quad (7.58)$$

These solutions show that the position of the particle varies **harmonically** (that is, as $\sin \omega t$) with a frequency $\nu = \omega/2\pi$. They also show that the particle is stationary ($p = 0$) when the displacement, x, has its maximum value, A, which is called the **amplitude** of the motion.

The total energy of a classical harmonic oscillator is proportional to the square of the amplitude of its motion. To confirm this remark we note that the kinetic energy is

$$E_k = \frac{p^2}{2m} = \frac{(m\omega A \cos \omega t)^2}{2m} = \tfrac{1}{2}m\omega^2A^2 \cos^2\omega t \quad (7.59)$$

Then, because $\omega = (k/m)^{1/2}$, this expression may be written

$$E_k = \tfrac{1}{2}kA^2 \cos^2\omega t \quad (7.60)$$

The force on the oscillator is $F = -kx$, so it follows from the relation $F = -dV/dx$ that the potential energy of a harmonic oscillator is

$$V = \tfrac{1}{2}kx^2 = \tfrac{1}{2}kA^2 \sin^2\omega t \quad (7.61)$$

The total energy is therefore

$$E = \tfrac{1}{2}kA^2 \cos^2\omega t + \tfrac{1}{2}kA^2 \sin^2\omega t = \tfrac{1}{2}kA^2 \quad (7.62)$$

(We have used $\cos^2\omega t + \sin^2\omega t = 1$.) That is, the energy of the oscillator is constant and, for a given force constant, is determined by its maximum displacement. It follows that the energy of an oscillating particle can be raised to any value by stretching the spring to any desired amplitude A. Note that the frequency of the motion depends only on the inherent properties of the oscillator (as represented by k and m) and is independent of the energy; the amplitude governs the energy, through $E = \tfrac{1}{2}kA^2$, and is independent of the frequency. In other words, the particle will oscillate at the same frequency regardless of the amplitude of its motion.

Discussion questions

7.1 Summarize the evidence that led to the introduction of quantum mechanics.

7.2 Explain why Planck's introduction of quantization accounted for the properties of black-body radiation.

7.3 Explain why Einstein's introduction of quantization accounted for the properties of heat capacities at low temperatures.

7.4 Explain the meaning and consequences of wave–particle duality.

7.5 Describe how a wavefunction determines the dynamical properties of a system and how those properties may be predicted.

7.6 Account for the uncertainty relation between position and linear momentum in terms of the shape of the wavefunction.

7.7 Suggest how the general shape of a wavefunction can be predicted without solving the Schrödinger equation explicitly.

Exercises

7.1(a) To what speed must an electron be accelerated for it to have a wavelength of 3.0 cm?

7.1(b) To what speed must a proton be accelerated for it to have a wavelength of 3.0 cm?

7.2(a) The fine-structure constant, α, plays a special role in the structure of matter; its approximate value is 1/137. What is the wavelength of an electron travelling at a speed αc, where c is the speed of light?

7.2(b) Calculate the linear momentum of photons of wavelength 350 nm. What speed does a hydrogen molecule need to travel to have the same linear momentum?

7.3(a) The speed of a certain proton is 0.45 Mm s^{-1}. If the uncertainty in its momentum is to be reduced to 0.0100 per cent, what uncertainty in its location must be tolerated?

7.3(b) The speed of a certain electron is 995 km s^{-1}. If the uncertainty in its momentum is to be reduced to 0.0010 per cent, what uncertainty in its location must be tolerated?

7.4(a) Calculate the energy per photon and the energy per mole of photons for radiation of wavelength (a) 600 nm (red), (b) 550 nm (yellow), (c) 400 nm (blue).

7.4(b) Calculate the energy per photon and the energy per mole of photons for radiation of wavelength (a) 200 nm (ultraviolet), (b) 150 pm (X-ray), (c) 1.00 cm (microwave).

7.5(a) Calculate the speed to which a stationary H atom would be accelerated if it absorbed each of the photons used in Exercise 7.4a.

7.5(b) Calculate the speed to which a stationary ^{4}He atom (mass 4.0026m_u) would be accelerated if it absorbed each of the photons used in Exercise 7.4b.

7.6(a) A glow-worm of mass 5.0 g emits red light (650 nm) with a power of 0.10 W entirely in the backward direction. To what speed will it have accelerated after 10 y if released into free space and assumed to live?

7.6(b) A photon-powered spacecraft of mass 10.0 kg emits radiation of wavelength 225 nm with a power of 1.50 kW entirely in the backward direction. To what speed will it have accelerated after 10.0 y if released into free space?

7.7(a) A sodium lamp emits yellow light (550 nm). How many photons does it emit each second if its power is (a) 1.0 W, (b) 100 W?

7.7(b) A laser used to read CDs emits red light of wavelength 700 nm. How many photons does it emit each second if its power is (a) 0.10 W, (b) 1.0 W?

7.8(a) The work function for metallic caesium is 2.14 eV. Calculate the kinetic energy and the speed of the electrons ejected by light of wavelength (a) 700 nm, (b) 300 nm.

7.8(b) The work function for metallic rubidium is 2.09 eV. Calculate the kinetic energy and the speed of the electrons ejected by light of wavelength (a) 650 nm, (b) 195 nm.

7.9(a) Calculate the size of the quantum involved in the excitation of (a) an electronic oscillation of period 1.0 fs, (b) a molecular vibration of period 10 fs, (c) a pendulum of period 1.0 s. Express the results in joules and kilojoules per mole.

7.9(b) Calculate the size of the quantum involved in the excitation of (a) an electronic oscillation of period 2.50 fs, (b) a molecular vibration of period 2.21 fs, (c) a balance wheel of period 1.0 ms. Express the results in joules and kilojoules per mole.

7.10(a) Calculate the de Broglie wavelength of (a) a mass of 1.0 g travelling at 1.0 cm s^{-1}, (b) the same, travelling at 100 km s^{-1}, (c) an He atom travelling at 1000 m s^{-1} (a typical speed at room temperature).

7.10(b) Calculate the de Broglie wavelength of an electron accelerated from rest through a potential difference of (a) 100 V, (b) 1.0 kV, (c) 100 kV.

7.11(a) An unnormalized wavefunction for a light atom rotating around a heavy atom to which it is bonded is $\psi(\phi) = e^{i\phi}$ with $0 \le \phi \le 2\pi$. Normalize this wavefunction.

7.11(b) An unnormalized wavefunction for an electron in a carbon nanotube of length L is $\sin(2\pi x/L)$. Normalize this wavefunction.

7.12(a) For the system described in Exercise 7.11a, what is the probability of finding the light atom in the volume element $d\phi$ at $\phi = \pi$?

7.12(b) For the system described in Exercise 7.11b, what is the probability of finding the electron in the range dx at $x = L/2$?

7.13(a) For the system described in Exercise 7.11a, what is the probability of finding the light atom between $\phi = \pi/2$ and $\phi = 3\pi/2$?

7.13(b) For the system described in Exercise 7.11b, what is the probability of finding the electron between $x = L/4$ and $x = L/2$?

7.14(a) Confirm that the operator $\hat{l}_z = (\hbar/i)d/d\phi$, where ϕ is an angle, is hermitian.

7.14(b) Show that the linear combinations $\hat{A} + i\hat{B}$ and $\hat{A} - i\hat{B}$ are not hermitian if $\hat{A}$ and $\hat{B}$ are hermitian operators.

7.15(a) Calculate the minimum uncertainty in the speed of a ball of mass 500 g that is known to be within 1.0 μm of a certain point on a bat. What is the minimum uncertainty in the position of a bullet of mass 5.0 g that is known to have a speed somewhere between 350.000 01 m s^{-1} and 350.000 00 m s^{-1}?

7.15(b) An electron is confined to a linear region with a length of the same order as the diameter of an atom (about 100 pm). Calculate the minimum uncertainties in its position and speed.

7.16(a) In an X-ray photoelectron experiment, a photon of wavelength 150 pm ejects an electron from the inner shell of an atom and it emerges with a speed of 21.4 Mm s^{-1}. Calculate the binding energy of the electron.

7.16(b) In an X-ray photoelectron experiment, a photon of wavelength 121 pm ejects an electron from the inner shell of an atom and it emerges with a speed of 56.9 Mm s^{-1}. Calculate the binding energy of the electron.

7.17(a) Determine the commutators of the operators (a) d/dx and $1/x$, (b) d/dx and x^2.

7.17(b) Determine the commutators of the operators a and $a^\dagger$, where $a = (\hat{x} + i\hat{p})/2^{1/2}$ and $a^\dagger = (\hat{x} - i\hat{p})/2^{1/2}$.

Problems*

Numerical problems

7.1 The Planck distribution gives the energy in the wavelength range $d\lambda$ at the wavelength λ. Calculate the energy density in the range 650 nm to 655 nm inside a cavity of volume 100 cm^3 when its temperature is (a) 25°C, (b) 3000°C.

7.2 For a black body, the temperature and the wavelength of emission maximum, λ_{max}, are related by Wien's law, $\lambda_{max}T = \frac{1}{5}c_2$, where $c_2 = hc/k$ (see Problem 7.12). Values of λ_{max} from a small pinhole in an electrically heated container were determined at a series of temperatures, and the results are given below. Deduce a value for Planck's constant.

θ/°C	1000	1500	2000	2500	3000	3500
λ_{max}/nm	2181	1600	1240	1035	878	763

7.3 The Einstein frequency is often expressed in terms of an equivalent temperature θ_E, where $\theta_E = h\nu/k$. Confirm that θ_E has the dimensions of temperature, and express the criterion for the validity of the high-temperature form of the Einstein equation in terms of it. Evaluate θ_E for (a) diamond, for which $\nu = 46.5$ THz and (b) for copper, for which $\nu = 7.15$ THz. What fraction of the Dulong and Petit value of the heat capacity does each substance reach at 25°C?

7.4 The ground-state wavefunction for a particle confined to a one-dimensional box of length L is

$$\psi = \left(\frac{2}{L}\right)^{1/2} \sin\left(\frac{\pi x}{L}\right)$$

Suppose the box is 10.0 nm long. Calculate the probability that the particle is (a) between $x = 4.95$ nm and 5.05 nm, (b) between $x = 1.95$ nm and 2.05 nm, (c) between $x = 9.90$ nm and 10.00 nm, (d) in the right half of the box, (e) in the central third of the box.

7.5 The ground-state wavefunction of a hydrogen atom is

$$\psi = \left(\frac{1}{\pi a_0^3}\right)^{1/2} e^{-r/a_0}$$

where $a_0 = 53$ pm (the Bohr radius). (a) Calculate the probability that the electron will be found somewhere within a small sphere of radius 1.0 pm centred on the nucleus. (b) Now suppose that the same sphere is located at $r = a_0$. What is the probability that the electron is inside it?

7.6 Atoms in a chemical bond vibrate around the equilibrium bond length. An atom undergoing vibrational motion is described by the wavefunction $\psi(x) = Ne^{-x^2/2a^2}$, where a is a constant and $-\infty < x < \infty$. (a) Normalize this function. (b) Calculate the probability of finding the particle in the range $-a \le x \le a$. *Hint.* The integral encountered in part (b) is the error function. It is defined and tabulated in M. Abramowitz and I.A. Stegun, *Handbook of mathematical functions*, Dover (1965) and is provided in most mathematical software packages.

7.7 Suppose that the state of the vibrating atom in Problem 7.6 is described by the wavefunction $\psi(x) = Nxe^{-x^2/2a^2}$. Where is the most probable location of the particle?

7.8 The normalized wavefunctions for a particle confined to move on a circle are $\psi(\phi) = (1/2\pi)^{1/2}e^{-im\phi}$, where $m = 0, \pm1, \pm2, \pm3, \ldots$ and $0 \le \phi \le 2\pi$. Determine $\langle \phi \rangle$.

7.9 A particle is in a state described by the wavefunction $\psi(x) = (2a/\pi)^{1/4}e^{-ax^2}$, where a is a constant and $-\infty \le x \le \infty$. Verify that the value of the product $\Delta p \Delta x$ is consistent with the predictions from the uncertainty principle.

7.10 A particle is in a state described by the wavefunction $\psi(x) = (2a)^{1/2}e^{-ax}$, where a is a constant and $0 \le x \le \infty$. Determine the expectation value of the commutator of the position and momentum operators.

Theoretical problems

7.11 Demonstrate that the Planck distribution reduces to the Rayleigh–Jeans law at long wavelengths.

7.12 Derive *Wien's law*, that $\lambda_{max}T$ is a constant, where λ_{max} is the wavelength corresponding to maximum in the Planck distribution at the temperature T, and deduce an expression for the constant as a multiple of the second radiation constant, $c_2 = hc/k$.

7.13 Use the Planck distribution to deduce the *Stefan–Boltzmann law* that the total energy density of black-body radiation is proportional to T^4, and find the constant of proportionality.

7.14‡ Prior to Planck's derivation of the distribution law for black-body radiation, Wien found empirically a closely related distribution function that is very nearly but not exactly in agreement with the experimental results, namely $\rho = (a/\lambda^5)e^{-b/\lambda kT}$. This formula shows small deviations from Planck's at long wavelengths. (a) By fitting Wien's empirical formula to Planck's at

* Problems denoted with the symbol ‡ were supplied by Charles Trapp, Carmen Giunta, and Marshall Cady.

short wavelengths determine the constants a and b. (b) Demonstrate that Wien's formula is consistent with Wien's law (Problem 7.12) and with the Stefan–Boltzmann law (Problem 7.13).

7.15 Normalize the following wavefunctions: (a) $\sin(n\pi x/L)$ in the range $0 \le x \le L$, where $n = 1, 2, 3, \ldots$, (b) a constant in the range $-L \le x \le L$, (c) $e^{-r/a}$ in three-dimensional space, (d) $re^{-r/2a}$ in three-dimensional space. *Hint.* The volume element in three dimensions is $d\tau = r^2 dr \sin\theta \, d\theta \, d\phi$, with $0 \le r < \infty$, $0 \le \theta \le \pi$, $0 \le \phi \le 2\pi$. Use the integral in Example 7.4.

7.16 (a) Two (unnormalized) excited state wavefunctions of the H atom are

$$\text{(i) } \psi = \left(2 - \frac{r}{a_0}\right) e^{-r/a_0} \qquad \text{(ii) } \psi = r \sin\theta \cos\phi \, e^{-r/2a_0}$$

Normalize both functions to 1. (b) Confirm that these two functions are mutually orthogonal.

7.17 Identify which of the following functions are eigenfunctions of the operator d/dx: (a) e^{ikx}, (b) $\cos kx$, (c) k, (d) kx, (e) e^{-ax^2}. Give the corresponding eigenvalue where appropriate.

7.18 Determine which of the following functions are eigenfunctions of the inversion operator $\hat{\imath}$ (which has the effect of making the replacement $x \to -x$): (a) $x^3 - kx$, (b) $\cos kx$, (c) $x^2 + 3x - 1$. State the eigenvalue of $\hat{\imath}$ when relevant.

7.19 Which of the functions in Problem 7.17 are (a) also eigenfunctions of d^2/dx^2 and (b) only eigenfunctions of d^2/dx^2? Give the eigenvalues where appropriate.

7.20 Construct quantum mechanical operators for the following observables: (a) kinetic energy in one and in three dimensions, (b) the inverse separation, $1/x$, (c) electric dipole moment in one dimension, (d) the mean square deviations of the position and momentum of a particle in one dimension from the mean values.

7.21 Write the time-independent Schrödinger equations for (a) an electron moving in one dimension about a stationary proton and subjected to a Coulombic potential, (b) a free particle, (c) a particle subjected to a constant, uniform force.

7.22 A particle is in a state described by the wavefunction $\psi = (\cos\chi)e^{ikx} + (\sin\chi)e^{-ikx}$, where χ (chi) is a parameter. What is the probability that the particle will be found with a linear momentum (a) $+k\hbar$, (b) $-k\hbar$? What form would the wavefunction have if it were 90 per cent certain that the particle had linear momentum $+k\hbar$?

7.23 Evaluate the kinetic energy of the particle with wavefunction given in Problem 7.22.

7.24 Calculate the average linear momentum of a particle described by the following wavefunctions: (a) e^{ikx}, (b) $\cos kx$, (c) e^{-ax^2}, where in each one x ranges from $-\infty$ to $+\infty$.

7.25 Evaluate the expectation values of r and r^2 for a hydrogen atom with wavefunctions given in Problem 7.16.

7.26 Calculate (a) the mean potential energy and (b) the mean kinetic energy of an electron in the ground state of a hydrogenic atom.

7.27 Use mathematical software to construct superpositions of cosine functions and determine the probability that a given momentum will be observed. If you plot the superposition (which you should), set $x = 0$ at the centre of the screen and build the superposition there. Evaluate the root mean square location of the packet, $\langle x^2 \rangle^{1/2}$.

7.28 Show that the expectation value of an operator that can be written as the square of an hermitian operator is positive.

7.29 (a) Given that any operators used to represent observables must satisfy the commutation relation in eqn 7.41, what would be the operator for position if the choice had been made to represent linear momentum parallel to the x-axis by multiplication by the linear momentum. These different choices are all valid 'representations' of quantum mechanics. (b) With the identification of $\hat{x}$ in this representation, what would be the operator for $1/x$? *Hint.* Think of $1/x$ as x^{-1}.

Applications: to nanoscience, environmental science, and astrophysics

7.30‡ The temperature of the Sun's surface is approximately 5800 K. On the assumption that the human eye evolved to be most sensitive at the wavelength of light corresponding to the maximum in the Sun's radiant energy distribution, determine the colour of light to which the eye is the most sensitive.

7.31 We saw in *Impact I7.1* that electron microscopes can obtain images with several hundredfold higher resolution than optical microscopes because of the short wavelength obtainable from a beam of electrons. For electrons moving at speeds close to c, the speed of light, the expression for the de Broglie wavelength (eqn 7.16) needs to be corrected for relativistic effects:

$$\lambda = \frac{h}{\left\{2m_e e\Delta\phi\left(1 + \dfrac{e\Delta\phi}{2m_e c^2}\right)\right\}^{1/2}}$$

where c is the speed of light in vacuum and $\Delta\phi$ is the potential difference through which the electrons are accelerated. (a) Use the expression above to calculate the de Broglie wavelength of electrons accelerated through 50 kV. (b) Is the relativistic correction important?

7.32‡ Solar energy strikes the top of the Earth's atmosphere at a rate of 343 W m^{-2}. About 30 per cent of this energy is reflected directly back into space by the Earth or the atmosphere. The Earth–atmosphere system absorbs the remaining energy and re-radiates it into space as black-body radiation. What is the average black-body temperature of the Earth? What is the wavelength of the most plentiful of the Earth's black-body radiation? *Hint.* Use Wien's law, Problem 7.12.

7.33‡ A star too small and cold to shine has been found by S. Kulkarni *et al.* (*Science* **270**, 1478 (1995)). The spectrum of the object shows the presence of methane, which, according to the authors, would not exist at temperatures much above 1000 K. The mass of the star, as determined from its gravitational effect on a companion star, is roughly 20 times the mass of Jupiter. The star is considered to be a brown dwarf, the coolest ever found. (a) From available thermodynamic data, test the stability of methane at temperatures above 1000 K. (b) What is λ_{max} for this star? (c) What is the energy density of the star relative to that of the Sun (6000 K)? (d) To determine whether the star will shine, estimate the fraction of the energy density of the star in the visible region of the spectrum.

7.34 Suppose that the wavefunction of an electron in a carbon nanotube is a linear combination of $\cos(nx)$ functions. Use mathematical software to construct superpositions of cosine functions and determine the probability that a given momentum will be observed. If you plot the superposition (which you should), set $x = 0$ at the centre of the screen and build the superposition there. Evaluate the root mean square location of the packet, $\langle x^2 \rangle^{1/2}$.

MATHEMATICAL BACKGROUND 3

Complex numbers

We describe here general properties of complex numbers and functions, which are mathematical constructs frequently encountered in quantum mechanics.

MB3.1 Definitions

Complex numbers have the general form

$$z = x + iy \quad \text{General form of a complex number} \quad \text{(MB3.1)}$$

where $i = (-1)^{1/2}$. The real numbers x and y are, respectively, the real and imaginary parts of z, denoted $\text{Re}(z)$ and $\text{Im}(z)$. When $y = 0$, $z = x$ is a real number; when $x = 0$, $z = iy$ is a pure imaginary number. Two complex numbers $z_1 = x_1 + iy_1$ and $z_2 = x_2 + iy_2$ are equal when $x_1 = x_2$ and $y_1 = y_2$. Although the general form of the imaginary part of a complex number is written iy, a specific numerical value is typically written in the reverse order; for instance, as 3i.

The **complex conjugate** of z, denoted z^*, is formed by replacing i by −i

$$z^* = x - iy \quad \text{Definition of the complex conjugate} \quad \text{(MB3.2)}$$

The product of z^* and z is denoted $|z|^2$ and is called the **square modulus** of z. From eqns MB3.1 and MB3.2,

$$|z|^2 = (x + iy)(x - iy) = x^2 + y^2 \quad \text{Square modulus} \quad \text{(MB3.3)}$$

since $i^2 = -1$. The square modulus is a real number. The **absolute value** or **modulus** is itself denoted $|z|$ and is given by:

$$|z| = (z^*z)^{1/2} = (x^2 + y^2)^{1/2} \quad \text{Absolute value or modulus} \quad \text{(MB3.4)}$$

Since $zz^* = |z|^2$ it follows that $z \times (z^*/|z|^2) = 1$, from which we can identify the (multiplicative) inverse of z (which exists for all nonzero complex numbers):

$$z^{-1} = \frac{z^*}{|z|^2} \quad \text{Inverse of a complex number} \quad \text{(MB3.5)}$$

• A brief illustration

Consider the complex number $z = 8 - 3i$. Its square modulus is

$$|z|^2 = z^*z = (8 - 3i)^*(8 - 3i) = (8 + 3i)(8 - 3i) = 64 + 9 = 73$$

The modulus is therefore $|z| = 73^{1/2}$. From eqn MB3.5, the inverse of z is

$$z^{-1} = \frac{8 + 3i}{73} = \frac{8}{73} + \frac{3}{73}i \quad \bullet$$

MB3.2 Polar representation

The complex number $z = x + iy$ can be represented as a point in a plane, the **complex plane**, with $\text{Re}(z)$ along the x-axis and $\text{Im}(z)$ along the y-axis (Fig. MB3.1). If, as shown in the figure, r and ϕ denote the polar coordinates of the point, then since $x = r\cos\phi$ and $y = r\sin\phi$, we can express the complex number in **polar form** as

$$z = r(\cos\phi + i\sin\phi) \quad \text{Polar form of a complex number} \quad \text{(MB3.6)}$$

The angle ϕ, called the **argument** of z, is the angle that z makes with the x-axis. Because $y/x = \tan\phi$, it follows that the polar form can be constructed from

$$r = (x^2 + y^2)^{1/2} = |z| \qquad \phi = \arctan\frac{y}{x} \quad \text{(MB3.7a)}$$

To convert from polar to Cartesian form, use

$$x = r\cos\phi \text{ and } y = r\sin\phi \text{ to form } z = x + iy \quad \text{(MB3.7b)}$$

One of the most useful relations involving complex numbers is **Euler's formula:**

$$e^{i\phi} = \cos\phi + i\sin\phi \quad \text{Euler's formula} \quad \text{(MB3.8a)}$$

The simplest proof of this relation is to expand the exponential function as a power series and to collect real and imaginary terms. It follows that

$$\cos\phi = \tfrac{1}{2}(e^{i\phi} + e^{-i\phi}) \qquad \sin\phi = -\tfrac{1}{2}i(e^{i\phi} - e^{-i\phi}) \quad \text{(MB3.8b)}$$

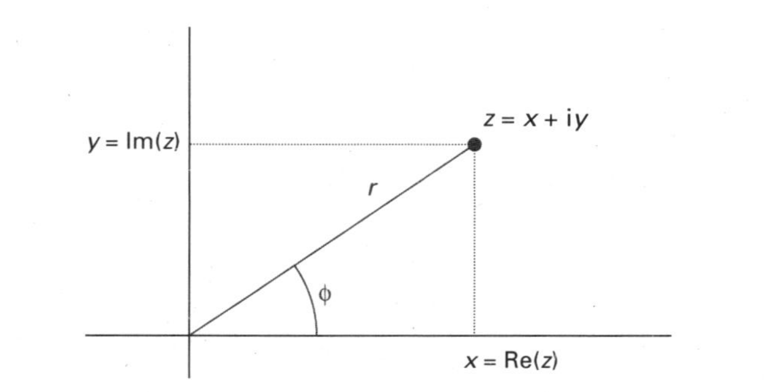

Fig. MB3.1 The representation of a complex number z as a point in the complex plane using cartesian coordinates (x,y) or polar coordinates (r,ϕ).

The polar form in eqn MB3.6 then becomes

$$z = re^{i\phi} \qquad \text{(MB3.7)}$$

- **A brief illustration**

Consider the complex number $z = 8 - 3i$. From the previous *brief illustration*, $r = |z| = 73^{1/2}$. The argument of z is

$$\phi = \arctan\frac{-3}{8} = -0.359 \text{ rad} \qquad \text{or} \qquad -20.6°$$

The polar form of the number is therefore

$$z = 73^{1/2}e^{-0.359i}$$ •

MB3.3 Operations

The following rules apply for arithmetic operations for the complex numbers $z_1 = x_1 + iy_1$ and $z_2 = x_2 + iy_2$.

1. Addition: $z_1 + z_2 = (x_1 + x_2) + i(y_1 + y_2)$ (MB3.10a)
2. Subtraction: $z_1 - z_2 = (x_1 - x_2) + i(y_1 - y_2)$ (MB3.10b)
3. Multiplication: $z_1 z_2 = (x_1 + iy_1)(x_2 + iy_2) = (x_1x_2 - y_1y_2) + i(x_1 y_2 + y_1x_2)$ (MB3.10c)
4. Division: We interpret z_1/z_2 as $z_1z_2^{-1}$ and use eqn MB3.5 for the inverse:

$$\frac{z_1}{z_2} = z_1z_2^{-1} = \frac{z_1z_2^{*}}{|z_2|^2} \qquad \text{(MB3.10d)}$$

- **A brief illustration**

Consider the complex numbers $z_1 = 6 + 2i$ and $z_2 = -4 - 3i$. Then

$$z_1 + z_2 = (6 - 4) + (2 - 3)i = 2 - i$$
$$z_1 - z_2 = 10 + 5i$$
$$z_1z_2 = \{6(-4) - 2(-3)\} + \{6(-3) + 2(-4)\}i = -18 - 26i$$
$$\frac{z_1}{z_2} = (6 + 2i)\left(\frac{-4 + 3i}{25}\right) = -\frac{6}{5} + \frac{2}{5}i$$ •

The polar form of a complex number is commonly used to perform arithmetical operations. For instance, the product of two complex numbers in polar form is

$$z_1z_2 = (r_1e^{i\phi_1})(r_2e^{i\phi_2}) = r_1r_2e^{i(\phi_1+\phi_2)} \qquad \text{(MB3.11)}$$

This multiplication is depicted in the complex plane as shown in Fig. MB3.2. The nth power and the nth root of a complex number are

$$z^n = (re^{i\phi})^n = r^ne^{in\phi} \qquad z^{1/n} = (re^{i\phi})^{1/n} = r^{1/n}e^{i\phi/n} \qquad \text{(MB3.12)}$$

The depictions in the complex plane are shown in Fig. MB3.3.

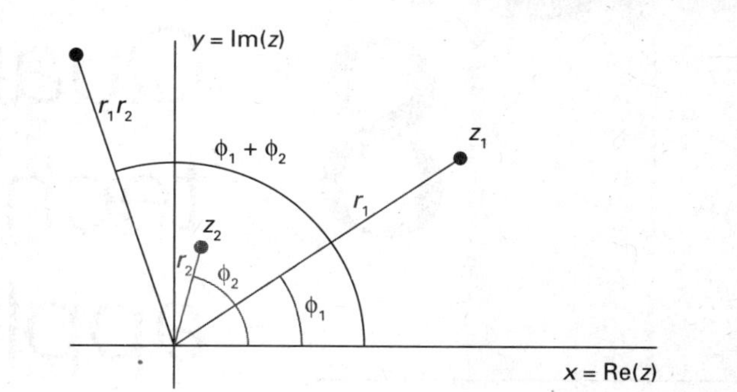

Fig. MB3.2 The multiplication of two complex numbers depicted in the complex plane.

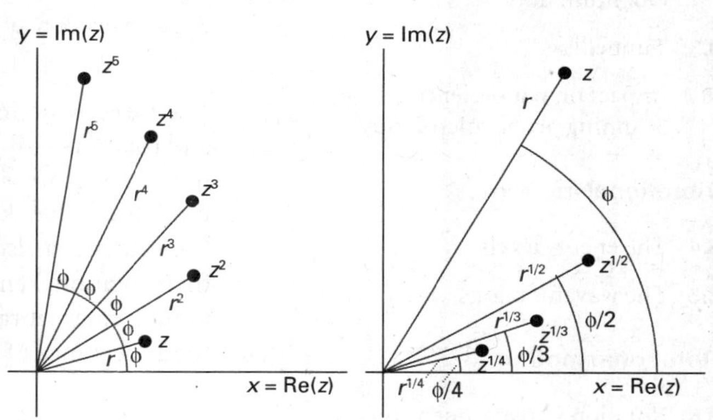

Fig. MB3.3 (a) The nth powers and (b) the nth roots ($n = 1, 2, 3, 4$) of a complex number depicted in the complex plane.

- **A brief illustration**

To determine the 5th root of $z = 8 - 3i$, we note that from the second *brief illustration* its polar form is

$$z = 73^{1/2}e^{-0.359i} = 8.544e^{-0.359i}$$

The 5th root is therefore

$$z^{1/5} = (8.544e^{-0.359i})^{1/5} = 8.544^{1/5}e^{-0.359i/5} = 1.536e^{-0.0718i}$$

It follows that $x = 1.536\cos(-0.0718) = 1.532$ and $y = 1.536\sin(-0.0718) = -0.110$ (note that we work in radians), so

$$(8 - 3i)^{1/5} = 1.532 - 0.110i$$ •

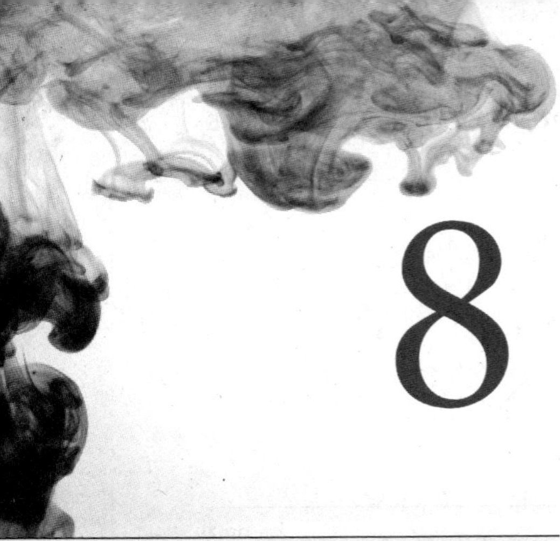

8 Quantum theory: techniques and applications

To calculate the properties of systems according to quantum mechanics we need to solve the appropriate Schrödinger equation. This chapter presents the essentials of the solutions for three basic types of motion: translation, vibration, and rotation. We shall see that only certain wavefunctions and their corresponding energies are acceptable. Hence, quantization emerges as a natural consequence of the equation and the conditions imposed on it. The solutions bring to light a number of nonclassical features of particles, especially their ability to tunnel into and through regions where classical physics would forbid them to be found. We also encounter a property of the electron, its spin, that has no classical counterpart.

The three basic modes of motion—translation (motion through space), vibration, and rotation—all play an important role in chemistry because they are ways in which molecules store energy. Gas-phase molecules, for instance, undergo translational motion and their kinetic energy is a contribution to the total internal energy of a sample. Molecules can also store energy as rotational kinetic energy and transitions between their rotational energy states can be observed spectroscopically. Energy is also stored as molecular vibration, and transitions between vibrational states are responsible for the appearance of infrared and Raman spectra.

Translational motion

Section 7.5 introduced the quantum mechanical description of free motion in one dimension. We saw there that the Schrödinger equation is

$$-\frac{\hbar^2}{2m}\frac{d^2\psi}{dx^2}=E\psi \qquad (8.1a)$$

or more succinctly

$$\hat{H}\psi=E\psi \qquad \hat{H}=-\frac{\hbar^2}{2m}\frac{d^2}{dx^2} \qquad (8.1b)$$

The general solutions of eqn 8.1 are (see *Mathematical background 4* following this chapter):

$$\psi_k=Ae^{ikx}+Be^{-ikx} \qquad E_k=\frac{k^2\hbar^2}{2m} \qquad (8.2)$$

Wavefunctions and energies of a free particle

Note that we are now labelling both the wavefunctions and the energies (that is, the eigenfunctions and eigenvalues of $\hat{H}$, with the index k. We can verify that these functions are solutions by substituting ψ_k into the left-hand side of eqn 8.1a and showing that the result is equal to $E_k\psi_k$. In this case, all values of k, and therefore all values of the energy, are permitted. It follows that the translational energy of a free particle is not quantized.

We saw in Section 7.5c that a wavefunction of the form e^{ikx} describes a particle with linear momentum $p_x = +k\hbar$, corresponding to motion towards positive x (to the right), and that a wavefunction of the form e^{-ikx} describes a particle with the same magnitude of linear momentum but travelling towards negative x (to the left). That is, e^{ikx} is an eigenfunction of the operator $\hat{p}_x$ with eigenvalue $+k\hbar$, and e^{-ikx} is an eigenfunction with eigenvalue $-k\hbar$. In either state, $|\psi|^2$ is independent of x, which implies that the position of the particle is completely unpredictable. This conclusion is consistent with the uncertainty principle, because, if the momentum is certain, then the position cannot be specified (the operators for x and p_x do not commute, Section 7.6).

8.1 A particle in a box

Key points **(a) The energies of a particle constrained to move in a finite region of space are quantized. (b) The energies and wavefunctions for a particle moving in a box are labelled by quantum numbers. The wavefunctions of a particle constrained to move in a one-dimensional box are mutually orthogonal sine functions with the same amplitude but different wavelengths. The zero point energy is the lowest, irremovable energy of a particle in a box. The correspondence principle states that classical mechanics emerges from quantum mechanics as high quantum numbers are reached.**

In this section, we consider a **particle in a box**, in which a particle of mass m is confined between two walls at $x = 0$ and $x = L$: the potential energy is zero inside the box but rises abruptly to infinity at the walls (Fig. 8.1). This model is an idealization of the potential energy of a gas-phase molecule that is free to move in a one-dimensional container or a bead confined to a wire. However, it is also the basis of the treatment of the electronic structure of metals (Chapter 19) and of a primitive treatment of conjugated molecules. The particle in a box is also used in statistical thermodynamics in assessing the contribution of the translational motion of molecules to their thermodynamic properties (Chapter 16).

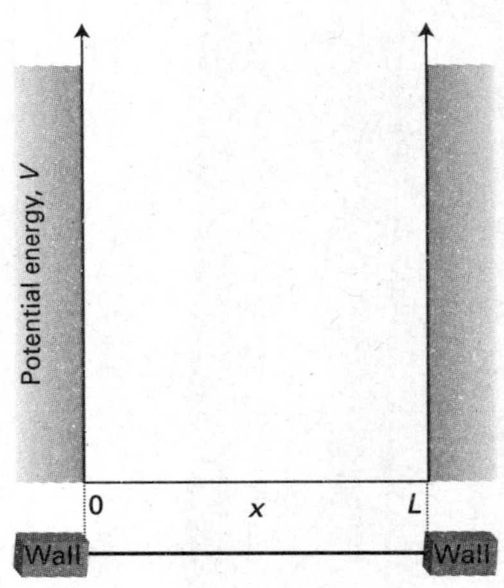

Fig. 8.1 A particle in a one-dimensional region with impenetrable walls. Its potential energy is zero between $x = 0$ and $x = L$, and rises abruptly to infinity as soon as it touches the walls.

(a) The acceptable solutions

The Schrödinger equation for the region between the walls (where $V = 0$) is the same as for a free particle (eqn 8.1), so the general solutions given in eqn 8.2 are also the same. However, it will turn out to be an advantage to use the relation $e^{\pm ix} = \cos x \pm i \sin x$ to write

$$\psi_k = Ae^{ikx} + Be^{-ikx} = A(\cos kx + i \sin kx) + B(\cos kx - i \sin kx)$$
$$= (A + B)\cos kx + (A - B)i \sin kx$$

If we absorb all numerical factors into two new coefficients C and D, the general solutions take the form

$$\psi_k(x) = C \sin kx + D \cos kx \qquad E_k = \frac{k^2\hbar^2}{2m} \tag{8.3}$$

For a free particle, any value of E_k corresponds to an acceptable solution. However, when the particle is confined within a region, the acceptable wavefunctions must satisfy certain **boundary conditions**, or constraints on the function at certain locations. As we shall see when we discuss penetration into barriers, a wavefunction decays exponentially with distance inside a barrier, such as a wall, and the decay is infinitely fast when the potential energy is infinite. This behaviour is consistent with the fact that it is physically impossible for the particle to be found with an infinite potential energy. We conclude that the wavefunction must be zero where V is infinite, at $x < 0$ and $x > L$. The continuity of the wavefunction then requires it to vanish just inside the well at $x = 0$ and $x = L$. That is, the boundary conditions are $\psi_k(0) = 0$ and $\psi_k(L) = 0$. These boundary conditions imply energy quantization, as we show in the following *Justification*.

Justification 8.1 *The energy levels and wavefunctions of a particle in a one-dimensional box*

Consider the wall at $x = 0$. According to eqn 8.3, $\psi(0) = D$ (because $\sin 0 = 0$ and $\cos 0 = 1$). However, because $\psi(0) = 0$ we must have $D = 0$. It follows that the wavefunction must be of the form $\psi_k(x) = C \sin kx$. The value of ψ at the other wall (at $x = L$) is $\psi_k(L) = C \sin kL$, which must also be zero. Taking $C = 0$ would give $\psi_k(x) = 0$ for all x, which would conflict with the Born interpretation (the particle must be somewhere). Therefore, kL must be chosen so that $\sin kL = 0$, which is satisfied by

$$kL = n\pi \qquad n = 1, 2, \ldots$$

The value $n = 0$ is ruled out, because it implies $k = 0$ and $\psi_k(x) = 0$ everywhere (because $\sin 0 = 0$), which is unacceptable. Negative values of n merely change the sign of $\sin kL$ (because $\sin(-x) = -\sin x$) and do not give rise to a new wavefunction. The wavefunctions are therefore

$$\psi_n(x) = C \sin(n\pi x/L) \qquad n = 1, 2, \ldots$$

(At this point we have started to label the solutions with the index n instead of k.) Because $E_k = k^2\hbar^2/2m$, and $k = n\pi/L$, it follows that the energy of the particle is limited to the values $n^2h^2/8mL^2$ with $n = 1, 2, \ldots$.

We conclude that the energy of the particle in a one-dimensional box is quantized and that this quantization arises from the boundary conditions that ψ must satisfy if it is to be an acceptable wavefunction. This is a general conclusion: *the need to satisfy boundary conditions implies that only certain wavefunctions are acceptable, and hence restricts observables to discrete values.* So far, only energy has been quantized; shortly we shall see that other physical observables may also be quantized.

(b) The properties of the solutions

We complete the derivation of the wavefunctions by finding the normalization constant (here written C and regarded as real; that is, does not contain $i = \sqrt{(-1)}$). To do so, we look for the value of C that ensures that the integral of ψ^2 over all the space available to the particle (that is, from $x = 0$ to $x = L$) is equal to 1

$$\int_0^L \psi^2\,dx = C^2 \int_0^L \sin^2 \frac{n\pi x}{L}\,dx = C^2 \times \frac{L}{2} = 1, \qquad \text{so } C = \left(\frac{2}{L}\right)^{1/2}$$

for all n. Therefore, the complete solution to the problem is

$$E_n = \frac{n^2h^2}{8mL^2} \qquad n = 1, 2, \ldots \qquad \text{Energies of a particle in a box} \qquad (8.4a)$$

$$\psi_n(x) = \left(\frac{2}{L}\right)^{1/2} \sin\left(\frac{n\pi x}{L}\right) \qquad \text{for } 0 \le x \le L \qquad \text{Wavefunctions of a particle in a box} \qquad (8.4b)$$

Self-test 8.1 Provide the intermediate steps for the determination of the normalization constant C. *Hint.* Use the standard integral $\int \sin^2 ax\,dx = \frac{1}{2}x - (1/4a)\sin 2ax$ + constant and the fact that $\sin 2m\pi = 0$, with $m = 0, 1, 2, \ldots$.

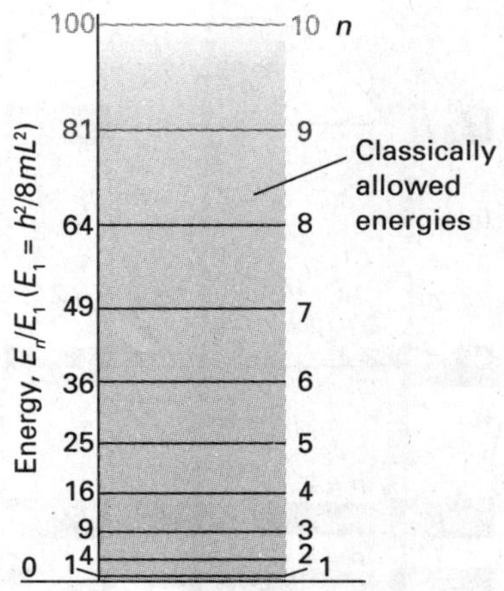

Fig. 8.2 The allowed energy levels for a particle in a box. Note that the energy levels increase as n^2, and that their separation increases as the quantum number increases.

The energies and wavefunctions are labelled with the 'quantum number' n. A **quantum number** is an integer (in some cases, as we shall see, a half-integer; that is, half an odd integer) that labels the state of the system. For a particle in a box there is an infinite number of acceptable solutions, and the quantum number n specifies the one of interest (Fig. 8.2). As well as acting as a label, a quantum number can often be used to calculate the energy corresponding to the state and to write down the wavefunction explicitly (in the present example, by using eqn 8.4).

Figure 8.3 shows some of the wavefunctions of a particle in a box: they are all sine functions with the same maximum amplitude but different wavelengths. Shortening the wavelength results in a sharper average curvature of the wavefunction and therefore an increase in the kinetic energy of the particle. Note that the number of nodes (points where the wavefunction passes through zero) also increases as n increases, and that the wavefunction ψ_n has $n-1$ nodes. Increasing the number of nodes between walls of a given separation increases the average curvature of the wavefunction and hence the kinetic energy of the particle.

The linear momentum of a particle in a box is not well-defined because the wavefunction $\sin kx$ (like $\cos kx$) is not an eigenfunction of the linear momentum operator. However, each wavefunction is a superposition of momentum eigenfunctions:

$$\psi_n = \left(\frac{2}{L}\right)^{1/2} \sin\frac{n\pi x}{L} = \frac{1}{2\mathrm{i}}\left(\frac{2}{L}\right)^{1/2}(\mathrm{e}^{\mathrm{i}kx} - \mathrm{e}^{-\mathrm{i}kx}) \qquad k = \frac{n\pi}{L} \tag{8.5}$$

A brief comment

It is often useful to write $\cos x = (\mathrm{e}^{\mathrm{i}x} + \mathrm{e}^{-\mathrm{i}x})/2$ and $\sin x = (\mathrm{e}^{\mathrm{i}x} - \mathrm{e}^{-\mathrm{i}x})/2\mathrm{i}$.

It follows that measurement of the linear momentum will give the value $+k\hbar$ for half the measurements of momentum and $-k\hbar$ for the other half. This detection of opposite directions of travel with equal probability is the quantum mechanical version of the classical picture that a particle in a box rattles from wall to wall, and in any given period spends half its time travelling to the left and half travelling to the right.

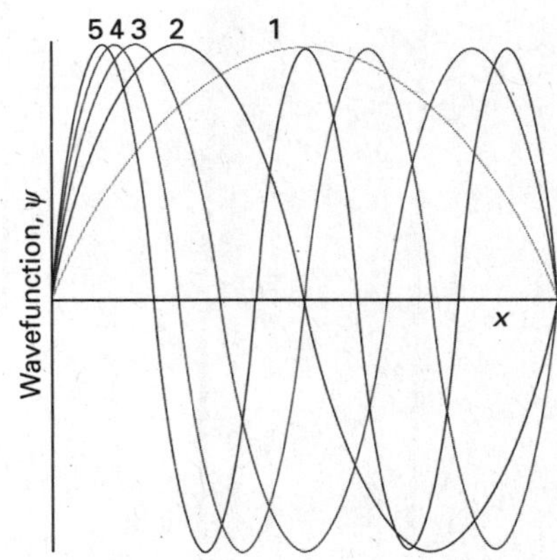

Fig. 8.3 The first five normalized wavefunctions of a particle in a box. Each wavefunction is a standing wave, and successive functions possess one more half wave and a correspondingly shorter wavelength.

interActivity Plot the probability density for a particle in a box with $n = 1, 2, \ldots 5$ and $n = 50$. How do your plots illustrate the correspondence principle?

Self-test 8.2 What is (a) the average value of the linear momentum of a particle in a box with quantum number n, (b) the average value of p^2? *Hint.* Compute expectation values. [(a) $\langle p\rangle = 0$, (b) $\langle p^2\rangle = n^2h^2/4L^2$]

Because n cannot be zero, the lowest energy that the particle may possess is not zero (as would be allowed by classical mechanics, corresponding to a stationary particle) but

$$E_1 = \frac{h^2}{8mL^2} \qquad \text{Zero-point energy of a particle in a box} \tag{8.6}$$

This lowest, irremovable energy is called the **zero-point energy**. The physical origin of the zero-point energy can be explained in two ways. First, the uncertainty principle requires a particle to possess kinetic energy if it is confined to a finite region: the location of the particle is not completely indefinite, so its momentum cannot be precisely zero. Hence it has nonzero kinetic energy. Second, if the wavefunction is to be zero at the walls, but smooth, continuous, and not zero everywhere, then it must be curved, and curvature in a wavefunction implies the possession of kinetic energy.

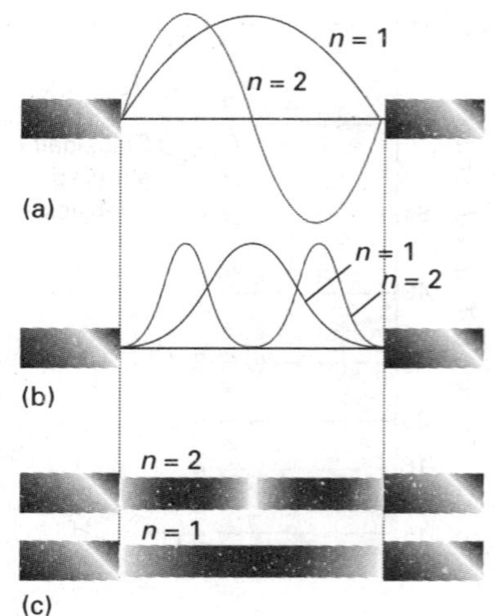

Fig. 8.4 (a) The first two wavefunctions, (b) the corresponding probability distributions, and (c) a representation of the probability distribution in terms of the darkness of shading.

The separation between adjacent energy levels with quantum numbers n and $n+1$ is

$$E_{n+1} - E_n = \frac{(n+1)^2h^2}{8mL^2} - \frac{n^2h^2}{8mL^2} = (2n+1)\frac{h^2}{8mL^2} \quad (8.7)$$

This separation decreases as the length of the container increases, and is very small when the container has macroscopic dimensions. The separation of adjacent levels becomes zero when the walls are infinitely far apart. Atoms and molecules free to move in normal laboratory-sized vessels may therefore be treated as though their translational energy is not quantized. The translational energy of completely free particles (those not confined by walls) is not quantized.

Self-test 8.3 Estimate a typical nuclear excitation energy in electronvolts (eV) by calculating the first excitation energy of a proton confined to a square well with a length equal to the diameter of a nucleus (approximately 1 fm). [0.6 GeV]

The probability density for a particle in a box is

$$\psi^2(x) = \frac{2}{L}\sin^2\frac{n\pi x}{L} \quad (8.8)$$

and varies with position. The non-uniformity is pronounced when n is small (Fig. 8.4), but—provided we take averages over a small region—$\psi^2(x)$ becomes more uniform as n increases. The distribution at high quantum numbers reflects the classical result that a particle bouncing between the walls spends, on the average, equal times at all points. That the quantum result corresponds to the classical prediction at high quantum numbers is an illustration of the **correspondence principle**, which states that classical mechanics emerges from quantum mechanics as high quantum numbers are reached.

Example 8.1 *Using the particle in a box solutions*

What is the probability, P, of locating a particle between $x=0$ (the left-hand end of a box) and $x=0.2$ nm in its lowest energy state in a box of length 1.0 nm?

Method The value of $\psi^2 dx$ is the probability of finding the particle in the small region dx located at x; therefore, the total probability of finding the particle in the specified region is the integral of $\psi^2 dx$ over that region. The wavefunction of the particle is given in eqn 8.4b with $n=1$.

Answer The probability of finding the particle in a region between $x=0$ and $x=l$ is

$$P = \int_0^l \psi_n^2\, dx = \frac{2}{L}\int_0^l \sin^2\frac{n\pi x}{L}dx = \frac{l}{L} - \frac{1}{2n\pi}\sin\frac{2\pi nl}{L}$$

We then set $n=1$ and $l=0.2$ nm, which gives $P=0.05$. The result corresponds to a chance of 1 in 20 of finding the particle in the region. As n becomes infinite, the sine term, which is multiplied by $1/n$, makes no contribution to P and the classical result, $P=l/L$, is obtained.

Self-test 8.4 Calculate the probability that a particle in the state with $n=1$ will be found between $x = 0.25L$ and $x = 0.75L$ in a box of length L (with $x = 0$ at the left-hand end of the box). [0.82]

8.2 Motion in two and more dimensions

Key points (a) The separation of variables technique can be used to solve the Schrödinger equation in multiple dimensions. The energies of a particle constrained to move in two or three dimensions are quantized. (b) Degeneracy occurs when different wavefunctions correspond to the same energy. Many of the states of a particle in a square or cubic box are degenerate.

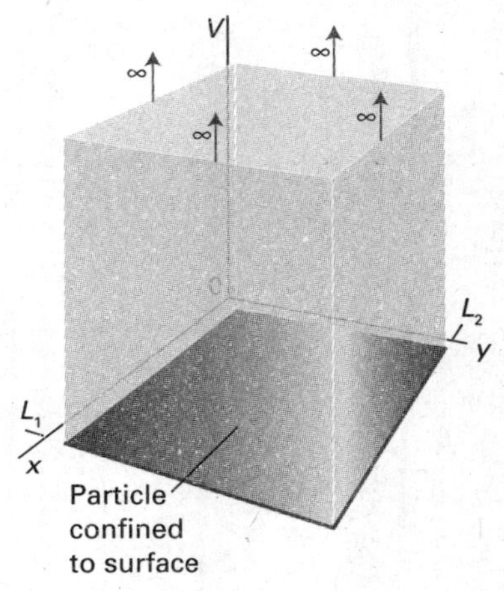

Fig. 8.5 A two-dimensional square well. The particle is confined to the plane bounded by impenetrable walls. As soon as it touches the walls, its potential energy rises to infinity.

Next, we consider a two-dimensional version of the particle in a box. Now the particle is confined to a rectangular surface of length L_1 in the x-direction and L_2 in the y-direction; the potential energy is zero everywhere except at the walls, where it is infinite (Fig. 8.5). The wavefunction is now a function of both x and y and the Schrödinger equation is

$$-\frac{\hbar^2}{2m}\left(\frac{\partial^2\psi}{\partial x^2}+\frac{\partial^2\psi}{\partial y^2}\right)=E\psi \tag{8.9}$$

We need to see how to solve this partial differential equation, a differential equation in more than one variable.

(a) Separation of variables

Some partial differential equations can be simplified by the **separation of variables technique** (*Mathematical background 4* following this chapter), which divides the equation into two or more ordinary differential equations, one for each variable. An important application of this procedure, as we shall see, is the separation of the Schrödinger equation for the hydrogen atom into equations that describe the radial and angular variation of the wavefunction. The technique is particularly simple for a two-dimensional square well, as can be seen by testing whether a solution of eqn 8.9 can be found by writing the wavefunction as a product of functions, one depending only on x and the other only on y:

$$\psi(x,y)=X(x)Y(y)$$

Separation of variables

With this substitution, we show in the following *Justification* that eqn 8.9 separates into two ordinary differential equations, one for each coordinate:

$$-\frac{\hbar^2}{2m}\frac{\mathrm{d}^2X}{\mathrm{d}x^2}=E_XX \qquad -\frac{\hbar^2}{2m}\frac{\mathrm{d}^2Y}{\mathrm{d}y^2}=E_YY \qquad E=E_X+E_Y \tag{8.10}$$

The quantity E_X is the energy associated with the motion of the particle parallel to the x-axis, and likewise for E_Y and motion parallel to the y-axis. Similarly, $X(x)$ is the wavefunction associated with the particle's freedom to move parallel to the x-axis and likewise for $Y(y)$ and motion parallel to the y-axis.

Justification 8.2 *The separation of variables technique applied to the particle in a two-dimensional box*

We follow the procedure in *Mathematical background 4* and apply it to eqn 8.9. The first step in the justification of the separability of the wavefunction into the product of two functions X and Y is to note that, because X is independent of y and Y is independent of x, we can write

$$\frac{\partial^2\psi}{\partial x^2}=\frac{\partial^2XY}{\partial x^2}=Y\frac{\mathrm{d}^2X}{\mathrm{d}x^2} \qquad \frac{\partial^2\psi}{\partial y^2}=\frac{\partial^2XY}{\partial y^2}=X\frac{\mathrm{d}^2Y}{\mathrm{d}y^2}$$

Then eqn 8.9 becomes

$$-\frac{\hbar^2}{2m}\left(Y\frac{d^2X}{dx^2}+X\frac{d^2Y}{dy^2}\right)=EXY$$

When both sides are divided by XY, we can rearrange the resulting equation into

$$\frac{1}{X}\frac{d^2X}{dx^2}+\frac{1}{Y}\frac{d^2Y}{dy^2}=-\frac{2mE}{\hbar^2}$$

The first term on the left is independent of y, so if y is varied only the second term can change. However, the sum of these two terms is a constant given by the right-hand side of the equation; therefore, even the second term cannot change when y is changed. In other words, the second term is a constant. By a similar argument, the first term is a constant when x changes. If we write these two constants as $-2mE_Y/\hbar^2$ and $-2mE_X/\hbar^2$ (because that captures the form of the original equation), we can write

$$\frac{1}{X}\frac{d^2X}{dx^2}=-\frac{2mE_X}{\hbar^2} \qquad \frac{1}{Y}\frac{d^2Y}{dy^2}=-\frac{2mE_Y}{\hbar^2}$$

Because the sum of the terms on the left of each equation is equal to $-2mE/\hbar^2$ it follows that $E_X+E_Y=E$. These two equations rearrange into the two ordinary (that is, single variable) differential equations in eqn 8.10.

Each of the two ordinary differential equations in eqn 8.10 is the same as the one-dimensional square-well Schrödinger equation. We can therefore adapt the results in eqn 8.4 without further calculation:

$$X_{n_1}(x)=\left(\frac{2}{L_1}\right)^{1/2}\sin\frac{n_1\pi x}{L_1} \qquad Y_{n_2}(y)=\left(\frac{2}{L_2}\right)^{1/2}\sin\frac{n_2\pi y}{L_2}$$

Then, because $\psi=XY$ and $E=E_X+E_Y$, we obtain

$$\psi_{n_1,n_2}(x,y)=\frac{2}{(L_1L_2)^{1/2}}\sin\frac{n_1\pi x}{L_1}\sin\frac{n_2\pi y}{L_2} \qquad (8.11a)$$

Wavefunctions and energies of a particle in a two-dimensional box

$$E_{n_1,n_2}=\left(\frac{n_1^2}{L_1^2}+\frac{n_2^2}{L_2^2}\right)\frac{h^2}{8m} \qquad 0\le x\le L_1,\ 0\le y\le L_2$$

with the quantum numbers taking the values $n_1=1, 2, \ldots$ and $n_2=1, 2, \ldots$ independently. Some of these functions are plotted in Fig. 8.6. They are the two-dimensional versions of the wavefunctions shown in Fig. 8.3. Note that two quantum numbers are needed in this two-dimensional problem.

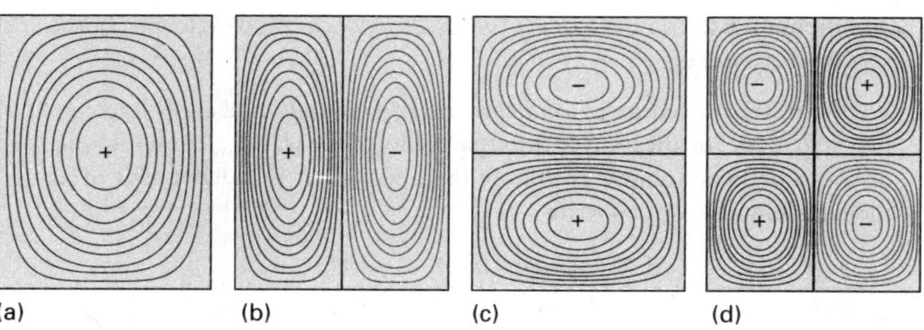

Fig. 8.6 The wavefunctions for a particle confined to a rectangular surface depicted as contours of equal amplitude. (a) $n_1=1$, $n_2=1$, the state of lowest energy, (b) $n_1=1$, $n_2=2$, (c) $n_1=2$, $n_2=1$, and (d) $n_1=2$, $n_2=2$.

interActivity Use mathematical software to generate three-dimensional plots of the functions in this illustration. Deduce a rule for the number of nodal lines in a wavefunction as a function of the values of n_x and n_y.

We treat a particle in a three-dimensional box in the same way. The wavefunctions have another factor (for the z-dependence), and the energy has an additional term in n_3^2/L_3^2. Solution of the Schrödinger equation by the separation of variables technique then gives

$$\psi_{n_1,n_2,n_3}(x,y,z) = \left(\frac{8}{L_1L_2L_3}\right)^{1/2} \sin\frac{n_1\pi x}{L_1}\sin\frac{n_2\pi y}{L_2}\sin\frac{n_3\pi z}{L_3}$$

Wavefunctions and energies of a particle in a three-dimensional box

$$E_{n_1,n_2,n_3} = \left(\frac{n_1^2}{L_1^2}+\frac{n_2^2}{L_2^2}+\frac{n_3^2}{L_3^2}\right)\frac{h^2}{8m} \qquad 0\le x\le L_1, 0\le y\le L_2, 0\le z\le L_3 \qquad (8.11b)$$

with the quantum numbers taking the values $n_1 = 1, 2, \ldots, n_2 = 1, 2, \ldots,$ and $n_3 = 1, 2, \ldots,$ independently.

(b) Degeneracy

An interesting feature of the solutions for a particle in a two-dimensional box is obtained when the plane surface is square, with $L_1 = L_2 = L$. Then eqn 8.11a becomes

$$\psi_{n_1,n_2}(x,y) = \frac{2}{L}\sin\frac{n_1\pi x}{L}\sin\frac{n_2\pi y}{L} \qquad E_{n_1,n_2} = (n_1^2+n_2^2)\frac{h^2}{8mL^2} \qquad (8.12)$$

Consider the cases $n_1 = 1, n_2 = 2$ and $n_1 = 2, n_2 = 1$:

$$\psi_{1,2} = \frac{2}{L}\sin\frac{\pi x}{L}\sin\frac{2\pi y}{L} \qquad E_{1,2} = \frac{5h^2}{8mL^2}$$

$$\psi_{2,1} = \frac{2}{L}\sin\frac{2\pi x}{L}\sin\frac{\pi y}{L} \qquad E_{2,1} = \frac{5h^2}{8mL^2}$$

We see that, although the wavefunctions are different, they are **degenerate**, meaning that they correspond to the same energy. In this case, in which there are two degenerate wavefunctions, we say that the energy level $5(h^2/8mL^2)$ is 'doubly degenerate'.

The occurrence of degeneracy is related to the symmetry of the system. Figure 8.7 shows contour diagrams of the two degenerate functions $\psi_{1,2}$ and $\psi_{2,1}$. As the box is square, we can convert one wavefunction into the other simply by rotating the plane by 90°. Interconversion by rotation through 90° is not possible when the plane is not square, and $\psi_{1,2}$ and $\psi_{2,1}$ are then not degenerate. Similar arguments account for the degeneracy of states in a cubic box. We shall see many other examples of degeneracy in the pages that follow (for instance, in the hydrogen atom), and all of them can be traced to the symmetry properties of the system (see Section 11.6).

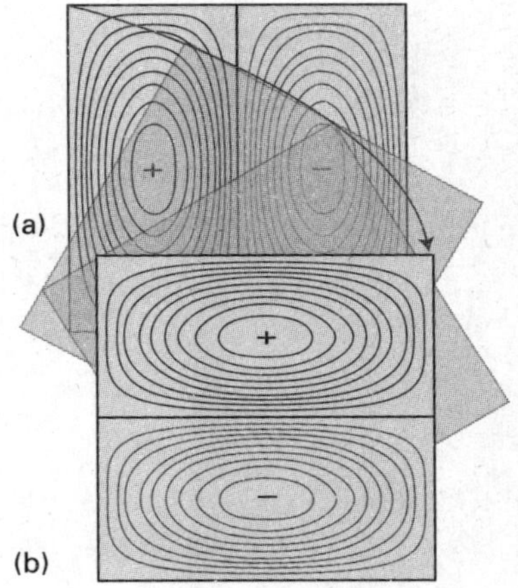

Fig. 8.7 The wavefunctions for a particle confined to a square surface. Note that one wavefunction can be converted into the other by a rotation of the box by 90°. The two functions correspond to the same energy. Degeneracy and symmetry are closely related.

IMPACT ON NANOSCIENCE
I8.1 Quantum dots

Nanoscience is the study of atomic and molecular assemblies with dimensions ranging from 1 nm to about 100 nm and nanotechnology is concerned with the incorporation of such assemblies into devices. The future economic impact of nanotechnology could be very significant. For example, increased demand for very small digital electronic devices has driven the design of ever smaller and more powerful microprocessors. However, there is an upper limit on the density of electronic circuits that can be incorporated into silicon-based chips with current fabrication technologies. As the ability to process data increases with the number of components in a chip, it follows that soon chips and the devices that use them will have to become bigger if processing power is to increase indefinitely. One way to circumvent this problem is to fabricate devices from nanometre-sized components.

We shall encounter several concepts of nanoscience throughout the text. Here we explore the possibility of using quantum mechanical effects that render the properties of an assembly dependent on its size.

Ordinary bulk metals conduct electricity because, in the presence of an electric field, electrons become mobile when they are easily excited into closely lying empty energy levels. By ignoring all the electrostatic interactions, we can treat the electrons as occupying the energy levels characteristic of independent particles in a three-dimensional box. Because the box has macroscopic dimensions, we know from eqn 8.7 that the separation between neighbouring levels is so small that they form a virtual continuum. Consequently, we are justified in neglecting energy quantization on the properties of the material. However, in a nanocrystal, a small cluster of atoms with dimensions in the nanometre scale, eqn 8.4a predicts that quantization of energy is significant and affects the properties of the sample. This quantum mechanical effect can be observed in 'boxes' of any shape. For example, you are invited to show in Problem 8.38 that the energy levels corresponding to spherically symmetrical wavefunctions of an electron in a spherical cavity of radius R are given by[1]

$$E_n = \frac{n^2h^2}{8m_eR^2}$$

The quantization of energy in nanocrystals has important technological implications when the material is a semiconductor, in which electrical conductivity increases with increasing temperature or upon excitation by light. That is, transfer of energy to a semiconductor increases the mobility of electrons in the material (see Chapter 19 for a more detailed discussion). Three-dimensional nanocrystals of semiconducting materials containing 10 to 10^5 atoms are called **quantum dots**. They can be made in solution or by depositing atoms on a surface, with the size of the nanocrystal being determined by the details of the synthesis.

First, we see that the energy required to induce electronic transitions from lower to higher energy levels, thereby increasing the mobility of electrons and inducing electrical conductivity, depends on the size of the quantum dot. The electrical properties of large, macroscopic samples of semiconductors cannot be tuned in this way. Second, in many quantum dots, such as the nearly spherical nanocrystals of cadmium selenide (CdSe), mobile electrons can be generated by absorption of visible light and, as the radius of the quantum dot decreases, the excitation wavelength decreases. That is, as the size of the quantum dot varies, so does the colour of the material. This phenomenon is indeed observed in suspensions of CdSe quantum dots of different sizes.

Because quantum dots are semiconductors with tunable electrical properties, there are many uses for these materials in the manufacture of transistors. The special optical properties of quantum dots can also be exploited. Just as the generation of an electron–hole pair requires absorption of light of a specific wavelength, so does recombination of the pair result in the emission of light of a specific wavelength. This property forms the basis for the use of quantum dots in the visualization of biological cells at work. For example, a CdSe quantum dot can be modified by covalent attachment of an organic spacer to its surface. When the other end of the spacer reacts specifically with a cellular component, such as a protein, nucleic acid, or membrane, the cell becomes labelled with a light-emitting quantum dot. The spatial distribution of emission intensity and, consequently, of the labelled molecule can then be measured

[1] There are solutions that are not spherically symmetrical and to which this expression does not apply.

with a microscope. Though this technique has been used extensively with organic molecules as labels, quantum dots are more stable and are stronger light emitters.

8.3 Tunnelling

Key points **Tunnelling is the penetration into or through classically forbidden regions. The transmission probability decreases exponentially with the thickness of the barrier and with the square-root of the mass of the particle.**

If the potential energy of a particle does not rise to infinity when it is in the walls of the container, and $E < V$, the wavefunction does not decay abruptly to zero. If the walls are thin (so that the potential energy falls to zero again after a finite distance), then the wavefunction oscillates inside the box, varies smoothly inside the region representing the wall, and oscillates again on the other side of the wall outside the box (Fig. 8.8). Hence the particle might be found on the outside of a container even though according to classical mechanics it has insufficient energy to escape. Such leakage by penetration through a classically forbidden region is called **tunnelling**.

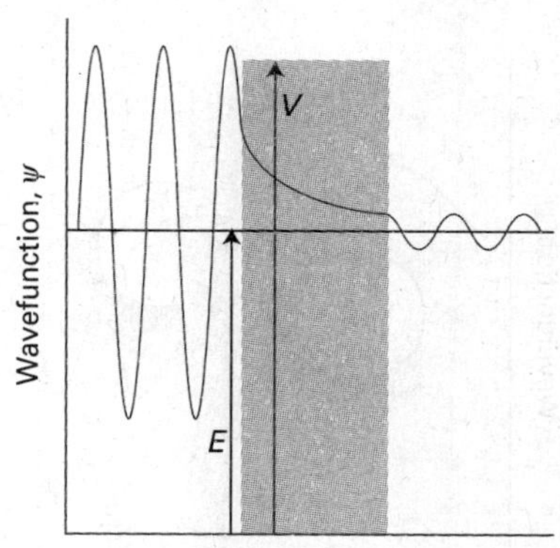

Fig. 8.8 A particle incident on a barrier from the left has an oscillating wavefunction, but inside the barrier there are no oscillations (for $E < V$). If the barrier is not too thick, the wavefunction is nonzero at its opposite face, and so oscillations begin again there. (Only the real component of the wavefunction is shown.)

The Schrödinger equation can be used to calculate the probability of tunnelling of a particle of mass m incident on a finite barrier from the left. On the left of the barrier (for $x < 0$) the wavefunctions are those of a particle with $V = 0$, so from eqn 8.2 we can write

$$\psi = Ae^{ikx} + Be^{-ikx} \qquad k\hbar = (2mE_k)^{1/2} \tag{8.13}$$

The Schrödinger equation for the region representing the barrier (for $0 \leq x \leq L$), where the potential energy has the constant value V, is

$$-\frac{\hbar^2}{2m}\frac{d^2\psi}{dx^2} + V\psi = E\psi \tag{8.14}$$

We shall consider particles that have $E < V$ (so, according to classical physics, the particle has insufficient energy to pass over the barrier), and therefore $V - E$ is positive. The general solutions of this equation are

$$\psi = Ce^{\kappa x} + De^{-\kappa x} \qquad \kappa\hbar = \{2m(V - E)\}^{1/2} \tag{8.15}$$

as we can readily verify by differentiating ψ twice with respect to x. The important feature to note is that the two exponentials are now real functions, as distinct from the complex, oscillating functions for the region where $V = 0$ (oscillating functions would be obtained if $E > V$). To the right of the barrier ($x > L$), where $V = 0$ again, the wavefunctions are

$$\psi = A'e^{ikx} + B'e^{-ikx} \qquad k\hbar = (2mE)^{1/2} \tag{8.16}$$

The complete wavefunction for a particle incident from the left consists of an incident wave, a wave reflected from the barrier, the exponentially changing amplitudes inside the barrier, and an oscillating wave representing the propagation of the particle to the right after tunnelling through the barrier successfully (Fig. 8.9). The acceptable wavefunctions must obey the conditions set out in Section 7.4b. In particular, they must be continuous at the edges of the barrier (at $x = 0$ and $x = L$, remembering that $e^0 = 1$):

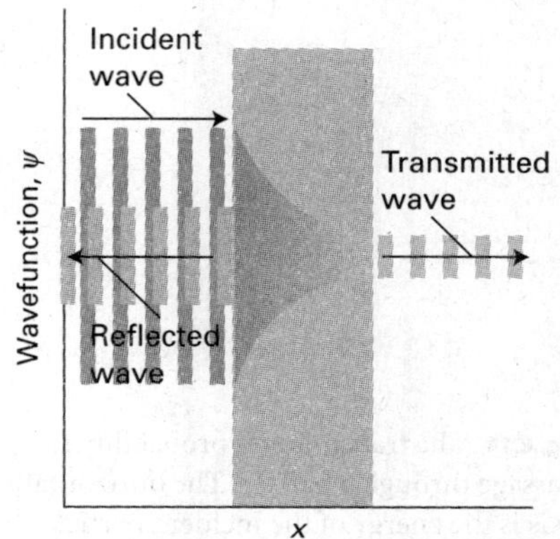

Fig. 8.9 When a particle is incident on a barrier from the left, the wavefunction consists of a wave representing linear momentum to the right, a reflected component representing momentum to the left, a varying but not oscillating component inside the barrier, and a (weak) wave representing motion to the right on the far side of the barrier.

$$A + B = C + D \qquad Ce^{\kappa L} + De^{-\kappa L} = A'e^{ikL} + B'e^{-ikL} \tag{8.17}$$

Their slopes (their first derivatives) must also be continuous there (Fig. 8.10):

$$ikA - ikB = \kappa C - \kappa D \qquad \kappa Ce^{\kappa L} - \kappa De^{-\kappa L} = ikA'e^{ikL} - ikB'e^{-ikL} \tag{8.18}$$

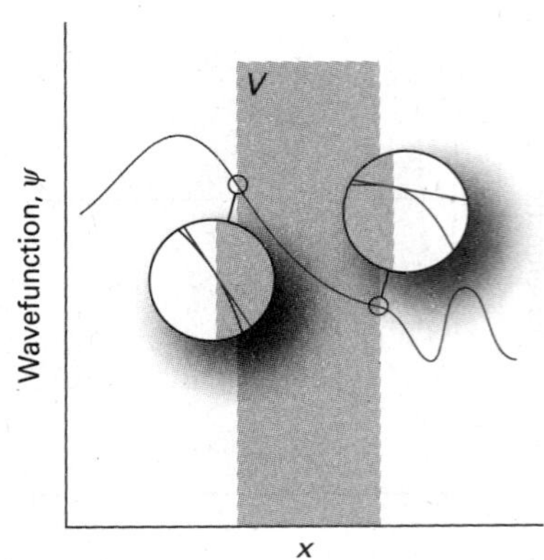

Fig. 8.10 The wavefunction and its slope must be continuous at the edges of the barrier. The conditions for continuity enable us to connect the wavefunctions in the three zones and hence to obtain relations between the coefficients that appear in the solutions of the Schrödinger equation.

At this stage, we have four equations for the six unknown coefficients. If the particles are shot towards the barrier from the left, there can be no particles travelling to the left on the right of the barrier. Therefore, we can set $B' = 0$, which removes one more unknown. We cannot set $B = 0$ because some particles may be reflected back from the barrier toward negative x.

The probability that a particle is travelling towards positive x (to the right) on the left of the barrier is proportional to $|A|^2$, and the probability that it is travelling to the right on the right of the barrier is $|A'|^2$. The ratio of these two probabilities is called the **transmission probability**, T. After some algebra (see Problem 8.8) we find

$$T = \left\{1 + \frac{(e^{\kappa L} - e^{-\kappa L})^2}{16\varepsilon(1-\varepsilon)}\right\}^{-1} \quad \text{Transmission probability} \quad (8.19a)$$

where $\varepsilon = E/V$. This function is plotted in Fig. 8.11; the transmission coefficient for $E > V$ is shown there too. For high, wide barriers (in the sense that $\kappa L \gg 1$), eqn 8.19a simplifies to

$$T \approx 16\varepsilon(1-\varepsilon)e^{-2\kappa L} \quad \text{Transmission probability for } \kappa L \gg 1 \quad (8.19b)$$

The transmission probability decreases exponentially with the thickness of the barrier and with $m^{1/2}$. It follows that particles of low mass are more able to tunnel through barriers than heavy ones (Fig. 8.12). Tunnelling is very important for electrons and muons (elementary particles with mass of about $207m_e$), and moderately important for protons (of mass $1840m_e$); for heavier particles it is less important. A number of effects in chemistry (for example, the very rapid equilibration of proton transfer reactions) is a manifestation of the ability of particles to tunnel through barriers. As we shall see in Chapter 22, electron tunnelling is one of the factors that determine the rates of electron transfer reactions at electrodes and in biological systems.

A problem related to tunnelling is that of a particle in a square-well potential of finite depth (Fig. 8.13). In this kind of potential, the wavefunction penetrates into the walls, where it decays exponentially towards zero, and oscillates within the well. The wavefunctions are found by ensuring, as in the discussion of tunnelling, that they and their slopes are continuous at the edges of the potential. Some of the lowest energy solutions are shown in Fig. 8.14. A further difference from the solutions for an infinitely deep well is that there is only a finite number of bound states. Regardless of the depth

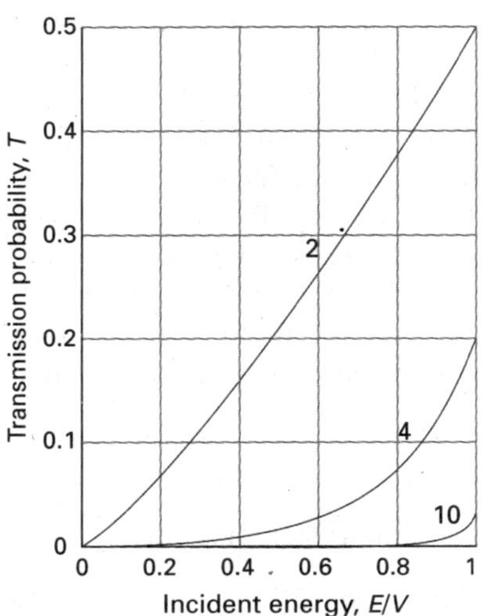

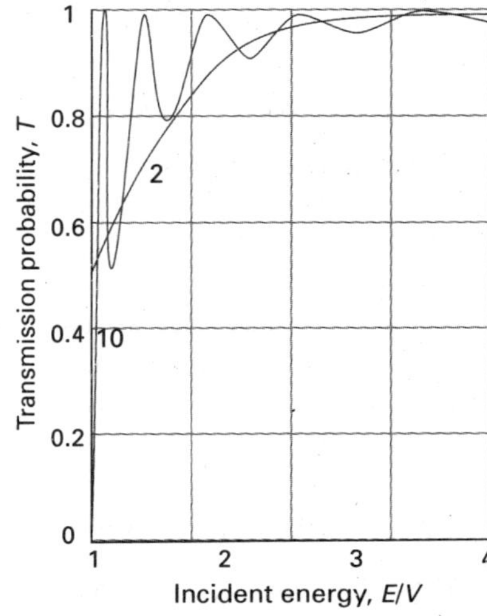

Fig. 8.11 The transmission probability for passage through a barrier. The horizontal axis is the energy of the incident particle expressed as a multiple of the barrier height. The curves are labelled with the value of $L(2mV)^{1/2}/\hbar$. The graph on the left is for $E < V$ and that on the right for $E > V$. Note that $T > 0$ for $E < V$, whereas classically T would be zero. However, $T < 1$ for $E > V$, whereas classically T would be 1.

interActivity Plot T against ε for a hydrogen molecule, a proton, and an electron.

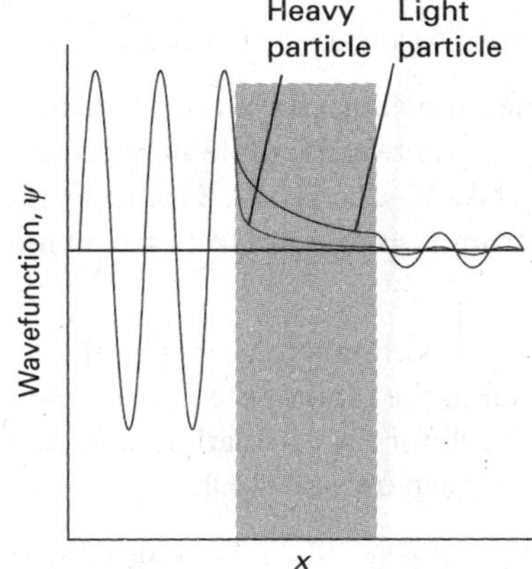

Fig. 8.12 The wavefunction of a heavy particle decays more rapidly inside a barrier than that of a light particle. Consequently, a light particle has a greater probability of tunnelling through the barrier.

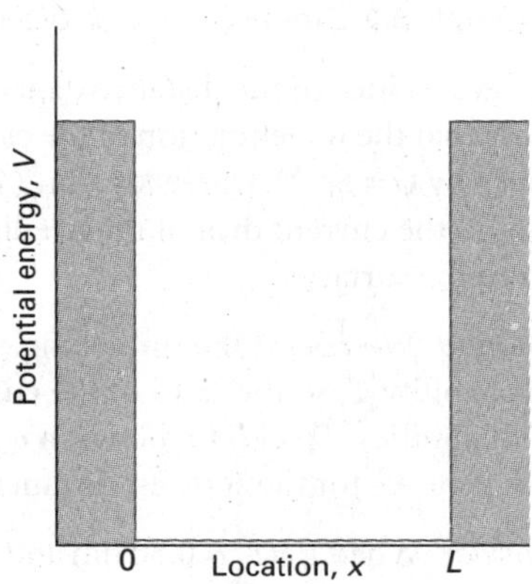

Fig. 8.13 A potential well with a finite depth.

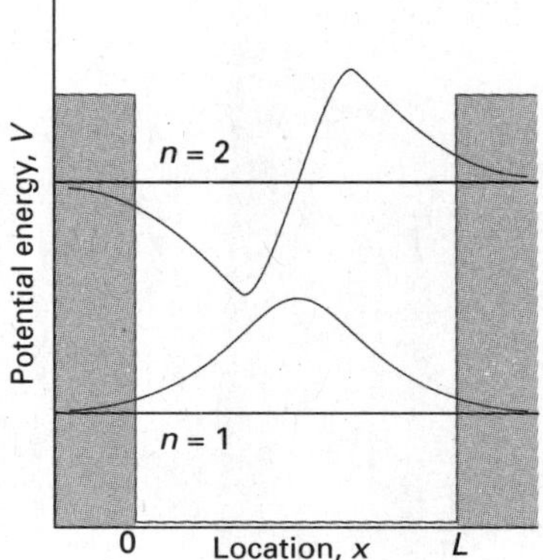

Fig. 8.14 The lowest two bound-state wavefunctions for a particle in the well shown in Fig. 8.13.

and length of the well, however, there is always at least one bound state. Detailed consideration of the Schrödinger equation for the problem shows that in general the number of levels is equal to N, with

$$N-1<\frac{(8mVL)^{1/2}}{h}<N \tag{8.20}$$

where V is the depth of the well and L is its length. We see that, the deeper and wider the well, the greater the number of bound states. As the depth becomes infinite, so the number of bound states also becomes infinite, as we have already seen.

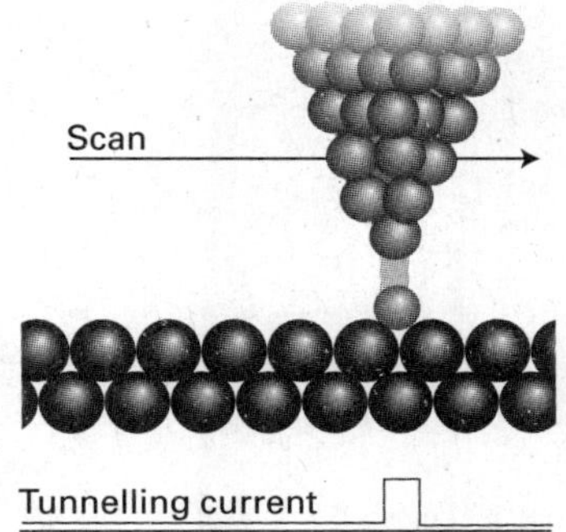

Fig. 8.15 A scanning tunnelling microscope makes use of the current of electrons that tunnel between the surface and the tip. That current is very sensitive to the distance of the tip above the surface.

IMPACT ON NANOSCIENCE

I8.2 Scanning probe microscopy

In *Impact I8.1* we outlined some advantages of working in the nanometre regime. Here we describe *scanning probe microscopy* (SPM), a collection of techniques that can be used to visualize and manipulate objects as small as atoms on surfaces.

One version of SPM is scanning tunnelling microscopy (STM), in which a platinum–rhodium or tungsten needle is scanned across the surface of a conducting solid. When the tip of the needle is brought very close to the surface, electrons tunnel across the intervening space (Fig. 8.15). In the constant-current mode of operation, the stylus moves up and down corresponding to the form of the surface, and the topography of the surface, including any adsorbates, can therefore be mapped on an atomic scale. The vertical motion of the stylus is achieved by fixing it to a piezoelectric cylinder, which contracts or expands according to the potential difference it experiences. In the constant-z mode, the vertical position of the stylus is held constant and the current is monitored. Because the tunnelling probability is very sensitive to the size of the gap, the microscope can detect tiny, atom-scale variations in the height of the surface.

Figure 8.16 shows an example of the kind of image obtained with a surface, in this case of gallium arsenide, that has been modified by addition of atoms, in this case caesium atoms. Each 'bump' on the surface corresponds to an atom. In a further variation of the STM technique, the tip may be used to nudge single atoms around on the surface, making possible the fabrication of complex and yet very tiny nanometre-sized structures.

Fig. 8.16 An STM image of caesium atoms on a gallium arsenide surface.

Example 8.2 *Exploring the origin of the current in scanning tunnelling microscopy*

To get an idea of the distance dependence of the tunnelling current in STM, suppose that the wavefunction of the electron in the gap between sample and needle is given by $\psi = Be^{-\kappa x}$, where $\kappa = \{2m_e(V-E)/\hbar^2\}^{1/2}$; take $V-E = 2.0$ eV. By what factor would the current drop if the needle is moved from $L_1 = 0.50$ nm to $L_2 = 0.60$ nm from the surface?

Method We regard the tunnelling current to be proportional to the transmission probability T, so the ratio of the currents is equal to the ratio of the transmission probabilities. To choose between eqn 8.19a or 8.19b for the calculation of T, first calculate κL for the shortest distance L_1: if $\kappa L_1 > 1$, then use eqn 8.19b.

Answer When $L = L_1 = 0.50$ nm and $V - E = 2.0\ \text{eV} = 3.20 \times 10^{-19}$ J the value of κL is

$$\kappa L_1 = \left\{\frac{2m_e(V-E)}{\hbar^2}\right\}^{1/2} L_1$$

$$= \left\{\frac{2 \times (9.109 \times 10^{-31}\ \text{kg}) \times (3.20 \times 10^{-19}\ \text{J})}{(1.054 \times 10^{-34}\ \text{J s})^2}\right\}^{1/2} \times (5.0 \times 10^{-10}\ \text{m})$$

$$= (7.25 \times 10^{9}\ \text{m}^{-1}) \times (5.0 \times 10^{-10}\ \text{m}) = 3.6$$

Because $\kappa L_1 > 1$, we use eqn 8.19b to calculate the transmission probabilities at the two distances. It follows that

$$\frac{\text{current at } L_2}{\text{current at } L_1} = \frac{T(L_2)}{T(L_1)} = \frac{16\varepsilon(1-\varepsilon)e^{-2\kappa L_2}}{16\varepsilon(1-\varepsilon)e^{-2\kappa L_1}} = e^{-2\kappa(L_2 - L_1)}$$

$$= e^{-2\times(7.25\times10^{-9}\ \text{m}^{-1})\times(1.0\times10^{-10}\ \text{m})} = 0.23$$

We conclude that, at a distance of 0.60 nm between the surface and the needle, the current is 23 per cent of the value measured when the distance is 0.50 nm.

Self-test 8.5 The ability of a proton to tunnel through a barrier contributes to the rapidity of proton transfer reactions in solution and therefore to the properties of acids and bases. Estimate the relative probabilities that a proton and a deuteron ($m_d = 3.342 \times 10^{-27}$ kg) can tunnel through the same barrier of height 1.0 eV (1.6×10^{-19} J) and length 100 pm when their energy is 0.9 eV. Comment on your answer.

[$T_H/T_D = 3.1 \times 10^2$; proton transfer reactions are expected to be much faster than deuteron transfer reactions.]

Vibrational motion

A particle undergoes **harmonic motion** if it experiences a 'Hooke's law' restoring force, in which the force is proportional to the displacement from the equilibrium position:

$$F = -k_f x \qquad \text{Hooke's law} \qquad (8.21)$$

Here, k_f is the **force constant**: the stiffer the 'spring', the greater the value of k_f. Because force is related to potential energy by $F = -dV/dx$, the force in eqn 8.21 corresponds to a potential energy

$$V = \tfrac{1}{2}k_f x^2 \qquad \text{Parabolic potential energy} \qquad (8.22)$$

This expression, which is the equation of a parabola (Fig. 8.17), is the origin of the term 'parabolic potential energy' for the potential energy characteristic of a harmonic oscillator. The Schrödinger equation for the particle is therefore

$$-\frac{\hbar^2}{2m}\frac{d^2\psi}{dx^2}+\tfrac{1}{2}k_f x^2\psi=E\psi \qquad (8.23)$$

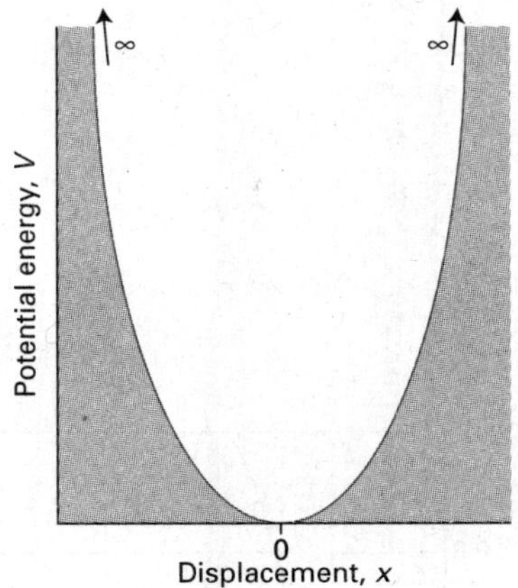

Fig. 8.17 The parabolic potential energy $V=\frac{1}{2}k_f x^2$ of a harmonic oscillator, where x is the displacement from equilibrium. The narrowness of the curve depends on the force constant k: the larger the value of k, the narrower the well.

8.4 The energy levels

Key point **The energies of a quantum mechanical harmonic oscillator are quantized with energies that form an equally spaced ladder.**

Equation 8.23 is a standard equation in the theory of differential equations and its solutions are well known to mathematicians. Quantization of energy levels arises from the boundary conditions: the oscillator will not be found with infinitely large displacements from equilibrium, so the only allowed solutions are those for which $\psi=0$ at $x=\pm\infty$. The permitted energy levels are

$$E_v=(v+\tfrac{1}{2})\hbar\omega \qquad \omega=\left(\frac{k_f}{m}\right)^{1/2} \qquad v=0,1,2,\ldots \qquad (8.24)$$

Energy levels of a harmonic oscillator

Note that ω (omega) increases with increasing force constant and decreasing mass. It follows from eqn 8.24 that the separation between adjacent levels is

$$E_{v+1}-E_v=\hbar\omega \qquad (8.25)$$

which is the same for all v. Therefore, the energy levels form a uniform ladder of spacing $\hbar\omega$ (Fig. 8.18). The energy separation $\hbar\omega$ is negligibly small for macroscopic objects (with large mass), but is of great importance for objects with mass similar to that of atoms.

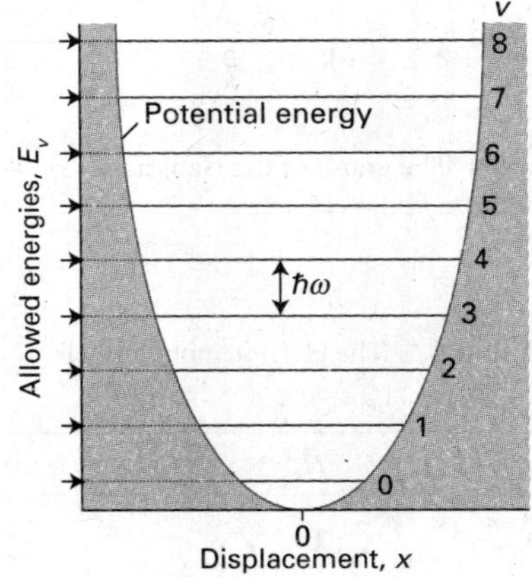

Fig. 8.18 The energy levels of a harmonic oscillator are evenly spaced with separation $\hbar\omega$, with $\omega=(k_f/m)^{1/2}$. Even in its lowest state, an oscillator has an energy greater than zero.

Because the smallest permitted value of v is 0, it follows from eqn 8.24 that a harmonic oscillator has a zero-point energy

$$E_0=\tfrac{1}{2}\hbar\omega \qquad (8.26)$$

Zero-point energy of a harmonic oscillator

The mathematical reason for the zero-point energy is that v cannot take negative values, for if it did the wavefunction would be ill-behaved. The physical reason is the same as for the particle in a square well: the particle is confined, its position is not completely uncertain, and therefore its momentum, and hence its kinetic energy, cannot be exactly zero. We can picture this zero-point state as one in which the particle fluctuates incessantly around its equilibrium position; classical mechanics would allow the particle to be perfectly still.

• A brief illustration

Atoms vibrate relative to one another in molecules with the bond acting like a spring. Consider an X–H bond, where a heavy X atom forms a stationary anchor for the very light H atom. That is, only the H atom moves, vibrating as a simple harmonic oscillator. Equation 8.24 describes the allowed vibrational energy levels of the bond. The force constant of a typical X–H chemical bond is around 500 N m^{-1}. For example, $k_f=516.3$ N m^{-1} for the ^{1}H^{35}Cl bond. Because the mass of a proton is about 1.7×10^{-27} kg, using $k_f=500$ N m^{-1} in eqn 8.24 gives $\omega\approx5.4\times10^{14}$ s^{-1} (5.4×10^2 THz). It follows from eqn 8.25 that the separation of adjacent levels is $\hbar\omega\approx5.7\times10^{-20}$ J (57 zJ, about 0.36 eV). This energy separation corresponds to 34 kJ mol^{-1}, which is chemically significant. From eqn 8.26, the zero-point energy of this molecular oscillator is about 28 zJ, which corresponds to 0.18 eV, or 17 kJ mol^{-1}. •

8.5 The wavefunctions

Key points (a) The wavefunctions of a harmonic oscillator have the form $\psi(x) = N \times$ (Hermite polynomial in x) $\times$ (bell-shaped Gaussian function). (b) The virial theorem states that, if the potential energy of a particle has the form $V = ax^b$, then its mean potential and kinetic energies are related by $2\langle E_k \rangle = b\langle V \rangle$. A quantum mechanical oscillator may be found at extensions that are forbidden by classical physics.

It is helpful at the outset to identify the similarities between the harmonic oscillator and the particle in a box, for then we shall be able to anticipate the form of the oscillator wavefunctions without detailed calculation. Like the particle in a box, a particle undergoing harmonic motion is trapped in a symmetrical well in which the potential energy rises to large values (and ultimately to infinity) for sufficiently large displacements (compare Figs. 8.1 and 8.17). However, there are two important differences. First, because the potential energy climbs towards infinity only as x^2 and not abruptly, the wavefunction approaches zero more slowly at large displacements than for the particle in a box. Second, as the kinetic energy of the oscillator depends on the displacement in a more complex way (on account of the variation of the potential energy), the curvature of the wavefunction also varies in a more complex way.

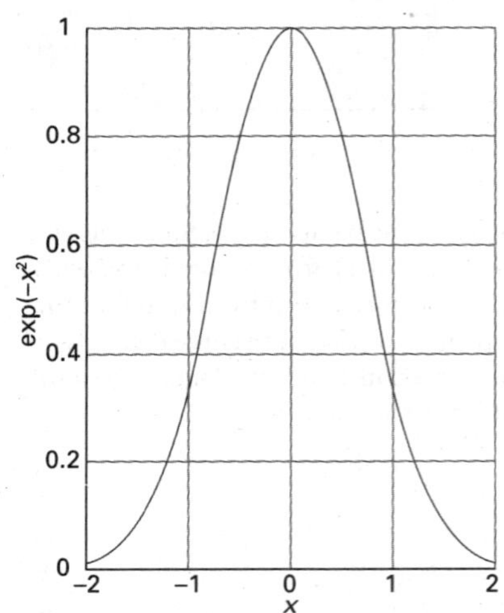

Fig. 8.19 The graph of the Gaussian function, $f(x) = e^{-x^2}$.

(a) The form of the wavefunctions

The detailed solution of eqn 8.23 shows that the wavefunction for a harmonic oscillator has the form

$$\psi(x) = N \times (\text{polynomial in } x) \times (\text{bell-shaped Gaussian function})$$

where N is a normalization constant. A Gaussian function is a function of the form e^{-x^2} (Fig. 8.19). The precise form of the wavefunctions is

$$\psi_v(x) = N_v H_v(y) e^{-y^2/2} \qquad y = \frac{x}{\alpha} \qquad \alpha = \left(\frac{\hbar^2}{mk_f}\right)^{1/4} \qquad \text{Wavefunctions of a harmonic oscillator} \qquad (8.27)$$

The factor $H_v(y)$ is a **Hermite polynomial** (Table 8.1). Hermite polynomials are members of a class of functions called orthogonal polynomials. These polynomials have a wide range of important properties, which allow a number of quantum mechanical calculations to be done with relative ease.

Because $H_0(y) = 1$, the wavefunction for the ground state (the lowest energy state, with $v = 0$) of the harmonic oscillator is

$$\psi_0(x) = N_0 e^{-y^2/2} = N_0 e^{-x^2/2\alpha^2} \qquad (8.28)$$

It follows that the probability density is the bell-shaped Gaussian function

$$\psi_0^2(x) = N_0^2 e^{-x^2/\alpha^2} \qquad (8.29)$$

The wavefunction and the probability distribution are shown in Fig. 8.20. Both curves have their largest values at zero displacement (at $x = 0$), so they capture the classical picture of the zero-point energy as arising from the ceaseless fluctuation of the particle about its equilibrium position.

Table 8.1 The Hermite polynomials $H_v(y)$

v	$H_v(y)$
0	1
1	$2y$
2	$4y^2 - 2$
3	$8y^3 - 12y$
4	$16y^4 - 48y^2 + 12$
5	$32y^5 - 160y^3 + 120y$
6	$64y^6 - 480y^4 + 720y^2 - 120$

The Hermite polynomials are solutions of the differential equation

$$H_v'' - 2yH_v' + 2vH_v = 0$$

where primes denote differentiation. They satisfy the recursion relation

$$H_{v+1} - 2yH_v + 2vH_{v-1} = 0$$

An important integral is

$$\int_{-\infty}^{\infty} H_{v'} H_v e^{-y^2} dy = \begin{cases} 0 & \text{if } v' \neq v \\ \pi^{1/2} 2^v v! & \text{if } v' = v \end{cases}$$

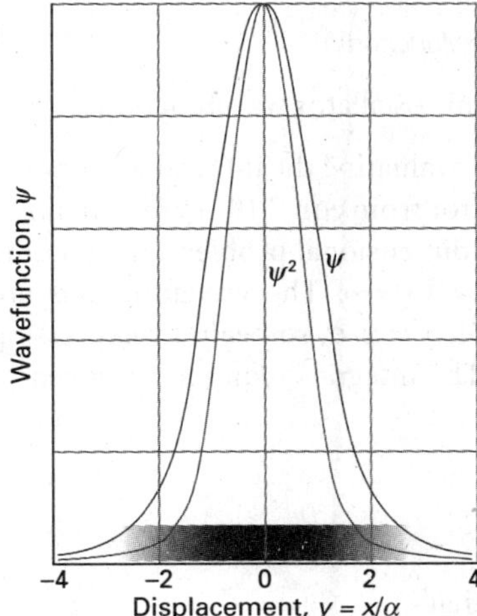

Fig. 8.20 The normalized wavefunction and probability distribution (shown also by shading) for the lowest energy state of a harmonic oscillator.

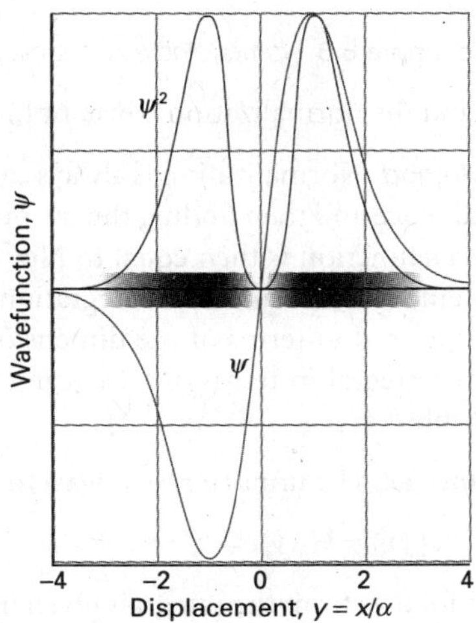

Fig. 8.21 The normalized wavefunction and probability distribution (shown also by shading) for the first excited state of a harmonic oscillator.

• A brief illustration

The wavefunction for the first excited state of the oscillator, the state with $v = 1$, is obtained by noting that $H_1(y) = 2y$ (note that some of the Hermite polynomials are very simple functions!):

$$\psi_1(x) = N_1 \times 2ye^{-y^2/2} \tag{8.30}$$

This function has a node at zero displacement ($x = 0$), and the probability density has maxima at $x = \pm\alpha$, corresponding to $y = \pm 1$ (Fig. 8.21). •

Once again, we should interpret the mathematical expressions we have derived. In the case of the harmonic oscillator wavefunctions in eqn 8.27, we should note the following.

1. The Gaussian function goes quickly to zero as the displacement increases (in either direction), so all the wavefunctions approach zero at large displacements.
2. The exponent y^2 is proportional to $x^2 \times (mk_f)^{1/2}$, so the wavefunctions decay more rapidly for large masses and large force constants (stiff springs).
3. As v increases, the Hermite polynomials become larger at large displacements (as x^v), so the wavefunctions grow large before the Gaussian function damps them down to zero: as a result, the wavefunctions spread over a wider range as v increases.

The shapes of several of the wavefunctions are shown in Fig. 8.22. At high quantum numbers, harmonic oscillator wavefunctions have their largest amplitudes near the turning points of the classical motion (the locations at which $V = E$, so the kinetic energy is zero). We see classical properties emerging in the correspondence limit of high quantum numbers, for a classical particle is most likely to be found at the turning points (where it is briefly stationary) and is least likely to be found at zero displacement (where it travels most rapidly).

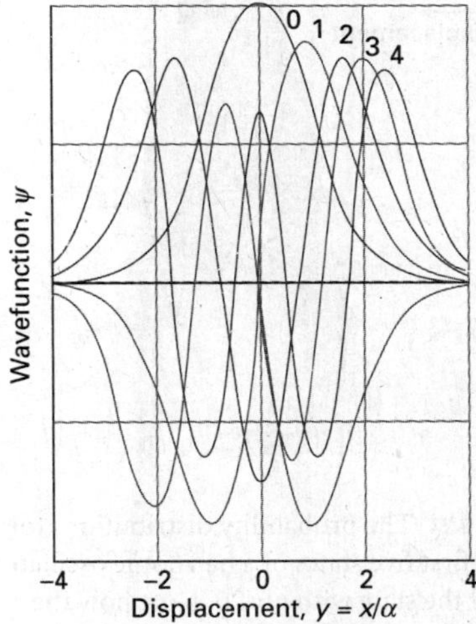

Fig. 8.22 The normalized wavefunctions for the first five states of a harmonic oscillator. Even values of v are purple; odd values are blue. Note that the number of nodes is equal to v and that alternate wavefunctions are symmetrical or antisymmetrical about $y = 0$ (zero displacement).

Example 8.3 *Normalizing a harmonic oscillator wavefunction*

Find the normalization constant for the harmonic oscillator wavefunctions.

Method Normalization is always carried out by evaluating the integral of $|\psi|^2$ over all space and then finding the normalization factor from eqn 7.19. The normalized wavefunction is then equal to $N\psi$. In this one-dimensional problem, the volume element is dx and the integration is from $-\infty$ to $+\infty$. The wavefunctions are expressed in terms of the dimensionless variable $y = x/\alpha$, so begin by expressing the integral in terms of y by using $dx = \alpha dy$. The integrals required are given in Table 8.1.

Answer The unnormalized wavefunction is

$$\psi_v(x) = H_v(y)e^{-y^2/2}$$

It follows from the integrals given in Table 8.1 that

$$\int_{-\infty}^{\infty} \psi_v^* \psi_v dx = \alpha \int_{-\infty}^{\infty} \psi_v^* \psi_v dy = \alpha \int_{-\infty}^{\infty} H_v^2(y) e^{-y^2} dy = \alpha \pi^{1/2} 2^v v!$$

where $v! = v(v-1)(v-2)\ldots 1$. Therefore,

$$N_v = \left(\frac{1}{\alpha \pi^{1/2} 2^v v!}\right)^{1/2}$$

Note that for a harmonic oscillator N_v is different for each value of v.

Self-test 8.6 Confirm, by explicit evaluation of the integral, that ψ_0 and ψ_1 are orthogonal.

[Evaluate the integral $\int_{-\infty}^{\infty} \psi_0^* \psi_1 dx$ by using the information in Table 8.1]

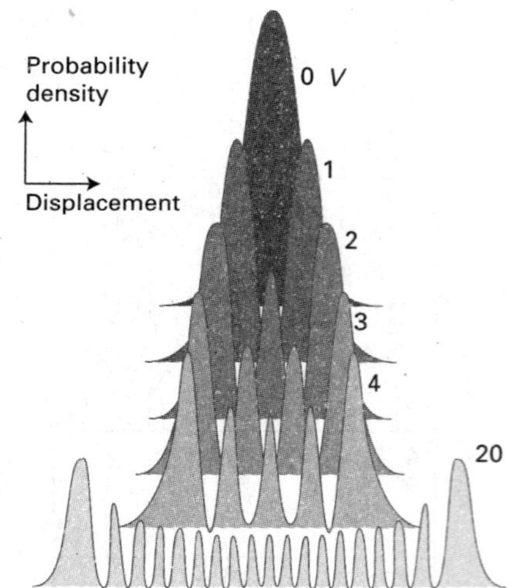

Fig. 8.23 The probability distributions for the first five states of a harmonic oscillator and the state with $v = 20$. Note how the regions of highest probability move towards the turning points of the classical motion as v increases.

interActivity To gain some insight into the origins of the nodes in the harmonic oscillator wavefunctions, plot the Hermite polynomials $H_v(y)$ for $v = 0$ through 5.

(b) The properties of oscillators

With the wavefunctions that are available, we can start calculating the properties of a harmonic oscillator. For instance, we can calculate the expectation values of an observable Ω by evaluating integrals of the type

$$\langle \Omega \rangle = \int_{-\infty}^{\infty} \psi_v^* \hat{\Omega} \psi_v dx \qquad (8.31)$$

(Here and henceforth, the wavefunctions are all taken as being normalized to 1.) When the explicit wavefunctions are substituted, the integrals look fearsome, but the Hermite polynomials have many simplifying features. For instance, we show in the following example that the mean displacement, $\langle x \rangle$, and the mean square displacement, $\langle x^2 \rangle$, of the oscillator when it is in the state with quantum number v are

$$\langle x \rangle = 0 \qquad \langle x^2 \rangle = (v + \tfrac{1}{2}) \frac{\hbar}{(mk_f)^{1/2}} \qquad (8.32)$$

The result for $\langle x \rangle$ shows that the oscillator is equally likely to be found on either side of $x = 0$ (like a classical oscillator). The result for $\langle x^2 \rangle$ shows that the mean square displacement increases with v. This increase is apparent from the probability densities in Fig. 8.23, and corresponds to the classical amplitude of swing increasing as the oscillator becomes more highly excited.

Example 8.4 *Calculating properties of a harmonic oscillator*

We can imagine the bending motion of a CO_2 molecule as a harmonic oscillation relative to the linear conformation of the molecule. We may be interested in the extent to which the molecule bends. Calculate the mean displacement of the oscillator when it is in a quantum state v.

Method Normalized wavefunctions must be used to calculate the expectation value. The operator for position along x is multiplication by the value of x (Section 7.5c). The resulting integral can be evaluated either by inspection (the integrand is the product of an odd and an even function), or by explicit evaluation using the formulas in Table 8.1. To give practice in this type of calculation, we illustrate the latter procedure. We shall need the relation $x = \alpha y$, which implies that $dx = \alpha dy$.

A brief comment

An even function is one for which $f(-x) = f(x)$; an odd function is one for which $f(-x) = -f(x)$. The product of an odd and even function is itself odd, and the integral of an odd function over a symmetrical range about $x = 0$ is zero.

Answer The integral we require is

$$\langle x\rangle = \int_{-\infty}^{\infty} \psi_v^* x \psi_v \mathrm{d}x = N_v^2 \int_{-\infty}^{\infty} (H_v \mathrm{e}^{-y^2/2}) x (H_v \mathrm{e}^{-y^2/2}) \mathrm{d}x$$

$$= \alpha^2 N_v^2 \int_{-\infty}^{\infty} (H_v \mathrm{e}^{-y^2/2}) y (H_v \mathrm{e}^{-y^2/2}) \mathrm{d}y$$

$$= \alpha^2 N_v^2 \int_{-\infty}^{\infty} H_v y H_v \mathrm{e}^{-y^2} \mathrm{d}y$$

Now use the recursion relation (see Table 8.1) to form

$$yH_v = vH_{v-1} + \tfrac{1}{2}H_{v+1}$$

which turns the integral into

$$\int_{-\infty}^{\infty} H_v y H_v \mathrm{e}^{-y^2} \mathrm{d}y = v \int_{-\infty}^{\infty} H_{v-1} H_v \mathrm{e}^{-y^2} \mathrm{d}y + \tfrac{1}{2} \int_{-\infty}^{\infty} H_{v+1} H_v \mathrm{e}^{-y^2} \mathrm{d}y$$

Both integrals are zero (see Table 8.1), so $\langle x\rangle = 0$. As remarked in the text, the mean displacement is zero because the displacement occurs equally on either side of the equilibrium position. The following *Self-test* extends this calculation by examining the mean square displacement, which we can expect to be non-zero and to increase with increasing v.

Self-test 8.7 Calculate the mean square displacement $\langle x^2\rangle$ of the particle from its equilibrium position. (Use the recursion relation twice.) [eqn 8.32]

The mean potential energy of an oscillator, the expectation value of $V = \tfrac{1}{2}kx^2$, can now be calculated very easily:

$$\langle V\rangle = \langle \tfrac{1}{2}k_f x^2\rangle = \tfrac{1}{2}(v + \tfrac{1}{2})\hbar \left(\frac{k_f}{m}\right)^{1/2} = \tfrac{1}{2}(v + \tfrac{1}{2})\hbar\omega \qquad (8.33)$$

Because the total energy in the state with quantum number v is $(v + \tfrac{1}{2})\hbar\omega$, it follows that

$$\langle V\rangle = \tfrac{1}{2}E_v \qquad (8.34a)$$

The total energy is the sum of the potential and kinetic energies, so it follows at once that the mean kinetic energy of the oscillator is

$$\langle E_k\rangle = \tfrac{1}{2}E_v \qquad (8.34b)$$

The result that the mean potential and kinetic energies of a harmonic oscillator are equal (and therefore that both are equal to half the total energy) is a special case of the **virial theorem**:

If the potential energy of a particle has the form $V = ax^b$, then its mean potential and kinetic energies are related by  (8.35)

$$2\langle E_k \rangle = b\langle V \rangle$$

For a harmonic oscillator $b = 2$, so $\langle E_k \rangle = \langle V \rangle$, as we have found. The virial theorem is a short cut to the establishment of a number of useful results, and we shall use it again.

An oscillator may be found at extensions with $V > E$ that are forbidden by classical physics, because they correspond to negative kinetic energy. For example, it follows from the shape of the wavefunction (see Problem 8.15) that in its lowest energy state there is about an 8 per cent chance of finding an oscillator stretched beyond its classical limit and an 8 per cent chance of finding it with a classically forbidden compression. These tunnelling probabilities are independent of the force constant and mass of the oscillator. The probability of being found in classically forbidden regions decreases quickly with increasing v, and vanishes entirely as v approaches infinity, as we would expect from the correspondence principle. Macroscopic oscillators (such as pendulums) are in states with very high quantum numbers, so the probability that they will be found in a classically forbidden region is wholly negligible. Molecules, however, are normally in their vibrational ground states, and for them the probability is very significant.

Rotational motion

The treatment of rotational motion can be broken down into two parts. The first deals with motion in two dimensions and the second with rotation in three dimensions.

8.6 Rotation in two dimensions: a particle on a ring

Key points **(a) The wavefunction of a particle on a ring must satisfy a cyclic boundary condition, and match at points separated by a complete revolution. (b) The energy and angular momentum of a particle on a ring are quantized.**

We consider a particle of mass m constrained to move in a circular path of radius r in the xy-plane with constant potential energy, which may be taken to be zero (Fig. 8.24). The total energy is equal to the kinetic energy, because $V = 0$ everywhere. We can therefore write $E = p^2/2m$. According to classical mechanics, the **angular momentum**, J_z, around the z-axis (which lies perpendicular to the xy-plane) is $J_z = \pm pr$, so the energy can be expressed as $J_z^2/2mr^2$. Because mr^2 is the **moment of inertia**, I, of the mass on its path, it follows that

$$E = \frac{J_z^2}{2I} \tag{8.36}$$

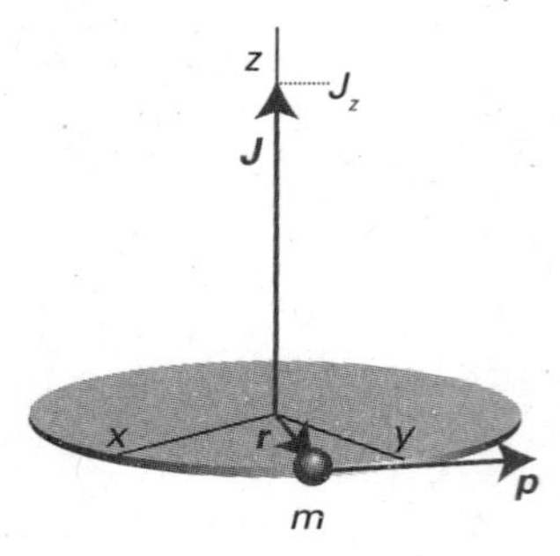

Fig. 8.24 The angular momentum of a particle of mass m on a circular path of radius r in the xy-plane is represented by a vector $\boldsymbol{J}$ with the single nonzero component, J_z, of magnitude pr perpendicular to the plane.

We shall now see that not all the values of the angular momentum are permitted in quantum mechanics, and therefore that both angular momentum and rotational energy are quantized.

(a) The qualitative origin of quantized rotation

Because $J_z = \pm pr$, and since the de Broglie relation gives $p = h/\lambda$, the angular momentum about the z-axis is

$$J_z = \pm\frac{hr}{\lambda}$$

Opposite signs correspond to opposite directions of travel. This equation shows that, the shorter the wavelength of the particle on a circular path of given radius, the greater the angular momentum of the particle. It follows that, if we can see why the wavelength is restricted to discrete values, then we shall understand why the angular momentum is quantized.

Suppose for the moment that λ can take an arbitrary value. In that case, the wavefunction depends on the azimuthal angle ϕ as shown in Fig. 8.25a. When ϕ increases beyond 2π, the wavefunction continues to change, but for an arbitrary wavelength it gives rise to a different value at each point, which is unacceptable (Section 7.4b). An acceptable solution is obtained only if the wavefunction reproduces itself on successive circuits, as in Fig. 8.25b. Because only some wavefunctions have this property, it follows that only some angular momenta are acceptable, and therefore that only certain rotational energies exist. Hence, the energy of the particle is quantized. Specifically, the only allowed wavelengths are

$$\lambda = \frac{2\pi r}{m_l}$$

with m_l, the conventional notation for this quantum number, taking integral values including 0. The value $m_l = 0$ corresponds to $\lambda = \infty$; a 'wave' of infinite wavelength has a constant height at all values of ϕ. The angular momentum is therefore limited to the values

$$J_z = \pm\frac{hr}{\lambda} = \frac{m_l hr}{2\pi r} = \frac{m_l h}{2\pi}$$

where we have allowed m_l to have positive or negative values. That is,

$$J_z = m_l\hbar \qquad m_l = 0, \pm1, \pm2, \ldots \qquad \text{Angular momentum of a particle on a ring} \qquad (8.37)$$

Positive values of m_l correspond to rotation in a clockwise sense around the z-axis (as viewed in the direction of z, Fig. 8.26) and negative values of m_l correspond to counterclockwise rotation around z. It then follows from eqn 8.36 that the energy is limited to the values

$$E = \frac{J_z^2}{2I} = \frac{m_l^2\hbar^2}{2I} \qquad \text{Energy levels of a particle on a ring} \qquad (8.38a)$$

We shall see shortly that the corresponding normalized wavefunctions are

$$\psi_{m_l}(\phi) = \frac{e^{im_l\phi}}{(2\pi)^{1/2}} \qquad \text{Wavefunctions of a particle on a ring} \qquad (8.38b)$$

The wavefunction with $m_l = 0$ is $\psi_0(\phi) = 1/(2\pi)^{1/2}$, and has the same value at all points on the circle.

We have arrived at a number of conclusions about rotational motion by combining some classical notions with the de Broglie relation. Such a procedure can be very useful for establishing the general form (and, as in this case, the exact energies) for a quantum mechanical system. However, to be sure that the correct solutions have been obtained, and to obtain practice for more complex problems where this less formal approach is inadequate, we need to solve the Schrödinger equation explicitly. The formal solution is described in the *Justification* that follows.

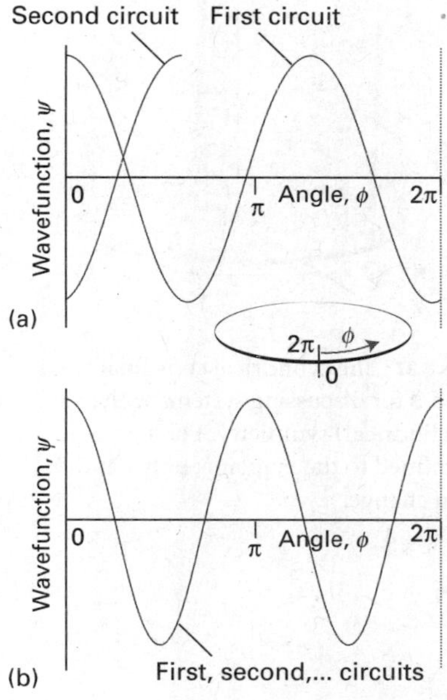

Fig. 8.25 Two solutions of the Schrödinger equation for a particle on a ring. The circumference has been opened out into a straight line; the points at $\phi = 0$ and 2π are identical. The solution in (a) is unacceptable because it is not single-valued. Moreover, on successive circuits it interferes destructively with itself, and does not survive. The solution in (b) is acceptable: it is single-valued, and on successive circuits it reproduces itself.

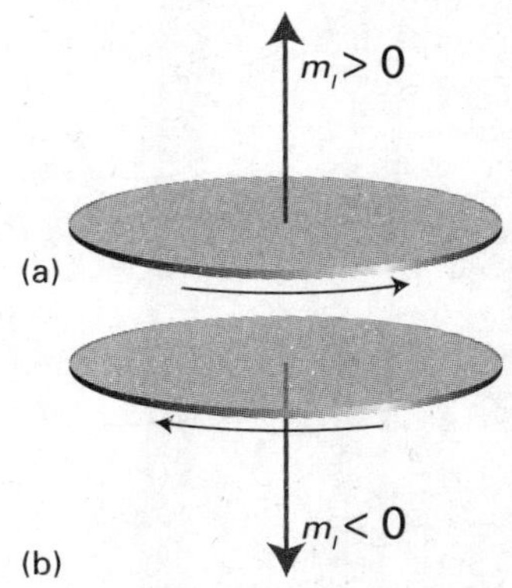

Fig. 8.26 The angular momentum of a particle confined to a plane can be represented by a vector of length $|m_l|$ units along the z-axis and with an orientation that indicates the direction of motion of the particle. The direction is given by the right-hand screw rule.

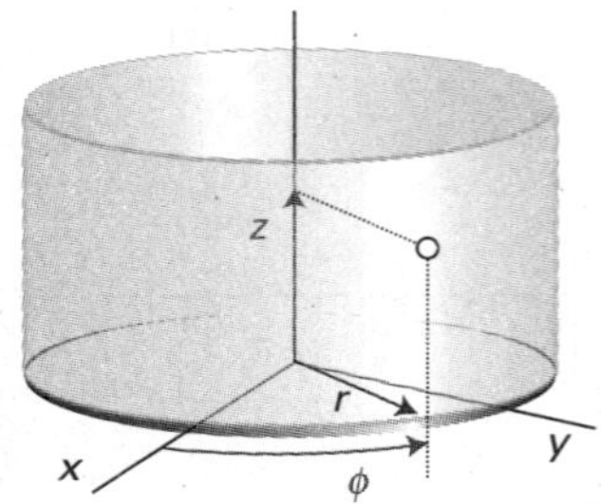

Fig. 8.27 The cylindrical coordinates z, r, and ϕ for discussing systems with axial (cylindrical) symmetry. For a particle confined to the xy-plane, only r and ϕ can change.

Justification 8.3 *The energies and wavefunctions of a particle on a ring*

The hamiltonian for a particle of mass m in a plane (with $V=0$) is the same as that given in eqn 8.9:

$$\hat{H}=-\frac{\hbar^2}{2m}\left(\frac{\partial^2}{\partial x^2}+\frac{\partial^2}{\partial y^2}\right)$$

and the Schrödinger equation is $\hat{H}\psi=E\psi$, with the wavefunction a function of the angle ϕ. It is always a good idea to use coordinates that reflect the full symmetry of the system, so we introduce the coordinates r and ϕ (Fig. 8.27), where $x=r\cos\phi$ and $y=r\sin\phi$. By standard manipulations we can write

$$\frac{\partial^2}{\partial x^2}+\frac{\partial^2}{\partial y^2}=\frac{\partial^2}{\partial r^2}+\frac{1}{r}\frac{\partial}{\partial r}+\frac{1}{r^2}\frac{\partial^2}{\partial\phi^2} \tag{8.39}$$

However, because the radius of the path is fixed, the derivatives with respect to r can be discarded. The hamiltonian then becomes

$$\hat{H}=-\frac{\hbar^2}{2mr^2}\frac{\mathrm{d}^2}{\mathrm{d}\phi^2}$$

The moment of inertia $I=mr^2$ has appeared automatically, so $\hat{H}$ may be written

$$\hat{H}=-\frac{\hbar^2}{2I}\frac{\mathrm{d}^2}{\mathrm{d}\phi^2} \tag{8.40}$$

and the Schrödinger equation is

$$\frac{\mathrm{d}^2\psi}{\mathrm{d}\phi^2}=-\frac{2IE}{\hbar^2}\psi \tag{8.41}$$

The normalized general solutions of the equation are

$$\psi_{m_l}(\phi)=\frac{\mathrm{e}^{\mathrm{i}m_l\phi}}{(2\pi)^{1/2}}\qquad m_l=\pm\frac{(2IE)^{1/2}}{\hbar} \tag{8.42}$$

The quantity m_l is just a dimensionless number at this stage.

We now select the acceptable solutions from among these general solutions by imposing the condition that the wavefunction should be single-valued. That is, the wavefunction ψ must satisfy a **cyclic boundary condition**, and match at points separated by a complete revolution: $\psi(\phi+2\pi)=\psi(\phi)$. On substituting the general wavefunction into this condition, we find

$$\psi_{m_l}(\phi+2\pi)=\frac{\mathrm{e}^{\mathrm{i}m_l(\phi+2\pi)}}{(2\pi)^{1/2}}=\frac{\mathrm{e}^{\mathrm{i}m_l\phi}\mathrm{e}^{2\pi\mathrm{i}m_l}}{(2\pi)^{1/2}}=\psi_{m_l}(\phi)\mathrm{e}^{2\pi\mathrm{i}m_l}$$

As $\mathrm{e}^{\mathrm{i}\pi}=-1$, this relation is equivalent to

$$\psi_{m_l}(\phi+2\pi)=(-1)^{2m_l}\psi_{m_l}(\phi) \tag{8.43}$$

Because we require $(-1)^{2m_l}=1$, $2m_l$ must be a positive or a negative even integer (including 0), and therefore m_l must be an integer: $m_l=0, \pm1, \pm2, \ldots$. The corresponding energies are therefore those given by eqn 8.38a with $m_l=0, \pm1, \pm2, \ldots$.

(b) Quantization of rotation

We can summarize the conclusions so far as follows. The energy is quantized and restricted to the values given in eqn 8.38a ($E=m_l^2\hbar^2/2I$). The occurrence of m_l as its square means that the energy of rotation is independent of the sense of rotation (the sign of m_l), as we expect physically. In other words, states with a given value of $|m_l|$ are doubly degenerate, except for $m_l = 0$, which is non-degenerate. Although the result has been derived for the rotation of a single mass point, it also applies to any body of moment of inertia I constrained to rotate about one axis.

We have also seen that the angular momentum is quantized and confined to the values given in eqn 8.37 ($J_z = m_l\hbar$). The increasing angular momentum is associated with the increasing number of nodes in the real and imaginary parts of the wavefunction: the wavelength decreases stepwise as $|m_l|$ increases, so the momentum with which the particle travels round the ring increases (Fig. 8.28). As shown in the following *Justification*, we can come to the same conclusion more formally by using the argument about the relation between eigenvalues and the values of observables established in Section 7.5.

A brief comment

The complex function $e^{im_l\phi}$ does not have nodes; however, it may be written as $\cos m_l\phi + i \sin m_l\phi$, and the real ($\cos m_l\phi$) and imaginary ($\sin m_l\phi$) components do have nodes.

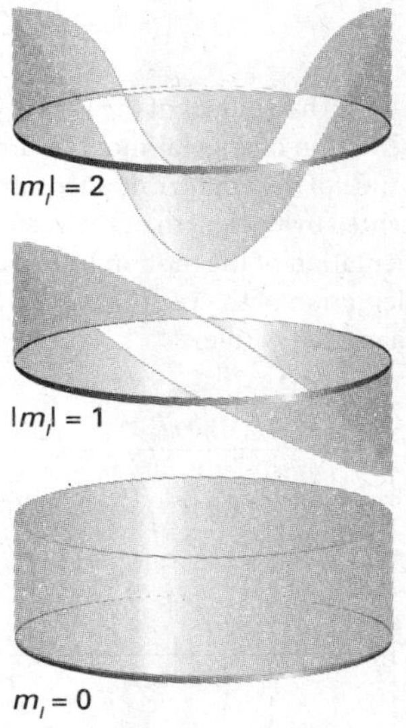

Fig. 8.28 The real parts of the wavefunctions of a particle on a ring. As shorter wavelengths are achieved, the magnitude of the angular momentum around the z-axis grows in steps of $\hbar$.

Justification 8.4 *The quantization of angular momentum*

In the discussion of translational motion in one dimension, we saw that the opposite signs in the wavefunctions e^{ikx} and e^{-ikx} correspond to opposite directions of travel, and that the linear momentum is given by the eigenvalue of the linear momentum operator. The same conclusions can be drawn here, but now we need the eigenvalues of the angular momentum operator. In classical mechanics the orbital angular momentum l_z about the z-axis is defined as

$$l_z = xp_y - yp_x \qquad \text{Definition of angular momentum} \qquad [8.44]$$

where p_x is the component of linear motion parallel to the x-axis and p_y is the component parallel to the y-axis.

The operators for the two linear momentum components are given in eqn 7.29, so the operator for angular momentum about the z-axis, which we denote $\hat{l}_z$, is

$$\hat{l}_z = \frac{\hbar}{i}\left(x\frac{\partial}{\partial y} - y\frac{\partial}{\partial x}\right) \qquad \text{Angular momentum operator} \qquad (8.45)$$

When expressed in terms of the coordinates r and ϕ, by standard manipulations this equation becomes

$$\hat{l}_z = \frac{\hbar}{i}\frac{\partial}{\partial \phi} \qquad \text{Angular momentum operator (polar form)} \qquad (8.46)$$

With the angular momentum operator available, we can test the wavefunction in eqn 8.42. Disregarding the normalization constant, we find

$$\hat{l}_z\psi_{m_l} = \frac{\hbar}{i}\frac{d\psi_{m_l}}{d\phi} = im_l\frac{\hbar}{i}e^{im_l\phi} = m_l\hbar\psi_{m_l} \qquad (8.47)$$

That is, ψ_{m_l} is an eigenfunction of $\hat{l}_z$, and corresponds to an angular momentum $m_l\hbar$. When m_l is positive, the angular momentum is positive (clockwise when seen from below); when m_l is negative, the angular momentum is negative (counterclockwise when seen from below). These features are the origin of the vector representation of angular momentum, in which the magnitude is represented by the length of a vector and the direction of motion by its orientation (Fig. 8.29).

A brief comment

The angular momentum in three dimensions is defined as

$$\boldsymbol{l} = \boldsymbol{r} \times \boldsymbol{p} = \begin{vmatrix} \boldsymbol{i} & \boldsymbol{j} & \boldsymbol{k} \\ x & y & z \\ p_x & p_y & p_z \end{vmatrix}$$
$$= (yp_z - zp_y)\boldsymbol{i} - (xp_z - zp_x)\boldsymbol{j} + (xp_y - yp_x)\boldsymbol{k}$$

where $\boldsymbol{i}$, $\boldsymbol{j}$, and $\boldsymbol{k}$ are unit vectors pointing along the positive directions on the x-, y-, and z-axes. It follows that the z-component of the angular momentum has a magnitude given by eqn 8.44. For more information on vectors, see *Mathematical background 5* following Chapter 9.

To locate the particle given its wavefunction in eqn 8.42, we form the probability density:

$$\psi^*_{m_l}\psi_{m_l} = \left(\frac{e^{im_l\phi}}{(2\pi)^{1/2}}\right)^*\left(\frac{e^{im_l\phi}}{(2\pi)^{1/2}}\right) = \left(\frac{e^{-im_l\phi}}{(2\pi)^{1/2}}\right)\left(\frac{e^{im_l\phi}}{(2\pi)^{1/2}}\right) = \frac{1}{2\pi}$$

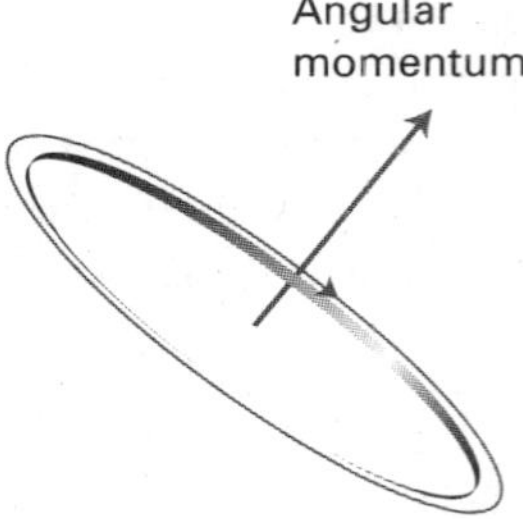

Fig. 8.29 The basic ideas of the vector representation of angular momentum: the magnitude of the angular momentum is represented by the length of the vector, and the orientation of the motion in space by the orientation of the vector (using the right-hand screw rule).

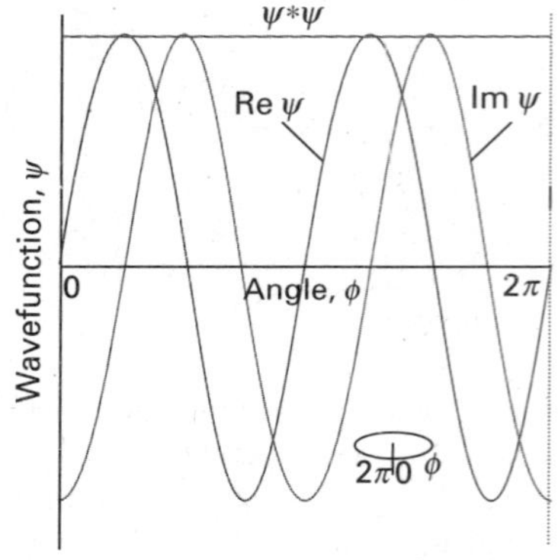

Fig. 8.30 The probability density for a particle in a definite state of angular momentum is uniform, so there is an equal probability of finding the particle anywhere on the ring.

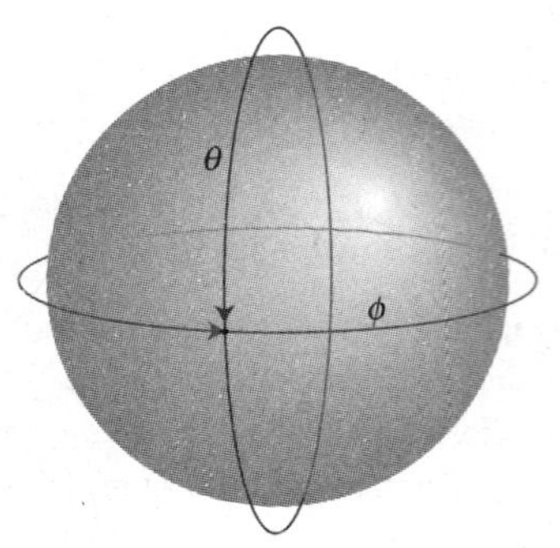

Fig. 8.31 The wavefunction of a particle on the surface of a sphere must satisfy two cyclic boundary conditions; this requirement leads to two quantum numbers for its state of angular momentum.

Because this probability density is independent of ϕ, the probability of locating the particle somewhere on the ring is also independent of ϕ (Fig. 8.30). Hence the location of the particle is completely indefinite, and knowing the angular momentum precisely eliminates the possibility of specifying the location of the particle. Angular momentum and angle are a pair of complementary observables (in the sense defined in Section 7.6), and the inability to specify them simultaneously with arbitrary precision is another example of the uncertainty principle.

8.7 Rotation in three dimensions: the particle on a sphere

Key points **(a) The wavefunction of a particle on a spherical surface must satisfy simultaneously two cyclic boundary conditions. (b) The energy and angular momentum of a particle on a sphere are quantized. (c) Space quantization is the restriction of the component of angular momentum around an axis to discrete values. (d) The vector model of angular momentum uses diagrams to represent the state of angular momentum of a rotating particle.**

We now consider a particle of mass m that is free to move anywhere on the surface of a sphere of radius r. We shall need the results of this calculation when we come to describe rotating molecules and the states of electrons in atoms. The requirement that the wavefunction should match as a path is traced over the poles as well as around the equator of the sphere surrounding the central point introduces a second cyclic boundary condition and therefore a second quantum number (Fig. 8.31).

(a) The Schrödinger equation

The hamiltonian for motion in three dimensions (Table 7.1) is

$$\hat{H}=-\frac{\hbar^2}{2m}\nabla^2+V \qquad \nabla^2=\frac{\partial^2}{\partial x^2}+\frac{\partial^2}{\partial y^2}+\frac{\partial^2}{\partial z^2} \qquad (8.48)$$

The symbol ∇^2 is a convenient abbreviation for the sum of the three second derivatives; it is called the **laplacian**, and read either 'del squared' or 'nabla squared'. For the particle confined to a spherical surface, $V=0$ wherever it is free to travel, and the radius r is a constant. The wavefunction is therefore a function of the **colatitude**, θ, and the **azimuth**, ϕ (Fig. 8.32), and so we write it as $\psi(\theta,\phi)$. The Schrödinger equation is

$$-\frac{\hbar^2}{2m}\nabla^2\psi=E\psi \qquad (8.49)$$

As shown in the following *Justification*, this partial differential equation can be simplified by the separation of variables procedure (*Mathematical background 4*) by expressing the wavefunction (for constant r) as the product

$$\psi(\theta,\phi)=\Theta(\theta)\Phi(\phi) \qquad \text{Separation of variables} \qquad (8.50)$$

where Θ is a function only of θ and Φ is a function only of ϕ.

Justification 8.5 *The separation of variables technique applied to the particle on a sphere*

The laplacian in spherical polar coordinates is

$$\nabla^2=\frac{\partial^2}{\partial r^2}+\frac{2}{r}\frac{\partial}{\partial r}+\frac{1}{r^2}\Lambda^2 \qquad \text{laplacian} \qquad (8.51a)$$

where the **legendrian**, Λ^2, is

$$\Lambda^2=\frac{1}{\sin^2\theta}\frac{\partial^2}{\partial\phi^2}+\frac{1}{\sin\theta}\frac{\partial}{\partial\theta}\sin\theta\frac{\partial}{\partial\theta} \quad \text{legendrian} \quad (8.51b)$$

Because r is constant, we can discard the part of the laplacian that involves differentiation with respect to r, and so write the Schrödinger equation as

$$\frac{1}{r^2}\Lambda^2\psi=-\frac{2mE}{\hbar^2}\psi$$

or, because $I=mr^2$, as

$$\Lambda^2\psi=-\varepsilon\psi \qquad \varepsilon=\frac{2IE}{\hbar^2}$$

To verify that this expression is separable, we substitute $\psi=\Theta\Phi$:

$$\frac{1}{\sin^2\theta}\frac{\partial^2(\Theta\Phi)}{\partial\phi^2}+\frac{1}{\sin\theta}\frac{\partial}{\partial\theta}\sin\theta\frac{\partial(\Theta\Phi)}{\partial\theta}=-\varepsilon\Theta\Phi$$

We now use the fact that Θ and Φ are each functions of one variable, so the partial derivatives become complete derivatives:

$$\frac{\Theta}{\sin^2\theta}\frac{d^2\Phi}{d\phi^2}+\frac{\Phi}{\sin\theta}\frac{d}{d\theta}\sin\theta\frac{d\Theta}{d\theta}=-\varepsilon\Theta\Phi$$

Division through by $\Theta\Phi$, multiplication by $\sin^2\theta$, and minor rearrangement gives

$$\frac{1}{\Phi}\frac{d^2\Phi}{d\phi^2}+\frac{\sin\theta}{\Theta}\frac{d}{d\theta}\sin\theta\frac{d\Theta}{d\theta}+\varepsilon\sin^2\theta=0$$

The first term on the left depends only on ϕ and the remaining two terms depend only on θ. We met a similar situation when discussing a particle on a rectangular surface (Justification 8.2), and by the same argument, the complete equation can be separated. Thus, if we set the first term equal to the numerical constant $-m_l^2$ (using a notation chosen with an eye to the future), the separated equations are

$$\frac{1}{\Phi}\frac{d^2\Phi}{d\phi^2}=-m_l^2 \qquad \frac{\sin\theta}{\Theta}\frac{d}{d\theta}\sin\theta\frac{d\Theta}{d\theta}+\varepsilon\sin^2\theta=m_l^2$$

The first of these two equations is the same as that in Justification 8.3, so it has the same solutions (eqn 8.42). The second is much more complicated to solve, but the solutions are tabulated as the associated Legendre functions. For reasons related to the behaviour of these functions, the cyclic boundary conditions on Θ arising from the need for the wavefunctions to match at $\theta=0$ and 2π (the North Pole) result in the introduction of a second quantum number, l, which identifies the acceptable solutions. The presence of the quantum number m_l in the second equation implies, as we see below, that the range of acceptable values of m_l is restricted by the value of l.

As indicated in Justification 8.5, solution of the Schrödinger equation shows that the acceptable wavefunctions are specified by two quantum numbers l and m_l which are restricted to the values

$$l=0, 1, 2, \ldots \qquad m_l=l, l-1, \ldots, -l \qquad (8.52)$$

Note that the **orbital angular momentum quantum number** l is non-negative and that, for a given value of l, there are $2l+1$ permitted values of the **magnetic quantum number**, m_l. The normalized wavefunctions are usually denoted $Y_{l,m_l}(\theta,\phi)$ and are called the **spherical harmonics** (Table 8.2).

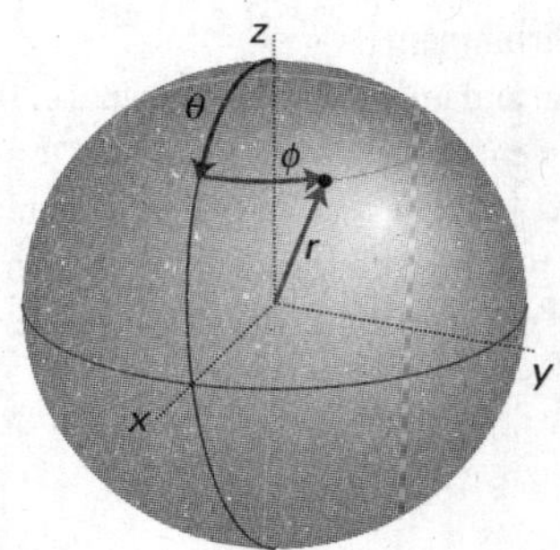

Fig. 8.32 Spherical polar coordinates. For a particle confined to the surface of a sphere, only the colatitude, θ, and the azimuth, ϕ, can change.

Table 8.2 The spherical harmonics

l	m_l	$Y_{l,m_l}(\theta,\phi)$
0	0	$\left(\frac{1}{4\pi}\right)^{1/2}$
1	0	$\left(\frac{3}{4\pi}\right)^{1/2}\cos\theta$
	±1	$\mp\left(\frac{3}{8\pi}\right)^{1/2}\sin\theta\, e^{\pm i\phi}$
2	0	$\left(\frac{5}{16\pi}\right)^{1/2}(3\cos^2\theta-1)$
	±1	$\mp\left(\frac{15}{8\pi}\right)^{1/2}\cos\theta\sin\theta\, e^{\pm i\phi}$
	±2	$\left(\frac{15}{32\pi}\right)^{1/2}\sin^2\theta\, e^{\pm 2i\phi}$
3	0	$\left(\frac{7}{16\pi}\right)^{1/2}(5\cos^3\theta-3\cos\theta)$
	±1	$\mp\left(\frac{21}{64\pi}\right)^{1/2}(5\cos^2\theta-1)\sin\theta\, e^{\pm i\phi}$
	±2	$\left(\frac{105}{32\pi}\right)^{1/2}\sin^2\theta\cos\theta\, e^{\pm 2i\phi}$
	±3	$\mp\left(\frac{35}{64\pi}\right)^{1/2}\sin^3\theta\, e^{\pm 3i\phi}$

The spherical harmonics are orthogonal and normalized in the following sense:

$$\int_0^{\pi}\int_0^{2\pi} Y_{l',m_l'}(\theta,\phi)^* Y_{l,m_l}(\theta,\phi)\sin\theta\, d\theta\, d\phi=\delta_{l'l}\delta_{m_l'm_l}$$

An important 'triple integral' is

$$\int_0^{\pi}\int_0^{2\pi} Y_{l'',m_l''}(\theta,\phi)^* Y_{l',m_l'}(\theta,\phi) Y_{l,m_l}(\theta,\phi)\sin\theta\, d\theta\, d\phi$$

$$=0 \quad \text{unless} \quad m_l''=m_l'+m_l$$

and we can form a triangle with sides of lengths l'', l', and l (such as 1, 2, and 3 or 1, 1, and 1, but not 1, 2, and 4).

A brief comment

The real and imaginary components of the Φ component of the wavefunctions, $e^{im_l\phi} = \cos m_l\phi + i \sin m_l\phi$, each have $|m_l|$ angular nodes, but these nodes are not seen when we plot the probability density, because $|e^{im_l\phi}|^2 = 1$.

Figure 8.33 is a representation of the spherical harmonics for $l = 0$ to 4 and $m_l = 0$, which emphasizes how the number of angular nodes (the angles at which the wavefunction passes through zero) increases as the value of l increases. There are no angular nodes around the z-axis for functions with $m_l = 0$, which corresponds to there being no component of orbital angular momentum about that axis. Figure 8.34 shows the distribution of the particle of a given angular momentum in more detail. In this representation, the value of $|Y_{l,m_l}|^2$ at each value of θ and ϕ is proportional to the distance of the surface from the origin. Note how, for a given value of l, the most probable location of the particle migrates towards the xy-plane as the value of $|m_l|$ increases.

It also follows from the solution of the Schrödinger equation that the energy E of the particle is restricted to the values

$$E = l(l+1)\frac{\hbar^2}{2I} \qquad l = 0, 1, 2, \ldots \qquad \text{Energy levels of a particle on a sphere} \qquad (8.53)$$

We see that the energy is quantized, and that it is independent of m_l. Because there are $2l + 1$ different wavefunctions (one for each value of m_l) that correspond to the same energy, it follows that a level with quantum number l is $(2l + 1)$-fold degenerate.

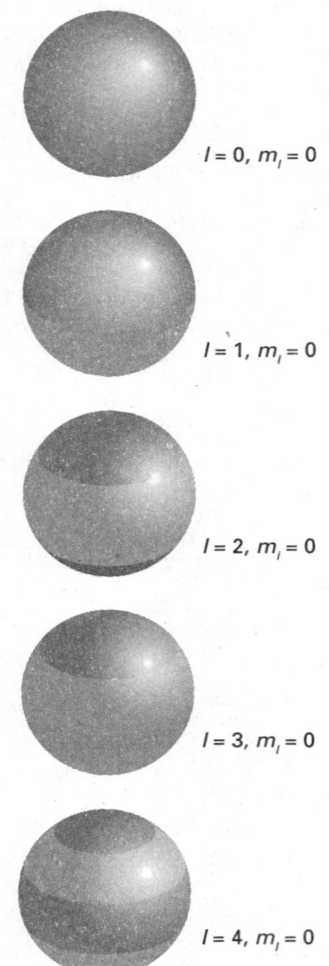

Fig. 8.33 A representation of the wavefunctions of a particle on the surface of a sphere that emphasizes the location of angular nodes: dark and light shading correspond to different signs of the wavefunction. Note that the number of nodes increases as the value of l increases. All these wavefunctions correspond to $m_l = 0$; a path round the vertical z-axis of the sphere does not cut through any nodes.

(b) Angular momentum

The energy of a rotating particle is related classically to its angular momentum J by $E = J^2/2I$. Therefore, by comparing this equation with eqn 8.53, we can deduce that, because the energy is quantized, then so too is the magnitude of the angular momentum, and confined to the values

$$\{l(l+1)\}^{1/2}\hbar \qquad l = 0, 1, 2 \ldots \qquad \text{Magnitude of angular momentum} \qquad (8.54a)$$

We have already seen (in the context of rotation in a plane) that the angular momentum about the z-axis is quantized, and that it has the values

$$m_l\hbar \qquad m_l = l, l-1, \ldots, -l \qquad \text{z-Component of angular momentum} \qquad (8.54b)$$

The fact that the number of nodes in $\psi_{l,m_l}(\theta,\phi)$ increases with l reflects the fact that higher angular momentum implies higher kinetic energy, and therefore a more sharply curved wavefunction. We can also see that the states corresponding to high angular momentum around the z-axis are those in which the most nodal lines cut the equator: a high kinetic energy now arises from motion parallel to the equator because the curvature is greatest in that direction.

• A brief illustration

Under certain circumstances, the particle on a sphere is a reasonable model for the description of the rotation of diatomic molecules. Consider, for example, the rotation of a $^1H^{127}I$ molecule: because of the large difference in atomic masses, it is appropriate to picture the ^1H atom as orbiting a stationary ^{127}I atom at a distance $r = 160$ pm, the equilibrium bond distance. The moment of inertia of $^1H^{127}I$ is then $I = m_H r^2 = 4.288 \times 10^{-47}$ kg m^2. It follows that

$$\frac{\hbar^2}{2I} = \frac{(1.054\,57 \times 10^{-34}\ \mathrm{J\ s})^2}{2 \times (4.288 \times 10^{-47}\ \mathrm{kg\ m^2})} = 1.297 \times 10^{-22}\ \mathrm{J}$$

or 0.1297 zJ. This energy corresponds to 78.09 J mol^{-1}. From eqn 8.53, the first few rotational energy levels are therefore 0 ($l = 0$), 0.2594 zJ ($l = 1$), 0.7782 zJ ($l = 2$), and 1.556 zJ ($l = 3$). The degeneracies of these levels are 1, 3, 5, and 7, respectively (from $2l + 1$) and

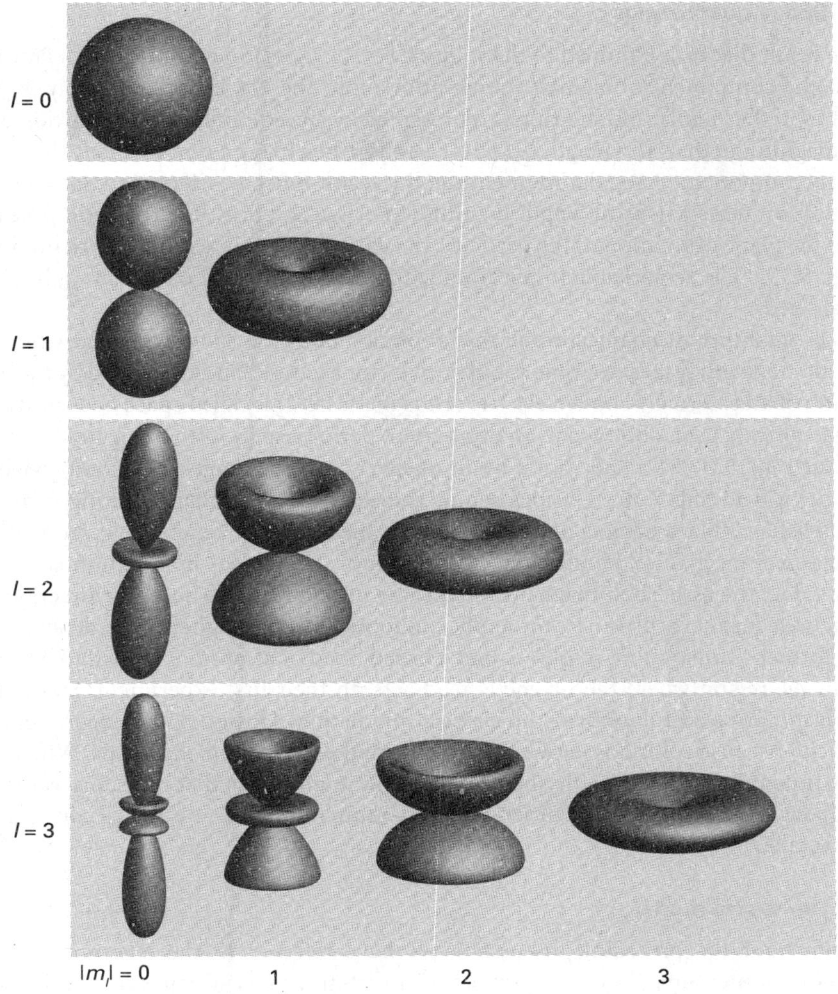

Fig. 8.34 A more complete representation of the wavefunctions for $l = 0, 1, 2$, and 3. The distance of a point on the surface from the origin is proportional to the square modulus of the amplitude of the wavefunction at that point.

interActivity Plot the first ten energy levels of a particle on spheres of different radius r. Which of the following statements are true: (a) for a given value of r, the energy separation between adjacent levels decreases with increasing l, (b) increasing r leads to an decrease in the value of the energy for each level, (c) the energy difference between adjacent levels increases as r increases?

the magnitudes of the angular momentum of the molecule are 0, $2^{1/2}\hbar$, $6^{1/2}\hbar$, and $(12)^{1/2}\hbar$ (from eqn 8.54a). It follows from our calculations that the $l = 0$ and $l = 1$ levels are separated by $\Delta E = 0.2594$ zJ. A transition between these two rotational levels of the molecule can be brought about by the emission or absorption of a photon with a frequency given by the Bohr frequency condition (eqn 7.14):

$$\nu = \frac{\Delta E}{h} = \frac{2.594 \times 10^{-22}\ \text{J}}{6.626 \times 10^{-34}\ \text{J s}} = 3.915 \times 10^{11}\ \text{Hz} = 391.5\ \text{GHz}$$

Radiation with this frequency belongs to the microwave region of the electromagnetic spectrum, so microwave spectroscopy is a convenient method for the study of molecular rotations . Because the transition energies depend on the moment of inertia, microwave spectroscopy is a very accurate technique for the determination of bond lengths. We discuss rotational spectra further in Chapter 12. ●

Self-test 8.8 Repeat the calculation for a $^2H^{127}I$ molecule (same bond length as $^1H^{127}I$).

[Energies are smaller by a factor of two; same angular momenta and numbers of components]

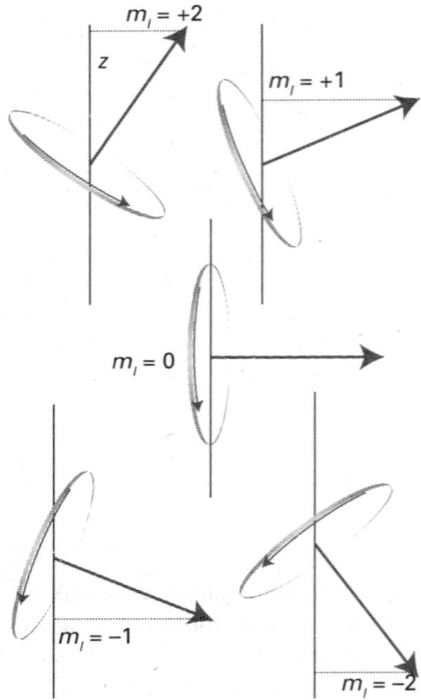

Fig. 8.35 The permitted orientations of angular momentum when $l = 2$. We shall see soon that this representation is too specific because the azimuthal orientation of the vector (its angle around z) is indeterminate.

(c) Space quantization

The result that m_l is confined to the values $l, l-1, \ldots, -l$ for a given value of l means that the component of angular momentum about the z-axis may take only $2l+1$ values. If the angular momentum is represented by a vector of length proportional to its magnitude (that is, of length $\{l(l+1)\}^{1/2}$ units), then to represent correctly the value of the component of angular momentum, the vector must be oriented so that its projection on the z-axis is of length m_l units. In classical terms, this restriction means that the plane of rotation of the particle can take only a discrete range of orientations (Fig. 8.35). The remarkable implication is that the orientation of a rotating body is quantized.

The quantum mechanical result that a rotating body may not take up an arbitrary orientation with respect to some specified axis (for example, an axis defined by the direction of an externally applied electric or magnetic field) is called **space quantization.** It had already been observed in an experiment performed by Otto Stern and Walther Gerlach in 1921, who had shot a beam of silver atoms through an inhomogeneous magnetic field (Fig. 8.36). The idea behind the experiment was that a rotating, charged body behaves like a magnet and interacts with the applied magnetic field. According to classical mechanics, because the orientation of the angular momentum can take any value, the associated magnet can take any orientation. Because the direction in which the magnet is driven by the applied inhomogeneous magnetic field depends on the former's orientation, it follows that a broad band of atoms is expected to emerge from the region where the magnetic field acts. In their first experiment, Stern and Gerlach appeared to confirm the classical prediction. However, the experiment is difficult because collisions between the atoms in the beam blurs the bands. When the experiment was repeated with a beam of very low intensity (so that collisions were less frequent) they observed discrete bands, as quantum mechanics was in due course able to explain.

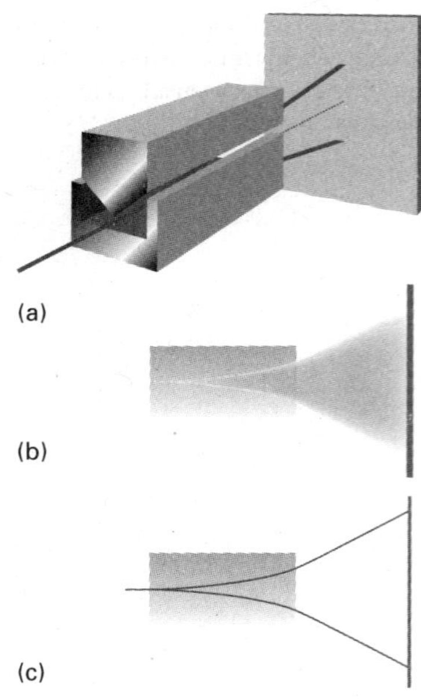

Fig. 8.36 (a) The experimental arrangement for the Stern–Gerlach experiment: the magnet provides an inhomogeneous field. (b) The classically expected result. (c) The observed outcome using silver atoms.

(d) The vector model

Throughout the preceding discussion, we have referred to the z-component of angular momentum (the component about an arbitrary axis, which is conventionally denoted z), and have made no reference to the x- and y-components (the components about the two axes perpendicular to z). The reason for this omission is found by examining the operators for the three components, each one being given by a term like that in eqn 8.45:

$$\hat{l}_x = \frac{\hbar}{\mathrm{i}}\left(y\frac{\partial}{\partial z} - z\frac{\partial}{\partial y}\right) \qquad \hat{l}_y = \frac{\hbar}{\mathrm{i}}\left(z\frac{\partial}{\partial x} - x\frac{\partial}{\partial z}\right) \qquad \hat{l}_z = \frac{\hbar}{\mathrm{i}}\left(x\frac{\partial}{\partial y} - y\frac{\partial}{\partial x}\right)$$

Angular momentum operators (8.55)

As you are invited to show in Problem 8.27, these three operators do not commute with one another:

$$[\hat{l}_x,\hat{l}_y] = \mathrm{i}\hbar\hat{l}_z \qquad [\hat{l}_y,\hat{l}_z] = \mathrm{i}\hbar\hat{l}_x \qquad [\hat{l}_z,\hat{l}_x] = \mathrm{i}\hbar\hat{l}_y$$

Angular momentum commutation relations (8.56a)

Therefore, we cannot specify more than one component (unless $l = 0$). In other words, l_x, l_y, and l_z are complementary observables. On the other hand, the operator for the square of the magnitude of the angular momentum is

$$\hat{l}^2 = \hat{l}_x^2 + \hat{l}_y^2 + \hat{l}_z^2 = \hbar^2\Lambda^2 \qquad (8.56b)$$

where Λ^2 is the legendrian in eqn 8.51b. This operator does commute with all three components:

$$[\hat{l}^2, \hat{l}_q] = 0 \qquad q = x, y, \text{and } z \tag{8.56c}$$

(See Problem 8.29.) Therefore, although we may specify the magnitude of the angular momentum and any of its components if l_z is known, then it is impossible to ascribe values to the other two components. It follows that the illustration in Fig. 8.35, which is summarized in Fig. 8.37a, gives a false impression of the state of the system, because it suggests definite values for the x- and y-components. A better picture must reflect the impossibility of specifying l_x and l_y if l_z is known.

The **vector model** of angular momentum uses pictures like that in Fig. 8.37b. The cones are drawn with side $\{l(l+1)\}^{1/2}$ units, and represent the magnitude of the angular momentum. Each cone has a definite projection (of m_l units) on the z-axis, representing the system's precise value of l_z. The l_x and l_y projections, however, are indefinite. The vector representing the state of angular momentum can be thought of as lying with its tip on any point on the mouth of the cone. At this stage it should not be thought of as sweeping round the cone; that aspect of the model will be added later when we allow the picture to convey more information.

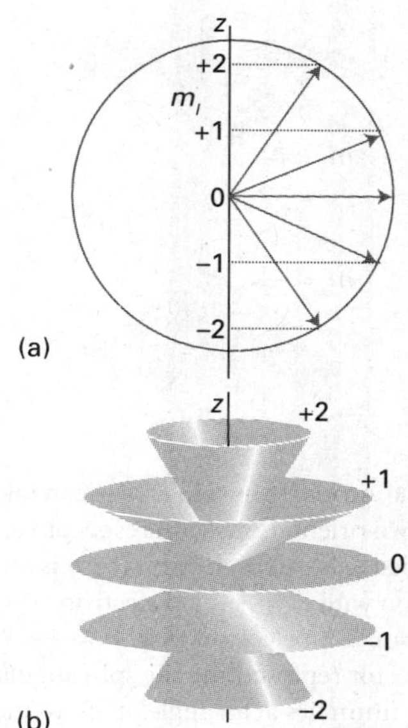

Fig. 8.37 (a) A summary of Fig. 8.35. However, because the azimuthal angle of the vector around the z-axis is indeterminate, a better representation is as in (b), where each vector lies at an unspecified azimuthal angle on its cone.

8.8 Spin

Key points **Spin is an intrinsic angular momentum of a fundamental particle. A fermion is a particle with a half-integral spin quantum number; a boson is a particle with an integral spin quantum number. For an electron, the spin quantum number is $s = \frac{1}{2}$. The spin magnetic quantum number is $m_s = s, s-1, \ldots, -s$; for an electron, $m_s = \pm\frac{1}{2}$.**

Stern and Gerlach observed two bands of Ag atoms in their experiment. This observation seems to conflict with one of the conclusions from quantum mechanics, because an angular momentum l gives rise to $2l + 1$ orientations, which is equal to 2 only if $l = \frac{1}{2}$, contrary to the conclusion that l must be an integer. The conflict was resolved by the suggestion that the angular momentum they were observing was not due to orbital angular momentum (the motion of an electron around the atomic nucleus) but arose instead from the motion of the electron about its own axis. This intrinsic angular momentum of the electron is called its **spin**. The explanation of the existence of spin emerged when Dirac combined quantum mechanics with special relativity and established the theory of relativistic quantum mechanics.

The spin of an electron about its own axis does not have to satisfy the same boundary conditions as those for a particle circulating around a central point, so the quantum number for spin angular momentum is subject to different restrictions. To distinguish this spin angular momentum from orbital angular momentum we use the **spin quantum number** s (in place of l; like l, s is a non-negative number) and m_s, the **spin magnetic quantum number**, for the projection on the z-axis. The magnitude of the spin angular momentum is $\{s(s+1)\}^{1/2}\hbar$ and the component $m_s\hbar$ is restricted to the $2s + 1$ values with

$$m_s = s, s-1, \ldots -s \tag{8.57}$$

The detailed analysis of the spin of a particle is sophisticated and shows that the property should not be taken to be an actual spinning motion. It is better to regard 'spin' as an intrinsic property like mass and charge. However, the picture of an actual spinning motion can be very useful when used with care. For an electron it turns out that only one value of s is allowed, namely, $s = \frac{1}{2}$, corresponding to an angular momentum of magnitude $(\frac{3}{4})^{1/2}\hbar = 0.866\hbar$. This spin angular momentum is an intrinsic

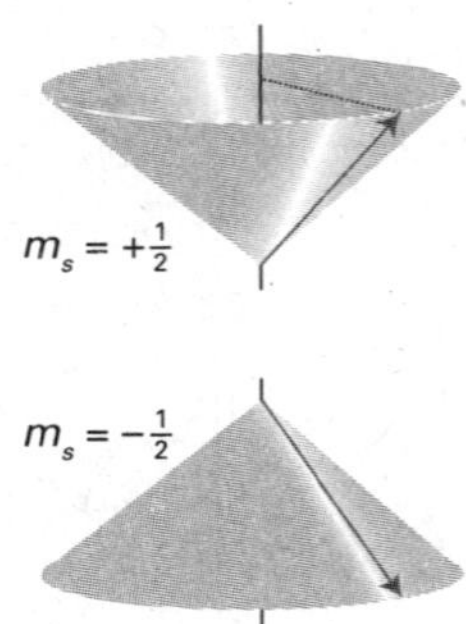

Fig. 8.38 An electron spin ($s = \frac{1}{2}$) can take only two orientations with respect to a specified axis. An α electron (top) is an electron with $m_s = +\frac{1}{2}$; a β electron (bottom) is an electron with $m_s = -\frac{1}{2}$. The vector representing the spin angular momentum lies at an angle of 55° to the z-axis (more precisely, the half-angle of the cones is $\arccos(\frac{1}{3}^{1/2})$).

property of the electron, like its rest mass and its charge, and every electron has exactly the same value: the magnitude of the spin angular momentum of an electron cannot be changed. The spin may lie in $2s + 1 = 2$ different orientations (Fig. 8.38). One orientation corresponds to $m_s = +\frac{1}{2}$ (this state is often denoted α or ↑); the other orientation corresponds to $m_s = -\frac{1}{2}$ (this state is denoted β or ↓).

The outcome of the Stern–Gerlach experiment can now be explained if we suppose that each Ag atom possesses an angular momentum due to the spin of a single electron, because the two bands of atoms then correspond to the two spin orientations. Why the atoms behave like this is explained in Chapter 9 (but it is already probably familiar from introductory chemistry that the ground-state configuration of a silver atom is $[Kr]4d^{10}5s^1$, a single unpaired electron outside a closed shell).

Like the electron, other elementary particles have characteristic spin. For example, protons and neutrons are spin-$\frac{1}{2}$ particles (that is, $s = \frac{1}{2}$) and invariably spin with angular momentum $(\frac{3}{4})^{1/2}\hbar = 0.866\hbar$. Because the masses of a proton and a neutron are so much greater than the mass of an electron, yet they all have the same spin angular momentum, the classical picture would be of these two particles spinning much more slowly than an electron. Some elementary particles have $s = 1$, and so have an intrinsic angular momentum of magnitude $2^{1/2}\hbar$. Some mesons are spin-1 particles (as are some atomic nuclei), but for our purposes the most important spin-1 particle is the photon. From the discussion in this chapter, we see that the photon has zero rest mass, zero charge, an energy $h\nu$, a linear momentum h/λ or $h\nu/c$, an intrinsic angular momentum of $2^{1/2}\hbar$, and travels at the speed c. We shall see the importance of photon spin in the next chapter.

Particles with half-integral spin are called **fermions** and those with integral spin (including 0) are called **bosons**. Thus, electrons and protons are fermions and photons are bosons. It is a very deep feature of nature that all the elementary particles that constitute matter are fermions, whereas the elementary particles that are responsible for the forces that bind fermions together are all bosons. Photons, for example, transmit the electromagnetic force that binds together electrically charged particles. Matter, therefore, is an assembly of fermions held together by forces conveyed by bosons.

The properties of angular momentum that we have developed are set out in Table 8.3. As mentioned there, when we use the quantum numbers l and m_l we shall mean orbital angular momentum (circulation in space). When we use s and m_s we shall mean spin angular momentum (intrinsic angular momentum). When we use j and m_j we shall mean either (or, in some contexts to be described in Chapter 9, a combination of orbital and spin momenta).

Table 8.3 Properties of the angular momentum of an electron

Quantum number	Symbol†	Values	Specifies
Orbital angular momentum	l	0, 1, 2, . . .‡	Magnitude, $\{l(l+1)\}^{1/2}\hbar$
Magnetic	m_l	$l, l-1, \ldots, -l$	Component on z-axis, $m_l\hbar$
Spin	s	$\frac{1}{2}$	Magnitude, $\{s(s+1)\}^{1/2}\hbar$
Spin magnetic	m_s	$\pm\frac{1}{2}$	Component on z-axis, $m_s\hbar$
Total*	j	$l+s, l+s-1, \ldots, \lvert l-s\rvert$	Magnitude, $\{j(j+1)\}^{1/2}\hbar$
Total magnetic	m_j	$j, j-1, \ldots, -j$	Component on z-axis, $m_j\hbar$

* To combine two angular momenta, use the Clebsch–Gordan series (see Section 9.10a):

$$j = j_1 + j_2, j_1 + j_2 - 2, \ldots, |j_1 - j_2|$$

† For many-electron systems, the quantum numbers are designated by upper-case letters (L, M_L, S, M_S, etc.).

‡ Note that the quantum numbers for magnitude (l, s, j, etc.) are never negative.

Checklist of key equations

Property	Equation	Comment
Wavefunctions of a free particle in one dimension	$\psi_k = Ae^{ikx} + Be^{-ikx}$	k continuously variable
Energies of a free particle	$E_k = k^2\hbar^2/2m$	k continuously variable
Wavefunctions of a particle in a one-dimensional box of length L	$\psi_n(x) = (2/L)^{1/2}\sin(n\pi x/L)$	$n = 1, 2, \ldots$
Energies of a particle in a one-dimensional box of length L	$E_n = n^2h^2/8mL^2$	$n = 1, 2, \ldots$
Wavefunctions of a particle in a two-dimensional box	$\psi_{n_1,n_2}(x,y) = \{2/(L_1L_2)^{1/2}\}\sin(n_1\pi x/L_1)\sin(n_2\pi y/L_2)$	$n_1 = 1, 2, \ldots, n_2 = 1, 2, \ldots$ $0 \le x \le L_1, 0 \le y \le L_2$
Energies of a particle in a two-dimensional box	$E_{n_1,n_2} = (n_1^2/L_1^2 + n_2^2/L_2^2)(h^2/8m)$	$n_1 = 1, 2, \ldots, n_2 = 1, 2, \ldots$ $0 \le x \le L_1, 0 \le y \le L_2$
Wavefunctions of a harmonic oscillator	$\psi_v(x) = N_vH_v(y)e^{-y^2/2}$, $y = x/\alpha$, $\alpha = (\hbar^2/mk)^{1/4}$	The Hermite polynomials $H_v(y)$ are listed in Table 8.1
Energies of a harmonic oscillator	$E_v = (v + 1/2)\hbar\omega$, $\omega = (k_f/m)^{1/2}$	$v = 0, 1, 2, \ldots$
Wavefunctions of a particle on a ring	$\psi_{m_l}(\phi) = (1/2\pi)^{1/2}e^{im_l\phi}$	$m_l = 0, \pm1, \pm2, \ldots$
Energies of a particle on a ring	$E = m_l^2\hbar^2/2I$	$I = mr^2$ and $m_l = 0, \pm1, \pm2, \ldots$
Angular momentum of a particle on a ring	$J_z = m_l\hbar$	$m_l = 0, \pm1, \pm2, \ldots$
Wavefunctions of a particle on a sphere	Spherical harmonics: $Y_{l,m_l}(\theta,\phi)$	See Table 8.2
Energies of a particle on a sphere	$E = l(l+1)\hbar^2/2I$	$l = 0, 1, 2, \ldots$
Magnitude of the angular momentum of a particle on a sphere	$\{l(l+1)\}^{1/2}\hbar$	$l = 0, 1, 2, \ldots$
z-component of the angular momentum of a particle on a sphere	$m_l\hbar$	$m_l = l, l-1, \ldots, -l$

Discussion questions

8.1 Discuss the physical origin of quantization energy for a particle confined to moving inside a one-dimensional box or on a ring.

8.2 In what ways does the quantum mechanical description of a harmonic oscillator merge with its classical description at high quantum numbers?

8.3 Define, justify, and provide examples of zero-point energy.

8.4 Discuss the physical origins of quantum mechanical tunnelling. Why is tunnelling more likely to contribute to the mechanisms of electron transfer and proton transfer processes than to mechanisms of group transfer reactions, such as AB + C → A + BC (where A, B, and C are large molecular groups)?

8.5 Distinguish between a fermion and a boson. Provide examples of each type of particle. What are the consequences of the difference between the types of particles?

8.6 Describe the features that stem from nanometre-scale dimensions that are not found in macroscopic objects.

Exercises

8.1(a) Calculate the energy separations in joules, kilojoules per mole, electronvolts, and reciprocal centimetres between the levels (a) $n = 2$ and $n = 1$, (b) $n = 6$ and $n = 5$ of an electron in a box of length 1.0 nm.

8.1(b) Calculate the energy separations in joules, kilojoules per mole, electronvolts, and reciprocal centimetres between the levels (a) $n = 3$ and $n = 1$, (b) $n = 7$ and $n = 6$ of an electron in a box of length 1.50 nm.

8.2(a) Calculate the probability that a particle will be found between $0.49L$ and $0.51L$ in a box of length L when it has (a) $n = 1$, (b) $n = 2$. Take the wavefunction to be a constant in this range.

8.2(b) Calculate the probability that a particle will be found between $0.65L$ and $0.67L$ in a box of length L when it has (a) $n = 1$, (b) $n = 2$. Take the wavefunction to be a constant in this range.

8.3(a) Calculate the expectation values of p and p^2 for a particle in the state $n = 1$ in a square-well potential.

8.3(b) Calculate the expectation values of p and p^2 for a particle in the state $n = 2$ in a square-well potential.

8.4(a) Calculate the expectation values of x and x^2 for a particle in the state $n = 1$ in a square-well potential.

8.4(b) Calculate the expectation values of x and x^2 for a particle in the state $n = 2$ in a square-well potential.

8.5(a) An electron is confined to a a square well of length L. What would be the length of the box such that the zero-point energy of the electron is equal to its rest mass energy, $m_e c^2$? Express your answer in terms of the parameter $\lambda_C = h/m_e c$, the 'Compton wavelength' of the electron.

8.5(b) Repeat Exercise 8.5a for a general particle of mass m in a cubic box.

8.6(a) What are the most likely locations of a particle in a box of length L in the state $n = 3$?

8.6(b) What are the most likely locations of a particle in a box of length L in the state $n = 5$?

8.7(a) Calculate the percentage change in a given energy level of a particle in a one-dimensional box when the length of the box is increased by 10 per cent.

8.7(b) Calculate the percentage change in a given energy level of a particle in a cubic box when the length of the edge of the cube is decreased by 10 per cent in each direction.

8.8(a) What is the value of n of a particle in a one-dimensional box such that the separation between neighbouring levels is equal to the energy of thermal motion ($\frac{1}{2}kT$).

8.8(b) A nitrogen molecule is confined in a cubic box of volume 1.00 m^3. Assuming that the molecule has an energy equal to $\frac{3}{2}kT$ at $T = 300$ K, what is the value of $n = (n_x^2 + n_y^2 + n_z^2)^{1/2}$ for this molecule? What is the energy separation between the levels n and $n + 1$? What is its de Broglie wavelength?

8.9(a) Calculate the zero-point energy of a harmonic oscillator consisting of a particle of mass 2.33×10^{-26} kg and force constant 155 N m^{-1}.

8.9(b) Calculate the zero-point energy of a harmonic oscillator consisting of a particle of mass 5.16×10^{-26} kg and force constant 285 N m^{-1}.

8.10(a) For a certain harmonic oscillator of effective mass 1.33×10^{-25} kg, the difference in adjacent energy levels is 4.82 zJ. Calculate the force constant of the oscillator.

8.10(b) For a certain harmonic oscillator of effective mass 2.88×10^{-25} kg, the difference in adjacent energy levels is 3.17 zJ. Calculate the force constant of the oscillator.

8.11(a) Calculate the wavelength of a photon needed to excite a transition between neighbouring energy levels of a harmonic oscillator of effective mass equal to that of a proton ($1.0078m_u$) and force constant 855 N m^{-1}.

8.11(b) Calculate the wavelength of a photon needed to excite a transition between neighbouring energy levels of a harmonic oscillator of effective mass equal to that of an oxygen atom ($15.9949m_u$) and force constant 544 N m^{-1}.

8.12(a) The vibrational frequency of H_2 is 131.9 THz. What is the vibrational frequency of D_2 ($D = {}^2H$)?

8.12(b) The vibrational frequency of H_2 is 131.9 THz. What is the vibrational frequency of T_2 ($T = {}^3H$)?

8.13(a) Calculate the minimum excitation energies of (a) a pendulum of length 1.0 m on the surface of the Earth, (b) the balance-wheel of a clockwork watch ($\nu = 5$ Hz).

8.13(b) Calculate the minimum excitation energies of (a) the 33 kHz quartz crystal of a watch, (b) the bond between two O atoms in O_2, for which $k_f = 1177$ N m^{-1}.

8.14(a) Confirm that the wavefunction for the ground state of a one-dimensional linear harmonic oscillator given in Table 8.1 is a solution of the Schrödinger equation for the oscillator and that its energy is $\frac{1}{2}\hbar\omega$.

8.14(b) Confirm that the wavefunction for the first excited state of a one-dimensional linear harmonic oscillator given in Table 8.1 is a solution of the Schrödinger equation for the oscillator and that its energy is $\frac{3}{2}\hbar\omega$.

8.15(a) Locate the nodes of the harmonic oscillator wavefunction with $v = 4$.

8.15(b) Locate the nodes of the harmonic oscillator wavefunction with $v = 5$.

8.16(a) What are the most probable displacements of a harmonic oscillator with $v = 1$?

8.16(b) What are the most probable displacements of a harmonic oscillator with $v = 3$?

8.17(a) Assuming that the vibrations of a ${}^{35}Cl_2$ molecule are equivalent to those of a harmonic oscillator with a force constant $k = 329$ N m^{-1}, what is the zero-point energy of vibration of this molecule? The effective mass of a homonuclear diatomic molecule is half its total mass, and $m({}^{35}Cl) = 34.9688m_u$.

8.17(b) Assuming that the vibrations of a ${}^{14}N_2$ molecule are equivalent to those of a harmonic oscillator with a force constant $k = 2293.8$ N m^{-1}, what is the zero-point energy of vibration of this molecule? The effective mass of a homonuclear diatomic molecule is half its total mass, and $m({}^{14}N) = 14.0031m_u$.

8.18(a) The wavefunction, $\psi(\phi)$, for the motion of a particle in a ring is of the form $\psi = Ne^{im\phi}$. Determine the normalization constant, N.

8.18(b) Confirm that wavefunctions for a particle in a ring with different values of the quantum number m_l are mutually orthogonal.

8.19(a) Calculate the minimum excitation energy of a proton constrained to rotate in a circle of radius 100 pm around a fixed point.

8.19(b) Calculate the value of $|m_l|$ for the system described in the preceding exercise corresponding to a rotational energy equal to the classical average energy at 25°C (which is equal to $\frac{1}{2}kT$).

8.20(a) Estimate the rotational quantum number of a bicycle wheel of diameter 60 cm and mass 1.0 kg when the bicycle is travelling at 20 km h^{-1}.

8.20(b) The mass of a vinyl gramophone record is 130 g and its diameter is 30 cm. Given that the moment of inertia of a solid uniform disc of mass m and radius r is $I = \frac{1}{2}mr^2$, estimate the rotational quantum number when the disc is rotating at 33 r.p.m.

8.21(a) The moment of inertia of a CH_4 molecule is 5.27×10^{-47} kg m^2. What is the minimum energy needed to start it rotating?

8.21(b) The moment of inertia of an SF_6 molecule is 3.07×10^{-45} kg m^2. What is the minimum energy needed to start it rotating?

8.22(a) Use the data in Exercise 8.21a to calculate the energy needed to excite a CH_4 molecule from a state with $l = 1$ to a state with $l = 2$.

8.22(b) Use the data in Exercise 8.21b to calculate the energy needed to excite an SF_6 molecule from a state with $l = 2$ to a state with $l = 3$.

8.23(a) What is the magnitude of the angular momentum of a CH_4 molecule when it is rotating with its minimum energy?

8.23(b) What is the magnitude of the angular momentum of an SF_6 molecule when it is rotating with its minimum energy?

8.24(a) Draw scale vector diagrams to represent the states (a) $s = \frac{1}{2}$, $m_s = +\frac{1}{2}$, (b) $l = 1$, $m_l = +1$, (c) $l = 2$, $m_l = 0$.

8.24(b) Draw the vector diagram for all the permitted states of a particle with $l = 6$.

Problems*

Numerical problems

8.1 Calculate the separation between the two lowest levels for an O_2 molecule in a one-dimensional container of length 5.0 cm. At what value of n does the energy of the molecule reach $\frac{1}{2}kT$ at 300 K, and what is the separation of this level from the one immediately below?

8.2 The mass to use in the expression for the vibrational frequency of a diatomic molecule is the effective mass $\mu = m_A m_B/(m_A + m_B)$, where m_A and m_B are the masses of the individual atoms. The following data on the infrared absorption wavenumbers (wavenumbers in cm^{-1}) of molecules are taken from *Spectra of diatomic molecules*, G. Herzberg, van Nostrand (1950):

$H^{35}Cl$	$H^{81}Br$	HI	CO	NO
2990	2650	2310	2170	1904

Calculate the force constants of the bonds and arrange them in order of increasing stiffness.

8.3 The rotation of an $^1H^{127}I$ molecule can be pictured as the orbital motion of an H atom at a distance 160 pm from a stationary I atom. (This picture is quite good; to be precise, both atoms rotate around their common centre of mass, which is very close to the I nucleus.) Suppose that the molecule rotates only in a plane. Calculate the energy needed to excite the molecule into rotation. What, apart from 0, is the minimum angular momentum of the molecule?

8.4 Calculate the energies of the first four rotational levels of $^1H^{127}I$ free to rotate in three dimensions, using for its moment of inertia $I = \mu R^2$, with $\mu = m_H m_I/(m_H + m_I)$ and $R = 160$ pm.

8.5 Use mathematical software to construct a wavepacket for a particle rotating on a circle of the form

$$\Psi(\phi,t) = \sum_{m_l=0}^{m_{l,\max}} c_{m_l} e^{i(m_l\phi - E_{m_l}t/\hbar)} \qquad E_{m_l} = m_l^2\hbar^2/2I$$

with coefficients c of your choice (for example, all equal). Explore how the wavepacket migrates on the ring but spreads with time.

8.6 Use mathematical software to construct a harmonic oscillator wavepacket of the form

$$\Psi(x,t) = \sum_{v=0}^{N} c_v \psi_v(x) e^{iE_v t/\hbar}$$

where the wavefunctions and energies are those of a harmonic oscillator and with coefficients c of your choice (for example, all equal). Explore how the wavepacket oscillates to and fro.

Theoretical problems

8.7 Suppose that 1.0 mol perfect gas molecules all occupy the lowest energy level of a cubic box. How much work must be done to change the volume of the box by ΔV? Would the work be different if the molecules all occupied a state $n \neq 1$? What is the relevance of this discussion to the expression for the expansion work discussed in Chapter 2? Can you identify a distinction between adiabatic and isothermal expansion?

8.8 Derive eqn 8.19a, the expression for the transmission probability, and show that when $\kappa L \gg 1$ it reduces to eqn 8.19b.

8.9‡ Consider the one-dimensional space in which a particle can experience one of three potentials depending upon its position. They are: $V = 0$ for $-\infty < x \leq 0$, 0, $V = V_2$ for $0 \leq x \leq L$, and $V = V_3$ for $L \leq x < \infty$. The particle wavefunction is to have both a component e^{ik_1x} that is incident upon the barrier V_2 and a reflected component e^{-ik_1x} in region 1 ($-\infty < x \leq 0$). In region 3 the wavefunction has only a forward component, e^{ik_3x}, which represents a particle that has traversed the barrier. The energy of the particle, E, is somewhere in the range of the $V_2 > E > V_3$. The transmission probability, T, is the ratio of the square modulus of the region 3 amplitude to the square modulus of the incident amplitude. (a) Base your calculation on the continuity of the amplitudes and the slope of the wavefunction at the locations of the zone boundaries and derive a general equation for T. (b) Show that the general equation for T reduces to eqn 8.19b in the high, wide barrier limit when $V_1 = V_3 = 0$. (c) Draw a graph of the probability of proton tunnelling when $V_3 = 0$, $L = 50$ pm, and $E = 10$ kJ mol^{-1} in the barrier range $E < V_2 < 2E$.

8.10 The wavefunction inside a long barrier of height V is $\psi = Ne^{-\kappa x}$. Calculate (a) the probability that the particle is inside the barrier and (b) the average penetration depth of the particle into the barrier.

8.11 Confirm that a function of the form e^{-gx^2} is a solution of the Schrödinger equation for the ground state of a harmonic oscillator and find an expression for g in terms of the mass and force constant of the oscillator.

8.12 Calculate the mean kinetic energy of a harmonic oscillator by using the relations in Table 8.1.

8.13 Calculate the values of $\langle x^3\rangle$ and $\langle x^4\rangle$ for a harmonic oscillator by using the relations in Table 8.1.

8.14 Determine the values of $\Delta x = (\langle x^2\rangle - \langle x\rangle^2)^{1/2}$ and $\Delta p = (\langle p^2\rangle - \langle p\rangle^2)^{1/2}$ for (a) a particle in a box of length L and (b) a harmonic oscillator. Discuss these quantities with reference to the uncertainty principle.

8.15 According to classical mechanics, the turning point, x_{tp}, of an oscillator occurs when its kinetic energy is zero, which is when its potential energy $\frac{1}{2}kx^2$ is equal to its total energy E. This equality occurs when

$$x_{tp}^2 = \frac{2E}{k} \quad \text{or} \quad x_{tp} = \pm\left(\frac{2E}{k}\right)^{1/2}$$

with E given by eqn 8.24. The probability of finding the oscillator stretched beyond a displacement x_{tp} is the sum of the probabilities $\psi^2 dx$ of finding it in any of the intervals dx lying between x_{tp} and infinity:

$$P = \int_{x_{tp}}^{\infty} \psi_v^2 \, dx$$

The variable of integration is best expressed in terms of $y = x/\alpha$ with $\alpha = (\hbar^2/mk)^{1/4}$. (a) Show that the turning points lie at $y_{tp} = \pm(2v+1)^{1/2}$. (b) Go on to show that for the state of lowest energy ($v = 0$), $y_{tp} = 1$ and the probability is $P = \frac{1}{2}(1 - \text{erf}1)$, where the *error function*, erf z, is defined as

$$\text{erf}\, z = 1 - \frac{2}{\pi^{1/2}} \int_z^{\infty} e^{-y^2} dy$$

The values of this function are tabulated and available in mathematical software packages.

8.16 Extend the calculation in Problem 8.15 by using mathematical software to calculate the probability that a harmonic oscillator will be found outside

* Problems denoted with the symbol ‡ were supplied by Charles Trapp, Carmen Giunta, and Marshall Cady.

the classically allowed displacements for general v and plot the probability as a function of v.

8.17 The intensities of spectroscopic transitions between the vibrational states of a molecule are proportional to the square of the integral $\int \psi_{v'} x \psi_v \, dx$ over all space. Use the relations between Hermite polynomials given in Table 8.1 to show that the only permitted transitions are those for which $v' = v \pm 1$ and evaluate the integral in these cases.

8.18 The potential energy of the rotation of one CH_3 group relative to its neighbour in ethane can be expressed as $V(\phi) = V_0 \cos 3\phi$. Show that for small displacements the motion of the group is harmonic and calculate the energy of excitation from $v = 0$ to $v = 1$. What do you expect to happen to the energy levels and wavefunctions as the excitation increases?

8.19 Show that, whatever superposition of harmonic oscillator states is used to construct a wavepacket, it is localized at the same place at the times $0, T, 2T, \ldots$, where T is the classical period of the oscillator.

8.20 Use the virial theorem to obtain an expression for the relation between the mean kinetic and potential energies of an electron in a hydrogen atom.

8.21 Evaluate the z-component of the angular momentum and the kinetic energy of a particle on a ring that is described by the (unnormalized) wavefunctions (a) $e^{i\phi}$, (b) $e^{-2i\phi}$, (c) $\cos \phi$, and (d) $(\cos \chi)e^{i\phi} + (\sin \chi)e^{-i\phi}$.

8.22 Is the Schrödinger equation for a particle on an elliptical ring of semimajor axes a and b separable? *Hint.* Although r varies with angle ϕ, the two are related by $r^2 = a^2\sin^2\phi + b^2\cos^2\phi$.

8.23 Confirm that the spherical harmonics (a) $Y_{0,0}$, (b) $Y_{2,-1}$, and (c) $Y_{3,+3}$ satisfy the Schrödinger equation for a particle free to rotate in three dimensions, and find its energy and angular momentum in each case.

8.24 Confirm that $Y_{3,+3}$ is normalized to 1. (The integration required is over the surface of a sphere.)

8.25 Derive an expression in terms of l and m_l for the half-angle of the apex of the cone used to represent an angular momentum according to the vector model. Evaluate the expression for an α spin. Show that the minimum possible angle approaches 0 as $l \to \infty$.

8.26 Show that the function $f = \cos ax \cos by \cos cz$ is an eigenfunction of ∇^2, and determine its eigenvalue.

8.27 Derive (in Cartesian coordinates) the quantum mechanical operators for the three components of angular momentum starting from the classical definition of angular momentum, $\boldsymbol{l} = \boldsymbol{r} \times \boldsymbol{p}$. Show that any two of the components do not mutually commute, and find their commutator.

8.28 Starting from the operator $\hat{l}_z = xp_y - yp_x$, prove that in spherical polar coordinates $\hat{l}_z = -i\hbar\partial/\partial\phi$.

8.29 Show that the commutator $[l^2, l_z] = 0$, and then, without further calculation, justify the remark that $[l^2, l_q] = 0$ for all $q = x, y$, and z.

8.30‡ A particle is confined to move in a one-dimensional box of length L. (a) If the particle is classical, show that the average value of x is $\frac{1}{2}L$ and that the root-mean square value is $L/3^{1/2}$. (b) Show that for large values of n, a quantum particle approaches the classical values. This result is an example of the correspondence principle, which states that, for very large values of the quantum numbers, the predictions of quantum mechanics approach those of classical mechanics.

Applications: to biology and nanotechnology

8.31 When β-carotene (1) is oxidized *in vivo*, it breaks in half and forms two molecules of retinal (vitamin A), which is a precursor to the pigment in the retina responsible for vision (see *Impact I13.1*). The conjugated system of retinal consists of 11 C atoms and one O atom. In the ground state of retinal, each level up to $n = 6$ is occupied by two electrons. Assuming an average internuclear distance of 140 pm, calculate (a) the separation in energy between the ground state and the first excited state in which one electron occupies the state with $n = 7$, and (b) the frequency of the radiation required to produce a transition between these two states. (c) Using your results, choose among the words in parentheses to generate a rule for the prediction of frequency shifts in the absorption spectra of linear polyenes:

1 β-Carotene

> The absorption spectrum of a linear polyene shifts to (higher/lower) frequency as the number of conjugated atoms (increases/decreases).

8.32 Many biological electron transfer reactions, such as those associated with biological energy conversion, may be visualized as arising from electron tunnelling between protein-bound co-factors, such as cytochromes, quinones, flavins, and chlorophylls. This tunnelling occurs over distances that are often greater than 1.0 nm, with sections of protein separating electron donor from acceptor. For a specific combination of donor and acceptor, the rate of electron tunnelling is proportional to the transmission probability, with $\kappa \approx 7\ \mathrm{nm}^{-1}$ (eqn 8.19). By what factor does the rate of electron tunnelling between two co-factors increase as the distance between them changes from 2.0 nm to 1.0 nm?

8.33 Carbon monoxide binds strongly to the Fe^{2+} ion of the haem group of the protein myoglobin. Estimate the vibrational frequency of CO bound to myoglobin by using the data in Problem 8.2 and by making the following assumptions: the atom that binds to the haem group is immobilized, the protein is infinitely more massive than either the C or O atom, the C atom binds to the Fe^{2+} ion, and binding of CO to the protein does not alter the force constant of the C≡O bond.

8.34 Of the four assumptions made in Problem 8.33, the last two are questionable. Suppose that the first two assumptions are still reasonable and that you have at your disposal a supply of myoglobin, a suitable buffer in which to suspend the protein, $^{12}C^{16}O$, $^{13}C^{16}O$, $^{12}C^{18}O$, $^{13}C^{18}O$, and an infrared spectrometer. Assuming that isotopic substitution does not affect the force constant of the C≡O bond, describe a set of experiments that: (a) proves which atom, C or O, binds to the haem group of myoglobin, and (b) allows for the determination of the force constant of the C≡O bond for myoglobin-bound carbon monoxide.

8.35 The particle on a ring is a useful model for the motion of electrons around the porphine ring (2), the conjugated macrocycle that forms the structural basis of the haem group and the chlorophylls. We may treat the group as a circular ring of radius 440 pm, with 22 electrons in the conjugated system moving along the perimeter of the ring. In the ground state of the molecule each state is occupied by two electrons. (a) Calculate the energy and angular momentum of an electron in the highest occupied level. (b) Calculate the frequency of radiation that can induce a transition between the highest occupied and lowest unoccupied levels.

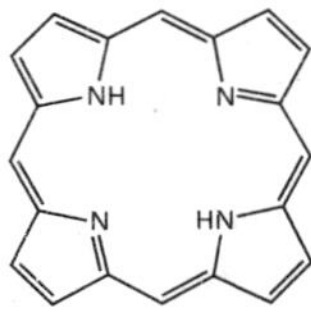

2 Porphine (free base form)

8.36 When in Chapter 18 we come to study macromolecules, such as synthetic polymers, proteins, and nucleic acids, we shall see that one conformation is that of a random coil. For a one-dimensional random coil of N units, the restoring force at small displacements and at a temperature T is

$$F = -\frac{kT}{2l}\ln\left(\frac{N+n}{N-n}\right)$$

where l is the length of each monomer unit and nl is the distance between the ends of the chain. Show that for small extensions ($n \ll N$) the restoring force is proportional to n and therefore the coil undergoes harmonic oscillation with force constant kT/Nl^2. Suppose that the mass to use for the vibrating chain is its total mass Nm, where m is the mass of one monomer unit, and deduce the root mean square separation of the ends of the chain due to quantum fluctuations in its vibrational ground state.

8.37 Here we explore further the idea introduced in *Impact I8.1* that quantum mechanical effects need to be invoked in the description of the electronic properties of metallic nanocrystals, here modelled as three-dimensional boxes. (a) Set up the Schrödinger equation for a particle of mass m in a three-dimensional rectangular box with sides L_1, L_2, and L_3. Show that the Schrödinger equation is separable. (b) Show that the wavefunction and the energy are defined by three quantum numbers. (c) Specialize the result from part (b) to an electron moving in a cubic box of side $L = 5$ nm and draw an energy diagram resembling Fig. 8.2 and showing the first 15 energy levels. Note that each energy level may consist of degenerate energy states. (d) Compare the energy level diagram from part (c) with the energy level diagram for an electron in a one-dimensional box of length $L = 5$ nm. Are the energy levels more or less sparsely distributed in the cubic box than in the one-dimensional box?

8.38 We remarked in *Impact I8.1* that the particle *in* a sphere is a reasonable starting point for the discussion of the electronic properties of spherical metal nanoparticles. Here, we justify the expression for the energy levels with $l = 0$. (a) The Hamiltonian for a particle free to move inside a sphere of radius R is

$$\hat{H} = -\frac{\hbar}{2m}\nabla^2$$

Show that the Schrödinger equation is separable into radial and angular components. That is, begin by writing $\psi(r,\theta,\phi) = u(r)Y(\theta,\phi)$, where $u(r)$ depends only on the distance of the particle away from the centre of the sphere, and $Y(\theta,\phi)$ is a spherical harmonic. Then show that the Schrödinger equation can be separated into two equations, one for u, the radial equation, and the other for Y, the angular equation:

$$-\frac{\hbar^2}{2m}\left(\frac{d^2u(r)}{dr^2} + \frac{2}{r}\frac{du(r)}{dr}\right) + \frac{l(l+1)\hbar^2}{2mr^2}u(r) = Eu(r)$$

$$\Lambda^2 Y = -l(l+1)Y$$

(b) Consider the case $l = 0$. Show by differentiation that the solution of the radial equation has the form

$$u(r) = (2\pi R)^{-1/2}\frac{\sin(n\pi r/R)}{r}$$

(c) Now go on to show that the allowed energies are given by:

$$E_n = \frac{n^2h^2}{8mR^2}$$

which is the expression given in *Impact I8.1* after substituting m_e for m.

8.39 The forces measured by atomic force microscopy (AFM) arise primarily from interactions between electrons of the stylus and on the surface. To get an idea of the magnitudes of these forces, calculate the force acting between two electrons separated by 2.0 nm. *Hint.* The Coulombic potential energy of a charge Q_1 at a distance r from another charge Q_2 is $V = Q_1Q_2/4\pi\varepsilon_0 r$, where $\varepsilon_0 = 8.854 \times 10^{-12}$ C^2 J^{-1} m^{-1} is the vacuum permittivity. To calculate the force between the electrons, note that $F = -dV/dr$.

MATHEMATICAL BACKGROUND 4

Differential equations

A **differential equation** is a relation between a function and its derivatives, as in

$$a\frac{d^2f}{dx^2}+b\frac{df}{dx}+cf=0 \qquad \text{(MB4.1)}$$

where f is a function of the variable x and the factors a, b, c may be either constants or functions of x. If the unknown function depends on only one variable, as in this example, the equation is called an **ordinary differential equation;** if it depends on more than one variable, as in

$$a\frac{\partial^2f}{\partial x^2}+b\frac{\partial^2f}{\partial y^2}+cf=0 \qquad \text{(MB4.2)}$$

it is called a **partial differential equation.** Here, f is a function of x and y, and the factors a, b, c may be either constants or functions of both variables. Note the change in symbol from d to ∂ to signify a *partial derivative* (see *Mathematical background 1*).

MB4.1 The structure of differential equations

The **order** of the differential equation is the order of the highest derivative that occurs in it: both examples above are second-order equations. Only rarely in science is a differential equation of order higher than 2 encountered.

A **linear differential equation** is one for which, if f is a solution, then so is constant $\times f$. Both examples above are linear. If the 0 on the right were replaced by a different number or a function other than f, then they would cease to be linear.

Solving a differential equation means something different from solving an algebraic equation. In the latter case, the solution is a value of the variable x (as in the solution $x = 2$ of the quadratic equation $x^2 - 4 = 0$). The solution of a differential equation is the entire function that satisfies the equation, as in

$$\frac{d^2f}{dx^2}+f=0 \quad \text{has the solution} \quad f=A\sin x+B\cos x \qquad \text{(MB4.3)}$$

with A and B constants. The process of finding a solution of a differential equation is called **integrating** the equation. The solution in eqn MB4.3 is an example of a **general solution** of a differential equation, that is, it is the most general solution of the equation and is expressed in terms of a number of constants (A and B in this case). When the constants are chosen to accord with certain specified **initial conditions** (if one variable is the time) or certain **boundary conditions** (to fulfil certain spatial restrictions on the solutions), we obtain the **particular solution** of the equation. The particular solution of a first-order differential equation requires one such condition; a second-order differential equation requires two.

• **A brief illustration**

If we are informed that $f(0) = 0$, then, because from eqn MB4.3 it follows that $f(0) = B$, we can conclude that $B = 0$. That still leaves A undetermined. If we are also told that $df/dx = 2$ at $x = 0$ (that is, $f'(0) = 2$, where the prime denotes a first derivative), then, because the general solution (but with $B = 0$) implies that $f'(x) = A\cos x$, we know that $f'(0) = A$, and therefore $A = 2$. The particular solution is therefore $f(x) = 2\sin x$. Figure MB4.1 shows a series of particular solutions corresponding to different boundary conditions. •

MB4.2 The solution of ordinary differential equations

The first-order linear differential equation

$$\frac{df}{dx}+af=0 \qquad \text{(MB4.4a)}$$

with a a function of x or a constant can be solved by direct integration. To proceed, we use the fact that the quantities df and dx (called *differentials*) can be treated algebraically like any quantity and rearrange the equation into

$$\frac{df}{f}=-a\,dx \qquad \text{(MB4.4b)}$$

and integrate both sides. For the left-hand side, we use the familiar result $\int dy/y = \ln y + \text{constant}$. After pooling all the constants into a single constant A, we obtain:

$$\ln f=-\int a\,dx+A \qquad \text{(MB4.4c)}$$

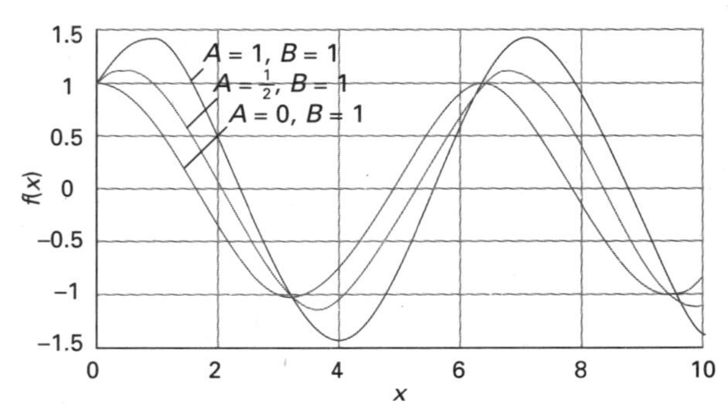

Fig. MB4.1 The solution of the differential equation in eqn MB4.3 with three different boundary conditions (as indicated by the resulting values of the constants A and B).

- **A brief illustration**

Suppose that in eqn MB4.4a the factor $a = 2x$; then the general solution, eqn MB4.4c, is

$$\ln f = -2\int x\mathrm{d}x + A = -x^2 + A$$

(We have absorbed the constant of integration into the constant A.) Therefore

$$f = \mathrm{e}^{A}\mathrm{e}^{-x^2}$$

If we are told that $f(0) = 1$, then we can infer that $A = 0$ and therefore that $f = \mathrm{e}^{-x^2}$. ●

The solution even of first-order differential equations quickly becomes more complicated. A nonlinear first-order equation of the form

$$\frac{\mathrm{d}f}{\mathrm{d}x} + af = b \qquad \text{(MB4.5a)}$$

with a and b functions of x (or constants) has a solution of the form

$$f\mathrm{e}^{\int a\mathrm{d}x} = \int \mathrm{e}^{\int a\mathrm{d}x}\, b\mathrm{d}x + A \qquad \text{(MB4.5b)}$$

as may be verified by differentiation. Mathematical software packages can often perform the required integrations.

Second-order differential equations are in general much more difficult to solve than first-order equations. One powerful approach commonly used to lay siege to second-order differential equations is to express the solution as a power series:

$$f(x) = \sum_{n=0}^{\infty} c_n x^n \qquad \text{(MB4.6)}$$

and then to use the differential equation to find a relation between the coefficients. This approach results, for instance, in the Hermite polynomials that form part of the solution of the Schrödinger equation for the harmonic oscillator (Section 8.5). Many of the second-order differential equations that occur in this text are tabulated in compilations of solutions or can be solved with mathematical software, and the specialized techniques that are needed to establish the form of the solutions may be found in mathematical texts.

MB4.3 The solution of partial differential equations

The only partial differential equations that we need to solve are those that can be separated into two or more ordinary differential equations by the technique known as **separation of variables.** To discover if the differential equation in eqn MB4.2 can be solved by this method we suppose that the full solution can be factored into functions that depend only on x or only on y, and write $f(x,y) = X(x)Y(y)$. At this stage there is no guarantee that the solution can be written in this way. Substituting this trial solution into the equation and recognizing that

$$\frac{\partial^2 XY}{\partial x^2} = Y\frac{\mathrm{d}^2 X}{\mathrm{d}x^2} \qquad \frac{\partial^2 XY}{\partial y^2} = X\frac{\mathrm{d}^2 Y}{\mathrm{d}y^2}$$

we obtain

$$aY\frac{\mathrm{d}^2 X}{\mathrm{d}x^2} + bX\frac{\mathrm{d}^2 Y}{\mathrm{d}y^2} + cXY = 0$$

We are using d instead of ∂ at this stage to denote differentials because each of the functions X and Y depends on one variable, x and y, respectively. Division through by XY turns this equation into

$$\frac{a}{X}\frac{\mathrm{d}^2 X}{\mathrm{d}x^2} + \frac{b}{Y}\frac{\mathrm{d}^2 Y}{\mathrm{d}y^2} + c = 0$$

Now suppose that a is a function only of x, b a function of y, and c a constant. (There are various other possibilities that permit the argument to continue.) Then the first term depends only on x and the second only on y. If x is varied, only the first term can change. But, as the other two terms do not change and the sum of the three terms is a constant (0), even that first term must be a constant. The same is true of the second term. Therefore because each term is equal to a constant, we can write

$$\frac{a}{X}\frac{\mathrm{d}^2 X}{\mathrm{d}x^2} = c_1 \qquad \frac{b}{Y}\frac{\mathrm{d}^2 Y}{\mathrm{d}y^2} = c_2 \qquad \text{with} \qquad c_1 + c_2 = -c$$

We now have two ordinary differential equations to solve by the techniques described in Section MB4.2. An example of this procedure is given in Section 8.2, for a particle in a two-dimensional region.

9 Atomic structure and spectra

We now use the principles of quantum mechanics introduced in the preceding two chapters to describe the internal structures of atoms. We see what experimental information is available from a study of the spectrum of atomic hydrogen. Then we set up the Schrödinger equation for an electron in an atom and separate it into angular and radial parts. The wavefunctions obtained are the 'atomic orbitals' of hydrogenic atoms. Next, we use these hydrogenic atomic orbitals to describe the structures of many-electron atoms. In conjunction with the Pauli exclusion principle, we account for the periodicity of atomic properties and the structure of the periodic table. The spectra of many-electron atoms are more complicated than those of hydrogen, but the same principles apply. We see in the closing sections of the chapter how such spectra are described by using term symbols, and the origin of the finer details of the appearance of spectra.

In this chapter we see how to use quantum mechanics to describe the **electronic structure** of an atom, the arrangement of electrons around a nucleus. The concepts we meet are of central importance for understanding the structures and reactions of atoms and molecules, and hence have extensive chemical applications. We need to distinguish between two types of atoms. A **hydrogenic atom** is a one-electron atom or ion of general atomic number Z; examples of hydrogenic atoms are H, He^+, Li^{2+}, O^{7+}, and even U^{91+}. A **many-electron atom** (or *polyelectronic atom*) is an atom or ion with more than one electron; examples include all neutral atoms other than H. So even He, with only two electrons, is a many-electron atom. Hydrogenic atoms are important because their Schrödinger equations can be solved exactly. They also provide a set of concepts that are used to describe the structures of many-electron atoms and, as we shall see in the next chapter, the structures of molecules too.

The structure and spectra of hydrogenic atoms

When an electric discharge is passed through gaseous hydrogen, the H_2 molecules are dissociated and the energetically excited H atoms that are produced emit light of discrete frequencies, producing a spectrum of a series of 'lines' (Fig. 9.1). The Swedish spectroscopist Johannes Rydberg noted (in 1890) that all the lines are described by the expression

$$\tilde{\nu} = R_{\mathrm{H}}\left(\frac{1}{n_1^2} - \frac{1}{n_2^2}\right) \qquad R_{\mathrm{H}} = 109\,677\ \mathrm{cm}^{-1} \qquad \text{Spectral lines of a hydrogen atom} \qquad (9.1)$$

with $n_1 = 1$ (the *Lyman series*), 2 (the *Balmer series*), and 3 (the *Paschen series*), and that in each case $n_2 = n_1 + 1, n_1 - 2, \ldots$ The constant R_{H} is now called the **Rydberg constant** for the hydrogen atom.

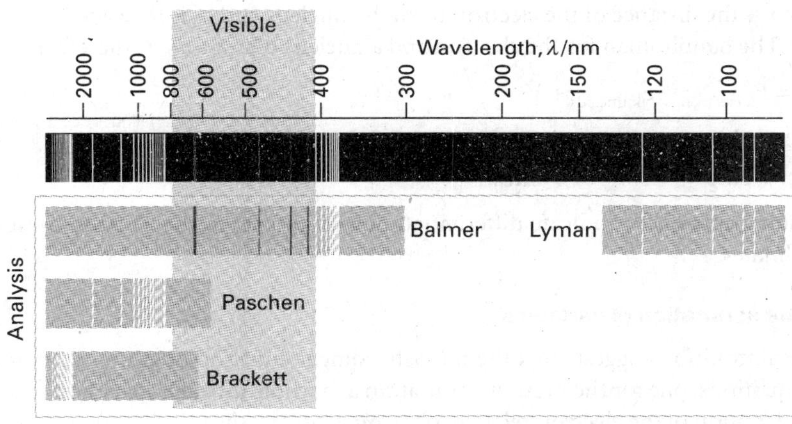

Fig. 9.1 The spectrum of atomic hydrogen. Both the observed spectrum and its resolution into overlapping series are shown. Note that the Balmer series lies in the visible region.

Self-test 9.1 Calculate the shortest wavelength line in the Paschen series.

[821 nm]

The form of eqn 9.1 strongly suggests that the wavenumber of each spectral line can be written as the difference of two **terms**, each of the form

$$T_n = \frac{R_H}{n^2} \tag{9.2}$$

The **Ritz combination principle** states that *the wavenumber of any spectral line (of any atom, not just hydrogenic atoms) is the difference between two terms.* We say that two terms T_1 and T_2 'combine' to produce a spectral line of wavenumber

$$\tilde{\nu} = T_1 - T_2 \tag{9.3}$$

Ritz combination principle

Thus, if each spectroscopic term represents an energy hcT, the difference in energy when the atom undergoes a transition between two terms is $\Delta E = hcT_1 - hcT_2$ and, according to the Bohr frequency condition ($\Delta E = h\nu$, Section 7.1c), the frequency of the radiation emitted is given by $\nu = cT_1 - cT_2$. This expression rearranges into the Ritz formula when expressed in terms of wavenumbers (on division by c; $\tilde{\nu} = \nu/c$). The Ritz combination principle applies to all types of atoms and molecules, but only for hydrogenic atoms do the terms have the simple form (constant)/n^2.

Because spectroscopic observations show that electromagnetic radiation is absorbed and emitted by atoms only at certain wavenumbers, it follows that only certain energy states of atoms are permitted. Our tasks in the first part of this chapter are to determine the origin of this energy quantization, to find the permitted energy levels, and to account for the value of R_H.

9.1 The structure of hydrogenic atoms

Key points (a) The Schrödinger equation for hydrogenic atoms separates into two equations: the solutions of one give the angular variation of the wavefunction and the solution of the other gives its radial dependence. (b) Close to the nucleus the radial wavefunction is proportional to r^l; far from the nucleus all wavefunctions approach zero exponentially.

The Coulomb potential energy of an electron in a hydrogenic atom of atomic number Z and therefore nuclear charge Ze is

$$V = -\frac{Ze^2}{4\pi\varepsilon_0 r} \tag{9.4}$$

where r is the distance of the electron from the nucleus and ε_0 is the vacuum permittivity. The hamiltonian for the electron and a nucleus of mass m_N is therefore

$$\hat{H} = \hat{E}_{k,\text{electron}} + \hat{E}_{k,\text{nucleus}} + \hat{V}$$
$$= -\frac{\hbar^2}{2m_e}\nabla_e^2 - \frac{\hbar^2}{2m_N}\nabla_N^2 - \frac{Ze^2}{4\pi\varepsilon_0 r} \qquad (9.5)$$

Hamiltonian for a hydrogenic atom

The subscripts on ∇^2 indicate differentiation with respect to the electron or nuclear coordinates.

(a) The separation of variables

Physical intuition suggests that the full Schrödinger equation ought to separate into two equations, one for the motion of the atom as a whole through space and the other for the motion of the electron relative to the nucleus. We show in *Further information 9.1* how this separation is achieved, and that the Schrödinger equation for the internal motion of the electron relative to the nucleus is

$$-\frac{\hbar^2}{2\mu}\nabla^2\psi - \frac{Ze^2}{4\pi\varepsilon_0 r}\psi = E\psi \qquad \frac{1}{\mu} = \frac{1}{m_e} + \frac{1}{m_N} \qquad (9.6)$$

Schrödinger equation for a hydrogenic atom

where differentiation is now with respect to the coordinates of the electron relative to the nucleus. The quantity μ is called the **reduced mass**. The reduced mass is very similar to the electron mass because m_N, the mass of the nucleus, is much larger than the mass of an electron, so $1/\mu \approx 1/m_e$ and therefore $\mu \approx m_e$. In all except the most precise work, the reduced mass can be replaced by m_e.

Because the potential energy is centrosymmetric (independent of angle), we can suspect that the equation for the wavefunction is separable into radial and angular components. Therefore, we write

$$\psi(r,\theta,\phi) = R(r)Y(\theta,\phi) \qquad (9.7)$$

and examine whether the Schrödinger equation can be separated into two equations, one for the **radial wavefunction** $R(r)$ and the other for the **angular wavefunction** $Y(\theta,\phi)$. As shown in *Further information 9.1*, the equation does separate, and the equations we have to solve are

$$\Lambda^2 Y = -l(l+1)Y \qquad (9.8a)$$

$$-\frac{\hbar^2}{2\mu}\frac{d^2u}{dr^2} + V_{\text{eff}}u = Eu \qquad (9.8b)$$

where $u(r) = rR(r)$ and

$$V_{\text{eff}} = -\frac{Ze^2}{4\pi\varepsilon_0 r} + \frac{l(l+1)\hbar^2}{2\mu r^2} \qquad (9.8c)$$

Equation 9.8a is the same as the Schrödinger equation for a particle free to move round a central point, and we considered it in Section 8.7. The solutions are the spherical harmonics (Table 8.2), and are specified by the quantum numbers l and m_l. We consider them in more detail shortly. Equation 9.8b is called the **radial wave equation**. The radial wave equation is the description of the motion of a particle of mass μ in a one-dimensional region $0 < r < \infty$ where the potential energy is $V_{\text{eff}}(r)$.

(b) The radial solutions

We can anticipate some features of the shapes of the radial wavefunctions by analysing the form of V_{eff}. The first term in eqn 9.8c is the Coulomb potential energy of the electron in the field of the nucleus. The second term stems from what in classical physics would be called the centrifugal force that arises from the angular momentum of the electron around the nucleus. When $l=0$, the electron has no angular momentum, and the effective potential energy is purely Coulombic and attractive at all radii (Fig. 9.2). When $l \neq 0$, the centrifugal term gives a positive (repulsive) contribution to the effective potential energy. When the electron is close to the nucleus ($r \approx 0$), this repulsive term, which is proportional to $1/r^2$, dominates the attractive Coulombic component, which is proportional to $1/r$, and the net result is an effective repulsion of the electron from the nucleus. The two effective potential energies, the one for $l=0$ and the one for $l \neq 0$, are therefore qualitatively very different close to the nucleus. However, they are similar at large distances because the centrifugal contribution tends to zero more rapidly (as $1/r^2$) than the Coulombic contribution (as $1/r$). Therefore, we can expect the solutions with $l=0$ and $l \neq 0$ to be quite different near the nucleus but similar far away from it. We show in the following *Justification* the following two important features of the radial wavefunction:

- Close to the nucleus the radial wavefunction is proportional to r^l, and the higher the orbital angular momentum, the less likely it is that the electron will be found there (Fig. 9.3).
- Far from the nucleus all radial wavefunctions approach zero exponentially.

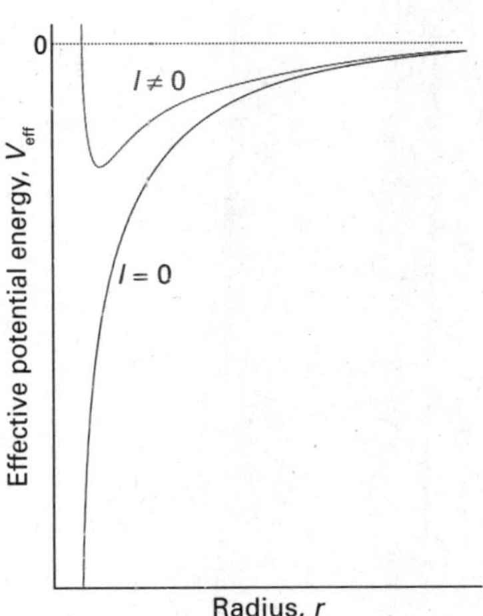

Fig. 9.2 The effective potential energy of an electron in the hydrogen atom. When the electron has zero orbital angular momentum, the effective potential energy is the Coulombic potential energy. When the electron has nonzero orbital angular momentum, the centrifugal effect gives rise to a positive contribution that is very large close to the nucleus. We can expect the $l=0$ and $l>0$ wavefunctions to be very different near the nucleus.

interActivity Plot the effective potential energy against r for several nonzero values of the orbital angular momentum l. How does the location of the minimum in the effective potential energy vary with l?

Justification 9.1 *The form of the radial wavefunction*

When r is very small (close to the nucleus), $u = rR \approx 0$, so the right-hand side of eqn 9.8b is zero; we can also ignore all but the largest terms (those depending on $1/r^2$) in eqn 9.8b and write

$$-\frac{d^2u}{dr^2}+\frac{l(l+1)}{r^2}u \approx 0$$

The solution of this equation (for $r \approx 0$) is

$$u \approx Ar^{l+1}+\frac{B}{r^l}$$

Because $R = u/r$, and R must be finite everywhere and in particular at $r=0$, we must set $B=0$, and hence obtain $R \approx Ar^l$.

Far from the nucleus, when r is very large, we can ignore terms in $1/r$ and $1/r^2$ and eqn 9.8b becomes

$$-\frac{\hbar^2}{2\mu}\frac{d^2u}{dr^2} \simeq Eu$$

where $\simeq$ means 'asymptotically equal to' in the sense that the values become equal as r becomes infinite (like an exponentially decaying function tending to zero). Because

$$\frac{d^2u}{dr^2}=\frac{d^2(rR)}{dr^2}=\frac{d}{dr}\frac{d(rR)}{dr}=\frac{d}{dr}\left(R+r\frac{dR}{dr}\right)$$

$$=2\frac{dR}{dr}+r\frac{d^2R}{dr^2} \simeq r\frac{d^2R}{dr^2}$$

as r becomes infinite, this equation has the form

$$-\frac{\hbar^2}{2\mu}\frac{d^2R}{dr^2} \simeq ER$$

The acceptable (finite) solution of this equation (for r large) is

$$R \simeq e^{-(2\mu|E|/\hbar^2)^{1/2}r}$$

and the wavefunction decays exponentially towards zero as r increases.

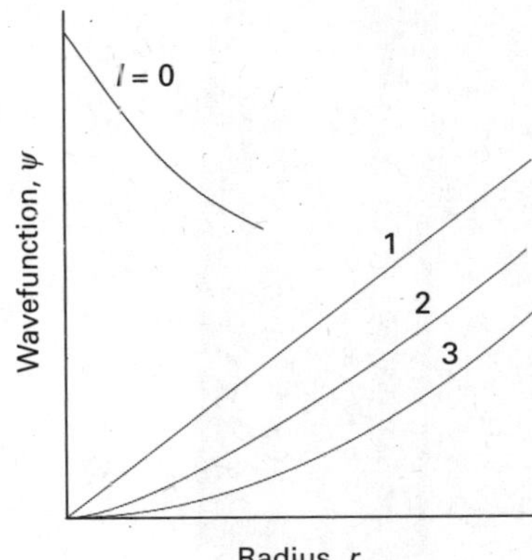

Fig. 9.3 Close to the nucleus, orbitals with $l=1$ are proportional to r, orbitals with $l=2$ are proportional to r^2, and orbitals with $l=3$ are proportional to r^3. Electrons are progressively excluded from the neighbourhood of the nucleus as l increases. An orbital with $l=0$ has a finite, nonzero value at the nucleus.

We shall not go through the technical steps of solving the radial equation for the full range of radii, and see how the form r^l close to the nucleus blends into the exponentially decaying form at great distances. It is sufficient to know that the two limits can be bridged only for integral values of a quantum number n, and that the allowed energies corresponding to the allowed solutions are

$$E_n = -\frac{Z^2\mu e^4}{32\pi^2\varepsilon_0^2\hbar^2 n^2} \quad \text{Bound state energies} \tag{9.9}$$

with $n = 1, 2, \ldots$ Likewise, the radial wavefunctions depend on the values of both n and l (but not on m_l because only l appears in the radial wave equation), and all of them have the form

$$R(r) = \overbrace{r^l}^{\text{Dominant close to the nucleus}} \times \overbrace{(\text{polynomial in } r)}^{\text{Bridges the two ends of the function}} \times \overbrace{(\text{decaying exponential in } r)}^{\text{Dominant far from the nucleus}} \tag{9.10}$$

These functions are most simply written in terms of the dimensionless quantity ρ (rho), where

$$\rho = \frac{2Zr}{na_0} \qquad a_0 = \frac{4\pi\varepsilon_0\hbar^2}{m_e e^2} \tag{9.11}$$

The **Bohr radius,** a_0, has the value 52.9 pm; it is so called because the same quantity appeared in Bohr's early model of the hydrogen atom as the radius of the electron orbit of lowest energy. Specifically, the radial wavefunctions for an electron with quantum numbers n and l are the (real) function

$$R_{n,l}(r) = N_{n,l}\rho^l L_{n+1}^{2l+1}(\rho)e^{-\rho/2} \quad \text{Radial wavefunctions} \tag{9.12}$$

where $L(\rho)$ is a polynomial called an *associated Laguerre polynomial*: it links the $r \approx 0$ solutions on its left (corresponding to $R \propto \rho^l$) to the exponentially decaying function on its right. The notation might look fearsome, but the polynomials have quite simple forms, such as 1, ρ, and $2 - \rho$ (they can be picked out in Table 9.1). The factor N ensures that the radial wavefunction is normalized to 1 in the sense that

Table 9.1 Hydrogenic radial wavefunctions

Orbital	n	l	$R_{n,l}$
1s	1	0	$2\left(\frac{Z}{a}\right)^{3/2} e^{-\rho/2}$
2s	2	0	$\frac{1}{8^{1/2}}\left(\frac{Z}{a}\right)^{3/2}(2-\rho)e^{-\rho/2}$
2p	2	1	$\frac{1}{24^{1/2}}\left(\frac{Z}{a}\right)^{3/2}\rho e^{-\rho/2}$
3s	3	0	$\frac{1}{243^{1/2}}\left(\frac{Z}{a}\right)^{3/2}(6-6\rho+\rho^2)e^{-\rho/2}$
3p	3	1	$\frac{1}{486^{1/2}}\left(\frac{Z}{a}\right)^{3/2}(4-\rho)\rho e^{-\rho/2}$
3d	3	2	$\frac{1}{2430^{1/2}}\left(\frac{Z}{a}\right)^{3/2}\rho^2 e^{-\rho/2}$

$\rho = (2Z/na)r$ with $a = 4\pi\varepsilon_0\hbar^2/\mu e^2$. For an infinitely heavy nucleus (or one that may be assumed to be so), $\mu = m_e$ and $a = a_0$, the Bohr radius. The full wavefunction is obtained by multiplying R by the appropriate Y given in Table 8.2.

$$\int_0^\infty R_{n,l}(r)^2 r^2 \mathrm{d}r = 1 \tag{9.13}$$

(The r^2 comes from the volume element in spherical coordinates, Section 7.4a.) Specifically, we can interpret the components of eqn 9.12 as follows:

1. The exponential factor ensures that the wavefunction approaches zero far from the nucleus.

2. The factor ρ^l ensures that (provided $l > 0$) the wavefunction vanishes at the nucleus.

3. The associated Laguerre polynomial is a function that in general oscillates from positive to negative values and accounts for the presence of radial nodes.

Expressions for some radial wavefunctions are given in Table 9.1 and illustrated in Fig. 9.4.

A brief comment

The zero at $r = 0$ is not a radial node because the radial wavefunction does not pass through zero at that point (because r cannot be negative). Nodes at the nucleus are all angular nodes.

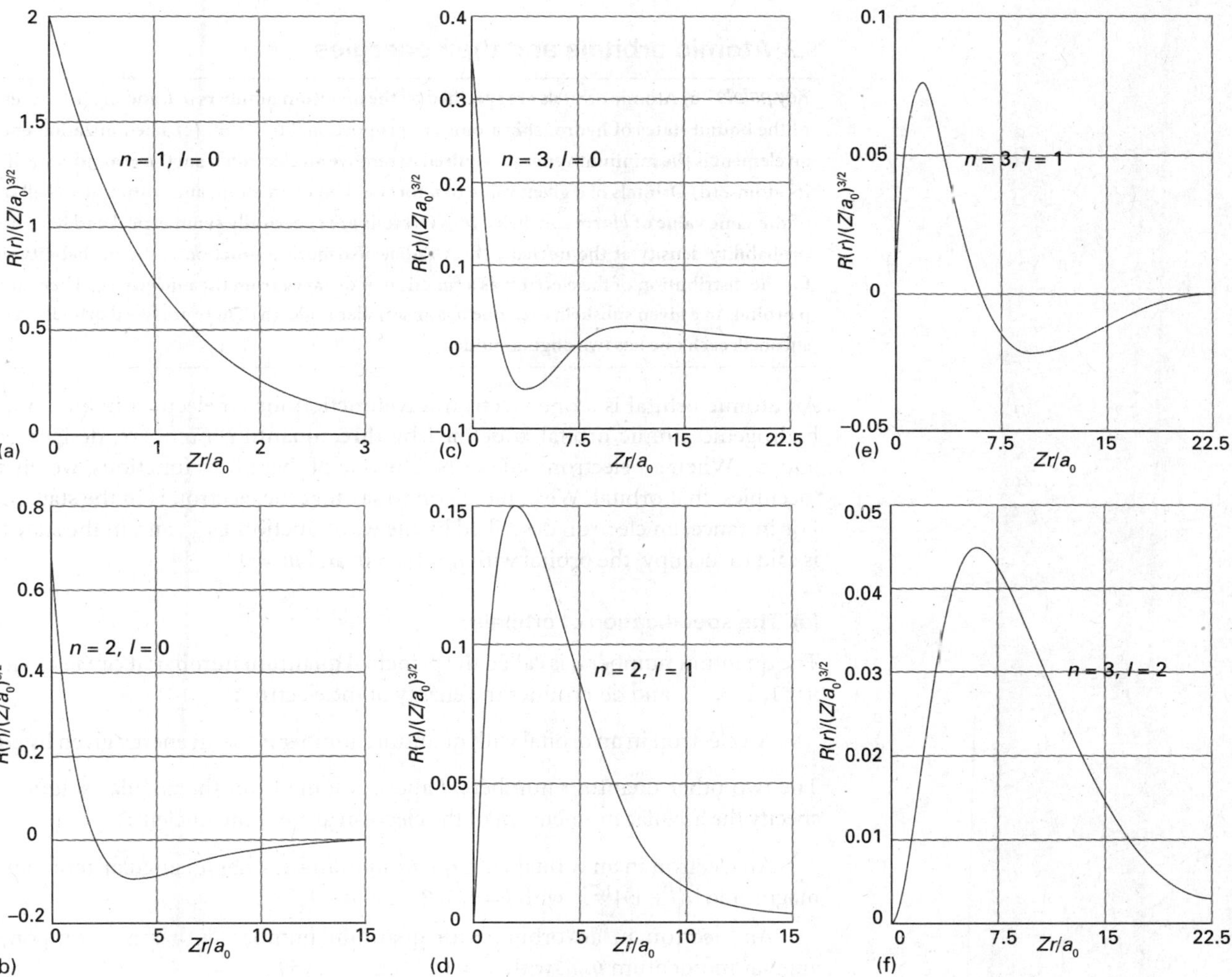

Fig. 9.4 The radial wavefunctions of the first few states of hydrogenic atoms of atomic number Z. Note that the orbitals with $l = 0$ have a nonzero and finite value at the nucleus. The horizontal scales are different in each case: orbitals with high principal quantum numbers are relatively distant from the nucleus.

interActivity Use mathematical software to find the locations of the radial nodes in hydrogenic wavefunctions with n up to 3.

• **A brief illustration**

To calculate the probability density at the nucleus for an electron with $n = 1$, $l = 0$, and $m_l = 0$, we evaluate ψ at $r = 0$:

$$\psi_{1,0,0}(0,\theta,\phi) = R_{1,0}(0)Y_{0,0}(\theta,\phi) = 2\left(\frac{Z}{a_0}\right)^{3/2}\left(\frac{1}{4\pi}\right)^{1/2}$$

The probability density is therefore

$$\psi_{1,0,0}(0,\theta,\phi)^2 = \frac{Z^3}{\pi a_0^3}$$

which evaluates to 2.15×10^{-6} pm^{-3} when $Z = 1$. •

Self-test 9.2 Evaluate the probability density at the nucleus of the electron for an electron with $n = 2$, $l = 0$, $m_l = 0$. $[(Z/a_0)^3/8\pi]$

9.2 Atomic orbitals and their energies

Key points (a) Atomic orbitals are specified by the quantum numbers n, l, and m_l. (b) The energies of the bound states of hydrogenic atoms are proportional to Z^2/n^2. (c) The ionization energy of an element is the minimum energy required to remove an electron from the ground state of one of its atoms. (d) Orbitals of a given value of n form a shell of an atom, and within that shell orbitals of the same value of l form subshells. (e) s Orbitals are spherically symmetrical and have nonzero probability density at the nucleus. (f) A radial distribution function is the probability density for the distribution of the electron as a function of distance from the nucleus. (g) There are three p orbitals in a given subshell; each one has an angular node. (h) There are five d orbitals in a given subshell; each one has two angular nodes.

An **atomic orbital** is a one-electron wavefunction for an electron in an atom. Each hydrogenic atomic orbital is defined by three quantum numbers, designated n, l, and m_l. When an electron is described by one of these wavefunctions, we say that it 'occupies' that orbital. We could go on to say that the electron is in the state $|n,l,m_l\rangle$. For instance, an electron described by the wavefunction $\psi_{1,0,0}$ and in the state $|1,0,0\rangle$ is said to 'occupy' the orbital with $n = 1$, $l = 0$, and $m_l = 0$.

(a) The specification of orbitals

The quantum number n is called the **principal quantum number**; it can take the value $n = 1, 2, 3, \ldots$ and determines the energy of the electron:

- An electron in an orbital with quantum number n has an energy given by eqn 9.9.

The two other quantum numbers, l and m_l, come from the angular solutions, and specify the angular momentum of the electron around the nucleus:

- An electron in an orbital with quantum number l has an angular momentum of magnitude $\{l(l+1)\}^{1/2}\hbar$, with $l = 0, 1, 2, \ldots, n-1$.
- An electron in an orbital with quantum number m_l has a z-component of angular momentum $m_l\hbar$, with $m_l = 0, \pm 1, \pm 2, \ldots, \pm l$.

Note how the value of the principal quantum number, n, controls the maximum value of l and l controls the range of values of m_l.

To define the state of an electron in a hydrogenic atom fully we need to specify not only the orbital it occupies but also its spin state. We saw in Section 8.8 that an electron possesses an intrinsic angular momentum that is described by the two quantum numbers s and m_s (the analogues of l and m_l). The value of s is fixed at $\frac{1}{2}$ for an electron, so we do not need to consider it further at this stage. However, m_s may be either $+\frac{1}{2}$ or $-\frac{1}{2}$, and to specify the state of an electron in a hydrogenic atom we need to specify which of these values describes it. It follows that, to specify the state of an electron in a hydrogenic atom, we need to give the values of four quantum numbers, namely n, l, m_l, and m_s.

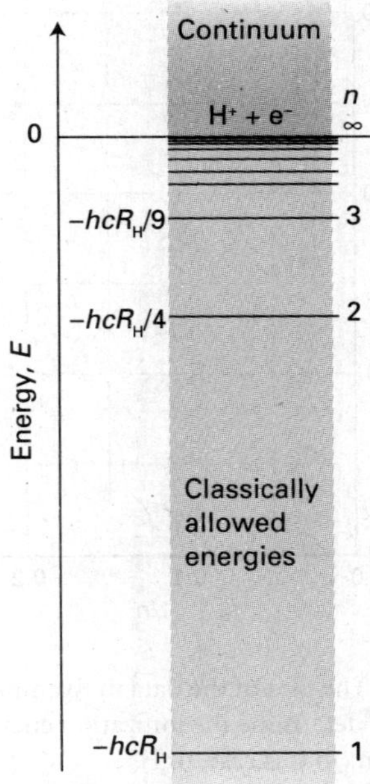

Fig. 9.5 The energy levels of a hydrogen atom. The values are relative to an infinitely separated, stationary electron and a proton.

(b) The energy levels

The energy levels predicted by eqn 9.9 are depicted in Fig. 9.5. The energies, and also the separation of neighbouring levels, are proportional to Z^2, so the levels are four times as wide apart (and the ground state four times deeper in energy) in He^+ ($Z = 2$) than in H ($Z = 1$). All the energies given by eqn 9.9 are negative. They refer to the **bound states** of the atom, in which the energy of the atom is lower than that of the infinitely separated, stationary electron and nucleus (which corresponds to the zero of energy). There are also solutions of the Schrödinger equation with positive energies. These solutions correspond to **unbound states** of the electron, the states to which an electron is raised when it is ejected from the atom by a high-energy collision or photon. The energies of the unbound electron are not quantized and form the continuum states of the atom.

Equation 9.9 is consistent with the spectroscopic result summarized by eqn 9.1, and we can identify the Rydberg constant for hydrogen ($Z = 1$) as

$$hcR_H = \frac{\mu_H e^4}{32\pi^2\varepsilon_0^2\hbar^2} \tag{9.14}$$

where μ_H is the reduced mass for hydrogen. The **Rydberg constant** itself, R_∞, is defined by the same expression except for the replacement of μ_H by the mass of an electron, m_e, corresponding to a nucleus of infinite mass:

$$R_H = \frac{\mu_H}{m_e}R_\infty \qquad R_\infty = \frac{m_e e^4}{8\varepsilon_0^2 h^3 c} \qquad \text{Rydberg constant} \qquad [9.15]$$

Insertion of the values of the fundamental constants into the expression for R_H gives almost exact agreement with the experimental value. The only discrepancies arise from the neglect of relativistic corrections (in simple terms, the increase of mass with speed), which the non-relativistic Schrödinger equation ignores.

(c) Ionization energies

The **ionization energy**, I, of an element is the minimum energy required to remove an electron from the ground state, the state of lowest energy, of one of its atoms in the gas phase. Because the ground state of hydrogen is the state with $n = 1$, with energy $E_1 = -hcR_H$ and the atom is ionized when the electron has been excited to the level corresponding to $n = \infty$ (see Fig. 9.5), the energy that must be supplied is

$$I = hcR_H \tag{9.16}$$

The value of I is 2.179 aJ (a, for atto, is the prefix that denotes 10^{-18}), which corresponds to 13.60 eV.

A note on good practice Ionization energies are sometimes referred to as *ionization potentials*. That is incorrect, but not uncommon. If the term is used at all, it should denote the potential difference through which an electron must be moved for its potential energy to change by an amount equal to the ionization energy, and reported in volts.

Example 9.1 *Measuring an ionization energy spectroscopically*

The emission spectrum of atomic hydrogen shows lines at 82 259, 97 492, 102 824, 105 292, 106 632, and 107 440 cm^{-1}, which correspond to transitions to the same lower state. Determine (a) the ionization energy of the lower state, (b) the value of the Rydberg constant.

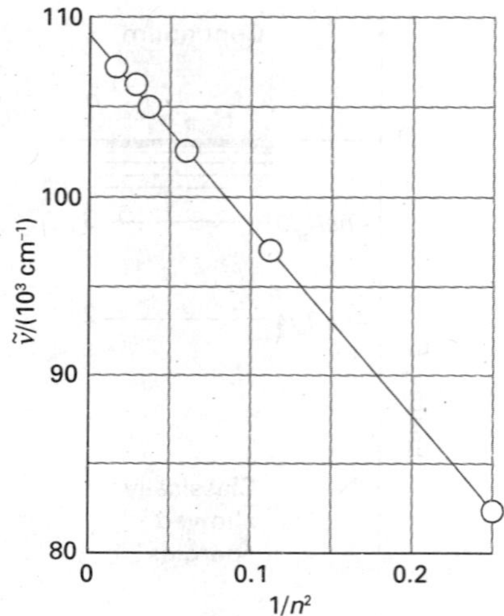

Fig. 9.6 The plot of the data in Example 9.1 used to determine the ionization energy of an atom (in this case, of H).

interActivity The initial value of n was not specified in Example 9.1. Show that the correct value can be determined by making several choices and selecting the one that leads to a straight line. The slope is, in this instance, numerically the same as R_H, so $R_H = 109\,679\ \text{cm}^{-1}$. A similar extrapolation procedure can be used for many-electron atoms (see Section 9.7).

Method The spectroscopic determination of ionization energies depends on the determination of the series limit, the wavenumber at which the series terminates and becomes a continuum. If the upper state lies at an energy $-hcR_H/n^2$, then, when the atom makes a transition to E_{lower}, a photon of wavenumber

$$\tilde{\nu} = -\frac{R_H}{n^2} - \frac{E_{\text{lower}}}{hc}$$

is emitted. However, because $I = -E_{\text{lower}}$, it follows that

$$\tilde{\nu} = \frac{I}{hc} - \frac{R_H}{n^2}$$

A plot of the wavenumbers against $1/n^2$ should give a straight line of slope $-R_H$ and intercept I/hc. Use a computer to make a least-squares fit of the data in order to obtain a result that reflects the precision of the data.

Answer The wavenumbers are plotted against $1/n^2$ in Fig. 9.6. The (least-squares) intercept lies at $109\,679\ \text{cm}^{-1}$, so the ionization energy is 2.1788 aJ (1312.1 kJ mol^{-1}).

Self-test 9.3 The emission spectrum of atomic deuterium shows lines at 15 238, 20 571, 23 039, and 24 380 cm^{-1}, which correspond to transitions to the same lower state. Determine (a) the ionization energy of the lower state, (b) the ionization energy of the ground state, (c) the mass of the deuteron (by expressing the Rydberg constant in terms of the reduced mass of the electron and the deuteron, and solving for the mass of the deuteron).

[(a) 328.1 kJ mol^{-1}, (b) 1312.4 kJ mol^{-1}, (c) 2.8×10^{-27} kg, a result very sensitive to R_D]

(d) Shells and subshells

All the orbitals of a given value of n are said to form a single **shell** of the atom. In a hydrogenic atom, all orbitals of given n, and therefore belonging to the same shell, have the same energy. It is common to refer to successive shells by letters:

$n =$	1	2	3	4 . . .
	K	L	M	N . . .

Specification of shells

Thus, all the orbitals of the shell with $n = 2$ form the L shell of the atom, and so on.

The orbitals with the same value of n but different values of l are said to form a **subshell** of a given shell. These subshells are generally referred to by letters:

$l =$	0	1	2	3	4	5	6 . . .
	s	p	d	f	g	h	i . . .

Specification of subshells

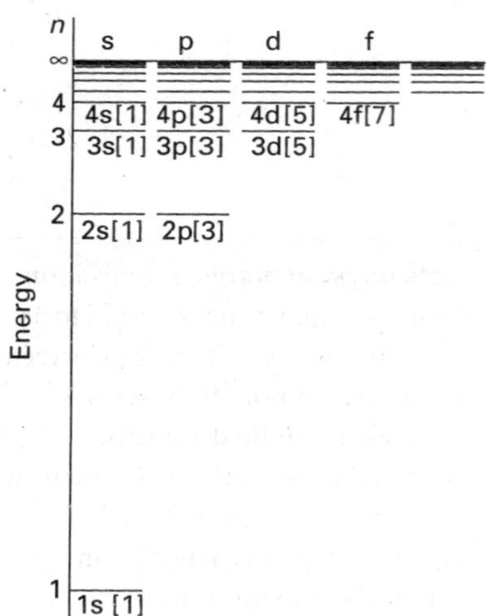

Fig. 9.7 The energy levels of the hydrogen atom showing the subshells and (in square brackets) the numbers of orbitals in each subshell. In hydrogenic atoms, all orbitals of a given shell have the same energy.

The letters then run alphabetically (j is not used because in some languages i and j are not distinguished). Figure 9.7 is a version of Fig. 9.5 that shows the subshells explicitly. Because l can range from 0 to $n - 1$, giving n values in all, it follows that there are n subshells of a shell with principal quantum number n. Thus, when $n = 1$, there is only one subshell, the one with $l = 0$. When $n = 2$, there are two subshells, the 2s subshell (with $l = 0$) and the 2p subshell (with $l = 1$).

When $n = 1$ there is only one subshell, that with $l = 0$, and that subshell contains only one orbital, with $m_l = 0$ (the only value of m_l permitted). When $n = 2$, there are four orbitals, one in the s subshell with $l = 0$ and $m_l = 0$, and three in the $l = 1$ subshell with $m_l = +1, 0, -1$. When $n = 3$ there are nine orbitals (one with $l = 0$, three with $l = 1$, and five with $l=2$). The organization of orbitals in the shells is summarized in Fig. 9.8. In general, the number of orbitals in a shell of principal quantum number n is n^2, so in a hydrogenic atom each energy level is n^2-fold degenerate.

(e) s Orbitals

The orbital occupied in the ground state is the one with $n=1$ (and therefore with $l=0$ and $m_l=0$, the only possible values of these quantum numbers when $n=1$). From Table 9.1 and $Y_{0,0}=1/2\pi^{1/2}$ we can write (for $Z=1$):

$$\psi=\frac{1}{(\pi a_0^3)^{1/2}}e^{-r/a_0} \tag{9.17}$$

This wavefunction is independent of angle and has the same value at all points of constant radius, that is, the 1s orbital is *spherically symmetrical.* The wavefunction decays exponentially from a maximum value of $1/(\pi a_0^3)^{1/2}$ at the nucleus (at $r=0$). It follows that the probability density of the electron is greatest at the nucleus itself.

We can understand the general form of the ground-state wavefunction by considering the contributions of the potential and kinetic energies to the total energy of the atom. The closer the electron is to the nucleus on average, the lower its average potential energy. This dependence suggests that the lowest potential energy should be obtained with a sharply peaked wavefunction that has a large amplitude at the nucleus and is zero everywhere else (Fig. 9.9). However, this shape implies a high kinetic energy, because such a wavefunction has a very high average curvature. The electron would have very low kinetic energy if its wavefunction had only a very low average curvature. However, such a wavefunction spreads to great distances from the nucleus and the average potential energy of the electron will be correspondingly high. The actual ground-state wavefunction is a compromise between these two extremes: the wavefunction spreads away from the nucleus (so the expectation value of the potential energy is not as low as in the first example, but nor is it very high) and has a

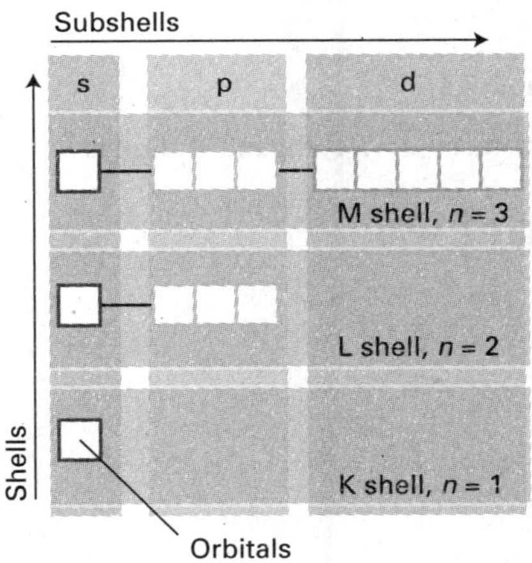

Fig. 9.8 The organization of orbitals (white squares) into subshells (characterized by l) and shells (characterized by n).

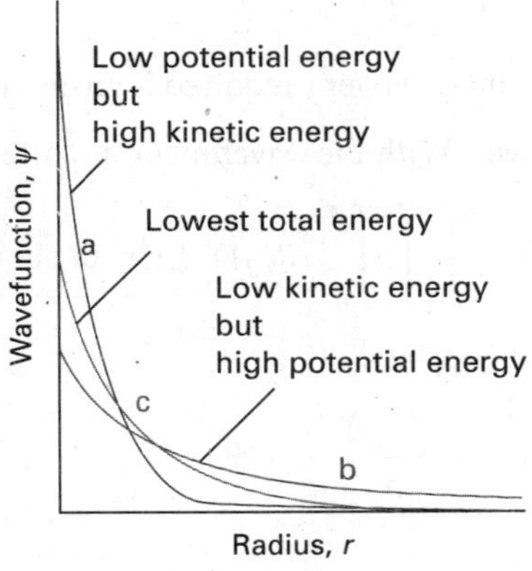

Fig. 9.9 The balance of kinetic and potential energies that accounts for the structure of the ground state of hydrogen (and similar atoms). (a) The sharply curved but localized orbital has high mean kinetic energy, but low mean potential energy; (b) the mean kinetic energy is low, but the potential energy is not very favourable; (c) the compromise of moderate kinetic energy and moderately favourable potential energy.

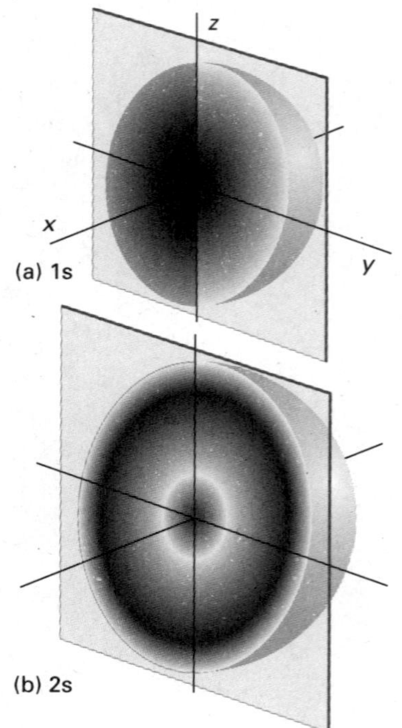

Fig. 9.10 Representations of the 1s and 2s hydrogenic atomic orbitals in terms of their electron densities (as represented by the density of shading).

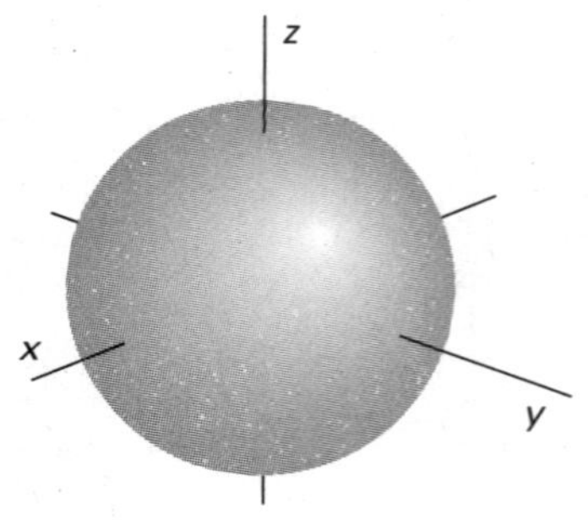

Fig. 9.11 The boundary surface of an s orbital, within which there is a 90 per cent probability of finding the electron.

reasonably low average curvature (so the expectation of the kinetic energy is not very low, but nor is it as high as in the first example).

By the virial theorem with $b = -1$ (eqn 8.35), $\langle E_k \rangle = -\frac{1}{2}\langle V \rangle$ and therefore $E = \langle E_k \rangle + \langle V \rangle = \frac{1}{2}\langle V \rangle$, so the total energy of an s electron becomes less negative as n increases and it is found at greater distances from the nucleus with a less negative potential energy. Thus, as n approaches infinity,

1. The kinetic energy becomes less positive and has fallen to zero when $n = \infty$.
2. The potential energy becomes less negative and has risen to zero when $n = \infty$.
3. The total energy becomes less negative and and has risen to zero when $n = \infty$.

One way of depicting the probability density of the electron is to represent $|\psi|^2$ by the density of shading (Fig. 9.10). A simpler procedure is to show only the **boundary surface**, the surface that captures a high proportion (typically about 90 per cent) of the electron probability. For the 1s orbital, the boundary surface is a sphere centred on the nucleus (Fig. 9.11).

Example 9.2 *Calculating the mean radius of an orbital*

Use hydrogenic orbitals to calculate the mean radius of a 1s orbital.

Method The mean radius is the expectation value

$$\langle r \rangle = \int \psi^* r \psi \, d\tau = \int r |\psi|^2 \, d\tau$$

We therefore need to evaluate the integral using the wavefunctions given in Table 9.1 and $d\tau = r^2 dr \sin\theta \, d\theta \, d\phi$. The angular parts of the wavefunction (Table 8.2) are normalized in the sense that

$$\int_0^{\pi}\int_0^{2\pi} |Y_{l,m_l}|^2 \sin\theta \, d\theta \, d\phi = 1$$

The integral over r required is given in Example 7.4.

Answer With the wavefunction written in the form $\psi = RY$, the integration is

$$\langle r \rangle = \int_0^{\infty}\int_0^{\pi}\int_0^{2\pi} rR_{n,l}^2 |Y_{l,m_l}|^2 r^2 \, dr \sin\theta \, d\theta \, d\phi = \int_0^{\infty} r^3 R_{n,l}^2 \, dr$$

For a 1s orbital

$$R_{1,0} = 2\left(\frac{Z}{a_0}\right)^{3/2} e^{-Zr/a_0}$$

Hence

$$\langle r \rangle = \frac{4Z^3}{a_0^3}\int_0^{\infty} r^3 e^{-2Zr/a_0} dr = \frac{3a_0}{2Z}$$

Self-test 9.4 Evaluate the mean radius of a 3s orbital by integration. [$27a_0/2Z$]

All s orbitals are spherically symmetric, but differ in the number of radial nodes. For example, the 1s, 2s, and 3s orbitals have 0, 1, and 2 radial nodes, respectively. In general, an ns orbital has $n-1$ radial nodes. As n increases, the radius of the spherical boundary surface that captures a given fraction of the probability also increases.

Self-test 9.5 (a) Use the fact that a 2s orbital has radial nodes where the polynomial factor (Table 9.1) is equal to zero, and locate the radial node at $2a_0/Z$ (see Fig. 9.4). (b) Similarly, locate the two nodes of a 3s orbital.

[(a) $2a_0/Z$; (b) $1.90a_0/Z$ and $7.10a_0/Z$]

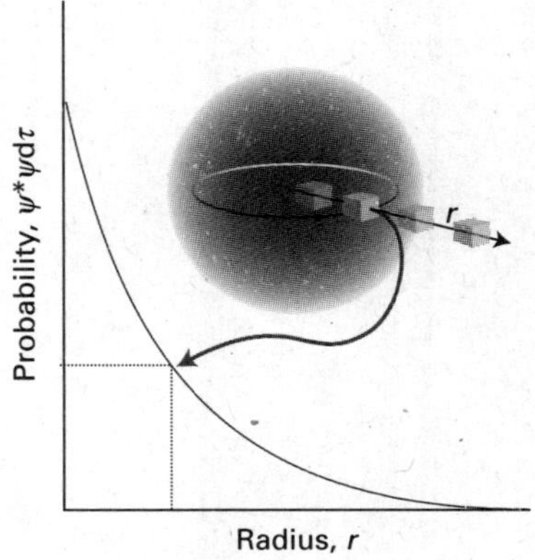

Fig. 9.12 A constant-volume electron-sensitive detector (the small cube) gives its greatest reading at the nucleus, and a smaller reading elsewhere. The same reading is obtained anywhere on a circle of given radius: the s orbital is spherically symmetrical.

(f) Radial distribution functions

The wavefunction tells us, through the value of $|\psi|^2$, the probability of finding an electron in any region. We can imagine a probe with a volume $d\tau$ and sensitive to electrons, and which we can move around near the nucleus of a hydrogen atom. Because the probability density in the ground state of the atom is $|\psi|^2 \propto e^{-2Zr/a_0}$, the reading from the detector decreases exponentially as the probe is moved out along any radius but is constant if the probe is moved on a circle of constant radius (Fig. 9.12).

Now consider the total probability of finding the electron *anywhere* between the two walls of a spherical shell of thickness dr at a radius r. The sensitive volume of the probe is now the volume of the shell (Fig. 9.13), which is $4\pi r^2 dr$ (the product of its surface area, $4\pi r^2$, and its thickness, dr). The probability that the electron will be found between the inner and outer surfaces of this shell is the probability density at the radius r multiplied by the volume of the probe, or $|\psi|^2 \times 4\pi r^2 dr$. This expression has the form $P(r)dr$, where

$$P(r) = 4\pi r^2 \psi^2 \tag{9.18a}$$

The more general expression, which also applies to orbitals that are not spherically symmetrical, is derived in the following *Justification*, and is

$$P(r) = r^2 R(r)^2 \quad \text{Radial distribution function} \tag{9.18b}$$

where $R(r)$ is the radial wavefunction for the orbital in question.

Justification 9.2 *The general form of the radial distribution function*

The probability of finding an electron in a volume element $d\tau$ when its wavefunction is $\psi = RY$ is $|RY|^2 d\tau$ with $d\tau = r^2 dr \sin\theta\, d\theta\, d\phi$. The total probability of finding the electron at any angle at a constant radius is the integral of this probability over the surface of a sphere of radius r, and is written $P(r)dr$, so

$$P(r)dr = \int_0^{\pi}\int_0^{2\pi} R(r)^2 |Y(\theta,\phi)|^2 r^2 dr \sin\theta\, d\theta\, d\phi$$

$$= r^2 R(r)^2 dr \overbrace{\int_0^{\pi}\int_0^{2\pi} |Y(\theta,\phi)|^2 \sin\theta\, d\theta\, d\phi}^{1} = r^2 R(r)^2 dr$$

The last equality follows from the fact that the spherical harmonics are normalized to 1 (see Example 9.2). It follows that $P(r) = r^2 R(r)^2$, as stated in the text.

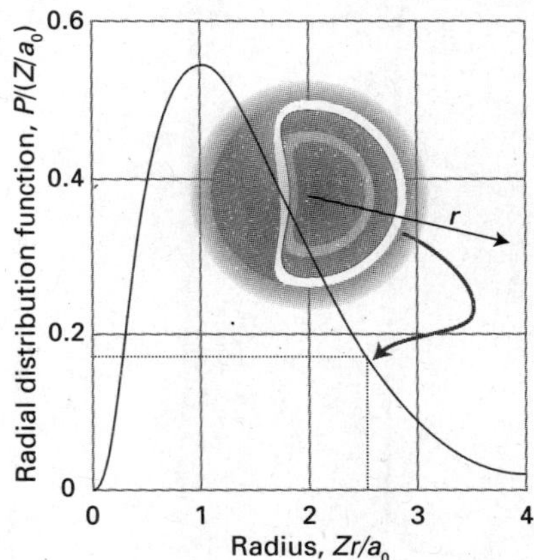

Fig. 9.13 The radial distribution function P gives the probability density that the electron will be found anywhere in a shell of radius r. For a 1s electron in hydrogen, P is a maximum when r is equal to the Bohr radius a_0. The value of P is equivalent to the reading that a detector shaped like a spherical shell would give as its radius is varied.

The **radial distribution function**, $P(r)$, is a probability density in the sense that, when it is multiplied by dr, it gives the probability of finding the electron anywhere between the two walls of a spherical shell of thickness dr at the radius r. For a 1s orbital,

$$P(r)=\frac{4Z^3}{a_0^3}r^2\mathrm{e}^{-2Zr/a_0} \qquad (9.19)$$

Let's interpret this expression:

1. Because $r^2=0$ at the nucleus, $P(0)=0$. The volume of the shell of inspection is zero when $r=0$.

2. As $r\to\infty$, $P(r)\to 0$ on account of the exponential term. The wavefunction has fallen to zero at great distances from the nucleus.

3. The increase in r^2 and the decrease in the exponential factor means that P passes through a maximum at an intermediate radius (see Fig. 9.13).

The maximum of $P(r)$, which can be found by differentiation, marks the most probable radius at which the electron will be found, and for a 1s orbital in hydrogen occurs at $r=a_0$, the Bohr radius. When we carry through the same calculation for the radial distribution function of the 2s orbital in hydrogen, we find that the most probable radius is $5.2a_0=275$ pm. This larger value reflects the expansion of the atom as its energy increases.

Example 9.3 *Calculating the most probable radius*

Calculate the most probable radius, r^*, at which an electron will be found when it occupies a 1s orbital of a hydrogenic atom of atomic number Z, and tabulate the values for the one-electron species from H to Ne^{9+}.

Method We find the radius at which the radial distribution function of the hydrogenic 1s orbital has a maximum value by solving $\mathrm{d}P/\mathrm{d}r=0$. If there are several maxima, then we choose the one corresponding to the greatest amplitude.

Answer The radial distribution function is given in eqn 9.19. It follows that

$$\frac{\mathrm{d}P}{\mathrm{d}r}=\frac{4Z^3}{a_0^3}\left(2r-\frac{2Zr^2}{a_0}\right)\mathrm{e}^{-2Zr/a_0}$$

This function is zero where the term in parentheses is zero, which (other than at $r=0$) is at

$$r^*=\frac{a_0}{Z}$$

Then, with $a_0=52.9$ pm, the most probable radius is

	H	He^+	Li^{2+}	Be^{3+}	B^{4+}	C^{5+}	N^{6+}	O^{7+}	F^{8+}	Ne^{9+}
r^*/pm	52.9	26.5	17.6	13.2	10.6	8.82	7.56	6.61	5.88	5.29

Notice how the 1s orbital is drawn towards the nucleus as the nuclear charge increases. At uranium the most probable radius is only 0.58 pm, almost 100 times closer than for hydrogen. (On a scale where $r^*=10$ cm for H, $r^*=1$ mm for U, Fig. 9.14.) We need to be cautious, though, in extending this result to very heavy atoms because relativistic effects are then important and complicate the calculation.

Self-test 9.6 Find the most probable distance of a 2s electron from the nucleus in a hydrogenic atom. $[(3+5^{1/2})a_0/Z]$

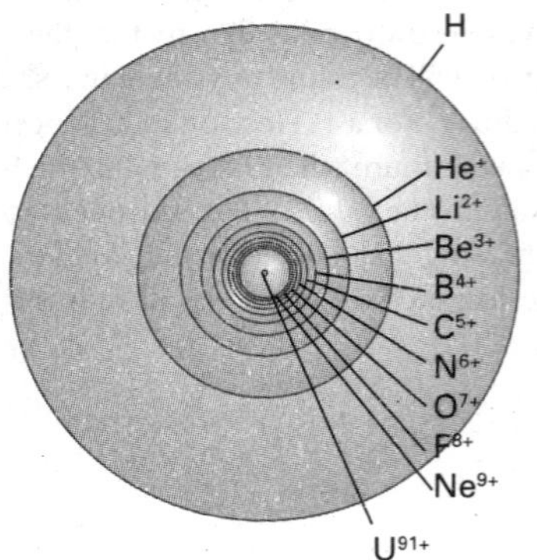

Fig. 9.14 A representation of the most probable radii of a variety of one-electron atoms and ions.

(g) p Orbitals

The three 2p orbitals are distinguished by the three different values that m_l can take when $l=1$. Because the quantum number m_l tells us the orbital angular momentum

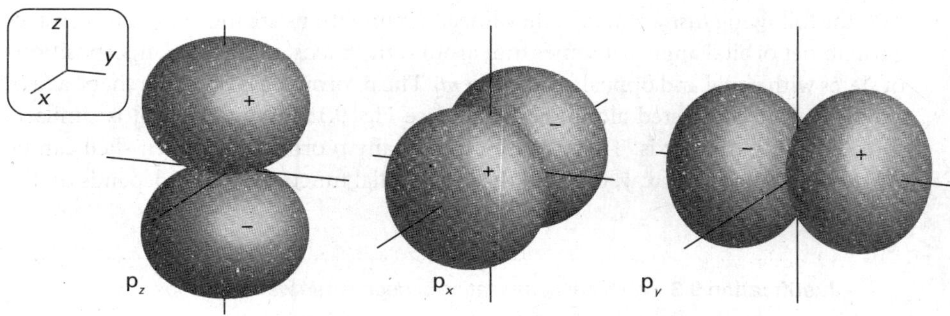

Fig. 9.15 The boundary surfaces of p orbitals. A nodal plane passes through the nucleus and separates the two lobes of each orbital. The dark and light areas denote regions of opposite sign of the wavefunction.

interActivity Use mathematical software to plot the boundary surfaces of the real parts of the spherical harmonics $Y_{1,m_l}(\theta,\phi)$. The resulting plots are not strictly the p orbital boundary surfaces, but sufficiently close to be reasonable representations of the shapes of hydrogenic orbitals.

around an axis, these different values of m_l denote orbitals in which the electron has different orbital angular momenta around an arbitrary z-axis but the same magnitude of that momentum (because l is the same for all three). The orbital with $m_l = 0$, for instance, has zero angular momentum around the z-axis. Its angular variation is proportional to $\cos\theta$, so the probability density, which is proportional to $\cos^2\theta$, has its maximum value on either side of the nucleus along the z-axis (at $\theta = 0$ and 180°). The wavefunction of a 2p orbital with $m_l = 0$ is

$$\psi_{\mathrm{p}_0} = R_{2,1}(r)Y_{1,0}(\theta,\phi) = \frac{1}{4(2\pi)^{1/2}}\left(\frac{Z}{a_0}\right)^{5/2} r\cos\theta\,\mathrm{e}^{-Zr/2a_0}$$
$$= r\cos\theta f(r) \qquad (9.20a)$$

where $f(r)$ is a function only of r. Because in spherical polar coordinates $z = r\cos\theta$, this wavefunction may also be written

$$\psi_{\mathrm{p}_0} = zf(r) \qquad (9.20b)$$

All p orbitals with $m_l = 0$ have wavefunctions of this form, but $f(r)$ depends on the value of n. This way of writing the orbital is the origin of the name 'p_z orbital': its boundary surface is shown in Fig. 9.15. The wavefunction is zero everywhere in the xy-plane, where $z = 0$, so the xy-plane is a **nodal plane** of the orbital: the wavefunction changes sign on going from one side of the plane to the other.

The wavefunctions of 2p orbitals with $m_l = \pm 1$ have the following form:

$$\psi_{\mathrm{p}_{\pm1}} = R_{2,1}(r)Y_{1,\pm1}(\theta,\phi) = \mp\frac{1}{8\pi^{1/2}}\left(\frac{Z}{a_0}\right)^{5/2} r\sin\theta\,\mathrm{e}^{\pm i\phi}\mathrm{e}^{-Zr/2a_0}$$
$$= \mp\frac{1}{2^{1/2}} r\sin\theta\,\mathrm{e}^{\pm i\phi} f(r) \qquad (9.21)$$

We saw in Chapter 8 that a particle that has net motion is described by a complex wavefunction. In the present case, the functions correspond to nonzero angular momentum about the z-axis: $\mathrm{e}^{+i\phi}$ corresponds to clockwise rotation when viewed from below, and $\mathrm{e}^{-i\phi}$ corresponds to counterclockwise rotation (from the same viewpoint). They have zero amplitude where $\theta = 0$ and 180° (along the z-axis) and maximum amplitude at 90°, which is in the xy-plane. To draw the functions it is usual to represent them as standing waves. To do so, we take the real linear combinations

$$\psi_{\mathrm{p}_x} = -\frac{1}{2^{1/2}}(\mathrm{p}_{+1} - \mathrm{p}_{-1}) = r\sin\theta\cos\phi f(r) = xf(r)$$
$$\psi_{\mathrm{p}_y} = \frac{i}{2^{1/2}}(\mathrm{p}_{+1} + \mathrm{p}_{-1}) = r\sin\theta\sin\phi f(r) = yf(r) \qquad (9.22)$$

(See the following *Justification*.) These linear combinations are indeed standing waves with no net orbital angular momentum around the z-axis, as they are superpositions of states with equal and opposite values of m_l. The p_x orbital has the same shape as a p_z orbital, but it is directed along the x-axis (see Fig. 9.15); the p_y orbital is similarly directed along the y-axis. The wavefunction of any p orbital of a given shell can be written as a product of x, y, or z and the same radial function (which depends on the value of n).

Justification 9.3 *The linear combination of degenerate wavefunctions*

We justify here the step of taking linear combinations of degenerate orbitals when we want to indicate a particular point. The freedom to do so rests on the fact that, whenever two or more wavefunctions correspond to the same energy, any linear combination of them is an equally valid solution of the Schrödinger equation.

Suppose ψ_1 and ψ_2 are both solutions of the Schrödinger equation with energy E; then we know that

$$\hat{H}\psi_1 = E\psi_1 \qquad \hat{H}\psi_2 = E\psi_2$$

Now consider the linear combination $\psi = c_1\psi_1 + c_2\psi_2$ where c_1 and c_2 are arbitrary coefficients. Then it follows that

$$\hat{H}\psi = \hat{H}(c_1\psi_1 + c_2\psi_2) = c_1\hat{H}\psi_1 + c_2\hat{H}\psi_2 = c_1E\psi_1 + c_2E\psi_2 = E\psi$$

Hence, the linear combination is also a solution corresponding to the same energy E.

(h) d Orbitals

When $n = 3$, l can be 0, 1, or 2. As a result, this shell consists of one 3s orbital, three 3p orbitals, and five 3d orbitals. Each value of the quantum number $m_l = +2, +1, 0, -1, -2$ corresponds to a different value for the component of the angular momentum about the z-axis. As for the p orbitals, d orbitals with opposite values of m_l (and hence opposite senses of motion around the z-axis) may be combined in pairs to give real standing waves, and the boundary surfaces of the resulting shapes are shown in Fig. 9.16. The real linear combinations have the following forms:

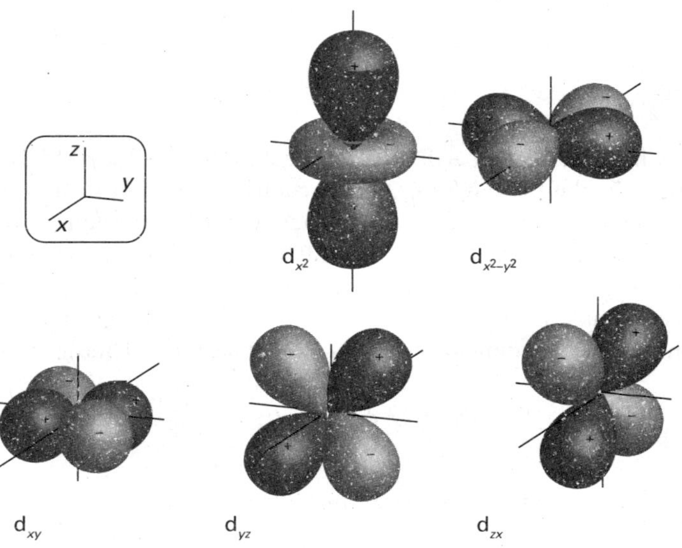

Fig. 9.16 The boundary surfaces of d orbitals. Two nodal planes in each orbital intersect at the nucleus and separate the lobes of each orbital. The dark and light areas denote regions of opposite sign of the wavefunction.

interActivity To gain insight into the shapes of the f orbitals, use mathematical software to plot the boundary surfaces of the spherical harmonics $Y_{3,m_l}(\theta,\phi)$.

$$d_{xy} = xyf(r) \qquad d_{yz} = yzf(r) \qquad d_{zx} = zxf(r)$$
$$d_{x^2-y^2} = \tfrac{1}{2}(x^2 - y^2)f(r) \qquad d_{z^2} = (\tfrac{1}{2}\sqrt{3})(3z^2 - r^2)f(r) \tag{9.23}$$

9.3 Spectroscopic transitions and selection rules

Key point **Allowed spectroscopic transitions of atoms are governed by selection rules that stem from the unit angular momentum of a photon and the conservation of angular momentum.**

The energies of the hydrogenic atoms are given by eqn 9.9. When the electron undergoes a **transition**, a change of state, from an orbital with quantum numbers n_1, l_1, m_{l1} to another (lower energy) orbital with quantum numbers n_2, l_2, m_{l2}, it undergoes a change of energy ΔE and discards the excess energy as a photon of electromagnetic radiation with a frequency ν given by the Bohr frequency condition (eqn 7.14).

It is tempting to think that all possible transitions are permissible, and that a spectrum arises from the transition of an electron from any initial orbital to any other orbital. However, this is not so, because a photon has an intrinsic spin angular momentum corresponding to $s = 1$ (Section 8.8). Because total angular momentum is conserved, the change in angular momentum of the electron must compensate for the angular momentum carried away by the photon. Thus, an electron in a d orbital ($l = 2$) cannot make a transition into an s orbital ($l = 0$) because the photon cannot carry away enough angular momentum. Similarly, an s electron cannot make a transition to another s orbital, because there would then be no change in the angular momentum of the electron to make up for the angular momentum carried away by the photon. It follows that some spectroscopic transitions are **allowed**, meaning that they can occur, whereas others are **forbidden**, meaning that they cannot occur.

A **selection rule** is a statement about which transitions are allowed. They are derived (for atoms) by identifying the transitions that conserve angular momentum when a photon is emitted or absorbed. We show in the following *Justification* that the selection rules for hydrogenic atoms are

$$\Delta l = \pm 1 \qquad \Delta m_l = 0, \pm 1 \tag{9.24}$$

Selection rules for hydrogenic atoms

The principal quantum number n can change by any amount consistent with the Δl for the transition, because it does not relate directly to the angular momentum.

Justification 9.4 *The identification of selection rules*

The underlying classical idea behind a spectroscopic transition is that, for an atom or molecule to be able to interact with the electromagnetic field and absorb or create a photon of frequency ν, it must possess, at least transiently, a dipole oscillating at that frequency. This transient dipole is expressed quantum mechanically in terms of the **transition dipole moment,** $\boldsymbol{\mu}_{fi}$, between the initial and final states, where[1]

$$\mu_{fi} = \int \psi_f^* \hat{\mu} \psi_i \, d\tau \tag{9.25}$$

and $\hat{\mu}$ is the electric dipole moment operator. For a one-electron atom $\hat{\mu}$ is multiplication by $-er$ with components $\mu_x = -ex$, $\mu_y = -ey$, and $\mu_z = -ez$. If the transition dipole moment is zero, then the transition is forbidden; the transition is allowed if the transition moment is nonzero.

[1] See our *Quanta, matter, and change* (2009) for a detailed development of the form of eqn 9.25.

To evaluate a transition dipole moment, we consider each component in turn. For example, for the z-component,

$$\mu_{z,\mathrm{fi}} = -e\int \psi_\mathrm{f}^* z \psi_\mathrm{i}\,\mathrm{d}\tau$$

To evaluate the integral, we note from Table 8.2 that $z = (4\pi/3)^{1/2} r Y_{1,0}$, so

$$\int \psi_\mathrm{f}^* z \psi_\mathrm{i}\,\mathrm{d}\tau = \int_0^\infty\int_0^\pi\int_0^{2\pi} \overbrace{R_{n_\mathrm{f},l_\mathrm{f}} Y^*_{l_\mathrm{f},m_{l,\mathrm{f}}}}^{\psi_\mathrm{f}^*} \overbrace{\left(\frac{4\pi}{3}\right)^{1/2} r Y_{1,0}}^{z} \overbrace{R_{n_\mathrm{i},l_\mathrm{i}} Y_{l_\mathrm{i},m_{l,\mathrm{i}}}}^{\psi_\mathrm{i}} \overbrace{r^2\mathrm{d}r \sin\theta\,\mathrm{d}\theta\mathrm{d}\phi}^{\mathrm{d}\tau}$$

This multiple integral is the product of three factors, an integral over r and two integrals over the angles, so the factors on the right can be grouped as follows:

$$\int \psi_\mathrm{f}^* z \psi_\mathrm{i}\,\mathrm{d}\tau = \left(\frac{4\pi}{3}\right)^{1/2} \int_0^\infty R_{n_\mathrm{f},l_\mathrm{f}} r R_{n_\mathrm{i},l_\mathrm{i}} r^2\mathrm{d}r \int_0^\pi\int_0^{2\pi} Y^*_{l_\mathrm{f},m_{l,\mathrm{f}}} Y_{1,0} Y_{l,m_{l,\mathrm{i}}} \sin\theta\,\mathrm{d}\theta\mathrm{d}\phi$$

It follows from the properties of the spherical harmonics (Table 8.2) that the integral

$$\int_0^\pi\int_0^{2\pi} Y^*_{l_\mathrm{f},m_{l,\mathrm{f}}} Y_{1,m} Y_{l_\mathrm{i},m_{l,\mathrm{i}}} \sin\theta\,\mathrm{d}\theta\mathrm{d}\phi$$

is zero unless $l_\mathrm{f} = l_\mathrm{i} \pm 1$ and $m_{l,\mathrm{f}} = m_{l,\mathrm{i}} + m$. Because $m = 0$ in the present case, the angular integral, and hence the z-component of the transition dipole moment, is zero unless $\Delta l = \pm 1$ and $\Delta m_l = 0$, which is a part of the set of selection rules. The same procedure, but considering the x- and y-components, results in the complete set of rules.

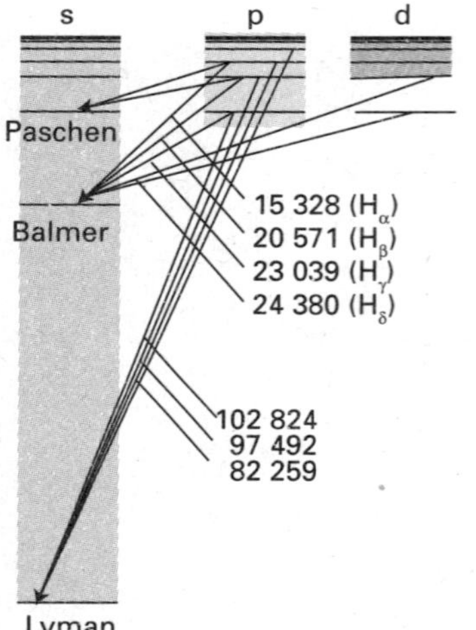

Fig. 9.17 A Grotrian diagram that summarizes the appearance and analysis of the spectrum of atomic hydrogen. The transitions are labelled with their wavenumbers (in cm^{-1}).

• A brief illustration

To identify the orbitals to which a 4d electron may make radiative transitions, we first identify the value of l and then apply the selection rule for this quantum number. Because $l = 2$, the final orbital must have $l = 1$ or 3. Thus, an electron may make a transition from a 4d orbital to any np orbital (subject to $\Delta m_l = 0, \pm 1$) and to any nf orbital (subject to the same rule). However, it cannot undergo a transition to any other orbital, so a transition to any ns orbital or to another nd orbital is forbidden. •

Self-test 9.7 To what orbitals may a 4s electron make electric-dipole allowed radiative transitions? [to np orbitals only]

The selection rules and the atomic energy levels jointly account for the structure of a **Grotrian diagram** (Fig. 9.17), which summarizes the energies of the states and the transitions between them. The thicknesses of the transition lines in the diagram denote their relative intensities in the spectrum; we see how to determine transition intensities in Section 13.2.

The structures of many-electron atoms

The Schrödinger equation for a many-electron atom is highly complicated because all the electrons interact with one another. One very important consequence of these interactions is that orbitals of the same value of n but different values of l are no longer degenerate in a many-electron atom. Moreover, even for a helium atom, with its two electrons, no analytical expression for the orbitals and energies can be given, and we are forced to make approximations. We shall adopt a simple approach based on what we already know about the structure of hydrogenic atoms. Later we shall see the kind of numerical computations that are currently used to obtain accurate wavefunctions and energies.

9.4 The orbital approximation

Key points **In the orbital approximation, each electron is regarded as occupying its own orbital. (a) A configuration is a statement of the occupied orbitals. (b) The Pauli exclusion principle, a special case of the Pauli principle, limits to two the number of electrons that can occupy a given orbital. (c) In many-electron atoms, s orbitals lie at a lower energy than p orbitals of the same shell due to the combined effects of penetration and shielding. (d) The building-up principle is an algorithm for predicting the ground-state electron configuration of an atom. (e) Ionization energies and electron affinities vary periodically through the periodic table.**

The wavefunction of a many-electron atom is a very complicated function of the coordinates of all the electrons, and we should write it $\Psi(r_1,r_2,\ldots)$, where r_i is the vector from the nucleus to electron i (upper-case Ψ is commonly used to denote a many-electron wavefunction). However, in the **orbital approximation** we suppose that a reasonable first approximation to this exact wavefunction is obtained by thinking of each electron as occupying its 'own' orbital, and write

$$\Psi(r_1,r_2,\ldots)=\psi(r_1)\psi(r_2)\ldots \qquad \text{Orbital approximation} \qquad (9.26)$$

We can think of the individual orbitals as resembling the hydrogenic orbitals, but corresponding to nuclear charges modified by the presence of all the other electrons in the atom. This description is only approximate, as the following *Justification* reveals, but it is a useful model for discussing the chemical properties of atoms, and is the starting point for more sophisticated descriptions of atomic structure.

Justification 9.5 *The orbital approximation*

The orbital approximation would be exact if there were no interactions between electrons. To demonstrate the validity of this remark, we need to consider a system in which the hamiltonian for the energy is the sum of two contributions, one for electron 1 and the other for electron 2:

$$\hat{H}=\hat{H}_1+\hat{H}_2$$

In an actual atom (such as helium atom), there is an additional term (proportional to $1/r_{12}$) corresponding to the interaction of the two electrons:

$$\hat{H}=\overbrace{-\frac{\hbar^2}{2m_e}\nabla_1^2-\frac{e^2}{4\pi\varepsilon_0 r_1}}^{\hat{H}_1}\overbrace{-\frac{\hbar^2}{2m_e}\nabla_2^2-\frac{e^2}{4\pi\varepsilon_0 r_2}}^{\hat{H}_2}+\frac{e^2}{4\pi r_{12}}$$

but we are ignoring that term. We shall now show that, if $\psi(r_1)$ is an eigenfunction of $\hat{H}_1$ with energy E_1, and $\psi(r_2)$ is an eigenfunction of $\hat{H}_2$ with energy E_2, then the product $\Psi(r_1,r_2)=\psi(r_1)\psi(r_2)$ is an eigenfunction of the combined hamiltonian $\hat{H}$. To do so we write

$$\begin{aligned}\hat{H}\Psi(r_1,r_2)&=(\hat{H}_1+\hat{H}_2)\psi(r_1)\psi(r_2)=\hat{H}_1\psi(r_1)\psi(r_2)+\psi(r_1)\hat{H}_2\psi(r_2)\\&=E_1\psi(r_1)\psi(r_2)+\psi(r_1)E_2\psi(r_2)=(E_1+E_2)\psi(r_1)\psi(r_2)\\&=E\Psi(r_1,r_2)\end{aligned}$$

where $E=E_1+E_2$. This is the result we need to prove. However, if the electrons interact (as they do in fact), then the proof fails.

(a) The helium atom

The orbital approximation allows us to express the electronic structure of an atom by reporting its **configuration**, a statement of its occupied orbitals (usually, but not necessarily, in its ground state). Thus, as the ground state of a hydrogenic atom consists of the single electron in a 1s orbital, we report its configuration as $1s^1$ (read 'one-ess-one').

A He atom has two electrons. We can imagine forming the atom by adding the electrons in succession to the orbitals of the bare nucleus (of charge $2e$). The first electron occupies a 1s hydrogenic orbital, but because $Z = 2$ that orbital is more compact than in H itself. The second electron joins the first in the 1s orbital, so the electron configuration of the ground state of He is $1s^2$.

(b) The Pauli principle

Lithium, with $Z = 3$, has three electrons. The first two occupy a 1s orbital drawn even more closely than in He around the more highly charged nucleus. The third electron, however, does not join the first two in the 1s orbital because that configuration is forbidden by the **Pauli exclusion principle**:

> No more than two electrons may occupy any given orbital, and if two do occupy one orbital, then their spins must be paired.

Pauli exclusion principle

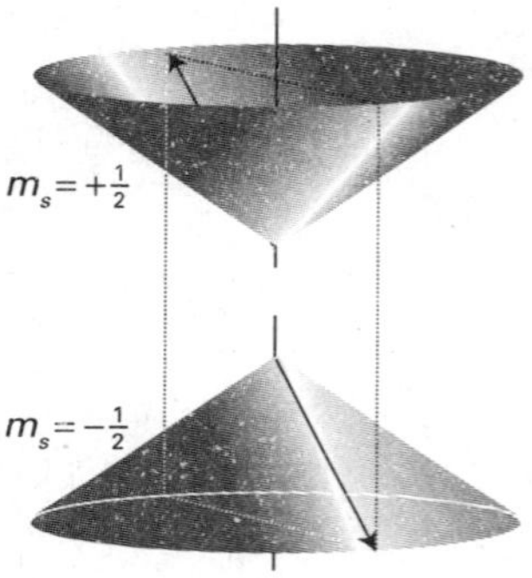

Fig. 9.18 Electrons with paired spins have zero resultant spin angular momentum. They can be represented by two vectors that lie at an indeterminate position on the cones shown here, but, wherever one lies on its cone, the other points in the opposite direction; their resultant is zero.

Electrons with paired spins, denoted ↑↓, have zero net spin angular momentum because the spin of one electron is cancelled by the spin of the other. Specifically, one electron has $m_s = +\frac{1}{2}$, the other has $m_s = -\frac{1}{2}$, and they are orientated on their respective cones so that the resultant spin is zero (Fig. 9.18). The exclusion principle is the key to the structure of complex atoms, to chemical periodicity, and to molecular structure. It was proposed by Wolfgang Pauli in 1924 when he was trying to account for the absence of some lines in the spectrum of helium. Later he was able to derive a very general form of the principle from theoretical considerations.

The Pauli exclusion principle in fact applies to any pair of identical fermions (particles with half integral spin). Thus it applies to protons, neutrons, and ^{13}C nuclei (all of which have spin $\frac{1}{2}$) and to ^{35}Cl nuclei (which have spin $\frac{3}{2}$). It does not apply to identical bosons (particles with integral spin), which include photons (spin 1), ^{12}C nuclei (spin 0). Any number of identical bosons may occupy the same state (that is, be described by the same wavefunction).

The Pauli *exclusion* principle is a special case of a general statement called the **Pauli principle**:

> When the labels of any two identical fermions are exchanged, the total wavefunction changes sign; when the labels of any two identical bosons are exchanged, the sign of the total wavefunction remains the same.

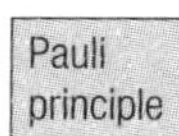

By 'total wavefunction' is meant the entire wavefunction, including the spin of the particles. To see that the Pauli principle implies the Pauli exclusion principle, we consider the wavefunction for two electrons $\psi(1,2)$. The Pauli principle implies that it is a fact of nature (which has its roots in the theory of relativity) that the wavefunction must change sign if we interchange the labels 1 and 2 wherever they occur in the function:

$$\Psi(2,1) = -\Psi(1,2) \tag{9.27}$$

Suppose the two electrons in an atom occupy an orbital ψ, then in the orbital approximation the overall wavefunction is $\psi(1)\psi(2)$. To apply the Pauli principle, we must deal with the total wavefunction, the wavefunction including spin. There are several possibilities for two spins: both α, denoted $\alpha(1)\alpha(2)$, both β, denoted $\beta(1)\beta(2)$, and one α the other β, denoted either $\alpha(1)\beta(2)$ or $\alpha(2)\beta(1)$. Because we cannot tell which electron is α and which is β, in the last case it is appropriate to express the spin states as the (normalized) linear combinations

$$\begin{aligned} \sigma_+(1,2) &= (1/2^{1/2})\{\alpha(1)\beta(2) + \beta(1)\alpha(2)\} \\ \sigma_-(1,2) &= (1/2^{1/2})\{\alpha(1)\beta(2) - \beta(1)\alpha(2)\} \end{aligned} \qquad (9.28)$$

A brief comment

A stronger justification for taking linear combinations in eqn 9.28 is that they correspond to eigenfunctions of the total spin operators S^2 and S_z, with $M_S = 0$ and, respectively, $S = 1$ and 0.

These combinations allow one spin to be α and the other β with equal probability. The total wavefunction of the system is therefore the product of the orbital part and one of the four spin states:

$$\begin{matrix} \psi(1)\psi(2)\alpha(1)\alpha(2) & \psi(1)\psi(2)\beta(1)\beta(2) \\ \psi(1)\psi(2)\sigma_+(1,2) & \psi(1)\psi(2)\sigma_-(1,2) \end{matrix} \qquad (9.29)$$

The Pauli principle says that, for a wavefunction to be acceptable (for electrons), it must change sign when the electrons are exchanged. In each case, exchanging the labels 1 and 2 converts the factor $\psi(1)\psi(2)$ into $\psi(2)\psi(1)$, which is the same, because the order of multiplying the functions does not change the value of the product. The same is true of $\alpha(1)\alpha(2)$ and $\beta(1)\beta(2)$. Therefore, the first two overall products are not allowed, because they do not change sign. The combination $\sigma_+(1,2)$ changes to

$$\sigma_+(2,1) = (1/2^{1/2})\{\alpha(2)\beta(1) + \beta(2)\alpha(1)\} = \sigma_+(1,2)$$

because it is simply the original function written in a different order. The third overall product is therefore also disallowed. Finally, consider $\sigma_-(1,2)$:

$$\begin{aligned} \sigma_-(2,1) &= (1/2^{1/2})\{\alpha(2)\beta(1) - \beta(2)\alpha(1)\} \\ &= -(1/2^{1/2})\{\alpha(1)\beta(2) - \beta(1)\alpha(2)\} = -\sigma_-(1,2) \end{aligned}$$

This combination does change sign (it is 'antisymmetric'). The product $\psi(1)\psi(2)\sigma_-(1,2)$ also changes sign under particle exchange, and therefore it is acceptable.

Now we see that only one of the four possible states is allowed by the Pauli principle, and the one that survives has paired α and β spins. This is the content of the Pauli exclusion principle. The exclusion principle is irrelevant when the orbitals occupied by the electrons are different, and both electrons may then have (but need not have) the same spin state. Nevertheless, even then the overall wavefunction must still be antisymmetric overall, and must still satisfy the Pauli principle itself.

A final point in this connection is that the acceptable product wavefunction $\psi(1)\psi(2)\sigma_-(1,2)$ can be expressed as a determinant:

$$\begin{aligned} \frac{1}{2^{1/2}}\begin{vmatrix} \psi(1)\alpha(1) & \psi(2)\alpha(2) \\ \psi(1)\beta(1) & \psi(2)\beta(2) \end{vmatrix} &= \frac{1}{2^{1/2}}\{\psi(1)\alpha(1)\psi(2)\beta(2) - \psi(2)\alpha(2)\psi(1)\beta(1)\} \\ &= \psi(1)\psi(2)\sigma_-(1,2) \end{aligned}$$

Any acceptable wavefunction for a closed-shell species can be expressed as a **Slater determinant**, as such determinants are known. In general, for N electrons in orbitals $\psi_a, \psi_b, \ldots$

$$\Psi(1,2,\ldots,N) = \frac{1}{(N!)^{1/2}}\begin{vmatrix} \psi_a(1)\alpha(1) & \psi_a(2)\alpha(2) & \psi_a(3)\alpha(3) & \cdots & \psi_a(N)\alpha(N) \\ \psi_a(1)\beta(1) & \psi_a(2)\beta(2) & \psi_a(3)\beta(3) & \cdots & \psi_a(N)\beta(N) \\ \psi_b(1)\alpha(1) & \psi_a(2)\alpha(2) & \psi_b(3)\alpha(3) & \cdots & \psi_b(N)\alpha(N) \\ \vdots & \vdots & \vdots & \vdots & \vdots \\ \psi_z(1)\beta(1) & \psi_z(2)\beta(2) & \psi_z(3)\beta(3) & \cdots & \psi_z(N)\beta(N) \end{vmatrix} \qquad [9.30a]$$

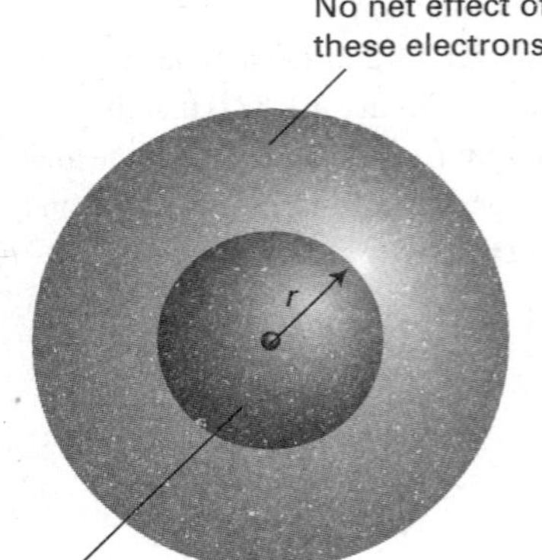

Fig. 9.19 An electron at a distance r from the nucleus experiences a Coulombic repulsion from all the electrons within a sphere of radius r and which is equivalent to a point negative charge located on the nucleus. The negative charge reduces the effective nuclear charge of the nucleus from Ze to $Z_{\text{eff}}e$.

Writing a many-electron wavefunction in this way ensures that it is antisymmetric under the interchange of any pair of electrons, as is explored in Problem 9.23. Because a Slater determinant takes up a lot of space, it is normally reported by writing only its diagonal elements, as in

$$\Psi(1,2,\ldots,N) = \left(\frac{1}{N!}\right)^{1/2} \det|\psi_a^{\alpha}(1)\psi_a^{\beta}(2)\psi_b^{\alpha}(3)\cdots\psi_z^{\beta}(N)| \qquad [9.30b]$$

Notation for a Slater determinant

Now we can return to lithium. In Li ($Z = 3$), the third electron cannot enter the 1s orbital because that orbital is already full: we say the K shell is *complete* and that the two electrons form a **closed shell**. Because a similar closed shell is characteristic of the He atom, we denote it [He]. The third electron is excluded from the K shell and must occupy the next available orbital, which is one with $n = 2$ and hence belonging to the L shell. However, we now have to decide whether the next available orbital is the 2s orbital or a 2p orbital, and therefore whether the lowest energy configuration of the atom is [He]$2s^1$ or [He]$2p^1$.

(c) Penetration and shielding

Unlike in hydrogenic atoms, the 2s and 2p orbitals (and, in general, all subshells of a given shell) are not degenerate in many-electron atoms. An electron in a many-electron atom experiences a Coulombic repulsion from all the other electrons present. If it is at a distance r from the nucleus, it experiences an average repulsion that can be represented by a point negative charge located at the nucleus and equal in magnitude to the total charge of the electrons within a sphere of radius r (Fig. 9.19). The effect of this point negative charge, when averaged over all the locations of the electron, is to reduce the full charge of the nucleus from Ze to $Z_{\text{eff}}e$, the **effective nuclear charge**. In everyday parlance, Z_{eff} itself is commonly referred to as the 'effective nuclear charge'. We say that the electron experiences a **shielded** nuclear charge, and the difference between Z and Z_{eff} is called the **shielding constant**, σ:

$$Z_{\text{eff}} = Z - \sigma \qquad [9.31]$$

Effective nuclear charge

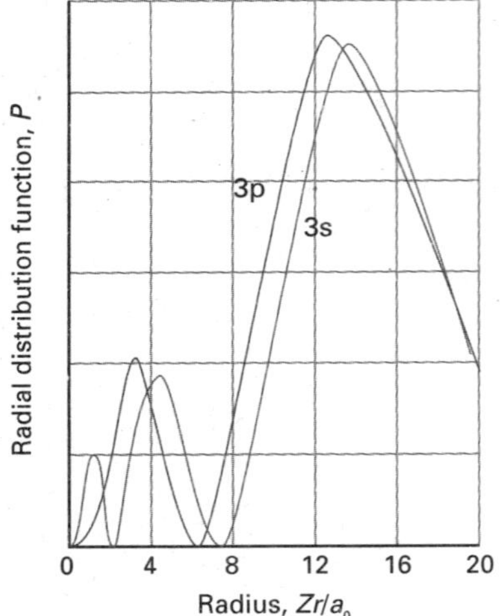

Fig. 9.20 An electron in an s orbital (here a 3s orbital) is more likely to be found close to the nucleus than an electron in a p orbital of the same shell (note the closeness of the innermost peak of the 3s orbital to the nucleus at $r = 0$). Hence an s electron experiences less shielding and is more tightly bound than a p electron.

interActivity Calculate and plot the graphs given above for $n = 4$.

The electrons do not actually 'block' the full Coulombic attraction of the nucleus: the shielding constant is simply a way of expressing the net outcome of the nuclear attraction and the electronic repulsions in terms of a single equivalent charge at the centre of the atom.

The shielding constant is different for s and p electrons because they have different radial distributions (Fig. 9.20). An s electron has a greater **penetration** through inner shells than a p electron, in the sense that it is more likely to be found close to the nucleus than a p electron of the same shell (the wavefunction of a p orbital, remember, is zero at the nucleus). Because only electrons inside the sphere defined by the location of the electron contribute to shielding, an s electron experiences less shielding than a p electron. Consequently, by the combined effects of penetration and shielding, an s electron is more tightly bound than a p electron of the same shell. Similarly, a d electron penetrates less than a p electron of the same shell (recall that the wavefunction of a d orbital varies as r^2 close to the nucleus, whereas a p orbital varies as r), and therefore experiences more shielding.

Shielding constants for different types of electrons in atoms have been calculated from their wavefunctions obtained by numerical solution of the Schrödinger equation for the atom (Table 9.2). We see that, in general, valence-shell s electrons do experience higher effective nuclear charges than p electrons, although there are some discrepancies. We return to this point shortly.

The consequence of penetration and shielding is that the energies of subshells of a shell in a many-electron atom (those with the same values of n but different values of l) in general lie in the order s < p < d < f. The individual orbitals of a given subshell (those with the same value of l but different values of m_l) remain degenerate because they all have the same radial characteristics and so experience the same effective nuclear charge.

We can now complete the Li story. Because the shell with $n = 2$ consists of two nondegenerate subshells, with the 2s orbital lower in energy than the three 2p orbitals, the third electron occupies the 2s orbital. This occupation results in the ground-state configuration $1s^22s^1$, with the central nucleus surrounded by a complete helium-like shell of two 1s electrons, and around that a more diffuse 2s electron. The electrons in the outermost shell of an atom in its ground state are called the **valence electrons** because they are largely responsible for the chemical bonds that the atom forms. Thus, the valence electron in Li is a 2s electron and its other two electrons belong to its core.

Table 9.2* Effective nuclear charge, $Z_{\text{eff}} = Z - \sigma$

Element	Z	Orbital	Z_{eff}
He	2	1s	1.6875
C	6	1s	5.6727
		2s	3.2166
		2p	3.1358

* More values are given in the *Data section*.

(d) The building-up principle

The extension of this argument is called the **building-up principle**, or the *Aufbau principle*, from the German word for building up, which will be familiar from introductory courses. In brief, we imagine the bare nucleus of atomic number Z, and then feed into the orbitals Z electrons in succession. The order of occupation is

1s 2s 2p 3s 3p 4s 3d 4p 5s 4d 5p 6s

and each orbital may accommodate up to two electrons. As an example, consider the carbon atom, for which $Z = 6$ and there are six electrons to accommodate. Two electrons enter and fill the 1s orbital, two enter and fill the 2s orbital, leaving two electrons to occupy the orbitals of the 2p subshell. Hence the ground-state configuration of C is $1s^22s^22p^2$, or more succinctly $[\text{He}]2s^22p^2$, with [He] the helium-like $1s^2$ core. However, we can be more precise: we can expect the last two electrons to occupy different 2p orbitals because they will then be further apart on average and repel each other less than if they were in the same orbital. Thus, one electron can be thought of as occupying the $2p_x$ orbital and the other the $2p_y$ orbital (the x, y, z designation is arbitrary, and it would be equally valid to use the complex forms of these orbitals), and the lowest energy configuration of the atom is $[\text{He}]2s^22p_x^12p_y^1$. The same rule applies whenever degenerate orbitals of a subshell are available for occupation. Thus, another rule of the building-up principle is:

Electrons occupy different orbitals of a given subshell before doubly occupying any one of them.

For instance, nitrogen ($Z = 7$) has the configuration $[\text{He}]2s^22p_x^12p_y^12p_z^1$, and only when we get to oxygen ($Z = 8$) is a 2p orbital doubly occupied, giving $[\text{He}]2s^22p_x^22p_y^12p_z^1$.

When electrons occupy orbitals singly we invoke **Hund's maximum multiplicity rule:**

An atom in its ground state adopts a configuration with the greatest number of unpaired electrons. [Hund's maximum multiplicity rule]

The explanation of Hund's rule is subtle, but it reflects the quantum mechanical property of **spin correlation**, that, as we demonstrate in the following *Justification*, electrons with parallel spins behave as if they have a tendency to stay well apart, and hence repel each other less. In essence, the effect of spin correlation is to allow the atom to shrink slightly, so the electron–nucleus interaction is improved when the spins are parallel. We can now conclude that, in the ground state of the carbon atom, the two 2p electrons have the same spin, that all three 2p electrons in the N atoms have the same spin (that is, they are parallel), and that the two 2p electrons in different orbitals in the O atom have the same spin (the two in the $2p_x$ orbital are necessarily paired).

Justification 9.6 *Spin correlation*

Suppose electron 1 is described by a wavefunction $\psi_a(r_1)$ and electron 2 is described by a wavefunction $\psi_b(r_2)$; then, in the orbital approximation, the joint wavefunction of the electrons is the product $\Psi = \psi_a(r_1)\psi_b(r_2)$. However, this wavefunction is not acceptable, because it suggests that we know which electron is in which orbital, whereas we cannot keep track of electrons. According to quantum mechanics, the correct description is either of the two following wavefunctions:

$$\Psi_{\pm} = (1/2^{1/2})\{\psi_a(r_1)\psi_b(r_2) \pm \psi_b(r_1)\psi_a(r_2)\}$$

According to the Pauli principle, because Ψ_+ is symmetrical under particle interchange, it must be multiplied by an antisymmetric spin function (the one denoted σ_-). That combination corresponds to a spin-paired state. Conversely, Ψ_- is antisymmetric, so it must be multiplied by one of the three symmetric spin states. These three symmetric states correspond to electrons with parallel spins (see Section 9.8 for an explanation).

Now consider the values of the two combinations when one electron approaches another, and $r_1 = r_2$. We see that Ψ_- vanishes, which means that there is zero probability of finding the two electrons at the same point in space when they have parallel spins. The other combination does not vanish when the two electrons are at the same point in space. Because the two electrons have different relative spatial distributions depending on whether their spins are parallel or not, it follows that their Coulombic interaction is different, and hence that the two states have different energies.

Neon, with $Z = 10$, has the configuration [He]2s^{2}2p^6, which completes the L shell. This closed-shell configuration is denoted [Ne], and acts as a core for subsequent elements. The next electron must enter the 3s orbital and begin a new shell, so an Na atom, with $Z = 11$, has the configuration [Ne]3s^1. Like lithium with the configuration [He]2s^1, sodium has a single s electron outside a complete core. This analysis has brought us to the origin of chemical periodicity. The L shell is completed by eight electrons, so the element with $Z = 3$ (Li) should have similar properties to the element with $Z = 11$ (Na). Likewise, Be ($Z = 4$) should be similar to $Z = 12$ (Mg), and so on, up to the noble gases He ($Z = 2$), Ne ($Z = 10$), and Ar ($Z = 18$).

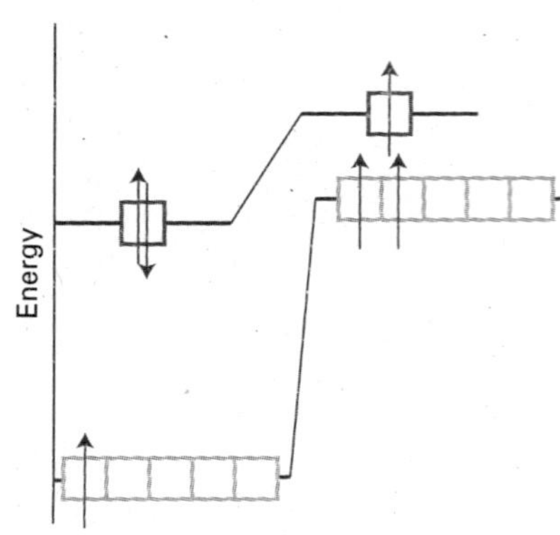

Fig. 9.21 Strong electron–electron repulsions in the 3d orbitals are minimized in the ground state of Sc if the atom has the configuration [Ar]3d^{1}4s^2 (shown on the left) instead of [Ar]3d^{2}4s^1 (shown on the right). The total energy of the atom is lower when it has the [Ar]3d^{1}4s^2 configuration despite the cost of populating the high energy 4s orbital.

Ten electrons can be accommodated in the five 3d orbitals, which accounts for the electron configurations of scandium to zinc. Calculations of the type discussed in Section 9.5 show that for these atoms the energies of the 3d orbitals are always lower than the energy of the 4s orbital. However, spectroscopic results show that Sc has the configuration [Ar]3d^{1}4s^2, instead of [Ar]3d^3 or [Ar]3d^{2}4s^1. To understand this observation, we have to consider the nature of electron–electron repulsions in 3d and 4s orbitals. The most probable distance of a 3d electron from the nucleus is less than that for a 4s electron, so two 3d electrons repel each other more strongly than two 4s electrons. As a result, Sc has the configuration [Ar]3d^{1}4s^2 rather than the two alternatives, for then the strong electron–electron repulsions in the 3d orbitals are minimized. The total energy of the atom is least despite the cost of allowing electrons to populate the high energy 4s orbital (Fig. 9.21). The effect just described is generally true for scandium through zinc, so their electron configurations are of the form [Ar]3d^{n}4s^2, where $n = 1$ for scandium and $n = 10$ for zinc. Two notable exceptions, which are observed experimentally, are Cr, with electron configuration [Ar]3d^{5}4s^1, and Cu, with electron configuration [Ar]3d^{10}4s^1.

At gallium, the building-up principle is used in the same way as in preceding periods. Now the 4s and 4p subshells constitute the valence shell, and the period terminates with krypton. Because 18 electrons have intervened since argon, this period is the first 'long period' of the periodic table. The existence of the d-block elements (the 'transition metals') reflects the stepwise occupation of the 3d orbitals, and the subtle shades of energy differences and effects of electron–electron repulsion along this series give rise to the rich complexity of inorganic d-metal chemistry. A similar intrusion of the f orbitals in Periods 6 and 7 accounts for the existence of the f block of the periodic table (the lanthanoids and actinoids).

We derive the configurations of cations of elements in the s, p, and d blocks of the periodic table by removing electrons from the ground-state configuration of the neutral atom in a specific order. First, we remove valence p electrons, then valence s electrons, and then as many d electrons as are necessary to achieve the specified charge. For instance, because the configuration of V is [Ar]3d^{3}4s^2, the V^{2+} cation has the configuration [Ar]3d^3. It is reasonable that we remove the more energetic 4s electrons in order to form the cation, but it is not obvious why the [Ar]3d^3 configuration is preferred in V^{2+} over the [Ar]3d^{1}4s^2 configuration, which is found in the isoelectronic Sc atom. Calculations show that the energy difference between [Ar]3d^3 and [Ar]3d^{1}4s^2 depends on Z_{eff}. As Z_{eff} increases, transfer of a 4s electron to a 3d orbital becomes more favourable because the electron–electron repulsions are compensated by attractive interactions between the nucleus and the electrons in the spatially compact 3d orbital. Indeed, calculations reveal that, for a sufficiently large Z_{eff}, [Ar]3d^3 is lower in energy than [Ar]3d^{1}4s^2. This conclusion explains why V^{2+} has a [Ar]3d^3 configuration and also accounts for the observed [Ar]4s^{0}3d^n configurations of the M^{2+} cations of Sc through Zn.

The configurations of anions of the p-block elements are derived by continuing the building-up procedure and adding electrons to the neutral atom until the configuration of the next noble gas has been reached. Thus, the configuration of the O^{2-} ion is achieved by adding two electrons to [He]2s^{2}2p^4, giving [He]2s^{2}2p^6, the same as the configuration of neon.

(e) Ionization energies and electron affinities

The minimum energy necessary to remove an electron from a many-electron atom in the gas phase is the **first ionization energy**, I_1, of the element. The **second ionization energy**, I_2, is the minimum energy needed to remove a second electron (from the singly charged cation). The variation of the first ionization energy through the periodic table is shown in Fig. 9.22 and some numerical values are given in Table 9.3. In thermodynamic calculations we often need the **standard enthalpy of ionization**, $\Delta_{\text{ion}}H^{\ominus}$. As shown in the following *Justification*, the two are related by

$$\Delta_{\text{ion}}H^{\ominus}(T) = I_1 + \tfrac{5}{2}RT \qquad (9.32)$$

Table 9.3* First and second ionization energies

Element	I_1/(kJ mol^{-1})	I_2/(kJ mol^{-1})
H	1312	
He	2372	5251
Mg	738	1451
Na	496	4562

* More values are given in the *Data section*.

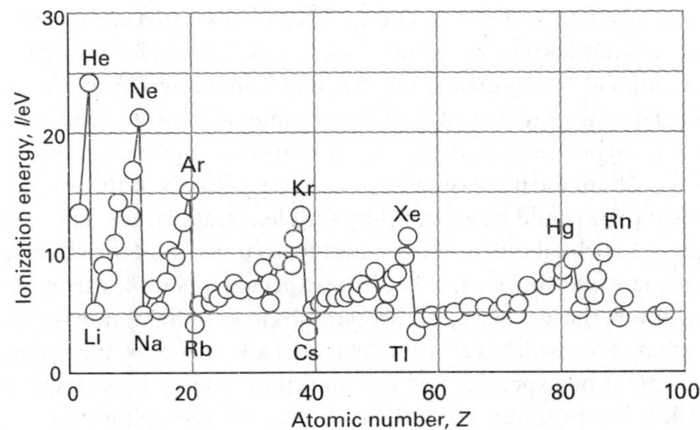

Fig. 9.22 The first ionization energies of the elements plotted against atomic number.

At 298 K, the difference between the ionization enthalpy and the corresponding ionization energy is 6.20 kJ mol^{-1}.

Justification 9.7 *The ionization enthalpy and the ionization energy*

It follows from Kirchhoff's law (Section 2.9 and eqn 2.36) that the reaction enthalpy for

$$M(g) \rightarrow M^+(g) + e^-(g)$$

at a temperature T is related to the value at $T = 0$ by

$$\Delta_r H^\ominus(T) = \Delta_r H^\ominus(0) + \int_0^T \Delta_r C_p^\ominus dT$$

The molar constant-pressure heat capacity of each species in the reaction is $\frac{5}{2}R$, so $\Delta_r C_p^\ominus = +\frac{5}{2}R$. The integral in this expression therefore evaluates to $+\frac{5}{2}RT$. The reaction enthalpy at $T = 0$ is the same as the (molar) ionization energy, I_1. Equation 9.32 then follows. The same expression applies to each successive ionization step, so the overall ionization enthalpy for the formation of M^{2+} is

$$\Delta_r H^\ominus(T) = I_1 + I_2 + 5RT$$

Table 9.4* Electron affinities, $E_a/(\text{kJ mol}^{-1})$

Cl	349		
F	322		
H	73		
O	141	O^-	−844

* More values are given in the *Data section*.

The **electron affinity**, E_{ea}, is the energy released when an electron attaches to a gas-phase atom (Table 9.4). In a common, logical (given its name), but not universal convention (which we adopt), the electron affinity is positive if energy is released when the electron attaches to the atom (that is, $E_{ea} > 0$ implies that electron attachment is exothermic). It follows from a similar argument to that given in the *Justification* above that the **standard enthalpy of electron gain**, $\Delta_{eg}H^\ominus$, at a temperature T is related to the electron affinity by

$$\Delta_{eg}H^\ominus(T) = -E_{ea} - \tfrac{5}{2}RT \qquad (9.33)$$

Note the change of sign. In typical thermodynamic cycles the $\frac{5}{2}RT$ that appears in eqn 9.32 cancels that in eqn 9.33, so ionization energies and electron affinities can be used directly. A final preliminary point is that the electron-gain enthalpy of a species X is the negative of the ionization enthalpy of its negative ion:

$$\Delta_{eg}H^\ominus(X) = -\Delta_{ion}H^\ominus(X^-) \qquad (9.34)$$

As ionization energy is often easier to measure than electron affinity; this relation can be used to determine numerical values of the latter.

As will be familiar from introductory chemistry, ionization energies and electron affinities show periodicities. The former is more regular and we concentrate on it. Lithium has a low first ionization energy because its outermost electron is well shielded from the nucleus by the core ($Z_{eff} = 1.3$, compared with $Z = 3$). The ionization energy of beryllium ($Z = 4$) is greater but that of boron is lower than that of beryllium because in the latter the outermost electron occupies a 2p orbital and is less strongly bound than if it had been a 2s electron. The ionization energy increases from boron to nitrogen on account of the increasing nuclear charge. However, the ionization energy of oxygen is less than would be expected by simple extrapolation. The explanation is that at oxygen a 2p orbital must become doubly occupied, and the electron–electron repulsions are increased above what would be expected by simple extrapolation along the row. In addition, the loss of a 2p electron results in a configuration with a half-filled subshell (like that of N), which is an arrangement of low energy, so the energy of $O^+ + e^-$ is lower than might be expected, and the ionization energy is correspondingly low too. (The kink is less pronounced in the next row, between phosphorus and sulfur, because their orbitals are more diffuse.) The values for oxygen, fluorine, and neon fall roughly on the same line, the increase of their ionization energies reflecting the increasing attraction of the more highly charged nuclei for the outermost electrons.

The outermost electron in sodium ($Z = 11$) is 3s. It is far from the nucleus, and the latter's charge is shielded by the compact, complete neon-like core, with the result that $Z_{\text{eff}} \approx 2.5$. As a result, the ionization energy of sodium is substantially lower than that of neon ($Z = 10$, $Z_{\text{eff}} \approx 5.8$). The periodic cycle starts again along this row, and the variation of the ionization energy can be traced to similar reasons.

Electron affinities are greatest close to fluorine, for the incoming electron enters a vacancy in a compact valence shell and can interact strongly with the nucleus. The attachment of an electron to an anion (as in the formation of O^{2-} from O^-) is invariably endothermic, so E_{ea} is negative. The incoming electron is repelled by the charge already present. Electron affinities are also small, and may be negative, when an electron enters an orbital that is far from the nucleus (as in the heavier alkali metal atoms) or is forced by the Pauli principle to occupy a new shell (as in the noble gas atoms).

9.5 Self-consistent field orbitals

Key point **The Schrödinger equation for many-electron atoms is solved numerically and iteratively until the solutions are self-consistent.**

The central difficulty of the Schrödinger equation is the presence of the electron–electron interaction terms. The potential energy of the electrons is

$$V = -\sum_i \frac{Ze^2}{4\pi\varepsilon_0 r_i} + \frac{1}{2}\sum_{i,j}{}' \frac{e^2}{4\pi\varepsilon_0 r_{ij}} \tag{9.35}$$

The prime on the second sum indicates that $i \neq j$, and the factor of one-half prevents double-counting of electron pair repulsions (1 interacting with 2 is the same as 2 interacting with 1). The first term is the total attractive interaction between the electrons and the nucleus. The second term is the total repulsive interaction between the electrons; r_{ij} is the distance between electrons i and j. It is hopeless to expect to find analytical solutions of a Schrödinger equation with such a complicated potential energy term, but computational techniques are available that give very detailed and reliable numerical solutions for the wavefunctions and energies. The techniques were originally introduced by D.R. Hartree (before computers were available) and then modified by V. Fock to take into account the Pauli principle correctly. In broad outline, the **Hartree–Fock self-consistent field** (HF-SCF) procedure is as follows.

Imagine that we have a rough idea of the structure of the atom. In the Ne atom, for instance, the orbital approximation suggests the configuration $1s^2 2s^2 2p^6$ with the orbitals approximated by hydrogenic atomic orbitals. Now consider one of the 2p electrons. A Schrödinger equation can be written for this electron by ascribing to it a potential energy due to the nuclear attraction and the repulsion from the other electrons. This equation has the form

$$\begin{aligned} &\hat{H}(1)\psi_{2p}(1) + V(\text{other electrons})\psi_{2p}(1) \\ &\quad - V(\text{exchange correction})\psi_{2p}(1) = E_{2p}\psi_{2p}(1) \end{aligned} \tag{9.36}$$

Although the equation is for the 2p orbital in neon, it depends on the wavefunctions of all the other occupied orbitals in the atom. A similar equation can be written for the 1s and 2s orbitals in the atom. The various terms are as follows:

- The first term on the left is the contribution of the kinetic energy and the attraction of the electron to the nucleus, just as in a hydrogenic atom.

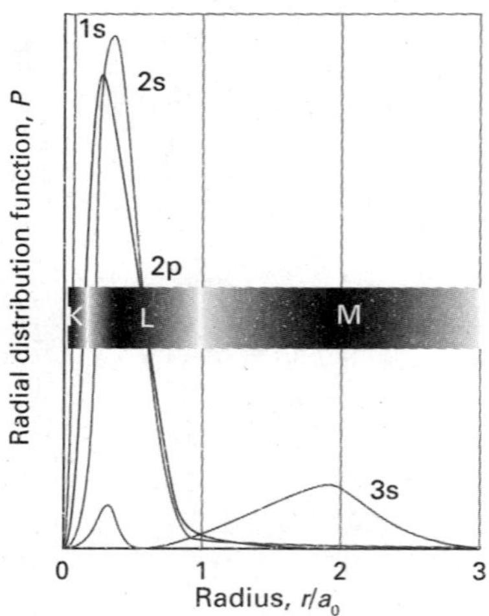

Fig. 9.23 The radial distribution functions for the orbitals of Na based on SCF calculations. Note the shell-like structure, with the 3s orbital outside the inner K and L shells.

- The second term takes into account the potential energy of the electron of interest due to the electrons in the other occupied orbitals.
- The third term is an *exchange correction* that takes into account the spin correlation effects discussed earlier.

There is no hope of solving eqn 9.36 analytically. However, it can be solved numerically if we guess an approximate form of the wavefunctions of all the orbitals except 2p. The procedure is then repeated for the other orbitals in the atom, the 1s and 2s orbitals. This sequence of calculations gives the form of the 2p, 2s, and 1s orbitals, and in general they will differ from the set used initially to start the calculation. These improved orbitals can be used in another cycle of calculation, and a second improved set of orbitals is obtained. The recycling continues until the orbitals and energies obtained are insignificantly different from those used at the start of the current cycle. The solutions are then self-consistent and accepted as solutions of the problem.

Figure 9.23 shows plots of some of the HF-SCF radial distribution functions for sodium. They show the grouping of electron density into shells, as was anticipated by the early chemists, and the differences of penetration as discussed above. These SCF calculations therefore support the qualitative discussions that are used to explain chemical periodicity. They also considerably extend that discussion by providing detailed wavefunctions and precise energies.

The spectra of complex atoms

The spectra of atoms rapidly become very complicated as the number of electrons increases, but there are some important and moderately simple features that make atomic spectroscopy useful in the study of the composition of samples as large and as complex as stars. The general idea is straightforward: lines in the spectrum (in either emission or absorption) occur when the atom undergoes a transition with a change of energy $|\Delta E|$, and emits or absorbs a photon of frequency $\nu = |\Delta E|/h$ and $\tilde{\nu} = |\Delta E|/hc$. Hence, we can expect the spectrum to give information about the energies of electrons in atoms. However, the actual energy levels are not given solely by the energies of the orbitals, because the electrons interact with one another in various ways, and there are contributions to the energy in addition to those we have already considered.

9.6 Linewidths

Key points **(a) Doppler broadening of a spectral line is caused by the distribution of molecular and atomic speeds in a sample. (b) Lifetime broadening arises from the finite lifetime of an excited state and a consequent blurring of energy levels. Collisions between atoms can affect excited state lifetimes and spectral linewidths. The natural linewidth of a transition is an intrinsic property that depends on the rate of spontaneous emission at the transition frequency.**

A number of effects contribute to the widths of spectroscopic lines. Some contributions to linewidths can be modified by changing the conditions, and to achieve high resolutions we need to know how to minimize these contributions. Other contributions cannot be changed, and represent an inherent limitation on resolution.

(a) Doppler broadening

One important broadening process in gaseous samples is the **Doppler effect**, in which radiation is shifted in frequency when the source is moving towards or away from the observer. When a source emitting electromagnetic radiation of frequency ν moves with a speed s relative to an observer, the observer detects radiation of frequency

$$\nu_{\text{receding}} = \nu\left(\frac{1-s/c}{1+s/c}\right)^{1/2} \qquad \nu_{\text{approaching}} = \left(\frac{1+s/c}{1-s/c}\right)^{1/2} \qquad \text{Doppler shifts} \qquad (9.37\text{a})$$

where c is the speed of light. For nonrelativistic speeds ($s \ll c$), these expressions simplify to

$$\nu_{\text{receding}} \approx \frac{\nu}{1+s/c} \qquad \nu_{\text{approaching}} \approx \frac{\nu}{1-s/c} \qquad (9.37\text{b})$$

Atoms reach high speeds in all directions in a gas, and a stationary observer detects the corresponding Doppler-shifted range of frequencies. Some atoms approach the observer, some move away; some move quickly, others slowly. The detected spectral 'line' is the absorption or emission profile arising from all the resulting Doppler shifts. As shown in the following *Justification*, the profile reflects the distribution of velocities parallel to the line of sight, which is a bell-shaped Gaussian curve. The Doppler line shape is therefore also a Gaussian (Fig. 9.24), and we show in the *Justification* that, when the temperature is T and the mass of the atom is m, then the observed width of the line at half-height (in terms of frequency or wavelength) is

$$\delta\nu_{\text{obs}} = \frac{2\nu}{c}\left(\frac{2kT\ln 2}{m}\right)^{1/2} \qquad \delta\lambda_{\text{obs}} = \frac{2\lambda}{c}\left(\frac{2kT\ln 2}{m}\right)^{1/2} \qquad \text{Doppler broadening} \qquad (9.38)$$

For an atom like Si at room temperature ($T \approx 300$ K), $\delta\nu/\nu \approx 2.3 \times 10^{-6}$. Doppler broadening increases with temperature because the molecules acquire a wider range of speeds. Therefore, to obtain spectra of maximum sharpness, it is best to work with cool samples.

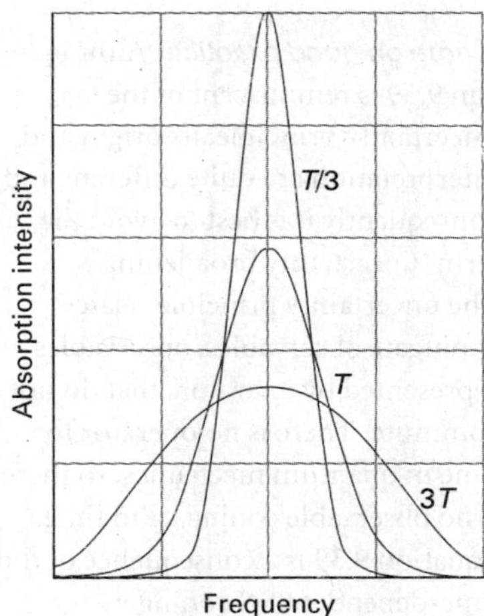

Fig. 9.24 The Gaussian shape of a Doppler-broadened spectral line reflects the Maxwell distribution of speeds in the sample at the temperature of the experiment. Notice that the line broadens as the temperature is increased.

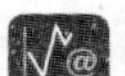
interActivity In a spectrometer that makes use of phase-sensitive detection, the output signal is proportional to the first derivative of the signal intensity, $dI/d\nu$. Plot the resulting line shape for various temperatures. How is the separation of the peaks related to the temperature?

Justification 9.8 *Doppler broadening*

We know from the Boltzmann distribution (*Fundamentals F.5a*) that the probability that an atom of mass m and speed s in a gas phase sample at a temperature T has kinetic energy $E_k = \frac{1}{2}ms^2$ is proportional to $e^{-ms^2/2kT}$. The observed frequencies, ν_{obs}, emitted or absorbed by the molecule are related to its speed by eqn 9.37b. When $s \ll c$, the Doppler shift in the frequency is

$$\nu_{\text{obs}} - \nu \approx \pm\nu s/c$$

which implies a symmetrical distribution of observed frequencies with respect to atomic speeds. More specifically, the intensity I of a transition at ν_{obs} is proportional to the probability of finding the atom that emits or absorbs at ν_{obs}, so it follows from the Boltzmann distribution and the expression for the Doppler shift that

$$I(\nu_{\text{obs}}) \propto e^{-mc^2(\nu_{\text{obs}}-\nu)^2/2\nu^2kT}$$

which has the form of a Gaussian function. The width at half-height can be calculated directly from the exponent to give eqn 9.38.

A brief comment

A Gaussian function of the general form $y(x) = ae^{-(x-b)^2/2\sigma^2}$, where a, b, and σ are constants, has a maximum $y(b) = a$ and a width at half-height $\delta x = 2\sigma(2\ln 2)^{1/2}$.

(b) Lifetime broadening

It is found that spectroscopic lines from gas-phase samples are not infinitely sharp even when Doppler broadening has been largely eliminated by working at low temperatures. This residual broadening is due to quantum mechanical effects. Specifically, when the Schrödinger equation is solved for a system that is changing with time,

A note on good practice Although eqn 9.39 is reminiscent of the uncertainty principle, its origin and interpretation are quite different and consequently it is best to avoid the term 'uncertainty broadening'. The uncertainty principle relates conjugate observables, observables represented by operators that do not commute. There is no operator for time in quantum mechanics, so there is no observable conjugate to time. Equation 9.39 is a consequence of the time-dependent Schrödinger equation.

it is found that it is impossible to specify the energy levels exactly. If on average a system survives in a state for a time τ (tau), the **lifetime** of the state, then its energy levels are blurred to an extent of order δE, where

$$\delta E \approx \frac{\hbar}{\tau} \qquad \text{Lifetime broadening} \qquad (9.39)$$

This expression is reminiscent of the Heisenberg uncertainty principle (eqn 7.39), and consequently this **lifetime broadening** is often called 'uncertainty broadening'. No excited state has an infinite lifetime; therefore, all states are subject to some lifetime broadening and the shorter the lifetimes of the states involved in a transition the broader the corresponding spectral lines.

• A brief illustration

When the energy spread is expressed as a wavenumber through $\delta E = hc\delta\tilde{\nu}$, and the values of the fundamental constants introduced, this relation becomes

$$\delta\tilde{\nu} \approx \frac{5.3\ \text{cm}^{-1}}{\tau/\text{ps}}$$

A typical electronic excited state natural lifetime is about 10^{-8} s (10 ns), corresponding to a natural width of about 5×10^{-4} cm^{-1} (15 MHz). A typical natural lifetime of a molecular rotation is about 10^3 s, corresponding to a natural linewidth of only 5×10^{-15} cm^{-1} (of the order of 10^{-4} Hz). •

Two processes are responsible for the finite lifetimes of excited states. The dominant one for low frequency transitions is **collisional deactivation**, which arises from collisions between atoms or with the walls of the container. If the **collisional lifetime**, the mean time between collisions, is τ_{col}, the resulting collisional linewidth is $\delta E_{col} \approx \hbar/\tau_{col}$. Because $\tau_{col} = 1/z$, where z is the collision frequency, and from the kinetic model of gases (Section 1.2) we know that z is proportional to the pressure, we see that the collisional linewidth is proportional to the pressure. The collisional linewidth can therefore be minimized by working at low pressures.

The rate of spontaneous emission cannot be changed. Hence it is a natural limit to the lifetime of an excited state, and the resulting lifetime broadening is the **natural linewidth** of the transition. The natural linewidth is an intrinsic property of the transition, and cannot be changed by modifying the conditions. Natural linewidths depend strongly on the transition frequency (as explained in Section 13.4, they increase as ν^3), so low frequency transitions have smaller natural linewidths than high frequency transitions.

9.7 Quantum defects and ionization limits

Key point **The general form of the expression for the energy of a level in a many-electron atom can be preserved by introducing an empirical quantum defect.**

One application of atomic spectroscopy is to the determination of ionization energies. However, we cannot use the procedure illustrated in Example 9.1 indiscriminately because the energy levels of a many-electron atom do not in general vary as $1/n^2$. If we confine attention to the outermost electrons, then we know that, as a result of penetration and shielding, they experience a nuclear charge of slightly more than $1e$ because in a neutral atom the other $Z - 1$ electrons cancel all but about one unit of nuclear charge. Typical values of Z_{eff} are a little more than 1, so we expect binding energies to be given by a term of the form $-hcR/n^2$, but lying slightly lower in energy than this formula predicts. We therefore introduce a **quantum defect**, δ, and write the energy as $-hcR/(n-\delta)^2$. The quantum defect is best regarded as a purely empirical quantity.

There are some excited states that are so diffuse that $\delta \to 0$ and the $1/n^2$ variation is valid: these states are called **Rydberg states.** In such cases we can write

$$\tilde{\nu} = \frac{I}{hc} - \frac{R}{n^2} \quad (9.40)$$

and a plot of wavenumber against $1/n^2$ can be used to obtain I by extrapolation; in practice, one would use a linear regression fit using a computer. If the lower state is not the ground state (a possibility if we wish to generalize the concept of ionization energy), the ionization energy of the ground state can be determined by adding the appropriate energy difference to the ionization energy obtained as described here.

9.8 Singlet and triplet states

Key points **Two electrons with paired spins form a singlet state; if their spins are parallel, they form a triplet state.**

Suppose we were interested in the energy levels of a He atom, with its two electrons. We know that the ground-state configuration is $1s^2$, and can anticipate that an excited configuration will be one in which one of the electrons has been promoted into a 2s orbital, giving the configuration $1s^12s^1$. The two electrons need not be paired because they occupy different orbitals. According to Hund's maximum multiplicity rule, the state of the atom with the spins parallel lies lower in energy than the state in which they are paired. Both states are permissible, and can contribute to the spectrum of the atom.

Parallel and antiparallel (paired) spins differ in their overall spin angular momentum. In the paired case, the two spin momenta cancel each other, and there is zero net spin (as was depicted in Fig. 9.18). The paired-spin arrangement is called a **singlet**. Its spin state is the one we denoted σ_- in the discussion of the Pauli principle:

$$\sigma_-(1,2) = (1/2^{1/2})\{\alpha(1)\beta(2) - \beta(1)\alpha(2)\} \quad \text{Singlet spin function} \quad (9.41a)$$

The angular momenta of two parallel spins add together to give a nonzero total spin, and the resulting state is called a **triplet**. As illustrated in Fig. 9.25, there are three ways of achieving a nonzero total spin, but only one way to achieve zero spin. The three spin states are the symmetric combinations introduced earlier:

$$\begin{aligned} &\alpha(1)\alpha(2) \\ &\sigma_+(1,2) = (1/2^{1/2})\{\alpha(1)\beta(2) + \beta(1)\alpha(2)\} \\ &\beta(1)\beta(2) \end{aligned} \quad \text{Triplet spin functions} \quad (9.41b)$$

The fact that the parallel arrangement of spins in the $1s^12s^1$ configuration of the He atom lies lower in energy than the antiparallel arrangement can now be expressed by saying that the triplet state of the $1s^12s^1$ configuration of He lies lower in energy than the singlet state. This is a general conclusion that applies to other atoms (and molecules) and, *for states arising from the same configuration, the triplet state generally lies lower than the singlet state.* The origin of the energy difference lies in the effect of spin correlation on the Coulombic interactions between electrons, as we saw in the case of Hund's rule for ground-state configurations. Because the Coulombic interaction between electrons in an atom is strong, the difference in energies between singlet

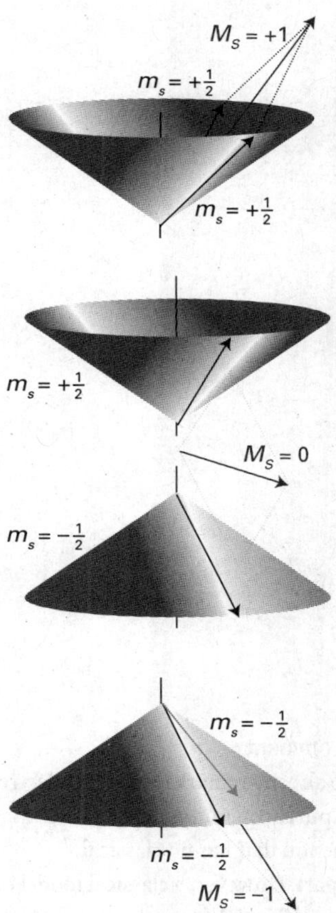

Fig. 9.25 When two electrons have parallel spins, they have a nonzero total spin angular momentum. There are three ways of achieving this resultant, which are shown by these vector representations. Note that, although we cannot know the orientation of the spin vectors on the cones, the angle between the vectors is the same in all three cases, for all three arrangements have the same total spin angular momentum (that is, the resultant of the two vectors has the same length in each case, but points in different directions). Compare this diagram with Fig. 9.18, which shows the antiparallel case. Note that, whereas two paired spins are precisely antiparallel, two 'parallel' spins are not strictly parallel.

and triplet states of the same configuration can be large. The two states of $1s^12s^1$ He, for instance, differ by 6421 cm^{-1} (corresponding to 0.80 eV).

The spectrum of atomic helium is more complicated than that of atomic hydrogen, but there are two simplifying features. One is that the only excited configurations it is necessary to consider are of the form $1s^1nl^1$: that is, only one electron is excited. Excitation of two electrons requires an energy greater than the ionization energy of the atom, so the He^+ ion is formed instead of the doubly excited atom. Second, no radiative transitions take place between singlet and triplet states because the relative orientation of the two electron spins cannot change during a transition. Thus, there is a spectrum arising from transitions between singlet states (including the ground state) and between triplet states, but not between the two. Spectroscopically, helium behaves like two distinct species, and the early spectroscopists actually thought of helium as consisting of 'parahelium' and 'orthohelium'. The Grotrian diagram for helium in Fig. 9.26 shows the two sets of transitions.

9.9 Spin–orbit coupling

Key points **The orbital and spin angular momenta interact magnetically. (a) Spin–orbit coupling results in the levels of a term having different energies. (b) Fine structure in a spectrum is due to transitions to different levels of a term.**

A brief comment

We have already remarked that the electron's spin is a purely quantum mechanical phenomenon that has no classical counterpart. However, a classical model can give us partial insight into the origin of an electron's magnetic moment. Namely, the magnetic field generated by a spinning electron, regarded classically as a moving charge, induces a magnetic moment. This model is merely a visualization aid and cannot be used to explain the magnitude of the magnetic moment of the electron or the origin of spin magnetic moments in electrically neutral particles, such as the neutron.

An electron has a magnetic moment that arises from its spin (Fig. 9.27). Similarly, an electron with orbital angular momentum (that is, an electron in an orbital with $l > 0$) is in effect a circulating current, and possesses a magnetic moment that arises from its orbital momentum. The interaction of the spin magnetic moment with the magnetic field arising from the orbital angular momentum is called **spin–orbit coupling**. The strength of the coupling, and its effect on the energy levels of the atom, depend on the relative orientations of the spin and orbital magnetic moments, and therefore on the relative orientations of the two angular momenta (Fig. 9.28).

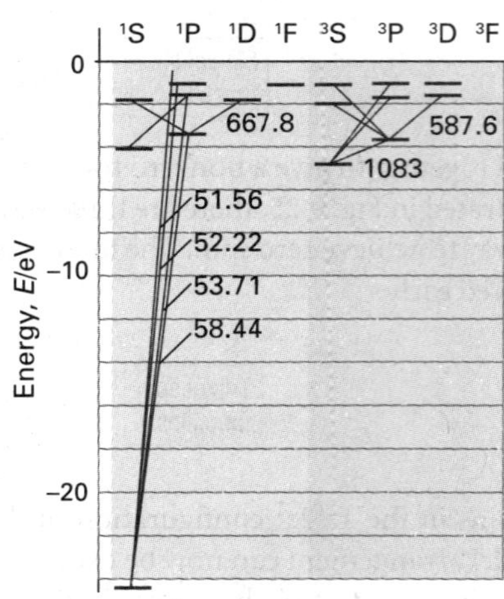

Fig. 9.26 Part of the Grotrian diagram for a helium atom. Note that there are no transitions between the singlet and triplet levels, denoted respectively by the left superscripts 1 and 3.

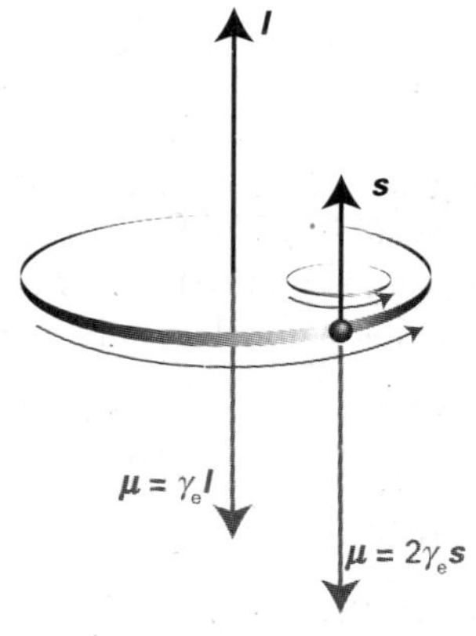

Fig. 9.27 Angular momentum gives rise to a magnetic moment (μ). For an electron, the magnetic moment is antiparallel to the orbital angular momentum, but proportional to it. For spin angular momentum, there is a factor of 2, which increases the magnetic moment to twice its expected value (see Section 9.10).

(a) The total angular momentum

One way of expressing the dependence of the spin–orbit interaction on the relative orientation of the spin and orbital momenta is to say that it depends on the total angular momentum of the electron, the vector sum of its spin and orbital momenta. Thus, when the spin and orbital angular momenta are nearly parallel, the total angular momentum is high; when the two angular momenta are opposed, the total angular momentum is low.

The total angular momentum of an electron is described by the quantum numbers j and m_j, with $j = l + \frac{1}{2}$ (when the two angular momenta are in the same direction) or $j = l - \frac{1}{2}$ (when they are opposed, Fig. 9.29). The different values of j that can arise for a given value of l label **levels** of a term. For $l = 0$, the only permitted value is $j = \frac{1}{2}$ (the total angular momentum is the same as the spin angular momentum because there is no other source of angular momentum in the atom). When $l = 1$, j may be either $\frac{3}{2}$ (the spin and orbital angular momenta are in the same sense) or $\frac{1}{2}$ (the spin and angular momenta are in opposite senses).

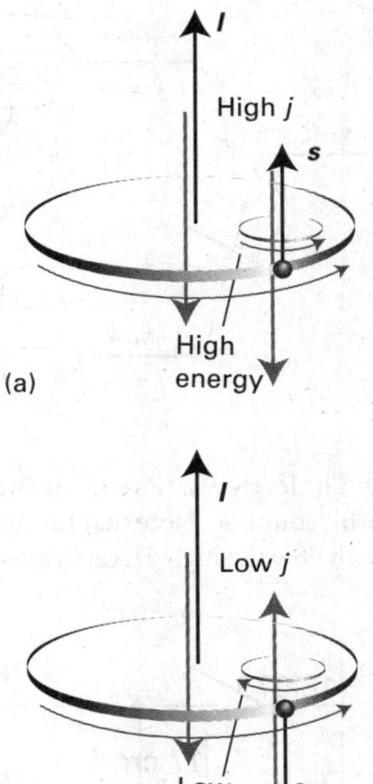

Fig. 9.28 Spin–orbit coupling is a magnetic interaction between spin and orbital magnetic moments. When the angular momenta are parallel, as in (a), the magnetic moments are aligned unfavourably; when they are opposed, as in (b), the interaction is favourable. This magnetic coupling is the cause of the splitting of a configuration into levels.

Example 9.4 *Identifying the levels of a configuration*

Identify the levels that may arise from the configurations (a) d^1, (b) s^1.

Method In each case, identify the value of l and then the possible values of j. For these one-electron systems, the total angular momentum is the sum and difference of the orbital and spin momenta.

Answer (a) For a d electron, $l = 2$ and there are two levels in the configuration, one with $j = 2 + \frac{1}{2} = \frac{5}{2}$ and the other with $j = 2 - \frac{1}{2} = \frac{3}{2}$. (b) For an s electron $l = 0$, so only one level is possible, and $j = \frac{1}{2}$.

Self-test 9.8 Identify the levels of the configurations (a) p^1 and (b) f^1.
[(a) $\frac{3}{2}, \frac{1}{2}$; (b) $\frac{7}{2}, \frac{5}{2}$]

The dependence of the spin–orbit interaction on the value of j is expressed in terms of the **spin–orbit coupling constant**, $\tilde{A}$ (which is typically expressed as a wavenumber). The quantum mechanical calculation outlined in *Further information 9.2* leads to the result that the energies of the levels with quantum numbers s, l, and j are given by

$$E_{l,s,j} = \tfrac{1}{2}hc\tilde{A}\{j(j+1) - l(l+1) - s(s+1)\} \tag{9.42}$$

• A brief illustration

The unpaired electron in the ground state of an alkali metal atom has $l = 0$, so $j = \frac{1}{2}$. Because the orbital angular momentum is zero in this state, the spin–orbit coupling energy is zero (as is confirmed by setting $j = s$ and $l = 0$ in eqn 9.42). When the electron is excited to an orbital with $l = 1$, it has orbital angular momentum and can give rise to a magnetic field that interacts with its spin. In this configuration the electron can have $j = \frac{3}{2}$ or $j = \frac{1}{2}$, and the energies of these levels are

$$E_{3/2} = \tfrac{1}{2}hc\tilde{A}\{\tfrac{3}{2}\times\tfrac{5}{2} - 1\times 2 - \tfrac{1}{2}\times\tfrac{3}{2}\} = \tfrac{1}{2}hc\tilde{A}$$

$$E_{1/2} = \tfrac{1}{2}hc\tilde{A}\{\tfrac{1}{2}\times\tfrac{3}{2} - 1\times 2 - \tfrac{1}{2}\times\tfrac{3}{2}\} = -hc\tilde{A}$$

The corresponding energies are shown in Fig. 9.30. Note that the baricentre (the 'centre of gravity') of the levels is unchanged, because there are four states of energy $\frac{1}{2}hc\tilde{A}$ and two of energy $-hc\tilde{A}$. •

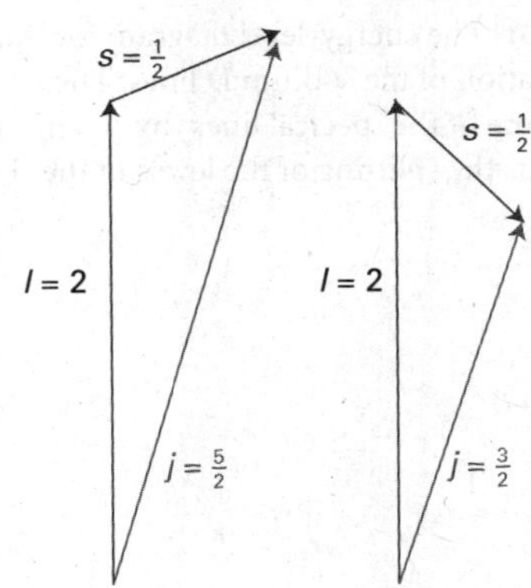

Fig. 9.29 The coupling of the spin and orbital angular momenta of a d electron ($l = 2$) gives two possible values of j depending on the relative orientations of the spin and orbital angular momenta of the electron.

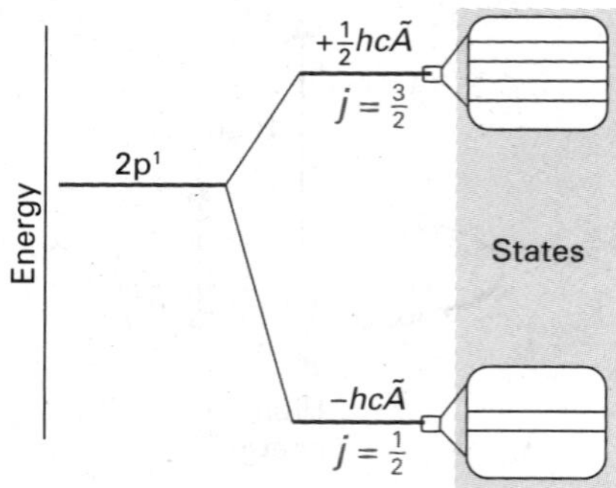

Fig. 9.30 The levels of a ^{2}P term arising from spin–orbit coupling. Note that the low-j level lies below the high-j level in energy.

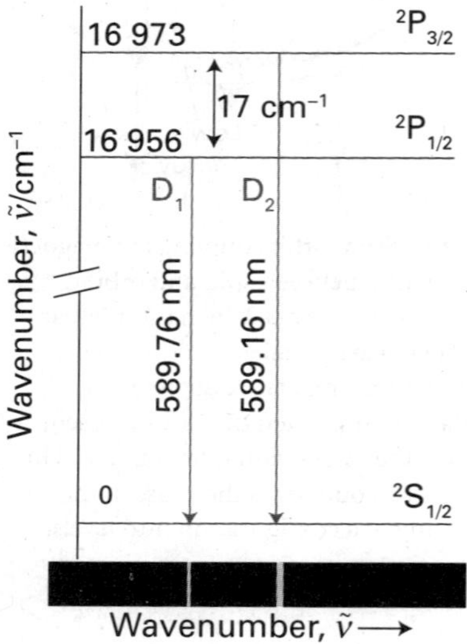

Fig. 9.31 The energy-level diagram for the formation of the sodium D lines. The splitting of the spectral lines (by 17 cm^{-1}) reflects the splitting of the levels of the ^{2}P term.

The strength of the spin–orbit coupling depends on the nuclear charge. To understand why this is so, imagine riding on the orbiting electron and seeing a charged nucleus apparently orbiting around us (like the Sun rising and setting). As a result, we find ourselves at the centre of a ring of current. The greater the nuclear charge, the greater this current, and therefore the stronger the magnetic field we detect. Because the spin magnetic moment of the electron interacts with this orbital magnetic field, it follows that the greater the nuclear charge, the stronger the spin–orbit interaction. The coupling increases sharply with atomic number (as Z^4). Whereas it is only small in H (giving rise to shifts of energy levels of no more than about 0.4 cm^{-1}), in heavy atoms like Pb it is very large (giving shifts of the order of thousands of reciprocal centimetres).

(b) Fine structure

Two spectral lines are observed when the p electron of an electronically excited alkali metal atom undergoes a transition and falls into a lower *s* orbital. One line is due to a transition starting in a $j=\frac{3}{2}$ level and the other line is due to a transition starting in the $j=\frac{1}{2}$ level of the same configuration. The two lines are an example of the **fine structure** of a spectrum, the structure in a spectrum due to spin–orbit coupling. Fine structure can be clearly seen in the emission spectrum from sodium vapour excited by an electric discharge (for example, in one kind of street lighting). The yellow line at 589 nm (close to 17 000 cm^{-1}) is actually a doublet composed of one line at 589.76 nm (16 956.2 cm^{-1}) and another at 589.16 nm (16 973.4 cm^{-1}); the components of this doublet are the 'D lines' of the spectrum (Fig. 9.31). Therefore, in Na, the spin–orbit coupling affects the energies by about 17 cm^{-1}.

Example 9.5 *Analysing a spectrum for the spin–orbit coupling constant*

The origin of the D lines in the spectrum of atomic sodium is shown in Fig. 9.31. Calculate the spin–orbit coupling constant for the upper configuration of the Na atom.

Method We see from Fig. 9.31 that the splitting of the lines is equal to the energy separation of the $j=\frac{3}{2}$ and $\frac{1}{2}$ levels of the excited configuration. This separation can be expressed in terms of $\tilde{A}$ by using eqn 9.42. Therefore, set the observed splitting equal to the energy separation calculated from eqn 9.42 and solve the equation for $\tilde{A}$.

Answer The two levels are split by

$$\Delta\tilde{\nu}=\tilde{A}\tfrac{1}{2}\{\tfrac{3}{2}(\tfrac{3}{2}+1)-\tfrac{1}{2}(\tfrac{1}{2}+1)\}=\tfrac{3}{2}\tilde{A}$$

The experimental value of $\Delta\tilde{\nu}$ is 17.2 cm^{-1}; therefore

$$\tilde{A}=\tfrac{2}{3}\times(17.2\ \text{cm}^{-1})=11.5\ \text{cm}^{-1}$$

The same calculation repeated for the other alkali metal atoms gives Li: 0.23 cm^{-1}, K: 38.5 cm^{-1}, Rb: 158 cm^{-1}, Cs: 370 cm^{-1}. Note the increase of $\tilde{A}$ with atomic number (but more slowly than Z^4 for these many-electron atoms).

Self-test 9.9 The configuration . . . 4p^{6}5d^1 of rubidium has two levels at 25 700.56 cm^{-1} and 25 703.52 cm^{-1} above the ground state. What is the spin–orbit coupling constant in this excited state? [1.18 cm^{-1}]

9.10 Term symbols and selection rules

Key points **A term symbol specifies the angular momentum states of an atom. (a) Angular momenta are combined into a resultant by using the Clebsch–Gordan series. (b) The multiplicity of a term is the value of $2S + 1$. (c) The total angular momentum in light atoms is obtained on the basis of Russell–Saunders coupling; in heavy atoms, *jj*-coupling is used. (d) Selection rules for light atoms include the fact that changes of total spin do not occur.**

We have used expressions such as 'the $j=\frac{3}{2}$ level of a configuration'. A **term symbol**, which is a symbol looking like $^2P_{3/2}$ or $^3D_2$, conveys this information much more succinctly. The convention of using lower-case letters to label orbitals and upper-case letters to label overall states applies throughout spectroscopy, not just to atoms.

A term symbol gives three pieces of information:

- The letter (P or D in the examples) indicates the total orbital angular momentum quantum number, L.
- The left superscript in the term symbol (the 2 in $^2P_{3/2}$) gives the multiplicity of the term.
- The right subscript in the term symbol (the $\frac{3}{2}$ in $^2P_{3/2}$) is the value of the total angular momentum quantum number, J.

We shall now say what each of these statements means; the contributions to the energies that we are about to discuss are summarized in Fig. 9.32.

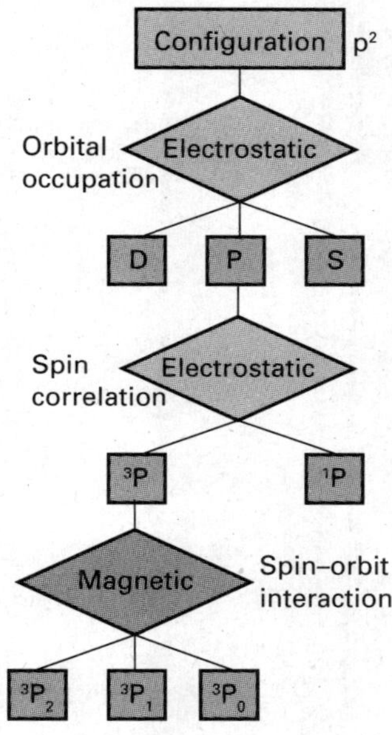

Fig. 9.32 A summary of the types of interaction that are responsible for the various kinds of splitting of energy levels in atoms. For light atoms, magnetic interactions are small, but in heavy atoms they may dominate the electrostatic (charge–charge) interactions.

(a) The total orbital angular momentum

When several electrons are present, it is necessary to judge how their individual orbital angular momenta add together or oppose each other. The **total orbital angular momentum quantum number**, L, tells us the magnitude of the angular momentum through $\{L(L+1)\}^{1/2}\hbar$. It has $2L+1$ orientations distinguished by the quantum number M_L, which can take the values $L, L-1, \ldots, -L$. Similar remarks apply to the **total spin quantum number**, S, and the quantum number M_S, and the **total angular momentum quantum number**, J, and the quantum number M_J.

The value of L (a non-negative integer) is obtained by coupling the individual orbital angular momenta by using the **Clebsch–Gordan series:**

$$L = l_1 + l_2, l_1 + l_2 - 1, \ldots, |l_1 - l_2| \qquad \text{Clebsch–Gordan series} \qquad (9.43)$$

The modulus signs are attached to $l_1 - l_2$ because L is non-negative. The maximum value, $L = l_1 + l_2$, is obtained when the two orbital angular momenta are in the same direction; the lowest value, $|l_1 - l_2|$, is obtained when they are in opposite directions. The intermediate values represent possible intermediate relative orientations of the two momenta (Fig. 9.33). For two p electrons (for which $l_1 = l_2 = 1$), $L = 2, 1, 0$. The code for converting the value of L into a letter is the same as for the s, p, d, f, . . . designation of orbitals, but uses upper-case Roman letters:

L:	0	1	2	3	4	5	6 . . .
	S	P	D	F	G	H	I . . .

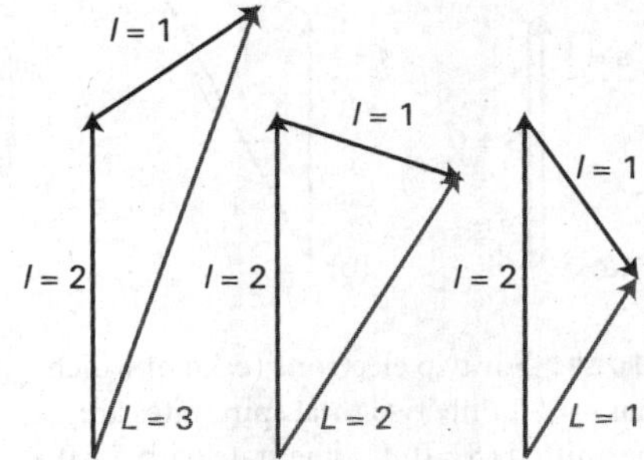

Fig. 9.33 The total orbital angular momenta of a p electron and a d electron correspond to $L = 3, 2$, and 1 and reflect the different relative orientations of the two momenta.

Thus, a p^2 configuration can give rise to D, P, and S terms. The terms differ in energy on account of the different spatial distribution of the electrons and the consequent differences in repulsion between them.

A closed shell has zero orbital angular momentum because all the individual orbital angular momenta sum to zero. Therefore, when working out term symbols, we need consider only the electrons of the unfilled shell. In the case of a single electron outside a closed shell, the value of L is the same as the value of l; so the configuration [Ne]$3s^1$ has only an S term.

Example 9.6 *Deriving the total orbital angular momentum of a configuration*

Find the terms that can arise from the configurations (a) d^2, (b) p^3.

Method Use the Clebsch–Gordan series and begin by finding the minimum value of L (so that we know where the series terminates). When there are more than two electrons to couple together, use two series in succession: first couple two electrons, and then couple the third to each combined state, and so on.

Answer (a) The minimum value is $|l_1 - l_2| = |2-2| = 0$. Therefore,

$$L = 2+2, 2+2-1, \ldots, 0 = 4, 3, 2, 1, 0$$

corresponding to G, F, D, P, S terms, respectively. (b) Coupling two electrons gives a minimum value of $|1-1| = 0$. Therefore,

$$L' = 1+1, 1+1-1, \ldots, 0 = 2, 1, 0$$

Now couple l_3 with $L' = 2$, to give $L = 3, 2, 1$; with $L' = 1$, to give $L = 2, 1, 0$; and with $L' = 0$, to give $L = 1$. The overall result is

$$L = 3, 2, 2, 1, 1, 1, 0$$

giving one F, two D, three P, and one S term.

Self-test 9.10 Repeat the question for the configurations (a) f^1d^1 and (b) d^3.
[(a) H, G, F, D, P; (b) I, 2H, 3G, 4F, 5D, 3P, S]

A note on good practice
Throughout our discussion of atomic spectroscopy, distinguish italic S, the total spin quantum number, from roman S, the term label.

(b) The multiplicity

When there are several electrons to be taken into account, we must assess their total spin angular momentum quantum number, S (a non-negative integer or half integer). Once again, we use the Clebsch–Gordan series in the form

$$S = s_1 + s_2, s_1 + s_2 - 1, \ldots, |s_1 - s_2| \tag{9.44}$$

to decide on the value of S, noting that each electron has $s = \frac{1}{2}$, which gives $S = 1, 0$ for two electrons (Fig. 9.34). If there are three electrons, the total spin angular momentum is obtained by coupling the third spin to each of the values of S for the first two spins, which results in $S = \frac{3}{2}$ and $S = \frac{1}{2}$.

The **multiplicity** of a term is the value of $2S + 1$. When $S = 0$ (as for a closed shell, like $1s^2$) the electrons are all paired and there is no net spin: this arrangement gives a singlet term, ^{1}S. A single electron has $S = s = \frac{1}{2}$, so a configuration such as [Ne]$3s^1$ can give rise to a doublet term, ^{2}S. Likewise, the configuration [Ne]$3p^1$ is a doublet, ^{2}P. When there are two unpaired electrons $S = 1$, so $2S + 1 = 3$, giving a triplet term, such as ^{3}D. We discussed the relative energies of singlets and triplets in Section 9.8 and saw that their energies differ on account of the different effects of spin correlation.

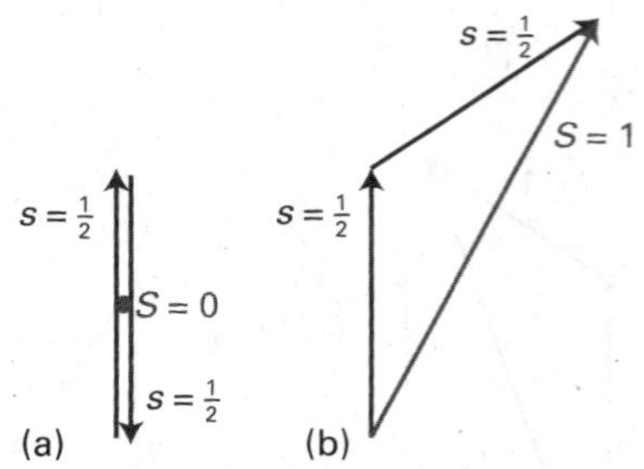

Fig. 9.34 For two electrons (each of which has $s = \frac{1}{2}$), only two total spin states are permitted ($S = 0, 1$). The state with $S = 0$ can have only one value of M_S ($M_S = 0$) and is a singlet; the state with $S = 1$ can have any of three values of M_S (+1, 0, −1) and is a triplet. The vector representations of the singlet and triplet states are shown in Figs. 9.18 and 9.25, respectively.

(c) The total angular momentum

As we have seen, the quantum number j tells us the relative orientation of the spin and orbital angular momenta of a single electron. The **total angular momentum quantum**

number, J (a non-negative integer or half integer), does the same for several electrons. If there is a single electron outside a closed shell, $J=j$, with j either $l+\frac{1}{2}$ or $|l-\frac{1}{2}|$. The $[Ne]3s^1$ configuration has $j=\frac{1}{2}$ (because $l=0$ and $s=\frac{1}{2}$), so the ^2S term has a single level, which we denote $^2S_{1/2}$. The $[Ne]3p^1$ configuration has $l=1$; therefore $j=\frac{3}{2}$ and $\frac{1}{2}$; the ^2P term therefore has two levels, $^2P_{3/2}$ and $^2P_{1/2}$. These levels lie at different energies on account of the magnetic spin–orbit interaction.

If there are several electrons outside a closed shell we have to consider the coupling of all the spins and all the orbital angular momenta. This complicated problem can be simplified when the spin–orbit coupling is weak (for atoms of low atomic number), for then we can use the **Russell–Saunders coupling** scheme. This scheme is based on the view that, if spin–orbit coupling is weak, then it is effective only when all the orbital momenta are operating cooperatively. We therefore imagine that all the orbital angular momenta of the electrons couple to give a total L, and that all the spins are similarly coupled to give a total S. Only at this stage do we imagine the two kinds of momenta coupling through the spin–orbit interaction to give a total J. The permitted values of J are given by the Clebsch–Gordan series

$$J=L+S, L+S-1, \ldots, |L-S| \qquad (9.45)$$

For example, in the case of the ^3D term of the configuration $[Ne]2p^13p^1$, the permitted values of J are 3, 2, 1 (because ^3D has $L=2$ and $S=1$), so the term has three levels, $^3D_3$, $^3D_2$, and $^3D_1$.

When $L \geq S$, the multiplicity is equal to the number of levels. For example, a ^2P term has the two levels $^2P_{3/2}$ and $^2P_{1/2}$, and ^3D has the three levels $^3D_3$, $^3D_2$, and $^3D_1$. However, this is not the case when $L < S$: the term ^2S, for example, has only the one level $^2S_{1/2}$.

Example 9.7 *Deriving term symbols*

Write the term symbols arising from the ground-state configurations of (a) Na, (b) F, and (c) the excited configuration $1s^22s^22p^13p^1$ of C.

Method Begin by writing the configurations, but ignore inner closed shells. Then couple the orbital momenta to find L and the spins to find S. Next, couple L and S to find J. Finally, express the term as $^{2S+1}\{L\}_J$, where $\{L\}$ is the appropriate letter. For F, for which the valence configuration is $2p^5$, treat the single gap in the closed-shell $2p^6$ configuration as a single particle.

Answer (a) For Na, the configuration is $[Ne]3s^1$, and we consider the single 3s electron. Because $L=l=0$ and $S=s=\frac{1}{2}$, it is possible for $J=j=s=\frac{1}{2}$ only. Hence the term symbol is $^2S_{1/2}$. (b) For F, the configuration is $[He]2s^22p^5$, which we can treat as $[Ne]2p^{-1}$ (where the notation $2p^{-1}$ signifies the absence of a 2p electron). Hence $L=1$, and $S=s=\frac{1}{2}$. Two values of $J=j$ are allowed: $J=\frac{3}{2}, \frac{1}{2}$. Hence, the term symbols for the two levels are $^2P_{3/2}$, $^2P_{1/2}$. (c) We are treating an excited configuration of carbon because, in the ground configuration, $2p^2$, the Pauli principle forbids some terms, and deciding which survive (^1D, ^3P, ^1S, in fact) is quite complicated. That is, there is a distinction between 'equivalent electrons', which are electrons that occupy the same orbitals, and 'inequivalent electrons', which are electrons that occupy different orbitals. The excited configuration of C under consideration is effectively $2p^13p^1$. This is a two-electron problem, and $l_1=l_2=1$, $s_1=s_2=\frac{1}{2}$. It follows that $L=2, 1, 0$ and $S=1, 0$. The terms are therefore ^3D and ^1D, ^3P and ^1P, and ^3S and ^1S. For ^3D, $L=2$ and $S=1$; hence $J=3, 2, 1$ and the levels are $^3D_3$, $^3D_2$, and $^3D_1$. For ^1D, $L=2$ and $S=0$, so the single level is $^1D_2$. The triplet of levels of ^3P is $^3P_2$, $^3P_1$, and $^3P_0$, and the singlet is $^1P_1$. For the ^3S term there is only one level, $^3S_1$ (because $J=1$ only), and the singlet term is $^1S_0$.

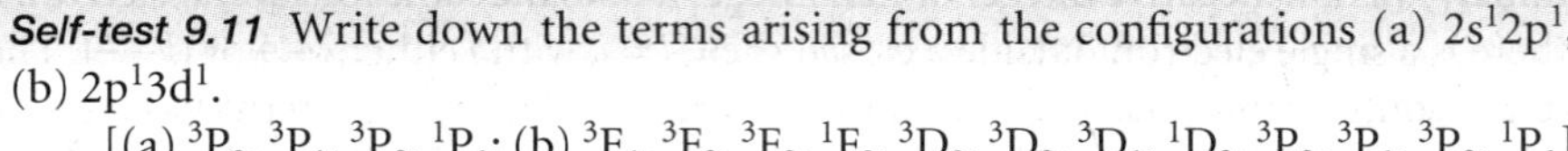

Self-test 9.11 Write down the terms arising from the configurations (a) $2s^1 2p^1$, (b) $2p^1 3d^1$.

[(a) $^3P_2$, $^3P_1$, $^3P_0$, $^1P_1$; (b) $^3F_4$, $^3F_3$, $^3F_2$, $^1F_3$, $^3D_3$, $^3D_2$, $^3D_1$, $^1D_2$, $^3P_2$, $^3P_1$, $^3P_0$, $^1P_1$]

Russell–Saunders coupling fails when the spin–orbit coupling is large (in heavy atoms, those with high Z). In that case, the individual spin and orbital momenta of the electrons are coupled into individual j values; then these momenta are combined into a grand total, J. This scheme is called ***jj*-coupling**. For example, in a p^2 configuration, the individual values of j are $\frac{3}{2}$ and $\frac{1}{2}$ for each electron. If the spin and the orbital angular momentum of each electron are coupled together strongly, it is best to consider each electron as a particle with angular momentum $j = \frac{3}{2}$ or $\frac{1}{2}$. These individual total momenta then couple as follows:

$$j_1 = \tfrac{3}{2} \text{ and } j_2 = \tfrac{3}{2} \qquad J = 3, 2, 1, 0$$
$$j_1 = \tfrac{3}{2} \text{ and } j_2 = \tfrac{1}{2} \qquad J = 2, 1$$
$$j_1 = \tfrac{1}{2} \text{ and } j_2 = \tfrac{3}{2} \qquad J = 2, 1$$
$$j_1 = \tfrac{1}{2} \text{ and } j_2 = \tfrac{1}{2} \qquad J = 1, 0$$

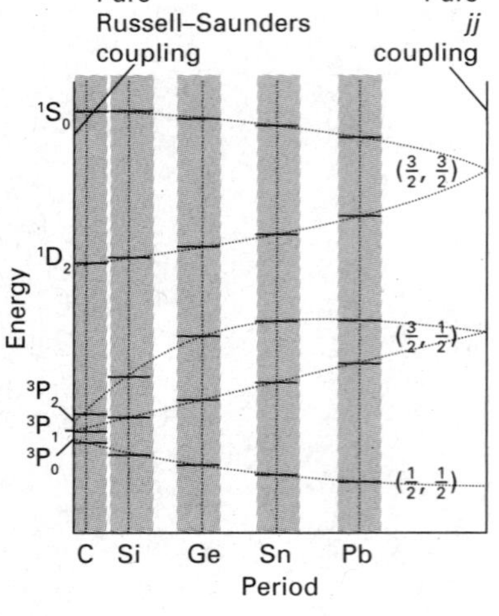

Fig. 9.35 The correlation diagram for some of the states of a two-electron system. All atoms lie between the two extremes, but, the heavier the atom, the closer it lies to the pure *jj*-coupling case.

For heavy atoms, in which *jj*-coupling is appropriate, it is best to discuss their energies using these quantum numbers.

Although *jj*-coupling should be used for assessing the energies of heavy atoms, the term symbols derived from Russell–Saunders coupling can still be used as labels. To see why this procedure is valid, we need to examine how the energies of the atomic states change as the spin–orbit coupling increases in strength. Such a **correlation diagram** is shown in Fig. 9.35. It shows that there is a correspondence between the low spin–orbit coupling (Russell–Saunders coupling) and high spin–orbit coupling (*jj*-coupling) schemes, so the labels derived by using the Russell–Saunders scheme can be used to label the states of the *jj*-coupling scheme.

(d) Selection rules

Any state of the atom, and any spectral transition, can be specified by using term symbols. For example, the transitions giving rise to the yellow sodium doublet (which were shown in Fig. 9.31) are

$$3p^1\,{}^2P_{3/2} \rightarrow 3s^1\,{}^2S_{1/2} \qquad 3p^1\,{}^2P_{1/2} \rightarrow 3s^1\,{}^2S_{1/2}$$

By convention, the upper term precedes the lower. The corresponding absorptions are therefore denoted

$${}^2P_{3/2} \leftarrow {}^2S_{1/2} \qquad {}^2P_{1/2} \leftarrow {}^2S_{1/2}$$

(The configurations have been omitted.)

We have seen that selection rules arise from the conservation of angular momentum during a transition and from the fact that a photon has a spin of 1. They can therefore be expressed in terms of the term symbols, because the latter carry information about angular momentum. A detailed analysis leads to the following rules:

$$\Delta S = 0 \quad \Delta L = 0, \pm 1 \quad \Delta l = \pm 1 \quad \Delta J = 0, \pm 1, \quad \text{but} \quad J = 0 \leftarrow\!|\!\rightarrow J = 0 \qquad (9.46)$$

Selection rules for atoms

where the symbol $\leftarrow\!|\!\rightarrow$ denotes a forbidden transition. The rule about ΔS (no change of overall spin) stems from the fact that the light does not affect the spin directly. The rules about ΔL and Δl express the fact that the orbital angular momentum of an individual electron must change (so $\Delta l = \pm 1$), but whether or not this results in an overall change of orbital momentum depends on the coupling.

The selection rules given above apply when Russell–Saunders coupling is valid (in light atoms, those of low Z). If we insist on labelling the terms of heavy atoms with symbols like ^{3}D, then we shall find that the selection rules progressively fail as the atomic number increases because the quantum numbers S and L become ill defined as *jj*-coupling becomes more appropriate. As explained above, Russell–Saunders term symbols are only a convenient way of labelling the terms of heavy atoms: they do not bear any direct relation to the actual angular momenta of the electrons in a heavy atom. For this reason, transitions between singlet and triplet states (for which $\Delta S = \pm 1$), while forbidden in light atoms, are allowed in heavy atoms.

IMPACT ON ASTROPHYSICS

I9.1 Spectroscopy of stars

The bulk of stellar material consists of neutral and ionized forms of hydrogen and helium atoms, with helium being the product of 'hydrogen burning' by nuclear fusion. However, nuclear fusion also makes heavier elements. It is generally accepted that the outer layers of stars are composed of lighter elements, such as H, He, C, N, O, and Ne in both neutral and ionized forms. Heavier elements, including neutral and ionized forms of Si, Mg, Ca, S, and Ar, are found closer to the stellar core. The core itself contains the heaviest elements and ^{56}Fe is particularly abundant because it is a very stable nuclide. All these elements are in the gas phase on account of the very high temperatures in stellar interiors. For example, the temperature is estimated to be 3.6 MK halfway to the centre of the Sun.

Astronomers use spectroscopic techniques to determine the chemical composition of stars because each element, and indeed each isotope of an element, has a characteristic spectral signature that is transmitted through space by the star's light. To understand the spectra of stars, we must first know why they shine. Nuclear reactions in the dense stellar interior generate radiation that travels to less dense outer layers. Absorption and re-emission of photons by the atoms and ions in the interior give rise to a quasi-continuum of radiation energy that is emitted into space by a thin layer of gas called the *photosphere*. To a good approximation, the distribution of energy emitted from a star's photosphere resembles the Planck distribution for a very hot black body (Section 7.1). For example, the energy distribution of our Sun's photosphere may be modelled by a Planck distribution with an effective temperature of 5.8 kK. Superimposed on the black-body radiation continuum are sharp absorption and emission lines from neutral atoms and ions present in the photosphere. Analysis of stellar radiation with a spectrometer mounted on to a telescope yields the chemical composition of the star's photosphere by comparison with known spectra of the elements. The data can also reveal the presence of small molecules, such as CN, C_2, TiO, and ZrO, in certain 'cold' stars, which are stars with relatively low effective temperatures.

The two outermost layers of a star are the *chromosphere*, a region just above the photosphere, and the *corona*, a region above the chromosphere that can be seen (with proper care) during eclipses. The photosphere, chromosphere, and corona comprise a star's 'atmosphere'. Our Sun's chromosphere is much less dense than its photosphere and its temperature is much higher, rising to about 10 kK. The reasons for this increase in temperature are not fully understood. The temperature of our Sun's corona is very high, rising up to 1.5 MK, so black-body emission is strong from the X-ray to the radiofrequency region of the spectrum. The spectrum of the Sun's corona is dominated by emission lines from electronically excited species, such as neutral atoms and a number of highly ionized species. The most intense emission lines in the visible range are from the Fe^{13+} ion at 530.3 nm, the Fe^{9+} ion at 637.4 nm, and the Ca^{4+} ion at 569.4 nm.

Because light from only the photosphere reaches our telescopes, the overall chemical composition of a star must be inferred from theoretical work on its interior and from spectral analysis of its atmosphere. Data on the Sun indicate that it is 92 per cent hydrogen and 7.8 per cent helium. The remaining 0.2 per cent is due to heavier elements, among which C, N, O, Ne, and Fe are the most abundant. More advanced analysis of spectra also permits the determination of other properties of stars, such as their relative speeds (Problem 9.29) and their effective temperatures (Problem 9.30).

Checklist of key equations

Property	Equation	Comment
Wavenumbers of the spectral lines of a hydrogen atom	$\tilde{\nu} = R_H\{(1/n_1^2) - (1/n_2^2)\}$	R_H is the Rydberg constant for hydrogen
Wavefunctions of hydrogenic atoms	$\psi(r,\theta,\phi) = R(r)Y(\theta,\phi)$	Y are spherical harmonics
Energies of hydrogenic atoms	$E_n = -Z^2\mu e^4/32\pi^2\varepsilon_0^2\hbar^2 n^2$	
Radial distribution function	$P(r) = r^2R(r)^2$	$P(r) = 4\pi r^2\psi^2$ for s orbitals
Orbital approximation	$\Psi(r_1,r_2,\ldots) = \psi(r_1)\psi(r_2)\ldots$	
Clebsch–Gordan series	$J = j_1+j_2, j_1+j_2-1, \ldots \lvert j_1-j_2\rvert$	J, j denote any kind of angular momenta
Selection rules	$\Delta S = 0, \Delta L = 0, \pm1, \Delta l = \pm1, \Delta J = 0, \pm1$, but $J=0 \leftarrow\!\!\mid\!\!\rightarrow J=0$	Light atoms
Lifetime broadening	$\delta E \approx \hbar/\tau$	

Further information

Further information 9.1 *The separation of motion*

(a) The separation of internal and external motion

Consider a one-dimensional system in which the potential energy depends only on the separation of the two particles. The total energy is

$$E = \frac{p_1^2}{2m_1} + \frac{p_2^2}{2m_2} + V \tag{9.47}$$

Where $p_1 = m_1\dot{x}_1$ and $p_2 = m_2\dot{x}_2$, the dot signifying differentiation with respect to time. The centre of mass (Fig. 9.36) is located at

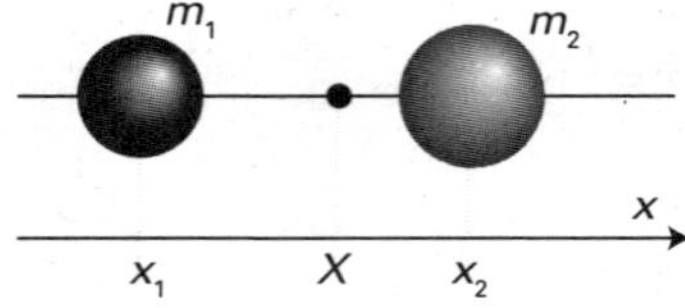

Fig. 9.36 The coordinates used for discussing the separation of the relative motion of two particles from the motion of the centre of mass.

$$X = \frac{m_1}{m}x_1 + \frac{m_2}{m}x_2 \qquad m = m_1 + m_2 \tag{9.48}$$

and the separation of the particles is $x = x_1 - x_2$. It follows that

$$x_1 = X + \frac{m_2}{m}x \qquad x_2 = X - \frac{m_1}{m}x \tag{9.49}$$

The linear momenta of the particles can be expressed in terms of the rates of change of x and X:

$$p_1 = m_1\dot{x}_1 = m_1\dot{X} + \frac{m_1m_2}{m}\dot{x} \qquad p_2 = m_2\dot{x}_2 = m_2\dot{X} - \frac{m_1m_2}{m}\dot{x} \tag{9.50}$$

Then it follows that

$$\frac{p_1^2}{2m_1} + \frac{p_2^2}{2m_2} = \tfrac{1}{2}m\dot{X}^2 + \tfrac{1}{2}\mu\dot{x}^2 \tag{9.51}$$

where μ is given in eqn 9.6. By writing $P = m\dot{X}$ for the linear momentum of the system as a whole and defining p as $\mu\dot{x}$, we find

$$E = \frac{P^2}{2m} + \frac{p^2}{2\mu} + V \tag{9.52a}$$

The corresponding hamiltonian (generalized to three dimensions) is therefore

$$\hat{H} = -\frac{\hbar^2}{2m}\nabla^2_{\text{c.m.}} - \frac{\hbar^2}{2\mu}\nabla^2 + V \qquad (9.52b)$$

where the first term differentiates with respect to the centre of mass coordinates and the second with respect to the relative coordinates.

Now we write the overall wavefunction as the product $\psi_{\text{total}} = \psi_{\text{c.m.}}\psi$, where the first factor is a function of only the centre of mass coordinates and the second is a function of only the relative coordinates. The overall Schrödinger equation, $\hat{H}\psi_{\text{total}} = E_{\text{total}}\psi_{\text{total}}$, then separates by the argument that we have used in Sections 8.2a and 8.7, with $E_{\text{total}} = E_{\text{c.m.}} + E$.

(b) The separation of angular and radial motion

The laplacian in three dimensions is given in eqn 8.51a. It follows that the Schrödinger equation in eqn 9.6 is

$$-\frac{\hbar^2}{2\mu}\left(\frac{\partial^2}{\partial r^2} + \frac{2}{r}\frac{\partial}{\partial r} + \frac{1}{r^2}\Lambda^2\right)RY + VRY = ERY \qquad (9.53)$$

Because R depends only on r and Y depends only on the angular coordinates, this equation becomes

$$-\frac{\hbar^2}{2\mu}\left(Y\frac{\mathrm{d}^2R}{\mathrm{d}r^2} + \frac{2Y}{r}\frac{\mathrm{d}R}{\mathrm{d}r} + \frac{R}{r^2}\Lambda^2 Y\right) + VRY = ERY \qquad (9.54)$$

If we multiply through by r^2/RY, we obtain

$$-\frac{\hbar^2}{2\mu R}\left(r^2\frac{\mathrm{d}^2R}{\mathrm{d}r^2} + 2r\frac{\mathrm{d}R}{\mathrm{d}r}\right) + Vr^2 - \frac{\hbar^2}{2\mu Y}\Lambda^2 Y = Er^2 \qquad (9.55)$$

At this point we employ the usual argument. The term in Y is the only one that depends on the angular variables, so it must be a constant. When we write this constant as $\hbar^2 l(l+1)/2\mu$, eqn 9.8b follows immediately.

Further information 9.2 *The energy of spin–orbit interaction*

The energy of a magnetic moment $\boldsymbol{\mu}$ in a magnetic field $\boldsymbol{\mathcal{B}}$ is equal to their scalar product $-\boldsymbol{\mu}\cdot\boldsymbol{\mathcal{B}}$. If the magnetic field arises from the orbital angular momentum of the electron, it is proportional to $\boldsymbol{l}$; if the magnetic moment $\boldsymbol{\mu}$ is that of the electron spin, then it is proportional to $\boldsymbol{s}$. It then follows that the energy of interaction is proportional to the scalar product $\boldsymbol{s}\cdot\boldsymbol{l}$:

$$\text{Energy of interaction} = -\boldsymbol{\mu}\cdot\boldsymbol{\mathcal{B}} \propto \boldsymbol{s}\cdot\boldsymbol{l}$$

(For the various vector manipulations used in this section, see *Mathematical background* 5 following this chapter.) Next, we note that the total angular momentum is the vector sum of the spin and orbital momenta: $\boldsymbol{j} = \boldsymbol{l} + \boldsymbol{s}$. The magnitude of the vector $\boldsymbol{j}$ is calculated by evaluating

$$\boldsymbol{j}\cdot\boldsymbol{j} = (\boldsymbol{l}+\boldsymbol{s})\cdot(\boldsymbol{l}+\boldsymbol{s}) = \boldsymbol{l}\cdot\boldsymbol{l} + \boldsymbol{s}\cdot\boldsymbol{s} + 2\boldsymbol{s}\cdot\boldsymbol{l}$$

so that

$$j^2 = l^2 + s^2 + 2\boldsymbol{s}\cdot\boldsymbol{l}$$

That is,

$$\boldsymbol{s}\cdot\boldsymbol{l} = \tfrac{1}{2}\{j^2 - l^2 - s^2\}$$

where we have used the fact that the scalar product of two vectors $\boldsymbol{u}$ and $\boldsymbol{v}$ is $\boldsymbol{u}\cdot\boldsymbol{v} = uv\cos\theta$, from which it follows that $\boldsymbol{u}\cdot\boldsymbol{u} = u^2$.

The preceding equation is a classical result. To make the transition to quantum mechanics, we treat all the quantities as operators, and write

$$\hat{\boldsymbol{s}}\cdot\hat{\boldsymbol{l}} = \tfrac{1}{2}\{\hat{j}^2 - \hat{l}^2 - \hat{s}^2\}$$

At this point, we evaluate the expectation value:

$$\begin{aligned}\langle j,l,s|\hat{\boldsymbol{s}}\cdot\hat{\boldsymbol{l}}|j,l,s\rangle &= \tfrac{1}{2}\langle j,l,s|\hat{j}^2 - \hat{l}^2 - \hat{s}^2|j,l,s\rangle \\ &= \tfrac{1}{2}\{j(j+1) - l(l+1) - s(s+1)\}\hbar^2\end{aligned} \qquad (9.56)$$

Then, by inserting this expression into the formula for the energy of interaction ($E \propto \boldsymbol{s}\cdot\boldsymbol{l}$, and writing the constant of proportionality as $hc\tilde{A}/\hbar^2$, we obtain eqn 9.42.

Discussion questions

9.1 Discuss the origin of the series of lines in the emission spectra of hydrogen. What region of the electromagnetic spectrum is associated with each of the series shown in Fig. 9.1?

9.2 Describe the separation of variables procedure as it is applied to simplify the description of a hydrogenic atom free to move through space.

9.3 List and describe the significance of the quantum numbers needed to specify the internal state of a hydrogenic atom.

9.4 Specify and account for the selection rules for transitions in hydrogenic atoms.

9.5 Explain the significance of (a) a boundary surface and (b) the radial distribution function for hydrogenic orbitals.

9.6 Outline the electron configurations of many-electron atoms in terms of their location in the periodic table.

9.7 Describe and account for the variation of first ionization energies along Period 2 of the periodic table. Would you expect the same variation in Period 3?

9.8 Describe the orbital approximation for the wavefunction of a many-electron atom. What are the limitations of the approximation?

9.9 Explain the origin of spin–orbit coupling and how it affects the appearance of a spectrum.

9.10 Describe the physical origins of linewidths in absorption and emission spectra. Do you expect the same contributions for species in condensed and gas phases?

Exercises

9.1(a) Determine the shortest and longest wavelength lines in the Lyman series.

9.1(b) The Pfund series has $n_1 = 5$. Determine the shortest and longest wavelength lines in the Pfund series.

9.2(a) Compute the wavelength, frequency, and wavenumber of the $n = 2 \rightarrow n = 1$ transition in He^+.

9.2(b) Compute the wavelength, frequency, and wavenumber of the $n = 5 \rightarrow n = 4$ transition in Li^{+2}.

9.3(a) When ultraviolet radiation of wavelength 58.4 nm from a helium lamp is directed on to a sample of krypton, electrons are ejected with a speed of 1.59 Mm s^{-1}. Calculate the ionization energy of krypton.

9.3(b) When ultraviolet radiation of wavelength 58.4 nm from a helium lamp is directed on to a sample of xenon, electrons are ejected with a speed of 1.79 Mm s^{-1}. Calculate the ionization energy of xenon.

9.4(a) State the orbital degeneracy of the levels in a hydrogen atom that have energy (a) $-hcR_H$; (b) $-\frac{1}{9}hcR_H$; (c) $-\frac{1}{25}hcR_H$.

9.4(b) State the orbital degeneracy of the levels in a hydrogenic atom (Z in parentheses) that have energy (a) $-4hcR_{atom}$ (2); (b) $-\frac{1}{4}hcR_{atom}$ (4), and (c) $-hcR_{atom}$ (5).

9.5(a) The wavefunction for the ground state of a hydrogen atom is Ne^{-r/a_0}. Determine the normalization constant N.

9.5(b) The wavefunction for the 2s orbital of a hydrogen atom is $N(2 - r/a_0)e^{-r/2a_0}$. Determine the normalization constant N.

9.6(a) By differentiation of the 2s radial wavefunction, show that it has two extrema in its amplitude, and locate them.

9.6(b) By differentiation of the 3s radial wavefunction, show that it has three extrema in its amplitude, and locate them.

9.7(a) At what radius does the probability of finding an electron at a point in the H atom fall to 50 per cent of its maximum value?

9.7(b) At what radius in the H atom does the radial distribution function of the ground state have (a) 50 per cent, (b) 75 per cent of its maximum value.

9.8(a) Locate the radial nodes in the 3s orbital of an H atom.

9.8(b) Locate the radial nodes in the 4p orbital of an H atom.

9.9(a) Calculate the average kinetic and potential energies of an electron in the ground state of a hydrogen atom.

9.9(b) Calculate the average kinetic and potential energies of a 2s electron in a hydrogenic atom of atomic number Z.

9.10(a) Write down the expression for the radial distribution function of a 2s electron in a hydrogenic atom and determine the radius at which the electron is most likely to be found.

9.10(b) Write down the expression for the radial distribution function of a 3s electron in a hydrogenic atom and determine the radius at which the electron is most likely to be found.

9.11(a) Write down the expression for the radial distribution function of a 2p electron in a hydrogenic atom and determine the radius at which the electron is most likely to be found.

9.11(b) Write down the expression for the radial distribution function of a 3p electron in a hydrogenic atom and determine the radius at which the electron is most likely to be found.

9.12(a) What is the orbital angular momentum of an electron in the orbitals (a) 1s, (b) 3s, (c) 3d? Give the numbers of angular and radial nodes in each case.

9.12(b) What is the orbital angular momentum of an electron in the orbitals (a) 4d, (b) 2p, (c) 3p? Give the numbers of angular and radial nodes in each case.

9.13(a) Locate the angular nodes and nodal planes of each of the 2p orbitals of a hydrogenic atom of atomic number Z. To locate the angular nodes, give the angle that the plane makes with the z-axis.

9.13(b) Locate the angular nodes and nodal planes of each of the 3d orbitals of a hydrogenic atom of atomic number Z. To locate the angular nodes, give the angle that the plane makes with the z-axis.

9.14(a) Which of the following transitions are allowed in the normal electronic emission spectrum of an atom: (a) 2s $\rightarrow$ 1s, (b) 2p $\rightarrow$ 1s, (c) 3d $\rightarrow$ 2p?

9.14(b) Which of the following transitions are allowed in the normal electronic emission spectrum of an atom: (a) 5d $\rightarrow$ 2s, (b) 5p $\rightarrow$ 3s, (c) 6p $\rightarrow$ 4f?

9.15(a) What is the Doppler-shifted wavelength of a red (680 nm) traffic light approached at 60 km h^{-1}?

9.15(b) At what speed of approach would a red (680 nm) traffic light appear green (530 nm)?

9.16(a) Estimate the lifetime of a state that gives rise to a line of width (a) 0.20 cm^{-1}, (b) 2.0 cm^{-1}.

9.16(b) Estimate the lifetime of a state that gives rise to a line of width (a) 200 MHz, (b) 2.45 cm^{-1}.

9.17(a) A molecule in a liquid undergoes about 1.0×10^{13} collisions in each second. Suppose that (a) every collision is effective in deactivating the molecule vibrationally and (b) that one collision in 100 is effective. Calculate the width (in cm^{-1}) of vibrational transitions in the molecule.

9.17(b) A molecule in a gas undergoes about 1.0×10^9 collisions in each second. Suppose that (a) every collision is effective in deactivating the molecule rotationally and (b) that one collision in 10 is effective. Calculate the width (in hertz) of rotational transitions in the molecule.

9.18(a) Write the ground-state electron configurations of the d-metals from scandium to zinc.

9.18(b) Write the ground-state electron configurations of the d-metals from yttrium to cadmium.

9.19(a) (a) Write the electronic configuration of the Ni^{2+} ion. (b) What are the possible values of the total spin quantum numbers S and M_S for this ion?

9.19(b) (a) Write the electronic configuration of the V^{2+} ion. (b) What are the possible values of the total spin quantum numbers S and M_S for this ion?

9.20(a) Calculate the permitted values of j for (a) a d electron, (b) an f electron.

9.20(b) Calculate the permitted values of j for (a) a p electron, (b) an h electron.

9.21(a) An electron in two different states of an atom is known to have $j = \frac{3}{2}$ and $\frac{1}{2}$. What is its orbital angular momentum quantum number in each case?

9.21(b) What are the allowed total angular momentum quantum numbers of a composite system in which $j_1 = 5$ and $j_2 = 3$?

9.22(a) What information does the term symbol $^1D_2$ provide about the angular momentum of an atom?

9.22(b) What information does the term symbol $^3F_4$ provide about the angular momentum of an atom?

9.23(a) Suppose that an atom has (a) 2, (b) 3 electrons in different orbitals. What are the possible values of the total spin quantum number *S*? What is the multiplicity in each case?

9.23(b) Suppose that an atom has (a) 4, (b) 5, electrons in different orbitals. What are the possible values of the total spin quantum number *S*? What is the multiplicity in each case?

9.24(a) What atomic terms are possible for the electron configuration ns^1nd^1? Which term is likely to lie lowest in energy?

9.24(b) What atomic terms are possible for the electron configuration np^1nd^1? Which term is likely to lie lowest in energy?

9.25(a) What values of *J* may occur in the terms (a) ^1S, (b) ^2P, (c) ^3P? How many states (distinguished by the quantum number M_J) belong to each level?

9.25(b) What values of *J* may occur in the terms (a) ^3D, (b) ^4D, (c) ^2G? How many states (distinguished by the quantum number M_J) belong to each level?

9.26(a) Give the possible term symbols for (a) Li $[He]2s^1$, (b) Na $[Ne]3p^1$.

9.26(b) Give the possible term symbols for (a) Sc $[Ar]3d^14s^2$, (b) Br $[Ar]3d^{10}4s^24p^5$.

9.27(a) Which of the following transitions between terms are allowed in the normal electronic emission spectrum of a many-electron atom: (a) $^3D_2 \rightarrow {}^3P_1$, (b) $^3P_2 \rightarrow {}^1S_0$, (c) $^3F_4 \rightarrow {}^3D_3$?

9.27(b) Which of the following transitions between terms are allowed in the normal electronic emission spectrum of a many-electron atom: (a) $^2P_{3/2} \rightarrow {}^2S_{1/2}$, (b) $^3P_0 \rightarrow {}^3S_1$, (c) $^3D_3 \rightarrow {}^1P_1$?

Problems*

Numerical problems

9.1 The *Humphreys series* is a group of lines in the spectrum of atomic hydrogen. It begins at 12 368 nm and has been traced to 3281.4 nm. What are the transitions involved? What are the wavelengths of the intermediate transitions?

9.2 A series of lines in the spectrum of atomic hydrogen lies at 656.46 nm, 486.27 nm, 434.17 nm, and 410.29 nm. What is the wavelength of the next line in the series? What is the ionization energy of the atom when it is in the lower state of the transitions?

9.3 The Li^{2+} ion is hydrogenic and has a Lyman series at 740 747 cm^{-1}, 877 924 cm^{-1}, 925 933 cm^{-1}, and beyond. Show that the energy levels are of the form $-hcR/n^2$ and find the value of *R* for this ion. Go on to predict the wavenumbers of the two longest-wavelength transitions of the Balmer series of the ion and find the ionization energy of the ion.

9.4 A series of lines in the spectrum of neutral Li atoms rise from combinations of $1s^22p^1\ {}^2P$ with $1s^2nd^1\ {}^2D$ and occur at 610.36 nm, 460.29 nm, and 413.23 nm. The d orbitals are hydrogenic. It is known that the ^2P term lies at 670.78 nm above the ground state, which is $1s^22s^1\ {}^2S$. Calculate the ionization energy of the ground-state atom.

9.5‡ W.P. Wijesundera *et al.* (*Phys. Rev.* A **51**, 278 (1995)) attempted to determine the electron configuration of the ground state of lawrencium, element 103. The two contending configurations are $[Rn]5f^{14}7s^27p^1$ and $[Rn]5f^{14}6d7s^2$. Write down the term symbols for each of these configurations, and identify the lowest level within each configuration. Which level would be lowest according to a simple estimate of spin–orbit coupling?

9.6 An emission line from K atoms is found to have two closely spaced components, one at 766.70 nm and the other at 770.11 nm. Account for this observation, and deduce what information you can.

9.7 Calculate the mass of the deuteron given that the first line in the Lyman series of H lies at 82 259.098 cm^{-1} whereas that of D lies at 82 281.476 cm^{-1}. Calculate the ratio of the ionization energies of H and D.

9.8 Positronium consists of an electron and a positron (same mass, opposite charge) orbiting round their common centre of mass. The broad features of the spectrum are therefore expected to be hydrogen-like, the differences arising largely from the mass differences. Predict the wavenumbers of the first three lines of the Balmer series of positronium. What is the binding energy of the ground state of positronium?

9.9 The *Zeeman effect* is the modification of an atomic spectrum by the application of a strong magnetic field. It arises from the interaction between applied magnetic fields and the magnetic moments due to orbital and spin angular momenta (recall the evidence provided for electron spin by the Stern–Gerlach experiment, Section 8.8). To gain some appreciation for the so-called *normal Zeeman effect*, which is observed in transitions involving singlet states, consider a p electron, with $l = 1$ and $m_l = 0, \pm 1$. In the absence of a magnetic field, these three states are degenerate. When a field of magnitude $\mathcal{B}$ is present, the degeneracy is removed and it is observed that the state with $m_l = +1$ moves up in energy by $\mu_B\mathcal{B}$, the state with $m_l = 0$ is unchanged, and the state with $m_l = -1$ moves down in energy by $\mu_B\mathcal{B}$, where $\mu_B = e\hbar/2m_e = 9.274 \times 10^{-24}$ J T^{-1} is the Bohr magneton (see Section 13.1). Therefore, a transition between a $^1S_0$ term and a $^1P_1$ term consists of three spectral lines in the presence of a magnetic field where, in the absence of the magnetic field, there is only one. (a) Calculate the splitting in reciprocal centimetres between the three spectral lines of a transition between a $^1S_0$ term and a $^1P_1$ term in the presence of a magnetic field of 2 T (where 1 T = 1 kg s^{-2} A^{-1}). (b) Compare the value you calculated in (a) with typical optical transition wavenumbers, such as those for the Balmer series of the H atom. Is the line splitting caused by the normal Zeeman effect relatively small or relatively large?

9.10 In 1976 it was mistakenly believed that the first of the 'superheavy' elements had been discovered in a sample of mica. Its atomic number was believed to be 126. What is the most probable distance of the innermost electrons from the nucleus of an atom of this element? (In such elements, relativistic effects are very important, but ignore them here.)

9.11 An electron in the ground-state He^+ ion undergoes a transition to a state described by the wavefunction $R_{4,1}(r)Y_{1,1}(\theta,\phi)$. (a) Describe the transition using term symbols. (b) Compute the wavelength, frequency, and wavenumber of the transition. (c) By how much does the mean radius of the electron change due to the transition?

9.12 The collision frequency *z* of a molecule of mass *m* in a gas at a pressure *p* is $z = 4\sigma(kT/\pi m)^{1/2}p/kT$, where σ is the collision cross-section. Find an

* Problems denoted with the symbol ‡ were supplied by Charles Trapp, Carmen Giunta, and Marshall Cady.

expression for the collision-limited lifetime of an excited state assuming that every collision is effective. Estimate the width of a rotational transition in HCl ($\sigma = 0.30$ nm^2) at 25°C and 1.0 atm. To what value must the pressure of the gas be reduced in order to ensure that collision broadening is less important than Doppler broadening?

Theoretical problems

9.13 What is the most probable point (not radius) at which a 2p electron will be found in the hydrogen atom?

9.14 Show by explicit integration that (a) hydrogenic 1s and 2s orbitals, (b) $2p_x$ and $2p_y$ orbitals are mutually orthogonal.

9.15‡ Explicit expressions for hydrogenic orbitals are given in Tables 9.1 and 8.2. (a) Verify both that the $3p_x$ orbital is normalized (to 1) and that $3p_x$ and $3d_{xy}$ are mutually orthogonal. (b) Determine the positions of both the radial nodes and nodal planes of the 3s, $3p_x$, and $3d_{xy}$ orbitals. (c) Determine the mean radius of the 3s orbital. (d) Draw a graph of the radial distribution function for the three orbitals (of part (b)) and discuss the significance of the graphs for interpreting the properties of many-electron atoms. (e) Create both *xy*-plane polar plots and boundary surface plots for these orbitals. Construct the boundary plots so that the distance from the origin to the surface is the absolute value of the angular part of the wavefunction. Compare the s, p, and d boundary surface plots with that of an f orbital, e.g. $\psi_f \propto x(5z^2 - r^2) \propto \sin\theta(5\cos^2\theta - 1)\cos\phi$.

9.16 Determine whether the p_x and p_y orbitals are eigenfunctions of l_z. If not, does a linear combination exist that is an eigenfunction of l_z?

9.17 Show that l_z and l^2 both commute with the hamiltonian for a hydrogen atom. What is the significance of this result?

9.18 The 'size' of an atom is sometimes considered to be measured by the radius of a sphere that contains 90 per cent of the charge density of the electrons in the outermost occupied orbital. Calculate the 'size' of a hydrogen atom in its ground state according to this definition. Go on to explore how the 'size' varies as the definition is changed to other percentages, and plot your conclusion.

9.19 Some atomic properties depend on the average value of $1/r$ rather than the average value of r itself. Evaluate the expectation value of $1/r$ for (a) a hydrogen 1s orbital, (b) a hydrogenic 2s orbital, (c) a hydrogenic 2p orbital.

9.20 One of the most famous of the obsolete theories of the hydrogen atom was proposed by Bohr. It has been replaced by quantum mechanics but, by a remarkable coincidence (not the only one where the Coulomb potential is concerned), the energies it predicts agree exactly with those obtained from the Schrödinger equation. In the Bohr atom, an electron travels in a circle around the nucleus. The Coulombic force of attraction ($Ze^2/4\pi\varepsilon_0 r^2$) is balanced by the centrifugal effect of the orbital motion. Bohr proposed that the angular momentum is limited to integral values of $\hbar$. When the two forces are balanced, the atom remains in a stationary state until it makes a spectral transition. Calculate the energies of a hydrogenic atom using the Bohr model.

9.21 The Bohr model of the atom is specified in Problem 9.20. What features of it are untenable according to quantum mechanics? How does the Bohr ground state differ from the actual ground state. Is there an experimental distinction between the Bohr and quantum mechanical models of the ground state?

9.22 Atomic units of length and energy may be based on the properties of a particular atom. The usual choice is that of a hydrogen atom, with the unit of length being the Bohr radius, a_0, and the unit of energy being the (negative of the) energy of the 1s orbital. If the positronium atom (e^+, e^-) were used instead, with analogous definitions of units of length and energy, what would be the relation between these two sets of atomic units?

9.23 Some of the selection rules for hydrogenic atoms were derived in *Justification* 9.4. Complete the derivation by considering the *x*- and *y*-components of the electric dipole moment operator.

9.24‡ Stern–Gerlach splittings of atomic beams are small and require either large magnetic field gradients or long magnets for their observation. For a beam of atoms with zero orbital angular momentum, such as H or Ag, the deflection is given by $x = \pm(\mu_B L^2/4E_k)\mathrm{d}\mathcal{B}/\mathrm{d}z$, where μ_B is the Bohr magneton (Problem 9.9), L is the length of the magnet, E_k is the average kinetic energy of the atoms in the beam, and $\mathrm{d}\mathcal{B}/\mathrm{d}z$ is the magnetic field gradient across the beam. (a) Use the Maxwell–Boltzmann velocity distribution to show that the average translational kinetic energy of the atoms emerging as a beam from a pinhole in an oven at temperature T is $2kT$. (b) Calculate the magnetic field gradient required to produce a splitting of 1.00 mm in a beam of Ag atoms from an oven at 1000 K with a magnet of length 50 cm.

9.25 The wavefunction of a many-electron closed-shell atom can be expressed as a Slater determinant (Section 9.4b). A useful property of determinants is that interchanging any two rows or columns changes their sign and therefore, if any two rows or columns are identical, then the determinant vanishes. Use this property to show that (a) the wavefunction is antisymmetric under particle exchange, (b) no two electrons can occupy the same orbital with the same spin.

Applications: to astrophysics and biochemistry

9.26 Hydrogen is the most abundant element in all stars. However, neither absorption nor emission lines due to neutral hydrogen are found in the spectra of stars with effective temperatures higher than 25 000 K. Account for this observation.

9.27 The distribution of isotopes of an element may yield clues about the nuclear reactions that occur in the interior of a star. Show that it is possible to use spectroscopy to confirm the presence of both $^4He^+$ and $^3He^+$ in a star by calculating the wavenumbers of the $n = 3 \rightarrow n = 2$ and of the $n = 2 \rightarrow n = 1$ transitions for each isotope.

9.28‡ Highly excited atoms have electrons with large principal quantum numbers. Such *Rydberg atoms* have unique properties and are of interest to astrophysicists. For hydrogen atoms with large n, derive a relation for the separation of energy levels. Calculate this separation for $n = 100$; also calculate the average radius, the geometric cross-section, and the ionization energy. Could a thermal collision with another hydrogen atom ionize this Rydberg atom? What minimum velocity of the second atom is required? Could a normal-sized neutral H atom simply pass through the Rydberg atom leaving it undisturbed? What might the radial wavefunction for a 100s orbital be like?

9.29 The spectrum of a star is used to measure its *radial velocity* with respect to the Sun, the component of the star's velocity vector that is parallel to a vector connecting the star's centre to the centre of the Sun. The measurement relies on the Doppler effect. When a star emitting electromagnetic radiation of frequency ν moves with a speed s relative to an observer, the observer detects radiation of frequency $\nu_{\text{receding}} = \nu f$ or $\nu_{\text{approaching}} = \nu/f$, where $f = \{(1 - s/c)/(1 + s/c)\}^{1/2}$ and c is the speed of light. It is easy to see that $\nu_{\text{receding}} < \nu$ and a receding star is characterized by a *red shift* of its spectrum with respect to the spectrum of an identical, but stationary source. Furthermore, $\nu_{\text{approaching}} > \nu$ and an approaching star is characterized by a *blue shift* of its spectrum with respect to the spectrum of an identical, but stationary source. In a typical experiment, ν is the frequency of a spectral line of an element measured in a stationary Earth-bound laboratory from a calibration source, such as an arc lamp. Measurement of the same spectral line in a star gives ν_{star} and the speed of recession or approach may be calculated from the value of ν and the equations above. (a) Three Fe I lines of the star HDE 271 182, which belongs to the Large Magellanic Cloud, occur at 438.882 nm, 441.000 nm, and 442.020 nm. The same lines occur at 438.392 nm, 440.510 nm, and 441.510 nm in the spectrum of an Earth-bound iron arc. Determine whether HDE 271 182 is receding from or approaching the Earth and estimate the star's radial speed with respect to the Earth. (b) What additional information would you need to calculate the radial velocity of HDE 271 182 with respect to the Sun?

9.30 In Problem 9.29, we saw that Doppler shifts of atomic spectral lines are used to estimate the speed of recession or approach of a star. From the discussion in Section 9.6a, it can be inferred that Doppler broadening of an atomic spectral line depends on the temperature of the star that emits the radiation. A spectral line of $^{48}Ti^{8+}$ (of mass $47.95m_u$) in a distant star was found to be shifted from 654.2 nm to 706.5 nm and to be broadened to 61.8 pm. What is the speed of recession and the surface temperature of the star?

9.31 The d-metals iron, copper, and manganese form cations with different oxidation states. For this reason, they are found in many oxidoreductases and in several proteins of oxidative phosphorylation and photosynthesis. Explain why many d-metals form cations with different oxidation states.

9.32 Thallium, a neurotoxin, is the heaviest member of Group 13 of the periodic table and is found most usually in the +1 oxidation state. Aluminium, which causes anaemia and dementia, is also a member of the group but its chemical properties are dominated by the +3 oxidation state. Examine this issue by plotting the first, second, and third ionization energies for the Group 13 elements against atomic number. Explain the trends you observe. *Hints.* The third ionization energy, I_3, is the minimum energy needed to remove an electron from the doubly charged cation: $E^{2+}(g) \rightarrow E^{3+}(g) + e^-(g)$, $I_3 = E(E^{3+}) - E(E^{2+})$. For data, see the links to databases of atomic properties provided in the text's web site.

MATHEMATICAL BACKGROUND 5

Vectors

A vector quantity has both magnitude and direction. The vector shown in Fig. MB5.1 has components on the x, y, and z axes with magnitudes v_x, v_y, and v_z, respectively. The vector may be represented as

$$\boldsymbol{v} = v_x\boldsymbol{i} + v_y\boldsymbol{j} + v_z\boldsymbol{k} \qquad \text{(MB5.1)}$$

where $\boldsymbol{i}$, $\boldsymbol{j}$, and $\boldsymbol{k}$ are **unit vectors**, vectors of magnitude 1, pointing along the positive directions on the x-, y-, and z-axes. The magnitude of the vector is denoted v or $|\boldsymbol{v}|$ and is given by

$$v = (v_x^2 + v_y^2 + v_z^2)^{1/2} \qquad \text{(MB5.2)}$$

MB5.1 Addition and subtraction

If $\boldsymbol{v} = v_x\boldsymbol{i} + v_y\boldsymbol{j} + v_z\boldsymbol{k}$ and $\boldsymbol{u} = u_x\boldsymbol{i} + u_y\boldsymbol{j} + u_z\boldsymbol{k}$, then

$$\boldsymbol{v} \pm \boldsymbol{u} = (v_x \pm u_x)\boldsymbol{i} + (v_y \pm u_y)\boldsymbol{j} + (v_z \pm u_z)\boldsymbol{k} \qquad \text{(MB5.3)}$$

A graphical method for adding and subtracting vectors is sometimes desirable. Consider two vectors $\boldsymbol{v}$ and $\boldsymbol{u}$ making an angle θ (Fig. MB5.2a). The first step in the addition of $\boldsymbol{v}$ to $\boldsymbol{u}$ consists of joining the tail of $\boldsymbol{v}$ to the head of $\boldsymbol{u}$, as shown in Fig. MB5.2b. In the second step, we draw a vector from the tail of $\boldsymbol{u}$ to the head of $\boldsymbol{v}$, as shown in Fig. MB5.2c. Reversing the order of addition leads to the same result. That is, we obtain the same resultant whether we add $\boldsymbol{u}$ to $\boldsymbol{v}$ or $\boldsymbol{v}$ to $\boldsymbol{u}$ (Fig. MB5.3).

To calculate the magnitude of the resultant $\boldsymbol{w} = \boldsymbol{u} + \boldsymbol{v}$ we note that $\boldsymbol{v}$, $\boldsymbol{u}$, and $\boldsymbol{w}$ form a triangle and that we know the magnitudes of two of its sides (u and v) and of the angle between them ($180° - \theta$; see Fig. MB5.2c). To calculate the magnitude of the third side, w, we make use of the *law of cosines*, which states that:

For a triangle with sides a, b, and c, and angle C facing side c:

$$c^2 = a^2 + b^2 - 2ab\cos C$$

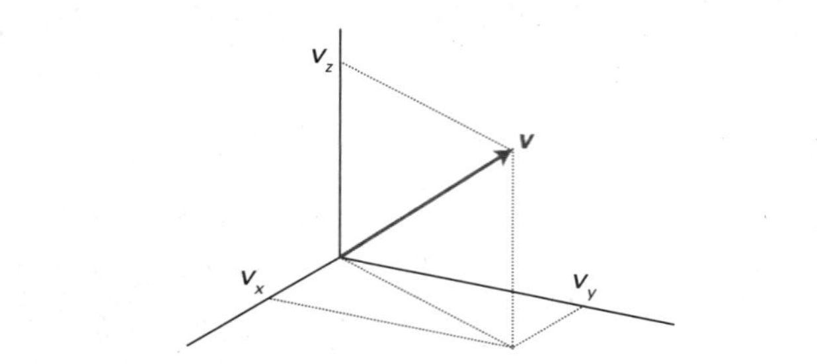

Fig. MB5.1 The vector $\boldsymbol{v}$ has components v_x, v_y, and v_z on the x-, y-, and z-axes, respectively. It has a magnitude v.

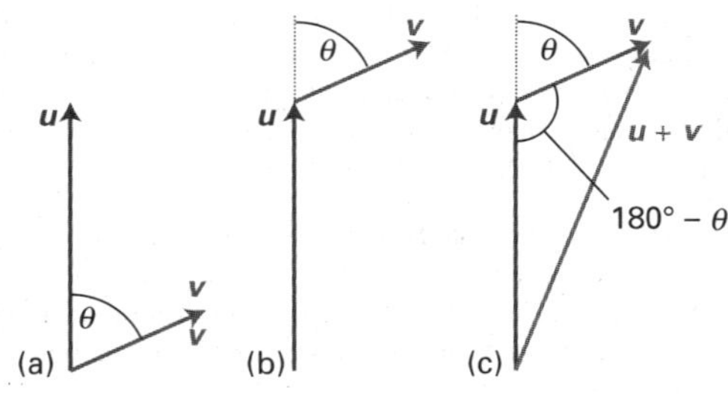

Fig. MB5.2 (a) The vectors $\boldsymbol{u}$ and $\boldsymbol{v}$ make an angle θ. (b) To add $\boldsymbol{v}$ to $\boldsymbol{u}$, we first join the tail of $\boldsymbol{v}$ to the head of $\boldsymbol{u}$, making sure that the angle θ between the vectors remains unchanged. (c) To finish the process, we draw the resultant vector by joining the tail of $\boldsymbol{u}$ to the head of $\boldsymbol{v}$.

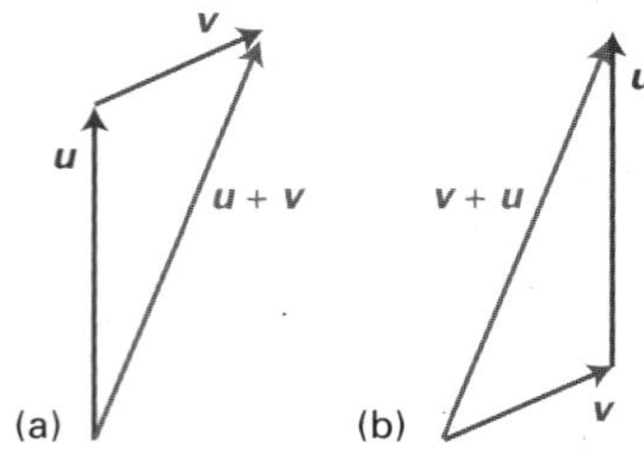

Fig. MB5.3 The addition of (a) $\boldsymbol{v}$ to $\boldsymbol{u}$ gives the same resultant as the addition of (b) $\boldsymbol{u}$ to $\boldsymbol{v}$.

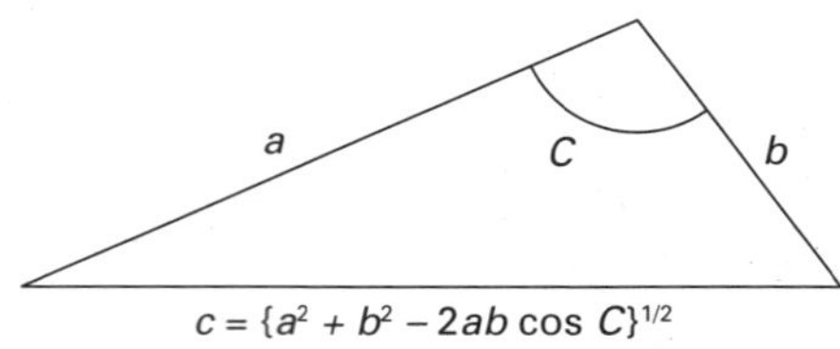

Fig. MB5.4 The graphical representation of the law of cosines.

This law is summarized graphically in Fig. MB5.4 and its application to the case shown in Fig. MB5.2c leads to the expression

$$w^2 = u^2 + v^2 - 2uv\cos(180° - \theta)$$

Because $\cos(180° - \theta) = -\cos\theta$, it follows after taking the square-root of both sides of the preceding expression that

$$w = (u^2 + v^2 + 2uv\cos\theta)^{1/2} \qquad \text{(MB5.4)}$$

The subtraction of vectors follows the same principles outlined above for addition. Consider again the vectors shown in

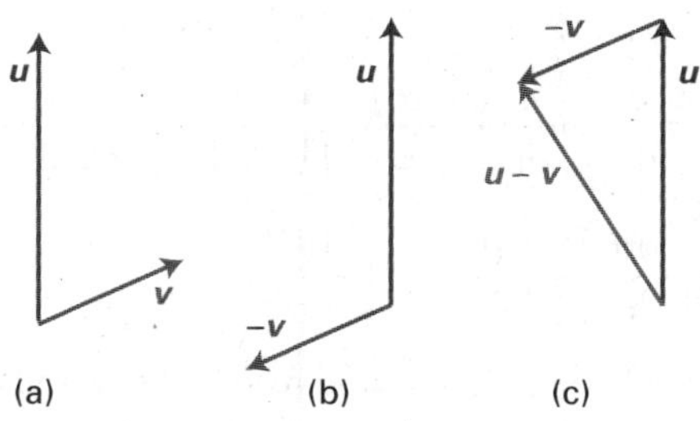

Fig. MB5.5 The graphical method for subtraction of the vector $\boldsymbol{v}$ from the vector $\boldsymbol{u}$ shown in (a) consists of (b) reversing the direction of $\boldsymbol{v}$ to form $-\boldsymbol{v}$, (c) moving the origin of $-\boldsymbol{v}$ to the tip of $\boldsymbol{u}$, and adding $-\boldsymbol{v}$ to $\boldsymbol{u}$.

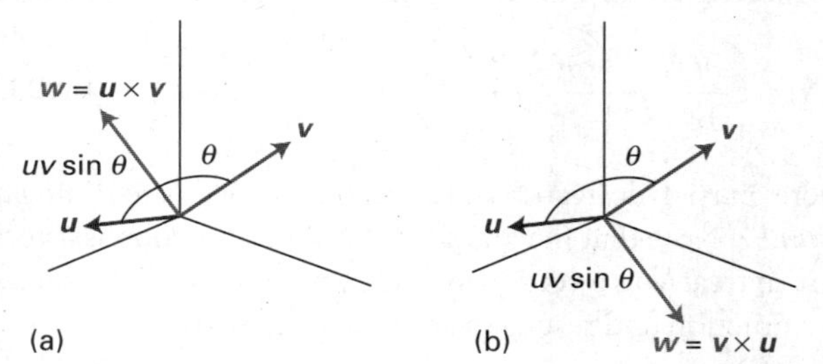

Fig. MB5.6 The direction of the cross-products of two vectors $\boldsymbol{u}$ and $\boldsymbol{v}$ with an angle θ between them: (a) $\boldsymbol{u}\times\boldsymbol{v}$ and (b) $\boldsymbol{v}\times\boldsymbol{u}$. Note that the resultant vector $\boldsymbol{w}$ is perpendicular to both $\boldsymbol{u}$ and $\boldsymbol{v}$ but the direction depends on the order in which the product is taken.

Fig. MB5.2a. We note that subtraction of $\boldsymbol{v}$ from $\boldsymbol{u}$ amounts to addition of $-\boldsymbol{v}$ to $\boldsymbol{u}$. It follows that in the first step of subtraction we draw $-\boldsymbol{v}$ by reversing the direction of $\boldsymbol{v}$ (Fig. MB5.5). Then, the second step consists of adding the $-\boldsymbol{v}$ to $\boldsymbol{u}$ by using the same strategy as in Fig. MB5.2.

MB5.2 Multiplication

There are two ways to multiply vectors. In one procedure, the **scalar product** (or **dot product**) of two vectors $\boldsymbol{u}$ and $\boldsymbol{v}$ is defined as

$$\boldsymbol{u}\cdot\boldsymbol{v} = uv\cos\theta \qquad \text{Scalar product} \qquad \text{(MB5.5)}$$

As its name suggests, the scalar product of two vectors is a scalar.

• **A brief illustration**

The energy of interaction between a magnetic moment $\boldsymbol{\mu}$ (which might be due to the orbital angular momentum, $\boldsymbol{l}$, of an electron, $\boldsymbol{\mu}=\gamma\boldsymbol{l}$) and a magnetic field $\boldsymbol{\mathcal{B}}$ is $E=-\boldsymbol{\mu}\cdot\boldsymbol{\mathcal{B}}$. Suppose the field is applied in the z-direction; then $\boldsymbol{\mathcal{B}}=\mathcal{B}\boldsymbol{k}$. The energy of interaction is then

$$E=-\boldsymbol{\mu}\cdot\boldsymbol{\mathcal{B}}=-\boldsymbol{\mu}\cdot\mathcal{B}\boldsymbol{k}=-\mu_z\mathcal{B}=-\mu\mathcal{B}\cos\theta$$

where θ is the angle between the magnetic moment and the field direction. •

The second type of vector multiplication is the **vector product** (or **cross-product**) of two vectors $\boldsymbol{u}$ and $\boldsymbol{v}$ to give a vector $\boldsymbol{w}$:

$$\boldsymbol{u}\times\boldsymbol{v}=\boldsymbol{w} \qquad \text{Vector product} \qquad \text{(MB5.6)}$$

where the length of $\boldsymbol{w}$ is $uv\sin\theta$. where θ is the angle between $\boldsymbol{u}$ and $\boldsymbol{v}$. The direction of $\boldsymbol{w}$ is determined by the 'right-hand rule' (Fig. MB5.6). An equivalent definition is

$$\boldsymbol{u}\times\boldsymbol{v}=\begin{vmatrix}\boldsymbol{i} & \boldsymbol{j} & \boldsymbol{k}\\ u_x & u_y & u_z\\ v_x & v_y & v_z\end{vmatrix} \qquad \text{Vector product} \qquad \text{(MB5.7)}$$
$$=(u_yv_z-u_zv_y)\boldsymbol{i}-(u_xv_z-u_zv_x)\boldsymbol{j}+(u_xv_y-u_yv_x)\boldsymbol{k}$$

where the structure in the middle is a determinant (see *Mathematical background* 6 following Chapter 10).

• **A brief illustration**

The angular momentum $\boldsymbol{l}$ is defined as the vector product of the position $\boldsymbol{r}=(x,y,z)$ and linear momentum: $\boldsymbol{p}=(p_x,p_y,p_z)$:

$$\boldsymbol{r}\times\boldsymbol{p}=\begin{vmatrix}\boldsymbol{i} & \boldsymbol{j} & \boldsymbol{k}\\ x & y & z\\ p_x & p_y & p_z\end{vmatrix}=(yp_z-zp_y)\boldsymbol{i}-(xp_z-zp_x)\boldsymbol{j}+(xp_y-yp_x)\boldsymbol{k}$$

We can now pick out the x-component as $l_x=yp_z-zp_y$, and likewise for the remaining two components. •

MB5.3 Differentiation

The derivative $\mathrm{d}\boldsymbol{v}/\mathrm{d}t$, where the components v_x, v_y, and v_z are themselves functions of t, is

$$\frac{\mathrm{d}\boldsymbol{v}}{\mathrm{d}t}=\left(\frac{\mathrm{d}v_x}{\mathrm{d}t}\right)\boldsymbol{i}+\left(\frac{\mathrm{d}v_y}{\mathrm{d}t}\right)\boldsymbol{j}+\left(\frac{\mathrm{d}v_z}{\mathrm{d}t}\right)\boldsymbol{k} \qquad \text{(MB5.8)}$$

The derivatives of scalar and vector products are obtained using the rules of differentiating a product:

$$\frac{\mathrm{d}(\boldsymbol{u}\cdot\boldsymbol{v})}{\mathrm{d}t}=\boldsymbol{u}\cdot\frac{\mathrm{d}\boldsymbol{v}}{\mathrm{d}t}+\boldsymbol{v}\cdot\frac{\mathrm{d}\boldsymbol{u}}{\mathrm{d}t} \qquad \text{(MB5.9a)}$$

$$\frac{\mathrm{d}(\boldsymbol{u}\times\boldsymbol{v})}{\mathrm{d}t}=\boldsymbol{u}\times\frac{\mathrm{d}\boldsymbol{v}}{\mathrm{d}t}+\frac{\mathrm{d}\boldsymbol{u}}{\mathrm{d}t}\times\boldsymbol{v} \qquad \text{(MB5.9b)}$$

In the latter, note the importance of preserving the order of vectors.

The **gradient** of a function $f(x,y,z)$, denoted grad f or ∇f, is

$$\nabla f = \left(\frac{\partial f}{\partial x}\right)i + \left(\frac{\partial f}{\partial y}\right)j + \left(\frac{\partial f}{\partial z}\right)k \quad \text{Gradient} \quad \text{(MB5.10)}$$

where partial derivatives are treated in *Mathematical background* 2. Note that the gradient of a scalar function is a vector. We can treat ∇ as a vector operator (in the sense that it operates on a function and results in a vector), and write

$$\nabla = i\frac{\partial}{\partial x} + j\frac{\partial}{\partial y} + k\frac{\partial}{\partial z} \quad \text{(MB5.11)}$$

The scalar product of ∇ and ∇f, using eqns MB5.10 and MB5.11, is

$$\nabla \cdot \nabla f = \left\{i\frac{\partial}{\partial x} + j\frac{\partial}{\partial y} + k\frac{\partial}{\partial z}\right\} \cdot \left\{\left(\frac{\partial f}{\partial x}\right)i + \left(\frac{\partial f}{\partial y}\right)j + \left(\frac{\partial f}{\partial z}\right)k\right\}$$

$$= \left(\frac{\partial^2 f}{\partial x^2}\right) + \left(\frac{\partial^2 f}{\partial y^2}\right) + \left(\frac{\partial^2 f}{\partial z^2}\right) \quad \text{(MB5.12)}$$

$\nabla \cdot \nabla f$ is normally denoted $\nabla^2 f$ and read 'del squared f'. Its form in polar coordinates is given in Table 7.1.

Molecular structure

10

The concepts developed in Chapter 9, particularly those of orbitals, can be extended to a description of the electronic structures of molecules. There are two principal quantum mechanical theories of molecular electronic structure. In valence-bond theory, the starting point is the concept of the shared electron pair. We see how to write the wavefunction for such a pair, and how it may be extended to account for the structures of a wide variety of molecules. The theory introduces the concepts of σ and π bonds, promotion, and hybridization that are used widely in chemistry. In molecular orbital theory (with which the bulk of the chapter is concerned), the concept of atomic orbital is extended to that of molecular orbital, which is a wavefunction that spreads over all the atoms in a molecule.

In this chapter we consider the origin of the strengths, numbers, and three-dimensional arrangement of chemical bonds between atoms. As we shall see, all chemical bonding can be traced to the interplay between the attraction of opposite charges, the repulsion of like charges, and the effect of changing kinetic energy as the electrons are confined to various regions when bonds form.

The quantum mechanical description of chemical bonding has become highly developed through the use of computers, and it is now possible to consider the structures of molecules of almost any complexity. We shall concentrate on the quantum mechanical description of the **covalent bond**, which was identified by G.N. Lewis (in 1916, before quantum mechanics was fully established) as an electron pair shared between two neighbouring atoms and denoted A–B. We shall see, however, that the other principal type of bond, an **ionic bond**, in which the cohesion arises from the Coulombic attraction between ions of opposite charge, is also captured as a limiting case of a covalent bond between dissimilar atoms.

There are two major approaches to the calculation of molecular structure, **valence-bond theory** (VB theory) and **molecular orbital theory** (MO theory). Almost all modern computational work makes use of MO theory, and we concentrate on that theory in this chapter. Valence-bond theory, though, has left its imprint on the language of chemistry, and it is important to know the significance of terms that chemists use every day. Therefore, our discussion is organized as follows. First, we set out the concepts common to all levels of description. Then we present VB theory, which gives us a simple qualitative understanding of bond formation and its associated language. Next, we present the basic ideas of MO theory. Finally, we see how computational techniques pervade all current discussions of molecular structure, including the prediction of chemical reactivity.

The Born–Oppenheimer approximation

Key point **The nuclei of atoms in a molecule are regarded as fixed at selected locations, and the Schrödinger equation is then solved for the wavefunction of the electrons alone.**

All theories of molecular structure make the same simplification at the outset. Whereas the Schrödinger equation for a hydrogen atom can be solved exactly, an exact solution is not possible for any molecule because even the simplest molecule consists of three particles (two nuclei and one electron). We therefore adopt the **Born–Oppenheimer approximation** in which it is supposed that the nuclei, being so much heavier than an electron, move relatively slowly and may be treated as stationary while the electrons move in their field. That is, we think of the nuclei as fixed at arbitrary locations, and then solve the Schrödinger equation for the wavefunction of the electrons alone.

The approximation is quite good for ground-state molecules, for calculations suggest that the nuclei in H_2 move through only about 1 pm while the electron speeds through 1000 pm, so even in this case the error of assuming that the nuclei are stationary is small. Exceptions to the approximation's validity include certain excited states of polyatomic molecules and the ground states of cations; both types of species are important when considering photoelectron spectroscopy (Section 10.4) and mass spectrometry.

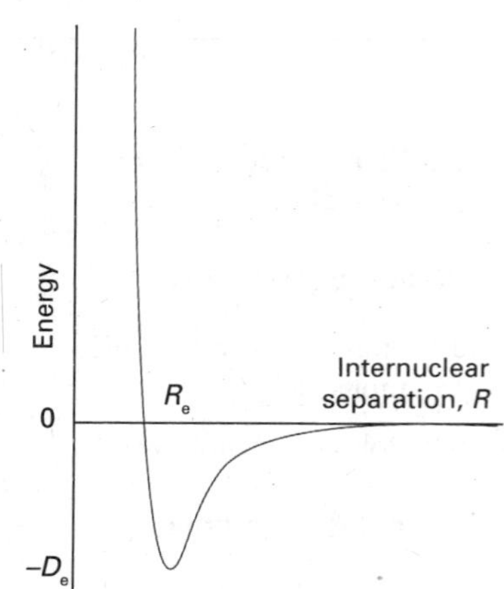

Fig. 10.1 A molecular potential energy curve. The equilibrium bond length R_e corresponds to the energy minimum.

The Born–Oppenheimer approximation allows us to select an internuclear separation in a diatomic molecule and then to solve the Schrödinger equation for the electrons at that nuclear separation. Then we choose a different separation and repeat the calculation, and so on. In this way we can explore how the energy of the molecule varies with bond length and obtain a **molecular potential energy curve** (Fig. 10.1). It is called a *potential* energy curve because the kinetic energy of the stationary nuclei is zero. Once the curve has been calculated or determined experimentally (by using the spectroscopic techniques described in Chapters 11 and 12), we can identify the **equilibrium bond length**, R_e, the internuclear separation at the minimum of the curve, and the **bond dissociation energy**, D_0, which is closely related to the depth, D_e, of the minimum below the energy of the infinitely widely separated and stationary atoms. When more than one molecular parameter is changed in a polyatomic molecule, such as its various bond lengths and angles, we obtain a potential energy *surface*; the overall equilibrium shape of the molecule corresponds to the global minimum of the surface.

A brief comment

The dissociation energy differs from the depth of the well by an energy equal to the zero-point vibrational energy of the bonded atoms: $D_0 = D_e - \frac{1}{2}h\nu$, where ν is the vibrational frequency of the bond (Section 12.8).

Valence-bond theory

Valence-bond theory was the first quantum mechanical theory of bonding to be developed. The language it introduced, which includes concepts such as spin pairing, σ and π bonds, and hybridization, is widely used throughout chemistry, especially in the description of the properties and reactions of organic compounds. Here we summarize essential topics of VB theory that should be familiar from introductory chemistry and set the stage for the development of MO theory.

10.1 Homonuclear diatomic molecules

Key point **In VB theory, a bond forms when an electron in an atomic orbital on one atom pairs its spin with that of an electron in an atomic orbital on another atom.**

We begin the account of VB theory by considering the simplest possible chemical bond, the one in molecular hydrogen, H_2. The spatial wavefunction for an electron on each of two widely separated H atoms is

$$\psi = \chi_{H1s_A}(r_1)\chi_{H1s_B}(r_2) \tag{10.1}$$

if electron 1 is on atom A and electron 2 is on atom B; in this chapter we use χ (chi) to denote atomic orbitals. For simplicity, we shall write this wavefunction as $\psi = A(1)B(2)$. When the atoms are close, it is not possible to know whether it is electron 1 or electron 2 that is on A. An equally valid description is therefore $\psi = A(2)B(1)$, in which electron 2 is on A and electron 1 is on B. When two outcomes are equally probable, quantum mechanics instructs us to describe the true state of the system as a superposition of the wavefunctions for each possibility (Section 7.5e), so a better description of the molecule than either wavefunction alone is one of the (unnormalized) linear combinations $\psi = A(1)B(2) \pm A(2)B(1)$. The combination with lower energy is the one with a + sign, so the valence-bond wavefunction of the electrons in an H_2 molecule is

A valence-bond wavefunction

$$\psi = A(1)B(2) + A(2)B(1) \tag{10.2}$$

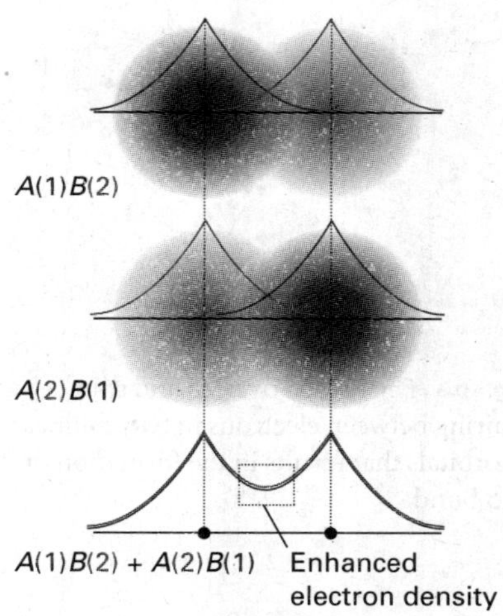

Fig. 10.2 It is very difficult to represent valence-bond wavefunctions because they refer to two electrons simultaneously. However, this illustration is an attempt. The top illustration represents $A(1)B(2)$, and the middle illustration represents the contribution $A(2)B(1)$. When the two contributions are superimposed, there is interference between the various contributions, resulting in an enhanced (two-electron) density in the internuclear region.

The reason why this linear combination has a lower energy than either the separate atoms or the linear combination with a negative sign can be traced to the constructive interference between the wave patterns represented by the terms $A(1)B(2)$ and $A(2)B(1)$, and the resulting enhancement of the probability density of the electrons in the internuclear region (Fig. 10.2).

The electron distribution described by the wavefunction in eqn 10.2 is called a **σ bond.** A σ bond has cylindrical symmetry around the internuclear axis, and is so called because, when viewed along the internuclear axis, it resembles a pair of electrons in an s orbital (and σ is the Greek equivalent of s).

A chemist's picture of a covalent bond is one in which the spins of two electrons pair as the atomic orbitals overlap. The origin of the role of spin, as we show in the following *Justification*, is that the wavefunction in eqn 10.2 can be formed only by a pair of spin-paired electrons. Spin pairing is not an end in itself: it is a means of achieving a wavefunction and the probability distribution implies that it corresponds to a low energy.

Justification 10.1 *Electron pairing in VB theory*

The Pauli principle requires the overall wavefunction of two electrons, the wavefunction including spin, to change sign when the labels of the electrons are interchanged (Section 9.4b). The overall VB wavefunction for two electrons is

$$\psi(1,2) = \{A(1)B(2) + A(2)B(1)\}\sigma(1,2)$$

where σ represents the spin component of the wavefunction. When the labels 1 and 2 are interchanged, this wavefunction becomes

$$\psi(2,1) = \{A(2)B(1) + A(1)B(2)\}\sigma(2,1) = \{A(1)B(2) + A(2)B(1)\}\sigma(2,1)$$

The Pauli principle requires that $\psi(2,1) = -\psi(1,2)$, which is satisfied only if $\sigma(2,1) = -\sigma(1,2)$. The combination of two spins that has this property is

$$\sigma_-(1,2) = (1/2^{1/2})\{\alpha(1)\beta(2) - \alpha(2)\beta(1)\}$$

which corresponds to paired electron spins (Section 9.8). Therefore, we conclude that the state of lower energy (and hence the formation of a chemical bond) is achieved if the electron spins are paired.

The VB description of H_2 can be applied to other homonuclear diatomic molecules. For N_2, for instance, we consider the valence electron configuration of each atom,

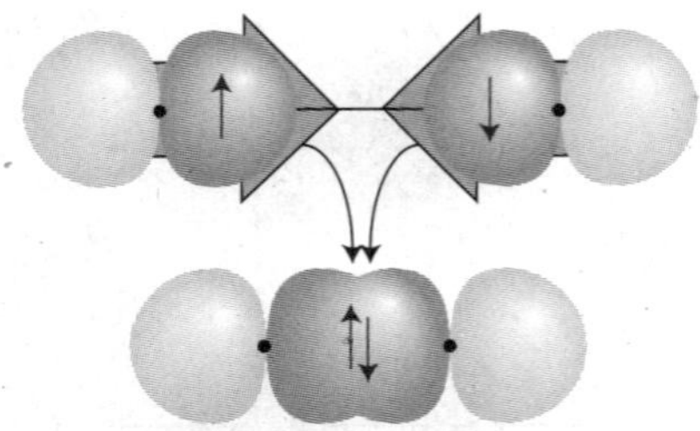

Fig. 10.3 The orbital overlap and spin pairing between electrons in two collinear p orbitals that results in the formation of a σ bond.

which is $2s^22p_x^12p_y^12p_z^1$. It is conventional to take the z-axis to be the internuclear axis, so we can imagine each atom as having a $2p_z$ orbital pointing towards a $2p_z$ orbital on the other atom (Fig. 10.3), with the $2p_x$ and $2p_y$ orbitals perpendicular to the axis. A σ bond is then formed by spin pairing between the two electrons in the two $2p_z$ orbitals. Its spatial wavefunction is given by eqn 10.2, but now A and B stand for the two $2p_z$ orbitals.

The remaining N2p orbitals cannot merge to give σ bonds as they do not have cylindrical symmetry around the internuclear axis. Instead, they merge to form two π bonds. A **π bond** arises from the spin pairing of electrons in two p orbitals that approach side-by-side (Fig. 10.4). It is so called because, viewed along the internuclear axis, a π bond resembles a pair of electrons in a p orbital (and π is the Greek equivalent of p).

There are two π bonds in N_2, one formed by spin pairing in two neighbouring $2p_x$ orbitals and the other by spin pairing in two neighbouring $2p_y$ orbitals. The overall bonding pattern in N_2 is therefore a σ bond plus two π bonds (Fig. 10.5), which is consistent with the Lewis structure :N≡N: for nitrogen.

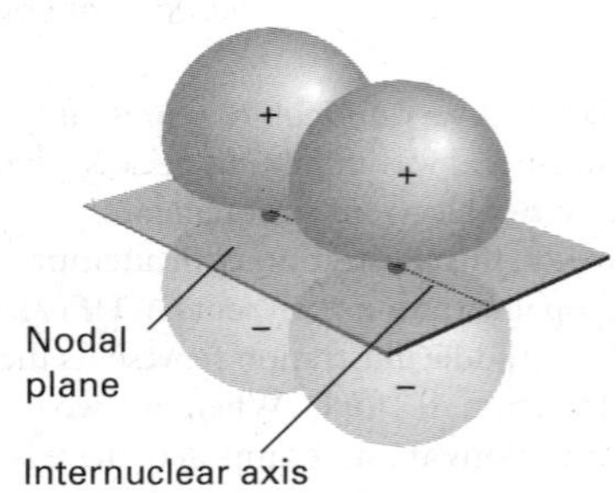

Fig. 10.4 A π bond results from orbital overlap and spin pairing between electrons in p orbitals with their axes perpendicular to the internuclear axis. The bond has two lobes of electron density separated by a nodal plane.

10.2 Polyatomic molecules

Key point To accommodate the shapes of polyatomic molecules, VB theory introduces the concepts of promotion and hybridization.

Each σ bond in a polyatomic molecule is formed by the spin pairing of electrons in atomic orbitals with cylindrical symmetry around the relevant internuclear axis. Likewise, π bonds are formed by pairing electrons that occupy atomic orbitals of the appropriate symmetry.

The VB description of H_2O will make this clear. The valence-electron configuration of an O atom is $2s^22p_x^22p_y^12p_z^1$. The two unpaired electrons in the O2p orbitals can each pair with an electron in an H1s orbital, and each combination results in the formation of a σ bond (each bond has cylindrical symmetry about the respective O–H internuclear axis). Because the $2p_y$ and $2p_z$ orbitals lie at 90° to each other, the two σ bonds also lie at 90° to each other (Fig. 10.6). We can predict, therefore, that H_2O should be an angular molecule, which it is. However, the theory predicts a bond angle of 90°, whereas the actual bond angle is 104.5°.

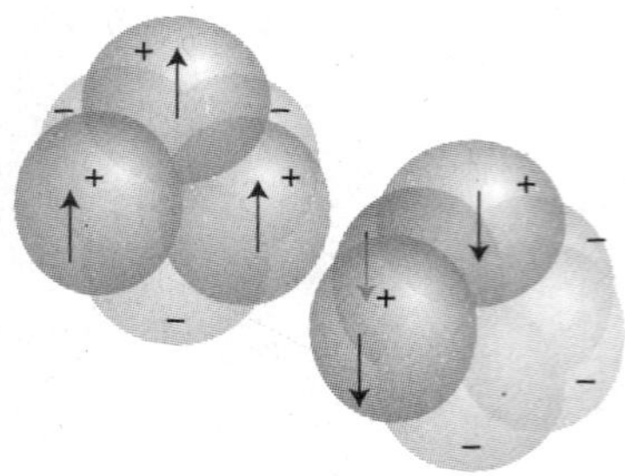

Fig. 10.5 The structure of bonds in a nitrogen molecule: there is one σ bond and two π bonds. As explained later, the overall electron density has cylindrical symmetry around the internuclear axis.

Self-test 10.1 Use VB theory to suggest a shape for the ammonia molecule, NH_3.

[Trigonal pyramidal with HNH bond angle 90°; experimental: 107°]

Another deficiency of this initial formulation of VB theory is its inability to account for carbon's tetravalence (its ability to form four bonds). The ground-state configuration of C is $2s^22p_x^12p_y^1$, which suggests that a carbon atom should be capable of forming only two bonds, not four. This deficiency is overcome by allowing for **promotion**, the excitation of an electron to an orbital of higher energy. In carbon, for example, the promotion of a 2s electron to a 2p orbital can be thought of as leading to the configuration $2s^12p_x^12p_y^12p_z^1$, with four unpaired electrons in separate orbitals. These electrons may pair with four electrons in orbitals provided by four other atoms (such as four H1s orbitals if the molecule is CH_4), and hence form four σ bonds. Although energy was required to promote the electron, it is more than recovered by the promoted atom's ability to form four bonds in place of the two bonds of the unpromoted atom. Promotion, and the formation of four bonds, is a characteristic feature of carbon because the promotion energy is quite small: the promoted electron leaves a doubly occupied 2s orbital and enters a vacant 2p orbital, hence significantly relieving the electron–electron repulsion it experiences in the former. However, we need to remember that promotion is not a 'real' process in which an atom somehow becomes excited and then forms bonds: it is a notional contribution to the overall energy change that occurs when bonds form.

The description of the bonding in CH_4 (and other alkanes) is still incomplete because it implies the presence of three σ bonds of one type (formed from H1s and C2p orbitals) and a fourth σ bond of a distinctly different character (formed from H1s and C2s). This problem is overcome by realizing that the electron density distribution in the promoted atom is equivalent to the electron density in which each electron occupies a **hybrid orbital** formed by interference between the C2s and C2p orbitals of the same atom. The origin of the hybridization can be appreciated by thinking of the four atomic orbitals centred on a nucleus as waves that interfere destructively and constructively in different regions, and give rise to four new shapes.

As we show in the following *Justification*, the specific linear combinations that give rise to four equivalent hybrid orbitals are

$$h_1 = s + p_x + p_y + p_z \qquad h_2 = s - p_x - p_y + p_z$$
$$h_3 = s - p_x + p_y - p_z \qquad h_4 = s + p_x - p_y - p_z \qquad \text{sp}^3 \text{ hybrid orbitals} \qquad (10.3)$$

As a result of the interference between the component orbitals, each hybrid orbital consists of a large lobe pointing in the direction of one corner of a regular tetrahedron (Fig. 10.7). The angle between the axes of the hybrid orbitals is the tetrahedral angle, arccos(−1/3) = 109.47°. Because each hybrid is built from one s orbital and three p orbitals, it is called an **sp³ hybrid orbital.**

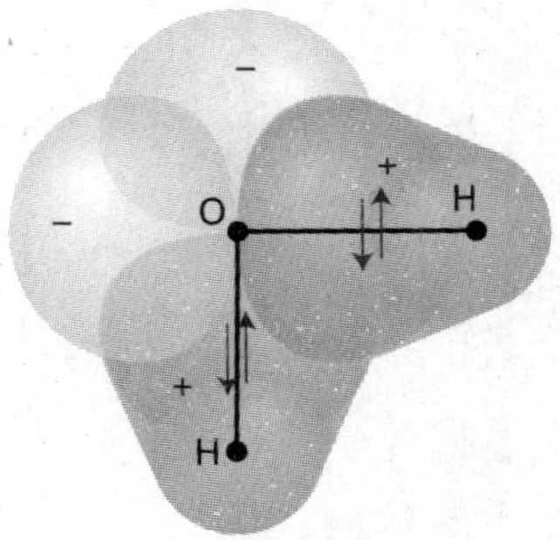

Fig. 10.6 A first approximation to the valence-bond description of bonding in an H_2O molecule. Each σ bond arises from the overlap of an H1s orbital with one of the O2p orbitals. This model suggests that the bond angle should be 90°, which is significantly different from the experimental value.

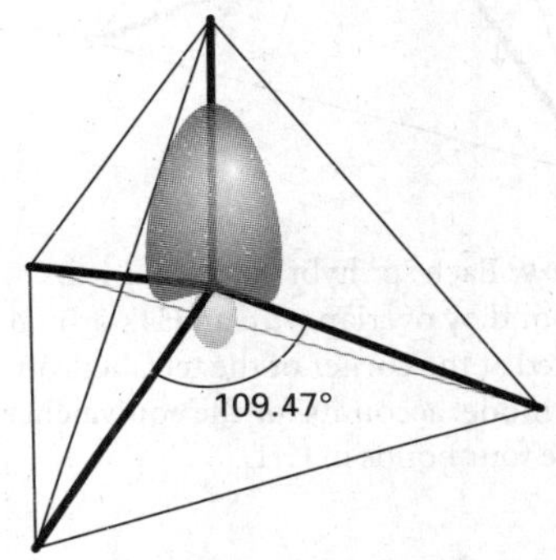

Fig. 10.7 An sp³ hybrid orbital formed from the superposition of s and p orbitals on the same atom. There are four such hybrids: each one points towards the corner of a regular tetrahedron. The overall electron density remains spherically symmetrical.

Justification 10.2 *Determining the form of tetrahedral hybrids*

We begin by supposing that each hybrid can be written in the form $h = as + b_xp_x + b_yp_y + b_zp_z$. The hybrid h_1 that points to the (1,1,1) corner of a cube (Fig. 10.8) must have equal contributions from all three p orbitals, so we can set the three b coefficients equal to each other and write $h_1 = as + b(p_x + p_y + p_z)$. The other three hybrids have the same composition (they are equivalent, apart from their direction in space), but are orthogonal to h_1. This orthogonality is achieved by choosing different signs for the p orbitals but the same overall composition. For instance, we might choose $h_2 = as + b(-p_x - p_y + p_z)$, in which case the orthogonality condition is

$$\int h_1h_2\,d\tau = \int (as + b(p_x + p_y + p_z))(as + b(-p_x - p_y + p_z))\,d\tau$$

$$= a^2\overbrace{\int s^2 d\tau}^{1} - b^2\overbrace{\int p_x^2 d\tau}^{1} - b^2\overbrace{\int p_y^2 d\tau}^{1} + b^2\overbrace{\int p_z^2 d\tau}^{1} - ab\overbrace{\int sp_x d\tau}^{0} - \cdots - b^2\overbrace{\int p_xp_y d\tau}^{0} + \cdots$$

$$= a^2 - b^2 - b^2 + b^2 = a^2 - b^2 = 0$$

We conclude that a solution is $a = b$ (the alternative solution, $a = -b$, simply corresponds to choosing different absolute phases for the p-orbitals) and the two hybrid orbitals are the h_1 and h_2 in eqn 10.3. A similar argument but with $h_3 = as + b(-p_x + p_y - p_z)$ or $h_4 = as + b(p_x - p_y - p_z)$ leads to the other two hybrids in eqn 10.3.

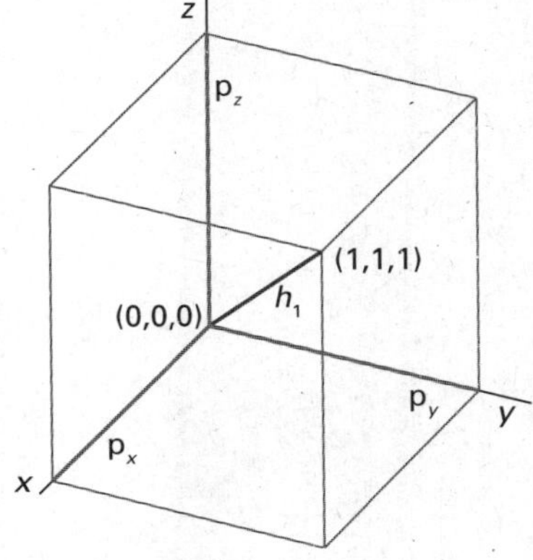

Fig. 10.8 One sp³ hybrid is constructed by supposing that it points to the (1,1,1) corner of a cube: it has equal contributions from all three p orbitals.

It is now easy to see how the valence-bond description of the CH_4 molecule leads to a tetrahedral molecule containing four equivalent C–H bonds. Each hybrid orbital of the promoted C atom contains a single unpaired electron; an H1s electron can pair with each one, giving rise to a σ bond pointing in a tetrahedral direction. For example, the (unnormalized) wavefunction for the bond formed by the hybrid orbital h_1 and the $1s_A$ orbital (with wavefunction that we shall denote A) is

$$\psi = h_1(1)A(2) + h_1(2)A(1) \tag{10.4}$$

As for H_2, to achieve this wavefunction, the two electrons it describes must be paired. Because each sp^3 hybrid orbital has the same composition, all four σ bonds are identical apart from their orientation in space (Fig. 10.9).

A hybrid orbital has enhanced amplitude in the internuclear region, which arises from the constructive interference between the s orbital and the positive lobes of the p orbitals (Fig. 10.10). As a result, the bond strength is greater than for a bond formed from an s or p orbital alone. This increased bond strength is another factor that helps to repay the promotion energy.

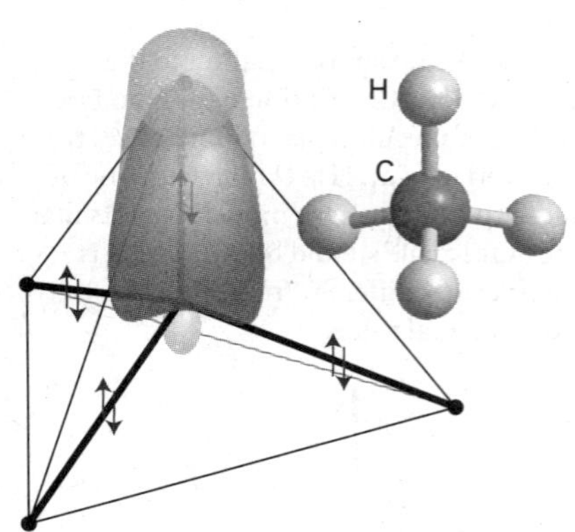

Fig. 10.9 Each sp^3 hybrid orbital forms a σ bond by overlap with an H1s orbital located at the corner of the tetrahedron. This model accounts for the equivalence of the four bonds in CH_4.

Hybridization is used to describe the structure of an ethene molecule, $H_2C{=}CH_2$, and the torsional rigidity of double bonds. An ethene molecule is planar, with HCH and HCC bond angles close to 120°. To reproduce the σ bonding structure, we promote each C atom to a $2s^12p^3$ configuration. However, instead of using all four orbitals to form hybrids, we form **sp^2 hybrid orbitals**:

$$\begin{aligned} h_1 &= s + 2^{1/2}p_y \\ h_2 &= s + (\tfrac{3}{2})^{1/2}p_x - (\tfrac{1}{2})^{1/2}p_y \\ h_3 &= s - (\tfrac{3}{2})^{1/2}p_x - (\tfrac{1}{2})^{1/2}p_y \end{aligned} \quad \boxed{sp^2 \text{ hybrid orbitals}} \tag{10.5}$$

These hybrids lie in a plane and point towards the corners of an equilateral triangle at 120° to each other (Fig. 10.11 and Problem 10.17). The third 2p orbital ($2p_z$) is not included in the hybridization; its axis is perpendicular to the plane in which the hybrids lie. The different signs of the coefficients ensure that constructive interference takes place in different regions of space, so giving the patterns in the illustration. The sp^2-hybridized C atoms each form three σ bonds by spin pairing with either the h_1

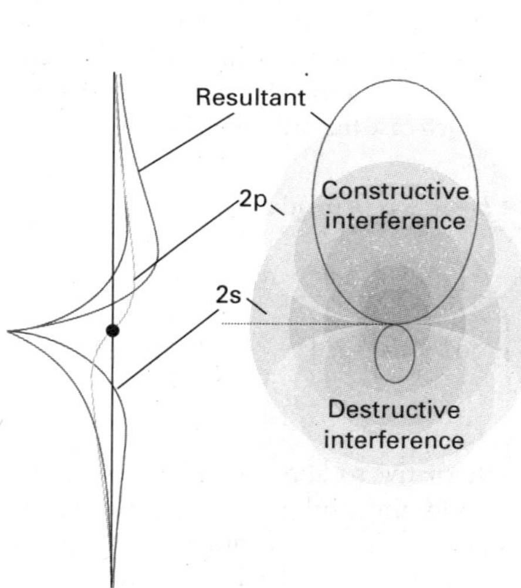

Fig. 10.10 A more detailed representation of the formation of an sp^3 hybrid by interference between wavefunctions centred on the same atomic nucleus. (To simplify the representation, we have ignored the radial node of the 2s orbital.)

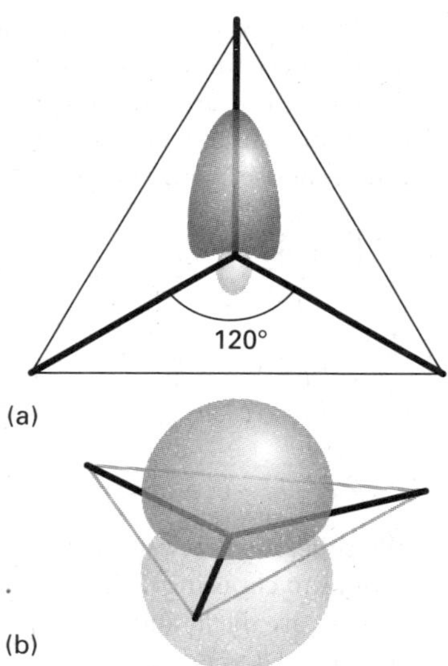

Fig. 10.11 (a) An s orbital and two p orbitals can be hybridized to form three equivalent orbitals that point towards the corners of an equilateral triangle. (b) The remaining unhybridized p orbital is perpendicular to the plane.

hybrid of the other C atom or with H1s orbitals. The σ framework therefore consists of C–H and C–C σ bonds at 120° to each other. When the two CH_2 groups lie in the same plane, the two electrons in the unhybridized p orbitals can pair and form a π bond (Fig. 10.12). The formation of this π bond locks the framework into the planar arrangement, for any rotation of one CH_2 group relative to the other leads to a weakening of the π bond (and consequently an increase in energy of the molecule).

A similar description applies to ethyne, HC≡CH, a linear molecule. Now the C atoms are **sp hybridized**, and the σ bonds are formed using hybrid atomic orbitals of the form

$$h_1 = s + p_z \qquad h_2 = s - p_z \qquad \text{sp hybrid orbitals} \qquad (10.6)$$

These two hybrids lie along the internuclear axis. The electrons in them pair either with an electron in the corresponding hybrid orbital on the other C atom or with an electron in one of the H1s orbitals. Electrons in the two remaining p orbitals on each atom, which are perpendicular to the molecular axis, pair to form two perpendicular π bonds (Fig. 10.13).

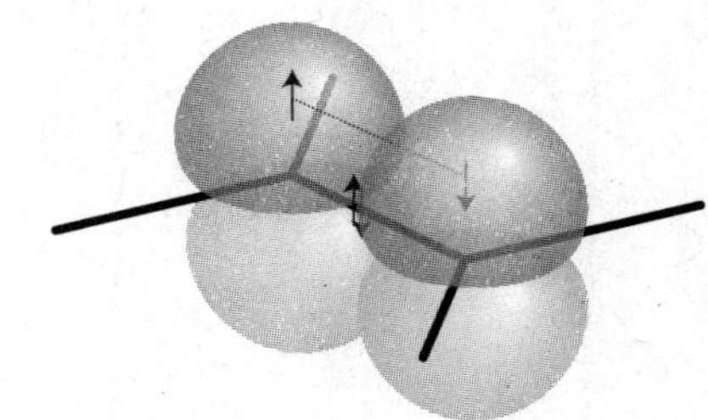

Fig. 10.12 A representation of the structure of a double bond in ethene; only the π bond is shown explicitly.

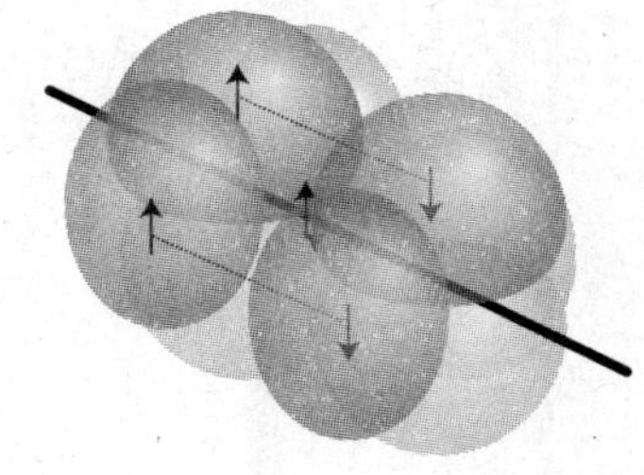

Fig. 10.13 A representation of the structure of a triple bond in ethyne; only the π bonds are shown explicitly. The overall electron density has cylindrical symmetry around the axis of the molecule.

Self-test 10.2 Hybrid orbitals do not always form bonds: they may also contain lone pairs of electrons. Use VB theory to suggest possible shapes for the hydrogen peroxide molecule, H_2O_2.

[Each H–O–O bond angle is predicted to be approximately 109° (experimental: 94.8°); rotation around the O–O bond is possible, so the molecule is conformationally mobile]

Other hybridization schemes, particularly those involving d orbitals, are often invoked in elementary descriptions of molecular structure to be consistent with other molecular geometries (Table 10.1). The hybridization of *N* atomic orbitals always results in the formation of *N* hybrid orbitals, which may either form bonds or may contain lone pairs of electrons. For example, sp^3d^2 hybridization results in six equivalent hybrid orbitals pointing towards the corners of a regular octahedron; it is sometimes invoked to account for the structure of octahedral molecules, such as SF_6.

Table 10.1* Some hybridization schemes

Coordination number	Arrangement	Composition
2	Linear	sp, pd, sd
	Angular	sd
3	Trigonal planar	sp^2, p^2d
	Unsymmetrical planar	spd
	Trigonal pyramidal	pd^2
4	Tetrahedral	sp^3, sd^3
	Irregular tetrahedral	spd^2, p^3d, dp^3
	Square planar	p^2d^2, sp^2d
5	Trigonal bipyramidal	sp^3d, spd^3
	Tetragonal pyramidal	sp^2d^2, sd^4, pd^4, p^3d^2
	Pentagonal planar	p^2d^3
6	Octahedral	sp^3d^2
	Trigonal prismatic	spd^4, pd^5
	Trigonal antiprismatic	p^3d^3

* Source: H. Eyring, J. Walter, and G.E. Kimball, *Quantum chemistry*, Wiley (1944).

Molecular orbital theory

In MO theory, electrons do not belong to particular bonds but spread throughout the entire molecule. This theory has been more fully developed than VB theory and provides the language that is widely used in modern discussions of bonding. To introduce it, we follow the same strategy as in Chapter 9, where the one-electron H atom was taken as the fundamental species for discussing atomic structure and then developed into a description of many-electron atoms. In this chapter we use the simplest molecular species of all, the hydrogen molecule-ion, H_2^+, to introduce the essential features of bonding and then use it to describe the structures of more complex systems.

10.3 The hydrogen molecule-ion

Key points **(a) A molecular orbital is constructed as a linear combination of atomic orbitals. (b) A bonding orbital arises from the constructive overlap of neighbouring atomic orbitals. (c) An antibonding orbital arises from the destructive overlap of neighbouring atomic orbitals.**

The hamiltonian for the single electron in H_2^+ is

$$\hat{H} = -\frac{\hbar^2}{2m_e}\nabla_1^2 + V \qquad V = -\frac{e^2}{4\pi\varepsilon_0}\left(\frac{1}{r_{A1}} + \frac{1}{r_{B1}} - \frac{1}{R}\right) \tag{10.7}$$

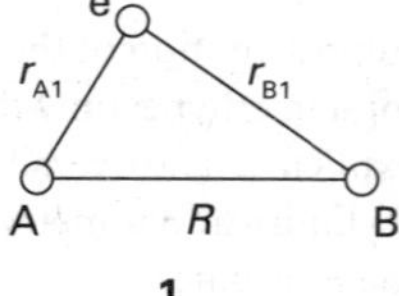

1

where r_{A1} and r_{B1} are the distances of the electron from the two nuclei A and B (1) and R is the distance between the two nuclei. In the expression for V, the first two terms in parentheses are the attractive contribution from the interaction between the electron and the nuclei; the remaining term is the repulsive interaction between the nuclei. The collection of fundamental constants $e^2/4\pi\varepsilon_0$ occurs widely throughout this chapter, and we shall denote it j_0.

The one-electron wavefunctions obtained by solving the Schrödinger equation $\hat{H}\psi = E\psi$ are called **molecular orbitals** (MO). A molecular orbital ψ gives, through the value of $|\psi|^2$, the distribution of the electron in the molecule. A molecular orbital is like an atomic orbital, but spreads throughout the molecule.

The Schrödinger equation can be solved analytically for H_2^+ (within the Born–Oppenheimer approximation), but the wavefunctions are very complicated functions; moreover, the solution cannot be extended to polyatomic systems. Therefore, we adopt a simpler procedure that, while more approximate, can be extended readily to other molecules.

(a) Linear combinations of atomic orbitals

If an electron can be found in an atomic orbital belonging to atom A and also in an atomic orbital belonging to atom B, then the overall wavefunction is a superposition of the two atomic orbitals:

$$\psi_{\pm} = N(A \pm B) \tag{10.8}$$

Linear combination of atomic orbitals

where, for H_2^+, A denotes χ_{H1s_A}, B denotes χ_{H1s_B}, and N is a normalization factor. The technical term for the superposition in eqn 10.8 is a **linear combination of atomic orbitals** (LCAO). An approximate molecular orbital formed from a linear combination of atomic orbitals is called an **LCAO-MO**. A molecular orbital that has cylindrical symmetry around the internuclear axis, such as the one we are discussing, is called a **σ orbital** because it resembles an s orbital when viewed along the axis and, more precisely, because it has zero orbital angular momentum around the internuclear axis.

Example 10.1 *Normalizing a molecular orbital*

Normalize the molecular orbital ψ_+ in eqn 10.8.

Method We need to find the factor N such that $\int\psi^*\psi\,d\tau = 1$. To proceed, substitute the LCAO into this integral, and make use of the fact that the atomic orbitals are individually normalized.

Answer Substitution of the wavefunction gives

$$\int\psi^*\psi\,d\tau = N^2\left\{\int A^2 d\tau + \int B^2 d\tau + 2\int AB\,d\tau\right\} = N^2(1+1+2S)$$

where $S = \int AB\,d\tau$ and has a value that depends on the nuclear separation (this 'overlap integral' will play a significant role later). For the integral to be equal to 1, we require

$$N = \frac{1}{\{2(1+S)\}^{1/2}}$$

In H_2^+, $S \approx 0.59$, so $N = 0.56$.

Self-test 10.3 Normalize the orbital ψ_- in eqn 10.8.

[$N = 1/\{2(1-S)\}^{1/2}$, so $N = 1.10$]

Figure 10.14 shows the contours of constant amplitude for the molecular orbital ψ_+ in eqn 10.8, and Fig. 10.15 shows its boundary surface. Plots like these are readily obtained using commercially available software. The calculation is quite straightforward, because all we need do is feed in the mathematical forms of the two atomic orbitals and then let the program do the rest. In this case, we use

$$A = \frac{e^{-r_A/a_0}}{(\pi a_0^3)^{1/2}} \qquad B = \frac{e^{-r_B/a_0}}{(\pi a_0^3)^{1/2}} \tag{10.9}$$

and note that r_A and r_B are not independent (2), but related by

$$r_B = \{r_A^2 + R^2 - 2r_A R\cos\theta\}^{1/2} \tag{10.10}$$

(b) Bonding orbitals

According to the Born interpretation, the probability density of the electron at each point in H_2^+ is proportional to the square modulus of its wavefunction at that point. The probability density corresponding to the (real) wavefunction ψ_+ in eqn 10.8 is

$$\psi_+^2 = N^2(A^2 + B^2 + 2AB) \tag{10.11}$$

This probability density is plotted in Fig. 10.16 and an important feature becomes apparent when we examine the internuclear region, where both atomic orbitals have similar amplitudes. According to eqn 10.11, the total probability density is proportional to the sum of:

- A^2, the probability density if the electron were confined to the atomic orbital A.
- B^2, the probability density if the electron were confined to the atomic orbital B.
- $2AB$, an extra contribution to the density from both atomic orbitals.

This last contribution, the **overlap density**, is crucial, because it represents an enhancement of the probability of finding the electron in the internuclear region. The

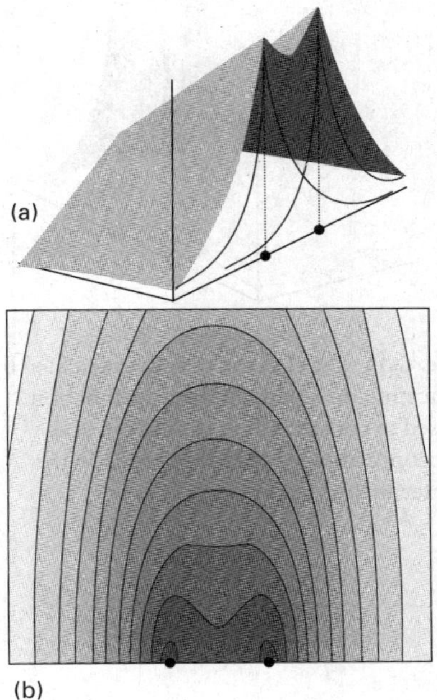

Fig. 10.14 (a) The amplitude of the bonding molecular orbital in a hydrogen molecule-ion in a plane containing the two nuclei and (b) a contour representation of the amplitude. To make this plot, we have taken $N^2 = 0.31$ (Example 10.1).

interActivity Plot the 1σ orbital for different values of the internuclear distance. Point to the features of the 1σ orbital that lead to bonding.

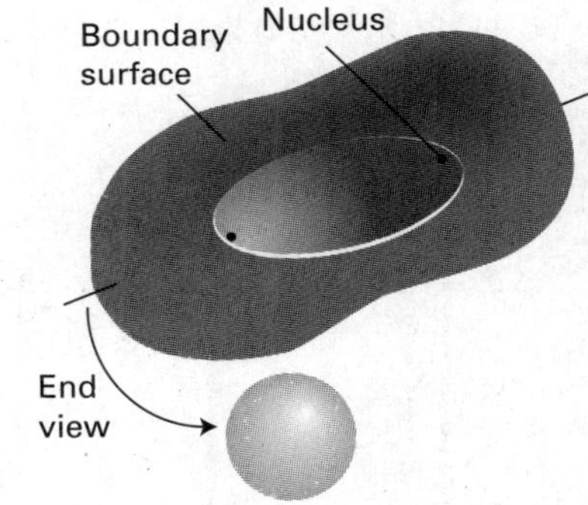

Fig. 10.15 A general indication of the shape of the boundary surface of a σ orbital.

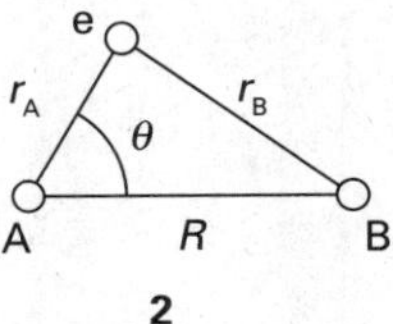

2

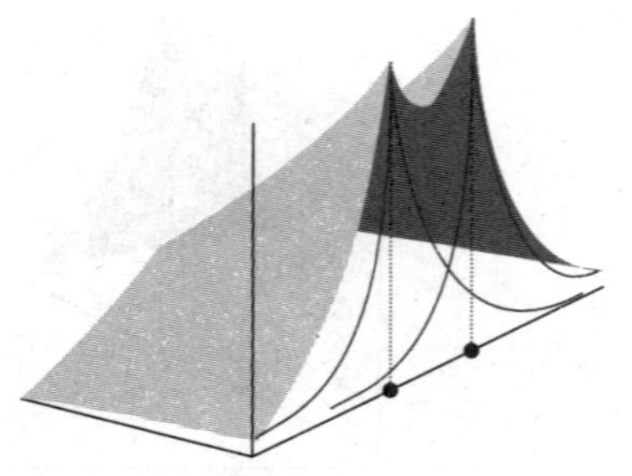

Fig. 10.16 The electron density calculated by forming the square of the wavefunction used to construct Fig. 10.14. Note the accumulation of electron density in the internuclear region.

enhancement can be traced to the constructive interference of the two atomic orbitals: each has a positive amplitude in the internuclear region, so the total amplitude is greater there than if the electron were confined to a single atomic orbital.

We shall frequently make use of the observation that *bonds form when electrons accumulate in regions where atomic orbitals overlap and interfere constructively*. The conventional explanation of this observation is based on the notion that accumulation of electron density between the nuclei puts the electron in a position where it interacts strongly with both nuclei. Hence, the energy of the molecule is lower than that of the separate atoms, where each electron can interact strongly with only one nucleus. This conventional explanation, however, has been called into question, because shifting an electron away from a nucleus into the internuclear region *raises* its potential energy. The modern (and still controversial) explanation does not emerge from the simple LCAO treatment given here. It seems that, at the same time as the electron shifts into the internuclear region, the atomic orbitals shrink. This orbital shrinkage improves the electron–nucleus attraction more than it is decreased by the migration to the internuclear region, so there is a net lowering of potential energy. The kinetic energy of the electron is also modified because the curvature of the wavefunction is changed, but the change in kinetic energy is dominated by the change in potential energy. Throughout the following discussion we ascribe the strength of chemical bonds to the accumulation of electron density in the internuclear region. We leave open the question whether in molecules more complicated than H_2^+ the true source of energy lowering is that accumulation itself or some indirect but related effect.

The σ orbital we have described is an example of a **bonding orbital**, an orbital which, if occupied, helps to bind two atoms together. Specifically, we label it 1σ as it is the σ orbital of lowest energy. An electron that occupies a σ orbital is called a **σ electron** and, if that is the only electron present in the molecule (as in the ground state of H_2^+), then we report the configuration of the molecule as $1\sigma^1$.

The energy $E_{1\sigma}$ of the 1σ orbital is (see Problem 10.18):

$$E_{1\sigma} = E_{\mathrm{H1s}} + \frac{j_0}{R} - \frac{j+k}{1+S} \qquad (10.12)$$

where E_{H1s} is the energy of a H1s orbital, j_0/R is the potential energy of repulsion between the two nuclei, and

$$S = \int AB\,\mathrm{d}\tau = \left\{1 + \frac{R}{a_0} + \frac{1}{3}\left(\frac{R}{a_0}\right)^2\right\}\mathrm{e}^{-R/a_0} \qquad (10.13a)$$

$$j = j_0\int \frac{A^2}{r_\mathrm{B}}\mathrm{d}\tau = \frac{j_0}{R}\left\{1 - \left(1 + \frac{R}{a_0}\right)\mathrm{e}^{-2R/a_0}\right\} \qquad (10.13b)$$

$$k = j_0\int \frac{AB}{r_\mathrm{B}}\mathrm{d}\tau = \frac{j_0}{a_0}\left(1 + \frac{R}{a_0}\right)\mathrm{e}^{-R/a_0} \qquad (10.13c)$$

We can interpret these three integrals as follows:

- All three integrals are positive and decline towards zero at large internuclear separations (S and k on account of the exponential term; j on account of the factor $1/R$). The integral S is discussed in more detail in Section 10.4c.
- The integral j is a measure of the interaction between a nucleus and the electron density centred on the other nucleus.
- The integral k is a measure of the interaction between a nucleus and the excess electron density in the internuclear region arising from overlap.

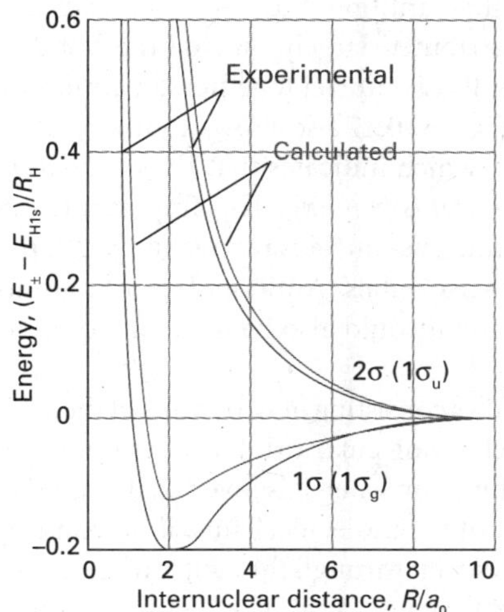

Fig. 10.17 The calculated and experimental molecular potential energy curves for a hydrogen molecule-ion showing the variation of the energy of the molecule as the bond length is changed. The alternative g,u notation is introduced in Section 10.3c.

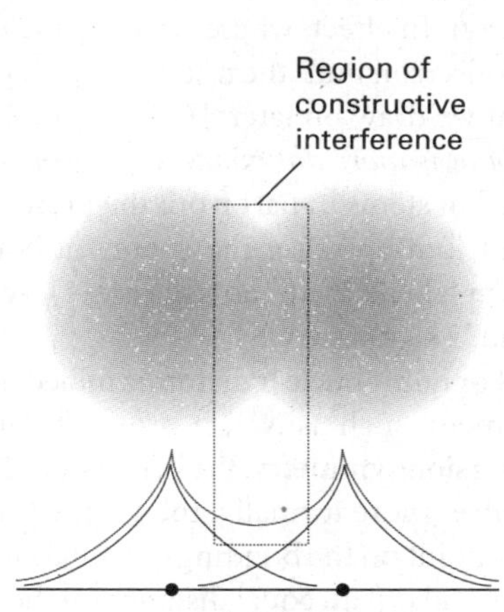

Fig. 10.18 A representation of the constructive interference that occurs when two H1s orbitals overlap and form a bonding σ orbital.

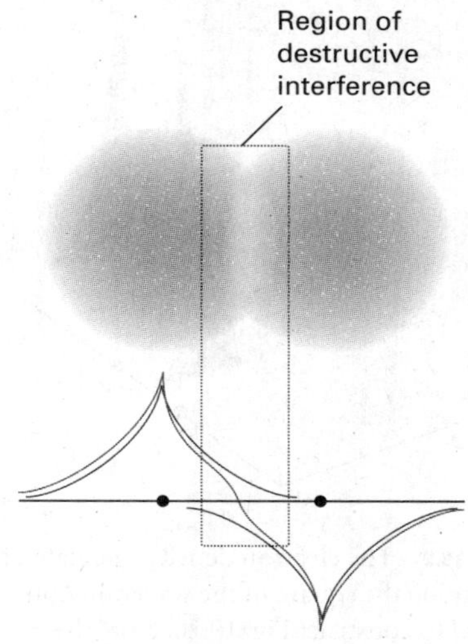

Fig. 10.19 A representation of the destructive interference that occurs when two H1s orbitals overlap and form an antibonding 2σ orbital.

Figure 10.17 is a plot of $E_{1\sigma}$ against R relative to the energy of the separated atoms. The energy of the 1σ orbital decreases as the internuclear separation decreases from large values because electron density accumulates in the internuclear region as the constructive interference between the atomic orbitals increases (Fig. 10.18). However, at small separations there is too little space between the nuclei for significant accumulation of electron density there. In addition, the nucleus–nucleus repulsion (which is proportional to $1/R$) becomes large. As a result, the energy of the molecule rises at short distances, and there is a minimum in the potential energy curve. Calculations on H_2^+ give $R_e = 130$ pm and $D_e = 1.77$ eV (171 kJ mol^{-1}); the experimental values are 106 pm and 2.6 eV, so this simple LCAO-MO description of the molecule, while inaccurate, is not absurdly wrong.

(c) Antibonding orbitals

The linear combination ψ_- in eqn 10.8 corresponds to a higher energy than that of ψ_+. Because it is also a σ orbital we label it 2σ. This orbital has an internuclear nodal plane where A and B cancel exactly (Figs. 10.19 and 10.20). The probability density is

$$\psi_-^2 = N^2(A^2 + B^2 - 2AB) \tag{10.14}$$

There is a reduction in probability density between the nuclei due to the $-2AB$ term (Fig. 10.21); in physical terms, there is destructive interference where the two atomic orbitals overlap. The 2σ orbital is an example of an **antibonding orbital**, an orbital that, if occupied, contributes to a reduction in the cohesion between two atoms and helps to raise the energy of the molecule relative to the separated atoms.

The energy $E_{2\sigma}$ of the 2σ antibonding orbital is given by (see Problem 10.18)

$$E_{2\sigma} = E_{\mathrm{H1s}} + \frac{j_0}{R} - \frac{j-k}{1-S} \tag{10.15}$$

where the integrals S, j, and k are the same as before (eqn 10.13). The variation of $E_{2\sigma}$ with R is shown in Fig. 10.17, where we see the destabilizing effect of an antibonding

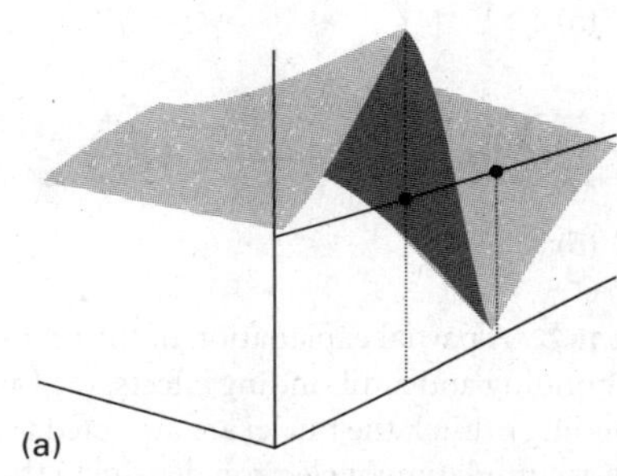

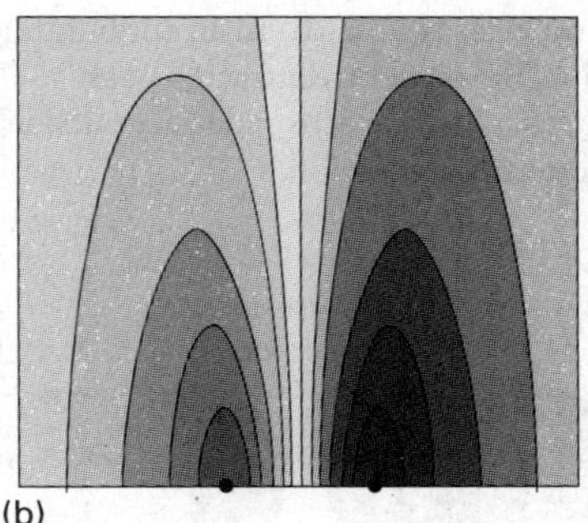

Fig. 10.20 (a) The amplitude of the antibonding molecular orbital in a hydrogen molecule-ion in a plane containing the two nuclei and (b) a contour representation of the amplitude. Note the internuclear node.

interActivity Plot the 2σ orbital for different values of the internuclear distance. Point to the features of the 2σ orbital that lead to antibonding.

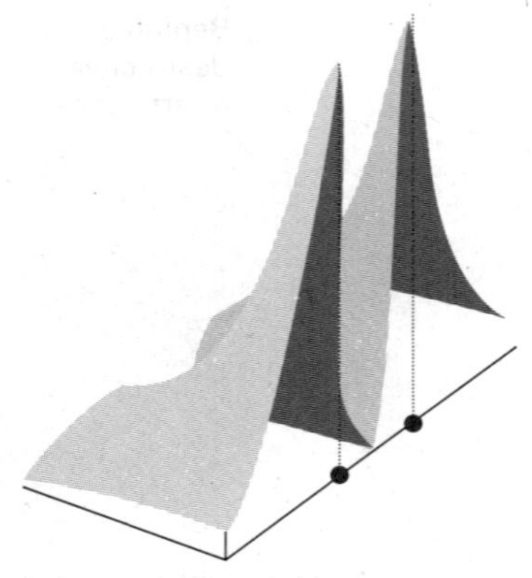

Fig. 10.21 The electron density calculated by forming the square of the wavefunction used to construct Fig. 10.20. Note the elimination of electron density from the internuclear region.

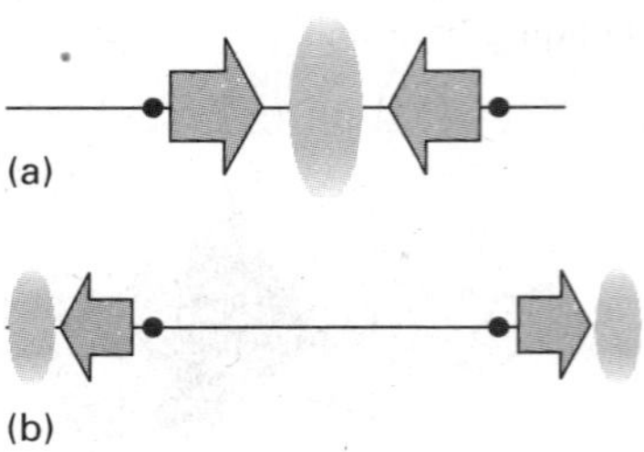

Fig. 10.22 A partial explanation of the origin of bonding and antibonding effects. (a) In a bonding orbital, the nuclei are attracted to the accumulation of electron density in the internuclear region. (b) In an antibonding orbital, the nuclei are attracted to an accumulation of electron density outside the internuclear region.

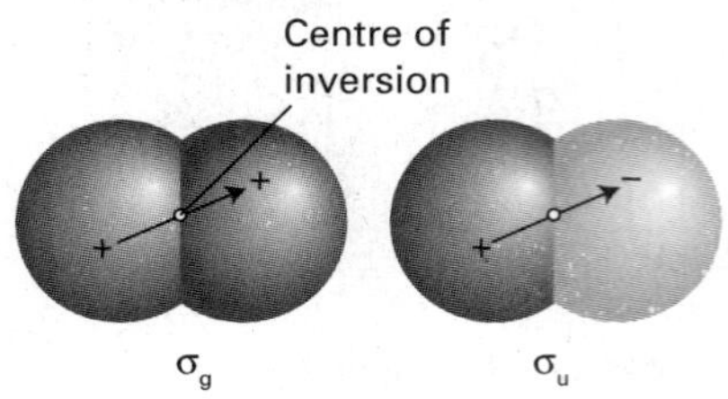

Fig. 10.23 The parity of an orbital is even (g) if its wavefunction is unchanged under inversion through the centre of symmetry of the molecule, but odd (u) if the wavefunction changes sign. Heteronuclear diatomic molecules do not have a centre of inversion, so for them the g, u classification is irrelevant.

electron. The effect is partly due to the fact that an antibonding electron is excluded from the internuclear region and hence is distributed largely outside the bonding region. In effect, whereas a bonding electron pulls two nuclei together, an antibonding electron pulls the nuclei apart (Fig. 10.22). Figure 10.17 also shows another feature that we draw on later: $|E_- - E_{H1s}| > |E_+ - E_{H1s}|$, which indicates that *the antibonding orbital is more antibonding than the bonding orbital is bonding*. This important conclusion stems in part from the presence of the nucleus–nucleus repulsion (j_0/R): this contribution raises the energy of both molecular orbitals. Antibonding orbitals are often labelled with an asterisk (*), so the 2σ orbital could also be denoted 2σ* (and read '2 sigma star').

For homonuclear diatomic molecules (molecules consisting of two atoms of the same element, such as N_2), it proves helpful to label a molecular orbital according to its **inversion symmetry**, the behaviour of the wavefunction when it is inverted through the centre (more formally, the centre of inversion) of the molecule. Thus, if we consider any point on the bonding σ orbital, and then project it through the centre of the molecule and out an equal distance on the other side, then we arrive at an identical value of the wavefunction (Fig. 10.23). This so-called **gerade symmetry** (from the German word for 'even') is denoted by a subscript g, as in σ_g. The same procedure applied to the antibonding 2σ orbital results in the same amplitude but opposite sign of the wavefunction. This **ungerade symmetry** ('odd symmetry') is denoted by a subscript u, as in σ_u. This inversion symmetry classification is not applicable to heteronuclear diatomic molecules (diatomic molecules formed by atoms from two different elements, such as CO) because these molecules do not have a centre of inversion. When using the g, u notation, each set of orbitals of the same inversion symmetry are labelled separately so, whereas 1σ becomes $1\sigma_g$, its antibonding partner, which so far we have called 2σ, is the first orbital of a different symmetry, and is denoted $1\sigma_u$. The general rule is that *each set of orbitals of the same symmetry designation is labelled separately.*

10.4 Homonuclear diatomic molecules

Key points **Electrons are added to available molecular orbitals in a manner that achieves the lowest overall energy. (a) As a first approximation, σ orbitals are constructed separately from valence s and p orbitals. (b) π Orbitals are constructed from the side-by-side overlap of p orbitals of the appropriate symmetry. (c) The overlap integral is a measure of the extent of orbital overlap. (d) The ground-state electron configurations of diatomic molecules are predicted by using the building up principle, and the bond order is a measure of the resulting net bonding character. (e) Photoelectron spectroscopy is a technique for determining the energies of electrons in molecular orbitals.**

In Chapter 9 we used the hydrogenic atomic orbitals and the building-up principle to deduce the ground electronic configurations of many-electron atoms. We now do the same for many-electron diatomic molecules by using the H_2^+ molecular orbitals as a basis for their discussion. The general procedure is to construct molecular orbitals by combining the available atomic orbitals:

Building-up principle for molecules

1. The electrons supplied by the atoms are accommodated in the orbitals so as to achieve the lowest overall energy subject to the constraint of the Pauli exclusion principle, that no more than two electrons may occupy a single orbital (and then must be paired).

2. If several degenerate molecular orbitals are available, electrons are added singly to each individual orbital before doubly occupying any one orbital (because that minimizes electron–electron repulsions).

3. According to Hund's maximum multiplicity rule (Section 9.4d), if two electrons do occupy different degenerate orbitals, then a lower energy is obtained if they do so with parallel spins.

(a) σ Orbitals

Consider H_2, the simplest many-electron diatomic molecule. Each H atom contributes a 1s orbital (as in H_2^+), so we can form the $1\sigma_g$ and $1\sigma_u$ orbitals from them, as we have seen already. At the experimental internuclear separation these orbitals will have the energies shown in Fig. 10.24, which is called a **molecular orbital energy level diagram**. Note that from two atomic orbitals we can build two molecular orbitals. In general, from *N* atomic orbitals we can build *N* molecular orbitals.

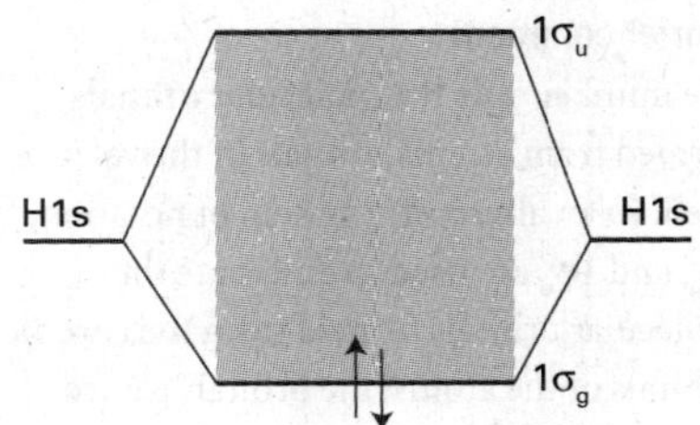

Fig. 10.24 A molecular orbital energy level diagram for orbitals constructed from the overlap of H1s orbitals; the separation of the levels corresponds to that found at the equilibrium bond length. The ground electronic configuration of H_2 is obtained by accommodating the two electrons in the lowest available orbital (the bonding orbital).

There are two electrons to accommodate, and both can enter $1\sigma_g$ by pairing their spins, as required by the Pauli principle (just as for atoms, Section 9.4b). The ground-state configuration is therefore $1\sigma_g^2$ and the atoms are joined by a bond consisting of an electron pair in a bonding σ orbital. This approach shows that an electron pair, which was the focus of Lewis's account of chemical bonding, represents the maximum number of electrons that can enter a bonding molecular orbital.

The same argument explains why He does not form diatomic molecules. Each He atom contributes a 1s orbital, so $1\sigma_g$ and $1\sigma_u$ molecular orbitals can be constructed. Although these orbitals differ in detail from those in H_2, their general shapes are the same and we can use the same qualitative energy level diagram in the discussion. There are four electrons to accommodate. Two can enter the $1\sigma_g$ orbital, but then it is full, and the next two must enter the $1\sigma_u$ orbital (Fig. 10.25). The ground electronic configuration of He_2 is therefore $1\sigma_g^2 1\sigma_u^2$. We see that there is one bond and one antibond. Because $1\sigma_u$ is raised in energy relative to the separate atoms more than $1\sigma_g$ is lowered, an He_2 molecule has a higher energy than the separated atoms, so it is unstable relative to them.

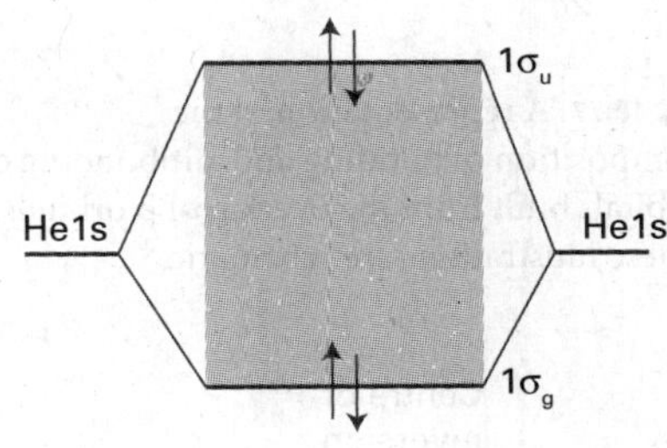

Fig. 10.25 The ground electronic configuration of the hypothetical four-electron molecule He_2 has two bonding electrons and two antibonding electrons. It has a higher energy than the separated atoms, and so is unstable.

We shall now see how the concepts we have introduced apply to homonuclear diatomic molecules in general. In elementary treatments, only the orbitals of the valence shell are used to form molecular orbitals so, for molecules formed with atoms from Period 2 elements, only the 2s and 2p atomic orbitals are considered. We shall make that approximation here too.

A general principle of molecular orbital theory is that *all orbitals of the appropriate symmetry* contribute to a molecular orbital. Thus, to build σ orbitals, we form linear combinations of all atomic orbitals that have cylindrical symmetry about the internuclear axis. These orbitals include the 2s orbitals on each atom and the $2p_z$ orbitals on the two atoms (Fig. 10.26). The general form of the σ orbitals that may be formed is therefore

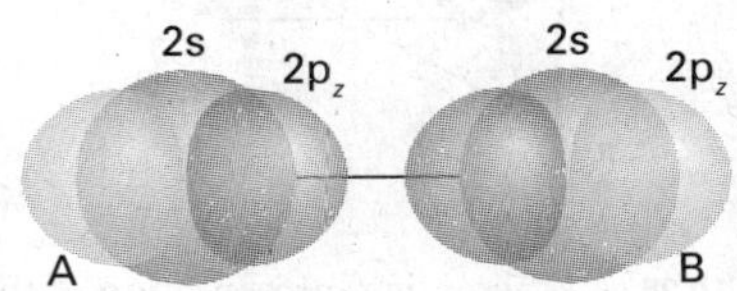

Fig. 10.26 According to molecular orbital theory, σ orbitals are built from all orbitals that have the appropriate symmetry. In homonuclear diatomic molecules of Period 2, that means that two 2s and two $2p_z$ orbitals should be used. From these four orbitals, four molecular orbitals can be built.

$$\psi = c_{A2s}\chi_{A2s} + c_{B2s}\chi_{B2s} + c_{A2p_z}\chi_{A2p_z} + c_{B2p_z}\chi_{B2p_z} \tag{10.16}$$

From these four atomic orbitals we can form four molecular orbitals of σ symmetry by an appropriate choice of the coefficients *c*.

The procedure for calculating the coefficients will be described in Section 10.6. At this stage we adopt a simpler route, and suppose that, because the 2s and $2p_z$ orbitals have distinctly different energies, they may be treated separately. That is, the four σ orbitals fall approximately into two sets, one consisting of two molecular orbitals of the form

$$\psi = c_{A2s}\chi_{A2s} + c_{B2s}\chi_{B2s} \tag{10.17a}$$

and another consisting of two orbitals of the form

$$\psi = c_{A2p_z}\chi_{A2p_z} + c_{B2p_z}\chi_{B2p_z} \tag{10.17b}$$

A brief comment

We number only the molecular orbitals formed from atomic orbitals in the valence shell. In an alternative system of notation, $1\sigma_g$ and $1\sigma_u$ are used to designate the molecular orbitals formed from the core 1s orbitals of the atoms; the orbitals we are considering would then be labelled starting from 2.

Because atoms A and B are identical, the energies of their 2s orbitals are the same, so the coefficients are equal (apart from a possible difference in sign); the same is true of the $2p_z$ orbitals. Therefore, the two sets of orbitals have the form $\chi_{A2s} \pm \chi_{B2s}$ and $\chi_{A2p_z} \pm \chi_{B2p_z}$.

The 2s orbitals on the two atoms overlap to give a bonding and an antibonding σ orbital ($1\sigma_g$ and $1\sigma_u$, respectively) in exactly the same way as we have already seen for 1s orbitals. The two $2p_z$ orbitals directed along the internuclear axis overlap strongly. They may interfere either constructively or destructively, and give a bonding or antibonding σ orbital (Fig. 10.27). These two σ orbitals are labelled $2\sigma_g$ and $2\sigma_u$, respectively. In general, note how the numbering follows the order of increasing energy.

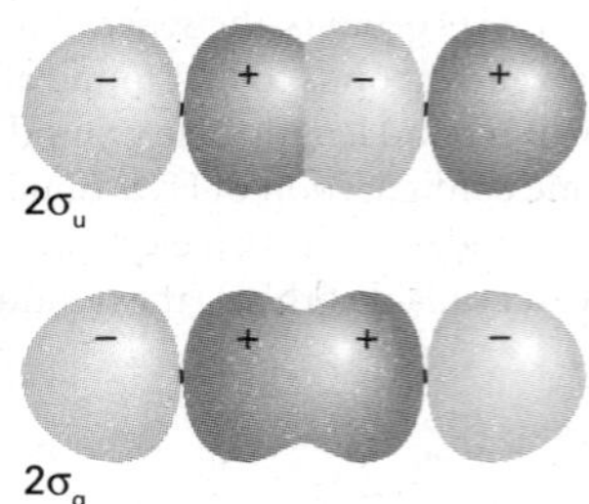

Fig. 10.27 A representation of the composition of bonding and antibonding σ orbitals built from the overlap of p orbitals. These illustrations are schematic.

(b) π Orbitals

Now consider the $2p_x$ and $2p_y$ orbitals of each atom. These orbitals are perpendicular to the internuclear axis and may overlap broadside-on. This overlap may be constructive or destructive and results in a bonding or an antibonding **π orbital** (Fig. 10.28). The notation π is the analogue of p in atoms, for when viewed along the axis of the molecule, a π orbital looks like a p orbital and has one unit of orbital angular momentum around the internuclear axis. The two neighbouring $2p_x$ orbitals overlap to give a bonding and antibonding π_x orbital, and the two $2p_y$ orbitals overlap to give two π_y orbitals. The π_x and π_y bonding orbitals are degenerate; so too are their antibonding partners. We also see from Fig. 10.28 that a bonding π orbital has odd parity and is denoted π_u and an antibonding π orbital has even parity, denoted π_g.

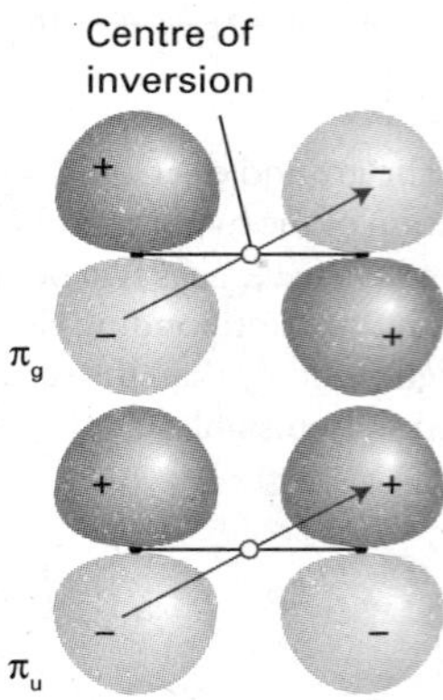

Fig. 10.28 A schematic representation of the structure of π bonding and antibonding molecular orbitals. The figure also shows that the bonding π orbital has odd parity, whereas the antibonding π orbital has even parity.

(c) The overlap integral

The extent to which two atomic orbitals on different atoms overlap is measured by the **overlap integral**, S:

$$S = \int \chi_A^* \chi_B \, d\tau \qquad \text{Definition of overlap integral} \qquad [10.18]$$

We have already met this integral (in Example 10.1 and eqn 10.13). If the atomic orbital χ_A on A is small wherever the orbital χ_B on B is large, or vice versa, then the product of their amplitudes is everywhere small and the integral—the sum of these products—is small (Fig. 10.29). If χ_A and χ_B are both large in some region of space, then S may be large. If the two normalized atomic orbitals are identical (for instance, 1s orbitals on the same nucleus), then $S = 1$. In some cases, simple formulas can be given for overlap integrals. For instance, the variation of S with internuclear separation for hydrogenic 1s orbitals on atoms of atomic number Z is given by

$$S(1s, 1s) = \left\{1 + \frac{ZR}{a_0} + \frac{1}{3}\left(\frac{ZR}{a_0}\right)^2\right\} e^{-ZR/a_0} \qquad (10.19)$$

and is plotted in Fig. 10.30 (eqn 10.19 is a generalization of eqn 10.13a, which was for H1s orbitals). It follows that $S = 0.59$ (an unusually large value) for two H1s orbitals at the equilibrium bond length in H_2^+. Typical values of S for orbitals with $n = 2$ are in the range 0.2 to 0.3.

Now consider the arrangement in which an s orbital is superimposed on a p_x orbital of a different atom (Fig. 10.31). The integral over the region where the product of orbitals is positive exactly cancels the integral over the region where the product of orbitals is negative, so overall $S = 0$ exactly. Therefore, there is no net overlap between the s and p orbitals in this arrangement.

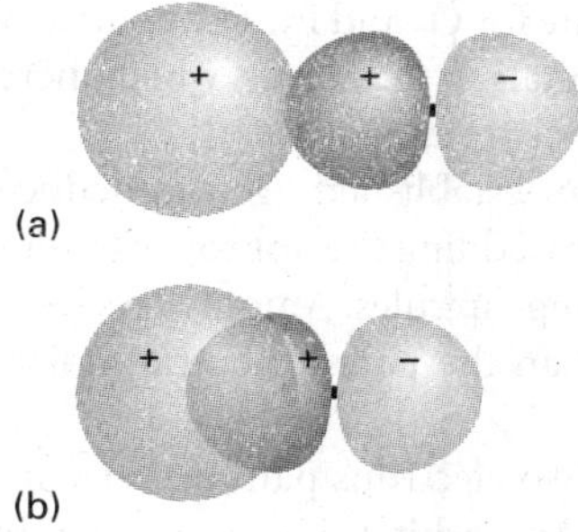

Fig. 10.29 (a) When two orbitals are on atoms that are far apart, the wavefunctions are small where they overlap, so S is small. (b) When the atoms are closer, both orbitals have significant amplitudes where they overlap, and S may approach 1. Note that S will decrease again as the two atoms approach more closely than shown here, because the region of negative amplitude of the p orbital starts to overlap the positive overlap of the s orbital. When the centres of the atoms coincide, $S = 0$.

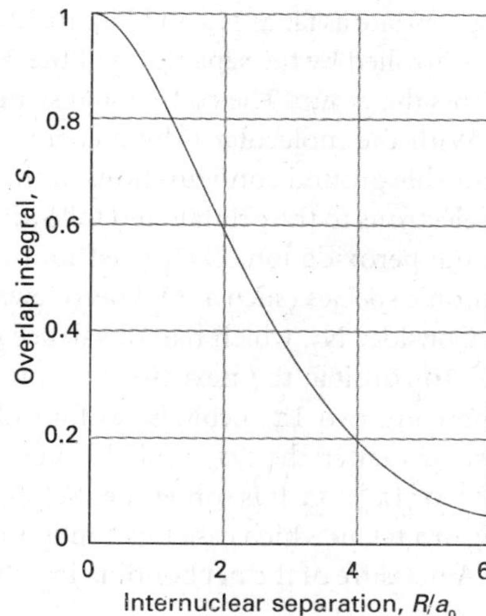

Fig. 10.30 The overlap integral, S, between two H1s orbitals as a function of their separation, R.

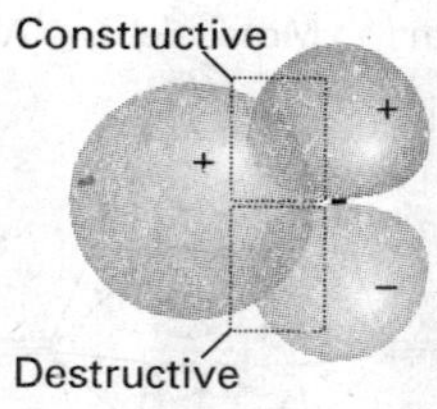

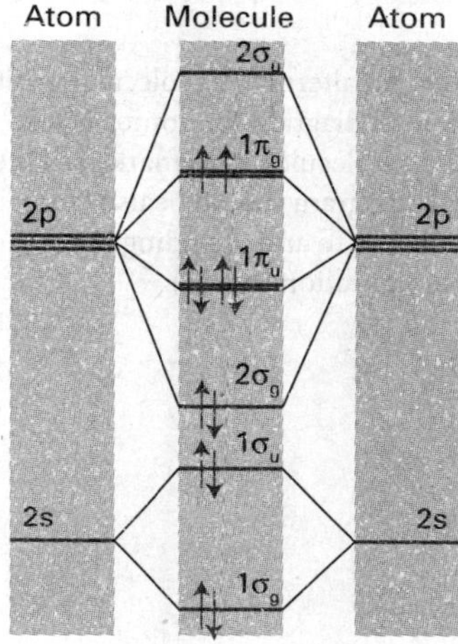

Fig. 10.31 A p orbital in the orientation shown here has zero net overlap ($S = 0$) with the s orbital at all internuclear separations.

(d) The electronic structures of homonuclear diatomic molecules

To construct the molecular orbital energy level diagram for Period 2 homonuclear diatomic molecules, we form eight molecular orbitals from the eight valence shell orbitals (four from each atom). In some cases, π orbitals are less strongly bonding than σ orbitals because their maximum overlap occurs off-axis. This relative weakness suggests that the molecular orbital energy level diagram ought to be as shown in Fig. 10.32. However, we must remember that we have assumed that 2s and $2p_z$ orbitals contribute to different sets of molecular orbitals, whereas in fact all four atomic orbitals have the same symmetry around the internuclear axis and contribute jointly to the four σ orbitals. Hence, there is no guarantee that this order of energies should prevail, and it is found experimentally (by spectroscopy) and by detailed calculation that the order varies along Period 2 (Fig. 10.33). The order shown in Fig. 10.34 is

Fig. 10.32 The molecular orbital energy level diagram for homonuclear diatomic molecules. The lines in the middle are an indication of the energies of the molecular orbitals that can be formed by overlap of atomic orbitals. As remarked in the text, this diagram should be used for O_2 (the configuration shown) and F_2.

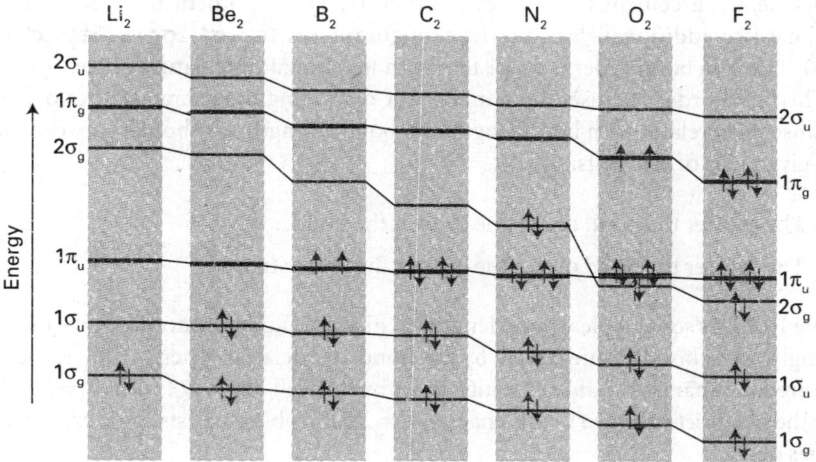

Fig. 10.33 The variation of the orbital energies as calculated for Period 2 homonuclear diatomics.

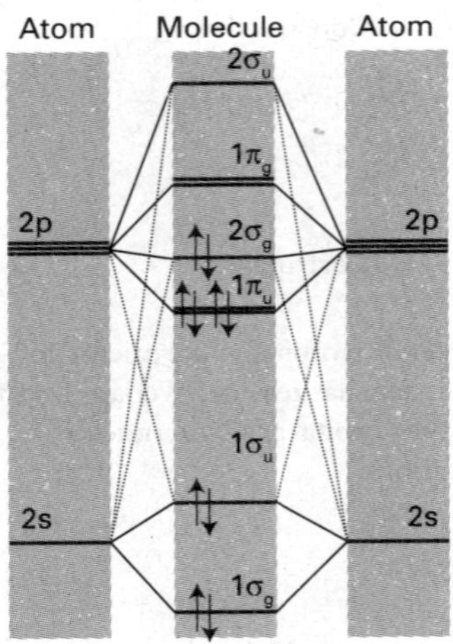

Fig. 10.34 An alternative molecular orbital energy level diagram for homonuclear diatomic molecules. As remarked in the text, this diagram should be used for diatomics up to and including N_2 (the configuration shown).

appropriate as far as N_2, and Fig. 10.32 is appropriate for O_2 and F_2. The relative order is controlled by the separation of the 2s and 2p orbitals in the atoms, which increases across the group. The consequent switch in order occurs at about N_2.

With the molecular orbital energy level diagram established, we can deduce the probable ground configurations of the molecules by adding the appropriate number of electrons to the orbitals and following the building-up rules. Anionic species (such as the peroxide ion, O_2^{2-}) need more electrons than the parent neutral molecules; cationic species (such as O_2^+) need fewer.

Consider N_2, which has 10 valence electrons. Two electrons pair, occupy, and fill the $1\sigma_g$ orbital; the next two occupy and fill the $1\sigma_u$ orbital. Six electrons remain. There are two $1\pi_u$ orbitals, so four electrons can be accommodated in them. The last two enter the $2\sigma_g$ orbital. Therefore, the ground-state configuration of N_2 is $1\sigma_g^2 1\sigma_u^2 1\pi_u^4 2\sigma_g^2$. It is sometimes helpful to include an asterisk to denote an antibonding orbital, in which case this configuration would be denoted $1\sigma_g^2 1\sigma_u^{*2} 1\pi_u^4 2\sigma_g^2$.

A measure of the net bonding in a diatomic molecule is its **bond order**, b:

$$b = \tfrac{1}{2}(N - N^*) \qquad \text{Definition of bond order} \qquad [10.20]$$

where N is the number of electrons in bonding orbitals and N^* is the number of electrons in antibonding orbitals. Thus, each electron pair in a bonding orbital increases the bond order by 1 and each pair in an antibonding orbital decreases b by 1. For H_2, $b = 1$, corresponding to a single bond, H–H, between the two atoms. In He_2, $b = 0$, and there is no bond. In N_2, $b = \tfrac{1}{2}(8 - 2) = 3$. This bond order accords with the Lewis structure of the molecule (:N≡N:).

The ground-state electron configuration of O_2, with 12 valence electrons, is based on Fig. 10.32, and is $1\sigma_g^2 1\sigma_u^2 2\sigma_g^2 1\pi_u^4 1\pi_g^2$ (or $1\sigma_g^2 1\sigma_u^{*2} 2\sigma_g^2 1\pi_u^4 1\pi_g^{*2}$). Its bond order is 2. According to the building-up principle, however, the two $1\pi_g$ electrons occupy different orbitals: one will enter $1\pi_{g,x}$ and the other will enter $1\pi_{g,y}$. Because the electrons are in different orbitals, they will have parallel spins. Therefore, we can predict that an O_2 molecule will have a net spin angular momentum $S = 1$ and, in the language introduced in Section 9.8, be in a triplet state. As electron spin is the source of a magnetic moment, we can go on to predict that oxygen should be paramagnetic, a substance that tends to move into a magnetic field (see Chapter 19). This prediction, which VB theory does not make, is confirmed by experiment.

An F_2 molecule has two more electrons than an O_2 molecule. Its configuration is therefore $1\sigma_g^2 1\sigma_u^{*2} 2\sigma_g^2 1\pi_u^4 1\pi_g^{*4}$ and $b = 1$. We conclude that F_2 is a singly bonded molecule, in agreement with its Lewis structure. The hypothetical molecule dineon, Ne_2, has two additional electrons: its configuration is $1\sigma_g^2 1\sigma_u^{*2} 2\sigma_g^2 1\pi_u^4 1\pi_g^{*4} 2\sigma_u^{*2}$ and $b = 0$. The zero bond order is consistent with the monatomic nature of Ne.

The bond order is a useful parameter for discussing the characteristics of bonds, because it correlates with bond length and bond strength. For bonds between atoms of a given pair of elements:

- The greater the bond order, the shorter the bond.
- The greater the bond order, the greater the bond strength.

Table 10.2 lists some typical bond lengths in diatomic and polyatomic molecules. The strength of a bond is measured by its bond dissociation energy, D_0, the energy required to separate the atoms to infinity or by the well depth D_e, with $D_0 = D_e - \tfrac{1}{2}\hbar\omega$ (see the first *brief comment* in this chapter; $\omega = 2\pi\nu$). Table 10.3 lists some experimental values of D_0.

A brief comment

Bond dissociation energies are commonly used in thermodynamic cycles, where bond enthalpies, $\Delta_{bond}H^{\ominus}$, should be used instead. It follows from the same kind of argument used in *Justification 9.7* concerning ionization enthalpies that

$$X_2(g) \rightarrow 2X(g) \qquad \Delta_{bond}H^{\ominus}(T) = D_0 + \tfrac{3}{2}RT$$

To derive this relation, we have supposed that the molar constant-pressure heat capacity of X_2 is $\tfrac{7}{2}R$ (Section 2.4 and eqn 2.26), for there is a contribution from two rotational modes as well as three translational modes.

Example 10.2 *Judging the relative bond strengths of molecules and ions*

Predict whether N_2^+ is likely to have a larger or smaller dissociation energy than N_2.

Method Because the molecule with the higher bond order is likely to have the higher dissociation energy, compare their electronic configurations and assess their bond orders.

Answer From Fig. 10.34, the electron configurations and bond orders are

$$N_2 \quad 1\sigma_g^2 1\sigma_u^{*2} 1\pi_u^4 2\sigma_g^2 \quad b=3$$

$$N_2^+ \quad 1\sigma_g^2 1\sigma_u^{*2} 1\pi_u^4 2\sigma_g^1 \quad b=2\tfrac{1}{2}$$

Because the cation has the smaller bond order, we expect it to have the smaller dissociation energy. The experimental dissociation energies are 945 kJ mol^{-1} for N_2 and 842 kJ mol^{-1} for N_2^+.

Self-test 10.4 Which can be expected to have the higher dissociation energy, F_2 or F_2^+? [F_2^+]

Table 10.2* Bond lengths

Bond	Order	R_e/pm
HH	1	74.14
NN	3	109.76
HCl	1	127.45
CH	1	*114*
CC	1	*154*
CC	2	*134*
CC	3	*120*

* More values will be found in the *Data section*. Numbers in italics are mean values for polyatomic molecules.

Table 10.3* Bond dissociation energies

Bond	Order	D_0/(kJ mol^{-1})
HH	1	432.1
NN	3	941.7
HCl	1	427.7
CH	1	*435*
CC	1	*368*
CC	2	*720*
CC	3	*962*

* More values will be found in the *Data section*. Numbers in italics are mean values for polyatomic molecules.

(e) Photoelectron spectroscopy

So far we have treated molecular orbitals as purely theoretical constructs, but is there experimental evidence for their existence? **Photoelectron spectroscopy** (PES) measures the ionization energies of molecules when electrons are ejected from different orbitals by absorption of a photon of known energy, and uses the information to infer the energies of molecular orbitals. The technique is also used to study solids, and in Chapter 22 we shall see the important information that it gives about species at or on surfaces.

Because energy is conserved when a photon ionizes a sample, the sum of the ionization energy, I, of the sample and the kinetic energy of the **photoelectron**, the ejected electron, must be equal to the energy of the incident photon $h\nu$ (Fig. 10.35):

$$h\nu = \tfrac{1}{2}m_e v^2 + I \qquad (10.21a)$$

This equation (which is like the one used for the photoelectric effect, eqn 7.15) can be refined in two ways. First, photoelectrons may originate from one of a number of different orbitals, and each one has a different ionization energy. Hence, a series of different kinetic energies of the photoelectrons will be obtained, each one satisfying

$$h\nu = \tfrac{1}{2}m_e v^2 + I_i \qquad (10.21b)$$

where I_i is the ionization energy for ejection of an electron from an orbital i. Therefore, by measuring the kinetic energies of the photoelectrons, and knowing ν, these ionization energies can be determined. Photoelectron spectra are interpreted in terms of an approximation called **Koopmans' theorem**, which states that the ionization energy I_i is equal to the orbital energy of the ejected electron (formally: $I_i = -\varepsilon_i$). That is, we can identify the ionization energy with the energy of the orbital from which it is ejected. The theorem is only an approximation because it ignores the fact that the remaining electrons adjust their distributions when ionization occurs.

The ionization energies of molecules are several electronvolts even for valence electrons, so it is essential to work in at least the ultraviolet region of the spectrum and with wavelengths of less than about 200 nm. Much work has been done with radiation generated by a discharge through helium: the He(I) line ($1s^12p^1 \rightarrow 1s^2$) lies at 58.43 nm, corresponding to a photon energy of 21.22 eV. Its use gives rise to the technique of **ultraviolet photoelectron spectroscopy** (UPS). When core electrons are being studied,

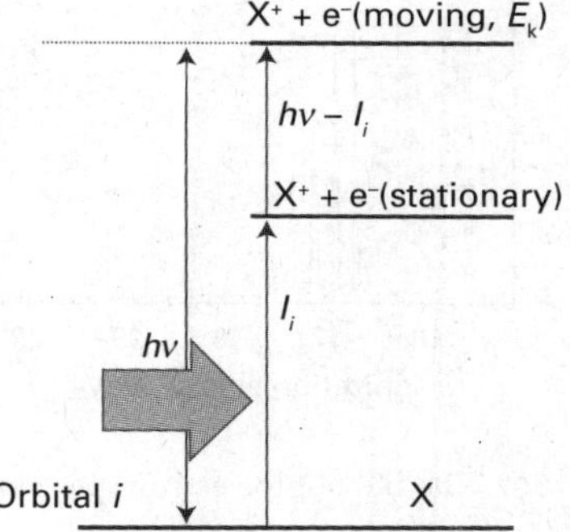

Fig. 10.35 An incoming photon carries an energy $h\nu$; an energy I_i is needed to remove an electron from an orbital i, and the difference appears as the kinetic energy of the electron.

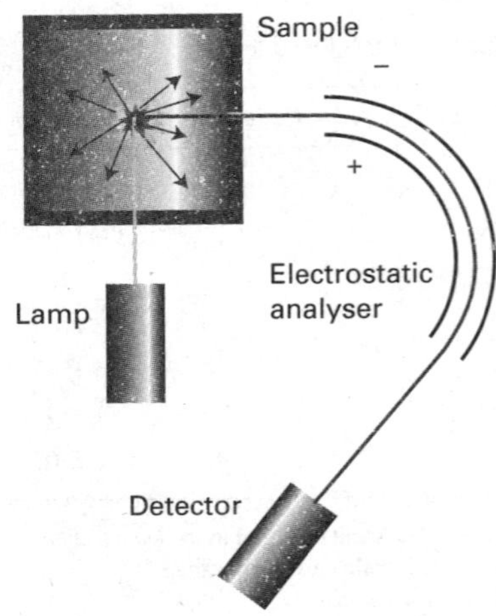

Fig. 10.36 A photoelectron spectrometer consists of a source of ionizing radiation (such as a helium discharge lamp for UPS and an X-ray source for XPS), an electrostatic analyser, and an electron detector. The deflection of the electron path caused by the analyser depends on their speed.

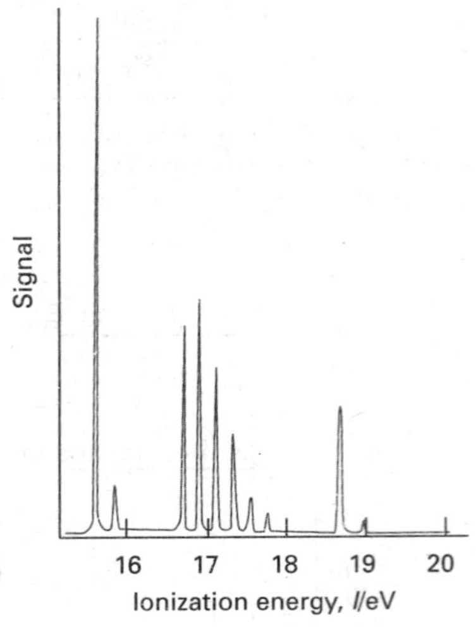

Fig. 10.37 The UV photoelectron spectrum of N_2.

photons of even higher energy are needed to expel them: X-rays are used, and the technique is denoted XPS.

The kinetic energies of the photoelectrons are measured using an electrostatic deflector that produces different deflections in the paths of the photoelectrons as they pass between charged plates (Fig. 10.36). As the field strength is increased, electrons of different speeds, and therefore kinetic energies, reach the detector. The electron flux can be recorded and plotted against kinetic energy to obtain the photoelectron spectrum.

• A brief illustration

Photoelectrons ejected from N_2 with He(I) radiation have kinetic energies of 5.63 eV (1 eV = 8065.5 cm^{-1}, Fig. 10.37). Helium(I) radiation of wavelength 58.43 nm has wavenumber 1.711×10^5 cm^{-1} and therefore corresponds to an energy of 21.22 eV. Then, from eqn 10.21, 21.22 eV = 5.63 eV + I_i, so I_i = 15.59 eV. This ionization energy is the energy needed to remove an electron from the occupied molecular orbital with the highest energy of the N_2 molecule, the $2\sigma_g$ bonding orbital. •

Self-test 10.5 Under the same circumstances, photoelectrons are also detected at 4.53 eV. To what ionization energy does that correspond? Suggest an origin.

[16.7 eV, $1\pi_u$]

It is often observed that photoejection results in cations that are excited vibrationally. Because different energies are needed to excite different vibrational states of the ion, the photoelectrons appear with different kinetic energies. The result is **vibrational fine structure**, a progression of lines with a frequency spacing that corresponds to the vibrational frequency of the molecule. Figure 10.38 shows an example of vibrational fine structure in the photoelectron spectrum of Br_2.

10.5 Heteronuclear diatomic molecules

Key points (a) A polar bond can be regarded as arising from a molecular orbital that is concentrated more on one atom than its partner. (b) The electronegativity of an element is a measure of the power of an atom to attract electrons to itself when it is part of a compound. (c) The variation principle provides a criterion of acceptability of an approximate wavefunction.

The electron distribution in a covalent bond in a heteronuclear diatomic molecule is not shared equally by the atoms because it is energetically favourable for the electron pair to be found closer to one atom than the other. This imbalance results in a **polar bond**, a covalent bond in which the electron pair is shared unequally by the two atoms. The bond in HF, for instance, is polar, with the electron pair closer to the F atom. The accumulation of the electron pair near the F atom results in that atom having a net negative charge, which is called a **partial negative charge** and denoted $\delta-$. There is a matching **partial positive charge**, $\delta+$, on the H atom.

(a) Polar bonds

A polar bond consists of two electrons in a bonding molecular orbital of the form

$$\psi = c_A A + c_B B \qquad \text{Form of wavefunction of a polar bond} \qquad (10.22)$$

with unequal coefficients. The proportion of the atomic orbital A in the bond is $|c_A|^2$ and that of B is $|c_B|^2$. A nonpolar bond has $|c_A|^2 = |c_B|^2$ and a pure ionic bond has one coefficient zero (so the species A^+B^- would have $c_A = 0$ and $c_B = 1$). The atomic orbital with the lower energy makes the larger contribution to the bonding molecular orbital. The opposite is true of the antibonding orbital, for which the dominant component comes from the atomic orbital with higher energy.

These points can be illustrated by considering HF, and judging the energies of the atomic orbitals from the ionization energies of the atoms. The general form of the molecular orbitals is

$$\psi = c_H\chi_H + c_F\chi_F \tag{10.23}$$

where χ_H is an H1s orbital and χ_F is an F2p$_z$ orbital (with z along the internuclear axis, the convention for linear molecules). The H1s orbital lies 13.6 eV below the zero of energy (the separated proton and electron) and the F2p$_z$ orbital lies at 17.4 eV (Fig. 10.39). Hence, the bonding σ orbital in HF is mainly F2p$_z$ and the antibonding σ orbital is mainly H1s orbital in character. The two electrons in the bonding orbital are most likely to be found in the F2p$_z$ orbital, so there is a partial negative charge on the F atom and a partial positive charge on the H atom.

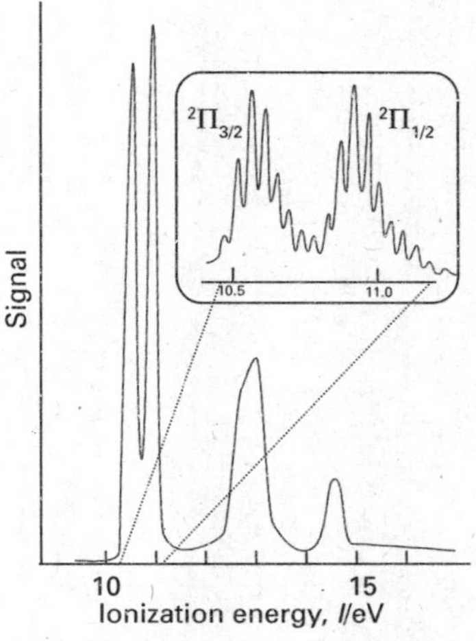

Fig. 10.38 The UV photoelectron spectrum of Br_2.

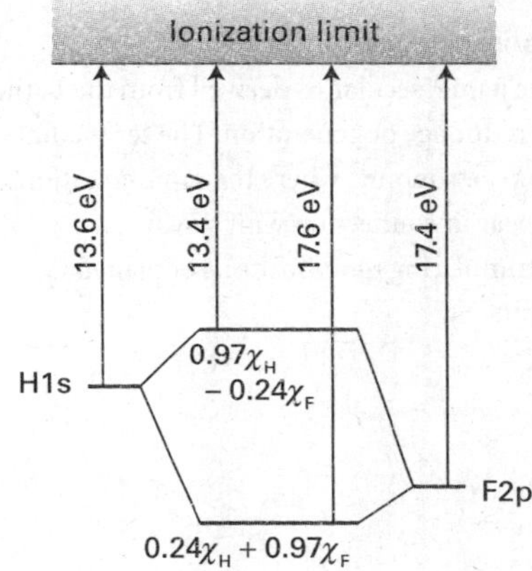

Fig. 10.39 The molecular orbital energy level diagram for HF showing the energy levels calculated for the basis H1s and F2p$_z$. The energies are shown relative to the ionization limit.

(b) Electronegativity

The charge distribution in bonds is commonly discussed in terms of the **electronegativity**, χ (chi), of the elements involved (there should be little danger of confusing this use of χ with its use to denote an atomic orbital, which is another common convention). The electronegativity is a parameter introduced by Linus Pauling as a measure of the power of an atom to attract electrons to itself when it is part of a compound. Pauling used valence-bond arguments to suggest that an appropriate numerical scale of electronegativities could be defined in terms of bond dissociation energies, D_0, and proposed that the difference in electronegativities could be expressed as

$$|\chi_A - \chi_B| = \{D_0(AB) - \tfrac{1}{2}[D_0(AA) + D_0(BB)]\}^{1/2} \qquad \text{Definition of Pauling electronegativity} \qquad [10.24]$$

where $D_0(AA)$ and $D_0(BB)$ are the dissociation energies of A–A and B–B bonds and $D_0(AB)$ is the dissociation energy of an A–B bond, all in electronvolts. (In later work Pauling used the geometrical mean of dissociation energies in place of the arithmetic mean.) This expression gives differences of electronegativities; to establish an absolute scale Pauling chose individual values that gave the best match to the values obtained from eqn 10.24. Electronegativities based on this definition are called **Pauling electronegativities** (Table 10.4). The most electronegative elements are those close to F (excluding the noble gases); the least are those close to Cs. It is found that the greater the difference in electronegativities, the greater the polar character of the bond. The difference for HF, for instance, is 1.78; a C–H bond, which is commonly regarded as almost nonpolar, has an electronegativity difference of 0.35.

The spectroscopist Robert Mulliken proposed an alternative definition of electronegativity. He argued that an element is likely to be highly electronegative if it has a high ionization energy (so it will not release electrons readily) and a high electron affinity (so it is energetically favorable to acquire electrons). The **Mulliken electronegativity scale** is therefore based on the definition

$$\chi_M = \tfrac{1}{2}(I + E_{ea}) \qquad \text{Definition of Mulliken electronegativity} \qquad [10.25]$$

where I is the ionization energy of the element and E_{ea} is its electron affinity (both in electronvolts). The Mulliken and Pauling scales are approximately in line with each other. A reasonably reliable conversion between the two is

$$\chi_P = 1.35\chi_M^{1/2} - 1.37 \tag{10.26}$$

Table 10.4* Pauling electronegativities

Element	χ_P
H	2.2
C	2.6
N	3.0
O	3.4
F	4.0
Cl	3.2
Cs	0.79

* More values will be found in the *Data section*.

(c) The variation principle

A more systematic way of discussing bond polarity and finding the coefficients in the linear combinations used to build molecular orbitals is provided by the **variation principle**:

> If an arbitrary wavefunction is used to calculate the energy, the value calculated is never less than the true energy.

Variation principle

This principle is the basis of all modern molecular structure calculations (Section 10.7). The arbitrary wavefunction is called the **trial wavefunction**. The principle implies that, if we vary the coefficients in the trial wavefunction until the lowest energy is achieved (by evaluating the expectation value of the hamiltonian for each wavefunction), then those coefficients will be the best. We might get a lower energy if we use a more complicated wavefunction (for example, by taking a linear combination of several atomic orbitals on each atom), but we shall have the optimum (minimum energy) molecular orbital that can be built from the chosen **basis set**, the given set of atomic orbitals.

The method can be illustrated by the trial wavefunction in eqn 10.23. We show in the following *Justification* that the coefficients are given by the solutions of the two **secular equations**

$$(\alpha_A - E)c_A + (\beta - ES)c_B = 0 \tag{10.27a}$$

$$(\beta - ES)c_A + (\alpha_B - E)c_B = 0 \tag{10.27b}$$

A brief comment

The name 'secular' is derived from the Latin word for age or generation. The term comes from astronomy, where the same equations appear in connection with slowly accumulating modifications of planetary orbits.

The parameter α is called a **Coulomb integral**. It is negative and can be interpreted as the energy of the electron when it occupies A (for α_A) or B (for α_B). In a homonuclear diatomic molecule, $\alpha_A = \alpha_B$. The parameter β is called a **resonance integral** (for classical reasons). It vanishes when the orbitals do not overlap, and at equilibrium bond lengths it is normally negative.

Justification 10.3 *The variation principle applied to a heteronuclear diatomic molecule*

The trial wavefunction in eqn 10.23 is real but not normalized because at this stage the coefficients can take arbitrary values. Therefore, we can write $\psi^* = \psi$ but do not assume that $\int \psi^2 \, d\tau = 1$. When a wavefunction is not normalized, we replace the expression

$$\langle \hat{\Omega} \rangle = \int \psi^* \hat{\Omega} \psi \, d\tau$$

by

$$\langle \hat{\Omega} \rangle = \int (N\psi^* \hat{\Omega} N\psi) d\tau = \frac{\int \psi^* \hat{\Omega} \psi \, d\tau}{\int \psi^* \psi \, d\tau}$$

(For the second equality, we have used eqn 7.19 for each N.) In this case, the energy of the trial wavefunction is the expectation value of the energy operator (the hamiltonian, $\hat{H}$) and we write:

$$E = \frac{\int \psi^* \hat{H} \psi \, d\tau}{\int \psi^* \psi \, d\tau} \tag{10.28}$$

We now search for values of the coefficients in the trial function that minimize the value of E. This is a standard problem in calculus, and is solved by finding the coefficients for which

$$\frac{\partial E}{\partial c_A}=0 \qquad \frac{\partial E}{\partial c_B}=0$$

The first step is to express the two integrals in eqn 10.28 in terms of the coefficients. The denominator is

$$\int \psi^2 \mathrm{d}\tau=\int (c_A A+c_B B)^2 \,\mathrm{d}\tau=c_A^2\int A^2\,\mathrm{d}\tau+c_B^2\int B^2\,\mathrm{d}\tau+2c_Ac_B\int AB\,\mathrm{d}\tau$$

$$=c_A^2+c_B^2+2c_Ac_BS$$

because the individual atomic orbitals are normalized and the third integral is the overlap integral S (eqn 10.18). The numerator is

$$\int \psi\hat{H}\psi\,\mathrm{d}\tau=\int (c_AA+c_BB)\hat{H}(c_AA+c_BB)\,\mathrm{d}\tau$$

$$=c_A^2\int A\hat{H}A\,\mathrm{d}\tau+c_B^2\int B\hat{H}B\,\mathrm{d}\tau+c_Ac_B\int A\hat{H}B\,\mathrm{d}\tau+c_Ac_B\int B\hat{H}A\,\mathrm{d}\tau$$

There are some complicated integrals in this expression, but we can combine them all into the parameters

$$\alpha_A=\int A\hat{H}A\,\mathrm{d}\tau \qquad \alpha_B=\int B\hat{H}B\,\mathrm{d}\tau \qquad [10.29]$$

$$\beta=\int A\hat{H}B\,\mathrm{d}\tau=\int B\hat{H}A\,\mathrm{d}\tau \text{ (by the hermiticity of } \hat{H})$$

Then

$$\int \psi\hat{H}\psi\,\mathrm{d}\tau=c_A^2\alpha_A+c_B^2\alpha_B+2c_Ac_B\beta$$

The complete expression for E is

$$E=\frac{c_A^2\alpha_A+c_B^2\alpha_B+2c_Ac_B\beta}{c_A^2+c_B^2+2c_Ac_BS} \qquad (10.30)$$

Its minimum is found by differentiation with respect to the two coefficients and setting the results equal to 0. After some straightforward work we obtain

$$\frac{\partial E}{\partial c_A}=\frac{2\times(c_A\alpha_A-c_AE+c_B\beta-c_BSE)}{c_A^2+c_B^2+2c_Ac_BS}$$

$$\frac{\partial E}{\partial c_B}=\frac{2\times(c_B\alpha_B-c_BE+c_A\beta-c_ASE)}{c_A^2+c_B^2+2c_Ac_BS}$$

For the derivatives to be equal to 0, the numerators of these expressions must vanish. That is, we must find values of c_A and c_B that satisfy the conditions

$$c_A\alpha_A-c_AE+c_B\beta-c_BSE=(\alpha_A-E)c_A+(\beta-ES)c_B=0$$
$$c_A\beta-c_ASE+c_B\alpha_B-c_BE=(\beta-ES)c_A+(\alpha_B-E)c_B=0$$

which are the secular equations (eqn 10.27).

To solve the secular equations for the coefficients we need to know the energy E of the orbital. As for any set of simultaneous equations, the secular equations have a solution if the **secular determinant**, the determinant of the coefficients, is zero; that is, if

$$\begin{vmatrix} \alpha_A - E & \beta - ES \\ \beta - ES & \alpha_B - E \end{vmatrix} = (\alpha_A - E)(\alpha_B - E) - (\beta - ES)^2 = 0 \qquad (10.31)$$

This quadratic equation, which expands to

$$(1 - S^2)E^2 + \{2\beta S - (\alpha_A + \alpha_B)\}E + (\alpha_A\alpha_B - \beta^2) = 0$$

has two roots that give the energies of the bonding and antibonding molecular orbitals formed from the atomic orbitals:

$$E_{\pm} = \frac{\alpha_A + \alpha_B - 2\beta S \pm \{(\alpha_A + \alpha_B - 2\beta S)^2 - 4(1 - S^2)(\alpha_A\alpha_B - \beta^2)\}^{1/2}}{2(1 - S^2)} \qquad (10.32a)$$

This expression becomes more transparent in two cases. For a *homonuclear diatomic molecule* we can set $\alpha_A = \alpha_B = \alpha$ and obtain

$$E_+ = \frac{\alpha + \beta}{1 + S} \qquad E_- = \frac{\alpha - \beta}{1 - S} \qquad (10.32b)$$

Homonuclear diatomic molecules

For $\beta < 0$, E_+ is the lower energy solution. For *heteronuclear diatomic molecules* we can make the approximation that $S = 0$ (simply to get a more transparent expression), and find

$$E_{\pm} = \tfrac{1}{2}(\alpha_A + \alpha_B) \pm \tfrac{1}{2}(\alpha_A - \alpha_B)\left\{1 + \left(\frac{2\beta}{\alpha_A - \alpha_B}\right)^2\right\}^{1/2} \qquad (10.32c)$$

Zero overlap approximation

- **A brief illustration**

The ionization energies of H1s and F2p electrons are 13.6 eV and 17.4 eV, respectively. Therefore, to calculate the energies of the bonding and antibonding orbitals in HF (using H1s and F2p$_z$ orbitals as a basis) we set $\alpha_H = -13.6$ eV and $\alpha_F = -17.4$ eV. We take $\beta = -1.0$ eV as a typical value and $S = 0$. Substituting these values into eqn 10.32c gives $E_+ = -17.6$ eV and $E_- = -13.4$ eV (as shown in Fig. 10.39). Had we used $S = 0.2$ (another typical value), then eqn 10.32a would have given $E_+ = -18.9$ eV and $E_- = -13.0$ eV. •

Self-test 10.6 The ionization energy of Cl is 13.1 eV; find the energies of the σ orbitals in the HCl molecule using $\beta = -1.0$ eV and $S = 0$.

[$E_- = -12.3$ eV, $E_+ = -14.4$ eV]

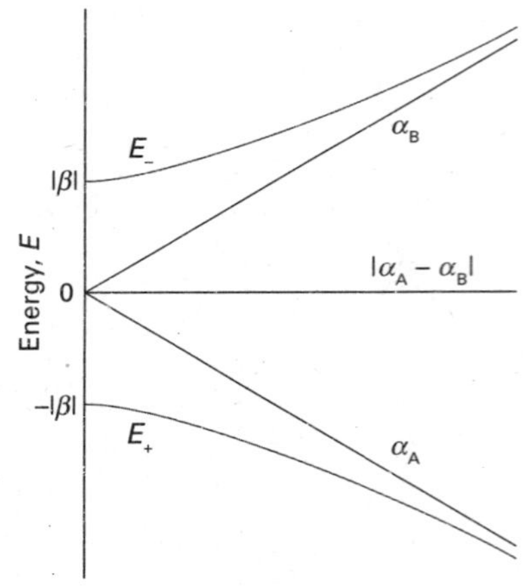

Fig. 10.40 The variation of the energies of molecular orbitals with the energy separation of the contributing atomic orbitals.

An important feature of eqn 10.32c is that as the energy difference $|\alpha_A - \alpha_B|$ between the interacting atomic orbitals increases, the bonding and antibonding effects decrease (Fig. 10.40). Thus, when $|\alpha_B - \alpha_A| \gg 2|\beta|$ we can make the approximation $(1 + x)^{1/2} \approx 1 + \tfrac{1}{2}x$ and obtain

$$E_+ \approx \alpha_A + \frac{\beta^2}{\alpha_A - \alpha_B} \qquad E_- \approx \alpha_B - \frac{\beta^2}{\alpha_A - \alpha_B} \qquad (10.33)$$

As these expressions show, and as can be seen from the graph, when the energy difference is very large, the energies of the resulting molecular orbitals differ only slightly from those of the atomic orbitals, which implies in turn that the bonding and antibonding effects are small. That is:

The strongest bonding and antibonding effects are obtained when the two contributing orbitals have closely similar energies.

Orbital contribution criterion

The difference in energy between core and valence orbitals is the justification for neglecting the contribution of core orbitals to bonding. The core orbitals of one atom have a similar energy to the core orbitals of the other atom; but core–core interaction is largely negligible because the overlap between them (and hence the value of β) is so small.

The values of the coefficients in the linear combination in eqn 10.22 are obtained by solving the secular equations using the two energies obtained from the secular determinant. The lower energy, E_+, gives the coefficients for the bonding molecular orbital, the upper energy, E_-, the coefficients for the antibonding molecular orbital. The secular equations give expressions for the ratio of the coefficients. Thus, the first of the two secular equations in eqn 10.27a, $(\alpha_A - E)c_A + (\beta - ES)c_B = 0$, gives

$$c_B = -\left(\frac{\alpha_A - E}{\beta - ES}\right)c_A \tag{10.34}$$

The wavefunction should also be normalized. This condition means that we must also ensure that

$$\int \psi^2\,d\tau = \int (c_A A + c_B B)^2\,d\tau = c_A^2\int A^2 d\tau + c_B^2\int B^2 d\tau + 2c_A c_B\int AB\,d\tau \tag{10.35}$$

$$= c_A^2 + c_B^2 + 2c_A c_B S = 1$$

When the preceding relation is substituted into this expression, we find

$$c_A = \frac{1}{\left\{1 + \left(\dfrac{\alpha_A - E}{\beta - ES}\right)^2 - 2S\left(\dfrac{\alpha_A - E}{\beta - ES}\right)\right\}^{1/2}} \tag{10.36}$$

which, together with eqn 10.32a, gives explicit expressions for the coefficients once we substitute the appropriate values of $E = E_\pm$ found previously. As before, this expression becomes more transparent in two cases. First, for a homonuclear diatomic molecule, with $\alpha_A = \alpha_B = \alpha$ and $E_\pm$ given in eqn 10.32b we find

$$E_+ = \frac{\alpha + \beta}{1 + S} \qquad c_A = \frac{1}{\{2(1+S)\}^{1/2}} \qquad c_B = c_A \tag{10.37a}$$

Homonuclear diatomic molecule

$$E_- = \frac{\alpha - \beta}{1 - S} \qquad c_A = \frac{1}{\{2(1-S)\}^{1/2}} \qquad c_B = -c_A \tag{10.37b}$$

For a heteronuclear diatomic molecule with $S = 0$, the coefficients are given by

$$c_A = \frac{1}{\left\{1 + \left(\dfrac{\alpha_A - E}{\beta}\right)^2\right\}^{1/2}} \qquad c_B = -\left(\frac{\alpha_A - E}{\beta}\right)c_A \tag{10.38}$$

Zero overlap approximation

with the appropriate values of $E = E_\pm$ taken from eqn 10.32c.

• **A brief illustration**

Here we continue the previous *brief illustration* using HF. With $\alpha_H = -13.6$ eV, $\alpha_F = -17.4$ eV, $\beta = -1.0$ eV, and $S = 0$ the two orbital energies were found to be $E_+ = -17.6$ eV and $E_- = -13.4$ eV. When these values are substituted into eqn 10.37 we find the following coefficients:

$$E_- = -13.4 \text{ eV} \qquad \psi_- = 0.97\chi_H - 0.24\chi_F$$
$$E_+ = -17.6 \text{ eV} \qquad \psi_+ = 0.24\chi_H + 0.97\chi_F$$

Notice how the lower energy orbital (the one with energy −17.6 eV) has a composition that is more F2p orbital than H1s, and that the opposite is true of the higher energy, antibonding orbital. Had we taken $S = 0.2$, then we would have found

$$E_- = -13.0 \text{ eV} \qquad \psi_- = 0.88\chi_H + 0.32\chi_F$$
$$E_+ = -18.9 \text{ eV} \qquad \psi_+ = 0.51\chi_H - 0.97\chi_F$$

It is no longer possible to interpret the coefficients as occupation probabilities of individual atomic orbitals or even their relative signs because now the basis orbitals are not orthogonal. •

Self-test 10.7 The ionization energy of Cl is 13.1 eV; find the form of the σ orbitals in the HCl molecule using $\beta = -1.0$ eV and $S = 0$.

[$\psi_- = -0.62\chi_H + 0.79\chi_{Cl}$; $\psi_+ = 0.79\chi_H + 0.62\chi_{Cl}$]

IMPACT ON BIOCHEMISTRY

I10.1 The biochemical reactivity of O_2, N_2, and NO

We can now see how some of these concepts are applied to diatomic molecules that play a vital biochemical role. At sea level, air contains approximately 23.1 per cent O_2 and 75.5 per cent N_2 by mass. Molecular orbital theory predicts correctly that O_2 has unpaired electron spins. It is a reactive component of the Earth's atmosphere; its most important biological role is as an oxidizing agent. By contrast N_2, the major component of the air we breathe, is so stable (on account of the triple bond connecting the atoms) and unreactive that *nitrogen fixation*, the reduction of atmospheric N_2 to NH_3, is among the most thermodynamically demanding of biochemical reactions, in the sense that it requires a great deal of energy derived from metabolism. So taxing is the process that only certain bacteria and archaea are capable of carrying it out, making nitrogen available first to plants and other micro-organisms in the form of ammonia. Only after incorporation into amino acids by plants does nitrogen adopt a chemical form that, when consumed, can be used by animals in the synthesis of proteins and other molecules that contain nitrogen.

The reactivity of O_2, while important for biological energy conversion, also poses serious physiological problems. During the course of metabolism, some electrons reduce O_2 to superoxide ion, O_2^-, which must be scavenged to prevent damage to cellular components. There is growing evidence for the involvement of the damage caused by reactive oxygen species (ROS), such as O_2^-, H_2O_2, and ·OH (the hydroxyl radical), in the mechanism of ageing and in the development of cardiovascular disease, cancer, stroke, inflammatory disease, and other conditions. For this reason, much effort has been expended on studies of the biochemistry of *antioxidants*, substances that can either deactivate ROS directly or halt the progress of cellular damage through reactions with radicals formed by processes initiated by ROS. Important examples of antioxidants are vitamin C (ascorbic acid), vitamin E (α-tocopherol), and uric acid.

Nitric oxide (nitrogen monoxide, NO) is a small molecule that diffuses quickly between cells, carrying chemical messages that help initiate a variety of processes, such as regulation of blood pressure, inhibition of platelet aggregation, and defence against inflammation and attacks to the immune system. Figure 10.41 shows the bonding scheme in NO and illustrates a number of points we have made about heteronuclear diatomic molecules. The ground configuration is $1\sigma^2 2\sigma^2 3\sigma^2 1\pi^4 2\pi^1$. The 3σ and 1π orbitals are predominantly of O character as that is the more electronegative element. The highest-energy occupied orbital is 2π; it is occupied by one electron and has more N character than O character. It follows that NO is a radical with an unpaired electron that can be regarded as localized more on the N atom than on the O atom. The lowest-energy unoccupied orbital is 4σ, which is also localized predominantly on N. Because NO is a radical, we expect it to be reactive. Its half-life is estimated as 1–5 s, so it needs to be synthesized often in the cell. As we saw above, there is a biochemical price to be paid for the reactivity of biological radicals.

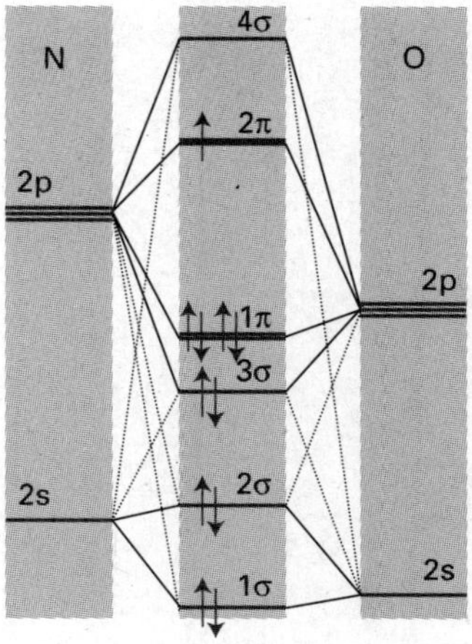

Fig. 10.41 The molecular orbital energy level diagram for NO.

Molecular orbitals for polyatomic systems

The molecular orbitals of polyatomic molecules are built in the same way as in diatomic molecules, the only difference being that we use more atomic orbitals to construct them. As for diatomic molecules, polyatomic molecular orbitals spread over the entire molecule. A molecular orbital has the general form

$$\psi = \sum_o c_o \chi_o \qquad (10.39)$$

General form of LCAO

where χ_o is an atomic orbital and the sum extends over all the valence orbitals of all the atoms in the molecule. To find the coefficients, we set up the secular equations and the secular determinant, just as for diatomic molecules, solve the latter for the energies, and then use these energies in the secular equations to find the coefficients of the atomic orbitals for each molecular orbital.

The principal difference between diatomic and polyatomic molecules lies in the greater range of shapes that are possible: a diatomic molecule is necessarily linear, but a triatomic molecule, for instance, may be either linear or angular (bent) with a characteristic bond angle. The shape of a polyatomic molecule—the specification of its bond lengths and its bond angles—can be predicted by calculating the total energy of the molecule for a variety of nuclear positions, and then identifying the conformation that corresponds to the lowest energy.

10.6 The Hückel approximation

Key points **(a) The Hückel method neglects overlap and interactions between atoms that are not neighbours. (b) It may be expressed in a compact manner by introducing matrices. (c) The strength of π bonding in conjugated systems is expressed by the π-binding energy, the delocalization energy, and the π-bond formation energy. (d) The stability of benzene arises from the geometry of the ring and the high delocalization energy.**

Molecular orbital theory takes large molecules and extended aggregates of atoms, such as solid materials, in its stride. First we consider conjugated molecules, in which there is an alternation of single and double bonds along a chain of carbon atoms.

Although the classification of an orbital as σ or π is strictly valid only in linear molecules, as will be familiar from introductory chemistry courses, it is also used to denote the local symmetry with respect to a given A–B bond axis.

The π molecular orbital energy level diagrams of conjugated molecules can be constructed using a set of approximations suggested by Erich Hückel in 1931. In his approach, the π orbitals are treated separately from the σ orbitals, and the latter form a rigid framework that determines the general shape of the molecule. All the C atoms are treated identically, so all the Coulomb integrals α for the atomic orbitals that contribute to the π orbitals are set equal. For example, in ethene, we take the σ bonds as fixed, and concentrate on finding the energies of the single π bond and its companion antibond.

(a) Ethene and frontier orbitals

We express the π orbitals as LCAOs of the C2p orbitals that lie perpendicular to the molecular plane. In ethene, for instance, we would write

$$\psi = c_A A + c_B B \qquad (10.40)$$

where the A is a C2p orbital on atom A, and so on. Next, the optimum coefficients and energies are found by the variation principle as explained in Section 10.5. That is, we solve the secular determinant, which in the case of ethene is eqn 10.31 with $\alpha_A = \alpha_B = \alpha$:

$$\begin{vmatrix} \alpha - E & \beta - ES \\ \beta - ES & \alpha - E \end{vmatrix} = 0 \qquad (10.41)$$

The roots of this determinant were given in eqn 10.32b. In a modern computation all the resonance integrals and overlap integrals would be included, but an indication of the molecular orbital energy level diagram can be obtained very readily if we make the following additional **Hückel approximations**:

1. All overlap integrals are set equal to zero.
2. All resonance integrals between non-neighbours are set equal to zero.
3. All remaining resonance integrals are set equal (to β).

Hückel approximations

These approximations are obviously very severe, but they let us calculate at least a general picture of the molecular orbital energy levels with very little work. The assumptions result in the following structure of the secular determinant:

1. All diagonal elements: $\alpha - E$.
2. Off-diagonal elements between neighbouring atoms: β.
3. All other elements: 0.

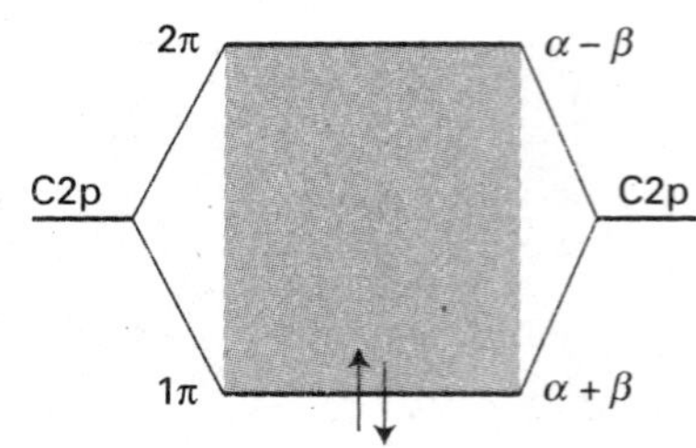

Fig. 10.42 The Hückel molecular orbital energy level diagram for the π orbitals of ethene. Two electrons occupy the lower π orbital.

These approximations convert eqn 10.41 to

$$\begin{vmatrix} \alpha - E & \beta \\ \beta & \alpha - E \end{vmatrix} = (\alpha - E)^2 - \beta^2 = 0 \qquad (10.42)$$

The roots of the equation are

$$E_{\pm} = \alpha \pm \beta \qquad (10.43)$$

The + sign corresponds to the bonding combination (β is negative) and the − sign corresponds to the antibonding combination (Fig. 10.42). We see the effect of neglecting overlap by comparing this result with eqn 10.32b.

The building-up principle leads to the configuration $1\pi^2$, because each carbon atom supplies one electron to the π system. The **highest occupied molecular orbital** in ethene, its HOMO, is the 1π orbital; the **lowest unoccupied molecular orbital**, its LUMO, is the 2π orbital (or, as it is sometimes denoted, the $2\pi^*$ orbital). These two orbitals jointly form the **frontier orbitals** of the molecule. The frontier orbitals are important because they are largely responsible for many of the chemical and spectroscopic properties of the molecule. For example, we can estimate that $2|\beta|$ is the $\pi^* \leftarrow \pi$ excitation energy of ethene, the energy required to excite an electron from the 1π to the 2π orbital. The constant β is often left as an adjustable parameter; an approximate value for π bonds formed from overlap of two C2p atomic orbitals is about −2.4 eV (−230 kJ mol^{-1}).

(b) The matrix formulation of the Hückel method

In preparation for making Hückel theory more sophisticated and readily applicable to bigger molecules, we need to reformulate it in terms of matrices and vectors (see *Mathematical background* 6 following this chapter). We have seen that the secular equations that we have to solve for a two-atom system have the form

$$(H_{AA} - E_iS_{AA})c_{i,A} + (H_{AB} - E_iS_{AB})c_{i,B} = 0 \qquad (10.44a)$$

$$(H_{BA} - E_iS_{BA})c_{i,A} + (H_{BB} - E_iS_{BB})c_{i,B} = 0 \qquad (10.44b)$$

where the eigenvalue E_i corresponds to a wavefunction of the form $\psi_i = c_{i,A}A + c_{i,B}B$. (These expressions generalize eqn 10.27.) There are two atomic orbitals, two eigenvalues, and two wavefunctions, so there are two pairs of secular equations, with the first corresponding to E_1 and ψ_1:

$$(H_{AA} - E_1S_{AA})c_{1,A} + (H_{AB} - E_1S_{AB})c_{1,B} = 0 \qquad (10.45a)$$

$$(H_{BA} - E_1S_{BA})c_{1,A} + (H_{BB} - E_1S_{BB})c_{1,B} = 0 \qquad (10.45b)$$

and another pair corresponding to E_2 and ψ_2:

$$(H_{AA} - E_2S_{AA})c_{2,A} + (H_{AB} - E_2S_{AB})c_{2,B} = 0 \qquad (10.45c)$$

$$(H_{BA} - E_2S_{BA})c_{2,A} + (H_{BB} - E_2S_{BB})c_{2,B} = 0 \qquad (10.45d)$$

If we introduce the following matrices and column vectors

$$\boldsymbol{H} = \begin{pmatrix} H_{AA} & H_{AB} \\ H_{BA} & H_{BB} \end{pmatrix} \quad \boldsymbol{S} = \begin{pmatrix} S_{AA} & S_{AB} \\ S_{BA} & S_{BB} \end{pmatrix} \quad \boldsymbol{c}_i = \begin{pmatrix} c_{i,A} \\ c_{i,B} \end{pmatrix} \qquad (10.46)$$

then each pair of equations may be written more succinctly as

$$(\boldsymbol{H} - E_i\boldsymbol{S})\boldsymbol{c}_i = 0 \quad \text{or} \quad \boldsymbol{H}\boldsymbol{c}_i = \boldsymbol{S}\boldsymbol{c}_iE_i \qquad (10.47)$$

The two sets of equations like these (with i = 1 and 2) can be combined into a single matrix equation by introducing the matrices

$$\boldsymbol{c} = (\boldsymbol{c}_1 \quad \boldsymbol{c}_2) = \begin{pmatrix} c_{1,A} & c_{2,A} \\ c_{1,B} & c_{2,B} \end{pmatrix} \quad \boldsymbol{E} = \begin{pmatrix} E_1 & 0 \\ 0 & E_2 \end{pmatrix} \qquad (10.48)$$

for then all four equation in eqn 10.45 are summarized by the single expression

$$\boldsymbol{H}\boldsymbol{c} = \boldsymbol{S}\boldsymbol{c}\boldsymbol{E} \qquad (10.49)$$

Self-test 10.8 Show by carrying out the necessary matrix operations that eqn 10.49 is a representation of all four equations in eqn 10.45.

In the Hückel approximation, $H_{AA} = H_{BB} = \alpha$, $H_{AB} = H_{BA} = \beta$, and we neglect overlap, setting $S = 1$, the unit matrix (with 1 on the diagonal and 0 elsewhere). Then

$$Hc = cE$$

At this point, we multiply from the left by the inverse matrix c^{-1}, use $c^{-1}c = 1$, and find

$$c^{-1}Hc = E \tag{10.50}$$

In other words, to find the eigenvalues E_i, we have to find a transformation of H that makes it diagonal. This procedure is called **matrix diagonalization.** The diagonal elements then correspond to the eigenvalues E_i and the columns of the matrix c that brings about this diagonalization are the coefficients of the members of the **basis set,** the set of atomic orbitals used in the calculation, and hence give us the composition of the molecular orbitals.

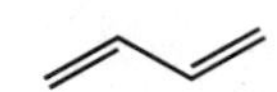

3 Butadiene

Example 10.3 *Finding molecular orbitals by matrix diagonalization*

Set up and solve the matrix equations within the Hückel approximation for the π orbitals of butadiene (**3**).

Method The matrices will be four-dimensional for this four-atom system. Ignore overlap, and construct the matrix H by using the Hückel approximation and the parameters α and β. Find the matrix c that diagonalizes H: for this step, use mathematical software. Full details are given in *Mathematical background 6.*

Answer

$$H = \begin{pmatrix} H_{11} & H_{12} & H_{13} & H_{14} \\ H_{21} & H_{22} & H_{23} & H_{24} \\ H_{31} & H_{32} & H_{33} & H_{34} \\ H_{41} & H_{42} & H_{43} & H_{44} \end{pmatrix} \overset{\text{Hückel approximation}}{=} \begin{pmatrix} \alpha & \beta & 0 & 0 \\ \beta & \alpha & \beta & 0 \\ 0 & \beta & \alpha & \beta \\ 0 & 0 & \beta & \alpha \end{pmatrix}$$

We write this matrix as

$$H = \alpha 1 + \beta \begin{pmatrix} 0 & 1 & 0 & 0 \\ 1 & 0 & 1 & 0 \\ 0 & 1 & 0 & 1 \\ 0 & 0 & 1 & 0 \end{pmatrix}$$

because most mathematical software can deal only with numerical matrices. The diagonalized form of the second matrix is

$$\beta \begin{pmatrix} +1.62 & 0 & 0 & 0 \\ 0 & +0.62 & 0 & 0 \\ 0 & 0 & -0.62 & 0 \\ 0 & 0 & 0 & -1.62 \end{pmatrix}$$

so we can infer that the diagonalized Hamiltonian matrix is

$$E = \begin{pmatrix} \alpha + 1.62\beta & 0 & 0 & 0 \\ 0 & \alpha + 0.62\beta & 0 & 0 \\ 0 & 0 & \alpha - 0.62\beta & 0 \\ 0 & 0 & 0 & \alpha - 1.62\beta \end{pmatrix}$$

The matrix that achieves the diagonalization is

$$c = \begin{pmatrix} 0.372 & 0.602 & 0.602 & -0.372 \\ 0.602 & 0.372 & -0.372 & 0.602 \\ 0.602 & -0.372 & -0.372 & -0.602 \\ 0.372 & -0.602 & 0.602 & 0.372 \end{pmatrix}$$

with each column giving the coefficients of the atomic orbitals for the corresponding molecular orbital. We can conclude that the energies and molecular orbitals are

$$E_1 = \alpha + 1.62\beta \qquad \psi_1 = 0.372\chi_A + 0.602\chi_B + 0.602\chi_C + 0.372\chi_D$$
$$E_2 = \alpha + 0.62\beta \qquad \psi_2 = 0.602\chi_A + 0.372\chi_B - 0.372\chi_C - 0.602\chi_D$$
$$E_3 = \alpha - 0.62\beta \qquad \psi_3 = 0.602\chi_A - 0.372\chi_B - 0.372\chi_C + 0.602\chi_D$$
$$E_4 = \alpha - 1.62\beta \qquad \psi_4 = -0.372\chi_A + 0.602\chi_B - 0.602\chi_C + 0.372\chi_D$$

where the C2p atomic orbitals are denoted by $\chi_A, \ldots, \chi_D$. Note that the molecular orbitals are mutually orthogonal and, with overlap neglected, normalized.

Self-test 10.9 Repeat the exercise for the allyl radical, $\cdot CH_2$–CH=CH_2.

[$E = \alpha + 1.41\beta, \alpha, \alpha - 1.41\beta$; $\psi_1 = 0.500\chi_A + 0.707\chi_B + 0.500\chi_C$, $\psi_2 = 0.707\chi_A - 0.707\chi_C$, $\psi_3 = 0.500\chi_A - 0.707\chi_B + 0.500\chi_C$]

(c) Butadiene and π-electron binding energy

As we saw in Example 10.3, the energies of the four LCAO-MOs for butadiene are

$$E = \alpha \pm 1.62\beta, \qquad \alpha \pm 0.62\beta \qquad (10.51)$$

These orbitals and their energies are drawn in Fig. 10.43. Note that the greater the number of internuclear nodes, the higher the energy of the orbital. There are four electrons to accommodate, so the ground-state configuration is $1\pi^2 2\pi^2$. The frontier orbitals of butadiene are the 2π orbital (the HOMO, which is largely bonding) and the 3π orbital (the LUMO, which is largely antibonding). 'Largely' bonding means that an orbital has both bonding and antibonding interactions between various neighbours, but the bonding effects dominate. 'Largely antibonding' indicates that the antibonding effects dominate.

An important point emerges when we calculate the total **π-electron binding energy**, E_π, the sum of the energies of each π electron, and compare it with what we find in ethene. In ethene the total energy is

$$E_\pi = 2(\alpha + \beta) = 2\alpha + 2\beta$$

In butadiene it is

$$E_\pi = 2(\alpha + 1.62\beta) + 2(\alpha + 0.62\beta) = 4\alpha + 4.48\beta$$

Therefore, the energy of the butadiene molecule lies lower by 0.48β (about 110 kJ mol^{-1}) than the sum of two individual π bonds. This extra stabilization of a conjugated system compared with a set of localized π bonds is called the **delocalization energy** of the molecule.

A closely related quantity is the **π-bond formation energy**, E_{bf}, the energy released when a π bond is formed. Because the contribution of α is the same in the molecule as in the atoms, we can find the π-bond formation energy from the π-electron binding energy by writing

$$E_{bf} = E_\pi - N_C\alpha \qquad \text{Definition of π-bond formation energy} \qquad [10.52]$$

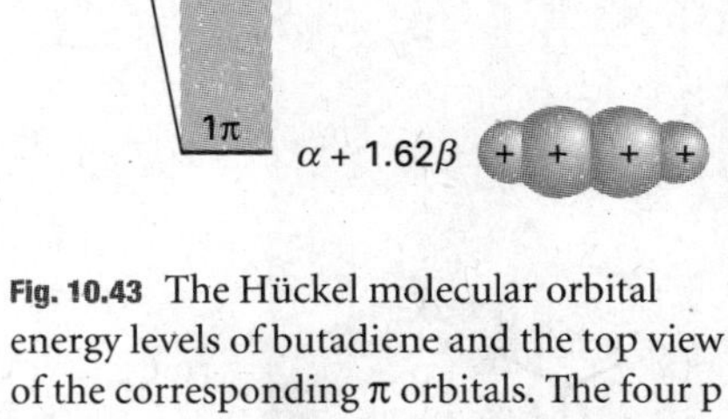

Fig. 10.43 The Hückel molecular orbital energy levels of butadiene and the top view of the corresponding π orbitals. The four p electrons (one supplied by each C) occupy the two lower π orbitals. Note that the orbitals are delocalized.

where N_C is the number of carbon atoms in the molecule. The π-bond formation energy in butadiene, for instance, is 4.48β.

Example 10.4 *Estimating the delocalization energy*

Use the Hückel approximation to find the energies of the π orbitals of cyclobutadiene, and estimate the delocalization energy.

Method Set up the secular determinant using the same basis as for butadiene, but note that atoms A and D are also now neighbours. Then solve for the roots of the secular equation and assess the total π-electron binding energy. For the delocalization energy, subtract from the total π-bond energy the energy of two π bonds.

Answer The hamiltonian matrix is

$$H=\begin{pmatrix}\alpha & \beta & 0 & \beta\\ \beta & \alpha & \beta & 0\\ 0 & \beta & \alpha & \beta\\ \beta & 0 & \beta & \alpha\end{pmatrix}$$

Diagonalization gives the energies of the orbitals as

$$E=\alpha+2\beta, \qquad \alpha, \qquad \alpha, \qquad \alpha-2\beta$$

Four electrons must be accommodated. Two occupy the lowest orbital (of energy $\alpha+2\beta$), and two occupy the doubly degenerate orbitals (of energy α). The total energy is therefore $4\alpha+4\beta$. Two isolated π bonds would have an energy $4\alpha+4\beta$; therefore, in this case, the delocalization energy is zero.

Self-test 10.10 Repeat the calculation for benzene (use software!).

[See next subsection]

(d) Benzene and aromatic stability

The most notable example of delocalization conferring extra stability is benzene and the aromatic molecules based on its structure. Benzene is often expressed in a mixture of valence-bond and molecular orbital terms, with typically valence-bond language used for its σ framework and molecular orbital language used to describe its π electrons.

First, the valence-bond component. The six C atoms are regarded as sp^2 hybridized, with a single unhydridized perpendicular 2p orbital. One H atom is bonded by (Csp^2,H1s) overlap to each C carbon, and the remaining hybrids overlap to give a regular hexagon of atoms (Fig. 10.44). The internal angle of a regular hexagon is 120°, so sp^2 hybridization is ideally suited for forming σ bonds. We see that the hexagonal shape of benzene permits strain-free σ bonding.

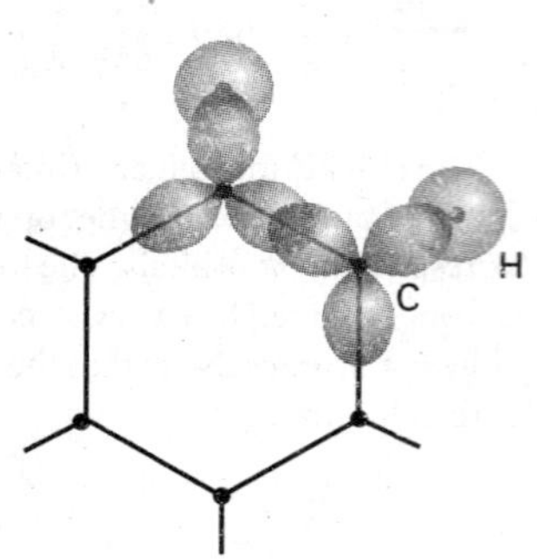

Fig. 10.44 The σ framework of benzene is formed by the overlap of Csp^2 hybrids, which fit without strain into a hexagonal arrangement.

Now consider the molecular orbital component of the description. The six C2p orbitals overlap to give six π orbitals that spread all round the ring. Their energies are calculated within the Hückel approximation by diagonalizing the hamiltonian matrix

$$H=\begin{pmatrix}\alpha & \beta & 0 & 0 & 0 & \beta\\ \beta & \alpha & \beta & 0 & 0 & 0\\ 0 & \beta & \alpha & \beta & 0 & 0\\ 0 & 0 & \beta & \alpha & \beta & 0\\ 0 & 0 & 0 & \beta & \alpha & \beta\\ \beta & 0 & 0 & 0 & \beta & \alpha\end{pmatrix}=\alpha 1+\beta\begin{pmatrix}0 & 1 & 0 & 0 & 0 & 1\\ 1 & 0 & 1 & 0 & 0 & 0\\ 0 & 1 & 0 & 1 & 0 & 0\\ 0 & 0 & 1 & 0 & 1 & 0\\ 0 & 0 & 0 & 1 & 0 & 1\\ 1 & 0 & 0 & 0 & 1 & 0\end{pmatrix}$$

The MO energies, the eigenvalues of this matrix, are simply

$$E = \alpha \pm 2\beta, \alpha \pm \beta, \alpha \pm \beta \qquad (10.53)$$

as shown in Fig. 10.45. The orbitals there have been given symmetry labels that we explain in Chapter 11. Note that the lowest energy orbital is bonding between all neighbouring atoms, the highest energy orbital is antibonding between each pair of neighbours, and the intermediate orbitals are a mixture of bonding, non-bonding, and antibonding character between adjacent atoms.

We now apply the building-up principle to the π system. There are six electrons to accommodate (one from each C atom), so the three lowest orbitals (a_{2u} and the doubly degenerate pair e_{1g}) are fully occupied, giving the ground-state configuration $a_{2u}^2e_{1g}^4$. A significant point is that the only molecular orbitals occupied are those with net bonding character.

The π-electron energy of benzene is

$$E_\pi = 2(\alpha + 2\beta) + 4(\alpha + \beta) = 6\alpha + 8\beta$$

If we ignored delocalization and thought of the molecule as having three isolated π bonds, it would be ascribed a π-electron energy of only $3(2\alpha + 2\beta) = 6\alpha + 6\beta$. The delocalization energy is therefore $2\beta \approx -460$ kJ mol^{-1}, which is considerably more than for butadiene. The π-bond formation energy in benzene is 8β.

This discussion suggests that aromatic stability can be traced to two main contributions. First, the shape of the regular hexagon is ideal for the formation of strong σ bonds: the σ framework is relaxed and without strain. Second, the π orbitals are such as to be able to accommodate all the electrons in bonding orbitals, and the delocalization energy is large.

A brief comment

The simple form of the eigenvalues in eqn 10.53 suggests that there is a more direct way of determining them than by using mathematical software. That is in fact the case, for symmetry arguments of the kind described in Chapter 11 show that the 6 × 6 matrix can be factorized into two 1 × 1 matrices and two 2 × 2 matrices, which are very easy to deal with.

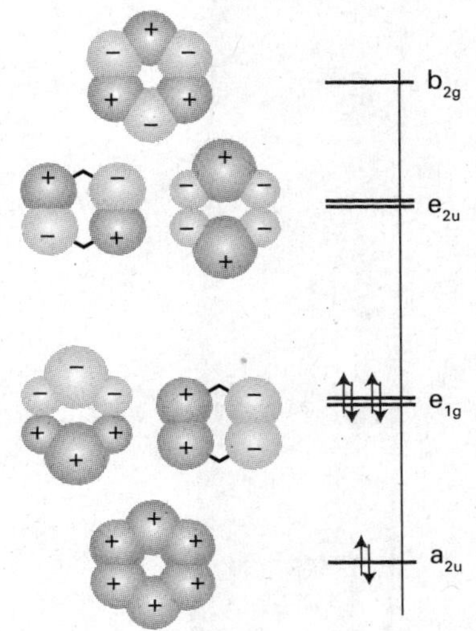

Fig. 10.45 The Hückel orbitals of benzene and the corresponding energy levels. The symmetry labels are explained in Chapter 11. The bonding and antibonding character of the delocalized orbitals reflects the numbers of nodes between the atoms. In the ground state, only the bonding orbitals are occupied.

10.7 Computational chemistry

Key points **(a) The Hartree–Fock equations are versions of the Schrödinger equation based on the occupation of individual molecular orbitals by electrons. The Roothaan equations are versions of these equations that are based on the molecular orbitals being expressed as linear combinations of molecular orbitals. (b) Semi-empirical calculations approximate integrals by estimating integrals using empirical data; *ab initio* methods evaluate all integrals numerically. (c) Density functional theories develop equations based on the electron density rather than the wavefunction itself.**

The severe assumptions of the Hückel method are now easy to avoid by using a variety of software packages that can be used not only to calculate the shapes and energies of molecular orbitals but also predict with reasonable accuracy the structure and reactivity of molecules. The full treatment of molecular electronic structure has received an enormous amount of attention by chemists and has become a keystone of modern chemical research.

A brief comment

The following sections provide a brief introduction. A more complete account with detailed examples will be found in Chapter 6 of our *Quanta, matter, and change* (2009). That chapter is also available in the online resource centre for this book.

(a) The Hartree–Fock equations

The starting point is to write down the many-electron wavefunction as a product of one-electron wavefunctions:

$$\Psi = \psi_a^\alpha(1)\psi_a^\beta(2)\ldots\psi_z^\beta(N_e)$$

This is the wavefunction for an N_e-electron closed-shell molecule in which electron 1 occupies molecular orbital ψ_a with spin α, electron 2 occupies molecular orbital ψ_a with spin β, and so on. We shall consider only closed-shell species. The wavefunction must satisfy the Pauli principle and change sign under the permutation of any pair of electrons. To achieve this behaviour, we write the wavefunction as a sum of all possible permutations with the appropriate sign:

$$\Psi = \psi_a^{\alpha}(1)\psi_a^{\beta}(2)\ldots\psi_z^{\beta}(N_e) - \psi_a^{\alpha}(2)\psi_a^{\beta}(1)\ldots\psi_z^{\beta}(N_e) + \cdots$$

There are $N_e!$ terms in this sum, and the entire sum can be written as a Slater determinant like that used in the description of many-electron atoms (Section 9.4b):

$$\Psi = \frac{1}{\sqrt{N_e!}}\begin{vmatrix} \psi_a^{\alpha}(1) & \psi_a^{\beta}(1) & \cdots & \cdots & \psi_z^{\beta}(1) \\ \psi_a^{\alpha}(2) & \psi_a^{\beta}(2) & \cdots & \cdots & \psi_z^{\beta}(2) \\ \vdots & \vdots & & & \vdots \\ \vdots & \vdots & & & \vdots \\ \psi_a^{\alpha}(N_e) & \psi_a^{\beta}(N_e) & \cdots & \cdots & \psi_z^{\beta}(N_e) \end{vmatrix} \qquad \text{A Slater determinant} \qquad (10.54)$$

where the initial factor ensures that the wavefunction is normalized if the component molecular orbitals are normalized.

When the determinantal wavefunction is combined with the variation principle (Section 10.5c), the optimum wavefunctions, in the sense of corresponding to the lowest total energy, must satisfy a modified version of the Schrödinger equation, which is written as a set of **Hartree–Fock equations**:

$$f_1\psi_m(1) = \varepsilon_m\psi_m(1) \qquad \text{Hartree–Fock equations} \qquad (10.55)$$

for each molecular orbital ψ_m. The **Fock operator** f_1 has terms that express mathematically (see *Further information 10.1*):

- the kinetic energy of the electron in ψ_m;
- the potential energy of interaction between the electron in ψ_m and the nuclei in the molecule;
- repulsive interactions between the electron in ψ_m and other electrons in the molecule;
- the effects of spin correlation between electrons in the molecule.

Because the Fock operator includes the effects of all the other electrons on electron 1, its detailed form depends on the wavefunctions of those electrons. To proceed, we have to guess the initial form of those wavefunctions, use them in the definition of the Fock operator, and solve the Hartree–Fock equations. That process is then continued using the newly found wavefunctions until each cycle of calculation leaves the energies and wavefunctions unchanged to within a chosen criterion. This is the origin of the term **self-consistent field** (SCF) for this type of procedure.

To solve the Hartree–Fock equations the molecular orbitals are expressed as linear combinations of N_b atomic orbitals χ_o (that is, N_b is the size of the basis set), which for simplicity we shall take to be real, and write

$$\psi_m = \sum_{o=1}^{N_b} c_{om}\chi_o \qquad \text{A general LCAO} \qquad (10.56)$$

For a given basis set, 'solving the Hartree–Fock equations for ψ_m' now corresponds to determining the values of the coefficients c_{om}. As we show in *Further information 10.1* the use of a linear combination like this leads to a set of equations that can be expressed in a matrix form known as the **Roothaan equations**:

$$\boldsymbol{Fc} = \boldsymbol{Sc\varepsilon} \qquad \text{Roothaan equations} \qquad (10.57)$$

where F is a matrix formed from the Fock operator with elements $F_{ab}=\int\chi_a(1)f_1\chi_b(1)\mathrm{d}\tau_1$, S is the matrix of overlap integrals with elements $S_{ab}=\int\chi_a(1)\chi_b(1)\mathrm{d}\tau_1$, and c and $\boldsymbol{\varepsilon}$ are matrices formed from the orbital coefficients c_{om} and molecular orbital energies ε_m, respectively. The resemblance of eqn 10.57 to eqn 10.49 ($Hc = ScE$) should be noted.

(b) Semi-empirical and *ab initio* methods

There are two main strategies for continuing the calculation from this point. In the **semi-empirical methods**, many of the integrals are estimated by appealing to spectroscopic data or physical properties such as ionization energies, and using a series of rules to set certain integrals equal to zero. We saw this procedure in a primitive form when we identified the integral α in eqn 10.32 with the negative of the ionization energy of an atom (see the *brief illustration* following that equation). In the ***ab initio* methods**, an attempt is made to calculate all the integrals that appear in the Fock and overlap matrices. Both procedures employ a great deal of computational effort and, along with cryptanalysts and meteorologists, theoretical chemists are among the heaviest users of the fastest computers.

We show in *Further information 10.1* that the Fock matrix includes integrals of the form

$$(AB|CD)=j_0\int A(1)B(1)\frac{1}{r_{12}}C(2)D(2)\mathrm{d}\tau_1\mathrm{d}\tau_2 \qquad (10.58)$$

where A, B, C, and D are atomic orbitals that in general may be centred on different nuclei. It can be appreciated that, if there are several dozen atomic orbitals used to build the molecular orbitals, then there will be tens of thousands of integrals of this form to evaluate (the number of integrals increases as the fourth power of the number of atomic orbitals in the basis). Some kind of approximation scheme is necessary.

One severe approximation used in the early days of computational chemistry was called **complete neglect of differential overlap** (CNDO), in which all integrals are set to zero unless A and B are the same orbitals centred on the same nucleus, and likewise for C and D. The surviving integrals are then adjusted until the energy levels are in good agreement with experiment or the computed enthalpy of formation of the compound is in agreement with experiment. More recent semi-empirical methods make less draconian decisions about which integrals are to be ignored, but they are all descendants of the early CNDO technique. These procedures are now readily available in commercial software packages and can be used with very little detailed knowledge of their mode of calculation. The packages also have sophisticated graphical output procedures, which enable one to analyse the shapes of orbitals and the distribution of electric charge in molecules. The latter is important when assessing, for instance, the likelihood that a given molecule will bind to an active site in an enzyme.

Commercial packages are also available for *ab initio* calculations. Here the problem is to evaluate as efficiently as possible thousands of integrals of the form $(AB|CD)$. This task is greatly facilitated by expressing the atomic orbitals used in the LCAOs as linear combinations of Gaussian orbitals. A **Gaussian type orbital** (GTO) is a function of the form $e^{-\zeta r^2}$. The advantage of GTOs over the correct orbitals (which for hydrogenic systems are proportional to $e^{-\zeta r}$) is that the product of two Gaussian functions is itself a Gaussian function that lies between the centres of the two contributing functions (Fig. 10.46). In this way, the four-centre integrals like that in eqn 10.58 become two-centre integrals of the form

$$(AB|CD)=j_0\int X(1)\frac{1}{r_{12}}Y(2)\mathrm{d}\tau_1\mathrm{d}\tau_2 \qquad (10.59)$$

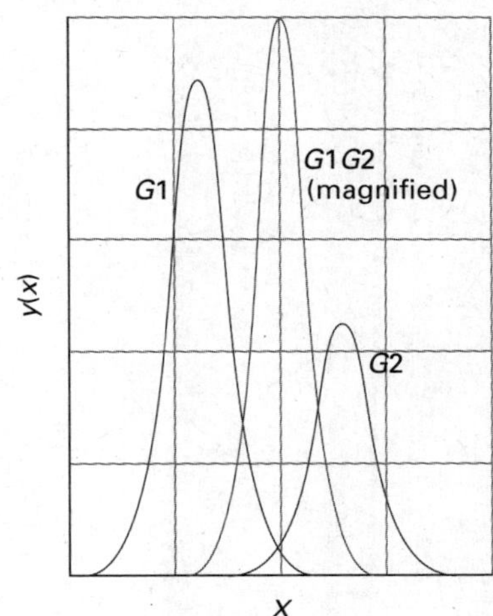

Fig. 10.46 The product of two Gaussian functions (the purple curves) is itself a Gaussian function located between the two contributing Gaussians.

where X is the Gaussian corresponding to the product AB and Y is the corresponding Gaussian from CD. Integrals of this form are much easier and faster to evaluate numerically than the original four-centre integrals. Although more GTOs have to be used to simulate the atomic orbitals, there is an overall increase in speed of computation.

• **A brief illustration**

Suppose we consider a one-dimensional 'homonuclear' system, with Gaussians of the form e^{-ax^2} located at 0 and R. Then one of the integrals that would have to be evaluated would include the term

$$\chi_A(1)\chi_B(1) = e^{-ax^2}e^{-a(x-R)^2} = e^{-2ax^2+2axR-aR^2}$$

Next we note that $-2a(x-\frac{1}{2}R)^2 = -2ax^2 + 2axR - \frac{1}{2}aR^2$, so we can write

$$\chi_A(1)\chi_B(1) = e^{-2a(x-\frac{1}{2}R)^2-\frac{1}{2}aR^2} = e^{-2a(x-\frac{1}{2}R)^2}e^{-\frac{1}{2}aR^2}$$

which is proportional to a single Gaussian centred on the midpoint of the internuclear distance. •

(c) Density functional theory

A technique that has gained considerable ground in recent years to become one of the most widely used techniques for the calculation of molecular structure is **density functional theory** (DFT). Its advantages include less demanding computational effort, less computer time, and—in some cases (particularly d-metal complexes)—better agreement with experimental values than is obtained from Hartree–Fock procedures.

The central focus of DFT is the electron density, ρ, rather than the wavefunction ψ. The 'functional' part of the name comes from the fact that the energy of the molecule is a function of the electron density, written $E[\rho]$, and the electron density is itself a function of position, $\rho(\boldsymbol{r})$, and in mathematics a function of a function is called a *functional*. The occupied orbitals are used to construct the electron density from

$$\rho(\boldsymbol{r}) = \sum_m |\psi_m(\boldsymbol{r})|^2 \qquad \text{Electron probability density} \qquad (10.60)$$

and are calculated from the **Kohn–Sham equations**, which are like the Hartree–Fock equations except for a term V_{XC}, called the **exchange–correlation potential**:

$$\left\{h_1 + j_0\int\frac{\rho(2)}{r_{12}}d\tau_2 + V_{XC}(1)\right\}\psi_m(1) = \varepsilon_m\psi_m(1) \qquad \text{Kohn–Sham equations} \qquad (10.61)$$

The first term on the left is the usual one-electron kinetic and potential energy contribution and the second term is the potential energy of repulsion between electrons 1 and 2. The challenge in DFT is to construct the exchange–correlation potential and computational chemists use several approximate expressions for V_{XC}.

The Kohn–Sham equations are solved iteratively and self-consistently. First, we guess the electron density. For this step it is common to use a superposition of atomic electron densities. Next, the Kohn–Sham equations are solved to obtain an initial set of orbitals. This set of orbitals is used to obtain a better approximation to the electron density and the process is repeated until the density and the exchange–correlation energy are constant to within some tolerance.

10.8 The prediction of molecular properties

Key points **(a) Graphical techniques plot a variety of surfaces based on electronic structure calculations. (b) Computational techniques are used to estimate enthalpies of formation and standard potentials. Electronic absorption spectra of conjugated systems correlate with the HOMO–LUMO energy gap.**

The results of molecular orbital calculations are only approximate, with deviations from experimental values increasing with the size of the molecule. Therefore, one goal of computational chemistry is to gain insight into trends in properties of molecules, without necessarily striving for ultimate accuracy. In the next sections we give a brief summary of strategies used by computational chemists for the prediction of molecular properties.

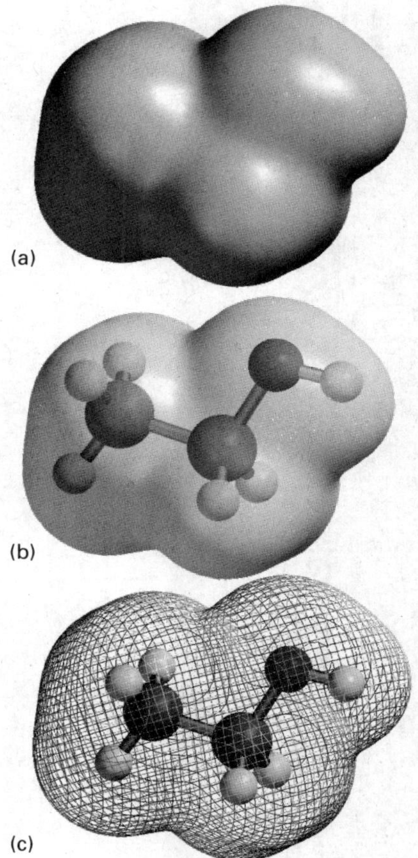

Fig. 10.47 Various representations of an isodensity surface of ethanol (a) solid surface, (b) transparent surface, and (c) mesh surface.

(a) Electron density and the electrostatic potential surfaces

One of the most significant developments in computational chemistry has been the introduction of graphical representations of molecular orbitals and electron densities. The raw output of a molecular structure calculation is a list of the coefficients of the atomic orbitals in each molecular orbital and the energies of these orbitals. The graphical representation of a molecular orbital uses stylized shapes to represent the basis set, and then scales their size to indicate the coefficient in the linear combination. Different signs of the wavefunctions are represented by different colours.

Once the coefficients are known, it is possible to construct a representation of the electron density in the molecule by noting which orbitals are occupied and then forming the squares of those orbitals. The total electron density at any point is then the sum of the squares of the wavefunctions evaluated at that point. The outcome is commonly represented by an **isodensity surface**, a surface of constant total electron density (Fig. 10.47). As shown in the illustration, there are several styles of representing an isodensity surface, as a solid form, as a transparent form with a ball-and-stick representation of the molecule within, or as a mesh. A related representation is a **solvent-accessible surface** in which the shape represents the shape of the molecule by imagining a sphere representing a solvent molecule rolling across the surface and plotting the locations of the centre of that sphere.

One of the most important aspects of a molecule other than its geometrical shape is the distribution of charge over its surface. The net charge at each point on an isodensity surface can be calculated by subtracting the charge due to the electron density at that point from the charge due to the nuclei: the result is an **electrostatic potential surface** (an 'elpot surface') in which net positive charge is shown in one colour and net negative charge is shown in another, with intermediate gradations of colour (Fig. 10.48).

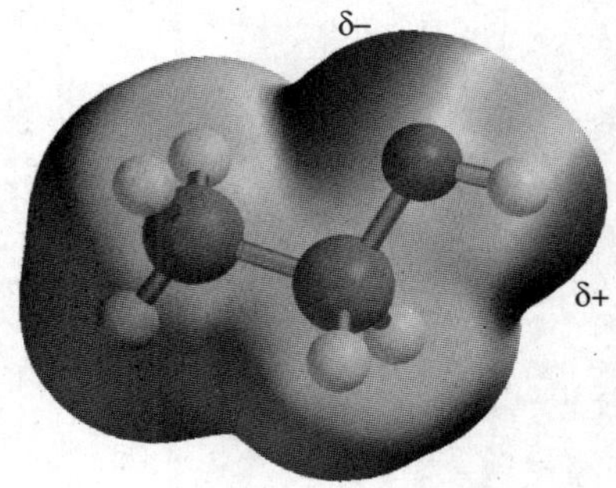

Fig. 10.48 An elpot diagram of ethanol; the molecule has the same orientation as in Fig. 10.47. Red denotes regions of relative negative potential and blue regions of relative positive potential (as in $^{\delta-}$O–H$^{\delta+}$).

Representations such as those we have illustrated are of critical importance in a number of fields. For instance, they may be used to identify an electron-poor region of a molecule that is susceptible to association with or chemical attack by an electron-rich region of another molecule. Such considerations are important for assessing the pharmacological activity of potential drugs.

(b) Thermodynamic and spectroscopic properties

We saw in Section 2.8b that computational chemistry is commonly used to estimate standard enthalpies of formation of molecules with complex three-dimensional structures. The computational approach also makes it possible to gain insight into the effect of solvation on the enthalpy of formation without conducting experiments. A calculation performed in the absence of solvent molecules estimates the properties of the molecule of interest in the gas phase. Computational methods are available that allow for the inclusion of several solvent molecules around a solute molecule, thereby taking into account the effect of molecular interactions with the solvent on the enthalpy of formation of the solute. Again, the numerical results are only estimates and the primary purpose of the calculation is to predict whether interactions with the solvent increase or decrease the enthalpy of formation. As an example, consider the amino acid glycine, which can exist in a neutral (**4**) or zwitterionic (**5**) form, in which the amino group is protonated and the carboxyl group is deprotonated. It is possible to show computationally that in the gas phase the neutral form has a lower enthalpy of formation than the zwitterionic form. However, in water the opposite is true because of strong interactions between the polar solvent and the charges in the zwitterion.

4 Glycine

5 Glycine (zwitterion)

Molecular orbital calculations can also be used to predict trends in electrochemical properties, such as standard potentials (Chapter 6). Several experimental and computational studies of aromatic hydrocarbons indicate that decreasing the energy of the LUMO enhances the ability of a molecule to accept an electron into the LUMO, with an attendant increase in the value of the standard potential of the molecule. The effect is also observed in quinones and flavins, which are co-factors involved in biological electron transfer reactions. For example, stepwise substitution of the hydrogen atoms in *p*-benzoquinone by methyl groups ($-CH_3$) results in a systematic increase in the energy of the LUMO and a decrease in the standard potential for formation of the semiquinone radical (**6**).

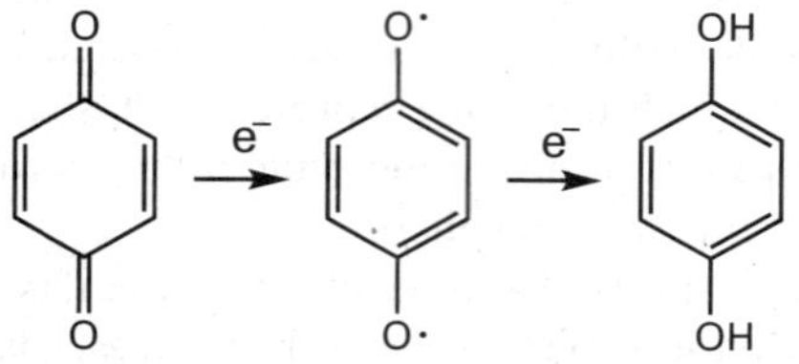

6

The standard potentials of naturally occurring quinones are also modified by the presence of different substituents, a strategy that imparts specific functions to specific quinones. For example, the substituents in coenzyme Q are largely responsible for positioning its standard potential so that the molecule can function as an electron shuttle between specific electroactive proteins in the respiratory chain (*Impact I6.1*).

Calculations based on semi-empirical, *ab initio*, and DFT methods are used to correlate the HOMO–LUMO energy gaps with the wavelengths of spectroscopic absorptions. For example, consider the linear polyenes shown in Table 10.5: ethene (C_2H_4), butadiene (C_4H_6), hexatriene (C_6H_8), and octatetraene (C_8H_{10}), all of which absorb in the ultraviolet region of the spectrum. The table also shows that, as expected, the wavelength of the lowest-energy electronic transition decreases as the energy separation between the HOMO and LUMO increases. We also see that the smallest HOMO–LUMO gap and longest transition wavelength correspond to octatetraene, the longest polyene in the group. It follows that the wavelength of the transition increases with increasing number of conjugated double bonds in linear polyenes.

Table 10.5 *Ab initio* calculations and spectroscopic data

Polyene	$\{E(\text{HOMO}) - E(\text{LUMO})\}/\text{eV}$	λ/nm
(C_2H_4)	18.1	163
	14.5	217
	12.7	252
	11.8	304

Extrapolation of the trend suggests that a sufficiently long linear polyene should absorb light in the visible region of the electromagnetic spectrum. This is indeed the case for β-carotene (7), which absorbs light with $\lambda \approx 450$ nm. The ability of β-carotene to absorb visible light is part of the strategy employed by plants to harvest solar energy for use in photosynthesis (Chapter 22).

7 *β*-Carotene

Checklist of key equations

Property	Equation	Comment
Valence-bond wavefunction	$\psi = A(1)B(2) + A(2)B(1)$	
Linear combination of atomic orbitals	$\psi_{\pm} = N(A \pm B)$	Homonuclear diatomic molecule
Overlap integral	$S = \int \chi_A^* \chi_B \, d\tau$	
Bond order	$b = \frac{1}{2}(N - N^*)$	
Photoelectron spectroscopy	$h\nu = \frac{1}{2}m_e v^2 + I_i$	I_i is the ionization energy from orbital i.
Linear combination of atomic orbitals	$\psi = \sum_i c_i \chi_i$	General case
Hückel equations	$\boldsymbol{Hc} = \boldsymbol{ScE}$	
π-Bond formation energy	$E_{bf} = E_\pi - N_C\alpha$	
Hartee–Fock equation	$f_1 \psi_m(1) = \varepsilon_m \psi_m(1)$	
Roothaan equations	$\boldsymbol{Fc} = \boldsymbol{Sc\varepsilon}$	

Further information

Further information 10.1 *Details of the Hartree–Fock method*

The Fock operator has the form

$$f_1 = h_1 + \sum_m \{2J_m(1) - K_m(1)\} \quad (10.62)$$

where the sum is over all occupied orbitals and h, J, and K are all operators. The first of the three terms in this expression is the **core Hamiltonian**

$$h_1 = -\frac{\hbar^2}{2m_e}\nabla_1^2 - j_0\sum_I \frac{Z_I}{r_{Ii}} \quad \text{Core hamiltonian} \quad (10.63a)$$

where I labels the nuclei in the molecule and $j_0 = e^2/4\pi\varepsilon_0$ (as in Section 10.3). The **Coulomb operator,** J, is

$$J_m(1)\psi_a(1) = j_0\int \psi_a(1)\frac{1}{r_{12}}\psi_m^*(2)\psi_m(2)\mathrm{d}\tau_2 \quad \text{Coulomb operator} \quad (10.63b)$$

and represents the repulsion experienced by electron 1 in orbital ψ_a from electron 2 in orbital ψ_m. The **exchange operator,** K, is

$$K_m(1)\psi_a(1) = j_0\int \psi_m(1)\frac{1}{r_{12}}\psi_m^*(2)\psi_a(2)\mathrm{d}\tau_2 \quad \text{Exchange operator} \quad (10.63c)$$

This integral represents the modification of the electron–electron repulsion that is due to spin correlation (Section 9.4d).

To construct the Roothaan equations we substitute the linear combination of atomic orbitals into the Hartree–Fock equations (eqn 10.55, $f_1\psi_m(1) = \varepsilon_m\psi_m(1)$), which gives

$$f_1\sum_{o=1}^{N_b} c_{oa}\chi_o(1) = \varepsilon_a\sum_{o=1}^{N_b} c_{oa}\chi_o(1)$$

Now multiply from the left by $\chi_{o'}(1)$ and integrate over the coordinates of electron 1:

$$\sum_{o=1}^{N_b} c_{oa}\overbrace{\int \chi_{o'}(1)f(1)\chi_o(1)\mathrm{d}r_1}^{F_{o'o}} = \varepsilon_a\sum_{o=1}^{N_b} c_{oa}\overbrace{\int \chi_{o'}(1)\chi_o(1)\mathrm{d}r_1}^{S_{o'o}}$$

That is,

$$\sum_{o=1}^{N_b} F_{o'o}c_{oa} = \varepsilon_a\sum_{o=1}^{N_b} S_{o'o}c_{oa}$$

This expression has the form of a relation between matrix elements of the product matrices $\boldsymbol{Fc}$ and $\boldsymbol{Sc}$:

$$(\boldsymbol{Fc})_{o'a} = (\boldsymbol{Sc})_{o'a}\varepsilon_a$$

If we now introduce the diagonal matrix $\boldsymbol{\varepsilon}$ with the values of ε_a along its diagonal, this relation can be written as the matrix equality $\boldsymbol{Fc} = \boldsymbol{Sc\varepsilon}$, as in eqn 10.57.

• A brief illustration

To set up the Roothaan equations for the HF molecule using the $N_b = 2$ basis set H1s (χ_A) and F2p$_z$ (χ_B) we write the two molecular orbitals ($m = a, b$) as

$$\psi_a = c_{Aa}\chi_A + c_{Ba}\chi_B \qquad \psi_b = c_{Ab}\chi_A + c_{Bb}\chi_B$$

The matrix $\boldsymbol{c}$ is then $\boldsymbol{c} = \begin{pmatrix} c_{Aa} & c_{Ab} \\ c_{Ba} & c_{Bb}\end{pmatrix}$ and the overlap matrix is $\boldsymbol{S} = \begin{pmatrix} 1 & S \\ S & 1\end{pmatrix}$. The Fock matrix is

$$\boldsymbol{F} = \begin{pmatrix} F_{AA} & F_{AB} \\ F_{BA} & F_{BB}\end{pmatrix} \quad \text{with} \quad F_{o'o} = \int \chi_{o'}f_1\chi_o\mathrm{d}\tau_1$$

Then the Roothaan equations ($\boldsymbol{Fc} = \boldsymbol{Sc\varepsilon}$) are

$$\begin{pmatrix} F_{AA} & F_{AB} \\ F_{BA} & F_{BB}\end{pmatrix}\begin{pmatrix} c_{Aa} & c_{Ab} \\ c_{Ba} & c_{Bb}\end{pmatrix} = \begin{pmatrix} 1 & S \\ S & 1\end{pmatrix}\begin{pmatrix} c_{Aa} & c_{Ab} \\ c_{Ba} & c_{Bb}\end{pmatrix}\begin{pmatrix} \varepsilon_a & 0 \\ 0 & \varepsilon_b\end{pmatrix}$$

This matrix equation expands to four individual equations, one of which is

$$F_{AA}c_{Aa} + F_{AB}c_{Ba} = \varepsilon_a c_{Aa} + S\varepsilon_a c_{Ba}$$

and which constitute four simultaneous equations for the coefficients c just like the secular equations developed earlier (such as in eqns 10.27 or 10.44). One major difference, though, is that, because f_1 is defined in terms of the molecular orbitals, the F factors depend on the coefficients we are trying to find. We develop this expression below. •

A quick look at the form of the Fock matrix gives us an idea of the magnitude of the challenges associated with implementation of the Hartree–Fock method. It follows from eqn 10.62 that

$$F_{o'o} = \int \chi_{o'}(1)\left\{h_1 + \sum_m [2J_m(1) - K_m(1)]\right\}\chi_o(1)\mathrm{d}\tau_1$$

Suppose we focus on the term involving K; then from eqn 10.63 it follows that one contribution to F is

$$j_0\int \chi_{o'}(1)\left\{\int \psi_m(1)\frac{1}{r_{12}}\psi_m(2)\chi_o(2)\mathrm{d}\tau_2\right\}\mathrm{d}\tau_1$$

$$= j_0\int \chi_{o'}(1)\psi_m(1)\frac{1}{r_{12}}\psi_m(2)\chi_o(2)\mathrm{d}\tau_1\mathrm{d}\tau_2$$

where to get the term on the right we have simply rearranged some factors. Each molecular orbital ψ is a linear combination of atomic orbitals χ, so even this single contribution is a sum of terms that have the form

$$(AB|CD) = j_0\int A(1)B(1)\frac{1}{r_{12}}C(2)D(2)\,\mathrm{d}\tau_1\mathrm{d}\tau_2$$

where A, B, C, and D are atomic orbitals, as we encountered in eqn 10.58.

• **A brief illustration**

The term F_{AB} in the hydrogen fluoride calculation that we have been developing is

$$F_{AB}=\int \chi_A(1)h_1\chi_B(1)d\tau_1+2\int \chi_A(1)J_a(1)\chi_B(1)d\tau_1 - \int \chi_A(1)K_a(1)\chi_B(1)d\tau_1$$

because only ψ_a is occupied, so only $m=a$ contributes to the sum over m. We use the definition of J_m in eqn 10.63b to write the second term on the right as follows:

$$\begin{aligned}\int \chi_A(1)J_a(1)\chi_B(1)d\tau_1 &= j_0\int \chi_A(1)\int \chi_B(1)\frac{1}{r_{12}}(c_{Aa}\chi_A(2)\\ &\quad + c_{Ba}\chi_B(2))(c_{Aa}\chi_A(2)+c_{Ba}\chi_B(2))d\tau_2 d\tau_1\\ &= j_0c_{Aa}^2\int \chi_A(1)\chi_B(1)\frac{1}{r_{12}}\chi_A(2)\chi_A(2)d\tau_1 d\tau_2+\cdots\\ &= j_0c_{Aa}^2(AB|AA)+\cdots\end{aligned}$$

There are four such terms, and four more from K. We now see how the coefficients c also appear in the Fs that appear in the Roothaan equations, which makes them so difficult to solve and forces us to use self-consistent numerical methods. •

Discussion questions

10.1 Compare the approximations built into valence-bond theory and molecular orbital theory.

10.2 Discuss the steps involved in the construction of sp^3, sp^2, and sp hybrid orbitals.

10.3 Distinguish between the Pauling and Mulliken electronegativity scales.

10.4 Why is spin-pairing associated with bond formation? Discuss the concept in the context of valence-bond and molecular-orbital methods.

10.5 Discuss the approximations built into the Hückel method.

10.6 Distinguish between delocalization energy, π-electron binding energy, and π-bond formation energy.

10.7 Use concepts of molecular orbital theory to describe the biochemical reactivity of O_2, N_2, and NO.

10.8 Outline the steps involved in the Hartree–Fock method method for the calculation of molecular electronic structure.

10.9 Why are self-consistent field procedures used in computational chemistry?

Exercises

10.1(a) Write the VB spatial wavefunction for the bonds in H_2O using the basis H1s and O2p.

10.1(b) Write the VB spatial wavefunction for the bonds in H_2O_2 using the basis H1s and O2p.

10.2(a) Write the total VB wavefunction (including spin) for the bond in OH^- using the basis H1s and $O2p_z$.

10.2(b) Write the total VB wavefunction (including spin) for the bond in HF using the basis H1s and $F2p_z$.

10.3(a) Write the VB wavefunction for a CH_4 molecule using the sp^3 hybrid orbitals h on C and the four H1s orbitals.

10.3(b) Write the VB wavefunction for a BF_3 molecule using the sp^2 hybrid orbitals h on C and the three F2p orbitals.

10.4(a) Show that the sp^3 hybrid orbitals h_3 and h_4 in eqn 10.3 are mutually orthogonal.

10.4(b) Show that the sp^2 hybrid orbitals h_2 and h_3 in eqn 10.5 are mutually orthogonal.

10.5(a) Give the ground-state electron configurations and bond orders of (a) Li_2, (b) Be_2, and (c) C_2.

10.5(b) Give the ground-state electron configurations of (a) H_2^-, (b) N_2, and (c) O_2.

10.6(a) Give the ground-state electron configurations of (a) CO, (b) NO, and (c) CN^-.

10.6(b) Give the ground-state electron configurations of (a) ClF, (b) CS, and (c) O_2^-.

10.7(a) From the ground-state electron configurations of B_2 and C_2, predict which molecule should have the greater bond dissociation energy.

10.7(b) Which of the molecules N_2, NO, O_2, C_2, F_2, and CN would you expect to be stabilized by (a) the addition of an electron to form AB^-, (b) the removal of an electron to form AB^+?

10.8(a) Sketch the molecular orbital energy level diagram for XeF and deduce its ground-state electron configurations. Is XeF likely to have a shorter bond length than XeF^+?

10.8(b) Sketch the molecular orbital energy level diagrams for BrCl and deduce its ground-state electron configurations. Is BrCl likely to have a shorter bond length than $BrCl^-$?

10.9(a) Use the electron configurations of NO and N_2 to predict which is likely to have the shorter bond length.

10.9(b) Arrange the species O_2^+, O_2, O_2^-, O_2^{2-} in order of increasing bond length.

10.10(a) Show that a molecular orbital of the form $A\sin\theta+B\cos\theta$ is normalized to 1 if the orbitals A and B are each normalized to 1 and $S=0$. What linear combination of A and B is orthogonal to this combination?

10.10(b) Normalize the molecular orbital $\psi_A + \lambda\psi_B$ in terms of the parameter λ and the overlap integral S.

10.11(a) Confirm that the bonding and antibonding combinations $\psi_A \pm \psi_B$ are mutually orthogonal in the sense that their mutual overlap is zero.

10.11(b) Suppose that a molecular orbital has the form $N(0.145A + 0.844B)$. Find a linear combination of the orbitals A and B that is orthogonal to this combination.

10.12(a) What is the energy of an electron that has been ejected from an orbital of ionization energy 11.0 eV by a photon of radiation of wavelength 100 nm?

10.12(b) What is the energy of an electron that has been ejected from an orbital of ionization energy 4.69 eV by a photon of radiation of wavelength 584 pm?

10.13(a) An electron ejected from an orbital of a diatomic molecule by 21.22 eV radiation was found to have a speed of 1.90 Mm s^{-1}. To what ionization energy does that correspond?

10.13(b) An electron ejected from an orbital of a diatomic molecule by He(I) radiation was found to have a speed of 0.501 per cent the speed of light, c. To what ionization energy does that correspond?

10.14(a) The ionization energy of Xe5p and F2p electrons are 12.1 eV and 17.4 eV, respectively. Calculate the energies and composition of the bonding and antibonding orbitals of XeF. Use $\beta = -1.5$ eV and $S = 0$.

10.14(b) The ionization energy of Xe5p and O2p electrons are 12.1 eV and 13.6 eV, respectively. Calculate the energies and composition of the bonding and antibonding orbitals of XeO. Use $\beta = -1.2$ eV and $S = 0$.

10.15(a) Repeat Exercise 10.14a but with $S = 0.20$.

10.15(b) Repeat Exercise 10.14b but with $S = 0.20$.

10.16(a) Construct the molecular orbital energy level diagrams of ethene on the basis that the molecule is formed from the appropriately hybridized CH_2 or CH fragments.

10.16(b) Construct the molecular orbital energy level diagrams of ethyne (acetylene) on the basis that the molecule is formed from the appropriately hybridized CH_2 or CH fragments.

10.17(a) Write down the secular determinants for (a) linear H_3, (b) cyclic H_3 within the Hückel approximation. Estimate the binding energy in each case.

10.17(b) Write down the secular determinant for the allyl radical, CH_2=CH–CH_2 and estimate the π-binding energy.

10.18(a) Predict the electronic configurations of (a) the benzene anion, (b) the benzene cation. Estimate the π-electron binding energy in each case within the Hückel approximation. *Hint.* Use mathematical software.

10.18(b) Predict the electronic configurations of (a) the naphthalene anion, (b) the naphthalene cation. Estimate the π-electron binding energy in each case within the Hückel approximation. *Hint.* Use mathematical software.

10.19(a) Use mathematical software to estimate the π-electron binding energy of (a) anthracene (**8**), (b) phenanthrene (**9**) within the Hückel approximation.

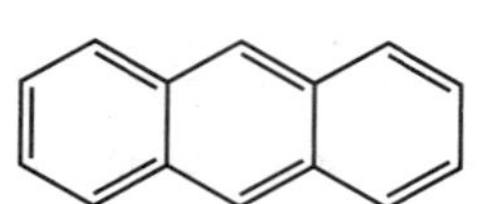

8 Anthracene

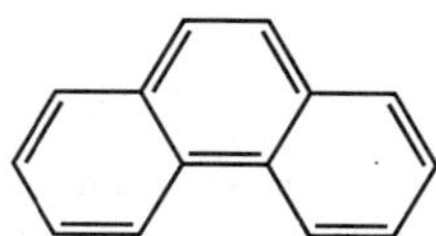

9 Phenanthrene

10.19(b) Use mathematical software to estimate the π-electron binding energy of azulene (**10**) within the Hückel approximation.

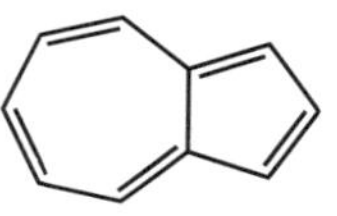

10 Azulene

Problems*

Numerical and graphical problems

10.1 Show graphically that, if a wave cos kx centred on A (so that x is measured from A) interferes with a similar wave cos $k'x$ centred on B (with x measured from B) a distance R away, then constructive interference occurs in the intermediate region when $k = k' = \pi/2R$ and destructive interference if $kR = \frac{1}{2}\pi$ and $k'R = \frac{3}{2}\pi$.

10.2 Before doing the calculation below, sketch how the overlap between a 1s orbital and a 2p orbital can be expected to depend on their separation. The overlap integral between an H1s orbital and an H2p orbital on nuclei separated by a distance R and forming a σ orbital is $S = \frac{1}{2}(R/a_0)\{1 + (R/a_0) + \frac{1}{3}(R/a_0)^2\}e^{-R/a_0}$. Plot this function, and find the separation for which the overlap is a maximum.

10.3 Calculate the total amplitude of the normalized bonding and antibonding LCAO-MOs that may be formed from two H1s orbitals at a separation of 106 pm. Plot the two amplitudes for positions along the molecular axis both inside and outside the internuclear region.

10.4 Repeat the calculation in Problem 10.3 but plot the probability densities of the two orbitals. Then form the difference density, the difference between ψ^2 and $\frac{1}{2}\{\psi_A^2 + \psi_B^2\}$.

10.5‡ Use the $2p_x$ and $2p_z$ hydrogenic atomic orbitals to construct simple LCAO descriptions of $2p\sigma$ and $2p\pi$ molecular orbitals. (a) Make a probability density plot, and both surface and contour plots of the xz-plane amplitudes of the $2p_z\sigma$ and $2p_z\sigma^*$ molecular orbitals. (b) Make surface and contour plots of the xz-plane amplitudes of the $2p_x\pi$ and $2p_x\pi^*$ molecular orbitals. Include plots for both internuclear distances, R, of $10a_0$ and $3a_0$, where $a_0 = 52.9$ pm. Interpret the graphs, and describe why this graphical information is useful.

10.6 Imagine a small electron-sensitive probe of volume 1.00 pm^3 inserted into an H_2^+ molecule-ion in its ground state. Calculate the probability that it

* Problems denoted with the symbol ‡ were supplied by Charles Trapp, Carmen Giunta, and Marshall Cady.

will register the presence of an electron at the following positions: (a) at nucleus A, (b) at nucleus B, (c) halfway between A and B, (c) at a point 20 pm along the bond from A and 10 pm perpendicularly. Do the same for the molecule-ion the instant after the electron has been excited into the antibonding LCAO-MO.

10.7‡ J.G. Dojahn *et al.* (*J. Phys. Chem.* **100**, 9649 (1996)) characterized the potential energy curves of homonuclear diatomic halogen molecules and molecular anions. Among the properties they report are the equilibrium internuclear distance R_e, the vibrational wavenumber, $\tilde{\nu}$, and the dissociation energy, D_e:

Species	R_e	$\tilde{\nu}/\text{cm}^{-1}$	D_e/eV
F_2	1.411	916.6	1.60
F_2^-	1.900	450.0	1.31

Rationalize these data in terms of molecular orbital configurations.

10.8 In a particular photoelectron spectrum using 21.21 eV photons, electrons were ejected with kinetic energies of 10.01 eV, 8.23 eV, and 5.22 eV. Sketch the molecular orbital energy level diagram for the species, showing the ionization energies of the three identifiable orbitals.

10.9‡ Set up and solve the Hückel secular equations for the π electrons of NO_3^-. Express the energies in terms of the Coulomb integrals α_O and α_N and the resonance integral β. Determine the delocalization energy of the ion.

10.10 In the 'free electron molecular orbital' (FEMO) theory, the electrons in a conjugated molecule are treated as independent particles in a box of length L. Sketch the form of the two occupied orbitals in butadiene predicted by this model and predict the minimum excitation energy of the molecule. The tetraene $CH_2{=}CHCH{=}CHCH{=}CHCH{=}CH_2$ can be treated as a box of length $8R$, where $R \approx 140$ pm (as in this case, an extra half bond-length is often added at each end of the box). Calculate the minimum excitation energy of the molecule and sketch the HOMO and LUMO. Estimate the colour a sample of the compound is likely to appear in white light.

10.11 The FEMO theory (Problem 10.10) of conjugated molecules is rather crude and better results are obtained with simple Hückel theory. (a) For a linear conjugated polyene with each of N_C carbon atoms contributing an electron in a 2p orbital, the energies E_k of the resulting π molecular orbitals are given by

$$E_k = \alpha + 2\beta\cos\frac{k\pi}{N_C+1} \qquad k = 1, 2, 3, \ldots, N_C$$

Use this expression to determine a reasonable empirical estimate of the resonance integral β for the homologous series consisting of ethene, butadiene, hexatriene, and octatetraene given that $\pi^* \leftarrow \pi$ ultraviolet absorptions from the HOMO to the LUMO occur at 61 500, 46 080, 39 750, and 32 900 cm^{-1}, respectively. (b) Calculate the π-electron delocalization energy, $E_{\text{deloc}} = E_\pi - N_\pi(\alpha + \beta)$, of octatetraene, where E_π is the total π-electron binding energy and N_π is the total number of π electrons. (c) In the context of this Hückel model, the π molecular orbitals are written as linear combinations of the carbon 2p orbitals. The coefficient of the jth atomic orbital in the kth molecular orbital is given by

$$c_{kj} = \left(\frac{2}{N_C+1}\right)^{1/2}\sin\frac{jk\pi}{N_C+1} \qquad j = 1, 2, 3, \ldots, N_C$$

Determine the values of the coefficients of each of the six 2p orbitals in each of the six π molecular orbitals of hexatriene. Match each set of coefficients (that is, each molecular orbital) with a value of the energy calculated with the expression given in part (a) of the molecular orbital. Comment on trends that relate the energy of a molecular orbital with its 'shape', which can be inferred from the magnitudes and signs of the coefficients in the linear combination that describes the molecular orbital.

10.12 For monocyclic conjugated polyenes (such as cyclobutadiene and benzene) with each of N carbon atoms contributing an electron in a 2p orbital, simple Hückel theory gives the following expression for the energies E_k of the resulting π molecular orbitals:

$$E_k = \alpha + 2\beta\cos\frac{2k\pi}{N_C} \qquad k = 0, \pm1, \pm2, \ldots, \pm N_C/2 \text{ (even } N\text{)}$$
$$k = 0, \pm1, \pm2, \ldots, \pm(N_C-1)/2 \text{ (odd } N\text{)}$$

(a) Calculate the energies of the π molecular orbitals of benzene and cyclooctatetraene. Comment on the presence or absence of degenerate energy levels. (b) Calculate and compare the delocalization energies of benzene (using the expression above) and hexatriene (see Problem 10.11a). What do you conclude from your results? (c) Calculate and compare the delocalization energies of cyclooctaene and octatetraene. Are your conclusions for this pair of molecules the same as for the pair of molecules investigated in part (b)?

10.13 Molecular orbital calculations based on semi-empirical, *ab initio*, and DFT methods describe the spectroscopic properties of conjugated molecules better than simple Hückel theory. (a) Using molecular modelling software and the computational method of your choice (semi-empirical, *ab initio*, or density functional methods), calculate the energy separation between the HOMO and LUMO of ethene, butadiene, hexatriene, and octatetraene. (b) Plot the HOMO–LUMO energy separations against the experimental frequencies for $\pi^* \leftarrow \pi$ ultraviolet absorptions for these molecules (Problem 10.11). Use mathematical software to find the polynomial equation that best fits the data. (c) Use your polynomial fit from part (b) to estimate the frequency of the $\pi^* \leftarrow \pi$ ultraviolet absorption of decapentaene from the calculated HOMO–LUMO energy separation. (d) Discuss why the calibration procedure of part (b) is necessary.

10.14 Electronic excitation of a molecule may weaken or strengthen some bonds because bonding and antibonding characteristics differ between the HOMO and the LUMO. For example, a carbon–carbon bond in a linear polyene may have bonding character in the HOMO and antibonding character in the LUMO. Therefore, promotion of an electron from the HOMO to the LUMO weakens this carbon–carbon bond in the excited electronic state, relative to the ground electronic state. Display the HOMO and LUMO of each molecule in Problem 10.13 and discuss in detail any changes in bond order that accompany the $\pi^* \leftarrow \pi$ ultraviolet absorptions in these molecules.

10.15 As mentioned in Section 2.8b, computational chemistry may be used to estimate the standard enthalpy of formation of molecules in the gas phase. (a) Using molecular modelling software and a semi-empirical method of your choice, calculate the standard enthalpy of formation of ethene, butadiene, hexatriene, and octatetraene in the gas phase. (b) Consult a database of thermochemical data, such as the online sources listed in this textbook's web site, and, for each molecule in part (a), calculate the relative error between the calculated and experimental values of the standard enthalpy of formation. (c) A good thermochemical database will also report the uncertainty in the experimental value of the standard enthalpy of formation. Compare experimental uncertainties with the relative errors calculated in part (b) and discuss the reliability of your chosen semi-empirical method for the estimation of thermochemical properties of linear polyenes.

Theoretical problems

10.16 Use hydrogenic atomic orbitals to write the explicit form of the sp^2 hybrid orbital h_2 in eqn 10.5. Determine the angle to the x-axis at which it has maximum amplitude.

10.17 Show that the sp^2 hybrids in eqn 10.5 make 120° to each other.

10.18 Derive eqns 10.12 and 10.15 by working with the normalized LCAO-MOs for the H_2^+ molecule-ion (Section 10.3a). Proceed by evaluating the expectation value of the hamiltonian for the ion. Make use of the fact that A and B each individually satisfy the Schrödinger equation for an isolated H atom.

10.19 Show that eqns 10.12 and 10.15 produce the result that

$$\Delta E = E_{2\sigma} - E_{1\sigma} = \frac{2k - 2Sj}{1 - S^2}$$

and go on to use the explicit expressions in eqn 10.13 to explore the range of internuclear separations over which $\Delta E > 0$.

10.20 Confirm the expressions for $\partial E/\partial c_A$ and $\partial E/\partial c_B$ derived in *Justification 10.3* (following eqn 10.30).

10.21 Show that if a matrix M can be written as $M = a\mathbf{1} + O$, where $\mathbf{1}$ is the unit matrix and O has off-diagonal elements, then to diagonalize M it is sufficient to diagonalize O. This result was used in Section 10.6b.

10.22 Show that the solutions of the secular determinant expression

$$\begin{vmatrix} \alpha_A - E & \beta \\ \beta & \alpha_B - E \end{vmatrix} = 0$$

for the orbital basis A, B can be written in terms of an angle θ, with

$$E_- = \alpha_B - \beta \tan\theta \qquad \psi_- = -A \sin\theta + B \cos\theta$$

$$E_+ = \alpha_A + \beta \tan\theta \qquad \psi_+ = A \cos\theta + B \sin\theta$$

and $\theta = \frac{1}{2}\arctan\{2\beta/(\alpha_B - \alpha_A)\}$.

10.23 We saw in the *brief illustration* in Section 10.7b that the product of two equivalent one-dimensional Gaussian functions is proportional to a Gaussian function. Repeat the calculation for a one-dimensional heteronuclear system.

10.24 Derive the three other equations for the HF molecule, the first of which is derived in the first *brief illustration* in *Further information* 10.1.

10.25 Derive the remaining terms for F_{AB}, the first of which is derived in the second brief illustration in *Further information* 10.1. Go on to identify equalities between the various integrals $(AB|CD)$ that you derive.

Applications: to astrophysics and biology

10.26‡ In Exercise 10.17a you were invited to set up the Hückel secular determinant for linear and cyclic H_3. The same secular determinant applies to the molecular ions H_3^+ and D_3^+. The molecular ion H_3^+ was discovered as long ago as 1912 by J.J. Thomson, but only more recently has the equivalent equilateral triangular structure been confirmed by M.J. Gaillard *et al.* (*Phys. Rev.* **A17**, 1797 (1978)). The molecular ion H_3^+ is the simplest polyatomic species with a confirmed existence and plays an important role in chemical reactions occurring in interstellar clouds that may lead to the formation of water, carbon monoxide, and ethyl alcohol. The H_3^+ ion has also been found in the atmospheres of Jupiter, Saturn, and Uranus. (a) Solve the Hückel secular equations for the energies of the H_3 system in terms of the parameters α and β, draw an energy level diagram for the orbitals, and determine the binding energies of H_3^+, H_3, and H_3^-. (b) Accurate quantum mechanical calculations by G.D. Carney and R.N. Porter (*J. Chem. Phys.* **65**, 3547 (1976)) give the dissociation energy for the process $H_3^+ \rightarrow H + H + H^+$ as 849 kJ mol^{-1}. From this information and data in Table 10.3, calculate the enthalpy of the reaction $H^+(g) + H_2(g) \rightarrow H_3^+(g)$. (c) From your equations and the information given, calculate a value for the resonance integral β in H_3^+. Then go on to calculate the binding energies of the other H_3 species in (a).

10.27‡ There is some indication that other hydrogen ring compounds and ions in addition to H_3 and D_3 species may play a role in interstellar chemistry. According to J.S. Wright and G.A. DiLabio (*J. Phys. Chem.* **96**, 10793 (1992)), H_5^-, H_6, and H_7^+ are particularly stable whereas H_4 and H_5^+ are not. Confirm these statements by Hückel calculations.

10.28 Here we develop a molecular orbital theory treatment of the peptide group –CONH–, which links amino acids in proteins. Specifically, we shall describe the factors that stabilize the planar conformation of the peptide group. (a) It will be familiar from introductory chemistry the planar conformation of the peptide group is explained by invoking delocalization of the π bond between the oxygen, carbon, and nitrogen atoms. It follows that we can model the peptide group with molecular orbital theory by making LCAO-MOs from 2p orbitals perpendicular to the plane defined by the O, C, and N atoms. The three combinations have the form:

$$\psi_1 = a\chi_O + b\chi_C + c\chi_N \qquad \psi_2 = d\chi_O - e\chi_N \qquad \psi_3 = f\chi_O - g\chi_C + h\chi_N$$

where the coefficients a through h are all positive. Sketch the orbitals ψ_1, ψ_2, and ψ_3 and characterize them as bonding, non-bonding, or antibonding molecular orbitals. In a non-bonding molecular orbital, a pair of electrons resides in an orbital confined largely to one atom and not appreciably involved in bond formation. (b) Show that this treatment is consistent only with a planar conformation of the peptide link. (c) Draw a diagram showing the relative energies of these molecular orbitals and determine the occupancy of the orbitals. *Hint.* Convince yourself that there are four electrons to be distributed among the molecular orbitals. (d) Now consider a non-planar conformation of the peptide link, in which the O2p and C2p orbitals are perpendicular to the plane defined by the O, C, and N atoms, but the N2p orbital lies on that plane. The LCAO-MOs are given by

$$\psi_4 = a\chi_O + b\chi_C \qquad \psi_5 = e\chi_N \qquad \psi_6 = f\chi_O - g\chi_C$$

Just as before, sketch these molecular orbitals and characterize them as bonding, non-bonding, or antibonding. Also, draw an energy level diagram and determine the occupancy of the orbitals. (e) Why is this arrangement of atomic orbitals consistent with a non-planar conformation for the peptide link? (f) Does the bonding MO associated with the planar conformation have the same energy as the bonding MO associated with the non-planar conformation? If not, which bonding MO is lower in energy? Repeat the analysis for the non-bonding and anti-bonding molecular orbitals. (g) Use your results from parts (a)–(f) to construct arguments that support the planar model for the peptide link.

10.29 Molecular orbital calculations may be used to predict trends in the standard potentials of conjugated molecules, such as the quinones and flavins, that are involved in biological electron transfer reactions. It is commonly assumed that decreasing the energy of the LUMO enhances the ability of a molecule to accept an electron into the LUMO, with an attendant increase in the value of the molecule's standard potential. Furthermore, a number of studies indicate that there is a linear correlation between the LUMO energy and the reduction potential of aromatic hydrocarbons. (a) The standard potentials at pH = 7 for the one-electron reduction of methyl-substituted 1,4-benzoquinones (**11**) to their respective semiquinone radical anions are:

11

R_2	R_3	R_5	R_6	$E^{\ominus}$/V
H	H	H	H	0.078
CH_3	H	H	H	0.023
CH_3	H	CH_3	H	−0.067
CH_3	CH_3	CH_3	H	−0.165
CH_3	CH_3	CH_3	CH_3	−0.260

Using molecular modelling software and the computational method of your choice (semi-empirical, *ab initio*, or density functional theory methods), calculate E_{LUMO}, the energy of the LUMO of each substituted 1,4-benzoquinone, and plot E_{LUMO} against $E^{\ominus}$. Do your calculations support a linear relation between E_{LUMO} and $E^{\ominus}$? (b) The 1,4-benzoquinone for which $R_2 = R_3 = CH_3$ and $R_5 = R_6 = OCH_3$ is a suitable model of ubiquinone, a component of the respiratory electron transport chain. Determine E_{LUMO} of this quinone and then use your results from part (a) to estimate its standard potential. (c) The 1,4-benzoquinone for which $R_2 = R_3 = R_5 = CH_3$ and $R_6 = H$ is a suitable model of plastoquinone, a component of the photosynthetic electron transport chain. Determine E_{LUMO} of this quinone and then use your results from part (a) to estimate its standard potential. Is plastoquinone expected to be a better or worse oxidizing agent than ubiquinone? (d) Based on your predictions and on basic concepts of biological electron transport, suggest a reason why ubiquinone is used in respiration and plastoquinone is used in photosynthesis.

MATHEMATICAL BACKGROUND 6

Matrices

A **matrix** is an array of numbers that are generalizations of ordinary numbers. We shall consider only square matrices, which have the numbers arranged in the same number of rows and columns By using matrices, we can manipulate large numbers of ordinary numbers simultaneously. A **determinant** is a particular combination of the numbers that appear in a matrix and is used to manipulate the matrix.

Matrices may be combined together by addition or multiplication according to generalizations of the rules for ordinary numbers. Although we describe below the key algebraic procedures involving matrices, it is important to note that most numerical matrix manipulations are now carried out with mathematical software. You are encouraged to use such software, if it is available to you.

MB6.1 Definitions

Consider a square matrix M of n^2 numbers arranged in n columns and n rows. These n^2 numbers are the **elements** of the matrix, and may be specified by stating the row, r, and column, c, at which they occur. Each element is therefore denoted M_{rc}. A **diagonal matrix** is a matrix in which the only nonzero elements lie on the major diagonal (the diagonal from M_{11} to M_{nn}). Thus, the matrix

$$M=\begin{pmatrix}1 & 0 & 0\\ 0 & 2 & 0\\ 0 & 0 & 1\end{pmatrix}$$

is a 3 × 3 diagonal square matrix. The condition may be written

$$M_{rc}=m_r\delta_{rc} \qquad \text{(MB6.1)}$$

where δ_{rc} is the **Kronecker delta**, which is equal to 1 for $r=c$ and to 0 for $r\neq c$. In the above example, $m_1=1$, $m_2=2$, and $m_3=1$. The **unit matrix**, **1** (and occasionally **I**), is a special case of a diagonal matrix in which all nonzero elements are 1.

The **transpose** of a matrix M is denoted M^{T} and is defined by

$$M^{\mathrm{T}}_{mn}=M_{nm} \qquad \boxed{\text{Transpose}} \qquad \text{(MB6.2)}$$

That is, the element in row n, column m of the original matrix becomes the element in row m, column n of the transpose (in effect, the elements are reflected across the diagonal). The **determinant**, $|M|$, of the matrix M is a real number arising from a specific procedure for taking sums and differences of products of matrix elements. For example, a 2 × 2 determinant is evaluated as

$$\begin{vmatrix}a & b\\ c & d\end{vmatrix}=ad-bc \qquad \boxed{\text{Determinant}} \qquad \text{(MB6.3a)}$$

and a 3 × 3 determinant is evaluated by expanding it as a sum of 2 × 2 determinants:

$$\begin{vmatrix}a & b & c\\ d & e & f\\ g & h & i\end{vmatrix}=a\begin{vmatrix}e & f\\ h & i\end{vmatrix}-b\begin{vmatrix}d & f\\ g & i\end{vmatrix}+c\begin{vmatrix}d & e\\ g & h\end{vmatrix} \qquad \text{(MB6.3b)}$$

$$=a(ei-fh)-b(di-fg)+c(dh-eg)$$

Note the sign change in alternate columns (b occurs with a negative sign in the expansion). An important property of a determinant is that, if any two rows or any two columns are interchanged, then the determinant changes sign.

- **A brief illustration**

The matrix

$$M=\begin{pmatrix}1 & 2\\ 3 & 4\end{pmatrix}$$

is a 2 × 2 matrix with the elements $M_{11}=1$, $M_{12}=2$, $M_{21}=3$, and $M_{22}=4$. Its transpose is

$$M^{\mathrm{T}}=\begin{pmatrix}1 & 3\\ 2 & 4\end{pmatrix}$$

and its determinant is

$$|M|=\begin{vmatrix}1 & 2\\ 3 & 4\end{vmatrix}=1\times 4-2\times 3=-2$$ •

MB6.2 Matrix addition and multiplication

Two matrices M and N may be added to give the sum $S=M+N$, according to the rule

$$S_{rc}=M_{rc}+N_{rc} \qquad \boxed{\text{Matrix addition}} \qquad \text{(MB6.4)}$$

That is, corresponding elements are added.

Two matrices may also be multiplied to give the product $P=MN$ according to the rule

$$P_{rc}=\sum_n M_{rn}N_{nc} \qquad \boxed{\text{Matrix multiplication}} \qquad \text{(MB6.5)}$$

These procedures are illustrated in Fig. MB6.1. It should be noticed that in general $MN\neq NM$, and matrix multiplication is in general non-commutative (that is, depends on the order of multiplication).

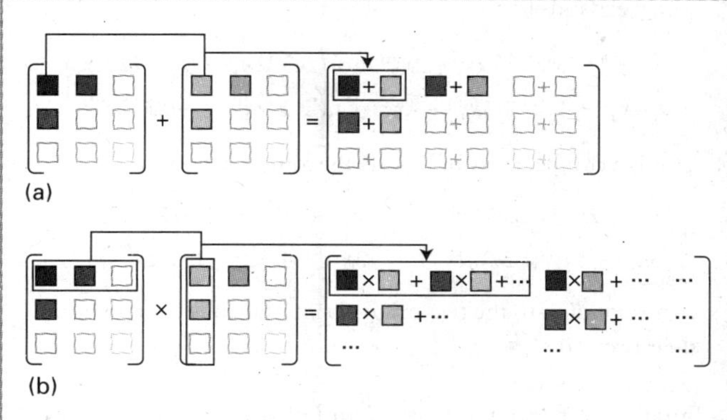

Fig. MB6.1 A diagrammatic representation of (a) matrix addition, (b) matrix multiplication.

• **A brief illustration**

Consider the matrices

$$M=\begin{pmatrix}1 & 2\\ 3 & 4\end{pmatrix} \quad \text{and} \quad N=\begin{pmatrix}5 & 6\\ 7 & 8\end{pmatrix}$$

Their sum is

$$S=\begin{pmatrix}1 & 2\\ 3 & 4\end{pmatrix}+\begin{pmatrix}5 & 6\\ 7 & 8\end{pmatrix}=\begin{pmatrix}6 & 8\\ 10 & 12\end{pmatrix}$$

and their product is

$$P=\begin{pmatrix}1 & 2\\ 3 & 4\end{pmatrix}\begin{pmatrix}5 & 6\\ 7 & 8\end{pmatrix}=\begin{pmatrix}1\times5+2\times7 & 1\times6+2\times8\\ 3\times5+4\times7 & 3\times6+4\times8\end{pmatrix}$$

$$=\begin{pmatrix}19 & 22\\ 43 & 50\end{pmatrix} \bullet$$

The **inverse** of a matrix M is denoted M^{-1}, and is defined so that

$$MM^{-1}=M^{-1}M=1 \qquad \text{Inverse} \qquad \text{(MB6.6)}$$

The inverse of a matrix is best constructed by using mathematical software and the tedious analytical approach is rarely necessary.

• **A brief illustration**

Consider the matrix M from the first *brief illustration* in this section. Mathematical software gives the following result:

$$M^{-1}=\begin{pmatrix}-2 & 1\\ \frac{3}{2} & -\frac{1}{2}\end{pmatrix} \bullet$$

MB6.3 Eigenvalue equations

An **eigenvalue equation** is an equation of the form

$$Mx=\lambda x \qquad \text{Eigenvalue equation} \qquad \text{(MB6.7a)}$$

where M is a square matrix with n rows and n columns, λ is a constant, the **eigenvalue**, and x is the **eigenvector**, an $n \times 1$ (column) matrix that satisfies the conditions of the eigenvalue equation and has the form:

$$x=\begin{pmatrix}x_1\\ x_2\\ \vdots\\ x_n\end{pmatrix}$$

In general, there are n eigenvalues $\lambda^{(i)}$, $i = 1, 2, \ldots n$, and n corresponding eigenvectors $x^{(i)}$. We write eqn MB6.7a as (noting that $1x=x$)

$$(M-\lambda 1)x=0 \qquad \text{(MB6.7b)}$$

Equation MB6.7b has a solution only if the determinant $|M - \lambda 1|$ of the coefficients of the matrix $M - \lambda 1$ is zero. It follows that the n eigenvalues may be found from the solution of the **secular equation:**

$$|M-\lambda 1|=0 \qquad \text{(MB6.8)}$$

A brief comment

If the inverse of the matrix $M-\lambda 1$ exists, then, from eqn MB6.7b, $(M-\lambda 1)^{-1}(M-\lambda 1)x=x=0$, a trivial solution. For a nontrivial solution, $(M-\lambda 1)^{-1}$ must not exist, which is the case if eqn MB6.8 holds.

• **A brief illustration**

Once again we use the matrix M in the first *brief illustration*, and write eqn MB6.7 as

$$\begin{pmatrix}1 & 2\\ 3 & 4\end{pmatrix}\begin{pmatrix}x_1\\ x_2\end{pmatrix}=\lambda\begin{pmatrix}x_1\\ x_2\end{pmatrix} \qquad \text{rearranged into}$$

$$\begin{pmatrix}1-\lambda & 2\\ 3 & 4-\lambda\end{pmatrix}\begin{pmatrix}x_1\\ x_2\end{pmatrix}=0$$

From the rules of matrix multiplication, the latter form expands into

$$\begin{pmatrix}(1-\lambda)x_1+2x_2\\ 3x_1+(4-\lambda)x_2\end{pmatrix}=0$$

which is simply a statement of the two simultaneous equations

$$(1-\lambda)x_1+2x_2=0 \text{ and } 3x_1+(4-\lambda)x_2=0$$

The condition for these two equations to have solutions is

$$|M-\lambda 1|=\begin{vmatrix}1-\lambda & 2\\ 3 & 4-\lambda\end{vmatrix}=(1-\lambda)(4-\lambda)-6=0$$

This condition corresponds to the quadratic equation

$$\lambda^2-5\lambda-2=0$$

with solutions $\lambda=+5.372$ and $\lambda=-0.372$, the two eigenvalues of the original equation. •

The n eigenvalues found by solving the secular equations are used to find the corresponding eigenvectors. To do so, we begin by considering an $n \times n$ matrix X which will be formed from the eigenvectors corresponding to all the eigenvalues. Thus, if the eigenvalues are $\lambda_1, \lambda_2, \ldots$, and the corresponding eigenvectors are

$$\boldsymbol{x}^{(1)} = \begin{pmatrix} x_1^{(1)} \\ x_2^{(1)} \\ \vdots \\ x_n^{(1)} \end{pmatrix} \qquad \boldsymbol{x}^{(2)} = \begin{pmatrix} x_1^{(2)} \\ x_2^{(2)} \\ \vdots \\ x_n^{(2)} \end{pmatrix}, \text{ etc.} \tag{MB6.9a}$$

the matrix X is

$$X = (\boldsymbol{x}^{(1)}, \boldsymbol{x}^{(2)}, \ldots, \boldsymbol{x}^{(n)}) = \begin{pmatrix} x_1^{(1)} & x_1^{(2)} & \cdots & x_1^{(n)} \\ x_2^{(1)} & x_2^{(2)} & \cdots & x_2^{(n)} \\ \vdots & \vdots & & \vdots \\ x_n^{(1)} & x_n^{(2)} & \cdots & x_n^{(n)} \end{pmatrix} \tag{MB6.9b}$$

Similarly, we form an $n \times n$ matrix $\boldsymbol{\Lambda}$ with the eigenvalues λ along the diagonal and zeroes elsewhere:

$$\boldsymbol{\Lambda} = \begin{pmatrix} \lambda_1 & 0 & \cdots & 0 \\ 0 & \lambda_2 & \cdots & 0 \\ \vdots & \vdots & & \vdots \\ 0 & 0 & \cdots & \lambda_n \end{pmatrix} \tag{MB6.10}$$

Now all the eigenvalue equations $M\boldsymbol{x}^{(i)} = \lambda_i \boldsymbol{x}^{(i)}$ may be confined into the single matrix equation

$$MX = X\Lambda \tag{MB6.11}$$

- **A brief illustration**

In the preceding *brief illustration* we established that if $M = \begin{pmatrix} 1 & 2 \\ 3 & 4 \end{pmatrix}$ then $\lambda_1 = +5.372$ and $\lambda_2 = -0.372$, with eigenvectors

$$\boldsymbol{x}^{(1)} = \begin{pmatrix} x_1^{(1)} \\ x_2^{(1)} \end{pmatrix} \quad \text{and} \quad \boldsymbol{x}^{(2)} = \begin{pmatrix} x_1^{(2)} \\ x_2^{(2)} \end{pmatrix}, \quad \text{respectively.}$$

We form

$$X = \begin{pmatrix} x_1^{(1)} & x_1^{(2)} \\ x_2^{(1)} & x_2^{(2)} \end{pmatrix} \qquad \Lambda = \begin{pmatrix} 5.372 & 0 \\ 0 & -0.372 \end{pmatrix}$$

The expression $MX = X\Lambda$ becomes

$$\begin{pmatrix} 1 & 2 \\ 3 & 4 \end{pmatrix} \begin{pmatrix} x_1^{(1)} & x_1^{(2)} \\ x_2^{(1)} & x_2^{(2)} \end{pmatrix} = \begin{pmatrix} x_1^{(1)} & x_1^{(2)} \\ x_2^{(1)} & x_2^{(2)} \end{pmatrix} \begin{pmatrix} 5.372 & 0 \\ 0 & -0.372 \end{pmatrix}$$

which expands to

$$\begin{pmatrix} x_1^{(1)} + 2x_2^{(1)} & x_1^{(2)} + 2x_2^{(2)} \\ 3x_1^{(1)} + 4x_2^{(1)} & 3x_1^{(2)} + 4x_2^{(2)} \end{pmatrix} = \begin{pmatrix} 5.372x_1^{(1)} & -0.372x_1^{(2)} \\ 5.372x_2^{(1)} & -0.372x_2^{(2)} \end{pmatrix}$$

This is a compact way of writing the four equations

$$x_1^{(1)} + 2x_2^{(1)} = 5.372x_1^{(1)} \qquad x_1^{(2)} + 2x_2^{(2)} = -0.372x_1^{(2)}$$
$$3x_1^{(1)} + 4x_2^{(1)} = 5.372x_2^{(1)} \qquad 3x_1^{(2)} + 4x_2^{(2)} = -0.372x_2^{(2)}$$

corresponding to the two original simultaneous equations and their two roots. •

Finally, we form X^{-1} from X and multiply eqn MB6.11 by it from the left:

$$X^{-1}MX = X^{-1}X\Lambda = \Lambda \tag{MB6.12}$$

A structure of the form $X^{-1}MX$ is called a **similarity transformation**. In this case the similarity transformation $X^{-1}MX$ makes M diagonal (because Λ is diagonal). It follows that, if the matrix X that causes $X^{-1}MX$ to be diagonal is known, then the problem is solved: the diagonal matrix so produced has the eigenvalues as its only nonzero elements, and the matrix X used to bring about the transformation has the corresponding eigenvectors as its columns. As will be appreciated once again, the solutions of eigenvalue equations are best found by using mathematical software.

- **A brief illustration**

To apply the similarity transformation, eqn MB6.12, to the matrix $\begin{pmatrix} 1 & 2 \\ 3 & 4 \end{pmatrix}$ from the preceding *brief illustration* it is best to use mathematical software to find the form of X. The result is

$$X = \begin{pmatrix} 0.416 & 0.825 \\ 0.909 & -0.566 \end{pmatrix}$$

This result can be verified by carrying out the multiplication

$$X^{-1}MX = \begin{pmatrix} 0.574 & 0.837 \\ 0.922 & -0.422 \end{pmatrix} \begin{pmatrix} 1 & 2 \\ 3 & 4 \end{pmatrix} \begin{pmatrix} 0.416 & 0.825 \\ 0.909 & -0.566 \end{pmatrix}$$
$$= \begin{pmatrix} 5.372 & 0 \\ 0 & -0.372 \end{pmatrix}$$

The result is indeed the diagonal matrix Λ calculated in the preceding *brief illustration*. It follows that the eigenvectors $\boldsymbol{x}^{(1)}$ and $\boldsymbol{x}^{(2)}$ are

$$\boldsymbol{x}^{(1)} = \begin{pmatrix} 0.416 \\ 0.909 \end{pmatrix} \quad \text{and} \quad \boldsymbol{x}^{(2)} = \begin{pmatrix} 0.825 \\ -0.566 \end{pmatrix} \quad \bullet$$

Molecular symmetry

11

In this chapter we sharpen the concept of 'shape' into a precise definition of 'symmetry', and show that symmetry may be discussed systematically. We see how to classify any molecule according to its symmetry and how to use this classification to discuss molecular properties. After describing the symmetry properties of molecules themselves, we turn to a consideration of the effect of symmetry transformations on orbitals and see that their transformation properties can be used to set up a labelling scheme. These symmetry labels are used to identify integrals that necessarily vanish. One important integral is the overlap integral between two orbitals. By knowing which atomic orbitals may have nonzero overlap, we can decide which ones can contribute to molecular orbitals. We also see how to select linear combinations of atomic orbitals that match the symmetry of the nuclear framework. Finally, by considering the symmetry properties of integrals, we see that it is possible to derive the selection rules that govern spectroscopic transitions.

The systematic discussion of symmetry is called **group theory**. Much of group theory is a summary of common sense about the symmetries of objects. However, because group theory is systematic, its rules can be applied in a straightforward, mechanical way. In most cases the theory gives a simple, direct method for arriving at useful conclusions with the minimum of calculation, and this is the aspect we stress here. In some cases, though, it leads to unexpected results.

The symmetry elements of objects

Some objects are 'more symmetrical' than others. A sphere is more symmetrical than a cube because it looks the same after it has been rotated through any angle about any diameter. A cube looks the same only if it is rotated through certain angles about specific axes, such as 90°, 180°, or 270° about an axis passing through the centres of any of its opposite faces (Fig. 11.1), or by 120° or 240° about an axis passing through any of its opposite corners. Similarly, an NH_3 molecule is 'more symmetrical' than an H_2O molecule because NH_3 looks the same after rotations of 120° or 240° about the axis shown in Fig. 11.2, whereas H_2O looks the same only after a rotation of 180°.

An action that leaves an object looking the same after it has been carried out is called a **symmetry operation**. Typical symmetry operations include rotations, reflections, and inversions. There is a corresponding **symmetry element** for each symmetry operation, which is the point, line, or plane with respect to which the symmetry operation is performed. For instance, a rotation (a symmetry operation) is carried out around an axis (the corresponding symmetry element). We shall see that we can classify molecules by identifying all their symmetry elements, and grouping together

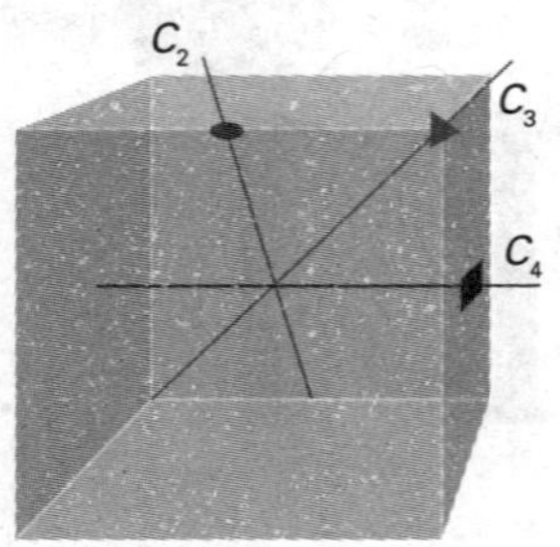

Fig. 11.1 Some of the symmetry elements of a cube. The twofold, threefold, and fourfold axes are labelled with the conventional symbols.

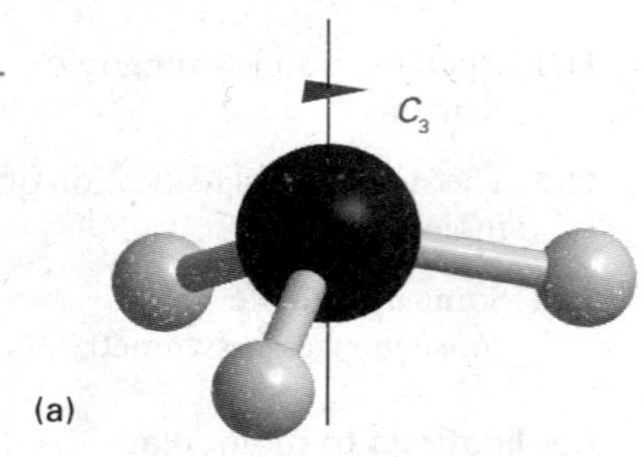

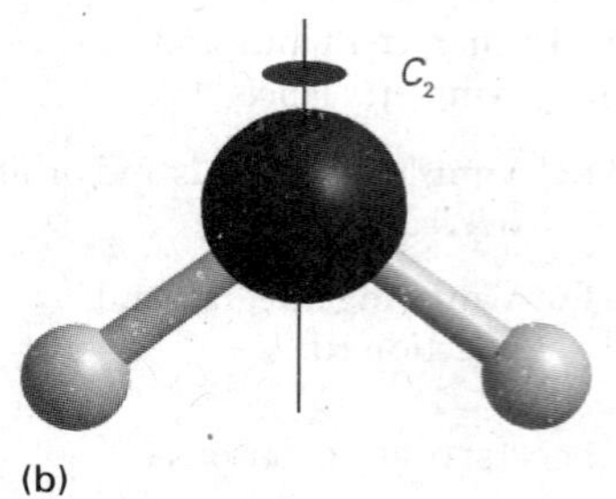

Fig. 11.2 (a) An NH_3 molecule has a threefold (C_3) axis and (b) an H_2O molecule has a twofold (C_2) axis. Both have other symmetry elements too.

molecules that possess the same set of symmetry elements. This procedure, for example, puts the trigonal pyramidal species NH_3 and SO_3^{2-} into one group and the angular species H_2O and SO_2 into another group.

11.1 Operations and symmetry elements

Key points **(a) Group theory is concerned with symmetry operations and the symmetry elements with which they are associated; point groups are composed of symmetry operations that preserve a single point. (b) A set of operations form a group if they satisfy certain criteria.**

The classification of objects according to symmetry elements corresponding to operations that leave at least one common point unchanged gives rise to the **point groups**. There are five kinds of symmetry operation (and five kinds of symmetry element) of this kind. When we consider crystals (Chapter 19), we shall meet symmetries arising from translation through space. These more extensive groups are called **space groups**.

(a) Notation

The **identity**, E, consists of doing nothing; the corresponding symmetry element is the entire object. Because every molecule is indistinguishable from itself if nothing is done to it, every object possesses at least the identity element. One reason for including the identity is that some molecules have only this symmetry element (**1**); another reason is technical and connected with the detailed formulation of group theory.

An ***n*-fold rotation** (the operation) about an ***n*-fold axis of symmetry**, C_n (the corresponding element) is a rotation through $360°/n$. The operation C_1 is a rotation through 360°, and is equivalent to the identity operation E. An H_2O molecule has one twofold axis, C_2. There is only one twofold rotation associated with a C_2 axis because clockwise and counterclockwise 180° rotations have an identical outcome. An NH_3 molecule has one threefold axis, C_3, with which is associated two symmetry operations, one being 120° rotation in a clockwise sense and the other 120° rotation in a counterclockwise sense. A pentagon has a C_5 axis, with two (clockwise and counterclockwise) rotations through 72° associated with it. It also has an axis denoted C_5^2, corresponding to two successive C_5 rotations; there are two such operations, one through 144° in a clockwise sense and the other through 144° in a counterclockwise sense. A cube has three C_4 axes, four C_3 axes, and six C_2 axes. However, even this high symmetry is exceeded by a sphere, which possesses an infinite number of symmetry axes (along any diameter) of all possible integral values of n. If a molecule possesses several rotation axes, then the one (or more) with the greatest value of n is called the **principal axis**. The principal axis of a benzene molecule is the sixfold axis perpendicular to the hexagonal ring (**2**).

A **reflection** (the operation) in a **mirror plane**, σ (the element), may contain the principal axis of a molecule or be perpendicular to it. If the plane is parallel to the principal axis, it is called 'vertical' and denoted σ_v. An H_2O molecule has two vertical

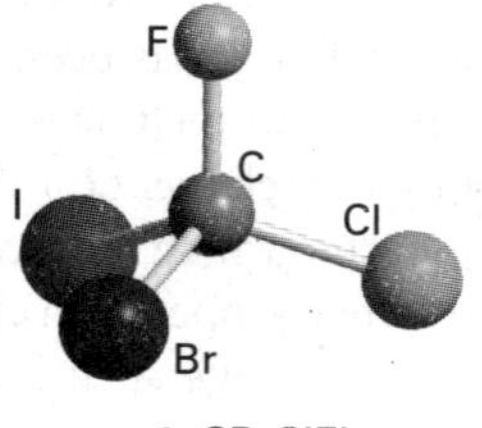

1 CBrClFI

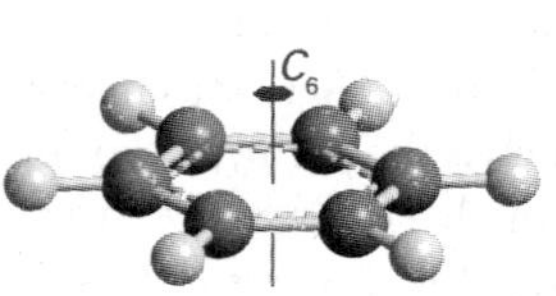

2 Benzene, C_6H_6

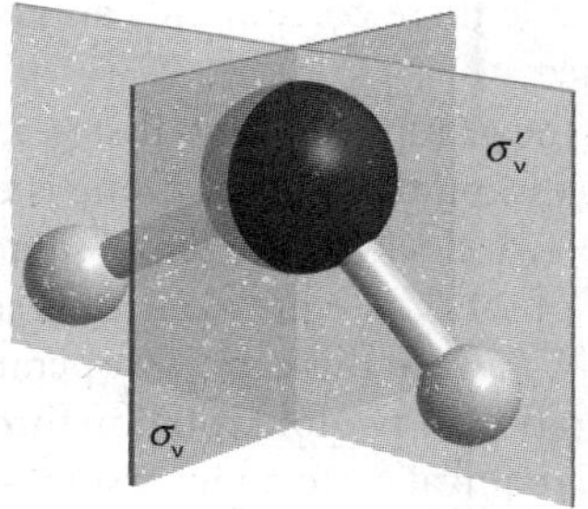

Fig. 11.3 An H_2O molecule has two mirror planes. They are both vertical (i.e. contain the principal axis), so are denoted σ_v and σ'_v.

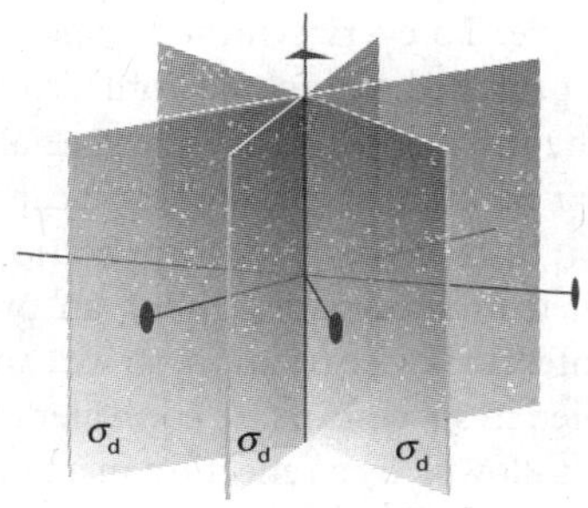

Fig. 11.4 Dihedral mirror planes (σ_d) bisect the C_2 axes perpendicular to the principal axis.

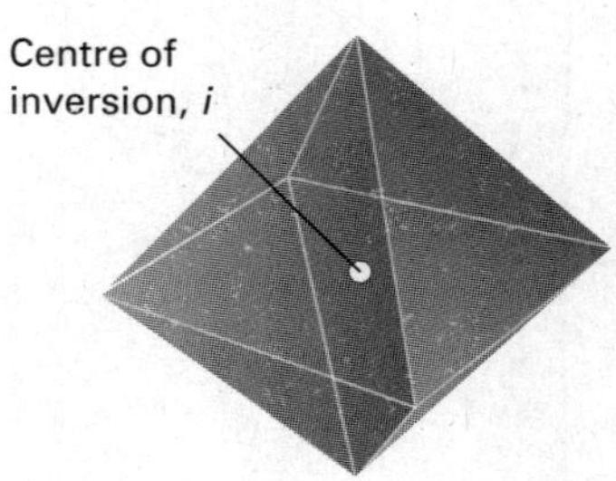

Fig. 11.5 A regular octahedron has a centre of inversion (i).

planes of symmetry (Fig. 11.3) and an NH_3 molecule has three. A vertical mirror plane that bisects the angle between two C_2 axes is called a 'dihedral plane' and is denoted σ_d (Fig. 11.4). When the plane of symmetry is perpendicular to the principal axis it is called 'horizontal' and denoted σ_h. A C_6H_6 molecule has a C_6 principal axis and a horizontal mirror plane (as well as several other symmetry elements).

In an **inversion** (the operation) through a **centre of symmetry**, i (the element), we imagine taking each point in a molecule, moving it to the centre of the molecule, and then moving it out the same distance on the other side; that is, the point (x, y, z) is taken into the point $(-x, -y, -z)$. Neither an H_2O molecule nor an NH_3 molecule has a centre of inversion, but a sphere and a cube do have one. A C_6H_6 molecule does have a centre of inversion, as does a regular octahedron (Fig. 11.5); a regular tetrahedron and a CH_4 molecule do not.

An ***n*-fold improper rotation** (the operation) about an ***n*-fold axis of improper rotation** or an ***n*-fold improper rotation axis**, S_n (the symmetry element), is composed of two successive transformations, neither of which alone is necessarily a symmetry operation. The first component is a rotation through $360°/n$, and the second is a reflection through a plane perpendicular to the axis of that rotation; neither operation alone needs to be a symmetry operation. A CH_4 molecule has three S_4 axes (Fig. 11.6).

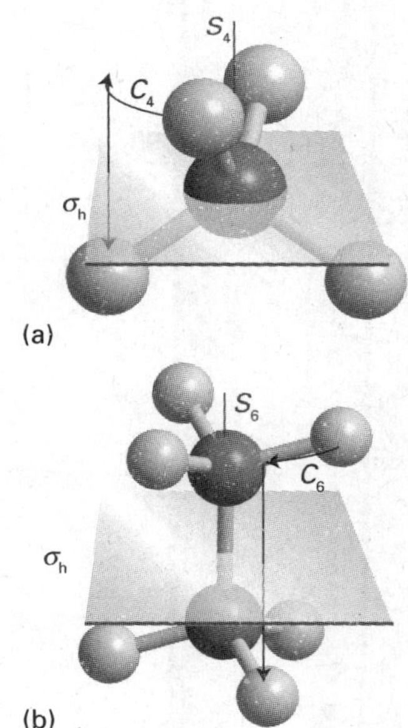

Fig. 11.6 (a) A CH_4 molecule has a fourfold improper rotation axis (S_4): the molecule is indistinguishable after a 90° rotation followed by a reflection across the horizontal plane, but neither operation alone is a symmetry operation. (b) The staggered form of ethane has an S_6 axis composed of a 60° rotation followed by a reflection.

(b) The criteria for being a group

In mathematics, a 'group' has a special meaning and is the basis of the name 'group theory' for the quantitative description of symmetry. A set of operations constitute a **group** if they satisfy the following criteria:

- The identity operation is a member of the set.
- The inverse of each operation is a member of the set.
- If R and S are members of the set, then the operation RS is also a member.

These criteria are satisfied by a large number of objects, but our concern is with symmetry operations, and we confine our remarks to them.

It is quite easy to see that the symmetry operations of a molecule fulfil the criteria that let them qualify as a group. First, we have seen that every molecule possesses the identity operation E. To judge whether the inverse of a symmetry operation is always present we need to note whether for each operation we can find another operation (or the same operation) that brings the molecule back to its original state. A reflection applied twice in succession (which we denote $\sigma\sigma$) is one example. A clockwise n-fold rotation followed by a counterclockwise n-fold rotation (denoted $C_n^-C_n^+$) is another example. To every symmetry operation of a molecule there corresponds an inverse and, provided we include both, criterion 2 is satisfied.

The third criterion is very special, and is called the **group property**. It states that, if two symmetry operations are carried out in succession, then the outcome is equivalent to a *single* symmetry operation. For example, two clockwise threefold rotations applied in succession, giving an overall rotation of 240°, is equivalent to a single counterclockwise rotation, so we can write $C_3^+C_3^+ = C_3^-$ and in this case two operations applied in succession are equivalent to a single operation. A twofold rotation through 180° followed by a reflection in a horizontal plane is equivalent to an inversion, so we can write $\sigma_h C_2 = i$. Once again, we see that successive operations are equivalent to a single operation, as criterion 3 requires.

All the symmetry operations of molecules satisfy the three criteria for them constituting a group, so we are justified in calling the theory of symmetry 'group theory' and using the powerful apparatus that mathematicians have assembled.

11.2 The symmetry classification of molecules

Key point **Molecules are classified according to the symmetry elements they possess.**

To classify molecules according to their symmetries, we list their symmetry elements and collect together molecules with the same list of elements. This procedure puts CH_4 and CCl_4, which both possess the same symmetry elements as a regular tetrahedron, into the same group, and H_2O into another group.

The name of the group to which a molecule belongs is determined by the symmetry elements it possesses. There are two systems of notation (Table 11.1). The **Schoenflies system** (in which a name looks like C_{4v}) is more common for the discussion of individual molecules, and the **Hermann–Mauguin system**, or **International system** (in which a name looks like $4mm$), is used almost exclusively in the discussion of crystal symmetry. The identification of a molecule's point group according to the Schoenflies system, which we outline below, is simplified by referring to the flow diagram in Fig. 11.7 and the shapes shown in Fig. 11.8.

Table 11.1 The notation for point groups*

C_i	$\bar{1}$										
C_s	m										
C_1	1	C_2	2	C_3	3	C_4	4	C_6	6		
		C_{2v}	$2mm$	C_{3v}	$3m$	C_{4v}	$4mm$	C_{6v}	$6mm$		
		C_{2h}	$2/m$	C_{3h}	$\bar{6}$	C_{4h}	$4/m$	C_{6h}	$6/m$		
		D_2	222	D_3	32	D_4	422	D_6	622		
		D_{2h}	mmm	D_{3h}	$\bar{6}2m$	D_{4h}	$4/mmm$	D_{6h}	$6/mmm$		
		D_{2d}	$\bar{4}2m$	D_{3d}	$\bar{3}m$	S_4	$\bar{4}$	S_6	$\bar{3}$		
T	23	T_d	$\bar{4}3m$	T_h	$m3$						
O	432	O_h	$m3m$								

* In the International system (or Hermann–Mauguin system) for point groups, a number n denotes the presence of an n-fold axis and m denotes a mirror plane. A slash (/) indicates that the mirror plane is perpendicular to the symmetry axis. It is important to distinguish symmetry elements of the same type but of different classes, as in $4/mmm$, in which there are three classes of mirror plane. A bar over a number indicates that the element is combined with an inversion. The only groups listed here are the so-called 'crystallographic point groups' (Section 19.1).

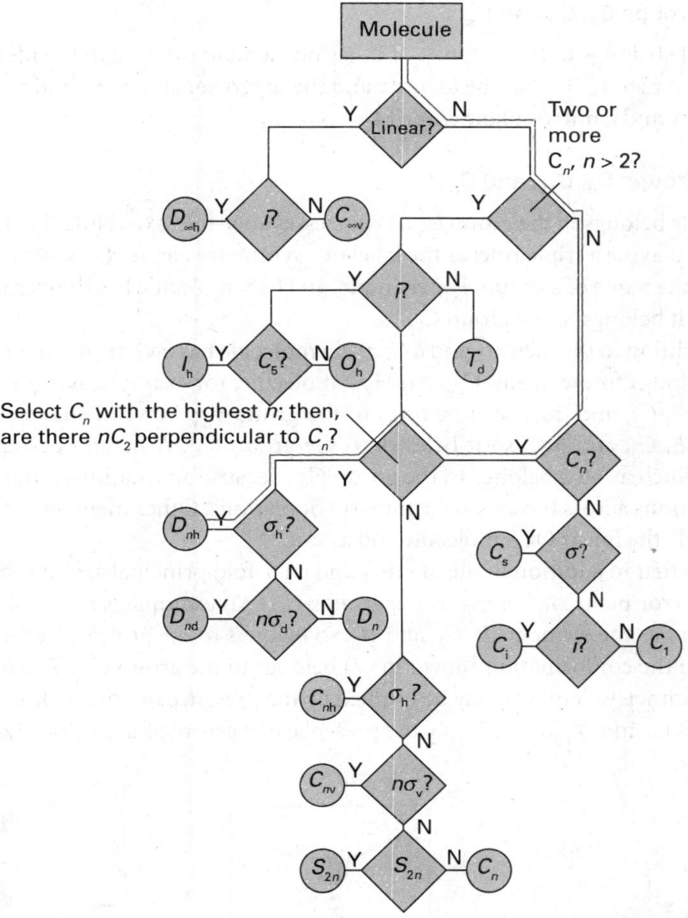

Fig. 11.7 A flow diagram for determining the point group of a molecule. Start at the top and answer the question posed in each diamond (Y = yes, N = no).

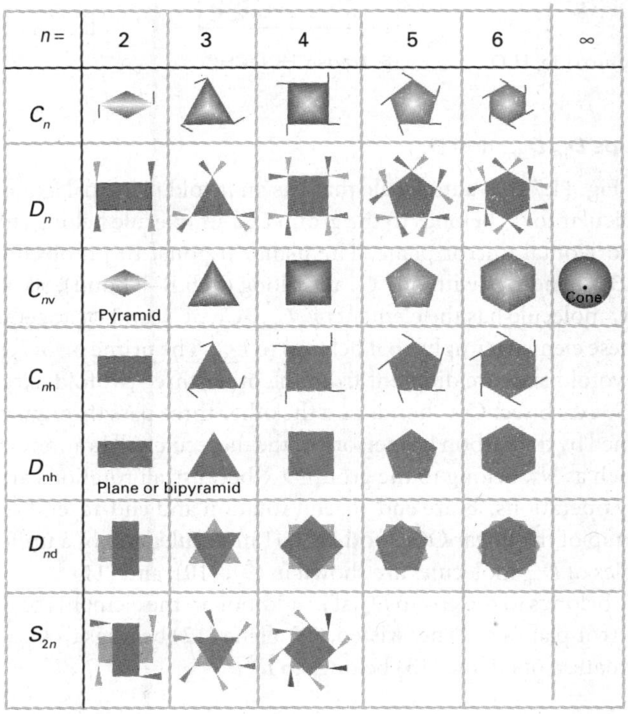

$n=$	2	3	4	5	6	∞
C_n						
D_n						
C_{nv}	Pyramid					Cone
C_{nh}						
D_{nh}	Plane or bipyramid					
D_{nd}						
S_{2n}						

Fig. 11.8 A summary of the shapes corresponding to different point groups. The group to which a molecule belongs can often be identified from this diagram without going through the formal procedure in Fig. 11.7.

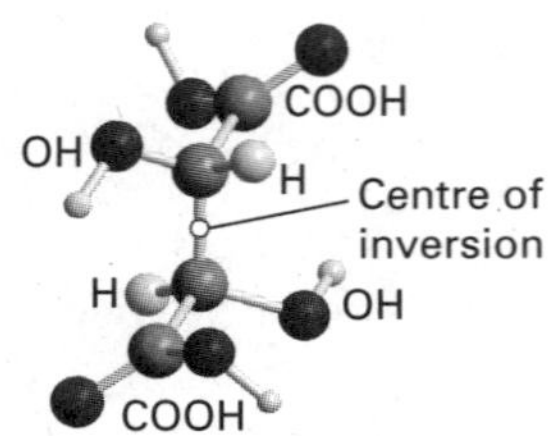

3 Meso-tartaric acid, HOOCCH(OH)CH(OH)COOH

4 Quinoline, C_9H_7N

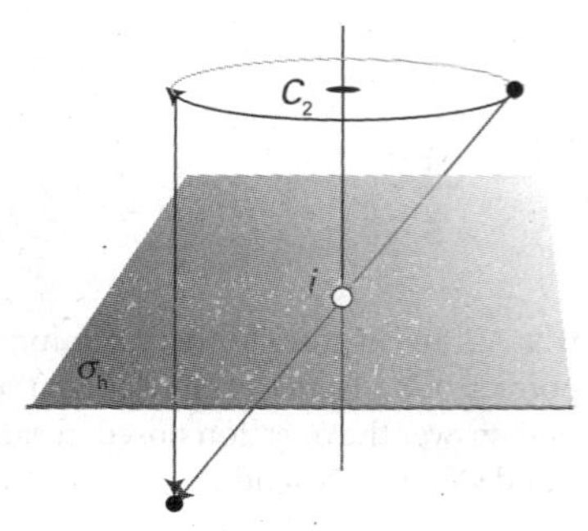

Fig. 11.9 The presence of a twofold axis and a horizontal mirror plane jointly imply the presence of a centre of inversion in the molecule.

(a) The groups C_1, C_i, and C_s

A molecule belongs to the group C_1 if it has no element other than the identity, as in (**1**). It belongs to C_i if it has the identity and the inversion alone (**3**), and to C_s if it has the identity and a mirror plane alone (**4**).

(b) The groups C_n, C_{nv}, and C_{nh}

A molecule belongs to the group C_n if it possesses an n-fold axis. Note that the symbol C_n is now playing a triple role: as the label of a symmetry element, a symmetry operation, and the name of a group. For example, an H_2O_2 molecule has the elements E and C_2 (**5**), so it belongs to the group C_2.

If in addition to the identity and a C_n axis a molecule has n vertical mirror planes σ_v, then it belongs to the group C_{nv}. An H_2O molecule, for example, has the symmetry elements E, C_2, and $2\sigma_v$, so it belongs to the group C_{2v}. An NH_3 molecule has the elements E, C_3, and $3\sigma_v$, so it belongs to the group C_{3v}. A heteronuclear diatomic molecule such as HCl belongs to the group $C_{\infty v}$ because all rotations around the axis and reflections across the axis are symmetry operations. Other members of the group $C_{\infty v}$ include the linear OCS molecule and a cone.

Objects that in addition to the identity and an n-fold principal axis also have a horizontal mirror plane σ_h belong to the groups C_{nh}. An example is *trans*-CHCl=CHCl (**6**), which has the elements E, C_2, and σ_h, so belongs to the group C_{2h}; the molecule $B(OH)_3$ in the conformation shown in (**7**) belongs to the group C_{3h}. The presence of certain symmetry elements may be implied by the presence of others: thus, in C_{2h} the operations C_2 and σ_h jointly imply the presence of a centre of inversion (Fig. 11.9).

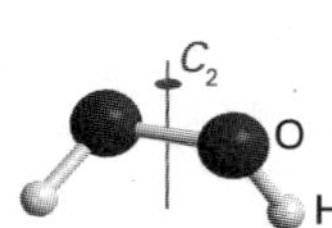

5 Hydrogen peroxide, H_2O_2

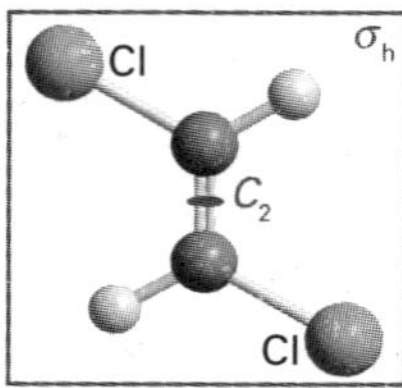

6 *trans*-CHCl=CHCl

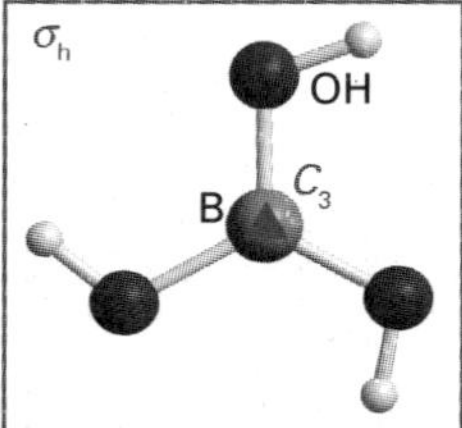

7 $B(OH)_3$

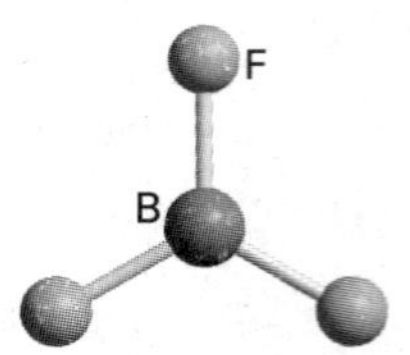

8 Boron trifluoride, BF_3

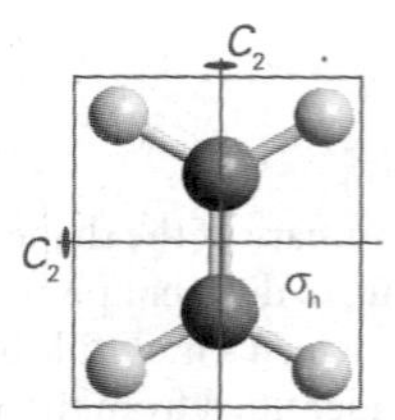

9 Ethene, CH_2=CH_2 (D_{2h})

(c) The groups D_n, D_{nh}, and D_{nd}

We see from Fig. 11.7 that a molecule that has an n-fold principal axis and n twofold axes perpendicular to C_n belongs to the group D_n. A molecule belongs to D_{nh} if it also possesses a horizontal mirror plane. The planar trigonal BF_3 molecule has the elements E, C_3, $3C_2$, and σ_h (with one C_2 axis along each B–F bond), so belongs to D_{3h} (**8**). The C_6H_6 molecule has the elements E, C_6, $3C_2$, $3C_2'$, and σ_h together with some others that these elements imply, so it belongs to D_{6h}. The prime on $3C_2'$ indicates that these three twofold axes are different from the other three twofold axes. In benzene, three of the C_2 axes bisect C–C bonds and the other three pass through vertices of the hexagon formed by the carbon framework of the molecule. All homonuclear diatomic molecules, such as N_2, belong to the group $D_{\infty h}$ because all rotations around the axis are symmetry operations, as are end-to-end rotation and end-to-end reflection; $D_{\infty h}$ is also the group of the linear OCO and HCCH molecules and of a uniform cylinder. Other examples of D_{nh} molecules are shown in (**9**), (**10**), and (**11**).

A molecule belongs to the group D_{nd} if in addition to the elements of D_n it possesses n dihedral mirror planes σ_d. The twisted, 90° allene (**12**) belongs to D_{2d}, and the staggered conformation of ethane (**13**) belongs to D_{3d}.

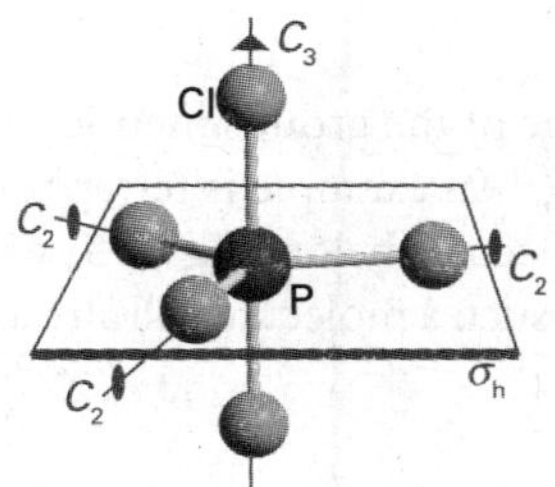

10 Phosphorus pentachloride, PCl_5 (D_{3h})

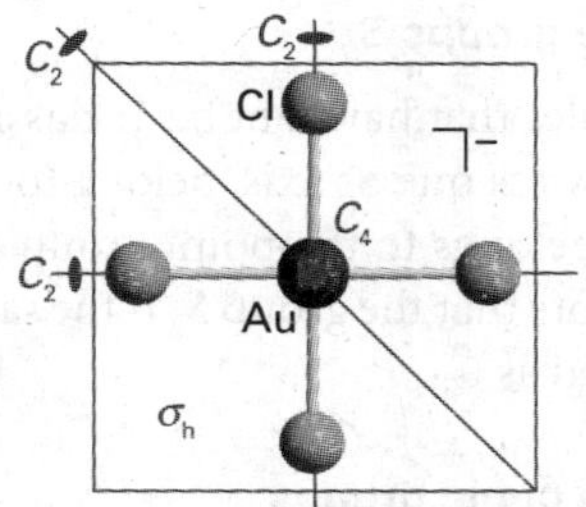

11 Tetrachloroaurate(III) ion, $[AuCl_4]^-$, (D_{4h})

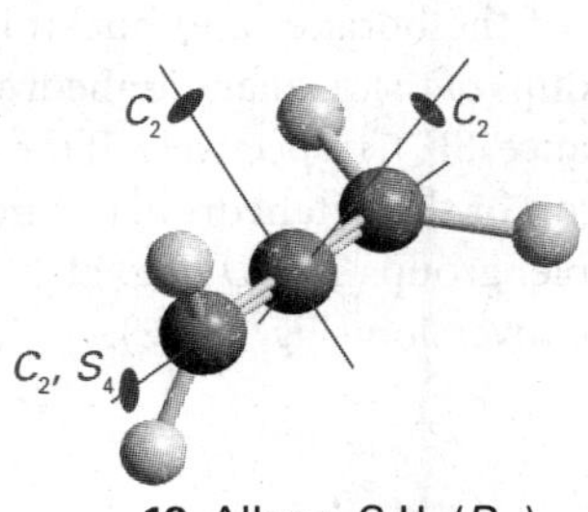

12 Allene, C_3H_4 (D_{2d})

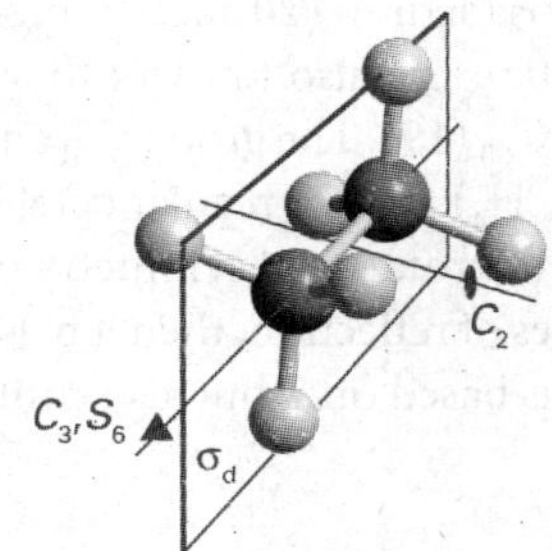

13 Ethane, C_2H_6 (D_{3d})

• **A brief illustration**

'Host' molecules, such as the bowl-shaped cryptophans, that encapsulate smaller 'guest' molecules have become a focus of interest for a wide variety of applications. Host–guest complexes are an important means of constructing nanoscale devices, selectively separating mixtures of small molecules on the basis of chemical and physical properties, delivering biologically active molecules to target cells, and providing unique environments to catalyse reactions. The shape of the host can influence both the encapsulation of guest molecules and the potential application of the complex. The anti and syn cryptophan isomers (14) and (15), for instance, belong to the groups D_3 and C_{3h}, respectively. •

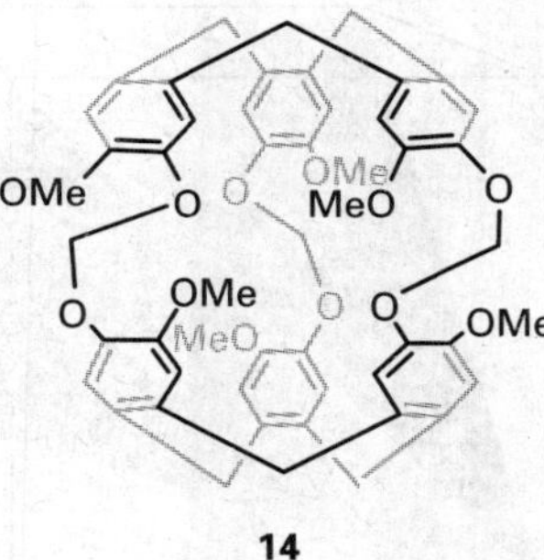

14

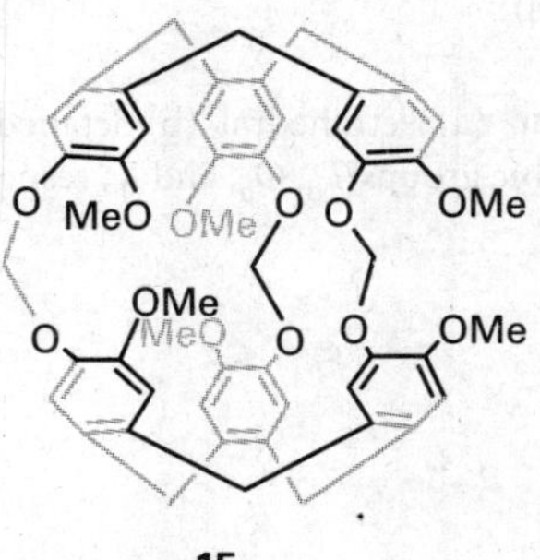

15

• **Another brief illustration**

Cucurbiturils are pumpkin-shaped water-soluble compounds composed of six, seven, or eight glycouril (16) units with a hydrophilic exterior and a hydrophobic interior cavity. With six glycouril units, for example, the host (17) belongs to the group D_{6h}. •

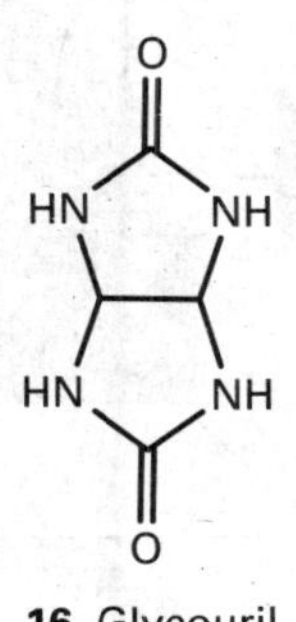

16 Glycouril

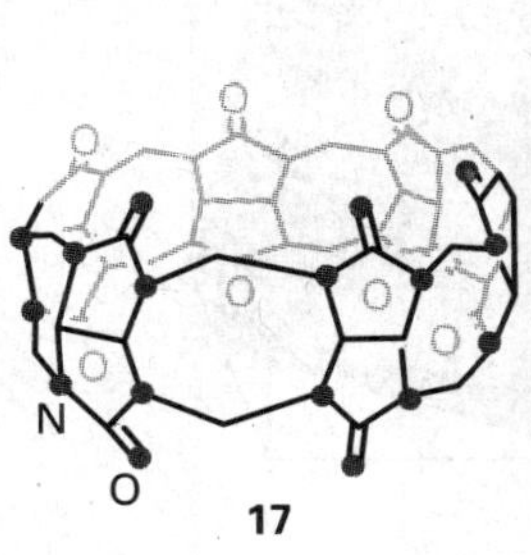

17

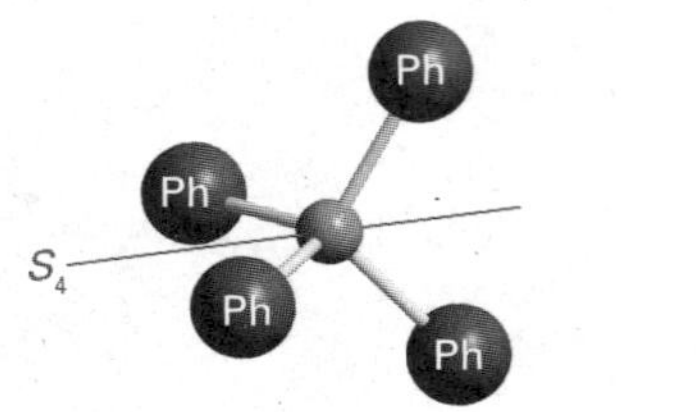

18 Tetraphenylmethane, $C(C_6H_5)_4$ (S_4)

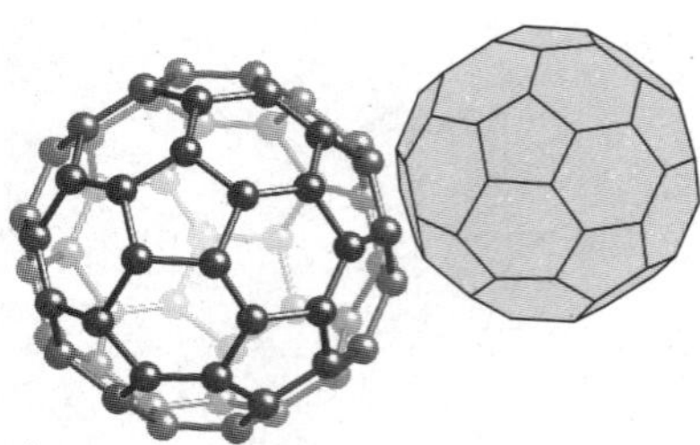

19 Buckminsterfullerene, C_{60} (I)

(d) The groups S_n

Molecules that have not been classified into one of the groups mentioned so far, but that possess one S_n axis, belong to the group S_n. An example is tetraphenylmethane, which belongs to the point group S_4 (**18**). Molecules belonging to S_n with $n > 4$ are rare. Note that the group S_2 is the same as C_i, so such a molecule will already have been classified as C_i.

(e) The cubic groups

A number of very important molecules (e.g. CH_4 and SF_6) possess more than one principal axis. Most belong to the **cubic groups**, and in particular to the **tetrahedral groups** T, T_d, and T_h (Fig. 11.10a) or to the **octahedral groups** O and O_h (Fig. 11.10b). A few icosahedral (20-faced) molecules belonging to the **icosahedral group**, I (Fig. 11.10c), are also known: they include some of the boranes and buckminsterfullerene, C_{60} (**19**). The groups T_d and O_h are the groups of the regular tetrahedron (for instance, CH_4) and the regular octahedron (for instance, SF_6), respectively. If the object possesses the rotational symmetry of the tetrahedron or the octahedron, but none of their planes of reflection, then it belongs to the simpler groups T or O (Fig. 11.11). The group T_h is based on T but also contains a centre of inversion (Fig. 11.12).

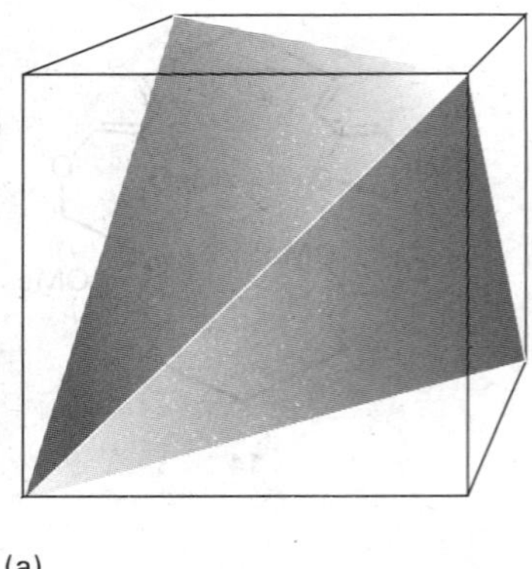

(a)

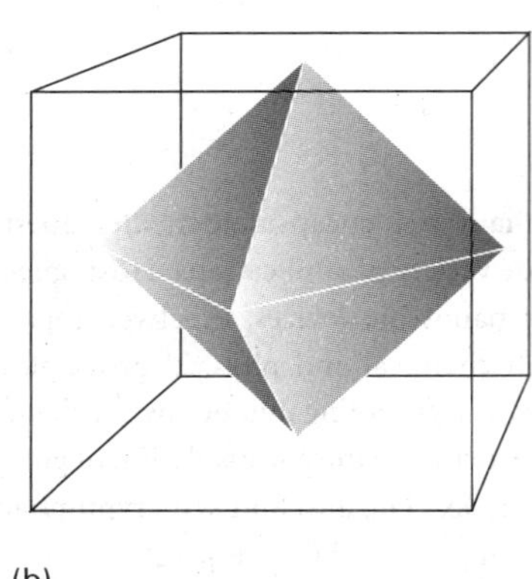

(b)

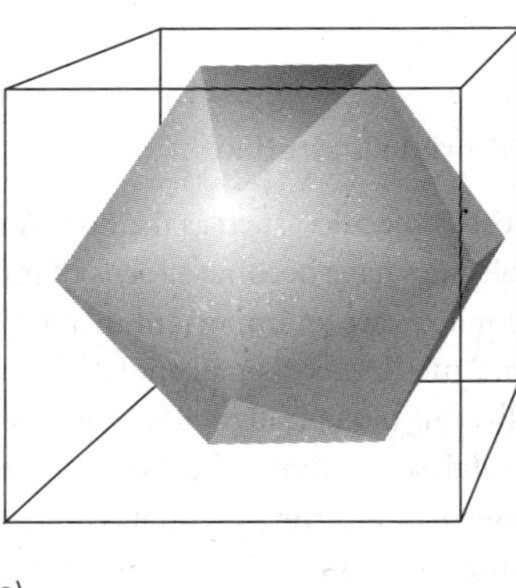

(c)

Fig. 11.10 (a) Tetrahedral, (b) octahedral, and (c) icosahedral molecules are drawn in a way that shows their relation to a cube: they belong to the cubic groups T_d, O_h, and I_h, respectively.

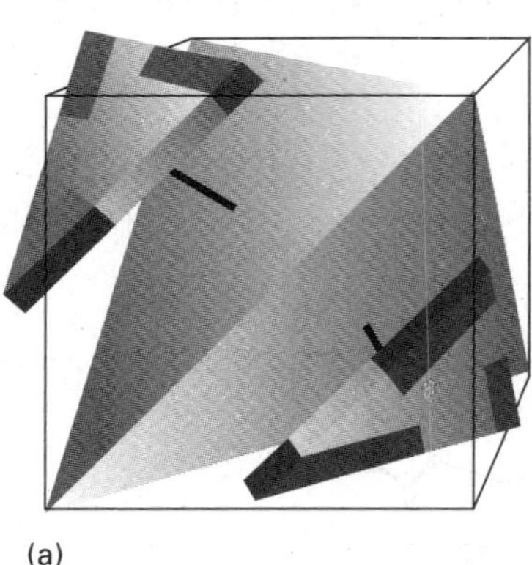

(a)

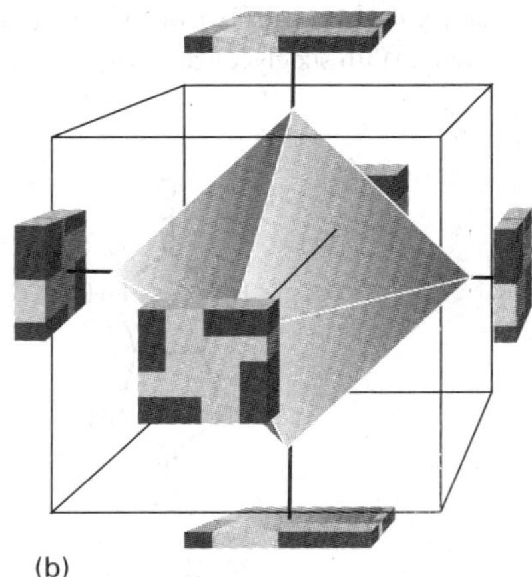

(b)

Fig. 11.11 Shapes corresponding to the point groups (a) T and (b) O. The presence of the decorated slabs reduces the symmetry of the object from T_d and O_h, respectively.

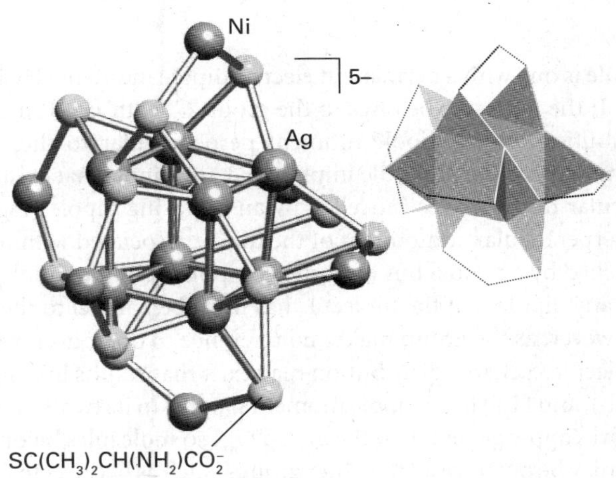

20

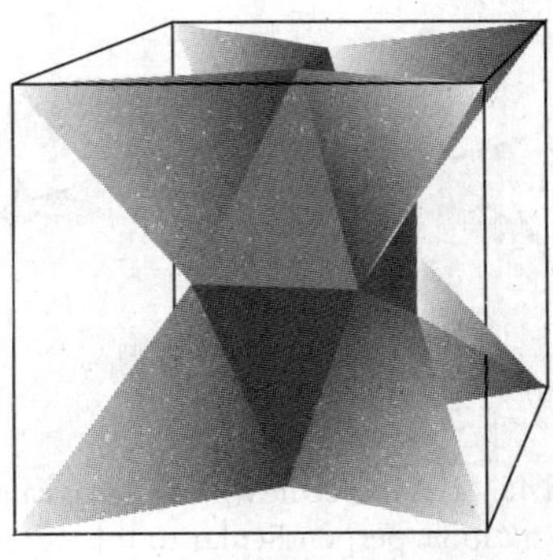

Fig. 11.12 The shape of an object belonging to the group T_h.

• A brief illustration

The ion $[Ag_8Ni_6\{SC(Me)_2CH(NH_2)CO_2\}_{12}Cl]^{5-}$ (**20**) is a tetrahedral host belonging to the group T_h. •

(f) The full rotation group

The **full rotation group**, R_3 (the 3 refers to rotation in three dimensions), consists of an infinite number of rotation axes with all possible values of *n*. A sphere and an atom belong to R_3, but no molecule does. Exploring the consequences of R_3 is a very important way of applying symmetry arguments to atoms, and is an alternative approach to the theory of orbital angular momentum.

Example 11.1 *Identifying a point group of a molecule*

Identify the point group to which a ruthenocene molecule (**21**) belongs.

Method Use the flow diagram in Fig. 11.7.

Answer The path to trace through the flow diagram in Fig. 11.7 is shown by a green line; it ends at D_{nh}. Because the molecule has a fivefold axis, it belongs to the group D_{5h}. If the rings were staggered, as they are in an excited state of ferrocene that lies 4 kJ mol^{-1} above the ground state (**22**), the horizontal reflection plane would be absent, but dihedral planes would be present.

Self-test 11.1 Classify the pentagonal antiprismatic excited state of ferrocene (**22**). [D_{5d}]

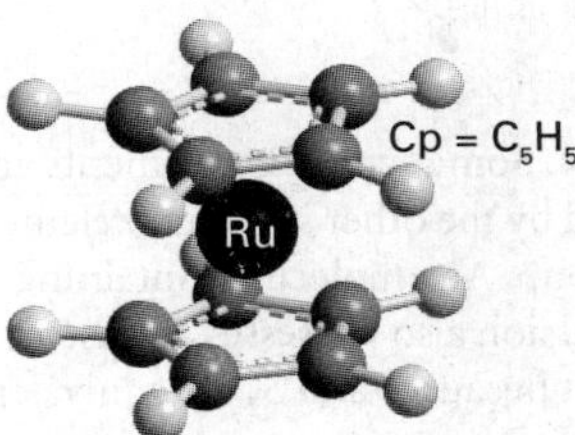

21 Ruthenocene, $Ru(Cp)_2$

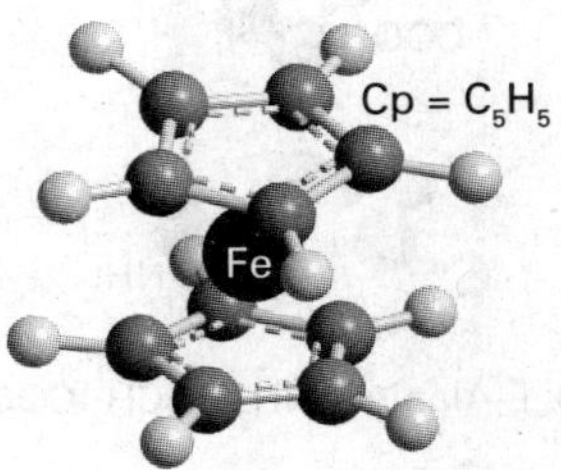

22 Ferrocene, $Fe(Cp)_2$

11.3 Some immediate consequences of symmetry

Key points **(a) Only molecules belonging to the groups C_n, C_{nv}, and C_s may have a permanent electric dipole moment. (b) A molecule may be chiral, and therefore optically active, only if it does not possess an axis of improper rotation, S_n.**

Some statements about the properties of a molecule can be made as soon as its point group has been identified.

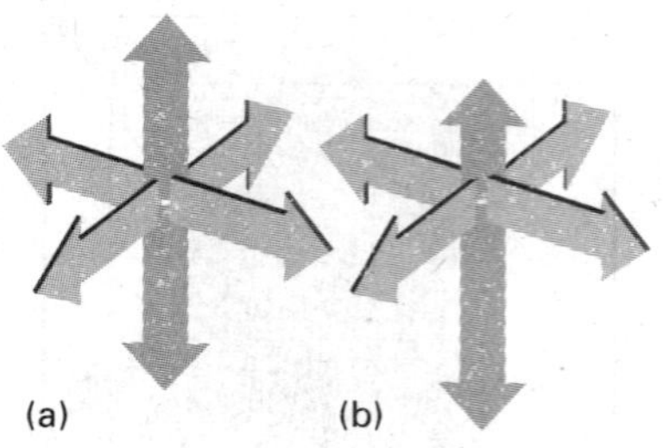

Fig. 11.13 (a) A molecule with a C_n axis cannot have a dipole perpendicular to the axis, but (b) it may have one parallel to the axis. The arrows represent local contributions to the overall electric dipole, such as may arise from bonds between pairs of neighbouring atoms with different electronegativities.

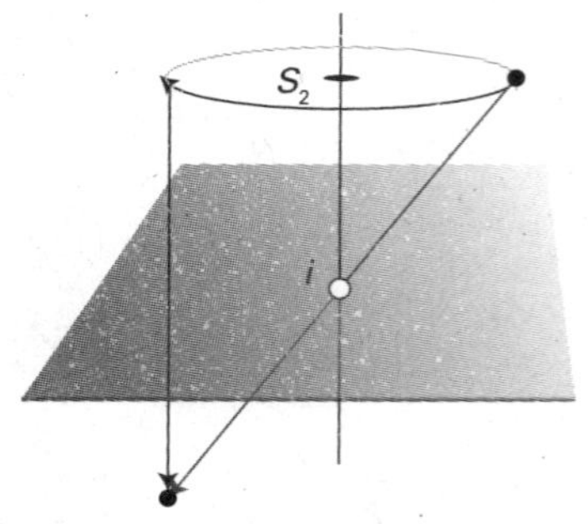

Fig. 11.14 Some symmetry elements are implied by the other symmetry elements in a group. Any molecule containing an inversion also possesses at least an S_2 element because i and S_2 are equivalent.

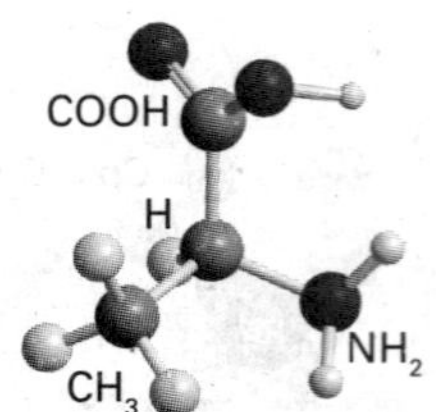

23 L-Alanine, $NH_2CH(CH_3)COOH$

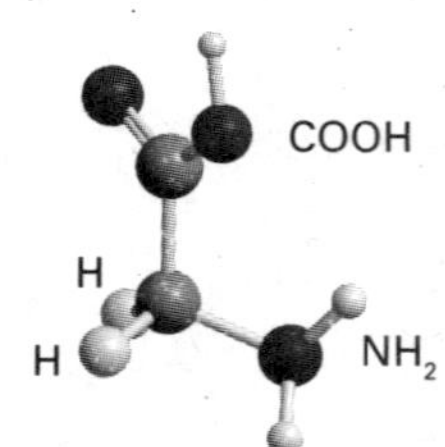

24 Glycine, NH_2CH_2COOH

(a) Polarity

A **polar molecule** is one with a permanent electric dipole moment (HCl, O_3, and NH_3 are examples). If the molecule belongs to the group C_n with $n > 1$, it cannot possess a charge distribution with a dipole moment perpendicular to the symmetry axis because the symmetry of the molecule implies that any dipole that exists in one direction perpendicular to the axis is cancelled by an opposing dipole (Fig. 11.13a). For example, the perpendicular component of the dipole associated with one O–H bond in H_2O is cancelled by an equal but opposite component of the dipole of the second O–H bond, so any dipole that the molecule has must be parallel to the twofold symmetry axis. However, as the group makes no reference to operations relating the two ends of the molecule, a charge distribution may exist that results in a dipole along the axis (Fig. 11.13b), and H_2O has a dipole moment parallel to its twofold symmetry axis. The same remarks apply generally to the group C_{nv}, so molecules belonging to any of the C_{nv} groups may be polar. In all the other groups, such as C_{3h}, D, etc., there are symmetry operations that take one end of the molecule into the other. Therefore, as well as having no dipole perpendicular to the axis, such molecules can have none along the axis, for otherwise these additional operations would not be symmetry operations. We can conclude that

> Only molecules belonging to the groups C_n, C_{nv}, and C_s may have a permanent electric dipole moment.

Criterion for being polar

For C_n and C_{nv}, that dipole moment must lie along the symmetry axis. Thus ozone, O_3, which is angular and belongs to the group C_{2v}, may be polar (and is), but carbon dioxide, CO_2, which is linear and belongs to the group $D_{\infty h}$, is not.

(b) Chirality

A **chiral molecule** (from the Greek word for 'hand') is a molecule that cannot be superimposed on its mirror image. An **achiral molecule** is a molecule that can be superimposed on its mirror image. Chiral molecules are **optically active** in the sense that they rotate the plane of polarized light. A chiral molecule and its mirror-image partner constitute an **enantiomeric pair** of optical isomers and rotate the plane of polarization in equal but opposite directions.

> A molecule may be chiral, and therefore optically active, only if it does not possess an axis of improper rotation, S_n.

Criterion for being chiral

However, we need to be aware that such an axis may be present under a different name, and be implied by other symmetry elements that are present. For example, molecules belonging to the groups C_{nh} possess an S_n axis implicitly because they possess both C_n and σ_h, which are the two components of an improper rotation axis. Any molecule containing a centre of inversion, i, also possesses an S_2 axis, because i is equivalent to C_2 in conjunction with σ_h, and that combination of elements is S_2 (Fig. 11.14). It follows that all molecules with centres of inversion are achiral and hence optically inactive. Similarly, because $S_1 = \sigma$, it follows that any molecule with a mirror plane is achiral.

A molecule may be chiral if it does not have a centre of inversion or a mirror plane, which is the case with the amino acid alanine (**23**), but not with glycine (**24**). However, a molecule may be achiral even though it does not have a centre of inversion. For example, the S_4 species (**25**) is achiral and optically inactive: though it lacks i (that is, S_2) it does have an S_4 axis.

Applications to molecular orbital theory and spectroscopy

We shall now turn our attention away from the symmetries of molecules themselves and direct it towards the symmetry characteristics of orbitals that belong to the various atoms in a molecule. This material will enable us to discuss the formulation and labelling of molecular orbitals and selection rules in spectroscopy.

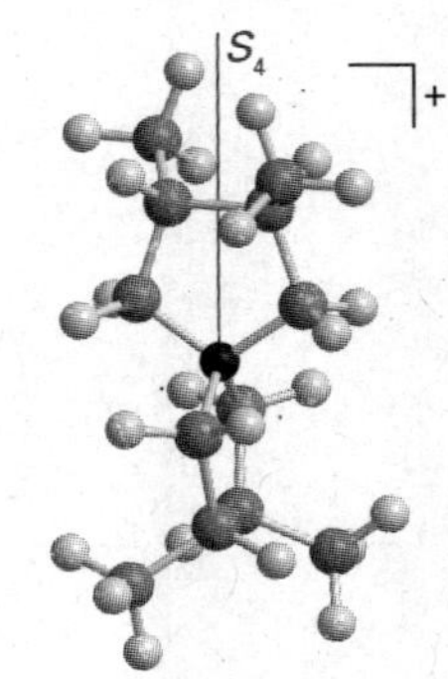

25 $N(CH_2CH(CH_3)CH(CH_3)CH_2)_2^+$

11.4 Character tables and symmetry labels

Key points (a) The character of an operation summarizes the effect of a symmetry operation on a function; it is the sum of the diagonal elements of a matrix that represents the effect of the operation. (b) The rows under the labels for the operations in a character table express the symmetry properties of the basis and are labelled with the symmetry species. (c) The character of the identity operation E is the degeneracy of the orbitals that form a basis. (d) The entries in a character table indicate how the basis functions transform under the symmetry operations. (e) Linear combinations of orbitals are also classified according to their symmetry.

We saw in Chapter 10 that molecular orbitals of diatomic and linear polyatomic molecules are labelled σ, π, etc. These labels refer to the symmetries of the orbitals with respect to rotations around the principal symmetry axis of the molecule. Thus, a σ orbital does not change sign under a rotation through any angle, a π orbital changes sign when rotated by 180°, and so on (Fig. 11.15). The symmetry classifications σ and π can also be assigned to individual atomic orbitals in a linear molecule. For example, we can speak of an individual p_z orbital as having σ symmetry if the z-axis lies along the bond, because p_z is cylindrically symmetrical about the bond. This labelling of orbitals according to their behaviour under rotations can be generalized and extended to nonlinear polyatomic molecules, where there may be reflections and inversions to take into account as well as rotations.

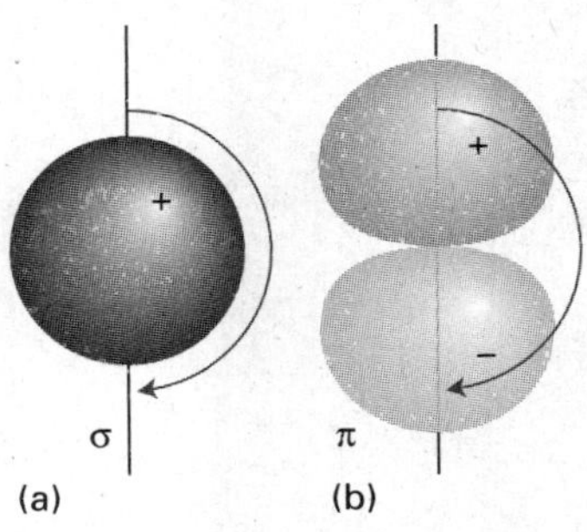

Fig. 11.15 A rotation through 180° about the internuclear axis (perpendicular to the page) (a) leaves the sign of a σ orbital unchanged but (b) the sign of a π orbital is changed. In the language introduced in this chapter, the characters of the C_2 rotation are +1 and −1 for the σ and π orbitals, respectively.

(a) Representations and characters

Labels analogous to σ and π are used to denote the symmetries of orbitals in polyatomic molecules. These labels look like a, a_1, e, e_g, and we first encountered them in Fig. 10.45 in connection with the molecular orbitals of benzene. As we shall see, these labels indicate the behaviour of the orbitals under the symmetry operations of the relevant point group of the molecule.

	C_2	(i.e. rotation by 180°)
σ	+1	(i.e. no change of sign)
π	−1	(i.e. change of sign)

A label is assigned to an orbital by referring to the **character table** of the group, a table that characterizes the different symmetry types possible in the point group. Thus, to assign the labels σ and π, we use the table shown in the margin. This table is a fragment of the full character table for a linear molecule. The entry +1 shows that the orbital remains the same and the entry −1 shows that the orbital changes sign under the operation C_2 at the head of the column (as illustrated in Fig. 11.15). So, to assign the label σ or π to a particular orbital, we compare the orbital's behaviour with the information in the character table.

The entries in a complete character table are derived by using the formal techniques of group theory and are called **characters**, χ (chi). These numbers characterize the essential features of each symmetry type in a way that we can illustrate by considering the C_{2v} molecule SO_2 and the valence p_x orbitals on each atom, which we shall denote p_S, p_A, and p_B (Fig. 11.16).

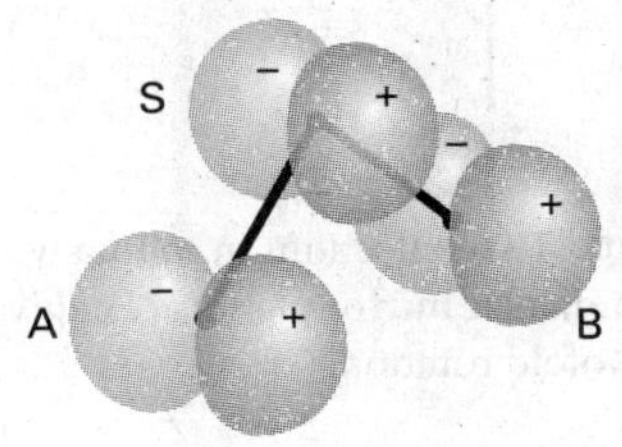

Fig. 11.16 The three p_x orbitals that are used to illustrate the construction of a matrix representation in a C_{2v} molecule (SO_2).

Under σ_v, the change $(p_S, p_B, p_A) \leftarrow (p_S, p_A, p_B)$ takes place. We can express this transformation by using matrix multiplication (see *Mathematical background 6* following Chapter 10 for a summary of the rules of matrix algebra):

$$(p_S, p_B, p_A) = (p_S, p_A, p_B)\begin{pmatrix} 1 & 0 & 0 \\ 0 & 0 & 1 \\ 0 & 1 & 0 \end{pmatrix} = (p_S, p_A, p_B)\boldsymbol{D}(\sigma_V) \qquad (11.1)$$

The matrix $D(\sigma_v)$ is called a **representative** of the operation σ_v. Representatives take different forms according to the **basis**, the set of orbitals that has been adopted.

We can use the same technique to find matrices that reproduce the other symmetry operations. For instance, C_2 has the effect $(-p_S, -p_B, -p_A) \leftarrow (p_S, p_A, p_B)$, and its representative is

$$D(C_2) = \begin{pmatrix} -1 & 0 & 0 \\ 0 & 0 & -1 \\ 0 & -1 & 0 \end{pmatrix} \qquad (11.2)$$

The effect of σ'_v is $(-p_S, -p_A, -p_B) \leftarrow (p_S, p_A, p_B)$, and its representative is

$$D(\sigma'_V) = \begin{pmatrix} -1 & 0 & 0 \\ 0 & -1 & 0 \\ 0 & 0 & -1 \end{pmatrix} \qquad (11.3)$$

The identity operation leaves the basis unchanged, so its representative is the 3×3 unit matrix:

$$D(E) = \begin{pmatrix} 1 & 0 & 0 \\ 0 & 1 & 0 \\ 0 & 0 & 1 \end{pmatrix} \qquad (11.4)$$

The set of matrices that represents *all* the operations of the group is called a **matrix representation**, **Γ** (uppercase gamma), of the group for the particular basis we have chosen. We denote this three-dimensional representation by $\Gamma^{(3)}$. The discovery of a matrix representation of the group means that we have found a link between symbolic manipulations of operations and algebraic manipulations of numbers. The following *Justification* explains why 'representation' is an accurate term.

Justification 11.1 *The representation of symmetry operations*

We saw in Section 11.1 that symmetry operations form a group if certain criteria are satisfied. Among them is the group property that, if R and S are symmetry operations, then RS is also a symmetry operation. The crucial point in this *Justification* is that the matrices used to reproduce the effect of symmetry operations on a given basis also satisfy the same group property. That is, if the operation S followed by the operation R is equivalent to the single operation RS, then the matrices also satisfy

$$D(R)D(S) = D(RS)$$

We can demonstrate this relation for the relation $\sigma_v\sigma'_v = C_2$ for the group C_{2v}, that is, a reflection in one plane followed by a reflection in a perpendicular plane is equivalent to a 180° rotation (Fig. 11.17). We use the matrices developed in the text:

$$D(\sigma_v)D(\sigma'_v) = \begin{pmatrix} 1 & 0 & 0 \\ 0 & 0 & 1 \\ 0 & 1 & 0 \end{pmatrix}\begin{pmatrix} -1 & 0 & 0 \\ 0 & -1 & 0 \\ 0 & 0 & -1 \end{pmatrix} = \begin{pmatrix} -1 & 0 & 0 \\ 0 & 0 & -1 \\ 0 & -1 & 0 \end{pmatrix} = D(C_2)$$

The same conclusion may be drawn for all combinations of the matrices listed above, so they do in fact 'represent' in a concrete way structure of the group of symmetry operations in this case.

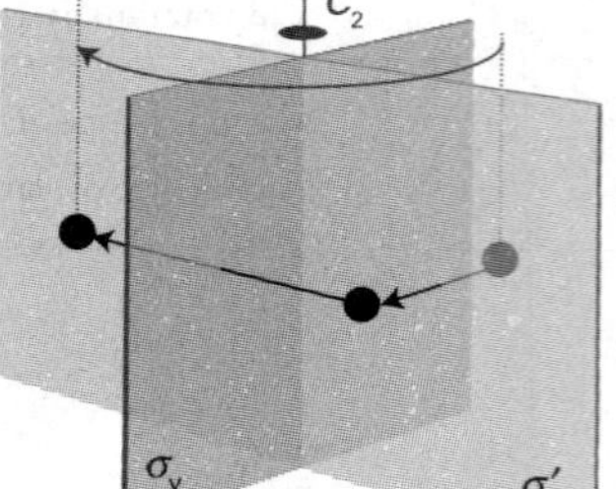

Fig. 11.17 Two reflections in mutually perpendicular mirror planes are equivalent to a twofold rotation.

The character of an operation in a particular matrix representation is the sum of the diagonal elements of the representative of that operation. Thus, in the basis we are illustrating, the characters of the representatives are

$D(E)$	$D(C_2)$	$D(\sigma_V)$	$D(\sigma_V')$
3	−1	1	−3

The character of an operation depends on the basis.

Inspection of the representatives shows that they are all of **block-diagonal form:**

$$D=\begin{pmatrix}[\bullet] & 0 & 0\\ 0 & [\bullet] & [\bullet]\\ 0 & [\bullet] & [\bullet]\end{pmatrix}$$

Block-diagonal matrix

The block-diagonal form of the representatives shows us that the symmetry operations of C_{2v} never mix p_S with the other two functions. Consequently, the basis can be cut into two parts, one consisting of p_S alone and the other of (p_A, p_B). It is readily verified that the p_S orbital itself is a basis for the one-dimensional representation

$$D(E)=1 \qquad D(C_2)=-1 \qquad D(\sigma_v)=1 \qquad D(\sigma_v')=-1$$

which we shall call $\Gamma^{(1)}$. The functions (p_A, p_B) are jointly a basis for the two-dimensional representation $\Gamma^{(2)}$:

$$D(E)=\begin{pmatrix}1 & 0\\ 0 & 1\end{pmatrix} \quad D(C_2)=\begin{pmatrix}0 & -1\\ -1 & 0\end{pmatrix} \quad D(\sigma_v)=\begin{pmatrix}0 & 1\\ 1 & 0\end{pmatrix} \quad D(\sigma_v')=\begin{pmatrix}-1 & 0\\ 0 & -1\end{pmatrix}$$

These matrices are the same as those of the original three-dimensional representation, except for the loss of the first row and column. We say that the original three-dimensional representation has been **reduced** to the 'direct sum' of a one-dimensional representation 'spanned' by p_S, and a two-dimensional representation spanned by (p_A, p_B). This reduction is consistent with the common sense view that the central orbital plays a role different from the other two. We denote the reduction symbolically by writing

$$\Gamma^{(3)}=\Gamma^{(1)}+\Gamma^{(2)} \tag{11.5}$$

The one-dimensional representation $\Gamma^{(1)}$ cannot be reduced any further, and is called an **irreducible representation** of the group (an 'irrep'). We can demonstrate that the two-dimensional representation $\Gamma^{(2)}$ is reducible (for this basis in this group) by switching attention to the linear combinations $p_1=p_A+p_B$ and $p_2=p_A-p_B$. These combinations are sketched in Fig. 11.18. The representatives in the new basis can be constructed from the old by noting, for example, that because, under σ_v, $(p_B, p_A) \leftarrow (p_A, p_B)$ it follows that $(p_1, -p_2) \leftarrow (p_1, p_2)$. In this way we find the following representation in the new basis:

$$D(E)=\begin{pmatrix}1 & 0\\ 0 & 1\end{pmatrix} \quad D(C_2)=\begin{pmatrix}-1 & 0\\ 0 & 1\end{pmatrix} \quad D(\sigma_v)=\begin{pmatrix}1 & 0\\ 0 & -1\end{pmatrix} \quad D(\sigma_v')=\begin{pmatrix}-1 & 0\\ 0 & -1\end{pmatrix}$$

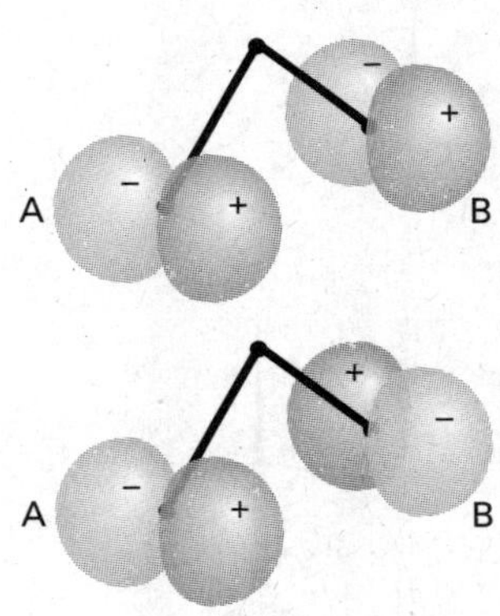

Fig. 11.18 Two symmetry-adapted linear combinations of the basis orbitals shown in Fig. 11.16. The two combinations each span a one-dimensional irreducible representation, and their symmetry species are different.

The new representatives are all in block-diagonal form (in this case, all the blocks are 1×1), and the two combinations are not mixed with each other by any operation of the group. We have therefore achieved the reduction of $\Gamma^{(2)}$ to the sum of two one-dimensional representations. Thus, p_1 spans

$$D(E)=1 \qquad D(C_2)=-1 \qquad D(\sigma_v)=1 \qquad D(\sigma_v')=-1$$

which is the same one-dimensional representation as that spanned by p_S, and p_2 spans

$$D(E)=1 \qquad D(C_2)=1 \qquad D(\sigma_v)=-1 \qquad D(\sigma_v')=-1$$

which is a different one-dimensional representation; we shall denote it $\Gamma^{(1)\prime}$.

Table 11.2* The C_{2v} character table

C_{2v}, $2mm$	E	C_2	σ_v	σ'_v	$h=4$	
A_1	1	1	1	1	z	z^2, y^2, x^2
A_2	1	1	−1	−1		xy
B_1	1	−1	1	−1	x	zx
B_2	1	−1	−1	1	y	yz

* More character tables are given at the end of the *Resource section*.

At this point we have found two irreducible representations of the group C_{2v} (Table 11.2). The two irreducible representations are normally labelled B_1 and A_2, respectively. An A or a B is used to denote a one-dimensional representation; A is used if the character under the principal rotation is +1, and B is used if the character is −1. Subscripts are used to distinguish the irreducible representations if there is more than one of the same type: A_1 is reserved for the representation with character 1 for all operations. When higher dimensional irreducible representations are permitted, E denotes a two-dimensional irreducible representation and T a three-dimensional irreducible representation; all the irreducible representations of C_{2v} are one-dimensional.

There are in fact only two more species of irreducible representations of this group, for a surprising theorem of group theory states that

$$\text{Number of symmetry species} = \text{number of classes} \tag{11.6}$$

Symmetry operations fall into the same **class** if they are of the same type (for example, rotations) and can be transformed into one another by a symmetry operation of the group. In C_{2v}, for instance, there are four classes (four columns in the character table), so there are only four species of irreducible representation. The character table in Table 11.2 therefore shows the characters of all the irreducible representations of this group.

(b) The structure of character tables

In general, the columns in a character table are labelled with the symmetry operations of the group. For instance, for the group C_{3v} the columns are headed E, C_3, and σ_v (Table 11.3). The numbers multiplying each operation are the numbers of members of each class. In the C_{3v} character table we see that the two threefold rotations (clockwise and counterclockwise rotations by 120°) belong to the same class: they are related by a reflection (Fig. 11.19). The three reflections (one through each of the three

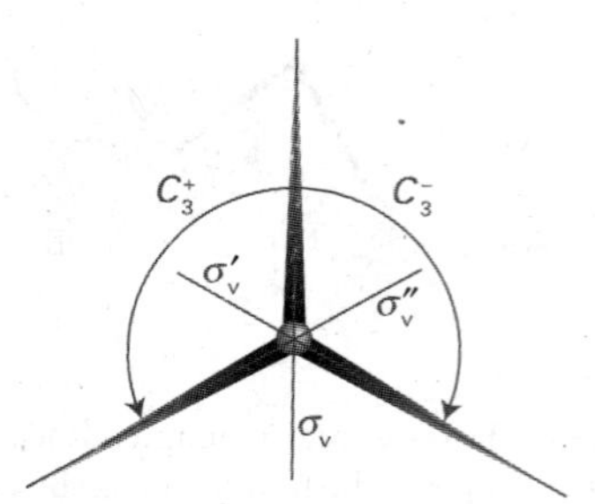

Fig. 11.19 Symmetry operations in the same class are related to one another by the symmetry operations of the group. Thus, the three mirror planes shown here are related by threefold rotations, and the two rotations shown here are related by reflection in σ_v.

Table 11.3* The C_{3v} character table

C_{3v}, $3m$	E	$2C_3$	$3\sigma_v$	$h=6$	
A_1	1	1	1	z	z^2, x^2+y^2
A_2	1	1	−1		
E	2	−1	0	(x, y)	$(xy, x^2-y^2), (yz, zx)$

* More character tables are given at the end of the *Resource section*.

vertical mirror planes) also lie in the same class: they are related by the threefold rotations. The two reflections of the group C_{2v} fall into different classes: although they are both reflections, one cannot be transformed into the other by any symmetry operation of the group.

The total number of operations in a group is called the **order**, h, of the group. The order of the group C_{3v}, for instance, is 6.

The rows under the labels for the operations summarize the symmetry properties of the orbitals. They are labelled with the **symmetry species** (the analogues of the labels σ and π). More formally, the symmetry species label the irreducible representations of the group, which are the basic types of behaviour that orbitals may show when subjected to the symmetry operations of the group, as we have illustrated for the group C_{2v}. By convention, irreducible representations are labelled with upper-case roman letters (such as $\mathrm{A_1}$ and E) and the orbitals to which they apply are labelled with the lower-case equivalents (so an orbital of symmetry species $\mathrm{A_1}$ is called an $\mathrm{a_1}$ orbital). Examples of each type of orbital are shown in Fig. 11.20.

A brief comment

Note that care must be taken to distinguish the identity element E (italic, a column heading) from the symmetry label E (roman, a row label).

(c) Character tables and orbital degeneracy

The character of the identity operation E tells us the degeneracy of the orbitals. Thus, in a C_{3v} molecule, any orbital with a symmetry label $\mathrm{a_1}$ or $\mathrm{a_2}$ is nondegenerate. Any doubly degenerate pair of orbitals in C_{3v} must be labelled e because, in this group, only E symmetry species have characters greater than 1.

Because there are no characters greater than 2 in the column headed E in C_{3v}, we know that there can be no triply degenerate orbitals in a C_{3v} molecule. This last point is a powerful result of group theory, for it means that, with a glance at the character table of a molecule, we can state the maximum possible degeneracy of its orbitals.

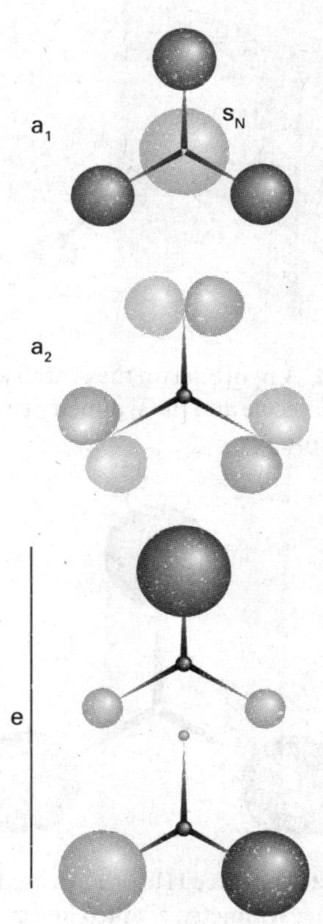

Fig. 11.20 Typical symmetry-adapted linear combinations of orbitals in a C_{3v} molecule.

Example 11.2 *Using a character table to judge degeneracy*

Can a trigonal planar molecule such as BF_3 have triply degenerate orbitals? What is the minimum number of atoms from which a molecule can be built that does display triple degeneracy?

Method First, identify the point group, and then refer to the corresponding character table in the *Resource section*. The maximum number in the column headed by the identity E is the maximum orbital degeneracy possible in a molecule of that point group. For the second part, consider the shapes that can be built from two, three, etc. atoms, and decide which number can be used to form a molecule that can have orbitals of symmetry species T.

Answer Trigonal planar molecules belong to the point group D_{3h}. Reference to the character table for this group shows that the maximum degeneracy is 2, as no character exceeds 2 in the column headed E. Therefore, the orbitals cannot be triply degenerate. A tetrahedral molecule (symmetry group T) has an irreducible representation with a T symmetry species. The minimum number of atoms needed to build such a molecule is four (as in P_4, for instance).

Self-test 11.2 A buckminsterfullerene molecule, C_{60} (**19**), belongs to the icosahedral point group. What is the maximum possible degree of degeneracy of its orbitals? [5]

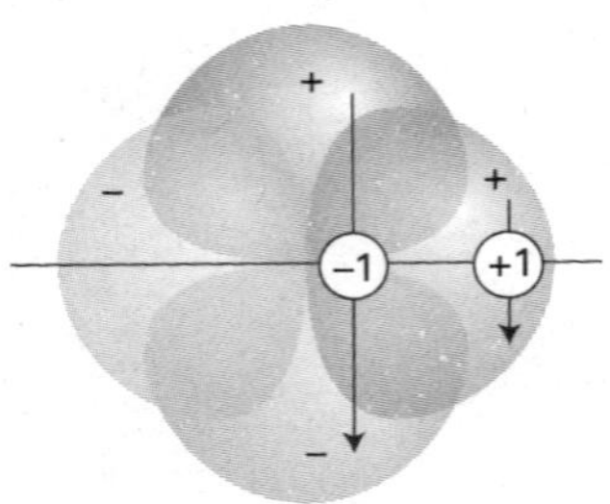

Fig. 11.21 The two orbitals shown here have different properties under reflection through the mirror plane: one changes sign (character −1), the other does not (character +1).

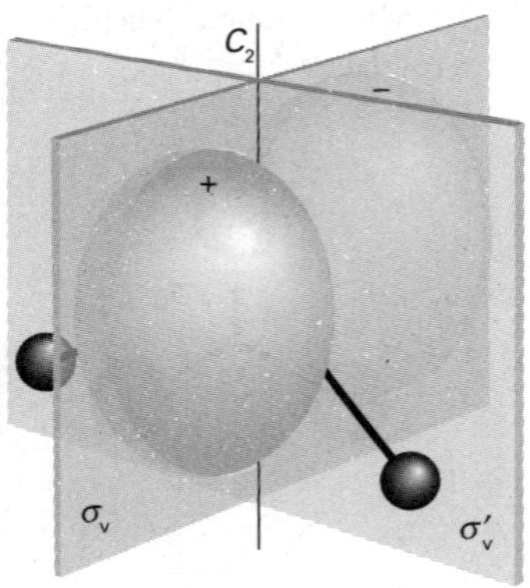

Fig. 11.22 A p_x orbital on the central atom of a C_{2v} molecule and the symmetry elements of the group.

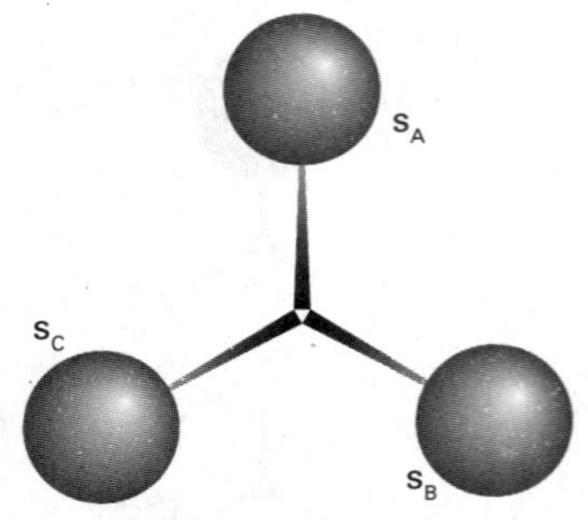

Fig. 11.23 The three H1s orbitals used to construct symmetry-adapted linear combinations in a C_{3v} molecule such as NH_3.

A brief comment

In previous chapters we used the Greek letter χ (chi) to denote atomic orbitals that comprise a basis set for molecular orbital calculations. To avoid confusion with our use of χ for characters in this chapter, atomic orbitals will be denoted by the Greek letter ψ (psi).

(d) Characters and operations

The characters in the rows labelled A and B and in the columns headed by symmetry operations other than the identity E indicate the behaviour of an orbital under the corresponding operations: a +1 indicates that an orbital is unchanged, and a −1 indicates that it changes sign. It follows that we can identify the symmetry label of the orbital by comparing the changes that occur to an orbital under each operation, and then comparing the resulting +1 or −1 with the entries in a row of the character table for the point group concerned.

For the rows labelled E or T (which refer to the behaviour of sets of doubly and triply degenerate orbitals, respectively), the characters in a row of the table are the sums of the characters summarizing the behaviour of the individual orbitals in the basis. Thus, if one member of a doubly degenerate pair remains unchanged under a symmetry operation but the other changes sign (Fig. 11.21), then the entry is reported as $\chi = 1 - 1 = 0$. Care must be exercised with these characters because the transformations of orbitals can be quite complicated; nevertheless, the sums of the individual characters are integers.

As an example, consider the O2p_x orbital in H_2O. Because H_2O belongs to the point group C_{2v}, we know by referring to the C_{2v} character table (Table 11.2) that the labels available for the orbitals are a_1, a_2, b_1, and b_2. We can decide the appropriate label for O2p_x by noting that under a 180° rotation (C_2) the orbital changes sign (Fig. 11.22), so it must be either B_1 or B_2, as only these two symmetry types have character −1 under C_2. The O2p_x orbital also changes sign under the reflection σ'_v, which identifies it as B_1. As we shall see, any molecular orbital built from this atomic orbital will also be a b_1 orbital. Similarly, O2p_y changes sign under C_2 but not under σ'_v; therefore, it can contribute to b_2 orbitals.

The behaviour of s, p, and d orbitals on a central atom under the symmetry operations of the molecule is so important that the symmetry species of these orbitals are generally indicated in a character table. To make these allocations, we look at the symmetry species of x, y, and z, which appear on the right-hand side of the character table. Thus, the position of z in Table 11.3 shows that p_z (which is proportional to $zf(r)$), has symmetry species A_1 in C_{3v}, whereas p_x and p_y (which are proportional to $xf(r)$ and $yf(r)$, respectively) are jointly of E symmetry. In technical terms, we say that p_x and p_y jointly **span** an irreducible representation of symmetry species E. An s orbital on the central atom always spans the fully symmetrical irreducible representation (typically labelled A_1 but sometimes A'_1) of a group as it is unchanged under all symmetry operations.

The five d orbitals of a shell are represented by xy for d_{xy}, etc., and are also listed on the right of the character table. We can see at a glance that in C_{3v}, d_{xy} and $d_{x^2-y^2}$ on a central atom jointly belong to E and hence form a doubly degenerate pair.

(e) The classification of linear combinations of orbitals

So far, we have dealt with the symmetry classification of individual orbitals. The same technique may be applied to linear combinations of orbitals on atoms that are related by symmetry transformations of the molecule, such as the combination $\psi_1 = \psi_A + \psi_B + \psi_C$ of the three H1s orbitals in the C_{3v} molecule NH_3 (Fig. 11.23). This combination remains unchanged under a C_3 rotation and under any of the three vertical reflections of the group, so its characters are

$$\chi(E) = 1 \qquad \chi(C_3) = 1 \qquad \chi(\sigma_v) = 1$$

Comparison with the C_{3v} character table shows that ψ_1 is of symmetry species A_1, and therefore that it contributes to a_1 molecular orbitals in NH_3.

Example 11.3 *Identifying the symmetry species of orbitals*

Identify the symmetry species of the orbital $\psi = \psi_A - \psi_B$ in a C_{2v} NO_2 molecule, where ψ_A is an O2p$_x$ orbital on one O atom and ψ_B that on the other O atom.

Method The negative sign in ψ indicates that the sign of ψ_B is opposite to that of ψ_A. We need to consider how the combination changes under each operation of the group, and then write the character as +1, −1, or 0 as specified above. Then we compare the resulting characters with each row in the character table for the point group, and hence identify the symmetry species.

Answer The combination is shown in Fig. 11.24. Under C_2, ψ changes into itself, implying a character of +1. Under the reflection σ_v, both orbitals change sign, so $\psi \rightarrow -\psi$, implying a character of −1. Under σ'_v, $\psi \rightarrow -\psi$, so the character for this operation is also −1. The characters are therefore

$$\chi(E)=1 \qquad \chi(C_2)=1 \qquad \chi(\sigma_v)=-1 \qquad \chi(\sigma'_v)=-1$$

These values match the characters of the A_2 symmetry species, so ψ can contribute to an a_2 orbital.

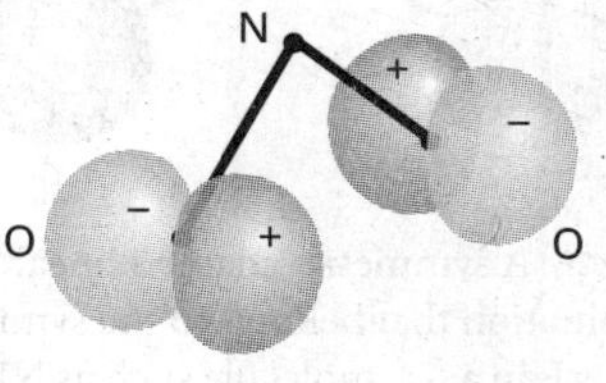

Fig. 11.24 One symmetry-adapted linear combination of O2p$_x$ orbitals in the C_{2v} NO_2^- molecule.

Self-test 11.3 Consider $PtCl_4^-$, in which the Cl ligands form a square planar array of point group D_{4h} (**26**). Identify the symmetry type of the combination $\psi_A - \psi_B + \psi_C - \psi_D$ where each ψ is a Cl3s orbital. [B_{2g}]

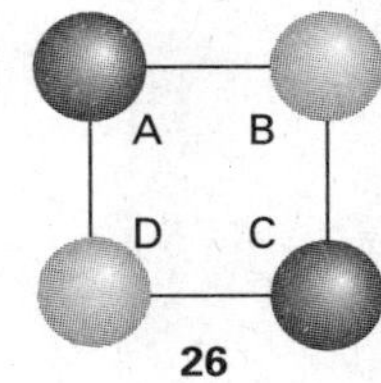

11.5 Vanishing integrals and orbital overlap

Key points **Character tables provide a basis for making various judgements based on symmetry. (a) They are used to decide whether an integral is necessarily zero: it must include a component that is a basis for the totally symmetric representation. (b) Only orbitals of the same symmetry species may have nonzero overlap. (c) Symmetry-adapted linear combinations are the building blocks of LCAO molecular orbitals.**

Suppose we had to evaluate the integral

$$I = \int f_1 f_2 \, d\tau \qquad (11.7)$$

where f_1 and f_2 are functions. For example, f_1 might be an atomic orbital A on one atom and f_2 an atomic orbital B on another atom, in which case I would be their overlap integral. If we knew that the integral is zero, we could say at once that a molecular orbital does not result from (A,B) overlap in that molecule. We shall now see that character tables provide a quick way of judging whether an integral is necessarily zero.

(a) The criteria for vanishing integrals

The key point in dealing with the integral I is that the value of any integral, and of an overlap integral in particular, is independent of the orientation of the molecule (Fig. 11.25). In group theory we express this point by saying that *I is invariant under any symmetry operation of the molecule*, and that each operation brings about the trivial transformation $I \rightarrow I$. Because the volume element $d\tau$ is invariant under any symmetry operation, it follows that the integral is nonzero only if the integrand itself, the product $f_1 f_2$, is unchanged by any symmetry operation of the molecular point group. If the integrand changed sign under a symmetry operation, the integral would

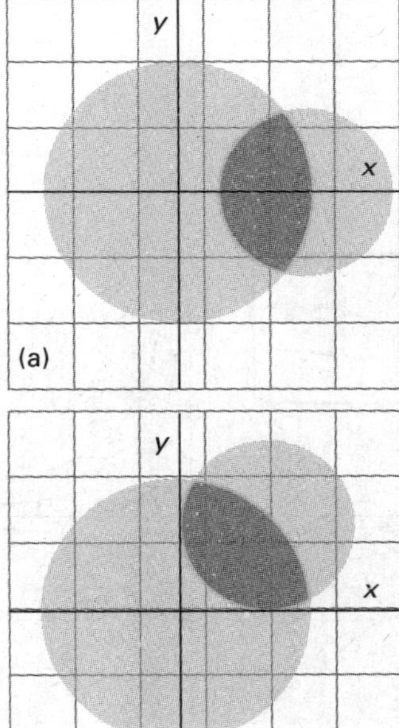

Fig. 11.25 The value of an integral I (for example, an area) is independent of the coordinate system used to evaluate it. That is, I is a basis of a representation of symmetry species A_1 (or its equivalent).

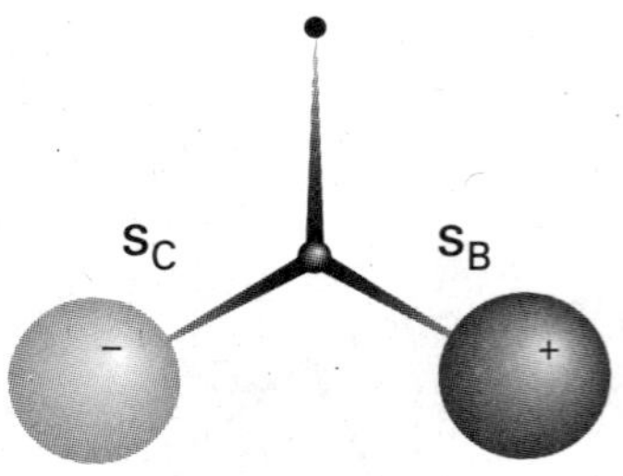

Fig. 11.26 A symmetry-adapted linear combination that belongs to the symmetry species E in a C_{3v} molecule such as NH_3. This combination can form a molecular orbital by overlapping with the p_x orbital on the central atom (the orbital with its axis parallel to the width of the page; see Fig. 11.29c).

be the sum of equal and opposite contributions, and hence would be zero. It follows that the only contribution to a nonzero integral comes from functions for which under any symmetry operation of the molecular point group $f_1f_2 \rightarrow f_1f_2$, and hence for which the characters of the operations are all equal to +1. Therefore, for I not to be zero, *the integrand f_1f_2 must have symmetry species A_1* (or its equivalent in the specific molecular point group).

We use the following procedure to deduce the symmetry species spanned by the product f_1f_2 and hence to see whether it does indeed span A_1.

1. Decide on the symmetry species of the individual functions f_1 and f_2 by reference to the character table, and write their characters in two rows in the same order as in the table.
2. Multiply the numbers in each column, writing the results in the same order.
3. Inspect the row so produced, and see if it can be expressed as a sum of characters from each column of the group. The integral must be zero if this sum does not contain A_1.

For example, if f_1 is the s_N orbital in NH_3 and f_2 is the linear combination $s_3 = s_B - s_C$ (Fig. 11.26), then, because s_N spans A_1 and s_3 is a member of the basis spanning E, we write

f_1:	1	1	1
f_2:	2	−1	0
f_1f_2:	2	−1	0

The characters 2, −1, 0 are those of E alone, so the integrand does not span A_1. It follows that the integral must be zero. Inspection of the form of the functions (see Fig. 11.26) shows why this is so: s_3 has a node running through s_N. Had we taken $f_1 = s_N$ and $f_2 = s_1$ instead, where $s_1 = s_A + s_B + s_C$, then because each spans A_1 with characters 1,1,1:

f_1:	1	1	1
f_2:	1	1	1
f_1f_2:	1	1	1

The characters of the product are those of A_1 itself. Therefore, s_1 and s_N may have nonzero overlap. A short cut that works when f_1 and f_2 are bases for irreducible representations of a group is to note their symmetry species: if they are different, then the integral of their product must vanish; if they are the same, then the integral may be nonzero.

It is important to note that group theory is specific about when an integral must be zero, but integrals that it allows to be nonzero may be zero for reasons unrelated to symmetry. For example, the N–H distance in ammonia may be so great that the (s_1, s_N) overlap integral is zero simply because the orbitals are so far apart.

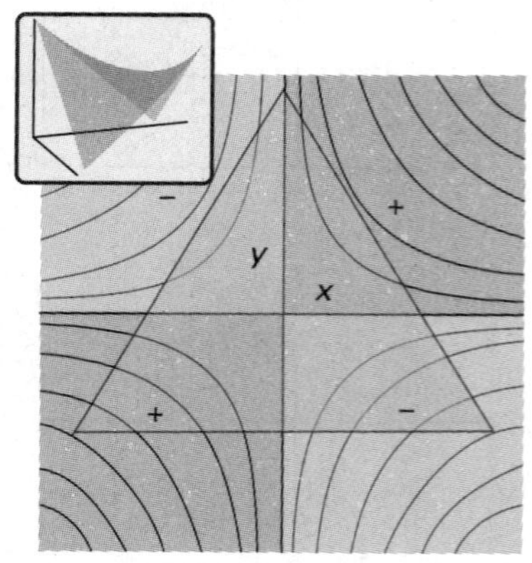

Fig. 11.27 The integral of the function $f = xy$ over the tinted region is zero. In this case, the result is obvious by inspection, but group theory can be used to establish similar results in less obvious cases. The insert shows the shape of the function in three dimensions.

Example 11.4 *Deciding if an integral must be zero (1)*

May the integral of the function $f = xy$ be nonzero when evaluated over a region the shape of an equilateral triangle centred on the origin (Fig. 11.27)?

Method First, note that an integral over a single function f is included in the previous discussion if we take $f_1 = f$ and $f_2 = 1$ in eqn 11.7. Therefore, we need to judge whether f alone belongs to the symmetry species A_1 (or its equivalent) in the point group of the system. To decide that, we identify the point group and then examine the character table to see whether f belongs to A_1 (or its equivalent).

Answer An equilateral triangle has the point-group symmetry D_{3h}. If we refer to the character table of the group, we see that xy is a member of a basis that spans the irreducible representation E″. Therefore, its integral must be zero, because the integrand has no component that spans A_1'.

Self-test 11.4 Can the function $x^2 + y^2$ have a nonzero integral when integrated over a regular pentagon centred on the origin? [Yes, Fig. 11.28]

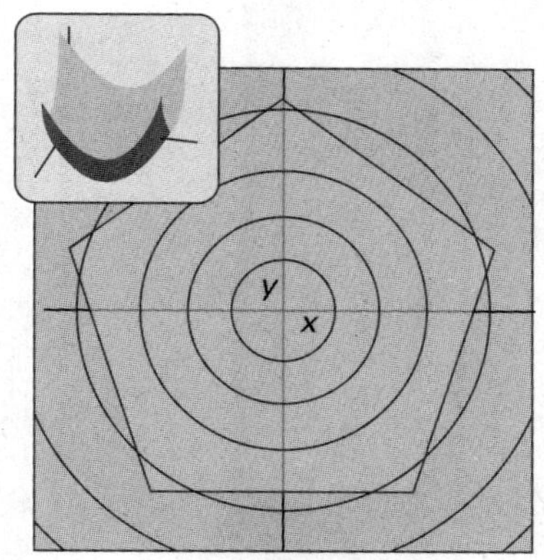

Fig. 11.28 The integration of a function over a pentagonal region. The insert shows the shape of the function in three dimensions.

In many cases, the product of functions f_1 and f_2 spans a sum of irreducible representations. For instance, in C_{2v} we may find the characters 2, 0, 0, −2 when we multiply the characters of f_1 and f_2 together. In this case, we note that these characters are the sum of the characters for A_2 and B_1:

	E	C_{2v}	σ_v	σ_v'
A_2	1	1	−1	−1
B_1	1	−1	1	−1
$A_2 + B_1$	2	0	0	−2

To summarize this result we write the symbolic expression $A_2 \times B_1 = A_2 + B_1$, which is called the **decomposition of a direct product**. This expression is symbolic. The × and + signs in this expression are not ordinary multiplication and addition signs: formally, they denote technical procedures with matrices called a 'direct product' and a 'direct sum'. Because the sum on the right does not include a component that is a basis for an irreducible representation of symmetry species A_1, we can conclude that the integral of $f_1 f_2$ over all space is zero in a C_{2v} molecule.

Whereas the decomposition of the characters 2, 0, 0, −2 can be done by inspection in this simple case, in other cases and more complex groups the decomposition is often far from obvious. For example, if we found the characters 8, −2, −6, 4, it would not be obvious that the sum contains A_1. Group theory, however, provides a systematic way of using the characters of the representation spanned by a product to find the symmetry species of the irreducible representations. The formal statement of the approach is as follows. We write the reduction of the representation as

$$\Gamma = \sum_n N_n \Gamma^{(n)} \qquad (11.8a)$$

where N_n is the number of times that the irreducible representation $\Gamma^{(n)}$ occurs in the reducible representation Γ; then

$$N_n = \frac{1}{h}\sum_R \chi^{(n)}(R)^* \chi(R) \qquad \text{Reduction of a representation} \qquad (11.8b)$$

where h is the order of the group, $\chi(R)$ the characters we are analysing for each operation R, and $\chi^{(n)}(R)$ the corresponding characters for the irreducible representation $\Gamma^{(n)}$. We have allowed for the possibility that the characters are complex, but in most cases they are real. The verbal interpretation of this recipe is as follows:

1. Write down a table with columns headed by the symmetry operations of the group.

2. In the first row write down the characters of the symmetry species we want to analyse.

3. In the second row, write down the characters of the irreducible representation Γ we are interested in.

4. Multiply the two rows together, add the products together, and divide by the order of the group.

The resulting number is the number of times $\Gamma^{(n)}$ occurs in the decomposition.

• A brief illustration

To find whether A_1 does indeed occur in the product with characters 8, −2, −6, 4 in C_{2v}, we draw up the following table:

	E	C_{2v}	σ_v	σ'_v	
					$h = 4$ (the order of the group)
f_1f_2	8	−2	−6	4	(the characters of the product)
A_1	1	1	1	1	(the symmetry species we are interested in)
	8	−2	−6	4	(the product of the two sets of characters)

The sum of the numbers in the last line is 4; when that number is divided by the order of the group, we get 1, so A_1 occurs once in the decomposition. When the procedure is repeated for all four symmetry species, we find that f_1f_2 spans $A_1 + 2A_2 + 5B_2$. •

Self-test 11.5 Does A_2 occur among the symmetry species of the irreducible representations spanned by a product with characters 7, −3, −1, 5 in the group C_{2v}? [No]

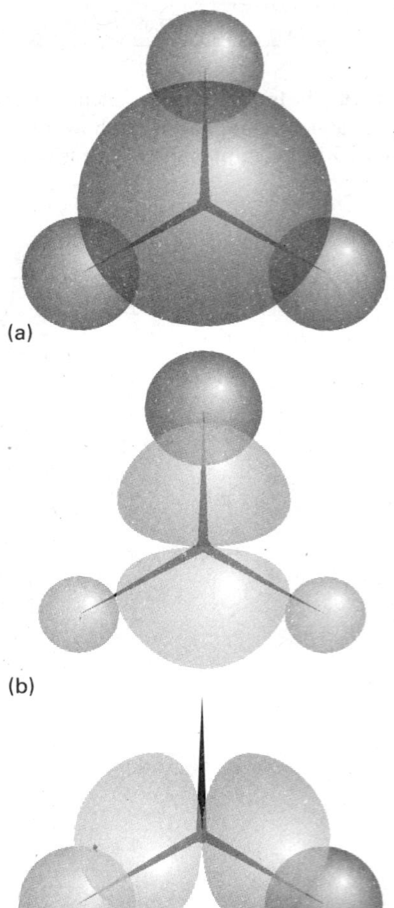

Fig. 11.29 Orbitals of the same symmetry species may have non-vanishing overlap. This diagram illustrates the three bonding orbitals that may be constructed from (N2s, H1s) and (N2p, H1s) overlap in a C_{3v} molecule. (a) a_1; (b) and (c) the two components of the doubly degenerate e orbitals. (There are also three antibonding orbitals of the same species.)

(b) Orbitals with nonzero overlap

The rules just given let us decide which atomic orbitals may have nonzero overlap in a molecule. We have seen that s_N may have nonzero overlap with s_1 (the combination $s_A + s_B + s_C$), so bonding and antibonding molecular orbitals can form from (s_N, s_1) overlap (Fig. 11.29). The general rule is that *only orbitals of the same symmetry species may have nonzero overlap*, so only orbitals of the same symmetry species form bonding and antibonding combinations. It should be recalled from Chapter 10 that the selection of atomic orbitals that had mutual nonzero overlap is the central and initial step in the construction of molecular orbitals by the LCAO procedure. We are therefore at the point of contact between group theory and the material introduced in that chapter. The molecular orbitals formed from a particular set of atomic orbitals with nonzero overlap are labelled with the lower-case letter corresponding to the symmetry species. Thus, the (s_N, s_1)-overlap orbitals are called a_1 orbitals (or a_1^* if we wish to emphasize that they are antibonding).

The linear combinations $s_2 = 2s_A - s_B - s_C$ and $s_3 = s_B - s_C$ have symmetry species E. Does the N atom have orbitals that have nonzero overlap with them (and give rise to *e* molecular orbitals)? Intuition (as supported by Figs. 11.29b and c) suggests that N2p_x and N2p_y should be suitable. We can confirm this conclusion by noting that the character table shows that, in C_{3v}, the functions x and y jointly belong to the symmetry species E. Therefore, N2p_x and N2p_y also belong to E, so may have nonzero overlap with s_2 and s_3. This conclusion can be verified by multiplying the characters and finding that the product of characters can be expressed as the decomposition $E \times E = A_1 + A_2 + E$. The two e orbitals that result are shown in Fig. 11.29 (there are also two antibonding e orbitals).

We can see the power of the method by exploring whether any d orbitals on the central atom can take part in bonding. As explained earlier, reference to the C_{3v} character table shows that d_{z^2} has A_1 symmetry and that the pairs ($d_{x^2-y^2}$, d_{xy}) and (d_{yz}, d_{zx}) each transform as E. It follows that molecular orbitals may be formed by (s_1, d_{z^2}) overlap and by overlap of the (s_2, s_3) combinations with the E d orbitals. Whether or not the d orbitals are in fact important is a question group theory cannot answer because the extent of their involvement depends on energy considerations, not symmetry.

Example 11.5 *Determining which orbitals can contribute to bonding*

The four H1s orbitals of methane span $A_1 + T_2$. With which of the C atom orbitals can they overlap? What bonding pattern would be possible if the C atom had d orbitals available?

Method Refer to the T_d character table (in the *Resource section*) and look for s, p, and d orbitals spanning A_1 or T_2.

Answer An s orbital spans A_1, so it may have nonzero overlap with the A_1 combination of H1s orbitals. The C2p orbitals span T_2, so they may have nonzero overlap with the T_2 combination. The d_{xy}, d_{yz}, and d_{zx} orbitals span T_2, so they may overlap the same combination. Neither of the other two d orbitals span A_1 (they span E), so they remain nonbonding orbitals. It follows that in methane there are (C2s,H1s)-overlap a_1 orbitals and (C2p,H1s)-overlap t_2 orbitals. The C3d orbitals might contribute to the latter. The lowest energy configuration is probably $a_1^2t_2^6$, with all bonding orbitals occupied.

Self-test 11.6 Consider the octahedral SF_6 molecule, with the bonding arising from overlap of S orbitals and a 2p orbital on each F directed towards the central S atom. The latter span $A_{1g} + E_g + T_{1u}$. What S orbitals have nonzero overlap? Suggest what the ground-state configuration is likely to be.

[3s(A_{1g}), 3p(T_{1u}), 3d(E_g); $a_{1g}^2t_{1u}^6e_g^4$]

(c) Symmetry-adapted linear combinations

So far, we have only asserted the forms of the linear combinations (such as s_1, etc.) that have a particular symmetry. Group theory also provides machinery that takes an arbitrary **basis**, or set of atomic orbitals (s_A, etc.), as input and generates combinations of the specified symmetry. Because these combinations are adapted to the symmetry of the molecule, they are called **symmetry-adapted linear combinations** (SALC). Symmetry-adapted linear combinations are the building blocks of LCAO molecular orbitals, for they include combinations such as those used to construct molecular orbitals in benzene. The construction of SALCs is the first step in any molecular orbital treatment of molecules.

The technique for building SALCs is derived by using the full power of group theory. We shall not show the derivation, which is very lengthy, but present the main conclusions as a set of rules. The formal expression is

$$\psi^{(n)} = \frac{1}{h}\sum_R \chi^{(n)}(R)^* R\phi_i \qquad \text{Generation of SALC} \qquad (11.9)$$

where $\psi^{(n)}$ is the symmetry-adapted linear combination we want to develop for the symmetry species $\Gamma^{(n)}$, h is the order of the group, R is an operation of the group, $\chi^{(n)}(R)$ is the character for that operation, and ϕ_i is one of the basis functions. As before, we have allowed for the possibility that a character is complex, but most are real. The verbal interpretation of this expression is:

1. Construct a table showing the effect of each operation on each orbital of the original basis.

2. To generate the combination of a specified symmetry species, take each column in turn and:

(i) Multiply each member of the column by the character of the corresponding operation.

(ii) Add together all the orbitals in each column with the factors as determined in (i).

(iii) Divide the sum by the order of the group.

	s_N	s_A	s_B	s_C
E	s_N	s_A	s_B	s_C
C_3^+	s_N	s_B	s_C	s_A
C_3^-	s_N	s_C	s_A	s_B
σ_v	s_N	s_A	s_C	s_B
σ_v'	s_N	s_B	s_A	s_C
σ_v''	s_N	s_C	s_B	s_A

• **A brief illustration**

From the (s_N,s_A,s_B,s_C) basis in NH_3 we form the table shown in the margin. To generate the A_1 combination, we take the characters for A_1 (1,1,1,1,1,1); then rules (i) and (ii) lead to

$$\psi \propto s_N + s_N + \cdots = 6s_N$$

The order of the group (the number of elements) is 6, so the combination of A_1 symmetry that can be generated from s_N is s_N itself. Applying the same technique to the column under s_A gives

$$\psi = \tfrac{1}{6}(s_A + s_B + s_C + s_A + s_B + s_C) = \tfrac{1}{3}(s_A + s_B + s_C)$$

The same combination is built from the other two columns, so they give no further information. The combination we have just formed is the s_1 combination we used before (apart from the numerical factor). •

We now form the overall molecular orbital by forming a linear combination of all the SALCs of the specified symmetry species. In this case, therefore, the a_1 molecular orbital is

$$\psi = c_N s_N + c_1 s_1$$

This is as far as group theory can take us. The coefficients are found by solving the Schrödinger equation by using the techniques outlined in Chapter 10; they do not come directly from the symmetry of the system.

We run into a problem when we try to generate an SALC of symmetry species E, because, for representations of dimension 2 or more, the rules generate sums of SALCs. This problem can be illustrated as follows. In C_{3v}, the E characters are 2, −1, −1, 0, 0, 0, so the column under s_N gives

$$\psi = \tfrac{1}{6}(2s_N - s_N - s_N + 0 + 0 + 0) = 0$$

The other columns give

$$\tfrac{1}{6}(2s_A - s_B - s_C) \qquad \tfrac{1}{6}(2s_B - s_A - s_C) \qquad \tfrac{1}{6}(2s_C - s_B - s_A)$$

However, any one of these three expressions can be expressed as a sum of the other two (they are not 'linearly independent'). The difference of the second and third gives $\tfrac{1}{2}(s_B - s_C)$, and this combination and the first, $\tfrac{1}{6}(2s_A - s_B - s_C)$, are the two (now linearly independent) SALCs we have used in the discussion of e orbitals.

11.6 Vanishing integrals and selection rules

Key points **A transition dipole moment is nonzero only if the direct product of its three components includes the totally symmetric representation.**

Integrals of the form

$$I = \int f_1 f_2 f_3 \,\mathrm{d}\tau \qquad (11.10)$$

are also common in quantum mechanics for they include matrix elements of operators (Section 7.5e), and it is important to know when they are necessarily zero. For the integral to be nonzero:

The product $f_1 f_2 f_3$ must span A_1 (or its equivalent) or contain a component that spans A_1.

Criterion for not necessarily vanishing

To test whether this is so, the characters of all three functions are multiplied together in the same way as in the rules set out above.

Example 11.6 *Deciding if an integral must be zero (2)*

Does the integral $\int(3d_{z^2})x(3d_{xy})\,\mathrm{d}\tau$ vanish in a C_{2v} molecule?

Method We must refer to the C_{2v} character table (Table 11.2) and the characters of the irreducible representations spanned by $3z^2 - r^2$ (the form of the d_{z^2} orbital), x, and xy; then we can use the procedure set out above (with one more row of multiplication).

Answer We draw up the following table:

	E	C_2	σ_v	σ'_v	
$f_3 = d_{xy}$	1	1	−1	−1	A_2
$f_2 = x$	1	−1	1	−1	B_1
$f_1 = d_{z^2}$	1	1	1	1	A_1
$f_1 f_2 f_3$	1	−1	−1	1	

The characters are those of B_2. Therefore, the integral is necessarily zero.

Self-test 11.7 Does the integral $\int(2p_x)(2p_y)(2p_z)\mathrm{d}\tau$ necessarily vanish in an octahedral O_h environment? [Yes]

We saw in Chapter 9 (*Justification* 9.4), and will see in more detail in Chapters 12 and 13, that the intensity of a spectral line arising from a molecular transition between some initial state with wavefunction ψ_i and a final state with wavefunction ψ_f depends on the (electric) transition dipole moment, μ_{fi}. The z-component of this vector is defined through

$$\mu_{z,fi} = -e\int \psi_f^* z \psi_i \,\mathrm{d}\tau \qquad [11.11]$$

where $-e$ is the charge of the electron. The transition moment has the form of the integral in eqn 11.10, so, once we know the symmetry species of the states, we can use group theory to formulate the selection rules for the transitions.

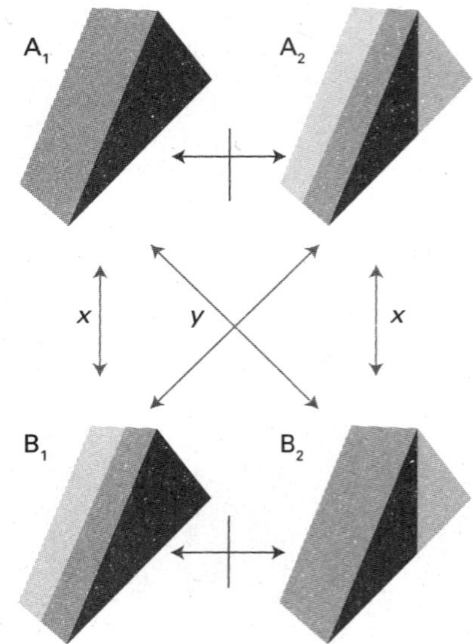

Fig. 11.30 The polarizations of the allowed transitions in a C_{2v} molecule. The shading indicates the structure of the orbitals of the specified symmetry species. The perspective view of the molecule makes it look rather like a door stop; however, from the side, each 'door stop' is in fact an isosceles triangle.

As an example, we investigate whether an electron in an a_1 orbital in H_2O (which belongs to the group C_{2v}) can make an electric dipole transition to a b_1 orbital (Fig. 11.30). We must examine all three components of the transition dipole moment, and take f_2 in eqn 11.10 as x, y, and z in turn. Reference to the C_{2v} character table shows that these components transform as B_1, B_2, and A_1, respectively. The three calculations run as follows:

	z-component				*y*-component				*z*-component				
	E	C_2	σ_v	σ_v'	E	C_2	σ_v	σ_v'	E	C_2	σ_v	σ_v'	
f_3	1	−1	1	−1	1	−1	1	−1	1	−1	1	−1	B_1
f_2	1	−1	1	−1	1	−1	−1	1	1	1	1	1	
f_1	1	1	1	1	1	1	1	1	1	1	1	1	A_1
$f_1f_2f_3$	1	1	1	1	1	1	−1	−1	1	−1	1	−1	

Only the first product (with $f_2 = x$) spans A_1, so only the x-component of the transition dipole moment may be nonzero. Therefore, we conclude that the electric dipole transitions between a_1 and b_1 are allowed. We can go on to state that the radiation emitted (or absorbed) is x-polarized and has its electric field vector in the x-direction, because that form of radiation couples with the x-component of a transition dipole.

Example 11.7 *Deducing a selection rule*

Is $p_x \rightarrow p_y$ an allowed transition in a tetrahedral environment?

Method We must decide whether the product $p_y q p_x$, with $q = x, y,$ or z, spans A_1 by using the T_d character table.

Answer The procedure works out as follows:

	E	$8C_3$	$3C_2$	$6\sigma_d$	$6S_4$	
$f_3(p_y)$	3	0	−1	1	−1	T_2
$f_2(q)$	3	0	−1	1	−1	T_2
$f_1(p_x)$	3	0	−1	1	−1	T_2
$f_1f_2f_3$	27	0	−1	1	−1	

We can use the decomposition procedure described in Section 11.5a to deduce that A_1 occurs (once) in this set of characters, so $p_x \rightarrow p_y$ is allowed.

A more detailed analysis (using the matrix representatives rather than the characters) shows that only $q = z$ gives an integral that may be nonzero, so the transition is z-polarized. That is, the electromagnetic radiation involved in the transition has its electric vector aligned in the z-direction.

Self-test 11.8 What are the allowed transitions, and their polarizations, of a b_1 electron in a C_{4v} molecule? [$b_1 \rightarrow b_1(z)$; $b_1 \rightarrow e(x,y)$]

The following chapters will show many more examples of the systematic use of symmetry. We shall see that the techniques of group theory greatly simplify the analysis of molecular structure and spectra.

Checklist of key equations

Property	Equation	Comment
Group property	If R and S are members of a group, then RS is also a member of the group	A criterion for being considered a group
Decomposition of a direct product	$\Gamma \times \Gamma' = \Gamma^{(1)} + \Gamma^{(2)} + \cdots$	
Reduction of a representation	$N_n = \frac{1}{h}\sum_R \chi^{(n)}(R)^*\chi(R)$	
Generation of a SALC	$\psi^{(n)} = \frac{1}{h}\sum_R \chi^{(n)}(R)^*R\phi_i$	
Typical integral	$I = \int f_1 f_2 f_3 \, d\tau$	Necessarily zero if integrand does not form a basis for the totally symmetric representation

Discussion questions

11.1 Explain what is meant by a 'group'.

11.2 Explain how a molecule is assigned to a point group.

11.3 List the symmetry operations and the corresponding symmetry elements of the point groups.

11.4 Explain the symmetry criteria that allow a molecule to be polar.

11.5 Explain the symmetry criteria that allow a molecule to be optically active.

11.6 Explain what is meant by (a) a representative and (b) a representation in the context of group theory.

11.7 Explain the construction and content of a character table.

11.8 Explain how spectroscopic selection rules arise and how they are formulated by using group theory.

11.9 Outline how a direct product is expressed as a direct sum and how to decide whether the totally symmetric irreducible representation is present in the direct product.

11.10 Identify and list four applications of character tables.

Exercises

11.1(a) The CH_3Cl molecule belongs to the point group C_{3v}. List the symmetry elements of the group and locate them in the molecule.

11.1(b) The CCl_4 molecule belongs to the point group T_d. List the symmetry elements of the group and locate them in the molecule.

11.2(a) Identify the point groups to which the following objects belong: (a) a sphere, (b) an isosceles triangle, (c) an equilateral triangle, (d) an unsharpened cylindrical pencil.

11.2(b) Identify the point groups to which the following objects belong: (a) a sharpened cylindrical pencil, (b) a three-bladed propellor, (c) a four-legged table, (d) yourself (approximately).

11.3(a) List the symmetry elements of the following molecules and name the point groups to which they belong: (a) NO_2, (b) N_2O, (c) $CHCl_3$, (d) $CH_2{=}CH_2$.

11.3(b) List the symmetry elements of the following molecules and name the point groups to which they belong: (a) naphthalene, (b) anthracene, (c) the three dichlorobenzenes.

11.4(a) Assign (a) *cis*-dichloroethene and (b) *trans*-dichloroethene to point groups.

11.4(b) Assign the following molecules to point groups: (a) HF, (b) IF_7 (pentagonal bipyramid), (c) XeO_2F_2 (see-saw), (d) $Fe_2(CO)_9$ (**27**), (e) cubane, C_8H_8, (f) tetrafluorocubane, $C_8H_4F_4$ (**28**).

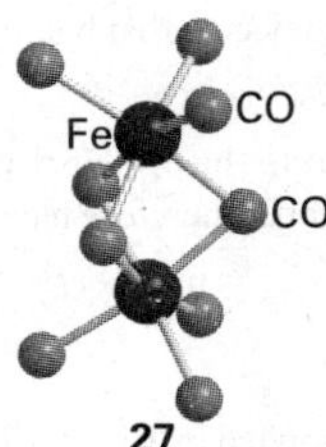

27

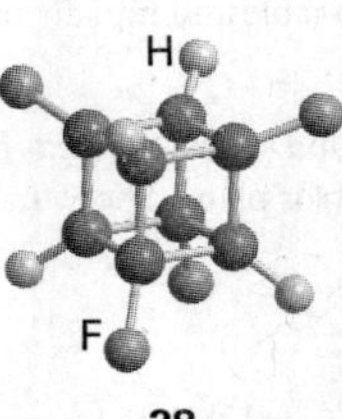

28

11.5(a) Which of the following molecules may be polar? (a) pyridine (C_{2v}), (b) nitroethane (C_s), (c) gas-phase $HgBr_2$ ($D_{\infty h}$), (d) $B_3N_3N_6$ (D_{3h}).

11.5(b) Which of the following molecules may be polar? (a) CH_3Cl (C_{3v}), (b) $HW_2(CO)_{10}$ (D_{4h}), (c) $SnCl_4$ (T_d).

11.6(a) Which of the molecules in Exercises 11.3a and 11.4a can be chiral?

11.6(b) Which of the molecules in Exercises 11.3b and 11.4b can be chiral?

11.7(a) Molecules belonging to the point groups D_{2h} or C_{3h} cannot be chiral. Which elements of these groups rule out chirality?

11.7(b) Molecules belonging to the point groups T_h or T_d cannot be chiral. Which elements of these groups rule out chirality?

11.8(a) The group D_2 consists of the elements E, C_2, C_2', and C_2'', where the three twofold rotations are around mutually perpendicular axes. Construct the group multiplication table.

11.8(b) The group C_{4v} consists of the elements E, $2C_4$, C_2, and $2\sigma_v$, $2\sigma_d$. Construct the group multiplication table.

11.9(a) Use symmetry properties to determine whether or not the integral $\int p_x z p_z \, d\tau$ is necessarily zero in a molecule with symmetry C_{4v}.

11.9(b) Use symmetry properties to determine whether or not the integral $\int p_x z p_z \, d\tau$ is necessarily zero in a molecule with symmetry D_{6h}.

11.10(a) Show that the transition $A_1 \rightarrow A_2$ is forbidden for electric dipole transitions in a C_{3v} molecule.

11.10(b) Is the transition $A_{1g} \rightarrow E_{2u}$ forbidden for electric dipole transitions in a D_{6h} molecule?

11.11(a) Show that the function xy has symmetry species B_2 in the group C_{4v}.

11.11(b) Show that the function xyz has symmetry species A_1 in the group D_2.

11.12(a) Consider the C_{2v} molecule NO_2. The combination $p_x(A) - p_x(B)$ of the two O atoms (with x perpendicular to the plane) spans A_2. Is there any orbital of the central N atom that can have a nonzero overlap with that combination of O orbitals? What would be the case in SO_2, where 3d orbitals might be available?

11.12(b) Consider the D_{3h} ion NO_3^-. Is there any orbital of the central N atom that can have a nonzero overlap with the combination $2p_z(A) - p_z(B) - p_z(C)$ of the three O atoms (with z perpendicular to the plane). What would be the case in SO_3, where 3d orbitals might be available?

11.13(a) The ground state of NO_2 is A_1 in the group C_{2v}. To what excited states may it be excited by electric dipole transitions, and what polarization of light is it necessary to use?

11.13(b) The ClO_2 molecule (which belongs to the group C_{2v}) was trapped in a solid. Its ground state is known to be B_1. Light polarized parallel to the y-axis (parallel to the OO separation) excited the molecule to an upper state. What is the symmetry of that state?

11.14(a) A set of basis functions is found to span a reducible representation of the group C_{4v} with characters 5,1,1,3,1 (in the order of operations in the character table in the *Resource section*). What irreducible representations does it span?

11.14(b) A set of basis functions is found to span a reducible representation of the group D_2 with characters 6,–2,0,0 (in the order of operations in the character table in the *Resource section*). What irreducible representations does it span?

11.15(a) What states of (a) benzene, (b) naphthalene may be reached by electric dipole transitions from their (totally symmetrical) ground states?

11.15(b) What states of (a) anthracene, (b) coronene (**29**) may be reached by electric dipole transitions from their (totally symmetrical) ground states?

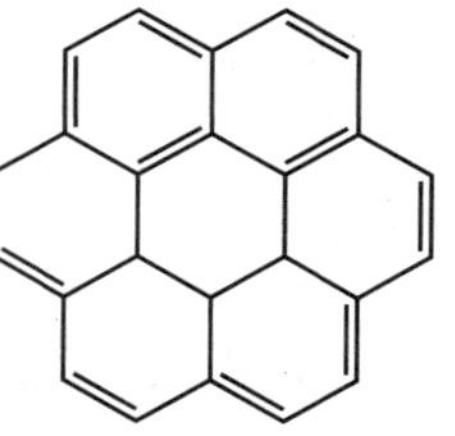

29 Coronene

11.16(a) Write $f_1 = \sin\theta$ and $f_2 = \cos\theta$, and show by symmetry arguments using the group C_s that the integral of their product over a symmetrical range around $\theta = 0$ is zero.

11.16(b) Write $f_1 = x$ and $f_2 = 3x^2 - 1$, and show by symmetry arguments using the group C_s that the integral of their product over a symmetrical range around $x = 0$ is zero.

Problems*

11.1 List the symmetry elements of the following molecules and name the point groups to which they belong: (a) staggered CH_3CH_3, (b) chair and boat cyclohexane, (c) B_2H_6, (d) $[Co(en)_3]^{3+}$, where en is ethylenediamine (ignore its detailed structure), (e) crown-shaped S_8. Which of these molecules can be (i) polar, (ii) chiral?

11.2 The group C_{2h} consists of the elements E, C_2, σ_h, i. Construct the group multiplication table and find an example of a molecule that belongs to the group.

11.3 The group D_{2h} has a C_2 axis perpendicular to the principal axis and a horizontal mirror plane. Show that the group must therefore have a centre of inversion.

11.4 Consider the H_2O molecule, which belongs to the group C_{2v}. Take as a basis the two H1s orbitals and the four valence orbital of the O atom and set up the 6 × 6 matrices that represent the group in this basis. Confirm by explicit matrix multiplication that the group multiplications (a) $C_2\sigma_v = \sigma_v'$ and (b) $\sigma_v\sigma_v' = C_2$. Confirm, by calculating the traces of the matrices, (a) that symmetry elements in the same class have the same character, (b) that the representation is reducible, and (c) that the basis spans $3A_1 + B_1 + 2B_2$.

11.5 Confirm that the z-component of orbital angular momentum is a basis for an irreducible representation of A_2 symmetry in C_{3v}.

11.6 The (one-dimensional) matrices $D(C_3) = 1$ and $D(C_2) = 1$, and $D(C_3) = 1$ and $D(C_2) = -1$ both represent the group multiplication $C_3C_2 = C_6$ in the

* Problems denoted with the symbol ‡ were supplied by Charles Trapp and Carmen Giunta.

group C_{6v} with $D(C_6) = +1$ and -1, respectively. Use the character table to confirm these remarks. What are the representatives of σ_v and σ_d in each case?

11.7 Construct the multiplication table of the Pauli spin matrices, $\boldsymbol{\sigma}$, and the 2×2 unit matrix:

$$\sigma_x = \begin{pmatrix} 0 & 1 \\ 1 & 0 \end{pmatrix} \quad \sigma_y = \begin{pmatrix} 0 & -i \\ i & 0 \end{pmatrix} \quad \sigma_z = \begin{pmatrix} 1 & 0 \\ 0 & -1 \end{pmatrix} \quad \sigma_0 = \begin{pmatrix} 1 & 0 \\ 0 & 1 \end{pmatrix}$$

Do the four matrices from a group under multiplication?

11.8 What irreducible representations do the four H1s orbitals of CH_4 span? Arè there s and p orbitals of the central C atom that may form molecular orbitals with them? Could d orbitals, even if they were present on the C atom, play a role in orbital formation in CH_4?

11.9 Suppose that a methane molecule became distorted to (a) C_{3v} symmetry by the lengthening of one bond, (b) C_{2v} symmetry, by a kind of scissors action in which one bond angle opened and another closed slightly. Would more d orbitals become available for bonding?

11.10‡ B.A. Bovenzi and G.A. Pearse, Jr. (*J. Chem. Soc. Dalton Trans.*, 2763 (1997)) synthesized coordination compounds of the tridentate ligand pyridine-2,6-diamidoxime ($C_7H_9N_5O_2$, **30**). Reaction with $NiSO_4$ produced a complex in which two of the essentially planar ligands are bonded at right angles to a single Ni atom. Name the point group and the symmetry operations of the resulting $[Ni(C_7H_9N_5O_2)_2]^{2+}$ complex cation.

30

11.11‡ R. Eujen *et al.* (*Inorg. Chem.* **36**, 1464 (1997)) prepared and characterized several square-planar Ag(III) complex anions. In the complex anion $[trans\text{-}Ag(CF_3)_2(CN)_2]^-$, the Ag–CN groups are collinear. (a) Assuming free rotation of the CF_3 groups (that is, disregarding the AgCF angles), name the point group of this complex anion. (b) Now suppose the CF_3 groups cannot rotate freely (because the ion was in a solid, for example). Structure (**31**) shows a plane that bisects the NC–Ag–CN axis and is perpendicular to it. Name the point group of the complex if each CF_3 group has a CF bond in that plane (so the CF_3 groups do not point to either CN group preferentially) and the CF_3 groups are (i) staggered (ii) eclipsed.

31

11.12‡ A computational study by C.J. Marsden (*Chem. Phys. Letts.* **245**, 475 (1995)) of AM_x compounds, where A is in Group 14 of the periodic table and M is an alkali metal, shows several deviations from the most symmetric structures for each formula. For example, most of the AM_4 structures were not tetrahedral but had two distinct values for MAM bond angles. They could be derived from a tetrahedron by a distortion shown in (**32**). (a) What is the point group of the distorted tetrahedron? (b) What is the symmetry species of the distortion considered as a vibration in the new, less symmetric group? Some AM_6 structures are not octahedral, but could be derived from an octahedron by translating a C–M–C axis as in (**33**). (c) What is the point group of the distorted octahedron? (d) What is the symmetry species of the distortion considered as a vibration in the new, less symmetric group?

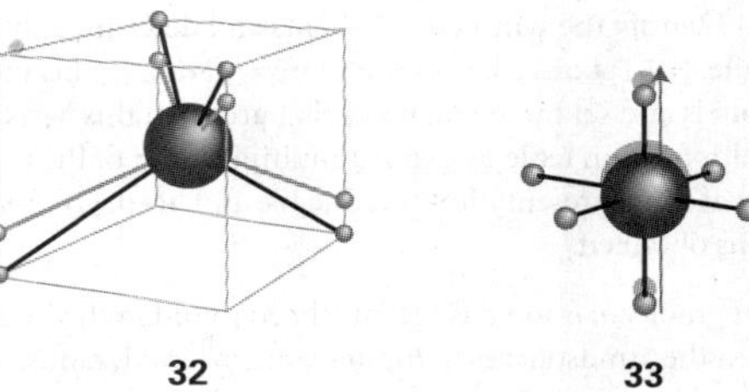

32 **33**

11.13 The algebraic forms of the f orbitals are a radial function multiplied by one of the factors (a) $z(5z^2 - 3r^2)$, (b) $y(5y^2 - 3r^2)$, (c) $x(5x^2 - 3r^2)$, (d) $z(x^2 - y^2)$, (e) $y(x^2 - z^2)$, (f) $x(z^2 - y^2)$, (g) xyz. Identify the irreducible representations spanned by these orbitals in (a) C_{2v}, (b) C_{3v}, (c) T_d, (d) O_h. Consider a lanthanoid ion at the centre of (a) a tetrahedral complex, (b) an octahedral complex. What sets of orbitals do the seven f orbitals split into?

11.14 Does the product xyz necessarily vanish when integrated over (a) a cube, (b) a tetrahedron, (c) a hexagonal prism, each centred on the origin?

11.15 The NO_2 molecule belongs to the group C_{2v}, with the C_2 axis bisecting the ONO angle. Taking as a basis the N2s, N2p, and O2p orbitals, identify the irreducible representations they span, and construct the symmetry-adapted linear combinations.

11.16 Construct the symmetry-adapted linear combinations of C2p$_z$ orbitals for benzene, and use them to calculate the Hückel secular determinant. This procedure leads to equations that are much easier to solve than using the original orbitals and show that the Hückel orbitals are those specified in Section 10.6d.

11.17 The phenanthrene molecule (**34**) belongs to the group C_{2v} with the C_2 axis perpendicular to the molecular plane. (a) Classify the irreducible representations spanned by the carbon 2p$_z$ orbitals and find their symmetry-adapted linear combinations. (b) Use your results from part (a) to calculate the Hückel secular determinant. (c) What states of phenanthrene may be reached by electric dipole transitions from its (totally symmetrical) ground state?

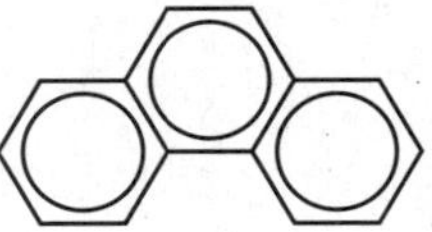

34 Phenanthrene

11.18‡ In a spectroscopic study of C_{60}, F. Negri *et al.* (*J. Phys. Chem.* **100**, 10849 (1996)) assigned peaks in the fluorescence spectrum. The molecule has icosahedral symmetry (I_h). The ground electronic state is A_{1g}, and the lowest-lying excited states are T_{1g} and G_g. (a) Are photon-induced transitions allowed from the ground state to either of these excited states? Explain your answer. (b) What if the transition is accompanied by a vibration that breaks the parity?

11.19 In the square-planar XeF_4 molecule, consider the symmetry-adapted linear combination $p_1 = p_A - p_B + p_C - p_D$ where p_A, p_B, p_C, and p_D are $2p_z$ atomic orbitals on the fluorine atoms (clockwise labelling of the F atoms). Using the reduced point group D_4 rather than the full symmetry point group of the molecule, determine which of the various s, p, and d atomic orbitals on the central Xe atom can form molecular orbitals with p_1.

Applications: to astrophysics and biology

11.20‡ The H_3^+ molecular ion, which plays an important role in chemical reactions occurring in interstellar clouds, is known to be equilateral triangular. (a) Identify the symmetry elements and determine the point group of this molecule. (b) Take as a basis for a representation of this molecule the three H1s orbitals and set up the matrices that group in this basis. (c) Obtain the group multiplication table by explicit multiplication of the matrices. (d) Determine if the representation is reducible and, if so, give the irreducible representations obtained.

11.21‡ The H_3^+ molecular ion has recently been found in the interstellar medium and in the atmospheres of Jupiter, Saturn, and Uranus. The H_4 analogues have not yet been found, and the square-planar structure is thought to be unstable with respect to vibration. Take as a basis for a representation of the point group of this molecule the four H1s orbitals and determine if this representation is reducible.

11.22 Some linear polyenes, of which β-carotene is an example, are important biological co-factors that participate in processes as diverse as the absorption of solar energy in photosynthesis (*Impact I21.1*) and protection against harmful biological oxidations. Use as a model of β-carotene a linear polyene containing 22 conjugated C atoms. (a) To what point group does this model of β-carotene belong? (b) Classify the irreducible representations spanned by the carbon $2p_z$ orbitals and find their symmetry-adapted linear combinations. (c) Use your results from part (b) to calculate the Hückel secular determinant. (d) What states of this model of β-carotene may be reached by electric dipole transitions from its (totally symmetrical) ground state?

11.23 The chlorophylls that participate in photosynthesis (*Impact I21.1*) and the haem groups of cytochromes (*Impact I6.1*) are derived from the porphine dianion group (**35**), which belongs to the D_{4h} point group. The ground electronic state is A_{1g} and the lowest-lying excited state is E_u. Is a photon-induced transition allowed from the ground state to the excited state? Explain your answer.

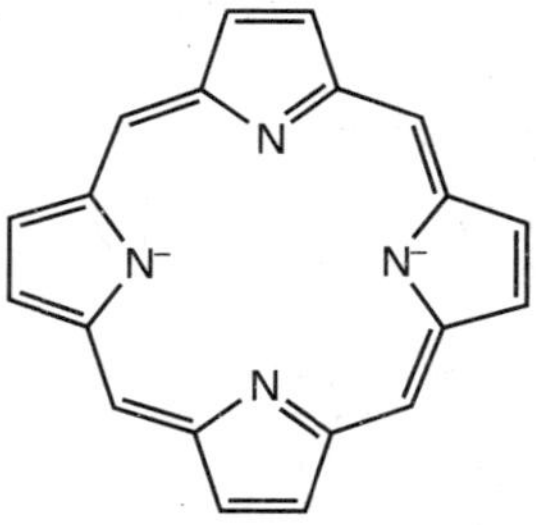

35 Porphine dianion

Molecular spectroscopy 1: rotational and vibrational spectra

12

The general strategy we adopt in the chapter is to set up expressions for the energy levels of molecules and then apply selection rules and considerations of populations to infer the form of spectra. Rotational energy levels are considered first: we see how to derive expressions for their values and how to interpret rotational spectra in terms of molecular dimensions. Not all molecules can occupy all rotational states: we see the experimental evidence for this restriction and its explanation in terms of nuclear spin and the Pauli principle. Next, we consider the vibrational energy levels of diatomic molecules and see that we can use the properties of harmonic oscillators developed in Chapter 8. Then we consider polyatomic molecules and find that their vibrations may be discussed as though they consisted of a set of independent harmonic oscillators, so the same approach as employed for diatomic molecules may be used. We also see that the symmetry properties of the vibrations of polyatomic molecules are helpful for deciding which modes of vibration can be studied spectroscopically.

The origin of spectral lines in molecular spectroscopy is the absorption, emission, or scattering of a photon when the energy of a molecule changes. The difference from atomic spectroscopy is that the energy of a molecule can change not only as a result of electronic transitions but also because it can undergo changes of rotational and vibrational state. Molecular spectra are therefore more complex than atomic spectra. However, they also contain information relating to more properties, and their analysis leads to values of bond strengths, lengths, and angles. They also provide a way of determining a variety of molecular properties, such as dipole moments. Molecular spectroscopy is also useful to astrophysicists and environmental scientists, for the chemical composition of interstellar space and of planetary atmospheres can be inferred from the rotational, vibrational, and electronic spectra of their constituents.

Pure rotational spectra, in which only the rotational state of a molecule changes, can be observed in the gas phase. Vibrational spectra of gaseous samples show features that arise from rotational transitions that accompany the excitation of vibration. Electronic spectra, which are described in Chapter 13, show features arising from simultaneous vibrational and rotational transitions. The simplest way of dealing with these complexities is to tackle each type of transition in turn, and then to see how simultaneous changes affect the appearance of spectra.

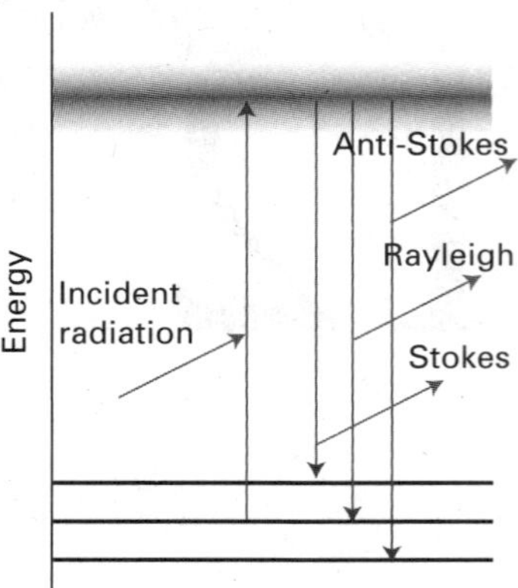

Fig. 12.1 In Raman spectroscopy, an incident photon is scattered from a molecule with either an increase in frequency (if the radiation collects energy from the molecule) or with a lower frequency if it loses energy to the molecule to give the anti-Stokes and Stokes lines, respectively. Scattering without change of frequency results in the Rayleigh line. The process can be regarded as taking place by an excitation of the molecule to a wide range of states (represented by the shaded band), and the subsequent return of the molecule to a lower state; the net energy change is then carried away by the photon.

General features of molecular spectroscopy

All types of spectra have some features in common, and we examine these first. In **emission spectroscopy**, a molecule undergoes a transition from a state of high energy E_1 to a state of lower energy E_2 and emits the excess energy as a photon. In **absorption spectroscopy**, the net absorption of incident radiation is monitored as its frequency is varied. We say *net* absorption, because it will become clear that, when a sample is irradiated, both absorption and emission at a given frequency are stimulated, and the detector measures the difference, the net absorption. In **Raman spectroscopy,** changes in molecular state are explored by examining the frequencies present in the radiation scattered by molecules. In Raman spectroscopy, about 1 in 10^7 of the incident photons collide with the molecules, give up some of their energy, and emerge with a lower energy. These scattered photons constitute the lower-frequency **Stokes radiation** from the sample (Fig. 12.1). Other incident photons may collect energy from the molecules (if they are already excited), and emerge as higher-frequency **anti-Stokes radiation.** The component of radiation scattered without change of frequency is called **Rayleigh radiation.**

The energy, $h\nu$, of the photon emitted or absorbed, and therefore the frequency ν of the radiation emitted or absorbed, is given by the Bohr frequency condition, $h\nu=|E_1-E_2|$ (eqn 7.14). Emission and absorption spectroscopy give the same information about energy level separations, but practical considerations generally determine which technique is employed. In Raman spectroscopy the difference between the frequencies of the scattered and incident radiation is determined by the transitions that take place within the molecule; this technique is used to study molecular vibrations and rotations. We discuss emission spectroscopy in Chapter 13, for it is more important for electronic transitions; here we focus on absorption and Raman spectroscopy, which are widely employed in studies of molecular rotations and vibrations.

12.1 Experimental techniques

Key points **Vibrational transitions are detected by monitoring the net absorption of infrared radiation; rotational transitions are detected by monitoring the net absorption of microwave radiation. In Raman spectroscopy, rotational and vibrational transitions are observed through analysis of radiation scattered by molecules.**

Common to all spectroscopic techniques is a **spectrometer,** an instrument that detects the characteristics of radiation scattered, emitted, or absorbed by atoms and molecules (see *Further information* 12.1). Figure 12.2 shows the general layout of an absorption spectrometer. Radiation from an appropriate source is directed toward a sample and the radiation transmitted strikes a **dispersing element** that separates it into different frequencies. The intensity of radiation at each frequency is then analysed by a suitable detector. In a typical Raman spectrometer, a monochromatic incident laser beam is passed through the sample and the radiation scattered from the front face of the sample is monitored (Fig. 12.3). This detection geometry allows for the study of gases, pure liquids, solutions, suspensions, and solids.

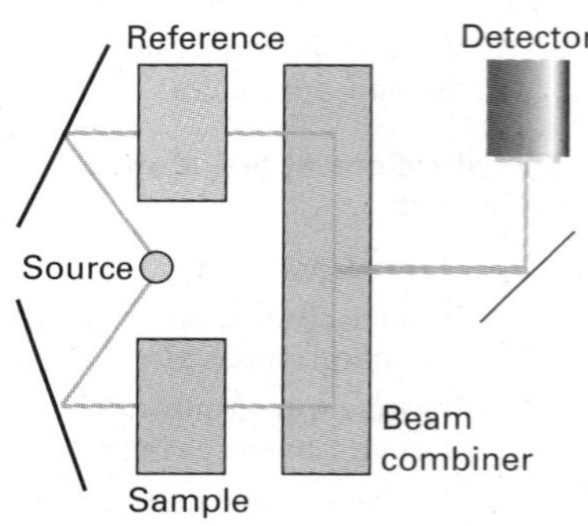

Fig. 12.2 The layout of a typical absorption spectrometer, in which the exciting beams of radiation pass alternately through a sample and a reference cell, and the detector is synchronized with them so that the relative absorption can be determined.

Modern spectrometers, particularly those operating in the infrared and near-infrared, now almost always use **Fourier transform techniques** of spectral detection and analysis. The heart of a Fourier transform spectrometer is a *Michelson interferometer*, a device for analysing the frequencies present in a composite signal. The total signal from a sample is like a chord played on a piano, and the Fourier transform of the signal is equivalent to the separation of the chord into its individual notes, its spectrum. The technique is described more fully in *Further information* 12.1.

The factors that contribute to the linewidths of the spectroscopic transitions of atoms (Section 9.6) apply to molecular spectra too. Thus, the linewidths of rotational spectra are minimized by working with cool samples and minimizing molecular collisions (to increase the collisional lifetimes). All linewidths have a natural limit determined by the lifetime of the upper state, which (as we show in Section 13.4a) increases as ν^3. Thus, rotational (microwave) transitions occur at much lower frequencies than vibrational (infrared) transitions and consequently have much longer lifetimes and hence much smaller natural linewidths: at low pressures rotational linewidths are due principally to Doppler broadening.

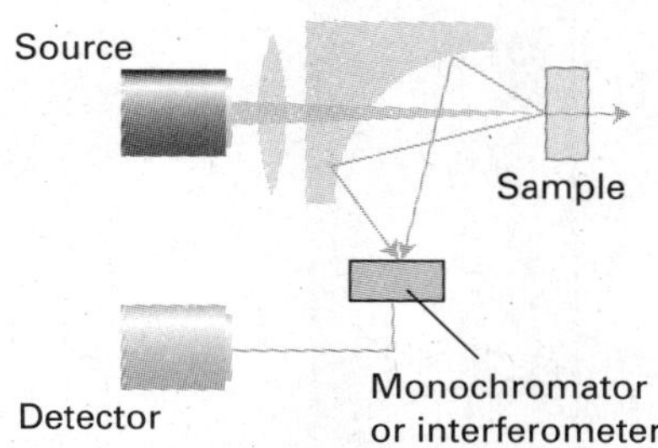

Fig. 12.3 A common arrangement adopted in Raman spectroscopy. A laser beam first passes through a lens and then through a small hole in a mirror with a curved reflecting surface. The focused beam strikes the sample and scattered light is both deflected and focused by the mirror. The spectrum is analysed by a monochromator or an interferometer.

12.2 Selection rules and transition moments

Key points **A gross selection rule specifies the general features a molecule must have if it is to have a spectrum of a given kind. Specific selection rules express the allowed transitions in terms of the changes in quantum numbers.**

We first met the concept of a 'selection rule' in Section 9.3 as a statement about whether a transition is forbidden or allowed. Selection rules also apply to molecular spectra, and the form they take depends on the type of transition. The underlying classical idea is that, for the molecule to be able to interact with the electromagnetic field and absorb or create a photon of frequency ν, it must possess, at least transiently, a dipole oscillating at that frequency. We saw in *Justification* 9.4 in Section 9.3 that this transient dipole is expressed quantum mechanically in terms of the transition dipole moment, $\boldsymbol{\mu}_{\text{fi}}$, between states ψ_{i} and ψ_{f}:

$$\boldsymbol{\mu}_{\text{fi}} = \int \psi_{\text{f}}^* \hat{\boldsymbol{\mu}} \psi_{\text{i}} \, \text{d}\tau \qquad \text{Definition of transition dipole moment} \qquad [12.1]$$

where $\hat{\boldsymbol{\mu}}$ is the electric dipole moment operator. The size of the transition dipole can be regarded as a measure of the charge redistribution that accompanies a transition: a transition will be active (and generate or absorb photons) only if the accompanying charge redistribution is dipolar (Fig. 12.4). Only if the transition dipole moment is nonzero does the transition contribute to the spectrum. It follows that, to identify the selection rules, we must establish the conditions for which $\boldsymbol{\mu}_{\text{fi}} \neq 0$.

A **gross selection rule** specifies the general features a molecule must have if it is to have a spectrum of a given kind. For instance, we shall see that a molecule gives a rotational spectrum only if it has a permanent electric dipole moment. This rule, and others like it for other types of transition, will be explained in the relevant sections of the chapter. A detailed study of the transition moment leads to the **specific selection rules** that express the allowed transitions in terms of the changes in quantum numbers. We have already encountered examples of specific selection rules when discussing atomic spectra (Sections 9.3 and 9.10), such as the rule $\Delta l = \pm 1$ for the angular momentum quantum number.

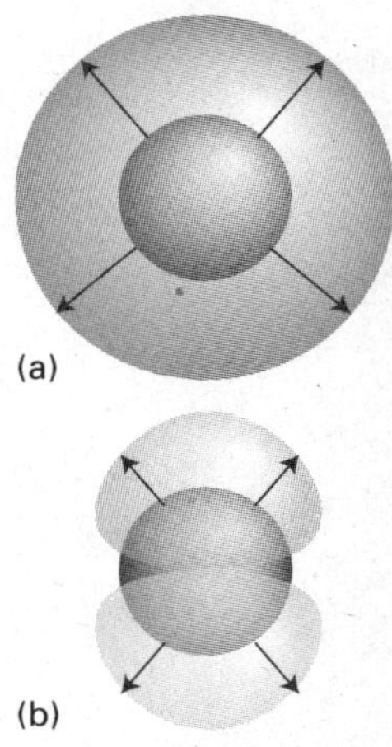

Fig. 12.4 (a) When a 1s electron becomes a 2s electron, there is a spherical migration of charge; there is no dipole moment associated with this migration of charge; this transition is electric-dipole forbidden. (b) In contrast, when a 1s electron becomes a 2p electron, there is a dipole associated with the charge migration; this transition is allowed. (There are subtle effects arising from the sign of the wavefunction that give the charge migration a dipolar character, which this diagram does not attempt to convey.) A similar dipolar redistribution of charge occurs in the active rotational and vibrational transitions of molecules, but is not always easy to visualize.

IMPACT ON ASTROPHYSICS

I12.1 Rotational and vibrational spectroscopy of interstellar species

Observations by the Cosmic Background Explorer (COBE) satellite support the long-held hypothesis that the distribution of energy in the current Universe can be modelled by a Planck distribution (eqn 7.8) with $T = 2.726 \pm 0.001$ K, the bulk of the radiation spanning the microwave region of the spectrum. This *cosmic microwave background radiation* is the residue of energy released during the Big Bang, the event that brought the Universe into existence. Very small fluctuations in the background temperature are believed to account for the large-scale structure of the Universe.

The interstellar space in our galaxy is a little warmer than the cosmic background and consists largely of dust grains and gas clouds. The dust grains are carbon-based compounds and silicates of aluminium, magnesium, and iron, in which are embedded trace amounts of methane, water, and ammonia. Interstellar clouds are significant because it is from them that new stars, and consequently new planets, are formed. The hottest clouds are plasmas with temperatures of up to 10^6 K and densities of only about 3×10^3 particles m^{-3}. Colder clouds range from 0.1 to 1000 solar masses (1 solar mass = 2×10^{30} kg), have a density of about 5×10^5 particles m^{-3}, consist largely of hydrogen atoms, and have a temperature of about 80 K. There are also colder and denser clouds, some with masses greater than 500 000 solar masses, densities greater than 10^9 particles m^{-3}, and temperatures that can be lower than 10 K. They are also called *molecular clouds*, because they are composed primarily of H_2 and CO gas in a ratio of about 10^5 to 1. There are also trace amounts of larger molecules. To place the densities in context, the density of liquid water at 298 K and 1 bar is about 3×10^{28} particles m^{-3}.

It follows from the Boltzmann distribution and the low temperature of a molecular cloud that the vast majority of a cloud's molecules are in their vibrational and electronic ground states. However, rotational excited states are populated at 10–100 K and decay by the emission of radiation. As a result, the spectrum of the cloud in the radiofrequency and microwave regions consists of sharp lines corresponding to rotational transitions (Fig. 12.5). The emitted radiation is collected by Earth-bound or space-borne radiotelescopes, telescopes with antennas and detectors optimized for the collection and analysis of radiation in this range. Earth-bound radiotelescopes are often located at the tops of high mountains, as atmospheric water vapour can reabsorb microwave radiation from space and hence interfere with the measurement.

Over 100 interstellar molecules have been identified by their rotational spectra, often by comparing radiotelescope data with spectra obtained in the laboratory or calculated by computational methods. The experiments have revealed the presence of trace amounts (with abundances of less than 10^{-8} relative to hydrogen) of neutral molecules, ions, and radicals. Examples of neutral molecules include hydrides, oxides (including water), sulfides, halogenated compounds, nitriles, hydrocarbons, aldehydes, alcohols, ethers, ketones, and amides. The largest molecule detected by rotational spectroscopy is the nitrile $HC_{11}N$.

Interstellar space can also be investigated with vibrational spectroscopy by using a combination of telescopes and infrared detectors. The experiments are conducted primarily in space-borne telescopes because the Earth's atmosphere absorbs a great deal of infrared radiation (see *Impact* I12.2). In most cases, absorption by an interstellar

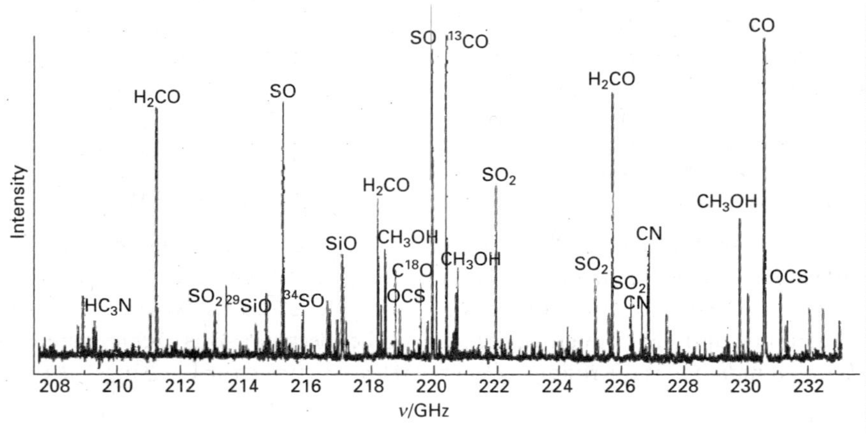

Fig. 12.5 Rotational spectrum of the Orion nebula, showing spectral fingerprints of diatomic and polyatomic molecules present in the interstellar cloud. (Adapted from G.A. Blake *et al.*, *Astrophys. J.* 315, 621 (1987).)

species is detected against the background of infrared radiation emitted by a nearby star. The data can detect the presence of gaseous and solid water, CO, and CO_2 in molecular clouds. In certain cases, infrared emission can be detected, but these events are rare because interstellar space is too cold and does not provide enough energy to promote a significant number of molecules to vibrationally excited states. However, infrared emissions can be observed if molecules are occasionally excited by high-energy photons emitted by hot stars in the vicinity of the cloud. For example, the polycyclic aromatic hydrocarbons hexabenzocoronene ($C_{42}H_{18}$, 1) and circumcoronene ($C_{54}H_{18}$, 2) have been identified from their characteristic infrared emissions.

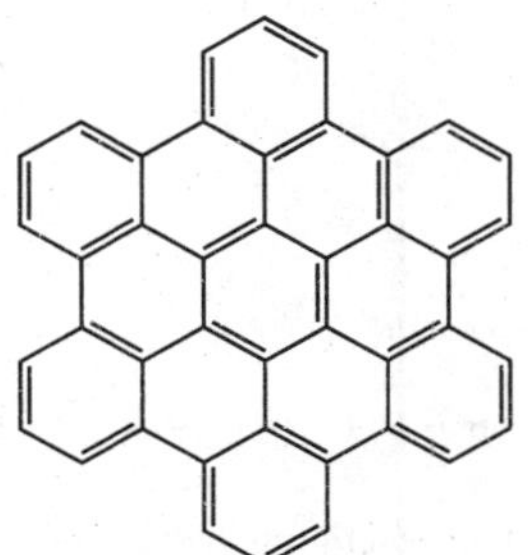

1 Hexabenzocoronene

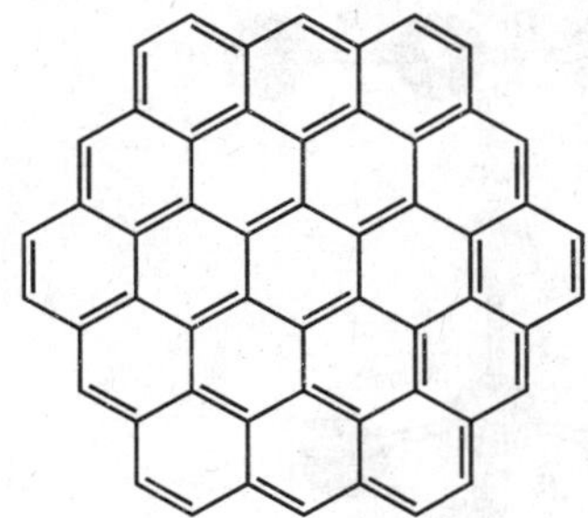

2 Circumcoronene

Pure rotation spectra

The general strategy we adopt for discussing molecular rotational and vibrational spectra and the information they contain is to find expressions for the energy levels of molecules and then to calculate the transition frequencies by applying the selection rules. We then predict the appearance of the spectrum by taking into account the transition moments and the populations of the states. In this section we illustrate the strategy by considering the rotational states of molecules.

12.3 Moments of inertia

Key points **A rigid rotor is a body that does not distort under the stress of rotation. Rigid rotors are classified by noting the number of equal principal moments of inertia.**

The key molecular parameter we shall need is the **moment of inertia**, I, of the molecule. The moment of inertia of a molecule is defined as the mass of each atom multiplied by the square of its distance from the rotational axis passing through the centre of mass of the molecule (Fig. 12.6):

$$I = \sum_i m_i x_i^2 \qquad \text{Definition of moment of inertia} \qquad [12.2]$$

where x_i is the perpendicular distance of the atom i from the axis of rotation. The moment of inertia depends on the masses of the atoms present and the molecular geometry, so we can suspect (and later shall see explicitly) that rotational spectroscopy will give information about bond lengths and bond angles.

In general, the rotational properties of any molecule can be expressed in terms of the moments of inertia about three perpendicular axes set in the molecule (Fig. 12.7). The convention is to label the moments of inertia I_a, I_b, and I_c, with the axes chosen so that $I_c \geq I_b \geq I_a$. For linear molecules, the moment of inertia around the internuclear axis is zero (because $x_i = 0$ for all the atoms). The explicit expressions for the moments of inertia of some symmetrical molecules are given in Table 12.1.

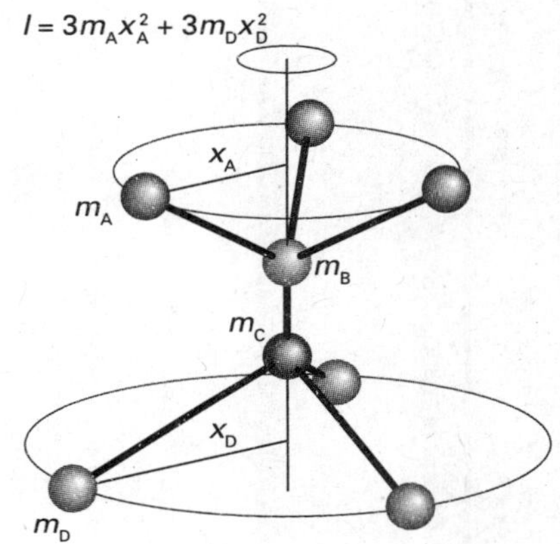

Fig. 12.6 The definition of moment of inertia. In this molecule there are three identical atoms attached to the B atom and three different but mutually identical atoms attached to the C atom. In this example, the centre of mass lies on an axis passing through the B and C atoms, and the perpendicular distances are measured from this axis.

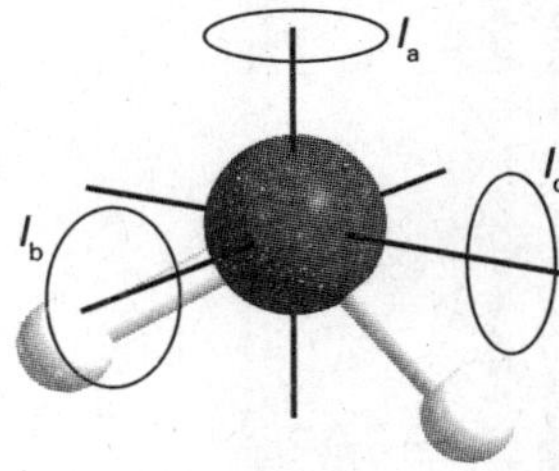

Fig. 12.7 An asymmetric rotor has three different moments of inertia; all three rotational axes coincide at the centre of mass of the molecule.

Table 12.1 Moments of inertia*

1. *Diatomic molecules*

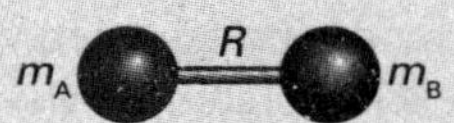

$$I=\mu R^2 \qquad \mu=\frac{m_A m_B}{m}$$

2. *Triatomic linear rotors*

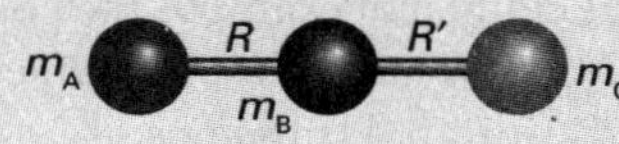

$$I=m_A R^2+m_C R'^2-\frac{(m_A R-m_C R')^2}{m}$$

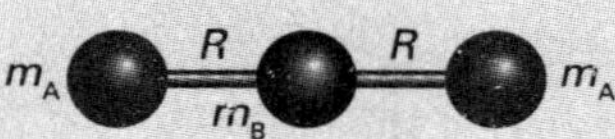

$$I=2m_A R^2$$

3. *Symmetric rotors*

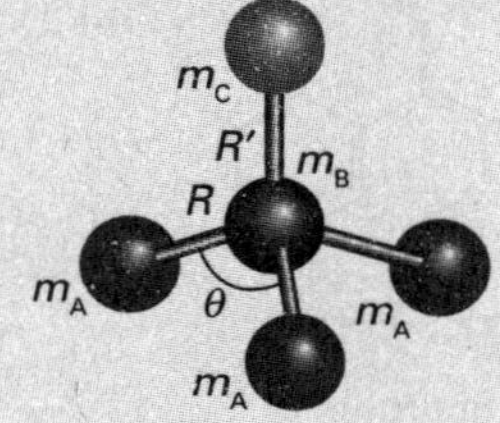

$$I_{\parallel}=2m_A(1-\cos\theta)R^2$$

$$I_{\perp}=m_A(1-\cos\theta)R^2+\frac{m_A}{m}(m_B+m_C)(1+2\cos\theta)R^2$$

$$+\frac{m_C}{m}\{(3m_A+m_B)R'+6m_A R[\tfrac{1}{3}(1+2\cos\theta)]^{1/2}\}R'$$

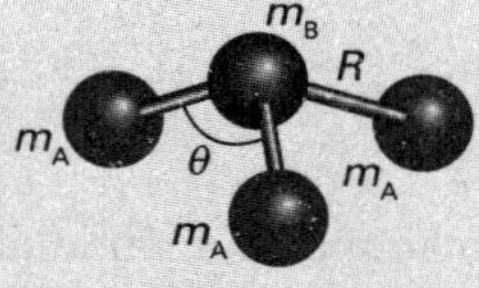

$$I_{\parallel}=2m_A(1-\cos\theta)R^2$$

$$I_{\perp}=m_A(1-\cos\theta)R^2+\frac{m_A m_B}{m}(1+2\cos\theta)R^2$$

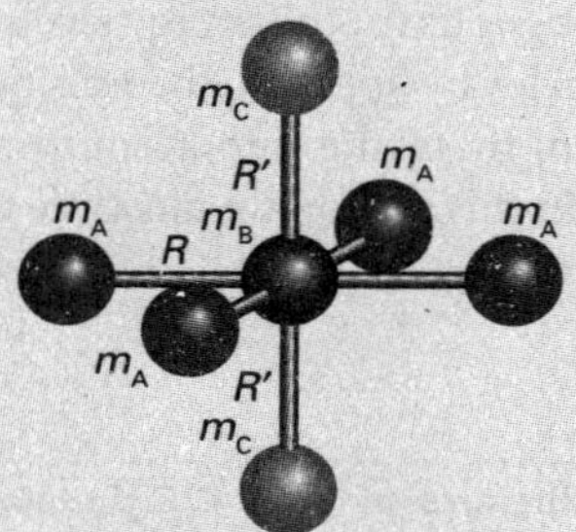

$$I_{\parallel}=4m_A R^2$$

$$I_{\perp}=2m_A R^2+2m_C R'^2$$

4. *Spherical rotors*

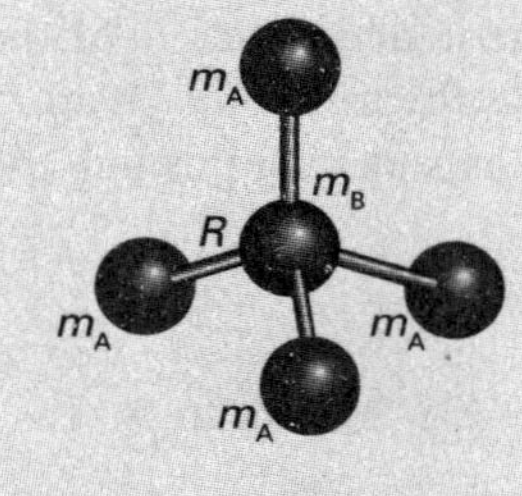

$$I=\tfrac{8}{3}m_A R^2$$

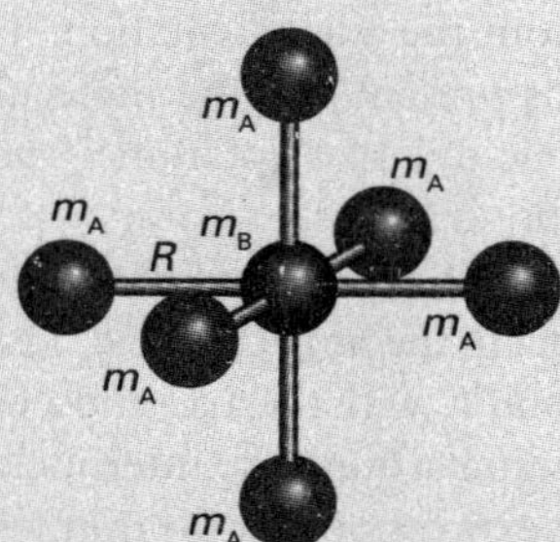

$$I=4m_A R^2$$

* In each case, m is the total mass of the molecule.

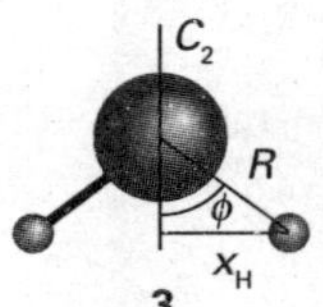

3

Example 12.1 *Calculating the moment of inertia of a molecule*

Calculate the moment of inertia of an H_2O molecule around the axis defined by the bisector of the HOH angle (3). The HOH bond angle is 104.5° and the bond length is 95.7 pm.

Method According to eqn 12.2, the moment of inertia is the sum of the masses multiplied by the squares of their distances from the axis of rotation. The latter can be expressed by using trigonometry and the bond angle and bond length.

Answer From eqn 12.2,

$$I = \sum_i m_i x_i^2 = m_H x_H^2 + 0 + m_H x_H^2 = 2m_H x_H^2$$

If the bond angle of the molecule is denoted 2ϕ and the bond length is R, trigonometry gives $x_H = R\sin\phi$. It follows that

$$I = 2m_H R^2 \sin^2\phi$$

Substitution of the data gives

$$I = 2\times(1.67\times10^{-27}\,\text{kg})\times(9.57\times10^{-11}\,\text{m})^2\times\sin^2(\tfrac{1}{2}\times104.5°)$$
$$= 1.91\times10^{-47}\,\text{kg m}^2$$

Note that the mass of the O atom makes no contribution to the moment of inertia for this mode of rotation as the atom is immobile while the H atoms circulate around it.

Self-test 12.1 Calculate the moment of inertia of a $CH^{35}Cl_3$ molecule around a rotational axis that contains the C–H bond. The C–Cl bond length is 177 pm and the HCCl angle is 107°; $m(^{35}Cl) = 34.97m_u$. [4.99×10^{-45} kg m^2]

A note on good practice The mass to use in the calculation of the moment of inertia is the actual atomic mass, not the element's molar mass; don't forget to convert from relative masses to actual masses by using the atomic mass constant m_u.

We shall suppose initially that molecules are **rigid rotors**, bodies that do not distort under the stress of rotation. Rigid rotors can be classified into four types (Fig. 12.8):

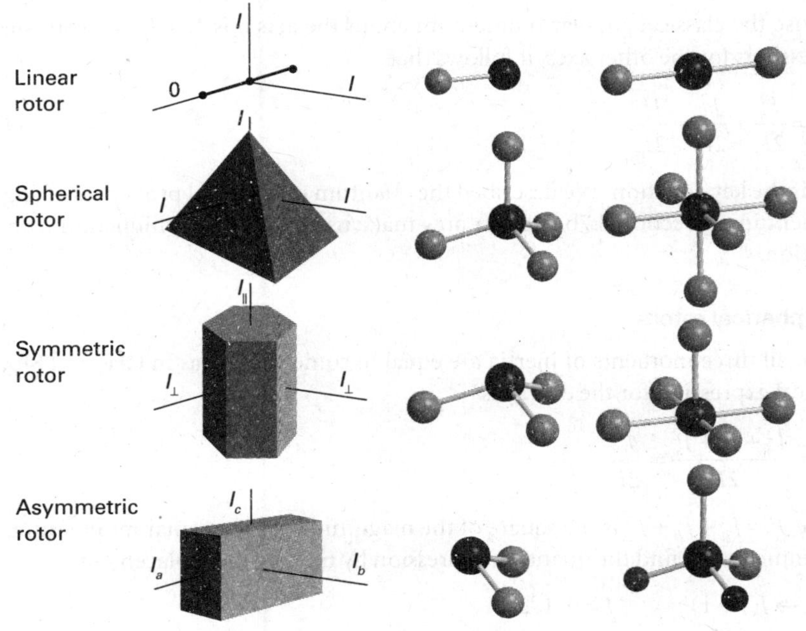

Fig. 12.8 A schematic illustration of the classification of rigid rotors and some typical molecules.

Spherical rotors have three equal moments of inertia (examples: CH_4, SiH_4, and SF_6).

Symmetric rotors have two equal moments of inertia and a third that is nonzero (examples: NH_3, CH_3Cl, and CH_3CN).

Linear rotors have two equal moments of inertia and a third that is zero (examples: CO_2, HCl, OCS, and HC≡CH).

Asymmetric rotors have three different and nonzero moments of inertia (examples: H_2O, H_2CO, and CH_3OH).

Spherical, symmetric, and asymmetric rotors are also called *spherical tops*, etc.

12.4 The rotational energy levels

Key points **(a) The energy levels of a rotor may be expressed in terms of the quantum numbers J, K, and M_J and rotational constants that are related to its moments of inertia. (b) Symmetric rotors are classified as prolate or oblate. (c) For a linear rotor rotation occurs only about an axis perpendicular to the line of atoms. (d) The degeneracies of spherical, symmetric, and linear rotors are $(2J+1)^2$, $2(2J+1)$, and $2J+1$, respectively. (e) Centrifugal distortion arises from forces that change the geometry of a molecule.**

The rotational energy levels of a rigid rotor may be obtained by solving the appropriate Schrödinger equation. Fortunately, however, there is a much less onerous short cut to the exact expressions that depends on noting the classical expression for the energy of a rotating body, expressing it in terms of the angular momentum, and then importing the quantum mechanical properties of angular momentum into the equations.

The classical expression for the energy of a body rotating about an axis a is

$$E_a = \tfrac{1}{2}I_a\omega_a^2 \tag{12.3}$$

where ω_a is the angular velocity (in radians per second, rad s^{-1}) about that axis and I_a is the corresponding moment of inertia. A body free to rotate about three axes has an energy

$$E = \tfrac{1}{2}I_a\omega_a^2 + \tfrac{1}{2}I_b\omega_b^2 + \tfrac{1}{2}I_c\omega_c^2 \tag{12.4}$$

Because the classical angular momentum about the axis a is $J_a = I_\alpha\omega_\alpha$, with similar expressions for the other axes, it follows that

$$E = \frac{J_a^2}{2I_a} + \frac{J_b^2}{2I_b} + \frac{J_c^2}{2I_c} \tag{12.5}$$

This is the key equation. We described the quantum mechanical properties of angular momentum in Section 8.7b and can now make use of them in conjunction with this equation.

(a) Spherical rotors

When all three moments of inertia are equal to some value I, as in CH_4 and SF_6, the classical expression for the energy is

$$E = \frac{J_a^2 + J_b^2 + J_c^2}{2I} = \frac{\mathcal{J}^2}{2I}$$

where $\mathcal{J}^2 = J_a^2 + J_b^2 + J_c^2$ is the square of the magnitude of the angular momentum. We can immediately find the quantum expression by making the replacement

$$\mathcal{J}^2 \rightarrow J(J+1)\hbar^2 \qquad J = 0, 1, 2, \ldots$$

Therefore, the energy of a spherical rotor is confined to the values

$$E_J = J(J+1)\frac{\hbar^2}{2I} \qquad J = 0, 1, 2, \ldots \tag{12.6}$$

The resulting ladder of energy levels is illustrated in Fig. 12.9. The energy is normally expressed in terms of the **rotational constant**, $\tilde{B}$, of the molecule, where

$$hc\tilde{B} = \frac{\hbar^2}{2I} \quad \text{so} \quad \tilde{B} = \frac{\hbar}{4\pi cI} \qquad \text{Definition of rotational constant} \quad [12.7]$$

The expression for the energy is then

$$E_J = hc\tilde{B}J(J+1) \qquad J = 0, 1, 2, \ldots \qquad \text{Energy levels of a spherical rotor} \quad (12.8)$$

The definition of $\tilde{B}$ as a wavenumber is convenient when we come to vibration–rotation spectra (Section 12.11). For pure rotational spectroscopy it is more common to define the rotational constant as a frequency and to denote it simply B. Then $B = \hbar/4\pi I$ and the energy is $E = hBJ(J+1)$. The two quantities are related by $B = c\tilde{B}$.

The energy of a rotational state is normally reported as the **rotational term**, $\tilde{F}(J)$, a wavenumber, by division of both sides of eqn 12.8 by hc:

$$\tilde{F}(J) = \tilde{B}J(J+1) \qquad \text{Rotational terms of a spherical rotor} \quad (12.9)$$

The separation of adjacent levels is

$$\tilde{F}(J+1) - \tilde{F}(J) = \tilde{B}(J+1)(J+2) - \tilde{B}J(J+1) = 2\tilde{B}(J+1) \tag{12.10}$$

Because the rotational constant is inversely proportional to I, large molecules have closely spaced rotational energy levels. We can estimate the magnitude of the separation by considering $C^{35}Cl_4$: from the bond lengths and masses of the atoms we find $I = 4.85 \times 10^{-45}$ kg m^2, and hence $\tilde{B} = 0.0577$ cm^{-1}.

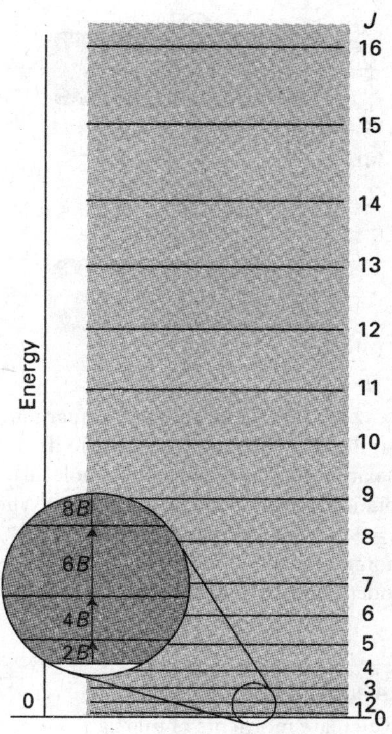

Fig. 12.9 The rotational energy levels of a linear or spherical rotor. Note that the energy separation between neighbouring levels increases as J increases.

(b) Symmetric rotors

In symmetric rotors, two moments of inertia are equal but different from the third (as in CH_3Cl, NH_3, and C_6H_6); the unique axis of the molecule is its **principal axis** (or *figure axis*). We shall write the unique moment of inertia (that about the principal axis) as $I_{\parallel}$ and the other two as $I_{\perp}$. If $I_{\parallel} > I_{\perp}$, the rotor is classified as **oblate** (like a pancake, and C_6H_6); if $I_{\parallel} < I_{\perp}$ it is classified as **prolate** (like a cigar, and CH_3Cl). The classical expression for the energy, eqn 12.5, becomes

$$E = \frac{J_b^2 + J_c^2}{2I_{\perp}} + \frac{J_a^2}{2I_{\parallel}}$$

Again, this expression can be written in terms of $\mathcal{J}^2 = J_a^2 + J_b^2 + J_c^2$

$$E = \frac{\mathcal{J}^2 - J_a^2}{2I_{\perp}} + \frac{J_a^2}{2I_{\parallel}} = \frac{\mathcal{J}^2}{2I_{\perp}} + \left(\frac{1}{2I_{\parallel}} - \frac{1}{2I_{\perp}}\right)J_a^2 \tag{12.11}$$

Now we generate the quantum expression by replacing $\mathcal{J}^2$ by $J(J+1)\hbar^2$, where J is the angular momentum quantum number. We also know from the quantum theory of angular momentum (Section 8.7) that the component of angular momentum about any axis is restricted to the values $K\hbar$, with $K = 0, \pm 1, \ldots, \pm J$. ($K$ is the quantum number used to signify a component on the principal axis; M_J is reserved for a component

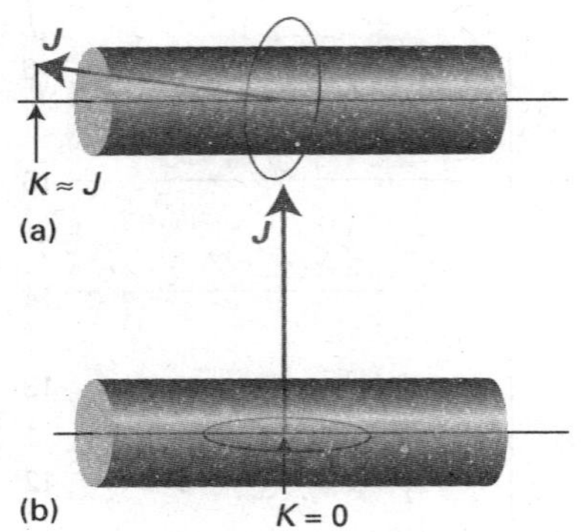

Fig. 12.10 The significance of the quantum number K. (a) When $|K|$ is close to its maximum value, J, most of the molecular rotation is around the figure axis. (b) When $K=0$ the molecule has no angular momentum about its principal axis: it is undergoing end-over-end rotation.

A note on good practice To calculate moments of inertia precisely, it is necessary to specify the nuclide.

on an externally defined axis.) Therefore, we also replace J_a^2 by $K^2\hbar^2$. It follows that the rotational terms are

$$\tilde{F}(J,K) = \tilde{B}J(J+1) + (\tilde{A}-\tilde{B})K^2 \quad J=0,1,2,\ldots \quad K=0,\pm 1,\ldots,\pm J \qquad (12.12)$$

Rotational terms of a symmetric rotor

with

$$\tilde{A} = \frac{\hbar}{4\pi c I_{\parallel}} \qquad \tilde{B} = \frac{\hbar}{4\pi c I_{\perp}} \qquad [12.13]$$

Equation 12.12 matches what we should expect for the dependence of the energy levels on the two distinct moments of inertia of the molecule. When $K = 0$, there is no component of angular momentum about the principal axis, and the energy levels depend only on $I_{\perp}$ (Fig. 12.10). When $K=\pm J$, almost all the angular momentum arises from rotation around the principal axis, and the energy levels are determined largely by $I_{\parallel}$. The sign of K does not affect the energy because opposite values of K correspond to opposite senses of rotation, and the energy does not depend on the sense of rotation.

Example 12.2 *Calculating the rotational energy levels of a molecule*

A $^{14}NH_3$ molecule is a symmetric rotor with bond length 101.2 pm and HNH bond angle 106.7°. Calculate its rotational terms.

Method Begin by calculating the rotational constants $\tilde{A}$ and $\tilde{B}$ by using the expressions for moments of inertia given in Table 12.1. Then use eqn 12.12 to find the rotational terms.

Answer Substitution of $m_A = 1.0078m_u$, $m_B = 14.0031m_u$, $R = 101.2$ pm, and $\theta = 106.7°$ into the second set of symmetric rotor expressions in Table 12.1 gives $I_{\parallel} = 4.4128\times10^{-47}$ kg m^2 and $I_{\perp} = 2.8059\times10^{-47}$ kg m^2. Hence, $\tilde{A} = 6.344$ cm^{-1} and $\tilde{B} = 9.977$ cm^{-1}. It follows from eqn 12.12 that

$$\tilde{F}(J,K)/\text{cm}^{-1} = 9.977J(J+1) - 3.633K^2$$

Upon multiplication by c, $\tilde{F}(J,K)$ acquires units of frequency and is denoted $F(J,K)$:

$$F(J,K)/\text{GHz} = 299.1J(J+1) - 108.9K^2$$

For $J = 1$, the energy needed for the molecule to rotate mainly about its figure axis ($K=\pm J$) is equivalent to 16.32 cm^{-1} (489.3 GHz), but end-over-end rotation ($K=0$) corresponds to 19.95 cm^{-1} (598.1 GHz).

Self-test 12.2 A $CH_3{}^{35}Cl$ molecule has a C–Cl bond length of 178 pm, a C–H bond length of 111 pm, and an HCH angle of 110.5°. Calculate its rotational energy terms.
[$\tilde{F}(J,K)/\text{cm}^{-1} = 0.472J(J+1) + 4.56K^2$; also $F(J,K)/\text{GHz} = 14.1J(J+1) + 137K^2$]

(c) Linear rotors

For a linear rotor (such as CO_2, HCl, and C_2H_2), in which the nuclei are regarded as mass points, the rotation occurs only about an axis perpendicular to the line of atoms and there is zero angular momentum around the line. Therefore, the component of angular momentum around the figure axis of a linear rotor is identically zero, and $K\equiv 0$ in eqn 12.12. The rotational terms of a linear molecule are therefore

$$\tilde{F}(J) = \tilde{B}J(J+1) \qquad J = 0, 1, 2, \ldots \qquad (12.14)$$

Rotational terms of a linear rotor

This expression is the same as eqn 12.9 but we have arrived at it in a significantly different way: here $K \equiv 0$ but for a spherical rotor $\tilde{A} = \tilde{B}$. Note that it is important to set K identically equal to 0 in eqn 12.12 so that the second term vanishes identically; there is then no need to worry about the consequences of $\tilde{A} \propto 1/I_{\parallel}$ approaching infinity as $I_{\parallel}$ approaches 0.

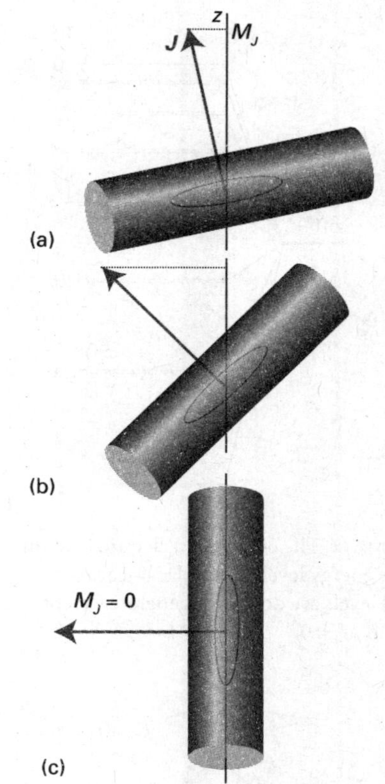

Fig. 12.11 The significance of the quantum number M_J. (a) When M_J is close to its maximum value, J, most of the molecular rotation is around the laboratory z-axis. (b) An intermediate value of M_J. (c) When $M_J = 0$ the molecule has no angular momentum about the z-axis. All three diagrams correspond to a state with $K = 0$; there are corresponding diagrams for different values of K, in which the angular momentum makes a different angle to the molecule's principal axis.

(d) Degeneracies and the Stark effect

The energy of a symmetric rotor depends on J and K, and each level except those with $K = 0$ is doubly degenerate: the states with K and $-K$ have the same energy. However, we must not forget that the angular momentum of the molecule has a component on an external, laboratory-fixed axis. This component is quantized, and its permitted values are $M_J\hbar$, with $M_J = 0, \pm 1, \ldots, \pm J$, giving $2J + 1$ values in all (Fig. 12.11). The quantum number M_J does not appear in the expression for the energy, but it is necessary for a complete specification of the state of the rotor. Consequently, all $2J + 1$ orientations of the rotating molecule have the same energy. It follows that a symmetric rotor level is $2(2J + 1)$-fold degenerate for $K \neq 0$ and $(2J + 1)$-fold degenerate for $K = 0$. A linear rotor has K fixed at 0, but the angular momentum may still have $2J + 1$ components on the laboratory axis, so its degeneracy is $2J + 1$.

A spherical rotor can be regarded as a version of a symmetric rotor in which $\tilde{A} = \tilde{B}$. The quantum number K may still take any one of $2J + 1$ values, but the energy is independent of which value it takes. Therefore, as well as having a $(2J + 1)$-fold degeneracy arising from its orientation in space, the rotor also has a $(2J + 1)$-fold degeneracy arising from its orientation with respect to an arbitrary axis in the molecule. The overall degeneracy of a symmetric rotor with quantum number J is therefore $(2J + 1)^2$. This degeneracy increases very rapidly: when $J = 10$, for instance, there are 441 states of the same energy.

The degeneracy associated with the quantum number M_J (the orientation of the rotation in space) is partly removed when an electric field is applied to a polar molecule (for example, HCl or NH_3), as illustrated in Fig. 12.12. The splitting of states by an electric field is called the **Stark effect**. The energy shift depends on the square of the permanent electric dipole moment, $\boldsymbol{\mu}$, because it depends on the distortion of the rotational wavefunction (a first-order term in $\mathcal{E}$), which favours low-energy orientations of $\boldsymbol{\mu}$, and also on the interaction of that distorted distribution with the applied field (another first-order term in $\mathcal{E}$). Thus we can write

$$E(J,M_J) = hc\tilde{B}J(J+1) + a\mu^2\mathcal{E}^2 \qquad (12.15)$$

Stark effect on the energy of a linear rotor

where a is a constant that depends on J and M_J. The observation of the Stark effect can therefore be used to measure the magnitudes (not the sign) of electric dipole moments, but the technique is limited to molecules that are sufficiently volatile to be studied by rotational spectroscopy. However, as spectra can be recorded for samples at pressures of only about 1 Pa and special techniques (such as using an intense laser beam or an electrical discharge) can be used to vaporize even some quite nonvolatile substances, a wide variety of samples may be studied. Sodium chloride, for example, can be studied as diatomic NaCl molecules at high temperatures.

(e) Centrifugal distortion

We have treated molecules as rigid rotors. However, the atoms of rotating molecules are subject to centrifugal forces that tend to distort the molecular geometry and

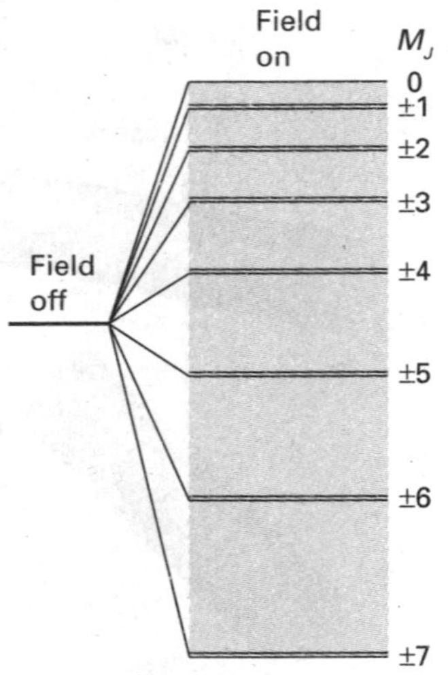

Fig. 12.12 The effect of an electric field on the energy levels of a polar linear rotor. All levels are doubly degenerate except that with $M_J = 0$.

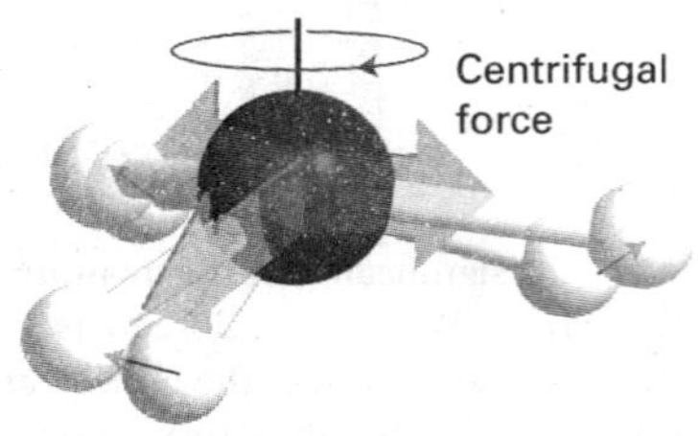

Fig. 12.13 The effect of rotation on a molecule. The centrifugal force arising from rotation distorts the molecule, opening out bond angles and stretching bonds slightly. The effect is to increase the moment of inertia of the molecule and hence to decrease its rotational constant.

change the moments of inertia (Fig. 12.13). The effect of centrifugal distortion on a diatomic molecule is to stretch the bond and hence to increase the moment of inertia. As a result, centrifugal distortion reduces the rotational constant and consequently the energy levels are slightly closer than the rigid-rotor expressions predict. The effect is usually taken into account largely empirically by subtracting a term from the energy and writing

$$\tilde{F}(J) = \tilde{B}J(J+1) - \tilde{D}_J J^2(J+1)^2 \qquad \text{Rotational terms affected by centrifugal distortion} \qquad (12.16)$$

The parameter $\tilde{D}_J$ is the **centrifugal distortion constant**. It is large when the bond is easily stretched. The centrifugal distortion constant of a diatomic molecule is related to the vibrational wavenumber of the bond, $\tilde{\nu}$ (which, as we shall see later, is a measure of its stiffness), through the approximate relation (see Problem 12.21)

$$\tilde{D}_J = \frac{4\tilde{B}^3}{\tilde{\nu}^2} \qquad \text{Centrifugal distortion constant} \qquad (12.17)$$

Hence the observation of the convergence of the rotational levels as J increases can be interpreted in terms of the rigidity of the bond.

12.5 Rotational transitions

Key points (a) For a molecule to give a pure rotational spectrum, it must be polar. The specific rotational selection rules are $\Delta J = \pm 1$, $\Delta M_J = 0, \pm 1$, $\Delta K = 0$. (b) Bond lengths may be obtained from analysis of microwave spectra.

Typical values of $\tilde{B}$ for small molecules are in the region of 0.1–10 cm^{-1} (for example, 0.356 cm^{-1} for NF_3 and 10.59 cm^{-1} for HCl), so rotational transitions lie in the microwave region of the spectrum. The transitions are detected by monitoring the net absorption of microwave radiation. Modulation of the transmitted intensity, which is used to facilitate detection and amplification of the absorption, can be achieved by varying the energy levels with an oscillating electric field. In this **Stark modulation**, an electric field of about 10^5 V m^{-1} and a frequency of 10–100 kHz is applied to the sample.

(a) Rotational selection rules

We have already remarked (Section 12.2) that the gross selection rule for the observation of a pure rotational spectrum is that a molecule must have a permanent electric dipole moment. That is, *for a molecule to give a pure rotational spectrum, it must be polar*. The classical basis of this rule is that a polar molecule appears to possess a fluctuating dipole when rotating but a nonpolar molecule does not (Fig. 12.14). The permanent dipole can be regarded as a handle with which the molecule stirs the electromagnetic field into oscillation (and vice versa for absorption). Homonuclear diatomic molecules and symmetrical linear molecules such as CO_2 are rotationally inactive. Spherical rotors cannot have electric dipole moments unless they become distorted by rotation, so they are also inactive except in special cases. An example of a spherical rotor that does become sufficiently distorted for it to acquire a dipole moment is SiH_4, which has a dipole moment of about 8.3 μD by virtue of its rotation when $J \approx 10$ (for comparison, HCl has a permanent dipole moment of 1.1 D; molecular dipole moments and their units are discussed in Section 17.1). The pure rotational spectrum of SiH_4 has been detected by using long path lengths (10 m) through high-pressure (4 atm) samples.

• **A brief illustration**

Of the molecules N_2, CO_2, OCS, H_2O, $CH_2{=}CH_2$, and C_6H_6, only OCS and H_2O are polar, so only these two molecules have microwave spectra. •

Self-test 12.3 Which of the molecules H_2, NO, N_2O, and CH_4 can have a pure rotational spectrum? [NO, N_2O]

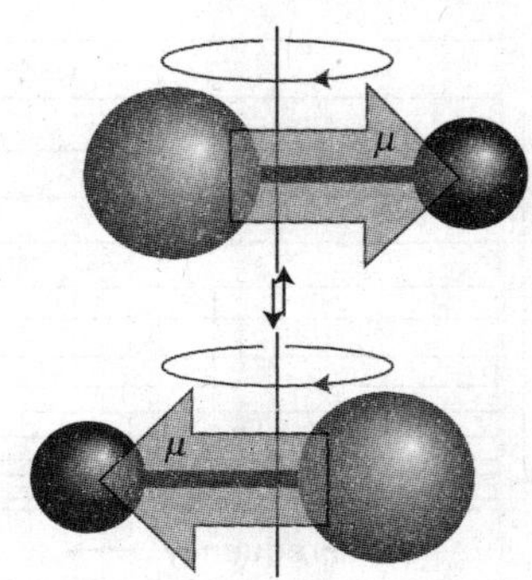

Fig. 12.14 To a stationary observer, a rotating polar molecule looks like an oscillating dipole that can stir the electromagnetic field into oscillation (and vice versa for absorption). This picture is the classical origin of the gross selection rule for rotational transitions.

The specific rotational selection rules are found by evaluating the transition dipole moment between rotational states. We show in *Further information 12.2* that, for a linear molecule, the transition moment vanishes unless the following conditions are fulfilled:

$$\Delta J = \pm 1 \qquad \Delta M_J = 0, \pm 1 \qquad \text{Rotational selection rules for linear rotors} \qquad (12.18)$$

The transition $\Delta J = +1$ corresponds to absorption and the transition $\Delta J = -1$ corresponds to emission. The allowed change in J in each case arises from the conservation of angular momentum when a photon, a spin-1 particle, is emitted or absorbed (Fig. 12.15).

When the transition moment is evaluated for all possible relative orientations of the molecule to the line of flight of the photon, it is found that the total $J+1 \leftrightarrow J$ transition intensity is proportional to

$$|\mu_{J+1,J}|^2 = \left(\frac{J+1}{2J+1}\right)\mu_0^2 \qquad (12.19)$$

where μ_0 is the permanent electric dipole moment of the molecule. The intensity is proportional to the square of the permanent electric dipole moment, so strongly polar molecules give rise to much more intense rotational lines than less polar molecules.

For symmetric rotors, an additional selection rule states that $\Delta K = 0$. To understand this rule, consider the symmetric rotor NH_3, where the electric dipole moment lies parallel to the figure axis. Such a molecule cannot be accelerated into different states of rotation around the figure axis by the absorption of radiation, so $\Delta K = 0$. Therefore, for symmetric rotors the selection rules are:

$$\Delta J = \pm 1 \qquad \Delta M_J = 0, \pm 1 \qquad \Delta K = 0 \qquad \text{Rotational selection rules for symmetric rotors} \qquad (12.20)$$

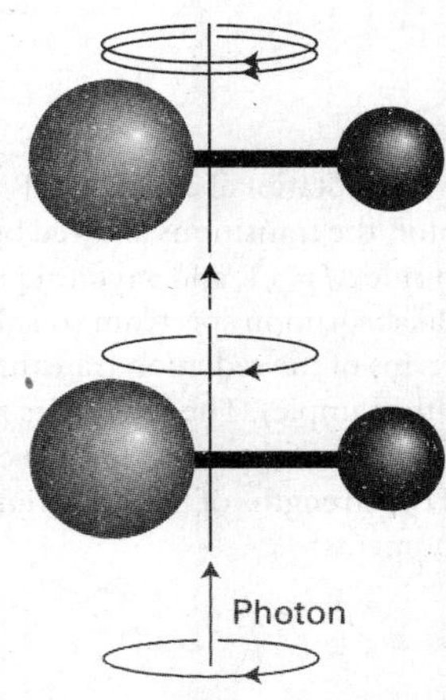

Fig. 12.15 When a photon is absorbed by a molecule, the angular momentum of the combined system is conserved. If the molecule is rotating in the same sense as the spin of the incoming photon, then J increases by 1.

(b) The appearance of rotational spectra

When these selection rules are applied to the expressions for the energy levels of a rigid spherical or linear rotor, it follows that the wavenumbers of the allowed $J+1 \leftarrow J$ absorptions are

$$\tilde{\nu}(J+1 \leftarrow J) = \tilde{F}(J+1) - \tilde{F}(J) = 2\tilde{B}(J+1) \qquad J = 0, 1, 2, \ldots \qquad (12.21a)$$

When centrifugal distortion is taken into account, the corresponding expression obtained from eqn 12.16 is

$$\tilde{\nu}(J+1 \leftarrow J) = 2\tilde{B}(J+1) - 4\tilde{D}_J(J+1)^3 \qquad (12.21b)$$

However, because the second term is typically very small compared with the first, the appearance of the spectrum closely resembles that predicted from eqn 12.21a.

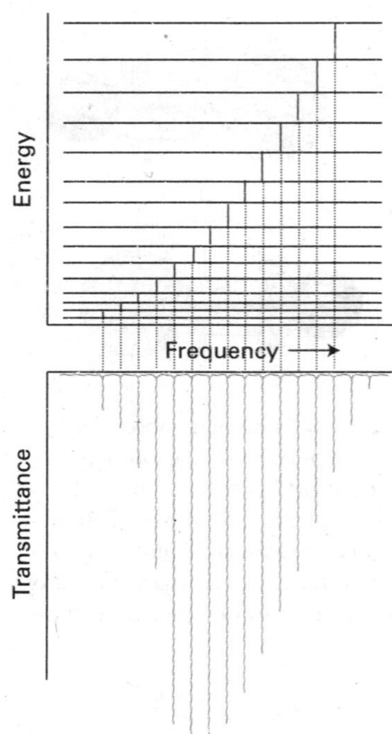

Fig. 12.16 The rotational energy levels of a linear rotor, the transitions allowed by the selection rule $\Delta J = \pm 1$, and a typical pure rotational absorption spectrum (displayed here in terms of the radiation transmitted through the sample). The intensities reflect the populations of the initial level in each case and the strengths of the transition dipole moments.

Example 12.3 *Predicting the appearance of a rotational spectrum*

Predict the form of the rotational spectrum of $^{14}NH_3$.

Method We calculated the energy levels in Example 12.2. The $^{14}NH_3$ molecule is a polar symmetric rotor, so the selection rules $\Delta J = \pm 1$ and $\Delta K = 0$ apply. For absorption, $\Delta J = +1$ and we can use eqn 12.21a.

Answer Because $\tilde{B} = 9.977$ cm^{-1}, we can draw up the following table for the $J + 1 \leftarrow J$ transitions.

J	0	1	2	3	...
$\tilde{\nu}$/cm^{-1}	19.95	39.91	59.86	79.82	...
ν/GHz	598.1	1197	1795	2393	...

The line spacing is 19.95 cm^{-1} (598.1 GHz).

Self-test 12.4 Repeat the problem for $C^{35}ClH_3$ (see Self-test 12.2 for details). [Lines of separation 0.944 cm^{-1} (28.3 GHz)]

The form of the spectrum predicted by eqn 12.21 is shown in Fig. 12.16. The most significant feature is that it consists of a series of lines with wavenumbers $2\tilde{B}$, $4\tilde{B}$, $6\tilde{B}$, ... and of separation $2\tilde{B}$. The measurement of the line spacing gives $\tilde{B}$, and hence the moment of inertia perpendicular to the principal axis of the molecule. Because the masses of the atoms are known, it is a simple matter to deduce the bond length of a diatomic molecule. However, in the case of a polyatomic molecule such as OCS or NH_3, the analysis gives only a single quantity, $I_\perp$, and it is not possible to infer both bond lengths (in OCS) or the bond length and bond angle (in NH_3). This difficulty can be overcome by using isotopically substituted molecules, such as ABC and A′BC; then, by assuming that $R(\text{A–B}) = R(\text{A}'\text{–B})$, both A–B and B–C bond lengths can be extracted from the two moments of inertia. A famous example of this procedure is the study of OCS; the actual calculation is worked through in Problem 12.7. The assumption that bond lengths are unchanged by isotopic substitution is only an approximation, but it is a good approximation in most cases. Nuclear spin, which differs from one isotope to another, also affects the appearance of high-resolution rotational spectra because spin is a source of angular momentum and can couple with the rotation of the molecule itself and hence affect the rotational energy levels.

The intensities of spectral lines increase with increasing J and pass through a maximum before tailing off as J becomes large. The most important reason for the maximum in intensity is the existence of a maximum in the population of rotational levels. The Boltzmann distribution (*Fundamentals F.5*) implies that the population of each state decays exponentially with increasing J, but the degeneracy of the levels increases, and these two opposite trends result in the population of the energy levels (as distinct from the individual states) passing through a maximum. Specifically, the population of a rotational energy level J is given by the Boltzmann expression

$$N_J \propto N g_J e^{-E_J/kT}$$

where N is the total number of molecules and g_J is the degeneracy of the level J. The value of J corresponding to a maximum of this expression is found by treating J as a continuous variable, differentiating with respect to J, and then setting the result equal to zero. The result is (see Problem 12.26)

$$J_{max} \approx \left(\frac{kT}{2hc\tilde{B}}\right)^{1/2} - \tfrac{1}{2} \qquad (12.22)$$

For a typical molecule (for example, OCS, with $\tilde{B} = 0.2\ \text{cm}^{-1}$) at room temperature, $kT \approx 1000hc\tilde{B}$, so $J_{max} \approx 30$. However, it must be recalled that the intensity of each transition also depends on the value of J (eqn 12.19) and on the population difference between the two states involved in the transition. Hence the value of J corresponding to the most intense line is not quite the same as the value of J for the most highly populated level.

12.6 Rotational Raman spectra

Key points **A molecule must be anisotropically polarizable for it to be rotationally Raman active. The specific selection rules are: (i) linear rotors, $\Delta J = 0, \pm 2$; (ii) symmetric rotors, $\Delta J = 0, \pm 1, \pm 2$; $\Delta K = 0$.**

The gross selection rule for rotational Raman transitions is *that the molecule must be anisotropically polarizable*. We begin by explaining what this means. A formal derivation of this rule is given in *Further information 12.2*.

The distortion of a molecule in an electric field is determined by its polarizability, α (Section 17.2). More precisely, if the strength of the field is $\mathcal{E}$, then the molecule acquires an induced dipole moment of magnitude

$$\mu = \alpha\mathcal{E} \tag{12.23}$$

in addition to any permanent dipole moment it may have. An atom is isotropically polarizable. That is, the same distortion is induced whatever the direction of the applied field. The polarizability of a spherical rotor is also isotropic. However, non-spherical rotors have polarizabilities that do depend on the direction of the field relative to the molecule, so these molecules are anisotropically polarizable (Fig. 12.17). The electron distribution in H_2, for example, is more distorted when the field is applied parallel to the bond than when it is applied perpendicular to it, and we write $\alpha_{\parallel} > \alpha_{\perp}$.

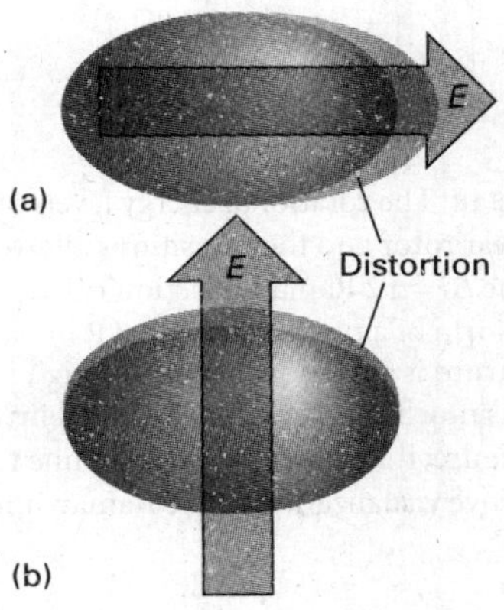

Fig. 12.17 An electric field applied to a molecule results in its distortion, and the distorted molecule acquires a contribution to its dipole moment (even if it is nonpolar initially). The polarizability may be different when the field is applied (a) parallel or (b) perpendicular to the molecular axis (or, in general, in different directions relative to the molecule); if that is so, then the molecule has an anisotropic polarizability.

All linear molecules and diatomics (whether homonuclear or heteronuclear) have anisotropic polarizabilities, and so are rotationally Raman active. This activity is one reason for the importance of rotational Raman spectroscopy, for the technique can be used to study many of the molecules that are inaccessible to microwave spectroscopy. Spherical rotors such as CH_4 and SF_6, however, are rotationally Raman inactive as well as microwave inactive. This inactivity does not mean that such molecules are never found in rotationally excited states. Molecular collisions do not have to obey such restrictive selection rules, and hence collisions between molecules can result in the population of any rotational state.

We show in *Further information 12.2* that the specific rotational Raman selection rules are

Linear rotors: $\Delta J = 0, \pm 2$

Symmetric rotors: $\Delta J = 0, \pm 1, \pm 2; \quad \Delta K = 0$ (12.24)

Rotational Raman selection rules

The $\Delta J = 0$ transitions do not lead to a shift in frequency of the scattered photon in pure rotational Raman spectroscopy, and contribute to the unshifted Rayleigh radiation.

We can predict the form of the Raman spectrum of a linear rotor by applying the selection rule $\Delta J = \pm 2$ to the rotational energy levels (Fig. 12.18). When the molecule makes a transition with $\Delta J = +2$, the scattered radiation leaves the molecule in a higher rotational state, so the wavenumber of the incident radiation, initially $\tilde{\nu}_i$, is decreased. These transitions account for the Stokes lines in the spectrum:

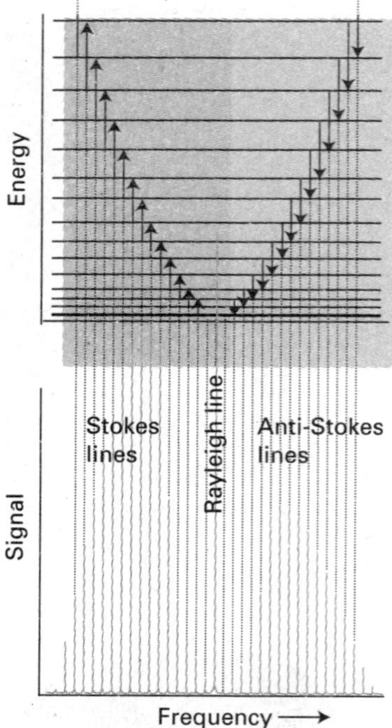

Fig. 12.18 The rotational energy levels of a linear rotor and the transitions allowed by the $\Delta J = \pm 2$ Raman selection rules. The form of a typical rotational Raman spectrum is also shown. The Rayleigh line is much stronger than depicted in the figure; it is shown as a weaker line to improve visualization of the Raman lines.

$$\tilde{\nu}(J+2 \leftarrow J) = \tilde{\nu}_i - \{\tilde{F}(J+2) - \tilde{F}(J)\} = \tilde{\nu}_i - 2\tilde{B}(2J+3) \tag{12.25a}$$

The Stokes lines appear to low frequency of the incident radiation and at displacements $6\tilde{B}$, $10\tilde{B}$, $14\tilde{B}$, ... from $\tilde{\nu}_i$ for $J = 0, 1, 2, \ldots$. When the molecule makes a transition with $\Delta J = -2$, the scattered photon emerges with increased energy. These transitions account for the anti-Stokes lines of the spectrum:

$$\tilde{\nu}(J-2 \leftarrow J) = \tilde{\nu}_i - \{\tilde{F}(J) - \tilde{F}(J-2)\} = \tilde{\nu}_i + 2\tilde{B}(2J-1) \tag{12.25b}$$

The anti-Stokes lines occur at displacements of $6\tilde{B}$, $10\tilde{B}$, $14\tilde{B}$, ... (for $J = 2, 3, 4, \ldots$; $J = 2$ is the lowest state that can contribute under the selection rule $\Delta J = -2$) to high frequency of the incident radiation. The separation of adjacent lines in both the Stokes and the anti-Stokes regions is $4\tilde{B}$, so from its measurement $I_\perp$ can be determined and then used to find the bond lengths exactly as in the case of microwave spectroscopy.

Example 12.4 *Predicting the form of a Raman spectrum*

Predict the form of the rotational Raman spectrum of $^{14}N_2$, for which $\tilde{B} = 1.99\ \text{cm}^{-1}$, when it is exposed to 336.732 nm laser radiation.

Method The molecule is rotationally Raman active because end-over-end rotation modulates its polarizability as viewed by a stationary observer. The Stokes and anti-Stokes lines are given by eqn 12.25.

Answer Because $\lambda_i = 336.732$ nm corresponds to $\tilde{\nu}_i = 29\,697.2\ \text{cm}^{-1}$, eqns 12.25a and 12.25b give the following line positions:

J	0	1	2	3
Stokes lines				
$\tilde{\nu}/\text{cm}^{-1}$	29 685.3	29 677.3	29 669.3	29 661.4
λ/nm	336.868	336.958	337.048	337.139
Anti-Stokes lines				
$\tilde{\nu}/\text{cm}^{-1}$			29 709.1	29 717.1
λ/nm			336.597	336.507

There will be a strong central line at 336.732 nm accompanied on either side by lines of increasing and then decreasing intensity (as a result of transition moment and population effects). The spread of the entire spectrum is very small, so the incident light must be highly monochromatic.

Self-test 12.5 Repeat the calculation for the rotational Raman spectrum of NH_3 ($\tilde{B} = 9.977\ \text{cm}^{-1}$).

[Stokes lines at 29 637.3, 29 597.4, 29 557.5, 29 517.6 cm^{-1}, anti-Stokes lines at 29 757.1, 29 797.0 cm^{-1}]

12.7 Nuclear statistics and rotational states

Key point **The appearance of rotational spectra is affected by nuclear statistics, the selective occupation of rotational states that stems from the Pauli principle.**

If eqn 12.25 is used in conjunction with the rotational Raman spectrum of CO_2, the rotational constant is inconsistent with other measurements of C–O bond lengths. The results are consistent only if it is supposed that the molecule can exist in states with even values of J, so the Stokes lines are $2 \leftarrow 0$, $4 \leftarrow 2$, ... and not $5 \leftarrow 3$, $3 \leftarrow 1$, ...

The explanation of the missing lines is the Pauli principle and the fact that ^{16}O nuclei are spin-0 bosons: just as the Pauli principle excludes certain electronic states, so too does it exclude certain molecular rotational states. The form of the Pauli principle given in Section 9.4b states that, when two identical bosons are exchanged, the overall wavefunction must remain unchanged in every respect, including sign. When a CO_2 molecule rotates through 180°, two identical O nuclei are interchanged, so the overall wavefunction of the molecule must remain unchanged. However, inspection of the form of the rotational wavefunctions (which have the same form as the s, p, etc. orbitals of atoms) shows that they change sign by $(-1)^J$ under such a rotation (Fig. 12.19). Therefore, only even values of J are permissible for CO_2, and hence the Raman spectrum shows only alternate lines.

The selective occupation of rotational states that stems from the Pauli principle is termed **nuclear statistics**. Nuclear statistics must be taken into account whenever a rotation interchanges equivalent nuclei. However, the consequences are not always as simple as for CO_2 because there are complicating features when the nuclei have nonzero spin: there may be several different relative nuclear spin orientations consistent with even values of J and a different number of spin orientations consistent with odd values of J. For molecular hydrogen and fluorine, for instance, with their two identical spin-$\frac{1}{2}$ nuclei, we show in the following *Justification* that there are three times as many ways of achieving a state with odd J than with even J, and there is a corresponding 3:1 alternation in intensity in their rotational Raman spectra (Fig. 12.20). In general, for a homonuclear diatomic molecule with nuclei of spin I, the numbers of ways of achieving states of odd and even J are in the ratio

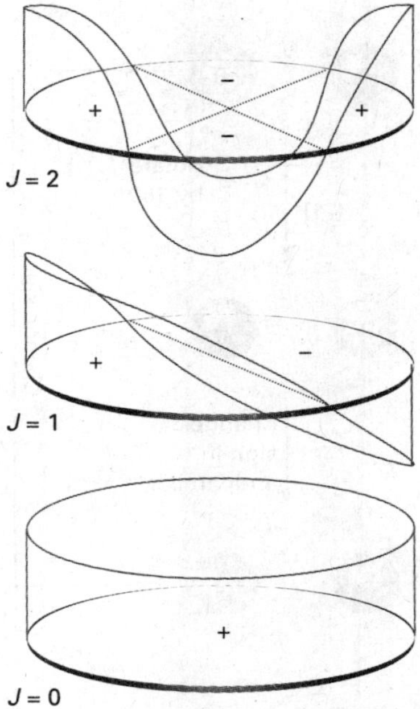

Fig. 12.19 The symmetries of rotational wavefunctions (shown here, for simplicity as a two-dimensional rotor) under a rotation through 180°. Wavefunctions with J even do not change sign; those with J odd do change sign.

$$\frac{\text{Number of ways of achieving odd } J}{\text{Number of ways of achieving even } J}=\begin{cases}(I+1)/I \text{ for half-integral spin nuclei}\\ I/(I+1) \text{ for integral spin nuclei}\end{cases} \tag{12.26}$$

For hydrogen, $I=\frac{1}{2}$, and the ratio is 3:1. For N_2, with $I=1$, the ratio is 1:2.

Justification 12.1 *The effect of nuclear statistics on rotational spectra*

Hydrogen nuclei are fermions, so the Pauli principle requires the overall wavefunction to change sign under particle interchange. However, the rotation of an H_2 molecule through 180° has a more complicated effect than merely relabelling the nuclei, because it interchanges their spin states too if the nuclear spins are paired ($\uparrow\downarrow$; $I_{total}=0$) but not if they are parallel ($\uparrow\uparrow$, $I_{total}=1$).

First, consider the case when the spins are parallel and their state is $\alpha(A)\alpha(B)$, $\alpha(A)\beta(B)+\alpha(B)\beta(A)$, or $\beta(A)\beta(B)$. The $\alpha(A)\alpha(B)$ and $\beta(A)\beta(B)$ combinations are unchanged when the molecule rotates through 180° so the rotational wavefunction must change sign to achieve an overall change of sign. Hence, only odd values of J are allowed. Although at first sight the spins must be interchanged in the combination $\alpha(A)\beta(B)+\alpha(B)\beta(A)$ so as to achieve a simple $A\leftrightarrow B$ interchange of labels (Fig. 12.21), $\beta(A)\alpha(B)+\beta(B)\alpha(A)$ is the same as $\alpha(A)\beta(B)+\alpha(B)\beta(A)$ apart from the order of terms, so only odd values of J are allowed for it too. In contrast, if the nuclear spins are paired, their wavefunction is $\alpha(A)\beta(B)-\alpha(B)\beta(A)$. This combination changes sign when α and β are exchanged (in order to achieve a simple $A\leftrightarrow B$ interchange overall). Therefore, for the overall wavefunction to change sign in this case requires the rotational wavefunction *not* to change sign. Hence, only even values of J are allowed if the nuclear spins are paired. In accord with the prediction of eqn 12.26, there are three ways of achieving odd J but only one of achieving even J.

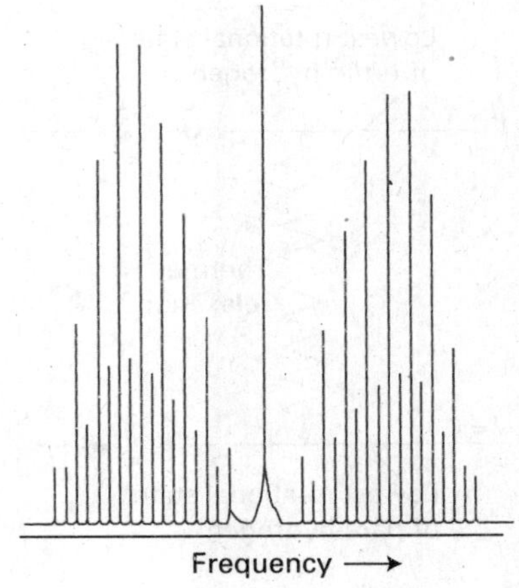

Fig. 12.20 The rotational Raman spectrum of a diatomic molecule with two identical spin-$\frac{1}{2}$ nuclei shows an alternation in intensity as a result of nuclear statistics. The Rayleigh line is much stronger than depicted in the figure; it is shown as a weaker line to improve visualization of the Raman lines.

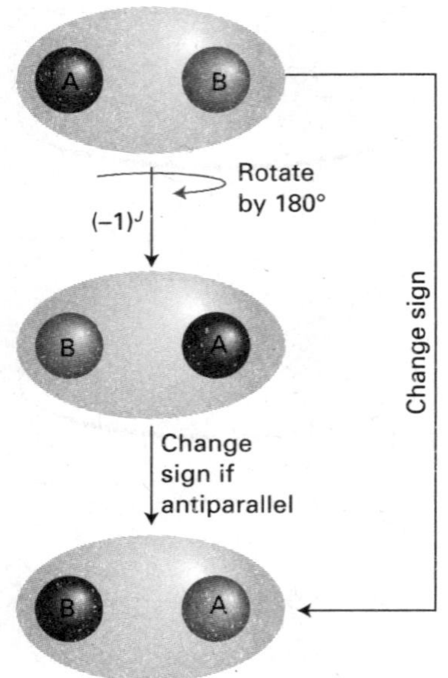

Fig. 12.21 The interchange of two identical fermion nuclei results in the change in sign of the overall wavefunction. The relabelling can be thought of as occurring in two steps: the first is a rotation of the molecule; the second is the interchange of unlike spins (represented by the different colours of the nuclei). The wavefunction changes sign in the second step if the nuclei have antiparallel spins.

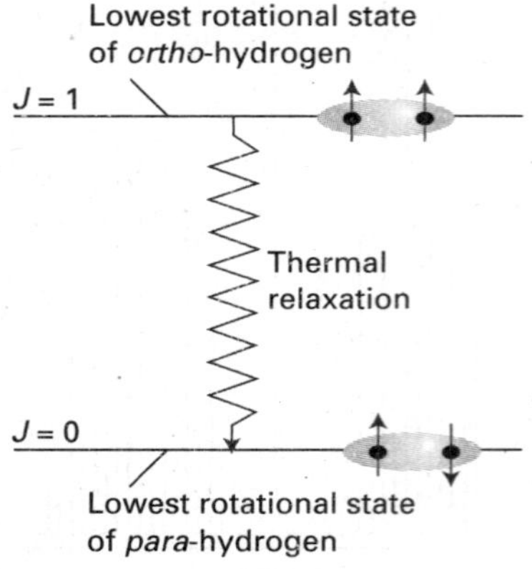

Fig. 12.22 When hydrogen is cooled, the molecules with parallel nuclear spins accumulate in their lowest available rotational state, the one with $J = 1$. They can enter the lowest rotational state ($J = 0$) only if the spins change their relative orientation and become antiparallel. This is a slow process under normal circumstances, so energy is slowly released.

Different relative nuclear spin orientations change into one another only very slowly, so an H_2 molecule with parallel nuclear spins remains distinct from one with paired nuclear spins for long periods. The two forms of hydrogen can be separated by physical techniques, and stored. The form with parallel nuclear spins is called ***ortho*-hydrogen** and the form with paired nuclear spins is called ***para*-hydrogen**. Because *ortho*-hydrogen cannot exist in a state with $J = 0$, it continues to rotate at very low temperatures and has an effective rotational zero-point energy (Fig. 12.22). This energy is of some concern to manufacturers of liquid hydrogen, for the slow conversion of *ortho*-hydrogen into *para*-hydrogen (which can exist with $J = 0$) as nuclear spins slowly realign releases rotational energy, which vaporizes the liquid. Techniques are used to accelerate the conversion of *ortho*-hydrogen to *para*-hydrogen to avoid this problem. One such technique is to pass hydrogen over a metal surface: the molecules adsorb on the surface as atoms, which then recombine in the lower energy *para*-hydrogen form.

The vibrations of diatomic molecules

In this section, we adopt the same strategy of finding expressions for the energy levels, establishing the selection rules, and then discussing the form of the spectrum. We shall also see how the simultaneous excitation of rotation modifies the appearance of a vibrational spectrum.

12.8 Molecular vibrations

Key point **The vibrational energy levels of a diatomic molecule modelled as a harmonic oscillator depend on a force constant k_f (a measure of the bond's stiffness) and the molecule's effective mass.**

We base our discussion on Fig. 12.23, which shows a typical potential energy curve (as in Fig. 10.1) of a diatomic molecule. In regions close to R_e (at the minimum of the curve) the potential energy can be approximated by a parabola, so we can write

$$V = \tfrac{1}{2}k_f x^2 \qquad x = R - R_e \qquad \text{Parabolic potential energy} \qquad (12.27)$$

where k_f is the **force constant** of the bond. The steeper the walls of the potential (the stiffer the bond), the greater the force constant.

To see the connection between the shape of the molecular potential energy curve and the value of k_f, note that we can expand the potential energy around its minimum by using a Taylor series, which is a common way of expressing how a function varies near a selected point (in this case, the minimum of the curve at $x = 0$):

$$V(x) = V(0) + \left(\frac{dV}{dx}\right)_0 x + \tfrac{1}{2}\left(\frac{d^2V}{dx^2}\right)_0 x^2 + \cdots \qquad (12.28)$$

The notation $(\ldots)_0$ means that the derivatives are first evaluated and then x is set equal to 0. The term $V(0)$ can be set arbitrarily to zero. The first derivative of V is zero at the minimum. Therefore, the first surviving term is proportional to the square of the displacement. For small displacements we can ignore all the higher terms, and so write

$$V(x) \approx \tfrac{1}{2}\left(\frac{d^2V}{dx^2}\right)_0 x^2 \qquad (12.29)$$

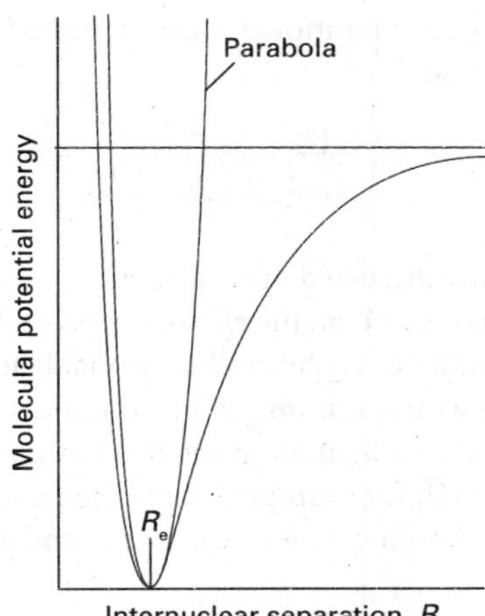

Fig. 12.23 A molecular potential energy curve can be approximated by a parabola near the bottom of the well. The parabolic potential leads to harmonic oscillations. At high excitation energies the parabolic approximation is poor (the true potential is less confining), and it is totally wrong near the dissociation limit.

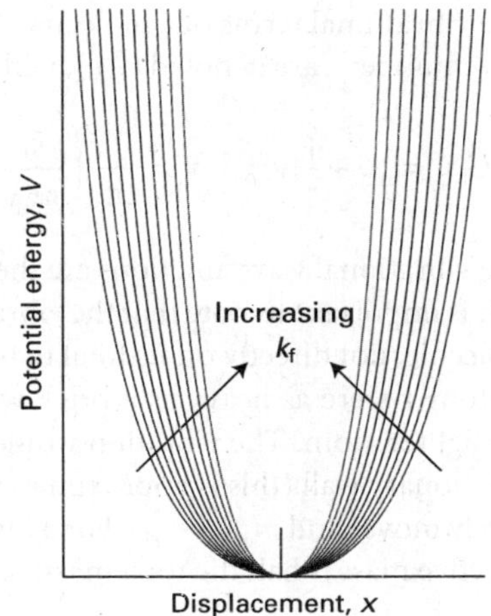

Fig. 12.24 The force constant is a measure of the curvature of the potential energy close to the equilibrium extension of the bond. A strongly confining well (one with steep sides, a stiff bond) corresponds to high values of k_f.

Therefore, the first approximation to a molecular potential energy curve is a parabolic potential, and we can identify the force constant as

$$k_f = \left(\frac{d^2V}{dx^2}\right)_0 \qquad \text{Formal definition of the force constant} \qquad [12.30]$$

We see that, if the potential energy curve is sharply curved close to its minimum, then k_f will be large. Conversely, if the potential energy curve is wide and shallow, then k_f will be small (Fig. 12.24).

The Schrödinger equation for the relative motion of two atoms of masses m_1 and m_2 with a parabolic potential energy is

$$-\frac{\hbar^2}{2m_{\text{eff}}}\frac{d^2\psi}{dx^2} + \tfrac{1}{2}k_f x^2\psi = E\psi \qquad (12.31)$$

where m_{eff} is the **effective mass**:

$$m_{\text{eff}} = \frac{m_1 m_2}{m_1 + m_2} \qquad \text{Effective mass} \qquad (12.32)$$

These equations are derived in the same way as in *Further information 9.1*, but here the separation of variables procedure is used to separate the relative motion of the atoms from the motion of the molecule as a whole.

The Schrödinger equation in eqn 12.31 is the same as eqn 8.23 for a particle of mass m undergoing harmonic motion. Therefore, we can use the results of Section 8.4 to write down the permitted vibrational energy levels:

$$E_v = (v + \tfrac{1}{2})\hbar\omega \qquad \omega = \left(\frac{k_f}{m_{\text{eff}}}\right)^{1/2} \qquad v = 0, 1, 2, \ldots \qquad \text{Vibrational energy levels of a diatomic molecule} \qquad (12.33)$$

A note on good practice Distinguish *effective mass* from *reduced mass*. The former is a measure of the mass that is moved during a vibration. The latter is the quantity that emerges from the separation of relative internal and overall translational motion. For a diatomic molecule the two are the same, but that is not true in general for vibrations of polyatomic molecules. Many, however, do not make this distinction and refer to both quantities as the 'reduced mass'.

The **vibrational terms** of a molecule, the energies of its vibrational states expressed as wavenumbers, are denoted $\tilde{G}(v)$, with $E_v = hc\tilde{G}(v)$, so

$$\tilde{G}(v) = (v + \tfrac{1}{2})\tilde{\nu} \qquad \tilde{\nu} = \frac{1}{2\pi c}\left(\frac{k_f}{m_{\text{eff}}}\right)^{1/2} \qquad \text{Vibrational terms of a diatomic molecule} \qquad (12.34)$$

The vibrational wavefunctions are the same as those discussed in Section 8.5.

It is important to note that the vibrational terms depend on the *effective* mass of the molecule, not directly on its total mass. This dependence is physically reasonable for, if atom 1 were as heavy as a brick wall, then we would find $m_{\text{eff}} \approx m_2$, the mass of the lighter atom. The vibration would then be that of a light atom relative to that of a stationary wall (this is approximately the case in HI, for example, where the I atom barely moves and $m_{\text{eff}} \approx m_{\text{H}}$). For a homonuclear diatomic molecule $m_1 = m_2$, and the effective mass is half the total mass: $m_{\text{eff}} = \frac{1}{2}m$.

• A brief illustration

An HCl molecule has a force constant of 516 N m^{-1}, a reasonably typical value for a single bond. The effective mass of $^1\text{H}^{35}\text{Cl}$ is 1.63×10^{-27} kg (note that this mass is very close to the mass of the hydrogen atom, 1.67×10^{-27} kg, so the Cl atom is acting like a brick wall). These values imply $\omega = 5.63 \times 10^{14}\ \text{s}^{-1}$, $\nu = 89.5$ THz (1 THz = 10^{12} Hz), $\tilde{\nu} = 2987\ \text{cm}^{-1}$, $\lambda = 3.35\ \mu\text{m}$. These characteristics correspond to electromagnetic radiation in the infrared region. •

12.9 Selection rules

Key points **The gross selection rule for infrared spectra is that the electric dipole moment of the molecule must change when the atoms are displaced relative to one another. The specific selection rule is $\Delta v = \pm 1$.**

The gross selection rule for a change in vibrational state brought about by absorption or emission of radiation is that *the electric dipole moment of the molecule must change when the atoms are displaced relative to one another*. Such vibrations are said to be **infrared active**. The classical basis of this rule is that the molecule can shake the electromagnetic field into oscillation if its dipole changes as it vibrates, and vice versa (Fig. 12.25); its formal basis is given in *Further information 12.2*. Note that the molecule need not have a permanent dipole: the rule requires only a change in dipole moment, possibly from zero. Some vibrations do not affect the molecule's dipole moment (for instance, the stretching motion of a homonuclear diatomic molecule), so they neither absorb nor generate radiation: such vibrations are said to be **infrared inactive**. Homonuclear diatomic molecules are infrared inactive because their dipole moments remain zero however long the bond; heteronuclear diatomic molecules are infrared active.

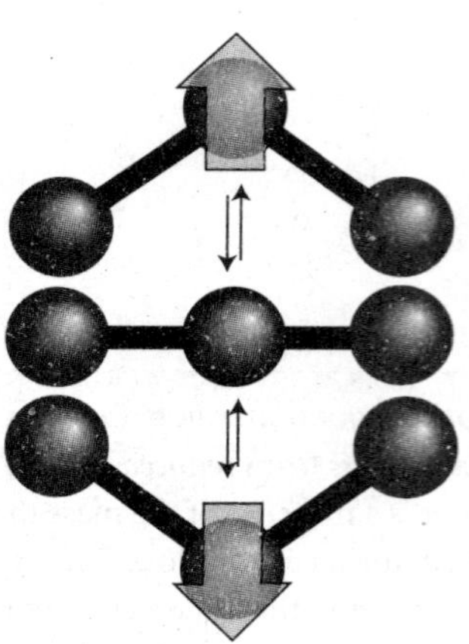

Fig. 12.25 The oscillation of a molecule, even if it is nonpolar, may result in an oscillating dipole that can interact with the electromagnetic field.

• A brief illustration

Of the molecules N_2, CO_2, OCS, H_2O, $CH_2{=}CH_2$, and C_6H_6, all except N_2 possess at least one vibrational mode that results in a change of dipole moment, so all except N_2 can show a vibrational absorption spectrum. Not all the modes of complex molecules are vibrationally active. For example, the symmetric stretch of CO_2, in which the O–C–O bonds stretch and contract symmetrically, is inactive because it leaves the dipole moment unchanged (at zero). •

Weak infrared transitions can be observed from homonuclear diatomic molecules trapped within various nanomaterials. For instance, when incorporated into solid C_{60}, H_2 molecules interact through van der Waals forces with the surrounding C_{60} molecules and acquire dipole moments, with the result that they have observable infrared spectra.

Self-test 12.6 Which of the molecules H_2, NO, N_2O, and CH_4 have infrared active vibrations? [NO, N_2O, CH_4]

The specific selection rule, which is obtained from an analysis of the expression for the transition moment and the properties of integrals over harmonic oscillator wavefunctions (as shown in *Further information 12.2*), is

$$\Delta v = \pm 1 \qquad \text{Specific vibrational selection rule} \qquad (12.35)$$

Transitions for which $\Delta v = +1$ correspond to absorption and those with $\Delta v = -1$ correspond to emission. It follows that the wavenumbers of allowed vibrational transitions, which are denoted $\Delta\tilde{G}_{v+\frac{1}{2}}$ for the transition $v+1 \leftarrow v$, are

$$\Delta\tilde{G}_{v+\frac{1}{2}} = \tilde{G}(v+1) - \tilde{G}(v) = \tilde{\nu} \qquad (12.36)$$

As we have seen, $\tilde{\nu}$ lies in the infrared region of the electromagnetic spectrum, so vibrational transitions absorb and generate infrared radiation.

At room temperature $kT/hc \approx 200\ \text{cm}^{-1}$, and most vibrational wavenumbers are significantly greater than $200\ \text{cm}^{-1}$. It follows from the Boltzmann distribution that almost all the molecules will be in their vibrational ground states initially. Hence, the dominant spectral transition will be the **fundamental transition**, $1 \leftarrow 0$. As a result, the spectrum is expected to consist of a single absorption line. If the molecules are formed in a vibrationally excited state, such as when vibrationally excited HF molecules are formed in the reaction $H_2 + F_2 \rightarrow 2\ HF^*$, the transitions $5 \rightarrow 4$, $4 \rightarrow 3, \ldots$ may also appear (in emission). In the harmonic approximation, all these lines lie at the same frequency, and the spectrum is also a single line. However, as we shall now show, the breakdown of the harmonic approximation causes the transitions to lie at slightly different frequencies, so several lines are observed.

12.10 **Anharmonicity**

Key points **(a) The Morse potential energy function can be used to describe anharmonic motion. (b) A Birge–Sponer plot may be used to determine the dissociation energy of the bond in a diatomic molecule.**

The vibrational terms in eqn 12.34 are only approximate because they are based on a parabolic approximation to the actual potential energy curve. A parabola cannot be correct at all extensions because it does not allow a bond to dissociate. At high vibrational excitations the swing of the atoms (more precisely, the spread of the vibrational wavefunction) allows the molecule to explore regions of the potential energy curve where the parabolic approximation is poor and additional terms in the Taylor expansion of V (eqn 12.28) must be retained. The motion then becomes **anharmonic**, in the sense that the restoring force is no longer proportional to the displacement. Because the actual curve is less confining than a parabola, we can anticipate that the energy levels become less widely spaced at high excitations.

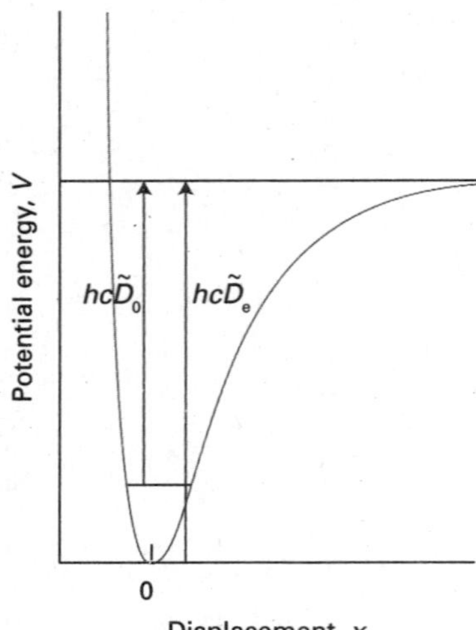

Fig. 12.26 The dissociation energy of a molecule, $\tilde{D}_0$, differs from the depth of the potential well, $\tilde{D}_e$, on account of the zero-point energy of the vibrations of the bond.

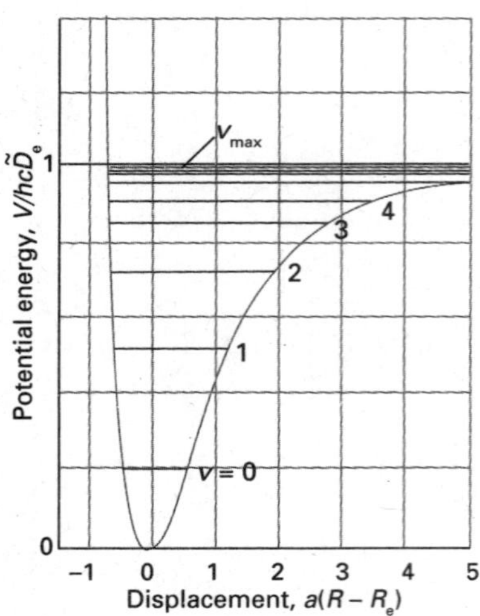

Fig. 12.27 The Morse potential energy curve reproduces the general shape of a molecular potential energy curve. The corresponding Schrödinger equation can be solved, and the values of the energies obtained. The number of bound levels is finite.

(a) The convergence of energy levels

One approach to the calculation of the energy levels in the presence of anharmonicity is to use a function that resembles the true potential energy more closely. The **Morse potential energy** is

$$V = hc\tilde{D}_e\{1 - e^{-a(R-R_e)}\}^2 \qquad a = \left(\frac{m_{\text{eff}}\omega^2}{2hc\tilde{D}_e}\right)^{1/2} \qquad \text{Morse potential energy} \qquad (12.37)$$

where $\tilde{D}_e$ is the depth of the potential minimum (Fig. 12.26). Near the well minimum the variation of V with displacement resembles a parabola (as can be checked by expanding the exponential as far as the first term) but, unlike a parabola, eqn 12.37 allows for dissociation at large displacements. The Schrödinger equation can be solved for the Morse potential and the permitted energy levels are

$$\tilde{G}(v) = (v+\tfrac{1}{2})\tilde{\nu} - (v+\tfrac{1}{2})^2 x_e\tilde{\nu} \qquad x_e = \frac{a^2\hbar}{2m_{\text{eff}}\omega} = \frac{\tilde{\nu}}{4\tilde{D}_e} \qquad (12.38)$$

The parameter x_e is called the **anharmonicity constant**. The number of vibrational levels of a Morse oscillator is finite, and $v = 0, 1, 2, \ldots, v_{\text{max}}$, as shown in Fig. 12.27 (see also Problem 12.24). The second term in the expression for $\tilde{G}$ subtracts from the first with increasing effect as v increases, and hence gives rise to the convergence of the levels at high quantum numbers.

Although the Morse oscillator is quite useful theoretically, in practice the more general expression

$$\tilde{G}(v) = (v+\tfrac{1}{2})\tilde{\nu} - (v+\tfrac{1}{2})^2 x_e\tilde{\nu} + (v+\tfrac{1}{2})^3 y_e\tilde{\nu} + \cdots \qquad (12.39)$$

where $x_e, y_e, \ldots$ are empirical dimensionless constants characteristic of the molecule, is used to fit the experimental data and to find the dissociation energy of the molecule. When anharmonicities are present, the wavenumbers of transitions with $\Delta v = +1$ are

$$\Delta\tilde{G}_{v+\frac{1}{2}} = \tilde{G}(v+1) - \tilde{G}(v) = \tilde{\nu} - 2(v+1)x_e\tilde{\nu} + \cdots \qquad (12.40)$$

Equation 12.40 shows that, when $x_e > 0$, the transitions move to lower wavenumbers as v increases.

Anharmonicity also accounts for the appearance of additional weak absorption lines corresponding to the transitions $2 \leftarrow 0$, $3 \leftarrow 0, \ldots$, even though these first, second, ... **overtones** are forbidden by the selection rule $\Delta v = \pm 1$. The first overtone, for example, gives rise to an absorption at

$$\tilde{G}(v+2) - \tilde{G}(v) = 2\tilde{\nu} - 2(2v+3)x_e\tilde{\nu} + \cdots \qquad (12.41)$$

The reason for the appearance of overtones is that the selection rule is derived from the properties of harmonic oscillator wavefunctions, which are only approximately valid when anharmonicity is present. Therefore, the selection rule is also only an approximation. For an anharmonic oscillator, all values of Δv are allowed, but transitions with $\Delta v > 1$ are allowed only weakly if the anharmonicity is slight.

(b) The Birge–Sponer plot

When several vibrational transitions are detectable, a graphical technique called a **Birge–Sponer plot** may be used to determine the dissociation energy, $hc\tilde{D}_0$, of the bond. The basis of the Birge–Sponer plot is that the sum of successive intervals $\Delta\tilde{G}_{v+\frac{1}{2}}$ from the zero-point level to the dissociation limit is the dissociation energy:

$$\tilde{D}_0 = \Delta\tilde{G}_{1/2} + \Delta\tilde{G}_{3/2} + \cdots = \sum_v \Delta\tilde{G}_{v+\frac{1}{2}} \qquad (12.42)$$

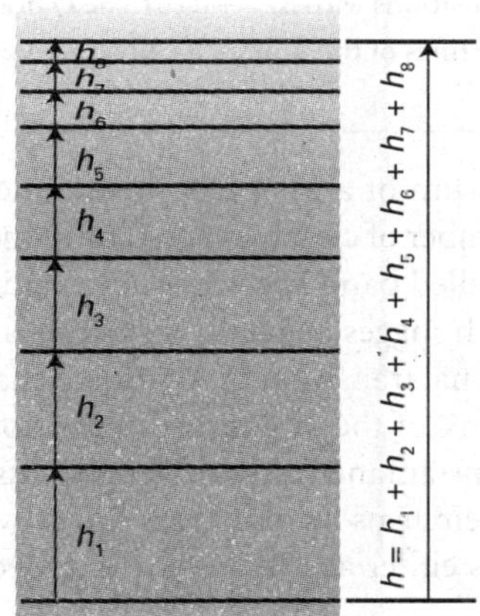

Fig. 12.28 The dissociation energy is the sum of the separations of the vibrational energy levels up to the dissociation limit just as the length of a ladder is the sum of the separations of its rungs.

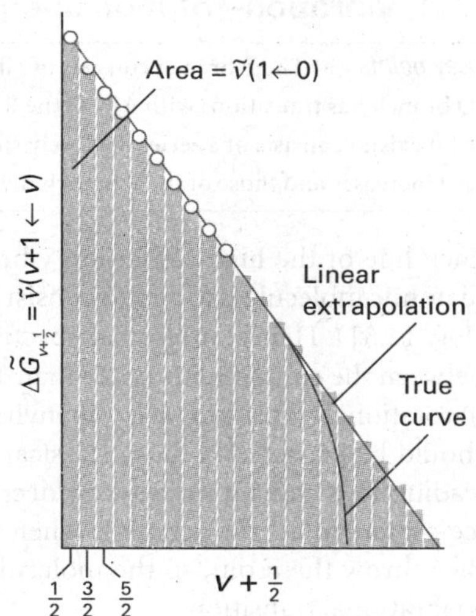

Fig. 12.29 The area under a plot of transition wavenumber against vibrational quantum number is equal to the dissociation energy of the molecule. The assumption that the differences approach zero linearly is the basis of the Birge–Sponer extrapolation.

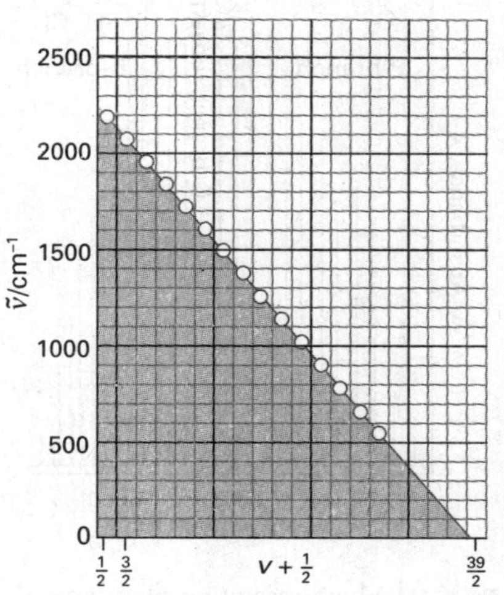

Fig. 12.30 The Birge–Sponer plot used in Example 12.5. The area is obtained simply by counting the squares beneath the line or using the formula for the area of a right triangle (area $= \frac{1}{2} \times$ base $\times$ height).

just as the height of the ladder is the sum of the separations of its rungs (Fig. 12.28). The construction in Fig. 12.29 shows that the area under the plot of $\Delta\tilde{G}_{v+\frac{1}{2}}$ against $v+\frac{1}{2}$ is equal to the sum, and therefore to $\tilde{D}_0$. The successive terms decrease linearly when only the x_e anharmonicity constant is taken into account and the inaccessible part of the spectrum can be estimated by linear extrapolation. Most actual plots differ from the linear plot as shown in Fig. 12.29, so the value of $\tilde{D}_0$ obtained in this way is usually an overestimate of the true value.

Example 12.5 *Using a Birge–Sponer plot*

The observed vibrational intervals of H_2^+ lie at the following values for $1 \leftarrow 0$, $2 \leftarrow 1, \ldots$, respectively (in cm^{-1}): 2191, 2064, 1941, 1821, 1705, 1591, 1479, 1368, 1257, 1145, 1033, 918, 800, 677, 548, 411. Determine the dissociation energy of the molecule.

Method Plot the separations against $v+\frac{1}{2}$, extrapolate linearly to the point cutting the horizontal axis, and then measure the area under the curve.

Answer The points are plotted in Fig. 12.30, and a linear extrapolation is shown. The area under the curve (use the formula for the area of a triangle or count the squares) is 214. Each square corresponds to 100 cm^{-1} (refer to the scale of the vertical axis); hence the dissociation energy is 21 400 cm^{-1} (corresponding to 256 kJ mol^{-1}).

Self-test 12.7 The vibrational levels of HgH converge rapidly, and successive intervals are 1203.7 (which corresponds to the $1 \leftarrow 0$ transition), 965.6, 632.4, and 172 cm^{-1}. Estimate the dissociation energy. [35.6 kJ mol^{-1}]

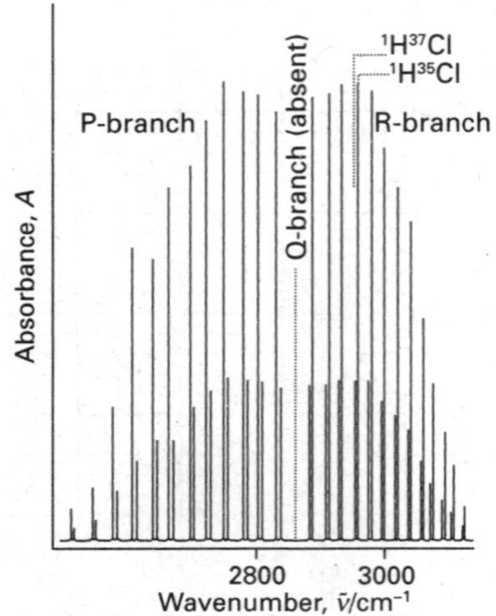

Fig. 12.31 A high-resolution vibration–rotation spectrum of HCl. The lines appear in pairs because $H^{35}Cl$ and $H^{37}Cl$ both contribute (their abundance ratio is 3:1). There is no Q branch, because $\Delta J = 0$ is forbidden for this molecule.

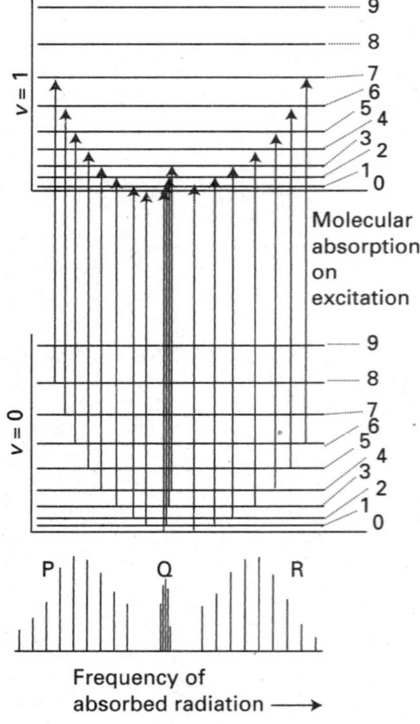

Fig. 12.32 The formation of P, Q, and R branches in a vibration–rotation spectrum. The intensities reflect the populations of the initial rotational levels.

12.11 Vibration–rotation spectra

Key points **(a) The P branch consists of vibration–rotation infrared transitions with $\Delta J = -1$; the Q branch has transitions with $\Delta J = 0$; the R branch has transitions with $\Delta J = +1$. (b) The Q branch (if it exists) consists of a series of closely spaced lines. The lines of the R branch converge slightly as J increases and those of the P branch diverge.**

Each line of the high resolution vibrational spectrum of a gas-phase heteronuclear diatomic molecule is found to consist of a large number of closely spaced components (Fig. 12.31). Hence, molecular spectra are often called **band spectra**. The separation between the components is less than 10 cm^{-1}, which suggests that the structure is due to rotational transitions accompanying the vibrational transition. A rotational change should be expected because classically we can think of the vibrational transition as leading to a sudden increase or decrease in the instantaneous bond length. Just as ice-skaters rotate more rapidly when they bring their arms in, and more slowly when they throw them out, so the molecular rotation is either accelerated or retarded by a vibrational transition.

(a) Spectral branches

A detailed analysis of the quantum mechanics of simultaneous vibrational and rotational changes shows that the rotational quantum number J changes by ± 1 during the vibrational transition of a diatomic molecule. If the molecule also possesses angular momentum about its axis, as in the case of the electronic orbital angular momentum of the paramagnetic molecule NO, then the selection rules also allow $\Delta J = 0$.

The appearance of the vibration–rotation spectrum of a diatomic molecule can be discussed in terms of the combined vibration–rotation terms, $\tilde{S}$:

$$\tilde{S}(v,J) = \tilde{G}(v) + \tilde{F}(J) \qquad (12.43)$$

If we ignore anharmonicity and centrifugal distortion we can use eqn 12.34 for the first term on the right and eqn 12.14 for the second, and obtain

$$\tilde{S}(v,J) = (v + \tfrac{1}{2})\tilde{\nu} + \tilde{B}J(J+1) \qquad (12.44)$$

In a more detailed treatment, $\tilde{B}$ is allowed to depend on the vibrational state because, as v increases, the molecule swells slightly and the moment of inertia changes. We shall continue with the simple expression initially.

When the vibrational transition $v+1 \leftarrow v$ occurs, J changes by ± 1 and in some cases by 0 (when $\Delta J = 0$ is allowed). The absorptions then fall into three groups called **branches** of the spectrum. The **P branch** consists of all transitions with $\Delta J = -1$:

$$\tilde{\nu}_P(J) = \tilde{S}(v+1, J-1) - \tilde{S}(v,J) = \tilde{\nu} - 2\tilde{B}J \qquad \text{P branch transitions} \qquad (12.45a)$$

This branch consists of lines at $\tilde{\nu} - 2\tilde{B}, \tilde{\nu} - 4\tilde{B}, \ldots$ with an intensity distribution reflecting both the populations of the rotational levels and the magnitude of the $J-1 \leftarrow J$ transition moment (Fig. 12.32). The **Q branch** consists of all lines with $\Delta J = 0$, and its wavenumbers are all

$$\tilde{\nu}_Q(J) = \tilde{S}(v+1, J) - \tilde{S}(v,J) = \tilde{\nu} \qquad \text{Q branch transitions} \qquad (12.45b)$$

for all values of J. This branch, when it is allowed (as in NO), appears at the vibrational transition wavenumber. In Fig. 12.31 there is a gap at the expected location of the Q branch because it is forbidden in HCl. The **R branch** consists of lines with $\Delta J = +1$:

$$\tilde{\nu}_R(J) = \tilde{S}(v+1, J+1) - \tilde{S}(v,J) = \tilde{\nu} + 2\tilde{B}(J+1) \qquad \text{R branch transitions} \qquad (12.45c)$$

This branch consists of lines displaced from $\tilde{\nu}$ to high wavenumber by $2\tilde{B}, 4\tilde{B}, \ldots$.

The separation between the lines in the P and R branches of a vibrational transition gives the value of $\tilde{B}$. Therefore, the bond length can be deduced without needing to take a pure rotational microwave spectrum. However, the latter is more precise because microwave frequencies can be measured with greater precision than infrared frequencies.

(b) Combination differences

The rotational constant of the vibrationally excited state, $\tilde{B}_1$ (in general, $\tilde{B}_v$), is different from that of the ground vibrational state, $\tilde{B}_0$. One contribution to the difference is the anharmonicity of the vibration, which results in a slightly extended bond in the upper state. However, even in the absence of anharmonicity, the average value of $1/R^2$ ($\langle 1/R^2 \rangle$, which is not the same as $1/\langle R^2 \rangle$) varies with the vibrational state (see Problems 12.19 and 12.20). As a result, the Q branch (if it exists) consists of a series of closely spaced lines. The lines of the R branch converge slightly as J increases; and those of the P branch diverge:

$$\begin{aligned} \tilde{\nu}_P(J) &= \tilde{\nu} - (\tilde{B}_1 + \tilde{B}_0)J + (\tilde{B}_1 - \tilde{B}_0)J^2 \\ \tilde{\nu}_Q(J) &= \tilde{\nu} + (\tilde{B}_1 - \tilde{B}_0)J(J+1) \\ \tilde{\nu}_R(J) &= \tilde{\nu} + (\tilde{B}_1 + \tilde{B}_0)(J+1) + (\tilde{B}_1 - \tilde{B}_0)(J+1)^2 \end{aligned} \tag{12.46}$$

To determine the two rotational constants individually, we use the method of **combination differences.** This procedure is used widely in spectroscopy to extract information about a particular state. It involves setting up expressions for the difference in the wavenumbers of transitions to a common state; the resulting expression then depends solely on properties of the other state.

As can be seen from Fig. 12.33, the transitions $\tilde{\nu}_R(J-1)$ and $\tilde{\nu}_P(J+1)$ have a common upper state, and hence can be anticipated to depend on $\tilde{B}_0$. Indeed, it is easy to show from eqn 12.46 that

$$\tilde{\nu}_R(J-1) - \tilde{\nu}_P(J+1) = 4\tilde{B}_0(J+\tfrac{1}{2}) \tag{12.47a}$$

Therefore, a plot of the combination difference against $J+\frac{1}{2}$ should be a straight line of slope $4\tilde{B}_0$, so the rotational constant of the molecule in the state $v=0$ can be determined. (Any deviation from a straight line is a consequence of centrifugal distortion, so that effect can be investigated too.) Similarly, $\tilde{\nu}_R(J)$ and $\tilde{\nu}_P(J)$ have a common lower state, and hence their combination difference gives information about the upper state:

$$\tilde{\nu}_R(J) - \tilde{\nu}_P(J) = 4\tilde{B}_1(J+\tfrac{1}{2}) \tag{12.47b}$$

The two rotational constants of $^1H^{35}Cl$ found in this way are $\tilde{B}_0 = 10.440\ \text{cm}^{-1}$ and $\tilde{B}_1 = 10.136\ \text{cm}^{-1}$.

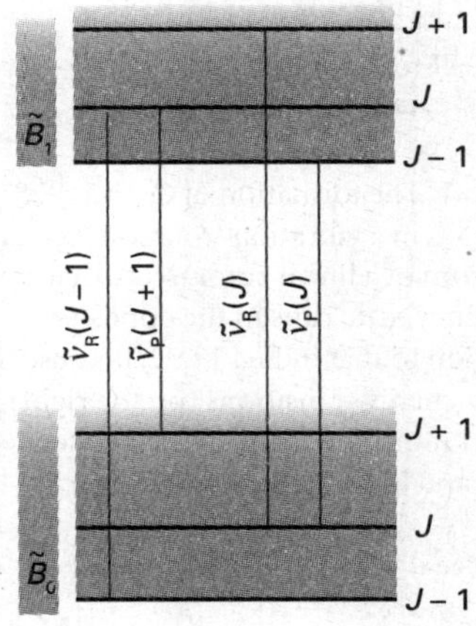

Fig. 12.33 The method of combination differences makes use of the fact that some transitions share a common level.

12.12 Vibrational Raman spectra of diatomic molecules

Key points **For a vibration to be Raman active, the polarizability must change as the molecule vibrates. The specific selection rule is $\Delta v = \pm 1$. In gas-phase spectra, the Stokes and anti-Stokes lines have a branch structure: the O branch ($\Delta J = -2$), the Q branch ($\Delta J = 0$), and the S branch ($\Delta J = +2$).**

The gross selection rule for vibrational Raman transitions is that *the polarizability should change as the molecule vibrates.* As homonuclear and heteronuclear diatomic molecules swell and contract during a vibration, the control of the nuclei over the electrons varies, and hence the molecular polarizability changes. Both types of diatomic molecule are therefore vibrationally Raman active. The specific selection rule for

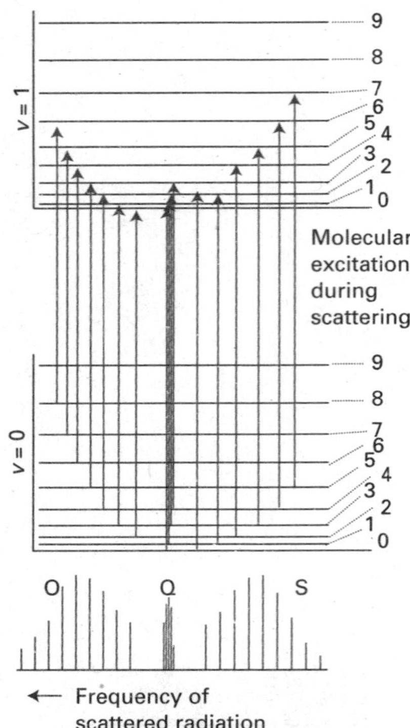

Fig. 12.34 The formation of O, Q, and S branches in a vibration–rotation Raman spectrum of a linear rotor. Note that the frequency scale runs in the opposite direction to that in Fig. 12.32, because the higher energy transitions (on the right) extract more energy from the incident beam and leave it at lower frequency.

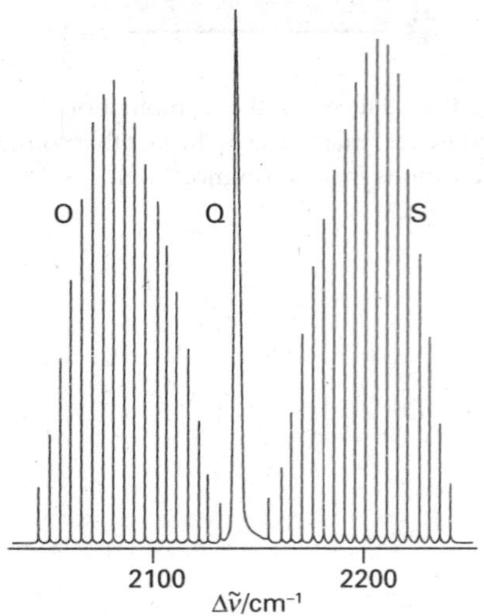

Fig. 12.35 The structure of a vibrational line in the vibrational Raman spectrum of carbon monoxide, showing the O, Q, and S branches.

vibrational Raman transitions in the harmonic approximation is $\Delta v = \pm 1$. The formal basis for the gross and specific selection rules is given in *Further information 12.2*.

The lines to high frequency of the incident radiation, the anti-Stokes lines, are those for which $\Delta v = -1$. The lines to low frequency, the Stokes lines, correspond to $\Delta v = +1$. The intensities of the anti-Stokes and Stokes lines are governed largely by the Boltzmann populations of the vibrational states involved in the transition. It follows that anti-Stokes lines are usually weak because very few molecules are in an excited vibrational state initially.

In gas-phase spectra, the Stokes and anti-Stokes lines have a branch structure arising from the simultaneous rotational transitions that accompany the vibrational excitation (Fig. 12.34). The selection rules are $\Delta J = 0, \pm 2$ (as in pure rotational Raman spectroscopy), and give rise to the **O branch** ($\Delta J = -2$), the **Q branch** ($\Delta J = 0$), and the **S branch** ($\Delta J = +2$):

$$\tilde{\nu}_O(J) = \tilde{\nu}_i - \tilde{\nu} - 2\tilde{B} + 4\tilde{B}J \qquad \text{O branch transitions}$$

$$\tilde{\nu}_Q(J) = \tilde{\nu}_i - \tilde{\nu} \qquad \text{Q branch transitions} \qquad (12.48)$$

$$\tilde{\nu}_S(J) = \tilde{\nu}_i - \tilde{\nu} - 6\tilde{B} - 4\tilde{B}J \qquad \text{S branch transitions}$$

where $\tilde{\nu}_i$ is the wavenumber of the incident radiation. Note that, unlike in infrared spectroscopy, a Q branch is obtained for all linear molecules. The spectrum of CO, for instance, is shown in Fig. 12.35: the structure of the Q branch arises from the differences in rotational constants of the upper and lower vibrational states.

The information available from vibrational Raman spectra adds to that from infrared spectroscopy because homonuclear diatomics can also be studied. The spectra can be interpreted in terms of the force constants, dissociation energies, and bond lengths, and some of the information obtained is included in Table 12.2.

The vibrations of polyatomic molecules

There is only one mode of vibration for a diatomic molecule, the bond stretch. In polyatomic molecules there are several modes of vibration because all the bond lengths and angles may change and the vibrational spectra are very complex. Nonetheless, we shall see that infrared and Raman spectroscopy can be used to obtain information about the structure of systems as large as animal and plant tissues. Raman spectroscopy is particularly useful for characterizing nanomaterials, especially carbon nanotubes.

Table 12.2* Properties of diatomic molecules

	$\tilde{\nu}$/cm^{-1}	R_e/pm	$\tilde{B}$/cm^{-1}	k/(N m^{-1})	D_0/(kJ mol^{-1})
$^{1}H_2$	4400	74	60.86	575	432
$^{1}H^{35}Cl$	2991	127	10.59	516	428
$^{1}H^{127}I$	2308	161	6.51	314	295
$^{35}Cl_2$	560	199	0.244	323	239

* More values are given in the *Data section*.

12.13 Normal modes

Key points **A normal mode is an independent, synchronous motion of atoms or groups of atoms that may be excited without leading to the excitation of any other normal mode. The number of normal modes is $3N-6$ (for nonlinear molecules) or $3N-5$ (linear molecules).**

We begin by calculating the total number of vibrational modes of a polyatomic molecule. We then see that we can choose combinations of these atomic displacements that give the simplest description of the vibrations.

As shown in the following *Justification*, for a nonlinear molecule that consists of N atoms, there are $3N-6$ independent modes of vibration. If the molecule is linear, there are $3N-5$ independent vibrational modes.

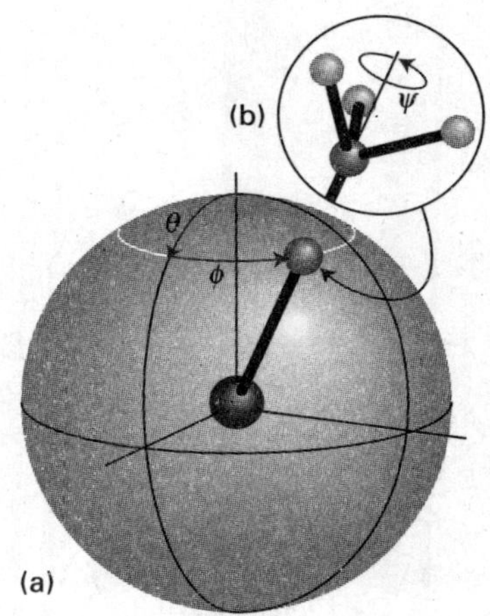

Fig. 12.36 (a) The orientation of a linear molecule requires the specification of two angles. (b) The orientation of a nonlinear molecule requires the specification of three angles.

• A brief illustration

Water, H_2O, is a nonlinear triatomic molecule, and has three modes of vibration (and three modes of rotation); CO_2 is a linear triatomic molecule, and has four modes of vibration (and only two modes of rotation). Even a middle-sized molecule such as naphthalene ($C_{10}H_8$) has 48 distinct modes of vibration. •

Justification 12.2 *The number of vibrational modes*

The total number of coordinates needed to specify the locations of N atoms is $3N$. Each atom may change its location by varying one of its three coordinates (x, y, and z), so the total number of displacements available is $3N$. These displacements can be grouped together in a physically sensible way. For example, three coordinates are needed to specify the location of the centre of mass of the molecule, so three of the $3N$ displacements correspond to the translational motion of the molecule as a whole. The remaining $3N-3$ are non-translational 'internal' modes of the molecule.

Two angles are needed to specify the orientation of a linear molecule in space: in effect, we need to give only the latitude and longitude of the direction in which the molecular axis is pointing (Fig. 12.36a). However, three angles are needed for a nonlinear molecule because we also need to specify the orientation of the molecule around the direction defined by the latitude and longitude (Fig. 12.36b). Therefore, two (linear) or three (nonlinear) of the $3N-3$ internal displacements are rotational. This leaves $3N-5$ (linear) or $3N-6$ (nonlinear) displacements of the atoms relative to one another: these are the vibrational modes. It follows that the number of modes of vibration N_{vib} is $3N-5$ for linear molecules and $3N-6$ for nonlinear molecules.

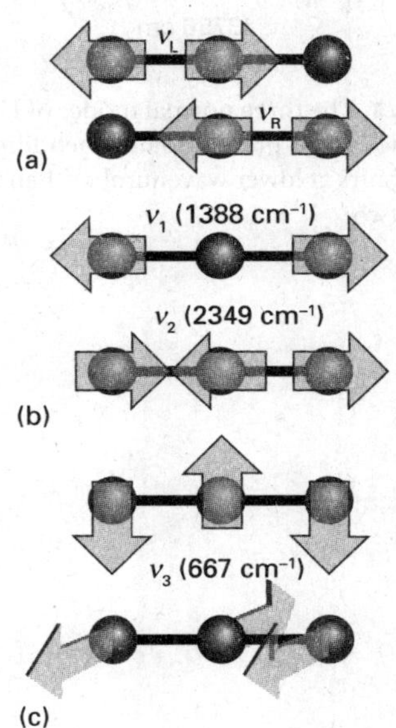

Fig. 12.37 Alternative descriptions of the vibrations of CO_2. (a) The stretching modes are not independent and, if one C–O group is excited, the other begins to vibrate. They are not normal modes of vibration of the molecule. (b) The symmetric and antisymmetric stretches are independent, and one can be excited without affecting the other: they are normal modes. (c) The two perpendicular bending motions are also normal modes.

The next step is to find the best description of the modes. One choice for the four modes of CO_2, for example, might be the ones in Fig. 12.37. This illustration shows the stretching of one bond (the mode ν_L), the stretching of the other (ν_R), and the two perpendicular bending modes (ν_2). The description, while permissible, has a disadvantage: when one CO bond vibration is excited, the motion of the C atom sets the other CO bond in motion, so energy flows backwards and forwards between ν_L and ν_R. Moreover, the position of the centre of mass of the molecule varies in the course of either vibration.

The description of the vibrational motion is much simpler if linear combinations of ν_L and ν_R are taken. For example, one combination is ν_1 in Fig. 12.37b: this mode is the **symmetric stretch**. In this mode, the C atom is buffeted simultaneously from each side and the motion continues indefinitely. Another mode is ν_3, the **antisymmetric stretch**, in which the two O atoms always move in the same direction as each other and opposite to that of the C atom. Both modes are independent in the sense that, if one is excited, then it does not excite the other. They are two of the 'normal modes' of the molecule, its independent, collective vibrational displacements. The two other normal modes are the bending modes ν_3. In general, a **normal mode** is an independent, synchronous motion of atoms or groups of atoms that may be excited without leading to the excitation of any other normal mode and without involving translation or rotation of the molecule as a whole.

The four normal modes of CO_2, and the N_{vib} normal modes of polyatomics in general, are the key to the description of molecular vibrations. Each normal mode, q, behaves like an independent harmonic oscillator (if anharmonicities are neglected), so each has a series of terms

$$\tilde{G}_q(v) = (v + \tfrac{1}{2})\tilde{\nu}_q \qquad \tilde{\nu}_q = \frac{1}{2\pi c}\left(\frac{k_q}{m_q}\right)^{1/2} \qquad \text{Vibrational terms of normal modes} \qquad (12.49)$$

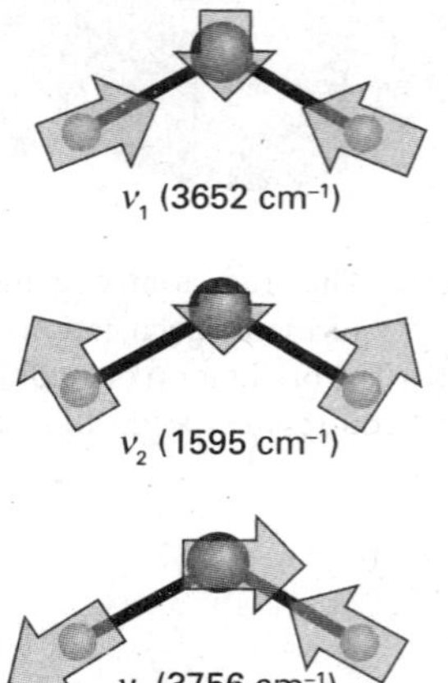

Fig. 12.38 The three normal modes of H_2O. The mode ν_2 is predominantly bending, and occurs at lower wavenumber than the other two.

where $\tilde{\nu}_q$ is the wavenumber of mode q and depends on the force constant k_q for the mode and on the effective mass m_q of the mode. The effective mass of the mode is a measure of the mass that is swung about by the vibration and in general is a complicated function of the masses of the atoms. For example, in the symmetric stretch of CO_2, the C atom is stationary, and the effective mass depends on the masses of only the O atoms. In the antisymmetric stretch and in the bends, all three atoms move, so all contribute to the effective mass. The three normal modes of H_2O are shown in Fig. 12.38: note that the predominantly bending mode (ν_2) has a lower frequency than the others, which are predominantly stretching modes. It is generally the case that the frequencies of bending motions are lower than those of stretching modes. One point that must be appreciated is that only in special cases (such as the CO_2 molecule) are the normal modes purely stretches or purely bends. In general, a normal mode is a composite motion of simultaneous stretching and bending of bonds. Another point in this connection is that heavy atoms generally move less than light atoms in normal modes.

12.14 Infrared absorption spectra of polyatomic molecules

Key points A normal mode is infrared active if it is accompanied by a change of dipole moment. The specific selection rule is $\Delta v_q = \pm 1$.

The gross selection rule for infrared activity is that *the motion corresponding to a normal mode should be accompanied by a change of dipole moment.* Deciding whether this is so can sometimes be done by inspection. For example, the symmetric stretch of CO_2 leaves the dipole moment unchanged (at zero, see Fig. 12.37), so this mode is infrared inactive. The antisymmetric stretch, however, changes the dipole moment because the molecule becomes unsymmetrical as it vibrates, so this mode is infrared active. Because the dipole moment change is parallel to the principal axis, the transitions arising from this mode are classified as **parallel bands** in the spectrum. Both bending modes are infrared active: they are accompanied by a changing dipole perpendicular to the principal axis, so transitions involving them lead to a **perpendicular band** in the spectrum. The latter bands eliminate the linearity of the molecule, and as a result a Q branch is observed; a parallel band does not have a Q branch.

The active modes are subject to the specific selection rule $\Delta v_q = \pm 1$ in the harmonic approximation, so the wavenumber of the fundamental transition (the 'first harmonic') of each active mode is $\tilde{\nu}_q$. From the analysis of the spectrum, a picture may be constructed of the stiffness of various parts of the molecule, that is, we can establish its **force field**, the set of force constants corresponding to all the displacements of the atoms. The force field may also be estimated by using the semi-empirical, *ab initio*, and DFT computational techniques described in Section 10.7. Superimposed on the simple force field scheme are the complications arising from anharmonicities and the effects of molecular rotation. Very often the sample is a liquid or a solid, and the molecules are unable to rotate freely. In a liquid, for example, a molecule may be able to rotate through only a few degrees before it is struck by another, so it changes its rotational state frequently. This random changing of orientation is called **tumbling**.

The lifetimes of rotational states in liquids are very short, so in most cases the rotational energies are ill-defined. Collisions occur at a rate of about 10^{13} s^{-1} and, even allowing for only a 10 per cent success rate in knocking the molecule into another rotational state, a lifetime broadening (eqn 9.39, in the form $\delta\tilde{\nu} \approx 1/2\pi c\tau$) of more than 1 cm^{-1} can easily result. The rotational structure of the vibrational spectrum is blurred by this effect, so the infrared spectra of molecules in condensed phases usually consist of broad lines spanning the entire range of the resolved gas-phase spectrum, and showing no branch structure.

One very important application of infrared spectroscopy to condensed phase samples, and one for which the blurring of the rotational structure by random collisions is a welcome simplification, is to chemical analysis. The vibrational spectra of different groups in a molecule give rise to absorptions at characteristic frequencies because a normal mode of even a very large molecule is often dominated by the motion of a small group of atoms. The intensities of the vibrational bands that can be identified with the motions of small groups are also transferable between molecules. Consequently, the molecules in a sample can often be identified by examining its infrared spectrum and referring to a table of characteristic frequencies and intensities (Table 12.3).

Table 12.3* Typical vibrational wavenumbers

Vibration type	$\tilde{\nu}$/cm^{-1}
C—H stretch	2850–2960
C—H bend	1340–1465
C—C stretch, bend	700–1250
C=C stretch	1620–1680

* More values are given in the *Data section*.

IMPACT ON ENVIRONMENTAL SCIENCE

I12.2 Climate change[1]

Solar energy strikes the top of the Earth's atmosphere at a rate of 343 W m^{-2}. About 30 per cent of this energy is reflected back into space by the Earth or the atmosphere. The Earth–atmosphere system absorbs the remaining energy and re-emits it into space as black-body radiation, with most of the intensity being carried by infrared radiation in the range 200–2500 cm^{-1} (4–50 μm). The Earth's average temperature is maintained by an energy balance between solar radiation absorbed by the Earth and black-body radiation emitted by the Earth.

The trapping of infrared radiation by certain gases in the atmosphere is known as the *greenhouse effect*, so called because it warms the Earth as if the planet were enclosed in a huge greenhouse. The result is that the natural greenhouse effect raises the average surface temperature well above the freezing point of water and creates an environment in which life is possible. The major constituents to the Earth's atmosphere, O_2 and N_2, do not contribute to the greenhouse effect because homonuclear diatomic molecules cannot absorb infrared radiation. However, the minor atmospheric gases, water vapour and CO_2, do absorb infrared radiation and hence are responsible for the greenhouse effect (Fig. 12.39). Water vapour absorbs strongly in the ranges 1300–1900 cm^{-1} (5.3–7.7 μm) and 3550–3900 cm^{-1} (2.6–2.8 μm), whereas CO_2 shows strong absorption in the ranges 500–725 cm^{-1} (14–20 μm) and 2250–2400 cm^{-1} (4.2–4.4 μm).

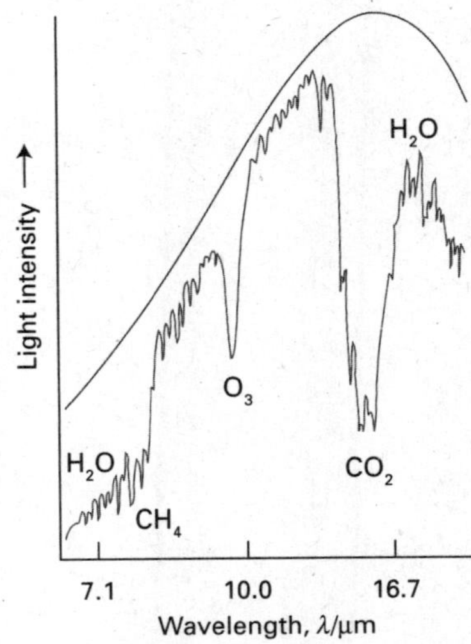

Fig. 12.39 The intensity of infrared radiation that would be lost from Earth in the absence of greenhouse gases is shown by the brown line. The blue line is the intensity of the radiation actually emitted. The maximum wavelength of radiation absorbed by each greenhouse gas is indicated.

[1] This section is based on a similar contribution initially prepared by Loretta Jones and appearing in *Chemical principles*, Peter Atkins and Loretta Jones, W.H. Freeman and Co., New York (2010).

Increases in the levels of greenhouse gases, which also include methane, dinitrogen oxide, ozone, and certain chlorofluorocarbons, as a result of human activity have the potential to enhance the natural greenhouse effect, leading to significant warming of the planet. This problem is referred to as *global warming*, and more generally as *climate change*, which we now explore in some detail.

The concentration of water vapour in the atmosphere has remained steady over time, but concentrations of some other greenhouse gases are rising. From about the year 1000 until about 1750, the CO_2 concentration remained fairly stable, but, since then, it has increased by 28 per cent. The concentration of methane, CH_4, has more than doubled during this time and is now at its highest level for 160 000 years (160 ka; a is the SI unit denoting 1 year). Studies of air pockets in ice cores taken from Antarctica show that increases in the concentration of both atmospheric CO_2 and CH_4 over the past 160 ka correlate well with increases in the global surface temperature.

Human activities are primarily responsible for the rising concentrations of atmospheric CO_2 and CH_4. Most of the atmospheric CO_2 comes from the burning of hydrocarbon fuels, which began on a large scale with the Industrial Revolution in the middle of the nineteenth century. The additional methane comes mainly from the petroleum industry and from agriculture.

The temperature of the surface of the Earth has increased by about 0.8 K since the middle of the nineteenth century (Fig. 12.40). In 2007 the Intergovernmental Panel on Climate Change (IPCC) estimated that our continued reliance on hydrocarbon fuels, coupled to current trends in population growth, could result in an additional increase of 1–3 K in the temperature of the Earth by 2100, relative to the surface temperature in 2000. Furthermore, the rate of temperature change is likely to be greater than at any time in the last 10 ka. To place a temperature rise of 3 K in perspective, it is useful to consider that the average temperature of the Earth during the last ice age was only 6 K colder than at present. Just as cooling the planet (for example, during an ice age) can lead to detrimental effects on ecosystems, so too can a dramatic warming of the globe. One example of a significant change in the environment caused by a temperature increase of 3 K is a rise in sea level by about 0.5 m, which is sufficient to alter weather patterns and submerge coastal ecosystems.

Computer projections for the next 200 years predict further increases in atmospheric CO_2 levels and suggest that, to maintain CO_2 at its current concentration, we would have to reduce hydrocarbon fuel consumption immediately by about 50 per cent. Clearly, in order to reverse global warming trends, we need to develop alternatives to fossil fuels, such as hydrogen (which can be used in fuel cells) and solar energy technologies.

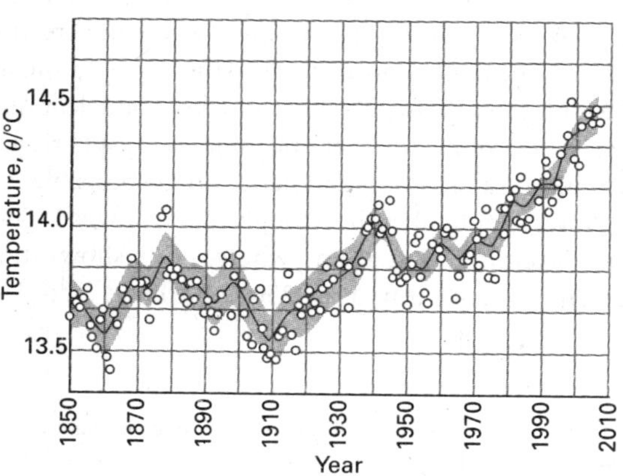

Fig. 12.40 The average change in surface temperature of the Earth from 1855 to 2002.

12.15 Vibrational Raman spectra of polyatomic molecules

Key points **The exclusion rule states that, if the molecule has a centre of symmetry, then no modes can be both infrared and Raman active. (a) Totally symmetrical vibrations give rise to polarized lines. (b) In resonance Raman spectroscopy the frequency of the incident radiation nearly coincides with the frequency of an electronic transition of the sample. (c) Coherent anti-Stokes Raman spectroscopy (CARS) is a Raman technique that relies on the use of two incident beams of radiation.**

The normal modes of vibration of molecules are Raman active if they are accompanied by a changing polarizability. It is sometimes quite difficult to judge by inspection when this is so. The symmetric stretch of CO_2, for example, alternately swells and contracts the molecule: this motion changes the polarizability of the molecule, so the mode is Raman active. The other modes of CO_2 leave the polarizability unchanged, so they are Raman inactive.

A more exact treatment of infrared and Raman activity of normal modes leads to the **exclusion rule**:

If the molecule has a centre of symmetry then no modes can be both infrared and Raman active. Exclusion rule

(A mode may be inactive in both.) Because it is often possible to judge intuitively if a mode changes the molecular dipole moment, we can use this rule to identify modes that are not Raman active. The rule applies to CO_2 but to neither H_2O nor CH_4 because they have no centre of symmetry. In general, it is necessary to use group theory to predict whether a mode is infrared or Raman active (Section 12.16).

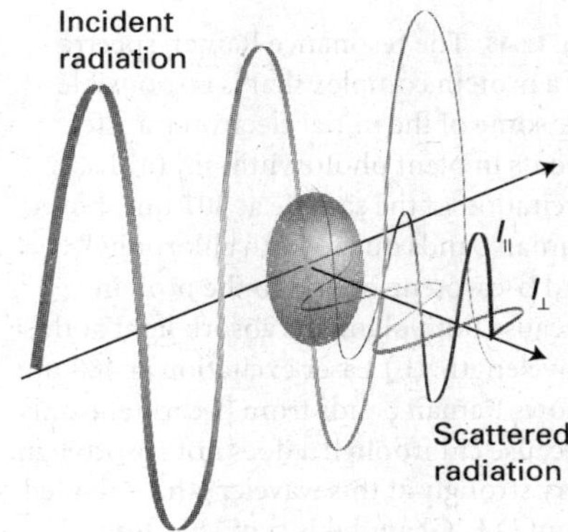

Fig. 12.41 The definition of the planes used for the specification of the depolarization ratio, ρ, in Raman scattering.

(a) Depolarization

The assignment of Raman lines to particular vibrational modes is aided by noting the state of polarization of the scattered light. The **depolarization ratio**, ρ, of a line is the ratio of the intensities, I, of the scattered light with polarizations perpendicular and parallel to the plane of polarization of the incident radiation:

$$\rho = \frac{I_{\perp}}{I_{\parallel}} \qquad \text{Definition of depolarization ratio} \qquad [12.50]$$

To measure ρ, the intensity of a Raman line is measured with a polarizing filter (a 'half-wave plate') first parallel and then perpendicular to the polarization of the incident beam. If the emergent light is not polarized, then both intensities are the same and ρ is close to 1; if the light retains its initial polarization, then $I_{\perp} = 0$, so $\rho = 0$ (Fig. 12.41). A line is classified as **depolarized** if it has ρ close to or greater than 0.75 and as **polarized** if $\rho < 0.75$. Only totally symmetrical vibrations give rise to polarized lines in which the incident polarization is largely preserved. Vibrations that are not totally symmetrical give rise to depolarized lines because the incident radiation can give rise to radiation in the perpendicular direction too.

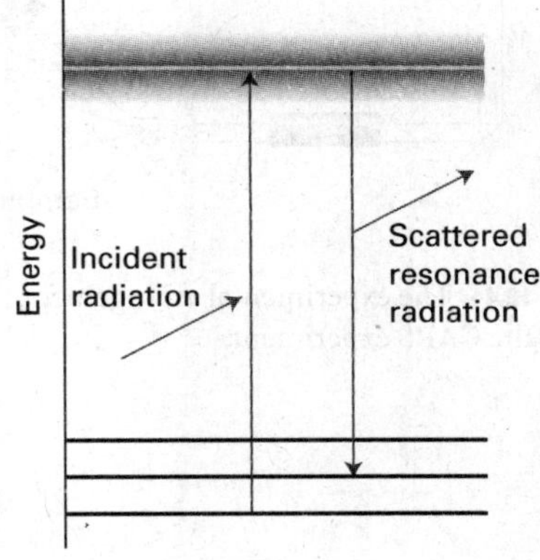

Fig. 12.42 In the resonance Raman effect the incident radiation has a frequency close to an actual electronic excitation of the molecule. A photon is emitted when the excited state returns to a state close to the ground state.

(b) Resonance Raman spectra

A modification of the basic Raman effect involves using incident radiation that nearly coincides with the frequency of an electronic transition of the sample (Fig. 12.42). The technique is then called **resonance Raman spectroscopy**. It is characterized by a much greater intensity in the scattered radiation. Furthermore, because it is often the case that only a few vibrational modes contribute to the more intense scattering, the spectrum is greatly simplified.

Resonance Raman spectroscopy is used to study biological molecules that absorb strongly in the ultraviolet and visible regions of the spectrum. Examples include the

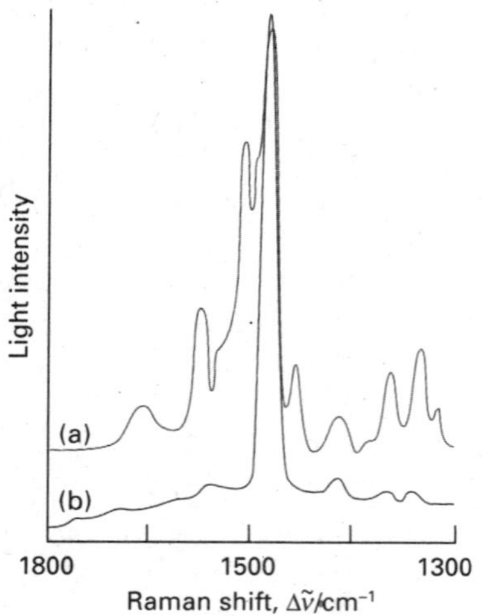

Fig. 12.43 The resonance Raman spectra of a protein complex that is responsible for some of the initial electron transfer events in plant photosynthesis. (a) Laser excitation of the sample at 407 nm shows Raman bands due to both chlorophyll *a* and β-carotene bound to the protein because both pigments absorb light at this wavelength. (b) Laser excitation at 488 nm shows Raman bands from β-carotene only because chlorophyll *a* does not absorb light very strongly at this wavelength. (Adapted from D.F. Ghanotakis *et al.*, *Biochim. Biophys. Acta* **974**, 44 (1989).)

pigments β-carotene and chlorophyll, which capture solar energy during plant photosynthesis. The resonance Raman spectra of Fig. 12.43 show vibrational transitions from only the few pigment molecules that are bound to very large proteins dissolved in an aqueous buffer solution. This selectivity arises from the fact that water (the solvent), amino acid residues, and the peptide group do not have electronic transitions at the laser wavelengths used in the experiment, so their conventional Raman spectra are weak compared to the enhanced spectra of the pigments. Comparison of the spectra in Figs. 12.43a and 12.43b also shows that, with proper choice of excitation wavelength, it is possible to examine individual classes of pigments bound to the same protein: excitation at 488 nm, where β-carotene absorbs strongly, shows vibrational bands from β-carotene only, whereas excitation at 407 nm, where chlorophyll *a* and β-carotene absorb, reveals features from both types of pigments.

(c) Coherent anti-Stokes Raman spectroscopy

The intensity of Raman transitions may be enhanced by **coherent anti-Stokes Raman spectroscopy** (CARS, Fig. 12.44). The technique relies on the fact that, if two laser beams of frequencies ν_1 and ν_2 pass through a sample, then they may mix together and give rise to coherent radiation of several different frequencies, one of which is

$$\nu' = 2\nu_1 - \nu_2 \tag{12.51}$$

Suppose that ν_2 is varied until it matches any Stokes line from the sample, such as the one with frequency $\nu_1 - \Delta\nu$; then the coherent emission will have frequency

$$\nu' = 2\nu_1 - (\nu_1 - \Delta\nu) = \nu_1 + \Delta\nu \tag{12.52}$$

which is the frequency of the corresponding anti-Stokes line. This coherent radiation forms a narrow beam of high intensity.

An advantage of CARS is that it can be used to study Raman transitions in the presence of competing incoherent background radiation, and so can be used to observe the Raman spectra of species in flames. One example is the vibration–rotation CARS spectrum of N_2 gas in a methane–air flame shown in Fig. 12.45.

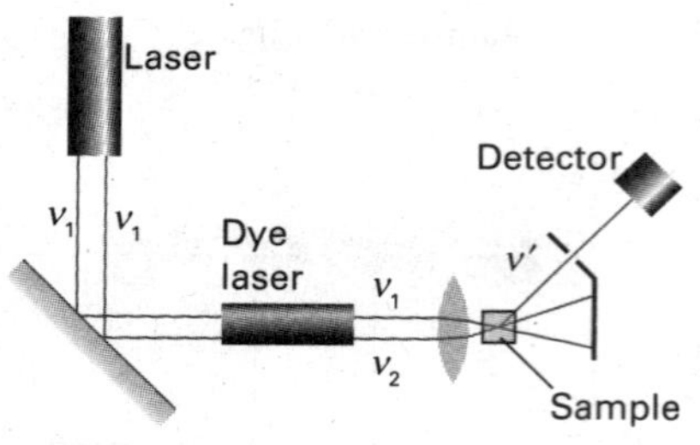

Fig. 12.44 The experimental arrangement for the CARS experiment.

12.16 Symmetry aspects of molecular vibrations

Key points **(a) A normal mode is infrared active if its symmetry species is the same as any of the symmetry species of *x*, *y*, or *z*. (b) A normal mode is Raman active if its symmetry species is the same as the symmetry species of a quadratic form.**

One of the most powerful ways of dealing with normal modes, especially of complex molecules, is to classify them according to their symmetries. Each normal mode must belong to one of the symmetry species of the molecular point group, as discussed in Chapter 11.

Example 12.6 *Identifying the symmetry species of a normal mode*

Establish the symmetry species of the normal mode vibrations of CH_4, which belongs to the group T_d.

Method The first step in the procedure is to identify the symmetry species of the irreducible representations spanned by all the 3*N* displacements of the atoms, using the characters of the molecular point group. Find these characters by counting 1 if the displacement is unchanged under a symmetry operation, −1 if it changes sign, and 0 if it is changed into some other displacement. Next, subtract the symmetry species of the translations. Translational displacements span the same symmetry species as *x*, *y*, and *z*, so they can be obtained from the rightmost column of the

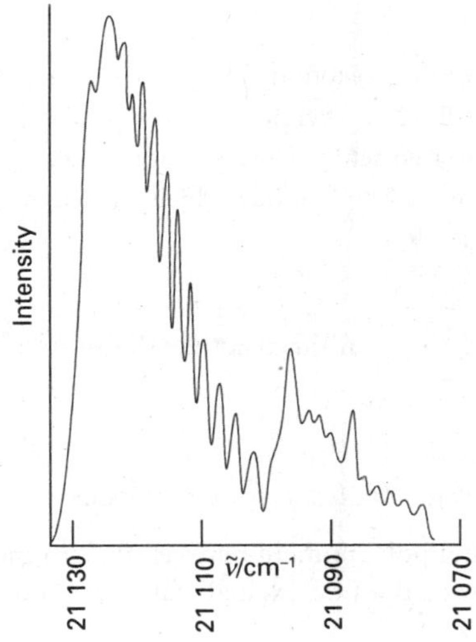

Fig. 12.45 CARS spectrum of a methane–air flame at 2104 K. The peaks correspond to the Q branch of the vibration–rotation spectrum of N_2 gas. (Adapted from J.F. Verdieck *et al.*, *J. Chem. Ed.* **59**, 495 (1982).)

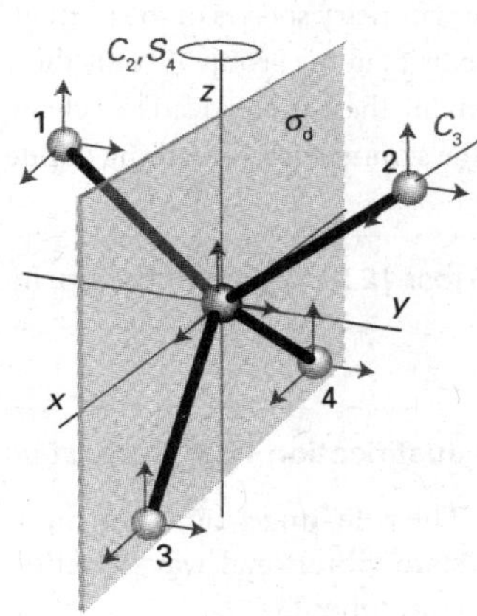

Fig. 12.46 The atomic displacements of CH_4 and the symmetry elements used to calculate the characters.

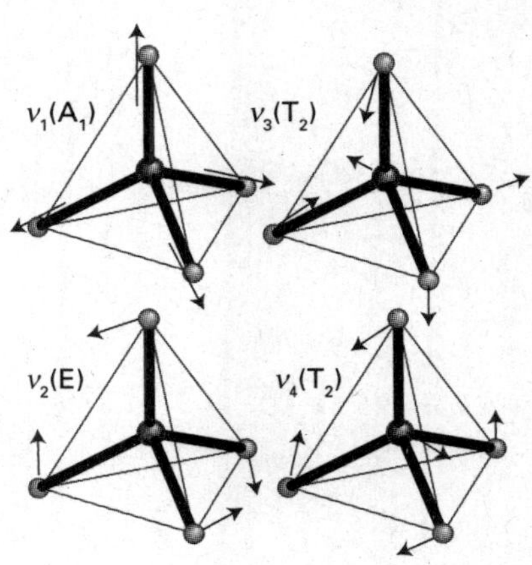

Fig. 12.47 Typical normal modes of vibration of a tetrahedral molecule. There are in fact two modes of symmetry species E and three modes of each T_2 symmetry species.

character table. Finally, subtract the symmetry species of the rotations, which are also given in the character table (and denoted there by R_x, R_y, or R_z).

Answer There are $3 \times 5 = 15$ degrees of freedom, of which $(3 \times 5) - 6 = 9$ are vibrations. Refer to Fig. 12.46. Under E, no displacement coordinates are changed, so the character is 15. Under C_3, no displacements are left unchanged, so the character is 0. Under the C_2 indicated, the z-displacement of the central atom is left unchanged, whereas its x- and y-components both change sign. Therefore $\chi(C_2) = 1 - 1 - 1 + 0 + 0 + \ldots = -1$. Under the S_4 indicated, the z-displacement of the central atom is reversed, so $\chi(S_4) = -1$. Under σ_d, the x- and z-displacements of C, H_3, and H_4 are left unchanged and the y-displacements are reversed; hence $\chi(\sigma_d) = 3 + 3 - 3 = 3$. The characters are therefore 15, 0, −1, −1, 3. By decomposing the direct product (Section 11.5a), we find that this representation spans $A_1 + E + T_1 + 3T_2$. The translations span T_2; the rotations span T_1. Hence, the nine vibrations span $A_1 + E + 2T_2$. The modes are shown in Fig. 12.47. We shall see in the next subsection that symmetry analysis gives a quick way of deciding which modes are active.

Self-test 12.8 Establish the symmetry species of the normal modes of H_2O.

[$2A_1 + B_2$]

(a) Infrared activity of normal modes

It is best to use group theory to judge the activities of more complex modes of vibration. This is easily done by checking the character table of the molecular point group for the symmetry species of the irreducible representations spanned by x, y, and z, for their species are also the symmetry species of the components of the electric dipole moment. Then apply the following rule:

If the symmetry species of a normal mode is the same as any of the symmetry species of x, y, or z, then the mode is infrared active.

Symmetry test for IR activity

• **A brief illustration**

To decide which modes of CH_4 are IR active, we note that we found in Example 12.6 that the symmetry species of the normal modes are $A_1 + E + 2T_2$. Therefore, because x, y, and z span T_2 in the group T_d, only the T_2 modes are infrared active. The distortions accompanying these modes lead to a changing dipole moment. The A_1 mode, which is inactive, is the symmetrical 'breathing' mode of the molecule. •

Self-test 12.9 Which of the normal modes of H_2O are infrared active? [All three]

Justification 12.3 *Using group theory to identify infrared active normal modes*

The rule hinges on the form of the transition dipole moment between the ground-state vibrational wavefunction, ψ_0, and that of the first excited state, ψ_1. The x-component is

$$\mu_{x,10} = -e\int \psi_1^* x \psi_0 \,\mathrm{d}\tau \tag{12.53}$$

with similar expressions for the two other components of the transition moment. The ground-state vibrational wavefunction is a Gaussian function of the form e^{-x^2}, so it is symmetrical in x. The wavefunction for the first excited state gives a non-vanishing integral only if it is proportional to x, for then the integrand is proportional to x^2 rather than to xy or xz. Consequently, the excited state wavefunction must have the same symmetry as the displacement x.

(b) Raman activity of normal modes

Group theory provides an explicit recipe for judging the Raman activity of a normal mode. In this case, the symmetry species of the quadratic forms (x^2, xy, etc.) listed in the character table are noted (they transform in the same way as the polarizability), and then we use the following rule:

If the symmetry species of a normal mode is the same as the symmetry species of a quadratic form, then the mode is Raman active.

Symmetry test for Raman activity

• **A brief illustration**

To decide which of the vibrations of CH_4 are Raman active, refer to the T_d character table. It was established in Example 12.6 that the symmetry species of the normal modes are $A_1 + E + 2T_2$. Because the quadratic forms span $A_1 + E + T_2$, all the normal modes are Raman active. By combining this information with that in Example 12.6, we see how the infrared and Raman spectra of CH_4 are assigned. The assignment of spectral features to the T_2 modes is straightforward because these are the only modes that are both infrared and Raman active. This leaves the A_1 and E modes to be assigned in the Raman spectrum. Measurement of the depolarization ratio distinguishes between these modes because the A_1 mode, being totally symmetric, is polarized and the E mode is depolarized. •

Self-test 12.10 Which of the vibrational modes of H_2O are Raman active? [All three]

Checklist of key equations

Property	Equation	Comment
Moment of inertia	$I = \sum_i m_i x_i^2$	x_i is perpendicular distance of atom i from the axis of rotation
Rotational terms of a spherical or linear rotor	$\tilde{F}(J) = \tilde{B}J(J+1)$	$J = 0, 1, 2, \ldots$; $\tilde{B} = \hbar/4\pi cI$
Rotational terms of a symmetric rotor	$\tilde{F}(J,K) = \tilde{B}J(J+1) + (\tilde{A} - \tilde{B})K^2$	$J = 0, 1, 2, \ldots$; $K = 0, \pm 1, \ldots, \pm J$ $\tilde{A} = \hbar/4\pi cI_{\parallel}$ $\tilde{B} = \hbar/4\pi cI_{\perp}$
Rotational terms of a spherical or linear rotor affected by centrifugal distortion	$\tilde{F}(J) = \tilde{B}J(J+1) - \tilde{D}_J J^2(J+1)^2$	$\tilde{D}_J = 4\tilde{B}^3/\tilde{\nu}^2$
Wavenumbers of rotational transitions of linear rotors	$\tilde{\nu}(J+1 \leftarrow J) = 2\tilde{B}(J+1)$	$J = 0, 1, 2, \ldots$
Wavenumbers of (i) Stokes and (ii) anti-Stokes lines in the rotational Raman spectrum of linear rotors	(i) $\tilde{\nu}(J+2 \leftarrow J) = \tilde{\nu}_i - 2\tilde{B}(2J+3)$ (ii) $\tilde{\nu}(J-2 \leftarrow J) = \tilde{\nu}_i + 2\tilde{B}(2J-1)$	$J = 0, 1, 2, \ldots$
Vibrational terms of a diatomic molecule	$\tilde{G}(v) = (v + \tfrac{1}{2})\tilde{\nu}$	$\tilde{\nu} = (1/2\pi c)(k_f/m_{\text{eff}})^{1/2}$ $m_{\text{eff}} = m_1 m_2/(m_1 + m_2)$
Wavenumbers of vibrational transitions of a diatomic molecule	$\Delta\tilde{G}_{v+\frac{1}{2}} = \tilde{\nu}$	$v = 0, 1, 2, \ldots$
Morse potential energy	$V = hc\tilde{D}_e\{1 - e^{-a(R-R_e)}\}^2$	$a = (m_{\text{eff}}\omega^2/2hc\tilde{D}_e)^{1/2}$
(i) Vibrational terms and (ii) wavenumbers of transitions of a diatomic molecule modelled with the Morse potential	(i) $\tilde{G}(v) = (v + \tfrac{1}{2})\tilde{\nu} - (v + \tfrac{1}{2})^2 x_e\tilde{\nu}$ (ii) $\Delta\tilde{G}_{v+\frac{1}{2}} = \tilde{\nu} - 2(v+1)x_e\tilde{\nu} + \cdots$	$x_e = \tilde{\nu}/4\tilde{D}_e$ In (ii), for a pure Morse potential, the series terminates after the second term
Vibration–rotation infrared transitions of a diatomic molecule	$\tilde{\nu}_P(J) = \tilde{\nu} - 2\tilde{B}J$ $\tilde{\nu}_Q(J) = \tilde{\nu}$ $\tilde{\nu}_R(J) = \tilde{\nu} + 2\tilde{B}(J+1)$	P ($J-1 \leftarrow J$), Q ($J \leftarrow J$), and R($J+1 \leftarrow J$) branches
Vibration-rotation Raman transitions of a diatomic molecule	$\tilde{\nu}_O(J) = \tilde{\nu}_i - \tilde{\nu} - 2\tilde{B} + 4\tilde{B}J$ $\tilde{\nu}_Q(J) = \tilde{\nu}_i - \tilde{\nu}$ $\tilde{\nu}_S(J) = \tilde{\nu}_i - \tilde{\nu} - 6\tilde{B} - 4\tilde{B}J$	O ($J-2 \leftarrow J$), Q ($J \leftarrow J$), and S($J+2 \leftarrow J$) branches
Depolarization ratio of a Raman line	$\rho = I_{\perp}/I_{\parallel}$	Polarized lines: $\rho < 0.75$ Depolarized lines: $\rho \geq 0.75$

Further information

Further information 12.1 *Spectrometers*

Here we provide additional brief details of the principles of operation of spectrometers, describing radiation sources, dispersing elements, detectors, and Fourier transform techniques. The information here is also relevant to the electronic transitions discussed in Chapter 13, where the radiation absorbed lies in the visible and ultraviolet regions of the spectrum.

(a) Sources of radiation

Sources of radiation are either *monochromatic*, those spanning a very narrow range of frequencies around a central value, or *polychromatic*, those spanning a wide range of frequencies. Monochromatic sources that can be tuned over a range of frequencies include the *klystron* and the *Gunn diode*, which operate in the microwave range, and lasers, which are discussed in Chapter 13.

Polychromatic sources that take advantage of black-body radiation from hot materials can be used from the infrared to the ultraviolet regions of the electromagnetic spectrum. Examples include mercury arcs inside a quartz envelope ($35\ \text{cm}^{-1} < \tilde{\nu} < 200\ \text{cm}^{-1}$), *Nernst filaments* and *globars* ($200\ \text{cm}^{-1} < \tilde{\nu} < 4000\ \text{cm}^{-1}$), and *quartz–tungsten–halogen lamps* ($320\ \text{nm} < \lambda < 2500\ \text{nm}$).

A *gas discharge lamp* is a common source of ultraviolet and visible radiation. In a *xenon discharge lamp*, an electrical discharge excites xenon atoms to excited states, which then emit ultraviolet radiation. In a *deuterium lamp*, excited D_2 molecules dissociate into electronically excited D atoms, which emit intense radiation between 200 nm and 400 nm.

For certain applications, synchrotron radiation is generated in a *synchrotron storage ring*, which consists of an electron beam travelling in a circular path with circumferences of up to several hundred metres. As electrons travelling in a circle are constantly accelerated by

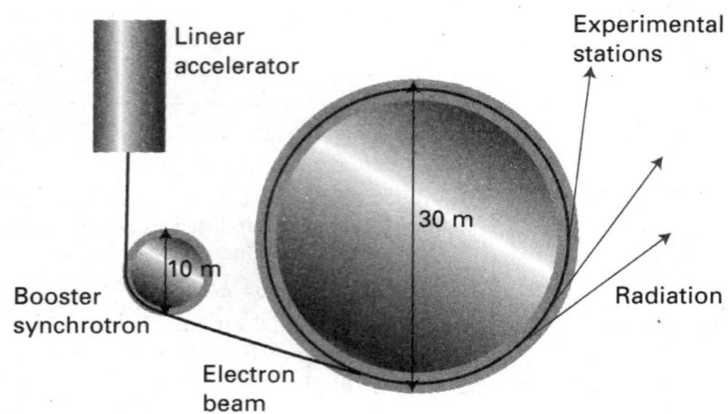

Fig. 12.48 A synchrotron storage ring. The electrons injected into the ring from the linear accelerator and booster synchrotron are accelerated to high speed in the main ring. An electron in a curved path is subject to constant acceleration, and an accelerated charge radiates electromagnetic energy.

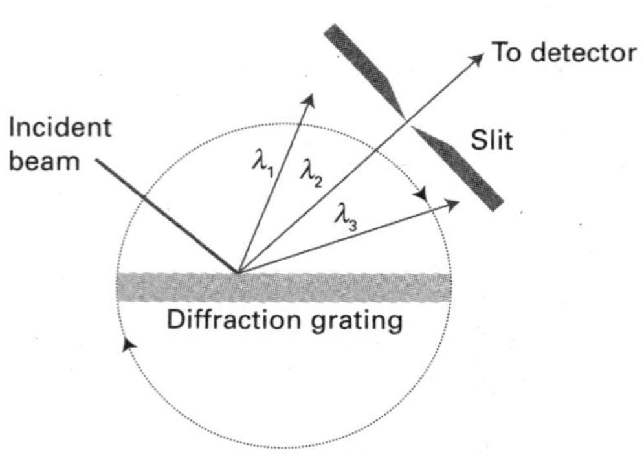

Fig. 12.50 A polychromatic beam is dispersed by a diffraction grating into three component wavelengths λ_1, λ_2, and λ_3. In the configuration shown, only radiation with λ_2 passes through a narrow slit and reaches the detector. Rotating the diffraction grating (as shown by the arrows on the dotted circle) allows λ_1 or λ_3 to reach the detector.

the forces that constrain them to their path, they generate radiation (Fig. 12.48). Synchrotron radiation spans a wide range of frequencies, including the infrared and X-rays. Except in the microwave region, synchrotron radiation is much more intense than can be obtained by most conventional sources.

(b) The dispersing element

The dispersing element in most absorption spectrometers operating in the ultraviolet to near-infrared region of the spectrum is a **diffraction grating**, which consists of a glass or ceramic plate into which fine grooves have been cut and covered with a reflective aluminium coating. The grating causes interference between waves reflected from its surface, and constructive interference occurs when

$$n\lambda = d(\sin\theta - \sin\phi) \tag{12.54}$$

where $n = 1, 2, \ldots$ is the *diffraction order*, λ is the wavelength of the diffracted radiation, d is the distance between grooves, θ is the angle of incidence of the beam, and ϕ is the angle of emergence of the beam (Fig. 12.49). For given values of n and θ, larger differences in ϕ are observed for different wavelengths when d is similar to the wavelength of radiation being analysed. Wide angular separation results in wide spatial separation between wavelengths some distance away from the grating, where a detector is placed.

In a **monochromator**, a narrow exit slit allows only a narrow range of wavelengths to reach the detector (Fig. 12.50). Turning the grating around an axis perpendicular to the incident and diffracted beams allows different wavelengths to be analysed; in this way, the absorption spectrum is built up one narrow wavelength range at a time. Typically, the grating is swept through an angle that investigates only the first order of diffraction ($n = 1$). In a **polychromator**, there is no slit and a broad range of wavelengths can be analysed simultaneously by *array detectors*, such as those discussed below.

(c) Fourier transform techniques

In a Fourier transform instrument, the diffraction grating is replaced by a Michelson interferometer, which works by splitting the beam from the sample into two and introducing a varying path difference, p, into one of them (Fig. 12.51). When the two components recombine, there is a phase difference between them, and they interfere either constructively or destructively depending on the

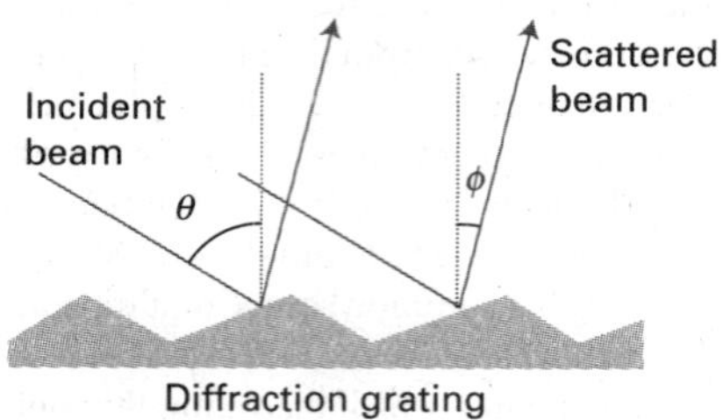

Fig. 12.49 One common dispersing element is a diffraction grating, which separates wavelengths spatially as a result of the scattering of light by fine grooves cut into a coated piece of glass. When a polychromatic light beam strikes the surface at an angle θ, several light beams of different wavelengths emerge at different angles ϕ (eqn 12.54).

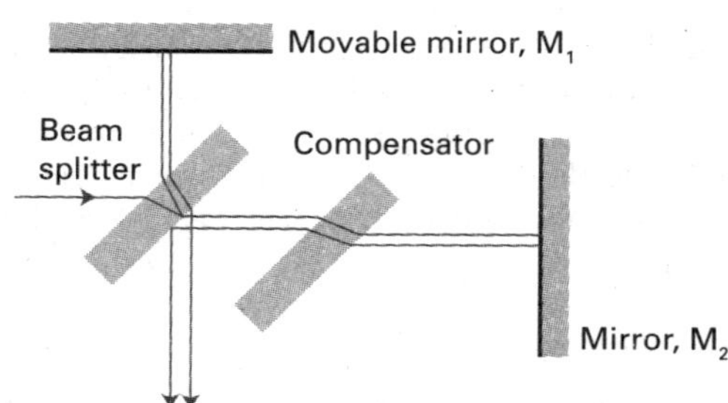

Fig. 12.51 A Michelson interferometer. The beam-splitting element divides the incident beam into two beams with a path difference that depends on the location of the mirror M_1. The compensator ensures that both beams pass through the same thickness of material.

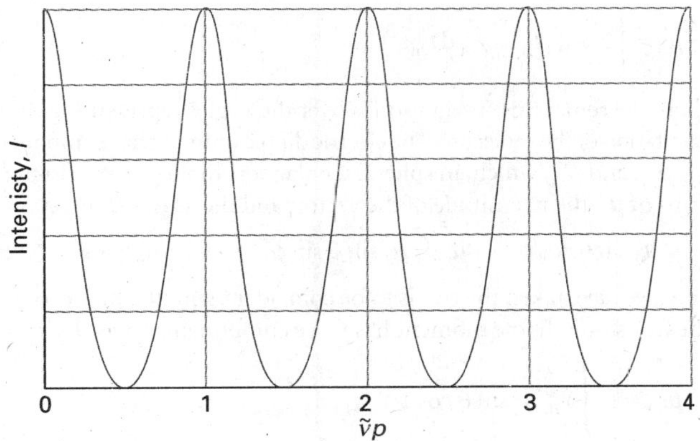

Fig. 12.52 An interferogram produced as the path length p is changed in the interferometer shown in Fig. 12.51. Only a single frequency component is present in the signal, so the graph is a plot of the function $I(p) = I_0(1 + \cos 2\pi\tilde{\nu}p)$, where I_0 is the intensity of the radiation.

interActivity Referring to Fig. 12.51, the mirror M_1 moves in finite distance increments, so the path difference p is also incremented in finite steps. Explore the effect of increasing the step size on the shape of the interferogram for a monochromatic beam of wavenumber $\tilde{\nu}$ and intensity I_0. That is, draw plots of $I(p)/I_0$ against $\tilde{\nu}p$, each with a different number of data points spanning the same total distance path taken by the movable mirror M_1.

difference in path lengths. The detected signal oscillates as the two components alternately come into and out of phase as the path difference is changed (Fig. 12.52). If the radiation has wavenumber $\tilde{\nu}$, the intensity of the detected signal due to radiation in the range of wavenumbers $\tilde{\nu}$ to $\tilde{\nu} + d\tilde{\nu}$, which we denote $I(p,\tilde{\nu})d\tilde{\nu}$, varies with p as

$$I(p,\tilde{\nu})d\tilde{\nu} = I(\tilde{\nu})(1 + \cos 2\pi\tilde{\nu}p)d\tilde{\nu} \qquad (12.55)$$

Hence, the interferometer converts the presence of a particular wavenumber component in the signal into a variation in intensity of the radiation reaching the detector. An actual signal consists of radiation spanning a large number of wavenumbers, and the total intensity at the detector, which we write $I(p)$, is the sum of contributions from all the wavenumbers present in the signal (Fig. 12.53):

$$I(p) = \int_0^\infty I(p,\tilde{\nu})d\tilde{\nu} = \int_0^\infty I(\tilde{\nu})(1 + \cos 2\pi\tilde{\nu}p)d\tilde{\nu} \qquad (12.56)$$

The problem is to find $I(\tilde{\nu})$, the variation of intensity with wavenumber, which is the spectrum we require, from the record of values of $I(p)$. This step is a standard technique of mathematics, and is the 'Fourier transformation' step from which this form of spectroscopy takes its name (see *Mathematical background 7* following Chapter 19). Specifically:

$$I(\tilde{\nu}) = 4\int_0^\infty \{I(p) - \tfrac{1}{2}I(0)\} \cos 2\pi\tilde{\nu}p \, dp \qquad (12.57)$$

where $I(0)$ is given by eqn 12.56 with $p = 0$. This integration is carried out numerically in a computer connected to the spectrometer, and the output, $I(\tilde{\nu})$, is the transmission spectrum of the sample (Fig. 12.54).

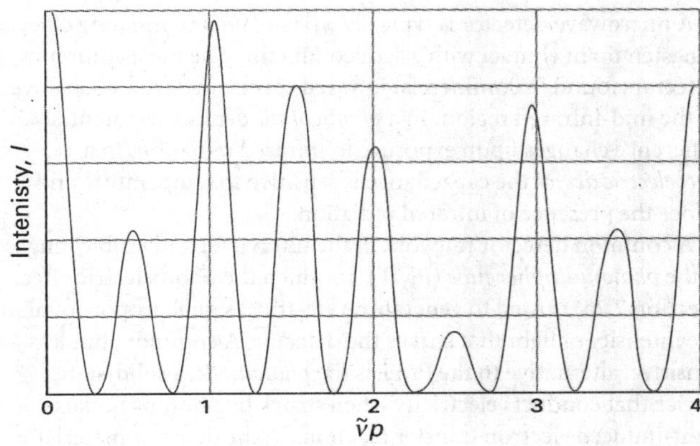

Fig. 12.53 An interferogram obtained when several (in this case, three) frequencies are present in the radiation.

interActivity For a signal consisting of only a few monochromatic beams, the integral in eqn 12.56 can be replaced by a sum over the finite number of wavenumbers. Use this information to draw your own version of Fig. 12.53. Then, go on to explore the effect of varying the wavenumbers and intensities of the three components of the radiation on the shape of the interferogram.

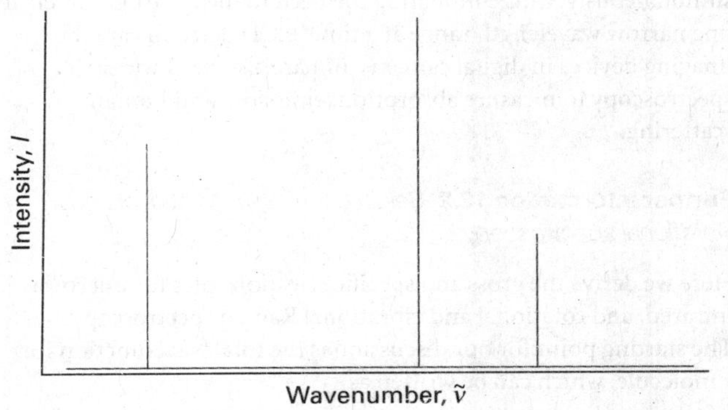

Fig. 12.54 The three frequency components and their intensities that account for the appearance of the interferogram in Fig. 12.53. This spectrum is the Fourier transform of the interferogram, and is a depiction of the contributing frequencies.

interActivity Calculate the Fourier transforms of the functions you generated in the previous *interActivity*.

A major advantage of the Fourier transform procedure is that all the radiation emitted by the source is monitored continuously. This is in contrast to a spectrometer in which a monochromator discards most of the generated radiation. As a result, Fourier transform spectrometers have a higher sensitivity than conventional spectrometers.

(d) Detectors

A **detector** is a device that converts radiation into an electric current or voltage for appropriate signal processing and display. Detectors may consist of a single radiation-sensing element or of several small elements arranged in one- or two-dimensional arrays.

A microwave detector is typically a *crystal diode* consisting of a tungsten tip in contact with a semiconductor. The most common detectors found in commercial infrared spectrometers are sensitive in the mid-infrared region. In a *photovoltaic device* the potential difference changes upon exposure to infrared radiation. In a *pyroelectric device* the capacitance is sensitive to temperature and hence the presence of infrared radiation.

A common detector for work in the ultraviolet and visible ranges is the *photomultiplier tube* (PMT), in which the photoelectric effect (Section 7.2a) is used to generate an electrical signal proportional to the intensity of light that strikes the detector. A common, but less sensitive, alternative to the PMT is the *photodiode*, a solid-state device that conducts electricity when struck by photons because light-induced electron transfer reactions in the detector material create mobile charge carriers (negatively charged electrons and positively charged 'holes'). In an *avalanche photodiode*, the photo-generated electrons are accelerated through a very large electrical potential difference. The high-energy electrons then collide with other atoms in the solid and ionize them, thus creating an avalanche of secondary charge carriers and increasing the sensitivity of the device toward photons.

The *charge-coupled device* (CCD) is a two-dimensional array of several million small photodiode detectors. With a CCD, a wide range of wavelengths that emerge from a polychromator are detected simultaneously, thus eliminating the need to measure light intensity one narrow wavelength range at a time. CCD detectors are the imaging devices in digital cameras, but are also used widely in spectroscopy to measure absorption, emission, and Raman scattering.

Further information 12.2 *Selection rules for rotational and vibrational spectroscopy*

Here we derive the gross and specific selection rules for microwave, infrared, and rotational and vibrational Raman spectroscopy. The starting point for our discussion is the total wavefunction for a molecule, which can be written as

$$\psi_{\text{total}} = \psi_{\text{c.m.}}\psi$$

where $\psi_{\text{c.m.}}$ describes the motion of the centre of mass and ψ describes the internal motion of the molecule. If we neglect the effect of electron spin, the Born–Oppenheimer approximation allows us to write ψ as the product of an electronic part, ψ_{ε}, a vibrational part, ψ_v, and a rotational part, which for a diatomic molecule can be represented by the spherical harmonics $Y_{J,M_J}(\theta,\phi)$ (Section 8.7). The transition dipole moment for a spectroscopic transition can now be written

$$\mu_{\text{fi}} = \int \psi_{\varepsilon\text{f}}^{*}\psi_{v\text{f}}^{*}Y_{J\text{f},M_J\text{f}}^{*}\hat{\mu}\psi_{\varepsilon\text{i}}\psi_{v\text{i}}Y_{J\text{i},M_J\text{i}}\text{d}\tau \qquad (12.58)$$

and our task is to explore conditions for which this integral vanishes or has a nonzero value.

(a) Microwave spectra

During a pure rotational transition the molecule does not change electronic or vibrational states. We identify $\mu_{\text{i}} = \int\psi_{\varepsilon\text{i}}^{*}\psi_{v\text{i}}^{*}\hat{\mu}\psi_{\varepsilon\text{i}}\psi_{v\text{i}}\text{d}\tau$ with the *permanent* electric dipole moment of the molecule in the state i. Equation 12.58 becomes

$$\mu_{\text{fi}} = \int Y_{J\text{f},M_J\text{f}}^{*}\mu_{\text{i}}Y_{J\text{i},M_J\text{i}}\,\text{d}\tau_{\text{angles}} \qquad (12.59)$$

where the remaining integration is over the angles representing the orientation of the molecule. The electric dipole moment has components $\mu_{\text{i},x}$, $\mu_{\text{i},y}$, and $\mu_{\text{i},z}$, which, in spherical polar coordinates, are written in terms of μ_0, the magnitude of the vector, and the angles θ and ϕ as

$$\mu_{\text{i},x} = \mu_0 \sin\theta\cos\phi \qquad \mu_{\text{i},y} = \mu_0\sin\theta\sin\phi \qquad \mu_{\text{i},z} = \mu_0\cos\theta \qquad (12.60)$$

Here, we have taken the z-axis to be coincident with the figure axis. The transition dipole moment has three components, given by

$$\mu_{\text{fi},x} = \mu_0\int Y_{J\text{f},M_J\text{f}}^{*}\sin\theta\cos\phi\,Y_{J\text{i},M_J\text{i}}\,\text{d}\tau_{\text{angles}}$$

$$\mu_{\text{fi},y} = \mu_0\int Y_{J\text{f},M_J\text{f}}^{*}\sin\theta\sin\phi\,Y_{J\text{i},M_J\text{i}}\,\text{d}\tau_{\text{angles}}$$

$$\mu_{\text{fi},z} = \mu_0\int Y_{J\text{f},M_J\text{f}}^{*}\cos\theta\,Y_{J\text{i},M_J\text{i}}\,\text{d}\tau_{\text{angles}} \qquad (12.61)$$

We see immediately that the molecule must have a permanent dipole moment in order to have a microwave spectrum. This is the gross selection rule for microwave spectroscopy.

For the specific selection rules we need to examine the conditions for which the integrals do not vanish, and we must consider each component. For the z-component, we simplify the integral by using $\cos\theta \propto Y_{1,0}$ (Table 8.2). It follows that

$$\mu_{\text{fi},z} \propto \int Y_{J\text{f},M_J\text{f}}^{*}Y_{1,0}Y_{J\text{i},M_J\text{i}}\,\text{d}\tau_{\text{angles}} \qquad (12.62\text{a})$$

According to the properties of the spherical harmonics (Table 8.2), this integral vanishes unless $J_{\text{f}} - J_{\text{i}} = \pm1$ and $M_{J,\text{f}} - M_{J,\text{i}} = 0$. These are two of the selection rules stated in eqn 12.18.

For the x- and y-components, we use $\cos\phi = \frac{1}{2}(\text{e}^{\text{i}\phi} + \text{e}^{-\text{i}\phi})$ to write $\sin\phi = -\frac{1}{2}\text{i}(\text{e}^{\text{i}\phi} - \text{e}^{-\text{i}\phi})$ to write $\sin\theta\cos\phi \propto Y_{1,1} + Y_{1,-1}$ and $\sin\theta\sin\phi \propto Y_{1,1} - Y_{1,-1}$. It follows that

$$\mu_{\text{fi},x} \propto \int Y_{J\text{f},M_J\text{f}}^{*}(Y_{1,+1} + Y_{1,-1})Y_{J\text{i},M_J\text{i}}\,\text{d}\tau_{\text{angles}}$$

$$\mu_{\text{fi},y} \propto \int Y_{J\text{f},M_J\text{f}}^{*}(Y_{1,+1} - Y_{1,-1})Y_{J\text{i},M_J\text{i}}\,\text{d}\tau_{\text{angles}} \qquad (12.62\text{b})$$

According to the properties of the spherical harmonics, these integrals vanish unless $J_{\text{f}} - J_{\text{i}} = \pm1$ and $M_{J,\text{f}} - M_{J,\text{i}} = \pm1$. This completes the selection rules of eqn 12.18.

(b) Rotational Raman spectra

We can understand the origin of the gross and specific selection rules for rotational Raman spectroscopy by using a diatomic molecule as an example. The incident electric field, $\mathcal{E}$, of a wave of electromagnetic radiation of frequency ω_{i} induces a molecular dipole moment that is given by

$$\mu_{\text{ind}} = \alpha\mathcal{E}(t) = \alpha\mathcal{E}\cos\omega_{\text{i}}t \qquad (12.63)$$

If the molecule is rotating at a circular frequency ω_{R}, to an external observer its polarizability is also time-dependent (if it is anisotropic), and we can write

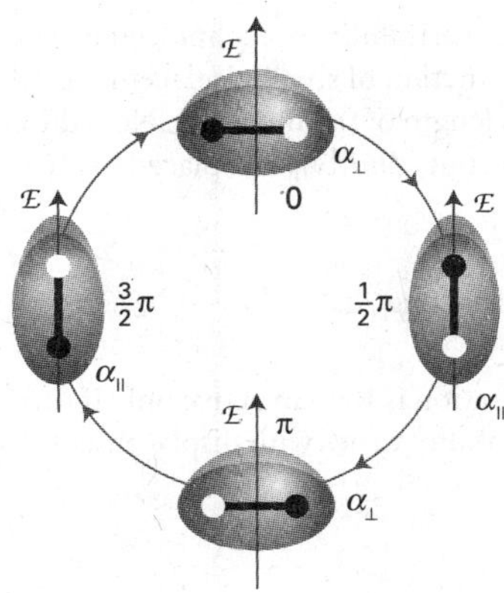

Fig. 12.55 The distortion induced in a molecule by an applied electric field returns to its initial value after a rotation of only 180° (that is, twice a revolution). This is the origin of the $\Delta J = \pm 2$ selection rule in rotational Raman spectroscopy.

$$\alpha = \alpha_0 + \Delta\alpha \cos 2\omega_R t \qquad (12.64)$$

where $\Delta\alpha = \alpha_\parallel - \alpha_\perp$ and α ranges from $\alpha_0 + \Delta\alpha$ to $\alpha_0 - \Delta\alpha$ as the molecule rotates. The 2 appears because the polarizability returns to its initial value twice each revolution (Fig. 12.55). Substituting this expression into the expression for the induced dipole moment gives

$$\begin{aligned}\mu_{\text{ind}} &= (\alpha_0 + \Delta\alpha \cos 2\omega_R t) \times (\mathcal{E} \cos \omega_i t)\\ &= \alpha_0 \mathcal{E} \cos \omega_i t + \mathcal{E}\Delta\alpha \cos 2\omega_R t \cos \omega_i t\\ &= \alpha_0 \mathcal{E} \cos \omega_i t + \tfrac{1}{2}\mathcal{E}\Delta\alpha\{\cos(\omega_i + 2\omega_R)t + \cos(\omega_i - 2\omega_R)t\}\end{aligned} \qquad (12.65)$$

This calculation shows that the induced dipole has a component oscillating at the incident frequency (which generates Rayleigh radiation), and that it also has two components at $\omega_i \pm 2\omega_R$, which give rise to the shifted Raman lines. These lines appear only if $\Delta\alpha \neq 0$; hence the polarizability must be anisotropic for there to be Raman lines. This is the gross selection rule for rotational Raman spectroscopy. We also see that the distortion induced in the molecule by the incident electric field returns to its initial value after a rotation of 180° (that is, twice a revolution). This is the classical origin of the specific selection rule $\Delta J = \pm 2$. The complete quantum mechanical calculation proceeds like that for microwave transitions but is too involved to include here.[2]

(c) Infrared spectra

The gross selection rule for infrared spectroscopy is based on an analysis of the transition dipole moment $\mu_{fi} = \int \psi_{vf}^* \hat{\mu} \psi_{vi}\, d\tau$, which arises from eqn 12.58 when the molecule does not change electronic or rotational states. For simplicity, we shall consider a one-dimensional oscillator (like a diatomic molecule). The electric dipole moment operator depends on the location of all the electrons and all the nuclei in the molecule, so it varies as the internuclear separation changes (Fig. 12.56). We can write its variation with displacement from the equilibrium separation, x, as

$$\mu = \mu_0 + \left(\frac{d\mu}{dx}\right)_0 x + \cdots \qquad (12.66)$$

[2] See our *Quanta, matter, and change* (2009).

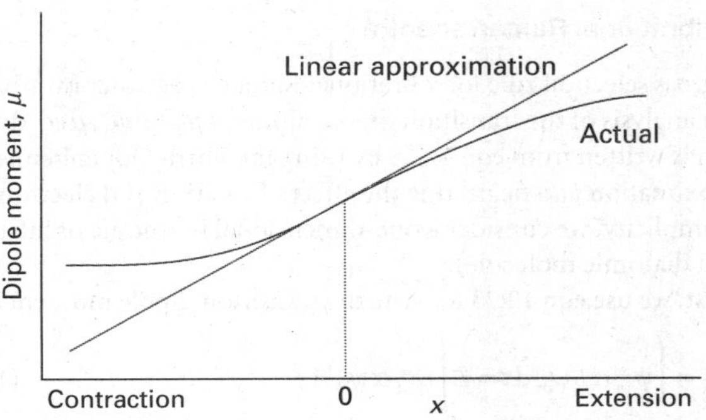

Fig. 12.56 The electric dipole moment of a heteronuclear diatomic molecule varies as shown by the purple curve. For small displacements the change in dipole moment is proportional to the displacement.

where μ_0 is the electric dipole moment operator when the nuclei have their equilibrium separation. It then follows that, with $f \neq i$ and keeping only the term linear in the small displacement x,

$$\mu_{fi} = \int \psi_{vf}^* \hat{\mu} \psi_{vi}\, dx = \mu_0 \int \psi_{vf}^* \psi_{vi}\, dx + \left(\frac{d\mu}{dx}\right)_0 \int \psi_{vf}^* x \psi_{vi}\, dx$$

The term multiplying μ_0 is zero because the states with different values of v are orthogonal. It follows that the transition dipole moment is

$$\mu_{fi} = \left(\frac{d\mu}{dx}\right)_0 \int \psi_{vf}^* x \psi_{vi}\, dx \qquad (12.67)$$

We see that the right-hand side is zero unless the dipole moment varies with displacement. This is the gross selection rule for infrared spectroscopy.

The specific selection rule is determined by considering the value of $\int \psi_{vf}^* x \psi_{vi}\, dx$. We need to write out the wavefunctions in terms of the Hermite polynomials given in Section 8.5 and then to use their properties (Example 8.4 should be reviewed, for it gives further details of the calculation). We note that $x = \alpha y$ with $\alpha = (\hbar^2/m_{\text{eff}} k_f)^{1/4}$ (eqn 8.27; note that in this context α is not the polarizability). Then we write

$$\int \psi_{vf}^* x \psi_{vi}\, dx = N_{v_f} N_{v_i} \int_{-\infty}^{\infty} H_{v_f} x H_{v_i} e^{-y^2} dx = \alpha^2 N_{v_f} N_{v_i} \int_{-\infty}^{\infty} H_{v_f} y H_{v_i} e^{-y^2} dy$$

To evaluate the integral we use the recursion relation

$$yH_v = vH_{v-1} + \tfrac{1}{2}H_{v+1}$$

which turns the matrix element into

$$\int \psi_{vf}^* x \psi_{vi}\, dx = \alpha^2 N_{v_f} N_{v_i} \left\{ v_i \int_{-\infty}^{\infty} H_{v_f} H_{v_i - 1}\, e^{-y^2} dy + \tfrac{1}{2} \int_{-\infty}^{\infty} H_{v_f} H_{v_i+1} e^{-y^2} dy \right\} \qquad (12.68)$$

The first integral is zero unless $v_f = v_i - 1$ and that the second is zero unless $v_f = v_i + 1$ (Table 8.1). It follows that the transition dipole moment is zero unless $\Delta v = \pm 1$.

(d) Vibrational Raman spectra

The gross selection rule for vibrational Raman spectroscopy is based on an analysis of the transition dipole moment $\mu_{fi} = \int \psi_{vf}^* \hat{\mu} \psi_{vi} \, d\tau$, which is written from eqn 12.58 by using the Born–Oppenheimer approximation and neglecting the effect of rotation and electron spin. For simplicity, we consider a one-dimensional harmonic oscillator (like a diatomic molecule).

First, we use eqn 12.23 to write the transition dipole moment as

$$\mu_{fi} = \int \psi_{vf}^* \alpha \mathcal{E} \psi_{vi} \, d\tau = \mathcal{E} \int \psi_{vf}^* \alpha \psi_{vi} \, d\tau \quad (12.69)$$

where $\alpha(x)$ is the polarizability of the molecule, which we expect to be a function of small displacements x from the equilibrium bond length of the molecule. Now the calculation proceeds as before, but $(dm/dx)_0$ is replaced by $\mathcal{E}(d\alpha/dx)_0$ in eqn 12.67. For $f \neq i$,

$$\mu_f = \mathcal{E}\left(\frac{d\alpha}{dx}\right)_0 \int \psi_{vf}^* x \psi_{vi} \, dx \quad (12.70)$$

Therefore, the vibration is Raman active only if $(d\alpha/dx)_0 \neq 0$, that is, the polarizability varies with displacement, and if $v_f - v_i = \pm 1$.

Discussion questions

12.1 Describe the physical origins of linewidths in the absorption and emission spectra of gases, liquids, and solids.

12.2 Discuss the physical origins of the gross and specific selection rules for microwave and infrared spectroscopy.

12.3 Discuss the physical origins of the gross and specific selection rules for rotational and vibrational Raman spectroscopy.

12.4 Explain how nuclear spin can influence the appearance of molecular spectra.

12.5 Consider a diatomic molecule that is highly susceptible to centrifugal distortion in its ground vibrational state. Do you expect excitation to high rotational energy levels to change the equilibrium bond length of this molecule? Justify your answer.

12.6 In what ways may the rotational and vibrational spectra of molecules change as a result of isotopic substitution?

12.7 Suppose that you wish to characterize the normal modes of benzene in the gas phase. Why is it important to obtain both infrared absorption and Raman spectra of your sample?

Exercises

12.1(a) Which of the following molecules may show a pure rotational microwave absorption spectrum: (a) H_2, (b) HCl, (c) CH_4, (d) CH_3Cl, (e) CH_2Cl_2?

12.1(b) Which of the following molecules may show a pure rotational microwave absorption spectrum: (a) H_2O, (b) H_2O_2, (c) NH_3, (d) N_2O?

12.2(a) Which of the following molecules may show a pure rotational Raman spectrum: (a) H_2, (b) HCl, (c) CH_4, (d) CH_3Cl?

12.2(b) Which of the following molecules may show a pure rotational Raman spectrum: (a) CH_2Cl_2, (b) CH_3CH_3, (c) SF_6, (d) N_2O?

12.3(a) Calculate the moment of inertia of an $^{31}PH_3$ molecule for rotation about its threefold axis. By how much does that moment of inertia change when ^{32}P replaces ^{31}P? ($m(^{31}P) = 30.97m_u$; $R_e = 142$ pm; HPH angle = 93.6°.)

12.3(b) Calculate the moment of inertia of a SiH_4 (bond length 147.98 pm) ion. By how much does that moment of inertia change when ^2H replaces ^1H?

12.4(a) Use the information in Table 12.1 to calculate the moments of inertia and the rotational constants (as frequencies and wavenumbers) of $^{35}Cl^{12}CH_3$. ($m(^{35}Cl) = 34.9688m_u$; R(C–H) = 111 pm; R(C–Cl) = 178 pm; HCH angle = 111°.)

12.4(b) Use the information in Table 12.1 to calculate the moments of inertia and the rotational constants (as frequencies and wavenumbers) of $H^{12}C^{35}Cl_3$. ($m(^{35}Cl) = 34.9688m_u$; R(C–H) = 107 pm; R(C–Cl) = 177 pm; ClCCl angle = 110°.)

12.5(a) Calculate the frequency of the $J = 4 \leftarrow 3$ transition in the pure rotational spectrum of $^{14}N^{16}O$. The equilibrium bond length is 115 pm.

12.5(b) Calculate the frequency of the $J = 3 \leftarrow 2$ transition in the pure rotational spectrum of $^{12}C^{16}O$. The equilibrium bond length is 112.81 pm.

12.6(a) If the wavenumber of the $J = 3 \leftarrow 2$ rotational transition of $^1H^{35}Cl$ considered as a rigid rotator is 63.56 cm^{-1}, what is (a) the moment of inertia of the molecule, (b) the bond length?

12.6(b) If the wavenumber of the $J = 1 \leftarrow 0$ rotational transition of $^1H^{81}Br$ considered as a rigid rotator is 16.93 cm^{-1}, what is (a) the moment of inertia of the molecule, (b) the bond length?

12.7(a) Given that the spacing of lines in the microwave spectrum of $^{27}Al^1H$ is constant at 12.604 cm^{-1}, calculate the moment of inertia and bond length of the molecule. ($m(^{27}Al) = 26.9815m_u$.)

12.7(b) Given that the spacing of lines in the microwave spectrum of $^{35}Cl^{19}F$ is constant at 1.033 cm^{-1}, calculate the moment of inertia and bond length of the molecule. ($m(^{35}Cl) = 34.9688m_u$, $m(^{19}F) = 18.9984m_u$.)

12.8(a) The rotational constant of $^{127}I^{35}Cl$ is 0.1142 cm^{-1}. Calculate the ICl bond length. ($m(^{35}Cl) = 34.9688m_u$, $m(^{127}I) = 126.9045m_u$.)

12.8(b) The rotational constant of $^{12}C^{16}O_2$ is 0.39021 cm^{-1}. Calculate the bond length of the molecule. ($m(^{12}C) = 12m_u$ exactly, $m(^{16}O) = 15.9949m_u$.)

12.9(a) Determine the HC and CN bond lengths in HCN from the rotational constants $B(^1H^{12}C^{14}N) = 44.316$ GHz and $B(^2H^{12}C^{14}N) = 36.208$ GHz.

12.9(b) Determine the CO and CS bond lengths in OCS from the rotational constants $B(^{16}O^{12}C^{32}S) = 6081.5$ MHz, $B(^{16}O^{12}C^{34}S) = 5932.8$ MHz.

12.10(a) The wavenumber of the incident radiation in a Raman spectrometer is 20 487 cm^{-1}. What is the wavenumber of the scattered Stokes radiation for the $J=2 \leftarrow 0$ transition of $^{14}N_2$?

12.10(b) The wavenumber of the incident radiation in a Raman spectrometer is 20 623 cm^{-1}. What is the wavenumber of the scattered Stokes radiation for the $J=4 \leftarrow 2$ transition of $^{16}O_2$?

12.11(a) The rotational Raman spectrum of $^{35}Cl_2$ ($m(^{35}Cl)=34.9688m_u$) shows a series of Stokes lines separated by 0.9752 cm^{-1} and a similar series of anti-Stokes lines. Calculate the bond length of the molecule.

12.11(b) The rotational Raman spectrum of $^{19}F_2$ ($m(^{19}F)=18.9984m_u$) shows a series of Stokes lines separated by 3.5312 cm^{-1} and a similar series of anti-Stokes lines. Calculate the bond length of the molecule.

12.12(a) Estimate the centrifugal distortion constant for $^1H^{127}I$, for which $\tilde{B}=6.511$ cm^{-1} and $\tilde{\nu}=2308$ cm^{-1}. By what factor would the constant change when ^2H is substituted for ^1H?

12.12(b) Estimate the centrifugal distortion constant for $^{79}Br^{81}Br$, for which $\tilde{B}=0.0809$ cm^{-1} and $\tilde{\nu}=323.2$ cm^{-1}. By what factor would the constant change when the ^{79}Br is replaced by ^{81}Br?

12.13(a) What is the most highly populated rotational level of Cl_2 at (a) 25°C, (b) 100°C? Take $\tilde{B}=0.244$ cm^{-1}.

12.13(b) What is the most highly populated rotational level of Br_2 at (a) 25°C, (b) 100°C? Take $\tilde{B}=0.0809$ cm^{-1}.

12.14(a) An object of mass 1.0 kg suspended from the end of a rubber band has a vibrational frequency of 2.0 Hz. Calculate the force constant of the rubber band.

12.14(b) An object of mass 2.0 g suspended from the end of a spring has a vibrational frequency of 3.0 Hz. Calculate the force constant of the spring.

12.15(a) Calculate the percentage difference in the fundamental vibration wavenumber of $^{23}Na^{35}Cl$ and $^{23}Na^{37}Cl$ on the assumption that their force constants are the same.

12.15(b) Calculate the percentage difference in the fundamental vibration wavenumber of $^1H^{35}Cl$ and $^2H^{37}Cl$ on the assumption that their force constants are the same.

12.16(a) The wavenumber of the fundamental vibrational transition of $^{35}Cl_2$ is 564.9 cm^{-1}. Calculate the force constant of the bond ($m(^{35}Cl)=34.9688m_u$).

12.16(b) The wavenumber of the fundamental vibrational transition of $^{79}Br^{81}Br$ is 323.2 cm^{-1}. Calculate the force constant of the bond ($m(^{79}Br)=78.9183m_u$, $m(^{81}Br)=80.9163m_u$).

12.17(a) Calculate the relative numbers of Cl_2 molecules ($\tilde{\nu}=559.7$ cm^{-1}) in the ground and first excited vibrational states at (a) 298 K, (b) 500 K.

12.17(b) Calculate the relative numbers of Br_2 molecules ($\tilde{\nu}=321$ cm^{-1}) in the second and first excited vibrational states at (a) 298 K, (b) 800 K.

12.18(a) The hydrogen halides have the following fundamental vibrational wavenumbers: 4141.3 cm^{-1} (HF); 2988.9 cm^{-1} ($H^{35}Cl$); 2649.7 cm^{-1} ($H^{81}Br$); 2309.5 cm^{-1} ($H^{127}I$). Calculate the force constants of the hydrogen–halogen bonds.

12.18(b) From the data in Exercise 12.18a, predict the fundamental vibrational wavenumbers of the deuterium halides.

12.19(a) For $^{16}O_2$, $\Delta\tilde{G}$ values for the transitions $v=1\leftarrow 0$, $2\leftarrow 0$, and $3\leftarrow 0$ are, respectively, 1556.22, 3088.28, and 4596.21 cm^{-1}. Calculate $\tilde{\nu}$ and x_e. Assume y_e to be zero.

12.19(b) For $^{14}N_2$, $\Delta\tilde{G}$ values for the transitions $v=1\leftarrow 0$, $2\leftarrow 0$, and $3\leftarrow 0$ are, respectively, 2345.15, 4661.40, and 6983.73 cm^{-1}. Calculate $\tilde{\nu}$ and x_e. Assume y_e to be zero.

12.20(a) The first five vibrational energy levels of HCl are at 1481.86, 4367.50, 7149.04, 9826.48, and 12 399.8 cm^{-1}. Calculate the dissociation energy of the molecule in reciprocal centimetres and electronvolts.

12.20(b) The first five vibrational energy levels of HI are at 1144.83, 3374.90, 5525.51, 7596.66, and 9588.35 cm^{-1}. Calculate the dissociation energy of the molecule in reciprocal centimetres and electronvolts.

12.21(a) Estimate the anharmonicity constant x_e for $^1H^{19}F$ from the data in Table 12.2. By what factor does x_e change when ^1H is replaced by ^2H? Assume a Morse potential.

12.21(b) Estimate the anharmonicity constant x_e for $^1H^{81}Br$ from the data in Table 12.2. By what factor does x_e change when ^1H is replaced by ^2H? Assume a Morse potential.

12.22(a) Infrared absorption by $^1H^{81}Br$ gives rise to an R branch from $v=0$. What is the wavenumber of the line originating from the rotational state with $J=2$? Use the information in Table 12.2.

12.22(b) Infrared absorption by $^1H^{127}I$ gives rise to an R branch from $v=0$. What is the wavenumber of the line originating from the rotational state with $J=2$? Use the information in Table 12.2.

12.23(a) Which of the following molecules may show infrared absorption spectra: (a) H_2, (b) HCl, (c) CO_2, (d) H_2O?

12.23(b) Which of the following molecules may show infrared absorption spectra: (a) CH_3CH_3, (b) CH_4, (c) CH_3Cl, (d) N_2?

12.24(a) How many normal modes of vibration are there for the following molecules: (a) H_2O, (b) H_2O_2, (c) C_2H_4?

12.24(b) How many normal modes of vibration are there for the following molecules: (a) C_6H_6, (b) $C_6H_6CH_3$, (c) HC≡C–C≡CH.

12.25(a) Which of the three vibrations of an AB_2 molecule are infrared or Raman active when it is (a) angular (bent), (b) linear?

12.25(b) Which of the vibrations of an AB_3 molecule are infrared or Raman active when it is (a) trigonal planar, (b) trigonal pyramidal?

12.26(a) Consider the vibrational mode that corresponds to the uniform expansion of the benzene ring. Is it (a) Raman, (b) infrared active?

12.26(b) Consider the vibrational mode that corresponds to the boat-like bending of a benzene ring. Is it (a) Raman, (b) infrared active?

12.27(a) The molecule CH_2Cl_2 belongs to the point group C_{2v}. The displacements of the atoms span $5A_1+2A_2+4B_1+4B_2$. What are the symmetries of the normal modes of vibration?

12.27(b) A carbon disulfide molecule belongs to the point group $D_{\infty h}$. The nine displacements of the three atoms span $A_{1g}+2A_{1u}+2E_{1u}+E_{1g}$. What are the symmetries of the normal modes of vibration?

Problems*

Numerical problems

12.1 The rotational constant of NH_3 is equivalent to 298 GHz. Compute the separation of the pure rotational spectrum lines in gigahertz (for the frequency), reciprocal centimetres (for the wavenumber), and millimetres (for the wavelength), and show that the value of B is consistent with an N–H bond length of 101.4 pm and a bond angle of 106.78°.

12.2 The rotational constant for CO is 1.9314 cm^{-1} and 1.6116 cm^{-1} in the ground and first excited vibrational states, respectively. By how much does the internuclear distance change as a result of this transition?

12.3 Pure rotational Raman spectra of gaseous C_6H_6 and C_6D_6 yield the following rotational constants: $\tilde{B}(C_6H_6) = 0.189\ 60\ cm^{-1}$, $\tilde{B}(C_6D_6) = 0.156\ 81\ cm^{-1}$. The moments of inertia of the molecules about any axis perpendicular to the C_6 axis were calculated from these data as $I(C_6H_6) = 1.4759 \times 10^{-45}$ kg m^2, $I(C_6D_6) = 1.7845 \times 10^{-45}$ kg m^2. Calculate the CC, CH, and CD bond lengths.

12.4 Rotational absorption lines from $^1H^{35}Cl$ gas were found at the following wavenumbers (R.L. Hausler and R.A. Oetjen, *J. Chem. Phys.* **21**, 1340 (1953)): 83.32, 104.13, 124.73, 145.37, 165.89, 186.23, 206.60, 226.86 cm^{-1}. Calculate the moment of inertia and the bond length of the molecule. Predict the positions of the corresponding lines in $^2H^{35}Cl$.

12.5 Is the bond length in HCl the same as that in DCl? The wavenumbers of the $J = 1 \leftarrow 0$ rotational transitions for $H^{35}Cl$ and $^2H^{35}Cl$ are 20.8784 and 10.7840 cm^{-1}, respectively. Accurate atomic masses are $1.007825m_u$ and $2.0140m_u$ for ^1H and ^2H, respectively. The mass of ^{35}Cl is $34.96885m_u$. Based on this information alone, can you conclude that the bond lengths are the same or different in the two molecules?

12.6 Thermodynamic considerations suggest that the copper monohalides CuX should exist mainly as polymers in the gas phase, and indeed it proved difficult to obtain the monomers in sufficient abundance to detect spectroscopically. This problem was overcome by flowing the halogen gas over copper heated to 1100 K (E.L. Manson *et al.*, *J. Chem. Phys.* **63**, 2724 (1975)). For CuBr the $J = 13 \rightarrow 14$, $14 \rightarrow 15$, and $15 \rightarrow 16$ transitions occurred at 84 421.34, 90 449.25, and 96 476.72 MHz, respectively. Calculate the rotational constant and bond length of CuBr.

12.7 The microwave spectrum of $^{16}O^{12}CS$ (C.H. Townes *et al.*, *Phys. Rev.* **74**, 1113 (1948)) gave absorption lines (in GHz) as follows:

J	1	2	3	4
^{32}S	24.325 92	36.488 82	48.651 64	60.814 08
^{34}S	23.732 33		47.462 40	

Use the expressions for moments of inertia in Table 12.1 and assume that the bond lengths are unchanged by substitution; calculate the CO and CS bond lengths in OCS.

12.8‡ In a study of the rotational spectrum of the linear FeCO radical, K. Tanaka *et al.* (*J. Chem. Phys.* **106**, 6820 (1997)) report the following $J + 1 \leftarrow J$ transitions:

J	24	25	26	27	28	29
ν/MHz	214 777.7	223 379.0	231 981.2	240 584.4	249 188.5	257 793.5

Evaluate the rotational constant of the molecule. Also, estimate the value of J for the most highly populated rotational energy level at 298 K and at 100 K.

12.9 The vibrational energy levels of NaI lie at the wavenumbers 142.81, 427.31, 710.31, and 991.81 cm^{-1}. Show that they fit the expression $(v + \frac{1}{2})\tilde{\nu} - (v + \frac{1}{2})^2 x\tilde{\nu}$, and deduce the force constant, zero-point energy, and dissociation energy of the molecule.

12.10 Predict the shape of the nitronium ion, NO_2^+, from its Lewis structure and the VSEPR model. It has one Raman active vibrational mode at 1400 cm^{-1}, two strong IR active modes at 2360 and 540 cm^{-1}, and one weak IR mode at 3735 cm^{-1}. Are these data consistent with the predicted shape of the molecule? Assign the vibrational wavenumbers to the modes from which they arise.

12.11 At low resolution, the strongest absorption band in the infrared absorption spectrum of $^{12}C^{16}O$ is centred at 2150 cm^{-1}. Upon closer examination at higher resolution, this band is observed to be split into two sets of closely spaced peaks, one on each side of the centre of the spectrum at 2143.26 cm^{-1}. The separation between the peaks immediately to the right and left of the centre is 7.655 cm^{-1}. Make the harmonic oscillator and rigid rotor approximations and calculate from these data: (a) the vibrational wavenumber of a CO molecule, (b) its molar zero-point vibrational energy, (c) the force constant of the CO bond, (d) the rotational constant $\tilde{B}$, and (e) the bond length of CO.

12.12 The HCl molecule is quite well described by the Morse potential with $hc\tilde{D}_e = 5.33$ eV, $\tilde{\nu} = 2989.7\ cm^{-1}$, and $x\tilde{\nu} = 52.05\ cm^{-1}$. Assuming that the potential is unchanged on deuteration, predict the dissociation energies ($hc\tilde{D}_0$) of (a) HCl, (b) DCl.

12.13 The Morse potential (eqn 12.37) is very useful as a simple representation of the actual molecular potential energy. When RbH was studied, it was found that $\tilde{\nu} = 936.8\ cm^{-1}$ and $x_e\tilde{\nu} = 14.15\ cm^{-1}$. Plot the potential energy curve from 50 pm to 800 pm around $R_e = 236.7$ pm. Then go on to explore how the rotation of a molecule may weaken its bond by allowing for the kinetic energy of rotation of a molecule and plotting $V^* = V + hc\tilde{B}J(J+1)$ with $\tilde{B} = \hbar/4\pi c\mu R^2$. Plot these curves on the same diagram for $J = 40$, 80, and 100, and observe how the dissociation energy is affected by the rotation. (Taking $\tilde{B} = 3.020\ cm^{-1}$ at the equilibrium bond length will greatly simplify the calculation.)

12.14‡ F. Luo, *et al.* (*J. Chem. Phys.* **98**, 3564 (1993)) observed He_2, a species that had escaped detection for a long time. The fact that the observation required temperatures in the neighbourhood of 1 mK is consistent with computational studies that suggest that $hc\tilde{D}_e$ for He_2 is about 15.1 yJ, $hc\tilde{D}_0$ about 0.02 yJ (1 yJ = 10^{-24} J), and R_e about 297 pm. (a) Estimate the fundamental vibrational wavenumber, force constant, moment of inertia, and rotational constant based on the harmonic oscillator and rigid-rotor approximations. (b) Such a weakly bound complex is hardly likely to be rigid. Estimate the vibrational wavenumber and anharmonicity constant based on the Morse potential.

12.15 As mentioned in Section 12.15, the semi-empirical, *ab initio*, and DFT methods discussed in Chapter 10 can be used to estimate the force field of a molecule. The molecule's vibrational spectrum can be simulated, and it is then possible to determine the correspondence between a vibrational frequency and the atomic displacements that give rise to a normal mode. (a) Using molecular modelling software[3] and the computational method of your choice (semi-empirical, *ab initio*, or DFT methods), calculate the fundamental vibrational wavenumbers and visualize the vibrational normal modes of SO_2 in the gas phase. (b) The experimental values of the fundamental vibrational wavenumbers of SO_2 in the gas phase are 525 cm^{-1}, 1151 cm^{-1}, and 1336 cm^{-1}.

* Problems denoted with the symbol ‡ were supplied by Charles Trapp, Carmen Giunta, and Marshall Cady.

[3] The web site contains links to molecular modelling freeware and to other sites where you may perform molecular orbital calculations directly from your web browser.

Compare the calculated and experimental values. Even if agreement is poor, is it possible to establish a correlation between an experimental value of the vibrational wavenumber with a specific vibrational normal mode?

12.16 Consider the molecule CH_3Cl. (a) To what point group does the molecule belong? (b) How many normal modes of vibration does the molecule have? (c) What are the symmetries of the normal modes of vibration for this molecule? (d) Which of the vibrational modes of this molecule are infrared active? (e) Which of the vibrational modes of this molecule are Raman active?

12.17 Suppose that three conformations are proposed for the nonlinear molecule H_2O_2 (**4**, **5**, and **6**). The infrared absorption spectrum of gaseous H_2O_2 has bands at 870, 1370, 2869, and 3417 cm^{-1}. The Raman spectrum of the same sample has bands at 877, 1408, 1435, and 3407 cm^{-1}. All bands correspond to fundamental vibrational wavenumbers and you may assume that: (i) the 870 and 877 cm^{-1} bands arise from the same normal mode, and (ii) the 3417 and 3407 cm^{-1} bands arise from the same normal mode. (a) If H_2O_2 were linear, how many normal modes of vibration would it have? (b) Give the symmetry point group of each of the three proposed conformations of nonlinear H_2O_2. (c) Determine which of the proposed conformations is inconsistent with the spectroscopic data. Explain your reasoning.

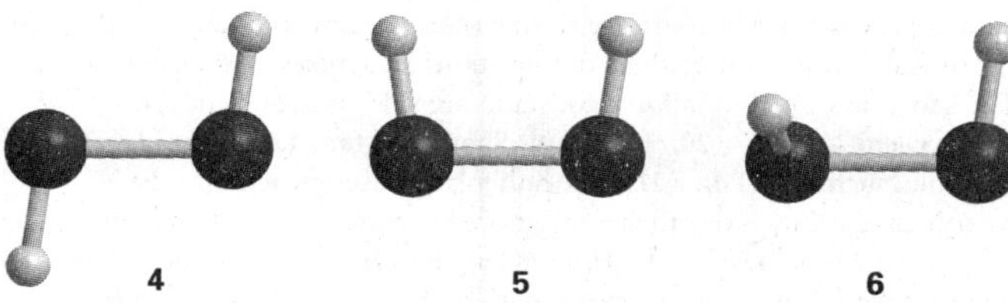

Theoretical problems

12.18 Show that the moment of inertia of a diatomic molecule composed of atoms of masses m_A and m_B and bond length R is equal to $m_{eff}R^2$, where $m_{eff} = m_A m_B/(m_A + m_B)$.

12.19 Suppose that the internuclear distance may be written $R = R_e + x$ where R_e is the equilibrium bond length. Also suppose that the potential well is symmetrical and confines the oscillator to small displacements. Deduce expressions for $1/\langle R\rangle^2$, $1/\langle R^2\rangle$, and $\langle 1/R^2\rangle$ to the lowest nonzero power of $\langle x^2\rangle/R_e^2$ and confirm that values are not the same.

12.20 Continue the development of Problem 12.19 by using the virial expression to relate $\langle x^2\rangle$ to the vibrational quantum number. Does your result imply that the rotational constant increases or decreases as the oscillator becomes excited to higher quantum states. What would be the effect of anharmonicity?

12.21 Derive eqn 12.17 for the centrifugal distortion constant $\tilde{D}_J$ of a diatomic molecule of effective mass m_{eff}. Treat the bond as an elastic spring with force constant k and equilibrium length r_e that is subjected to a centrifugal distortion to a new length r_c. Begin the derivation by letting the particles experience a restoring force of magnitude $k(r_c - r_e)$ that is countered perfectly by a centrifugal force $m_{eff}\omega^2 r_c$, where ω is the angular velocity of the rotating molecule. Then introduce quantum mechanical effects by writing the angular momentum as $\{J(J+1)\}^{1/2}\hbar$. Finally, write an expression for the energy of the rotating molecule, compare it with eqn 12.16, and write an expression for $\tilde{D}_J$.

12.22 Derive an expression for the force constant of an oscillator that can be modelled by a Morse potential (eqn 12.37).

12.23 Suppose a particle confined to a cavity in a microporous material has a potential energy of the form $V(x) = V_0(e^{-a^2/x^2} - 1)$. Sketch the form of the potential energy. What is the value of the force constant corresponding to this potential energy? Would the particle undergo simple harmonic motion? Sketch the likely form of the first two vibrational wavefunctions.

12.24 Show that there are a finite number of bound states of a Morse oscillator and find an expression for the maximum value of the vibrational quantum number. *Hint.* Show that the vibrational terms (eqn 12.38) pass through a maximum as v increases.

12.25 In the group theoretical language developed in Chapter 11, a spherical rotor is a molecule that belongs to a cubic or icosahedral point group, a symmetric rotor is a molecule with at least a threefold axis of symmetry, and an asymmetric rotor is a molecule without a threefold (or higher) axis. Linear molecules are linear rotors. Classify each of the following molecules as a spherical, symmetric, linear, or asymmetric rotor and justify your answers with group theoretical arguments: (a) CH_4, (b) CH_3CN, (c) CO_2, (d) CH_3OH, (e) benzene, (f) pyridine.

12.26 Derive an expression for the value of J corresponding to the most highly populated rotational energy level of a diatomic rotor at a temperature T remembering that the degeneracy of each level is $2J + 1$. Evaluate the expression for ICl (for which $\tilde{B} = 0.1142\ cm^{-1}$) at 25°C. Repeat the problem for the most highly populated level of a spherical rotor, taking note of the fact that each level is $(2J+1)^2$-fold degenerate. Evaluate the expression for CH_4 (for which $\tilde{B} = 5.24\ cm^{-1}$) at 25°C.

12.27 The moments of inertia of the linear mercury(II) halides are very large, so the O and S branches of their vibrational Raman spectra show little rotational structure. Nevertheless, the peaks of both branches can be identified and have been used to measure the rotational constants of the molecules (R.J.H. Clark and D.M. Rippon, *J. Chem. Soc. Faraday Soc. II*, **69**, 1496 (1973)). Show, from a knowledge of the value of J corresponding to the intensity maximum, that the separation of the peaks of the O and S branches is given by the Placzek–Teller relation $\delta\tilde{\nu} = (32\tilde{B}kT/hc)^{1/2}$. The following widths were obtained at the temperatures stated:

	$HgCl_2$	$HgBr_2$	HgI_2
θ/°C	282	292	292
$\delta\tilde{\nu}/cm^{-1}$	23.8	15.2	11.4

Calculate the bond lengths in the three molecules.

Applications: to biology, environmental science, and astrophysics

12.28 The protein haemerythrin is responsible for binding and carrying O_2 in some invertebrates. Each protein molecule has two Fe^{2+} ions that are in very close proximity and work together to bind one molecule of O_2. The Fe_2O_2 group of oxygenated haemerythrin is coloured and has an electronic absorption band at 500 nm. The resonance Raman spectrum of oxygenated haemerythrin obtained with laser excitation at 500 nm has a band at 844 cm^{-1} that has been attributed to the O–O stretching mode of bound $^{16}O_2$. (a) Why is resonance Raman spectroscopy and not infrared spectroscopy the method of choice for the study of the binding of O_2 to haemerythrin? (b) Proof that the 844 cm^{-1} band arises from a bound O_2 species may be obtained by conducting experiments on samples of haemerythrin that have been mixed with $^{18}O_2$, instead of $^{16}O_2$. Predict the fundamental vibrational wavenumber of the ^{18}O–^{18}O stretching mode in a sample of haemerythrin that has been treated with $^{18}O_2$. (c) The fundamental vibrational wavenumbers for the O–O stretching modes of O_2, O_2^- (superoxide anion), and O_2^{2-} (peroxide anion) are 1555, 1107, and 878 cm^{-1}, respectively. Explain this trend in terms of the electronic structures of O_2, O_2^-, and O_2^{2-} *Hint.* Review Section 10.4. What are the bond orders of O_2, O_2^-, and O_2^{2-}? (d) Based on the data given above, which of the following species best describes the Fe_2O_2 group of haemerythrin: $Fe_2^{2+}O_2$, $Fe^{2+}Fe^{3+}O_2^-$, or $Fe_2^{3+}O_2^{2-}$? Explain your reasoning. (e) The resonance Raman spectrum of haemerythrin mixed with $^{16}O^{18}O$ has two bands that can be attributed to the O–O stretching mode of bound oxygen. Discuss how this observation may be used to exclude one or more of the four proposed schemes (7–10) for binding of O_2 to the Fe_2 site of haemerythrin.

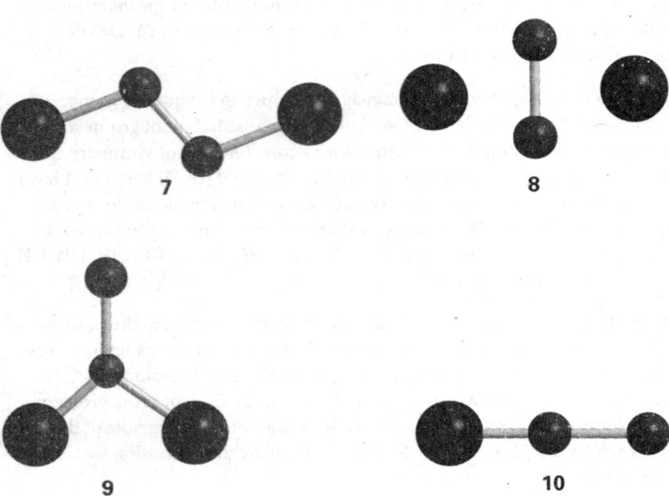

12.29‡ A mixture of carbon dioxide (2.1 per cent) and helium, at 1.00 bar and 298 K in a gas cell of length 10 cm has an infrared absorption band centred at 2349 cm^{-1} with an intensity of absorption, $A(\tilde{\nu})$, described by:

$$A(\tilde{\nu}) = \frac{a_1}{1 + a_2(\tilde{\nu} - a_3)^2} + \frac{a_4}{1 + a_5(\tilde{\nu} - a_6)^2}$$

where the coefficients are $a_1 = 0.932$, $a_2 = 0.005050\ cm^2$, $a_3 = 2333\ cm^{-1}$, $a_4 = 1.504$, $a_5 = 0.01521\ cm^2$, $a_6 = 2362\ cm^{-1}$. (a) Draw a graph of $A(\tilde{\nu})$. What is the origin of both the band and the band width? What are the allowed and forbidden transitions of this band? (b) Calculate the transition wavenumbers and intensity of absorption of the band with a simple harmonic oscillator–rigid rotor model and compare the result with the experimental spectra. The CO bond length is 116.2 pm. (c) Within what height, h, is basically all the infrared emission from the Earth in this band absorbed by atmospheric carbon dioxide? The mole fraction of CO_2 in the atmosphere is 3.3×10^{-4} and $T/K = 288 - 0.0065(h/m)$ below 10 km. Draw a surface plot of the atmospheric absorption of the band as a function of both height and wavenumber.

12.30 A. Dalgarno, in 'Chemistry in the interstellar medium', *Frontiers of Astrophysics*, E.H. Avrett (ed.), Harvard University Press, Cambridge (1976), notes that, although both CH and CN spectra show up strongly in the interstellar medium in the constellation Ophiuchus, the CN spectrum has become the standard for the determination of the temperature of the cosmic microwave background radiation. Demonstrate through a calculation why CH would not be as useful for this purpose as CN. The rotational constant $\tilde{B}_0$ for CH is 14.190 cm^{-1}.

12.31‡ There is a gaseous interstellar cloud in the constellation Ophiuchus that is illuminated from behind by the star ζ-Ophiuci. Analysis of the electronic–vibrational–rotational absorption lines obtained by H.S. Uhler and R.A. Patterson (*Astrophys. J.* **42**, 434 (1915)) shows the presence of CN molecules in the interstellar medium. A strong absorption line in the ultraviolet region at $\lambda = 387.5$ nm was observed corresponding to the transition $J = 0 - 1$. Unexpectedly, a second strong absorption line with 25 per cent of the intensity of the first was found at a slightly longer wavelength ($\Delta\lambda = 0.061$ nm) corresponding to the transition $J = 1 - 1$ (here allowed). Calculate the temperature of the CN molecules. Gerhard Herzberg, who was later to receive the Nobel Prize for his contributions to spectroscopy, calculated the temperature as 2.3 K. Although puzzled by this result, he did not realize its full significance. If he had, his prize might have been for the discovery of the cosmic microwave background radiation.

12.32‡ The H_3^+ ion has recently been found in the interstellar medium and in the atmospheres of Jupiter, Saturn, and Uranus. The rotational energy levels of H_3^+, an oblate symmetric rotor, are given by eqn 12.12, with $\tilde{C}$ replacing A, when centrifugal distortion and other complications are ignored. Experimental values for vibrational–rotational constants are $\tilde{\nu}(E') = 2521.6\ cm^{-1}$, $\tilde{B} = 43.55\ cm^{-1}$, and $\tilde{C} = 20.71\ cm^{-1}$. (a) Show that, for a nonlinear planar molecule (such as H_3^+), $I_C = 2I_B$. The rather large discrepancy with the experimental values is due to factors ignored in eqn 12.12. (b) Calculate an approximate value of the H–H bond length in H_3^+. (c) The value of R_e obtained from the best quantum mechanical calculations by J.B. Anderson (*J. Chem. Phys.* **96**, 3702 (1991)) is 87.32 pm. Use this result to calculate the values of the rotational constants $\tilde{B}$ and $\tilde{C}$. (d) Assuming that the geometry and force constants are the same in D_3^+ and H_3^+, calculate the spectroscopic constants of D_3^+. The molecular ion D_3^+ was first produced by J.T. Shy *et al.* (*Phys. Rev. Lett* **45**, 535 (1980)) who observed the $\nu_2(E')$ band in the infrared.

12.33 The space immediately surrounding stars, also called the *circumstellar space*, is significantly warmer because stars are very intense black-body emitters with temperatures of several kilokelvin. Discuss how such factors as cloud temperature, particle density, and particle velocity may affect the rotational spectrum of CO in an interstellar cloud. What new features in the spectrum of CO can be observed in gas ejected from and still near a star with temperatures of about 1000 K, relative to gas in a cloud with temperature of about 10 K? Explain how these features may be used to distinguish between circumstellar and interstellar material on the basis of the rotational spectrum of CO.

Molecular spectroscopy 2: electronic transitions

13

Simple analytical expressions for the electronic energy levels of molecules cannot be given, so this chapter concentrates on the qualitative features of electronic transitions. A common theme throughout the chapter is that electronic transitions occur within a stationary nuclear framework. We pay particular attention to spontaneous radiative decay processes, which include fluorescence and phosphorescence. A specially important example of stimulated radiative decay is that responsible for the action of lasers, and we see how this stimulated emission may be achieved and employed.

The energies needed to change the electron distributions of molecules are of the order of several electronvolts (1 eV is equivalent to about 8000 cm^{-1} or 100 kJ mol^{-1}). Consequently, the photons emitted or absorbed when such changes occur lie in the visible and ultraviolet regions of the spectrum (Table 13.1).

Considerable information can be obtained from the radiation emitted when excited electronic states decay radiatively back to the ground state. For instance, lasers have brought unprecedented precision to spectroscopy, made Raman spectroscopy a widely useful technique, and have made it possible to study chemical reactions on a femtosecond timescale. We shall see the principles of their action in this chapter and encounter their applications throughout the rest of the book.

The characteristics of electronic transitions

In the lowest vibrational state of the ground electronic state of a molecule the nuclei are at their equilibrium locations and experience no net force from the electrons and other nuclei in the molecule. The electron distribution is changed when an electronic

Table 13.1* Colour, frequency, and energy of light

Colour	λ/nm	$\nu/(10^{14}$ Hz)	E/(kJ mol^{-1})
Infrared	>1000	<3.0	<120
Red	700	4.3	170
Yellow	580	5.2	210
Blue	470	6.4	250
Ultraviolet	<300	>10	>400

* More values are given in the *Data section*.

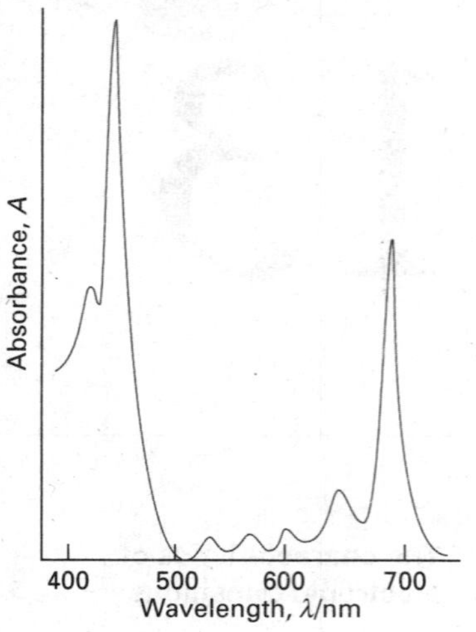

Fig. 13.1 The absorption spectrum of chlorophyll in the visible region. Note that it absorbs in the red and blue regions, and that green light is not absorbed.

transition occurs and the nuclei become subjected to different forces. They start to vibrate around their new equilibrium locations and the vibrational transitions that accompany the electronic transition give rise to the **vibrational structure** of the electronic transition. This structure can be resolved for gaseous samples, but in a liquid or solid the lines usually merge together and result in a broad, almost featureless band (Fig. 13.1). Superimposed on the vibrational transitions that accompany the electronic transition of a molecule in the gas phase is an additional structure that arises from rotational transitions. The electronic spectra of gaseous samples are therefore very complicated but rich in information.

13.1 Measurements of intensity

Key point **The intensity of absorption is reported as the molar absorption coefficient by using the Beer–Lambert law; the total absorption is reported as the integrated absorption coefficient.**

It is found empirically that the transmitted intensity I varies with the length, L, of the sample and the molar concentration, [J], of the absorbing species J in accord with the **Beer–Lambert law:**

$$I = I_0 10^{-\varepsilon[\mathrm{J}]L} \qquad (13.1)$$

Beer–Lambert law

where I_0 is the incident intensity. The quantity ε (epsilon) is called the **molar absorption coefficient** (formerly, and still widely, the 'extinction coefficient'). The molar absorption coefficient depends on the frequency of the incident radiation and is greatest where the absorption is most intense. Its dimensions are 1/(concentration × length), and it is normally convenient to express it in cubic decimetres per mole per centimetre ($\mathrm{dm^3\,mol^{-1}\,cm^{-1}}$); in SI base units it is expressed in metres-squared per mole ($\mathrm{m^2\,mol^{-1}}$). The latter units imply that ε may be regarded as a (molar) cross-section for absorption and that, the greater the cross-sectional area of the molecule for absorption, the greater is its ability to block the passage of the incident radiation at a given frequency. The Beer–Lambert law is an empirical result. However, it is simple to account for its form as we show in the following *Justification*.

Justification 13.1 *The Beer–Lambert law*

The change in intensity, dI, that occurs when light passes through a layer of thickness dL containing an absorbing species J at a molar concentration [J] is proportional to the thickness of the layer, the concentration of J, and the intensity, I, incident on the layer. We can therefore write

$$\mathrm{d}I = -\kappa[\mathrm{J}]I\,\mathrm{d}L$$

where κ (kappa) is the proportionality coefficient, or equivalently

$$\frac{\mathrm{d}I}{I} = -\kappa[\mathrm{J}]\mathrm{d}L$$

This expression applies to each successive layer into which the sample can be regarded as being divided. Therefore, to obtain the intensity that emerges from a sample of thickness L when the intensity incident on one face of the sample is I_0, we sum all the successive changes:

$$\int_{I_0}^{I} \frac{\mathrm{d}I}{I} = -\kappa\int_0^L [\mathrm{J}]\mathrm{d}L$$

If the concentration is uniform, [J] is independent of location, and the expression integrates to

$$\ln\frac{I}{I_0} = -\kappa[\mathrm{J}]L$$

This expression gives the Beer–Lambert law when the logarithm is converted to base 10 by using $\ln x = (\ln 10)\log x$ and replacing κ by $\varepsilon \ln 10$.

The spectral characteristics of a sample are commonly reported as the **transmittance**, T, of the sample at a given frequency:

$$T = \frac{I}{I_0} \qquad \text{Definition of transmittance} \qquad [13.2]$$

and the **absorbance**, A, of the sample:

$$A = \log\frac{I_0}{I} \qquad \text{Definition of absorbance} \qquad [13.3]$$

The two quantities are related by $A = -\log T$ (note the common logarithm) and the Beer–Lambert law becomes

$$A = \varepsilon[\mathrm{J}]L \qquad (13.4)$$

The product $\varepsilon[\mathrm{J}]L$ was known formerly as the *optical density* of the sample.

• **A brief illustration**

The Beer–Lambert law implies that the intensity of electromagnetic radiation transmitted through a sample at a given wavenumber decreases exponentially with the sample thickness and the molar concentration. If the transmittance is 0.1 for a path length of 1 cm (corresponding to a 90 per cent reduction in intensity), then it would be $(0.1)^2 = 0.01$ for a path of double the length (corresponding to a 99 per cent reduction in intensity overall). •

The maximum value of the molar absorption coefficient, ε_{max}, is an indication of the intensity of a transition. However, as absorption bands generally spread over a range of wavenumbers, quoting the absorption coefficient at a single wavenumber might not give a true indication of the intensity of a transition. The **integrated absorption coefficient**, $\mathcal{A}$, is the sum of the absorption coefficients over the entire band (Fig. 13.2), and corresponds to the area under the plot of the molar absorption coefficient against wavenumber:

$$\mathcal{A} = \int_{\text{band}} \varepsilon(\tilde{\nu})\mathrm{d}\tilde{\nu} \qquad \text{Definition of integrated absorption coefficient} \qquad [13.5]$$

For lines of similar widths, the integrated absorption coefficients are proportional to the heights of the lines.

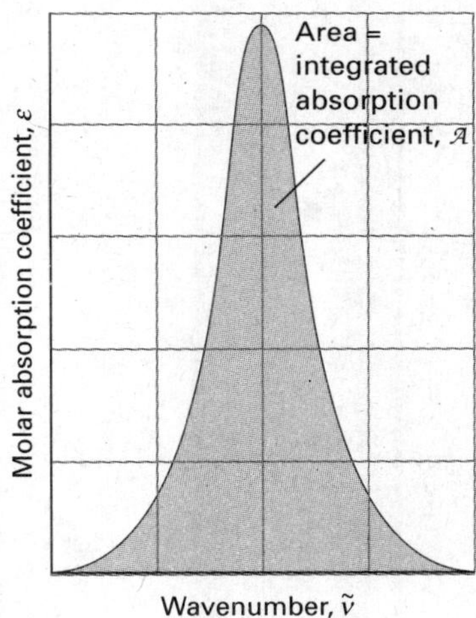

Fig. 13.2 The integrated absorption coefficient of a transition is the area under a plot of the molar absorption coefficient against the wavenumber of the incident radiation.

13.2 The electronic spectra of diatomic molecules

Key points (a) The term symbols of diatomic molecules express the components of electronic angular momentum around the internuclear axis. (b) Selection rules for electronic transitions are based on considerations of angular momentum and symmetry. (c) The Franck–Condon principle provides a basis for explaining the vibrational structure of electronic transitions. (d) In gas-phase samples, rotational structure is present too and can give rise to band heads.

We saw in Section 9.10 how the states of atoms are expressed by using term symbols and that the selection rules for electronic transitions could be expressed in terms of these term symbols. Much the same is true of diatomic molecules, one principal difference being the replacement of full spherical symmetry of atoms by the cylindrical symmetry defined by the axis of the molecule. The second principal difference is the fact that a diatomic molecule can vibrate and rotate.

(a) Term symbols

The term symbols of linear molecules (the analogues of the symbols ^{2}P, etc. for atoms) are constructed in a similar way to those for atoms, with the Roman upper-case letter (the P in this instance) representing the total orbital angular momentum of the electrons around the nucleus. In a linear molecule, and specifically a diatomic molecule, a Greek upper-case letter represents the total orbital angular momentum of the electrons around the internuclear axis. If this component of orbital angular momentum is $\Lambda\hbar$ with $\Lambda = 0, \pm 1, \pm 2 \ldots$, we use the following designation:

$\lvert\Lambda\rvert$	0	1	2	...
	Σ	Π	Δ	...

These labels are the analogues of S, P, D, . . . for atoms for states with $L = 0, 1, 2, \ldots$. To decide on the value of L for atoms we had to use the Clebsch–Gordan series to couple the individual angular momenta. The procedure to determine Λ is much simpler in a diatomic molecule because we simply add the values of the individual components of each electron, $\lambda\hbar$:

$$\Lambda = \lambda_1 + \lambda_2 + \cdots \tag{13.6}$$

A single electron in a σ orbital has $\lambda = 0$: the orbital is cylindrically symmetrical and has no angular nodes when viewed along the internuclear axis. Therefore, if that is the only type of electron present, $\Lambda = 0$. The term symbol for the ground state of H_2 with electron configuration $1\sigma_g^2$ is therefore Σ. A π electron in a diatomic molecule has one unit of orbital angular momentum about the internuclear axis ($\lambda = \pm 1$) and, if it is the only electron outside a closed shell, gives rise to a Π term. If there are two π electrons (as in the ground state of O_2, with configuration . . . $1\pi_g^2$, there are two possible outcomes. If the electrons are travelling in opposite directions, then $\lambda_1 = +1$ and $\lambda_2 = -1$ (or vice versa) and $\Lambda = 0$, corresponding to a Σ term. Alternatively, the electrons might occupy the same π orbital and $\lambda_1 = \lambda_2 = +1$ (or -1), and $\Lambda = \pm 2$, corresponding to a Δ term. In O_2 it is energetically favourable for the electrons to occupy different orbitals, so the ground term is Σ.

As in atoms, we use a left superscript with the value of $2S + 1$ to denote the multiplicity of the term, where S is the total spin quantum number of the electrons. For H_2^+, because there is only one electron, $S = s = \frac{1}{2}$ and the term symbol is $^2\Sigma$, a doublet term. For H_2, with no net spin, $S = 0$ and the ground state is a singlet term, $^1\Sigma$. In O_2, because in the ground state the two π electrons occupy different orbitals (as we saw above), they may have either parallel or antiparallel spins; the lower energy is obtained (as in atoms) if the spins are parallel, so $S = 1$ and the ground state is $^3\Sigma$.

The overall parity of the state (its symmetry under inversion through the centre of the molecule, if it has one) is added as a right subscript to the term symbol. For H_2^+ in its ground state, the parity of the only occupied orbital ($1\sigma_g$) is g, so the term itself is also g, and in full dress is $^2\Sigma_g$. If there are several electrons, the overall parity is calculated by noting the parity of each occupied orbital and using

$$\mathrm{g} \times \mathrm{g} = \mathrm{g} \qquad \mathrm{u} \times \mathrm{u} = \mathrm{g} \qquad \mathrm{u} \times \mathrm{g} = \mathrm{u} \tag{13.7}$$

These rules are generated by interpreting g as +1 and u as −1. The term symbol for the ground state of any closed-shell homonuclear diatomic molecule is $^1\Sigma_g$ because the spin is zero (a singlet term in which all electrons paired), there is no orbital angular momentum from a closed shell, and the overall parity is g. The parity of the ground state of O_2 is also g × g = g, so it is denoted $^3\Sigma_g$. If the molecule is heteronuclear, parity is irrelevant and the ground state of a closed-shell species, such as CO, is $^1\Sigma$.

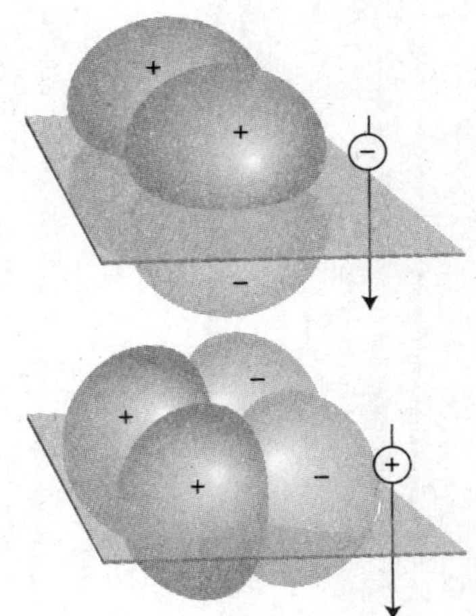

Fig. 13.3 The + or − on a term symbol refers to the overall symmetry of a configuration under reflection in a plane containing the two nuclei.

• A brief illustration

An excited configuration of O_2 is . . . $1\pi_g^2$ with both π electrons in the same orbital. As we have seen, $|\Lambda|=2$, represented by Δ. The two electrons must be paired if they occupy the same orbital, so $S=0$. The overall parity is g×g=g. Therefore, the term symbol is $^1\Delta_g$. •

We saw in Chapter 11 that angular momentum is an aspect of the symmetry of states. That remains true for linear molecules, and the term symbols can also be thought of as denoting various aspects of rotational and inversion symmetry of the electronic wavefunction of the molecule. With that in mind, there is an additional symmetry operation that distinguishes different types of Σ term: reflection in a plane containing the internuclear axis. A + superscript on Σ is used to denote a wavefunction that does not change sign under this reflection and a – sign is used if the wavefunction changes sign (Fig. 13.3).

• A brief illustration

If we think of O_2 in its ground state as having one electron in $1\pi_{g,x}$, which changes sign under reflection in the *yz*-plane, and the other electron in $1\pi_{g,y}$, which does not change sign under reflection in the same plane, then the overall reflection symmetry is (closed shell) × (+) × (−) = (−), and the full term symbol of the ground electronic state of O_2 is $^3\Sigma_g^-$. •

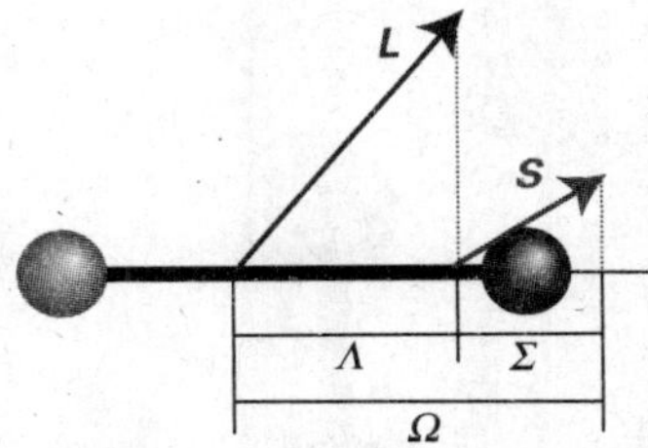

Fig. 13.4 The coupling of spin and orbital angular momenta in a linear molecule: only the components along the internuclear axis are conserved.

As we saw in connection with atoms, another aspect of angular momentum that it is sometimes necessary to denote is the total angular momentum. In atoms that is denoted by the value of *J* and appears as a right subscript in the term symbol, as in $^2P_{1/2}$, with different values of *J* corresponding to different *levels* of a term. In a linear molecule, only the angular momentum about the internuclear axis is well defined, and has the value Ω. For light molecules, where the spin–orbit coupling is weak, Ω is obtained by adding together the components of orbital angular momentum around the axis (the value of Λ) and the component of the electron spin on that axis (Fig. 13.4). The latter is denoted Σ, where $\Sigma = S, S-1, S-2, \ldots, -S$. Then

$$\Omega = \Lambda + \Sigma \tag{13.8}$$

The value of $|\Omega|$ may then be attached to the term symbol as a right subscript (just like *J* is used in atoms) to denote the different levels. These levels differ in energy, as in atoms, as a result of spin–orbit coupling.

A brief comment

It is important to distinguish between the (upright) term symbol Σ and the (sloping) quantum number Σ.

• A brief illustration

The ground-state configuration of NO is . . . π_g^1, so it is a $^2\Pi$ term with $\Lambda = \pm 1$ and $\Sigma = \pm\frac{1}{2}$. Therefore, there are two levels of the term, one with $\Omega = \pm\frac{1}{2}$ and the other with $\pm\frac{3}{2}$, denoted $^2\Pi_{1/2}$ and $^2\Pi_{3/2}$, respectively. Each level is doubly degenerate (corresponding to the opposite signs of Ω). In NO, $^2\Pi_{1/2}$ lies slightly lower than $^2\Pi_{3/2}$. •

(b) Selection rules

A number of selection rules govern which transitions will be observed in the electronic spectrum of a molecule. The selection rules concerned with changes in angular momentum are

$$\Delta\Lambda = 0, \pm 1 \qquad \Delta S = 0 \qquad \Delta\Sigma = 0 \qquad \Delta\Omega = 0, \pm 1$$

Selection rules for linear molecules

As in atoms (Section 9.3), the origins of these rules are conservation of angular momentum during a transition and the fact that a photon has a spin of 1.

There are two selection rules concerned with changes in symmetry. First, as we show in the following *Justification*,

For Σ terms, only $\Sigma^+ \leftrightarrow \Sigma^+$ and $\Sigma^- \leftrightarrow \Sigma^-$ are allowed

Second, the **Laporte selection rule** for centrosymmetric molecules (those with a centre of inversion) and atoms states that *the only allowed transitions are transitions that are accompanied by a change of parity*. That is,

For centrosymmetric molecules, only u $\rightarrow$ g and g $\rightarrow$ u are allowed

Laporte selection rule

Justification 13.2 *Symmetry-based selection rules*

The last two selection rules result from the fact that the electric-dipole transition moment introduced in *Justification* 9.4, $\mu_{fi} = \int \psi_f^* \hat{\mu} \psi_i \, d\tau$, vanishes unless the integrand is invariant under all symmetry operations of the molecule.

The *z*-component of the dipole moment operator is the component of μ responsible for $\Sigma \leftrightarrow \Sigma$ transitions (the other components have Π symmetry and cannot make a contribution). The *z*-component of μ has (+) symmetry with respect to reflection in a plane containing the internuclear axis. Therefore, for a $(+) \leftrightarrow (-)$ transition, the overall symmetry of the transition dipole moment is $(+) \times (+) \times (-) = (-)$, so it must be zero and hence $\Sigma^+ \leftrightarrow \Sigma^-$ transitions are not allowed. The integrals for $\Sigma^+ \leftrightarrow \Sigma^+$ and $\Sigma^- \leftrightarrow \Sigma^-$ transform as $(+) \times (+) \times (+) = (+)$ and $(-) \times (+) \times (-) = (+)$, respectively, and so both transitions are allowed.

The three components of the dipole moment operator transform like *x*, *y*, and *z*, and in a centrosymmetric molecule are all u. Therefore, for a g $\rightarrow$ g transition, the overall parity of the transition dipole moment is g $\times$ u $\times$ g = u, so it must be zero. Likewise, for a u $\rightarrow$ u transition, the overall parity is u $\times$ u $\times$ u = u, so the transition dipole moment must also vanish. Hence, transitions without a change of parity are forbidden. For a g $\leftrightarrow$ u transition the integral transforms as g $\times$ u $\times$ u = g, and is allowed.

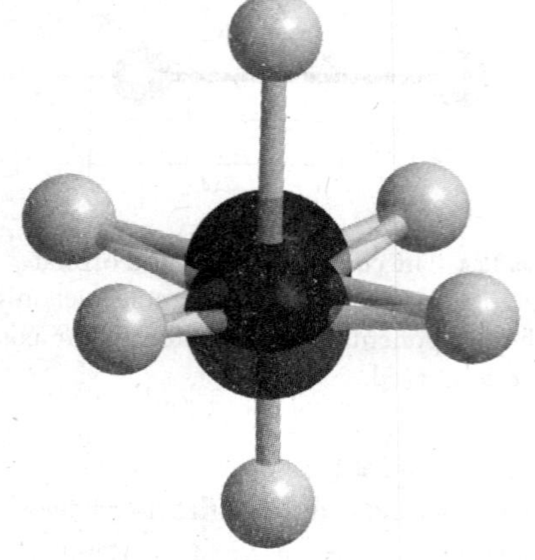

Fig. 13.5 A d–d transition is parity-forbidden because it corresponds to a g–g transition. However, a vibration of the molecule can destroy the inversion symmetry of the molecule and the g,u classification no longer applies. The removal of the centre of symmetry gives rise to a vibronically allowed transition.

A forbidden g $\rightarrow$ g transition can become allowed if the centre of symmetry is eliminated by an asymmetrical vibration, such as the one shown in Fig. 13.5. When the centre of symmetry is lost, g $\rightarrow$ g and u $\rightarrow$ u transitions are no longer parity-forbidden and become weakly allowed. A transition that derives its intensity from an asymmetrical vibration of a molecule is called a **vibronic transition**.

Self-test 13.1 Which of the following electronic transitions are allowed in O_2?

${}^3\Sigma_g^- \leftrightarrow {}^1\Delta_g$, ${}^3\Sigma_g^- \leftrightarrow {}^1\Sigma_g^+$, ${}^3\Sigma_g^- \leftrightarrow {}^3\Delta_u$, ${}^3\Sigma_g^- \leftrightarrow {}^3\Sigma_u^+$, ${}^3\Sigma_g^- \leftrightarrow {}^3\Sigma_u^-$ [${}^3\Sigma_g^- \leftrightarrow {}^3\Sigma_u^-$]

(c) Vibrational structure

To account for the vibrational structure in electronic spectra of molecules (Fig. 13.6), we apply the **Franck–Condon principle**:

> Because the nuclei are so much more massive than the electrons, an electronic transition takes place very much faster than the nuclei can respond.

Franck–Condon principle

As a result of the transition, electron density is rapidly built up in new regions of the molecule and removed from others. In classical terms, the initially stationary nuclei suddenly experience a new force field, to which they respond by beginning to vibrate and (in classical terms) swing backwards and forwards from their original separation (which was maintained during the rapid electronic excitation). The stationary equilibrium separation of the nuclei in the initial electronic state therefore becomes a turning point in the final electronic state (Fig. 13.7). We can imagine the transition as taking place up the vertical line in Fig. 13.7. This interpretation is the origin of the expression **vertical transition**, which is used to denote an electronic transition that occurs without change of nuclear geometry.

The vibrational structure of the spectrum depends on the relative horizontal position of the two potential energy curves, and a long **vibrational progression**, a lot of vibrational structure, is stimulated if the upper potential energy curve is appreciably displaced horizontally from the lower. The upper curve is usually displaced to greater equilibrium bond lengths because electronically excited states usually have more antibonding character than electronic ground states. The separation of the vibrational lines depends on the vibrational energies of the *upper* electronic state. Hence, electronic absorption spectra may be used to assess the force fields and dissociation energies of electronically excited molecules.

The quantum mechanical version of the Franck–Condon principle refines this picture. Instead of saying that the nuclei stay at the same locations and are stationary during the transition, we say that *they retain their initial dynamic state*. In quantum mechanics, the dynamical state is expressed by the wavefunction, so an equivalent statement is that the nuclear wavefunction does not change during the electronic transition. Initially the molecule is in the lowest vibrational state of its ground electronic state with a bell-shaped wavefunction centred on the equilibrium bond length (Fig. 13.8). To find the nuclear state to which the transition takes place, we look for the vibrational wavefunction that most closely resembles this initial wavefunction, for that corresponds to the nuclear dynamical state that is least changed in the transition. Intuitively, we can see that the final wavefunction is the one with a large peak close to the position of the initial bell-shaped function. As we saw in Section 8.5, provided the vibrational quantum number is not zero, the biggest peaks of vibrational wavefunctions occur close to the edges of the confining potential, so we can expect the transition to occur to those vibrational states, in accord with the classical description. However, several vibrational states have their major peaks in similar positions, so we should expect transitions to occur to a range of vibrational states, as is observed.

The quantitative form of the Franck–Condon principle and the justification of the preceding description is derived from the expression for the transition dipole moment (as in *Justification* 13.2). The dipole moment operator is a sum over all nuclei and electrons in the molecule:

$$\hat{\mu} = -e\sum_i \boldsymbol{r}_i + e\sum_I Z_I \boldsymbol{R}_I \tag{13.9}$$

where the vectors are the distances from the centre of charge of the molecule. The intensity of the transition is proportional to the square modulus, $|\mu_{fi}|^2$, of the magnitude

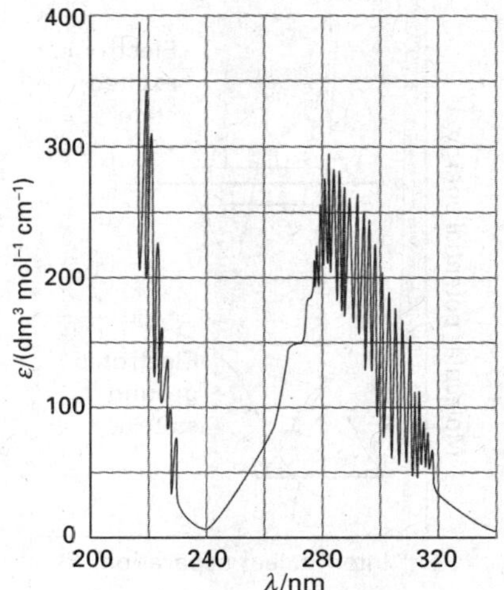

Fig. 13.6 The electronic spectra of some molecules show significant vibrational structure. Shown here is the ultraviolet spectrum of gaseous SO_2 at 298 K. As explained in the text, the sharp lines in this spectrum are due to transitions from a lower electronic state to different vibrational levels of a higher electronic state.

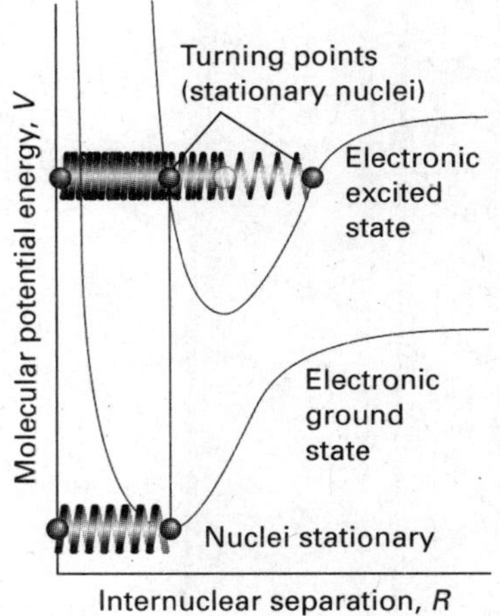

Fig. 13.7 According to the Franck–Condon principle, the most intense vibronic transition is from the ground vibrational state to the vibrational state lying vertically above it. Transitions to other vibrational levels also occur, but with lower intensity.

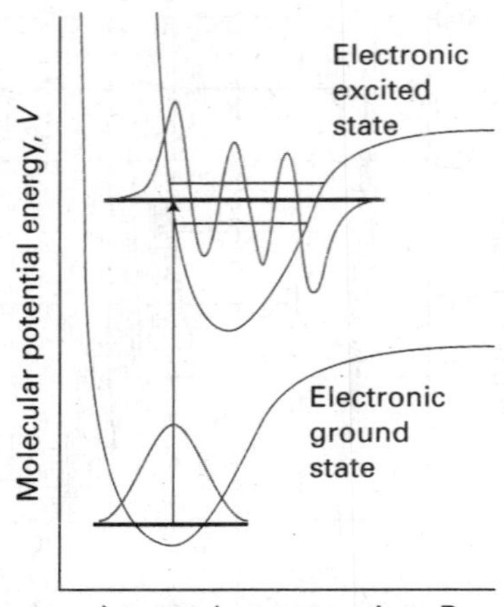

Fig. 13.8 In the quantum mechanical version of the Franck–Condon principle, the molecule undergoes a transition to the upper vibrational state that most closely resembles the vibrational wavefunction of the vibrational ground state of the lower electronic state. The two wavefunctions shown here have the greatest overlap integral of all the vibrational states of the upper electronic state and hence are most closely similar.

of the transition dipole moment, and we show in the following *Justification* that this intensity is proportional to the square modulus of the overlap integral, $S(v_f,v_i)$, between the vibrational states of the initial and final electronic states. This overlap integral is a measure of the match between the vibrational wavefunctions in the upper and lower electronic states: $S=1$ for a perfect match and $S=0$ when there is no similarity.

Justification 13.3 *The Franck–Condon approximation*

The overall state of the molecule consists of an electronic part, ψ_ε, and a vibrational part, ψ_v. Therefore, within the Born–Oppenheimer approximation, the transition dipole moment factorizes as follows:

$$\mu_{\text{fi}} = \int \psi^*_{\varepsilon_\text{f}}\psi^*_{v_\text{f}}\left\{-e\sum_i \boldsymbol{r}_i + e\sum_I Z_I\boldsymbol{R}_I\right\}\psi_{\varepsilon_\text{i}}\psi_{v_\text{i}}\,\text{d}\tau$$

$$= -e\sum_i\int\psi^*_{\varepsilon_\text{f}}\boldsymbol{r}_i\psi_{\varepsilon_\text{i}}\,\text{d}\tau_\varepsilon\int\psi^*_{v_\text{f}}\psi_{v_\text{i}}\,\text{d}\tau_v + e\sum_I Z_I\int\psi^*_{\varepsilon_\text{f}}\psi_{\varepsilon_\text{i}}\,\text{d}\tau_\varepsilon\int\psi^*_{v_\text{f}}\boldsymbol{R}_I\psi_{v_\text{i}}\,\text{d}\tau_v$$

The second term on the right of the second row is zero, because two different electronic states are orthogonal. Therefore,

$$\mu_{\text{fi}} = -e\sum_i\int\psi^*_{\varepsilon_\text{f}}\boldsymbol{r}_i\psi_{\varepsilon_\text{i}}\,\text{d}\tau_\varepsilon\int\psi^*_{v_\text{f}}\psi_{v_\text{i}}\,\text{d}\tau_v = \mu_{\varepsilon_\text{f}\varepsilon_\text{i}}S(v_\text{f},v_\text{i})$$

where

$$\mu_{\varepsilon_\text{f}\varepsilon_\text{i}} = -e\sum_i\int\psi^*_{\varepsilon_\text{f}}\boldsymbol{r}_i\psi_{\varepsilon_\text{i}}\,\text{d}\tau_\varepsilon \qquad \text{and} \qquad S(v_\text{f},v_\text{i}) = \int\psi^*_{v_\text{f}}\psi_{v_\text{i}}\,\text{d}\tau_v$$

The matrix element $\mu_{\varepsilon_\text{f}\varepsilon_\text{i}}$ is the electric-dipole transition moment arising from the redistribution of electrons (and a measure of the 'kick' this redistribution gives to the electromagnetic field, and vice versa for absorption). The factor $S(v_\text{f},v_\text{i})$, is the overlap integral between the vibrational state ψ_{v_i} in the initial electronic state of the molecule, and the vibrational state ψ_{v_f} in the final electronic state of the molecule.

Because the transition intensity is proportional to the square of the magnitude of the transition dipole moment, the intensity of an absorption is proportional to $|S(v_\text{f},v_\text{i})|^2$, which is known as the **Franck–Condon factor** for the transition. It follows that, the greater the overlap of the vibrational state wavefunction in the upper electronic state with the vibrational wavefunction in the lower electronic state, the greater the absorption intensity of that particular simultaneous electronic and vibrational transition.

Example 13.1 *Calculating a Franck–Condon factor*

Consider the transition from one electronic state to another, their bond lengths being R_e and R'_e and their force constants equal. Calculate the Franck–Condon factor for the 0–0 transition and show that the transition is most intense when the bond lengths are equal.

Method We need to calculate $S(0,0)$, the overlap integral of the two ground-state vibrational wavefunctions, and then take its square. The difference between harmonic and anharmonic vibrational wavefunctions is negligible for $n=0$, so harmonic oscillator wavefunctions can be used (Table 8.1).

Answer We use the (real) wavefunctions

$$\psi_0 = \left(\frac{1}{\alpha\pi^{1/2}}\right)^{1/2}\text{e}^{-x^2/2\alpha^2} \qquad \psi'_0 = \left(\frac{1}{\alpha\pi^{1/2}}\right)^{1/2}\text{e}^{-x'^2/2\alpha^2}$$

where $x = R - R_e$ and $x' = R - R'_e$ with $\alpha = (\hbar^2/mk)^{1/4}$ (Section 8.5a). The overlap integral is

$$S(0,0) = \langle 0|0\rangle = \int_{-\infty}^{\infty} \psi'_0\psi_0 \mathrm{d}R = \frac{1}{\alpha\pi^{1/2}}\int_{-\infty}^{\infty} \mathrm{e}^{-(x^2+x'^2)/2\alpha^2}\mathrm{d}x$$

We now write $\alpha z = R - \frac{1}{2}(R_e + R'_e)$, and manipulate this expression into

$$S(0,0) = \frac{1}{\pi^{1/2}}\mathrm{e}^{-(R_e-R'_e)^2/4\alpha^2}\int_{-\infty}^{\infty} \mathrm{e}^{-z^2}\mathrm{d}z$$

The value of the integral is $\pi^{1/2}$. Therefore, the overlap integral is

$$S(0,0) = \mathrm{e}^{-(R_e-R'_e)^2/4\alpha^2}$$

and the Franck–Condon factor is

$$S(0,0)^2 = \mathrm{e}^{-(R_e-R'_e)^2/2\alpha^2}$$

This factor is equal to 1 when $R'_e = R_e$ and decreases as the equilibrium bond lengths diverge from each other (Fig. 13.9).

For Br_2, $R_e = 228$ pm and there is an upper state with $R'_e = 266$ pm. Taking the vibrational wavenumber as 250 cm^{-1} gives $S(0,0)^2 = 5.1 \times 10^{-10}$, so the intensity of the 0–0 transition is only 5.1×10^{-10} of what it would have been if the potential curves had been directly above each other.

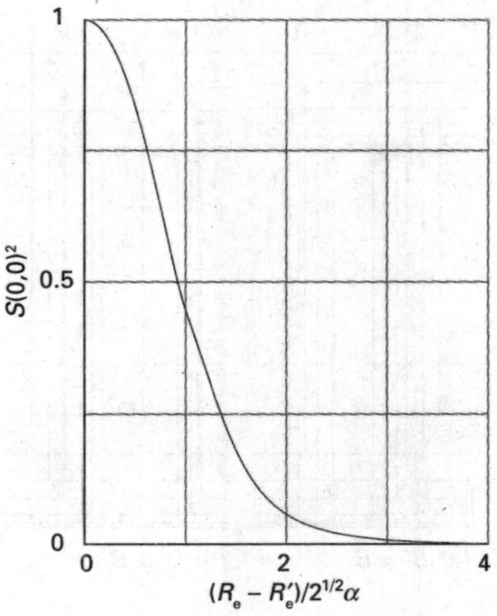

Fig. 13.9 The Franck–Condon factor for the arrangement discussed in Example 13.1.

Self-test 13.2 Suppose the vibrational wavefunctions can be approximated by rectangular functions of width W and W', centred on the equilibrium bond lengths (Fig. 13.10). Find the corresponding Franck–Condon factors when the centres are coincident and $W' < W$. $[S^2 = W'/W]$

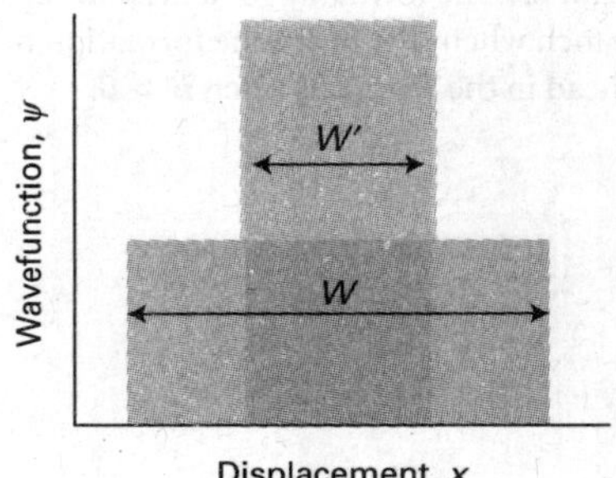

Fig. 13.10 The model wavefunctions used in Self-test 13.2.

(d) Rotational structure

Just as in vibrational spectroscopy, where a vibrational transition is accompanied by rotational excitation, so rotational transitions accompany the excitation of the vibrational excitation that accompanies electronic excitation. We therefore see P, Q, and R branches for each vibrational transition, and the electronic transition has a very rich structure. However, the principal difference is that electronic excitation can result in much larger changes in bond length than vibrational excitation causes alone, and the rotational branches have a more complex structure than in vibration–rotation spectra.

We suppose that the rotational constants of the electronic ground and excited states are $\tilde{B}$ and $\tilde{B}'$, respectively. The rotational energy levels of the initial and final states are

$$E(J) = hc\tilde{B}J(J+1) \qquad E(J') = hc\tilde{B}'J'(J'+1) \tag{13.10}$$

When a transition occurs with $\Delta J = -1$ the wavenumber of the vibrational component of the electronic transition is shifted from $\tilde{\nu}$ to

$$\tilde{\nu} + \tilde{B}'(J-1)J - \tilde{B}J(J+1) = \tilde{\nu} - (\tilde{B}' + \tilde{B})J + (\tilde{B}' - \tilde{B})J^2$$

This transition is a contribution to the P branch (just as in Section 12.11). There are corresponding transitions for the Q and R branches with wavenumbers that may be calculated in a similar way. All three branches are:

P branch ($\Delta J = -1$): $\quad \tilde{\nu}_P(J) = \tilde{\nu} - (\tilde{B}' + \tilde{B})J + (\tilde{B}' - \tilde{B})J^2 \quad$ Branch structure $\quad$ (13.11a)

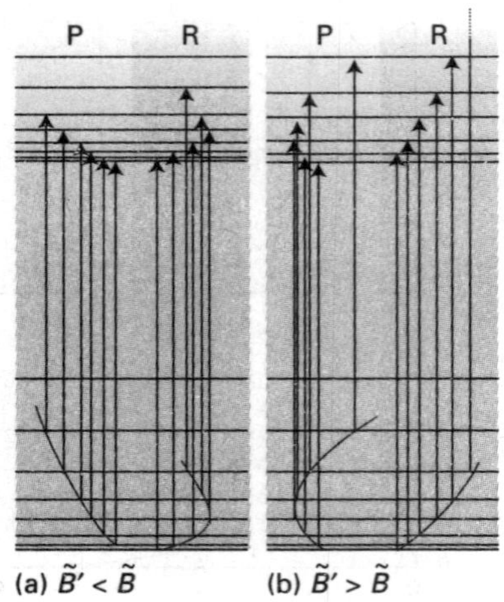

Fig. 13.11 When the rotational constants of a diatomic molecule differ significantly in the initial and final states of an electronic transition, the P and R branches show a head. (a) The formation of a head in the R branch when $\tilde{B}' < \tilde{B}$; (b) the formation of a head in the P branch when $\tilde{B}' > \tilde{B}$.

Q branch ($\Delta J = 0$): $$\tilde{\nu}_Q(J) = \tilde{\nu} + (\tilde{B}' - \tilde{B})J(J+1) \qquad (13.11b)$$

R branch ($\Delta J = +1$): $$\tilde{\nu}_R(J) = \tilde{\nu} + (\tilde{B}' + \tilde{B})(J+1) + (\tilde{B}' - \tilde{B})(J+1)^2 \qquad (13.11c)$$

These expressions are the analogues of eqn 12.46.

First, suppose that the bond length in the electronically excited state is greater than that in the ground state; then $\tilde{B}' < \tilde{B}$ and $\tilde{B}' - \tilde{B}$ is negative. In this case the lines of the R branch converge with increasing J and when J is such that $|\tilde{B}' - \tilde{B}|(J+1) > \tilde{B}' + \tilde{B}$ the lines start to appear at successively decreasing wavenumbers. That is, the R branch has a **band head** (Fig. 13.11a). When the bond is shorter in the excited state than in the ground state, $\tilde{B}' > \tilde{B}$ and $\tilde{B}' - \tilde{B}$ is positive. In this case, the lines of the P branch begin to converge and go through a head when J is such that $|\tilde{B}' - \tilde{B}|J > \tilde{B}' + \tilde{B}$ (Fig. 13.11b).

13.3 The electronic spectra of polyatomic molecules

Key points (a) In d-metal complexes, the presence of ligands removes the degeneracy of d orbitals and vibrationally allowed transitions can occur between them. (b) Charge-transfer transitions typically involve the migration of electrons between the ligands and the central metal atom. (c) Other chromophores include double bonds ($\pi^* \leftarrow \pi$ transitions) and carbonyl groups ($\pi^* \leftarrow$ n transitions). (d) Circular dichroism is the differential absorption of light with opposite circular polarizations.

The absorption of a photon can often be traced to the excitation of specific types of electrons or to electrons that belong to a small group of atoms in a polyatomic molecule. For example, when a carbonyl group (>C=O) is present, an absorption at about 290 nm is normally observed, although its precise location depends on the nature of the rest of the molecule. Groups with characteristic optical absorptions are called **chromophores** (from the Greek for 'colour bringer'), and their presence often accounts for the colours of substances (Table 13.2).

(a) d–d transitions

In a free atom, all five d orbitals of a given shell are degenerate. In a d-metal complex, where the immediate environment of the atom is no longer spherical, the d orbitals are not all degenerate, and electrons can absorb energy by making transitions between them.

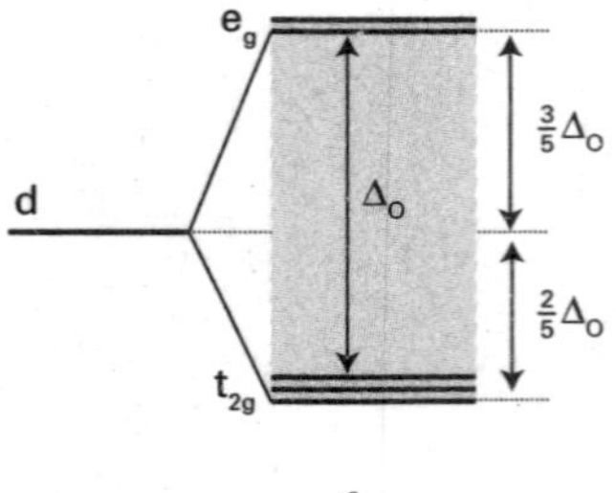

1

To see the origin of this splitting, we regard the six ligands as point negative charges that repel the d electrons of the central ion (Fig. 13.12). As a result, the orbitals fall into two groups, with $d_{x^2-y^2}$ and d_{z^2} pointing directly towards the ligand positions, and d_{xy}, d_{yz}, and d_{zx} pointing between them. An electron occupying an orbital of the former group has a less favourable potential energy than when it occupies any of the three orbitals of the other group, and so the d orbitals split into the two sets shown in (1) with an energy difference Δ_O: a triply degenerate set comprising the d_{xy}, d_{yz}, and d_{zx}

Table 13.2* Absorption characteristics of some groups and molecules

Group	$\tilde{\nu}$/cm^{-1}	λ_{max}/nm	ε/(dm^3 mol^{-1} cm^{-1})
C=C ($\pi^* \leftarrow \pi$)	61 000	163	15 000
	57 300	174	5 500
C=O ($\pi^* \leftarrow$ n)	35 000–37 000	270–290	10–20
H_2O ($\pi^* \leftarrow$ n)	60 000	167	7 000

* More values are given in the *Data section.*

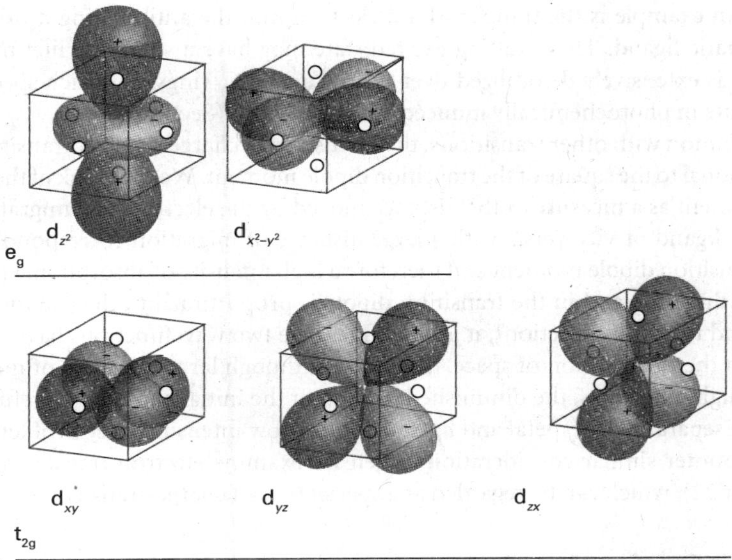

Fig. 13.12 The classification of d orbitals in an octahedral environment.

orbitals and labelled t_{2g}, and a doubly degenerate set comprising the $d_{x^2-y^2}$ and d_{z^2} orbitals and labelled e_g. The three t_{2g} orbitals lie below the two e_g orbitals in energy; the difference in energy is denoted Δ_O and called the **ligand-field splitting parameter** (the O denoting octahedral symmetry). The ligand field splitting is typically about 10 per cent of the overall energy of interaction between the ligands and the central metal atom, which is largely responsible for the existence of the complex. The d orbitals also divide into two sets in a tetrahedral complex, but in this case the e orbitals lie below the t_2 orbitals (the g,u classification is no longer relevant as a tetrahedral complex has no centre of inversion) and their separation is written Δ_T.

Neither Δ_O nor Δ_T is large, so transitions between the two sets of orbitals typically occur in the visible region of the spectrum. The transitions are responsible for many of the colours that are so characteristic of d-metal complexes. As an example, the spectrum of $[Ti(OH_2)_6]^{3+}$ (2) near 20 000 cm^{-1} (500 nm) is shown in Fig. 13.13, and can be ascribed to the promotion of its single d electron from a t_{2g} orbital to an e_g orbital. The wavenumber of the absorption maximum suggests that $\Delta_O \approx 20\,000\ cm^{-1}$ for this complex, which corresponds to about 2.5 eV.

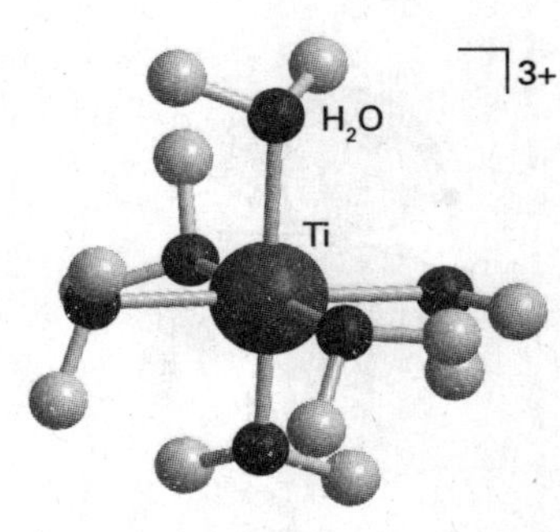

2 $[Ti(OH_2)_6]^{3+}$

According to the Laporte rule (Section 13.2b), d–d transitions are parity-forbidden in octahedral complexes because they are $g \rightarrow g$ transitions (more specifically $e_g \leftarrow t_{2g}$ transitions). However, d–d transitions become weakly allowed as vibronic transitions as a result of coupling to asymmetrical vibrations such as that shown in Fig. 13.5.

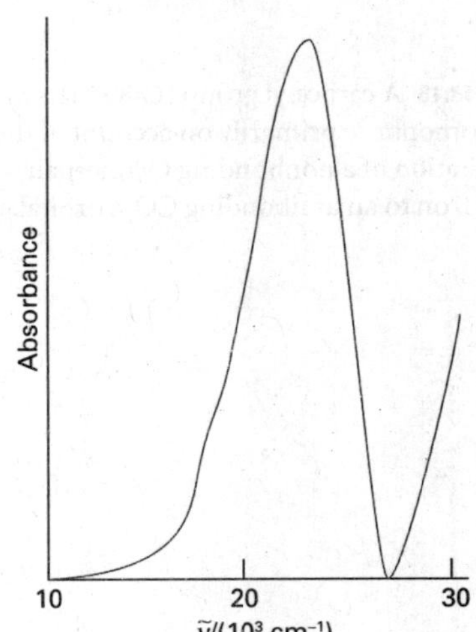

Fig. 13.13 The electronic absorption spectrum of $[Ti(OH_2)_6]^{+3}$ in aqueous solution.

(b) Charge-transfer transitions

A d-metal complex may absorb radiation as a result of the transfer of an electron from the ligands into the d orbitals of the central atom, or vice versa. In such **charge-transfer transitions** the electron moves through a considerable distance, which means that the transition dipole moment may be large and the absorption correspondingly intense. This mode of chromophore activity accounts for the intense violet colour (which arises from strong absorption within the range 420–700 nm) of the permanganate ion, MnO_4^-. In this oxoanion, the electron migrates from an orbital that is largely confined to the O atom ligands to an orbital that is largely confined to the Mn atom. It is therefore an example of a **ligand-to-metal charge-transfer transition** (LMCT). The reverse migration, a **metal-to-ligand charge-transfer transition** (MLCT), can also occur. An example is the transfer of a d electron into the antibonding π orbitals of an aromatic ligand. The resulting excited state may have a very long lifetime if the electron is extensively delocalized over several aromatic rings, and such species can participate in photochemically induced redox reactions (Section 21.10).

In common with other transitions, the intensities of charge-transfer transitions are proportional to the square of the transition dipole moment. We can think of the transition moment as a measure of the distance moved by the electron as it migrates from metal to ligand or vice versa, with a large distance of migration corresponding to a large transition dipole moment and therefore a high intensity of absorption. However, because the integrand in the transition dipole is proportional to the product of the initial and final wavefunctions, it is zero unless the two wavefunctions have nonzero values in the same region of space. Therefore, although large distances of migration favour high intensities, the diminished overlap of the initial and final wavefunctions for large separations of metal and ligands favours low intensities (see Problem 13.8). We encounter similar considerations when we examine electron transfer reactions (Chapter 22), which can be regarded as a special type of charge-transfer transition.

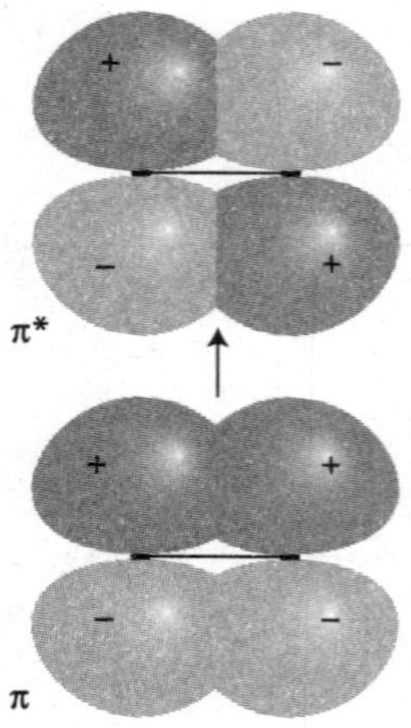

Fig. 13.14 A C=C double bond acts as a chromophore. One of its important transitions is the $\pi^* \leftarrow \pi$ transition illustrated here, in which an electron is promoted from a π orbital to the corresponding antibonding orbital.

(c) $\pi^* \leftarrow \pi$ and $\pi^* \leftarrow n$ transitions

Absorption by a C=C double bond results in the excitation of a π electron into an antibonding π* orbital (Fig. 13.14). The chromophore activity is therefore due to a $\pi^* \leftarrow \pi$ transition (which is normally read 'π to π-star transition'). Its energy is about 7 eV for an unconjugated double bond, which corresponds to an absorption at 180 nm (in the ultraviolet). When the double bond is part of a conjugated chain, the energies of the molecular orbitals lie closer together and the $\pi^* \leftarrow \pi$ transition moves to longer wavelength; it may even lie in the visible region if the conjugated system is long enough. An important example of a $\pi^* \leftarrow \pi$ transition is provided by the photochemical mechanism of vision (*Impact I13.1*).

The transition responsible for absorption in carbonyl compounds can be traced to the lone pairs of electrons on the O atom. The Lewis concept of a 'lone pair' of electrons is represented in molecular orbital theory by a pair of electrons in an orbital confined largely to one atom and not appreciably involved in bond formation. One of these electrons may be excited into an empty π* orbital of the carbonyl group (Fig. 13.15), which gives rise to an $\pi^* \leftarrow n$ transition (an 'n to π-star transition'). Typical absorption energies are about 4 eV (290 nm). Because $\pi^* \leftarrow n$ transitions in carbonyls are symmetry forbidden, the absorptions are weak.

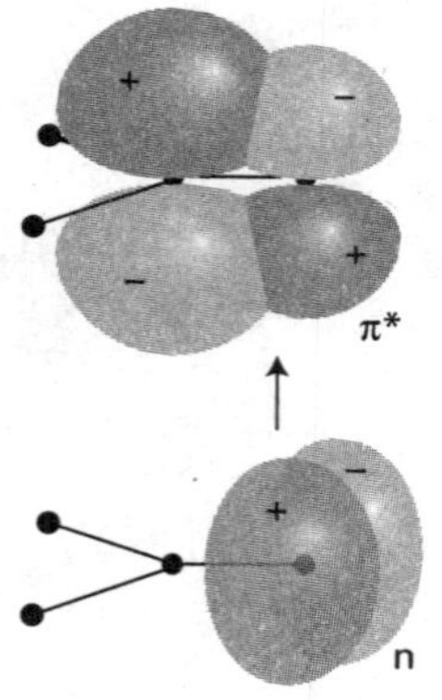

Fig. 13.15 A carbonyl group (C=O) acts as a chromophore primarily on account of the excitation of a nonbonding O lone-pair electron to an antibonding CO π orbital.

(d) Circular dichroism

Electronic spectra can reveal additional details of molecular structure when experiments are conducted with **polarized light**, electromagnetic radiation with electric and magnetic fields that oscillate only in certain directions. Light is **plane polarized** when the electric and magnetic fields each oscillate in a single plane (Fig. 13.16). The plane of polarization may be oriented in any direction around the direction of propagation (the y-direction in Fig. 13.16), with the electric and magnetic fields perpendicular to that direction (and perpendicular to each other). An alternative mode of polarization is **circular polarization**, in which the electric and magnetic fields rotate around the direction of propagation in either a clockwise or a counterclockwise sense but remain perpendicular to it and each other.

When plane-polarized radiation passes through samples of certain kinds of matter, the plane of polarization is rotated around the direction of propagation. This rotation is the familiar phenomenon of optical activity, observed when the molecules in the sample are chiral (Section 11.3b). Chiral molecules have a second characteristic: they

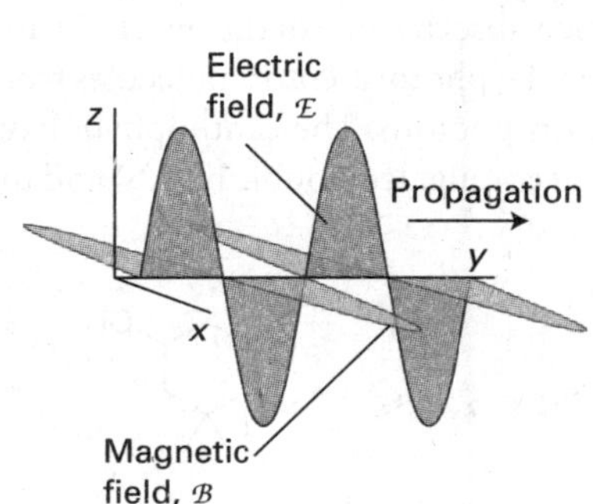

Fig. 13.16 Electromagnetic radiation consists of a wave of electric and magnetic fields perpendicular to the direction of propagation (in this case the y-direction), and mutually perpendicular to each other. This illustration shows a plane-polarized wave, with the electric and magnetic fields oscillating in the yz- and xy-planes, respectively.

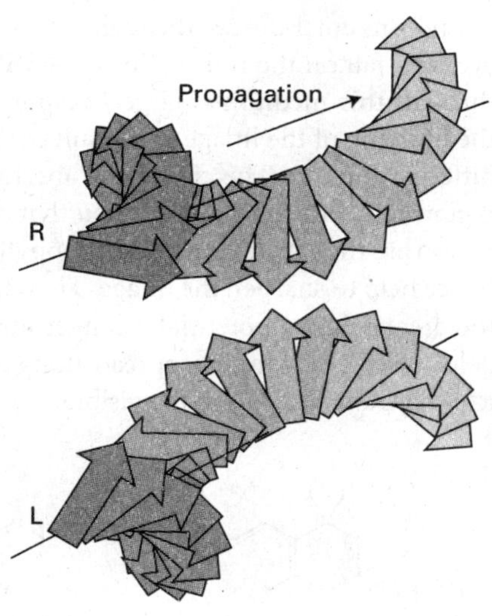

Fig. 13.17 In circularly polarized light, the electric field at different points along the direction of propagation rotates. The arrays of arrows in these illustrations show the view of the electric field: (a) right-circularly polarized, (b) left-circularly polarized light.

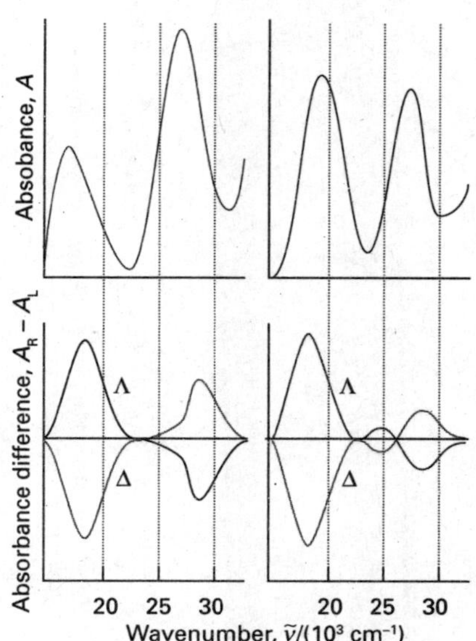

Fig. 13.18 (a) The absorption spectra of two isomers, denoted mer and fac, of $[Co(ala)_3]$, where ala is the conjugate base of alanine, and (b) the corresponding CD spectra. The left- and right-handed forms of these isomers give similar absorption spectra. However, the CD spectra are distinctly different, and the absolute configurations (denoted Λ and Δ) have been assigned by comparison with the CD spectra of a complex of known absolute configuration.

absorb left and right circularly polarized light to different extents. In a circularly polarized ray of light, the electric field describes a helical path as the wave travels through space (Fig. 13.17), and the rotation may be either clockwise or counterclockwise. The differential absorption of left- and right-circularly polarized light is called **circular dichroism**. In terms of the absorbances for the two components, A_L and A_R, the circular dichroism of a sample of molar concentration [J] and path-length L is reported as

$$\Delta\varepsilon = \varepsilon_L - \varepsilon_R = \frac{A_L - A_R}{[J]L} \tag{13.12}$$

Circular dichroism is a useful adjunct to visible and UV spectroscopy. For example, the CD spectra of the enantiomeric pairs of chiral d-metal complexes are distinctly different, whereas there is little difference between their absorption spectra (Fig. 13.18). Moreover, CD spectra can be used to assign the absolute configuration of complexes by comparing the observed spectrum with the CD spectrum of a similar complex of known handedness.

IMPACT ON BIOCHEMISTRY

I13.1 Vision

The eye is an exquisite photochemical organ that acts as a transducer, converting radiant energy into electrical signals that travel along neurons. Here we concentrate on the events taking place in the human eye, but similar processes occur in all animals. Indeed, a single type of protein, rhodopsin, is the primary receptor for light throughout the animal kingdom, which indicates that vision emerged very early in evolutionary history, no doubt because of its enormous value for survival.

Photons enter the eye through the cornea, pass through the ocular fluid that fills the eye, and fall on the retina. The ocular fluid is principally water, and passage of light through this medium is largely responsible for the *chromatic aberration* of the eye, the blurring of the image as a result of different frequencies being brought to slightly different focuses. The chromatic aberration is reduced to some extent by the tinted region called the *macular pigment* that covers part of the retina. The pigments in this region are the carotene-like xanthophylls (**3**), which absorb some of the blue light and hence help to sharpen the image. They also protect the photoreceptor molecules from too great a flux of potentially dangerous high energy photons. The xanthophylls have delocalized electrons that spread along the chain of conjugated double bonds, and the $\pi^* \leftarrow \pi$ transition lies in the visible.

3 A xanthophyll

4 11-*cis*-retinal

5 All-*trans*-retinal

About 57 per cent of the photons that enter the eye reach the retina; the rest are scattered or absorbed by the ocular fluid. Here the primary act of vision takes place, in which the chromophore of a rhodopsin molecule absorbs a photon in another $\pi^* \leftarrow \pi$ transition. A rhodopsin molecule consists of an opsin protein molecule to which is attached a 11-*cis*-retinal molecule (**4**). The latter resembles half a carotene molecule, showing Nature's economy in its use of available materials. The attachment is by the formation of a protonated Schiff's base, utilizing the –CHO group of the chromophore and the terminal NH_2 group of the sidechain, a lysine residue from opsin. The free 11-*cis*-retinal molecule absorbs in the ultraviolet, but attachment to the opsin protein molecule shifts the absorption into the visible region. The rhodopsin molecules are situated in the membranes of special cells (the 'rods' and the 'cones') that cover the retina. The opsin molecule is anchored into the cell membrane by two hydrophobic groups and largely surrounds the chromophore (Fig. 13.19).

Immediately after the absorption of a photon, the 11-*cis*-retinal molecule undergoes photoisomerization into all-*trans*-retinal (**5**). Photoisomerization takes about 200 fs and about 67 pigment molecules isomerize for every 100 photons that are absorbed. The process occurs because the $\pi^* \leftarrow \pi$ excitation of an electron loosens one of the π bonds (the one indicated by the arrow in **4**), its torsional rigidity is lost, and one part of the molecule swings round into its new position. At that point, the molecule returns to its ground state, but is now trapped in its new conformation. The straightened tail of all-*trans*-retinal results in the molecule taking up more space than 11-*cis*-retinal did, so the molecule presses against the coils of the opsin molecule that surrounds it. In about 0.25–0.50 ms from the initial absorption event, the rhodopsin molecule is activated both by the isomerization of retinal and deprotonation of its Schiff's base tether to opsin, forming an intermediate known as *metarhodopsin II*.

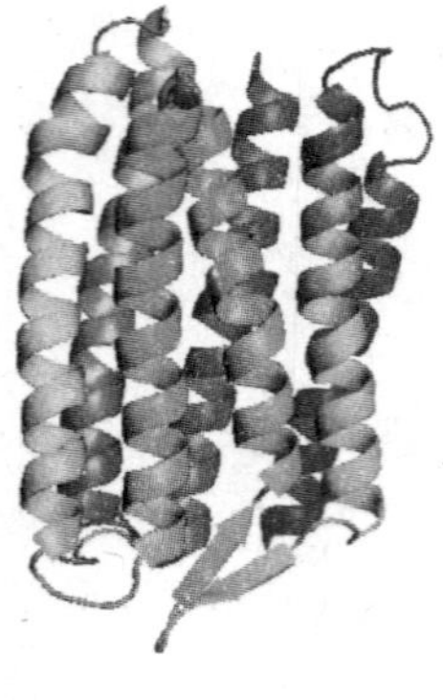

Fig. 13.19 The structure of the rhodopsin molecule, consisting of an opsin protein to which is attached an 11-*cis*-retinal molecule embedded in the space surrounded by the helical regions. Only the protein is shown.

In a sequence of biochemical events known as the *biochemical cascade*, metarhodopsin II activates the protein transducin, which in turn activates a phosphodiesterase enzyme that hydrolyses cyclic guanine monophosphate (cGMP) to GMP. The reduction in the concentration of cGMP causes ion channels, proteins that mediate the movement of ions across biological membranes (*Impact I20.2*), to close. The result is an imbalance of charge that in turn creates an electrical potential across the membrane. The pulse of electric potential travels through the optical nerve and into the optical cortex, where it is interpreted as a signal and incorporated into the web of events we call 'vision'.

The resting state of the rhodopsin molecule is restored by a series of nonradiative chemical events powered by ATP. The process involves the escape of all-*trans*-retinal as all-*trans*-retinol (in which –CHO has been reduced to $-CH_2OH$) from the opsin molecule by a process catalysed by the enzyme rhodopsin kinase and the attachment of another protein molecule, arrestin. The free all-*trans*-retinol molecule now undergoes enzyme-catalysed isomerization into 11-*cis*-retinol followed by dehydrogenation to form 11-*cis*-retinal, which is then delivered back into an opsin molecule. At this point, the cycle of excitation, photoisomerization, and regeneration is ready to begin again.

The fates of electronically excited states

A **radiative decay process** is a process in which a molecule discards its excitation energy as a photon. A more common fate is **nonradiative decay**, in which the excess energy is transferred into the vibration, rotation, and translation of the surrounding molecules. This thermal degradation converts the excitation energy completely into thermal motion of the environment (that is, to 'heat'). An excited molecule may also take part in a chemical reaction, as we discuss in Chapter 22.

13.4 Fluorescence and phosphorescence

Key points (a) The rates of radiative transitions are summarized by the Einstein coefficients of stimulated and spontaneous processes. (b) Fluorescence is radiative decay between states of the same multiplicity. (c) Phosphorescence is radiative decay between states of different multiplicity and persists after the exciting radiation is removed.

In **fluorescence**, spontaneous emission of radiation occurs within a few nanoseconds after the exciting radiation is extinguished (Fig. 13.20). In **phosphorescence**, the spontaneous emission may persist for long periods (even hours, but characteristically seconds or fractions of seconds). The difference suggests that fluorescence is a fast conversion of absorbed radiation into re-emitted energy, and that phosphorescence involves the storage of energy in a reservoir from which it slowly leaks.

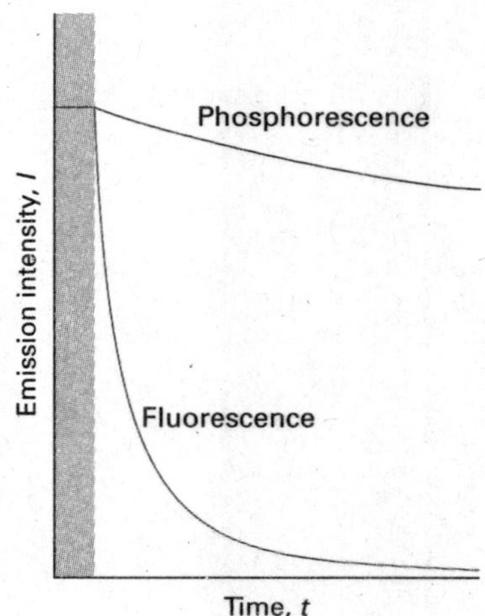

Fig. 13.20 The empirical (observation-based) distinction between fluorescence and phosphorescence is that the former is extinguished very quickly after the exciting source is removed, whereas the latter continues with relatively slowly diminishing intensity.

(a) Stimulated and spontaneous radiative processes

Einstein identified three contributions to the transitions between states. **Stimulated absorption** is the transition from a low energy state to one of higher energy that is driven by the electromagnetic field oscillating at the transition frequency. The transition rate, w, is the rate of change of probability of the molecule being found in the upper state. The more intense the electromagnetic field (the more intense the incident radiation), the greater the rate at which transitions are induced and hence the stronger the absorption by the sample. Einstein wrote the transition rate as

$$w = B\rho \qquad \text{Rate of stimulated absorption} \qquad (13.13)$$

The constant B is the **Einstein coefficient of stimulated absorption** and $\rho d\nu$ is the energy density of radiation in the frequency range ν to $\nu + d\nu$, where ν is the frequency of the transition. When the molecule is exposed to black-body radiation from a source of temperature T, ρ is given by the Planck distribution (eqn 7.8):

$$\rho = \frac{8\pi h\nu^3/c^3}{e^{h\nu/kT} - 1} \qquad (13.14)$$

where the slight difference between the forms of the Planck distribution shown here and in eqn 7.8 stems from the fact that it is written here as $\rho d\nu$, and $d\lambda = (c/\nu^2)d\nu$.

For the time being, we can treat B as an empirical parameter that characterizes the transition: if B is large, then a given intensity of incident radiation will induce transitions strongly and the sample will be strongly absorbing. The **total rate of absorption**, W, the number of molecules excited during an interval divided by the duration of the interval, is the transition rate of a single molecule multiplied by the number of molecules N in the lower state: $W = Nw$.

Einstein considered that the radiation was also able to induce the molecule in the upper state to undergo a transition to the lower state, and hence to generate a photon of frequency ν. Thus, he wrote the rate of this stimulated emission as

$$w' = B'\rho \qquad (13.15)$$

Rate of stimulated emission

where B' is the **Einstein coefficient of stimulated emission.** Note that only radiation of the same frequency as the transition can stimulate an excited state to fall to a lower state. However, he realized that stimulated emission was not the only means by which the excited state could generate radiation and return to the lower state, and suggested that an excited state could undergo **spontaneous emission** at a rate that was independent of the intensity of the radiation (of any frequency) that is already present. Einstein therefore wrote the total rate of transition from the upper to the lower state as

$$w' = A + B'\rho \qquad (13.16)$$

Total rate of emission

The constant A is the **Einstein coefficient of spontaneous emission.**

As we demonstrate in the following *Justification*, Einstein was able to show that the two coefficients of stimulated absorption and emission are equal, and that the coefficient of spontaneous emission is related to them by

$$A = \left(\frac{8\pi h\nu^3}{c^3}\right)B \qquad B' = B \qquad (13.17)$$

Relation between the Einstein coefficients

The important features of these equations are

- The coefficient of spontaneous emission increases as the third power of the frequency and therefore the separation in energy of the upper and lower states.
- The rates of stimulated absorption and emission between two states are the same for a given intensity of incident radiation at the transition frequency.

Justification 13.4 *The relation between the Einstein coefficients*

The expressions for the rates w and w' are for the transitions of individual molecules. The total rates of emission and absorption depend on the numbers of molecules in the two states involved in the transition. That is, the total rate of absorption is Nw and the total rate of emission is $N'w'$, where N is the population of the lower state and N' is the population of the upper state. At thermal equilibrium the total rates of emission and absorption are equal, so

$$NB\rho = N'(A + B'\rho)$$

This expression rearranges into

$$\rho = \frac{N'A}{NB - N'B'} = \frac{A/B}{N/N' - B'/B} = \frac{A/B}{e^{h\nu/kT} - B'/B}$$

We have used the Boltzmann expression (*Fundamentals F.5*) for the ratio of populations of states of energies E and E' in the last step:

$$\frac{N'}{N} = e^{-h\nu/kT} \qquad h\nu = E' - E$$

This result has the same form as the Planck distribution (eqn 13.14), which describes the radiation density at thermal equilibrium. Indeed, when we compare the two expressions for ρ, we can conclude that the coefficients are related by eqn 13.17.

(b) Fluorescence

Figure 13.21 shows the sequence of steps involved in fluorescence. The initial stimulated absorption takes the molecule to an excited electronic state, and if the absorption spectrum were monitored it would look like the one shown in Fig. 13.22a. The excited molecule is subjected to collisions with the surrounding molecules, and as it gives up energy nonradiatively it steps down the ladder of vibrational levels to the lowest vibrational level of the electronically excited molecular state. The surrounding molecules, however, might now be unable to accept the larger energy difference needed to lower the molecule to the ground electronic state. It might therefore survive long enough to undergo spontaneous emission and emit the remaining excess energy as radiation. The downward electronic transition is vertical (in accord with the Franck–Condon principle) and the fluorescence spectrum has a vibrational structure characteristic of the *lower* electronic state (Fig. 13.22b).

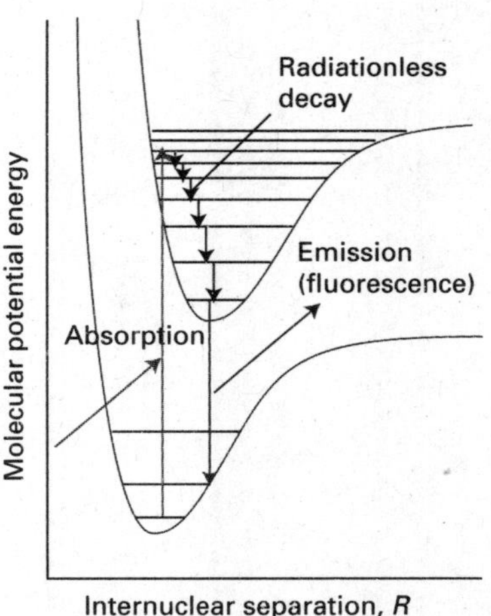

Fig. 13.21 The sequence of steps leading to fluorescence. After the initial absorption, the upper vibrational states undergo radiationless decay by giving up energy to the surroundings. A radiative transition then occurs from the vibrational ground state of the upper electronic state.

Provided they can be seen, the 0–0 absorption and fluorescence transitions can be expected to be coincident. The absorption spectrum arises from 1–0, 2–0, . . . transitions that occur at progressively higher wavenumber and with intensities governed by the Franck–Condon principle. The fluorescence spectrum arises from 0–0, 0–1, . . . *downward* transitions that occur with decreasing wavenumbers. The 0–0 absorption and fluorescence peaks are not always exactly coincident, however, because the solvent may interact differently with the solute in the ground and excited states (for instance, the hydrogen bonding pattern might differ). Because the solvent molecules do not have time to rearrange during the transition, the absorption occurs in an environment characteristic of the solvated ground state; however, the fluorescence occurs in an environment characteristic of the solvated excited state (Fig. 13.23).

Fluorescence occurs at lower frequencies (longer wavelengths) than that of the incident radiation because the emissive transition occurs after some vibrational energy has been discarded into the surroundings. The vivid oranges and greens of fluorescent dyes are an everyday manifestation of this effect: they absorb in the ultraviolet and blue, and fluoresce in the visible. The mechanism also suggests that the intensity of the fluorescence ought to depend on the ability of the solvent molecules to accept the electronic and vibrational quanta. It is indeed found that a solvent composed of molecules with widely spaced vibrational levels (such as water) can in some cases accept the large quantum of electronic energy and so extinguish, or 'quench', the fluorescence. The rate at which fluorescence is quenched by other molecules also gives valuable kinetic information; this important aspect of fluorescence is taken further in Section 21.10.

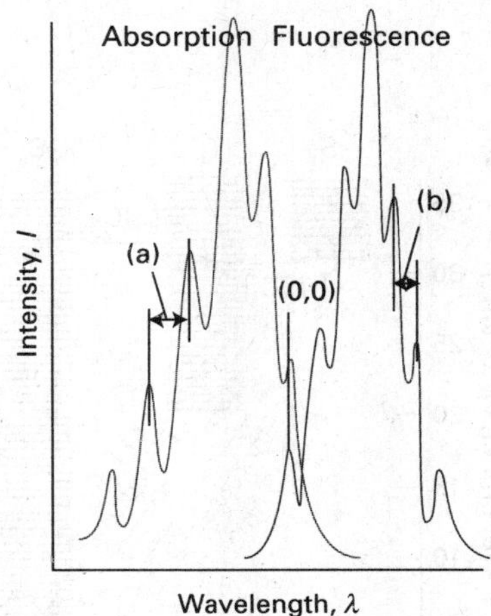

Fig. 13.22 An absorption spectrum (a) shows a vibrational structure characteristic of the upper state. A fluorescence spectrum (b) shows a structure characteristic of the lower state; it is also displaced to lower frequencies (but the 0–0 transitions are coincident) and resembles a mirror image of the absorption.

(c) Phosphorescence

Figure 13.24 shows the sequence of events leading to phosphorescence for a molecule with a singlet ground state. The first steps are the same as in fluorescence, but the presence of a triplet excited state plays a decisive role. The singlet and triplet excited states share a common geometry at the point where their potential energy curves intersect. Hence, if there is a mechanism for unpairing two electron spins (and achieving the conversion of ↑↓ to ↑↑), the molecule may undergo **intersystem crossing**, a nonradiative

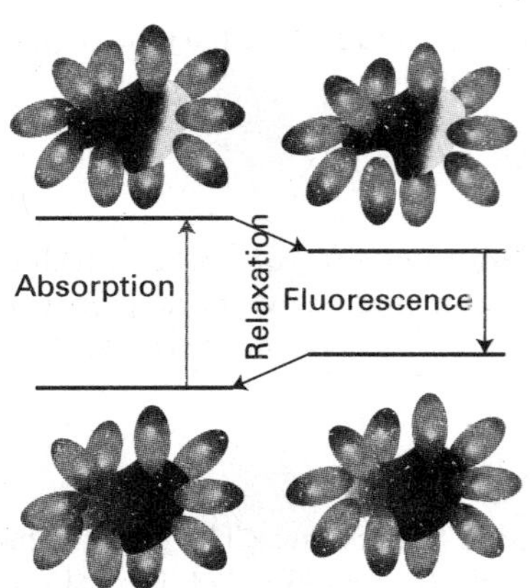

Fig. 13.23 The solvent can shift the fluorescence spectrum relative to the absorption spectrum. On the left we see that the absorption occurs with the solvent (the ellipses) in the arrangement characteristic of the ground electronic state of the molecule (the sphere). However, before fluorescence occurs, the solvent molecules relax into a new arrangement, and that arrangement is preserved during the subsequent radiative transition.

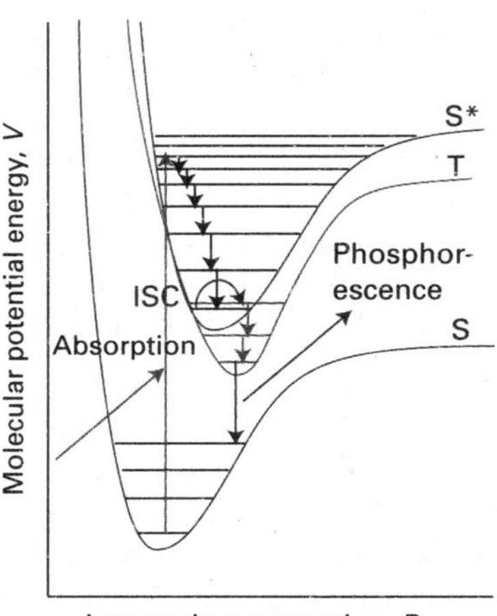

Fig. 13.24 The sequence of steps leading to phosphorescence. The important step is the intersystem crossing (ISC), the switch from a singlet state to a triplet state brought about by spin–orbit coupling. The triplet state acts as a slowly radiating reservoir because the return to the ground state is spin-forbidden.

transition between states of different multiplicity, and become a triplet state. We saw in the discussion of atomic spectra (Section 9.10d) that singlet–triplet transitions may occur in the presence of spin–orbit coupling, and the same is true in molecules. We can expect intersystem crossing to be important when a molecule contains a moderately heavy atom (such as sulfur), because then the spin–orbit coupling is large.

If an excited molecule crosses into a triplet state, it continues to deposit energy into the surroundings. However, it is now stepping down the triplet's vibrational ladder, and at the lowest energy level it is trapped because the triplet state is at a lower energy than the corresponding singlet (recall Hund's rule, Section 9.4d). The solvent cannot absorb the final, large quantum of electronic excitation energy, and the molecule cannot radiate its energy because return to the ground state is spin-forbidden. The radiative transition, however, is not totally forbidden because the spin–orbit coupling that was responsible for the intersystem crossing also breaks the selection rule. The molecules are therefore able to emit weakly, and the emission may continue long after the original excited state was formed.

The mechanism accounts for the observation that the excitation energy seems to get trapped in a slowly leaking reservoir. It also suggests (as is confirmed experimentally) that phosphorescence should be most intense from solid samples: energy transfer is then less efficient and intersystem crossing has time to occur as the singlet excited state steps slowly past the intersection point. The mechanism also suggests that the phosphorescence efficiency should depend on the presence of a moderately heavy atom (with strong spin–orbit coupling), which is in fact the case. The confirmation of the mechanism is the experimental observation (using the sensitive magnetic resonance techniques described in Chapter 14) that the sample is paramagnetic while the reservoir state, with its unpaired electron spins, is populated.

The various types of nonradiative and radiative transitions that can occur in molecules are often represented on a schematic **Jablonski diagram** of the type shown in Fig. 13.25.

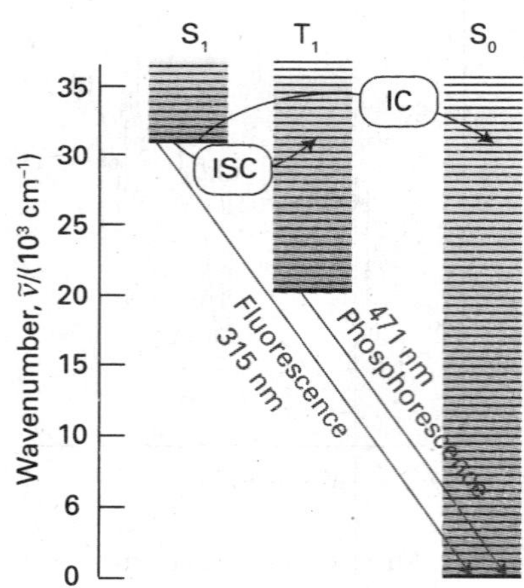

Fig. 13.25 A Jablonski diagram (here, for naphthalene) is a simplified portrayal of the relative positions of the electronic energy levels of a molecule. Vibrational levels of a given electronic state lie above each other, but the relative horizontal locations of the columns bear no relation to the nuclear separations in the states. The ground vibrational states of each electronic state are correctly located vertically but the other vibrational states are shown only schematically. (IC: internal conversion; ISC: intersystem crossing.)

IMPACT ON BIOCHEMISTRY

I13.2 Fluorescence microscopy

Fluorescence is a very important technique for the study of biological molecules. In **fluorescence microscopy**, images of biological cells at work are obtained by attaching a large number of fluorescent molecules to proteins, nucleic acids, and membranes and then measuring the distribution of fluorescence intensity within the illuminated area. Apart from a small number of co-factors, such as the chlorophylls and flavins, the majority of the building blocks of proteins and nucleic acids do not fluoresce strongly. Four notable exceptions are the amino acids tryptophan ($\lambda_{abs} \approx 280$ nm and $\lambda_{fluor} \approx 348$ nm in water), tyrosine ($\lambda_{abs} \approx 274$ nm and $\lambda_{fluor} \approx 303$ nm in water), and phenylalanine ($\lambda_{abs} \approx 257$ nm and $\lambda_{fluor} \approx 282$ nm in water), and the oxidized form of the sequence serine–tyrosine–glycine (**6**) found in the green fluorescent protein (GFP) of certain jellyfish. The wild type of GFP from *Aequora victoria* absorbs strongly at 395 nm and emits maximally at 509 nm and is commonly used as a fluorescent label.

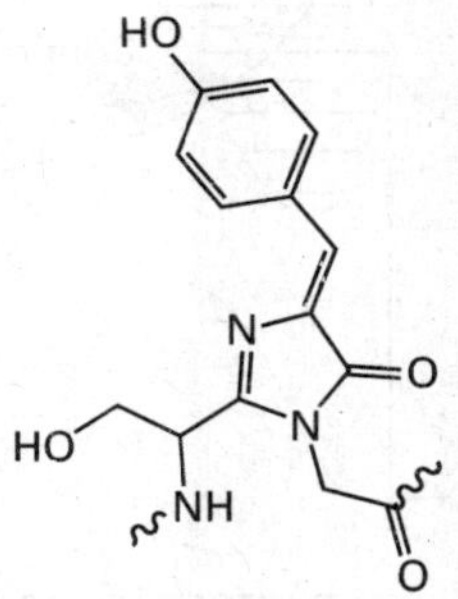

6 The chromophore of GFP

Fluorescence microscopy has been used for many years to image biological cells, but the visualization of molecules requires creative strategies. In a conventional light microscope, an image is constructed from a pattern of diffracted light waves that emanate from the illuminated object. As a result, some information about the specimen is lost by destructive interference of scattered light waves. Ultimately, this *diffraction limit* prevents the study of samples that are much smaller than the wavelength of light used as a probe. In practice, two objects will appear as distinct images under a microscope if the distance between their centres is greater than the *Airy radius*, $r_{Airy} = 0.61\lambda/a$, where λ is the wavelength of the incident beam of radiation and a is the numerical aperture of the objective lens, the lens that collects light scattered by the object. The numerical aperture of the objective lens is defined as $a = n_r \sin \alpha$, where n_r is the refractive index of the lens material (the greater the refractive index, the greater the bending of a ray of light by the lens) and the angle α is the half-angle of the widest cone of scattered light that can be collected by the lens (so the lens collects light beams sweeping a cone with angle 2α).

Most molecules—including biological polymers—have dimensions that are much smaller than visible wavelengths, so special techniques had to be developed to make single-molecule spectroscopy possible. In **near-field scanning optical microscopy** (NSOM), a very thin metal-coated optical fibre is used to deliver light to a small area. It is possible to construct fibres with tip diameters in the range of 50 to 100 nm, which are indeed smaller than visible wavelengths. The fibre tip is placed very close to the sample, in a region known as the *near field*, where, according to classical physics, waves do not undergo diffraction. In **far-field confocal microscopy**, laser light focused by an objective lens is used to illuminate about 1 μm^3 of a very dilute sample placed beyond the near field. This illumination scheme is limited by diffraction and, as a result, data from far-field microscopy have less structural detail than data from NSOM. However, far-field microscopes are very easy to construct and the technique can be used to probe single molecules as long as there is one molecule, on average, in the illuminated area.

13.5 Dissociation and predissociation

Key point **Two further fates of an electronically excited species are dissociation and internal conversion to a dissociative state.**

Another fate for an electronically excited molecule is **dissociation**, the breaking of bonds (Fig. 13.26). The onset of dissociation can be detected in an absorption spectrum by seeing that the vibrational structure of a band terminates at a certain energy. Absorption occurs in a continuous band above this **dissociation limit** because the

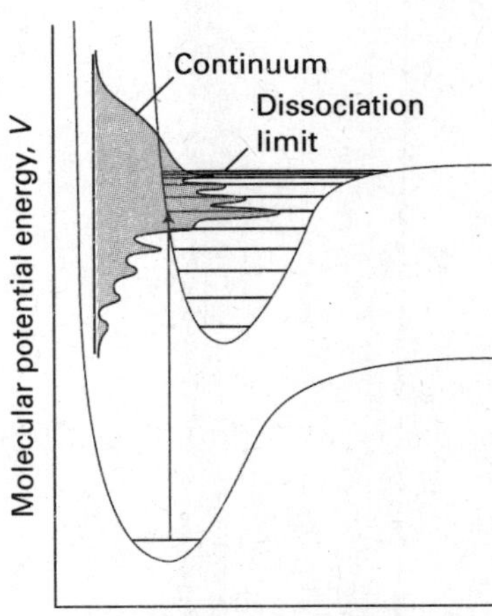

Fig. 13.26 When absorption occurs to unbound states of the upper electronic state, the molecule dissociates and the absorption is a continuum. Below the dissociation limit the electronic spectrum shows a normal vibrational structure.

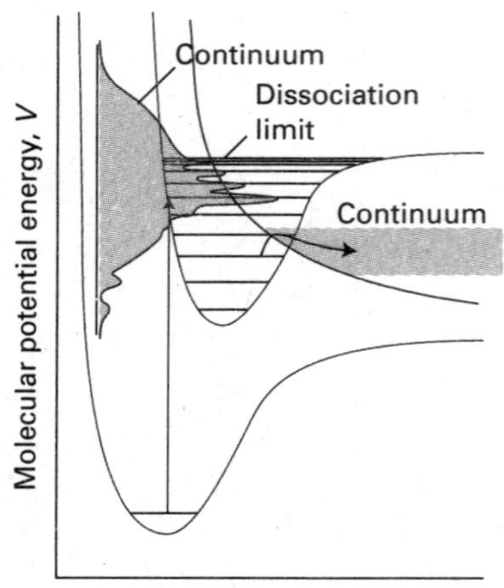

Fig. 13.27 When a dissociative state crosses a bound state, as in the upper part of the illustration, molecules excited to levels near the crossing may dissociate. This process is called predissociation, and is detected in the spectrum as a loss of vibrational structure that resumes at higher frequencies.

final state is an unquantized translational motion of the fragments. Locating the dissociation limit is a valuable way of determining the bond dissociation energy.

In some cases, the vibrational structure disappears but resumes at higher photon energies. This **predissociation** can be interpreted in terms of the molecular potential energy curves shown in Fig. 13.27. When a molecule is excited to a vibrational level, its electrons may undergo a redistribution that results in it undergoing an **internal conversion**, a radiationless conversion to another state of the same multiplicity. An internal conversion occurs most readily at the point of intersection of the two molecular potential energy curves, because there the nuclear geometries of the two states are the same. The state into which the molecule converts may be dissociative, so the states near the intersection have a finite lifetime and hence their energies are imprecisely defined. As a result, the absorption spectrum is blurred in the vicinity of the intersection. When the incoming photon brings enough energy to excite the molecule to a vibrational level high above the intersection, the internal conversion does not occur (the nuclei are unlikely to have the same geometry). Consequently, the levels resume their well-defined, vibrational character with correspondingly well-defined energies, and the line structure resumes on the high-frequency side of the blurred region.

13.6 Laser action

Key points **(a) To achieve laser action, it is necessary to generate a population inversion. (b) The characteristics of the cavity determine the resonant modes of a laser. (c) Pulses are generated by the techniques of Q-switching and mode locking.**

The word laser is an acronym formed from light amplification by stimulated emission of radiation. In stimulated emission (Section 13.4), an excited state is stimulated to emit a photon by radiation of the same frequency: the more photons that are present, the greater the probability of the emission. The essential feature of laser action is positive-feedback: the more photons present of the appropriate frequency, the more photons of that frequency that will be stimulated to form.

Laser radiation has a number of striking characteristics (Table 13.3). Each of them (sometimes in combination with the others) opens up interesting opportunities in physical chemistry. As we have seen, Raman spectroscopy has flourished on account of the high intensity monochromatic radiation available from lasers and photochemistry

Table 13.3 Characteristics of laser radiation and their chemical applications

Characteristic	Advantage	Application
High power	Multiphoton process Low detector noise High scattering intensity	Spectroscopy Improved sensitivity Raman spectroscopy (Chapter 12)
Monochromatic	High resolution State selection	Spectroscopy Photochemical studies (Chapter 21) State-to-state reaction dynamics (Chapter 22)
Collimated beam	Long path lengths Forward-scattering observable	Improved sensitivity Raman spectroscopy (Chapter 12)
Coherent	Interference between separate beams	CARS (Chapter 12)
Pulsed	Precise timing of excitation	Fast reactions (Chapters 21 and 22) Relaxation (Chapter 21) Energy transfer (Chapter 21)

has enabled reactions to be studied on timescales of femtosecond and even attoseconds on account of the ultrashort pulses that lasers can generate (Section 22.4e).

Lasers lie very much on the frontier of physics and chemistry, for their operation depends on details of optics and, in some cases, of solid-state processes. In this section, we discuss the mechanisms of laser action, and then explore their applications in chemistry. We discuss the modes of operation of a number of some commonly available laser systems in *Further information 13.1*.

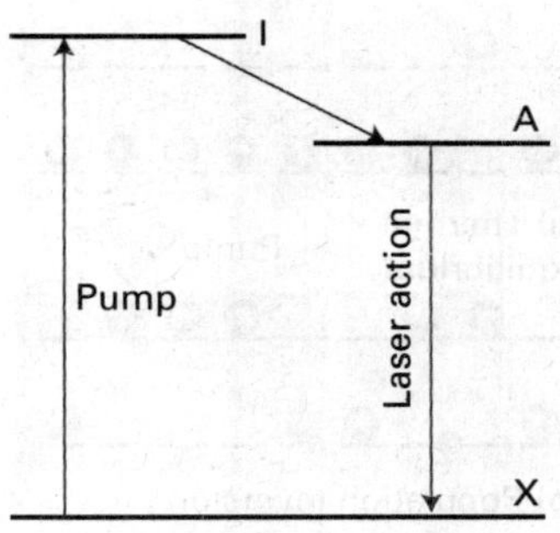

Fig. 13.28 The transitions involved in one kind of three-level laser. The pumping pulse populates the intermediate state I, which in turn populates the laser state A. The laser transition is the stimulated emission A → X.

(a) Population inversion

One requirement of laser action is the existence of a **metastable excited state**, an excited state with a long enough lifetime for it to participate in stimulated emission. Another requirement is the existence of a greater population in the metastable state than in the lower state where the transition terminates, for then there will be a net emission of radiation. Because at thermal equilibrium the opposite is true, it is necessary to achieve a **population inversion** in which there are more molecules in the upper state than in the lower.

One way of achieving population inversion is illustrated in Fig. 13.28. The molecule is excited to an intermediate state I, which then gives up some of its energy nonradiatively and changes into a lower state A; the laser transition is the return of A to the ground state X. Because three energy levels are involved overall, this arrangement leads to a **three-level laser**. In practice, I consists of many states, all of which can convert to the upper of the two laser states A. The I ← X transition is stimulated with an intense flash of light in the process called **pumping**. The pumping is often achieved with an electric discharge through xenon or with the light of another laser. The conversion of I to A should be rapid, and the laser transitions from A to X should be relatively slow.

The disadvantage of this three-level arrangement is that it is difficult to achieve population inversion, because so many ground-state molecules must be converted to the excited state by the pumping action. The arrangement adopted in a **four-level laser** simplifies this task by having the laser transition terminate in a state A′ other than the ground state (Fig. 13.29). Because A′ is unpopulated initially, any population in A corresponds to a population inversion, and we can expect laser action if A is sufficiently metastable. Moreover, this population inversion can be maintained if the X ← A′ transitions are rapid, for these transitions will deplete any population in A′ that stems from the laser transition, and keep the state A′ relatively empty.

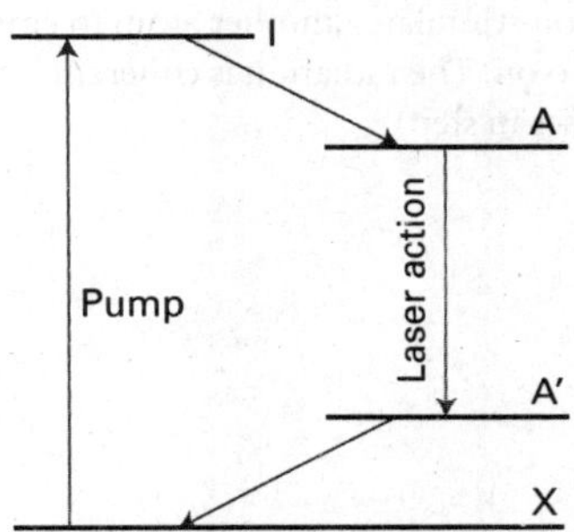

Fig. 13.29 The transitions involved in a four-level laser. Because the laser transition terminates in an excited state (A′), the population inversion between A and A′ is much easier to achieve.

(b) Cavity and mode characteristics

The laser medium is confined to a cavity that ensures that only certain photons of a particular frequency, direction of travel, and state of polarization are generated abundantly. The cavity is essentially a region between two mirrors that reflect the light back and forth. This arrangement can be regarded as a version of the particle in a box, with the particle now being a photon. As in the treatment of a particle in a box (Section 8.1), the only wavelengths that can be sustained satisfy

$$n \times \tfrac{1}{2}\lambda = L \qquad (13.18)$$

where n is an integer and L is the length of the cavity. That is, only an integral number of half-wavelengths fit into the cavity; all other waves undergo destructive interference with themselves. In addition, not all wavelengths that can be sustained by the cavity are amplified by the laser medium (many fall outside the range of frequencies of the laser transitions), so only a few contribute to the laser radiation. These wavelengths are the **resonant modes** of the laser.

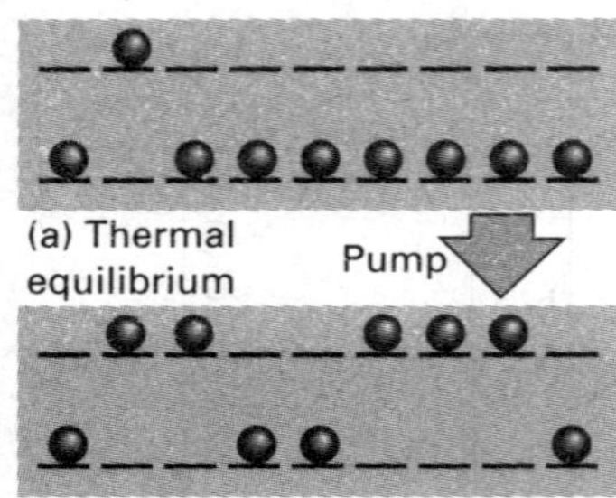

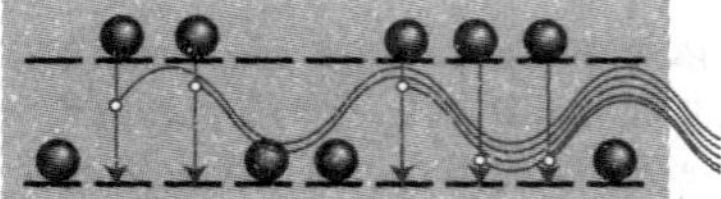

Fig. 13.30 A schematic illustration of the steps leading to laser action. (a) The Boltzmann population of states, with more atoms (or molecules or ions) in the ground state. (b) When the initial state absorbs, the populations are inverted (the atoms are pumped to the excited state). (c) A cascade of radiation then occurs, as one emitted photon stimulates another atom to emit, and so on. The radiation is coherent (phases in step).

Photons with the correct wavelength for the resonant modes of the cavity and the correct frequency to stimulate the laser transition are highly amplified. One photon might be generated spontaneously and travel through the medium. It stimulates the emission of another photon, which in turn stimulates more (Fig. 13.30). The cascade of energy builds up rapidly, and soon the cavity is an intense reservoir of radiation at all the resonant modes it can sustain. Some of this radiation can be withdrawn if one of the mirrors is partially transmitting.

The resonant modes of the cavity have various natural characteristics, and to some extent may be selected. Only photons that are travelling strictly parallel to the axis of the cavity undergo more than a couple of reflections, so only they are amplified, all others simply vanishing into the surroundings. Hence, laser light generally forms a beam with very low divergence. It may also be polarized, with its electric vector in a particular plane (or in some other state of polarization), by including a polarizing filter into the cavity or by making use of polarized transitions in a solid medium.

Laser radiation is **coherent** in the sense that the electromagnetic waves are all in step. In **spatial coherence** the waves are in step across the cross-section of the beam emerging from the cavity. In **temporal coherence** the waves remain in step along the beam. The latter is normally expressed in terms of a **coherence length**, l_C, the distance over which the waves remain coherent, and is related to the range of wavelengths, $\Delta\lambda$ present in the beam:

$$l_C = \frac{\lambda^2}{2\Delta\lambda} \qquad (13.19)$$

If the beam were perfectly monochromatic, with strictly one wavelength present, $\Delta\lambda$ would be zero and the waves would remain in step for an infinite distance. When many wavelengths are present, the waves get out of step in a short distance and the coherence length is small. A typical light bulb gives out light with a coherence length of only about 400 nm; a He–Ne laser with $\Delta\lambda \approx 2$ pm has a coherence length of about 10 cm.

(c) Pulsed lasers

A laser can generate radiation for as long as the population inversion is maintained. A laser can operate continuously when heat is easily dissipated, for then the population of the upper level can be replenished by pumping. When overheating is a problem, the laser can be operated only in pulses, perhaps of microsecond or millisecond duration, so that the medium has a chance to cool or the lower state discard its population. However, it is sometimes desirable to have pulses of radiation rather than a continuous output, with a lot of power concentrated into a brief pulse. One way of achieving pulses is by **Q-switching**, the modification of the resonance characteristics of the laser cavity. The name comes from the 'Q-factor' used as a measure of the quality of a resonance cavity in microwave engineering.

Example 13.2 *Relating the power and energy of a laser*

A laser rated at 0.10 J can generate radiation in 3.0 ns pulses at a pulse repetition rate of 10 Hz. Assuming that the pulses are rectangular, calculate the peak power output and the average power output of this laser.

Method The power output is the energy released in an interval divided by the duration of the interval, and is expressed in watts ($1\ \mathrm{W} = 1\ \mathrm{J\ s^{-1}}$). To calculate the peak power output, P_{peak}, we divide the energy released during the pulse divided by the duration of the pulse. The average power output, P_{average}, is the total energy released by a large number of pulses divided by the duration of the time interval over which the total energy was measured. So, the average power is simply the energy released by one pulse multiplied by the pulse repetition rate.

Answer From the data,

$$P_{\text{peak}} = \frac{0.10\ \text{J}}{3.0 \times 10^{-9}\ \text{s}} = 3.3 \times 10^{7}\ \text{J s}^{-1}$$

That is, the peak power output is 33 MW. The pulse repetition rate is 10 Hz, so ten pulses are emitted by the laser in every second of operation. It follows that the average power output is

$$P_{\text{average}} = 0.10\ \text{J} \times 10\ \text{s}^{-1} = 1.0\ \text{J s}^{-1} = 1.0\ \text{W}$$

The peak power is much higher than the average power because this laser emits light for only 30 ns during each second of operation.

Self-test 13.3 Calculate the peak power and average power output of a laser with a pulse energy of 2.0 mJ, a pulse duration of 30 ps, and a pulse repetition rate of 38 MHz. [$P_{\text{peak}} = 67$ MW, $P_{\text{average}} = 76$ kW]

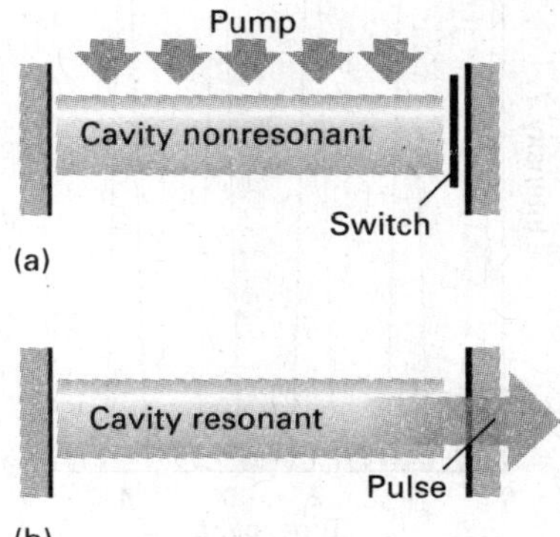

Fig. 13.31 The principle of *Q*-switching. The excited state is populated while the cavity is nonresonant. Then the resonance characteristics are suddenly restored, and the stimulated emission emerges in a giant pulse.

The aim of *Q*-switching is to achieve a healthy population inversion in the absence of the resonant cavity, then to plunge the population-inverted medium into a cavity and hence to obtain a sudden pulse of radiation. The switching may be achieved by impairing the resonance characteristics of the cavity in some way while the pumping pulse is active and then suddenly to improve them (Fig. 13.31). One technique is to use the ability of some crystals, such as those of potassium dihydrogenphosphate (KH_2PO_4), to change their optical properties when an electrical potential difference is applied. Switching the potential on and off can store and then release energy in a laser cavity, resulting in an intense pulse of stimulated emission.

The technique of **mode locking** can produce pulses of picosecond duration and less. A laser radiates at a number of different frequencies, depending on the precise details of the resonance characteristics of the cavity and in particular on the number of half-wavelengths of radiation that can be trapped between the mirrors (the cavity modes). The resonant modes differ in frequency by multiples of $c/2L$ (as can be inferred from eqn 13.18 with $\nu = c/\lambda$). Normally, these modes have random phases relative to each other. However, it is possible to lock their phases together. As we show in the following *Justification*, interference then occurs to give a series of sharp peaks, and the energy of the laser is obtained in short bursts (Fig. 13.32). The sharpness of the peaks depends on the range of modes superimposed and, the wider the range, the narrower the pulses. In a laser with a cavity of length 30 cm, the peaks are separated by 2 ns. If 1000 modes contribute, the width of the pulses is 4 ps.

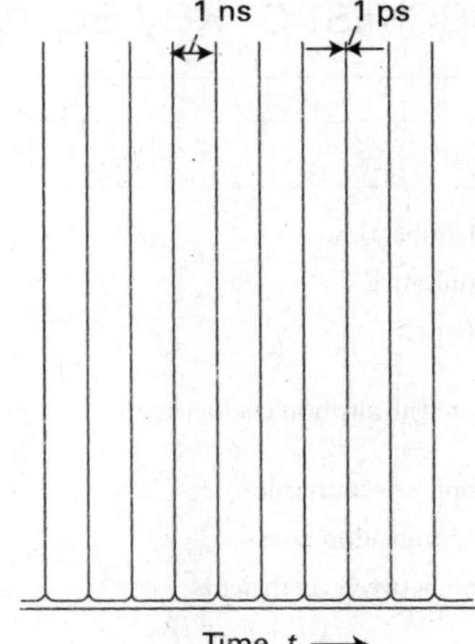

Fig. 13.32 The output of a mode-locked laser consists of a stream of very narrow pulses separated by an interval equal to the time it takes for light to make a round trip inside the cavity.

Justification 13.5 *The origin of mode locking*

The general expression for a (complex) wave of amplitude $\mathcal{E}_0$ and frequency ω is $\mathcal{E}_0 e^{i\omega t}$. Therefore, each wave that can be supported by a cavity of length L has the form

$$\mathcal{E}_n(t) = \mathcal{E}_0 e^{2\pi i(\nu + nc/2L)t}$$

where ν is the lowest frequency. A wave formed by superimposing N modes with $n = 0, 1, \ldots, N-1$ has the form

$$\mathcal{E}(t) = \sum_{n=0}^{N-1} \mathcal{E}_n(t) = \mathcal{E}_0 e^{2\pi i\nu t} \sum_{n=0}^{N-1} e^{i\pi nct/L}$$

A brief comment

The sum of a geometrical progression of N terms is

$$S = 1 + x + x^2 + \cdots + x^{N-1} = \frac{1-x^N}{1-x}$$

Note also that $e^{ix} - e^{-ix} = 2i \sin x$.

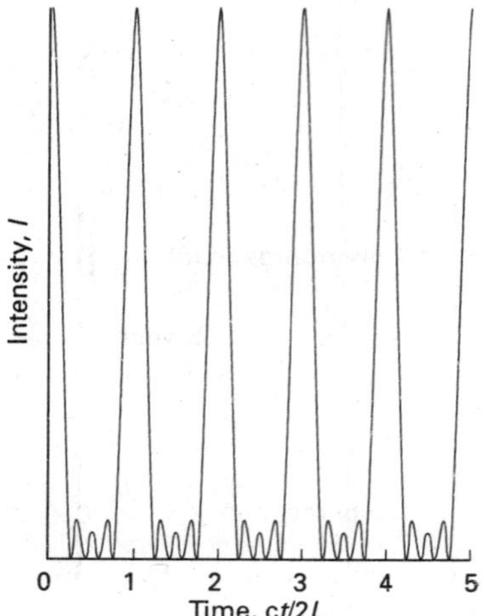

Fig. 13.33 The function derived in *Justification 13.5* showing in more detail the structure of the pulses generated by a mode-locked laser.

The sum is a geometrical progression of N terms:

$$\sum_{n=0}^{N-1} e^{i\pi nct/L} = 1 + e^{i\pi ct/L} + e^{2i\pi ct/L} + \cdots e^{(N-1)i\pi ct/L}$$

$$= \frac{\sin(N\pi ct/2L)}{\sin(\pi ct/2L)} \times e^{(N-1)i\pi ct/2L}$$

The intensity, I, of the radiation is proportional to the square modulus of the total amplitude, so

$$I \propto \mathcal{E}^*\mathcal{E} = \mathcal{E}_0^2 \frac{\sin^2(N\pi ct/2L)}{\sin^2(\pi ct/2L)}$$

This function is shown in Fig. 13.33. We see that it is a series of peaks with maxima separated by $t = 2L/c$, the round-trip transit time of the light in the cavity, and that the peaks become sharper as N is increased.

Mode locking is achieved by varying the Q-factor of the cavity periodically at the frequency $c/2L$. The modulation can be pictured as the opening of a shutter in synchrony with the round-trip travel time of the photons in the cavity, so only photons making the journey in that time are amplified. The modulation can be achieved by linking a prism in the cavity to a transducer driven by a radiofrequency source at a frequency $c/2L$. The transducer sets up standing-wave vibrations in the prism and modulates the loss it introduces into the cavity. We also see in Section 19.10c that the unique optical properties of some materials can be exploited to bring about mode-locking.

Checklist of key equations

Property	Equation	Comment
Beer–Lambert law	$I = I_0 10^{-\varepsilon[J]L}$	Uniform sample
Transmittance	$T = I/I_0$	Definition
Absorbance	$A = \log(I_0/I)$	Definition
Integrated absorption coefficient	$\mathcal{A} = \int_{\text{band}} \varepsilon(\tilde{\nu})\mathrm{d}\tilde{\nu}$	ε is the molar absorption coefficient
Electronic selection rules	$\Delta\Lambda = 0, \pm 1 \quad S = 0 \quad \Delta\Sigma = 0 \quad \Delta\Omega = 0, \pm 1$	Linear molecules
Einstein transition rates	$w = B\rho \quad w' = A + B'\rho$	A: spontaneous; B and B': stimulated
Relation between coefficients	$A = (8\pi h\nu^3/c^3)B \quad B' = B$	

Further information

Further information 13.1 *Examples of practical lasers*

Figure 13.34 summarizes the requirements for an efficient laser. In practice, the requirements can be satisfied by using a variety of different systems, and this section reviews some that are commonly available. We also include some lasers that operate by using other than electronic transitions. Noticeably absent from this discussion are solid state lasers (including the ubiquitous diode lasers), which we discuss in Chapter 19.

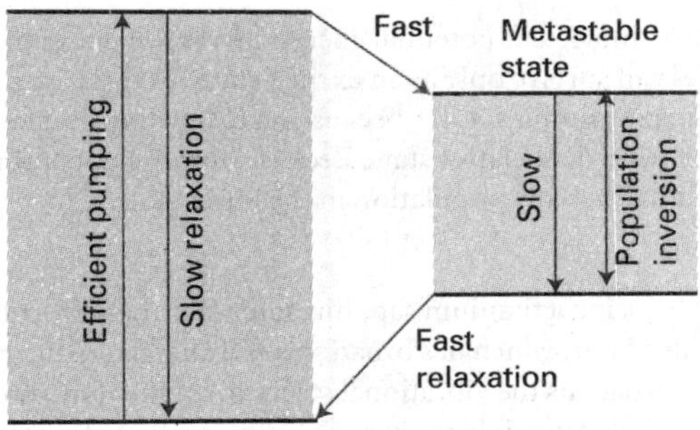

Fig. 13.34 A summary of the features needed for efficient laser action.

(a) Gas lasers

Because gas lasers can be cooled by a rapid flow of the gas through the cavity, they can be used to generate high powers. The pumping is normally achieved using a gas that is different from the gas responsible for the laser emission itself.

In the **helium–neon laser** the active medium is a mixture of helium and neon in a mole ratio of about 5:1 (Fig. 13.35). The initial step is the excitation of an He atom to the metastable $1s^12s^1$ configuration by using an electric discharge (the collisions of electrons and ions cause transitions that are not restricted by electric-dipole selection rules). The excitation energy of this transition happens to match

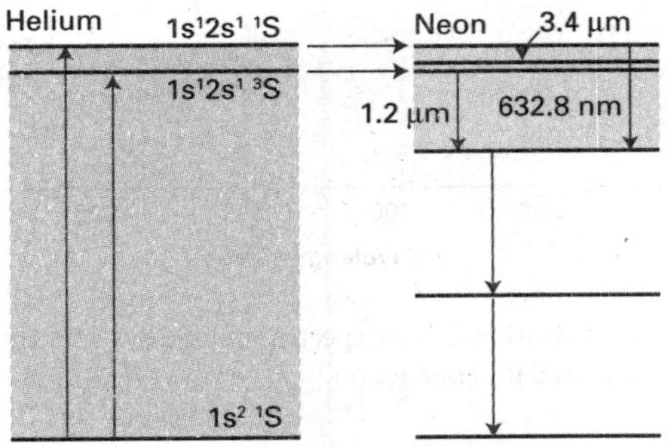

Fig. 13.35 The transitions involved in a helium–neon laser. The pumping (of the neon) depends on a coincidental matching of the helium and neon energy separations, so excited He atoms can transfer their excess energy to Ne atoms during a collision.

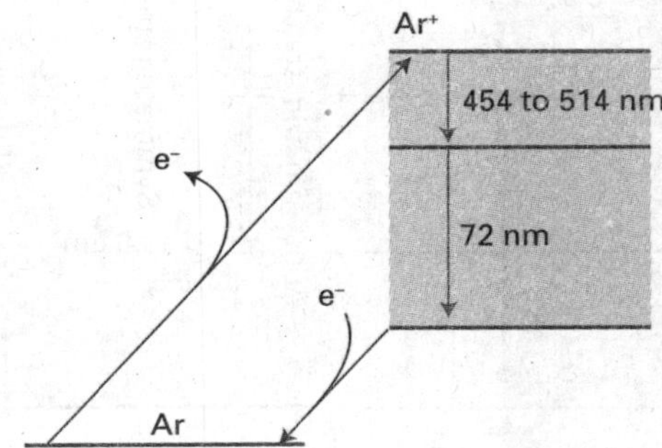

Fig. 13.36 The transitions involved in an argon-ion laser.

an excitation energy of neon, and during an He–Ne collision efficient transfer of energy may occur, leading to the production of highly excited, metastable Ne atoms with unpopulated intermediate states. Laser action generating 633 nm radiation (among about 100 other lines) then occurs.

The **argon-ion laser** (Fig. 13.36), one of a number of 'ion lasers', consists of argon at about 1 Torr, through which is passed an electric discharge. The discharge results in the formation of Ar^+ and Ar^{2+} ions in excited states, which undergo a laser transition to a lower state. These ions then revert to their ground states by emitting hard ultraviolet radiation (at 72 nm), and are then neutralized by a series of electrodes in the laser cavity. One of the design problems is to find materials that can withstand this damaging residual radiation. There are many lines in the laser transition because the excited ions may make transitions to many lower states, but two strong emissions from Ar^+ are at 488 nm (blue) and 514 nm (green); other transitions occur elsewhere in the visible region, in the infrared, and in the ultraviolet. The **krypton-ion laser** works similarly. It is less efficient, but gives a wider range of wavelengths, the most intense being at 647 nm (red), but it can also generate yellow, green, and violet lines.

The **carbon dioxide laser** works on a slightly different principle (Fig. 13.37), for its radiation (between 9.2 μm and 10.8 μm, with the strongest emission at 10.6 μm, in the infrared) arises from vibrational transitions. Most of the working gas is nitrogen, which becomes vibrationally excited by electronic and ionic collisions in an electric discharge. The vibrational levels happen to coincide with the ladder of antisymmetric stretch (ν_2, see Fig. 12.37) energy levels of CO_2, which pick up the energy during a collision. Laser action then occurs from the lowest excited level of ν_2 to the lowest excited level of the symmetric stretch (ν_1), which has remained unpopulated during the collisions. This transition is allowed by anharmonicities in the molecular potential energy. Some helium is included in the gas to help remove energy from this state and maintain the population inversion.

In a **nitrogen laser**, the efficiency of the stimulated transition (at 337 nm, in the ultraviolet, the transition $C^3\Pi_u \rightarrow B^3\Pi_g$) is so great that a single passage of a pulse of radiation is enough to generate laser radiation and mirrors are unnecessary: such lasers are said to be **superradiant**.

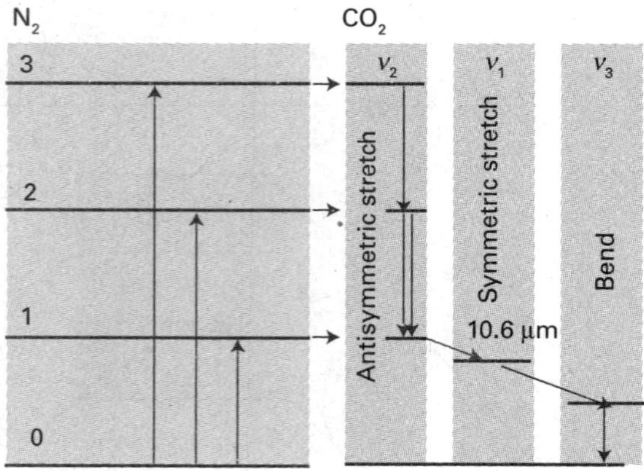

Fig. 13.37 The transitions involved in a carbon dioxide laser. The pumping also depends on the coincidental matching of energy separations; in this case the vibrationally excited N_2 molecules have excess energies that correspond to a vibrational excitation of the antisymmetric stretch of CO_2. The laser transition is from $\nu_2 = 1$ to $\nu_1 = 1$.

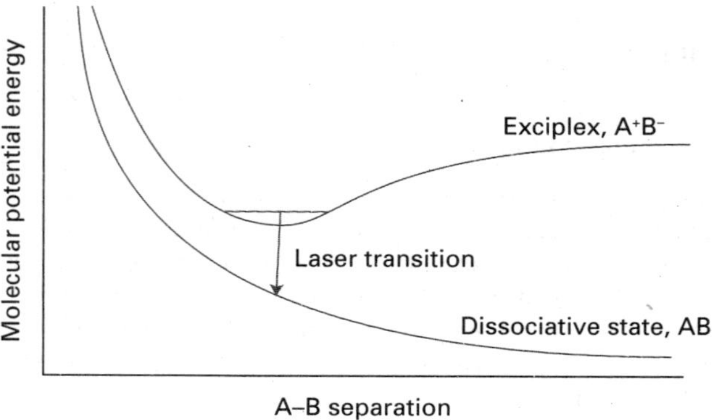

Fig. 13.38 The molecular potential energy curves for an exciplex. The species can survive only as an excited state (in this case a charge-transfer complex A^+B^-, because on discarding its energy it enters the lower, dissociative state. Because only the upper state can exist, there is never any population in the lower state.

(b) Exciplex lasers

The population inversion needed for laser action is achieved in an underhand way in **exciplex lasers**, for in these (as we shall see) the lower state does not effectively exist. This odd situation is achieved by forming an **exciplex**, a combination of two atoms that survives only in an excited state and which dissociates as soon as the excitation energy has been discarded. An exciplex can be formed in a mixture of xenon, chlorine, and neon (which acts as a buffer gas). An electric discharge through the mixture produces excited Cl atoms, which attach to the Xe atoms to give the exciplex XeCl*. The exciplex survives for about 10 ns, which is time for it to participate in laser action at 308 nm (in the ultraviolet). As soon as XeCl* has discarded a photon, the atoms separate because the molecular potential energy curve of the ground state is dissociative, and the ground state of the exciplex cannot become populated (Fig. 13.38). The KrF* exciplex laser is another example: it produces radiation at 249 nm.

A brief comment

The term 'excimer laser' is also widely encountered and used loosely when 'exciplex laser' is more appropriate. An exciplex has the form AB*, whereas an excimer, an excited dimer, is AA*.

(c) Dye lasers

Gas lasers and most solid state lasers operate at discrete frequencies and, although the frequency required may be selected by suitable optics, the laser cannot be tuned continuously. The tuning problem is overcome by using a titanium sapphire laser (Further information 19.1) or a **dye laser**, which has broad spectral characteristics because the solvent broadens the vibrational structure of the transitions into bands. Hence, it is possible to scan the wavelength continuously (by rotating the diffraction grating in the cavity) and achieve laser action at any chosen wavelength. A commonly used dye is Rhodamine 6G in methanol (Fig. 13.39). As the gain is very high, only a short length of the optical path need be through the dye. The excited states of the active medium, the dye, are sustained by another laser or a flash lamp, and the dye solution is flowed through the laser cavity to avoid thermal degradation.

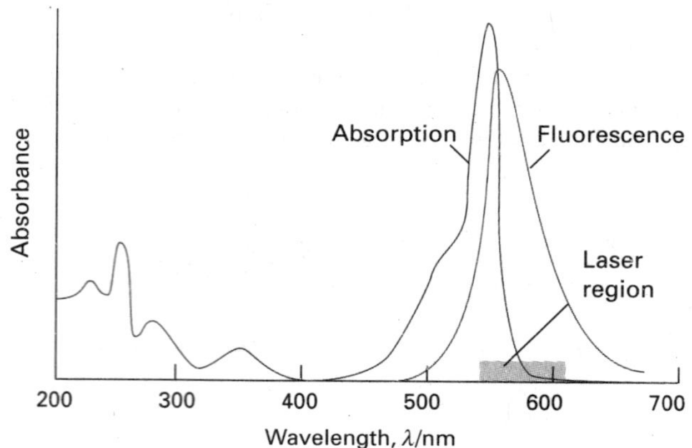

Fig. 13.39 The optical absorption spectrum of the dye Rhodamine 6G and the region used for laser action.

Discussion questions

13.1 Explain the origin of the term symbol $^3\Sigma_g^-$ for the ground state of dioxygen.

13.2 Explain the basis of the Franck–Condon principle and how it leads to the formation of a vibrational progression.

13.3 How do the band heads in P and R branches arise? Could the Q branch show a head?

13.4 Explain how colour can arise from molecules.

13.5 Suppose that you are a colour chemist and had been asked to intensify the colour of a dye without changing the type of compound, and that the dye in question was a polyene. Would you choose to lengthen or to shorten the chain? Would the modification to the length shift the apparent colour of the dye towards the red or the blue?

13.6 Describe the mechanism of fluorescence. In what respects is a fluorescence spectrum not the exact mirror image of the corresponding absorption spectrum?

13.7 The oxygen molecule absorbs ultraviolet radiation in a transition from its $^3\Sigma_g^-$ ground electronic state to an excited state that is energetically close to a dissociative $^5\Pi_u$ state. The absorption band has a relatively large experimental linewidth. Account for this observation.

13.8 Describe the principles of (a) continuous-wave and (b) pulsed laser action.

Exercises

13.1(a) The molar absorption coefficient of a substance dissolved in hexane is known to be 855 $dm^3\ mol^{-1}\ cm^{-1}$ at 270 nm. Calculate the percentage reduction in intensity when light of that wavelength passes through 2.5 mm of a solution of concentration 3.25 $mmol\ dm^{-3}$.

13.1(b) The molar absorption coefficient of a substance dissolved in hexane is known to be 327 $dm^3\ mol^{-1}\ cm^{-1}$ at 300 nm. Calculate the percentage reduction in intensity when light of that wavelength passes through 1.50 mm of a solution of concentration 2.22 $mmol\ dm^{-3}$.

13.2(a) A solution of an unknown component of a biological sample when placed in an absorption cell of path length 1.00 cm transmits 20.1 per cent of light of 340 nm incident upon it. If the concentration of the component is 0.111 $mmol\ dm^{-3}$, what is the molar absorption coefficient?

13.2(b) When light of wavelength 400 nm passes through 3.5 mm of a solution of an absorbing substance at a concentration 0.667 $mmol\ dm^{-3}$, the transmission is 65.5 per cent. Calculate the molar absorption coefficient of the solute at this wavelength and express the answer in $cm^2\ mol^{-1}$.

13.3(a) The molar absorption coefficient of a solute at 540 nm is 286 $dm^3\ mol^{-1}\ cm^{-1}$. When light of that wavelength passes through a 6.5 mm cell containing a solution of the solute, 46.5 per cent of the light was absorbed. What is the concentration of the solution?

13.3(b) The molar absorption coefficient of a solute at 440 nm is 323 $dm^3\ mol^{-1}\ cm^{-1}$. When light of that wavelength passes through a 7.50 mm cell containing a solution of the solute, 52.3 per cent of the light was absorbed. What is the concentration of the solution?

13.4(a) The absorption associated with a particular transition begins at 230 nm, peaks sharply at 260 nm, and ends at 290 nm. The maximum value of the molar absorption coefficient is 1.21×10^4 $dm^3\ mol^{-1}\ cm^{-1}$. Estimate the integrated absorption coefficient of the transition assuming a triangular lineshape.

13.4(b) The absorption associated with a certain transition begins at 199 nm, peaks sharply at 220 nm, and ends at 275 nm. The maximum value of the molar absorption coefficient is 2.25×10^4 $dm^3\ mol^{-1}\ cm^{-1}$. Estimate the integrated absorption coefficient of the transition assuming an inverted parabolic lineshape (Fig. 13.40).

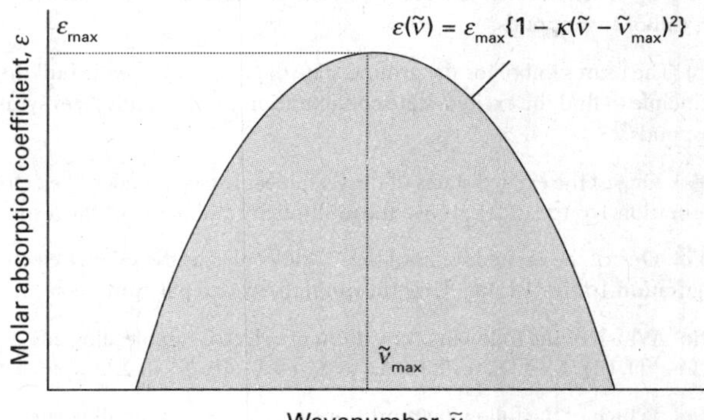

Fig. 13.40

13.5(a) The following data were obtained for the absorption by Br_2 in carbon tetrachloride using a 2.0 mm cell. Calculate the molar absorption coefficient of bromine at the wavelength employed:

$[Br_2]/(mol\ dm^{-3})$	0.0010	0.0050	0.0100	0.0500
T/(per cent)	81.4	35.6	12.7	3.0×10^{-3}

13.5(b) The following data were obtained for the absorption by a dye dissolved in methylbenzene using a 2.50 mm cell. Calculate the molar absorption coefficient of the dye at the wavelength employed:

$[dye]/(mol\ dm^{-3})$	0.0010	0.0050	0.0100	0.0500
T/(per cent)	73	21	4.2	1.33×10^{-5}

13.6(a) A 2.0-mm cell was filled with a solution of benzene in a non-absorbing solvent. The concentration of the benzene was 0.010 $mol\ dm^{-3}$ and the wavelength of the radiation was 256 nm (where there is a maximum in the absorption). Calculate the molar absorption coefficient of benzene at this wavelength given that the transmission was 48 per cent. What will the transmittance be in a 4.0-mm cell at the same wavelength?

13.6(b) A 2.50-mm cell was filled with a solution of a dye. The concentration of the dye was 15.5 $mmol\ dm^{-3}$. Calculate the molar absorption coefficient of benzene at this wavelength given that the transmission was 32 per cent. What will the transmittance be in a 4.50-mm cell at the same wavelength?

13.7(a) A swimmer enters a gloomier world (in one sense) on diving to greater depths. Given that the mean molar absorption coefficient of sea water in the visible region is 6.2×10^{-3} dm^3 mol^{-1} cm^{-1}, calculate the depth at which a diver will experience (a) half the surface intensity of light, (b) one-tenth the surface intensity.

13.7(b) Given that the maximum molar absorption coefficient of a molecule containing a carbonyl group is 30 dm^3 mol^{-1} cm^{-1} near 280 nm, calculate the thickness of a sample that will result in (a) half the initial intensity of radiation, (b) one-tenth the initial intensity.

13.8(a) The electronic absorption bands of many molecules in solution have half-widths at half-height of about 5000 cm^{-1}. Estimate the integrated absorption coefficients of bands for which (a) $\varepsilon_{max} \approx 1 \times 10^4$ dm^3 mol^{-1} cm^{-1}, (b) $\varepsilon_{max} \approx 5 \times 10^2$ dm^3 mol^{-1} cm^{-1}.

13.8(b) The electronic absorption band of a compound in solution had a Gaussian lineshape and a half-width at half-height of 4233 cm^{-1} and $\varepsilon_{max} = 1.54 \times 10^4$ dm^3 mol^{-1} cm^{-1}. Estimate the integrated absorption coefficient.

13.9(a) The term symbol for one of the excited states of H_2 is $^3\Pi_u$. Use the building-up principle to find the excited-state configuration to which this term symbol corresponds.

13.9(b) The term symbol for the ground state of N_2^+ is $^2\Pi_g$. Use the building-up principle to find the excited-state configuration to which this term symbol corresponds.

13.10(a) One of the excited states of the C_2 molecule has the valence electron configuration $1\sigma_g^2 1\sigma_u^2 1\pi_u^3 1\pi_g^1$. Give the multiplicity and parity of the term.

13.10(b) One of the excited states of the C_2 molecule has the valence electron configuration $1\sigma_g^2 1\sigma_u^2 1\pi_u^2 1\pi_g^2$. Give the multiplicity and parity of the term.

13.11(a) Which of the following transitions are electric-dipole allowed? (a) $^2\Pi \leftrightarrow {}^2\Pi$, (b) $^1\Sigma \leftrightarrow {}^1\Sigma$, (c) $\Sigma \leftrightarrow \Delta$, (d) $\Sigma^+ \leftrightarrow \Sigma^-$, (e) $\Sigma^+ \leftrightarrow \Sigma^+$.

13.11(b) Which of the following transitions are electric-dipole allowed? (a) $^1\Sigma_g^+ \leftrightarrow {}^1\Sigma_u^+$, (b) $^3\Sigma_g^+ \leftrightarrow {}^3\Sigma_u^+$, (c) $t_{2g} \leftrightarrow e_g$, (d) $\pi^* \leftrightarrow n$.

13.12(a) The ground-state wavefunction of a certain molecule is described by the vibrational wavefunction $\psi_0 = N_0 e^{-ax^2}$. Calculate the Franck–Condon factor for a transition to a vibrational state described by the wavefunction $\psi_0' = N_0' e^{-b(x-x_0)^2}$, with $b = a/2$.

13.12(b) The ground-state wavefunction of a certain molecule is described by the vibrational wavefunction $\psi_0 = N_0 e^{-ax^2}$. Calculate the Franck–Condon factor for a transition to a vibrational state described by the wavefunction $\psi_1' = N_1' x e^{-b(x-x_0)^2}$, with $b = a/2$.

13.13(a) The following parameters describe the electronic ground state and an excited electronic state of SnO: $\tilde{B} = 0.3540$ cm^{-1}, $\tilde{B}' = 0.3101$ cm^{-1}. Which branch of the transition between them shows a head? At what value of J will it occur?

13.13(b) The following parameters describe the electronic ground state and an excited electronic state of BeH: $\tilde{B} = 10.308$ cm^{-1}, $\tilde{B}' = 10.470$ cm^{-1}. Which branch of the transition between them shows a head? At what value of J will it occur?

13.14(a) The R-branch of the $^1\Pi_u \leftarrow {}^1\Sigma_g^+$ transition of H_2 shows a band head at the very low value of $J = 1$. The rotational constant of the ground state is 60.80 cm^{-1}. What is the rotational constant of the upper state? Has the bond length increased or decreased in the transition?

13.14(b) The P-branch of the $^2\Pi \leftarrow {}^2\Sigma^+$ transition of CdH shows a band head at $J = 25$. The rotational constant of the ground state is 5.437 cm^{-1}. What is the rotational constant of the upper state? Has the bond length increased or decreased in the transition?

13.15(a) The two compounds 2,3-dimethyl-2-butene (**7**) and 2,5-dimethyl-2,4-hexadiene (**8**) are to be distinguished by their ultraviolet absorption spectra. The maximum absorption in one compound occurs at 192 nm and in the other at 243 nm. Match the maxima to the compounds and justify the assignment.

7 2,3-Dimethyl-2-butene

8 2,5-Dimethyl-2,4-hexadiene

13.15(b) 1,3,5-hexatriene (a kind of 'linear' benzene) was converted into benzene itself. On the basis of a free-electron molecular orbital model (in which hexatriene is treated as a linear box and benzene as a ring), would you expect the lowest energy absorption to rise or fall in energy?

13.16(a) The compound $CH_3CH{=}CHCHO$ has a strong absorption in the ultraviolet at 46 950 cm^{-1} and a weak absorption at 30 000 cm^{-1}. Justify these features and assign the ultraviolet absorption transitions.

13.16(b) 3-Buten-2-one (**9**) has a strong absorption at 213 nm and a weaker absorption at 320 nm. Justify these features and assign the ultraviolet absorption transitions.

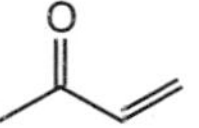

9 3-Butene-2-one

13.17(a) The complex ion $[Fe(OH_2)_6]^{3+}$ has an electronic absorption spectrum with a maximum at 700 nm. Estimate a value of Δ_O for the complex.

13.17(b) The complex ion $[Fe(CN)_6]^{3-}$ has an electronic absorption spectrum with a maximum at 305 nm. Estimate a value of Δ_O for the complex.

13.18(a) The line marked A in Fig. 13.41 is the fluorescence spectrum of benzophenone in solid solution in ethanol at low temperatures observed when the sample is illuminated with 360 nm light. What can be said about the vibrational energy levels of the carbonyl group in (a) its ground electronic state and (b) its excited electronic state?

13.18(b) When naphthalene is illuminated with 360 nm light it does not absorb, but the line marked B in Fig 13.41 is the phosphorescence spectrum of a solid solution of a mixture of naphthalene and benzophenone in ethanol.

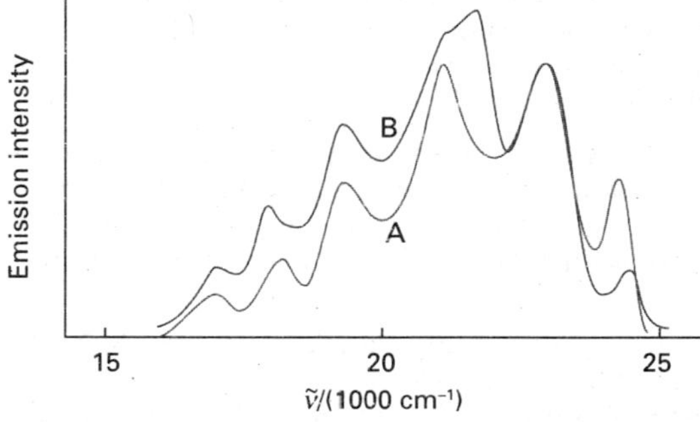

Fig. 13.41

Now a component of fluorescence from naphthalene can be detected. Account for this observation.

13.19(a) Consider a laser cavity of length 30 cm. What are the allowed wavelengths and frequencies of the resonant modes?

13.19(b) Consider a laser cavity of length 1.0 m. What are the allowed wavelengths and frequencies of the resonant modes?

13.20(a) A pulsed laser rated at 0.10 mJ can generate radiation with peak power output of 5.0 MW and average power output of 7.0 kW. What are the pulse duration and repetition rate?

13.20(b) A pulsed laser rated at 20.0 μJ can generate radiation with peak power output of 100 kW and average power output of 0.40 mW. What are the pulse duration and repetition rate?

13.21(a) Use mathematical software or an electronic spreadsheet to simulate the output of a mode-locked laser (that is, plots such as that shown in Fig. 13.33) for $L = 30$ cm and $N = 100$ and 1000.

13.21(b) Use mathematical software or an electronic spreadsheet to simulate the output of a mode-locked laser (that is, plots such as that shown in Fig. 13.33) for $L = 1.0$ cm and $N = 50$ and 500.

Problems*

Numerical problems

13.1 The vibrational wavenumber of the oxygen molecule in its electronic ground state is 1580 cm^{-1}, whereas that in the first excited state (B $^3\Sigma_u^-$), to which there is an allowed electronic transition, is 700 cm^{-1}. Given that the separation in energy between the minima in their respective potential energy curves of these two electronic states is 6.175 eV, what is the wavenumber of the lowest energy transition in the band of transitions originating from the $v = 0$ vibrational state of the electronic ground state to this excited state? Ignore any rotational structure or anharmonicity.

13.2 We are now ready to understand more deeply the features of photoelectron spectra (Section 10.4e). The highest kinetic energy electrons in the photoelectron spectrum of H_2O using 21.22 eV radiation are at about 12–13 eV and show a large vibrational spacing of 0.41 eV. The symmetric stretching mode of the neutral H_2O molecule lies at 3652 cm^{-1}. (a) What conclusions can be drawn from the nature of the orbital from which the electron is ejected? (b) In the same spectrum of H_2O, the band near 7.0 eV shows a long vibrational series with spacing 0.125 eV. The bending mode of H_2O lies at 1596 cm^{-1}. What conclusions can you draw about the characteristics of the orbital occupied by the photoelectron?

13.3 The electronic spectrum of the IBr molecule shows two low-lying, well-defined convergence limits at 14 660 and 18 345 cm^{-1}. Energy levels for the iodine and bromine atoms occur at 0, 7598; and 0, 3685 cm^{-1}, respectively. Other atomic levels are at much higher energies. What possibilities exist for the numerical value of the dissociation energy of IBr? Decide which is the correct possibility by calculating this quantity from $\Delta_f H^{\ominus}$(IBr,g) = +40.79 kJ mol^{-1} and the dissociation energies of I_2(g) and Br_2(g) which are 146 and 190 kJ mol^{-1}, respectively.

13.4 In many cases it is possible to assume that an absorption band has a Gaussian lineshape (one proportional to e^{-x^2}) centred on the band maximum. Assume such a lineshape, and show that $\mathcal{A} \approx 1.0645\varepsilon_{max}\Delta\tilde{\nu}_{1/2}$, where $\Delta\tilde{\nu}_{1/2}$ is the width at half-height. The absorption spectrum of azoethane ($CH_3CH_2N_2$) between 24 000 cm^{-1} and 34 000 cm^{-1} is shown in Fig. 13.42. First, estimate $\mathcal{A}$ for the band by assuming that it is Gaussian. Then integrate the absorption band graphically. The latter can be done either by ruling and counting squares, or by tracing the line shape on to paper and weighing. A more sophisticated procedure would be to use mathematical software to fit a polynomial to the absorption band (or a Gaussian), and then to integrate the result analytically.

13.5 A lot of information about the energy levels and wavefunctions of small inorganic molecules can be obtained from their ultraviolet spectra. An example of a spectrum with considerable vibrational structure, that of gaseous SO_2 at 25°C, is shown in Fig. 13.6. Estimate the integrated absorption coefficient for the transition. What electronic states are accessible from the A_1 ground state of this C_{2v} molecule by electric dipole transitions?

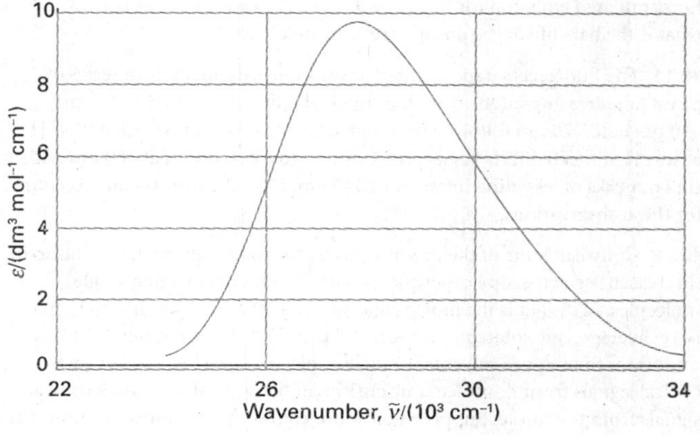

Fig. 13.42

13.6‡ J.G. Dojahn *et al.* (*J. Phys. Chem.* **100**, 9649 (1996)) characterized the potential energy curves of the ground and electronic states of homonuclear diatomic halogen anions. These anions have a $^2\Sigma_u^+$ ground state and $^2\Pi_g$, $^2\Pi_u$, and $^2\Sigma_g^+$ excited states. To which of the excited states are transitions by absorption of photons allowed? Explain.

13.7 A transition of particular importance in O_2 gives rise to the 'Schumann–Runge band' in the ultraviolet region. The wavenumbers (in cm^{-1}) of transitions from the ground state to the vibrational levels of the first excited state ($^3\Sigma_u^-$) are 50 062.6, 50 725.4, 51 369.0, 51 988.6, 52 579.0, 53 143.4, 53 679.6, 54 177.0, 54 641.8, 55 078.2, 55 460.0, 55 803.1, 56 107.3, 56 360.3, 56 570.6. What is the dissociation energy of the upper electronic state? (Use a Birge–Sponer plot.) The same excited state is known to dissociate into one ground state O atom and one excited state atom with an energy 190 kJ mol^{-1} above the ground state. (This excited atom is responsible for a great deal of photochemical mischief in the atmosphere.) Ground state O_2 dissociates into two ground-state atoms. Use this information to calculate the dissociation energy of ground-state O_2 from the Schumann–Runge data.

13.8 Suppose that we can model a charge-transfer transition in a one-dimensional system as a process in which a rectangular wavefunction that is nonzero in the range $0 \le x \le a$ makes a transition to another rectangular

* Problems denoted with the symbol ‡ were supplied by Charles Trapp and Carmen Giunta.

wavefunction that is nonzero in the range $\frac{1}{2}a \leq x \leq b$. Evaluate the transition moment $\int \psi_{\rm f}^{*} x \psi_{\rm i}\,{\rm d}x$.

13.9 Aromatic hydrocarbons and I_2 form complexes from which charge-transfer electronic transitions are observed. The hydrocarbon acts an electron donor and I_2 as an electron acceptor. The energies $h\nu_{max}$ of the charge transfer transitions for a number of hydrocarbon–I_2 complexes are given below:

Hydrocarbon	benzene	biphenyl	naphthalene	phenanthrene	pyrene	anthracene
$h\nu_{max}$/eV	4.184	3.654	3.452	3.288	2.989	2.890

Investigate the hypothesis that there is a correlation between the energy of the HOMO of the hydrocarbon (from which the electron comes in the charge-transfer transition) and $h\nu_{max}$. Use one of the molecular electronic structure methods discussed in Chapter 10 to determine the energy of the HOMO of each hydrocarbon in the data set.

13.10 A certain molecule fluoresces at a wavelength of 400 nm with a half-life of 1.0 ns. It phosphoresces at 500 nm. If the ratio of the transition probabilities for stimulated emission for the $S^* \rightarrow S$ to the $T \rightarrow S$ transitions is 1.0×10^5, what is the half-life of the phosphorescent state?

13.11 The fluorescence spectrum of anthracene vapour shows a series of peaks of increasing intensity with individual maxima at 440 nm, 410 nm, 390 nm, and 370 nm followed by a sharp cut-off at shorter wavelengths. The absorption spectrum rises sharply from zero to a maximum at 360 nm with a trail of peaks of lessening intensity at 345 nm, 330 nm, and 305 nm. Account for these observations.

13.12 Consider some of the precautions that must be taken when conducting fluorescence microscopy experiments with the aim of detecting single molecules. (a) What is the molar concentration of a solution in which there is, on average, one solute molecule in 1.0 μm^3 (1.0 fL) of solution? (b) It is important to use pure solvents in single-molecule spectroscopy because optical signals from fluorescent impurities in the solvent may mask optical signals from the solute. Suppose that water containing a fluorescent impurity of molar mass 100 g mol^{-1} is used as solvent and that analysis indicates the presence of 0.10 mg of impurity per 1.0 kg of solvent. On average, how many impurity molecules will be present in 1.0 μm^3 of solution? You may take the density of water as 1.0 g cm^{-3}. Comment on the suitability of this solvent for single-molecule spectroscopy experiments.

13.13 Light-induced degradation of molecules, also called *photobleaching*, is a serious problem in fluorescence microscopy. A molecule of a fluorescent dye commonly used to label biopolymers can withstand about 10^6 excitations by photons before light-induced reactions destroy its π system and the molecule no longer fluoresces. For how long will a single dye molecule fluoresce while being excited by 1.0 mW of 488 nm radiation from a continuous-wave argon ion laser? You may assume that the dye has an absorption spectrum that peaks at 488 nm and that every photon delivered by the laser is absorbed by the molecule.

Theoretical problems

13.14 It is common to make measurements of absorbance at two wavelengths and use them to find the individual concentrations of two components A and B in a mixture. Show that the molar concentrations of A and B are

$$[A] = \frac{\varepsilon_{B2}A_1 - \varepsilon_{B1}A_2}{(\varepsilon_{A1}\varepsilon_{B2} - \varepsilon_{A2}\varepsilon_{B1})l} \qquad [B] = \frac{\varepsilon_{A1}A_2 - \varepsilon_{A2}A_1}{(\varepsilon_{A1}\varepsilon_{B2} - \varepsilon_{A2}\varepsilon_{B1})l}$$

where A_1 and A_2 are absorbances of the mixture at wavelengths λ_1 and λ_2, and the molar extinction coefficients of A (and B) at these wavelengths are ε_{A1} and ε_{A2} (and ε_{B1} and ε_{B2}).

13.15 When pyridine is added to a solution of iodine in carbon tetrachloride the 520 nm band of absorption shifts toward 450 nm. However, the absorbance of the solution at 490 nm remains constant: this feature is called an *isosbestic point*. Show that an isosbestic point should occur when two absorbing species are in equilibrium.

13.16 Spin angular momentum is conserved when a molecule dissociates into atoms. What atom multiplicities are permitted when (a) an O_2 molecule, (b) an N_2 molecule dissociates into atoms?

13.17 Assume that the electronic states of the π electrons of a conjugated molecule can be approximated by the wavefunctions of a particle in a one-dimensional box, and that the dipole moment can be related to the displacement along this length by $\mu = -ex$. Show that the transition probability for the transition $n = 1 \rightarrow n = 2$ is nonzero, whereas that for $n = 1 \rightarrow n = 3$ is zero. *Hint.* The following relations will be useful:

$$\sin x \sin y = \tfrac{1}{2}\cos(x-y) - \tfrac{1}{2}\cos(x+y)$$

$$\int x \cos ax\,{\rm d}x = \frac{1}{a^2}\cos ax + \frac{x}{a}\sin ax$$

13.18 Use a group theoretical argument to decide which of the following transitions are electric-dipole allowed: (a) the $\pi^* \leftarrow \pi$ transition in ethene, (b) the $\pi^* \leftarrow$ n transition in a carbonyl group in a C_{2v} environment.

13.19 Estimate the transition dipole moment of a charge-transfer transition modelled as the migration of an electron from a H1s orbital on one atom to another H1s orbital on an atom a distance R away. Approximate the transition moment by $-eRS$ where S is the overlap integral of the two orbitals. Sketch the oscillator strength as a function of R using the curve for S given in Fig. 10.29. Why does the intensity fall to zero as R approaches zero and infinity?

13.20 The Beer–Lambert law states that the absorbance of a sample at a wavenumber $\tilde{\nu}$ is proportional to the molar concentration [J] of the absorbing species J and to the length L of the sample (eqn 13.4). In this problem you will show that the intensity of fluorescence emission from a sample of J is also proportional to [J] and L. Consider a sample of J that is illuminated with a beam of intensity $I_0(\tilde{\nu})$ at the wavenumber $\tilde{\nu}$. Before fluorescence can occur, a fraction of $I_0(\tilde{\nu})$ must be absorbed and an intensity $I(\tilde{\nu})$ will be transmitted. However, not all of the absorbed intensity is emitted and the intensity of fluorescence depends on the fluorescence quantum yield, ϕ_f, the efficiency of photon emission. The fluorescence quantum yield ranges from 0 to 1 and is proportional to the ratio of the integral of the fluorescence spectrum over the integrated absorption coefficient. Because of a Stokes shift of magnitude $\Delta\tilde{\nu}_{\rm Stokes}$, fluorescence occurs at a wavenumber $\tilde{\nu}_f$, with $\tilde{\nu}_f + \Delta\tilde{\nu}_{\rm Stokes} = \tilde{\nu}$. It follows that the fluorescence intensity at $\tilde{\nu}_f$, $I_f(\tilde{\nu}_f)$, is proportional to ϕ_f and to the intensity of exciting radiation that is absorbed by J, $I_{abs}(\tilde{\nu}) = I_0(\tilde{\nu}) - I(\tilde{\nu})$. (a) Use the Beer–Lambert law to express $I_{abs}(\tilde{\nu})$ in terms of $I_0(\tilde{\nu})$, [J], L, and $\varepsilon(\tilde{\nu})$, the molar absorption coefficient of J at $\tilde{\nu}$. (b) Use your result from part (a) to show that $I_f(\tilde{\nu}_f) \propto I_0(\tilde{\nu})\varepsilon(\tilde{\nu})\phi_f[{\rm J}]L$.

Applications: to biochemistry, environmental science, and astrophysics

13.21 The protein haemerythrin (Her) is responsible for binding and carrying O_2 in some invertebrates. Each protein molecule has two Fe^{2+} ions that are in very close proximity and work together to bind one molecule of O_2. The Fe_2O_2 group of oxygenated haemerythrin is coloured and has an electronic absorption band at 500 nm. Figure 13.43 shows the UV-visible absorption spectrum of a derivative of haemerythrin in the presence of different concentrations of CNS^- ions. What may be inferred from the spectrum?

13.22 The flux of visible photons reaching Earth from the North Star is about 4×10^3 mm^{-2} s^{-1}. Of these photons, 30 per cent are absorbed or scattered by the atmosphere and 25 per cent of the surviving photons are scattered by the surface of the cornea of the eye. A further 9 per cent are absorbed inside the cornea. The area of the pupil at night is about 40 mm^2 and the response time of the eye is about 0.1 s. Of the photons passing through the pupil, about 43 per cent are absorbed in the ocular medium. How many photons from the

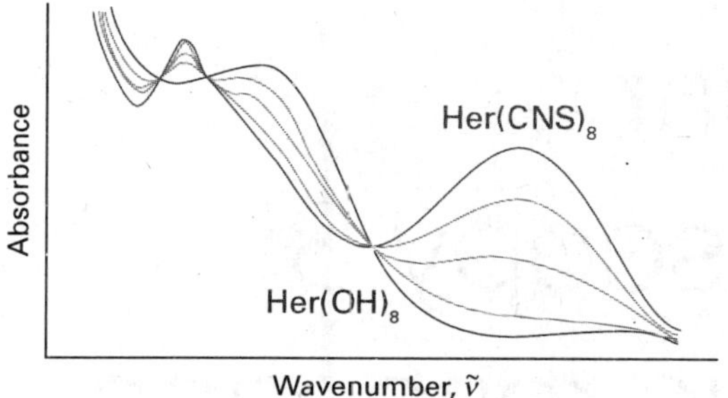

Fig. 13.43

North Star are focused on to the retina in 0.1 s? For a continuation of this story, see R.W. Rodieck, *The first steps in seeing*, Sinauer, Sunderland (1998).

13.23 Use molecule (**10**) as a model of the *trans* conformation of the chromophore found in rhodopsin. In this model, the methyl group bound to the nitrogen atom of the protonated Schiff's base replaces the protein. (a) Using molecular modelling software and the computational method of your instructor's choice, calculate the energy separation between the HOMO and LUMO of (**10**). (b) Repeat the calculation for the 11-*cis* form of (**10**). (c) Based on your results from parts (a) and (b), do you expect the experimental frequency for the $\pi^* \leftarrow \pi$ visible absorption of the *trans* form of (**10**) to be higher or lower than that for the 11-*cis* form of (**10**)?

C_{11} NH^+

10

13.24‡ Ozone absorbs ultraviolet radiation in a part of the electromagnetic spectrum energetic enough to disrupt DNA in biological organisms and that is absorbed by no other abundant atmospheric constituent. This spectral range, denoted UV-B, spans the wavelengths of about 290 nm to 320 nm. The molar extinction coefficient of ozone over this range is given in the table below (W.B. DeMore *et al.*, *Chemical kinetics and photochemical data for use in stratospheric modeling: Evaluation Number 11*, JPL Publication 94–26 (1994).)

λ/nm	292.0	296.3	300.8	305.4	310.1	315.0	320.0
$\varepsilon/(\text{dm}^3\,\text{mol}^{-1}\,\text{cm}^{-1})$	1512	865	477	257	135.9	69.5	34.5

Compute the integrated absorption coefficient of ozone over the wavelength range 290–320 nm. (*Hint*: $\varepsilon(\tilde{\nu})$ can be fitted to an exponential function quite well.)

13.25‡ The abundance of ozone is typically inferred from measurements of UV absorption and is often expressed in terms of *Dobson units* (DU): 1 DU is equivalent to a layer of pure ozone 10^{-3} cm thick at 1 atm and 0°C. Compute the absorbance of UV radiation at 300 nm expected for an ozone abundance of 300 DU (a typical value) and 100 DU (a value reached during seasonal Antarctic ozone depletions) given a molar absorption coefficient of 476 $\text{dm}^3\,\text{mol}^{-1}\,\text{cm}^{-1}$.

13.26‡ G.C.G. Wachewsky *et al.* (*J. Phys. Chem.* **100**, 11559 (1996)) examined the UV absorption spectrum of CH_3I, a species of interest in connection with stratospheric ozone chemistry. They found the integrated absorption coefficient to be dependent on temperature and pressure to an extent inconsistent with internal structural changes in isolated CH_3I molecules; they explained the changes as due to dimerization of a substantial fraction of the CH_3I, a process which would naturally be pressure- and temperature-dependent. (a) Compute the integrated absorption coefficient over a triangular lineshape in the range 31 250 to 34 483 cm^{-1} and a maximal molar absorption coefficient of 150 $\text{dm}^3\,\text{mol}^{-1}\,\text{cm}^{-1}$ at 31 250 cm^{-1}. (b) Suppose 1 per cent of the CH_3I units in a sample at 2.4 Torr and 373 K exists as dimers. Compute the absorbance expected at 31 250 cm^{-1} in a sample cell of length 12.0 cm. (c) Suppose 18 per cent of the CH_3I units in a sample at 100 Torr and 373 K exists as dimers. Compute the absorbance expected at 31 250 cm^{-1} in a sample cell of length 12.0 cm; compute the molar absorption coefficient that would be inferred from this absorbance if dimerization was not considered.

13.27‡ The molecule Cl_2O_2 is believed to participate in the seasonal depletion of ozone over Antarctica. M. Schwell *et al.* (*J. Phys. Chem.* **100**, 10070 (1996)) measured the ionization energies of Cl_2O_2 by photoelectron spectroscopy in which the ionized fragments were detected using a mass spectrometer. From their data, we can infer that the ionization enthalpy of Cl_2O_2 is 11.05 eV and the enthalpy of the dissociative ionization $Cl_2O_2 \rightarrow Cl + OClO^+ + e^-$ is 10.95 eV. They used this information to make some inferences about the structure of Cl_2O_2. Computational studies had suggested that the lowest energy isomer is ClOOCl, but that $ClClO_2$ (C_{2v}) and ClOClO are not very much higher in energy. The Cl_2O_2 in the photoionization step is the lowest energy isomer, whatever its structure may be, and its enthalpy of formation had previously been reported as +133 kJ mol^{-1}. The Cl_2O_2 in the dissociative ionization step is unlikely to be ClOOCl, for the product can be derived from it only with substantial rearrangement. Given $\Delta_f H^{\ominus}(OClO^+) = +1096$ kJ mol^{-1} and $\Delta_f H^{\ominus}(e^-) = 0$, determine whether the Cl_2O_2 in the dissociative ionization is the same as that in the photoionization. If different, how much greater is its $\Delta_f H^{\ominus}$? Are these results consistent with or contradictory to the computational studies?

13.28‡ One of the principal methods for obtaining the electronic spectra of unstable radicals is to study the spectra of comets, which are almost entirely due to radicals. Many radical spectra have been found in comets, including that due to CN. These radicals are produced in comets by the absorption of far ultraviolet solar radiation by their parent compounds. Subsequently, their fluorescence is excited by sunlight of longer wavelength. The spectra of comet Hale–Bopp (C/1995 O1) have been the subject of many recent studies. One such study is that of the fluorescence spectrum of CN in the comet at large heliocentric distances by R.M. Wagner and D.G. Schleicher (*Science* 275, 1918 (1997)), in which the authors determine the spatial distribution and rate of production of CN in the coma. The (0–0) vibrational band is centred on 387.6 nm and the weaker (1–1) band with relative intensity 0.1 is centred on 386.4 nm. The band heads for (0–0) and (0–1) are known to be 388.3 and 421.6 nm, respectively. From these data, calculate the energy of the excited S_1 state relative to the ground S_0 state, the vibrational wavenumbers and the difference in the vibrational wavenumbers of the two states, and the relative populations of the $v = 0$ and $v = 1$ vibrational levels of the S_1 state. Also estimate the effective temperature of the molecule in the excited S_1 state. Only eight rotational levels of the S_1 state are thought to be populated. Is that observation consistent with the effective temperature of the S_1 state?

14 Molecular spectroscopy 3: magnetic resonance

One of the most widely used spectroscopic procedures in chemistry makes use of the classical concept of resonance. The chapter begins with an account of conventional nuclear magnetic resonance, which shows how the resonance frequency of a magnetic nucleus is affected by its electronic environment and the presence of magnetic nuclei in its vicinity. Then we turn to the modern versions of NMR, which are based on the use of pulses of electromagnetic radiation and the processing of the resulting signal by Fourier transform techniques. The experimental techniques for electron paramagnetic resonance resemble those used in the early days of NMR. The information obtained is used to investigate species with unpaired electrons.

When two pendulums share a slightly flexible support and one is set in motion, the other is forced into oscillation by the motion of the common axle. As a result, energy flows between the two pendulums. The energy transfer occurs most efficiently when the frequencies of the two pendulums are identical. The condition of strong effective coupling when the frequencies of two oscillators are identical is called **resonance**. Resonance is the basis of a number of everyday phenomena, including the response of radios to the weak oscillations of the electromagnetic field generated by a distant transmitter. Historically, spectroscopic techniques that measure transitions between nuclear and electron spin states have carried the term 'resonance' in their names because they have depended on matching a set of energy levels to a source of monochromatic radiation and observing the strong absorption that occurs at resonance.

The effect of magnetic fields on electrons and nuclei

The Stern–Gerlach experiment (Section 8.8) provided evidence for electron spin. It turns out that many nuclei also possess spin angular momentum. Orbital and spin angular momenta give rise to magnetic moments, and to say that electrons and nuclei have magnetic moments means that, to some extent, they behave like small bar magnets with energies that depend on their orientation in an applied magnetic field. First, we establish how the energies of electrons and nuclei depend on the applied field. Then we see how to use this dependence to study the structure and dynamics of complex molecules.

14.1 The energies of electrons in magnetic fields

Key points **Electrons interact with magnetic fields, which remove the degeneracy of the quantized m_s states. The different energies can be represented on the vector model as vectors precessing at the Larmor frequency, ν_L.**

Classically, the energy of a magnetic moment $\boldsymbol{\mu}$ in a magnetic field $\mathcal{B}$ is equal to the scalar product

$$E = -\boldsymbol{\mu}\cdot\mathcal{B} \tag{14.1}$$

More formally, $\mathcal{B}$ is the magnetic induction and is measured in tesla, T; 1 T = 1 kg s^{-2}A^{-1}. The (non-SI) unit gauss, G, is also occasionally used: 1 T = 10^4 G.

Quantum mechanically, we write the hamiltonian as

$$\hat{H} = -\hat{\boldsymbol{\mu}}\cdot\mathcal{B} \tag{14.2}$$

To write an expression for $\hat{\boldsymbol{\mu}}$, we recall from *Further information 9.2* (on spin–orbit coupling in atoms) that the magnetic moment of an electron is proportional to its angular momentum. For an electron possessing orbital angular momentum we write

$$\hat{\boldsymbol{\mu}} = \gamma_e\hat{\boldsymbol{l}} \quad \text{and} \quad \hat{H} = -\gamma_e\mathcal{B}\cdot\hat{\boldsymbol{l}} \tag{14.3}$$

where $\hat{\boldsymbol{l}}$ is the orbital angular momentum operator and from classical electrodynamics

$$\gamma_e = -\frac{e}{2m_e} \qquad \text{Definition of magnetogyric ratio of an electron} \qquad [14.4]$$

γ_e is called the **magnetogyric ratio** of the electron. Its negative sign (arising from the sign of the electron's charge) shows that the orbital moment is opposite in direction to the orbital angular momentum vector (as is depicted in Fig. 9.27).

For a magnetic field of magnitude $\mathcal{B}_0$ along the z-direction, the hamiltonian in eqn 14.3 becomes

$$\hat{H} = -\gamma_e\mathcal{B}_0\hat{l}_z \tag{14.5a}$$

Because the eigenvalues of the operator $\hat{l}_z$ are $m_l\hbar$ the eigenvalues of this hamiltonian are

$$E_{m_l} = -\gamma_e m_l\hbar\mathcal{B}_0 = \mu_B m_l\mathcal{B}_0 \tag{14.5b}$$

The combination $-\gamma_e\hbar$ occurs widely and, as in this equation, is expressed as the **Bohr magneton, μ_B**:

$$\mu_B = -\gamma_e\hbar = \frac{e\hbar}{2m_e} = 9.274\times10^{-24}\ \text{J T}^{-1} \qquad \text{Definition of the Bohr magneton} \qquad [14.6]$$

The Bohr magneton, a positive quantity, is often regarded as the fundamental quantum of magnetic moment.

The spin magnetic moment of an electron, which has a spin quantum number $s=\frac{1}{2}$ (Section 8.8), is also proportional to its spin angular momentum. However, instead of eqn 14.3, the spin magnetic moment and hamiltonian operators are, respectively,

$$\hat{\boldsymbol{\mu}} = g_e\gamma_e\hat{\boldsymbol{s}} \quad \text{and} \quad \hat{H} = -g_e\gamma_e\mathcal{B}\cdot\hat{\boldsymbol{s}} \tag{14.7}$$

where $\hat{\boldsymbol{s}}$ is the spin angular momentum operator and the extra factor g_e is called the **g-value** of the electron: $g_e = 2.002\,319\ldots$. Dirac's relativistic theory (his modification of the Schrödinger equation to make it consistent with Einstein's special relativity) gives $g_e = 2$; the additional $0.002\,319\ldots$ arises from interactions of the electron with the electromagnetic fluctuations of the vacuum that surrounds the electron. For a magnetic field of magnitude $\mathcal{B}_0$ in the z-direction

A brief comment

Scalar products (or 'dot products') are explained in *Mathematical background 5* following Chapter 9.

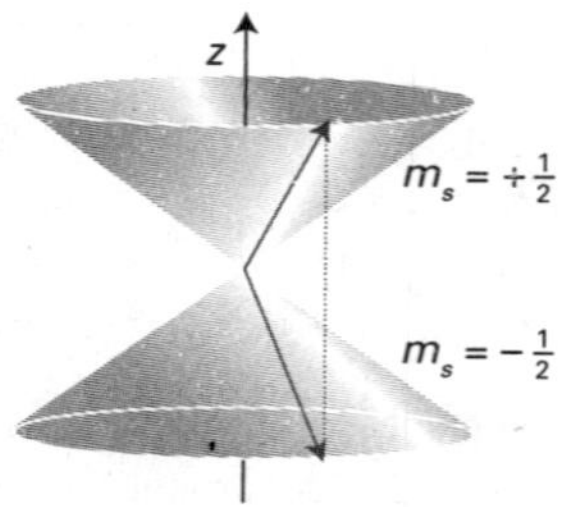

Fig. 14.1 The interactions between the m_s states of an electron and an external magnetic field may be visualized as the precession of the vectors representing the angular momentum.

$$\hat{H} = -g_e\gamma_e\mathcal{B}_0\hat{s}_z \tag{14.8a}$$

Because the eigenvalues of the operator $\hat{s}_z$ are $m_s\hbar$ with $m_s = +\frac{1}{2}$ (α) and $m_s = -\frac{1}{2}$ (β), it follows that the energies of an electron spin in a magnetic field are

$$E_{m_s} = -g_e\gamma_e m_s\hbar\mathcal{B}_0 = g_e\mu_B m_s\mathcal{B}_0 \tag{14.8b}$$

Energies of an electron spin in a magnetic field

In the absence of a magnetic field, the states with different values of m_s are degenerate. When a field is present, the degeneracy is removed: the state with $m_s = +\frac{1}{2}$ moves up in energy by $\frac{1}{2}g_e\mu_B\mathcal{B}_0$ and the state with $m_s = -\frac{1}{2}$ moves down by $\frac{1}{2}g_e\mu_B\mathcal{B}_0$. The different energies arising from an interaction with an external field are sometimes represented on the vector model by picturing the vectors as **precessing**, or sweeping round their cones (Fig. 14.1), with the rate of precession equal to the **Larmor frequency**, ν_L:

$$\nu_L = \frac{|\gamma_e\mathcal{B}_0|}{2\pi} \tag*{[14.9]}$$

Definition of Larmor frequency

Equation 14.9 shows that the Larmor frequency increases with the strength of the magnetic field. For a field of 1 T, the Larmor frequency is 30 GHz.

14.2 The energies of nuclei in magnetic fields

Key points **The spin quantum number, *I*, of a nucleus is either an integer or a half-integer. Nuclei interact with magnetic fields, which remove the degeneracy of the quantized m_I states.**

The nuclear spin quantum number, *I*, is a fixed characteristic property of a nucleus and, depending on the nuclide, is either an integer or a half-integer (Table 14.1). A nucleus with spin quantum number *I* has the following properties:

1. An angular momentum of magnitude $\{I(I+1)\}^{1/2}\hbar$.
2. A component of angular momentum $m_I\hbar$ on a specified axis ('the *z*-axis'), where $m_I = I, I-1, \ldots, -I$.
3. If $I > 0$, a magnetic moment with a constant magnitude and an orientation that is determined by the value of m_I.

According to the second property, the spin, and hence the magnetic moment, of the nucleus may lie in $2I+1$ different orientations relative to an axis. A proton has $I = \frac{1}{2}$ and its spin may adopt either of two orientations; a ^{14}N nucleus has $I = 1$ and its spin may adopt any of three orientations; both ^{12}C and ^{16}O have $I = 0$ and hence zero magnetic moment.

Table 14.1 Nuclear constitution and the nuclear spin quantum number*

Number of protons	Number of neutrons	*I*
even	even	0
odd	odd	integer (1, 2, 3, . . .)
even	odd	half-integer ($\frac{1}{2}, \frac{3}{2}, \frac{5}{2}, \ldots$)
odd	even	half-integer ($\frac{1}{2}, \frac{3}{2}, \frac{5}{2}, \ldots$)

* The spin of a nucleus may be different if it is in an excited state; throughout this chapter we deal only with the ground state of nuclei.

Table 14.2* Nuclear spin properties

Nuclide	Natural abundance/%	Spin I	g-factor, g_I	Magnetogyric ratio, $\gamma/(10^7\ \mathrm{T^{-1}\,s^{-1}})$	NMR frequency at 1 T, ν/MHz
^{1}n		$\frac{1}{2}$	−3.826	−18.32	29.164
^{1}H	99.98	$\frac{1}{2}$	5.586	26.75	42.576
^{2}H	0.02	1	0.857	4.11	6.536
^{13}C	1.11	$\frac{1}{2}$	1.405	6.73	10.708
^{14}N	99.64	1	0.404	1.93	3.078

* More values are given in the *Data section.*

The energy of interaction between a nucleus with a magnetic moment $\boldsymbol{\mu}$ and an external magnetic field $\mathcal{B}$ may be calculated by using operators analogous to those of eqn 14.3:

$$\hat{\mu} = \gamma\hat{I} \qquad \text{and} \qquad \hat{H} = -\gamma\mathcal{B}\cdot\hat{I} \tag{14.10a}$$

where γ is the magnetogyric ratio of the specified nucleus, an empirically determined characteristic arising from its internal structure (Table 14.2). The corresponding energies when the magnetic field of magnitude $\mathcal{B}_0$ is applied along the z-axis are

$$E_{m_I} = -\gamma\hbar\mathcal{B}_0 m_I \tag{14.10b}$$

Energies of a nuclear spin in a magnetic field

As for electrons, the nuclear spin may be pictured as precessing around the direction of the applied field at a rate proportional to the applied field. For protons, a field of 1 T corresponds to a Larmor frequency (eqn 14.9, with γ_e replaced by γ) of about 40 MHz.

The magnetic moment of a nucleus is sometimes expressed in terms of the **nuclear g-factor**, g_I, a characteristic of the nucleus, and the **nuclear magneton**, μ_N, a quantity independent of the nucleus, by using

$$\gamma\hbar = g_I\mu_N \qquad \mu_N = \frac{e\hbar}{2m_p} = 5.051 \times 10^{-27}\ \mathrm{J\,T^{-1}} \tag*{[14.11]}$$

Definitions of nuclear g-factor and nuclear magneton

where m_p is the mass of the proton. The nuclear magneton is about 2000 times smaller than the Bohr magneton, so nuclear magnetic moments—and consequently the energies of interaction with magnetic fields—are about 2000 times weaker than the electron spin magnetic moment. Nuclear g-factors vary between −6 and +6 (Table 14.2): positive values of g_I and γ denote a magnetic moment that lies in the same direction as the spin angular momentum vector; negative values indicate that the magnetic moment and spin lie in opposite directions. For the remainder of this chapter we shall assume that γ is positive, as is the case for the majority of nuclei. In such cases, it follows from eqn 14.10b that states with $m_I < 0$ lie above states with $m_I > 0$.

14.3 Magnetic resonance spectroscopy

Key points **Electron paramagnetic resonance (EPR), a microwave technique, is the observation of the frequency at which an electron spin comes into resonance with an electromagnetic field when the molecule is exposed to a strong magnetic field. Nuclear magnetic resonance (NMR), a radiofrequency technique, is the analogous observation for nuclei.**

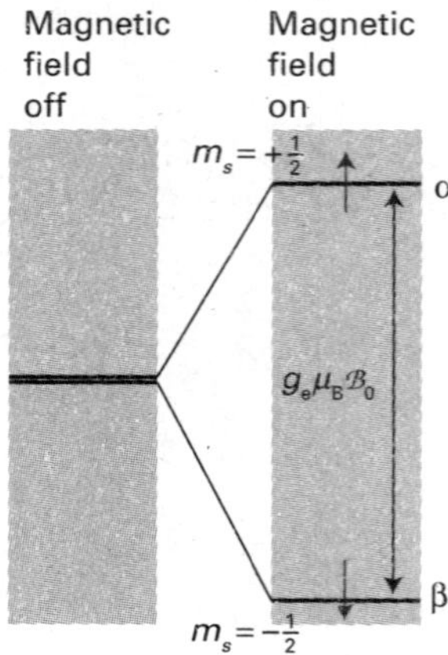

Fig. 14.2 Electron spin levels in a magnetic field. Note that the β state is lower in energy than the α state (because the magnetogyric ratio of an electron is negative). Resonance is achieved when the frequency of the incident radiation matches the frequency corresponding to the energy separation.

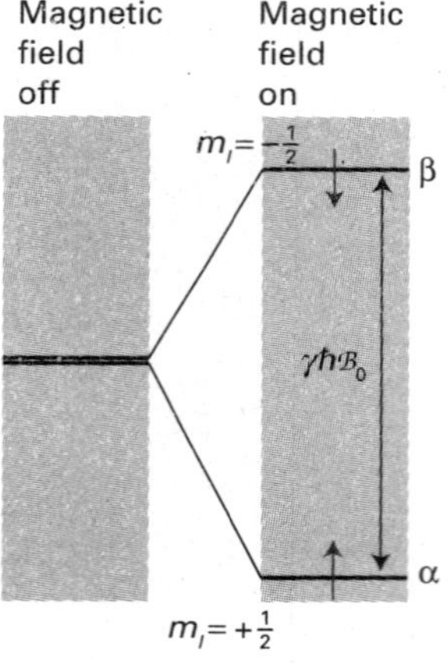

Fig. 14.3 The nuclear spin energy levels of a spin-$\frac{1}{2}$ nucleus with positive magnetogyric ratio (for example, ^{1}H or ^{13}C) in a magnetic field. Resonance occurs when the energy separation of the levels matches the energy of the photons in the electromagnetic field.

In its original form, the magnetic resonance experiment is the resonant absorption of radiation by nuclei or unpaired electrons in a magnetic field. From eqn 14.8b, the separation between the (upper) $m_s = +\frac{1}{2}$ and (lower) $m_s = -\frac{1}{2}$ levels of an electron spin in a magnetic field of magnitude $\mathcal{B}_0$ in the z-direction is

$$\Delta E = E_{1/2} - E_{-1/2} = \tfrac{1}{2}g_e\mu_B\mathcal{B}_0 - (-\tfrac{1}{2}g_e\mu_B\mathcal{B}_0) = g_e\mu_B\mathcal{B}_0 \qquad (14.12a)$$

If the sample is exposed to radiation of frequency ν, the energy separations come into resonance with the radiation when the frequency satisfies the **resonance condition** (Fig. 14.2):

$$h\nu = g_e\mu_B\mathcal{B}_0 \qquad \text{Resonance condition for electrons} \qquad (14.12b)$$

At resonance there is strong coupling between the electron spins and the radiation, and strong absorption occurs as the spins make the transition β → α. **Electron paramagnetic resonance** (EPR), or **electron spin resonance** (ESR), is the study of molecules and ions containing unpaired electrons by observing the magnetic field at which they come into resonance with radiation of known frequency. Magnetic fields of about 0.3 T (the value used in most commercial EPR spectrometers) correspond to resonance with an electromagnetic field of frequency 10 GHz (10^{10} Hz) and wavelength 3 cm. Because 3 cm radiation falls in the microwave region of the electromagnetic spectrum, EPR is a microwave technique.

The energy separation between the (lower, for $\gamma > 0$) $m_I = +\frac{1}{2}$ and (upper) $m_I = -\frac{1}{2}$ states of a **spin-$\frac{1}{2}$ nucleus,** a nucleus with $I = \frac{1}{2}$, is

$$\Delta E = E_{-1/2} - E_{+1/2} = \tfrac{1}{2}\gamma\hbar\mathcal{B}_0 - (-\tfrac{1}{2}\gamma\hbar\mathcal{B}_0) = \gamma\hbar\mathcal{B}_0 \qquad (14.13a)$$

and resonant absorption occurs when the resonance condition (Fig. 14.3)

$$h\nu = \gamma\hbar\mathcal{B}_0 \qquad \text{Resonance condition for spin-}\tfrac{1}{2}\text{ nuclei} \qquad (14.13b)$$

is fulfilled. Because $\gamma\hbar\mathcal{B}_0/h$ is the Larmor frequency of the nucleus, this resonance occurs when the frequency of the electromagnetic field matches the Larmor frequency ($\nu = \nu_L$). In its simplest form, **nuclear magnetic resonance** (NMR) is the study of the properties of molecules containing magnetic nuclei by applying a magnetic field and observing the frequency of the resonant electromagnetic field. Larmor frequencies of nuclei at the fields normally employed (about 12 T) typically lie in the radiofrequency region of the electromagnetic spectrum (close to 500 MHz), so NMR is a radiofrequency technique.

For much of this chapter we consider spin-$\frac{1}{2}$ nuclei, but NMR is applicable to nuclei with any nonzero spin. As well as protons, which are the most common nuclei studied by NMR, spin-$\frac{1}{2}$ nuclei include ^{13}C, ^{19}F, and ^{31}P. Nuclear magnetic resonance is far more important than EPR, and so we consider it first and at greater length.

Nuclear magnetic resonance

Although the NMR technique is simple in concept, NMR spectra can be highly complex. However, they have proved invaluable in chemistry, for they reveal so much structural information. A magnetic nucleus is a very sensitive, non-invasive probe of the surrounding electronic structure.

14.4 The NMR spectrometer

Key points **NMR spectrometers consist of a source of radiofrequency radiation and a superconducting magnet. The resonance absorption intensity increases with the strength of the applied magnetic field (as $\mathcal{B}_0^2$).**

An NMR spectrometer consists of the appropriate sources of radiofrequency radiation and a magnet that can produce a uniform, intense field. Most modern instruments use a superconducting magnet capable of producing fields of the order of 10 T and more (Fig. 14.4). The sample is rotated rapidly to average out magnetic inhomogeneities; however, although sample spinning is essential for the investigation of small molecules, for large molecules it can lead to irreproducible results and is often avoided. Although a superconducting magnet operates at the temperature of liquid helium (4 K), the sample itself is normally at room temperature or held in a variable temperature enclosure between, typically, −150 to +100°C.

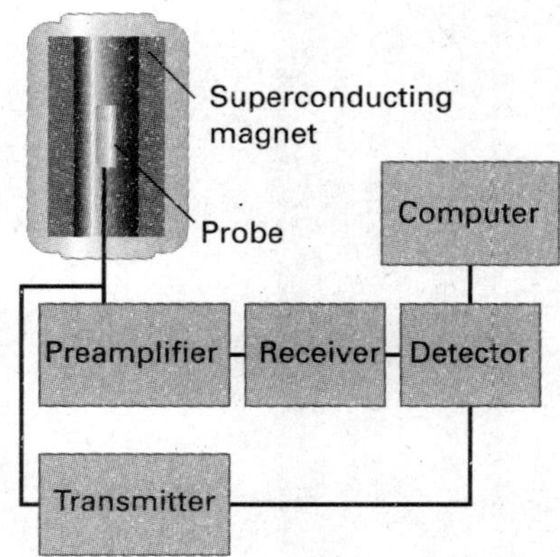

Fig. 14.4 The layout of a typical NMR spectrometer. The link from the transmitter to the detector indicates that the high frequency of the transmitter is subtracted from the high frequency signal detected to give a low frequency signal for processing.

The intensity of an NMR transition depends on a number of factors. We show in the following *Justification* that

$$\text{Intensity} \propto (N_\alpha - N_\beta)\mathcal{B}_0 \qquad (14.14\text{a})$$

where

$$N_\alpha - N_\beta \approx \frac{N\gamma\hbar\mathcal{B}_0}{2kT} \qquad (14.14\text{b})$$

with N the total number of spins ($N = N_\alpha + N_\beta$). It follows that decreasing the temperature increases the intensity by increasing the population difference. By combining these two equations we see that the intensity is proportional to $\mathcal{B}_0^2$, so NMR transitions can be enhanced significantly by increasing the strength of the applied magnetic field. We shall also see (Section 14.6) that the use of high magnetic fields simplifies the appearance of spectra and so allows them to be interpreted more readily. We also conclude that absorptions of nuclei with large magnetogyric ratios (^{1}H, for instance) are more intense than those with small magnetogyric ratios (^{13}C, for instance)

Justification 14.1 *Intensities in NMR spectra*

From the general considerations of transition intensities in *Justification 13.4*, we know that the rate of absorption of electromagnetic radiation is proportional to the population of the lower energy state (N_α in the case of a proton NMR transition) and the rate of stimulated emission is proportional to the population of the upper state (N_β). At the low frequencies typical of magnetic resonance, we can neglect spontaneous emission as it is very slow. Therefore, the net rate of absorption is proportional to the difference in populations, and we can write

$$\text{Rate of absorption} \propto N_\alpha - N_\beta$$

The intensity of absorption, the rate at which energy is absorbed, is proportional to the product of the rate of absorption (the rate at which photons are absorbed) and the energy of each photon, and the latter is proportional to the frequency ν of the incident radiation (through $E = h\nu$). At resonance, this frequency is proportional to the applied magnetic field (through $\nu = \nu_L = \gamma\mathcal{B}_0/2\pi$), so we can write

$$\text{Intensity of absorption} \propto (N_\alpha - N_\beta)\mathcal{B}_0$$

as in eqn 14.14a. To write an expression for the population difference, we use the Boltzmann distribution (*Fundamentals F.5*) to write the ratio of populations as

$$\frac{N_\beta}{N_\alpha} = e^{-\Delta E/kT} \approx 1 - \frac{\Delta E}{kT} = 1 - \frac{\gamma\hbar\mathcal{B}_0}{kT}$$

where we have used $e^{-x} \approx 1 - x$ (which is valid for $x \ll 1$) and $\Delta E = E_\beta - E_\alpha$. The expansion of the exponential term is appropriate for $\Delta E \ll kT$, a condition usually met for nuclear spins. It follows that

$$N_\alpha - N_\beta = N_\alpha\left(1 - \frac{N_\beta}{N_\alpha}\right) \approx \frac{\gamma\hbar\mathcal{B}_0 N_\alpha}{kT}$$

Because $N_\alpha \approx \frac{1}{2}N$ (the two spin states are nearly equally populated),

$$N_\alpha - N_\beta = N_\alpha\left(1 - \frac{N_\beta}{N_\alpha}\right) \approx \frac{\gamma\hbar\mathcal{B}_0 N}{2kT}$$

which is eqn 14.14b.

14.5 The chemical shift

Key points **(a) The chemical shift of a nucleus is the difference between its resonance frequency and that of a reference standard. (b) The shielding constant is the sum of a local contribution, a neighbouring group contribution, and a solvent contribution. (c) The local contribution is the sum of a diamagnetic contribution and a paramagnetic contribution. (d) The neighbouring group contribution arises from the currents induced in nearby groups of atoms. (e) The solvent contribution can arise from specific molecular interactions between the solute and the solvent.**

Nuclear magnetic moments interact with the *local* magnetic field. The local field may differ from the applied field because the latter induces electronic orbital angular momentum (that is, the circulation of electronic currents) which gives rise to a small additional magnetic field $\delta\mathcal{B}$ at the nuclei. This additional field is proportional to the applied field, and it is conventional to write

$$\delta\mathcal{B} = -\sigma\mathcal{B}_0 \qquad \text{Definition of shielding constant} \qquad [14.15]$$

where the dimensionless quantity σ is called the **shielding constant** of the nucleus (σ is usually positive but may be negative). The ability of the applied field to induce an electronic current in the molecule, and hence affect the strength of the resulting local magnetic field experienced by the nucleus, depends on the details of the electronic structure near the magnetic nucleus of interest, so nuclei in different chemical groups have different shielding constants. The calculation of reliable values of the shielding constant is very difficult, but trends in it are quite well understood and we concentrate on them.

(a) The δ scale of chemical shifts

Because the total local field is

$$\mathcal{B}_{loc} = \mathcal{B}_0 + \delta\mathcal{B} = (1 - \sigma)\mathcal{B}_0 \qquad (14.16)$$

the nuclear Larmor frequency is

$$\nu_L = \frac{\gamma\mathcal{B}_{loc}}{2\pi} = (1 - \sigma)\frac{\gamma\mathcal{B}_0}{2\pi} \qquad (14.17)$$

This frequency is different for nuclei in different environments. Hence, different nuclei, even of the same element, come into resonance at different frequencies if they are in different molecular environments.

Resonance frequencies are expressed in terms of an empirical quantity called the **chemical shift**, which is related to the difference between the resonance frequency, ν, of the nucleus in question and that of a reference standard, ν°:

$$\delta = \frac{\nu - \nu^\circ}{\nu^\circ} \times 10^6 \qquad [14.18]$$

Definition of chemical shift

The standard for protons is the proton resonance in tetramethylsilane ($Si(CH_3)_4$, commonly referred to as TMS), which bristles with protons and dissolves without reaction in many liquids. Other references are used for other nuclei. For ^{13}C, the reference frequency is the ^{13}C resonance in TMS; for ^{31}P it is the ^{31}P resonance in 85 per cent H_3PO_4(aq). The advantage of the δ-scale is that shifts reported on it are independent of the applied field (because both numerator and denominator are proportional to the applied field).

- **A brief illustration**

From eqn 14.18,

$$\nu - \nu^\circ = \nu^\circ \delta \times 10^{-6}$$

A nucleus with $\delta = 1.00$ in a spectrometer operating at 500 MHz will have a shift relative to the reference equal to

$$\nu - \nu^\circ = (500\ \text{MHz}) \times 1.00 \times 10^{-6} = 500\ \text{Hz}$$

In a spectrometer operating at 100 MHz, the shift relative to the reference would be only 100 Hz. ●

A note on good practice In much of the literature, chemical shifts are reported in 'parts per million', ppm, in recognition of the factor of 10^6 in the definition. This practice is unnecessary.

The relation between δ and σ is obtained by substituting eqn 14.17 into eqn 14.18:

$$\delta = \frac{(1-\sigma)\mathcal{B}_0 - (1-\sigma^\circ)\mathcal{B}_0}{(1-\sigma^\circ)\mathcal{B}_0} \times 10^6$$
$$= \frac{\sigma^\circ - \sigma}{1-\sigma^\circ} \times 10^6 \approx (\sigma^\circ - \sigma) \times 10^6 \qquad (14.19)$$

Relation between δ and σ

As the shielding σ, gets smaller, δ increases. Therefore, we speak of nuclei with large chemical shift as being strongly **deshielded**. Some typical chemical shifts are given in Fig. 14.5. As can be seen from the illustration, the nuclei of different elements have very different ranges of chemical shifts. The ranges exhibit the variety of electronic environments of the nuclei in molecules: the higher the atomic number of the element, the greater the number of electrons around the nucleus and hence the greater the range of shieldings. By convention, NMR spectra are plotted with δ increasing from right to left.

The existence of a chemical shift explains the general features of the spectrum of ethanol shown in Fig.14.6. The CH_3 protons form one group of nuclei with $\delta \approx 1.2$. The two CH_2 protons are in a different part of the molecule, experience a different local magnetic field, and resonate at $\delta \approx 3.6$. Finally, the OH proton is in another environment, and has a chemical shift of $\delta \approx 4$. The increasing value of δ (that is, the decrease in shielding) is consistent with the electron-withdrawing power of the O atom: it reduces the electron density of the OH proton most, and that proton is strongly deshielded. It reduces the electron density of the distant methyl protons least, and those nuclei are least deshielded.

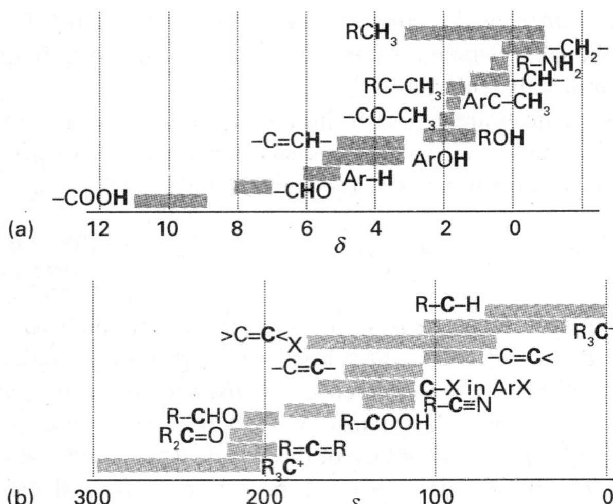

Fig. 14.5 The range of typical chemical shifts for (a) ^{1}H resonances and (b) ^{13}C resonances.

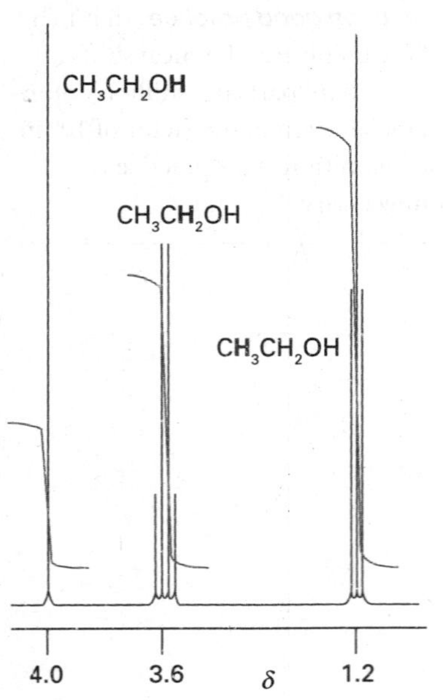

Fig. 14.6 The ^{1}H-NMR spectrum of ethanol. The bold letters denote the protons giving rise to the resonance peak, and the step-like curve is the integrated signal.

The relative intensities of the signals (the areas under the absorption lines) can be used to help distinguish which group of lines corresponds to which chemical group. The determination of the area under an absorption line is referred to as the **integration** of the signal (just as any area under a curve may be determined by mathematical integration). Data analysis software performs this integration and the values are represented as the height of step-like curves superimposed on the spectrum, as in Fig. 14.6. In ethanol the group intensities are in the ratio 3:2:1 because there are three CH_3 protons, two CH_2 protons, and one OH proton in each molecule. Counting the number of magnetic nuclei as well as noting their chemical shifts helps in the identification of the sample.

(b) The origin of shielding constants

The calculation of shielding constants is difficult, even for small molecules, for it requires detailed information (using the techniques outlined in Chapter 10) about the distribution of electron density in the ground and excited states and the excitation energies of the molecule. Nevertheless, considerable success has been achieved with small molecules such as H_2O and CH_4 and even large molecules, such as proteins, are within the scope of some types of calculation. However, it is easier to understand the different contributions to chemical shifts by studying the large body of empirical information now available for large molecules.

The empirical approach supposes that the observed shielding constant is the sum of three contributions:

$$\sigma = \sigma(\text{local}) + \sigma(\text{neighbour}) + \sigma(\text{solvent}) \tag{14.20}$$

The **local contribution**, $\sigma(\text{local})$, is essentially the contribution of the electrons of the atom that contains the nucleus in question. The **neighbouring group contribution**, $\sigma(\text{neighbour})$, is the contribution from the groups of atoms that form the rest of the molecule. The **solvent contribution**, $\sigma(\text{solvent})$, is the contribution from the solvent molecules.

(c) The local contribution

It is convenient to regard the local contribution to the shielding constant as the sum of a **diamagnetic contribution**, σ_d, and a **paramagnetic contribution**, σ_p:

$$\sigma(\text{local}) = \sigma_d + \sigma_p \qquad \text{Local contribution to the shielding constant} \qquad (14.21)$$

A diamagnetic contribution to σ(local) opposes the applied magnetic field and shields the nucleus in question. A paramagnetic contribution to σ(local) reinforces the applied magnetic field and deshields the nucleus in question. Therefore, $\sigma_d > 0$ and $\sigma_p < 0$. The total local contribution is positive if the diamagnetic contribution dominates, and is negative if the paramagnetic contribution dominates.

The diamagnetic contribution arises from the ability of the applied field to generate a circulation of charge in the ground-state electron distribution of the atom. The circulation generates a magnetic field that opposes the applied field and hence shields the nucleus. The magnitude of σ_d depends on the electron density close to the nucleus and can be calculated from the **Lamb formula:**[1]

$$\sigma_d = \frac{e^2\mu_0}{12\pi m_e}\left\langle \frac{1}{r} \right\rangle \qquad \text{Lamb formula} \qquad (14.22)$$

where μ_0 is the vacuum permeability (a fundamental constant, see inside the front cover) and r is the electron–nucleus distance.

• A brief illustration

To calculate σ_d for the proton in a free H atom, we need to calculate the expectation value of $1/r$ for a hydrogen 1s orbital. Wavefunctions are given in Table 9.1, and the integral we need is given in Example 7.4. Because $d\tau = r^2\, dr \sin\theta\, d\theta\, d\phi$, we can write

$$\left\langle \frac{1}{r} \right\rangle = \int \frac{\psi^*\psi}{r} d\tau = \frac{1}{\pi a_0^3}\int_0^{2\pi} d\phi \int_0^{\pi} \sin\theta\, d\theta \int_0^{\infty} r e^{-2r/a_0} dr = \frac{4}{a_0^3}\int_0^{\infty} r e^{-2r/a_0} dr = \frac{1}{a_0}$$

Therefore,

$$\sigma_d = \frac{e^2\mu_0}{12\pi m_e a_0}$$

With the values of the fundamental constants inside the front cover, this expression evaluates to 1.78×10^{-5}. •

The diamagnetic contribution is the only contribution in closed-shell free atoms. It is also the only contribution to the local shielding for electron distributions that have spherical or cylindrical symmetry. Thus, it is the only contribution to the local shielding from inner cores of atoms, for cores remain nearly spherical even though the atom may be a component of a molecule and its valence electron distribution highly distorted. The diamagnetic contribution is broadly proportional to the electron density of the atom containing the nucleus of interest. It follows that the shielding is decreased if the electron density on the atom is reduced by the influence of an electronegative atom nearby. That reduction in shielding as the electronegativity of a neighbouring atom increases translates into an increase in the chemical shift δ (Fig. 14.7).

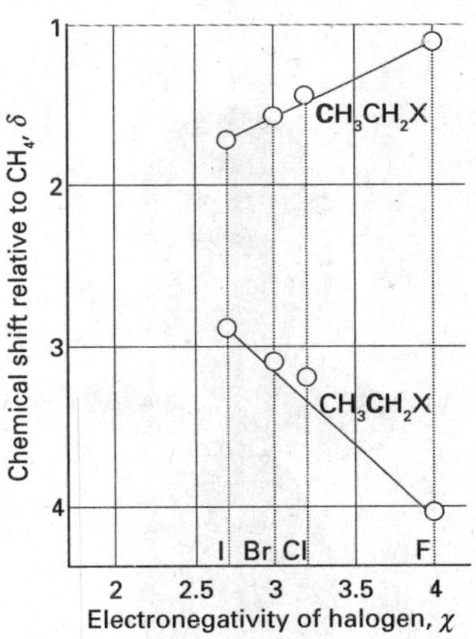

Fig. 14.7 The variation of chemical shielding with electronegativity. The shifts for the methylene protons agree with the trend expected with increasing electronegativity. However, to emphasize that chemical shifts are subtle phenomena, notice that the trend for the methyl protons is opposite to that expected. For these protons another contribution (the magnetic anisotropy of C–H and C–X bonds) is dominant.

The local paramagnetic contribution, σ_p, arises from the ability of the applied field to force electrons to circulate through the molecule by making use of orbitals that are unoccupied in the ground state. It is zero in free atoms and around the axes of linear molecules (such as ethyne, HC≡CH) where the electrons can circulate freely and a field applied along the internuclear axis is unable to force them into other orbitals. We can expect large paramagnetic contributions from small atoms (because the induced

[1] For a derivation, see our *Molecular quantum mechanics* (2005).

currents are then close to the nucleus) in molecules with low lying excited states (because an applied field can then induce significant currents). In fact, the paramagnetic contribution is the dominant local contribution for atoms other than hydrogen.

(d) Neighbouring group contributions

The neighbouring group contribution arises from the currents induced in nearby groups of atoms. Consider the influence of the neighbouring group X on the proton H in a molecule such as H–X. The applied field generates currents in the electron distribution of X and gives rise to an induced magnetic moment proportional to the applied field; the constant of proportionality is the magnetic susceptibility, χ (chi), of the group X: $\mu_{\text{induced}} = \chi\mathcal{B}_0$. The susceptibility is negative for a diamagnetic group because the induced moment is opposite to the direction of the applied field. As we show in the following *Justification*, the induced moment gives rise to a magnetic field with a component parallel to the applied field and at a distance r and angle θ (1) that has the form

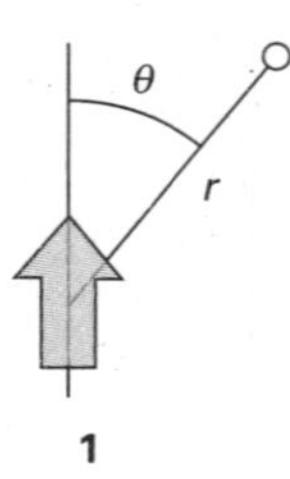

1

$$\mathcal{B}_{\text{local}} \propto \frac{\mu_{\text{induced}}(1-3\cos^2\theta)}{r^3} \qquad \text{Local dipolar field} \qquad (14.23a)$$

We see that the strength of the additional magnetic field experienced by the proton is inversely proportional to the cube of the distance r between H and X. Second, if the magnetic susceptibility is independent of the orientation of the molecule (is 'isotropic'), because $1 - 3\cos^2\theta$ is zero when averaged over a sphere (see Problem 14.17), the local field averages to zero. To a good approximation, the shielding constant σ(neighbour) depends on the distance r and the difference $\chi_{\parallel} - \chi_{\perp}$ as

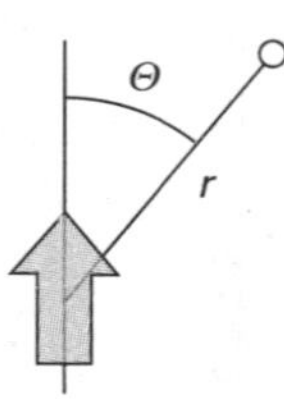

2

$$\sigma(\text{neighbour}) \propto (\chi_{\parallel} - \chi_{\perp})\left(\frac{1-3\cos^2\Theta}{r^3}\right) \qquad \text{Neighbouring group contribution to the shielding constant} \qquad (14.23b)$$

where Θ (upper-case theta) is the angle between the X–H axis and the symmetry axis of the neighbouring group (2). Equation 14.23 shows that the neighbouring group contribution may be positive or negative according to the relative magnitudes of the two magnetic susceptibilities and the relative orientation of the nucleus with respect to X. If $54.7° < \Theta < 125.3°$, then $1 - 3\cos^2\Theta$ is positive, but it is negative otherwise (Figs. 14.8 and 14.9).

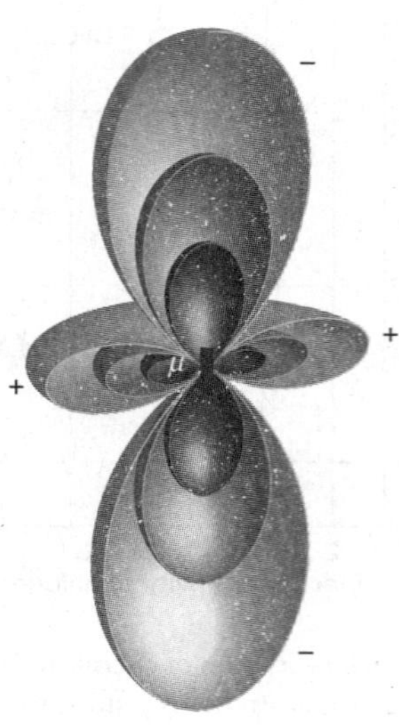

Fig. 14.8 A depiction of the field arising from a point magnetic dipole. The three shades of colour represent the strength of field declining with distance (as $1/r^3$), and each surface shows the angle dependence of the z-component of the field for each distance.

Justification 14.2 *Dipolar fields*

Standard electromagnetic theory gives the magnetic field at a point $\boldsymbol{r}$ from a point magnetic dipole $\boldsymbol{\mu}$ as

$$\boldsymbol{\mathcal{B}} = \frac{\mu_0}{4\pi r^3}\left(\boldsymbol{\mu} - \frac{3(\boldsymbol{\mu}\cdot\boldsymbol{r})\boldsymbol{r}}{r^2}\right)$$

where μ_0 is the vacuum permeability (a fundamental constant with the defined value $4\pi \times 10^{-7}\ \text{T}^2\ \text{J}^{-1}\ \text{m}^3$). The electric field due to a point electric dipole is given by a similar expression:

$$\boldsymbol{\mathcal{E}} = \frac{1}{4\pi\varepsilon_0 r^3}\left(\boldsymbol{\mu} - \frac{3(\boldsymbol{\mu}\cdot\boldsymbol{r})\boldsymbol{r}}{r^2}\right)$$

where ε_0 is the vacuum permittivity, which is related to μ_0 by $\varepsilon_0 = 1/\mu_0 c^2$. The component of magnetic field in the z-direction is

$$\mathcal{B}_z = \frac{\mu_0}{4\pi r^3}\left(\mu_z - \frac{3(\boldsymbol{\mu}\cdot\boldsymbol{r})z}{r^2}\right)$$

with $z = r\cos\theta$, the z-component of the distance vector $\boldsymbol{r}$. If the magnetic dipole is also parallel to the z-direction, $\mu_z = \mu$ and $\boldsymbol{\mu}\cdot\boldsymbol{r} = \mu r\cos\theta$. It follows that

$$\mathcal{B}_z = \frac{\mu_0}{4\pi r^3}\left(\mu - \frac{3(\mu r\cos\theta)(r\cos\theta)}{r^2}\right) = \frac{\mu\mu_0(1-3\cos^2\theta)}{4\pi r^3}$$

as in eqn 14.23a.

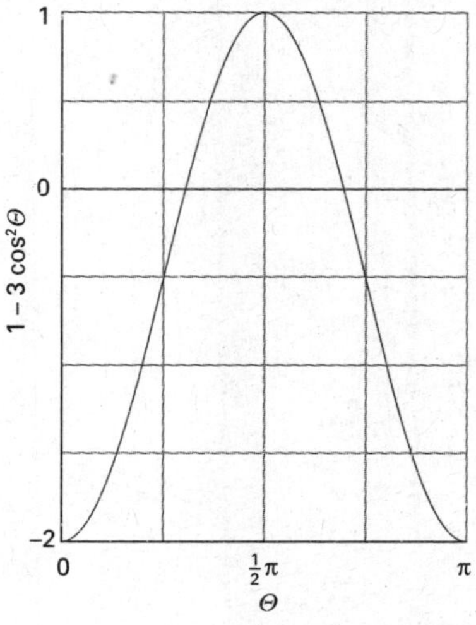

Fig. 14.9 The variation of the function $1 - 3\cos^2\Theta$ with the angle Θ.

A special case of a neighbouring group effect is found in aromatic compounds. The strong anisotropy of the magnetic susceptibility of the benzene ring is ascribed to the ability of the field to induce a **ring current**, a circulation of electrons around the ring, when it is applied perpendicular to the molecular plane. Protons in the plane are deshielded (Fig. 14.10), but any that happen to lie above or below the plane (as members of substituents of the ring) are shielded.

(e) The solvent contribution

A solvent can influence the local magnetic field experienced by a nucleus in a variety of ways. Some of these effects arise from specific interactions between the solute and the solvent (such as hydrogen-bond formation and other forms of Lewis acid–base complex formation). The anisotropy of the magnetic susceptibility of the solvent molecules, especially if they are aromatic, can also be the source of a local magnetic field. Moreover, if there are steric interactions that result in a loose but specific interaction between a solute molecule and a solvent molecule, then protons in the solute molecule may experience shielding or deshielding effects according to their location relative to the solvent molecule (Fig. 14.11). We shall see that the NMR spectra of species that contain protons with widely different chemical shifts are easier to interpret than those in which the shifts are similar, so the appropriate choice of solvent may help to simplify the appearance and interpretation of a spectrum.

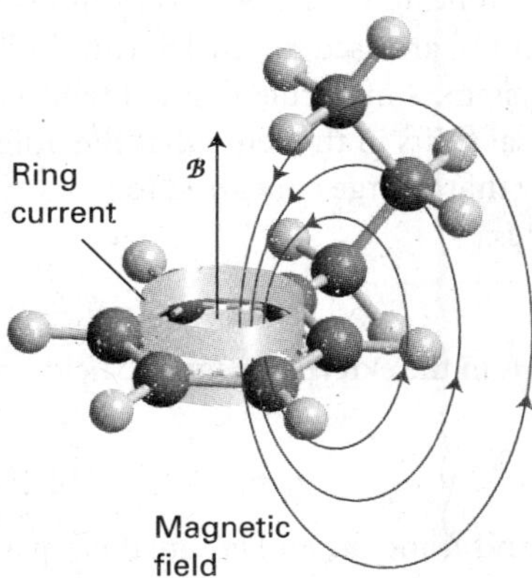

Fig. 14.10 The shielding and deshielding effects of the ring current induced in the benzene ring by the applied field. Protons attached to the ring are deshielded but a proton attached to a substituent that projects above the ring is shielded.

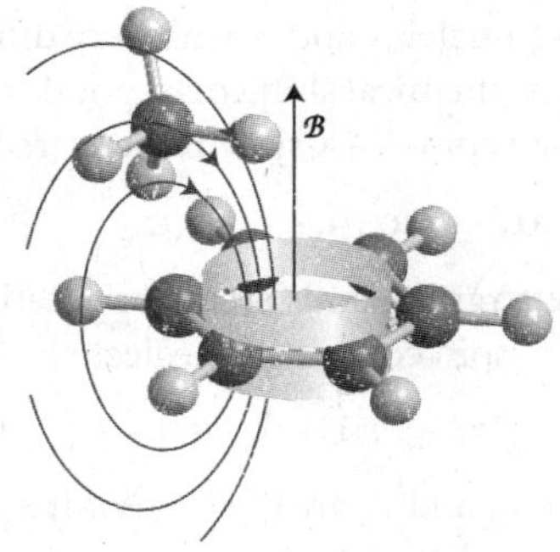

Fig. 14.11 An aromatic solvent (benzene here) can give rise to local currents that shield or deshield a proton in a solute molecule. In this relative orientation of the solvent and solute, the proton on the solute molecule is shielded.

14.6 The fine structure

Key points **(a) Spin–spin coupling is expressed in terms of the spin–spin coupling constant *J* and depends on the relative orientation of two nuclear spins. (b) *N* equivalent spin-$\frac{1}{2}$ nuclei split the resonance of a nearby spin or group of equivalent spins into *N* + 1 lines with an intensity distribution given by Pascal's triangle. (c) The coupling constant decreases as the number of bonds separating two nuclei increases. (d) Spin–spin coupling can be explained in terms of the polarization mechanism and the Fermi contact interaction. (e) Chemically and magnetically equivalent nuclei have the same chemical shifts. (f) In strongly coupled spectra, transitions cannot be allocated to definite groups.**

The splitting of resonances into individual lines by spin–spin coupling in Fig. 14.6 is called the **fine structure** of the spectrum. It arises because each magnetic nucleus may contribute to the local field experienced by the other nuclei and so modify their resonance frequencies. The strength of the interaction is expressed in terms of the **scalar coupling constant**, *J*, and reported in hertz (Hz). The scalar coupling constant is so called because the energy of interaction it describes is proportional to the scalar product of the two interacting spins: $E \propto \boldsymbol{I}_1 \cdot \boldsymbol{I}_2$. As explained in *Mathematical background 5*, a scalar product depends on the angle between the two vectors, so writing the energy in this way is simply a way of saying that the energy of interaction between two spins depends on their relative orientation. The constant of proportionality in this expression is written $hJ/\hbar^2$ (so $E = (hJ/\hbar^2)\boldsymbol{I}_1 \cdot \boldsymbol{I}_2$); because each spin angular momentum is proportional to $\hbar$, E is then proportional to hJ and J is a frequency (with units hertz). For nuclei that are constrained to align with the applied field in the *z*-direction, the only contribution to $\boldsymbol{I}_1 \cdot \boldsymbol{I}_2$ is $I_{1z}I_{2z}$, with eigenvalues $m_1m_2\hbar^2$, so the energy due to spin–spin coupling is

$$E_{m_1m_2} = hJm_1m_2 \qquad \text{Spin–spin coupling energy} \qquad (14.24)$$

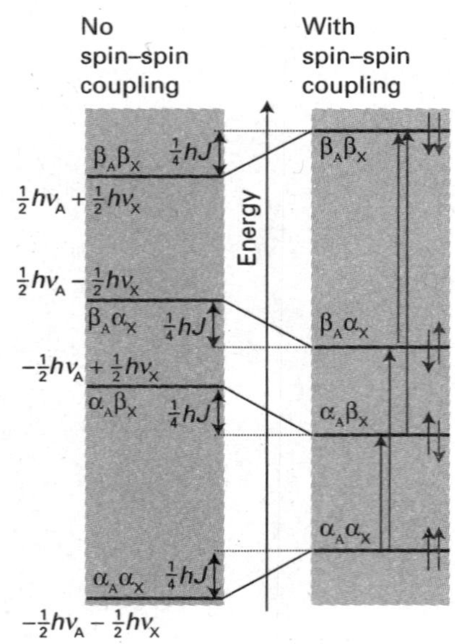

Fig. 14.12 The energy levels of an AX system. The four levels on the left are those of the two spins in the absence of spin–spin coupling. The four levels on the right show how a positive spin–spin coupling constant affects the energies. The transitions shown are for $\beta \leftarrow \alpha$ of A or X, the other nucleus (X or A, respectively) remaining unchanged. We have exaggerated the effect for clarity in practice, the splitting caused by spin–spin coupling is much smaller than that caused by the applied field.

(a) The energy levels of coupled systems

It will be useful for later discussions to consider an NMR spectrum in terms of the energy levels of the nuclei and the transitions between them. In NMR, letters far apart in the alphabet (typically A and X) are used to indicate nuclei with very different chemical shifts; letters close together (such as A and B) are used for nuclei with similar chemical shifts. We shall consider first an AX system, a molecule that contains two spin-$\frac{1}{2}$ nuclei A and X with very different chemical shifts in the sense that the difference in chemical shift corresponds to a frequency that is large compared to *J*.

For a spin-$\frac{1}{2}$ AX system there are four spin states:

$$\alpha_A\alpha_X \qquad \alpha_A\beta_X \qquad \beta_A\alpha_X \qquad \beta_A\beta_X$$

The energy depends on the orientation of the spins in the external magnetic field, and if spin–spin coupling is neglected

$$E_{m_Am_X} = -\gamma\hbar(1-\sigma_A)\mathcal{B}_0 m_A - \gamma\hbar(1-\sigma_X)\mathcal{B}_0 m_X = -h\nu_A m_A - h\nu_X m_X \qquad (14.25a)$$

where ν_A and ν_X are the Larmor frequencies of A and X and m_A and m_X are their quantum numbers ($m_A = \pm\frac{1}{2}$, $m_X = \pm\frac{1}{2}$). This expression gives the four lines on the left of Fig. 14.12. When spin–spin coupling is included (by using eqn 14.24), the energy levels are

$$E_{m_Am_X} = -h\nu_A m_A - h\nu_X m_X + hJm_A m_X \qquad (14.25b)$$

If $J > 0$, a lower energy is obtained when $m_A m_X < 0$, which is the case if one spin is α and the other is β. A higher energy is obtained if both spins are α or both spins are β.

The opposite is true if $J<0$. The resulting energy level diagram (for $J>0$) is shown on the right of Fig. 14.12. We see that the αα and ββ states are both raised by $\frac{1}{4}hJ$ and that the αβ and βα states are both lowered by $\frac{1}{4}hJ$.

When a transition of nucleus A occurs, nucleus X remains unchanged. Therefore, the A resonance is a transition for which $\Delta m_A=+1$ and $\Delta m_X=0$ There are two such transitions, one in which $\beta_A \leftarrow \alpha_A$ occurs when the X nucleus is α, and the other in which $\beta_A \leftarrow \alpha_A$ occurs when the X nucleus is β. They are shown in Fig. 14.12 and in a slightly different form in Fig. 14.13. The energies of the transitions are

$$\Delta E = h\nu_A \pm \tfrac{1}{2}hJ \qquad (14.26a)$$

Therefore, the A resonance consists of a doublet of separation J centred on the chemical shift of A (Fig. 14.14). Similar remarks apply to the X resonance, which consists of two transitions according to whether the A nucleus is α or β (as shown in Fig. 14.13). The transition energies are

$$\Delta E = h\nu_X \pm \tfrac{1}{2}hJ \qquad (14.26b)$$

It follows that the X resonance also consists of two lines of the same separation J, but they are centred on the chemical shift of X (as shown in Fig. 14.14).

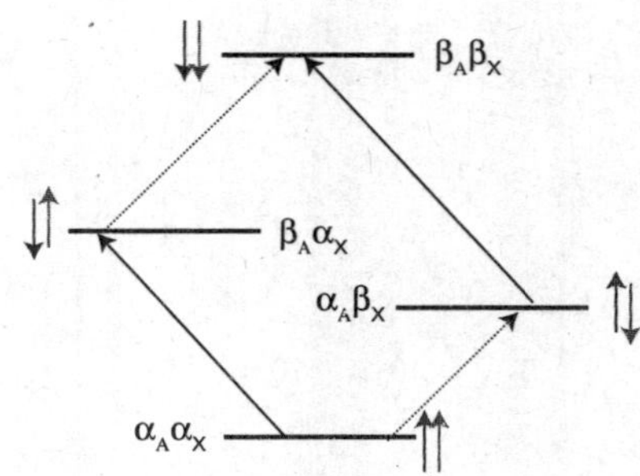

Fig. 14.13 An alternative depiction of the energy levels and transitions shown in Fig. 14.11. Once again, we have exaggerated the effect of spin–spin coupling.

(b) Patterns of coupling

We have seen that, in an AX system, spin–spin coupling results in a doublet of lines for the A resonance and a doublet of lines for the X resonance of the same separation. The X resonance in an AX_n species (such as an AX_2 or AX_3 species) is also a doublet with splitting J. As we shall explain below, *a group of equivalent nuclei resonates like a single nucleus.* The only difference for the X resonance of an AX_n species is that the intensity is n times as great as that of an AX species (Fig. 14.15). The A resonance in an AX_n species, though, is quite different from the A resonance in an AX species. For example, consider an AX_2 species with two equivalent X nuclei. The A resonance is split into a doublet of separation J by one X, and each line of that doublet is split again by the same amount by the second X (Fig. 14.16). This splitting results in three lines in

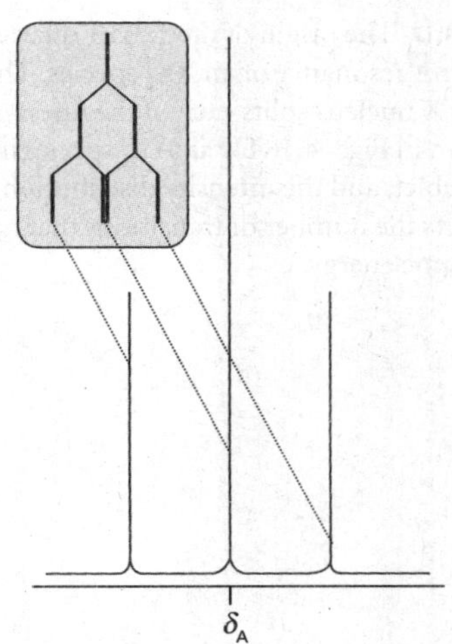

Fig. 14.16 The origin of the 1:2:1 triplet in the A resonance of an AX_2 species. The resonance of A is split into two by coupling with one X nucleus (as shown in the inset), and then each of those two lines is split into two by coupling to the second X nucleus. Because each X nucleus causes the same splitting, the two central transitions are coincident and give rise to an absorption line of double the intensity of the outer lines.

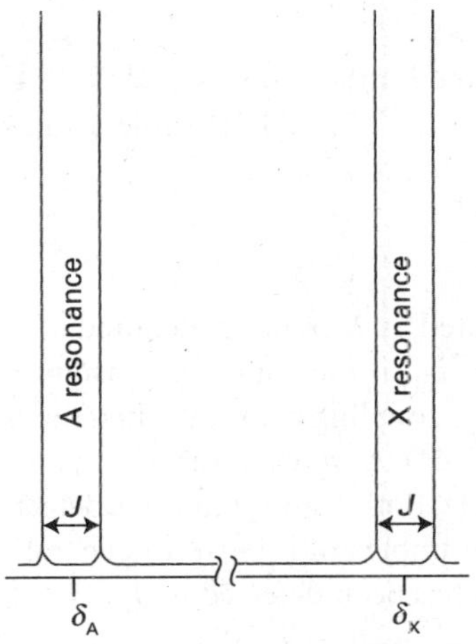

Fig. 14.14 The effect of spin–spin coupling on an AX spectrum. Each resonance is split into two lines separated by J. The pairs of resonances are centred on the chemical shifts of the protons in the absence of spin–spin coupling.

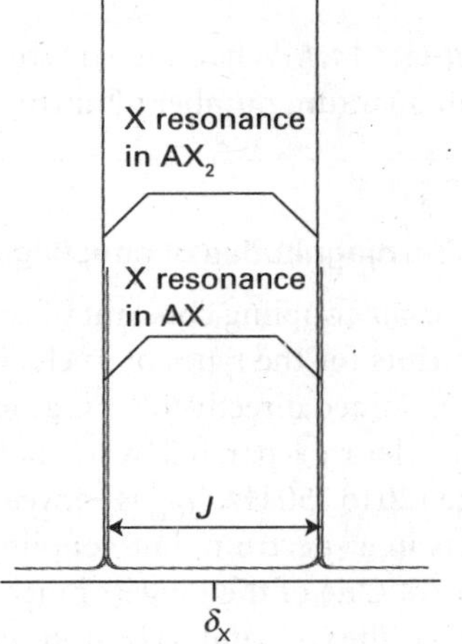

Fig. 14.15 The X resonance of an AX_2 species is also a doublet, because the two equivalent X nuclei behave like a single nucleus; however, the overall absorption is twice as intense as that of an AX species.

the intensity ratio 1:2:1 (because the central frequency can be obtained in two ways). The A resonance of an A_nX_2 species would also be a 1:2:1 triplet of splitting J, the only difference being that the intensity of the A resonance would be n times as great as that of AX_2.

Three equivalent X nuclei (an AX_3 species) split the resonance of A into four lines of intensity ratio 1:3:3:1 and separation J (Fig. 14.17). The X resonance, though, is still a doublet of separation J. In general, n equivalent spin-$\frac{1}{2}$ nuclei split the resonance of a nearby spin or group of equivalent spins into $n+1$ lines with an intensity distribution given by 'Pascal's triangle' in which each entry is the sum of the two entries immediately above (3). The easiest way of constructing the pattern of fine structure is to draw a diagram in which each successive row shows the splitting due to an additional proton. The procedure is illustrated in Fig. 14.18 and was used in Figs. 14.16 and 14.17. It is easily extended to molecules containing nuclei with $I>\frac{1}{2}$ (Fig. 14.19).

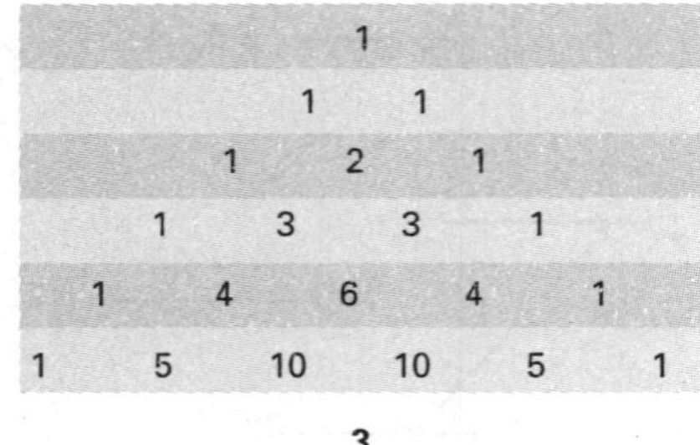

3

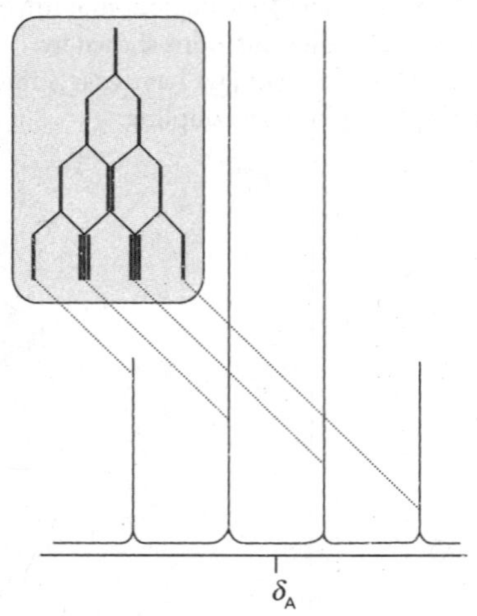

Fig. 14.17 The origin of the 1:3:3:1 quartet in the A resonance of an AX_3 species. The third X nucleus splits each of the lines shown in Fig. 14.16 for an AX_2 species into a doublet, and the intensity distribution reflects the number of transitions that have the same energy.

Example 14.1 *Accounting for the fine structure in a spectrum*

Account for the fine structure in the NMR spectrum of the C–H protons of ethanol.

Method Consider how each group of equivalent protons (for instance, three methyl protons) split the resonances of the other groups of protons. There is no splitting within groups of equivalent protons. Each splitting pattern can be decided by referring to Pascal's triangle.

Answer The three protons of the CH_3 group split the resonance of the CH_2 protons into a 1:3:3:1 quartet with a splitting J. Likewise, the two protons of the CH_2 group split the resonance of the CH_3 protons into a 1:2:1 triplet with the same splitting J. The OH resonance is not split because the OH protons migrate rapidly from molecule to molecule (including molecules of impurities in the sample) and their effect averages to zero. In gaseous ethanol, where this migration does not occur, the OH resonance appears as a triplet, showing that the CH_2 protons interact with the OH proton.

Self-test 14.1 What fine-structure can be expected for the protons in $^{14}NH_4^+$? The spin quantum number of nitrogen-14 is 1. [1:1:1 triplet from N]

(c) The magnitudes of coupling constants

The scalar coupling constant of two nuclei separated by N bonds is denoted ^NJ, with subscripts for the types of nuclei involved. Thus, $^1J_{CH}$ is the coupling constant for a proton joined directly to a ^{13}C atom, and $^2J_{CH}$ is the coupling constant when the same two nuclei are separated by two bonds (as in ^{13}C–C–H). A typical value of $^1J_{CH}$ is in the range 120 to 250 Hz; $^2J_{CH}$ is between −10 and +20 Hz. Both ^3J and ^4J can give detectable effects in a spectrum, but couplings over larger numbers of bonds can generally be ignored. One of the longest range couplings that has been detected is $^9J_{HH} = 0.4$ Hz between the CH_3 and CH_2 protons in $CH_3C{\equiv}C{-}C{\equiv}C{-}C{\equiv}C{-}CH_2OH$.

As we have remarked (in the discussion following eqn 14.25b), the sign of J_{XY} indicates whether the energy of two spins is lower when they are parallel ($J<0$) or when they are antiparallel ($J>0$). It is found that $^1J_{CH}$ is often positive, $^2J_{HH}$ is often negative, $^3J_{HH}$ is often positive, and so on. An additional point is that J varies with the angle between the bonds (Fig. 14.20). Thus, a $^3J_{HH}$ coupling constant is often found to depend on the dihedral angle ϕ (4) according to the **Karplus equation:**

H ϕ H

4

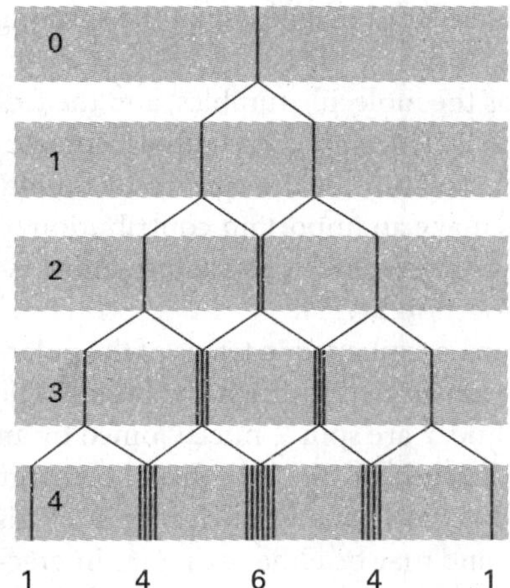

Fig. 14.18 The intensity distribution of the A resonance of an AX_n resonance can be constructed by considering the splitting caused by 1, 2, . . . n protons, as in Figs. 14.16 and 14.17. The resulting intensity distribution has a binomial distribution and is given by the integers in the corresponding row of Pascal's triangle. Note that, although the lines have been drawn side-by-side for clarity, the members of each group are coincident. Four protons, in AX_4, split the A resonance into a 1:4:6:4:1 quintet.

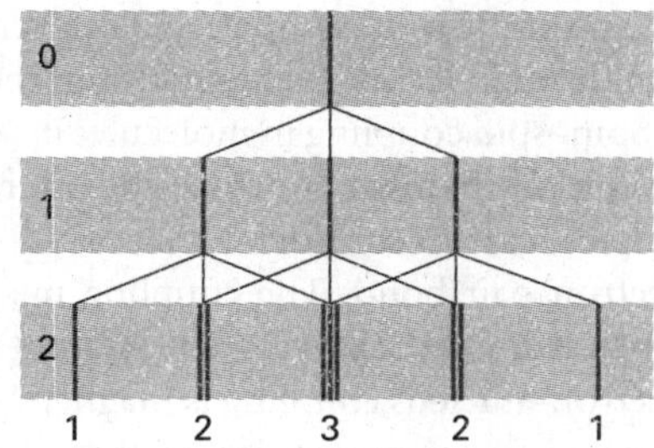

Fig. 14.19 The intensity distribution arising from spin–spin interaction with nuclei with $I=1$ can be constructed similarly, but each successive nucleus splits the lines into three equal intensity components. Two equivalent spin-1 nuclei give rise to a 1:2:3:2:1 quintet.

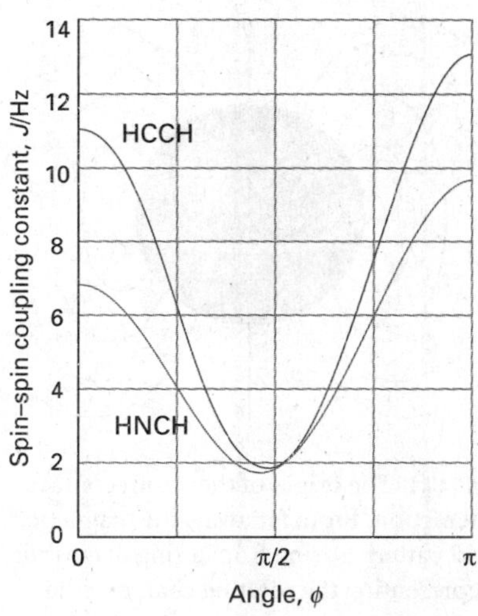

Fig. 14.20 The variation of the spin–spin coupling constant with angle predicted by the Karplus equation for an HCCH group and an HNCH group.

interActivity Draw a family of curves showing the variation of $^3J_{HH}$ with ϕ for which $A=+7.0$ Hz, $B=-1.0$ Hz, and C varies slightly from a typical value of +5.0 Hz. What is the effect of changing the value of the parameter C on the shape of the curve? In a similar fashion, explore the effect of the values of A and B on the shape of the curve.

$$^3J_{HH}=A+B\cos\phi+C\cos 2\phi \quad \text{Karplus equation} \quad (14.27)$$

with A, B, and C empirical constants with values close to +7 Hz, −1 Hz, and +5 Hz, respectively, for an HCCH fragment. It follows that the measurement of $^3J_{HH}$ in a series of related compounds can be used to determine their conformations. The coupling constant $^1J_{CH}$ also depends on the hybridization of the C atom, as the following values indicate:

	sp	sp^2	sp^3
$^1J_{CH}$/Hz	250	160	125

(d) The origin of spin–spin coupling

Spin–spin coupling is a very subtle phenomenon and it is better to treat J as an empirical parameter than to use calculated values. However, we can get some insight into its origins, if not its precise magnitude—or always reliably its sign—by considering the magnetic interactions within molecules.

A nucleus with spin projection m_I gives rise to a magnetic field with z-component $\mathcal{B}_{nuc}$ at a distance R, where, to a good approximation,

$$\mathcal{B}_{nuc}=-\frac{\gamma\hbar\mu_0}{4\pi R^3}(1-3\cos^2\theta)m_I \quad (14.28)$$

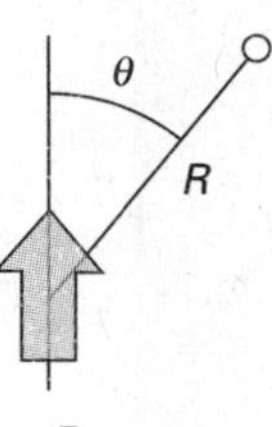

5

The angle θ is defined in (5); we saw a version of this expression in eqn 14.23a. The magnitude of this field is about 0.1 mT when $R=0.3$ nm, corresponding to a splitting

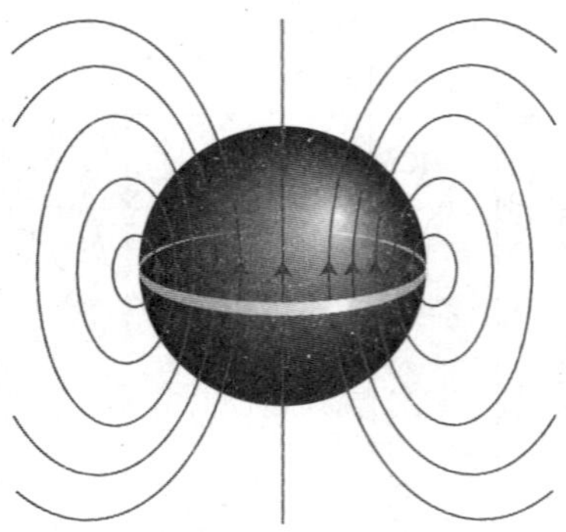

Fig. 14.21 The origin of the Fermi contact interaction. From far away, the magnetic field pattern arising from a ring of current (representing the rotating charge of the nucleus, the pale grey sphere) is that of a point dipole. However, if an electron can sample the field close to the region indicated by the sphere, the field distribution differs significantly from that of a point dipole. For example, if the electron can penetrate the sphere, then the spherical average of the field it experiences is not zero.

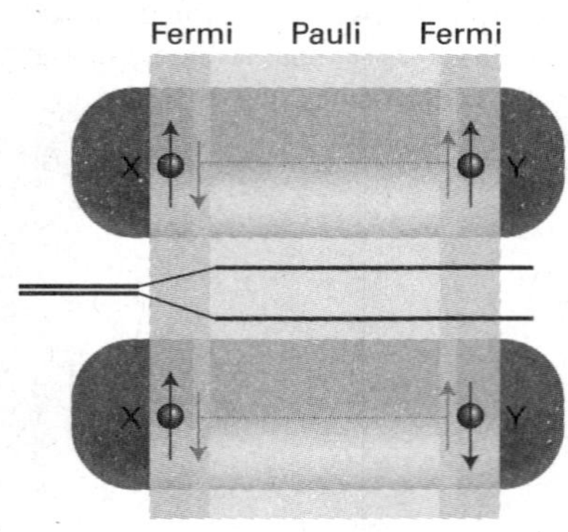

Fig. 14.22 The polarization mechanism for spin–spin coupling ($^1J_{HH}$). The two arrangements have slightly different energies. In this case, J is positive, corresponding to a lower energy when the nuclear spins are antiparallel.

of resonance signal of about 10^4 Hz, and is of the order of magnitude of the splitting observed in solid samples (see Section 14.13a).

In a liquid, the angle θ sweeps over all values as the molecule tumbles, and the factor $1 - 3\cos^2\theta$ averages to zero (see Problem 14.17). Hence the direct dipolar interaction between spins cannot account for the fine structure of the spectra of rapidly tumbling molecules. The direct interaction does make an important contribution to the spectra of solid samples and is a very useful indirect source of structure information through its involvement in spin relaxation (Section 14.11).

Spin–spin coupling in molecules in solution can be explained in terms of the **polarization mechanism**, in which the interaction is transmitted through the bonds. The simplest case to consider is that of $^1J_{XY}$ where X and Y are spin-$\frac{1}{2}$ nuclei joined by an electron-pair bond. The coupling mechanism depends on the fact that the energy depends on the relative orientation of the bonding electron and nuclear spins. This electron–nucleus coupling is magnetic in origin, and may be either a dipolar interaction or a **Fermi contact interaction**. A pictorial description of the latter is as follows. First, we regard the magnetic moment of the nucleus as arising from the circulation of a current in a tiny loop with a radius similar to that of the nucleus (Fig. 14.21). Far from the nucleus the field generated by this loop is indistinguishable from the field generated by a point magnetic dipole. Close to the loop, however, the field differs from that of a point dipole. The magnetic interaction between this non-dipolar field and the electron's magnetic moment is the contact interaction. The contact interaction—essentially the failure of the point-dipole approximation—depends on the very close approach of an electron to the nucleus and hence can occur only if the electron occupies an s orbital (which is the reason why $^1J_{CH}$ depends on the hybridization ratio). We shall suppose that it is energetically favourable for an electron spin and a nuclear spin to be antiparallel (as is the case for a proton and an electron in a hydrogen atom).

If the X nucleus is α, a β electron of the bonding pair will tend to be found nearby, because that is an energetically favourable arrangement (Fig. 14.22). The second electron in the bond, which must have α spin if the other is β (by the Pauli principle), will be found mainly at the far end of the bond because electrons tend to stay apart to reduce their mutual repulsion. Because it is energetically favourable for the spin of Y to be antiparallel to an electron spin, a Y nucleus with β spin has a lower energy than when it has α spin. The opposite is true when X is β, for now the α spin of Y has the lower energy. In other words, the antiparallel arrangement of nuclear spins lies lower in energy than the parallel arrangement as a result of their magnetic coupling with the bond electrons. That is, $^1J_{CH}$ is positive.

To account for the value of $^2J_{XY}$, as in H–C–H, we need a mechanism that can transmit the spin alignments through the central C atom (which may be ^{12}C, with no nuclear spin of its own). In this case (Fig. 14.23), an X nucleus with α spin polarizes the electrons in its bond, and the α electron is likely to be found closer to the C nucleus. The more favourable arrangement of two electrons on the same atom is with their spins parallel (Hund's rule, Section 9.4), so the more favourable arrangement is for the α electron of the neighbouring bond to be close to the C nucleus. Consequently, the β electron of that bond is more likely to be found close to the Y nucleus, and therefore that nucleus will have a lower energy if it is α. Hence, according to this mechanism, the lower energy will be obtained if the Y spin is parallel to that of X. That is, $^2J_{HH}$ is negative.

The coupling of nuclear spin to electron spin by the Fermi contact interaction is most important for proton spins, but it is not necessarily the most important mechanism for other nuclei. These nuclei may also interact by a dipolar mechanism with the electron magnetic moments and with their orbital motion, and there is no simple way of specifying whether J will be positive or negative.

(e) Equivalent nuclei

A group of nuclei are **chemically equivalent** if they are related by a symmetry operation of the molecule and have the same chemical shifts. Chemically equivalent nuclei are nuclei that would be regarded as 'equivalent' according to ordinary chemical criteria. Nuclei are **magnetically equivalent** if, as well as being chemically equivalent, they also have identical spin–spin interactions with any other magnetic nuclei in the molecule.

The difference between chemical and magnetic equivalence is illustrated by CH_2F_2 and $H_2C{=}CF_2$. In each of these molecules the protons are chemically equivalent: they are related by symmetry and undergo the same chemical reactions. However, although the protons in CH_2F_2 are magnetically equivalent, those in $CH_2{=}CF_2$ are not. One proton in the latter has a *cis* spin-coupling interaction with a given F nucleus whereas the other proton has a *trans* interaction with it. In contrast, in CH_2F_2 both protons are connected to a given F nucleus by identical bonds, so there is no distinction between them. Strictly speaking, the CH_3 protons in ethanol (and other compounds) are magnetically inequivalent on account of their different interactions with the CH_2 protons in the next group. However, they are in practice made magnetically equivalent by the rapid rotation of the CH_3 group, which averages out any differences. Magnetically inequivalent species can give very complicated spectra (for instance, the proton and ^{19}F spectra of $H_2C{=}CF_2$ each consist of 12 lines), and we shall not consider them further.

An important feature of chemically equivalent magnetic nuclei is that, although they do couple together, the coupling has no effect on the appearance of the spectrum. The reason for the invisibility of the coupling is set out in the following *Justification*, but qualitatively it is that all allowed nuclear spin transitions are *collective* reorientations of groups of equivalent nuclear spins that do not change the relative orientations of the spins within the group (Fig. 14.24). Then, because the relative orientations of nuclear spins are not changed in any transition, the magnitude of the coupling between them is undetectable. Hence, an isolated CH_3 group gives a single, unsplit line because all the allowed transitions of the group of three protons occur without change of their relative orientations.

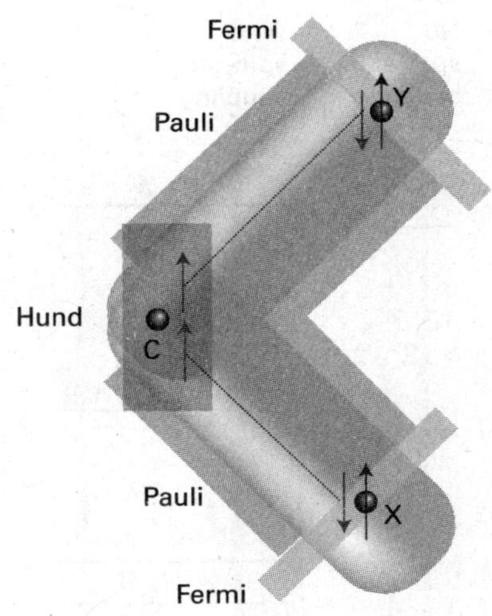

Fig. 14.23 The polarization mechanism for $^2J_{HH}$ spin–spin coupling. The spin information is transmitted from one bond to the next by a version of the mechanism that accounts for the lower energy of electrons with parallel spins in different atomic orbitals (Hund's rule of maximum multiplicity). In this case, $J < 0$, corresponding to a lower energy when the nuclear spins are parallel.

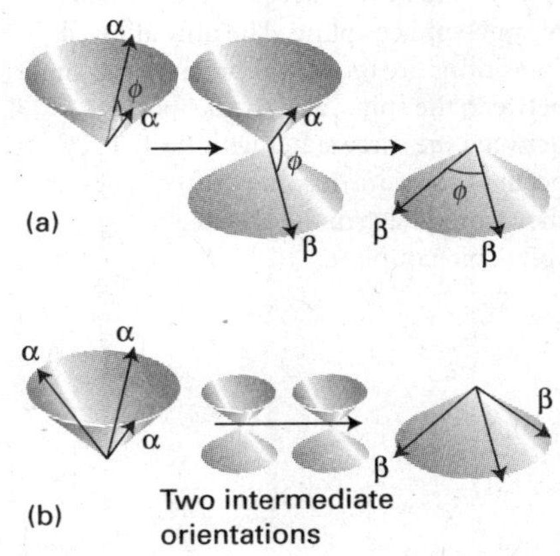

Fig. 14.24 (a) A group of two equivalent nuclei realigns as a group, without change of angle between the spins, when a resonant absorption occurs. Hence it behaves like a single nucleus and the spin–spin coupling between the individual spins of the group is undetectable. (b) Three equivalent nuclei also realign as a group without change of their relative orientations.

Justification 14.3 *The energy levels of an A_2 system*

Consider an A_2 system of two spin-$\frac{1}{2}$ nuclei. First, consider the energy levels in the absence of spin–spin coupling. There are four spin states that (just as for two electrons) can be classified according to their total spin I (the analogue of S for two electrons) and their total projection M_I on the z-axis. The states are analogous to those we developed for two electrons in singlet and triplet states (eqn 9.41):

Spins parallel, $I = 1$:	$M_I = +1$	$\alpha\alpha$
	$M_I = 0$	$(1/2^{1/2})\{\alpha\beta + \beta\alpha\}$
	$M_I = -1$	$\beta\beta$
Spins paired, $I = 0$:	$M_I = 0$	$(1/2^{1/2})\{\alpha\beta - \beta\alpha\}$

The sign in $\alpha\beta + \beta\alpha$ signifies an in-phase alignment of spins and $I = 1$; the – sign in $\alpha\beta - \beta\alpha$ signifies an alignment out of phase by π, and hence $I = 0$ (see Fig. 9.18). The effect of a magnetic field on these four states is shown in Fig. 14.25: the energies of the two states with $M_I = 0$ are unchanged by the field because they are composed of equal proportions of α and β spins.

As remarked in Section 14.6, the spin–spin coupling energy is proportional to the scalar product of the vectors representing the spins, $E = (hJ/\hbar^2)\boldsymbol{I}_1 \cdot \boldsymbol{I}_2$. The scalar product can be expressed in terms of the total nuclear spin by noting that

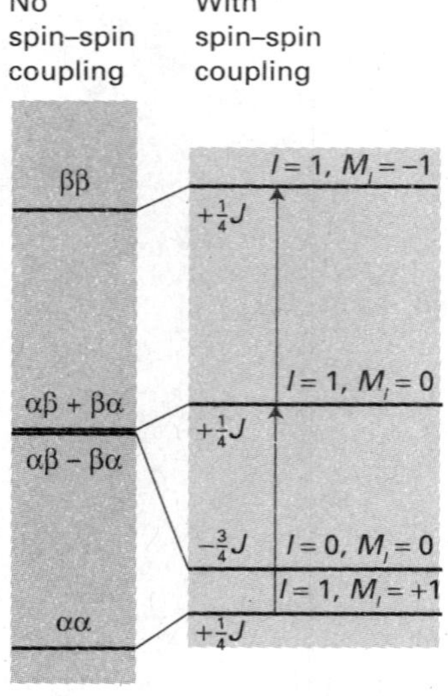

Fig. 14.25 The energy levels of an A_2 system in the absence of spin–spin coupling are shown on the left. When spin–spin coupling is taken into account, the energy levels on the right are obtained. Note that the three states with total nuclear spin $I = 1$ correspond to parallel spins and give rise to the same increase in energy (J is positive); the one state with $I = 0$ (antiparallel nuclear spins) has a lower energy in the presence of spin–spin coupling. The only allowed transitions are those that preserve the angle between the spins, and so take place between the three states with $I = 1$. They occur at the same resonance frequency as they would have in the absence of spin–spin coupling.

$$I^2 = (\boldsymbol{I}_1 + \boldsymbol{I}_2)\cdot(\boldsymbol{I}_1 + \boldsymbol{I}_2) = I_1^2 + I_2^2 + 2\boldsymbol{I}_1\cdot\boldsymbol{I}_2$$

rearranging this expression to

$$\boldsymbol{I}_1\cdot\boldsymbol{I}_2 = \tfrac{1}{2}\{I^2 - I_1^2 - I_2^2\}$$

and replacing the magnitudes by their quantum mechanical values:

$$\boldsymbol{I}_1\cdot\boldsymbol{I}_2 = \tfrac{1}{2}\{I(I+1) - I_1(I_1+1) - I_2(I_2+1)\}\hbar^2$$

Then, because $I_1 = I_2 = \frac{1}{2}$, it follows that

$$E = \tfrac{1}{2}hJ\{I(I+1) - \tfrac{3}{2}\}$$

For parallel spins, $I = 1$ and $E = +\frac{1}{4}hJ$; for antiparallel spins $I = 0$ and $E = -\frac{3}{4}hJ$, as in Fig. 14.25. We see that three of the states move in energy in one direction and the fourth (the one with antiparallel spins) moves three times as much in the opposite direction. The resulting energy levels are shown on the right in Fig. 14.25.

The NMR spectrum of the A_2 species arises from transitions between the levels. However, the radiofrequency field affects the two equivalent protons equally, so it cannot change the orientation of one proton relative to the other; therefore, the transitions take place within the set of states that correspond to parallel spin (those labelled $I = 1$), and no spin-parallel state can change to a spin-antiparallel state (the state with $I = 0$). Put another way, the allowed transitions are subject to the selection rule $\Delta I = 0$.This selection rule is in addition to the rule $\Delta M_I = \pm 1$ that arises from the conservation of angular momentum and the unit spin of the photon. The allowed transitions are shown in Fig. 14.25: we see that there are only two transitions, and that they occur at the same resonance frequency that the nuclei would have in the absence of spin–spin coupling. Hence, the spin–spin coupling interaction does not affect the appearance of the spectrum.

(f) Strongly coupled nuclei

NMR spectra are usually much more complex than the foregoing simple analysis suggests. We have described the extreme case in which the differences in chemical shifts are much greater than the spin–spin coupling constants. In such cases it is simple to identify groups of magnetically equivalent nuclei and to think of the groups of nuclear spins as reorientating relative to each other. The spectra that result are called **first-order spectra**.

Transitions cannot be allocated to definite groups when the differences in their chemical shifts are comparable to their spin–spin coupling interactions. The complicated spectra that are then obtained are called **strongly coupled spectra** (or 'second-order spectra') and are much more difficult to analyse (Fig. 14.26). Because the difference in resonance frequencies increases with field, but spin–spin coupling constants are independent of it, a second-order spectrum may become simpler (and first-order) at high fields and individual groups of nuclei become identifiable again.

A clue to the type of analysis that is appropriate is given by the notation for the types of spins involved. Thus, an AX spin system (which consists of two nuclei with a large chemical shift difference) has a first-order spectrum. An AB system, on the other hand (with two nuclei of similar chemical shifts), gives a spectrum typical of a strongly coupled system. An AX system may have widely different Larmor frequencies because A and X are nuclei of different elements (such as ^{13}C and ^{1}H), in which case they form a **heteronuclear spin system**. AX may also denote a **homonuclear spin system** in which the nuclei are of the same element but in markedly different environments.

14.7 Conformational conversion and exchange processes

Key point **Coalescence of two NMR lines occurs when a conformational interchange or chemical exchange of nuclei is fast; the spectrum shows a single line at the mean of the two chemical shifts.**

The appearance of an NMR spectrum is changed if magnetic nuclei can jump rapidly between different environments. Consider a molecule, such as *N,N*-dimethylformamide, that can jump between conformations; in its case, the methyl shifts depend on whether they are *cis* or *trans* to the carbonyl group (Fig. 14.27). When the jumping rate is low, the spectrum shows two sets of lines, one each from molecules in each conformation. When the interconversion is fast, the spectrum shows a single line at the mean of the two chemical shifts. At intermediate inversion rates, the line is very broad. This maximum broadening occurs when the lifetime, τ, of a conformation gives rise to a linewidth that is comparable to the difference of resonance frequencies, $\delta\nu$ and both broadened lines blend together into a very broad line. Coalescence of the two lines occurs when

$$\tau = \frac{\sqrt{2}}{\pi \delta\nu} \qquad \text{Condition for coalescence of two NMR lines} \qquad (14.29)$$

• **A brief illustration**

The NO group in *N,N*-dimethylnitrosamine, $(CH_3)_2N{-}NO$ (**6**), rotates about the N–N bond and, as a result, the magnetic environments of the two CH_3 groups are interchanged. The two CH_3 resonances are separated by 390 Hz in a 600 MHz spectrometer. According to eqn 14.29,

$$\tau = \frac{\sqrt{2}}{\pi \times (390\ \text{s}^{-1})} = 1.2\ \text{ms}$$

It follows that the signal will collapse to a single line when the interconversion rate exceeds about $1/\tau = 830\ \text{s}^{-1}$. •

Self-test 14.2 What would you deduce from the observation of a single line from the same molecule in a 300 MHz spectrometer?

[Conformation lifetime less than 2.3 ms]

A similar explanation accounts for the loss of fine structure in solvents able to exchange protons with the sample. For example, hydroxyl protons are able to exchange with water protons. When this **chemical exchange** occurs, a molecule ROH with an α-spin proton (we write this ROH_α) rapidly converts to ROH_β and then perhaps to ROH_α again because the protons provided by the solvent molecules in successive exchanges have random spin orientations. Therefore, instead of seeing a spectrum composed of contributions from both ROH_α and ROH_β molecules (that is, a spectrum showing a doublet structure due to the OH proton) we see a spectrum that shows no splitting caused by coupling of the OH proton (as in Fig. 14.6). The effect is observed when the lifetime of a molecule due to this chemical exchange is so short that the lifetime broadening is greater than the doublet splitting. Because this splitting is often very small (a few hertz), a proton must remain attached to the same molecule for longer than about 0.1 s for the splitting to be observable. In water, the exchange rate is much faster than that, so alcohols show no splitting from the OH protons. In dry dimethylsulfoxide (DMSO), the exchange rate may be slow enough for the splitting to be detected.

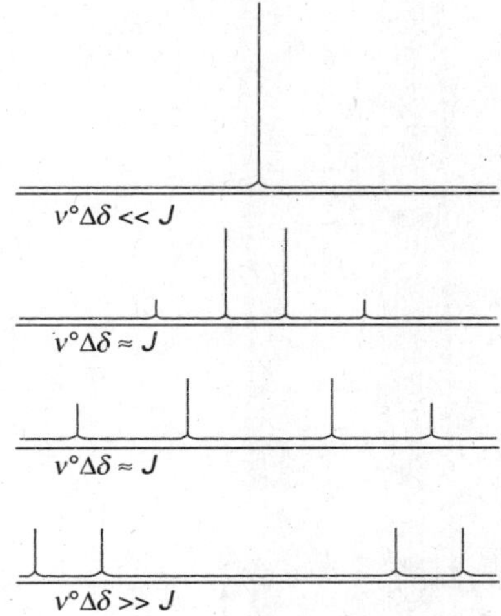

Fig. 14.26 The NMR spectra of an A_2 system (top) and an AX system (bottom) are simple 'first-order' spectra. At intermediate relative values of the chemical shift difference and the spin–spin coupling, complex 'strongly coupled' spectra are obtained. Note how the inner two lines of the bottom spectrum move together, grow in intensity, and form the single central line of the top spectrum. The two outer lines diminish in intensity and are absent in the top spectrum.

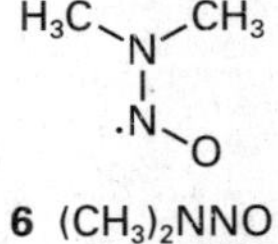

6 $(CH_3)_2NNO$

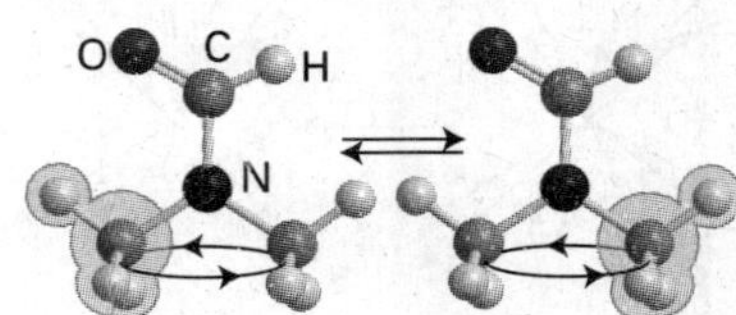

Fig. 14.27 When a molecule changes from one conformation to another, the positions of its protons are interchanged and jump between magnetically distinct environments.

Pulse techniques in NMR

The common method of detecting the energy separation between nuclear spin states is more sophisticated than simply looking for the frequency at which resonance occurs. One of the best analogies that has been suggested to illustrate the preferred way of observing an NMR spectrum is that of detecting the spectrum of vibrations of a bell. We could stimulate the bell with a gentle vibration at a gradually increasing frequency, and note the frequencies at which it resonated with the stimulation. A lot of time would be spent getting zero response when the stimulating frequency was between the bell's vibrational modes. However, if we were simply to hit the bell with a hammer, we would immediately obtain a clang composed of all the frequencies that the bell can produce. The equivalent in NMR is to monitor the radiation nuclear spins emit as they return to equilibrium after the appropriate stimulation. The resulting **Fourier-transform NMR** gives greatly increased sensitivity, so opening up much of the periodic table to the technique. Moreover, multiple-pulse FTNMR gives chemists unparalleled control over the information content and display of spectra. We need to understand how the equivalent of the hammer blow is delivered and how the signal is monitored and interpreted. These features are generally expressed in terms of the vector model of angular momentum introduced in Section 8.7d; the mathematical basis of Fourier transform techniques in general is discussed in *Mathematical background 7* following Chapter 19.

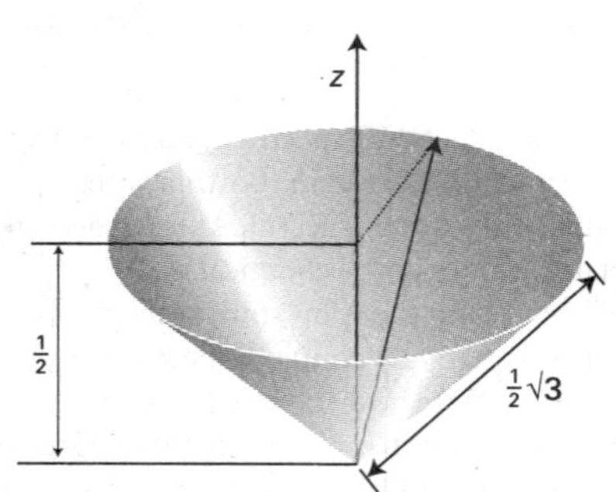

Fig. 14.28 The vector model of angular momentum for a single spin-$\frac{1}{2}$ nucleus. The angle around the z-axis is indeterminate.

14.8 The magnetization vector

Key points **(a) In the presence of a magnetic field, the magnetization vector grows in magnitude and precesses at the Larmor frequency. (b) When a radiofrequency pulse is applied, the magnetization vector tips and rotates in a different plane. Free-induction decay (FID) is the decay of the magnetization after the pulse. (c) Fourier transformation of the FID curve gives the NMR spectrum.**

Consider a sample composed of many identical spin-$\frac{1}{2}$ nuclei. By analogy with the discussion of angular momenta in Section 8.7d, a nuclear spin can be represented by a vector of length $\{I(I+1)\}^{1/2}$ units with a component of length m_I units along the z-axis. As the uncertainty principle does not allow us to specify the x- and y-components of the angular momentum, all we know is that the vector lies somewhere on a cone around the z-axis. For $I=\frac{1}{2}$, the length of the vector is $\frac{1}{2}\sqrt{3}$ and it makes an angle of 55° to the z-axis (Fig. 14.28).

In the absence of a magnetic field, the sample consists of equal numbers of α and β nuclear spins with their vectors lying at random angles on the cones. These angles are unpredictable, and at this stage we picture the spin vectors as stationary. The **magnetization**, M, of the sample, its net nuclear magnetic moment, is zero (Fig. 14.29a).

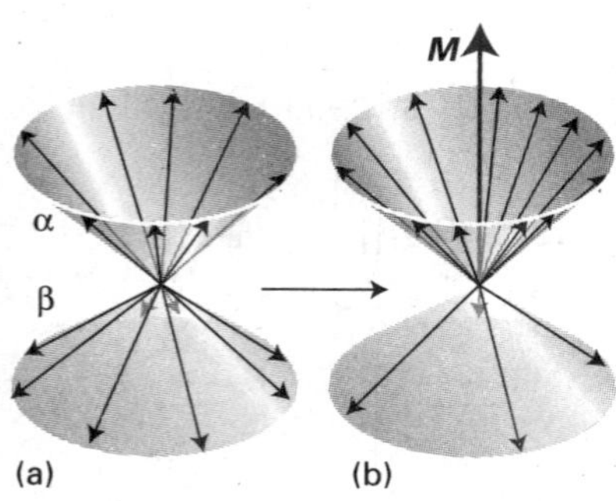

Fig. 14.29 The magnetization of a sample of spin-$\frac{1}{2}$ nuclei is the resultant of all their magnetic moments. (a) In the absence of an externally applied field, there are equal numbers of α and β spins at random angles around the z-axis (the field direction) and the magnetization is zero. (b) In the presence of a field, the spins precess around their cones (that is, there is an energy difference between the α and β states) and there are slightly more α spins than β spins. As a result, there is a net magnetization along the z-axis.

(a) The effect of the static field

Two changes occur in the magnetization when a magnetic field of magnitude $\mathcal{B}_0$ is present and aligned in the z-direction. First, the energies of the two orientations change, the α spins moving to low energy and the β spins to high energy (provided $\gamma > 0$). At 10 T, the Larmor frequency for protons is 427 MHz, and in the vector model the individual vectors are pictured as precessing at this rate. This motion is a pictorial representation of the difference in energy of the spin states (it is not an actual representation of reality but is inspired by the actual motion of a classical bar magnet in a magnetic field). As the field is increased, the Larmor frequency increases and the precession becomes faster. Secondly, the populations of the two spin states (the numbers of α and β spins) at thermal equilibrium change, and there will be more α spins than β spins. Because $h\nu_L/kT \approx 7\times10^{-5}$ for protons at 300 K and 10 T, it follows from the Boltzmann distribution that $N_\beta/N_\alpha = e^{-h\nu_L/kT}$ is only slightly less than 1. That is, there is only a tiny imbalance of populations, and it is even smaller for other nuclei with their smaller magnetogyric ratios. However, despite its smallness, the imbalance means that there is a net magnetization that we can represent by a vector M pointing in the z-direction and with a length proportional to the population difference (Fig. 14.29b).

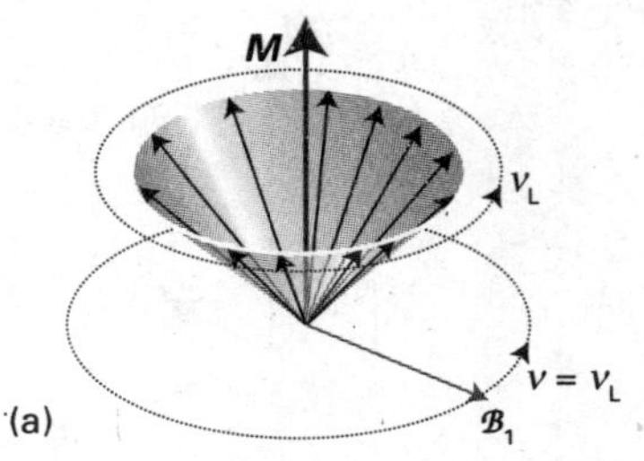

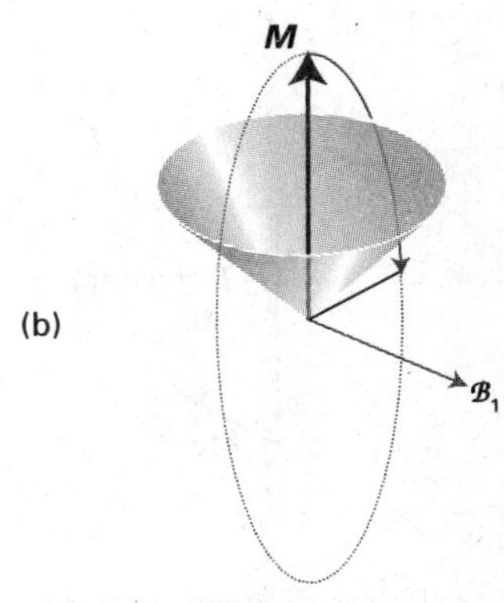

Fig. 14.30 (a) In a resonance experiment, a circularly polarized radiofrequency magnetic field $\mathcal{B}_1$ is applied in the xy-plane (the magnetization vector lies along the z-axis). (b) If we step into a frame rotating at the radiofrequency, $\mathcal{B}_1$ appears to be stationary, as does the magnetization M if the Larmor frequency is equal to the radiofrequency. When the two frequencies coincide, the magnetization vector of the sample rotates around the direction of the $\mathcal{B}_1$ field.

(b) The effect of the radiofrequency field

We now consider the effect of a radiofrequency field circularly polarized in the xy-plane, so that the magnetic component of the electromagnetic field (the only component we need to consider) is rotating around the z-direction in the same sense as the Larmor precession of the nuclei. The strength of the rotating magnetic field is $\mathcal{B}_1$.

To interpret the effects of radiofrequency pulses on the magnetization, it is useful to imagine stepping on to a platform, a so-called **rotating frame**, that rotates around the direction of the applied field. Suppose we choose the frequency of the radiofrequency field to be equal to the Larmor frequency of the spins, $\nu_L = \gamma\mathcal{B}_0/2\pi$; this choice is equivalent to selecting the resonance condition in the conventional experiment. The rotating magnetic field is in step with the precessing spins, the nuclei experience a steady $\mathcal{B}_1$ field, and precess about it at a frequency $\gamma\mathcal{B}_1/2\pi$ (Fig. 14.30). Now suppose that the $\mathcal{B}_1$ field is applied in a pulse of duration $\frac{1}{4}\times(2\pi/\gamma\mathcal{B}_1)$, the magnetization tips through an angle of $\frac{1}{4}\times 2\pi = \pi/2$ (90°) in the rotating frame and we say that we have applied a **90° pulse**, or a 'π/2 pulse' (Fig. 14.31a). The duration of the pulse depends on the strength of the $\mathcal{B}_1$ field, but is typically of the order of microseconds.

Now imagine stepping out of the rotating frame. To a fixed external observer (the role played by a radiofrequency coil), the magnetization vector is rotating at the Larmor frequency in the xy-plane (Fig. 14.31b). The rotating magnetization induces in the coil a signal that oscillates at the Larmor frequency and that can be amplified and processed. In practice, the processing takes place after subtraction of a constant high frequency component (the radiofrequency used for $\mathcal{B}_1$), so that all the signal manipulation takes place at frequencies of a few kilohertz.

As time passes, the individual spins move out of step (partly because they are precessing at slightly different rates, as we shall explain later), so the magnetization vector shrinks exponentially with a time constant T_2 and induces an ever weaker signal in the detector coil. The form of the signal that we can expect is therefore the oscillating-decaying **free-induction decay** (FID) shown in Fig. 14.32. The y-component of the magnetization varies as

$$M_y(t) = M_0\cos(2\pi\nu_L t)e^{-t/T_2} \qquad \text{Free induction decay} \qquad (14.30)$$

We have considered the effect of a pulse applied at exactly the Larmor frequency. However, virtually the same effect is obtained off resonance, provided that the pulse is applied close to ν_L. If the difference in frequency is small compared to the inverse of the duration of the 90° pulse, the magnetization will end up in the xy-plane. Note that we do not need to know the Larmor frequency beforehand: the short pulse is the analogue of the hammer blow on the bell, exciting a range of frequencies. The detected signal shows that a particular resonant frequency is present.

(c) Time- and frequency-domain signals

We can think of the magnetization vector of a homonuclear AX spin system with $J=0$ as consisting of two parts, one formed by the A spins and the other by the X spins. When the 90° pulse is applied, both magnetization vectors are rotated into the xy-plane. However, because the A and X nuclei precess at different frequencies, they

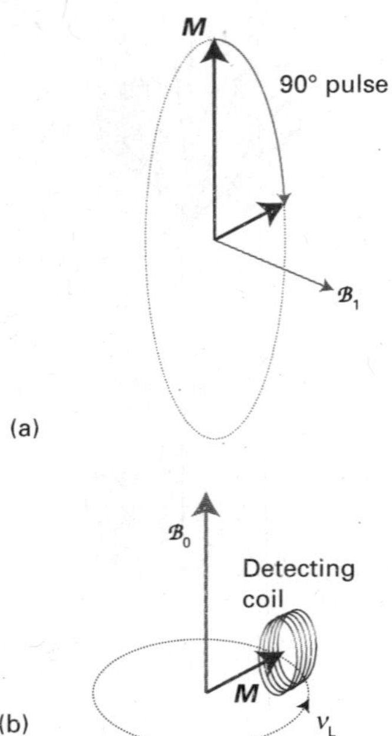

Fig. 14.31 (a) If the radiofrequency field is applied for a certain time, the magnetization vector is rotated into the *xy*-plane. (b) To an external stationary observer (the coil), the magnetization vector is rotating at the Larmor frequency, and can induce a signal in the coil.

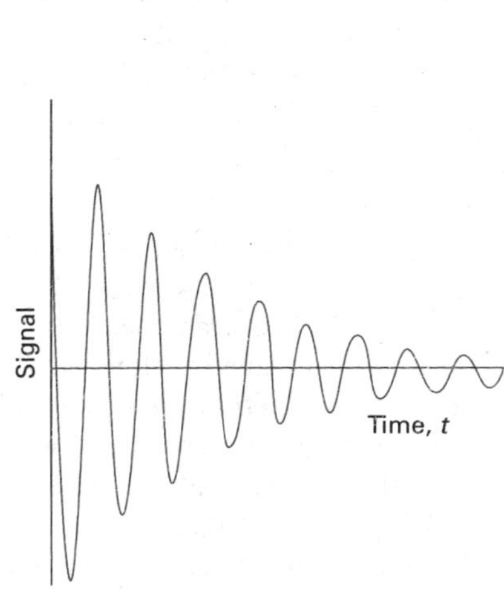

Fig. 14.32 A simple free-induction decay of a sample of spins with a single resonance frequency.

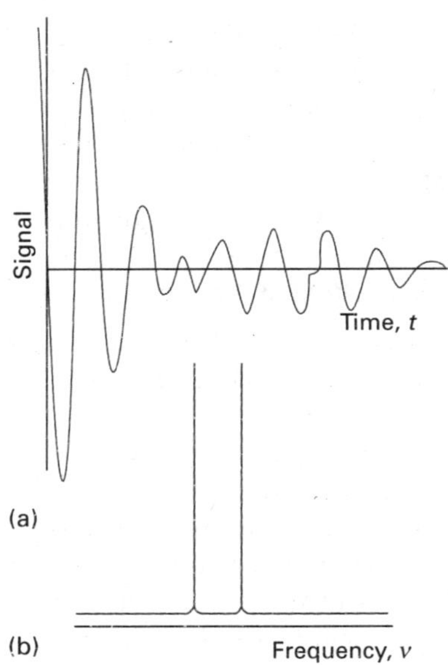

Fig. 14.33 (a) A free induction decay signal of a sample of AX species and (b) its analysis into its frequency components.

interActivity The *Living graphs* section of the text's web site has an applet that allows you to calculate and display the FID curve from an AX system. Explore the effect on the shape of the FID curve of changing the chemical shifts (and therefore the Larmor frequencies) of the A and X nuclei.

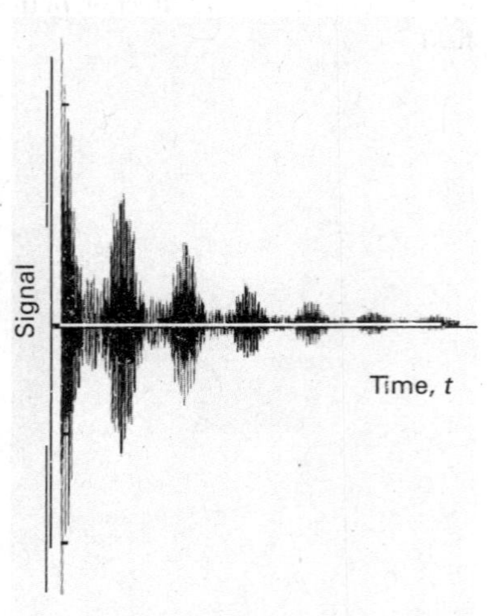

Fig. 14.34 A free induction decay signal of a sample of ethanol. Its Fourier transform is the frequency-domain spectrum shown in Fig. 14.6. The total length of the image corresponds to about 1 s.

induce two signals in the detector coils, and the overall FID curve may resemble that in Fig. 14.33a. The composite FID curve is the analogue of the struck bell emitting a rich tone composed of all the frequencies (in this case, just the two resonance frequencies of the uncoupled A and X nuclei) at which it can vibrate.

The problem we must address is how to recover the resonance frequencies present in a free-induction decay. We know that the FID curve is a sum of decaying oscillating functions, so the problem is to analyse it into its components by carrying out a Fourier transformation (*Further information 14.1* and *Mathematical background 7*). When the signal in Fig. 14.33a is transformed in this way, we get the frequency-domain spectrum shown in Fig. 14.33b. One line represents the Larmor frequency of the A nuclei and the other that of the X nuclei.

The FID curve in Fig. 14.34 is obtained from a sample of ethanol. The frequency-domain spectrum obtained from it by Fourier transformation is the one that we have already discussed (Fig. 14.6). We can now see why the FID curve in Fig. 14.34 is so complex: it arises from the precession of a magnetization vector that is composed of eight components, each with a characteristic frequency.

14.9 Spin relaxation

Key points **Spin relaxation is the return of a spin system to equilibrium. (a) During longitudinal (or spin–lattice) relaxation, β spins revert to α spins. Transverse (or spin–spin) relaxation is the randomization of spin directions. (b) The longitudinal relaxation time T_1 can be measured by the inversion recovery technique. (c) The transverse relaxation time T_2 can be measured by observing spin echoes.**

There are two reasons why the component of the magnetization vector in the xy-plane shrinks. Both reflect the fact that the nuclear spins are not in thermal equilibrium with their surroundings (for then M lies parallel to z). At thermal equilibrium the spins have a Boltzmann distribution, with more α spins than β spins. The return to equilibrium is the process called **spin relaxation.**

(a) Longitudinal and transverse relaxation

Consider the effect of a 180° pulse, which may be visualized in the rotating frame as a flip of the net magnetization vector from one direction along the z-axis (with more α spins than β spins) to the opposite direction (with more β spins than α spins). After the pulse, the populations revert to their thermal equilibrium values exponentially. As they do so, the z-component of magnetization reverts to its equilibrium value M_0 with a time constant called the **longitudinal relaxation time,** T_1 (Fig. 14.35):

$$M_z(t) - M_0 \propto e^{-t/T_1} \qquad \text{Definition of longitudinal relaxation time} \qquad (14.31)$$

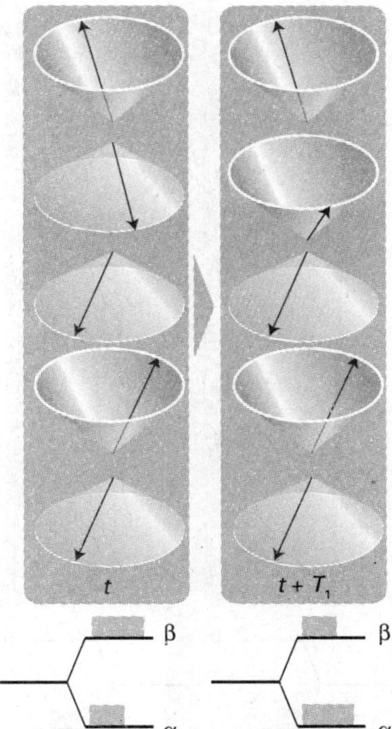

Fig. 14.35 In longitudinal relaxation the spins relax back towards their thermal equilibrium populations. On the left we see the precessional cones representing spin-$\frac{1}{2}$ angular momenta, and they do not have their thermal equilibrium populations (there are more β-spins than α-spins). On the right, which represents the sample a long time after a time T_1 has elapsed, the populations are those characteristic of a Boltzmann distribution. In actuality, T_1 is the time constant for relaxation to the arrangement on the right and T_1 ln 2 is the half-life of the arrangement on the left.

Because this relaxation process involves giving up energy to the surroundings (the 'lattice') as β spins revert to α spins, the time constant T_1 is also called the **spin–lattice relaxation time.** Spin–lattice relaxation is caused by local magnetic fields that fluctuate at a frequency close to the resonance frequency of the β → α transition. Such fields can arise from the tumbling motion of molecules in a fluid sample. If molecular tumbling is too slow or too fast compared to the resonance frequency, it will give rise to a fluctuating magnetic field with a frequency that is either too low or too high to stimulate a spin change from β to α, so T_1 will be long. Only if the molecule tumbles at about the resonance frequency will the fluctuating magnetic field be able to induce spin changes effectively, and only then will T_1 be short. The rate of molecular tumbling increases with temperature and with reducing viscosity of the solvent, so we can expect a dependence like that shown in Fig. 14.36. The quantitative treatment of relaxation times depends on setting up models of molecular motion and using, for instance, the diffusion equation (Section 20.9).

Now consider the events following a 90° pulse. The magnetization vector in the xy-plane is large when the spins are bunched together immediately after the pulse. However, this orderly bunching of spins is not at equilibrium and, even if there were no spin–lattice relaxation, we would expect the individual spins to spread out until they were uniformly distributed with all possible angles around the z-axis (Fig. 14.37). At that stage, the component of magnetization vector in the plane would be zero. The randomization of the spin directions occurs exponentially with a time constant called the **transverse relaxation time,** T_2:

$$M_y(t) \propto e^{-t/T_2} \qquad \text{Definition of transverse relaxation time} \qquad (14.32)$$

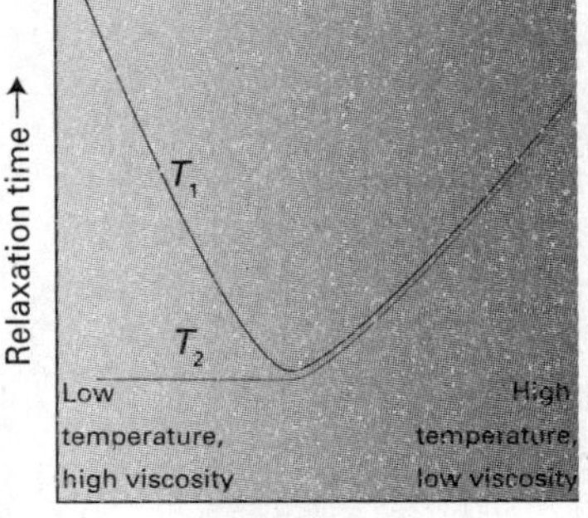

Fig. 14.36 The variation of the two relaxation times with the rate at which the molecules move (either by tumbling or migrating through the solution). The horizontal axis can be interpreted as representing temperature or viscosity. Note that, at rapid rates of motion, the two relaxation times coincide.

Because the relaxation involves the relative orientation of the spins around their respective cones, T_2 is also known as the **spin–spin relaxation time.** Any relaxation process that changes the balance between α and β spins will also contribute to this randomization, so the time constant T_2 is almost always less than or equal to T_1.

Local magnetic fields also affect spin–spin relaxation. When the fluctuations are slow, each molecule lingers in its local magnetic environment and the spin orientations randomize quickly around their cones. If the molecules move rapidly from one magnetic environment to another, the effects of differences in local magnetic field average to zero: individual spins do not precess at very different rates, they can remain bunched for longer, and spin–spin relaxation does not take place as quickly. In other words, slow molecular motion corresponds to short T_2 and fast motion corresponds

Fig. 14.37 The transverse relaxation time, T_2, is the time constant for the phases of the spins to become randomized (another condition for equilibrium) and to change from the orderly arrangement shown on the left to the disorderly arrangement on the right (long after a time T_2 has elapsed). Note that the populations of the states remain the same; only the relative phase of the spins relaxes. In actuality, T_2 is the time constant for relaxation to the arrangement on the right and $T_2 \ln 2$ is the half-life of the arrangement on the left.

to long T_2 (as shown in Fig. 14.36). Calculations show that, when the motion is fast, the main randomizing effect arises from $\beta \rightarrow \alpha$ transitions rather than different precession rates on the cones, and then $T_2 \approx T_1$.

If the y-component of magnetization decays with a time constant T_2, the spectral line is broadened (Fig. 14.38), and its width at half-height becomes

$$\Delta \nu_{1/2} = \frac{1}{\pi T_2} \qquad (14.33)$$

Width at half-height of an NMR line

This connection between decay rate and spectral width emerges naturally from a Fourier analysis (*Mathematical background 7*). Typical values of T_2 in proton NMR are of the order of seconds, so linewidths of around 0.1 Hz can be anticipated, in broad agreement with observation.

So far, we have assumed that the equipment, and in particular the magnet, is perfect, and that the differences in Larmor frequencies arise solely from interactions within the sample. In practice, the magnet is not perfect, and the field is different at different locations in the sample. The inhomogeneity broadens the resonance, and in most cases this **inhomogeneous broadening** dominates the broadening we have discussed so far. It is common to express the extent of inhomogeneous broadening in terms of an **effective transverse relaxation time,** T_2^*, by using a relation like eqn 14.33, but writing

$$T_2^* = \frac{1}{\pi \Delta \nu_{1/2}} \qquad [14.34]$$

Definition of effective transverse relaxation time

where $\Delta \nu_{1/2}$ is the observed width at half-height of a line with a Lorenztian shape of the form $I \propto 1/(1 + \nu^2)$.

• A brief illustration

Consider a line in a spectrum with a width of 10 Hz. It follows from eqn 14.34 that the effective transverse relaxation time is

$$T_2^* = \frac{1}{\pi \times (10\ \mathrm{s}^{-1})} = 32\ \mathrm{ms} \quad \bullet$$

(b) The measurement of T_1

The longitudinal relaxation time T_1 can be measured by the **inversion recovery technique.** The first step is to apply a 180° pulse to the sample. A 180° pulse is achieved by applying the $\mathcal{B}_1$ field for twice as long as for a 90° pulse, so the magnetization vector precesses through 180° and points in the z-direction (Fig. 14.39). No signal can be seen at this stage because there is no component of magnetization in the xy-plane (where the coil can detect it). The β spins begin to relax back into α spins, and the magnetization vector first shrinks exponentially, falling through zero to its thermal equilibrium value, M_0. After an interval τ, a 90° pulse is applied that rotates the remaining magnetization into the xy-plane, where it generates an FID signal. The frequency-domain spectrum is then obtained by Fourier transformation.

The intensity of the spectrum obtained in this way depends on the length of the magnetization vector that is rotated into the xy-plane. The length of that vector changes exponentially as the interval between the two pulses is increased, so the intensity of the spectrum also changes exponentially with increasing τ. We can therefore measure T_1 by fitting an exponential curve to the series of spectra obtained with different values of τ.

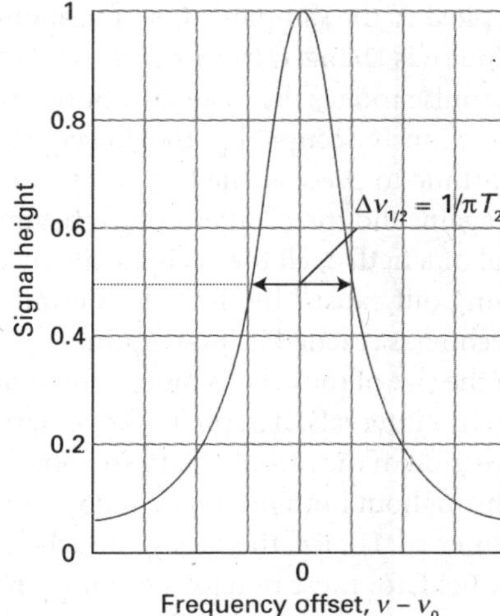

Fig. 14.38 A Lorentzian absorption line. The width at half-height is inversely proportional to the parameter T_2 and, the longer the transverse relaxation time, the narrower the line.

interActivity The *Living graphs* section of the text's web site has an applet that allows you to calculate and display Lorenztian absorption lines. Explore the effect of the parameter T_2 on the width and the maximal intensity of a Lorentzian line. Rationalize your observations.

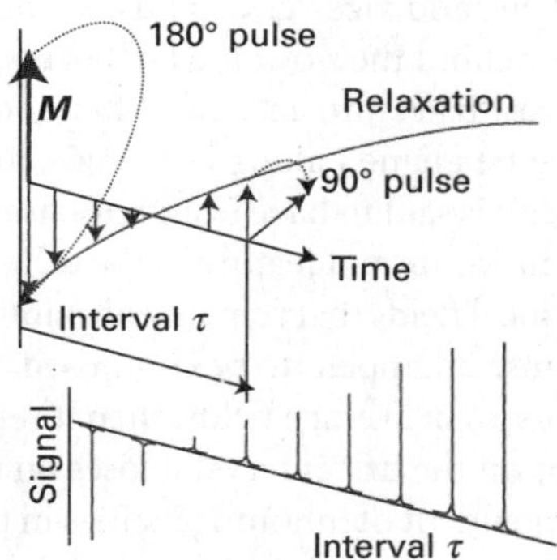

Fig. 14.39 The result of applying a 180° pulse to the magnetization in the rotating frame and the effect of a subsequent 90° pulse. The amplitude of the frequency-domain spectrum varies with the interval between the two pulses because spin–lattice relaxation has time to occur.

(c) Spin echoes

The measurement of T_2 (as distinct from T_2^*) depends on being able to eliminate the effects of inhomogeneous broadening. The cunning required is at the root of some of the most important advances that have been made in NMR since its introduction.

A **spin echo** is the magnetic analogue of an audible echo: transverse magnetization is created by a radiofrequency pulse, decays away, is reflected by a second pulse, and grows back to form an echo. The sequence of events is shown in Fig. 14.40. We can consider the overall magnetization as being made up of a number of different magnetizations, each of which arises from a **spin packet** of nuclei with very similar precession frequencies. The spread in these frequencies arises because the applied field $\mathcal{B}_0$ is inhomogeneous, so different parts of the sample experience different fields. The precession frequencies also differ if there is more than one chemical shift present. As will be seen, the importance of a spin echo is that it can suppress the effects of both field inhomogeneities and chemical shifts.

First, a 90° pulse is applied to the sample. We follow events by using the rotating frame, in which $\mathcal{B}_1$ is stationary along the x-axis and causes the magnetization to rotate into the xy-plane. The spin packets now begin to fan out because they have different Larmor frequencies, with some above the radiofrequency and some below. The detected signal depends on the resultant of the spin-packet magnetization vectors, and decays with a time-constant T_2^* because of the combined effects of field inhomogeneity and spin–spin relaxation.

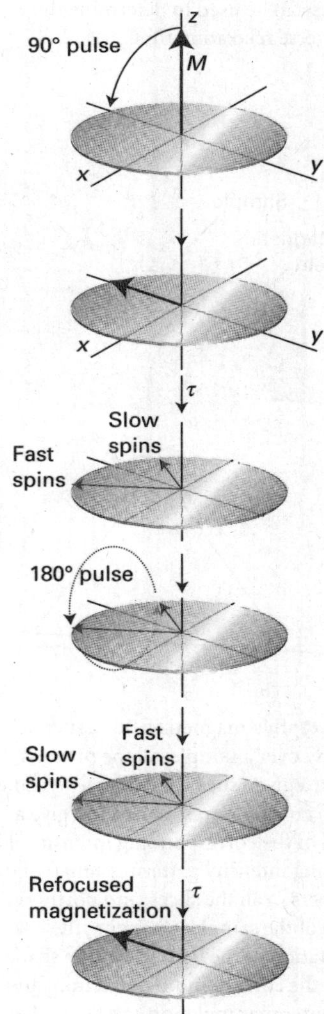

Fig. 14.40 The sequence of pulses leading to the observation of a spin echo.

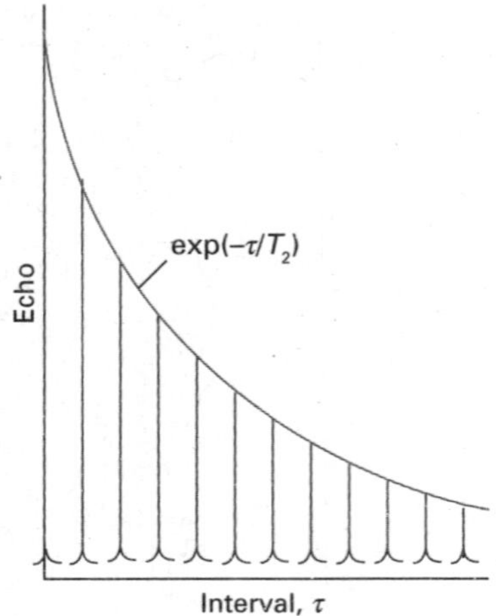

Fig. 14.41 The exponential decay of spin echoes can be used to determine the transverse relaxation time.

After an evolution period τ, a 180° pulse is applied to the sample—this time about the y-axis of the rotating frame (the axis of the pulse is changed from x to y by a 90° phase shift of the radiofrequency radiation). The pulse rotates the magnetization vectors of the faster spin packets into the positions previously occupied by the slower spin packets, and vice versa. Thus, as the vectors continue to precess, the fast vectors are now behind the slow; the fan begins to close up again, and the resultant signal begins to grow back into an echo. After another interval of length τ, all the vectors will once more be aligned along the y-axis, and the fanning out caused by the field inhomogeneity is said to have been **refocused**: the spin echo has reached its maximum.

The important feature of the technique is that the size of the echo is independent of any local fields that remain constant during the two τ intervals. If a spin packet is 'fast' because it happens to be composed of spins in a region of the sample that experiences higher than average fields, then it remains fast throughout both intervals, and what it gains on the first interval it loses on the second interval. Hence, the size of the echo is independent of inhomogeneities in the magnetic field, for these remain constant. The true transverse relaxation arises from fields that vary on a molecular distance scale, and there is no guarantee that an individual 'fast' spin will remain 'fast' in the refocusing phase: the spins within the packets therefore spread with a time constant T_2. Hence, the effects of the true relaxation are not refocused, and the size of the echo decays with the time constant T_2 (Fig. 14.41).

IMPACT ON MEDICINE

I14.1 Magnetic resonance imaging

One of the most striking applications of nuclear magnetic resonance is in medicine. *Magnetic resonance imaging* (MRI) is a portrayal of the concentrations of protons in a solid object. The technique relies on the application of specific pulse sequences to an object in an inhomogeneous magnetic field.

If an object containing hydrogen nuclei (a tube of water or a human body) is placed in an NMR spectrometer and exposed to a *homogeneous* magnetic field, then a single resonance signal will be detected. Now consider a flask of water in a magnetic field that varies linearly in the z-direction according to $\mathcal{B}_0 + \mathcal{G}_z z$, where $\mathcal{G}_z$ is the field gradient along the z-direction (Fig. 14.42). Then the water protons will be resonant at the frequencies

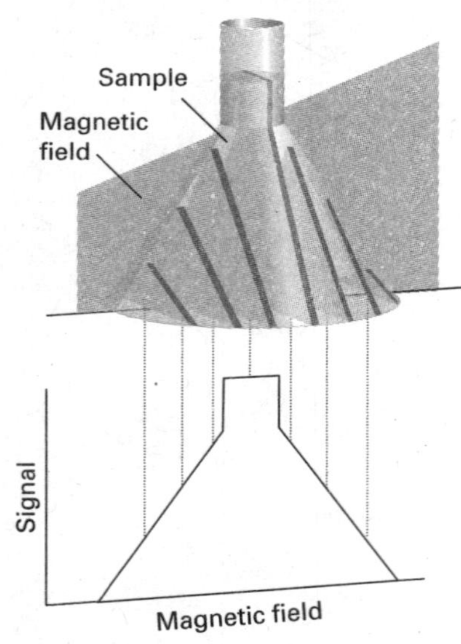

Fig. 14.42 In a magnetic field that varies linearly over a sample, all the protons within a given slice (that is, at a given field value) come into resonance and give a signal of the corresponding intensity. The resulting intensity pattern is a map of the numbers in all the slices, and portrays the shape of the sample. Changing the orientation of the field shows the shape along the corresponding direction, and computer manipulation can be used to build up the three-dimensional shape of the sample.

$$\nu_L(z) = \frac{\gamma}{2\pi}(\mathcal{B}_0 + \mathcal{G}_z z) \tag{14.35}$$

(Similar equations may be written for gradients along the x- and y-directions.) Application of a 90° radiofrequency pulse with $\nu = \nu_L(z)$ will result in a signal with an intensity that is proportional to the numbers of protons at the position z. This is an example of *slice selection*, the application of a selective 90° pulse that excites nuclei in a specific region, or slice, of the sample. It follows that the intensity of the NMR signal will be a projection of the numbers of protons on a line parallel to the field gradient. The image of a three-dimensional object such as a flask of water can be obtained if the slice selection technique is applied at different orientations (see Fig. 14.43). In *projection reconstruction*, the projections can be analysed on a computer to reconstruct the three-dimensional distribution of protons in the object.

In practice, the NMR signal is not obtained by direct analysis of the FID curve after application of a single 90° pulse. Instead, spin echoes are often detected with several variations of the 90°–τ–180° pulse sequence (Section 14.9c). In *phase encoding*, field gradients are applied during the evolution period and the detection period of a spin-echo pulse sequence. The first step consists of a 90° pulse that results in slice selection along the z-direction. The second step consists of application of a *phase gradient*, a field gradient along the y-direction, during the evolution period. At each position along the gradient, a spin packet will precess at a different Larmor frequency due to chemical shift effects and the field inhomogeneity, so each packet will dephase to a different extent by the end of the evolution period. We can control the extent of dephasing by changing the duration of the evolution period, so Fourier transformation on τ gives information about the location of a proton along the y-direction.[2] For each value of τ, the next steps are application of the 180° pulse and then of a *read gradient*, a field gradient along the x-direction, during detection of the echo. Protons at different positions along x experience different fields and will resonate at different frequencies. Therefore Fourier transformation of the FID gives different signals for protons at different positions along x.

A common problem with the techniques described above is image contrast, which must be optimized in order to show spatial variations in water content in the sample. One strategy for solving this problem takes advantage of the fact that the relaxation times of water protons are shorter for water in biological tissues than for the pure liquid. Furthermore, relaxation times from water protons are also different in healthy and diseased tissues. A *T_1-weighted image* is obtained by repeating the spin-echo sequence before spin–lattice relaxation can return the spins in the sample to equilibrium. Under these conditions, differences in signal intensities are directly related to differences in T_1. A *T_2-weighted image* is obtained by using an evolution period τ that is relatively long. Each point on the image is an echo signal that behaves in the manner shown in Fig. 14.41, so signal intensities are strongly dependent on variations in T_2. However, allowing so much of the decay to occur leads to weak signals even for those protons with long spin–spin relaxation times. Another strategy involves the use of *contrast agents*, paramagnetic compounds that shorten the relaxation times of nearby protons. The technique is particularly useful in enhancing image contrast and in diagnosing disease if the contrast agent is distributed differently in healthy and diseased tissues.

The MRI technique is used widely to detect physiological abnormalities and to observe metabolic processes. With *functional MRI*, blood flow in different regions of the brain can be studied and related to the mental activities of the subject. The technique is based on differences in the magnetic properties of deoxygenated and oxygenated haemoglobin, the iron-containing protein that transports O_2 in red blood cells. The more paramagnetic deoxygenated haemoglobin affects the proton resonances of tissue differently from the oxygenated protein. Because there is greater blood flow in active regions of the brain than in inactive regions, changes in the intensities of proton resonances due to changes in levels of oxygenated haemoglobin can be related to brain activity.

The special advantage of MRI is that it can image *soft* tissues (Fig. 14.43), whereas X-rays are largely used for imaging hard, bony structures and abnormally dense regions, such as tumours. In fact, the invisibility of hard structures in MRI is an advantage, as it allows the imaging of structures encased by bone, such as the brain and the spinal cord. X-rays are known to be dangerous on account of the ionization they cause; the high magnetic fields used in MRI may also be dangerous but, apart from anecdotes about the extraction of loose fillings from teeth, there is no convincing evidence of their harmfulness, and the technique is considered safe.

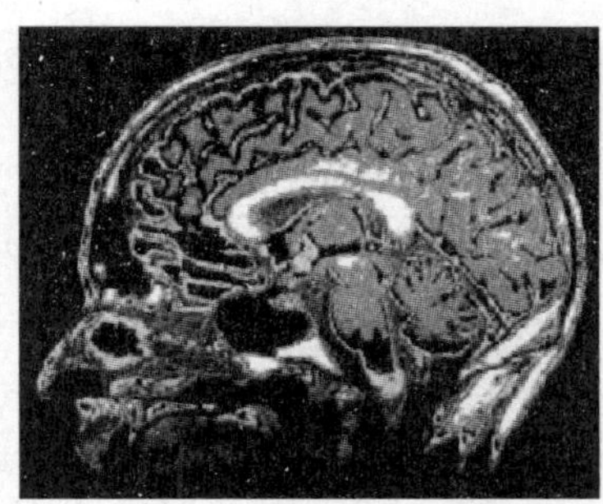

Fig. 14.43 The great advantage of MRI is that it can display soft tissue, such as in this cross-section through a patient's head. (Courtesy of the University of Manitoba.)

[2] For technical reasons, it is more common to vary the magnitude of the phase gradient.

14.10 Spin decoupling

Key point **In proton decoupling of ^{13}C-NMR spectra, protons are made to undergo rapid spin reorientations and the ^{13}C nucleus senses an average orientation. As a result, its resonance is a single line and not a group of lines.**

Carbon-13 is a **dilute-spin species** in the sense that it is unlikely that more than one ^{13}C nucleus will be found in any given small molecule (provided the sample has not been enriched with that isotope; the natural abundance of ^{13}C is only 1.1 per cent). Even in large molecules, although more than one ^{13}C nucleus may be present, it is unlikely that they will be close enough to give an observable splitting. Hence, it is not normally necessary to take into account ^{13}C–^{13}C spin–spin coupling within a molecule.

Protons are **abundant-spin species** in the sense that a molecule is likely to contain many of them. If we were observing a ^{13}C-NMR spectrum, we would obtain a very complex spectrum on account of the coupling of the one ^{13}C nucleus with many of the protons that are present. To avoid this difficulty, ^{13}C-NMR spectra are normally observed using the technique of **proton decoupling**. Thus, if the CH_3 protons of ethanol are irradiated with a second, strong, resonant radiofrequency pulse, they undergo rapid spin reorientations and the ^{13}C nucleus senses an average orientation. As a result, its resonance is a single line and not a 1:3:3:1 quartet. Proton decoupling has the additional advantage of enhancing sensitivity, because the intensity is concentrated into a single transition frequency instead of being spread over several transition frequencies (see Section 14.11). If care is taken to ensure that the other parameters on which the strength of the signal depends are kept constant, the intensities of proton-decoupled spectra are proportional to the number of ^{13}C nuclei present. The technique is widely used to characterize synthetic polymers.

14.11 The nuclear Overhauser effect

Key point **The nuclear Overhauser effect is the modification of one resonance by the saturation of another.**

We have seen already that one advantage of protons in NMR is their high magnetogyric ratio, which results in relatively large Boltzmann population differences and hence greater resonance intensities than for most other nuclei. In the steady-state **nuclear Overhauser effect** (NOE), spin relaxation processes involving internuclear dipole–dipole interactions are used to transfer this population advantage to another nucleus (such as ^{13}C or another proton), so that the latter's resonances are modified.

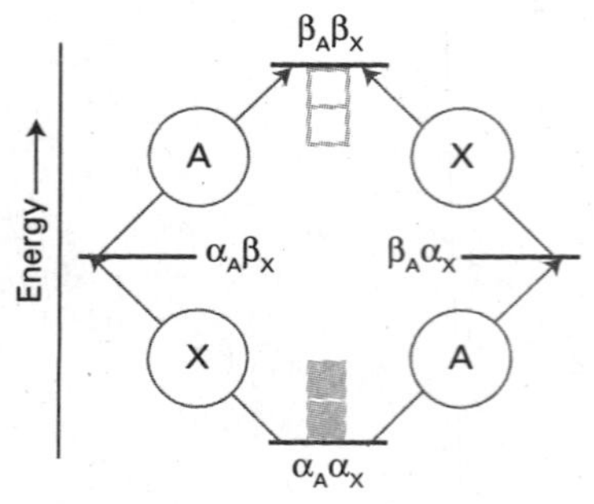

Fig. 14.44 The energy levels of an AX system and an indication of their relative populations. Each grey square above the line represents an excess population and each white square below the line represents a population deficit. The transitions of A and X are marked.

To understand the effect, we consider the populations of the four levels of a homonuclear (for instance, proton) AX system; these levels were shown in Fig. 14.13. At thermal equilibrium, the population of the $\alpha_A\alpha_X$ level is the greatest, and that of the $\beta_A\beta_X$ level is the least; the other two levels have the same energy and an intermediate population. The thermal equilibrium absorption intensities reflect these populations as shown in Fig. 14.44. Now consider the combined effect of spin relaxation and keeping the X spins saturated (that is, their populations equalized). When we saturate the X transition, the populations of the X levels are equalized ($N_{\alpha X} = N_{\beta X}$) and all transitions involving $\alpha_X \leftrightarrow \beta_X$ spin flips are no longer observed. At this stage there is no change in the populations of the A levels. If that were all there were to happen, all we would see would be the loss of the X resonance and no effect on the A resonance.

Now consider the effect of spin relaxation. Relaxation can occur in a variety of ways if there is a dipolar interaction between the A and X spins. One possibility is for the magnetic field acting between the two spins to cause them *both* to flop simultaneously

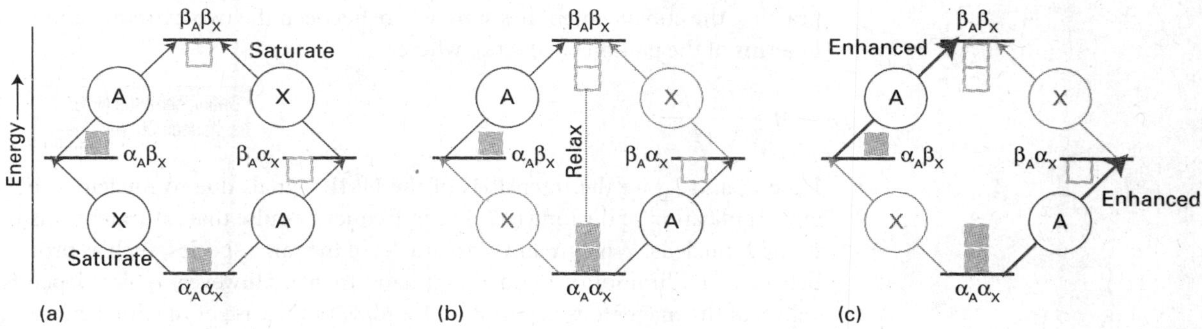

Fig. 14.45 (a) When the X transition is saturated, the populations of its two states are equalized and the population excess and deficit become as shown (using the same symbols as in Fig. 14.44). (b) Dipole–dipole relaxation relaxes the populations of the highest and lowest states, and they regain their original populations. (c) The A transitions reflect the difference in populations resulting from the preceding changes, and are enhanced compared with those shown in Fig. 14.44.

from β to α, so the $\alpha_A\alpha_X$ and $\beta_A\beta_X$ states regain their thermal equilibrium populations. However, the populations of the $\alpha_A\beta_X$ and $\beta_A\alpha_X$ levels remain unchanged at the values characteristic of saturation. As we see from Fig. 14.45, the population difference between the states joined by transitions of A is now greater than at equilibrium, so the resonance absorption is enhanced. Another possibility is for the dipolar interaction between the two spins to cause α_A to flip to β_A and simultaneously β_X to flop to α_X (or vice versa). This transition equilibrates the populations of $\alpha_A\beta_X$ and $\beta_A\alpha_X$ but leaves the $\alpha_A\alpha_X$ and $\beta_A\beta_X$ populations unchanged. Now we see from the illustration that the population differences in the states involved in the A transitions are decreased, so the resonance absorption is diminished.

Which effect wins? Does the NOE enhance the A absorption or does it diminish it? As in the discussion of relaxation times in Section 14.9, the efficiency of the intensity-enhancing $\beta_A\beta_X \leftrightarrow \alpha_A\alpha_X$ relaxation is high if the dipole field oscillates at a frequency close to the transition frequency, which in this case is about 2ν; likewise, the efficiency of the intensity-diminishing $\alpha_A\beta_X \leftrightarrow \beta_A\alpha_X$ relaxation is high if the dipole field is stationary (as there is no frequency difference between the initial and final states). A large molecule rotates so slowly that there is very little motion at 2ν, so we expect an intensity decrease (Fig. 14.46). A small molecule rotating rapidly can be expected to have substantial motion at 2ν, and a consequent enhancement of the signal. In

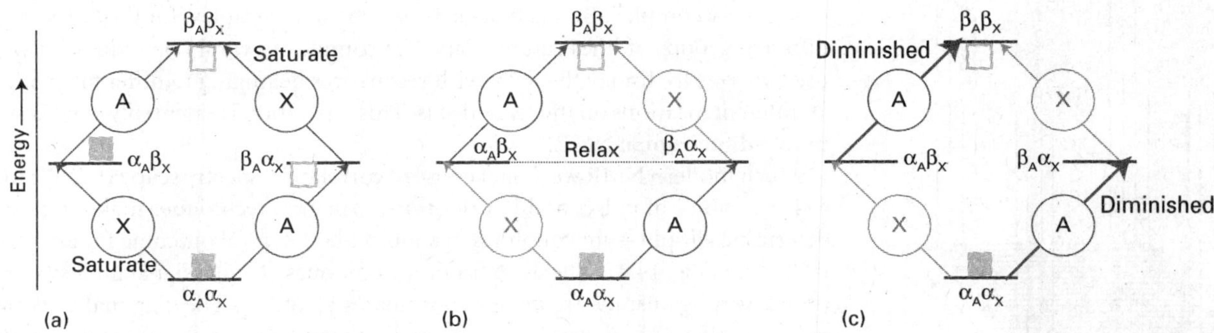

Fig. 14.46 (a) When the X transition is saturated, just as in Fig. 14.45 the populations of its two states are equalized and the population excess and deficit become as shown. (b) Dipole–dipole relaxation relaxes the populations of the two intermediate states, and they regain their original populations. (c) The A transitions reflect the difference in populations resulting from the preceding changes, and are diminished compared with those shown in Fig. 14.44.

practice, the enhancement lies somewhere between the two extremes and is reported in terms of the parameter η (eta), where

$$\eta = \frac{I_A - I_A^\circ}{I_A^\circ} \qquad \text{Definition of the NOE enhancement parameter} \qquad [14.36]$$

Here I_A° and I_A are the intensities of the NMR signals due to nucleus A before and after application of the long ($>T_1$) radiofrequency pulse that saturates transitions due to the X nucleus. When A and X are nuclei of the same species, such as protons, η lies between -1 (diminution) and $+\frac{1}{2}$ (enhancement). However, η also depends on the values of the magnetogyric ratios of A and X. In the case of maximal enhancement it is possible to show that

$$\eta = \frac{\gamma_X}{2\gamma_A} \qquad (14.37)$$

where γ_A and γ_X are the magnetogyric ratios of nuclei A and X, respectively. For ^{13}C close to a saturated proton, the ratio evaluates to 1.99, which shows that an enhancement of about a factor of 2 can be achieved.

The NOE is also used to determine interproton distances. The Overhauser enhancement of a proton A generated by saturating a spin X depends on the fraction of A's spin–lattice relaxation that is caused by its dipolar interaction with X. Because the dipolar field is proportional to r^{-3}, where r is the internuclear distance, and the relaxation effect is proportional to the square of the field, and therefore to r^{-6}, the NOE may be used to determine the geometries of molecules in solution. The determination of the structure of a small protein in solution involves the use of several hundred NOE measurements, effectively casting a net over the protons present. The enormous importance of this procedure is that we can determine the conformation of biological macromolecules in an aqueous environment and do not need to try to make the single crystals that are essential for an X-ray diffraction investigation (Chapter 19).

14.12 Two-dimensional NMR

Key points **In two-dimensional NMR, spectra are displayed in two axes, with resonances belonging to different groups lying at different locations on the second axis. In correlation spectroscopy (COSY), all spin–spin couplings in a molecule are determined. In nuclear Overhauser effect spectroscopy (NOESY), internuclear distances up to about 0.5 nm are determined.**

An NMR spectrum contains a great deal of information and, if many protons are present, is very complex. Even a first-order spectrum is complex, for the fine structure of different groups of lines can overlap. The complexity would be reduced if we could use two axes to display the data, with resonances belonging to different groups lying at different locations on the second axis. This separation is essentially what is achieved in **two-dimensional NMR**.

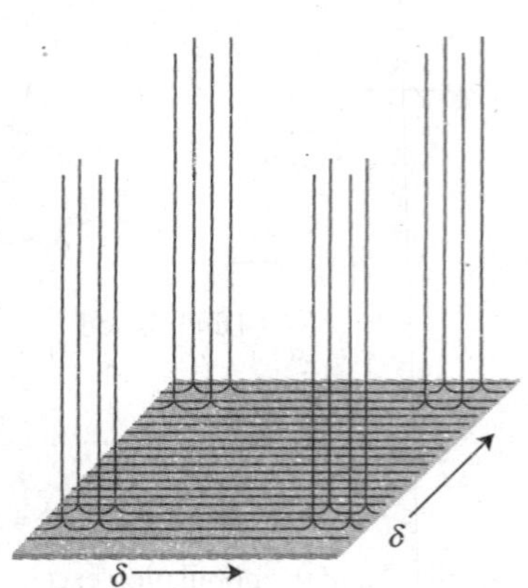

Fig. 14.47 An idealization of the COSY spectrum of an AX spin system.

Much modern NMR work makes use of **correlation spectroscopy** (COSY) in which a clever choice of pulses and Fourier transformation techniques makes it possible to determine all spin–spin couplings in a molecule. A typical outcome for an AX system is shown in Fig. 14.47. The diagram shows contours of equal signal intensity on a plot of intensity against the frequency coordinates ν_1 and ν_2. The **diagonal peaks** are signals centred on (δ_A,δ_A) and (δ_X,δ_X) and lie along the diagonal where $\nu_1 = \nu_2$. That is, the spectrum along the diagonal is equivalent to the one-dimensional spectrum obtained with the conventional NMR technique (Fig. 14.14). The **cross-peaks** (or *off-diagonal peaks*) are signals centred on (δ_A,δ_X) and (δ_X,δ_A) and owe their existence to the coupling between the A and X nuclei.

Although information from two-dimensional NMR spectroscopy is trivial in an AX system, it can be of enormous help in the interpretation of more complex spectra, leading to a map of the couplings between spins and to the determination of the bonding network in complex molecules. Indeed, the spectrum of a synthetic or biological polymer that would be impossible to interpret in one-dimensional NMR can often be interpreted reasonably rapidly by two-dimensional NMR.

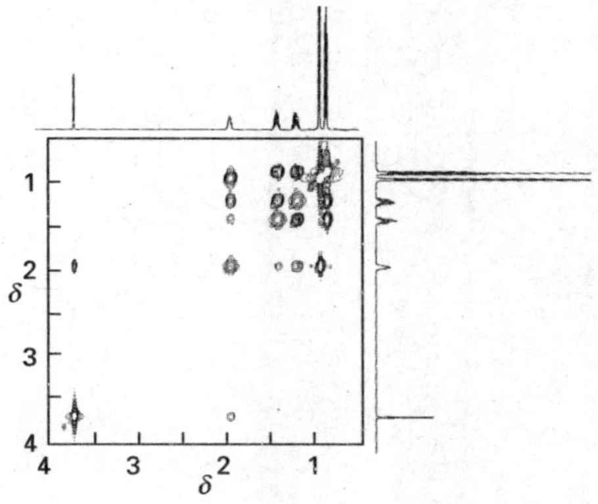

Fig. 14.48 Proton COSY spectrum of isoleucine. (The *brief illustration* and corresponding spectrum are adapted from K.E. van Holde, *et al.*, *Principles of physical biochemistry*, Prentice Hall, Upper Saddle River (1998).)

• A brief illustration

Figure 14.48 is a portion of the COSY spectrum of the amino acid isoleucine (7), showing the resonances associated with the protons bound to the carbon atoms. From the molecular structure, we expect that: (i) the C_a–H proton is coupled only to the C_b–H proton, (ii) the C_b–H protons are coupled to the C_a–H, C_c–H, and C_d–H protons, and (iii) the inequivalent C_d–H protons are coupled to the C_b–H and C_e–H protons. We now note that:

• The resonance with $\delta = 3.6$ shares a cross-peak with only one other resonance at $\delta = 1.9$, which in turn shares cross-peaks with resonances at $\delta = 1.4$, 1.2, and 0.9. We conclude that the resonances at $\delta = 3.6$ and 1.9 correspond to the C_a–H and C_b–H protons, respectively.

• The proton with resonance at $\delta = 0.8$ is not coupled to the C_b–H protons, so we assign the resonance at $\delta = 0.8$ to the C_e–H protons.

• The resonances at $\delta = 1.4$ and 1.2 do not share cross-peaks with the resonance at $\delta = 0.9$.

• In the light of the expected couplings, we assign the resonance at $\delta = 0.9$ to the C_c–H protons and the resonances at $\delta = 1.4$ and 1.2 to the inequivalent C_d–H protons. •

H_2N–$\overset{a}{C}H$(–$\overset{b}{C}H(\overset{c}{C}H_3)$–$\overset{d}{C}H_2$–$\overset{e}{C}H_3$)–C(=O)–OH

7 Isoleucine

We have seen that the nuclear Overhauser effect can provide information about internuclear distances through analysis of enhancement patterns in the NMR spectrum before and after saturation of selected resonances. In **nuclear Overhauser effect spectroscopy** (NOESY) a map of all possible NOE interactions is obtained by again using a proper choice of radiofrequency pulses and Fourier transformation techniques. Like a COSY spectrum, a NOESY spectrum consists of a series of diagonal peaks that correspond to the one-dimensional NMR spectrum of the sample. The off-diagonal peaks indicate which nuclei are close enough to each other to give rise to a nuclear Overhauser effect. NOESY data reveal internuclear distances up to about 0.5 nm.

14.13 Solid-state NMR

Key points **(a) Broad NMR linewidths in solid samples are determined by magnetic interactions between nuclear spins and chemical shift anisotropy. (b) Magic-angle spinning (MAS) is a technique in which the NMR linewidths in a solid sample are reduced by spinning at an angle of 54.74° to the applied magnetic field.**

The principal difficulty with the application of NMR to solids is the low resolution characteristic of solid samples. Nevertheless, there are good reasons for seeking to overcome these difficulties. They include the possibility that a compound of interest is unstable in solution or that it is insoluble, so conventional solution NMR cannot be employed. Moreover, many species are intrinsically interesting as solids, and it is important to determine their structures and dynamics. Synthetic polymers are particularly interesting in this regard, and information can be obtained about the arrangement of molecules, their conformations, and the motion of different parts of the chain. This kind of information is crucial to an interpretation of the bulk properties of the polymer in terms of its molecular characteristics. Similarly, inorganic substances, such as the zeolites that are used as molecular sieves and shape-selective catalysts, can be studied using solid-state NMR, and structural problems can be resolved that cannot be tackled by X-ray diffraction. The recent surge of interest in inorganic nanomaterials has also contributed to the development of solid-state NMR studies.

Problems of resolution and linewidth are not the only features that plague NMR studies of solids, but the rewards are so great that considerable efforts have been made to overcome them and have achieved notable success. Because molecular rotation has almost ceased (except in special cases, including 'plastic crystals' in which the molecules continue to tumble), spin–lattice relaxation times are very long but spin–spin relaxation times are very short. Hence, in a pulse experiment, there is a need for lengthy delays—of several seconds—between successive pulses so that the spin system has time to revert to equilibrium. Even gathering the murky information may therefore be a lengthy process. Moreover, because lines are so broad, very high powers of radiofrequency radiation may be required to achieve saturation. Whereas solution pulse NMR uses transmitters of a few tens of watts, solid-state NMR may require transmitters rated at several hundreds of watts.

(a) The origins of linewidths in solids

There are three principal contributions to the linewidths of solids. One is the direct magnetic dipolar interaction between nuclear spins. As we saw in the discussion of spin–spin coupling, a nuclear magnetic moment will give rise to a local magnetic field, which points in different directions at different locations around the nucleus. If we are interested only in the component parallel to the direction of the applied magnetic field (because only this component has a significant effect), then we can use a classical expression in *Justification* 14.2 to write the magnitude of the local magnetic field as

$$\mathcal{B}_{\text{loc}} = -\frac{\gamma\hbar\mu_0 m_I}{4\pi R^3}(1-3\cos^2\theta) \tag{14.38}$$

Unlike in solution, this field is not motionally averaged to zero. Many nuclei may contribute to the total local field experienced by a nucleus of interest, and different nuclei in a sample may experience a wide range of fields. Typical dipole fields are of the order of 1 mT, which corresponds to splittings and linewidths of the order of 10 kHz.

A second source of linewidth is the anisotropy of the chemical shift. We have seen that chemical shifts arise from the ability of the applied field to generate electron currents in molecules. In general, this ability depends on the orientation of the molecule relative to the applied field. In solution, when the molecule is tumbling rapidly, only the average value of the chemical shift is relevant. However, the anisotropy is not averaged to zero for stationary molecules in a solid, and molecules in different orientations have resonances at different frequencies. The chemical shift anisotropy also varies with the angle between the applied field and the principal axis of the molecule as $1-3\cos^2\theta$.

The third contribution is the electric quadrupole interaction. Nuclei with $I>\frac{1}{2}$ have a distribution of charge that gives rise to an electric quadrupole moment (for instance, the positive charge may be concentrated around the equator or at the poles). An electric quadrupole interacts with an electric field gradient, such as may arise from a non-spherical distribution of charge around the nucleus. This interaction also varies as $1-3\cos^2\theta$.

(b) The reduction of linewidths

Fortunately, there are techniques available for reducing the linewidths of solid samples. One technique, **magic-angle spinning** (MAS), takes note of the $1-3\cos^2\theta$

dependence of the dipole–dipole interaction, the chemical shift anisotropy, and the electric quadrupole interaction. The 'magic angle' is the angle at which $1 - 3\cos^2\theta = 0$, and corresponds to 54.74°. In the technique, the sample is spun at high speed at the magic angle to the applied field (Fig. 14.49). All the dipolar interactions and the anisotropies average to the value they would have at the magic angle, but at that angle they are zero. The difficulty with MAS is that the spinning frequency must not be less than the width of the spectrum, which is of the order of kilohertz. However, gas-driven sample spinners that can be rotated at up to 25 kHz are now routinely available, and a considerable body of work has been done.

Pulsed techniques similar to those described in the previous section may also be used to reduce linewidths. The dipolar field of protons, for instance, may be reduced by a decoupling procedure. However, because the range of coupling strengths is so large, radiofrequency power of the order of 1 kW is required. Elaborate pulse sequences have also been devised that reduce linewidths by averaging procedures that make use of twisting the magnetization vector through an elaborate series of angles.

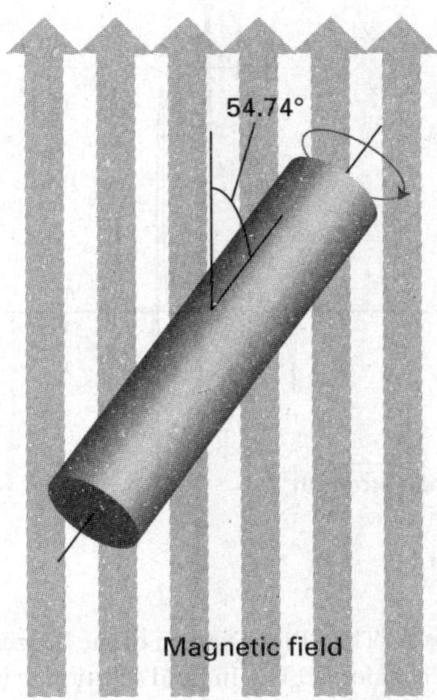

Fig. 14.49 In magic angle spinning, the sample spins at 54.74° (that is, arccos $1/3^{1/2}$) to the applied magnetic field. Rapid motion at this angle averages dipole–dipole interactions and chemical shift anisotropies to zero.

Electron paramagnetic resonance

Electron paramagnetic resonance (EPR) is less widely applicable than NMR because it cannot be detected in normal, spin-paired molecules and the sample must possess unpaired electron spins. It is used to study radicals formed during chemical reactions or by radiation, radicals that act as probes of biological structure, many d-metal complexes, and molecules in triplet states (such as those involved in phosphorescence, Section 13.4). The sample may be a gas, a liquid, or a solid, but the free rotation of molecules in the gas phase gives rise to complications.

14.14 The EPR spectrometer

Key point **EPR spectrometers consist of a microwave source, a cavity in which the sample is inserted, a microwave detector, and an electromagnet.**

Both Fourier-transform (FT) and continuous wave (CW) EPR spectrometers are available. The FT-EPR instrument is based on the concepts developed in Section 14.8, except that pulses of microwaves are used to excite electron spins in the sample. The layout of the more common CW-EPR spectrometer is shown in Fig. 14.50. It consists of a microwave source (a klystron or a Gunn oscillator), a cavity in which the sample is inserted in a glass or quartz container, a microwave detector, and an electromagnet with a field that can be varied in the region of 0.3 T. The EPR spectrum is obtained by monitoring the microwave absorption as the field is changed, and a typical spectrum (of the benzene radical anion, $C_6H_6^-$) is shown in Fig. 14.51. The peculiar appearance of the spectrum, which is in fact the first derivative of the absorption, arises from the detection technique, which is sensitive to the slope of the absorption curve (Fig. 14.52).

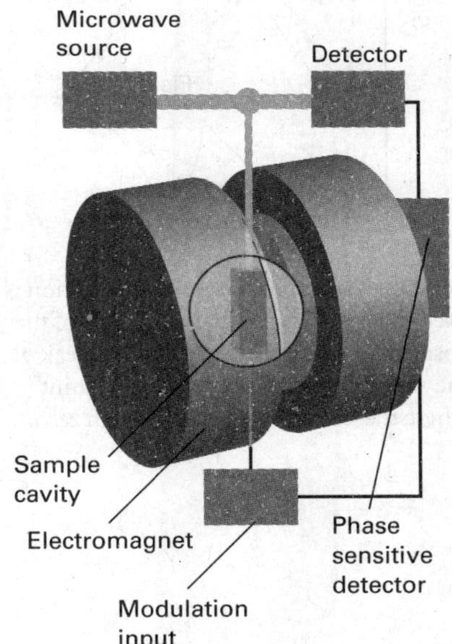

Fig. 14.50 The layout of a continuous-wave EPR spectrometer. A typical magnetic field is 0.3 T, which requires 9 GHz (3 cm) microwaves for resonance.

14.15 The *g*-value

Key point **The EPR resonance condition is written in terms of the *g*-value of the radical, *g*; the deviation of *g* from $g_e = 2.0023$ depends on the ability of the applied field to induce local electron currents in the radical.**

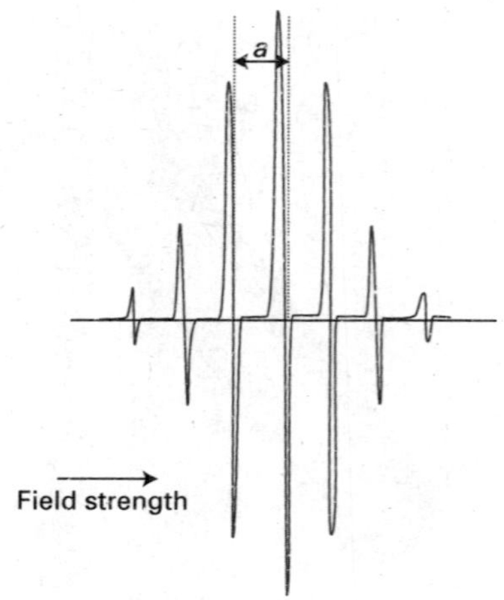

Fig. 14.51 The EPR spectrum of the benzene radical anion, $C_6H_6^-$, in fluid solution. *a* is the hyperfine splitting of the spectrum; the centre of the spectrum is determined by the *g*-value of the radical.

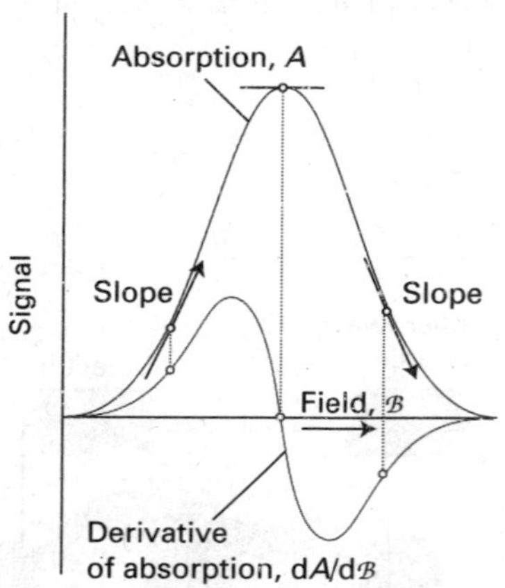

Fig. 14.52 When phase-sensitive detection is used, the signal is the first derivative of the absorption intensity. Note that the peak of the absorption corresponds to the point where the derivative passes through zero.

Equation 14.12b gives the resonance frequency for a transition between the $m_s = -\frac{1}{2}$ and the $m_s = +\frac{1}{2}$ levels of a 'free' electron in terms of the *g*-value $g_e \approx 2.0023$. The magnetic moment of an unpaired electron in a radical also interacts with an external field, but the *g*-value is different from that for a free electron because of local magnetic fields induced by the molecular framework of the radical. Consequently, the resonance condition is normally written as

$$h\nu = g\mu_B \mathcal{B}_0 \qquad \text{EPR resonance condition} \qquad (14.39)$$

where *g* is the **g-value** of the radical.

- **A brief illustration**

The centre of the EPR spectrum of the methyl radical occurred at 329.40 mT in a spectrometer operating at 9.2330 GHz (radiation belonging to the X band of the microwave region). Its *g*-value is therefore

$$g = \frac{h\nu}{\mu_B \mathcal{B}_0} = \frac{(6.626\,08 \times 10^{-34}\,\text{J s}) \times (9.2330 \times 10^9\,\text{s}^{-1})}{(9.2740 \times 10^{-24}\,\text{J T}^{-1}) \times (0.329\,40\,\text{T})} = 2.0027 \bullet$$

Self-test 14.3 At what magnetic field would the methyl radical come into resonance in a spectrometer operating at 34.000 GHz (radiation belonging to the Q band of the microwave region)? [1.213 T]

The *g*-value in a molecular environment (a radical or a d-metal complex) is related to the ease with which the applied field can stir up currents through the molecular framework and the strength of the magnetic field the currents generate. Therefore, the *g*-value gives some information about electronic structure and plays a similar role in EPR to that played by shielding constants in NMR.

Electrons can migrate through the molecular framework by making use of excited states (Fig. 14.53). This additional path for circulation of electrons gives rise to a local magnetic field that adds to the applied field. Therefore, we expect the ease of stirring up currents to be inversely proportional to the separation of energy levels, ΔE, in the molecule. As we saw in Section 9.9, the strength of the field generated by electronic currents in atoms (and analogously in molecules) is related to the extent of coupling between spin and orbital angular momenta. That is, the local field strength is proportional to the molecular spin–orbit coupling constant, ξ.

We can conclude from the discussion above that the *g*-value of a radical or d-metal complex differs from g_e, the 'free-electron' *g*-value, by an amount that is proportional to $\xi/\Delta E$. This proportionality is widely observed. Many organic radicals have *g*-values close to 2.0027 and inorganic radicals have *g*-values typically in the range 1.9 to 2.1. The *g*-values of paramagnetic d-metal complexes often differ considerably from g_e, varying from 0 to 6, because in them ΔE is small (on account of the splitting of d orbitals brought about by interactions with ligands, as we saw in Section 13.3).

Just as in the case of the chemical shift in NMR spectroscopy, the *g*-value is anisotropic, that is, its magnitude depends on the orientation of the radical with respect to the applied field. In solution, when the molecule is tumbling rapidly, only the average value of the *g*-value is observed. Therefore, anisotropy of the *g*-value is observed only for radicals trapped in solids.

14.16 Hyperfine structure

Key points **The hyperfine structure of an EPR spectrum is its splitting of individual resonance lines into components by the magnetic interaction between the electron and nuclei with spin. (a) If a radical contains N equivalent nuclei with spin quantum number I, then there are $2NI+1$ hyperfine lines with an intensity distribution given by a modified version of Pascal's triangle. (b) Hyperfine structure can be explained by dipole–dipole interactions, Fermi contact interactions, and the polarization mechanism.**

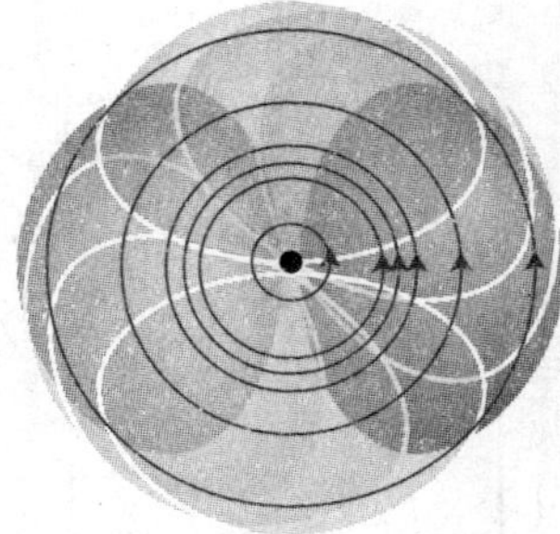

Fig. 14.53 An applied magnetic field can induce circulation of electrons that makes use of excited state orbitals.

The most important feature of EPR spectra is their **hyperfine structure**, the splitting of individual resonance lines into components. In general in spectroscopy, the term 'hyperfine structure' means the structure of a spectrum that can be traced to interactions of the electrons with nuclei other than as a result of the latter's point electric charge. The source of the hyperfine structure in EPR is the magnetic interaction between the electron spin and the magnetic dipole moments of the nuclei present in the radical.

(a) The effects of nuclear spin

Consider the effect on the EPR spectrum of a single H nucleus located somewhere in a radical. The proton spin is a source of magnetic field and, depending on the orientation of the nuclear spin, the field it generates adds to or subtracts from the applied field. The total local field is therefore

$$\mathcal{B}_{\rm loc} = \mathcal{B} + am_I \qquad m_I = \pm\tfrac{1}{2} \tag{14.40}$$

where a is the **hyperfine coupling constant.** Half the radicals in a sample have $m_I = +\frac{1}{2}$, so half resonate when the applied field satisfies the condition

$$h\nu = g\mu_{\rm B}(\mathcal{B} + \tfrac{1}{2}a), \qquad \text{or} \qquad \mathcal{B} = \frac{h\nu}{g\mu_{\rm B}} - \tfrac{1}{2}a \tag{14.41a}$$

The other half (which have $m_I = -\frac{1}{2}$) resonate when

$$h\nu = g\mu_{\rm B}(\mathcal{B} - \tfrac{1}{2}a), \qquad \text{or} \qquad \mathcal{B} = \frac{h\nu}{g\mu_{\rm B}} + \tfrac{1}{2}a \tag{14.41b}$$

Therefore, instead of a single line, the spectrum shows two lines of half the original intensity separated by a and centred on the field determined by g (Fig. 14.54).

If the radical contains an ^{14}N atom ($I = 1$), its EPR spectrum consists of three lines of equal intensity, because the ^{14}N nucleus has three possible spin orientations, and each spin orientation is possessed by one-third of all the radicals in the sample. In general, a spin-I nucleus splits the spectrum into $2I + 1$ hyperfine lines of equal intensity.

When there are several magnetic nuclei present in the radical, each one contributes to the hyperfine structure. In the case of equivalent protons (for example, the two CH_2 protons in the radical CH_3CH_2) some of the hyperfine lines are coincident. It is not hard to show that, if the radical contains N equivalent protons, then there are $N+1$ hyperfine lines with a binomial intensity distribution (the intensity distribution given by Pascal's triangle). The spectrum of the benzene radical anion in Fig. 14.51, which has seven lines with intensity ratio 1:6:15:20:15:6:1, is consistent with a radical containing six equivalent protons. More generally, if the radical contains N equivalent nuclei with spin quantum number I, then there are $2NI+1$ hyperfine lines with an intensity distribution based on a modified version of Pascal's triangle as shown in the following *Example*.

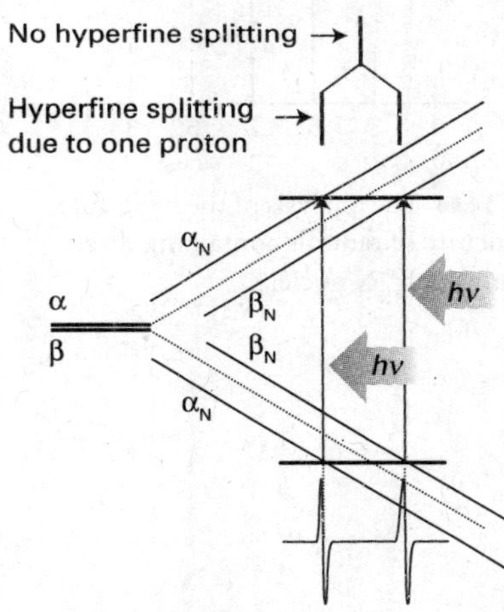

Fig. 14.54 The hyperfine interaction between an electron and a spin-$\frac{1}{2}$ nucleus results in four energy levels in place of the original two. As a result, the spectrum consists of two lines (of equal intensity) instead of one. The intensity distribution can be summarized by a simple stick diagram. The diagonal lines show the energies of the states as the applied field is increased, and resonance occurs when the separation of states matches the fixed energy of the microwave photon.

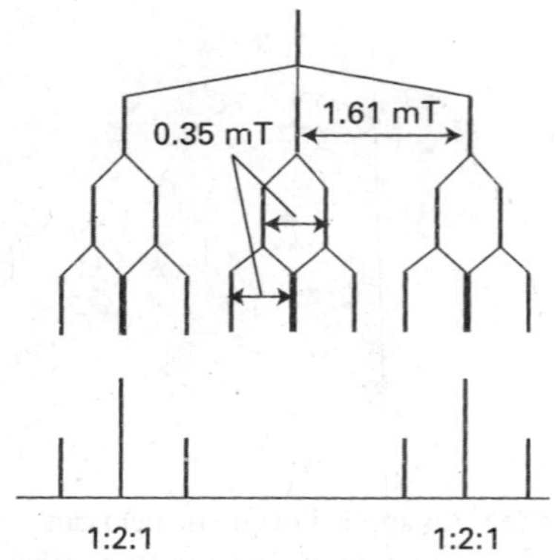

Fig. 14.55 The analysis of the hyperfine structure of radicals containing one ^{14}N nucleus ($I = 1$) and two equivalent protons.

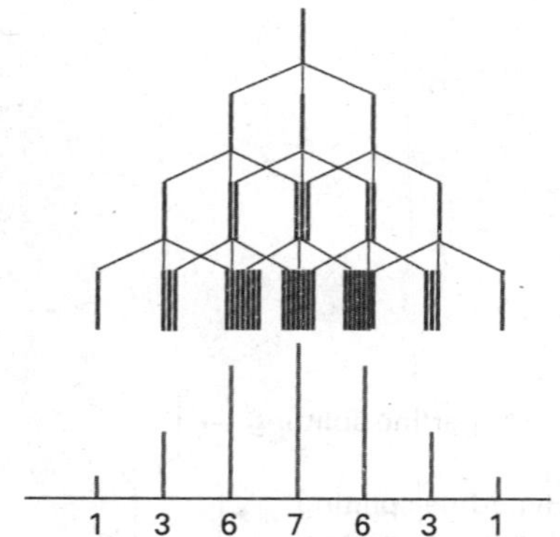

Fig. 14.56 The analysis of the hyperfine structure of radicals containing three equivalent ^{14}N nuclei.

Example 14.2 *Predicting the hyperfine structure of an EPR spectrum*

A radical contains one ^{14}N nucleus ($I = 1$) with hyperfine constant 1.61 mT and two equivalent protons ($I = \frac{1}{2}$) with hyperfine constant 0.35 mT. Predict the form of the EPR spectrum.

Method We should consider the hyperfine structure that arises from each type of nucleus or group of equivalent nuclei in succession. So, split a line with one nucleus, then each of those lines is split by a second nucleus (or group of nuclei), and so on. It is best to start with the nucleus with the largest hyperfine splitting; however, any choice could be made, and the order in which nuclei are considered does not affect the conclusion.

Answer The ^{14}N nucleus gives three hyperfine lines of equal intensity separated by 1.61 mT. Each line is split into doublets of spacing 0.35 mT by the first proton, and each line of these doublets is split into doublets with the same 0.35 mT splitting (Fig. 14.55). The central lines of each split doublet coincide, so the proton splitting gives 1:2:1 triplets of internal splitting 0.35 mT. Therefore, the spectrum consists of three equivalent 1:2:1 triplets.

Self-test 14.4 Predict the form of the EPR spectrum of a radical containing three equivalent ^{14}N nuclei. [Fig. 14.56]

The hyperfine structure of an EPR spectrum is a kind of fingerprint that helps to identify the radicals present in a sample. Moreover, because the magnitude of the splitting depends on the distribution of the unpaired electron near the magnetic nuclei present, the spectrum can be used to map the molecular orbital occupied by the unpaired electron. For example, because the hyperfine splitting in $C_6H_6^-$ is 0.375 mT, and one proton is close to a C atom with one-sixth the unpaired electron spin density (because the electron is spread uniformly around the ring), the hyperfine splitting caused by a proton in the electron spin entirely confined to a single adjacent C atom should be 6×0.375 mT $= 2.25$ mT. If in another aromatic radical we find a hyperfine splitting constant a, then the **spin density**, ρ, the probability that an unpaired electron is on the atom, can be calculated from the **McConnell equation**:

$$a = Q\rho \qquad \text{McConnell equation} \qquad (14.42)$$

with $Q = 2.25$ mT. In this equation, ρ is the spin density on a C atom and a is the hyperfine splitting observed for the H atom to which it is attached.

- **A brief illustration**

The hyperfine structure of the EPR spectrum of the radical anion (naphthalene)$^-$ can be interpreted as arising from two groups of four equivalent protons. Those at the 1, 4, 5, and 8 positions in the ring have $a = 0.490$ mT and those in the 2, 3, 6, and 7 positions have $a = 0.183$ mT. The densities obtained by using the McConnell equation are 0.22 and 0.08, respectively (**8**). •

0.22 0.08

8

Self-test 14.5 The spin density in (anthracene)$^-$ is shown in (**9**). Predict the form of its EPR spectrum.

[A 1:2:1 triplet of splitting 0.43 mT split into a 1:4:6:4:1 quintet of splitting 0.22 mT, split into a 1:4:6:4:1 quintet of splitting 0.11 mT, $3 \times 5 \times 5 = 75$ lines in all]

0.097 0.193 0.048

9

(b) The origin of the hyperfine interaction

The hyperfine interaction is an interaction between the magnetic moments of the unpaired electron and the nuclei. There are two contributions to the interaction.

An electron in a p orbital does not approach the nucleus very closely, so it experiences a field that appears to arise from a point magnetic dipole. The resulting interaction is called the **dipole–dipole interaction.** The contribution of a magnetic nucleus to the local field experienced by the unpaired electron is given by an expression like that in eqn 14.28. A characteristic of this type of interaction is that it is anisotropic. Furthermore, just as in the case of NMR, the dipole–dipole interaction averages to zero when the radical is free to tumble. Therefore, hyperfine structure due to the dipole–dipole interaction is observed only for radicals trapped in solids.

An s electron is spherically distributed around a nucleus and so has zero average dipole–dipole interaction with the nucleus even in a solid sample. However, because an s electron has a nonzero probability of being at the nucleus, it is incorrect to treat the interaction as one between two point dipoles. An s electron has a Fermi contact interaction with the nucleus, which as we saw in Section 14.6d is a magnetic interaction that occurs when the point dipole approximation fails. The contact interaction is isotropic (that is, independent of the radical's orientation), and consequently is shown even by rapidly tumbling molecules in fluids (provided the spin density has some s character).

The dipole–dipole interactions of p electrons and the Fermi contact interaction of s electrons can be quite large. For example, a 2p electron in a nitrogen atom experiences an average field of about 4.8 mT from the ^{14}N nucleus. A 1s electron in a hydrogen atom experiences a field of about 50 mT as a result of its Fermi contact interaction with the central proton. More values are listed in Table 14.3. The magnitudes of the contact interactions in radicals can be interpreted in terms of the s orbital character of the molecular orbital occupied by the unpaired electron, and the dipole–dipole interaction can be interpreted in terms of the p character. The analysis of hyperfine structure therefore gives information about the composition of the orbital, and especially the hybridization of the atomic orbitals (see Problem 14.13).

We still have the source of the hyperfine structure of the $C_6H_6^-$ anion and other aromatic radical anions to explain. The sample is fluid, and as the radicals are tumbling the hyperfine structure cannot be due to the dipole–dipole interaction. Moreover, the protons lie in the nodal plane of the π orbital occupied by the unpaired electron, so the structure cannot be due to a Fermi contact interaction. The explanation lies in a **polarization mechanism** similar to the one responsible for spin–spin coupling in NMR. There is a magnetic interaction between a proton and the α electrons ($m_s = +\frac{1}{2}$) which results in one of the electrons tending to be found with a greater probability nearby (Fig. 14.57). The electron with opposite spin is therefore more likely to be close to the C atom at the other end of the bond. The unpaired electron on the C atom has a lower energy if it is parallel to that electron (Hund's rule favours parallel electrons on atoms), so the unpaired electron can detect the spin of the proton indirectly. Calculation using this model leads to a hyperfine interaction in agreement with the observed value of 2.25 mT.

Table 14.3* Hyperfine coupling constants for atoms, a/mT

Nuclide	Isotropic coupling	Anisotropic coupling
^{1}H	50.8 (1s)	
^{2}H	7.8 (1s)	
^{14}N	55.2 (2s)	4.8 (2p)
^{19}F	1720 (2s)	108.4 (2p)

* More values are given in the *Data section*.

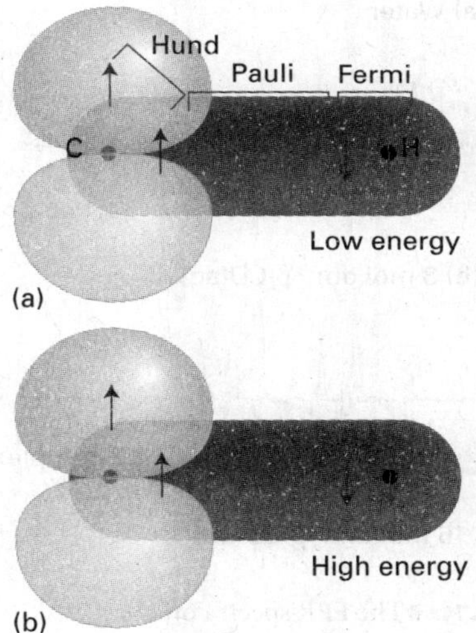

Fig. 14.57 The polarization mechanism for the hyperfine interaction in π-electron radicals. The arrangement in (a) is lower in energy than that in (b), so there is an effective coupling between the unpaired electron and the proton.

IMPACT ON BIOCHEMISTRY AND NANOSCIENCE

I14.2 Spin probes

We saw in Sections 14.15 and 14.16 that anisotropy of the g-value and of the nuclear hyperfine interactions can be observed when a radical is immobilized in a solid. Figure 14.58 shows the variation of the lineshape of the EPR spectrum of the di-*tert*-butyl nitroxide radical (**10**) with temperature. At 292 K, the radical tumbles freely and isotropic hyperfine coupling to the ^{14}N nucleus gives rise to three sharp peaks. At

N
O·
10

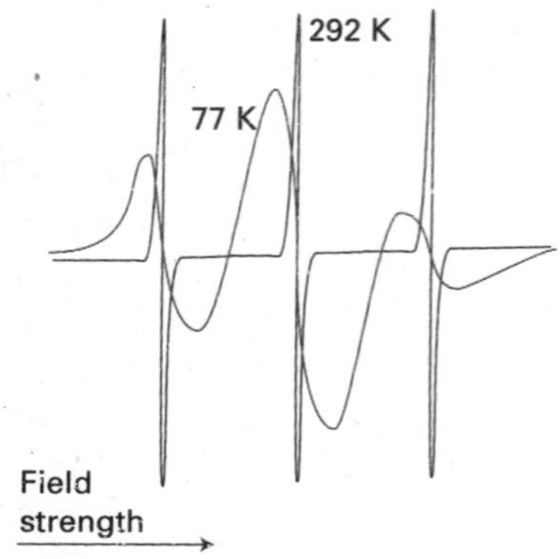

Fig. 14.58 EPR spectra of the di-*tert*-butyl nitroxide radical at 292 K and 77 K. Adapted from J.R. Bolton, in *Biological applications of electron spin resonance*, H.M. Swartz, J.R. Bolton, and D.C. Borg (ed.), Wiley, New York (1972).

77 K, motion of the radical is restricted. Both isotropic and anisotropic hyperfine couplings determine the appearance of the spectrum, which now consists of three broad peaks.

A *spin probe* (or *spin label*) is a radical that interacts with a molecular assembly (a biopolymer or a nanostructure) and with an EPR spectrum that reports on the structural and dynamical properties of the assembly. The ideal spin probe is one with a spectrum that broadens significantly as its motion is restricted to a relatively small extent. Nitroxide spin probes have been used to show that the hydrophobic interiors of biological membranes, once thought to be rigid, are in fact very fluid and individual lipid molecules move laterally through the sheet-like structure of the membrane. The EPR spectrum also can reveal whether a nitroxide spin probe is free in solution, positioned as a guest within a macromolecular host, or intercalated within micelles (see Chapter 18). For example, hyperfine coupling constants to the ^{14}N nucleus can change if the N–O group is exposed to the solvent or buried in the assembly.

Benzyl *tert*-butyl nitroxide (11) and dibenzylnitroxide (12) are particularly well-suited spin probes for supramolecular systems, such as those formed with the host β-cyclodextrin (13). As the concentration of the host system is increased, the EPR spectrum shifts from that of the free nitroxide to that of the 1:1 complexed radical (Fig. 14.59). The variations in the nitrogen hyperfine coupling are attributed to the extent of exposure of the N–O group to water, with the lowest value for β-cyclodextrin and its hydrophobic cavity. The hyperfine coupling constant for the benzyl hydrogens two bonds from the unpaired electron reflects the conformation of the nitroxide radical in the various macromolecular host systems, particularly with regard to rotation of the benzyl group about the C–N bond. The symmetric nitroxide spin probe in (12) can be incorporated into two β-cyclodextrin cavities. This 1:2 inclusion complex exhibits reduced nitrogen hyperfine splitting, which is consistent with the less polar environment achieved by the complete shielding of the nitroxide from solvent.

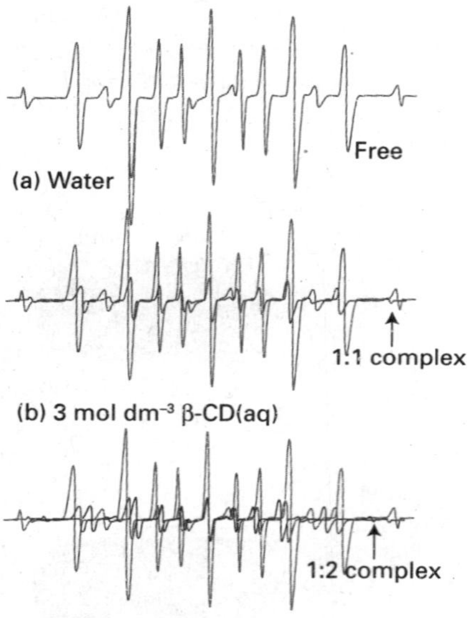

Fig. 14.59 The EPR spectra of dibenzylnitroxide in water with different concentrations of β-cyclodextrin. Based on P. Franchi *et al.*, *Current Organic Chemistry*, 1831, 8 (2004).

11

12

13 β-Cyclodextrin

Checklist of key equations

Property	Equation	Comment
Resonance condition	$h\nu = g_e\mu_B\mathcal{B}_0$	For electrons
	$h\nu = \gamma\hbar\mathcal{B}_0$	For spin-$\frac{1}{2}$ nuclei
δ-Scale of chemical shifts	$\delta = \{(\nu - \nu^\circ)/\nu^\circ\} \times 10^6$	
Relation between chemical shift and shielding constant	$\delta \approx (\sigma^\circ - \sigma) \times 10^6$	
Local contribution to the shielding constant	$\sigma(\text{local}) = \sigma_d + \sigma_p$	
Lamb formula	$\sigma_d = (e^2\mu_0/12\pi m_e)\langle 1/r\rangle$	
Neighbouring group contribution to the shielding constant	$\sigma(\text{neighbour}) \propto (\chi_\parallel - \chi_\perp)(1 - 3\cos^2\Theta)/r^3$	The angle Θ is defined in (2)
Karplus equation	$J = A + B\cos\phi + C\cos 2\phi$	A, B, and C are empirical constants
Condition for coalescence of two NMR lines	$\tau = \sqrt{2}/\pi\delta\nu$	Conformational conversions and exchange processes
Free-induction decay	$M_y(t) = M_0\cos(2\pi\nu_L t)e^{-t/T_2}$	
Width at half-height of an NMR line	$\Delta\nu_{1/2} = 1/\pi T_2$	Inhomogeneous broadening is treated by using T_2^*
NOE enhancement parameter	$\eta = (I_A - I_A^\circ)/I_A^\circ$	
McConnell equation	$a = Q\rho$	$Q = 2.25$ mT

Further information

Further information 14.1 *Fourier transformation of the FID curve*

The analysis of the FID curve is achieved by the standard mathematical technique of Fourier transformation, which is explained more fully in *Mathematical background 7* following Chapter 19). We start by noting that the signal $S(t)$ in the time domain, the total FID curve, is the sum (more precisely, the integral) over all the contributing frequencies

$$S(t) = \int_{-\infty}^{\infty} I(\nu)e^{2\pi i\nu t}d\nu \qquad (14.43)$$

Because $e^{2\pi i\nu t} = \cos(2\pi\nu t) + i\sin(2\pi\nu t)$, the expression above is a sum over harmonically oscillating functions, with each one weighted by the intensity $I(\nu)$.

We need $I(\nu)$, the spectrum in the frequency domain; it is obtained by evaluating the integral

$$I(\nu) = 2\,\mathrm{Re}\int_0^{\infty} S(t)e^{-2\pi i\nu t}dt \qquad (14.44)$$

where Re means take the real part of the following expression. This integral is very much like an overlap integral: it gives a nonzero value if $S(t)$ contains a component that matches the oscillating function $e^{2\pi i\nu t}$. The integration is carried out at a series of frequencies ν on a computer that is built into the spectrometer.

Discussion questions

14.1 To what extent are all spectroscopic techniques resonance techniques, and are magnetic resonance techniques best so-called?

14.2 Discuss in detail the origins of the local, neighbouring group, and solvent contributions to the shielding constant.

14.3 Describe the significance of the chemical shift in relation to the terms 'high-field' and 'low-field'.

14.4 Explain why groups of equivalent protons do not exhibit the spin–spin coupling that exists between them.

14.5 Explain the difference between magnetically equivalent and chemically equivalent nuclei, and give two examples of each.

14.6 Discuss in detail the effects of a 90° pulse and of a 180° pulse on a system of spin-$\frac{1}{2}$ nuclei in a static magnetic field.

14.7 Suggest a reason why the relaxation times of ^{13}C nuclei are typically much longer than those of ^{1}H nuclei.

14.8 Discuss how the Fermi contact interaction and the polarization mechanism contribute to spin–spin couplings in NMR and hyperfine interactions in EPR.

14.9 Suggest how spin probes could be used to estimate the depth of a crevice in a biopolymer, such as the active site of an enzyme.

Exercises

14.1(a) Calculate the Larmor frequency of an electron in a magnetic field of 1.0 T.

14.1(b) Calculate the Larmor frequency of a proton in a magnetic field of 1.0 T.

14.2(a) For how long must a magnetic field of 1.0 T be applied to rotate the angular momentum vector of an electron through 90°?

14.2(b) For how long must a magnetic field of 1.0 T be applied to rotate the angular momentum vector of a proton through 90°?

14.3(a) What is the resonance frequency of a proton in a magnetic field of 14.1 T?

14.3(b) What is the resonance frequency of a ^{19}F nucleus in a magnetic field of 16.2 T?

14.4(a) Calculate the frequency separation of the nuclear spin levels of a ^{13}C nucleus in a magnetic field of 14.4 T given that the magnetogyric ratio is $6.73 \times 10^{7}\ T^{-1}\ s^{-1}$.

14.4(b) Calculate the frequency separation of the nuclear spin levels of a ^{14}N nucleus in a magnetic field of 15.4 T given that the magnetogyric ratio is $1.93 \times 10^{7}\ T^{-1}\ s^{-1}$.

14.5(a) Which has the greater energy level separation in a 600 MHz NMR spectrometer, a proton or a deuteron?

14.5(b) Which has the greater energy level separation, a ^{14}N nucleus in an NMR spectrometer operating at 14 T or an electron in an EPR spectrometer operating at 0.30 T?

14.6(a) Use Table 14.2 to predict the magnetic fields at which (a) ^{1}H, (b) ^{2}H, (c) ^{13}C come into resonance at (i) 250 MHz, (ii) 500 MHz.

14.6(b) Use Table 14.2 to predict the magnetic fields at which (a) ^{14}N, (b) ^{19}F, and (c) ^{31}P come into resonance at (i) 300 MHz, (ii) 750 MHz.

14.7(a) Calculate the relative population differences ($\delta N/N$) for protons in fields of (a) 0.30 T, (b) 1.5 T, and (c) 10 T at 25°C.

14.7(b) Calculate the relative population differences ($\delta N/N$) for ^{13}C nuclei in fields of (a) 0.50 T, (b) 2.5 T, and (c) 15.5 T at 25°C.

14.8(a) Evaluate the strength of the z-component of a magnetic field at 100 pm from an electron spin when θ is (a) 0, (b) 90°.

14.8(b) Evaluate the strength of the z-component of a magnetic field at 100 pm from a proton spin when θ is (a) 0, (b) 90°.

14.9(a) The first generally available NMR spectrometers operated at a frequency of 60 MHz; today it is not uncommon to use a spectrometer that operates at 800 MHz. What are the relative population differences ($\delta N/N$) of ^{13}C spin states in these two spectrometers at 25°C?

14.9(b) What are the relative population differences ($\delta N/N$) of electron spins in an EPR spectrometer operating at 0.33 T at (a) 25°C, (b) 77 K?

14.10(a) The chemical shift of the CH_3 protons in acetaldehyde (ethanal) is $\delta = 2.20$ and that of the CHO proton is 9.80. What is the difference in local magnetic field between the two regions of the molecule when the applied field is (a) 1.5 T, (b) 15 T?

14.10(b) The chemical shift of the CH_3 protons in diethyl ether is $\delta = 1.16$ and that of the CH_2 protons is 3.36. What is the difference in local magnetic field between the two regions of the molecule when the applied field is (a) 1.9 T, (b) 16.5 T?

14.11(a) Sketch the appearance of the ^{1}H-NMR spectrum of acetaldehyde (ethanal) using $J = 2.90$ Hz and the data in Exercise 14.10a in a spectrometer operating at (a) 250 MHz, (b) 500 MHz.

14.11(b) Sketch the appearance of the ^{1}H-NMR spectrum of diethyl ether using $J = 6.97$ Hz and the data in Exercise 14.10b in a spectrometer operating at (a) 350 MHz, (b) 650 MHz.

14.12(a) Construct a version of Pascal's triangle to show the fine structure that might arise from spin–spin coupling to a group of four spin-$\frac{3}{2}$ nuclei.

14.12(b) Construct a version of Pascal's triangle to show the fine structure that might arise from spin–spin coupling to a group of three spin-$\frac{5}{2}$ nuclei.

14.13(a) Two groups of protons are made equivalent by the isomerization of a fluxional molecule. At low temperatures, where the interconversion is slow, one group has $\delta = 4.0$ and the other has $\delta = 5.2$. At what rate of interconversion will the two signals merge in a spectrometer operating at 250 MHz?

14.13(b) Two groups of protons are made equivalent by the isomerization of a fluxional molecule. At low temperatures, where the interconversion is slow, one group has $\delta = 5.5$ and the other has $\delta = 6.8$. At what rate of interconversion will the two signals merge in a spectrometer operating at 350 MHz?

14.14(a) Sketch the form of the ^{19}F-NMR spectra of a natural sample of tetrafluoroborate ions, BF_4^-, allowing for the relative abundances of ^{10}B and ^{11}B.

14.14(b) From the data in Table 14.2, predict the frequency needed for ^{31}P-NMR in an NMR spectrometer designed to observe proton resonance at 500 MHz. Sketch the proton and ^{31}P resonances in the NMR spectrum of PH_4^+.

14.15(a) Sketch the form of an $A_3M_2X_4$ spectrum, where A, M, and X are protons with distinctly different chemical shifts and $J_{AM} > J_{AX} > J_{MX}$.

14.15(b) Sketch the form of an $A_2M_2X_5$ spectrum, where A, M, and X are protons with distinctly different chemical shifts and $J_{AM} > J_{AX} > J_{MX}$.

14.16(a) Which of the following molecules have sets of nuclei that are chemically but not magnetically equivalent? (a) CH_3CH_3, (b) $CH_2{=}CH_2$.

14.16(b) Which of the following molecules have sets of nuclei that are chemically but not magnetically equivalent? (a) $CH_2{=}C{=}CF_2$, (b) *cis*- and *trans*-$[Mo(CO)_4(PH_3)_2]$.

14.17(a) What is the effective transverse relaxation time when the width of a resonance line is 1.5 Hz?

14.17(b) What is the effective transverse relaxation time when the width of a resonance line is 12 Hz?

14.18(a) Predict the maximum enhancement (as the value of η) that could be obtained in a NOE observation in which ^{31}P is coupled to protons.

14.18(b) Predict the maximum enhancement (as the value of η) that could be obtained in a NOE observation in which ^{19}F is coupled to protons.

14.19(a) The duration of a 90° or 180° pulse depends on the strength of the $\mathcal{B}_1$ field. If a 90° pulse requires 10 μs, what is the strength of the $\mathcal{B}_1$ field? How long would the corresponding 180° pulse require?

14.19(b) The duration of a 90° or 180° pulse depends on the strength of the $\mathcal{B}_1$ field. If a 180° pulse requires 12.5 μs, what is the strength of the $\mathcal{B}_1$ field? How long would the corresponding 90° pulse require?

14.20(a) What magnetic field would be required in order to use an EPR X-band spectrometer (9 GHz) to observe ^{1}H-NMR and a 300 MHz spectrometer to observe EPR?

14.20(b) Some commercial EPR spectrometers use 8 mm microwave radiation (the Q band). What magnetic field is needed to satisfy the resonance condition?

14.21(a) The centre of the EPR spectrum of atomic hydrogen lies at 329.12 mT in a spectrometer operating at 9.2231 GHz. What is the g-value of the electron in the atom?

14.21(b) The centre of the EPR spectrum of atomic deuterium lies at 330.02 mT in a spectrometer operating at 9.2482 GHz. What is the g-value of the electron in the atom?

14.22(a) A radical containing two equivalent protons shows a three-line spectrum with an intensity distribution 1:2:1. The lines occur at 330.2 mT, 332.5 mT, and 334.8 mT. What is the hyperfine coupling constant for each proton? What is the g-value of the radical given that the spectrometer is operating at 9.319 GHz?

14.22(b) A radical containing three equivalent protons shows a four-line spectrum with an intensity distribution 1:3:3:1. The lines occur at 331.4 mT, 333.6 mT, 335.8 mT, and 338.0 mT. What is the hyperfine coupling constant for each proton? What is the g-value of the radical given that the spectrometer is operating at 9.332 GHz?

14.23(a) A radical containing two inequivalent protons with hyperfine constants 2.0 mT and 2.6 mT gives a spectrum centred on 332.5 mT. At what fields do the hyperfine lines occur and what are their relative intensities?

14.23(b) A radical containing three inequivalent protons with hyperfine constants 2.11 mT, 2.87 mT, and 2.89 mT gives a spectrum centred on 332.8 mT. At what fields do the hyperfine lines occur and what are their relative intensities?

14.24(a) Predict the intensity distribution in the hyperfine lines of the EPR spectra of (a) $\cdot CH_3$, (b) $\cdot CD_3$.

14.24(b) Predict the intensity distribution in the hyperfine lines of the EPR spectra of (a) $\cdot CH_2CH_3$, (b) $\cdot CD_2CD_3$.

14.25(a) The benzene radical anion has $g = 2.0025$. At what field should you search for resonance in a spectrometer operating at (a) 9.302 GHz, (b) 33.67 GHz?

14.25(b) The naphthalene radical anion has $g = 2.0024$. At what field should you search for resonance in a spectrometer operating at (a) 9.312 GHz, (b) 33.88 GHz?

14.26(a) The EPR spectrum of a radical with a single magnetic nucleus is split into four lines of equal intensity. What is the nuclear spin of the nucleus?

14.26(b) The EPR spectrum of a radical with two equivalent nuclei of a particular kind is split into five lines of intensity ratio 1:2:3:2:1. What is the spin of the nuclei?

14.27(a) Sketch the form of the hyperfine structures of radicals XH_2 and XD_2, where the nucleus X has $I = \frac{5}{2}$.

14.27(b) Sketch the form of the hyperfine structures of radicals XH_3 and XD_3, where the nucleus X has $I = \frac{3}{2}$.

14.28(a) A fluxional radical has EPR resonances at $g_{\parallel} = 2.012$ and $g_{\perp} = 2.032$ parallel and perpendicular to its molecular axis, respectively. At what tumbling rate (in rotations per second) would the two resonances merge in a spectrometer operating at 0.30 T?

14.28(b) A fluxional radical has EPR resonances at $g_{\parallel} = 2.022$ and $g_{\perp} = 2.023$ parallel and perpendicular to its molecular axis, respectively. At what tumbling rate (in rotations per second) would the two resonances merge in a spectrometer operating at 1.0 T?

Problems*

Numerical problems

14.1 A scientist investigates the possibility of neutron spin resonance, and has available a commercial NMR spectrometer operating at 300 MHz. What field is required for resonance? What is the relative population difference at room temperature? Which is the lower energy spin state of the neutron?

14.2 Two groups of protons have $\delta = 4.0$ and $\delta = 5.2$ and are interconverted by a conformational change of a fluxional molecule. In a 60 MHz spectrometer the spectrum collapsed into a single line at 280 K but at 300 MHz the collapse did not occur until the temperature had been raised to 300 K. What is the activation energy of the interconversion?

14.3‡ Suppose that the FID in Fig. 14.32 was recorded in a 300 MHz spectrometer, and that the interval between maxima in the oscillations in the FID is 0.10 s. What is the Larmor frequency of the nuclei and the spin–spin relaxation time?

14.4 Use mathematical software to construct the FID curve for a set of three nuclei with resonances at $\delta = 3.2$, 4.1, and 5.0 in a spectrometer operating at 800 MHz. Suppose that $T_2 = 1.0$ s. Go on to plot FID curves that show how they vary as the frequency of the spectrometer is changed from 200 MHz to 800 MHz.

14.5‡ In a classic study of the application of NMR to the measurement of rotational barriers in molecules, P.M. Nair and J.D. Roberts (*J. Am. Chem. Soc.* 79, 4565 (1957)) obtained the 40 MHz ^{19}F-NMR spectrum of $F_2BrCCBrCl_2$. Their spectra are reproduced in Fig. 14.60. At 193 K the spectrum shows five resonance peaks. Peaks I and III are separated by 160 Hz, as are IV and V. The ratio of the integrated intensities of peak II to peaks I, III, IV, and V is approximately 10 to 1. At 273 K, the five peaks have collapsed into one.

* Problems denoted with the symbol ‡ were supplied by Charles Trapp and Carmen Giunta.

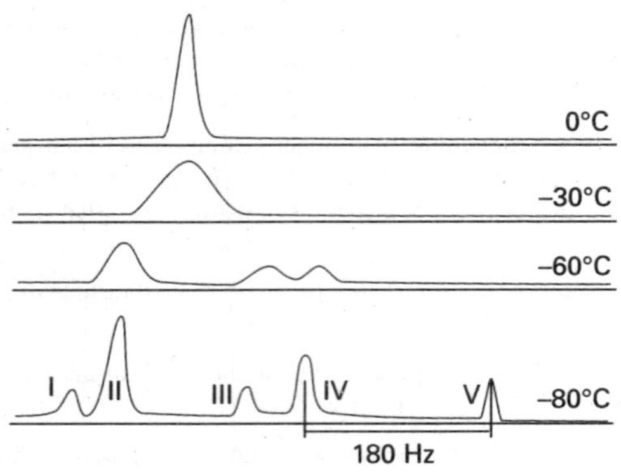

Fig. 14.60

Explain the spectrum and its change with temperature. At what rate of interconversion will the spectrum collapse to a single line? Calculate the rotational energy barrier between the rotational isomers on the assumption that it is related to the rate of interconversion between the isomers.

14.6‡ Various versions of the Karplus equation (eqn 14.27) have been used to correlate data on vicinal proton coupling constants in systems of the type $R_1R_2CHCHR_3R_4$. The original version (M. Karplus, *J. Am. Chem. Soc.* **85**, 2870 (1963)) is ${}^3J_{HH} = A\cos^2\phi_{HH} + B$. When $R_3 = R_4 = H$, ${}^3J_{HH} = 7.3$ Hz; when $R_3 = CH_3$ and $R_4 = H$, ${}^3J_{HH} = 8.0$ Hz; when $R_3 = R_4 = CH_3$, ${}^3J_{HH} = 11.2$ Hz. Assume that only staggered conformations are important and determine which version of the Karplus equation fits the data better.

14.7‡ It might be unexpected that the Karplus equation, which was first derived for ${}^3J_{HH}$ coupling constants, should also apply to vicinal coupling between the nuclei of metals such as tin. T.N. Mitchell and B. Kowall (*Magn. Reson. Chem.* **33**, 325 (1995)) have studied the relation between ${}^3J_{HH}$ and ${}^3J_{SnSn}$ in compounds of the type $Me_3SnCH_2CHRSnMe_3$ and find that ${}^3J_{SnSn} = 78.86\,{}^3J_{HH} + 27.84$ Hz. (a) Does this result support a Karplus-type equation for tin? Explain your reasoning. (b) Obtain the Karplus equation for ${}^3J_{SnSn}$ and plot it as a function of the dihedral angle. (c) Draw the preferred conformation.

14.8 Figure 14.61 shows the proton COSY spectrum of 1-nitropropane ($NO_2CH_2CH_2CH_3$). The circles show enhanced views of the spectral features. Account for the appearance of off-diagonal peaks in the spectrum. (Spectrum provided by Prof. G. Morris.)

14.9 The z-component of the magnetic field at a distance R from a magnetic moment parallel to the z-axis is given by eqn 14.28. In a solid, a proton at a distance R from another can experience such a field and the measurement of the splitting it causes in the spectrum can be used to calculate R. In gypsum, for instance, the splitting in the H_2O resonance can be interpreted in terms of a magnetic field of 0.715 mT generated by one proton and experienced by the other. What is the separation of the protons in the H_2O molåecule?

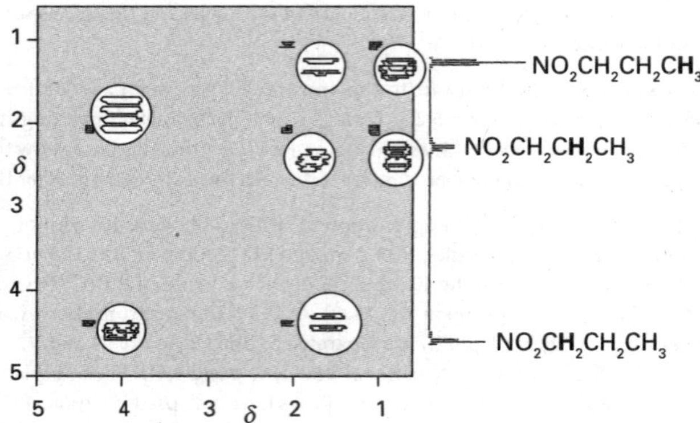

Fig. 14.61

14.10 The angular NO_2 molecule has a single unpaired electron and can be trapped in a solid matrix or prepared inside a nitrite crystal by radiation damage of NO_2^- ions. When the applied field is parallel to the OO direction the centre of the spectrum lies at 333.64 mT in a spectrometer operating at 9.302 GHz. When the field lies along the bisector of the ONO angle, the resonance lies at 331.94 mT. What are the g-values in the two orientations?

14.11 The hyperfine coupling constant in $\cdot CH_3$ is 2.3 mT. Use the information in Table 14.3 to predict the splitting between the hyperfine lines of the spectrum of $\cdot CD_3$. What are the overall widths of the hyperfine spectra in each case?

14.12 The p-dinitrobenzene radical anion can be prepared by reduction of p-dinitrobenzene. The radical anion has two equivalent N nuclei ($I = 1$) and four equivalent protons. Predict the form of the EPR spectrum using $a(N) = 0.148$ mT and $a(H) = 0.112$ mT.

14.13 When an electron occupies a 2s orbital on an N atom it has a hyperfine interaction of 55.2 mT with the nucleus. The spectrum of NO_2 shows an isotropic hyperfine interaction of 5.7 mT. For what proportion of its time is the unpaired electron of NO_2 occupying a 2s orbital? The hyperfine coupling constant for an electron in a 2p orbital of an N atom is 4.8 mT. In NO_2 the anisotropic part of the hyperfine coupling is 1.3 mT. What proportion of its time does the unpaired electron spend in the 2p orbital of the N atom in NO_2? What is the total probability that the electron will be found on (a) the N atoms, (b) the O atoms? What is the hybridization ratio of the N atom? Does the hybridization support the view that NO_2 is angular?

14.14 The hyperfine coupling constants observed in the radical anions (**14**), (**15**), and (**16**) are shown (in millitesla, mT). Use the value for the benzene radical anion to map the probability of finding the unpaired electron in the π orbital on each C atom.

14: NO_2, NO_2; 0.011, 0.172, 0.011, 0.172

15: NO_2, NO_2; 0.450, 0.272, 0.106, 0.450

16: NO_2, NO_2; 0.112, 0.112, 0.112, 0.112

Theoretical problems

14.15 Derive an expression for the diamagnetic shielding arising from (a) an electron in the 1s orbital of a hydrogenic atom of atomic number Z, (b) an electron in a H2s orbital. *Hint*. Use eqn 14.22 and the information in the *brief illustration* that follows it.

14.16 In this problem you will use the molecular electronic structure methods described in Chapter 10 to investigate the hypothesis that the magnitude of the ^{13}C chemical shift correlates with the net charge on a ^{13}C atom. (a) Using molecular modelling software[3] and the computational method of your choice, calculate the net charge at the C atom *para* to the substituents in this series of molecules: benzene, phenol, toluene, trifluorotoluene, benzonitrile, and nitrobenzene. (b) The ^{13}C chemical shifts of the *para* C atoms in each of the molecules that you examined in part (a) are given below:

[3] The web site contains links to molecular modelling freeware and to other sites where you may perform molecular orbital calculations directly from your web browser.

Substituent	OH	CH_3	H	CF_3	CN	NO_2
δ	130.1	128.4	128.5	128.9	129.1	129.4

Is there a linear correlation between net charge and ^{13}C chemical shift of the *para* C atom in this series of molecules? (c) If you did find a correlation in part (b), use the concepts developed in this chapter to explain the physical origins of the correlation.

14.17 In a liquid, the dipolar magnetic field averages to zero: show this result by evaluating the average of the field given in eqn 14.28. *Hint.* The volume element in polar coordinates is $\sin\theta\, d\theta \cdot d\phi$.

14.18 When interacting with a large biopolymer or even larger organelle, a small molecule might not rotate freely in all directions and the dipolar interaction might not average to zero. Suppose a molecule is bound so that, although the vector separating two protons may rotate freely around the *z*-axis, the colatitude may vary only between 0 and θ'. Average the dipolar field over this restricted range of orientations and confirm that the average vanishes when $\theta' = \pi$ (corresponding to rotation over an entire sphere). What is the average value of the local dipolar field for the H_2O molecule in Problem 14.9 if it is bound to a biopolymer that enables it to rotate up to $\theta' = 30°$?

14.19 The shape of a spectral line, $I(\omega)$, is related to the free induction decay signal $S(t)$ by eqn 14.44, where 'Re' means take the real part of what follows. Calculate the lineshape corresponding to an oscillating, decaying function $S(t) = \cos \omega_0 t\, e^{-t/\tau}$.

14.20 In the language of Problem 14.19, show that, if $S(t) = (a \cos \omega_1 t + b \cos \omega_2 t)e^{-t/\tau}$, then the spectrum consists of two lines with intensities proportional to a and b and located at $\omega = \omega_1$ and ω_2, respectively.

14.21 Suppose that a signal is (a) a decaying exponential function proportional to $e^{-t/\tau}$, (b) a Gaussian function proportional to e^{-t^2/τ^2}. To what linewidth (at half-height) does each process lead?

Applications: to biochemistry and medicine

14.22 Interpret the following features of the NMR spectra of hen lysozyme: (a) saturation of a proton resonance assigned to the side chain of methionine-105 changes the intensities of proton resonances assigned to the side chains of tryptophan-28 and tyrosine-23; (b) saturation of proton resonances assigned to tryptophan-28 did not affect the spectrum of tyrosine-23.

14.23 Suggest a reason why the spin–lattice relaxation time of benzene (a small molecule) in a mobile, deuterated hydrocarbon solvent increases with temperature, whereas that of an oligonucleotide (a large molecule) decreases.

14.24 NMR spectroscopy may be used to determine the equilibrium constant for dissociation of a complex between a small molecule, such as an enzyme inhibitor I, and a protein, such as an enzyme E:

$$EI \rightleftharpoons E + I \qquad K = [E][I]/[EI]$$

In the limit of slow chemical exchange, the NMR spectrum of a proton in I would consist of two resonances: one at ν_I for free I and another at ν_{EI} for bound I. When chemical exchange is fast, the NMR spectrum of the same proton in I consists of a single peak with a resonance frequency ν given by $\nu = f_I\nu_I + f_{EI}\nu_{EI}$, where $f_I = [I]/([I] + [EI])$ and $f_{EI} = [EI]/([I] + [EI])$ are, respectively, the fractions of free I and bound I. For the purposes of analysing the data, it is also useful to define the frequency differences $\delta\nu = \nu - \nu_I$ and $\delta\nu = \nu_{EI} - \nu_I$. Show that, when the initial concentration of I, $[I]_0$, is much greater than the initial concentration of E, $[E]_0$, a plot of $[I]_0$ against $\delta\nu^{-1}$ is a straight line with slope $[E]_0\Delta\nu$ and y-intercept K.

14.25 The molecular electronic structure methods described in Chapter 10 may be used to predict the spin density distribution in a radical. Recent EPR studies have shown that the amino acid tyrosine participates in a number of biological electron-transfer reactions, including the processes of water oxidation to O_2 in plant photosystem II (*Impact I21.1*). During the course of these electron-transfer reactions, a tyrosine radical forms, with spin density delocalized over the side chain of the amino acid. (a) The phenoxy radical shown in (**17**) is a suitable model of the tyrosine radical. Using molecular modelling software and the computational method of your choice (semi-empirical or *ab initio* methods), calculate the spin densities at the O atom and at all of the C atoms in (**17**). (b) Predict the form of the EPR spectrum of (**17**).

·O–C_6H_4–CH_3

17

14.26 Sketch the EPR spectra of the di-*tert*-butyl nitroxide radical (**10**) at 292 K in the limits of very low concentration (at which electron exchange is negligible), moderate concentration (at which electron exchange effects begin to be observed), and high concentration (at which electron exchange effects predominate). Discuss how the observation of electron exchange between nitroxide spin probes can inform the study of lateral mobility of lipids in a biological membrane.

14.27 You are designing an MRI spectrometer. What field gradient (in microtesla per metre, $\mu T\, m^{-1}$) is required to produce a separation of 100 Hz between two protons separated by the long diameter of a human kidney (taken as 8 cm) given that they are in environments with $\delta = 3.4$? The radiofrequency field of the spectrometer is at 400 MHz and the applied field is 9.4 T.

14.28 Suppose a uniform disc-shaped organ is in a linear field gradient, and that the MRI signal is proportional to the number of protons in a slice of width δx at each horizontal distance x from the centre of the disc. Sketch the shape of the absorption intensity for the MRI image of the disc before any computer manipulation has been carried out.

15 Statistical thermodynamics 1: the concepts

Statistical thermodynamics provides the link between the microscopic properties of matter and its bulk properties. Two key ideas are introduced in this chapter. The first is the Boltzmann distribution, which is used to predict the populations of states in systems at thermal equilibrium. In this chapter we see its derivation in terms of the distribution of particles over available states. The derivation leads naturally to the introduction of the partition function, which is the central mathematical concept of this and the next chapter. We see how to interpret the partition function and how to calculate it in a number of simple cases. We then see how to extract thermodynamic information from the partition function. In the final part of the chapter, we generalize the discussion to include systems that are composed of assemblies of interacting particles. Very similar equations are developed to those in the first part of the chapter, but they are much more widely applicable.

The preceding chapters of this part of the text have shown how the energy levels of molecules can be calculated, determined spectroscopically, and related to their structures. The next major step is to see how knowledge of these energy levels can be used to account for the properties of matter in bulk. To do so, we now introduce the concepts of **statistical thermodynamics**, the link between individual molecular properties and bulk thermodynamic properties.

The crucial step in going from the quantum mechanics of individual molecules to the thermodynamics of bulk samples is to recognize that the latter deals with the *average* behaviour of large numbers of molecules. For example, the pressure of a gas depends on the average force exerted by its molecules, and there is no need to specify which molecules happen to be striking the wall at any instant. Nor is it necessary to consider the fluctuations in the pressure as different numbers of molecules collide with the wall at different moments. The fluctuations in pressure are very small compared with the steady pressure: it is highly improbable that there will be a sudden lull in the number of collisions, or a sudden surge. Fluctuations in other thermodynamic properties also occur, but for large numbers of particles they are negligible compared to the mean values.

This chapter introduces statistical thermodynamics in two stages. The first, the derivation of the Boltzmann distribution for individual particles, is of restricted applicability, but it has the advantage of taking us directly to a result of central importance in a straightforward and elementary way. We can *use* statistical thermodynamics once we have deduced the Boltzmann distribution. Then (in Section 15.5) we extend the arguments to systems composed of interacting particles.

The distribution of molecular states

We consider a closed system composed of N molecules. Although the total energy is constant at E, it is not possible to be definite about how that energy is shared between the molecules. Collisions result in the ceaseless redistribution of energy not only between the molecules but also among their different modes of motion. The closest we can come to a description of the distribution of energy is to report the **population** of a state, the average number of molecules that occupy it, and to say that on average there are n_i molecules in a state of energy ε_i. The populations of the states remain almost constant, but the precise identities of the molecules in each state may change at every collision.

The problem we address in this section is the calculation of the populations of states for any type of molecule in any mode of motion at any temperature. The only restriction is that the molecules should be independent, in the sense that the total energy of the system is a sum of their individual energies. We are discounting (at this stage) the possibility that in a real system a contribution to the total energy may arise from interactions between molecules. We also adopt the **principle of equal *a priori* probabilities**, the assumption that all possibilities for the distribution of energy are equally probable. *A priori* means in this context loosely 'as far as one knows'. We have no reason to presume otherwise than that, for a collection of molecules at thermal equilibrium, vibrational states of a certain energy, for instance, are as likely to be populated as rotational states of the same energy.

One very important conclusion that will emerge from the following analysis is that the populations of states depend on a single parameter, the 'temperature'. That is, statistical thermodynamics provides a molecular justification for the concept of temperature and some insight into this crucially important quantity.

15.1 Configurations and weights

Key points **(a) The weight of a configuration is the number of ways that molecules can be distributed over the available states. (b) The most probable distribution, that of the greatest weight, is the Boltzmann distribution.**

Any individual molecule may exist in states with energies ε_0, ε_1 . . . We shall always take ε_0, the lowest state, as the zero of energy ($\varepsilon_0 = 0$), and measure all other energies relative to that state. To obtain the actual internal energy, U, we may have to add a constant to the calculated energy of the system. For example, if we are considering the vibrational contribution to the internal energy, then we must add the total zero-point energy of any oscillators in the sample.

(a) Instantaneous configurations

At any instant there will be N_0 molecules in the state with energy ε_0, N_1 with ε_1, and so on. The specification of the set of populations N_0, N_1, . . . in the form $\{N_0, N_1, \ldots\}$ is a statement of the instantaneous **configuration** of the system. The instantaneous configuration fluctuates with time because the populations change. We can picture a large number of different instantaneous configurations. One, for example, might be $\{N,0,0,\ldots\}$, corresponding to every molecule being in its ground state. Another might be $\{N-2,2,0,0,\ldots\}$, in which two of the molecules are in the first excited state. The latter configuration is intrinsically more likely to be found than the former because it can be achieved in more ways: $\{N,0,0,\ldots\}$ can be achieved in only one way, but $\{N-2,2,\ldots\}$ can be achieved in $\frac{1}{2}N(N-1)$ different ways (Fig. 15.1; see the

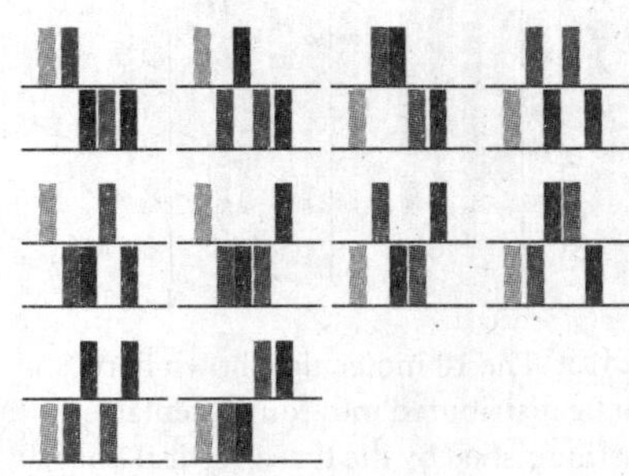

Fig. 15.1 Whereas a configuration $\{5,0,0,\ldots\}$ can be achieved in only one way, a configuration $\{3,2,0,\ldots\}$ can be achieved in the ten different ways shown here, where the tinted blocks represent different molecules.

following *Justification*). At this stage in the argument, we are ignoring the requirement that the total energy of the system must be constant (the second configuration has a higher energy than the first); the constraint of total energy will be imposed later in this section.

If, as a result of collisions, the system were to fluctuate between the configurations $\{N,0,0,\ldots\}$ and $\{N-2,2,0,\ldots\}$, it would almost always be found in the second, more likely state (especially if N were large). In other words, a system free to switch between the two configurations would show properties characteristic almost exclusively of the second configuration. A general configuration $\{N_0,N_1,\ldots\}$ can be achieved in $\mathcal{W}$ different ways, where $\mathcal{W}$ is called the **weight** of the configuration. The weight of the configuration $\{N_0,N_1,\ldots\}$ is given by the expression

A brief comment

More formally, $\mathcal{W}$ is called the *multinomial coefficient*. In eqn 15.1, $x!$, x factorial, denotes $x(x-1)(x-2)\ldots 1$ and, by definition, $0! = 1$.

$$\mathcal{W} = \frac{N!}{N_0!N_1!N_2!\ldots} \qquad \text{The weight of a configuration} \qquad (15.1)$$

Equation 15.1 is a generalization of the formula in $\mathcal{W} = \frac{1}{2}N(N-1)$, and reduces to it for the configuration $\{N-2,2,0,\ldots\}$.

• **A brief illustration**

To calculate the number of ways of distributing 20 identical objects with the arrangement 1, 0, 3, 5, 10, 1, we note that the configuration is $\{1,0,3,5,10,1\}$ with $N = 20$; therefore the weight is

$$\mathcal{W} = \frac{20!}{1!0!3!5!10!1!} = 9.31 \times 10^8 \quad \bullet$$

Self-test 15.1 Calculate the weight of the configuration in which 20 objects are distributed in the arrangement 0, 1, 5, 0, 8, 0, 3, 2, 0, 1. $[4.19 \times 10^{10}]$

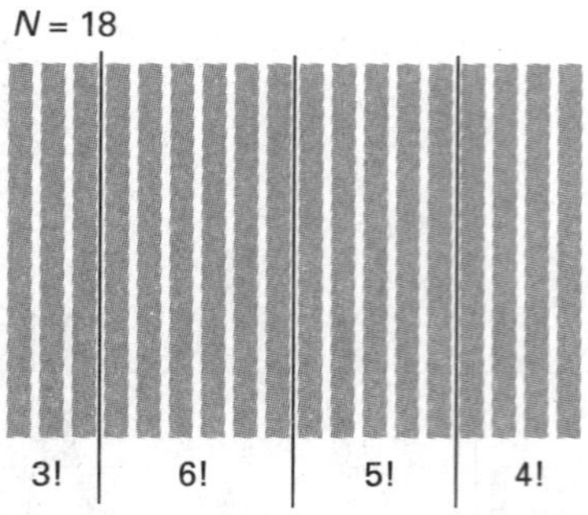

Justification 15.1 *The weight of a configuration*

First, consider the weight of the configuration $\{N-2,2,0,0,\ldots\}$. One candidate for promotion to an upper state can be selected in N ways. There are $N-1$ candidates for the second choice, so the total number of choices is $N(N-1)$. However, we should not distinguish the choice (Jack, Jill) from the choice (Jill, Jack) because they lead to the same configurations. Therefore, only half the choices lead to distinguishable configurations, and the total number of distinguishable choices is $\frac{1}{2}N(N-1)$.

Now we generalize this remark. Consider the number of ways of distributing N balls into bins. The first ball can be selected in N different ways, the next ball in $N-1$ different ways for the balls remaining, and so on. Therefore, there are $N(N-1)\ldots 1 = N!$ ways of selecting the balls for distribution over the bins. However, if there are N_0 balls in the bin labelled ε_0, there would be $N_0!$ different ways in which the same balls could have been chosen (Fig. 15.2). Similarly, there are $N_1!$ ways in which the N_1 balls in the bin labelled ε_1 can be chosen, and so on. Therefore, the total number of distinguishable ways of distributing the balls so that there are N_0 in bin ε_0, N_1 in bin ε_1, etc. regardless of the order in which the balls were chosen is $N!/N_0!N_1!\ldots$, which is the content of eqn 15.1.

Fig. 15.2 The 18 molecules shown here can be distributed into four receptacles (distinguished by the three vertical lines) in 18! different ways. However, 3! of the selections that put three molecules in the first receptacle are equivalent, 6! that put six molecules into the second receptacle are equivalent, and so on. Hence the number of distinguishable arrangements is 18!/3!6!5!4!.

It will turn out to be more convenient to deal with the natural logarithm of the weight, ln $\mathcal{W}$, rather than with the weight itself. We shall therefore need the expression

$$\ln \mathcal{W} = \ln \frac{N!}{N_0!N_1!N_2!\ldots!} = \ln N! - \ln(N_0!N_1!N_2!\cdots!)$$
$$= \ln N! - (\ln N_0! + \ln N_1! + \ln N_2! + \cdots)$$
$$= \ln N! - \sum_i \ln N_i!$$

where in the first line we have used $\ln(x/y) = \ln x - \ln y$ and in the second $\ln xy = \ln x + \ln y$. One reason for introducing $\ln \mathcal{W}$ is that it is easier to make approximations. In particular, we can simplify the factorials by using **Stirling's approximation** in the form

$$\ln x! \approx x \ln x - x \qquad \text{Stirling's approximation} \qquad (15.2)$$

A brief comment

A more accurate form of Stirling's approximation is

$$x! \approx (2\pi)^{1/2} x^{x+\frac{1}{2}} e^{-x}$$

and is in error by less than 1 per cent when x is greater than about 10. We deal with far larger values of x, and the simplified version in eqn 15.2 is adequate.

Then the approximate expression for the weight is

$$\ln \mathcal{W} = (N \ln N - N) - \sum_i (N_i \ln N_i - N_i) = N \ln N - \sum_i N_i \ln N_i \qquad (15.3)$$

The final form of eqn 15.3 is derived by noting that the sum of N_i is equal to N, so the second and fourth terms in the second expression cancel.

(b) The Boltzmann distribution

We have seen that the configuration $\{N-2,2,0,\ldots\}$ dominates $\{N,0,0,\ldots\}$, and it should be easy to believe that there may be other configurations that have a much greater weight than both. We shall see, in fact, that there is a configuration with so great a weight that it overwhelms all the rest in importance to such an extent that the system will almost always be found in it. The properties of the system will therefore be characteristic of that particular dominating configuration. This dominating configuration can be found by looking for the values of N_i that lead to a maximum value of $\mathcal{W}$. Because $\mathcal{W}$ is a function of all the N_i, we can do this search by varying the N_i and looking for the values that correspond to $d\mathcal{W} = 0$ (just as in the search for the maximum of any function), or equivalently a maximum value of $\ln \mathcal{W}$. However, there are two difficulties with this procedure.

The first difficulty is that the only permitted configurations are those corresponding to the specified, constant, total energy of the system. This requirement rules out many configurations; $\{N,0,0,\ldots\}$ and $\{N-2,2,0,\ldots\}$, for instance, have different energies, so both cannot occur in the same isolated system. It follows that, in looking for the configuration with the greatest weight, we must ensure that the configuration also satisfies the condition

$$\text{Constant total energy:} \qquad \sum_i N_i \varepsilon_i = E \qquad (15.4)$$

where E is the total energy of the system.

The second constraint is that, because the total number of molecules present is also fixed (at N), we cannot arbitrarily vary all the populations simultaneously. Thus, increasing the population of one state by 1 demands that the population of another state must be reduced by 1. Therefore, the search for the maximum value of $\mathcal{W}$ is also subject to the condition

$$\text{Constant total number of molecules:} \qquad \sum_i N_i = N \qquad (15.5)$$

We show in *Further information 15.1* that the populations in the configuration of greatest weight, subject to the two constraints in eqns 15.4 and 15.5, depend on the energy of the state according to the **Boltzmann distribution**:

$$\frac{N_i}{N}=\frac{e^{-\beta\varepsilon_i}}{\sum_i e^{-\beta\varepsilon_i}} \qquad \text{Boltzmann distribution} \qquad (15.6a)$$

where $\varepsilon_0 \le \varepsilon_1 \le \varepsilon_2 \ldots$. Equation 15.6a is the justification of the remark that a single parameter, here denoted β, determines the most probable populations of the states of the system. We shall see in Section 15.3b that

$$\beta=\frac{1}{kT} \qquad (15.6b)$$

where T is the thermodynamic temperature and k is Boltzmann's constant. In other words, *the thermodynamic temperature is the unique parameter that governs the most probable populations of states of a system at thermal equilibrium*. In *Further information 15.1*, moreover, we see that β is a more natural measure of temperature than T itself.

15.2 The molecular partition function

Key points **(a) The molecular partition function indicates the number of thermally accessible states of a collection of molecules at a temperature *T*. (b) The translational partition function is calculated by noting that translational states form a near continuum. When the energy is a sum of contributions from independent modes of motion, the partition function is a product of partition functions for each mode of motion.**

The Boltzmann distribution is hugely important throughout physical chemistry (and science in general). From now on we write it as

$$p_i=\frac{e^{-\beta\varepsilon_i}}{q} \qquad \text{Population of a state} \qquad (15.7)$$

where p_i is the fraction of molecules in the state i, $p_i = N_i/N$, and q is the **molecular partition function**:

$$q=\sum_i e^{-\beta\varepsilon_i} \qquad \text{Definition of the molecular partition function} \qquad [15.8]$$

The sum in q is sometimes expressed slightly differently. It may happen that several states have the same energy, and so give the same contribution to the sum. If, for example, g_i states have the same energy ε_i (so the level is g_i-fold degenerate), we could write

$$q=\sum_{\text{levels } I} g_I e^{-\beta\varepsilon_I} \qquad (15.9)$$

where the sum is now over energy levels (sets of states with the same energy), not individual states. We use the letter i to label individual states and I to label levels; when appropriate, we replace these labels by the appropriate quantum numbers.

Example 15.1 *Writing a partition function*

Write an expression for the partition function of a linear molecule (such as HCl) treated as a rigid rotor.

Method To use eqn 15.9 we need to know (a) the energies of the levels, (b) the degeneracies, the number of states that belong to each level. Whenever calculating a partition function, the energies of the levels are expressed relative to 0 for the state of lowest energy. The energy levels of a rigid linear rotor were derived in Section 12.4c.

Answer From eqn 12.14, the energy levels of a linear rotor are $hc\tilde{B}J(J+1)$, with $J = 0, 1, 2, \ldots$. Therefore label the levels with this quantum number. The state of lowest energy has zero energy, so no adjustment need be made to the energies given by this expression. Each level consists of $2J+1$ degenerate states. Therefore,

$$q = \sum_{J=0}^{\infty} \overbrace{(2J+1)}^{g_J} e^{-\beta \overbrace{hc\tilde{B}J(J+1)}^{\varepsilon_J}}$$

The sum can be evaluated numerically by supplying the value of $\tilde{B}$ (from spectroscopy or calculation) and the temperature. For reasons explained in Section 16.2b, this expression applies only to unsymmetrical linear rotors (for instance, HCl, not CO_2).

Self-test 15.2 Write the partition function for a two-level system, the lower state (at energy 0) being nondegenerate, and the upper state (at an energy ε) doubly degenerate. $[q = 1 + 2e^{-\beta\varepsilon}]$

(a) An interpretation of the partition function

Some insight into the significance of a partition function can be obtained by considering how q depends on the temperature. When T is close to zero, the parameter $\beta = 1/kT$ is close to infinity. Then every term except one in the sum defining q is zero because each one has the form e^{-x} with $x \to \infty$. The exception is the term with $\varepsilon_0 \equiv 0$ (or the g_0 terms at zero energy if the ground state is g_0-fold degenerate), because then $\varepsilon_0/kT \equiv 0$ whatever the temperature, including zero. As there is only one surviving term when $T = 0$, and its value is g_0, it follows that

$$\lim_{T\to 0} q = g_0 \tag{15.10}$$

That is, at $T = 0$, the partition function is equal to the degeneracy of the ground state.

Now consider the case when T is so high that for each term in the sum $\varepsilon_j/kT \approx 0$. Because $e^{-x} = 1$ when $x = 0$ each term in the sum now contributes 1. It follows that the sum is equal to the number of molecular states, which in general is infinite:

$$\lim_{T\to\infty} q = \infty \tag{15.11}$$

In some idealized cases, the molecule may have only a finite number of states; then the upper limit of q is equal to the number of states. For example, if we were considering only the spin energy levels of a doublet ($S = \frac{1}{2}$) radical in a magnetic field, then there would be only two states ($M_S = \pm\frac{1}{2}$). The partition function for such a system can therefore be expected to rise towards 2 as T is increased towards infinity.

We see that the molecular partition function gives an indication of the number of states that are thermally accessible to a molecule at the temperature of the system.

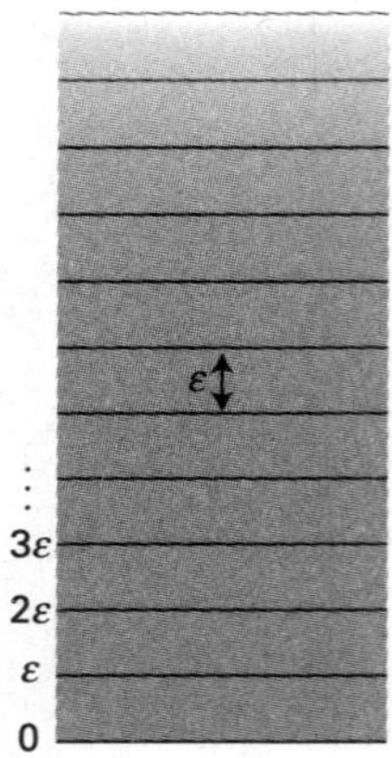

Fig. 15.3 The equally spaced infinite array of energy levels used in the calculation of the partition function. A harmonic oscillator has the same spectrum of levels.

At $T = 0$, only the ground level is accessible and $q = g_0$. At very high temperatures, virtually all states are accessible, and q is correspondingly large.

Example 15.2 *Evaluating the partition function for a uniform ladder of energy levels*

Evaluate the partition function for a molecule with an infinite number of equally spaced nondegenerate energy levels (Fig. 15.3). These levels can be thought of as the vibrational energy levels of a diatomic molecule in the harmonic approximation.

Method We expect the partition function to increase from 1 at $T = 0$ and approach infinity as T goes to ∞. To evaluate eqn 15.8 explicitly, note that

$$1 + x + x^2 + \cdots = \frac{1}{1-x}$$

Answer If the separation of neighbouring levels is ε, the partition function is

$$q = 1 + e^{-\beta\varepsilon} + e^{-2\beta\varepsilon} + \cdots = 1 + e^{-\beta\varepsilon} + (e^{-\beta\varepsilon})^2 + \cdots = \frac{1}{1 - e^{-\beta\varepsilon}}$$

This expression is plotted in Fig. 15.4: notice that, as anticipated, q rises from 1 to infinity as the temperature is raised.

Self-test 15.3 Find and plot an expression for the partition function of a system with one state at zero energy and another state at the energy ε.

[$q = 1 + e^{-\beta\varepsilon}$, Fig. 15.5]

A brief comment

The sum of the infinite series $S = 1 + x + x^2 \cdots$ is obtained by multiplying both sides by x, which gives $xS = x + x^2 + x^3 \cdots = S - 1$ and hence $S = 1/(1 - x)$.

It follows from eqn 15.7 and the expression for q derived in Example 15.2 for a uniform ladder of states of spacing ε,

$$q = \frac{1}{1 - e^{-\beta\varepsilon}} \qquad \text{Partition function for a uniform array of states} \qquad (15.12)$$

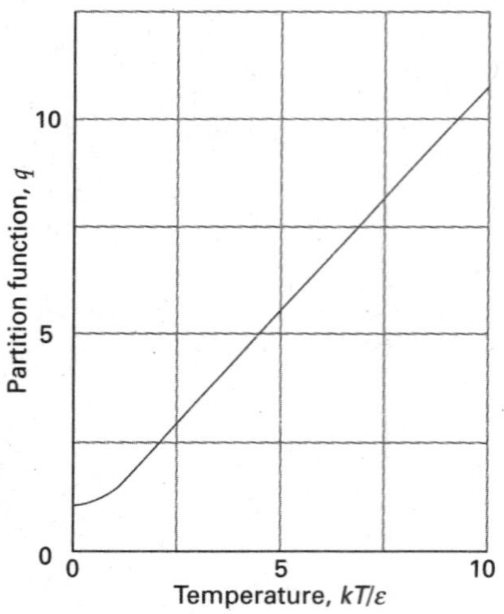

Fig. 15.4 The partition function for the system shown in Fig. 15.3 (a harmonic oscillator) as a function of temperature.

interActivity Plot the partition function of a harmonic oscillator against temperature for several values of the energy separation ε. How does q vary with temperature when T is high, in the sense that $kT \gg \varepsilon$ (or $\beta\varepsilon \ll 1$)?

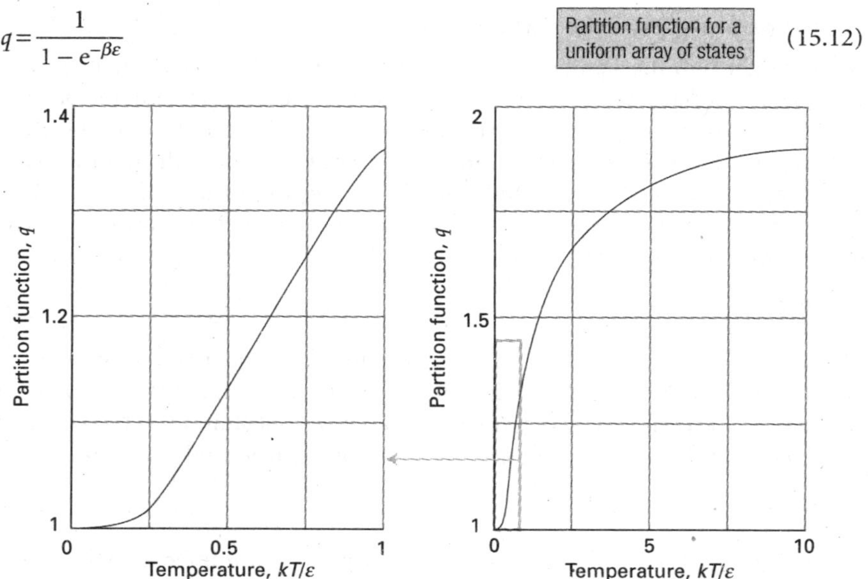

Fig. 15.5 The partition function for a two-level system as a function of temperature. The two graphs differ in the scale of the temperature axis to show the approach to 1 as $T \to 0$ and the slow approach to 2 as $T \to \infty$.

interActivity Consider a three-level system with levels 0, ε, and 2ε. Plot the partition function against kT/ε.

that the fraction of molecules in the state with energy ε_i is

$$p_i = \frac{e^{-\beta\varepsilon_i}}{q} = (1 - e^{-\beta\varepsilon})e^{-\beta\varepsilon_i} \qquad (15.13)$$

Figure 15.6 shows how p_i varies with temperature. At very low temperatures, where q is close to 1, only the lowest state is significantly populated. As the temperature is raised, the population breaks out of the lowest state, and the upper states become progressively more highly populated. At the same time, the partition function rises from 1 and its value gives an indication of the range of states populated. The name 'partition function' reflects the sense in which q measures how the total number of molecules is distributed—partitioned—over the available states.

The corresponding expressions for a two-level system derived in Self-test 15.3 are

$$p_0 = \frac{1}{1+e^{-\beta\varepsilon}} \qquad p_1 = \frac{e^{-\beta\varepsilon}}{1+e^{-\beta\varepsilon}} \qquad (15.14)$$

Populations of a two-state system

These functions are plotted in Fig. 15.7. Notice how the populations tend towards equality ($p_0 = \frac{1}{2}$, $p_1 = \frac{1}{2}$) as $T \to \infty$. A common error is to suppose that all the molecules in the system will be found in the upper energy state when $T = \infty$; however, we see from eqn 15.14 that, as $T \to \infty$ the populations of states become equal. The same conclusion is true of multi-level systems too: as $T \to \infty$, all states become equally populated.

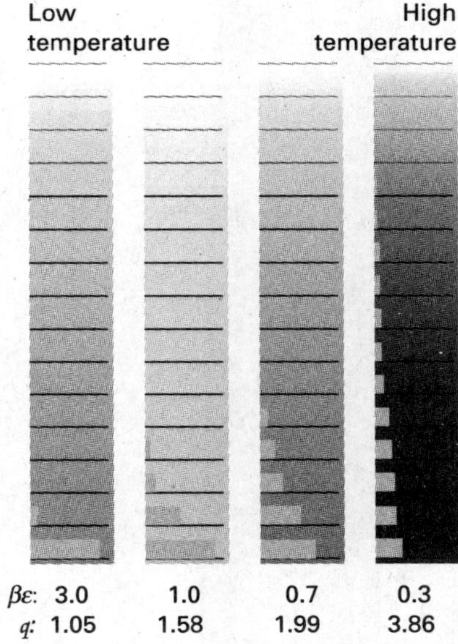

Fig. 15.6 The populations of the energy levels of the system shown in Fig. 15.3 at different temperatures, and the corresponding values of the partition function calculated in Example 15.2. Note that $\beta = 1/kT$.

interActivity To visualize the content of Fig. 15.6 in a different way, plot the functions p_0, p_1, p_2, and p_3 against kT/ε.

Example 15.3 *Using the partition function to calculate a population*

Calculate the proportion of I_2 molecules in their ground, first excited, and second excited vibrational states at 25°C. The vibrational wavenumber is 214.6 cm^{-1}.

Method Vibrational energy levels have a constant separation (in the harmonic approximation, Section 12.8), so the partition function is given by eqn 15.12 and the populations by eqn 15.13. To use the latter equation, we identify the index i with the quantum number v, and calculate p_v for $v = 0$, 1, and 2. At 298.15 K, $kT/hc = 207.226$ cm^{-1}.

Answer First, we note that

$$\beta\varepsilon = \frac{hc\tilde{\nu}}{kT} = \frac{214.6 \text{ cm}^{-1}}{207.226 \text{ cm}^{-1}} = 1.036$$

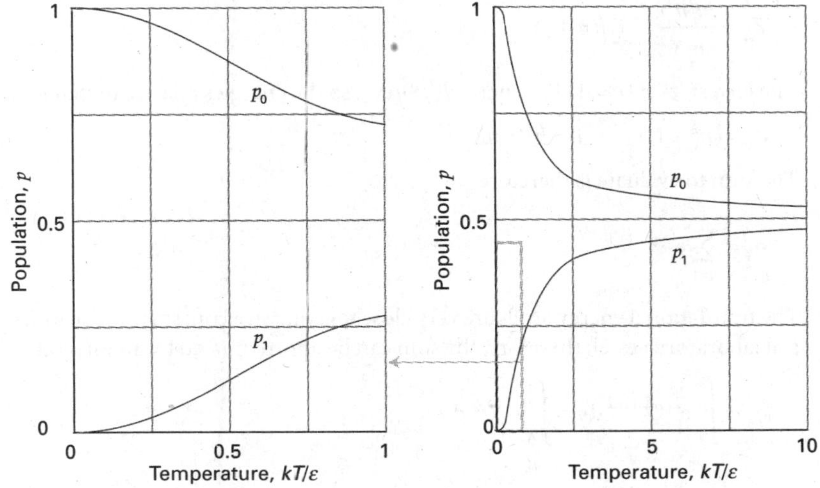

Fig. 15.7 The fraction of populations of the two states of a two-level system as a function of temperature (eqn 15.14). Note that, as the temperature approaches infinity, the populations of the two states become equal (and the fractions both approach 0.5).

interActivity Consider a three-level system with levels 0, ε, and 2ε. Plot the functions p_0, p_1, and p_2 against kT/ε.

Then it follows from eqn 15.13 that the populations are

$$p_v = (1 - e^{-\beta\varepsilon})e^{-v\beta\varepsilon} = 0.645e^{-1.036v}$$

Therefore, $p_0 = 0.645$, $p_1 = 0.229$, $p_2 = 0.081$. The I–I bond is not stiff and the atoms are heavy: as a result, the vibrational energy separations are small and at room temperature several vibrational levels are significantly populated. The value of the partition function, $q = 1.55$, reflects this small but significant spread of populations.

Self-test 15.4 At what temperature would the $v = 1$ level of I_2 have (a) half the population of the ground state, (b) the same population as the ground state?

[(a) 445 K, (b) infinite]

(b) Approximations and factorizations

In general, exact analytical expressions for partition functions cannot be obtained. However, closed approximate expressions can often be found and prove to be very important in a number of chemical applications. For instance, the expression for the partition function for a particle of mass m free to move in a one-dimensional container of length X can be evaluated by making use of the fact that the separation of energy levels is very small and that large numbers of states are accessible at normal temperatures. As shown in the following *Justification*, in this case

$$q_X = \left(\frac{2\pi m}{h^2\beta}\right)^{1/2} X = \left(\frac{2\pi mkT}{h^2}\right)^{1/2} X \qquad (15.15)$$

Partition function for translation in one dimension

This expression shows that the partition function for translational motion increases with the length of the box and the mass of the particle, for in each case the separation of the energy levels becomes smaller and more levels become thermally accessible. For a given mass and length of the box, the partition function also increases with increasing temperature, because more states become accessible.

Justification 15.2 *The partition function for a particle in a one-dimensional box*

The energy levels of a molecule of mass m in a container of length X are given by eqn 8.4a with $L = X$:

$$E_n = \frac{n^2h^2}{8mX^2} \qquad n = 1, 2, \ldots$$

The lowest level ($n = 1$) has energy $h^2/8mX^2$, so the energies relative to that level are

$$\varepsilon_n = (n^2 - 1)\varepsilon \qquad \varepsilon = h^2/8mX^2$$

The sum to evaluate is therefore

$$q_X = \sum_{n=1}^{\infty} e^{-(n^2-1)\beta\varepsilon}$$

The translational energy levels are very close together in a container the size of a typical laboratory vessel; therefore, the sum can be approximated by an integral:

$$q_X = \int_1^{\infty} e^{-(n^2-1)\beta\varepsilon}\,dn = \int_0^{\infty} e^{-n^2\beta\varepsilon}\,dn$$

The extension of the lower limit to $n = 0$ and the replacement of $n^2 - 1$ by n^2 introduces negligible error but turns the integral into standard form. We make the substitution $x^2 = n^2\beta\varepsilon$, implying $dn = dx/(\beta\varepsilon)^{1/2}$, and therefore that

$$q_X = \left(\frac{1}{\beta\varepsilon}\right)^{1/2} \overbrace{\int_0^\infty e^{-x^2}dx}^{\pi^{1/2}/2} = \left(\frac{1}{\beta\varepsilon}\right)^{1/2}\left(\frac{\pi^{1/2}}{2}\right) = \left(\frac{2\pi m}{h^2\beta}\right)^{1/2} X$$

Another useful feature of partition functions is used to derive expressions when the energy of a molecule arises from several different, independent sources: if the energy is a sum of contributions from independent modes of motion, then the partition function is a product of partition functions for each mode of motion. For instance, suppose the molecule we are considering is free to move in three dimensions. We take the length of the container in the y-direction to be Y and that in the z-direction to be Z. The total energy of a molecule ε is the sum of its translational energies in all three directions:

$$\varepsilon_{n_1n_2n_3} = \varepsilon_{n_1}^{(X)} + \varepsilon_{n_2}^{(Y)} + \varepsilon_{n_3}^{(Z)} \qquad (15.16)$$

where n_1, n_2, and n_3 are the quantum numbers for motion in the x-, y-, and z-directions, respectively. Therefore, because $e^{a+b+c} = e^a e^b e^c$, the partition function factorizes as follows:

$$q = \sum_{\text{all } n} e^{-\beta\varepsilon_{n_1}^{(X)} - \beta\varepsilon_{n_2}^{(Y)} - \beta\varepsilon_{n_3}^{(Z)}} = \sum_{\text{all } n} e^{-\beta\varepsilon_{n_1}^{(X)}} e^{-\beta\varepsilon_{n_2}^{(Y)}} e^{-\beta\varepsilon_{n_3}^{(Z)}}$$

$$= \left(\sum_{n_1} e^{-\beta\varepsilon_{n_1}^{(X)}}\right)\left(\sum_{n_2} e^{-\beta\varepsilon_{n_2}^{(Y)}}\right)\left(\sum_{n_3} e^{-\beta\varepsilon_{n_3}^{(Z)}}\right) = q_X q_Y q_Z \qquad (15.17)$$

It is generally true that, if the energy of a molecule can be written as the sum of independent terms, then the partition function is the corresponding product of individual contributions.

Equation 15.15 gives the partition function for translational motion in the x-direction. The only change for the other two directions is to replace the length X by the lengths Y or Z. Hence the partition function for motion in three dimensions is

$$q = \left(\frac{2\pi m}{h^2\beta}\right)^{3/2} XYZ \qquad (15.18)$$

The product of lengths XYZ is the volume, V, of the container, so we can write

$$q = \frac{V}{\Lambda^3} \qquad \Lambda = h\left(\frac{\beta}{2\pi m}\right)^{1/2} = \frac{h}{(2\pi mkT)^{1/2}} \qquad (15.19)$$

Partition function for translation in three dimensions

The quantity Λ has the dimensions of length and is called the **thermal wavelength** (sometimes the *thermal de Broglie wavelength*) of the molecule. The thermal wavelength decreases with increasing mass and temperature. As in the one-dimensional case, the partition function increases with the mass of the particle (as $m^{3/2}$) and the volume of the container (as V); for a given mass and volume, the partition function increases with temperature (as $T^{3/2}$).

• A brief illustration

To calculate the translational partition function of an H_2 molecule confined to a 100 cm^3 vessel at 25°C we use $m = 2.016m_u$; then

$$\Lambda = \frac{6.626\times10^{-34}\ \text{J s}}{\{2\pi\times(2.016\times1.6605\times10^{-27}\ \text{kg})\times(1.38\times10^{-23}\ \text{J K}^{-1})\times(298\ \text{K})\}^{1/2}}$$
$$= 7.12\times10^{-11}\ \text{m}$$

where we have used 1 J = 1 kg m^2 s^{-2}. Therefore,

$$q = \frac{1.00\times10^{-4}\ \text{m}^3}{(7.12\times10^{-11}\ \text{m})^3} = 2.77\times10^{26}$$

About 10^{26} quantum states are thermally accessible, even at room temperature and for this light molecule. Many states are occupied if the thermal wavelength (which in this case is 71.2 pm) is small compared with the linear dimensions of the container. •

Self-test 15.5 Calculate the translational partition function for a D_2 molecule under the same conditions. [$q = 7.8\times10^{26}$, $2^{3/2}$ times larger]

The validity of the approximations that led to eqn 15.19 can be expressed in terms of the average separation of the particles in the container, *d*. We do not have to worry about the role of the Pauli principle in the occupation of states if there are many states available for each molecule. Because q is the total number of accessible states, the average number of states per molecule is q/N. For this quantity to be large, $q \gg 1$, we require $V/N\Lambda^3 \gg 1$. However, V/N is the volume occupied by a single particle, and therefore the average separation of the particles is $d = (V/N)^{1/3}$. The condition for there being many states available per molecule is therefore $d^3/\Lambda^3 \gg 1$, and therefore $d \gg \Lambda$. That is, for eqn 15.19 to be valid, *the average separation of the particles must be much greater than their thermal wavelength*. For H_2 molecules at 1 bar and 298 K, the average separation is 3 nm, which is significantly larger than their thermal wavelength (71.2 pm).

The internal energy and the entropy

The importance of the molecular partition function is that it contains all the information needed to calculate the thermodynamic properties of a system of independent particles. In this respect, q plays a role in statistical thermodynamics very similar to that played by the wavefunction in quantum mechanics: q is a kind of thermal wavefunction. Here we start to see how this information can be extracted.

15.3 The internal energy

Key points **(a) The internal energy is proportional to the derivative of the partition function with respect to temperature. (b) The parameter $\beta = 1/kT$.**

We shall begin to unfold the importance of q by showing how to derive an expression for the internal energy of the system.

(a) The relation between U and q

The total energy of the system relative to the energy of the lowest state is

$$E(T)=\sum_i N_i\varepsilon_i \tag{15.20}$$

The energy depends on the temperature because the populations of the states depend on the temperature. Because the most probable configuration is so strongly dominating, we can use the Boltzmann distribution for the populations and write

$$E(T)=\frac{N}{q}\sum_i \varepsilon_i e^{-\beta\varepsilon_i} \tag{15.21}$$

To manipulate this expression into a form involving only q we note that

$$\varepsilon_i e^{-\beta\varepsilon_i}=-\frac{d}{d\beta}e^{-\beta\varepsilon_i}$$

It follows that

$$E(T)=-\frac{N}{q}\sum_i \frac{d}{d\beta}e^{-\beta\varepsilon_i}=-\frac{N}{q}\frac{d}{d\beta}\sum_i e^{-\beta\varepsilon_i}=-\frac{N}{q}\frac{dq}{d\beta} \tag{15.22}$$

- **A brief illustration**

From the two-level partition function $q=1+e^{-\beta\varepsilon}$ we can deduce that the total energy of N two-level systems is

$$E(T)=-\left(\frac{N}{1+e^{-\beta\varepsilon}}\right)\frac{d}{d\beta}(1+e^{-\beta\varepsilon})=\frac{N\varepsilon e^{-\beta\varepsilon}}{1+e^{-\beta\varepsilon}}=\frac{N\varepsilon}{1+e^{\beta\varepsilon}}$$

This function is plotted in Fig. 15.8. Notice how the energy is zero at $T=0$, when only the lower state (at the zero of energy) is occupied, and rises to $\frac{1}{2}N\varepsilon$ as $T\to\infty$, when the two levels become equally populated. •

There are several points in relation to eqn 15.22 that need to be made. Because $\varepsilon_0=0$ (remember that we measure all energies from the lowest available level), $E(T)$ should be interpreted as the value of the internal energy relative to its value at $T=0$, $U(0)$. Therefore, to obtain the conventional internal energy U, we must add the internal energy at $T=0$:

$$U(T)=U(0)+E(T) \tag{15.23}$$

Secondly, because the partition function may depend on variables other than the temperature (for example, the volume), the derivative with respect to β in eqn 15.22 is actually a *partial* derivative with these other variables held constant. The complete expression relating the molecular partition function to the thermodynamic internal energy of a system of independent molecules is therefore

$$U(T)=U(0)-\frac{N}{q}\left(\frac{\partial q}{\partial\beta}\right)_V \tag{15.24a}$$

Internal energy in terms of the partition function . . .

An equivalent form is obtained by noting that $dx/x=d\ln x$:

$$U(T)=U(0)-N\left(\frac{\partial \ln q}{\partial\beta}\right)_V \tag{15.24b}$$

. . . and an alternative version

These two equations confirm that we need know only the partition function (as a function of temperature) to calculate the internal energy relative to its value at $T=0$.

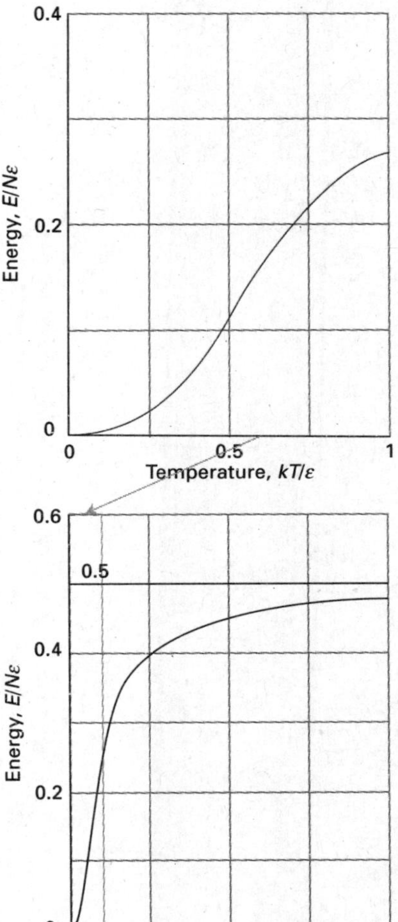

Fig. 15.8 The total energy of a two-level system (expressed as a multiple of $N\varepsilon$) as a function of temperature, on two temperature scales. The graph at the top shows the slow rise away from zero energy at low temperatures; the slope of the graph at $T=0$ is 0 (that is, the heat capacity is zero at $T=0$). The graph below shows the slow rise to 0.5 as $T\to\infty$ as both states become equally populated (see Fig. 15.7).

interActivity Draw graphs similar to those in Fig. 15.8 for a three-level system with levels 0, ε, and 2ε.

(b) The value of β

We now confirm that the parameter β, which we have anticipated is equal to $1/kT$, does indeed have that value. To do so, we compare the equipartition expression for the internal energy of a monatomic perfect gas, which from *Fundamentals F.5* we know to be

$$U(T) = U(0) + \tfrac{3}{2}nRT \tag{15.25a}$$

with the value calculated from the translational partition function (see the following *Justification*), which is

$$U(T) = U(0) + \frac{3N}{2\beta} \tag{15.25b}$$

It follows by comparing these two expressions that

$$\beta = \frac{N}{nRT} = \frac{nN_A}{nN_AkT} = \frac{1}{kT} \tag{15.26}$$

We have used $N = nN_A$, where n is the amount of gas molecules, N_A is Avogadro's constant, and $R = N_Ak$. Although we have proved that $\beta = 1/kT$ by examining a very specific example, the translational motion of a perfect monatomic gas, the result is general.

Justification 15.3 *The internal energy of a perfect gas*

To use eqn 15.24, we introduce the translational partition function from eqn 15.19:

$$\left(\frac{\partial q}{\partial \beta}\right)_V = \left(\frac{\partial}{\partial \beta}\frac{V}{\Lambda^3}\right)_V = V\frac{\mathrm{d}}{\mathrm{d}\beta}\frac{1}{\Lambda^3} = -3\frac{V}{\Lambda^4}\frac{\mathrm{d}\Lambda}{\mathrm{d}\beta}$$

Then we note from the formula for Λ in eqn 15.19 that

$$\frac{\mathrm{d}\Lambda}{\mathrm{d}\beta} = \frac{\mathrm{d}}{\mathrm{d}\beta}\left\{\frac{h\beta^{1/2}}{(2\pi m)^{1/2}}\right\} = \frac{1}{2\beta^{1/2}} \times \frac{h}{(2\pi m)^{1/2}} = \frac{\Lambda}{2\beta}$$

and so obtain

$$\left(\frac{\partial q}{\partial \beta}\right)_V = -\frac{3V}{2\beta\Lambda^3}$$

By eqn 15.24a,

$$U(T) = U(0) - N\left(\frac{\Lambda^3}{V}\right)\left(-\frac{3V}{2\beta\Lambda^3}\right) = U(0) + \frac{3N}{2\beta}$$

as in eqn 15.25b.

15.4 The statistical entropy

Key point **The statistical entropy is defined by the Boltzmann formula but may be expressed in terms of the molecular partition function.**

If it is true that the partition function contains all thermodynamic information, then it must be possible to use it to calculate the entropy as well as the internal energy. Because we know (from Section 3.2) that entropy is related to the dispersal of energy and that the partition function is a measure of the number of thermally accessible states, we can be confident that the two are indeed related.

We shall develop the relation between the entropy and the partition function in two stages. In *Further information 15.2*, we justify one of the most celebrated equations in statistical thermodynamics, the **Boltzmann formula** for the entropy:

$$S = k \ln \mathcal{W} \quad \text{Boltzmann formula for the entropy} \quad [15.27]$$

In this expression, $\mathcal{W}$ is the weight of the most probable configuration of the system. In the second stage, we express $\mathcal{W}$ in terms of the partition function.

The statistical entropy behaves in exactly the same way as the thermodynamic entropy. Thus, as the temperature is lowered, the value of $\mathcal{W}$, and hence of S, decreases because fewer configurations are consistent with the total energy. In the limit $T \to 0$, $\mathcal{W} = 1$, so $\ln \mathcal{W} = 0$, because only one configuration (every molecule in the lowest level) is compatible with $E = 0$. It follows that $S \to 0$ as $T \to 0$, which is compatible with the Third Law of thermodynamics, that the entropies of all perfect crystals approach the same value as $T \to 0$ (Section 3.4).

Now we relate the Boltzmann formula for the entropy to the partition function. To do so, we substitute the expression for $\ln \mathcal{W}$ given in eqn 15.3 into eqn 15.27 and, as shown in the following *Justification*, obtain

$$S(T) = \frac{U(T) - U(0)}{T} + Nk \ln q \quad \text{Entropy in terms of the partition function} \quad (15.28)$$

Justification 15.4 *The statistical entropy*

The first stage is to use eqn 15.3 ($\ln \mathcal{W} = N \ln N - \sum_i N_i \ln N_i$) and $N = \sum_i N_i$ to write

$$S(T) = k\left\{N \ln N - \sum_i N_i \ln N_i\right\} = k\left\{\left(\sum_i N_i\right) \ln N - \sum_i N_i \ln N_i\right\}$$
$$= k \sum_i N_i \{\ln N - \ln N_i\}$$

Next, we use $\ln x - \ln y = \ln(x/y) = -\ln(y/x)$ to write this expression as

$$S(T) = -k \sum_i N_i \ln \frac{N_i}{N} = -Nk \sum_i p_i \ln p_i$$

where $p_i = N_i/N$, the fraction of molecules in state i. It follows from eqn 15.7 that

$$\ln p_i = -\beta\varepsilon_i - \ln q$$

and therefore that

$$S(T) = -Nk\left(-\beta \sum_i p_i\varepsilon - \sum_i p_i \ln q\right) = k\beta\{U(T) - U(0)\} + Nk \ln q$$

We have used the fact that the sum over the p_i is equal to 1 and that (from eqns 15.20 and 15.23)

$$N \sum_i p_i\varepsilon_i = \sum_i N_i p_i \varepsilon_i = \sum_i N_i \varepsilon_i = E(T) = U(T) - U(0)$$

We have already established that $\beta = 1/kT$, so eqn 15.28 immediately follows.

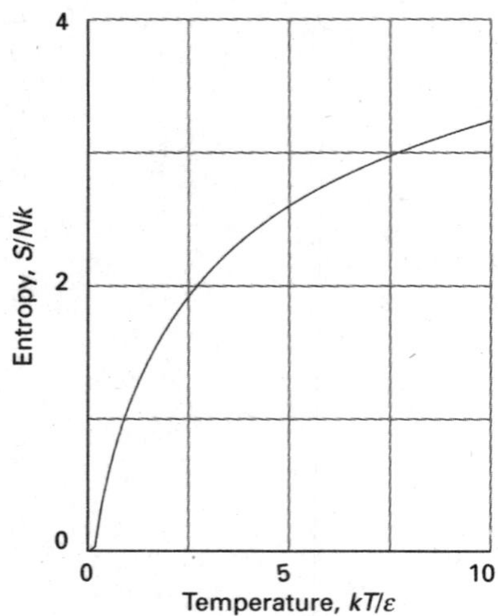

Fig. 15.9 The temperature variation of the entropy of the system shown in Fig. 15.3 (expressed here as a multiple of Nk). The entropy approaches zero as $T \rightarrow 0$, and increases without limit as $T \rightarrow \infty$.

interActivity Plot the function dS/dT, the temperature coefficient of the entropy, against kT/ε. Is there a temperature at which this coefficient passes through a maximum? If you find a maximum, explain its physical origins.

Example 15.4 *Calculating the entropy of a collection of oscillators*

Calculate the entropy of a collection of N independent harmonic oscillators, and evaluate it using vibrational data for I_2 vapour at 25°C (Example 15.3).

Method To use eqn 15.28, we use the partition function for a molecule with evenly spaced vibrational energy levels, eqn 15.12. With the partition function available, the internal energy can be found by differentiation (as in eqn 15.24a), and the two expressions then combined to give S.

Answer The molecular partition function as given in eqn 15.12 is

$$q = \frac{1}{1 - e^{-\beta\varepsilon}}$$

The internal energy is obtained by using eqn 15.24a:

$$U(T) - U(0) = -\frac{N}{q}\left(\frac{\partial q}{\partial \beta}\right)_V = \frac{N\varepsilon e^{-\beta\varepsilon}}{1 - e^{-\beta\varepsilon}} = \frac{N\varepsilon}{e^{\beta\varepsilon} - 1}$$

The entropy is therefore

$$S(T) = Nk\left\{\frac{\beta\varepsilon}{e^{\beta\varepsilon} - 1} - \ln(1 - e^{-\beta\varepsilon})\right\}$$

This function is plotted in Fig. 15.9. For I_2 at 25°C, $\beta\varepsilon = 1.036$ (Example 15.3), so $S_m = 8.38\ \mathrm{J\,K^{-1}\,mol^{-1}}$.

Self-test 15.6 Evaluate the molar entropy of N two-level systems and plot the resulting expression. What is the entropy when the two states are equally thermally accessible? [$S(T)/Nk = \beta\varepsilon/(1 + e^{\beta\varepsilon}) + \ln(1 + e^{-\beta\varepsilon})$; see Fig. 15.10; $S = Nk \ln 2$]

IMPACT ON TECHNOLOGY

I15.1 Reaching very low temperatures

Common refrigerators do not need to reach temperatures too far below the melting point of water, but the study of physical and chemical phenomena at very low temperatures requires more sophisticated technology. The world record low temperature stands at about 100 pK for solids and at about 500 pK for gases, where molecules move so slowly it takes them about 10 s to travel 1 cm. Gases may be cooled by Joule–Thomson expansion below their inversion temperature (Section 2.12), and temperatures lower than 4 K (the boiling point of helium) can be reached by the evaporation of liquid helium by pumping rapidly through large diameter pipes. Temperatures as low as about 1 K can be reached in this way, but at lower temperatures helium is insufficiently volatile for this procedure to be effective; moreover, the superfluid phase begins to interfere with the cooling process by creeping round the apparatus.

Common methods used to reach very low temperatures include *laser cooling* and *adiabatic demagnetization*. In laser cooling, also called *optical trapping*, atoms in the gas phase are cooled by inelastic collisions with photons from intense laser beams, which act as walls of a very small container. For example, the technique can be used to cool a group of 2000 rubidium atoms to 20 nK. Adiabatic demagnetization relies on the fact that, in the absence of a magnetic field, the unpaired electrons of a paramagnetic material are orientated at random, but in the presence of a magnetic field there are more β spins ($m_s = -\frac{1}{2}$) than α spins ($m_s = +\frac{1}{2}$). In thermodynamic terms, the application of a magnetic field lowers the entropy of a sample (Fig. 15.11), and at a given temperature, the entropy of a sample is lower when the field is on than when it is off.

A sample of paramagnetic material, such as a d- or f-metal complex, is cooled to about 1 K by using helium. Gadolinium(III) sulfate octahydrate, $Gd_2(SO_4)_3 \cdot 8H_2O$, has been used because each gadolinium ion carries several unpaired electrons but is separated from its neighbours by a coordination sphere of hydrating H_2O molecules. The sample is then exposed to a strong magnetic field while it is surrounded by helium, which provides thermal contact with the cold reservoir. This magnetization step is isothermal, and heat leaves the sample as the electron spins adopt the lower energy state (AB in Fig. 15.11). Thermal contact between the sample and the surroundings is now broken by pumping away the helium and the magnetic field is reduced to zero. This step is adiabatic and effectively reversible, so the state of the sample changes from B to C. At the end of this step the sample is the same as it was at A except that it now has a lower entropy. That lower entropy in the absence of a magnetic field corresponds to a lower temperature. That is, adiabatic demagnetization has cooled the sample.

Even lower temperatures can be reached if nuclear spins (which also behave like small magnets) are used instead of electron spins in the technique of *adiabatic nuclear demagnetization*. This technique was used to reach the current world record (in silver) of 280 pK.

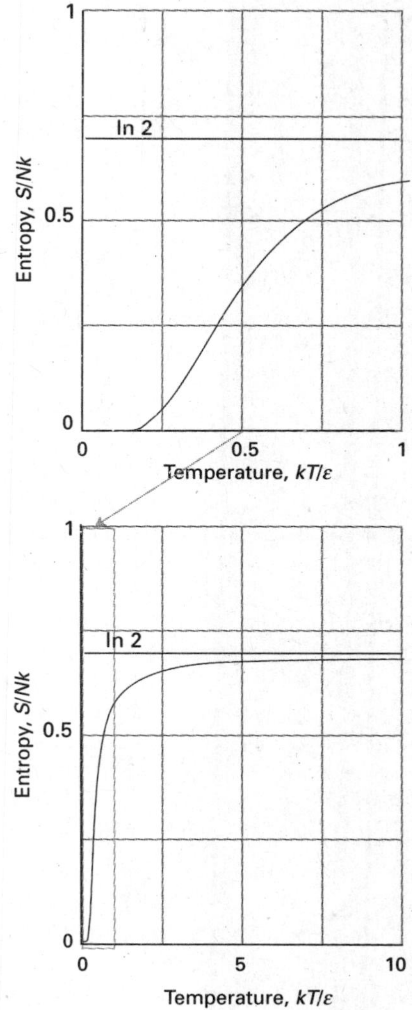

Fig. 15.10 The temperature variation of the entropy of a two-level system (expressed as a multiple of Nk). As $T \to \infty$ the two states become equally populated and S approaches $Nk \ln 2$.

interActivity Draw graphs similar to those in Fig. 15.10 for a three-level system with levels 0, ε, and 2ε.

The canonical partition function

In this section we see how to generalize our conclusions to include systems composed of interacting molecules. We shall also see how to obtain the molecular partition function from the more general form of the partition function developed here.

15.5 The canonical ensemble

Key points **(a) A canonical ensemble is an imaginary collection of replications of the actual system with a common temperature. It is used to extend statistical thermodynamics to include interacting molecules. (b) The thermodynamic limit is reached when the number of replications becomes infinite. (c) Most members of the ensemble have an energy very close to the mean value.**

The crucial new concept we need when treating systems of interacting particles is the 'ensemble'. Like so many scientific terms, the term has basically its normal meaning of 'collection', but it has been sharpened and refined into a precise significance.

(a) The concept of ensemble

To set up an ensemble, we take a closed system of specified volume, composition, and temperature, and think of it as replicated $\tilde{N}$ times (Fig. 15.12). All the identical closed systems are regarded as being in thermal contact with one another, so they can exchange energy. The total energy of all the systems is $\tilde{E}$ and, because they are in thermal equilibrium with one another, they all have the same temperature, T. This imaginary collection of replications of the actual system with a common temperature is called the **canonical ensemble**. The word 'canon' means 'according to a rule'.

There are two other important ensembles. In the **microcanonical ensemble** the condition of constant temperature is replaced by the requirement that all the systems should have exactly the same energy: each system is individually isolated. In the **grand canonical ensemble** the volume and temperature of each system is the same, but they

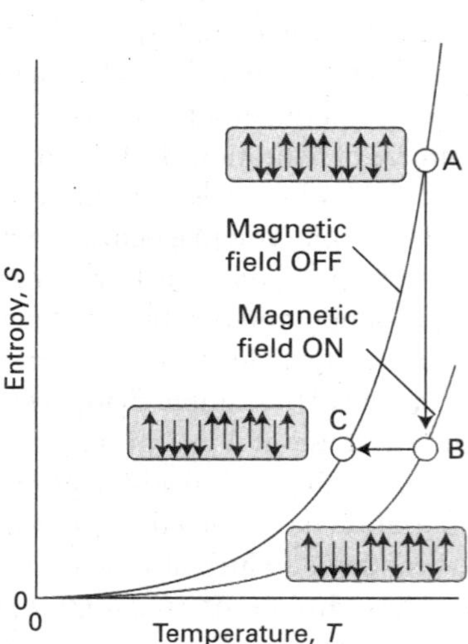

Fig. 15.11 The technique of adiabatic demagnetization is used to attain very low temperatures. The upper curve shows the variation in the entropy of a paramagnetic system in the absence of an applied field. The lower curve shows the variation in entropy when a field is applied and has made the electron spins more orderly. The isothermal magnetization step is from A to B; the adiabatic demagnetization step (at constant entropy) is from B to C.

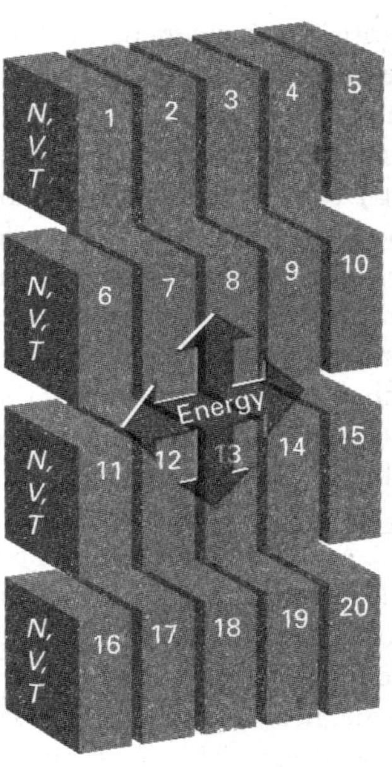

Fig. 15.12 A representation of the canonical ensemble, in this case for $\tilde{N} = 20$. The individual replications of the actual system all have the same composition and volume. They are all in mutual thermal contact, and so all have the same temperature. Energy may be transferred between them as heat, and so they do not all have the same energy. The total energy $\tilde{E}$ of all 20 replications is a constant because the ensemble is isolated overall.

are open, which means that matter can be imagined as able to pass between the systems; the composition of each one may fluctuate, but now the chemical potential is the same in each system:

Microcanonical ensemble: N, V, E common

Canonical ensemble: N, V, T common   Definitions of ensembles

Grand canonical ensemble: μ, V, T common

The important point about an ensemble is that it is a collection of *imaginary* replications of the system, so we are free to let the number of members be as large as we like; when appropriate, we can let $\tilde{N}$ become infinite. The number of members of the ensemble in a state with energy E_i is denoted $\tilde{N}_i$, and we can speak of the configuration of the ensemble (by analogy with the configuration of the system used in Section 15.1) and its weight, $\tilde{\mathcal{W}}$. Note that $\tilde{N}$ is unrelated to N, the number of molecules in the actual system; $\tilde{N}$ is the number of imaginary replications of that system.

(b) Dominating configurations

Just as in Section 15.1, some of the configurations of the ensemble will be very much more probable than others. For instance, it is very unlikely that the whole of the total energy, $\tilde{E}$, will accumulate in one system. By analogy with the earlier discussion, we can anticipate that there will be a dominating configuration, and that we can evaluate the thermodynamic properties by taking the average over the ensemble using that single, most probable, configuration. In the **thermodynamic limit** of $\tilde{N} \to \infty$, this dominating configuration is overwhelmingly the most probable, and it dominates the properties of the system virtually completely.

The quantitative discussion follows the argument in Section 15.1 with the modification that N and N_i are replaced by $\tilde{N}$ and $\tilde{N}_i$. The weight of a configuration $\{\tilde{N}_0, \tilde{N}_1, \ldots\}$ is

$$\tilde{\mathcal{W}} = \frac{\tilde{N}!}{\tilde{N}_0!\tilde{N}_1!\ldots} \qquad (15.29)$$

The configuration of greatest weight, subject to the constraints that the total energy of the ensemble is constant at $\tilde{E}$ and that the total number of members is fixed at $\tilde{N}$, is given by the **canonical distribution**:

$$\frac{\tilde{N}_i}{\tilde{N}} = \frac{e^{-\beta E_i}}{Q} \qquad Q = \sum_i e^{-\beta E_i} \qquad (15.30)$$

Definition of canonical partition function

The quantity Q, which is a function of the temperature, is called the **canonical partition function**.

(c) Fluctuations from the most probable distribution

The canonical distribution in eqn 15.30 is only apparently an exponentially decreasing function of the energy of the system. We must appreciate that eqn 15.30 gives the probability of occurrence of members in a single state i of the entire system of energy E_i. There may in fact be numerous states with almost identical energies. For example, in a gas the identities of the molecules moving slowly or quickly can change without necessarily affecting the total energy. The density of states, the number of states in an energy range divided by the width of the range (Fig. 15.13), is a very sharply increasing function of energy. It follows that the probability of a member of an ensemble having a specified energy (as distinct from being in a specified state) is given by eqn 15.30, a sharply decreasing function, multiplied by a sharply increasing function (Fig. 15.14). Therefore, the overall distribution is a sharply peaked function. We conclude that most members of the ensemble have an energy very close to the mean value.

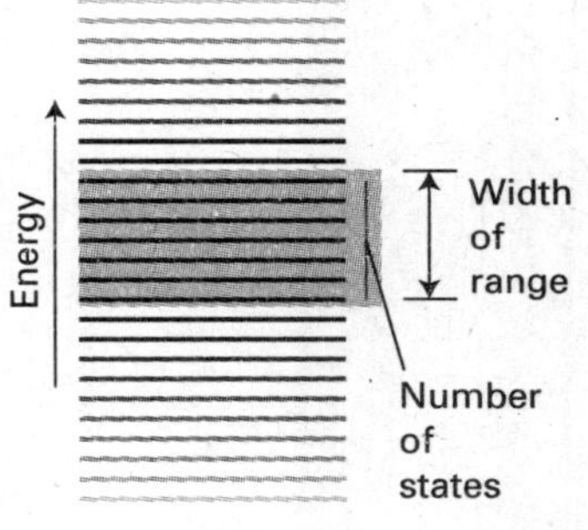

Fig. 15.13 The energy density of states is the number of states in an energy range divided by the width of the range.

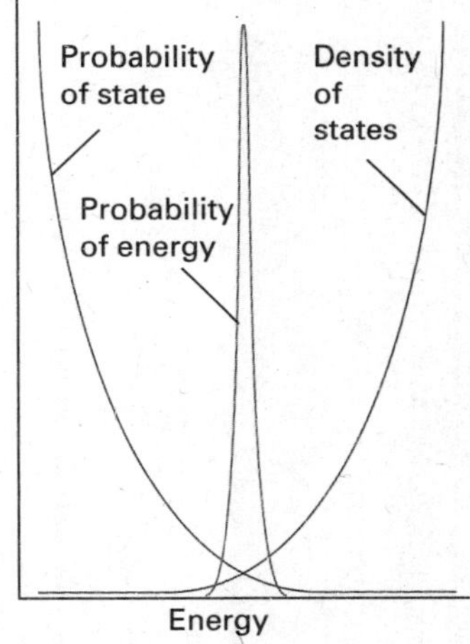

Fig. 15.14 To construct the form of the distribution of members of the canonical ensemble in terms of their energies, we multiply the probability that any one is in a state of given energy, eqn 15.32, by the density of states corresponding to that energy (a steeply rising function). The product is a sharply peaked function at the mean energy, which shows that almost all the members of the ensemble have that energy.

15.6 The thermodynamic information in the partition function

Key points **(a) The internal energy of a system composed of interacting molecules is proportional to the derivative of the canonical partition function with respect to temperature. (b) The entropy of an interacting system can be calculated from the canonical partition function.**

Like the molecular partition function, the canonical partition function carries all the thermodynamic information about a system. However, Q is more general than q because it does not assume that the molecules are independent. We can therefore use Q to discuss the properties of condensed phases and real gases where molecular interactions are important.

(a) The internal energy

If the total energy of the ensemble is $\tilde{E}$, and there are $\tilde{N}$ members, the average energy of a member is $E = \tilde{E}/\tilde{N}$. We use this quantity to calculate the internal energy of the system in the thermodynamic limit of $\tilde{N}$ (and $\tilde{E}$) approaching infinity:

$$U(T) = U(0) + E(T) = U(0) + \tilde{E}(T)/\tilde{N} \quad \text{as} \quad \tilde{N} \to \infty \qquad (15.31)$$

The fraction, $\tilde{p}_i$, of members of the ensemble in a state i with energy E_i is given by the analogue of eqn 15.7 as

$$\tilde{p}_i = \frac{e^{-\beta E_i}}{Q} \tag{15.32}$$

It follows that the internal energy is given by

$$U(T) = U(0) + \sum_i \tilde{p}_i E_i = U(0) + \frac{1}{Q}\sum_i E_i e^{-\beta E_i} \tag{15.33}$$

By the same argument that led to eqn 15.24,

$$U(T) = U(0) - \frac{1}{Q}\left(\frac{\partial Q}{\partial \beta}\right)_V = U(0) - \left(\frac{\partial \ln Q}{\partial \beta}\right)_V \tag{15.34}$$

Internal energy in terms of the canonical partition function

(b) The entropy

The total weight, $\tilde{\mathcal{W}}$, of a configuration of the ensemble is the product of the average weight $\mathcal{W}$ of each member of the ensemble, $\tilde{\mathcal{W}} = \mathcal{W}^{\tilde{N}}$. Hence, we can calculate S from

$$S = k \ln \mathcal{W} = k \ln \tilde{\mathcal{W}}^{1/\tilde{N}} = \frac{k}{\tilde{N}} \ln \tilde{\mathcal{W}} \tag{15.35}$$

It follows, by the same argument used in Section 15.4, that

$$S(T) = \frac{U(T) - U(0)}{T} + k \ln Q \tag{15.36}$$

Entropy in terms of the canonical partition function

15.7 Independent molecules

Key points **(a) For distinguishable independent molecules, $Q = q^N$; for indistinguishable independent molecules, $Q = q^N/N!$. (b) The entropy of a perfect gas is given by the Sackur–Tetrode equation.**

We shall now see how to recover the molecular partition function from the more general canonical partition function when the molecules are independent. When the molecules are independent and distinguishable (in the sense to be described), we show in the following *Justification* that the relation between Q and q is

$$Q = q^N \tag{15.37}$$

Justification 15.5 *The relation between Q and q*

The total energy of a collection of N independent molecules is the sum of the energies of the molecules. Therefore, we can write the total energy of a state i of the system as

$$E_i = \varepsilon_i(1) + \varepsilon_i(2) + \cdots + \varepsilon_i(N)$$

In this expression, $\varepsilon_i(1)$ is the energy of molecule 1 when the system is in the state i, $\varepsilon_i(2)$ the energy of molecule 2 when the system is in the same state i, and so on. The canonical partition function is then

$$Q = \sum_i e^{-\beta\varepsilon_i(1) - \beta\varepsilon_i(2) - \cdots - \beta\varepsilon_i(N)}$$

The sum over the states of the system can be reproduced by letting each molecule enter all its own individual states (although we meet an important proviso shortly). Therefore, instead of summing over the states i of the system, we can sum over all the individual states i of molecule 1, all the states i of molecule 2, and so on. This rewriting of the original expression leads to

$$Q=\left(\sum_i \mathrm{e}^{-\beta\varepsilon_i}\right)\left(\sum_i \mathrm{e}^{-\beta\varepsilon_i}\right)\cdots\left(\sum_i \mathrm{e}^{-\beta\varepsilon_i}\right)=\left(\sum_i \mathrm{e}^{-\beta\varepsilon_i}\right)^N=q^N$$

(a) Distinguishable and indistinguishable molecules

If all the molecules are identical and free to move through space, we cannot distinguish them and the relation $Q = q^N$ is not valid. Suppose that molecule 1 is in some state a, molecule 2 is in b, and molecule 3 is in c, then one member of the ensemble has an energy $E = \varepsilon_a + \varepsilon_b + \varepsilon_c$. This member, however, is indistinguishable from one formed by putting molecule 1 in state b, molecule 2 in state c, and molecule 3 in state a, or some other permutation. There are six such permutations in all, and $N!$ in general. In the case of indistinguishable molecules, it follows that we have counted too many states in going from the sum over system states to the sum over molecular states, so writing $Q = q^N$ overestimates the value of Q. The detailed argument is quite involved, but at all except very low temperatures it turns out that the correction factor is $1/N!$. Therefore:

- For indistinguishable independent molecules: $Q = q^N/N!$ (15.38a)
- For distinguishable independent molecules: $Q = q^N$ (15.38b)

Relation between Q and q

For molecules to be indistinguishable, they must be of the same kind: an Ar atom is never indistinguishable from a Ne atom. Their identity, however, is not the only criterion. Each identical molecule in a crystal lattice, for instance, can be 'named' with a set of coordinates. Identical molecules in a lattice can therefore be treated as distinguishable because their sites are distinguishable, and we use eqn 15.38b. On the other hand, identical molecules in a gas are free to move to different locations, and there is no way of keeping track of the identity of a given molecule; we therefore use eqn 15.38a.

(b) The entropy of a monatomic gas

An important application of the previous material is the derivation (as shown in the following *Justification*) of the **Sackur–Tetrode equation** for the entropy of a monatomic gas:

$$S(T)=nR\ln\left(\frac{\mathrm{e}^{5/2}V}{nN_\mathrm{A}\Lambda^3}\right)\qquad \Lambda=\frac{h}{(2\pi mkT)^{1/2}} \tag{15.39a}$$

Sackur–Tetrode equation . . .

This equation implies that the molar entropy of a perfect gas of high molar mass is greater than one of low molar mass under the same conditions (because the former has more thermally accessible translational states). Because the gas is perfect, we can use the relation $V = nRT/p$ to express the entropy in terms of the pressure as

$$S(T)=nR\ln\left(\frac{\mathrm{e}^{5/2}kT}{p\Lambda^3}\right) \tag{15.39b}$$

. . . in terms of pressure

Justification 15.6 *The Sackur–Tetrode equation*

For a gas of independent molecules, Q may be replaced by $q^N/N!$, with the result that eqn 15.36 becomes

$$S(T)=\frac{U(T)-U(0)}{T}+Nk\ln q-k\ln N!$$

Because the number of molecules ($N=nN_A$) in a typical sample is large, we can use Stirling's approximation (eqn 15.2) to write

$$S(T)=\frac{U(T)-U(0)}{T}+nR\ln q-nR\ln N+nR$$

The only mode of motion for a gas of atoms is translation, and the partition function is $q=V/\Lambda^3$ (eqn 15.19), where Λ is the thermal wavelength. The internal energy is given by eqn 15.25a, so the entropy is

$$S(T)=\tfrac{3}{2}nR+nR\left(\ln\frac{V}{\Lambda^3}-\ln nN_A+1\right)=nR\left(\ln e^{3/2}+\ln\frac{V}{\Lambda^3}-\ln nN_A+\ln e\right)$$

which rearranges into eqn 15.39.

Example 15.5 *Using the Sackur–Tetrode equation*

Calculate the standard molar entropy of gaseous argon at 25°C.

Method To calculate the molar entropy, S_m, from eqn 15.39b, divide both sides by n. To calculate the standard molar entropy, $S_m^{\ominus}$, set $p=p^{\ominus}$ in the expression for S_m:

$$S_m^{\ominus}=R\ln\left(\frac{e^{5/2}kT}{p^{\ominus}\Lambda^3}\right)$$

Answer The mass of an Ar atom is $m=39.95m_u$. At 25°C, its thermal wavelength is 16.0 pm (by the same kind of calculation as in the *brief illustration* in Section 15.2b). Therefore,

$$S_m^{\ominus}=R\ln\left\{\frac{e^{5/2}\times(4.12\times10^{-21}\text{ J})}{(10^5\text{ N m}^{-2})\times(1.60\times10^{-11}\text{ m})^3}\right\}=18.6R=155\text{ J K}^{-1}\text{ mol}^{-1}$$

We can anticipate, on the basis of the number of accessible states for a lighter molecule, that the standard molar entropy of Ne is likely to be smaller than for Ar; its actual value is 17.60R at 298 K.

Self-test 15.7 Calculate the translational contribution to the standard molar entropy of H_2 at 25°C. [14.2R]

The Sackur–Tetrode equation implies that, when a monatomic perfect gas expands isothermally from V_i to V_f, its entropy changes by

$$\Delta S=nR\ln(aV_f)-nR\ln(aV_i)=nR\ln\frac{V_f}{V_i} \qquad (15.40)$$

where aV is the collection of quantities inside the logarithm of eqn 15.39a. This is exactly the expression we obtained by using classical thermodynamics (Example 3.1). Now, though, we see that that classical expression is in fact a consequence of the increase in the number of accessible translational states when the volume of the container is increased (Fig. 15.15).

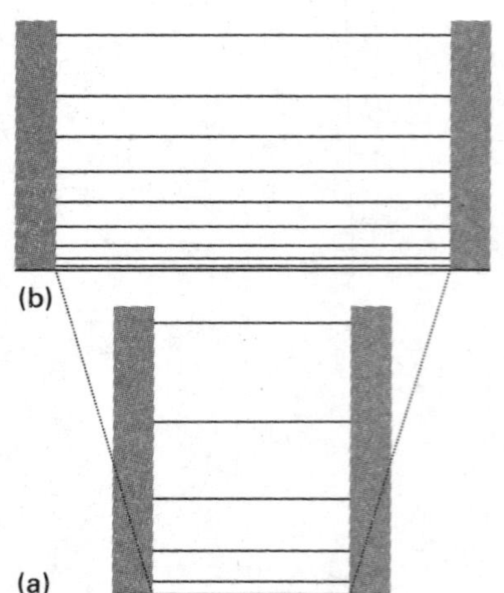

Fig. 15.15 As the width of a container is increased (going from (a) to (b)), the energy levels become closer together (as $1/L^2$), and as a result more are thermally accessible at a given temperature. Consequently, the entropy of the system rises as the container expands.

Checklist of key equations

Property	Equation	Comment
Weight of the configuration $\{N_0,N_1,\ldots\}$	$\mathcal{W}=N!/N_0!N_1!\ldots$	
Boltzmann distribution	$N_i=Ne^{-\beta\varepsilon_i}/q$	$\beta=1/kT$
Molecular partition function	$q=\sum_i e^{-\beta\varepsilon_i}$	
Translational partition function	$q=V/\Lambda^3$	Perfect gas
Thermal wavelength	$\Lambda=h/(2\pi mkT)^{1/2}$	
Mean energy	$E(T)=-(N/q)(\partial q/\partial\beta)_V=-N(\partial \ln q/\partial\beta)_V$	
Internal energy	$U(T)=U(0)+E(T)$	
Boltzmann formula for the entropy	$S=k\ln\mathcal{W}$	
The entropy in terms of the partition function	$S=\{U-U(0)\}/T+Nk\ln q$	Distinguishable molecules
	$S=\{U-U(0)\}/T+Nk\ln q-Nk(\ln N-1)$	Indistinguishable molecules
Canonical partition function	$Q=\sum_i e^{-\beta E_i}$	
Internal energy of an ensemble	$U=U(0)-(\partial\ln Q/\partial\beta)_V$	
Entropy of an ensemble	$S=\{U-U(0)\}/T+k\ln Q$	
Canonical partition function in terms of the molecular partition function	$Q=q^N$	Distinguishable independent molecules
	$Q=q^N/N!$	Indistinguishable independent molecules
Sackur–Tetrode equation	$S(T)=nR\ln(e^{5/2}V/nN_A\Lambda^3)$	Entropy of a monatomic perfect gas

➔ For a chart of the relations between principal equations, see the Road map section of the Resource section.

Further information

Further information 15.1 *The Boltzmann distribution*

We remarked in Section 15.1 that ln $\mathcal{W}$ is easier to handle than $\mathcal{W}$. Therefore, to find the form of the Boltzmann distribution, we look for the condition for ln $\mathcal{W}$ being a maximum rather than dealing directly with $\mathcal{W}$. If you are interested in the outline of the derivation, you need go no further than Section FI15.1a. However, if you wish to learn about some of the mathematical details of the calculation, go on to Section FI15.1b.

(a) The derivation

Because ln $\mathcal{W}$ depends on all the N_i, when a configuration changes and the N_i change to N_i+dN_i, the function ln $\mathcal{W}$ changes to ln $\mathcal{W}$ + d ln $\mathcal{W}$, where

$$d\ln\mathcal{W}=\sum_i\left(\frac{\partial\ln\mathcal{W}}{\partial N_i}\right)dN_i$$

All this expression states is that a change in ln $\mathcal{W}$ is the sum of contributions arising from changes in each value of N_i. At a maximum, d ln $\mathcal{W}=0$. However, when the N_i change, they do so subject to the two constraints

$$\sum_i\varepsilon_i dN_i=0 \qquad \sum_i dN_i=0 \tag{15.41}$$

The first constraint recognizes that the total energy must not change, and the second recognizes that the total number of molecules must not change. These two constraints prevent us from solving d ln $\mathcal{W}=0$ simply by setting all $(\partial\ln\mathcal{W}/\partial N_i)=0$ because the dN_i are not all independent.

The way to take constraints into account was devised by the French mathematician Lagrange, and is called the **method of undetermined multipliers** (see below). All we need here is the rule that a constraint should be multiplied by a constant and then added to the main variation equation. The variables are then treated as though they were all independent, and the constants are evaluated at the end of the calculation.

We employ the technique as follows. The two constraints in eqn 15.41 are multiplied by the constants $-\beta$ and α, respectively (the minus sign in $-\beta$ has been included for future convenience), and then added to the expression for d ln $\mathcal{W}$:

$$\begin{aligned}d\ln\mathcal{W}&=\sum_i\left(\frac{\partial\ln\mathcal{W}}{\partial N_i}\right)dN_i+\alpha\sum_i dN_i-\beta\sum_i\varepsilon_i dN_i\\&=\sum_i\left\{\left(\frac{\partial\ln\mathcal{W}}{\partial N_i}\right)+\alpha-\beta\varepsilon_i\right\}dN_i\end{aligned}$$

All the dN_i are now treated as independent. Hence the only way of satisfying ln $\mathcal{W}=0$ is to require that, for each i,

$$\frac{\partial \ln \mathcal{W}}{\partial N_i}+\alpha-\beta\varepsilon_i=0 \qquad (15.42)$$

when the N_i have their most probable values.

Differentiation of $\ln \mathcal{W}$ as given in eqn 15.3 with respect to N_i gives

$$\frac{\partial \ln \mathcal{W}}{\partial N_i}=\frac{\partial(N\ln N)}{\partial N_i}-\sum_j\frac{\partial(N_j\ln N_j)}{\partial N_i}$$

Note that we have had to change the summation index (from i to j) to avoid confusion with the index on N_i. The derivative of the first term is obtained as follows:

$$\frac{\partial(N\ln N)}{\partial N_i}=\left(\frac{\partial N}{\partial N_i}\right)\ln N+N\left(\frac{\partial \ln N}{\partial N_i}\right)$$
$$=\ln N+\frac{\partial N}{\partial N_i}=\ln N+1$$

The $\ln N$ in the first term on the right in the second line arises because $N=N_1+N_2+\cdots$ and so the derivative of N with respect to any of the N_i is 1: that is, $\partial N/\partial N_i=1$. The second term on the right in the second line arises because $\partial(\ln N)/\partial N_i=(1/N)\partial N/\partial N_i$. The final 1 is then obtained in the same way as in the preceding remark, by using $\partial N/\partial N_i=1$.

For the derivative of the second term we first note that

$$\frac{\partial \ln N_j}{\partial N_i}=\frac{1}{N_j}\left(\frac{\partial N_j}{\partial N_i}\right)$$

If $i\neq j$, N_j is independent of N_i, so $\partial N_j/\partial N_i=0$. However, if $i=j$,

$$\frac{\partial N_j}{\partial N_i}=\frac{\partial N_j}{\partial N_j}=1$$

Therefore,

$$\frac{\partial N_j}{\partial N_i}=\delta_{ij}$$

with δ_{ij} the Kronecker delta ($\delta_{ij}=1$ if $i=j$; $\delta_{ij}=0$ otherwise). Then

$$\sum_j\frac{\partial(N_j\ln N_j)}{\partial N_i}=\sum_j\left\{\left(\frac{\partial N_j}{\partial N_i}\right)\ln N_j+N_j\left(\frac{\partial \ln N_j}{\partial N_i}\right)\right\}$$
$$=\sum_j\left\{\left(\frac{\partial N_j}{\partial N_i}\right)\ln N_j+\left(\frac{\partial N_j}{\partial N_i}\right)\right\}$$
$$=\sum_j\left(\frac{\partial N_j}{\partial N_i}\right)(\ln N_j+1)$$
$$=\sum_j\delta_{ij}(\ln N_j+1)=\ln N_i+1$$

and therefore

$$\frac{\partial \ln \mathcal{W}}{\partial N_i}=-(\ln N_i+1)+(\ln N+1)=-\ln\frac{N_i}{N}$$

It follows from eqn 15.42 that

$$-\ln\frac{N_i}{N}+\alpha-\beta\varepsilon_i=0$$

and therefore that

$$\frac{N_i}{N}=\mathrm{e}^{\alpha-\beta\varepsilon_i}$$

At this stage we note that

$$N=\sum_i N_i=\sum_i N\mathrm{e}^{\alpha-\beta\varepsilon_i}=N\mathrm{e}^{\alpha}\sum_i\mathrm{e}^{-\beta\varepsilon_i}$$

Because the N cancels on each side of this equality, it follows that

$$\mathrm{e}^{\alpha}=\frac{1}{\sum_j\mathrm{e}^{-\beta\varepsilon_j}} \qquad (15.43)$$

and

$$\frac{N_i}{N}=\mathrm{e}^{\alpha-\beta\varepsilon_i}=\mathrm{e}^{\alpha}\mathrm{e}^{-\beta\varepsilon_i}=\frac{1}{\sum_j\mathrm{e}^{-\beta\varepsilon_j}}\mathrm{e}^{-\beta\varepsilon_i}$$

which is eqn 15.6a (because at this stage we are free to replace the summation index j by i).

(b) The method of undetermined multipliers

To understand the derivation above more fully we need to see how we take constraints into account. Suppose we need to find the maximum (or minimum) value of some function f that depends on several variables $x_1, x_2, \ldots, x_n$. When the variables undergo a small change from x_i to $x_i+\delta x_i$ the function changes from f to $f+\delta f$, where

$$\delta f=\sum_i^n\left(\frac{\partial f}{\partial x_i}\right)\delta x_i \qquad (15.44)$$

At a minimum or maximum, $\delta f=0$, so then

$$\sum_i^n\left(\frac{\partial f}{\partial x_i}\right)\delta x_i=0 \qquad (15.45)$$

If the x_i were all independent, all the δx_i would be arbitrary, and this equation could be solved by setting each $(\partial f/\partial x_i)=0$ individually. When the x_i are not all independent, the δx_i are not all independent, and the simple solution is no longer valid. We proceed as follows.

Let the constraint connecting the variables be an equation of the form $g=0$. For example, in the preceding section one constraint was $n_0+n_1+\cdots=N$, which can be written

$$g=0, \text{ with } g=(n_0+n_1+\cdots)-N$$

The constraint $g=0$ is always valid, so g remains unchanged when the x_i are varied:

$$\delta g=\sum_i\left(\frac{\partial g}{\partial x_i}\right)\delta x_i=0 \qquad (15.46)$$

Because δg is zero, we can multiply it by a parameter, λ, and add it to eqn 15.45:

$$\sum_i^n\left\{\left(\frac{\partial f}{\partial x_i}\right)+\lambda\left(\frac{\partial g}{\partial x_i}\right)\right\}\delta x_i=0 \qquad (15.47)$$

This equation can be solved for one of the δx, δx_n for instance, in terms of all the other δx_i. All those other δx_i ($i = 1, 2, \ldots n-1$) are independent, because there is only one constraint on the system. But here is the trick: λ is arbitrary; therefore we can choose it so that the coefficient of δx_n in eqn 15.47 is zero. That is, we choose λ so that

$$\left(\frac{\partial f}{\partial x_n}\right) + \lambda\left(\frac{\partial g}{\partial x_n}\right) = 0 \qquad (15.48)$$

Then eqn 15.47 becomes

$$\sum_i^{n-1}\left\{\left(\frac{\partial f}{\partial x_i}\right) + \lambda\left(\frac{\partial g}{\partial x_i}\right)\right\}\delta x_i = 0 \qquad (15.49)$$

Now the $n-1$ variations δx_i *are* independent, so the solution of this equation is

$$\left(\frac{\partial f}{\partial x_i}\right) + \lambda\left(\frac{\partial g}{\partial x_i}\right) = 0 \qquad i = 1, 2, \ldots, n-1 \qquad (15.50)$$

However, eqn 15.48 has exactly the same form as this equation, so the maximum or minimum of f can be found by solving

$$\left(\frac{\partial f}{\partial x_i}\right) + \lambda\left(\frac{\partial g}{\partial x_i}\right) = 0 \qquad i = 1, 2, \ldots, n \qquad (15.51)$$

The use of this approach was illustrated in Section FI15.1a for two constraints and therefore two undetermined multipliers λ_1 and λ_2 (α and $-\beta$).

The multipliers λ cannot always remain undetermined. One approach is to solve eqn 15.48 instead of incorporating it into the minimization scheme. In Section FI15.1a we used the alternative procedure of keeping λ undetermined until a property was calculated for which the value was already known. Thus, we found that $\beta = 1/kT$ by calculating the internal energy of a perfect gas.

Further information 15.2 *The Boltzmann formula*

A change in the internal energy

$$U(T) = U(0) + \sum_i N_i\varepsilon_i \qquad (15.52)$$

may arise from either a modification of the energy levels of a system (when ε_i changes to $\varepsilon_i + \mathrm{d}\varepsilon$) or from a modification of the populations (when N_i changes to $N_i + \mathrm{d}N_i$). The most general change is therefore

$$\mathrm{d}U = \mathrm{d}U(0) + \sum_i N_i\mathrm{d}\varepsilon_i + \sum_i \varepsilon_i\mathrm{d}N_i \qquad (15.53)$$

Because the energy levels do not change when a system is heated at constant volume (Fig. 15.16), in the absence of all changes other than heating

$$\mathrm{d}U = \sum_i \varepsilon_i\mathrm{d}N_i$$

We know from thermodynamics (and specifically from eqn 3.46) that under the same conditions

$$\mathrm{d}U = \mathrm{d}q_{\mathrm{rev}} = T\,\mathrm{d}S$$

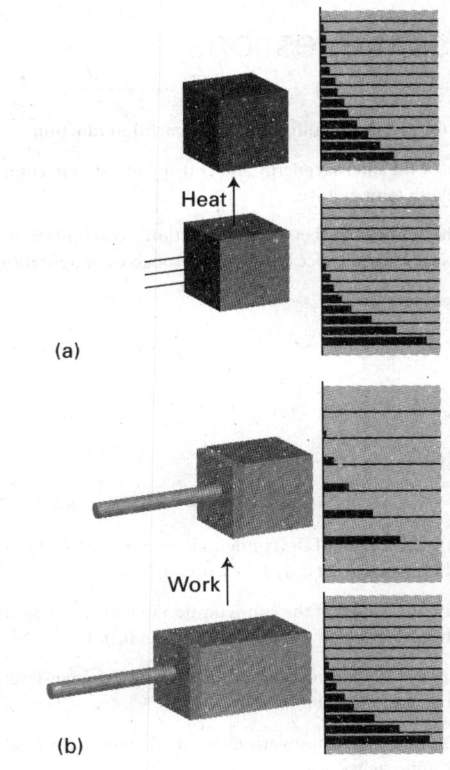

Fig. 15.16 (a) When a system is heated, the energy levels are unchanged but their populations are changed. (b) When work is done on a system, the energy levels themselves are changed. The levels in this case are the one-dimensional particle-in-a-box energy levels of Chapter 8: they depend on the size of the container and move apart as its length is decreased.

Therefore,

$$\mathrm{d}S = \frac{\mathrm{d}U}{T} = k\beta\sum_i \varepsilon_i\mathrm{d}N_i \qquad (15.54)$$

For changes in the most probable configuration (the only one we need consider), we rearrange eqn 15.42 to

$$\beta\varepsilon_i = \frac{\partial \ln \mathcal{W}}{\partial N_i} + \alpha$$

and find that

$$\mathrm{d}S = k\sum_i\left(\frac{\partial \ln \mathcal{W}}{\partial N_i}\right)\mathrm{d}N_i + k\alpha\sum_i \mathrm{d}N_i$$

Because the number of molecules is constant, the sum over the $\mathrm{d}N_i$ is zero. Hence

$$\mathrm{d}S = k\sum_i\left(\frac{\partial \ln \mathcal{W}}{\partial N_i}\right)\mathrm{d}N_i = k(\mathrm{d}\ln \mathcal{W})$$

This relation strongly suggests the definition $S = k\ln \mathcal{W}$, as in eqn 15.27.

Discussion questions

15.1 Describe the physical significance of the partition function.

15.2 Explain how the internal energy and entropy of a system composed of two levels vary with temperature.

15.3 Discuss the relationship between 'population', 'configuration', and 'weight'. What is the significance of the most probable configuration?

15.4 What is temperature?

15.5 What is the difference between a 'state' and an 'energy level'? Why is it important to make this distinction?

15.6 Explain what is meant by an *ensemble* and why it is useful in statistical thermodynamics.

15.7 Under what circumstances may identical particles be regarded as distinguishable?

Exercises

15.1(a) Calculate the weight of the configuration in which 16 objects are distributed in the arrangement 0, 1, 2, 3, 8, 0, 0, 0, 0, 2.

15.1(b) Calculate the weight of the configuration in which 21 objects are distributed in the arrangement 6, 0, 5, 0, 4, 0, 3, 0, 2, 0, 0, 1.

15.2(a) What are the relative populations of the states of a two-level system when the temperature is infinite?

15.2(b) What are the relative populations of the states of a two-level system as the temperature approaches zero?

15.3(a) What is the temperature of a two-level system of energy separation equivalent to 400 cm^{-1} when the population of the upper state is one-third that of the lower state?

15.3(b) What is the temperature of a two-level system of energy separation equivalent to 300 cm^{-1} when the population of the upper state is one-half that of the lower state?

15.4(a) A certain molecule has a nondegenerate excited state lying at 540 cm^{-1} above the nondegenerate ground state. At what temperature will 10 per cent of the molecules be in the upper state?

15.4(b) A certain molecule has a doubly degenerate excited state lying at 360 cm^{-1} above the nondegenerate ground state. At what temperature will 15 per cent of the molecules be in the upper level?

15.5(a) Calculate (a) the thermal wavelength, (b) the translational partition function at (i) 300 K and (ii) 3000 K of a molecule of molar mass 150 g mol^{-1} in a container of volume 1.00 cm^3.

15.5(b) Calculate (a) the thermal wavelength, (b) the translational partition function of a Ne atom in a cubic box of side 1.00 cm at (i) 300 K and (ii) 3000 K.

15.6(a) Calculate the ratio of the translational partition functions of D_2 and H_2 at the same temperature and volume.

15.6(b) Calculate the ratio of the translational partition functions of xenon and helium at the same temperature and volume.

15.7(a) By what factor does the number of available configurations increase when 100 J of energy is added to a system containing 1.00 mol of particles at constant volume at 298 K?

15.7(b) By what factor does the number of available configurations increase when 20 m^3 of air at 1.00 atm and 300 K is allowed to expand by 0.0010 per cent at constant temperature?

15.8(a) The bond length of O_2 is 120.75 pm. Use the high-temperature approximation to calculate the rotational partition function of the molecule at 300 K.

15.8(b) The bond length of N_2 is 109.75 pm. Use the high-temperature approximation to calculate the rotational partition function of the molecule at 300 K.

15.9(a) The NOF molecule is an asymmetric rotor with rotational constants 3.1752 cm^{-1}, 0.3951 cm^{-1}, and 0.3505 cm^{-1}. Calculate the rotational partition function of the molecule at (a) 25°C, (b) 100°C.

15.9(b) The H_2O molecule is an asymmetric rotor with rotational constants 27.877 cm^{-1}, 14.512 cm^{-1}, and 9.285 cm^{-1}. Calculate the rotational partition function of the molecule at (a) 25°C, (b) 100°C.

15.10(a) The rotational constant of CO is 1.931 cm^{-1}. Evaluate the rotational partition function explicitly (without approximation) and plot its value as a function of temperature. At what temperature is the value within 5 per cent of the value calculated from the approximate formula?

15.10(b) The rotational constant of HI is 6.511 cm^{-1}. Evaluate the rotational partition function explicitly (without approximation) and plot its value as a function of temperature. At what temperature is the value within 5 per cent of the value calculated from the approximate formula?

15.11(a) The rotational constant of CH_4 is 5.241 cm^{-1}. Evaluate the rotational partition function explicitly (without approximation but ignoring the role of nuclear statistics) and plot its value as a function of temperature. At what temperature is the value within 5 per cent of the value calculated from the approximate formula?

15.11(b) The rotational constant of CCl_4 is 0.0572 cm^{-1}. Evaluate the rotational partition function explicitly (without approximation but ignoring the role of nuclear statistics) and plot its value as a function of temperature. At what temperature is the value within 5 per cent of the value calculated from the approximate formula?

15.12(a) The rotational constants of CH_3Cl are $\tilde{A} = 5.097$ cm^{-1} and $\tilde{B} = 0.443$ cm^{-1}. Evaluate the rotational partition function explicitly (without approximation but ignoring the role of nuclear statistics) and plot its value as a function of temperature. At what temperature is the value within 5 per cent of the value calculated from the approximate formula?

15.12(b) The rotational constants of NH_3 are $\tilde{A} = 6.196$ cm^{-1} and $\tilde{B} = 9.444$ cm^{-1}. Evaluate the rotational partition function explicitly (without approximation but ignoring the role of nuclear statistics) and plot its value as a function of temperature. At what temperature is the value within 5 per cent of the value calculated from the approximate formula?

15.13(a) Give the symmetry number for each of the following molecules: (a) CO, (b) O_2, (c) H_2S, (d) SiH_4, and (e) $CHCl_3$.

15.13(b) Give the symmetry number for each of the following molecules: (a) CO_2, (b) O_3, (c) SO_3, (d) SF_6, and (e) Al_2Cl_6.

15.14(a) Estimate the rotational partition function of ethene at 25°C given that $\tilde{A} = 4.828\ \text{cm}^{-1}$, $\tilde{B} = 1.0012\ \text{cm}^{-1}$, and $\tilde{C} = 0.8282\ \text{cm}^{-1}$. Take the symmetry number into account.

15.14(b) Evaluate the rotational partition function of pyridine, C_5H_5N, at room temperature given that $\tilde{A} = 0.2014\ \text{cm}^{-1}$, $\tilde{B} = 0.1936\ \text{cm}^{-1}$, and $\tilde{C} = 0.0987\ \text{cm}^{-1}$. Take the symmetry number into account.

15.15(a) The vibrational wavenumber of Br_2 is 323.2 cm^{-1}. Evaluate the vibrational partition function explicitly (without approximation) and plot its value as a function of temperature. At what temperature is the value within 5 per cent of the value calculated from the approximate formula?

15.15(b) The vibrational wavenumber of I_2 is 214.5 cm^{-1}. Evaluate the vibrational partition function explicitly (without approximation) and plot its value as a function of temperature. At what temperature is the value within 5 per cent of the value calculated from the approximate formula?

15.16(a) Calculate the vibrational partition function of CS_2 at 500 K given the wavenumbers 658 cm^{-1} (symmetric stretch), 397 cm^{-1} (bend; two modes), and 1535 cm^{-1} (asymmetric stretch).

15.16(b) Calculate the vibrational partition function of HCN at 900 K given the wavenumbers 3311 cm^{-1} (symmetric stretch), 712 cm^{-1} (bend; two modes), and 2097 cm^{-1} (asymmetric stretch).

15.17(a) Calculate the vibrational partition function of CCl_4 at 500 K given the wavenumbers 459 cm^{-1} (symmetric stretch, A), 217 cm^{-1} (deformation, E), 776 cm^{-1} (deformation, T), and 314 cm^{-1} (deformation, T).

15.17(b) Calculate the vibrational partition function of CI_4 at 500 K given the wavenumbers 178 cm^{-1} (symmetric stretch, A), 90 cm^{-1} (deformation, E), 555 cm^{-1} (deformation, T), and 125 cm^{-1} (deformation, T).

15.18(a) A certain atom has a threefold degenerate ground level, a nondegenerate electronically excited level at 3500 cm^{-1}, and a threefold degenerate level at 4700 cm^{-1}. Calculate the partition function of these electronic states at 1900 K.

15.18(b) A certain atom has a doubly degenerate ground level, a triply degenerate electronically excited level at 1250 cm^{-1}, and a doubly degenerate level at 1300 cm^{-1}. Calculate the partition function of these electronic states at 2000 K.

15.19(a) Calculate the electronic contribution to the molar internal energy at 1900 K for a sample composed of the atoms specified in Exercise 15.18a.

15.19(b) Calculate the electronic contribution to the molar internal energy at 2000 K for a sample composed of the atoms specified in Exercise 15.18b.

15.20(a) An electron spin can adopt either of two orientations in a magnetic field, and its energies are $\pm\mu_B\mathcal{B}$, where μ_B is the Bohr magneton. Deduce an expression for the partition function and mean energy of the electron and sketch the variation of the functions with $\mathcal{B}$. Calculate the relative populations of the spin states at (a) 4.0 K, (b) 298 K when $\mathcal{B} = 1.0$ T.

15.20(b) A nitrogen nucleus spin can adopt any of three orientations in a magnetic field, and its energies are 0, $\pm\gamma_N\hbar\mathcal{B}$, where γ_N is the magnetogyric ratio of the nucleus. Deduce an expression for the partition function and mean energy of the nucleus and sketch the variation of the functions with $\mathcal{B}$. Calculate the relative populations of the spin states at (a) 1.0 K, (b) 298 K when $\mathcal{B} = 20.0$ T.

15.21(a) Consider a system of distinguishable particles having only two nondegenerate energy levels separated by an energy that is equal to the value of kT at 10 K. Calculate (a) the ratio of populations in the two states at (1) 1.0 K, (2) 10 K, and (3) 100 K, (b) the molecular partition function at 10 K, (c) the molar energy at 10 K, (d) the molar heat capacity at 10 K, (e) the molar entropy at 10 K.

15.21(b) Consider a system of distinguishable particles having only three nondegenerate energy levels separated by an energy that is equal to the value of kT at 25.0 K. Calculate (a) the ratio of populations in the states at (1) 1.00 K, (2) 25.0 K, and (3) 100 K, (b) the molecular partition function at 25.0 K, (c) the molar energy at 25.0 K, (d) the molar heat capacity at 25.0 K, (e) the molar entropy at 25.0 K.

15.22(a) At what temperature would the population of the first excited vibrational state of HCl be $1/e$ times its population of the ground state?

15.22(b) At what temperature would the population of the first excited rotational level of HCl be $1/e$ times its population of the ground state?

15.23(a) Calculate the standard molar entropy of neon gas at (a) 200 K, (b) 298.15 K.

15.23(b) Calculate the standard molar entropy of xenon gas at (a) 100 K, (b) 298.15 K.

15.24(a) Calculate the vibrational contribution to the entropy of Cl_2 at 500 K given that the wavenumber of the vibration is 560 cm^{-1}.

15.24(b) Calculate the vibrational contribution to the entropy of Br_2 at 600 K given that the wavenumber of the vibration is 321 cm^{-1}.

15.25(a) Identify the systems for which it is essential to include a factor of $1/N!$ on going from Q to q: (a) a sample of helium gas, (b) a sample of carbon monoxide gas, (c) a solid sample of carbon monoxide, (d) water vapour.

15.25(b) Identify the systems for which it is essential to include a factor of $1/N!$ on going from Q to q: (a) a sample of carbon dioxide gas, (b) a sample of graphite, (c) a sample of diamond, (d) ice.

Problems*

Numerical problems

15.1 Use mathematical software to evaluate $\mathcal{W}$ for $N = 20$ for a series of distributions over a uniform ladder of energy levels, ensuring that the total energy is constant. Identify the configuration of greatest weight and compare it to the distribution predicted by the Boltzmann expression. Explore what happens as the value of the total energy is changed.

15.2‡ Consider a system A consisting of subsystems A_1 and A_2, for which $\mathcal{W}_1 = 1 \times 10^{20}$ and $\mathcal{W}_2 = 2 \times 10^{20}$. What is the number of configurations available to the combined system? Also, compute the entropies S, S_1, and S_2. What is the significance of this result?

15.3‡ Consider 1.00×10^{22} ^{4}He atoms in a box of dimensions 1.0 cm × 1.0 cm × 1.0 cm. Calculate the occupancy of the first excited level at 1.0 mK, 2.0 K, and 4.0 K. Do the same for ^{3}He. What conclusions might you draw from the results of your calculations?

15.4 This problem is also best done using mathematical software. Equation 15.12 is the partition function for a harmonic oscillator. Consider a Morse oscillator (Section 12.10) in which the energy levels are given by eqn 12.38.

$$E_v = (v + \tfrac{1}{2})hc\tilde{\nu} - (v + \tfrac{1}{2})^2 hcx_e\tilde{\nu}$$

Evaluate the partition function for this oscillator, remembering (1) to measure energies from the lowest level and (2) to note that there is only a finite number of levels. Plot the partition function against temperature for a variety of values of x_e, and—on the same graph—compare your results with that for a harmonic oscillator.

15.5 Explore the conditions under which the 'integral' approximation for the translational partition function is not valid by considering the translational partition function of an Ar atom in a cubic box of side 1.00 cm. Estimate the temperature at which, according to the integral approximation, $q = 10$ and evaluate the exact partition function at that temperature.

15.6 A certain atom has a doubly degenerate ground level pair and an upper level of four degenerate states at 450 cm^{-1} above the ground level. In an atomic beam study of the atoms it was observed that 30 per cent of the atoms were in the upper level, and the translational temperature of the beam was 300 K. Are the electronic states of the atoms in thermal equilibrium with the translational states?

15.7 (a) Calculate the electronic partition function of a tellurium atom at (i) 298 K, (ii) 5000 K by direct summation using the following data:

Term	Degeneracy	Wavenumber/cm^{-1}
Ground	5	0
1	1	4 707
2	3	4 751
3	5	10 559

(b) What proportion of the Te atoms are in the ground term and in the term labelled 2 at the two temperatures? (c) Calculate the electronic contribution to the standard molar entropy of gaseous Te atoms.

15.8 The four lowest electronic levels of a Ti atom are: $^3F_2$, $^3F_3$, $^3F_4$, and $^5F_1$, at 0, 170, 387, and 6557 cm^{-1}, respectively. There are many other electronic states at higher energies. The boiling point of titanium is 3287°C. What are the relative populations of these levels at the boiling point? (*Hint.* The degeneracies of the levels are $2J + 1$.)

15.9 The NO molecule has a doubly degenerate excited electronic level 121.1 cm^{-1} above the doubly degenerate electronic ground term. Calculate and plot the electronic partition function of NO from $T = 0$ to 1000 K. Evaluate (a) the term populations and (b) the electronic contribution to the molar internal energy at 300 K. Calculate the electronic contribution to the molar entropy of the NO molecule at 300 K and 500 K.

15.10‡ J. Sugar and A. Musgrove (*J. Phys. Chem. Ref. Data* 22, 1213 (1993)) have published tables of energy levels for germanium atoms and cations from Ge$^+$ to Ge^{+31}. The lowest-lying energy levels in neutral Ge are as follows:

	$^3P_0$	$^3P_1$	$^3P_2$	$^1D_2$	$^1S_0$
(E/hc)/cm^{-1}	0	557.1	1410.0	7125.3	16 367.3

Calculate the electronic partition function at 298 K and 1000 K by direct summation. *Hint.* The degeneracy of a level is $2J + 1$.

15.11 Calculate, by explicit summation, the vibrational partition function and the vibrational contribution to the molar internal energy of I_2 molecules at (a) 100 K, (b) 298 K given that its vibrational energy levels lie at the following wavenumbers above the zero-point energy level: 0, 213.30, 425.39, 636.27, 845.93 cm^{-1}. What proportion of I_2 molecules are in the ground and first two excited levels at the two temperatures? Calculate the vibrational contribution to the molar entropy of I_2 at the two temperatures.

15.12‡ (a) The standard molar entropy of graphite at 298, 410, and 498 K is 5.69, 9.03, and 11.63 J K^{-1} mol^{-1}, respectively. If 1.00 mol C(graphite) at 298 K is surrounded by thermal insulation and placed next to 1.00 mol C(graphite) at 498 K, also insulated, how many configurations are there altogether for the combined but independent systems? (b) If the same two samples are now placed in thermal contact and brought to thermal equilibrium, the final temperature will be 410 K. (Why might the final temperature not be the average? It isn't.) How many configurations are there now in the combined system? Neglect any volume changes. (c) Demonstrate that this process is spontaneous.

Theoretical problems

15.13 Explore the consequences of using the full version of Stirling's approximation, $x! \approx (2\pi)^{1/2}x^{x+1/2}e^{-x}$, in the development of the expression for the configuration of greatest weight. Does the more accurate approximation have a significant effect on the form of the Boltzmann distribution?

15.14 A sample consisting of five molecules has a total energy 5ε. Each molecule is able to occupy states of energy $j\varepsilon$, with $j = 0, 1, 2, \ldots$. (a) Calculate the weight of the configuration in which the molecules are distributed evenly over the available states. (b) Draw up a table with columns headed by the energy of the states and write beneath them all configurations that are consistent with the total energy. Calculate the weights of each configuration and identify the most probable configurations.

15.15 A sample of nine molecules is numerically tractable but on the verge of being thermodynamically significant. Draw up a table of configurations for $N = 9$, total energy 9ε in a system with energy levels $j\varepsilon$ (as in Problem 15.14). Before evaluating the weights of the configurations, guess (by looking for the most 'exponential' distribution of populations) which of the configurations will turn out to be the most probable. Go on to calculate the weights and identify the most probable configuration.

* Problems denoted with the symbol ‡ were supplied by Charles Trapp and Carmen Giunta.

15.16 The most probable configuration is characterized by a parameter we know as the 'temperature'. The temperatures of the system specified in Problems 15.14 and 15.15 must be such as to give a mean value of ε for the energy of each molecule and a total energy $N\varepsilon$ for the system. (a) Show that the temperature can be obtained by plotting p_j against j, where p_j is the (most probable) fraction of molecules in the state with energy $j\varepsilon$. Apply the procedure to the system in Problem 15.15. What is the temperature of the system when ε corresponds to 50 cm^{-1}? (b) Choose configurations other than the most probable, and show that the same procedure gives a worse straight line, indicating that a temperature is not well-defined for them.

15.17 A certain molecule can exist in either a nondegenerate singlet state or a triplet state (with degeneracy 3). The energy of the triplet exceeds that of the singlet by ε. Assuming that the molecules are distinguishable (localized) and independent, (a) obtain the expression for the molecular partition function. (b) Find expressions in terms of ε for the molar energy, molar heat capacity, and molar entropy of such molecules and calculate their values at $T = \varepsilon/k$.

15.18 Consider a system with energy levels $\varepsilon_j = j\varepsilon$ and N molecules. (a) Show that, if the mean energy per molecule is $a\varepsilon$, then the temperature is given by

$$\beta = \frac{1}{\varepsilon}\ln\left(1+\frac{1}{a}\right)$$

Evaluate the temperature for a system in which the mean energy is ε, taking ε equivalent to 50 cm^{-1}. (b) Calculate the molecular partition function q for the system when its mean energy is $a\varepsilon$. (c) Show that the entropy of the system is

$$S/k = (1+a)\ln(1+a) - a\ln a$$

and evaluate this expression for a mean energy ε.

15.19‡ For gases, the canonical partition function, Q, is related to the molecular partition function q by $Q = q^N/N!$. Use the expression for q and general thermodynamic relations to derive the perfect gas law $pV = nRT$.

15.20 In the following pair of problems we explore the concept of negative absolute temperature ($T < 0$). Show that for a two-level system (energy separation ε) that the temperature is formally negative when the population of the upper state exceeds that of the lower state. Use the partition function for this system to derive and plot expressions for the internal energy and the entropy (and the partition function itself) as a function of (a) kT/ε, (b) $\varepsilon\beta$ from -10 to $+10$ in each case.

15.21 The thermodynamic relation $(\partial U/\partial S)_V = T$ applies formally to $T < 0$ as well as to $T > 0$. Plot the U calculated in Problem 15.20 against S and confirm that $(\partial U/\partial S)_V < 0$ and $(\partial U/\partial S)_V > 0$ over the appropriate ranges of temperature.

Applications: to atmospheric science and astrophysics

15.22‡ The variation of the atmospheric pressure p with altitude h is predicted by the *barometric formula* to be $p = p_0\,e^{-h/H}$ where p_0 is the pressure at sea level and $H = RT/Mg$ with M the average molar mass of air and T the average temperature. Obtain the barometric formula from the Boltzmann distribution. Recall that the potential energy of a particle at height h above the surface of the Earth is mgh. Convert the barometric formula from pressure to number density, $\mathcal{N}$. Compare the relative number densities, $\mathcal{N}(h)/\mathcal{N}(0)$, for O_2 and H_2O at $h = 8.0$ km, a typical cruising altitude for commercial aircraft.

15.23‡ Planets lose their atmospheres over time unless they are replenished. A complete analysis of the overall process is very complicated and depends upon the radius of the planet, temperature, atmospheric composition, and other factors. Prove that the atmosphere of planets cannot be in an equilibrium state by demonstrating that the Boltzmann distribution leads to a uniform finite number density as $r \to \infty$. *Hint.* Recall that in a gravitational field the potential energy is $V(r) = -GMm/r$, where G is the gravitational constant, M is the mass of the planet, and m the mass of the particle.

15.24‡ Consider the electronic partition function of a perfect atomic hydrogen gas at a density of 1.99×10^{-4} kg m^{-3} and 5780 K. These are the mean conditions within the Sun's photosphere, the surface layer of the Sun that is about 190 km thick. (a) Show that this partition function, which involves a sum over an infinite number of quantum states that are solutions to the Schrödinger equation for an isolated atomic hydrogen atom, is infinite. (b) Develop a theoretical argument for truncating the sum and estimate the maximum number of quantum states that contribute to the sum. (c) Calculate the equilibrium probability that an atomic hydrogen electron is in each quantum state. Are there any general implications concerning electronic states that will be observed for other atoms and molecules? Is it wise to apply these calculations in the study of the Sun's photosphere?

16 Statistical thermodynamics 2: applications

In this chapter we apply the concepts of statistical thermodynamics to the calculation of chemically significant quantities. First, we establish the relations between thermodynamic functions and partition functions. Next, we show that the molecular partition function can be factorized into contributions from each mode of motion and establish the formulas for the partition functions for translational, rotational, and vibrational modes of motion and the contribution of electronic excitation. These contributions can be calculated from spectroscopic data. Finally, we turn to specific applications, which include the mean energies of modes of motion, the heat capacities of substances, and residual entropies. In the final section, we see how to calculate the equilibrium constant of a reaction and through that calculation understand some of the molecular features that determine the magnitudes of equilibrium constants and their variation with temperature.

A partition function is the bridge between thermodynamics, spectroscopy, and quantum mechanics. Once it is known, a partition function can be used to calculate thermodynamic functions, heat capacities, entropies, and equilibrium constants. It also sheds light on the significance of these properties.

Fundamental relations

In this section we see how to obtain any thermodynamic function once we know the partition function. Then we see how to calculate the molecular partition function, and through that the thermodynamic functions, from spectroscopic data.

16.1 The thermodynamic functions

Key point **The following functions are written in terms of the canonical partition function: (a) the Helmoltz energy, (b) the pressure, (c) the enthalpy, (d) the Gibbs energy.**

We have already derived (in Chapter 15) the two expressions for calculating the internal energy and the entropy of a system from its canonical partition function, Q:

$$U-U(0)=-\left(\frac{\partial \ln Q}{\partial \beta}\right)_V \qquad S=\frac{U-U(0)}{T}+k\ln Q \tag{16.1}$$

where $\beta=1/kT$. If the molecules are independent, we can go on to make the substitutions $Q=q^N$ (for distinguishable molecules, as in a solid) or $Q=q^N/N!$ (for indistinguishable molecules, as in a gas). All the thermodynamic functions introduced in Part 1 are related to U and S, so we have a route to their calculation from Q.

(a) Helmholtz energy

The Helmholtz energy, A, is defined as $A = U - TS$. This relation implies that $A(0) = U(0)$, so substitution for U and S by using eqn 16.1 leads to the very simple expression

$$A - A(0) = -kT \ln Q \qquad \text{Helmholtz energy in terms of } Q \qquad (16.2)$$

(b) The pressure

By an argument like that leading to eqn 3.35, it follows from $A = U - TS$ that $dA = -pdV - SdT$. Therefore, on imposing constant temperature, the pressure and the Helmholtz energy are related by $p = -(\partial A/\partial V)_T$. It then follows from eqn 16.2 that

$$p = kT\left(\frac{\partial \ln Q}{\partial V}\right)_T \qquad \text{Pressure in terms of } Q \qquad (16.3)$$

This relation is entirely general, and may be used for any type of substance, including perfect gases, real gases, and liquids. Because Q is in general a function of the volume, temperature, and amount of substance, eqn 16.3 is an equation of state.

Example 16.1 *Deriving an equation of state*

Derive an expression for the pressure of a gas of independent particles.

Method We should suspect that the pressure is that given by the perfect gas law. To proceed systematically, substitute the explicit formula for Q for a gas of independent, indistinguishable molecules (see eqn 15.38 and the *Checklist of key equations* at the end of Chapter 15) into eqn 16.3.

Answer For a gas of independent molecules, $Q = q^N/N!$ with $q = V/\Lambda^3$:

$$p = kT\left(\frac{\partial \ln Q}{\partial V}\right)_T = \frac{kT}{Q}\left(\frac{\partial Q}{\partial V}\right)_T = \frac{NkT}{q}\left(\frac{\partial q}{\partial V}\right)_T$$

$$= \frac{NkT\Lambda^3}{V} \times \frac{1}{\Lambda^3} = \frac{NkT}{V} = \frac{nRT}{V}$$

To derive this relation, we have used

$$\left(\frac{\partial q}{\partial V}\right)_T = \left(\frac{\partial (V/\Lambda^3)}{\partial V}\right)_T = \frac{1}{\Lambda^3}$$

and $NkT = nN_AkT = nRT$. The calculation shows that the equation of state of a gas of independent particles is indeed the perfect gas law.

Self-test 16.1 Derive the equation of state of a sample for which $Q = q^N f/N!$, with $q = V/\Lambda^3$, where f depends on the volume. [$p = nRT/V + kT(\partial \ln f/\partial V)_T$]

(c) The enthalpy

At this stage we can use the expressions for U and p in the definition $H = U + pV$ to obtain an expression for the enthalpy, H, of any substance:

$$H - H(0) = -\left(\frac{\partial \ln Q}{\partial \beta}\right)_V + kTV\left(\frac{\partial \ln Q}{\partial V}\right)_T \qquad \text{Enthalpy in terms of } Q \qquad (16.4)$$

We have already seen that $U - U(0) = \frac{3}{2}nRT$ for a gas of independent particles (eqn 15.25a), and have just shown that $pV = nRT$. Therefore, for such a gas,

$$H - H(0) = \tfrac{5}{2}nRT \tag{16.5}$$

(d) The Gibbs energy

One of the most important thermodynamic functions for chemistry is the Gibbs energy, $G = H - TS = A + pV$. We can now express this function in terms of the partition function by combining the expressions for A and p:

$$G - G(0) = -kT \ln Q + kTV\left(\frac{\partial \ln Q}{\partial V}\right)_T \tag{16.6}$$

Gibbs energy in terms of Q

This expression takes a simple form for a gas of independent molecules because pV in the expression $G = A + pV$ can be replaced by nRT:

$$G - G(0) = -kT \ln Q + nRT \tag{16.7}^\circ$$

Furthermore, because $Q = q^N/N!$, and therefore $\ln Q = N \ln q - \ln N!$, it follows by using Stirling's approximation ($\ln N! = N \ln N - N$) that we can write

$$\begin{aligned} G - G(0) &= -NkT \ln q + kT \ln N! + nRT \\ &= -nRT \ln q + kT(N \ln N - N) + nRT \\ &= -nRT \ln \frac{q}{N} \end{aligned} \tag{16.8}^\circ$$

with $N = nN_A$. Now we see another interpretation of the Gibbs energy: it is proportional to the logarithm of the average number of thermally accessible states per molecule.

It will turn out to be convenient to define the **molar partition function**, $q_m = q/n$ (with units mol^{-1}), for then

$$G - G(0) = -nRT \ln \frac{q_m}{N_A} \tag{16.9}^\circ$$

Gibbs energy of independent molecules

16.2 The molecular partition function

Key points **The molecular partition function factorizes into a product of: (a) translational, (b) rotational, (c) vibrational, and (d) electronic contributions. (e) The contributions to the overall partition function are summarized in the *Checklist of key equations*.**

The energy of a molecule is the sum of contributions from its different modes of motion:

$$\varepsilon_i = \varepsilon_i^T + \varepsilon_i^R + \varepsilon_i^V + \varepsilon_i^E \tag{16.10}$$

where T denotes translation, R rotation, V vibration, and E the electronic contribution. The electronic contribution is not actually a 'mode of motion', but it is convenient to include it here. The separation of terms in eqn 16.10 is only approximate (except for translation) because the modes are not completely independent, but in most cases it is satisfactory. The separation of the electronic and vibrational motions is justified provided only the ground electronic state is occupied (for otherwise the vibrational characteristics depend on the electronic state) and, for the electronic ground state, that the Born–Oppenheimer approximation is valid (Chapter 10). The separation of the vibrational and rotational modes is justified to the extent that the rotational constant is independent of the vibrational state.

Given that the energy is a sum of independent contributions, the partition function factorizes into a product of contributions (recall Section 15.2b):

$$q = \sum_i e^{-\beta\varepsilon_i} = \sum_{i\,(\text{all states})} e^{-\beta\varepsilon_i^{T}-\beta\varepsilon_i^{R}-\beta\varepsilon_i^{V}-\beta\varepsilon_i^{E}}$$

$$= \sum_{i\,(\text{translational})}\ \sum_{i\,(\text{rotational})}\ \sum_{i\,(\text{vibrational})}\ \sum_{i\,(\text{electronic})} e^{-\beta\varepsilon_i^{T}-\beta\varepsilon_i^{R}-\beta\varepsilon_i^{V}-\beta\varepsilon_i^{E}} \qquad \text{Factorization of the partition function} \qquad (16.11)$$

$$= \left(\sum_{i\,(\text{translational})} e^{-\beta\varepsilon_i^{T}}\right)\left(\sum_{i\,(\text{rotational})} e^{-\beta\varepsilon_i^{R}}\right)\left(\sum_{i\,(\text{vibrational})} e^{-\beta\varepsilon_i^{V}}\right)\left(\sum_{i\,(\text{electronic})} e^{-\beta\varepsilon_i^{E}}\right)$$

$$= q^{T}q^{R}q^{V}q^{E}$$

This factorization allows us to investigate each contribution separately.

(a) The translational contribution

The translational partition function of a molecule of mass m in a container of volume V was derived in Section 15.2:

$$q^{T} = \frac{V}{\Lambda^3} \qquad \Lambda = h\left(\frac{\beta}{2\pi m}\right)^{1/2} = \frac{h}{(2\pi mkT)^{1/2}} \qquad \text{Translational contribution to } q \qquad (16.12)$$

Notice that $q^{T} \to \infty$ as $T \to \infty$ because an infinite number of states becomes accessible as the temperature is raised. Even at room temperature $q^{T} \approx 2 \times 10^{28}$ for an O_2 molecule in a vessel of volume 100 cm^3.

The thermal wavelength, Λ, lets us judge whether the approximations that led to the expression for q^{T} are valid. The approximations are valid if many states are occupied, which requires V/Λ^3 to be large. That will be so if Λ is small compared with the linear dimensions of the container. For H_2 at 25°C, $\Lambda = 71$ pm, which is far smaller than any conventional container is likely to be (but comparable to pores in zeolites or cavities in clathrates). For O_2, a heavier molecule, $\Lambda = 18$ pm. We saw in Section 15.2 that an equivalent criterion of validity is that Λ should be much less than the average separation of the molecules in the sample.

(b) The rotational contribution

As demonstrated in Example 15.1, the partition function of a nonsymmetrical (AB) linear rotor is

$$q^{R} = \sum_J (2J+1)e^{-\beta hc\tilde{B}J(J+1)} \qquad (16.13)$$

The direct method of calculating q^{R} is to substitute the experimental values of the rotational energy levels into this expression and to sum the series numerically.

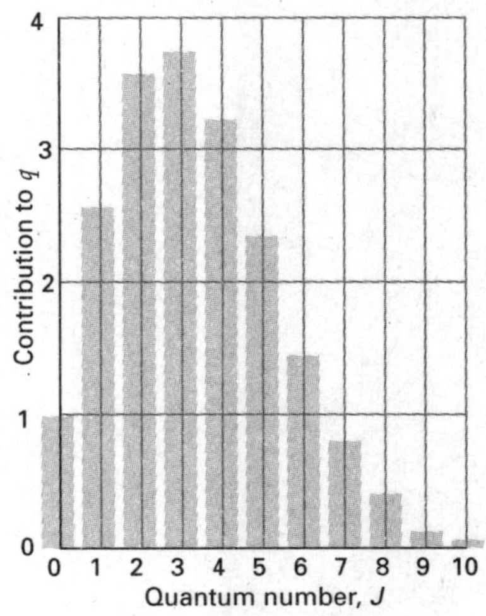

Fig. 16.1 The contributions to the rotational partition function of an HCl molecule at 25°C. The vertical axis is the value of $(2J+1)e^{-\beta hcBJ(J+1)}$. Successive terms (which are proportional to the populations of the levels) pass through a maximum because the population of individual states decreases exponentially, but the degeneracy of the levels increases with J.

Example 16.2 *Evaluating the rotational partition function explicitly*

Evaluate the rotational partition function of $^{1}H^{35}Cl$ at 25°C, given that $\tilde{B}$ = 10.591 cm^{-1}.

Method We use eqn 16.13 and evaluate it term by term. A useful relation is kT/hc = 207.22 cm^{-1} at 298.15 K. The sum is readily evaluated by using mathematical software.

Answer To show how successive terms contribute, we draw up the following table by using $hc\tilde{B}/kT = 0.051\,11$ (Fig. 16.1):

J	0	1	2	3	4	...	10
$(2J+1)e^{-0.05111J(J+1)}$	1	2.71	3.68	3.79	3.24	...	0.08

The sum required by eqn 16.13 (the sum of the numbers in the second row of the table) is 19.9; hence $q^{R} = 19.9$ at this temperature. Taking J up to 50 gives $q^{R} = 19.902$. Notice that about ten J-levels are significantly populated but the number of populated *states* is larger on account of the $(2J+1)$-fold degeneracy of each level. We shall shortly encounter the approximation that $q^{R} \approx kT/hc\tilde{B}$, which in the present case gives $q^{R} = 19.6$, in good agreement with the exact value and with much less work.

Self-test 16.2 Evaluate the rotational partition function for HCl at 0°C. [18.26]

At room temperature $kT/hc \approx 200\ \text{cm}^{-1}$. The rotational constants of many molecules are close to 1 cm^{-1} (Table 12.2) and often smaller (though the very light H_2 molecule, for which $\tilde{B} = 60.9\ \text{cm}^{-1}$, is one exception). It follows that many rotational levels are populated at normal temperatures. When this is the case, we show in the following *Justification* that the partition function may be approximated by

$$q^{R} = \frac{kT}{hc\tilde{B}} \qquad \text{for linear rotors} \tag{16.14a}$$

$$q^{R} = \left(\frac{kT}{hc}\right)^{3/2}\left(\frac{\pi}{\tilde{A}\tilde{B}\tilde{C}}\right)^{1/2} \qquad \text{for non-linear rotors} \tag{16.14b}$$

where $\tilde{A}$, $\tilde{B}$, and $\tilde{C}$ are the rotational constants of the molecule. However, before using these expressions, read on (to eqns 16.15 and 16.16).

Justification 16.1 *The rotational contribution to the molecular partition function*

When many rotational states are occupied and kT is much larger than the separation between neighbouring states, the sum in the partition function can be approximated by an integral, much as we did for translational motion in *Justification 15.2*:

$$q^{R} = \int_0^{\infty} (2J+1)e^{-\beta hc\tilde{B}J(J+1)}\,\text{d}J$$

Although this integral looks complicated, it can be evaluated without much effort by noticing that because

$$\frac{\text{d}}{\text{d}J}e^{aJ(J+1)} = \left\{\frac{\text{d}}{\text{d}J}aJ(J+1)\right\}e^{aJ(J+1)} = a(2J+1)e^{aJ(J+1)}$$

it can also be written as

$$q^{R} = -\frac{1}{\beta hc\tilde{B}}\int_0^{\infty}\left(\frac{\text{d}}{\text{d}J}e^{-\beta hc\tilde{B}J(J+1)}\right)\text{d}J$$

Then, because the integral of a derivative of a function is the function itself, we obtain

$$q^{R} = -\frac{1}{\beta hc\tilde{B}}e^{-\beta hc\tilde{B}J(J+1)}\Bigg|_0^{\infty} = \frac{1}{\beta hc\tilde{B}}$$

which (because $\beta = 1/kT$) is eqn 16.14a. The calculation for a nonlinear molecule is along the same lines, but slightly trickier: it is presented in *Further information 16.1*.

A useful way of expressing the temperature above which the rotational approximation is valid is to introduce the **characteristic rotational temperature**, $\theta_R = hc\tilde{B}/k$. Then 'high temperature' means $T \gg \theta_R$ and under these conditions the rotational partition function of a linear molecule is simply T/θ_R. Some typical values of θ_R are shown in Table 16.1. The value for H_2 is abnormally high and we must be careful with the approximation for this molecule.

Table 16.1* Rotational and vibrational temperatures

Molecule	Mode	θ_V/K	θ_R/K
H_2		6330	88
HCl		4300	15.2
I_2		39	0.053
CO_2	ν_1	1997	0.561
	ν_2	3380	
	ν_3	960	

* For more values, see Table 12.2 in the *Data section* and use $hc/k = 1.439$ K cm.

The general conclusion at this stage is that molecules with large moments of inertia (and hence small rotational constants and low characteristic rotational temperatures) have large rotational partition functions. The large value of q^R reflects the closeness in energy (compared with kT) of the rotational states in large, heavy molecules, and the large number of them that are accessible at normal temperatures.

We must take care, however, not to include too many rotational states in the sum. For a homonuclear diatomic molecule or a symmetrical linear molecule (such as CO_2 or HC≡CH), a rotation through 180° results in an indistinguishable state of the molecule. Hence, the number of thermally accessible states is only half the number that can be occupied by a heteronuclear diatomic molecule, where rotation through 180° does result in a distinguishable state. Therefore, for a symmetrical linear molecule

$$q^R = \frac{kT}{2hc\tilde{B}} = \frac{T}{2\theta_R} \quad (16.15a)$$

The equations for symmetrical and nonsymmetrical molecules can be combined into a single expression by introducing the **symmetry number**, σ, which is the number of indistinguishable orientations of the molecule. Then

$$q^R = \frac{kT}{\sigma hc\tilde{B}} = \frac{T}{\sigma\theta_R} \quad (16.15b)$$

Rotational contribution to q in the high temperature limit (linear rotors)

For a heteronuclear diatomic molecule $\sigma = 1$; for a homonuclear diatomic molecule or a symmetrical linear molecule, $\sigma = 2$.

Justification 16.2 *The origin of the symmetry number*

The quantum mechanical origin of the symmetry number is the Pauli principle, which forbids the occupation of certain states. We saw in Section 12.7, for example, that H_2 may occupy rotational states with even J only if its nuclear spins are paired (*para*-hydrogen), and odd J states only if its nuclear spins are parallel (*ortho*-hydrogen). There are three states of *ortho*-H_2 to each value of J (because there are three parallel spin states of the two nuclei).

To set up the rotational partition function we note that 'ordinary' molecular hydrogen is a mixture of one part *para*-H_2 (with only its even-J rotational states occupied) and three parts *ortho*-H_2 (with only its odd-J rotational states occupied). Therefore, the average partition function per molecule is

$$q^R = \tfrac{1}{4}\sum_{\text{even } J}(2J+1)e^{-\beta hc\tilde{B}J(J+1)} + \tfrac{3}{4}\sum_{\text{odd } J}(2J+1)e^{-\beta hc\tilde{B}J(J+1)}$$

The odd-J states are more heavily weighted than the even-J states (Fig. 16.2). From the illustration we see that we would obtain approximately the same answer for the partition function (the sum of all the populations) if each J term contributed half its normal value to the sum. That is, the last equation can be approximated as

$$q^R = \tfrac{1}{2}\sum_J (2J+1)e^{-\beta hc\tilde{B}J(J+1)}$$

This approximation is very good when many terms contribute (at high temperatures).

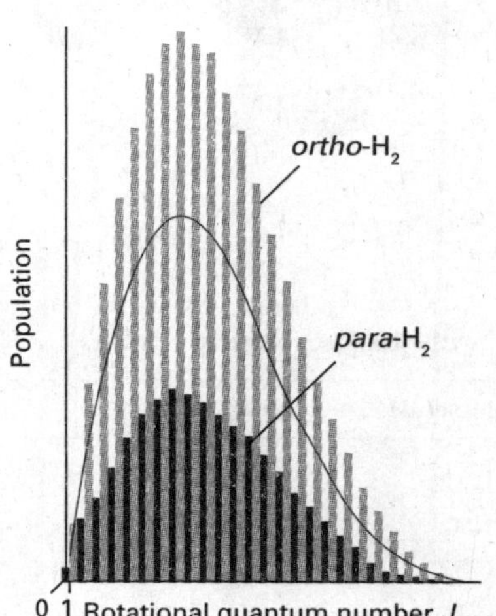

Fig. 16.2 The values of the individual terms $(2J+1)e^{-\beta hcBJ(J+1)}$ contributing to the mean partition function of a 3:1 mixture of *ortho*- and *para*-H_2. The partition function is the sum of all these terms. At high temperatures, the sum is approximately equal to the sum of the terms over all values of J, each with a weight of $\frac{1}{2}$. This is the sum of the contributions indicated by the curve.

The same type of argument may be used for linear symmetrical molecules in which identical bosons are interchanged by rotation (such as CO_2). As pointed out in Section 12.7, if the nuclear spin of the bosons is 0, then only even-J states are admissible. Because only half the rotational states are occupied, the rotational partition function is only half the value of the sum obtained by allowing all values of J to contribute (Fig. 16.3).

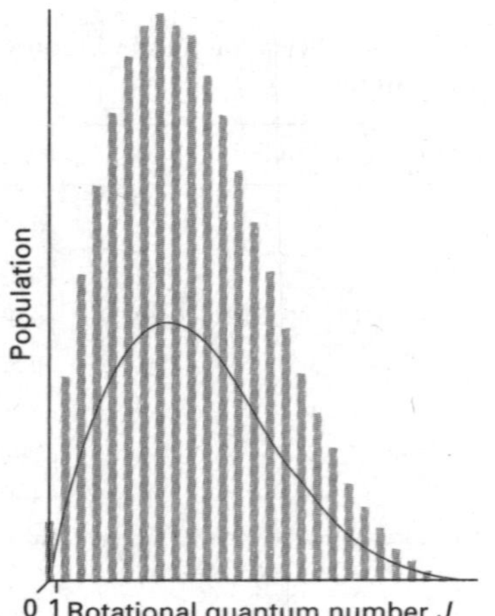

Fig. 16.3 The relative populations of the rotational energy levels of CO_2. Only states with even J values are occupied. The full line shows the smoothed, averaged population of levels.

The same care must be exercised for other types of symmetrical molecule, and for a nonlinear molecule we write

$$q^{R} = \frac{1}{\sigma}\left(\frac{kT}{hc}\right)^{3/2}\left(\frac{\pi}{\tilde{A}\tilde{B}\tilde{C}}\right)^{1/2} \qquad (16.16)$$

Rotational contribution to q in the high temperature limit (nonlinear molecules)

Some typical values of the symmetry numbers required are given in Table 16.2. The value $\sigma(H_2O) = 2$ reflects the fact that a 180° rotation about the bisector of the H–O–H angle interchanges two indistinguishable atoms. In NH_3, there are three indistinguishable orientations around the axis shown in (1). For CH_4, any of three 120° rotations about any of its four C–H bonds leaves the molecule in an indistinguishable state, so the symmetry number is $3 \times 4 = 12$. For benzene, any of six orientations around the axis perpendicular to the plane of the molecule leaves it apparently unchanged, as does a rotation of 180° around any of six axes in the plane of the molecule (three of which pass along each C–H bond and the remaining three pass through each C–C bond in the plane of the molecule). For the way that group theory is used to identify the value of the symmetry number, see Problem 16.18.

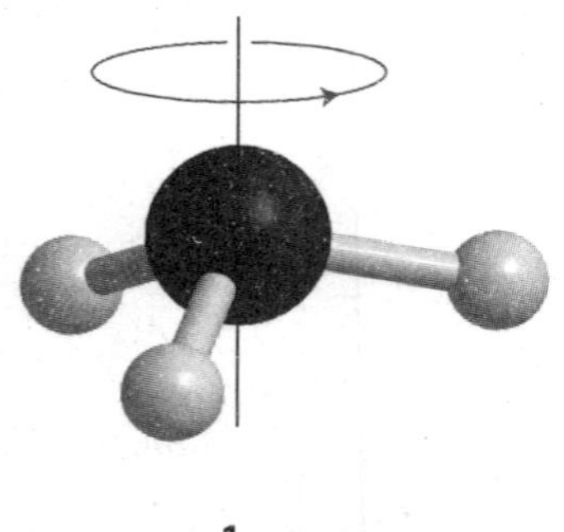

1

Table 16.2* Symmetry numbers

Molecule	σ
H_2O	2
NH_3	3
CH_4	12
C_6H_6	12

* For more values, see Table 12.2 in the *Data section*.

(c) The vibrational contribution

The vibrational partition function of a molecule is calculated by substituting the measured vibrational energy levels into the exponentials appearing in the definition of q^{V}, and summing them numerically. In a polyatomic molecule each normal mode (Section 12.13) has its own partition function (provided the anharmonicities are so small that the modes are independent). The overall vibrational partition function is the product of the individual partition functions, and we can write $q^{V} = q^{V}(1)q^{V}(2)\ldots$, where $q^{V}(K)$ is the partition function for the Kth normal mode and is calculated by direct summation of the observed spectroscopic levels.

If the vibrational excitation is not too great, the harmonic approximation may be made, and the vibrational energy levels written as

$$E_{v} = (v + \tfrac{1}{2})hc\tilde{\nu} \qquad v = 0, 1, 2, \ldots \qquad (16.17)$$

If, as usual, we measure energies from the zero-point level, then the permitted values are $\varepsilon_{v} = vhc\tilde{\nu}$ and the partition function is

$$q^{V} = \sum_{v} e^{-\beta vhc\tilde{\nu}} = \sum_{v} (e^{-\beta hc\tilde{\nu}})^{v} \qquad (16.18)$$

(because $e^{ax} = (e^{x})^{a}$). We met this sum in Example 15.2 (which is no accident: the ladder-like array of levels in Fig. 15.3 is exactly the same as that of a harmonic oscillator). The series can be summed in the same way, and gives

$$q^{V} = \frac{1}{1 - e^{-\beta hc\tilde{\nu}}} \qquad (16.19)$$

Vibrational contribution to q

This function is plotted in Fig. 16.4. In a polyatomic molecule, each normal mode gives rise to a partition function of this form.

Example 16.3 *Calculating a vibrational partition function*

The wavenumbers of the three normal modes of H_2O are 3656.7 cm^{-1}, 1594.8 cm^{-1}, and 3755.8 cm^{-1}. Evaluate the vibrational partition function at 1500 K.

Method Use eqn 16.19 for each mode, and then form the product of the three contributions. At 1500 K, $kT/hc = 1042.6$ cm^{-1}.

Answer We draw up the following table displaying the contributions of each mode:

Mode:	1	2	3
$\tilde{\nu}$/cm^{-1}	3656.7	1594.8	3755.8
$hc\tilde{\nu}/kT$	3.507	1.530	3.602
q^V	1.031	1.276	1.028

The overall vibrational partition function is therefore

$$q^V = 1.031 \times 1.276 \times 1.028 = 1.353$$

The three normal modes of H_2O are at such high wavenumbers that even at 1500 K most of the molecules are in their vibrational ground state. However, there may be so many normal modes in a large molecule that their excitation may be significant even though each mode is not appreciably excited. For example, a nonlinear molecule containing 10 atoms has $3N - 6 = 24$ normal modes (Section 12.13). If we assume a value of about 1.1 for the vibrational partition function of one normal mode, the overall vibrational partition function is about $q^V \approx (1.1)^{24} = 9.8$, which indicates significant vibrational excitation relative to a smaller molecule, such as H_2O.

Self-test 16.3 Repeat the calculation for CO_2, where the vibrational wavenumbers are 1388 cm^{-1}, 667.4 cm^{-1}, and 2349 cm^{-1}, the second being the doubly degenerate bending mode. [6.79]

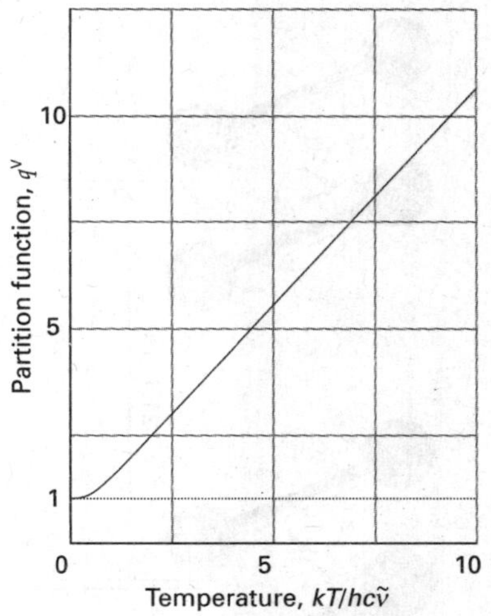

Fig. 16.4 The vibrational partition function of a molecule in the harmonic approximation. Note that the partition function is linearly proportional to the temperature when the temperature is high ($T \gg \theta_V$).

interActivity Plot the temperature dependence of the vibrational contribution to the molecular partition function for several values of the vibrational wavenumber. Estimate from your plots the temperature above which the harmonic oscillator is in the 'high temperature' limit.

In many molecules the vibrational wavenumbers are so great that $\beta hc\tilde{\nu} > 1$. For example, the lowest vibrational wavenumber of CH_4 is 1306 cm^{-1}, so $\beta hc\tilde{\nu} = 6.3$ at room temperature. C–H stretches normally lie in the range 2850 to 2960 cm^{-1}, so for them $\beta hc\tilde{\nu} \approx 14$. In these cases, $e^{-\beta hc\tilde{\nu}}$ in the denominator of q^V is very close to zero (for example, $e^{-6.3} = 0.002$), and the vibrational partition function for a single mode is very close to 1 ($q^V = 1.002$ when $\beta hc\tilde{\nu} = 6.3$, implying that only the zero-point level is significantly occupied.

Now consider the case of bonds so weak that $\beta hc\tilde{\nu} \ll kT$. When this condition is satisfied, the partition function may be approximated by expanding the exponential ($e^x = 1 + x + \cdots$):

$$q^V = \frac{1}{1-(1-\beta hc\tilde{\nu}+\cdots)} \tag{16.20}$$

That is, for weak bonds at high temperatures,

$$q^V = \frac{1}{\beta hc\tilde{\nu}} = \frac{kT}{hc\tilde{\nu}} \qquad \text{Vibrational contribution to } q \text{ in the high temperature limit} \tag{16.21}$$

The temperatures for which eqn 16.21 is valid can be expressed in terms of the **characteristic vibrational temperature**, $\theta_V = hc\tilde{\nu}/k$ (Table 16.1). The value for H_2 is abnormally high because the atoms are so light and the vibrational frequency is correspondingly high. In terms of the vibrational temperature, 'high temperature' means

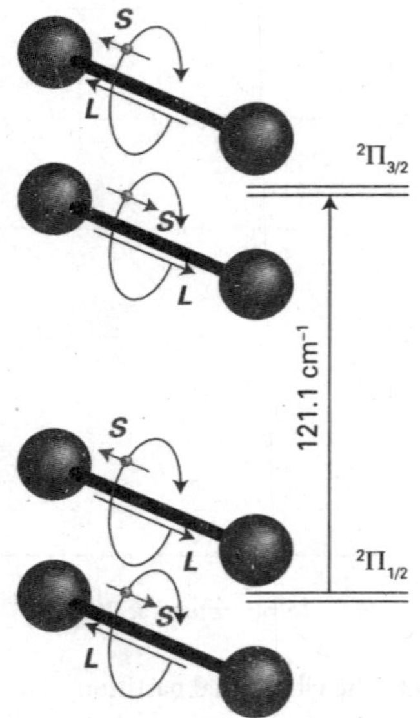

Fig. 16.5 The doubly degenerate ground electronic level of NO (with the spin and orbital angular momentum around the axis in opposite directions) and the doubly degenerate first excited level (with the spin and orbital momenta parallel). The upper level is thermally accessible at room temperature.

$T \gg \theta_V$ and, when this condition is satisfied, $q^V = T/\theta_V$ (the analogue of the rotational expression).

(d) The electronic contribution

Electronic energy separations from the ground state are usually very large, so for most cases $q^E = 1$. An important exception arises in the case of atoms and molecules having electronically degenerate ground states, in which case $q^E = g^E$, where g^E is the degeneracy of the electronic ground state. Alkali metal atoms, for example, have doubly degenerate ground states (corresponding to the two orientations of their electron spin), so $q^E = 2$.

Some atoms and molecules have low-lying electronically excited states. (At high enough temperatures, all atoms and molecules have thermally accessible excited states.) An example is NO, which has the configuration . . . π^1. The orbital angular momentum may take two orientations with respect to the molecular axis (corresponding to circulation clockwise or counterclockwise around the axis), and the spin angular momentum may also take two orientations with respect to the axis, giving four states in all (Fig. 16.5). The energy of the two states in which the orbital and spin momenta are parallel (giving the $^2\Pi_{3/2}$ term) is slightly greater than that of the two other states in which they are antiparallel (giving the $^2\Pi_{1/2}$ term). The separation, which arises from spin–orbit coupling (Section 9.9), is only 121 cm^{-1}. Hence, at normal temperatures, all four states are thermally accessible. If we denote the energies of the two levels as $E_{1/2} = 0$ and $E_{3/2} = \varepsilon$, the partition function is

$$q^E = \sum_{\text{energy levels}} g_j e^{-\beta\varepsilon_j} = 2 + 2e^{-\beta\varepsilon} \tag{16.22}$$

Figure 16.6 shows the variation of this function with temperature. At $T = 0$, $q^E = 2$, because only the doubly degenerate ground state is accessible. At high temperatures, $q^E \to 4$ because all four states are accessible. At 25°C, $q^E = 3.1$.

(e) The overall partition function

The partition functions for each mode of motion of a molecule are collected in the *Checklist* at the end of the chapter. The overall partition function is the product of each contribution. For a diatomic molecule with no low-lying electronically excited states and $T \gg \theta_R$

$$q = g^E\left(\frac{V}{\Lambda^3}\right)\left(\frac{T}{\sigma\theta_R}\right)\left(\frac{1}{1-e^{-\theta_V/T}}\right) \tag{16.23}$$

Example 16.4 *Calculating a thermodynamic function from spectroscopic data*

Calculate the value of $G_m^\ominus - G_m^\ominus(0)$ for H_2O(g) at 1500 K given that $\tilde{A} = 27.8778$ cm^{-1}, $\tilde{B} = 14.5092$ cm^{-1}, and $\tilde{C} = 9.2869$ cm^{-1} and the information in Example 16.3.

Method The starting point is eqn 16.9. For the standard value, we evaluate the translational partition function at $p^\ominus$ (that is, at 10^5 Pa exactly). The vibrational partition function was calculated in Example 16.3. Use the expressions in the *Checklist* for the other contributions.

Answer Because $m = 18.015m_u$, it follows that $q_m^{T\ominus}/N_A = 1.729 \times 10^8$. For the vibrational contribution we have already found that $q^V = 1.353$. From Table 16.2 we see that $\sigma = 2$, so the rotational contribution is $q^R = 486.7$. Therefore,

$$\begin{aligned} G_m^\ominus - G_m^\ominus(0) &= -(8.3145 \text{ J K}^{-1}\text{ mol}^{-1}) \times (1500 \text{ K}) \times \ln\{(1.706 \times 10^8) \times 486.7 \times 1.352\} \\ &= -317.5 \text{ kJ mol}^{-1} \end{aligned}$$

Self-test 16.4 Repeat the calculation for CO_2. The vibrational data are given in Self-test 16.3; $\tilde{B} = 0.3902\ \text{cm}^{-1}$. [$-366.9\ \text{kJ mol}^{-1}$]

Overall partition functions obtained from eqn 16.23 are approximate because they assume that the rotational levels are very close together and that the vibrational levels are harmonic. These approximations are avoided by using the energy levels identified spectroscopically and evaluating the sums explicitly.

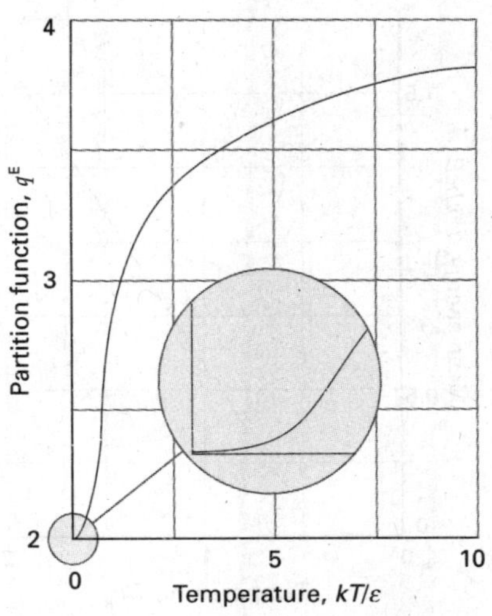

Fig. 16.6 The variation with temperature of the electronic partition function of an NO molecule. Note that the curve resembles that for a two-level system (Fig. 15.5), but rises from 2 (the degeneracy of the lower level) and approaches 4 (the total number of states) at high temperatures.

interActivity Plot the temperature dependence of the electronic partition function for several values of the energy separation ε between two doubly degenerate levels. From your plots, estimate the temperature at which the population of the excited level begins to increase sharply.

Using statistical thermodynamics

We can now calculate partition functions and, from them, any thermodynamic quantity, so gaining insight into a variety of physical, chemical, and biological processes. In this section, we indicate how to do the calculations for four important properties.

16.3 Mean energies

Key points **The mean energy of a mode of motion can be calculated from the contribution of that mode to the molecular partition function. The mean energy is the sum of contributions from: (a) translation, (b) rotation, and (c) vibration.**

It is often useful to know the mean energy, $\langle\varepsilon\rangle$, of various modes of motion. When the molecular partition function can be factorized into contributions from each mode, the mean energy of each mode M (from eqn 15.22) is

$$\langle\varepsilon^{M}\rangle = -\frac{1}{q^{M}}\left(\frac{\partial q^{M}}{\partial\beta}\right)_V \qquad M = T, R, V, \text{or } E \qquad \text{Mean energy of a mode of motion} \qquad (16.24)$$

(a) The mean translational energy

To see a pattern emerging, we consider first a one-dimensional system of length X, for which $q^{T} = X/\Lambda$, with $\Lambda = h(\beta/2\pi m)^{1/2}$. Then, if we note that Λ is a constant times $\beta^{1/2}$,

$$\langle\varepsilon^{T}\rangle = -\frac{\Lambda}{X}\left(\frac{\partial}{\partial\beta}\frac{X}{\Lambda}\right)_V = -\beta^{1/2}\frac{d}{d\beta}\left(\frac{1}{\beta^{1/2}}\right) = \frac{1}{2\beta} = \tfrac{1}{2}kT \qquad (16.25a)$$

For a molecule free to move in three dimensions, the analogous calculation leads to

$$\langle\varepsilon^{T}\rangle = \tfrac{3}{2}kT \qquad \text{Mean translational energy} \qquad (16.25b)$$

Both conclusions are in agreement with the classical equipartition theorem (see *Fundamentals F.5*) that the mean energy of each quadratic contribution to the energy is $\tfrac{1}{2}kT$. Furthermore, the fact that the mean energy is independent of the size of the container is consistent with the thermodynamic result that the internal energy of a perfect gas is independent of its volume (Section 2.11).

(b) The mean rotational energy

The mean rotational energy of a linear molecule is obtained from the partition function given in eqn 16.13:

$$q^{R} = 1 + 3e^{-2\beta hc\tilde{B}} + 5e^{-6\beta hc\tilde{B}} + \cdots$$

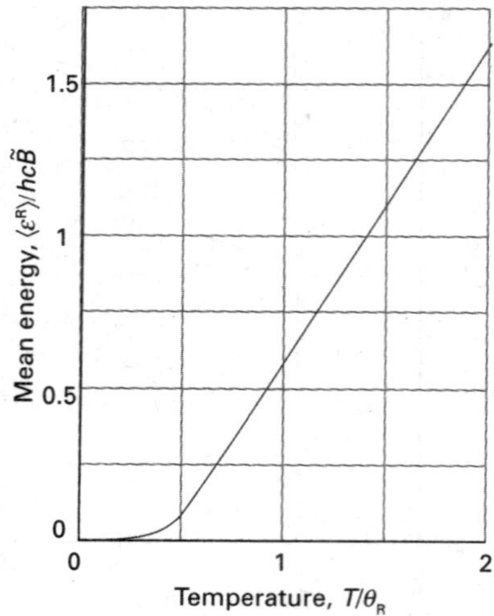

Fig. 16.7 The mean rotational energy of a nonsymmetrical linear rotor as a function of temperature. At high temperatures ($T \gg \theta_R$), the energy is linearly proportional to the temperature, in accord with the equipartition theorem.

interActivity Plot the temperature dependence of the mean rotational energy for several values of the rotational constant (for reasonable values of the rotational constant, see the *Data section*). From your plots, estimate the temperature at which the mean rotational energy begins to increase sharply.

Hence

$$\langle \varepsilon^{\mathrm{R}} \rangle = \frac{hc\tilde{B}(6e^{-2\beta hc\tilde{B}} + 30e^{-6\beta hc\tilde{B}} + \cdots)}{1 + 3e^{-2\beta hc\tilde{B}} + 5e^{-6\beta hc\tilde{B}} + \cdots} \quad \text{Mean rotational energy} \quad (16.26a)$$

This function is plotted in Fig. 16.7. At high temperatures ($T \gg \theta_R$), q^R is given by eqn 16.15, and

$$\langle \varepsilon^{\mathrm{R}} \rangle = -\frac{1}{q^{\mathrm{R}}}\frac{\mathrm{d}q^{\mathrm{R}}}{\mathrm{d}\beta} = -\sigma hc\beta\tilde{B}\frac{\mathrm{d}}{\mathrm{d}\beta}\frac{1}{\sigma hc\beta\tilde{B}} = \frac{1}{\beta} = kT \quad \text{Mean rotational energy (high temperature limit)} \quad (16.26b)$$

(q^R is independent of V, so the partial derivatives have been replaced by complete derivatives.) The high-temperature result is also in agreement with the equipartition theorem, for the classical expression for the energy of a linear rotor is $E_k = \frac{1}{2}I_\perp\omega_a^2 + \frac{1}{2}I_\perp\omega_b^2$. (There is no rotation around the line of atoms.) It follows from the equipartition theorem that the mean rotational energy is $2 \times \frac{1}{2}kT = kT$.

(c) The mean vibrational energy

The vibrational partition function in the harmonic approximation is given in eqn 16.19. Because q^V is independent of the volume, it follows that

$$\frac{\mathrm{d}q^{\mathrm{V}}}{\mathrm{d}\beta} = \frac{\mathrm{d}}{\mathrm{d}\beta}\left(\frac{1}{1-e^{-\beta hc\tilde{\nu}}}\right) = -\frac{hc\tilde{\nu}e^{-\beta hc\tilde{\nu}}}{(1-e^{-\beta hc\tilde{\nu}})^2} \quad (16.27)$$

and hence from

$$\langle \varepsilon^{\mathrm{V}} \rangle = -\frac{1}{q^{\mathrm{V}}}\frac{\mathrm{d}q^{\mathrm{V}}}{\mathrm{d}\beta} = -(1-e^{-\beta hc\tilde{\nu}})\left\{-\frac{hc\tilde{\nu}e^{-\beta hc\tilde{\nu}}}{(1-e^{-\beta hc\tilde{\nu}})^2}\right\} = \frac{hc\tilde{\nu}e^{-\beta hc\tilde{\nu}}}{1-e^{-\beta hc\tilde{\nu}}}$$

that

$$\langle \varepsilon^{\mathrm{V}} \rangle = \frac{hc\tilde{\nu}}{e^{\beta hc\tilde{\nu}} - 1} \quad \text{Mean vibrational energy} \quad (16.28)$$

The zero-point energy, $\frac{1}{2}hc\tilde{\nu}$, can be added to the right-hand side if the mean energy is to be measured from 0 rather than the lowest attainable level (the zero-point level). The variation of the mean energy with temperature is illustrated in Fig. 16.8. At high temperatures, when $T \gg \theta_V$, or $\beta hc\tilde{\nu} \ll 1$, the exponential functions can be expanded ($e^x = 1 + x + \cdots$) and all but the leading terms discarded. This approximation leads to

$$\langle \varepsilon^{\mathrm{V}} \rangle = \frac{hc\tilde{\nu}}{(1 + \beta hc\tilde{\nu} + \cdots) - 1} \approx \frac{1}{\beta} = kT \quad \text{Mean vibrational energy (high temperature limit)} \quad (16.29)$$

This result is in agreement with the value predicted by the classical equipartition theorem, because the energy of a one-dimensional oscillator is $E = \frac{1}{2}mv_x^2 + \frac{1}{2}k_f x^2$ and the mean energy of each quadratic term is $\frac{1}{2}kT$.

16.4 Heat capacities

Key points **(a) The constant-volume heat capacity can be calculated from the molecular partition function. (b) The total heat capacity of a molecular substance is the sum of the contributions of each mode.**

The constant-volume heat capacity is defined as $C_V = (\partial U/\partial T)_V$. The derivative with respect to T is converted into a derivative with respect to β by using

$$\frac{\mathrm{d}}{\mathrm{d}T}=\frac{\mathrm{d}\beta}{\mathrm{d}T}\frac{\mathrm{d}}{\mathrm{d}\beta}=-\frac{1}{kT^2}\frac{\mathrm{d}}{\mathrm{d}\beta}=-k\beta^2\frac{\mathrm{d}}{\mathrm{d}\beta} \tag{16.30}$$

It follows that

$$C_V=-k\beta^2\left(\frac{\partial U}{\partial \beta}\right)_V \tag{16.31a}$$

Because the internal energy of a perfect gas is a sum of contributions, the heat capacity is also a sum of contributions from each mode. The contribution of mode M is

$$C_V^{\mathrm{M}}=N\left(\frac{\partial\langle\varepsilon^{\mathrm{M}}\rangle}{\partial T}\right)_V=-Nk\beta^2\left(\frac{\partial\langle\varepsilon^{\mathrm{M}}\rangle}{\partial \beta}\right)_V \tag{16.31b}$$

Contribution of a mode to the constant-volume heat capacity

(a) The individual contributions

The temperature is always high enough (provided the gas is above its condensation temperature) for the mean translational energy to be $\frac{3}{2}kT$, the equipartition value. Therefore, the translational contribution to the molar constant-volume heat capacity is

$$C_{V,\mathrm{m}}^{\mathrm{T}}=N_{\mathrm{A}}\frac{\mathrm{d}(\frac{3}{2}kT)}{\mathrm{d}T}=\tfrac{3}{2}R \tag{16.32}$$

Translational contribution to C_V

Translation is the only mode of motion for a monatomic gas, so for such a gas $C_{V,\mathrm{m}}=\frac{3}{2}R=12.47\ \mathrm{J\,K^{-1}\,mol^{-1}}$. This result is very reliable: helium, for example, has this value over a range of 2000 K. We saw in Section 2.5c that $C_{p,\mathrm{m}}-C_{V,\mathrm{m}}=R$, so for a monatomic perfect gas $C_{p,\mathrm{m}}=\frac{5}{2}R$ and therefore

$$\gamma=\frac{C_p}{C_V}=\tfrac{5}{3} \tag{16.33}°$$

Heat capacity ratio for a monatomic gas

When the temperature is high enough for the rotations of the molecules to be highly excited (when $T\gg\theta_{\mathrm{R}}$), we can use the equipartition value kT for the mean rotational energy (for a linear rotor) to obtain $C_{V,\mathrm{m}}=R$. For nonlinear molecules, the mean rotational energy rises to $\frac{3}{2}kT$, so the molar rotational heat capacity rises to $\frac{3}{2}R$ when $T\gg\theta_{\mathrm{R}}$. Only the lowest rotational state is occupied when the temperature is very low, and then rotation does not contribute to the heat capacity. We can calculate the rotational heat capacity at intermediate temperatures by differentiating the equation for the mean rotational energy (eqn 16.26a). The resulting (untidy) expression, which is plotted in Fig. 16.9, shows that the contribution rises from zero (when $T=0$) to the equipartition value (when $T\gg\theta_{\mathrm{R}}$). Because the translational contribution is always present, we can expect the molar heat capacity of a gas of diatomic molecules $C_{V,\mathrm{m}}^{\mathrm{T}}+C_{V,\mathrm{m}}^{\mathrm{R}}$ to rise from $\frac{3}{2}R$ to $\frac{5}{2}R$ as the temperature is increased above θ_{R}. Problem 16.20 explores how the overall shape of the curve can be traced to the sum of thermal excitations between all the available rotational energy levels (Fig. 16.10).

Molecular vibrations contribute to the heat capacity, but only when the temperature is high enough for them to be significantly excited. The equipartition mean energy is kT for each mode, so the maximum contribution to the molar heat capacity is R. However, it is very unusual for the vibrations to be so highly excited that equipartition is valid, and it is more appropriate to use the full expression for the vibrational heat capacity, which is obtained by differentiating eqn 16.28:

$$C_{V,\mathrm{m}}^{\mathrm{V}}=Rf(T)\qquad f(T)=\left(\frac{\theta_{\mathrm{V}}}{T}\right)^2\left(\frac{\mathrm{e}^{-\theta_{\mathrm{V}}/2T}}{1-\mathrm{e}^{-\theta_{\mathrm{V}}/T}}\right)^2 \tag{16.34}$$

Vibrational contribution to C_V

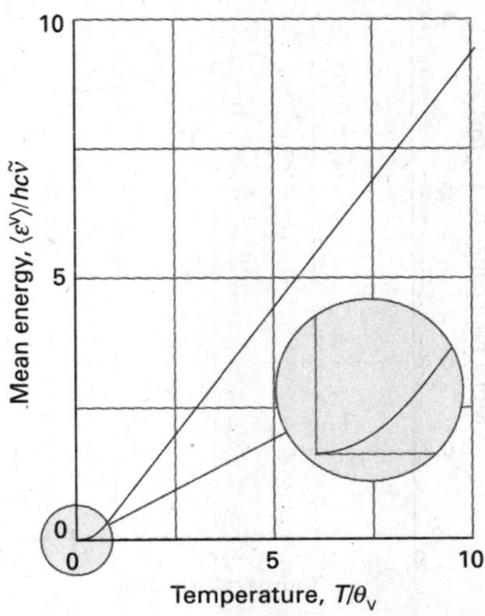

Fig. 16.8 The mean vibrational energy of a molecule in the harmonic approximation as a function of temperature. At high temperatures ($T\gg\theta_{\mathrm{V}}$), the energy is linearly proportional to the temperature, in accord with the equipartition theorem.

interActivity Plot the temperature dependence of the mean vibrational energy for several values of the vibrational wavenumber (for reasonable values of the vibrational wavenumber, see the *Data section*). From your plots, estimate the temperature at which the mean vibrational energy begins to increase sharply.

A brief comment

Equation 16.34 is essentially the same as the Einstein formula for the heat capacity of a solid (eqn 7.11) with θ_{V} the Einstein temperature, θ_{E}. The only difference is that vibrations can take place in three dimensions in a solid.

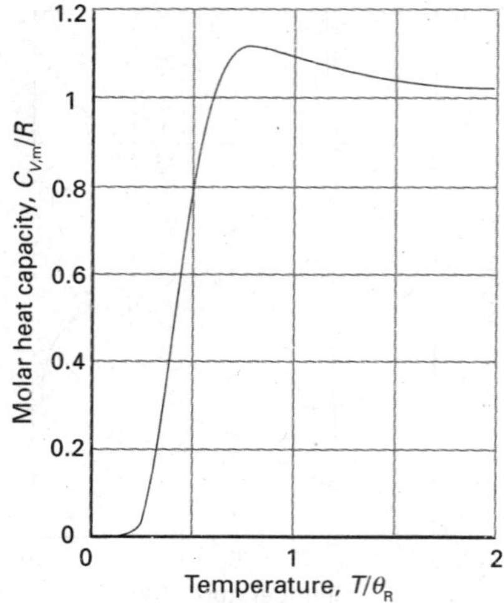

Fig. 16.9 The temperature dependence of the rotational contribution to the heat capacity of a linear molecule.

interActivity The *Living graphs* section of the text's web site has applets for the calculation of the temperature dependence of the rotational contribution to the heat capacity. Explore the effect of the rotational constant on the plot of $C^R_{V,m}$ against T.

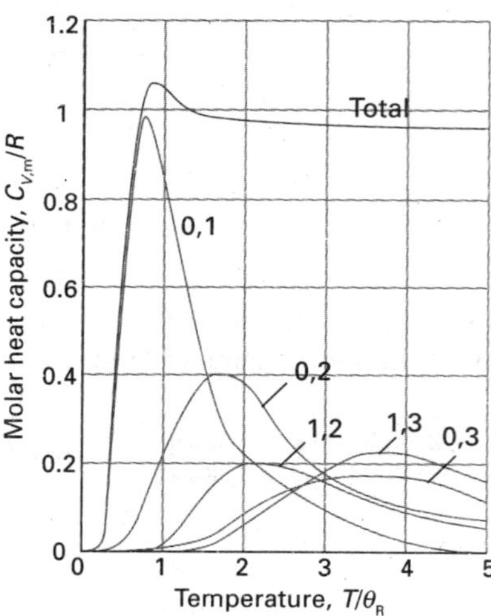

Fig. 16.10 The rotational heat capacity of a linear molecule can be regarded as the sum of contributions from a collection of two-level systems, in which the rise in temperature stimulates transitions between J levels, some of which are shown here. The calculation on which this illustration is based is sketched in Problem 16.20.

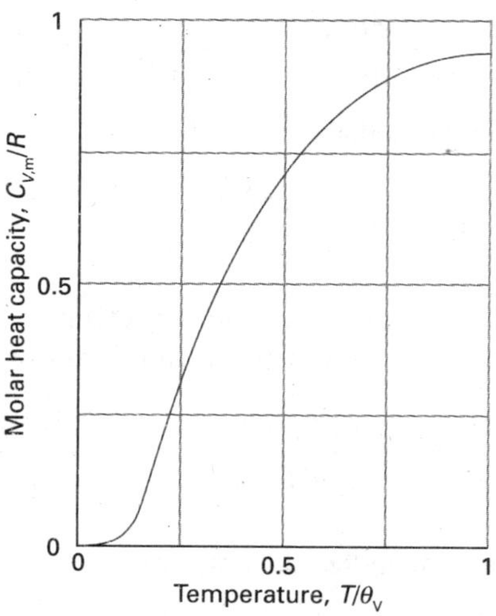

Fig. 16.11 The temperature dependence of the vibrational heat capacity of a molecule in the harmonic approximation calculated by using eqn 16.34. Note that the heat capacity is within 10 per cent of its classical value for temperatures greater than θ_V.

interActivity The *Living graphs* section of the text's web site has applets for the calculation of the temperature dependence of the vibrational contribution to the heat capacity. Explore the effect of the vibrational wavenumber on the plot of $C^V_{V,m}$ against T.

where $\theta_V = hc\tilde{\nu}/k$ is the characteristic vibrational temperature. The curve in Fig. 16.11 shows how the vibrational heat capacity depends on temperature. Note that, even when the temperature is only slightly above θ_V, the heat capacity is close to its equipartition value.

(b) The overall heat capacity

The total heat capacity of a molecular substance is the sum of each contribution (Fig. 16.12). When equipartition is valid (when the temperature is well above the characteristic temperature of the mode, $T \gg \theta_M$) we can estimate the heat capacity by counting the numbers of modes that are active. In gases, all three translational modes are always active and contribute $\frac{3}{2}R$ to the molar heat capacity. If we denote the number of active rotational modes by ν^*_R (so for most molecules at normal temperatures $\nu^*_R = 2$ for linear molecules, and 3 for nonlinear molecules), then the rotational contribution is $\frac{1}{2}\nu^*_R R$. If the temperature is high enough for ν^*_V vibrational modes to be active, the vibrational contribution to the molar heat capacity is $\nu^*_V R$. In most cases $\nu^*_V \approx 0$. It follows that the total molar heat capacity is

$$C_{V,m} = \tfrac{1}{2}(3 + \nu^*_R + 2\nu^*_V)R \qquad \text{Total heat capacity (at high temperatures)} \qquad (16.35)$$

• A brief illustration

The characteristic temperatures (in round numbers) of the vibrations of H_2O are 5300 K, 2300 K, and 5400 K; the vibrations are therefore not excited at 373 K. The three rotational modes of H_2O have characteristic temperatures 40 K, 21 K, and 13 K, so they are fully excited, like the three translational modes. The translational contribution is $\frac{3}{2}R$ = 12.5 J K^{-1} mol^{-1}. Fully excited rotations contribute a further 12.5 J K^{-1} mol^{-1}. Therefore, a value close to 25 J K^{-1} mol^{-1} is predicted. The experimental value is 26.1 J K^{-1} mol^{-1}. The discrepancy is probably due to deviations from perfect gas behaviour. •

Self-test 16.5 Estimate the molar constant-volume heat capacity of gaseous I_2 at 25°C ($\tilde{B}$ = 0.037 cm^{-1}; $\tilde{\nu}$ = 214.5 cm^{-1}). [29 J K^{-1} mol^{-1}]

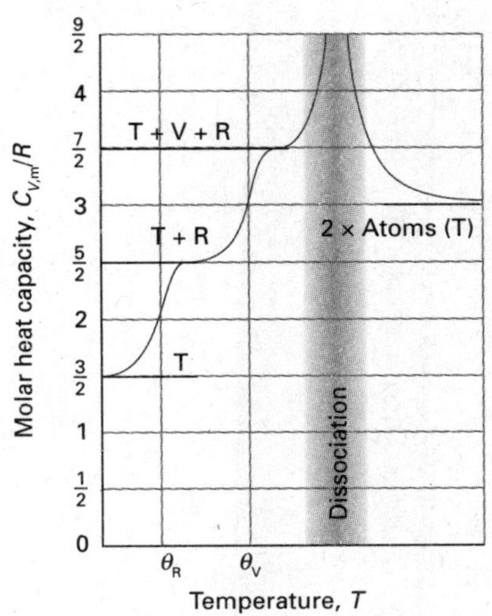

Fig. 16.12 The general features of the temperature dependence of the heat capacity of diatomic molecules are as shown here. Each mode becomes active when its characteristic temperature is exceeded. The heat capacity becomes very large when the molecule dissociates because the energy is used to cause dissociation and not to raise the temperature. Then it falls back to the translation-only value of the atoms.

16.5 Equations of state

Key point **The canonical partition function of a gas factorizes into a part arising from the kinetic energy and a configuration integral, Z, which depends on the intermolecular interactions.**

The relation between p and Q in eqn 16.3 is a very important route to the equations of state of real gases in terms of intermolecular forces, for the latter can be built into Q. We have already seen (Example 16.1) that the partition function for a gas of independent particles leads to the perfect gas equation of state, $pV = nRT$. Real gases differ from perfect gases in their equations of state and we saw in Section 1.3 that their equations of state may be written

$$\frac{pV_m}{RT} = 1 + \frac{B}{V_m} + \frac{C}{V_m^2} + \cdots \qquad (16.36)$$

where B is the second virial coefficient and C is the third virial coefficient.

The total kinetic energy of a gas is the sum of the kinetic energies of the individual molecules. Therefore, even in a real gas the canonical partition function factorizes into a part arising from the kinetic energy, which is the same as for the perfect gas, and a factor called the **configuration integral**, Z, which depends on the intermolecular potentials. We therefore write

$$Q = \frac{Z}{\Lambda^{3N}} \qquad \text{Q in terms of the configuration integral} \qquad (16.37)$$

By comparing this equation with eqn 15.38 ($Q = q^N/N!$, with $q = V/\Lambda^3$), we see that for a perfect gas of atoms (with no contributions from rotational or vibrational modes)

$$Z = \frac{V^N}{N!} \qquad \text{Configuration integral (perfect monatomic gas)} \qquad (16.38)°$$

For a real monatomic gas (for which the intermolecular interactions are isotropic), Z is related to the total potential energy E_p of interaction of all the particles by

$$Z = \frac{1}{N!}\int e^{-\beta E_p} d\tau_1 d\tau_2 \cdots d\tau_N \qquad \text{Configuration integral (real monatomic gas)} \qquad (16.39)$$

where $d\tau_i$ is the volume element for atom i. The physical origin of this term is that the probability of occurrence of each arrangement of molecules possible in the sample is given by a Boltzmann distribution in which the exponent is given by the potential energy corresponding to that arrangement.

• **A brief illustration**

When the molecules do not interact with one another, $E_p = 0$ and hence $e^{-\beta E_p} = 1$. Then

$$Z = \frac{1}{N!}\int d\tau_1 d\tau_2 \cdots d\tau_N = \frac{V^N}{N!}$$

because $\int d\tau = V$, where V is the volume of the container. This result coincides with eqn 16.38. •

When interactions between pairs of particles are significant and we can ignore three-body interactions, etc., the configuration integral simplifies to

$$Z = \frac{V^{N-2}}{N!}\int e^{-\beta E_p} d\tau_1 d\tau_2 \qquad (16.40)$$

The second virial coefficient then turns out to be

$$B = -\frac{N_A}{2V}\int f d\tau_1 d\tau_2 \qquad (16.41)$$

Second virial coefficient

The quantity f is the **Mayer f-function**: it goes to zero when the two particles are so far apart that $E_p = 0$. When the intermolecular interaction depends only on the separation r of the particles and not on their relative orientation or their absolute position in space, as in the interaction of closed-shell atoms in a uniform sample, the volume element simplifies to $4\pi r^2 dr$ (because the integrals over the angular variables in $d\tau = r^2\, dr \sin\theta\, d\theta d\phi$ give a factor of 4π) and eqn 16.41 becomes

$$B = -2\pi N_A \int_0^\infty f r^2 dr \qquad f = e^{-\beta E_p} - 1 \qquad (16.42)$$

The integral can be evaluated (usually numerically) by substituting an expression for the intermolecular potential energy.

Intermolecular potential energies are discussed in more detail in Chapter 17, where several expressions are developed for them. At this stage, we can illustrate how eqn 16.42 is used by considering the **hard-sphere potential**, which is infinite when the separation of the two molecules, r, is less than or equal to a certain value σ, and is zero for greater separations. Then

$$e^{-\beta E_p} = 0 \qquad f = -1 \quad \text{when} \quad r \le \sigma \quad (\text{and } E_p = \infty) \qquad (16.43a)$$

$$e^{-\beta E_p} = 1 \qquad f = 0 \quad \text{when} \quad r > \sigma \quad (\text{and } E_p = 0) \qquad (16.43b)$$

It follows from eqn 16.42 that the second virial coefficient is

$$B = 2\pi N_A \int_0^\sigma r^2 dr = \tfrac{2}{3}\pi N_A \sigma^3 \qquad (16.44)$$

This calculation of B raises the question as to whether a potential can be found that, when the virial coefficients are evaluated, gives the van der Waals equation of state. Such a potential can be found for weak attractive interactions ($a \ll RT$): it consists of a hard-sphere repulsive core and a long-range, shallow attractive region (see Problem 16.22). A further point is that, once a second virial coefficient has been calculated for a given intermolecular potential, it is possible to calculate other thermodynamic properties that depend on the form of the potential. For example, it is possible to calculate the isothermal Joule–Thomson coefficient, μ_T (Section 2.12a), from the thermodynamic relation

$$\lim_{p\to 0} \mu_T = B - T\frac{dB}{dT} \qquad (16.45)$$

(see Problem 16.17) and from the result calculate the Joule–Thomson coefficient itself by using eqn 2.53.

16.6 Molecular interactions in liquids

Key points **(a) The radial distribution function, $g(r)$, is the probability that a molecule will be found in the range dr at a distance r from another molecule. (b) The radial distribution function may be calculated with Monte Carlo and molecular dynamics techniques. (c) The internal energy and pressure of a fluid may be expressed in terms of the radial distribution function.**

The starting point for the discussion of solids is the well ordered structure of a perfect crystal, which will be discussed in Chapter 19. The starting point for the discussion of gases is the completely disordered distribution of the molecules of a perfect gas, as we saw in Chapter 1. Liquids lie between these two extremes. We shall see that the structural and thermodynamic properties of liquids depend on the nature of intermolecular interactions and that an equation of state can be built in a similar way to that just demonstrated for real gases.

(a) The radial distribution function

The average relative locations of the particles of a liquid are expressed in terms of the **radial distribution function,** $g(r)$. This function is defined so that $g(r)r^2dr$ is the probability that a molecule will be found in the range dr at a distance r from another molecule. In a perfect crystal, $g(r)$ is a periodic array of sharp spikes, representing the certainty (in the absence of defects and thermal motion) that molecules (or ions) lie at definite locations. This regularity continues out to the edges of the crystal, so we say that crystals have **long-range order**. When the crystal melts, the long-range order is lost and, wherever we look at long distances from a given molecule, there is equal probability of finding a second molecule. Close to the first molecule, though, the nearest neighbours might still adopt approximately their original relative positions and, even if they are displaced by newcomers, the new particles might adopt their vacated positions. It is still possible to detect a sphere of nearest neighbours at a distance r_1, and perhaps beyond them a sphere of next-nearest neighbours at r_2. The existence of this **short-range order** means that the radial distribution function can be expected to oscillate at short distances, with a peak at r_1, a smaller peak at r_2, and perhaps some more structure beyond that.

The radial distribution function of the oxygen atoms in liquid water is shown in Fig. 16.13. Closer analysis shows that any given H_2O molecule is surrounded by other molecules at the corners of a tetrahedron. The form of $g(r)$ at 100°C shows that the intermolecular interactions (in this case, principally by hydrogen bonds) are strong enough to affect the local structure right up to the boiling point. Raman spectra indicate that in liquid water most molecules participate in either three or four hydrogen bonds. Infrared spectra show that about 90 per cent of hydrogen bonds are intact at the melting point of ice, falling to about 20 per cent at the boiling point.

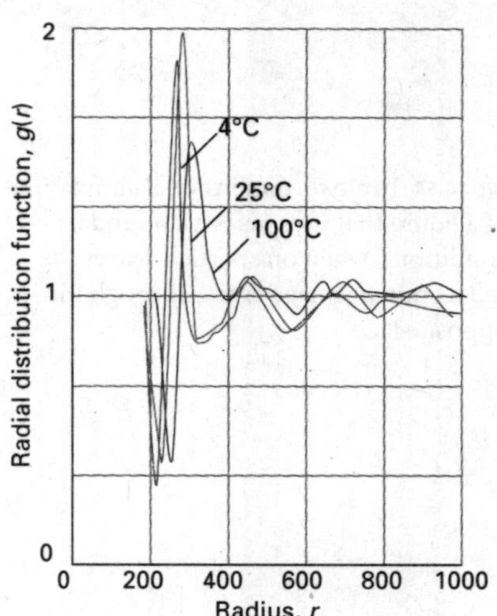

Fig. 16.13 The radial distribution function of the oxygen atoms in liquid water at three temperatures. Note the expansion as the temperature is raised. (Based on A.H. Narten, M.D. Danford, and H.A. Levy, *Discuss. Faraday. Soc.* **43**, 97 (1967).)

The formal expression for the radial distribution function for molecules 1 and 2 in a fluid consisting of N particles is the somewhat fearsome equation

$$g(r_{12}) = \frac{\displaystyle\iint\cdots\int e^{-\beta V_N} d\tau_3 d\tau_4 \cdots d\tau_N}{\displaystyle N^2 \iint\cdots\int e^{-\beta V_N} d\tau_1 d\tau_2 \cdots d\tau_N} \qquad \text{Radial distribution function} \qquad (16.46)$$

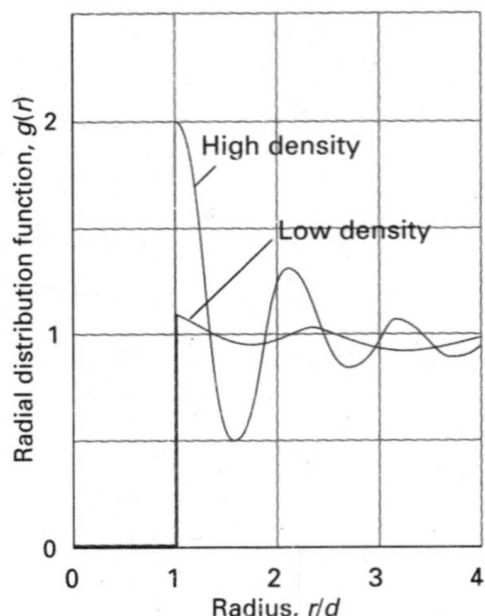

Fig. 16.14 The radial distribution function for a simulation of a liquid using impenetrable hard spheres (ball bearings) of diameter d.

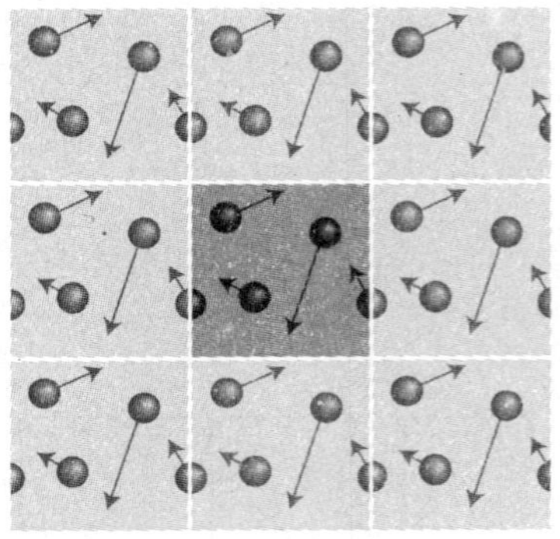

Fig. 16.15 In a two-dimensional simulation of a liquid that uses periodic boundary conditions, when one particle leaves the cell its mirror image enters through the opposite face.

where $\beta = 1/kT$ and V_N is the N-particle potential energy. Although fearsome, this expression is nothing more than the Boltzmann distribution for the relative locations of two molecules in a field provided by all the other molecules in the system.

(b) The calculation of *g*(*r*)

Because the radial distribution function can be calculated by making assumptions about the intermolecular interactions, it can be used to test theories of liquid structure. However, even a fluid of hard spheres without attractive interactions (a collection of ball-bearings in a container) gives a function that oscillates near the origin (Fig. 16.14), and one of the factors influencing, and sometimes dominating, the structure of a liquid is the geometrical problem of stacking together reasonably hard spheres. Indeed, the radial distribution function of a liquid of hard spheres shows more pronounced oscillations at a given temperature than that of any other type of liquid. The attractive part of the potential modifies this basic structure, but sometimes only quite weakly. One of the reasons behind the difficulty of describing liquids theoretically is the similar importance of both the attractive and repulsive (hard core) components of the potential.

There are several ways of building the intermolecular potential into the calculation of $g(r)$. Numerical methods take a box of about 10^3 particles (the number increases as computers grow more powerful), and the rest of the liquid is simulated by surrounding the box with replications of the original box (Fig. 16.15). Then, whenever a particle leaves the box through one of its faces, its image arrives through the opposite face. When calculating the interactions of a molecule in a box, it interacts with all the molecules in the box and all the periodic replications of those molecules and itself in the other boxes.

In the **Monte Carlo method**, the particles in the box are moved through small but otherwise random distances, and the change in total potential energy of the N particles in the box, ΔV_N, is calculated using one of the intermolecular potentials discussed in Section 17.5. Whether or not this new configuration is accepted is then judged from the following rules:

1 If the potential energy is not greater than before the change, then the configuration is accepted.

If the potential energy is greater than before the change, then it is necessary to check if the new configuration is reasonable and can exist in equilibrium with configurations of lower potential energy at a given temperature. To make progress, we use the result that, at equilibrium, the ratio of populations of two states with energy separation ΔV_N is $e^{-\Delta V_N/kT}$. Because we are testing the viability of a configuration with a higher potential energy than the previous configuration in the calculation, $\Delta V_N > 0$ and the exponential factor varies between 0 and 1. In the Monte Carlo method, the second rule, therefore, is:

2 The exponential factor is compared with a random number between 0 and 1; if the factor is larger than the random number, then the configuration is accepted; if the factor is not larger, the configuration is rejected.

The configurations generated with Monte Carlo calculations can be used to construct $g(r)$ simply by counting the number of pairs of particles with a separation r and averaging the result over the whole collection of configurations.

In the **molecular dynamics** approach, the history of an initial arrangement is followed by calculating the trajectories of all the particles under the influence of the intermolecular potentials and the forces they exert. The calculation gives a series of snapshots of the liquid, and $g(r)$ can be calculated as before. The temperature of the system is inferred by computing the mean kinetic energy of the particles and using the equipartition result that

$$\langle \tfrac{1}{2}mv_q^2 \rangle = \tfrac{1}{2}kT \qquad (16.47)$$

for each coordinate q.

(c) The thermodynamic properties of liquids

Once $g(r)$ is known it can be used to calculate the thermodynamic properties of liquids. For example, the contribution of the pairwise additive intermolecular potential, V_2, to the internal energy is given by the integral

$$U_{\text{interaction}}(T) = \frac{2\pi N^2}{V}\int_0^\infty g(r)V_2 r^2 \mathrm{d}r \qquad (16.48)$$

Contribution of pairwise interactions to the internal energy

That is, $U_{\text{interaction}}$ is essentially the average two-particle potential energy weighted by $g(r)r^2\mathrm{d}r$, which is the probability that the pair of particles have a separation between r and $r+\mathrm{d}r$. Likewise, the contribution that pairwise interactions make to the pressure is

$$\frac{pV}{nRT} = 1 - \frac{2\pi N}{3kTV}\int_0^\infty g(r)v_2 r^2 \mathrm{d}r \qquad v_2 = r\frac{\mathrm{d}V_2}{\mathrm{d}r} \qquad (16.49\text{a})$$

The quantity v_2 is called the **virial** (hence the term 'virial equation of state'). To understand the physical content of this expression, we rewrite it as

$$p = \frac{nRT}{V} - \frac{2\pi}{3}\left(\frac{N}{V}\right)^2\int_0^\infty g(r)v_2 r^2 \mathrm{d}r \qquad (16.49\text{b})$$

Pressure in terms of $g(r)$

The first term on the right is the **kinetic pressure**, the contribution to the pressure from the impact of the molecules in free flight. The second term is essentially the internal pressure, $\pi_T = (\partial U/\partial V)_T$ (Section 2.11), representing the contribution to the pressure from the intermolecular forces. To see the connection, we should recognize $-\mathrm{d}V_2/\mathrm{d}r$ (in v_2) as the force required to move two molecules apart, and therefore $-r(\mathrm{d}V_2/\mathrm{d}r)$ as the work required to separate the molecules through a distance r. The second term is therefore the average of this work over the range of pairwise separations in the liquid as represented by the probability of finding two molecules at separations between r and $r+\mathrm{d}r$, which is $g(r)r^2\mathrm{d}r$. In brief, the integral, when multiplied by the square of the number density, is the change in internal energy of the system as it expands, and therefore is equal to the internal pressure.

16.7 Residual entropies

Key point **The residual entropy is a nonzero entropy at $T=0$ arising from molecular disorder.**

Entropies may be calculated from spectroscopic data; they may also be measured experimentally (Section 3.3d). In many cases there is good agreement, but in some the experimental entropy is less than the calculated value. One possibility is that the experimental determination failed to take a phase transition into account and a contribution of the form $\Delta_{\text{trs}}H/T_{\text{trs}}$ was incorrectly omitted from the sum. Another possibility is that some disorder is present in the solid even at $T=0$. The entropy at $T=0$ is then greater than zero and is called the **residual entropy**.

The origin and magnitude of the residual entropy can be explained by considering a crystal composed of AB molecules, where A and B are similar atoms (such as CO, with its very small electric dipole moment). There may be so little energy difference between . . . AB AB AB AB . . . , . . . AB BA BA AB . . . , and other arrangements that the molecules adopt the orientations AB and BA at random in the solid. We can readily calculate the entropy arising from residual disorder by using the Boltzmann formula $S = k \ln \mathcal{W}$. To do so, we suppose that two orientations are equally probable, and that the sample consists of N molecules. Because the same energy can be achieved in 2^N different ways (because each molecule can take either of two orientations), the total number of ways of achieving the same energy is $\mathcal{W} = 2^N$. It follows that

$$S = k \ln 2^N = Nk \ln 2 = nR \ln 2 \qquad (16.50a)$$

We can therefore expect a residual molar entropy of $R \ln 2 = 5.8\ \mathrm{J\,K^{-1}\,mol^{-1}}$ for solids composed of molecules that can adopt either of two orientations at $T = 0$. If s orientations are possible, the residual molar entropy will be

$$S_m(0) = R \ln s \qquad \text{Residual entropy} \qquad (16.50b)$$

An $FClO_3$ molecule, for example, can adopt four orientations with about the same energy (with the F atom at any of the four corners of a tetrahedron), and the calculated residual molar entropy of $R \ln 4 = 11.5\ \mathrm{J\,K^{-1}\,mol^{-1}}$ is in good agreement with the experimental value ($10.1\ \mathrm{J\,K^{-1}\,mol^{-1}}$). For CO, the measured residual entropy is $5\ \mathrm{J\,K^{-1}\,mol^{-1}}$, which is close to $R \ln 2$, the value expected for a random structure of the form . . . CO CO OC CO OC OC . . . .

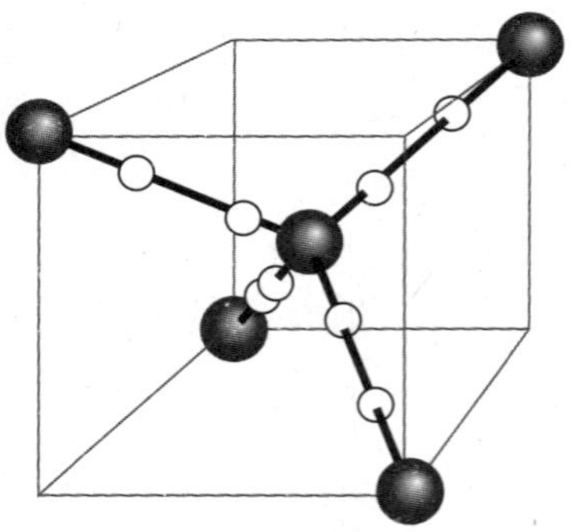

Fig. 16.16 The possible locations of H atoms around a central O atom in an ice crystal are shown by the white spheres. Only one of the locations on each bond may be occupied by an atom, and two H atoms must be close to the O atom and two H atoms must be distant from it.

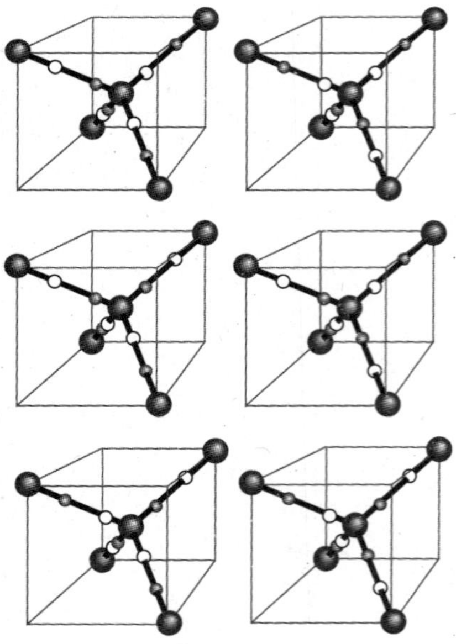

Fig. 16.17 The six possible arrangements of H atoms in the locations identified in Fig. 16.16. Occupied locations are denoted by grey spheres and unoccupied locations by white spheres.

- **A brief illustration**

Consider a sample of ice with N H_2O molecules. Each O atom is surrounded tetrahedrally by four H atoms, two of which are attached by short σ bonds, the other two being attached by long hydrogen bonds (Fig. 16.16). It follows that each of the $2N$ H atoms can be in one of two positions (either close to or far from an O atom as shown in Fig. 16.17), resulting in 2^{2N} possible arrangements. However, not all these arrangements are acceptable. Indeed, of the $2^4 = 16$ ways of arranging four H atoms around one O atom, only 6 have two short and two long OH distances and hence are acceptable. Therefore, the number of permitted arrangements is

$$\mathcal{W} = 2^{2N}\left(\tfrac{6}{16}\right)^N = \left(\tfrac{3}{2}\right)^N$$

It then follows that the residual molar entropy is

$$S_m(0) \approx k \ln\left(\tfrac{3}{2}\right)^{N_A} = N_A k \ln\left(\tfrac{3}{2}\right) = R \ln\left(\tfrac{3}{2}\right) = 3.4\ \mathrm{J\,K^{-1}\,mol^{-1}}$$

which is in good agreement with the experimental value of $3.4\ \mathrm{J\,K^{-1}\,mol^{-1}}$. The model, however, is not exact because it ignores the possibility that next-nearest neighbours and those beyond can influence the local arrangement of bonds. •

16.8 Equilibrium constants

Key points **(a) The equilibrium constant can be written in terms of the partition function. (b) The equilibrium constant for dissociation of a diatomic molecule in the gas phase may be calculated from spectroscopic data. (c) The physical basis of equilibrium can be understood by using the principles of statistical thermodynamics.**

The Gibbs energy of a gas of independent molecules is given by eqn 16.9 in terms of the molar partition function, $q_m = q/n$. The equilibrium constant K of a reaction is related to the standard Gibbs energy of reaction by $\Delta_r G^\ominus = -RT \ln K$. To calculate the equilibrium constant, we need to combine these two equations. We shall consider gas phase reactions in which the equilibrium constant is expressed in terms of the partial pressures of the reactants and products.

(a) The relation between K and the partition function

To find an expression for the standard reaction Gibbs energy we need expressions for the standard molar Gibbs energies, $G^{\ominus}/n$, of each species. For these expressions, we need the value of the molar partition function when $p = p^{\ominus}$ (where $p^{\ominus} = 1$ bar): we denote this **standard molar partition function** $q_m^{\ominus}$. Because only the translational component depends on the pressure, we can find $q_m^{\ominus}$ by evaluating the partition function with V replaced by $V_m^{\ominus}$, where $V_m^{\ominus} = RT/p^{\ominus}$. For a species J it follows that

$$G_m^{\ominus}(\mathrm{J}) = G_m^{\ominus}(\mathrm{J},0) - RT\ln\frac{q_{\mathrm{J,m}}^{\ominus}}{N_A} \qquad (16.51)^{\circ}$$

where $q_{\mathrm{J,m}}^{\ominus}$ is the standard molar partition function of J. By combining expressions like this one (as shown in the following *Justification*), the equilibrium constant for the reaction $a\mathrm{A} + b\mathrm{B} \rightarrow c\mathrm{C} + d\mathrm{D}$ is given by the expression

$$K = \frac{(q_{\mathrm{C,m}}^{\ominus}/N_A)^c(q_{\mathrm{D,m}}^{\ominus}/N_A)^d}{(q_{\mathrm{A,m}}^{\ominus}/N_A)^a(q_{\mathrm{B,m}}^{\ominus}/N_A)^b}\mathrm{e}^{-\Delta_r E_0/RT} \qquad (16.52a)$$

where $\Delta_r E_0$ is the difference in molar energies of the ground states of the products and reactants (this term is defined more precisely in the *Justification*), and is calculated from the bond dissociation energies of the species (Fig. 16.18). In terms of the stoichiometric numbers introduced in Section 2.8a, we would write

$$K = \left\{\prod_J \left(\frac{q_{\mathrm{J,m}}^{\ominus}}{N_A}\right)^{\nu_J}\right\}\mathrm{e}^{-\Delta_r E_0/RT} \qquad (16.52b)$$

Equilibrium constant in terms of partition functions

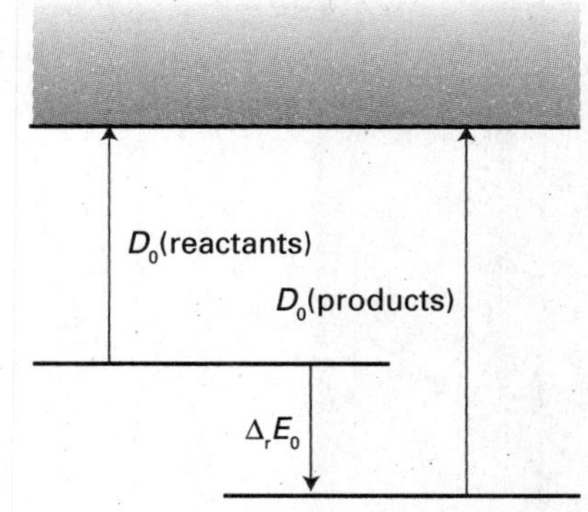

Fig. 16.18 The definition of $\Delta_r E_0$ for the calculation of equilibrium constants.

Justification 16.3 *The equilibrium constant in terms of the partition function 1*

The standard molar reaction Gibbs energy for the reaction is

$$\begin{aligned}\Delta_r G^{\ominus} &= cG_m^{\ominus}(\mathrm{C}) + dG_m^{\ominus}(\mathrm{D}) - aG_m^{\ominus}(\mathrm{A}) - bG_m^{\ominus}(\mathrm{B})\\ &= cG_m^{\ominus}(\mathrm{C},0) + dG_m^{\ominus}(\mathrm{D},0) - aG_m^{\ominus}(\mathrm{A},0) - bG_m^{\ominus}(\mathrm{B},0)\\ &\quad - RT\left\{c\ln\frac{q_{\mathrm{C,m}}^{\ominus}}{N_A} + d\ln\frac{q_{\mathrm{D,m}}^{\ominus}}{N_A} - a\ln\frac{q_{\mathrm{A,m}}^{\ominus}}{N_A} - b\ln\frac{q_{\mathrm{B,m}}^{\ominus}}{N_A}\right\}\end{aligned}$$

Because $G(0) = U(0)$, the first term on the right is

$$\Delta_r E_0 = cU_m^{\ominus}(\mathrm{C},0) + dU_m^{\ominus}(\mathrm{D},0) - aU_m^{\ominus}(\mathrm{A},0) - bU_m^{\ominus}(\mathrm{B},0) \qquad (16.53)$$

the reaction internal energy at $T = 0$ (a molar quantity).

Now we can write

$$\begin{aligned}\Delta_r G^{\ominus} &= \Delta_r E_0 - RT\left\{\ln\left(\frac{q_{\mathrm{C,m}}^{\ominus}}{N_A}\right)^c + \ln\left(\frac{q_{\mathrm{D,m}}^{\ominus}}{N_A}\right)^d - \ln\left(\frac{q_{\mathrm{A,m}}^{\ominus}}{N_A}\right)^a - \ln\left(\frac{q_{\mathrm{B,m}}^{\ominus}}{N_A}\right)^b\right\}\\ &= \Delta_r E_0 - RT\ln\frac{(q_{\mathrm{C,m}}^{\ominus}/N_A)^c(q_{\mathrm{D,m}}^{\ominus}/N_A)^d}{(q_{\mathrm{A,m}}^{\ominus}/N_A)^a(q_{\mathrm{B,m}}^{\ominus}/N_A)^b}\\ &= -RT\left\{-\frac{\Delta_r E_0}{RT} + \ln\frac{(q_{\mathrm{C,m}}^{\ominus}/N_A)^c(q_{\mathrm{D,m}}^{\ominus}/N_A)^d}{(q_{\mathrm{A,m}}^{\ominus}/N_A)^a(q_{\mathrm{B,m}}^{\ominus}/N_A)^b}\right\}\end{aligned}$$

At this stage we can pick out an expression for K by comparing this equation with $\Delta_r G^{\ominus} = -RT\ln K$, which gives

$$\ln K = -\frac{\Delta_r E_0}{RT} + \ln\frac{(q_{\mathrm{C,m}}^{\ominus}/N_A)^c(q_{\mathrm{D,m}}^{\ominus}/N_A)^d}{(q_{\mathrm{A,m}}^{\ominus}/N_A)^a(q_{\mathrm{B,m}}^{\ominus}/N_A)^b}$$

This expression is easily rearranged into eqn 16.52a by forming the exponential of both sides.

(b) A dissociation equilibrium

We shall illustrate the application of eqn 16.52 to an equilibrium in which a diatomic molecule X_2 dissociates into its atoms:

$$X_2(g) \rightleftharpoons 2X(g) \qquad K = \frac{p_X^2}{p_{X_2}p^{\ominus}}$$

According to eqn 16.52 (with $a=1$, $b=0$, $c=2$, and $d=0$):

$$K = \frac{(q^{\ominus}_{X,m}/N_A)^2}{q^{\ominus}_{X_2,m}/N_A} e^{-\Delta_r E_0/RT} = \frac{(q^{\ominus}_{X,m})^2}{q^{\ominus}_{X_2,m}N_A} e^{-\Delta_r E_0/RT} \tag{16.54a}$$

with

$$\Delta_r E_0 = 2U^{\ominus}_m(X,0) - U^{\ominus}_m(X_2,0) = D_0(X\text{–}X) \tag{16.54b}$$

where $D_0(X\text{–}X)$ is the dissociation energy of the X–X bond. The standard molar partition functions of the atoms X are

$$q^{\ominus}_{X,m} = g_X\left(\frac{V^{\ominus}_m}{\Lambda_X^3}\right) = \frac{RTg_X}{p^{\ominus}\Lambda_X^3}$$

where g_X is the degeneracy of the electronic ground state of X and we have used $V^{\ominus}_m = RT/p^{\ominus}$. The diatomic molecule X_2 also has rotational and vibrational degrees of freedom, so its standard molar partition function is

$$q^{\ominus}_{X_2,m} = g_{X_2}\left(\frac{V^{\ominus}_m}{\Lambda_{X_2}^3}\right)q^R_{X_2}q^V_{X_2} = \frac{RTg_{X_2}q^R_{X_2}q^V_{X_2}}{p^{\ominus}\Lambda_{X_2}^3}$$

where g_{X_2} is the degeneracy of the electronic ground state of X_2. It follows from eqn 16.52 that the equilibrium constant is

$$K = \frac{kTg_X^2\Lambda_{X_2}^3}{p^{\ominus}g_{X_2}q^R_{X_2}q^V_{X_2}\Lambda_X^6} e^{-D_0/RT} \tag{16.55}$$

where we have used $R/N_A = k$. All the quantities in this expression can be calculated from spectroscopic data. The Λs are defined in the *Checklist* and depend on the masses of the species and the temperature; the expressions for the rotational and vibrational partition functions are also available in the *Checklist* and depend on the rotational constant and vibrational wavenumber of the molecule.

• A brief illustration

To evaluate the equilibrium constant for the dissociation $Na_2(g) \rightleftharpoons 2\,Na(g)$ at 1000 K we use the following data: $\tilde{B} = 0.1547\ cm^{-1}$, $\tilde{\nu} = 159.2\ cm^{-1}$, $D_0 = 70.4\ kJ\ mol^{-1}$. Then, noting that the Na atoms have doublet ground terms, the partition functions and other quantities required are as follows:

$\Lambda(Na_2) = 8.14$ pm $\qquad \Lambda(Na) = 11.5$ pm
$q^R(Na_2) = 2246 \qquad q^V(Na_2) = 4.885$
$g(Na) = 2 \qquad g(Na_2) = 1$

Then, from eqn 16.55,

$$K = \frac{(1.38\times10^{-23}\ J\ K^{-1})\times(1000\ K)\times4\times(8.14\times10^{-12}\ m)^3}{(10^5\ Pa)\times2246\times4.885\times(1.15\times10^{-11}\ m)^6}\times e^{-8.47} = 2.46$$

where we have used $1\ J = 1\ kg\ m^2\ s^{-2}$ and $1\ Pa = 1\ kg\ m^{-1}\ s^{-1}$. •

(c) Contributions to the equilibrium constant

We are now in a position to appreciate the physical basis of equilibrium constants. To see what is involved, consider a simple R $\rightleftharpoons$ P gas-phase equilibrium (R for reactants, P for products).

Figure 16.19 shows two sets of energy levels: one set of states belongs to R, and the other belongs to P. The populations of the states are given by the Boltzmann distribution, and are independent of whether any given state happens to belong to R or to P. We can therefore imagine a single Boltzmann distribution spreading, without distinction, over the two sets of states. If the spacings of R and P are similar (as in Fig. 16.19), and P lies above R, the diagram indicates that R will dominate in the equilibrium mixture. However, if P has a high density of states (a large number of states in a given energy range, as in Fig. 16.20), then, even though its zero-point energy lies above that of R, the species P might still dominate at equilibrium.

It is quite easy to show (see the following *Justification*) that the ratio of numbers of R and P molecules at equilibrium is given by

$$\frac{N_P}{N_R} = \frac{q_P}{q_R} e^{-\Delta_r E_0/RT} \qquad (16.56a)$$

and therefore that the equilibrium constant for the reaction is

$$K = \frac{q_P}{q_R} e^{-\Delta_r E_0/RT} \qquad (16.56b)$$

just as would be obtained from eqn 16.52.

A brief comment

For an R $\rightleftharpoons$ P equilibrium, the V factors in the partition functions cancel, so the appearance of q in place of $q^\ominus$ has no effect. In the case of a more general reaction, the conversion from q to $q^\ominus$ comes about at the stage of converting the pressures that occur in K to numbers of molecules.

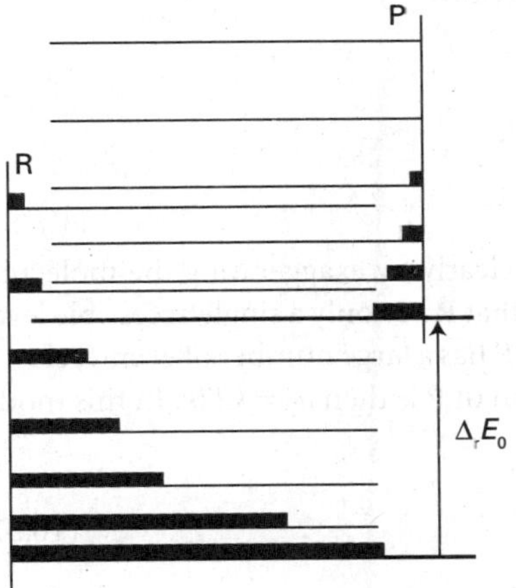

Fig. 16.19 The array of R(eactants) and P(roducts) energy levels. At equilibrium all are accessible (to differing extents, depending on the temperature), and the equilibrium composition of the system reflects the overall Boltzmann distribution of populations. As $\Delta_r E_0$ increases, R becomes dominant.

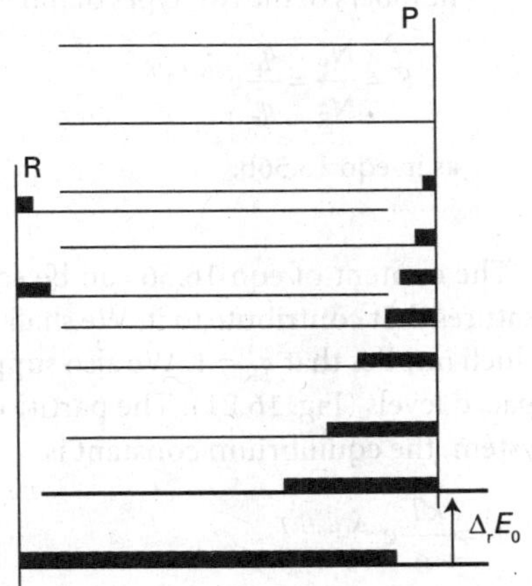

Fig. 16.20 It is important to take into account the densities of states of the molecules. Even though P might lie well above R in energy (that is, $\Delta_r E_0$ is large and positive), P might have so many states that its total population dominates in the mixture. In classical thermodynamic terms, we have to take entropies into account as well as enthalpies when considering equilibria.

Justification 16.4 *The equilibrium constant in terms of the partition function 2*

The population in a state i of the composite (R,P) system is

$$n_i = \frac{N\mathrm{e}^{-\beta\varepsilon_i}}{q}$$

where N is the total number of molecules. The total number of R molecules is the sum of these populations taken over the states belonging to R; these states we label r with energies ε_r. The total number of P molecules is the sum over the states belonging to P; these states we label p with energies ε'_p (the prime is explained in a moment):

$$N_\mathrm{R} = \sum_\mathrm{r} n_\mathrm{r} = \frac{N}{q}\sum_\mathrm{r} \mathrm{e}^{-\beta\varepsilon_\mathrm{r}} \qquad N_\mathrm{P} = \sum_\mathrm{p} n_\mathrm{p} = \frac{N}{q}\sum_\mathrm{p} \mathrm{e}^{-\beta\varepsilon'_\mathrm{p}}$$

The sum over the states of R is its partition function, q_R, so

$$N_\mathrm{R} = \frac{Nq_\mathrm{R}}{q}$$

The sum over the states of P is also a partition function, but the energies are measured from the ground state of the combined system, which is the ground state of R. However, because $\varepsilon'_\mathrm{p} = \varepsilon_\mathrm{p} + \Delta\varepsilon_0$ where $\Delta\varepsilon_0$ is the separation of zero-point energies,

$$N_\mathrm{P} = \frac{N}{q}\sum_\mathrm{p} \mathrm{e}^{-\beta(\varepsilon_\mathrm{p}+\Delta\varepsilon_0)} = \frac{N}{q}\left(\sum_\mathrm{p} \mathrm{e}^{-\beta\varepsilon_\mathrm{p}}\right)\mathrm{e}^{-\beta\Delta\varepsilon_0} = \frac{Nq_\mathrm{P}}{q}\mathrm{e}^{-\Delta_\mathrm{r}E_0/RT}$$

The switch from $\Delta\varepsilon_0/k$ to $\Delta_\mathrm{r}E_0/R$ in the last step is the conversion of molecular energies to molar energies.

The equilibrium constant of the R $\rightleftharpoons$ P reaction is proportional to the ratio of the numbers of the two types of molecule. Therefore,

$$K = \frac{N_\mathrm{P}}{N_\mathrm{R}} = \frac{q_\mathrm{P}}{q_\mathrm{r}}\mathrm{e}^{-\Delta_\mathrm{r}E_0/RT}$$

as in eqn 16.56b.

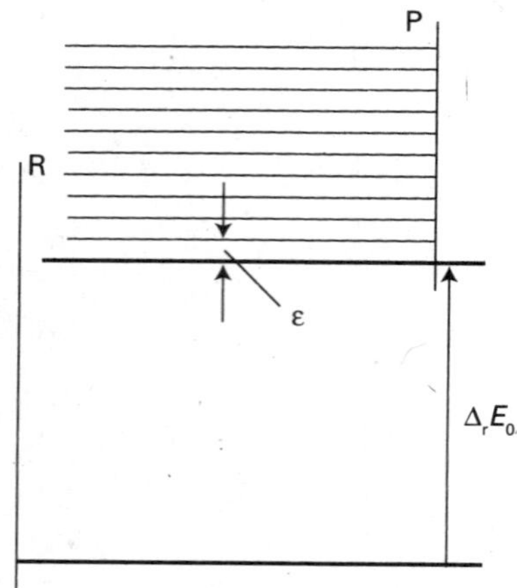

Fig. 16.21 The model used in the text for exploring the effects of energy separations and densities of states on equilibria. The products P can dominate provided ΔE_0 is not too large and P has an appreciable density of states.

The content of eqn 16.56 can be seen most clearly by exaggerating the molecular features that contribute to it. We shall suppose that R has only a single accessible level, which implies that $q_\mathrm{R} = 1$. We also suppose that P has a large number of evenly, closely spaced levels (Fig. 16.21). The partition function of P is then $q_\mathrm{P} = kT/\varepsilon$. In this model system, the equilibrium constant is

$$K = \frac{kT}{\varepsilon}\mathrm{e}^{-\Delta_\mathrm{r}E_0/RT} \qquad (16.57)$$

When $\Delta_\mathrm{r}E_0$ is very large, the exponential term dominates and $K \ll 1$, which implies that very little P is present at equilibrium. When $\Delta_\mathrm{r}E_0$ is small but still positive, K can exceed 1 because the factor kT/ε may be large enough to overcome the small size of the exponential term. The size of K then reflects the predominance of P at equilibrium on account of its high density of states. At low temperatures $K \ll 1$ and the system consists entirely of R. At high temperatures the exponential function approaches 1 and the pre-exponential factor is large. Hence P becomes dominant. We see that, in this endothermic reaction (endothermic because P lies above R), a rise in temperature favours P, because its states become accessible. This behaviour is what we saw, from the outside, in Chapter 6.

The model also shows why the Gibbs energy, G, and not just the enthalpy, determines the position of equilibrium. It shows that the density of states (and hence the entropy) of each species as well as their relative energies controls the distribution of populations and hence the value of the equilibrium constant.

IMPACT ON BIOCHEMISTRY

I16.1 The helix–coil transition in polypeptides

The hydrogen bonds between amino acids of a polypeptide give rise to stable helical or sheet structures, which may collapse into a random coil when certain conditions are changed. The unwinding of a helix into a random coil is a *cooperative transition*, in which the polymer becomes increasingly more susceptible to structural changes once the process has begun. We examine here a model based on the principles of statistical thermodynamics that accounts for the cooperativity of the helix–coil transition in polypeptides.

To calculate the fraction of polypeptide molecules present as helix or coil we need to set up the partition function for the various states of the molecule. To illustrate the approach, consider a short polypeptide with four amino acid residues, each labelled *h* if it contributes to a helical region and *c* if it contributes to a random coil region. We suppose that conformations *hhhh* and *cccc* contribute terms q_0 and q_4, respectively, to the partition function q. Then we assume that each of the four conformations with one *c* amino acid (such as *hchh*) contributes q_1. Similarly, each of the six states with two *c* amino acids contributes a term q_2, and each of the four states with three *c* amino acids contributes a term q_3. The partition function is then

$$q = q_0 + 4q_1 + 6q_2 + 4q_3 + q_4 = q_0\left(1 + \frac{4q_1}{q_0} + \frac{6q_2}{q_0} + \frac{4q_3}{q_0} + \frac{q_4}{q_0}\right)$$

We shall now suppose that each partition function differs from q_0 only by the energy of each conformation relative to *hhhh*, and write

$$\frac{q_i}{q_0} = e^{-(\varepsilon_i - \varepsilon_0)/kT}$$

Next, we suppose that the conformational transformations are non-cooperative, in the sense that the energy associated with changing one *h* amino acid into one *c* amino acid has the same value regardless of how many *h* or *c* amino acid residues are in the reactant or product state and regardless of where in the chain the conversion occurs. That is, we suppose that the difference in energy between c^ih^{4-i} and $c^{i+1}h^{3-i}$ has the same value γ for all i. This assumption implies that $\varepsilon_i - \varepsilon_0 = i\gamma$ and therefore that

$$q/q_0 = 1 + 4s + 6s^2 + 4s^3 + s^4 = (1+s)^4 \qquad s = e^{-\gamma/kT} \tag{16.58}$$

where s is called the *stability parameter*. The extension of this treatment to take into account a longer chain of residues is now straightforward: we simply replace the 4 in the sum by N:

$$\frac{q}{q_0} = (1+s)^N \tag{16.59}$$

A cooperative transformation is more difficult to accommodate, and depends on building a model of how neighbours facilitate each other's conformational change. In the simple *zipper model*, conversion from *h* to *c* is allowed only if a residue adjacent to the one undergoing the conversion is already a *c* residue. Thus, the zipper model allows a transition of the type . . . *hhhch* . . . → . . . *hhhcc* . . . , but not a transition of the type . . . *hhhch* . . . → . . . *hchch*. . . . The only exception to this rule is, of course,

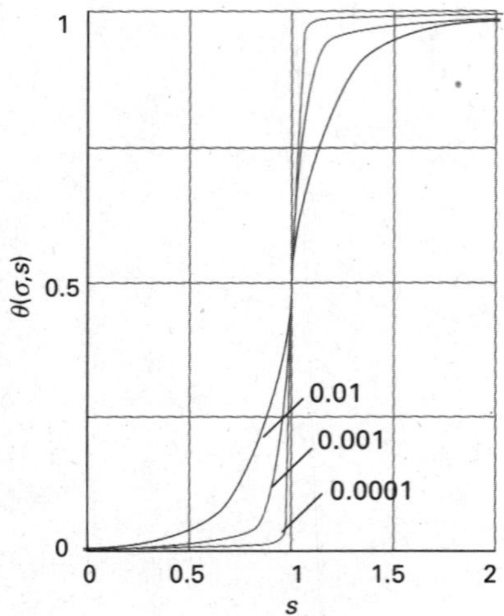

Fig. 16.22 Plots of the degree of conversion θ, against s for several values of σ. The curves show the sigmoidal shape characteristics of cooperative behaviour.

the very first conversion from h to c in a fully helical chain. Cooperativity is included in the zipper model by assuming that the first conversion from h to c, called the *nucleation step*, is less favourable than the remaining conversions and replacing s for that step by ss, where $s \ll 1$. Each subsequent step is called a *propagation step* and has a stability parameter s.

A more sophisticated model for the helix–coil transition must allow for helical segments to form in different regions of a long polypeptide chain, with the nascent helices being separated by shrinking coil segments. Calculations based on this more complete *Zimm–Bragg model* give

$$\theta = \tfrac{1}{2}\left(1 + \frac{(s-1)+2\sigma}{[(s-1)^2+4s\sigma]^{1/2}}\right) \qquad (16.60)$$

where θ = (mean number of coil units)/(total units) is the *degree of conversion* of a polypeptide to a random coil. Figure 16.22 shows plots of θ against s for several values of σ. The curves show the sigmoidal shape characteristic of cooperative behaviour. There is a sudden surge of transition to a random coil as s passes through 1 and, the smaller the parameter σ, the greater the sharpness and hence the greater the cooperativity of the transition. That is, the harder it is to get coil formation started, the sharper the transition from helix to coil.

Checklist of key equations

Property	Equation	Comment
Helmholtz energy	$A - A(0) = -kT\ln Q$	
Pressure	$p = kT(\partial \ln Q/\partial V)_T$	
Enthalpy	$H - H(0) = -(\partial \ln Q/\partial \beta)_V + kTV(\partial \ln Q/\partial V)_T$	
Gibbs energy	$G - G(0) = -kT\ln Q + kTV(\partial \ln Q/\partial V)_T$	
Molecular energy	$\varepsilon = \varepsilon^T + \varepsilon^R + \varepsilon^V + \varepsilon^E$	Assumes that R, V, E modes are independent
Molecular partition function	$q = q^T q^R q^V q^E$	Assumes that R, V, E modes are independent
Contributions to the partition function:		
Translational	$q^T = V/\Lambda^3$	$\Lambda = h/(2\pi mkT)^{1/2}$
	$q_m^{T\ominus}/N_A = kT/p^\ominus\Lambda^3$	
Rotational		
linear molecules	$q^R = T/\sigma\theta_R$	High temperature limit; $\theta_R = hc\tilde{B}/k$
nonlinear molecules	$q^R = (1/\sigma)(kT/hc)^{3/2}(\pi/\tilde{A}\tilde{B}\tilde{C})^{1/2}$	High temperature limit
Vibrational	$q^V = (1 - e^{-\theta_V/T})^{-1}$	Diatomic molecule in the
	For $T \gg \theta_V$, $q^V = T/\theta_V$	harmonic approximation; $\theta_V = hc\tilde{\nu}/k = h\nu/k$
Electronic	$q^E = g_0$ [+ higher terms]	
Mean energy of a mode of motion	$\langle\varepsilon^M\rangle = -(1/q^M)(\partial q^M/\partial\beta)_V$	M = T, R, V, or E
Contribution of a mode to the constant-volume heat capacity	$C_V^M = -Nk\beta^2(\partial\langle\varepsilon^M\rangle/\partial\beta)_V$	M = T, R, V, or E
Residual entropy	$S_m(0) = R\ln s$	
Equilibrium constant in terms of the partition function	$K = \left\{\prod_J (q_{J,m}^\ominus/N_A)^{\nu_J}\right\} e^{-\Delta_r E_0/RT}$	Gas phase reaction

➔ For a chart of the relations between principal equations, see the Road map section of the Resource section.

Further information

Further information 16.1 *The rotational partition function of a symmetric rotor*

The energies of a symmetric rotor are

$$E_{J,K,M_J} = hc\tilde{B}J(J+1) + hc(\tilde{A}-\tilde{B})K^2$$

with $J = 0, 1, 2, \ldots, K = J, J-1, \ldots, -J$, and $M_J = J, J-1, \ldots, -J$. Instead of considering these ranges, we can cover the same values by allowing K to range from $-\infty$ to ∞, with J confined to $|K|, |K|+1, \ldots, \infty$ for each value of K (Fig. 16.23). Because the energy is independent of M_J, and there are $2J+1$ values of M_J for each value of J, each value of J is $(2J+1)$-fold degenerate. It follows that the partition function

$$q = \sum_{J=0}^{\infty} \sum_{K=-J}^{J} \sum_{M_J=-J}^{J} \mathrm{e}^{-E_{J,K,M_J}/kT} \qquad (16.61)$$

can be written equivalently as

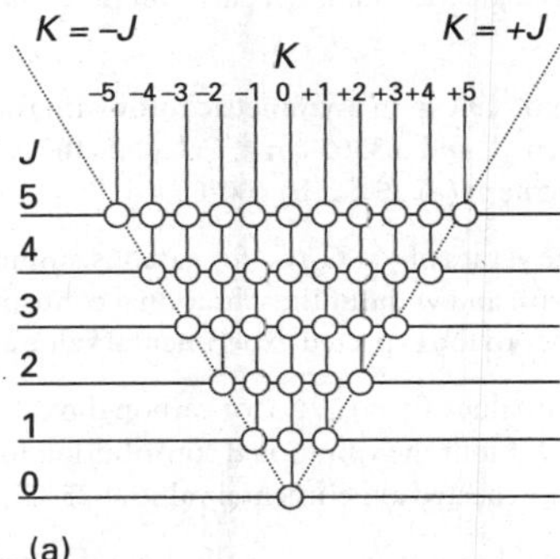

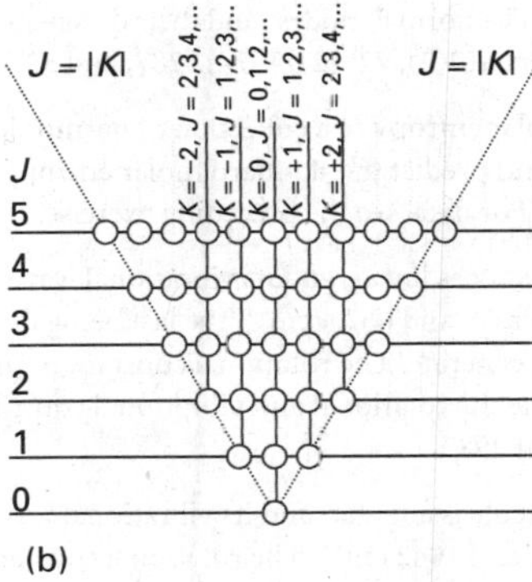

Fig. 16.23 (a) The sum over $J = 0, 1, 2, \ldots$ and $K = J, J-1, \ldots, -J$ (depicted by the circles) can be covered (b) by allowing K to range from $-\infty$ to ∞, with J confined to $|K|, |K|+1, \ldots, \infty$ for each value of K.

$$\begin{aligned} q &= \sum_{K=-\infty}^{\infty} \sum_{J=|K|}^{\infty} (2J+1)\mathrm{e}^{-E_{J,K,M_J}/kT} \\ &= \sum_{K=-\infty}^{\infty} \sum_{J=|K|}^{\infty} (2J+1)\mathrm{e}^{-hc\{\tilde{B}J(J+1)+(\tilde{A}-\tilde{B})K^2\}/kT} \\ &= \sum_{K=-\infty}^{\infty} \mathrm{e}^{\{-hc\{(\tilde{A}-\tilde{B})kT\}K^2} \sum_{J=|K|}^{\infty} (2J+1)\mathrm{e}^{-hc\tilde{B}J(J+1)/kT} \end{aligned}$$

Now we assume that the temperature is so high that numerous states are occupied and that the sums may be approximated by integrals. Then

$$q = \int_{-\infty}^{\infty} \mathrm{e}^{-\{hc(\tilde{A}-\tilde{B})/kT\}K^2} \int_{|K|}^{\infty} (2J+1)\mathrm{e}^{-hc\tilde{B}J(J+1)/kT}\,\mathrm{d}J\,\mathrm{d}K \qquad (16.62)$$

As in *Justification 16.1*, the integral over J can be recognized as the integral of the derivative of a function, which is the function itself, so

$$\begin{aligned} \int_{|K|}^{\infty} (2J+1)\mathrm{e}^{-hc\tilde{B}J(J+1)/kT}\mathrm{d}J &= \int_{|K|}^{\infty} \left(-\frac{kT}{hc\tilde{B}}\right)\frac{\mathrm{d}}{\mathrm{d}J}\mathrm{e}^{-hc\tilde{B}J(J+1)/kT}\mathrm{d}J \\ &= \left(-\frac{kT}{hc\tilde{B}}\right)\mathrm{e}^{-hc\tilde{B}J(J+1)/kT}\Big|_{|K|}^{\infty} \\ &= \left(\frac{kT}{hc\tilde{B}}\right)\mathrm{e}^{-hc\tilde{B}|K|(|K|+1)/kT} \qquad (16.63) \\ &\approx \left(\frac{kT}{hc\tilde{B}}\right)\mathrm{e}^{-hc\tilde{B}K^2/kT} \end{aligned}$$

In the last line we have supposed that $|K| \gg 1$ for most contributions. Now we can write eqn 16.62 as

$$\begin{aligned} q &= \frac{kT}{hc\tilde{B}} \int_{-\infty}^{\infty} \mathrm{e}^{-\{hc(\tilde{A}-\tilde{B})/kT\}K^2}\mathrm{e}^{-hc\tilde{B}K^2/kT}\,\mathrm{d}K \\ &= \frac{kT}{hc\tilde{B}} \int_{-\infty}^{\infty} \mathrm{e}^{-\{hc\tilde{A}/kT\}K^2}\,\mathrm{d}K = \left(\frac{kT}{hc\tilde{B}}\right)\left(\frac{kT}{hc\tilde{A}}\right)^{1/2} \overbrace{\int_{-\infty}^{\infty} \mathrm{e}^{-x^2}\,\mathrm{d}x}^{\pi^{1/2}} \qquad (16.64) \\ &= \left(\frac{kT}{hc}\right)^{3/2}\left(\frac{\pi}{\tilde{A}\tilde{B}^2}\right)^{1/2} \end{aligned}$$

For an asymmetric rotor, one of the $\tilde{B}$s is replaced by $\tilde{C}$, to give eqn 16.14b.

Discussion questions

16.1 Discuss the limitations of the expressions $q^{R}=kT/hc\tilde{B}$, $q^{V}=kT/hc\tilde{\nu}$, and $q^{E}=g^{E}$.

16.2 Explain the origin of the symmetry number.

16.3 Explain the origin of residual entropy.

16.4 Describe the molecular features that determine the magnitudes of the constant-volume molar heat capacity of a molecular substance.

16.5 Describe the features that determine whether particles should be regarded as indistinguishable or not.

16.6 Discuss and illustrate the proposition that $1/T$ is a more natural measurement of temperature than T itself.

16.7 Describe the molecular features that determine the magnitudes of equilibrium constants and their variation with temperature.

Exercises

16.1(a) Evaluate the thermal wavelength of CO_2 at 25°C.

16.1(b) Evaluate the thermal wavelength of SO_2 at 25°C.

16.2(a) Evaluate the translational partition function of CO_2 at 25°C in a container of volume 1.0 cm^3.

16.2(b) Evaluate the translational partition function wavelength of SO_2 at 25°C in a container of volume 1.0 cm^3.

16.3(a) Use the equipartition theorem to estimate the constant-volume molar heat capacity of (a) I_2, (b) CH_4, (c) C_6H_6 in the gas phase at 25°C.

16.3(b) Use the equipartition theorem to estimate the constant-volume molar heat capacity of (a) O_3, (b) C_2H_6, (c) CO_2 in the gas phase at 25°C.

16.4(a) Evaluate the rotational temperature of $H^{35}Cl$.

16.4(b) Evaluate the rotational temperature of H_2 and D_2.

16.5(a) Estimate the rotational partition function of HCl at (a) 25°C and (b) 250°C.

16.5(b) Estimate the rotational partition function of O_2 at (a) 25°C and (b) 250°C.

16.6(a) Give the symmetry number for each of the following molecules: (a) CO, (b) O_2, (c) H_2S, (d) SiH_4, and (e) $CHCl_3$.

16.6(b) Give the symmetry number for each of the following molecules: (a) CO_2, (b) O_3, (c) SO_3, (d) SF_6, and (e) Al_2Cl_6.

16.7(a) Calculate the rotational partition function of H_2O at 298 K from its rotational constants 27.878 cm^{-1}, 14.509 cm^{-1}, and 9.287 cm^{-1}. Above what temperature is the high-temperature approximation valid to within 10 per cent of the true value?

16.7(b) Calculate the rotational partition function of SO_2 at 298 K from its rotational constants 2.027 36 cm^{-1}, 0.344 17 cm^{-1}, and 0.293 535 cm^{-1}. Above what temperature is the high-temperature approximation valid to within 10 per cent of the true value?

16.8(a) From the results of Exercise 16.7a, calculate the rotational contribution to the molar entropy of gaseous water at 25°C.

16.8(b) From the results of Exercise 16.7b, calculate the rotational contribution to the molar entropy of sulfur dioxide at 25°C.

16.9(a) Calculate the rotational partition function of CH_4 (a) by direct summation of the energy levels at 298 K and 500 K, and (b) by the high-temperature approximation. Take $\tilde{B}=5.2412$ cm^{-1}.

16.9(b) Calculate the rotational partition function of CH_3CN (a) by direct summation of the energy levels at 298 K and 500 K, and (b) by the high-temperature approximation. Take $\tilde{A}=5.28$ cm^{-1} and $\tilde{B}=0.307$ cm^{-1}.

16.10(a) The bond length of O_2 is 120.75 pm. Use the high-temperature approximation to calculate the rotational partition function of the molecule at 300 K.

16.10(b) The NOF molecule is an asymmetric rotor with rotational constants 3.1752 cm^{-1}, 0.3951 cm^{-1}, and 0.3505 cm^{-1}. Calculate the rotational partition function of the molecule at (a) 25°C, (b) 100°C.

16.11(a) Estimate the values of $\gamma=C_p/C_V$ for gaseous ammonia and methane. Do this calculation with and without the vibrational contribution to the energy. Which is closer to the expected experimental value at 25°C?

16.11(b) Estimate the value of $\gamma=C_p/C_V$ for carbon dioxide. Do this calculation with and without the vibrational contribution to the energy. Which is closer to the expected experimental value at 25°C?

16.12(a) Plot the molar heat capacity of a collection of harmonic oscillators as a function of T/θ_V, and predict the vibrational heat capacity of ethyne at (a) 298 K, (b) 500 K. The normal modes (and their degeneracies in parentheses) occur at wavenumbers 612(2), 729(2), 1974, 3287, and 3374 cm^{-1}.

16.12(b) Plot the molar entropy of a collection of harmonic oscillators as a function of T/θ_V, and predict the standard molar entropy of ethyne at (a) 298 K, (b) 500 K. For data, see the preceding exercise.

16.13(a) A CO_2 molecule is linear, and its vibrational wavenumbers are 1388.2 cm^{-1}, 2349.2 cm^{-1}, and 667.4 cm^{-1}, the last being doubly degenerate and the others nondegenerate. The rotational constant of the molecule is 0.3902 cm^{-1}. Calculate the rotational and vibrational contributions to the molar Gibbs energy at 298 K.

16.13(b) An O_3 molecule is angular, and its vibrational wavenumbers are 1110 cm^{-1}, 705 cm^{-1}, and 1042 cm^{-1}. The rotational constants of the molecule are 3.553 cm^{-1}, 0.4452 cm^{-1}, and 0.3948 cm^{-1}. Calculate the rotational and vibrational contributions to the molar Gibbs energy at 298 K.

16.14(a) The ground level of Cl is $^2P_{3/2}$ and a $^2P_{1/2}$ level lies 881 cm^{-1} above it. Calculate the electronic partition function of Cl atoms at (a) 500 K and (b) 900 K.

16.14(b) The first electronically excited state of O_2 is $^1\Delta_g$ and lies 7918.1 cm^{-1} above the ground state, which is $^3\Sigma_g^-$. Calculate the electronic partition function of O_2 molecules at (a) 500 K and (b) 900 K.

16.15(a) Use the information in Exercise 16.14a to calculate the electronic contribution to the heat capacity of Cl atoms at (a) 500 K and (b) 900 K.

16.15(b) Use the information in Exercise 16.14b to calculate the electronic contribution to the heat capacity of of O_2 at 400 K.

16.16(a) Use the information in Exercise 16.14a to calculate the electronic contribution to the molar Gibbs energy of Cl atoms at (a) 500 K and (b) 900 K.

16.16(b) Use the information in Exercise 16.14a to calculate the electronic contribution to the molar Gibbs energy of O_2 at 400 K.

16.17(a) The ground state of the Co^{2+} ion in $CoSO_4{\cdot}7H_2O$ may be regarded as $^4T_{9/2}$. The entropy of the solid at temperatures below 1 K is derived almost entirely from the electron spin. Estimate the molar entropy of the solid at these temperatures.

16.17(b) Estimate the contribution of the spin to the molar entropy of a solid sample of a d-metal complex with $S = \frac{5}{2}$.

16.18(a) Sketch the form of the Mayer *f*-function for the hard-sphere potential specified in eqn 16.43.

16.18(b) Sketch the form of the Mayer *f*-function for an intermolecular potential energy of the form $E_p = -\varepsilon(\sigma^6/r^6 - \sigma^{12}/r^{12})$.

16.19(a) Calculate the residual molar entropy of a solid in which the molecules can adopt (a) three, (b) five, (c) six orientations of equal energy at $T = 0$.

16.19(b) Suppose that the hexagonal molecule $C_6H_nF_{6-n}$ has a residual entropy on account of the similarity of the H and F atoms. Calculate the residual for each value of n.

16.20(a) Calculate the equilibrium constant of the reaction $I_2(g) \rightleftharpoons 2\,I(g)$ at 1000 K from the following data for I_2: $\tilde{\nu} = 214.36\ \mathrm{cm^{-1}}$, $\tilde{B} = 0.0373\ \mathrm{cm^{-1}}$, $D_e = 1.5422$ eV. The ground state of the I atoms is $^2P_{3/2}$, implying fourfold degeneracy.

16.20(b) Calculate the equilibrium constant at 298 K for the gas-phase isotopic exchange reaction $2\,^{79}Br^{81}Br \rightleftharpoons {}^{79}Br^{79}Br + {}^{81}Br^{81}Br$. The Br_2 molecule has a nondegenerate ground state, with no other electronic states nearby. Base the calculation on the wavenumber of the vibration of $^{79}Br^{81}Br$, which is 323.33 $\mathrm{cm^{-1}}$.

Problems*

Numerical problems

16.1 The NO molecule has a doubly degenerate electronic ground state and a doubly degenerate excited state at 121.1 $\mathrm{cm^{-1}}$. Calculate and plot the electronic contribution to the molar heat capacity of the molecule up to 500 K.

16.2 Explore whether a magnetic field can influence the heat capacity of a paramagnetic molecule by calculating the electronic contribution to the heat capacity of an NO_2 molecule in a magnetic field. Estimate the total constant-volume heat capacity using equipartition, and calculate the percentage change in heat capacity brought about by a 5.0 T magnetic field at (a) 50 K, (b) 298 K.

16.3 The energy levels of a CH_3 group attached to a larger fragment are given by the expression for a particle on a ring, provided the group is rotating freely. What is the high-temperature contribution to the heat capacity and entropy of such a freely rotating group at 25°C? The moment of inertia of CH_3 about its threefold rotation axis (the axis that passes through the C atom and the centre of the equilateral triangle formed by the H atoms) is $5.341 \times 10^{-47}\ \mathrm{kg\ m^2}$.

16.4 Calculate the temperature dependence of the heat capacity of *p*-H_2 (in which only rotational states with even values of J are populated) at low temperatures on the basis that its rotational levels $J = 0$ and $J = 2$ constitute a system that resembles a two-level system except for the degeneracy of the upper level. Use $\tilde{B} = 60.864\ \mathrm{cm^{-1}}$ and sketch the heat capacity curve. The experimental heat capacity of *p*-H_2 does in fact show a peak at low temperatures.

16.5 The pure rotational microwave spectrum of $H^{35}Cl$ has absorption lines at the following wavenumbers (in $\mathrm{cm^{-1}}$): 21.19, 42.37, 63.56, 84.75, 105.93, 127.12 148.31 169.49, 190.68, 211.87, 233.06, 254.24, 275.43, 296.62, 317.80, 338.99, 360.18, 381.36, 402.55, 423.74, 444.92, 466.11, 487.30, 508.48. Calculate the rotational partition function at 25°C by direct summation.

16.6 Calculate the standard molar entropy of $N_2(g)$ at 298 K from its rotational constant $\tilde{B} = 1.9987\ \mathrm{cm^{-1}}$ and its vibrational wavenumber $\tilde{\nu} = 2358\ \mathrm{cm^{-1}}$. The thermochemical value is 192.1 $\mathrm{J\ K^{-1}\ mol^{-1}}$. What does this suggest about the solid at $T = 0$?

16.7‡ J.G. Dojahn *et al.* (*J. Phys. Chem.* **100**, 9649 (1996)) characterized the potential energy curves of the ground and electronic states of homonuclear diatomic halogen anions. The ground state of F_2^- is $^2\Sigma_u^+$ with a fundamental vibrational wavenumber of 450.0 $\mathrm{cm^{-1}}$ and equilibrium internuclear distance of 190.0 pm. The first two excited states are at 1.609 and 1.702 eV above the ground state. Compute the standard molar entropy of F_2^- at 298 K.

16.8‡ In a spectroscopic study of buckminsterfullerene C_{60}, F. Negri *et al.* (*J. Phys. Chem.* **100**, 10849 (1996)) reviewed the wavenumbers of all the vibrational modes of the molecule:

Mode	Number	Degeneracy	Wavenumber/cm^{-1}
A_u	1	1	976
T_{1u}	4	3	525, 578, 1180, and 1430
T_{2u}	5	3	354, 715, 1037, 1190, 1540
G_u	6	4	345, 757, 776, 963, 1315, 1410
H_u	7	5	403, 525, 667, 738, 1215, 1342, 1566

How many modes have a vibrational temperature θ_V below 1000 K? Estimate the molar constant-volume heat capacity of C_{60} at 1000 K, counting as active all modes with θ_V below this temperature.

16.9‡ Treat carbon monoxide as a perfect gas and apply equilibrium statistical thermodynamics to the study of its properties, as specified below, in the temperature range 100–1000 K at 1 bar. $\tilde{\nu} = 2169.8\ \mathrm{cm^{-1}}$, $\tilde{B} = 1.931\ \mathrm{cm^{-1}}$, and $D_0 = 11.09$ eV; neglect anharmonicity and centrifugal distortion. (a) Examine the probability distribution of molecules over available rotational and vibrational states. (b) Explore numerically the differences, if any, between the rotational molecular partition function as calculated with the discrete energy distribution and that calculated with the classical, continuous energy distribution. (c) Calculate the individual contributions to $U_m(T) - U_m(100\ \mathrm{K})$, $C_{V,m}(T)$, and $S_m(T) - S_m(100\ \mathrm{K})$ made by the translational, rotational, and vibrational degrees of freedom.

16.10 Use mathematical software to evaluate the second virial coefficient in eqn 16.42 for a intermolecular potential energy of the form $E_p = -\varepsilon(\sigma^6/r^6 - \sigma^{12}/r^{12})$ and plot it as a function of temperature. Discuss how changing the range (as expressed by σ) and the depth of the potential well (as expressed by ε) affect the value of B.

* Problems denoted with the symbol ‡ were supplied by Charles Trapp, Carmen Giunta, and Marshall Cady.

16.11 Calculate and plot as a function of temperature, in the range 300 K to 1000 K, the equilibrium constant for the reaction $CD_4(g) + HCl(g) \rightleftharpoons CHD_3(g) + DCl(g)$ using the following data (numbers in parentheses are degeneracies):

Molecule	$\tilde{\nu}/cm^{-1}$	$\tilde{B}/cm^{-1}$	$\tilde{A}/cm^{-1}$
CHD_3	2993(1), 2142(1), 1003(3), 1291(2), 1036(2)	3.28	2.63
CD_4	2109(1), 1092(2), 2259(3), 996(3)	2.63	
HCl	2991(1)	10.59	
DCl	2145(1)	5.445	

16.12 The exchange of deuterium between acid and water is an important type of equilibrium, and we can examine it using spectroscopic data on the molecules. Calculate the equilibrium constant at (a) 298 K and (b) 800 K for the gas-phase exchange reaction $H_2O + DCl \rightleftharpoons HDO + HCl$ from the following data:

Molecule	$\tilde{\nu}/cm^{-1}$	$\tilde{A}$ cm^{-1}	$\tilde{B}$ cm^{-1}	$\tilde{C}$ cm^{-1}
H_2O	3656.7, 1594.8, 3755.8	27.88	14.51	9.29
HDO	2726.7, 1402.2, 3707.5	23.38	9.102	6.417
HCl	2991		10.59	
DCl	2145		5.449	

Theoretical problems

16.13 Derive the Sackur–Tetrode equation for a monatomic gas confined to a two-dimensional surface, and hence derive an expression for the standard molar entropy of condensation to form a mobile surface film.

16.14‡ For H_2 at very low temperatures, only translational motion contributes to the heat capacity. At temperatures above $\theta_R = hc\tilde{B}/k$, the rotational contribution to the heat capacity becomes significant. At still higher temperatures, above $\theta_V = h\nu/k$, the vibrations contribute. But at this latter temperature, dissociation of the molecule into the atoms must be considered. (a) Explain the origin of the expressions for θ_R and θ_V, and calculate their values for hydrogen. (b) Obtain an expression for the molar constant-pressure heat capacity of hydrogen at all temperatures taking into account the dissociation of hydrogen. (c) Make a plot of the molar constant-pressure heat capacity as a function of temperature in the high-temperature region where dissociation of the molecule is significant.

16.15 Derive expressions for the internal energy, heat capacity, entropy, Helmholtz energy, and Gibbs energy of a harmonic oscillator. Express the results in terms of the vibrational temperature, θ_V, and plot graphs of each property against T/θ_V.

16.16 Use mathematical software to evaluate the heat capacity of the bound states of a Morse oscillator (Section 12.10) in which the energy levels are given by eqn 12.38:

$$E_v = (v+\tfrac{1}{2})hc\tilde{\nu} - (v+\tfrac{1}{2})^2 hcx_e\tilde{\nu}$$

Plot the heat capacity as a function of temperature. Can you devise a way to include the unbound states that lie above the dissociation limit? Use the parameters for HCl (Exercise 12.12).

16.17 Derive eqn 16.45, that $\mu_T = B - TdB/dT$ in the limit $p \to 0$. *Hint*: Start by writing $\mu = (V_m/C_{p,m})(\alpha T - 1)$ and $\mu_T = -C_p\mu$ (see Sections 2.11 and 2.12 for definitions of these terms) and the virial equation in eqn 1.19.

16.18 A formal way of arriving at the value of the symmetry number is to note that σ is the order (the number of elements) of the *rotational subgroup* of the molecule, the point group of the molecule with all but the identity and the rotations removed. The rotational subgroup of H_2O is $\{E, C_2\}$, so $\sigma = 2$. The rotational subgroup of NH_3 is $\{E, 2C_3\}$, so $\sigma = 3$. This recipe makes it easy to find the symmetry numbers for more complicated molecules. The rotational subgroup of CH_4 is obtained from the T character table as $\{E, 8C_3, 3C_2\}$, so $\sigma = 12$. For benzene, the rotational subgroup of D_{6h} is $\{E, 2C_6, 2C_3, C_2, 3C_2', 3C_2''\}$, so $\sigma = 12$. (a) Estimate the rotational partition function of ethene at 25°C given that $\tilde{A} = 4.828$ cm^{-1}, $\tilde{B} = 1.0012$ cm^{-1}, and $\tilde{C} = 0.8282$ cm^{-1}. (b) Evaluate the rotational partition function of pyridine, C_5H_5N, at room temperature ($\tilde{A} = 0.2014$ cm^{-1}, $\tilde{B} = 0.1936$ cm^{-1}, $\tilde{C} = 0.0987$ cm^{-1}).

16.19 Although expressions like $\langle\varepsilon\rangle = -d\ln q/d\beta$ are useful for formal manipulations in statistical thermodynamics, and for expressing thermodynamic functions in neat formulas, they are sometimes more trouble than they are worth in practical applications. When presented with a table of energy levels, it is often much more convenient to evaluate the following sums directly:

$$q = \sum_j e^{-\beta\varepsilon_j} \qquad \dot{q} = \sum_j \beta\varepsilon_j e^{-\beta\varepsilon_j} \qquad \ddot{q} = \sum_j (\beta\varepsilon_j)^2 e^{-\beta\varepsilon_j}$$

(a) Derive expressions for the internal energy, heat capacity, and entropy in terms of these three functions. (b) Apply the technique to the calculation of the electronic contribution to the constant-volume molar heat capacity of magnesium vapour at 5000 K using the following data:

Term	^1S	$^3P_0$	$^3P_1$	$^3P_2$	$^1P_1$	$^3S_1$
Degeneracy	1	1	3	5	3	3
$\tilde{\nu}/cm^{-1}$	0	21 850	21 870	21 911	35 051	41 197

16.20 Show how the heat capacity of a linear rotor is related to the following sum:

$$\zeta(\beta) = \frac{1}{q^2}\sum_{J,J'} \{\varepsilon(J) - \varepsilon(J')\}^2 g(J')e^{-\beta\{\varepsilon(J)+\varepsilon(J')\}}$$

by

$$C = \tfrac{1}{2}Nk\beta^2\zeta(\beta)$$

where the $\varepsilon(J)$ are the rotational energy levels and $g(J)$ their degeneracies. Then go on to show graphically that the total contribution to the heat capacity of a linear rotor can be regarded as a sum of contributions due to transitions $0 \to 1, 0 \to 2, 1 \to 2, 1 \to 3$, etc. In this way, construct Fig. 16.10 for the rotational heat capacities of a linear molecule.

16.21 Set up a calculation like that in Problem 16.20 to analyse the vibrational contribution to the heat capacity in terms of excitations between levels and illustrate your results graphically in terms of a diagram like that in Fig. 16.10.

16.22 Suppose that an intermolecular potential has a hard-sphere core of radius r_1 and a shallow attractive well of uniform depth e out to a distance r_2. Show, by using eqn 16.41 and the condition $\varepsilon \ll kT$, that such a model is approximately consistent with a van der Waals equation of state when $b \ll V_m$, and relate the van der Waals parameters and the Joule–Thomson coefficient to the parameters in this model.

16.23 Explore the consequences of modelling the pair distribution function in eqn 16.49a as

$$g(r) = 1 + \cos\left(\frac{4r}{d} - 4\right)e^{-(r/d-1)}$$

for $r \geq d$ and $g(r) = 0$ for $r < d$ and the intermolecular potential energy specified in Problem 16.10 ($E_p = -\varepsilon(\sigma^6/r^6 - \sigma^{12}/r^{12})$). Begin by plotting $g(r)$ to verify that it resembles the form shown in Fig. 16.15. Then evaluate the virial for the potential energy (eqn 16.49a with V_2 identified with E_p). Finally, explore the internal pressure of the fluid and discuss how it varies with temperature and the parameters in the intermolecular potential energy.

16.24 Determine whether a magnetic field can influence the value of an equilibrium constant. Consider the equilibrium $I_2(g) \rightleftharpoons 2\,I(g)$ at 1000 K, and calculate the ratio of equilibrium constants $K(\mathcal{B})/K$, where $K(\mathcal{B})$ is the equilibrium constant when a magnetic field $\mathcal{B}$ is present and removes the degeneracy of the four states of the $^2P_{3/2}$ level. Data on the species are given in Exercise 16.20a. The electronic g-value of the atoms is $\frac{4}{3}$. Calculate the field required to change the equilibrium constant by 1 per cent.

16.25 The heat capacity ratio of a gas determines the speed of sound in it through the formula $c_s = (\gamma RT/M)^{1/2}$, where $\gamma = C_p/C_V$ and M is the molar mass of the gas. Deduce an expression for the speed of sound in a perfect gas of (a) diatomic, (b) linear triatomic, (c) nonlinear triatomic molecules at high temperatures (with translation and rotation active). Estimate the speed of sound in air at 25°C.

Applications: to biology, materials science, environmental science, and astrophysics

16.26 An average human DNA molecule has 5×10^8 binucleotides (rungs on the DNA ladder) of four different kinds. If each rung were a random choice of one of these four possibilities, what would be the residual entropy associated with this typical DNA molecule?

16.27 It is possible to write an approximate expression for the partition function of a protein molecule by including contributions from only two states: the native and denatured forms of the polymer. Proceeding with this crude model gives us insight into the contribution of denaturation to the heat capacity of a protein. According to this model, the total energy of a system of N protein molecules is

$$E = \frac{N\varepsilon e^{-\varepsilon/kT}}{1+e^{-\varepsilon/kT}}$$

where ε is the energy separation between the denatured and native forms. (a) Show that the constant-volume molar heat capacity is

$$C_{V,m} = \frac{R(\varepsilon_m/RT)^2 e^{-\varepsilon_m/RT}}{(1+e^{-\varepsilon_m/RT})^2}$$

(b) Plot the variation of $C_{V,m}$ with temperature. (c) If the function $C_{V,m}(T)$ has a maximum or minimum, derive an expression for the temperature at which it occurs.

16.28‡ R. Viswanathan *et al.* (*J. Phys. Chem.* **100**, 10784 (1996)) studied thermodynamic properties of several boron–silicon gas-phase species experimentally and theoretically. These species can occur in the high-temperature chemical vapour deposition (CVD) of silicon-based semiconductors. Among the computations they reported was computation of the Gibbs energy of BSi(g) at several temperatures based on a $^4\Sigma^-$ ground state with equilibrium internuclear distance of 190.5 pm and fundamental vibrational wavenumber of 772 cm^{-1} and a $^2P_0$ first excited level 8000 cm^{-1} above the ground level. Compute the standard molar Gibbs energy $G_m^{\ominus}(2000\text{ K}) - G_m^{\ominus}(0)$.

16.29‡ The molecule Cl_2O_2, which is believed to participate in the seasonal depletion of ozone over Antarctica, has been studied by several means. M. Birk *et al.* (*J. Chem. Phys.* **91**, 6588 (1989)) report its rotational constants (B) as 13 109.4, 2409.8, and 2139.7 MHz. They also report that its rotational spectrum indicates a molecule with a symmetry number of 2. J. Jacobs *et al.* (*J. Amer. Chem. Soc.* **116**, 1106 (1994)) report its vibrational wavenumbers as 753, 542, 310, 127, 646, and 419 cm^{-1}. Compute $G_m^{\ominus}(200\text{ K}) - G_m^{\ominus}(0)$ of Cl_2O_2.

16.30‡ J. Hutter *et al.* (*J. Amer. Chem. Soc.* **116**, 750 (1994)) examined the geometric and vibrational structure of several carbon molecules of formula C_n. Given that the ground state of C_3, a molecule found in interstellar space and in flames, is an angular singlet with moments of inertia 39.340, 39.032, and $0.3082m_u$ Å^2 (where 1 Å $= 10^{-10}$ m) and with vibrational wavenumbers of 63.4, 1224.5, and 2040 cm^{-1}, compute $G_m^{\ominus}(10.00\text{ K}) - G_m^{\ominus}(0)$ and $G_m^{\ominus}(1000\text{ K}) - G_m^{\ominus}(0)$ for C_3.

17 Molecular interactions

In this chapter we examine molecular interactions in gases and liquids and interpret them in terms of electric properties of molecules, such as electric dipole moments and polarizabilities. All these properties reflect the degree to which the nuclei of atoms exert control over the electrons in a molecule, either by causing electrons to accumulate in particular regions, or by permitting them to respond more or less strongly to the effects of external electric fields. We shall see here and in Chapter 18 that molecular interactions govern the structures and functions of molecular assemblies.

Molecular interactions are responsible for the unique properties of substances as simple as water and as complex as biological and synthetic macromolecules. The shapes and chemical properties of molecular assemblies also result from specific patterns of interactions between two or more atoms, molecules, or macromolecules. **Supramolecular chemistry** is the field of chemistry that studies the relationships between structure and function in molecular assemblies, such as drug–receptor complexes (*Impact I17.1*) and nanoscale catalysts. Molecular assemblies are treated lightly in this chapter and more extensively in Chapter 18. The interaction between ions is treated in Chapter 5 (for solutions) and Chapter 19 (for solids).

We begin our examination of molecular interactions by describing the electric properties of molecules, which may be interpreted in terms of concepts of electronic structure introduced in Chapter 10. We shall see that small imbalances of charge distributions in molecules allow them to interact with one another. This interaction results in the cohesion of molecules to form supramolecular assemblies and the bulk phases of matter.

Electric properties of molecules

Many of the electric properties of molecules can be traced to the competing influences of nuclei with different charges or to the competition between the control exercised by a nucleus and the influence of an externally applied field. The former competition may result in an electric dipole moment. The latter may result in properties such as refractive index and optical activity.

17.1 Electric dipole moments

Key points **A polar molecule is a molecule with a permanent electric dipole moment. The magnitude of a dipole moment is the product of the partial charge and the separation.**

An **electric dipole** consists of two electric charges $+Q$ and $-Q$ separated by a distance R. This arrangement of charges is represented by a vector $\boldsymbol{\mu}$ (1). The magnitude of $\boldsymbol{\mu}$ is

$\mu = QR$ and, although the SI unit of dipole moment is coulomb metre (C m), it is still commonly reported in the non-SI unit debye, D, named after Peter Debye, a pioneer in the study of dipole moments of molecules, where

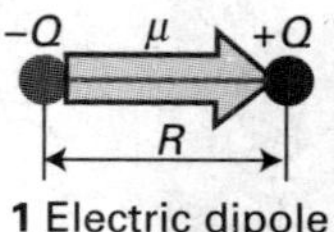

1 Electric dipole

$$1\ \mathrm{D} = 3.335\,64 \times 10^{-30}\ \mathrm{C\,m} \qquad (17.1)$$

The dipole moment of a pair of charges $+e$ and $-e$ separated by 100 pm is 1.6×10^{-29} C m, corresponding to 4.8 D. Dipole moments of small molecules are typically about 1 D. The conversion factor in eqn 17.1 stems from the original definition of the debye in terms of c.g.s. units: 1 D is the dipole moment of two equal and opposite charges of magnitude 1 e.s.u. separated by 1 Å.

A **polar molecule** is a molecule with a permanent electric dipole moment. The permanent dipole moment stems from the partial charges on the atoms in the molecule that arise from differences in electronegativity or other features of bonding (Sections 10.6–10.8). Nonpolar molecules acquire an induced dipole moment in an electric field on account of the distortion the field causes in their electronic distributions and nuclear positions; however, this induced moment is only temporary, and disappears as soon as the perturbing field is removed. Polar molecules also have their existing dipole moments temporarily modified by an applied field.

A brief comment

In elementary chemistry, an electric dipole moment is often represented by the arrow +⟶ added to the Lewis structure for the molecule, with the + marking the positive end. Note that the direction of the arrow is opposite to that of μ.

Microwave spectroscopy (Section 12.5) is used to measure the electric dipole moments of molecules for which a rotational spectrum can be observed. Measurements on a liquid or solid bulk sample are made with a method explained later. Computational software is now widely available, and typically computes electric dipole moments by assessing the electron density at each point in the molecule and its coordinates relative to the centroid of the molecule. However, it is still important to be able to formulate simple models of the origin of these moments and to understand how they arise. The following paragraphs focus on this aspect.

Table 17.1* Dipole moments (μ) and polarizability volumes (α')

	μ/D	$\alpha'/(10^{-30}\ \mathrm{m}^3)$
CCl_4	0	10.5
H_2	0	0.819
H_2O	1.85	1.48
HCl	1.08	2.63
HI	0.42	5.45

* More values are given in the *Data section*.

All heteronuclear diatomic molecules are polar, and typical values of μ include 1.08 D for HCl and 0.42 D for HI (Table 17.1). Molecular symmetry is of the greatest importance in deciding whether a polyatomic molecule is polar or not. Indeed, molecular symmetry is more important than the question of whether or not the atoms in the molecule belong to the same element. Homonuclear polyatomic molecules may be polar if they have low symmetry and the atoms are in inequivalent positions. For instance, the angular molecule ozone, O_3 (2), is homonuclear; however, it is polar because the central O atom is different from the outer two (it is bonded to two atoms; they are bonded only to one); moreover, the dipole moments associated with each bond make an angle to each other and do not cancel. Heteronuclear polyatomic molecules may be nonpolar if they have high symmetry, because individual bond dipoles may then cancel. The heteronuclear linear triatomic molecule CO_2, for example, is nonpolar because, although there are partial charges on all three atoms, the dipole moment associated with the OC bond points in the opposite direction to the dipole moment associated with the CO bond, and the two cancel (3).

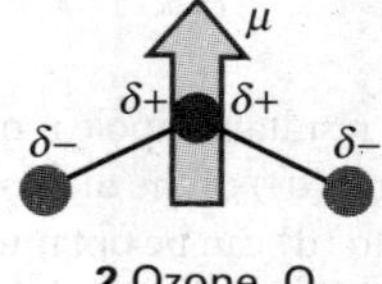

2 Ozone, O_3

3 Carbon dioxide, CO_2

To a first approximation, it is possible to resolve the dipole moment of a polyatomic molecule into contributions from various groups of atoms in the molecule and the directions in which these individual contributions lie (Fig. 17.1). Thus, 1,4-dichlorobenzene is nonpolar by symmetry on account of the cancellation of two equal but opposing C–Cl moments (exactly as in carbon dioxide). 1,2-Dichlorobenzene, however, has a dipole moment that is approximately the resultant of two chlorobenzene dipole moments arranged at 60° to each other. This technique of 'vector addition' can be applied with fair success to other series of related molecules, and the resultant μ_{res} of two dipole moments μ_1 and μ_2 that make an angle θ to each other (4) is approximately (see *Mathematical background* 4)

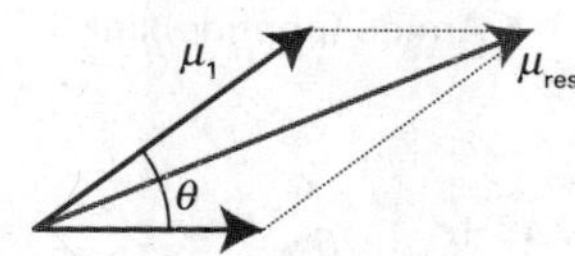

4 Addition of dipole moments

$$\mu_{res} \approx (\mu_1^2 + \mu_2^2 + 2\mu_1\mu_2 \cos\theta)^{1/2} \qquad (17.2a)$$

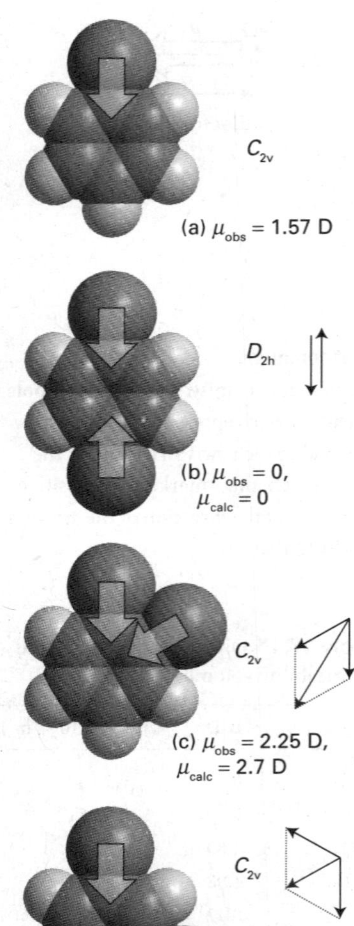

Fig. 17.1 The resultant dipole moments (red in (c) and (d)) of the dichlorobenzene isomers (b) to (d) can be obtained approximately by vectorial addition of two chlorobenzene dipole moments (1.57 D).

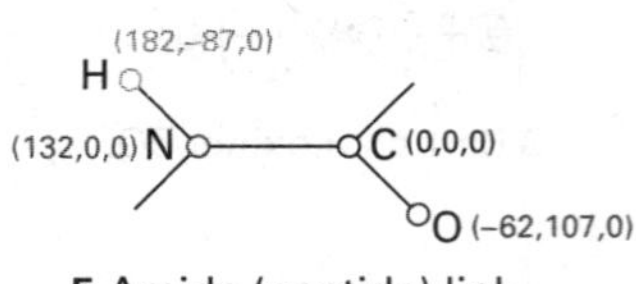

5 Amide (peptide) link

(182,–87,0) +0.18 H
(132,0,0) N –0.36
+0.45 C (0,0,0)
O (–62,107,0) –0.38

6

When the two dipole moments have the same magnitude (as in the dichlorobenzenes), this equation simplifies to

$$\mu_{res} \approx 2\mu_1 \cos \tfrac{1}{2}\theta \tag{17.2b}$$

Self-test 17.1 Estimate the ratio of the electric dipole moments of *ortho* (1,2-) and *meta* (1,3-) disubstituted benzenes. [μ (*ortho*)/μ (*meta*) $= 3^{1/2} \approx 1.7$]

A better approach to the calculation of dipole moments is to take into account the locations and magnitudes of the partial charges on all the atoms. These partial charges are included in the output of many molecular structure software packages. To calculate the x-component, for instance, we need to know the partial charge on each atom and the atom's x-coordinate relative to a point in the molecule and form the sum

$$\mu_x = \sum_J Q_J x_J \tag{17.3a}$$

Here Q_J is the partial charge of atom J, x_J is the x-coordinate of atom J, and the sum is over all the atoms in the molecule. Analogous expressions are used for the y- and z-components. For an electrically neutral molecule, the origin of the coordinates is arbitrary, so it is best chosen to simplify the measurements. In common with all vectors, the magnitude of $\boldsymbol{\mu}$ is related to the three components μ_x, μ_y, and μ_z by

$$\mu = (\mu_x^2 + \mu_y^2 + \mu_z^2)^{1/2} \tag{17.3b}$$

Example 17.1 *Calculating a molecular dipole moment*

Estimate the electric dipole moment of the amide group shown in (**5**) by using the partial charges (as multiples of e) in Table 17.2 and the locations of the atoms shown.

Method We use eqn 17.3a to calculate each of the components of the dipole moment and then eqn 17.3b to assemble the three components into the magnitude of the dipole moment. Note that the partial charges are multiples of the fundamental charge, $e = 1.609 \times 10^{-19}$ C.

Answer The expression for μ_x is

$$\begin{aligned}\mu_x &= (-0.36e) \times (132\text{ pm}) + (0.45e) \times (0\text{ pm}) + (0.18e) \times (182\text{ pm}) \\ &\quad + (-0.38e) \times (-62.0\text{ pm}) \\ &= 8.8e\text{ pm} \\ &= 8.8 \times (1.609 \times 10^{-19}\text{ C}) \times (10^{-12}\text{ m}) = 1.4 \times 10^{-30}\text{ C m}\end{aligned}$$

corresponding to $\mu_x = +0.42$ D. The expression for μ_y is:

$$\begin{aligned}\mu_y &= (-0.36e) \times (0\text{ pm}) + (0.45e) \times (0\text{ pm}) + (0.18e) \times (-87\text{ pm}) \\ &\quad + (-0.38e) \times (107\text{ pm}) \\ &= -56e\text{ pm} = -9.0 \times 10^{-30}\text{ C m}\end{aligned}$$

It follows that $\mu_y = -2.7$ D. The amide group is planar, so $\mu_z = 0$ and

$$\mu = \{(0.42\text{ D})^2 + (-2.7\text{ D})^2\}^{1/2} = 2.7\text{ D}$$

We can find the orientation of the dipole moment by arranging an arrow of length 2.7 units of length to have x-, y-, and z-components of 0.42, –2.7, and 0 units; the orientation is superimposed on (**6**).

Self-test 17.2 Calculate the electric dipole moment of formaldehyde by using the information in (7). [2.3 D]

17.2 Polarizabilities

Key point **The polarizability is a measure of the ability of a molecule to undergo a redistribution of charge in response to the application of an electric field, resulting in the induction of a dipole moment.**

An applied electric field can distort a molecule as well as align its permanent electric dipole moment. The **induced dipole moment**, μ^*, is generally proportional to the field strength, $\mathcal{E}$, and we write

$$\mu^* = \alpha\mathcal{E} \qquad \text{Definition of polarizability} \qquad (17.4)$$

The constant of proportionality α is the **polarizability** of the molecule. The greater the polarizability, the larger is the induced dipole moment for a given applied field. In a formal treatment, we should use vector quantities and allow for the possibility that the induced dipole moment might not lie parallel to the applied field, but for simplicity we discuss polarizabilities in terms of (scalar) magnitudes.

Polarizability has the units (coulomb metre)^2 per joule ($C^2\ m^2\ J^{-1}$). That collection of units is awkward, so α is often expressed as a **polarizability volume,** α', by using the relation

$$\alpha' = \frac{\alpha}{4\pi\varepsilon_0} \qquad \text{Definition of the polarizability volume} \qquad [17.5]$$

where ε_0 is the vacuum permittivity. Because the units of $4\pi\varepsilon_0$ are coulomb-squared per joule per metre ($C^2\ J^{-1}\ m^{-1}$), it follows that α' has the dimensions of volume (hence its name). Polarizability volumes are similar in magnitude to actual molecular volumes (of the order of $10^{-30}\ m^3$, $10^{-3}\ nm^3$, 1 Å^3).

Some experimental polarizability volumes of molecules are given in Table 17.1. As shown in the following *Justification*, polarizability volumes correlate with the HOMO–LUMO separations in atoms and molecules. The electron distribution can be distorted readily if the LUMO lies close to the HOMO in energy, so the polarizability is then large. If the LUMO lies high above the HOMO, an applied field cannot perturb the electron distribution significantly, and the polarizability is low. Molecules with small HOMO–LUMO gaps are typically large, with numerous electrons.

Justification 17.1 *Polarizabilities and molecular structures*

The quantum mechanical expression for the molecular polarizability in the z-direction is[1]

$$\alpha = 2\sum_{n\neq 0}\frac{|\mu_{z,0n}|^2}{E_n - E_0} \qquad (17.6)$$

where $\mu_{z,0n}$ is the *transition* electric dipole moment in the z-direction, a measure of the extent to which electric charge is shifted when an electron migrates from

[1] For a derivation of eqn 17.6 see our *Quanta, matter and change—A molecular approach to physical chemistry* (2009).

Table 17.2 Partial charges in polypeptides

Atom	Partial charge/e
C(=O)	+0.45
C(–CO)	+0.06
H(–C)	+0.02
H(–N)	+0.18
H(–O)	+0.42
N	−0.36
O	−0.38

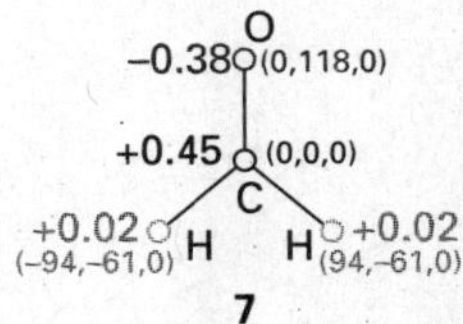

7

A brief comment

When using older compilations of data, it is useful to note that polarizability volumes have the same numerical values as the 'polarizabilities' reported using c.g.s. electrical units, so the tabulated values previously called 'polarizabilities' can be used directly.

the ground state to create an excited state. The sum is over the excited states, with energies E_n. The content of eqn 17.6 can be appreciated by approximating the excitation energies by a mean value ΔE (an indication of the HOMO–LUMO separation) and supposing that the most important transition dipole moment is approximately equal to the charge of an electron multiplied by the molecular radius R. Then

$$\alpha \approx \frac{2e^2R^2}{\Delta E}$$

This expression shows that α increases with the size of the molecule and with the ease with which it can be excited (the smaller the value of ΔE).

If the excitation energy is approximated by the energy needed to remove an electron to infinity from a distance R from a single positive charge, we can write $\Delta E \approx e^2/4\pi\varepsilon_0 R$. When this expression is substituted into the equation above, both sides are divided by $4\pi\varepsilon_0$, and the factor of 2 ignored in this approximation, we obtain $\alpha' \approx R^3$, which is of the same order of magnitude as the molecular volume.

For most molecules, the polarizability is anisotropic, by which is meant that its value depends on the orientation of the molecule relative to the field. The polarizability volume of benzene when the field is applied perpendicular to the ring is 0.0067 nm^3 and it is 0.0123 nm^3 when the field is applied in the plane of the ring. The anisotropy of the polarizability determines whether a molecule is rotationally Raman active (Section 12.6).

17.3 Polarization

Key points **The polarization is the electric dipole moment density. Orientation polarization is the polarization arising from the permanent dipole moments. Distortion polarization is the polarization arising from the distortion of the positions of the nuclei by the applied field. Electronic polarizability is the polarizability due to the distortion of the electron distribution.**

The **polarization**, P, of a sample is the electric dipole moment density, the mean electric dipole moment of the molecules, $\langle\mu\rangle$, multiplied by the number density, $\mathcal{N}$:

$$P = \langle\mu\rangle\mathcal{N} \qquad \text{Definition of polarization} \qquad (17.7)$$

In the following pages we refer to the sample as a **dielectric**, by which is meant a polarizable, nonconducting medium.

The polarization of an isotropic fluid sample is zero in the absence of an applied field because the molecules adopt ceaselessly changing random orientations due to thermal motion, so $\langle\mu\rangle = 0$. In the presence of a weak electric field, the orientations of the molecular dipoles fluctuate but we show in the following *Justification* that the mean value of the dipole moment for the sample at a temperature T is

$$\langle\mu_z\rangle = \frac{\mu^2\mathcal{E}}{3kT} \qquad \text{Mean dipole moment in the presence of a weak electric field} \qquad (17.8)$$

where z is the direction of the applied field $\mathcal{E}$. At very high electric fields the orientations of molecular dipoles fluctuate about the field direction to a lesser extent and the mean dipole moment approaches its maximum value of $\langle\mu_z\rangle = \mu$.

Justification 17.2 *The thermally averaged dipole moment*

The probability dp that a dipole has an orientation in the range θ to $\theta + d\theta$ is given by the Boltzmann distribution (Section 15.1b), which in this case is

$$dp = \frac{e^{-E(\theta)/kT} \sin\theta\, d\theta}{\int_0^{\pi} e^{-E(\theta)/kT} \sin\theta\, d\theta}$$

where $E(\theta)$ is the energy of the dipole in the field: $E(\theta) = -\mu\mathcal{E}\cos\theta$, with $0 \le \theta \le \pi$. The average value of the component of the dipole moment parallel to the applied electric field is therefore

$$\langle\mu_z\rangle = \int \mu\cos\theta\, dp = \mu\int\cos\theta\, dp = \frac{\mu\int_0^{\pi} e^{x\cos\theta}\cos\theta\sin\theta\, d\theta}{\int_0^{\pi} e^{x\cos\theta}\sin\theta\, d\theta}$$

with $x = \mu\mathcal{E}/kT$. The integral takes on a simpler appearance when we write $y = \cos\theta$ and $dy = -\sin\theta\, d\theta$, and change the limits of integration to $y = -1$ (at $\theta = \pi$) and $y = 1$ (at $\theta = 0$):

$$\langle\mu_z\rangle = \frac{\mu\int_{-1}^{1} y e^{xy}\, dy}{\int_{-1}^{1} e^{xy}\, dy}$$

At this point we use

$$\int_{-1}^{1} e^{xy}\, dy = \frac{e^x - e^{-x}}{x} \qquad \int_{-1}^{1} y e^{xy}\, dy = \frac{e^x + e^{-x}}{x} - \frac{e^x - e^{-x}}{x^2}$$

It is now straightforward algebra to combine these two results and to obtain

$$\langle\mu_z\rangle = \mu L(x) \qquad L(x) = \frac{e^x + e^{-x}}{e^x - e^{-x}} - \frac{1}{x} \qquad x = \frac{\mu\mathcal{E}}{kT} \qquad (17.9)$$

$L(x)$ is called the **Langevin function.**

Under most circumstances, x is very small (for example, if $\mu = 1$ D and $T = 300$ K, then x exceeds 0.01 only if the field strength exceeds 100 kV cm^{-1}, and most measurements are done at much lower strengths). The exponentials in the Langevin function can be expanded as $e^x = 1 + x + \frac{1}{2}x^2 + \frac{1}{6}x^3 + \cdots$ when the field is so weak that $x \ll 1$, and the largest term that survives is

$$L(x) = \tfrac{1}{3}x + \cdots \qquad (17.10)$$

Therefore, the average molecular dipole moment is given by eqn 17.8.

When the applied field changes direction slowly, the permanent dipole moment has time to reorient—the whole molecule rotates into a new direction—and follows the field. However, when the frequency of the field is high, a molecule cannot change direction fast enough to follow the change in direction of the applied field and the permanent dipole moment then makes no contribution to the polarization of the sample. Because a molecule takes about 1 ps to turn through about 1 radian in a fluid, the loss of this contribution to the polarization occurs when measurements are made at frequencies greater than about 10^{11} Hz (in the microwave region). We say that the **orientation polarization**, the polarization arising from the permanent dipole moments, is lost at such high frequencies.

The next contribution to the polarization to be lost as the frequency is raised is the **distortion polarization**, the polarization that arises from the distortion of the positions of the nuclei by the applied field. The molecule is bent and stretched by the applied field, and the molecular dipole moment changes accordingly. The time taken for a molecule to bend is approximately the inverse of the molecular vibrational frequency, so the distortion polarization disappears when the frequency of the radiation is increased through the infrared. The disappearance of polarization occurs in stages: as shown in the following *Justification*, each successive stage occurs as the incident frequency rises above the frequency of a particular mode of vibration. At even higher frequencies, in the visible region, only the electrons are mobile enough to respond to the rapidly changing direction of the applied field. The polarization that remains is now due entirely to the distortion of the electron distribution, and the surviving contribution to the molecular polarizability is called the **electronic polarizability**.

Justification 17.3 *The frequency dependence of polarizabilities*

The quantum mechanical expression for the polarizability of a molecule in the presence of an electric field that is oscillating at a frequency ω in the z-direction is[2]

$$\alpha(\omega)=\frac{2}{\hbar}\sum_n \frac{\omega_{n0}|\mu_{z,0n}|^2}{\omega_{n0}^2-\omega^2} \qquad \text{Frequency dependence of the polarizability} \qquad (17.11)$$

The quantities in this expression (which is valid provided that ω is not close to ω_{n0}) are the same as those in *Justification* 17.1, with $\hbar\omega_{n0}=E_n-E_0$. As $\omega\to 0$, the equation reduces to eqn 17.6 for the static polarizability. As ω becomes very high (and much higher than any excitation frequency of the molecule so that the ω_{n0}^2 in the denominator can be ignored), the polarizability becomes

$$\alpha(\omega)=-\frac{2}{\hbar\omega^2}\sum_n \omega_{n0}|\mu_{z,0n}|^2\to 0 \qquad \text{as} \qquad \omega\to\infty$$

That is, when the incident frequency is much higher than any excitation frequency, the polarizability becomes zero. The argument applies to each type of excitation, vibrational as well as electronic, and accounts for the successive decreases in polarizability as the frequency is increased.

17.4 Relative permittivities

Key points **The permittivity is the quantity ε in the Coulomb potential energy, $V=Q_1Q_2/4\pi\varepsilon r$. The relative permittivity is given by $\varepsilon_r=\varepsilon/\varepsilon_0$ and may be calculated from electric properties by using the Debye equation or the Clausius–Mossotti equation.**

When two charges Q_1 and Q_2 are separated by a distance r in a vacuum, the Coulomb potential energy of their interaction is

$$V=\frac{Q_1Q_2}{4\pi\varepsilon_0 r} \qquad (17.12a)$$

[2] For a derivation of eqn 17.11 see our *Quanta, matter and change—A molecular approach to physical chemistry* (2009).

When the same two charges are immersed in a medium (such as air or a liquid), their potential energy is reduced to

$$V=\frac{Q_1Q_2}{4\pi\varepsilon r} \qquad (17.12b)$$

where ε is the **permittivity** of the medium. The permittivity is normally expressed in terms of the dimensionless **relative permittivity**, ε_r, (formerly and still widely called the dielectric constant) of the medium:

$$\varepsilon_r=\frac{\varepsilon}{\varepsilon_0} \qquad [17.13]$$

Definition of relative permittivity

The relative permittivity can have a very significant effect on the strength of the interactions between ions in solution. For instance, water has a relative permittivity of 78 at 25°C, so the interionic Coulombic interaction energy is reduced by nearly two orders of magnitude from its vacuum value. Some of the consequences of this reduction for electrolyte solutions were explored in Chapter 5.

The relative permittivity of a substance is large if its molecules are polar or highly polarizable. The quantitative relation between the relative permittivity and the electric properties of the molecules is obtained by considering the polarization of a medium, and is expressed by the **Debye equation**:

$$\frac{\varepsilon_r-1}{\varepsilon_r+2}=\frac{\rho P_m}{M} \qquad (17.14)$$

Debye equation

where ρ is the mass density of the sample, M is the molar mass of the molecules, and P_m is the **molar polarization**, which is defined as

$$P_m=\frac{N_A}{3\varepsilon_0}\left(\alpha+\frac{\mu^2}{3kT}\right) \qquad [17.15]$$

Definition of molar polarization

(where α is the polarizability, not the polarizability volume α'). The term $\mu^2/3kT$ stems from the thermal averaging of the electric dipole moment in the presence of the applied field (eqn 17.8). The corresponding expression without the contribution from the permanent dipole moment is called the **Clausius–Mossotti equation:**

$$\frac{\varepsilon_r-1}{\varepsilon_r+2}=\frac{\rho N_A\alpha}{3M\varepsilon_0} \qquad (17.16)$$

Clausius–Mossotti equation

The Clausius–Mossotti equation is used when there is no contribution from permanent electric dipole moments to the polarization, either because the molecules are nonpolar or because the frequency of the applied field is so high that the molecules cannot orientate quickly enough to follow the change in direction of the field.

Example 17.2 *Determining dipole moment and polarizability*

The relative permittivity of a substance is measured by comparing the capacitance of a capacitor with and without the sample present (C and C_0, respectively) and using $\varepsilon_r=C/C_0$. The relative permittivity of camphor (**8**) was measured at a series of temperatures with the results given below. Determine the dipole moment and the polarizability volume of the molecule.

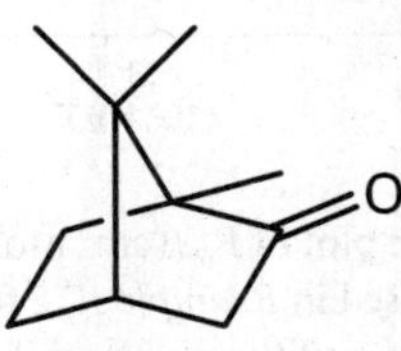

8 Camphor

θ/°C	ρ/(g cm^{-3})	ε_r
0	0.99	12.5
20	0.99	11.4
40	0.99	10.8
60	0.99	10.0
80	0.99	9.50
100	0.99	8.90
120	0.97	8.10
140	0.96	7.60
160	0.95	7.11
200	0.91	6.21

Method Equations 17.14 and 17.15 imply that the polarizability and permanent electric dipole moment of the molecules in a sample can be determined by measuring ε_r at a series of temperatures, calculating P_m, and plotting it against $1/T$. The slope of the graph is $N_A\mu^2/9\varepsilon_0 k$ and its intercept at $1/T=0$ is $N_A\alpha/3\varepsilon_0$. We need to calculate $(\varepsilon_r-1)/(\varepsilon_r+2)$ at each temperature, and then multiply by M/ρ to form P_m.

Answer For camphor, $M = 152.23$ g mol^{-1}. We can therefore use the data to draw up the following table:

θ/°C	(10^3 K)/T	ε_r	$(\varepsilon_r-1)/(\varepsilon_r+2)$	P_m/(cm^3 mol^{-1})
0	3.66	12.5	0.793	122
20	3.41	11.4	0.776	119
40	3.19	10.8	0.766	118
60	3.00	10.0	0.750	115
80	2.83	9.50	0.739	114
100	2.68	8.90	0.725	111
120	2.54	8.10	0.703	110
140	2.42	7.60	0.688	109
160	2.31	7.11	0.670	107
200	2.11	6.21	0.634	106

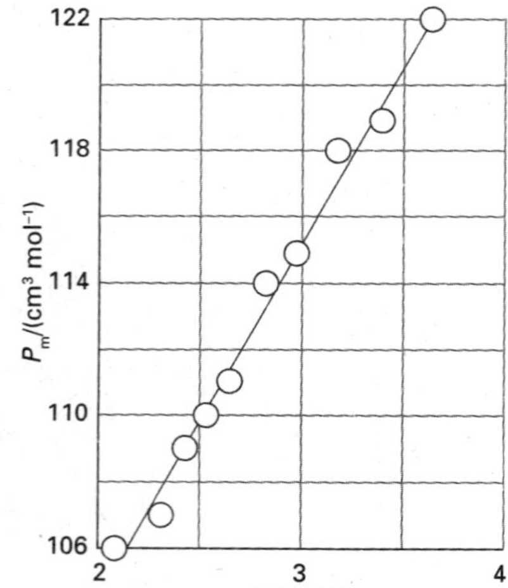

Fig. 17.2 The plot of P_m/(cm^3 mol^{-1}) against (10^3 K)/T used in *Example 17.2* for the determination of the polarizability and dipole moment of camphor.

The points are plotted in Fig. 17.2. The intercept lies at 82.9, so $\alpha' = 3.3\times10^{-23}$ cm^3. The slope is 10.7, so $\mu = 4.42\times10^{-30}$ C m, corresponding to 1.33 D. Because the Debye equation describes molecules that are free to rotate, the data show that camphor, which does not melt until 175°C, is rotating even in the solid. It is an approximately spherical molecule.

Self-test 17.3 The relative permittivity of chlorobenzene is 5.71 at 20°C and 5.62 at 25°C. Assuming a constant density (1.11 g cm^{-3}), estimate its polarizability volume and dipole moment. [1.4×10^{-23} cm^3, 1.1 D]

The Maxwell equations, which describe the properties of electromagnetic radiation, relate the refractive index at a (visible or ultraviolet) specified wavelength to the relative permittivity at that frequency:

$$n_r = \varepsilon_r^{1/2} \quad \text{Relation between refractive index and relative permittivity} \qquad (17.17)$$

where the refractive index, n_r, of the medium is the ratio of the speed of light in a vacuum, c, to its speed c' in the medium: $n_r = c/c'$. (A beam of light changes direction ('bends') when it passes from a region of one refractive index to a region with a different refractive index.) Therefore, the molar polarization, P_m, and the molecular polarizability, α, can be measured at frequencies typical of visible light (about 10^{15} to 10^{16} Hz) by measuring the refractive index of the sample and using the Clausius–Mossotti equation.

Interactions between molecules

A **van der Waals interaction** is the attractive interaction between closed-shell molecules; some strict definitions also require that it be an interaction with a potential energy that depends on the distance between the molecules as $1/r^6$. In addition, there are interactions between ions and the partial charges of polar molecules and repulsive interactions that prevent the complete collapse of matter to nuclear densities. The repulsive interactions arise from Coulombic repulsions and, indirectly, from the Pauli principle and the exclusion of electrons from regions of space where the orbitals of neighbouring species overlap.

17.5 Interactions between dipoles

Key points A van der Waals interaction between closed-shell molecules is an interaction that gives rise to a potential energy that is inversely proportional to the sixth power of their separation. The potential energy of each type of intermolecular interaction has a characteristic distance dependence: (a) point-dipole–point-charge ($1/r^2$); (b) dipole–dipole ($1/r^3$, non-rotating molecules; $1/r^6$, rotating molecules); (c) dipole–induced-dipole interaction ($1/r^6$); (d) dispersion (or London) interaction ($1/r^6$). (e) A hydrogen bond is an interaction of the form A–H···B, where A and B are N, O, or F. (f) A hydrophobic interaction favours the clustering of hydrophobic groups in aqueous environments. (g) The total attractive interaction energy between rotating molecules is then the sum of the three van der Waals contributions discussed above.

Most of the discussion in this section is based on the Coulombic potential energy of interaction between two charges (eqn 17.12a). We can easily adapt this expression to find the potential energy of a point charge and a dipole and extend it to the interaction between two dipoles.

(a) The potential energy of interaction

We show in the *Justification* below that the potential energy of interaction between a point dipole $\mu_1 = Q_1 l$ and the point charge Q_2 in the arrangement shown in (9) is

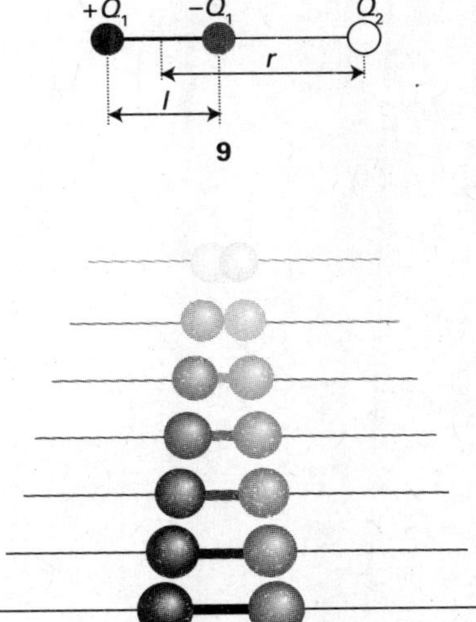

$$V = -\frac{\mu_1 Q_2}{4\pi\varepsilon_0 r^2} \qquad (17.18)$$

Energy of interaction between a point dipole and a point charge

With μ in coulomb metres, Q_2 in coulombs, and r in metres, V is obtained in joules. A **point dipole** is a dipole in which the separation between the charges is much smaller than the distance at which the dipole is being observed, ($l \ll r$). The potential energy rises towards zero (the value at infinite separation of the charge and the dipole) more rapidly (as $1/r^2$) than that between two point charges (which varies as $1/r$) because, from the viewpoint of the point charge, the partial charges of the dipole seem to merge and cancel as the distance r increases (Fig. 17.3).

Fig. 17.3 There are two contributions to the diminishing field of an electric dipole with distance (here seen from the side). The potentials of the charges decrease (shown here by a fading intensity) and the two charges appear to merge, so their combined effect approaches zero more rapidly than by the distance effect alone.

Justification 17.4 *The interaction between a point charge and a point dipole*

The sum of the potential energies of repulsion between like charges and attraction between opposite charges in the orientation shown in (**9**) is

$$V=\frac{1}{4\pi\varepsilon_0}\left(-\frac{Q_1Q_2}{r-\frac{1}{2}l}+\frac{Q_1Q_2}{r+\frac{1}{2}l}\right)=\frac{Q_1Q_2}{4\pi\varepsilon_0 r}\left(-\frac{1}{1-x}+\frac{1}{1+x}\right)$$

where $x=l/2r$ Because $l\ll r$ for a point dipole, this expression can be simplified by expanding the terms in x by using

$$\frac{1}{1+x}=1-x+x^2-\cdots \qquad \frac{1}{1-x}=1+x+x^2+\cdots$$

and retaining only the leading surviving term:

$$V=\frac{Q_1Q_2}{4\pi\varepsilon_0 r}\{-(1+x+\cdots)+(1-x+\cdots)\}\approx-\frac{2xQ_1Q_2}{4\pi\varepsilon_0 r}=-\frac{Q_1Q_2l}{4\pi\varepsilon_0 r^2}$$

With $\mu_1=Q_1l$, this expression becomes eqn 17.18. This expression should be multiplied by cos θ when the point charge lies at an angle θ to the axis of the dipole.

10

Example 17.3 *Calculating the interaction energy of two dipoles*

Calculate the potential energy of interaction of two dipoles in the arrangement shown in (**10**) when their separation is r.

Method We proceed in exactly the same way as in *Justification 17.4*, but now the total interaction energy is the sum of four pairwise terms, two attractions between opposite charges, which contribute negative terms to the potential energy, and two repulsions between like charges, which contribute positive terms.

Answer The sum of the four contributions is

$$V=\frac{1}{4\pi\varepsilon_0}\left(-\frac{Q_1Q_2}{r+l}+\frac{Q_1Q_2}{r}+\frac{Q_1Q_2}{r}-\frac{Q_1Q_2}{r-l}\right)=-\frac{Q_1Q_2}{4\pi\varepsilon_0 r}\left(\frac{1}{1+x}-2+\frac{1}{1-x}\right)$$

with $x=l/r$. As before, provided $l\ll r$ we can expand the two terms in x and retain only the first surviving term, which is equal to $2x^2$. This step results in the expression

$$V=-\frac{2x^2Q_1Q_2}{4\pi\varepsilon_0 r}$$

Therefore, because $\mu_1=Q_1l$ and $\mu_2=Q_2l$, the potential energy of interaction in the alignment shown in the illustration is

$$V=-\frac{\mu_1\mu_2}{2\pi\varepsilon_0 r^3}$$

This interaction energy approaches zero more rapidly (as $1/r^3$) than for the previous case: now both interacting entities appear neutral to each other at large separations. See *Further information 17.1* for the general expression.

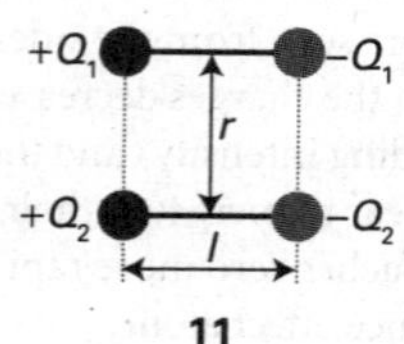

11

Self-test 17.4 Derive an expression for the potential energy when the dipoles are in the arrangement shown in (**11**). $[V=\mu_1\mu_2/4\pi\varepsilon_0 r^3]$

Table 17.3 Multipole interaction potential energies

Interaction type	Distance dependence of potential energy	Typical energy/ (kJ mol^{-1})	Comment
Ion–ion	$1/r$	250	Only between ions*
Ion–dipole	$1/r^2$	15	
Dipole–dipole	$1/r^3$	2	Between stationary polar molecules
	$1/r^6$	0.6	Between rotating polar molecules
London (dispersion)	$1/r^6$	2	Between all types of molecules
Hydrogen bond		20	Interaction of the type A–H···B, with A, B = O, N, or F

* Electrolyte solutions are treated in Chapter 5; ionic solids in Chapter 19.

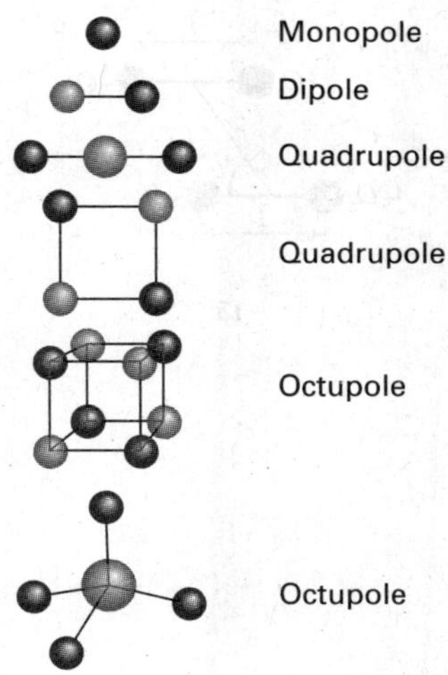

Fig. 17.4 Typical charge arrays corresponding to electric multipoles. The field arising from an arbitrary finite charge distribution can be expressed as the superposition of the fields arising from a superposition of multipoles.

Table 17.3 summarizes the various expressions for the interaction of charges and dipoles. It is quite easy to extend the formulas given there to obtain expressions for the energy of interaction of higher **multipoles**, or arrays of point charges (Fig. 17.4). Specifically, an ***n*-pole** is an array of point charges with an n-pole moment but no lower moment. Thus, a **monopole** ($n = 1$) is a point charge, and the monopole moment is what we normally call the overall charge. A dipole ($n = 2$), as we have seen, is an array of charges that has no monopole moment (no net charge). A **quadrupole** ($n = 3$) consists of an array of point charges that has neither net charge nor dipole moment (as for CO_2 molecules, 3). An **octupole** ($n = 4$) consists of an array of point charges that sum to zero and which has neither a dipole moment nor a quadrupole moment (as for CH_4 molecules, 12). The feature to remember is that the interaction energy falls off more rapidly the higher the order of the multipole. For the interaction of an n-pole with an m-pole, the potential energy varies with distance as

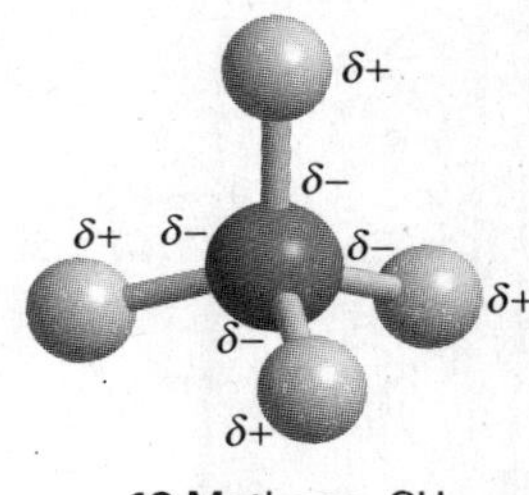

12 Methane, CH_4

$$V \propto \frac{1}{r^{n+m-1}} \qquad \text{Energy of interaction between multipoles} \qquad (17.19)$$

The reason for the even steeper decrease with distance is the same as before: the array of charges appears to blend together into neutrality more rapidly with distance the higher the number of individual charges that contribute to the multipole. Note that a given molecule may have a charge distribution that corresponds to a superposition of several different multipoles.

The same kind of argument as that used to derive expressions for the potential energy can be used to establish the distance dependence of the strength of the electric field generated by a dipole. We shall need this expression when we calculate the dipole moment induced in one molecule by another.

A brief comment

The electric field is actually a vector, and we cannot simply add and subtract magnitudes without taking into account the directions of the fields. In the cases we consider, this will not be a complication because the two charges of the dipoles will be collinear and give rise to fields in the same direction. Be careful, though, with more general arrangements of charges.

The starting point for the calculation is the strength of the electric field generated by a point electric charge:

$$\mathcal{E} = \frac{Q}{4\pi\varepsilon_0 r^2} \qquad \text{Electric field generated by a point charge} \qquad (17.20)$$

The field generated by a dipole is the sum of the fields generated by each partial charge. For the point-dipole arrangement shown in Fig. 17.5, the same procedure that was used to derive the potential energy gives

$$\mathcal{E} = \frac{\mu}{2\pi\varepsilon_0 r^3} \qquad \text{Electric field generated by a point dipole} \qquad (17.21)$$

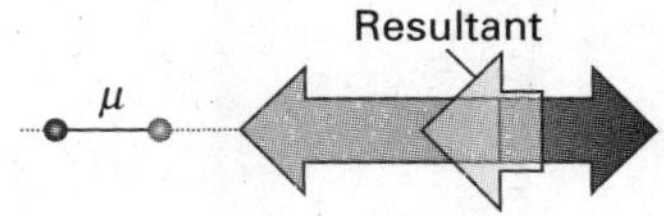

Fig. 17.5 The electric field of a dipole is the sum of the opposing fields from the positive and negative charges, each of which is proportional to $1/r^2$. The difference, the net field, is proportional to $1/r^3$.

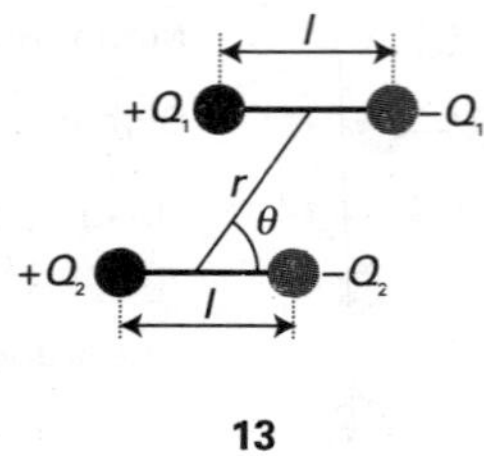

13

The electric field of a multipole (in this case a dipole) decreases more rapidly with distance (as $1/r^3$ for a dipole) than a monopole (a point charge).

(b) Dipole–dipole interactions

The potential energy of interaction between two polar molecules is a complicated function of their relative orientation. When the two dipoles are parallel (as in **13**), the potential energy is simply (see *Further information 17.1*)

$$V=\frac{\mu_1\mu_2 f(\theta)}{4\pi\varepsilon_0 r^3} \qquad f(\theta)=1-3\cos^2\theta \tag{17.22}$$

Energy of interaction between two fixed parallel dipoles

This expression applies to polar molecules in a fixed, parallel, orientation in a solid. In a fluid of freely rotating molecules, the interaction between dipoles averages to zero because $f(\theta)$ changes sign as the orientation changes, and its average value is zero. Physically, the like partial charges of two freely rotating molecules are close together as much as the two opposite charges, and the repulsion of the former is cancelled by the attraction of the latter. Mathematically, this result arises from the fact that the average (or mean value) of the function $1-3\cos^2\theta$ is

$$(1/\pi)\int_0^\pi (1-3\cos^2\theta)\sin\theta\,d\theta=0$$

The interaction energy of two *freely* rotating dipoles is zero. However, because their mutual potential energy depends on their relative orientation, the molecules do not in fact rotate completely freely, even in a gas. In fact, the lower energy orientations are marginally favoured, so there is a nonzero average interaction between polar molecules. We show in the following *Justification* that the average potential energy of two rotating molecules that are separated by a distance r is

$$\langle V\rangle=-\frac{C}{r^6} \qquad C=\frac{2\mu_1^2\mu_2^2}{3(4\pi\varepsilon_0)^2kT} \tag{17.23}$$

Average energy of interaction between two rotating polar molecules

This expression describes the **Keesom interaction**, and is the first of the contributions to the van der Waals interaction.

Justification 17.5 *The Keesom interaction*

The detailed calculation of the Keesom interaction energy is quite complicated, but the form of the final answer can be constructed quite simply. First, we note that the average interaction energy of two polar molecules rotating at a fixed separation r is given by

$$\langle V\rangle=\frac{\mu_1\mu_2\langle f\rangle}{4\pi\varepsilon_0 r^3}$$

where $\langle f\rangle$ now includes a weighting factor in the averaging that is equal to the probability that a particular orientation will be adopted. This probability is given by the Boltzmann distribution $p\propto e^{-E/kT}$, with E interpreted as the potential energy of interaction of the two dipoles in that orientation. That is,

$$p\propto e^{-V/kT} \qquad V=\frac{\mu_1\mu_2 f}{4\pi\varepsilon_0 r^3}$$

When the potential energy of interaction of the two dipoles is very small compared with the energy of thermal motion, we can use $V \ll kT$, expand the exponential function in p, and retain only the first two terms:

$$p \propto 1 - V/kT + \cdots$$

We now write the weighted average of f as

$$\langle f \rangle = \frac{\int_0^\pi fp \, d\theta}{\int_0^\pi d\theta} = \frac{1}{\pi}\int_0^\pi f e^{-V/kT} d\theta = \frac{1}{\pi}\int_0^\pi f(1 - V/kT) d\theta + \cdots$$

It follows that

$$\langle f \rangle = \frac{1}{\pi}\int_0^\pi f \, d\theta - \frac{1}{\pi}\int_0^\pi f(V/kT) d\theta + \cdots = \frac{1}{\pi}\int_0^\pi f d\theta - \frac{1}{\pi}\int_0^\pi \frac{\mu_1\mu_2}{4\pi\varepsilon_0 kTr^3} f^2 d\theta + \cdots$$

$$= \overbrace{\frac{1}{\pi}\int_0^\pi f \, d\theta}^{\langle f \rangle_0} - \frac{\mu_1\mu_2}{4\pi\varepsilon_0 kTr^3}\overbrace{\left(\frac{1}{\pi}\int_0^\pi f^2 \, d\theta\right)}^{\langle f^2 \rangle_0} + \cdots$$

$$= \langle f \rangle_0 - \frac{\mu_1\mu_2}{4\pi\varepsilon_0 kTr^3}\langle f^2 \rangle_0 + \cdots$$

where $\langle \cdots \rangle_0$ denotes an unweighted spherical average. The spherical average of f is

$$\langle f \rangle_0 = \frac{1}{\pi}\int_0^\pi (1 - 3\cos^2\theta)\sin\theta \, d\theta = 0$$

so the first term in the expression for $\langle f \rangle$ vanishes. However, the average value of f^2 is nonzero because f^2 is positive at all orientations, so we can write

$$\langle V \rangle = -\frac{\mu_1^2\mu_2^2\langle f^2 \rangle_0}{(4\pi\varepsilon_0)^2 kTr^6}$$

The average value $\langle f^2 \rangle_0$ turns out to be $\frac{2}{3}$ when the calculation is carried through in detail. The final result is that quoted in eqn 17.23.

The important features of eqn 17.23 are:

- The negative sign shows that the average interaction is attractive.
- The dependence of the average interaction energy on the inverse sixth power of the separation identifies it as a van der Waals interaction.
- The inverse dependence on the temperature reflects the way in which the greater thermal motion overcomes the mutual orientating effects of the dipoles at higher temperatures.
- The inverse sixth power arises from the inverse third power of the interaction potential energy that is weighted by the energy in the Boltzmann term, which is also proportional to the inverse third power of the separation.

At 25°C the average interaction energy for pairs of molecules with $\mu = 1$ D is about -0.06 kJ mol^{-1} when the separation is 0.5 nm. This energy should be compared with the average molar kinetic energy of $\frac{3}{2}RT = 3.7$ kJ mol^{-1} at the same temperature. The interaction energy is also much smaller than the energies involved in the making and breaking of chemical bonds.

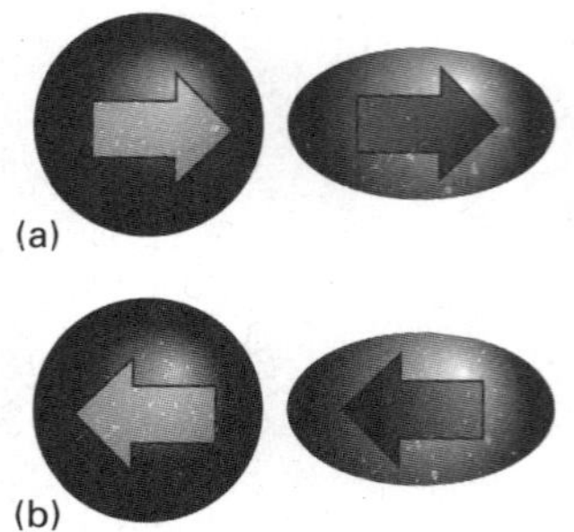

Fig. 17.6 (a) A polar molecule can induce a dipole in a nonpolar molecule, and (b) the latter's orientation follows the former's, so the interaction does not average to zero.

(c) Dipole–induced-dipole interactions

A polar molecule with dipole moment μ_1 can induce a dipole μ_2^* in a neighbouring polarizable molecule (Fig. 17.6). The induced dipole interacts with the permanent dipole of the first molecule, and the two are attracted together. The average interaction energy when the separation of the molecules is r is

$$V = -\frac{C}{r^6} \qquad C = \frac{\mu_1^2 \alpha_2'}{4\pi\varepsilon_0}$$ Energy of interaction between a polar molecule and a polarizable molecule (17.24)

where α_2' is the polarizability volume of molecule 2 and μ_1 is the permanent dipole moment of molecule 1. Note that the C in this expression is different from the C in eqn 17.23 and other expressions below: we are using the same symbol in C/r^6 to emphasize the similarity of form of each expression.

The dipole–induced-dipole interaction energy is independent of the temperature because thermal motion has no effect on the averaging process. Moreover, like the dipole–dipole interaction, the potential energy depends on $1/r^6$: this distance dependence stems from the $1/r^3$ dependence of the field (and hence the magnitude of the induced dipole) and the $1/r^3$ dependence of the potential energy of interaction between the permanent and induced dipoles. For a molecule with $\mu = 1$ D (such as HCl) near a molecule of polarizability volume $\alpha' = 10 \times 10^{-30}$ m^3, the average interaction energy is about -0.8 kJ mol^{-1} when the separation is 0.3 nm.

(d) Induced-dipole–induced-dipole interactions

Nonpolar molecules (including closed-shell atoms, such as Ar) attract one another even though neither has a permanent dipole moment. The abundant evidence for the existence of interactions between them is the formation of condensed phases of nonpolar substances, such as the condensation of hydrogen or argon to a liquid at low temperatures and the fact that benzene is a liquid at normal temperatures.

The interaction between nonpolar molecules arises from the transient dipoles that all molecules possess as a result of fluctuations in the instantaneous positions of electrons. To appreciate the origin of the interaction, suppose that the electrons in one molecule flicker into an arrangement that gives the molecule an instantaneous dipole moment μ_1^*. This dipole generates an electric field that polarizes the other molecule, and induces in that molecule an instantaneous dipole moment μ_2^*. The two dipoles attract each other and the potential energy of the pair is lowered. Although the first molecule will go on to change the size and direction of its instantaneous dipole, the electron distribution of the second molecule will follow, that is, the two dipoles are correlated in direction (Fig. 17.7). Because of this correlation, the attraction between the two instantaneous dipoles does not average to zero, and gives rise to an induced-dipole–induced-dipole interaction. This interaction is called either the **dispersion interaction** or the **London interaction** (for Fritz London, who first described it).

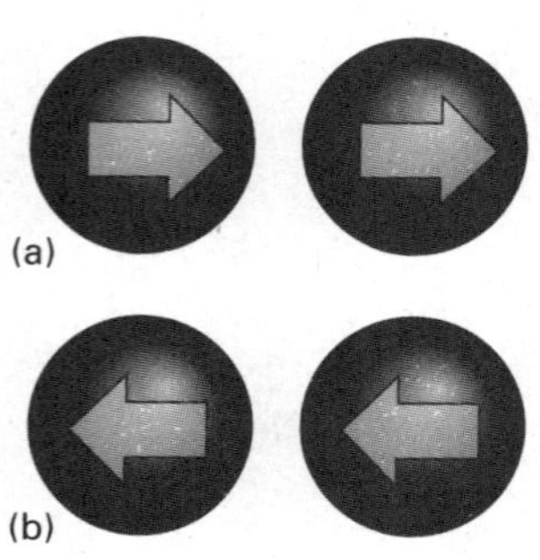

Fig. 17.7 (a) In the dispersion interaction, an instantaneous dipole on one molecule induces a dipole on another molecule, and the two dipoles then interact to lower the energy. (b) The two instantaneous dipoles are correlated and, although they occur in different orientations at different instants, the interaction does not average to zero.

The strength of the dispersion interaction depends on the polarizability of the first molecule because the instantaneous dipole moment μ_1^* depends on the looseness of the control that the nuclear charge exercises over the outer electrons. The strength of the interaction also depends on the polarizability of the second molecule, for that polarizability determines how readily a dipole can be induced by another molecule. The actual calculation of the dispersion interaction is quite involved, but a reasonable approximation to the interaction energy is given by the **London formula**:

$$V = -\frac{C}{r^6} \qquad C = \tfrac{3}{2}\alpha_1'\alpha_2'\frac{I_1 I_2}{I_1 + I_2}$$ London formula (17.25)

where I_1 and I_2 are the ionization energies of the two molecules (Table 9.3). This interaction energy is also proportional to the inverse sixth power of the separation of the molecules, which identifies it as a third contribution to the van der Waals interaction. The dispersion interaction generally dominates all the interactions between molecules other than hydrogen bonds.

• A brief illustration

For two CH_4 molecules, we can substitute $\alpha' = 2.6 \times 10^{-30}$ m^3 and $I \approx 700$ kJ mol^{-1} to obtain $V = -5$ kJ mol^{-1} for $r = 0.3$ nm. A very rough check on this figure is the enthalpy of vaporization of methane, which is 8.2 kJ mol^{-1}. However, this comparison is insecure, partly because the enthalpy of vaporization is a many-body quantity and partly because the long-distance assumption breaks down. •

(e) Hydrogen bonding

The interactions described so far are universal in the sense that they are possessed by all molecules independent of their specific identity. However, there is a type of interaction possessed by molecules that have a particular constitution. A **hydrogen bond** is an attractive interaction between two species that arises from a link of the form A–H···B, where A and B are highly electronegative elements and B possesses a lone pair of electrons. Hydrogen bonding is conventionally regarded as being limited to N, O, and F but, if B is an anionic species (such as Cl^-), it may also participate in hydrogen bonding. There is no strict cut-off for an ability to participate in hydrogen bonding, but N, O, and F participate most effectively.

The formation of a hydrogen bond can be regarded either as the approach between a partial positive charge of H and a partial negative charge of B or as a particular example of delocalized molecular orbital formation in which A, H, and B each supply one atomic orbital from which three molecular orbitals are constructed (Fig. 17.8). Experimental evidence and theoretical arguments have been presented in favour of both views and the matter has not yet been resolved. The electrostatic interaction model can be understood readily in terms of the discussion in Section 17.5b. Here we develop the molecular orbital model.

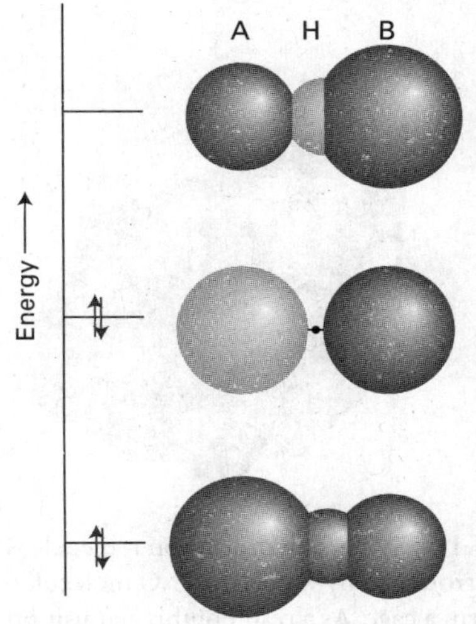

Fig. 17.8 The molecular orbital interpretation of the formation of an A–H···B hydrogen bond. From the three A, H, and B orbitals, three molecular orbitals can be formed (their relative contributions are represented by the sizes of the spheres. Only the two lower energy orbitals are occupied, and there may therefore be a net lowering of energy compared with the separate AH and B species.

Thus, if the A–H bond is regarded as formed from the overlap of an orbital on A, ψ_A, and a hydrogen 1s orbital, ψ_H, and the lone pair on B occupies an orbital on B, ψ_B, then, when the two molecules are close together, we can build three molecular orbitals from the three basis orbitals:

$$\psi = c_1\psi_A + c_2\psi_H + c_3\psi_B$$

One of the molecular orbitals is bonding, one almost nonbonding, and the third anti-bonding. These three orbitals need to accommodate four electrons (two from the original A–H bond and two from the lone pair of B), so two enter the bonding orbital and two enter the nonbonding orbital. Because the anti-bonding orbital remains empty, the net effect—depending on the precise location of the almost nonbonding orbital—may be a lowering of energy.

In practice, the strength of the bond is found to be about 20 kJ mol^{-1}. Because the bonding depends on orbital overlap, it is virtually a contact-like interaction that is turned on when AH touches B and is zero as soon as the contact is broken. If hydrogen bonding is present, it dominates the other intermolecular interactions. The properties of liquid and solid water, for example, are dominated by the hydrogen bonding between H_2O molecules. The structure of DNA and hence the transmission of genetic information is crucially dependent on the strength of hydrogen bonds between base pairs. The structural evidence for hydrogen bonding comes from noting that the internuclear distance between formally nonbonded atoms is less than their van der Waals contact distance, which suggests that a dominating attractive interaction is present. For example, the O–O distance in O–H···O is expected to be 280 pm on the basis of van der Waals radii, but is found to be 270 pm in typical compounds. Moreover, the H···O distance is expected to be 260 pm but is found to be only 170 pm.

Hydrogen bonds may be either symmetric or unsymmetric. In a symmetric hydrogen bond, the H atom lies midway between the two other atoms. This arrangement is rare, but occurs in $F–H···F^-$, where both bond lengths are 120 pm. More common is the unsymmetrical arrangement, where the A–H bond is shorter than the H···B bond. Simple electrostatic arguments, treating A–H···B as an array of point charges (partial negative charges on A and B, partial positive on H) suggest that the lowest energy is achieved when the bond is linear, because then the two partial negative charges are furthest apart. The experimental evidence from structural studies supports a linear or near-linear arrangement.

(f) The hydrophobic interaction

Nonpolar molecules do dissolve slightly in polar solvents, but strong interactions between solute and solvent are not possible and as a result it is found that each individual solute molecule is surrounded by a solvent cage (Fig. 17.9). To understand the consequences of this effect, consider the thermodynamics of transfer of a nonpolar hydrocarbon solute from a nonpolar solvent to water, a polar solvent. Experiments indicate that the process is endergonic ($\Delta_{\text{transfer}}G > 0$), as expected on the basis of the increase in polarity of the solvent, but exothermic ($\Delta_{\text{transfer}}H < 0$). Therefore, it is a large decrease in the entropy of the system ($\Delta_{\text{transfer}}S < 0$) that accounts for the positive Gibbs energy of transfer. For example, the process

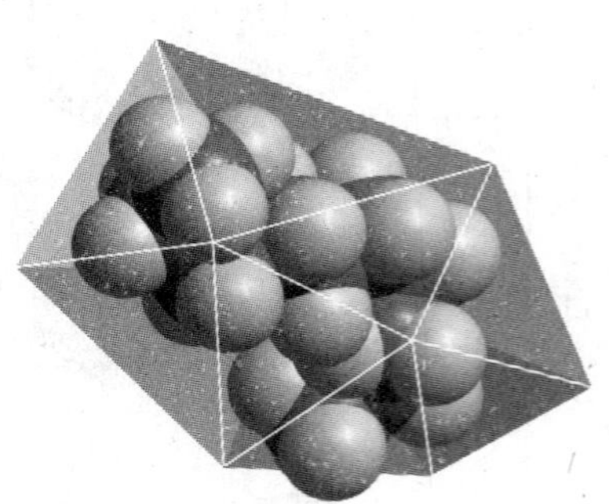

Fig. 17.9 When a hydrocarbon molecule is surrounded by water, the H_2O molecules form a cage. As a result of this acquisition of structure, the entropy of the water decreases, so the dispersal of the hydrocarbon into the water is entropy-opposed; its coalescence is entropy-favoured.

$$CH_4(\text{in } CCl_4) \rightarrow CH_4(\text{aq})$$

has $\Delta_{\text{transfer}}G = +12\text{ kJ mol}^{-1}$, $\Delta_{\text{transfer}}H = -10\text{ kJ mol}^{-1}$, and $\Delta_{\text{transfer}}S = -75\text{ J K}^{-1}\text{ mol}^{-1}$ at 298 K. Substances characterized by a positive Gibbs energy of transfer from a nonpolar to a polar solvent are called **hydrophobic**.

It is possible to quantify the hydrophobicity of a small molecular group R by defining the **hydrophobicity constant**, π, as

$$\pi = \log\frac{S}{S_0} \qquad \text{Definition of hydrophobicity constant} \qquad [17.26]$$

where S is the ratio of the molar solubility of the compound R–A in octanol, a nonpolar solvent, to that in water, and S_0 is the ratio of the molar solubility of the compound H–A in octanol to that in water. Therefore, positive values of π indicate hydrophobicity and negative values of π indicate hydrophilicity, the thermodynamic preference for water as a solvent. It is observed experimentally that the π values of most groups do not depend on the nature of A. However, measurements do suggest group additivity of π values, as the following data show:

R	CH_3	CH_3CH_2	$CH_3(CH_2)_2$	$CH_3(CH_2)_3$	$CH_3(CH_2)_4$
π	0.5	1.0	1.5	2.0	2.5

Thus, acyclic saturated hydrocarbons become more hydrophobic as the carbon chain length increases. This trend can be rationalized by $\Delta_{\text{transfer}}H$ becoming more positive and $\Delta_{\text{transfer}}S$ more negative as the number of carbon atoms in the chain increases.

At the molecular level, formation of a solvent cage around a hydrophobic molecule involves the formation of new hydrogen bonds among solvent molecules. This process is exothermic and accounts for the negative values of $\Delta_{\text{transfer}}H$. On the other hand, the increase in order associated with formation of a very large number of small solvent cages decreases the entropy of the system and accounts for the negative values of $\Delta_{\text{transfer}}S$. However, when many solute molecules cluster together, fewer (albeit larger) cages are required and more solvent molecules are free to move. The net effect of formation of large clusters of hydrophobic molecules is then a decrease in the organization of the solvent and therefore a net *increase* in entropy of the system. This increase in entropy of the solvent is large enough to render spontaneous the association of hydrophobic molecules in a polar solvent.

The increase in entropy that results from fewer structural demands on the solvent placed by the clustering of nonpolar molecules is the origin of the **hydrophobic interaction**, which tends to stabilize aggregation of hydrophobic groups in micelles and biopolymers (Chapter 18). The hydrophobic interaction is an example of an ordering process that is driven by a tendency toward greater disorder of the solvent.

(g) The total attractive interaction

We shall consider molecules that are unable to participate in hydrogen bond formation. The total attractive interaction energy between rotating molecules is then the sum of the dipole–dipole, dipole–induced-dipole, and dispersion interactions. Only the dispersion interaction contributes if both molecules are nonpolar. In a fluid phase, all three contributions to the potential energy vary as the inverse sixth power of the separation of the molecules, so we may write

$$V = -\frac{C_6}{r^6} \qquad (17.27)$$

where C_6 is a coefficient that depends on the identity of the molecules.

Although attractive interactions between molecules are often expressed as in eqn 17.27, we must remember that this equation has only limited validity. First, we have taken into account only dipolar interactions of various kinds, for they have the longest range and are dominant if the average separation of the molecules is large. However, in a complete treatment we should also consider quadrupolar and higher-order multipole interactions, particularly if the molecules do not have permanent dipole moments. Secondly, the expressions have been derived by assuming that the molecules can rotate reasonably freely. That is not the case in most solids, and in rigid media the dipole–dipole interaction is proportional to $1/r^3$ because the Boltzmann averaging procedure is irrelevant when the molecules are trapped into a fixed orientation.

A different kind of limitation is that eqn 17.27 relates to the interactions of pairs of molecules. There is no reason to suppose that the energy of interaction of three (or more) molecules is the sum of the pairwise interaction energies alone. The total dispersion energy of three closed-shell atoms, for instance, is given approximately by the **Axilrod–Teller formula:**

$$V = -\frac{C_6}{r_{AB}^6} - \frac{C_6}{r_{BC}^6} - \frac{C_6}{r_{CA}^6} + \frac{C'}{(r_{AB}r_{BC}r_{CA})^3} \qquad (17.28a)$$

Axilrod–Teller formula

where

$$C' = a(3\cos\theta_A \cos\theta_B \cos\theta_C + 1) \qquad (17.28b)$$

The parameter a is approximately equal to $\frac{3}{4}\alpha' C_6$; the angles θ are the internal angles of the triangle formed by the three atoms (**14**). The term in C' (which represents the non-additivity of the pairwise interactions) is negative for a linear arrangement of atoms (so that arrangement is stabilized) and positive for an equilateral triangular cluster (so that arrangement is destabilized). It is found that the three-body term contributes about 10 per cent of the total interaction energy in liquid argon.

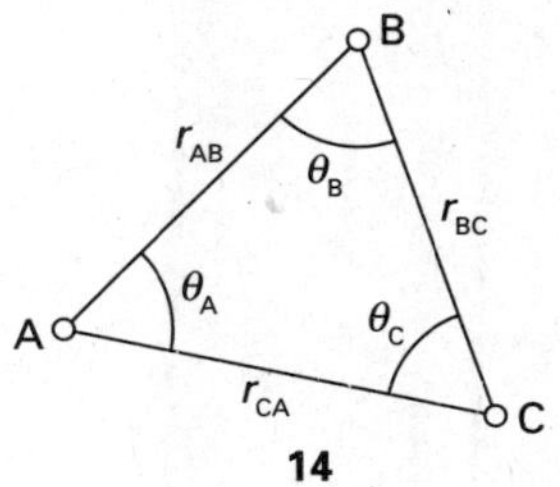

14

IMPACT ON MEDICINE

I17.1 Molecular recognition and drug design

Here we encounter the first of a series of examples of work being done in the area of supramolecular chemistry that illustrate some of the roles of intermolecular interactions in chemistry. Our focus here is on the binding of a drug, a small molecule or protein, to a specific receptor site of a target molecule, such as a larger protein or nucleic acid. The chemical result of the formation of this supramolecular assembly is the inhibition of the progress of disease.

The binding of a ligand, or *guest*, to a biopolymer, or *host*, is governed by molecular interactions. To devise efficient therapies, we need to know how to characterize and optimize molecular interactions between the host and the guest. Examples of biological *host–guest complexes* include enzyme–substrate complexes, antigen–antibody complexes, and drug–receptor complexes. In all these cases, a site on the guest contains functional groups that can interact with complementary functional groups of the host. For example, a hydrogen bond donor group of the guest must be positioned near a hydrogen bond acceptor group of the host for tight binding to occur. It is generally true that many specific intermolecular contacts must be made in a biological host–guest complex and, as a result, a guest binds only hosts that are chemically similar. The strict rules governing molecular recognition of a guest by a host control every biological process, from metabolism to immunological response, and provide important clues for the design of effective drugs for the treatment of disease.

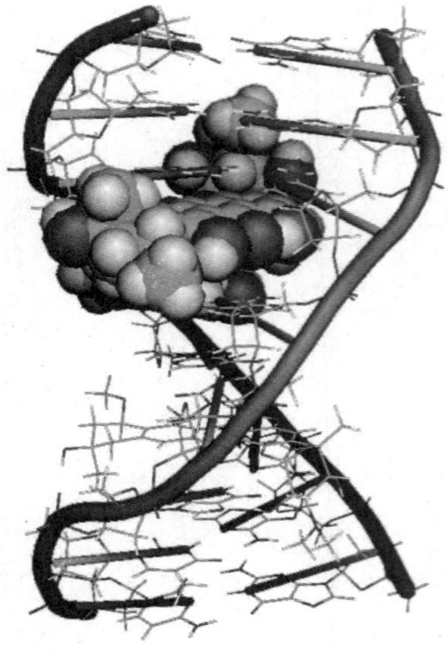

Interactions between nonpolar groups can be important in the binding of a guest to a host. For example, many enzyme active sites have hydrophobic pockets that bind nonpolar groups of a substrate. In addition to dispersion, repulsive, and hydrophobic interactions, **π-stacking interactions** are also possible, in which the planar π systems of aromatic macrocycles lie one on top of the other, in a nearly parallel orientation. Such interactions are responsible for the stacking of hydrogen-bonded base pairs in DNA (Fig. 17.10). Some drugs with planar π systems (for example, the molecule shown in Fig. 17.10 as a space-filling model) are effective because they intercalate between base pairs through π stacking interactions, causing the helix to unwind slightly and altering the function of DNA.

Fig. 17.10 Some drugs with planar π systems intercalate between base pairs of DNA.

Coulombic interactions can be important in the interior of a biopolymer host, where the relative permittivity can be much lower than that of the aqueous exterior. For example, at physiological pH, amino acid side chains containing carboxylic acid or amine groups are negatively and positively charged, respectively, and can attract each other. Dipole–dipole interactions are also possible because many of the building blocks of biopolymers are polar, including the peptide link, –CONH– (see *Example 17.1*). However, hydrogen bonding interactions are by far the most prevalent in a biological host–guest complexes. Many effective drugs bind tightly and inhibit the action of enzymes that are associated with the progress of a disease. In many cases, a successful inhibitor will be able to form the same hydrogen bonds with the binding site that the normal substrate of the enzyme can form, except that the drug is chemically inert towards the enzyme.

There are two main strategies for the discovery of a drug. In *structure-based design*, new drugs are developed on the basis of the known structure of the receptor site of a known target. However, in many cases a number of so-called *lead compounds* are known to have some biological activity but little information is available about the target. To design a molecule with improved pharmacological efficacy, **quantitative structure–activity relationships** (QSAR) are often established by correlating data on activity of lead compounds with molecular properties, also called *molecular descriptors*, which can be determined either experimentally or computationally.

In broad terms, the first stage of the QSAR method consists of compiling molecular descriptors for a very large number of lead compounds. Descriptors such as molar mass, molecular dimensions and volume, and relative solubility in water and nonpolar solvents are available from routine experimental procedures. Quantum mechanical descriptors determined by semi-empirical and *ab initio* calculations include bond orders and HOMO and LUMO energies.

In the second stage of the process, biological activity is expressed as a function of the molecular descriptors. An example of a QSAR equation is:

$$Activity = c_0 + c_1 d_1 + c_2 d_1^2 + c_3 d_2 + c_4 d_2^2 + \cdots \qquad (17.29)$$

A QSAR equation

where d_i is the value of the descriptor and c_i is a coefficient calculated by fitting the data by regression analysis. The quadratic terms account for the fact that biological activity can have a maximum or minimum value at a specific descriptor value. For example, a molecule might not cross a biological membrane and become available for binding to targets in the interior of the cell if it is too hydrophilic, in which case it will not partition into the hydrophobic layer of the cell membrane (see Section 18.7 for details of membrane structure), or too hydrophobic, for then it may bind too tightly to the membrane. It follows that the activity will peak at some intermediate value of a parameter that measures the relative solubility of the drug in water and organic solvents.

In the final stage of the QSAR process, the activity of a drug candidate can be estimated from its molecular descriptors and the QSAR equation either by interpolation or extrapolation of the data. The predictions are more reliable when a large number of lead compounds and molecular descriptors are used to generate the QSAR equation.

The traditional QSAR technique has been refined into **3D QSAR**, in which sophisticated computational methods are used to gain further insight into the three-dimensional features of drug candidates that lead to tight binding to the receptor site of a target. The process begins by using a computer to superimpose three-dimensional structural models of lead compounds and looking for common features, such as similarities in shape, location of functional groups, and electrostatic potential plots, which can be obtained from molecular orbital calculations. The key assumption of the method is that common structural features are indicative of molecular properties that enhance binding of the drug to the receptor. The collection of superimposed molecules is then placed inside a three-dimensional grid of points. An atomic probe, typically an sp^3-hybridized carbon atom, visits each grid point and two energies of interaction are calculated: E_{steric}, the steric energy reflecting interactions between the probe and electrons in uncharged regions of the drug, and E_{elec}, the electrostatic energy arising from interactions between the probe and a region of the molecule carrying a partial charge. The measured equilibrium constant for binding of the drug to the target, K_{bind}, is then assumed to be related to the interaction energies at each point r by the 3D QSAR equation

$$\log K_{bind} = c_0 + \sum_r \{c_S(r)E_{steric}(r) + c_E(r)E_{elec}(r)\} \qquad (17.30)$$

A 3D QSAR equation

where the $c(r)$ are coefficients calculated by regression analysis, with the coefficients c_S and c_E reflecting the relative importance of steric and electrostatic interactions, respectively, at the grid point r. Visualization of the regression analysis is facilitated by colouring each grid point according to the magnitude of the coefficients. Figure 17.11 shows results of a 3D QSAR analysis of the binding of steroids, molecules with the carbon skeleton shown, to human corticosteroid-binding globulin (CBG). Indeed, we

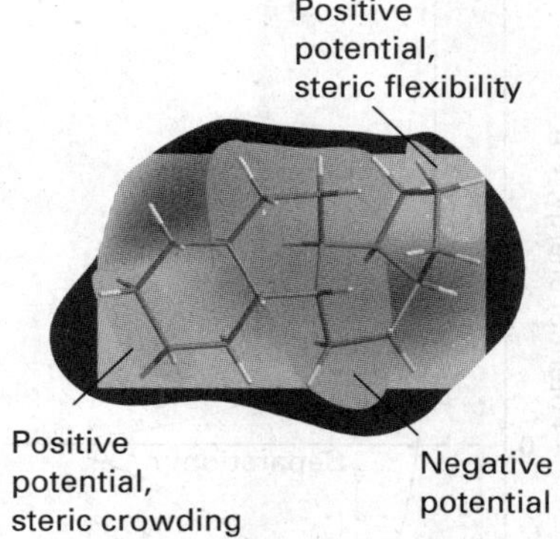

Fig. 17.11 A 3D QSAR analysis of the binding of steroids, molecules with the carbon skeleton shown, to human corticosteroid-binding globulin (CBG). The ellipses indicate areas in the protein's binding site with positive or negative electrostatic potentials and with little or much steric crowding. It follows from the calculations that addition of large substituents near the left-hand side of the molecule (as it is drawn on the page) leads to poor affinity of the drug to the binding site. Also, substituents that lead to the accumulation of negative electrostatic potential at either end of the drug are likely to show enhanced affinity for the binding site. (Adapted from P. Krogsgaard-Larsen, T. Liljefors, U. Madsen (ed.), *Textbook of drug design and discovery*, Taylor & Francis, London (2002).)

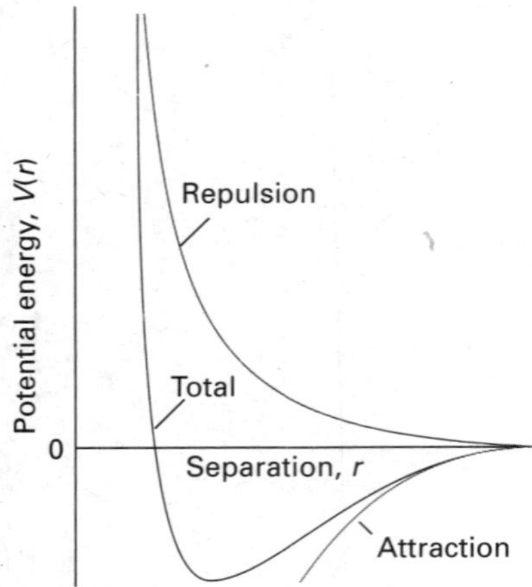

Fig. 17.12 The general form of an intermolecular potential energy curve. At long range the interaction is attractive, but at close range the repulsions dominate.

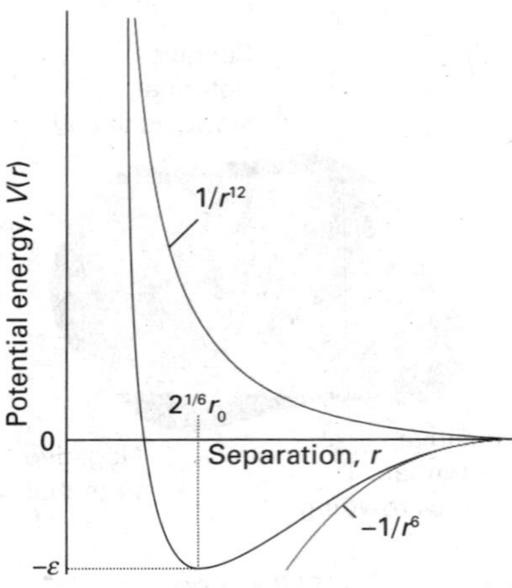

Fig. 17.13 The Lennard-Jones potential, and the relation of the parameters to the features of the curve. The green and purple lines are the two contributions.

Table 17.4* Lennard-Jones (12,6) parameters

	(ε/k)/K	r_0/pm
Ar	111.84	362.3
CCl_4	378.86	624.1
N_2	91.85	391.9
Xe	213.96	426.0

* More values are given in the *Data section.*

see that the technique lives up to the promise of opening a window into the chemical nature of the binding site even when its structure is not known.

The QSAR and 3D QSAR methods, though powerful, have limited power: the predictions are only as good as the data used in the correlations are both reliable and abundant. However, the techniques have been used successfully to identify compounds that deserve further synthetic elaboration, such as addition or removal of functional groups, and testing.

17.6 Repulsive and total interactions

Key point **The Lennard-Jones (12,6) potential is a model of the total intermolecular potential energy, with a term proportional to r^{-6} that represents attractive interactions and a term proportional to r^{-12} that represents repulsive interactions.**

When molecules are squeezed together, the nuclear and electronic repulsions and the rising electronic kinetic energy begin to dominate the attractive forces. The repulsions increase steeply with decreasing separation in a way that can be deduced only by very extensive, complicated molecular structure calculations of the kind described in Chapter 10 (Fig. 17.12).

In many cases, however, progress can be made by using a greatly simplified representation of the potential energy, where the details are ignored and the general features expressed by a few adjustable parameters. One such approximation is the **hard-sphere potential**, in which it is assumed that the potential energy rises abruptly to infinity as soon as the particles come within a separation d:

$$V = \infty \quad \text{for} \quad r \le d \qquad V = 0 \quad \text{for} \quad r > d \qquad \text{Hard-sphere potential} \qquad (17.31)$$

This very simple potential is surprisingly useful for assessing a number of properties. Another widely used approximation is the **Mie potential**:

$$V = \frac{C_n}{r^n} - \frac{C_m}{r^m} \qquad \text{Mie potential} \qquad (17.32)$$

with $n > m$. The first term represents repulsions and the second term attractions. The **Lennard-Jones potential** is a special case of the Mie potential with $n = 12$ and $m = 6$ (Fig. 17.13); it is often written in the form

$$V = 4\varepsilon\left\{\left(\frac{r_0}{r}\right)^{12} - \left(\frac{r_0}{r}\right)^{6}\right\} \qquad \text{Lennard-Jones potential} \qquad (17.33)$$

The two parameters are ε, the depth of the well (not to be confused with the symbol of the permittivity of a medium used in Section 17.4), and r_0, the separation at which $V = 0$ (Table 17.4). The well minimum occurs at $r_e = 2^{1/6}r_0$. Although the Lennard-Jones potential has been used in many calculations, there is plenty of evidence to show that $1/r^{12}$ is a very poor representation of the repulsive potential, and that an exponential form, e^{-r/r_0}, is greatly superior. An exponential function is more faithful to the exponential decay of atomic wavefunctions at large distances, and hence to the overlap that is responsible for repulsion. The potential with an exponential repulsive term and a $1/r^6$ attractive term is known as an **exp-6 potential**. These potentials can be used to calculate the virial coefficients of gases, as explained in Section 16.5, and through them various properties of real gases, such as the Joule–Thompson coefficient. The potentials are also used to model the structures of condensed fluids.

With the advent of **atomic force microscopy** (AFM), in which the force between a molecular sized probe and a surface is monitored (see *Impact I8.2*), it has become possible to measure directly the forces acting between molecules. The force, F, is the negative slope of potential, so for a Lennard-Jones potential between individual molecules we write

$$F=-\frac{\mathrm{d}V}{\mathrm{d}r}=\frac{24\varepsilon}{r_0}\left\{2\left(\frac{r_0}{r}\right)^{13}-\left(\frac{r_0}{r}\right)^{7}\right\} \qquad (17.34)$$

The net attractive force is greatest (from $\mathrm{d}F/\mathrm{d}r = 0$) at $r = (26/7)^{1/6}r_0$, or $1.244r_0$, and at that distance is equal to $-144(7/26)^{7/6}\varepsilon/13r_0$, or $-2.396\varepsilon/r_0$. For typical parameters, the magnitude of this force is about 10 pN.

IMPACT ON MATERIALS SCIENCE
I17.2 Hydrogen storage in molecular clathrates

Another example of supramolecular chemistry is the work leading to technologies that will make hydrogen gas a widely used alternative fuel. Hydrogen gas is efficient and environmentally clean in the sense that it is possible to use it in fuel cells without generating carbon dioxide, a greenhouse gas. Effective storage and delivery of hydrogen gas is key to the commercial development of devices that use it as a fuel. However, because H_2 molecules interact only weakly with one another, the liquefaction of hydrogen for storage and transport requires very high pressures, very low temperatures, or both. For example, at 1 atm hydrogen gas condenses only at 20 K. Whereas adsorption of hydrogen on solid surfaces, a process discussed in more detail in Chapter 23, is one way to solve the storage problem, more recent solutions involve insertion of H_2 molecules as guests in cage-like structures called **clathrates**.

Water is a common host, leading to solid materials known as *solid hydrogen gas hydrates*. One such clathrate forms at 2.5–6.0 MPa (25–60 atm) and 249 K. Hydrogen molecules are encapsulated by weak van der Waals interactions with host molecules and can be released either by increasing the temperature or by decreasing the pressure on the material.

The tetrahedral coordination geometry of the O atom in a water molecule and hydrogen bonding between water molecules lead to a variety of structures for water clathrates. Figure 17.14 shows that these structures, typically represented by adjoining polyhedra with vertices denoting the positions of O atoms, possess cages of different sizes. The so-called T cavity is a tetrakaidecahedron cage consisting of 12 pentagonal faces and two hexagonal faces. The smaller D cavity (shown in Fig. 17.14) is a pentagonal dodecahedron cage consisting of 12 pentagonal faces. The H cavity (also shown in Fig. 17.14) is a hexakaidecahedron structure consisting of 12 pentagonal and four hexagonal faces.

Spectroscopic and diffraction studies of hydrogen gas hydrates indicate that this type of material can encapsulate 5.2 per cent H_2 by mass. This percentage is a promising level for hydrogen storage, but more research is needed to result in a material with reasonable commercial utility.

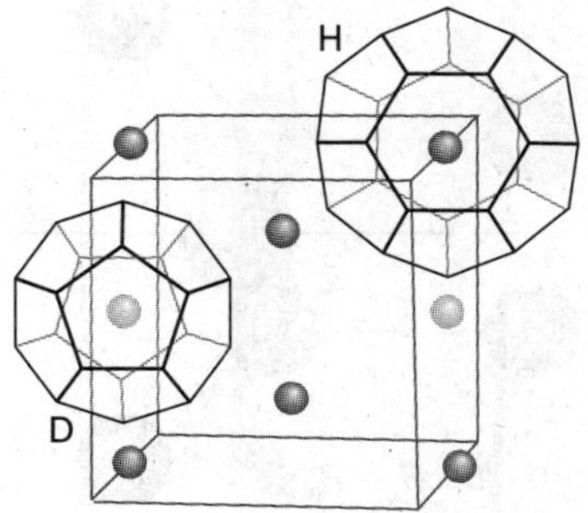

Fig. 17.14 A gas hydrate showing representative D and H cavities in a partial representation of the structure. The D cavity is a pentagonal dodecahedron cage consisting of 12 pentagonal faces. The H cavity is a hexakaidecahedron structure consisting of 12 pentagonal and four hexagonal faces.

Gases and liquids

The form of matter with the least order is a gas. In a perfect gas there are no intermolecular interactions and the distribution of molecules is completely random. In a real gas there are weak attractions and repulsions that have minimal effect on the

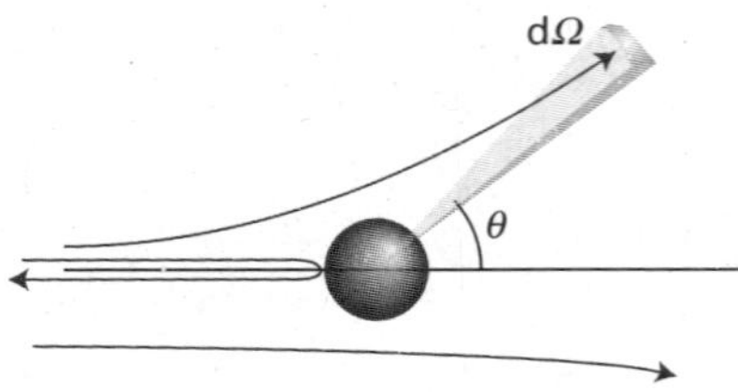

Fig. 17.15 The definition of the solid angle, $d\Omega$, for scattering.

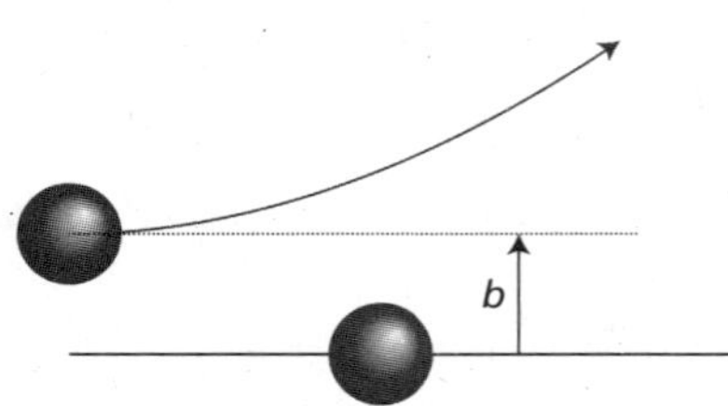

Fig. 17.16 The definition of the impact parameter, b, as the perpendicular separation of the initial paths of the particles.

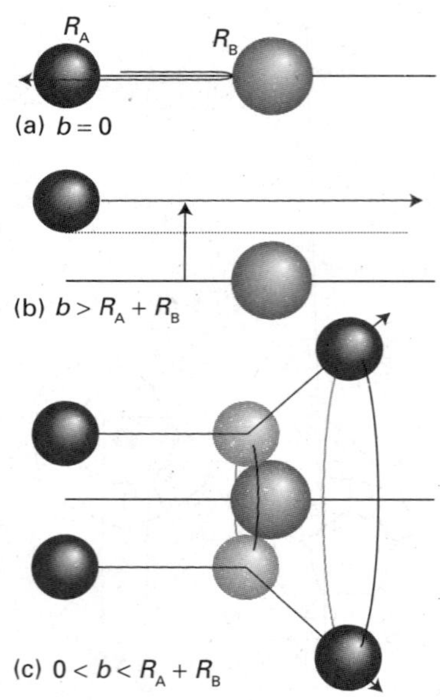

Fig. 17.17 Three typical cases for the collisions of two hard spheres: (a) $b = 0$, giving backward scattering; (b) $b > R_A + R_B$, giving forward scattering; (c) $0 < b < R_A + R_B$, leading to scattering into one direction on a ring of possibilities. (The target molecule is taken to be so heavy that it remains virtually stationary.)

relative locations of the molecules but that cause deviations from the perfect gas law for the dependence of pressure on the volume, temperature, and amount (Section 1.3).

The attractions between molecules are responsible for the condensation of gases into liquids at low temperatures. First, at low enough temperatures the molecules of a gas have insufficient kinetic energy to escape from each other's attraction and they stick together. Second, although molecules attract each other when they are a few diameters apart, as soon as they come into contact they repel each other. This repulsion is responsible for the fact that liquids and solids have a definite bulk and do not collapse to an infinitesimal point. The molecules are held together by molecular interactions, but their kinetic energies are comparable to their potential energies. As a result, we saw in Section 16.6 that, although the molecules of a liquid are not free to escape completely from the bulk, the whole structure is very mobile and we can speak only of the *average* relative locations of molecules. In the following sections we build on those concepts and add thermodynamic arguments to describe the surface of a liquid and the condensation of a gas into a liquid.

17.7 Molecular interactions in gases

Key points **A molecular beam is a collimated, narrow stream of molecules travelling though an evacuated vessel. Molecular beam techniques are used to investigate molecular interactions in gases. van der Waals molecules are complexes of the form AB in which A and B are held together by van der Waals forces.**

Molecular interactions in the gas phase can be studied in **molecular beams**, which consist of a collimated, narrow stream of molecules travelling though an evacuated vessel. The beam is directed towards other molecules, and the scattering that occurs on impact is related to the intermolecular interactions.

The primary experimental information from a molecular beam experiment is the fraction of the molecules in the incident beam that are scattered into a particular direction. The fraction is normally expressed in terms of dI, the rate at which molecules are scattered into a cone (described by a solid angle $d\Omega$) that represents the area covered by the 'eye' of the detector (Fig. 17.15). This rate is reported as the **differential scattering cross-section**, σ, the constant of proportionality between the value of dI and the intensity, I, of the incident beam, the number density of target molecules, $\mathcal{N}$, and the infinitesimal path length dx through the sample:

$$dI = \sigma I \mathcal{N} dx \qquad (17.35)$$

The value of σ (which has the dimensions of area) depends on the **impact parameter**, b, the initial perpendicular separation of the paths of the colliding molecules (Fig. 17.16), and the details of the intermolecular potential. The role of the impact parameter is most easily seen by considering the impact of two hard spheres (Fig. 17.17). If $b = 0$, the lighter projectile is on a trajectory that leads to a head-on collision, so the only scattering intensity is detected when the detector is at $\theta = \pi$. When the impact parameter is so great that the spheres do not make contact ($b > R_A + R_B$), there is no scattering and the scattering cross-section is zero at all angles except $\theta = 0$. Glancing blows, with $0 < b \leq R_A + R_B$, lead to scattering intensity in cones around the forward direction.

The scattering pattern of real molecules, which are not hard spheres, depends on the details of the intermolecular potential, including the anisotropy that is present when the molecules are non-spherical. The scattering also depends on the relative speed of approach of the two particles: a very fast particle might pass through the interaction region without much deflection, whereas a slower one on the same path might be temporarily captured and undergo considerable deflection (Fig. 17.18). The variation of the scattering cross-section with the relative speed of approach should therefore give information about the strength and range of the intermolecular potential.

A further point is that the outcome of collisions is determined by quantum, not classical, mechanics. The wave nature of the particles can be taken into account, at least to some extent, by drawing all classical trajectories that take the projectile particle from source to detector, and then considering the effects of interference between them.

Two quantum mechanical effects are of great importance. A particle with a certain impact parameter might approach the attractive region of the potential in such a way that the particle is deflected towards the repulsive core (Fig. 17.19), which then repels it out through the attractive region to continue its flight in the forward direction. Some molecules, however, also travel in the forward direction because they have impact parameters so large that they are undeflected. The wavefunctions of the particles that take the two types of path interfere, and the intensity in the forward direction is modified. The effect is called **quantum oscillation.** The same phenomenon accounts for the optical 'glory effect', in which a bright halo can sometimes be seen surrounding an illuminated object. (The coloured rings around the shadow of an aircraft cast on clouds by the Sun, and often seen in flight, is an example of an optical glory.)

The second quantum effect we need consider is the observation of a strongly enhanced scattering in a non-forward direction. This effect is called **rainbow scattering** because the same mechanism accounts for the appearance of an optical rainbow. The origin of the phenomenon is illustrated in Fig. 17.20. As the impact parameter decreases, there comes a stage at which the scattering angle passes through a maximum and the interference between the paths results in a strongly scattered beam. The **rainbow angle,** θ_r, is the angle for which $d\theta/db = 0$ and the scattering is strong.

Another phenomenon that can occur in certain beams is the capturing of one species by another. The vibrational temperature in supersonic beams is so low that **van der Waals molecules** may be formed, which are complexes of the form AB in which A and B are held together by van der Waals forces or hydrogen bonds. Large numbers of such molecules have been studied spectroscopically, including ArHCl, $(HCl)_2$, $ArCO_2$, and $(H_2O)_2$. More recently, van der Waals clusters of water molecules have been pursued as far as $(H_2O)_6$. The study of their spectroscopic properties gives detailed information about the intermolecular potentials involved.

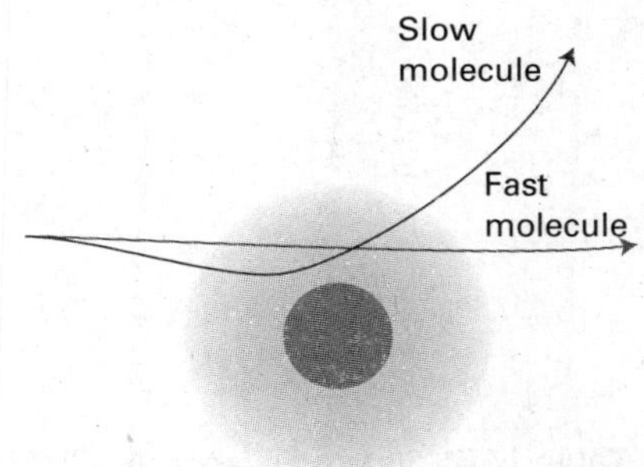

Fig. 17.18 The extent of scattering may depend on the relative speed of approach as well as the impact parameter. The dark central zone represents the repulsive core; the fuzzy outer zone represents the long-range attractive potential.

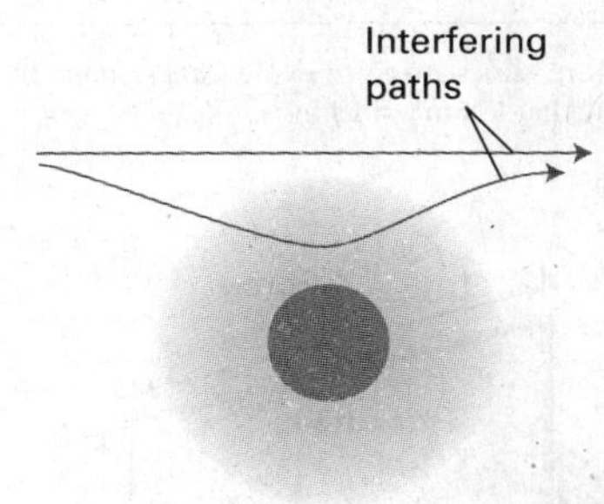

Fig. 17.19 Two paths leading to the same destination will interfere quantum mechanically; in this case they give rise to quantum oscillations in the forward direction.

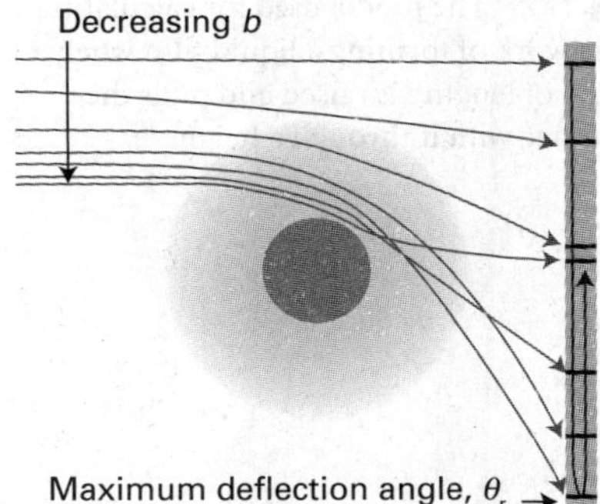

Fig. 17.20 The interference of paths leading to rainbow scattering. The rainbow angle, θ_r, is the maximum scattering angle reached as b is decreased. Interference between the numerous paths at that angle modifies the scattering intensity markedly.

17.8 The liquid–vapour interface

Key points (a) Liquids tend to adopt shapes that minimize their surface area. (b) The minimization of surface area results in the formation of bubbles, cavities, and droplets. (c) Capillary action is the tendency of liquids to rise up narrow tubes.

So far, we have concentrated on the properties of gases. In Section 16.6, we described the structure of liquids. Now we turn our attention to the physical boundary between phases, such as the surface where solid is in contact with liquid or liquid is in contact with its vapour. In this section we concentrate on the liquid–vapour interface, which is interesting because it is so mobile. Chapter 19 deals with solid surfaces and their important role in catalysis.

(a) Surface tension

Liquids tend to adopt shapes that minimize their surface area, for then the maximum number of molecules are in the bulk and hence surrounded by and interacting with neighbours. Droplets of liquids therefore tend to be spherical, because a sphere is the shape with the smallest surface-to-volume ratio. However, there may be other forces present that compete against the tendency to form this ideal shape and, in particular, gravity may flatten spheres into puddles or oceans.

Surface effects may be expressed in the language of Helmholtz and Gibbs energies (Chapter 3). The link between these quantities and the surface area is the work needed to change the area by a given amount, and the fact that dA and dG are equal (under different conditions) to the work done in changing the energy of a system. The work needed to change the surface area, σ, of a sample by an infinitesimal amount dσ is proportional to dσ, and we write

$$\mathrm{d}w = \gamma \mathrm{d}\sigma \qquad \text{Definition of surface tension} \qquad [17.36]$$

Table 17.5* Surface tensions of liquids at 293 K

	$\gamma/(\mathrm{mN\,m^{-1}})$
Benzene	28.88
Mercury	472
Methanol	22.6
Water	72.75

* More values are given in the *Data section*. Note that $1\ \mathrm{N\,m^{-1}} = 1\ \mathrm{J\,m^{-2}}$.

The constant of proportionality, γ, is called the **surface tension**; its dimensions are energy/area and its units are typically joules per metre squared ($\mathrm{J\,m^{-2}}$). However, as in Table 17.5, values of γ are usually reported in newtons per metre ($\mathrm{N\,m^{-1}}$ because $1\ \mathrm{J} = 1\ \mathrm{N\,m}$). The work of surface formation at constant volume and temperature can be identified with the change in the Helmholtz energy, and we can write

$$\mathrm{d}A = \gamma \mathrm{d}\sigma \qquad (17.37)$$

Because the Helmholtz energy decreases (d$A < 0$) if the surface area decreases (d$\sigma < 0$), surfaces have a natural tendency to contract. This is a more formal way of expressing what we have already described.

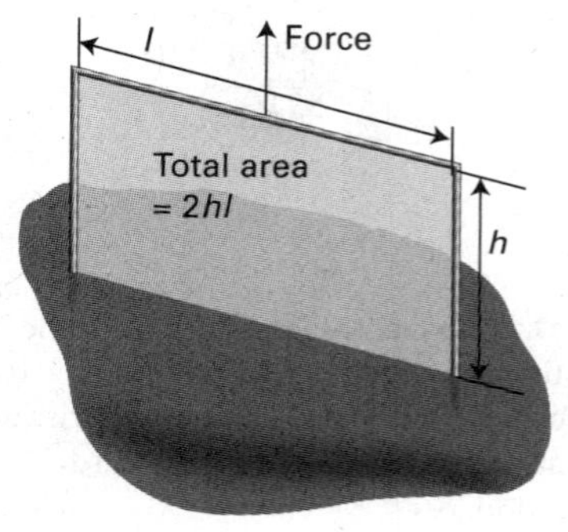

Fig. 17.21 The model used for calculating the work of forming a liquid film when a wire of length l is raised and pulls the surface with it through a height h.

Example 17.4 *Using the surface tension*

Calculate the work needed to raise a wire of length l and to stretch the surface of a liquid through a height h in the arrangement shown in Fig. 17.21. Disregard gravitational potential energy.

Method According to eqn 17.36, the work required to create a surface of area σ given that the surface tension does not vary as the surface is formed is $w = \gamma\sigma$. Therefore, all we need do is to calculate the surface area of the two-sided rectangle formed as the frame is withdrawn from the liquid.

Answer When the wire of length l is raised through a height h it increases the area of the liquid by twice the area of the rectangle (because there is a surface on each side). The total increase is therefore $2lh$ and the work done is $2\gamma lh$.

The expression $2\gamma lh$ can be expressed as force × distance by writing it as $2\gamma l \times h$, and identifying γl as the opposing force on the wire of length l. This interpretation is why γ is called a tension and why its units are often chosen to be newtons per metre ($\mathrm{N\,m^{-1}}$, so γl is a force in newtons).

Self-test 17.5 Calculate the work of creating a spherical cavity of radius r in a liquid of surface tension γ. [$4\pi r^2\gamma$]

(b) Curved surfaces

The minimization of the surface area of a liquid may result in the formation of a curved surface. A **bubble** is a region in which vapour (and possibly air too) is trapped by a thin film; a **cavity** is a vapour-filled hole in a liquid. What are widely called 'bubbles' in liquids are therefore strictly cavities. True bubbles have two surfaces (one on each side of the film); cavities have only one. The treatments of both are similar, but a factor of 2 is required for bubbles to take into account the doubled surface area.

A **droplet** is a small volume of liquid at equilibrium surrounded by its vapour (and possibly also air).

The pressure on the concave side of an interface, p_{in}, is always greater than the pressure on the convex side, p_{out}. This relation is expressed by the **Laplace equation**, which is derived in the following *Justification*:

$$p_{\text{in}} = p_{\text{out}} + \frac{2\gamma}{r} \qquad \text{Laplace equation} \qquad (17.38)$$

Justification 17.6 *The Laplace equation*

The cavities in a liquid are at equilibrium when the tendency for their surface area to decrease is balanced by the rise of internal pressure which would then result. When the pressure inside a cavity is p_{in} and its radius is r, the outward force is

$$\text{pressure} \times \text{area} = 4\pi r^2 p_{\text{in}}$$

The force inwards arises from the external pressure and the surface tension. The former has magnitude $4\pi r^2 p_{\text{out}}$. The latter is calculated as follows. The change in surface area when the radius of a sphere changes from r to $r + \mathrm{d}r$ is

$$\mathrm{d}\sigma = 4\pi(r + \mathrm{d}r)^2 - 4\pi r^2 = 8\pi r\,\mathrm{d}r$$

(The second-order infinitesimal, $(\mathrm{d}r)^2$, is ignored.) The work done when the surface is stretched by this amount is therefore

$$\mathrm{d}w = 8\pi\gamma r\,\mathrm{d}r$$

As force × distance is work, the force opposing stretching through a distance $\mathrm{d}r$ when the radius is r is

$$F = 8\pi\gamma r$$

The total inward force is therefore $4\pi r^2 p_{\text{out}} + 8\pi\gamma r$. At equilibrium, the outward and inward forces are balanced, so we can write

$$4\pi r^2 p_{\text{in}} = 4\pi r^2 p_{\text{out}} + 8\pi\gamma r$$

which rearranges into eqn 17.38.

The Laplace equation shows that the difference in pressure decreases to zero as the radius of curvature becomes infinite (when the surface is flat, Fig. 17.22). Small cavities have small radii of curvature, so the pressure difference across their surface is quite large. For instance, a 'bubble' (actually, a cavity) of radius 0.10 mm in champagne implies a pressure difference of 1.5 kPa, which is enough to sustain a column of water of height 15 cm.

(c) Capillary action

The tendency of liquids to rise up capillary tubes (tubes of narrow bore; the name comes from the Latin word for 'hair'), which is called **capillary action**, is a consequence of surface tension. Consider what happens when a glass capillary tube is first immersed in water or any liquid that has a tendency to adhere to the walls. The energy is lowest when a thin film covers as much of the glass as possible. As this film creeps up the inside wall it has the effect of curving the surface of the liquid inside the tube. This curvature implies that the pressure just beneath the curving meniscus is less than the atmospheric pressure by approximately $2\gamma/r$, where r is the radius of the tube and we assume a hemispherical surface. The pressure immediately under the flat surface outside the tube is P, the atmospheric pressure; but inside the tube under the curved

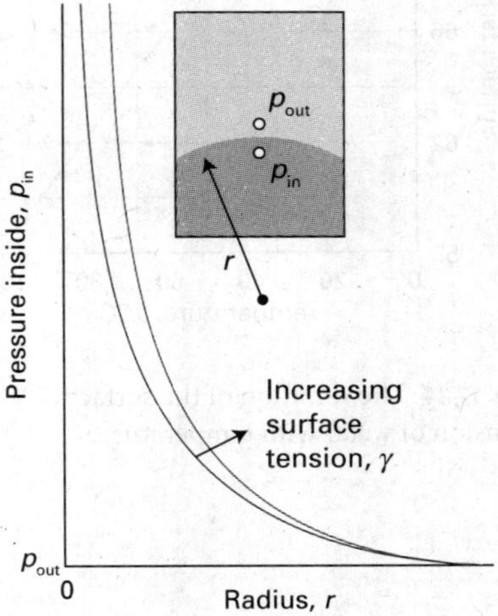

Fig. 17.22 The dependence of the pressure inside a curved surface on the radius of the surface, for two different values of the surface tension.

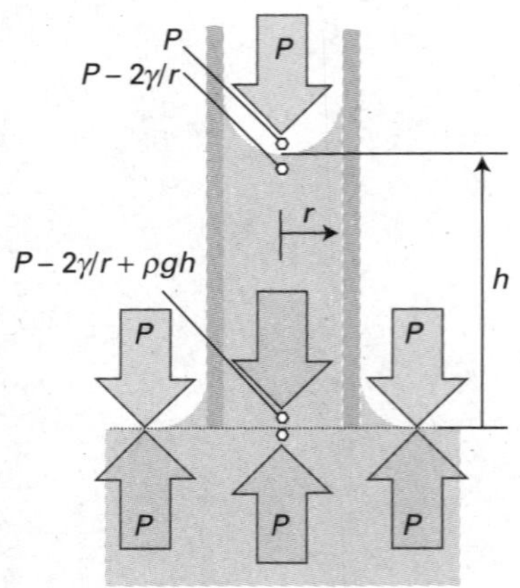

Fig. 17.23 When a capillary tube is first stood in a liquid, the latter climbs up the walls, so curving the surface. The pressure just under the meniscus is less than that arising from the atmosphere by $2\gamma/r$. The pressure is equal at equal heights throughout the liquid provided the hydrostatic pressure (which is equal to ρgh) cancels the pressure difference arising from the curvature.

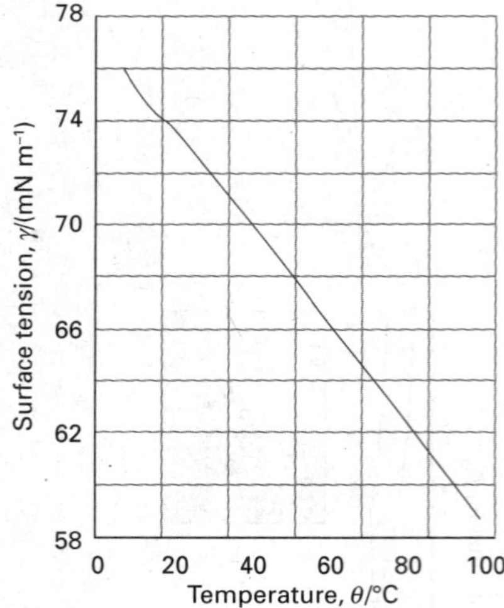

Fig. 17.24 The variation of the surface tension of water with temperature.

surface it is only $P - 2\gamma/r$. The excess external pressure presses the liquid up the tube until hydrostatic equilibrium (equal pressures at equal depths) has been reached (Fig. 17.23).

To calculate the height to which the liquid rises, we note that the pressure exerted by a column of liquid of mass density ρ and height h is

$$p = \rho gh \tag{17.39}$$

This hydrostatic pressure matches the pressure difference $2\gamma/r$ at equilibrium. Therefore, the height of the column at equilibrium is obtained by equating $2\gamma/r$ and ρgh, which gives

$$h = \frac{2\gamma}{\rho gr} \tag{17.40}$$

This simple expression provides a reasonably accurate way of measuring the surface tension of liquids. Surface tension decreases with increasing temperature (Fig. 17.24).

• A brief illustration

If water at 25°C (and density 997.1 kg m^{-3}) rises through 7.36 cm in a capillary of radius 0.20 mm, its surface tension at that temperature is

$$\begin{aligned}\gamma &= \tfrac{1}{2}\rho ghr \\ &= \tfrac{1}{2}\times(997.1\text{ kg m}^{-3})\times(9.81\text{ m s}^{-2})\times(7.36\times10^{-2}\text{ m})\times(2.0\times10^{-4}\text{ m}) \\ &= 72\text{ mN m}^{-1}\end{aligned}$$

where we have used 1 kg m s^{-2} = 1 N. •

When the adhesive forces between the liquid and the material of the capillary wall are weaker than the cohesive forces within the liquid (as for mercury in glass), the liquid in the tube retracts from the walls. This retraction curves the surface with the concave, high pressure side downwards. To equalize the pressure at the same depth throughout the liquid the surface must fall to compensate for the heightened pressure arising from its curvature. This compensation results in a capillary depression.

In many cases there is a nonzero angle between the edge of the meniscus and the wall. If this contact angle is θ_c, then eqn 17.40 should be modified by multiplying the right-hand side by $\cos\theta_c$. The origin of the contact angle can be traced to the balance of forces at the line of contact between the liquid and the solid (Fig. 17.25). If the solid–gas, solid–liquid, and liquid–gas surface tensions (essentially the energy needed to create unit area of each of the interfaces) are denoted γ_{sg}, γ_{sl}, and γ_{lg}, respectively, then the vertical forces are in balance if

$$\gamma_{sg} = \gamma_{sl} + \gamma_{lg}\cos\theta_c \tag{17.41}$$

This expression solves to

$$\cos\theta_c = \frac{\gamma_{sg} - \gamma_{sl}}{\gamma_{lg}} \tag{17.42}$$

If we note that the superficial work of adhesion of the liquid to the solid (the work of adhesion divided by the area of contact) is

$$w_{ad} = \gamma_{sg} + \gamma_{lg} - \gamma_{sl} \tag{17.43}$$

eqn 17.42 can be written

$$\cos\theta_c = \frac{w_{ad}}{\gamma_{lg}} - 1 \tag{17.44}$$

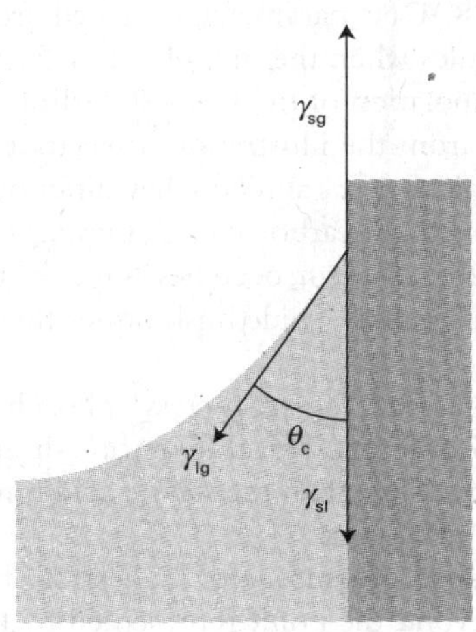

Fig. 17.25 The balance of forces that results in a contact angle, θ_c.

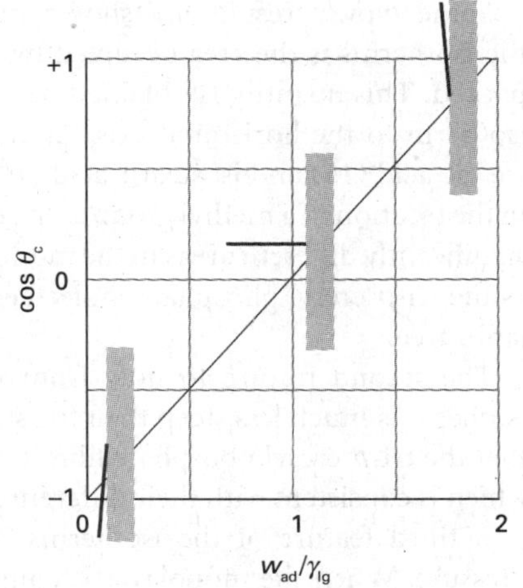

Fig. 17.26 The variation of contact angle (shown by the semaphore-like object) as the ratio w_{ad}/γ_{lg} changes.

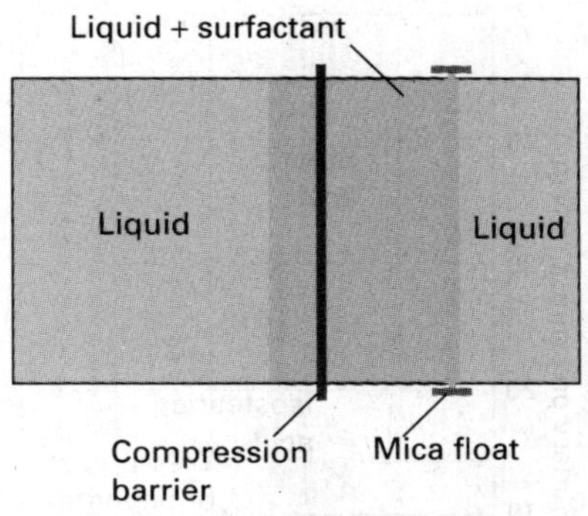

Fig. 17.27 A schematic diagram of the apparatus used to measure the surface pressure and other characteristics of a surface film. The surfactant is spread on the surface of the liquid in the trough, and then compressed horizontally by moving the compression barrier towards the mica float. The latter is connected to a torsion wire, so the difference in force on either side of the float can be monitored.

We now see that the liquid 'wets' (spreads over) the surface, corresponding to $0 < \theta_c < 90°$, when $1 < w_{ad}/\gamma_{lg} < 2$ (Fig. 17.26). The liquid does not wet the surface, corresponding to $90° < \theta_c < 180°$, when $0 < w_{ad}/\gamma_{lg} < 1$. For mercury in contact with glass, $\theta_c = 140°$, which corresponds to $w_{ad}/\gamma_{lg} = 0.23$, indicating a relatively low work of adhesion of the mercury to glass on account of the strong cohesive forces within mercury.

17.9 Surface films

Key points **(a) The surface pressure is the difference between the surface tension of the pure solvent and the solution. The collapse pressure is the highest surface pressure that a surface film can sustain. (b) A surfactant modifies the surface tension and surface pressure.**

The compositions of surface layers have been investigated by the simple but technically elegant procedure of slicing thin layers off the surfaces of solutions and analysing their compositions. The physical properties of surface films have also been investigated. Surface films one molecule thick are called **monolayers**. When a monolayer has been transferred to a solid support, it is called a **Langmuir–Blodgett film**, after Irving Langmuir and Katherine Blodgett, who developed experimental techniques for studying them.

(a) Surface pressure

The principal apparatus used for the study of surface monolayers is a **surface film balance** (Fig. 17.27). This device consists of a shallow trough and a barrier that can be moved along the surface of the liquid in the trough, and hence compress any monolayer on the surface. The **surface pressure**, π, the difference between the surface tension of the pure solvent and the solution ($\pi = \gamma^* - \gamma$) is measured by using a torsion wire attached to a strip of mica that rests on the surface and pressing against one edge of the monolayer. The parts of the apparatus that are in touch with liquids are coated in polytetrafluoroethene to eliminate effects arising from the liquid–solid interface. In an actual experiment, a small amount (about 0.01 mg) of the surfactant under investigation is dissolved in a volatile solvent and then poured on to the surface of the water; the compression barrier is then moved across the surface and the surface pressure exerted on the mica bar is monitored.

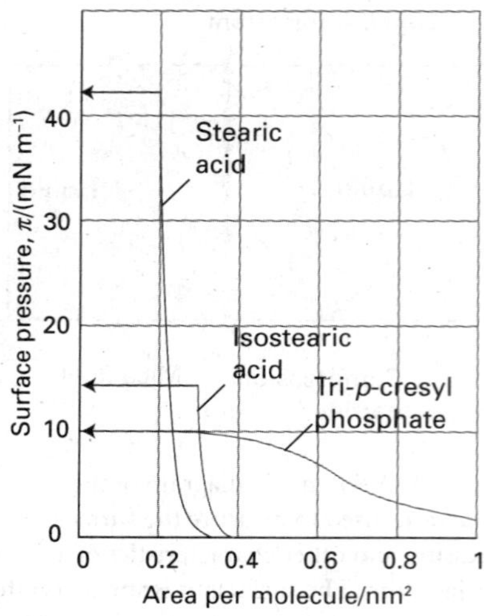

Fig. 17.28 The variation of surface pressure with the area occupied by each surfactant molecule. The collapse pressures are indicated by the horizontal arrow.

Some typical results are shown in Fig. 17.28. One parameter obtained from the isotherms is the area occupied by the molecules when the monolayer is closely packed. This quantity is obtained from the extrapolation of the steepest part of the isotherm to the horizontal axis. As can be seen from the illustration, even though stearic acid (**15**) and isostearic acid (**16**) are chemically very similar (they differ only in the location of a methyl group at the end of a long hydrocarbon chain), they occupy significantly different areas in the monolayer. Neither, though, occupies as much area as the tri-*p*-cresyl phosphate molecule (**17**), which is like a wide bush rather than a lanky tree.

The second feature to note from Fig. 17.28 is that the tri-*p*-cresyl phosphate isotherm is much less steep than the stearic acid isotherms. This difference indicates that the tri-*p*-cresyl phosphate film is more compressible than the stearic acid films, which is consistent with their different molecular structures.

A third feature of the isotherms is the **collapse pressure**, the highest surface pressure. When the monolayer is compressed beyond the point represented by the collapse pressure, the monolayer buckles and collapses into a film several molecules thick. As can be seen from the isotherms in Fig. 17.28, stearic acid has a high collapse pressure, but that of tri-*p*-cresyl phosphate is significantly smaller, indicating a much weaker film.

(b) The thermodynamics of surface layers

A **surfactant** is a species that is active at the interface between two phases, such as at the interface between hydrophilic and hydrophobic phases. A surfactant accumulates at the interface, and modifies its surface tension and hence the surface pressure. To establish the relation between the concentration of surfactant at a surface and the change in surface tension it brings about, we consider two phases α and β in contact and suppose that the system consists of several components J, each one present in an overall amount n_J. If the components were distributed uniformly through the two phases right up to the interface, which is taken to be a plane of surface area σ, the total Gibbs energy, G, would be the sum of the Gibbs energies of both phases, $G = G(\alpha) + G(\beta)$. However, the components are not uniformly distributed because one may accumulate at the interface. As a result, the sum of the two Gibbs energies differs from G by an amount called the **surface Gibbs energy**, $G(\sigma)$:

$$G(\sigma) = G - \{G(\alpha) + G(\beta)\} \qquad \text{Definition of surface Gibbs energy} \qquad [17.45]$$

Similarly, if it is supposed that the concentration of a species J is uniform right up to the interface, then from its volume we would conclude that it contains an amount $n_J(\alpha)$ of J in phase α and an amount $n_J(\beta)$ in phase β. However, because a species may accumulate at the interface, the total amount of J differs from the sum of these two

15 Stearic acid, $C_{17}H_{35}COOH$

16 Isostearic acid, $C_{17}H_{35}COOH$

17 Tri-*p*-cresylphosphate

amounts by $n_J(\sigma) = n_J - \{n_J(\alpha) + n_J(\beta)\}$. This difference is expressed in terms of the surface excess, Γ_J:

$$\Gamma_J = \frac{n_J(\sigma)}{\sigma} \qquad \text{Definition of surface excess} \qquad [17.46]$$

The surface excess may be either positive (an accumulation of J at the interface) or negative (a deficiency there).

The relation between the change in surface tension and the composition of a surface (as expressed by the surface excess) was derived by Gibbs. In the following *Justification* we derive the **Gibbs isotherm**, between the changes in the chemical potentials of the substances present in the interface and the change in surface tension:

$$d\gamma = -\sum_J \Gamma_J d\mu_J \qquad \text{Gibbs isotherm} \qquad (17.47)$$

Justification 17.7 *The Gibbs isotherm*

A general change in G is brought about by changes in T, p, and the n_J:

$$dG = -SdT + Vdp + \gamma d\sigma + \sum_J \mu_J dn_J$$

When this relation is applied to G, $G(\alpha)$, and $G(\beta)$ we find

$$dG(\sigma) = -S(\sigma)dT + \gamma d\sigma + \sum_J \mu_J dn_J(\sigma)$$

because at equilibrium the chemical potential of each component is the same in every phase, $\mu_J(\alpha) = \mu_J(\beta) = \mu_J(\sigma)$. Just as in the discussion of partial molar quantities (Section 5.1), the last equation integrates at constant temperature to

$$G(\sigma) = \gamma\sigma + \sum_J \mu_J n_J(\sigma)$$

We are seeking a connection between the change of surface tension $d\gamma$ and the change of composition at the interface. Therefore, we use the argument that in Section 5.1 led to the Gibbs–Duhem equation (eqn 5.12b), but this time we compare the expression

$$dG(\sigma) = \gamma d\sigma + \sum_J \mu_J dn_J(\sigma)$$

(which is valid at constant temperature) with the expression for the same quantity but derived from the preceding equation:

$$dG(\sigma) = \gamma d\sigma + \sigma d\gamma + \sum_J \mu_J dn_J(\sigma) + \sum_J n_J(\sigma) d\mu_J$$

The comparison implies that, at constant temperature,

$$\sigma d\gamma + \sum_J n_J(\sigma) d\mu_J = 0$$

Division by σ then gives eqn 17.47.

Now consider a simplified model of the interface in which the 'oil' and 'water' phases are separated by a geometrically flat surface. This approximation implies that only the surfactant, S, accumulates at the surface, and hence that Γ_{oil} and Γ_{water} are both zero. Then the Gibbs isotherm equation becomes

$$d\gamma = -\Gamma_S d\mu_S \qquad (17.48)$$

For dilute solutions,

$$d\mu_S = RT \ln c \tag{17.49}$$

where c is the molar concentration of the surfactant. It follows that

$$d\gamma = -RT\Gamma_S \frac{dc}{c}$$

at constant temperature, or

$$\left(\frac{\partial \gamma}{\partial c}\right)_T = -\frac{RT\Gamma_S}{c} \tag{17.50}$$

If the surfactant accumulates at the interface, its surface excess is positive and eqn 17.50 implies that $(\partial\gamma/\partial c)_T < 0$. That is, the surface tension decreases when a solute accumulates at a surface. Conversely, if the concentration dependence of γ is known, then the surface excess may be predicted and used to infer the area occupied by each surfactant molecule on the surface.

17.10 Condensation

Key point **Nucleation provides surfaces to which molecules can attach and thereby induce condensation.**

We now bring together concepts from this chapter and Chapter 4 to explain the condensation of a gas to a liquid. We saw in Section 4.4 that the vapour pressure of a liquid depends on the pressure applied to the liquid. Because curving a surface gives rise to a pressure differential of $2\gamma/r$, we can expect the vapour pressure above a curved surface to be different from that above a flat surface. By substituting this value of the pressure difference into eqn 4.4 ($p = p^*e^{V_m\Delta P/RT}$, where p^* is the vapour pressure when the pressure difference is zero) we obtain the **Kelvin equation** for the vapour pressure of a liquid when it is dispersed as droplets of radius r:

$$p = p^*e^{2\gamma V_m/rRT} \qquad \text{Kelvin equation} \tag{17.51}$$

The analogous expression for the vapour pressure inside a cavity can be written at once. The pressure of the liquid outside the cavity is less than the pressure inside, so the only change is in the sign of the exponent in the last expression.

For droplets of water of radius 1 μm and 1 nm the ratios p/p^* at 25°C are about 1.001 and 3, respectively. The second figure, although quite large, is unreliable because at that radius the droplet is less than about 10 molecules in diameter and the basis of the calculation is suspect. The first figure shows that the effect is usually small; nevertheless it may have important consequences.

Consider, for example, the formation of a cloud. Warm, moist air rises into the cooler regions higher in the atmosphere. At some altitude the temperature is so low that the vapour becomes thermodynamically unstable with respect to the liquid and we expect it to condense into a cloud of liquid droplets. The initial step can be imagined as a swarm of water molecules congregating into a microscopic droplet. Because the initial droplet is so small it has an enhanced vapour pressure. Therefore, instead of growing it evaporates. This effect stabilizes the vapour because an initial tendency to condense is overcome by a heightened tendency to evaporate. The vapour phase is then said to be **supersaturated**. It is thermodynamically unstable with respect to the liquid but not unstable with respect to the small droplets that need to form before the bulk liquid phase can appear, so the formation of the latter by a simple, direct mechanism is hindered.

Clouds do form, so there must be a mechanism. Two processes are responsible. The first is that a sufficiently large number of molecules might congregate into a droplet so big that the enhanced evaporative effect is unimportant. The chance of one of these **spontaneous nucleation centres** forming is low, and in rain formation it is not a dominant mechanism. The more important process depends on the presence of minute dust particles or other kinds of foreign matter. These **nucleate** the condensation (that is, provide centres at which it can occur) by providing surfaces to which the water molecules can attach.

Liquids may be **superheated** above their boiling temperatures and **supercooled** below their freezing temperatures. In each case the thermodynamically stable phase is not achieved on account of the kinetic stabilization that occurs in the absence of nucleation centres. For example, superheating occurs because the vapour pressure inside a cavity is artificially low, so any cavity that does form tends to collapse. This instability is encountered when an unstirred beaker of water is heated, for its temperature may be raised above its boiling point. Violent bumping often ensues as spontaneous nucleation leads to bubbles big enough to survive. To ensure smooth boiling at the true boiling temperature, nucleation centres, such as small pieces of sharp-edged glass or bubbles (cavities) of air, should be introduced.

Checklist of key equations

Property	Equation	Comment
Magnitude of the dipole moment	$\mu = QR$	Definition
Magnitude of the induced dipole moment	$\mu^* = \alpha E$	Linear approximation
Polarization of a sample	$P = \langle \mu \rangle \mathcal{N}$	
Potential energy of interaction between two point charges in a medium	$V = Q_1Q_2/4\pi\varepsilon r$	The relative permittivity of the medium is $\varepsilon_r = \varepsilon/\varepsilon_0$
Debye equation	$(\varepsilon_r - 1)/(\varepsilon_r + 2) = \rho P_m/M$	
Clausius–Mossoti equation	$(\varepsilon_r - 1)/(\varepsilon_r + 2) = \rho N_A\alpha/3M\varepsilon_0$	
Energy of interaction between a point dipole and a point charge	$V = -\mu_1 Q_2/4\pi\varepsilon_0 r^2$	
Energy of interaction between two fixed dipoles	$V = \mu_1\mu_2 f(\theta)/4\pi\varepsilon_0 r^3, f(\theta) = 1 - 3\cos^2\theta$	Parallel dipoles
Energy of interaction between two rotating dipoles	$V = -2\mu_1^2\mu_2^2/3(4\pi\varepsilon_0)^2kTr^6$	
Energy of interaction between a polar molecule and a polarizable molecule	$V = -\mu_1^2\alpha_2'/4\pi\varepsilon_0 r^6$	
London formula	$V = -\frac{3}{2}\alpha_1'\alpha_2' I_1 I_2/(I_1 + I_2)r^6$	
Axilrod–Teller formula	$V = -C_6/r_{AB}^6 - C_6/r_{BC}^6 - C_6/r_{CA}^6 + C'/(r_{AB}r_{BC}r_{CA})^3$	Applies to closed shell atoms
Lennard–Jones potential	$V = 4\varepsilon\{(r_0/r)^{12} - (r_0/r)^6\}$	
Laplace equation	$p_{in} = p_{out} + 2\gamma/r$	
Surface Gibbs energy	$G(\sigma) = G - \{G(\alpha) + G(\beta)\}$	Definition
Surface excess	$\Gamma_J = n_J(\sigma)/\sigma$	Definition
Gibbs isotherm	$d\gamma = -\sum_J \Gamma_J d\mu_J$	
Kelvin equation	$p = p^* e^{2\gamma V_m/rRT}$	

Further information

Further information 17.1 *The dipole–dipole interaction*

An important problem in physical chemistry is the calculation of the potential energy of interaction between two point dipoles with moments $\boldsymbol{\mu}_1$ and $\boldsymbol{\mu}_2$, separated by a vector $\boldsymbol{r}$. From classical electromagnetic theory, the potential energy of $\boldsymbol{\mu}_2$ in the electric field $\boldsymbol{\mathcal{E}}_1$ generated by $\boldsymbol{\mu}_1$ is given by the dot (scalar) product

$$V = -\boldsymbol{\mathcal{E}}_1 \cdot \boldsymbol{\mu}_2 \qquad (17.52)$$

To calculate $\boldsymbol{\mathcal{E}}_1$, we consider a distribution of point charges Q_i located at x_i, y_i, and z_i from the origin. The Coulomb potential ϕ due to this distribution at a point with coordinates x, y, and z is:

$$\phi = \sum_i \frac{Q_i}{4\pi\varepsilon_0} \frac{1}{\{(x-x_i)^2 + (y-y_i)^2 + (z-z_i)^2\}^{1/2}} \qquad (17.53)$$

where $\boldsymbol{r}$ is the location of the point of interest and the $\boldsymbol{r}_i$ are the locations of the charges Q_i.

A brief comment

The potential energy of a charge Q_1 in the presence of another charge Q_2 may be written as $V = Q_1\phi$ where $\phi = Q_2/4\pi\varepsilon_0 r$ is the Coulomb potential (*Fundamentals* F.6). If there are several charges $Q_2, Q_3, \ldots$ present in the system, then the total potential experienced by the charge Q_1 is the sum of the potential generated by each charge: $\phi = \phi_2 + \phi_3 + \cdots$. The electric field strength is the negative gradient of the electric potential: $\boldsymbol{\mathcal{E}} = -\nabla\phi$.

If we suppose that all the charges are close to the origin (in the sense that $r_i \ll r$), we can use a Taylor expansion to write

$$\phi(\boldsymbol{r}) = \sum_i \frac{Q_i}{4\pi\varepsilon_0}\left\{\frac{1}{r} + \left(\frac{\partial\{(x-x_i)^2+(y-y_i)^2+(z-z_i)^2\}^{1/2}}{\partial x_i}\right)_{x_i=0} x_i + \cdots\right\}$$

$$= \sum_i \frac{Q_i}{4\pi\varepsilon_0}\left\{\frac{1}{r} + \frac{xx_i}{r^3} + \cdots\right\} \qquad (17.54)$$

where the ellipses include the terms arising from derivatives with respect to y_i and z_i and higher derivatives. If the charge distribution is electrically neutral, the first term disappears because $\sum_i Q_i = 0$. Next we note that, $\sum_i Q_i x_i = \mu_x$ and likewise for the y- and z-components. That is,

$$\phi = \frac{1}{4\pi\varepsilon_0 r^3}(\mu_x x + \mu_y y + \mu_z z) = \frac{1}{4\pi\varepsilon_0 r^3}\boldsymbol{\mu}_1 \cdot \boldsymbol{r} \qquad (17.55)$$

The electric field strength is

$$\boldsymbol{\mathcal{E}}_1 = \frac{1}{4\pi\varepsilon_0}\nabla\frac{\boldsymbol{\mu}_1\cdot\boldsymbol{r}}{r^3} = -\frac{\boldsymbol{\mu}_1}{4\pi\varepsilon_0 r^3} - \frac{\boldsymbol{\mu}_1\cdot\boldsymbol{r}}{4\pi\varepsilon_0}\nabla\frac{1}{r^3} \qquad (17.56)$$

It follows from eqns 17.52 and 17.56 that

$$V = \frac{\boldsymbol{\mu}_1\cdot\boldsymbol{\mu}_2}{4\pi\varepsilon_0 r^3} - 3\frac{(\boldsymbol{\mu}_1\cdot\boldsymbol{r})(\boldsymbol{\mu}_2\cdot\boldsymbol{r})}{4\pi\varepsilon_0 r^5} \qquad (17.57)$$

For the arrangement shown in (13), in which $\boldsymbol{\mu}_1\cdot\boldsymbol{r} = \mu_1 r\cos\theta$ and $\boldsymbol{\mu}_2\cdot\boldsymbol{r} = \mu_2 r\cos\theta$, eqn 17.57 becomes:

$$V = \frac{\mu_1\mu_2 f(\theta)}{4\pi\varepsilon_0 r^3} \qquad f(\theta) = 1 - 3\cos^2\theta \qquad (17.58)$$

which is eqn 17.22.

Further information 17.2 *The basic principles of molecular beams*

The basic arrangement for a molecular beam experiment is shown in Fig. 17.29. If the pressure of vapour in the source is increased so that the mean free path of the molecules in the emerging beam is much shorter than the diameter of the pinhole, many collisions take place even outside the source. The net effect of these collisions, which give rise to **hydrodynamic flow**, is to transfer momentum into the direction of the beam. The molecules in the beam then travel with very similar speeds, so further downstream few collisions take place between them. This condition is called **molecular flow**. Because the spread in speeds is so small, the molecules are effectively in a state of very low translational temperature (Fig. 17.30). The translational

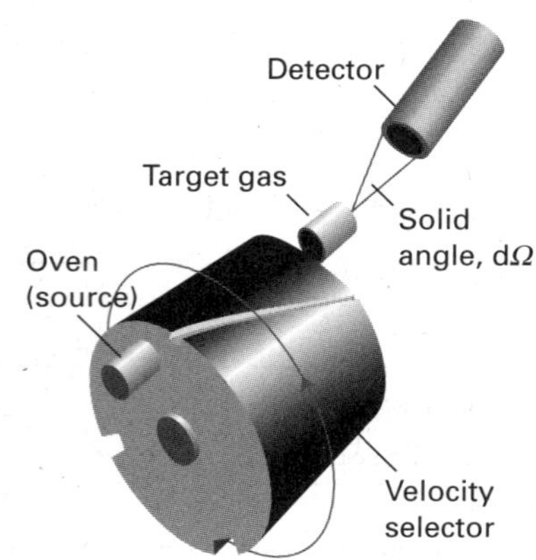

Fig. 17.29 The basic arrangement of a molecular beam apparatus. The atoms or molecules emerge from a heated source, and pass through the velocity selector, a rotating slotted cylinder. The scattering occurs from the target gas (which might take the form of another beam), and the flux of particles entering the detector set at some angle is recorded.

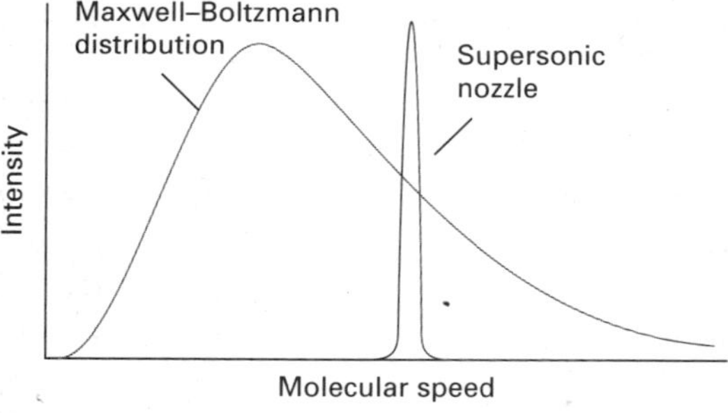

Fig. 17.30 The shift in the mean speed and the width of the distribution brought about by use of a supersonic nozzle.

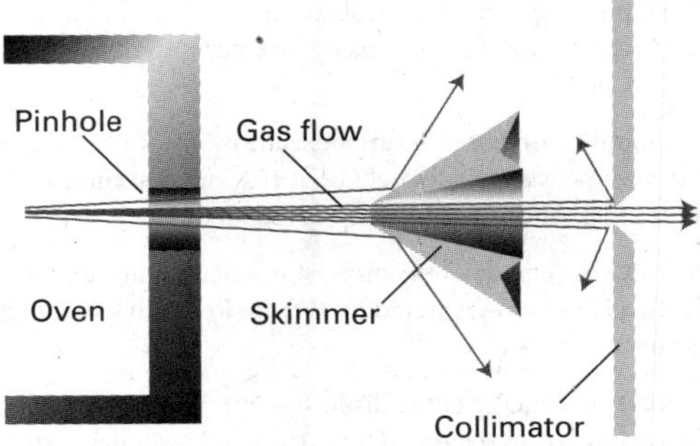

Fig. 17.31 A supersonic nozzle skims off some of the molecules of the beam and leads to a beam with well defined velocity.

temperature may reach as low as 1 K. Such jets are called **supersonic** because the average speed of the molecules in the jet is much greater than the speed of sound in the jet.

A supersonic jet can be converted into a more parallel **supersonic beam** if it is 'skimmed' in the region of hydrodynamic flow and the excess gas pumped away. A skimmer consists of a conical nozzle shaped to avoid any supersonic shock waves spreading back into the gas and so increasing the translational temperature (Fig. 17.31). A jet or beam may also be formed by using helium or neon as the principal gas, and injecting molecules of interest into it in the hydrodynamic region of flow.

The low translational temperature of the molecules is reflected in the low rotational and vibrational temperatures of the molecules. In this context, a rotational or vibrational temperature means the temperature that should be used in the Boltzmann distribution to reproduce the observed populations of the states. However, as rotational modes equilibrate more slowly, and vibrational modes equilibrate even more slowly, the rotational and vibrational populations of the species correspond to somewhat higher temperatures, of the order of 10 K for rotation and 100 K for vibrations.

The target gas may be either a bulk sample or another molecular beam. The latter **crossed beam technique** gives a lot of information because the states of both the target and projectile molecules may be controlled. The intensity of the incident beam is measured by the **incident beam flux**, *I*, which is the number of particles passing through a given area in a given interval divided by the area and the duration of the interval.

The detectors may consist of a chamber fitted with a sensitive pressure gauge, a bolometer (a detector that responds to the incident energy by making use of the temperature dependence of resistance), or an ionization detector, in which the incoming molecule is first ionized and then detected electronically. The state of the scattered molecules may also be determined spectroscopically, and is of interest when the collisions change their vibrational or rotational states.

Discussion questions

17.1 Explain how the permanent dipole moment and the polarizability of a molecule arise.

17.2 Explain why the polarizability of a molecule decreases at high frequencies.

17.3 Describe the experimental procedures available for determining the electric dipole moment of a molecule.

17.4 Identify the terms in and limit the generality of the following expressions: (a) $V=-Q_2\mu_1/4\pi\varepsilon_0 r^2$, (b) $V=-Q_2\mu_1\cos\theta/4\pi\varepsilon_0 r^2$, and (c) $V=\mu_2\mu_1(1-3\cos^2\theta)/4\pi\varepsilon_0 r^3$.

17.5 Draw examples of the arrangement of electrical charges that correspond to a monopole, dipole, quadrupole, and octupole and suggest a reason for the different distance dependencies of their electric fields.

17.6 Account for the theoretical conclusion that many attractive interactions between molecules vary with their separation as $1/r^6$.

17.7 Describe the formation of a hydrogen bond in terms of molecular orbitals. How does this description relate to one in terms of electrostatic interactions between partial charges?

17.8 Account for the hydrophobic interaction and discuss its manifestations.

17.9 Describe the process of condensation.

17.10 Describe how molecular beams are used to investigate intermolecular potentials.

Exercises

17.1(a) Which of the following molecules may be polar: ClF_3, O_3, H_2O_2?

17.1(b) Which of the following molecules may be polar: SO_3, XeF_4, SF_4?

17.2(a) Calculate the resultant of two dipole moments of magnitude 1.5 D and 0.80 D that make an angle of 109.5° to each other.

17.2(b) Calculate the resultant of two dipole moments of magnitude 2.5 D and 0.50 D that make an angle of 120° to each other.

17.3(a) Calculate the magnitude and direction of the dipole moment of the following arrangement of charges in the *xy*-plane: 3*e* at (0,0), −*e* at (0.32 nm, 0), and −2*e* at an angle of 20° from the *x*-axis and a distance of 0.23 nm from the origin.

17.3(b) Calculate the magnitude and direction of the dipole moment of the following arrangement of charges in the *xy*-plane: 4*e* at (0,0), −2*e* at (162 pm, 0), and −2*e* at an angle of 30° from the *x*-axis and a distance of 143 pm from the origin.

17.4(a) Calculate the molar energy required to reverse the direction of an H_2O molecule located 100 pm from a Li^+ ion. Take the dipole moment of water as 1.85 D.

17.4(b) Calculate the molar energy required to reverse the direction of an HCl molecule located 300 pm from a Mg^{2+} ion. Take the dipole moment of HCl as 1.08 D.

17.5(a) The polarizability volume of H_2O is 1.48×10^{-24} cm^3; calculate the dipole moment of the molecule (in addition to the permanent dipole moment) induced by an applied electric field of strength 1.0 kV cm^{-1}.

17.5(b) The polarizability volume of NH_3 is 2.22×10^{-30} m^3; calculate the dipole moment of the molecule (in addition to the permanent dipole moment) induced by an applied electric field of strength 15.0 kV m^{-1}.

17.6(a) The molar polarization of fluorobenzene vapour varies linearly with T^{-1}, and is 70.62 cm^3 mol^{-1} at 351.0 K and 62.47 cm^3 mol^{-1} at 423.2 K. Calculate the polarizability and dipole moment of the molecule.

17.6(b) The molar polarization of the vapour of a compound was found to vary linearly with T^{-1}, and is 75.74 cm^3 mol^{-1} at 320.0 K and 71.43 cm^3 mol^{-1} at 421.7 K. Calculate the polarizability and dipole moment of the molecule.

17.7(a) At 0°C, the molar polarization of liquid chlorine trifluoride is 27.18 cm^3 mol^{-1} and its density is 1.89 g cm^{-3}. Calculate the relative permittivity of the liquid.

17.7(b) At 0°C, the molar polarization of a liquid is 32.16 cm^3 mol^{-1} and its density is 1.92 g cm^{-3}. Calculate the relative permittivity of the liquid. Take $M = 55.0$ g mol^{-1}.

17.8(a) The refractive index of CH_2I_2 is 1.732 for 656 nm light. Its density at 20°C is 3.32 g cm^{-3}. Calculate the polarizability of the molecule at this wavelength.

17.8(b) The refractive index of a compound is 1.622 for 643 nm light. Its density at 20°C is 2.99 g cm^{-3}. Calculate the polarizability of the molecule at this wavelength. Take $M = 65.5$ g mol^{-1}.

17.9(a) The polarizability volume of H_2O at optical frequencies is 1.5×10^{-24} cm^3: estimate the refractive index of water. The experimental value is 1.33; what may be the origin of the discrepancy?

17.9(b) The polarizability volume of a liquid of molar mass 72.3 g mol^{-1} and density 865 kg mol^{-1} at optical frequencies is 2.2×10^{-30} m^3: estimate the refractive index of the liquid.

17.10(a) The dipole moment of chlorobenzene is 1.57 D and its polarizability volume is 1.23×10^{-23} cm^3. Estimate its relative permittivity at 25°C, when its density is 1.173 g cm^{-3}.

17.10(b) The dipole moment of bromobenzene is 5.17×10^{-30} C m and its polarizability volume is approximately 1.5×10^{-19} m^3. Estimate its relative permittivity at 25°C, when its density is 1491 kg m^{-3}.

17.11(a) Estimate the energy of the dispersion interaction (use the London formula) for two He atoms separated by 1.0 nm. Relevant data can be found in the *Data section*.

17.11(b) Estimate the energy of the dispersion interaction (use the London formula) for two Ar atoms separated by 1.0 nm. Relevant data can be found in the *Data section*.

17.12(a) How much energy (in kJ mol^{-1}) is required to break the hydrogen bond in a vacuum ($\varepsilon_r = 1$)? (Use the electrostatic model of the hydrogen bond.)

17.12(b) How much energy (in kJ mol^{-1}) is required to break the hydrogen bond in water ($\varepsilon_r \approx 80.0$)? Use the electrostatic model of the hydrogen bond.

17.13(a) Calculate the vapour pressure of a spherical droplet of water of radius 10 nm at 20°C. The vapour pressure of bulk water at that temperature is 2.3 kPa and its density is 0.9982 g cm^{-3}.

17.13(b) Calculate the vapour pressure of a spherical droplet of water of radius 20.0 nm at 35.0°C. The vapour pressure of bulk water at that temperature is 5.623 kPa and its density is 994.0 kg m^{-3}.

17.14(a) The contact angle for water on clean glass is close to zero. Calculate the surface tension of water at 20°C given that at that temperature water climbs to a height of 4.96 cm in a clean glass capillary tube of internal radius 0.300 mm. The density of water at 20°C is 998.2 kg m^{-3}.

17.14(b) The contact angle for water on clean glass is close to zero. Calculate the surface tension of water at 30°C given that at that temperature water climbs to a height of 9.11 cm in a clean glass capillary tube of internal radius 0.320 mm. The density of water at 30°C is 0.9956 g cm^{-3}.

17.15(a) Calculate the pressure differential of water across the surface of a spherical droplet of radius 200 nm at 20°C.

17.15(b) Calculate the pressure differential of ethanol across the surface of a spherical droplet of radius 220 nm at 20°C. The surface tension of ethanol at that temperature is 22.39 mN m^{-1}.

Problems*

Numerical problems

17.1 Suppose an H_2O molecule ($\mu = 1.85$ D) approaches an anion. What is the favourable orientation of the molecule? Calculate the electric field (in volts per metre) experienced by the anion when the water dipole is (a) 1.0 nm, (b) 0.3 nm, (c) 30 nm from the ion.

17.2 The electric dipole moment of toluene (methylbenzene) is 0.4 D. Estimate the dipole moments of the three xylenes (dimethylbenzene). Which answer can you be sure about?

17.3 Plot the magnitude of the electric dipole moment of hydrogen peroxide as the H–O–O–H (azimuthal) angle ϕ changes from 0 to 2π. Use the dimensions shown in (**18**).

18

17.4 An H_2O molecule is aligned by an external electric field of strength 1.0 kV m^{-1} and an Ar atom ($\alpha' = 1.66 \times 10^{-24}$ cm^3) is brought up slowly from one side. At what separation is it energetically favourable for the H_2O molecule to flip over and point towards the approaching Ar atom?

* Problems denoted with the symbol ‡ were supplied by Charles Trapp and Carmen Giunta.

17.5 The relative permittivity of chloroform was measured over a range of temperatures with the following results:

θ/°C	−80	−70	−60	−40	−20	0	20
ε	3.1	3.1	7.0	6.5	6.0	5.5	5.0
ρ/(g cm^{-3})	1.65	1.64	1.64	1.61	1.57	1.53	1.50

The freezing point of chloroform is −64°C. Account for these results and calculate the dipole moment and polarizability volume of the molecule.

17.6 The relative permittivities of methanol (m.p. −95°C) corrected for density variation are given below. What molecular information can be deduced from these values? Take $\rho = 0.791$ g cm^{-3} at 20°C.

θ/°C	−185	−170	−150	−140	−110	−80	−50	−20	0	20
ε_r	3.2	3.6	4.0	5.1	67	57	49	43	38	34

17.7 In his classic book *Polar molecules*, Debye reports some early measurements of the polarizability of ammonia. From the selection below, determine the dipole moment and the polarizability volume of the molecule.

T/K	292.2	309.0	333.0	387.0	413.0	446.0
P_m/(cm^3 mol^{-1})	57.57	55.01	51.22	44.99	42.51	39.59

The refractive index of ammonia at 273 K and 100 kPa is 1.000 379 (for yellow sodium light). Calculate the molar polarizability of the gas at this temperature and at 292.2 K. Combine the value calculated with the static molar polarizability at 292.2 K and deduce from this information alone the molecular dipole moment.

17.8 Values of the molar polarization of gaseous water at 100 kPa as determined from capacitance measurements are given below as a function of temperature.

T/K	384.3	420.1	444.7	484.1	522.0
P_m/(cm^3 mol^{-1})	57.4	53.5	50.1	46.8	43.1

Calculate the dipole moment of H_2O and its polarizability volume.

17.9‡ F. Luo *et al.* (*J. Chem. Phys.* **98**, 3564 (1993)) reported experimental observation of the He_2 complex, a species that had escaped detection for a long time. The fact that the observation required temperatures in the neighbourhood of 1 mK is consistent with computational studies that suggest that $hc\tilde{D}_e$ for He_2 is about 1.51×10^{-23} J, $hc\tilde{D}_0$ about 2×10^{-26} J, and R about 297 pm. (a) Determine the Lennard-Jones parameters, r_0 and ε, and plot the Lennard-Jones potential for He–He interactions. (b) Plot the Morse potential given that $a = 5.79 \times 10^{10}$ m^{-1}.

17.10‡ D.D. Nelson *et al.* (*Science* **238**, 1670 (1987)) examined several weakly bound gas-phase complexes of ammonia in search of examples in which the H atoms in NH_3 formed hydrogen bonds, but found none. For example, they found that the complex of NH_3 and CO_2 has the carbon atom nearest the nitrogen (299 pm away): the CO_2 molecule is at right angles to the C–N 'bond', and the H atoms of NH_3 are pointing away from the CO_2. The permanent dipole moment of this complex is reported as 1.77 D. If the N and C atoms are the centres of the negative and positive charge distributions, respectively, what is the magnitude of those partial charges (as multiples of e)?

17.11‡ From data in Table 17.1 calculate the molar polarization, relative permittivity, and refractive index of methanol at 20°C. Its density at that temperature is 0.7914 g cm^{-3}.

17.12 The surface tensions of a series of aqueous solutions of a surfactant A were measured at 20 °C, with the following results:

[A]/(mol dm^{-3})	0	0.10	0.20	0.30	0.40	0.50
γ/(mN m^{-1})	72.8	70.2	67.7	65.1	62.8	59.8

Calculate the surface excess concentration.

Theoretical problems

17.13 Calculate the potential energy of the interaction between two linear quadrupoles when they are (a) collinear, (b) parallel and separated by a distance r.

17.14 Show that, in a gas (for which the refractive index is close to 1), the refractive index depends on the pressure as $n_r = 1 + \text{const} \times p$, and find the constant of proportionality. Go on to show how to deduce the polarizability volume of a molecule from measurements of the refractive index of a gaseous sample.

17.15 Acetic acid vapour contains a proportion of planar, hydrogen-bonded dimers. The relative permittivity of pure liquid acetic acid is 7.14 at 290 K and increases with increasing temperature. Suggest an interpretation of the latter observation. What effect should isothermal dilution have on the relative permittivity of solutions of acetic acid in benzene?

17.16 Show that the mean interaction energy of N atoms of diameter d interacting with a potential energy of the form C_6/R^6 is given by $U = -2N^2C_6/3Vd^3$, where V is the volume in which the molecules are confined and all effects of clustering are ignored. Hence, find a connection between the van der Waals parameter a and C_6, from $n^2a/V^2 = (\partial U/\partial V)_T$.

17.17 Suppose the repulsive term in a Lennard-Jones (12,6)-potential is replaced by an exponential function of the form $e^{-r/d}$. Sketch the form of the potential energy and locate the distance at which it is a minimum.

17.18 The *cohesive energy density*, $\mathcal{U}$, is defined as U/V, where U is the mean potential energy of attraction within the sample and V its volume. Show that $\mathcal{U} = \frac{1}{2}\mathcal{N}\int V(R)\mathrm{d}\tau$, where $\mathcal{N}$ is the number density of the molecules and $V(R)$ is their attractive potential energy and where the integration ranges from d to infinity and over all angles. Go on to show that the cohesive energy density of a uniform distribution of molecules that interact by a van der Waals attraction of the form $-C_6/R^6$ is equal to $(2\pi/3)(N_A^2/d^3M^2)\rho^2C_6$, where ρ is the mass density of the solid sample and M is the molar mass of the molecules.

17.19 Consider the collision between a hard-sphere molecule of radius R_1 and mass m, and an infinitely massive impenetrable sphere of radius R_2. Plot the scattering angle θ as a function of the impact parameter b. Carry out the calculation using simple geometrical considerations.

17.20 The dependence of the scattering characteristics of atoms on the energy of the collision can be modelled as follows. We suppose that the two colliding atoms behave as impenetrable spheres, as in Problem 17.19, but that the effective radius of the heavy atoms depends on the speed v of the light atom. Suppose its effective radius depends on v as R_2e^{-v/v^*}, where v^* is a constant. Take $R_1 = \frac{1}{2}R_2$ for simplicity and an impact parameter $b = \frac{1}{2}R_2$, and plot the scattering angle as a function of (a) speed, (b) kinetic energy of approach.

Applications: to biochemistry

17.21 Phenylalanine (Phe, **19**) is a naturally occurring amino acid. What is the energy of interaction between its phenyl group and the electric dipole moment of a neighbouring peptide group? Take the distance between the groups as 4.0 nm and treat the phenyl group as a benzene molecules. The dipole moment of the peptide group is $\mu = 2.7$ D and the polarizability volume of benzene is $\alpha' = 1.04 \times 10^{-29}$ m^3.

COOH
NH$_2$

19

17.22 Now consider the London interaction between the phenyl groups of two Phe residues (see Problem 17.21). (a) Estimate the potential energy of interaction between two such rings (treated as benzene molecules) separated by 4.0 nm. For the ionization energy, use $I = 5.0$ eV. (b) Given that force is the negative slope of the potential, calculate the distance dependence of the force acting between two nonbonded groups of atoms, such as the phenyl groups of Phe, in a polypeptide chain that can have a London dispersion interaction with each other. What is the separation at which the force between the phenyl groups (treated as benzene molecules) of two Phe residues is zero? (*Hint.* Calculate the slope by considering the potential energy at r and $r + \delta r$, with $\delta r \ll r$, and evaluating $\{V(r + \delta r) - V(r)\}/\delta r$. At the end of the calculation, let δr become vanishingly small).

17.23 Molecular orbital calculations may be used to predict structures of intermolecular complexes. Hydrogen bonds between purine and pyrimidine bases are responsible for the double helix structure of DNA (see Chapter 18). Consider methyl-adenine (**20**, with $R = CH_3$) and methyl-thymine (**21**, with $R = CH_3$) as models of two bases that can form hydrogen bonds in DNA. (a) Using molecular modelling software and the computational method of your choice, calculate the atomic charges of all atoms in methyl-adenine and methyl-thymine. (b) Based on your tabulation of atomic charges, identify the atoms in methyl-adenine and methyl-thymine that are likely to participate in hydrogen bonds. (c) Draw all possible adenine–thymine pairs that can be linked by hydrogen bonds, keeping in mind that linear arrangements of the A–H···B fragments are preferred in DNA. For this step, you may want to use your molecular modelling software to align the molecules properly. (d) Consult Chapter 18 and determine which of the pairs that you drew in part (c) occur naturally in DNA molecules. (e) Repeat parts (a)–(d) for cytosine and guanine, which also form base pairs in DNA (see Chapter 18 for the structures of these bases).

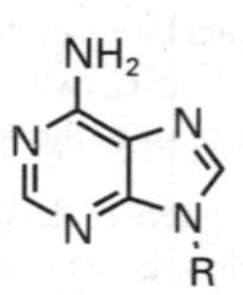

20 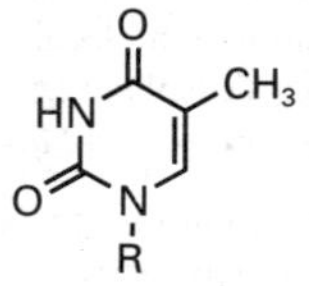 **21**

17.24 Molecular orbital calculations may be used to predict the dipole moments of molecules. (a) Using molecular modelling software and the computational method of your choice, calculate the dipole moment of the peptide link, modelled as a *trans*-*N*-methylacetamide (**22**). Plot the energy of interaction between these dipoles against the angle θ for $r = 3.0$ nm (see eqn 17.22). (b) Compare the maximum value of the dipole–dipole interaction energy from part (a) to 20 kJ mol^{-1}, a typical value for the energy of a hydrogen bonding interaction in biological systems.

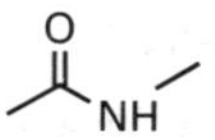

22

17.25 This problem gives a simple example of a quantitative structure–activity relation (QSAR). The binding of nonpolar groups of amino acid to hydrophobic sites in the interior of proteins is governed largely by hydrophobic interactions. (a) Consider a family of hydrocarbons R–H. The hydrophobicity constants, π, for $R = CH_3$, CH_2CH_3, $(CH_2)_2CH_3$, $(CH_2)_3CH_3$, and $(CH_2)_4CH_3$ are, respectively, 0.5, 1.0, 1.5, 2.0, and 2.5. Use these data to predict the π value for $(CH_2)_6CH_3$. (b) The equilibrium constants K_I for the dissociation of inhibitors (**23**) from the enzyme chymotrypsin were measured for different substituents R:

R	CH_3CO	CN	NO_2	CH_3	Cl
π	−0.20	−0.025	0.33	0.5	0.9
$\log K_I$	−1.73	−1.90	−2.43	−2.55	−3.40

Plot $\log K_I$ against π. Does the plot suggest a linear relationship? If so, what are the slope and intercept to the $\log K_I$ axis of the line that best fits the data? (c) Predict the value of K_I for the case $R = H$.

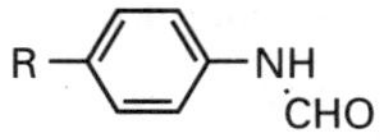

23

17.26 Derivatives of the compound TIBO (**24**) inhibit the enzyme reverse transcriptase, which catalyses the conversion of retroviral RNA to DNA. A QSAR analysis of the activity A of a number of TIBO derivatives suggests the following equation:

$$\log A = b_0 + b_1 S + b_2 W$$

where S is a parameter related to the drug's solubility in water and W is a parameter related to the width of the first atom in a substituent X shown in **24**. (a) Use the following data to determine the values of b_0, b_1, and b_2. *Hint.* The QSAR equation relates one dependent variable, $\log A$, to two independent variables, S and W. To fit the data, you must use the mathematical procedure of *multiple regression*, which can be performed with mathematical software or an electronic spreadsheet.

X	H	Cl	SCH_3	OCH_3	CN	CHO	Br	CH_3	CCH
$\log A$	7.36	8.37	8.3	7.47	7.25	6.73	8.52	7.87	7.53
S	3.53	4.24	4.09	3.45	2.96	2.89	4.39	4.03	3.80
W	1.00	1.80	1.70	1.35	1.60	1.60	1.95	1.60	1.60

(b) What should be the value of W for a drug with $S = 4.84$ and $\log A = 7.60$?

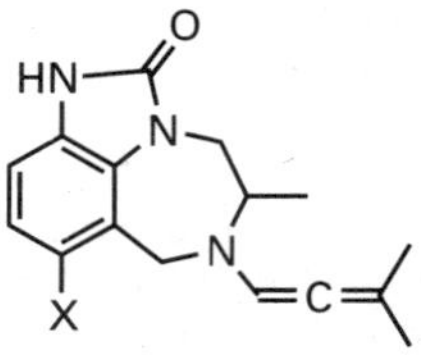

24

Materials 1: macromolecules and self-assembly

18

Atoms, small molecules, and macromolecules can form large assemblies, sometimes by processes involving self-assembly, that are held together by one or more of the molecular interactions described in Chapter 17. Macromolecules, although built from covalently linked components, adopt shapes that are governed by these interactions. We consider a range of structures in this chapter, beginning with a structureless random coil, partially structured coils, and then the structurally precise forces that operate in polypeptides and nucleic acids. We go on to explore colloids, micelles, and biological membranes, which are assemblies with some of the typical properties of molecules but also with their own characteristic features. Macromolecules, whether natural or synthetic, need to be characterized in terms of their molar mass, their size, and their shape and we conclude the chapter by considering how these features are determined experimentally.

There are macromolecules everywhere, inside us and outside us. Some are natural: they include polysaccharides such as cellulose, polypeptides such as protein enzymes, and polynucleotides such as deoxyribonucleic acid (DNA). Others are synthetic: they include **polymers** such as nylon and polystyrene that are manufactured by stringing together and (in some cases) cross-linking smaller units known as **monomers**. Molecules both large and small may also gather together in a process that is called 'self-assembly' and give rise to aggregates that to some extent behave like macromolecules. One example is the assembly of the protein actin into filaments in muscle tissue. Solutions of macromolecules are examples of a large class of systems known as 'colloids' that consist of a stable dispersion of one finely divided material in another.

Macromolecules and aggregates give rise to special problems that include the investigation and description of their shapes, the determination of their sizes, and the large deviations from ideality of their solutions. Natural macromolecules differ in certain respects from synthetic macromolecules, particularly in their composition and the resulting structure, but the two share a number of common properties. We concentrate on these common properties here.

Structure and dynamics

The concept of the 'structure' of a macromolecule takes on different meanings at the different levels at which we think about the arrangement of the chain or network of monomers. The term **configuration** refers to the structural features that can be changed only by breaking chemical bonds and forming new ones. Thus, the chains –A–B–C– and –A–C–B– have different configurations. The term **conformation** refers to the spatial arrangement of the different parts of a chain, and one conformation can be changed into another by rotating one part of a chain around a bond.

18.1 The different levels of structure

Key points The primary structure of a macromolecule is the sequence of small molecular residues making up the polymer. The secondary structure is the spatial arrangement of a chain of residues. The tertiary structure is the overall three-dimensional structure of a macromolecule. The quaternary structure is the manner in which large molecules are formed by the aggregation of others.

The **primary structure** of a macromolecule is the sequence of small molecular residues making up the polymer. The residues may form either a chain, as in polyethene, or a more complex network in which cross-links connect different chains, as in cross-linked polyacrylamide. In a synthetic polymer, virtually all the residues are identical and it is sufficient to name the monomer used in the synthesis. Thus, the repeating unit of polyethene and its derivatives is $-CHXCH_2-$, and the primary structure of the chain is specified by denoting it as $-(CHXCH_2)_n-$.

The concept of primary structure ceases to be trivial in the case of synthetic copolymers and biological macromolecules, for in general these substances are chains formed from different molecules. For example, proteins are **polypeptides** formed from different amino acids (about twenty occur naturally) strung together by the **peptide link**, –CONH–. The determination of the primary structure is then a highly complex problem of chemical analysis called **sequencing**. The **degradation** of a polymer is a disruption of its primary structure, when the chain breaks into shorter components.

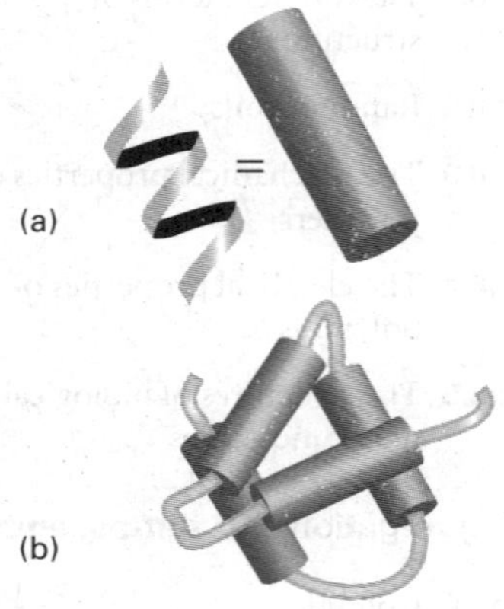

Fig. 18.1 (a) A polymer adopts a highly organized helical conformation, an example of a secondary structure. The helix is represented as a cylinder. (b) Several helical segments connected by short random coils pack together, providing an example of tertiary structure.

The **secondary structure** of a macromolecule is the (often local) spatial arrangement of a chain. The secondary structure of a molecule of polyethene in a good solvent is typically a random coil; in the absence of a solvent polyethene forms lamellar crystals with a hairpin-like bend about every 100 monomer units, presumably because for that number of monomers the intermolecular (in this case *intra*molecular) potential energy is sufficient to overcome thermal disordering. The secondary structure of a protein is a highly organized arrangement determined largely by hydrogen bonds, and taking the form of random coils, helices (Fig. 18.1a), or sheets in various segments of the molecule. The loss of secondary structure is called **denaturation**. When the hydrogen bonds in a protein are destroyed (for instance, by heating, as when cooking an egg) the structure denatures into a random coil.

The **tertiary structure** is the overall three-dimensional structure of a macromolecule. For instance, the hypothetical protein shown in Fig. 18.1b has helical regions connected by short random-coil sections. The helices interact to form a compact tertiary structure. Denaturation may also occur at this level.

The **quaternary structure** of a macromolecule is the manner in which large molecules are formed by the aggregation of others. Figure 18.2 shows how four molecular subunits, each with a specific tertiary structure, aggregate together. Quaternary structure can be very important in biology. For example, the oxygen-transport protein haemoglobin consists of four subunits that work together to take up and release O_2.

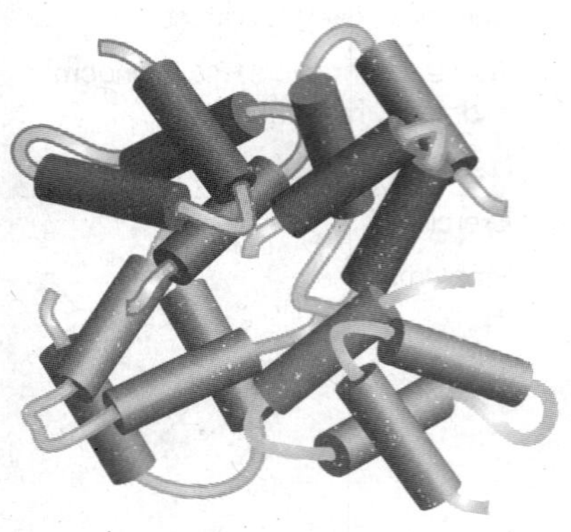

Fig. 18.2 Several subunits with specific tertiary structures pack together, providing an example of quaternary structure.

18.2 Random coils

Key points The least structured conformation of a macromolecule is a random coil, which can be modelled as a freely jointed chain. (a) The root mean square separation between the ends of a chain and the radius of gyration are useful measures of the size of a random coil. (b) The conformational entropy is the statistical entropy arising from the arrangement of bonds in a random coil. (c) The freely jointed chain model is improved by removing the freedom of bond angles to take any value. (d) The persistence length is a measure of the rigidity of a region of a polymer chain.

The most likely conformation of a chain of identical units not capable of forming hydrogen bonds or any other type of specific bond is a **random coil**. Polyethene is a simple example. The random coil model is a helpful starting point for estimating the orders of magnitude of the hydrodynamic properties of polymers and denatured proteins in solution.

The simplest model of a random coil is a **freely jointed chain**, in which any bond is free to make any angle with respect to the preceding one (Fig. 18.3). We assume that the residues occupy zero volume, so different parts of the chain can occupy the same region of space. The model is obviously an oversimplification because a bond is actually constrained to a cone of angles around a direction defined by its neighbour (Fig. 18.4) and real chains are self-avoiding in the sense that distant parts of the same chain cannot occupy the same space.

In a hypothetical one-dimensional freely jointed chain all the residues lie in a straight line, and the angle between neighbours is either 0° or 180°. The residues in a three-dimensional freely jointed chain are not restricted to lie in a line or a plane.

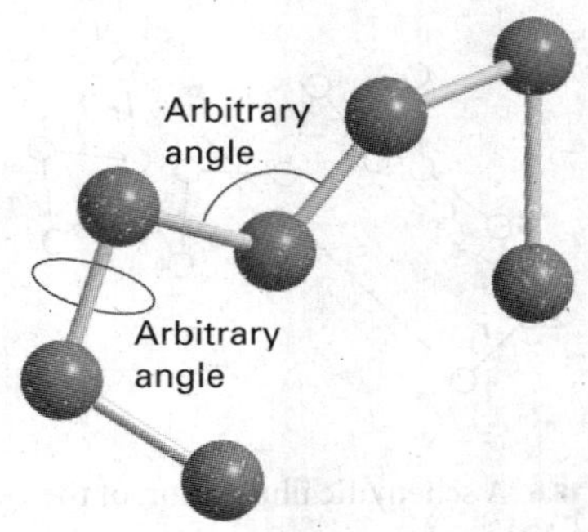

Fig. 18.3 A freely jointed chain is like a three-dimensional random walk, each step being in an arbitrary direction but of the same length.

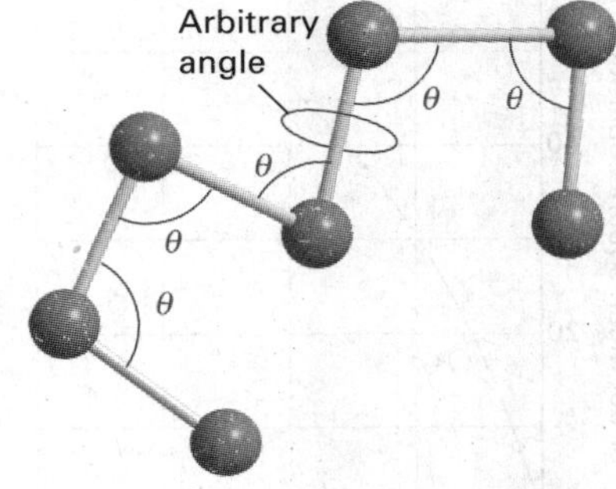

Fig. 18.4 A better description is obtained by fixing the bond angle (for example, at the tetrahedral angle) and allowing free rotation about a bond direction.

(a) Measures of size

As shown in *Further information 18.1*, the probability, P, that the ends of a long one-dimensional freely jointed chain composed of N units of length l (and therefore of total length Nl) are a distance nl apart is

$$P=\left(\frac{2}{\pi N}\right)^{1/2} e^{-n^2/2N} \qquad \text{Probability distribution for a one-dimensional random coil} \qquad (18.1)$$

This function is plotted in Fig. 18.5. We also show in *Further information 18.1* that eqn 18.1 can be used to calculate the probability that the ends of a long three-dimensional freely jointed chain lie in the range r to $r+\mathrm{d}r$. We write this probability as $f(r)\mathrm{d}r$, where

$$f(r)=4\pi\left(\frac{a}{\pi^{1/2}}\right)^3 r^2 e^{-a^2r^2} \qquad a=\left(\frac{3}{2Nl^2}\right)^{1/2} \qquad \text{Probability distribution for a three-dimensional random coil} \qquad (18.2)$$

In some coils, the ends may be far apart whereas in others their separation is small. Here and elsewhere we are ignoring the fact that the chain cannot be longer than Nl. Although eqn 18.2 gives a nonzero probability for $r > Nl$, the values are so small that the errors in pretending that r can range up to infinity are negligible. An alternative interpretation of eqn 18.2 is to regard each coil in a sample as ceaselessly writhing from one conformation to another; then $f(r)\mathrm{d}r$ is the probability that at any instant the chain will be found with the separation of its ends between r and $r+\mathrm{d}r$.

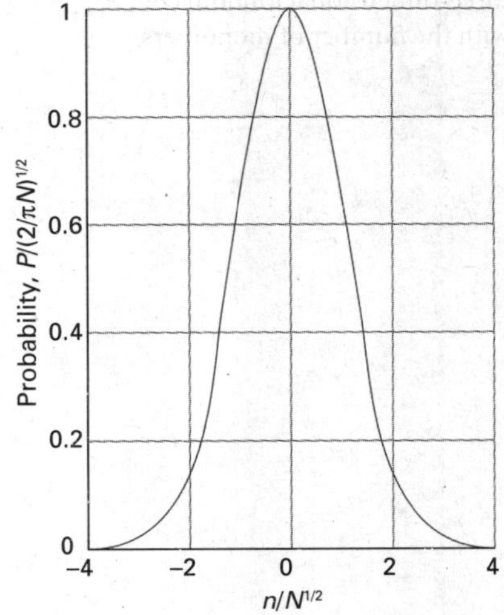

Fig. 18.5 The probability distribution for the separation of the ends of a one-dimensional random coil. The separation of the ends is nl, where l is the bond length.

• **A brief illustration**

Suppose that $N = 1000$ and $l = 150$ pm; then the probability that the ends of a one-dimensional random coil are 3 nm apart is given by eqn 18.1 by setting $n = (3000\text{ pm})/(150\text{ pm}) = 20$ and is 0.0207 (1 in 48 chance of being found there). If the coil is three-dimensional, we set $a = 2.58 \times 10^{-4}\text{ pm}^{-1}$. Then the probability density at $r = 3$ nm is given by eqn 18.2 as $f = 1.92 \times 10^{-4}\text{ pm}^{-1}$. The probability that the ends will be found in a shell of radius 3 nm and thickness 10 pm (regardless of direction) is therefore 1.92×10^{-3} (1 in 520). •

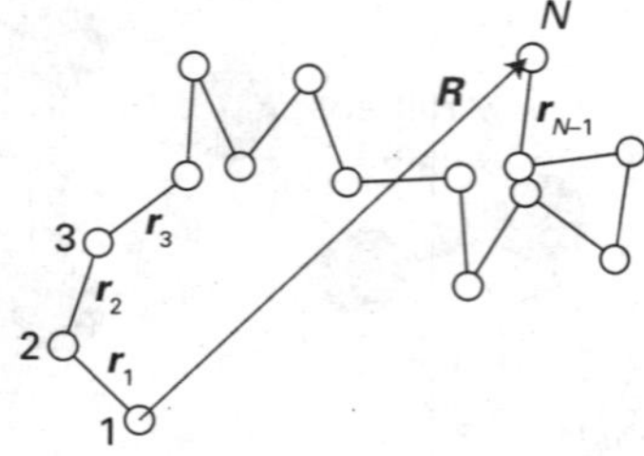

Fig. 18.6 A schematic illustration of the calculation of the root mean square separation of the ends of a random coil.

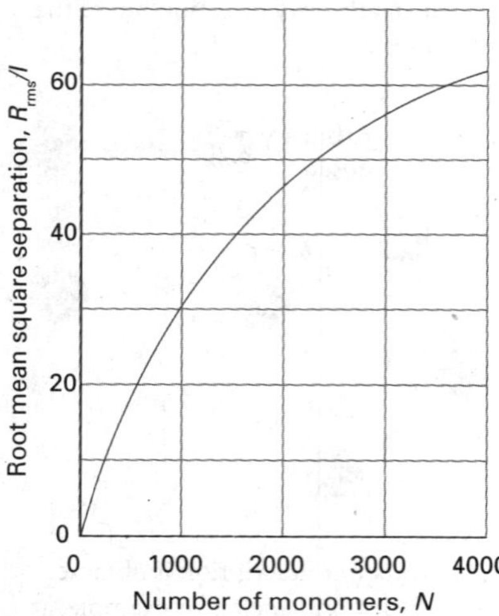

Fig. 18.7 The variation of the root mean square separation of the ends of a three-dimensional random coil, R_{rms}, with the number of monomers.

There are several measures of the geometrical size of a random coil. The **contour length**, R_c, is the length of the macromolecule measured along its backbone from atom to atom. For a polymer of N monomer units each of length l, the contour length is

$$R_c = Nl \qquad \text{Contour length} \qquad [18.3]$$

The **root mean square separation**, R_{rms}, is a measure of the average separation of the ends of a random coil: it is the square root of the mean value of R^2. To determine its value we note that the vector joining the two ends of the chain is the vector sum of the vectors joining neighbouring monomers: $\boldsymbol{R} = \sum_{i=1}^{N} \boldsymbol{r}_i$ (Fig. 18.6). The mean square separation of the ends of the chain is therefore

$$\langle R^2 \rangle = \langle \boldsymbol{R} \cdot \boldsymbol{R} \rangle = \sum_{i,j}^{N} \langle \boldsymbol{r}_i \cdot \boldsymbol{r}_j \rangle = \sum_{i}^{N} \langle r_i^2 \rangle + \sum_{i \neq j}^{N} \langle \boldsymbol{r}_i \cdot \boldsymbol{r}_j \rangle$$

When N is large (which we assume throughout) the second sum vanishes because the individual vectors all lie in random directions. The first sum is Nl^2 as all bond lengths are the same (and equal to l); so, after taking square roots, we conclude that

$$R_{rms} = N^{1/2} l \qquad \text{Root mean square separation} \qquad (18.4)$$

We see that, as the number of monomer units increases, the root mean square separation of its end increases as $N^{1/2}$ (Fig. 18.7), and consequently its volume increases as $N^{3/2}$. The result must be multiplied by a factor when the chain is not freely jointed (see below).

- **A brief illustration**

For $N = 1000$ and $l = 150$ pm, the contour length is $R_c = 1000 \times 150$ pm $= 150$ nm. The root mean square separation of the ends of the coil is $R_{rms} = (1000)^{1/2} \times 150$ pm $= 4.74$ nm. •

Another convenient measure of size is the **radius of gyration**, R_g, which is the radius of a hollow sphere of mass m that has the same moment of inertia (and therefore rotational characteristics) as the actual molecule of the same mass. We show in the following *Justification* that

$$R_g = N^{1/2} l \qquad \text{Radius of gyration of a one-dimensional random coil} \qquad (18.5)$$

A similar calculation for a three-dimensional random coil (Problem 18.17) gives

$$R_g = \left(\frac{N}{6}\right)^{1/2} l \qquad \text{Radius of gyration of a three-dimensional random coil} \qquad (18.6)$$

The radius of gyration is smaller in this case because the extra dimensions enable the coil to be more compact. The radius of gyration may also be calculated for other geometries. For example, a solid uniform sphere of radius R has $R_g = (\frac{3}{5})^{1/2} R$, and a long thin uniform rod of length l has $R_g = l/(12)^{1/2}$ for rotation about an axis perpendicular to the long axis. A solid sphere with the same radius and mass as a random coil will have a greater radius of gyration as it is entirely dense throughout.

Justification 18.1 *The radius of gyration*

For a one-dimensional random coil with $N+1$ identical monomers (and therefore N bonds) each of mass m, the moment of inertia around the centre of the chain (which is also at the first monomer, because steps occur in equal numbers to left and right) is

$$I=\sum_{i=0}^{N} m_i r_i^2 = m\sum_{i=0}^{N} r_i^2$$

This moment of inertia is set equal to $m_{\text{tot}}R_{\text{g}}^2$, where m_{tot} is the total mass of the polymer, $m_{\text{tot}}=(N+1)m$. Therefore, after averaging over all conformations,

$$R_{\text{g}}^2=\frac{1}{N+1}\sum_{i=0}^{N}\langle r_i^2\rangle$$

For a linear random chain, $\langle r_i^2\rangle = Nl^2$ (see Problem 18.16) and, as there are $N+1$ such terms in the sum, we find

$$R_{\text{g}}^2=Nl^2$$

Equation 18.5 then follows after taking the square root of each side.

The random coil model ignores the role of the solvent: a poor solvent will tend to cause the coil to tighten so that solute–solvent contacts are minimized; a good solvent does the opposite. Therefore, calculations based on this model are better regarded as lower bounds to the dimensions for a polymer in a good solvent and as an upper bound for a polymer in a poor solvent. The model is most reliable for a polymer in a bulk solid sample, where the coil is likely to have its natural dimensions.

(b) Conformational entropy

The random coil is the least structured conformation of a polymer chain and corresponds to the state of greatest entropy. Any stretching of the coil introduces order and reduces the entropy. Conversely, the formation of a random coil from a more extended form is a spontaneous process (provided enthalpy contributions do not interfere). As shown in the following *Justification*, we can use the same model to deduce that the change in **conformational entropy**, the statistical entropy arising from the arrangement of bonds, when a one-dimensional chain containing N bonds of length l is stretched or compressed by nl is

$$\Delta S=-\tfrac{1}{2}kN\ln\{(1+\nu)^{1+\nu}(1-\nu)^{1-\nu}\}\qquad \nu=n/N \qquad \text{Conformational entropy of a random coil} \qquad (18.7)$$

This function is plotted in Fig. 18.8, and we see that minimum extension corresponds to maximum entropy.

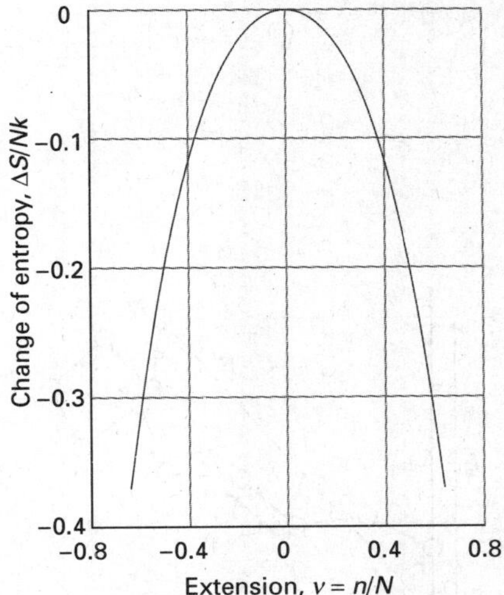

Fig. 18.8 The change in molar entropy of a perfect elastomer as its extension changes; $\nu=1$ corresponds to complete extension; $\nu=0$, the conformation of highest entropy, corresponds to the random coil.

• A brief illustration

As before, suppose $N=1000$ and $l=150$ pm. The change in entropy when the (one-dimensional) random coil is stretched through 1.5 nm (corresponding to $n=10$ and $\nu=1/100$) is $\Delta S=-0.050k$. The change in molar entropy is therefore $\Delta S_{\text{m}}=-0.050R$, or -0.42 J K^{-1} mol^{-1} (we have used $R=N_{\text{A}}k$). •

Justification 18.2 *The conformational entropy of a freely jointed chain*

The conformational entropy of the chain is $S = k \ln W$, where W is given by eqn 18.38 in *Further information 18.1*. Therefore,

$$S/k = \ln N! - \ln\{\tfrac{1}{2}(N+n)\}! - \ln\{\tfrac{1}{2}(N-n)\}!$$

Because the factorials are large (except for large extensions), we can use Stirling's approximation to obtain

$$S/k = -\ln(2\pi)^{1/2} + (N+1)\ln 2 + (N+\tfrac{1}{2})\ln N - \tfrac{1}{2}\ln\{(N+n)^{N+n+1}(N-n)^{N-n+1}\}$$

We have seen that the most probable conformation of a one-dimensional chain is the one with the ends close together ($n = 0$). This conformation also corresponds to maximum entropy, as may be confirmed by differentiation. Therefore, the maximum entropy is

$$S/k = -\ln(2\pi)^{1/2} + (N+1)\ln 2 + \tfrac{1}{2}\ln N$$

The change in entropy when the chain is stretched or compressed by nl is therefore the difference of these two quantities, and the resulting expression is eqn 18.7.

(c) Constrained chains

The freely jointed chain model is improved by removing the freedom of bond angles to take any value. For long chains, we can simply take groups of neighbouring bonds and consider the direction of their resultant. Although each successive individual bond is constrained to a single cone of angle θ relative to its neighbour, the resultant of several bonds lies in a random direction. By concentrating on such groups rather than individuals, it turns out that for long chains the expressions for the root mean square separation and the radius of gyration given above should be multiplied by

$$F = \left(\frac{1-\cos\theta}{1+\cos\theta}\right)^{1/2} \tag{18.8}$$

For tetrahedral bonds, for which $\cos\theta = \tfrac{1}{3}$ (that is, $\theta = 109.5°$), $F = 2^{1/2}$. Therefore:

$$R_{\text{rms}} = (2N)^{1/2} l \qquad R_{\text{g}} = \left(\frac{N}{3}\right)^{1/2} l \tag{18.9}$$

Dimensions of a tetrahedrally constrained chain

The model of a randomly coiled molecule is still an approximation, even after the bond angles have been restricted, because it does not take into account the impossibility of two or more atoms occupying the same place. Such self-avoidance tends to swell the coil, so (in the absence of solvent effects) it is better to regard R_{rms} and R_{g} as lower bounds to the actual values.

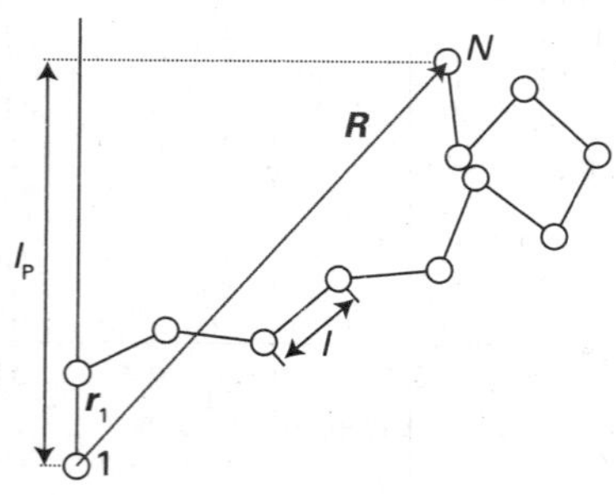

Fig. 18.9 The persistence length is defined as the average value of the projection of the end-to-end vector on the first bond of the chain.

(d) Partly rigid coils

An important measure of the flexibility of a chain is the **persistence length**, l_{p}, a measure of the length over which the direction of the first monomer–monomer direction is sustained. If the chain is a rigid rod, then the persistence length is the same as the contour length. For a freely jointed random coil, the persistence length is just the length of the monomer–monomer bond. Therefore, the persistence length can be regarded as a measure of the stiffness of the chain. In general, the persistence length of a chain of identical monomers of length l is defined as the average value of the projection of the end-to-end vector on the first bond of the chain (Fig. 18.9):

$$l_p = \left\langle \frac{r_1}{l} \cdot R \right\rangle = \frac{1}{l} \sum_{i=0}^{N-1} \langle r_1 \cdot r_i \rangle$$ Definition of persistence length [18.10]

(The sum ends at $N-1$ because the last atom is atom N and the last bond is from atom $N-1$ to atom N.) Experimental values of persistence lengths are as follows:

poly(glycine)	poly(L-alanine)	poly(L-proline)
0.6 nm	2 nm	22 nm

These values suggest that the stiffness of the chain increases from left to right along the series.

The mean square distance between the ends of a chain that has a nonzero persistence length can be expected to be greater than for a random coil because the partial rigidity of the coil does not let it roll up so tightly. We show in *Further information 18.1* that

$$R_{rms} = N^{1/2} l F \qquad \text{where} \qquad F = \left(\frac{2l_p}{l} - 1 \right)^{1/2} \tag{18.11}$$

For a random coil, $l_p = l$, so $R_{rms} = N^{1/2} l$, as we have already found. For $l_p > l$, $F > 1$, so the coil has swollen, as we anticipated.

18.3 The mechanical properties of polymers

Key points **The elastic properties of a material are summarized by a stress–strain curve. A perfect elastomer is a polymer for which the internal energy is independent of the extension. The disruption of long-range order in a polymer occurs at a melting temperature. Synthetic polymers undergo a transition from a state of high to low chain mobility at the glass transition temperature.**

The stress–strain curve shown in Fig. 18.10 shows how a material responds to stress. The region of **elastic deformation** is where the strain is proportional to the stress and is reversible: when the stress is removed, the sample returns to its initial shape. As we shall see in more detail in Section 19.8, the slope of the stress–strain curve in this region is 'Young's modulus', E, for the material. At the **yield point**, the reversible, linear deformation gives way to **plastic deformation**, where the strain is no longer linearly proportional to the stress and the initial shape of the sample is not recovered when the stress is removed. Thermosetting plastics have only a very short elastic range; thermoplastics typically (but not universally) have a long plastic range. An **elastomer** is specifically a polymer with a long elastic range. They typically have numerous cross-links (such as the sulfur links in vulcanized rubber) that pull them back into their original shape when the stress is removed.

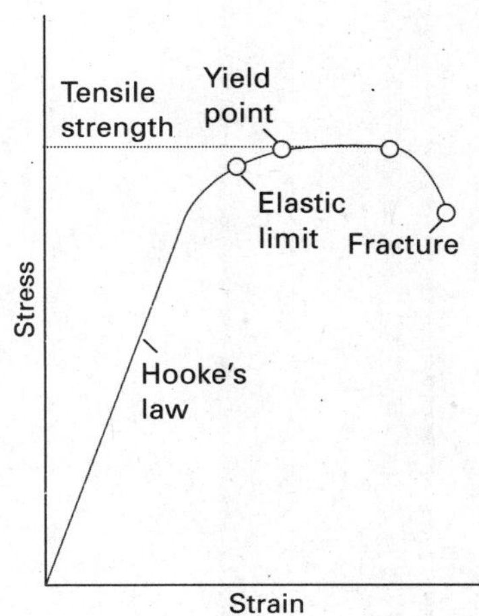

Fig. 18.10 A typical stress–strain curve.

Although practical elastomers are typically extensively cross-linked, even a freely jointed chain behaves as an elastomer for small extensions. It is a model of a **perfect elastomer**, a polymer in which the internal energy is independent of the extension. We saw in Section 18.2b that the contraction of an extended chain to a random coil is spontaneous in the sense that it corresponds to an increase in entropy; the entropy change of the surroundings is zero because no energy is released when the coil forms. In the following *Justification* we also see that the restoring force, $\mathcal{F}$, of a one-dimensional random coil when the chain is stretched or compressed by nl is

$$\mathcal{F} = \frac{kT}{2l} \ln\left(\frac{1+\nu}{1-\nu} \right) \qquad \nu = n/N \tag{18.12a}$$

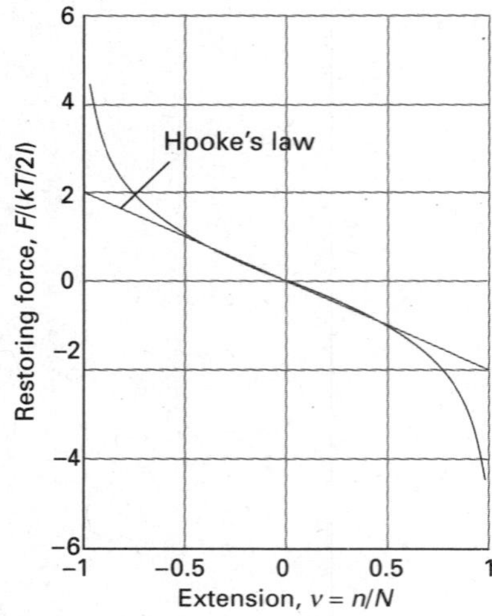

Fig. 18.11 The restoring force, $\mathcal{F}$, of a one-dimensional perfect elastomer. For small strains, $\mathcal{F}$ is linearly proportional to the extension, corresponding to Hooke's law.

where N is the total number of bonds of length l. This function is plotted in Fig. 18.11. At low extensions, when $\nu \ll 1$ we can use $\ln(1+x) = x - \frac{1}{2}x^2 + \cdots$ and find (retaining only linear terms) that

$$\mathcal{F} \approx \frac{\nu kT}{l} = \frac{nkT}{Nl} \qquad (18.12b)$$

That is, for small displacements the sample obeys Hooke's law: the restoring force is proportional to the displacement (which is proportional to n). For small displacements, therefore, the whole coil shakes with simple harmonic motion. When this equation is rearranged to

$$nl = \left(\frac{Nl^2}{kT}\right)\mathcal{F} \qquad (18.12c)$$

we see that for small displacements, the strain, as measured by the extension nl, is proportional to the applied force, as is characteristic of the elastic deformation region of an elastomer.

Justification 18.3 *Hooke's law*

The work done on an elastomer when it is extended through a distance dx is $\mathcal{F}dx$, where $\mathcal{F}$ is the restoring force. The change in internal energy is therefore

$$dU = T\,dS + \mathcal{F}\,dx$$

It follows that

$$\left(\frac{\partial U}{\partial x}\right)_T = T\left(\frac{\partial S}{\partial x}\right)_T + \mathcal{F}$$

In a perfect elastomer, as in a perfect gas, the internal energy is independent of the dimensions (at constant temperature), so $(\partial U/\partial x)_T = 0$. The restoring force is therefore

$$\mathcal{F} = -T\left(\frac{\partial S}{\partial x}\right)_T$$

If now we substitute eqn 18.7 into this expression , we obtain

$$\mathcal{F} = -\frac{T}{l}\left(\frac{\partial S}{\partial n}\right)_T = \frac{T}{Nl}\left(\frac{\partial S}{\partial \nu}\right)_T = \frac{kT}{2l}\ln\left(\frac{1+\nu}{1-\nu}\right)$$

as in eqn 18.12a.

The crystallinity of synthetic polymers can be destroyed by thermal motion at sufficiently high temperatures. This change in crystallinity may be thought of as a kind of intramolecular melting from a crystalline solid to a more fluid random coil. Polymer melting also occurs at a specific **melting temperature**, T_m, which increases with the strength and number of intermolecular interactions in the material. Thus, polyethene, which has chains that interact only weakly in the solid, has $T_m = 414$ K and nylon-66 fibres, in which there are strong hydrogen bonds between chains, has $T_m = 530$ K. High melting temperatures are desirable in most practical applications involving fibres and plastics.

All synthetic polymers undergo a transition from a state of high to low chain mobility at the **glass transition temperature**, T_g. To visualize the glass transition, we consider what happens to an elastomer as we lower its temperature. There is sufficient energy available at normal temperatures for limited bond rotation to occur and the flexible chains writhe. At lower temperatures, the amplitudes of the writhing motion decrease until a specific temperature, T_g, is reached at which motion is frozen completely and the sample forms a glass. Glass transition temperatures well below 300 K are desirable in elastomers that are to be used at normal temperatures. Both the glass transition temperature and the melting temperature of a polymer may be measured by differential scanning calorimetry (*Impact I2.1*). Because the motion of the segments of a polymer chain increases at the glass transition temperature, T_g may also be determined from a plot of the specific volume of a polymer (the reciprocal of its mass density) against temperature (Fig. 18.12).

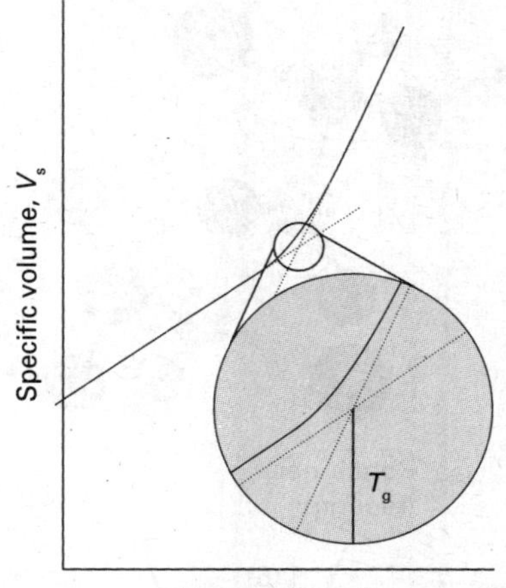

Fig. 18.12 The variation of specific volume with temperature of a synthetic polymer. The glass transition temperature, T_g, is at the point of intersection of extrapolations of the two linear parts of the curve.

18.4 The electrical properties of polymers

Key points In conducting polymers conjugated double bonds facilitate electron conduction along the chain. These polymers are slightly better electrical conductors than silicon semiconductors but are worse than metallic conductors.

Most of the macromolecules and self-assembled structures considered in this chapter are insulators, or very poor electrical conductors. However, a variety of newly developed macromolecular materials have electrical conductivities that rival those of silicon-based semiconductors and even metallic conductors. We examine one example in detail: **conducting polymers**, in which extensively conjugated double bonds facilitate electron conduction along the polymer chain.

One example of a conducting polymer is polyacetylene (polyethyne, Fig. 18.13). Whereas the delocalized π bonds do suggest that electrons can move up and down the chain, the electrical conductivity of polyacetylene increases significantly when it is partially oxidized by I_2 and other strong oxidants. The product is a **polaron**, a partially localized cation radical that travels through the chain, as shown in Fig. 18.13. Oxidation of the polymer by one more equivalent forms either **bipolarons**, a di-cation that moves as a unit through the chain, or **solitons**, two separate cation radicals that move independently. Polarons and solitons contribute to the mechanism of charge conduction in polyacetylene.

Conducting polymers are slightly better electrical conductors than silicon semiconductors but are far worse than metallic conductors. They are currently used in a number of devices, such as electrodes in batteries, electrolytic capacitors, and sensors. Recent studies of photon emission by conducting polymers may lead to new technologies for light-emitting diodes and flat-panel displays. Conducting polymers also show promise as molecular wires that can be incorporated into nanometre-sized electronic devices.

A brief comment

The 2000 Nobel Prize in chemistry was awarded to A.J. Heeger, A.G. MacDiarmid, and H. Shirakawa for their pioneering work in the synthesis and characterization of conducting polymers.

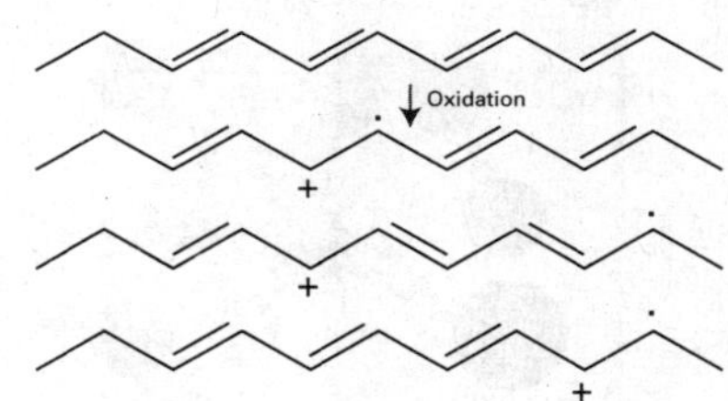

Fig. 18.13 The mechanism of migration of a partially localized cation radical, or polaron, in polyacetylene.

18.5 The structures of biological macromolecules

Key points (a) The secondary structure of a protein is the spatial arrangement of the polypeptide chain and includes the α-helix and β-sheet. Helical and sheet-like polypeptide chains are folded into a tertiary structure by bonding influences between the residues of the chain. Some proteins have a quaternary structure as aggregates of two or more polypeptide chains. (b) In DNA, two polynucleotide chains held together by hydrogen-bonded base pairs wind around each other to form a double helix. In RNA, single chains fold into complex structures by formation of specific base pairs.

A protein is a polypeptide composed of linked α-amino acids, $NH_2CHRCOOH$, where R is one of about 20 groups. For a protein to function correctly, it needs to have

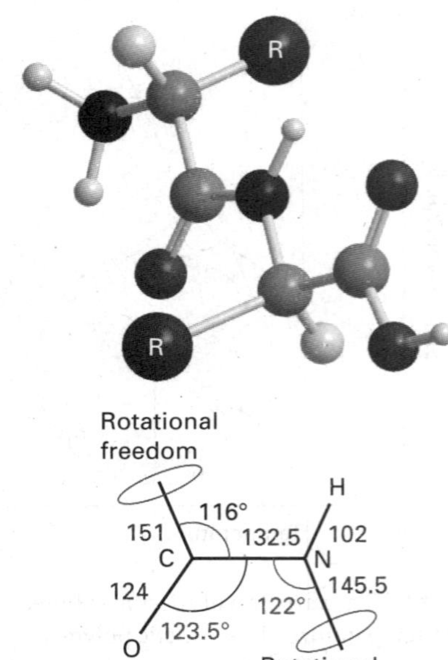

Fig. 18.14 The dimensions that characterize the peptide link. The C–NH–CO–C atoms define a plane (the C–N bond has partial double-bond character), but there is rotational freedom around the C–CO and N–C bonds.

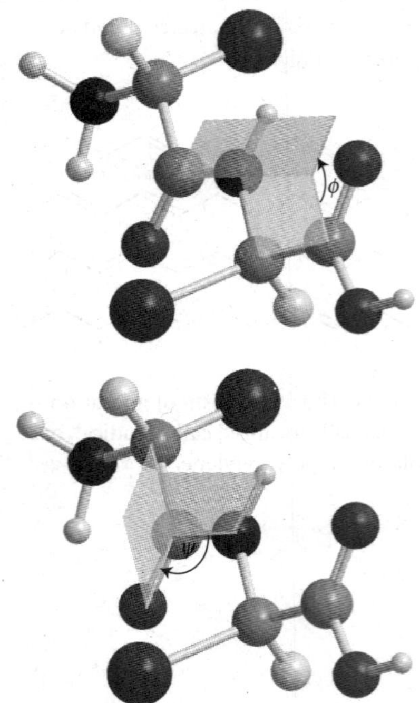

Fig. 18.15 The definition of the torsional angles ψ and ϕ between two peptide units. In this case (an α-L-polypeptide) the chain has been drawn in its all-*trans* form, with $\psi = \phi = 180°$.

a well defined conformation. For example, an enzyme has its greatest catalytic efficiency only when it is in a specific conformation. The amino acid sequence of a protein contains the necessary information to create the active conformation of the protein as it is formed. However, the prediction of the observed conformation from the primary structure, the so-called *protein folding problem*, is extraordinarily difficult and is still the focus of much research. Nucleic acids are key components of the mechanism of storage and transfer of genetic information in biological cells. Deoxyribonucleic acid (DNA) contains the instructions for protein synthesis, which is carried out by different forms of ribonucleic acid (RNA).

(a) Proteins

The origin of the secondary structures of proteins is found in the rules formulated by Linus Pauling and Robert Corey in 1951 that seek to identify the principal contributions to the lowering of energy of the molecule by focusing on the role of hydrogen bonds and the peptide link, –CONH–. The latter can act both as a donor of the H atom (the NH part of the link) and as an acceptor (the CO part). The **Corey–Pauling rules** are as follows (Fig. 18.14):

1. The four atoms of the peptide link lie in a relatively rigid plane.

The planarity of the link is due to delocalization of π electrons over the O, C, and N atoms and the maintenance of maximum overlap of their p orbitals.

2. The N, H, and O atoms of a hydrogen bond lie in a straight line (with displacements of H tolerated up to not more than 30° from the N–O vector).
3. All NH and CO groups are engaged in hydrogen bonding.

The rules are satisfied by two structures. One, in which hydrogen bonding between peptide links leads to a helical structure, is a *helix*, which can be arranged as either a right- or a left-handed screw. The other, in which hydrogen bonding between peptide links leads to a planar structure, is a *sheet*; this form is the secondary structure of the protein fibroin, the constituent of silk.

Because the planar peptide link is relatively rigid, the geometry of a polypeptide chain can be specified by the two angles that two neighbouring planar peptide links make to each other. Figure 18.15 shows the two angles ϕ and ψ commonly used to specify this relative orientation. The sign convention is that a positive angle means that the front atom must be rotated clockwise to bring it into an eclipsed position relative to the rear atom. For an all-*trans* form of the chain, all ϕ and ψ are 180°. A helix is obtained when all the ϕ are equal and when all the ψ are equal. For a right-handed helix (Fig. 18.16), all $\phi = 57°$ and all $\psi = -47°$. For a left-handed helix, both angles are positive. The torsional contribution to the total potential energy is

$$V_{\text{torsion}} = A(1 + \cos 3\phi) + B(1 + \cos 3\psi) \tag{18.13}$$

in which A and B are constants of the order of 1 kJ mol^{-1}. Because only two angles are needed to specify the conformation of a helix, and they range from −180° to +180°, the torsional potential energy of the entire molecule can be represented on a **Ramachandran plot**, a contour diagram in which one axis represents ϕ and the other represents ψ.

Figure 18.17 shows the Ramachandran plots for the helical form of polypeptide chains formed from the nonchiral amino acid glycine (R = H) and the chiral amino acid L-alanine (R = CH_3). The glycine map is almost symmetrical, with minima of equal depth at $\phi = -80°$, $\psi = -60°$ and at $\phi = +80°$, $\psi = -0°$. In contrast, the map for L-alanine is unsymmetrical, and there are three distinct low-energy conformations (marked I, II, III). The minima of regions I and II lie close to the angles typical of

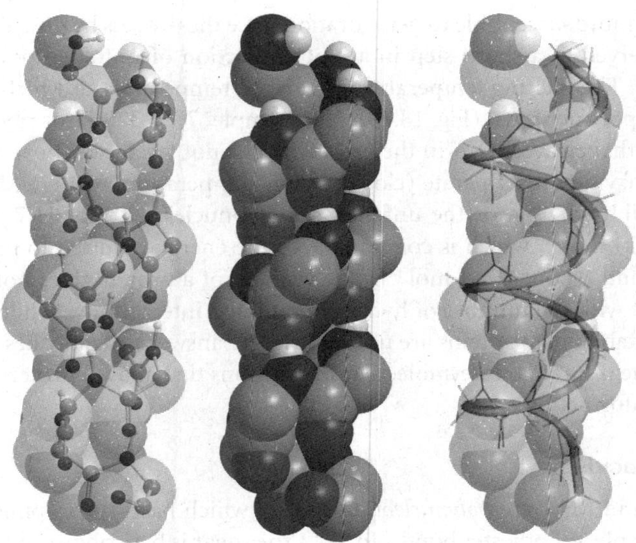

Fig 18.16 The polypeptide α helix, with poly-L-glycine as an example. There are 3.6 residues per turn, and a translation along the helix of 150 pm per residue, giving a pitch of 540 pm. The diameter (ignoring side chains) is about 600 pm.

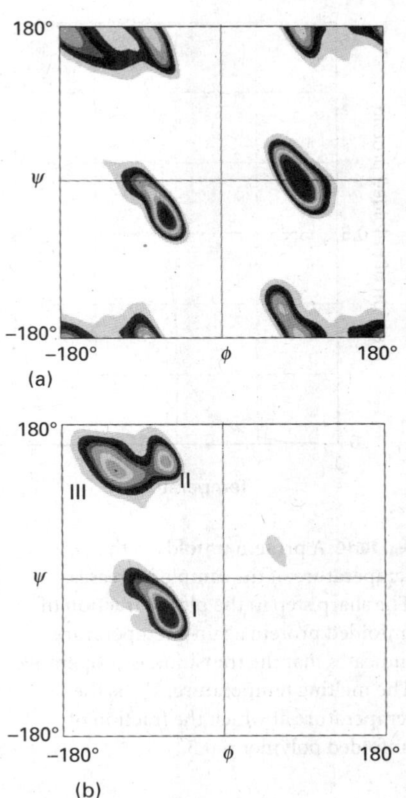

Fig. 18.17 Contour plots of potential energy against the angles ψ and ϕ, also known as a Ramachandran diagram, for (a) a glycyl residue of a polypeptide chain and (b) an alanyl residue. The glycyl diagram is symmetrical, but that for alanyl is unsymmetrical and the global minimum corresponds to an α-helix. (Reproduced with permission, T. Hovmöller *et al.*, *Acta Cryst.* **D58**, 768 (2002).)

right- and left-handed helices, but the former has a lower minimum. This result is consistent with the observation that polypeptides of the naturally occurring L-amino acids tend to form right-handed helices.

A **β-sheet** (also called the **β-pleated sheet**) is formed by hydrogen bonding between two extended polypeptide chains (large absolute values of the torsion angles ϕ and ψ). In an **antiparallel β-sheet** (Fig. 18.18a), $\phi = -139°$, $\psi = 113°$, and the N–H···O atoms of the hydrogen bonds form a straight line. This arrangement is a consequence of the antiparallel arrangement of the chains: every N–H bond on one chain is aligned with a C–O bond from another chain. Antiparallel β-sheets are very common in proteins. In a **parallel β-sheet** (Fig. 18.18b), $\phi = -119°$, $\psi = 113°$, and the N–H···O atoms of the hydrogen bonds are not perfectly aligned. This arrangement is a result of the parallel arrangement of the chains: each N–H bond on one chain is aligned with a N–H bond of another chain and, as a result, each C–O bond of one chain is aligned with a C–O bond of another chain. These structures are not common in proteins.

Covalent and non-covalent interactions may cause polypeptide chains with well defined secondary structures to fold into tertiary structures. Although the rules that govern protein folding are still being elucidated, a few general conclusions may be drawn from X-ray diffraction studies of water-soluble natural proteins and synthetic polypeptides. In an aqueous environment, the chains fold in such a way as to place nonpolar R groups in the interior (which is often not very accessible to solvent) and charged R groups on the surface (in direct contact with the polar solvent). Other factors that promote the folding of proteins include covalent disulfide (–S–S–) links, Coulombic interactions between ions (which depend on the degree of protonation of groups and therefore on the pH), van der Waals interactions, and hydrophobic interactions (Section 17.5f). The clustering of nonpolar, hydrophobic, amino acids into the interior of a protein is driven primarily by hydrophobic interactions.

Proteins are relatively unstable towards chemical and thermal **denaturation**, the loss of structure. Thermal denaturation is similar to the melting of synthetic polymers. Denaturation is a **cooperative process** in the sense that the biopolymer becomes

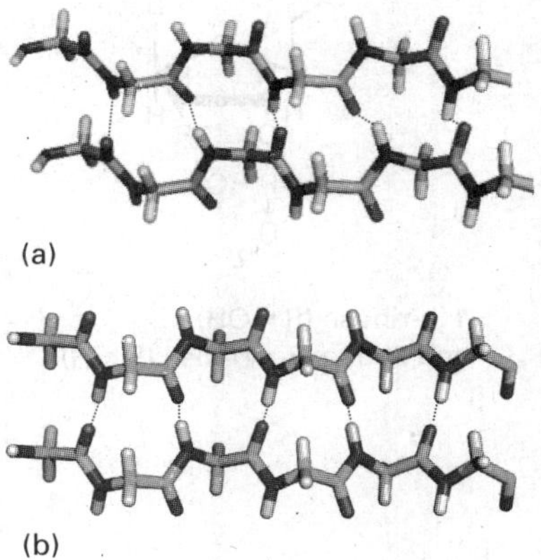

Fig. 18.18 The two types of β-sheets: (a) antiparallel ($\phi = -139°$, $\psi = 113°$), in which the N–H–O atoms of the hydrogen bonds form a straight line; (b) parallel ($\phi = -119°$, $\psi = 113°$ in which the N–H–O atoms of the hydrogen bonds are not perfectly aligned.

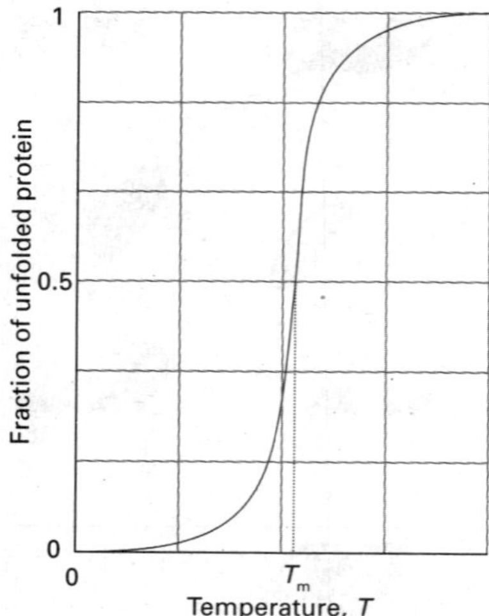

Fig. 18.19 A protein unfolds as the temperature of the sample increases. The sharp step in the plot of fraction of unfolded protein against temperature indicates that the transition is cooperative. The melting temperature, T_m, is the temperature at which the fraction of unfolded polymer is 0.5.

increasingly more susceptible to denaturation once the process begins. This cooperativity is observed as a sharp step in a plot of fraction of unfolded polymer versus temperature. The **melting temperature**, T_m, is the temperature at which the fraction of unfolded polymer is 0.5 (Fig. 18.19). For example, $T_m = 320$ K for ribonuclease T_1 (an enzyme that cleaves RNA in the cell), which is not far above the temperature at which the enzyme must operate (close to body temperature, 310 K). More surprisingly, the Gibbs energy for the unfolding of ribonuclease T_1 at pH 7.0 and 298 K is only 19.5 kJ mol^{-1}, which is comparable to the energy required to break a single hydrogen bond (about 20 kJ mol^{-1}). The stability of a protein does not increase in a simple way with the number of hydrogen bonding interactions. While the reasons for the low stability of proteins are not known, the answer probably lies in a delicate balance of the intra- and intermolecular interactions that allow a protein to fold into its active conformation.

(b) Nucleic acids

Both DNA and RNA are *polynucleotides* (1), in which base–sugar–phosphate units are linked by phosphodiester bonds. In RNA the sugar is β-D-ribose and in DNA it is β-D-2-deoxyribose (as shown in 1). The most common bases are adenine (A, 2), cytosine (C, 3), guanine (G, 4), thymine (T, found in DNA only, 5), and uracil (U, found in RNA only, 6). At physiological pH, each phosphate group of the chain carries a negative charge and the bases are deprotonated and neutral. This charge distribution leads to two important properties. One is that the polynucleotide chain is a **polyelectrolyte**, a macromolecule with many different charged sites, with a large and negative overall surface charge. The second is that the bases can interact by hydrogen bonding, as shown for A–T (7) and C–G base pairs (8). The secondary and tertiary structures of DNA and RNA arise primarily from the pattern of this hydrogen bonding between bases of one or more chains.

In DNA, two polynucleotide chains wind around each other to form a double helix (Fig. 18.20). The chains are held together by links involving A–T and C–G base pairs that lie parallel to each other and perpendicular to the major axis of the helix. The structure is stabilized further by interactions between the planar π systems of the bases. In B-DNA, the most common form of DNA found in biological cells, the helix is right-handed with a diameter of 2.0 nm and a pitch of 3.4 nm.

1 D-ribose (R = OH) and 2′-deoxy-D-ribose (R = H)

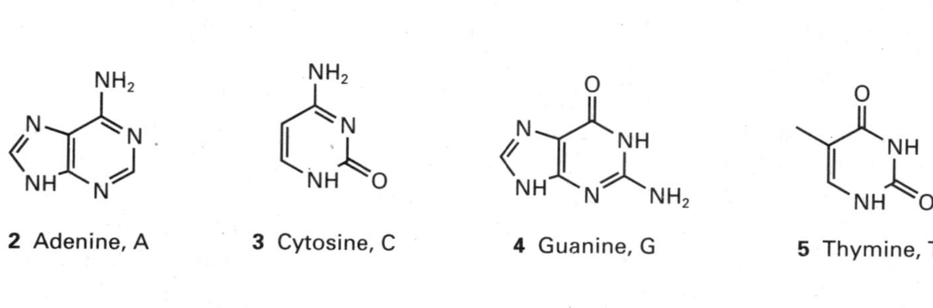

2 Adenine, A **3** Cytosine, C **4** Guanine, G **5** Thymine, T

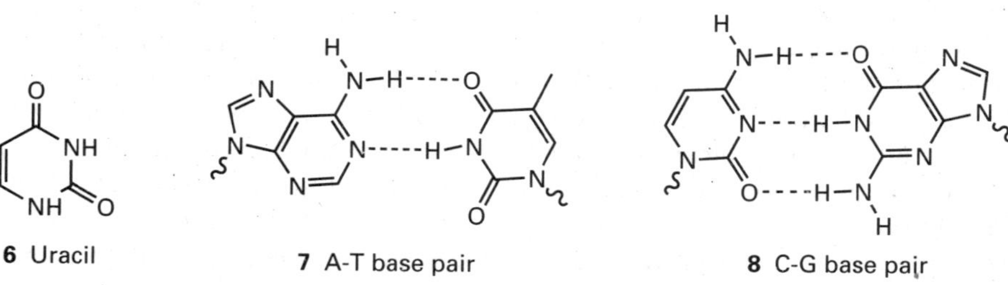

6 Uracil **7** A-T base pair **8** C-G base pair

Aggregation and self-assembly

Much of the material discussed in this chapter also applies to aggregates of particles that form by **self-assembly**, the spontaneous formation of complex structures of molecules or macromolecules held together by molecular interactions, such as Coulombic, dispersion, hydrogen bonding, or hydrophobic interactions. We have already encountered a few examples of self-assembly, such as the formation of liquid crystals (*Impact I5.2*), of protein quaternary structures from two or more polypeptide chains, and (by implication) of a DNA double helix from two polynucleotide chains. Now we concentrate on the specific properties of additional self-assembled systems, including small aggregates that are at the heart of detergent action and extended sheets like those forming biological cell membranes. We also consider examples in which the controlled design of new materials with enhanced properties is informed by an understanding of the principles underlying self-assembly.

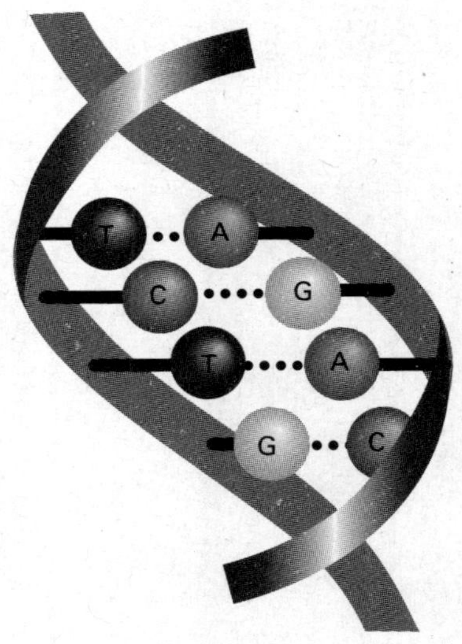

Fig. 18.20 DNA double helix, in which two polynucleotide chains are linked together by hydrogen bonds between adenine (A) and thymine (T) and between cytosine (C) and guanine (G).

18.6 Colloids

Key points **A disperse system is a dispersion of small particles of one material in another. (a) Colloids are classified as lyophilic and lyophobic. A surfactant is a species that accumulates at the interface of two phases or substances. (b) Many colloid particles are thermodynamically unstable but kinetically nonlabile. (c) The radius of shear is the radius of the sphere that captures the rigid layer of charge attached to a colloid particle. The zeta potential is the electric potential at the radius of shear relative to its value in the distant, bulk medium. The inner shell of charge and the outer atmosphere jointly constitute the electric double layer. Flocculation is the reversible aggregation of colloidal particles; coagulation is the irreversible aggregation of colloidal particles. The Schultze–Hardy rule states that hydrophobic colloids are flocculated most efficiently by ions of opposite charge type and high charge number.**

A **colloid**, or **disperse phase**, is a dispersion of small particles of one material in another that does not settle out under gravity. In this context, 'small' means that one dimension at least is smaller than about 500 nm in diameter (about the wavelength of visible light). Many colloids are suspensions of nanoparticles (particles of diameter up to about 100 nm). In general, colloidal particles are aggregates of numerous atoms or molecules, but are commonly but not universally too small to be seen with an ordinary optical microscope. They pass through most filter papers, but can be detected by light-scattering and sedimentation.

(a) Classification and preparation

The name given to the colloid depends on the two phases involved. A **sol** is a dispersion of a solid in a liquid (such as clusters of gold atoms in water) or of a solid in a solid (such as ruby glass, which is a gold-in-glass sol, and achieves its colour by light scattering). An **aerosol** is a dispersion of a liquid in a gas (like fog and many sprays) or a solid in a gas (such as smoke): the particles are often large enough to be seen with a microscope. An **emulsion** is a dispersion of a liquid in a liquid (such as milk). A **foam** is a dispersion of a gas in a liquid.

A further classification of colloids is as **lyophilic**, or solvent attracting, and **lyophobic**, solvent repelling. If the solvent is water, the terms **hydrophilic** and **hydrophobic**, respectively, are used instead. Lyophobic colloids include the metal sols. Lyophilic colloids generally have some chemical similarity to the solvent, such as –OH groups able to form hydrogen bonds. A **gel** is a semirigid mass of a lyophilic sol.

The preparation of aerosols can be as simple as sneezing (which produces an imperfect aerosol). Laboratory and commercial methods make use of several techniques. Material (for example, quartz) may be ground in the presence of the dispersion medium. Passing a heavy electric current through a cell may lead to the sputtering (crumbling) of an electrode into colloidal particles. Arcing between electrodes immersed in the support medium also produces a colloid. Chemical precipitation sometimes results in a colloid. A precipitate (for example, silver iodide) already formed may be dispersed by the addition of a peptizing agent (for example, potassium iodide). Clays may be peptized by alkalis, the OH ion being the active agent.

Emulsions are normally prepared by shaking the two components together vigorously, although some kind of emulsifying agent usually has to be added to stabilize the product. This emulsifying agent may be a soap (the salt of a long-chain carboxylic acid) or other **surfactant** (surface active) species, or a lyophilic sol that forms a protective film around the dispersed phase. In milk, which is an emulsion of fats in water, the emulsifying agent is casein, a protein containing phosphate groups. It is clear from the formation of cream on the surface of milk that casein is not completely successful in stabilizing milk: the dispersed fats coalesce into oily droplets which float to the surface. This coagulation may be prevented by ensuring that the emulsion is dispersed very finely initially: intense agitation with ultrasonics brings this dispersion about, the product being 'homogenized' milk.

One way to form an aerosol is to tear apart a spray of liquid with a jet of gas. The dispersal is aided if a charge is applied to the liquid, for then electrostatic repulsions help to blast it apart into droplets. This procedure may also be used to produce emulsions, for the charged liquid phase may be directed into another liquid.

Colloids are often purified by dialysis (*Impact I5.1*). The aim is to remove much (but not all, for reasons explained later) of the ionic material that may have accompanied their formation. A membrane (for example, cellulose) is selected that is permeable to solvent and ions, but not to the colloid particles. Dialysis is very slow, and is normally accelerated by applying an electric field and making use of the charges carried by many colloidal particles; the technique is then called **electrodialysis**.

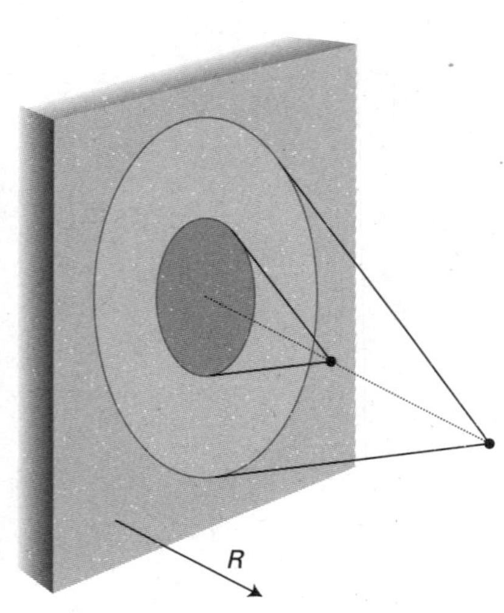

Fig. 18.21 Although the attraction between individual molecules is proportional to $1/R^6$, more molecules are within range at large separations (pale region) than at small separation (dark region), so the total interaction energy declines more slowly and is proportional to a lower power of R.

(b) Structure and stability

Colloids are thermodynamically unstable with respect to the bulk. This instability can be expressed thermodynamically by noting that because the change in Helmholtz energy, dA, when the surface area of the sample changes by $d\sigma$ at constant temperature and pressure is $dA = \gamma d\sigma$, where γ is the interfacial surface tension (Section 17.8a), it follows that $dA < 0$ if $d\sigma < 0$. The survival of colloids must therefore be a consequence of the kinetics of collapse: colloids are thermodynamically unstable but kinetically nonlabile.

At first sight, even the kinetic argument seems to fail: colloidal particles attract each other over large distances, so there is a long-range force that tends to condense them into a single blob. The reasoning behind this remark is as follows. The energy of attraction between two individual atoms i and j separated by a distance R_{ij}, one in each colloidal particle, varies with their separation as $1/R_{ij}^6$ (Section 17.5). The sum of all these pairwise interactions, however, decreases only as approximately $1/R^2$ (the precise variation depending on the shape of the particles and their closeness), where R is the separation of the centres of the particles. The change in the power from 6 to 2 stems from the fact that at short distances only a few molecules interact but at large distances many individual molecules are at about the same distance from one another, and contribute equally to the sum (Fig. 18.21), so the total interaction does not fall off as fast as the single molecule–molecule interaction.

Several factors oppose the long-range dispersion attraction. For example, there may be a protective film at the surface of the colloid particles that stabilizes the interface and cannot be penetrated when two particles touch. Thus the surface atoms of a platinum sol in water react chemically and are turned into $-Pt(OH)_3H_3$, and this layer encases the particle like a shell. A fat can be emulsified by a soap because the long hydrocarbon tails penetrate the oil droplet but the carboxylate head groups (or other hydrophilic groups in synthetic detergents) surround the surface, form hydrogen bonds with water, and give rise to a shell of negative charge that repels a possible approach from another similarly charged particle.

(c) The electrical double layer

A major source of kinetic nonlability of colloids is the existence of an electric charge on the surfaces of the particles. On account of this charge, ions of opposite charge tend to cluster nearby, and an ionic atmosphere is formed, just as for ions (Section 5.13).

We need to distinguish two regions of charge. First, there is a fairly immobile layer of ions that adhere tightly to the surface of the colloidal particle, and which may include water molecules (if that is the support medium). The radius of the sphere that captures this rigid layer is called the **radius of shear** and is the major factor determining the mobility of the particles. The electric potential at the radius of shear relative to its value in the distant, bulk medium is called the **zeta potential**, ζ, or the **electrokinetic potential**. Second, the charged unit attracts an oppositely charged atmosphere of mobile ions. The inner shell of charge and the outer ionic atmosphere is called the **electrical double layer**.

The theory of the stability of lyophobic dispersions was developed by B. Derjaguin and L. Landau and independently by E. Verwey and J.T.G. Overbeek, and is known as the **DLVO theory**.[1] It assumes that there is a balance between the repulsive interaction between the charges of the electrical double layers on neighbouring particles and the attractive interactions arising from van der Waals interactions between the molecules in the particles. The potential energy arising from the repulsion of double layers on particles of radius a has the form

$$V_{\text{repulsion}} = +\frac{Aa^2\zeta^2}{R}e^{-s/r_D} \tag{18.14}$$

where A is a constant, ζ is the zeta potential, R is the separation of centres, s is the separation of the surfaces of the two particles ($s = R - 2a$ for spherical particles of radius a), and r_D is the thickness of the double layer. This expression is valid for small particles with a thick double layer ($a \ll r_D$). When the double layer is thin ($r_D \ll a$), the expression is replaced by

$$V_{\text{repulsion}} = \tfrac{1}{2}Aa\zeta^2 \ln(1 + e^{-s/r_D}) \tag{18.15}$$

In each case, the thickness of the double layer can be estimated from an expression like that derived for the thickness of the ionic atmosphere in the Debye–Hückel theory (eqn 5.91) in which there is a competition between the assembling influences of the attraction between opposite charges and the disruptive effect of thermal motion:

$$r_D = \left(\frac{\varepsilon RT}{2\rho F^2 I b^{\ominus}}\right)^{1/2} \quad \text{Thickness of the electrical double layer} \tag{18.16}$$

[1] The derivation of the expressions quoted here is too complicated to include here. For a full description, see Volume 1 of R.J. Hunter, *Foundations of colloid science*, Oxford University Press (1987).

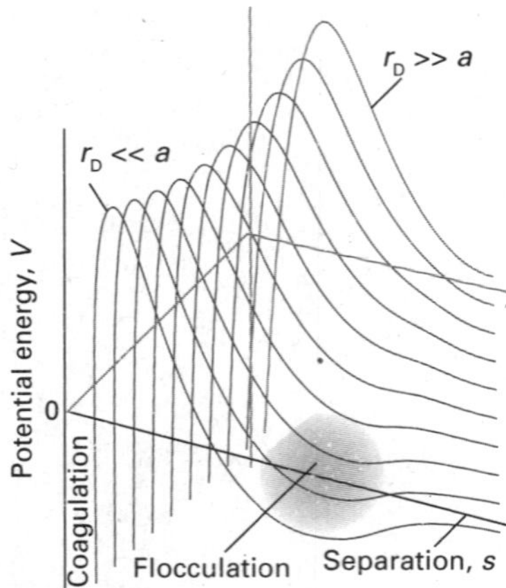

Fig. 18.22 The potential energy of interaction as a function of the separation of the centres of the two particles and its variation with the ratio of the particle size (radius a for spherical particles) to the thickness of the electrical double layer, r_D. The regions labelled coagulation and flocculation show the dips in the potential energy curves where these processes occur.

where I is the ionic strength of the solution, ρ its mass density, and $b^{\ominus} = 1$ mol kg^{-1} (F is Faraday's constant and ε is the permittivity, $\varepsilon = \varepsilon_r\varepsilon_0$). The potential energy arising from the attractive interaction has the form

$$V_{\text{attraction}} = -\frac{B}{s} \tag{18.17}$$

where B is another constant. The variation of the total potential energy with separation is shown in Fig. 18.22.

At high ionic strengths, the ionic atmosphere is dense and the potential shows a secondary minimum at large separations. Aggregation of the particles arising from the stabilizing effect of this secondary minimum is called **flocculation.** The flocculated material can often be redispersed by agitation because the well is so shallow. **Coagulation**, the irreversible aggregation of distinct particles into large particles, occurs when the separation of the particles is so small that they enter the primary minimum of the potential energy curve and van der Waals forces are dominant.

The ionic strength is increased by the addition of ions, particularly those of high charge type, so such ions act as flocculating agents. This increase is the basis of the empirical **Schulze–Hardy rule**, that hydrophobic colloids are flocculated most efficiently by ions of opposite charge type and high charge number. The Al^{3+} ions in alum are very effective, and are used to induce the congealing of blood. When river water containing colloidal clay flows into the sea, the salt water induces flocculation and coagulation, and is a major cause of silting in estuaries. Metal oxide sols tend to be positively charged, whereas sulfur and the noble metals tend to be negatively charged.

The primary role of the electric double layer is to confer kinetic non-lability. Colliding colloidal particles break through the double layer and coalesce only if the collision is sufficiently energetic to disrupt the layers of ions and solvating molecules, or if thermal motion has stirred away the surface accumulation of charge. This disruption may occur at high temperatures, which is one reason why sols precipitate when they are heated.

18.7 Micelles and biological membranes

Key points **(a) A micelle is a colloid-sized cluster of molecules that forms at the critical micelle concentration and at the Krafft temperature. Micelles can assume a number of shapes, depending on temperature, shape, and concentration of constituent molecules. (b) Some micelles exist as parallel sheets two molecules thick that are either extended (planar bilayers) or fold back on to themselves (unilamellar vesicles). (c) Self-assembled monolayers are ordered molecular aggregates that form a single layer of material on a surface.**

In aqueous solutions surfactant molecules or ions can cluster together as **micelles,** which are colloid-sized clusters of molecules, for their hydrophobic tails tend to congregate (through hydrophobic interactions—see Section 17.5f), and their hydrophilic head groups provide protection (Fig. 18.23).

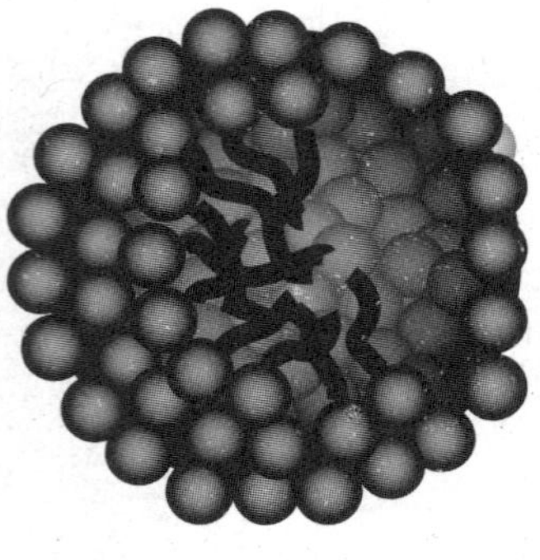

Fig. 18.23 A schematic version of a spherical micelle. The hydrophilic groups are represented by spheres and the hydrophobic hydrocarbon chains are represented by the stalks; these stalks are mobile.

(a) Micelle formation

Micelles form only above the **critical micelle concentration** (CMC) and above the **Krafft temperature.** The CMC is detected by noting a pronounced change in physical properties of the solution, particularly the molar conductivity (Fig. 18.24). There is no abrupt change in properties at the CMC; rather, there is a transition region corresponding to a range of concentrations around the CMC where physical properties vary smoothly but nonlinearly with the concentration. The hydrocarbon interior of a micelle is like a droplet of oil. Nuclear magnetic resonance shows that the hydrocarbon tails are mobile, but slightly more restricted than in the bulk. Micelles are important in industry and biology on account of their solubilizing function: matter can be transported by water after it has been dissolved in their hydrocarbon interiors. For this reason, micellar systems are used as detergents, for organic synthesis, froth flotation, and petroleum recovery.

Non-ionic surfactant molecules may cluster together in clumps of 1000 or more, but ionic species tend to be disrupted by the electrostatic repulsions between head groups and are normally limited to groups of less than about 100. However, the disruptive effect depends more on the effective size of the head group than the charge. For example, ionic surfactants such as sodium dodecyl sulfate (SDS) and cetyl trimethylammonium bromide (CTAB) form rods at moderate concentrations, whereas sugar surfactants form small, approximately spherical micelles. The micelle population is often polydisperse, and the shapes of the individual micelles vary with shape of the constituent surfactant molecules, surfactant concentration, and temperature. A useful predictor of the shape of the micelle is the **surfactant parameter,** N_s, defined as

$$N_s = \frac{V}{Al} \qquad \text{Definition of the surfactant parameter} \qquad [18.18]$$

where V is the volume of the hydrophobic surfactant tail, A is the area of the hydrophilic surfactant head group, and l is the maximum length of the surfactant tail. Table 18.1 summarizes the dependence of aggregate structure on the surfactant parameter.

In aqueous solutions spherical micelles form, as shown in Fig. 18.23, with the polar head groups of the surfactant molecules on the micellar surface and interacting favorably with solvent and ions in solution. Hydrophobic interactions stabilize the aggregation of the hydrophobic surfactant tails in the micellar core. Under certain experimental conditions, a **liposome** may form, with an inward pointing inner surface of molecules surrounded by an outward pointing outer layer (Fig. 18.25). Liposomes may be used to carry nonpolar drug molecules in blood.

Increasing the ionic strength of the aqueous solution reduces repulsions between surface head groups, and cylindrical micelles can form. These cylinders may stack together in reasonably close-packed (hexagonal) arrays, forming **lyotropic mesomorphs** and, more colloquially, 'liquid crystalline phases'.

Reverse micelles form in nonpolar solvents, with small polar surfactant head groups in a micellar core and more voluminous hydrophobic surfactant tails extending into the organic bulk phase. These spherical aggregates can solubilize water in organic solvents by creating a pool of trapped water molecules in the micellar core. As aggregates arrange at high surfactant concentrations to yield long-range positional order, many other types of structures are possible including cubic and hexagonal shapes.

The enthalpy of micelle formation reflects the contributions of interactions between micelle chains within the micelles and between the polar head groups and the

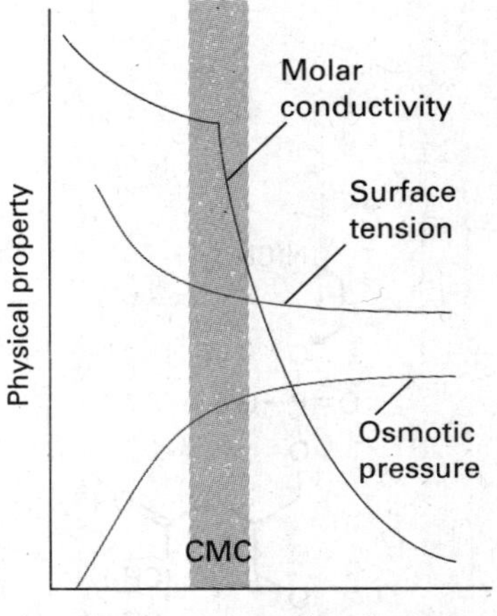

Fig. 18.24 The typical variation of some physical properties of an aqueous solution of sodium dodecylsulfate close to the critical micelle concentration (CMC).

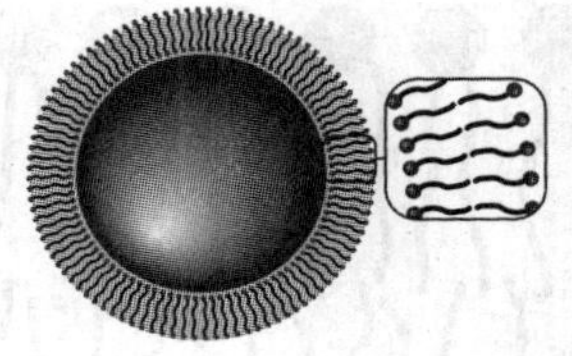

Fig. 18.25 The cross-sectional structure of a spherical liposome.

Table 18.1 Variation of micelle shape with the surfactant parameter

Value or range of the surfactant parameter, N_s	Micelle shape
< 0.33	Spherical
0.33 to 0.50	Cylindrical rods
0.50 to 1.00	Vesicles
1.00	Planar bilayers
> 1.00	Reverse micelles and other shapes

surrounding medium. Consequently, enthalpies of micelle formation display no readily discernible pattern and may be positive (endothermic) or negative (exothermic). Many non-ionic micelles form endothermically, with ΔH of the order of 10 kJ per mole of surfactant molecules. That such micelles do form above the CMC indicates that the entropy change accompanying their formation must then be positive, and measurements suggest a value of about $+140\ \mathrm{J\ K^{-1}\ mol^{-1}}$ at room temperature. The fact that the entropy change is positive even though the molecules are clustering together shows that hydrophobic interactions are important in the formation of micelles.

9 Phosphatidyl choline

(b) Bilayers, vesicles, and membranes

Some micelles at concentrations well above the CMC form extended parallel sheets two molecules thick, called **planar bilayers.** The individual molecules lie perpendicular to the sheets, with hydrophilic groups on the outside in aqueous solution and on the inside in nonpolar media. When segments of planar bilayers fold back on themselves, **unilamellar vesicles** may form where the spherical hydrophobic bilayer shell separates an inner aqueous compartment from the external aqueous environment.

Bilayers show a close resemblance to biological membranes, and are often a useful model on which to base investigations of biological structures. However, actual membranes are highly sophisticated structures. The basic structural element of a membrane is a phospholipid, such as phosphatidyl choline (**9**), which contains long hydrocarbon chains (typically in the range C_{14}–C_{24}) and a variety of polar groups, such as $-CH_2CH_2N(CH_3)_3^+$. The hydrophobic chains stack together to form an extensive layer about 5 nm across. The lipid molecules form layers instead of micelles because the hydrocarbon chains are too bulky to allow packing into nearly spherical clusters.

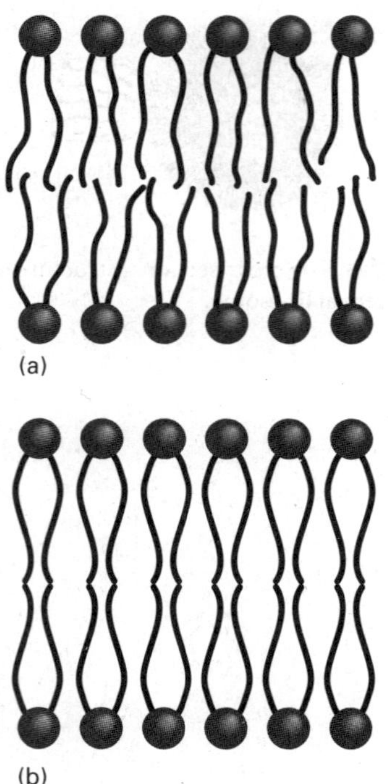

Fig. 18.26 A depiction of the variation with temperature of the flexibility of hydrocarbon chains in a lipid bilayer. (a) At physiological temperature, the bilayer exists as a liquid crystal, in which some order exists but the chains writhe. (b) At a specific temperature, the chains are largely frozen and the bilayer is said to exist as a gel.

The bilayer is a highly mobile structure, as shown by EPR studies with spin-labelled phospholipids (*Impact I14.2*). Not only are the hydrocarbon chains ceaselessly twisting and turning in the region between the polar groups, but the phospholipid and cholesterol molecules migrate over the surface. It is better to think of the membrane as a viscous fluid rather than a permanent structure, with a viscosity about 100 times that of water. In common with diffusional behaviour in general (Section 20.8), the average distance a phospholipid molecule diffuses is proportional to the square-root of the time; more precisely, for a molecule confined to a two-dimensional plane, the average distance travelled in a time t is equal to $(4Dt)^{1/2}$. Typically, a phospholipid molecule migrates through about 1 μm in about 1 min.

All lipid bilayers undergo a transition from a state of high to low chain mobility at a temperature that depends on the structure of the lipid. To visualize the transition, we consider what happens to a membrane as we lower its temperature (Fig. 18.26). There is sufficient energy available at normal temperatures for limited bond rotation to occur and the flexible chains writhe. However, the membrane is still highly organized in the sense that the bilayer structure does not come apart and the system is best described as a liquid crystal. At lower temperatures, the amplitudes of the writhing motion decrease until a specific temperature is reached at which motion is largely frozen. The membrane is said to exist as a gel. Biological membranes exist as liquid crystals at physiological temperatures.

Phase transitions in membranes are often observed as 'melting' from gel to liquid crystal by differential scanning calorimetry (*Impact I2.1*). The data show relations between the structure of the lipid and the melting temperature. For example, the melting temperature increases with the length of the hydrophobic chain of the lipid. This correlation is reasonable, as we expect longer chains to be held together more strongly by hydrophobic interactions than shorter chains. It follows that stabilization of the gel phase in membranes of lipids with long chains results in relatively high melting temperatures. On the other hand, any structural elements that prevent alignment of the hydrophobic chains in the gel phase lead to low melting temperatures. Indeed, lipids containing unsaturated chains, those containing some C=C bonds, form membranes with lower melting temperatures than those formed from lipids with fully saturated chains, those consisting of C–C bonds only.

Interspersed among the phospholipids of biological membranes are sterols, such as cholesterol (**10**), which is largely hydrophobic but does contain a hydrophilic –OH group. Sterols, which are present in different proportions in different types of cells, prevent the hydrophobic chains of lipids from 'freezing' into a gel and, by disrupting the packing of the chains, spread the melting point of the membrane over a range of temperatures.

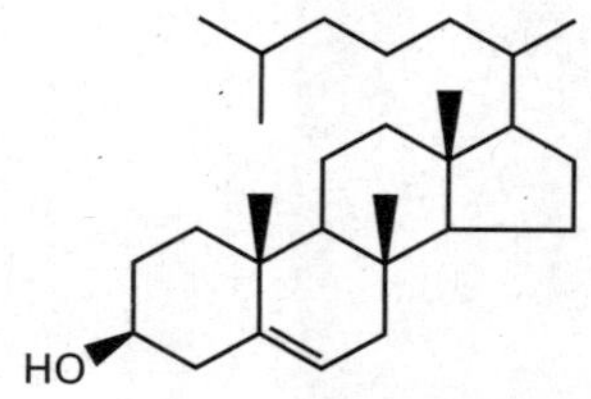

10 Cholesterol

(c) Self-assembled monolayers

Molecular self-assembly can be used as the basis for manipulation of surfaces on the nanometre scale. Of current interest are **self-assembled monolayers** (SAMs), ordered molecular aggregates that form a single layer of material on a surface. To understand the formation of SAMs, consider exposing molecules such as alkyl thiols RSH, where R represents an alkyl chain, to an Au(0) surface. The thiols react with the surface, forming $RS^-Au(I)$ adducts:

$$RSH + Au(0)_n \rightarrow RS^-Au(I) \cdot Au(0)_{n-1} + \tfrac{1}{2}H_2$$

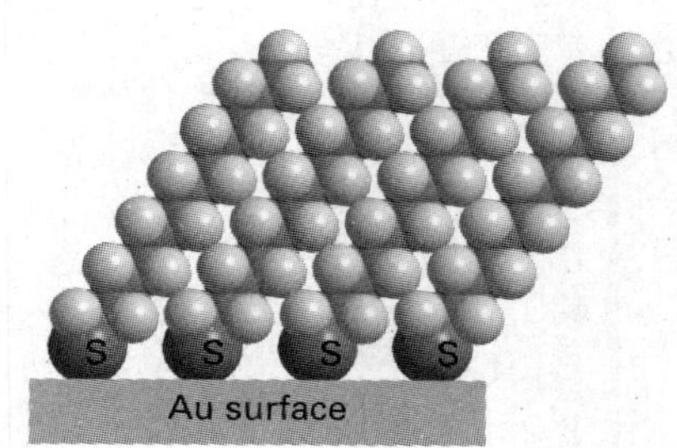

Fig. 18.27 Self-assembled monolayers of alkylthiols formed on to a gold surface by reaction of the thiol groups with the surface and aggregation of the alkyl chains.

If R is a sufficiently long chain, van der Waals interactions between the adsorbed RS units lead to the formation of a highly ordered monolayer on the surface, as shown in Fig. 18.27. It is observed that the Gibbs energy of formation of SAMs increases with the length of the alkyl chain, with each methylene group contributing 400–4000 J mol^{-1} to the overall Gibbs energy of formation.

A self-assembled monolayer alters the properties of the surface. For example, a hydrophilic surface may be rendered hydrophobic once covered with a SAM. Furthermore, attaching functional groups to the exposed ends of the alkyl groups may impart specific chemical reactivity or ligand-binding properties to the surface, leading to applications in chemical (or biochemical) sensors and reactors.

Determination of size and shape

We have seen the importance of knowing the sizes of macromolecules. X-ray diffraction (which is described in detail in Chapter 19) can reveal the position of almost every atom other than hydrogen even in very large molecules. However, there are several reasons why other techniques must also be used. In the first place, the sample might be a mixture of molecules with different chain lengths and extents of cross-linking, in which case sharp X-ray images are not obtained. Even if all the molecules in the sample are identical, it might prove impossible to obtain a single crystal, which is essential for diffraction studies because only then does the electron density (which is responsible for the scattering) have a large-scale periodic variation. Furthermore, although work on proteins and DNA has shown how immensely interesting and motivating the data can be, the information is incomplete. For instance, what can be said about the shape of the molecule in its natural environment, a biological cell? What can be said about the response of its shape to changes in its environment?

18.8 Mean molar masses

Key points Macromolecules can be monodisperse, with a single molar mass, or polydisperse, with various molar masses. Depending on the measurement technique, the mean values of molar masses of polydisperse systems are obtained as the number-average, viscosity-average, weight-average, and Z-average molar masses.

A pure protein is **monodisperse**, meaning that it has a single, definite molar mass. There may be small variations, such as one amino acid replacing another, depending on the source of the sample. A synthetic polymer, however, is **polydisperse**, in the sense that a sample is a mixture of molecules with various chain lengths and molar masses. The various techniques that are used to measure molar mass result in different types of mean values of polydisperse systems.

The mean obtained from the determination of molar mass by osmometry (Section 5.5e) is the **number-average molar mass**, $\bar{M}_n$, which is the value obtained by weighting each molar mass by the number of molecules of that mass present in the sample:

$$\bar{M}_n = \frac{1}{N}\sum_i N_i M_i = \langle M \rangle \qquad [18.19]$$

Definition of the number-average molar mass

where N_i is the number of molecules with molar mass M_i and there are N molecules in all. The notation $\langle X \rangle$ denotes the usual (number) average of a property X, and we shall use it again below. For reasons related to the ways in which macromolecules contribute to physical properties, viscosity measurements give the **viscosity-average molar mass**, $\bar{M}_v$, light-scattering experiments give the **weight-average molar mass**, $\bar{M}_w$, and sedimentation experiments give the ***Z*-average molar mass**, $\bar{M}_Z$. (The name is derived from the z-coordinate used to depict data in a procedure for determining the average.) Although such averages are often best left as empirical quantities, some may be interpreted in terms of the composition of the sample. Thus, the weight-average molar mass is the average calculated by weighting the molar masses of the molecules by the mass of each one present in the sample:

$$\bar{M}_w = \frac{1}{m}\sum_i m_i M_i \qquad [18.20a]$$

Definition of the weight-average molar mass

In this expression, m_i is the total mass of molecules of molar mass M_i and m is the total mass of the sample. Because $m_i = N_i M_i / N_A$, we can also express this average as

$$\bar{M}_w = \frac{\sum_i N_i M_i^2}{\sum_i N_i M_i} = \frac{\langle M^2 \rangle}{\langle M \rangle} \qquad (18.20b)$$

Interpretation of the weight-average molar mass

This expression shows that the weight-average molar mass is proportional to the mean square molar mass. Similarly, the Z-average molar mass turns out to be proportional to the mean cubic molar mass:

$$\bar{M}_Z = \frac{\sum_i N_i M_i^3}{\sum_i N_i M_i^2} = \frac{\langle M^3 \rangle}{\langle M^2 \rangle} \qquad (18.20c)$$

Interpretation of the Z-average molar mass

Example 18.1 *Calculating number and mass averages*

Determine the number-average and the weight-average molar masses of a sample of poly(vinyl chloride) from the following data:

Molar mass interval/ (kg mol^{-1})	Average molar mass within interval/(kg mol^{-1})	Mass of sample within interval/g
5–10	7.5	9.6
10–15	12.5	8.7
15–20	17.5	8.9
20–25	22.5	5.6
25–30	27.5	3.1
30–35	32.5	1.7

Method The relevant equations are eqns 18.19 and 18.20a. Calculate the two averages by weighting the molar mass within each interval by the number and mass, respectively, of the molecules in each interval. Obtain the numbers in each interval by dividing the mass of the sample in each interval by the average molar mass for that interval. Because the number of molecules is proportional to the amount of substance (the number of moles), the number-weighted average can be obtained directly from the amounts in each interval.

Answer The amounts in each interval are as follows:

Interval	5–10	10–15	15–20	20–25	25–30	30–35
Molar mass/(kg mol^{-1})	7.5	12.5	17.5	22.5	27.5	32.5
Amount/mmol	1.3	0.70	0.51	0.25	0.11	0.052
					Total:	2.92

The number-average molar mass is therefore

$$\bar{M}_n/(\text{kg mol}^{-1}) = \frac{1}{2.92}(1.3 \times 7.5 + 0.70 \times 12.5 + 0.51 \times 17.5 + 0.25 \times 22.5 + 0.11 \times 27.5 + 0.052 \times 32.5) = 13$$

The weight-average molar mass is calculated directly from the data after noting that the total mass of the sample is 37.6 g:

$$\bar{M}_w/(\text{kg mol}^{-1}) = \frac{1}{37.6}(9.6 \times 7.5 + 8.7 \times 12.5 + 8.9 \times 17.5 + 5.6 \times 22.5 + 3.1 \times 27.5 + 1.7 \times 32.5) = 16$$

Note the different values of the two averages. In this instance, $\bar{M}_w/\bar{M}_n = 1.2$.

Self-test 18.1 Evaluate the *Z*-average molar mass of the sample. [19 kg mol^{-1}]

The ratio $\bar{M}_w/\bar{M}_n$ is called the **heterogeneity index** (or 'polydispersity index'). It follows from eqns 18.19 and 18.20b that

$$\frac{\bar{M}_w}{\bar{M}_n} = \frac{\langle M^2 \rangle}{\langle M \rangle^2} \qquad [18.21]$$

Definition of the heterogeneity index

That is, the index is proportional to the ratio of the mean square molar mass to the square of the mean molar mass. In the determination of protein molar masses we expect the various averages to be the same because the sample is monodisperse (unless there has been degradation). A synthetic polymer normally spans a range of molar masses and the different averages yield different values. Typical synthetic materials have $\bar{M}_w/\bar{M}_n \approx 4$ but much recent research has been devoted to developing methods that give much lower polydisperities. The term 'monodisperse' is conventionally applied to synthetic polymers in which this index is less than 1.1; commercial polyethene samples might be much more heterogeneous, with a ratio close to 30. One consequence of a narrow molar mass distribution for synthetic polymers is often a higher degree of three-dimensional long-range order in the solid and therefore higher density and melting point. The spread of values is controlled by the choice of catalyst and reaction conditions. In practice, it is found that long-range order is determined more by structural factors (branching, for instance) than by molar mass.

A note on good practice The masses of macromolecules are often reported in daltons (Da), where 1 Da $= m_u$ (with $m_u = 1.661 \times 10^{-27}$ kg). Note that 1 Da is a measure of *molecular* mass not of *molar* mass. We might say that the mass (not the molar mass) of a certain macromolecule is 100 kDa (that is, its mass is $100 \times 10^3 \times m_u$); we could also say that its molar mass is 100 kg mol^{-1}; we should not say (even though it is common practice) that its molar mass is 100 kDa.

18.9 The techniques

Average molar masses may be determined by osmotic pressure of polymer solutions. The upper limit for the reliability of membrane osmometry is about 1000 kg mol^{-1}. A major problem for macromolecules of relatively low molar mass (less than about 10 kg mol^{-1}) is their ability to percolate through the membrane. One consequence of this partial permeability is that membrane osmometry tends to overestimate the average molar mass of a polydisperse mixture. Several techniques for the determination of molar mass and polydispersity that are not so limited include mass spectrometry, laser light scattering, ultracentrifugation, electrophoresis, and viscosity measurements.

(a) Mass spectrometry

Key point **In the MALDI-TOF technique, matrix-assisted laser desorption/ionization is coupled with a time-of-flight mass spectrometer to measure the molar masses of macromolecules.**

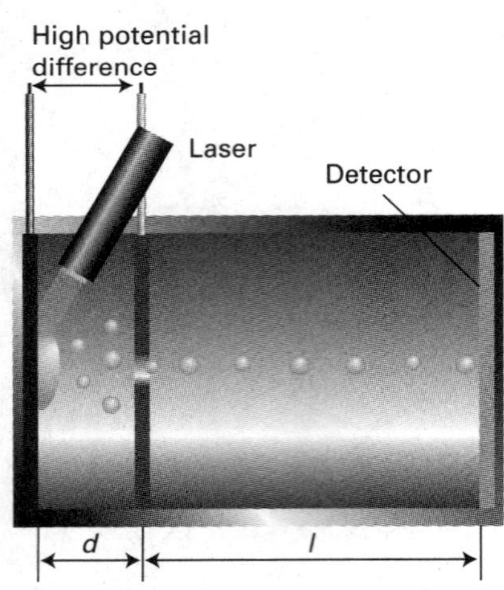

Fig. 18.28 A matrix-assisted laser desorption/ionization time-of-flight (MALDI-TOF) mass spectrometer. A laser beam ejects macromolecules and ions from the solid matrix. The ionized macromolecules are accelerated by an electrical potential difference over a distance *d* and then travel through a drift region of length *l*. Ions with the smallest mass to charge ratio (*m*/*z*) reach the detector first.

Mass spectrometry is among the most accurate techniques for the determination of molar masses. The procedure consists of ionizing the sample in the gas phase and then measuring the mass-to-charge number ratio (m/z; more precisely, the dimensionless ratio m/zm_u) of all ions. Macromolecules present a challenge because it is difficult to produce gaseous ions of large species without fragmentation. However, two new techniques have emerged that circumvent this problem: **matrix-assisted laser desorption/ionization** (MALDI) and **electrospray ionization**. We shall discuss **MALDI-TOF mass spectrometry**, so called because the MALDI technique is coupled to a time-of-flight (TOF) ion detector.

Figure 18.28 shows a schematic view of a MALDI-TOF mass spectrometer. The macromolecule is first embedded in a solid matrix that often consists of an organic material such as *trans*-3-indoleacrylic acid and inorganic salts such as sodium chloride or silver trifluoroacetate. This sample is then irradiated with a pulsed laser. The laser beam ejects electronically excited matrix ions, cations, and neutral macromolecules, thus creating a dense gas plume above the sample surface. The macromolecule is ionized by collisions and complexation with small cations, such as H^+, Na^+, and Ag^+.

Figure 18.29 shows the MALDI-TOF mass spectrum of a polydisperse sample of poly(butylene adipate) (PBA, **11**). The MALDI technique produces mostly singly charged molecular ions that are not fragmented. Therefore, the multiple peaks in the spectrum arise from polymers of different lengths, with the intensity of each peak being proportional to the abundance of each polymer in the sample. Values of $\bar{M}_n$, $\bar{M}_w$, and the heterogeneity index can be calculated from the data. It is also possible

11

to use the mass spectrum to verify the structure of a polymer, as shown in the following example.

Example 18.2 *Interpreting the mass spectrum of a polymer*

The mass spectrum in Fig. 18.29 consists of peaks spaced by 200 g mol^{-1}. The peak at 4113 g mol^{-1} corresponds to the polymer for which $n = 20$. From these data, verify that the sample consists of polymers with the general structure given by (**11**).

Method Because each peak corresponds to a different value of n, the molar mass difference, ΔM, between peaks corresponds to the molar mass, M, of the repeating unit (the group inside the brackets in **11**). Furthermore, the molar mass of the terminal groups (the groups outside the brackets in **11**) may be obtained from the molar mass of any peak by using

$$M(\text{terminal groups}) = M(\text{polymer with } n \text{ repeating units}) - n\Delta M - M(\text{cation})$$

where the last term corresponds to the molar mass of the cation that attaches to the macromolecule during ionization.

Answer The value of ΔM is consistent with the molar mass of the repeating unit shown in (**11**), which is 200 g mol^{-1}. The molar mass of the terminal group is calculated by recalling that Na^+ is the cation in the matrix:

$$M(\text{terminal group}) = 4113\text{ g mol}^{-1} - 20(200\text{ g mol}^{-1}) - 23\text{ g mol}^{-1} = 90\text{ g mol}^{-1}$$

The result is consistent with the molar mass of the $-O(CH_2)_4OH$ terminal group (89 g mol^{-1}) plus the molar mass of the –H terminal group (1 g mol^{-1}).

Self-test 18.2 What would be the molar mass of the $n = 20$ polymer if silver trifluoroacetate were used instead of NaCl in the preparation of the matrix?

[4198 g mol^{-1}]

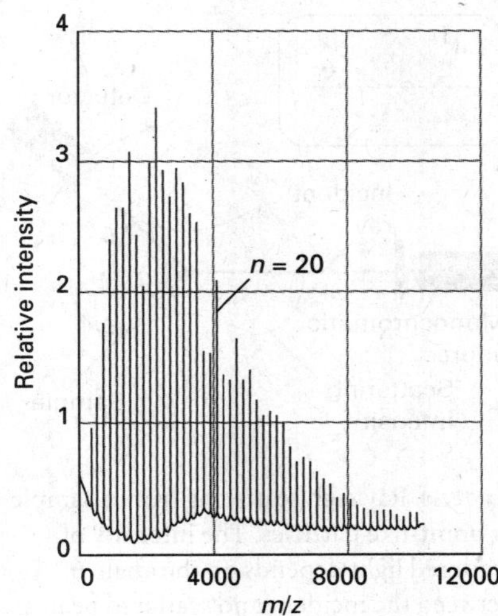

Fig. 18.29 MALDI-TOF spectrum of a sample of poly(butylene adipate) with $\bar{M}_n = 4525$ g mol^{-1} (Adapted from Mudiman *et al.*, *J. Chem. Educ.*, 74, 1288 (1997).)

(b) Laser light scattering

Key points **(a) The intensity of Rayleigh light scattering by a sample increases with decreasing wavelength of the incident radiation and increasing size of the particles in the sample. Analysis of Rayleigh scattering leads to the determination of the molar mass of a macromolecule or aggregate. (b) The analysis of Rayleigh scattering needs to take into account the non-ideality of solutions of macromolecules. (c) Dynamic light scattering is a technique for the determination of the diffusion properties and molar masses of macromolecules and aggregates.**

The intensity of light scattered from a dilute solution excited by plane-polarized light measured at a distance r from the solution is proportional to the intensity of incident light, I_0, and to $(1/r^2)\sin^2\phi$ where r is the distance of the detector from the sample (the factor $1/r^2$ occurs because the wave is spreading out over a sphere of radius r and surface area $4\pi r^2$, so any sample of the radiation is diluted by a factor proportional to r^2), and we write

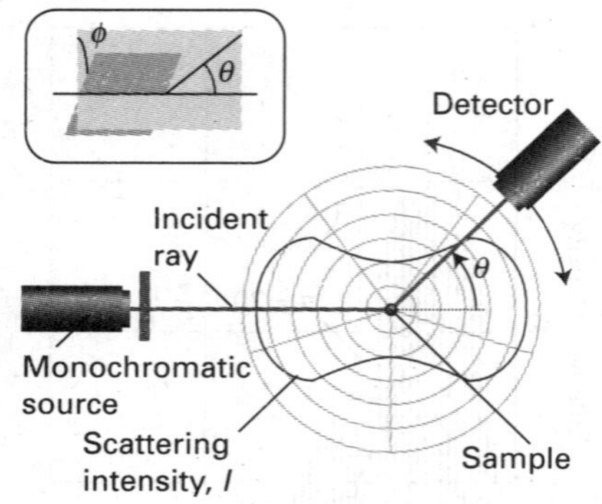

Fig. 18.30 Rayleigh scattering from a sample of point-like particles. The intensity of scattered light depends on the angle θ between the incident and scattered beams. The inset shows the angle ϕ between the plane of polarization of the incident beam and the plane defined by the incident and scattered beams. In a typical experimental arrangement, $\phi = 90°$.

$$I(\theta,\phi,r) = R(\theta)I_0 \times \frac{\sin^2\phi}{r^2} \quad \text{Definition of the Rayleigh ratio} \qquad [18.22a]$$

where the coefficient of proportionality $R(\theta)$, which depends on the angle θ, is called the **Rayleigh ratio** (Fig. 18.30). The value of $R(\theta)$ is found by inverting this expression and inserting the measured value of the intensity at the location θ, ϕ, and r:

$$R(\theta) = \frac{I(\theta,\phi,r)}{I_0} \times \frac{r^2}{\sin^2\phi} \qquad (18.22b)$$

The intensity of scattered radiation is proportional to the concentration of scattering molecules and for a solution of a polymer of mass concentration c_P, the Rayleigh ratio may be written[2]

$$R(\theta) = Kc_P\bar{M}_w \quad \text{Relation between the Rayleigh ratio and the polymer molar mass} \qquad (18.23)$$

The proportionality constant K recognizes that the scattering by the macromolecules depends on the difference of refractive index (n_r) between them and the solvent (if the refractive indexes were the same, the solute would be invisible) and in a calculation that we do not reproduce here, it is found that

$$K = \frac{2\pi^2 n_r^2}{N_A\lambda^4}\frac{dn_r}{dc_p} \qquad (18.24)$$

All the quantities in this expression can be measured in separate experiments, so K is, in principle, known.

When the size of the molecule is not negligible compared with the wavelength of the incident radiation, interference between different parts of the same molecule must be taken into account. This effect is taken into account by multiplying the right-hand side of eqn 8.22b by a structure factor, $P(\theta)$:[2]

$$P(\theta) \approx 1 - p(\theta) \quad \text{with} \quad p(\theta) = \frac{16\pi^2 R_g^2 \sin^2\frac{1}{2}\theta}{3\lambda^2} \quad \text{Structure factor of a small macromolecule} \qquad (18.25)$$

where R_g is the radius of gyration of the macromolecule. (The radius of gyration plays a role not through its normal appearance in expressions relating to hydrodynamic behaviour but because it turns out to be the appropriate measure of the size of the molecule.) As we show in the following *Justification*, we should expect a straight line when $1/R(\theta)$ is plotted against $(\sin^2\frac{1}{2}\theta)/R(\theta)$, from which the weight-average molar mass and the radius of gyration can be determined from the intercept and slope, respectively. Table 18.2 lists some experimental values of R_g obtained in this way.

Table 18.2* Radius of gyration

	$M/(\text{kg mol}^{-1})$	R_g/nm
Serum albumin	66.2	2.98
Polystyrene	3.2×10^3	50‡
DNA	4×10^3	117

* More values are given in the *Data section*.
‡ In a poor solvent.

Justification 18.4 *The analysis of scattering intensity*

Equations 18.22b and 18.25 can be combined as follows:

$$\frac{1}{R(\theta)} = \frac{1}{KP(\theta)c_p\bar{M}_w} \approx \frac{1}{Kc_p\bar{M}_w(1-p(\theta))}$$

Then we use $(1-x)^{-1} \approx 1 + x$ (which is valid when $x \ll 1$) to write

$$\frac{1}{R(\theta)} \approx \frac{1}{Kc_p\bar{M}_w}(1+p(\theta)) = \frac{1}{Kc_p\bar{M}_w} + \frac{p(\theta)}{Kc_p\bar{M}_w}$$

[2] For a derivation of this and the following equation, see C.A. Johnson and D.A. Gabriel, *Laser light scattering*, Dover, New York (1995) and references cited therein. See also R.J. Hunter, cited in footnote 1.

As we are ignoring terms of order x^2, we can replace the factor $Kc_p\bar{M}_w$ in the second term by $R(\theta)$, and so obtain

$$\frac{1}{R(\theta)} \approx \frac{1}{Kc_p\bar{M}_w} + \frac{p(\theta)}{R(\theta)} \qquad (18.26a)$$

Because $p(\theta) \propto \sin^2\frac{1}{2}\theta$, this expression has the form

$$\frac{1}{R(\theta)} \approx \frac{1}{Kc_p\bar{M}_w} + a \times \frac{\sin^2\frac{1}{2}\theta}{R(\theta)} \qquad a = \frac{16\pi^2 R_g^2}{3\lambda^2} \qquad (18.26b)$$

where the weight-average molar mass can be determined from the intercept and the radius of gyration can be determined from the slope a.

Example 18.3 *Determining the size of a polymer by light scattering*

The following data for a sample of polystyrene in butanone were obtained at 20°C with plane-polarized light at $\lambda = 546$ nm.

θ/°	26.0	36.9	66.4	90.0	113.6
$R(\theta)$/m^2	19.7	18.8	17.1	16.0	14.4

In separate experiments, it was determined that $K = 6.42 \times 10^{-5}$ mol m^5 kg^{-2}. From this information, calculate R_g and $\bar{M}_w$ for the sample. Assume that the polymer is small enough that eqn 18.25 holds. Take $c_P = 311$ kg m^{-3}.

Method As shown in the text, a plot of $1/R(\theta)$ against $(\sin^2\frac{1}{2}\theta)/R(\theta)$ should be a straight line with slope $16\pi^2R_g^2/3\lambda^2$ and y-intercept $1/Kc_p\bar{M}_w$.

Answer We construct a table of values of $1/R(\theta)$ and $(\sin^2\frac{1}{2}\theta)/R(\theta)$ and plot the data (Fig. 18.31).

θ/°	26.0	36.9	66.4	90.0	113.6
$\{10^2/R(\theta)\}$/m^{-2}	5.08	5.32	5.85	6.25	6.94
$\{10^3 \times (\sin^2\frac{1}{2}\theta)/R(\theta)\}$/m^{-2}	2.57	5.33	17.5	31.3	48.6

The best straight line through the data has a slope of 0.388 and a y-intercept of 5.07×10^{-2}. From these values and the value of K, we calculate $R_g = 4.69 \times 10^{-8}$ m $=$ 46.9 nm and $\bar{M}_w = 987$ kg mol^{-1}.

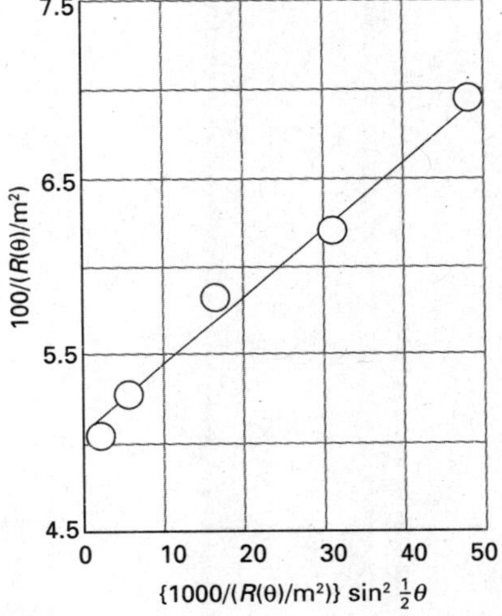

Fig. 18.31 Plot of the data for Example 18.3.

Self-test 18.3 The following data for an aqueous solution of a protein with $c_p =$ 2.0 kg m^{-3} were obtained at 20°C with laser light at $\lambda = 532$ nm:

θ/°	15.0	45.0	70.0	85.0	90.0
$R(\theta)$/m^2	23.8	22.9	21.6	20.7	20.4

In a separate experiment, it was determined that $K = 2.40 \times 10^{-2}$ mol m^5 kg^{-2}. From this information, calculate the radius of gyration and the molar mass of the protein. Assume the protein is small enough that eqn 18.25 holds.

[$R_g = 39.8$ nm; $M = 498$ kg mol^{-1}]

(c) Sedimentation

Key points **(a) The rate of sedimentation in an ultracentrifuge depends on molar masses and shapes of the macromolecules in the sample. (b) The weight-average and Z-average molar mass of a sample of macromolecules can be determined from equilibrium measurements of sedimentation in an ultracentrifuge.**

In a gravitational field, heavy particles settle towards the foot of a column of solution by the process called **sedimentation**. The rate of sedimentation depends on the strength of the field and on the masses and shapes of the particles. Spherical molecules (and compact molecules in general) sediment faster than rod-like and extended molecules. When the sample is at equilibrium, the particles are dispersed over a range of heights in accord with the Boltzmann distribution (because the gravitational field competes with the stirring effect of thermal motion). The spread of heights depends on the masses of the molecules, so the equilibrium distribution is another way to determine molar mass.

Sedimentation is normally very slow, but it can be accelerated by **ultracentrifugation**, a technique that replaces the gravitational field with a centrifugal field. The effect can be achieved in an ultracentrifuge, which is essentially a cylinder that can be rotated at high speed about its axis with a sample in a cell near its outer edge. Modern ultracentrifuges can produce accelerations equivalent to about 10^5 that of gravity ('10^5 g'). Initially the sample is uniform, but the 'top' (innermost) boundary of the solute moves outwards as sedimentation proceeds.

A solute particle of mass m has an effective mass $m_{\text{eff}} = bm$ on account of the buoyancy of the medium, with

$$b = 1 - \rho v_{\text{s}} \tag{18.27}$$

where ρ is the solution density, v_{s} is the partial specific volume of the solute ($v_{\text{s}} = (\partial V/\partial m_{\text{B}})_T$, with m_{B} the total mass of solute), and ρv_{s} is the mass of solvent displaced per gram of solute. The solute particles at a distance r from the axis of a rotor spinning at an angular velocity ω experience a centrifugal force of magnitude $m_{\text{eff}} r\omega^2$. The acceleration outwards is countered by a frictional force proportional to the speed, $s = \mathrm{d}r/\mathrm{d}t$, of the particles through the medium. This force is written fs, where f is the **frictional coefficient**. The particles therefore adopt a **drift speed**, a constant speed through the medium, which is found by equating the two forces $m_{\text{eff}} r\omega^2$ and fs. The forces are equal when

$$s = \frac{m_{\text{eff}} r\omega^2}{f} = \frac{bmr\omega^2}{f} \tag{18.28}$$

The drift speed depends on the angular velocity and the radius, and it is convenient to define the **sedimentation constant**, S, as

$$S = \frac{s}{r\omega^2} \qquad \text{Definition of the sedimentation constant} \qquad [18.29]$$

Then, because the average molecular mass is related to the average molar mass $\bar{M}_{\text{n}}$ through $m = \bar{M}_{\text{n}}/N_{\text{A}}$

$$S = \frac{b\bar{M}_{\text{n}}}{fN_{\text{A}}} \tag{18.30}$$

For a spherical particle of radius a in a solvent of viscosity η, the frictional coefficient f is given by **Stokes' relation**:

$$f = 6\pi a\eta \tag{18.31}$$

On substituting this expression into eqn 18.30, we obtain

$$S = \frac{b\bar{M}_{\text{n}}}{6\pi a\eta N_{\text{A}}} \qquad \text{Relation between } S \text{ and the molar mass of a spherical polymer} \tag{18.32}$$

and S may be used to determine either $\bar{M}_n$ or a. Again, if the molecules are not spherical, we use the appropriate value of f given in Table 18.3. As always when dealing with macromolecules, the measurements must be carried out at a series of concentrations and then extrapolated to zero concentration to avoid the complications that arise from the interference between bulky molecules.

Table 18.3* Frictional coefficients and molecular geometry†

a/b	Prolate	Oblate
2	1.04	1.04
3	1.18	1.17
6	1.31	1.28
8	1.43	1.37
10	1.54	1.46

* More values and analytical expressions are given in the *Data section*.
† Entries are the ratio f/f_0, where $f_0 = 6\pi\eta c$, where $c = (ab^2)^{1/3}$ for prolate ellipsoids and $c = (a^2b)^{1/3}$ for oblate ellipsoids; $2a$ is the major axis and $2b$ is the minor axis.

Example 18.4 *Determining a sedimentation constant*

The sedimentation of the protein bovine serum albumin (BSA) was monitored at 25°C. The initial location of the solute surface was at 5.50 cm from the axis of rotation, and during centrifugation at 56 850 r.p.m. it receded as follows:

t/s	0	500	1000	2000	3000	4000	5000
r/cm	5.50	5.55	5.60	5.70	5.80	5.91	6.01

Calculate the sedimentation coefficient.

Method Equation 18.29 can be interpreted as a differential equation for $s = dr/dt$ in terms of r; so integrate it to obtain a formula for r in terms of t. The integrated expression, an expression for r as a function of t, will suggest how to plot the data and obtain from it the sedimentation constant.

Answer Equation 18.29 may be written

$$\frac{dr}{dt} = r\omega^2 S$$

This equation integrates to

$$\ln\frac{r}{r_0} = \omega^2 St$$

It follows that a plot of $\ln(r/r_0)$ against t should be a straight line of slope $\omega^2 S$. Use $\omega = 2\pi\nu$, where ν is in cycles per second, and draw up the following table:

t/s	0	500	1000	2000	3000	4000	5000
$10^2 \ln(r/r_0)$	0	0.905	1.80	3.57	5.31	7.19	8.87

The straight-line graph (Fig. 18.32) has a slope of 1.78×10^{-5}; so $\omega^2 S = 1.78 \times 10^{-5}\ \mathrm{s^{-1}}$. Because $\omega = 2\pi \times (56\,850/60)\ \mathrm{s^{-1}} = 5.95 \times 10^3\ \mathrm{s^{-1}}$, it follows that $S = 5.02 \times 10^{-13}$ s. The unit 10^{-13} s is sometimes called a 'svedberg' and denoted Sv; in this case $S = 5.02$ Sv.

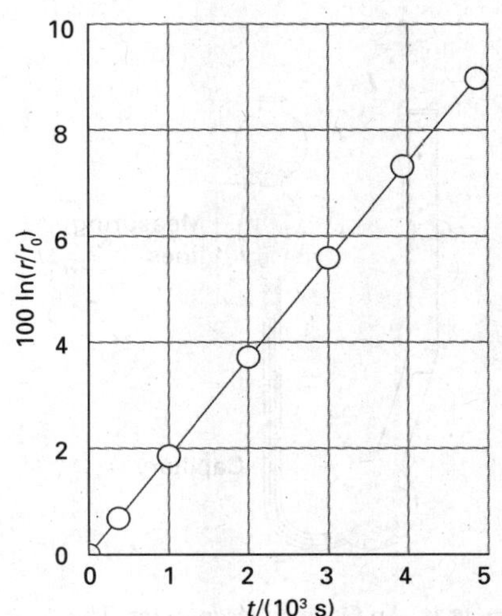

Fig. 18.32 A plot of the data for Example 18.4.

Self-test 18.4 Calculate the sedimentation constant given the following data (the other conditions being the same as above):

t/s	0	500	1000	2000	3000	4000	5000
r/cm	5.65	5.68	5.71	5.77	5.84	5.9	5.97

[3.11 Sv]

The difficulty with using sedimentation rates to measure molar masses lies in the inaccuracies inherent in the determination of diffusion coefficients of polydisperse systems. This problem can be avoided by allowing the system to reach equilibrium, for the transport property D is then no longer relevant. As we show in the following *Justification*, the weight-average molar mass can be obtained from the ratio of concentrations of the macromolecules at two different radii in a centrifuge operating at angular frequency ω:

$$\bar{M}_w = \frac{2RT}{(r_2^2 - r_1^2)b\omega^2}\ln\frac{c_2}{c_1} \quad (18.33)$$

An alternative treatment of the data leads to the Z-average molar mass. The centrifuge is run more slowly in this technique than in the sedimentation rate method to avoid having all the solute pressed in a thin film against the bottom of the cell. At these slower speeds, several days may be needed for equilibrium to be reached.

Justification 18.5 *The weight-average molar mass from sedimentation experiments*

The centrifugal force acting on a molecule at a radius r when it is rotating around the axis of the centrifuge at a frequency ω is $m\omega^2 r$. This force corresponds to a difference in potential energy (using $F = -dV/dr$) of $-\frac{1}{2}m\omega^2 r^2$. The difference in potential energy between r_1 and r_2 (with $r_2 > r_1$) is therefore $\frac{1}{2}m\omega^2(r_1^2 - r_2^2)$. According to the Boltzmann distribution, the ratio of concentrations of molecules at these two radii should therefore be

$$\frac{c_2}{c_1} = e^{-\frac{1}{2}m_{\text{eff}}\omega^2(r_1^2 - r_2^2)/kT}$$

The effective mass, m_{eff}, which allows for buoyancy effects, is $m(1 - v\rho)$, and m/k can be replaced by M/R, where $R = N_A k$ is the gas constant. Then, by taking logarithms of both sides, the last equation becomes

$$\ln\frac{c_2}{c_1} = \frac{M(1 - v\rho)\omega^2(r_2^2 - r_1^2)}{2RT}$$

which rearranges into eqn 18.33.

(d) Viscosity

Key point **The viscosity-average molar mass can be determined from measurements of the viscosity of solutions of macromolecules.**

The formal definition of viscosity is given in Section 20.4; for now, we need to know that highly viscous liquids flow slowly and retard the motion of objects through them. The presence of a macromolecular solute increases the viscosity of a solution. The effect is large even at low concentration, because big molecules affect the fluid flow over an extensive region surrounding them. At low concentrations the viscosity, η, of the solution is related to the viscosity of the pure solvent, η_0, by

$$\eta = \eta_0(1 + [\eta]c + [\eta]'c^2 + \cdots) \quad (18.34)$$

The **intrinsic viscosity**, $[\eta]$, is the analogue of a virial coefficient (and has dimensions of 1/concentration). It follows from eqn 18.34 that

$$[\eta] = \lim_{c\to 0}\left(\frac{\eta - \eta_0}{c\eta_0}\right) = \lim_{c\to 0}\left(\frac{\eta/\eta_0 - 1}{c}\right) \quad \text{Definition of the intrinsic viscosity} \quad [18.35]$$

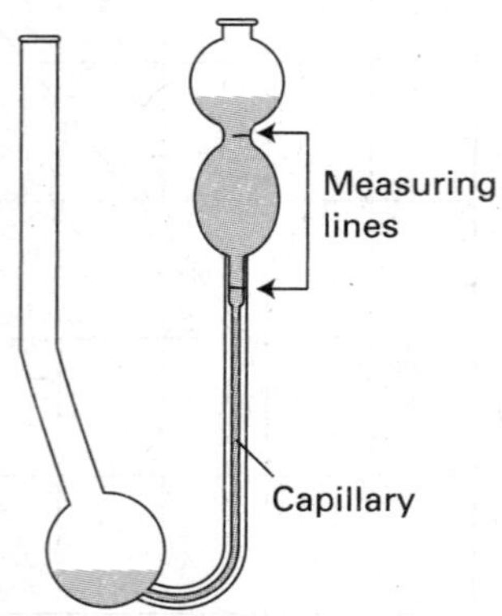

Fig. 18.33 An Ostwald viscometer. The viscosity is measured by noting the time required for the liquid to drain between the two marks.

Viscosities are measured in several ways. In the **Ostwald viscometer** shown in Fig. 18.33, the time taken for a solution to flow through the capillary is noted, and compared with a standard sample. The method is well suited to the determination of $[\eta]$ because the ratio of the viscosities of the solution and the pure solvent is proportional to the drainage time t and t_0 after correcting for different densities ρ and ρ_0:

$$\frac{\eta}{\eta_0} = \frac{t}{t_0} \times \frac{\rho}{\rho_0} \quad (18.36)$$

Table 18.4* Intrinsic viscosity

	Solvent	θ/°C	K/(cm³ g⁻¹)	a
Polystyrene	Benzene	25	9.5×10^{-3}	0.74
Poly(methylpropene)	Benzene	23	8.3×10^{-2}	0.50
Various proteins	Guanidine hydrochloride + $HSCH_2CH_2OH$		7.2×10^{-3}	0.66

* More values are given in the *Data section*.

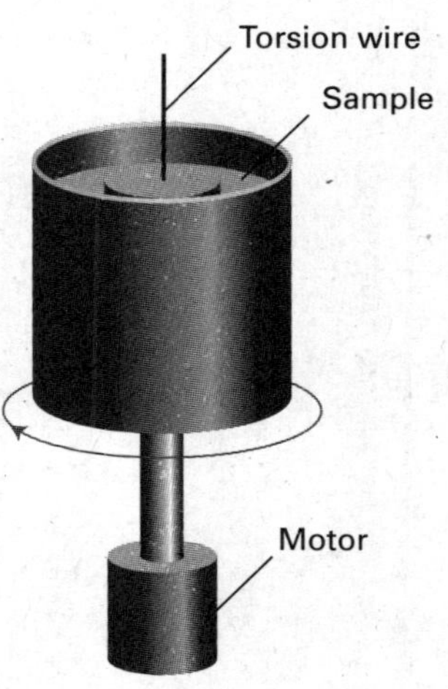

Fig. 18.34 A rotating rheometer. The torque on the inner drum is observed when the outer container is rotated.

This ratio can be used directly in eqn 18.35. Viscometers in the form of rotating concentric cylinders are also used (Fig. 18.34), and the torque on the inner cylinder is monitored while the outer one is rotated. Such **rotating rheometers** (some instruments for the measurement of viscosity are also called rheometers, from the Greek word for 'flow') have the advantage over the Ostwald viscometer that the shear gradient between the cylinders is simpler than in the capillary and effects of the kind discussed shortly can be studied more easily.

There are many complications in the interpretation of viscosity measurements. Much of the work is based on empirical observations, and the determination of molar mass is usually based on comparisons with a standard, nearly monodisperse sample. Some regularities are observed that help in the determination. For example, it is found that some solutions of macromolecules often fit the **Mark–Kuhn–Houwink–Sakurada equation:**

$$[\eta] = K\bar{M}_v^a \qquad \text{Mark–Kuhn–Houwink–Sakurada equation} \qquad (18.37)$$

where K and a are constants that depend on the solvent and type of macromolecule (Table 18.4); the viscosity-average molar mass, $\bar{M}_v$, appears in this expression.

Example 18.5 *Using intrinsic viscosity to measure molar mass*

The viscosities of a series of solutions of polystyrene in toluene were measured at 25°C with the following results:

c/(g dm⁻³)	0	2	4	6	8	10
η/(10⁻⁴ kg m⁻¹ s⁻¹)	5.58	6.15	6.74	7.35	7.98	8.64

Calculate the intrinsic viscosity and estimate the molar mass of the polymer by using eqn 18.37 with $K = 3.80\times10^{-5}$ dm³ g⁻¹ and $a = 0.63$.

Method The intrinsic viscosity is defined in eqn 18.35; therefore, form this ratio at the series of data points and extrapolate to $c = 0$. Interpret $\bar{M}_v$ as $\bar{M}_v$/(g mol⁻¹) in eqn 18.37.

Answer We draw up the following table:

c/(g dm⁻³)	0	2	4	6	8	10
η/η_0	1	1.102	1.208	1.317	1.43	1.549
$100[(\eta/\eta_0) - 1]/(c/\text{g dm}^{-3})$		5.11	5.20	5.28	5.38	5.49

The points are plotted in Fig. 18.35. The extrapolated intercept at $c = 0$ is 0.0504, so $[\eta] = 0.0504$ dm³ g⁻¹. Therefore

$$\bar{M}_v = \left(\frac{[\eta]}{K}\right)^{1/a} = 9.0\times10^4 \text{ g mol}^{-1}$$

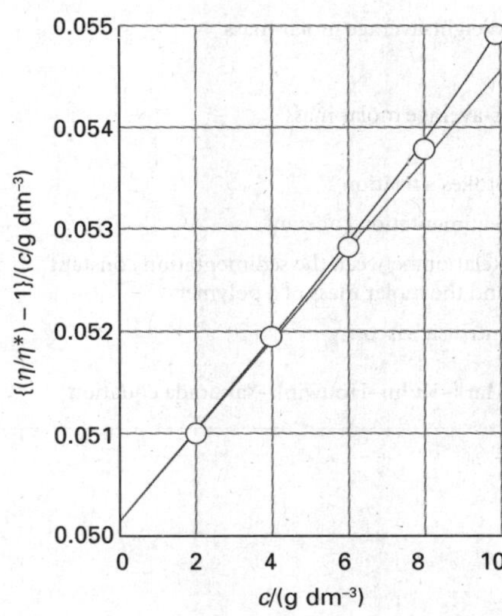

Fig. 18.35 The plot used for the determination of intrinsic viscosity, which is taken from the intercept at $c = 0$; see Example 18.5.

Self-test 18.5 Show that the intrinsic viscosity may also be obtained as $[\eta] = \lim_{c\to 0}(\eta/\eta_0)$ and evaluate the viscosity-average molar mass by using this relation. [$90\ \text{kg mol}^{-1}$]

In some cases, the flow is non-Newtonian in the sense that the viscosity of the solution changes as the rate of flow increases. A decrease in viscosity with increasing rate of flow indicates the presence of long rod-like molecules that are orientated by the flow and hence slide past each other more freely. In some somewhat rare cases the stresses set up by the flow are so great that long molecules are broken up, with further consequences for the viscosity.

Checklist of key equations

Property	Equation	Comment
Contour length of a random coil	$R_c = Nl$	
Root-mean-square separation of a random coil	$R_{rms} = N^{1/2}l$	Unconstrained chain
	$R_{rms} = (2N)^{1/2}l$	Constrained tetrahedral chain
Radius of gyration of a 3D random coil	$R_g = (N/6)^{1/2}l$	Unconstrained chain
	$R_g = (N/3)^{1/2}l$	Constrained tetrahedral chain
Conformational entropy of a random coil	$\Delta S = -\frac{1}{2}kN\ln\{(1+\nu)^{1+\nu}(1-\nu)^{1-\nu}\}$	
	$\nu = n/N$	
Thickness of the electrical double layer	$r_D \propto (T/I)^{1/2}$	Debye–Hückel theory
Number-average molar mass	$\bar{M}_n = (1/N)\sum_i N_i M_i$	Definition
Weight-average molar mass	$\bar{M}_w = (1/m)\sum_i m_i M_i$	Definition
Z-average molar mass	$\bar{M}_Z = \left(\sum_i N_i M_i^3\right)\Big/\left(\sum_i N_i M_i^2\right)$	Interpretation
Stokes's relation	$f = 6\pi a\eta$	
Sedimentation constant	$S = s/r\omega^2$	Definition
Relation between the sedimentation constant and the molar mass of a polymer	$S = b\bar{M}_n/6\pi a\eta N_A$	Spherical polymer
Intrinsic viscosity	$[\eta] = \lim_{c\to 0}((\eta-\eta_0)/c\eta_0)$	Definition
Mark–Kuhn–Houwink–Sakurada equation	$[\eta] = K\bar{M}_v^a$	θ-solution

Further information

Further information 18.1 *Random and nearly random coils*

In this section, we consider various statistical aspects of the structures of random and nearly random coils in one and three dimensions and derive the expressions quoted in the text.

(a) A one-dimensional freely jointed random coil

Consider a one-dimensional freely jointed polymer. We can specify the conformation of a molecule by stating the number of bonds pointing to the right (N_R) and the number pointing to the left (N_L). The distance between the ends of the chain is $(N_R - N_L)l$, where l is the length of an individual bond. We write $n = N_R - N_L$ and the total number of bonds as $N = N_R + N_L$.

The number of ways W of forming a chain with a given end-to-end distance nl is the number of ways of having N_R right-pointing and N_L left-pointing bonds. There are $N(N-1)(N-2)\cdots 1 = N!$ ways of selecting whether a step should be to the right or the left. If N_L steps are to the left, $N_R = N - N_L$ will be to the right. However, we end up at the same point for all $N_L!$ and $N_R!$ choices of which step is to the left and which to the right. Therefore

$$W = \frac{N!}{N_L!N_R!} = \frac{N!}{\{\tfrac{1}{2}(N+n)\}!\{\tfrac{1}{2}(N-n)\}!} \qquad (18.38)$$

The probability that the separation is nl is

$$P = \frac{\text{number of polymers with } N_R \text{ bonds to the right}}{\text{total number of arrangements of bonds}}$$

$$= \frac{N!/N_R!(N-N_R)!}{2^N} = \frac{N!}{\{\tfrac{1}{2}(N+n)\}!\{\tfrac{1}{2}(N-n)\}!2^N}$$

When the chain is compact in the sense that $n \ll N$, it is more convenient to evaluate $\ln P$: the factorials are then large and we can use Stirling's approximation (Section 15.1a) in the form

$$\ln x! \approx \ln(2\pi)^{1/2} + (x + \tfrac{1}{2})\ln x - x$$

The result, after quite a lot of algebra (see Problem 18.20), is

$$\ln P = \ln\left(\frac{2}{\pi N}\right)^{1/2} - \tfrac{1}{2}(N+n+1)\ln(1+\nu) - \tfrac{1}{2}(N-n+1)\ln(1-\nu) \qquad (18.39)$$

where $\nu = n/N$. For a compact coil ($\nu \ll 1$) we use the approximation $\ln(1 \pm \nu) \approx \pm\nu - \tfrac{1}{2}\nu^2$ and so obtain

$$\ln P \approx \ln\left(\frac{2}{\pi N}\right)^{1/2} - \tfrac{1}{2}N\nu^2$$

which rearranges into eqn 18.1:

$$P = \left(\frac{2}{\pi N}\right)^{1/2} e^{-n^2/2N} \qquad (18.40)$$

To show that the total probability of the chain ends being at any separation, we integrate P over all values of n. However, because n can change only in steps of 2, the integration step size is $\tfrac{1}{2}dn$, not dn itself. Then (with N allowed to become infinite),

$$\sum_{n=-N}^{N} P \rightarrow \int_{-\infty}^{\infty} P(n)(\tfrac{1}{2}dn) = \tfrac{1}{2}\left(\frac{2}{\pi N}\right)^{1/2}\int_{-\infty}^{\infty} e^{-n^2/2N}dn = 1$$

(b) A three-dimensional freely jointed random coil

The length of a step in three dimensions, l, can be expressed in terms of its projections on each of three orthogonal axes as $l^2 = l_x^2 + l_y^2 + l_z^2$. The average values of l_x^2, l_y^2, and l_z^2 are all the same in a spherically symmetric environment, so the average length of a step in the x-direction (or any of the other two directions) can be obtained by writing $l^2 = 3\langle l_x^2\rangle$, and is $x = \langle l_x^2\rangle^{1/2} = l/3^{1/2}$. The probability that the random walk will end up at a distance x from the origin is given by eqn 18.1 with $n = x/(l/3^{1/2}) = 3^{1/2}x/l$:

$$P(x) = \left(\frac{2}{\pi N}\right)^{1/2} e^{-3x^2/2Nl^2} \qquad (18.41)$$

If x is regarded as continuously variable, we need to replace this probability by a probability density $f(x)$ such that $f(x)dx$ is the probability that the ends of the chain will be found between x and $x + dx$. Because $dx = 2(l/3^{1/2})dn$ (for the factor 2, see the remark at the end of the preceding section), $dn = (3^{1/2}/2l)dx$, so

$$f(x) = \frac{1}{2l}\left(\frac{6}{\pi N}\right)^{1/2} e^{-3x^2/2Nl^2} \qquad (18.42)$$

Because the probabilities of making steps along all three coordinates are independent, the probability of finding the ends of the chain in a region of volume $dV = dx\,dy\,dz$ at a distance r is the product of these densities:

$$f(x,y,z)dV = f(x)f(y)f(z)dx\,dy\,dz = \frac{1}{8l^3}\left(\frac{6}{\pi N}\right)^{3/2} e^{-3r^2/2Nl^2}dV \qquad (18.43)$$

The volume of a spherical shell at a distance r is $4\pi r^2$, so the total probability of finding the ends at a separation between r and $r + dr$, regardless of orientation, is

$$f(r)dr = \frac{4\pi}{8l^3}\left(\frac{6}{\pi N}\right)^{3/2} r^2 e^{-3r^2/2Nl^2}dr \qquad (18.44)$$

from which $f(r)$ can be identified, as in eqn 18.2.

(c) A partially rigid coil

In each of the following steps we use $N \rightarrow \infty$ when necessary. We start from

$$\langle R^2\rangle = \sum_i \langle r_i^2\rangle + \sum_{i \neq j}^{N} \langle \boldsymbol{r}_i \cdot \boldsymbol{r}_j\rangle \qquad (18.45)$$

The first term is Nl^2 regardless of the rigidity of the coil. The second term can be written as follows:

$$\sum_{i\neq j}^{N}\langle r_i\cdot r_j\rangle = 2\sum_{i=2}^{N}\langle r_1\cdot r_i\rangle + 2\sum_{i=3}^{N}\langle r_2\cdot r_i\rangle + \cdots$$

There are $N-1$ such terms, and provided we allow N to become infinite, all the sums on the right have the same value, so

$$\sum_{i\neq j}^{N}\langle r_i\cdot r_j\rangle = 2(N-1)\sum_{i=2}^{N}\langle r_1\cdot r_j\rangle \approx 2N\sum_{i=2}^{N}\langle r_1\cdot r_j\rangle$$

The sum on the right is essentially the persistence length. Specifically

$$\sum_{j=2}^{N}\langle r_1\cdot r_j\rangle = \sum_{j=1}^{N}\langle r_1\cdot r_j\rangle - \langle r_1^2\rangle = ll_p - l^2 \qquad (18.46)$$

Now we bring the three pieces of the calculation together:

$$\langle R^2\rangle = Nl^2 + 2N(ll_p - l^2) = 2Nll_p - Nl^2 \qquad (18.47)$$

which, on taking the square root of both sides, is eqn 18.11.

Discussion questions

18.1 Describe the various measures of the size of a random coil and identify how they depend on the number of units.

18.2 What are the consequences of there being partial rigidity in an otherwise random coil?

18.3 It is observed that the critical micelle concentration of sodium dodecyl sulfate in aqueous solution decreases as the concentration of added sodium chloride increases. Explain this effect.

18.4 Distinguish between number-average, weight-average, and Z-average molar masses. Identify experimental techniques that can measure each of these properties.

18.5 Suggest reasons why different techniques produce different mass averages.

18.6 Why is the protein-folding problem so difficult to resolve?

Exercises

18.1(a) A polymer chain consists of 700 segments, each 0.90 nm long. If the chain were ideally flexible, what would be the r.m.s. separation of the ends of the chain?

18.1(b) A polymer chain consists of 1200 segments, each 1.125 nm long. If the chain were ideally flexible, what would be the r.m.s. separation of the ends of the chain?

18.2(a) Calculate the contour length (the length of the extended chain) and the root mean square separation (the end-to-end distance) for polyethylene with a molar mass of 280 kg mol^{-1}.

18.2(b) Calculate the contour length (the length of the extended chain) and the root mean square separation (the end-to-end distance) for polypropylene of molar mass 174 kg mol^{-1}.

18.3(a) The radius of gyration of a long chain molecule is found to be 7.3 nm. The chain consists of C–C links. Assume the chain is randomly coiled and estimate the number of links in the chain.

18.3(b) The radius of gyration of a long chain molecule is found to be 18.9 nm. The chain consists of links of length 450 pm. Assume the chain is randomly coiled and estimate the number of links in the chain.

18.4(a) What is the probability that the ends of a polyethene chain of molar mass 65 kg mol^{-1} are 10 nm apart when the polymer is treated as a one-dimensional freely jointed chain?

18.4(b) What is the probability that the ends of a polyethene chain of molar mass 85 kg mol^{-1} are 15 nm apart when the polymer is treated as a one-dimensional freely jointed chain?

18.5(a) What is the probability that the ends of a polyethene chain of molar mass 65 kg mol^{-1} are between 10 nm and 10.1 nm apart when the polymer is treated as a three-dimensional freely jointed chain?

18.5(b) What is the probability that the ends of a polyethene chain of molar mass 85 kg mol^{-1} are between 15 nm and 15.1 nm apart when the polymer is treated as a three-dimensional freely jointed chain?

18.6(a) Calculate the change in molar entropy when the ends of a one-dimensional polyethene chain of molar mass 65 kg mol^{-1} are moved apart by 1.0 nm.

18.6(b) Calculate the change in molar entropy when the ends of a one-dimensional polyethene chain of molar mass 85 kg mol^{-1} are moved apart by 2.0 nm.

18.7(a) By what percentage does the radius of gyration of a polymer chain increase (+) or decrease (−) when the bond angle between units is limited to 109°? What is the percentage change in volume of the coil?

18.7(b) By what percentage does the root mean square separation of the ends of a polymer chain increase (+) or decrease (−) when the bond angle between units is limited to 120°? What is the percentage change in volume of the coil?

18.8(a) By what percentage does the radius of gyration of a polymer chain increase (+) or decrease (−) when the persistence length changes from l (the bond length) to 5.0 per cent of the contour length? What is the percentage change in volume of the coil?

18.8(b) By what percentage does the root mean square separation of the ends of a polymer chain increase (+) or decrease (−) when the persistence length changes from l (the bond length) to 2.5 per cent of the contour length? What is the percentage change in volume of the coil?

18.9(a) The radius of gyration of a three-dimensional partially rigid polymer of 1000 units each of length 150 pm was measured as 2.1 nm. What is the persistence length of the polymer?

18.9(b) The radius of gyration of a three-dimensional partially rigid polymer of 1500 units each of length 164 pm was measured as 3.0 nm. What is the persistence length of the polymer?

18.10(a) Calculate the restoring force when the ends of a one-dimensional polyethene chain of molar mass 65 kg mol^{-1} are moved apart by 1.0 nm at 20°C.

18.10(b) Calculate the restoring force when the ends of a one-dimensional polyethene chain of molar mass 85 kg mol^{-1} are moved apart by 2.0 nm at 25°C.

18.11(a) Calculate the number-average molar mass and the mass-average molar mass of a mixture of equal amounts of two polymers, one having $M = 62$ kg mol^{-1} and the other $M = 78$ kg mol^{-1}.

18.11(b) Calculate the number-average molar mass and the mass-average molar mass of a mixture of two polymers, one having $M = 62$ kg mol^{-1} and the other $M = 78$ kg mol^{-1}, with their amounts (numbers of moles) in the ratio 3:2.

18.12(a) A solution consists of solvent, 30 per cent by mass, of a dimer with $M = 30$ kg mol^{-1} and its monomer. What average molar mass would be obtained from measurement of (a) osmotic pressure, (b) light scattering?

18.12(b) A solution consists of 25 per cent by mass of a trimer with $M = 22$ kg mol^{-1} and its monomer. What average molar mass would be obtained from measurement of: (a) osmotic pressure, (b) light scattering?

18.13(a) What is the relative rate of sedimentation for two spherical particles of the same density, but which differ in radius by a factor of 10?

18.13(b) What is the relative rate of sedimentation for two spherical particles with densities 1.10 g cm^{-3} and 1.18 g cm^{-3} and which differ in radius by a factor of 8.4, the former being the larger? Use $\rho = 0.794$ g cm^{-3} for the density of the solution.

18.14(a) Human haemoglobin has a specific volume of 0.749×10^{3} m^3 kg^{-1}, a sedimentation constant of 4.48 Sv, and a diffusion coefficient of 6.9×10^{-11} m^2 s^{-1}. Determine its molar mass from this information.

18.14(b) A synthetic polymer has a specific volume of 8.01×10^{-4} m^3 kg^{-1}, a sedimentation constant of 7.46 Sv, and a diffusion coefficient of 7.72×10^{-11} m^2 s^{-1}. Determine its molar mass from this information.

18.15(a) Find the drift speed of a particle of radius 20 μm and density 1750 kg m^{-3} which is settling from suspension in water (density 1000 kg m^{-3}) under the influence of gravity alone. The viscosity of water is 8.9×10^{-4} kg m^{-1} s^{-1}.

18.15(b) Find the drift speed of a particle of radius 15.5 μm and density 1250 kg m^{-3} which is settling from suspension in water (density 1000 kg m^{-3}) under the influence of gravity alone. The viscosity of water is 8.9×10^{-4} kg m^{-1} s^{-1}.

18.16(a) At 20°C the diffusion coefficient of a macromolecule is found to be 8.3×10^{-11} m^2 s^{-1}. Its sedimentation constant is 3.2 Sv in a solution of density 1.06 g cm^{-3}. The specific volume of the macromolecule is 0.656 cm^{-3} g^{-1}. Determine the molar mass of the macromolecule.

18.16(b) At 20°C the diffusion coefficient of a macromolecule is found to be 7.9×10^{-11} m^2 s^{-1}. Its sedimentation constant is 5.1 Sv in a solution of density 997 kg m^{-1}. The specific volume of the macromolecule is 0.721 cm^{-3} g^{-1}. Determine the molar mass of the macromolecule.

18.17(a) The data from a sedimentation equilibrium experiment performed at 300 K on a macromolecular solute in aqueous solution show that a graph of $\ln c$ against r^2 is a straight line with a slope of 729 cm^{-2}. The rotational rate of the centrifuge was 50 000 r.p.m. The specific volume of the solute is 0.61 cm^3 g^{-1}. Calculate the molar mass of the solute.

18.17(b) The data from a sedimentation equilibrium experiment performed at 293 K on a macromolecular solute in aqueous solution show that a graph of $\ln c$ against $(r/\text{cm})^2$ is a straight line with a slope of 821. The rotation rate of the centrifuge was 1080 Hz. The specific volume of the solute is 7.2×10^{-4} m^3 kg^{-1}. Calculate the molar mass of the solute.

Problems*

Numerical problems

18.1 The following table lists the glass transition temperatures, T_g, of several polymers. Discuss the reasons why the structure of the monomer unit has an effect on the value of T_g.

Polymer	Poly(oxymethylene)	Polyethene	Poly(vinyl chloride)	Polystyrene
Structure	$-(OCH_2)_n-$	$-(CH_2CH_2)_n-$	$-(CH_2-CHCl)_n-$	$-(CH_2-CH(C_6H_5))_n-$
T_g/K	198	253	354	381

18.2 In a sedimentation experiment the position of the boundary as a function of time was found to be as follows:

t/min	15.5	29.1	36.4	58.2
r/cm	5.05	5.09	5.12	5.19

The rotation rate of the centrifuge was 45 000 r.p.m. Calculate the sedimentation constant of the solute.

18.3 Evaluate the radius of gyration, R_g, of (a) a solid sphere of radius a, (b) a long straight rod of radius a and length l. Show that, in the case of a solid sphere of specific volume v_s, $R_g/\text{nm} \approx 0.056902 \times \{(v_s/\text{cm}^3\,\text{g}^{-1})(M/\text{g mol}^{-1})\}^{1/3}$. Evaluate R_g for a species with $M = 100$ kg mol^{-1}, $v_s = 0.750$ cm^3 g^{-1}, and, in the case of the rod, of radius 0.50 nm.

18.4 Calculate the speed of operation (in r.p.m.) of an ultracentrifuge needed to obtain a readily measurable concentration gradient in a sedimentation equilibrium experiment. Take that gradient to be a concentration at the bottom of the cell about five times greater than at the top. Use $r_{top} = 5.0$ cm, $r_{bott} = 7.0$ cm, $M \approx 10^5$ g mol^{-1}, $\rho v_s \approx 0.75$, $T = 298$ K.

18.5 The concentration dependence of the viscosity of a polymer solution is found to be as follows:

c/(g dm^{-3})	1.32	2.89	5.73	9.17
η/(g m^{-1} s^{-1})	1.08	1.20	1.42	1.73

The viscosity of the solvent is 0.985 g m^{-1} s^{-1}. What is the intrinsic viscosity of the polymer?

18.6 The times of flow of dilute solutions of polystyrene in benzene through a viscometer at 25°C are given in the table below. From these data, calculate the molar mass of the polystyrene samples. Since the solutions are dilute, assume that the densities of the solutions are the same as those of pure benzene. η(benzene) $= 0.601 \times 10^{-3}$ kg m^{-1} s^{-1} (0.601 cP) at 25°C.

c/(g dm^{-3})	0	2.22	5.00	8.00	10.00
t/s	208.2	248.1	303.4	371.8	421.3

* Problems denoted with the symbol ‡ were supplied by Charles Trapp and Carmen Giunta.

18.7 The viscosities of solutions of polyisobutene in benzene were measured at 24°C (the θ temperature for the system) with the following results:

$c/(\text{g}/10^2\ \text{cm}^3)$	0	0.2	0.4	0.6	0.8	1.0
$\eta/(10^{-3}\ \text{kg m}^{-1}\ \text{s}^{-1})$	0.647	0.690	0.733	0.777	0.821	0.865

Use the information in Table 18.4 to deduce the molar mass of the polymer.

18.8‡ Polystyrene in cyclohexane at 34.5°C forms a θ solution, with an intrinsic viscosity related to the molar mass by $[\eta] = KM^a$. The following data on polystyrene in cyclohexane are taken from L.J. Fetters *et al.*, (*J. Phys. Chem. Ref. Data* **23**, 619 (1994)):

$M/(\text{kg mol}^{-1})$	10.0	19.8	106	249	359	860	1800	5470	9720	56 800
$[\eta]/(\text{cm}^3\ \text{g}^{-1})$	8.90	11.9	28.1	44.0	51.2	77.6	113.9	195	275	667

Determine the parameters K and a. What is the molar mass of a polystyrene that forms a θ solution in cyclohexane with $[\eta] = 100\ \text{cm}^3\ \text{g}^{-1}$?

18.9‡ Standard polystyrene solutions of known average molar masses continue to be used as for the calibration of many methods of characterizing polymer solutions. M. Kolinsky and J. Janca (*J. Polym. Sci., Polym. Chem.* **12**, 1181 (1974)) studied polystyrene in tetrahydrofuran (THF) for use in calibrating a gel permeation chromatograph. Their results for the intrinsic viscosity, $[\eta]$, as a function of average molar mass at 25°C are given in the table below. (a) Obtain the Mark–Houwink constants that fit these data. (b) Compare your values to those in Table 18.4 and Example 18.5. How might you explain the differences?

$\bar{M}_v/(\text{kg mol}^{-1})$	5.0	10.3	19.85	51	98.2	173	411	867
$[\eta]/(\text{cm}^3\ \text{g}^{-1})$	5.2	8.8	14.0	27.6	43.6	67.0	125.0	206.7

18.10 The concentration dependence of the osmotic pressure of solutions of a macromolecule at 20°C was found to be as follows:

$c/(\text{g dm}^{-3})$	1.21	2.72	5.08	6.60
Π/Pa	134	321	655	898

Determine the molar mass of the macromolecule and the osmotic virial coefficient.

18.11 The osmotic pressure of a fraction of poly(vinyl chloride) in a ketone solvent was measured at 25°C. The density of the solvent (which is virtually equal to the density of the solution) was 0.798 g cm^{-3}. Calculate the molar mass and the osmotic virial coefficient, B, of the fraction from the following data:

$c/(\text{g}/10^2\ \text{cm}^3)$	0.200	0.400	0.600	0.088	1.000
h/cm	0.48	1.2	1.86	2.76	3.88

Theoretical problems

18.12 Derive an expression for the fundamental vibrational frequency of a one-dimensional random coil that has been slightly stretched and then released. Evaluate this frequency for a sample of polyethene of molar mass 65 kg mol^{-1} at 20°C. Account physically for the dependence of frequency on temperature and molar mass.

18.13 In formamide as solvent, poly(γ-benzyl-L-glutamate) is found by light scattering experiments to have a radius of gyration proportional to M; in contrast, polystyrene in butanone has R_g proportional to $M^{1/2}$. Present arguments to show that the first polymer is a rigid rod whereas the second is a random coil.

18.14 A polymerization process produced a Gaussian distribution of polymers in the sense that the proportion of molecules having a molar mass in the range M to $M + \text{d}M$ was proportional to $e^{-(M-\bar{M})^2/2\gamma}$. What is the number average molar mass when the distribution is narrow?

18.15 Use eqn 18.2 to deduce expressions for (a) the root mean square separation of the ends of the chain, (b) the mean separation of the ends, and (c) their most probable separation. Evaluate these three quantities for a fully flexible chain with $N = 4000$ and $l = 154$ pm.

18.16 Deduce the relation $\langle r_i^2 \rangle = Nl^2$ for the mean square distance of a monomer from the origin in a freely jointed chain of N units each of length l. *Hint.* Use the distribution in eqn 18.2.

18.17 Deduce an expression for the radius of gyration of a three-dimensional freely jointed chain (eqn 18.6).

18.18 Derive expressions for the moments of inertia and hence the radii of gyration of (a) a uniform thin disc, (b) a long uniform rod, (c) a uniform sphere.

18.19 Construct a two-dimensional random walk by using a random number generating routine with mathematical software or electronic spreadsheet. Construct a walk of 50 and 100 steps. If there are many people working on the problem, investigate the mean and most probable separations in the plots by direct measurement. Do they vary as $N^{1/2}$?

18.20 Confirm the expression for ln P in eqn 18.39.

18.21 The effective radius, a, of a random coil is related to its radius of gyration, R_g, by $a = \gamma R_g$, with $\gamma = 0.85$. Deduce an expression for the osmotic virial coefficient, B, in terms of the number of chain units for (a) a freely jointed chain, (b) a chain with tetrahedral bond angles. Evaluate B for $l = 154$ pm and $N = 4000$. Estimate B for a randomly coiled polyethylene chain of arbitrary molar mass, M, and evaluate it for $M = 56$ kg mol^{-1}. Use $B = \frac{1}{2}N_A v_P$, where v_P is the excluded volume due to a single molecule.

18.22 Radius of gyration is defined in *Justification* 18.1. Show that an equivalent definition is that R_g is the average root mean square distance of the atoms or groups (all assumed to be of the same mass), that is, that $R_g^2 = (1/N)\sum_j R_j^2$, where R_j is the distance of atom j from the centre of mass.

18.23 Consider the thermodynamic description of stretching rubber. The observables are the tension, t, and length, l (the analogues of p and V for gases). Because $\text{d}w = t\text{d}l$, the basic equation is $\text{d}U = T\text{d}S + t\text{d}l$. If $G = U - TS - tl$, find expressions for dG and dA, and deduce the Maxwell relations

$$\left(\frac{\partial S}{\partial l}\right)_T = -\left(\frac{\partial t}{\partial T}\right)_l \qquad \left(\frac{\partial S}{\partial t}\right)_T = -\left(\frac{\partial l}{\partial T}\right)_t$$

Go on to deduce the equation of state for rubber

$$\left(\frac{\partial U}{\partial l}\right)_T = t - \left(\frac{\partial t}{\partial T}\right)_t$$

18.24 On the assumption that the tension required to keep a sample at a constant length is proportional to the temperature ($t = aT$, the analogue of $p \propto T$), show that the tension can be ascribed to the dependence of the entropy on the length of the sample. Account for this result in terms of the molecular nature of the sample.

Applications: to biochemistry and technology

18.25 Commercial software (more specifically 'molecular mechanics' or 'conformational search' software) automate the calculations that lead to Ramachandran plots, such as those in Fig. 18.17. In this problem our model for the protein is the dipeptide (**12**) in which the terminal methyl groups replace the rest of the polypeptide chain. (a) Draw three initial conformers of the dipeptide with R = H: one with $\phi = +75°$, $\psi = -65°$, a second with $\phi = \psi = +180°$, and a third with $\phi = +65°$, $\psi = +35°$. Use software of your instructor's choice to optimize the geometry of each conformer and find the final ϕ and ψ angles in each case. Did all of the initial conformers converge to the same final conformation? If not, what do these final conformers represent?

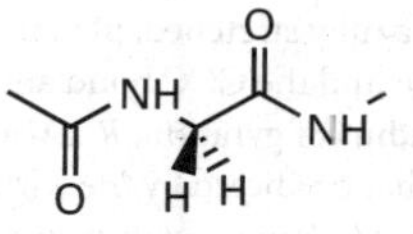

12

(b) Use the approach in part (a) to investigate the case R = CH_3, with the same three initial conformers as starting points for the calculations. Rationalize any similarities and differences between the final conformers of the dipeptides with R = H and R = CH_3.

18.26 Calculate the excluded volume in terms of the molecular volume on the basis that the molecules are spheres of radius a. Evaluate the osmotic virial coefficient in the case of bushy stunt virus, $a = 14.0$ nm, and haemoglobin, $a = 3.2$ nm (see Problem 18.21). Evaluate the percentage deviation of the Rayleigh ratios of 1.00 g/(100 cm^3) solutions of bushy stunt virus ($M = 1.07 \times 10^4$ kg mol^{-1}) and haemoglobin ($M = 66.5$ kg mol^{-1}) from the ideal solution values. In eqn 18.8, let $P_\theta = 1$ and assume that both solutions have the same K value.

18.27 Use the information below and the expression for R_g of a solid sphere quoted in the text (following eqn 18.6), to classify the species below as globular or rod-like.

	M/(g mol^{-1})	v_s/(cm^3 g^{-1})	R_g/nm
Serum albumin	66×10^3	0.752	2.98
Bushy stunt virus	10.6×10^6	0.741	12.0
DNA	4×10^6	0.556	117.0

18.28 Suppose that a rod-like DNA molecule of length 250 nm undergoes a conformational change to a closed-circular (cc) form. (a) Use the information in Problem 18.27 and an incident wavelength $\lambda = 488$ nm to calculate the ratio of scattering intensities by each of these conformations, I_{rod}/I_{cc}, when $\theta = 20°$, 45°, and 90°. (b) Suppose that you wish to use light scattering as a technique for the study of conformational changes in DNA molecules. Based on your answer to part (a), at which angle would you conduct the experiments? Justify your choice.

18.29 In an ultracentrifugation experiment at 20°C on bovine serum albumin the following data were obtained: $\rho = 1.001$ g cm^{-3}, $v_s = 1.112$ cm^{-3} g^{-1}, $\omega/2\pi = 322$ Hz,

r/cm	5.0	5.1	5.2	5.3	5.4
c/(mg cm^{-3})	0.536	0.284	0.148	0.077	0.039

Evaluate the molar mass of the sample.

18.30 Sedimentation studies on haemoglobin in water gave a sedimentation constant $S = 4.5$ Sv at 20°C. The diffusion coefficient is 6.3×10^{-11} m^2 s^{-1} at the same temperature. Calculate the molar mass of haemoglobin using $v_s = 0.75$ cm^3 g^{-1} for its partial specific volume and $\rho = 0.998$ g cm^{-3} for the density of the solution. Estimate the effective radius of the haemoglobin molecule given that the viscosity of the solution is 1.00×10^3 kg m^{-1} s^{-1}.

18.31 The rate of sedimentation of a recently isolated protein was monitored at 20°C and with a rotor speed of 50 000 r.p.m. The boundary receded as follows:

t/s	0	300	600	900	1200	1500	1800
r/cm	6.127	6.153	6.179	6.206	6.232	6.258	6.284

Calculate the sedimentation constant and the molar mass of the protein on the basis that its partial specific volume is 0.728 cm^3 g^{-1} and its diffusion coefficient is 7.62×10^{-11} m^2 s^{-1} at 20°C, the density of the solution then being 0.9981 g cm^{-3}. Suggest a shape for the protein given that the viscosity of the solution is 1.00×10^3 kg m^{-1} s^{-1} at 20°C.

18.32 For some proteins, the isoelectric point must be obtained by extrapolation because the macromolecule might not be stable over a very wide pH range. Estimate the pH of the isoelectric point from the following data for a protein:

pH	4.5	5.0	5.5	6.0
Drift speed/(μm s^{-1})	−0.10	−0.20	−0.30	−0.35

18.33 Here we use concepts developed in Chapter 15 and this chapter to enhance our understanding of closed-circular and supercoiled DNA. (a) The average end-to-end distance of a flexible polymer (such as a fully denatured polypeptide or a strand of DNA) is $N^{1/2}l$, where N is the number of groups (residues or bases) and l is the length of each group. Initially, therefore, one end of the polymer can be found anywhere within a sphere of radius $N^{1/2}l$ centred on the other end. When the ends join to form a circle, they are confined to a volume of radius l. What is the change in molar entropy? Plot the function you derive as a function of N. (b) The energy necessary to twist ccDNA by i turns is $\varepsilon_i = ki^2$, with k an empirical constant and i being negative or positive depending on the sense of the twist. For example, one twist ($i = \pm 1$) makes ccDNA resemble the number 8. (i) Show that the distribution of the populations $p_i = n_i/N$ of ccDNA molecules with i turns at a specified temperature has the form of a Gaussian function. (ii) Plot the expression you derived in part (a) for several values of the temperature. Does the curve has a maximum? If so, at what value of i? Comment on variations of the shape of the curve with temperature. (iii) Calculate p_0, p_1, p_5, and p_{10} at 298 K.

18.34 The melting temperature of a DNA molecule can be determined by differential scanning calorimetry (*Impact I2.1*). The following data were obtained in aqueous solutions containing the specified concentration c_{salt} of an soluble ionic solid for a series of DNA molecules with varying base pair composition, with f the fraction of GC base pairs:

$c_{salt} = 1.0 \times 10^{-2}$ mol dm^{-3}

f	0.375	0.509	0.589	0.688	0.750
T_m/K	339	344	348	351	354

$c_{salt} = 0.15$ mol dm^{-3}

f	0.375	0.509	0.589	0.688	0.750
T_m/K	359	364	368	371	374

(a) Estimate the melting temperature of a DNA molecule containing 40.0 per cent GC base pairs in both samples. *Hint.* Begin by plotting T_m against fraction of GC base pairs and examining the shape of the curve. (b) Do the data show an effect of concentration of ions in solution on the melting temperature of DNA? If so, provide a molecular interpretation for the effect you observe.

18.35 The fluidity of a lipid bilayer dispersed in aqueous solution depends on temperature and there are two important melting transitions. One transition is from a 'solid crystalline' state in which the hydrophobic chains are packed together tightly (hence move very little) to a 'liquid crystalline state', in which there is increased but still limited movement of the of the chains. The second transition, which occurs at a higher temperature than the first, is from the liquid crystalline state to a liquid state, in which the hydrophobic interactions holding the aggregate together are largely disrupted. (a) It is observed that the transition temperatures increase with the hydrophobic chain length and decrease with the number of C=C bonds in the chain. Explain these observations. (b) What effect is the inclusion of cholesterol likely to have on the transition temperatures of a lipid bilayer? Justify your answer.

18.36 Polystyrene is a synthetic polymer with the structure $-(CH_2-CH(C_6H_5))_n-$. A batch of polydisperse polystyrene was prepared by initiating the polymerization with t-butyl radicals. As a result, the t-butyl group is expected to be covalently attached to the end of the final products. A sample from this batch was embedded in an organic matrix containing silver trifluoroacetate and the resulting MALDI-TOF spectrum consisted of a large number of peaks separated by 104 g mol^{-1}, with the most intense peak at 25 578 g mol^{-1}. Comment on the purity of this sample and determine the number of (CH_2–$CH(C_6H_5)$) units in the species that gives rise to the most intense peak in the spectrum.

18.37 A manufacturer of polystyrene beads claims that they have an average molar mass of 250 kg mol^{-1}. Solutions of these beads are studied by a physical chemistry student by dilute solution viscometry with an Ostwald viscometer in both the 'good' solvent toluene and the theta solvent cyclohexane. The drainage times, t_D, as a function of concentration for the two solvents are given in the table below. (a) Fit the data to the virial equation for viscosity,

$$\eta = \eta^*(1 + [\eta]c + k'[\eta]^2c^2 + \cdots)$$

where k' is called the *Huggins constant* and is typically in the range 0.35–0.40. From the fit, determine the intrinsic viscosity and the Huggins constant. (b) Use the empirical Mark–Kuhn–Houwink–Sakurada equation (eqn 18.37) to determine the molar mass of polystyrene in the two solvents. For theta solvents, $a = 0.5$ and $K = 8.2 \times 10^{-5}$ dm^{-3} g^{-1} for cyclohexane; for the good solvent toluene $a = 0.72$ and $K = 1.15 \times 10^{-5}$ dm^{-3} g^{-1}. (c) According to a general theory proposed by Kirkwood and Riseman, the root mean square end-to-end distance of a polymer chain in solution is related to $[\eta]$ by $\phi\langle r^2\rangle^{3/2}/M$, where ϕ is a universal constant with the value 2.84×10^{26} when $[\eta]$ is expressed in cubic decimetres per gram and the distance is in metres. Calculate this quantity for each solvent. (d) From the molar masses calculate the average number of styrene ($C_6H_5CH{=}CH_2$) monomer units, $\langle n\rangle$. (e) Calculate the length of a fully stretched, planar zigzag configuration, taking the C–C distance as 154 pm and the CCC bond angle to be 109°. (f) Use eqn 18.6 to calculate the radius of gyration, R_g. Also calculate $\langle r^2\rangle^{1/2} = n^{1/2}l$. Compare this result with that predicted by the Kirkwood–Riseman theory: which gives the better fit? (g) Compare your values for M to the results of Problem 18.36. Is there any reason why they should or should not agree? Is the manufacturer's claim valid?

$c/(\text{g dm}^{-3}\text{ toluene})$	0	1.0	3.0	5.0
t_D/s	8.37	9.11	10.72	12.52
$c/(\text{g dm}^{-3}\text{ cyclohexane})$	0	1.0	1.5	2.0
t_D/s	8.32	8.67	8.85	9.03

18.38‡ The determination of the average molar masses of conducting polymers is an important part of their characterization. S. Holdcroft (*J. Polym. Sci., Polym. Phys.* **29**, 1585 (1991)) has determined the molar mases and Mark–Houwink constants for the electronically conducting polymer, poly(3-hexylthiophene) (P3HT) in tetrahydrofuran (THF) at 25°C by methods similar to those used for nonconducting polymers. The values for molar mass and intrinsic viscosity in the table below are adapted from their data. Determine the constants in the Mark–Kuhn–Houwink–Sakurada equation from these results and compare to the values obtained in your solution to Problem 18.9.

$\bar{M}_v/(\text{kg mol}^{-1})$	3.8	11.1	15.3	58.8
$[\eta]/(\text{cm}^3\text{ g}^{-1})$	6.23	17.44	23.73	85.28

Materials 2: solids

19

First, we see how to describe the regular arrangement of atoms in crystals and the symmetry of their arrangement. Then we consider the basic principles of X-ray diffraction and see how the diffraction pattern can be interpreted in terms of the distribution of electron density in a unit cell. X-ray diffraction leads to information about the structures of metallic, ionic, and molecular solids, and we review some typical results and their rationalization in terms of atomic and ionic radii. With structures established, we move on to the properties of solids, and see how their mechanical, electrical, optical, and magnetic properties stem from the properties of their constituent atoms and molecules.

The solid state includes most of the materials that make modern technology possible. It includes the wide varieties of steel that are used in architecture and engineering, the semiconductors and metallic conductors that are used in information technology and power distribution, the ceramics that increasingly are replacing metals, and the synthetic and natural polymers discussed in Chapter 18 that are used in the textile industry and in the fabrication of many of the common objects of the modern world. The properties of solids stem, of course, from the arrangement and properties of the constituent atoms, and one of the challenges of this chapter is to see how a wide range of bulk properties, including rigidity, electrical conductivity, and optical and magnetic properties, stem from the properties of atoms. One crucial aspect of this link is the pattern in which the atoms (and molecules) are stacked together, and we start this chapter with an examination of how the structures of solids are described and determined.

Crystallography

Early in the history of modern science it was suggested that the regular external form of crystals implied an internal regularity of their constituents. In this section we see how to describe and determine the arrangement of atoms inside crystals.

19.1 Lattices and unit cells

Key points **A space lattice is the pattern formed by points representing the locations of structural motifs. A unit cell is an imaginary parallelepiped that contains one unit of a translationally repeating pattern. Unit cells are classified into seven crystal systems according to their rotational symmetries. The Bravais lattices are the 14 distinct space lattices in three dimensions.**

A crystal is built up from regularly repeating 'structural motifs', which may be atoms, molecules, or groups of atoms, molecules, or ions. A **space lattice** is the pattern

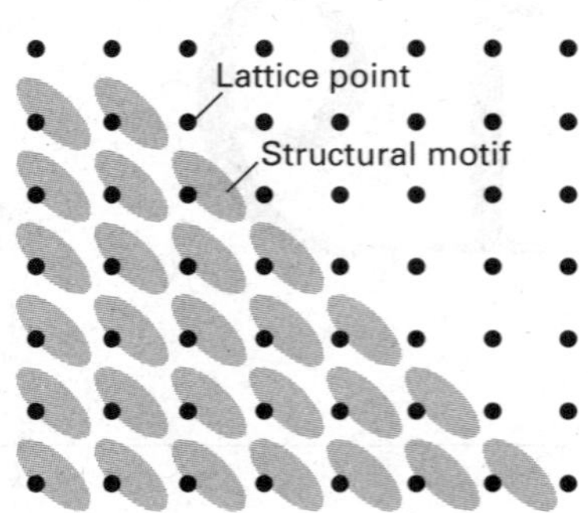

Fig. 19.1 Each lattice point specifies the location of a structural motif (for example, a molecule or a group of molecules). The crystal lattice is the array of lattice points; the crystal structure is the collection of structural motifs arranged according to the lattice.

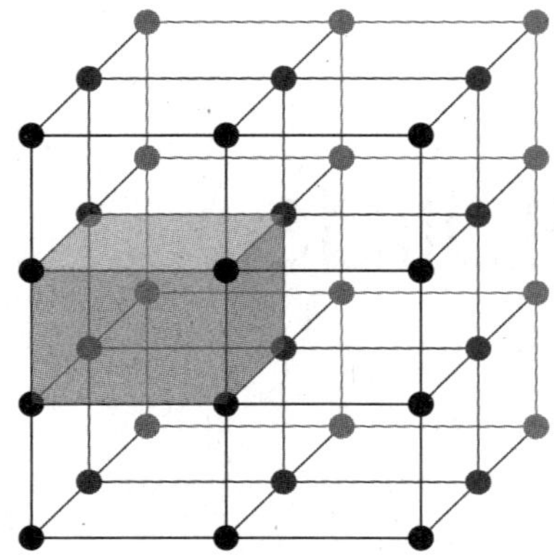

Fig. 19.2 A unit cell is a parallel-sided (but not necessarily rectangular) figure from which the entire crystal structure can be constructed by using only translations (not reflections, rotations, or inversions).

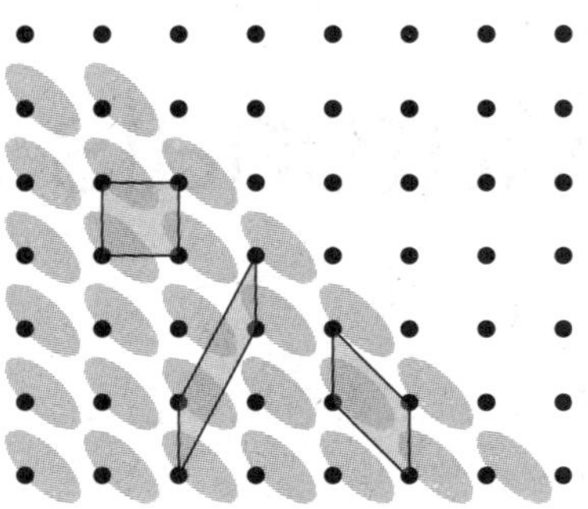

Fig. 19.3 A unit cell can be chosen in a variety of ways, as shown here. It is conventional to choose the cell that represents the full symmetry of the lattice. In this rectangular lattice, the rectangular unit cell would normally be adopted.

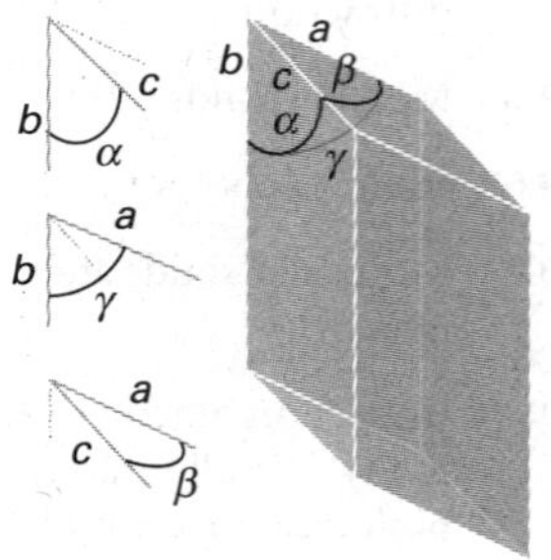

Fig. 19.4 The notation for the sides and angles of a unit cell. Note that the angle α lies in the plane (b,c) and perpendicular to the axis a.

A brief comment

A *symmetry operation* is an action (such as a rotation, reflection, or inversion) that leaves an object looking the same after it has been carried out. There is a corresponding *symmetry element* for each symmetry operation, which is the point, line, or plane with respect to which the symmetry operation is performed. For instance, an *n-fold rotation* (the symmetry operation) about an *n-fold axis of symmetry* (the corresponding symmetry element) is a rotation through $360°/n$. See Chapter 11 for a more detailed discussion of symmetry.

formed by points representing the locations of these motifs (Fig. 19.1). The space lattice is, in effect, an abstract scaffolding for the crystal structure. More formally, a space lattice is a three-dimensional, infinite array of points, each of which is surrounded in an identical way by its neighbours, and which defines the basic structure of the crystal. In some cases there may be a structural motif centred on each lattice point, but that is not necessary. The crystal structure itself is obtained by associating with each lattice point an identical structural motif.

The **unit cell** is an imaginary parallelepiped (parallel-sided figure) that contains one unit of the translationally repeating pattern (Fig. 19.2). A unit cell can be thought of as the fundamental region from which the entire crystal may be constructed by purely translational displacements (like bricks in a wall). A unit cell is commonly formed by joining neighbouring lattice points by straight lines (Fig. 19.3). Such unit cells are called **primitive**. It is sometimes more convenient to draw larger **non-primitive unit cells** that also have lattice points at their centres or on pairs of opposite faces. An infinite number of different unit cells can describe the same lattice, but the one with sides that have the shortest lengths and that are most nearly perpendicular to one another is normally chosen. The lengths of the sides of a unit cell are denoted a, b, and c, and the angles between them are denoted α, β, and γ (Fig. 19.4).

Unit cells are classified into seven **crystal systems** by noting the rotational symmetry elements they possess. A *cubic unit cell*, for example, has four threefold axes in a tetrahedral array (Fig. 19.5). A *monoclinic unit cell* has one twofold axis; the unique axis is by convention the b axis (Fig. 19.6). A *triclinic unit cell* has no rotational symmetry, and typically all three sides and angles are different (Fig. 19.7). Table 19.1 lists the **essential symmetries**, the elements that must be present for the unit cell to belong to a particular crystal system.

There are only 14 distinct space lattices in three dimensions. These **Bravais lattices** are illustrated in Fig. 19.8. It is conventional to portray these lattices by primitive unit cells in some cases and by non-primitive unit cells in others. A **primitive unit cell** (with lattice points only at the corners) is denoted P. A **body-centred unit cell** (I) also has a lattice point at its centre. A **face-centred unit cell** (F) has lattice points at its corners and also at the centres of its six faces. A **side-centred unit cell** (A, B, or C) has

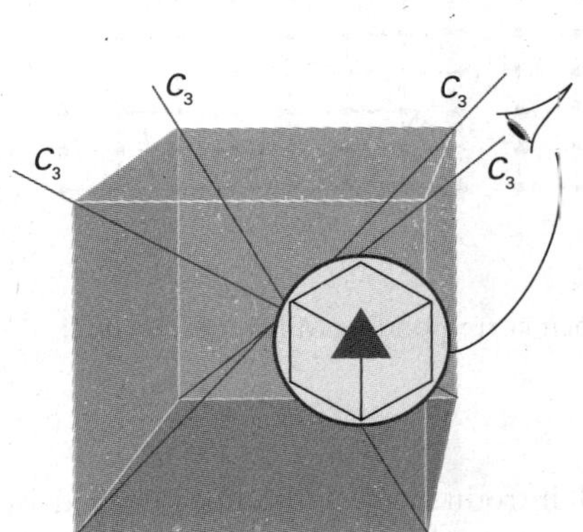

Fig. 19.5 A unit cell belonging to the cubic system has four threefold axes, denoted C_3, arranged tetrahedrally. The insert shows the threefold symmetry.

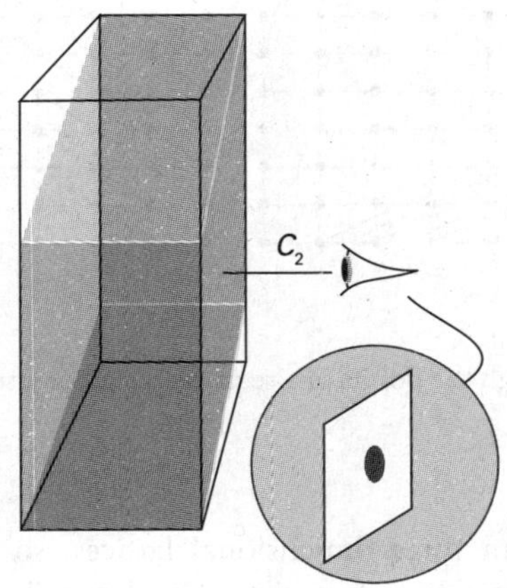

Fig. 19.6 A unit belonging to the monoclinic system has a twofold axis (denoted C_2 and shown in more detail in the insert).

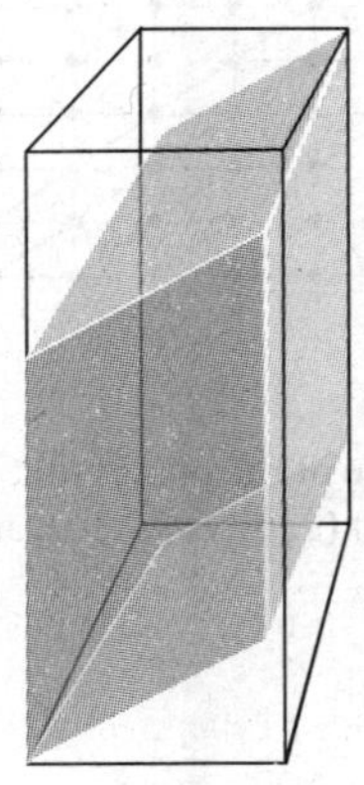

Fig. 19.7 A triclinic unit cell has no axes of rotational symmetry.

Table 19.1 The seven crystal systems

System	Essential symmetries
Triclinic	None
Monoclinic	One C_2 axis
Orthorhombic	Three perpendicular C_2 axes
Rhombohedral	One C_3 axis
Tetragonal	One C_4 axis
Hexagonal	One C_6 axis
Cubic	Four C_3 axes in a tetrahedral arrangement

lattice points at its corners and at the centres of two opposite faces. For simple structures, it is often convenient to choose an atom belonging to the structural motif, or the centre of a molecule, as the location of a lattice point or the vertex of a unit cell, but that is not a necessary requirement.

19.2 The identification of lattice planes

Key point **Crystal planes are specified by a set of Miller indices.**

The spacing of the planes of lattice points in a crystal is an important quantitative aspect of its structure. However, there are many different sets of planes (Fig. 19.9), and we need to be able to label them. Two-dimensional lattices are easier to visualize

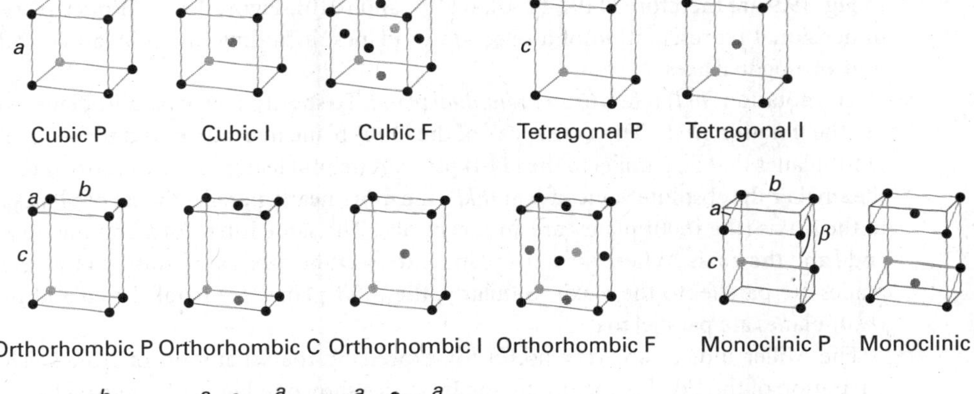

Fig. 19.8 The fourteen Bravais lattices. The points are lattice points and are not necessarily occupied by atoms. P denotes a primitive unit cell (R is used for a trigonal lattice), I a body-centred unit cell, F a face-centred unit cell, and C (or A or B) a cell with lattice points on two opposite faces.

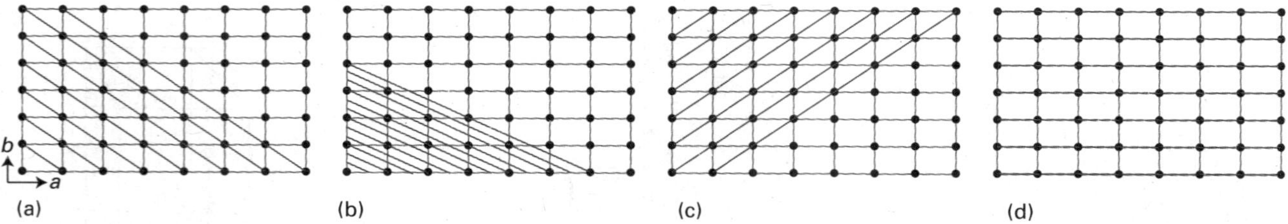

Fig. 19.9 Some of the planes that can be drawn through the points of a rectangular space lattice and their corresponding Miller indices (*hkl*): (a) (110), (b) (230), (c) ($\bar{1}$10), and (d) (010).

than three-dimensional lattices, so we shall introduce the concepts involved by referring to two dimensions initially and then extend the conclusions by analogy to three dimensions.

Consider a two-dimensional rectangular lattice formed from a unit cell of sides a, b (as in Fig. 19.9). Each plane in the illustration (except the plane passing through the origin) can be distinguished by the distances at which it intersects the a and b axes. One way to label each set of parallel planes would therefore be to quote the smallest intersection distances. For example, we could denote the four sets in the illustration as $(1a, 1b)$, $(\frac{1}{2}a, \frac{1}{3}b)$, $(-1a, 1b)$, and $(\infty a, 1b)$. However, if we agree to quote distances along the axes as multiples of the lengths of the unit cell, then we can label the planes more simply as $(1, 1)$, $(\frac{1}{2}, \frac{1}{3})$, $(-1, 1)$, and $(\infty, 1)$. If the lattice in Fig. 19.9 is the top view of a three-dimensional orthorhombic lattice in which the unit cell has a length c in the z-direction, all four sets of planes intersect the z-axis at infinity. Therefore, the full labels are $(1, 1, \infty)$, $(\frac{1}{2}, \frac{1}{3}, \infty)$, $(-1, 1, \infty)$, and $(\infty, 1, \infty)$.

The presence of fractions and infinity in the labels is inconvenient. They can be eliminated by taking the reciprocals of the labels. As we shall see, taking reciprocals turns out to have further advantages. The **Miller indices**, (*hkl*), are the reciprocals of intersection distances (with fractions cleared by multiplying through by an appropriate factor, if taking the reciprocal results in a fraction). For example, the $(1, 1, \infty)$ planes in Fig. 19.9a are the (110) planes in the Miller notation. Similarly, the $(\frac{1}{2}, \frac{1}{3}, \infty)$ planes are denoted (230). Negative indices are written with a bar over the number, and Fig. 19.9c shows the ($\bar{1}$10) planes. The Miller indices for the four sets of planes in Fig. 19.9 are therefore (110), (230), ($\bar{1}$10), and (010). Figure 19.10 shows a three-dimensional representation of a selection of planes, including one in a lattice with non-orthogonal axes.

The notation (*hkl*) refers to an *individual* plane. To specify a *set* of parallel planes we use the notation {*hkl*}. Thus, we speak of the (110) plane in a lattice, and the set of all {110} planes that lie parallel to the (110) plane. A helpful feature to remember is that, the smaller the absolute value of h in {*hkl*}, the more nearly parallel the set of planes is to the a axis (the {*h*00} planes are an exception). The same is true of k and the b axis and l and the c axis. When $h = 0$, the planes intersect the a axis at infinity, so the {0*kl*} planes are parallel to the a axis. Similarly, the {*h*0*l*} planes are parallel to b and the {*hk*0} planes are parallel to c.

The Miller indices are very useful for expressing the separation of planes. The separation of the {*hk*0} planes in the square lattice shown in Fig. 19.11 is given by

$$\frac{1}{d_{hk0}^2} = \frac{h^2 + k^2}{a^2} \quad \text{or} \quad d_{hk0} = \frac{a}{(h^2 + k^2)^{1/2}} \qquad (19.1)$$

By extension to three dimensions, the separation of the {*hkl*} planes of a cubic lattice is given by

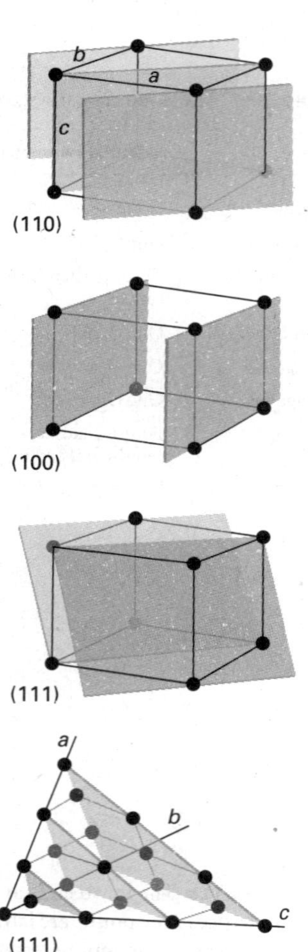

Fig. 19.10 Some representative planes in three dimensions and their Miller indices. Note that a 0 indicates that a plane is parallel to the corresponding axis, and that the indexing may also be used for unit cells with non-orthogonal axes.

$$\frac{1}{d_{hkl}^2}=\frac{h^2+k^2+l^2}{a^2} \quad \text{or} \quad d_{hkl}=\frac{a}{(h^2+k^2+l^2)^{1/2}} \tag{19.2}$$

The corresponding expression for a general orthorhombic lattice is the generalization of this expression:

$$\frac{1}{d_{hkl}^2}=\frac{h^2}{a^2}+\frac{k^2}{b^2}+\frac{l^2}{c^2} \tag{19.3}$$

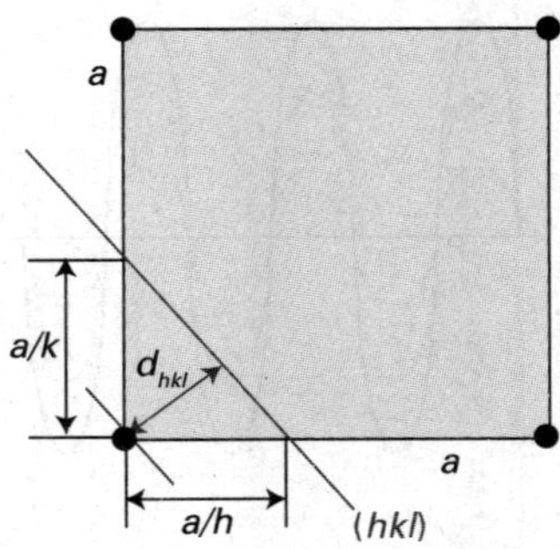

Fig. 19.11 The dimensions of a unit cell and their relation to the plane passing through the lattice points.

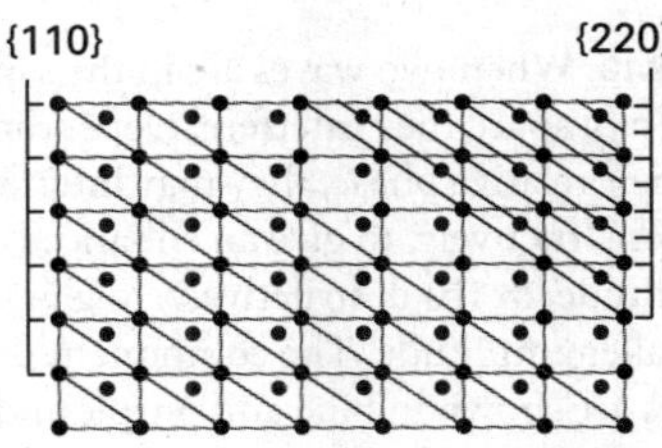

Fig. 19.12 The separation of the {220} planes is half that of the {110} planes. In general, the separation of the planes {*nh*,*nk*,*nl*} is *n* times smaller than the separation of the {*hkl*} planes.

Example 19.1 *Using the Miller indices*

Calculate the separation of (a) the {123} planes and (b) the {246} planes of an orthorhombic unit cell with $a = 0.82$ nm, $b = 0.94$ nm, and $c = 0.75$ nm.

Method For the first part, simply substitute the information into eqn 19.3. For the second part, instead of repeating the calculation, note that, if all three Miller indices are multiplied by n, then their separation is reduced by that factor (Fig. 19.12):

$$\frac{1}{d_{nh,nk,nl}^2}=\frac{(nh)^2}{a^2}+\frac{(nk)^2}{b^2}+\frac{(nl)^2}{c^2}=n^2\left(\frac{h^2}{a^2}+\frac{k^2}{b^2}+\frac{l^2}{c^2}\right)=\frac{n^2}{d_{hkl}^2}$$

which implies that

$$d_{nh,nk,nl}=\frac{d_{hkl}}{n}$$

Answer Substituting the indices into eqn 19.3 gives

$$\frac{1}{d_{123}^2}=\frac{1^2}{(0.82\text{ nm})^2}+\frac{2^2}{(0.94\text{ nm})^2}+\frac{3^2}{(0.75\text{ nm})^2}=0.22\text{ nm}^{-2}$$

Hence, $d_{123} = 0.21$ nm. It then follows immediately that d_{246} is one-half this value, or 0.11 nm.

Self-test 19.1 Calculate the separation of (a) the {133} planes and (b) the {399} planes in the same lattice. [0.19 nm, 0.063 nm]

A note on good practice It is always sensible to look for analytical relations between quantities rather than to evaluate expressions numerically each time for that emphasizes the relations between quantities (and avoids unnecessary work).

19.3 The investigation of structure

Key points **(a) The positions of atoms in a solid may be revealed by analysing the pattern of diffraction of X-rays by a single crystal or powder (a collection of crystallites). (b) Bragg's law relates the glancing angle of incidence of X-rays and the separation of lattice planes. (c) The scattering factor is a measure of the ability of an atom to diffract radiation. (d) The structure factor is the overall amplitude of a wave diffracted by the {*hkl*} planes. Fourier synthesis is the construction of the electron density distribution from structure factors. (e) A Patterson synthesis is a map of interatomic vectors obtained by Fourier analysis of diffraction intensities. (f) Structure refinement is the adjustment of structural parameters to give the best fit between the observed intensities and those calculated from the model of the structure deduced from the diffraction pattern.**

A characteristic property of waves is that they interfere with one another, giving a greater displacement where peaks coincide with peaks or troughs coincide with troughs, and a smaller displacement where peaks coincide with troughs (Fig. 19.13). According to classical electromagnetic theory, the intensity of electromagnetic radiation is

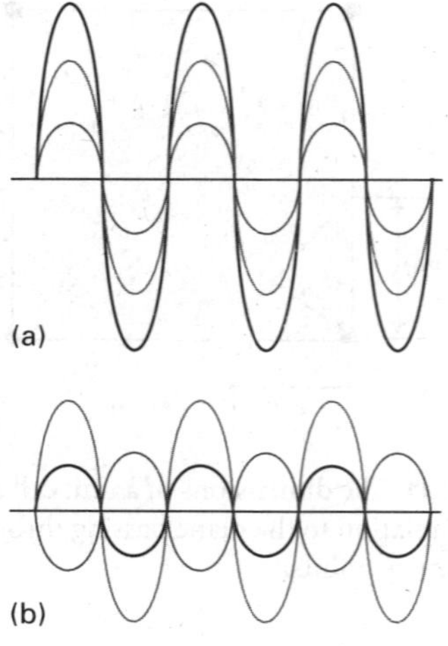

Fig. 19.13 When two waves are in the same region of space they interfere. Depending on their relative phase, they may interfere (a) constructively, to give an enhanced amplitude, or (b) destructively, to give a smaller amplitude. The component waves are shown in blue and purple and the resultant in black.

proportional to the square of the amplitude of the waves. Therefore, the regions of constructive or destructive interference show up as regions of enhanced or diminished intensities. The phenomenon of **diffraction** is the interference caused by an object in the path of waves, and the pattern of varying intensity that results is called the **diffraction pattern**. Diffraction occurs when the dimensions of the diffracting object are comparable to the wavelength of the radiation.

(a) X-ray diffraction

Wilhelm Röntgen discovered X-rays in 1895. Seventeen years later, Max von Laue suggested that they might be diffracted when passed through a crystal, for by then he had realized that their wavelengths are comparable to the separation of lattice planes. This suggestion was confirmed almost immediately by Walter Friedrich and Paul Knipping and has grown since then into a technique of extraordinary power. The bulk of this section will deal with the determination of structures using X-ray diffraction. The mathematical procedures necessary for the determination of structure from X-ray diffraction data are enormously complex, but such is the degree of integration of computers into the experimental apparatus that the technique is almost fully automated, even for large molecules and complex solids. The analysis is aided by molecular modelling techniques, which can guide the investigation towards a plausible structure.

X-rays are electromagnetic radiation with wavelengths of the order of 10^{-10} m. They are typically generated by bombarding a metal with high-energy electrons (Fig. 19.14). The electrons decelerate as they plunge into the metal and generate radiation with a continuous range of wavelengths called **Bremsstrahlung** (*Bremse* is German for deceleration, *Strahlung* for ray.) Superimposed on the continuum are a few high-intensity, sharp peaks (Fig. 19.15). These peaks arise from collisions of the incoming electrons with the electrons in the inner shells of the atoms. A collision expels an electron from an inner shell, and an electron of higher energy drops into the vacancy, emitting the excess energy as an X-ray photon (Fig. 19.16). If the electron

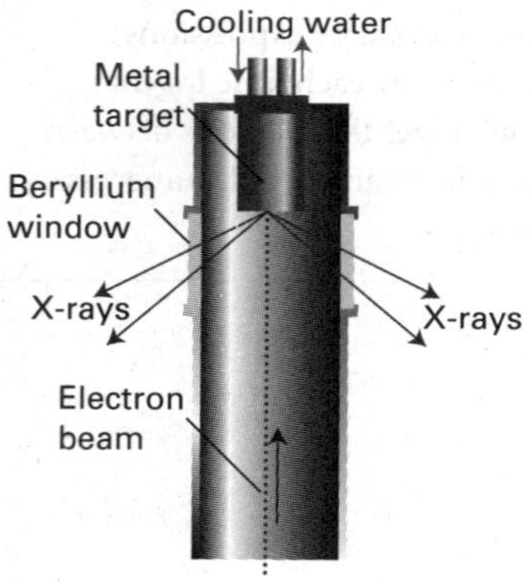

Cooling water
Metal target
Beryllium window
X-rays
X-rays
Electron beam

Fig. 19.14 X-rays are generated by directing an electron beam on to a cooled metal target. Beryllium is transparent to X-rays (on account of the small number of electrons in each atom) and is used for the windows.

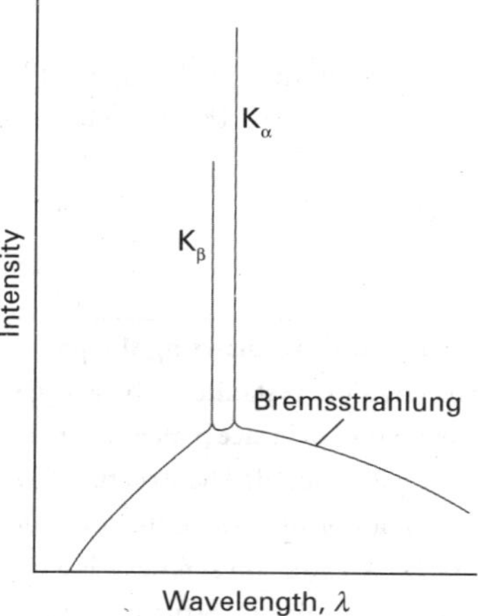

Fig. 19.15 The X-ray emission from a metal consists of a broad, featureless Bremsstrahlung background, with sharp transitions superimposed on it. The label K indicates that the radiation comes from a transition in which an electron falls into a vacancy in the K shell of the atom.

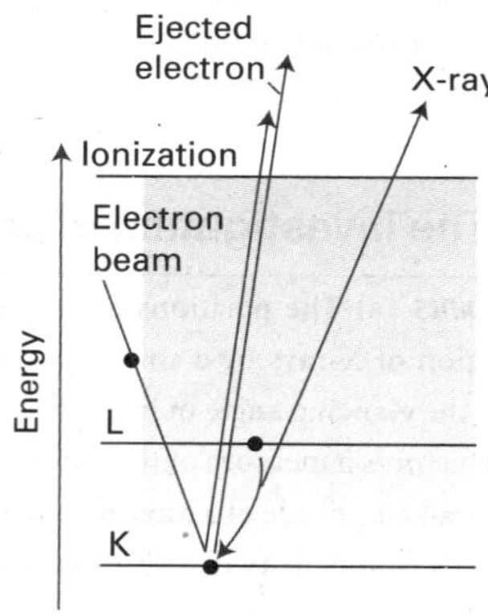

Fig. 19.16 The processes that contribute to the generation of X-rays. An incoming electron collides with an electron (in the K shell), and ejects it. Another electron (from the L shell in this illustration) falls into the vacancy and emits its excess energy as an X-ray photon.

falls into a K shell (a shell with $n = 1$), the X-rays are classified as **K-radiation**, and similarly for transitions into the L ($n = 2$) and M ($n = 3$) shells. Strong, distinct lines are labelled K_α, K_β, and so on. Increasingly, X-ray diffraction makes use of the radiation available from synchrotron sources (*Further information 12.1*), for its high intensity greatly enhances the sensitivity of the technique.

von Laue's original method consisted of passing a broad-band beam of X-rays into a single crystal, and recording the diffraction pattern photographically. The idea behind the approach was that a crystal might not be suitably orientated to act as a diffraction grating for a single wavelength but, whatever its orientation, diffraction would be achieved for at least one of the wavelengths if a range of wavelengths was used. There is currently a resurgence of interest in this approach because synchrotron radiation spans a range of X-ray wavelengths.

An alternative technique was developed by Peter Debye and Paul Scherrer and independently by Albert Hull. They used monochromatic radiation and a powdered sample. When the sample is a powder, at least some of the crystallites will be orientated so as to give rise to diffraction. In modern powder diffractometers the intensities of the reflections are monitored electronically as the detector is rotated around the sample in a plane containing the incident ray (Fig. 19.17). Powder diffraction techniques are used to identify a sample of a solid substance by comparison of the positions of the diffraction lines and their intensities with diffraction patterns stored in a large data bank. Powder diffraction data are also used to help determine phase diagrams, for different crystalline phases result in different diffraction patterns, and to determine the relative amounts of each phase present in a mixture. The technique is also used for the initial determination of the dimensions and symmetries of unit cells.

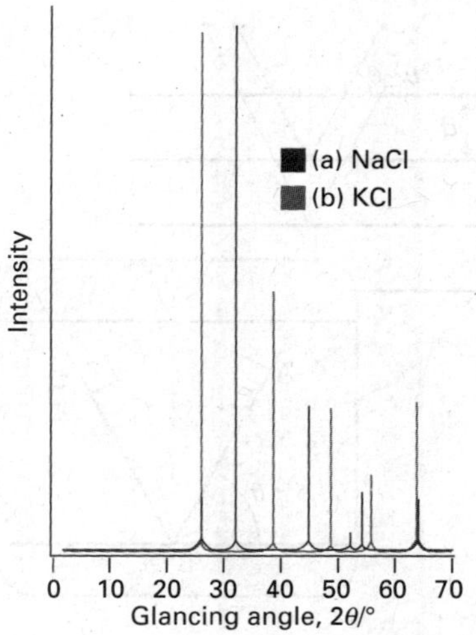

Fig. 19.17 X-ray powder diffraction patterns of (a) NaCl, (b) KCl. The smaller number of lines in (b) is a consequence of the similarity of the K^+ and Cl^- scattering factors, as discussed later in the chapter.

The method developed by the Braggs (William and his son Lawrence, who later jointly won the Nobel Prize) is the foundation of almost all modern work in X-ray crystallography. They used a single crystal and a monochromatic beam of X-rays, and rotated the crystal until a reflection was detected. There are many different sets of planes in a crystal, so there are many angles at which a reflection occurs. The complete set of data consists of the list of angles at which reflections are observed and their intensities.

Single-crystal diffraction patterns are measured by using a **four-circle diffractometer** (Fig. 19.18). The computer linked to the diffractometer determines the unit cell dimensions and the angular settings of the diffractometer's four circles that are needed to observe any particular intensity peak in the diffraction pattern. The computer controls the settings, and moves the crystal and the detector for each one in turn. At each setting, the diffraction intensity is measured, and background intensities are assessed by making measurements at slightly different settings. Computing techniques are now available that lead not only to automatic indexing but also to the automated determination of the shape, symmetry, and size of the unit cell. Moreover, several techniques are now available for sampling large amounts of data, including area detectors and image plates, which sample whole regions of diffraction patterns simultaneously.

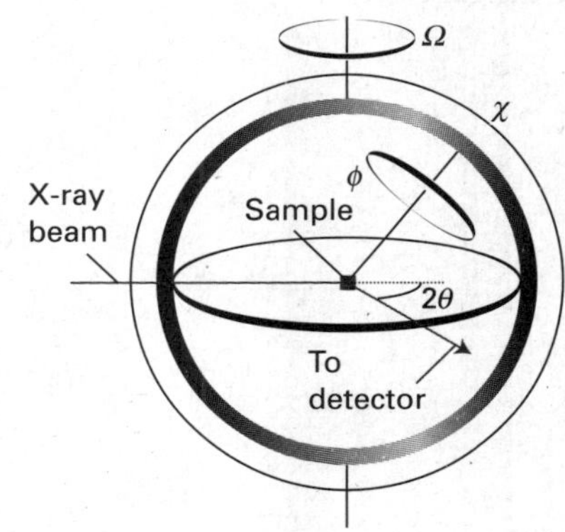

Fig. 19.18 A four-circle diffractometer. The settings of the orientations (ϕ, χ, θ, and Ω) of the components are controlled by computer; each (hkl) reflection is monitored in turn, and their intensities are recorded.

(b) Bragg's law

An early approach to the analysis of diffraction patterns produced by crystals was to regard a lattice plane as a semi-transparent mirror, and to model a crystal as stacks of reflecting lattice planes of separation d (Fig. 19.19). The model makes it easy to calculate the angle the crystal must make to the incoming beam of X-rays for constructive interference to occur. It has also given rise to the name **reflection** to denote an intense beam arising from constructive interference.

Consider the reflection of two parallel rays of the same wavelength by two adjacent planes of a lattice, as shown in Fig. 19.19. One ray strikes a point on the upper plane

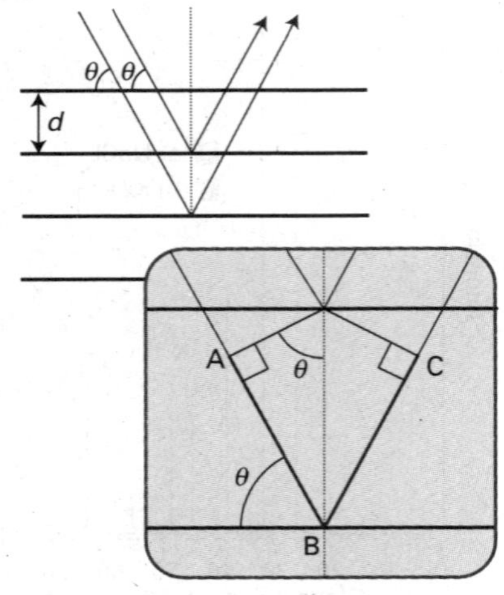

Fig. 19.19 The conventional derivation of Bragg's law treats each lattice plane as a reflecting the incident radiation. The path lengths differ by AB + BC, which depends on the glancing angle, θ. Constructive interference (a 'reflection') occurs when AB + BC is equal to an integer number of wavelengths.

but the other ray must travel an additional distance AB before striking the plane immediately below. Similarly, the reflected rays will differ in path length by a distance BC. The net path length difference of the two rays is then

$$AB + BC = 2d \sin \theta$$

where θ is the **glancing angle.** For many glancing angles the path-length difference is not an integer number of wavelengths, and the waves interfere largely destructively. However, when the path-length difference is an integer number of wavelengths ($AB + BC = n\lambda$), the reflected waves are in phase and interfere constructively. It follows that a reflection should be observed when the glancing angle satisfies **Bragg's law:**

$$n\lambda = 2d \sin \theta \quad \text{Bragg's law} \quad (19.4)$$

Reflections with $n = 2, 3, \ldots$ are called second-order, third-order, and so on; they correspond to path-length differences of 2, 3, . . . wavelengths. In modern work it is normal to absorb the n into d, to write Bragg's law as

$$\lambda = 2d \sin \theta \quad \text{Alternative form of Bragg's law} \quad (19.5)$$

and to regard the nth-order reflection as arising from the $\{nh,nk,nl\}$ planes (see Example 19.1).

The primary use of Bragg's law is in the determination of the spacing between the layers in the lattice for, once the angle θ corresponding to a reflection has been determined, d may readily be calculated.

• A brief illustration

A first-order reflection from the {111} planes of a cubic crystal was observed at a glancing angle of 11.2° when Cu(K_α) X-rays of wavelength 154 pm were used. According to eqn 19.5, the {111} planes responsible for the diffraction have separation $d_{111} = \lambda/2 \sin \theta$. The separation of the {111} planes of a cubic lattice of side a is given by eqn 19.2 as $d_{111} = a/3^{1/2}$. Therefore,

$$a = \frac{3^{1/2}\lambda}{2 \sin \theta} = \frac{3^{1/2} \times (154\ \text{pm})}{2 \sin 11.2°} = 687\ \text{pm} \quad \bullet$$

Self-test 19.2 Calculate the angle at which the same crystal will give a reflection from the {123} planes. [24.8°]

Some types of unit cell give characteristic and easily recognizable patterns of lines. For example, in a cubic lattice of unit cell dimension a the spacing is given by eqn 19.2, so the angles at which the $\{hkl\}$ planes give first-order reflections are given by

$$\sin \theta = (h^2 + k^2 + l^2)^{1/2} \frac{\lambda}{2a}$$

The reflections are then predicted by substituting the values of h, k, and l:

$\{hkl\}$	{100}	{110}	{111}	{200}	{210}	{211}	{220}	{300}	{221}	{310} . . .
$h^2+k^2+l^2$	1	2	3	4	5	6	8	9	9	10 . . .

Notice that 7 (and 15, . . .) is missing because the sum of the squares of three integers cannot equal 7 (or 15, . . .). Therefore the pattern has absences that are characteristic of the cubic P lattice.

Self-test 19.3 Normally, experimental procedures measure 2θ rather than θ itself. A diffraction examination of the element polonium gave lines at the following values of 2θ (in degrees) when 71.0 pm Mo X-rays were used: 12.1, 17.1, 21.0, 24.3, 27.2, 29.9, 34.7, 36.9, 38.9, 40.9, 42.8. Identify the unit cell and determine its dimensions. [cubic P; $a = 337$ pm]

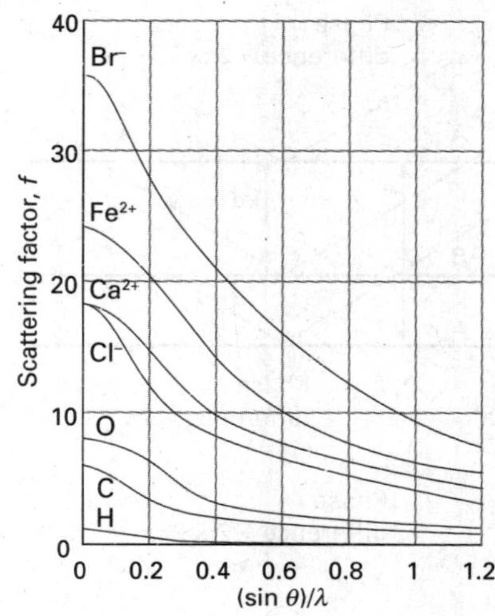

Fig. 19.20 The variation of the scattering factor of atoms and ions with atomic number and angle. The scattering factor in the forward direction (at $\theta = 0$, and hence at $(\sin\theta)/\lambda = 0$) is equal to the number of electrons present in the species.

(c) Scattering factors

To prepare the way to discussing modern methods of structural analysis we need to note that the scattering of X-rays is caused by the oscillations an incoming electromagnetic wave generates in the electrons of atoms, and heavy atoms give rise to stronger scattering than light atoms. This dependence on the number of electrons is expressed in terms of the **scattering factor**, f, of the element. If the scattering factor is large, then the atoms scatter X-rays strongly. The scattering factor of an atom is related to the electron density distribution in the atom, $\rho(r)$, by

$$f = 4\pi\int_0^\infty \rho(r)\frac{\sin kr}{kr}r^2\mathrm{d}r \qquad k = \frac{4\pi}{\lambda}\sin\theta \qquad \text{Scattering factor} \qquad (19.6)$$

The value of f is greatest in the forward direction and smaller for directions away from the forward direction (Fig. 19.20). The detailed analysis of the intensities of reflections must take this dependence on direction into account (in single crystal studies as well as for powders). We show in the following *Justification* that, in the forward direction (for $\theta = 0$), f is equal to the total number of electrons in the atom.

Justification 19.1 *The forward scattering factor*

As $\theta \to 0$, so $k \to 0$. Because $\sin x = x - \frac{1}{6}x^3 + \cdots$,

$$\lim_{x\to 0}\frac{\sin x}{x} = \lim_{x\to 0}\frac{x - \frac{1}{6}x^3 + \cdots}{x} = \lim_{x\to 0}(1 - \tfrac{1}{6}x^2 + \cdots) = 1$$

The factor $(\sin kr)/kr$ is therefore equal to 1 for forward scattering. It follows that in the forward direction

$$f = 4\pi\int_0^\infty \rho(r)r^2\mathrm{d}r$$

The integral over the electron density ρ (the number of electrons in an infinitesimal region divided by the volume of the region) multiplied by the volume element $4\pi r^2\mathrm{d}r$ is the total number of electrons, N_e, in the atom. Hence, in the forward direction, $f = N_e$. For example, the scattering factors of Na^+, K^+, and Cl^- are 10, 18, and 18, respectively. The scattering factor is smaller in non-forward directions because $(\sin kr)/kr < 1$ for $\theta > 0$, so the integral is smaller than the value calculated here.

(d) The electron density

The problem we now address is how to interpret the data from a diffractometer in terms of the detailed structure of a crystal. To do so, we must go beyond Bragg's law.

If a unit cell contains several atoms with scattering factors f_j and coordinates (x_ja, y_jb, z_jc), then we show in the following *Justification* that the overall amplitude of a wave diffracted by the $\{hkl\}$ planes is given by

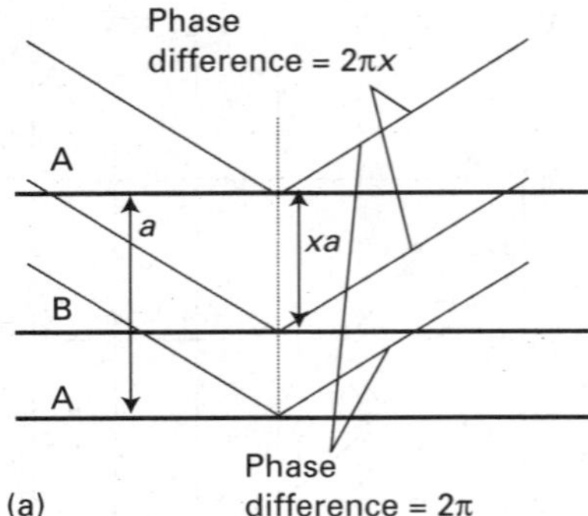

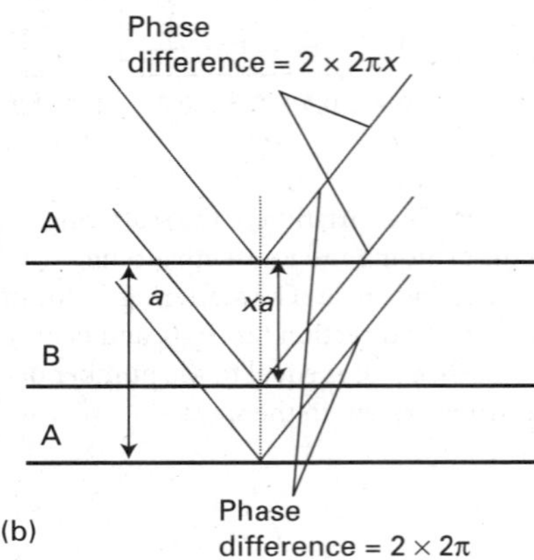

Fig. 19.21 Diffraction from a crystal containing two kinds of atoms. (a) For a (100) reflection from the A planes, there is a phase difference of 2π between waves reflected by neighbouring planes. (b) For a (200) reflection, the phase difference is 4π. The reflection from a B plane at a fractional distance *xa* from an A plane has a phase that is *x* times these phase differences.

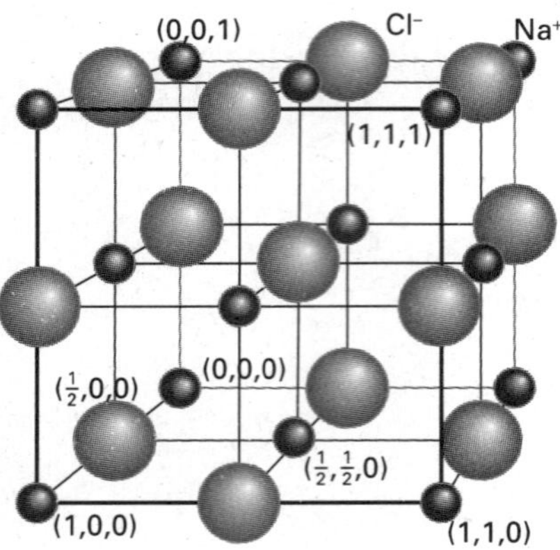

Fig. 19.22 The location of the atoms for the structure factor calculation in Example 19.2. The purple circles are Na^+, the green circles are Cl^-.

$$F_{hkl}=\sum_j f_j e^{i\phi_{hkl}(j)} \quad \text{where} \quad \phi_{hkl}(j)=2\pi(hx_j+ky_j+lz_j)$$ Definition of structure factor [19.7]

where $i=(-1)^{1/2}$. The sum is over all the atoms in the unit cell. The quantity F_{hkl} is called the **structure factor**.

Justification 19.2 *The structure factor*

We begin by showing that, if in the unit cell there is an A atom at the origin and a B atom at the coordinates (xa, yb, zc), where x, y, and z lie in the range 0 to 1, then the phase difference between the *hkl* reflections of the A and B atoms is $\phi_{hkl}=2\pi(hx+ky+lz)$.

Consider the crystal shown schematically in Fig. 19.21. The reflection corresponds to two waves from adjacent A planes, the phase difference of the waves being 2π. If there is a B atom at a fraction x of the distance between the two A planes, then it gives rise to a wave with a phase difference $2\pi x$ relative to an A reflection. To see this conclusion, note that, if $x=0$, there is no phase difference; if $x=\frac{1}{2}$ the phase difference is π; if $x=1$, the B atom lies where the upper A atom is and the phase difference is 2π. Now consider a (200) reflection. There is now a $2\times 2\pi$ difference between the waves from the two A layers, and if B were to lie at $x=0.5$ it would give rise to a wave that differed in phase by 2π from the wave from the lower A layer. Thus, for a general fractional position x, the phase difference for a (200) reflection is $2\times 2\pi x$. For a general $(h00)$ reflection, the phase difference is therefore $h\times 2\pi x$. For three dimensions, this result generalizes to $f_{hkl}=2\pi(hx+ky+lz)$.

If the amplitude of the waves scattered from A is f_A at the detector, that of the waves scattered from B is $f_B e^{i\phi_{hkl}}$, with ϕ_{hkl} as the phase difference given in eqn 19.7. The total amplitude at the detector is therefore

$$F_{hkl}=f_A+f_B e^{i\phi_{hkl}}$$

This expression generalizes to eqn 19.7 when there are several atoms present each with scattering factor f_j.

Example 19.2 *Calculating a structure factor*

Calculate the structure factors for the unit cell in Fig. 19.22.

Method The structure factor is defined by eqn 19.7. To use this equation, consider the ions at the locations specified in Fig. 19.22. Write f^+ for the Na^+ scattering factor and f^- for the Cl^- scattering factor. Note that ions in the body of the cell contribute to the scattering with a strength f. However, ions on faces are shared between two cells (use $\frac{1}{2}f$), those on edges by four cells (use $\frac{1}{4}f$) and those at corners by eight cells (use $\frac{1}{8}f$). Two useful relations are

$$e^{i\pi}=-1 \qquad \cos\phi=\tfrac{1}{2}(e^{i\phi}+e^{-i\phi})$$

Answer From eqn 19.7, and summing over the coordinates of all 27 atoms in the illustration:

$$F_{hkl}=f^+(\tfrac{1}{8}+\tfrac{1}{8}e^{2\pi il}+\cdots+\tfrac{1}{2}e^{2\pi i(\frac{1}{2}h+\frac{1}{2}k+l)})$$
$$+f^-(e^{2\pi i(\frac{1}{2}h+\frac{1}{2}k+\frac{1}{2}l)}+\tfrac{1}{4}e^{2\pi i(\frac{1}{2}h)}+\cdots+\tfrac{1}{4}e^{2\pi i(\frac{1}{2}h+l)})$$

To simplify this 27-term expression, we use

$$e^{2\pi ih}=e^{2\pi ik}=e^{2\pi il}=1$$

because h, k, and l are all integers:

$$F_{hkl} = f^+\{1 + \cos(h+k)\pi + \cos(h+l)\pi + \cos(k+l)\pi\} + f^-\{(-1)^{h+k+l} + \cos k\pi + \cos l\pi + \cos h\pi\}$$

Then, because $\cos h\pi = (-1)^h$

$$F_{hkl} = f^+\{1 + (-1)^{h+k} + (-1)^{h+l} + (-1)^{l+k}\} + f^-\{(-1)^{h+k+l} + (-1)^h + (-1)^k + (-1)^l\}$$

Now note that:

- if h, k, and l are all even, $F_{hkl} = f^+\{1+1+1+1\} + f^-\{1+1+1+1\} = 4(f^+ + f^-)$
- if h, k, and l are all odd, $F_{hkl} = 4(f^+ - f^-)$
- if one index is odd and two are even, or vice versa, $F_{hkl} = 0$

The hkl all-odd reflections are less intense than the hkl all-even. For $f^+ = f^-$, which is the case for identical atoms in a cubic P arrangement, the hkl all-odd have zero intensity.

Self-test 19.4 Which reflections cannot be observed for a cubic I lattice?
[for $h+k+l$ odd, $F_{hkl} = 0$]

Because the intensity is proportional to the square modulus of the amplitude of the wave, the intensity, I_{hkl}, at the detector is

$$I_{hkl} \propto F^\star_{hkl} F_{hkl} = (f_A + f_B e^{-i\phi_{hkl}})(f_A + f_B e^{i\phi_{hkl}})$$

This expression expands to

$$I_{hkl} \propto f_A^2 + f_B^2 + f_A f_B(e^{i\phi_{hkl}} + e^{-i\phi_{hkl}}) = f_A^2 + f_B^2 + 2f_A f_B \cos\phi_{hkl}$$

The cosine term either adds to or subtracts from $f_A^2 + f_B^2$ depending on the value of ϕ_{hkl}, which in turn depends on h, k, and l and x, y, and z. Hence, there is a variation in the intensities of the lines with different hkl. The A and B reflections interfere destructively when the phase difference is π, and the total intensity is zero if the atoms have the same scattering power. For example, if the unit cells are cubic I with a B atom at $x = y = z = \frac{1}{2}$, then the A,B phase difference is $(h+k+l)\pi$. Therefore, all reflections for odd values of $h+k+l$ vanish (as we saw in Self-test 19.4) because the waves are displaced in phase by π. Hence the diffraction pattern for a cubic I lattice can be constructed from that for the cubic P lattice (a cubic lattice without points at the centre of its unit cells) by striking out all reflections with odd values of $h+k+l$. Recognition of these **systematic absences** in a powder spectrum immediately indicates a cubic I lattice (Fig. 19.23).

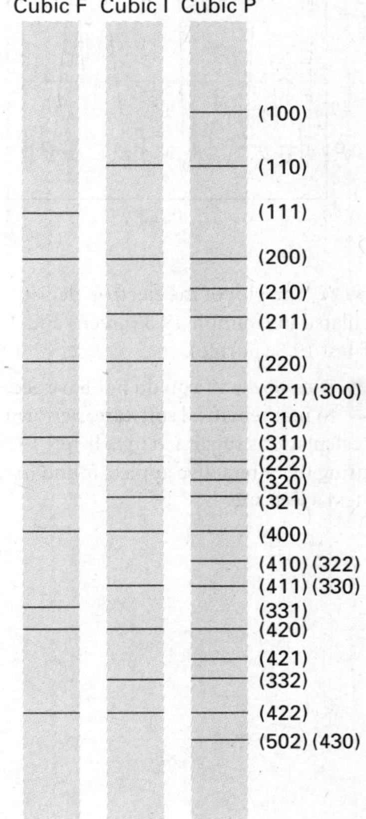

Fig. 19.23 The powder diffraction patterns and the systematic absences of three versions of a cubic cell: cubic F (fcc; h, k, l all even or all odd are present), cubic I (bcc; $h+k+l$ = odd are absent), cubic P. Comparison of the observed pattern with patterns like these enables the unit cell to be identified. The locations of the lines give the cell dimensions.

Because the intensity of the (hkl) reflection is proportional to $|F_{hkl}|^2$, in principle we can determine the structure factors experimentally by taking the square root of the corresponding intensities (but see below). Then, once we know all the structure factors F_{hkl}, we can calculate the electron density distribution, $\rho(r)$, in the unit cell by using the expression

$$\rho(r) = \frac{1}{V}\sum_{hkl} F_{hkl} e^{-2\pi i(hx+ky+lz)} \quad \text{Fourier synthesis of the electron density} \quad (19.8)$$

where V is the volume of the unit cell. Equation 19.8 is called a **Fourier synthesis** of the electron density. (See *Mathematical background 7* which follows this chapter for more information on Fourier series and transforms.)

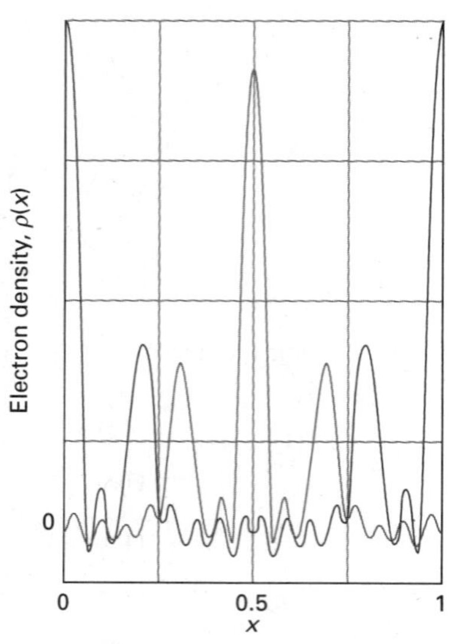

Fig. 19.24 The plot of the electron density calculated in Example 19.3 (green) and Self-test 19.5 (purple).

interActivity If you do not have access to mathematical software, perform the calculations suggested in Self-test 19.5 by using the interactive applets found in the text's web site.

Example 19.3 *Calculating an electron density by Fourier synthesis*

Consider the $\{h00\}$ planes of a crystal extending indefinitely in the x-direction. In an X-ray analysis the structure factors were found as follows:

h:	0	1	2	3	4	5	6	7	8	9
F_h	16	−10	2	−1	7	−10	8	−3	2	−3
h:	10	11	12	13	14	15				
F_h	6	−5	3	−2	2	−3				

(and $F_{-h} = F_h$). Construct a plot of the electron density projected on to the x-axis of the unit cell.

Method Because $F_{-h} = F_h$, it follows from eqn 19.8 that

$$V\rho(x) = \sum_{h=-\infty}^{\infty} F_h e^{-2\pi i h x} = F_0 + \sum_{h=1}^{\infty} (F_h e^{-2\pi i h x} + F_{-h} e^{2\pi i h x})$$

$$= F_0 + \sum_{h=1}^{\infty} F_h (e^{-2\pi i h x} + e^{2\pi i h x}) = F_0 + 2\sum_{h=1}^{\infty} F_h \cos 2\pi h x$$

and we evaluate the sum (truncated at $h = 15$) for points $0 < x < 1$ using mathematical software.

Answer The results are plotted in Fig. 19.24 (green line). The positions of three atoms can be discerned very readily. The more terms there are included, the more accurate the density plot. Terms corresponding to high values of h (short wavelength cosine terms in the sum) account for the finer details of the electron density; low values of h account for the broad features.

Self-test 19.5 Use mathematical software to experiment with different structure factors (including changing signs as well as amplitudes). For example, use the same values of F_h as above, but with positive signs for all values of h.

[Fig. 19.24 (purple line)]

(e) The phase problem

A problem with the procedure outlined so far is that, as we have seen, the observed intensity I_{hkl} is proportional to the square modulus $|F_{hkl}|^2$, so we cannot say whether we should use $+|F_{hkl}|$ or $-|F_{hkl}|$ in the sum in eqn 19.8. In fact, the difficulty is more severe for non-centrosymmetric unit cells because, if we write F_{hkl} as the complex number $|F_{hkl}|e^{i\alpha}$ where α is the phase of F_{hkl} and $|F_{hkl}|$ is its magnitude, then the intensity lets us determine $|F_{hkl}|$ but tells us nothing of its phase, which may lie anywhere from 0 to 2π. This ambiguity is called the **phase problem**; its consequences are illustrated by comparing the two plots in Fig. 19.24. Some way must be found to assign phases to the structure factors, for otherwise the sum for ρ cannot be evaluated and the method would be useless.

The phase problem can be overcome to some extent by a variety of methods. One procedure that is widely used for inorganic materials with a reasonably small number of atoms in a unit cell and for organic molecules with a small number of heavy atoms is the **Patterson synthesis**. Instead of the structure factors F_{hkl}, the values of $|F_{hkl}|^2$, which can be obtained without ambiguity from the intensities, are used in an expression that resembles eqn 19.8:

$$P(\boldsymbol{r}) = \frac{1}{V}\sum_{hkl} |F_{hkl}|^2 e^{-2\pi i(hx+ky+lz)}$$ Patterson synthesis (19.9)

The outcome of a Patterson synthesis is a map of the vector separations of the atoms (the distances and directions between atoms) in the unit cell. Thus, if atom A is at the coordinates (x_A, y_A, z_A) and atom B is at (x_B, y_B, z_B), then there will be a peak at $(x_A - x_B, y_A - y_B, z_A - z_B)$ in the Patterson map. There will also be a peak at the negative of these coordinates, because there is a vector from B to A as well as a vector from A to B. The height of the peak in the map is proportional to the product of the atomic numbers of the two atoms, Z_AZ_B. For example, if the unit cell has the structure shown in Fig. 19.25a, the Patterson synthesis would be the map shown in Fig. 19.25b, where the location of each spot relative to the origin gives the separation and relative orientation of each pair of atoms in the original structure.

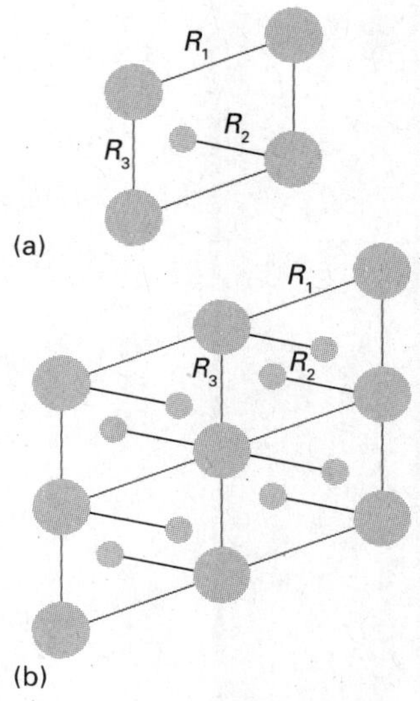

Fig. 19.25 The Patterson synthesis corresponding to the pattern in (a) is the pattern in (b). The distance and orientation of each spot from the origin gives the orientation and separation of one atom–atom separation in (a). Some of the typical distances and their contribution to (b) are shown as R_1, etc.

Heavy atoms dominate the scattering because their scattering factors are large, of the order of their atomic numbers, and their locations may be deduced quite readily. The sign of F_{hkl} can then be calculated from the locations of the heavy atoms in the unit cell, and to a high probability the phase calculated for them will be the same as the phase for the entire unit cell. To see why this is so, we have to note that a structure factor of a centrosymmetric cell has the form

$$F = (\pm)f_{\text{heavy}} + (\pm)f_{\text{light}} + (\pm)f_{\text{light}} + \cdots \qquad (19.10)$$

where f_{heavy} is the scattering factor of the heavy atom and f_{light} the scattering factors of the light atoms. The f_{light} are all much smaller than f_{heavy}, and their phases are more or less random if the atoms are distributed throughout the unit cell. Therefore, the net effect of the f_{light} is to change F only slightly from f_{heavy}, and we can be reasonably confident that F will have the same sign as that calculated from the location of the heavy atom. This phase can then be combined with the observed $|F|$ (from the reflection intensity) to perform a Fourier synthesis of the full electron density in the unit cell, and hence to locate the light atoms as well as the heavy atoms.

Modern structural analyses make extensive use of **direct methods**. Direct methods are based on the possibility of treating the atoms in a unit cell as being virtually randomly distributed (from the radiation's point of view), and then using statistical techniques to compute the probabilities that the phases have a particular value. It is possible to deduce relations between some structure factors and sums (and sums of squares) of others, which have the effect of constraining the phases to particular values (with high probability, so long as the structure factors are large). For example, the **Sayre probability relation** has the form

sign of $F_{h+h',k+k',l+l'}$ is probably equal to (sign of F_{hkl}) × (sign of $F_{h'k'l'}$) Sayre probability relation (19.11)

For example, if F_{122} and F_{232} are both large and negative, then it is highly likely that F_{354}, provided it is large, will be positive.

(f) Structure refinement

In the final stages of the determination of a crystal structure, the parameters describing the structure (atom positions, for instance) are adjusted systematically to give the best fit between the observed intensities and those calculated from the model of the structure deduced from the diffraction pattern. This process is called **structure refinement**. Not only does the procedure give accurate positions for all the atoms in the unit cell, but it also gives an estimate of the errors in those positions and in the bond lengths and angles derived from them. The procedure also provides information on the vibrational amplitudes of the atoms.

19.4 Neutron and electron diffraction

Key point **Neutrons generated in a nuclear reactor and then slowed to thermal velocities have wavelengths similar to those of X-rays and may also be used for diffraction studies of solids.**

According to the de Broglie relation (eqn 7.16, $\lambda = h/p$), particles have wavelengths and may therefore undergo diffraction. Neutrons generated in a nuclear reactor and then slowed to thermal velocities have wavelengths similar to those of X-rays and may also be used for diffraction studies. For instance, a neutron generated in a reactor and slowed to thermal velocities by repeated collisions with a moderator (such as graphite) until it is travelling at about 4 km s^{-1} has a wavelength of about 100 pm. In practice, a range of wavelengths occurs in a neutron beam, but a monochromatic beam can be selected by diffraction from a crystal, such as a single crystal of germanium.

Example 19.4 *Calculating the typical wavelength of thermal neutrons*

Calculate the typical wavelength of neutrons that have reached thermal equilibrium with their surroundings at 373 K.

Method We need to relate the wavelength to the temperature. There are two linking steps. First, the de Broglie relation expresses the wavelength in terms of the linear momentum. Then the linear momentum can be expressed in terms of the kinetic energy, the mean value of which is given in terms of the temperature by the equipartition theorem (see Section 16.3).

Answer From the equipartition principle, we know that the mean translational kinetic energy of a neutron at a temperature T travelling in the x-direction is $E_k = \frac{1}{2}kT$. The kinetic energy is also equal to $p^2/2m$, where p is the momentum of the neutron and m is its mass. Hence, $p = (mkT)^{1/2}$. It follows from the de Broglie relation $\lambda = h/p$ that the neutron's wavelength is

$$\lambda = \frac{h}{(mkT)^{1/2}}$$

Therefore, at 373 K,

$$\lambda = \frac{6.626 \times 10^{-34}\ \text{J s}}{\{(1.675 \times 10^{-27}\ \text{kg}) \times (1.381 \times 10^{-23}\ \text{J K}^{-1}) \times (373\ \text{K})\}^{1/2}}$$

$$= \frac{6.626 \times 10^{-34}\ \text{J s}}{(1.675 \times 1.381 \times 373 \times 10^{-50})^{1/2}(\text{kg}^2\ \text{m}^2\ \text{s}^{-2})^{1/2}}$$

$$= 2.26 \times 10^{-10}\ \text{m} = 226\ \text{pm}$$

where we have used 1 J = 1 kg m^2 s^{-2}.

Self-test 19.6 Calculate the temperature needed for the average wavelength of the neutrons to be 100 pm. [1.90×10^3 K]

Neutron diffraction differs from X-ray diffraction in two main respects. First, the scattering of neutrons is a nuclear phenomenon. Neutrons pass through the extranuclear electrons of atoms and interact with the nuclei through the 'strong force' that is responsible for binding nucleons together. As a result, the intensity with which neutrons are scattered is independent of the number of electrons and neighbouring elements in the periodic table may scatter neutrons with markedly different intensities. Neutron diffraction can be used to distinguish atoms of elements such as Ni and Co that are present in the same compound and to study order–disorder phase transitions in FeCo. A second difference is that neutrons possess a magnetic moment due to their spin. This magnetic moment can couple to the magnetic fields of atoms or ions in a crystal (if the ions have unpaired electrons) and modify the diffraction pattern. One consequence is that neutron diffraction is well suited to the investigation of magnetically ordered lattices in which neighbouring atoms may be of the same element but have different orientations of their electronic spin (Fig. 19.26).

Electrons accelerated from rest through a potential difference of 40 kV have wavelengths of about 6 pm, and so are also suitable for diffraction studies. However, their main application is to the study of surfaces.

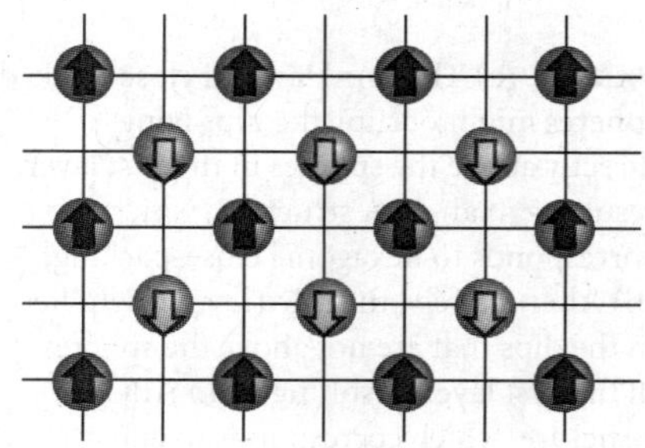

Fig. 19.26 If the spins of atoms at lattice points are orderly, as in this material, where the spins of one set of atoms are aligned antiparallel to those of the other set, neutron diffraction detects two interpenetrating simple cubic lattices on account of the magnetic interaction of the neutron with the atoms, but X-ray diffraction would see only a single bcc lattice.

19.5 Metallic solids

Key points **(a) Many elemental metals have close-packed structures with coordination number 12; close-packed structures are commonly either cubic (ccp) or hexagonal (hcp). (b) The departure from close packing suggests that such factors as specific covalent bonding between neighbouring atoms influence the structure of a solid.**

We now turn to the information that has been obtained from these various diffraction measurements. The bonding within a solid may be of various kinds. Simplest of all (in principle) are elemental metals, where electrons are delocalized over arrays of identical cations and bind them together into a rigid but ductile and malleable whole. Most metallic elements crystallize in one of three simple forms, two of which can be explained in terms of hard spheres packing together in the closest possible arrangement.

(a) Close packing

Figure 19.27 shows a **close-packed** layer of identical spheres, one with maximum utilization of space. A close-packed three-dimensional structure is obtained by stacking such close-packed layers on top of one another. However, this stacking can be done in different ways, which result in close-packed **polytypes**, or structures that are identical in two dimensions (the close-packed layers) but differ in the third dimension.

In all polytypes, the spheres of the second close-packed layer lie in the depressions of the first layer (Fig. 19.28). The third layer may be added in either of two ways. In one, the spheres are placed so that they reproduce the first layer (Fig. 19.29a), to give an ABA pattern of layers. Alternatively, the spheres may be placed over the gaps in the

Fig. 19.27 The first layer of close-packed spheres used to build a three-dimensional close-packed structure.

Fig. 19.28 The second layer of close-packed spheres occupies the dips of the first layer. The two layers are the AB component of the close-packed structure.

Fig. 19.29 (a) The third layer of close-packed spheres might occupy the dips lying directly above the spheres in the first layer, resulting in an ABA structure, which corresponds to hexagonal close-packing. (b) Alternatively, the third layer might lie in the dips that are not above the spheres in the first layer, resulting in an ABC structure, which corresponds to cubic close-packing.

Table 19.2 The crystal structures of some elements

Structure	Element
hcp*	Be, Cd, Co, He, Mg, Sc, Ti, Zn
fcc* (ccp, cubic F)	Ag, Al, Ar, Au, Ca, Cu, Kr, Ne, Ni, Pd, Pb, Pt, Rh, Rn, Sr, Xe
bcc (cubic I)	Ba, Cs, Cr, Fe, K, Li, Mn, Mo, Rb, Na, Ta, W, V
cubic P	Po

* Close-packed structures.

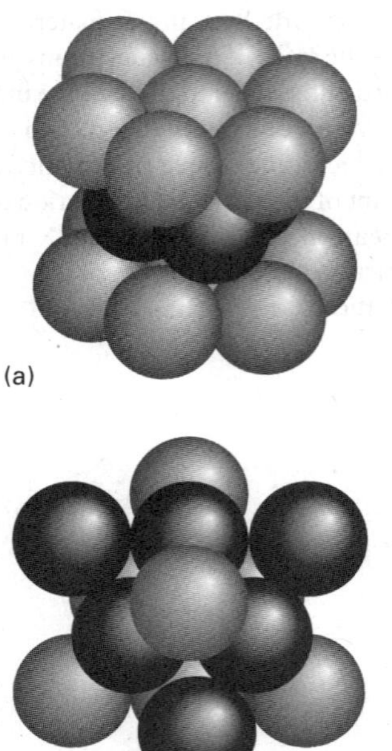

Fig. 19.30 A fragment of the structure shown in Fig. 19.29 revealing the (a) hexagonal (b) cubic symmetry. The tints on the spheres are the same as for the layers in Fig. 19.29.

first layer (Fig. 19.29b), so giving an ABC pattern. Two polytypes are formed if the two stacking patterns are repeated in the vertical direction. If the ABA pattern is repeated, to give the sequence of layers ABABAB . . ., the spheres are **hexagonally close-packed** (hcp). Alternatively, if the ABC pattern is repeated, to give the sequence ABCABC . . . , the spheres are **cubic close-packed** (ccp). We can see the origins of these names by referring to Fig. 19.30. The cubic close-packed (ccp) structure belongs to the cubic P Bravais lattice, from which one can construct a face-centred cubic (fcc) unit cell. It is also possible to have random sequences of layers; however, the hcp and ccp polytypes are the most important. Table 19.2 lists some elements possessing these structures.

The compactness of close-packed structures is indicated by their **coordination number**, the number of atoms immediately surrounding any selected atom, which is 12 in all cases. Another measure of their compactness is the **packing fraction**, the fraction of space occupied by the spheres, which is 0.740 (see the following *Justification*). That is, in a close-packed solid of identical hard spheres, only 26.0 per cent of the volume is empty space. The fact that many metals are close-packed accounts for their high densities.

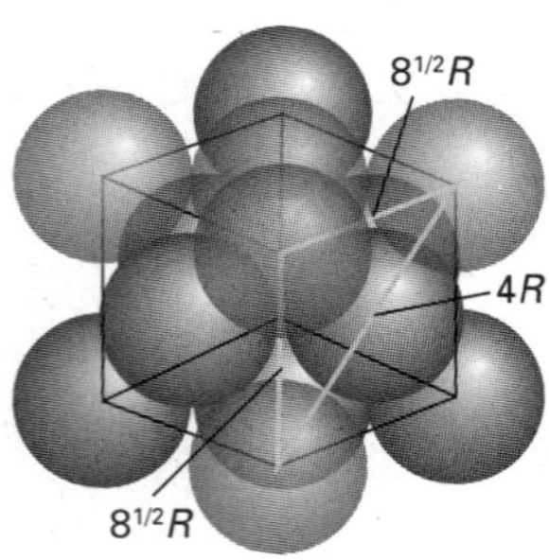

Fig. 19.31 The calculation of the packing fraction of a ccp unit cell.

Justification 19.3 *The packing fraction*

To calculate a packing fraction of a ccp structure, we first calculate the volume of a unit cell, and then calculate the total volume of the spheres that fully or partially occupy it. The first part of the calculation is a straightforward exercise in geometry. The second part involves counting the fraction of spheres that occupy the cell.

Refer to Fig. 19.31. Because a diagonal of any face passes completely through one sphere and halfway through two other spheres, its length is $4R$. The length of a side is therefore $8^{1/2}R$ and the volume of the unit cell is $8^{3/2}R^3$. Because each cell contains the equivalent of $6\times\frac{1}{2}+8\times\frac{1}{8}=4$ spheres, and the volume of each sphere is $\frac{4}{3}\pi R^3$, the total occupied volume is $\frac{16}{3}\pi R^3$. The fraction of space occupied is therefore $\frac{16}{3}\pi R^3/8^{3/2}R^3=\frac{16}{3}\pi/8^{3/2}$, or 0.740. Because an hcp structure has the same coordination number, its packing fraction is the same. The packing fractions of structures that are not close-packed are calculated similarly (see Exercises 19.20 and 19.21).

(b) Less closely packed structures

As shown in Table 19.2, a number of common metals adopt structures that are less than close-packed. The departure from close packing suggests that factors such as specific covalent bonding between neighbouring atoms are beginning to influence the structure and impose a specific geometrical arrangement. One such arrangement results in a cubic I (bcc, for body-centred cubic) structure, with one sphere at the centre of a cube formed by eight others. The coordination number of a bcc structure is only 8, but there are six more atoms not much further away than the eight nearest neighbours. The packing fraction of 0.68 is not much smaller than the value for a close-packed structure (0.74), and shows that about two-thirds of the available space is actually occupied.

19.6 Ionic solids

Key points **(a) Representative ionic structures include the caesium-chloride, rock-salt, and zinc-blende structures. The radius-ratio rule may be used cautiously to predict which of these three structures is likely. (b) The lattice enthalpy is the change in enthalpy accompanying the complete separation of the components of the solid. The lattice enthalpy is expressed by the Born–Mayer equation. Experimental values of the lattice enthalpy are obtained by using a Born–Haber cycle.**

Two questions arise when we consider ionic solids: the relative locations adopted by the ions and the energetics of the resulting structure.

(a) Structure

When crystals of compounds of monatomic ions (such as NaCl and MgO) are modelled by stacks of hard spheres it is essential to allow for the different ionic radii (typically with the cations smaller than the anions) and different charges. The **coordination number** of an ion is the number of nearest neighbours of opposite charge; the structure itself is characterized as having **(n_+,n_-)-coordination**, where n_+ is the coordination number of the cation and n_- that of the anion.

Even if, by chance, the ions have the same size, the problems of ensuring that the unit cells are electrically neutral makes it impossible to achieve 12-coordinate close-packed ionic structures. As a result, ionic solids are generally less dense than metals. The best packing that can be achieved is the (8,8)-coordinate **caesium-chloride structure** in which each cation is surrounded by eight anions and each anion is surrounded by eight cations (Fig. 19.32). In this structure, an ion of one charge occupies the centre of a cubic unit cell with eight counter ions at its corners. The structure is adopted by CsCl itself and also by CaS, CsCN (with some distortion), and CsAu.

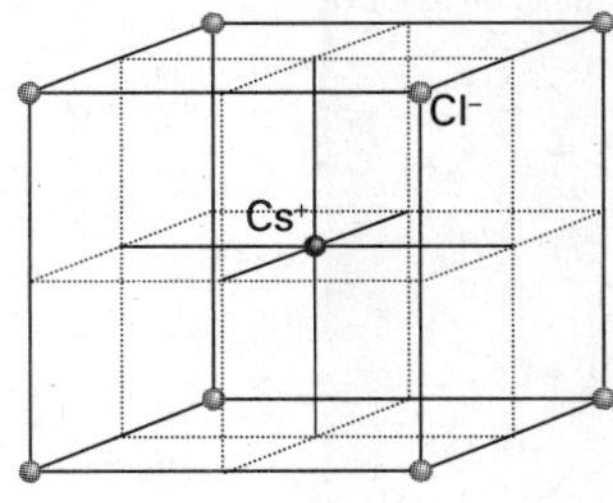

Fig. 19.32 The caesium-chloride structure consists of two interpenetrating simple cubic arrays of ions, one of cations and the other of anions, so that each cube of ions of one kind has a counter ion at its centre.

When the radii of the ions differ more than in CsCl, even eight-coordinate packing cannot be achieved. One common structure adopted is the (6,6)-coordinate **rock-salt structure** typified by NaCl (Fig. 19.33). In this structure, each cation is surrounded by six anions and each anion is surrounded by six cations. The rock-salt structure can be pictured as consisting of two interpenetrating slightly expanded cubic F (fcc) arrays, one composed of cations and the other of anions. This structure is adopted by NaCl itself and also by several other MX compounds, including KBr, AgCl, MgO, and ScN.

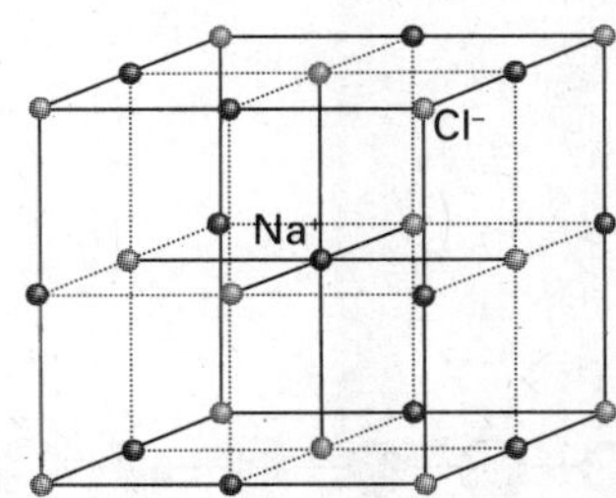

Fig. 19.33 The rock-salt (NaCl) structure consists of two mutually interpenetrating slightly expanded face-centred cubic arrays of ions. The entire assembly shown here is the unit cell.

The switch from the caesium-chloride structure to the rock-salt structure is related to the value of the **radius ratio**, γ:

$$\gamma = \frac{r_{\text{smaller}}}{r_{\text{larger}}} \quad \text{Definition of the radius ratio} \quad [19.12]$$

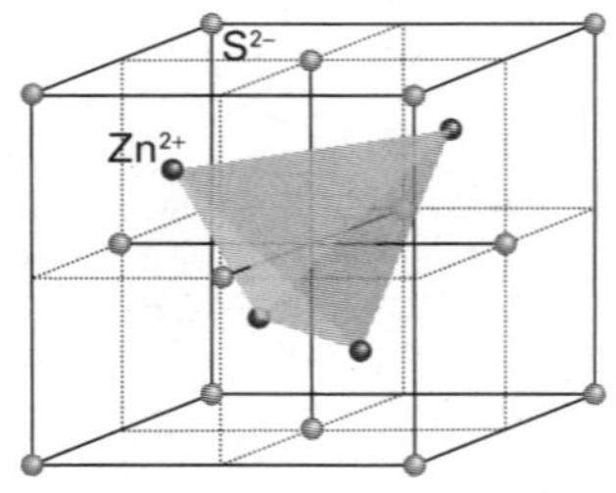

Fig. 19.34 The structure of the sphalerite form of ZnS showing the location of the Zn ions in the tetrahedral holes formed by the array of S ions. (There is an S ion at the centre of the cube inside the tetrahedron of Zn ions.)

Table 19.3* Ionic radii, r/pm

Na^+	102 (6†), 116 (8)
K^+	138 (6), 151 (8)
F^-	128 (2), 131 (4)
Cl^-	181 (close packing)

* More values are given in the *Data section*.
† Coordination number.

The two radii are those of the larger and smaller ions in the crystal. The **radius-ratio rule**, which is derived by considering the geometrical problem of packing the maximum number of hard spheres of one radius around a hard sphere of a different radius, can be summarized as follows:

Radius ratio	**Structural type**
$\gamma < 2^{1/2} - 1 = 0.414$	Sphalerite
$2^{1/2} - 1 = 0.414 < \gamma < 0.732$	Rock-salt
$\gamma > 3^{1/2} - 1 = 0.732$	Caesium-chloride

The sphalerite (or zinc-blende) structure is shown in Fig. 19.34. The deviation of a structure from that expected on the basis of the radius-ratio rule is often taken to be an indication of a shift from ionic towards covalent bonding; however, a major source of unreliability is the arbitrariness of ionic radii and their variation with coordination number.

Ionic radii are derived from the distance between centres of adjacent ions in a crystal. However, we need to apportion the total distance between the two ions by defining the radius of one ion and then inferring the radius of the other ion. One scale that is widely used is based on the value 140 pm for the radius of the O^{2-} ion (Table 19.3). Other scales are also available (such as one based on F^- for discussing halides), and it is essential not to mix values from different scales. Because ionic radii are so arbitrary, predictions based on them must be viewed cautiously.

(b) Energetics

The **lattice energy** of a solid is the difference in potential energy of the ions packed together in a solid and widely separated as a gas. The lattice energy is always positive; a high lattice energy indicates that the ions interact strongly with one another to give a tightly bonded solid. The **lattice enthalpy**, ΔH_L, is the change in standard molar enthalpy for the process

$$MX(s) \rightarrow M^+(g) + X^-(g)$$

and its equivalent for other charge types and stoichiometries. The lattice enthalpy is equal to the lattice energy at $T = 0$; at normal temperatures they differ by only a few kilojoules per mole, and the difference is normally neglected.

Each ion in a solid experiences electrostatic attractions from all the other oppositely charged ions and repulsions from all the other like-charged ions. The total Coulombic potential energy is the sum of all the electrostatic contributions. Each cation is surrounded by anions, and there is a large negative contribution from the attraction of the opposite charges. Beyond those nearest neighbours, there are cations that contribute a positive term to the total potential energy of the central cation. There is also a negative contribution from the anions beyond those cations, a positive contribution from the cations beyond them, and so on to the edge of the solid. These repulsions and attractions become progressively weaker as the distance from the central ion increases, but the net outcome of all these contributions is a lowering of energy.

First, consider a simple one-dimensional model of a solid consisting of a long line of uniformly spaced alternating cations and anions, with d the distance between their centres, the sum of the ionic radii (Fig. 19.35). If the charge numbers of the ions have the same absolute value (+1 and −1, or +2 and −2, for instance), then $z_1 = +z$, $z_2 = -z$, and $z_1z_2 = -z^2$. The potential energy of the central ion is calculated by summing all the terms, with negative terms representing attractions to oppositely charged ions and positive terms representing repulsions from like-charged ions. Suppose the central ion is a cation; then the potential energy of its interaction with ions extending in a line to the right is

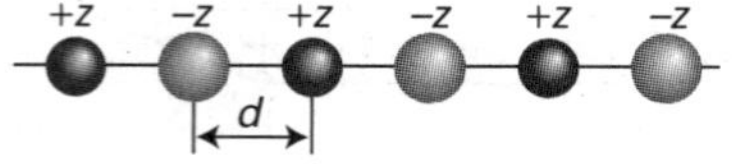

Fig. 19.35 A line of alternating cations and ions used in the calculation of the Madelung constant in one dimension.

$$E_p(\text{cation}) = \frac{1}{4\pi\varepsilon_0} \times \left(-\frac{z^2e^2}{d} + \frac{z^2e^2}{2d} - \frac{z^2e^2}{3d} + \frac{z^2e^2}{4d} - \cdots \right)$$

$$= -\frac{z^2e^2}{4\pi\varepsilon_0 d}\left(1 - \frac{1}{2} + \frac{1}{3} - \frac{1}{4} + \cdots\right)$$

$$= -\frac{z^2e^2}{4\pi\varepsilon_0 d} \times \ln 2$$

We have used the relation $1 - \frac{1}{2} + \frac{1}{3} - \frac{1}{4} + \cdots = \ln 2$. Next, we multiply E_p by 2 to obtain the total energy arising from interactions on each side of the ion to obtain

$$E_p(\text{cation}) = -2 \ln 2 \times \frac{z^2e^2}{4\pi\varepsilon_0 d}$$

with $d = r_{\text{cation}} + r_{\text{anion}}$. This energy is negative, corresponding to a net attraction. The same expression applies to the potential energy of the neighbouring anion, $E_p(\text{anion})$. The total potential energy per mole of ion pairs is therefore

$$E_p = \tfrac{1}{2}N_A\{E_p(\text{cation}) + E_p(\text{anion})\} = -2 \ln 2 \times \frac{z^2N_Ae^2}{4\pi\varepsilon_0 d}$$

We have introduced the factor of $\frac{1}{2}$ to avoid double counting the interaction (Jack with Jill, Jill with Jack, etc.). This calculation can be extended to three-dimensional arrays of ions with different charges:

$$E_p = -A \times \frac{|z_Az_B|N_Ae^2}{4\pi\varepsilon_0 d} \qquad (19.13)$$

Potential energy in terms of the Madelung constant

The factor A is a positive numerical constant called the **Madelung constant**; its value depends on how the ions are arranged about one another. For ions arranged in the same way as in sodium chloride, $A = 1.748$. Table 19.4 lists Madelung constants for other common structures.

Table 19.4 Madelung constants

Structural type	A
Caesium chloride	1.763
Fluorite	2.519
Rock salt	1.748
Rutile	2.408
Sphalerite	1.638
Wurtzite	1.641

There are also repulsions arising from the overlap of the atomic orbitals of the ions and the role of the Pauli principle. These repulsions are taken into account by supposing that, because wavefunctions decay exponentially with distance at large distances from the nucleus, and repulsive interactions depend on the overlap of orbitals, the repulsive contribution to the potential energy has the form

$$E_p^* = N_AC'e^{-d/d^*} \qquad (19.14)$$

with C' and d^* constants; the latter is commonly taken to be 34.5 pm. The total potential energy is the sum of E_p and E_p^* and passes through a minimum when $d(E_p + E_p^*)/dd = 0$ (Fig. 19.36). A short calculation leads to the following expression for the minimum total potential energy (see Problem 19.27):

$$E_{p,\min} = -\frac{N_A|z_Az_B|e^2}{4\pi\varepsilon_0 d}\left(1 - \frac{d^*}{d}\right)A \qquad (19.15)$$

Born–Mayer equation

This expression is called the **Born–Mayer equation**. Provided we ignore zero-point contributions to the energy, we can identify the negative of this potential energy with the lattice energy. We see that large lattice energies are expected when the ions are highly charged (so $|z_Az_B|$ is large) and small (so d is small).

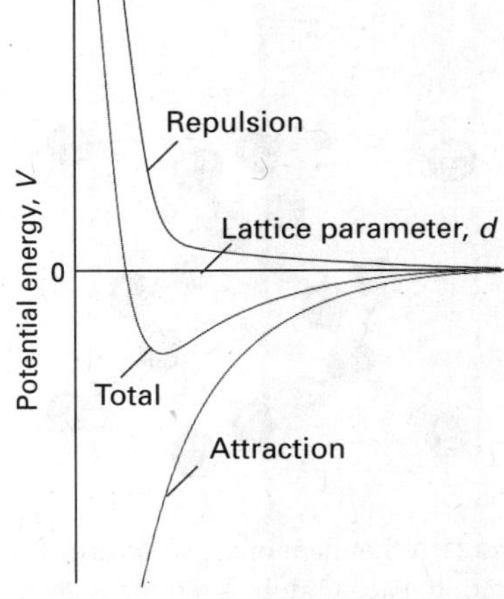

Fig. 19.36 The contributions to the total potential energy of an ionic crystal.

Experimental values of the lattice enthalpy (the enthalpy, rather than the energy) are obtained by using a **Born–Haber cycle**, a closed path of transformations starting and ending at the same point, one step of which is the formation of the solid compound

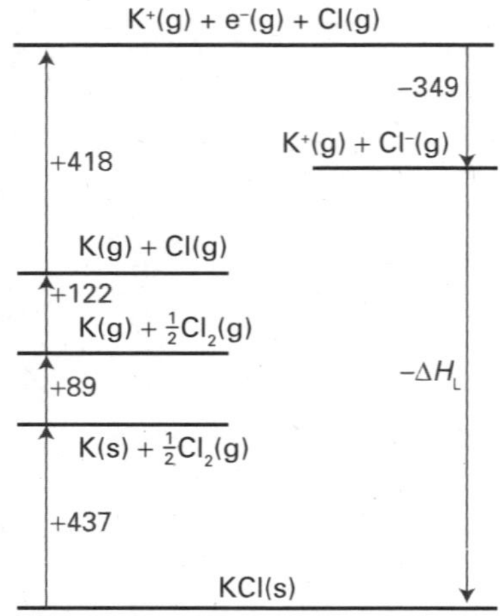

Fig. 19.37 The Born–Haber cycle for KCl at 298 K. Enthalpy changes are in kilojoules per mole.

Table 19.5* Lattice enthalpies at 298 K, $\Delta H_L/(\text{kJ mol}^{-1})$

NaF	787
NaBr	751
MgO	3850
MgS	3406

* More values are given in the *Data section*.

from a gas of widely separated ions. A typical cycle, for potassium chloride, is shown in Fig. 19.37. It consists of the following steps (for convenience, starting at the elements):

	$\Delta H/(\text{kJ mol}^{-1})$	
1. Sublimation of K(s)	+89	[dissociation enthalpy of K(s)]
2. Dissociation of $\frac{1}{2}$ $Cl_2(g)$	+122	[$\frac{1}{2}\times$ dissociation enthalpy of $Cl_2(g)$]
3. Ionization of K(g)	+418	[ionization enthalpy of K(g)]
4. Electron attachment to Cl(g)	−349	[electron gain enthalpy of Cl(g)]
5. Formation of solid from gas	$-\Delta H_L/(\text{kJ mol}^{-1})$	
6. Decomposition of compound	+437	[negative of enthalpy of formation of KCl(s)]

Because the sum of these enthalpy changes is equal to zero, we can infer from

$$89 + 122 + 418 - 349 - \Delta H_L/(\text{kJ mol}^{-1}) + 437 = 0$$

that $\Delta H_L = +717$ kJ mol^{-1}. Some lattice enthalpies obtained in this way are listed in Table 19.5. As can be seen from the data, the trends in values are in general accord with the predictions of the Born–Mayer equation. Agreement is typically taken to imply that the ionic model of bonding is valid for the substance; disagreement implies that there is a covalent contribution to the bonding. It is important, though, to be cautious, because numerical agreement might be coincidental.

19.7 Molecular solids and covalent networks

Key points **A covalent network solid is a solid in which covalent bonds in a definite spatial orientation link the atoms in a network extending through the crystal. A molecular solid is a solid consisting of discrete molecules held together by van der Waals interactions and, in certain cases, hydrogen bonding.**

X-ray diffraction studies of solids reveal a huge amount of information, including interatomic distances, bond angles, stereochemistry, and vibrational parameters. In this section we can do no more than hint at the diversity of types of solids found when molecules pack together or atoms link together in extended networks.

In **covalent network solids**, covalent bonds in a definite spatial orientation link the atoms in a network extending through the crystal. The demands of directional bonding, which have only a small effect on the structures of many metals, now override the geometrical problem of packing spheres together, and elaborate and extensive structures may be formed. Examples include silicon, red phosphorus, boron nitride, and—very importantly—diamond, graphite, and carbon nanotubes, which we discuss in detail.

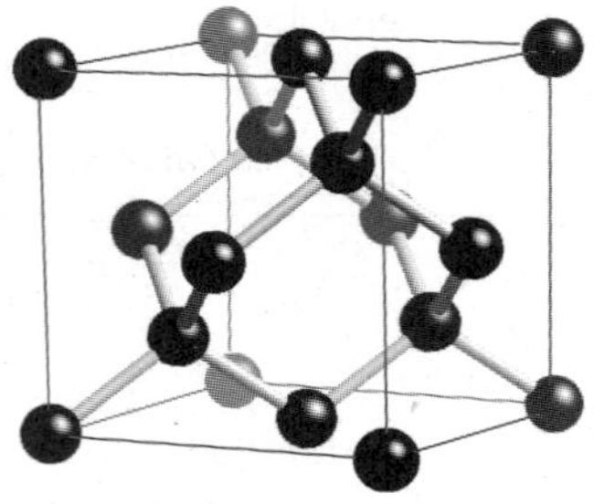

Fig. 19.38 A fragment of the structure of diamond. Each C atom is tetrahedrally bonded to four neighbours. This framework-like structure results in a rigid crystal.

Diamond and graphite are two *allotropes*, distinct forms of an element that differ in the way that atoms are linked, of carbon. In diamond each sp^3-hybridized carbon is bonded tetrahedrally to its four neighbours (Fig. 19.38). The network of strong C–C bonds is repeated throughout the crystal and, as a result, diamond is the hardest known substance.

In graphite, σ bonds between sp^2-hybridized carbon atoms form hexagonal rings which, when repeated throughout a plane, give rise to graphene sheets (Fig. 19.39). Because the sheets can slide against each other when impurities are present, graphite is used widely as a lubricant.

Carbon nanotubes are thin cylinders of carbon atoms that are both mechanically strong and highly conducting (see *Impact I19.2*). They are synthesized by condensing

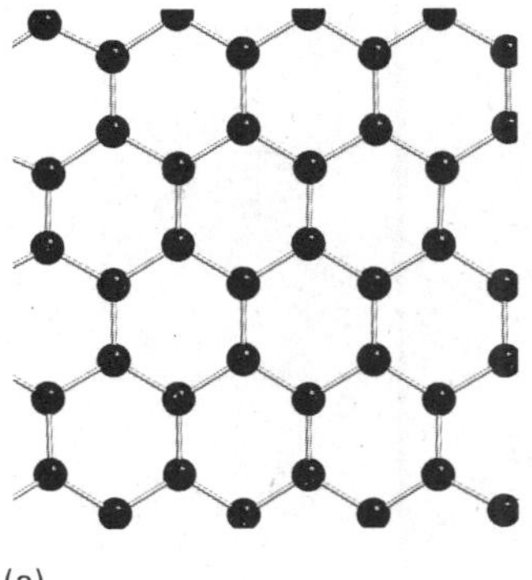

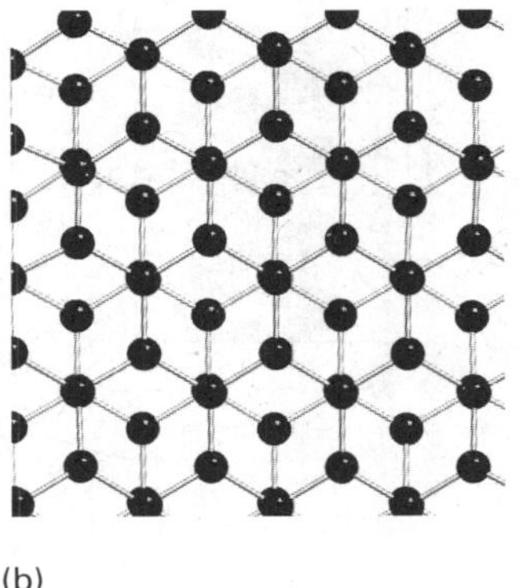

Fig. 19.39 Graphite consists of flat planes of hexagons of carbon atoms lying above one another. (a) The arrangement of carbon atoms in a sheet; (b) the relative arrangement of neighbouring sheets. When impurities are present, the planes can slide over one another easily.

a carbon plasma either in the presence or absence of a catalyst. The simplest structural motif is called a *single-walled nanotube* (SWNT) and is shown in Fig. 19.40. In a SWNT, sp^2-hybridized carbon atoms form hexagonal rings reminiscent of the structure of the carbon sheets found in graphite. The tubes have diameters between 1 and 2 nm and lengths of several micrometres. The features shown in Fig. 19.40 have been confirmed by direct visualization with scanning tunnelling microscopy (*Impact I8.2*). A *multi-walled nanotube* (MWNT) consists of several concentric SWNTs and its diameter varies between 2 and 25 nm.

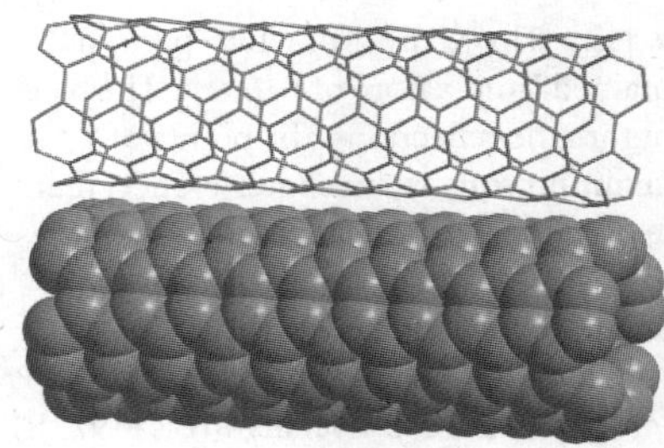

Fig. 19.40 In a single-walled nanotube (SWNT), sp^2-hybridized carbon atoms form hexagonal rings that grow as tubes with diameters between 1 and 2 nm and lengths of several micrometres.

Molecular solids, which are the subject of the overwhelming majority of modern structural determinations, are held together by van der Waals interactions and, in certain cases, hydrogen bonding (Chapter 17). The observed crystal structure is Nature's solution to the problem of condensing objects of various shapes into an aggregate of minimum energy (actually, for $T > 0$, of minimum Gibbs energy). The prediction of the structure is a very difficult task, but software specifically designed to explore interaction energies can now make reasonably reliable predictions. The problem is made more complicated by the role of hydrogen bonds, which in some cases dominate the crystal structure, as in ice (Fig. 19.41), but in others (for example, in phenol) distort a structure that is determined largely by the van der Waals interactions.

IMPACT ON BIOCHEMISTRY
I19.1 X-ray crystallography of biological macromolecules

X-ray crystallography is the deployment of X-ray diffraction techniques for the determination of the location of all the atoms in molecules as complicated as biopolymers. Bragg's law helps us understand the features of one of the most seminal X-ray images of all, the characteristic X-shaped pattern obtained by Rosalind Franklin and Maurice Wilkins from strands of DNA and used by James Watson and Francis Crick in their construction of the double-helix model of DNA (Fig. 19.42). To interpret this image by using Bragg's law we have to be aware that it was obtained by using a fibre consisting of many DNA molecules oriented with their axes parallel to the axis of the fibre, with X-rays incident from a perpendicular direction. All the molecules in the fibre are parallel (or nearly so), but are randomly distributed in the perpendicular directions; as a result, the diffraction pattern exhibits the periodic structure parallel to the fibre axis superimposed on a general background of scattering from the distribution of molecules in the perpendicular directions.

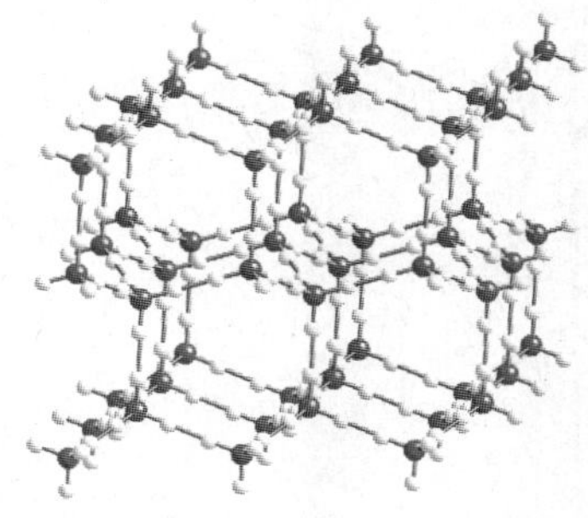

Fig. 19.41 A fragment of the crystal structure of ice (ice-I). Each O atom is at the centre of a tetrahedron of four O atoms at a distance of 276 pm. The central O atom is attached by two short O–H bonds to two H atoms and by two long hydrogen bonds to the H atoms of two of the neighbouring molecules. Overall, the structure consists of planes of hexagonal puckered rings of H_2O molecules (like the chair form of cyclohexane).

There are two principal features in Fig. 19.42: the strong 'meridional' scattering upward and downward by the fibre and the X-shaped distribution at smaller scattering angles. Because scattering through large angles occurs for closely spaced features (from $\lambda = 2d \sin\theta$, if d is small, then θ must be large to preserve the equality), we can

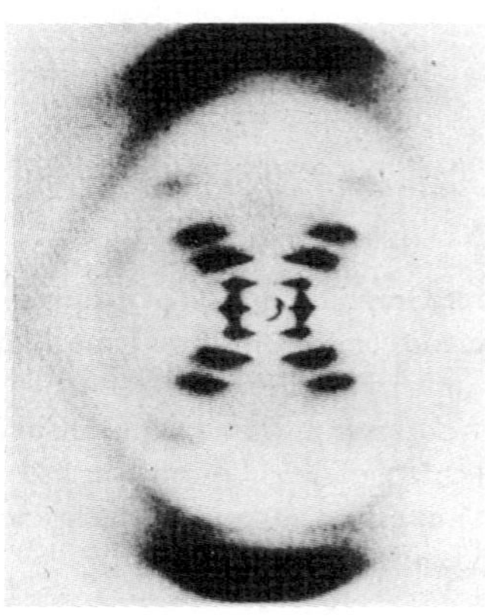

Fig. 19.42 The X-ray diffraction pattern obtained from a fibre of B-DNA. The black dots are the reflections, the points of maximum constructive interference, that are used to determine the structure of the molecule. (Adapted from an illustration that appears in J.P. Glusker and K.N. Trueblood, *Crystal structure analysis: A primer*. Oxford University Press (1972).)

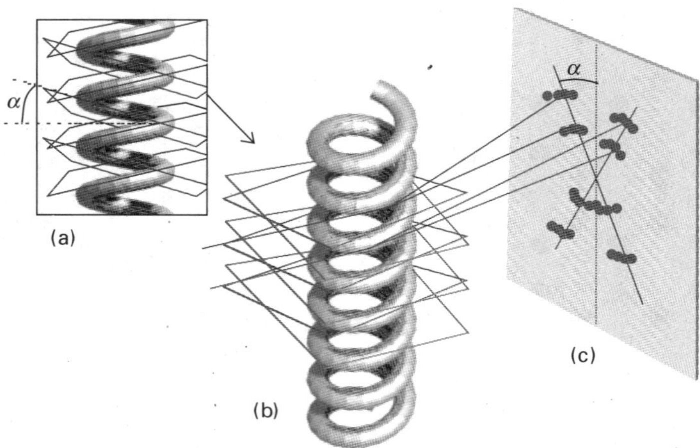

Fig. 19.43 The origin of the X pattern characteristic of diffraction by a helix. (a) A helix can be thought of as consisting of an array of planes at an angle α together with an array of planes at an angle $-\alpha$. (b) The diffraction spots from one set of planes appear at an angle α to the vertical, giving one leg of the X, and those of the other set appear at an angle $-\alpha$, giving rise to the other leg of the X. The lower half of the X appears because the helix has up–down symmetry in this arrangement. (c) The sequence of spots outward along a leg of the X corresponds to first-, second-, . . . order diffraction ($n = 1, 2, \ldots$).

infer that the meridional scattering arises from closely spaced components and that the inner X-shaped pattern arises from features with a longer periodicity. Because the meridional pattern occurs at a distance of about 10 times that of the innermost spots of the X-pattern, the large-scale structure is about 10 times bigger than the small-scale structure. From the geometry of the instrument, the wavelength of the radiation, and Bragg's law, we can infer that the periodicity of the small-scale feature is 340 pm whereas that of the large-scale feature is 3400 pm (that is, 3.4 nm).

To see that the cross is characteristic of a helix, look at Fig. 19.43. Each turn of the helix defines two planes, one orientated at an angle α to the horizontal and the other at $-\alpha$. As a result, to a first approximation, a helix can be thought of as consisting of an array of planes at an angle α together with an array of planes at an angle $-\alpha$ with a separation within each set determined by the pitch of the helix. Thus, a DNA molecule is like two arrays of planes, each set corresponding to those treated in the derivation of Bragg's law, with a perpendicular separation $d = p \cos \alpha$, where p is the pitch of the helix, each canted at the angles $\pm\alpha$ to the horizontal. The diffraction spots from one set of planes therefore occur at an angle α to the vertical, giving one leg of the X, and those of the other set occur at an angle $-\alpha$, giving rise to the other leg of the X. The experimental arrangement has up–down symmetry, so the diffraction pattern repeats to produce the lower half of the X. The sequence of spots outward along a leg corresponds to first-, second-, . . . order diffraction ($n = 1, 2, \ldots$ in eqn 19.4). Therefore from the X-ray pattern, we see at once that the molecule is helical and we can measure the angle α directly, and find $\alpha = 40°$. Finally, with the angle α and the pitch p determined, we can determine the radius r of the helix from $\tan \alpha = p/r$, from which it follows that $r = (3.4\ \text{nm})/(\tan 40°) = 4.1\ \text{nm}$.

To derive the relation between the helix and the cross-like pattern we have ignored the detailed structure of the helix, the fact that it is a periodic array of nucleotide bases, not a smooth wire. In Fig. 19.44 we represent the bases by points, and see that there is an additional periodicity of separation h, forming planes that are perpendicular to the axis to the molecule (and the fibre). These planes give rise to the strong meridional

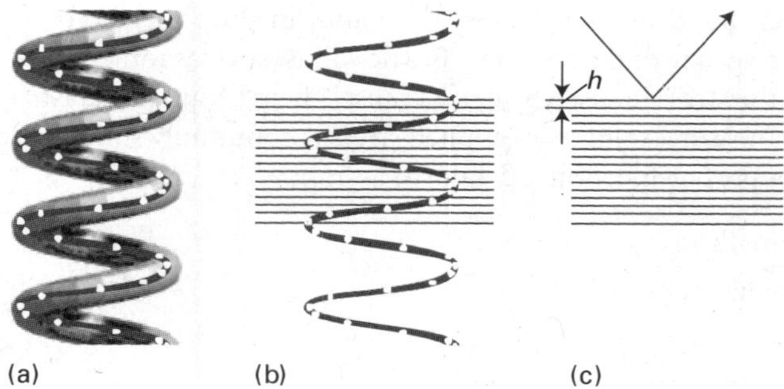

Fig. 19.44 The effect of the internal structure of the helix on the X-ray diffraction pattern. (a) The residues of the macromolecule are represented by points. (b) Parallel planes passing through the residues are perpendicular to the axis of the molecule. (c) The planes give rise to strong diffraction with an angle that allows us to determine the layer spacing h from $\lambda = 2h \sin \theta$.

diffraction with an angle that allows us to determine the layer spacing from Bragg's law in the form $\lambda = 2h \sin \theta$ as $h = 340$ pm.

The success of modern biochemistry in explaining such processes as DNA replication, protein biosynthesis, and enzyme catalysis is a direct result of developments in preparatory, instrumental, and computational procedures that have led to the determination of large numbers of structures of biological macromolecules by techniques based on X-ray diffraction. Most work is now done not on fibres but on crystals, in which the large molecules lie in orderly ranks. But even so crystallography yields only a static picture of biological structure and does not lend insight into changes that accompany biological processes. Therefore, information from crystallographic and spectroscopic studies is considered together to describe biochemical reactions.

The properties of solids

In this section we consider how the bulk properties of solids, particularly their mechanical, electrical, optical, and magnetic properties, stem from the properties of their constituent atoms. The rational fabrication of modern materials depends crucially on an understanding of this link.

19.8 Mechanical properties

Key points **The mechanical properties of a solid are discussed in terms of the relationship between stress, the applied force divided by the area to which it is applied, and strain, the distortion of a sample resulting from an applied stress. The response of a solid to an applied stress is summarized by the Young's modulus, the bulk modulus, the shear modulus, and Poisson's ratio.**

The fundamental concepts for the discussion of the mechanical properties of solids are stress and strain. The **stress** on an object is the applied force divided by the area to which it is applied. The **strain** is the resulting distortion of the sample. The general field of the relations between stress and strain is called **rheology**.

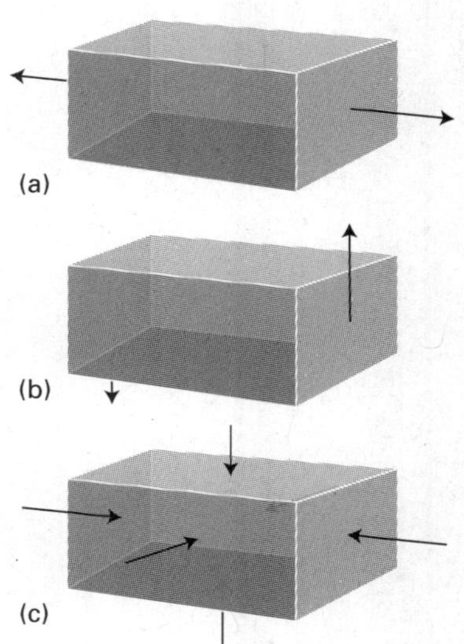

Fig. 19.45 Types of stress applied to a body. (a) Uniaxial stress, (b) shear stress, (c) hydrostatic pressure.

Stress may be applied in a number of different ways. Thus, **uniaxial stress** is a simple compression or extension in one direction (Fig. 19.45); **hydrostatic stress** is a stress applied simultaneously in all directions, as in a body immersed in a fluid. A **pure shear** is a stress that tends to push opposite faces of the sample in opposite directions. A sample subjected to a small stress typically undergoes **elastic deformation** in the sense that it recovers its original shape when the stress is removed. For low stresses, the strain is linearly proportional to the stress. The response becomes nonlinear at high stresses but may remain elastic. Above a certain threshold, the strain becomes **plastic** in the sense that recovery does not occur when the stress is removed. Plastic deformation

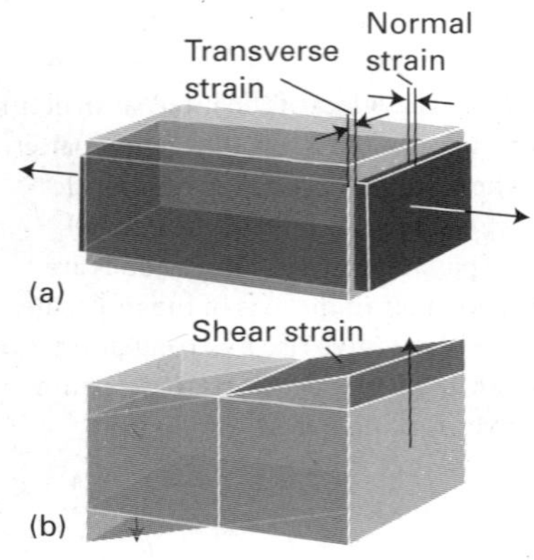

Fig. 19.46 (a) Normal stress and the resulting strain. (b) Shear stress. Poisson's ratio indicates the extent to which a body changes shape when subjected to a uniaxial stress.

occurs when bond breaking takes place and, in pure metals, typically takes place through the agency of dislocations. Brittle solids, such as ionic solids, exhibit sudden fracture as the stress focused by cracks causes them to spread catastrophically.

The response of a solid to an applied stress is commonly summarized by a number of coefficients of proportionality known as 'moduli':

$$E = \frac{\text{normal stress}}{\text{normal strain}} \qquad \text{Definition of Young's modulus} \qquad [19.16a]$$

$$K = \frac{\text{pressure}}{\text{fractional change in volume}} \qquad \text{Definition of bulk modulus} \qquad [19.16b]$$

$$G = \frac{\text{shear stress}}{\text{shear strain}} \qquad \text{Definition of shear modulus} \qquad [19.16c]$$

where 'normal stress' refers to stretching and compression of the material, as shown in Fig. 19.46a and 'shear stress' refers to the stress depicted in Fig. 19.46b. The bulk modulus is the inverse of the isothermal compressibility, κ_T, first encountered in Section 2.11 (eqn 2.43, $\kappa_T = -(\partial V/\partial p)_T/V$). A third ratio, called **Poisson's ratio,** indicates how the sample changes its shape:

$$\nu_P = \frac{\text{transverse strain}}{\text{normal strain}} \qquad \text{Definition of Poisson's ratio} \qquad [19.17]$$

The moduli are interrelated:

$$G = \frac{E}{2(1+\nu_P)} \qquad K = \frac{E}{3(1-2\nu_P)} \qquad \text{Relations between the moduli} \qquad (19.18)$$

We can use thermodynamic arguments to discover the relation of the moduli to the molecular properties of the solid. Thus, in the following *Justification*, we show that, if neighbouring molecules interact by a Lennard-Jones potential, then the bulk modulus and the compressibility of the solid are related to the Lennard-Jones parameter ε (the depth of the potential well) by

$$K = \frac{8N_A\varepsilon}{V_m} \qquad \kappa_T = \frac{V_m}{8N_A\varepsilon} \qquad (19.19)$$

We see that the bulk modulus is large (the solid stiff) if the potential well represented by the Lennard-Jones potential is deep and the solid is dense (its molar volume small).

Justification 19.4 *The relation between compressibility and molecular interactions*

We begin by writing an expression for K from the definition of κ_T (eqn 2.43, $\kappa_T = -(\partial U/\partial p)_T/V$), but in terms of the variation of the internal energy U with the volume V. To do so, we note that the thermodynamic relation $p = -(\partial A/\partial V)_T$, which comes from the relation $dA = -pdV - SdT$ at constant temperature, becomes $p = -(\partial U/\partial V)_T$ at $T = 0$ (because $A = U - TS$). Therefore, at $T = 0$,

$$K = \frac{1}{\kappa_T} = -\frac{V}{(\partial V/\partial p)_T} = -V\left(\frac{\partial p}{\partial V}\right)_T = V\left(\frac{\partial^2 U}{\partial V^2}\right)_T$$

This expression shows that the bulk modulus (and through eqn 19.18, the other two moduli) depends on the curvature of a plot of the internal energy against volume. To develop this conclusion, we note that the variation of internal energy with volume can be expressed in terms of its variation with a lattice parameter, R, such as the length of the side of a unit cell:

$$\frac{\partial U}{\partial V} = \frac{\partial U}{\partial R}\frac{\partial R}{\partial V}$$

and so

$$\frac{\partial^2 U}{\partial V^2} = \frac{\partial U}{\partial R}\frac{\partial^2 R}{\partial V^2} + \frac{\partial^2 U}{\partial V \partial R}\frac{\partial R}{\partial V} = \frac{\partial U}{\partial R}\frac{\partial^2 R}{\partial V^2} + \frac{\partial^2 U}{\partial R^2}\left(\frac{\partial R}{\partial V}\right)^2$$

To calculate K at the equilibrium volume of the sample, we set $R = R_0$ and recognize that $\partial U/\partial R = 0$ at equilibrium, so

$$K = V\left(\frac{\partial^2 U}{\partial R^2}\right)_{T,0}\left(\frac{\partial R}{\partial V}\right)^2_{T,0}$$

where the 0 denotes that the derivatives are evaluated at the equilibrium dimensions of the unit cell by setting $R = R_0$ after the derivative has been calculated. At this point we can write $V = aR^3$, where a is a constant that depends on the crystal structure, which implies that $\partial R/\partial V = 1/(3aR^2)$. Then, if the internal energy is given by a pairwise Lennard-Jones (12,6)-potential (eqn 17.33) we can write

$$\left(\frac{\partial^2 U}{\partial R^2}\right)_{T,0} = \frac{72nN_A\varepsilon}{R_0^2} \qquad (19.20)$$

where n is the amount of substance in the sample of volume V_0. It then follows that

$$K = \frac{72nN_A\varepsilon}{9aR_0^3} = \frac{8nN_A\varepsilon}{V_0} = \frac{8N_A\varepsilon}{V_m}$$

where we have used $V_m = V_0/n$, which is the first of eqn 19.19. Its reciprocal is κ_T. To obtain the result in eqn 19.20, we have used the fact that, at equilibrium, $R = R_0$ and $\sigma^6/R_0^6 = \frac{1}{2}$ where σ is the scale parameter for the intermolecular potential (r_0 in eqn 17.33).

The typical behaviour of a solid under stress is illustrated in Fig. 19.47. For small strains, the stress–strain relation is a Hooke's law of force, with the strain directly proportional to the stress. For larger strains, though, dislocations begin to play a major role and the strain becomes plastic in the sense that the sample does not recover its original shape when the stress is removed (recall Fig. 18.10).

The differing rheological characteristics of metals can be traced to the presence of **slip planes**, which are planes of atoms that under stress may slip or slide relative to one another. The slip planes of a ccp structure are the close-packed planes, and careful inspection of a unit cell shows that there are eight sets of slip planes in different directions. As a result, metals with cubic close-packed structures, like copper, are malleable: they can easily be bent, flattened, or pounded into shape. In contrast, a hexagonal close-packed structure has only one set of slip planes; and metals with hexagonal close packing, like zinc or cadmium, tend to be brittle.

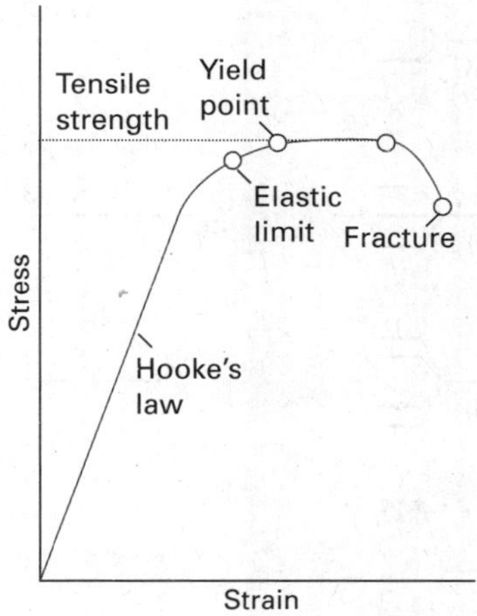

Fig. 19.47 At small strains, a body obeys Hooke's law (stress proportional to strain) and is elastic (recovers its shape when the stress is removed). At high strains, the body is no longer elastic, may yield and become plastic. At even higher strains, the solid fails (at its limiting tensile strength) and finally fractures.

19.9 Electrical properties

Key points **Electronic conductors are classified as metallic conductors or semiconductors according to the temperature dependence of their conductivities. An insulator is a semiconductor with a very low electrical conductivity. (a) According to the band theory, electrons occupy molecular orbitals formed from the overlap of atomic orbitals. Full bands are called valence bands and empty bands are called conduction bands. (b) The occupation of the orbitals in a solid is given by the Fermi–Dirac distribution. (c) Semiconductors are classified as p-type or n-type according to whether conduction is due to holes in the valence band or electrons in the conduction band.**

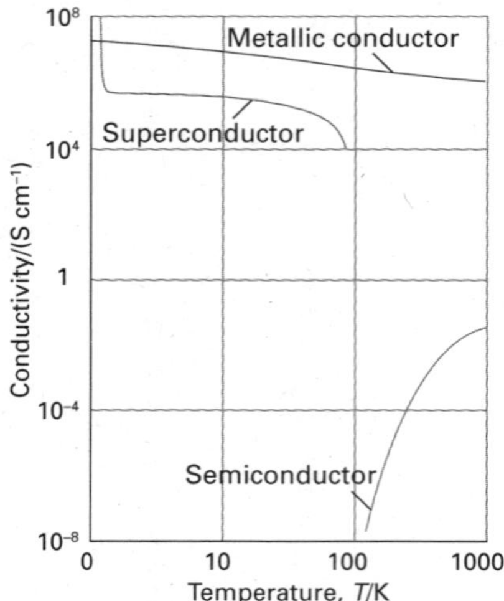

Fig. 19.48 The variation of the electrical conductivity of a substance with temperature is the basis of its classification as a metallic conductor, a semiconductor, or a superconductor. We shall see in Chapter 20 that conductivity is expressed in siemens per metre ($S\,m^{-1}$ or, as here, $S\,cm^{-1}$), where $1\,S = 1\,\Omega^{-1}$ (the resistance is expressed in ohms, Ω).

We shall confine attention to electronic conductivity, but note that some ionic solids display ionic conductivity. Two types of solid are distinguished by the temperature dependence of their electrical conductivity (Fig. 19.48):

A **metallic conductor** is a substance with a conductivity that decreases as the temperature is raised.

A **semiconductor** is a substance with a conductivity that increases as the temperature is raised.

A semiconductor generally has a lower conductivity than that typical of metals, but the magnitude of the conductivity is not the criterion of the distinction. It is conventional to classify semiconductors with very low electrical conductivities, such as most synthetic polymers, as **insulators.** We shall use this term, but it should be appreciated that it is one of convenience rather than one of fundamental significance. A **superconductor** is a solid that conducts electricity without resistance.

(a) The formation of bands

The central aspect of solids that determines their electrical properties is the distribution of their electrons. There are two models of this distribution. In one, the **nearly free-electron approximation**, the valence electrons are assumed to be trapped in a box with a periodic potential, with low energy corresponding to the locations of cations. In the **tight-binding approximation**, the valence electrons are assumed to occupy molecular orbitals delocalized throughout the solid. The latter model is more in accord with the discussion in the foregoing chapters, and we confine our attention to it.

We shall consider a one-dimensional solid, which consists of a single, infinitely long line of atoms. At first sight, this model may seem too restrictive and unrealistic. However, not only does it give us the concepts we need to understand conductivity in three-dimensional, macroscopic samples of metals and semiconductors, it is also the starting point for the description of long and thin structures, such as the carbon nanotubes discussed earlier in the chapter.

Suppose that each atom has one s orbital available for forming molecular orbitals. We can construct the LCAO-MOs of the solid by adding N atoms in succession to a line, and then infer the electronic structure using the building-up principle. One atom contributes one s orbital at a certain energy (Fig. 19.49). When a second atom is brought up it overlaps the first and forms bonding and antibonding orbitals. The third atom overlaps its nearest neighbour (and only slightly the next-nearest) and, from these three atomic orbitals, three molecular orbitals are formed: one is fully bonding, one fully antibonding, and the intermediate orbital is nonbonding between neighbours. The fourth atom leads to the formation of a fourth molecular orbital. At this stage, we can begin to see that the general effect of bringing up successive atoms is to spread the range of energies covered by the molecular orbitals, and also to fill in the range of energies with more and more orbitals (one more for each atom). When N atoms have been added to the line, there are N molecular orbitals covering a band of energies of finite width, and the Hückel secular determinant (Section 10.6) is

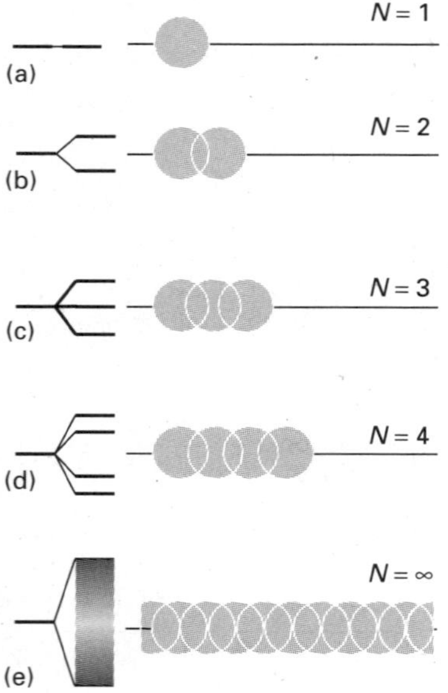

Fig. 19.49 The formation of a band of N molecular orbitals by successive addition of N atoms to a line. Note that the band remains of finite width as N becomes infinite and, although it looks continuous, it consists of N different orbitals.

$$\begin{vmatrix} \alpha-E & \beta & 0 & 0 & 0 & \cdots & 0 \\ \beta & \alpha-E & \beta & 0 & 0 & \cdots & 0 \\ 0 & \beta & \alpha-E & \beta & 0 & \cdots & 0 \\ 0 & 0 & \beta & \alpha-E & \beta & \cdots & 0 \\ 0 & 0 & 0 & \beta & \alpha-E & \cdots & 0 \\ \vdots & \vdots & \vdots & \vdots & \vdots & \cdots & \vdots \\ 0 & 0 & 0 & 0 & 0 & \cdots & \alpha-E \end{vmatrix} = 0$$

where β is now the (s,s) resonance integral. The theory of determinants applied to such a symmetrical example as this (technically a 'tridiagonal determinant') leads to the following expression for the roots:

$$E_k = \alpha + 2\beta \cos\frac{k\pi}{N+1} \qquad k = 1, 2, \ldots, N \tag{19.21}$$

When N is infinitely large, the difference between neighbouring energy levels (the energies corresponding to k and $k + 1$) is infinitely small, but, as we show in the following *Justification*, the band still has finite width overall:

$$E_N - E_1 \rightarrow -4\beta \qquad \text{as} \qquad N \rightarrow \infty \tag{19.22}$$

We can think of this band as consisting of N different molecular orbitals, the lowest-energy orbital ($k = 1$) being fully bonding, and the highest-energy orbital ($k = N$) being fully antibonding between adjacent atoms (Fig. 19.50). Similar bands form in three-dimensional solids.

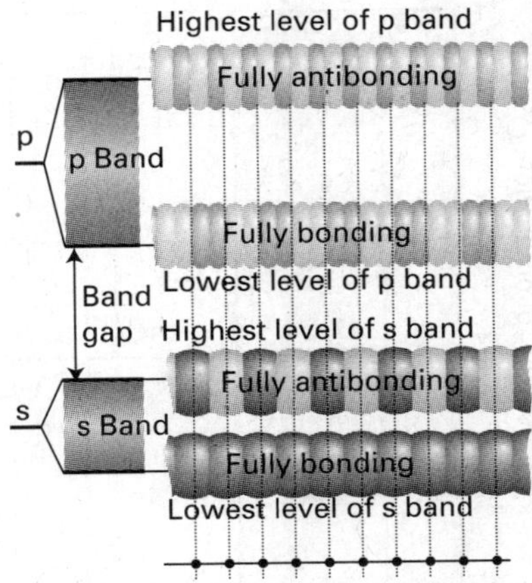

Fig. 19.50 The overlap of s orbitals gives rise to an s band and the overlap of p orbitals gives rise to a p band. In this case, the s and p orbitals of the atoms are so widely spaced that there is a band gap. In many cases the separation is less and the bands overlap.

Justification 19.5 *The width of a band*

The energy of the level with $k = 1$ is

$$E_1 = \alpha + 2\beta \cos\frac{\pi}{N+1}$$

As N becomes infinite, the cosine term becomes $\cos 0 = 1$. Therefore, in this limit

$$E_1 = \alpha + 2\beta$$

When k has its maximum value of N,

$$E_N = \alpha + 2\beta \cos\frac{N\pi}{N+1}$$

As N approaches infinity, we can ignore the 1 in the denominator, and the cosine term becomes $\cos \pi = -1$. Therefore, in this limit $E_N = \alpha - 2\beta$. The difference between the upper and lower energies of the band is therefore 4β.

The band formed from overlap of s orbitals is called the **s band**. If the atoms have p orbitals available, the same procedure leads to a **p band** (as shown in the upper half of Fig. 19.50). If the atomic p orbitals lie higher in energy than the s orbitals, then the p band lies higher than the s band, and there may be a **band gap**, a range of energies to which no orbital corresponds. However, the s and p bands may also be contiguous or even overlap (as is the case for the 3s and 3p bands in magnesium).

(b) The occupation of orbitals

Now consider the electronic structure of a solid formed from atoms each able to contribute one electron (for example, the alkali metals). There are N atomic orbitals and therefore N molecular orbitals packed into an apparently continuous band. There are N electrons to accommodate.

At $T = 0$, only the lowest $\frac{1}{2}N$ molecular orbitals are occupied (Fig. 19.51), and the HOMO is called the **Fermi level**. However, unlike in molecules, there are empty orbitals very close in energy to the Fermi level, so it requires hardly any energy to excite the uppermost electrons. Some of the electrons are therefore very mobile and give rise to electrical conductivity.

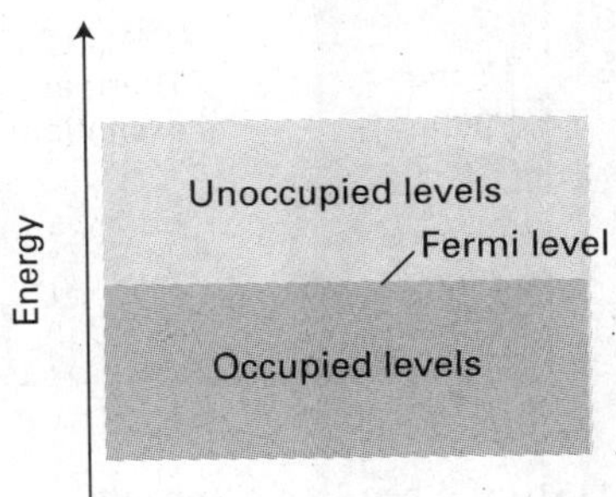

Fig. 19.51 When N electrons occupy a band of N orbitals, it is only half full and the electrons near the Fermi level (the top of the filled levels) are mobile.

At temperatures above absolute zero, electrons can be excited by the thermal motion of the atoms. The population, P, of the orbitals is given by the **Fermi–Dirac**

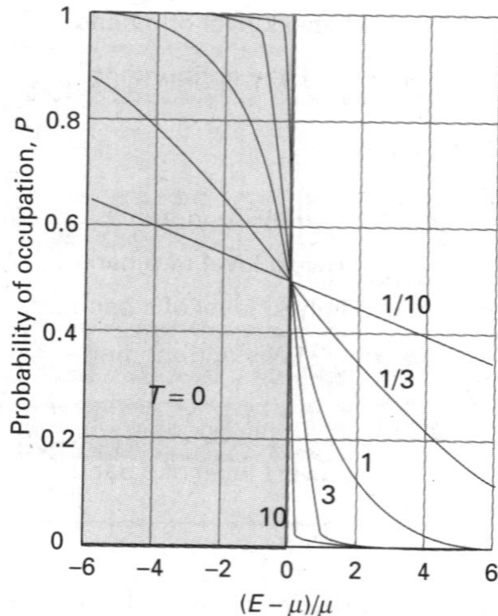

Fig. 19.52 The Fermi–Dirac distribution, which gives the population of the levels at a temperature T. The high-energy tail decays exponentially towards zero. The curves are labelled with the value of μ/kT. The tinted grey region shows the occupation of levels at $T=0$.

interActivity Express the population P as a function of the variables $(E-\mu)/\mu$ and μ/kT and then display the set of curves shown in Fig. 19.52 as a single surface.

distribution, a version of the Boltzmann distribution that takes into account the effect of the Pauli principle:

$$P = \frac{1}{e^{(E-\mu)/kT}+1} \quad \text{The Fermi–Dirac distribution} \qquad (19.23)$$

The quantity μ is the **chemical potential**, which in this context is the energy of the level for which $P=\frac{1}{2}$ (note that the chemical potential decreases as the temperature increases). The chemical potential in eqn 19.23 has the dimensions of energy, not energy per mole. The shape of the Fermi–Dirac distribution is shown in Fig. 19.52. For energies well above μ, the 1 in the denominator can be neglected, and then

$$P \approx e^{-(E-\mu)/kT} \qquad (19.24)$$

The population now resembles a Boltzmann distribution, decaying exponentially with increasing energy. The higher the temperature, the longer the exponential tail.

The electrical conductivity of a metallic solid decreases with increasing temperature even though more electrons are excited into empty orbitals. This apparent paradox is resolved by noting that the increase in temperature causes more vigorous thermal motion of the atoms, so collisions between the moving electrons and an atom are more likely. That is, the electrons are scattered out of their paths through the solid, and are less efficient at transporting charge.

(c) Insulators and semiconductors

When each atom provides two electrons, the $2N$ electrons fill the N orbitals of the s band. The Fermi level now lies at the top of the band (at $T=0$), and there is a gap before the next band begins (Fig. 19.53). As the temperature is increased, the tail of the Fermi–Dirac distribution extends across the gap, and electrons leave the lower band, which is called the **valence band**, and populate the empty orbitals of the upper band, which is called the **conduction band**. As a consequence of electron promotion, positively charged 'holes' are left in in the valence band. The holes and promoted electrons are now mobile, and the solid is an electrical conductor. In fact, it is a semiconductor, because the electrical conductivity depends on the number of electrons that are promoted across the gap, and that number increases as the temperature is raised. If the gap is large, though, very few electrons will be promoted at ordinary temperatures and the conductivity will remain close to zero, resulting in an insulator. Thus, the conventional distinction between an insulator and a semiconductor is related to the size of the band gap and is not an absolute distinction like that between a metal (incomplete bands at $T=0$) and a semiconductor (full bands at $T=0$).

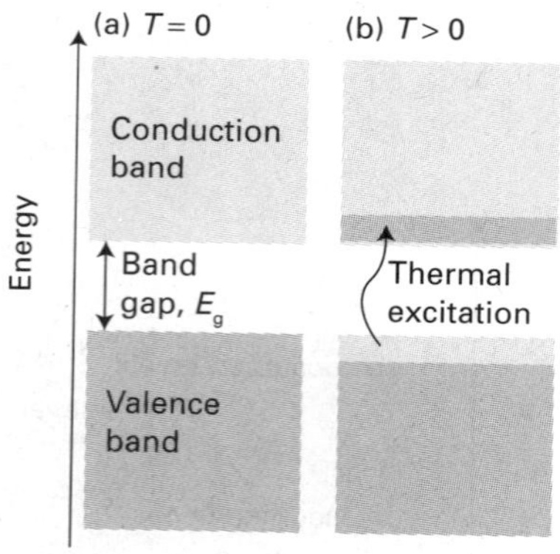

Fig. 19.53 (a) When $2N$ electrons are present, the band is full and the material is an insulator at $T=0$. (b) At temperatures above $T=0$, electrons populate the levels of the upper *conduction band* and the solid is a semiconductor.

Figure 19.53 depicts conduction in an **intrinsic semiconductor**, in which semiconduction is a property of the band structure of the pure material. Examples of intrinsic semiconductors include silicon and germanium. A **compound semiconductor** is an intrinsic semiconductor that is a combination of different elements, such as GaN, CdS, and many d-metal oxides. An **extrinsic semiconductor** is one in which charge carriers are present as a result of the replacement of some atoms (to the extent of about 1 in 10^9) by **dopant** atoms, the atoms of another element. If the dopants can trap electrons, they withdraw electrons from the filled band, leaving holes which allow the remaining electrons to move (Fig. 19.54a). This procedure gives rise to **p-type semiconductivity**, the p indicating that the holes are positive relative to the electrons in the band. An example is silicon doped with indium. We can picture the semiconduction as arising from the transfer of an electron from a Si atom to a neighbouring In atom. The electrons at the top of the silicon valence band are now mobile, and carry current through the solid. Alternatively, a dopant might carry excess electrons (for example, phosphorus atoms introduced into germanium), and these additional electrons occupy otherwise empty bands, giving **n-type semiconductivity**, where n denotes the negative charge of the carriers (Fig. 19.54b).

Now we consider the properties of a **p–n junction**, the interface of a p-type and n-type semiconductor. Consider the application of a 'reverse bias' to the junction, in the sense that a negative electrode is attached to the p-type semiconductor and a positive electrode is attached to the n-type semiconductor (Fig. 19.55a). Under these conditions, the positively charged holes in the p-type semicondutor are attracted to the negative electrode and the negatively charged electrons in the n-type semiconductor are attracted to the positive electrode. As a consequence, charge does not flow across the junction. Now consider the application of a 'forward bias' to the junction, in the sense that the positive electrode is attached to the p-type semiconductor and the negative electrode is attached to the n-type semiconductor (Fig. 19.55b). Now charge flows across the junction, with electrons in the n-type semiconductor moving toward the positive electrode and holes moving in the opposite direction. It follows that a p–n junction affords a great deal of control over the magnitude and direction of current through a material. This control is essential for the operation of transistors and diodes, which are key components of modern electronic devices.

As electrons and holes move across a p–n junction under forward bias, they recombine and release energy. However, as long as the forward bias continues to be applied, the flow of charge from the electrodes to the semiconductors will replenish them with electrons and holes, so the junction will sustain a current. In some solids, the energy of electron–hole recombination is released as heat and the device becomes warm. This is the case for silicon semiconductors, and is one reason why computers need efficient cooling systems.

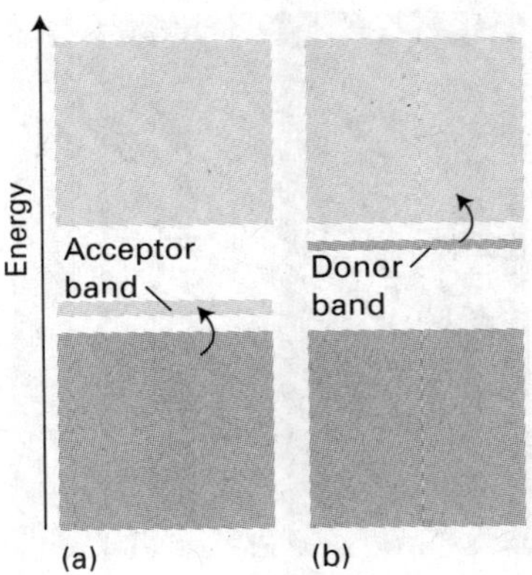

Fig. 19.54 (a) A dopant with fewer electrons than its host can form a narrow band that accepts electrons from the valence band. The holes in the band are mobile and the substance is a p-type semiconductor. (b) A dopant with more electrons than its host forms a narrow band that can supply electrons to the conduction band. The electrons it supplies are mobile and the substance is an n-type semiconductor.

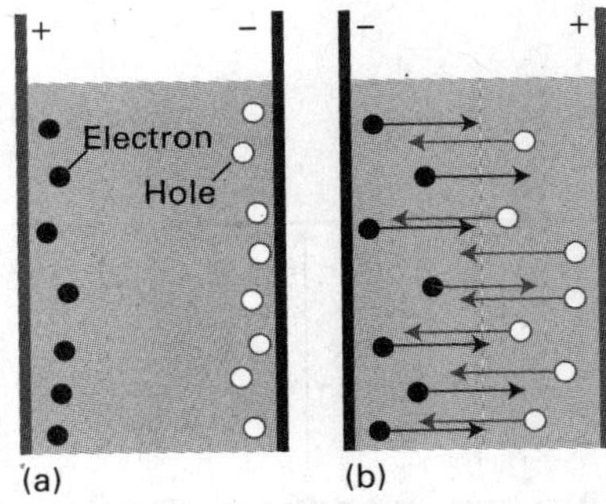

Fig. 19.55 A p–n junction under (a) reverse bias, (b) forward bias.

IMPACT ON NANOSCIENCE
I19.2 Nanowires

We have already remarked throughout the text that research on nanometre-sized materials is motivated by the possibility that they will form the basis for cheaper and smaller electronic devices. The synthesis of *nanowires*, nanometre-sized atomic assemblies that conduct electricity, is a major step in the fabrication of nanodevices. An important type of nanowire is based on carbon nanotubes, which, like graphite, can conduct electrons through delocalized π molecular orbitals that form from unhybridized 2p orbitals on carbon. Recent studies have shown a correlation between structure and conductivity in single-walled nanotubes (SWNTs) that does not occur in graphite. The SWNT in Fig. 19.40 is a semiconductor. If the hexagons are rotated by 90° about their sixfold axis, the resulting SWNT is a metallic conductor.

Carbon nanotubes are promising building blocks not only because they have useful electrical properties but also because they have unusual mechanical properties. For example, an SWNT has a Young's modulus that is approximately five times larger and a tensile strength that is approximately 375 times larger than that of steel.

Silicon nanowires can be made by focusing a pulsed laser beam on to a solid target composed of silicon and iron. The laser ejects Fe and Si atoms from the surface of the target, forming a vapour that can condense into liquid $FeSi_n$ nanoclusters at sufficiently low temperatures. The phase diagram for this complex mixture shows that solid silicon and liquid $FeSi_n$ coexist at temperatures higher than 1473 K. Hence, it is possible to precipitate solid silicon from the mixture if the experimental conditions are controlled to maintain the $FeSi_n$ nanoclusters in a liquid state that is supersaturated with silicon. It is observed that the silicon precipitate consists of nanowires with diameters of about 10 nm and lengths greater than 1 μm.

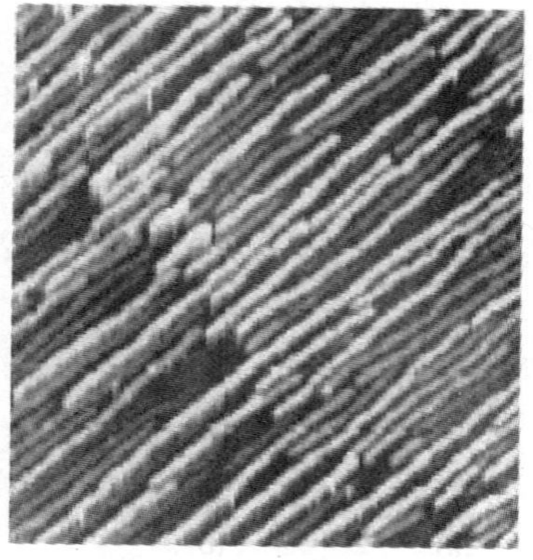

Fig. 19.56 Germanium nanowires fabricated on to a silicon surface by molecular beam epitaxy. (Reproduced with permission from T. Ogino *et al.*, *Acc. Chem. Res.* **32**, 447 (1999).)

Nanowires are also fabricated by *molecular beam epitaxy* (MBE), in which gaseous atoms or molecules are sprayed on to a crystalline surface in an ultra-high vacuum chamber. The result is formation of highly ordered structures. Through careful control of the chamber temperature and of the spraying process, it is possible to deposit thin films on to a surface or to create nanometre-sized assemblies with specific shapes. For example, Fig. 19.56 shows an AFM image of germanium nanowires on a silicon surface. The wires are about 2 nm high, 10–32 nm wide, and 10–600 nm long.

Direct manipulation of atoms on a surface also leads to the formation of nanowires. The Coulomb attraction between an atom and the tip of an STM can be exploited to move atoms along a surface, arranging them into patterns, such as wires.

19.10 Optical properties

Key points **(a) The optical properties of molecular solids can be understood in terms of the formation and migration of excitons. (b) The spectroscopic properties of metallic conductors and semiconductors can be understood in terms of the light-induced promotion of electrons from valence bands to conduction band. (c) Nonlinear optical phenomena arise from changes in the optical properties of a material in the presence of intense electromagnetic radiation.**

In this section, we explore the consequences of interactions between electromagnetic radiation and solids. Our focus will be on the origins of phenomena that inform the design of useful devices, such as lasers and light-emitting diodes.

(a) Light absorption by excitons in molecular solids

From the discussion in earlier chapters, we are already familiar with the factors that determine the energy and intensity of light absorbed by atoms and molecules in the gas phase and in solution. Now we consider the effects on the electronic absorption spectrum of bringing atoms or molecules together into a solid.

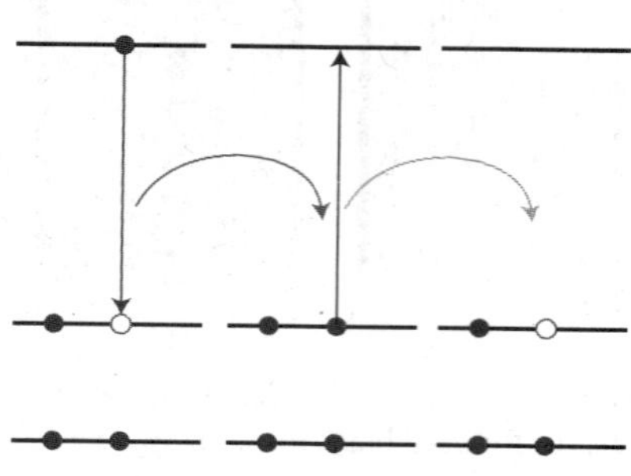

Fig. 19.57 The electron–hole pair shown on the left can migrate through a solid lattice as the excitation hops from molecule to molecule. The mobile excitation is called an exciton.

Consider an electronic excitation of a molecule (or an ion) in a crystal. If the excitation corresponds to the removal of an electron from one orbital of a molecule and its elevation to an orbital of higher energy, then the excited state of the molecule can be envisaged as the coexistence of an electron and a hole. This electron–hole pair, the particle-like **exciton**, migrates from molecule to molecule in the crystal (Fig. 19.57). Exciton formation causes spectral lines to shift, split, and change intensity.

The electron and the hole jump together from molecule to molecule as they migrate. A migrating excitation of this kind is called a **Frenkel exciton**. The electron and hole can also be on different molecules, but in each other's vicinity. A migrating excitation of this kind, which is now spread over several molecules (more usually ions), is a **Wannier exciton**.

Frenkel excitons are more common in molecular solids. Their migration implies that there is an interaction between the species that constitute the crystal, for otherwise the excitation on one unit could not move to another. This interaction affects the energy levels of the system. The strength of the interaction governs the rate at which an exciton moves through the crystal: a strong interaction results in fast migration, and a vanishingly small interaction leaves the exciton localized on its original molecule. The specific mechanism of interaction that leads to exciton migration is the interaction between the transition dipole moments of the excitation. Thus, an electric dipole transition in a molecule is accompanied by a shift of charge, and the transient dipole exerts a force on an adjacent molecule. The latter responds by shifting its charge. This process continues and the excitation migrates through the crystal.

The energy shift arising from the interaction between transition dipoles can be understood in terms of their electrostatic interaction. An all-parallel arrangement of the dipoles (Fig. 19.58a) is energetically unfavourable, so the absorption occurs at a higher frequency than in the isolated molecule. Conversely, a head-to-tail alignment of transient dipoles (Fig. 19.58b) is energetically favourable, and the transition occurs at a lower frequency than in the isolated molecules.

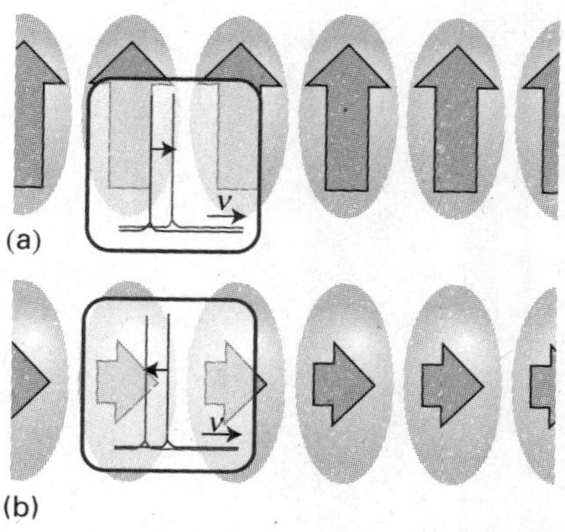

Fig. 19.58 (a) The alignment of transition dipoles (the gold arrows) is energetically unfavourable, and the exciton absorption is shifted to higher energy (higher frequency). (b) The alignment is energetically favourable for a transition in this orientation, and the exciton band occurs at lower frequency than in the isolated molecules.

• A brief illustration

Recall from Section 17.5 that the potential energy of interaction between two parallel dipoles μ_1 and μ_2 separated by a distance r is $V = \mu_1\mu_2(1 - 3\cos^2\theta)/4\pi\varepsilon_0 r^3$, where the angle θ is defined in (1). We see that $\theta = 0°$ for a head-to-tail alignment and $\theta = 90°$ for a parallel alignment. It follows that $V < 0$ (an attractive interaction) for $0° \leq \theta < 54.74°$, $V = 0$ when $\theta = 54.74°$ (for then $1 - 3\cos^2\theta = 0$), and $V > 0$ (a repulsive interaction) for $54.74° < \theta \leq 90°$. This result is expected on the basis of qualitative arguments. In a head-to-tail arrangement, the interaction between the region of partial positive charge in one molecule with the region of partial negative charge in the other molecule is attractive. By contrast, in a parallel arrangement, the molecular interaction is repulsive because of the close approach of regions of partial charge with the same sign. •

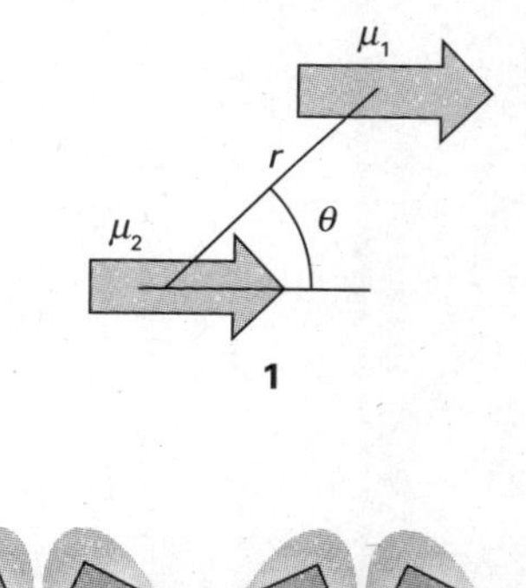

1

It follows from this discussion that, when $0° \leq \theta < 54.74°$, the frequency of exciton absorption is lower than the corresponding absorption frequency for the isolated molecule (a *red shift* in the spectrum of the solid with respect to that of the isolated molecule). Conversely, when $54.74° < \theta \leq 90°$, the frequency of exciton absorption is higher than the corresponding absorption frequency for the isolated molecule (a *blue shift* in the spectrum of the solid with respect to that of the isolated molecule). In the special case $\theta = 54.74°$, the solid and the isolated molecule have absorption lines at the same frequency.

If there are N molecules per unit cell, there are N **exciton bands** in the spectrum (if all of them are allowed). The splitting between the bands is the **Davydov splitting**. To understand the origin of the splitting, consider the case $N = 2$ with the molecules arranged as in Fig. 19.59 and suppose that the transition dipoles are along the length of the molecules. The radiation stimulates the collective excitation of the transition dipoles that are in-phase between neighbouring unit cells. *Within* each unit cell the transition dipoles may be arrayed in the two different ways shown in the illustration. Since the two orientations correspond to different interaction energies, with interaction being repulsive in one and attractive in the other, the two transitions appear in the spectrum at two bands of different frequencies. The Davydov splitting is determined by the energy of interaction between the transition dipoles within the unit cell.

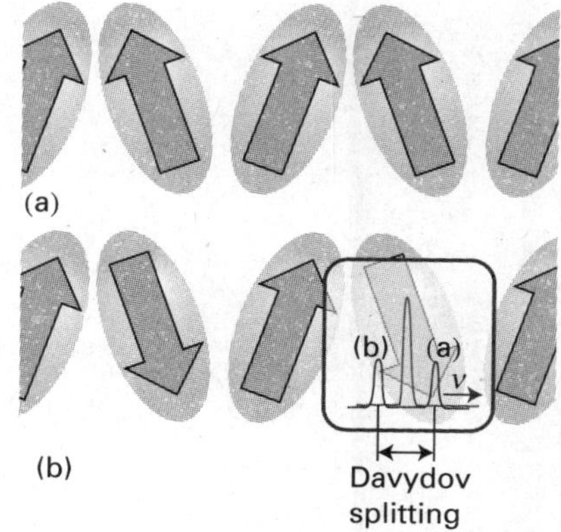

Fig. 19.59 When the transition moments within a unit cell may lie in different relative directions, as depicted in (a) and (b), the energies of the transitions are shifted and give rise to the two bands labelled (a) and (b) in the spectrum. The separation of the bands is the Davydov splitting.

(b) Light absorption by metals and semiconductors

Now we turn our attention to metallic conductors and semiconductors. Again we need to consider the consequences of interactions between particles, in this case atoms, which are now so strong that we need to abandon arguments based primarily on van der Waals interactions in favour of a full molecular orbital treatment, the band model of Section 19.9.

Consider Fig. 19.51, which shows bands in an idealized metallic conductor. The absorption of light can excite electrons from the occupied levels to the unoccupied levels. There is a near continuum of unoccupied energy levels above the Fermi level, so we expect to observe absorption over a wide range of frequencies. In metals, the bands are sufficiently wide that radiation from the radiofrequency to the middle of the ultraviolet region of the electromagnetic spectrum is absorbed (many metals are transparent to very high-frequency radiation, such as X-rays and γ-rays). Because this range of absorbed frequencies includes the entire visible spectrum, we expect that all metals should appear black. However, we know that metals are shiny (that is, they reflect light) and some are coloured (that is, they absorb light of only certain wavelengths), so we need to extend our model.

To explain the shiny appearance of a smooth metal surface, we need to realize that the absorbed energy can be re-emitted very efficiently as light, with only a small fraction of the energy being released to the surroundings as heat. Because the atoms near the surface of the material absorb most of the radiation, emission also occurs primarily from the surface. In essence, if the sample is excited with visible light, then visible light will be reflected from the surface, accounting for the lustre of the material.

The perceived colour of a metal depends on the frequency range of reflected light which, in turn, depends on the frequency range of light that can be absorbed and, by extension, on the band structure. Silver reflects light with nearly equal efficiency across the visible spectrum because its band structure has many unoccupied energy levels that can be populated by absorption of, and depopulated by emission of, visible light. On the other hand, copper has its characteristic colour because it has relatively fewer unoccupied energy levels that can be excited with violet, blue, and green light. The material reflects at all wavelengths, but more light is emitted at lower frequencies (corresponding to yellow, orange, and red). Similar arguments account for the colours of other metals, such as the yellow of gold.

Finally, consider semiconductors. We have already seen that promotion of electrons from the valence to the conduction band of a semiconductor can be the result of thermal excitation, if the band gap E_g is comparable to the energy that can be supplied by heating. In some materials, the band gap is very large and electron promotion can occur only by excitation with electromagnetic radiation. However, we see from Fig. 19.53 that there is a frequency $\nu_{min} = E_g/h$ below which light absorption cannot occur. Above this frequency threshold, a wide range of frequencies can be absorbed by the material, as in a metal.

• A brief illustration

The semiconductor cadmium sulfide (CdS) has a band gap energy of 2.4 eV (equivalent to 0.38 aJ). It follows that the minimum electronic absorption frequency is

$$\nu_{min} = \frac{3.8 \times 10^{-19}\ \text{J}}{6.626 \times 10^{-34}\ \text{J s}} = 5.8 \times 10^{14}\ \text{s}^{-1}$$

This frequency, 5.8×10^{14} Hz, corresponds to a wavelength of 517 nm (green light; see Table 13.1). Lower frequencies, corresponding to yellow, orange, and red, are not absorbed and consequently CdS appears yellow-orange. •

Self-test 19.7 Predict the colours of the following materials, given their band-gap energies (in parentheses): GaAs (1.43 eV), HgS (2.1 eV), and ZnS (3.6 eV).

[Black, red, and colourless]

(c) Nonlinear optical phenomena

Nonlinear optical phenomena arise from changes in the optical properties of a material in the presence of an intense electric field from electromagnetic radiation. Here we explore two phenomena that not only can be studied conveniently with intense laser beams but are commonly used in the laboratory to modify the output of lasers for specific experiments, such as those described in Section 13.6.

In **frequency doubling**, or **second harmonic generation**, an intense laser beam is converted to radiation with twice (and in general a multiple) of its initial frequency as it passes though a suitable material. It follows that frequency doubling and tripling of a Nd–YAG laser, which emits radiation at 1064 nm (see *Further information 19.1*), produce green light at 532 nm and ultraviolet radiation at 355 nm, respectively.

We can account for frequency doubling by examining how a substance responds nonlinearly to incident radiation of frequency $\omega = 2\pi\nu$. Radiation of a particular frequency arises from oscillations of an electric dipole at that frequency and the incident electric field $\mathcal{E}$ induces an electric dipole of magnitude μ, in the substance. At low light intensity, most materials respond linearly, in the sense that $\mu = \alpha\mathcal{E}$, where α is the polarizability (see Section 17.2). To allow for nonlinear response by some materials at high light intensity, we can write

$$\mu = \alpha\mathcal{E} + \tfrac{1}{2}\beta\mathcal{E}^2 + \cdots \qquad (19.25)$$

The induced dipole moment in terms of the hyperpolarizability

where the coefficient β is the **hyperpolarizability** of the material. The nonlinear term $\beta\mathcal{E}^2$ can be expanded as follows if we suppose that the incident electric field is $\mathcal{E}_0 \cos \omega t$:

$$\beta\mathcal{E}^2 = \beta\mathcal{E}_0^2 \cos^2\omega t = \tfrac{1}{2}\beta\mathcal{E}_0^2(1 + \cos 2\omega t) \qquad (19.26)$$

Hence, the nonlinear term contributes an induced electric dipole that oscillates at the frequency 2ω and that can act as a source of radiation of that frequency. Common materials that can be used for frequency doubling in laser systems include crystals of potassium dihydrogenphosphate (KH_2PO_4), lithium niobate ($LiNbO_3$), and β-barium borate (β-BaB_2O_4).

Another important nonlinear optical phenomenon is the **optical Kerr effect**, which arises from a change in refractive index of a well chosen medium, the **Kerr medium**, when it is exposed to intense laser pulses. Because a beam of light changes direction when it passes from a region of one refractive index to a region with a different refractive index, changes in refractive index result in the self-focusing of an intense laser pulse as it travels through the Kerr medium (Fig. 19.60).

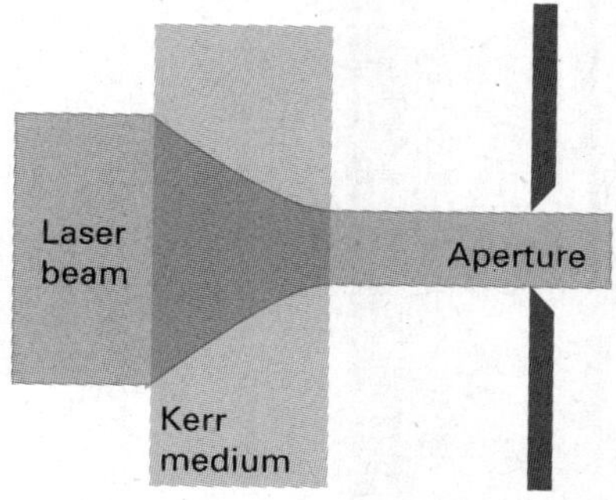

Fig. 19.60 An illustration of the Kerr effect. An intense laser beam is focused inside a Kerr medium and passes through a small aperture in the laser cavity. This effect may be used to mode-lock a laser, as explained in the text.

The optical Kerr effect is used as a mechanism of mode-locking lasers (Section 13.6c). A Kerr medium is included in the cavity and next to it is a small aperture. The procedure makes use of the fact that the **gain**, the growth in intensity, of a frequency component of the radiation in the cavity is very sensitive to amplification and, once a particular frequency begins to grow, it can quickly dominate. When the power inside the cavity is low, a portion of the photons will be blocked by the aperture, creating a significant loss. A spontaneous fluctuation in intensity—a bunching of photons—may begin to turn on the optical Kerr effect and the changes in the refractive index of the Kerr medium will result in a **Kerr lens**, which is the self-focusing of the laser beam. The bunch of photons can pass through and travel to the far end of the cavity, amplifying as it goes. The Kerr lens immediately disappears (if the medium is well chosen), but is re-created when the intense pulse returns from the mirror at the far end. In this way, that particular bunch of photons may grow to considerable intensity because it alone is stimulating emission in the cavity. Sapphire is an example of a Kerr medium that facilitates the mode-locking of titanium sapphire lasers, resulting in very short laser pulses of duration in the femtosecond range.

In addition to being useful laboratory tools, nonlinear optical materials are also finding many applications in the telecommunications industry, which is becoming ever more reliant on optical signals transmitted through optical fibres to carry voice and data. Judicious use of nonlinear phenomena leads to more ways in which the properties of optical signals, and hence the information they carry, can be manipulated.

A brief comment

The refractive index, n_r, of the medium, the ratio of the speed of light in a vacuum, c, to its speed c' in the medium: $n_r = c/c'$. A beam of light changes direction ('bends') when it passes from a region of one refractive index to a region with a different refractive index.

19.11 Magnetic properties

Key points (a) A diamagnetic material moves out of a magnetic field; a paramagnetic material moves into a magnetic field. The Curie law describes the temperature dependence of the molar magnetic susceptibility. (b) Ferromagnetism is the cooperative alignment of electron spins in a material; antiferromagnetism results from alternating spin orientations in a material. (c) Temperature-independent paramagnetism arises from induced electron currents in a molecule.

The magnetic properties of metallic solids and semiconductors depend strongly on the band structures of the material. Here we confine our attention largely to magnetic properties that stem from collections of individual centres (molecules or ions, such as d-metal complexes). Much of the discussion applies to liquid and gas phase samples as well as to solids.

(a) Magnetic susceptibility

The magnetic and electric properties of molecules and solids are analogous. For instance, some molecules and ions possess permanent magnetic dipole moments, and an applied magnetic field can induce a magnetic moment, with the result that the entire solid sample becomes magnetized. The analogue of the electric polarization, P, is the **magnetization**, $\mathcal{M}$, the average molecular magnetic dipole moment multiplied by the number density of magnetic centres in the sample. The magnetization induced by a field of strength $\mathcal{H}$ is proportional to $\mathcal{H}$, and we write

$$\mathcal{M} = \chi\mathcal{H} \qquad \text{Definition of the magnetization} \qquad [19.27]$$

where χ is the dimensionless **volume magnetic susceptibility**. A closely related quantity is the **molar magnetic susceptibility**, χ_m:

$$\chi_m = \chi V_m \qquad \text{Definition of the molar magnetic susceptibility} \qquad [19.28]$$

where V_m is the molar volume of the substance (we shall soon see why it is sensible to introduce this quantity). The **magnetic flux density**, $\mathcal{B}$, is related to the applied field strength and the magnetization by

$$\mathcal{B} = \mu_0(\mathcal{H} + \mathcal{M}) = \mu_0(1+\chi)\mathcal{H} \qquad \text{Definition of the magnetic flux density} \qquad [19.29]$$

where μ_0 is the vacuum permeability, $\mu_0 = 4\pi \times 10^{-7}\ \text{J C}^{-2}\ \text{m}^{-1}\ \text{s}^2$. The magnetic flux density can be thought of as the density of magnetic lines of force permeating the medium. This density is increased if $\mathcal{M}$ adds to $\mathcal{H}$ (when $\chi > 0$), but the density is decreased if $\mathcal{M}$ opposes $\mathcal{H}$ (when $\chi < 0$). Materials for which χ is positive are called **paramagnetic**. Those for which χ is negative are called **diamagnetic**.

Just as polar molecules with a permanent electric dipole moment of magnitude μ in fluid phases contribute a term proportional to $\mu^2/3kT$ to the electric polarization of a medium (eqn 17.15), so molecules and ions with a permanent magnetic dipole moment of magnitude m contribute to the magnetization an amount proportional to $m^2/3kT$. However, unlike for polar molecules, this contribution to the magnetization is obtained even for paramagnetic species trapped in solids, because the direction of the spin of the electrons is typically not coupled to the orientation of the molecular framework and so contributes even when the nuclei are stationary. An applied field can also induce a magnetic moment by stirring up currents in the electron distribution like those responsible for the chemical shift in NMR (Section 14.5). The constant of proportionality between the induced moment and the applied field is called the **magnetizability**, ξ (xi), and the magnetic analogue of eqn 17.15 is

$$\chi = \mathcal{N}\mu_0\left(\xi + \frac{m^2}{3kT}\right) \qquad (19.30)$$

We can now see why it is convenient to introduce χ_m, because the product of the number density $\mathcal{N}$ and the molar volume is Avogadro's constant, N_A:

$$\mathcal{N}V_m = \frac{NV_m}{V} = \frac{nN_AV_m}{nV_m} = N_A \qquad (19.31)$$

Hence

$$\chi_m = N_A\mu_0\left(\xi + \frac{m^2}{3kT}\right) \qquad (19.32)$$

and the density dependence of the susceptibility (which occurs in eqn 19.30 via $\mathcal{N} = N_A\chi/M$) has been eliminated. The expression for χ_m is in agreement with the empirical **Curie law**:

$$\chi_m = A + \frac{C}{T} \qquad \text{Curie law} \qquad (19.33)$$

with $A = N_A\mu_0\xi$ and $C = N_A\mu_0 m^2/3k$. As indicated above, and in contrast to electric moments, this expression applies to solids as well as fluid phases.

The magnetic susceptibility is traditionally measured with a **Gouy balance**. This instrument consists of a sensitive balance from which the sample hangs in the form of a narrow cylinder and lies between the poles of a magnet. If the sample is paramagnetic, it is drawn into the field, and its apparent weight is greater than when the field is off. A diamagnetic sample tends to be expelled from the field and appears to weigh less when the field is turned on. The balance is normally calibrated against a sample of known susceptibility. The modern version of the determination makes use of a **superconducting quantum interference device** (SQUID, Fig. 19.61). A SQUID takes advantage of the quantization of magnetic flux and the property of current loops in superconductors that, as part of the circuit, include a weakly conducting link through which electrons must tunnel. The current that flows in the loop in a magnetic field depends on the value of the magnetic flux, and a SQUID can be exploited as a very sensitive magnetometer.

Table 19.6 lists some experimental values. A typical paramagnetic volume susceptibility is about 10^{-3}, and a typical diamagnetic volume susceptibility is about $(-)10^{-5}$. The permanent magnetic moment can be extracted from susceptibility measurements by plotting χ against $1/T$.

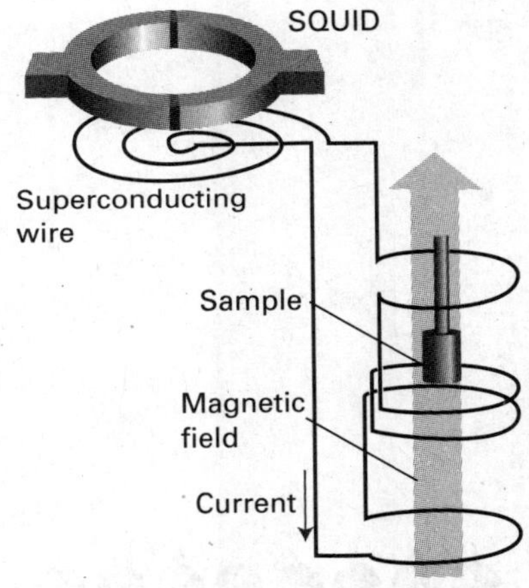

Fig. 19.61 The arrangement used to measure magnetic susceptibility with a SQUID. The sample is moved upwards in small increments and the potential difference across the SQUID is measured.

Table 19.6* Magnetic susceptibilities at 298 K

	$\chi/10^{-6}$	$\chi_m/(10^{-10}\ m^3\ mol^{-1})$
$H_2O(l)$	−9.02	−1.63
NaCl(s)	−16	−3.8
Cu(s)	−9.7	−0.69
$CuSO_4 \cdot 5H_2O(s)$	+167	+183

* More values are given in the *Data section*.

(b) The permanent magnetic moment

The permanent magnetic moment of a magnetic centre arises from any unpaired electron spins. We saw in Section 14.1 that the magnitude of the magnetic moment of an electron is proportional to the magnitude of the spin angular momentum, $\{s(s+1)\}^{1/2}\hbar$

$$\mu = g_e\{s(s+1)\}^{1/2}\mu_B \qquad \mu_B = \frac{e\hbar}{2m_e} \tag{19.34}$$

where $g_e = 2.0023$ (see Section 14.1). If there are several unpaired electron spins in each molecule or ion, they combine to a total spin S, and then $s(s+1)$ should be replaced by $S(S+1)$. It follows that the spin contribution to the molar magnetic susceptibility is

$$\chi_m = \frac{N_A g_e^2 \mu_0 \mu_B^2 S(S+1)}{3kT} \qquad \text{Spin contribution} \tag{19.35}$$

This expression shows that the susceptibility is positive, so the spin magnetic moments contribute to the paramagnetic susceptibilities of materials. The contribution decreases with increasing temperature because the thermal motion randomizes the spin orientations. In practice, a contribution to the paramagnetism also arises from the orbital angular momenta of electrons: we have discussed the spin-only contribution.

• A brief illustration

Consider a complex salt with three unpaired electrons per complex cation at 298 K, of mass density 3.24 g cm^{-3}, and molar mass 200 g mol^{-1}. First note that

$$\frac{N_A g_e^2 \mu_0 \mu_B^2}{3k} = 6.3001 \times 10^{-6}\ \text{m}^3\ \text{K}^{-1}\ \text{mol}^{-1}$$

Consequently,

$$\chi_m = 6.3001 \times 10^{-6} \times \frac{S(S+1)}{T/\text{K}}\ \text{m}^3\ \text{mol}^{-1}$$

Substitution of the data with $S = \frac{3}{2}$ gives $\chi_m = 7.9 \times 10^{-8}\ \text{m}^3\ \text{mol}^{-1}$. Note that the density is not needed at this stage. To obtain the volume magnetic susceptibility, the molar susceptibility is divided by the molar volume $V_m = M/\rho$, where ρ is the mass density. In this illustration, $V_m = 61.7\ \text{cm}^3\ \text{mol}^{-1}$, so $\chi = 1.3 \times 10^{-3}$. •

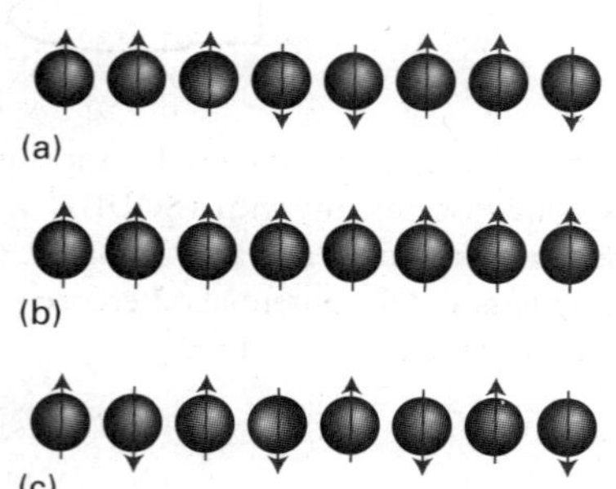

Fig. 19.62 (a) In a paramagnetic material, the electron spins are aligned at random in the absence of an applied magnetic field. (b) In a ferromagnetic material, the electron spins are locked into a parallel alignment over large domains. (c) In an antiferromagnetic material, the electron spins are locked into an antiparallel arrangement. The latter two arrangements survive even in the absence of an applied field.

At low temperatures, some paramagnetic solids make a phase transition to a state in which large domains of spins align with parallel orientations. This cooperative alignment gives rise to a very strong magnetization and is called **ferromagnetism** (Fig. 19.62). In other cases, the cooperative effect leads to alternating spin orientations: the spins are locked into a low-magnetization arrangement to give an **antiferromagnetic phase.** The ferromagnetic phase has a nonzero magnetization in the absence of an applied field, but the antiferromagnetic phase has a zero magnetization because the spin magnetic moments cancel. The ferromagnetic transition occurs at the **Curie temperature**, and the antiferromagnetic transition occurs at the **Néel temperature**.

(c) Induced magnetic moments

An applied magnetic field induces the circulation of electronic currents. These currents give rise to a magnetic field that usually opposes the applied field, so the substance is diamagnetic. In a few cases the induced field augments the applied field, and the substance is then paramagnetic.

The great majority of molecules and ions with no unpaired electron spins are diamagnetic. In these cases, the induced electron currents occur within the orbitals that are occupied in its ground state. In the few cases in which species are paramagnetic despite having no unpaired electrons, the induced electron currents flow in the opposite direction because they can make use of unoccupied orbitals that lie close to the HOMO in energy. This orbital paramagnetism can be distinguished from spin paramagnetism by the fact that it is temperature-independent: this is why it is called **temperature-independent paramagnetism** (TIP).

We can summarize these remarks as follows. All molecules and ions have a diamagnetic component to their susceptibility, but it is dominated by spin paramagnetism if unpaired electrons are present. In a few cases (where there are low-lying excited states) TIP is strong enough to make the species paramagnetic even though their electrons are paired.

19.12 Superconductors

Key points **Superconductors conduct electricity without resistance below a critical temperature T_c. Type I superconductors show abrupt loss of superconductivity when an applied magnetic field exceeds a critical value $\mathcal{H}_c$. Type II superconductors show a gradual loss of superconductivity and diamagnetism with increasing magnetic field.**

The resistance to flow of electrical current of a normal metallic conductor decreases smoothly with temperature but never vanishes. However, certain solids known as **superconductors** conduct electricity without resistance below a critical temperature, T_c. Following the discovery in 1911 that mercury is a superconductor below 4.2 K, the boiling point of liquid helium, physicists and chemists made slow but steady progress in the discovery of superconductors with higher values of T_c. Metals, such as tungsten, mercury, and lead, tend to have T_c values below about 10 K. Intermetallic compounds, such as Nb_3X (X = Sn, Al, or Ge), and alloys, such as Nb/Ti and Nb/Zr, have intermediate T_c values ranging between 10 K and 23 K. In 1986, **high-temperature superconductors** (HTSC) were discovered. Several **ceramics**, inorganic powders that have been fused and hardened by heating to a high temperature, containing oxocuprate motifs, Cu_mO_n, are now known with T_c values well above 77 K, the boiling point of the inexpensive refrigerant liquid nitrogen. For example, $HgBa_2Ca_2Cu_2O_8$ has $T_c = 153$ K.

Superconductors have unique magnetic properties as well. Some superconductors, classed as **Type I**, show abrupt loss of superconductivity when an applied magnetic field exceeds a critical value $\mathcal{H}_c$ characteristic of the material. It is observed that the value of $\mathcal{H}_c$ depends on temperature and T_c as

$$\mathcal{H}_c(T) = \mathcal{H}_c(0)\left(1 - \frac{T^2}{T_c^2}\right) \qquad (19.36)$$

where $\mathcal{H}_c(0)$ is the value of $\mathcal{H}_c$ as $T \to 0$. Type I superconductors are also completely diamagnetic below $\mathcal{H}_c$, meaning that no magnetic field lines penetrate into the material. This complete exclusion of a magnetic field in a material is known as the **Meissner effect**, which can be visualized by the levitation of a superconductor above a magnet. **Type II** superconductors, which include the HTSCs, show a gradual loss of superconductivity and diamagnetism with increasing magnetic field.

There is a degree of periodicity in the elements that exhibit superconductivity. The metals iron, cobalt, nickel, copper, silver, and gold do not display superconductivity, nor do the alkali metals. It is observed that, for simple metals, ferromagnetism and superconductivity never coexist, but in some of the oxocuprate superconductors ferromagnetism and superconductivity can coexist. One of the most widely studied oxocuprate superconductors $YBa_2Cu_3O_7$ (informally known as '123' on account of the proportions of the metal atoms in the compound) has the structure shown in Fig. 19.63. The square-pyramidal CuO_5 units arranged as two-dimensional layers and

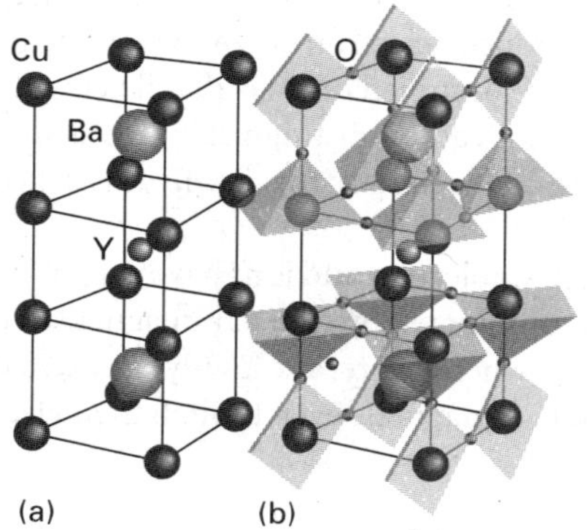

Fig. 19.63 Structure of the $YBa_2Cu_3O_7$ superconductor. (a) Metal atom positions. (b) The polyhedra show the positions of oxygen atoms and indicate that the metal ions are in square-planar and square-pyramidal coordination environments.

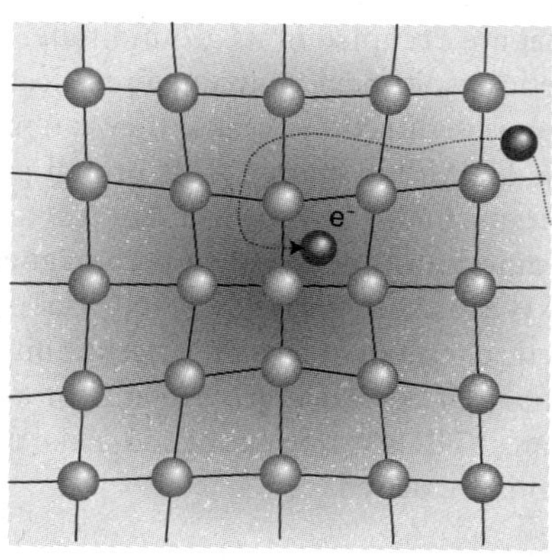

Fig. 19.64 The formation of a Cooper pair. One electron distorts the crystal lattice and the second electron has a lower energy if it goes to that region. These electron–lattice interactions effectively bind the two electrons into a pair.

the square planar CuO_4 units arranged in sheets are common structural features of oxocuprate HTSCs.

The mechanism of superconduction is well-understood for low-temperature materials but there is as yet no settled explanation of high-temperature superconductivity. The central concept of low-temperature superconduction is the existence of a **Cooper pair**, a pair of electrons that exists on account of the indirect electron–electron interactions fostered by the nuclei of the atoms in the lattice. Thus, if one electron is in a particular region of a solid, the nuclei there move toward it to give a distorted local structure (Fig. 19.64). Because that local distortion is rich in positive charge, it is favourable for a second electron to join the first. Hence, there is a virtual attraction between the two electrons, and they move together as a pair. The local distortion can be easily disrupted by thermal motion of the ions in the solid, so the virtual attraction occurs only at very low temperatures. A Cooper pair undergoes less scattering than an individual electron as it travels through the solid because the distortion caused by one electron can attract back the other electron should it be scattered out of its path in a collision. Because the Cooper pair is stable against scattering, it can carry charge freely through the solid, and hence give rise to superconduction.

The Cooper pairs responsible for low-temperature superconductivity are likely to be important in HTSCs, but the mechanism for pairing is hotly debated. There is evidence implicating the arrangement of CuO_5 layers and CuO_4 sheets in the mechanism of high-temperature superconduction. It is believed that movement of electrons along the linked CuO_4 units accounts for superconductivity, whereas the linked CuO_5 units act as 'charge reservoirs' that maintain an appropriate number of electrons in the superconducting layers.

Superconductors can sustain large currents and, consequently, are excellent materials for the high-field magnets used in modern NMR spectroscopy (Chapter 14). However, the potential uses of superconducting materials are not limited to the field to chemical instrumentation. For example, HTSCs with T_c values near ambient temperature would be very efficient components of an electrical power transmission system, in which energy loss due to electrical resistance would be minimized. The appropriate technology is not yet available, but research in this area of materials science is active.

Checklist of key equations

Property	Equation	Comment
Separation of neighbouring planes in a rectangular lattice	$1/d_{hkl}^2 = (h^2/a^2) + (k^2/b^2) + (l^2/c^2)$	
Bragg's law	$\lambda = 2d \sin\theta$	
Scattering factor	$f = 4\pi \int_0^\infty \rho(r)\{(\sin kr)/kr\}r^2\,dr$	$k = (4\pi/\lambda)\sin\theta$
Structure factor	$F_{hkl} = \sum_j f_j e^{i\phi_{hkl}(j)}$	$\phi_{hkl}(j) = 2\pi(hx_j + ky_j + lz_j)$
Fourier synthesis	$\rho(r) = (1/V)\sum_{hkl} F_{hkl} e^{-2\pi i(hx+ky+lz)}$	
Patterson synthesis	$P(r) = (1/V)\sum_{hkl} \lvert F_{hkl}\rvert^2 e^{-2\pi i(hx+ky+lz)}$	
Radius ratio	$\gamma = r_{smaller}/r_{larger}$	Definition
Born–Mayer equation	$E_{p,min} = -(N_A \lvert z_A z_B\rvert e^2/4\pi\varepsilon_0 d)(1 - d^*/d)A$	
Young's modulus	E = normal stress/normal strain	Definition
Bulk modulus	K = pressure/fractional change in volume	Definition
Shear modulus	G = shear stress/shear strain	Definition
Poisson's ratio	ν_p = transverse strain/normal strain	Definition
Fermi–Dirac distribution	$P = (e^{(E-\mu)/kT} + 1)^{-1}$	μ is the chemical potential
Magnetization of a material	$\mathcal{M} = \chi\mathcal{H}$	Definition
Curie law	$\chi_m = A + C/T$	$A = N_A\mu_0\xi$ $C = N_A\mu_0 m^2/3k$
Spin contribution to the molar magnetic susceptibility	$\chi_m = N_A g_e^2 \mu_0 \mu_B^2 S(S+1)/3kT$	

Further information

Further information 19.1 *Solid state lasers and light-emitting diodes*

Here we explore the further consequences of light emission in solids, focusing our attention on ionic crystals and semiconductors used in the design of lasers and light-emitting diodes. In Chapter 13 we discussed the conditions under which a material can become a laser and it would be helpful to review those concepts.

The **neodymium laser** is an example of a four-level laser, in which the laser transition terminates in a state other than the ground state of the laser material (Fig. 19.65). In one form it consists of Nd^{3+} ions at low concentration in yttrium aluminium garnet (YAG, specifically $Y_3Al_5O_{12}$), and is then known as a **Nd-YAG laser**. The population inversion results from pumping a majority of the Nd^{3+} ions into an excited state by using an intense flash from another source, followed by a radiationless transition to another excited state. The pumping flash need not be monochromatic because the upper level actually consists of several states spanning a band of frequencies. A neodymium laser operates at a number of wavelengths in the

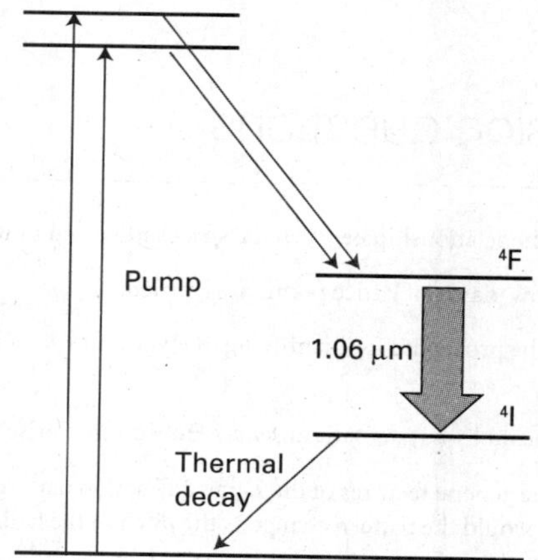

Fig. 19.65 The transitions involved in a neodymium laser. The laser action takes place between the ^{4}F and ^{4}I excited states.

infrared, the band at 1064 nm being most common. The transition at 1064 nm is very efficient and the laser is capable of substantial power output, either in continuous or pulsed (by Q-switching or mode-locking as discussed in Section 13.6c) modes of operation.

The **titanium sapphire laser** consists of Ti^{3+} ions at low concentration in a crystal of sapphire (Al_2O_3). The electronic absorption spectrum of Ti^{3+} ion in sapphire is very similar to that shown in Fig. 13.13, with a broad absorption band centred at around 500 nm that arises from vibronically allowed d–d transitions of the Ti^{3+} ion in an octahedral environment provided by oxygen atoms of the host lattice. As a result, the emission spectrum of Ti^{3+} in sapphire is also broad and laser action occurs over a wide range of wavelengths (Fig. 19.66). Therefore, the titanium sapphire laser is an example of a **vibronic laser**, in which the laser transitions originate from vibronic transitions in the laser medium. The titanium sapphire laser is usually pumped by another laser, such as a Nd–YAG laser or an argon-ion laser (*Further information 13.1*), and can be operated in either a continuous or pulsed fashion. Mode-locked titanium sapphire lasers produce energetic (20 mJ to 1 J) and very short (20–100 fs, 1 fs = 10^{-15} s) pulses. When considered together with broad wavelength tunability (700–1000 nm), these features of the titanium sapphire laser justify its wide use in modern spectroscopy and photochemistry.

The unique electrical properties of p–n junctions between semiconductors can be put to good use in optical devices. In some materials, most notably gallium arsenide, GaAs, energy from electron–hole recombination is released not as heat but is carried away by photons as electrons move across the junction under forward bias. Practical **light-emitting diodes** of this kind are widely used in electronic displays. The wavelength of emitted light depends on the band gap of the semiconductor. Gallium arsenide itself emits infrared light, but the band gap is widened by incorporating phosphorus, and a material of composition approximately $GaAs_{0.6}P_{0.4}$ emits light in the red region of the spectrum.

A light-emitting diode is not a laser, because no resonance cavity and stimulated emission are involved. In **diode lasers**, light emission due to electron–hole recombination is employed as the basis of laser action. The population inversion can be sustained by sweeping away the electrons that fall into the holes of the p-type semiconductor, and a resonant cavity can be formed by using the high refractive index of the semiconducting material and cleaving single crystals so that the light is trapped by the abrupt variation of refractive index. One widely used material is $Ga_{1-x}Al_xAs$, which produces infrared laser radiation and is widely used in compact-disc (CD) players.

High-power diode lasers are also used to pump other lasers. One example is the pumping of Nd:YAG lasers by $Ga_{0.91}Al_{0.09}As$/$Ga_{0.7}Al_{0.3}As$ diode lasers. The Nd:YAG laser is often used to pump yet another laser, such as a Ti:sapphire laser. As a result, it is now possible to construct a laser system for steady-state or time-resolved spectroscopy entirely out of solid-state components.

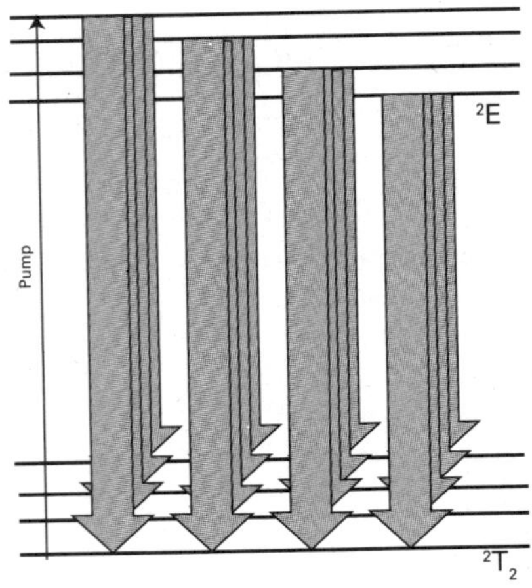

Fig. 19.66 The transitions involved in a titanium sapphire laser. The laser medium consists of sapphire (Al_2O_3) doped with Ti^{3+} ions. Monochromatic light from a pump laser induces a $^2E \leftarrow {}^2T_2$ transition in a Ti^{3+} ion that resides in a site with octahedral symmetry. After radiationless vibrational excitation in the ^2E state, laser emission occurs from a very large number of closely spaced vibronic states of the medium. As a result, the titanium sapphire laser emits radiation over a broad spectrum that spans from about 700 nm to about 1000 nm.

Discussion questions

19.1 Describe the relationship between the space lattice and unit cell.

19.2 Explain how planes of lattice points are labelled.

19.3 Describe the procedure for identifying the type and size of a cubic unit cell.

19.4 What is meant by a systematic absence? How do they arise?

19.5 Explain the general features of the X-ray diffraction pattern of a helical molecule. How would the pattern change as the pitch of the helix is increased?

19.6 Describe what is meant by 'scattering factor'. How is it related to the number of electrons in the atoms scattering X-rays?

19.7 Describe the phase problem and explain how it may be overcome.

19.8 Describe the structures of elemental metallic solids in terms of the packing of hard spheres. To what extent is the hard-sphere model inaccurate?

19.9 Describe the caesium-chloride and rock-salt structures in terms of the occupation of holes in expanded close-packed lattices.

19.10 Explain how metallic conductors and semiconductors are identified and explain their electrical and optical properties in terms of band theory.

19.11 Describe the characteristics of the Fermi–Dirac distribution. Why is it appropriate to call the parameter μ a chemical potential?

19.12 Explain the origin of Davydov splitting in the exciton bands of a crystal.

19.13 To what extent are the electric and magnetic properties of molecules analogous? How do they differ?

Exercises

19.1(a) Equivalent lattice points within the unit cell of a Bravais lattice have identical surroundings. What points within a face-centred cubic unit cell are equivalent to the point $(\frac{1}{2}, 0, 0)$?

19.1(b) Equivalent lattice points within the unit cell of a Bravais lattice have identical surroundings. What points within a body-centred cubic unit cell are equivalent to the point $(\frac{1}{2}, 0, \frac{1}{2})$?

19.2(a) Find the Miller indices of the planes that intersect the crystallographic axes at the distances $(2a, 3b, 2c)$ and $(2a, 2b, \infty c)$.

19.2(b) Find the Miller indices of the planes that intersect the crystallographic axes at the distances $(1a, 3b, -c)$ and $(2a, 3b, 4c)$.

19.3(a) Calculate the separations of the planes {111}, {211}, and {100} in a crystal in which the cubic unit cell has side 432 pm.

19.3(b) Calculate the separations of the planes {121}, {221}, and {244} in a crystal in which the cubic unit cell has side 523 pm.

19.4(a) The glancing angle of a Bragg reflection from a set of crystal planes separated by 99.3 pm is 20.85°. Calculate the wavelength of the X-rays.

19.4(b) The glancing angle of a Bragg reflection from a set of crystal planes separated by 128.2 pm is 19.76°. Calculate the wavelength of the X-rays.

19.5(a) What are the values of 2θ of the first three diffraction lines of bcc iron (atomic radius 126 pm) when the X-ray wavelength is 58 pm?

19.5(b) What are the values of 2θ of the first three diffraction lines of fcc gold (atomic radius 144 pm) when the X-ray wavelength is 154 pm?

19.6(a) Copper K_α radiation consists of two components of wavelengths 154.433 pm and 154.051 pm. Calculate the separation of the diffraction lines arising from the two components in a powder diffraction pattern recorded in a circular camera of radius 5.74 cm (with the sample at the centre) from planes of separation 77.8 pm.

19.6(b) A synchrotron source produces X-radiation at a range of wavelengths. Consider two components of wavelengths 95.401 and 96.035 pm. Calculate the separation of the diffraction lines arising from the two components in a powder diffraction pattern recorded in a circular camera of radius 5.74 cm (with the sample at the centre) from planes of separation 82.3 pm.

19.7(a) What is the value of the scattering factor in the forward direction for Br^-?

19.7(b) What is the value of the scattering factor in the forward direction for Mg^{2+}?

19.8(a) The compound Rb_3TlF_6 has a tetragonal unit cell with dimensions $a = 651$ pm and $c = 934$ pm. Calculate the volume of the unit cell.

19.8(b) Calculate the volume of the hexagonal unit cell of sodium nitrate, for which the dimensions are $a = 1692.9$ pm and $c = 506.96$ pm.

19.9(a) The orthorhombic unit cell of $NiSO_4$ has the dimensions $a = 634$ pm, $b = 784$ pm, and $c = 516$ pm, and the density of the solid is estimated as 3.9 g cm^{-3}. Determine the number of formula units per unit cell and calculate a more precise value of the density.

19.9(b) An orthorhombic unit cell of a compound of molar mass 135.01 g mol^{-1} has the dimensions $a = 589$ pm, $b = 822$ pm, and $c = 798$ pm. The density of the solid is estimated as 2.9 g cm^{-3}. Determine the number of formula units per unit cell and calculate a more precise value of the density.

19.10(a) The unit cells of $SbCl_3$ are orthorhombic with dimensions $a = 812$ pm, $b = 947$ pm, and $c = 637$ pm. Calculate the spacing, d, of the (411) planes.

19.10(b) An orthorhombic unit cell has dimensions $a = 679$ pm, $b = 879$ pm, and $c = 860$ pm. Calculate the spacing, d, of the (322) planes.

19.11(a) A substance known to have a cubic unit cell gives reflections with Cu K_α radiation (wavelength 154 pm) at glancing angles 19.4°, 22.5°, 32.6°, and 39.4°. The reflection at 32.6° is known to be due to the (220) planes. Index the other reflections.

19.11(b) A substance known to have a cubic unit cell gives reflections with radiation of wavelength 137 pm at the glancing angles 10.7°, 13.6°, 17.7°, and 21.9°. The reflection at 17.7° is known to be due to the (111) planes. Index the other reflections.

19.12(a) Potassium nitrate crystals have orthorhombic unit cells of dimensions $a = 542$ pm, $b = 917$ pm, and $c = 645$ pm. Calculate the glancing angles for the (100), (010), and (111) reflections using Cu K_α radiation (154 pm).

19.12(b) Calcium carbonate crystals in the form of aragonite have orthorhombic unit cells of dimensions $a = 574.1$ pm, $b = 796.8$ pm, and $c = 495.9$ pm. Calculate the glancing angles for the (100), (010), and (111) reflections using radiation of wavelength 83.42 pm (from aluminium).

19.13(a) Copper(I) chloride forms cubic crystals with four formula units per unit cell. The only reflections present in a powder photograph are those with either all even indices or all odd indices. What is the (Bravais) lattice type of the unit cell?

19.13(b) A powder diffraction photograph from tungsten shows lines which index as (110), (200), (211), (220), (310), (222), (321), (400), ... Identify the (Bravais) lattice type of the unit cell.

19.14(a) The coordinates, in units of a, of the atoms in a body-centred cubic lattice are (0,0,0), (0,1,0), (0,0,1), (0,1,1), (1,0,0), (1,1,0), (1,0,1), and (1,1,1). Calculate the structure factors F_{hkl} when all the atoms are identical.

19.14(b) The coordinates, in units of a, of the atoms in a body-centred cubic lattice are (0,0,0), (0,1,0), (0,0,1), (0,1,1), (1,0,0), (1,1,0), (1,0,1), (1,1,1), and $(\frac{1}{2},\frac{1}{2},\frac{1}{2})$. Calculate the structure factors F_{hkl} when all the atoms are identical.

19.15(a) In an X-ray investigation, the following structure factors were determined (with $F_{-h00} = F_{h00}$)

h	0	1	2	3	4	5	6	7	8	9
F_{h00}	10	−10	8	−8	6	−6	4	−4	2	−2

Construct the electron density along the corresponding direction.

19.15(b) In an X-ray investigation, the following structure factors were determined (with $F_{-h00} = F_{h00}$)

h	0	1	2	3	4	5	6	7	8	9
F_{h00}	10	10	4	4	6	6	8	8	10	10

Construct the electron density along the corresponding direction.

19.16(a) Construct the Patterson synthesis from the information in Exercise 19.15a.

19.16(b) Construct the Patterson synthesis from the information in Exercise 19.15b.

19.17(a) In a Patterson synthesis, the spots correspond to the lengths and directions of the vectors joining the atoms in a unit cell. Sketch the pattern that would be obtained for a planar, triangular isolated BF_3 molecule.

19.17(b) In a Patterson synthesis, the spots correspond to the lengths and directions of the vectors joining the atoms in a unit cell. Sketch the pattern that would be obtained from the C atoms in an isolated benzene molecule.

19.18(a) What velocity should neutrons have if they are to have wavelength 50 pm?

19.18(b) What velocity should neutrons have if they are to have wavelength 105 pm?

19.19(a) Calculate the wavelength of neutrons that have reached thermal equilibrium by collision with a moderator at 300 K.

19.19(b) Calculate the wavelength of neutrons that have reached thermal equilibrium by collision with a moderator at 380 K.

19.20(a) Calculate the packing fraction for close packed cylinders.

19.20(b) Calculate the packing fraction for equilateral triangular rods stacked as shown in (2).

2

19.21(a) Calculate the packing fractions of (a) a primitive cubic unit cell, (b) a bcc unit cell, (c) an fcc unit cell composed of identical hard spheres.

19.21(b) Calculate the atomic packing fraction for a side-centred C cubic unit cell.

19.22(a) Verify that the radius ratios for sixfold coordination is 0.414.

19.22(b) Verify that the radius ratios for eightfold coordination is 0.732.

19.23(a) From the data in Table 19.3 determine the radius of the smallest cation that can have (a) sixfold and (b) eightfold coordination with the O^{2-} ion.

19.23(b) From the data in Table 19.3 determine the radius of the smallest cation that can have (a) sixfold and (b) eightfold coordination with the K^+ ion.

19.24(a) Is there an expansion or a contraction as titanium transforms from hcp to body-centred cubic? The atomic radius of titanium is 145.8 pm in hcp but 142.5 pm in bcc.

19.24(b) Is there an expansion or a contraction as iron transforms from hcp to bcc? The atomic radius of iron is 126 in hcp but 122 pm in bcc.

19.25(a) Calculate the lattice enthalpy of CaO from the following data:

	$\Delta H/(\text{kJ mol}^{-1})$
Sublimation of Ca(s)	+178
Ionization of Ca(g) to Ca^{2+}(g)	+1735
Dissociation of O_2(g)	+249
Electron attachment to O(g)	−141
Electron attachment to O^-(g)	+844
Formation of CaO(s) from Ca(s) and O_2(g)	−635

19.25(b) Calculate the lattice enthalpy of $MgBr_2$ from the following data:

	$\Delta H/(\text{kJ mol}^{-1})$
Sublimation of Mg(s)	+148
Ionization of Mg(g) to Mg^{2+}(g)	+2187
Vaporization of Br_2(l)	+31
Dissociation of Br_2(g)	+193
Electron attachment to Br(g)	−331
Formation of $MgBr_2$(s) from Mg(s) and Br_2(l)	−524

19.26(a) Young's modulus for polyethene at room temperature is 1.2 GPa. What strain will be produced when a mass of 1.0 kg is suspended from a polyethene thread of diameter 1.0 mm?

19.26(b) Young's modulus for iron at room temperature is 215 GPa. What strain will be produced when a mass of 10.0 kg is suspended from an iron wire of diameter 0.10 mm?

19.27(a) Poisson's ratio for polyethene is 0.45. What change in volume takes place when a cube of polyethene of volume 1.0 cm^3 is subjected to a uniaxial stress that produces a strain of 1.0 per cent?

19.27(b) Poisson's ratio for lead is 0.41. What change in volume takes place when a cube of lead of volume 1.0 dm^3 is subjected to a uniaxial stress that produces a strain of 2.0 per cent?

19.28(a) Is arsenic-doped germanium a p-type or n-type semiconductor?

19.28(b) Is gallium-doped germanium a p-type or n-type semiconductor?

19.29(a) The promotion of an electron from the valence band into the conduction band in pure TiO_2 by light absorption requires a wavelength of less than 350 nm. Calculate the energy gap in electronvolts between the valence and conduction bands.

19.29(b) The band gap in silicon is 1.12 eV. Calculate the minimum frequency of electromagnetic radiation that results in promotion of electrons from the valence to the conduction band.

19.30(a) The magnetic moment of $CrCl_3$ is $3.81\mu_B$. How many unpaired electrons does the Cr atom possess?

19.30(b) The magnetic moment of Mn^{2+} in its complexes is typically $5.3\mu_B$. How many unpaired electrons does the ion possess?

19.31(a) Calculate the molar susceptibility of benzene given that its volume susceptibility is -7.2×10^{-7} and its density 0.879 g cm^{-3} at 25°C.

19.31(b) Calculate the molar susceptibility of cyclohexane given that its volume susceptibility is -7.9×10^{-7} and its density 811 kg m^{-3} at 25°C.

19.32(a) Data on a single crystal of MnF_2 give $\chi_m = 0.1463$ cm^3 mol^{-1} at 294.53 K. Determine the effective number of unpaired electrons in this compound and compare your result with the theoretical value.

19.32(b) Data on a single crystal of $NiSO_4{\cdot}7H_2O$ give $\chi_m = 6.00\times10^{-8}$ m^3 mol^{-1} at 298 K. Determine the effective number of unpaired electrons in this compound and compare your result with the theoretical value.

19.33(a) Estimate the spin-only molar susceptibility of $CuSO_4{\cdot}5H_2O$ at 25°C.

19.33(b) Estimate the spin-only molar susceptibility of $MnSO_4{\cdot}4H_2O$ at 298 K.

19.34(a) Lead has $T_c = 7.19$ K and $\mathcal{H}_c = 63.9$ kA m^{-1}. At what temperature does lead become superconducting in a magnetic field of 20 kA m^{-1}?

19.34(b) Tin has $T_c = 3.72$ K and $\mathcal{H}_c = 25$ kA m^{-1}. At what temperature does tin become superconducting in a magnetic field of 15 kA m^{-1}?

Problems*

Numerical problems

19.1 In the early days of X-ray crystallography there was an urgent need to know the wavelengths of X-rays. One technique was to measure the diffraction angle from a mechanically ruled grating. Another method was to estimate the separation of lattice planes from the measured density of a crystal. The density of NaCl is 2.17 g cm^{-3} and the (100) reflection using Pd K_α radiation occurred at 6.0°. Calculate the wavelength of the X-rays.

19.2 The element polonium crystallizes in a cubic system. Bragg reflections, with X-rays of wavelength 154 pm, occur at $\sin\theta = 0.225$, 0.316, and 0.388 from the (100), (110), and (111) sets of planes. The separation between the sixth and seventh lines observed in the powder diffraction pattern is larger than between the fifth and sixth lines. Is the unit cell simple, body-centred, or face-centred? Calculate the unit cell dimension.

19.3 Elemental silver reflects X-rays of wavelength 154.18 pm at angles of 19.076°, 22.171°, and 32.256°. However, there are no other reflections at angles of less than 33°. Assuming a cubic unit cell, determine its type and dimension. Calculate the density of silver.

19.4 In their book *X-rays and crystal structures* (which begins 'It is now two years since Dr. Laue conceived the idea . . .') the Braggs give a number of simple examples of X-ray analysis. For instance, they report that the reflection from (100) planes in KCl occurs at 5° 23′, but for NaCl it occurs at 6° 0′ for X-rays of the same wavelength. If the side of the NaCl unit cell is 564 pm, what is the side of the KCl unit cell? The densities of KCl and NaCl are 1.99 g cm^{-3} and 2.17 g cm^{-3}, respectively. Do these values support the X-ray analysis?

19.5 Calculate the coefficient of thermal expansion of diamond given that the (111) reflection shifts from 22° 2′ 25″ to 21° 57′ 59″ on heating a crystal from 100 K to 300 K and 154.0562 pm X-rays are used.

19.6 The carbon–carbon bond length in diamond is 154.45 pm. If diamond were considered to be a close-packed structure of hard spheres with radii equal to half the bond length, what would be its expected density? The diamond lattice is face-centred cubic and its actual density is 3.516 g cm^{-3}. Can you explain the discrepancy?

19.7 The volume of a monoclinic unit cell is $abc \sin\beta$. Naphthalene has a monoclinic unit cell with two molecules per cell and sides in the ratio 1.377:1:1.436. The angle β is 122° 49′ and the density of the solid is 1.152 g cm^{-3}. Calculate the dimensions of the cell.

19.8‡ B.A. Bovenzi and G.A. Pearse, Jr. (*J. Chem. Soc. Dalton Trans.* 2793–8 (1997)) synthesized coordination compounds of the tridentate ligand pyridine-2,6-diamidoxime ($C_7H_9N_5O_2$). The compound that they isolated from the reaction of the ligand with $CuSO_4$(aq) did not contain a $[Cu(C_7H_9N_5O_2)_2]^{2+}$ complex cation as expected. Instead, X-ray diffraction analysis revealed a linear polymer of formula $[Cu(Cu(C_7H_9N_5O_2)(SO_4)\cdot 2H_2O]_n$, which features bridging sulfate groups. The unit cell was primitive monoclinic with $a = 1.0427$ nm, $b = 0.8876$ nm, $c = 1.3777$ nm, and $\beta = 93.254°$. The mass density of the crystals is 2.024 g cm^{-3}. How many monomer units are there per unit cell?

19.9‡ D. Sellmann *et al.* (*Inorg. Chem.* **36**, 1397 (1997)) describe the synthesis and reactivity of the ruthenium nitrido compound $[N(C_4H_9)_4][Ru(N)(S_2C_6H_4)_2]$. The ruthenium complex anion has the two 1,2-benzenedithiolate ligands (**3**) at the base of a rectangular pyramid and the nitrido ligand at the apex. Compute the mass density of the compound given that it crystallizes into an orthorhombic unit cell with $a = 3.6881$ nm, $b = 0.9402$ nm, and $c = 1.7652$ nm and eight formula units per cell. Replacing the ruthenium with an osmium results in a compound with the same crystal structure and a unit cell with a volume less than 1 per cent larger. Estimate the mass density of the osmium analogue.

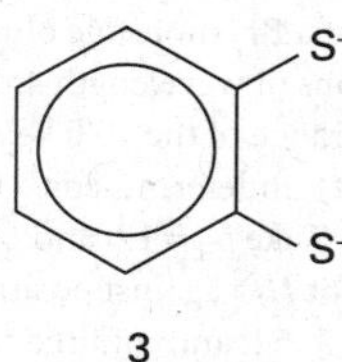

3

19.10 The unit cell dimensions of NaCl, KCl, NaBr, and KBr, all of which crystallize in face-centred cubic lattices, are 562.8 pm, 627.7 pm, 596.2 pm, and 658.6 pm, respectively. In each case, anion and cation are in contact along an edge of the unit cell. Do the data support the contention that ionic radii are constants independent of the counterion?

19.11 The powder diffraction patterns of (a) tungsten, (b) copper obtained in a camera of radius 28.7 mm are shown in Fig. 19.67. Both were obtained with 154 pm X-rays and the scales are marked. Identify the unit cell in each case, and calculate the lattice spacing. Estimate the metallic radii of W and Cu.

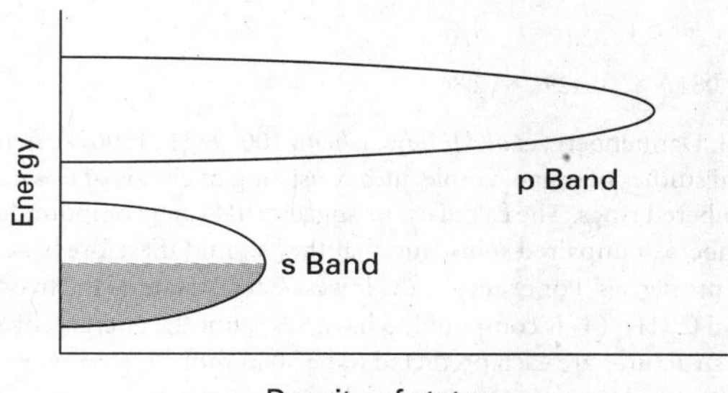

Fig. 19.67

19.12 Genuine pearls consist of concentric layers of calcite crystals ($CaCO_3$) in which the trigonal axes are oriented along the radii. The nucleus of a cultured pearl is a piece of mother-of-pearl that has been worked into a sphere on a lathe. The oyster then deposits concentric layers of calcite on the central seed. Suggest an X-ray method for distinguishing between real and cultured pearls.

19.13 The structures of crystalline macromolecules may be determined by X-ray diffraction techniques by methods similar to those for smaller molecules. Fully crystalline polyethene has its chains aligned in an orthorhombic unit cell of dimensions 740 pm × 493 pm × 253 pm. There are two repeating CH_2CH_2 units per unit cell. Calculate the theoretical density of fully crystalline polyethene. The actual density ranges from 0.92 to 0.95 g cm^{-3}.

19.14 The scattering of electrons or neutrons from a pair of nuclei separated by a distance R_{ij} and orientated at a definite angle to the incident beam can be calculated. When the molecule consists of a number of atoms, we sum over the contribution from all pairs, and find that the total intensity has an angular variation given by the *Wierl equation*:

$$I(\theta) = \sum_{i,j} f_i f_j \frac{\sin sR_{ij}}{sR_{ij}} \qquad s = \frac{4\pi}{\lambda}\sin\tfrac{1}{2}\theta$$

* Problems denoted with the symbol ‡ were supplied by Charles Trapp and Carmen Giunta.

where λ is the wavelength of the electrons in the beam and θ is the scattering angle. The *electron scattering factor*, f, is a measure of the intensity of the electron scattering powers of the atoms. (a) Predict from the Wierl equation the positions of the first maximum and first minimum in the neutron and electron diffraction patterns of a Br_2 molecule obtained with neutrons of wavelength 78 pm and electrons of wavelength 4.0 pm. (b) Use the Wierl equation to predict the appearance of the 10.0 keV electron diffraction pattern of CCl_4 with an (as yet) undetermined C–Cl bond length but of known tetrahedral symmetry. Take $f_{Cl} = 17f$ and $f_C = 6f$ and note that $R(Cl,Cl) = (8/3)^{1/2}R(C,Cl)$. Plot I/f^2 against positions of the maxima that occurred at 3° 10′, 5° 22′, and 7° 54′ and minima that occurred at 1° 46′, 4° 6′, 6° 40′, and 9° 10′. What is the C–Cl bond length in CCl_4?

19.15 Aided by the Born–Mayer equation for the lattice enthalpy and a Born–Haber cycle, show that formation of CaCl is an exothermic process (the sublimation enthalpy of Ca(s) is 176 kJ mol^{-1}. Show that an explanation for the nonexistence of CaCl can be found in the reaction enthalpy for the reaction $2CaCl(s) \rightarrow Ca(s) + CaCl_2$.

19.16 In an intrinsic semiconductor, the band gap is so small that the Fermi–Dirac distribution results in some electrons populating the conduction band. It follows from the exponential form of the Fermi–Dirac distribution that the conductance G, the inverse of the resistance (with units of siemens, $1\ S = 1\ \Omega^{-1}$), of an intrinsic semiconductor should have an Arrhenius-like temperature dependence, shown in practice to have the form $G = G_0e^{-E_g/2kT}$, where E_g is the band gap. The conductance of a sample of germanium varied with temperature as indicated below. Estimate the value of E_g.

T/K	312	354	420
G/S	0.0847	0.429	2.86

19.17‡ J.J. Dannenberg, *et al.* (*J. Phys. Chem.* **100**, 9631 (1996)) carried out theoretical studies of organic molecules consisting of chains of unsaturated four-membered rings. The calculations suggest that such compounds have large numbers of unpaired spins, and that they should therefore have unusual magnetic properties. For example, the lowest-energy state of the five-ring compound $C_{22}H_{14}$ (4) is computed to have $S = 3$, but the energies of $S = 2$ and $S = 4$ structures are each predicted to be 50 kJ mol^{-1} higher in energy. Compute the molar magnetic susceptibility of these three low-lying levels at 298 K. Estimate the molar susceptibility at 298 K if each level is present in proportion to its Boltzmann factor (effectively assuming that the degeneracy is the same for all three of these levels).

4

19.18‡ P.G. Radaelli *et al.* (*Science* **265**, 380 (1994)) report the synthesis and structure of a material that becomes superconducting at temperatures below 45 K. The compound is based on a layered compound $Hg_2Ba_2YCu_2O_{8-\delta}$, which has a tetragonal unit cell with $a = 0.38606$ nm and $c = 2.8915$ nm; each unit cell contains two formula units. The compound is made superconducting by partially replacing Y by Ca, accompanied by a change in unit cell volume by less than 1 per cent. Estimate the Ca content x in superconducting $Hg_2Ba_2Y_{1-x}Ca_xCu_2O_{7.55}$ given that the mass density of the compound is 7.651 g cm^{-3}.

Theoretical problems

19.19 Show that the separation of the (hkl) planes in an orthorhombic crystal with sides a, b, and c is given by eqn 19.3.

19.20 Show that the volume of a triclinic unit cell of sides a, b, and c and angles α, β, and γ is

$$V = abc(1 - \cos^2\alpha - \cos^2\beta - \cos^2\gamma + 2\cos\alpha\cos\beta\cos\gamma)^{1/2}$$

Use this expression to derive expressions for monoclinic and orthorhombic unit cells. For the derivation, it may be helpful to use the result from vector analysis that $V = \boldsymbol{a}\cdot\boldsymbol{b}\times\boldsymbol{c}$ and to calculate V^2 initially.

19.21 Use mathematical software to draw a graph of the scattering factor f agains $(\sin\theta)/\lambda$ for an atom of atomic number Z for which $\rho(r) = 3Z/4\pi R^3$ for $0 \le r \le R$ and $\rho(r) = 0$ for $r > R$, with R a parameter that represents the radius of the atom. Explore how f varies with Z and R.

19.22 Calculate the scattering factor for a hydrogenic atom of atomic number Z in which the single electron occupies (a) the 1s orbital, (b) the 2s orbital. Radial wavefunctions are given in Table 9.1. Plot f as a function of $(\sin\theta)/\lambda$. *Hint*. Interpret $4\pi\rho(r)r^2$ as the radial distribution function $P(r)$ of eqn 9.18.

19.23 Explore how the scattering factor of Problem 19.22 changes when the actual 1s wavefunction of a hydrogenic atom is replaced by a Gaussian function.

19.24 Rods of elliptical cross-section with semi-major and -minor axes a and b are close-packed as shown in (5). What is the packing fraction? Draw a graph of the packing fraction against the eccentricity ε of the ellipse. For an ellipse with semi-major axis a and semi-minor axis b, $\varepsilon = (1 - b^2/a^2)^{1/2}$.

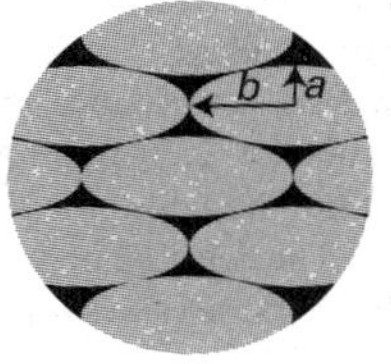

5

19.25 The coordinates of the four I atoms in the unit cell of KIO_4 are (0,0,0), $(0,\frac{1}{2},\frac{1}{2})$, $(\frac{1}{2},\frac{1}{2},\frac{1}{2})$, $(\frac{1}{2},0,\frac{3}{4})$. By calculating the phase of the I reflection in the structure factor, show that the I atoms contribute no net intensity to the (114) reflection.

19.26 The coordinates, in units of a, of the A atoms, with scattering factor f_A, in a cubic lattice are (0,0,0), (0,1,0), (0,0,1), (0,1,1), (1,0,0), (1,1,0), (1,0,1), and (1,1,1). There is also a B atom, with scattering factor f_B, at $(\frac{1}{2},\frac{1}{2},\frac{1}{2})$. Calculate the structure factors F_{hkl} and predict the form of the powder diffraction pattern when (a) $f_A = f, f_B = 0$, (b) $f_B = \frac{1}{2}f_A$, and (c) $f_A = f_B = f$.

19.27 Derive the Born–Mayer equation (eqn 19.15) by calculating the energy at which $d(E_p + E_p^*)/dd = 0$, with E_p and E_p^* given by eqns 19.13 and 19.14, respectively.

19.28 For an isotropic substance, the moduli and Poisson's ratio may be expressed in terms of two parameters λ and μ called the *Lamé constants*:

$$E = \frac{\mu(3\lambda + 2\mu)}{\lambda + \mu} \qquad K = \frac{3\lambda + 2\mu}{3} \qquad G = \mu \qquad \nu_P = \frac{\lambda}{2(\lambda + \mu)}$$

Use the Lamé constants to confirm the relations between G, K, and E given in eqn 19.18.

19.29 When energy levels in a band form a continuum, the density of states $\rho(E)$, the number of levels in an energy range divided by the width of the range, may be written as $\rho(E) = dk/dE$, where dk is the change in the quantum number k and dE is the energy change. (a) Use eqn 19.21 to show that

$$\rho(E) = -\frac{(N+1)/2\pi\beta}{\left\{1 - \left(\dfrac{E - \alpha}{2\beta}\right)^2\right\}^{1/2}}$$

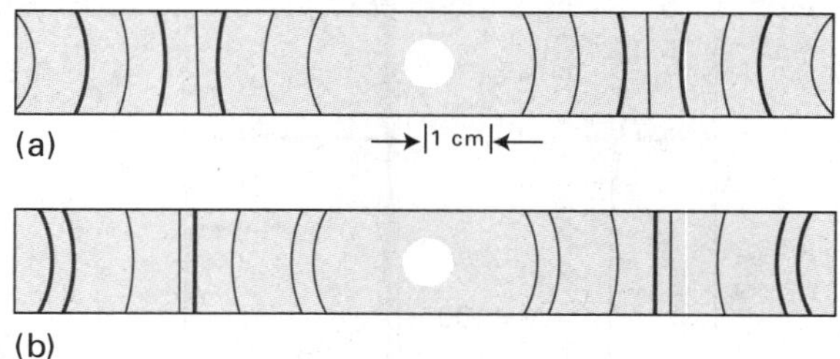

Fig. 19.68

where k, N, α, and β have the meanings described in Section 19.9. (b) Use the expression above to show that $\rho(E)$ becomes infinite as E approaches $\alpha \pm 2\beta$. That is, show that the density of states increases towards the edges of the bands in a one-dimensional metallic conductor.

19.30 The treatment in Problem 19.29 applies only to one-dimensional solids. In three dimensions, the variation of density of states is more like that shown in Fig. 19.68. Account for the fact that in a three-dimensional solid the greatest density of states lies near the centre of the band and the lowest density at the edges.

19.31 Here we investigate quantitatively the spectra of molecular solids. We begin by considering a dimer, with each monomer having a single transition with transition dipole moment μ_{mon} and wavenumber $\tilde{\nu}_{\text{mon}}$. We assume that the ground state wavefunctions are not perturbed as a result of dimerization. and then write the dimer excited state wavefunctions ψ_i as linear combinations of the excited state wavefunctions ψ_1 and ψ_2 of the monomer: $\psi_i = c_j\psi_1 + c_k\psi_2$. Now we write the hamiltonian matrix with diagonal elements set to the energy between the excited and ground state of the monomer (which, expressed as a wavenumber, is simply $\tilde{\nu}_{\text{mon}}$), and off-diagonal elements corresponding to the energy of interaction between the transition dipoles. Using the arrangement discussed in (1), we write this interaction energy (as a wavenumber) as:

$$\beta = \frac{\mu_{\text{mon}}^2}{4\pi\varepsilon_0 hcr^3}(1 - 3\cos^2\theta)$$

It follows that the hamiltonian matrix is

$$\hat{H} = \begin{pmatrix} \tilde{\nu}_{\text{mon}} & \beta \\ \beta & \tilde{\nu}_{\text{mon}} \end{pmatrix}$$

The eigenvalues of the matrix are the dimer transition wavenumbers $\tilde{\nu}_1$ and $\tilde{\nu}_2$. Theeigenvectors are the wavefunctions for the excited states of the dimer and have the form $\begin{pmatrix} c_j \\ c_k \end{pmatrix}$. (a) The intensity of absorption of incident radiation is proportional to the square of the transition dipole moment (Section 9.3). The monomer transition dipole moment is $\mu_{\text{mon}} = \int \psi_1^* \hat{\mu} \psi_0 \, d\tau = \int \psi_2^* \hat{\mu} \psi_0 \, d\tau$, where ψ_0 is the wavefunction of the monomer ground state. Assume that the dimer ground state may also be described by ψ_0 and show that the transition dipole moment μ_i of each dimer transition is given by $\mu_i = \mu_{\text{mon}}(c_j + c_k)$.

19.32 (a) Consider a dimer of monomers with $\mu_{\text{mon}} = 4.00$ D, $\tilde{\nu}_{\text{mon}} = 25\,000$ cm^{-1}, and $r = 0.5$ nm. How do the transition wavenumbers $\tilde{\nu}_1$ and $\tilde{\nu}_2$ vary with the angle θ? The relative intensities of the dimer transitions may be estimated by calculating the ratio μ_2^2/μ_1^2. How does this ratio vary with the angle θ? (c) Now expand the treatment given above to a chain of N monomers ($N = 5$, 10, 15, and 20), with $\mu_{\text{mon}} = 4.00$ D, $\tilde{\nu}_{\text{mon}} = 25\,000$ cm^{-1}, and $r = 0.5$ nm. For simplicity, assume that $\theta = 0$ and that only nearest neighbours interact with interaction energy V. For example the hamiltonian matrix for the case $N = 4$ is

$$\hat{H} = \begin{pmatrix} \tilde{\nu}_{\text{mon}} & V & 0 & 0 \\ V & \tilde{\nu}_{\text{mon}} & V & 0 \\ 0 & V & \tilde{\nu}_{\text{mon}} & V \\ 0 & 0 & V & \tilde{\nu}_{\text{mon}} \end{pmatrix}$$

How does the wavenumber of the lowest energy transition vary with size of the chain? How does the transition dipole moment of the lowest energy transition vary with the size of the chain?

19.33 Show that if a substance responds nonlinearly to two sources of radiation, one of frequency ω_1 and the other of frequency ω_2, then it may give rise to radiation of the sum and difference of the two frequencies. This nonlinear optical phenomenon is known as *frequency mixing* and is used to expand the wavelength range of lasers in laboratory applications, such as spectroscopy and photochemistry.

19.34 The magnetizability, ξ, and the volume and molar magnetic susceptibilities can be calculated from the wavefunctions of molecules. For instance, the magnetizability of a hydrogenic atom is given by the expression $\xi = -(e^2/6m_e)\langle r^2 \rangle$, where $\langle r^2 \rangle$ is the (expectation) mean value of r^2 in the atom. Calculate ξ and χ_m for the ground state of a hydrogenic atom.

19.35 Nitrogen dioxide, a paramagnetic compound, is in equilibrium with its dimer, dinitrogen tetroxide, a diamagnetic compound. Derive an expression in terms of the equilibrium constant, K, for the dimerization to show how the molar susceptibility varies with the pressure of the sample. Suggest how the susceptibility might be expected to vary as the temperature is changed at constant pressure.

19.36 An NO molecule has thermally accessible electronically excited states. It also has an unpaired electron, and so may be expected to be paramagnetic. However, its ground state is not paramagnetic because the magnetic moment of the orbital motion of the unpaired electron almost exactly cancels the spin magnetic moment. The first excited state (at 121 cm^{-1}) is paramagnetic because the orbital magnetic moment adds to, rather than cancels, the spin magnetic moment. The upper state has a magnetic moment of $2\mu_B$. Because the upper state is thermally accessible, the paramagnetic susceptibility of NO shows a pronounced temperature dependence even near room temperature. Calculate the molar paramagnetic susceptibility of NO and plot it as a function of temperature.

Applications to: biochemistry and nanoscience

19.37 Although the crystallization of large biological molecules may not be as readily accomplished as that of small molecules, their crystal lattices are no different. Tobacco seed globulin forms face-centred cubic crystals with unit cell dimension of 12.3 nm and a density of 1.287 g cm^{-3}. Determine its molar mass.

19.38 What features in an X-ray diffraction pattern suggest a helical conformation for a biological macromolecule? Use Fig. 19.42 to deduce as much quantitative information as you can about the shape and size of a DNA molecule.

19.39 A transistor is a semiconducting device that is commonly used either as a switch or an amplifier of electrical signals. Prepare a brief report on the design of a nanometre-sized transistor that uses a carbon nanotube as a component. A useful starting point is the work summarized by S.J. Tans *et al.* (*Nature* **393**, 49 (1998)).

19.40 The tip of a scanning tunnelling microscope can be used to move atoms on a surface. The movement of atoms and ions depends on their ability to leave one position and stick to another, and therefore on the energy changes that occur. As an illustration, consider a two-dimensional square lattice of univalent positive and negative ions separated by 200 pm, and consider a cation on top of this array. Calculate, by direct summation, its Coulombic interaction when it is in an empty lattice point directly above an anion.

MATHEMATICAL BACKGROUND 7

Fourier series and Fourier transforms

Some of the most versatile mathematical functions are the trigonometric functions sine and cosine. As a result, it is often very helpful to express a general function as a linear combination of these functions and then to carry out manipulations on the resulting series. Because sines and cosines have the form of waves, the linear combinations often have a straightforward physical interpretation. Throughout this discussion, the function $f(x)$ is real.

MB7.1 Fourier series

A *Fourier series* is a linear combination of sines and cosines that replicates a periodic function:

$$f(x)=\tfrac{1}{2}a_0+\sum_{n=1}^{\infty}\left\{a_n\cos\frac{2n\pi x}{L}+b_n\sin\frac{2n\pi x}{L}\right\} \qquad \text{(MB7.1)}$$

A periodic function is one that repeats periodically, such that $f(x+2L)=f(x)$ where $2L$ is the period. Although it is perhaps not surprising that sines and cosines can be used to replicate continuous functions, it turns out that—with certain limitations—they can also be used to replicate discontinuous functions. The coefficients in eqn MB7.1 are found by making use of the orthogonality of the sine and cosine functions

$$\int_{-L}^{L}\sin\frac{m\pi x}{L}\cos\frac{m\pi x}{L}\mathrm{d}x=0 \qquad \text{(MB7.2a)}$$

and the integrals

$$\int_{-L}^{L}\sin\frac{m\pi x}{L}\sin\frac{n\pi x}{L}\mathrm{d}x=\int_{-L}^{L}\cos\frac{m\pi x}{L}\cos\frac{n\pi x}{L}\mathrm{d}x=L\delta_{mn} \qquad \text{(MB7.2b)}$$

where $\delta_{mn}=1$ if $m=n$ and 0 if $m\neq n$. Thus, multiplication of both sides of eqn MB7.1 by $\cos(k\pi x/L)$ and integration from $-L$ to L gives an expression for the coefficient a_k, and multiplication by $\sin(k\pi x/L)$ and integration likewise gives an expression for b_k:

$$a_k=\frac{1}{L}\int_{-L}^{L}f(x)\cos\frac{k\pi x}{L}\mathrm{d}x \qquad k=0,1,2,\ldots$$
$$b_k=\frac{1}{L}\int_{-L}^{L}f(x)\sin\frac{k\pi x}{L}\mathrm{d}x \qquad k=0,1,2,\ldots \qquad \text{(MB7.3)}$$

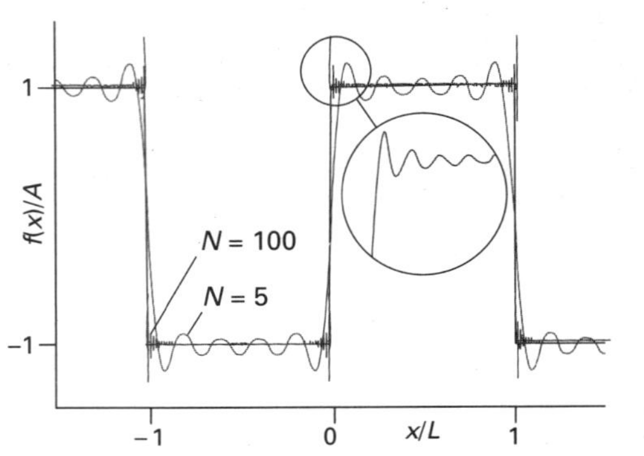

Fig. MB7.1 A square wave and two successive approximations by Fourier series ($N=5$ and $N=100$). The inset shows a magnification of the $N=100$ approximation.

• A brief illustration

Figure MB7.1 shows a graph of a square wave of amplitude A that is periodic between $-L$ and L. The mathematical form of the wave is

$$f(x)=\begin{cases}-A & -L\le x\le 0\\ +A & 0\le x\le L\end{cases}$$

The coefficients a are all zero because $f(x)$ is antisymmetric ($f(-x)=-f(x)$) whereas all the cosine functions are symmetric ($\cos(-x)=\cos(x)$) and so cosine waves make no contribution to the sum. The coefficients b are obtained from

$$b_k=\frac{1}{L}\int_{-L}^{L}f(x)\sin\frac{k\pi x}{L}\mathrm{d}x$$
$$=\frac{1}{L}\int_{-L}^{0}(-A)\sin\frac{k\pi x}{L}\mathrm{d}x+\frac{1}{L}\int_{0}^{L}A\sin\frac{k\pi x}{L}\mathrm{d}x=\frac{2A}{k\pi}\{1-(-1)^k\}$$

The final expression has been formulated to acknowledge that the two integrals cancel when k is even but add together when k is odd. Therefore,

$$f(x)=\frac{2A}{\pi}\sum_{k=1}^{N}\frac{1-(-1)^k}{k}\sin\frac{2k\pi x}{L}=\frac{4A}{\pi}\sum_{n=1}^{N}\frac{1}{2n-1}\sin\frac{2(2n-1)\pi x}{L}=$$

with $N\to\infty$. The sum over n is the same as the sum over k; in the latter, terms with k even are all zero. This function is plotted in Fig. MB7.1 for two values of N to show how the series becomes more faithful to the original function as N increases. •

Self-test MB7.1 Repeat the analysis for a saw-tooth wave, $f(x)=Ax$ in the range $-L\le x<L$ and $f(x+2L)=f(x)$ elsewhere. Use graphing software to depict the result.

$$[f(x)=(2AL/\pi)\sum_{n=1}^{\infty}\{(-1)^{n+1}/n\}\sin(n\pi x/L),\ \text{Fig. MB7.2}]$$

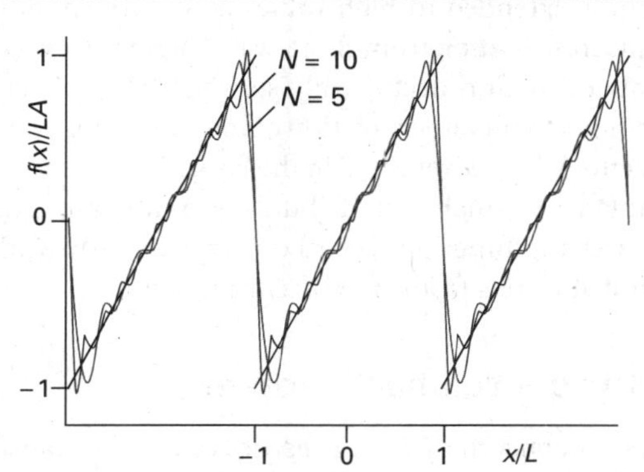

Fig. MB7.2 A saw-tooth function and its representation as a Fourier series with two successive approximations ($N = 5$ and $N = 10$).

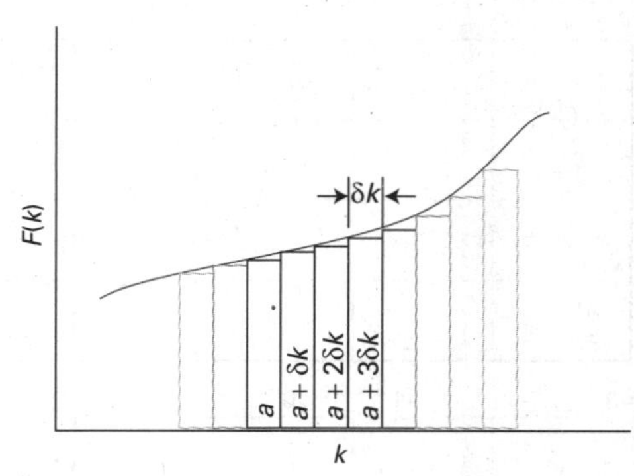

Fig. MB7.3 The formal definition of an integral as the sum of the value of a function at a series of infinitely spaced points multiplied by the separation of each point.

MB7.2 Fourier transforms

The Fourier series in eqn MB7.1 can be expressed in a more succinct manner if we allow the coefficients to be complex numbers and make use of *de Moivre's relation*

$$e^{in\pi x/L} = \cos\frac{n\pi x}{L} + i\sin\frac{n\pi x}{L} \quad \text{(MB7.4)}$$

for then we may write

$$f(x) = \sum_{n=-\infty}^{\infty} c_n e^{in\pi x/L} \qquad c_n = \frac{1}{2L}\int_{-L}^{L} f(x)e^{-in\pi x/L}dx \quad \text{(MB7.5)}$$

This complex formalism is well suited to the extension of this discussion to functions with periods that become infinite. If a period is infinite, we are effectively dealing with a non-periodic function, such as the decaying exponential function e^{-x}.

We write $\delta k = \pi/L$ and consider the limit as $L \to \infty$ and therefore $\delta k \to 0$: that is, eqn MB7.5 becomes

$$\begin{aligned} f(x) &= \lim_{L\to\infty} \sum_{n=-\infty}^{\infty} \left\{ \frac{1}{2L}\int_{-L}^{L} f(x')e^{-in\pi x'/L}dx' \right\} e^{in\pi x/L} \\ &= \lim_{\delta k\to 0} \sum_{n=-\infty}^{\infty} \left\{ \frac{\delta k}{2\pi}\int_{-\pi/\delta k}^{\pi/\delta k} f(x')e^{-in\delta k x'}dx' \right\} e^{in\delta k x} \quad \text{(MB7.6)} \\ &= \frac{1}{2\pi}\lim_{\delta k\to 0} \sum_{n=-\infty}^{\infty} \left\{ \int_{-\infty}^{\infty} f(x')e^{-in\delta k(x'-x)}dx' \right\} \delta k \end{aligned}$$

In the last line we have anticipated that the limits of the integral will become infinite. At this point we should recognize that a formal definition of an integral is the sum of the value of a function at a series of infinitely spaced points multiplied by the separation of each point (Fig. MB7.3):

$$\int_a^b F(k)dk = \lim_{\delta k\to 0} \sum_{n=-\infty}^{\infty} F(n\delta k)\delta k \quad \text{(MB7.7)}$$

Exactly this form appears on the right-hand side of eqn MB7.6, so we can write that equation as

$$f(x) = \frac{1}{2\pi}\int_{-\infty}^{\infty} \tilde{f}(k)e^{ikx}dk \quad \text{where} \quad \tilde{f}(k) = \int_{-\infty}^{\infty} f(x')e^{-ikx'}dx' \quad \text{(MB7.8)}$$

(At this stage we can drop the prime on x.) We call the function $\tilde{f}(k)$ the *Fourier transform* of $f(x)$; the original function $f(x)$ is the *inverse Fourier transform* of $\tilde{f}(k)$.

• A brief illustration

The Fourier transform of the symmetrical exponential function $f(x) = e^{-a|x|}$ is

$$\begin{aligned} \tilde{f}(k) &= \int_{-\infty}^{\infty} e^{-a|x|-ikx}dx \\ &= \int_{-\infty}^{0} e^{ax-ikx}dx + \int_{0}^{\infty} e^{-ax-ikx}dx \\ &= \frac{1}{a-ik} + \frac{1}{a+ik} = \frac{2a}{a^2+k^2} \end{aligned}$$

The original function and its Fourier transform are drawn in Fig. MB7.4. •

Self-test MB7.2 Evaluate the Fourier transform of the Gaussian function $e^{-a^2x^2}$.

$[\tilde{f}(k) = (\pi/a^2)^{1/2} e^{-k^2/4a^2}]$

The physical interpretation of eqn MB7.8 is that $f(x)$ is expressed as a superposition of harmonic (sine and cosine) functions of wavelength $\lambda = 2\pi/k$, and that the weight of each

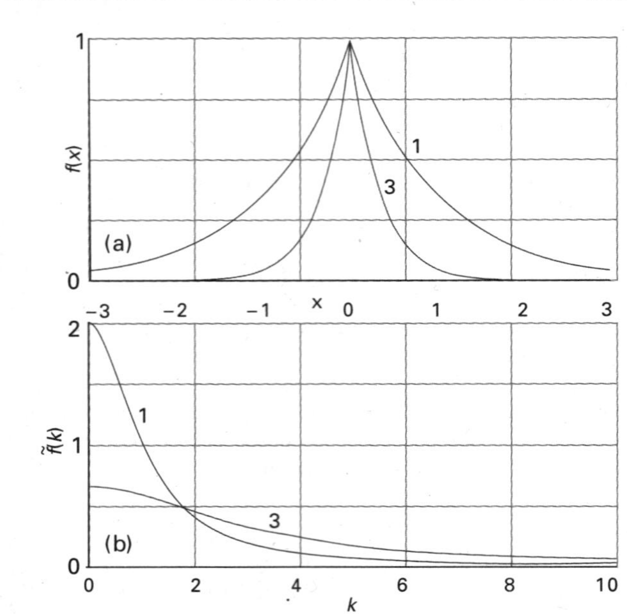

Fig. MB7.4 (a) The symmetrical exponential function $f(x) = e^{-a|x|}$ and (b) its Fourier transform for two values of the decay constant a. Note how the function with the more rapid decay has a Fourier transform richer in short-wavelength (high k) components.

constituent function is given by the Fourier transform at the corresponding value of k. This interpretation is consistent with the calculation in the *brief illustration*. As we see from Fig. MB7.4, when the exponential function falls away rapidly, the Fourier transform is extended to high values of k, corresponding to a significant contribution from short-wavelength waves. When the exponential function decays only slowly, the most significant contributions to the superposition come from long-wavelength components, which is reflected in the Fourier transform, with its predominance of small-k contributions in this case. In general, a slowly varying function has a Fourier transform with significant contributions from small-k components.

MB7.3 The convolution theorem

A final point concerning the properties of Fourier transforms is the *convolution theorem*, which states that, if a function is the 'convolution' of two other functions, that is if

$$F(x) = \int_{-\infty}^{\infty} f_1(x')f_2(x-x')\mathrm{d}x' \tag{MB7.9a}$$

then the Fourier transform of $F(x)$ is the product of the Fourier transforms of its component functions:

$$\tilde{F}(k) = \tilde{f}_1(k)\tilde{f}_2(k) \tag{MB7.9b}$$

• **A brief illustration**

If $F(x)$ is the convolution of two Gaussian functions,

$$F(x) = \int_{-\infty}^{\infty} e^{-a^2x'^2}e^{-b^2(x-x')^2}\mathrm{d}x'$$

then from Self-test MB7.2 we can immediately write its transform as

$$\tilde{F}(k) = \left(\frac{\pi}{a^2}\right)^{1/2} e^{-k^2/4a^2}\left(\frac{\pi}{b^2}\right)^{1/2} e^{-k^2/4b^2} = \frac{\pi}{ab}e^{-(k^2/4)(1/a^2+1/b^2)}$$ •

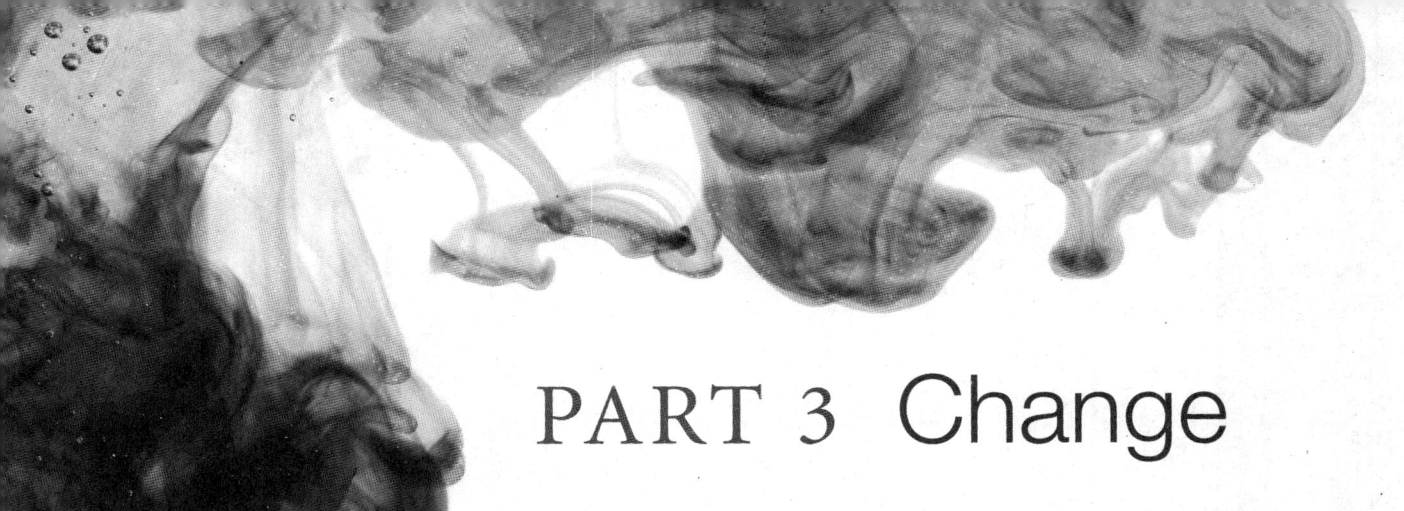

PART 3 Change

Part 3 considers the processes by which change occurs. We prepare the ground for a discussion of the rates of reactions by considering the motion of molecules in gases and in liquids. Then we establish the precise meaning of reaction rate, and see how the overall rate, and the complex behaviour of some reactions, may be expressed in terms of elementary steps and the atomic events that take place when molecules meet. Of enormous importance in both industry and biology is the control of reaction rates by catalysis, which we discuss in the last chapter of the text.

Molecules in motion

20

One of the simplest types of molecular motion to describe is the random motion of molecules of a perfect gas. We see that a simple theory accounts for the pressure of a gas and the rates at which molecules and energy migrate through gases. Molecular mobility is particularly important in liquids. Another simple kind of motion is the largely uniform motion of ions in solution in the presence of an electric field. Molecular and ionic motion have common features and, by considering them from a more general viewpoint, we derive expressions that govern the migration of properties through matter. One of the most useful consequences of this general approach is the formulation of the diffusion equation, which is an equation that shows how matter and energy spread through media of various kinds. Finally, we build a simple model for all types of molecular motion, in which the molecules migrate in a series of small steps, and see that it accounts for many of the properties of migrating molecules in both gases and condensed phases.

This chapter provides techniques for discussing the motion of all kinds of particles in all kinds of fluids. We set the scene by considering a simple type of motion, that of molecules in a perfect gas, and go on to see that molecular motion in liquids shows a number of similarities. We shall concentrate on the **transport properties** of a substance, its ability to transfer matter, energy, or some other property from one place to another. Four examples of transport properties are

Diffusion, the migration of matter down a concentration gradient.

Thermal conduction, the migration of energy down a temperature gradient.

Electric conduction, the migration of electric charge up or down an electrical potential gradient.

Viscosity, the migration of linear momentum down a velocity gradient.

It is convenient to include in the discussion **effusion**, the emergence of a gas from a container through a small hole.

Molecular motion in gases

Here we present the kinetic model of a perfect gas as a starting point for the discussion of its transport properties. In the **kinetic model** of gases we assume that the only contribution to the energy of the gas is from the kinetic energies of the molecules. The kinetic model is one of the most remarkable—and arguably most beautiful—models in physical chemistry for, from a set of very slender assumptions, powerful quantitative conclusions can be deduced.

20.1 The kinetic model of gases

Key points **The kinetic model of a gas considers only the contribution to the energy from the kinetic energies of the molecules. (a) Important results from the model include expressions for the pressure and the root mean square speed. The Maxwell distribution of speeds gives the fraction of molecules that have speeds in a specified range. (b) The collision frequency is the number of collisions made by a molecule in an interval divided by the length of the interval. (c) The mean free path is the average distance a molecule travels between collisions.**

The kinetic model is based on three assumptions:

1. The gas consists of molecules of mass m in ceaseless random motion.
2. The size of the molecules is negligible, in the sense that their diameters are much smaller than the average distance travelled between collisions.
3. The molecules interact only through brief, infrequent, and elastic collisions.

An **elastic collision** is a collision in which the total translational kinetic energy of the molecules is conserved.

(a) Pressure and molecular speeds

From the very economical assumptions of the kinetic model, we show in the following *Justification* that the pressure and volume of the gas are related by

$$pV = \tfrac{1}{3}nMc^2 \qquad \text{(20.1)}^\circ$$

The pressure of a perfect gas according to the kinetic model

where $M = mN_A$, the molar mass of the molecules, and c is the **root mean square speed** of the molecules, the square root of the mean of the squares of the speeds, v, of the molecules:

$$c = \langle v^2 \rangle^{1/2} \qquad [20.2]$$

Definition of the root mean square speed

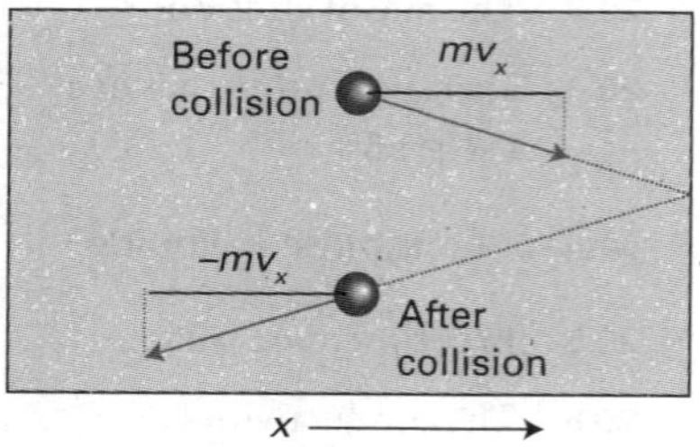

Fig. 20.1 The pressure of a gas arises from the impact of its molecules on the walls. In an elastic collision of a molecule with a wall perpendicular to the x-axis, the x-component of velocity is reversed but the y- and z-components are unchanged.

Justification 20.1 *The pressure of a gas according to the kinetic model*

Consider the arrangement in Fig. 20.1. When a particle of mass m that is travelling with a component of velocity v_x parallel to the x-axis collides with the wall on the right and is reflected, its linear momentum (the product of its mass and its velocity) changes from mv_x before the collision to $-mv_x$ after the collision (when it is travelling in the opposite direction). The x-component of momentum therefore changes by $2mv_x$ on each collision (the y- and z-components are unchanged). Many molecules collide with the wall in an interval Δt, and the total change of momentum is the product of the change in momentum of each molecule multiplied by the number of molecules that reach the wall during the interval.

Because a molecule with velocity component v_x can travel a distance $v_x\Delta t$ along the x-axis in an interval Δt, all the molecules within a distance $v_x\Delta t$ of the wall will strike it if they are travelling towards it (Fig. 20.2). It follows that, if the wall has area A, then all the particles in a volume $A \times v_x\Delta t$ will reach the wall (if they are travelling towards it). The number density of particles is nN_A/V, where n is the total amount of molecules in the container of volume V and N_A is Avogadro's constant, so the number of molecules in the volume $Av_x\Delta t$ is $(nN_A/V) \times Av_x\Delta t$.

At any instant, half the particles are moving to the right and half are moving to the left. Therefore, the average number of collisions with the wall during the interval Δt

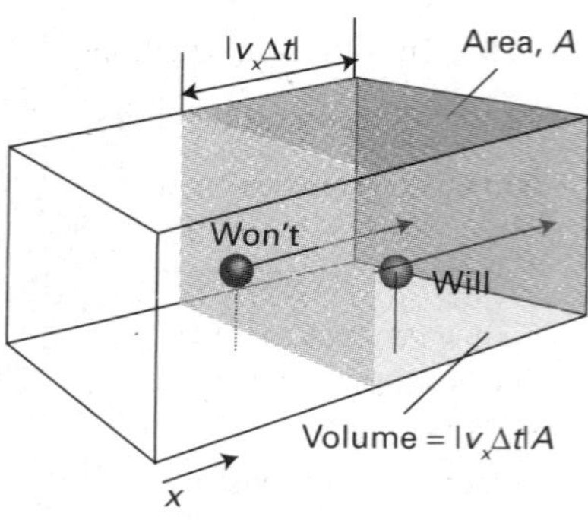

Fig. 20.2 A molecule will reach the wall on the right within an interval Δt if it is within a distance $v_x\Delta t$ of the wall and travelling to the right.

is $\frac{1}{2}nN_AAv_x\Delta t/V$. The total momentum change in that interval is the product of this number and the change $2mv_x$:

$$\text{Momentum change} = \frac{nN_AAv_x\Delta t}{2V} \times 2mv_x = \frac{nmAN_Av_x^2\Delta t}{V} = \frac{nMAv_x^2\Delta t}{V}$$

where $M = mN_A$.

Next, to find the force, we calculate the rate of change of momentum, which is this change of momentum divided by the interval Δt during which it occurs:

$$\text{Rate of change of momentum} = \frac{nMAv_x^2}{V}$$

This rate of change of momentum is equal to the force (by Newton's second law of motion). It follows that the pressure, the force divided by the area, is

$$\text{Pressure} = \frac{nMv_x^2}{V}$$

Not all the molecules travel with the same velocity, so the detected pressure, p, is the average (denoted $\langle\cdots\rangle$) of the quantity just calculated:

$$p = \frac{nM\langle v_x^2\rangle}{V}$$

This expression already resembles the perfect gas equation of state.

To write an expression of the pressure in terms of the root mean square speed, c, we begin by writing the speed of a single molecule, v, as $v = v_x^2 + v_y^2 + v_z^2$. Because the root-mean-square speed, c, is defined as $c = \langle v^2\rangle^{1/2}$ (eqn 20.2), it follows that

$$c^2 = \langle v^2\rangle = \langle v_x^2\rangle + \langle v_y^2\rangle + \langle v_z^2\rangle$$

However, because the molecules are moving randomly, all three averages are the same. It follows that $c^2 = 3\langle v_x^2\rangle$. Equation 20.1 follows immediately by substituting $\langle v_x^2\rangle = \frac{1}{3}c^2$ into $p = nM\langle v_x^2\rangle/V$.

Equation 20.1 is one of the key results of the kinetic model. We see that, if the root mean square speed of the molecules depends only on the temperature, then at constant temperature

$$pV = \text{constant}$$

which is the content of Boyle's law (Section 1.2). Moreover, for eqn 20.1 to be the equation of state of a perfect gas, its right-hand side must be equal to nRT. It follows that the root mean square speed of the molecules in a gas at a temperature T must be

$$c = \left(\frac{3RT}{M}\right)^{1/2} \qquad \text{Root mean square speed in a perfect gas} \qquad (20.3)°$$

We can conclude that the root mean square speed of the molecules of a gas is proportional to the square root of the temperature and inversely proportional to the square root of the molar mass. That is, the higher the temperature, the higher the root mean square speed of the molecules, and, at a given temperature, heavy molecules travel more slowly than light molecules.

• A brief illustration

The root mean square speed of N_2 molecules ($M = 28.02\ \text{g mol}^{-1}$) at 298 K is found from eqn 20.3 to be

$$c = \left(\frac{3 \times 8.3145\ \text{J K}^{-1}\ \text{mol}^{-1} \times 298\ \text{K}}{28.02 \times 10^{-3}\ \text{kg mol}^{-1}}\right)^{1/2} = 515\ \text{m s}^{-1}$$

Sound waves are pressure waves, and for them to propagate the molecules of the gas must move to form regions of high and low pressure. Therefore, we should expect the speed of sound in air to be approximately 500 m s^{-1}. The experimental value is 340 m s^{-1}. •

Equation 20.3 is an expression for the mean square speed of molecules. However, in an actual gas the speeds of individual molecules span a wide range, and the collisions in the gas continually redistribute the speeds among the molecules. Before a collision, a molecule may be travelling rapidly, but after a collision it may be accelerated to a very high speed, only to be slowed again by the next collision. The fraction of molecules that have speeds in the range v to $v + dv$ is proportional to the width of the range, and is written $f(v)dv$, where $f(v)$ is called the **distribution of speeds**. Note that, in common with other distribution functions, $f(v)$ acquires physical significance only after it is multiplied by the range of speeds of interest.

The precise form of f for molecules of a gas at a temperature T was derived by J.C. Maxwell, and is

$$f(v) = 4\pi\left(\frac{M}{2\pi RT}\right)^{3/2} v^2 e^{-Mv^2/2RT} \qquad \text{Maxwell distribution of speeds} \qquad (20.4)$$

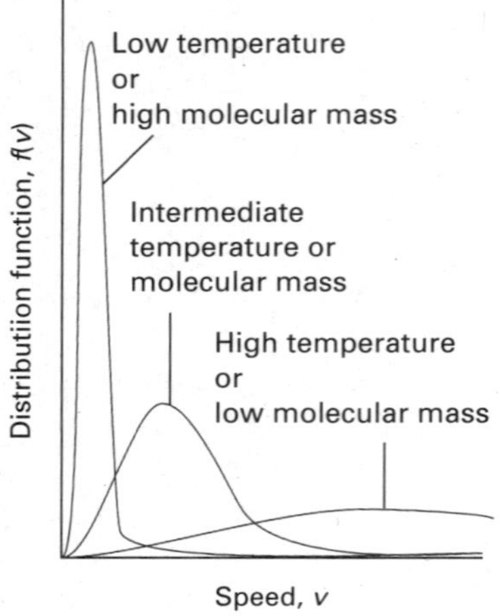

Fig. 20.3 The distribution of molecular speeds with temperature and molar mass. Note that the most probable speed (corresponding to the peak of the distribution) increases with temperature and with decreasing molar mass, and simultaneously the distribution becomes broader.

interActivity (a) Plot different distributions by keeping the molar mass constant at 100 g mol^{-1} and varying the temperature of the sample between 200 K and 2000 K. (b) Use mathematical software or the *Living graph* applet from the text's web site to evaluate numerically the fraction of molecules with speeds in the range 100 m s^{-1} to 200 m s^{-1} at 300 K and 1000 K. (c) Based on your observations, provide a molecular interpretation of temperature.

This expression is called the **Maxwell distribution of speeds** and is derived in the following *Justification*. Let's consider its features, which are also shown pictorially in Fig. 20.3:

1. Equation 20.4 includes a decaying exponential function, the term $e^{-Mv^2/2RT}$. Its presence implies that the fraction of molecules with very high speeds will be very small because e^{-x^2} becomes very small when x^2 is large.

2. The factor $M/2RT$ multiplying v^2 in the exponent is large when the molar mass, M, is large, so the exponential factor goes most rapidly towards zero when M is large. That is, heavy molecules are unlikely to be found with very high speeds.

3. The opposite is true when the temperature, T, is high: then the factor $M/2RT$ in the exponent is small, so the exponential factor falls towards zero relatively slowly as v increases. In other words, a greater fraction of the molecules can be expected to have high speeds at high temperatures than at low temperatures.

4. A factor v^2 (the term before the e) multiplies the exponential. This factor goes to zero as v goes to zero, so the fraction of molecules with very low speeds will also be very small.

5. The remaining factors (the term in parentheses in eqn 20.4 and the 4π) simply ensure that, when we add together the fractions over the entire range of speeds from zero to infinity, then we get 1.

To use eqn 20.4 to calculate the fraction of molecules in a given narrow range of speeds, Δv, we evaluate $f(v)$ at the speed of interest, then multiply it by the width of the range of speeds of interest, that is, we form $f(v)\Delta v$. To use the distribution to calculate the fraction in a range of speeds that is too wide to be treated as infinitesimal, we evaluate the integral:

$$\text{Fraction in the range } v_1 \text{ to } v_2 = \int_{v_1}^{v_2} f(v)\,dv \qquad (20.5)$$

This integral is the area under the graph of f as a function of v and, except in special cases, has to be evaluated numerically by using mathematical software (Fig. 20.4).

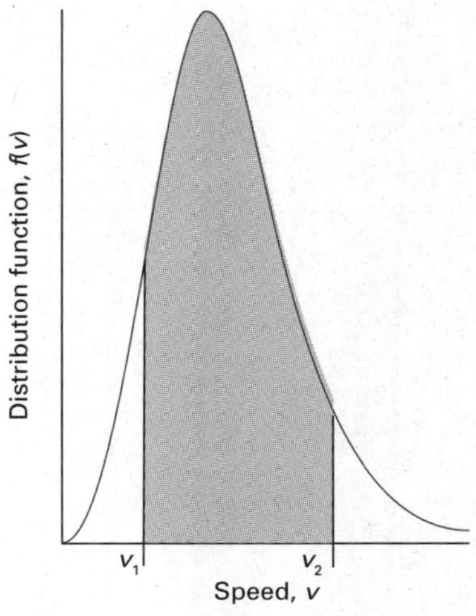

Fig. 20.4 To calculate the probability that a molecule will have a speed in the range v_1 to v_2, we integrate the distribution between those two limits; the integral is equal to the area of the curve between the limits, as shown shaded here.

Justification 20.2 *The Maxwell distribution of speeds*

The Boltzmann distribution is a key result of physical chemistry; it was introduced in *Fundamentals F.5* and treated fully in Section 15.1. It implies that the fraction of molecules with velocity components v_x, v_y, v_z is proportional to an exponential function of their kinetic energy, E_k, which is

$$E_k = \tfrac{1}{2}mv_x^2 + \tfrac{1}{2}mv_y^2 + \tfrac{1}{2}mv_z^2$$

Therefore, we can use the relation $a^{x+y+z+\cdots} = a^x a^y a^z \ldots$ to write

$$f = Ke^{-E_k/kT} = Ke^{-(\frac{1}{2}mv_x^2+\frac{1}{2}mv_y^2+\frac{1}{2}mv_z^2)/kT} = Ke^{-mv_x^2/2kT}e^{-mv_y^2/2kT}e^{-mv_z^2/2kT}$$

where K is a constant of proportionality (at constant temperature) and $f\mathrm{d}v_x\mathrm{d}v_y\mathrm{d}v_z$ is the fraction of molecules in the velocity range v_x to $v_x + \mathrm{d}v_x$, v_y to $v_y + \mathrm{d}v_y$, and v_z to $v_z + \mathrm{d}v_z$. We see that the fraction factorizes into three factors, one for each axis, and we can write $f = f(v_x)f(v_y)f(v_z)$ with

$$f(v_x) = K^{1/3}e^{-mv_x^2/2kT}$$

and likewise for the two other directions.

To determine the constant K, we note that a molecule must have a velocity somewhere in the range $-\infty < v_x < \infty$, so

$$\int_{-\infty}^{\infty} f(v_x)\mathrm{d}v_x = 1$$

Substitution of the expression for $f(v_x)$ then gives

$$1 = K^{1/3}\int_{-\infty}^{\infty} e^{-mv_x^2/2kT}\mathrm{d}v_x = K^{1/3}\left(\frac{2\pi kT}{m}\right)^{1/2}$$

where we have used the standard integral

$$\int_{-\infty}^{\infty} e^{-ax^2}\mathrm{d}x = \left(\frac{\pi}{a}\right)^{1/2}$$

Therefore, $K = (m/2\pi kT)^{3/2} = (M/2\pi RT)^{3/2}$, where M is the molar mass of the molecules. At this stage we know that

$$f(v_x) = \left(\frac{M}{2\pi RT}\right)^{1/2} e^{-Mv_x^2/2RT} \qquad (20.6)$$

The probability that a molecule has a velocity in the range v_x to $v_x + \mathrm{d}v_x$, v_y to $v_y + \mathrm{d}v_y$, v_z to $v_z + \mathrm{d}v_z$ is the product of these individual probabilities:

$$f(v_x)f(v_y)f(v_z)\mathrm{d}v_x\mathrm{d}v_y\mathrm{d}v_z = \left(\frac{M}{2\pi RT}\right)^{3/2} e^{-Mv^2/2RT}\mathrm{d}v_x\mathrm{d}v_y\mathrm{d}v_z$$

where $v^2 = v_x^2 + v_y^2 + v_z^2$. The probability $f(v)\mathrm{d}v$ that the molecules have a speed in the range v to $v + \mathrm{d}v$ regardless of direction is the sum of the probabilities that the velocity lies in any of the volume elements $\mathrm{d}v_x\mathrm{d}v_y\mathrm{d}v_z$ forming a spherical shell of radius v and thickness $\mathrm{d}v$ (Fig. 20.5). The sum of the volume elements on the right-hand side of the last equation is the volume of this shell, $4\pi v^2\mathrm{d}v$. Therefore, the probability that it is in a volume element $\mathrm{d}v_x\mathrm{d}v_y\mathrm{d}v_z$ at a distance v from the origin

$$f(v) = 4\pi\left(\frac{M}{2\pi RT}\right)^{3/2} v^2 e^{-Mv^2/2RT}$$

as given in eqn 20.4.

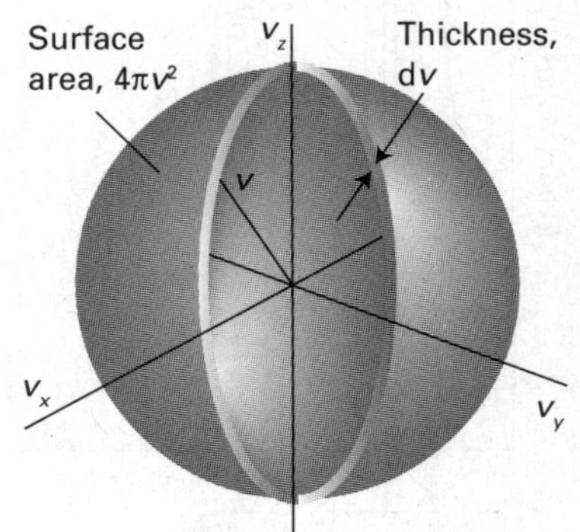

Fig. 20.5 To evaluate the probability that a molecule has a speed in the range v to $v + \mathrm{d}v$, we evaluate the total probability that the molecule will have a speed that is anywhere on the surface of a sphere of radius $v = (v_x^2 + v_y^2 + v_z^2)^{1/2}$ by summing the probabilities that it is in a volume element $\mathrm{d}v_x\mathrm{d}v_y\mathrm{d}v_z$ at a distance v from the origin.

Example 20.1 *Calculating the mean speed of molecules in a gas*

What is the mean speed, $\bar{c}$, of N_2 molecules in air at 25°C?

Method A mean speed is calculated by multiplying each speed by the fraction of molecules that have that speed, and then adding all the products together. When the speed varies over a continuous range, the sum is replaced by an integral. To employ this approach here, we note that the fraction of molecules with a speed in the range v to $v + dv$ is $f(v)dv$, so the product of this fraction and the speed is $vf(v)dv$. The mean speed, $\bar{c}$, is obtained by evaluating the integral

$$\bar{c} = \int_{-\infty}^{\infty} vf(v)\mathrm{d}v$$

with $f(v)$ given in eqn 20.4.

Answer The integral required is

$$\bar{c} = 4\pi\left(\frac{M}{2\pi RT}\right)^{3/2}\int_0^{\infty} v^3 \mathrm{e}^{-Mv^2/2RT}\mathrm{d}v$$

$$= 4\pi\left(\frac{M}{2\pi RT}\right)^{3/2} \times \tfrac{1}{2}\left(\frac{2RT}{M}\right)^2 = \left(\frac{8RT}{\pi M}\right)^{1/2}$$

where we have used the standard result from tables of integrals (or software) that

$$\int_0^{\infty} x^3 \mathrm{e}^{-ax^2}\mathrm{d}x = \frac{1}{2a^2}$$

Substitution of the data then gives

$$\bar{c} = \left(\frac{8 \times (8.3141\ \mathrm{J\ K^{-1}\ mol^{-1}}) \times (298\ \mathrm{K})}{\pi \times (28.02 \times 10^{-3}\ \mathrm{kg\ mol^{-1}})}\right)^{1/2} = 475\ \mathrm{m\ s^{-1}}$$

where we have used $1\ \mathrm{J} = 1\ \mathrm{kg\ m^2\ s^{-2}}$.

Self-test 20.1 Evaluate the root mean square speed of the molecules by integration. You will need the integral

$$\int_0^{\infty} x^4 \mathrm{e}^{-ax^2}\mathrm{d}x = \tfrac{3}{8}\left(\frac{\pi}{a^5}\right)^{1/2}$$

$[c = (3RT/M)^{1/2}, 515\ \mathrm{m\ s^{-1}}]$

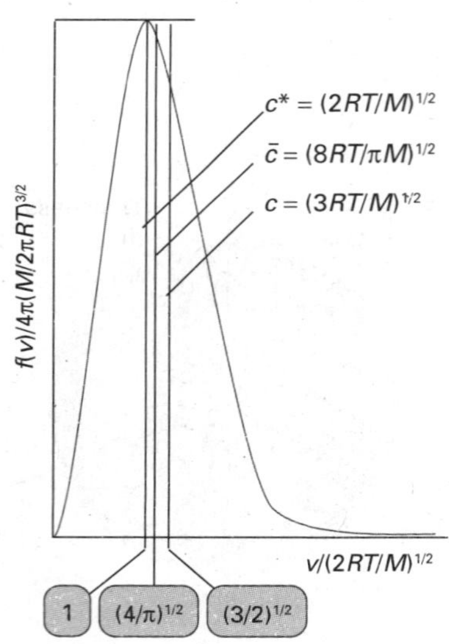

Fig. 20.6 A summary of the conclusions that can be deduced from the Maxwell distribution for molecules of molar mass M at a temperature T: c^* is the most probable speed, $\bar{c}$ is the mean speed, and c is the root mean square speed.

As shown in Example 20.1, we can use the Maxwell distribution to evaluate the **mean speed**, $\bar{c}$, of the molecules in a gas:

$$\bar{c} = \left(\frac{8RT}{\pi M}\right)^{1/2} \qquad \text{Mean speed} \qquad (20.7)$$

We can identify the **most probable speed**, c^*, by differentiating f with respect to v and looking for the value of v at which the derivative is zero (other than at $v = 0$ and $v = \infty$):

$$c^* = \left(\frac{2RT}{M}\right)^{1/2} \qquad \text{Most probable speed} \qquad (20.8)$$

Figure 20.6 summarizes these results. Note that the mean speed is the value of v that divides the distribution into two equal areas.

The **relative mean speed**, $\bar{c}_{rel}$, the mean speed with which one molecule approaches another, can also be calculated from the distribution:

$$\bar{c}_{rel} = 2^{1/2}\bar{c} \qquad \text{Relative mean speed} \qquad (20.9)$$

This result is much harder to derive, but the diagram in Fig. 20.7 should help to show that it is plausible. The last result can also be generalized to the relative mean speed of two dissimilar molecules of masses m_A and m_B:

$$\bar{c}_{rel} = \left(\frac{8kT}{\pi\mu}\right)^{1/2} \qquad \mu = \frac{m_A m_B}{m_A + m_B} \qquad \text{Relative mean speed} \qquad (20.10)$$

Note that the molecular masses (not the molar masses) and Boltzmann's constant, $k = R/N_A$, appear in this expression; the quantity μ is called the **reduced mass** of the molecules. Equation 20.10 turns into eqn 20.9 when the molecules are identical (that is, $m_A = m_B = m$, so $\mu = \frac{1}{2}m$).

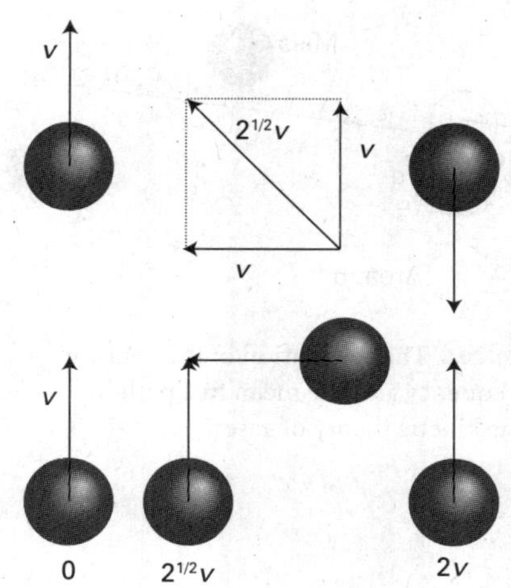

Fig. 20.7 A simplified version of the argument to show that the relative mean speed of molecules in a gas is related to the mean speed. When the molecules are moving in the same direction, the relative mean speed is zero; it is $2v$ when the molecules are approaching each other. A typical mean direction of approach is from the side, and the mean speed of approach is then $2^{1/2}v$. The last direction of approach is the most characteristic, so the mean speed of approach can be expected to be about $2^{1/2}v$. This value is confirmed by more detailed calculation.

(b) The collision frequency

A qualitative picture of the events taking place in a gas was first described in Section 1.2. The kinetic model enables us to make that picture more quantitative. In particular, it enables us to calculate the frequency with which molecular collisions occur and the distance a molecule travels on average between collisions.

We count a 'hit' whenever the centres of two molecules come within a distance d of each other, where d, the **collision diameter**, is of the order of the actual diameters of the molecules (for impenetrable hard spheres d is the diameter). As we show in the following *Justification*, we can use kinetic model to deduce that the **collision frequency**, z, the number of collisions made by one molecule divided by the time interval during which the collisions are counted, when there are N molecules in a volume V is

$$z = \sigma\bar{c}_{rel}\mathcal{N} \qquad \text{Collision frequency} \qquad (20.11a)^{\circ}$$

with $\mathcal{N} = N/V$ and $\bar{c}_{rel}$ given in eqn 20.10. The area $\sigma = \pi d^2$ is called the **collision cross-section** of the molecules. Some typical collision cross-sections are given in Table 20.1. In terms of the pressure

$$z = \frac{\sigma\bar{c}_{rel}p}{kT} \qquad \text{Collision frequency in terms of the pressure} \qquad (20.11b)^{\circ}$$

A brief comment

The reduced mass arises whenever relative motion of two particles is encountered. It also occurs in the hydrogen atom when considering the relative motion of the electron and nucleus (Section 9.1) and in the description of the vibration of a diatomic molecule (Section 12.8).

Justification 20.3 *Using the kinetic model to calculate the collision frequency*

When a molecule travels through a gas it sweeps out a 'collision tube' of area $\sigma = \pi d^2$ and length $\lambda = \bar{c}_{rel}\Delta t$ where $\bar{c}_{rel}$ is the relative velocity and Δt is the interval before the first collision (Fig. 20.8). There is one molecule in this tube of volume $\sigma\lambda$, so the number density is $1/\sigma\lambda = 1/\sigma\bar{c}_{rel}\Delta t$. This number density must be equal to the bulk number density, $\mathcal{N} = N/V = p/kT$, so from $p/kT = 1/\sigma\bar{c}_{rel}\Delta t$ we can infer that $\Delta t = kT/\sigma\bar{c}_{rel}p$. The collision frequency, z, is the inverse of the time between collisions, so $z = 1/\Delta t = \sigma\bar{c}_{rel}p/kT$, as in eqn 20.11b.

Table 20.1* Collision cross-sections

	σ/nm^2
C_6H_6	0.88
CO_2	0.52
He	0.21
N_2	0.43

* More values are given in the *Data section*.

Equation 20.11a shows that, at constant volume (and therefore constant number density), the collision frequency increases with increasing temperature. The reason

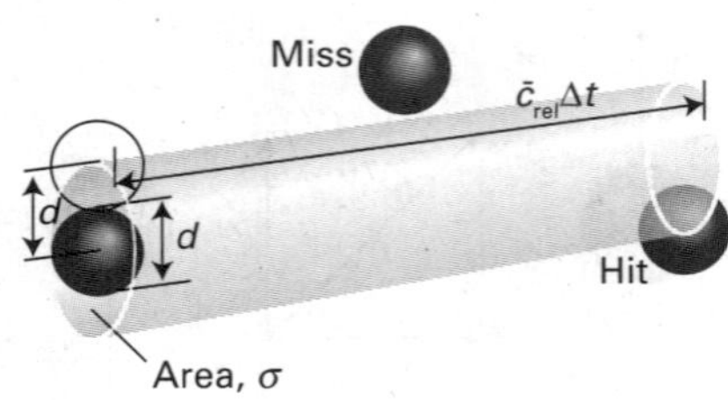

Fig. 20.8 The calculation of the collision frequency and the mean free path in the kinetic theory of gases.

for this increase is that the relative mean speed increases with temperature (eqns 20.9 and 20.10). Equation 20.11b shows that, at constant temperature, the collision frequency is proportional to the pressure. Such a proportionality is plausible for, the greater the pressure, the greater the number density of molecules in the sample, and the rate at which they encounter one another is greater even though their average speed remains the same. For an N_2 molecule in a sample at 1 atm and 25°C, $z \approx 5 \times 10^9\ \mathrm{s}^{-1}$, so a given molecule collides about 5×10^9 times each second. We are beginning to appreciate the timescale of events in gases.

(c) The mean free path

Once we have the collision frequency, we can calculate the **mean free path**, λ (lambda), the average distance a molecule travels between collisions. As implied by the derivation in *Justification 20.3*

$$\lambda = \bar{c}_{rel}\Delta t = \frac{\bar{c}_{rel}}{z} \quad \text{Mean free path} \quad (20.12)$$

Substitution of the expression for z in eqn 20.11b gives

$$\lambda = \frac{kT}{\sigma p} \quad \text{Mean free path in terms of the pressure} \quad (20.13)$$

Doubling the pressure reduces the mean free path by half. A typical mean free path in nitrogen gas at 1 atm is 70 nm, or about 10^3 molecular diameters. Although the temperature appears in eqn 20.13, in a sample of constant volume, the pressure is proportional to T, so T/p remains constant when the temperature is increased. Therefore, the mean free path is independent of the temperature in a sample of gas in a container of fixed volume. The distance between collisions is determined by the number of molecules present in the given volume, not by the speed at which they travel.

In summary, a typical gas (N_2 or O_2) at 1 atm and 25°C can be thought of as a collection of molecules travelling with a mean speed of about 500 m s^{-1}. Each molecule makes a collision within about 1 ns, and between collisions it travels about 10^3 molecular diameters. The kinetic model of gases is valid (and the gas behaves nearly perfectly) if the diameter of the molecules is much smaller than the mean free path ($d \ll \lambda$), for then the molecules spend most of their time far from one another.

IMPACT ON ASTROPHYSICS

I20.1 The Sun as a ball of perfect gas

The kinetic model of gases is valid when the size of the particles is negligible compared with their mean free path. It may seem absurd, therefore, to expect the kinetic model and, as a consequence, the perfect gas law, to be applicable to the dense matter of stellar interiors. In the Sun, for instance, the density at its centre is 1.50 times that of liquid water and comparable to that of water about halfway to its surface. However, we have to realize that the state of matter is that of a *plasma*, in which the electrons have been stripped from the atoms of hydrogen and helium that make up the bulk of the matter of stars. As a result, the particles making up the plasma have diameters comparable to those of nuclei, or about 10 fm. Therefore, a mean free path of only 0.1 pm satisfies the criterion for the validity of the kinetic theory and the perfect gas law. We can therefore use $pV = nRT$ as the equation of state for the stellar interior. Although the Coulombic interaction between charged particles is strong, at the high temperatures of stellar interiors the kinetic energy of the charged particles is very much greater and so 'kinetic-energy only' is a tolerable approximation.

As for any perfect gas, the pressure in the interior of the Sun is related to the mass density, $\rho = m/V$, by $p = \rho RT/M$. Atoms are stripped of their electrons in the interior of stars so, if we suppose that the interior consists of ionized hydrogen atoms, the mean molar mass is one-half the molar mass of hydrogen, or 0.5 g mol^{-1} (the mean of the molar mass of H^+ and e^-, the latter being virtually 0). Halfway to the centre of the Sun, the temperature is 3.6 MK and the mass density is 1.20 g cm^{-3} (slightly denser than water); so the pressure there works out as 7.2×10^{13} Pa, or about 720 million atmospheres.

We can combine this result with the expression for the pressure from the kinetic model (eqn 20.1). Because the total kinetic energy of the particles is $E_k = \frac{1}{2}Nmc^2$, we can write $p = \frac{2}{3}E_k/V$. That is, the pressure of the plasma is related to the *kinetic energy density*, $\rho_k = E_k/V$, the kinetic energy of the molecules in a region divided by the volume of the region, by $p = \frac{2}{3}\rho_k$. It follows that the kinetic energy density halfway to the centre of the Sun is about 0.11 GJ cm^{-3}. In contrast, on a warm day (25°C) on Earth, the (translational) kinetic energy density of our atmosphere is only 0.15 J cm^{-3}.

20.2 Collisions with walls and surfaces

Key point **The collision flux, Z_W, is the number of collisions with an area in a given time interval divided by the area and the duration of the interval.**

The key result for accounting for transport in the gas phase (and in Chapter 23 for the discussion of surface chemistry) is the rate at which molecules strike an area, which may be an imaginary area embedded in the gas, or part of a real wall. The **collision flux**, Z_W, is the number of collisions with the area in a given time interval divided by the area and the duration of the interval. The **collision frequency**, the number of hits per second, is obtained by multiplication of the collision flux by the area of interest. We show in the following *Justification* that the collision flux is

$$Z_W = \frac{p}{(2\pi mkT)^{1/2}} \qquad \text{Collision flux} \qquad (20.14)^\circ$$

When $p = 100$ kPa (1.00 bar) and $T = 300$ K, $Z_W \approx 3 \times 10^{23}$ cm^{-2} s^{-1} for O_2.

Justification 20.4 *The collision flux*

Consider a wall of area A perpendicular to the x-axis (as in Fig. 20.2). If a molecule has $v_x > 0$ (that is, it is travelling in the direction of positive x), then it will strike the wall within an interval Δt if it lies within a distance $v_x\Delta t$ of the wall. Therefore, all molecules in the volume $Av_x\Delta t$, and with positive x-component of velocities, will strike the wall in the interval Δt. The total number of collisions in this interval is therefore the volume $Av_x\Delta t$ multiplied by the number density, $\mathcal{N}$, of molecules. However, to take account of the presence of a range of velocities in the sample, we must sum the result over all the positive values of v_x weighted by the probability distribution of velocities (eqn 20.6):

$$\text{Number of collisions} = \mathcal{N}A\Delta t\int_0^\infty v_x f(v_x)\mathrm{d}x$$

The collision flux is the number of collisions divided by A and Δt, so

$$Z_W = \mathcal{N}\int_0^\infty v_x f(v_x)\mathrm{d}x$$

Then, using the velocity distribution in eqn 20.6,

$$\int_0^\infty v_x f(v_x)\mathrm{d}v_x = \left(\frac{m}{2\pi kT}\right)^{1/2}\int_0^\infty v_x \mathrm{e}^{-mv_x^2/2kT}\mathrm{d}v_x = \left(\frac{kT}{2\pi m}\right)^{1/2}$$

where we have used the standard integral

$$\int_0^\infty x\mathrm{e}^{-ax^2}\mathrm{d}x = \frac{1}{2a}$$

Therefore,

$$Z_\mathrm{W} = \mathcal{N}\left(\frac{kT}{2\pi m}\right)^{1/2} = \tfrac{1}{4}\bar{c}\mathcal{N} \qquad (20.15)^\circ$$

where we have used eqn 20.7 in the form $\bar{c} = (8kT/\pi m)^{1/2}$, which implies that $\frac{1}{4}\bar{c} = (kT/2\pi m)^{1/2}$. Substitution of $\mathcal{N} = nN_\mathrm{A}/V = p/kT$ gives eqn 20.14.

20.3 The rate of effusion

Key points **Effusion is the emergence of a gas from a container through a small hole. Graham's law of effusion states that the rate of effusion is inversely proportional to the square root of the molar mass.**

The essential empirical observations on effusion are summarized by **Graham's law of effusion**, which states that the rate of effusion is inversely proportional to the square root of the molar mass. The basis of this result is that, as remarked above, the mean speed of molecules is inversely proportional to $M^{1/2}$, so the rate at which they strike the area of the hole is also inversely proportional to $M^{1/2}$. However, by using the expression for the rate of collisions, we can obtain a more detailed expression for the rate of effusion and hence use effusion data more effectively.

When a gas at a pressure p and temperature T is separated from a vacuum by a small hole, the rate of escape of its molecules is equal to the rate at which they strike the area of the hole (which is given by eqn 20.14). Therefore, for a hole of area A_0,

$$\text{Rate of effusion} = Z_\mathrm{W}A_0 = \frac{pA_0}{(2\pi mkT)^{1/2}} = \frac{pA_0N_\mathrm{A}}{(2\pi MRT)^{1/2}} \qquad \text{Rate of effusion} \qquad (20.16)^\circ$$

where, in the last step, we have used $R = N_\mathrm{A}k$ and $M = mN_\mathrm{A}$. This rate is inversely proportional to $M^{1/2}$, in accord with Graham's law.

Equation 20.16 is the basis of the **Knudsen method** for the determination of the vapour pressures of liquids and solids, particularly of substances with very low vapour pressures. Thus, if the vapour pressure of a sample is p, and it is enclosed in a cavity with a small hole, then the rate of loss of mass from the container is proportional to p.

Example 20.2 *Calculating the vapour pressure from a mass loss*

Caesium (m.p. 29°C, b.p. 686°C) was introduced into a container and heated to 500°C. When a hole of diameter 0.50 mm was opened in the container for 100 s, a mass loss of 385 mg was measured. Calculate the vapour pressure of liquid caesium at 500 K.

Method The pressure of vapour is constant inside the container despite the effusion of atoms because the hot liquid metal replenishes the vapour. The rate of effusion is therefore constant, and given by eqn 20.16. To express the rate in terms of mass, multiply the number of atoms that escape by the mass of each atom.

Answer The mass loss Δm in an interval Δt is related to the collision flux by

$$\Delta m = Z_W A_0 m \Delta t$$

where A_0 is the area of the hole and m is the mass of one atom. It follows that

$$Z_W = \frac{\Delta m}{A_0 m \Delta t}$$

Because Z_W is related to the pressure by eqn 20.14, we can write

$$p = \left(\frac{2\pi RT}{M}\right)^{1/2} \frac{\Delta m}{A_0 \Delta t}$$

Because $M = 132.9 \text{ g mol}^{-1}$, substitution of the data gives $p = 8.7$ kPa (using $1 \text{ Pa} = 1 \text{ N m}^{-2} = 1 \text{ J m}^{-1}$), or 65 Torr.

Self-test 20.2 How long would it take 1.0 g of Cs atoms to effuse out of the oven under the same conditions? [260 s]

20.4 Transport properties of a perfect gas

Key points **(a) Flux is the quantity of a property passing through a given area in a given time interval divided by the area and the duration of the interval. Diffusion is the migration of matter down a concentration gradient. Fick's first law of diffusion states that the flux of matter is proportional to the concentration gradient. Thermal conduction is the migration of energy down a temperature gradient and the flux of energy is proportional to the temperature gradient. Viscosity is the migration of linear momentum down a velocity gradient and the flux of momentum is proportional to the velocity gradient. (b) The coefficients of diffusion, thermal conductivity, and viscosity of a perfect gas are proportional to the product of the mean free path and mean speed.**

Transport properties are commonly expressed in terms of a number of 'phenomenological' equations, or equations that are empirical summaries of experimental observations. These phenomenological equations apply to all kinds of properties and media. In the following sections, we introduce the equations for the general case and then show how to calculate the parameters that appear in them.

(a) The phenomenological equations

The rate of migration of a property is measured by its **flux**, J, the quantity of that property passing through a given area in a given time interval divided by the area and the duration of the interval. If matter is flowing (as in diffusion), we speak of a **matter flux** of so many molecules per square metre per second; if the property is energy (as in thermal conduction), then we speak of the **energy flux** and express it in joules per square metre per second, and so on. To calculate the total quantity of each property transferred through a given area A in a given time interval Δt, we multiply the flux by the area and the time interval, and form $JA\Delta t$.

Experimental observations on transport properties show that the flux of a property is usually proportional to the first derivative of some other related property. For example, the flux of matter diffusing parallel to the z-axis of a container is found to be proportional to the first derivative of the concentration:

$$J(\text{matter}) \propto \frac{d\mathcal{N}}{dz} \qquad \text{Fick's first law of diffusion} \qquad (20.17)$$

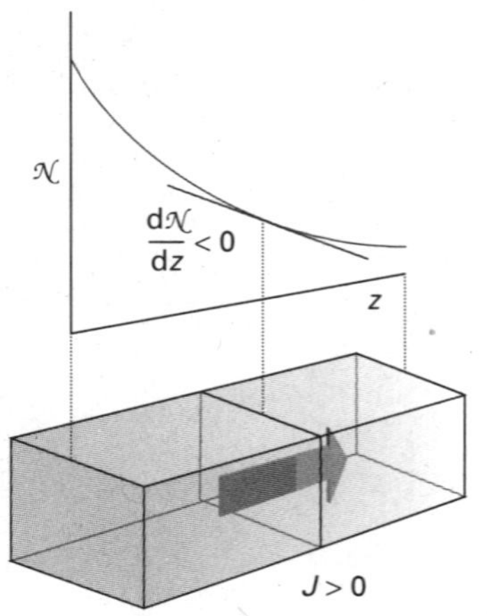

Fig. 20.9 The flux of particles down a concentration gradient. Fick's first law states that the flux of matter (the number of particles passing through an imaginary window in a given interval divided by the area of the window and the length of the interval) is proportional to the density gradient at that point.

Table 20.2* Transport properties of gases at 1 atm

	κ/(J K^{-1} m^{-1} s^{-1})	η/μP†	
	273 K	273 K	293 K
Ar	0.0163	210	223
CO_2	0.0145	136	147
He	0.1442	187	196
N_2	0.0240	166	176

* More values are given in the *Data section*.
† 1 μP = 10^{-7} kg m^{-1} s^{-1}.

where $\mathcal{N}$ is the number density of particles with units number per metre cubed (m^{-3}). The SI units of J are number per metre squared per second (m^{-2} s^{-1}). The proportionality of the flux of matter to the concentration gradient is sometimes called **Fick's first law of diffusion**: the law implies that diffusion will be faster when the concentration varies steeply with position than when the concentration is nearly uniform. There is no net flux if the concentration is uniform ($d\mathcal{N}/dz = 0$). Similarly, the rate of thermal conduction (the flux of the energy associated with thermal motion) is found to be proportional to the temperature gradient:

$$J(\text{energy}) \propto \frac{dT}{dz} \qquad \text{Flux of energy} \qquad (20.18)$$

The SI units of this flux are joules per metre squared per second (J m^{-2} s^{-1}).

A positive value of J signifies a flux towards positive z; a negative value of J signifies a flux towards negative z. Because matter flows down a concentration gradient, from high concentration to low concentration, J is positive if $d\mathcal{N}/dz$ is negative (Fig. 20.9). Therefore, the coefficient of proportionality in eqn 20.17 must be negative, and we write it $-D$:

$$J(\text{matter}) = -D\frac{d\mathcal{N}}{dz} \qquad \text{Fick's first law in terms of the diffusion coefficient} \qquad (20.19)$$

The constant D is the called the **diffusion coefficient**; its SI units are metre squared per second (m^2 s^{-1}). Energy migrates down a temperature gradient, and the same reasoning leads to

$$J(\text{energy}) = -\kappa\frac{dT}{dz} \qquad \text{Flux of energy in terms of the coefficient of thermal conductivity} \qquad (20.20)$$

where κ is the **coefficient of thermal conductivity**. The SI units of κ are joules per kelvin per metre per second (J K^{-1} m^{-1} s^{-1}). Some experimental values are given in Table 20.2.

To see the connection between the flux of momentum and the viscosity, consider a fluid in a state of **Newtonian flow**, which can be imagined as occurring by a series of layers moving past one another (Fig. 20.10). The layer next to the wall of the vessel is stationary, and the velocity of successive layers varies linearly with distance, z, from the wall. Molecules ceaselessly move between the layers and bring with them the x-component of linear momentum they possessed in their original layer. A layer is retarded by molecules arriving from a more slowly moving layer because they have a low momentum in the x-direction. A layer is accelerated by molecules arriving from a more rapidly moving layer. We interpret the net retarding effect as the fluid's viscosity.

Because the retarding effect depends on the transfer of the x-component of linear momentum into the layer of interest, the viscosity depends on the flux of this x-component in the z-direction. The flux of the x-component of momentum is proportional to dv_x/dz because there is no net flux when all the layers move at the same velocity. We can therefore write

$$J(x\text{-component of momentum}) = -\eta\frac{dv_x}{dz} \qquad \text{Momentum flux in terms of the coefficient of viscosity} \qquad (20.21)$$

The constant of proportionality, η, is the **coefficient of viscosity** (or simply 'the viscosity'). Its units are kilograms per metre per second (kg m^{-1} s^{-1}). Viscosities are often reported in poise (P), where 1 P = 10^{-1} kg m^{-1} s^{-1}. Some experimental values are given in Table 20.2.

Table 20.3 Transport properties of perfect gases

Property	Transported quantity	Simple kinetic theory	Units
Diffusion	Matter	$D=\frac{1}{3}\lambda\bar{c}$	$m^2\ s^{-1}$
Thermal conductivity	Energy	$\kappa=\frac{1}{3}\lambda\bar{c}C_{V,m}[A]$ $=\dfrac{\bar{c}C_{V,m}}{3\sqrt{2}\sigma N_A}$	$J\ K^{-1}\ m^{-1}\ s^{-1}$
Viscosity	Linear momentum	$\eta=\frac{1}{3}\lambda\bar{c}m\mathcal{N}$ $=\dfrac{m\bar{c}}{3\sqrt{2}\sigma}$	$kg\ m^{-1}\ s^{-1}$

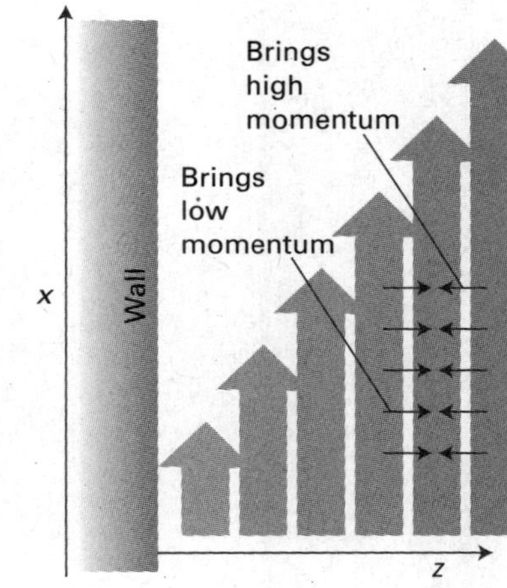

Fig. 20.10 The viscosity of a fluid arises from the transport of linear momentum. In this illustration the fluid is undergoing Newtonian (laminar) flow, and particles bring their initial momentum when they enter a new layer. If they arrive with high x-component of momentum they accelerate the layer; if with low x-component of momentum they retard the layer.

(b) The transport parameters

As shown in *Further information 20.1* and summarized in Table 20.3, the kinetic model leads to expressions for the diffusional parameters of a perfect gas. The diffusion coefficient, for instance, is

$$D=\tfrac{1}{3}\lambda\bar{c} \quad \text{Diffusion coefficient of a perfect gas} \quad (20.22)°$$

As usual, we need to consider the significance of this expression:

1. The mean free path, λ, decreases as the pressure is increased (eqn 20.13), so D decreases with increasing pressure and, as a result, the gas molecules diffuse more slowly.

2. The mean speed, $\bar{c}$, increases with the temperature (eqn 20.7), so D also increases with temperature. As a result, molecules in a hot sample diffuse more quickly than those in a cool sample (for a given concentration gradient).

3. Because the mean free path increases when the collision cross-section of the molecules decreases (eqn 20.13), the diffusion coefficient is greater for small molecules than for large molecules.

Similarly, according to the kinetic model of gases, the thermal conductivity of a perfect gas A having molar concentration [A] is given by the expression

$$\kappa=\tfrac{1}{3}\lambda\bar{c}C_{V,m}[A] \quad \text{Coefficient of thermal conductivity of a perfect gas} \quad (20.23)°$$

where $C_{V,m}$ is the molar heat capacity at constant volume. To interpret this expression, we note that:

1. Because λ is inversely proportional to the pressure, and hence inversely proportional to the molar concentration of the gas, the thermal conductivity is independent of the pressure.

2. The thermal conductivity is greater for gases with a high heat capacity because a given temperature gradient then corresponds to a greater energy gradient.

The physical reason for the pressure independence of κ is that the thermal conductivity can be expected to be large when many molecules are available to transport the energy, but the presence of so many molecules limits their mean free path and they cannot carry the energy over a great distance. These two effects balance. The thermal conductivity is indeed found experimentally to be independent of the pressure, except when the pressure is very low, when $\kappa \propto p$. At low pressures λ exceeds the dimensions of the apparatus, and the distance over which the energy is transported is determined by the size of the container and not by the other molecules present. The flux is still proportional to the number of carriers, but the length of the journey no longer depends on λ, so $\kappa \propto$ [A], which implies that $\kappa \propto p$.

Finally, the kinetic model leads to the following expression for the viscosity (see *Further information 20.1*):

$$\eta = \tfrac{1}{3}M\lambda\bar{c}[\mathrm{A}] \qquad \text{Coefficient of viscosity of a perfect gas} \qquad (20.24)^\circ$$

where [A] is the molar concentration of the gas molecules and M is their molar mass. We can interpret this expression as follows:

1. Because $\lambda \propto 1/p$ (eqn 20.13) and $[\mathrm{A}] \propto p$, it follows that $\eta \propto \bar{c}$, independent of p. That is, the viscosity is independent of the pressure.

2. Because $\bar{c} \propto T^{1/2}$ (eqn 20.7), $\eta \propto T^{1/2}$. That is, the viscosity of a gas *increases* with temperature.

The physical reason for the pressure independence of the viscosity is the same as for the thermal conductivity: more molecules are available to transport the momentum, but they carry it less far on account of the decrease in mean free path. The increase of viscosity with temperature is explained when we remember that at high temperatures the molecules travel more quickly, so the flux of momentum is greater. By contrast, as we shall see in Section 20.5, the viscosity of a liquid *decreases* with increase in temperature because intermolecular interactions must be overcome.

Molecular motion in liquids

We outlined what is currently known about the structure of simple liquids in Section 16.6. Here we consider a particularly simple type of motion through a liquid, that of an ion, and see that the information that motion provides can be used to infer the behaviour of uncharged species too.

20.5 Experimental results

Key point **Molecular motion in liquids can be studied by NMR, EPR, inelastic neutron scattering, and viscosity measurements.**

The motion of molecules in liquids can be studied experimentally by a variety of methods. Relaxation time measurements in NMR and EPR (Chapter 14) can be interpreted in terms of the mobilities of the molecules, and have been used to show that big molecules in viscous fluids typically rotate in a series of small (about 5°) steps, whereas small molecules in nonviscous fluids typically jump through about 1 radian (57°) in each step. Another important technique is **inelastic neutron scattering**, in which the energy neutrons collect or discard as they pass through a sample is interpreted in terms of the motion of its particles. The same technique is used to examine the internal dynamics of macromolecules.

More mundane than these experiments are viscosity measurements (Table 20.4). For a molecule to move in a liquid, it must acquire at least a minimum energy to escape from its neighbours. The probability that a molecule has at least an energy E_a is proportional to $e^{-E_a/RT}$, so the mobility of the molecules in the liquid should follow

Table 20.4* Viscosities of liquids at 298 K

	$\eta/(10^{-3}\ \mathrm{kg\ m^{-1}\ s^{-1}})$
Benzene	0.601
Mercury	1.55
Pentane	0.224
Water†	0.891

* More values are given in the *Data section*.
† The viscosity of water corresponds to 0.891 cP.

this type of temperature dependence. Because the coefficient of viscosity, η, is inversely proportional to the mobility of the particles, we should expect that

$$\eta \propto e^{E_a/RT} \tag{20.25}$$

(Note the positive sign of the exponent.) This expression implies that the viscosity should decrease sharply with increasing temperature. Such a variation is found experimentally, at least over reasonably small temperature ranges (Fig. 20.11). The activation energy typical of viscosity is comparable to the mean potential energy of intermolecular interactions.

One problem with the interpretation of viscosity measurements is that the change in density of the liquid as it is heated makes a pronounced contribution to the temperature variation of the viscosity. Thus, the temperature dependence of viscosity at constant volume, when the density is constant, is much less than that at constant pressure. The intermolecular interactions between the molecules of the liquid govern the magnitude of E_a, but the problem of calculating it is immensely difficult and still largely unsolved. At low temperatures, the viscosity of water decreases as the pressure is increased. This behaviour is consistent with the rupture of hydrogen bonds.

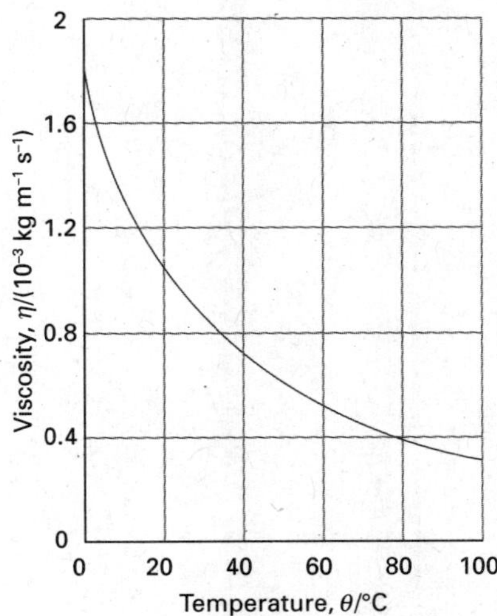

Fig. 20.11 The experimental temperature dependence of the viscosity of water. As the temperature is increased, more molecules are able to escape from the potential wells provided by their neighbours, and so the liquid becomes more fluid. A plot of ln η against $1/T$ is a straight line (over a small range) with positive slope.

20.6 The conductivities of electrolyte solutions

Key points **The conductance is the inverse of resistance. Kohlrausch's law describes the concentration dependence of the molar conductivity of a strong electrolyte (a substance that dissociates fully into ions in solution).**

Further insight into the nature of molecular motion can be obtained by studying the motion of ions in solution, for ions can be dragged through the solvent by the application of a potential difference between two electrodes immersed in the sample. By studying the transport of charge through electrolyte solutions it is possible to build up a picture of the events that occur in them and, in some cases, to extrapolate the conclusions to species that have zero charge, that is, to neutral molecules.

The fundamental measurement used to study the motion of ions is that of the electrical resistance, R, of the solution. The **conductance**, G, of a solution is the inverse of its resistance R: $G = 1/R$. As resistance is expressed in ohms, Ω, the conductance of a sample is expressed in Ω^{-1}. The reciprocal ohm used to be called the mho, but its official designation is now the siemens, S, and $1\ \text{S} = 1\ \Omega^{-1} = 1\ \text{C V}^{-1}\ \text{s}^{-1}$. The conductance of a sample decreases with its length l and increases with its cross-sectional area A. We therefore write

$$G = \frac{\kappa A}{l} \qquad \text{Conductance of a solution} \tag{20.26}$$

where κ is the **conductivity**. With the conductance in siemens and the dimensions in metres, it follows that the SI units of κ are siemens per metre (S m^{-1}).

The conductivity of a solution depends on the number of ions present, and it is normal to introduce the **molar conductivity**, Λ_m, which is defined as

$$\Lambda_m = \frac{\kappa}{c} \qquad \text{Definition of molar conductivity of a solution} \quad [20.27]$$

where c is the molar concentration of the added electrolyte. The SI unit of molar conductivity is siemens metre-squared per mole (S m^2 mol^{-1}), and typical values are about 10 mS m^2 mol^{-1} (where 1 mS = 10^{-3} S).

The molar conductivity is found to vary with the concentration. One reason for this variation is that the number of ions in the solution might not be proportional to the concentration of the electrolyte. For instance, the concentration of ions in a solution of a weak acid depends on the concentration of the acid in a complicated way, and doubling the concentration of the acid added does not double the number of ions. Secondly, because ions interact strongly with one another, the conductivity of a solution is not exactly proportional to the number of ions present.

In an extensive series of measurements during the nineteenth century, Friedrich Kohlrausch showed that at low concentrations the molar conductivities of strong electrolytes (substances that are fully dissociated into ions in solution) vary linearly with the square root of the concentration:

$$\Lambda_m = \Lambda_m^\circ - \mathcal{K}c^{1/2} \qquad \text{Kohlrausch's law} \qquad (20.28)$$

This variation is called **Kohlrausch's law**. The constant Λ_m° is the **limiting molar conductivity**, the molar conductivity in the limit of zero concentration (when the ions are effectively infinitely far apart and do not interact with one another). The constant $\mathcal{K}$ is found to depend more on the stoichiometry of the electrolyte (that is, whether it is of the form MA, or M_2A, etc.) than on its specific identity. In due course we shall see that the $c^{1/2}$ dependence arises from interactions between ions: when charge is conducted ionically, ions of one charge are moving past the ions of interest and retard its progress.

Kohlrausch was also able to establish experimentally that Λ_m° can be expressed as the sum of contributions from its individual ions. If the limiting molar conductivity of the cations is denoted λ_+ and that of the anions λ_-, then his **law of the independent migration of ions** states that

$$\Lambda_m^\circ = \nu_+\lambda_+ + \nu_-\lambda_- \qquad \text{Law of independent migration of ions} \qquad (20.29)^\circ$$

where ν_+ and ν_- are the numbers of cations and anions per formula unit of electrolyte (for example, $\nu_+ = \nu_- = 1$ for HCl, NaCl, and $CuSO_4$, but $\nu_+ = 1$, $\nu_- = 2$ for $MgCl_2$).

20.7 The mobilities of ions

Key points **(a) The drift speed is the terminal speed when an accelerating force is balanced by the viscous drag. The Grotthuss mechanism describes the motion of a proton in water as resulting from rearrangement of bonds in a group of water molecules. (b) The ionic conductivity is the contribution of ions of one type to the molar conductivity of a solution. (c) The Debye–Hückel–Onsager theory explains the concentration dependence of the molar conductivity of a strong electrolyte in terms of ionic interactions.**

To interpret conductivity measurements we need to know why ions move at different rates, why they have different molar conductivities, and why the molar conductivities of strong electrolytes decrease with the square root of the molar concentration. The central idea in this section is that, although the motion of an ion remains largely random, the presence of an electric field biases its motion, and the ion undergoes net migration through the solution.

(a) The drift speed

When the potential difference between two electrodes a distance l apart is $\Delta\phi$, the ions in the solution between them experience a uniform electric field of magnitude

$$\mathcal{E} = \frac{\Delta\phi}{l} \qquad (20.30)$$

Table 20.5* Ionic mobilities in water at 298 K

	$u/(10^{-8}\ m^2\ s^{-1}\ V^{-1})$		$u/(10^{-8}\ m^2\ s^{-1}\ V^{-1})$
H^+	36.23	OH^-	20.64
Na^+	5.19	Cl^-	7.91
K^+	7.62	Br^-	8.09
Zn^{2+}	5.47	SO_4^{2-}	8.29

* More values are given in the *Data section*.

In such a field, an ion of charge ze experiences a force of magnitude

$$\mathcal{F} = ze\mathcal{E} = \frac{ze\Delta\phi}{l} \tag{20.31}$$

(In this chapter we disregard the sign of the charge number and so avoid notational complications.) A cation responds to the application of the field by accelerating towards the negative electrode and an anion responds by accelerating towards the positive electrode. However, this acceleration is short-lived. As the ion moves through the solvent it experiences a frictional retarding force, $\mathcal{F}_{fric}$, proportional to its speed. If we assume that the Stokes's relation formula (eqn 18.31) for a sphere of radius a and speed s applies even on a microscopic scale (and independent evidence from magnetic resonance suggests that it often gives at least the right order of magnitude), then we can write this retarding force as

$$\mathcal{F}_{fric} = fs \qquad f = 6\pi\eta a \qquad \text{Frictional retarding force} \tag{20.32}$$

The two forces act in opposite directions, and the ions quickly reach a terminal speed, the **drift speed**, when the accelerating force is balanced by the viscous drag. The net force is zero when

$$s = \frac{ze\mathcal{E}}{f} \qquad \text{Drift speed} \tag{20.33}$$

It follows that the drift speed of an ion is proportional to the strength of the applied field. We write

$$s = u\mathcal{E} \qquad \text{Definition of ionic mobility} \qquad [20.34]$$

where u is called the **mobility** of the ion (Table 20.5). Comparison of eqns 20.33 and 20.34 and use of eqn 20.32 shows that

$$u = \frac{ze}{f} = \frac{ze}{6\pi\eta a} \qquad \text{Ionic mobility in terms of viscosity} \tag{20.35}$$

• A brief illustration

For an order of magnitude estimate we can take $z = 1$ and a the radius of an ion such as Cs^+ (which might be typical of a smaller ion plus its hydration sphere), which is 170 pm. For the viscosity, we use $\eta = 1.0$ cP (1.0×10^{-3} kg m^{-1} s^{-1}, Table 20.4). Then $u \approx 5 \times 10^{-8}$ m^2 V^{-1} s^{-1}. This value means that, when there is a potential difference of 1 V across a solution of length 1 cm (so $\mathcal{E} = 100$ V m^{-1}), the drift speed is typically about 5 μm s^{-1}. That speed might seem slow, but not when expressed on a molecular scale, for it corresponds to an ion passing about 10^4 solvent molecules per second. •

Because the drift speed governs the rate at which charge is transported, we might expect the conductivity to decrease with increasing solution viscosity and ion size. Experiments confirm these predictions for bulky ions (such as R_4N^+ and RCO_2^-) but not for small ions. For example, the molar conductivities of the alkali metal ions increase from Li^+ to Cs^+ (Table 20.5) even though the ionic radii increase. The paradox is resolved when we realize that the radius a in the Stokes formula is the **hydrodynamic radius** (or 'Stokes radius') of the ion, its effective radius in the solution taking into account all the H_2O molecules it carries in its hydration sphere. Small ions give rise to stronger electric fields than large ones (the electric field at the surface of a sphere of radius r is proportional to ze/r^2 and it follows that the smaller the radius the stronger the field), so small ions are more extensively solvated than big ions. Thus, an ion of small ionic radius may have a large hydrodynamic radius because it drags many solvent molecules through the solution as it migrates. The hydrating H_2O molecules are often very labile, however, and NMR and isotope studies have shown that the exchange between the coordination sphere of the ion and the bulk solvent is very rapid.

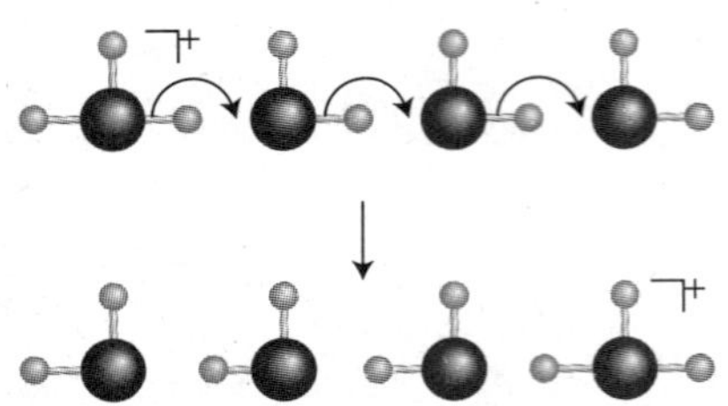
Fig. 20.12 A highly schematic diagram showing the effective motion of a proton in water.

The proton, although it is very small, has a very high molar conductivity (Table 20.5)! Proton and ^{17}O-NMR show that the times characteristic of protons hopping from one molecule to the next are about 1.5 ps, which is comparable to the time that inelastic neutron scattering shows it takes a water molecule to reorientate through about 1 rad (1 to 2 ps). According to the **Grotthuss mechanism**, there is an effective motion of a proton that involves the rearrangement of bonds in a group of water molecules (Fig. 20.12). The model is consistent with the observation that the molar conductivity of protons increases as the pressure is raised, for increasing pressure ruptures the hydrogen bonds in water. The mobility of NH_4^+ is also anomalous and presumably occurs by an analogous mechanism.

(b) Mobility and conductivity

Ionic mobilities provide a link between measurable and theoretical quantities. As a first step we establish in the following *Justification* the following relation between an ion's mobility and its molar conductivity:

$$\lambda_{\pm} = zu_{\pm}F \qquad \text{Relation between ionic mobility and molar conductivity} \qquad (20.36)°$$

where F is Faraday's constant ($F = N_Ae$).

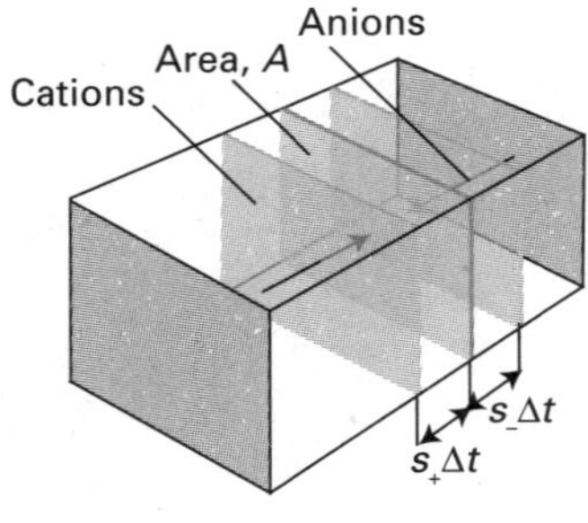

Fig. 20.13 In the calculation of the current, all the cations within a distance $s_+\Delta t$ (that is, those in the volume $s_+A\Delta t$) will pass through the area A. The anions in the corresponding volume on the other side of the window will also contribute to the current similarly.

Justification 20.5 *The relation between ionic mobility and molar conductivity*

To keep the calculation simple, we ignore signs in the following, and concentrate on the magnitudes of quantities: the direction of ion flux can always be decided by common sense.

Consider a solution of a fully dissociated strong electrolyte at a molar concentration c. Let each formula unit give rise to ν_+ cations of charge z_+e and ν_- anions of charge z_-e. The molar concentration of each type of ion is therefore νc (with $\nu = \nu_+$ or ν_-), and the number density of each type is νcN_A. The number of ions of one kind that pass through an imaginary window of area A during an interval Δt is equal to the number within the distance $s\Delta t$ (Fig. 20.13), and therefore to the number in the volume $s\Delta tA$. (The same argument was used in Section 20.1 in the discussion of the pressure of a gas.) The number of ions of that kind in this volume is equal to $s\Delta tA\nu cN_A$. The flux through the window (the number of this type of ion passing through the window divided by the area of the window and the duration of the interval) is therefore

$$J(\text{ions}) = \frac{s\Delta tA\nu cN_A}{A\Delta t} = s\nu cN_A$$

Each ion carries a charge ze, so the flux of charge is

$$J(\text{charge}) = zsvceN_A = szvcF$$

Because $s = u\mathcal{E}$, the flux is

$$J(\text{charge}) = zuvcF\mathcal{E}$$

The current, I, through the window due to the ions we are considering is the charge flux times the area:

$$I = JA = zuvcF\mathcal{E}A$$

Because the electric field is the potential gradient, $\Delta\phi/l$, we can write

$$I = \frac{zuvcFA\Delta\phi}{l} \qquad (20.37)$$

Current and potential difference are related by Ohm's law, $\Delta\phi = IR$, so it follows that

$$I = \frac{\Delta\phi}{R} = G\Delta\phi = \frac{\kappa A\Delta\phi}{l}$$

where we have used eqn 20.26 in the form $\kappa = Gl/A$. Note that the proportionality of current to potential difference ($I \propto \Delta\phi$) is another example of a phenomenological flux equation like those introduced in Section 20.4. Comparison of the last two expressions gives $\kappa = zuvcF$. Division by the molar concentration of ions, vc, then results in eqn 20.36 for cations (u_+) and anions (u_-).

Equation 20.36 applies to the cations and to the anions. Therefore, for the solution itself in the limit of zero concentration (when there are no interionic interactions),

$$\Lambda_m^\circ = (z_+u_+v_+ + z_-u_-v_-)F \qquad (20.38)^\circ$$

For a symmetrical z:z electrolyte (for example, $CuSO_4$ with $z = 2$), this equation simplifies to

$$\Lambda_m^\circ = z(u_+ + u_-)F \qquad (20.39)^\circ$$

• **A brief illustration**

Earlier, we estimated the typical ionic mobility as $5 \times 10^{-8}\ \text{m}^2\ \text{V}^{-1}\ \text{s}^{-1}$; so, with $z = 1$ for both the cation and anion, we can estimate that a typical limiting molar conductivity should be about $10\ \text{mS m}^2\ \text{mol}^{-1}$, in accord with experiment. The experimental value for KCl, for instance, is $15\ \text{mS m}^2\ \text{mol}^{-1}$. •

(c) Ion–ion interactions

The remaining problem is to account for the $c^{1/2}$ dependence of the Kohlrausch law (eqn 20.28). In Section 5.13 we saw something similar: the activity coefficients of ions at low concentrations also depend on $c^{1/2}$ and depend on their charge type rather than their specific identities. That $c^{1/2}$ dependence was explained in terms of the properties of the ionic atmosphere around each ion, and we can suspect that the same explanation applies here too.

To accommodate the effect of motion, we need to modify the picture of an ionic atmosphere as a spherical haze of charge. Because the ions forming the atmosphere do not adjust to the moving ion immediately, the atmosphere is incompletely formed in front of the moving ion and incompletely decayed behind the ion (Fig. 20.14). The overall effect is the displacement of the centre of charge of the atmosphere a short distance behind the moving ion. Because the two charges are opposite, the result is a retardation of the moving ion. This reduction of the ions' mobility is called the

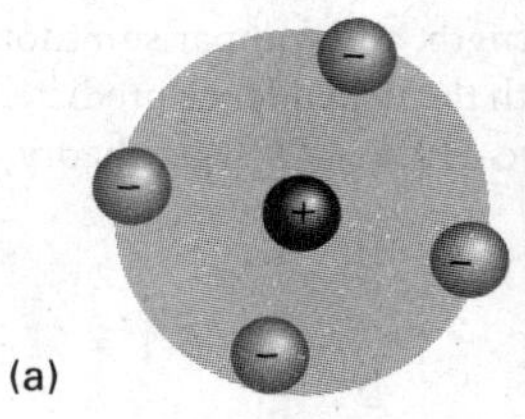

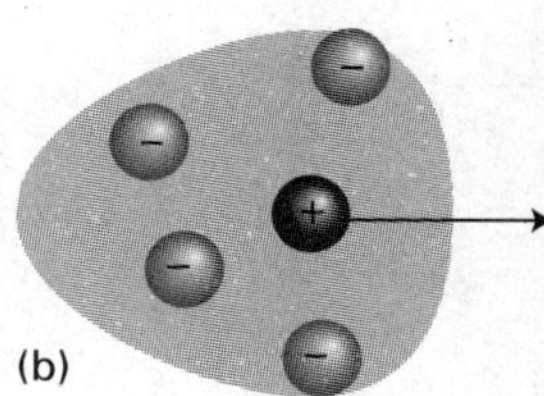

Fig. 20.14 (a) In the absence of an applied field, the ionic atmosphere is spherically symmetric, but (b) when a field is present it is distorted and the centres of negative and positive charge no longer coincide. The attraction between the opposite charges retards the motion of the central ion.

Table 20.6* Debye–Hückel–Onsager coefficients for (1,1)-electrolytes at 298 K

Solvent	$A/(\text{mS m}^2\,\text{mol}^{-1}/(\text{mol dm}^{-3})^{1/2})$	$B/(\text{mol dm}^{-3})^{-1/2}$
Methanol	15.61	0.923
Propanone	32.8	1.63
Water	6.02	0.229

* More values are given in the *Data section*.

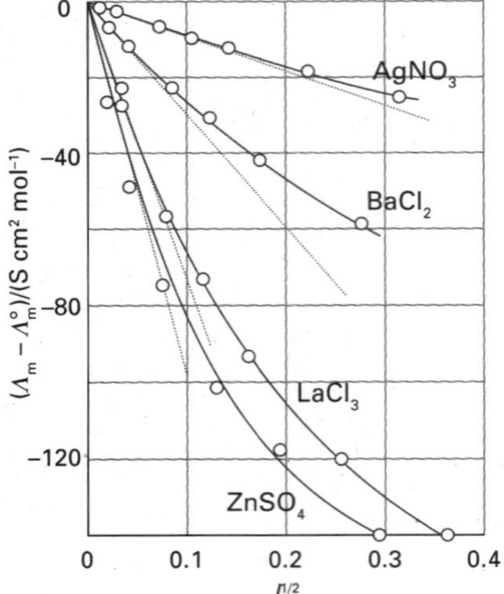

Fig. 20.15 The dependence of molar conductivities on the square root of the ionic strength, and comparison (dotted lines) with the dependence predicted by the Debye–Hückel–Onsager theory.

relaxation effect. A confirmation of the picture is obtained by observing the conductivities of ions at high frequencies, which are greater than at low frequencies: the atmosphere does not have time to follow the rapidly changing direction of motion of the ion, and its effect averages to zero.

The ionic atmosphere has another effect on the motion of the ions. We have seen that the moving ion experiences a viscous drag. When the ionic atmosphere is present this drag is enhanced because the ionic atmosphere moves in an opposite direction to the central ion. The enhanced viscous drag, which is called the **electrophoretic effect**, reduces the mobility of the ions, and hence also reduces their conductivities.

The quantitative formulation of these effects is far from simple, but the **Debye–Hückel–Onsager theory** is an attempt to obtain quantitative expressions at about the same level of sophistication as the Debye–Hückel theory itself. The theory leads to a Kohlrausch-like expression in which

$$\mathcal{K} = A + B\Lambda_m^\circ \qquad (20.40a)$$

with

$$A \propto \frac{z^2}{\eta T^{1/2}} \qquad B \propto \frac{z^3}{T^{3/2}} \qquad (20.40b)$$

See Table 20.6 for some values of A and B. The slopes of the conductivity curves are predicted to depend on the charge type of the electrolyte, in accord with the Kohlrausch law, and some comparisons between theory and experiment are shown in Fig. 20.15. The agreement is quite good at very low ionic strengths, corresponding to very low molar concentrations (less than about 10^{-3} M, depending on the charge type).

IMPACT ON BIOCHEMISTRY

I20.2 Ion channels

Controlled transport of molecules and ions across biological membranes is at the heart of a number of key cellular processes, such as the transmission of nerve impulses, the transfer of glucose into red blood cells, and the synthesis of ATP by oxidative phosphorylation (*Impact I6.1*). Here we examine some of the ways in which ions cross the alien environment of the lipid bilayer.

The thermodynamic tendency to transport an ion through the membrane is partially determined by a concentration gradient (more precisely, an activity gradient) across the membrane, which results in a difference in molar Gibbs energy between the inside and the outside of the cell, and a transmembrane potential gradient, which is due to the different potential energy of the ions on each side of the bilayer. There is a tendency, called **passive transport**, for a species to move spontaneously down concentration and membrane potential gradients. It is also possible to move a species against these gradients, but now the flow is not spontaneous and must be driven by an exergonic process, such as the hydrolysis of ATP. This process is called **active transport**.

The transport of ions into or out of a cell needs to be mediated (that is, facilitated by other species) because the hydrophobic environment of the membrane is inhospitable to ions. There are two mechanisms for ion transport: mediation by a carrier molecule and transport through a 'channel former', a protein that creates a hydrophilic pore through which the ion can pass. An example of a channel former is the polypeptide gramicidin A, which increases the membrane permeability to cations such as H^+, K^+, and Na^+.

Ion channels are proteins that effect the movement of specific ions down a membrane potential gradient. They are highly selective, so there is a channel protein for Ca^{2+}, another for Cl^-, and so on. The opening of the gate may be triggered by potential differences between the two sides of the membrane or by the binding of an 'effector molecule' to a specific receptor site on the channel.

Ions such as H^+, Na^+, K^+, and Ca^{2+} are often transported actively across membranes by integral proteins called **ion pumps**. Ion pumps are molecular machines that work by adopting conformations that are permeable to one ion but not others depending on the state of phosphorylation of the protein. Because protein phosphorylation requires dephosphorylation of ATP, the conformational change that opens or closes the pump is endergonic and requires the use of energy stored during metabolism.

The structures of a number of channel proteins have been obtained by the now traditional X-ray diffraction techniques described in Chapter 19. Information about the flow of ions across channels and pumps is supplied by the **patch clamp technique.** One of many possible experimental arrangements is shown in Fig. 20.16. With mild suction, a 'patch' of membrane from a whole cell or a small section of a broken cell can be attached tightly to the tip of a micropipette filled with an electrolyte solution and containing an electronic conductor, the so-called 'patch electrode'. A potential difference (the 'clamp') is applied between the patch electrode and an intracellular electronic conductor in contact with the cytosol of the cell. If the membrane is permeable to ions at the applied potential difference, a current flows through the completed circuit. Using narrow micropipette tips with diameters of less than 1 μm, ion currents of a few picoamperes (1 pA = 10^{-12} A) have been measured across sections of membranes containing only one ion channel protein.

A detailed picture of the mechanism of action of ion channels has emerged from analysis of patch clamp data and structural data. Here we focus on the K^+ ion channel protein, which, like all other mediators of ion transport, spans the membrane bilayer (Fig. 20.17). The pore through which ions move has a length of 3.4 nm and is divided into two regions: a wide region with a length of 2.2 nm and diameter of 1.0 nm and a narrow region with a length of 1.2 nm and diameter of 0.3 nm. The narrow region is called the 'selectivity filter' of the K^+ ion channel because it allows only K^+ ions to pass.

Filtering is a subtle process that depends on ionic size and the thermodynamic tendency of an ion to lose its hydrating water molecules. Upon entering the selectivity filter, the K^+ ion is stripped of its hydrating shell and is then gripped by carbonyl groups of the protein. Dehydration of the K^+ ion is endergonic ($\Delta_{dehyd}G^{\ominus} = +203\ kJ\ mol^{-1}$), but is driven by the energy of interaction between the ion and the protein. The Na^+ ion, though smaller than the K^+ ion, does not pass through the selectivity filter of the K^+ ion channel because interactions with the protein are not sufficient to compensate for the high Gibbs energy of dehydration of Na^+ ($\Delta_{dehyd}G^{\ominus} = +301\ kJ\ mol^{-1}$). More specifically, a dehydrated Na^+ ion is too small and cannot be held tightly by the protein carbonyl groups, which are positioned for ideal interactions with the larger K^+ ion. In its hydrated form, the Na^+ ion is too large (larger than a dehydrated K^+ ion), does not fit in the selectivity filter, and does not cross the membrane.

Though very selective, a K^+ ion channel can still let other ions pass through. For example, K^+ and Tl^+ ions have similar radii and Gibbs energies of dehydration, so Tl^+

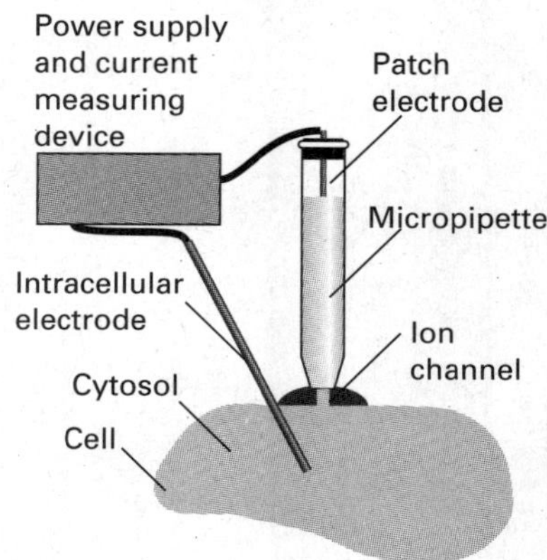

Fig. 20.16 A representation of the patch clamp technique for the measurement of ionic currents through membranes in intact cells. A section of membrane containing an ion channel is in tight contact with the tip of a micropipette containing an electrolyte solution and the patch electrode. An intracellular electronic conductor is inserted into the cytosol of the cell and the two conductors are connected to a power supply and current measuring device.

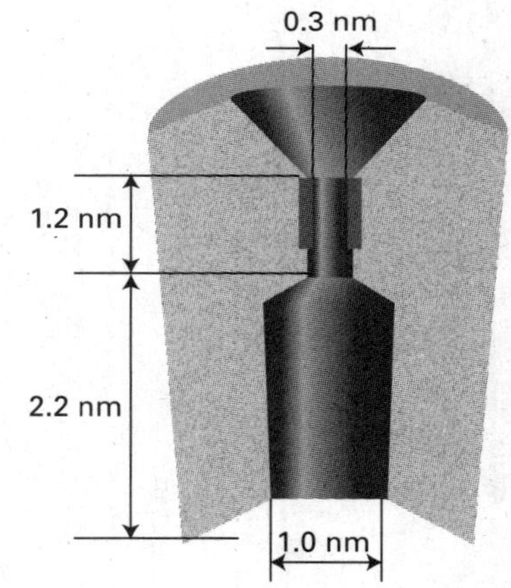

Fig. 20.17 A schematic representation of the cross-section of a membrane-spanning K^+ ion channel protein. The bulk of the protein is shown in light shades of grey. The pore through which ions move is divided into two regions: a wide region with a length of 2.2 nm and diameter of 1.0 nm, and a narrow region, the *selectivity filter*, with a length of 1.2 nm and diameter of 0.3 nm. The selectivity filter has a number of carbonyl groups (shown in dark green) that grip K^+ ions. As explained in the text, electrostatic repulsions between two bound K^+ ions 'encourage' ionic movement through the selectivity filter and across the membrane.

can cross the membrane. As a result, Tl^+ is a neurotoxin because it replaces K^+ in many neuronal functions.

The efficiency of transfer of K^+ ions through the channel can also be explained by structural features of the protein. For efficient transport to occur, a K^+ ion must enter the protein, but then must not be allowed to remain inside for very long so that, as one K^+ ion enters the channel from one side, another K^+ ion leaves from the opposite side. An ion is lured into the channel by water molecules about halfway through the length of the membrane. Consequently, the thermodynamic cost of moving an ion from an aqueous environment to the less hydrophilic interior of the protein is minimized. The ion is 'encouraged' to leave the protein by electrostatic interactions in the selectivity filter, which can bind two K^+ ions simultaneously, usually with a bridging water molecule. Electrostatic repulsion prevents the ions from binding too tightly, minimizing the residence time of an ion in the selectivity filter, and maximizing the transport rate.

Diffusion

We are now in a position to extend the discussion of ionic motion to cover the migration of neutral molecules and of ions in the absence of an applied electric field. We shall do this by expressing ion motion in a more general way than hitherto, and will then discover that the same equations apply even when the charge on the particles is zero.

20.8 The thermodynamic view

Key points **The thermodynamic force represents the spontaneous tendency of molecules to disperse as a consequence of the Second Law. (a) Fick's first law of diffusion can be deduced by considering the thermodynamic force and viscous drag of a solution. (b) The diffusion coefficient and the ionic mobility are related by the Einstein relation. (c) The Stokes–Einstein equation relates the diffusion coefficient to the frictional force.**

We saw in Part 1 that, at constant temperature and pressure, the maximum non-expansion work that can be done per mole when a substance moves from a location where its chemical potential is μ to a location where its chemical potential is $\mu + \mathrm{d}\mu$ is $\mathrm{d}w = \mathrm{d}\mu$. In a system in which the chemical potential depends on the position x,

$$\mathrm{d}w = \mathrm{d}\mu = \left(\frac{\partial \mu}{\partial x}\right)_{p,T} \mathrm{d}x \qquad (20.41)$$

We also saw in Chapter 2 (Table 2.1) that, in general, work can always be expressed in terms of an opposing force (which here we write $\mathcal{F}$), and that

$$\mathrm{d}w = -\mathcal{F}\,\mathrm{d}x \qquad (20.42)$$

By comparing these two expressions, we see that the slope of the chemical potential can be interpreted as an effective force per mole of molecules. We write this **thermodynamic force** as

$$\mathcal{F} = -\left(\frac{\partial \mu}{\partial x}\right)_{p,T} \qquad \text{Definition of the thermodynamic force} \qquad [20.43]$$

There is not necessarily a real force pushing the particles down the slope of the chemical potential. As we shall see, the force may represent the spontaneous tendency of the molecules to disperse as a consequence of the Second Law and the hunt for maximum entropy.

(a) Fick's first law of diffusion

In a solution in which the activity of the solute is a, the chemical potential is

$$\mu = \mu^{\ominus} + RT \ln a$$

If the solution is not uniform the activity depends on the position and we can write

$$\mathcal{F} = -RT\left(\frac{\partial \ln a}{\partial x}\right)_{p,T} \qquad (20.44)$$

If the solution is ideal, a may be replaced by the molar concentration c, and then

$$\mathcal{F} = -\frac{RT}{c}\left(\frac{\partial c}{\partial x}\right)_{p,T} \qquad (20.45)^{\circ}$$

where we have also used the relation $\mathrm{d}\ln y/\mathrm{d}x = (1/y)(\mathrm{d}y/\mathrm{d}x)$.

Example 20.3 *Calculating the thermodynamic force*

Suppose the concentration of a solute decays exponentially along the length of a container. Calculate the thermodynamic force on the solute at 25°C given that the concentration falls to half its value in 10 cm.

Method According to eqn 20.45, the thermodynamic force is calculated by differentiating the concentration with respect to distance. Therefore, write an expression for the variation of the concentration with distance, and then differentiate it.

Answer The concentration varies with position as

$$c = c_0 \mathrm{e}^{-x/\lambda}$$

where λ is the decay constant. Therefore,

$$\frac{\mathrm{d}c}{\mathrm{d}x} = -\frac{c}{\lambda}$$

Equation 20.45 then implies that

$$\mathcal{F} = \frac{RT}{\lambda}$$

We know that the concentration falls to $\frac{1}{2}c_0$ at $x = 10$ cm, so we can find λ from $\frac{1}{2} = \mathrm{e}^{-(10\ \mathrm{cm})/\lambda}$. That is $\lambda = (10\ \mathrm{cm}/\ln 2)$. It follows that

$$\mathcal{F} = (8.3145\ \mathrm{J\,K^{-1}\,mol^{-1}}) \times (298\ \mathrm{K}) \times \ln 2/(1.0 \times 10^{-1}\ \mathrm{m}) = 17\ \mathrm{kN\,mol^{-1}}$$

where we have used $1\ \mathrm{J} = 1\ \mathrm{N\,m}$.

Self-test 20.3 Calculate the thermodynamic force on the molecules of molar mass M in a vertical tube in a gravitational field on the surface of the Earth, and evaluate $\mathcal{F}$ for molecules of molar mass 100 g mol^{-1}. Comment on its magnitude relative to that just calculated.

[$\mathcal{F} = -Mg$, -0.98 N mol^{-1}; the force arising from the concentration gradient greatly dominates that arising from the gravitational gradient.]

In Section 20.4 we saw that Fick's first law of diffusion (that the particle flux is proportional to the concentration gradient) could be deduced from the kinetic model of gases. We shall now show that it can be deduced more generally and that it applies to the diffusion of species in condensed phases too.

We suppose that the flux of diffusing particles is motion in response to a thermodynamic force arising from a concentration gradient. The particles reach a steady drift speed, s, when the thermodynamic force, $\mathcal{F}$, is matched by the viscous drag. This drift speed is proportional to the thermodynamic force, and we write $s \propto \mathcal{F}$. However, the particle flux, J, is proportional to the drift speed, and the thermodynamic force is proportional to the concentration gradient, $\mathrm{d}c/\mathrm{d}x$. The chain of proportionalities ($J \propto s$, $s \propto \mathcal{F}$, and $\mathcal{F} \propto \mathrm{d}c/\mathrm{d}x$) implies that $J \propto \mathrm{d}c/\mathrm{d}x$, which is the content of Fick's law.

(b) The Einstein relation

If we divide both sides of eqn 20.19 by Avogadro's constant, thereby converting numbers into amounts (numbers of moles), then Fick's law becomes

$$J = -D\frac{\mathrm{d}c}{\mathrm{d}x} \qquad \text{Fick's first law in terms of the concentration gradient} \qquad (20.46)$$

In this expression, D is the diffusion coefficient and $\mathrm{d}c/\mathrm{d}x$ is the slope of the molar concentration. The flux is related to the drift speed by

$$J = sc \qquad (20.47)$$

This relation follows from the argument that we have used several times before. Thus, all particles within a distance $s\Delta t$, and therefore in a volume $s\Delta tA$, can pass through a window of area A in an interval Δt. Hence, the amount of substance that can pass through the window in that interval is $s\Delta tAc$. Therefore,

$$sc = -D\frac{\mathrm{d}c}{\mathrm{d}x}$$

If now we express $\mathrm{d}c/\mathrm{d}x$ in terms of $\mathcal{F}$ by using eqn 20.45, we find

$$s = -\frac{D}{c}\frac{\mathrm{d}c}{\mathrm{d}x} = \frac{D\mathcal{F}}{RT} \qquad (20.48)$$

Therefore, once we know the effective force and the diffusion coefficient, D, we can calculate the drift speed of the particles (and vice versa) whatever the origin of the force.

There is one case where we already know the drift speed and the effective force acting on a particle: an ion in solution has a drift speed $s = u\mathcal{E}$ when it experiences a force $ez\mathcal{E}$ from an electric field of strength $\mathcal{E}$ (so $\mathcal{F} = N_A ez\mathcal{E} = zF\mathcal{E}$). Therefore, substituting these known values into eqn 20.48 gives

$$u\mathcal{E} = \frac{zF\mathcal{E}D}{RT}$$

and hence

$$u = \frac{zFD}{RT} \qquad (20.49)$$

This equation rearranges into the very important result known as the **Einstein relation** between the diffusion coefficient and the ionic mobility:

$$D = \frac{uRT}{zF} \qquad \text{Einstein relation} \qquad (20.50)^{\circ}$$

On inserting the typical value $u = 5 \times 10^{-8}\ \mathrm{m^2\ s^{-1}\ V^{-1}}$, we find $D \approx 1 \times 10^{-9}\ \mathrm{m^2\ s^{-1}}$ at 25°C as a typical value of the diffusion coefficient of an ion in water.

(c) The Stokes–Einstein equation

Equations 20.35 ($u = ez/f$) and 20.49 relate the mobility of an ion to the frictional force and to the diffusion coefficient, respectively. We can combine the two expressions into the **Stokes–Einstein equation**:

$$D = \frac{kT}{f} \qquad \text{Stokes–Einstein equation} \qquad (20.51)$$

If the frictional force is described by Stokes's relation (eqn 18.31), then we also obtain a relation between the diffusion coefficient and the viscosity of the medium:

$$D = \frac{kT}{6\pi\eta a} \qquad (20.52)$$

An important feature of eqn 20.51 (and of its special case, eqn 20.52) is that it makes no reference to the charge of the diffusing species. Therefore, the equation also applies in the limit of vanishingly small charge, that is, it also applies to neutral molecules. Consequently, we may use viscosity measurements to estimate the diffusion coefficients for electrically neutral molecules in solution (Table 20.7). It must not be forgotten, however, that both equations depend on the assumption that the viscous drag is proportional to the speed.

Table 20.7* Diffusion coefficients at 298 K

	$D/(10^{-9}\ \mathrm{m^2\ s^{-1}})$
H^+ in water	9.31
I_2 in hexane	4.05
Na^+ in water	1.33
Sucrose in water	0.522

* More values are given in the *Data section*.

Example 20.4 *Interpreting the mobility of an ion*

Use the experimental value of the mobility to evaluate the diffusion coefficient, the limiting molar conductivity, and the hydrodynamic radius of a sulfate ion in aqueous solution.

Method The starting point is the mobility of the ion, which is given in Table 20.5. The diffusion coefficient can then be determined from the Einstein relation, eqn 20.50. The ionic conductivity is related to the mobility by eqn 20.36. To estimate the hydrodynamic radius, a, of the ion, use the Stokes–Einstein relation to find f and the Stokes law to relate f to a.

Answer From Table 20.5, the mobility of SO_4^{2-} is $8.29 \times 10^{-8}\ \mathrm{m^2\ s^{-1}\ V^{-1}}$. It follows from eqn 20.50 that

$$D = \frac{uRT}{zF} = 1.1 \times 10^{-9}\ \mathrm{m^2\ s^{-1}}$$

From eqn 20.36 it follows that

$$\lambda_- = zu_-F = 16\ \mathrm{mS\ m^2\ mol^{-1}}$$

Finally, from $f = 6\pi\eta a$ using 0.891 cP (or $8.91 \times 10^{-4}\ \mathrm{kg\ m^{-1}\ s^{-1}}$) for the viscosity of water (Table 20.4):

$$a = \frac{kT}{6\pi\eta D} = 220\ \mathrm{pm}$$

The bond length in SO_4^{2-} is 144 pm, so the radius calculated here is plausible and consistent with a small degree of solvation.

Self-test 20.4 Repeat the calculation for the NH_4^+ ion.

[$1.96 \times 10^{-9}\ \mathrm{m^2\ s^{-1}}$, $7.4\ \mathrm{mS\ m^2\ mol^{-1}}$, 125 pm]

20.9 The diffusion equation

Key points **The diffusion equation is a relation between the rate of change of concentration at a point and the spatial variation of the concentration at that point. (a) The generalized diffusion equation takes into account the combined effects of diffusion and convection. (b) The diffusion equation is a second-order differential equation with respect to space and a first-order differential equation with respect to time. Its solution requires specification of two boundary conditions for the spatial dependence and an initial condition for the time dependence.**

We now turn to the discussion of time-dependent diffusion processes, where we are interested in the spreading of inhomogeneities with time. One example is the temperature of a metal bar that has been heated at one end: if the source of heat is removed, then the bar gradually settles down into a state of uniform temperature. When the source of heat is maintained and the bar is connected at the far end to a thermal sink, it settles down into a steady state of nonuniform temperature. Another example (and one more relevant to chemistry) is the concentration distribution in a solvent to which a solute is added. We shall focus on the description of the diffusion of particles, but similar arguments apply to the diffusion of physical properties, such as temperature. Our aim is to obtain an equation for the rate of change of the concentration of particles in an inhomogeneous region.

The central equation of this section is the **diffusion equation**, also called 'Fick's second law of diffusion', which relates the rate of change of concentration at a point to the spatial variation of the concentration at that point:

$$\frac{\partial c}{\partial t}=D\frac{\partial^2 c}{\partial x^2} \qquad \text{Diffusion equation} \qquad (20.53)$$

We show in the following *Justification* that the diffusion equation follows from Fick's first law of diffusion.

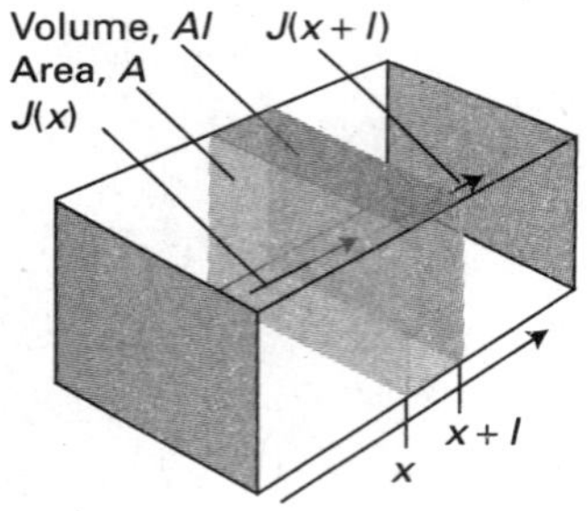

Fig. 20.18 The net flux in a region is the difference between the flux entering from the region of high concentration (on the left) and the flux leaving to the region of low concentration (on the right).

Justification 20.6 *The diffusion equation*

Consider a thin slab of cross-sectional area A that extends from x to $x+l$ (Fig. 20.18). Let the concentration at x be c at the time t. The amount (number of moles) of particles that enter the slab in the infinitesimal interval $\mathrm{d}t$ is $JA\mathrm{d}t$, so the rate of increase in molar concentration inside the slab (which has volume Al) on account of the flux from the left is

$$\frac{\partial c}{\partial t}=\frac{JA\mathrm{d}t}{Al\mathrm{d}t}=\frac{J}{l}$$

There is also an outflow through the right-hand window. The flux through that window is J', and the rate of change of concentration that results is

$$\frac{\partial c}{\partial t}=-\frac{J'A\mathrm{d}t}{Al\mathrm{d}t}=-\frac{J'}{l}$$

The net rate of change of concentration is therefore

$$\frac{\partial c}{\partial t}=\frac{J-J'}{l}$$

Each flux is proportional to the concentration gradient at the window. So, by using Fick's first law, we can write

$$J-J'=-D\frac{\partial c}{\partial x}+D\frac{\partial c'}{\partial x}=-D\frac{\partial c}{\partial x}+D\frac{\partial}{\partial x}\left\{c+\left(\frac{\partial c}{\partial x}\right)l\right\}=Dl\frac{\partial^2 c}{\partial x^2}$$

When this relation is substituted into the expression for the rate of change of concentration in the slab, we get eqn 20.53.

The diffusion equation shows that the rate of change of concentration is proportional to the curvature (more precisely, to the second derivative) of the concentration with respect to distance. If the concentration changes sharply from point to point (if the distribution is highly wrinkled) then the concentration changes rapidly with time. Where the curvature is positive (a dip, Fig. 20.19), the change in concentration is positive; the dip tends to fill. Where the curvature is negative (a heap), the change in concentration is negative; the heap tends to spread. If the curvature is zero, then the concentration is constant in time. If the concentration decreases linearly with distance, then the concentration at any point is constant because the inflow of particles is exactly balanced by the outflow.

The diffusion equation can be regarded as a mathematical formulation of the intuitive notion that there is a natural tendency for the wrinkles in a distribution to disappear. More succinctly: Nature abhors a wrinkle.

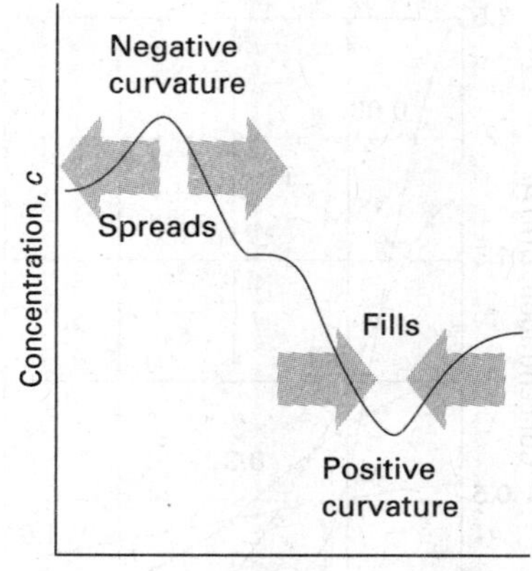

Fig. 20.19 Nature abhors a wrinkle. The diffusion equation tells us that peaks in a distribution (regions of negative curvature) spread and troughs (regions of positive curvature) fill in.

(a) Diffusion with convection

The transport of particles arising from the motion of a streaming fluid is called **convection.** If for the moment we ignore diffusion, then the flux of particles through an area A in an interval Δt when the fluid is flowing at a velocity v can be calculated in the way we have used several times before (by counting the particles within a distance $v\Delta t$), and is

$$J=\frac{cAv\Delta t}{A\Delta t}=cv \qquad \text{Convective flux} \qquad (20.54)$$

This J is called the **convective flux**. The rate of change of concentration in a slab of thickness l and area A is, by the same argument as before and assuming that the velocity does not depend on the position,

$$\frac{\partial c}{\partial t}=\frac{J-J'}{l}=\left\{c-\left[c+\left(\frac{\partial c}{\partial x}\right)l\right]\right\}\frac{v}{l}=-v\frac{\partial c}{\partial x} \qquad (20.55)$$

When both diffusion and convection occur, the total change of concentration in a region is the sum of the two effects, and the **generalized diffusion equation** is

$$\frac{\partial c}{\partial t}=D\frac{\partial^2 c}{\partial x^2}-v\frac{\partial c}{\partial x} \qquad \text{Generalized diffusion equation} \qquad (20.56)$$

A further refinement, which is important in chemistry, is the possibility that the concentrations of particles may change as a result of reaction. When reactions are included in eqn 20.56 (Section 22.2), we get a powerful differential equation for discussing the properties of reacting, diffusing, convecting systems and which is the basis of reactor design in chemical industry and of the utilization of resources in living cells.

(b) Solutions of the diffusion equation

The diffusion equation, eqn 20.53, is a second-order differential equation with respect to space and a first-order differential equation with respect to time. Therefore, we must specify two boundary conditions for the spatial dependence and a single initial condition for the time dependence.

As an illustration, consider a solvent in which the solute is initially coated on one surface of the container (for example, a layer of sugar on the bottom of a deep beaker of water). The single initial condition is that at $t=0$ all N_0 particles are concentrated on the yz-plane (of area A) at $x=0$. The two boundary conditions are derived from the requirements (1) that the concentration must everywhere be finite and (2) that the total amount (number of moles) of particles present is n_0 (with $n_0=N_0/N_A$) at all

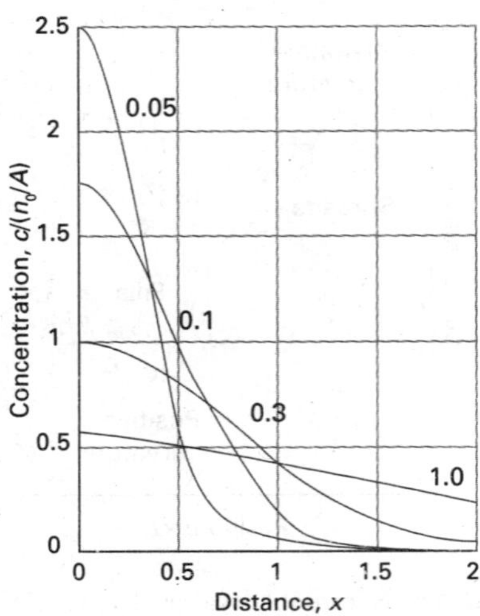

Fig. 20.20 The concentration profiles above a plane from which a solute is diffusing. The curves are plots of eqn 20.57 and are labelled with different values of Dt. The units of Dt and x are arbitrary, but are related so that Dt/x^2 is dimensionless. For example, if x is in metres, Dt would be in metres^2; so, for $D = 10^{-9}\ \mathrm{m^2\ s^{-1}}$, $Dt = 0.1\ \mathrm{m^2}$ corresponds to $t = 10^8$ s.

interActivity Generate a family of curves similar to that shown in Fig. 20.20 but by using eqn 20.58, which describes diffusion in three dimensions.

times. These requirements imply that the flux of particles is zero at the top and bottom surfaces of the system. Under these conditions it is found that

$$c(x,t) = \frac{n_0}{A(\pi Dt)^{1/2}} e^{-x^2/4Dt} \tag{20.57}$$

as may be verified by direct substitution. Figure 20.20 shows the shape of the concentration distribution at various times, and it is clear that the concentration spreads and tends to uniformity.

Another useful result is for a localized concentration of solute in a three-dimensional solvent (a sugar lump suspended in a large flask of water). The concentration of diffused solute is spherically symmetrical and at a radius r is

$$c(r,t) = \frac{n_0}{8(\pi Dt)^{3/2}} e^{-r^2/4Dt} \tag{20.58}$$

Other chemically (and physically) interesting arrangements, such as transport of substances across biological membranes can be treated. In many cases the solutions are more cumbersome.

The solutions of the diffusion equation are useful for experimental determinations of diffusion coefficients. In the **capillary technique**, a capillary tube, open at one end and containing a solution, is immersed in a well-stirred larger quantity of solvent, and the change of concentration in the tube is monitored. The solute diffuses from the open end of the capillary at a rate that can be calculated by solving the diffusion equation with the appropriate boundary conditions, so D may be determined. In the **diaphragm technique**, the diffusion occurs through the capillary pores of a sintered glass diaphragm separating the well-stirred solution and solvent. The concentrations are monitored and then related to the solutions of the diffusion equation corresponding to this arrangement. Diffusion coefficients may also be measured by laser light scattering techniques and by NMR.

20.10 Diffusion probabilities

Key point **Diffusion is a very slow process.**

The solutions of the diffusion equation can be used to predict the concentration of particles (or the value of some other physical quantity, such as the temperature in a nonuniform system) at any location. We can also use them to calculate the net distance through which the particles diffuse in a given time.

Example 20.5 *Calculating the net distance of diffusion*

Calculate the net distance travelled on average by particles in a time t if they have a diffusion constant D.

Method We need to calculate the probability that a particle will be found at a certain distance from the origin, and then calculate the average distance travelled by weighting each distance by that probability.

Answer The number of particles in a slab of thickness dx and area A at x, where the molar concentration is c, is $cAN_A\mathrm{d}x$. The probability that any of the $N_0 = n_0N_A$ particles is in the slab is therefore $cAN_A\mathrm{d}x/N_0$. If the particle is in the slab, it has travelled a distance x from the origin. Therefore, the mean distance travelled by all the particles is the sum of each x weighted by the probability of its occurrence:

$$\langle x\rangle = \int_0^\infty \frac{xcAN_A}{N_0}\mathrm{d}x = \frac{1}{(\pi Dt)^{1/2}}\int_0^\infty x e^{-x^2/4Dt}\mathrm{d}x = 2\left(\frac{Dt}{\pi}\right)^{1/2}$$

where we have used the same standard integral as that used in *Justification 20.4*.

Self-test 20.5 Derive an expression for the root mean square distance travelled by diffusing particles in a time t. $[\langle x^2\rangle^{1/2} = (2Dt)^{1/2}]$

As shown in Example 20.5, the average distance travelled by a diffusing particle in a time t is

$$\langle x\rangle = 2\left(\frac{Dt}{\pi}\right)^{1/2} \qquad (20.59)$$

and the root mean square distance travelled in the same time is

$$\langle x^2\rangle^{1/2} = (2Dt)^{1/2} \qquad (20.60)$$

The latter is a valuable measure of the spread of particles when they can diffuse in both directions from the origin (for then $\langle x\rangle = 0$ at all times). The root mean square distance travelled by particles with a typical diffusion coefficient ($D = 5\times10^{-10}\ \mathrm{m^2\,s^{-1}}$) is illustrated in Fig. 20.21, which shows how long it takes for diffusion to increase the net distance travelled on average to about 1 cm in an unstirred solution. The graph shows that diffusion is a very slow process (which is why solutions are stirred, to encourage mixing by convection).

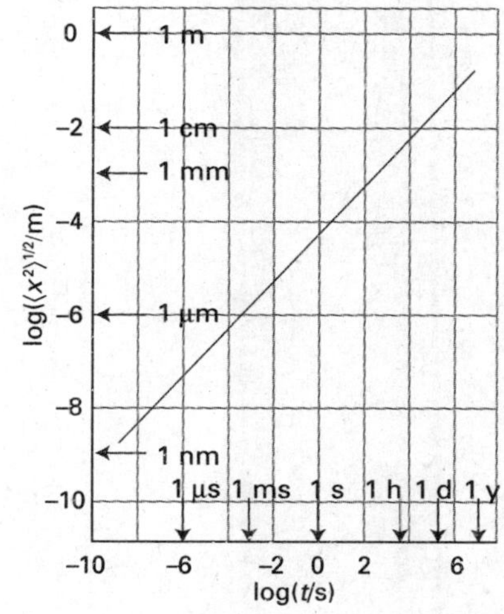

Fig. 20.21 The root mean square distance covered by particles with $D = 5\times10^{-10}\ \mathrm{m^2\,s^{-1}}$. Note the great slowness of diffusion.

20.11 The statistical view

Key points **Diffusion can be described mathematically with a random walk model. The Einstein–Smoluchowski equation relates the diffusion coefficient to the parameters used in the formulation of the random walk model.**

An intuitive picture of diffusion is of the particles moving in a series of small steps and gradually migrating from their original positions. We shall explore this idea using a model in which the particles can jump through a distance λ in a time τ. The total distance travelled by a particle in a time t is therefore $t\lambda/\tau$. However, the particle will not necessarily be found at that distance from the origin. The direction of each step may be different, and the net distance travelled must take the changing directions into account.

If we simplify the discussion by allowing the particles to travel only along a straight line (the x-axis), and for each step (to the left or the right) to be through the same distance λ, then we obtain the **one-dimensional random walk.** The same model was used in the discussion of a one-dimensional random coil in Section 18.1. We can use the result (eqn 18.1) established there by substituting $N = t/\tau$ for the total number of steps and $n = x/\lambda$ for the displacement from the origin, and obtain

$$P = \left(\frac{2\tau}{\pi t}\right)^{1/2} e^{-x^2\tau/2t\lambda^2} \qquad (20.61)$$

The differences of detail between eqns 20.57 and 20.61 arise from the fact that in the present calculation the particles can migrate in either direction from the origin. Moreover, they can be found only at discrete points separated by λ instead of being anywhere on a continuous line. The fact that the two expressions are so similar suggests that diffusion can indeed be interpreted as the outcome of a large number of steps in random directions.

We can now relate the coefficient D to the step length λ and the rate at which the jumps occur. Thus, by comparing the two exponents in eqns 20.59 and 20.63 we can immediately write down the **Einstein–Smoluchowski equation:**

$$D = \frac{\lambda^2}{2\tau} \qquad \text{Einstein–Smoluchowski equation} \qquad (20.62)$$

• **A brief illustration**

Suppose that a SO_4^{2-} ion jumps through its own diameter each time it makes a move in an aqueous solution; then, because $D = 1.1 \times 10^{-9}\ m^2\ s^{-1}$ and $a = 220$ pm (as deduced from mobility measurements), it follows from $\lambda = 2a$ that $\tau = 88$ ps. Because τ is the time for one jump, the ion makes 1×10^{10} jumps per second. •

The Einstein–Smoluchowski equation is the central connection between the microscopic details of particle motion and the macroscopic parameters relating to diffusion (for example, the diffusion coefficient and, through the Stokes–Einstein relation, the viscosity). It also brings us back full circle to the properties of the perfect gas. For if we interpret λ/τ as $\bar{c}$, the mean speed of the molecules, and interpret λ as a mean free path, then we can recognize in the Einstein–Smoluchowski equation exactly the same expression as we obtained from the kinetic model of gases, eqn 20.22. That is, the diffusion of a perfect gas is a random walk with an average step size equal to the mean free path.

Checklist of key equations

Property	Equation	Comment
Pressure of a perfect gas from the kinetic model	$pV = \frac{1}{3}nMc^2$	Kinetic model
Maxwell distribution of speeds	$f(v) = 4\pi(M/2\pi RT)^{3/2}v^2e^{-Mv^2/2RT}$	
Root mean square speed in a perfect gas	$c = \langle v^2\rangle^{1/2} = (3RT/M)^{1/2}$	Kinetic model
Mean speed in a perfect gas	$\bar{c} = (8RT/\pi M)^{1/2}$	Kinetic model
Most probable speed in a perfect gas	$c^* = (2RT/M)^{1/2}$	Kinetic model
Relative mean speed in a perfect gas	$\bar{c}_{rel} = 2^{1/2}\bar{c}$	Kinetic model
The collision frequency in a perfect gas	$z = \sigma\bar{c}_{rel}\mathcal{N}$, $\sigma = \pi d^2$	Kinetic model
Mean free path in a perfect gas	$\lambda = \bar{c}_{rel}/z$	
Collision flux	$Z_W = p/(2\pi mkT)^{1/2}$	
Rate of effusion	$Z_W A_0 = pA_0N_A/(2\pi MRT)^{1/2}$	
Fick's first law of diffusion	$J(\text{matter}) = -D\,d\mathcal{N}/dz$	
Flux of energy	$J(\text{energy}) = -\kappa dT/dz$	
Flux of momentum	$J(x\text{-component of momentum}) = -\eta\,dv_x/dz$	
Diffusion coefficient of a perfect gas	$D = \frac{1}{3}\lambda\bar{c}$	
Coefficient of thermal conductivity of a perfect gas	$\kappa = \frac{1}{3}\lambda\bar{c}C_{V,m}[A]$	
Coefficient of viscosity of a perfect gas	$\eta = \frac{1}{3}M\lambda\bar{c}[A]$	
Conductance	$G = \kappa A/l$	
Molar conductivity	$\Lambda_m = \kappa/c$	
Kohlrausch's law	$\Lambda_m = \Lambda_m^\circ - \mathcal{K}c^{1/2}$	
Law of independent migration of ions	$\Lambda_m^\circ = \nu_+\lambda_+ + \nu_-\lambda_-$	
Drift speed	$s = uE$, $u = ze/6\pi\eta a$	
Ionic conductivity	$\lambda_\pm = zu_\pm F$	
Einstein relation	$D = uRT/zF$	
Stokes–Einstein equation	$D = kT/f$	
Diffusion equation	$\partial c/\partial t = D\partial^2 c/\partial x^2$	
Generalized diffusion equation	$\partial c/\partial t = D\partial^2 c/\partial x^2 - v\partial c/\partial x$	
Einstein–Smoluchowski equation	$D = \lambda^2/2\tau$	

Further information

Further information 20.1 *The transport characteristics of a perfect gas*

In this *Further information* section, we derive expressions for the diffusion characteristics (specifically, the diffusion coefficient, the thermal conductivity, and the viscosity) of a perfect gas on the basis of the kinetic molecular theory.

(a) The diffusion coefficient, *D*

Consider the arrangement depicted in Fig. 20.22. On average, the molecules passing through the area A at $z = 0$ have travelled about one mean free path λ since their last collision. Therefore, the number density where they originated is $\mathcal{N}(z)$ evaluated at $z = -\lambda$. This number density is approximately

$$\mathcal{N}(-\lambda) = \mathcal{N}(0) - \lambda\left(\frac{\mathrm{d}\mathcal{N}}{\mathrm{d}z}\right)_0 \qquad (20.63)$$

where we have used a Taylor expansion of the form $f(x) = f(0) + (\mathrm{d}f/\mathrm{d}x)_0 x + \cdots$ truncated after the second term. The average number of impacts on the imaginary window of area A_0 during an interval Δt is $Z_W A_0 \Delta t$, with $Z_W = \frac{1}{4}\mathcal{N}\bar{c}$ (eqn 20.15). Therefore, the flux from left to right, $J(\mathrm{L} \rightarrow \mathrm{R})$, arising from the supply of molecules on the left, is

$$J(\mathrm{L} \rightarrow \mathrm{R}) = \frac{\frac{1}{4}A_0\mathcal{N}(-\lambda)\bar{c}\Delta t}{A_0\Delta t} = \tfrac{1}{4}\mathcal{N}(-\lambda)\bar{c} \qquad (20.64)$$

There is also a flux of molecules from right to left. On average, the molecules making the journey have originated from $z = +\lambda$ where the number density is $\mathcal{N}(\lambda)$. Therefore,

$$J(\mathrm{L} \leftarrow \mathrm{R}) = -\tfrac{1}{4}\mathcal{N}(\lambda)\bar{c} \qquad (20.65)$$

The average number density at $z = +\lambda$ is approximately

$$\mathcal{N}(\lambda) = \mathcal{N}(0) + \lambda\left(\frac{\mathrm{d}\mathcal{N}}{\mathrm{d}z}\right)_0 \qquad (20.66)$$

The net flux is

$$\begin{aligned} J_z &= J(\mathrm{L} \rightarrow \mathrm{R}) + J(\mathrm{L} \leftarrow \mathrm{R}) \\ &= \tfrac{1}{4}\bar{c}\left\{\left[\mathcal{N}(0) - \lambda\left(\frac{\mathrm{d}\mathcal{N}}{\mathrm{d}z}\right)_0\right] - \left[\mathcal{N}(0) + \lambda\left(\frac{\mathrm{d}\mathcal{N}}{\mathrm{d}z}\right)_0\right]\right\} \\ &= -\tfrac{1}{2}\bar{c}\lambda\left(\frac{\mathrm{d}\mathcal{N}}{\mathrm{d}z}\right)_0 \end{aligned} \qquad (20.67)$$

This equation shows that the flux is proportional to the first derivative of the concentration, in agreement with Fick's law.

At this stage it looks as though we can pick out a value of the diffusion coefficient by comparing eqns 20.19 and 20.67, so obtaining $D = \frac{1}{2}\lambda\bar{c}$. It must be remembered, however, that the calculation is quite crude, and is little more than an assessment of the order of magnitude of D. One aspect that has not been taken into account is illustrated in Fig. 20.23, which shows that, although a molecule may have begun its journey very close to the window, it could have a long flight before it gets there. Because the path is long, the molecule is likely to collide before reaching the window, so it ought to be added to the graveyard of other molecules that have collided. To take this effect into account involves a lot of work, but the end result is the appearance of a factor of $\frac{2}{3}$ representing the lower flux. The modification results in eqn 20.22.

(b) Thermal conductivity

According to the equipartition theorem (Section 16.3), each molecule carries an average energy $\varepsilon = \nu kT$, where ν is a number of the order of 1.

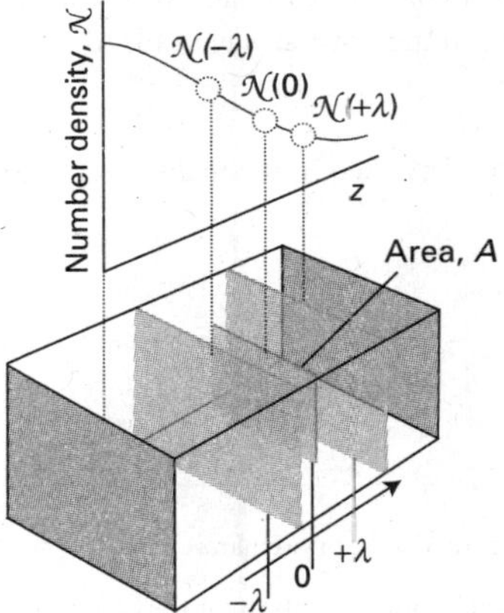

Fig. 20.22 The calculation of the rate of diffusion of a gas considers the net flux of molecules through a plane of area A as a result of arrivals from on average a distance λ away in each direction, where λ is the mean free path.

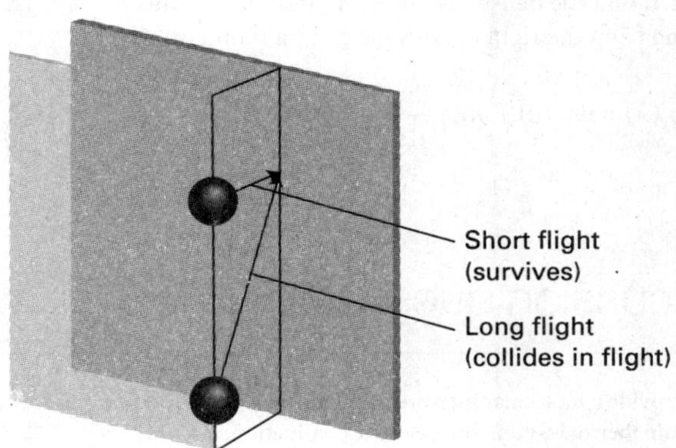

Fig. 20.23 One approximation ignored in the simple treatment is that some particles might make a long flight to the plane even though they are only a short perpendicular distance away, and therefore they have a higher chance of colliding during their journey.

For monatomic particles, $\nu=\frac{3}{2}$. When one molecule passes through the imaginary window, it transports that energy on average. We suppose that the number density is uniform but that the temperature is not. On average, molecules arrive from the left after travelling a mean free path from their last collision in a hotter region, and therefore with a higher energy. Molecules also arrive from the right after travelling a mean free path from a cooler region. The two opposing energy fluxes are therefore

$$J(\mathrm{L}\rightarrow\mathrm{R})=\tfrac{1}{4}\bar{c}\mathcal{N}\varepsilon(-\lambda)\qquad \varepsilon(-\lambda)=\nu k\left\{T-\lambda\left(\frac{\mathrm{d}T}{\mathrm{d}z}\right)_0\right\}$$

$$J(\mathrm{L}\leftarrow\mathrm{R})=\tfrac{1}{4}\bar{c}\mathcal{N}\varepsilon(+\lambda)\qquad \varepsilon(+\lambda)=\nu k\left\{T+\lambda\left(\frac{\mathrm{d}T}{\mathrm{d}z}\right)_0\right\}\tag{20.68}$$

and the net flux is

$$J_z=J(\mathrm{L}\rightarrow\mathrm{R})-J(\mathrm{L}\leftarrow\mathrm{R})=-\tfrac{1}{2}\nu k\lambda\bar{c}\mathcal{N}\left(\frac{\mathrm{d}T}{\mathrm{d}z}\right)_0\tag{20.69}$$

As before, we multiply by $\frac{2}{3}$ to take long flight paths into account, and so arrive at

$$J_z=-\tfrac{1}{3}\nu k\lambda\bar{c}\mathcal{N}\left(\frac{\mathrm{d}T}{\mathrm{d}z}\right)_0\tag{20.70}$$

The energy flux is proportional to the temperature gradient, as we wanted to show. Comparison of this equation with eqn 20.20 shows that

$$\kappa=\tfrac{1}{3}\nu k\lambda\bar{c}\mathcal{N}\tag{20.71}$$

Equation 20.23 then follows from $C_{V,\mathrm{m}}=\nu kN_{\mathrm{A}}$ for a perfect gas, where [A] is the molar concentration of A. For this step, we use $\mathcal{N}=N/V=nN_{\mathrm{A}}/V=N_{\mathrm{A}}[\mathrm{A}]$.

(c) Viscosity

Molecules travelling from the right in Fig. 20.24 (from a fast layer to a slower one) transport a momentum $mv_x(\lambda)$ to their new layer at $z=0$; those travelling from the left transport $mv_x(-\lambda)$ to it. If it is assumed that the density is uniform, the collision flux is $\frac{1}{4}\mathcal{N}\bar{c}$. Those arriving from the right on average carry a momentum

$$mv_x(\lambda)=mv_x(0)+m\lambda\left(\frac{\mathrm{d}v_x}{\mathrm{d}z}\right)_0\tag{20.72a}$$

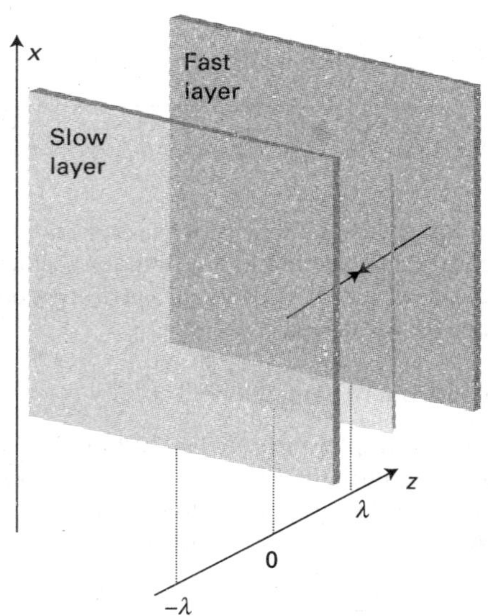

Fig. 20.24 The calculation of the viscosity of a gas examines the net x-component of momentum brought to a plane from faster and slower layers on average a mean free path away in each direction.

Those arriving from the left bring a momentum

$$mv_x(-\lambda)=mv_x(0)-m\lambda\left(\frac{\mathrm{d}v_x}{\mathrm{d}z}\right)_0\tag{20.72b}$$

The net flux of x-momentum in the z-direction is therefore

$$J=\tfrac{1}{4}\mathcal{N}\bar{c}\left\{\left[mv_x(0)-m\lambda\left(\frac{\mathrm{d}v_x}{\mathrm{d}z}\right)_0\right]-\left[mv_x(0)+m\lambda\left(\frac{\mathrm{d}v_x}{\mathrm{d}z}\right)_0\right]\right\}$$

$$=-\tfrac{1}{2}\mathcal{N}m\lambda\bar{c}\left(\frac{\mathrm{d}v_x}{\mathrm{d}z}\right)_0\tag{20.73}$$

The flux is proportional to the velocity gradient, as we wished to show. Comparison of this expression with eqn 20.21, and multiplication by $\frac{2}{3}$ in the normal way, leads to

$$\eta=\tfrac{1}{3}\mathcal{N}m\lambda\bar{c}\tag{20.74}$$

which can easily be converted into eqn 20.24 by using $Nm=nM$ and $[\mathrm{A}]=n/V$.

Discussion questions

20.1 Provide a molecular interpretation for each of the following processes: diffusion, thermal conduction, electric conduction, viscosity.

20.2 Provide a molecular interpretation for the observation that the viscosity of a gas increases with temperature whereas the viscosity of a liquid decreases with increasing temperature.

20.3 Discuss the mechanism of proton conduction in liquid water. How could this mechanism be tested?

20.4 What is the nature of a thermodynamic force?

20.5 Provide a molecular interpretation for the observation that mediated transport across a biological membrane leads to a maximum flux $J_{\max}$ when the concentration of the transported species becomes very large.

20.6 Discuss how nuclear magnetic resonance spectroscopy, inelastic neutron scattering, and dynamic light scattering may be used to measure the mobility of molecules in liquids.

Exercises

20.1(a) Determine the ratios of (a) the mean speeds, (b) the mean kinetic energies of H_2 molecules and Hg atoms at 20°C.

20.1(b) Determine the ratios of (a) the mean speeds, (b) the mean kinetic energies of He atoms and Hg atoms at 25°C.

20.2(a) A 1.0 dm^3 glass bulb contains 1.0×10^{23} H_2 molecules. If the pressure exerted by the gas is 100 kPa, what are (a) the temperature of the gas, (b) the root mean square speeds of the molecules? (c) Would the temperature be different if they were O_2 molecules?

20.2(b) The best laboratory vacuum pump can generate a vacuum of about 1 nTorr. At 25°C and assuming that air consists of N_2 molecules with a collision diameter of 395 pm, calculate (a) the mean speed of the molecules, (b) the mean free path, (c) the collision frequency in the gas.

20.3(a) Use the Maxwell distribution of speeds to estimate the fraction of N_2 molecules at 500 K that have speeds in the range 290 to 300 m s^{-1}.

20.3(b) Use the Maxwell distribution of speeds to estimate the fraction of CO_2 molecules at 300 K that have speeds in the range 200 to 250 m s^{-1}.

20.4(a) Find an expression for the root mean square deviation of the speed of molecules in a gas from its mean value, $\Delta c = \{\langle c^2\rangle - \langle c\rangle^2\}^{1/2}$.

20.4(b) Find a relation between $\langle c^2\rangle^{1/2}$ and $\langle c^4\rangle^{1/4}$ for molecules in a gas at a temperature T.

20.5(a) At what pressure does the mean free path of argon at 25°C become comparable to the size of a 1 dm^3 vessel that contains it? Take $\sigma = 0.36$ nm^2.

20.5(b) At what pressure does the mean free path of argon at 25°C become comparable to the diameters of the atoms themselves?

20.6(a) At an altitude of 20 km the temperature is 217 K and the pressure 0.050 atm. What is the mean free path of N_2 molecules? ($\sigma = 0.43$ nm^2.)

20.6(b) At an altitude of 15 km the temperature is 217 K and the pressure 12.1 kPa. What is the mean free path of N_2 molecules? ($\sigma = 0.43$ nm^2.)

20.7(a) How many collisions does a single Ar atom make in 1.0 s when the temperature is 25°C and the pressure is (a) 10 atm, (b) 1.0 atm, (c) 1.0 μatm?

20.7(b) How many collisions per second does an N_2 molecule make at an altitude of 15 km? (See Exercise 20.6b for data.)

20.8(a) Calculate the mean free path of molecules in air using $\sigma = 0.43$ nm^2 at 25°C and (a) 10 atm, (b) 1.0 atm, (c) 1.0 μatm.

20.8(b) Calculate the mean free path of carbon dioxide molecules using $\sigma = 0.52$ nm^2 at 25°C and (a) 15 atm, (b) 1.0 bar, (c) 1.0 Torr.

20.9(a) A solid surface with dimensions 2.5 mm × 3.0 mm is exposed to argon gas at 90 Pa and 500 K. How many collisions do the Ar atoms make with this surface in 15 s?

20.9(b) A solid surface with dimensions 3.5 mm × 4.0 cm is exposed to helium gas at 111 Pa and 1500 K. How many collisions do the He atoms make with this surface in 10 s?

20.10(a) An effusion cell has a circular hole of diameter 2.50 mm. If the molar mass of the solid in the cell is 260 g mol^{-1} and its vapour pressure is 0.835 Pa at 400 K, by how much will the mass of the solid decrease in a period of 2.00 h?

20.10(b) An effusion cell has a circular hole of diameter 3.00 mm. If the molar mass of the solid in the cell is 300 g mol^{-1} and its vapour pressure is 0.224 Pa at 450 K, by how much will the mass of the solid decrease in a period of 24.00 h?

20.11(a) A solid compound of molar mass 100 g mol^{-1} was introduced into a container and heated to 400°C. When a hole of diameter 0.50 mm was opened in the container for 400 s, a mass loss of 285 mg was measured. Calculate the vapour pressure of the compound at 400°C.

20.11(b) A solid compound of molar mass 200 g mol^{-1} was introduced into a container and heated to 300°C. When a hole of diameter 0.50 mm was opened in the container for 500 s, a mass loss of 277 mg was measured. Calculate the vapour pressure of the compound at 300°C.

20.12(a) A manometer was connected to a bulb containing carbon dioxide under slight pressure. The gas was allowed to escape through a small pinhole, and the time for the manometer reading to drop from 75 cm to 50 cm was 52 s. When the experiment was repeated using nitrogen (for which $M = 28.02$ g mol^{-1}) the same fall took place in 42 s. Calculate the molar mass of carbon dioxide.

20.12(b) A manometer was connected to a bulb containing nitrogen under slight pressure. The gas was allowed to escape through a small pinhole, and the time for the manometer reading to drop from 65.1 cm to 42.1 cm was 18.5 s. When the experiment was repeated using a fluorocarbon gas, the same fall took place in 82.3 s. Calculate the molar mass of the fluorocarbon.

20.13(a) A space vehicle of internal volume 3.0 m^3 is struck by a meteor and a hole of radius 0.10 mm is formed. If the oxygen pressure within the vehicle is initially 80 kPa and its temperature 298 K, how long will the pressure take to fall to 70 kPa?

20.13(b) A container of internal volume 22.0 m^3 was punctured, and a hole of radius 0.050 mm was formed. If the nitrogen pressure within the vehicle is initially 122 kPa and its temperature 293 K, how long will the pressure take to fall to 105 kPa?

20.14(a) Calculate the flux of energy arising from a temperature gradient of 2.5 K m^{-1} in a sample of argon in which the mean temperature is 273 K.

20.14(b) Calculate the flux of energy arising from a temperature gradient of 3.5 K m^{-1} in a sample of hydrogen in which the mean temperature is 260 K.

20.15(a) Use the experimental value of the thermal conductivity of neon (Table 20.2) to estimate the collision cross-section of Ne atoms at 273 K.

20.15(b) Use the experimental value of the thermal conductivity of nitrogen (Table 20.2) to estimate the collision cross-section of N_2 molecules at 298 K.

20.16(a) In a double-glazed window, the panes of glass are separated by 5.0 cm. What is the rate of transfer of heat by conduction from the warm room (25°C) to the cold exterior (−10°C) through a window of area 1.0 m^2? What power of heater is required to make good the loss of heat?

20.16(b) Two sheets of copper of area 1.50 m^2 are separated by 10.0 cm. What is the rate of transfer of heat by conduction from the warm sheet (50°C) to the cold sheet (−10°C). What is the rate of loss of heat?

20.17(a) Use the experimental value of the coefficient of viscosity for neon (Table 20.2) to estimate the collision cross-section of Ne atoms at 273 K.

20.17(b) Use the experimental value of the coefficient of viscosity for nitrogen (Table 20.2) to estimate the collision cross-section of the molecules at 273 K.

20.18(a) Calculate the inlet pressure required to maintain a flow rate of 9.5×10^5 dm^3 h^{-1} of nitrogen at 293 K flowing through a pipe of length 8.50 m and diameter 1.00 cm. The pressure of gas as it leaves the tube is 1.00 bar. The volume of the gas is measured at that pressure.

20.18(b) Calculate the inlet pressure required to maintain a flow rate of $8.70\ cm^3\ s^{-1}$ of nitrogen at 300 K flowing through a pipe of length 10.5 m and diameter 15 mm. The pressure of gas as it leaves the tube is 1.00 bar. The volume of the gas is measured at that pressure.

20.19(a) Calculate the viscosity of air at (a) 273 K, (b) 298 K, (c) 1000 K. Take $\sigma \approx 0.40\ nm^2$. (The experimental values are 173 μP at 273 K, 182 μP at 20°C, and 394 μP at 600°C.)

20.19(b) Calculate the viscosity of benzene vapour at (a) 273 K, (b) 298 K, (c) 1000 K. Take $\sigma \approx 0.88\ nm^2$.

20.20(a) Calculate the thermal conductivities of (a) argon, (b) helium at 300 K and 1.0 mbar. Each gas is confined in a cubic vessel of side 10 cm, one wall being at 310 K and the one opposite at 295 K. What is the rate of flow of energy as heat from one wall to the other in each case?

20.20(b) Calculate the thermal conductivities of (a) neon, (b) nitrogen at 300 K and 15 mbar. Each gas is confined in a cubic vessel of side 15 cm, one wall being at 305 K and the one opposite at 295 K. What is the rate of flow of energy as heat from one wall to the other in each case?

20.21(a) Calculate the thermal conductivity of argon ($C_{V,m} = 12.5\ J\ K^{-1}\ mol^{-1}$, $\sigma = 0.36\ nm^2$) at room temperature (20°C).

20.21(b) Calculate the thermal conductivity of nitrogen ($C_{V,m} = 20.8\ J\ K^{-1}\ mol^{-1}$, $\sigma = 0.43\ nm^2$) at room temperature (20°C).

20.22(a) Calculate the diffusion constant of argon at 25°C and (a) 1.00 Pa, (b) 100 kPa, (c) 10.0 MPa. If a pressure gradient of $0.10\ atm\ cm^{-1}$ is established in a pipe, what is the flow of gas due to diffusion?

20.22(b) Calculate the diffusion constant of nitrogen at 25°C and (a) 10.0 Pa, (b) 100 kPa, (c) 15.0 MPa. If a pressure gradient of $0.20\ bar\ m^{-1}$ is established in a pipe, what is the flow of gas due to diffusion?

20.23(a) The mobility of a chloride ion in aqueous solution at 25°C is $7.91 \times 10^{-8}\ m^2\ s^{-1}\ V^{-1}$. Calculate the molar ionic conductivity.

20.23(b) The mobility of an acetate ion in aqueous solution at 25°C is $4.24 \times 10^{-8}\ m^2\ s^{-1}\ V^{-1}$. Calculate the molar ionic conductivity.

20.24(a) The mobility of a Rb^+ ion in aqueous solution is $7.92 \times 10^{-8}\ m^2\ s^{-1}\ V^{-1}$ at 25°C. The potential difference between two electrodes placed in the solution is 35.0 V. If the electrodes are 8.00 mm apart, what is the drift speed of the Rb^+ ion?

20.24(b) The mobility of a Li^+ ion in aqueous solution is $4.01 \times 10^{-8}\ m^2\ s^{-1}\ V^{-1}$ at 25°C. The potential difference between two electrodes placed in the solution is 12.0 V. If the electrodes are 1.00 cm apart, what is the drift speed of the ion?

20.25(a) The limiting molar conductivities of KCl, KNO_3, and $AgNO_3$ are $14.99\ mS\ m^2\ mol^{-1}$, $14.50\ mS\ m^2\ mol^{-1}$, and $13.34\ mS\ m^2\ mol^{-1}$, respectively (all at 25°C). What is the limiting molar conductivity of AgCl at this temperature?

20.25(b) The limiting molar conductivities of NaI, $NaCH_3CO_2$, and $Mg(CH_3CO_2)_2$ are $12.69\ mS\ m^2\ mol^{-1}$, $9.10\ mS\ m^2\ mol^{-1}$, and $18.78\ mS\ m^2\ mol^{-1}$, respectively (all at 25°C). What is the limiting molar conductivity of MgI_2 at this temperature?

20.26(a) At 25°C the molar ionic conductivities of Li^+, Na^+, and K^+ are $3.87\ mS\ m^2\ mol^{-1}$, $5.01\ mS\ m^2\ mol^{-1}$, and $7.35\ mS\ m^2\ mol^{-1}$, respectively. What are their mobilities?

20.26(b) At 25°C the molar ionic conductivities of F^-, Cl^-, and Br^- are $5.54\ mS\ m^2\ mol^{-1}$, $7.635\ mS\ m^2\ mol^{-1}$, and $7.81\ mS\ m^2\ mol^{-1}$, respectively. What are their mobilities?

20.27(a) The mobility of a NO_3^- ion in aqueous solution at 25°C is $7.40 \times 10^{-8}\ m^2\ s^{-1}\ V^{-1}$. Calculate its diffusion coefficient in water at 25°C.

20.27(b) The mobility of a $CH_3CO_2^-$ ion in aqueous solution at 25°C is $4.24 \times 10^{-8}\ m^2\ s^{-1}\ V^{-1}$. Calculate its diffusion coefficient in water at 25°C.

20.28(a) Suppose the concentration of a solute decays linearly along the length of a container. Calculate the thermodynamic force on the solute at 25°C and 10 cm and 20 cm given that the concentration falls to half its value in 10 cm.

20.28(b) Suppose the concentration of a solute increases as x^2 along the length of a container. Calculate the thermodynamic force on the solute at 25°C and 8 cm and 16 cm given that the concentration falls to half its value in 8 cm.

20.29(a) Suppose the concentration of a solute follows a Gaussian distribution (proportional to e^{-x^2}) along the length of a container. Calculate the thermodynamic force on the solute at 20°C and 5.0 cm given that the concentration falls to half its value in 5.0 cm.

20.29(b) Suppose the concentration of a solute follows a Gaussian distribution (proportional to e^{-x^2}) along the length of a container. Calculate the thermodynamic force on the solute at 18°C and 10.0 cm given that the concentration falls to half its value in 10.0 cm.

20.30(a) The diffusion coefficient of CCl_4 in heptane at 25°C is $3.17 \times 10^{-9}\ m^2\ s^{-1}$. Estimate the time required for a CCl_4 molecule to have a root mean square displacement of 5.0 mm.

20.30(b) The diffusion coefficient of I_2 in hexane at 25°C is $4.05 \times 10^{-9}\ m^2\ s^{-1}$. Estimate the time required for an iodine molecule to have a root mean square displacement of 1.0 cm.

20.31(a) Estimate the effective radius of a sucrose molecule in water at 25°C given that its diffusion coefficient is $5.2 \times 10^{-10}\ m^2\ s^{-1}$ and that the viscosity of water is 1.00 cP.

20.31(b) Estimate the effective radius of a glycine molecule in water at 25°C given that its diffusion coefficient is $1.055 \times 10^{-9}\ m^2\ s^{-1}$ and that the viscosity of water is 1.00 cP.

20.32(a) The diffusion coefficient for molecular iodine in benzene is $2.13 \times 10^{-9}\ m^2\ s^{-1}$. How long does a molecule take to jump through about one molecular diameter (approximately the fundamental jump length for translational motion)?

20.32(b) The diffusion coefficient for CCl_4 in heptane is $3.17 \times 10^{-9}\ m^2\ s^{-1}$. How long does a molecule take to jump through about one molecular diameter (approximately the fundamental jump length for translational motion)?

20.33(a) What are the root mean square distances travelled by an iodine molecule in benzene and by a sucrose molecule in water at 25°C in 1.0 s?

20.33(b) About how long, on average, does it take for the molecules in Exercise 20.33a to drift to a point (a) 1.0 mm, (b) 1.0 cm from their starting points?

Problems*

Numerical problems

20.1 The speed of molecules can be measured with a rotating slotted-disc apparatus, which consists of five coaxial 5.0 cm diameter discs separated by 1.0 cm, the slots in their rims being displaced by 2.0° between neighbours. The relative intensities, I, of the detected beam of Kr atoms for two different temperatures and at a series of rotation rates were as follows:

ν/Hz	20	40	80	100	120
I (40 K)	0.846	0.513	0.069	0.015	0.002
I (100 K)	0.592	0.485	0.217	0.119	0.057

Find the distributions of molecular velocities, $f(v_x)$, at these temperatures, and check that they conform to the theoretical prediction for a one-dimensional system.

20.2 Cars were timed by police radar as they passed in both directions below a bridge. Their velocities (kilometres per hour, numbers of cars in parentheses) to the east and west were as follows: 80 E (40), 85 E (62), 90 E (53), 95 E (12), 100 E (2); 80 W (38), 85 W (59), 90 W (50), 95 W (10), 100 W (2). What are (a) the mean velocity, (b) the mean speed, (c) the root mean square speed?

20.3 A population consists of people of the following heights (in metres, numbers of individuals in brackets): 1.80 (1), 1.82 (2), 1.84 (4), 1.86 (7), 1.88 (10), 1.90 (15), 1.92 (9), 1.94 (4), 1.96 (0), 1.98 (1). What are (a) the mean height, (b) the root mean square height of the population?

20.4 Calculate the ratio of the thermal conductivities of gaseous hydrogen at 300 K to gaseous hydrogen at 10 K. Be circumspect, and think about the modes of motion that are thermally active at the two temperatures.

20.5 A Knudsen cell was used to determine the vapour pressure of germanium at 1000°C. During an interval of 7200 s the mass loss through a hole of radius 0.50 mm amounted to 43 μg. What is the vapour pressure of germanium at 1000°C? Assume the gas to be monatomic.

20.6 An atomic beam is designed to function with (a) cadmium, (b) mercury. The source is an oven maintained at 380 K, there being a small slit of dimensions $1.0\ \text{cm} \times 1.0 \times 10^{-3}$ cm. The vapour pressure of cadmium is 0.13 Pa and that of mercury is 12 Pa at this temperature. What is the atomic current (the number of atoms per second) in the beams?

20.7 Conductivities are often measured by comparing the resistance of a cell filled with the sample to its resistance when filled with some standard solution, such as aqueous potassium chloride. The conductivity of water is 76 mS m^{-1} at 25°C and the conductivity of 0.100 mol dm^{-3} KCl(aq) is 1.1639 S m^{-1}. A cell had a resistance of 33.21 Ω when filled with 0.100 mol dm^{-3} KCl(aq) and 300.0 Ω when filled with 0.100 mol dm^{-3} CH_3COOH(aq). What is the molar conductivity of acetic acid at that concentration and temperature?

20.8 The resistances of a series of aqueous NaCl solutions, formed by successive dilution of a sample, were measured in a cell with cell constant (the constant C in the relation $\kappa = C/R$) equal to 0.2063 cm^{-1}. The following values were found:

c/(mol dm^{-3})	0.00050	0.0010	0.0050	0.010	0.020	0.050
R/Ω	3314	1669	342.1	174.1	89.08	37.14

Verify that the molar conductivity follows the Kohlrausch law and find the limiting molar conductivity. Determine the coefficient $\mathcal{K}$. Use the value of $\mathcal{K}$ (which should depend only on the nature, not the identity of the ions) and the information that $\lambda(Na^+) = 5.01$ mS m^2 mol^{-1} and $\lambda(I^-) = 7.68$ mS m^2 mol^{-1} to predict (a) the molar conductivity, (b) the conductivity, (c) the resistance it would show in the cell, of 0.010 mol dm^{-3} NaI(aq) at 25°C.

20.9 After correction for the water conductivity, the conductivity of a saturated aqueous solution of AgCl at 25°C was found to be 0.1887 mS m^{-1}. What is the solubility of silver chloride at this temperature?

20.10 What are the drift speeds of Li^+, Na^+, and K^+ in water when a potential difference of 10 V is applied across a 1.00-cm conductivity cell? How long would it take an ion to move from one electrode to the other? In conductivity measurements it is normal to use alternating current: what are the displacements of the ions in (a) centimetres, (b) solvent diameters, about 300 pm, during a half cycle of 1.0 kHz applied potential?

20.11 The mobilities of H^+ and Cl^- at 25°C in water are 3.623×10^{-7} $m^2\ s^{-1}\ V^{-1}$ and 7.91×10^{-8} $m^2\ s^{-1}\ V^{-1}$, respectively. What proportion of the current is carried by the protons in 10^{-3} M HCl(aq)? What fraction do they carry when the NaCl is added to the acid so that the solution is 1.0 mol dm^{-3} in the salt? Note how concentration as well as mobility governs the transport of current.

20.12 A dilute solution of potassium permanganate in water at 25°C was prepared. The solution was in a horizontal tube of length 10 cm, and at first there was a linear gradation of intensity of the purple solution from the left (where the concentration was 0.100 mol dm^{-3}) to the right (where the concentration was 0.050 mol dm^{-3}). What are the magnitude and sign of the thermodynamic force acting on the solute (a) close to the left face of the container, (b) in the middle, (c) close to the right face? Give the force per mole and force per molecule in each case.

20.13 Estimate the diffusion coefficients and the effective hydrodynamic radii of the alkali metal cations in water from their mobilities at 25°C. Estimate the approximate number of water molecules that are dragged along by the cations. Ionic radii are given in Table 20.3.

20.14 Nuclear magnetic resonance can be used to determine the mobility of molecules in liquids. A set of measurements on methane in carbon tetrachloride showed that its diffusion coefficient is 2.05×10^{-9} $m^2\ s^{-1}$ at 0°C and 2.89×10^{-9} $m^2\ s^{-1}$ at 25°C. Deduce what information you can about the mobility of methane in carbon tetrachloride.

20.15 A concentrated sucrose solution is poured into a cylinder of diameter 5.0 cm. The solution consisted of 10 g of sugar in 5.0 cm^3 of water. A further 1.0 dm^3 of water is then poured very carefully on top of the layer, without disturbing the layer. Ignore gravitational effects, and pay attention only to diffusional processes. Find the concentration at 5.0 cm above the lower layer after a lapse of (a) 10 s, (b) 1.0 years.

20.16 In a series of observations on the displacement of rubber latex spheres of radius 0.212 μm, the mean square displacements after selected time intervals were on average as follows:

t/s	30	60	90	120
$10^{12}\langle x^2\rangle/m^2$	88.2	113.5	128	144

These results were originally used to find the value of Avogadro's constant, but there are now better ways of determining N_A, so the data can be used to find another quantity. Find the effective viscosity of water at the temperature of this experiment (25°C).

* Problems denoted with the symbol ‡ were supplied by Charles Trapp, Carmen Giunta, and Marshall Cady.

20.17‡ A.K. Srivastava *et al.* (*J. Chem. Eng. Data* **41**, 431 (1996)) measured the conductance of several salts in a binary solvent mixture of water and a dipolar aprotic solvent 1,3-dioxolan-2-one. They report the following conductances at 25°C in a solvent 80 per cent 1,3-dioxolan-2-one by mass:

NaI

$c/(\text{mmol dm}^{-3})$	32.02	20.28	12.06	8.64	2.85	1.24	0.83
$\Lambda_m/(\text{S cm}^2\ \text{mol}^{-1})$	50.26	51.99	54.01	55.75	57.99	58.44	58.67

KI

$c/(\text{mmol dm}^{-3})$	17.68	10.8	87.19	2.67	1.28	0.83	0.19
$\Lambda_m/(\text{S cm}^2\ \text{mol}^{-1})$	42.45	45.91	47.53	51.81	54.09	55.78	57.42

Calculate Λ_m° for NaI and KI in this solvent and $\lambda^\circ(\text{Na}) - \lambda^\circ(\text{K})$. Compare your results to the analogous quantities in aqueous solution using Table 20.5 in the *Data section*.

20.18‡ A. Fenghour *et al.* (*J. Phys. Chem. Ref. Data* **24**, 1649 (1995)) have compiled an extensive table of viscosity coefficients for ammonia in the liquid and vapour phases. Deduce the effective molecular diameter of NH_3 based on each of the following vapour-phase viscosity coefficients: (a) $\eta = 9.08 \times 10^{-6}$ kg m^{-1} s^{-1} at 270 K and 1.00 bar; (b) $\eta = 1.749 \times 10^{-5}$ kg m^{-1} s^{-1} at 490 K and 10.0 bar.

20.19‡ G. Bakale *et al.* (*J. Phys. Chem.* **100**, 12477 (1996)) measured the mobility of singly charged C_{60}^- ions in a variety of nonpolar solvents. In cyclohexane at 22°C, the mobility is 1.1 cm^2 V^{-1} s^{-1}. Estimate the effective radius of the C_{60}^- ion. The viscosity of the solvent is 0.93×10^{-3} kg m^{-1} s^{-1}. *Comment.* The researchers interpreted the substantial difference between this number and the van der Waals radius of neutral C_{60} in terms of a solvation layer around the ion.

Theoretical problems

20.20 Start from the Maxwell–Boltzmann distribution and derive an expression for the most probable speed of a gas of molecules at a temperature *T*. Go on to demonstrate the validity of the equipartition conclusion that the average translational kinetic energy of molecules free to move in three dimensions is $\frac{3}{2}kT$.

20.21 Consider molecules that are confined to move in a plane (a two-dimensional gas). Calculate the distribution of speeds and determine the mean speed of the molecules at a temperature *T*.

20.22 A specially constructed velocity-selector accepts a beam of molecules from an oven at a temperature *T* but blocks the passage of molecules with a speed greater than the mean. What is the mean speed of the emerging beam, relative to the initial value, treated as a one-dimensional problem?

20.23 What is the proportion of gas molecules having (a) more than, (b) less than the root mean square speed? (c) What are the proportions having speeds greater and smaller than the mean speed?

20.24 Calculate the fractions of molecules in a gas that have a speed in a range Δv at the speed nc^* relative to those in the same range at c^* itself? This calculation can be used to estimate the fraction of very energetic molecules (which is important for reactions). Evaluate the ratio for $n = 3$ and $n = 4$.

20.25 Derive an expression that shows how the pressure of a gas inside an effusion oven (a heated chamber with a small hole in one wall) varies with time if the oven is not replenished as the gas escapes. Then show that $t_{1/2}$, the time required for the pressure to decrease to half its initial value, is independent of the initial pressure. *Hint.* Begin by setting up a differential equation relating dp/dt to $p = NkT/V$, and then integrating it.

20.26 Confirm that eqn 20.57 is a solution of the diffusion equation with the correct initial value.

20.27 Calculate the relation between $\langle x^2\rangle^{1/2}$ and $\langle x^4\rangle^{1/4}$ for diffusing particles at a time *t* if they have a diffusion constant *D*.

20.28 The diffusion equation is valid when many elementary steps are taken in the time interval of interest, but the random walk calculation lets us discuss distributions for short times as well as for long. Use eqn 20.61 to calculate the probability of being six paces from the origin (that is, at $x = 6\lambda$) after (a) four, (b) six, (c) twelve steps.

20.29‡ A dilute solution of a weak (1,1)-electrolyte contains both neutral ion pairs and ions in equilibrium ($AB \rightleftharpoons A^+ + B^-$). Prove that molar conductivities are related to the degree of ionization by the equations:

$$\frac{1}{\Lambda_m} = \frac{1}{\Lambda_m(\alpha)} + \frac{(1-\alpha)\Lambda_m^\circ}{\alpha^2 \Lambda_m(\alpha)^2} \qquad \Lambda_m(\alpha) = \lambda_+ + \lambda_- = \Lambda_m^\circ - \mathcal{K}(\alpha c)^{1/2}$$

where Λ_m° is the molar conductivity at infinite dilution and $\mathcal{K}$ is the constant in Kohlrausch's law (eqn 20.28).

Applications: to astrophysics and biochemistry

20.30 Calculate the escape velocity (the minimum initial velocity that will take an object to infinity) from the surface of a planet of radius *R*. What is the value for (a) the Earth, $R = 6.37$ Mm, $g = 9.81$ m s^{-2}, (b) Mars, $R = 3.38$ Mm, $m_{\text{Mars}}/m_{\text{Earth}} = 0.108$. At what temperatures do H_2, He, and O_2 molecules have mean speeds equal to their escape speeds? What proportion of the molecules have enough speed to escape when the temperature is (a) 240 K, (b) 1500 K? Calculations of this kind are very important in considering the composition of planetary atmospheres.

20.31‡ Interstellar space is a medium quite different from the gaseous environments we commonly encounter on Earth. For instance, a typical density of the medium is about 1 atom cm^{-3} and that atom is typically H; the effective temperature due to stellar background radiation is about 10 000 K. Estimate the diffusion coefficient and thermal conductivity of H under these conditions. *Comment.* Energy is in fact transferred much more effectively by radiation.

20.32 The principal components of the atmosphere of the Earth are diatomic molecules, which can rotate as well as translate. Given that the translational kinetic energy density of the atmosphere is 0.15 J cm^{-3}, what is the total kinetic energy density, including rotation?

20.33‡ In the *standard model* of stellar structure (I. Nicholson, *The sun.* Rand McNally, New York (1982)), the interior of the Sun is thought to consist of 36 per cent H and 64 per cent He by mass, at a density of 158 g cm^{-3}. Both atoms are completely ionized. The approximate dimensions of the nuclei can be calculated from the formula $r_{\text{nucleus}} = 1.4A^{1/3}$ fm, where *A* is the mass number. The size of the free electron, $r_e \approx 10^{-18}$ m, is negligible compared to the size of the nuclei. (a) Calculate the excluded volume in 1.0 cm^3 of the stellar interior and on that basis decide upon the applicability of the perfect gas law to this system. (b) The standard model suggests that the pressure in the stellar interior is 2.5×10^{11} atm. Calculate the temperature of the Sun's interior based on the perfect gas model. The generally accepted standard model value is 16 MK. (c) Would a van der Waals type of equation (with $a = 0$) give a better value for *T*?

20.34 Enrico Fermi, the great Italian scientist, was a master at making good approximate calculations based on little or no actual data. Hence, such calculations are often called 'Fermi calculations'. Do a Fermi calculation on how long it would take for a gaseous air-borne cold virus of molar mass 100 kg

mol^{-1} to travel the distance between two conversing people 1.0 m apart by diffusion in still air.

20.35 The diffusion coefficient of a particular kind of t-RNA molecule is $D = 1.0 \times 10^{-11}\ \text{m}^2\ \text{s}^{-1}$ in the medium of a cell interior. How long does it take molecules produced in the cell nucleus to reach the walls of the cell at a distance 1.0 μm, corresponding to the radius of the cell?

20.36‡ In this problem, we examine a model for the transport of oxygen from air in the lungs to blood. First, show that, for the initial and boundary conditions $c(x,t) = c(x,0) = c_0$, $(0 < x < \infty)$ and $c(0,t) = c_s$, $(0 \le t \le \infty)$ where c_0 and c_s are constants, the concentration, $c(x,t)$, of a species is given by

$$c(x,t) = c_0 + (c_s - c_0)\{1 - \text{erf}\,\xi\} \qquad \xi(x,t) = \frac{x}{(4Dt)^{1/2}}$$

where erf ξ is the error function and the concentration $c(x,t)$ evolves by diffusion from the yz-plane of constant concentration, such as might occur if a condensed phase is absorbing a species from a gas phase. Now draw graphs of concentration profiles at several different times of your choice for the diffusion of oxygen into water at 298 K (when $D = 2.10 \times 10^{-9}\ \text{m}^2\ \text{s}^{-1}$) on a spatial scale comparable to passage of oxygen from lungs through alveoli into the blood. Use $c_0 = 0$ and set c_s equal to the solubility of oxygen in water. *Hint.* Use mathematical software.

21 The rates of chemical reactions

This chapter is the first of a sequence that explores the rates of chemical reactions. The chapter begins with a discussion of the definition of reaction rate and outlines the techniques for its measurement. The results of such measurements show that reaction rates depend on the concentration of reactants (and products) in characteristic ways that can be expressed in terms of differential equations known as rate laws. The solutions of these equations are used to predict the concentrations of species at any time after the start of the reaction. The form of the rate law also provides insight into the series of elementary steps by which a reaction takes place. The key task in this connection is the construction of a rate law from a proposed mechanism and its comparison with experiment. Simple elementary steps have simple rate laws, and these rate laws can be combined together by invoking one or more approximations. These approximations include the concept of the rate-determining stage of a reaction, the steady-state concentration of a reaction intermediate, and the existence of a pre-equilibrium. We go on to consider examples of reaction mechanisms, focusing on polymerization reactions and photochemistry, in which reactions are initiated by light.

This chapter introduces the principles of **chemical kinetics**, the study of reaction rates, by showing how the rates of reactions may be measured and interpreted. The remaining chapters of this part of the text then develop this material in more detail and apply it to more complicated or more specialized cases. The rate of a chemical reaction might depend on variables under our control, such as the pressure, the temperature, and the presence of a catalyst, and we may be able to optimize the rate by the appropriate choice of conditions. The study of reaction rates also leads to an understanding of the **mechanisms** of reactions, their analysis into a sequence of elementary steps.

Empirical chemical kinetics

The first steps in the kinetic analysis of reactions are to establish the stoichiometry of the reaction and identify any side reactions. The basic data of chemical kinetics are then the concentrations of the reactants and products at different times after a reaction has been initiated. The rates of most chemical reactions are sensitive to the temperature, so in conventional experiments the temperature of the reaction mixture must be held constant throughout the course of the reaction. This requirement puts severe demands on the design of an experiment. Gas-phase reactions, for instance, are often carried out in a vessel held in contact with a substantial block of metal. Liquid-phase reactions, including flow reactions, must be carried out in an efficient thermostat. Special efforts have to be made to study reactions at low temperatures, as in the study of the kinds of reactions that take place in interstellar clouds. Thus, supersonic expansion of the reaction gas can be used to attain temperatures as low as 10 K. For work in the liquid phase and the solid phase, very low temperatures are often reached by flowing cold liquid or cold gas around the reaction vessel. Alternatively, the entire reaction vessel is immersed in a thermally insulated container filled with a cryogenic liquid, such as liquid helium (for work at around 4 K) or liquid nitrogen (for work at around 77 K). Non-isothermal conditions are sometimes employed. For instance, the shelf-life of an expensive pharmaceutical may be explored by slowly raising the temperature of a single sample.

21.1 Experimental techniques

Key points **(a) The rates of chemical reactions are measured by using techniques that monitor the concentrations of species present in the reaction mixture. (b) Examples of experimental techniques include real-time and quenching procedures, flow and stopped-flow techniques, and flash photolysis.**

The method used to monitor concentrations depends on the species involved and the rapidity with which their concentrations change. Many reactions reach equilibrium over periods of minutes or hours, and several techniques may then be used to follow the changing concentrations.

(a) Monitoring the progress of a reaction

A reaction in which at least one component is a gas might result in an overall change in pressure in a system of constant volume, so its progress may be followed by recording the variation of pressure with time.

Example 21.1 *Monitoring the variation in pressure*

Predict how the total pressure varies during the gas-phase decomposition $2\,N_2O_5(g) \rightarrow 4\,NO_2(g) + O_2(g)$ in a constant-volume container.

Method The total pressure (at constant volume and temperature and assuming perfect gas behaviour) is proportional to the number of gas-phase molecules. Therefore, because each mole of N_2O_5 gives rise to $\frac{5}{2}$ mol of gas molecules, we can expect the pressure to rise to $\frac{5}{2}$ times its initial value. To confirm this conclusion, express the progress of the reaction in terms of the fraction, α, of N_2O_5 molecules that have reacted.

Answer Let the initial pressure be p_0 and the initial amount of N_2O_5 molecules present be n. When a fraction α of the N_2O_5 molecules has decomposed, the amounts of the components in the reaction mixture are:

	N_2O_5	NO_2	O_2	Total
Amount:	$n(1-\alpha)$	$2\alpha n$	$\frac{1}{2}\alpha n$	$n(1+\frac{3}{2}\alpha)$

When $\alpha = 0$ the pressure is p_0, so at any stage the total pressure is

$$p = (1 + \tfrac{3}{2}\alpha)p_0$$

When the reaction is complete, the pressure will have risen to $\frac{5}{2}$ times its initial value.

Self-test 21.1 Repeat the calculation for $2\,NOBr(g) \rightarrow 2\,NO(g) + Br_2(g)$.

$[p = (1 + \tfrac{1}{2}\alpha)p_0]$

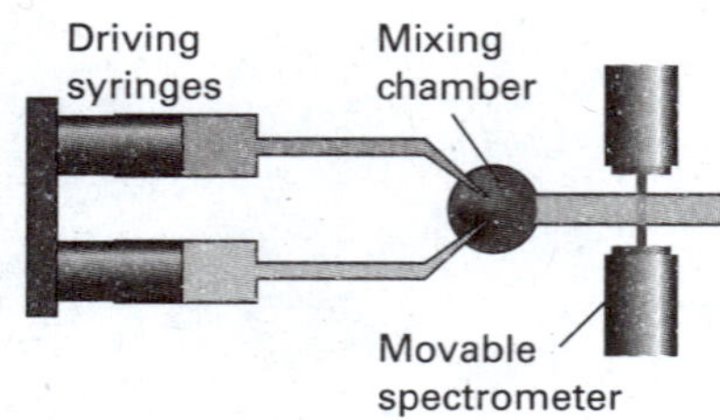

Fig. 21.1 The arrangement used in the flow technique for studying reaction rates. The reactants are injected into the mixing chamber at a steady rate. The location of the spectrometer corresponds to different times after initiation.

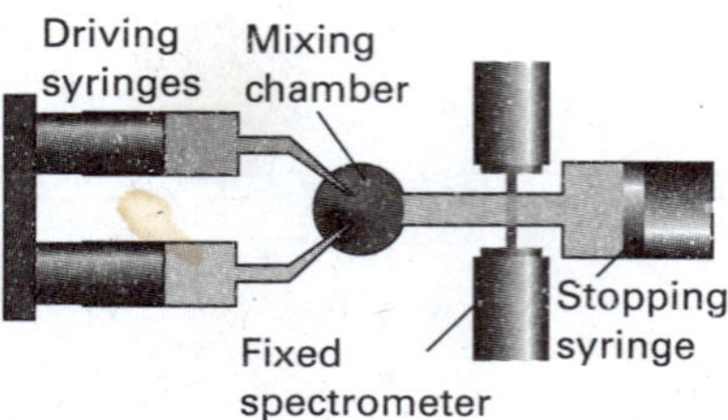

Fig. 21.2 In the stopped-flow technique the reagents are driven quickly into the mixing chamber by the driving syringes and then the time dependence of the concentrations is monitored.

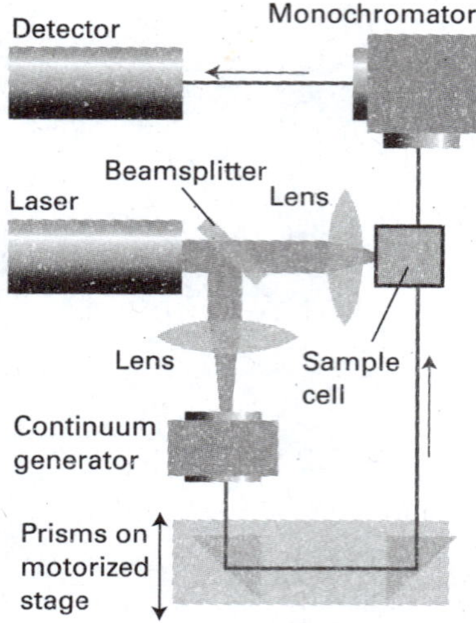

Fig. 21.3 A configuration used for time-resolved absorption spectroscopy, in which the same pulsed laser is used to generate a monochromatic pump pulse and, after continuum generation in a suitable liquid, a 'white' light probe pulse. The time delay between the pump and probe pulses may be varied by moving the motorized stage in the direction shown by the double arrow.

Spectrophotometry, the measurement of absorption of radiation in a particular spectral region, is widely applicable, and is especially useful when one substance in the reaction mixture has a strong characteristic absorption in a conveniently accessible region of the electromagnetic spectrum. For example, the progress of the reaction

$$H_2(g) + Br_2(g) \rightarrow 2\,HBr(g)$$

can be followed by measuring the absorption of visible light by bromine. A reaction that changes the number or type of ions present in a solution may be followed by monitoring the electrical conductivity of the solution. The replacement of neutral molecules by ionic products can result in dramatic changes in the conductivity, as in the reaction

$$(CH_3)_3CCl(aq) + H_2O(l) \rightarrow (CH_3)_3COH(aq) + H^+(aq) + Cl^-(aq)$$

If hydrogen ions are produced or consumed, the reaction may be followed by monitoring the pH of the solution.

Other methods of determining composition include emission spectroscopy, mass spectrometry, gas chromatography, nuclear magnetic resonance, and electron paramagnetic resonance (for reactions involving radicals or paramagnetic d-metal ions).

(b) Application of the techniques

In a **real-time analysis** the composition of the system is analysed while the reaction is in progress. Either a small sample is withdrawn or the bulk solution is monitored. In the **flow method** the reactants are mixed as they flow together in a chamber (Fig. 21.1). The reaction continues as the thoroughly mixed solutions flow through the outlet tube, and observation of the composition at different positions along the tube is equivalent to the observation of the reaction mixture at different times after mixing. The disadvantage of conventional flow techniques is that a large volume of reactant solution is necessary. This makes the study of fast reactions particularly difficult because to spread the reaction over a length of tube the flow must be rapid. This disadvantage is avoided by the **stopped-flow technique**, in which the reagents are mixed very quickly in a small chamber fitted with a syringe instead of an outlet tube (Fig. 21.2). The flow ceases when the plunger of the syringe reaches a stop, and the reaction continues in the mixed solutions. Observations, commonly using spectroscopic techniques such as ultraviolet–visible absorption, circular dichroism, and fluorescence emission, are made on the sample as a function of time. The technique allows for the study of reactions that occur on the millisecond to second timescale. The suitability of the stopped-flow method to the study of small samples means that it is appropriate for many biochemical reactions, and it has been widely used to study the kinetics of protein folding and enzyme action (see *Impact I16.1*).

Very fast reactions can be studied by **flash photolysis**, in which the sample is exposed to a brief flash of light that initiates the reaction and then the contents of the reaction chamber are monitored. The apparatus used for flash photolysis studies is based on the experimental design for **time-resolved spectroscopy**, in which reactions occurring on a picosecond or femtosecond timescale may be monitored by using electronic absorption or emission, infrared absorption, or Raman scattering. The spectra are recorded at a series of times following laser excitation. The laser pulse can initiate the reaction by forming a reactive species, such as an excited electronic state of a molecule, a radical, or an ion. We discuss examples of excited state reactions in Section 21.10.

The arrangement shown in Fig. 21.3 is often used to study ultrafast chemical reactions that can be initiated by light, such as the initial events of vision (*Impact I13.1*). A strong and short laser pulse, the *pump*, promotes a molecule A to an excited electronic state A^* that can either emit a photon (as fluorescence or phosphorescence) or react with another species B to yield a product C:

$$A + h\nu \rightarrow A^* \qquad \text{(absorption)}$$

$$A^* \rightarrow A \qquad \text{(emission)}$$

$$A^* + B \rightarrow [AB] \rightarrow C \qquad \text{(reaction)}$$

Here [AB] denotes either an intermediate or an activated complex (see Section 21.5). The rates of appearance and disappearance of the various species are determined by observing time-dependent changes in the absorption spectrum of the sample during the course of the reaction. This monitoring is done by passing a weak pulse of white light, the *probe*, through the sample at different times after the laser pulse. Pulsed 'white' light can be generated directly from the laser pulse by the phenomenon of **continuum generation**, in which focusing an ultrafast laser pulse on a sample containing a liquid (such as water or carbon tetrachloride) or a solid (such as sapphire) results in an outgoing beam with a wide distribution of frequencies. A time delay between the strong laser pulse and the 'white' light pulse can be introduced by allowing one of the beams to travel a longer distance before reaching the sample. For example, a difference in travel distance of $\Delta d = 3$ mm corresponds to a time delay $\Delta t = \Delta d/c \approx 10$ ps between two beams, where c is the speed of light. The relative distances travelled by the two beams in Fig. 21.3 are controlled by directing the 'white' light beam to a motorized stage carrying a pair of mirrors.

Variations of the arrangement in Fig. 21.3 allow for the observation of fluorescence decay kinetics of A^* and time-resolved Raman spectra during the course of the reaction. The fluorescence lifetime of A^* can be determined by exciting A as before and measuring the decay of the fluorescence intensity after the pulse with a fast photodetector system. In this case, continuum generation is not necessary. Time-resolved resonance Raman spectra of A, A^*, B, [AB], or C can be obtained by initiating the reaction with a strong laser pulse of a certain wavelength and then, some time later, irradiating the sample with another laser pulse that can excite the resonance Raman spectrum of the desired species. Also in this case continuum generation is not necessary. Instead, the Raman excitation beam may be generated in a dye laser (see *Further information 13.1*).

In contrast to real-time analysis, **quenching methods** are based on stopping, or quenching, the reaction after it has been allowed to proceed for a certain time. In this way the composition is analysed at leisure and reaction intermediates may be trapped. These methods are suitable only for reactions that are slow enough for there to be little reaction during the time it takes to quench the mixture. In the **chemical quench flow method**, the reactants are mixed in much the same way as in the flow method but the reaction is quenched by another reagent, such as solution of acid or base, after the mixture has travelled along a fixed length of the outlet tube. Different reaction times can be selected by varying the flow rate along the outlet tube. An advantage of the chemical quench flow method over the stopped-flow method is that spectroscopic fingerprints are not needed in order to measure the concentration of reactants and products. Once the reaction has been quenched, the solution may be examined by 'slow' techniques, such as gel electrophoresis, mass spectrometry, and chromatography. In the **freeze quench method**, the reaction is quenched by cooling the mixture within milliseconds and the concentrations of reactants, intermediates, and products are measured spectroscopically.

21.2 The rates of reactions

Key points (a) The instantaneous rate of a reaction is the slope of the tangent to the graph of concentration against time (expressed as a positive quantity). (b) A rate law is an expression for the reaction rate in terms of the concentrations of the species that occur in the overall chemical reaction. (c) For a rate law of the form $v = k_r[A]^a[B]^b \ldots$, the rate constant is k_r, the order with respect to A is a, and the overall order is $a + b + \cdots$. (d) The isolation method and the method of initial rates are often used in the determination of rate laws.

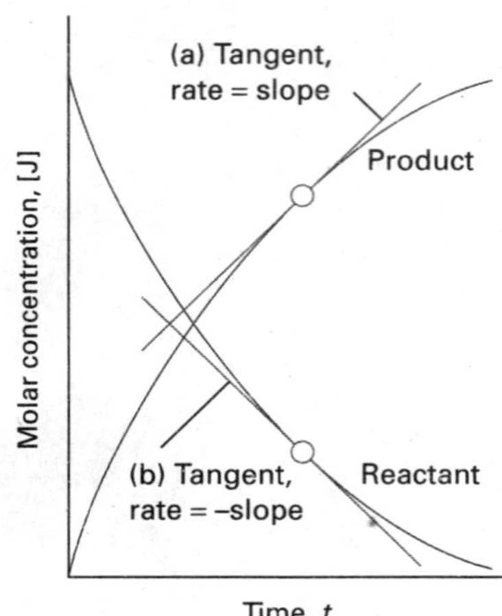

Fig. 21.4 The definition of (instantaneous) rate as the slope of the tangent drawn to the curve showing the variation of concentration with time. For negative slopes, the sign is changed when reporting the rate, so all reaction rates are positive.

Reaction rates depend on the composition and the temperature of the reaction mixture. The next few sections look at these observations in more detail.

(a) The definition of rate

Consider a reaction of the form $A + 2\,B \rightarrow 3\,C + D$, in which at some instant the molar concentration of a participant J is [J] and the volume of the system is constant. The instantaneous **rate of consumption** of one of the reactants at a given time is $d[R]/dt$, where R is A or B. This rate is a positive quantity (Fig. 21.4). The **rate of formation** of one of the products (C or D, which we denote P) is $d[P]/dt$ (note the difference in sign). This rate is also positive.

It follows from the stoichiometry for the reaction $A + 2\,B \rightarrow 3\,C + D$ that

$$\frac{d[D]}{dt} = \frac{1}{3}\frac{d[C]}{dt} = -\frac{d[A]}{dt} = -\frac{1}{2}\frac{d[B]}{dt}$$

so the rate of the reaction is related to the rates of change of concentration of products and reactants in several ways. The undesirability of having different rates to describe the same reaction is avoided by using the extent of reaction, ξ (xi, the quantity introduced in Section 6.1a):

$$\xi = \frac{n_J - n_{J,0}}{\nu_J} \qquad \text{Definition of the extent of reaction} \qquad [21.1]$$

where ν_J is the stoichiometric number of species J, and defining the unique **rate of reaction**, v, as the rate of change of the extent of reaction:

$$v = \frac{1}{V}\frac{d\xi}{dt} \qquad \text{Definition of the rate of reaction} \qquad [21.2]$$

where V is the volume of the system. It follows that

$$v = \frac{1}{\nu_J} \times \frac{1}{V}\frac{dn_J}{dt} \qquad (21.3a)$$

(Remember that ν_J is negative for reactants and positive for products.) For a homogeneous reaction in a constant-volume system the volume V can be taken inside the differential and we use $[J] = n_J/V$ to write

$$v = \frac{1}{\nu_J}\frac{d[J]}{dt} \qquad (21.3b)$$

For a heterogeneous reaction, we use the (constant) surface area, A, occupied by the species in place of V and use $\sigma_J = n_J/A$ to write

$$v = \frac{1}{\nu_J}\frac{d\sigma_J}{dt} \qquad (21.3c)$$

In each case there is now a single rate for the entire reaction (for the chemical equation as written). With molar concentrations in moles per cubic decimetre and time in seconds, reaction rates of homogeneous reactions are reported in moles per cubic decimetre per second (mol dm^{-3} s^{-1}) or related units. For gas-phase reactions, such as those taking place in the atmosphere, concentrations are often expressed in molecules per cubic centimetre (molecules cm^{-3}) and rates in molecules per cubic centimetre per second (molecules cm^{-3} s^{-1}). For heterogeneous reactions, rates are expressed in moles per square metre per second (mol m^{-2} s^{-1}) or related units.

• A brief illustration

If the rate of formation of NO in the reaction $2\,NOBr(g) \rightarrow 2\,NO(g) + Br_2(g)$ is reported as 0.16 mmol dm^{-3} s^{-1}, we use $\nu_{NO} = +2$ to report that $v = 0.080$ mmol dm^{-3} s^{-1}. Because $\nu_{NOBr} = -2$ it follows that $d[NOBr]/dt = -0.16$ mmol dm^{-3} s^{-1}. The rate of consumption of NOBr is therefore 0.16 mmol dm^{-3} s^{-1}, or 9.6×10^{16} molecules cm^{-3} s^{-1}. •

Self-test 21.2 The rate of change of molar concentration of CH_3 radicals in the reaction $2\,CH_3(g) \rightarrow CH_3CH_3(g)$ was reported as $d[CH_3]/dt = -1.2$ mol dm^{-3} s^{-1} under particular conditions. What is (a) the rate of reaction and (b) the rate of formation of CH_3CH_3? [(a) 0.60 mol dm^{-3} s^{-1}, (b) 0.60 mol dm^{-3} s^{-1}]

(b) Rate laws and rate constants

The rate of reaction is often found to be proportional to the concentrations of the reactants raised to a power. For example, the rate of a reaction may be proportional to the molar concentrations of two reactants A and B, so we write

$$v = k_r[A][B] \tag{21.4}$$

with each concentration raised to the first power. The coefficient k_r is called the **rate constant** for the reaction. The rate constant is independent of the concentrations but depends on the temperature. An experimentally determined equation of this kind is called the **rate law** of the reaction. More formally, a rate law is an equation that expresses the rate of reaction as a function of the concentrations of all the species present in the overall chemical equation for the reaction at some time:

$$v = f([A],[B],\ldots) \qquad \text{Definition of the rate law in terms of the concentration} \qquad [21.5a]$$

For homogeneous gas-phase reactions, it is often more convenient to express the rate law in terms of partial pressures. In this case, we write

$$v = f(p_A, p_B, \ldots) \qquad \text{Definition of the rate law in terms of the pressure} \qquad [21.5b]$$

The units of k_r are always such as to convert the product of concentrations into a rate expressed as a change in concentration divided by time. For example, if the rate law is the one shown in eqn 21.4, with concentrations expressed in mol dm^{-3}, then the units of k_r will be dm^3 mol^{-1} s^{-1} because

$$\text{dm}^3\,\text{mol}^{-1}\,\text{s}^{-1} \times \text{mol dm}^{-3} \times \text{mol dm}^{-3} = \text{mol dm}^{-3}\,\text{s}^{-1}$$

In gas-phase studies, including studies of the processes taking place in the atmosphere, concentrations are commonly expressed in molecules cm^{-3}, so the rate constant for the reaction above would be expressed in cm^3 $molecule^{-1}$ s^{-1}. We can use the approach just developed to determine the units of the rate constant from rate laws of any form. For example, the rate constant for a reaction with rate law of the form $k_r[A]$ is commonly expressed in s^{-1}.

• A brief illustration

The rate constant for the reaction $O(g) + O_3(g) \rightarrow 2\,O_2(g)$ is 8.0×10^{-15} cm^3 $molecule^{-1}$ s^{-1} at 298 K. To express this rate constant in dm^3 mol^{-1} s^{-1}, we make use of

$$1\text{ cm} = 10^{-2}\text{ m} = 10^{-2} \times 10\text{ dm} = 10^{-1}\text{ dm}$$

$$1\text{ mol} = 6.022 \times 10^{23}\text{ molecules, so 1 molecule} = \frac{1\text{ mol}}{6.022 \times 10^{23}}$$

It follows that

$$\begin{aligned} k_r &= 8.0 \times 10^{-15}\text{ cm}^3\text{ molecule}^{-1}\text{ s}^{-1} \\ &= 8.0 \times 10^{-15}\,(10^{-1}\text{ dm})^3 \left(\frac{1\text{ mol}}{6.022 \times 10^{23}}\right)^{-1}\text{ s}^{-1} \\ &= 8.0 \times 10^{-15} \times 10^{-3} \times 6.022 \times 10^{23}\text{ dm}^3\text{ mol}^{-1}\text{ s}^{-1} \\ &= 4.8 \times 10^{6}\text{ dm}^3\text{ mol}^{-1}\text{ s}^{-1} \end{aligned}$$ •

Self-test 21.3 A reaction has a rate law of the form $k_r[A]^2[B]$. What are the units of the rate constant if the reaction rate is measured in mol dm^{-3} s^{-1}?

[dm^6 mol^{-2} s^{-1}]

The rate law of a reaction is determined experimentally, and cannot in general be inferred from the stoichiometry of the balanced chemical equation for the reaction. The reaction of hydrogen and bromine, for example, has a very simple stoichiometry, $H_2(g) + Br_2(g) \rightarrow 2\,HBr(g)$, but its rate law is complicated:

$$v = \frac{k_a[H_2][Br_2]^{3/2}}{[Br_2] + k_b[HBr]} \qquad (21.6)$$

In certain cases the rate law does reflect the stoichiometry of the reaction, but that is either a coincidence or reflects a feature of the underlying reaction mechanism.

A practical application of a rate law is that, once we know the law and the value of the rate constant, we can predict the rate of reaction from the composition of the mixture. Moreover, as we shall see later, by knowing the rate law, we can go on to predict the composition of the reaction mixture at a later stage of the reaction. A rate law is also a guide to the mechanism of the reaction, for any proposed mechanism must be consistent with the observed rate law.

(c) Reaction order

Many reactions are found to have rate laws of the form

$$v = k_r[A]^a[B]^b \cdots \qquad (21.7)$$

The power to which the concentration of a species (a product or a reactant) is raised in a rate law of this kind is the **order** of the reaction with respect to that species. A reaction with the rate law in eqn 21.4 is **first-order** in A and first-order in B. The **overall order** of a reaction with a rate law like that in eqn 21.7 is the sum of the individual orders, $a + b + \cdots$. The rate law in eqn 21.4 is therefore second-order overall.

A reaction need not have an integral order, and many gas-phase reactions do not. For example, a reaction having the rate law

$$v = k[\mathrm{A}]^{1/2}[\mathrm{B}] \qquad (21.8)$$

is half-order in A, first-order in B, and three-halves-order overall. Some reactions obey a **zero-order rate law**, and therefore have a rate that is independent of the concentration of the reactant (so long as some is present). Thus, the catalytic decomposition of phosphine (PH_3) on hot tungsten at high pressures has the rate law

$$v = k_{\mathrm{r}} \qquad (21.9)$$

The PH_3 decomposes at a constant rate until it has almost entirely disappeared. Zero-order reactions typically occur when there is a bottle-neck of some kind in the mechanism, as in heterogeneous reactions when the surface is saturated regardless of how much reactant remains. Zero-order reactions are also found for a number of enzyme reactions when there is a large excess of reactant relative to the enzyme, and the amount of enzyme present governs the rate, not the amount of reactant.

When a rate law is not of the form in eqn 21.7, the reaction does not have an overall order and may not even have definite orders with respect to each participant. Thus, although eqn 21.6 shows that the reaction of hydrogen and bromine is first-order in H_2, the reaction has an indefinite order with respect to both Br_2 and HBr and has no overall order.

These remarks point to three important questions:

- How do we identify the rate law and obtain the rate constant from the experimental data? We concentrate on this aspect in this chapter.
- How do we construct reaction mechanisms that are consistent with the rate law? We shall develop the techniques of doing so in Sections 21.8–10 and in Chapter 23.
- How do we account for the values of the rate constants and their temperature dependence? We shall see a little of what is involved in this chapter, but leave the details until Chapter 22.

(d) The determination of the rate law

The determination of a rate law is simplified by the **isolation method** in which the concentrations of all the reactants except one are in large excess. If B is in large excess, for example, then to a good approximation its concentration is constant throughout the reaction. Although the true rate law might be $v = k_{\mathrm{r}}[\mathrm{A}][\mathrm{B}]$, we can approximate [B] by $[\mathrm{B}]_0$, its initial value, and write

$$v = k_{\mathrm{r}}'[\mathrm{A}] \qquad k_{\mathrm{r}}' = k_{\mathrm{r}}[\mathrm{B}]_0 \qquad (21.10)$$

which has the form of a first-order rate law. Because the true rate law has been forced into first-order form by assuming that the concentration of B is constant, eqn 21.10 is called a **pseudofirst-order rate law**. The dependence of the rate on the concentration of each of the reactants may be found by isolating them in turn (by having all the other substances present in large excess), and so constructing a picture of the overall rate law.

In the **method of initial rates**, which is often used in conjunction with the isolation method, the rate is measured at the beginning of the reaction for several different initial concentrations of reactants. We shall suppose that the rate law for a reaction with A isolated is $v = k_{\mathrm{r}}'[\mathrm{A}]^a$; then its initial rate, v_0, is given by the initial values of the concentration of A, and we write $v_0 = k_{\mathrm{r}}'[\mathrm{A}]_0^a$. Taking (common) logarithms gives:

$$\log v_0 = \log k_{\mathrm{r}}' + a \log[\mathrm{A}]_0 \qquad (21.11)$$

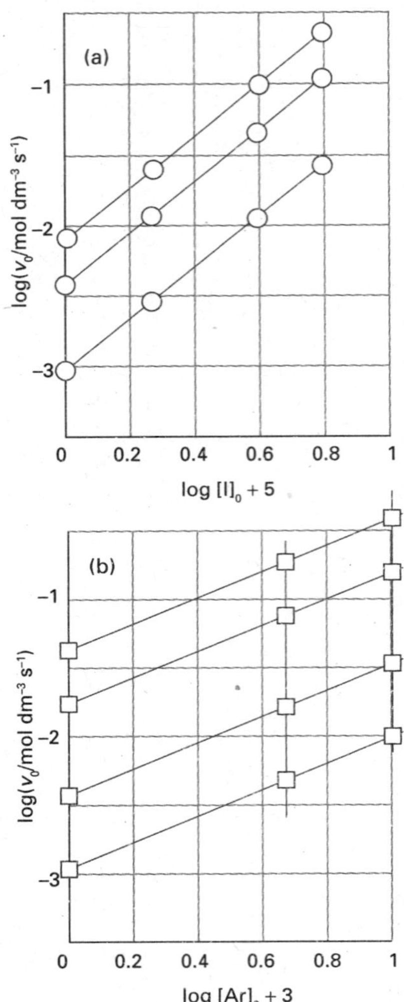

Fig. 21.5 The plot of log v_0 against (a) log$[\text{I}]_0$ for a given $[\text{Ar}]_0$, and (b) log$[\text{Ar}]_0$ for a given $[\text{I}]_0$.

A note on good practice The units of k_r come automatically from the calculation, and are always such as to convert the product of concentrations to a rate in concentration/time (for example, mol dm^{-3} s^{-1}).

For a series of initial concentrations, a plot of the logarithms of the initial rates against the logarithms of the initial concentrations of A should be a straight line with slope *a*.

Example 21.2 *Using the method of initial rates*

The recombination of iodine atoms in the gas phase in the presence of argon was investigated and the order of the reaction was determined by the method of initial rates. The initial rates of reaction of $2\,\text{I(g)} + \text{Ar(g)} \rightarrow \text{I}_2\text{(g)} + \text{Ar(g)}$ were as follows:

$[\text{I}]_0/(10^{-5}$ mol dm$^{-3})$		1.0	2.0	4.0	6.0
$v_0/(\text{mol dm}^{-3}\,\text{s}^{-1})$	(a)	8.70×10^{-4}	3.48×10^{-3}	1.39×10^{-2}	3.13×10^{-2}
	(b)	4.35×10^{-3}	1.74×10^{-2}	6.96×10^{-2}	1.57×10^{-1}
	(c)	8.69×10^{-3}	3.47×10^{-2}	1.38×10^{-1}	3.13×10^{-1}

The Ar concentrations are (a) 1.0 mmol dm^{-3}, (b) 5.0 mmol dm^{-3}, and (c) 10.0 mmol dm^{-3}. Determine the orders of reaction with respect to the I and Ar atom concentrations and the rate constant.

Method Plot the logarithm of the initial rate, $\log v_0$, against $\log[\text{I}]_0$ for a given concentration of Ar, and, separately, against $\log[\text{Ar}]_0$ for a given concentration of I. The slopes of the two lines are the orders of reaction with respect to I and Ar, respectively. The intercepts with the vertical axis give $\log k_r'$ and, by using eqn 21.10, k_r.

Answer The plots are shown in Fig. 21.5. The slopes are 2 and 1, respectively, so the (initial) rate law is $v_0 = k_r[\text{I}]_0^2[\text{Ar}]_0$. This rate law signifies that the reaction is second-order in [I], first-order in [Ar], and third-order overall. The intercept corresponds to $k_r = 9\times10^9$ mol^{-2} dm^6 s^{-1}.

Self-test 21.4 The initial rate of a reaction depended on concentration of a substance J as follows:

$[\text{J}]_0/(\text{mmol dm}^{-3})$	5.0	8.2	17	30
$v_0/(10^{-7}\,\text{mol dm}^{-3}\,\text{s}^{-1})$	3.6	9.6	41	130

Determine the order of the reaction with respect to J and calculate the rate constant.
[2, 1.4×10^{-2} dm^3 mol^{-1} s^{-1}]

The method of initial rates might not reveal the full rate law, for once the products have been generated they might participate in the reaction and affect its rate. For example, products participate in the synthesis of HBr, because eqn 21.6 shows that the full rate law depends on the concentration of HBr. To avoid this difficulty, the rate law should be fitted to the data throughout the reaction. The fitting may be done, in simple cases at least, by using a proposed rate law to predict the concentration of any component at any time, and comparing it with the data. A law should also be tested by observing whether the addition of products or, for gas-phase reactions, a change in the surface-to-volume ratio in the reaction chamber affects the rate.

21.3 Integrated rate laws

Key points **An integrated rate law is an expression for the concentration of a reactant or product as a function of time. The half-life $t_{1/2}$ of a reaction is the time it takes for the concentration of a species to fall to half its initial value. The time constant τ is the time required for the concentration of a reactant to fall to 1/e of its initial value.**

Because rate laws are differential equations, we must integrate them if we want to find the concentrations as a function of time. Even the most complex rate laws may be integrated numerically. However, in a number of simple cases analytical solutions, known as **integrated rate laws**, are easily obtained, and prove to be very useful. We examine a few of these simple cases here.

(a) First-order reactions

As shown in the following *Justification*, the integrated form of the first-order rate law

$$\frac{d[A]}{dt} = -k_r[A] \qquad (21.12a)$$

is

$$\ln\left(\frac{[A]}{[A]_0}\right) = -k_r t \qquad [A] = [A]_0 e^{-k_r t} \qquad (21.12b)$$

Integrated first-order rate law

where $[A]_0$ is the initial concentration of A (at $t = 0$).

Justification 21.1 *First-order integrated rate law*

First, we rearrange eqn 21.12a into

$$\frac{d[A]}{[A]} = -k_r dt$$

This expression can be integrated directly because k_r is a constant independent of t. Initially (at $t = 0$) the concentration of A is $[A]_0$, and at a later time t it is $[A]$, so we make these values the limits of the integrals and write

$$\int_{[A]_0}^{[A]} \frac{d[A]}{[A]} = -k_r \int_0^t dt$$

Because the integral of $1/x$ is $\ln x$, eqn 21.12b is obtained immediately.

Equation 21.12b shows that, if $\ln([A]/[A]_0)$ is plotted against t, then a first-order reaction will give a straight line of slope $-k_r$. Some rate constants determined in this way are given in Table 21.1. The second expression in eqn 21.12b shows that in a first-order reaction the reactant concentration decreases exponentially with time with a rate determined by k_r (Fig. 21.6).

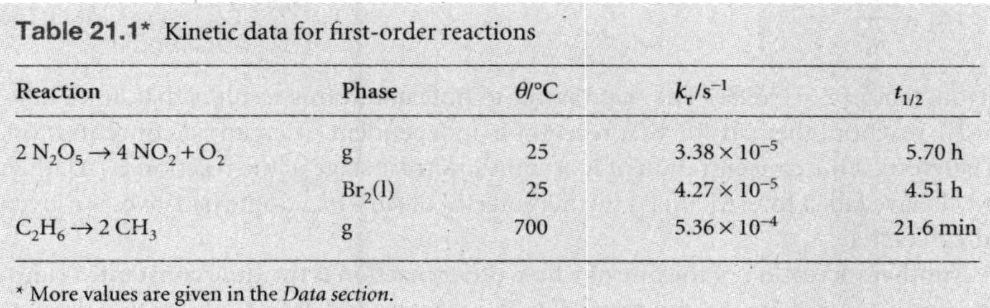

Table 21.1* Kinetic data for first-order reactions

Reaction	Phase	θ/°C	k_r/s^{-1}	$t_{1/2}$
$2 N_2O_5 \rightarrow 4 NO_2 + O_2$	g	25	3.38×10^{-5}	5.70 h
	$Br_2(l)$	25	4.27×10^{-5}	4.51 h
$C_2H_6 \rightarrow 2 CH_3$	g	700	5.36×10^{-4}	21.6 min

* More values are given in the *Data section*.

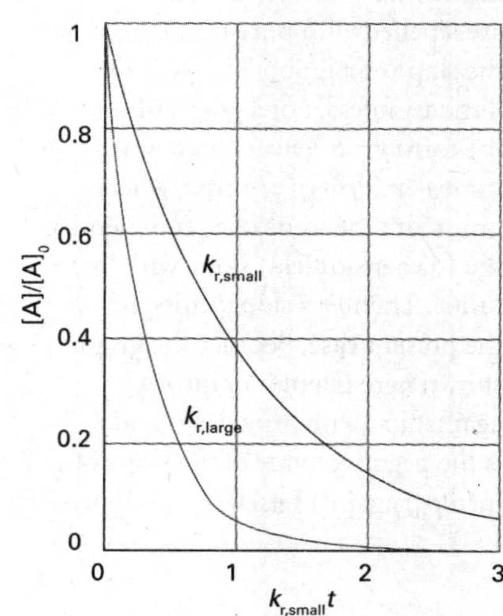

Fig. 21.6 The exponential decay of the reactant in a first-order reaction. The larger the rate constant, the more rapid the decay: here $k_{r,large} = 3k_{r,small}$.

interActivity For a first-order reaction of the form $A \rightarrow nB$ (with n possibly fractional), the concentration of the product varies with time as $[B] = n[B]_0(1 - e^{-k_r t})$. Plot the time dependence of [A] and [B] for the cases $n = 0.5$, 1, and 2.

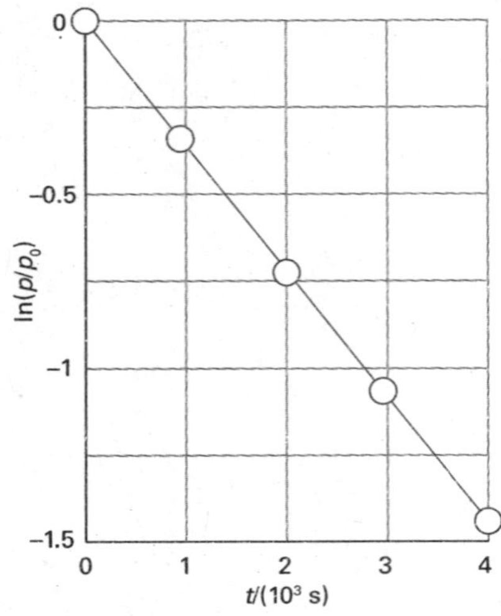

Fig. 21.7 The determination of the rate constant of a first-order reaction: a straight line is obtained when $\ln[A]/[A]_0$ (or, as here, $\ln p/p_0$) is plotted against t; the slope gives k_r.

Example 21.3 *Analysing a first-order reaction*

The variation in the partial pressure of azomethane with time was followed at 600 K, with the results given below. Confirm that the decomposition

$$CH_3N_2CH_3(g) \rightarrow CH_3CH_3(g) + N_2(g)$$

is first-order in azomethane, and find the rate constant at 600 K.

t/s	0	1000	2000	3000	4000
p/Pa	10.9	7.63	5.32	3.71	2.59

Method As indicated in the text, to confirm that a reaction is first-order, plot $\ln([A]/[A]_0)$ against time and expect a straight line. Because the partial pressure of a gas is proportional to its concentration (provided it can be treated as a perfect gas), an equivalent procedure is to plot $\ln(p/p_0)$ against t. If a straight line is obtained, its slope can be identified with k_r.

Answer We draw up the following table:

t/s	0	1000	2000	3000	4000
$\ln(p/p_0)$	0	−0.357	−0.717	−1.078	−1.437

Figure 21.7 shows the plot of $\ln(p/p_0)$ against t. The plot is straight, confirming a first-order reaction, and its slope is -3.6×10^{-4}. Therefore, $k_r = 3.6 \times 10^{-4}\ s^{-1}$.

Self-test 21.5 In a particular experiment, it was found that the concentration of N_2O_5 in liquid bromine varied with time as follows:

t/s	0	200	400	600	1000
$[N_2O_5]/(\text{mol dm}^{-3})$	0.110	0.073	0.048	0.032	0.014

Confirm that the reaction is first-order in N_2O_5 and determine the rate constant.
$[k_r = 2.1 \times 10^{-3}\ s^{-1}]$

A note on good practice Because the horizontal and vertical axes of graphs are labelled with pure numbers, the slope of a graph is always dimensionless. For a graph of the form $y = b + mx$ we can write $y = b + (m \text{ units})(x/\text{units})$, where 'units' are the units of x, and identify the (dimensionless) slope with 'm units'. Then m = slope/units. In the present case, because the graph shown here is a plot of $\ln(p/p_0)$ against t/s (with 'units' = s) and k_r is the negative value of the slope of $\ln(p/p_0)$ against t itself, k_r = −slope/s.

(b) Half-lives and time constants

A useful indication of the rate of a first-order chemical reaction is the **half-life,** $t_{1/2}$, of a substance, the time taken for the concentration of a reactant to fall to half its initial value. The time for [A] to decrease from $[A]_0$ to $\frac{1}{2}[A]_0$ in a first-order reaction is given by eqn 21.12b as

$$k_r t_{1/2} = -\ln\left(\frac{\frac{1}{2}[A]_0}{[A]_0}\right) = -\ln \tfrac{1}{2} = \ln 2$$

Hence

$$t_{1/2} = \frac{\ln 2}{k_r} \qquad \text{Half-life of a first-order reaction} \qquad (21.13)$$

(Note that ln 2 = 0.693.) The main point to note about this result is that, for a first-order reaction, the half-life of a reactant is independent of its initial concentration. Therefore, if the concentration of A at some *arbitrary* stage of the reaction is [A], then it will have fallen to $\frac{1}{2}$[A] after a further interval of $(\ln 2)/k_r$. Some half-lives are given in Table 21.1.

Another indication of the rate of a first-order reaction is the **time constant,** τ (tau), the time required for the concentration of a reactant to fall to 1/e of its initial value. From eqn 21.12b it follows that

$$k_r\tau=-\ln\left(\frac{[A]_0/e}{[A]_0}\right)=-\ln\frac{1}{e}=1$$

That is, the time constant of a first-order reaction is the reciprocal of the rate constant:

$$\tau=\frac{1}{k_r} \quad \text{Time constant of a first-order reaction} \quad (21.14)$$

(c) Second-order reactions

We show in the following *Justification* that the integrated form of the second-order rate law

$$\frac{d[A]}{dt}=-k_r[A]^2 \quad (21.15a)$$

is either of the following two forms:

$$\frac{1}{[A]}-\frac{1}{[A]_0}=k_rt \quad \text{Integrated second-order rate law} \quad (21.15b)$$

$$[A]=\frac{[A]_0}{1+k_rt[A]_0} \quad \text{Alternative form of the integrated rate law} \quad (21.15c)$$

where $[A]_0$ is the initial concentration of A (at $t=0$).

Justification 21.2 *Second-order integrated rate law*

To integrate eqn 21.15a we rearrange it into

$$\frac{d[A]}{[A]^2}=-k_r dt$$

The concentration of A is $[A_0]$ at $t=0$ and $[A]$ at a general time t later. Therefore,

$$-\int_{[A]_0}^{[A]}\frac{d[A]}{[A]^2}=k_r\int_0^t dt$$

Because the integral of $1/x^2$ is $-1/x$, we obtain eqn 21.15b by substitution of the limits

$$\frac{1}{[A]}\bigg|_{[A]_0}^{[A]}=\frac{1}{[A]}-\frac{1}{[A]_0}=k_rt$$

We can then rearrange this expression into eqn 21.15c.

Equation 21.15b shows that to test for a second-order reaction we should plot $1/[A]$ against t and expect a straight line. The slope of the graph is k_r. Some rate constants determined in this way are given in Table 21.2. The rearranged form, eqn 21.15c, lets us predict the concentration of A at any time after the start of the reaction. It shows that the concentration of A approaches zero more slowly than in a first-order reaction with the same initial rate (Fig. 21.8).

It follows from eqn 21.15b by substituting $t=t_{1/2}$ and $[A]=\frac{1}{2}[A]_0$ that the half-life of a species A that is consumed in a second-order reaction is

$$t_{1/2}=\frac{1}{k_r[A]_0} \quad \text{Half-life of a second-order reaction} \quad (21.16)$$

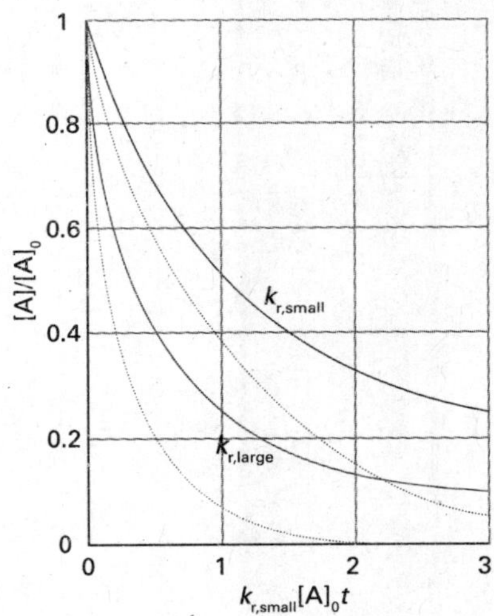

Fig. 21.8 The variation with time of the concentration of a reactant in a second-order reaction. The grey lines are the corresponding decays in a first-order reaction with the same initial rate. For this illustration, $k_{r,large}=3k_{r,small}$.

interActivity For a second-order reaction of the form $A\rightarrow nB$ (with n possibly fractional), the concentration of the product varies with time as $[B]=nk_rt[A]_0^2/(1+k_rt[A]_0)$. Plot the time dependence of [A] and [B] for the cases $n=0.5$, 1, and 2.

Table 21.2* Kinetic data for second-order reactions

Reaction	Phase	θ/°C	k_r/(dm^3 mol^{-1} s^{-1})
$2\,NOBr \rightarrow 2\,NO + Br_2$	g	10	0.80
$2\,I \rightarrow I_2$	g	23	7×10^9
$CH_3Cl + CH_3O^-$	$CH_3OH(l)$	20	2.29×10^{-6}

* More values are given in the *Data section*.

Therefore, unlike a first-order reaction, the half-life of a substance in a second-order reaction varies with the initial concentration. A practical consequence of this dependence is that species that decay by second-order reactions (which includes some environmentally harmful substances) may persist in low concentrations for long periods because their half-lives are long when their concentrations are low. In general, for an nth-order reaction (with $n > 1$) of the form A → products, the half-life is related to the rate constant and the initial concentration of A by

$$t_{1/2} = \frac{2^{n-1}-1}{(n-1)k_r[A]_0^{n-1}} \qquad (21.17)$$

Half-life of an nth-order reaction

(See Problem 21.22.)

Another type of second-order reaction is one that is first-order in each of two reactants A and B:

$$\frac{d[A]}{dt} = -k_r[A][B] \qquad (21.18)$$

Such a rate law cannot be integrated until we know how the concentration of B is related to that of A. For example, if the reaction is A + B → P, where P denotes products, and the initial concentrations are $[A]_0$ and $[B]_0$, then it is shown in the following *Justification* that, at a time t after the start of the reaction, the concentrations satisfy the relation

$$\ln\left(\frac{[B]/[B]_0}{[A]/[A]_0}\right) = ([B]_0 - [A]_0)k_r t \qquad (21.19)$$

Integrated rate law of a second-order reaction of the type A + B → P

Therefore, a plot of the expression on the left against t should be a straight line from which k_r can be obtained.

Justification 21.3 *Overall second-order rate law*

It follows from the reaction stoichiometry that, when the concentration of A has fallen to $[A]_0 - x$, the concentration of B will have fallen to $[B]_0 - x$ (because each A that disappears entails the disappearance of one B). It follows that

$$\frac{d[A]}{dt} = -k_r([A]_0 - x)([B]_0 - x)$$

Because $[A] = [A]_0 - x$, it follows that $d[A]/dt = -dx/dt$ and the rate law may be written as

$$\frac{dx}{dt} = k_r([A]_0 - x)([B]_0 - x)$$

The initial condition is that $x=0$ when $t=0$; so the integration required is

$$\int_0^x \frac{\mathrm{d}x}{([\mathrm{A}]_0-x)([\mathrm{B}]_0-x)}=k_\mathrm{r}\int_0^t \mathrm{d}t$$

The integral on the right is simply $k_\mathrm{r}t$. The integral on the left is evaluated by using the method of partial fractions in which we write

$$\frac{1}{(a-x)(b-x)}=\frac{1}{b-a}\left(\frac{1}{a-x}-\frac{1}{b-x}\right)$$

It follows that

$$\int\frac{\mathrm{d}x}{(a-x)(b-x)}=\frac{1}{b-a}\left[\int\frac{\mathrm{d}x}{a-x}-\int\frac{\mathrm{d}x}{b-x}\right]=\frac{1}{b-a}\left[\ln\frac{1}{a-x}-\ln\frac{1}{b-x}\right]+\text{constant}$$

and therefore that

$$\int_0^x\frac{\mathrm{d}x}{([\mathrm{A}]_0-x)([\mathrm{B}]_0-x)}=\frac{1}{[\mathrm{B}]_0-[\mathrm{A}]_0}\left\{\ln\left(\frac{[\mathrm{A}]_0}{[\mathrm{A}]_0-x}\right)-\ln\left(\frac{[\mathrm{B}]_0}{[\mathrm{B}]_0-x}\right)\right\}$$

This expression can be simplified and rearranged into eqn 21.19 by combining the two logarithms by using $\ln y-\ln z=\ln(y/z)$ and noting that $[\mathrm{A}]=[\mathrm{A}]_0-x$ and $[\mathrm{B}]=[\mathrm{B}]_0-x$. Similar calculations may be carried out to find the integrated rate laws for other orders, and some are listed in Table 21.3.

Table 21.3 Integrated rate laws

Order	Reaction	Rate law*	$t_{1/2}$
0	$\mathrm{A}\rightarrow\mathrm{P}$	$v=k_\mathrm{r}$ $k_\mathrm{r}t=x$ for $0\le x\le[\mathrm{A}]_0$	$[\mathrm{A}]_0/2k_\mathrm{r}$
1	$\mathrm{A}\rightarrow\mathrm{P}$	$v=k_\mathrm{r}[\mathrm{A}]$ $k_\mathrm{r}t=\ln\dfrac{[\mathrm{A}]_0}{[\mathrm{A}]_0-x}$	$(\ln 2)/k_\mathrm{r}$
2	$\mathrm{A}\rightarrow\mathrm{P}$	$v=k_\mathrm{r}[\mathrm{A}]^2$ $k_\mathrm{r}t=\dfrac{x}{[\mathrm{A}]_0([\mathrm{A}]_0-x)}$	$1/k_\mathrm{r}[\mathrm{A}]_0$
	$\mathrm{A}+\mathrm{B}\rightarrow\mathrm{P}$	$v=k_\mathrm{r}[\mathrm{A}][\mathrm{B}]$ $k_\mathrm{r}t=\dfrac{1}{[\mathrm{B}]_0-[\mathrm{A}]_0}\ln\dfrac{[\mathrm{A}]_0([\mathrm{B}]_0-x)}{([\mathrm{A}]_0-x)[\mathrm{B}]_0}$	
	$\mathrm{A}+2\,\mathrm{B}\rightarrow\mathrm{P}$	$v=k_\mathrm{r}[\mathrm{A}][\mathrm{B}]$ $k_\mathrm{r}t=\dfrac{1}{[\mathrm{B}]_0-2[\mathrm{A}]_0}\ln\dfrac{[\mathrm{A}]_0([\mathrm{B}]_0-2x)}{([\mathrm{A}]_0-x)[\mathrm{B}]_0}$	
	$\mathrm{A}\rightarrow\mathrm{P}$ with autocatalysis	$v=k_\mathrm{r}[\mathrm{A}][\mathrm{P}]$ $k_\mathrm{r}t=\dfrac{1}{[\mathrm{A}]_0+[\mathrm{P}]_0}\ln\dfrac{[\mathrm{A}]_0([\mathrm{P}]_0+x)}{([\mathrm{A}]_0-x)[\mathrm{P}]_0}$	
3	$\mathrm{A}+2\,\mathrm{B}\rightarrow\mathrm{P}$	$v=k_\mathrm{r}[\mathrm{A}][\mathrm{B}]^2$ $k_\mathrm{r}t=\dfrac{2x}{(2[\mathrm{A}]_0-[\mathrm{B}]_0)([\mathrm{B}]_0-2x)[\mathrm{B}]_0}+\dfrac{1}{(2[\mathrm{A}]_0-[\mathrm{B}]_0)^2}\ln\dfrac{[\mathrm{A}]_0([\mathrm{B}]_0-2x)}{([\mathrm{A}]_0-x)[\mathrm{B}]_0}$	
$n\ge 2$	$\mathrm{A}\rightarrow\mathrm{P}$	$v=k_\mathrm{r}[\mathrm{A}]^n$ $k_\mathrm{r}t=\dfrac{1}{n-1}\left\{\dfrac{1}{([\mathrm{A}]_0-x)^{n-1}}-\dfrac{1}{[\mathrm{A}]_0^{n-1}}\right\}$	$\dfrac{2^{n-1}-1}{(n-1)k_\mathrm{r}[\mathrm{A}]_0^{n-1}}$

* $x=[\mathrm{P}]$ and $v=\mathrm{d}x/\mathrm{d}t$.

21.4 Reactions approaching equilibrium

Key points (a) The equilibrium constant for a reaction is equal to the ratio of the forward and reverse rate constants. (b) In relaxation methods of kinetic analysis, the equilibrium position of a reaction is first shifted suddenly and then allowed to readjust to the equilibrium composition characteristic of the new conditions.

Because all the rate laws considered so far disregard the possibility that the reverse reaction is important, none of them describes the overall rate when the reaction is close to equilibrium. At that stage the products may be so abundant that the reverse reaction must be taken into account. In practice, however, most kinetic studies are made on reactions that are far from equilibrium, and the reverse reactions are unimportant.

(a) First-order reactions close to equilibrium

We can explore the variation of the composition with time close to chemical equilibrium by considering the reaction in which A forms B and both forward and reverse reactions are first-order (as in some isomerizations). The scheme we consider is

$$\begin{aligned} &A \rightarrow B \qquad v = k_r[A] \\ &B \rightarrow A \qquad v = k_r'[B] \end{aligned} \tag{21.20}$$

The concentration of A is reduced by the forward reaction (at a rate $k_r[A]$) but it is increased by the reverse reaction (at a rate $k_r'[B]$). The net rate of change is therefore

$$\frac{d[A]}{dt} = -k_r[A] + k_r'[B] \tag{21.21}$$

If the initial concentration of A is $[A]_0$, and no B is present initially, then at all times $[A] + [B] = [A]_0$. Therefore,

$$\frac{d[A]}{dt} = -k_r[A] + k_r'([A]_0 - [A]) = -(k_r + k_r')[A] + k_r'[A]_0 \tag{21.22}$$

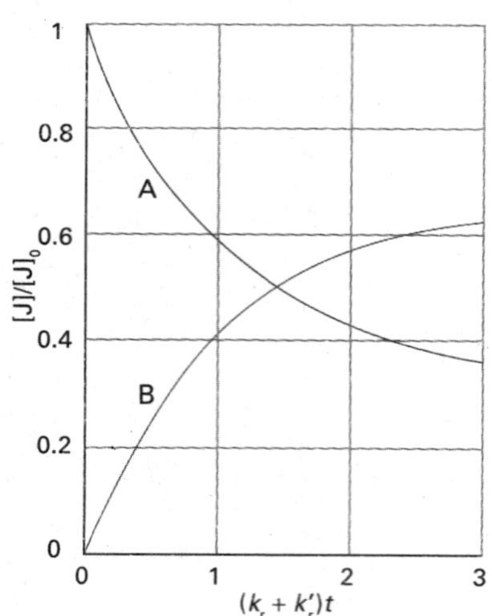

Fig. 21.9 The approach of concentrations to their equilibrium values as predicted by eqn 21.23 for a reaction A ⇌ B that is first-order in each direction, and for which $k_r = 2k_r'$.

interActivity Set up the rate equations and plot the corresponding graphs for the approach to an equilibrium of the form A ⇌ 2 B.

The solution of this first-order differential equation (as may be checked by differentiation) is

$$[A] = \frac{k_r' + k_r e^{-(k_r+k_r')t}}{k_r' + k_r}[A]_0 \tag{21.23}$$

Figure 21.9 shows the time dependence predicted by this equation.

As $t \rightarrow \infty$, the concentrations reach their equilibrium values, which are given by eqn 21.23 as:

$$[A]_{eq} = \frac{k_r'[A]_0}{k_r + k_r'} \qquad [B]_{eq} = [A]_0 - [A]_{eq} = \frac{k_r[A]_0}{k_r + k_r'} \tag{21.24}$$

It follows that the equilibrium constant of the reaction is

$$K = \frac{[B]_{eq}}{[A]_{eq}} = \frac{k_r}{k_r'} \tag{21.25}$$

(This expression is only approximate because thermodynamically precise equilibrium constants are expressed in terms of activities, not concentrations.) Exactly the same conclusion can be reached—more simply, in fact—by noting that, at equilibrium, the forward and reverse rates must be the same, so

$$k_r[A]_{eq} = k_r'[B]_{eq} \tag{21.26}$$

This relation rearranges into eqn 21.25. The theoretical importance of eqn 21.25 is that it relates a thermodynamic quantity, the equilibrium constant, to quantities relating to rates. Its practical importance is that, if one of the rate constants can be measured, then the other may be obtained if the equilibrium constant is known.

For a more general reaction, the overall equilibrium constant can be expressed in terms of the rate constants for all the intermediate stages of the reaction mechanism:

$$K=\frac{k_a}{k'_a}\times\frac{k_b}{k'_b}\times\cdots \qquad \text{The equilibrium constant in terms of the rate constants} \qquad (21.27)$$

where the *k*s are the rate constants for the individual steps and the *k*′s are those for the corresponding reverse steps.

(b) Relaxation methods

The term **relaxation** denotes the return of a system to equilibrium. It is used in chemical kinetics to indicate that an externally applied influence has shifted the equilibrium position of a reaction, normally suddenly, and that the reaction is adjusting to the equilibrium composition characteristic of the new conditions (Fig. 21.10). We shall consider the response of reaction rates to a **temperature jump**, a sudden change in temperature. We know from Section 6.4 that the equilibrium composition of a reaction depends on the temperature (provided $\Delta_r H^\ominus$ is nonzero), so a shift in temperature acts as a perturbation on the system. One way of achieving a temperature jump is to discharge a capacitor through a sample made conducting by the addition of ions, but laser or microwave discharges can also be used. Temperature jumps of between 5 and 10 K can be achieved in about 1 μs with electrical discharges. The high energy output of pulsed lasers (Section 13.6) is sufficient to generate temperature jumps of between 10 and 30 K within nanoseconds in aqueous samples. Some equilibria are also sensitive to pressure, and **pressure-jump techniques** may then also be used.

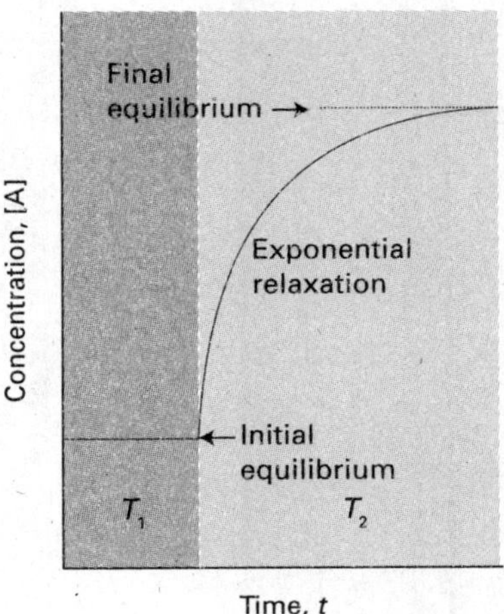

Fig. 21.10 The relaxation to the new equilibrium composition when a reaction initially at equilibrium at a temperature T_1 is subjected to a sudden change of temperature, which takes it to T_2.

When a sudden temperature increase is applied to a simple $A \rightleftharpoons B$ equilibrium that is first-order in each direction, we show in the following *Justification* that the composition relaxes exponentially to the new equilibrium composition:

$$x=x_0 e^{-t/\tau} \qquad \frac{1}{\tau}=k_r+k'_r \qquad \text{Relaxation after a temperature jump} \qquad (21.28)$$

where x_0 is the departure from equilibrium immediately after the temperature jump and x is the departure from equilibrium at the new temperature after a time t.

Justification 21.4 *Relaxation to equilibrium*

When the temperature of a system at equilibrium is increased suddenly, the rate constants change from their earlier values to the new values k_r and k'_r characteristic of that temperature, but the concentrations of A and B remain for an instant at their old equilibrium values. As the system is no longer at equilibrium, it readjusts to the new equilibrium concentrations, which are now given by

$$k_r[A]_{eq}=k'_r[B]_{eq}$$

and it does so at a rate that depends on the new rate constants. We write the deviation of [A] from its new equilibrium value as x, so $[A]=x+[A]_{eq}$ and $[B]=[B]_{eq}-x$. The concentration of A then changes as follows:

$$\frac{d[A]}{dt}=-k_r[A]+k'_r[B]=-k_r([A]_{eq}+x)+k'_r([B]_{eq}-x)=-(k_r+k'_r)x$$

because the two terms involving the equilibrium concentrations cancel. Because $d[A]/dt=dx/dt$, this equation is a first-order differential equation with the solution that resembles eqn 21.12b and is given in eqn 21.28.

Equation 21.28 shows that the concentrations of A and B relax into the new equilibrium at a rate determined by the sum of the two new rate constants. Because the equilibrium constant under the new conditions is $K \approx k_r/k_r'$, its value may be combined with the relaxation time measurement to find the individual k_r and k_r'.

Example 21.4 *Analysing a temperature-jump experiment*

The equilibrium constant for the autoprotolysis of water, $H_2O(l) \rightleftharpoons H^+(aq) + OH^-(aq)$, is $K_w = a(H^+)a(OH^-) = 1.008 \times 10^{-14}$ at 298 K. After a temperature-jump, the reaction returns to equilibrium with a relaxation time of 37 μs at 298 K and pH ≈ 7. Given that the forward reaction is first-order and the reverse is second-order overall, calculate the rate constants for the forward and reverse reactions.

Method We need to derive an expression for the relaxation time, τ (the time constant for return to equilibrium), in terms of k_r (forward, first-order reaction) and k_r' (reverse, second-order reaction). We can proceed as above, but it will be necessary to make the assumption that the deviation from equilibrium (x) is so small that terms in x^2 can be neglected. Relate k_r and k_r' through the equilibrium constant, but be careful with units because K_w is dimensionless.

Answer The forward rate at the final temperature is $k_r[H_2O]$ and the reverse rate is $k_r'[H^+][OH^-]$. The net rate of deprotonation of H_2O is

$$\frac{d[H_2O]}{dt} = -k_r[H_2O] + k_r'[H^+][OH^-]$$

We write $[H_2O] = [H_2O]_{eq} + x$, $[H^+] = [H^+]_{eq} - x$, and $[OH^-] = [OH^-]_{eq} - x$, and obtain

$$\begin{aligned}\frac{dx}{dt} &= -\{k_r + k_r'([H^+]_{eq} + [OH^-]_{eq})\}x - k_r[H_2O]_{eq} + k_r'[H^+]_{eq}[OH^-]_{eq} + k_r'x^2 \\ &\approx -\{k_r + k_r'([H^+]_{eq} + [OH^-]_{eq})\}x\end{aligned}$$

where we have neglected the term in x^2 because it is so small and have used the equilibrium condition $k_r[H_2O]_{eq} = k_r'[H^+]_{eq}[OH^-]_{eq}$ to eliminate the terms that are independent of x. It follows that

$$\frac{1}{\tau} = k_r + k_r'([H^+]_{eq} + [OH^-]_{eq})$$

At this point we note that

$$K_w = a(H^+)a(OH^-) \approx ([H^+]_{eq}/c^\ominus)([OH^-]_{eq}/c^\ominus) = [H^+]_{eq}[OH^-]_{eq}/c^{\ominus 2}$$

with $c^\ominus = 1$ mol dm^{-3}. For this electrically neutral system, $[H^+] = [OH^-]$, so the concentration of each type of ion is $K_w^{1/2}c^\ominus$, and hence

$$\frac{1}{\tau} = k_r + k_r'(K_w^{1/2}c^\ominus + K_w^{1/2}c^\ominus) = k_r'\left\{\frac{k_r}{k_r'} + 2K_w^{1/2}c^\ominus\right\}$$

A note on good practice Notice how we keep track of units through the use of $c^\ominus$: K and K_w are dimensionless, k_r' is expressed in dm^3 mol^{-1} s^{-1}, and k_r is expressed in s^{-1}.

At this point we note that

$$\frac{k_r}{k_r'} = \frac{[H^+]_{eq}[OH^-]_{eq}}{[H_2O]_{eq}} = \frac{K_w c^{\ominus 2}}{[H_2O]_{eq}}$$

The molar concentration of pure water is 55.6 mol dm^{-3}, so $[H_2O]_{eq}/c^\ominus = 55.6$. If we write $K = K_w/55.6 = 1.81 \times 10^{-16}$, we obtain

$$\frac{1}{\tau} = k_r'\{K + 2K_w^{1/2}\}c^\ominus$$

Hence,

$$k_r' = \frac{1}{\tau(K+2K_w^{1/2})c^{\ominus}}$$

$$= \frac{1}{(3.7\times10^{-5}\,\text{s})\times(2.0\times10^{-7})\times(1\,\text{mol dm}^{-3})} = 1.4\times10^{11}\,\text{dm}^3\,\text{mol}^{-1}\,\text{s}^{-1}$$

It follows that

$$k_r = k_r' K c^{\ominus} = 2.4\times10^{-5}\,\text{s}^{-1}$$

The reaction is faster in ice, where $k_r' = 8.6\times10^{12}\,\text{dm}^3\,\text{mol}^{-1}\,\text{s}^{-1}$.

Self-test 21.6 Derive an expression for the relaxation time of a concentration when the reaction $A + B \rightleftharpoons C + D$ is second-order in both directions.

$[1/\tau = k_r([A]+[B])_{eq} + k_r'([C]+[D])_{eq}]$

21.5 The temperature dependence of reaction rates

Key points **(a) The temperature dependence of the rate constant of a reaction typically follows the Arrhenius equation. (b) The activation energy E_a is the minimum kinetic energy required for reaction during a molecular encounter. The pre-exponential factor A is a measure of the rate at which collisions occur irrespective of their energy.**

The rate constants of most reactions increase as the temperature is raised. Many reactions in solution fall somewhere in the range spanned by the hydrolysis of methyl ethanoate (where the rate constant at 35°C is 1.82 times that at 25°C) and the hydrolysis of sucrose (where the factor is 4.13).

(a) The Arrhenius parameters

It is found experimentally for many reactions that a plot of ln k_r against $1/T$ gives a straight line. This behaviour is normally expressed mathematically by introducing two parameters, one representing the intercept and the other the slope of the straight line, and writing the **Arrhenius equation**

$$\ln k_r = \ln A - \frac{E_a}{RT} \qquad \text{Arrhenius equation} \qquad (21.29)$$

The parameter A, which corresponds to the intercept of the line at $1/T = 0$ (at infinite temperature, Fig. 21.11), is called the **pre-exponential factor** or the 'frequency factor'. The parameter E_a, which is obtained from the slope of the line ($-E_a/R$), is called the **activation energy**. Collectively the two quantities are called the **Arrhenius parameters** (Table 21.4).

Table 21.4* Arrhenius parameters

(1) First-order reactions	A/s^{-1}	E_a/(kJ mol^{-1})
$CH_3NC \rightarrow CH_3CN$	3.98×10^{13}	160
$2\,N_2O_5 \rightarrow 4\,NO_2 + O_2$	4.94×10^{13}	103.4
(2) Second-order reactions	A/(dm^3 mol^{-1} s^{-1})	E_a/(kJ mol^{-1})
$OH + H_2 \rightarrow H_2O + H$	8.0×10^{10}	42
$NaC_2H_5O + CH_3I$ in ethanol	2.42×10^{11}	81.6

* More values are given in the *Data section*.

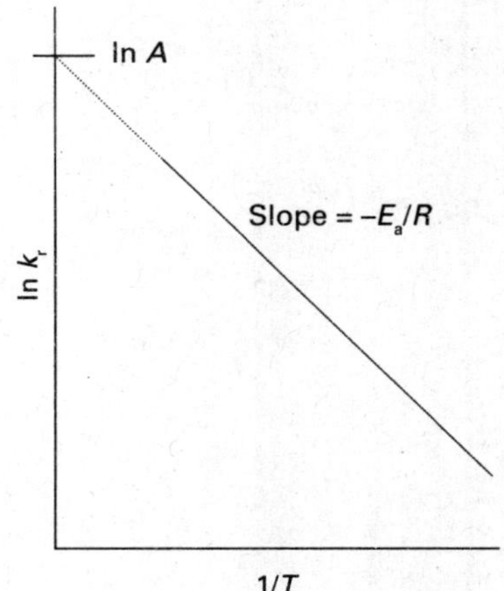

Fig. 21.11 A plot of ln k_r against $1/T$ is a straight line when the reaction follows the behaviour described by the Arrhenius equation (eqn 21.29). The slope gives $-E_a/R$ and the intercept at $1/T = 0$ gives ln A.

Example 21.5 *Determining the Arrhenius parameters*

The rate of the second-order decomposition of acetaldehyde (ethanal, CH_3CHO) was measured over the temperature range 700–1000 K, and the rate constants are reported below. Find E_a and A.

T/K	700	730	760	790	810	840	910	1000
$k_r/(\mathrm{dm^3\,mol^{-1}\,s^{-1}})$	0.011	0.035	0.105	0.343	0.789	2.17	20.0	145

Method According to eqn 21.29, the data can be analysed by plotting $\ln(k_r/\mathrm{dm^3\,mol^{-1}\,s^{-1}})$ against $1/(T/\mathrm{K})$, or more conveniently $(10^3\,\mathrm{K})/T$, and getting a straight line. As explained in Example 21.3, we obtain the activation energy from the dimensionless slope by writing $-E_a/R$ = slope/units, where in this case 'units' $= 1/(10^3\,\mathrm{K})$, so $E_a = -\mathrm{slope} \times R \times 10^3\,\mathrm{K}$. The intercept at $1/T = 0$ is $\ln(A/\mathrm{dm^3\,mol^{-1}\,s^{-1}})$.

Answer We draw up the following table:

$(10^3\,\mathrm{K})/T$	1.43	1.37	1.32	1.27	1.23	1.19	1.10	1.00
$\ln(k_r/\mathrm{dm^3\,mol^{-1}\,s^{-1}})$	−4.51	−3.35	−2.25	−1.07	−0.24	0.77	3.00	4.98

Now plot $\ln k_r$ against $1/T$ (Fig. 21.12). The least-squares fit is to a line with slope −22.7 and intercept 27.7. Therefore,

$$E_a = 22.7 \times (8.3145\,\mathrm{J\,K^{-1}\,mol^{-1}}) \times 10^3\,\mathrm{K} = 189\,\mathrm{kJ\,mol^{-1}}$$
$$A = e^{27.7}\,\mathrm{dm^3\,mol^{-1}\,s^{-1}} = 1.1 \times 10^{12}\,\mathrm{dm^3\,mol^{-1}\,s^{-1}}$$

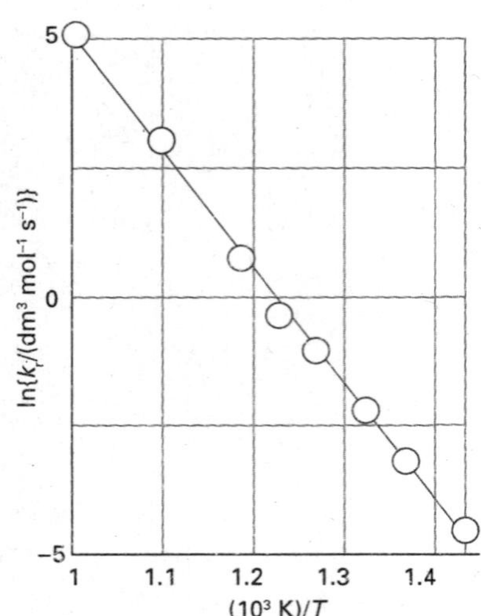

Fig. 21.12 The Arrhenius plot using the data in Example 21.5.

A note on good practice Note that A has the same units as k_r. In practice, A is obtained from one of the mid-range data values rather than using a lengthy extrapolation.

Self-test 21.7 Determine A and E_a from the following data:

T/K	300	350	400	450	500
$k_r/(\mathrm{dm^3\,mol^{-1}\,s^{-1}})$	7.9×10^6	3.0×10^7	7.9×10^7	1.7×10^8	3.2×10^8

[$8 \times 10^{10}\,\mathrm{dm^3\,mol^{-1}\,s^{-1}}$, $23\,\mathrm{kJ\,mol^{-1}}$]

The fact that E_a is given by the slope of the plot of $\ln k_r$ against $1/T$ means that, the higher the activation energy, the stronger the temperature dependence of the rate constant (that is, the steeper the slope). *A high activation energy signifies that the rate constant depends strongly on temperature.* If a reaction has zero activation energy, its rate is independent of temperature. In some cases the activation energy is negative, which indicates that the rate decreases as the temperature is raised. We shall see that such behaviour is a signal that the reaction has a complex mechanism.

The temperature dependence of some reactions is non-Arrhenius, in the sense that a straight line is not obtained when $\ln k$ is plotted against $1/T$. However, it is still possible to define an activation energy at any temperature as

$$E_a = RT^2\left(\frac{d\ln k_r}{dT}\right) \qquad \text{Definition of the activation energy} \qquad [21.30]$$

This definition reduces to the earlier one (as the slope of a straight line) for a temperature-independent activation energy. However, the definition in eqn 21.30 is more general than eqn 21.29, because it allows E_a to be obtained from the slope (at the temperature of interest) of a plot of $\ln k_r$ against $1/T$ even if the Arrhenius plot is not a straight line.

(b) The interpretation of the parameters

For the present chapter we shall regard the Arrhenius parameters as purely empirical quantities that enable us to discuss the variation of rate constants with temperature; however, it is useful to have an interpretation in mind. To find one, we begin by writing eqn 21.29 as

$$k_r = A\,e^{-E_a/RT} \qquad \text{Alternative form of the Arrhenius equation} \qquad (21.31)$$

Next, to interpret E_a we consider how the molecular potential energy changes in the course of a chemical reaction that begins with a collision between molecules of A and molecules of B (Fig. 21.13).

As the reaction event proceeds, A and B come into contact, distort, and begin to exchange or discard atoms. The **reaction coordinate** is the collection of motions, such as changes in interatomic distances and bond angles, that are directly involved in the formation of products from reactants. (The reaction coordinate is essentially a geometrical concept and quite distinct from the extent of reaction.) The potential energy rises to a maximum and the cluster of atoms that corresponds to the region close to the maximum is called the **activated complex**. After the maximum, the potential energy falls as the atoms rearrange in the cluster and reaches a value characteristic of the products. The climax of the reaction is at the peak of the potential energy, which corresponds to the activation energy E_a. Here two reactant molecules have come to such a degree of closeness and distortion that a small further distortion will send them in the direction of products. This crucial configuration is called the **transition state** of the reaction. Although some molecules entering the transition state might revert to reactants, if they pass through this configuration then it is inevitable that products will emerge from the encounter.

We also conclude from the preceding discussion that, for a reaction involving the collision of two molecules, *the activation energy is the minimum kinetic energy that reactants must have in order to form products*. For example, in a gas-phase reaction there are numerous collisions each second, but only a tiny proportion are sufficiently energetic to lead to reaction. The fraction of collisions with a kinetic energy in excess of an energy E_a is given by the Boltzmann distribution as $e^{-E_a/RT}$. We show in the following *Justification* that we can interpret the exponential factor in eqn 21.31 as the fraction of collisions that have enough kinetic energy to lead to reaction.

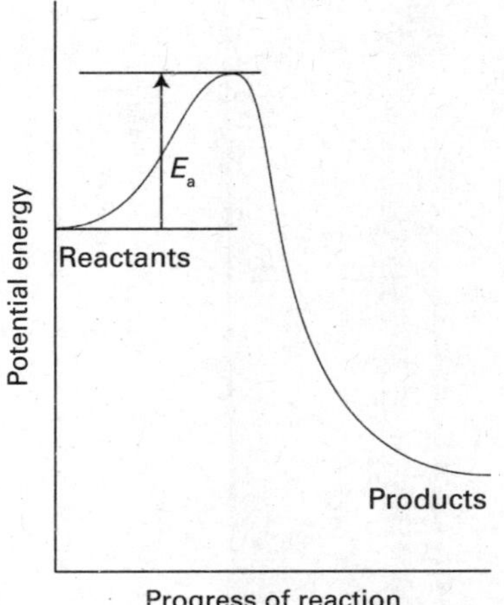

Fig. 21.13 A potential energy profile for an exothermic reaction. The height of the barrier between the reactants and products is the activation energy of the reaction.

A brief comment

The terms *activated complex* and *transition state* are often used as synonyms; however, we shall preserve a distinction.

Justification 21.5 *Interpreting the activation energy*

Suppose the energy levels available to the system form a uniform array of separation ε. The Boltzmann distribution is

$$\frac{N_i}{N} = \frac{e^{-i\varepsilon\beta}}{q} = (1-e^{-\varepsilon\beta})e^{-i\varepsilon\beta}$$

where $\beta = 1/kT$ and we have used the result in eqn 15.12 for the partition function q. The total number of molecules in states with energy of at least $i_{max}\varepsilon$ is

$$\sum_{i=i_{max}}^{\infty} N_i = \sum_{i=0}^{\infty} N_i - \sum_{i=0}^{i_{max}-1} N_i = N - \frac{N}{q}\sum_{i=0}^{i_{max}-1} e^{-i\varepsilon\beta}$$

The sum of the finite geometrical series is

$$\sum_{i=0}^{i_{max}-1} e^{-i\varepsilon\beta} = \frac{1-e^{-i_{max}\varepsilon\beta}}{1-e^{-\varepsilon\beta}} = q(1-e^{-i_{max}\varepsilon\beta})$$

Therefore, the fraction of molecules in states with energy of at least $\varepsilon_{\max} = i_{\max}\varepsilon$ is

$$\frac{\sum_{i=i_{\max}}^{\infty} N_i}{N} = 1-(1-e^{-i_{\max}\varepsilon\beta}) = e^{-i_{\max}\varepsilon\beta} = e^{-\varepsilon_{\max}/kT}$$

The calculation using translational states is more involved, and is presented in Section 22.4.

The pre-exponential factor is a measure of the rate at which collisions occur irrespective of their energy. Hence, the product of A and the exponential factor, $e^{-E_a/RT}$, gives the rate of *successful* collisions. We shall develop these remarks in Chapter 22 and see that they have their analogues for reactions that take place in liquids.

Accounting for the rate laws

We now move on to the second stage of the analysis of kinetic data, their explanation in terms of a postulated reaction mechanism.

21.6 Elementary reactions

Key points **The mechanism of reaction is the sequence of elementary steps involved in a reaction. The molecularity of an elementary reaction is the number of molecules coming together to react. An elementary unimolecular reaction has first-order kinetics; an elementary bimolecular reaction has second-order kinetics.**

Most reactions occur in a sequence of steps called **elementary reactions**, each of which involves only a small number of molecules or ions. A typical elementary reaction is

$$H + Br_2 \rightarrow HBr + Br$$

Note that the phase of the species is not specified in the chemical equation for an elementary reaction, and the equation represents the specific process occurring to individual molecules. This equation, for instance, signifies that an H atom attacks a Br_2 molecule to produce an HBr molecule and a Br atom. The **molecularity** of an elementary reaction is the number of molecules coming together to react in an elementary reaction. In a **unimolecular reaction**, a single molecule shakes itself apart or its atoms into a new arrangement, as in the isomerization of cyclopropane to propene. In a **bimolecular reaction**, a pair of molecules collide and exchange energy, atoms, or groups of atoms, or undergo some other kind of change. It is most important to distinguish molecularity from order:

- *reaction order* is an empirical quantity, and obtained from the experimental rate law;
- *molecularity* refers to an elementary reaction proposed as an individual step in a mechanism.

The rate law of a unimolecular elementary reaction is first-order in the reactant:

$$A \rightarrow P \qquad \frac{d[A]}{dt} = -k_r[A] \qquad (21.32)$$

where P denotes products (several different species may be formed). A unimolecular reaction is first-order because the number of A molecules that decay in a short interval is proportional to the number available to decay. (Ten times as many decay in the same interval when there are initially 1000 A molecules as when there are only 100 present.) Therefore, the rate of decomposition of A is proportional to its molar concentration at any moment during the reaction.

An elementary bimolecular reaction has a second-order rate law:

$$A + B \rightarrow P \qquad \frac{d[A]}{dt} = -k_r[A][B] \tag{21.33}$$

A bimolecular reaction is second-order because its rate is proportional to the rate at which the reactant species meet, which in turn is proportional to their concentrations. Therefore, if we have evidence that a reaction is a single-step, bimolecular process, we can write down the rate law (and then go on to test it). Bimolecular elementary reactions are believed to account for many homogeneous reactions, such as the dimerizations of alkenes and dienes and reactions such as

$$CH_3I(alc) + CH_3CH_2O^-(alc) \rightarrow CH_3OCH_2CH_3(alc) + I^-(alc)$$

(where 'alc' signifies alcohol solution). There is evidence that the mechanism of this reaction is a single elementary step

$$CH_3I + CH_3CH_2O^- \rightarrow CH_3OCH_2CH_3 + I^- \tag{21.34}$$

This mechanism is consistent with the observed rate law

$$v = k_r[CH_3I][CH_3CH_2O^-] \tag{21.35}$$

We shall see below how to combine a series of simple steps together into a mechanism and how to arrive at the corresponding rate law. For the present we emphasize that, *if the reaction is an elementary bimolecular process, then it has second-order kinetics, but if the kinetics are second-order, then the reaction might be complex.* The postulated mechanism can be explored only by detailed detective work on the system, and by investigating whether side products or intermediates appear during the course of the reaction. Detailed analysis of this kind was one of the ways, for example, in which the reaction $H_2(g) + I_2(g) \rightarrow 2\,HI(g)$ was shown to proceed by a complex mechanism. For many years the reaction had been accepted on good, but insufficiently meticulous evidence as a fine example of a simple bimolecular reaction, $H_2 + I_2 \rightarrow HI + HI$, in which atoms exchanged partners during a collision.

21.7 Consecutive elementary reactions

Key points **(a) The concentration of a reaction intermediate rises to a maximum and then falls to zero whilst the concentration of the product rises from zero. (b) The rate-determining step is the slowest step in a reaction mechanism that controls the rate of the overall reaction. (c) In the steady-state approximation, it is assumed that the concentrations of all reaction intermediates remain constant and small throughout the reaction. (d) Provided a reaction has not reached equilibrium, the products of competing reactions are controlled by kinetics. (e) Pre-equilibrium is a state in which an intermediate is in equilibrium with the reactants and which arises when the rates of formation of the intermediate and its decay back into reactants are much faster than its rate of formation of products.**

Some reactions proceed through the formation of an intermediate (I), as in the consecutive unimolecular reactions

$$A \xrightarrow{k_a} I \xrightarrow{k_b} P$$

An example is the decay of a radioactive family, such as

$$^{239}\mathrm{U} \xrightarrow{23.5\ \mathrm{min}} {}^{239}\mathrm{Np} \xrightarrow{2.35\ \mathrm{day}} {}^{239}\mathrm{Pu}$$

(The times are half-lives.) We can discover the characteristics of this type of reaction by setting up the rate laws for the net rate of change of the concentration of each substance.

(a) The variation of concentrations with time

The rate of unimolecular decomposition of A is

$$\frac{\mathrm{d}[\mathrm{A}]}{\mathrm{d}t} = -k_\mathrm{a}[\mathrm{A}] \qquad (21.36)$$

and A is not replenished. The intermediate I is formed from A (at a rate $k_\mathrm{a}[\mathrm{A}]$) but decays to P (at a rate $k_\mathrm{b}[\mathrm{I}]$). The net rate of formation of I is therefore

$$\frac{\mathrm{d}[\mathrm{I}]}{\mathrm{d}t} = k_\mathrm{a}[\mathrm{A}] - k_\mathrm{b}[\mathrm{I}] \qquad (21.37)$$

The product P is formed by the unimolecular decay of I:

$$\frac{\mathrm{d}[\mathrm{P}]}{\mathrm{d}t} = k_\mathrm{b}[\mathrm{I}] \qquad (21.38)$$

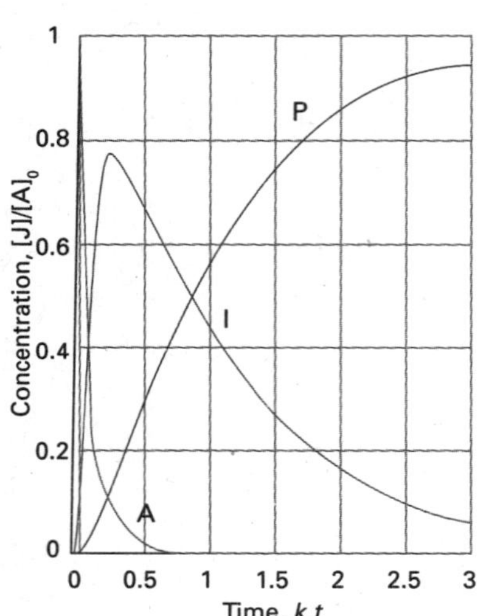

Fig. 21.14 The concentrations of A, I, and P in the consecutive reaction scheme A → I → P. The curves are plots of eqns 21.39, 21.41, and 21.42 with $k_\mathrm{a} = 10k_\mathrm{b}$. If the intermediate I is in fact the desired product, it is important to be able to predict when its concentration is greatest; see Example 21.6.

interActivity Use mathematical software, an electronic spreadsheet, or the applets found in the *Living graphs* section of the text's web site to investigate the effects on [A], [I], [P], and t_max of increasing the ratio $k_\mathrm{a}/k_\mathrm{b}$ from 10 (as in Fig. 21.14) to 0.01. Compare your results with those shown in Fig. 21.16.

We suppose that initially only A is present, and that its concentration is $[\mathrm{A}]_0$.

The first of the rate laws, eqn 21.36, is an ordinary first-order decay, so we can write

$$[\mathrm{A}] = [\mathrm{A}]_0 \mathrm{e}^{-k_\mathrm{a}t} \qquad (21.39)$$

When this equation is substituted into eqn 21.37, we obtain after rearrangement

$$\frac{\mathrm{d}[\mathrm{I}]}{\mathrm{d}t} + k_\mathrm{b}[\mathrm{I}] = k_\mathrm{a}[\mathrm{A}]_0 \mathrm{e}^{-k_\mathrm{a}t} \qquad (21.40)$$

This differential equation has a standard form (see *Mathematical background* 4) and, after setting $[\mathrm{I}]_0 = 0$, the solution is

$$[\mathrm{I}] = \frac{k_\mathrm{a}}{k_\mathrm{b} - k_\mathrm{a}}(\mathrm{e}^{-k_\mathrm{a}t} - \mathrm{e}^{-k_\mathrm{b}t})[\mathrm{A}]_0 \qquad (21.41)$$

At all times $[\mathrm{A}] + [\mathrm{I}] + [\mathrm{P}] = [\mathrm{A}]_0$, so it follows that

$$[\mathrm{P}] = \left\{1 + \frac{k_\mathrm{a}\mathrm{e}^{-k_\mathrm{b}t} - k_\mathrm{b}\mathrm{e}^{-k_\mathrm{a}t}}{k_\mathrm{b} - k_\mathrm{a}}\right\}[\mathrm{A}]_0 \qquad (21.42)$$

The concentration of the intermediate I rises to a maximum and then falls to zero (Fig. 21.14). The concentration of the product P rises from zero towards $[\mathrm{A}]_0$.

Example 21.6 *Analysing consecutive reactions*

Suppose that in an industrial batch process a substance A produces the desired compound I, which goes on to decay to a worthless product C, each step of the reaction being first-order. At what time will I be present in greatest concentration?

Method The time dependence of the concentration of I is given by eqn 21.41. We can find the time at which [I] passes through a maximum, t_max, by calculating $\mathrm{d}[\mathrm{I}]/\mathrm{d}t$ and setting the resulting rate equal to zero.

Answer It follows from eqn 21.41 that

$$\frac{d[I]}{dt} = -\frac{k_a[A]_0(k_a e^{-k_at} - k_b e^{-k_bt})}{k_b - k_a}$$

This rate is equal to zero when $k_a e^{-k_at} = k_b e^{-k_bt}$. Therefore,

$$t_{max} = \frac{1}{k_a - k_b} \ln \frac{k_a}{k_b}$$

For a given value of k_a, as k_b increases both the time at which [I] is a maximum and the yield of I decrease.

Self-test 21.8 Calculate the maximum concentration of I and justify the last remark.
$[[I]_{max}/[A]_0 = (k_a/k_b)^c, c = k_b/(k_b - k_a)]$

(b) The steady-state approximation

One feature of the calculation so far has probably not gone unnoticed: there is a considerable increase in mathematical complexity as soon as the reaction mechanism has more than a couple of steps. A reaction scheme involving many steps is nearly always unsolvable analytically, and alternative methods of solution are necessary. One approach is to integrate the rate laws numerically. An alternative approach, which continues to be widely used because it leads to convenient expressions and more readily digestible results, is to make an approximation.

The **steady-state approximation** (which is also widely called the **quasi-steady-state approximation**, QSSA, to distinguish it from a true steady state) assumes that, after an initial **induction period**, an interval during which the concentrations of intermediates, I, rise from zero, and during the major part of the reaction, the rates of change of concentrations of all reaction intermediates are negligibly small (Fig. 21.15):

$$\frac{d[I]}{dt} \approx 0 \qquad (21.43)$$

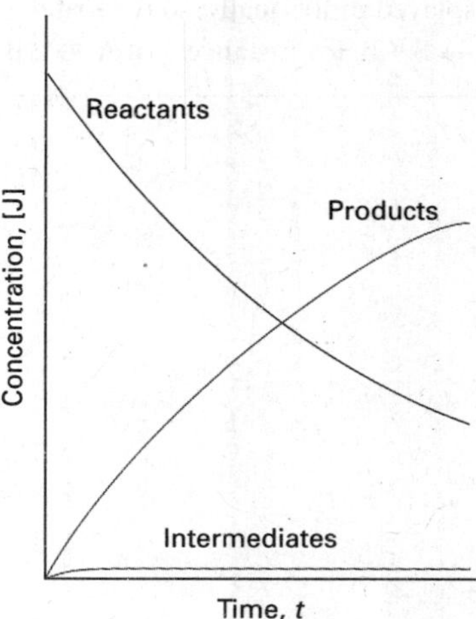

Fig. 21.15 The basis of the steady-state approximation. It is supposed that the concentrations of intermediates remain small and hardly change during most of the course of the reaction.

This approximation greatly simplifies the discussion of reaction schemes. For example, when we apply the approximation to the consecutive first-order mechanism, we set $d[I]/dt = 0$ in eqn 21.37, which then becomes $k_a[A] - k_b[I] \approx 0$. Then

$$[I] \approx (k_a/k_b)[A] \qquad (21.44)$$

For this expression to be consistent with eqn 21.41, we require $k_a/k_b \ll 1$ (so that, even though [A] does depend on the time, the dependence of [I] on the time is negligible). On substituting this value of [I] into eqn 21.38, that equation becomes

$$\frac{d[P]}{dt} = k_b[I] \approx k_a[A] \qquad (21.45)$$

and we see that P is formed by a first-order decay of A, with a rate constant k_a, the rate constant of the slower, rate-determining, step. We can write down the solution of this equation at once by substituting the solution for [A], eqn 21.39, and integrating:

$$[P] = k_a[A]_0 \int_0^t e^{-k_at} dt = (1 - e^{-k_at})[A]_0 \qquad (21.46)$$

This is the same (approximate) result as before, eqn 21.42 (when $k_b \gg k_a$), but much more quickly obtained. Figure 21.16 compares the approximate solutions found here with the exact solutions found earlier: k_b does not have to be very much bigger than k_a for the approach to be reasonably accurate.

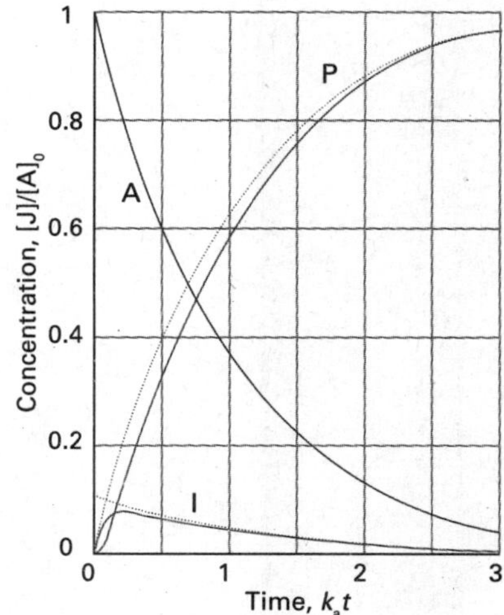

Fig. 21.16 A comparison of the exact result for the concentrations of a consecutive reaction and the concentrations obtained by using the steady-state approximation (dotted lines) for $k_b = 20k_a$. (The curve for [A] is unchanged.)

Example 21.7 *Using the steady-state approximation*

Devise the rate law for the decomposition of N_2O_5,

$$2\,N_2O_5(g) \rightarrow 4\,NO_2(g) + O_2(g)$$

on the basis of the following mechanism:

$$\begin{array}{ll} N_2O_5 \rightarrow NO_2 + NO_3 & k_a \\ NO_2 + NO_3 \rightarrow N_2O_5 & k_a' \\ NO_2 + NO_3 \rightarrow NO_2 + O_2 + NO & k_b \\ NO + N_2O_5 \rightarrow NO_2 + NO_2 + NO_2 & k_c \end{array}$$

A note on good practice Note that when writing the equation for an elementary reaction all the species are displayed individually; so we write $A \rightarrow B + B$, for instance, not $A \rightarrow 2\,B$.

Method First identify the intermediates (species that occur in the reaction steps but do not appear in the overall reaction) and write expressions for their net rates of formation. Then, all net rates of change of the concentrations of intermediates are set equal to zero and the resulting equations are solved algebraically.

Answer The intermediates are NO and NO_3; the net rates of change of their concentrations are

$$\frac{d[NO]}{dt} = k_b[NO_2][NO_3] - k_c[NO][N_2O_5] \approx 0$$

$$\frac{d[NO_3]}{dt} = k_a[N_2O_5] - k_a'[NO_2][NO_3] - k_b[NO_2][NO_3] \approx 0$$

The net rate of change of concentration of N_2O_5 is

$$\frac{d[N_2O_5]}{dt} = -k_a[N_2O_5] + k_a'[NO_2][NO_3] - k_c[NO][N_2O_5]$$

We use

$$k_b[NO_2][NO_3] - k_c[NO][N_2O_5] = 0$$

and

$$k_a[N_2O_5] - k_a'[NO_2][NO_3] - k_b[NO_2][NO_3] = 0$$

to write

$$[NO] = \frac{k_b[NO_2][NO_3]}{k_c[N_2O_5]}$$

$$[NO_3] = \frac{k_a[N_2O_5]}{(k_a' + k_b)[NO_2]}$$

and then substitute these expressions into that for $d[N_2O_5]/dt$ to obtain

$$\frac{d[N_2O_5]}{dt} = -\frac{2k_ak_b[N_2O_5]}{k_a' + k_b}$$

Self-test 21.9 Derive the rate law for the decomposition of ozone in the reaction $2\,O_3(g) \rightarrow 3\,O_2(g)$ on the basis of the (incomplete) mechanism

$$\begin{array}{ll} O_3 \rightarrow O_2 + O & k_a \\ O_2 + O \rightarrow O_3 & k_a' \\ O + O_3 \rightarrow O_2 + O_2 & k_b \end{array}$$

$[d[O_3]/dt = -2k_ak_b[O_3]^2/(k_a'[O_2] + k_b[O_3])]$

(c) The rate-determining step

Equation 21.46 shows that when $k_b \gg k_a$ the formation of the final product P depends on only the *smaller* of the two rate constants. That is, the rate of formation of P depends on the rate at which I is formed, not on the rate at which I changes into P. For this reason, the step A → I is called the 'rate-determining step' of the reaction. Its existence has been likened to building a six-lane highway up to a single-lane bridge: the traffic flow is governed by the rate of crossing the bridge. Similar remarks apply to more complicated reaction mechanisms, and in general the **rate-determining step** is the slowest step in a mechanism and controls the overall rate of the reaction. However, the rate-determining step is not just the slowest step: it must be slow *and* be a crucial gateway for the formation of products. If a faster reaction can also lead to products, then the slowest step is irrelevant because the slow reaction can then be sidestepped (Fig. 21.17).

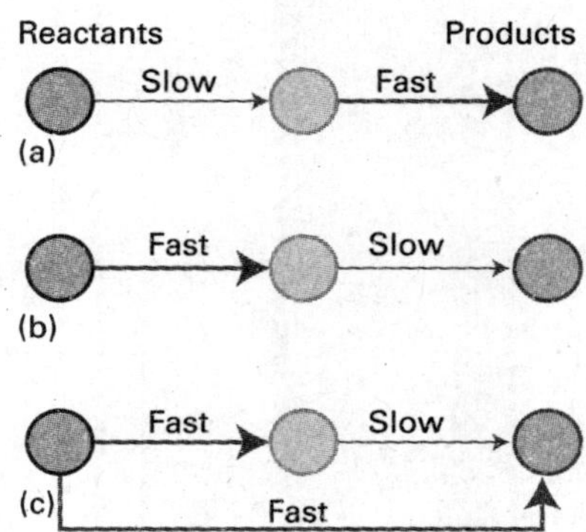

Fig. 21.17 In these diagrams of reaction schemes, heavy arrows represent fast steps and light arrows represent slow steps. (a) The first step is rate-determining; (b) the second step is rate-determining; (c) although one step is slow, it is not rate-determining because there is a fast route that circumvents it.

The rate law of a reaction that has a rate-determining step can often be written down almost by inspection. If the first step in a mechanism is rate-determining, then the rate of the overall reaction is equal to the rate of the first step because all subsequent steps are so fast that once the first intermediate is formed it results immediately in the formation of products. Figure 21.18 shows the reaction profile for a mechanism of this kind in which the slowest step is the one with the highest activation energy. Once over the initial barrier, the intermediates cascade into products. However, a rate-determining step may also stem from the low concentration of a crucial reactant and need not correspond to the step with highest activation barrier.

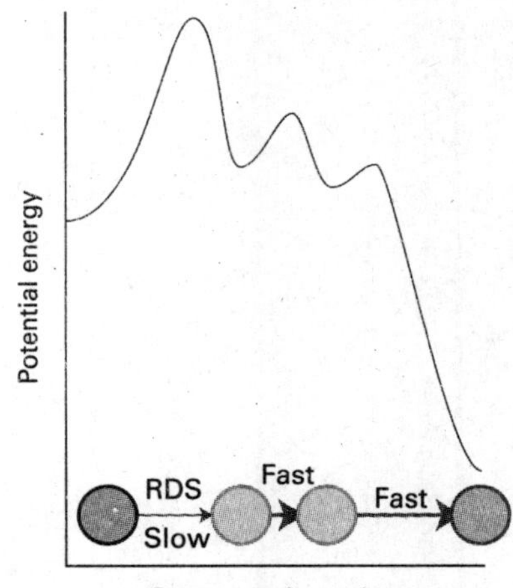

Fig. 21.18 The reaction profile for a mechanism in which the first step (RDS) is rate-determining.

(d) Pre-equilibria

From a simple sequence of consecutive reactions we now turn to a slightly more complicated mechanism in which an intermediate I reaches an equilibrium with the reactants A and B:

$$A + B \rightleftharpoons I \rightarrow P \tag{21.47}$$

The rate constants are k_a and k_a' for the forward and reverse reactions of the equilibrium and k_b for the final step. This scheme involves a **pre-equilibrium**, in which an intermediate is in equilibrium with the reactants. A pre-equilibrium can arise when the rate of decay of the intermediate back into reactants is much faster than the rate at which it forms products; thus, the condition is possible when $k_a' \gg k_b$ but not when $k_b \gg k_a'$. Because we assume that A, B, and I are in equilibrium, we can write

$$K = \frac{[I]}{[A][B]} \qquad K = \frac{k_a}{k_a'} \tag{21.48}$$

In writing these equations, we are presuming that the rate of reaction of I to form P is too slow to affect the maintenance of the pre-equilibrium (see the example below). We are also ignoring the fact, as is commonly done, that the standard concentration $c^{\ominus}$ should appear in the expression for K to ensure that it is dimensionless. The rate of formation of P may now be written:

$$\frac{d[P]}{dt} = k_b[I] = k_b K[A][B] \tag{21.49}$$

This rate law has the form of a second-order rate law with a composite rate constant:

$$\frac{d[P]}{dt} = k[A][B] \qquad k = k_b K = \frac{k_a k_b}{k_a'} \tag{21.50}$$

Example 21.8 *Analysing a pre-equilibrium*

Repeat the pre-equilibrium calculation but without ignoring the fact that I is slowly leaking away as it forms P.

Method Begin by writing the net rates of change of the concentrations of the substances and then invoke the steady-state approximation for the intermediate I. Use the resulting expression to obtain the rate of change of the concentration of P.

Answer The net rates of change of P and I are

$$\frac{d[P]}{dt} = k_b[I]$$

$$\frac{d[I]}{dt} = k_a[A][B] - k_a'[I] - k_b[I] \approx 0$$

The second equation solves to

$$[I] \approx \frac{k_a[A][B]}{k_a' + k_b}$$

When we substitute this result into the expression for the rate of formation of P, we obtain

$$\frac{d[P]}{dt} \approx k_r[A][B] \qquad k_r = \frac{k_a k_b}{k_a' + k_b}$$

This expression reduces to that in eqn 21.50 when the rate constant for the decay of I into products is much smaller than that for its decay into reactants, $k_b \ll k_a'$.

Self-test 21.10 Show that the pre-equilibrium mechanism in which $A + A \rightleftharpoons I$ (K) followed by $I + B \rightarrow P$ (k_b) results in an overall third-order reaction.
$[d[P]/dt = k_b K[A]^2[B]]$

(e) Kinetic and thermodynamic control of reactions

In some cases reactants can give rise to a variety of products, as in nitrations of mono-substituted benzene, when various proportions of the *ortho*-, *meta*-, and *para*-substituted products are obtained, depending on the directing power of the original substituent. Suppose two products, P_1 and P_2, are produced by the following competing reactions:

$$A + B \rightarrow P_1 \qquad \text{Rate of formation of } P_1 = k_1[A][B]$$

$$A + B \rightarrow P_2 \qquad \text{Rate of formation of } P_2 = k_2[A][B]$$

The relative proportion in which the two products have been produced at a given stage of the reaction (before it has reached equilibrium) is given by the ratio of the two rates, and therefore of the two rate constants:

$$\frac{[P_2]}{[P_1]} = \frac{k_2}{k_1} \tag{21.51}$$

This ratio represents the **kinetic control** over the proportions of products, and is a common feature of the reactions encountered in organic chemistry where reactants are chosen that facilitate pathways favouring the formation of a desired product. If a reaction is allowed to reach equilibrium, then the proportion of products is determined by thermodynamic rather than kinetic considerations, and the ratio of concentrations is controlled by considerations of the standard Gibbs energies of all the reactants and products.

Self-test 21.11 Two products are formed in reactions in which there is kinetic control of the ratio of products. The activation energy for the reaction leading to product 1 is greater than that leading to product 2. Will the ratio of product concentrations $[P_1]/[P_2]$ increase or decrease if the temperature is raised?

[The ratio $[P_1]/[P_2]$ will increase]

Examples of reaction mechanisms

Many reactions take place by mechanisms that involve several elementary steps. Some take place at a useful rate only after absorption of light or if a catalyst is present. In the following sections we begin to see how to develop the ideas introduced so far to deal with these special kinds of reactions. We leave the study of catalysis to Chapter 23 and focus here on the kinetic analysis of a special class of reactions in the gas phase, polymerization kinetics, and photochemical reactions.

21.8 Unimolecular reactions

Key points **(a) The Lindemann–Hinshelwood mechanism and the RRKM model of 'unimolecular' reactions account for the first-order kinetics of gas-phase reactions. (b) The overall activation energy of a reaction with a complex mechanism may be positive or negative.**

A number of gas-phase reactions follow first-order kinetics, as in the isomerization of cyclopropane mentioned earlier:

$$\textit{cyclo}\text{-}C_3H_6 \rightarrow CH_3CH{=}CH_2 \qquad v = k_r[\textit{cyclo}\text{-}C_3H_6] \tag{21.52}$$

The problem with the interpretation of first-order rate laws is that presumably a molecule acquires enough energy to react as a result of its collisions with other molecules. However, collisions are simple bimolecular events, so how can they result in a first-order rate law? First-order gas-phase reactions are widely called 'unimolecular reactions' because they also involve an elementary unimolecular step in which the reactant molecule changes into the product. This term must be used with caution, though, because the overall mechanism has bimolecular as well as unimolecular steps.

(a) The Lindemann–Hinshelwood mechanism

The first successful explanation of unimolecular reactions was provided by Frederick Lindemann in 1921 and then elaborated by Cyril Hinshelwood. In the **Lindemann–Hinshelwood mechanism** it is supposed that a reactant molecule A becomes energetically excited by collision with another A molecule in a bimolecular step (Fig. 21.19):

$$A + A \rightarrow A^* + A \qquad \frac{d[A^*]}{dt} = k_a[A]^2 \tag{21.53}$$

The energized molecule (A*) might lose its excess energy by collision with another molecule:

$$A + A^* \rightarrow A + A \qquad \frac{d[A^*]}{dt} = -k_a'[A][A^*] \tag{21.54}$$

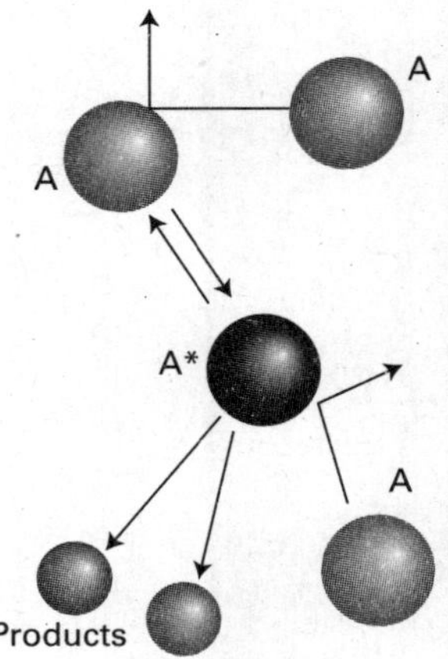

Fig. 21.19 A representation of the Lindemann–Hinshelwood mechanism of unimolecular reactions. The species A is excited by collision with A, and the excited A molecule (A*) may either be deactivated by a collision with A or go on to decay by a unimolecular process to form products.

Alternatively, the excited molecule might shake itself apart and form products P. That is, it might undergo the unimolecular decay

$$A^* \rightarrow P \qquad \frac{d[A^*]}{dt} = -k_b[A^*] \tag{21.55}$$

If the unimolecular step is slow enough to be the rate-determining step, the overall reaction will have first-order kinetics, as observed. This conclusion can be demonstrated explicitly by applying the steady-state approximation to the net rate of formation of A*:

$$\frac{d[A^*]}{dt} = k_a[A]^2 - k_a'[A][A^*] - k_b[A^*] \approx 0 \tag{21.56}$$

This equation solves to

$$[A^*] = \frac{k_a[A]^2}{k_b + k_a'[A]} \tag{21.57}$$

so the rate law for the formation of P is

$$\frac{d[P]}{dt} = k_b[A^*] = \frac{k_a k_b[A]^2}{k_b + k_a'[A]} \tag{21.58}$$

At this stage the rate law is not first-order. However, if the rate of deactivation by (A*,A) collisions is much greater than the rate of unimolecular decay, in the sense that

$$k_a'[A^*][A] \gg k_b[A^*] \qquad \text{or} \qquad k_a'[A] \gg k_b$$

then we can neglect k_b in the denominator and obtain

$$\frac{d[P]}{dt} = k_r[A] \qquad k_r = \frac{k_a k_b}{k_a'} \tag{21.59}$$

Equation 21.59 is a first-order rate law, as we set out to show.

The Lindemann–Hinshelwood mechanism can be tested because it predicts that, as the concentration (and therefore the partial pressure) of A is reduced, the reaction should switch to overall second-order kinetics. Thus, when $k_a'[A] \ll k_b$, the rate law in eqn 21.58 is

$$\frac{d[P]}{dt} \approx k_a[A]^2 \tag{21.60}$$

The physical reason for the change of order is that at low pressures the rate-determining step is the bimolecular formation of A*. If we write the full rate law in eqn 21.58 as

$$\frac{d[P]}{dt} = k_r[A] \qquad k_r = \frac{k_a k_b[A]}{k_b + k_a'[A]} \tag{21.61}$$

then the expression for the effective rate constant, k_r, can be rearranged to

$$\frac{1}{k_r} = \frac{k_a'}{k_a k_b} + \frac{1}{k_a[A]} \tag{21.62}$$

Hence, a test of the theory is to plot $1/k_r$ against $1/[A]$, and to expect a straight line. This behaviour is observed often at low concentrations but deviations are common at high concentrations. In Chapter 22 we develop the description of the mechanism to take into account experimental results over a range of concentrations and pressures.

(b) The activation energy of a composite reaction

Although the rate of each step of a complex mechanism might increase with temperature and show Arrhenius behaviour, is that true of a composite reaction? To answer this question, we consider the high-pressure limit of the Lindemann–Hinshelwood mechanism as expressed in eqn 21.59. If each of the rate constants has an Arrhenius-like temperature dependence, we can use eqn 21.31 for each of them, and write

$$k_r = \frac{k_a k_b}{k'_a} = \frac{(A_a e^{-E_a(a)/RT})(A_b e^{-E_a(b)/RT})}{(A'_a e^{-E'_a(a)/RT})} \tag{21.63}$$

$$= \frac{A_a A_b}{A'_a} e^{-\{E_a(a)+E_a(b)-E'_a(a)\}/RT}$$

That is, the composite rate constant k has an Arrhenius-like form with activation energy

$$E_a = E_a(a) + E_a(b) - E'_a(a) \tag{21.64}$$

Provided $E_a(a) + E_a(b) > E'_a(a)$, the activation energy is positive and the rate increases with temperature. However, it is conceivable that $E_a(a) + E_a(b) < E'_a(a)$ (Fig. 21.20), in which case the activation energy is negative and the rate will *decrease* as the temperature is raised. There is nothing remarkable about this behaviour: all it means is that the reverse reaction (corresponding to the deactivation of A*) is so sensitive to temperature that its rate increases sharply as the temperature is raised, and depletes the steady-state concentration of A*. The Lindemann–Hinshelwood mechanism is an unlikely candidate for this type of behaviour because the deactivation of A* has only a small activation energy, but there are reactions with analogous mechanisms in which a negative activation energy is observed.

When we examine the general rate law given in eqn 21.58, it is clear that the temperature dependence may be difficult to predict because each rate constant in the expression for k_r increases with temperature, and the outcome depends on whether the terms in the numerator dominate those in the denominator, or vice versa. The fact that so many reactions do show Arrhenius-like behaviour with positive activation energies suggests that their rate laws are in a 'simple' regime, like eqn 21.60 rather than eqn 21.58, and that the temperature dependence is dominated by the activation energy of the rate-determining stage.

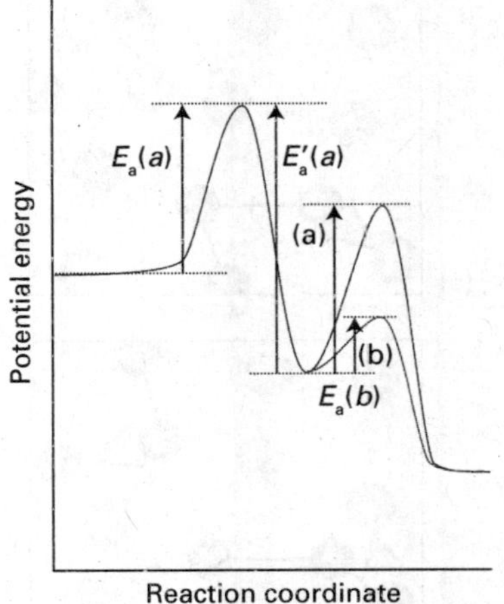

Fig. 21.20 For a reaction with a pre-equilibrium, there are three activation energies to take into account, two referring to the reversible steps of the pre-equilibrium and one for the final step. The relative magnitudes of the activation energies determine whether the overall activation energy is (a) positive or (b) negative.

21.9 Polymerization kinetics

Key points (a) In stepwise polymerization any two monomers in the reaction mixture can link together at any time. The longer a stepwise polymerization proceeds, the higher the average molar mass of the product. (b) In chain polymerization an activated monomer attacks another monomer and links to it. The slower the initiation of the chain, the higher the average molar mass of the polymer.

There are two major classes of polymerization processes and the average molar mass of the product varies with time in distinctive ways. In **stepwise polymerization** any two monomers present in the reaction mixture can link together at any time and growth of the polymer is not confined to chains that are already forming (Fig. 21.21). As a result, monomers are consumed early in the reaction and, as we shall see, the average molar mass of the product grows with time. In **chain polymerization** an activated monomer, M, attacks another monomer, links to it, then that unit attacks another monomer, and so on. The monomer is used up as it becomes linked to the growing chains (Fig. 21.22). High polymers are formed rapidly and only the yield, not the average molar mass, of the polymer is increased by allowing long reaction times.

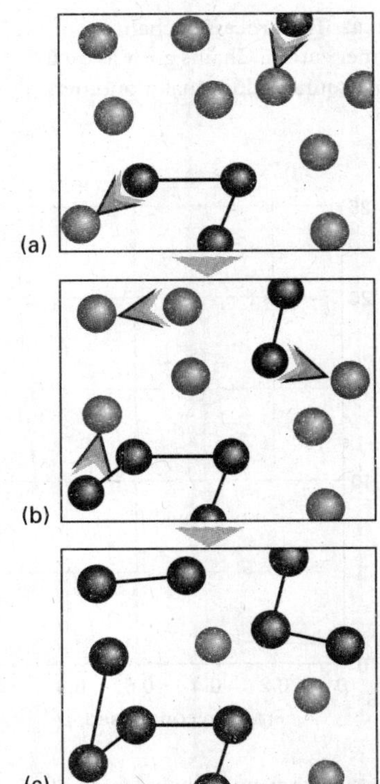

Fig. 21.21 In stepwise polymerization, growth can start at any pair of monomers, and so new chains begin to form throughout the reaction.

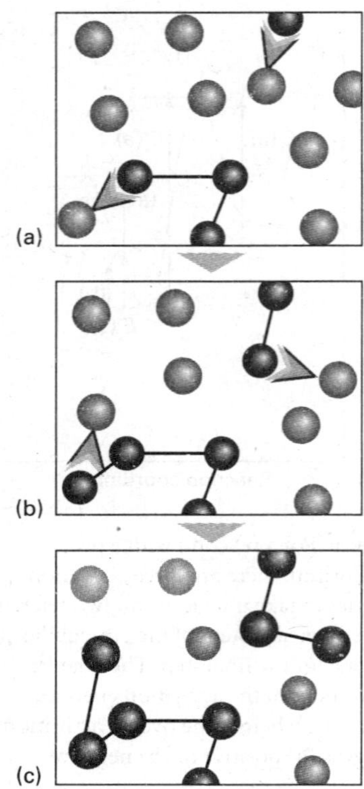

Fig. 21.22 The process of chain polymerization. Chains grow as each chain acquires additional monomers.

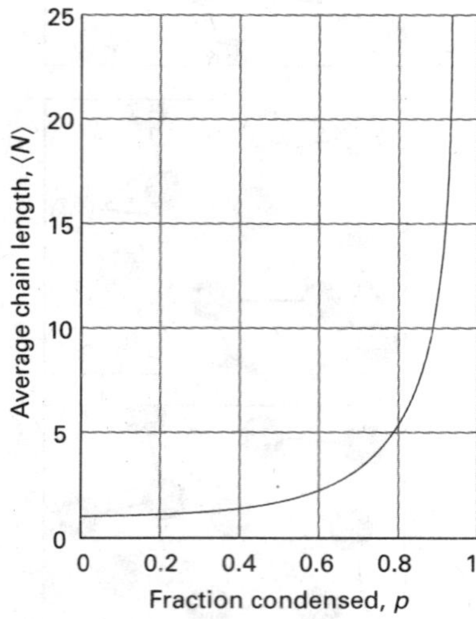

Fig. 21.23 The average chain length of a polymer as a function of the fraction of reacted monomers, p. Note that p must be very close to 1 for the chains to be long.

interActivity Plot the variation of p with time for a range of k_r values of your choosing (take $[A]_0 = 1.0\ mol\ dm^{-3}$).

(a) Stepwise polymerization

Stepwise polymerization commonly proceeds by a condensation reaction, in which a small molecule (typically H_2O) is eliminated in each step. Stepwise polymerization is the mechanism of production of polyamides, as in the formation of nylon-66:

$$\begin{aligned} &H_2N(CH_2)_6NH_2 + HOOC(CH_2)_4COOH \\ &\quad \rightarrow H_2N(CH_2)_6NHCO(CH_2)_4COOH + H_2O \\ &\quad \rightarrow H\text{–}[NH(CH_2)_6NHCO(CH_2)_4CO]_n\text{–}OH \end{aligned}$$

Polyesters and polyurethanes are formed similarly (the latter without elimination). A polyester, for example, can be regarded as the outcome of the stepwise condensation of a hydroxyacid HO–R–COOH. We shall consider the formation of a polyester from such a monomer, and measure its progress in terms of the concentration of the –COOH groups in the sample (which we denote A), for these groups gradually disappear as the condensation proceeds. Because the condensation reaction can occur between molecules containing any number of monomer units, chains of many different lengths can grow in the reaction mixture.

In the absence of a catalyst, we can expect the condensation to be overall second-order in the concentration of the –OH and –COOH (or A) groups, and write

$$\frac{d[A]}{dt} = -k_r[OH][A] \tag{21.65a}$$

However, because there is one –OH group for each –COOH group, this equation is the same as

$$\frac{d[A]}{dt} = -k_r[A]^2 \tag{21.65b}$$

If we assume that the rate constant for the condensation is independent of the chain length, then k_r remains constant throughout the reaction. The solution of this rate law is given by eqn 21.15c, and is

$$[A] = \frac{[A]_0}{1 + k_r t[A]_0} \tag{21.66}$$

The fraction, p, of –COOH groups that have condensed at time t is, after application of eqn 21.66:

$$p = \frac{[A]_0 - [A]}{[A]_0} = \frac{k_r t[A]_0}{1 + k_r t[A]_0} \qquad \text{Fraction of condensed groups} \tag{21.67}$$

Next, we calculate the **degree of polymerization**, which is defined as the average number of monomer residues per polymer molecule. This quantity is the ratio of the initial concentration of A, $[A]_0$, to the concentration of end groups, [A], at the time of interest, because there is one A group per polymer molecule. For example, if there were initially 1000 A groups and there are now only 10, each polymer must be 100 units long on average. Because we can express [A] in terms of p (eqn 21.67), the average number of monomers per polymer molecule, $\langle N \rangle$, is

$$\langle N \rangle = \frac{[A]_0}{[A]} = \frac{1}{1-p} \qquad \text{Degree of polymerization} \tag{21.68a}$$

This result is illustrated in Fig. 21.23. When we express p in terms of the rate constant k_r (eqn 21.67), we find

$$\langle N\rangle = 1 + k_r t[A]_0 \qquad (21.68b)$$

Degree of polymerization in terms of the rate constant

The average length grows linearly with time. Therefore, the longer a stepwise polymerization proceeds, the higher the average molar mass of the product.

(b) Chain polymerization

Many gas-phase reactions and liquid-phase polymerization reactions are **chain reactions.** In a chain reaction, a reaction intermediate produced in one step generates an intermediate in a subsequent step, then that intermediate generates another intermediate, and so on. The intermediates in a chain reaction are called **chain carriers.** In a **radical chain reaction** the chain carriers are radicals (species with unpaired electrons).

Chain polymerization occurs by addition of monomers to a growing polymer, often by a radical chain process. It results in the rapid growth of an individual polymer chain for each activated monomer. Examples include the addition polymerizations of ethene, methyl methacrylate, and styrene, as in

$$-CH_2CH_2X\cdot + CH_2{=}CHX \rightarrow -CH_2CHXCH_2CHX\cdot$$

and subsequent reactions. The central feature of the kinetic analysis (which is summarized in the following *Justification*) is that the rate of polymerization is proportional to the square root of the initiator concentration:

$$v = k_r[I]^{1/2}[M] \qquad (21.69)$$

Justification 21.6 *The rate of chain polymerization*

There are three basic types of reaction step in a chain polymerization process:

(a) Initiation:

$$I \rightarrow R\cdot + R\cdot \qquad v_i = k_i[I]$$

$$M + R\cdot \rightarrow \cdot M_1 \qquad \text{(fast)}$$

where I is the initiator, R· the radical I forms, and $\cdot M_1$ a monomer radical. We have shown a reaction in which a radical is produced, but in some polymerizations the initiation step leads to the formation of an ionic chain carrier. The rate-determining step is the formation of the radicals R· by homolysis of the initiator, so the rate of initiation is equal to the v_i given above.

(b) Propagation:

$$M + \cdot M_1 \rightarrow \cdot M_2$$

$$M + \cdot M_2 \rightarrow \cdot M_3$$

$$\vdots$$

$$M + \cdot M_{n-1} \rightarrow \cdot M_n \qquad v_p = k_p[M][\cdot M]$$

If we assume that the rate of propagation is independent of chain size for sufficiently large chains, then we can use only the equation given above to describe the propagation process. Consequently, for sufficiently large chains, the rate of propagation is equal to the overall rate of polymerization.

Because this chain of reactions propagates quickly, the rate at which the total concentration of radicals grows is equal to the rate of the rate-determining initiation step. It follows that

$$\left(\frac{d[\cdot M]}{dt}\right)_{\text{production}} = 2fk_i[I] \qquad (21.70)$$

where f is the fraction of radicals R· that successfully initiate a chain.

(c) Termination:

$$\cdot M_n + \cdot M_m \rightarrow M_{n+m} \qquad \text{(mutual termination)}$$
$$\cdot M_n + \cdot M_m \rightarrow M_n + M_m \qquad \text{(disproportionation)}$$
$$M + \cdot M_n \rightarrow \cdot M + M_n \qquad \text{(chain transfer)}$$

In **mutual termination** two growing radical chains combine. In termination by **disproportionation** a hydrogen atom transfers from one chain to another, corresponding to the oxidation of the donor and the reduction of the acceptor. In **chain transfer**, a new chain initiates at the expense of the one currently growing.

Here we suppose that only mutual termination occurs. If we assume that the rate of termination is independent of the length of the chain, the rate law for termination is

$$v_t = k_t[\cdot M]^2$$

and the rate of change of radical concentration by this process is

$$\left(\frac{d[\cdot M]}{dt}\right)_{\text{depletion}} = -2k_t[\cdot M]^2$$

The steady-state approximation gives:

$$\frac{d[\cdot M]}{dt} = 2fk_i[I] - 2k_t[\cdot M]^2 = 0$$

The steady-state concentration of radical chains is therefore

$$[\cdot M] = \left(\frac{fk_i}{k_t}\right)^{1/2}[I]^{1/2} \qquad (21.71)$$

Because the rate of propagation of the chains is the negative of the rate at which the monomer is consumed, we can write $v_p = -d[M]/dt$ and

$$v_p = k_p[\cdot M][M] = k_p\left(\frac{fk_i}{k_t}\right)^{1/2}[I]^{1/2}[M] \qquad (21.72)$$

This rate is also the rate of polymerization, which has the form of eqn 21.69.

The **kinetic chain length**, ν, is the ratio of the number of monomer units consumed per activated centre produced in the initiation step:

$$\nu = \frac{\text{number of monomer units consumed}}{\text{number of activated centres produced}} \qquad \text{Definition of the kinetic chain length} \qquad [21.73a]$$

The kinetic chain length can be expressed in terms of the rate expressions in *Justification* 21.6. To do so, we recognize that monomers are consumed at the rate that chains propagate. Then,

$$\nu = \frac{\text{rate of propagation of chains}}{\text{rate of production of radicals}} \qquad \text{Definition of the kinetic chain length in terms of reaction rates} \qquad [21.73b]$$

By making the steady-state approximation, we set the rate of production of radicals equal to the termination rate. Therefore, we can write the expression for the kinetic chain length as

$$\nu = \frac{k_p[\cdot M][M]}{2k_t[\cdot M]^2} = \frac{k_p[M]}{2k_t[\cdot M]}$$

When we substitute the steady-state expression, eqn 21.71, for the radical concentration, we obtain

$$\nu = k_r[M][I]^{-1/2} \qquad k_r = \tfrac{1}{2}k_p(fk_ik_t)^{-1/2} \tag{21.74}$$

Consider a polymer produced by a chain mechanism with mutual termination. In this case, the average number of monomers in a polymer molecule, $\langle N \rangle$, produced by the reaction is the sum of the numbers in the two combining polymer chains. The average number of units in each chain is ν. Therefore,

$$\langle N \rangle = 2\nu = 2k_r[M][I]^{-1/2} \tag{21.75}$$

Degree of polymerization in a chain process

with k_r given in eqn 21.74. We see that, the slower the initiation of the chain (the smaller the initiator concentration and the smaller the initiation rate constant), the greater the kinetic chain length, and therefore the higher the average molar mass of the polymer. Some of the consequences of molar mass for polymers were explored in Chapter 18: now we have seen how we can exercise kinetic control over them.

21.10 Photochemistry

Key points **(a) The primary quantum yield of a photochemical reaction is the number of reactant molecules producing specified primary products for each photon absorbed. (b) The observed lifetime of an excited state is related to the quantum yield and rate constant of emission. (c) A Stern–Volmer plot is used to analyse the kinetics of fluorescence quenching in solution. Collisional deactivation, electron transfer, and resonance energy transfer are common fluorescence quenching processes. (d) The efficiency of resonance energy transfer decreases with increasing separation between donor and acceptor molecules.**

Many reactions can be initiated by the absorption of electromagnetic radiation by one of the mechanisms described in Chapter 13. The most important of all are the photochemical processes that capture the radiant energy of the Sun. Some of these reactions lead to the heating of the atmosphere during the daytime by absorption of ultraviolet radiation. Others include the absorption of visible radiation during photosynthesis (*Impact I21.1*). Without photochemical processes, the Earth would be simply a warm, sterile, rock. Table 21.5 summarizes common photochemical reactions.

Photochemical processes are initiated by the absorption of radiation by at least one component of a reaction mixture. In a **primary process**, products are formed directly from the excited state of a reactant. Examples include fluorescence (Section 13.4) and the *cis–trans* photoisomerization of retinal (Table 21.5, see also *Impact I13.1*). Products of a **secondary process** originate from intermediates that are formed directly from the excited state of a reactant.

Competing with the formation of photochemical products is a host of primary photophysical processes that can deactivate the excited state (Table 21.6). Therefore, it is important to consider the timescales of excited state formation and decay before describing the mechanisms of photochemical reactions. Electronic transitions caused by absorption of ultraviolet and visible radiation occur within 10^{-16}–10^{-15} s. We expect, then, that the upper limit for the rate constant of a first-order photochemical reaction is about 10^{16} s^{-1}. Fluorescence is slower than absorption, with typical lifetimes of 10^{-12}–10^{-6} s. Therefore, the excited singlet state can initiate very fast photochemical reactions in the femtosecond (10^{-15} s) to picosecond (10^{-12} s) timescale. Examples of such ultrafast reactions are the initial events of vision (*Impact I13.1*) and of photosynthesis. Typical intersystem crossing (ISC) and phosphorescence times

Table 21.5 Examples of photochemical processes

Process	General form	Example
Ionization	$A^* \rightarrow A^+ + e^-$	$NO^* \xrightarrow{134\ nm} NO^+ + e^-$
Electron transfer	$A^* + B \rightarrow A^+ + B^-$ or $A^- + B^+$	$[Ru(bpy)_3^{2+}]^* + Fe^{3+} \xrightarrow{452\ nm} Ru(bpy)_3^{3+} + Fe^{2+}$
Dissociation	$A^* \rightarrow B + C$	$O_3^* \xrightarrow{1180\ nm} O_2 + O$
	$A^* + B\text{–}C \rightarrow A + B + C$	$Hg^*\ CH_4 \xrightarrow{254\ nm} Hg + CH_3 + H$
Addition	$2\ A^* \rightarrow B$	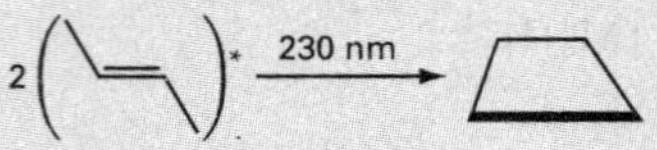 2 ( )* 230 nm →
	$A^* + B \rightarrow AB$	
Abstraction	$A^* + B\text{–}C \rightarrow A\text{–}B + C$	$Hg^* + H_2 \xrightarrow{254\ nm} HgH + H$
Isomerization or rearrangement	$A^* \rightarrow A'$	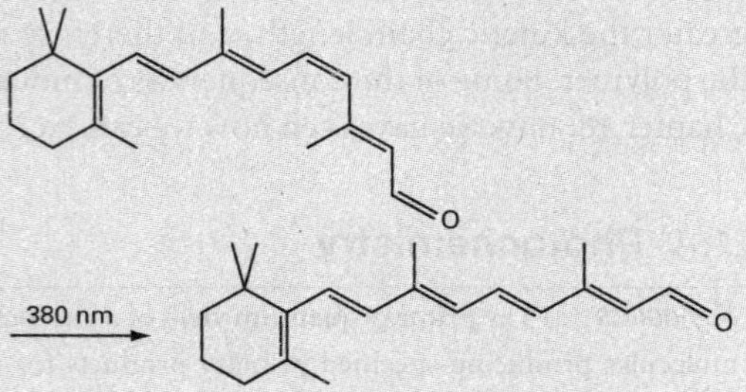 380 nm

* Excited state.

Table 21.6 Common photophysical processes†

Primary absorption	$S + h\nu \rightarrow S^*$
Excited-state absorption	$S^* + h\nu \rightarrow S^{**}$ $T^* + h\nu \rightarrow T^{**}$
Fluorescence	$S^* \rightarrow S + h\nu$
Stimulated emission	$S^* + h\nu \rightarrow S + 2h\nu$
Intersystem crossing (ISC)	$S^* \rightarrow T^*$
Phosphorescence	$T^* \rightarrow S + h\nu$
Internal conversion (IC)	$S^* \rightarrow S$
Collision-induced emission	$S^* + M \rightarrow S + M + h\nu$
Collisional deactivation	$S^* + M \rightarrow S + M$ $T^* + M \rightarrow S + M$
Electronic energy transfer:	
Singlet–singlet	$S^* + S \rightarrow S + S^*$
Triplet–triplet	$T^* + T \rightarrow T + T^*$
Excimer formation	$S^* + S \rightarrow (SS)^*$
Energy pooling	
Singlet–singlet	$S^* + S^* \rightarrow S^{**} + S$
Triplet–triplet	$T^* + T^* \rightarrow S^* + S$

† S denotes a singlet state, T a triplet state, and M is a third-body.

for large organic molecules are 10^{-12}–10^{-4} s and 10^{-6}–10^{-1} s, respectively. As a consequence, excited triplet states are photochemically important. Indeed, because phosphorescence decay is several orders of magnitude slower than most typical reactions, species in excited triplet states can undergo a very large number of collisions with other reactants before deactivation.

(a) The primary quantum yield

We shall see that the rates of deactivation of the excited state by radiative, non-radiative, and chemical processes determine the yield of product in a photochemical reaction. The **primary quantum yield**, ϕ, is defined as the number of photophysical or photochemical events that lead to primary products divided by the number of photons absorbed by the molecule in the same interval:

$$\phi = \frac{\text{number of events}}{\text{number of photons absorbed}} \qquad \text{Definition of the primary quantum yield} \qquad [21.76a]$$

When we divide both the numerator and denominator of this expression by the time interval over which the events occurred, we see that the primary quantum yield is also the rate of radiation-induced primary events divided by the rate of photon absorption, I_{abs}:

$$\phi = \frac{\text{rate of process}}{\text{intensity of light absorbed}} = \frac{v}{I_{abs}} \qquad \text{Definition of the primary quantum yield in terms of rates of processes} \qquad [21.76b]$$

A molecule in an excited state must either decay to the ground state or form a photochemical product. Therefore, the total number of molecules deactivated by radiative processes, non-radiative processes, and photochemical reactions must be equal to the number of excited species produced by absorption of light. We conclude that the sum of primary quantum yields ϕ_i for *all* photophysical and photochemical events i must be equal to 1, regardless of the number of reactions involving the excited state. It follows that

$$\sum_i \phi_i = \sum_i \frac{v_i}{I_{abs}} = 1 \qquad (21.77)$$

It follows that for an excited singlet state that decays to the ground state only via the photophysical processes described earlier in this section, we write

$$\phi_f + \phi_{IC} + \phi_p = 1$$

where ϕ_f, ϕ_{IC}, and ϕ_p are the quantum yields of fluorescence, internal conversion, and phosphorescence, respectively (intersystem crossing from the singlet to the triplet state is taken into account with the measurement of ϕ_p). The quantum yield of photon emission by fluorescence and phosphorescence is $\phi_{emission} = \phi_f + \phi_p$, which is less than 1. If the excited singlet state also participates in a primary photochemical reaction with quantum yield ϕ_r, we write

$$\phi_f + \phi_{IC} + \phi_p + \phi_r = 1$$

We can now strengthen the link between reaction rates and primary quantum yield already established by eqns 21.76 and 21.77. By taking the constant I_{abs} out of the summation in eqn 21.77 and rearranging, we obtain $I_{abs} = \sum_i v_i$. Substituting this result into eqn 21.76b gives the general result

$$\phi = \frac{v}{\sum_i v_i} \qquad (21.78)$$

Therefore, the primary quantum yield may be determined directly from the experimental rates of *all* photophysical and photochemical processes that deactivate the excited state.

(b) Mechanism of decay of excited singlet states

Consider the formation and decay of an excited singlet state in the absence of a chemical reaction:

Absorption:	$S + h\nu_i \rightarrow S^*$	$v_{abs} = I_{abs}$
Fluorescence:	$S^* \rightarrow S + h\nu_f$	$v_f = k_f[S^*]$
Internal conversion:	$S^* \rightarrow S$	$v_{IC} = k_{IC}[S^*]$
Intersystem crossing:	$S^* \rightarrow T^*$	$v_{ISC} = k_{ISC}[S^*]$

in which S is an absorbing species, S* an excited singlet state, T* an excited triplet state, and $h\nu_i$ and $h\nu_f$ are the energies of the incident and fluorescent photons, respectively. From the methods developed earlier in this chapter and the rates of the steps that form and destroy the excited singlet state S*, we write the rate of formation and decay of S* as:

Rate of formation of $[S^*] = I_{abs}$

Rate of decay of $[S^*] = -k_f[S^*] - k_{ISC}[S^*] - k_{IC}[S^*] = -(k_f + k_{ISC} + k_{IC})[S^*]$

It follows that the excited state decays by a first-order process so, when the light is turned off, the concentration of S* varies with time t as:

$$[S^*]_t = [S^*]_0 e^{-t/\tau_0} \qquad (21.79)$$

where the **observed lifetime**, τ_0, of the first excited singlet state is defined as:

$$\tau_0 = \frac{1}{k_f + k_{ISC} + k_{IC}} \qquad [21.80]$$

Definition of the observed lifetime of the excited singlet state

We show in the following *Justification* that the quantum yield of fluorescence is

$$\phi_f = \frac{k_f}{k_f + k_{ISC} + k_{IC}} \qquad (21.81)$$

Quantum yield of fluorescence

Justification 21.7 *The quantum yield of fluorescence*

Most fluorescence measurements are conducted by illuminating a relatively dilute sample with a continuous and intense beam of light. It follows that [S*] is small and constant, so we may invoke the steady-state approximation (Section 21.7) and write:

$$\frac{d[S^*]}{dt} = I_{abs} - k_f[S^*] - k_{ISC}[S^*] - k_{IC}[S^*] = I_{abs} - (k_f + k_{ISC} + k_{IC})[S^*] = 0$$

Consequently,

$$I_{abs} = (k_f + k_{ISC} + k_{IC})[S^*]$$

By using this expression and eqn 21.76b, the quantum yield of fluorescence is written as:

$$\phi_f = \frac{v_f}{I_{abs}} = \frac{k_f[S^*]}{(k_f + k_{ISC} + k_{IC})[S^*]}$$

which, by cancelling the [S*], simplifies to eqn 21.81.

The observed fluorescence lifetime can be measured by using a pulsed laser technique (Section 21.1). First, the sample is excited with a short light pulse from a laser using a wavelength at which S absorbs strongly. Then, the exponential decay of the fluorescence intensity after the pulse is monitored. From eqns 21.80 and 21.81, it follows that

$$\tau_0 = \frac{1}{k_f + k_{ISC} + k_{IC}} = \left(\frac{k_f}{k_f + k_{ISC} + k_{IC}}\right) \times \frac{1}{k_f} = \frac{\phi_f}{k_f} \qquad (21.82)$$

• **A brief illustration**

In water, the fluorescence quantum yield and observed fluorescence lifetime of tryptophan are $\phi_f = 0.20$ and $\tau_0 = 2.6$ ns, respectively. It follows from eqn 21.82 that the fluorescence rate constant k_f is

$$k_f = \frac{\phi_f}{\tau_0} = \frac{0.20}{2.6 \times 10^{-9}\ \text{s}} = 7.7 \times 10^7\ \text{s}^{-1}$$ •

(c) Quenching

The shortening of the lifetime of the excited state by the presence of another species is called **quenching**. Quenching may be either a desired process, such as in energy or electron transfer, or an undesired side reaction that can decrease the quantum yield of a desired photochemical process. Quenching effects may be studied by monitoring the emission from the excited state that is involved in the photochemical reaction.

The addition of a quencher, Q, opens an additional channel for deactivation of S*:

Quenching: $\quad S^* + Q \rightarrow S + Q \qquad v_Q = k_Q[Q][S^*]$

The **Stern–Volmer equation**, which is derived in the following *Justification*, relates the fluorescence quantum yields $\phi_{f,0}$ and ϕ_f measured in the absence and presence, respectively, of a quencher Q at a molar concentration [Q]:

$$\frac{\phi_{f,0}}{\phi_f} = 1 + \tau_0 k_Q[Q] \quad \text{Stern–Volmer equation} \quad (21.83)$$

This equation tells us that a plot of $\phi_{f,0}/\phi_f$ against [Q] should be a straight line with slope $\tau_0 k_Q$. Such a plot is called a **Stern–Volmer plot** (Fig. 21.24). The method may also be applied to the quenching of phosphorescence.

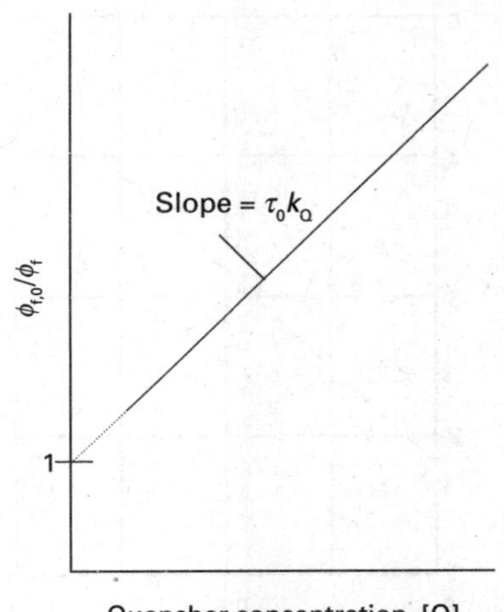

Fig. 21.24 The format of a Stern–Volmer plot and the interpretation of the slope in terms of the rate constant for quenching and the observed fluorescence lifetime in the absence of quenching.

Justification 21.8 *The Stern–Volmer equation*

With the addition of quenching, the steady-state approximation for [S*] now gives:

$$\frac{d[S^*]}{dt} = I_{abs} - (k_f + k_{IC} + k_{ISC} + k_Q[Q])[S^*] = 0$$

and the fluorescence quantum yield in the presence of the quencher is:

$$\phi_f = \frac{k_f}{k_f + k_{ISC} + k_{IC} + k_Q[Q]}$$

When [Q] = 0, the quantum yield is

$$\phi_{f,0} = \frac{k_f}{k_f + k_{ISC} + k_{IC}}$$

It follows that

$$\begin{aligned}\frac{\phi_{f,0}}{\phi_f} &= \left(\frac{k_f}{k_f + k_{ISC} + k_{IC}}\right) \times \left(\frac{k_f + k_{ISC} + k_{IC} + k_Q[Q]}{k_f}\right) \\ &= \frac{k_f + k_{ISC} + k_{IC} + k_Q[Q]}{k_f + k_{ISC} + k_{IC}} \\ &= 1 + \frac{k_Q}{k_f + k_{ISC} + k_{IC}}[Q]\end{aligned}$$

By using eqn 21.80, this expression simplifies to eqn 21.83.

Because the fluorescence intensity and lifetime are both proportional to the fluorescence quantum yield (specifically, from eqn 21.82, $\tau_0 = \phi_f/k_f$), plots of $I_{f,0}/I_f$ and τ_0/τ (where the subscript 0 indicates a measurement in the absence of quencher) against [Q] should also be linear with the same slope and intercept as those shown for eqn 21.83.

Example 21.9 *Determining the quenching rate constant*

The molecule 2,2′-bipyridine (1, bpy) forms a complex with the Ru^{2+} ion. Ruthenium(II) tris-(2,2′-bipyridyl), $Ru(bpy)_3^{2+}$ (2), has a strong metal-to-ligand charge transfer (MLCT) transition (Section 13.3) at 450 nm. The quenching of the $^*Ru(bpy)_3^{2+}$ excited state by $Fe(OH_2)_6^{3+}$ in acidic solution was monitored by measuring emission lifetimes at 600 nm. Determine the quenching rate constant for this reaction from the following data:

$[Fe(OH_2)_6^{3+}]/(10^{-4}\ mol\ dm^{-3})$	0	1.6	4.7	7	9.4
$\tau/(10^{-7}\ s)$	6	4.05	3.37	2.96	2.17

1 2,2′-Bipyridine (bpy)

2 $[Ru(bpy)_3]^{2+}$

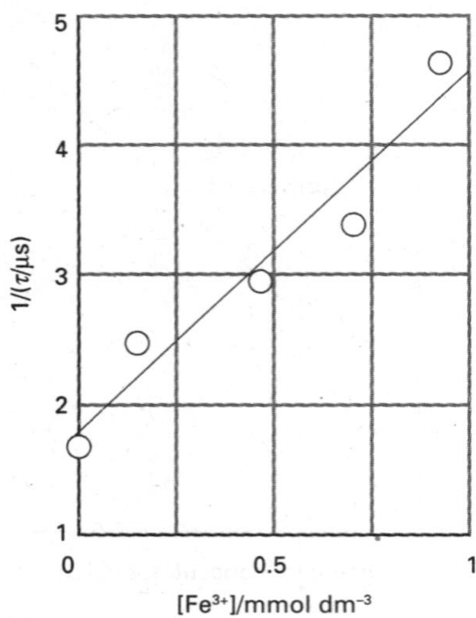

Fig. 21.25 The Stern–Volmer plot of the data for Example 21.9.

Method Re-write the Stern–Volmer equation (eqn 21.83) for use with lifetime data; then fit the data to a straight line.

Answer Upon substitution of τ_0/τ for $\phi_{f,0}/\phi_f$ in eqn 21.83 and after rearrangement, we obtain:

$$\frac{1}{\tau}=\frac{1}{\tau_0}+k_Q[Q] \qquad (21.84)$$

Figure 21.25 shows a plot of $1/\tau$ against $[Fe^{3+}]$ and the results of a fit to eqn 21.84. The slope of the line is 2.8×10^9, so $k_Q = 2.8\times10^9$ dm^3 mol^{-1} s^{-1}. This example shows that measurements of emission lifetimes are preferred because they yield the value of k_Q directly. To determine the value of k_Q from intensity or quantum yield measurements, we need to make an independent measurement of τ_0.

Self-test 21.12 The quenching of tryptophan fluorescence by dissolved O_2 gas was monitored by measuring emission lifetimes at 348 nm in aqueous solutions. Determine the quenching rate constant for this process from the following data:

$[O_2]/(10^{-2}$ mol dm$^{-3})$	0	2.3	5.5	8	10.8
$\tau/(10^{-9}$ s$)$	2.6	1.5	0.92	0.71	0.57

[1.3×10^{10} dm^3 mol^{-1} s^{-1}]

Three common mechanisms for bimolecular quenching of an excited singlet (or triplet) state are:

Collisional deactivation:	$S^* + Q \rightarrow S + Q$
Resonance energy transfer:	$S^* + Q \rightarrow S + Q^*$
Electron transfer:	$S^* + Q \rightarrow S^+ + Q^-$ or $S^- + Q^+$

The quenching rate constant itself does not give much insight into the mechanism of quenching. For the system of Example 21.9, it is known that the quenching of the excited state of $Ru(bpy)_3^{2+}$ is a result of electron transfer to Fe^{3+}, but the quenching data do not allow us to prove the mechanism.

There are, however, some criteria that govern the relative efficiencies of collisional quenching, energy transfer, and electron transfer. Collisional quenching is particularly efficient when Q is a heavy species, such as iodide ion, which receives energy from S* and then decays primarily by internal conversion to the ground state. As we show in detail in Section 22.9, according to the **Marcus theory** of electron transfer, which was proposed by R.A. Marcus in 1965, the rates of electron transfer (from ground or excited states) depend on:

- The distance between the donor and acceptor, with electron transfer becoming more efficient as the distance between donor and acceptor decreases.
- The reaction Gibbs energy, $\Delta_r G$, with electron transfer becoming more efficient as the reaction becomes more exergonic. For example, efficient photooxidation of S requires that the reduction potential of S* be lower than the reduction potential of Q.
- The reorganization energy, the energy cost incurred by molecular rearrangements of donor, acceptor, and medium during electron transfer. The electron transfer rate is predicted to increase as this reorganization energy is matched closely by the reaction Gibbs energy.

Electron transfer can also be studied by time-resolved spectroscopy (Section 21.1). The oxidized and reduced products often have electronic absorption spectra distinct from those of their neutral parent compounds. Therefore, the rapid appearance of such known features in the absorption spectrum after excitation by a laser pulse may be taken as indication of quenching by electron transfer. In the following section we explore energy transfer in detail.

(d) Resonance energy transfer

We visualize the process $S^* + Q \rightarrow S + Q^*$ as follows. The oscillating electric field of the incoming electromagnetic radiation induces an oscillating electric dipole moment in S. Energy is absorbed by S if the frequency of the incident radiation, ν, is such that $\nu = \Delta E_S/h$, where ΔE_S is the energy separation between the ground and excited electronic states of S and h is Planck's constant. This is the 'resonance condition' for absorption of radiation. The oscillating dipole on S now can affect electrons bound to a nearby Q molecule by inducing an oscillating dipole moment in the latter. If the frequency of oscillation of the electric dipole moment in S is such that $\nu = \Delta E_Q/h$ then Q will absorb energy from S.

The efficiency, η_T, of resonance energy transfer is defined as

$$\eta_T = 1 - \frac{\phi_f}{\phi_{f,0}} \qquad [21.85]$$

Definition of the efficiency of resonance energy transfer

According to the **Förster theory** of resonance energy transfer, energy transfer is efficient when:

- The energy donor and acceptor are separated by a short distance (of the order of nanometres).
- Photons emitted by the excited state of the donor can be absorbed directly by the acceptor.

For donor–acceptor systems that are held rigidly either by covalent bonds or by a protein 'scaffold', η_T increases with decreasing distance, R, according to

$$\eta_T = \frac{R_0^6}{R_0^6 + R^6} \qquad (21.86)$$

Efficiency of energy transfer in terms of the donor–acceptor distance

where R_0 is a parameter (with units of distance) that is characteristic of each donor–acceptor pair.[1] Equation 21.86 has been verified experimentally and values of R_0 are available for a number of donor–acceptor pairs (Table 21.7).

The emission and absorption spectra of molecules span a range of wavelengths, so the second requirement of the Förster theory is met when the emission spectrum of the donor molecule overlaps significantly with the absorption spectrum of the acceptor. In the overlap region, photons emitted by the donor have the proper energy to be absorbed by the acceptor (Fig. 21.26).

In many cases, it is possible to prove that energy transfer is the predominant mechanism of quenching if the excited state of the acceptor fluoresces or phosphoresces at a characteristic wavelength. In a pulsed laser experiment, the rise in fluorescence intensity from Q^* with a characteristic time that is the same as that for the decay of the fluorescence of S^* is often taken as indication of energy transfer from S to Q.

Equation 21.86 forms the basis of **fluorescence resonance energy transfer** (FRET), in which the dependence of the energy transfer efficiency, η_T, on the distance, R,

1 See our *Quanta, matter, and change* (2009) for a justification of eqn 21.86.

Table 21.7 Values of R_0 for some donor–acceptor pairs*

Donor†	Acceptor	R_0/nm
Naphthalene	Dansyl	2.2
Dansyl	ODR	4.3
Pyrene	Coumarin	3.9
IEDANS	FITC	4.9
Tryptophan	IEDANS	2.2
Tryptophan	Haem (heme)	2.9

* Additional values may be found in J.R. Lacowicz, *Principles of fluorescence spectroscopy*, Kluwer Academic/Plenum, New York (1999).

† Abbreviations:
Dansyl: 5-dimethylamino-1-naphthalenesulfonic acid;
FITC: fluorescein 5-isothiocyanate;
IEDANS: 5-((((2-iodoacetyl)amino)ethyl)amino)naphthalene-1-sulfonic acid;
ODR: octadecyl-rhodamine.

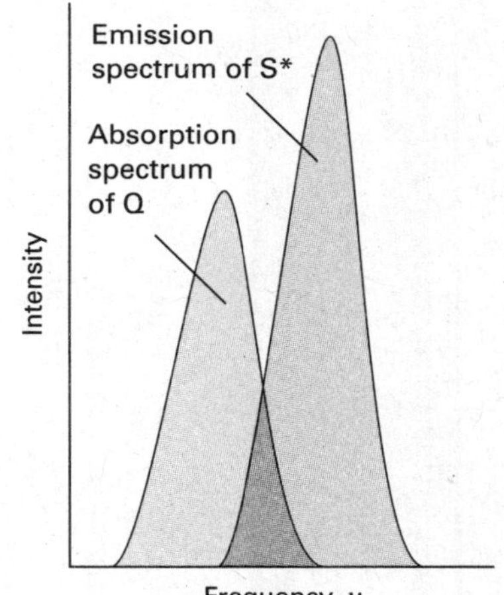

Fig. 21.26 According to the Förster theory, the rate of energy transfer from a molecule S^* in an excited state to a quencher molecule Q is optimized at radiation frequencies in which the emission spectrum of S^* overlaps with the absorption spectrum of Q, as shown in the shaded region.

between energy donor and acceptor can be used to measure distances in biological systems. In a typical FRET experiment, a site on a biopolymer or membrane is labelled covalently with an energy donor and another site is labelled covalently with an energy acceptor. In certain cases, the donor or acceptor may be natural constituents of the system, such as amino acid groups, co-factors, or enzyme substrates. The distance between the labels is then calculated from the known value of R_0 and eqn 21.86. Several tests have shown that the FRET technique is useful for measuring distances ranging from 1 to 9 nm.

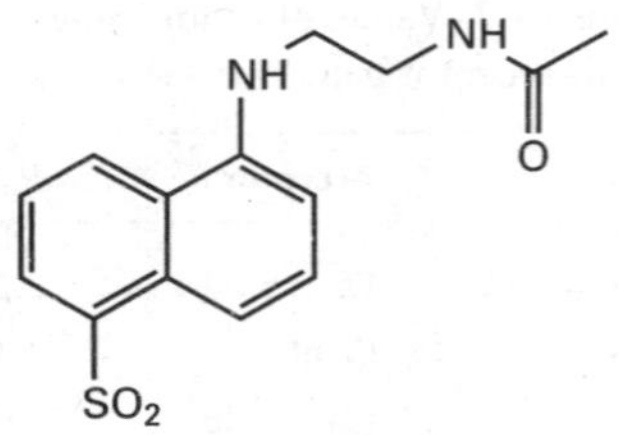

NH NH O SO2

3 1.5-I AEDANS

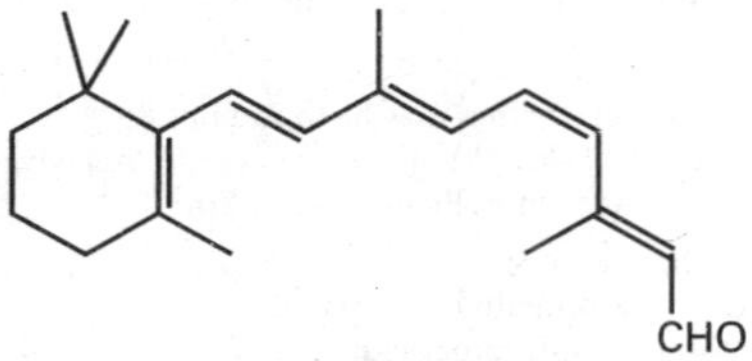

CHO

4 11-*cis*-Retinal

• A brief illustration

As an illustration of the FRET technique, consider a study of the protein rhodopsin (*Impact I13.1*). When an amino acid on the surface of rhodopsin was labelled covalently with the energy donor 1.5-I AEDANS (**3**), the fluorescence quantum yield of the label decreased from 0.75 to 0.68 due to quenching by the visual pigment 11-*cis*-retinal (**4**). From eqn 21.85, we calculate $\eta_T = 1 - (0.68/0.75) = 0.093$ and from eqn 21.86 and the known value of $R_0 = 5.4$ nm for the 1.5-I AEDANS/11-*cis*-retinal pair we calculate $R = 7.9$ nm. Therefore, we take 7.9 nm to be the distance between the surface of the protein and 11-*cis*-retinal. •

If donor and acceptor molecules diffuse in solution or in the gas phase, Förster theory predicts that the efficiency of quenching by energy transfer increases as the average distance travelled between collisions of donor and acceptor decreases. That is, the quenching efficiency increases with concentration of quencher, as predicted by the Stern–Volmer equation.

IMPACT ON BIOCHEMISTRY

I21.1 Harvesting of light during plant photosynthesis

A large proportion of solar radiation with wavelengths below 400 nm and above 1000 nm is absorbed by atmospheric gases such as ozone and O_2, which absorb ultraviolet radiation, and CO_2 and H_2O, which absorb infrared radiation (*Impact I12.2*). As a result, plants, algae, and some species of bacteria evolved photosynthetic apparatus that captures visible and near-infrared radiation. Plants use radiation in the wavelength range of 400–700 nm to drive the endergonic reduction of CO_2 to glucose, with concomitant oxidation of water to O_2 ($\Delta_r G^{\ominus} = +2880$ kJ mol^{-1}), in essence the reverse of glycolysis and the citric acid cycle (*Impact I6.1*):

$$6\,CO_2(g) + 6\,H_2O(l) \underset{\text{glycolysis and the citric acid cycle}}{\overset{\text{photosynthesis}}{\rightleftarrows}} C_6H_{12}O_6(s) + 6\,O_2(g)$$

Electrons flow from reductant to oxidant via a series of electrochemical reactions that are coupled to the synthesis of ATP. The process takes place in the *chloroplast*, a special organelle of the plant cell, where chlorophylls *a* and *b* (**5**) and carotenoids (of which β-carotene, **6**, is an example) bind to integral proteins called *light-harvesting complexes*, which absorb solar energy and transfer it to protein complexes known as *reaction centres*, where light-induced electron transfer reactions occur. The combination of a light-harvesting complex and a reaction centre complex is called a *photosystem*. Plants have two photosystems that drive the reduction of NADP$^+$ (**7**) by water:

$$2\,H_2O + 2\,NADP^+ \xrightarrow{\text{light, photosystems I and II}} O_2 + 2\,NADPH + 2\,H^+$$

It is clear that energy from light is required to drive this reaction because, in the dark, $E^{\ominus} = -1.135$V and $\Delta_r G^{\ominus} = +438.0$ kJ mol^{-1}.

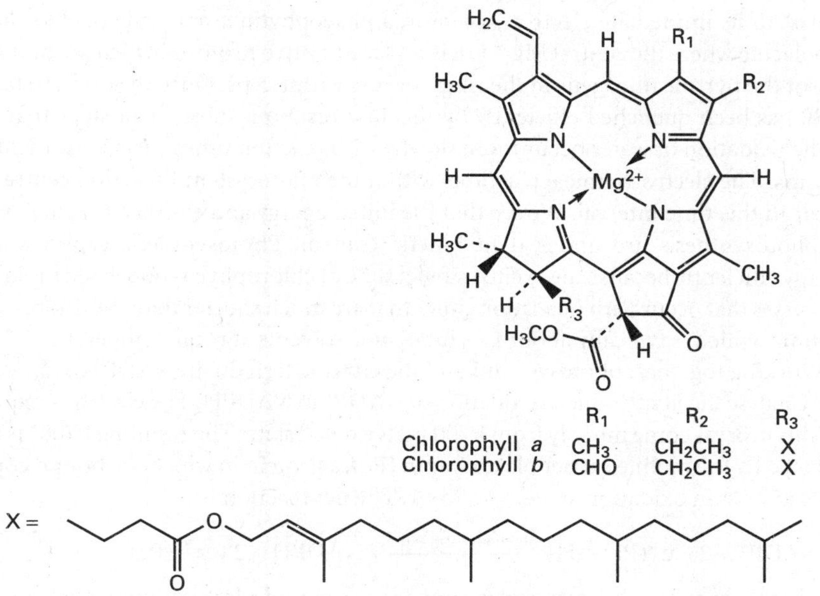

	R_1	R_2	R_3
Chlorophyll *a*	CH_3	CH_2CH_3	X
Chlorophyll *b*	CHO	CH_2CH_3	X

5 Chlorophyll *a* and *b*

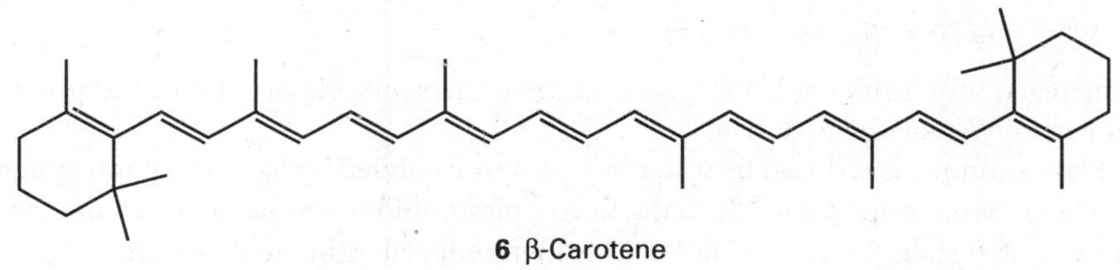

6 β-Carotene

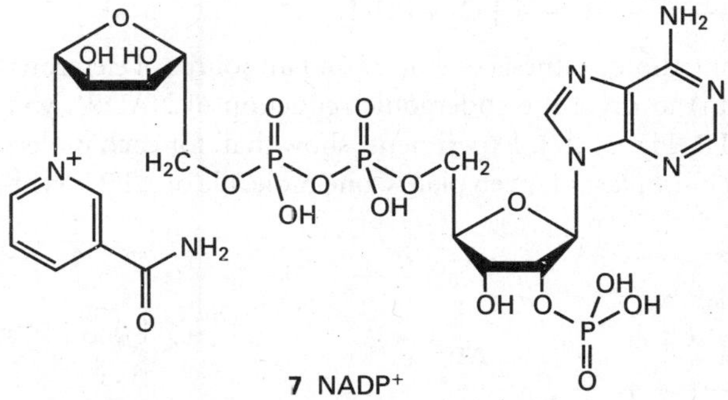

7 $NADP^+$

Light-harvesting complexes bind large numbers of pigments in order to provide a sufficiently large area for capture of radiation. In photosystems I and II, absorption of a photon *raises* a chlorophyll or carotenoid molecule to an excited singlet state and within 0.1–5 ps the energy hops to a nearby pigment via the Förster mechanism. About 100–200 ps later, which corresponds to thousands of hops within the light-harvesting complex, more than 90 per cent of the absorbed energy reaches the reaction centre. There, a chlorophyll *a* dimer becomes electronically excited and initiates ultrafast electron transfer reactions. For example, the transfer of an electron from the excited singlet state of P680, the chlorophyll dimer of the photosystem II reaction centre, to its immediate electron acceptor, a phaeophytin *a* molecule (a chlorophyll *a* molecule where the central Mg^{2+} ion is replaced by two protons, which are bound to two of the pyrrole nitrogens in the ring), occurs within 3 ps. Once the excited state of P680 has been quenched efficiently by this first reaction, subsequent steps that lead to the oxidation of water occur more slowly, with reaction times varying from 200 ps to 1 ms. The electrochemical reactions within the photosystem I reaction centre also occur in this time interval. We see that the initial energy and electron transfer events of photosynthesis are under tight kinetic control. Photosynthesis captures solar energy efficiently because the excited singlet state of chlorophyll is quenched rapidly by processes that occur with relaxation times that are much shorter than the fluorescence lifetime, which is typically about 1 ns in organic solvents at room temperature.

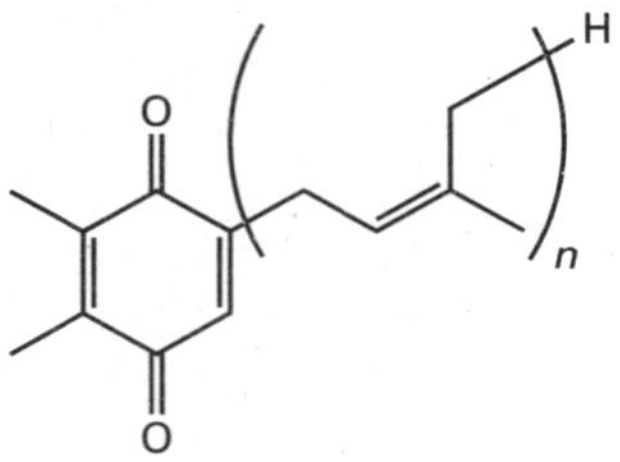

Working together, photosystem I and the enzyme ferredoxin:NADP$^+$ oxidoreductase catalyse the light-induced oxidation of NADP$^+$ to NADPH. The electrons required for this process come initially from P700 in its excited state. The resulting P700$^+$ is then reduced by the mobile carrier plastocyanin (Pc), a protein in which the bound copper ion can exist in oxidation states +2 and +1. The net reaction is

$$\mathrm{NADP^+ + 2\,Cu^+(Pc) + H^+ \xrightarrow{\text{light, photosystem I}} NADPH + 2\,Cu^{2+}(Pc)}$$

8 Plastoquinone

Oxidized plastocyanin accepts electrons from reduced plastoquinone (PQ, **8**). The process is catalysed by the cytochrome b_6f complex, a membrane protein complex:

$$\mathrm{PQH_2 + 2\,Cu^{2+}(Pc) \xrightarrow{\text{cyt}b_6f\text{ complex}} PQ + 2\,H^+ + 2\,Cu^+(Pc)}$$

$$E^{\ominus} = +0.370\ \mathrm{V},\ \Delta_r G^{\ominus} = -71.4\ \mathrm{kJ\ mol^{-1}}$$

This reaction is sufficiently exergonic to drive the synthesis of ATP in the process known as *photophosphorylation*.

Plastoquinone is reduced by water in a process catalysed by light and photosystem II. The electrons required for the reduction of plastoquinone come initially from P680 in its excited state. The resulting P680$^+$ is then reduced ultimately by water. The net reaction is

$$\mathrm{H_2O + PQ \xrightarrow{\text{light, photosystem II}} \tfrac{1}{2}O_2 + PQH_2}$$

In this way, plant photosynthesis uses an abundant source of electrons (water) and of energy (the Sun) to drive the endergonic reduction of NADP$^+$, with concomitant synthesis of ATP (Fig. 21.27). Experiments show that, for each molecule of NADPH formed in the chloroplast of green plants, one molecule of ATP is synthesized.

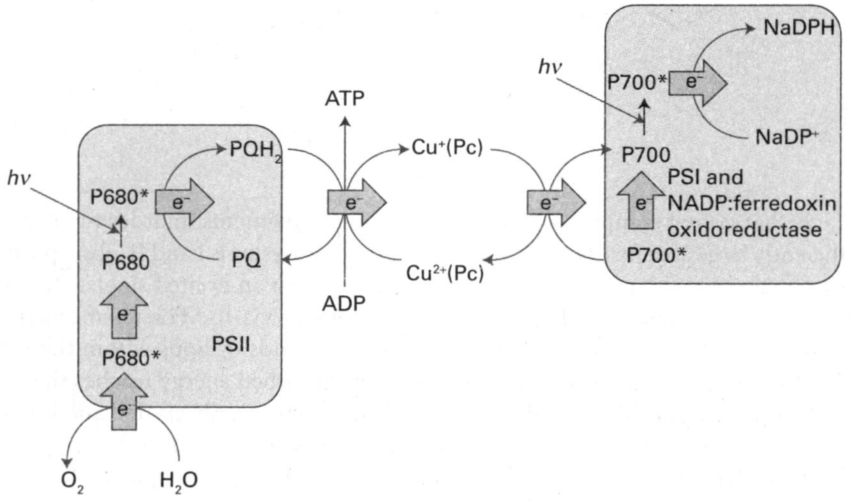

Fig. 21.27 In plant photosynthesis, light-induced electron transfer processes lead to the oxidation of water to O_2 and the reduction of NADP$^+$ to NADPH, with concomitant production of ATP. The energy stored in ATP and NADPH is used to reduce CO_2 to carbohydrate in a separate set of reactions. The scheme summarizes the general patterns of electron flow and does not show all the intermediate electron carriers in photosystems I and II, the cytochrome b_6f complex, and ferredoxin:NADP$^+$ oxidoreductase.

The ATP and NADPH molecules formed by the light-induced electron transfer reactions of plant photosynthesis participate directly in the reduction of CO_2 to glucose in the chloroplast:

$$6\,CO_2 + 12\,NADPH + 12\,ATP + 12\,H^+ \rightarrow C_6H_{12}O_6 + 12\,NADP^+ + 12\,ADP + 12\,P_i + 6\,H_2O$$

In summary, plant photosynthesis uses solar energy to transfer electrons from a poor reductant (water) to carbon dioxide. In the process, high energy molecules (carbohydrates, such as glucose) are synthesized in the cell. Animals feed on the carbohydrates derived from photosynthesis. During aerobic metabolism, the O_2 released by photosynthesis as a waste product is used to oxidize carbohydrates to CO_2, driving biological processes, such as biosynthesis, muscle contraction, cell division, and nerve conduction. Hence, the sustenance of life on Earth depends on a tightly regulated carbon–oxygen cycle that is driven by solar energy.

Checklist of key equations

Property or process	Equation	Comment
Rate of a reaction	$v = (1/V)(\mathrm{d}\xi/\mathrm{d}t),\ \xi = (n_J - n_{J,0})/v_J$	Definition
(a) Integrated rate law, (b) half-life, and (c) time constant of a first-order reaction of type A → P	(a) $\ln([A]/[A]_0) = -k_r t$ or $[A] = [A]_0 e^{-k_r t}$ (b) $t_{1/2} = \ln 2/k_r$ (c) $\tau = 1/k_r$	
(a) Integrated rate law and (b) half-life of a second-order reaction of type A → P	(a) $1/[A] - 1/[A]_0 = k_r t$ or $[A] = [A]_0/(1 + k_r t[A]_0)$ (b) $t_{1/2} = 1/k_r[A]_0$	
Integrated rate law of a second-order reaction of type A + B → P	$\ln\{([B]/[B]_0)/([A]/[A]_0)\} = ([B]_0 - [A]_0)k_r t$	
Half-life of an *n*th order reaction	$t_{1/2} = (2^{n-1} - 1)/(n-1)k_r[A]_0^{n-1}$	
Equilibrium constant in terms of rate constants	$K = (k_a/k_a') \times (k_b/k_b') \times \cdots$	
Relaxation of an equilibrium A ⇌ B after a temperature jump	$x = x_0 e^{-t/\tau} \quad 1/\tau = k_r + k_r'$	First-order in each direction
Arrhenius equation	$\ln k_r = \ln A - E_a/RT$	
Activation energy	$E_a = RT^2(\mathrm{d}\ln k_r/\mathrm{d}T)$	Definition
Degree of polymerization	$\langle N\rangle = 1/(1-p)$ or $\langle N\rangle = 1 + k_r t[A]_0$	Stepwise polymerization
Kinetic chain length	$v = k_r[M][I]^{-1/2} \quad k_r = \tfrac{1}{2}k_p(fk_ik_t)^{-1/2}$	Chain polymerization
Primary quantum yield	$\phi = v/I_{abs}$	
Quantum yield of fluorescence	$\phi_f = k_f/(k_f + k_{ISC} + k_{IC})$	
Observed excited state lifetime	$\tau_0 = \phi_f/k_f$	Absence of quencher
Stern–Volmer equation	$\phi_{f,0}/\phi_f = 1 + \tau_0 k_Q[Q]$	
Efficiency of resonance energy transfer	$\eta_T = 1 - (\phi_f/\phi_{f,0})$ or $\eta_T = R_0^6/(R_0^6 + R^6)$	

Discussion questions

21.1 Consult literature sources and list the observed ranges of timescales during which the following processes occur: radiative decay of excited electronic states, molecular rotational motion, molecular vibrational motion, proton transfer reactions, energy transfer between fluorescent molecules used in FRET analysis, electron transfer events between complex ions in solution, and collisions in liquids.

21.2 Describe the main features, including advantages and disadvantages, of the following experimental methods for determining the rate law of a reaction: the isolation method, the method of initial rates, and fitting data to integrated rate law expressions.

21.3 Why may reaction orders change under different circumstances?

21.4 When might deviations from the Arrhenius law be observed?

21.5 Is it possible for the activation energy of a reaction to be negative? Explain your conclusion and provide a molecular interpretation.

21.6 Assess the validity of the following statement: the rate-determining step is the slowest step in a reaction mechanism.

21.7 Distinguish between kinetic and thermodynamic control of a reaction. Suggest criteria for expecting one rather than the other.

21.8 Bearing in mind distinctions between the mechanisms of stepwise and chain polymerization, describe ways in which it is possible to control the molar mass of a polymer by manipulating the kinetic parameters of polymerization.

21.9 Distinguish between the primary quantum yield and overall quantum yield of a chemical reaction. Describe an experimental procedure for the determination of the quantum yield.

21.10 Summarize the factors that govern the rates of photo-induced electron transfer according to Marcus theory and that govern the rates of resonance energy transfer according to Förster theory. Can you find similarities between the two theories?

Exercises

21.1(a) Predict how the total pressure varies during the gas-phase reaction $2\,ICl(g) + H_2(g) \rightarrow I_2(g) + 2\,HCl(g)$ in a constant-volume container.

21.1(b) Predict how the total pressure varies during the gas-phase reaction $N_2(g) + 3\,H_2(g) \rightarrow 2\,NH_3(g)$ in a constant-volume container.

21.2(a) The rate of the reaction $A + 2\,B \rightarrow 3\,C + D$ was reported as $2.7\ mol\ dm^{-3}\ s^{-1}$. State the rates of formation and consumption of the participants.

21.2(b) The rate of the reaction $A + 3\,B \rightarrow C + 2\,D$ was reported as $2.7\ mol\ dm^{-3}\ s^{-1}$. State the rates of formation and consumption of the participants.

21.3(a) The rate of formation of C in the reaction $2\,A + B \rightarrow 2\,C + 3\,D$ is $2.7\ mol\ dm^{-3}\ s^{-1}$. State the reaction rate, and the rates of formation or consumption of A, C, and D.

21.3(b) The rate of consumption of B in the reaction $A + 3\,B \rightarrow C + 2\,D$ is $2.7\ mol\ dm^{-3}\ s^{-1}$. State the reaction rate, and the rates of formation or consumption of A, B, and D.

21.4(a) The rate law for the reaction in Exercise 21.2a was found to be $v = k_r[A][B]$. What are the units of k_r? Express the rate law in terms of the rates of formation and consumption of (a) A, (b) C.

21.4(b) The rate law for the reaction in Exercise 21.2b was found to be $v = k_r[A][B]^2$. What are the units of k_r? Express the rate law in terms of the rates of formation and consumption of (a) A, (b) C.

21.5(a) The rate law for the reaction in Exercise 21.3a was reported as $d[C]/dt = k_r[A][B][C]$. Express the rate law in terms of the reaction rate; what are the units for k_r in each case?

21.5(b) The rate law for the reaction in Exercise 21.3b was reported as $d[C]/dt = k_r[A][B][C]^{-1}$. Express the rate law in terms of the reaction rate; what are the units for k_r in each case?

21.6(a) If the rate laws are expressed with (a) concentrations in moles per decimetre cubed, (b) pressures in kilopascals, what are the units of the second-order and third-order rate constants?

21.6(b) If the rate laws are expressed with (a) concentrations in molecules per metre cubed, (b) pressures in pascals, what are the units of the second-order and third-order rate constants?

21.7(a) At 518°C, the rate of decomposition of a sample of gaseous acetaldehyde, initially at a pressure of 363 Torr, was $1.07\ Torr\ s^{-1}$ when 5.0 per cent had reacted and $0.76\ Torr\ s^{-1}$ when 20.0 per cent had reacted. Determine the order of the reaction.

21.7(b) At 400 K, the rate of decomposition of a gaseous compound initially at a pressure of 12.6 kPa, was $9.71\ Pa\ s^{-1}$ when 10.0 per cent had reacted and $7.67\ Pa\ s^{-1}$ when 20.0 per cent had reacted. Determine the order of the reaction.

21.8(a) At 518°C, the half-life for the decomposition of a sample of gaseous acetaldehyde (ethanal) initially at 363 Torr was 410 s. When the pressure was 169 Torr, the half-life was 880 s. Determine the order of the reaction.

21.8(b) At 400 K, the half-life for the decomposition of a sample of a gaseous compound initially at 55.5 kPa was 340 s. When the pressure was 28.9 kPa, the half-life was 178 s. Determine the order of the reaction.

21.9(a) The rate constant for the first-order decomposition of N_2O_5 in the reaction $2\,N_2O_5(g) \rightarrow 4\,NO_2(g) + O_2(g)$ is $k_r = 3.38 \times 10^{-5}\ s^{-1}$ at 25°C. What is the half-life of N_2O_5? What will be the pressure, initially 500 Torr, at (a) 50 s, (b) 20 min after initiation of the reaction?

21.9(b) The rate constant for the first-order decomposition of a compound A in the reaction $2\,A \rightarrow P$ is $k_r = 3.56 \times 10^{-7}\ s^{-1}$ at 25°C. What is the half-life of A? What will be the pressure, initially 33.0 kPa at (a) 50 s, (b) 20 min after initiation of the reaction?

21.10(a) A second-order reaction of the type $A + B \rightarrow P$ was carried out in a solution that was initially $0.075\ mol\ dm^{-3}$ in A and $0.050\ mol\ dm^{-3}$ in B. After 1.0 h the concentration of A had fallen to $0.020\ mol\ dm^{-3}$. (a) Calculate the rate constant. (b) What is the half-life of the reactants?

21.10(b) A second-order reaction of the type $A + 2\,B \rightarrow P$ was carried out in a solution that was initially $0.050\ mol\ dm^{-3}$ in A and $0.030\ mol\ dm^{-3}$ in B. After 1.0 h the concentration of A had fallen to $0.010\ mol\ dm^{-3}$. (a) Calculate the rate constant. (b) What is the half-life of the reactants?

21.11(a) The second-order rate constant for the reaction

$$CH_3COOC_2H_5(aq) + OH^-(aq) \rightarrow CH_3CO_2^-(aq) + CH_3CH_2OH(aq)$$

is $0.11\ dm^3\ mol^{-1}\ s^{-1}$. What is the concentration of ester ($CH_3COOC_2H_5$) after (a) 20 s, (b) 15 min when ethyl acetate is added to sodium hydroxide so that the initial concentrations are $[NaOH] = 0.060\ mol\ dm^{-3}$ and $[CH_3COOC_2H_5] = 0.110\ mol\ dm^{-3}$?

21.11(b) The second-order rate constant for the reaction $A + 2\,B \rightarrow C + D$ is $0.34\ dm^3\ mol^{-1}\ s^{-1}$. What is the concentration of C after (a) 20 s, (b) 15 min when the reactants are mixed with initial concentrations of $[A] = 0.027\ mol\ dm^{-1}$ and $[B] = 0.130\ mol\ dm^{-3}$?

21.12(a) A reaction $2\,A \rightarrow P$ has a second-order rate law with $k_r = 4.30 \times 10^{-4}\ dm^3\ mol^{-1}\ s^{-1}$. Calculate the time required for the concentration of A to change from $0.210\ mol\ dm^{-3}$ to $0.010\ mol\ dm^{-3}$.

21.12(b) A reaction $2\,A \rightarrow P$ has a third-order rate law with $k_r = 6.50 \times 10^{-4}\ dm^6\ mol^{-2}\ s^{-1}$. Calculate the time required for the concentration of A to change from $0.067\ mol\ dm^{-3}$ to $0.015\ mol\ dm^{-3}$.

21.13(a) The equilibrium $NH_3(aq) + H_2O(l) \rightleftharpoons NH_4^+(aq) + OH^-(aq)$ at 25°C is subjected to a temperature jump that slightly increases the concentration of $NH_4^+(aq)$ and $OH^-(aq)$. The measured relaxation time is 7.61 ns. The equilibrium constant for the system is 1.78×10^{-5} at 25°C, and the equilibrium concentration of $NH_3(aq)$ is $0.15\ mol\ dm^{-3}$. Calculate the rate constants for the forward and reversed steps.

21.13(b) The equilibrium $A \rightleftharpoons B + C$ at 25°C is subjected to a temperature jump that slightly increases the concentrations of B and C. The measured relaxation time is 3.0 μs. The equilibrium constant for the system is 2.0×10^{-16} at 25°C, and the equilibrium concentrations of B and C at 25°C are both $2.0 \times 10^{-4}\ mol\ dm^{-3}$. Calculate the rate constants for the forward and reverse steps.

21.14(a) The rate constant for the decomposition of a certain substance is $2.80 \times 10^{-3}\ dm^3\ mol^{-1}\ s^{-1}$ at 30°C and $1.38 \times 10^{-2}\ dm^3\ mol^{-1}\ s^{-1}$ at 50°C. Evaluate the Arrhenius parameters of the reaction.

21.14(b) The rate constant for the decomposition of a certain substance is $1.70 \times 10^{-2}\ dm^3\ mol^{-1}\ s^{-1}$ at 24°C and $2.01 \times 10^{-2}\ dm^3\ mol^{-1}\ s^{-1}$ at 37°C. Evaluate the Arrhenius parameters of the reaction.

21.15(a) The rate of a chemical reaction is found to triple when the temperature is raised from 24°C to 49°C. Determine the activation energy.

21.15(b) The rate of a chemical reaction is found to double when the temperature is raised from 25°C to 35°C. Determine the activation energy.

21.16(a) The reaction mechanism for the decomposition of A_2

$A_2 \rightleftharpoons A + A$ (fast)

$A + B \rightarrow P$ (slow)

involves an intermediate A. Deduce the rate law for the reaction in two ways by (a) assuming a pre-equilibrium and (b) making a steady-state approximation.

21.16(b) The reaction mechanism for renaturation of a double helix from its strands A and B:

$A + B \rightleftharpoons$ unstable helix (fast)

Unstable helix $\rightarrow$ stable double helix (slow)

involves an intermediate. Deduce the rate law for the reaction in two ways by (a) assuming a pre-equilibrium and (b) making a steady-state approximation.

21.17(a) The effective rate constant for a gaseous reaction that has a Lindemann–Hinshelwood mechanism is $2.50 \times 10^{-4}\ s^{-1}$ at 1.30 kPa and $2.10 \times 10^{-5}\ s^{-1}$ at 12 Pa. Calculate the rate constant for the activation step in the mechanism.

21.17(b) The effective rate constant for a gaseous reaction that has a Lindemann–Hinshelwood mechanism is $1.7 \times 10^{-3}\ s^{-1}$ at 1.09 kPa and $2.2 \times 10^{-4}\ s^{-1}$ at 25 Pa. Calculate the rate constant for the activation step in the mechanism.

21.18(a) The mechanism of a composite reaction consists of a fast pre-equilibrium step with forward and reverse activation energies of $25\ kJ\ mol^{-1}$ and $38\ kJ\ mol^{-1}$, respectively, followed by an elementary step of activation energy $10\ kJ\ mol^{-1}$. What is the activation energy of the composite reaction?

21.18(b) The mechanism of a composite reaction consists of a fast pre-equilibrium step with forward and reverse activation energies of $27\ kJ\ mol^{-1}$ and $35\ kJ\ mol^{-1}$, respectively, followed by an elementary step of activation energy $15\ kJ\ mol^{-1}$. What is the activation energy of the composite reaction?

21.19(a) Calculate the fraction condensed and the degree of polymerization at $t = 5.00$ h of a polymer formed by a stepwise process with $k_r = 1.39\ dm^3\ mol^{-1}\ s^{-1}$ and an initial monomer concentration of $1.00 \times 10^{-2}\ mol\ dm^{-3}$.

21.19(b) Calculate the fraction condensed and the degree of polymerization at $t = 10.00$ hr of a polymer formed by a stepwise process with $k_r = 2.80 \times 10^{-2}\ dm^3\ mol^{-1}\ s^{-1}$ and an initial monomer concentration of $5.00 \times 10^{-2}\ mol\ dm^{-3}$.

21.20(a) Consider a polymer formed by a chain process. By how much does the kinetic chain length change if the concentration of initiator increases by a factor of 3.6 and the concentration of monomer decreases by a factor of 4.2?

21.20(b) Consider a polymer formed by a chain process. By how much does the kinetic chain length change if the concentration of initiator decreases by a factor of 10.0 and the concentration of increases by a factor of 5.0?

21.21(a) In a photochemical reaction $A \rightarrow 2\,B + C$, the quantum yield with 500 nm light is $2.1 \times 10^2\ mol\ einstein^{-1}$ (1 einstein = 1 mol photons). After exposure of 300 mmol of A to the light, 2.28 mmol of B is formed. How many photons were absorbed by A?

21.21(b) In a photochemical reaction $A \rightarrow B + C$, the quantum yield with 500 nm light is $1.2 \times 10^2\ mol\ einstein^{-1}$. After exposure of 200 mmol A to the light, 1.77 mmol B is formed. How many photons were absorbed by A?

21.22(a) In an experiment to measure the quantum yield of a photochemical reaction, the absorbing substance was exposed to 490 nm light from a 100 W source for 45 min. The intensity of the transmitted light was 40 per cent of the intensity of the incident light. As a result of irradiation, 0.344 mol of the absorbing substance decomposed. Determine the quantum yield.

21.22(b) In an experiment to measure the quantum yield of a photochemical reaction, the absorbing substance was exposed to 320 nm radiation from a 87.5 W source for 28.0 min. The intensity of the transmitted light was 0.257 that of the incident light. As a result of irradiation, 0.324 mol of the absorbing substance decomposed. Determine the quantum yield.

21.23(a) Consider the quenching of an organic fluorescent species with $\tau_0 = 6.0$ ns by a d-metal ion with $k_Q = 3.0 \times 10^8\ dm^3\ mol^{-1}\ s^{-1}$. Predict the concentration of quencher required to decrease the fluorescence intensity of the organic species to 50 per cent of the unquenched value.

21.23(b) Consider the quenching of an organic fluorescent species with $\tau_0 = 3.5$ ns by a d-metal ion with $k_Q = 2.5 \times 10^9\ dm^3\ mol^{-1}\ s^{-1}$. Predict the concentration of quencher required to decrease the fluorescence intensity of the organic species to 75 per cent of the unquenched value.

21.24(a) An aminoacid on the surface of a protein was labelled covalently with 1.5-I AEDANS and another was labelled covalently with FITC. The fluorescence quantum yield of 1.5-IAEDANS decreased by 10 per cent due to quenching by FITC. What is the distance between the aminoacids? *Hint.* See Table 21.6.

21.24(b) An aminoacid on the surface of an enzyme was labelled covalently with 1.5-I AEDANS and it is known that the active site contains a tryptophan residue. The fluorescence quantum yield of tryptophan decreased by 15 per cent due to quenching by 1.5-IAEDANS. What is the distance between the active site and the surface of the enzyme?

Problems*

Numerical problems

21.1 The data below apply to the formation of urea from ammonium cyanate, $NH_4CNO \rightarrow NH_2CONH_2$. Initially 22.9 g of ammonium cyanate was dissolved in enough water to prepare 1.00 dm^3 of solution. Determine the order of the reaction, the rate constant, and the mass of ammonium cyanate left after 300 min.

t/min	0	20.0	50.0	65.0	150
m(urea)/g	0	7.0	12.1	13.8	17.7

21.2 The data below apply to the reaction $(CH_3)_3CBr + H_2O \rightarrow (CH_3)_3COH + HBr$. Determine the order of the reaction, the rate constant, and the molar concentration of $(CH_3)_3CBr$ after 43.8 h.

t	0	3.15	6.20	10.00	18.30	30.80
$[(CH_3)_3CBr]/(10^{-2}$ mol $dm^{-3})$	10.39	8.96	7.76	6.39	3.53	2.07

21.3 The thermal decomposition of an organic nitrile produced the following data:

$t/(10^3$ s)	0	2.00	4.00	6.00	8.00	10.00	12.00	∞
[nitrile]/(mol dm^{-3})	1.50	1.26	1.07	0.92	0.81	0.72	0.65	0.40

Determine the order of the reaction and the rate constant.

21.4 The following data have been obtained for the decomposition of $N_2O_5(g)$ at 67°C according to the reaction $2\,N_2O_5(g) \rightarrow 4\,NO_2(g) + O_2(g)$. Determine the order of the reaction, the rate constant, and the half-life. It is not necessary to obtain the result graphically; you may do a calculation using estimates of the rates of change of concentration.

t/min	0	1	2	3	4	5
$[N_2O_5]$/(mol dm^{-3})	1.000	0.705	0.497	0.349	0.246	0.173

21.5 The gas-phase decomposition of acetic acid at 1189 K proceeds by way of two parallel reactions:

(1) $CH_3COOH \rightarrow CH_4 + CO_2$ $\quad k_1 = 3.74\ s^{-1}$

(2) $CH_3COOH \rightarrow CH_2CO + H_2O$ $\quad k_2 = 4.65\ s^{-1}$

What is the maximum percentage yield of the ketene CH_2CO obtainable at this temperature?

21.6 Sucrose is readily hydrolysed to glucose and fructose in acidic solution. The hydrolysis is often monitored by measuring the angle of rotation of plane-polarized light passing through the solution. From the angle of rotation the concentration of sucrose can be determined. An experiment on the hydrolysis of sucrose in 0.50 M HCl(aq) produced the following data:

t/min	0	14	39	60	80	110	140	170	210
[sucrose]/(mol dm^{-3})	0.316	0.300	0.274	0.256	0.238	0.211	0.190	0.170	0.146

Determine the rate constant of the reaction and the half-life of a sucrose molecule.

21.7 The composition of a liquid-phase reaction $2\,A \rightarrow B$ was followed by a spectrophotometric method with the following results:

t/min	0	10	20	30	40	∞
[B]/(mol dm^{-3})	0	0.089	0.153	0.200	0.230	0.312

Determine the order of the reaction and its rate constant.

21.8 The ClO radical decays rapidly by way of the reaction, $2\,ClO \rightarrow Cl_2 + O_2$. The following data have been obtained:

$t/(10^{-3}$ s)	0.12	0.62	0.96	1.60	3.20	4.00	5.75
[ClO]/(10^{-6} mol dm^{-3})	8.49	8.09	7.10	5.79	5.20	4.77	3.95

Determine the rate constant of the reaction and the half-life of a ClO radical.

21.9 Cyclopropane isomerizes into propene when heated to 500°C in the gas phase. The extent of conversion for various initial pressures has been followed by gas chromatography by allowing the reaction to proceed for a time with various initial pressures:

p_0/Torr	200	200	400	400	600	600
t/s	100	200	100	200	100	200
p/Torr	186	173	373	347	559	520

where p_0 is the initial pressure and p is the final pressure of cyclopropane. What is the order and rate constant for the reaction under these conditions?

21.10 The addition of hydrogen halides to alkenes has played a fundamental role in the investigation of organic reaction mechanisms. In one study (M.J. Haugh and D.R. Dalton, *J. Amer. Chem. Soc.* **97**, 5674 (1975)), high pressures of hydrogen chloride (up to 25 atm) and propene (up to 5 atm) were examined over a range of temperatures and the amount of 2-chloropropane formed was determined by NMR. Show that, if the reaction $A + B \rightarrow P$ proceeds for a short time δt, the concentration of product follows $[P]/[A] = k_r[A]^{m-1}[B]^n\delta t$ if the reaction is mth-order in A and nth-order in B. In a series of runs the ratio of [chloropropane] to [propene] was independent of [propene] but the ratio of [chloropropane] to [HCl] for constant amounts of propene depended on [HCl]. For $\delta t \approx 100$ h (which is short on the timescale of the reaction) the latter ratio rose from zero to 0.05, 0.03, 0.01 for p(HCl) = 10 atm, 7.5 atm, 5.0 atm, respectively. What are the orders of the reaction with respect to each reactant?

21.11 Show that the following mechanism can account for the rate law of the reaction in Problem 21.10:

$HCl + HCl \rightleftharpoons (HCl)_2$ $\quad K_1$

$HCl + CH_3CH{=}CH_2 \rightleftharpoons \text{complex}$ $\quad K_2$

$(HCl)_2 + \text{complex} \rightarrow CH_3CHClCH_3 + 2\,HCl$ $\quad k_r$ (slow)

What further tests could you apply to verify this mechanism?

21.12 A first-order decomposition reaction is observed to have the following rate constants at the indicated temperatures. Estimate the activation energy.

$k_r/(10^{-3}\ s^{-1})$	2.46	45.1	576
θ/°C	0	20.0	40.0

21.13 The second-order rate constants for the reaction of oxygen atoms with aromatic hydrocarbons have been measured (R. Atkinson and J.N. Pitts, *J. Phys. Chem.* **79**, 295 (1975)). In the reaction with benzene the rate constants are 1.44×10^7 dm^3 mol^{-1} s^{-1} at 300.3 K, 3.03×10^7 dm^3 mol^{-1} s^{-1} at 341.2 K, and 6.9×10^7 dm^3 mol^{-1} s^{-1} at 392.2 K. Find the pre-exponential factor and activation energy of the reaction.

21.14 In the experiments described in Problems 21.10 and 21.11 an inverse temperature dependence of the reaction rate was observed, the overall rate of reaction at 70°C being approximately one-third that at 19°C. Estimate the apparent activation energy and the activation energy of the rate-determining

* Problems denoted with the symbol ‡ were supplied by Charles Trapp, Carmen Giunta, and Marshall Cady.

step given that the enthalpies of the two equilibria are both of the order of -14 kJ mol^{-1}.

21.15 Use mathematical software or an electronic spreadsheet to examine the time dependence of [I] in the reaction mechanism $A \rightarrow I \rightarrow P$ (k_a, k_b). In all of the following calculations, use $[A]_0 = 1$ mol dm^{-3} and a time range of 0 to 5 s. (a) Plot [I] against t for $k_a = 10$ s^{-1} and $k_b = 1$ s^{-1}. (b) Increase the ratio k_b/k_a steadily by decreasing the value of k_a and examine the plot of [I] against t at each turn. What approximation about d[I]/dt becomes increasingly valid?

21.16 Consider the dimerization $2\,A \rightleftharpoons A_2$, with forward rate constant k_a and reverse rate constant k_a'. (a) Derive the following expression for the relaxation time in terms of the total concentration of protein, $[A]_{tot} = [A] + 2[A_2]$:

$$\frac{1}{\tau^2} = k_a'^2 + 8k_a k_a'[A]_{tot}$$

(b) Describe the computational procedures that lead to the determination of the rate constants k_a and k_a' from measurements of τ for different values of $[A]_{tot}$. (c) Use the data provided below and the procedure you outlined in part (b) to calculate the rate constants k_a and k_a', and the equilibrium constant K for formation of hydrogen-bonded dimers of 2-pyridone:

[P]/(mol dm^{-3})	0.500	0.352	0.251	0.151	0.101
τ/ns	2.3	2.7	3.3	4.0	5.3

21.17 In Problem 21.9 the isomerization of cyclopropane over a limited pressure range was examined. If the Lindemann mechanism of first-order reactions is to be tested we also need data at low pressures. These have been obtained (H.O. Pritchard *et al.*, *Proc. R. Soc.* **A217**, 563 (1953)):

p/Torr	84.1	11.0	2.89	0.569	0.120	0.067
$10^4\,k_r$/s^{-1}	2.98	2.23	1.54	0.857	0.392	0.303

Test the Lindemann theory with these data.

21.18 Dansyl chloride, which absorbs maximally at 330 nm and fluoresces maximally at 510 nm, can be used to label aminoacids in fluorescence microscopy and FRET studies. Tabulated below is the variation of the fluorescence intensity of an aqueous solution of dansyl chloride with time after excitation by a short laser pulse (with I_0 the initial fluorescence intensity). The ratio of intensities is equal to the ratio of the rates of photon emission.

t/ns	5.0	10.0	15.0	20.0
I_f/I_0	0.45	0.21	0.11	0.05

(a) Calculate the observed fluorescence lifetime of dansyl chloride in water. (b) The fluorescence quantum yield of dansyl chloride in water is 0.70. What is the fluorescence rate constant?

21.19 When benzophenone is illuminated with ultraviolet radiation it is excited into a singlet state. This singlet changes rapidly into a triplet, which phosphoresces. Triethylamine acts as a quencher for the triplet. In an experiment in methanol as solvent, the phosphorescence intensity varied with amine concentration as shown below. A time-resolved laser spectroscopy experiment had also shown that the half-life of the fluorescence in the absence of quencher is 29 μs. What is the value of k_Q?

[Q]/(mol dm^{-3})	0.0010	0.0050	0.0100
I_f/(arbitrary units)	0.41	0.25	0.16

21.20 An electronically excited state of Hg can be quenched by N_2 according to

$$Hg^*\,(g) + N_2\,(g, v=0) \rightarrow Hg\,(g) + N_2\,(g, v=1)$$

in which energy transfer from Hg* excites N_2 vibrationally. Fluorescence lifetime measurements of samples of Hg with and without N_2 present are summarized below ($T = 300$ K):

$p_{N_2} = 0.0$ atm

Relative fluorescence intensity	1.000	0.606	0.360	0.22	0.135
t/μs	0.0	5.0	10.0	15.0	20.0

$p_{N_2} = 9.74 \times 10^{-4}$ atm

Relative fluorescence intensity	1.000	0.585	0.342	0.200	0.117
t/μs	0.0	3.0	6.0	9.0	12.0

You may assume that all gases are perfect. Determine the rate constant for the energy transfer process.

21.21 The Förster theory of resonance energy transfer and the basis for the FRET technique can be tested by performing fluorescence measurements on a series of compounds in which an energy donor and an energy acceptor are covalently linked by a rigid molecular linker of variable and known length. L. Stryer and R.P. Haugland, *Proc. Natl. Acad. Sci. USA* **58**, 719 (1967) collected the following data on a family of compounds with the general composition dansyl-(L-prolyl)$_n$-naphthyl, in which the distance R between the naphthyl donor and the dansyl acceptor was varied from 1.2 nm to 4.6 nm by increasing the number of prolyl units in the linker:

R/nm	1.2	1.5	1.8	2.8	3.1	3.4	3.7	4.0	4.3	4.6
η_T	0.99	0.94	0.97	0.82	0.74	0.65	0.40	0.28	0.24	0.16

Are the data described adequately by eqn 21.86? If so, what is the value of R_0 for the naphthyl–dansyl pair?

Theoretical problems

21.22 Show that $t_{1/2}$ is given by eqn 21.17 for a reaction that is nth-order in A. Then deduce an expression for the time it takes for the concentration of a substance to fall to one-third the initial value in an nth-order reaction.

21.23 The equilibrium $A \rightleftharpoons B$ is first-order in both directions. Derive an expression for the concentration of A as a function of time when the initial molar concentrations of A and B are $[A]_0$ and $[B]_0$. What is the final composition of the system?

21.24 Derive an integrated expression for a second-order rate law $v = k[A][B]$ for a reaction of stoichiometry $2\,A + 3\,B \rightarrow P$.

21.25 Derive the integrated form of a third-order rate law $v = k[A]^2[B]$ in which the stoichiometry is $2\,A + B \rightarrow P$ and the reactants are initially present in (a) their stoichiometric proportions, (b) with B present initially in twice the amount.

21.26 Show that the definition of E_a given in eqn 21.30 reduces to eqn 21.29 for a temperature-independent activation energy.

21.27 Set up the rate equations for the reaction mechanism:

$$A \underset{k_a'}{\overset{k_a}{\rightleftharpoons}} B \underset{k_b'}{\overset{k_b}{\rightleftharpoons}} CH$$

Show that the mechanism is equivalent to

$$A \underset{k_r'}{\overset{k_r}{\rightleftharpoons}} C$$

under specified circumstances.

21.28 Show that the ratio $t_{1/2}/t_{3/4}$, where $t_{1/2}$ is the half-life and $t_{3/4}$ is the time for the concentration of A to decrease to $\frac{3}{4}$ of its initial value (implying that $t_{3/4} < t_{1/2}$), can be written as a function of n alone, and can therefore be used as a rapid assessment of the order of a reaction.

21.29 Derive an equation for the steady state rate of the sequence of reactions $A \rightleftarrows B \rightleftarrows C \rightleftarrows D$, with [A] maintained at a fixed value and the product D removed as soon as it is formed.

21.30 Consider the dimerization $2\,A \rightleftarrows A_2$ with forward rate constant k_r and backward rate constant k_r'. Show that the relaxation time is:

$$\tau = \frac{1}{k_r' + 4k_r[A]_{eq}}$$

21.31 Express the root mean square deviation $\{\langle M^2\rangle - \langle M\rangle^2\}^{1/2}$ of the molar mass of a condensation polymer in terms of the fraction p, and deduce its time dependence.

21.32 Calculate the ratio of the mean cube molar mass to the mean square molar mass in terms of (a) the fraction p, (b) the chain length.

21.33 Calculate the average polymer length in a polymer produced by a chain mechanism in which termination occurs by a disproportionation reaction of the form $M\cdot + \cdot M \rightarrow M + {:}M$.

21.34 Derive an expression for the time dependence of the degree of polymerization for a stepwise polymerization in which the reaction is acid-catalysed by the –COOH acid functional group. The rate law is $d[A]/dt = -k_r[A]^2[OH]$.

21.35 Conventional equilibrium considerations do not apply when a reaction is being driven by light absorption. Thus the steady-state concentration of products and reactants might differ significantly from equilibrium values. For instance, suppose the reaction $A \rightarrow B$ is driven by light absorption, and that its rate is I_a, but that the reverse reaction $B \rightarrow A$ is bimolecular and second-order with a rate $k_r[B]^2$. What is the stationary state concentration of B? Why does this 'photostationary state' differ from the equilibrium state?

21.36 The photochemical chlorination of chloroform in the gas phase has been found to follow the rate law $d[CCl_4]/dt = k_r[Cl_2]^{1/2}I_a^{1/2}$. Devise a mechanism that leads to this rate law when the chlorine pressure is high.

Applications to: biochemistry and environmental science

21.37 Pharmacokinetics is the study of the rates of absorption and elimination of drugs by organisms. In most cases, elimination is slower than absorption and is a more important determinant of availability of a drug for binding to its target. A drug can be eliminated by many mechanisms, such as metabolism in the liver, intestine, or kidney followed by excretion of breakdown products through urine or faeces. As an example of pharmacokinetic analysis, consider the elimination of beta adrenergic blocking agents (beta blockers), drugs used in the treatment of hypertension. After intravenous administration of a beta blocker, the blood plasma of a patient was analysed for remaining drug and the data are shown below, where c is the drug concentration measured at a time t after the injection.

t/min	30	60	120	150	240	360	480
c/(ng cm^{-3})	699	622	413	292	152	60	24

(a) Is removal of the drug a first- or second-order process? (b) Calculate the rate constant and half-life of the process. *Comment.* An essential aspect of drug development is the optimization of the half-life of elimination, which needs to be long enough to allow the drug to find and act on its target organ but not so long that harmful side-effects become important.

21.38 Consider a mechanism for the helix–coil transition in polypeptides that begins in the middle of the chain:

$$hhhh\ldots \rightleftharpoons hchh\ldots$$

$$hchh\ldots \rightleftharpoons cccc\ldots$$

The first conversion from h to c, also called a nucleation step, is relatively slow, so neither step may be rate-determining. (a) Set up the rate equations for this mechanism. (b) Apply the steady-state approximation and show that, under these circumstances, the mechanism is equivalent to $hhhh\ldots \rightleftharpoons cccc\ldots$.

21.39‡ The oxidation of HSO_3^- by O_2 in aqueous solution is a reaction of importance to the processes of acid rain formation and flue gas desulfurization. R.E. Connick *et al.* (*Inorg. Chem.* **34**, 4543 (1995)) report that the reaction $2\,HSO_3^- + O_2 \rightarrow 2\,SO_4^{2-} + 2\,H^+$ follows the rate law $v = k_r[HSO_3^-]^2[H^+]^2$. Given pH = 5.6 and an oxygen molar concentration of 2.4×10^{-4} mol dm^{-3} (both presumed constant), an initial HSO_3^- molar concentration of 5×10^{-5} mol dm^{-3}, and a rate constant of 3.6×10^{6} dm^9 mol^{-3} s^{-1}, what is the initial rate of reaction? How long would it take for HSO_3^- to reach half its initial concentration?

21.40 In light-harvesting complexes, the fluorescence of a chlorophyll molecule is quenched by nearby chlorophyll molecules. Given that for a pair of chlorophyll *a* molecules $R_0 = 5.6$ nm, by what distance should two chlorophyll *a* molecules be separated to shorten the fluorescence lifetime from 1 ns (a typical value for monomeric chlorophyll *a* in organic solvents) to 10 ps?

21.41‡ Ultraviolet radiation photolyses O_3 to O_2 and O. Determine the rate at which ozone is consumed by 305 nm radiation in a layer of the stratosphere of thickness 1 km. The quantum yield is 0.94 at 220 K, the concentration about 8×10^{-9} mol dm^{-3}, the molar absorption coefficient 260 dm^3 mol^{-1} cm^{-1}, and the flux of 305 nm radiation about 1×10^{14} photons cm^{-2} s^{-1}. Data from W.B. DeMore *et al.*, *Chemical kinetics and photochemical data for use in stratospheric modeling: Evaluation Number 11*, JPL Publication 94-26 (1994).

Reaction dynamics

22

The simplest quantitative account of reaction rates is in terms of collision theory, which can be used only for the discussion of reactions between simple species in the gas phase. Reactions in solution are classified into two types: diffusion-controlled and activation-controlled. The former can be expressed quantitatively in terms of the diffusion equation. In transition state theory, it is assumed that the reactant molecules form a complex that can be discussed in terms of the population of its energy levels. Transition state theory inspires a thermodynamic approach to reaction rates, in which the rate constant is expressed in terms of thermodynamic parameters. This approach is useful for parametrizing the rates of reactions in solution. The highest level of sophistication is in terms of potential energy surfaces and the motion of molecules through these surfaces. As we shall see, such an approach gives an intimate picture of the events that occur when reactions occur and is open to experimental study. We also use transition state theory to examine the transfer of electrons in homogeneous systems and at electrodes.

Now we are at the heart of chemistry. Here we examine the details of what happens to molecules at the climax of reactions. Extensive changes of structure are taking place and energies the size of dissociation energies are being redistributed among bonds: old bonds are being ripped apart and new bonds are being formed.

As may be imagined, the calculation of the rates of such processes from first principles is very difficult. Nevertheless, like so many intricate problems, the broad features can be established quite simply. Only when we enquire more deeply do the complications emerge. In this chapter we look at several approaches to the calculation of a rate constant for elementary bimolecular processes, ranging from electron transfer to chemical reactions involving bond breakage and formation. Although a great deal of information can be obtained from gas-phase reactions, many reactions of interest take place in condensed phases, and we shall also see to what extent their rates can be predicted.

Reactive encounters

In this section we consider two elementary approaches to the calculation of reaction rates, one relating to gas-phase reactions and the other to reactions in solution. Both approaches are based on the view that reactant molecules must meet, and that reaction takes place only if the molecules have a certain minimum energy. In the collision theory of bimolecular gas-phase reactions, which we mentioned briefly in Section 21.5b, products are formed only if the collision is sufficiently energetic; otherwise the colliding reactant molecules separate again. In solution, the reactant molecules may simply diffuse together and then acquire energy from their immediate surroundings while they are in contact.

22.1 Collision theory

Key points In collision theory, it is supposed that the rate is proportional to (a) the collision frequency, (b) the fraction of collisions that occur with at least the kinetic energy E_a along their lines of centres, and (c) a steric factor. (d) The RRK model predicts the steric factor and rate constant of unimolecular reactions.

We shall consider the bimolecular elementary reaction

$$A + B \rightarrow P \qquad v = k_r[A][B] \tag{22.1}$$

where P denotes products, and aim to calculate the second-order rate constant k_r.

We can anticipate the general form of the expression for k_r by considering the physical requirements for reaction. We expect the rate v to be proportional to the rate of collisions, and therefore to the mean speed of the molecules, $\bar{c} \propto (T/M)^{1/2}$, where M is the molar mass of the molecules, their collision cross-section, σ, and the number densities $\mathcal{N}_A$ and $\mathcal{N}_B$ of A and B (and therefore to their molar concentrations):

$$v \propto \sigma(T/M)^{1/2}\mathcal{N}_A\mathcal{N}_B \propto \sigma(T/M)^{1/2}[A][B]$$

However, a collision will be successful only if the kinetic energy exceeds a minimum value, the activation energy, E_a, of the reaction. This requirement suggests that the rate constant should also be proportional to a Boltzmann factor of the form $e^{-E_a/RT}$. So we can anticipate, by writing the reaction rate in the form given in eqn 22.1, that

$$k_r \propto \sigma(T/M)^{1/2}e^{-E_a/RT}$$

Not every collision will lead to reaction even if the energy requirement is satisfied, because the reactants may need to collide in a certain relative orientation. This 'steric requirement' suggests that a further factor, P, should be introduced, and that

$$k_r \propto P\sigma(T/M)^{1/2}e^{-E_a/RT} \tag{22.2}$$

As we shall see in detail below, this expression (which resembles the Arrhenius expression for the rate constant) has the form predicted by collision theory. It reflects three aspects of a successful collision:

$$k_r \propto \text{steric requirement} \times \text{encounter rate} \times \text{minimum energy requirement}$$

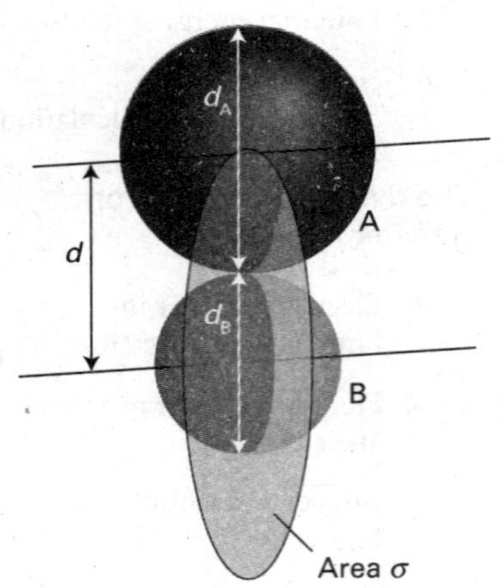

Fig. 22.1 The collision cross-section for two molecules can be regarded to be the area within which the projectile molecule (A) must enter around the target molecule (B) in order for a collision to occur. If the diameters of the two molecules are d_A and d_B, the radius of the target area is $d = \frac{1}{2}(d_A + d_B)$ and the cross-section is πd^2.

(a) Collision rates in gases

We have anticipated that the reaction rate, and hence k_r, depends on the frequency with which molecules collide. The **collision density**, Z_{AB}, is the number of collisions between molecules A and B in a region of the sample in an interval of time divided by the volume of the region and the duration of the interval. The frequency of collisions of a single molecule in a gas was calculated in Section 20.1. As shown in the following *Justification*, that result can be adapted to deduce that

$$Z_{AB} = \sigma\left(\frac{8kT}{\pi\mu}\right)^{1/2} N_A^2[A][B] \qquad \text{Collision density for two different molecules in a perfect gas} \tag{22.3a}$$

where N_A is Avogadro's constant, σ is the collision cross-section (Fig. 22.1):

$$\sigma = \pi d^2 \qquad d = \tfrac{1}{2}(d_A + d_B) \qquad \text{Collision cross-section} \tag{22.3b}$$

and μ is the reduced mass:

$$\mu = \frac{m_A m_B}{m_A + m_B}$$

Reduced mass of a two-particle system (22.3c)

Similarly, the collision density for like molecules at a molar concentration [A] is

$$Z_{AA} = \sigma\left(\frac{4kT}{\pi m_A}\right)^{1/2} N_A^2[A]^2$$

Collision density for two like molecules in a perfect gas (22.4)

Collision densities may be very large. For example, in nitrogen at room temperature and pressure, with $d = 280$ pm, $Z = 5 \times 10^{34}$ m^{-3} s^{-1}.

Justification 22.1 *The collision density*

It follows from eqn 20.11 that the collision frequency, z, for a single A molecule of mass m_A in a gas of other A molecules is

$$z = \sigma \bar{c}_{rel} \mathcal{N}_A \quad (22.5)$$

where $\mathcal{N}_A$ is the number density of A molecules, $\bar{c}_{rel}$ is their relative mean speed,

$$\bar{c}_{rel} = \left(\frac{8kT}{\pi\mu}\right)^{1/2} \quad (22.6)$$

and μ is the reduced mass (eqn 22.3c), which in this case is simply $\frac{1}{2}m_A$.

The total collision density is the collision frequency multiplied by the number density of A molecules:

$$Z_{AA} = \tfrac{1}{2} z \mathcal{N}_A = \tfrac{1}{2}\sigma \bar{c}_{rel} \mathcal{N}_A^2 \quad (22.7a)$$

The factor of $\frac{1}{2}$ has been introduced to avoid double counting of the collisions (so one A molecule colliding with another A molecule is counted as one collision regardless of their actual identities). For collisions of A and B molecules present at number densities $\mathcal{N}_A$ and $\mathcal{N}_B$, the collision density is

$$Z_{AB} = \sigma \bar{c}_{rel} \mathcal{N}_A \mathcal{N}_B \quad (22.7b)$$

Note that we have discarded the factor $\frac{1}{2}$ because now we are considering an A molecule colliding with any of the B molecules as a collision.

The number density of a species J is $\mathcal{N}_J = N_A[J]$, where [J] is the molar concentration and N_A is Avogadro's constant. Equations 22.3a and 22.4 then follow.

(b) The energy requirement

According to collision theory, the rate of change in the molar concentration of A molecules is the product of the collision density and the probability that a collision occurs with sufficient energy. The latter condition can be incorporated by writing the collision cross-section as a function of the kinetic energy of approach of the two colliding species, and setting the cross-section, $\sigma(\varepsilon)$, equal to zero if the kinetic energy of approach is below a certain threshold value, ε_a. Later, we shall identify $N_A\varepsilon_a$ as E_a, the (molar) activation energy of the reaction. Then, for a collision with a specific relative speed of approach v_{rel} (not, at this stage, a mean value),

$$\frac{d\mathcal{N}_A}{dt} = -\sigma(\varepsilon) v_{rel} \mathcal{N}_A \mathcal{N}_B \quad (22.8a)$$

Or, in terms of molar concentrations,

$$\frac{d[A]}{dt} = -\sigma(\varepsilon)v_{rel}N_A[A][B] \tag{22.8b}$$

The kinetic energy associated with the relative motion of the two particles is $\varepsilon = \frac{1}{2}\mu v_{rel}^2$; therefore the relative speed is $v_{rel} = (2\varepsilon/\mu)^{1/2}$. At this point we recognize that a wide range of approach energies ε is present in a sample, so we should average the expression just derived over a Boltzmann distribution of energies $f(\varepsilon)$ (Section 15.1b), and write

$$\frac{d[A]}{dt} = -\left\{\int_0^\infty \sigma(\varepsilon)v_{rel}f(\varepsilon)d\varepsilon\right\}N_A[A][B] \tag{22.9}$$

and hence recognize the rate constant as

$$k_r = N_A\int_0^\infty \sigma(\varepsilon)v_{rel}f(\varepsilon)d\varepsilon \tag{22.10}$$

Now suppose that the reactive collision cross-section is zero below ε_a. We show in the following *Justification* that, above ε_a, a plausible expression for $\sigma(\varepsilon)$ is

$$\sigma(\varepsilon) = \left(1 - \frac{\varepsilon_a}{\varepsilon}\right)\sigma \tag{22.11}$$

The collision cross-section

Note that, when $\varepsilon = \varepsilon_a$, $\sigma(\varepsilon) = 0$, so the cross-section rises smoothly from its value 0 below ε_a, and that, when $\varepsilon \gg \varepsilon_a$, it attains the constant value $\sigma(\varepsilon) = \sigma$.

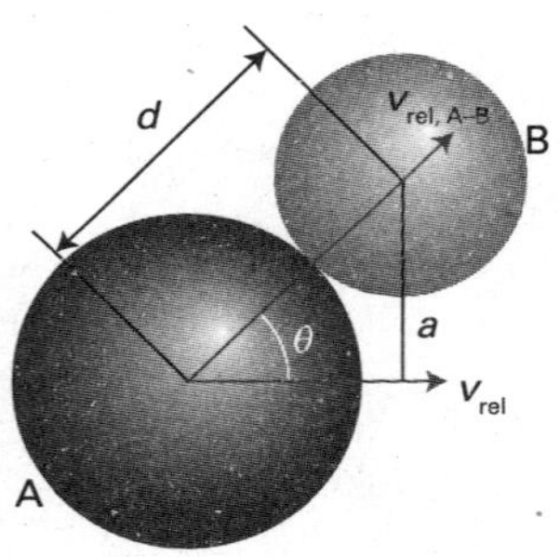

Fig. 22.2 The parameters used in the calculation of the dependence of the collision cross-section on the relative kinetic energy of two molecules A and B.

Justification 22.2 *The collision cross-section*

Consider two colliding molecules A and B with relative speed v_{rel} and relative kinetic energy $\varepsilon = \frac{1}{2}\mu v_{rel}^2$ (Fig. 22.2). Intuitively we expect that a head-on collision between A and B will be most effective in bringing about a chemical reaction. Therefore, $v_{rel,A-B}$, the magnitude of the relative velocity component parallel to an axis that contains the vector connecting the centres of A and B, must be large. From trigonometry and the definitions of the distances a and d, and the angle θ given in Fig. 22.2, it follows that

$$v_{rel,A-B} = v_{rel}/\cos\theta = v_{rel}\left(\frac{d^2 - a^2}{d^2}\right)^{1/2}$$

We assume that only the kinetic energy associated with the head-on component of the collision, ε_{A-B}, can lead to a chemical reaction. After squaring both sides of the equation above and multiplying by $\frac{1}{2}\mu$, it follows that

$$\varepsilon_{A-B} = \varepsilon\frac{d^2 - a^2}{d^2}$$

The existence of an energy threshold, ε_a, for the formation of products implies that there is a maximum value of a, a_{max}, above which reactions do not occur. Setting $a = a_{max}$ and $\varepsilon_{A-B} = \varepsilon_a$ in this expression gives

$$a_{max}^2 = \left(1 - \frac{\varepsilon_a}{\varepsilon}\right)d^2$$

Substitution of $\sigma(\varepsilon)$ for πa_{max}^2 and σ for πd^2 in the equation above gives eqn 22.11. Note that the equation can be used only when $\varepsilon > \varepsilon_a$.

With the energy dependence of the collision cross-section established, we can evaluate the integral in eqn 22.10. In the following *Justification* we show that

$$k_r = N_A\sigma\bar{c}_{rel}e^{-E_a/RT} \qquad (22.12)$$

The rate constant for a gas-phase bimolecular elementary reaction (excluding steric requirements)

Justification 22.3 *The rate constant*

The Maxwell distribution of molecular speeds given in Section 20.1 may be expressed in terms of the kinetic energy, ε, by writing $\varepsilon = \frac{1}{2}\mu v^2$, then $dv = d\varepsilon/(2\mu\varepsilon)^{1/2}$ and eqn 20.4 becomes

$$f(v)dv = 4\pi\left(\frac{\mu}{2\pi kT}\right)^{3/2}\left(\frac{2\varepsilon}{\mu}\right)e^{-\varepsilon/kT}\frac{d\varepsilon}{(2\mu\varepsilon)^{1/2}}$$
$$= 2\pi\left(\frac{1}{\pi kT}\right)^{3/2}\varepsilon^{1/2}e^{-\varepsilon/kT}d\varepsilon = f(\varepsilon)d\varepsilon$$

The integral we need to evaluate is therefore

$$\int_0^\infty \sigma(\varepsilon)v_{rel}f(\varepsilon)d\varepsilon = 2\pi\left(\frac{1}{\pi kT}\right)^{3/2}\int_0^\infty \sigma(\varepsilon)\left(\frac{2\varepsilon}{\mu}\right)^{1/2}\varepsilon^{1/2}e^{-\varepsilon/kT}d\varepsilon$$
$$= \left(\frac{8}{\pi\mu kT}\right)^{1/2}\left(\frac{1}{kT}\right)\int_0^\infty \varepsilon\sigma(\varepsilon)e^{-\varepsilon/kT}d\varepsilon$$

To proceed, we introduce the expression for $\sigma(\varepsilon)$ in eqn 22.11, and evaluate

$$\int_0^\infty \varepsilon\sigma(\varepsilon)e^{-\varepsilon/kT}d\varepsilon = \sigma\int_{\varepsilon_a}^\infty \varepsilon\left(1-\frac{\varepsilon_a}{\varepsilon}\right)e^{-\varepsilon/kT}d\varepsilon = (kT)^2\sigma e^{-\varepsilon_a/kT}$$

We have made use of the fact that $\sigma = 0$ for $\varepsilon < \varepsilon_a$ and have used the two integrals

$$\int e^{-ax}dx = -\frac{e^{-ax}}{a} + \text{constant} \qquad \text{and} \qquad \int xe^{-ax}dx = \frac{e^{-ax}}{a^2} + \frac{xe^{-ax}}{a} + \text{constant}$$

It follows that

$$\int_0^\infty \sigma(\varepsilon)v_{rel}f(\varepsilon)d\varepsilon = \sigma\left(\frac{8kT}{\pi\mu}\right)^{1/2}e^{-\varepsilon_a/kT}$$

as in eqn 22.12 (with $\varepsilon_a/kT = E_a/RT$).

Equation 22.12 has the Arrhenius form $k_r = Ae^{-E_a/RT}$ provided the exponential temperature dependence dominates the weak square-root temperature dependence of the pre-exponential factor. It follows that we can identify the activation energy, E_a, with the minimum kinetic energy along the line of approach that is needed for reaction, and that the pre-exponential factor is a measure of the rate at which collisions occur in the gas.

(c) The steric requirement

The simplest procedure for calculating k_r is to use for σ the values obtained for non-reactive collisions (for example, typically those obtained from viscosity measurements) or from tables of molecular radii. Table 22.1 compares some values of the pre-exponential factor calculated in this way with values obtained from Arrhenius

Table 22.1* Arrhenius parameters for gas-phase reactions

	$A/(\mathrm{dm^3\,mol^{-1}\,s^{-1}})$			
	Experiment	Theory	$E_a/(\mathrm{kJ\,mol^{-1}})$	P
$2\,NOCl \rightarrow 2\,NO + 2\,Cl$	9.4×10^{9}	5.9×10^{10}	102	0.16
$2\,ClO \rightarrow Cl_2 + O_2$	6.3×10^{7}	2.5×10^{10}	0	2.5×10^{-3}
$H_2 + C_2H_4 \rightarrow C_2H_6$	1.24×10^{6}	7.4×10^{11}	180	1.7×10^{-6}
$K + Br_2 \rightarrow KBr + Br$	1.0×10^{12}	2.1×10^{11}	0	4.8

* More values are given in the *Data section*.

plots (Section 21.5). One of the reactions shows fair agreement between theory and experiment, but for others there are major discrepancies. In some cases the experimental values are orders of magnitude smaller than those calculated, which suggests that the collision energy is not the only criterion for reaction and that some other feature, such as the relative orientation of the colliding species, is important. Moreover, one reaction in the table has a pre-exponential factor larger than theory, which seems to indicate that the reaction occurs more quickly than the particles collide!

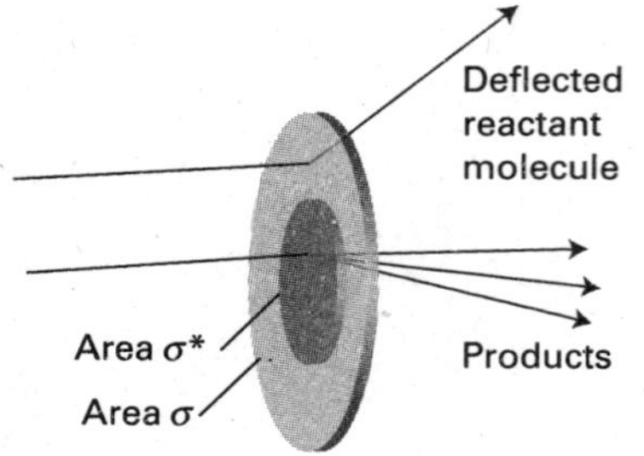

Fig. 22.3 The collision cross-section is the target area that results in simple deflection of the projectile molecule; the reactive cross-section is the corresponding area for chemical change to occur on collision.

We can accommodate the disagreement between experiment and theory by introducing a **steric factor**, P, and expressing the **reactive cross-section**, σ^*, as a multiple of the collision cross-section, $\sigma^* = P\sigma$ (Fig. 22.3). Then the rate constant becomes

$$k_r = P\sigma\left(\frac{8kT}{\pi\mu}\right)^{1/2} N_A e^{-E_a/RT} \qquad (22.13)$$

The rate constant for a gas-phase bimolecular elementary reaction (including steric requirements)

This expression has the form we anticipated in eqn 22.2. The steric factor is normally found to be several orders of magnitude smaller than 1.

Example 22.1 *Estimating a steric factor (1)*

Estimate the steric factor for the reaction $H_2 + C_2H_4 \rightarrow C_2H_6$ at 628 K given that the pre-exponential factor is $1.24\times10^{6}\,\mathrm{dm^3\,mol^{-1}\,s^{-1}}$.

Method To calculate P, we need to calculate the pre-exponential factor, A, by using eqn 22.12 and then compare the answer with experiment: the ratio is P. Table 20.1 lists collision cross-sections for non-reactive encounters. The best way to estimate the collision cross-section for dissimilar spherical species is to calculate the collision diameter for each one (from $\sigma = \pi d^2$), to calculate the mean of the two diameters, and then to calculate the cross-section for that mean diameter. However, as neither species is spherical, a simpler but more approximate procedure is just to take the average of the two collision cross-sections.

Answer The reduced mass of the colliding pair is

$$\mu = \frac{m_1 m_2}{m_1 + m_2} = 3.12\times10^{-27}\,\mathrm{kg}$$

because $m_1 = 2.016m_u$ for H_2 and $m_2 = 28.05m_u$ for C_2H_4 (the atomic mass constant, m_u, is defined inside the front cover). Hence at 628 K

$$\left(\frac{8kT}{\pi\mu}\right)^{1/2} = 2.66\times10^{3}\,\mathrm{m\,s^{-1}}$$

From Table 20.1, $\sigma(H_2) = 0.27\ nm^2$ and $\sigma(C_2H_4) = 0.64\ nm^2$, giving a mean collision cross-section of $\sigma = 0.46\ nm^2$. Therefore,

$$A = \sigma\left(\frac{8kT}{\pi\mu}\right)^{1/2} N_A = 7.37\times10^{11}\ dm^3\ mol^{-1}\ s^{-1}$$

Experimentally $A = 1.24\times10^6\ dm^3\ mol^{-1}\ s^{-1}$, so it follows that $P = 1.7\times10^{-6}$. The very small value of P is one reason why catalysts are needed to bring this reaction about at a reasonable rate. As a general guide, the more complex the molecules, the smaller the value of P.

Self-test 22.1 It is found for the reaction $NO + Cl_2 \rightarrow NOCl + Cl$ that $A = 4.0\times10^9\ dm^3\ mol^{-1}\ s^{-1}$ at 298 K. Use $\sigma(NO) = 0.42\ nm^2$ and $\sigma(Cl_2) = 0.93\ nm^2$ to estimate the P factor for the reaction. [0.018]

An example of a reaction for which it is possible to estimate the steric factor is $K + Br_2 \rightarrow KBr + Br$, for which $P = 4.8$. In this reaction, the distance of approach at which reaction occurs appears to be considerably larger than the distance needed for deflection of the path of the approaching molecules in a non-reactive collision. It has been proposed that the reaction proceeds by a **harpoon mechanism.** This brilliant name is based on a model of the reaction that pictures the K atom as approaching a Br_2 molecule, and when the two are close enough an electron (the harpoon) flips across from K to Br_2. In place of two neutral particles there are now two ions, so there is a Coulombic attraction between them: this attraction is the line on the harpoon. Under its influence the ions move together (the line is wound in), the reaction takes place, and KBr + Br emerge. The harpoon extends the cross-section for the reactive encounter, and the reaction rate is greatly underestimated by taking for the collision cross-section the value for simple mechanical contact between $K + Br_2$.

Example 22.2 *Estimating a steric factor (2)*

Estimate the value of P for the harpoon mechanism by calculating the distance at which it becomes energetically favourable for the electron to leap from K to Br_2.

Method We should begin by identifying all the contributions to the energy of interaction between the colliding species. There are three contributions to the energy of the process $K + Br_2 \rightarrow K^+ + Br_2^-$. The first is the ionization energy, I, of K. The second is the electron affinity, E_{ea}, of Br_2. The third is the Coulombic interaction energy between the ions when they have been formed: when their separation is R, this energy is $-e^2/4\pi\varepsilon_0 R$. The electron flips across when the sum of these three contributions changes from positive to negative (that is, when the sum is zero).

Answer The net change in energy when the transfer occurs at a separation R is

$$E = I - E_{ea} - \frac{e^2}{4\pi\varepsilon_0 R}$$

The ionization energy I is larger than E_{ea}, so E becomes negative only when R has decreased to less than some critical value R^* given by

$$\frac{e^2}{4\pi\varepsilon_0 R^*} = I - E_{ea}$$

When the particles are at this separation, the harpoon shoots across from K to Br_2, so we can identify the reactive cross-section as $\sigma^* = \pi R^{*2}$. This value of σ^* implies that the steric factor is

$$P=\frac{\sigma^*}{\sigma}=\frac{R^{*2}}{d^2}=\left\{\frac{e^2}{4\pi\varepsilon_0 d(I-E_{ea})}\right\}^2$$

where $d = R(K) + R(Br_2)$. With $I = 420$ kJ mol^{-1} (corresponding to 0.70 aJ), $E_{ea} \approx$ 250 kJ mol^{-1} (corresponding to 0.42 aJ), and $d = 400$ pm, we find $P = 4.2$, in good agreement with the experimental value (4.8).

Self-test 22.2 Estimate the value of P for the harpoon reaction between Na and Cl_2 for which $d \approx 350$ pm; take $E_{ea} \approx 230$ kJ mol^{-1}. [2.2]

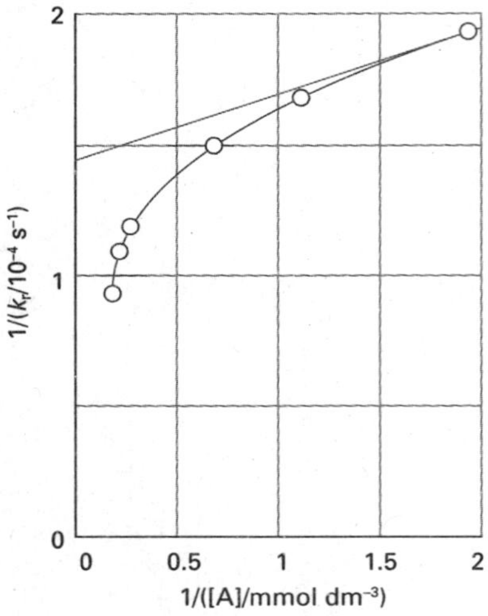

Fig. 22.4 The pressure dependence of the unimolecular isomerization of *trans*-CHD=CHD showing a pronounced departure from the straight line predicted by the Lindemann–Hinshelwood mechanism.

(d) The RRK model

Another instance where the steric factor P can be estimated is for unimolecular gas-phase reactions and its introduction brings the Lindemann–Hinshelwood mechanism into closer agreement with experiment. According to that theory (Section 21.8a), a plot of $1/k_r$ against $1/[A]$ should be linear. However, as Fig. 22.4 shows, a typical plot has a pronounced curvature, corresponding to a larger value of k_r (a smaller value of $1/k_r$) at high pressures (low $1/[A]$) than would be expected by extrapolation of the reasonably linear low pressure (high $1/[A]$) data.

The improved model was proposed in 1926 by O.K. Rice and H.C. Ramsperger and almost simultaneously by L.S. Kassel, and is now known as the **Rice–Ramsperger–Kassel model** (RRK model). The model has been elaborated, largely by R.A. Marcus, into the RRKM model. Here we outline Kassel's original approach to the RRK model. The essential feature of the model is that, although a molecule might have enough energy to react, that energy is distributed over all the modes of motion of the molecule, and reaction will occur only when enough of that energy has migrated into a particular location (such as a bond) in the molecule. This distribution leads to a P factor of the form[1]

$$P=\left(1-\frac{E^*}{E}\right)^{s-1} \qquad \text{The steric factor from the RRK model} \qquad (22.14a)$$

where s is the number of modes of motion over which the energy may be dissipated, E^* is the energy required for the bond of interest to break, and E is the energy available in the collision. We can then write the **Kassel form** of the unimolecular rate constant for the decay of A^* to products as

$$k_b(E)=\left(1-\frac{E^*}{E}\right)^{s-1}k_b \qquad \text{for} \qquad E\geq E^* \qquad \text{Kassel form of the unimolecular rate constant} \qquad (22.14b)$$

where k_b is the rate constant used in the original Lindemann theory.

• A brief illustration

Suppose that an energy of 250 kJ mol^{-1} is available in a collision but 200 kJ mol^{-1} is needed to break a particular bond in a molecule with $s = 10$. Then

$$P=\left(1-\frac{200\text{ kJ mol}^{-1}}{250\text{ kJ mol}^{-1}}\right)^9=5\times10^{-7}$$

If 500 kJ mol^{-1} is available,

$$P=\left(1-\frac{200\text{ kJ mol}^{-1}}{500\text{ kJ mol}^{-1}}\right)^9=1\times10^{-2}$$

and the collision is much more efficient. •

[1] The derivation is given in our *Quanta, matter, and change* (2009).

The energy dependence of the rate constant given by eqn 21.14b is shown in Fig. 22.5 for various values of s. We see that the rate constant is smaller at a given excitation energy if s is large, as it takes longer for the excitation energy to migrate through all the oscillators of a large molecule and accumulate in the critical mode. As E becomes very large, however, the term in parentheses approaches 1, and $k_b(E)$ becomes independent of the energy and the number of oscillators in the molecule, as there is now enough energy to accumulate immediately in the critical mode regardless of the size of the molecule.

Example 22.2 and the calculations summarized in this section illustrate two points about steric factors. First, the concept is not wholly useless because in some cases its numerical value can be estimated. Second (and more pessimistically) most reactions are much more complex than $K + Br_2$ and unimolecular gas-phase reactions and we cannot expect to obtain P so easily. It is clear that we need a more powerful theory that lets us calculate rate constants for a wider variety of reactions. We go part of the way toward describing such a theory in Section 22.4 after we have established some of the features of reactions in solution.

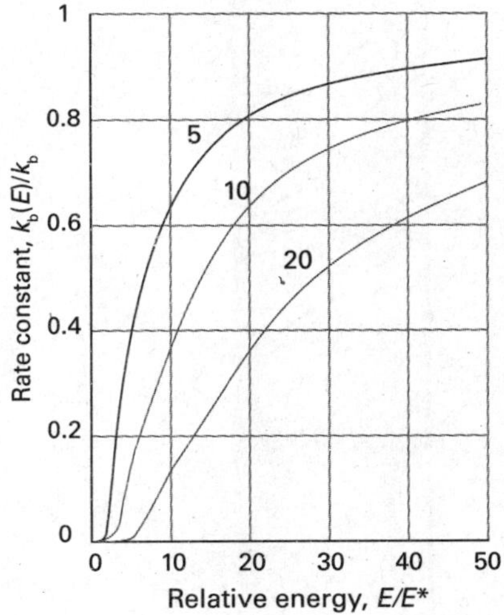

Fig. 22.5 The energy dependence of the rate constant given by eqn 22.14b for three values of s.

22.2 Diffusion-controlled reactions

Key points **(a) The rate of a diffusion-controlled reaction is controlled by the rate at which reactant molecules encounter each other in solution. The rate of an activation-controlled reaction is controlled by the rate of accumulating sufficient energy. (b) An indication that a reaction is diffusion-controlled is that its rate constant is of the order of 10^9 dm^3 mol^{-1} s^{-1} or greater.**

Encounters between reactants in solution occur in a very different manner from encounters in gases. Reactant molecules have to jostle their way through the solvent, so their encounter frequency is considerably less than in a gas. However, because a molecule also migrates only slowly away from a location, two reactant molecules that encounter each other stay near each other for much longer than in a gas. This lingering of one molecule near another on account of the hindering presence of solvent molecules is called the **cage effect**. Such an encounter pair may accumulate enough energy to react even though it does not have enough energy to do so when it first forms. The activation energy of a reaction is a much more complicated quantity in solution than in a gas because the encounter pair is surrounded by solvent and we need to consider the energy of the entire local assembly of reactant and solvent molecules. Some experimental data on Arrhenius parameters in solution are given in Table 22.2.

(a) Classes of reaction

The complicated overall process can be divided into simpler parts by setting up a simple kinetic scheme. We suppose that the rate of formation of an encounter pair AB is first-order in each of the reactants A and B:

$$A + B \rightarrow AB \qquad v = k_d[A][B]$$

Table 22.2* Arrhenius parameters for reactions in solution

	Solvent	$A/(\mathrm{dm^3\ mol^{-1}\ s^{-1}})$	$E_a/(\mathrm{kJ\ mol^{-1}})$
$(CH_3)_3CCl$ solvolysis	Water	7.1×10^{16}	100
	Ethanol	3.0×10^{13}	112
	Chloroform	1.4×10^{4}	45
$CH_3CH_2Br + OH^-$	Ethanol	4.3×10^{11}	90

* More values are given in the *Data section*.

As we shall see, k_d (where the d signifies diffusion) is determined by the diffusional characteristics of A and B. The encounter pair can break up without reaction with a rate constant k_d' or it can go on to form products P with a rate constant k_a (where the a signifies an activated process). If we suppose that both processes are pseudofirst-order reactions (with the solvent perhaps playing a role), then we can write

$$AB \rightarrow A + B \qquad v = k_d'[AB]$$

and

$$AB \rightarrow P \qquad v = k_a[AB]$$

The concentration of AB can now be found from the equation for the net rate of change of concentration of AB and using the steady-state approximation:

$$\frac{d[AB]}{dt} = k_d[A][B] - k_d'[AB] - k_a[AB] \approx 0$$

This expression solves to

$$[AB] = \frac{k_d[A][B]}{k_a + k_d'}$$

The rate of formation of products is therefore

$$\frac{d[P]}{dt} \approx k_a[AB] = k_r[A][B] \qquad k_r = \frac{k_a k_d}{k_a + k_d'} \tag{22.15}$$

Two limits can now be distinguished. If the rate of separation of the unreacted encounter pair is much slower than the rate at which it forms products, then $k_d' \ll k_a$ and the effective rate constant is

$$k_r \approx \frac{k_a k_d}{k_a} = k_d \tag{22.16}$$

Rate constant for a diffusion-controlled reaction

In this **diffusion-controlled limit**, the rate of reaction is governed by the rate at which the reactant molecules diffuse through the solvent. Because the combination of radicals involves very little activation energy, radical and atom recombination reactions are often diffusion-controlled.

An **activation-controlled reaction** arises when a substantial activation energy is involved in the reaction AB $\rightarrow$ P. Then $k_a \ll k_d'$ and

$$k_r \approx \frac{k_a k_d}{k_d'} = k_a K \tag{22.17}$$

Rate constant for an activation-controlled reaction

where K is the equilibrium constant for $A + B \rightleftharpoons AB$. In this limit, the reaction proceeds at the rate at which energy accumulates in the encounter pair from the surrounding solvent.

(b) Diffusion and reaction

The rate of a diffusion-controlled reaction is calculated by considering the rate at which the reactants diffuse together. As shown in the following *Justification*, the rate constant for a reaction in which the two reactant molecules react if they come within a distance R^* of one another is

$$k_d = 4\pi R^* D N_A \tag{22.18}$$

The rate constant of a diffusion-controlled reaction in terms of the diffusion coefficients where D is the sum of the diffusion coefficients of the two reactant species in the solution. It follows from this expression that an indication that a reaction is diffusion-controlled is that its rate constant is of the order of 10^9 dm^3 mol^{-1} s^{-1} or greater, as may be confirmed by taking values of R^* and D to be 100 nm and 10^{-9} m^2 s^{-1}, respectively.

Justification 22.4 *Solution of the radial diffusion equation*

From the form of the diffusion equation (Section 20.9) corresponding to motion in three dimensions, $D_B\nabla^2[B] = \partial[B]/\partial t$, the concentration of B when the system has reached a steady state ($\partial[B]/\partial t = 0$) satisfies $\nabla^2[B]_r = 0$, where the subscript r signifies a quantity that varies with the distance r. For a spherically symmetrical system, ∇^2 can be replaced by radial derivatives alone (see Table 7.1), so the equation satisfied by $[B]_r$ is

$$\frac{d^2[B]_r}{dr^2} + \frac{2}{r}\frac{d[B]_r}{dr} = 0$$

The general solution of this equation is

$$[B]_r = a + \frac{b}{r}$$

as may be verified by substitution. We need two boundary conditions to pin down the values of the two constants. One condition is that $[B]_r$ has its bulk value [B] as $r \to \infty$. The second condition is that the concentration of B is zero at $r = R^*$, the distance at which reaction occurs. It follows that $a = [B]$ and $b = -R^*[B]$, and hence that (for $r \geq R^*$)

$$[B]_r = \left(1 - \frac{R^*}{r}\right)[B] \qquad (22.19)$$

Figure 22.6 illustrates the variation of concentration expressed by this equation.

The rate of reaction is the (molar) flux, J, of the reactant B towards A multiplied by the area of the spherical surface of radius R^*:

$$\text{Rate of reaction} = 4\pi R^{*2}J$$

From Fick's first law (eqn 20.19), the flux towards A is proportional to the concentration gradient, so at a radius R^*:

$$J = D_B\left(\frac{d[B]_r}{dr}\right)_{r=R^*} = \frac{D_B[B]}{R^*}$$

(A sign change has been introduced because we are interested in the flux towards decreasing values of r.) When this condition is substituted into the previous equation we obtain

$$\text{Rate of reaction} = 4\pi R^* D_B[B]$$

The rate of the diffusion-controlled reaction is equal to the average flow of B molecules to all the A molecules in the sample. If the bulk concentration of A is [A], the number of A molecules in the sample of volume V is $N_A[A]V$; the global flow of all B to all A is therefore $4\pi R^* D_B N_A[A][B]V$. Because it is unrealistic to suppose that all A are stationary; we replace D_B by the sum of the diffusion coefficients of the two species and write $D = D_A + D_B$. Then the rate of formation of the encounter pair AB is

$$\frac{d[AB]}{dt} = 4\pi R^* D N_A[A][B]$$

Hence, the diffusion-controlled rate constant is as given in eqn 22.18.

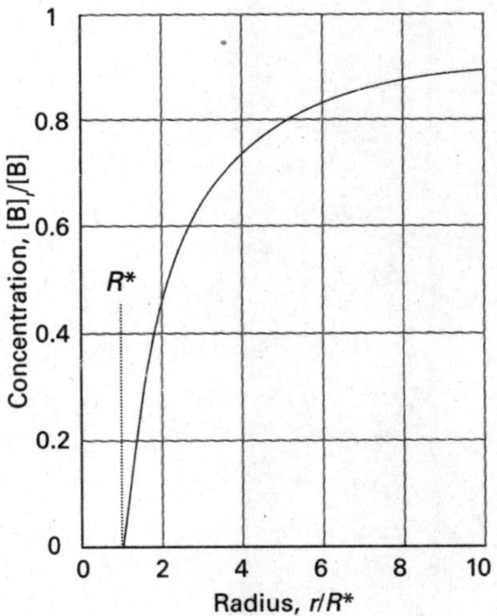

Fig. 22.6 The concentration profile for reaction in solution when a molecule B diffuses towards another reactant molecule and reacts if it reaches R^*.

We can take eqn 22.18 further by incorporating the Stokes–Einstein equation (eqn 20.51) relating the diffusion constant and the hydrodynamic radius R_A and R_B of each molecule in a medium of viscosity η:

$$D_A = \frac{kT}{6\pi\eta R_A} \qquad D_B = \frac{kT}{6\pi\eta R_B} \tag{22.20}$$

As these relations are approximate, little extra error is introduced if we write $R_A = R_B = \frac{1}{2}R^*$, which leads to

$$k_d = \frac{8RT}{3\eta} \tag{22.21}$$

The rate constant of a diffusion-controlled reaction in terms of the viscosity coefficient

(The R in this equation is the gas constant.) The radii have cancelled because, although the diffusion constants are smaller when the radii are large, the reactive collision radius is larger and the particles need to travel a shorter distance to meet. In this approximation, the rate constant is independent of the identities of the reactants, and depends only on the temperature and the viscosity of the solvent.

• A brief illustration

The rate constant for the recombination of I atoms in hexane at 298 K, when the viscosity of the solvent is 0.326 cP (with 1 P = 10^{-1} kg m^{-1} s^{-1}) is

$$k_d = \frac{8 \times (8.3145\ \mathrm{J\ K^{-1}\ mol^{-1}}) \times (298\ \mathrm{K})}{3 \times (3.26 \times 10^{-4}\ \mathrm{kg\ m^{-1}\ s^{-1}})} = 2.0 \times 10^{7}\ \mathrm{m^3\ mol^{-1}\ s^{-1}}$$

where we have used 1 J = 1 kg m^2 s^{-2}. Because 1 m^3 = 10^3 dm^3, this result corresponds to 2.0×10^{10} dm^3 mol^{-1} s^{-1}. The experimental value is 1.3×10^{10} dm^3 mol^{-1} s^{-1}, so the agreement is very good considering the approximations involved. •

22.3 The material balance equation

Key point **The material balance equation combines the effects of diffusion, convection, and reaction.**

The diffusion of reactants plays an important role in many chemical processes, such as the diffusion of O_2 molecules into red blood corpuscles and the diffusion of a gas towards a catalyst. We can have a glimpse of the kinds of calculations involved by considering the diffusion equation (Section 20.9) generalized to take into account the possibility that the diffusing, convecting molecules are also reacting.

Consider a small volume element in a chemical reactor (or a biological cell) modelled as a one-dimensional system. The net rate at which J molecules enter the region by diffusion and convection is given by eqn 20.56:

$$\frac{\partial[\mathrm{J}]}{\partial t} = D\frac{\partial^2[\mathrm{J}]}{\partial x^2} - v\frac{\partial[\mathrm{J}]}{\partial x} \tag{22.22}$$

The net rate of change of molar concentration due to chemical reaction is

$$\frac{\partial[\mathrm{J}]}{\partial t} = -k_r[\mathrm{J}] \tag{22.23}$$

if we suppose that J disappears by a pseudofirst-order reaction. Therefore, the overall rate of change of the concentration of J is

$$\frac{\partial[\mathrm{J}]}{\partial t}=\underbrace{D\frac{\partial^2[\mathrm{J}]}{\partial x^2}}_{\substack{\text{Spread due to}\\\text{non-uniform}\\\text{concentration}}}-\underbrace{v\frac{\partial[\mathrm{J}]}{\partial x}}_{\substack{\text{Change}\\\text{due to}\\\text{convection}}}-\underbrace{k_{\mathrm{r}}[\mathrm{J}]}_{\substack{\text{Loss}\\\text{due to}\\\text{reaction}}} \qquad (22.24)$$

Material balance equation

Equation 22.24 is called the **material balance equation.** If the rate constant is large, then [J] will decline rapidly. However, if the diffusion constant is large, then the decline can be replenished as J diffuses rapidly into the region. The convection term, which may represent the effects of stirring, can sweep material either into or out of the region according to the signs of v and the concentration gradient $\partial[\mathrm{J}]/\partial x$.

The material balance equation, even for a one-dimensional system, is a second-order partial differential equation and is far from easy to solve in general. Some idea of how it is solved can be obtained by considering the special case in which there is no convective motion (as in an unstirred reaction vessel):

$$\frac{\partial[\mathrm{J}]}{\partial t}=D\frac{\partial^2[\mathrm{J}]}{\partial x^2}-k_{\mathrm{r}}[\mathrm{J}] \qquad (22.25)$$

As may be verified by substitution, if the solution of this equation in the absence of reaction (that is, for $k_{\mathrm{r}}=0$) is [J], then the solution $[\mathrm{J}]^*$ in the presence of reaction ($k_{\mathrm{r}}>0$) is

$$[\mathrm{J}]^*=[\mathrm{J}]\mathrm{e}^{-k_{\mathrm{r}}t} \qquad (22.26)$$

We have already met one solution of the diffusion equation in the absence of reaction: eqn 20.57 is the solution for a system in which initially a layer of n_0N_{A} molecules is spread over a plane of area A:

$$[\mathrm{J}]=\frac{n_0\mathrm{e}^{-x^2/4Dt}}{A(\pi Dt)^{1/2}} \qquad (22.27)$$

Solution of the material balance equation

When this expression is substituted into eqn 22.26, we obtain the concentration of J as it diffuses away from its initial surface layer and undergoes reaction in the solution above (Fig. 22.7).

Only in some special cases can the full material balance equation be solved analytically. Most modern work on reactor design and cell kinetics uses numerical methods to solve the equation, and detailed solutions for realistic environments, such as vessels of different shapes (which influence the boundary conditions on the solutions) and with a variety of inhomogeneously distributed reactants, can be obtained reasonably easily.

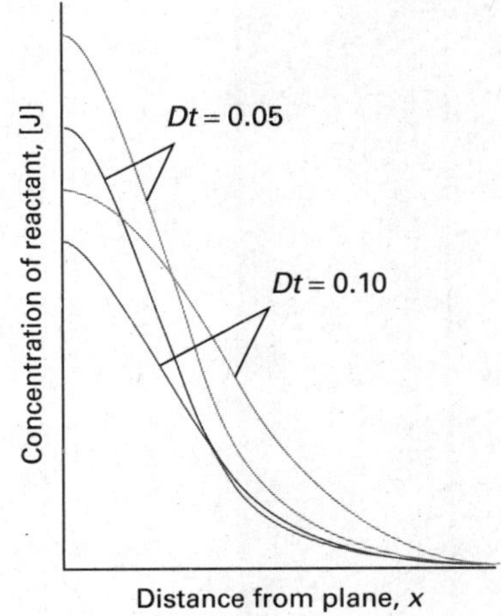

Fig. 22.7 The concentration profiles for a diffusing, reacting system (for example, a column of solution) in which one reactant is initially in a layer at $x=0$. In the absence of reaction (grey lines) the concentration profiles are the same as in Fig. 20.20.

interActivity Use the interactive applet found in the *Living graphs* section of the text's web site to explore the effect of varying the value of the rate constant k on the spatial variation of [J] for a constant value of the diffusion constant D.

Transition state theory

We saw in Section 21.5 that an activated complex forms between reactants as they collide and begin to assume the nuclear and electronic configurations characteristic of products. We also saw that the change in potential energy associated with formation of the activated complex accounts for the activation energy of the reaction. We now consider a more detailed calculation of rate constants which uses the concepts of statistical thermodynamics developed in Chapter 16. The approach we describe, which is called **transition state theory** (also widely referred to as *activated complex theory*), has the advantage that a quantity corresponding to the steric factor appears

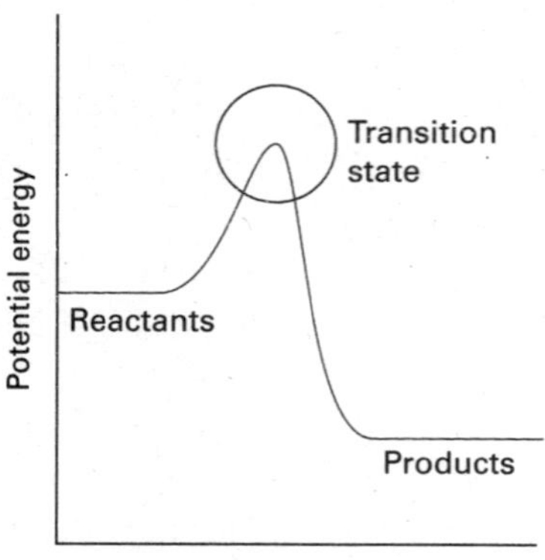

Fig. 22.8 A reaction profile (for an exothermic reaction). The horizontal axis is the reaction coordinate, and the vertical axis is potential energy. The activated complex is the region near the potential maximum, and the transition state corresponds to the maximum itself.

automatically and P does not need to be grafted on to an equation as an afterthought. Transition state theory is an attempt to identify the principal features governing the size of a rate constant in terms of a model of the events that take place during the reaction. There are several approaches to the calculation, all of which lead to the same final expression; here we present the simplest approach.

22.4 The Eyring equation

Key points **(a) In transition state theory, it is supposed that an activated complex is in equilibrium with the reactants, and that the rate at which that complex forms products depends on the rate at which it passes through a transition state. (b) The concentration of the activated complex can be calculated from the partition functions of the participating species. (c) The Eyring equation is an expression for the rate constant in transition state theory. (d) Results of collision and transition state theory agree when considering collisions of structureless particles. (e) Ultrafast laser spectroscopy can be used to observe and manipulate clusters that resemble activated complexes.**

Transition state theory pictures a reaction between A and B as proceeding through the formation of an activated complex, $C^{\ddagger}$, in a rapid pre-equilibrium (Fig. 22.8):

$$A + B \rightleftharpoons C^{\ddagger} \qquad K^{\ddagger} = \frac{p_{C}^{\ddagger} p^{\ominus}}{p_A p_B} \tag{22.28}$$

When we express the partial pressures, p_J, in terms of the molar concentrations, [J], by using $p_J = RT[J]$, the concentration of activated complex is related to the (dimensionless) equilibrium constant by

$$[C^{\ddagger}] = \frac{RT}{p^{\ominus}} K^{\ddagger}[A][B] \tag{22.29}$$

The activated complex falls apart by unimolecular decay into products, P, with a rate constant $k^{\ddagger}$:

$$C^{\ddagger} \rightarrow P \qquad v = k^{\ddagger}[C^{\ddagger}] \tag{22.30}$$

It follows that

$$v = k_r[A][B] \qquad k_r = \frac{RT}{p^{\ominus}} k^{\ddagger} K^{\ddagger} \tag{22.31}$$

Our task is to calculate the unimolecular rate constant $k^{\ddagger}$ and the equilibrium constant $K^{\ddagger}$.

(a) The rate of decay of the activated complex

An activated complex can form products if it passes through the **transition state**, the arrangement the atoms must achieve in order to convert to products (Section 21.5b). If its vibration-like motion along the reaction coordinate occurs with a frequency $\nu^{\ddagger}$, then the frequency with which the cluster of atoms forming the complex approaches the transition state is also $\nu^{\ddagger}$. However, it is possible that not every oscillation along the reaction coordinate takes the complex through the transition state. For instance, the centrifugal effect of rotations might also be an important contribution to the breakup of the complex, and in some cases the complex might be rotating too slowly, or rotating rapidly but about the wrong axis. Therefore, we suppose that the rate of passage of the complex through the transition state is only proportional to rather than equal to the vibrational frequency along the reaction coordinate, and write

A note on good practice We consider it appropriate to distinguish the transition state from the activated complex, but not everyone does so. The *activated complex* is a cluster of atoms formed from the reactants; the *transition state* is the specific configuration of those atoms that is the gateway to the formation of products.

$$k^{\ddagger} = \kappa\nu^{\ddagger} \tag{22.32}$$

Rate constant for passage through the transition state in terms of the transmission coefficient

where κ is the **transmission coefficient.** In the absence of information to the contrary, κ is assumed to be about 1, signifying that almost every visit to the transition state leads on to products.

(b) The concentration of the activated complex

We saw in Section 16.8 how to calculate equilibrium constants from structural data. Equation 16.52a of that section can be used directly, which in this case gives

$$K^{\ddagger} = \frac{N_A q_{C^{\ddagger}}^{\ominus}}{q_A^{\ominus} q_B^{\ominus}} e^{-\Delta_r E_0/RT} \tag{22.33}$$

where $p^{\ominus} = 1$ bar and

$$\Delta_r E_0 = E_0(C^{\ddagger}) - E_0(A) - E_0(B) \tag{22.34}$$

The $q_J^{\ominus}$ are the standard molar partition functions, as defined in Section 16.2. Note that the units of N_A and the $q_J^{\ominus}$ are mol^{-1}, so $K^{\ddagger}$ is dimensionless (as is appropriate for an equilibrium constant).

In the final step of this part of the calculation, we focus attention on the partition function of the activated complex. We have already assumed that a vibration of the activated complex $C^{\ddagger}$ tips it through the transition state. The partition function for this vibration is

$$q = \frac{1}{1 - e^{-h\nu^{\ddagger}/kT}}$$

where $\nu^{\ddagger}$ is its frequency (the same frequency that determines $k^{\ddagger}$). This frequency is much lower than for an ordinary molecular vibration because the oscillation corresponds to the complex falling apart (Fig. 22.9), so the force constant is very low. Therefore, provided that $h\nu^{\ddagger}/kT \ll 1$ the exponential may be expanded and the partition function reduces to

$$q = \frac{1}{1 - \left(1 - \dfrac{h\nu^{\ddagger}}{kT} + \cdots\right)} \approx \frac{kT}{h\nu^{\ddagger}}$$

We can therefore write

$$q_{C^{\ddagger}} \approx \frac{kT}{h\nu^{\ddagger}} \bar{q}_{C^{\ddagger}} \tag{22.35}$$

where $\bar{q}$ denotes the partition function for all the other modes of the complex. The constant $K^{\ddagger}$ is therefore

$$K^{\ddagger} = \frac{kT}{h\nu^{\ddagger}} \bar{K}^{\ddagger} \qquad \bar{K}^{\ddagger} = \frac{N_A \bar{q}_{C^{\ddagger}}^{\ominus}}{q_A^{\ominus} q_B^{\ominus}} e^{-\Delta_r E_0/RT} \tag{22.36}$$

with $\bar{K}^{\ddagger}$ a kind of equilibrium constant, but with one vibrational mode of $C^{\ddagger}$ discarded.

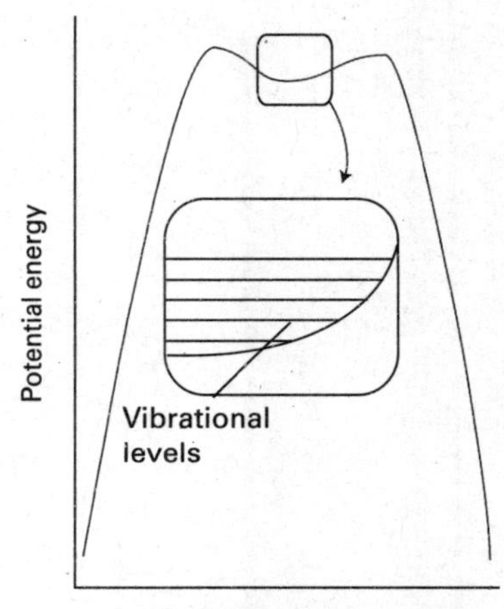

Fig. 22.9 In an elementary depiction of the activated complex close to the transition state, there is a broad, shallow dip in the potential energy surface along the reaction coordinate. The complex vibrates harmonically and almost classically in this well. However, this depiction is an oversimplification, for in many cases there is no dip at the top of the barrier, and the curvature of the potential energy, and therefore the force constant, is negative. Formally, the vibrational frequency is then imaginary. We ignore this problem here.

(c) The rate constant

We can now combine all the parts of the calculation into

$$k_r = k^{\ddagger} \frac{RT}{p^{\ominus}} K^{\ddagger} = \kappa\nu^{\ddagger} \frac{kT}{h\nu^{\ddagger}} \frac{RT}{p^{\ominus}} \bar{K}^{\ddagger} \tag{22.37a}$$

At this stage the unknown frequencies $\nu^{\ddagger}$ cancel and, after writing $\bar{K}_C^{\ddagger} = (RT/p^{\ominus})\bar{K}^{\ddagger}$, we obtain the **Eyring equation**:

$$k_r = \kappa \frac{kT}{h} \bar{K}_C^{\ddagger} \qquad \text{Eyring equation} \qquad (22.37b)$$

The factor $\bar{K}_C^{\ddagger}$ is given by eqn 22.36 and the definition $\bar{K}_C^{\ddagger} = (RT/p^{\ominus})\bar{K}^{\ddagger}$ in terms of the partition functions of A, B, and $C^{\ddagger}$, so in principle we now have an explicit expression for calculating the second-order rate constant for a bimolecular reaction in terms of the molecular parameters for the reactants and the activated complex and the quantity κ.

The partition functions for the reactants can normally be calculated quite readily, using either spectroscopic information about their energy levels or the approximate expressions set out in the Checklist at the end of Chapter 16. The difficulty with the Eyring equation, however, lies in the calculation of the partition function of the activated complex: $C^{\ddagger}$ is difficult to investigate spectroscopically (but see Section 22.4e), and in general we need to make assumptions about its size, shape, and structure. We shall illustrate what is involved in one simple but significant case.

(d) The collision of structureless particles

Consider the case of two structureless particles A and B colliding to give an activated complex that resembles a diatomic molecule. Because the reactants J = A, B are structureless 'atoms', the only contributions to their partition functions are the translational terms:

$$q_J^{\ominus} = \frac{V_m^{\ominus}}{\Lambda_J^3} \qquad \Lambda_J = \frac{h}{(2\pi m_J kT)^{1/2}} \qquad V_m^{\ominus} = \frac{RT}{p^{\ominus}} \qquad (22.38a)$$

The activated complex is a diatomic cluster of mass $m_{C^{\ddagger}} = m_A + m_B$ and moment of inertia I. It has one vibrational mode, but that mode corresponds to motion along the reaction coordinate and therefore does not appear in $\bar{q}_{C^{\ddagger}}$. It follows that the standard molar partition function of the activated complex is

$$\bar{q}_{C^{\ddagger}}^{\ominus} = \left(\frac{2IkT}{\hbar^2}\right)\frac{V_m^{\ominus}}{\Lambda_{C^{\ddagger}}^3} \qquad (22.38b)$$

The moment of inertia of a diatomic molecule of bond length r is μr^2, where $\mu = m_A m_B/(m_A + m_B)$ is the effective mass, so the expression for the rate constant is

$$\begin{aligned} k_r &= \kappa \frac{kT}{h}\frac{RT}{p^{\ominus}}\left(\frac{N_A \Lambda_A^3 \Lambda_B^3}{\Lambda_{C^{\ddagger}}^3 V_m^{\ominus}}\right)\left(\frac{2IkT}{\hbar^2}\right)e^{-\Delta_r E_0/RT} \\ &= \kappa \frac{kT}{h} N_A \left(\frac{\Lambda_A \Lambda_B}{\Lambda_{C^{\ddagger}}}\right)^3 \left(\frac{2IkT}{\hbar^2}\right)e^{-\Delta_r E_0/RT} \\ &= \kappa N_A \left(\frac{8kT}{\pi\mu}\right)^{1/2} \pi r^2 e^{-\Delta_r E_0/RT} \end{aligned} \qquad \text{The rate constant for a reaction between structureless particles} \qquad (22.39)$$

Finally, by identifying $\kappa\pi r^2$ as the reactive cross-section σ^*, we arrive at precisely the same expression as that obtained from simple collision theory (eqn 22.13 with $\Delta_r E_0 = E_a$).

(e) Observation and manipulation of the activated complex

The development of femtosecond pulsed lasers has made it possible to make observations on species that have such short lifetimes that in a number of respects they resemble an activated complex. In a typical experiment designed to detect an activated complex, a femtosecond laser pulse is used to excite a molecule to a dissociative state, and then a second femtosecond pulse is fired at an interval after the dissociating pulse. The frequency of the second pulse is set at an absorption of one of the free fragmentation products, so its absorption is a measure of the abundance of the dissociation product. For example, when ICN is dissociated by the first pulse, the emergence of CN from the photoactivated state can be monitored by watching the growth of the free CN absorption (or, more commonly, its laser-induced fluorescence). In this way it has been found that the CN signal remains zero until the fragments have separated by about 600 pm, which takes about 205 fs.

Some sense of the progress that has been made in the study of the intimate mechanism of chemical reactions can be obtained by considering the decay of the ion pair Na^+I^-. As shown in Fig. 22.10, excitation of the ionic species with a femtosecond laser pulse forms an excited state that corresponds to a covalently bonded NaI molecule. The system can be described with two potential energy surfaces, one largely 'ionic' and another 'covalent', which cross at an internuclear separation of 693 pm. A short laser pulse is composed of a wide range of frequencies, which excite many vibrational states of NaI simultaneously. Consequently, the electronically excited complex exists as a superposition of states, or a localized wavepacket (Section 7.6), which oscillates between the 'covalent' and 'ionic' potential energy surfaces, as shown in Fig. 22.10. The complex can also dissociate, shown as movement of the wavepacket toward very long internuclear separation along the dissociative surface. However, not every outward-going swing leads to dissociation because there is a chance that the I atom can be harpooned again, in which case it fails to make good its escape. The dynamics of the system is probed by a second laser pulse with a frequency that corresponds to the absorption frequency of the free Na product or to the frequency at which Na absorbs when it is a part of the complex. The latter frequency depends on the Na···I distance, so an absorption (in practice, a laser-induced fluorescence) is obtained each time the wavepacket returns to that separation.

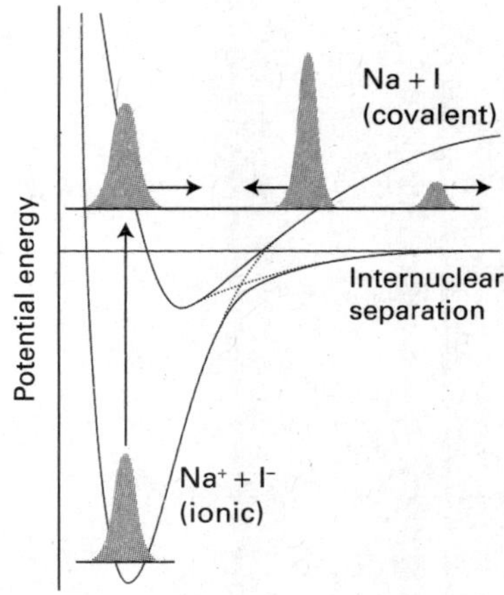

Fig. 22.10 Excitation of the ion pair Na^+I^- forms an excited state with covalent character. Also shown is movement between a 'covalent' surface and an 'ionic' surface of the wavepacket formed by laser excitation.

A typical set of results is shown in Fig. 22.11. The bound Na absorption intensity shows up as a series of pulses that recur in about 1 ps, showing that the wavepacket oscillates with about that period. The decline in intensity shows the rate at which the complex can dissociate as the two atoms swing away from each other. The free Na absorption also grows in an oscillating manner, showing the periodicity of wavepacket oscillation, each swing of which gives it a chance to dissociate. The precise period of the oscillation in NaI is 1.25 ps, corresponding to a vibrational wavenumber of 27 cm^{-1} (recall that the activated complex theory assumes that such a vibration has a very low frequency). The complex survives for about ten oscillations. In contrast, although the oscillation frequency of NaBr is similar, it barely survives one oscillation.

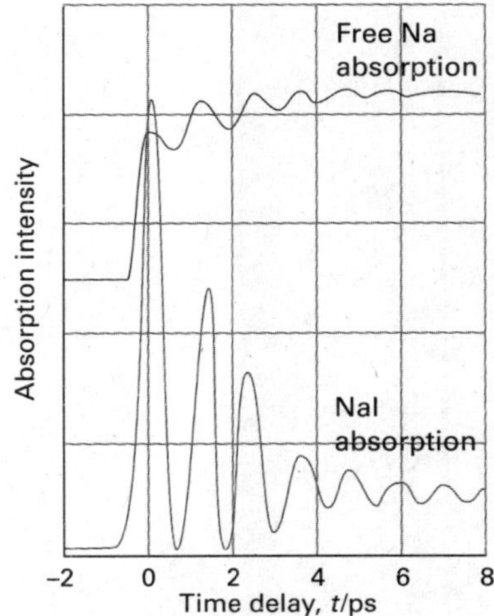

Fig. 22.11 Femtosecond spectroscopic results for the reaction in which sodium iodide separates into Na and I. The lower curve is the absorption of the electronically excited complex and the upper curve is the absorption of free Na atoms. (Adapted from A.H. Zewail, *Science* **242**, 1645 (1988).)

Femtosecond spectroscopy has also been used to examine analogues of the activated complex involved in bimolecular reactions. Thus, a molecular beam can be used to produce a van der Waals molecule (Section 17.7), such as IH···OCO. The HI bond can be dissociated by a femtosecond pulse, and the H atom is ejected towards the O atom of the neighbouring CO_2 molecule to form HOCO. Hence, the van der Waals molecule is a source of a species that resembles the activated complex of the reaction

$$H + CO_2 \rightarrow [HOCO]^\ddagger \rightarrow HO + CO$$

The probe pulse is tuned to the OH radical, which enables the evolution of $[HOCO]^\ddagger$ to be studied in real time. Femtosecond transition state spectroscopy has also been used to study more complex reactions, such as the Diels–Alder reaction, nucleophilic substitution reactions, and pericyclic addition and cleavage reactions. Biological processes that are open to study by femtosecond spectroscopy include the energy-converting processes of photosynthesis and the photostimulated processes of vision. In other experiments, the photoejection of carbon monoxide from myoglobin and the attachment of O_2 to the exposed site have been studied to obtain rate constants for the two processes.

22.5 Thermodynamic aspects

Key points **(a) The rate constant may be parametrized in terms of the Gibbs energy, entropy, and enthalpy of activation. (b) The kinetic salt effect is the effect of an added inert salt on the rate of a reaction between ions.**

The statistical thermodynamic version of transition state theory rapidly runs into difficulties because only in some cases is anything known about the structure of the activated complex. However, the concepts that it introduces, principally that of an equilibrium between the reactants and the activated complex, have motivated a more general, empirical approach in which the activation process is expressed in terms of thermodynamic functions.

(a) Activation parameters

If we accept that $\bar{K}^{\ddagger}$ is an equilibrium constant (despite one mode of $C^{\ddagger}$ having been discarded), we can express it in terms of a **Gibbs energy of activation**, $\Delta^{\ddagger}G$, through the definition

$$\Delta^{\ddagger}G = -RT\ln\bar{K}^{\ddagger} \qquad [22.40]$$

Definition of the Gibbs energy of activation in terms of $\bar{K}^{\ddagger}$

(All the $\Delta^{\ddagger}X$ in this section are *standard* thermodynamic quantities, $\Delta^{\ddagger}X^{\ominus}$, but we shall omit the standard state sign to avoid overburdening the notation.) Then the rate constant becomes

$$k_r = \kappa\frac{kT}{h}\frac{RT}{p^{\ominus}}e^{-\Delta^{\ddagger}G/RT} \qquad (22.41)$$

The Eyring equation in terms of the Gibbs energy of activation

Because $G = H - TS$, the Gibbs energy of activation can be divided into an **entropy of activation**, $\Delta^{\ddagger}S$, and an **enthalpy of activation**, $\Delta^{\ddagger}H$, by writing

$$\Delta^{\ddagger}G = \Delta^{\ddagger}H - T\Delta^{\ddagger}S \qquad (22.42)$$

Relation of the Gibbs energy of activation to $\Delta^{\ddagger}S$ and $\Delta^{\ddagger}H$

When eqn 22.42 is used in eqn 22.41 and κ is absorbed into the entropy term, we obtain

$$k_r = Be^{\Delta^{\ddagger}S/R}e^{-\Delta^{\ddagger}H/RT} \qquad B = \frac{kT}{h}\frac{RT}{p^{\ominus}} \qquad (22.43)$$

The Eyring equation in terms of the entropy and enthalpy of activation

The formal definition of activation energy, $E_a = RT^2(\partial\ln k_r/\partial T)$, then gives $E_a = \Delta^{\ddagger}H + 2RT$, so

$$k_r = e^2Be^{\Delta^{\ddagger}S/R}e^{-E_a/RT} \qquad (22.44)$$

The Eyring equation in terms of the entropy of activation and the activation energy

A brief comment

For reactions of the type $A + B \rightarrow P$ in the gas phase, $E_a = \Delta^{\ddagger}H + 2RT$. For these reactions in solution, $E_a = \Delta^{\ddagger}H + RT$.

from which it follows that the Arrhenius factor A can be identified as

$$A = e^2 B e^{\Delta^{\ddagger}S/R} \tag{22.45}$$

The pre-exponential factor in terms of the entropy of activation

The entropy of activation is negative because two reactant species come together to form one species. However, if there is a reduction in entropy below what would be expected for the simple encounter of A and B, then A will be smaller than that expected on the basis of simple collision theory. Indeed, we can identify that additional reduction in entropy, $\Delta^{\ddagger}S_{\text{steric}}$, as the origin of the steric factor of collision theory, and write

$$P = e^{\Delta^{\ddagger}S_{\text{steric}}/R} \tag{22.46}$$

The steric factor in terms of the entropy of activation

Thus, the more complex the steric requirements of the encounter, the more negative the value of $\Delta^{\ddagger}S_{\text{steric}}$, and the smaller the value of P.

Gibbs energies, enthalpies, entropies, volumes, and heat capacities of activation are widely used to report experimental reaction rates, especially for organic reactions in solution. They are encountered when relationships between equilibrium constants and rates of reaction are explored using **correlation analysis**, in which $\ln K$ (which is equal to $-\Delta_r G^{\ominus}/RT$) is plotted against $\ln k_r$ (which is proportional to $-\Delta^{\ddagger}G/RT$). In many cases the correlation is linear, signifying that, as the reaction becomes thermodynamically more favourable, its rate constant increases (Fig. 22.12). This linear correlation is the origin of the alternative name **linear free energy relation**.

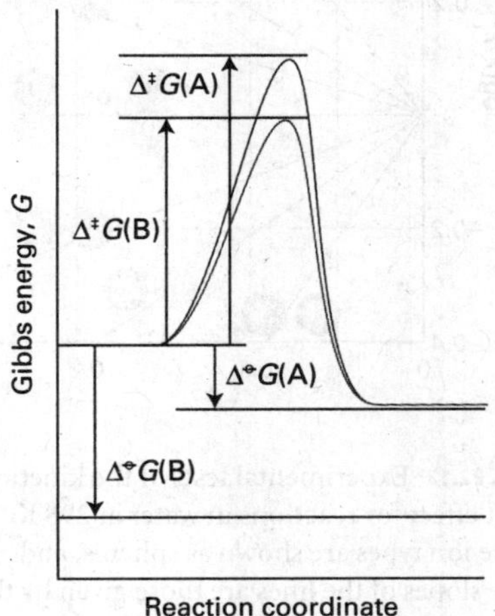

Fig. 22.12 For a related series of reactions, as the magnitude of the standard reaction Gibbs energy increases, so the activation barrier decreases. The approximate linear correlation between $\Delta^{\ddagger}G$ and $\Delta_r G^{\ominus}$ is the origin of linear free energy relations.

(b) Reactions between ions

The thermodynamic version of transition state theory simplifies the discussion of reactions in solution. The statistical thermodynamic theory is very complicated to apply because the solvent plays a role in the activated complex. In the thermodynamic approach we combine the rate law

$$\frac{d[P]}{dt} = k^{\ddagger}[C^{\ddagger}]$$

with the thermodynamic equilibrium constant

$$K = \frac{a_{C^{\ddagger}}}{a_A a_B} = K_{\gamma}\frac{[C^{\ddagger}]c^{\ominus}}{[A][B]} \qquad K_{\gamma} = \frac{\gamma_{C^{\ddagger}}}{\gamma_A \gamma_B}$$

Then

$$\frac{d[P]}{dt} = k_r[A][B] \qquad k_r = \frac{k^{\ddagger}K}{K_{\gamma}} \tag{22.47a}$$

If k_r° is the rate constant when the activity coefficients are 1 (that is, $k_r^{\circ} = k^{\ddagger}K$), we can write

$$k_r = \frac{k_r^{\circ}}{K_{\gamma}} \tag{22.47b}$$

At low concentrations the activity coefficients can be expressed in terms of the ionic strength, I, of the solution by using the Debye–Hückel limiting law (Section 5.13, particularly eqn 5.75.) Although eqn 5.75 is for the mean activity coefficient, eqn 5.95a shows that it is the outcome of two contributions for each type of ion present and that we may write

$$\log \gamma_A = -Az_A^2 I^{1/2} \qquad \log \gamma_B = -Az_B^2 I^{1/2} \tag{22.48a}$$

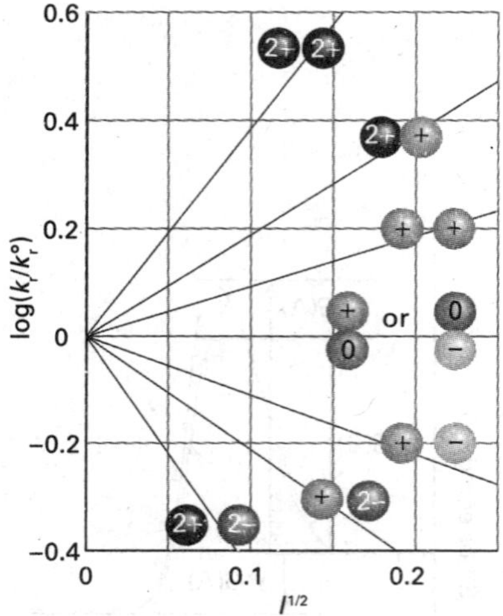

Fig. 22.13 Experimental tests of the kinetic salt effect for reactions in water at 298 K. The ion types are shown as spheres, and the slopes of the lines are those given by the Debye–Hückel limiting law and eqn 22.49.

with $A = 0.509$ in aqueous solution at 298 K and z_A and z_B the charge numbers of A and B, respectively. Since the activated complex forms from reaction of one of the ions of A with one of the ions of B, the charge number of the activated complex is $z_A + z_B$. Therefore

$$\log \gamma_{C^\ddagger} = -A(z_A + z_B)^2 I^{1/2} \tag{22.48b}$$

Inserting these relations into eqn 22.47b results in

$$\log k_r = \log k_r^\circ - A\{z_A^2 + z_B^2 - (z_A + z_B)^2\}I^{1/2} = \log k_r^\circ + 2Az_Az_B I^{1/2} \quad \text{Kinetic salt effect} \tag{22.49}$$

Equation 22.49 expresses the **kinetic salt effect**, the variation of the rate constant of a reaction between ions with the ionic strength of the solution (Fig. 22.13). If the reactant ions have the same sign (as in a reaction between cations or between anions), then increasing the ionic strength by the addition of inert ions increases the rate constant. The formation of a single, highly charged ionic complex from two less highly charged ions is favoured by a high ionic strength because the new ion has a denser ionic atmosphere and interacts with that atmosphere more strongly. Conversely, ions of opposite charge react more slowly in solutions of high ionic strength. Now the charges cancel and the complex has a less favourable interaction with its atmosphere than the separated ions.

Example 22.3 *Analysing the kinetic salt effect*

The rate constant for the base (OH^-) hydrolysis of $[CoBr(NH_3)_5]^{2+}$ varies with ionic strength as tabulated below. What can be deduced about the charge of the activated complex in the rate-determining stage?

I	0.0050	0.0100	0.0150	0.0200	0.0250	0.0300
k_r/k_r°	0.718	0.631	0.562	0.515	0.475	0.447

Method According to eqn 22.49, plot $\log(k_r/k_r^\circ)$ against $I^{1/2}$, when the slope will give $1.02z_Az_B$, from which we can infer the charges of the ions involved in the formation of the activated complex.

Answer Form the following table:

I	0.0050	0.0100	0.0150	0.0200	0.0250	0.0300
$I^{1/2}$	0.071	0.100	0.122	0.141	0.158	0.173
$\log(k_r/k_r^\circ)$	−0.14	−0.20	−0.25	−0.29	−0.32	−0.35

These points are plotted in Fig. 22.14. The slope of the (least squares) straight line is −2.04, indicating that $z_Az_B = -2$. Because $z_A = -1$ for the OH^- ion, if that ion is involved in the formation of the activated complex, then the charge number of the second ion is +2. This analysis suggests that the pentaamminebromocobalt(III) cation participates in the formation of the activated complex. The rate constant is also influenced by the relative permittivity of the medium.

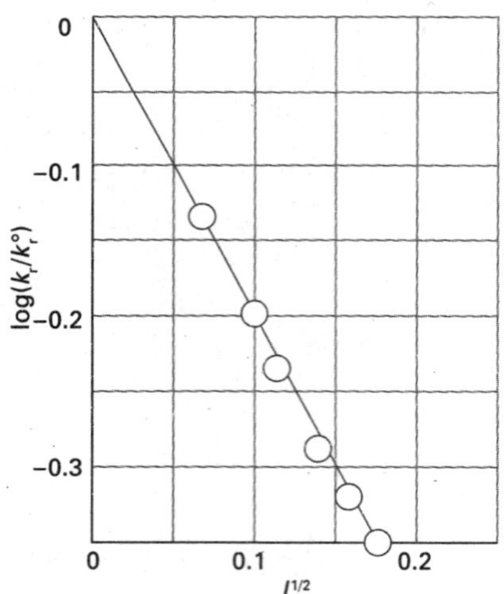

Fig. 22.14 The experimental ionic strength dependence of the rate constant of a hydrolysis reaction: the slope gives information about the charge types involved in the activated complex of the rate-determining step. See Example 22.3.

Self-test 22.3 An ion of charge number +1 is known to be involved in the activated complex of a reaction. Deduce the charge number of the other ion from the following data:

I	0.0015	0.0025	0.0035	0.0045	0.0055	0.0065
k_r/k_r°	0.930	0.902	0.884	0.867	0.853	0.841

[−1]

The dynamics of molecular collisions

We now come to the third and most detailed level of our examination of the factors that govern the rates of reactions.

22.6 Reactive collisions

Key points **(a) Techniques for the study of reactive collisions include infrared chemiluminescence, laser-induced fluorescence, multiphoton ionization, reaction product imaging, and resonant multiphoton ionization. (b) The rate constant of a reaction is the sum of state-to-state rate constants over all final states and over a Boltzmann-weighted sum of initial states.**

Molecular beams allow us to study collisions between molecules in preselected energy states, and can be used to determine the states of the products of a reactive collision. Information of this kind is essential if a full picture of the reaction is to be built, because the rate constant is an average over events in which reactants in different initial states evolve into products in their final states.

(a) Experimental probes of reactive collisions

Detailed experimental information about the intimate processes that occur during reactive encounters comes from molecular beams, especially crossed molecular beams (Fig. 22.15). The detector for the products of the collision of two beams can be moved to different angles, so the angular distribution of the products can be determined. Because the molecules in the incoming beams can be prepared with different energies (for example, with different translational energies by using rotating sectors and supersonic nozzles, with different vibrational energies by using selective excitation with lasers, as shown in Section 22.9b, and with different orientations by using electric fields), it is possible to study the dependence of the success of collisions on these variables and to study how they affect the properties of the outcoming product molecules.

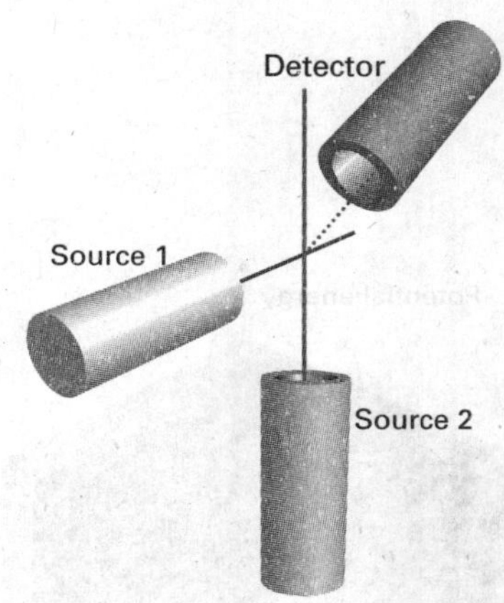

Fig. 22.15 In a crossed-beam experiment, state-selected molecules are generated in two separate sources, and are directed perpendicular to one another. The detector responds to molecules (which may be product molecules if chemical reaction occurs) scattered into a chosen direction.

One method for examining the energy distribution in the products is **infrared chemiluminescence**, in which vibrationally excited molecules emit infrared radiation as they return to their ground states. By studying the intensities of the infrared emission spectrum, the populations of the vibrational states may be determined (Fig. 22.16). Another method makes use of **laser-induced fluorescence**. In this technique, a laser is used to excite a product molecule from a specific vibration–rotation level; the intensity of the fluorescence from the upper state is monitored and interpreted in terms of the population of the initial vibration–rotation state. **Multiphoton ionization** (MPI) techniques are also good alternatives for the study of weakly fluorescing molecules. In MPI, the absorption of several photons by a molecule results in ionization if the total photon energy is greater than the ionization energy of the molecule. One or more pulsed lasers are used to generate the molecular ions, which are commonly detected by time-of-flight mass spectrometry (TOF-MS, Section 18.9). An important variant of MPI is **resonant multiphoton ionization** (REMPI), in which one or more photons promote a molecule to an electronically excited state and then additional photons are used to generate ions from the excited state. The power of REMPI lies in the fact that the experimenter can choose which reactant or product to study by tuning the laser frequency to the electronic absorption band of a specific molecule.

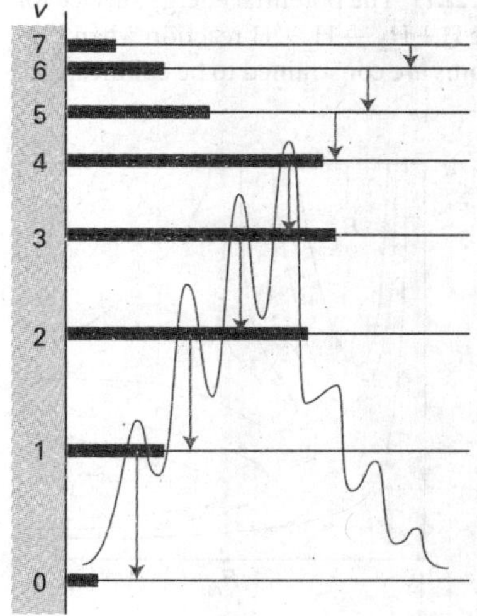

Fig. 22.16 Infrared chemiluminescence from CO produced in the reaction $O + CS \rightarrow CO + S$ arises from the non-equilibrium populations of the vibrational states of CO and the radiative relaxation to equilibrium.

(b) State-to-state dynamics

The concept of collision cross-section was introduced in connection with collision theory in Section 22.1, where we saw that the second-order rate constant, k_r, can be expressed as a Boltzmann-weighted average of the reactive collision cross-section and the relative speed of approach. We shall write eqn 22.10 as

$$k_r = \langle \sigma v_{rel} \rangle N_A \qquad (22.50)$$

where the angle brackets denote a Boltzmann average. Molecular beam studies provide a more sophisticated version of this quantity, for they provide the **state-to-state cross-section**, $\sigma_{nn'}$, and hence the **state-to-state rate constant**, $k_{nn'}$:

$$k_{nn'} = \langle \sigma_{nn'} v_{rel} \rangle N_A \qquad (22.51)$$

The state-to-state rate constant

The rate constant k_r is the sum of the state-to-state rate constants over all final states (because a reaction is successful whatever the final state of the products) and over a Boltzmann-weighted sum of initial states (because the reactants are initially present with a characteristic distribution of populations at a temperature T):

$$k_r = \sum_{n,n'} k_{nn'}(T) f_n(T) \qquad (22.52)$$

The reaction rate constant in terms of the state-to-state rate constant

where $f_n(T)$ is the Boltzmann factor at a temperature T. It follows that, if we can determine or calculate the state-to-state cross-sections for a wide range of approach speeds and initial and final states, then we have a route to the calculation of the rate constant for the reaction.

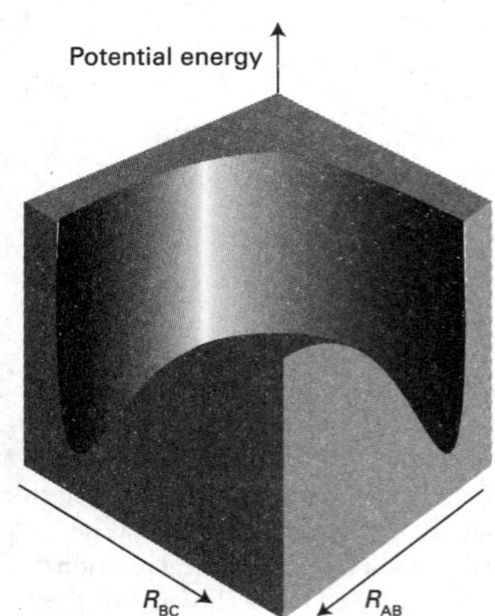

Fig. 22.17 The potential energy surface for the $H + H_2 \rightarrow H_2 + H$ reaction when the atoms are constrained to be collinear.

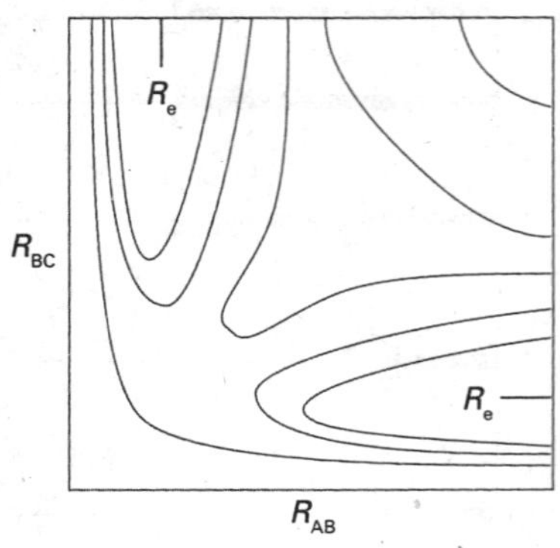

Fig. 22.18 The contour diagram (with contours of equal potential energy) corresponding to the surface in Fig. 22.17. R_e marks the equilibrium bond length of an H_2 molecule (strictly, it relates to the arrangement when the third atom is at infinity).

22.7 Potential energy surfaces

Key point **A potential energy surface maps the potential energy as a function of the relative positions of all the atoms taking part in a reaction.**

One of the most important concepts for discussing beam results and calculating the state-to-state collision cross-section is the **potential energy surface** of a reaction, the potential energy as a function of the relative positions of all the atoms taking part in the reaction. Potential energy surfaces may be constructed from experimental data, with the techniques described in Section 22.6, and from results of quantum chemical calculations. Density functional theory (DFT, Section 10.7c) provides a relatively cheap way of incorporating electron correlation and is increasingly being applied to reactive systems.

To illustrate the features of a potential energy surface we consider the collision between an H atom and an H_2 molecule. Detailed calculations show that the approach of an atom along the H–H axis requires less energy for reaction than any other approach, so initially we confine our attention to a collinear approach. Two parameters are required to define the nuclear separations: one is the H_A–H_B separation R_{AB}, and the other is the H_B–H_C separation R_{BC}.

At the start of the encounter R_{AB} is infinite and R_{BC} is the H_2 equilibrium bond length. At the end of a successful reactive encounter R_{AB} is equal to the equilibrium bond length and R_{BC} is infinite. The total energy of the three-atom system depends on their relative separations, and can be found by doing a molecular orbital calculation. The plot of the total energy of the system against R_{AB} and R_{BC} gives the potential energy surface of this collinear reaction (Fig. 22.17). This surface is normally depicted as a contour diagram (Fig. 22.18).

When R_{AB} is very large, the variations in potential energy represented by the surface as R_{BC} changes are those of an isolated H_2 molecule as its bond length is altered. A section through the surface at $R_{AB} = \infty$, for example, is the same as the H_2 bonding potential energy curve like that shown in Fig. 10.17. At the edge of the diagram where R_{BC} is very large, a section through the surface is the molecular potential energy curve of an isolated H_AH_B molecule.

The actual path of the atoms in the course of the encounter depends on their total energy, the sum of their kinetic and potential energies. However, we can obtain an initial idea of the paths available to the system for paths that correspond to least potential energy. For example, consider the changes in potential energy as H_A approaches H_BH_C. If the H_B–H_C bond length is constant during the initial approach of H_A, the potential energy of the H_3 cluster rises along the path marked A in Fig. 22.19. We see that the potential energy reaches a high value as H_A is pushed into the molecule and then decreases sharply as H_C breaks off and separates to a great distance. An alternative reaction path can be imagined (B) in which the H_B–H_C bond length increases while H_A is still far away. Both paths, although feasible if the molecules have sufficient initial kinetic energy, take the three atoms to regions of high potential energy in the course of the encounter.

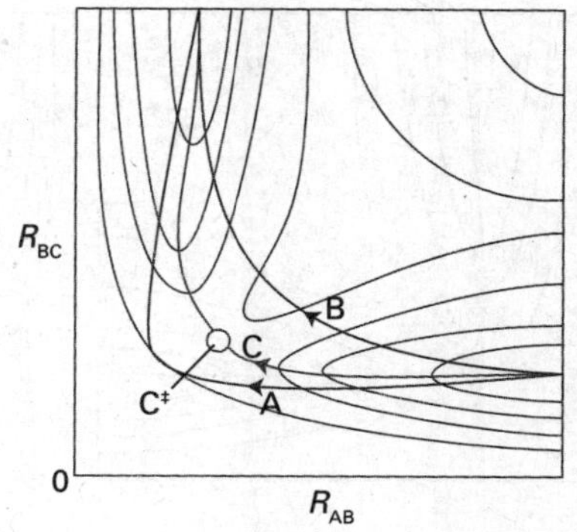

Fig. 22.19 Various trajectories through the potential energy surface shown in Fig. 22.18. Path A corresponds to a route in which R_{BC} is held constant as H_A approaches; path B corresponds to a route in which R_{BC} lengthens at an early stage during the approach of H_A; path C is the route along the floor of the potential valley.

The path of least potential energy is the one marked C, corresponding to R_{BC} lengthening as H_A approaches and begins to form a bond with H_B. The H_B–H_C bond relaxes at the demand of the incoming atom, and the potential energy climbs only as far as the saddle-shaped region of the surface, to the **saddle point** marked C‡. The encounter of least potential energy is one in which the atoms take route C up the floor of the valley, through the saddle point, and down the floor of the other valley as H_C recedes and the new H_A–H_B bond achieves its equilibrium length. This path is the reaction coordinate we met in Section 22.4.

We can now make contact with the transition state theory of reaction rates. In terms of trajectories on potential surfaces, the transition state can be identified with a critical geometry such that every trajectory that goes through this geometry goes on to react (Fig. 22.20).

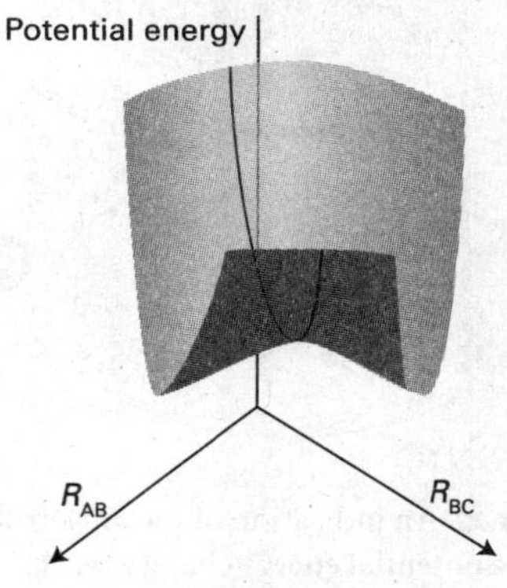

Fig. 22.20 The transition state is a set of configurations (here, marked by the line across the saddle point) through which successful reactive trajectories must pass.

22.8 Some results from experiments and calculations

Key points (a) The direction of approach between reactants affects the distribution of reaction products. (b) In an attractive surface, the saddle point (the highest point) occurs early on the reaction coordinate. In a repulsive surface, the saddle point occurs late on the reaction coordinate. (c) Classical mechanics may be used to map the trajectories of atoms during a reaction. (d) Quantum mechanical scattering theory focuses on the evolution of a wavefunction representing initially the reactants and finally the products.

To travel successfully from reactants to products the incoming molecules must possess enough kinetic energy to be able to climb to the saddle point of the potential surface. Therefore, the shape of the surface can be explored experimentally by changing the relative speed of approach (by selecting the beam velocity) and the degree of vibrational excitation and observing whether reaction occurs and whether the products emerge in a vibrationally excited state (Fig. 22.21). For example, one question that can be answered is whether it is better to smash the reactants together with a lot of translational kinetic energy or to ensure instead that they approach in highly excited vibrational states. Thus, is trajectory C_2^*, where the H_BH_C molecule is initially vibrationally excited, more efficient at leading to reaction than the trajectory C_1^*, in which the total energy is the same but has a high translational kinetic energy?

(a) The direction of attack and separation

Figure 22.22 shows the results of a calculation of the potential energy as an H atom approaches an H_2 molecule from different angles, the H_2 bond being allowed to relax to the optimum length in each case. The potential barrier is least for collinear attack, as

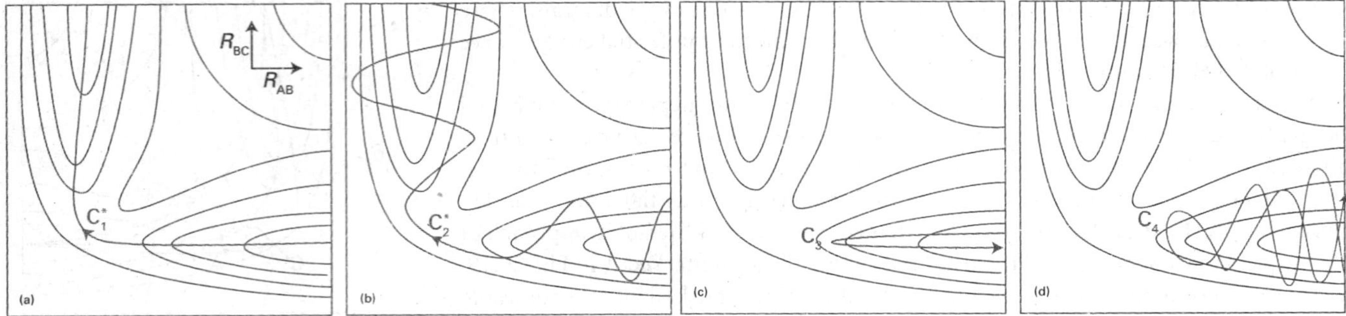

Fig. 22.21 Some successful (*) and unsuccessful encounters. (a) C_1^* corresponds to the path along the foot of the valley; (b) C_2^* corresponds to an approach of A to a vibrating BC molecule, and the formation of a vibrating AB molecule as C departs. (c) C_3 corresponds to A approaching a non-vibrating BC molecule, but with insufficient translational kinetic energy; (d) C_4 corresponds to A approaching a vibrating BC molecule, but still the energy, and the phase of the vibration, is insufficient for reaction.

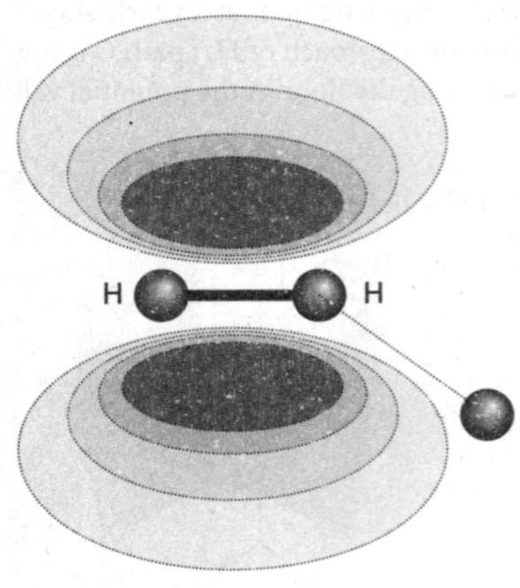

Fig. 22.22 An indication of the anisotropy of the potential energy changes as H approaches H_2 with different angles of attack. The collinear attack has the lowest potential barrier to reaction. The contours indicate the potential energy along the reaction coordinate for each configuration.

we assumed earlier. (But we must be aware that other lines of attack are feasible and contribute to the overall rate.) In contrast, Fig. 22.23 shows the potential energy changes that occur as a Cl atom approaches an HI molecule. The lowest barrier occurs for approaches within a cone of half-angle 30° surrounding the H atom. The relevance of this result to the calculation of the steric factor of collision theory should be noted: not every collision is successful, because not every one lies within the reactive cone.

If the collision is sticky, so that when the reactants collide they orbit around each other, the products can be expected to emerge in random directions because all memory of the approach direction has been lost. A rotation takes about 1 ps, so if the collision is over in less than that time the complex will not have had time to rotate and the products will be thrown off in a specific direction. In the collision of K and I_2, for example, most of the products are thrown off in the forward direction. This product distribution is consistent with the harpoon mechanism (Section 22.1c) because the transition takes place at long range. In contrast, the collision of K with CH_3I leads to reaction only if the molecules approach each other very closely. In this mechanism, K effectively bumps into a brick wall, and the KI product bounces out in the backward direction. The detection of this anisotropy in the angular distribution of products gives an indication of the distance and orientation of approach needed for reaction, as well as showing that the event is complete in less than 1 ps.

(b) Attractive and repulsive surfaces

Some reactions are very sensitive to whether the energy has been predigested into a vibrational mode or left as the relative translational kinetic energy of the colliding molecules. For example, if two HI molecules are hurled together with more than twice the activation energy of the reaction, then no reaction occurs if all the energy is translational. For F + HCl → Cl + HF, for example, the reaction is about five times as efficient when the HCl is in its first vibrational excited state than when, although HCl has the same total energy, it is in its vibrational ground state.

The origin of these requirements can be found by examining the potential energy surface. Figure 22.24 shows an **attractive surface** in which the saddle point occurs early in the reaction coordinate. Figure 22.25 shows a **repulsive surface** in which the saddle point occurs late. A surface that is attractive in one direction is repulsive in the reverse direction.

Consider first the attractive surface. If the original molecule is vibrationally excited, then a collision with an incoming molecule takes the system along *C*. This path is

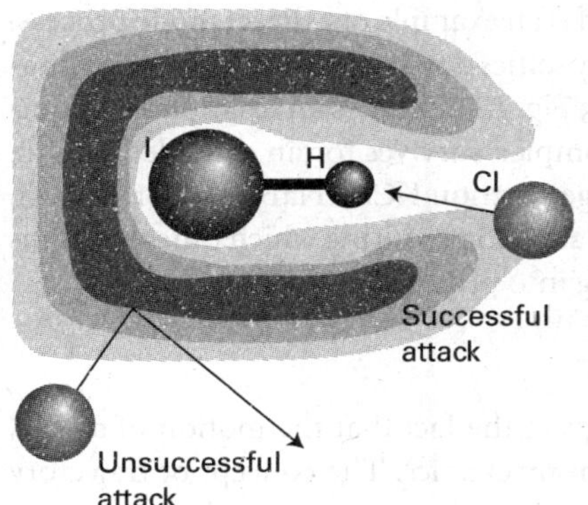

Fig. 22.23 The potential energy barrier for the approach of Cl to HI. In this case, successful encounters occur only when Cl approaches within a cone surrounding the H atom.

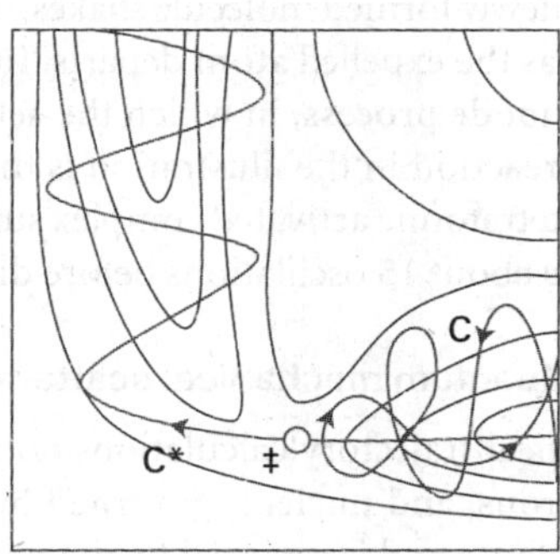

Fig. 22.24 An attractive potential energy surface. A successful encounter (C^*) involves high translational kinetic energy and results in a vibrationally excited product.

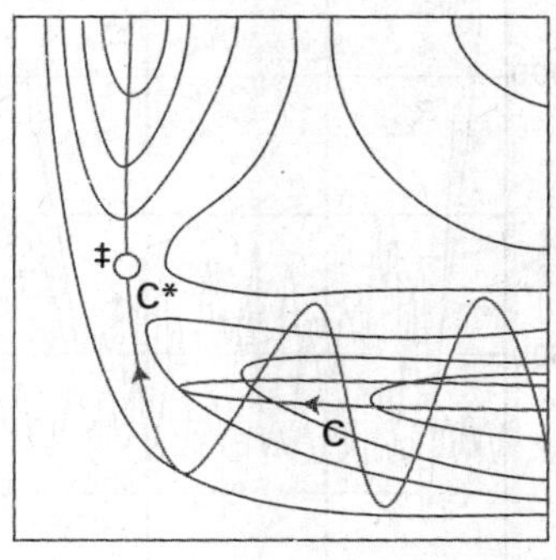

Fig. 22.25 A repulsive potential energy surface. A successful encounter (C^*) involves initial vibrational excitation and the products have high translational kinetic energy. A reaction that is attractive in one direction is repulsive in the reverse direction.

bottled up in the region of the reactants, and does not take the system to the saddle point. If, however, the same amount of energy is present solely as translational kinetic energy, then the system moves along C^* and travels smoothly over the saddle point into products. We can therefore conclude that reactions with attractive potential energy surfaces proceed more efficiently if the energy is in relative translational motion. Moreover, the potential surface shows that once past the saddle point the trajectory runs up the steep wall of the product valley, and then rolls from side to side as it falls to the foot of the valley as the products separate. In other words, the products emerge in a vibrationally excited state.

Now consider the repulsive surface (Fig. 22.25). On trajectory C the collisional energy is largely in translation. As the reactants approach, the potential energy rises. Their path takes them up the opposing face of the valley, and they are reflected back into the reactant region. This path corresponds to an unsuccessful encounter, even though the energy is sufficient for reaction. On C^* some of the energy is in the vibration of the reactant molecule and the motion causes the trajectory to weave from side to side up the valley as it approaches the saddle point. This motion may be sufficient to tip the system round the corner to the saddle point and then on to products. In this case, the product molecule is expected to be in an unexcited vibrational state. Reactions with repulsive potential surfaces can therefore be expected to proceed more efficiently if the excess energy is present as vibrations. This is the case with the $H + Cl_2 \rightarrow HCl + Cl$ reaction, for instance.

(c) Classical trajectories

A clear picture of the reaction event can be obtained by using classical mechanics to calculate the trajectories of the atoms taking place in a reaction from a set of initial conditions, such as velocities, relative orientations, and internal energies of the reacting particles. The initial values used for the internal energy reflect the quantization of electronic, vibrational, and rotational energies in molecules but the features of quantum mechanics are not used explicitly in the calculation of the trajectory.

Figure 22.26 shows the result of such a calculation of the positions of the three atoms in the reaction $H + H_2 \rightarrow H_2 + H$, the horizontal coordinate now being time and the vertical coordinate the separations. This illustration shows clearly the vibration of the original molecule and the approach of the attacking atom. The reaction itself, the

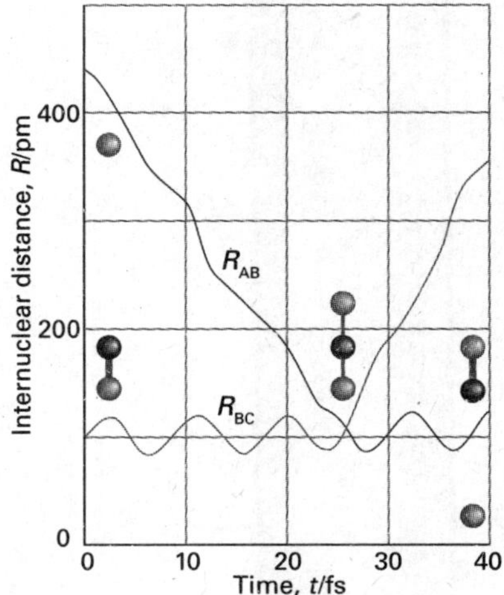

Fig. 22.26 The calculated trajectories for a reactive encounter between A and a vibrating BC molecule leading to the formation of a vibrating AB molecule. This direct-mode reaction is between H and H_2. (M. Karplus, R.N. Porter, and R.D. Sharma, *J. Chem. Phys.* 43, 3258 (1965).)

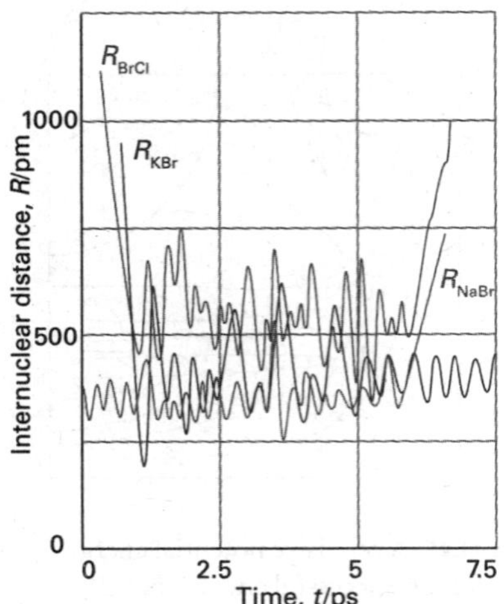

Fig. 22.27 An example of the trajectories calculated for a complex-mode reaction, $KCl + NaBr \rightarrow KBr + NaCl$, in which the collision cluster has a long lifetime. (P. Brumer and M. Karplus, *Faraday Disc. Chem. Soc.*, 55, 80 (1973).)

switch of partners, takes place very rapidly and is an example of a **direct mode process**. The newly formed molecule shakes, but quickly settles down to steady, harmonic vibration as the expelled atom departs. In contrast, Fig. 22.27 shows an example of a **complex mode process**, in which the activated complex survives for an extended period. The reaction in the illustration is the exchange reaction $KCl + NaBr \rightarrow KBr + NaCl$. The tetratomic activated complex survives for about 5 ps, during which time the atoms make about 15 oscillations before dissociating into products.

(d) Quantum mechanical scattering theory

Classical trajectory calculations do not recognize the fact that the motion of atoms, electrons, and nuclei is governed by quantum mechanics. The concept of trajectory then fades and is replaced by the unfolding in time of a wavefunction that represents initially the reactants and finally products.

Complete quantum mechanical calculations of time-dependent wavefunctions and rate constants are very onerous because it is necessary to take into account all the allowed electronic, vibrational, and rotational states populated by each atom and molecule in the system at a given temperature. It is common to define a 'channel' as a group of molecules in well-defined quantum mechanically allowed states. Then, at a given temperature, there are many channels that represent the reactants and many channels that represent possible products, with some transitions between channels being allowed but others not allowed. Furthermore, not every transition leads to a chemical reaction. For example, the process $H_2^* + OH \rightarrow H_2 + (OH)^*$, where the asterisk denotes an excited state, amounts to energy transfer between H_2 and OH, whereas the process $H_2^* + OH \rightarrow H_2O + H$ represents a chemical reaction. What complicates a quantum mechanical calculation of wavefunctions and rate constants even in this simple four-atom system is that many reacting channels present at a given temperature can lead to the desired products $H_2O + H$, which themselves may be formed as many distinct channels. The **cumulative reaction probability**, $\mathcal{P}(E)$, at a fixed total energy E is then written as

$$\mathcal{P}(E) = \sum_{i,j} P_{ij}(E) \qquad \text{Cumulative reaction probability} \qquad (22.53)$$

where $P_{ij}(E)$ is the probability for a transition between a reacting channel i and a product channel j and the summation is over all possible transitions that lead to product. It is then possible to show that the rate constant is given by

$$k_r(T) = \frac{\int_0^\infty \mathcal{P}(E)e^{-E/kT}\,dE}{hQ_R(T)} \qquad \text{The rate constant in terms of the cumulative reaction probability} \qquad (22.54)$$

where $Q_R(T)$ is the partition function density (the partition function divided by the volume) of the reactants at the temperature T. The significance of eqn 22.54 is that it provides a direct connection between an experimental quantity, the rate constant, and a theoretical quantity, $\mathcal{P}(E)$.

The dynamics of electron transfer

We end the chapter by applying the concepts of transition state theory and quantum theory to the study of a deceptively simple process, electron transfer. We begin by examining the features of a theory that describes the factors governing the rates of electron transfer in homogeneous systems. Then, we discuss electron transfer processes occurring on the surfaces of electrodes.

22.9 Electron transfer in homogeneous systems

Key points **(a) Electrons are transferred by tunnelling through a potential energy barrier. (b) The rate constant of electron transfer in a donor–acceptor complex depends on the distance between electron donor and acceptor, the standard reaction Gibbs energy, and the reorganization energy. (c) The key features of the theory of electron transfer in homogeneous systems have been verified experimentally.**

Consider electron transfer from a donor species D to an acceptor species A in solution. The net reaction is

$$D + A \rightleftarrows D^+ + A^- \qquad v = k_r[D][A] \tag{22.55}$$

In the first step of the mechanism, D and A must diffuse through the solution and collide to form a complex DA, in which the donor and acceptor are separated by a distance comparable to *r*, the distance between the edges of each species. The first (reversible) step is the formation of an encounter complex:

$$D + A \rightleftarrows DA \qquad (k_a, k_a') \tag{22.56a}$$

Next, (reversible) electron transfer occurs within the DA complex to yield D^+A^-:

$$DA \rightleftarrows D^+A^- \qquad (k_{et}, k_{et}') \tag{22.56b}$$

The D^+A^- complex can also break apart and the ions diffuse through the solution:

$$D^+A^- \rightarrow D^+ + A^- \qquad (k_d) \tag{22.56c}$$

We show in the following *Justification* that

$$\frac{1}{k_r} = \frac{1}{k_a} + \frac{k_a'}{k_a k_{et}}\left(1 + \frac{k_{et}'}{k_d}\right) \tag{22.57}$$

The rate constant for the overall electron transfer process

Justification 22.5 *The rate constant for electron transfer in solution*

We begin by identifying the rate of the overall reaction (eqn 22.55) to the rate of formation of separated ions, the reaction products:

$$v = k_r[D][A] = k_d[D^+A^-]$$

There are two reaction intermediates, DA and D^+A^-, and we apply the steady-state approximation to both. From

$$\frac{d[D^+A^-]}{dt} = k_{et}[DA] - k_{et}'[D^+A^-] - k_d[D^+A^-] = 0$$

it follows that

$$[DA] = \frac{k_{et}' + k_d}{k_{et}}[D^+A^-]$$

and from

$$\frac{d[DA]}{dt} = k_a[D][A] - k_a'[DA] - k_{et}[DA] + k_{et}'[D^+A^-]$$

$$= k_a[D][A] - \left\{\frac{(k_a' + k_{et})(k_{et}' + k_d)}{k_{et}} - k_{et}'\right\}[D^+A^-] = 0$$

it follows that

$$[D^+A^-] = \frac{k_a k_{et}}{k'_a k'_{et} + k'_a k_d + k_d k_{et}}[D][A]$$

When we multiply this expression by k_d, we see that the resulting equation has the form of the rate of electron transfer, $v = k_r[D][A]$, with k_r given by

$$k_r = \frac{k_d k_a k_{et}}{k'_a k'_{et} + k'_a k_d + k_d k_{et}}$$

To obtain eqn 22.57, we divide the numerator and denominator on the right-hand side of this expression by $k_d k_{et}$ and then form the reciprocal of both sides.

To gain insight into eqn 22.57 and the factors that determine the rate of electron transfer reactions in solution, we assume that the main decay route for D^+A^- is dissociation of the complex into separated ions, or $k_d \gg k'_{et}$. It follows that

$$\frac{1}{k_r} \approx \frac{1}{k_a}\left(1 + \frac{k'_a}{k_{et}}\right)$$

When $k_{et} \gg k'_a$, we see that $k_r \approx k_a$ and the rate of product formation is controlled by diffusion of D and A in solution, which results in the formation of the DA complex. When $k_{et} \ll k'_a$, we see that $k_r = (k_a/k'_a)k_{et}$ or, after using eqn 22.56a,

$$k_r \approx K_{DA} k_{et} \text{ with } K_{DA} = k_a/k'_a \qquad (22.58)$$

The overall rate constant in the limit $k_{et} \ll k'_a$

and the process is controlled by the activation energy of electron transfer in the DA complex. Using transition state theory (Section 22.5), we write

$$k_{et} = \kappa \nu^{\ddagger} e^{-\Delta^{\ddagger}G/RT} \qquad (22.59)$$

where κ is the transmission coefficient, $\nu^{\ddagger}$ is the vibrational frequency with which the activated complex approaches the transition state, and $\Delta^{\ddagger}G$ is the Gibbs energy of activation.

Our first task is to write theoretical expressions for $\kappa\nu^{\ddagger}$ and $\Delta^{\ddagger}G$. The discussion concentrates on the following two key aspects of the theory, which was developed independently by R.A. Marcus, N.S. Hush, V.G. Levich, and R.R. Dogonadze:

- Electrons are transferred by tunnelling through a potential energy barrier, the height of which is partly determined by the ionization energies of the DA and D^+A^- complexes. Electron tunnelling influences the magnitude of $\kappa\nu^{\ddagger}$.
- The complex DA and the solvent molecules surrounding it undergo structural rearrangements prior to electron transfer. The energy associated with these rearrangements and the standard reaction Gibbs energy determine $\Delta^{\ddagger}G$.

(a) The role of electron tunnelling

We saw in Section 13.2c that, according to the Franck–Condon principle, electronic transitions are so fast that they can be regarded as taking place in a stationary nuclear framework. This principle also applies to an electron transfer process in which an electron migrates from one energy surface, representing the dependence of the energy of DA on its geometry, to another representing the energy of D^+A^-. We can represent the potential energy (and the Gibbs energy) surfaces of the two complexes (the reactant complex, DA, and the product complex, D^+A^-) by the parabolas characteristic of harmonic oscillators, with the displacement coordinate q corresponding to the changing geometries (Fig. 22.28). This coordinate represents a collective mode of the donor, acceptor, and solvent.

According to the Franck–Condon principle, the nuclei do not have time to move when the system passes from the reactant to the product surface as a result of the transfer of an electron. Therefore, electron transfer can occur only after thermal fluctuations bring the geometry of DA to q^* in Fig. 22.28, the value of the nuclear coordinate at which the two parabolas intersect.

The factor $\kappa\nu^{\ddagger}$ is a measure of the probability that the system will convert from reactants (DA) to products (D^+A^-) at q^* by electron transfer within the thermally excited DA complex. To understand the process, we must turn our attention to the effect that the rearrangement of nuclear coordinates has on electronic energy levels of DA and D^+A^- for a given distance r between D and A (Fig. 22.29). Initially, the electron to be transferred occupies the HOMO of D, and the overall energy of DA is lower than that of D^+A^- (Fig. 22.29a). As the nuclei rearrange to a configuration represented by q^* in Fig. 22.29b, the highest occupied electronic level of DA and the lowest unoccupied electronic level of D^+A^- converge in energy and electron transfer becomes energetically feasible. Over reasonably short distances r, the main mechanism of electron transfer is tunnelling through the potential energy barrier depicted in Fig. 22.29b. The height of the barrier increases with the ionization energies of the DA and D^+A^- complexes. After an electron moves from the HOMO of D to the LUMO of A, the system relaxes to the configuration represented by $q_0^{\rm P}$ in Fig. 22.29c. As shown in the illustration, now the energy of D^+A^- is lower than that of DA, reflecting the thermodynamic tendency for A to remain reduced and for D to remain oxidized.

The tunnelling event responsible for electron transfer is similar to that described in Section 8.3 where we saw that the transmission probability decreases exponentially with the thickness of the barrier. So the theory supposes similarly that the probability of electron transfer from D to A decreases exponentially with increasing distance between D and A in the DA complex and that for fixed values of the temperature and the Gibbs energy of activation, the rate constant $k_{\rm et}$ varies with the edge-to-edge distance r as[2]

$$k_{\rm et} \propto e^{-\beta r} \qquad (22.60)$$

The value of β depends on the medium through which the electron must travel from donor to acceptor. In a vacuum, $28\ {\rm nm}^{-1} < \beta < 35\ {\rm nm}^{-1}$, whereas $\beta \approx 9\ {\rm nm}^{-1}$ when the intervening medium is a molecular link.

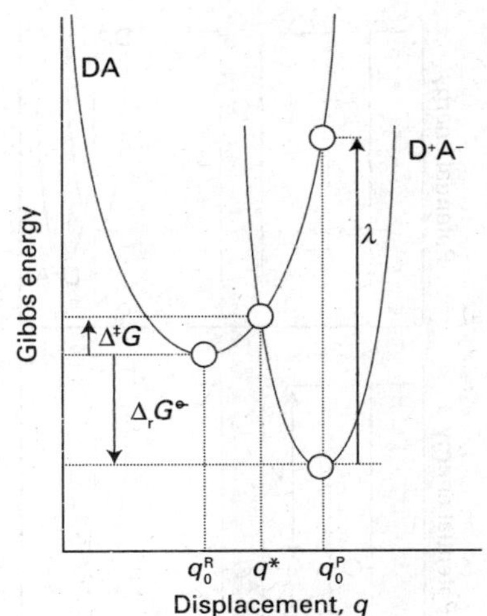

Fig. 22.28 The Gibbs energy surfaces of the complexes DA and D^+A^- involved in an electron transfer process are represented by parabolas characteristic of harmonic oscillators, with the displacement coordinate q corresponding to the changing geometries of the system. In the plot, $q_0^{\rm R}$ and $q_0^{\rm P}$ are the values of q at which the minima of the reactant and product parabolas occur, respectively. The parabolas intersect at $q = q^*$. The plots also portray the Gibbs energy of activation, $\Delta^{\ddagger}G$, the standard reaction Gibbs energy, $\Delta_{\rm r}G^{\ominus}$, and the reorganization energy, λ (discussed in Section 22.9b).

(b) The expression for the rate of electron transfer

Marcus noted that the DA complex and the medium surrounding it must rearrange spatially as charge is redistributed to form the ions D^+ and A^-. These molecular rearrangements, which contribute to the Gibbs energy of activation, $\Delta^{\ddagger}G$, include the relative reorientation of the D and A molecules in DA and the relative reorientation of the solvent molecules surrounding DA. We show in *Further information 22.1* that

$$\Delta^{\ddagger}G = \frac{(\Delta_{\rm r}G^{\ominus} + \lambda)^2}{4\lambda} \qquad (22.61)$$

The Gibbs energy of activation for electron transfer

where $\Delta_{\rm r}G^{\ominus}$ is the standard reaction Gibbs energy for the electron transfer process $DA \rightarrow D^+A^-$, and λ is the **reorganization energy**, the energy needed for the molecular

[2] For a full mathematical treatment of Marcus theory, see our *Quanta, matter, and change* (2009).

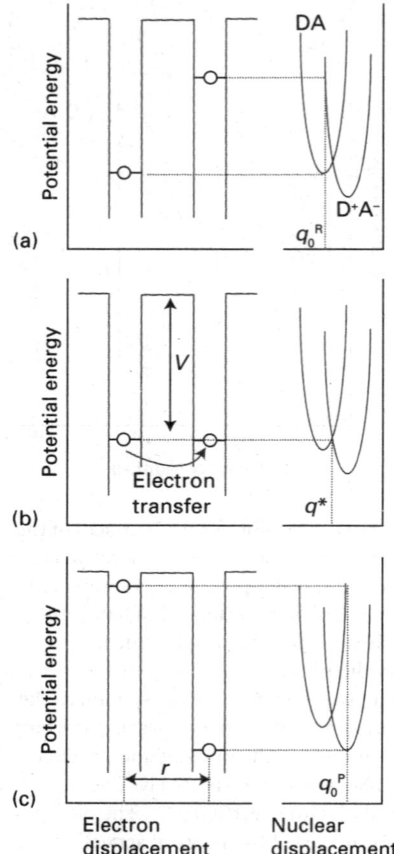

Fig. 22.29 Correspondence between the electronic energy levels (shown on the left) and the nuclear energy levels (shown on the right) for the DA and D^+A^- complexes involved in an electron transfer process. (a) At the nuclear configuration denoted by q_0^R, the electron to be transferred in DA is in an occupied electronic energy level and the lowest unoccupied energy level of D^+A^- is of too high an energy to be a good electron acceptor. (b) As the nuclei rearrange to a configuration represented by q^*, DA and D^+A^- become degenerate and electron transfer occurs by tunnelling through the barrier of height V and width r, the edge-to-edge distance between donor and acceptor. (c) The system relaxes to the equilibrium nuclear configuration of D^+A^- denoted by q_0^P, in which the lowest unoccupied electronic level of DA is higher in energy than the highest occupied electronic level of D^+A^-. (Adapted from R.A. Marcus and N. Sutin, *Biochim. Biophys. Acta* **811**, 265 (1985).)

rearrangement that must take place for DA to attain the equilibrium geometry of D^+A^-. Equation 22.61 shows that $\Delta^{\ddagger}G = 0$ when $\Delta_r G^{\ominus} = -\lambda$, with the implication that the reaction is not slowed down by an activation barrier when the reorganization energy is equal to the standard reaction Gibbs energy.

Equations 22.59 and 22.60 suggest that the expression for k_{et} has the form

$$k_{et} \propto e^{-\beta r} e^{-\Delta^{\ddagger}G/RT} \qquad \text{The rate constant of electron transfer} \qquad (22.62)$$

However, this expression has some limitations which are revealed by the detailed derivation. First, it describes only systems in which the electroactive species are so far apart that the wavefunctions for D and A do not overlap extensively. An example of such a 'weakly coupled system' is the cytochrome *c*/cytochrome b_5 complex, in which the electroactive haem-bound Fe atoms shuttle between oxidation states +2 and +3 during electron transfer and are about 1.7 nm apart. Examples of 'strongly coupled systems', in which wavefunctions for D and A do overlap extensively, are mixed-valence, binuclear d-metal complexes with the general structure $L_m M^{n+}\text{–B–}M^{p+}L_m$ with the electroactive metal ions separated by a bridging ligand B. In these systems, $r < 1.0$ nm. The weak coupling limit applies to a large number of electron transfer reactions, including those between proteins during metabolism.

A second limitation of eqn 22.62 is that it applies only at high temperatures. At low temperatures, thermal fluctuations alone cannot bring the reactants to the transition state and transition state theory, which is at the heart of the theory, fails to account for any observed electron transfer. Electron transfer can still occur, but by *nuclear tunnelling* from the reactant to the product surfaces. We saw in Section 8.5 that the wavefunctions for the lower levels of the quantum mechanical harmonic oscillator extend significantly beyond classically allowed regions, so an oscillator can tunnel into a region of space in which another oscillator may be found.

(c) Experimental results

It is difficult to measure the distance dependence of k_{et} when the reactants are ions or molecules that are free to move in solution. In such cases, electron transfer occurs after a donor–acceptor complex forms and it is not possible to exert control over r, the edge-to-edge distance. The most meaningful experimental tests of the dependence of k_{et} on r are those in which the same donor and acceptor are positioned at a variety of distances, perhaps by covalent attachment to molecular linkers (**1**, for example). Under these conditions, the term $e^{-\Delta^{\ddagger}G/RT}$ is a constant and, after taking the natural logarithm of eqn 22.60, we obtain

(a) (b) (c) (d) (e)

(f) $R_1 = R_2 = H$ (h) $R_1 = R_2 = Cl$

(g) $R_1 = H$, $R_2 = Cl$

1

$$\ln k_{et} = -\beta r + \text{constant} \tag{22.63}$$

Distance dependence of the rate constant of electron transfer

This expression implies that a plot of $\ln k_{et}$ against r should be a straight line with slope $-\beta$.

The dependence of k_{et} on the standard reaction Gibbs energy has been investigated in systems where the edge-to-edge distance, the reorganization energy, and $\kappa\nu^{\ddagger}$ are constant for a series of reactions. Then eqn 22.62 may be written after taking logarithms of both sides and using eqn 22.61 as

$$\ln k_{et} = -\frac{RT}{4\lambda}\left(\frac{\Delta_r G^{\ominus}}{RT}\right)^2 - \tfrac{1}{2}\left(\frac{\Delta_r G^{\ominus}}{RT}\right) + \text{constant} \tag{22.64}$$

Dependence of the rate constant of electron transfer on the reaction Gibbs energy

This expression implies that a plot of $\ln k_{et}$ (or $\log k_{et}$) against $\Delta_r G^{\ominus}$ (or $-\Delta_r G^{\ominus}$) should be shaped like a downward parabola. It also indicates that the rate constant increases as $\Delta_r G^{\ominus}$ decreases but only up to the stage when $-\Delta_r G^{\ominus} = \lambda$. Beyond that, the reaction enters the **inverted region**, in which the rate constant decreases as the reaction becomes more exergonic ($\Delta_r G^{\ominus}$ becomes more negative). The inverted region has been observed in a series of special compounds in which the electron donor and acceptor are linked covalently to a molecular spacer of known and fixed size (Fig. 22.30).

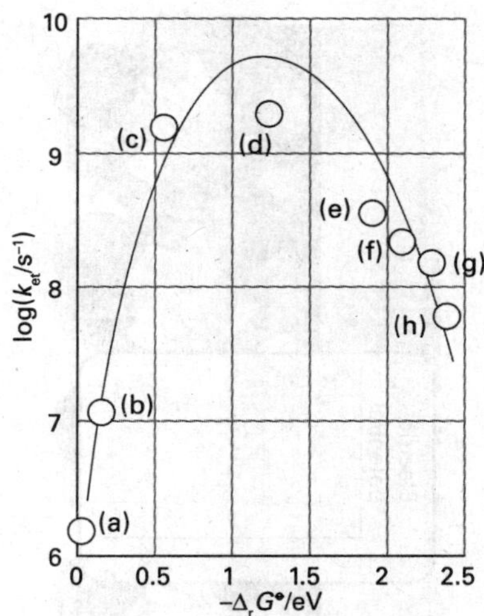

Fig. 22.30 Variation of $\log k_{et}$ with $-\Delta_r G^{\ominus}$ for a series of compounds with the structures given in (1). Kinetic measurements were conducted in 2-methyltetrahydrofuran and at 296 K. The distance between donor (the reduced biphenyl group) and the acceptor is constant for all compounds in the series because the molecular linker remains the same. Each acceptor has a characteristic standard reduction potential, so it follows that the standard Gibbs energy for the electron transfer process is different for each compound in the series. The line is a fit to a version of eqn 22.64 and the maximum of the parabola occurs at $-\Delta_r G^{\ominus} = \lambda = 1.2\ \text{eV} = 1.2 \times 10^2\ \text{kJ mol}^{-1}$. (Reproduced with permission from J.R. Miller *et al., J. Am. Chem. Soc.* **106**, 3047 (1984).)

22.10 Electron transfer processes at electrodes

Key points **(a) An electrical double layer consists of sheets of opposite charge at the surface of the electrode and next to it in the solution. Models of the double layer include the Helmholtz layer model and the Gouy–Chapman model. The Galvani potential difference is the potential difference between the bulk of the metal electrode and the bulk of the solution. (b) The current density at an electrode is expressed by the Butler–Volmer equation. (c) To induce current to flow through an electrolytic cell and bring about a nonspontaneous cell reaction, the applied potential difference must exceed the cell potential by at least the cell overpotential. (d) In working galvanic cells the overpotential leads to a smaller potential than under zero-current conditions and the cell potential decreases as current is generated.**

As for homogeneous systems (Section 22.9), electron transfer at the surface of an electrode involves electron tunnelling. However, the electrode possesses a nearly infinite number of closely spaced electronic energy levels rather than the small number of discrete levels of a typical complex. Furthermore, specific interactions with the electrode surface give the solute and solvent special properties that can be very different from those observed in the bulk of the solution. For this reason, we begin with a description of the electrode–solution interface. Then, we describe the kinetics of electrode processes by using a largely phenomenological (rather than strictly theoretical) approach that draws on the thermodynamic language inspired by transition state theory.

(a) The electrode–solution interface

The most primitive model of the boundary between the solid and liquid phases is as an **electrical double layer**, which consists of a sheet of positive charge at the surface of the electrode and a sheet of negative charge next to it in the solution (or vice versa). We shall see that this arrangement creates an electrical potential difference, called the **Galvani potential difference**, between the bulk of the metal electrode and the bulk of the solution.

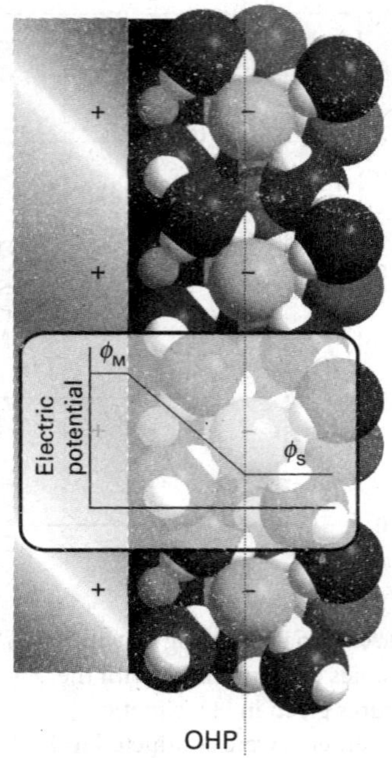

Fig. 22.31 A simple model of the electrode–solution interface treats it as two rigid planes of charge. One plane, the outer Helmholtz plane (OHP), is due to the ions with their solvating molecules and the other plane is that of the electrode itself. The plot shows the dependence of the electric potential with distance from the electrode surface according to this model. Between the electrode surface and the OHP, the potential varies linearly from ϕ_M, the value in the metal, to ϕ_S, the value in the bulk of the solution.

More sophisticated models for the electrode–solution interface attempt to describe the gradual changes in the structure of the solution between two extremes, one the charged electrode surface and the other the bulk solution. In the **Helmholtz layer model** of the interface the solvated ions arrange themselves along the surface of the electrode but are held away from it by their hydration spheres (Fig. 22.31). The location of the sheet of ionic charge, which is called the **outer Helmholtz plane** (OHP), is identified as the plane running through the solvated ions. In this simple model, the electrical potential changes linearly within the layer bounded by the electrode surface on one side and the OHP on the other. In a refinement of this model, ions that have discarded their solvating molecules and have become attached to the electrode surface by chemical bonds are regarded as forming the **inner Helmholtz plane** (IHP). The Helmholtz layer model ignores the disrupting effect of thermal motion, which tends to break up and disperse the rigid outer plane of charge. In the **Gouy-Chapman model** of the **diffuse double layer**, the disordering effect of thermal motion is taken into account in much the same way as the Debye–Hückel model describes the ionic atmosphere of an ion (Section 5.13) with the latter's single central ion replaced by an infinite, plane electrode.

Figure 22.32 shows how the local concentrations of cations and anions differ in the Gouy–Chapman model from their bulk concentrations. Ions of opposite charge cluster close to the electrode and ions of the same charge are repelled from it. The modification of the local concentrations near an electrode implies that it might be misleading to use activity coefficients characteristic of the bulk to discuss the thermodynamic properties of ions near the interface. This is one of the reasons why measurements of the dynamics of electrode processes are almost always done using a large excess of supporting electrolyte (for example, a 1 M solution of a salt, an acid, or a base). Under such conditions, the activity coefficients are almost constant because the inert ions dominate the effects of local changes caused by any reactions taking place.

Neither the Helmholtz nor the Gouy–Chapman model is a very good representation of the structure of the double layer. The former overemphasizes the rigidity of the local solution; the latter underemphasizes its structure. The two are combined in the **Stern model**, in which the ions closest to the electrode are constrained into a rigid Helmholtz plane while outside that plane the ions are dispersed as in the Gouy–Chapman model (Fig. 22.33). Yet another level of sophistication is found in the **Grahame model**, which adds an inner Helmholtz plane to the Stern model.

The potential difference between points in the bulk metal and the bulk solution is the **Galvani potential difference**, $\Delta\phi$. Apart from a constant, this Galvani potential difference is the electrode potential that was discussed in Chapter 6. We shall ignore the constant, which cannot be measured anyway, and identify changes in $\Delta\phi$ with changes in electrode potential.

(b) The Butler–Volmer equation

We shall consider a reaction at the electrode in which an ion is reduced by the transfer of a single electron in the rate-determining step. The quantity we focus on is the **current density**, j, the electric current flowing through a region of an electrode divided by the area of the region. We show in *Further information 22.2* that an analysis of the effect of the Galvani potential difference at the electrode on the current density leads to the **Butler–Volmer equation**:

$$j = j_0\{e^{(1-\alpha)f\eta} - e^{-\alpha f\eta}\} \quad \text{Butler–Volmer equation} \quad (22.65)$$

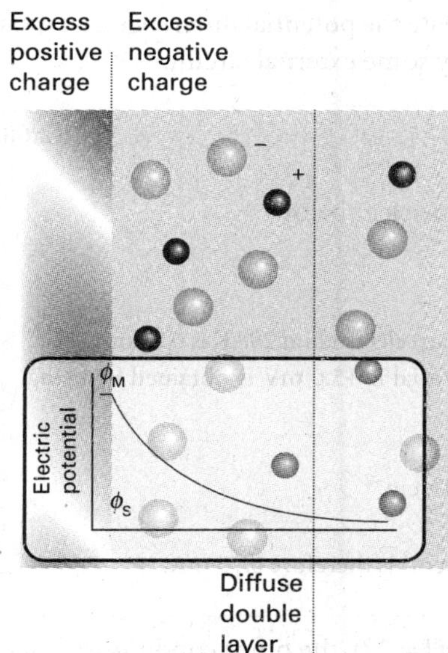

Fig. 22.32 The Gouy–Chapman model of the electrical double layer treats the outer region as an atmosphere of counter-charge, similar to the Debye–Hückel theory of ion atmospheres. The plot of electrical potential against distance from the electrode surface shows the meaning of the diffuse double layer (see text for details).

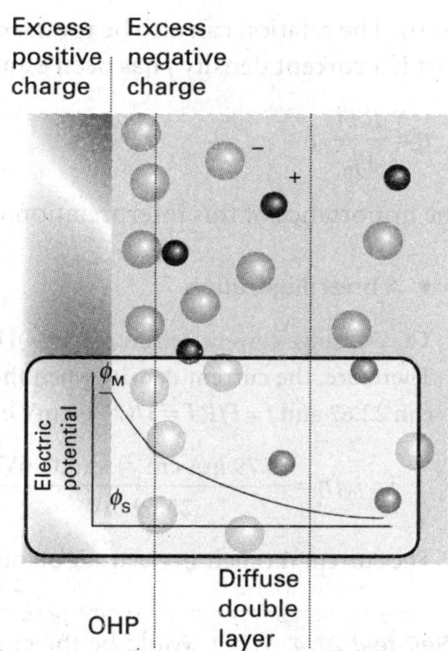

Fig. 22.33 A representation of the Stern model of the electrode–solution interface. The model incorporates the idea of an outer Helmholtz plane near the electrode surface and of a diffuse double layer further away from the surface.

where we have written $f = F/RT$, with F as Faraday's constant. The quantity η (eta) is the **overpotential**:

$$\eta = E' - E \qquad \text{Definition of the overpotential} \qquad [22.66]$$

where E is the electrode potential at equilibrium (when there is no net flow of current), and E' is the electrode potential when a current is being drawn from the cell. The quantity α is the **transfer coefficient**, and is an indication of whether the transition state between the reduced and oxidized forms of the electroactive species in solution is reactant-like ($\alpha = 0$) or product-like ($\alpha = 1$). The quantity j_0 is the **exchange-current density**, the magnitude of the equal but opposite current densities when the electrode is at equilibrium. Figure 22.34 shows how eqn 22.65 predicts the current density to depend on the overpotential for different values of the transfer coefficient.

When the overpotential is so small that $f\eta \ll 1$ (in practice, η less than about 0.01 V) the exponentials in eqn 22.65 can be expanded by using $e^x = 1 + x + \cdots$ to give

$$j = j_0\{1 + (1-\alpha)f\eta + \cdots - (1 - \alpha f\eta + \cdots)\} \approx j_0 f\eta \qquad (22.67)$$

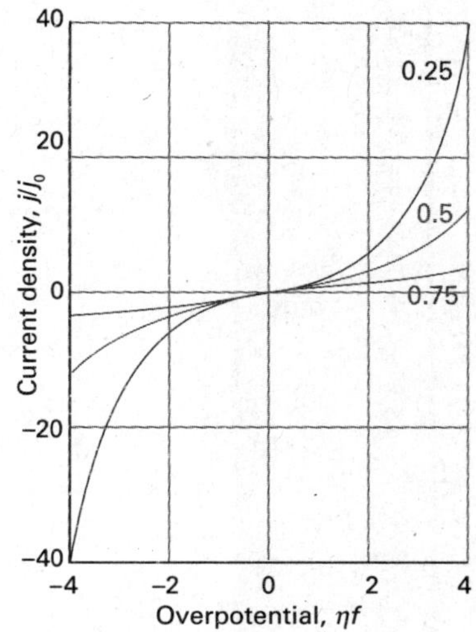

Fig. 22.34 The dependence of the current density on the overpotential for different values of the transfer coefficient.

This equation shows that the current density is proportional to the overpotential, so at low overpotentials the interface behaves like a conductor that obeys Ohm's law. When there is a small positive overpotential the current is anodic ($j > 0$ when $\eta > 0$), and when the overpotential is small and negative the current is cathodic ($j < 0$ when

$\eta < 0$). The relation can also be reversed to calculate the potential difference that must exist if a current density j has been established by some external circuit:

$$\eta = \frac{RTj}{Fj_0} \qquad (22.68)$$

The importance of this interpretation will become clear below.

• A brief illustration

The exchange current density of a $Pt(s)|H_2(g)|H^+(aq)$ electrode at 298 K is 0.79 mA cm^{-2}. Therefore, the current density when the overpotential is +5.0 mV is obtained by using eqn 22.67 and $f = F/RT = 1/(25.69\text{ mV})$:

$$j = j_0 f\eta = \frac{(0.79\text{ mA cm}^{-2}) \times (5.0\text{ mV})}{25.69\text{ mV}} = 0.15\text{ mA cm}^{-2}$$

The current through an electrode of total area 5.0 cm^2 is therefore 0.75 mA. •

Self-test 22.4 What would be the current at pH = 2.0, the other conditions being the same? [−18 mA (cathodic)]

When the overpotential is large and positive (in practice, $\eta \geq 0.12$ V), corresponding to the electrode being the anode in electrolysis, the second exponential in eqn 22.65 is much smaller than the first, and may be neglected. Then

$$j = j_0 e^{(1-\alpha)f\eta} \qquad (22.69)$$

so

$$\ln j = \ln j_0 + (1 - \alpha)f\eta \qquad (22.70)$$

When the overpotential is large but negative (in practice, $\eta < -0.12$ V), corresponding to the cathode in electrolysis, the first exponential in eqn 22.65 may be neglected. Then

$$j = -j_0 e^{-\alpha f\eta} \qquad (22.71)$$

so

$$\ln(-j) = \ln j_0 - \alpha f\eta \qquad (22.72)$$

Some experimental values for the Butler–Volmer parameters are given in Table 22.3. From them we can see that exchange current densities vary over a very wide range. Exchange currents are generally large when the redox process involves no bond breaking (as in the $[Fe(CN)_6]^{3-}$, $[Fe(CN)_6]^{4-}$ couple) or if only weak bonds are broken

Table 22.3* Exchange current densities and transfer coefficients at 298 K

Reaction	Electrode	$j_0/(\text{A cm}^{-2})$	α
$2H^+ + 2e^- \rightarrow H_2$	Pt	7.9×10^{-4}	
	Ni	6.3×10^{-6}	0.58
	Pb	5.0×10^{-12}	
$Fe^{3+} + e^- \rightarrow Fe^{2+}$	Pt	2.5×10^{-3}	0.58

* More values are given in the *Data section*.

(as in Cl_2, Cl^-). They are generally small when more than one electron needs to be transferred, or when multiple or strong bonds are broken, as in the N_2, N_3^- couple and in redox reactions of organic compounds.

(c) Electrolysis

To induce current to flow through an electrolytic cell and bring about a nonspontaneous cell reaction, the applied potential difference must exceed the zero-current potential by at least the **cell overpotential**, the sum of the overpotentials at the two electrodes and the ohmic drop (IR_s, where R_s is the internal resistance of the cell) due to the current through the electrolyte. The additional potential needed to achieve a detectable rate of reaction may need to be large when the exchange current density at the electrodes is small. For similar reasons, a working galvanic cell generates a smaller potential than under zero-current conditions. In this section we see how to cope with both aspects of the overpotential.

The relative rates of gas evolution or metal deposition during electrolysis can be estimated from the Butler–Volmer equation and tables of exchange current densities. From eqn 22.71 and assuming equal transfer coefficients, we write the ratio of the cathodic currents as

$$\frac{j'}{j} = \frac{j_0'}{j_0} e^{(\eta-\eta')\alpha f} \qquad (22.73)$$

where j' is the current density for electrodeposition and j is that for gas evolution, and j_0' and j_0 are the corresponding exchange current densities. This equation shows that metal deposition is favoured by a large exchange current density and relatively high gas evolution overpotential (so $\eta - \eta'$ is positive and large). Note that $\eta < 0$ for a cathodic process, so $-\eta' > 0$. The exchange current density depends strongly on the nature of the electrode surface, and changes in the course of the electrodeposition of one metal on another. A very crude criterion is that significant evolution or deposition occurs only if the overpotential exceeds about 0.6 V.

A glance at Table 22.3 shows the wide range of exchange current densities for a metal/hydrogen electrode. The most sluggish exchange currents occur for lead and mercury, and the value of 1 pA cm^{-2} corresponds to a monolayer of atoms being replaced in about 5 years. For such systems, a high overpotential is needed to induce significant hydrogen evolution. In contrast, the value for platinum (1 mA cm^{-2}) corresponds to a monolayer being replaced in 0.1 s, so gas evolution occurs for a much lower overpotential.

The exchange current density also depends on the crystal face exposed. For the deposition of copper on copper, the (100) face has $j_0 = 1$ mA cm^{-2}, so for the same overpotential the (100) face grows at 2.5 times the rate of the (111) face, for which $j_0 = 0.4$ mA cm^{-2}.

(d) Working galvanic cells

In working galvanic cells (those not balanced against an external potential), the overpotential results in a smaller generated potential than under zero-current conditions. Furthermore, we expect the cell potential to decrease as current is generated because it is then no longer working reversibly and can therefore do less than maximum work.

We shall consider the cell $M|M^+(aq)||M'^+(aq)|M'$ and ignore all the complications arising from liquid junctions. The potential of the cell is $E' = \Delta\phi_R - \Delta\phi_L$. Because the cell potential differences differ from their zero-current values by overpotentials, we can write $\Delta\phi_X = E_X + \eta_X$ where X is L or R for the left or right electrode, respectively. The cell potential is therefore

$$E' = E + \eta_R - \eta_L \tag{22.74a}$$

To avoid confusion about signs (η_R is negative; η_L is positive) and to emphasize that a working cell has a lower potential than a zero-current cell, we shall write this expression as

$$E' = E - |\eta_R| - |\eta_L| \tag{22.74b}$$

with E the cell potential. We should also subtract the ohmic potential difference IR_s, where R_s is the cell's internal resistance:

$$E' = E - |\eta_R| - |\eta_L| - IR_s \tag{22.74c}$$

The ohmic term is a contribution to the cell's irreversibility—it is a thermal dissipation term—so the sign of IR_s is always such as to reduce the potential in the direction of zero.

The overpotentials in eqn 22.74 can be calculated from the Butler–Volmer equation for a given current, I, being drawn. We shall simplify the equations by supposing that the areas, A, of the electrodes are the same, that only one electron is transferred in the rate-determining steps at the electrodes, that the transfer coefficients are both $\frac{1}{2}$, and that the high-overpotential limit of the Butler–Volmer equation may be used. Then from eqns 22.71 and 22.74c we find

$$E' = E - IR_s - \frac{4RT}{F}\ln\left(\frac{I}{A\bar{j}}\right) \qquad \bar{j} = (j_{0L}j_{0R})^{1/2} \tag{22.75}$$

where j_{0L} and j_{0R} are the exchange current densities for the two electrodes.

• A brief illustration

Suppose that a cell consists of two electrodes each of area 10 cm^2 with exchange current densities 5 μA cm^{-2} and has internal resistance 10 Ω. At 298 K RT/F = 25.7 mV. The zero-current cell potential is 1.5 V. If the cell is producing a current of 10 mA, its working potential will be

$$E' = 1.5\text{ V} - \overbrace{(10\text{ mA})\times(10\,\Omega)}^{0.10\text{ V}} - \overbrace{4(25.7\text{ mV})\ln\left(\frac{10\text{ mA}}{(10\text{ cm}^2)\times(5\,\mu\text{A cm}^{-2})}\right)}^{0.54\text{ V}} = 0.9\text{ V}$$

We have used 1 A Ω = 1 V. Note that we have ignored various other factors that reduce the cell potential, such as the inability of reactants to diffuse rapidly enough to the electrodes. •

Electric storage cells operate as galvanic cells while they are producing electricity but as electrolytic cells while they are being charged by an external supply. The lead–acid battery is an old device, but one well suited to the job of starting cars (and the only one available). During charging the cathode reaction is the reduction of Pb^{2+} and its deposition as lead on the lead electrode. Deposition occurs instead of the reduction of the acid to hydrogen because the latter has a low exchange current density on lead. The anode reaction during charging is the oxidation of Pb(II) to Pb(IV), which is deposited as the oxide PbO_2. On discharge, the two reactions run in reverse. Because they have such high exchange current densities the discharge can occur rapidly, which is why the lead battery can produce large currents on demand.

IMPACT ON TECHNOLOGY

I22.1 Fuel cells

A fuel cell operates like a conventional galvanic cell with the exception that the reactants are supplied from outside rather than forming an integral part of its construction. A fundamental and important example of a fuel cell is the hydrogen/oxygen cell, such as the ones used in space missions (Fig. 22.35). One of the electrolytes used is concentrated aqueous potassium hydroxide maintained at 200°C and 20–40 atm; the electrodes may be porous nickel in the form of sheets of compressed powder. The cathode reaction is the reduction

$$O_2(g) + 2\,H_2O(l) + 4\,e^- \rightarrow 4\,OH^-(aq) \qquad E^\ominus = +0.40\ \text{V}$$

and the anode reaction is the oxidation

$$H_2(g) + 2\,OH^-(aq) \rightarrow 2\,H_2O(l) + 2\,e^-$$

For the corresponding reduction, $E^\ominus = -0.83$ V. Because the overall reaction

$$2\,H_2(g) + O_2(g) \rightarrow 2\,H_2O(l) \qquad E^\ominus_{cell} = +1.23\ \text{V}$$

is exothermic as well as spontaneous, it is less favourable thermodynamically at 200°C than at 25°C, so the cell potential is lower at the higher temperature. However, the increased pressure compensates for the increased temperature, and $E \approx +1.2$ V at 200°C and 40 atm.

One advantage of the hydrogen/oxygen system is the large exchange current density of the hydrogen reaction. Unfortunately, the oxygen reaction has an exchange current density of only about 0.1 nA cm^{-2}, which limits the current available from the cell. One way round the difficulty is to use a catalytic surface (to increase j_0) with a large surface area. One type of highly developed fuel cell has phosphoric acid as the electrolyte and operates with hydrogen and air at about 200°C; the hydrogen is obtained from a reforming reaction on natural gas:

Anode: $2\,H_2(g) \rightarrow 4\,H^+(aq) + 4\,e^-$

Cathode: $O_2(g) + 4\,H^+(aq) + 4\,e^- \rightarrow 2\,H_2O(l)$

This fuel cell has shown promise for *combined heat and power systems* (CHP systems). In such systems, the waste heat is used to heat buildings or to do work. Efficiency in a CHP plant can reach 80 per cent. The power output of batteries of such cells has reached the order of 10 MW. Although hydrogen gas is an attractive fuel, it has disadvantages for mobile applications: it is difficult to store and dangerous to handle. One possibility for portable fuel cells is to store the hydrogen in carbon nanotubes. It has been shown that carbon nanofibres in herringbone patterns can store huge amounts of hydrogen and result in an energy density (the magnitude of the released energy divided by the volume of the material) twice that of gasoline.

Cells with molten carbonate electrolytes at about 600°C can make use of natural gas directly. Solid-state electrolytes are also used. They include one version in which the electrolyte is a solid polymeric ionic conductor at about 100°C, but in current versions it requires very pure hydrogen to operate successfully. Solid ionic conducting oxide cells operate at about 1000°C and can use hydrocarbons directly as fuel. Until these materials have been developed, one attractive fuel is methanol, which is easy to handle and is rich in hydrogen atoms:

Anode: $CH_3OH(l) + 6\,OH^-(aq) \rightarrow 5\,H_2O(l) + CO_2(g) + 6\,e^-$

Cathode: $O_2(g) + 4\,e^- + 2\,H_2O(l) \rightarrow 4\,OH^-(aq)$

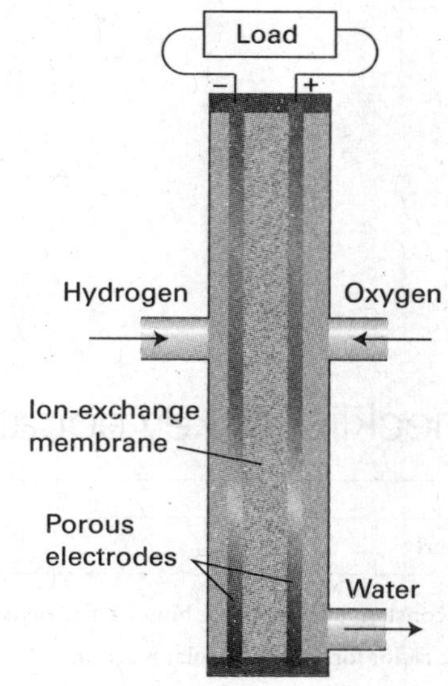

Fig. 22.35 A single cell of a hydrogen/oxygen fuel cell. In practice, a stack of many cells is used.

One disadvantage of methanol, however, is the phenomenon of 'electro-osmotic drag' in which protons moving through the polymer electrolyte membrane separating the anode and cathode carry water and methanol with them into the cathode compartment where the potential is sufficient to oxidize CH_3OH to CO_2, so reducing the efficiency of the cell.

Checklist of key equations

Property	Equation	Comment
Rate constant of a gas-phase bimolecular elementary reaction	$k_r = P\sigma(8kT/\pi\mu)^{1/2}N_A e^{-E_a/RT}$	Collision theory
Steric factor for a unimolecular reaction	$P = (1-(E^*/E))^{s-1}$	RRK model
Rate constant of a diffusion-controlled reaction	$k_d = 4\pi R^* DN_A = 8RT/3\eta$	
Material balance equation	$\partial[J]/\partial t = D\partial^2[J]/\partial x^2 - v\partial[J]/\partial x - k_r[J]$	
Eyring equation	$k_r = \kappa(kT/h)\bar{K}_C^{\ddagger}$	
	$k_r = e^2 B e^{\Delta^{\ddagger}S/R} e^{-E_a/RT}$	
Kinetic salt effect	$\log k_r = \log k_r^{\circ} + 2Az_Az_BI^{1/2}$	
State-to-state rate constant	$k_{nn'} = \langle\sigma_{nn'}v_{rel}\rangle N_A$	
Rate constant in terms of the cumulative	$k_r(T) = \left(\int_0^{\infty} \mathcal{P}(E)e^{-E/kT}\,dE\right)/hQ_R(T)$	
reaction probability	$\mathcal{P}(E) = \sum_{i,j} P_{ij}(E)$	
Rate constant of electron transfer in homogeneous systems	$k_{et} \propto e^{-\beta r}e^{-\Delta^{\ddagger}G/RT}$	Marcus theory
	$\Delta^{\ddagger}G = (\Delta_r G^{\ominus} + \lambda)^2/4\lambda$	
Butler–Volmer equation	$j = j_0\{e^{(1-\alpha)f\eta} - e^{-\alpha f\eta}\}$	

Further information

Further information 22.1 *The Gibbs energy of activation of electron transfer*

The simplest way to derive an expression for the Gibbs energy of activation of electron transfer processes is to construct a model in which the surfaces for DA (the 'reactant complex', denoted R) and for D^+A^- (the 'product complex', denoted P) are described by classical harmonic oscillators with identical effective masses μ and angular frequencies ω, but displaced minima, as shown in Fig. 22.28. The molar Gibbs energies $G_{m,R}(q)$ and $G_{m,P}(q)$ of the reactant and product complexes, respectively, may be written as

$$G_{m,R}(q) = \tfrac{1}{2}N_A\mu\omega^2(q-q_0^R)^2 + G_{m,R}(q_0^R) \qquad (22.76a)$$

$$G_{m,P}(q) = \tfrac{1}{2}N_A\mu\omega^2(q-q_0^P)^2 + G_{m,P}(q_0^P) \qquad (22.76b)$$

where q_0^R and q_0^P are the values of q at which the minima of the reactant and product parabolas occur, respectively. The standard reaction Gibbs energy for the electron transfer process $DA \rightarrow D^+A^-$ is $\Delta_r G^{\ominus} = G_{m,P}(q_0^P) - G_{m,R}(q_0^R)$, the difference in standard molar Gibbs energy between the minima of the parabolas. In Fig. 22.28, $\Delta_r G^{\ominus} < 0$.

We also note that q^*, the value of q corresponding to the transition state of the complex, may be written in terms of the parameter α, the fractional change in q:

$$q^* = q_0^R + \alpha(q_0^P - q_0^R) \qquad (22.77)$$

We see from Fig. 22.28 that $\Delta^{\ddagger}G = G_{m,R}(q^*) - G_{m,R}(q_0^R)$. It then follows from eqns 22.76a, 22.76b, and 22.77 that

$$\Delta^{\ddagger}G = \tfrac{1}{2}N_A\mu\omega^2(q^* - q_0^R)^2 = \tfrac{1}{2}N_A\mu\omega^2\{\alpha(q_0^P - q_0^R)\}^2 \qquad (22.78)$$

We now define the reorganization energy, λ, as

$$\lambda = \tfrac{1}{2}N_A\mu\omega^2(q_0^P - q_0^R)^2 \tag{22.79}$$

which can be interpreted as $G_{m,R}(q_0^P) - G_{m,R}(q_0^R)$ and, consequently, as the (Gibbs) energy required to deform the equilibrium configuration of DA to the equilibrium configuration of D^+A^- (as shown in Fig. 22.28). It follows from eqns 22.78 and 22.79 that

$$\Delta^{\ddagger}G = \alpha^2\lambda \tag{22.80}$$

Because $G_{m,R}(q^*) = G_{m,P}(q^*)$, it follows from eqns 22.76b, 22.77, 22.79, and 22.80 that

$$\begin{aligned}\alpha^2\lambda &= \tfrac{1}{2}N_A\mu\omega^2\{(\alpha-1)(q_0^P - q_0^R)\}^2 + \Delta_r G^{\ominus}\\ &= (\alpha-1)^2\lambda + \Delta_r G^{\ominus}\end{aligned} \tag{22.81}$$

which implies that

$$\alpha = \tfrac{1}{2}\left(\frac{\Delta_r G^{\ominus}}{\lambda} + 1\right) \tag{22.82}$$

By combining eqns 22.80 and 22.82, we obtain eqn 22.61. We can obtain an identical relation if we allow the harmonic oscillators to have different angular frequencies and hence different curvatures.

Further information 22.2 *The Butler–Volmer equation*

Because an electrode reaction is heterogeneous, we express the rate of charge transfer as the flux of products, the amount of material produced over a region of the electrode surface in an interval of time divided by the area of the region and the duration of the interval.

A first-order heterogeneous rate law has the form

$$\text{Product flux} = k_r[\text{species}] \tag{22.83}$$

where [species] is the molar concentration of the relevant species in solution close to the electrode, just outside the double layer. The rate constant has dimensions of length/time (with units, for example, of centimetres per second, cm s^{-1}). If the molar concentrations of the oxidized and reduced materials outside the double layer are [Ox] and [Red], respectively, then the rate of reduction of Ox, v_{Ox}, is

$$v_{Ox} = k_c[\text{Ox}] \tag{22.84a}$$

and the rate of oxidation of Red, v_{Red}, is

$$v_{Red} = k_a[\text{Red}] \tag{22.84b}$$

(The notation k_c and k_a is justified below.)

Consider a reaction at the electrode in which an ion is reduced by the transfer of a single electron in the rate-determining step. The net current density at the electrode is the difference between the current densities arising from the reduction of Ox and the oxidation of Red. Because the redox processes at the electrode involve the transfer of one electron per reaction event, the current densities, j, arising from the redox processes are the rates (as expressed above) multiplied by the charge transferred per mole of reaction, which is given by Faraday's constant. Therefore, there is a **cathodic current density** of magnitude

$$j_c = Fk_c[\text{Ox}] \quad \text{for} \quad \text{Ox} + e^- \rightarrow \text{Red} \tag{22.85a}$$

Cathodic current density

arising from the reduction (because, as we saw in Chapter 6, the cathode is the site of reduction). There is also an opposing **anodic current density** of magnitude

$$j_a = Fk_a[\text{Red}] \quad \text{for} \quad \text{Red} \rightarrow \text{Ox} + e^- \tag{22.85b}$$

Anodic current density

arising from the oxidation (because the anode is the site of oxidation). The net current density at the electrode is the difference

$$j = j_a - j_c = Fk_a[\text{Red}] - Fk_c[\text{Ox}] \tag{22.85c}$$

Note that, when $j_a > j_c$, so that $j > 0$, the current is anodic (Fig. 22.36a); when $j_c > j_a$, so that $j < 0$, the current is cathodic (Fig. 22.36b).

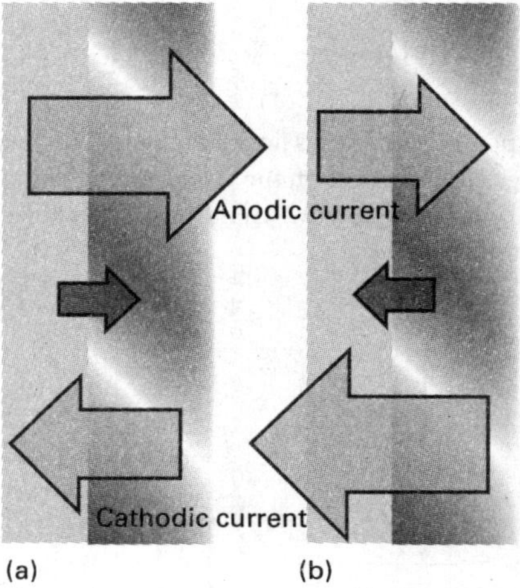

Fig. 22.36 The net current density is defined as the difference between the cathodic and anodic contributions. (a) When $j_a > j_c$, the net current is anodic, and there is a net oxidation of the species in solution. (b) When $j_c > j_a$, the net current is cathodic, and the net process is reduction.

If a species is to participate in reduction or oxidation at an electrode, it must discard any solvating molecules, migrate through the electrode–solution interface, and adjust its hydration sphere as it receives or discards electrons. Likewise, a species already at the inner plane must be detached and migrate into the bulk. Because both processes are activated, we can expect to write their rate constants in the form suggested by transition state theory (Section 22.5) as

$$k_r = Be^{-\Delta^{\ddagger}G/RT} \tag{22.86}$$

where $\Delta^{\ddagger}G$ is the activation Gibbs energy and B is a constant with the same dimensions as k_r.

When eqn 22.86 is inserted into eqn 22.85c we obtain

$$j = FB_a[\text{Red}]e^{-\Delta^{\ddagger}G_a/RT} - FB_c[\text{Ox}]e^{-\Delta^{\ddagger}G_c/RT} \tag{22.87}$$

This expression allows the activation Gibbs energies to be different for the cathodic and anodic processes. That they are different is the central feature of the remaining discussion.

Next, we relate j to the Galvani potential difference, which varies across the electrode–solution interface as shown schematically in Fig. 22.37. Consider the reduction reaction, $\text{Ox} + e^- \rightarrow \text{Red}$, and the corresponding reaction profile. If the transition state of the activated

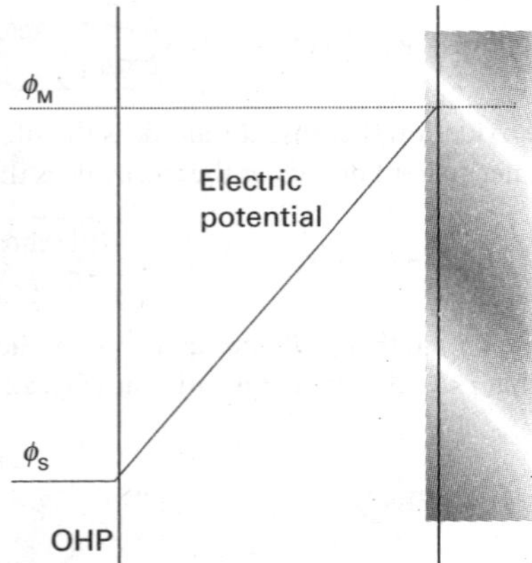

Fig. 22.37 The potential, ϕ, varies linearly between two plane parallel sheets of charge, and its effect on the Gibbs energy of the transition state depends on the extent to which the latter resembles the species at the inner or outer planes.

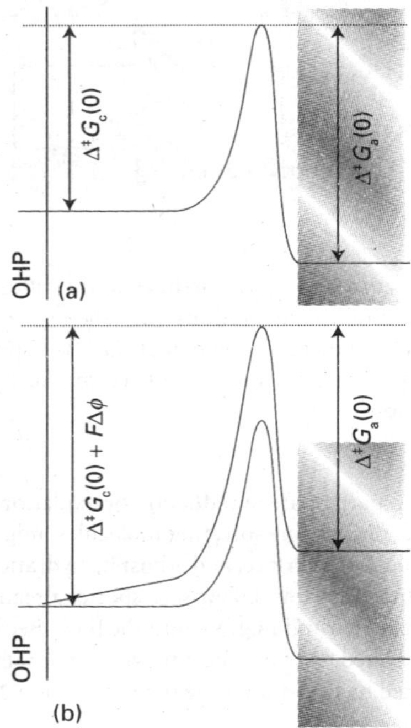

Fig. 22.38 When the transition state resembles a species that has undergone reduction, the activation Gibbs energy for the anodic current is almost unchanged, but the full effect applies to the cathodic current. (a) Zero potential difference; (b) nonzero potential difference.

complex is product-like (as represented by the peak of the reaction profile being close to the electrode in Fig. 22.38), the activation Gibbs energy is changed from $\Delta^{\ddagger}G_c(0)$, the value it has in the absence of a potential difference across the double layer, to

$$\Delta^{\ddagger}G_c = \Delta^{\ddagger}G_c(0) + F\Delta\phi \tag{22.88a}$$

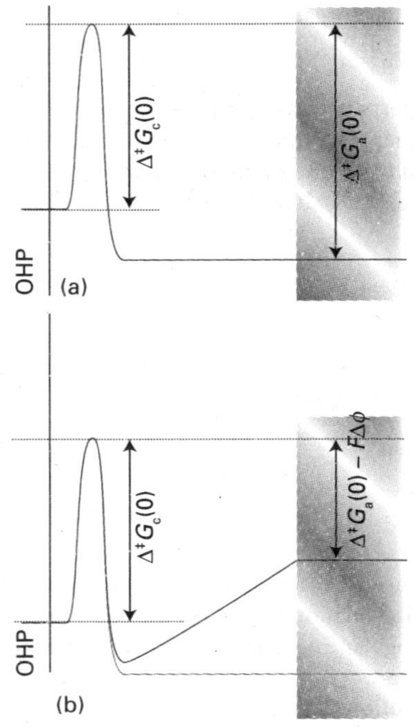

Fig. 22.39 When the transition state resembles a species that has undergone oxidation, the activation Gibbs energy for the cathodic current is almost unchanged but the activation Gibbs energy for the anodic current is strongly affected. (a) Zero potential difference; (b) nonzero potential difference.

Thus, if the electrode is more positive than the solution, $\Delta\phi > 0$, then more work has to be done to form an activated complex from Ox; in this case the activation Gibbs energy is increased. If the transition state is reactant-like (represented by the peak of the reaction profile being close to the outer plane of the double-layer in Fig. 22.39), then $\Delta^{\ddagger}G_c$ is independent of $\Delta\phi$. In a real system, the transition state has an intermediate resemblance to these extremes (Fig. 22.40) and the activation Gibbs energy for reduction may be written as

$$\Delta^{\ddagger}G_c = \Delta^{\ddagger}G_c(0) + \alpha F\Delta\phi \tag{22.88b}$$

The parameter α lies in the range 0 to 1. Experimentally, α is often found to be about 0.5.

Now consider the oxidation reaction, Red + $e^- \rightarrow$ Ox and its reaction profile. Similar remarks apply. In this case, Red discards an electron to the electrode, so the extra work is zero if the transition state is reactant-like (represented by a peak close to the electrode). The extra work is the full $-F\Delta\phi$ if it resembles the product (the peak close to the outer plane). In general, the activation Gibbs energy for this anodic process is

$$\Delta^{\ddagger}G_a = \Delta^{\ddagger}G_a(0) - (1-\alpha)F\Delta\phi \tag{22.89}$$

The two activation Gibbs energies can now be inserted in place of the values used in eqn 22.87 with the result that

$$j = FB_a[\mathrm{Red}]e^{-\Delta^{\ddagger}G_a(0)/RT}e^{(1-\alpha)F\Delta\phi/RT} - FB_c[\mathrm{Ox}]e^{-\Delta^{\ddagger}G_c(0)/RT}e^{-\alpha F\Delta\phi/RT} \tag{22.90}$$

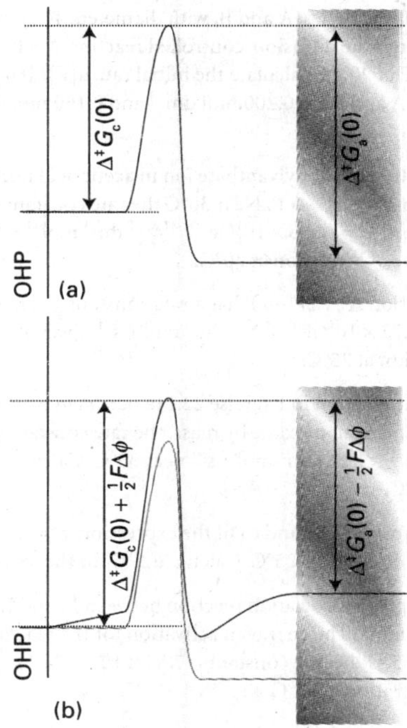

Fig. 22.40 When the transition state is intermediate in its resemblance to reduced and oxidized species, as represented here by a peak located at an intermediate position as measured by α (with $0<\alpha<1$), both activation Gibbs energies are affected; here, $\alpha \approx 0.5$. (a) Zero potential difference; (b) nonzero potential difference.

This is an explicit, if complicated, expression for the net current density in terms of the potential difference.

The appearance of eqn 22.90 can be simplified. First, in a purely cosmetic step we write $f=F/RT$. Next, we identify the individual cathodic and anodic current densities:

$$\left.\begin{aligned} j_a &= FB_a[\text{Red}]e^{-\Delta^{\ddagger}G_a(0)/RT}e^{(1-\alpha)f\Delta\phi} \\ j_c &= FB_c[\text{Ox}]e^{-\Delta^{\ddagger}G_c(0)/RT}e^{-\alpha f\Delta\phi} \end{aligned}\right\} j=j_a-j_c \tag{22.91}$$

If the cell is balanced against an external source, the Galvani potential difference, $\Delta\phi$, can be identified as the (zero-current) electrode potential, E, and we can write

$$\begin{aligned} j_a &= FB_a[\text{Red}]e^{-\Delta^{\ddagger}G_a(0)/RT}e^{(1-\alpha)fE} \\ j_c &= FB_c[\text{Ox}]e^{-\Delta^{\ddagger}G_c(0)/RT}e^{-\alpha fE} \end{aligned} \tag{22.92}$$

When these equations apply, there is no net current at the electrode (as the cell is balanced), so the two current densities must be equal. From now on we denote them both as j_0.

When the cell is producing current (that is, when a load is connected between the electrode being studied and a second counter electrode) the electrode potential changes from its zero-current value, E, to a new value, E', and the difference is the electrode's overpotential, $\eta=E'-E$. Hence, $\Delta\phi$ changes to $\Delta\phi=E+\eta$ and the two current densities become

$$j_a=j_0e^{(1-\alpha)f\eta} \qquad j_c=j_0e^{-\alpha f\eta} \tag{22.93}$$

Then from eqn 22.90 we obtain the Butler–Volmer equation, eqn 22.65.

Discussion questions

22.1 Describe the essential features of the harpoon mechanism.

22.2 In what senses are diffusion-controlled reactions and activation-controlled reactions activated processes?

22.3 Discuss critically the formulation of the Eyring equation.

22.4 What quantum mechanical effects play a role in molecular collisions?

22.5 Discuss the physical origin of the kinetic salt effect.

22.6 Describe how the shape of a potential energy surface governs the efficiencies of reactive collisions and the role of molecular vibration.

22.7 Describe how the distance between electron donor and acceptor, and the reorganization energy of redox active species and the surrounding medium determine the rate of electron transfer in homogeneous systems.

22.8 In what sense is electron transfer at an electrode an activated process?

Exercises

22.1(a) Calculate the collision frequency, z, and the collision density, Z, in ammonia, $R=190$ pm, at 25°C and 100 kPa. What is the percentage increase when the temperature is raised by 10 K at constant volume?

22.1(b) Calculate the collision frequency, z, and the collision density, Z, in carbon monoxide, $R=180$ pm at 25°C and 100 kPa. What is the percentage increase when the temperature is raised by 10 K at constant volume?

22.2(a) Collision theory demands knowing the fraction of molecular collisions having at least the kinetic energy E_a along the line of flight. What is this fraction when (a) $E_a=10$ kJ mol^{-1}, (b) $E_a=100$ kJ mol^{-1} at (i) 300 K and (ii) 1000 K?

22.2(b) Collision theory demands knowing the fraction of molecular collisions having at least the kinetic energy E_a along the line of flight. What is this fraction when (a) $E_a=15$ kJ mol^{-1}, (b) $E_a=150$ kJ mol^{-1} at (i) 300 K and (ii) 800 K?

22.3(a) Calculate the percentage increase in the fractions in Exercise 22.2a when the temperature is raised by 10 K.

22.3(b) Calculate the percentage increase in the fractions in Exercise 22.2b when the temperature is raised by 10 K.

22.4(a) Use the collision theory of gas-phase reactions to calculate the theoretical value of the second-order rate constant for the reaction $H_2(g) + I_2(g) \rightarrow 2\,HI(g)$ at 650 K, assuming that it is elementary bimolecular. The collision cross-section is 0.36 nm^2, the reduced mass is 3.32×10^{-27} kg, and the activation energy is 171 kJ mol^{-1}.

22.4(b) Use the collision theory of gas-phase reactions to calculate the theoretical value of the second-order rate constant for the reaction $D_2(g) + Br_2(g) \rightarrow 2\,DBr(g)$ at 450 K, assuming that it is elementary bimolecular. Take the collision cross-section as 0.30 nm^2, the reduced mass as $3.930m_u$, and the activation energy as 200 kJ mol^{-1}.

22.5(a) In a certain collision, an energy equivalent to 400 kJ mol^{-1} is delivered; the energy needed to break a bond is 350 kJ mol^{-1}; there are 5 relevant molecular modes. What is the value of the *P*-factor for the reactive encounter?

22.5(b) In a certain collision, an energy equivalent to 300 kJ mol^{-1} is delivered; the energy needed to break a bond is 250 kJ mol^{-1}; there are 6 relevant molecular modes. What is the value of the *P*-factor for the reactive encounter?

22.6(a) A typical diffusion coefficient for small molecules in aqueous solution at 25°C is 5×10^{-9} m^2 s^{-1}. If the critical reaction distance is 0.4 nm, what value is expected for the second-order rate constant for a diffusion-controlled reaction?

22.6(b) Suppose that the typical diffusion coefficient for a reactant in aqueous solution at 25°C is 4.2×10^{-9} m^2 s^{-1}. If the critical reaction distance is 0.50 nm, what value is expected for the second-order rate constant for the diffusion-controlled reaction?

22.7(a) Calculate the magnitude of the diffusion-controlled rate constant at 298 K for a species in (a) water, (b) pentane. The viscosities are 1.00×10^{-3} kg m^{-1} s^{-1}, and 2.2×10^{-4} kg m^{-1} s^{-1}, respectively.

22.7(b) Calculate the magnitude of the diffusion-controlled rate constant at 298 K for a species in (a) decylbenzene, (b) concentrated sulfuric acid. The viscosities are 3.36 cP and 27 cP, respectively.

22.8(a) Calculate the magnitude of the diffusion-controlled rate constant at 298 K for the recombination of two atoms in water, for which $\eta = 0.89$ cP. Assuming the concentration of the reacting species is 1.0 mmol dm^{-3} initially, how long does it take for the concentration of the atoms to fall to half that value? Assume the reaction is elementary.

22.8(b) Calculate the magnitude of the diffusion-controlled rate constant at 298 K for the recombination of two atoms in benzene, for which $\eta = 0.601$ cP. Assuming the concentration of the reacting species is 1.8 mmol dm^{-3} initially, how long does it take for the concentration of the atoms to fall to half that value? Assume the reaction is elementary.

22.9(a) For the gaseous reaction $A + B \rightarrow P$, the reactive cross-section obtained from the experimental value of the pre-exponential factor is 9.2×10^{-22} m^2. The collision cross-sections of A and B estimated from the transport properties are 0.95 and 0.65 nm^2, respectively. Calculate the *P*-factor for the reaction.

22.9(b) For the gaseous reaction $A + B \rightarrow P$, the reactive cross-section obtained from the experimental value of the pre-exponential factor is 8.7×10^{-22} m^2. The collision cross-sections of A and B estimated from the transport properties are 0.88 and 0.40 nm^2, respectively. Calculate the *P*-factor for the reaction.

22.10(a) Two neutral species, A and B, with diameters 588 pm and 1650 pm, respectively, undergo the diffusion-controlled reaction $A + B \rightarrow P$ in a solvent of viscosity 2.37×10^{-3} kg m^{-1} s^{-1} at 40°C. Calculate the initial rate $d[P]/dt$ if the initial concentrations of A and B are 0.150 mol dm^{-3} and 0.330 mol dm^{-3}, respectively.

22.10(b) Two neutral species, A and B, with diameters 442 pm and 885 pm, respectively, undergo the diffusion-controlled reaction $A + B \rightarrow P$ in a solvent of viscosity 1.27 cP at 20°C. Calculate the initial rate $d[P]/dt$ if the initial concentrations of A and B are 0.200 mol dm^{-3} and 0.150 mol dm^{-3}, respectively.

22.11(a) The reaction of propylxanthate ion in acetic acid buffer solutions has the mechanism $A^- + H^+ \rightarrow P$. Near 30°C the rate constant is given by the empirical expression $k_2 = (2.05 \times 10^{13})e^{-(8681\,K)/T}$ dm^3 mol^{-1} s^{-1}. Evaluate the energy and entropy of activation at 30°C.

22.11(b) The reaction $A^- + H^+ \rightarrow P$ has a rate constant given by the empirical expression $k_2 = (8.72 \times 10^{12})e^{(6134\,K)/T}$ dm^3 mol^{-1} s^{-1}. Evaluate the energy and entropy of activation at 25°C.

22.12(a) When the reaction in Exercise 22.11a occurs in a dioxane/water mixture that is 30 per cent dioxane by mass, the rate constant fits $k_2 = (7.78 \times 10^{14})e^{-(9134\,K)/T}$ dm^3 mol^{-1} s^{-1} near 30°C. Calculate $\Delta^{\ddagger}G$ for the reaction at 30°C.

22.12(b) A rate constant is found to fit the expression $k_2 = (6.45 \times 10^{13})$ $e^{-(5375\,K)/T}$ dm^3 mol^{-1} s^{-1} near 25°C. Calculate $\Delta^{\ddagger}G$ for the reaction at 25°C.

22.13(a) The gas-phase association reaction between F_2 and IF_5 is first-order in each of the reactants. The energy of activation for the reaction is 58.6 kJ mol^{-1}. At 65°C the rate constant is 7.84×10^{-3} kPa^{-1} s^{-1}. Calculate the entropy of activation at 65°C.

22.13(b) A gas-phase recombination reaction is first-order in each of the reactants. The energy of activation for the reaction is 49.6 kJ mol^{-1}. At 55°C the rate constant is 0.23 m^3 s^{-1}. Calculate the entropy of activation at 55°C.

22.14(a) Calculate the entropy of activation for a collision between two structureless particles at 300 K, taking $M = 50$ g mol^{-1} and $\sigma = 0.40$ nm^2.

22.14(b) Calculate the entropy of activation for a collision between two structureless particles at 500 K, taking $M = 78$ g mol^{-1} and $\sigma = 0.62$ nm^2.

22.15(a) The pre-exponential factor for the gas-phase decomposition of ozone at low pressures is 4.6×10^{12} dm^3 mol^{-1} s^{-1} and its activation energy is 10.0 kJ mol^{-1}. What are (a) the entropy of activation, (b) the enthalpy of activation, (c) the Gibbs energy of activation at 298 K?

22.15(b) The pre-exponential factor for a gas-phase decomposition of ozone at low pressures is 2.3×10^{13} dm^3 mol^{-1} s^{-1} and its activation energy is 30.0 kJ mol^{-1}. What are (a) the entropy of activation, (b) the enthalpy of activation, (c) the Gibbs energy of activation at 298 K?

22.16(a) The rate constant of the reaction $H_2O_2(aq) + I^-(aq) + H^+(aq) \rightarrow H_2O(l) + HIO(aq)$ is sensitive to the ionic strength of the aqueous solution in which the reaction occurs. At 25°C, $k = 12.2$ dm^6 mol^{-2} min^{-1} at an ionic strength of 0.0525. Use the Debye–Hückel limiting law to estimate the rate constant at zero ionic strength.

22.16(b) At 25°C, $k_r = 1.55$ dm^6 mol^{-2} min^{-1} at an ionic strength of 0.0241 for a reaction in which the rate-determining step involves the encounter of two singly charged cations. Use the Debye–Hückel limiting law to estimate the rate constant at zero ionic strength.

22.17(a) For an electron donor–acceptor pair, $H_{AB} = 0.03$ cm^{-1}, $\Delta_r G^{\circ} = -0.182$ eV, and $k_{et} = 30.5$ s^{-1} at 298 K. Estimate the value of the reorganization energy.

22.17(b) For an electron donor–acceptor pair, $k_{et} = 2.02 \times 10^5$ s^{-1} for $\Delta_r G^{\circ} = -0.665$ eV. The standard reaction Gibbs energy changes to $\Delta_r G^{\circ} = -0.975$ eV when a substituent is added to the electron acceptor and the rate constant for electron transfer changes to $k_{et} = 3.33 \times 10^6$ s^{-1}. The experiments were conducted at 298 K. Assume that the distance between donor and acceptor is the same in both experiments and estimate the values of H_{AB} and λ.

22.18(a) For an electron donor–acceptor pair, $k_{et} = 2.02 \times 10^5\ s^{-1}$ when $r = 1.11$ nm and $k_{et} = 4.51 \times 10^5\ s^{-1}$ when $r = 1.23$ nm. Assuming that $\Delta_r G^\ominus$ and λ are the same in both experiments, estimate the value of β.

22.18(b) Refer to Exercise 22.18a. Estimate the value of k_{et} when $r = 1.48$ nm.

22.19(a) The transfer coefficient of a certain electrode in contact with M^{3+} and M^{4+} in aqueous solution at 25°C is 0.39. The current density is found to be 55.0 mA cm^{-2} when the overpotential is 125 mV. What is the overpotential required for a current density of 75 mA cm^{-2}?

22.19(b) The transfer coefficient of a certain electrode in contact with M^{2+} and M^{3+} in aqueous solution at 25°C is 0.42. The current density is found to be 17.0 mA cm^{-2} when the overpotential is 105 mV. What is the overpotential required for a current density of 72 mA cm^{-2}?

22.20(a) Determine the exchange current density from the information given in Exercise 22.19a.

22.20(b) Determine the exchange current density from the information given in Exercise 22.19b.

22.21(a) To a first approximation, significant evolution or deposition occurs in electrolysis only if the overpotential exceeds about 0.6 V. To illustrate this criterion determine the effect that increasing the overpotential from 0.40 V to 0.60 V has on the current density in the electrolysis of a certain electrolyte solution, which is 1.0 mA cm^{-2} at 0.4 V and 25°C. Take $\alpha = 0.5$.

22.21(b) Determine the effect that increasing the overpotential from 0.50 V to 0.60 V has on the current density in the electrolysis of a certain electrolyte solution, which is 1.22 mA cm^{-2} at 0.50 V and 25°C. Take $\alpha = 0.50$.

22.22(a) Use the data in Table 22.3 for the exchange current density and transfer coefficient for the reaction $2\ H^+ + 2\ e^- \rightarrow H_2$ on nickel at 25°C to determine what current density would be needed to obtain an overpotential of 0.20 V as calculated from (a) the Butler–Volmer equation, and (b) the Tafel equation (eqn 22.69). Is the validity of the Tafel approximation affected at higher overpotentials (of 0.4 V and more)?

22.22(b) Use the data in Table 22.3 for the exchange current density and transfer coefficient for the reaction $Fe^{3+} + e^- \rightarrow Fe^{2+}$ on platinum at 25°C to determine what current density would be needed to obtain an overpotential of 0.30 V as calculated from (a) the Butler–Volmer equation, and (b) the Tafel equation (eqn 22.69). Is the validity of the Tafel approximation affected at higher overpotentials (of 0.4 V and more)?

22.23(a) A typical exchange current density, that for H^+ discharge at platinum, is 0.79 mA cm^{-2} at 25°C. What is the current density at an electrode when its overpotential is (a) 10 mV, (b) 100 mV, (c) −5.0 V? Take $\alpha = 0.5$.

22.23(b) The exchange current density for a Pt$|Fe^{3+},Fe^{2+}$ electrode is 2.5 mA cm^{-2}. The standard potential of the electrode is +0.77 V. Calculate the current flowing through an electrode of surface area 1.0 cm^2 as a function of the potential of the electrode. Take unit activity for both ions.

22.24(a) How many electrons or protons are transported through the double layer in each second when the Pt,$H_2|H^+$, Pt$|Fe^{3+},Fe^{2+}$, and Pb,$H_2|H^+$ electrodes are at equilibrium at 25°C? Take the area as 1.0 cm^2 in each case. Estimate the number of times each second a single atom on the surface takes part in a electron transfer event, assuming an electrode atom occupies about $(280\ pm)^2$ of the surface.

22.24(b) How many electrons or protons are transported through the double layer in each second when the Cu,$H_2|H^+$ and Pt$|Ce^{4+},Ce^{3+}$ electrodes are at equilibrium at 25°C? Take the area as 1.0 cm^2 in each case. Estimate the number of times each second a single atom on the surface takes part in a electron transfer event, assuming an electrode atom occupies about $(260\ pm)^2$ of the surface.

22.25(a) What is the effective resistance at 25°C of an electrode interface when the overpotential is small? Evaluate it for 1.0 cm^2 (a) Pt,$H_2|H^+$, (b) Hg,$H_2|H^+$ electrodes.

22.25(b) Evaluate the effective resistance at 25°C of an electrode interface for 1.0 cm^2 (a) Pb,$H_2|H^+$, (b) Pt$|Fe^{2+},Fe^{3+}$ electrodes.

22.26(a) The exchange current density for H^+ discharge at zinc is about 50 pA cm^{-2}. Can zinc be deposited from a unit activity aqueous solution of a zinc salt?

22.26(b) The standard potential of the $Zn^{2+}|Zn$ electrode is −0.76 V at 25°C. The exchange current density for H^+ discharge at platinum is 0.79 mA cm^{-2}. Can zinc be plated on to platinum at that temperature? (Take unit activities.)

Problems*

Numerical problems

22.1 In the dimerization of methyl radicals at 25°C, the experimental pre-exponential factor is $2.4 \times 10^{10}\ dm^3\ mol^{-1}\ s^{-1}$. What are (a) the reactive cross-section, (b) the P factor for the reaction if the C–H bond length is 154 pm?

22.2 Nitrogen dioxide reacts bimolecularly in the gas phase to give $2\ NO + O_2$. The temperature dependence of the second-order rate constant for the rate law $d[P]/dt = k[NO_2]^2$ is given below. What are the P factor and the reactive cross-section for the reaction?

T/K	600	700	800	1000
$k_r/(cm^3\ mol^{-1}\ s^{-1})$	4.6×10^2	9.7×10^3	1.3×10^5	3.1×10^6

Take $\sigma = 0.60\ nm^2$.

22.3 The diameter of the methyl radical is about 308 pm. What is the maximum rate constant in the expression $d[C_2H_6]/dt = k_r[CH_3]^2$ for second-order recombination of radicals at room temperature? 10 per cent of a 1.0-dm^3 sample of ethane at 298 K and 100 kPa is dissociated into methyl radicals. What is the minimum time for 90 per cent recombination?

22.4 The rates of thermolysis of a variety of *cis*- and *trans*-azoalkanes have been measured over a range of temperatures in order to settle a controversy concerning the mechanism of the reaction. In ethanol an unstable *cis*-azoalkane decomposed at a rate that was followed by observing the N_2 evolution, and this led to the rate constants listed below (P.S. Engel and D.J. Bishop, *J. Amer. Chem. Soc.* **97**, 6754 (1975)). Calculate the enthalpy, entropy, energy, and Gibbs energy of activation at −20°C.

θ/°C	−24.82	−20.73	−17.02	−13.00	−8.95
$10^4 \times k_r/s^{-1}$	1.22	2.31	4.39	8.50	14.3

22.5 In an experimental study of a bimolecular reaction in aqueous solution, the second-order rate constant was measured at 25°C and at a variety of ionic

* Problems denoted with the symbol ‡ were supplied by Charles Trapp, Carmen Giunta, and Marshall Cady.

strengths and the results are tabulated below. It is known that a singly charged ion is involved in the rate-determining step. What is the charge on the other ion involved?

I	0.0025	0.0037	0.0045	0.0065	0.0085
$k_r/(\text{dm}^3\ \text{mol}^{-1}\ \text{s}^{-1})$	1.05	1.12	1.16	1.18	1.26

22.6 The rate constant of the reaction $I^-(aq) + H_2O_2(aq) \rightarrow H_2O(l) + IO^-(aq)$ varies slowly with ionic strength, even though the Debye–Hückel limiting law predicts no effect. Use the following data from 25°C to find the dependence of $\log k_r$ on the ionic strength:

I	0.0207	0.0525	0.0925	0.1575
$k_r/(\text{dm}^3\ \text{mol}^{-1}\ \text{min}^{-1})$	0.663	0.670	0.679	0.694

Evaluate the limiting value of k_r at zero ionic strength. What does the result suggest for the dependence of $\log \gamma$ on ionic strength for a neutral molecule in an electrolyte solution?

22.7 The total cross-sections for reactions between alkali metal atoms and halogen molecules are given in the table below (R.D. Levine and R.B. Bernstein, *Molecular reaction dynamics*, Clarendon Press, Oxford, 72 (1974)). Assess the data in terms of the harpoon mechanism.

σ^*/nm^2	Cl_2	Br_2	I_2
Na	1.24	1.16	0.97
K	1.54	1.51	1.27
Rb	1.90	1.97	1.67
Cs	1.96	2.04	1.95

Electron affinities are approximately 1.3 eV (Cl_2), 1.2 eV (Br_2), and 1.7 eV (I_2), and ionization energies are 5.1 eV (Na), 4.3 eV (K), 4.2 eV (Rb), and 3.9 eV (Cs).

22.8‡ One of the most historically significant studies of chemical reaction rates was that by M. Bodenstein (*Z. physik. Chem.* **29**, 295 (1899)) of the gas-phase reaction $2\,HI(g) \rightarrow H_2(g) + I_2(g)$ and its reverse, with rate constants k_r and k_r', respectively. The measured rate constants as a function of temperature are

T/K	647	666	683	700	716	781
$k_r/(22.4\ \text{dm}^3\ \text{mol}^{-1}\ \text{min}^{-1})$	0.230	0.588	1.37	3.10	6.70	105.9
$k_r'/(22.4\ \text{dm}^3\ \text{mol}^{-1}\ \text{min}^{-1})$	0.0140	0.0379	0.0659	0.172	0.375	3.58

Demonstrate that these data are consistent with the collision theory of bimolecular gas-phase reactions.

22.9 In an experiment on the $Pt|H_2|H^+$ electrode in dilute H_2SO_4 the following current densities were observed at 25°C. Evaluate α and j_0 for the electrode.

η/mV	50	100	150	200	250
$j/(\text{mA cm}^{-2})$	2.66	8.91	29.9	100	335

How would the current density at this electrode depend on the overpotential of the same set of magnitudes but of opposite sign?

22.10 The standard potentials of lead and tin are −126 mV and −136 mV, respectively, at 25°C, and the overpotentials for their deposition are close to zero. What should their relative activities be in order to ensure simultaneous deposition from a mixture?

22.11‡ The rate of deposition of iron, v, on the surface of an iron electrode from an aqueous solution of Fe^{2+} has been studied as a function of potential, E, relative to the standard hydrogen electrode, by J. Kanya (*J. Electroanal. Chem.* **84**, 83 (1977)). The values in the table below are based on the data obtained with an electrode of surface area 9.1 cm² in contact with a solution of concentration 1.70 μmol dm⁻³ in Fe^{2+}. (a) Assuming unit activity coefficients, calculate the zero current potential of the Fe^{2+}/Fe cathode and the overpotential at each value of the working potential. (b) Calculate the cathodic current density, j_c, from the rate of deposition of Fe^{2+} for each value of E. (c) Examine the extent to which the data fit eqn 22.69 and calculate the exchange current density.

$v/(\text{pmol s}^{-1})$	1.47	2.18	3.11	7.26
$-E$/mV	702	727	752	812

22.12‡ The thickness of the diffuse double layer according to the Gouy–Chapman model is given by eqn 18.16. Use this equation to calculate and plot the thickness as a function of concentration and electrolyte type at 25°C. For examples, choose aqueous solutions of NaCl and Na_2SO_4 ranging in concentration from 0.1 to 100 mmol dm⁻³.

22.13‡ V.V. Losev and A.P. Pchel'nikov (*Soviet Electrochem.* **6**, 34 (1970)) obtained the following current–voltage data for an indium anode relative to a standard hydrogen electrode at 293 K:

$-E$/V	0.388	0.365	0.350	0.335
$j/(\text{A m}^{-2})$	0	0.590	1.438	3.507

Use these data to calculate the transfer coefficient and the exchange current density. What is the cathodic current density when the potential is 0.365 V?

22.14‡ An early study of the hydrogen overpotential is that of H. Bowden and T. Rideal (*Proc. Roy. Soc.* **A120**, 59 (1928)), who measured the overpotential for H_2 evolution with a mercury electrode in dilute aqueous solutions of H_2SO_4 at 25°C. Determine the exchange current density and transfer coefficient, α, from their data:

$j/(\text{mA m}^{-2})$	2.9	6.3	28	100	250	630	1650	3300
η/V	0.60	0.65	0.73	0.79	0.84	0.89	0.93	0.96

Explain any deviations from the result expected from eqn 22.69.

Theoretical problems

22.15 Confirm that eqn 22.26 is a solution of eqn 22.25, where [J] is a solution of the same equation but with $k_r = 0$ and for the same initial conditions.

22.16 Confirm that, if the initial condition is $[J] = 0$ at $t = 0$ everywhere, and the boundary condition is $[J] = [J]_0$ at $t > 0$ at all points on a surface, then the solutions $[J]^*$ in the presence of a first-order reaction that removed J are related to those in the absence of reaction, [J], by

$$[J]^* = k_r \int_0^t [J]e^{-k_r t}dt + [J]e^{-k_r t}$$

Base your answer on eqn 22.25.

22.17 Estimate the orders of magnitude of the partition functions involved in a rate expression. State the order of magnitude of q_m^T/N_A, q^R, q^V, q^E for typical molecules. Check that in the collision of two structureless molecules the order of magnitude of the pre-exponential factor is of the same order as that predicted by collision theory. Go on to estimate the P factor for a reaction in which $A + B \rightarrow P$, and A and B are nonlinear triatomic molecules.

22.18 Use the Debye–Hückel limiting law to show that changes in ionic strength can affect the rate of reaction catalysed by H^+ from the deprotonation of a weak acid. Consider the mechanism: $H^+(aq) + B(aq) \rightarrow P$, where H^+ comes from the deprotonation of the weak acid, HA. The weak acid has a fixed concentration. First show that $\log [H^+]$, derived from the ionization of HA, depends on the activity coefficients of ions and thus depends on the ionic strength. Then find the relationship between log(rate) and $\log [H^+]$ to show that the rate also depends on the ionic strength.

22.19 The Eyring equation can also be applied to physical processes. As an example, consider the rate of diffusion of an atom stuck to the surface of a solid. Suppose that in order to move from one site to another it has to reach the top of the barrier where it can vibrate classically in the vertical direction and in one horizontal direction, but vibration along the other horizontal direction takes it into the neighbouring site. Find an expression for the rate of diffusion, and evaluate it for W atoms on a tungsten surface ($E_a = 60$ kJ mol^{-1}). Suppose that the vibration frequencies at the transition state are (a) the same as, (b) one-half the value for the adsorbed atom. What is the value of the diffusion coefficient D at 500 K? (Take the site separation as 316 pm and $\nu = 1 \times 10^{11}$ Hz.)

22.20‡ Show that bimolecular reactions between nonlinear molecules are much slower than between atoms even when the activation energies of both reactions are equal. Use transition state theory and make the following assumptions. (1) All vibrational partition functions are close to 1; (2) all rotational partition functions are approximately $1 \times 10^{1.5}$, which is a reasonable order of magnitude number; (3) the translational partition function for each species is 1×10^{26}.

22.21 This exercise gives some familiarity with the difficulties involved in predicting the structure of activated complexes. It also demonstrates the importance of femtosecond spectroscopy to our understanding of chemical dynamics because direct experimental observation of the activated complex removes much of the ambiguity of theoretical predictions. Consider the attack of H on D_2, which is one step in the $H_2 + D_2$ reaction. (a) Suppose that the H approaches D_2 from the side and forms a complex in the form of an isosceles triangle. Take the H–D distance as 30 per cent greater than in H_2 (74 pm) and the D–D distance as 20 per cent greater than in H_2. Let the critical coordinate be the antisymmetric stretching vibration in which one H–D bond stretches as the other shortens. Let all the vibrations be at about 1000 cm^{-1}. Estimate k_2 for this reaction at 400 K using the experimental activation energy of about 35 kJ mol^{-1}. (b) Now change the model of the activated complex in part (a) and make it linear. Use the same estimated molecular bond lengths and vibrational frequencies to calculate k_2 for this choice of model. (c) Clearly, there is much scope for modifying the parameters of the models of the activated complex. Use mathematical software or write and run a program that allows you to vary the structure of the complex and the parameters in a plausible way, and look for a model (or more than one model) that gives a value of k close to the experimental value, 4×10^5 dm^3 mol^{-1} s^{-1}.

22.22 If $\alpha = \frac{1}{2}$, an electrode interface is unable to rectify alternating current because the current density curve is symmetrical about $\eta = 0$. When $\alpha \neq \frac{1}{2}$, the magnitude of the current density depends on the sign of the overpotential, and so some degree of 'faradaic rectification' may be obtained. Suppose that the overpotential varies as $\eta = \eta_0 \cos \omega t$. Derive an expression for the mean flow of current (averaged over a cycle) for general α, and confirm that the mean current is zero when $\alpha = \frac{1}{2}$. In each case work in the limit of small η_0 but to second order in $\eta_0 F/RT$. Calculate the mean direct current at 25°C for a 1.0 cm^2 hydrogen–platinum electrode with $\alpha = 0.38$ when the overpotential varies between ±10 mV at 50 Hz.

22.23 Now suppose that the overpotential is in the high overpotential region at all times even though it is oscillating. What waveform will the current across the interface show if it varies linearly and periodically (as a sawtooth waveform) between η_- and η_+ around η_0? Take $\alpha = \frac{1}{2}$.

Applications: to biochemistry and environmental science

22.24‡ R. Atkinson (*J. Phys. Chem. Ref. Data* 26, 215 (1997)) has reviewed a large set of rate constants relevant to the atmospheric chemistry of volatile organic compounds. The recommended rate constant for the bimolecular association of O_2 with an alkyl radical R at 298 K is 4.7×10^9 dm^3 mol^{-1} s^{-1} for R = C_2H_5 and 8.4×10^9 dm^3 mol^{-1} s^{-1} for R = cyclohexyl. Assuming no energy barrier, compute the steric factor, P, for each reaction. (*Hint.* Obtain collision diameters from collision cross-sections of similar molecules in the *Data section*.)

22.25‡ The compound α-tocopherol, a form of vitamin E, is a powerful antioxidant that may help to maintain the integrity of biological membranes. R.H. Bisby and A.W. Parker (*J. Amer. Chem. Soc.* 117, 5664 (1995)) studied the reaction of photochemically excited duroquinone with the antioxidant in ethanol. Once the duroquinone was photochemically excited, a bimolecular reaction took place at a rate described as diffusion-limited. (a) Estimate the rate constant for a diffusion-limited reaction in ethanol. (b) The reported rate constant was 2.77×10^9 dm^3 mol^{-1} s^{-1}; estimate the critical reaction distance if the sum of diffusion constants is 1×10^{-9} m^2 s^{-1}.

22.26 The study of conditions that optimize the association of proteins in solution guides the design of protocols for formation of large crystals that are amenable to analysis by the X-ray diffraction techniques discussed in Chapter 19. It is important to characterize protein dimerization because the process is considered to be the rate-determining step in the growth of crystals of many proteins. Consider the variation with ionic strength of the rate constant of dimerization in aqueous solution of a cationic protein P:

I	0.0100	0.0150	0.0200	0.0250	0.0300	0.0350
$k/k°$	8.10	13.30	20.50	27.80	38.10	52.00

What can be deduced about the charge of P?

22.27 A useful strategy for the study of electron transfer in proteins consists of attaching an electroactive species to the protein's surface and then measuring k_{et} between the attached species and an electroactive protein cofactor. J.W. Winkler and H.B. Gray (*Chem. Rev.* 92, 369 (1992)) summarize data for cytochrome *c* (*Impact I6.1*) modified by replacement of the haem iron by a zinc ion, resulting in a zinc-porphyrin (ZnP) moiety in the interior of the protein, and by attachment of a ruthenium ion complex to a surface histidine aminoacid. The edge-to-edge distance between the electroactive species was thus fixed at 1.23 nm. A variety of ruthenium ion complexes with different standard reduction potentials were used. For each ruthenium-modified protein, either the $Ru^{2} \rightarrow ZnP^{+}$ or the $ZnP^{*} \rightarrow Ru^{3+}$, in which the electron donor is an electronic excited state of the zinc-porphyrin formed by laser excitation, was monitored. This arrangement leads to different standard reaction Gibbs energies because the redox couples ZnP^+/ZnP and ZnP^+/ZnP^* have different standard potentials, with the electronically excited porphyrin being a more powerful reductant. Use the following data to estimate the reorganization energy for this system:

$-\Delta_r G^{\ominus}$/eV	0.665	0.705	0.745	0.975	1.015	1.055
$k_{et}/(10^6\ s^{-1})$	0.657	1.52	1.12	8.99	5.76	10.1

22.28 The rate constant for electron transfer between a cytochrome *c* and the bacteriochlorophyll dimer of the reaction centre of the purple bacterium *Rhodobacter sphaeroides* decreases with decreasing temperature in the range 300 K to 130 K. Below 130 K, the rate constant becomes independent of temperature. Account for these results.

22.29 Calculate the thermodynamic limit to the zero-current potential of fuel cells operating on (a) hydrogen and oxygen, (b) methane and air, and (c) propane and air. Use the Gibbs energy information in the *Data section*, and take the species to be in their standard states at 25°c.

23 Catalysis

This chapter extends the material introduced in Chapters 21 and 22 by showing how to deal with catalysis. We begin with a description of homogeneous catalysis and apply the associated concepts to enzyme-catalysed reactions. We go on to consider heterogeneous catalysis by exploring the extent to which a solid surface is covered and the variation of the extent of coverage with pressure and temperature. Then we use this material to discuss how surfaces affect the rate and course of chemical change by acting as the site of catalysis.

A **catalyst** is a substance that accelerates a reaction but undergoes no net chemical change. The catalyst lowers the activation energy of the reaction by providing an alternative path that avoids the slow, rate-determining step of the uncatalysed reaction (Fig. 23.1).

A **homogeneous catalyst** is a catalyst in the same phase as the reaction mixture. For example, the decomposition of hydrogen peroxide in aqueous solution is catalysed by iodide ion. **Enzymes**, which are biological catalysts, are very specific and can have a dramatic effect on the reactions they control. We shall examine enzyme catalysis in Section 23.2. A **heterogeneous catalyst** is a catalyst in a different phase from the reaction mixture. For example, the hydrogenation of ethene to ethane, a gas-phase reaction, is accelerated in the presence of a solid catalyst such as palladium, platinum, or nickel. The metal provides a surface upon which the reactants bind; this binding facilitates encounters between reactants and increases the rate of the reaction. Most of this chapter is an exploration of catalytic activity on surfaces.

Homogeneous catalysis

Homogeneous catalysts can be very effective. For instance, the activation energy for the decomposition of hydrogen peroxide in solution is 76 kJ mol^{-1}, and the reaction is slow at room temperature. When a little iodide ion is added, the activation energy falls to 57 kJ mol^{-1} and the rate constant increases by a factor of 2000. The enzyme catalase reduces the activation energy even further, to 8 kJ mol^{-1}, corresponding to an acceleration of the reaction by a factor of 10^{15} at 298 K.

23.1 Features of homogeneous catalysis

Key points **Catalysts are substances that accelerate reactions but undergo no net chemical change. A homogeneous catalyst is a catalyst in the same phase as the reaction mixture. Examples of homogeneous catalysis include acid and base catalysis.**

We can obtain some idea of the mode of action of homogeneous catalysts by examining the kinetics of the iodide-catalysed decomposition of hydrogen peroxide:

$$2\,H_2O_2(aq) \rightarrow 2\,H_2O(l) + O_2(g)$$

The reaction is believed to proceed through the following pre-equilibrium:

$$H_3O^+ + H_2O_2 \rightleftharpoons H_3O_2^+ + H_2O \qquad K = \frac{[H_3O_2^+]}{[H_2O_2][H_3O^+]}$$

$$H_3O_2^+ + I^- \rightarrow HOI + H_2O \qquad v = k_a[H_3O_2^+][I^-]$$

$$HOI + H_2O_2 \rightarrow H_3O^+ + O_2 + I^- \qquad \text{(fast)}$$

where we have set the activity of H_2O in the equilibrium constant equal to 1 and assumed that the thermodynamic properties of the other substances are ideal. The second step is rate-determining. Therefore, we can obtain the rate law of the overall reaction by setting the overall rate equal to the rate of the second step and using the equilibrium constant to express the concentration of $H_3O_2^+$ in terms of the reactants. The result is

$$\frac{d[O_2]}{dt} = k_r[H_2O_2][H_3O^+][I^-]$$

with $k_r = k_a K$, in agreement with the observed dependence of the rate on the I^- concentration and the pH of the solution. The observed activation energy is that of the effective rate constant $k_a K$.

In **acid catalysis** the crucial step is the transfer of a proton to the substrate:

$$X + HA \rightarrow HX^+ + A^- \qquad HX^+ \rightarrow \text{products}$$

Acid catalysis is the primary process in the solvolysis of esters and keto–enol tautomerism:

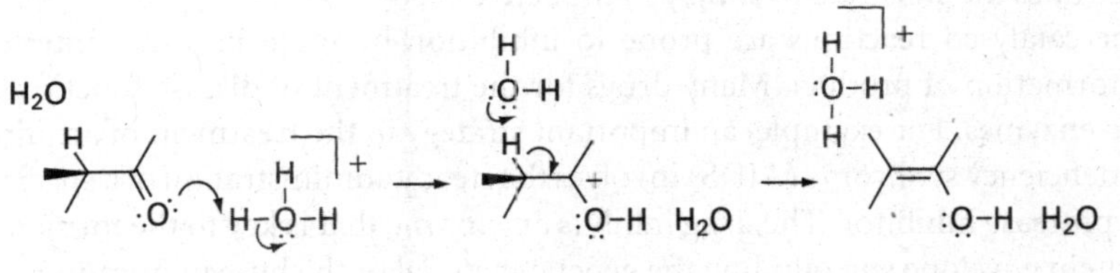

In **base catalysis**, a proton is transferred from the substrate to a base:

$$XH + B \rightarrow X^- + BH^+ \qquad X^- \rightarrow \text{products}$$

Base catalysis is the primary step in the isomerization and halogenation of organic compounds, and of the Claisen and aldol condensation reactions. The base-catalysed version of keto–enol tautomerism, for instance, is

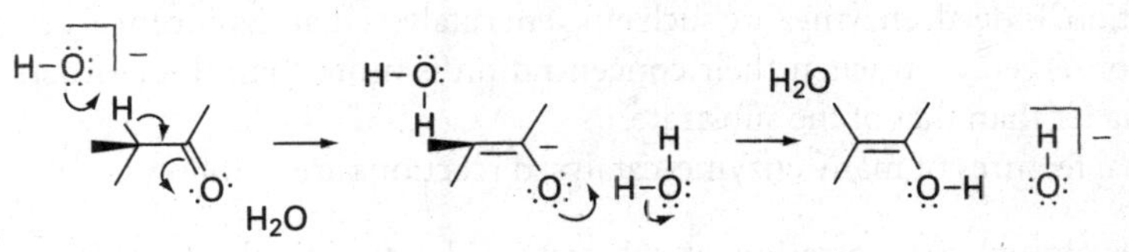

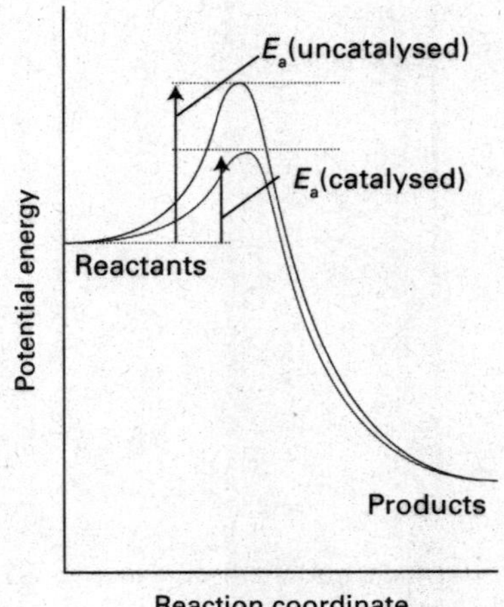

Fig. 23.1 A catalyst provides a different path with a lower activation energy. The result is an increase in the rate of formation of products.

23.2 Enzymes

Key points **Enzymes are homogeneous, biological catalysts. (a) The Michaelis–Menten mechanism of enzyme kinetics accounts for the dependence of rate on the concentration of the substrate. A Lineweaver–Burk plot is used to determine the parameters that occur in the mechanism. (b) The turnover frequency is the number of catalytic cycles performed by the active site of an enzyme in a given interval divided by the duration of the interval. The catalytic efficiency is the effective rate constant of the enzymatic reaction. (c) In competitive inhibition of an enzyme, the inhibitor binds only to the active site of the enzyme. In uncompetitive inhibition the inhibitor binds to a site of the enzyme that is removed from the active site, but only if the substrate is already present. In non-competitive inhibition, the inhibitor binds to a site other than the active site.**

Enzymes are homogeneous biological catalysts. These ubiquitous compounds are special proteins or nucleic acids that contain an **active site**, which is responsible for binding the **substrates**, the reactants, and processing them into products. As is true of any catalyst, the active site returns to its original state after the products are released. Many enzymes consist primarily of proteins, some featuring organic or inorganic co-factors in their active sites. However, certain RNA molecules can also be biological catalysts, forming *ribozymes*. A very important example of a ribozyme is the *ribosome*, a large assembly of proteins and catalytically active RNA molecules responsible for the synthesis of proteins in the cell.

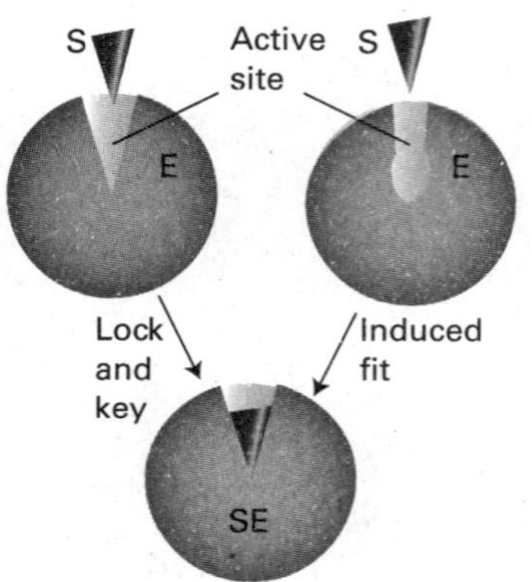

Fig. 23.2 Two models that explain the binding of a substrate to the active site of an enzyme. In the lock-and-key model, the active site and substrate have complementary three-dimensional structures and dock perfectly without the need for major atomic rearrangements. In the induced fit model, binding of the substrate induces a conformational change in the active site. The substrate fits well in the active site after the conformational change has taken place.

The structure of the active site is specific to the reaction that it catalyses, with groups in the substrate interacting with groups in the active site by intermolecular interactions, such as hydrogen bonding, electrostatic, or van der Waals interactions. Figure 23.2 shows two models that explain the binding of a substrate to the active site of an enzyme. In the **lock-and-key model**, the active site and substrate have complementary three-dimensional structures and dock perfectly without the need for major atomic rearrangements. Experimental evidence favours the **induced fit model**, in which binding of the substrate induces a conformational change in the active site. Only after the change does the substrate fit snugly in the active site.

Enzyme-catalysed reactions are prone to inhibition by molecules that interfere with the formation of product. Many drugs for the treatment of disease function by inhibiting enzymes. For example, an important strategy in the treatment of acquired immune deficiency syndrome (AIDS) involves the steady administration of a specially designed protease inhibitor. The drug inhibits an enzyme that is key to the formation of the protein envelope surrounding the genetic material of the human immunodeficiency virus (HIV). Without a properly formed envelope, HIV cannot replicate in the host organism.

(a) The Michaelis–Menten mechanism of enzyme catalysis

Experimental studies of enzyme kinetics are typically conducted by monitoring the initial rate of product formation in a solution in which the enzyme is present at very low concentration. Indeed, enzymes are such efficient catalysts that significant accelerations may be observed even when their concentration is more than three orders of magnitude smaller than that of the substrate.

The principal features of many enzyme-catalysed reactions are as follows:

- For a given initial concentration of substrate, $[S]_0$, the initial rate of product formation is proportional to the total concentration of enzyme, $[E]_0$.
- For a given $[E]_0$ and low values of $[S]_0$, the rate of product formation is proportional to $[S]_0$.

- For a given $[E]_0$ and high values of $[S]_0$, the rate of product formation becomes independent of $[S]_0$, reaching a maximum value known as the **maximum velocity**, v_{max}.

The **Michaelis–Menten mechanism** accounts for these features. According to this mechanism, an enzyme–substrate complex is formed in the first step and either the substrate is released unchanged or after modification to form products:

$$E + S \rightleftarrows ES \qquad k_a, k'_a$$

$$ES \rightarrow P + E \qquad k_b$$

Michaelis–Menten mechanism

We show in the following *Justification* that this mechanism leads to the **Michaelis–Menten equation** for the rate of product formation

$$v = \frac{k_b[E]_0}{1 + K_m/[S]_0} \qquad (23.1)$$

Michaelis–Menten equation

where $K_M = (k'_a + k_b)/k_a$ is the **Michaelis constant**, characteristic of a given enzyme acting on a given substrate and having the dimensions of a molar concentration.

Justification 23.1 *The Michaelis–Menten equation*

The rate of product formation according to the Michaelis–Menten mechanism is

$$v = k_b[ES]$$

We can obtain the concentration of the enzyme–substrate complex by invoking the steady-state approximation and writing

$$\frac{d[ES]}{dt} = k_a[E][S] - k'_a[ES] - k_b[ES] = 0$$

It follows that

$$[ES] = \left(\frac{k_a}{k'_a + k_b}\right)[E][S]$$

where [E] and [S] are the concentrations of *free* enzyme and substrate, respectively. Now we define the Michaelis constant as

$$K_M = \frac{k'_a + k_b}{k_a} = \frac{[E][S]}{[ES]}$$

To express the rate law in terms of the concentrations of enzyme and substrate added, we note that $[E]_0 = [E] + [ES]$. Moreover, because the substrate is typically in large excess relative to the enzyme, the free substrate concentration is approximately equal to the initial substrate concentration and we can write $[S] \approx [S]_0$. It then follows that:

$$[ES] = \frac{[E]_0}{1 + K_M/[S]_0}$$

Equation 23.1 is obtained when this expression for [ES] is substituted into that for the rate of product formation ($v = k_b[ES]$).

Equation 23.1 shows that, in accord with experimental observations (Fig. 23.3):

- When $[S]_0 \ll K_M$ the rate is proportional to $[S]_0$:

$$v = \frac{k_b}{K_M}[S]_0[E]_0 \qquad (23.2a)$$

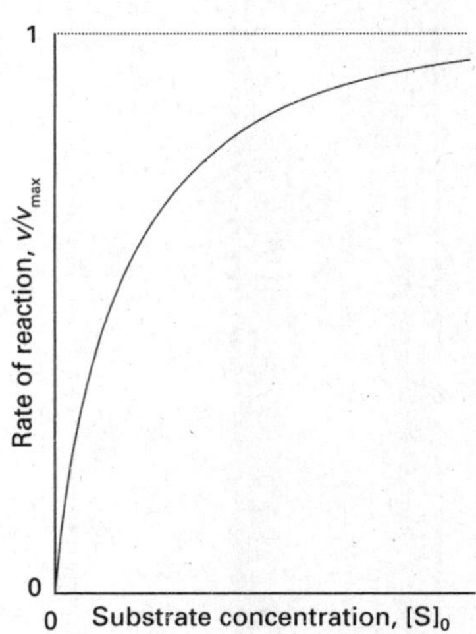

Fig. 23.3 The variation of the rate of an enzyme-catalysed reaction with substrate concentration. The approach to a maximum rate, v_{max}, for large $[S]_0$ is explained by the Michaelis–Menten mechanism.

interActivity Use the Michaelis–Menten equation to generate two families of curves showing the dependence of v on $[S]_0$: one in which K_M varies but v_{max} is constant, and another in which v_{max} varies but K_M is constant.

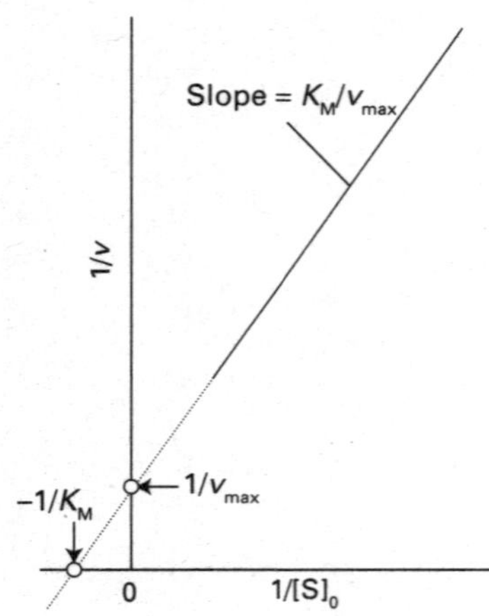

Fig. 23.4 A Lineweaver–Burk plot for the analysis of an enzyme-catalysed reaction that proceeds by a Michaelis-Menten mechanism and the significance of the intercepts and the slope.

- When $[S]_0 \gg K_M$ the rate reaches its maximum value and is independent of $[S]_0$:

$$v = v_{max} = k_b[E]_0 \qquad (23.2b)$$

Substitution of the definition of v_{max} into eqn 23.1 gives

$$v = \frac{v_{max}}{1 + K_M/[S]_0} \qquad (23.3a)$$

which can be rearranged into a form amenable to data analysis by linear regression by taking reciprocals of both sides:

$$\frac{1}{v} = \frac{1}{v_{max}} + \left(\frac{K_M}{v_{max}}\right)\frac{1}{[S]_0} \qquad (23.3b)$$

Form of the Michaelis–Menten equation used in a Lineweaver–Burk plot

A **Lineweaver–Burk plot** is a plot of $1/v$ against $1/[S]_0$, and according to eqn 23.3b it should yield a straight line with slope of K_M/v_{max}, a y-intercept at $1/v_{max}$, and an x-intercept at $-1/K_M$ (Fig. 23.4). The value of k_b is then calculated from the y-intercept and eqn 23.2b. However, the plot cannot give the individual rate constants k_a and k_a' that appear in the expression for K_M. The stopped-flow technique described in Section 21.1b can give the additional data needed, because we can find the rate of formation of the enzyme–substrate complex by monitoring the concentration after mixing the enzyme and substrate. This procedure gives a value for k_a, and k_a' is then found by combining this result with the values of k_b and K_M.

(b) The catalytic efficiency of enzymes

The **turnover frequency**, or **catalytic constant**, of an enzyme, k_{cat}, is the number of catalytic cycles (turnovers) performed by the active site in a given interval divided by the duration of the interval. This quantity has units of a first-order rate constant and, in terms of the Michaelis–Menten mechanism, is numerically equivalent to k_b, the rate constant for release of product from the enzyme–substrate complex. It follows from the identification of k_{cat} with k_b and from eqn 23.2b that

$$k_{cat} = k_b = \frac{v_{max}}{[E]_0} \qquad (23.4)$$

Turnover frequency

The **catalytic efficiency**, η (eta), of an enzyme is the ratio k_{cat}/K_M. The higher the value of η, the more efficient is the enzyme. We can think of the catalytic efficiency as the effective rate constant of the enzymatic reaction. From $K_M = (k_a' + k_b)/k_a$ and eqn 23.4, it follows that

$$\eta = \frac{k_{cat}}{K_M} = \frac{k_a k_b}{k_a' + k_b} \qquad (23.5)$$

Catalytic efficiency

The efficiency reaches its maximum value of k_a when $k_b \gg k_a'$. Because k_a is the rate constant for the formation of a complex from two species that are diffusing freely in solution, the maximum efficiency is related to the maximum rate of diffusion of E and S in solution. This limit (which is discussed further in Section 22.2) leads to rate constants of about 10^8–10^9 dm^3 mol^{-1} s^{-1} for molecules as large as enzymes at room temperature. The enzyme catalase has $\eta = 4.0 \times 10^8$ dm^3 mol^{-1} s^{-1} and is said to have attained 'catalytic perfection', in the sense that the rate of the reaction it catalyses is controlled only by diffusion: it acts as soon as a substrate makes contact.

Example 23.1 *Determining the catalytic efficiency of an enzyme*

The enzyme carbonic anhydrase catalyses the hydration of CO_2 in red blood cells to give bicarbonate (hydrogencarbonate) ion:

$$CO_2(g) + H_2O(l) \rightarrow HCO_3^-(aq) + H^+(aq)$$

The following data were obtained for the reaction at pH = 7.1, 273.5 K, and an enzyme concentration of 2.3 nmol dm^{-3}:

$[CO_2]/(\text{mmol dm}^{-3})$	1.25	2.5	5	20
rate/(mmol dm^{-3} s^{-1})	2.78×10^{-2}	5.00×10^{-2}	8.33×10^{-2}	1.67×10^{-1}

Determine the catalytic efficiency of carbonic anhydrase at 273.5 K.

Method Prepare a Lineweaver–Burk plot and determine the values of K_M and v_{max} by linear regression analysis. From eqn 23.4 and the enzyme concentration, calculate k_{cat} and the catalytic effciency from eqn 23.5.

Answer We draw up the following table:

$1/([CO_2]/(\text{mmol dm}^{-3}))$	0.800	0.400	0.200	0.0500
$1/(v/(\text{mmol dm}^{-3}\,\text{s}^{-1}))$	36.0	20.0	12.0	6.0

Figure 23.5 shows the Lineweaver–Burk plot for the data. The slope is 40.0 and the y-intercept is 4.00. Hence,

$$v_{max}/(\text{mmol dm}^{-3}\,\text{s}^{-1}) = \frac{1}{\text{intercept}} = \frac{1}{4.00} = 0.250$$

and

$$K_M/(\text{mmol dm}^{-3}) = \frac{\text{slope}}{\text{intercept}} = \frac{40.00}{4.00} = 10.0$$

It follows that

$$k_{cat} = \frac{v_{max}}{[E]_0} = \frac{2.5 \times 10^{-4}\,\text{mol dm}^{-3}\,\text{s}^{-1}}{2.3 \times 10^{-9}\,\text{mol dm}^{-3}} = 1.1 \times 10^5\,\text{s}^{-1}$$

and

$$\eta = \frac{k_{cat}}{K_M} = \frac{1.1 \times 10^5\,\text{s}^{-1}}{10.0 \times 10^{-3}\,\text{mol dm}^{-3}} = 1.1 \times 10^7\,\text{dm}^3\,\text{mol}^{-1}\,\text{s}^{-1}$$

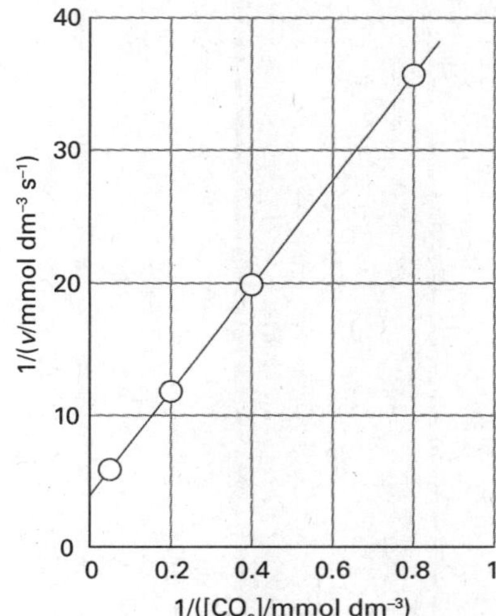

Fig. 23.5 The Lineweaver-Burk plot of the data for Example 23.1.

A note on good practice The slope and the intercept are unitless: we have remarked previously, that all graphs should be plotted as pure numbers.

Self-test 23.1 The enzyme α-chymotrypsin is secreted in the pancreas of mammals and cleaves peptide bonds made between certain amino acids. Several solutions containing the small peptide *N*-glutaryl-l-phenylalanine-*p*-nitroanilide at different concentrations were prepared and the same small amount of α-chymotrypsin was added to each one. The following data were obtained on the initial rates of the formation of product:

[S]/(mmol dm^{-3})	0.334	0.450	0.667	1.00	1.33	1.67
v/(mmol dm^{-3} s^{-1})	0.152	0.201	0.269	0.417	0.505	0.667

Determine the maximum velocity and the Michaelis constant for the reaction.

[$v_{max} = 2.76$ mmol dm^{-3} s^{-1}, $K_M = 5.77$ mmol dm^{-3}]

(c) Mechanisms of enzyme inhibition

An inhibitor, I, decreases the rate of product formation from the substrate by binding to the enzyme, to the ES complex, or to the enzyme and ES complex simultaneously. The most general kinetic scheme for enzyme inhibition is then:

$$\mathrm{E+S \rightleftarrows ES} \qquad k_a, k_a'$$

$$\mathrm{ES \rightarrow P+E} \qquad k_b$$

$$\mathrm{EI \rightleftharpoons E+I} \qquad K_I = \frac{[\mathrm{E}][\mathrm{I}]}{[\mathrm{EI}]} \tag{23.6a}$$

$$\mathrm{ESI \rightleftharpoons ES+I} \qquad K_I' = \frac{[\mathrm{ES}][\mathrm{I}]}{[\mathrm{ESI}]} \tag{23.6b}$$

The lower the values of K_I and K_I' the more efficient are the inhibitors. The rate of product formation is always given by $v = k_b[\mathrm{ES}]$, because only ES leads to product. As shown in the following *Justification*, the rate of reaction in the presence of an inhibitor is

$$v = \frac{v_{max}}{\alpha' + \alpha K_M/[\mathrm{S}]_0} \qquad \text{Effect of inhibition on the rate} \tag{23.7}$$

where $\alpha = 1 + [\mathrm{I}]/K_I$ and $\alpha' = 1 + [\mathrm{I}]/K_I'$. This equation is very similar to the Michaelis–Menten equation for the uninhibited enzyme (eqn 23.1) and is also amenable to analysis by a Lineweaver–Burk plot:

$$\frac{1}{v} = \frac{\alpha'}{v_{max}} + \left(\frac{\alpha K_M}{v_{max}}\right)\frac{1}{[\mathrm{S}]_0} \tag{23.8}$$

Justification 23.2 *Enzyme inhibition*

By mass balance, the total concentration of enzyme is:

$$[\mathrm{E}]_0 = [\mathrm{E}] + [\mathrm{EI}] + [\mathrm{ES}] + [\mathrm{ESI}]$$

By using eqns 23.6a and 23.6b and the definitions

$$\alpha = 1 + \frac{[\mathrm{I}]}{K_I} \quad \text{and} \quad \alpha' = 1 + \frac{[\mathrm{I}]}{K_I'}$$

it follows that

$$[\mathrm{E}]_0 = [\mathrm{E}]\alpha + [\mathrm{ES}]\alpha'$$

By using $K_M = [\mathrm{E}][\mathrm{S}]/[\mathrm{ES}]$ and replacing [S] with $[\mathrm{S}]_0$ we can write

$$[\mathrm{E}]_0 = \frac{K_M[\mathrm{ES}]}{[\mathrm{S}]_0}\alpha + [\mathrm{ES}]\alpha' = [\mathrm{ES}]\left(\frac{\alpha K_M}{[\mathrm{S}]_0} + \alpha'\right)$$

The expression for the rate of product formation is then:

$$v = k_b[\mathrm{ES}] = \frac{k_b[\mathrm{E}]_0}{\alpha K_M/[\mathrm{S}]_0 + \alpha'}$$

which, by using eqn 23.2b, gives eqn 23.7.

There are three major modes of inhibition that give rise to distinctly different kinetic behaviour (Fig. 23.6). In **competitive inhibition** the inhibitor binds only to the active site of the enzyme and thereby inhibits the attachment of the substrate.

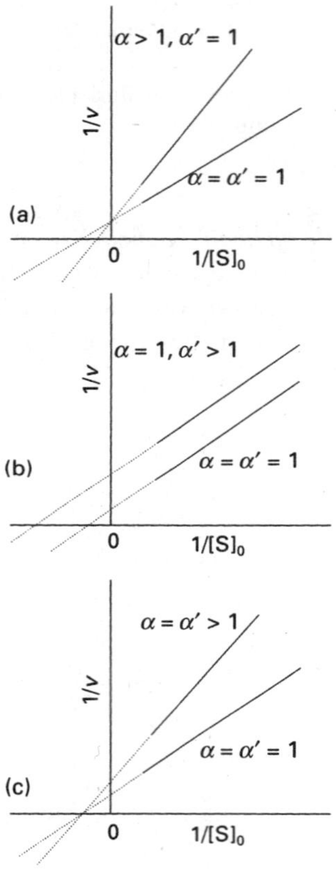

Fig. 23.6 Lineweaver–Burk plots characteristic of the three major modes of enzyme inhibition: (a) competitive inhibition, (b) uncompetitive inhibition, and (c) non-competitive inhibition, showing the special case $\alpha = \alpha' > 1$.

interActivity Use eqn 23.8 to explore the effect of competitive, uncompetitive, and non-competitive inhibition on the shapes of the plots of v against [S] for constant K_M and v_{max}.

This condition corresponds to $\alpha > 1$ and $\alpha' = 1$ (because ESI does not form). In this limit, eqn 23.8 becomes

$$\frac{1}{v} = \frac{1}{v_{max}} + \left(\frac{\alpha K_M}{v_{max}}\right)\frac{1}{[S]_0}$$

Competitive inhibition

The y-intercept is unchanged but the slope of the Lineweaver–Burk plot increases by a factor of α relative to the slope for data on the uninhibited enzyme (Fig. 23.6a). In **uncompetitive inhibition** the inhibitor binds to a site of the enzyme that is removed from the active site, but only if the substrate is already present. The inhibition occurs because ESI reduces the concentration of ES, the active type of complex. In this case $\alpha = 1$ (because EI does not form) and $\alpha' > 1$ and eqn 23.8 becomes

$$\frac{1}{v} = \frac{\alpha'}{v_{max}} + \left(\frac{K_M}{v_{max}}\right)\frac{1}{[S]_0}$$

Uncompetitive inhibition

The y-intercept of the Lineweaver–Burk plot increases by a factor of α' relative to the y-intercept for data on the uninhibited enzyme but the slope does not change (Fig. 23.6b). In **non-competitive inhibition** (also called **mixed inhibition**) the inhibitor binds to a site other than the active site, and its presence reduces the ability of the substrate to bind to the active site. Inhibition occurs at both the E and ES sites. This condition corresponds to $\alpha > 1$ and $\alpha' > 1$. Both the slope and y-intercept of the Lineweaver–Burk plot increase upon addition of the inhibitor. Figure 23.6c shows the special case of $K_I = K_I'$ and $\alpha = \alpha'$, which results in intersection of the lines at the x-axis.

In all cases, the efficiency of the inhibitor may be obtained by determining K_M and v_{max} from a control experiment with uninhibited enzyme and then repeating the experiment with a known concentration of inhibitor. From the slope and y-intercept of the Lineweaver-Burk plot for the inhibited enzyme (eqn 23.8), the mode of inhibition, the values of α or α', and the values of K_I and K_I' may be obtained.

Example 23.2 *Distinguishing between types of inhibition*

Five solutions of a substrate, S, were prepared with the concentrations given in the first column below and each one was divided into five equal volumes. The same concentration of enzyme was present in each one. An inhibitor, I, was then added in four different concentrations to the samples, and the initial rate of formation of product was determined with the results given below. Does the inhibitor act competitively or non-competitively? Determine K_I and K_M.

	$v/(\mu mol\ dm^{-3}\ s^{-1})$ for $[I]/(mmol\ dm^{-3})$ =				
$[S]_0/(mmol\ dm^{-3})$	0	0.20	0.40	0.60	0.80
0.050	0.033	0.026	0.021	0.018	0.016
0.10	0.055	0.045	0.038	0.033	0.029
0.20	0.083	0.071	0.062	0.055	0.050
0.40	0.111	0.100	0.091	0.084	0.077
0.60	0.126	0.116	0.108	0.101	0.094

Method We draw a series of Lineweaver–Burk plots for different inhibitor concentrations. If the plots resemble those in Fig. 23.6a, then the inhibition is

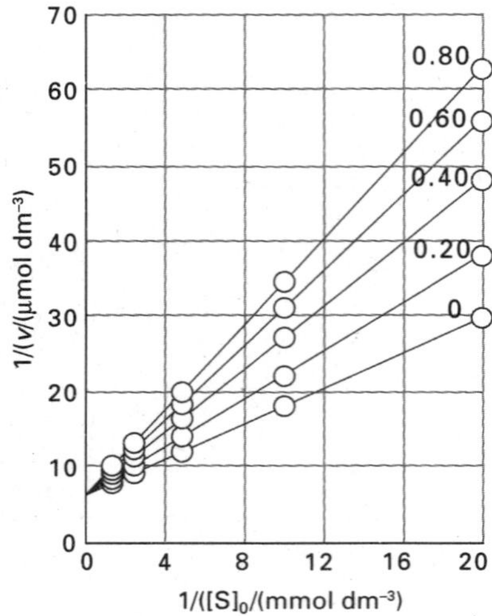

Fig. 23.7 Lineweaver–Burk plots for the data in Example 23.2. Each line corresponds to a different concentration of inhibitor.

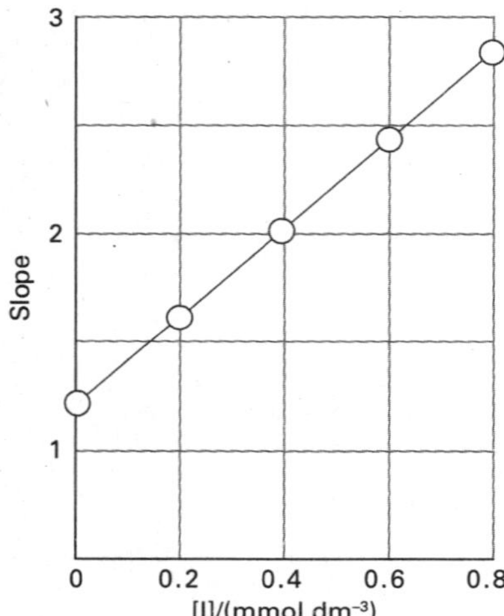

Fig. 23.8 Plot of the slopes of the plots in Fig. 23.7 against [I] based on the data in Example 23.2.

competitive. On the other hand, if the plots resemble those in Fig. 23.6c, then the inhibition is non-competitive. To find K_I, we need to determine the slope at each value of [I], which is equal to $\alpha K_M/v_{max}$, or $K_M/v_{max} + K_M[I]/K_I v_{max}$, then plot this slope against [I]: the intercept at [I] = 0 is the value of K_M/v_{max} and the slope is $K_M/K_I v_{max}$.

Answer First we draw up a table of $1/[S]_0$ and $1/v$ for each value of [I]:

	$1/(v/(\mu mol\ dm^{-3}\ s^{-1}))$ for $[I]/(mmol\ dm^{-3})$ =				
$1/([S]_0/(mmol\ dm^{-3}))$	0	0.20	0.40	0.60	0.80
20	30	38	48	56	62
10	18	22	26	30	34
5.0	12	14	16	18	20
2.5	9.01	10.0	11.0	11.9	13.0
1.7	7.94	8.62	9.26	9.90	10.6

The five plots (one for each [I]) are given in Fig. 23.7. We see that they pass through the same intercept on the vertical axis, so the inhibition is competitive. The mean of the (least squares) intercepts is 5.83, so $v_{max} = 0.172\ \mu mol\ dm^{-3}\ s^{-1}$ (note how it picks up the units for v in the data). The (least squares) slopes of the lines are as follows:

$[I]/(mmol\ dm^{-3})$	0	0.20	0.40	0.60	0.80
Slope	1.219	1.627	2.090	2.489	2.832

These values are plotted in Fig. 23.8. The intercept at [I] = 0 is 1.234, so K_M = 0.212 mmol dm^{-3}. The (least squares) slope of the line is 2.045, so

$$K_I/(mmol\ dm^{-3}) = \frac{K_M}{slope \times v_{max}} = \frac{0.212}{2.045 \times 0.172} = 0.603$$

Self-test 23.2 Repeat the question using the following data:

	$v/(\mu mol\ dm^{-3}\ s^{-1})$ for $[I]/(mmol\ dm^{-3})$ =				
$[S]_0/(mmol\ dm^{-3})$	0	0.20	0.40	0.60	0.80
0.050	0.020	0.015	0.012	0.0098	0.0084
0.10	0.035	0.026	0.021	0.017	0.015
0.20	0.056	0.042	0.033	0.028	0.024
0.40	0.080	0.059	0.047	0.039	0.034
0.60	0.093	0.069	0.055	0.046	0.039

[Non-competitive, $K_M = 0.30$ mmol dm^{-3}, $K_I = 0.57$ mmol dm^{-3}]

Heterogeneous catalysis

The remainder of this chapter is devoted to developing and applying concepts of structure and reactivity in heterogeneous catalysis. For simplicity, we consider only gas/solid systems. To understand the catalytic role of a solid surface we begin by describing its unique structural features. Then, because many reactions catalysed by surfaces involve reactants and products in the gas phase, we discuss **adsorption**, the attachment of particles to a solid surface, and **desorption**, the reverse process. Finally, we consider specific mechanisms of heterogeneous catalysis.

23.3 The growth and structure of solid surfaces

Key points **Adsorption is the attachment of molecules to a surface; the substance that adsorbs is the adsorbate and the underlying material is the adsorbent or substrate. The reverse of adsorption is desorption. (a) Surface defects play an important role in surface growth and catalysis. (b) Techniques for studying surface composition and structure include scanning electron microscopy, scanning probe microscopy, photoemission spectroscopy, Auger electron spectroscopy, and low energy electron diffraction.**

The substance that adsorbs is the **adsorbate** and the underlying material that we are concerned with in this section is the **adsorbent** or **substrate**.

(a) Surface growth

A simple picture of a perfect crystal surface is as a tray of oranges in a grocery store (Fig. 23.9). A gas molecule that collides with the surface can be imagined as a ping-pong ball bouncing erratically over the oranges. The molecule loses energy as it bounces, but it is likely to escape from the surface before it has lost enough kinetic energy to be trapped. The same is true, to some extent, of an ionic crystal in contact with a solution. There is little energy advantage for an ion in solution to discard some of its solvating molecules and stick at an exposed position on the surface.

The picture changes when the surface has defects, for then there are ridges of incomplete layers of atoms or ions. A common type of surface defect is a **step** between two otherwise flat layers of atoms called **terraces** (Fig. 23.10). A step defect might itself have defects, for it might have kinks. When an atom settles on a terrace it bounces across it under the influence of the intermolecular potential, and might come to a step or a corner formed by a kink. Instead of interacting with a single terrace atom, the molecule now interacts with several, and the interaction may be strong enough to trap it. Likewise, when ions deposit from solution, the loss of the solvation interaction is offset by a strong Coulombic interaction between the arriving ions and several ions at the surface defect.

The rapidity of growth depends on the crystal plane concerned, and the slowest growing faces dominate the appearance of the crystal. This feature is explained in Fig. 23.11, where we see that, although the horizontal face grows forward most rapidly, it grows itself out of existence, and the slower-growing faces survive.

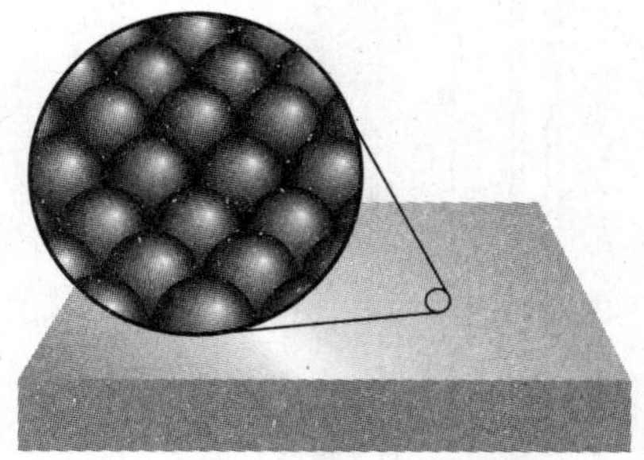

Fig. 23.9 A schematic diagram of the flat surface of a solid. This primitive model is largely supported by scanning tunnelling microscope images (see *Impact I8.2*).

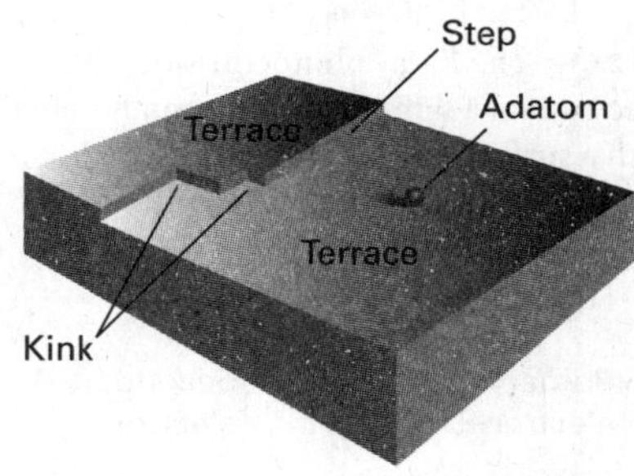

Fig. 23.10 Some of the kinds of defects that may occur on otherwise perfect terraces. Defects play an important role in surface growth and catalysis.

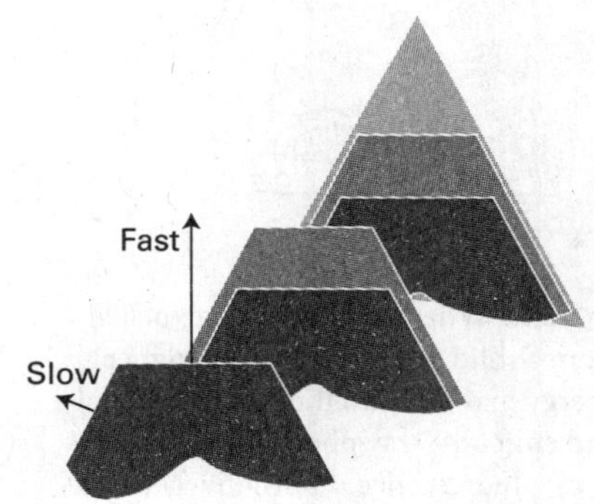

Fig. 23.11 The more slowly growing faces of a crystal dominate its final external appearance. Three successive stages of the growth are shown.

(b) Surface composition and structure

Under normal conditions, a surface exposed to a gas is constantly bombarded with molecules and a freshly prepared surface is covered very quickly. Just how quickly can be estimated using the kinetic model of gases and the following expression for the collision flux (eqn 20.14):

$$Z_W = \frac{p}{(2\pi mkT)^{1/2}} \qquad \text{Collision flux} \qquad (23.9)$$

A practical form of this equation is

$$Z_W = \frac{Z_0(p/\text{Pa})}{\{(T/\text{K})(M/(\text{g mol}^{-1}))\}^{1/2}} \qquad \text{with} \qquad Z_0 = 2.63 \times 10^{24}\ \text{m}^{-2}\ \text{s}^{-1}$$

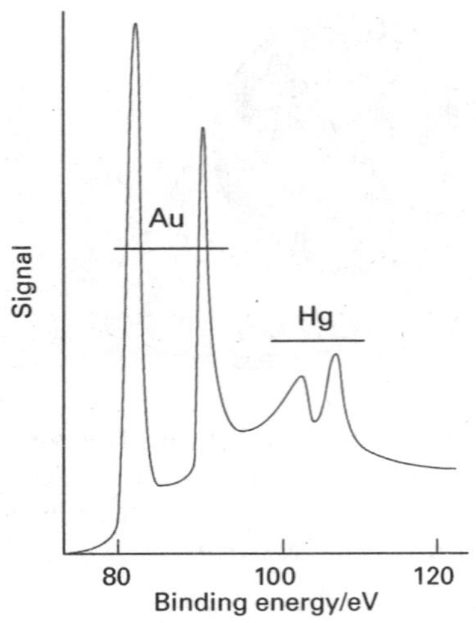

Fig. 23.12 The X-ray photoemission spectrum of a sample of gold contaminated with a surface layer of mercury. (M.W. Roberts and C.S. McKee, *Chemistry of the metal–gas interface*, Oxford (1978).)

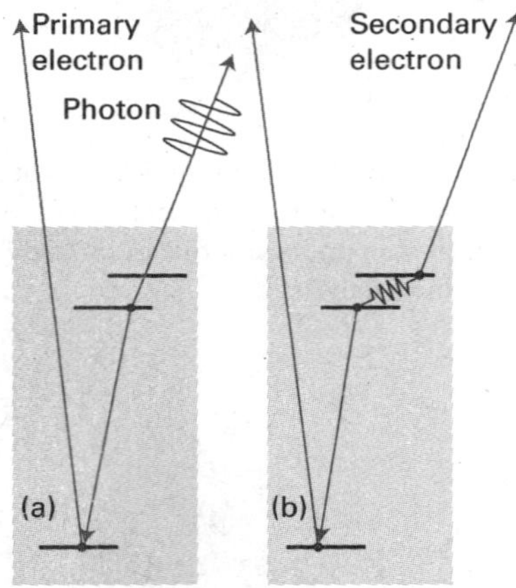

Fig. 23.13 When an electron is expelled from a solid (a) an electron of higher energy may fall into the vacated orbital and emit an X-ray photon to produce X-ray fluorescence. Alternatively (b) the electron falling into the orbital may give up its energy to another electron, which is ejected in the Auger effect.

where M is the molar mass of the gas. For air ($M \approx 29$ g mol^{-1}) at 1 atm and 25°C the collision flux is 3×10^{27} m^{-2} s^{-1}. Because 1 m^2 of metal surface consists of about 10^{19} atoms, each atom is struck about 10^8 times each second. Even if only a few collisions leave a molecule adsorbed to the surface, the time for which a freshly prepared surface remains clean is very short.

The obvious way to retain cleanliness is to reduce the pressure. When it is reduced to 10^{-4} Pa (as in a simple vacuum system) the collision flux falls to about 10^{18} m^{-2} s^{-1}, corresponding to one hit per surface atom in each 0.1 s. Even that is too brief in most experiments, and in **ultrahigh vacuum** (UHV) techniques pressures as low as 0.1 μPa (when $Z_W = 10^{15}$ m^{-2} s^{-1}) are reached on a routine basis and as low as 1 nPa (when $Z_W = 10^{13}$ m^{-2} s^{-1}) are reached with special care. These collision fluxes correspond to each surface atom being hit once every 10^5 to 10^6 s, or about once a day.

The chemical composition of a surface can be determined by a variety of ionization techniques. The same techniques can be used to detect any remaining contamination after cleaning and to detect layers of material adsorbed later in the experiment. One technique is **photoemission spectroscopy**, a derivative of the photoelectric effect, in which X-rays (for XPS) or hard (short wavelength) ultraviolet (for UPS) ionizing radiation is used, giving rise to ejected electrons from adsorbed species. The kinetic energies of the electrons ejected from their orbitals are measured and the pattern of energies is a fingerprint of the material present (Fig. 23.12). UPS, which examines electrons ejected from valence shells, is also used to establish the bonding characteristics and the details of valence shell electronic structures of substances on the surface. Its usefulness is its ability to reveal which orbitals of the adsorbate are involved in the bond to the substrate. For instance, the principal difference between the photoemission results on free benzene and benzene adsorbed on palladium is in the energies of the π electrons. This difference is interpreted as meaning that the C_6H_6 molecules lie parallel to the surface and are attached to it by their π orbitals. In contrast, pyridine (C_6H_5N) stands almost perpendicular to the surface, and is attached by a σ bond formed by the nitrogen lone pair.

A very important technique, which is widely used in the microelectronics industry, is **Auger electron spectroscopy** (AES). The **Auger effect** (pronounced oh-zhey) is the emission of a second electron after high energy radiation has expelled another. The first electron to depart leaves a hole in a low-lying orbital, and an upper electron falls into it. The energy this releases may result either in the generation of radiation, which is called **X-ray fluorescence** (Fig. 23.13a) or in the ejection of another electron (Fig. 23.13b). The latter is the secondary electron of the Auger effect. The energies of the secondary electrons are characteristic of the material present, so the Auger effect effectively takes a fingerprint of the sample. In practice, the Auger spectrum is normally obtained by irradiating the sample with an electron beam of energy in the range 1–5 keV rather than electromagnetic radiation. In **scanning Auger electron microscopy** (SAM), the finely focused electron beam is scanned over the surface and a map of composition is compiled; the resolution can reach below about 50 nm.

One of the most informative techniques for determining the arrangement of the atoms close to the surface is **low energy electron diffraction** (LEED). This technique is like X-ray diffraction (Chapter 19) but uses the wave character of electrons, and the sample is now the surface of a solid. The use of low energy electrons (with energies in the range 10–200 eV, corresponding to wavelengths in the range 100–400 pm) ensures that the diffraction is caused only by atoms on and close to the surface. The experimental arrangement is shown in Fig. 23.14, and typical LEED patterns, obtained by photographing the fluorescent screen through the viewing port, are shown in Fig. 23.15.

Example 23.3 *Interpreting a LEED pattern*

The LEED pattern from a clean unreconstructed (110) face of palladium is shown in (a) below. The reconstructed surface gives a LEED pattern shown as (b). What can be inferred about the structure of the reconstructed surface?

```
 • • •            • • •
                  • • •
 • • •            • • •
(a)          (b)  • • •
 • • •            • • •
```

Method Recall from Bragg's law (Section 19.3), $\lambda = 2d \sin \theta$, that, for a given wavelength, the smaller the separation d of the layers, the greater the scattering angle (so that $2d \sin \theta$ remains constant). In terms of the LEED pattern, the farther apart the atoms responsible for the pattern, the closer the spots appear in the pattern. Twice the separation between the atoms corresponds to half the separation between the spots, and vice versa. Therefore, inspect the two patterns and identify how the new pattern relates to the old.

Answer The horizontal separation between spots is unchanged, which indicates that the atoms remain in the same position in that dimension when reconstruction occurs. However, the vertical spacing is halved, which suggests that the atoms are twice as far apart in that direction as they are in the unreconstructed surface.

Self-test 23.3 Sketch the LEED pattern for a surface that was reconstructed from that shown in (a) above by tripling the vertical separation. [:::::::]

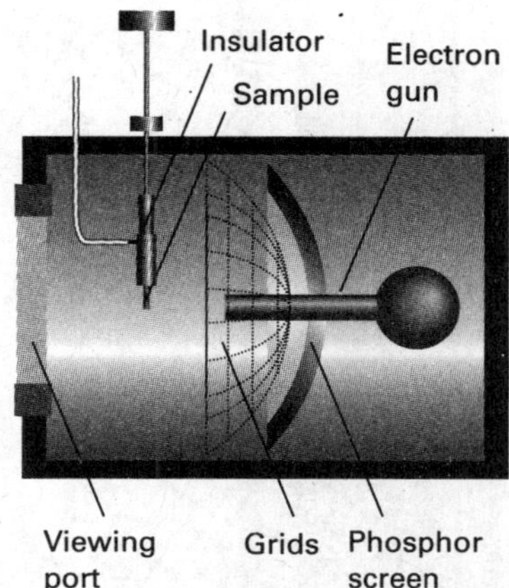

Fig. 23.14 A schematic diagram of the apparatus used for a LEED experiment. The electrons diffracted by the surface layers are detected by the fluorescence they cause on the phosphor screen.

Observations using LEED show that the surface of a crystal rarely has exactly the same form as a slice through the bulk. As a general rule, it is found that metal surfaces are simply truncations of the bulk lattice, but the distance between the top layer of atoms and the one below is contracted by around 5 per cent. Semiconductors generally have surfaces reconstructed to a depth of several layers. Reconstruction occurs in ionic solids. For example, in lithium fluoride the Li^+ and F^- ions close to the surface apparently lie on slightly different planes. An actual example of the detail that can now be obtained from refined LEED techniques is shown in Fig. 23.16 for CH_3C– adsorbed on a (111) plane of rhodium.

The presence of terraces, steps, and kinks in a surface shows up in LEED patterns, and their surface density (the number of defects in a region divided by the area of the

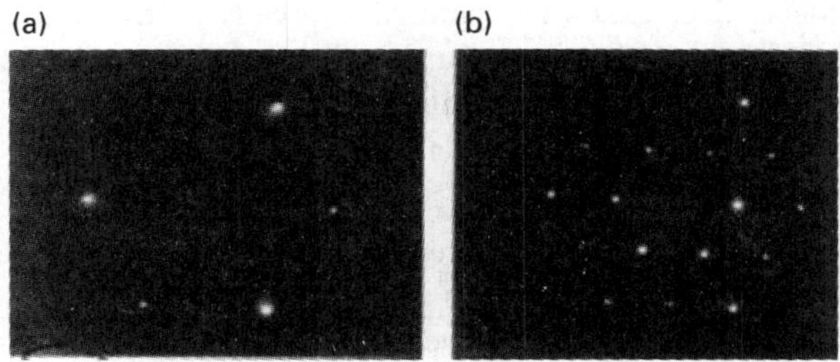

Fig. 23.15 LEED photographs of (a) a clean platinum surface and (b) after its exposure to propyne, $CH_3C{\equiv}CH$. (Photographs provided by Professor G.A. Somorjai.)

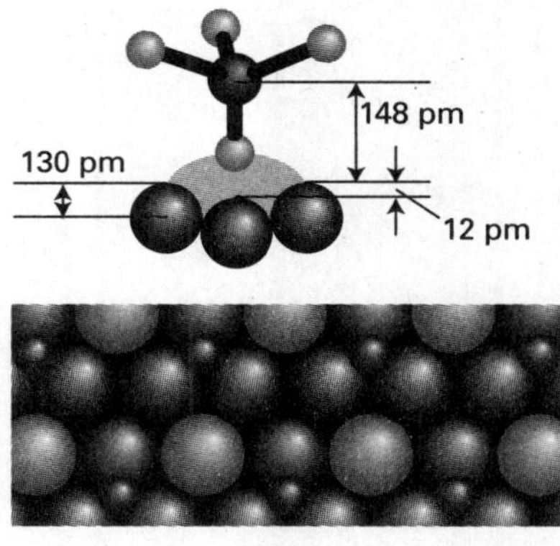

Fig. 23.16 The structure of a surface close to the point of attachment of CH_3C– to the (110) surface of rhodium at 300 K and the changes in positions of the metal atoms that accompany chemisorption.

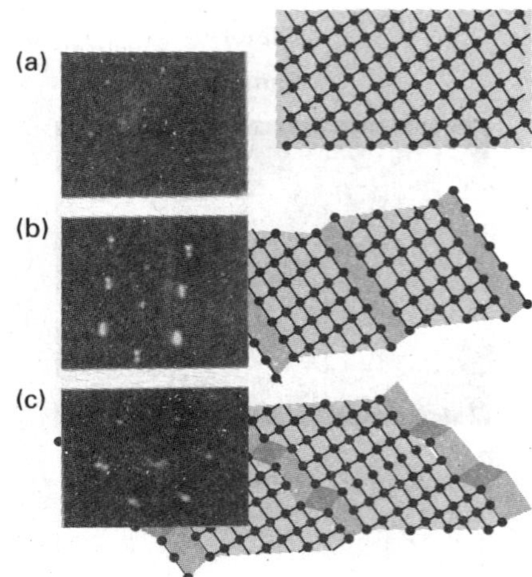

Fig. 23.17 LEED patterns may be used to assess the defect density of a surface. The photographs correspond to a platinum surface with (a) low defect density, (b) regular steps separated by about six atoms, and (c) regular steps with kinks. (Photographs provided by Professor G.A. Samorjai.)

region) can be estimated. The importance of this type of measurement will emerge later. Three examples of how steps and kinks affect the pattern are shown in Fig. 23.17. The samples used were obtained by cleaving a crystal at different angles to a plane of atoms. Only terraces are produced when the cut is parallel to the plane, and the density of steps increases as the angle of the cut increases. The observation of additional structure in the LEED patterns, rather than blurring, shows that the steps are arrayed regularly.

23.4 The extent of adsorption

Key points **The fractional coverage, θ, is the ratio of the number of occupied sites to the number of available sites. Techniques for studying the rates of surface processes include flash desorption, second harmonic generation, gravimetry, and molecular beam reactive scattering. (a) Physisorption is adsorption by a van der Waals interaction; chemisorption is adsorption by formation of a chemical bond. (b) An adsorption isotherm is the variation of θ with pressure at a chosen temperature. Examples include the Langmuir, BET, Temkin, and Freundlich isotherms.**

The extent of surface coverage is normally expressed as the **fractional coverage**, θ:

$$\theta = \frac{\text{number of adsorption sites occupied}}{\text{number of adsorption sites available}} \qquad \text{Definition of the fractional coverage} \qquad [23.10]$$

The fractional coverage is often expressed in terms of the volume of adsorbate adsorbed by $\theta = V/V_\infty$, where V_∞ is the volume of adsorbate corresponding to complete monolayer coverage. The **rate of adsorption**, $d\theta/dt$, is the rate of change of surface coverage, and can be determined by observing the change of fractional coverage with time.

Among the principal techniques for measuring $d\theta/dt$ are flow methods, in which the sample itself acts as a pump because adsorption removes particles from the gas. One commonly used technique is therefore to monitor the rates of flow of gas into and out of the system: the difference is the rate of gas uptake by the sample. Integration of this rate then gives the fractional coverage at any stage. In **flash desorption** the sample is suddenly heated (electrically) and the resulting rise of pressure is interpreted in terms of the amount of adsorbate originally on the sample. The interpretation may be confused by the desorption of a compound (for example, WO_3 from oxygen on tungsten). **Gravimetry**, in which the sample is weighed on a microbalance during the experiment, can also be used. A common instrument for gravimetric measurements is the **quartz crystal microbalance** (QCM), in which the mass of a sample laid on the surface of a quartz crystal is related to changes in the latter's mechanical properties. The key principle behind the operation of a QCM is the ability of a quartz crystal to vibrate at a characteristic frequency when an oscillating electric field is applied. The vibrational frequency decreases when material is spread over the surface of the crystal and the change in frequency is proportional to the mass of material. Masses as small as a few nanograms can be measured reliably in this way.

(a) Physisorption and chemisorptions

Molecules and atoms can attach to surfaces in two ways. In **physisorption** (an abbreviation of 'physical adsorption'), there is a van der Waals interaction (for example, a dispersion or a dipolar interaction) between the adsorbate and the substrate. van der Waals interactions have a long range but are weak, and the energy released when a particle is physisorbed is of the same order of magnitude as the enthalpy of condensation. Such small energies can be absorbed as vibrations of the lattice and dissipated as thermal motion, and a molecule bouncing across the surface will gradually lose its energy and finally adsorb to it in the process called **accommodation**. The enthalpy of physisorption can be measured by monitoring the rise in temperature of a sample of known heat capacity, and typical values are in the region of -20 kJ mol^{-1} (Table 23.1). This small enthalpy change is insufficient to lead to bond breaking, so a physisorbed molecule retains its identity, although it might be distorted by the presence of the surface.

In **chemisorption** (an abbreviation of 'chemical adsorption'), the molecules (or atoms) stick to the surface by forming a chemical (usually covalent) bond, and tend to find sites that maximize their coordination number with the substrate. The enthalpy of chemisorption is very much greater than that for physisorption, and typical values are in the region of -200 kJ mol^{-1} (Table 23.2). The distance between the surface and the closest adsorbate atom is also typically shorter for chemisorption than for physisorption. A chemisorbed molecule may be torn apart at the demand of the unsatisfied valencies of the surface atoms, and the existence of molecular fragments on the surface as a result of chemisorption is one reason why solid surfaces catalyse reactions.

Except in special cases, chemisorption must be exothermic. A spontaneous process requires $\Delta G < 0$ at constant pressure and temperature. Because the translational freedom of the adsorbate is reduced when it is adsorbed, ΔS is negative. Therefore, in order for $\Delta G = \Delta H - T\Delta S$ to be negative, ΔH must be negative (that is, the process is exothermic). Exceptions may occur if the adsorbate dissociates and has high translational mobility on the surface. For example, H_2 adsorbs endothermically on glass because there is a large increase of translational entropy accompanying the dissociation of the molecules into atoms that move quite freely over the surface. In this case, the entropy change in the process $H_2(g) \rightarrow 2\,H(\text{glass})$ is sufficiently positive to overcome the small positive enthalpy change.

The enthalpy of adsorption depends on the extent of surface coverage, mainly because the adsorbate particles interact. If the particles repel each other (as for CO on palladium) the adsorption becomes less exothermic (the enthalpy of adsorption less negative) as coverage increases. Moreover, LEED studies show that such species settle on the surface in a disordered way until packing requirements demand order. If the adsorbate particles attract one another (as for O_2 on tungsten), then they tend to cluster together in islands, and growth occurs at the borders. These adsorbates also show order–disorder transitions when they are heated enough for thermal motion to overcome the particle–particle interactions, but not so much that they are desorbed.

Table 23.1* Maximum observed enthalpies of physisorption

Adsorbate	$\Delta_{ad}H^{\ominus}/(\text{kJ mol}^{-1})$
CH_4	−21
H_2	−84
H_2O	−59
N_2	−21

* More values are given in the *Data section*.

Table 23.2* Enthalpies of chemisorption, $\Delta_{ad}H^{\ominus}/(\text{kJ mol}^{-1})$

Adsorbate	Adsorbent (substrate)		
	Cr	Fe	Ni
C_2H_4	−427	−285	−243
CO		−192	
H_2	−188	−134	
NH_3		−188	−155

* More values are given in the *Data section*.

(b) Adsorption isotherms

In chemisorption the free gas and the adsorbed gas are in dynamic equilibrium, and the fractional coverage of the surface depends on the pressure of the overlying gas. The variation of θ with pressure at a chosen temperature is called the **adsorption isotherm**.

The simplest physically plausible isotherm is based on three assumptions:

- Adsorption cannot proceed beyond monolayer coverage.
- All sites are equivalent and the surface is uniform (that is, the surface is perfectly flat on a microscopic scale).
- The ability of a molecule to adsorb at a given site is independent of the occupation of neighbouring sites (that is, there are no interactions between adsorbed molecules).

The dynamic equilibrium is

$$A(g) + M(\text{surface}) \rightleftharpoons AM(\text{surface})$$

with rate constants k_a for adsorption and k_d for desorption. The rate of change of surface coverage due to adsorption is proportional to the partial pressure p of A and the number of vacant sites $N(1-\theta)$, where N is the total number of sites:

$$\frac{d\theta}{dt} = k_a pN(1-\theta) \qquad \text{Rate of adsorption} \qquad (23.11a)$$

The rate of change of θ due to desorption is proportional to the number of adsorbed species, $N\theta$:

$$\frac{d\theta}{dt} = -k_d N\theta \qquad \text{Rate of desorption} \qquad (23.11b)$$

At equilibrium there is no net change (that is, the sum of these two rates is zero), and solving for θ gives the **Langmuir isotherm**:

$$\theta = \frac{Kp}{1+Kp} \qquad K = \frac{k_a}{k_d} \qquad \text{Langmuir isotherm} \qquad (23.12)$$

Example 23.4 *Using the Langmuir isotherm*

The data given below are for the adsorption of CO on charcoal at 273 K. Confirm that they fit the Langmuir isotherm, and find the constant K and the volume corresponding to complete coverage. In each case V has been corrected to 1.00 atm (101.325 kPa).

p/kPa	13.3	26.7	40.0	53.3	66.7	80.0	93.3
V/cm^3	10.2	18.6	25.5	31.5	36.9	41.6	46.1

Method From eqn 23.12,

$$Kp\theta + \theta = Kp$$

With $\theta = V/V_\infty$, where V_∞ is the volume corresponding to complete coverage, this expression can be rearranged into

$$\frac{p}{V} = \frac{p}{V_\infty} + \frac{1}{KV_\infty}$$

Hence, a plot of p/V against p should give a straight line of slope $1/V_\infty$ and intercept $1/KV_\infty$.

Answer The data for the plot are as follows:

p/kPa	13.3	26.7	40.0	53.3	66.7	80.0	93.3
$(p/\text{kPa})/(V/\text{cm}^3)$	1.30	1.44	1.57	1.69	1.81	1.92	2.02

The points are plotted in Fig. 23.18. The (least squares) slope is 0.00900, so $V_\infty = 111$ cm^3. The intercept at $p = 0$ is 1.20, so

$$K = \frac{1}{(111\ \text{cm}^3) \times (1.20\ \text{kPa cm}^{-3})} = 7.51 \times 10^{-3}\ \text{kPa}^{-1}$$

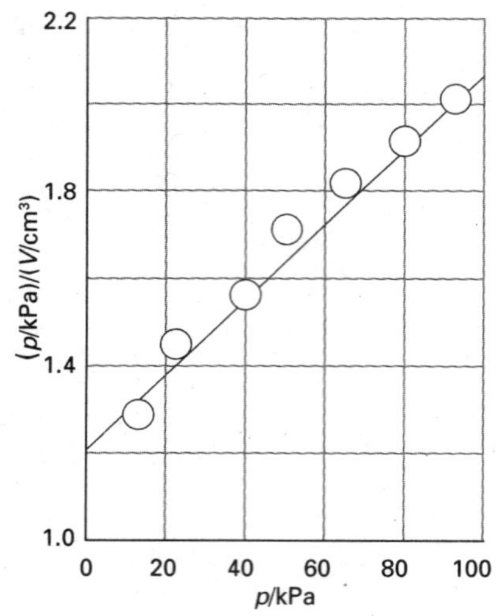

Fig. 23.18 The plot of the data in Example 23.4. As illustrated here, the Langmuir isotherm predicts that a straight line should be obtained when p/V is plotted against p.

Self-test 23.4 Repeat the calculation for the following data:

p/kPa	13.3	26.7	40.0	53.3	66.7	80.0	93.3
V/cm^3	10.3	19.3	27.3	34.1	40.0	45.5	48.0

[128 cm^3, 6.69×10^{-3} kPa^{-1}]

For adsorption with dissociation, the rate of adsorption is proportional to the pressure and to the probability that both atoms will find sites, which is proportional to the square of the number of vacant sites,

$$\frac{d\theta}{dt} = k_a p\{N(1-\theta)\}^2 \qquad (23.13a)$$

The rate of desorption is proportional to the frequency of encounters of atoms on the surface, and is therefore second-order in the number of atoms present:

$$\frac{d\theta}{dt} = -k_d(N\theta)^2 \qquad (23.13b)$$

The condition for no net change leads to the isotherm

$$\theta = \frac{(Kp)^{1/2}}{1+(Kp)^{1/2}} \qquad \text{Langmuir isotherm for adsorption with dissociation} \qquad (23.14)$$

The surface coverage now depends more weakly on pressure than for non-dissociative adsorption.

The shapes of the Langmuir isotherms with and without dissociation are shown in Figs. 23.19 and 23.20. The fractional coverage increases with increasing pressure, and approaches 1 only at very high pressure, when the gas is forced on to every available site of the surface. Different curves (and therefore different values of K) are obtained at different temperatures, and the temperature dependence of K can be used to determine the **isosteric enthalpy of adsorption**, $\Delta_{ad}H^{\ominus}$, the standard enthalpy of adsorption at a fixed surface coverage. To determine this quantity we recognize that K is essentially an equilibrium constant, and then use the van't Hoff equation (eqn 6.21) to write:

$$\left(\frac{\partial \ln K}{\partial T}\right)_\theta = \frac{\Delta_{ad}H^{\ominus}}{RT^2} \qquad \text{Isosteric enthalpy of adsorption from the equilibrium constant} \qquad (23.15)$$

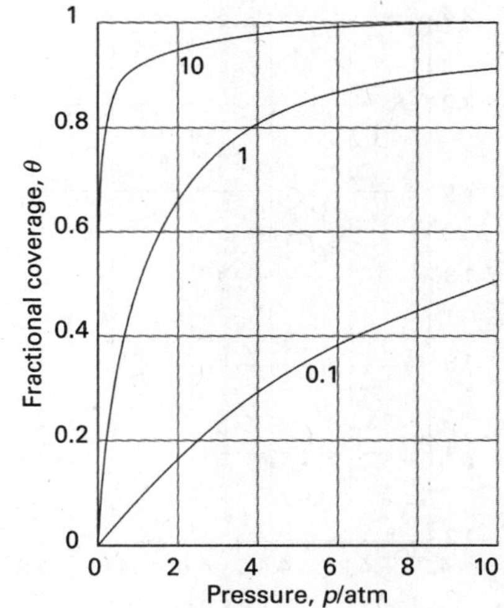

Fig. 23.19 The Langmuir isotherm for dissociative adsorption, $X_2(g) \rightarrow 2\,X(\text{surface})$, for different values of K.

interActivity Using eqn 23.14, generate a family of curves showing the dependence of $1/\theta$ on $1/p$ for several values of K.

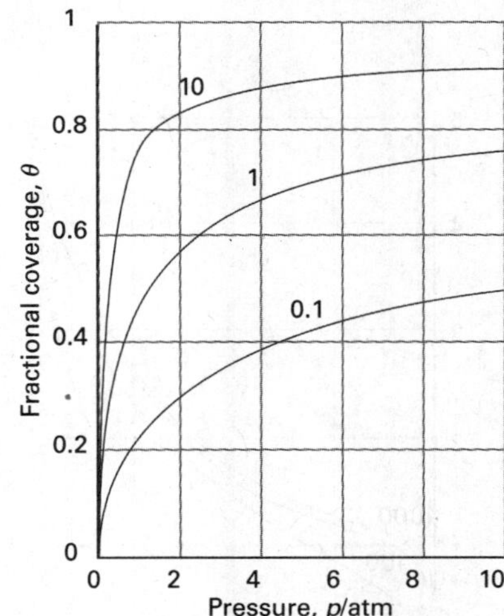

Fig. 23.20 The Langmuir isotherm for non-dissociative adsorption for different values of K.

interActivity Using eqn 23.12, generate a family of curves showing the dependence of $1/\theta$ on $1/p$ for several values of K. Taking these results together with those of the previous *interActivity*, discuss how plots of $1/\theta$ against $1/p$ can be used to distinguish between adsorption with and without dissociation.

Example 23.5 *Measuring the isosteric enthalpy of adsorption*

The data below show the pressures of CO needed for the volume of adsorption (corrected to 1.00 atm and 273 K) to be 10.0 cm^3 using the same sample as in Example 23.4. Calculate the adsorption enthalpy at this surface coverage.

T/K	200	210	220	230	240	250
p/kPa	4.00	4.95	6.03	7.20	8.47	9.85

Method The Langmuir isotherm can be rearranged to

$$Kp = \frac{\theta}{1-\theta}$$

Therefore, when θ is constant,

$$\ln K + \ln p = \text{constant}$$

It follows from eqn 23.15 that

$$\left(\frac{\partial \ln p}{\partial T}\right)_\theta = -\left(\frac{\partial \ln K}{\partial T}\right)_\theta = -\frac{\Delta_{ad}H^{\ominus}}{RT^2}$$

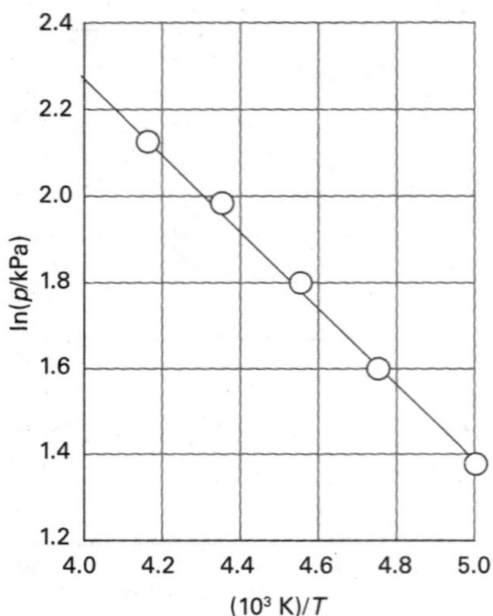

Fig. 23.21 The isosteric enthalpy of adsorption can be obtained from the slope of the plot of ln p against $1/T$, where p is the pressure needed to achieve the specified surface coverage. The data used are from Example 23.5.

With $\mathrm{d}(1/T)/\mathrm{d}T = -1/T^2$, this expression rearranges to

$$\left(\frac{\partial \ln p}{\partial(1/T)}\right)_\theta = \frac{\Delta_{\mathrm{ad}}H^\ominus}{R}$$

Therefore, a plot of ln p against $1/T$ should be a straight line of slope $\Delta_{\mathrm{ad}}H^\ominus/R$.

Answer We draw up the following table:

T/K	200	210	220	230	240	250
$10^3/(T/\mathrm{K})$	5.00	4.76	4.55	4.35	4.17	4.00
$\ln(p/\mathrm{kPa})$	1.39	1.60	1.80	1.97	2.14	2.29

The points are plotted in Fig. 23.21. The slope (of the least squares fitted line) is −0.904, so

$$\Delta_{\mathrm{ad}}H^\ominus = -(0.904 \times 10^3\ \mathrm{K}) \times R = -7.52\ \mathrm{kJ\ mol^{-1}}$$

The value of K can be used to obtain a value of $\Delta_{\mathrm{ad}}G^\ominus$, and then that value combined with $\Delta_{\mathrm{ad}}H^\ominus$ to obtain the standard entropy of adsorption. The expression for $(\partial \ln p/\partial T)_\theta$ in this example is independent of the model for the isotherm.

Self-test 23.5 Repeat the calculation using the following data:

T/K	200	210	220	230	240	250
p/kPa	4.32	5.59	7.07	8.80	10.67	12.80

[-9.0 kJ mol^{-1}]

If the initial adsorbed layer can act as a substrate for further (for example, physical) adsorption, then, instead of the isotherm levelling off to some saturated value at high pressures, it can be expected to rise indefinitely. The most widely used isotherm dealing with multilayer adsorption was derived by Stephen Brunauer, Paul Emmett, and Edward Teller (see *Further information 23.1*), and is called the **BET isotherm**:

$$\frac{V}{V_{\mathrm{mon}}} = \frac{cz}{(1-z)\{1-(1-c)z\}} \quad \text{with} \quad z = \frac{p}{p^*} \qquad \text{BET isotherm} \qquad (23.16)$$

In this expression, p^* is the vapour pressure above a layer of adsorbate that is more than one molecule thick and which resembles a pure bulk liquid, V_{mon} is the volume corresponding to monolayer coverage, and c is a constant that is large when the enthalpy of desorption from a monolayer is large compared with the enthalpy of vaporization of the liquid adsorbate:

$$c = \mathrm{e}^{(\Delta_{\mathrm{des}}H^\ominus - \Delta_{\mathrm{vap}}H^\ominus)/RT} \qquad (23.17)$$

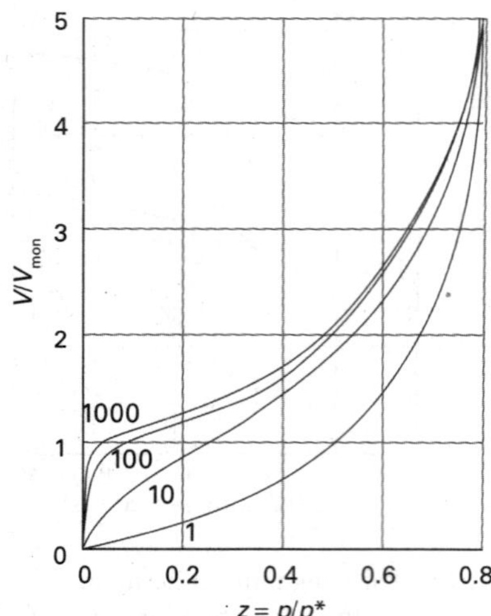

Fig. 23.22 Plots of the BET isotherm for different values of c. The value of V/V_{mon} rises indefinitely because the adsorbate may condense on the covered substrate surface.

interActivity Using eqn 23.16, generate a family of curves showing the dependence of $zV_{\mathrm{mon}}/(1-z)V$ on z for different values of c.

Figure 23.22 illustrates the shapes of BET isotherms. They rise indefinitely as the pressure is increased because there is no limit to the amount of material that may condense when multilayer coverage may occur. A BET isotherm is not accurate at all pressures, but it is widely used in industry to determine the surface areas of solids.

Example 23.6 *Using the BET isotherm*

The data below relate to the adsorption of N_2 on rutile (TiO_2) at 75 K. Confirm that they fit a BET isotherm in the range of pressures reported, and determine V_{mon} and c.

p/kPa	0.160	1.87	6.11	11.67	17.02	21.92	27.29
V/mm^3	601	720	822	935	1046	1146	1254

At 75 K, $p^* = 76.0$ kPa. The volumes have been corrected to 1.00 atm and 273 K and refer to 1.00 g of substrate.

Method Equation 23.16 can be reorganized into

$$\frac{z}{(1-z)V} = \frac{1}{cV_{\text{mon}}} + \frac{(c-1)z}{cV_{\text{mon}}}$$

It follows that $(c-1)/cV_{\text{mon}}$ can be obtained from the slope of a plot of the expression on the left against z, and cV_{mon} can be found from the intercept at $z = 0$. The results can then be combined to give c and V_{mon}.

Answer We draw up the following table:

p/kPa	0.160	1.87	6.11	11.67	17.02	21.92	27.29
10^3z	2.11	24.6	80.4	154	224	288	359
$10^4z/(1-z)(V/\text{mm}^3)$	0.035	0.350	1.06	1.95	2.76	3.53	4.47

These points are plotted in Fig. 23.23. The least squares best line has an intercept at 0.0398, so

$$\frac{1}{cV_{\text{mon}}} = 3.98 \times 10^{-6}\ \text{mm}^{-3}$$

The slope of the line is 1.23×10^{-2}, so

$$\frac{c-1}{cV_{\text{mon}}} = (1.23 \times 10^{-2}) \times 10^3 \times 10^{-4}\ \text{mm}^{-3} = 1.23 \times 10^{-3}\ \text{mm}^{-3}$$

The solutions of these equations are $c = 310$ and $V_{\text{mon}} = 811\ \text{mm}^3$. At 1.00 atm and 273 K, 811 mm^3 corresponds to 3.6×10^{-5} mol, or 2.2×10^{19} atoms. Because each atom occupies an area of about 0.16 nm^2, the surface area of the sample is about 3.5 m^2.

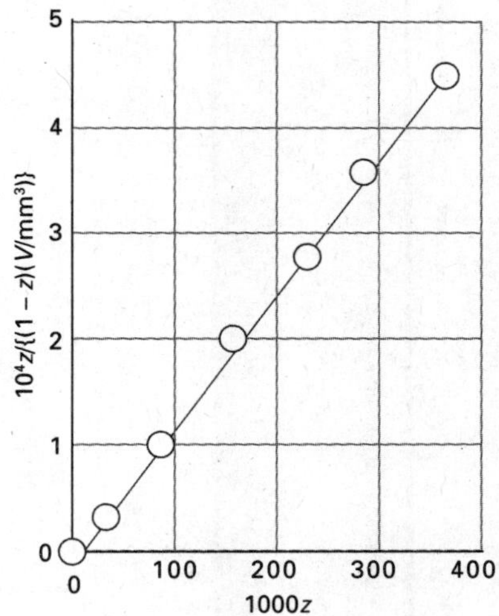

Fig. 23.23 The BET isotherm can be tested, and the parameters determined, by plotting $z/(1-z)V$ against $z = p/p^*$. The data are from Example 23.6.

Self-test 23.6 Repeat the calculation for the following data:

p/kPa	0.160	1.87	6.11	11.67	17.02	21.92	27.29
V/cm^3	235	559	649	719	790	860	950

[370, 615 cm^3]

When $c \gg 1$, the BET isotherm takes the simpler form

$$\frac{V}{V_{\text{mon}}} = \frac{1}{1-z} \qquad \text{BET isotherm when } c \gg 1 \qquad (23.18)$$

This expression is applicable to unreactive gases on polar surfaces, for which $c \approx 10^2$ because $\Delta_{\text{des}}H^{\ominus}$ is then significantly greater than $\Delta_{\text{vap}}H^{\ominus}$ (eqn 23.17). The BET isotherm fits experimental observations moderately well over restricted pressure ranges, but it errs by underestimating the extent of adsorption at low pressures and by overestimating it at high pressures.

An assumption of the Langmuir isotherm is the independence and equivalence of the adsorption sites. Deviations from the isotherm can often be traced to the failure of these assumptions. For example, the enthalpy of adsorption often becomes less negative as θ increases, which suggests that the energetically most favourable sites are occupied first. Various attempts have been made to take these variations into account. The **Temkin isotherm**,

$$\theta = c_1 \ln(c_2 p) \qquad \text{Temkin isotherm} \qquad (23.19)$$

where c_1 and c_2 are constants, corresponds to supposing that the adsorption enthalpy changes linearly with pressure. The **Freundlich isotherm**

$$\theta = c_1 p^{1/c_2} \qquad \text{Freundlich isotherm} \qquad (23.20)$$

corresponds to a logarithmic change. This isotherm attempts to incorporate the role of substrate–substrate interactions on the surface.

Different isotherms agree with experiment more or less well over restricted ranges of pressure, but they remain largely empirical. Empirical, however, does not mean useless for, if the parameters of a reasonably reliable isotherm are known, reasonably reliable results can be obtained for the extent of surface coverage under various conditions. This kind of information is essential for any discussion of heterogeneous catalysis.

23.5 The rates of surface processes

Key points **(a) The sticking probability is the proportion of collisions with the surface that lead to adsorption. (b) Desorption is an activated process; the desorption activation energy is measured by temperature-programmed desorption) or thermal desorption spectroscopy. (c) Diffusion characteristics of an adsorbate can be examined by using STM or field-ionization microscopy.**

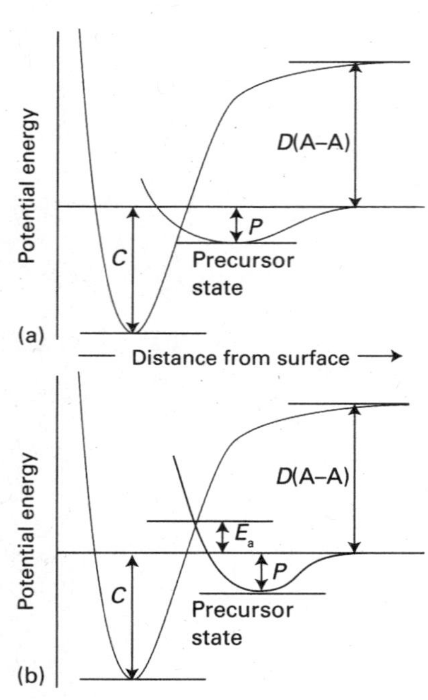

Fig. 23.24 The potential energy profiles for the dissociative chemisorption of an A_2 molecule. In each case, P is the enthalpy of (non-dissociative) physisorption and C that for chemisorption (at $T = 0$). The relative locations of the curves determines whether the chemisorption is (a) not activated or (b) activated.

The rates of surface processes may be studied by techniques already described in this chapter. Another technique, **second harmonic generation** (SHG), is very important for the study of all types of surfaces, including thin films and liquid–gas interfaces. We saw in Section 19.10 that second harmonic generation is the conversion of an intense, pulsed laser beam to radiation with twice its initial frequency as it passes though a material. In addition to a number of crystals, surfaces are also suitable materials for SHG. For example, adsorption of gas molecules on to a surface alters the intensity of the SHG signal, allowing for characterization of processes by the various isotherms discussed above. Because pulsed lasers are the excitation sources, time-resolved measurements of the kinetics and dynamics of surface processes are possible over timescales as short as femtoseconds.

Figure 23.24 shows how the potential energy of a molecule varies with its distance from the substrate surface. As the molecule approaches the surface its energy falls as it becomes physisorbed into the **precursor state** for chemisorption. Dissociation into fragments often takes place as a molecule moves into its chemisorbed state, and after an initial increase of energy as the bonds stretch there is a sharp decrease as the adsorbate–substrate bonds reach their full strength. Even if the molecule does not fragment, there is likely to be an initial increase of potential energy as the molecule approaches the surface and the bonds adjust.

In most cases, therefore, we can expect there to be a potential energy barrier separating the precursor and chemisorbed states. This barrier, though, might be low, and might not rise above the energy of a distant, stationary particle (as in Fig. 23.24a). In this case, chemisorption is not an activated process and can be expected to be rapid. Many gas adsorptions on clean metals appear to be non-activated. In some cases the barrier rises above the zero axis (as in Fig. 23.24b); such chemisorptions are activated and slower than the non-activated kind. An example is H_2 on copper, which has an activation energy in the region of 20–40 kJ mol^{-1}.

One point that emerges from this discussion is that rates are not good criteria for distinguishing between physisorption and chemisorption. Chemisorption can be fast if the activation energy is small or zero, but it may be slow if the activation energy is large. Physisorption is usually fast, but it can appear to be slow if adsorption is taking place on a porous medium.

(a) The rate of adsorption

The rate at which a surface is covered by adsorbate depends on the ability of the substrate to dissipate the energy of the incoming particle as thermal motion as it crashes on to the surface. If the energy is not dissipated quickly, the particle migrates over the surface until a vibration expels it into the overlying gas or it reaches an edge. The proportion of collisions with the surface that successfully lead to adsorption is called the **sticking probability**, *s*:

$$s = \frac{\text{rate of adsorption of particles by the surface}}{\text{rate of collision of particles with the surface}} \quad \text{Definition of the sticking probability} \quad [23.21]$$

The denominator can be calculated from the kinetic model, and the numerator can be measured by observing the rate of change of pressure.

Values of s vary widely. For example, at room temperature CO has s in the range 0.1–1.0 for several d-metal surfaces, but for N_2 on rhenium $s < 10^{-2}$, indicating that more than a hundred collisions are needed before one molecule sticks successfully. Beam studies on specific crystal planes show a pronounced specificity: for N_2 on tungsten, s ranges from 0.74 on the (320) faces down to less than 0.01 on the (110) faces at room temperature. The sticking probability decreases as the surface coverage increases (Fig. 23.25). A simple assumption is that s is proportional to $1 - \theta$, the fraction uncovered, and it is common to write

$$s = (1 - \theta)s_0 \quad (23.22)$$

where s_0 is the sticking probability on a perfectly clean surface. The results in the illustration do not fit this expression because they show that s remains close to s_0 until the coverage has risen to about 6×10^{13} molecules cm^{-2}, and then falls steeply. The explanation is probably that the colliding molecule does not enter the chemisorbed state at once, but moves over the surface until it encounters an empty site.

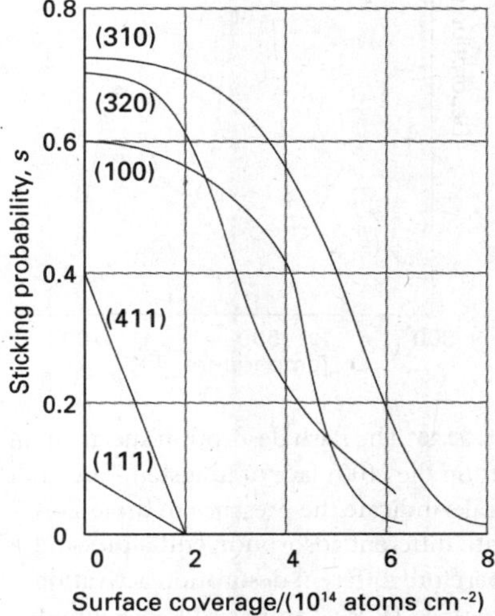

Fig. 23.25 The sticking probability of N_2 on various faces of a tungsten crystal and its dependence on surface coverage. Note the very low sticking probability for the (110) and (111) faces. (Data provided by Professor D.A. King.)

(b) The rate of desorption

Desorption is always activated because the particles have to be lifted from the foot of a potential well. A physisorbed particle vibrates in its shallow potential well, and might shake itself off the surface after a short time. The temperature dependence of the first-order rate of departure can be expected to be Arrhenius-like, with an activation energy for desorption, E_d, comparable to the enthalpy of physisorption:

$$k_d = Ae^{-E_d/RT} \quad (23.23)$$

Therefore, the half-life for remaining on the surface has a temperature dependence

$$t_{1/2} = \frac{\ln 2}{k_d} = \tau_0 e^{E_d/RT} \qquad \tau_0 = \frac{\ln 2}{A} \quad (23.24)$$

(Note the positive sign in the exponent.) If we suppose that $1/\tau_0$ is approximately the same as the vibrational frequency of the weak particle–surface bond (about 10^{12} Hz) and $E_d \approx 25$ kJ mol^{-1}, then residence half-lives of around 10 ns are predicted at room temperature. Lifetimes close to 1 s are obtained only by lowering the temperature to about 100 K. For chemisorption, with $E_d = 100$ kJ mol^{-1} and guessing that $\tau_0 = 10^{-14}$ s (because the adsorbate–substrate bond is quite stiff), we expect a residence half-life of about 3×10^3 s (about an hour) at room temperature, decreasing to 1 s at about 350 K.

The desorption activation energy can be measured in several ways. However, we must be guarded in its interpretation because it often depends on the fractional coverage, and so may change as desorption proceeds. Moreover, the transfer of concepts such as 'reaction order' and 'rate constant' from bulk studies to surfaces is hazardous, and there are few examples of strictly first-order or second-order desorption kinetics (just as there are few integral-order reactions in the gas phase too).

If we disregard these complications, one way of measuring the desorption activation energy is to monitor the rate of increase in pressure when the sample is maintained at a series of temperatures, and to attempt to make an Arrhenius plot. A more sophisticated technique is **temperature programmed desorption** (TPD) or **thermal desorption spectroscopy** (TDS). The basic observation is a surge in desorption rate (as monitored by a mass spectrometer) when the temperature is raised linearly to the temperature at which desorption occurs rapidly, but once the desorption has occurred there is no more adsorbate to escape from the surface, so the desorption flux falls again as the temperature continues to rise. The TPD spectrum, the plot of desorption flux against temperature, therefore shows a peak, the location of which depends on the desorption activation energy. There are three maxima in the example shown in Fig. 23.26, indicating the presence of three sites with different activation energies.

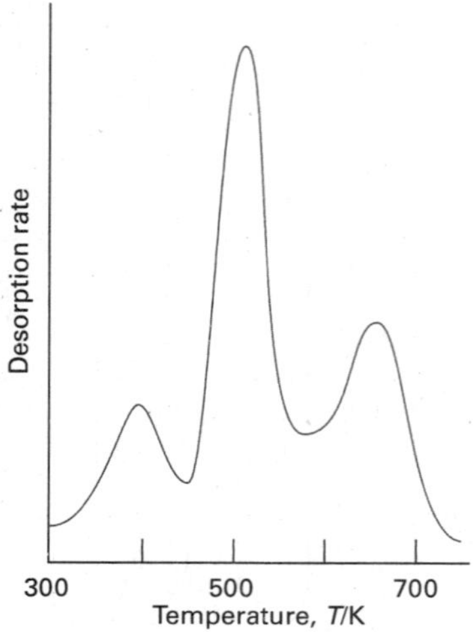

Fig. 23.26 The flash desorption spectrum of H_2 on the (100) face of tungsten. The three peaks indicate the presence of three sites with different adsorption enthalpies and therefore different desorption activation energies. (P.W. Tamm and L.D. Schmidt, *J. Chem. Phys.*, 51, 5352 (1969).)

In many cases only a single activation energy (and a single peak in the TPD spectrum) is observed. When several peaks are observed they might correspond to adsorption on different crystal planes or to multilayer adsorption. For instance, Cd atoms on tungsten show two activation energies, one of 18 kJ mol^{-1} and the other of 90 kJ mol^{-1}. The explanation is that the more tightly bound Cd atoms are attached directly to the substrate, and the less strongly bound are in a layer (or layers) above the primary overlayer. Another example of a system showing two desorption activation energies is CO on tungsten, the values being 120 kJ mol^{-1} and 300 kJ mol^{-1}. The explanation is believed to be the existence of two types of metal–adsorbate binding site, one involving a simple M–CO bond, the other adsorption with dissociation into individually adsorbed C and O atoms.

(c) Mobility on surfaces

A further aspect of the strength of the interactions between adsorbate and substrate is the mobility of the adsorbate. Mobility is often a vital feature of a catalyst's activity, because a catalyst might be impotent if the reactant molecules adsorb so strongly that they cannot migrate. The activation energy for diffusion over a surface need not be the same as for desorption because the particles may be able to move through valleys between potential peaks without leaving the surface completely. In general, the activation energy for migration is about 10–20 per cent of the energy of the surface–adsorbate bond, but the actual value depends on the extent of coverage. The defect structure of the sample (which depends on the temperature) may also play a dominant role because the adsorbed molecules might find it easier to skip across a terrace than to roll along the foot of a step, and these molecules might become trapped in vacancies in an otherwise flat terrace. Diffusion may also be easier across one crystal face than another, and so the surface mobility depends on which lattice planes are exposed.

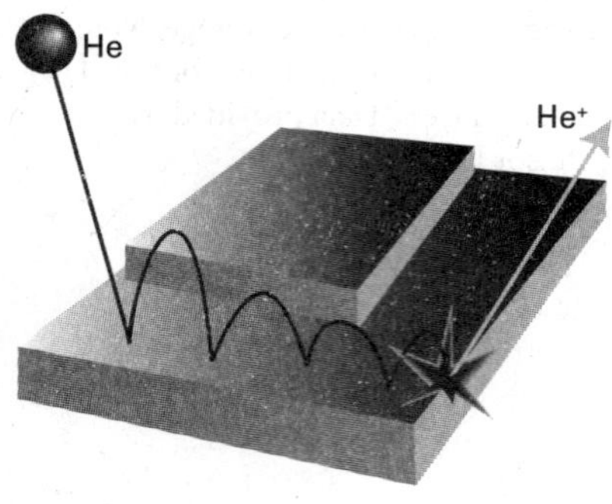

Fig. 23.27 The events leading to an FIM image of a surface. The He atom migrates across the surface until it is ionized at an exposed atom, when it is pulled off by the externally applied potential. (The bouncing motion is due to the intermolecular potential, not gravity!)

Diffusion characteristics of an adsorbate can be examined by using STM to follow the change in surface characteristics or by **field-ionization microscopy** (FIM), which portrays the electrical characteristics of a surface by using the ionization of noble gas atoms to probe the surface (Fig. 23.27). An individual atom is imaged, the temperature is raised, and then lowered after a definite interval. A new image is then recorded, and the new position of the atom measured (Fig. 23.28). A sequence of images shows

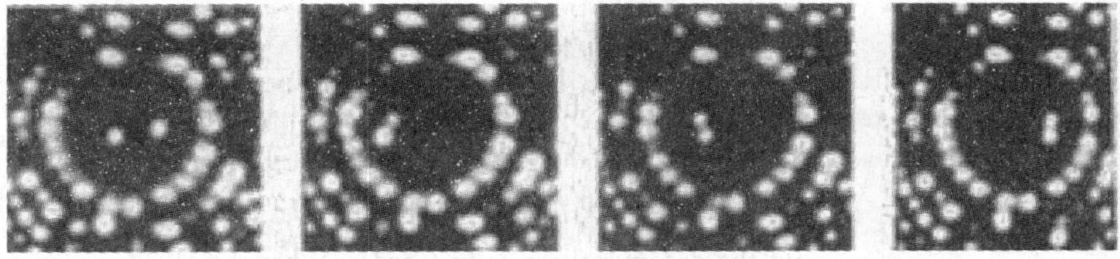

Fig. 23.28 FIM micrographs showing the migration of Re atoms on rhenium during 3 s intervals at 375 K. (Photographs provided by Professor G. Ehrlich.)

that the atom makes a random walk across the surface, and the diffusion coefficient, D, can be inferred from the mean distance, d, travelled in an interval τ by using the two-dimensional random walk expression $d = (D\tau)^{1/2}$. The value of D for different crystal planes at different temperatures can be determined directly in this way, and the activation energy for migration over each plane obtained from the Arrhenius-like expression

$$D = D_0 e^{-E_D/RT} \tag{23.25}$$

where E_D is the activation energy for diffusion. Typical values for W atoms on tungsten have E_D in the range 57–87 kJ mol^{-1} and $D_0 \approx 3.8 \times 10^{-11}$ m^2 s^{-1}. For CO on tungsten, the activation energy falls from 144 kJ mol^{-1} at low surface coverage to 88 kJ mol^{-1} when the coverage is high.

23.6 Mechanisms of heterogeneous catalysis

Key points **In the Langmuir–Hinshelwood mechanism of surface-catalysed reactions, the reaction takes place by encounters between molecular fragments and atoms adsorbed on the surface. In the Eley–Rideal mechanism of a surface-catalysed reaction, a gas-phase molecule collides with another molecule already adsorbed on the surface.**

Many catalysts depend on **co-adsorption**, the adsorption of two or more species. One consequence of the presence of a second species may be the modification of the electronic structure at the surface of a metal. For instance, partial coverage of d-metal surfaces by alkali metals has a pronounced effect on the electron distribution and reduces the work function of the metal. Such modifiers can act as promoters (to enhance the action of catalysts) or as poisons (to inhibit catalytic action).

Figure 23.29 shows the potential energy curve for a reaction influenced by the action of a heterogeneous catalyst. Differences between Fig. 23.29 and 23.1 arise from the fact that heterogeneous catalysis normally depends on at least one reactant being adsorbed (usually chemisorbed) and modified to a form in which it readily undergoes reaction, and desorption of products. Modification of the reactant often takes the form of a fragmentation of the reactant molecules. In practice, the active phase is dispersed as very small particles of linear dimension less than 2 nm on a porous oxide support. **Shape-selective catalysts**, such as the zeolites (*Impact I23.1*), which have a pore size that can distinguish shapes and sizes at a molecular scale, have high internal specific surface areas, in the range of 100–500 m^2 g^{-1}.

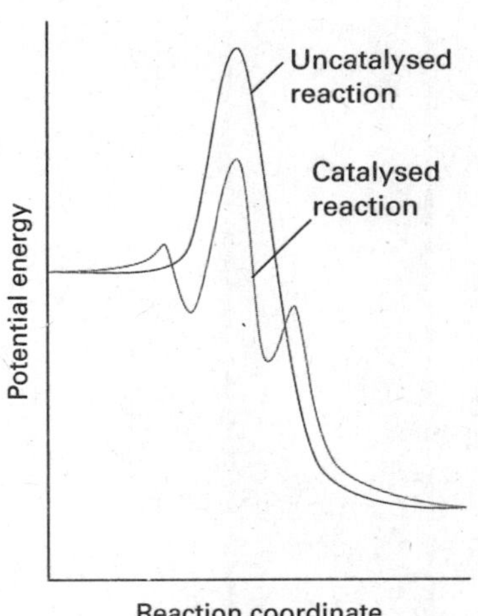

Fig. 23.29 The reaction profile for catalysed and uncatalysed reactions. The catalysed reaction path includes activation energies for adsorption and desorption as well as an overall lower activation energy for the process.

The decomposition of phosphine (PH_3) on tungsten is first-order at low pressures and zeroth-order at high pressures. To account for these observations, we write down a plausible rate law in terms of an adsorption isotherm and explore its form in the limits of high and low pressure. If the rate is supposed to be proportional to the surface coverage and we suppose that θ is given by the Langmuir isotherm, we would write

$$v = k_r\theta = \frac{k_r Kp}{1 + Kp} \qquad (23.26)$$

where p is the pressure of phosphine. When the pressure is so low that $Kp \ll 1$, we can neglect Kp in the denominator and obtain

$$v = k_r Kp \qquad (23.27a)$$

and the decomposition is first-order. When $Kp \gg 1$, we can neglect the 1 in the denominator, whereupon the Kp terms cancel and we are left with

$$v = k_r \qquad (23.27b)$$

and the decomposition is zeroth-order.

Self-test 23.7 Suggest the form of the rate law for the deuteration of NH_3 in which D_2 adsorbs dissociatively but not extensively (that is, $Kp \ll 1$, with p the partial pressure of D_2), and NH_3 (with partial pressure p') adsorbs at different sites.

[$v = k_r(Kp)^{1/2}K'p'/(1 + K'p')$]

In the **Langmuir–Hinshelwood mechanism** (LH mechanism) of surface-catalysed reactions, the reaction takes place by encounters between molecular fragments and atoms adsorbed on the surface. We therefore expect the rate law to be second-order in the extent of surface coverage:

$$\mathrm{A + B \rightarrow P} \qquad v = k_r\theta_A\theta_B \qquad (23.28)$$

Insertion of the appropriate isotherms for A and B then gives the reaction rate in terms of the partial pressures of the reactants. For example, if A and B follow Langmuir isotherms, and adsorb without dissociation, so that

$$\theta_A = \frac{K_A p_A}{1 + K_A p_A + K_B p_B} \qquad \theta_B = \frac{K_B p_B}{1 + K_A p_A + K_B p_B} \qquad (23.29)$$

then it follows that the rate law is

$$v = \frac{k_r K_A K_B p_A p_B}{(1 + K_A p_A + K_B p_B)^2} \qquad (23.30)$$

The rate law according to the Langmuir–Hinshelwood mechanism

The parameters K in the isotherms and the rate constant k_r are all temperature-dependent, so the overall temperature dependence of the rate may be strongly non-Arrhenius (in the sense that the reaction rate is unlikely to be proportional to $e^{-E_a/RT}$). The Langmuir–Hinshelwood mechanism is dominant for the catalytic oxidation of CO to CO_2.

In the **Eley–Rideal mechanism** (ER mechanism) of a surface-catalysed reaction, a gas-phase molecule collides with another molecule already adsorbed on the surface. The rate of formation of product is expected to be proportional to the partial pressure, p_B, of the non-adsorbed gas B and the extent of surface coverage, θ_A, of the adsorbed gas A. It follows that the rate law should be

$$\mathrm{A + B \rightarrow P} \qquad v = k_r p_B\theta_A \qquad (23.31)$$

The rate constant, k_r, might be much larger than for the uncatalysed gas-phase reaction because the reaction on the surface has a low activation energy and the adsorption itself is often not activated.

If we know the adsorption isotherm for A, we can express the rate law in terms of its partial pressure, p_A. For example, if the adsorption of A follows a Langmuir isotherm in the pressure range of interest, then the rate law would be

$$v = \frac{k_r K p_A p_B}{1 + K p_A} \qquad (23.32)$$

The rate law according to the Eley-Rideal mechanism

If A were a diatomic molecule that adsorbed as atoms, we would substitute the isotherm given in eqn 23.14 instead.

According to eqn 23.32, when the partial pressure of A is high (in the sense $Kp_A \gg 1$) there is almost complete surface coverage, and the rate is equal to $k_r p_B$. Now the rate-determining step is the collision of B with the adsorbed fragments. When the pressure of A is low $Kp_A \ll 1$, perhaps because of its reaction, the rate is equal to $k_r K p_A p_B$; now the extent of surface coverage is important in the determination of the rate.

Almost all thermal surface-catalysed reactions are thought to take place by the LH mechanism, but a number of reactions with an ER mechanism have also been identified from molecular beam investigations. For example, the reaction between H(g) and D(ad) to form HD(g) is thought to be by an ER mechanism involving the direct collision and pick-up of the adsorbed D atom by the incident H atom. However, the two mechanisms should really be thought of as ideal limits, and all reactions lie somewhere between the two and show features of each one.

23.7 Catalytic activity at surfaces

Key point **The activity of a catalyst depends on the strength of chemisorption.**

It has become possible to investigate how the catalytic activity of a surface depends on its structure as well as its composition. For instance, the cleavage of C–H and H–H bonds appears to depend on the presence of steps and kinks, and a terrace often has only minimal catalytic activity. The reaction $H_2 + D_2 \rightarrow 2\,HD$ has been studied in detail. For this reaction, terrace sites are inactive but one molecule in ten reacts when it strikes a step. Although the step itself might be the important feature, it may be that the presence of the step merely exposes a more reactive crystal face (the step face itself). Likewise, the dehydrogenation of hexane to hexene depends strongly on the kink density, and it appears that kinks are needed to cleave C–C bonds. These observations suggest a reason why even small amounts of impurities may poison a catalyst: they are likely to attach to step and kink sites, and so impair the activity of the catalyst entirely. A constructive outcome is that the extent of dehydrogenation may be controlled relative to other types of reactions by seeking impurities that adsorb at kinks and act as specific poisons.

The activity of a catalyst depends on the strength of chemisorption as indicated by the 'volcano' curve in Fig. 23.30 (which is so-called on account of its general shape). To be active, the catalyst should be extensively covered by adsorbate, which is the case if chemisorption is strong. On the other hand, if the strength of the substrate–adsorbate bond becomes too great, the activity declines either because the other reactant molecules cannot react with the adsorbate or because the adsorbate molecules are immobilized on the surface. This pattern of behaviour suggests that the activity of a catalyst should initially increase with strength of adsorption (as measured, for instance, by the enthalpy of adsorption) and then decline, and that the most active catalysts should be those lying near the summit of the volcano. Most active metals are those that lie close to the middle of the d block.

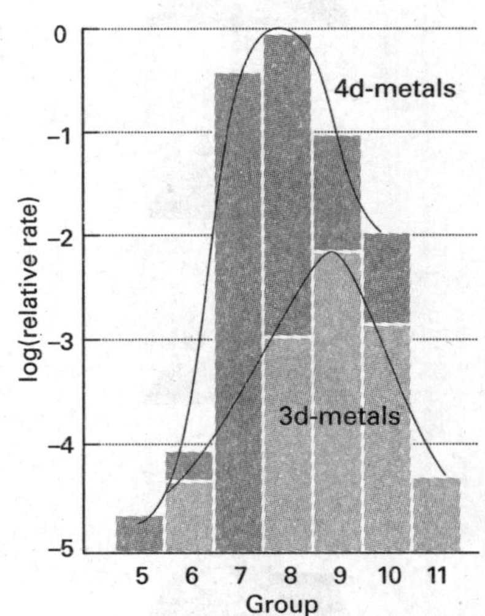

Fig. 23.30 A volcano curve of catalytic activity arises because, although the reactants must adsorb reasonably strongly, they must not adsorb so strongly that they are immobilized. The lower curve refers to the first series of d-block metals, the upper curve to the second and third series d-block metals. The group numbers relate to the periodic table inside the back cover.

Table 23.3 Chemisorption abilities*

	O_2	C_2H_2	C_2H_4	CO	H_2	CO_2	N_2
Ti, Cr, Mo, Fe	+	+	+	+	+	+	+
Ni, Co	+	+	+	+	+	+	−
Pd, Pt	+	+	+	+	+	−	−
Mn, Cu	+	+	+	+	±	−	−
Al, Au	+	+	+	−	−	−	−
Li, Na, K	+	+	−	−	−	−	−
Mg, Ag, Zn, Pb	+	−	−	−	−	−	−

* +, Strong chemisorption; ±, chemisorption; −, no chemisorption.

Many metals are suitable for adsorbing gases, and the general order of adsorption strengths decreases along the series O_2, C_2H_2, C_2H_4, CO, H_2, CO_2, N_2. Some of these molecules adsorb dissociatively (for example, H_2). Elements from the d block, such as iron, vanadium, and chromium, show a strong activity towards all these gases, but manganese and copper are unable to adsorb N_2 and CO_2. Metals towards the left of the periodic table (for example, magnesium and lithium) can adsorb (and, in fact, react with) only the most active gas (O_2). These trends are summarized in Table 23.3.

IMPACT ON TECHNOLOGY

I23.1 Catalysis in the chemical industry

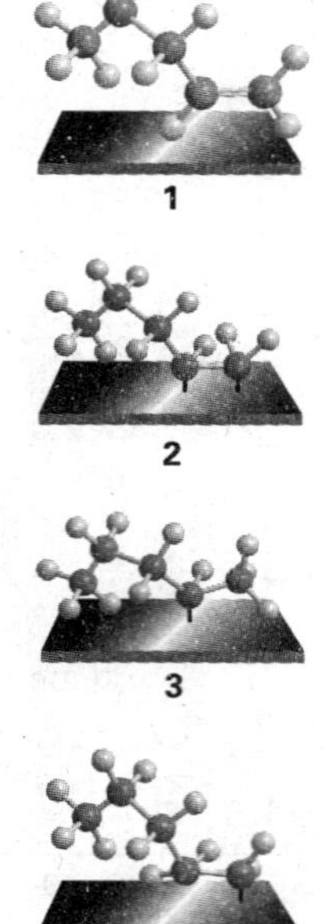

Almost the whole of modern chemical industry depends on the development, selection, and application of catalysts (Table 23.4). All we can hope to do in this section is to give a brief indication of some of the problems involved. Other than the ones we consider, these problems include the danger of the catalyst being poisoned by byproducts or impurities, and economic considerations relating to cost and lifetime.

An example of catalytic action is found in the hydrogenation of alkenes. The alkene (**1**) adsorbs by forming two bonds with the surface (**2**), and on the same surface there may be adsorbed H atoms. When an encounter occurs, one of the alkene–surface bonds is broken (forming **3** or **4**) and later an encounter with a second H atom releases the fully hydrogenated hydrocarbon, which is the thermodynamically more stable species. The evidence for a two-stage reaction is the appearance of different

Table 23.4 Properties of catalysts

Catalyst	Function	Examples
Metals	Hydrogenation Dehydrogenation	Fe, Ni, Pt, Ag
Semiconducting oxides and sulfides	Oxidation Desulfurization	NiO, ZnO, MgO, Bi_2O_3/MoO_3, MoS_2
Insulating oxides	Dehydration	Al_2O_3, SiO_2, MgO
Acids	Polymerization Isomerization Cracking Alkylation	H_3PO_4, H_2SO_4, SiO_3/Al_2O_3, zeolites

isomeric alkenes in the mixture. The formation of isomers comes about because, while the hydrocarbon chain is waving about over the surface of the metal, an atom in the chain might chemisorb again to form (5) and then desorb to (6), an isomer of the original molecule. The new alkene would not be formed if the two hydrogen atoms attached simultaneously.

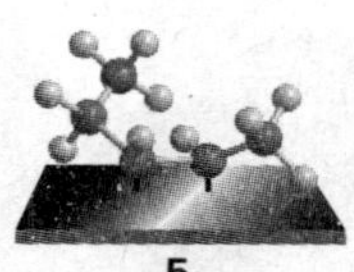

5

6

A major industrial application of catalytic hydrogenation is to the formation of edible fats from vegetable and animal oils. Raw oils obtained from sources such as the soya bean have the structure $CH_2(OOCR)CH(OOCR')CH_2(OOCR'')$, where R, R′, and R″ are long-chain hydrocarbons with several double bonds. One disadvantage of the presence of many double bonds is that the oils are susceptible to atmospheric oxidation, and therefore are liable to become rancid. The geometrical configuration of the chains is responsible for the liquid nature of the oil, and in many applications a solid fat is at least much better and often necessary. Controlled partial hydrogenation of an oil with a catalyst carefully selected so that hydrogenation is incomplete and so that the chains do not isomerize (finely divided nickel, in fact) is used on a wide scale to produce edible fats. The process, and the industry, are not made any easier by the seasonal variation of the number of double bonds in the oils.

Catalytic oxidation is also widely used in industry and in pollution control. Although in some cases it is desirable to achieve complete oxidation (as in the production of nitric acid from ammonia), in others partial oxidation is the aim. For example, the complete oxidation of propene to carbon dioxide and water is wasteful, but its partial oxidation to propenal (acrolein, $CH_2{=}CHCHO$) is the start of important industrial processes. Likewise, the controlled oxidations of ethene to ethanol, ethanal (acetaldehyde), and (in the presence of acetic acid or chlorine) to chloroethene (vinyl chloride, for the manufacture of PVC), are the initial stages of very important chemical industries.

Some of these oxidation reactions are catalysed by d-metal oxides of various kinds. The physical chemistry of oxide surfaces is very complex, as can be appreciated by considering what happens during the oxidation of propene to propenal on bismuth molybdate. The first stage is the adsorption of the propene molecule with loss of a hydrogen to form the propenyl (allyl) radical, $CH_2{=}CHCH_2$. An O atom in the surface can now transfer to this radical, leading to the formation of propenal and its desorption from the surface. The H atom also escapes with a surface O atom, and goes on to form H_2O, which leaves the surface. The surface is left with vacancies and metal ions in lower oxidation states. These vacancies are attacked by O_2 molecules in the overlying gas, which then chemisorb as O_2^- ions, so reforming the catalyst. This sequence of events, which is called the **Mars van Krevelen mechanism**, involves great upheavals of the surface, and some materials break up under the stress.

Many of the small organic molecules used in the preparation of all kinds of chemical products come from oil. These small building blocks of polymers, perfumes, and petrochemicals in general, are usually cut from the long-chain hydrocarbons drawn from the Earth as petroleum. The catalytically induced fragmentation of the long-chain hydrocarbons is called **cracking**, and is often brought about on silica–alumina catalysts. These catalysts act by forming unstable carbocations, which dissociate and rearrange to more highly branched isomers. These branched isomers burn more smoothly and efficiently in internal combustion engines, and are used to produce higher octane fuels.

Catalytic **reforming** uses a dual-function catalyst, such as a dispersion of platinum and acidic alumina. The platinum provides the metal function, and brings about dehydrogenation and hydrogenation. The alumina provides the acidic function, being able to form carbocations from alkenes. The sequence of events in catalytic reforming shows up very clearly the complications that must be unravelled if a reaction as

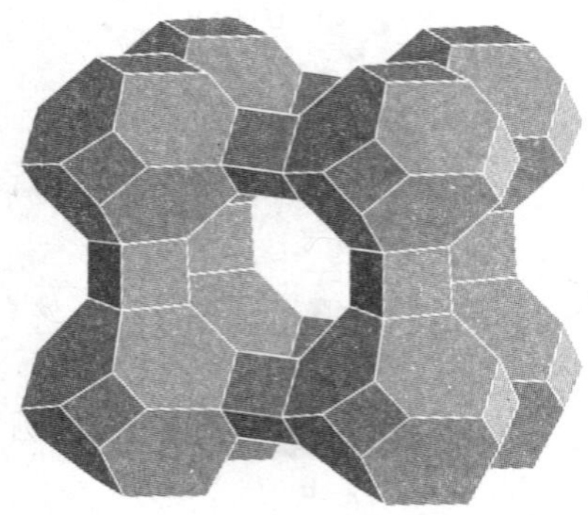

Fig. 23.31 A framework representation of the general layout of the Si, Al, and O atoms in a zeolite material. Each vertex corresponds to a Si or Al atom and each edge corresponds to the approximate location of a O atom. Note the large central pore, which can hold cations, water molecules, or other small molecules.

important as this is to be understood and improved. The first step is the attachment of the long-chain hydrocarbon by chemisorption to the platinum. In this process first one and then a second H atom is lost, and an alkene is formed. The alkene migrates to a Brønsted acid site, where it accepts a proton and attaches to the surface as a carbocation. This carbocation can undergo several different reactions. It can break into two, isomerize into a more highly branched form, or undergo varieties of ring-closure. Then the adsorbed molecule loses a proton, escapes from the surface, and migrates (possibly through the gas) as an alkene to a metal part of the catalyst where it is hydrogenated. We end up with a rich selection of smaller molecules which can be withdrawn, fractionated, and then used as raw materials for other products.

The concept of a solid surface has been extended with the availability of **microporous materials**, in which the surface effectively extends deep inside the solid. Zeolites are microporous aluminosilicates with the general formula $\{[M^{n+}]_{x/n}\cdot[H_2O]_m\}\{[AlO_2]_x[SiO_2]_y\}^{x-}$, where M^{n+} cations and H_2O molecules bind inside the cavities, or pores, of the Al–O–Si framework (Fig. 23.31). Small neutral molecules, such as CO_2, NH_3, and hydrocarbons (including aromatic compounds), can also adsorb to the internal surfaces, accounting partially for the utility of zeolites as catalysts.

Some zeolites for which $M = H^+$ are very strong acids and catalyse a variety of reactions that are of particular importance to the petrochemical industry. Examples include the dehydration of methanol to form hydrocarbons such as gasoline and other fuels:

$$x\,CH_3OH \xrightarrow{\text{zeolite}} (CH_2)_x + x\,H_2O$$

and the isomerization of *m*-xylene (**7**) to *p*-xylene (**8**). The catalytically important form of these acidic zeolites may be either a Brønsted acid (**9**) or a Lewis acid (**10**). Like enzymes, a zeolite catalyst with a specific compostion and structure is very selective toward certain reactants and products because only molecules of certain sizes can enter and exit the pores in which catalysis occurs. It is also possible that zeolites derive their selectivity from the ability to bind and to stabilize only transition states that fit properly in the pores. The analysis of the mechanism of zeolyte catalysis is greatly facilitated by computer simulation of microporous systems, which shows how molecules fit in the pores, migrate through the connecting tunnels, and react at the appropriate active sites.

7 $\xrightarrow{\text{Zeolite}}$ **8**

9 $\xrightarrow{-H_2O}$ **10**

Checklist of key equations

Property	Equation	Comment
Michaelis–Menten equation	$v = v_{max}[S]_0/([S]_0 + K_M)$	
Lineweaver–Burk plot	$1/v = 1/v_{max} + (K_M/v_{max})(1/[S]_0)$	
Turnover frequency	$k_{cat} = v_{max}/[E]_0$	Definition
Catalytic efficiency	$\eta = k_{cat}/K_M$	Definition
Effect of inhibition	$v = v_{max}[S]_0/(\alpha'[S]_0 + \alpha K_M)$	Assumes Michaelis–Menten mechanism
Langmuir isotherm	$\theta = Kp/(1 + Kp)$	Assumes independent sites, monolayer coverage
Isosteric enthalpy of adsorption	$(\partial \ln K/\partial T)_\theta = \Delta_{ad}H^{\ominus}/RT^2$	
BET isotherm	$V/V_{mon} = cz/(1 - z)\{1 - (1 - c)z\}$	Multilayer adsorption
Temkin isotherm	$\theta = c_1 \ln(c_2 p)$	
Freundlich isotherm	$\theta = c_1 p^{1/c_2}$	
Sticking probability	$s = (1 - \theta)s_0$	Approximate form
Langmuir–Hinshelwood mechanism	$v = k_r K_A K_B p_A p_B/(1 + K_A p_A + K_B p_B)^2$	
Eley–Rideal mechanism	$v = k_r K p_A p_B/(1 + K p_A)$	

Further information

Further information 23.1 *The BET isotherm*

We suppose that at equilibrium a fraction θ_0 of the surface sites are unoccupied, a fraction θ_1 is covered by a monolayer, a fraction θ_2 is covered by a bilayer, and so on. The number of adsorbed molecules is therefore

$$N = N_{sites}(\theta_1 + 2\theta_2 + 3\theta_3 + \cdots)$$

where N_{sites} is the total number of sites. We now follow the derivation that led to the Langmuir isotherm (eqn 23.12) but allow for different rates of desorption from the substrate and the various layers:

First layer: Rate of adsorption $= Nk_{a,0}p\theta_0$
Rate of desorption $= Nk_{d,0}\theta_1$
At equilibrium $k_{a,0}p\theta_0 = k_{d,0}\theta_1$

Second layer: Rate of adsorption $= Nk_{a,1}p\theta_1$
Rate of desorption $= Nk_{d,1}\theta_2$
At equilibrium $k_{a,1}p\theta_1 = k_{d,1}\theta_2$

Third layer: Rate of adsorption $= Nk_{a,2}p\theta_2$
Rate of desorption $= Nk_{d,2}\theta_3$
At equilibrium $k_{a,2}p\theta_2 = k_{d,2}\theta_3$

and so on. We now suppose that once a monolayer has been formed, all the rate constants involving adsorption and desorption from the physisorbed layers are the same, and write these equations as

$$k_{a,0}p\theta_0 = k_{d,0}\theta_1, \text{ so } \theta_1 = (k_{a,0}/k_{d,0})p\theta_0 = K_0 p\theta_0$$
$$k_{a,1}p\theta_1 = k_{d,1}\theta_2, \text{ so } \theta_2 = (k_{a,1}/k_{d,1})p\theta_1 = (k_{a,0}/k_{d,0})(k_{a,1}/k_{d,1})p^2\theta_0 = K_0K_1p^2\theta_0$$
$$k_{a,1}p\theta_2 = k_{d,1}\theta_3, \text{ so } \theta_3 = (k_{a,1}/k_{d,1})p\theta_2 = (k_{a,0}/k_{d,0})(k_{a,1}/k_{d,1})^2p^3\theta_0 = K_0K_1^2p^3\theta_0$$

and so on, with $K_0 = k_{a,0}/k_{d,0}$ and $K_1 = k_{a,1}/k_{d,1}$ the equilibrium constants for adsorption to the substrate and an overlayer, respectively. Now, because $\theta_0 + \theta_1 + \theta_2 + \cdots = 1$, it follows that with

$$\begin{aligned}\theta_0 + K_0p\theta_0 + K_0K_1p^2\theta_0 + K_0K_1^2p^3\theta_0 + \cdots &= \theta_0 + K_0p\theta_0\{1 + K_1p + K_1^2p^2 + \cdots\}\\ &= \left\{1 + \frac{K_0p}{1 - K_1p}\right\}\theta_0 = \left\{\frac{1 - K_1p + K_0p}{1 - K_1p}\right\}\theta_0\end{aligned}$$

then, because this expression is equal to 1,

$$\theta_0 = \frac{1 - K_1p}{1 - (K_1 - K_0)p}$$

In a similar way, we can write the number of adsorbed species as

$$\begin{aligned}N &= N_{sites}K_0p\theta_0 + 2N_{sites}K_0K_1p^2\theta_0 + \cdots\\ &= N_{sites}K_0p\theta_0(1 + 2K_1p + 3K_1^2p^2 + \cdots)\\ &= \frac{N_{sites}K_0p\theta_0}{(1 - K_1p)^2}\end{aligned}$$

By combining the last two expressions, we obtain

$$N = \frac{N_{\text{sites}}K_0 p}{(1-K_1 p)^2} \times \frac{1-K_1 p}{1-(K_1-K_0)p}$$

$$= \frac{N_{\text{sites}}K_0 p}{(1-K_1 p)\{1-(K_1-K_0)p\}}$$

The ratio N/N_{sites} is equal to the ratio V/V_{mon}, where V is the total volume adsorbed and V_{mon} the volume adsorbed had there been complete monolayer coverage. The term K_1 is the equilibrium constant for the 'reaction' in which the 'reactant' is a molecule physisorbed on to adsorbed layers and the 'product' is the molecule in the vapour. This process is very much like the equilibrium $M(g) \rightleftharpoons M(l)$, for which $K = 1/p^*$, where p^* is the vapour pressure of the liquid. Therefore, with $K_1 = 1/p^*$, $z = p/p^*$, and $c = K_0/K_1$, the last equation becomes

$$\frac{V}{V_{\text{mon}}} = \frac{K_0 p}{(1-p/p^*)\{1-(1-K_0/K_1)p/p^*\}}$$

$$= \frac{cz}{(1-z)\{1-(1-c)z\}}$$

as in eqn 23.16.

Discussion questions

23.1 Discuss the features, advantages, and limitations of the Michaelis–Menten mechanism of enzyme action.

23.2 Prepare a report on the application of the experimental strategies described in Chapter 21 to the study of enzyme-catalysed reactions. Refer to the following topics: (a) the determination of reaction rates over a large timescale; (b) the determination of the rate constants and equilibrium constant of binding of substrate to an enzyme; and (c) the characterization of intermediates in a catalytic cycle. Your report should be similar in content and extent to one of the *Impact* sections found throughout this text.

23.3 A plot of the rate of an enzyme-catalysed reaction against temperature has a maximum, in an apparent deviation from the behaviour predicted by the Arrhenius relation (eqn 21.29). Suggest a molecular interpretation for this effect.

23.4 Distinguish between competitive, non-competitive, and uncompetitive inhibition of enzymes. Discuss how these modes of inhibition may be detected experimentally.

23.5 Drawing from knowledge you have acquired through the text, describe the advantages and limitations of each of the microscopy, diffraction, and scattering techniques designated by the acronyms AFM, FIM, LEED, SAM, SEM, and STM.

23.6 Distinguish between the following adsorption isotherms: Langmuir, BET, Temkin, and Freundlich and indicate when and why they are likely to be appropriate.

23.7 Account for the dependence of catalytic activity of a surface on the strength of chemisorption, as shown in Fig. 23.24.

Exercises

23.1(a) Consider the base-catalysed reaction

(1) $AH + B \rightleftharpoons BH^+ + A^-$ $\quad k_a, k_a'$, both fast

(2) $A^- + AH \rightarrow$ product $\quad k_b$, slow

Deduce the rate law.

23.1(b) Consider the acid-catalysed reaction

(1) $HA + H^+ \rightleftharpoons HAH^+$ $\quad k_a, k_a'$, both fast

(2) $HAH^+ + B \rightarrow BH^+ + AH$ $\quad k_b$, slow

Deduce the rate law.

23.2(a) The enzyme-catalysed conversion of a substrate at 25°C has a Michaelis constant of 0.046 mol dm^{-3}. The rate of the reaction is 1.04×10^{-3} mol dm^{-3} s^{-1} when the substrate concentration is 0.105 mol dm^{-3}. What is the maximum velocity of this reaction?

23.2(b) The enzyme-catalysed conversion of a substrate at 25°C has a Michaelis constant of 0.032 mol dm^{-3}. The rate of the reaction is 2.05×10^{-4} mol dm^{-3} s^{-1} when the substrate concentration is 0.875 mol dm^{-3}. What is the maximum velocity of this reaction?

23.3(a) The enzyme-catalysed conversion of a substrate at 25°C has a Michaelis constant of 0.015 mol dm^{-3}, and a maximum velocity of 4.25×10^{-4} mol dm^{-3} s^{-1} when the enzyme concentration is 3.60×10^{-9} mol dm^{-3}. Calculate k_{cat} and η. Is the enzyme 'catalytically perfect'?

23.3(b) The enzyme-catalysed conversion of a substrate at 25°C has a Michaelis constant of 9.0×10^{5} mol dm^{-3} and a maximum velocity of 2.24×10^{-5} mol dm^{-3} s^{-1} when the enzyme concentration is 1.60×10^{-9} mol dm^{-3}. Calculate k_{cat} and η. Is the enzyme 'catalytically perfect'?

23.4(a) Consider an enzyme-catalysed reaction that follows Michaelis–Menten kinetics with $K_M = 3.0 \times 10^{-3}$ mol dm^{-3}. What concentration of a competitive inhibitor characterized by $K_I = 2.0 \times 10^{-5}$ mol dm^{-3} will reduce the rate of formation of product by 50 per cent when the substrate concentration is held at 1.0×10^{-4} mol dm^{-3}?

23.4(b) Consider an enzyme-catalysed reaction that follows Michaelis–Menten kinetics with $K_M = 7.5 \times 10^{-4}$ mol dm^{-3}. What concentration of a competitive inhibitor characterized by $K_I = 5.6 \times 10^{-4}$ mol dm^{-3} will reduce the rate of formation of product by 75 per cent when the substrate concentration is held at 1.0×10^{-4} mol dm^{-3}?

23.5(a) Calculate the frequency of molecular collisions per square centimetre of surface in a vessel containing (a) hydrogen, (b) propane at 25°C when the pressure is (i) 100 Pa, (ii) 0.10 μTorr.

23.5(b) Calculate the frequency of molecular collisions per square centimetre of surface in a vessel containing (a) nitrogen, (b) methane at 25°C when the pressure is (i) 10.0 Pa, (ii) 0.150 μTorr.

23.6(a) What pressure of argon gas is required to produce a collision rate of 4.5×10^{20} s^{-1} at 425 K on a circular surface of diameter 1.5 mm?

23.6(b) What pressure of nitrogen gas is required to produce a collision rate of 5.00×10^{19} s^{-1} at 525 K on a circular surface of diameter 2.0 mm?

23.7(a) The LEED pattern from a clean unreconstructed (110) face of a metal is shown below. Sketch the LEED pattern for a surface that was reconstructed by doubling the vertical separation between the atoms.

• • •

• • •

• • •

• • •

• • •

23.7(b) The LEED pattern from a clean unreconstructed (110) face of a metal is shown below. Sketch the LEED pattern for a surface that was reconstructed by tripling the horizontal separation between the atoms.

• • •

• • •

• • •

23.8(a) A monolayer of N_2 molecules is adsorbed on the surface of 1.00 g of an Fe/Al_2O_3 catalyst at 77 K, the boiling point of liquid nitrogen. Upon warming, the nitrogen occupies 3.86 cm^3 at 0°C and 760 Torr. What is the surface area of the catalyst?

23.8(b) A monolayer of CO molecules is adsorbed on the surface of 1.00 g of an Fe/Al_2O_3 catalyst at 77 K, the boiling point of liquid nitrogen. Upon warming, the carbon monoxide occupies 3.75 cm^3 at 0°C and 1.00 bar. What is the surface area of the catalyst?

23.9(a) The volume of oxygen gas at 0°C and 104 kPa adsorbed on the surface of 1.00 g of a sample of silica at 0°C was 0.286 cm^3 at 145.4 Torr and 1.443 cm^3 at 760 Torr. What is the value of V_{mon}?

23.9(b) The volume of gas at 20°C and 1.00 bar adsorbed on the surface of 1.50 g of a sample of silica at 0°C was 1.52 cm^3 at 56.4 kPa and 2.77 cm^3 at 108 kPa. What is the value of V_{mon}?

23.10(a) The enthalpy of adsorption of CO on a surface is found to be −120 kJ mol^{-1}. Estimate the mean lifetime of a CO molecule on the surface at 400 K.

23.10(b) The enthalpy of adsorption of ammonia on a nickel surface is found to be −155 kJ mol^{-1}. Estimate the mean lifetime of an NH_3 molecule on the surface at 500 K.

23.11(a) A certain solid sample adsorbs 0.44 mg of CO when the pressure of the gas is 26.0 kPa and the temperature is 300 K. The mass of gas adsorbed when the pressure is 3.0 kPa and the temperature is 300 K is 0.19 mg. The Langmuir isotherm is known to describe the adsorption. Find the fractional coverage of the surface at the two pressures.

23.11(b) A certain solid sample adsorbs 0.63 mg of CO when the pressure of the gas is 36.0 kPa and the temperature is 300 K. The mass of gas adsorbed when the pressure is 4.0 kPa and the temperature is 300 K is 0.21 mg. The Langmuir isotherm is known to describe the adsorption. Find the fractional coverage of the surface at the two pressures.

23.12(a) The adsorption of a gas is described by the Langmuir isotherm with $K = 0.75$ kPa^{-1} at 25°C. Calculate the pressure at which the fractional surface coverage is (a) 0.15, (b) 0.95.

23.12(b) The adsorption of a gas is described by the Langmuir isotherm with $K = 0.548$ kPa^{-1} at 25°C. Calculate the pressure at which the fractional surface coverage is (a) 0.20, (b) 0.75.

23.13(a) A solid in contact with a gas at 12 kPa and 25°C adsorbs 2.5 mg of the gas and obeys the Langmuir isotherm. The enthalpy change when 1.00 mmol of the adsorbed gas is desorbed is +10.2 kJ. What is the equilibrium pressure for the adsorption of 2.5 mg of gas at 40°C?

23.13(b) A solid in contact with a gas at 8.86 kPa and 25°C adsorbs 4.67 mg of the gas and obeys the Langmuir isotherm. The enthalpy change when 1.00 mmol of the adsorbed gas is desorbed is +12.2 kJ. What is the equilibrium pressure for the adsorption of the same mass of gas at 45°C?

23.14(a) Nitrogen gas adsorbed on charcoal to the extent of 0.921 cm^3 g^{-1} at 490 kPa and 190 K, but at 250 K the same amount of adsorption was achieved only when the pressure was increased to 3.2 MPa. What is the enthalpy of adsorption of nitrogen on charcoal?

23.14(b) Nitrogen gas adsorbed on a surface to the extent of 1.242 cm^3 g^{-1} at 350 kPa and 180 K, but at 240 K the same amount of adsorption was achieved only when the pressure was increased to 1.02 MPa. What is the enthalpy of adsorption of nitrogen on the surface?

23.15(a) In an experiment on the adsorption of oxygen on tungsten it was found that the same volume of oxygen was desorbed in 27 min at 1856 K and 2.0 min at 1978 K. What is the activation energy of desorption? How long would it take for the same amount to desorb at (a) 298 K, (b) 3000 K?

23.15(b) In an experiment on the adsorption of ethene on iron it was found that the same volume of the gas was desorbed in 1856 s at 873 K and 8.44 s at 1012 K. What is the activation energy of desorption? How long would it take for the same amount of ethene to desorb at (a) 298 K, (b) 1500 K?

23.16(a) The average time for which an oxygen atom remains adsorbed to a tungsten surface is 0.36 s at 2548 K and 3.49 s at 2362 K. What is the activation energy for chemisorption?

23.16(b) The average time for which a hydrogen atom remains adsorbed on a manganese surface is 35 per cent shorter at 1000 K than at 600 K. What is the activation energy for chemisorption?

23.17(a) For how long on average would an H atom remain on a surface at 400 K if its desorption activation energy is (a) 15 kJ mol^{-1}, (b) 150 kJ mol^{-1}? Take $\tau_0 = 0.10$ ps. For how long on average would the same atoms remain at 1000 K?

23.17(b) For how long on average would an atom remain on a surface at 298 K if its desorption activation energy is (a) 20 kJ mol^{-1}, (b) 200 kJ mol^{-1}? Take $\tau_0 = 0.12$ ps. For how long on average would the same atoms remain at 800 K?

23.18(a) Hydrogen iodide is very strongly adsorbed on gold but only slightly adsorbed on platinum. Assume the adsorption follows the Langmuir isotherm and predict the order of the HI decomposition reaction on each of the two metal surfaces.

23.18(b) Suppose it is known that ozone adsorbs on a particular surface in accord with a Langmuir isotherm. How could you use the pressure dependence of the fractional coverage to distinguish between adsorption (a) without dissociation, (b) with dissociation into $O + O_2$, (c) with dissociation into $O + O + O$?

Problems*

Numerical problems

23.1 The following results were obtained for the action of an ATPase on ATP at 20°C, when the concentration of the ATPase was 20 nmol dm^{-3}:

$[ATP]/(\mu mol\ dm^{-3})$	0.60	0.80	1.4	2.0	3.0
$v/(\mu mol\ dm^{-3}\ s^{-1})$	0.81	0.97	1.30	1.47	1.69

Determine the Michaelis constant, the maximum velocity of the reaction, the turnover number, and the catalytic efficiency of the enzyme.

23.2 There are different ways to represent and analyse data for enzyme-catalysed reactions. For example, in the *Eadie–Hofstee plot*, $v/[S]_0$ is plotted against v. Alternatively, in the *Hanes plot* $v/[S]_0$ is plotted against $[S]_0$. (a) Using the simple Michaelis–Menten mechanism, derive relations between $v/[S]_0$ and v and between $v/[S]_0$ and $[S]_0$. (b) Discuss how the values of K_M and v_{max} are obtained from analysis of the Eadie–Hofstee and Hanes plots. (c) Determine the Michaelis constant and the maximum velocity of the reaction from Problem 23.1 by using Eadie–Hofstee and Hanes plots to analyse the data.

23.3 In general, the catalytic efficiency of an enzyme depends on the pH of the medium in which it operates. One way to account for this behaviour is to propose that the enzyme and the enzyme–substrate complex are active only in specific protonation states. This proposition can be summarized by the following mechanism:

$$EH + S \rightleftharpoons ESH \qquad k_a, k'_a$$

$$ESH \rightarrow E + P \qquad k_b$$

$$EH \rightleftharpoons E^- + H^+ \qquad K_{E,a} = \frac{[E^-][H^+]}{[EH]}$$

$$EH_2^+ \rightleftharpoons EH + H^+ \qquad K_{E,b} = \frac{[EH][H^+]}{[EH_2^+]}$$

$$ESH \rightleftharpoons ES^- + H^+ \qquad K_{ES,a} = \frac{[ES^-][H^+]}{[ESH]}$$

$$ESH_2 \rightleftharpoons ESH + H^+ \qquad K_{ES,b} = \frac{[ESH][H^+]}{[ESH_2]}$$

in which only the EH and ESH forms are active. (a) For the mechanism above, show that

$$v = \frac{v'_{max}}{1 + K'_M[S]_0}$$

with

$$v'_{max} = \frac{v_{max}}{1 + \dfrac{[H^+]}{K_{ES,b}} + \dfrac{K_{ES,a}}{[H^+]}}$$

$$K'_M = K_M \frac{1 + \dfrac{[H^+]}{K_{E,b}} + \dfrac{K_{E,a}}{[H^+]}}{1 + \dfrac{[H^+]}{K_{ES,b}} + \dfrac{K_{ES,a}}{[H^+]}}$$

where v_{max} and K_M correspond to the form EH of the enzyme. (b) For pH values ranging from 0 to 14, plot v'_{max} against pH for a hypothetical reaction for which $v_{max} = 1.0 \times 10^{-6}$ mol dm^{-3} s^{-1}, $K_{ES,b} = 1.0 \times 10^{-6}$ mol dm^{-3} and $K_{ES,a} = 1.0 \times 10^{-8}$. Is there a pH at which v_{max} reaches a maximum value? If so, determine the pH. (c) Redraw the plot in part (b) by using the same value of v_{max}, but $K_{ES,b} = 1.0 \times 10^{-4}$ mol dm^{-3} and $K_{ES,a} = 1.0 \times 10^{-10}$ mol dm^{-3}. Account for any differences between this plot and the plot from part (b).

23.4 The enzyme carboxypeptidase catalyses the hydrolysis of polypeptides and here we consider its inhibition. The following results were obtained when the rate of the enzymolysis of carbobenzoxy-glycyl-D-phenylalanine (CBGP) was monitored without inhibitor:

$[CBGP]_0/(10^{-2}\ mol\ dm^{-3})$	1.25	3.84	5.81	7.13
Relative reaction rate	0.398	0.669	0.859	1.000

(All rates in this problem were measured with the same concentration of enzyme and are relative to the rate measured when $[CBGP]_0 = 0.0713$ mol dm^{-3} in the absence of inhibitor.) When 2.0×10^{-3} mol dm^{-3} phenylbutyrate ion was added to a solution containing the enzyme and substrate, the following results were obtained:

$[CBGP]_0/(10^{-2}\ mol\ dm^{-3})$	1.25	2.50	4.00	5.50
Relative reaction rate	0.172	0.301	0.344	0.548

In a separate experiment, the effect of 5.0×10^{-2} mol dm^{-3} benzoate ion was monitored and the results were:

$[CBGP]_0/(10^{-2}\ mol\ dm^{-3})$	1.75	2.50	5.00	10.00
Relative reaction rate	0.183	0.201	0.231	0.246

Determine the mode of inhibition of carboxypeptidase by the phenylbutyrate ion and benzoate ion.

23.5 The movement of atoms and ions on a surface depends on their ability to leave one position and stick to another, and therefore on the energy changes that occur. As an illustration, consider a two-dimensional square lattice of univalent positive and negative ions separated by 200 pm, and consider a cation on the upper terrace of this array. Calculate, by direct summation, its Coulombic interaction when it is in an empty lattice point directly above an anion. Now consider a high step in the same lattice, and let the cation move into the corner formed by the step and the terrace. Calculate the Coulombic energy for this position, and decide on the likely settling point for the cation.

23.6 In a study of the catalytic properties of a titanium surface it was necessary to maintain the surface free from contamination. Calculate the collision frequency per square centimetre of surface made by O_2 molecules at (a) 100 kPa, (b) 1.00 Pa and 300 K. Estimate the number of collisions made with a single surface atom in each second. The conclusions underline the importance of working at very low pressures (much lower than 1 Pa, in fact) in order to study the properties of uncontaminated surfaces. Take the nearest-neighbour distance as 291 pm.

23.7 Nickel is face-centred cubic with a unit cell of side 352 pm. What is the number of atoms per square centimetre exposed on a surface formed by (a) (100), (b) (110), (c) (111) planes? Calculate the frequency of molecular collisions per surface atom in a vessel containing (a) hydrogen, (b) propane at 25°C when the pressure is (i) 100 Pa, (ii) 0.10 μTorr.

23.8 The data below are for the chemisorption of hydrogen on copper powder at 25°C. Confirm that they fit the Langmuir isotherm at low coverages. Then find the value of K for the adsorption equilibrium and the adsorption volume corresponding to complete coverage.

p/Pa	25	129	253	540	1000	1593
V/cm^3	0.042	0.163	0.221	0.321	0.411	0.471

* Problems denoted with the symbol ‡ were supplied by Charles Trapp, Carmen Giunta, and Marshall Cady.

23.9 The data for the adsorption of ammonia on barium fluoride are reported below. Confirm that they fit a BET isotherm and find values of c and V_{mon}.

(a) $\theta = 0°C$, $p^* = 429.6$ kPa:

p/kPa	14.0	37.6	65.6	79.2	82.7	100.7	106.4
V/cm^3	11.1	13.5	14.9	16.0	15.5	17.3	16.5

(b) $\theta = 18.6°C$, $p^* = 819.7$ kPa:

p/kPa	5.3	8.4	14.4	29.2	62.1	74.0	80.1	102.0
V/cm^3	9.2	9.8	10.3	11.3	12.9	13.1	13.4	14.1

23.10 The following data have been obtained for the adsorption of H_2 on the surface of 1.00 g of copper at 0°C. The volume of H_2 below is the volume that the gas would occupy at STP (0°C and 1 atm).

p/atm	0.050	0.100	0.150	0.200	0.250
V/cm^3	23.8	13.3	8.70	6.80	5.71

Determine the volume of H_2 necessary to form a monolayer and estimate the surface area of the copper sample. The density of liquid hydrogen is 0.708 g cm^{-3}.

23.11 The adsorption of solutes on solids from liquids often follows a Freundlich isotherm. Check the applicability of this isotherm to the following data for the adsorption of acetic acid on charcoal at 25°C and find the values of the parameters c_1 and c_2.

[acid]/(mol dm^{-3})	0.05	0.10	0.50	1.0	1.5
w_a/g	0.04	0.06	0.12	0.16	0.19

w_a is the mass adsorbed per gram of charcoal.

23.12 In some catalytic reactions the products may adsorb more strongly than the reacting gas. This is the case, for instance, in the catalytic decomposition of ammonia on platinum at 1000°C. As a first step in examining the kinetics of this type of process, show that the rate of ammonia decomposition should follow

$$\frac{dp_{NH_3}}{dt} = -k_c\frac{p_{NH_3}}{p_{H_2}}$$

in the limit of very strong adsorption of hydrogen. Start by showing that, when a gas J adsorbs very strongly and its pressure is p_J, that the fraction of uncovered sites is approximately $1/Kp_J$. Solve the rate equation for the catalytic decomposition of NH_3 on platinum and show that a plot of $F(t) = (1/t)\ln(p/p_0)$ against $G(t) = (p - p_0)/t$, where p is the pressure of ammonia, should give a straight line from which k_c can be determined. Check the rate law on the basis of the data below, and find k_c for the reaction.

t/s	0	30	60	100	160	200	250
p/kPa	13.3	11.7	11.2	10.7	10.3	9.9	9.6

23.13‡ A. Akgerman and M. Zardkoohi (*J. Chem. Eng. Data* **41**, 185 (1996)) examined the adsorption of phenol from aqueous solution on to fly ash at 20°C. They fitted their observations to a Freundlich isotherm of the form $c_{ads} = Kc_{sol}^{1/n}$, where c_{ads} is the concentration of adsorbed phenol and c_{sol} is the concentration of aqueous phenol. Among the data reported are the following:

c_{sol}/(mg g^{-1})	8.26	15.65	25.43	31.74	40.00
c_{ads}/(mg g^{-1})	4.4	19.2	35.2	52.0	67.2

Determine the constants K and n. What further information would be necessary in order to express the data in terms of fractional coverage, θ?

23.14‡ C. Huang and W.P. Cheng (*J. Colloid Interface Sci.* **188**, 270 (1997)) examined the adsorption of the hexacyanoferrate(III) ion, $[Fe(CN)_6]^{3-}$, on γ-Al_2O_3 from aqueous solution. They modelled the adsorption with a modified Langmuir isotherm, obtaining the following values of K at pH = 6.5:

T/K	283	298	308	318
$10^{-11}K$	2.642	2.078	1.286	1.085

Determine the isosteric enthalpy of adsorption, $\Delta_{ads}H^\ominus$, at this pH. The researchers also reported $\Delta_{ads}S^\ominus = +146$ J mol^{-1} K^{-1} under these conditions. Determine $\Delta_{ads}G^\ominus$.

23.15‡ M.-G. Olivier and R. Jadot (*J. Chem. Eng. Data* **42**, 230 (1997)) studied the adsorption of butane on silica gel. They report the following amounts of absorption (in moles per kilogram of silica gel) at 303 K:

p/kPa	31.00	38.22	53.03	76.38	101.97
n/(mol kg^{-1})	1.00	1.17	1.54	2.04	2.49
p/kPa	130.47	165.06	182.41	205.75	219.91
n/(mol kg^{-1})	2.90	3.22	3.30	3.35	3.36

Fit these data to a Langmuir isotherm, and determine the value of n that corresponds to complete coverage and the constant K.

23.16‡ The following data were obtained for the extent of adsorption, s, of acetone on charcoal from an aqueous solution of molar concentration, c, at 18°C.

c/(mmol dm^{-3})	15.0	23.0	42.0	84.0	165	390	800
s/(mmol acetone/g charcoal)	0.60	0.75	1.05	1.50	2.15	3.50	5.10

Which isotherm fits this data best: Langmuir, Freundlich, or Temkin?

Theoretical problems

23.17 Autocatalysis is the catalysis of a reaction by the products. For example, for a reaction A → P it may be found that the rate law is $v = k_r[A][P]$ and the reaction rate is proportional to the concentration of P. The reaction gets started because there are usually other reaction routes for the formation of some P initially, which then takes part in the autocatalytic reaction proper. (a) Integrate the rate equation for an autocatalytic reaction of the form A → P, with rate law $v = k_r[A][P]$, and show that

$$\frac{[P]}{[P]_0} = (b+1)\frac{e^{at}}{1+be^{at}}$$

where $a = ([A]_0 + [P]_0)k_r$ and $b = [P]_0/[A]_0$. *Hint.* Starting with the expression $v = -d[A]/dt = k_r[A][P]$, write $[A] = [A]_0 - x$, $[P] = [P]_0 + x$ and then write the expression for the rate of change of either species in terms of x. To integrate the resulting expression, use

$$\frac{1}{([A]_0 - x)([P]_0 + x)} = \frac{1}{[A]_0 + [P]_0}\left(\frac{1}{[A]_0 - x} + \frac{1}{[P]_0 + x}\right)$$

(b) Plot $[P]/[P]_0$ against at for several values of b. Discuss the effect of autocatalysis on the shape of a plot of $[P]/[P]_0$ against t by comparing your results with those for a first-order process, in which $[P]/[P]_0 = 1 - e^{-k_rt}$. (c) Show that for the autocatalytic process discussed in parts (a) and (b), the reaction rate reaches a maximum at $t_{max} = -(1/a)\ln b$. (d) An autocatalytic reaction A → P is observed to have the rate law $d[P]/dt = k_r[A]^2[P]$. Solve the rate law for initial concentrations $[A]_0$ and $[P]_0$. Calculate the time at which the rate reaches a maximum. (e) Another reaction with the stoichiometry A → P has the rate law $d[P]/dt = k_r[A][P]^2$; integrate the rate law for initial concentrations $[A]_0$ and $[P]_0$. Calculate the time at which the rate reaches a maximum.

23.18 Many biological and biochemical processes involve autocatalytic steps (Problem 23.17). In the SIR model of the spread and decline of infectious diseases the population is divided into three classes; the susceptibles, S, who can catch the disease, the infectives, I, who have the disease and can transmit it, and the removed class, R, who have either had the disease and recovered, are dead, are immune or isolated. The model mechanism for this process implies the following rate laws:

$$\frac{dS}{dt} = -rSI \qquad \frac{dI}{dt} = rSI - aI \qquad \frac{dR}{dt} = aI$$

What are the autocatalytic steps of this mechanism? Find the conditions on the ratio a/r that decide whether the disease will spread (an epidemic) or die out. Show that a constant population is built into this system, namely that S + I + R = N, meaning that the timescales of births, deaths by other causes, and migration are assumed large compared to that of the spread of the disease.

23.19 Michaelis and Menten derived their rate law by assuming a rapid pre-equilibrium of E, S, and ES. Derive the rate law in this manner, and identify the conditions under which it becomes the same as that based on the steady-state approximation (eqn 23.1).

23.20 For many enzymes, the mechanism of action involves the formation of two intermediates:

$$\begin{array}{ll} E + S \rightarrow ES & v = k_a[E][S] \\ ES \rightarrow E + S & v = k_a'[ES] \\ ES \rightarrow ES' & v = k_b[ES] \\ ES' \rightarrow E + P & v = k_c[ES'] \end{array}$$

Show that the rate of formation of product has the same form as that shown in eqn 23.1

$$v = \frac{v_{max}}{1 + K_M/[S]_0}$$

but with v_{max} and K_M given by

$$v_{max} = \frac{k_b k_c[E]_0}{k_b + k_c} \quad \text{and} \quad K_M = \frac{k_c(k_a' + k_b)}{k_a(k_b + k_c)}$$

23.21 Some enzymes are inhibited by high concentrations of their own substrates. (a) Show that when substrate inhibition is important the reaction rate v is given by

$$v = \frac{v_{max}}{1 + K_M/[S]_0 + [S]_0/K_I}$$

where K_I is the equilibrium constant for dissociation of the inhibited enzyme–substrate complex. (b) What effect does substrate inhibition have on a plot of $1/v$ against $1/[S]_0$?

23.22 Although the attractive van der Waals interaction between individual molecules varies as R^{-6}, the interaction of a molecule with a nearby solid (a homogeneous collection of molecules) varies as R^{-3}, where R is its vertical distance above the surface. Confirm this assertion. Calculate the interaction energy between an Ar atom and the surface of solid argon on the basis of a Lennard-Jones (6,12)-potential. Estimate the equilibrium distance of an atom above the surface.

Applications to: chemical engineering and environmental science

23.23 The designers of a new industrial plant wanted to use a catalyst code-named CR-1 in a step involving the fluorination of butadiene. As a first step in the investigation they determined the form of the adsorption isotherm. The volume of butadiene adsorbed per gram of CR-1 at 15°C varied with pressure as given below. Is the Langmuir isotherm suitable at this pressure?

p/kPa	13.3	26.7	40.0	53.3	66.7	80.0
V/cm^3	17.9	33.0	47.0	60.8	75.3	91.3

Investigate whether the BET isotherm gives a better description of the adsorption of butadiene on CR-1. At 15°C, p^*(butadiene) = 200 kPa. Find V_{mon} and c.

23.24‡ In a study relevant to automobile catalytic converters, C.E. Wartnaby *et al.* (*J. Phys. Chem.* **100**, 12483 (1996)) measured the enthalpy of adsorption of CO, NO, and O_2 on initially clean platinum 110 surfaces. They report $\Delta_{ads}H^{\ominus}$ for NO to be −160 kJ mol^{-1}. How much more strongly adsorbed is NO at 500°C than at 400°C?

23.25‡ The removal or recovery of volatile organic compounds (VOCs) from exhaust gas streams is an important process in environmental engineering. Activated carbon has long been used as an adsorbent in this process, but the presence of moisture in the stream reduces its effectiveness. M.-S. Chou and J.-H. Chiou (*J. Envir. Engrg.* ASCE, **123**, 437 (1997)) have studied the effect of moisture content on the adsorption capacities of granular activated carbon (GAC) for normal hexane and cyclohexane in air streams. From their data for dry streams containing cyclohexane, shown in the table below, they conclude that GAC obeys a Langmuir-type model in which $q_{VOC,RH=0} = abc_{VOC}/(1 + bc_{VOC})$, where $q = m_{VOC}/m_{GAC}$, RH denotes relative humidity, a the maximum adsorption capacity, b is an affinity parameter, and p is the abundance in parts per million (ppm). The following table gives values of $q_{VOC,RH=0}$ for cyclohexane:

c/ppm	33.6°C	41.5°C	57.4°C	76.4°C	99°C
200	0.080	0.069	0.052	0.042	0.027
500	0.093	0.083	0.072	0.056	0.042
1000	0.101	0.088	0.076	0.063	0.045
2000	0.105	0.092	0.083	0.068	0.052
3000	0.112	0.102	0.087	0.072	0.058

(a) By linear regression of $1/q_{VOC,RH=0}$ against $1/c_{VOC}$, test the goodness of fit and determine values of a and b. (b) The parameters a and b can be related to $\Delta_{ads}H$, the enthalpy of adsorption, and Δ_bH, the difference in activation energy for adsorption and desorption of the VOC molecules, through Arrhenius-type equations of the form $a = k_a \exp(-\Delta_{ads}H/RT)$ and $b = k_b \exp(-\Delta_bH/RT)$. Test the goodness of fit of the data to these equations and obtain values for k_a, k_b, $\Delta_{ads}H$, and Δ_bH. (c) What interpretation might you give to k_a and k_b?

23.26‡ M.-S. Chou and J.-H. Chiou (*J. Envir. Engrg.*, ASCE, **123**, 437 (1997)) have studied the effect of moisture content on the adsorption capacities of granular activated carbon (GAC, Norit PK 1-3) for the volatile organic compounds (VOCs) normal hexane and cyclohexane in air streams. The following table shows the adsorption capacities ($q_{water} = m_{water}/m_{GAC}$) of GAC for pure water from moist air streams as a function of relative humidity (RH) in the absence of VOCs at 41.5°C.

RH	0.00	0.26	0.49	0.57	0.80	1.00
q_{water}	0.00	0.026	0.072	0.091	0.161	0.229

The authors conclude that the data at this and other temperatures obey a Freundlich-type isotherm, $q_{water} = k(\mathrm{RH})^{1/n}$. (a) Test this hypothesis for their data at 41.5°C and determine the constants k and n. (b) Why might VOCs obey the Langmuir model, but water the Freundlich model? (c) When both water vapour and cyclohexane were present in the stream the values given in the table below were determined for the ratio $r_{VOC} = q_{VOC}/q_{VOC,RH=0}$ at 41.5°C.

RH	0.00	0.10	0.25	0.40	0.53	0.76	0.81
r_{VOC}	1.00	0.98	0.91	0.84	0.79	0.67	0.61

The authors propose that these data fit the equation $r_{VOC} = 1 - q_{water}$. Test their proposal and determine values for k and n and compare to those obtained in part (b) for pure water. Suggest reasons for any differences.

23.27‡ The release of petroleum products by leaky underground storage tanks is a serious threat to clean ground water. BTEX compounds (benzene, toluene, ethylbenzene, and xylenes) are of primary concern due to their ability to cause health problems at low concentrations. D.S. Kershaw *et al.* (*J. Geotech. & Geoenvir. Engrg.* **123**, 324 (1997)) have studied the ability of ground tyre rubber to sorb (adsorb and absorb) benzene and *o*-xylene. Though sorption involves more than surface interactions, sorption data are usually found to fit one of the adsorption isotherms. In this study, the authors have tested how well their data fit the linear ($q = Kc_{eq}$), Freundlich ($q = K_F c_{eq}^{1/n}$), and Langmuir ($q = K_L M c_{eq}/(1 + K_L c_{eq})$) type isotherms, where q is the mass of solvent sorbed per gram of ground rubber (in milligrams per gram), the Ks and M are empirical constants, and c_{eq} the equilibrium concentration of contaminant in solution (in milligrams per litre). (a) Determine the units of the empirical constants. (b) Determine which of the isotherms best fits the data in the table below for the sorption of benzene on ground rubber.

c_{eq}/(mg dm^{-3})	97.10	36.10	10.40	6.51	6.21	2.48
q/(mg g^{-1})	7.13	4.60	1.80	1.10	0.55	0.31

(c) Compare the sorption efficiency of ground rubber to that of granulated activated charcoal, which for benzene has been shown to obey the Freundlich isotherm in the form $q = 1.0c_{eq}^{1.6}$ with coefficient of determination $R^2 = 0.94$.

Resource section

Contents

The following is a directory of all tables in the text, those included in this *Data section* are marked with an asterisk. The remainder will be found on the pages indicated.

Part 1 Road maps

Gas laws (Chapter 1)

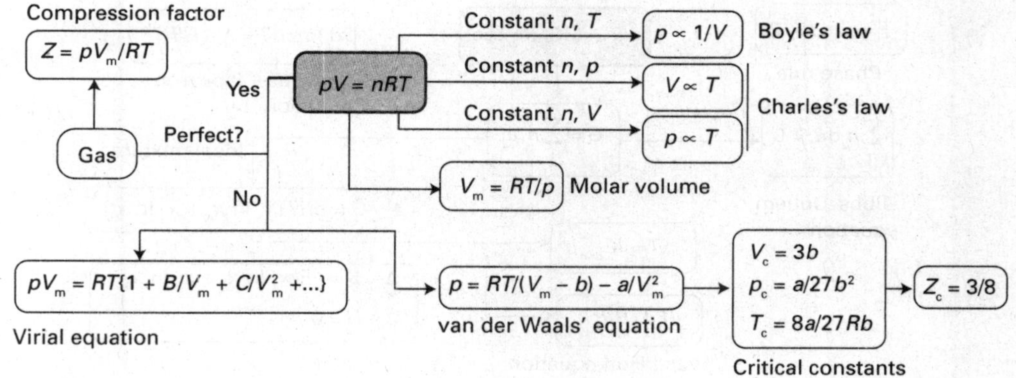

The First Law (Chapter 2)

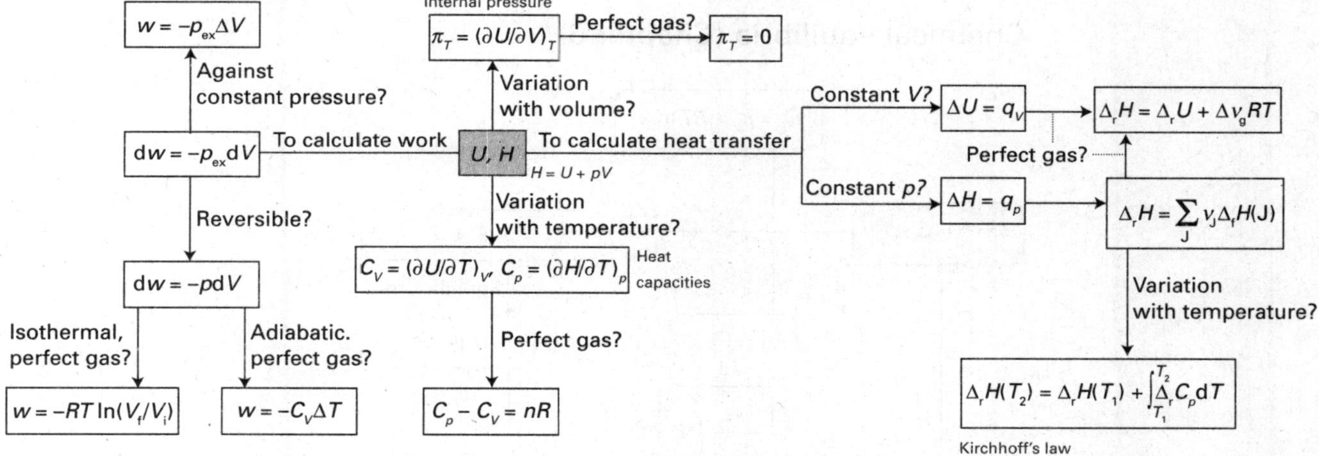

The Second Law (Chapter 3)

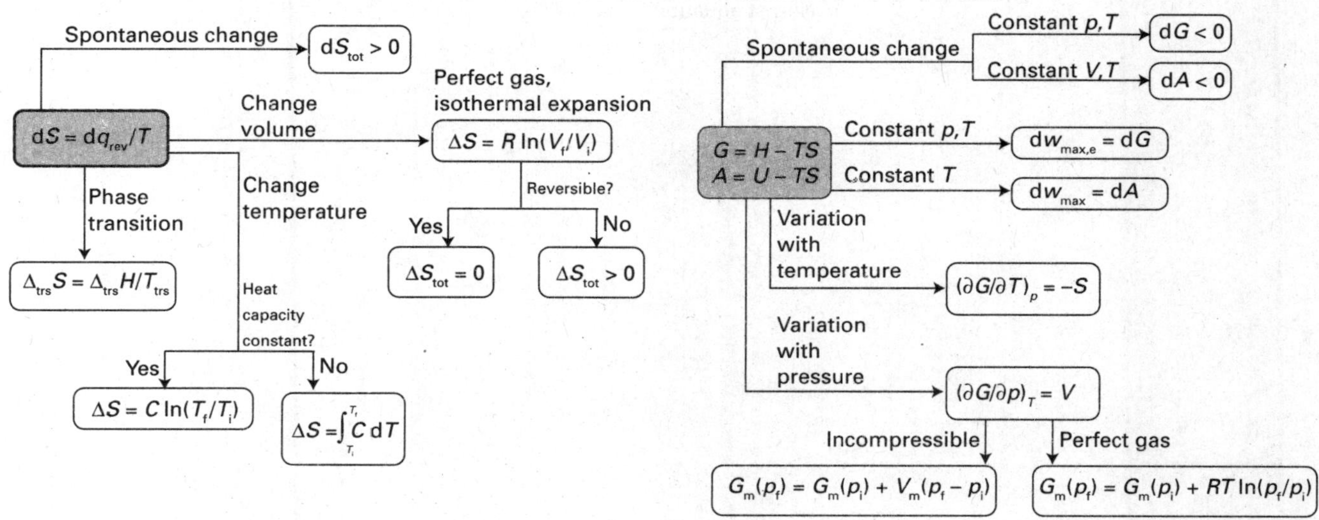

Physical equilibria (Chapters 4 and 5)

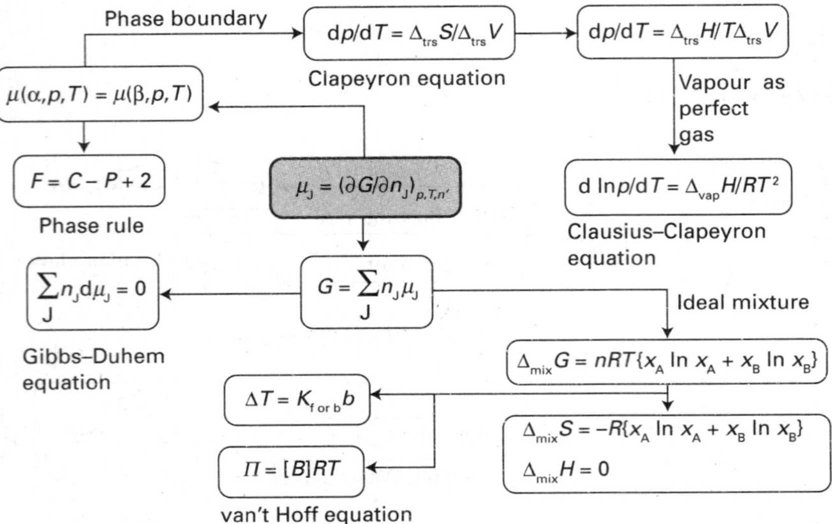

Chemical equilibria (Chapter 6)

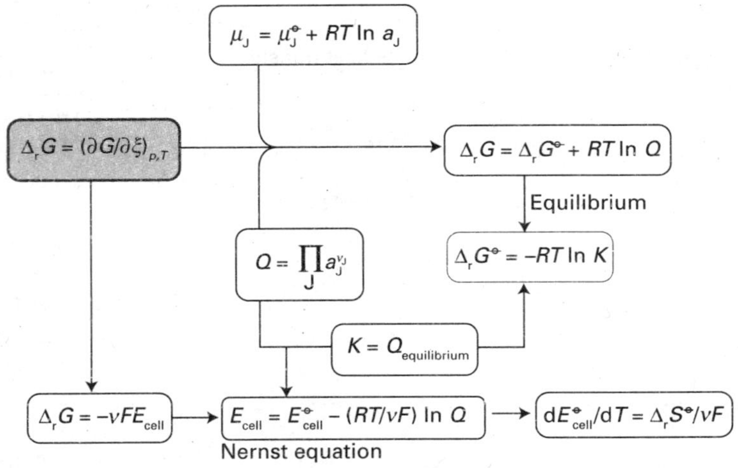

Statistical thermodynamics (Chapters 15 and 16)

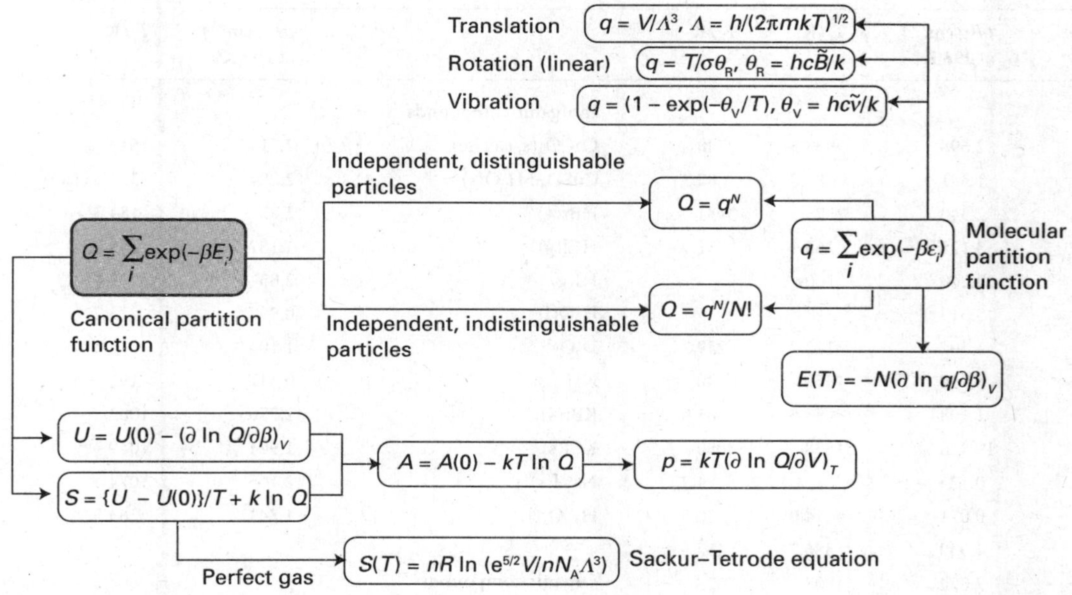

Part 2 Data section

The following tables reproduce and expand the data given in the short tables in the text, and follow their numbering. Standard states refer to a pressure of $p^\ominus = 1$ bar. The general references are as follows:

AIP: D.E. Gray (ed.), *American Institute of Physics handbook*. McGraw Hill, New York (1972).

E: J. Emsley, *The elements*. Oxford University Press (1991).

HCP: D.R. Lide (ed.), *Handbook of chemistry and physics*. CRC Press, Boca Raton (2000).

JL: A.M. James and M.P. Lord, *Macmillan's chemical and physical data*. Macmillan, London (1992).

KL: G.W.C. Kaye and T.H. Laby (ed.), *Tables of physical and chemical constants*. Longman, London (1973).

LR: G.N. Lewis and M. Randall, revised by K.S. Pitzer and L. Brewer, *Thermodynamics*. McGraw Hill, New York (1961).

NBS: NBS tables of chemical thermodynamic properties, published as *J. Phys. Chem. Reference Data*, **11**, Supplement 2 (1982).

RS: R.A. Robinson and R.H. Stokes, *Electrolyte solutions*, Butterworth, London (1959).

TDOC: J.B. Pedley, J.D. Naylor, and S.P. Kirby, *Thermochemical data of organic compounds*. Chapman & Hall, London (1986).

Physical properties of selected materials

	ρ/(g cm^{-3}) at 293 K†	T_f/K	T_b/K		ρ/(g cm^{-3}) at 293 K†	T_f/K	T_b/K
Elements				**Inorganic compounds**			
Aluminium(s)	2.698	933.5	2740	$CaCO_3$(s, calcite)	2.71	1612	1171d
Argon(g)	1.381	83.8	87.3	$CuSO_4\cdot 5H_2O$(s)	2.284	383($-H_2O$)	423($-5H_2O$)
Boron(s)	2.340	2573	3931	HBr(g)	2.77	184.3	206.4
Bromine(l)	3.123	265.9	331.9	HCl(g)	1.187	159.0	191.1
Carbon(s, gr)	2.260	3700s		HI(g)	2.85	222.4	237.8
Carbon(s, d)	3.513			H_2O(l)	0.997	273.2	373.2
Chlorine(g)	1.507	172.2	239.2	D_2O(l)	1.104	277.0	374.6
Copper(s)	8.960	1357	2840	NH_3(g)	0.817	195.4	238.8
Fluorine(g)	1.108	53.5	85.0	KBr(s)	2.750	1003	1708
Gold(s)	19.320	1338	3080	KCl(s)	1.984	1049	1773s
Helium(g)	0.125		4.22	NaCl(s)	2.165	1074	1686
Hydrogen(g)	0.071	14.0	20.3	H_2SO_4(l)	1.841	283.5	611.2
Iodine(s)	4.930	386.7	457.5				
Iron(s)	7.874	1808	3023	**Organic compounds**			
Krypton(g)	2.413	116.6	120.8	Acetaldehyde, CH_3CHO(l)	0.788	152	293
Lead(s)	11.350	600.6	2013	Acetic acid, CH_3COOH(l)	1.049	289.8	391
Lithium(s)	0.534	453.7	1620	Acetone, $(CH_3)_2CO$(l)	0.787	178	329
Magnesium(s)	1.738	922.0	1363	Aniline, $C_6H_5NH_2$(l)	1.026	267	457
Mercury(l)	13.546	234.3	629.7	Anthracene, $C_{14}H_{10}$(s)	1.243	490	615
Neon(g)	1.207	24.5	27.1	Benzene, C_6H_6(l)	0.879	278.6	353.2
Nitrogen(g)	0.880	63.3	77.4	Carbon tetrachloride, CCl_4(l)	1.63	250	349.9
Oxygen(g)	1.140	54.8	90.2	Chloroform, $CHCl_3$(l)	1.499	209.6	334
Phosphorus(s, wh)	1.820	317.3	553	Ethanol, C_2H_5OH(l)	0.789	156	351.4
Potassium(s)	0.862	336.8	1047	Formaldehyde, HCHO(g)		181	254.0
Silver(s)	10.500	1235	2485	Glucose, $C_6H_{12}O_6$(s)	1.544	415	
Sodium(s)	0.971	371.0	1156	Methane, CH_4(g)		90.6	111.6
Sulfur(s, α)	2.070	386.0	717.8	Methanol, CH_3OH(l)	0.791	179.2	337.6
Uranium(s)	18.950	1406	4018	Naphthalene, $C_{10}H_8$(s)	1.145	353.4	491
Xenon(g)	2.939	161.3	166.1	Octane, C_8H_{18}(l)	0.703	216.4	398.8
Zinc(s)	7.133	692.7	1180	Phenol, C_6H_5OH(s)	1.073	314.1	455.0
				Sucrose, $C_{12}H_{22}O_{11}$(s)	1.588	457d	

d: decomposes; s: sublimes; Data: AIP, E, HCP, KL. † For gases, at their boiling points.

Masses and natural abundances of selected nuclides

Nuclide		m/m_u	Abundance/%
H	^{1}H	1.0078	99.985
	^{2}H	2.0140	0.015
He	^{3}He	3.0160	0.000 13
	^{4}He	4.0026	100
Li	^{6}Li	6.0151	7.42
	^{7}Li	7.0160	92.58
B	^{10}B	10.0129	19.78
	^{11}B	11.0093	80.22
C	^{12}C	12*	98.89
	^{13}C	13.0034	1.11
N	^{14}N	14.0031	99.63
	^{15}N	15.0001	0.37
O	^{16}O	15.9949	99.76
	^{17}O	16.9991	0.037
	^{18}O	17.9992	0.204
F	^{19}F	18.9984	100
P	^{31}P	30.9738	100
S	^{32}S	31.9721	95.0
	^{33}S	32.9715	0.76
	^{34}S	33.9679	4.22
Cl	^{35}Cl	34.9688	75.53
	^{37}Cl	36.9651	24.4
Br	^{79}Br	78.9183	50.54
	^{81}Br	80.9163	49.46
I	^{127}I	126.9045	100

* Exact value.

Table 1.4 Second virial coefficients, $B/(\text{cm}^3\ \text{mol}^{-1})$

	100 K	273 K	373 K	600 K
Air	−167.3	−13.5	3.4	19.0
Ar	−187.0	−21.7	−4.2	11.9
CH_4		−53.6	−21.2	8.1
CO_2		−142	−72.2	−12.4
H_2	−2.0	13.7	15.6	
He	11.4	12.0	11.3	10.4
Kr		−62.9	−28.7	1.7
N_2	−160.0	−10.5	6.2	21.7
Ne	−6.0	10.4	12.3	13.8
O_2	−197.5	−22.0	−3.7	12.9
Xe		−153.7	−81.7	−19.6

Data: AIP, JL. The values relate to the expansion in eqn 1.19b of Section 1.3b; convert to eqn 1.19a using $B' = B/RT$.
For Ar at 273 K, $C = 1200\ \text{cm}^6\ \text{mol}^{-1}$.

Table 1.5 Critical constants of gases

	p_c/atm	$V_c/(\text{cm}^3\ \text{mol}^{-1})$	T_c/K	Z_c	T_B/K
Ar	48.00	75.25	150.72	0.292	411.5
Br_2	102	135	584	0.287	
C_2H_4	50.50	124	283.1	0.270	
C_2H_6	48.20	148	305.4	0.285	
C_6H_6	48.6	260	562.7	0.274	
CH_4	45.6	98.7	190.6	0.288	510.0
Cl_2	76.1	124	417.2	0.276	
CO_2	72.85	94.0	304.2	0.274	714.8
F_2	55	144			
H_2	12.8	64.99	33.23	0.305	110.0
H_2O	218.3	55.3	647.4	0.227	
HBr	84.0	363.0			
HCl	81.5	81.0	324.7	0.248	
He	2.26	57.76	5.21	0.305	22.64
HI	80.8	423.2			
Kr	54.27	92.24	209.39	0.291	575.0
N_2	33.54	90.10	126.3	0.292	327.2
Ne	26.86	41.74	44.44	0.307	122.1
NH_3	111.3	72.5	405.5	0.242	
O_2	50.14	78.0	154.8	0.308	405.9
Xe	58.0	118.8	289.75	0.290	768.0

Data: AIP, KL.

Table 1.6 van der Waals coefficients

	a/(atm dm^6 mol^{-2})	b/(10^{-2} dm^3 mol^{-1})		a/(atm dm^6 mol^{-2})	b/(10^{-2} dm^3 mol^{-1})
Ar	1.337	3.20	H_2S	4.484	4.34
C_2H_4	4.552	5.82	He	0.0341	2.38
C_2H_6	5.507	6.51	Kr	5.125	1.06
C_6H_6	18.57	11.93	N_2	1.352	3.87
CH_4	2.273	4.31	Ne	0.205	1.67
Cl_2	6.260	5.42	NH_3	4.169	3.71
CO	1.453	3.95	O_2	1.364	3.19
CO_2	3.610	4.29	SO_2	6.775	5.68
H_2	0.2420	2.65	Xe	4.137	5.16
H_2O	5.464	3.05			

Data: HCP.

Table 2.2 Temperature variation of molar heat capacities†

	a	b/(10^{-3} K^{-1})	c/(10^5 K^2)
Monatomic gases			
	20.78	0	0
Other gases			
Br_2	37.32	0.50	−1.26
Cl_2	37.03	0.67	−2.85
CO_2	44.22	8.79	−8.62
F_2	34.56	2.51	−3.51
H_2	27.28	3.26	0.50
I_2	37.40	0.59	−0.71
N_2	28.58	3.77	−0.50
NH_3	29.75	25.1	−1.55
O_2	29.96	4.18	−1.67
Liquids (from melting to boiling)			
$C_{10}H_8$, naphthalene	79.5	0.4075	0
I_2	80.33	0	0
H_2O	75.29	0	0
Solids			
Al	20.67	12.38	0
C (graphite)	16.86	4.77	−8.54
$C_{10}H_8$, naphthalene	−115.9	3.920×10^3	0
Cu	22.64	6.28	0
I_2	40.12	49.79	0
NaCl	45.94	16.32	0
Pb	22.13	11.72	0.96

† For $C_{p,m}/(\mathrm{J\,K^{-1}\,mol^{-1}}) = a + bT + c/T^2$.
Source: LR.

Table 2.3 Standard enthalpies of fusion and vaporization at the transition temperature, $\Delta_{trs}H^{\ominus}/(kJ\ mol^{-1})$

	T_f/K	Fusion	T_b/K	Vaporization		T_f/K	Fusion	T_b/K	Vaporization
Elements					**Inorganic compounds**				
Ag	1234	11.30	2436	250.6	CO_2	217.0	8.33	194.6	25.23s
Ar	83.81	1.188	87.29	6.506	CS_2	161.2	4.39	319.4	26.74
Br_2	265.9	10.57	332.4	29.45	H_2O	273.15	6.008	373.15	40.656
Cl_2	172.1	6.41	239.1	20.41					44.016 at 298 K
F_2	53.6	0.26	85.0	3.16	H_2S	187.6	2.377	212.8	18.67
H_2	13.96	0.117	20.38	0.916	H_2SO_4	283.5	2.56		
He	3.5	0.021	4.22	0.084	NH_3	195.4	5.652	239.7	23.35
Hg	234.3	2.292	629.7	59.30					
I_2	386.8	15.52	458.4	41.80	**Organic compounds**				
N_2	63.15	0.719	77.35	5.586	CH_4	90.68	0.941	111.7	8.18
Na	371.0	2.601	1156	98.01	CCl_4	250.3	2.47	349.9	30.00
O_2	54.36	0.444	90.18	6.820	C_2H_6	89.85	2.86	184.6	14.7
Xe	161	2.30	165	12.6	C_6H_6	278.61	10.59	353.2	30.8
K	336.4	2.35	1031	80.23	C_6H_{14}	178	13.08	342.1	28.85
					$C_{10}H_8$	354	18.80	490.9	51.51
					CH_3OH	175.2	3.16	337.2	35.27
									37.99 at 298 K
					C_2H_5OH	158.7	4.60	352	43.5

Data: AIP; s denotes sublimation.

Table 2.5 Lattice enthalpies at 298 K, $\Delta H_L^{\ominus}/(kJ\ mol^{-1})$

	F	Cl	Br	I
Halides				
Li	1037	852	815	761
Na	926	787	752	705
K	821	717	689	649
Rb	789	695	668	632
Cs	750	676	654	620
Ag	969	912	900	886
Be		3017		
Mg		2524		
Ca		2255		
Sr		2153		

Oxides							
MgO	3850	CaO	3461	SrO	3283	BaO	3114
Sulfides							
MgS	3406	CaS	3119	SrS	2974	BaS	2832

Entries refer to $MX(s) \rightarrow M^+(g) + X^-(g)$.
Data: Principally D. Cubicciotti, *J. Chem. Phys.* **31**, 1646 (1959).

Table 2.6 Thermodynamic data for organic compounds at 298 K

	M/(g mol^{-1})	$\Delta_f H^⦵$/(kJ mol^{-1})	$\Delta_f G^⦵$/(kJ mol^{-1})	$S_m^⦵$/(J K^{-1} mol^{-1})†	$C_{p,m}^⦵$/(J K^{-1} mol^{-1})	$\Delta_c H^⦵$/(kJ mol^{-1})
C(s) (graphite)	12.011	0	0	5.740	8.527	−393.51
C(s) (diamond)	12.011	+1.895	+2.900	2.377	6.113	−395.40
CO_2(g)	44.040	−393.51	−394.36	213.74	37.11	
Hydrocarbons						
CH_4(g), methane	16.04	−74.81	−50.72	186.26	35.31	−890
CH_3(g), methyl	15.04	+145.69	+147.92	194.2	38.70	
C_2H_2(g), ethyne	26.04	+226.73	+209.20	200.94	43.93	−1300
C_2H_4(g), ethene	28.05	+52.26	+68.15	219.56	43.56	−1411
C_2H_6(g), ethane	30.07	−84.68	−32.82	229.60	52.63	−1560
C_3H_6(g), propene	42.08	+20.42	+62.78	267.05	63.89	−2058
C_3H_6(g), cyclopropane	42.08	+53.30	+104.45	237.55	55.94	−2091
C_3H_8(g), propane	44.10	−103.85	−23.49	269.91	73.5	−2220
C_4H_8(g), 1-butene	56.11	−0.13	+71.39	305.71	85.65	−2717
C_4H_8(g), *cis*-2-butene	56.11	−6.99	+65.95	300.94	78.91	−2710
C_4H_8(g), *trans*-2-butene	56.11	−11.17	+63.06	296.59	87.82	−2707
C_4H_{10}(g), butane	58.13	−126.15	−17.03	310.23	97.45	−2878
C_5H_{12}(g), pentane	72.15	−146.44	−8.20	348.40	120.2	−3537
C_5H_{12}(l)	72.15	−173.1				
C_6H_6(l), benzene	78.12	+49.0	+124.3	173.3	136.1	−3268
C_6H_6(g)	78.12	+82.93	+129.72	269.31	81.67	−3302
C_6H_{12}(l), cyclohexane	84.16	−156	+26.8	204.4	156.5	−3920
C_6H_{14}(l), hexane	86.18	−198.7		204.3		−4163
$C_6H_5CH_3$(g), methylbenzene (toluene)	92.14	+50.0	+122.0	320.7	103.6	−3953
C_7H_{16}(l), heptane	100.21	−224.4	+1.0	328.6	224.3	
C_8H_{18}(l), octane	114.23	−249.9	+6.4	361.1		−5471
C_8H_{18}(l), iso-octane	114.23	−255.1				−5461
$C_{10}H_8$(s), naphthalene	128.18	+78.53				−5157
Alcohols and phenols						
CH_3OH(l), methanol	32.04	−238.66	−166.27	126.8	81.6	−726
CH_3OH(g)	32.04	−200.66	−161.96	239.81	43.89	−764
C_2H_5OH(l), ethanol	46.07	−277.69	−174.78	160.7	111.46	−1368
C_2H_5OH(g)	46.07	−235.10	−168.49	282.70	65.44	−1409
C_6H_5OH(s), phenol	94.12	−165.0	−50.9	146.0		−3054
Carboxylic acids, hydroxy acids, and esters						
HCOOH(l), formic	46.03	−424.72	−361.35	128.95	99.04	−255
CH_3COOH(l), acetic	60.05	−484.5	−389.9	159.8	124.3	−875
CH_3COOH(aq)	60.05	−485.76	−396.46	178.7		
$CH_3CO_2^-$(aq)	59.05	−486.01	−369.31	+86.6	−6.3	
$(COOH)_2$(s), oxalic	90.04	−827.2			117	−254
C_6H_5COOH(s), benzoic	122.13	−385.1	−245.3	167.6	146.8	−3227
$CH_3CH(OH)COOH$(s), lactic	90.08	−694.0				−1344
$CH_3COOC_2H_5$(l), ethyl acetate	88.11	−479.0	−332.7	259.4	170.1	−2231

Table 2.6 (Continued)

	$M/(\mathrm{g\ mol^{-1}})$	$\Delta_f H^{\ominus}/(\mathrm{kJ\ mol^{-1}})$	$\Delta_f G^{\ominus}/(\mathrm{kJ\ mol^{-1}})$	$S_m^{\ominus}/(\mathrm{J\ K^{-1}\ mol^{-1}})$†	$C_{p,m}^{\ominus}/(\mathrm{J\ K^{-1}\ mol^{-1}})$	$\Delta_c H^{\ominus}/(\mathrm{kJ\ mol^{-1}})$
Alkanals and alkanones						
HCHO(g), methanal	30.03	−108.57	−102.53	218.77	35.40	−571
CH_3CHO(l), ethanal	44.05	−192.30	−128.12	160.2		−1166
CH_3CHO(g)	44.05	−166.19	−128.86	250.3	57.3	−1192
CH_3COCH_3(l), propanone	58.08	−248.1	−155.4	200.4	124.7	−1790
Sugars						
$C_6H_{12}O_6$(s), α-D-glucose	180.16	−1274				−2808
$C_6H_{12}O_6$(s), β-D-glucose	180.16	−1268	−910	212		
$C_6H_{12}O_6$(s), β-D-fructose	180.16	−1266				−2810
$C_{12}H_{22}O_{11}$(s), sucrose	342.30	−2222	−1543	360.2		−5645
Nitrogen compounds						
$CO(NH_2)_2$(s), urea	60.06	−333.51	−197.33	104.60	93.14	−632
CH_3NH_2(g), methylamine	31.06	−22.97	+32.16	243.41	53.1	−1085
$C_6H_5NH_2$(l), aniline	93.13	+31.1				−3393
$CH_2(NH_2)COOH$(s), glycine	75.07	−532.9	−373.4	103.5	99.2	−969

Data: NBS, TDOC. † Standard entropies of ions may be either positive or negative because the values are relative to the entropy of the hydrogen ion.

Table 2.8 Thermodynamic data for elements and inorganic compounds at 298 K

	$M/(\mathrm{g\ mol^{-1}})$	$\Delta_f H^{\ominus}/(\mathrm{kJ\ mol^{-1}})$	$\Delta_f G^{\ominus}/(\mathrm{kJ\ mol^{-1}})$	$S_m^{\ominus}/(\mathrm{J\ K^{-1}\ mol^{-1}})$†	$C_{p,m}^{\ominus}/(\mathrm{J\ K^{-1}\ mol^{-1}})$
Aluminium (aluminum)					
Al(s)	26.98	0	0	28.33	24.35
Al(l)	26.98	+10.56	+7.20	39.55	24.21
Al(g)	26.98	+326.4	+285.7	164.54	21.38
Al^{3+}(g)	26.98	+5483.17			
Al^{3+}(aq)	26.98	−531	−485	−321.7	
Al_2O_3(s, α)	101.96	−1675.7	−1582.3	50.92	79.04
$AlCl_3$(s)	133.24	−704.2	−628.8	110.67	91.84
Argon					
Ar(g)	39.95	0	0	154.84	20.786
Antimony					
Sb(s)	121.75	0	0	45.69	25.23
SbH_3(g)	124.77	+145.11	+147.75	232.78	41.05
Arsenic					
As(s, α)	74.92	0	0	35.1	24.64
As(g)	74.92	+302.5	+261.0	174.21	20.79
As_4(g)	299.69	+143.9	+92.4	314	
AsH_3(g)	77.95	+66.44	+68.93	222.78	38.07

Table 2.8 (Continued)

	$M/(\mathrm{g\ mol^{-1}})$	$\Delta_f H^{\ominus}/(\mathrm{kJ\ mol^{-1}})$	$\Delta_f G^{\ominus}/(\mathrm{kJ\ mol^{-1}})$	$S_m^{\ominus}/(\mathrm{J\ K^{-1}\ mol^{-1}})$†	$C_{p,m}^{\ominus}/(\mathrm{J\ K^{-1}\ mol^{-1}})$
Barium					
Ba(s)	137.34	0	0	62.8	28.07
Ba(g)	137.34	+180	+146	170.24	20.79
Ba^{2+}(aq)	137.34	−537.64	−560.77	+9.6	
BaO(s)	153.34	−553.5	−525.1	70.43	47.78
$BaCl_2$(s)	208.25	−858.6	−810.4	123.68	75.14
Beryllium					
Be(s)	9.01	0	0	9.50	16.44
Be(g)	9.01	+324.3	+286.6	136.27	20.79
Bismuth					
Bi(s)	208.98	0	0	56.74	25.52
Bi(g)	208.98	+207.1	+168.2	187.00	20.79
Bromine					
Br_2(l)	159.82	0	0	152.23	75.689
Br_2(g)	159.82	+30.907	+3.110	245.46	36.02
Br(g)	79.91	+111.88	+82.396	175.02	20.786
Br^-(g)	79.91	−219.07			
Br^-(aq)	79.91	−121.55	−103.96	+82.4	−141.8
HBr(g)	90.92	−36.40	−53.45	198.70	29.142
Cadmium					
Cd(s, γ)	112.40	0	0	51.76	25.98
Cd(g)	112.40	+112.01	+77.41	167.75	20.79
Cd^{2+}(aq)	112.40	−75.90	−77.612	−73.2	
CdO(s)	128.40	−258.2	−228.4	54.8	43.43
$CdCO_3$(s)	172.41	−750.6	−669.4	92.5	
Caesium (cesium)					
Cs(s)	132.91	0	0	85.23	32.17
Cs(g)	132.91	+76.06	+49.12	175.60	20.79
Cs^+(aq)	132.91	−258.28	−292.02	+133.05	−10.5
Calcium					
Ca(s)	40.08	0	0	41.42	25.31
Ca(g)	40.08	+178.2	+144.3	154.88	20.786
Ca^{2+}(aq)	40.08	−542.83	−553.58	−53.1	
CaO(s)	56.08	−635.09	−604.03	39.75	42.80
$CaCO_3$(s) (calcite)	100.09	−1206.9	−1128.8	92.9	81.88
$CaCO_3$(s) (aragonite)	100.09	−1207.1	−1127.8	88.7	81.25
CaF_2(s)	78.08	−1219.6	−1167.3	68.87	67.03
$CaCl_2$(s)	110.99	−795.8	−748.1	104.6	72.59
$CaBr_2$(s)	199.90	−682.8	−663.6	130	
Carbon (for 'organic' compounds of carbon, see Table 14.5)					
C(s) (graphite)	12.011	0	0	5.740	8.527
C(s) (diamond)	12.011	+1.895	+2.900	2.377	6.113
C(g)	12.011	+716.68	+671.26	158.10	20.838

Table 2.8 (Continued)

	$M/(\mathrm{g\ mol^{-1}})$	$\Delta_f H^\ominus/(\mathrm{kJ\ mol^{-1}})$	$\Delta_f G^\ominus/(\mathrm{kJ\ mol^{-1}})$	$S_m^\ominus/(\mathrm{J\ K^{-1}\ mol^{-1}})$†	$C_{p,m}^\ominus/(\mathrm{J\ K^{-1}\ mol^{-1}})$
Carbon (Continued)					
$C_2(g)$	24.022	+831.90	+775.89	199.42	43.21
$CO(g)$	28.011	−110.53	−137.17	197.67	29.14
$CO_2(g)$	44.010	−393.51	−394.36	213.74	37.11
$CO_2(aq)$	44.010	−413.80	−385.98	117.6	
$H_2CO_3(aq)$	62.03	−699.65	−623.08	187.4	
$HCO_3^-(aq)$	61.02	−691.99	−586.77	+91.2	
$CO_3^{2-}(aq)$	60.01	−677.14	−527.81	−56.9	
$CCl_4(l)$	153.82	−135.44	−65.21	216.40	131.75
$CS_2(l)$	76.14	+89.70	+65.27	151.34	75.7
$HCN(g)$	27.03	+135.1	+124.7	201.78	35.86
$HCN(l)$	27.03	+108.87	+124.97	112.84	70.63
$CN^-(aq)$	26.02	+150.6	+172.4	+94.1	
Chlorine					
$Cl_2(g)$	70.91	0	0	223.07	33.91
$Cl(g)$	35.45	+121.68	+105.68	165.20	21.840
$Cl^-(g)$	34.45	−233.13			
$Cl^-(aq)$	35.45	−167.16	−131.23	+56.5	−136.4
$HCl(g)$	36.46	−92.31	−95.30	186.91	29.12
$HCl(aq)$	36.46	−167.16	−131.23	56.5	−136.4
Chromium					
$Cr(s)$	52.00	0	0	23.77	23.35
$Cr(g)$	52.00	+396.6	+351.8	174.50	20.79
$CrO_4^{2-}(aq)$	115.99	−881.15	−727.75	+50.21	
$Cr_2O_7^{2-}(aq)$	215.99	−1490.3	−1301.1	+261.9	
Copper					
$Cu(s)$	63.54	0	0	33.150	24.44
$Cu(g)$	63.54	+338.32	+298.58	166.38	20.79
$Cu^+(aq)$	63.54	+71.67	+49.98	+40.6	
$Cu^{2+}(aq)$	63.54	+64.77	+65.49	−99.6	
$Cu_2O(s)$	143.08	−168.6	−146.0	93.14	63.64
$CuO(s)$	79.54	−157.3	−129.7	42.63	42.30
$CuSO_4(s)$	159.60	−771.36	−661.8	109	100.0
$CuSO_4 \cdot H_2O(s)$	177.62	−1085.8	−918.11	146.0	134
$CuSO_4 \cdot 5H_2O(s)$	249.68	−2279.7	−1879.7	300.4	280
Deuterium					
$D_2(g)$	4.028	0	0	144.96	29.20
$HD(g)$	3.022	+0.318	−1.464	143.80	29.196
$D_2O(g)$	20.028	−249.20	−234.54	198.34	34.27
$D_2O(l)$	20.028	−294.60	−243.44	75.94	84.35
$HDO(g)$	19.022	−245.30	−233.11	199.51	33.81
$HDO(l)$	19.022	−289.89	−241.86	79.29	

Table 2.8 (Continued)

	$M/(\text{g mol}^{-1})$	$\Delta_f H^{\ominus}/(\text{kJ mol}^{-1})$	$\Delta_f G^{\ominus}/(\text{kJ mol}^{-1})$	$S_m^{\ominus}/(\text{J K}^{-1}\text{ mol}^{-1})$†	$C_{p,m}^{\ominus}/(\text{J K}^{-1}\text{ mol}^{-1})$
Fluorine					
F_2(g)	38.00	0	0	202.78	31.30
F(g)	19.00	+78.99	+61.91	158.75	22.74
F^-(aq)	19.00	−332.63	−278.79	−13.8	−106.7
HF(g)	20.01	−271.1	−273.2	173.78	29.13
Gold					
Au(s)	196.97	0	0	47.40	25.42
Au(g)	196.97	+366.1	+326.3	180.50	20.79
Helium					
He(g)	4.003	0	0	126.15	20.786
Hydrogen (see also deuterium)					
H_2(g)	2.016	0	0	130.684	28.824
H(g)	1.008	+217.97	+203.25	114.71	20.784
H^+(aq)	1.008	0	0	0	0
H^+(g)	1.008	+1536.20			
H_2O(s)	18.015			37.99	
H_2O(l)	18.015	−285.83	−237.13	69.91	75.291
H_2O(g)	18.015	−241.82	−228.57	188.83	33.58
H_2O_2(l)	34.015	−187.78	−120.35	109.6	89.1
Iodine					
I_2(s)	253.81	0	0	116.135	54.44
I_2(g)	253.81	+62.44	+19.33	260.69	36.90
I(g)	126.90	+106.84	+70.25	180.79	20.786
I^-(aq)	126.90	−55.19	−51.57	+111.3	−142.3
HI(g)	127.91	+26.48	+1.70	206.59	29.158
Iron					
Fe(s)	55.85	0	0	27.28	25.10
Fe(g)	55.85	+416.3	+370.7	180.49	25.68
Fe^{2+}(aq)	55.85	−89.1	−78.90	−137.7	
Fe^{3+}(aq)	55.85	−48.5	−4.7	−315.9	
Fe_3O_4(s) (magnetite)	231.54	−1118.4	−1015.4	146.4	143.43
Fe_2O_3(s) (haematite)	159.69	−824.2	−742.2	87.40	103.85
FeS(s, α)	87.91	−100.0	−100.4	60.29	50.54
FeS_2(s)	119.98	−178.2	−166.9	52.93	62.17
Krypton					
Kr(g)	83.80	0	0	164.08	20.786
Lead					
Pb(s)	207.19	0	0	64.81	26.44
Pb(g)	207.19	+195.0	+161.9	175.37	20.79
Pb^{2+}(aq)	207.19	−1.7	−24.43	+10.5	
PbO(s, yellow)	223.19	−217.32	−187.89	68.70	45.77
PbO(s, red)	223.19	−218.99	−188.93	66.5	45.81
PbO_2(s)	239.19	−277.4	−217.33	68.6	64.64

Table 2.8 (Continued)

	M/(g mol^{-1})	$\Delta_f H^{\ominus}$/(kJ mol^{-1})	$\Delta_f G^{\ominus}$/(kJ mol^{-1})	$S_m^{\ominus}$/(J K^{-1} mol^{-1})†	$C_{p,m}^{\ominus}$/(J K^{-1} mol^{-1})
Lithium					
Li(s)	6.94	0	0	29.12	24.77
Li(g)	6.94	+159.37	+126.66	138.77	20.79
Li^+(aq)	6.94	−278.49	−293.31	+13.4	68.6
Magnesium					
Mg(s)	24.31	0	0	32.68	24.89
Mg(g)	24.31	+147.70	+113.10	148.65	20.786
Mg^{2+}(aq)	24.31	−466.85	−454.8	−138.1	
MgO(s)	40.31	−601.70	−569.43	26.94	37.15
$MgCO_3$(s)	84.32	−1095.8	−1012.1	65.7	75.52
$MgCl_2$(s)	95.22	−641.32	−591.79	89.62	71.38
Mercury					
Hg(l)	200.59	0	0	76.02	27.983
Hg(g)	200.59	+61.32	+31.82	174.96	20.786
Hg^{2+}(aq)	200.59	+171.1	+164.40	−32.2	
Hg_2^{2+}(aq)	401.18	+172.4	+153.52	+84.5	
HgO(s)	216.59	−90.83	−58.54	70.29	44.06
Hg_2Cl_2(s)	472.09	−265.22	−210.75	192.5	102
$HgCl_2$(s)	271.50	−224.3	−178.6	146.0	
HgS(s, black)	232.65	−53.6	−47.7	88.3	
Neon					
Ne(g)	20.18	0	0	146.33	20.786
Nitrogen					
N_2(g)	28.013	0	0	191.61	29.125
N(g)	14.007	+472.70	+455.56	153.30	20.786
NO(g)	30.01	+90.25	+86.55	210.76	29.844
N_2O(g)	44.01	+82.05	+104.20	219.85	38.45
NO_2(g)	46.01	+33.18	+51.31	240.06	37.20
N_2O_4(g)	92.1	+9.16	+97.89	304.29	77.28
N_2O_5(s)	108.01	−43.1	+113.9	178.2	143.1
N_2O_5(g)	108.01	+11.3	+115.1	355.7	84.5
HNO_3(l)	63.01	−174.10	−80.71	155.60	109.87
HNO_3(aq)	63.01	−207.36	−111.25	146.4	−86.6
NO_3^-(aq)	62.01	−205.0	−108.74	+146.4	−86.6
NH_3(g)	17.03	−46.11	−16.45	192.45	35.06
NH_3(aq)	17.03	−80.29	−26.50	111.3	
NH_4^+(aq)	18.04	−132.51	−79.31	+113.4	79.9
NH_2OH(s)	33.03	−114.2			
HN_3(l)	43.03	+264.0	+327.3	140.6	
HN_3(g)	43.03	+294.1	+328.1	238.97	43.68
N_2H_4(l)	32.05	+50.63	+149.43	121.21	98.87
NH_4NO_3(s)	80.04	−365.56	−183.87	151.08	139.3
NH_4Cl(s)	53.49	−314.43	−202.87	94.6	84.1

Table 2.8 (Continued)

	$M/(g\ mol^{-1})$	$\Delta_f H^\ominus/(kJ\ mol^{-1})$	$\Delta_f G^\ominus/(kJ\ mol^{-1})$	$S_m^\ominus/(J\ K^{-1}\ mol^{-1})$†	$C_{p,m}^\ominus/(J\ K^{-1}\ mol^{-1})$
Oxygen					
$O_2(g)$	31.999	0	0	205.138	29.355
$O(g)$	15.999	+249.17	+231.73	161.06	21.912
$O_3(g)$	47.998	+142.7	+163.2	238.93	39.20
$OH^-(aq)$	17.007	−229.99	−157.24	−10.75	−148.5
Phosphorus					
P(s, wh)	30.97	0	0	41.09	23.840
P(g)	30.97	+314.64	+278.25	163.19	20.786
$P_2(g)$	61.95	+144.3	+103.7	218.13	32.05
$P_4(g)$	123.90	+58.91	+24.44	279.98	67.15
$PH_3(g)$	34.00	+5.4	+13.4	210.23	37.11
$PCl_3(g)$	137.33	−287.0	−267.8	311.78	71.84
$PCl_3(l)$	137.33	−319.7	−272.3	217.1	
$PCl_5(g)$	208.24	−374.9	−305.0	364.6	112.8
$PCl_5(s)$	208.24	−443.5			
$H_3PO_3(s)$	82.00	−964.4			
$H_3PO_3(aq)$	82.00	−964.8			
$H_3PO_4(s)$	94.97	−1279.0	−1119.1	110.50	106.06
$H_3PO_4(l)$	94.97	−1266.9			
$H_3PO_4(aq)$	94.97	−1277.4	−1018.7	−222	
$PO_4^{3-}(aq)$	94.97	−1277.4	−1018.7	−221.8	
$P_4O_{10}(s)$	283.89	−2984.0	−2697.0	228.86	211.71
$P_4O_6(s)$	219.89	−1640.1			
Potassium					
K(s)	39.10	0	0	64.18	29.58
K(g)	39.10	+89.24	+60.59	160.336	20.786
$K^+(g)$	39.10	+514.26			
$K^+(aq)$	39.10	−252.38	−283.27	+102.5	21.8
KOH(s)	56.11	−424.76	−379.08	78.9	64.9
KF(s)	58.10	−576.27	−537.75	66.57	49.04
KCl(s)	74.56	−436.75	−409.14	82.59	51.30
KBr(s)	119.01	−393.80	−380.66	95.90	52.30
KI(s)	166.01	−327.90	−324.89	106.32	52.93
Silicon					
Si(s)	28.09	0	0	18.83	20.00
Si(g)	28.09	+455.6	+411.3	167.97	22.25
$SiO_2(s, \alpha)$	60.09	−910.94	−856.64	41.84	44.43
Silver					
Ag(s)	107.87	0	0	42.55	25.351
Ag(g)	107.87	+284.55	+245.65	173.00	20.79
$Ag^+(aq)$	107.87	+105.58	+77.11	+72.68	21.8
AgBr(s)	187.78	−100.37	−96.90	107.1	52.38

Table 2.8 (Continued)

	$M/(\mathrm{g\ mol^{-1}})$	$\Delta_f H^\ominus/(\mathrm{kJ\ mol^{-1}})$	$\Delta_f G^\ominus/(\mathrm{kJ\ mol^{-1}})$	$S_m^\ominus/(\mathrm{J\ K^{-1}\ mol^{-1}})$†	$C_{p,m}^\ominus/(\mathrm{J\ K^{-1}\ mol^{-1}})$
Silver (Continued)					
AgCl(s)	143.32	−127.07	−109.79	96.2	50.79
Ag_2O(s)	231.74	−31.05	−11.20	121.3	65.86
$AgNO_3$(s)	169.88	−129.39	−33.41	140.92	93.05
Sodium					
Na(s)	22.99	0	0	51.21	28.24
Na(g)	22.99	+107.32	+76.76	153.71	20.79
Na^+(aq)	22.99	−240.12	−261.91	+59.0	46.4
NaOH(s)	40.00	−425.61	−379.49	64.46	59.54
NaCl(s)	58.44	−411.15	−384.14	72.13	50.50
NaBr(s)	102.90	−361.06	−348.98	86.82	51.38
NaI(s)	149.89	−287.78	−286.06	98.53	52.09
Sulfur					
S(s, α) (rhombic)	32.06	0	0	31.80	22.64
S(s, β) (monoclinic)	32.06	+0.33	+0.1	32.6	23.6
S(g)	32.06	+278.81	+238.25	167.82	23.673
S_2(g)	64.13	+128.37	+79.30	228.18	32.47
S^{2-}(aq)	32.06	+33.1	+85.8	−14.6	
SO_2(g)	64.06	−296.83	−300.19	248.22	39.87
SO_3(g)	80.06	−395.72	−371.06	256.76	50.67
H_2SO_4(l)	98.08	−813.99	−690.00	156.90	138.9
H_2SO_4(aq)	98.08	−909.27	−744.53	20.1	−293
SO_4^{2-}(aq)	96.06	−909.27	−744.53	+20.1	−293
HSO_4^-(aq)	97.07	−887.34	−755.91	+131.8	−84
H_2S(g)	34.08	−20.63	−33.56	205.79	34.23
H_2S(aq)	34.08	−39.7	−27.83	121	
HS^-(aq)	33.072	−17.6	+12.08	+62.08	
SF_6(g)	146.05	−1209	−1105.3	291.82	97.28
Tin					
Sn(s, β)	118.69	0	0	51.55	26.99
Sn(g)	118.69	+302.1	+267.3	168.49	20.26
Sn^{2+}(aq)	118.69	−8.8	−27.2	−17	
SnO(s)	134.69	−285.8	−256.9	56.5	44.31
SnO_2(s)	150.69	−580.7	−519.6	52.3	52.59
Xenon					
Xe(g)	131.30	0	0	169.68	20.786
Zinc					
Zn(s)	65.37	0	0	41.63	25.40
Zn(g)	65.37	+130.73	+95.14	160.98	20.79
Zn^{2+}(aq)	65.37	−153.89	−147.06	−112.1	46
ZnO(s)	81.37	−348.28	−318.30	43.64	40.25

Source: NBS. † Standard entropies of ions may be either positive or negative because the values are relative to the entropy of the hydrogen ion.

Table 2.9 Expansion coefficients, α, and isothermal compressibilities, κ_T

	$\alpha/(10^{-4}\,K^{-1})$	$\kappa_T/(10^{-6}\,atm^{-1})$
Liquids		
Benzene	12.4	92.1
Carbon tetrachloride	12.4	90.5
Ethanol	11.2	76.8
Mercury	1.82	38.7
Water	2.1	49.6
Solids		
Copper	0.501	0.735
Diamond	0.030	0.187
Iron	0.354	0.589
Lead	0.861	2.21

The values refer to 20°C.
Data: AIP(α), KL(κ_T).

Table 2.10 Inversion temperatures, normal freezing and boiling points, and Joule–Thomson coefficients at 1 atm and 298 K

	T_I/K	T_f/K	T_b/K	μ/(K atm^{-1})
Air	603			0.189 at 50°C
Argon	723	83.8	87.3	
Carbon dioxide	1500	194.7s		1.11 at 300 K
Helium	40		4.22	−0.062
Hydrogen	202	14.0	20.3	−0.03
Krypton	1090	116.6	120.8	
Methane	968	90.6	111.6	
Neon	231	24.5	27.1	
Nitrogen	621	63.3	77.4	0.27
Oxygen	764	54.8	90.2	0.31

s: sublimes.
Data: AIP, JL, and M.W. Zemansky, *Heat and thermodynamics*. McGraw-Hill, New York (1957).

Table 3.1 Standard entropies (and temperatures) of phase transitions, $\Delta_{trs}S^{\ominus}/(J\,K^{-1}\,mol^{-1})$

	Fusion (at T_f)	Vaporization (at T_b)
Ar	14.17 (at 83.8 K)	74.53 (at 87.3 K)
Br_2	39.76 (at 265.9 K)	88.61 (at 332.4 K)
C_6H_6	38.00 (at 278.6 K)	87.19 (at 353.2 K)
CH_3COOH	40.4 (at 289.8 K)	61.9 (at 391.4 K)
CH_3OH	18.03 (at 175.2 K)	104.6 (at 337.2 K)
Cl_2	37.22 (at 172.1 K)	85.38 (at 239.0 K)
H_2	8.38 (at 14.0 K)	44.96 (at 20.38 K)
H_2O	22.00 (at 273.2 K)	109.1 (at 373.2 K)
H_2S	12.67 (at 187.6 K)	87.75 (at 212.0 K)
He	4.8 (at 1.8 K and 30 bar)	19.9 (at 4.22 K)
N_2	11.39 (at 63.2 K)	75.22 (at 77.4 K)
NH_3	28.93 (at 195.4 K)	97.41 (at 239.73 K)
O_2	8.17 (at 54.4 K)	75.63 (at 90.2 K)

Data: AIP.

Table 3.2 Standard enthalpies and entropies of vaporization of liquids at their normal boiling points

	$\Delta_{vap}H^{\ominus}/(kJ\,mol^{-1})$	θ_b/°C	$\Delta_{vap}S^{\ominus}/(J\,K^{-1}\,mol^{-1})$
Benzene	30.8	80.1	+87.2
Carbon disulfide	26.74	46.25	+83.7
Carbon tetrachloride	30.00	76.7	+85.8
Cyclohexane	30.1	80.7	+85.1
Decane	38.75	174	+86.7
Dimethyl ether	21.51	−23	+86
Ethanol	38.6	78.3	+110.0
Hydrogen sulfide	18.7	−60.4	+87.9
Mercury	59.3	356.6	+94.2
Methane	8.18	−161.5	+73.2
Methanol	35.21	65.0	+104.1
Water	40.7	100.0	+109.1

Data: JL.

Table 3.3 Standard Third-Law entropies at 298 K: see Tables 2.6 and 2.8

Table 3.4 Standard Gibbs energies of formation (at 298 K): see Tables 2.6 and 2.8

Table 3.6 The fugacity coefficient of nitrogen at 273 K

p/atm	ϕ	p/atm	ϕ
1	0.999 55	300	1.0055
10	0.9956	400	1.062
50	0.9912	600	1.239
100	0.9703	800	1.495
150	0.9672	1000	1.839
200	0.9721		

Data: LR.

Table 5.1 Henry's law constants for gases in water at 298 K, K/(kPa kg mol^{-1})

	Water	Benzene
CH_4	7.55×10^4	44.4×10^3
CO_2	3.01×10^3	8.90×10^2
H_2	1.28×10^5	2.79×10^4
N_2	1.56×10^5	1.87×10^4
O_2	7.92×10^4	

Data: converted from R.J. Silbey and R.A. Alberty, *Physical chemistry*. Wiley, New York (2001).

Table 5.2 Freezing-point and boiling-point constants

	K_f/(K kg mol^{-1})	K_b/(K kg mol^{-1})
Acetic acid	3.90	3.07
Benzene	5.12	2.53
Camphor	40	
Carbon disulfide	3.8	2.37
Carbon tetrachloride	30	4.95
Naphthalene	6.94	5.8
Phenol	7.27	3.04
Water	1.86	0.51

Data: KL.

Table 5.5 Mean activity coefficients in water at 298 K

$b/b^\ominus$	HCl	KCl	$CaCl_2$	H_2SO_4	$LaCl_3$	$In_2(SO_4)_3$
0.001	0.966	0.966	0.888	0.830	0.790	
0.005	0.929	0.927	0.789	0.639	0.636	0.16
0.01	0.905	0.902	0.732	0.544	0.560	0.11
0.05	0.830	0.816	0.584	0.340	0.388	0.035
0.10	0.798	0.770	0.524	0.266	0.356	0.025
0.50	0.769	0.652	0.510	0.155	0.303	0.014
1.00	0.811	0.607	0.725	0.131	0.387	
2.00	1.011	0.577	1.554	0.125	0.954	

Data: RS, HCP, and S. Glasstone, *Introduction to electrochemistry*. Van Nostrand (1942).

Table 6.2 Standard potentials at 298 K. (a) In electrochemical order

Reduction half-reaction	$E^{\ominus}/V$
Strongly oxidizing	
$H_4XeO_6 + 2H^+ + 2e^- \rightarrow XeO_3 + 3H_2O$	+3.0
$F_2 + 2e^- \rightarrow 2F^-$	+2.87
$O_3 + 2H^+ + 2e^- \rightarrow O_2 + H_2O$	+2.07
$S_2O_8^{2-} + 2e^- \rightarrow 2SO_4^{2-}$	+2.05
$Ag^{2+} + e^- \rightarrow Ag^+$	+1.98
$Co^{3+} + e^- \rightarrow Co^{2+}$	+1.81
$H_2O_2 + 2H^+ + 2e^- \rightarrow 2H_2O$	+1.78
$Au^+ + e^- \rightarrow Au$	+1.69
$Pb^{4+} + 2e^- \rightarrow Pb^{2+}$	+1.67
$2HClO + 2H^+ + 2e^- \rightarrow Cl_2 + 2H_2O$	+1.63
$Ce^{4+} + e^- \rightarrow Ce^{3+}$	+1.61
$2HBrO + 2H^+ + 2e^- \rightarrow Br_2 + 2H_2O$	+1.60
$MnO_4^- + 8H^+ + 5e^- \rightarrow Mn^{2+} + 4H_2O$	+1.51
$Mn^{3+} + e^- \rightarrow Mn^{2+}$	+1.51
$Au^{3+} + 3e^- \rightarrow Au$	+1.40
$Cl_2 + 2e^- \rightarrow 2Cl^-$	+1.36
$Cr_2O_7^{2-} + 14H^+ + 6e^- \rightarrow 2Cr^{3+} + 7H_2O$	+1.33
$O_3 + H_2O + 2e^- \rightarrow O_2 + 2OH^-$	+1.24
$O_2 + 4H^+ + 4e^- \rightarrow 2H_2O$	+1.23
$ClO_4^- + 2H^+ + 2e^- \rightarrow ClO_3^- + H_2O$	+1.23
$MnO_2 + 4H^+ + 2e^- \rightarrow Mn^{2+} + 2H_2O$	+1.23
$Br_2 + 2e^- \rightarrow 2Br^-$	+1.09
$Pu^{4+} + e^- \rightarrow Pu^{3+}$	+0.97
$NO_3^- + 4H^+ + 3e^- \rightarrow NO + 2H_2O$	+0.96
$2Hg^{2+} + 2e^- \rightarrow Hg_2^{2+}$	+0.92
$ClO^- + H_2O + 2e^- \rightarrow Cl^- + 2OH^-$	+0.89
$Hg^{2+} + 2e^- \rightarrow Hg$	+0.86
$NO_3^- + 2H^+ + e^- \rightarrow NO_2 + H_2O$	+0.80
$Ag^+ + e^- \rightarrow Ag$	+0.80
$Hg_2^{2+} + 2e^- \rightarrow 2Hg$	+0.79
$Fe^{3+} + e^- \rightarrow Fe^{2+}$	+0.77
$BrO^- + H_2O + 2e^- \rightarrow Br^- + 2OH^-$	+0.76
$Hg_2SO_4 + 2e^- \rightarrow 2Hg + SO_4^{2-}$	+0.62
$MnO_4^{2-} + 2H_2O + 2e^- \rightarrow MnO_2 + 4OH^-$	+0.60
$MnO_4^- + e^- \rightarrow MnO_4^{2-}$	+0.56
$I_2 + 2e^- \rightarrow 2I^-$	+0.54
$Cu^+ + e^- \rightarrow Cu$	+0.52
$I_3^- + 2e^- \rightarrow 3I^-$	+0.53
$NiOOH + H_2O + e^- \rightarrow Ni(OH)_2 + OH^-$	+0.49
$Ag_2CrO_4 + 2e^- \rightarrow 2Ag + CrO_4^{2-}$	+0.45
$O_2 + 2H_2O + 4e^- \rightarrow 4OH^-$	+0.40
$ClO_4^- + H_2O + 2e^- \rightarrow ClO_3^- + 2OH^-$	+0.36
$[Fe(CN)_6]^{3-} + e^- \rightarrow [Fe(CN)_6]^{4-}$	+0.36
$Cu^{2+} + 2e^- \rightarrow Cu$	+0.34
$Hg_2Cl_2 + 2e^- \rightarrow 2Hg + 2Cl^-$	+0.27
$AgCl + e^- \rightarrow Ag + Cl^-$	+0.22
$Bi^{3+} + 3e^- \rightarrow Bi$	+0.20
$Cu^{2+} + e^- \rightarrow Cu^+$	+0.16
$Sn^{4+} + 2e^- \rightarrow Sn^{2+}$	+0.15
$AgBr + e^- \rightarrow Ag + Br^-$	+0.07
$Ti^{4+} + e^- \rightarrow Ti^{3+}$	0.00
$2H^+ + 2e^- \rightarrow H_2$	0, by definition
$Fe^{3+} + 3e^- \rightarrow Fe$	−0.04
$O_2 + H_2O + 2e^- \rightarrow HO_2^- + OH^-$	−0.08
$Pb^{2+} + 2e^- \rightarrow Pb$	−0.13
$In^+ + e^- \rightarrow In$	−0.14
$Sn^{2+} + 2e^- \rightarrow Sn$	−0.14
$AgI + e^- \rightarrow Ag + I^-$	−0.15
$Ni^{2+} + 2e^- \rightarrow Ni$	−0.23
$Co^{2+} + 2e^- \rightarrow Co$	−0.28
$In^{3+} + 3e^- \rightarrow In$	−0.34
$Tl^+ + e^- \rightarrow Tl$	−0.34
$PbSO_4 + 2e^- \rightarrow Pb + SO_4^{2-}$	−0.36
$Ti^{3+} + e^- \rightarrow Ti^{2+}$	−0.37
$Cd^{2+} + 2e^- \rightarrow Cd$	−0.40
$In^{2+} + e^- \rightarrow In^+$	−0.40
$Cr^{3+} + e^- \rightarrow Cr^{2+}$	−0.41
$Fe^{2+} + 2e^- \rightarrow Fe$	−0.44
$In^{3+} + 2e^- \rightarrow In^+$	−0.44
$S + 2e^- \rightarrow S^{2-}$	−0.48
$In^{3+} + e^- \rightarrow In^{2+}$	−0.49
$U^{4+} + e^- \rightarrow U^{3+}$	−0.61
$Cr^{3+} + 3e^- \rightarrow Cr$	−0.74
$Zn^{2+} + 2e^- \rightarrow Zn$	−0.76
$Cd(OH)_2 + 2e^- \rightarrow Cd + 2OH^-$	−0.81
$2H_2O + 2e^- \rightarrow H_2 + 2OH^-$	−0.83
$Cr^{2+} + 2e^- \rightarrow Cr$	−0.91
$Mn^{2+} + 2e^- \rightarrow Mn$	−1.18
$V^{2+} + 2e^- \rightarrow V$	−1.19
$Ti^{2+} + 2e^- \rightarrow Ti$	−1.63
$Al^{3+} + 3e^- \rightarrow Al$	−1.66
$U^{3+} + 3e^- \rightarrow U$	−1.79
$Sc^{3+} + 3e^- \rightarrow Sc$	−2.09
$Mg^{2+} + 2e^- \rightarrow Mg$	−2.36
$Ce^{3+} + 3e^- \rightarrow Ce$	−2.48
$La^{3+} + 3e^- \rightarrow La$	−2.52
$Na^+ + e^- \rightarrow Na$	−2.71
$Ca^{2+} + 2e^- \rightarrow Ca$	−2.87
$Sr^{2+} + 2e^- \rightarrow Sr$	−2.89
$Ba^{2+} + 2e^- \rightarrow Ba$	−2.91
$Ra^{2+} + 2e^- \rightarrow Ra$	−2.92
$Cs^+ + e^- \rightarrow Cs$	−2.92
$Rb^+ + e^- \rightarrow Rb$	−2.93
$K^+ + e^- \rightarrow K$	−2.93
$Li^+ + e^- \rightarrow Li$	−3.05

Table 6.2 Standard potentials at 298 K. (b) In alphabetical order

Reduction half-reaction	$E^{\ominus}$/V	Reduction half-reaction	$E^{\ominus}$/V
$Ag^+ + e^- \rightarrow Ag$	+0.80	$I_2 + 2e^- \rightarrow 2I^-$	+0.54
$Ag^{2+} + e^- \rightarrow Ag^+$	+1.98	$I_3^- + 2e^- \rightarrow 3I^-$	+0.53
$AgBr + e^- \rightarrow Ag + Br^-$	+0.0713	$In^+ + e^- \rightarrow In$	−0.14
$AgCl + e^- \rightarrow Ag + Cl^-$	+0.22	$In^{2+} + e^- \rightarrow In^+$	−0.40
$Ag_2CrO_4 + 2e^- \rightarrow 2Ag + CrO_4^{2-}$	+0.45	$In^{3+} + 2e^- \rightarrow In^+$	−0.44
$AgF + e^- \rightarrow Ag + F^-$	+0.78	$In^{3+} + 3e^- \rightarrow In$	−0.34
$AgI + e^- \rightarrow Ag + I^-$	−0.15	$In^{3+} + e^- \rightarrow In^{2+}$	−0.49
$Al^{3+} + 3e^- \rightarrow Al$	−1.66	$K^+ + e^- \rightarrow K$	−2.93
$Au^+ + e^- \rightarrow Au$	+1.69	$La^{3+} + 3e^- \rightarrow La$	−2.52
$Au^{3+} + 3e^- \rightarrow Au$	+1.40	$Li^+ + e^- \rightarrow Li$	−3.05
$Ba^{2+} + 2e^- \rightarrow Ba$	+2.91	$Mg^{2+} + 2e^- \rightarrow Mg$	−2.36
$Be^{2+} + 2e^- \rightarrow Be$	−1.85	$Mn^{2+} + 2e^- \rightarrow Mn$	−1.18
$Bi^{3+} + 3e^- \rightarrow Bi$	+0.20	$Mn^{3+} + e^- \rightarrow Mn^{2+}$	+1.51
$Br_2 + 2e^- \rightarrow 2Br^-$	+1.09	$MnO_2 + 4H^+ + 2e^- \rightarrow Mn^{2+} + 2H_2O$	+1.23
$BrO^- + H_2O + 2e^- \rightarrow Br^- + 2OH^-$	+0.76	$MnO_4^- + 8H^+ + 5e^- \rightarrow Mn^{2+} + 4H_2O$	+1.51
$Ca^{2+} + 2e^- \rightarrow Ca$	−2.87	$MnO_4^- + e^- \rightarrow MnO_4^{2-}$	+0.56
$Cd(OH)_2 + 2e^- \rightarrow Cd + 2OH^-$	−0.81	$MnO_4^{2-} + 2H_2O + 2e^- \rightarrow MnO_2 + 4OH^-$	+0.60
$Cd^{2+} + 2e^- \rightarrow Cd$	−0.40	$Na^+ + e^- \rightarrow Na$	−2.71
$Ce^{3+} + 3e^- \rightarrow Ce$	−2.48	$Ni^{2+} + 2e^- \rightarrow Ni$	−0.23
$Ce^{4+} + e^- \rightarrow Ce^{3+}$	+1.61	$NiOOH + H_2O + e^- \rightarrow Ni(OH)_2 + OH^-$	+0.49
$Cl_2 + 2e^- \rightarrow 2Cl^-$	+1.36	$NO_3^- + 2H^+ + e^- \rightarrow NO_2 + H_2O$	−0.80
$ClO^- + H_2O + 2e^- \rightarrow Cl^- + 2OH^-$	+0.89	$NO_3^- + 4H^+ + 3e^- \rightarrow NO + 2H_2O$	+0.96
$ClO_4^- + 2H^+ + 2e^- \rightarrow ClO_3^- + H_2O$	+1.23	$NO_3^- + H_2O + 2e^- \rightarrow NO_2^- + 2OH^-$	+0.10
$ClO_4^- + H_2O + 2e^- \rightarrow ClO_3^- + 2OH^-$	+0.36	$O_2 + 2H_2O + 4e^- \rightarrow 4OH^-$	+0.40
$Co^{2+} + 2e^- \rightarrow Co$	−0.28	$O_2 + 4H^+ + 4e^- \rightarrow 2H_2O$	+1.23
$Co^{3+} + e^- \rightarrow Co^{2+}$	+1.81	$O_2 + e^- \rightarrow O_2^-$	−0.56
$Cr^{2+} + 2e^- \rightarrow Cr$	−0.91	$O_2 + H_2O + 2e^- \rightarrow HO_2^- + OH^-$	−0.08
$Cr_2O_7^{2-} + 14H^+ + 6e^- \rightarrow 2Cr^{3+} + 7H_2O$	+1.33	$O_3 + 2H^+ + 2e^- \rightarrow O_2 + H_2O$	+2.07
$Cr^{3+} + 3e^- \rightarrow Cr$	−0.74	$O_3 + H_2O + 2e^- \rightarrow O_2 + 2OH^-$	+1.24
$Cr^{3+} + e^- \rightarrow Cr^{2+}$	−0.41	$Pb^{2+} + 2e^- \rightarrow Pb$	−0.13
$Cs^+ + e^- \rightarrow Cs$	−2.92	$Pb^{4+} + 2e^- \rightarrow Pb^{2+}$	+1.67
$Cu^+ + e^- \rightarrow Cu$	+0.52	$PbSO_4 + 2e^- \rightarrow Pb + SO_4^{2-}$	−0.36
$Cu^{2+} + 2e^- \rightarrow Cu$	+0.34	$Pt^{2+} + 2e^- \rightarrow Pt$	+1.20
$Cu^{2+} + e^- \rightarrow Cu^+$	+0.16	$Pu^{4+} + e^- \rightarrow Pu^{3+}$	+0.97
$F_2 + 2e^- \rightarrow 2F^-$	+2.87	$Ra^{2+} + 2e^- \rightarrow Ra$	−2.92
$Fe^{2+} + 2e^- \rightarrow Fe$	−0.44	$Rb^+ + e^- \rightarrow Rb$	−2.93
$Fe^{3+} + 3e^- \rightarrow Fe$	−0.04	$S + 2e^- \rightarrow S^{2-}$	−0.48
$Fe^{3+} + e^- \rightarrow Fe^{2+}$	+0.77	$S_2O_8^{2-} + 2e^- \rightarrow 2SO_4^{2-}$	+2.05
$[Fe(CN)_6]^{3-} + e^- \rightarrow [Fe(CN)_6]^{4-}$	+0.36	$Sc^{3+} + 3e^- \rightarrow Sc$	−2.09
$2H^+ + 2e^- \rightarrow H_2$	0, by definition	$Sn^{2+} + 2e^- \rightarrow Sn$	−0.14
$2H_2O + 2e^- \rightarrow H_2 + 2OH^-$	−0.83	$Sn^{4+} + 2e^- \rightarrow Sn^{2+}$	+0.15
$2HBrO + 2H^+ + 2e^- \rightarrow Br_2 + 2H_2O$	+1.60	$Sr^{2+} + 2e^- \rightarrow Sr$	−2.89
$2HClO + 2H^+ + 2e^- \rightarrow Cl_2 + 2H_2O$	+1.63	$Ti^{2+} + 2e^- \rightarrow Ti$	−1.63
$H_2O_2 + 2H^+ + 2e^- \rightarrow 2H_2O$	+1.78	$Ti^{3+} + e^- \rightarrow Ti^{2+}$	−0.37
$H_4XeO_6 + 2H^+ + 2e^- \rightarrow XeO_3 + 3H_2O$	+3.0	$Ti^{4+} + e^- \rightarrow Ti^{3+}$	0.00
$Hg_2^{2+} + 2e^- \rightarrow 2Hg$	+0.79	$Tl^+ + e^- \rightarrow Tl$	−0.34
$Hg_2Cl_2 + 2e^- \rightarrow 2Hg + 2Cl^-$	+0.27	$U^{3+} + 3e^- \rightarrow U$	−1.79
$Hg^{2+} + 2e^- \rightarrow Hg$	+0.86	$U^{4+} + e^- \rightarrow U^{3+}$	−0.61
$2Hg^{2+} + 2e^- \rightarrow Hg_2^{2+}$	+0.92	$V^{2+} + 2e^- \rightarrow V$	−1.19
$Hg_2SO_4 + 2e^- \rightarrow 2Hg + SO_4^{2-}$	+0.62	$V^{3+} + e^- \rightarrow V^{2+}$	−0.26
		$Zn^{2+} + 2e^- \rightarrow Zn$	−0.76

Table 9.2 Effective nuclear charge, $Z_{eff} = Z - \sigma$

	H							He
1s	1							1.6875
	Li	**Be**	**B**	**C**	**N**	**O**	**F**	**Ne**
1s	2.6906	3.6848	4.6795	5.6727	6.6651	7.6579	8.6501	9.6421
2s	1.2792	1.9120	2.5762	3.2166	3.8474	4.4916	5.1276	5.7584
2p			2.4214	3.1358	3.8340	4.4532	5.1000	5.7584
	Na	**Mg**	**Al**	**Si**	**P**	**S**	**Cl**	**Ar**
1s	10.6259	11.6089	12.5910	13.5745	14.5578	15.5409	16.5239	17.5075
2s	6.5714	7.3920	8.3736	9.0200	9.8250	10.6288	11.4304	12.2304
2p	6.8018	7.8258	8.9634	9.9450	10.9612	11.9770	12.9932	14.0082
3s	2.5074	3.3075	4.1172	4.9032	5.6418	6.3669	7.0683	7.7568
3p			4.0656	4.2852	4.8864	5.4819	6.1161	6.7641

Data: E. Clementi and D.L. Raimondi, *Atomic screening constants from SCF functions.* IBM Res. Note NJ-27 (1963). *J. Chem. Phys.* **38**, 2686 (1963).

Table 9.3 Ionization energies, $I/(\text{kJ mol}^{-1})$

H							He
1312.0							2372.3
							5250.4
Li	**Be**	**B**	**C**	**N**	**O**	**F**	**Ne**
513.3	899.4	800.6	1086.2	1402.3	1313.9	1681	2080.6
7298.0	1757.1	2427	2352	2856.1	3388.2	3374	3952.2
Na	**Mg**	**Al**	**Si**	**P**	**S**	**Cl**	**Ar**
495.8	737.7	577.4	786.5	1011.7	999.6	1251.1	1520.4
4562.4	1450.7	1816.6	1577.1	1903.2	2251	2297	2665.2
		2744.6		2912			
K	**Ca**	**Ga**	**Ge**	**As**	**Se**	**Br**	**Kr**
418.8	589.7	578.8	762.1	947.0	940.9	1139.9	1350.7
3051.4	1145	1979	1537	1798	2044	2104	2350
		2963	2735				
Rb	**Sr**	**In**	**Sn**	**Sb**	**Te**	**I**	**Xe**
403.0	549.5	558.3	708.6	833.7	869.2	1008.4	1170.4
2632	1064.2	1820.6	1411.8	1794	1795	1845.9	2046
		2704	2943.0	2443			
Cs	**Ba**	**Tl**	**Pb**	**Bi**	**Po**	**At**	**Rn**
375.5	502.8	589.3	715.5	703.2	812	930	1037
2420	965.1	1971.0	1450.4	1610			
		2878	3081.5	2466			

Data: E.

Table 9.4 Electron affinities, $E_{ea}/(\mathrm{kJ\ mol^{-1}})$

H							**He**
72.8							−21
Li	**Be**	**B**	**C**	**N**	**O**	**F**	**Ne**
59.8	≤0	23	122.5	−7	141 −844	322	−29
Na	**Mg**	**Al**	**Si**	**P**	**S**	**Cl**	**Ar**
52.9	≤0	44	133.6	71.7	200.4 −532	348.7	−35
K	**Ca**	**Ga**	**Ge**	**As**	**Se**	**Br**	**Kr**
48.3	2.37	36	116	77	195.0	324.5	−39
Rb	**Sr**	**In**	**Sn**	**Sb**	**Te**	**I**	**Xe**
46.9	5.03	34	121	101	190.2	295.3	−41
Cs	**Ba**	**Tl**	**Pb**	**Bi**	**Po**	**At**	**Rn**
45.5	13.95	30	35.2	101	186	270	−41

Data: E.

Table 10.2 Bond lengths, R_e/pm

(a) Bond lengths in specific molecules

Br_2	228.3
Cl_2	198.75
CO	112.81
F_2	141.78
H_2^+	106
H_2	74.138
HBr	141.44
HCl	127.45
HF	91.680
HI	160.92
N_2	109.76
O_2	120.75

(b) Mean bond lengths from covalent radii*

H	37						
C	77(1) 67(2) 60(3)	N	74(1) 65(2)	O	66(1) 57(2)	F	64
Si	118	P	110	S	104(1) 95(2)	Cl	99
Ge	122	As	121	Se	104	Br	114
		Sb	141	Te	137	I	133

* Values are for single bonds except where indicated otherwise (values in parentheses). The length of an A—B covalent bond (of given order) is the sum of the corresponding covalent radii.

Table 10.3a Bond dissociation enthalpies, $\Delta H^{\ominus}(\text{A–B})/(\text{kJ mol}^{-1})$ at 298 K*

Diatomic molecules

H–H	436	F–F	155	Cl–Cl	242	Br–Br	193	I–I	151
O=O	497	C=O	1076	N≡N	945				
H–O	428	H–F	565	H–Cl	431	H–Br	366	H–I	299

Polyatomic molecules

H–CH_3	435	H–NH_2	460	H–OH	492	H–C_6H_5	469
H_3C–CH_3	368	H_2C=CH_2	720	HC≡CH	962		
HO–CH_3	377	Cl–CH_3	352	Br–CH_3	293	I–CH_3	237
O=CO	531	HO–OH	213	O_2N–NO_2	54		

* To a good approximation bond dissociation enthalpies and dissociation energies are related by $\Delta H^{\ominus} = D_e + \frac{3}{2}RT$ with $D_e = D_0 + \frac{1}{2}\hbar\omega$. For precise values of D_0 for diatomic molecules, see Table 12.2.
Data: HCP, KL.

Table 10.3b Mean bond enthalpies, $\Delta H^{\ominus}(\text{A–B})/(\text{kJ mol}^{-1})$*

	H	C	N	O	F	Cl	Br	I	S	P	Si
H	436										
C	412	348(i)									
		612(ii)									
		838(iii)									
		518(a)									
N	388	305(i)	163(i)								
		613(ii)	409(ii)								
		890(iii)	946(iii)								
O	463	360(i)	157	146(i)							
		743(ii)		497(ii)							
F	565	484	270	185	155						
Cl	431	338	200	203	254	242					
Br	366	276				219	193				
I	299	238				210	178	151			
S	338	259			496	250	212		264		
P	322									201	
Si	318		374	466							226

* Mean bond enthalpies are such a crude measure of bond strength that they need not be distinguished from dissociation energies.
(i) Single bond, (ii) double bond, (iii) triple bond, (a) aromatic.
Data: HCP and L. Pauling, *The nature of the chemical bond*. Cornell University Press (1960).

Table 10.4 Pauling (*italics*) and Mulliken electronegativities

H							He
2.20							
3.06							
Li	Be	B	C	N	O	F	Ne
0.98	*1.57*	*2.04*	*2.55*	*3.04*	*3.44*	*3.98*	
1.28	1.99	1.83	2.67	3.08	3.22	4.43	4.60
Na	Mg	Al	Si	P	S	Cl	Ar
0.93	*1.31*	*1.61*	*1.90*	*2.19*	*2.58*	*3.16*	
1.21	1.63	1.37	2.03	2.39	2.65	3.54	3.36
K	Ca	Ga	Ge	As	Se	Br	Kr
0.82	*1.00*	*1.81*	*2.01*	*2.18*	*2.55*	*2.96*	*3.0*
1.03	1.30	1.34	1.95	2.26	2.51	3.24	2.98
Rb	Sr	In	Sn	Sb	Te	I	Xe
0.82	*0.95*	*1.78*	*1.96*	*2.05*	*2.10*	*2.66*	*2.6*
0.99	1.21	1.30	1.83	2.06	2.34	2.88	2.59
Cs	Ba	Tl	Pb	Bi			
0.79	*0.89*	*2.04*	*2.33*	*2.02*			

Data: Pauling values: A.L. Allred, *J. Inorg. Nucl. Chem.* **17**, 215 (1961); L.C. Allen and J.E. Huheey, *ibid.*, **42**, 1523 (1980). Mulliken values: L.C. Allen, *J. Am. Chem. Soc.* **111**, 9003 (1989). The Mulliken values have been scaled to the range of the Pauling values.

Table 12.2 Properties of diatomic molecules

	$\tilde{\nu}/\mathrm{cm}^{-1}$	θ_V/K	$\tilde{B}/\mathrm{cm}^{-1}$	θ_R/K	R_e/pm	$k_f/(\mathrm{N\,m^{-1}})$	$D_0/(\mathrm{kJ\,mol^{-1}})$	σ
$^{1}H_2^+$	2321.8	3341	29.8	42.9	106	160	255.8	2
$^{1}H_2$	4400.39	6332	60.864	87.6	74.138	574.9	432.1	2
$^{2}H_2$	3118.46	4487	30.442	43.8	74.154	577.0	439.6	2
$^{1}H^{19}F$	4138.32	5955	20.956	30.2	91.680	965.7	564.4	1
$^{1}H^{35}Cl$	2990.95	4304	10.593	15.2	127.45	516.3	427.7	1
$^{1}H^{81}Br$	2648.98	3812	8.465	12.2	141.44	411.5	362.7	1
$^{1}H^{127}I$	2308.09	3321	6.511	9.37	160.92	313.8	294.9	1
$^{14}N_2$	2358.07	3393	1.9987	2.88	109.76	2293.8	941.7	2
$^{16}O_2$	1580.36	2274	1.4457	2.08	120.75	1176.8	493.5	2
$^{19}F_2$	891.8	1283	0.8828	1.27	141.78	445.1	154.4	2
$^{35}Cl_2$	559.71	805	0.2441	0.351	198.75	322.7	239.3	2
$^{12}C^{16}O$	2170.21	3122	1.9313	2.78	112.81	1903.17	1071.8	1
$^{79}Br^{81}Br$	323.2	465	0.0809	10.116	283.3	245.9	190.2	1

Data: AIP.

Table 12.3 Typical vibrational wavenumbers, $\tilde{\nu}/\text{cm}^{-1}$

C—H stretch	2850–2960
C—H bend	1340–1465
C—C stretch, bend	700–1250
C=C stretch	1620–1680
C≡C stretch	2100–2260
O—H stretch	3590–3650
H-bonds	3200–3570
C=O stretch	1640–1780
C≡N stretch	2215–2275
N—H stretch	3200–3500
C—F stretch	1000–1400
C—Cl stretch	600–800
C—Br stretch	500–600
C—I stretch	500
CO_3^{2-}	1410–1450
NO_3^-	1350–1420
NO_2^-	1230–1250
SO_4^{2-}	1080–1130
Silicates	900–1100

Data: L.J. Bellamy, *The infrared spectra of complex molecules* and *Advances in infrared group frequencies*. Chapman and Hall.

Table 13.1 Colour, frequency, and energy of light

Colour	λ/nm	$\nu/(10^{14}\text{ Hz})$	$\tilde{\nu}/(10^4\text{ cm}^{-1})$	E/eV	$E/(\text{kJ mol}^{-1})$
Infrared	>1000	<3.00	<1.00	<1.24	<120
Red	700	4.28	1.43	1.77	171
Orange	620	4.84	1.61	2.00	193
Yellow	580	5.17	1.72	2.14	206
Green	530	5.66	1.89	2.34	226
Blue	470	6.38	2.13	2.64	254
Violet	420	7.14	2.38	2.95	285
Near ultraviolet	300	10.0	3.33	4.15	400
Far ultraviolet	<200	>15.0	>5.00	>6.20	>598

Data: J.G. Calvert and J.N. Pitts, *Photochemistry*. Wiley, New York (1966).

Table 13.2 Absorption characteristics of some groups and molecules

Group	$\tilde{\nu}_{max}/(10^4\text{ cm}^{-1})$	λ_{max}/nm	$\varepsilon_{max}/(\text{dm}^3\text{ mol}^{-1}\text{ cm}^{-1})$
C=C ($\pi^* \leftarrow \pi$)	6.10	163	1.5×10^4
	5.73	174	5.5×10^3
C=O ($\pi^* \leftarrow n$)	3.7–3.5	270–290	10–20
—N=N—	2.9	350	15
	>3.9	<260	Strong
$—NO_2$	3.6	280	10
	4.8	210	1.0×10^4
$C_6H_5—$	3.9	255	200
	5.0	200	6.3×10^3
	5.5	180	1.0×10^5
$[Cu(OH_2)_6]^{2+}(aq)$	1.2	810	10
$[Cu(NH_3)_4]^{2+}(aq)$	1.7	600	50
H_2O ($\pi^* \leftarrow n$)	6.0	167	7.0×10^3

Table 14.2 Nuclear spin properties

Nuclide	Natural abundance %	Spin I	Magnetic moment μ/μ_N	g-value	$\gamma/(10^7\ T^{-1}\ s^{-1})$	NMR frequency at 1 T, ν/MHz
^1n*		$\frac{1}{2}$	−1.9130	−3.8260	−18.324	29.164
^1H	99.9844	$\frac{1}{2}$	2.792 85	5.5857	26.752	42.576
^2H	0.0156	1	0.857 44	0.857 44	4.1067	6.536
^3H*		$\frac{1}{2}$	2.978 96	−4.2553	−20.380	45.414
^{10}B	19.6	3	1.8006	0.6002	2.875	4.575
^{11}B	80.4	$\frac{3}{2}$	2.6886	1.7923	8.5841	13.663
^{13}C	1.108	$\frac{1}{2}$	0.7024	1.4046	6.7272	10.708
^{14}N	99.635	1	0.403 56	0.403 56	1.9328	3.078
^{17}O	0.037	$\frac{5}{2}$	−1.893 79	−0.7572	−3.627	5.774
^{19}F	100	$\frac{1}{2}$	2.628 87	5.2567	25.177	40.077
^{31}P	100	$\frac{1}{2}$	1.1316	2.2634	10.840	17.251
^{33}S	0.74	$\frac{3}{2}$	0.6438	0.4289	2.054	3.272
^{35}Cl	75.4	$\frac{3}{2}$	0.8219	0.5479	2.624	4.176
^{37}Cl	24.6	$\frac{3}{2}$	0.6841	0.4561	2.184	3.476

* Radioactive.
μ is the magnetic moment of the spin state with the largest value of m_I: $\mu = g_I \mu_N I$ and μ_N is the nuclear magneton (see inside front cover).
Data: KL and HCP.

Table 14.3 Hyperfine coupling constants for atoms, a/mT

Nuclide	Spin	Isotropic coupling	Anisotropic coupling
^1H	$\frac{1}{2}$	50.8(1s)	
^2H	1	7.8(1s)	
^{13}C	$\frac{1}{2}$	113.0(2s)	6.6(2p)
^{14}N	1	55.2(2s)	4.8(2p)
^{19}F	$\frac{1}{2}$	1720(2s)	108.4(2p)
^{31}P	$\frac{1}{2}$	364(3s)	20.6(3p)
^{35}Cl	$\frac{3}{2}$	168(3s)	10.0(3p)
^{37}Cl	$\frac{3}{2}$	140(3s)	8.4(3p)

Data: P.W. Atkins and M.C.R. Symons, *The structure of inorganic radicals*. Elsevier, Amsterdam (1967).

Table 16.1 Rotational and vibrational temperatures: see Table 12.2

Table 16.2 Symmetry numbers: see Table 12.2

Table 17.1 Dipole moments (μ), polarizabilities (α), and polarizability volumes (α')

	$\mu/(10^{-30}$ C m)	μ/D	$\alpha'/(10^{-30}$ m^3)	$\alpha/(10^{-40}$ J^{-1} C^2 m^2)
Ar	0	0	1.66	1.85
C_2H_5OH	5.64	1.69		
$C_6H_5CH_3$	1.20	0.36		
C_6H_6	0	0	10.4	11.6
CCl_4	0	0	10.3	11.7
CH_2Cl_2	5.24	1.57	6.80	7.57
CH_3Cl	6.24	1.87	4.53	5.04
CH_3OH	5.70	1.71	3.23	3.59
CH_4	0	0	2.60	2.89
$CHCl_3$	3.37	1.01	8.50	9.46
CO	0.390	0.117	1.98	2.20
CO_2	0	0	2.63	2.93
H_2	0	0	0.819	0.911
H_2O	6.17	1.85	1.48	1.65
HBr	2.67	0.80	3.61	4.01
HCl	3.60	1.08	2.63	2.93
He	0	0	0.20	0.22
HF	6.37	1.91	0.51	0.57
HI	1.40	0.42	5.45	6.06
N_2	0	0	1.77	1.97
NH_3	4.90	1.47	2.22	2.47
1,2-$C_6H_4(CH_3)_2$	2.07	0.62		

Data: HCP and C.J.F. Böttcher and P. Bordewijk, *Theory of electric polarization*. Elsevier, Amsterdam (1978).

Table 17.4 Lennard-Jones (12,6) parameters

	(ε/k)/K	r_0/pm
Ar	111.84	362.3
C_2H_2	209.11	463.5
C_2H_4	200.78	458.9
C_2H_6	216.12	478.2
C_6H_6	377.46	617.4
CCl_4	378.86	624.1
Cl_2	296.27	448.5
CO_2	201.71	444.4
F_2	104.29	357.1
Kr	154.87	389.5
N_2	91.85	391.9
O_2	113.27	365.4
Xe	213.96	426.0

Source: F. Cuadros, I. Cachadiña, and W. Ahamuda, *Molec. Engineering*, **6**, 319 (1996).

Table 17.5 Surface tensions of liquids at 293 K

	$\gamma/(\mathrm{mN\ m^{-1}})$
Benzene	28.88
Carbon tetrachloride	27.0
Ethanol	22.8
Hexane	18.4
Mercury	472
Methanol	22.6
Water	72.75
	72.0 at 25°C
	58.0 at 100°C

Data: KL.

Table 18.2 Radius of gyration of some macromolecules

	$M/(\mathrm{kg\ mol^{-1}})$	R_g/nm
Serum albumin	66	2.98
Myosin	493	46.8
Polystyrene	3.2×10^3	50 (in poor solvent)
DNA	4×10^3	117.0
Tobacco mosaic virus	3.9×10^4	92.4

Data: C. Tanford, *Physical chemistry of macromolecules*. Wiley, New York (1961).

Table 18.3 Frictional coefficients and molecular geometry

Major axis/Minor axis	Prolate	Oblate
2	1.04	1.04
3	1.11	1.10
4	1.18	1.17
5	1.25	1.22
6	1.31	1.28
7	1.38	1.33
8	1.43	1.37
9	1.49	1.42
10	1.54	1.46
50	2.95	2.38
100	4.07	2.97

Data: K.E. Van Holde, *Physical biochemistry*. Prentice-Hall, Englewood Cliffs (1971).

Sphere; radius a, $c = af_0$

Prolate ellipsoid; major axis $2a$, minor axis $2b$, $c = (ab^2)^{1/3}$

$$f = \left\{ \frac{(1-b^2/a^2)^{1/2}}{(b/a)^{2/3}\ln\{[1+(1-b^2/a^2)^{1/2}]/(b/a)\}} \right\} f_0$$

Oblate ellipsoid; major axis $2a$, minor axis $2b$, $c = (a^2b)^{1/3}$

$$f = \left\{ \frac{(a^2/b^2-1)^{1/2}}{(a/b)^{2/3}\arctan[(a^2/b^2-1)^{1/2}]} \right\} f_0$$

Long rod; length l, radius a, $c = (3a^2/4)^{1/3}$

$$f = \left\{ \frac{(1/2a)^{2/3}}{(3/2)^{1/3}\{2\ln(l/a)-0.11\}} \right\} f_0$$

In each case $f_0 = 6\pi\eta c$ with the appropriate value of c.

Table 18.4 Intrinsic viscosity

Macromolecule	Solvent	$\theta/°\mathrm{C}$	$K/(10^{-3}\ \mathrm{cm^3\ g^{-1}})$	a
Polystyrene	Benzene	25	9.5	0.74
	Cyclohexane	34†	81	0.50
Polyisobutylene	Benzene	23†	83	0.50
	Cyclohexane	30	26	0.70
Amylose	0.33 M KCl(aq)	25†	113	0.50
Various proteins‡	Guanidine hydrochloride + $HSCH_2CH_2OH$		7.16	0.66

† The θ temperature.

‡ Use $[\eta] = KN^a$; N is the number of amino acid residues.

Data: K.E. Van Holde, *Physical biochemistry*. Prentice-Hall, Englewood Cliffs (1971).

Table 19.3 Ionic radii (r/pm)†

Li^+(4)	Be^{2+}(4)	B^{3+}(4)	N^{3-}	O^{2-}(6)	F^-(6)		
59	27	12	171	140	133		
Na^+(6)	Mg^{2+}(6)	Al^{3+}(6)	P^{3-}	S^{2-}(6)	Cl^-(6)		
102	72	53	212	184	181		
K^+(6)	Ca^{2+}(6)	Ga^{3+}(6)	As^{3-}(6)	Se^{2-}(6)	Br^-(6)		
138	100	62	222	198	196		
Rb^+(6)	Sr^{2+}(6)	In^{3+}(6)		Te^{2-}(6)	I^-(6)		
149	116	79		221	220		
Cs^+(6)	Ba^{2+}(6)	Tl^{3+}(6)					
167	136	88					
d-block elements (high-spin ions)							
Sc^{3+}(6)	Ti^{4+}(6)	Cr^{3+}(6)	Mn^{3+}(6)	Fe^{2+}(6)	Co^{3+}(6)	Cu^{2+}(6)	Zn^{2+}(6)
73	60	61	65	63	61	73	75

† Numbers in parentheses are the coordination numbers of the ions. Values for ions without a coordination number stated are estimates.
Data: R.D. Shannon and C.T. Prewitt, *Acta Cryst.* **B25**, 925 (1969).

Table 19.5 Lattice enthalpies at 298 K: see Table 2.5

Table 19.6 Magnetic susceptibilities at 298 K

	$\chi/10^{-6}$	$\chi_m/(10^{-10}\,m^3\,mol^{-1})$
H_2O(l)	−9.02	−1.63
C_6H_6(l)	−8.8	−7.8
C_6H_{12}(l)	−10.2	−11.1
CCl_4(l)	−5.4	−5.2
NaCl(s)	−16	−3.8
Cu(s)	−9.7	−0.69
S(rhombic)	−12.6	−1.95
Hg(l)	−28.4	−4.21
Al(s)	+20.7	+2.07
Pt(s)	+267.3	+24.25
Na(s)	+8.48	+2.01
K(s)	+5.94	+2.61
$CuSO_4\cdot 5H_2O$(s)	+167	+183
$MnSO_4\cdot 4H_2O$(s)	+1859	+1835
$NiSO_4\cdot 7H_2O$(s)	+355	+503
$FeSO_4$(s)	+3743	+1558

Source: Principally HCP, with $\chi_m = \chi V_m = \chi\rho/M$.

Table 20.1 Collision cross-sections, σ/nm^2

Ar	0.36
C_2H_4	0.64
C_6H_6	0.88
CH_4	0.46
Cl_2	0.93
CO_2	0.52
H_2	0.27
He	0.21
N_2	0.43
Ne	0.24
O_2	0.40
SO_2	0.58

Data: KL.

Table 20.2 Transport properties of gases at 1 atm

	$\kappa/(\mathrm{J\,K^{-1}\,m^{-1}\,s^{-1}})$	$\eta/\mu\mathrm{P}$	
	273 K	273 K	293 K
Air	0.0241	173	182
Ar	0.0163	210	223
C_2H_4	0.0164	97	103
CH_4	0.0302	103	110
Cl_2	0.079	123	132
CO_2	0.0145	136	147
H_2	0.1682	84	88
He	0.1442	187	196
Kr	0.0087	234	250
N_2	0.0240	166	176
Ne	0.0465	298	313
O_2	0.0245	195	204
Xe	0.0052	212	228

Data: KL.

Table 20.4 Viscosities of liquids at 298 K, $\eta/(10^{-3}\,\mathrm{kg\,m^{-1}\,s^{-1}})$

Benzene	0.601
Carbon tetrachloride	0.880
Ethanol	1.06
Mercury	1.55
Methanol	0.553
Pentane	0.224
Sulfuric acid	27
Water†	0.891

† The viscosity of water over its entire liquid range is represented with less than 1 per cent error by the expression

$$\log(\eta_{20}/\eta) = A/B,$$
$$A = 1.370\,23(t-20) + 8.36\times10^{-4}(t-20)^2$$
$$B = 109 + t \qquad t = \theta/{}^\circ\mathrm{C}$$

Convert $\mathrm{kg\,m^{-1}\,s^{-1}}$ to centipoise (cP) by multiplying by 10^3 (so $\eta \approx 1$ cP for water).
Data: AIP, KL.

Table 20.5 Ionic mobilities in water at 298 K, $u/(10^{-8}\,\mathrm{m^2\,s^{-1}\,V^{-1}})$

Cations		Anions	
Ag^+	6.24	Br^-	8.09
Ca^{2+}	6.17	$CH_3CO_2^-$	4.24
Cu^{2+}	5.56	Cl^-	7.91
H^+	36.23	CO_3^{2-}	7.46
K^+	7.62	F^-	5.70
Li^+	4.01	$[Fe(CN)_6]^{3-}$	10.5
Na^+	5.19	$[Fe(CN)_6]^{4-}$	11.4
NH_4^+	7.63	I^-	7.96
$[N(CH_3)_4]^+$	4.65	NO_3^-	7.40
Rb^+	7.92	OH^-	20.64
Zn^{2+}	5.47	SO_4^{2-}	8.29

Data: Principally Table 20.4 and $u = \lambda/zF$.

Table 20.6 Debye–Hückel–Onsager coefficients for (1,1)-electrolytes at 298 K

Solvent	$A/(\mathrm{mS\,m^2\,mol^{-1}}/(\mathrm{mol\,dm^{-3}})^{1/2})$	$B/(\mathrm{mol\,dm^{-3}})^{-1/2}$
Acetone (propanone)	3.28	1.63
Acetonitrile	2.29	0.716
Ethanol	8.97	1.83
Methanol	15.61	0.923
Nitrobenzene	4.42	0.776
Nitromethane	111	0.708
Water	6.020	0.229

Data: J.O'M. Bockris and A.K.N. Reddy, *Modern electrochemistry*. Plenum, New York (1970).

Table 20.7 Diffusion coefficients at 298 K, $D/(10^{-9}\ m^2\ s^{-1})$

Molecules in liquids				Ions in water			
I_2 in hexane	4.05	H_2 in $CCl_4(l)$	9.75	K^+	1.96	Br^-	2.08
in benzene	2.13	N_2 in $CCl_4(l)$	3.42	H^+	9.31	Cl^-	2.03
CCl_4 in heptane	3.17	O_2 in $CCl_4(l)$	3.82	Li^+	1.03	F^-	1.46
Glycine in water	1.055	Ar in $CCl_4(l)$	3.63	Na^+	1.33	I^-	2.05
Dextrose in water	0.673	CH_4 in $CCl_4(l)$	2.89			OH^-	5.03
Sucrose in water	0.5216	H_2O in water	2.26				
		CH_3OH in water	1.58				
		C_2H_5OH in water	1.24				

Data: AIP.

Table 21.1 Kinetic data for first-order reactions

	Phase	$\theta/°C$	k_r/s^{-1}	$t_{1/2}$
$2\,N_2O_5 \rightarrow 4\,NO_2 + O_2$	g	25	3.38×10^{-5}	5.70 h
	$HNO_3(l)$	25	1.47×10^{-6}	131 h
	$Br_2(l)$	25	4.27×10^{-5}	4.51 h
$C_2H_6 \rightarrow 2\,CH_3$	g	700	5.36×10^{-4}	21.6 min
Cyclopropane $\rightarrow$ propene	g	500	6.71×10^{-4}	17.2 min
$CH_3N_2CH_3 \rightarrow C_2H_6 + N_2$	g	327	3.4×10^{-4}	34 min
Sucrose $\rightarrow$ glucose + fructose	$aq(H^+)$	25	6.0×10^{-5}	3.2 h

g: High pressure gas-phase limit.
Data: Principally K.J. Laidler, *Chemical kinetics*. Harper & Row, New York (1987); M.J. Pilling and P.W. Seakins, *Reaction kinetics*. Oxford University Press (1995); J. Nicholas, *Chemical kinetics*. Harper & Row, New York (1976). See also JL.

Table 21.2 Kinetic data for second-order reactions

	Phase	$\theta/°C$	$k/(dm^3\ mol^{-1}\ s^{-1})$
$2\,NOBr \rightarrow 2\,NO + Br_2$	g	10	0.80
$2\,NO_2 \rightarrow 2\,NO + O_2$	g	300	0.54
$H_2 + I_2 \rightarrow 2\,HI$	g	400	2.42×10^{-2}
$D_2 + HCl \rightarrow DH + DCl$	g	600	0.141
$2\,I \rightarrow I_2$	g	23	7×10^9
	hexane	50	1.8×10^{10}
$CH_3Cl + CH_3O^-$	methanol	20	2.29×10^{-6}
$CH_3Br + CH_3O^-$	methanol	20	9.23×10^{-6}
$H^+ + OH^- \rightarrow H_2O$	water	25	1.35×10^{11}
	ice	−10	8.6×10^{12}

Data: Principally K.J. Laidler, *Chemical kinetics*. Harper & Row, New York (1987); M.J. Pilling and P.W. Seakins, *Reaction kinetics*. Oxford University Press (1995); J. Nicholas, *Chemical kinetics*. Harper & Row, New York (1976).

Table 21.4 Arrhenius parameters

First-order reactions	A/s^{-1}	$E_a/(kJ\ mol^{-1})$
Cyclopropane → propene	1.58×10^{15}	272
$CH_3NC \rightarrow CH_3CN$	3.98×10^{13}	160
cis-CHD=CHD → *trans*-CHD=CHD	3.16×10^{12}	256
Cyclobutane → $2\ C_2H_4$	3.98×10^{13}	261
$C_2H_5I \rightarrow C_2H_4 + HI$	2.51×10^{17}	209
$C_2H_6 \rightarrow 2\ CH_3$	2.51×10^{7}	384
$2\ N_2O_5 \rightarrow 4\ NO_2 + O_2$	4.94×10^{13}	103
$N_2O \rightarrow N_2 + O$	7.94×10^{11}	250
$C_2H_5 \rightarrow C_2H_4 + H$	1.0×10^{13}	167
Second-order, gas-phase	$A/(dm^3\ mol^{-1}\ s^{-1})$	$E_a/(kJ\ mol^{-1})$
$O + N_2 \rightarrow NO + N$	1×10^{11}	315
$OH + H_2 \rightarrow H_2O + H$	8×10^{10}	42
$Cl + H_2 \rightarrow HCl + H$	8×10^{10}	23
$2\ CH_3 \rightarrow C_2H_6$	2×10^{10}	ca. 0
$NO + Cl_2 \rightarrow NOCl + Cl$	4.0×10^{9}	85
$SO + O_2 \rightarrow SO_2 + O$	3×10^{8}	27
$CH_3 + C_2H_6 \rightarrow CH_4 + C_2H_5$	2×10^{8}	44
$C_6H_5 + H_2 \rightarrow C_6H_6 + H$	1×10^{8}	ca. 25
Second-order, solution	$A/(dm^3\ mol^{-1}\ s^{-1})$	$E_a/(kJ\ mol^{-1})$
$C_2H_5ONa + CH_3I$ in ethanol	2.42×10^{11}	81.6
$C_2H_5Br + OH^-$ in water	4.30×10^{11}	89.5
$C_2H_5I + C_2H_5O^-$ in ethanol	1.49×10^{11}	86.6
$CH_3I + C_2H_5O^-$ in ethanol	2.42×10^{11}	81.6
$C_2H_5Br + OH^-$ in ethanol	4.30×10^{11}	89.5
$CO_2 + OH^-$ in water	1.5×10^{10}	38
$CH_3I + S_2O_3^{2-}$ in water	2.19×10^{12}	78.7
Sucrose + H_2O in acidic water	1.50×10^{15}	107.9
$(CH_3)_3CCl$ solvolysis		
in water	7.1×10^{16}	100
in methanol	2.3×10^{13}	107
in ethanol	3.0×10^{13}	112
in acetic acid	4.3×10^{13}	111
in chloroform	1.4×10^{4}	45
$C_6H_5NH_2 + C_6H_5COCH_2Br$		
in benzene	91	34

Data: Principally J. Nicholas, *Chemical kinetics*. Harper & Row, New York (1976) and A.A. Frost and R.G. Pearson, *Kinetics and mechanism*. Wiley, New York (1961).

Table 22.1 Arrhenius parameters for gas-phase reactions

	$A/(\mathrm{dm^3\ mol^{-1}\ s^{-1}})$		$E_a/(\mathrm{kJ\ mol^{-1}})$	P
	Experiment	Theory		
$2\,NOCl \rightarrow 2\,NO + Cl_2$	9.4×10^{9}	5.9×10^{10}	102.0	0.16
$2\,NO_2 \rightarrow 2\,NO + O_2$	2.0×10^{9}	4.0×10^{10}	111.0	5.0×10^{-2}
$2\,ClO \rightarrow Cl_2 + O_2$	6.3×10^{7}	2.5×10^{10}	0.0	2.5×10^{-3}
$H_2 + C_2H_4 \rightarrow C_2H_6$	1.24×10^{6}	7.4×10^{11}	180	1.7×10^{-6}
$K + Br_2 \rightarrow KBr + Br$	1.0×10^{12}	2.1×10^{11}	0.0	4.8

Data: Principally M.J. Pilling and P.W. Seakins, *Reaction kinetics*. Oxford University Press (1995).

Table 22.2 Arrhenius parameters for reactions in solution: see Table 21.4

Table 22.3 Exchange current densities and transfer coefficients at 298 K

Reaction	Electrode	$j_0/(\mathrm{A\ cm^{-2}})$	α
$2\,H^+ + 2\,e^- \rightarrow H_2$	Pt	7.9×10^{-4}	
	Cu	1×10^{-6}	
	Ni	6.3×10^{-6}	0.58
	Hg	7.9×10^{-13}	0.50
	Pb	5.0×10^{-12}	
$Fe^{3+} + e^- \rightarrow Fe^{2+}$	Pt	2.5×10^{-3}	0.58
$Ce^{4+} + e^- \rightarrow Ce^{3+}$	Pt	4.0×10^{-5}	0.75

Data: Principally J.O'M. Bockris and A.K.N. Reddy, *Modern electrochemistry*. Plenum, New York (1970).

Table 23.1 Maximum observed enthalpies of physisorption, $\Delta_{ad}H^{\ominus}/(\mathrm{kJ\ mol^{-1}})$

C_2H_2	−38	H_2	−84
C_2H_4	−34	H_2O	−59
CH_4	−21	N_2	−21
Cl_2	−36	NH_3	−38
CO	−25	O_2	−21
CO_2	−25		

Data: D.O. Haywood and B.M.W. Trapnell, *Chemisorption*. Butterworth (1964).

Table 23.2 Enthalpies of chemisorption, $\Delta_{ad}H^{\ominus}/(\text{kJ mol}^{-1})$

Adsorbate	Adsorbent (substrate)											
	Ti	Ta	Nb	W	Cr	Mo	Mn	Fe	Co	Ni	Rh	Pt
H_2		−188			−188	−167	−71	−134			−117	
N_2		−586						−293				
O_2						−720					−494	−293
CO	−640							−192	−176			
CO_2	−682	−703	−552	−456	−339	−372	−222	−225	−146	−184		
NH_3				−301				−188		−155		
C_2H_4		−577		−427	−427			−285		−243	−209	

Data: D.O. Haywood and B.M.W. Trapnell, *Chemisorption*. Butterworth (1964).

Part 3 Character tables

The groups C_1, C_s, C_i

C_1 (1)	E	$h=1$
A	1	

$C_s = C_h$ (m)	E	σ_h	$h=2$	
A′	1	1	x, y, R_z	x^2, y^2, z^2, xy
A″	1	−1	z, R_x, R_y	yz, xz

$C_i = S_2$ ($\bar{1}$)	E	i	$h=2$	
A_g	1	1	R_x, R_y, R_z	$x^2, y^2, z^2, xy, xz, yz$
A_u	1	−1	x, y, z	

The groups C_{nv}

C_{2v}, $2mm$	E	C_2	σ_v	σ'_v	$h=4$	
A_1	1	1	1	1	z, z^2, x^2, y^2	
A_2	1	1	−1	−1	xy	R_z
B_1	1	−1	1	−1	x, xz	R_y
B_2	1	−1	−1	1	y, yz	R_x

C_{3v}, $3m$	E	$2C_3$	$3\sigma_v$	$h=6$	
A_1	1	1	1	z, z^2, x^2+y^2	
A_2	1	1	−1		R_z
E	2	−1	0	$(x, y), (xy, x^2-y^2)$ (xz, yz)	(R_x, R_y)

C_{4v}, $4mm$	E	C_2	$2C_4$	$2\sigma_v$	$2\sigma_d$	$h=8$	
A_1	1	1	1	1	1	z, z^2, x^2+y^2	
A_2	1	1	1	−1	−1		R_z
B_1	1	1	−1	1	−1	x^2-y^2	
B_2	1	1	−1	−1	1	xy	
E	2	−2	0	0	0	$(x, y), (xz, yz)$	(R_x, R_y)

C_{5v}	E	$2C_5$	$2C_5^2$	$5\sigma_v$	$h=10, \alpha=72°$	
A_1	1	1	1	1	z, z^2, x^2+y^2	
A_2	1	1	1	−1		R_z
E_1	2	$2\cos\alpha$	$2\cos 2\alpha$	0	$(x, y), (xz, yz)$	(R_x, R_y)
E_2	2	$2\cos 2\alpha$	$2\cos\alpha$	0	(xy, x^2-y^2)	

C_{6v}, $6mm$	E	C_2	$2C_3$	$2C_6$	$3\sigma_d$	$3\sigma_v$	$h=12$	
A_1	1	1	1	1	1	1	z, z^2, x^2+y^2	
A_2	1	1	1	1	−1	−1		R_z
B_1	1	−1	1	−1	−1	1		
B_2	1	−1	1	−1	1	−1		
E_1	2	−2	−1	1	0	0	$(x, y), (xz, yz)$	(R_x, R_y)
E_2	2	2	−1	−1	0	0	(xy, x^2-y^2)	

$C_{\infty v}$	E	$2C_\phi$†	$\infty\sigma_v$	$h=\infty$	
$A_1(\Sigma^+)$	1	1	1	z, z^2, x^2+y^2	
$A_2(\Sigma^-)$	1	1	−1		R_z
$E_1(\Pi)$	2	$2\cos\phi$	0	$(x, y), (xz, yz)$	(R_x, R_y)
$E_2(\Delta)$	2	$2\cos 2\phi$	0	(xy, x^2-y^2)	

† There is only one member of this class if $\phi=\pi$.

The groups D_n

D_2, 222	E	C_2^z	C_2^y	C_2^x	$h=4$	
A_1	1	1	1	1	x^2, y^2, z^2	
B_1	1	1	−1	−1	z, xy	R_z
B_2	1	−1	1	−1	y, xz	R_y
B_3	1	−1	−1	1	x, yz	R_x

D_3, 32	E	$2C_3$	$3C_2'$	$h=6$	
A_1	1	1	1	z^2, x^2+y^2	
A_2	1	1	−1	z	R_z
E	2	−1	0	$(x, y), (xz, yz), (xy, x^2-y^2)$	(R_x, R_y)

D_4, 422	E	C_2	$2C_4$	$2C_2'$	$2C_2''$	$h=8$	
A_1	1	1	1	1	1	z^2, x^2+y^2	
A_2	1	1	1	−1	−1	z	R_z
B_1	1	1	−1	1	−1	x^2-y^2	
B_2	1	1	−1	−1	1	xy	
E	2	−2	0	0	0	$(x, y), (xz, yz)$	(R_x, R_y)

The groups D_{nh}

D_{3h}, $\bar{6}2m$	E	σ_h	$2C_3$	$2S_3$	$3C_2'$	$3\sigma_v$	$h=12$	
A_1'	1	1	1	1	1	1	z^2, x^2+y^2	
A_2'	1	1	1	1	−1	−1		R_z
A_1''	1	−1	1	−1	1	−1		
A_2''	1	−1	1	−1	−1	1	z	
E′	2	2	−1	−1	0	0	$(x, y), (xy, x^2-y^2)$	
E″	2	−2	−1	1	0	0	(xz, yz)	(R_x, R_y)

D_{4h}, 4/*mmm*	E	$2C_4$	C_2	$2C_2'$	$2C_2''$	i	$2S_4$	σ_h	$2\sigma_v$	$2\sigma_d$	$h=16$	
A_{1g}	1	1	1	1	1	1	1	1	1	1	x^2+y^2, z^2	
A_{2g}	1	1	1	−1	−1	1	1	1	−1	−1		R_z
B_{1g}	1	−1	1	1	−1	1	−1	1	1	−1	x^2-y^2	
B_{2g}	1	−1	1	−1	1	1	−1	1	−1	1	xy	
E_g	2	0	−2	0	0	2	0	−2	0	0	(xz, yz)	(R_x, R_y)
A_{1u}	1	1	1	1	1	−1	−1	−1	−1	−1		
A_{2u}	1	1	1	−1	−1	−1	−1	−1	1	1	z	
B_{1u}	1	−1	1	1	−1	−1	1	−1	−1	1		
B_{2u}	1	−1	1	−1	1	−1	1	−1	1	−1		
E_u	2	0	−2	0	0	−2	0	2	0	0	(x, y)	

D_{5h}	E	$2C_5$	$2C_5^2$	$5C_2$	σ_h	$2S_5$	$2S_5^3$	$5\sigma_v$	$h=20$	$\alpha=72°$
A_1'	1	1	1	1	1	1	1	1	x^2+y^2, z^2	
A_2'	1	1	1	−1	1	1	1	−1		R_z
E_1'	2	$2\cos\alpha$	$2\cos 2\alpha$	0	2	$2\cos\alpha$	$2\cos 2\alpha$	0	(x, y)	
E_2'	2	$2\cos 2\alpha$	$2\cos\alpha$	0	2	$2\cos 2\alpha$	$2\cos\alpha$	0	(x^2-y^2, xy)	
A_1''	1	1	1	1	−1	−1	−1	−1		
A_2''	1	1	1	−1	−1	−1	−1	1	z	
E_1''	2	$2\cos\alpha$	$2\cos 2\alpha$	0	−2	$-2\cos\alpha$	$-2\cos 2\alpha$	0	(xz, yz)	(R_x, R_y)
E_2''	2	$2\cos 2\alpha$	$2\cos\alpha$	0	−2	$-2\cos 2\alpha$	$-2\cos\alpha$	0		

$D_{\infty h}$	E	$2C_\phi$	...	$\infty\sigma_v$	i	$2S_\infty$	...	$\infty C_2'$	$h=\infty$	
$A_{1g}(\Sigma_g^+)$	1	1	...	1	1	1	...	1	z^2, x^2+y^2	
$A_{1u}(\Sigma_u^+)$	1	1	...	1	−1	−1	...	−1	z	
$A_{2g}(\Sigma_g^-)$	1	1	...	−1	1	1	...	−1		R_z
$A_{2u}(\Sigma_u^-)$	1	1	...	−1	−1	−1	...	1		
$E_{1g}(\Pi_g)$	2	$2\cos\phi$	...	0	2	$-2\cos\phi$	...	0	(xz, yz)	(R_x, R_y)
$E_{1u}(\Pi_u)$	2	$2\cos\phi$	...	0	−2	$2\cos\phi$	...	0	(x, y)	
$E_{2g}(\Delta_g)$	2	$2\cos 2\phi$	...	0	2	$2\cos 2\phi$	...	0	(xy, x^2-y^2)	
$E_{2u}(\Delta_u)$	2	$2\cos 2\phi$	...	0	−2	$-2\cos 2\phi$	...	0		
⋮	⋮	⋮		⋮	⋮	⋮		⋮		

The cubic groups

T_d, $\bar{4}3m$	E	$8C_3$	$3C_2$	$6\sigma_d$	$6S_4$	$h=24$	
A_1	1	1	1	1	1	$x^2+y^2+z^2$	
A_2	1	1	1	−1	−1		
E	2	−1	2	0	0	$(3z^2-r^2, x^2-y^2)$	
T_1	3	0	−1	−1	1		(R_x, R_y, R_z)
T_2	3	0	−1	1	−1	$(x, y, z), (xy, xz, yz)$	

O_h ($m3m$)	E	$8C_3$	$6C_2$	$6C_4$	$3C_2$ $(=C_4^2)$	i	$6S_4$	$8S_6$	$3\sigma_h$	$6\sigma_d$	$h=48$	
A_{1g}	1	1	1	1	1	1	1	1	1	1	$x^2+y^2+z^2$	
A_{2g}	1	1	−1	−1	1	1	−1	1	1	−1		
E_g	2	−1	0	0	2	2	0	−1	2	0	$(2z^2-x^2-y^2, x^2-y^2)$	
T_{1g}	3	0	−1	1	−1	3	1	0	−1	−1		(R_x, R_y, R_z)
T_{2g}	3	0	1	−1	−1	3	−1	0	−1	1	(xy, yz, zx)	
A_{1u}	1	1	1	1	1	−1	−1	−1	−1	−1		
A_{2u}	1	1	−1	−1	1	−1	1	−1	−1	1		
E_u	2	−1	0	0	2	−2	0	1	−2	0		
T_{1u}	3	0	−1	1	−1	−3	−1	0	1	1	(x, y, z)	
T_{2u}	3	0	1	−1	−1	−3	1	0	1	−1		

The icosahedral group

I	E	$12C_5$	$12C_5^2$	$20C_3$	$15C_2$	$h=60$	
A	1	1	1	1	1	$x^2+y^2+z^2$	
T_1	3	$\frac{1}{2}(1+\sqrt{5})$	$\frac{1}{2}(1-\sqrt{5})$	0	−1	(x, y, z)	(R_x, R_y, R_z)
T_2	3	$\frac{1}{2}(1-\sqrt{5})$	$\frac{1}{2}(1+\sqrt{5})$	0	−1		
G	4	−1	−1	1	0		
H	5	0	0	−1	1	$(2z^2-x^2-y^2, x^2-y^2, xy, yz, zx)$	

Further information: P.W. Atkins, M.S. Child, and C.S.G. Phillips, *Tables for group theory*. Oxford University Press (1970). In this source, which is available on the web (see p. *xiv* for more details), other character tables such as D_2, D_4, D_{2d}, D_{3d}, and D_{5d} can be found.

Solutions to a) exercises

A horizontal bar over the last digit in some answers denotes an insignificant digit.

Chapter 1

E1.1(a) (a) 24 atm (b) 22 atm
E1.2(a) (a) 3.42 bar (b) 3.38 atm
E1.3(a) 30 lb in^{-2}
E1.4(a) 4.20×10^{-2} atm
E1.5(a) 0.50 m^3
E1.6(a) 102 kPa
E1.7(a) 8.3147 J K^{-1} mol^{-1}
E1.8(a) S_8
E1.9(a) 6.2 kg
E1.10(a) (a) (i) 0.762 (ii) 0.238 (iii) 0.752 bar (iv) 0.235 bar
(b) (i) 0.782 (ii) 0.208 (iii) 0.0099 bar (iv) 0.772 bar (v) 0.205
E1.11(a) 169 g mol^{-1}
E1.12(a) −273°C
E1.13(a) (a) (i) 1.0 atm (ii) 8.2×10^2 atm
(b) (i) 1.0 atm (ii) 1.8×10^3 atm
E1.14(a) $a = 7.61 \times 10^{-2}$ kg m^5 s^{-2} mol^{-2}, $b = 2.26 \times 10^{-5}$ m^3 mol^{-1}
E1.15(a) (a) 0.88 (b) 1.2 dm^3 mol^{-1}
E1.16(a) 140 atm
E1.17(a) (a) 50.7 atm (b) 35.2 atm, 0.695
E1.18(a) (a) 0.67, 0.33 (b) 2.0 atm, 1.0 atm (c) 3.0 atm
E1.19(a) 32.9 cm^3 mol^{-1}, 1.33 dm^6 atm mol^{-2}, 0.118 nm
E1.20(a) (a) 1.41×10^3 K (b) 0.139 nm
E1.21(a) (a) $T = 3.64 \times 10^3$ K, $p = 8.7$ atm (b) $T = 2.60 \times 10^3$ K, $p = 4.5$ atm
(c) $T = 46.7$ K, $p = 0.18$ atm
E1.22(a) 0.66

Chapter 2

E2.1(a) On Earth: 2.6×10^3 J needed, on the moon: 4.2×10^2 J needed
E2.2(a) -1.0×10^2 J
E2.3(a) (a) $w = -1.57$ kJ, $q = +1.57$ kJ (b) $w = -1.13$ kJ, $q = +1.13$ kJ (c) 0
E2.4(a) $p_2 = 1.33$ atm, $w = 0$, $q = \Delta U = +1.25$ kJ
E2.5(a) (a) −88 J (b) −167 J
E2.6(a) $\Delta H = q = -40.656$ kJ, $w = 3.10$ kJ, $\Delta U = -37.55$ kJ
E2.7(a) $w = -1.5$ kJ
E2.8(a) (a) $q = \Delta H = +2.83 \times 10^4$ J $= +28.3$ kJ, $w = -1.45$ kJ,
$\Delta U = +26.8$ kJ
(b) $\Delta H = +28.3$ kJ, $\Delta U = +26.8$ kJ, $w = 0$, $q = +26.8$ kJ
E2.9(a) $13\bar{1}$ K
E2.10(a) $w = -194$ J
E2.11(a) 22 kPa
E2.12(a) $C_{p,m} = 30$ J K^{-1} mol^{-1}, $C_{V,m} = 22$ J K^{-1} mol^{-1}
E2.13(a) $q_p = +2.2$ kJ, $\Delta H = +2.2$ kJ, $\Delta U = +1.6$ kJ
E2.14(a) $w = -3.2$ kJ, $\Delta U = -3.2$ kJ, $\Delta T = -38$ K, $\Delta H = -4.5$ kJ
E2.15(a) $V_f = 0.00944$ m^3, $T_f = 288$ K, $w = -4.6 \times 10^2$ J
E2.16(a) $q = +13.0$ kJ, $w = -1.0$ kJ, $\Delta U = 12.0$ kJ
E2.17(a) $\Delta_L H^\ominus(SrI_2,s) = 1953$ kJ mol^{-1}
E2.18(a) −4564.7 kJ mol^{-1}
E2.19(a) $\Delta_f H[(CH_2)_3,g] = +53$ kJ mol^{-1}, $\Delta_r H = -33$ kJ mol^{-1}
E2.20(a) $\Delta_c U^\ominus = -5152$ kJ mol^{-1}, $C = 1.58$ kJ K^{-1}, $\Delta T = 205$ K
E2.21(a) +65.49 kJ mol^{-1}
E2.22(a) −383 kJ mol^{-1}
E2.23(a) (a) $\Delta_r H^\ominus(3) = -114.40$ kJ mol^{-1}, $\Delta_r U = -111.92$ kJ mol^{-1}
(b) $\Delta_f H^\ominus(HCl,g) = -92.31$ kJ mol^{-1},
$\Delta_f H^\ominus(H_2O,g) = -241.82$ kJ mol^{-1}
E2.24(a) −1368 kJ mol^{-1}
E2.25(a) (a) −392.1 kJ mol^{-1} (b) −946.6 kJ mol^{-1}
E2.26(a) −56.98 kJ mol^{-1}
E2.27(a) (a) $\Delta_r H^\ominus(298\text{ K}) = +131.29$ kJ mol^{-1},
$\Delta_r U^\ominus(298\text{ K}) = +128.81$ kJ mol^{-1}
(b) $\Delta_r H^\ominus(378\text{ K}) = +132.56$ kJ mol^{-1},
$\Delta_r U^\ominus(378\text{ K}) = +129.42$ kJ mol^{-1}
E2.28(a) −218.66 kJ mol^{-1}
E2.29(a) −1892 kJ mol^{-1}
E2.30(a) 0.71 K atm^{-1}
E2.31(a) $\Delta U = 131$ J mol^{-1}, $q = +8.05 \times 10^3$ J mol^{-1}, $w = -7.92 \times 10^3$ J mol^{-1}
E2.32(a) 1.31×10^{-3} K^{-1}
E2.33(a) $1.\bar{1} \times 10^3$ atm
E2.34(a) $\left(\dfrac{\partial H_m}{\partial p}\right)_T = -7.2$ J atm^{-1} mol^{-1}, $q(\text{supplied}) = +8.1$ kJ

Chapter 3

E3.1(a) (a) 92 J K^{-1} (b) 67 J K^{-1}
E3.2(a) 152.67 J K^{-1} mol^{-1}
E3.3(a) −22.1 J K^{-1}
E3.4(a) $q = 0$, $\Delta S = 0$, $\Delta U = +4.1$ kJ, $\Delta H = +5.4$ kJ
E3.5(a) $\Delta H = 0$, $\Delta H_{tot} = 0$, $\Delta S_{tot} = +93.4$ J K^{-1}
E3.6(a) (a) $q = 0$ (b) −20 J (c) −20 J (d) $-0.34\bar{7}$ K (e) +0.60 J K^{-1}
E3.7(a) (a) +87.8 J K^{-1} mol^{-1} (b) −87.8 J K^{-1} mol^{-1}
E3.8(a) (a) −386.1 J K^{-1} mol^{-1} (b) +92.6 J K^{-1} mol^{-1}
(c) −153.1 J K^{-1} mol^{-1}
E3.9(a) (a) −521.5 kJ mol^{-1} (b) +25.8 kJ mol^{-1} (c) −178.7 kJ mol^{-1}
E3.10(a) (a) −522.1 kJ mol^{-1} (b) +25.78 kJ mol^{-1} (c) −178.6 kJ mol^{-1}
E3.11(a) −93.05 kJ mol^{-1}
E3.12(a) −50 kJ mol^{-1}
E3.13(a) (a) $\Delta S(\text{gas}) = +2.9$ J K^{-1}, $\Delta S(\text{surroundings}) = -2.9$ J K^{-1},
$\Delta S(\text{total}) = 0$
(b) $\Delta S(\text{gas}) = +2.9$ J K^{-1}, $\Delta S(\text{surroundings}) = 0$,
$\Delta S(\text{total}) = +2.9$ J K^{-1}
(c) $\Delta S(\text{gas}) = 0$, $\Delta S(\text{surroundings}) = 0$, $\Delta S(\text{total}) = 0$
E3.14(a) 817.90 kJ mol^{-1}
E3.15(a) $\eta = 1 - \dfrac{333\text{ K}}{373\text{ K}} = 0.11$, $\eta = 1 - \dfrac{353\text{ K}}{573\text{ K}} = 0.38$
E3.16(a) −3.8 J
E3.17(a) −36.5 J K^{-1}
E3.18(a) 12 kJ

E3.19(a) $+7.3\ \text{kJ mol}^{-1}$
E3.20(a) $-0.55\ \text{kJ mol}^{-1}$
E3.21(a) $+10\ \text{kJ}$
E3.22(a) $+11\ \text{kJ mol}^{-1}$

Chapter 4

E4.1(a) (a) single phase (b) three phases (c) two phases (d) two phases
E4.2(a) 0.71 J
E4.3(a) -1.0×10^{-4} K
E4.4(a) 4
E4.5(a) $5.2\ \text{kJ mol}^{-1}$
E4.6(a) $70\ \text{J mol}^{-1}$
E4.7(a) 2.71 kPa
E4.8(a) $\Delta_{\text{fus}}S=+45.2\bar{3}\ \text{J K}^{-1}\ \text{mol}^{-1}$, $\Delta_{\text{fus}}H=+16\ \text{kJ mol}^{-1}$
E4.9(a) 31°C
E4.10(a) $+20.80\ \text{kJ mol}^{-1}$
E4.11(a) (a) $+34.08\ \text{kJ mol}^{-1}$ (b) 350.5 K
E4.12(a) 281.8 K or 8.7°C
E4.13(a) $25\ \text{g s}^{-1}$
E4.14(a) (a) 1.7×10^{3} g (b) 31×10^{3} g (c) 1.4 g
E4.15(a) (a) $+49\ \text{kJ mol}^{-1}$ (b) $21\bar{5}$°C (c) $+101\ \text{J K}^{-1}\ \text{mol}^{-1}$
E4.16(a) 272.80 K
E4.17(a) 0.0763 = 7.63 per cent

Chapter 5

E5.1(a) $886.8\ \text{cm}^{3}$
E5.2(a) $56\ \text{cm}^{3}\ \text{mol}^{-1}$
E5.3(a) 6.4×10^{3} kPa
E5.4(a) 1.3×10^{2} kPa
E5.5(a) $85\ \text{g mol}^{-1}$
E5.6(a) $3.8\times10^{2}\ \text{g mol}^{-1}$
E5.7(a) −0.09°C
E5.8(a) $\Delta_{\text{mix}}G=-0.35\ \text{kJ}$, $\Delta_{\text{mix}}S=+1.2\ \text{J K}^{-1}$
E5.9(a) $+4.71\ \text{J K}^{-1}\ \text{mol}^{-1}$
E5.10(a) (a) $x_{\text{A}}=\frac{1}{2}$ (b) 0.8600
E5.11(a) (a) $3.4\times10^{-3}\ \text{mol kg}^{-1}$ (b) $3.37\times10^{-2}\ \text{mol kg}^{-1}$
E5.12(a) $0.17\ \text{mol dm}^{-3}$
E5.13(a) $0.135\ \text{mol kg}^{-1}$, 24.0 g anthracene
E5.14(a) $87\ \text{kg mol}^{-1}$
E5.15(a) $a_{\text{A}}=0.833$, $\gamma_{\text{A}}=0.93$, $a_{\text{B}}=0.125$, $\gamma_{\text{B}}=0.12\bar{5}$, $a_{\text{B}}=2.8$
E5.16(a) $p_{\text{A}}=32.2$ Torr, $p_{\text{B}}=x_{\text{B}}K_{\text{B}}=6.1$ Torr, $p_{\text{total}}=38.3$ Torr, $y_{\text{A}}=0.840$, $y_{\text{B}}=0.160$
E5.17(a) $a_{\text{A}}=0.498$, $a_{\text{M}}=0.667$, $\gamma_{\text{A}}=1.24$, $\gamma_{\text{M}}=1.11$
E5.18(a) 0.90
E5.19(a) (a) 2.73 g (b) 2.92 g
E5.20(a) $I=0.060$, CaCl_2: $\gamma_{\pm}=0.56$, $a(\text{Ca}^{2-})=0.0056$, $a(\text{Cl}^{+})=0.011$
E5.21(a) $K_{\text{B}}=2.01$
E5.22(a) $x_1=0.92$, $x_2=0.08$, $y_1=0.97$, $y_2=0.03$
E5.23(a) $x_{\text{A}}=0.267$, $x_{\text{B}}=0.733$, $p_{\text{total}}=58.6$ kPa
E5.24(a) (a) solution is ideal (b) $y_{\text{A}}=0.830$, $y_{\text{B}}=0.1703$
E5.25(a) (a) 20.6 kPa (b) 0.668 (c) 0.332
E5.26(a) (a) $y_{\text{M}}=0.36$ (b) $y_{\text{M}}=0.80$ (*i.e.*, $y_{\text{O}}=0.20$).
E5.29(a) $x_{\text{B}}\approx0.26$ and its melting point is labeled $T_2\approx200$°C.
E5.31(a) (a) 76 per cent (c) $\dfrac{n_{\text{c}}}{n_{\text{a}}}=1.11$, $\dfrac{n_{\text{c}}}{n_{\text{a}}}=1.46$

E5.32(a) (b) 620 Torr (c) 490 Torr (d) $x_{\text{Hex}}=0.50$ $y_{\text{Hex}}=0.72$ (e) $y_{\text{Hex}}=0.50$, $x_{\text{Hex}}=0.30$

Chapter 6

E6.1(a) $n_{\text{A}}=0.9$ mol, $n_{\text{B}}=1.2$ mol
E6.2(a) $\Delta G\approx-0.64$ kJ
E6.3(a) $K\approx6\times10^{5}$
E6.4(a) 2.85×10^{-6}
E6.5(a) (a) 0.141 (b) 13.5
E6.6(a) (a) $\Delta_{\text{r}}G^{\ominus}=-68.26\ \text{kJ mol}^{-1}$, $K=9.13\times10^{11}$ (b) $K_{400\ \text{K}}=1.32\times10^{9}$, $\Delta_{\text{r}}G^{\ominus}_{400\ \text{K}}=-69.8\ \text{kJ mol}^{-1}$
E6.7(a) $K=(0.0831451\ \text{K}^{-1})\times K_cT$
E6.8(a) (b) 0.33 (c) 0.33 (d) $+2.8\ \text{kJ mol}^{-1}$
E6.9(a) $K_1=0.045$, $T_2=15\overline{00}$ K
E6.10(a) $\Delta_{\text{r}}H^{\ominus}=+2.77\ \text{kJ mol}^{-1}$, $\Delta_{\text{r}}G^{\ominus}=-16.5\ \text{J K}^{-1}\ \text{mol}^{-1}$
E6.11(a) $K=(0.0831451\ \text{K}^{-1})\times K_cT$
E6.12(a) (a) $K(25°\text{C})=1.17\times10^{6}$, $K_c(25°\text{C})=4.72\times10^{4}$ (b) $K(100°\text{C})=9.95\times10^{5}$, $K_c(100°\text{C})=3.21\times10^{4}$
E6.13(a) $+12.3\ \text{kJ mol}^{-1}$
E6.14(a) 50 per cent
E6.15(a) $x_{\text{borneol}}=0.9663$, $x_{\text{iso}}=0.0337$
E6.16(a) (a) $\Delta_{\text{r}}H^{\ominus}=52.89\ \text{kJ mol}^{-1}$ (b) $\Delta_{\text{r}}H^{\ominus}=-52.89\ \text{kJ mol}^{-1}$
E6.17(a) $-14.4\ \text{kJ mol}^{-1}$
E6.18(a) 1110 K (837°C)
E6.19(a) $-1108\ \text{kJ mol}^{-1}$
E6.21(a) (a) +1.10 V (b) +0.22 V (c) +1.23 V
E6.22(a) (a) $\text{Cd}^{2+}(\text{aq})+2\ \text{Br}^{-}(\text{aq})+2\ \text{Ag(s)}\rightarrow\text{Cd(s)}+2\ \text{AgBr(s)}$ (c) −0.62 V
E6.23(a) (a) 6.5×10^{9} (b) 1.4×10^{12}
E6.24(a) (a) 8.5×10^{-17} (b) $9.2\times10^{-9}\ \text{mol dm}^{-3}$ or $2.2\ \mu\text{g dm}^{-3}$

Chapter 7

E7.1(a) $0.024\ \text{m s}^{-1}$
E7.2(a) 332 pm
E7.3(a) 700 pm

E7.4(a)

λ/nm	E/J	E_{m}/(kJ mol^{-1})
(a) 600	3.31×10^{-19}	199
(b) 550	3.61×10^{-19}	218
(c) 400	4.97×10^{-19}	299

E7.5(a)

λ/nm	E_{photon}/J	v/(km s^{-1})
(a) 600	3.31×10^{-19}	19.9
(b) 550	3.61×10^{-19}	20.8
(c) 400	4.97×10^{-19}	24.4

E7.6(a) $21\ \text{m s}^{-1}$
E7.7(a) (a) 2.77×10^{18} (b) 2.77×10^{20}
E7.8(a) (a) no electron ejection (b) $837\ \text{km s}^{-1}$
E7.9(a) (a) 6.6×10^{-19} J, $4.0\times10^{2}\ \text{kJ mol}^{-1}$ (b) 6.6×10^{-20} J, $40\ \text{kJ mol}^{-1}$ (c) 6.6×10^{-34} J, $4.0\times10^{-13}\ \text{kJ mol}^{-1}$
E7.10(a) (a) 6.6×10^{-29} m (b) 6.6×10^{-36} m (c) 99.7 pm

E7.11(a) $N=\left(\frac{1}{2\pi}\right)^{1/2}$

E7.12(a) $(1/2\pi)\,\mathrm{d}\phi$

E7.13(a) $\frac{1}{2}$

E7.15(a) $\Delta v_{\min}=1.1\times10^{-28}\ \mathrm{m\ s^{-1}}$, $\Delta q_{\min}=1\times10^{-27}$ m

E7.16(a) 6.96 keV

E7.17(a) (a) $\left[\frac{\mathrm{d}}{\mathrm{d}x},\frac{1}{x}\right]=-\frac{1}{x^2}$ (b) $\left[\frac{\mathrm{d}}{\mathrm{d}x},x^2\right]=2x$

Chapter 8

E8.1(a) (a) 1.81×10^{-19} J, 1.13 eV, 9100 $\mathrm{cm^{-1}}$, 109 kJ $\mathrm{mol^{-1}}$
(b) 6.6×10^{-19} J, 4.1 eV, 33 000 $\mathrm{cm^{-1}}$, 400 kJ $\mathrm{mol^{-1}}$

E8.2(a) (a) 0.04 (b) 0

E8.3(a) $\frac{h^2}{4L^2}$

E8.4(a) $\frac{L}{2}, L^2\left(\frac{1}{3}-\frac{1}{2\pi^2}\right)$

E8.5(a) $\frac{h}{8^{1/2}m_ec}=\frac{\lambda_C}{8^{1/2}}$

E8.6(a) $\frac{L}{6},\frac{L}{2}$ and $\frac{5L}{6}$

E8.7(a) −17.4 per cent

E8.8(a) $\frac{2kTmL^2}{h^2}-\frac{1}{2}$

E8.9(a) 4.30×10^{-21} J

E8.10(a) 278 N $\mathrm{m^{-1}}$

E8.11(a) 2.64 μm

E8.12(a) 8.3673×10^{-28} kg, 1.6722×10^{-27} kg, $\omega_D=93.3$ THz

E8.13(a) (a) 3.3×10^{-34} J (b) 3.3×10^{-33} J

E8.15(a) $\pm0.525\alpha$ or $\pm1.65\alpha$

E8.16(a) $\pm\alpha$

E8.17(a) 5.61×10^{-21} J

E8.18(a) $\left(\frac{1}{2\pi}\right)^{1/2}$

E8.19(a) 3.32×10^{-22} J

E8.20(a) 3.2×10^{34}

E8.21(a) 2.11×10^{-22} J

E8.22(a) 4.22×10^{-22} J

E8.23(a) 1.49×10^{-34} J s

Chapter 9

E9.1(a) 9.118×10^{-6} cm, 1.216×10^{-5} cm

E9.2(a) $\tilde{\nu}=3.292\times10^5\ \mathrm{cm^{-1}}$, $\lambda=3.038\times10^{-6}$ cm, $\nu=9.869\times10^{15}$ Hz

E9.3(a) 14.0 eV

E9.4(a) (a) 1 (b) 9 (c) $g=25$

E9.5(a) $N=\left(\frac{2}{a_0^{2/3}}\right)\left(\frac{TT}{4}\right)=\frac{TT}{2a_0^{2/3}}$

E9.6(a) $4a_0$, 0

E9.7(a) $r=0.35a_0$

E9.8(a) 101 pm and 376 pm

E9.9(a) $\langle V\rangle=2E_{1s}$, $\langle T\rangle=-E_{1s}$,

E9.10(a) $5.24\frac{a_0}{Z}$

E9.11(a) $r=2a_0/Z$

E9.13(a) $\theta=\pi/2$, $\theta=0$, $\theta=0$

E9.14(a) (a) forbidden (b) allowed (c) allowed

E9.15(a) $0.999\,999\,944\times680$ nm

E9.16(a) (a) 27 ps (b) 2.7 ps

E9.17(a) (a) 53 $\mathrm{cm^{-1}}$ (b) 0.53 $\mathrm{cm^{-1}}$

E9.19(a) (a) [Ar]3d^8 (b) $S=1,0$, $M_S=-1,0,+1$, $M_S=0$

E9.20(a) (a) $\frac{5}{2},\frac{3}{2}$ (b) $\frac{7}{2},\frac{5}{2}$

E9.21(a) $l=1$

E9.22(a) $L=2, S=0, J=2$

E9.23(a) (a) 1,0, 3,1 (b) $\frac{3}{2},\frac{1}{2}$, and $\frac{1}{2}$, 4, 2, 2

E9.24(a) $^3D_3$, $^3D_2$, $^3D_1$, $^1D_2$

E9.25(a) (a) $J=0$, (b) $J=0$, (c) $J=2,1,0$

E9.26(a) (a) $^2S_{1/2}$ (b) $^2P_{3/2}$ and $^2P_{1/2}$

E9.27(a) (a) allowed (b) forbidden (c) allowed

Chapter 10

E10.1(a) $\{A(1)p_x(2)+A(2)p_x(1)\}\times\{B(3)p_y(4)+B(4)p_x(3)\}$

E10.2(a) $\{s(1)p_z(2)+s(2)p_z(1)\}\times\{\alpha(1)\beta(2)-\alpha(2)\beta(1)\}$

E10.5(a) (a) $1\sigma_g^2$, $b=1$ (b) $1\sigma_g^21\sigma_u^2$, $b=0$ (c) $1\sigma_g^21\sigma_u^21\pi_u^4$, $b=2$

E10.6(a) (a) $1\sigma^22\sigma^21\pi^43\sigma^2$ (b) $1\sigma^22\sigma^23\sigma^21\pi^42\pi^1$ (c) $1\sigma^22\sigma^21\pi^43\sigma^2$

E10.7(a) C_2

E10.10(a) $A\cos\theta-B\sin\theta$

E10.11(a) 0

E10.12(a) 1.4 eV, 2.2×10^{-19} J

E10.13(a) 10.96 eV, 1.76×10^{-18} J

E10.14(a) $\psi_+=0.97\chi_F+0.25\chi_{Xe}$, $\psi_-=0.25\chi_F-0.97\chi_{Xe}$

E10.15(a) $E_{bond}=-18.1$ eV, $E_{anti}=-12.0$ eV, $\psi_{bond}=1.01\chi_F-0.36\chi_{Xe}$, $\psi_{anti}=0.16\chi_F+0.96\chi_{Xe}$

E10.17(a) (a) $3\alpha+2^{3/2}\beta$ (b) $3\alpha+3\beta$

E10.18(a) (a) $a_{2u}^2e_{1g}^4e_{2u}^1$, $7\alpha+7\beta$, $7\alpha+7\beta$ (b) $a_{2u}^2e_{1g}^3$, $5\alpha+7\beta$

E10.17(a) (a) $14\alpha+19.314\beta$ (b) $14\alpha+19.448\beta$

Chapter 11

E11.2(a) (a) R_3 (b) C_{2v} (c) D_{3h} (d) $D_{\infty h}$

E11.3(a) (a) C_{2v} (b) $C_{\infty v}$ (c) C_{3v} (d) D_{2h}

E11.4(a) (a) C_{2v} (b) C_{2h}

E11.7(a) σ_h, i

E11.12(a) d_{xy}

E11.14(a) $2A_1+B_1+E$

E11.15(a) (a) either E_{1u} or A_{2u} (b) B_{3u}(x-polarized), B_{2u}(y-polarized), B_{1u}(z-polarized)

Chapter 12

E12.2(a) (c) CH_4 is inactive

E12.3(a) 7.173×10^{-47} kg $\mathrm{m^2}$, $I_\parallel$ will not change

E12.4(a) $I_\parallel=5.60\times10^{-47}$ kg $\mathrm{m^2}$, $I_\perp=6.29\times10^{-46}$ kg $\mathrm{m^2}$, $\tilde{A}=5.00\ \mathrm{cm^{-1}}$, $\tilde{B}=0.445\ \mathrm{cm^{-1}}$, $A=1.50\times10^{11}$ Hz, $B=1.33\times10^{10}$ Hz

E12.5(a) 4.09×10^{11} Hz

E12.6(a) (a) 2.642×10^{-47} kg $\mathrm{m^2}$ (b) 127.4 pm

E12.7(a) $I=4.442\times10^{-47}$ kg $\mathrm{m^2}$, $R=165.9$ pm

E12.8(a) 232.1 pm

E12.9(a) $R = 106.5$ pm, $R' = 115.6$ pm
E12.10(a) 20 475 cm^{-1}
E12.11(a) 198.9 pm
E12.12(a) $\tilde{D}_J = 2.111 \times 10^{-4}$ cm^{-1}, 0.1253
E12.13(a) (a) 20 (b) 24
E12.14(a) 1.6×10^2 N m^{-1}
E12.15(a) 1.089 per cent
E12.16(a) 327.8 N m^{-1}
E12.17(a) (a) 0.067 (b) 0.20

E12.18(a)

	HF	HCl	HBr	HI
$\tilde{\nu}/cm^{-1}$	4141.3	2988.9	2649.7	2309.5
m_{eff}/m_u	0.9570	0.9697	0.9954	0.9999
$k/(N\ m^{-1})$	967.0	515.6	411.8	314.2

E12.19(a) $\tilde{\nu} = 1580.38$ cm^{-1}, $x_e = 7.644 \times 10^{-3}$
E12.20(a) 5.15 eV
E12.21(a) $x_e = 0.02101, \dfrac{x_e(^2H^{19}F)}{x_e(^1H^{19}F)} = 0.5256$
E12.22(a) 2699.77 cm^{-1}
E12.26(a) Raman active
E12.27(a) $4A_1 + A_2 + 2B_1 + 2B_2$

Chapter 13

E13.1(a) 80 per cent
E13.2(a) 6.28×10^3 $dm^3\ mol^{-1}\ cm^{-1}$
E13.3(a) 1.5 mmol dm^{-3}
E13.4(a) 5.4×10^7 $dm^3\ mol^{-1}\ cm^{-2}$
E13.5(a) 4.5×10^2 $dm^3\ mol^{-1}\ cm^{-2}$
E13.6(a) 23 per cent
E13.7(a) (a) 0.87 m (b) 2.9 m
E13.8(a) (a) 5×10^7 $dm^3\ mol^{-1}\ cm^{-2}$ (b) 2.5×10^6 $dm^3\ mol^{-1}\ cm^{-2}$
E13.9(a) $1\sigma_g^1 1\pi_u^1$
E13.10(a) 3, u
E13.11(a) (a) allowed (b) allowed (c) forbidden (d) forbidden (e) allowed
E13.12(a) $\dfrac{2\sqrt{2}}{3} e^{-2ax_0^2/3}$
E13.13(a) R branch has a band head, $J = 7$
E13.14(a) $30.4\ cm^{-1} < \tilde{B}' < 40.5\ cm^{-1}$
E13.17(a) $\Delta_O = P - \tilde{\nu}$, 14×10^3 cm^{-1}
E13.18(a) (a) $\tilde{\nu} \approx 1800$ cm^{-1}
E13.19(a) $\lambda = 60$ cm ($\nu = 500$ MHz)
E13.20(a) 20 ps, 70 MHz

Chapter 14

E14.1(a) 28 GHz
E14.2(a) 8.9×10^{-12} s
E14.3(a) 600 MHz
E14.4(a) 154 MHz
E14.5(a) $\Delta E = 3.98 \times 10^{-25}$ J, $\Delta E = 6.11 \times 10^{-26}$ J, larger for the proton
E14.7(a) (a) 1×10^{-6} (b) 5.1×10^{-6} (c) 3.4×10^{-5}
E14.8(a) (a) 1.86 T (b) 0
E14.9(a) 13
E14.10(a) (a) 11 μT (b) 110 μT
E14.13(a) 6.7×10^2 s^{-1}
E14.17(a) 0.21 s
E14.18(a) 1.234
E14.19(a) $\mathcal{B}_1 = 5.9 \times 10^{-4}$ T, 20 μs
E14.20(a) (a) 2×10^2 T (b) 10 mT
E14.21(a) 2.0022
E14.22(a) $a = 2.3$ mT, $g = 2.002\bar{5}$
E14.23(a) equal intensity, 330.2 mT, 332.2 mT, 332.8 mT, 334.8 mT
E14.25(a) (a) 331.9 mT (b) 1.201 T
E14.26(a) $I = \frac{3}{2}$
E14.28(a) 1.9×10^8 s^{-1}

Chapter 15

E15.1(a) 21621600
E15.2(a) 1
E15.3(a) 524 K
E15.4(a) $35\bar{4}$ K
E15.5(a) (a) (i) 8.23 pm (ii) 2.60 pm (b) (i) 1.79×10^{27} (ii) 5.67×10^{28}
E15.6(a) 2.83
E15.7(a) 2.4×10^{25}
E15.8(a) 72.2
E15.9(a) (a) 7.97×10^3 (b) 1.12×10^4
E15.10(a) 18 K
E15.11(a) 37 K
E15.12(a) 4.5 K
E15.13(a) (a) 1 (b) 2 (c) 2 (d) 12 (e) 3
E15.14(a) 660.6
E15.14(a) 4500 K
E15.16(a) 2.571
E15.17(a) 42.3
E15.18(a) 3.1561
E15.19(a) +2.46 kJ
E15.20(a) $1 + e^{-2\mu_B\beta\mathcal{B}}, \dfrac{2\mu_B\mathcal{B}e^{-2\mu_B\beta\mathcal{B}}}{1 + e^{-2\mu_B\beta\mathcal{B}}}, -\mu_B\mathcal{B} + \dfrac{2\mu_B\mathcal{B}e^{-2\mu_B\beta\mathcal{B}}}{1 + e^{-2\mu_B\beta\mathcal{B}}}$ (a) 0.71 (b) 0.996
E15.21(a) (a) (1) 5×10^{-5} (2) 0.4 (3) 0.905 (b) 1.4 (c) 22 J mol^{-1} (d) 1.6 J K^{-1} mol^{-1} (e) 4.8 J K^{-1} mol^{-1}
E15.22(a) 4303 K
E15.23(a) (a) 138 J K^{-1} mol^{-1} (b) 146 J K^{-1} mol^{-1}
E15.24(a) 5.20 J K^{-1} mol^{-1}
E15.25(a) (a) He gas (b) CO gas (d) H_2O vapour

Chapter 16

E16.1(a) 15.27 pm
E16.2(a) 2.8×10^{26}
E16.3(a) (a) $\frac{7}{2}R$ (b) $3R$ (c) $7R$
E16.4(a) 15.24 K
E16.5(a) (a) 19.6 (b) 34.3
E16.6(a) (a) 1 (b) 2 (c) 2 (d) 12 (e) 3
E16.7(a) $q^R = 43.1$, $\theta_R = 22.36$ K
E16.8(a) 43.76 J K^{-1} mol^{-1}
E16.9(a) (a) 36.95, 80.08 (b) 36.7, 79.7
E16.10(a) 72.5
E16.11(a) closer, closer
E16.12(a) (a) 14.93 J K^{-1} mol^{-1} (b) 25.65 J K^{-1} mol^{-1}
E16.13(a) -13.8 kJ mol^{-1}, -0.20 kJ mol^{-1}

E16.14(a) (a) 4.158 (b) 4.489
E16.15(a) (a) 0.236 (b) 0.193
E16.16(a) (a) -6.42 kJ mol^{-1} (b) -14.0 kJ mol^{-1}
E16.17(a) 11.5 J K^{-1} mol^{-1}
E16.19(a) (a) 9.13 J K^{-1} mol^{-1} (b) 13.4 J K^{-1} mol^{-1} (c) 14.9 J K^{-1} mol^{-1}
E16.20(a) 3.70×10^{-3}

Chapter 17

E17.1(a) ClF_3, O_3, H_2O_2
E17.2(a) 1.4 D
E17.3(a) 37D, 11.7°
E17.4(a) 1.07×10^{3} kJ mol^{-1}
E17.5(a) 5.0 μD
E17.6(a) 1.66 D, 1.01×10^{-39} J^{-1} C^2 m^2, 9.06×10^{-30} m^3
E17.7(a) 4.75
E17.8(a) 1.42×10^{-39} J^{-1} C^2 m^2
E17.9(a) 1.34
E17.10(a) 17.7
E17.11(a) 0.071 J mol^{-1}
E17.12(a) 289 kJ mol^{-1}
E17.13(a) 2.6 kPa
E17.14(a) 72.8 mN m^{-1}
E17.15(a) 728 kPa

Chapter 18

E18.1(a) 27 nm
E18.2(a) $R_c = 3.08$ μm, $R_{rms} = 30.8$ nm
E18.3(a) 1.4×10^{4}
E18.4(a) 0.017
E18.5(a) 6.4×10^{-3}
E18.6(a) -19 mJ mol^{-1} K^{-1}
E18.7(a) $\left(\dfrac{R_{g,\text{constrained coil}} - R_{g,\text{random coil}}}{R_{g,\text{random coil}}}\right)\times 100\% = +41.42\%$,

$\left(\dfrac{V_{\text{constrained coil}} - V_{\text{random coil}}}{V_{\text{random coil}}}\right)\times 100\% = +182.8\%$

E18.8(a) $\left(\dfrac{R_g - R_{g,\text{random coil}}}{R_{g,\text{random coil}}}\right)\times 100\% = +895\%$ when $N = 1000$,

$\left(\dfrac{V - V_{g,\text{random coil}}}{V_{g,\text{random coil}}}\right)\times 100\% = +9.84\times10^{4}\,\%$ when $N = 1000$

E18.9(a) 1.3×10^{4} pm
E18.10(a) 3.7×10^{-14} N
E18.11(a) $\bar{M}_n = 70$ kg mol^{-1}, $\bar{M}_w = 69$ kg mol^{-1}
E18.12(a) (a) 18 kg mol^{-1} (b) 20 kg mol^{-1}
E18.13(a) 100
E18.14(a) 64 kg mol^{-1}
E18.15(a) 0.73 mm s^{-1}
E18.16(a) 31 kg mol^{-1}
E18.17(a) 3.4×10^{3} kg mol^{-1}

Chapter 19

E19.1(a) $(1,\tfrac{1}{2},0)$, $(1,0,\tfrac{1}{2})$, $(\tfrac{1}{2},\tfrac{1}{2},\tfrac{1}{2})$
E19.2(a) (323) and (110)
E19.3(a) $d_{111} = 249$ pm, $d_{211} = 176$ pm, $d_{100} = 432$ pm
E19.4(a) 70.7 pm
E19.5(a) $\sin\theta_{110} = 16°$, $\sin\theta_{200} = 23°$, $\sin\theta_{211} = 28°$
E19.6(a) 0.214 cm
E19.7(a) $f_{Br^-} = 36$
E19.8(a) 0.396 nm^3
E19.9(a) $N = 4$, $\rho = 4.01$ g cm^{-3}
E19.10(a) 190 pm
E19.11(a) 111, 200, 311, cubic F
E19.12(a) $\theta_{100} = 8.17°$, $\theta_{010} = 4.82°$, $\theta_{111} = 11.75°$
E19.13(a) face-centred cubic
E19.14(a) f
E19.18(a) 7.9 km s^{-1}
E19.19(a) 252 pm
E19.20(a) 0.9069
E19.21(a) (a) 0.5236 (b) 0.6802 (c) 0.7405
E19.22(a) 0.41421
E19.23(a) (a) 58.0 pm (b) 102 pm
E19.24(a) expansion
E19.25(a) 3500 kJ mol^{-1}
E19.26(a) 0.010
E19.27(a) 9.3×10^{-4} cm^3
E19.29(a) 3.54 eV
E19.30(a) 3 unpaired electrons
E19.31(a) -6.4×10^{-5} cm^3 mol^{-1}
E19.32(a) 5
E19.33(a) $+1.6\times10^{-8}$ m^3 mol^{-1}
E19.34(a) 6.0 K

Chapter 20

E20.1(a) (a) 9.975
E20.2(a) (a) 72 K (b) 946 m s^{-1} (c) the temperature would not be different
E20.3(a) 9.06×10^{-3}
E20.4(a) $\left(3-\dfrac{8}{\pi}\right)^{1/2}\left(\dfrac{RT}{M}\right)^{1/2}$, $\left(1-\dfrac{8}{\pi}\right)^{1/2}c$
E20.5(a) 0.0652 Pa
E20.6(a) 0.97 μm
E20.7(a) 397 m s^{-1} (a) 5.0×10^{10} s^{-1} (b) 5.0×10^{9} s^{-1} (c) 5.0×10^{3} s^{-1}
E20.8(a) (a) 6.7 nm (b) 67 nm (c) 6.7 cm
E20.9(a) 1.9×10^{20}
E20.10(a) 104 mg
E20.11(a) 415 Pa
E20.12(a) 42.4 g mol^{-1}
E20.13(a) 1.3 days
E20.14(a) -0.013 J m^{-2} s^{-1}
E20.15(a) 0.0562 nm^2
E20.16(a) 17 W
E20.17(a) 0.142 nm^2
E20.18(a) $p_1 = 205$ kPa, $\dfrac{dV}{dt} = \dfrac{\pi R^4}{16\eta p_0 l}(p_1^2 - p_2^2)$ Poiseuille's formula
E20.19(a) (a) 127 μP (b) 132 μP (c) 243 μP
E20.20(a) (a) $\kappa = 5.4$ mJ K^{-1} m^{-1} s^{-1}, $J_{\text{energy}} = -0.81$ W m^{-2}, Rate of energy flow $= -8.1$ mW (b) $\kappa = 29$ mJ K^{-1} m^{-1} s^{-1}, $J_{\text{energy}} = -4.4$ W m^{-2}, Rate of energy flow $= -44$ mW
E20.21(a) 5.4 mJ m^{-1} s^{-1}

E20.22(a) (a) $D = 1.07\ \mathrm{m^2\,s^{-1}}$, $J = 438\ \mathrm{mol\,m^{-2}\,s^{-1}}$ (b) $D = 1.07 \times 10^{-5}\ \mathrm{m^2\,s^{-1}}$, $J = 4.38\ \mathrm{mmol\,m^{-2}\,s^{-1}}$, (c) $D = 1.07 \times 10^{-7}\ \mathrm{m^2\,s^{-1}}$, $J = 43.8\ \mathrm{\mu mol\,m^{-2}\,s^{-1}}$
E20.23(a) $7.63\ \mathrm{mS\,m^2\,mol^{-1}}$
E20.24(a) $347\ \mathrm{\mu m\,s^{-1}}$
E20.25(a) $13.83\ \mathrm{mS\,m^2\,mol^{-1}}$
E20.26(a) $u(\mathrm{Li^+}) = 4.01 \times 10^{-8}\ \mathrm{m^2\,V^{-1}\,s^{-1}}$, $u(\mathrm{Na^+}) = 5.19 \times 10^{-8}\ \mathrm{m^2\,V^{-1}\,s^{-1}}$, $u(\mathrm{K^+}) = 7.62 \times 10^{-8}\ \mathrm{m^2\,V^{-1}\,s^{-1}}$
E20.27(a) $1.90 \times 10^{-9}\ \mathrm{m^2\,s^{-1}}$
E20.28(a) $\mathcal{F}(10\ \mathrm{cm}) = 25\ \mathrm{kN\,mol^{-1}}$, $\mathcal{F}(20\ \mathrm{cm}) = \infty$
E20.29(a) $67.5\ \mathrm{kN\,mol^{-1}}$
E20.30(a) $t = 1.3 \times 10^3\ \mathrm{s}$
E20.31(a) $a = 0.42\ \mathrm{nm}$
E20.32(a) 27.3 ps
E20.33(a) $\langle r^2\rangle^{1/2} = 113\ \mathrm{\mu m}$, $\langle r^2\rangle^{1/2} = 56\ \mathrm{\mu m}$

Chapter 21

E21.1(a) no change in pressure
E21.2(a) $8.1\ \mathrm{mol\,dm^{-3}\,s^{-1}}$, $2.7\ \mathrm{mol\,dm^{-3}\,s^{-1}}$, $2.7\ \mathrm{mol\,dm^{-3}\,s^{-1}}$, $5.4\ \mathrm{mol\,dm^{-3}\,s^{-1}}$
E21.3(a) $v = 1.3\bar{5}\ \mathrm{mol\,dm^{-3}\,s^{-1}}$, $4.0\bar{5}\ \mathrm{mol\,dm^{-3}\,s^{-1}}$, $2.7\ \mathrm{mol\,dm^{-3}\,s^{-1}}$, $1.3\bar{5}\ \mathrm{mol\,dm^{-3}\,s^{-1}}$
E21.4(a) $\mathrm{dm^3\,mol^{-1}\,s^{-1}}$, (a) $k_r[\mathrm{A}][\mathrm{B}]$ (b) $3k_r[\mathrm{A}][\mathrm{B}]$
E21.5(a) $v = \frac{1}{2}k_r[\mathrm{A}][\mathrm{B}][\mathrm{C}]$, $[k_r] = \mathrm{dm^6\,mol^{-2}\,s^{-1}}$
E21.6(a) (a) $[k_r] = \mathrm{dm^3\,mol^{-1}\,s^{-1}}$, $[k_r] = \mathrm{dm^6\,mol^{-2}\,s^{-1}}$ (b) $[k_r] = \mathrm{kPa^{-1}\,s^{-1}}$ $[k_r] = \mathrm{kPa^{-2}\,s^{-1}}$
E21.7(a) second-order
E21.8(a) $n = 2$
E21.9(a) 1.03×10^4 s, (a) 498 Torr (b) 461 Torr
E21.10(a) (a) $16.\bar{2}\ \mathrm{dm^3\,mol^{-1}\,h^{-1}}$, $4.5 \times 10^{-3}\ \mathrm{dm^3\,mol^{-1}\,s^{-1}}$ (b) 5.1×10^3 s, 2.1×10^3 s
E21.11(a) (a) $0.098\ \mathrm{mol\,dm^{-3}}$ (b) $0.050\ \mathrm{mol\,dm^{-3}}$
E21.12(a) $1.11 \times 10^5\ \mathrm{s} = 128$ days
E21.13(a) $1.28 \times 10^4\ \mathrm{dm^3\,mol^{-1}\,s^{-1}}$, $4.0 \times 10^{10}\ \mathrm{dm^{-3}\,mol\,s^{-1}}$
E21.14(a) $64.9\ \mathrm{kJ\,mol^{-1}}$, $4.32 \times 10^8\ \mathrm{mol\,dm^{-3}\,s^{-1}}$
E21.15(a) $35\ \mathrm{kJ\,mol^{-1}}$
E21.16(a) (i) $k_3K^{1/2}[\mathrm{A_2}]^{1/2}[\mathrm{B}]$ (ii) $\dfrac{k_2^2[\mathrm{B}]^2}{4k_1'}\left(\sqrt{1+\dfrac{16k_1'k_1[\mathrm{A_2}]}{k_2^2[\mathrm{B}]^2}}-1\right)$, $k_2K^{1/2}[\mathrm{A_2}]^{1/2}[\mathrm{B}]$, $2k_1[\mathrm{A_2}]$
E21.17(a) $1.9 \times 10^{-6}\ \mathrm{Pa^{-1}\,s^{-1}}$, $1.9\ \mathrm{MPa^{-1}\,s^{-1}}$
E21.18(a) $-3\ \mathrm{kJ\,mol^{-1}}$
E21.19(a) 251, 0.996
E21.20(a) 0.125
E21.21(a) 3.3×10^{18}
E21.22(a) 0.52
E21.23(a) $0.56\ \mathrm{mol\,dm^{-3}}$
E21.24(a) 7.1 nm

Chapter 22

E22.1(a) $9.49 \times 10^9\ \mathrm{s^{-1}}$, $1.15 \times 10^{35}\ \mathrm{s^{-1}\,m^{-3}}$, 1.7 per cent
E22.2(a) (a) (i) 0.018, (ii) 0.30 (b) (i) 3.9×10^{-18} (ii) 6.0×10^{-6}
E22.3(a) (a) (i) 14% (ii) 1% (b) (i) 280%, 260 % (ii) 13%
E22.4(a) $1.03 \times 10^{-5}\ \mathrm{m^3\,mol^{-1}\,s^{-1}} = 1.03 \times 10^{-2}\ \mathrm{dm^3\,mol^{-1}\,s^{-1}}$
E22.5(a) 2.4×10^{-4}
E22.6(a) $3 \times 10^{10}\ \mathrm{dm^3\,mol^{-1}\,s^{-1}}$
E22.7(a) (a) $6.61 \times 10^9\ \mathrm{dm^3\,mol^{-1}\,s^{-1}}$ (b) $3.0 \times 10^{10}\ \mathrm{dm^3\,mol^{-1}\,s^{-1}}$
E22.8(a) $7.4 \times 10^9\ \mathrm{dm^3\,mol^{-1}\,s^{-1}}$, 6.7×10^{-8} s
E22.9(a) $0.79\ \mathrm{nm^2}$, 1.16×10^{-3}
E22.10(a) $1.87 \times 10^8\ \mathrm{mol\,dm^{-3}\,s^{-1}}$
E22.11(a) $+69.66\ \mathrm{kJ\,mol^{-1}}$, $-25.3\ \mathrm{J\,K^{-1}\,mol^{-1}}$
E22.12(a) $+73.4\ \mathrm{kJ\,mol^{-1}}$, $+71.9\ \mathrm{kJ\,mol^{-1}}$
E22.13(a) $-91\ \mathrm{J\,K^{-1}\,mol^{-1}}$
E22.14(a) $-72\ \mathrm{J\,K^{-1}\,mol^{-1}}$
E22.15(a) (a) $-46\ \mathrm{J\,K^{-1}\,mol^{-1}}$ (b) $+5.0\ \mathrm{kJ\,mol^{-1}}$ (c) $+18.7\ \mathrm{kJ\,mol^{-1}}$
E22.16(a) $7.1\ \mathrm{dm^6\,mol^{-2}\,min^{-1}}$
E22.17(a) $1.\bar{9} \times 10^{-19}$ J, 1.2 eV
E22.18(a) $12.\bar{5}\ \mathrm{nm^{-1}}$
E22.19(a) 0.138 V
E22.20(a) $2.82\ \mathrm{mA\,cm^{-2}}$
E22.21(a) increases, factor of 50
E22.22(a) (a) $1.7 \times 10^{-4}\ \mathrm{A\,cm^{-2}}$ (b) $1.7 \times 10^{-4}\ \mathrm{A\,cm^{-2}}$
E22.23(a) (a) $0.31\ \mathrm{mA\,cm^{-2}}$ (b) $5.44\ \mathrm{mA\,cm^{-2}}$ (c) $-2 \times 10^{42}\ \mathrm{mA\,cm^{-2}}$
E22.24(a) (a) $4.9 \times 10^{15}\ \mathrm{cm^{-2}\,s^{-1}}$ (b) $1.6 \times 10^{16}\ \mathrm{cm^{-2}\,s^{-1}}$ (c) $3.1 \times 10^7\ \mathrm{cm^{-2}\,s^{-1}}$, $3.9\ \mathrm{s^{-1}}$, $12\ \mathrm{s^{-1}}$, $2.4 \times 10^{-8}\ \mathrm{s^{-1}}$
E22.25(a) (a) $33\ \Omega$ (b) $3.3 \times 10^{10}\ \Omega$
E22.26(a) one can (barely) deposit zinc

Chapter 23

E23.1(a) $\dfrac{k_bK[\mathrm{AH}]^2[\mathrm{B}]}{[\mathrm{BH^+}]}$
E23.2(a) $1.50\ \mathrm{mmol\,dm^{-3}\,s^{-1}}$
E23.3(a) $k_{\mathrm{cat}} = 1.18 \times 10^5\ \mathrm{s^{-1}}$, $\eta = 7.9 \times 10^6\ \mathrm{dm^3\,mol^{-1}\,s^{-1}}$
E23.4(a) $2.0 \times 10^{-5}\ \mathrm{mol\,dm^{-3}}$
E23.5(a) (a) (i) $1.07 \times 10^{21}\ \mathrm{cm^{-2}\,s^{-1}}$ (ii) $1.4 \times 10^{14}\ \mathrm{cm^{-2}\,s^{-1}}$ (b) (i) $2.30 \times 10^{20}\ \mathrm{cm^{-2}\,s^{-1}}$ (ii) $3.1 \times 10^{13}\ \mathrm{cm^{-2}\,s^{-1}}$
E23.6(a) 0.13 bar
E23.8(a) $12\ \mathrm{m^2}$
E23.9(a) $33.6\ \mathrm{cm^3}$
E23.10(a) chemisorption, 50 s
E23.11(a) $\theta_1 = 0.83$, $\theta_2 = 0.36$
E23.12(a) (a) 0.24 kPa (b) 25 kPa
E23.13(a) 15 kPa
E23.14(a) $-12.\bar{4}\ \mathrm{kJ\,mol^{-1}}$
E23.15(a) $E_d = 65\bar{1}\ \mathrm{kJ\,mol^{-1}}$, (a) 1.6×10^{97} min (b) 2.8×10^{-6} min
E23.16(a) $61\bar{1}\ \mathrm{kJ\,mol^{-1}}$
E23.17(a) (a) $t_{1/2}(400\ \mathrm{K}) = 9.1$ ps, $t_{1/2}(1000\ \mathrm{K}) = 0.60$ ps (b) $t_{1/2}(400\ \mathrm{K}) = 4.1 \times 10^6$ s, $t_{1/2}(1000\ \mathrm{K}) = 6.6\ \mathrm{\mu s}$
E23.18(a) (a) zeroth-order (b) first-order

Solutions to odd-numbered problems

A horizontal bar over the last digit in some answers denotes an insignificant digit.

Chapter 1

P1.1 −233°N

P1.3 −272.95°C

P1.5 (a) 0.0245 kPa (b) 9.14 kPa (c) 0.0245 kPa

P1.7 (a) 12.5 $dm^3\,mol^{-1}$ (b) 12.3 $dm^3\,mol^{-1}$

P1.9 (a) 0.941 $dm^3\,mol^{-1}$ (b) 2.69 $dm^3\,mol^{-1}$, 2.67 $dm^3\,mol^{-1}$ (c) 5.11 $dm^3\,mol^{-1}$

P1.11 (a) 0.1353 $dm^3\,mol^{-1}$ (b) 0.6957 (c) 0.7158

P1.13 $a = 5.649$ $dm^6\,atm\,mol^{-2}$, $b = 59.4$ $cm^3\,mol^{-1}$, $p = 21$ atm

P1.15 1.26 $dm^6\,atm\,mol^{-2}$, 34.6 $cm^3\,mol^{-1}$

P1.17 $\frac{1}{3}$

P1.19 0.0866 atm^{-1}, 2.12 $dm^3\,mol^{-1}$

P1.23 0.011

P1.25 3.4×10^8 dm^3

P1.27 (a) 1.7×10^{-5} (b) 0.72

P1.31 0.0029 atm

Chapter 2

P2.1 *Total cycle*

State	p/atm	V/dm^3	T/K
1	1.00	22.44	273
2	1.00	44.8	546
3	0.50	44.8	273

Thermodynamic quantities calculated for reversible steps.

Step	Process	q/kJ	w/kJ	ΔU/kJ	ΔH/kJ
$1 \rightarrow 2$	p constant $= p_{ex}$	+5.67	−2.27	+3.40	+5.67
$2 \rightarrow 3$	V constant	−3.40	0	−3.40	−5.67
$3 \rightarrow 1$	Isothermal, reversible	−1.57	+1.57	0	0
Cycle		+0.70	−0.70	0	0

P2.3 $w = 0$, $\Delta U = +2.35$ kJ, $\Delta H = +3.03$ kJ

P2.5 (a) $w = 0$, $\Delta U = +6.19$ kJ, $q = +6.19$ kJ, $\Delta H = +8.67$ kJ
(b) $q = 0$, $\Delta U(b) = -6.19$ kJ, $\Delta H(b) = -8.67$ kJ, $w = -6.19$ kJ
(c) −4.29 kJ

P2.7 −89.03 $kJ\,mol^{-1}$

P2.9 $\Delta_r H^\ominus = +17.7$ $kJ\,mol^{-1}$, $\Delta_f H^\ominus$(metallocene, 583 K) = +116.0 $kJ\,mol^{-1}$

P2.11 $\Delta T = +37$ K, $m = 4.09$ kg

P2.13 $n = 0.903$, $k = -73.7$ $kJ\,mol^{-1}$

P2.15 $\Delta_c H^\ominus = -25\,968$ $kJ\,mol^{-1}$, $\Delta_f H^\ominus(C_{60}) = 2357$ $kJ\,mol^{-1}$

P2.17 (a) 240 $kJ\,mol^{-1}$ (b) 228 $kJ\,mol^{-1}$

P2.19 41.40 $J\,K^{-1}\,mol^{-1}$

P2.21 +3.60 kJ

P2.23 (a) $(2x-2y+2)dx + (4y-2x-4)dy$ (b) −2 (c) $\left(y+\frac{1}{x}\right)dx + (x-1)dy$

P2.27 (a) $1 + \frac{p}{(\partial U/\partial V)_p}$ (b) $1 + p\left(\frac{\partial V}{\partial U}\right)_p$

P2.31 (a) −1.5 kJ (b) −1.6 kJ

P2.33 increase

P2.35 $T = \left(\frac{p}{nR}\right) \times (V - nb) + \left(\frac{na}{RV^2}\right) \times (V - nb)$, $\left(\frac{\partial T}{\partial p}\right)_V = \frac{V - nb}{nR}$

P2.37 $\mu C_p = \left(\frac{1 - \frac{nb\zeta}{V}}{\zeta - 1}\right)V$, $\mu = 1.41$ K atm^{-1}, $T_I = \frac{27}{4}T_c\left(1 - \frac{b}{V_m}\right)^2$, $T_I = 1946$ K

P2.39 $\frac{1}{\lambda} = 1 - \frac{(3V_r - 1)^2}{4T_rV_r^3}$, $C_{p,m} - C_{V,m} = 9.2$ $J\,K^{-1}\,mol^{-1}$

P2.41 (a) $\mu = \frac{aT^2}{C_p}$ (b) $C_V = C_p - R\left(1 + \frac{2apT}{R}\right)^2$

P2.43 (a) 16.2 $kJ\,mol^{-1}$ (b) 114.8 $kJ\,mol^{-1}$ (c) 122.0 $kJ\,mol^{-1}$

P2.45 (a) 29.9 K MPa^{-1} (b) −2.99 K

Chapter 3

P3.1 (a) $\Delta_{trs}S(l \rightarrow s, -5°C) = -21.3$ $J\,K^{-1}\,mol^{-1}$, $\Delta S_{sur} = +21.7$ $J\,K^{-1}\,mol^{-1}$, $\Delta S_{total} = +0.4$ $J\,K^{-1}\,mol^{-1}$
(b) $\Delta_{trs}S(l \rightarrow g, T) = +109.7$ $J\,K^{-1}\,mol^{-1}$, $\Delta S_{sur} = -111.2$ $J\,K^{-1}\,mol^{-1}$, $\Delta S_{total} = -1.5$ $J\,K^{-1}\,mol^{-1}$

P3.3 (a) $q(Cu) = 43.9$ kJ, $q(H_2O) = -43.9$ kJ, $\Delta S(H_2O) = -118.\bar{1}$ $J\,K^{-1}$, $\Delta S(Cu) = 145.\bar{9}$ $J\,K^{-1}$, ΔS(total) = 28 $J\,K^{-1}$
(b) $\theta = 49.9°C = 323.1$ K, $q(Cu) = 38.4$ kJ, $\Delta S(H_2O) = -119.\bar{8}$ $J\,K^{-1}$, $\Delta S(Cu) = 129.\bar{2}$ $J\,K^{-1}$, ΔS(total) = 9 $J\,K^{-1}$

P3.5

	Step1	Step2	Step3	Step4	Cycle
q	+11.5 kJ	0	−5.74 kJ	0	−5.8 kJ
w	−11.5 kJ	−3.74 kJ	+5.74 kJ	+3.74 kJ	−5.8 kJ
ΔU	0	−3.74 kJ	0	+3.74 kJ	0
ΔH	0	−6.23 kJ	0	+6.23 kJ	0
ΔS	+19.1 $J\,K^{-1}$	0	−19.1 $J\,K^{-1}$	0	0
ΔS_{tot}	0	0	0	0	0
ΔG	−11.5 kJ	?	+5.73 kJ	?	0

P3.7 (a) 200.7 $J\,K^{-1}\,mol^{-1}$ (b) 232.0 $J\,K^{-1}\,mol^{-1}$

P3.9 +22.6 $J\,K^{-1}$

P3.11 (a) 63.88 $J\,K^{-1}\,mol^{-1}$ (b) 66.08 $J\,K^{-1}\,mol^{-1}$

P3.13 32.1 $kJ\,mol^{-1}$

P3.15 46.60 $J\,K^{-1}\,mol^{-1}$

P3.17 (a) 7 mol^{-1} (b) +107 $kJ\,mol^{-1}$

P3.23 (a) $\dfrac{R}{V_m - b}$ (b) $\dfrac{R\left(1+\dfrac{a}{RV_mT}\right)e^{-a/RV_mT}}{V_m - b}$

P3.25 $\left(\dfrac{\partial V}{\partial S}\right)_p = \left(\dfrac{\partial T}{\partial p}\right)_S, \left(\dfrac{\partial S}{\partial V}\right)_T = \left(\dfrac{\partial p}{\partial T}\right)_V$

P3.29 $\pi_T \approx \dfrac{p^2}{R} \times \dfrac{\Delta B}{\Delta T}$ (a) 3.0×10^{-3} atm (b) 0.30 atm

P3.31 $\pi_T = \dfrac{nap}{RTV}$

P3.33 $T\,dS = C_p dT - \alpha TV\,dp$, $q_{rev} = -\alpha TV\,\Delta p$, $q_{rev} = -0.50$ kJ

P3.35 $f = 0.9974$ atm

P3.37 -21 kJ mol^{-1}

P3.39 13 per cent

P3.43 $\eta = 1 - \left(\dfrac{V_B}{V_A}\right)^{1/c}$, $\Delta S_2 = +33$ J K^{-1}, $\Delta S_{sur,2} = -33$ J K^{-1}, $\Delta S_4 = -33$ J K^{-1}, $\Delta S_{sur,4} = +33$ J K^{-1}

P3.45 (a) 1.00 kJ (b) 8.4 kJ

Chapter 4

P4.1 $T_3 = 196.0$ K, $p_3 = 11.1$ Torr

P4.3 (a) $+5.56\times10^3$ Pa K^{-1} (b) 2.6 per cent

P4.5 (a) -1.63 cm^3 mol^{-1} (b) $+30.1$ dm^3 mol^{-1} (c) $+6\times10^2$ J mol^{-1}

P4.7 22°C

P4.9 (a) 227.5°C (b) +53 kJ mol^{-1}

P4.11 (b) 178.18 K (c) $T = 383.6$ K, $\Delta_{vap}H = 33.0$ kJ mol^{-1}

P4.15 9.8 Torr

P4.17 $T_h = 363$ K (90°C)

P4.19 (1) $\dfrac{dp}{dT} = \dfrac{\alpha_2 - \alpha_1}{\kappa_{T,2} - \kappa_{T,1}}$ (2) $\dfrac{dp}{dT} = \dfrac{C_{p,m2} - C_{p,m1}}{TV_m(\alpha_2 - \alpha_1)}$

P4.21 $n = 17$

P4.23 (b) 112 K (c) 8.07 kJ mol^{-1}

Chapter 5

P5.1 $K_A = 15.58$ kPa, $K_B = 47.03$ kPa

P5.3 $V_B = -1.4$ cm^3 mol^{-1}, $V_A = 18.\overline{04}$ cm^3 mol^{-1}

P5.5 $V_E^* = 57.6$ cm^3, $V_W^* = 45.6$ cm^3, $\Delta V \approx +0.95$ cm^3

P5.7 4 ions

P5.11 (a) $V_1 = V_1^* + a_0x_2^2 + a_1(3x_1 - x_2)x_2^2$, $V_2 = V_2^* + a_0x_1^2 + a_1(x_1 - 3x_2)x_1^2$
(b) $V_1 = 75.63$ cm^3 mol^{-1}, $V_2 = 99.06$ cm^3 mol^{-1}

P5.13 371 bar

P5.15 -4.64 kJ mol^{-1}

P5.17 (b) 391.0 K (c) 0.532

P5.23 Mg = 16, Mg_2Cu = 43

P5.25 (b) $x_{Si} = 0.13$, $\dfrac{n_{Ca_2Si}}{n_{liq}} = 0.5$ (c) $\dfrac{n_{Si}}{n_{liq}} = 0.53$, $\dfrac{n_{Si}}{n_{CaSi_2}} = 0.67$

P5.27 $\mu_A = \mu_A^* + RT\ln x_A + gRTx_B^2$

P5.29 73.96 cm^3 mol^{-1}

P5.31 $\phi - \phi(0) + \displaystyle\int_0^r \left(\dfrac{\phi - 1}{r}\right) dr$

P5.33 $\phi = 1 - \dfrac{1}{3}A'b^{1/2}$, $\phi = \dfrac{\Delta T}{2bK_f}$

P5.35 (1) 56 μg N_2 (2) 14 μg N_2 (3) 1.7×10^2 μg N_2

P5.39 (a) $R' = 84\,784.0$ g cm K^{-1} mol^{-1} (b) $M = 1.1\times10^5$ g mol^{-1}
(d) $B' = 21.4$ cm^3 g^{-1}, $C' = 211$ cm^6 g^{-2} (e) $19\bar{6}$ cm^6 g^{-2}

Chapter 6

P6.1 (a) +4.48 kJ mol^{-1} (b) $p_{IBr} = 0.101$ atm

P6.3 $\Delta_f H^\ominus = \frac{3}{2}R \times (B - CT)$, $\Delta_r C_p^\ominus = 70.5$ J K^{-1} mol^{-1}

P6.5 $\Delta_r G^\ominus(T)/(\text{kJ mol}^{-1}) = 78 - 0.161 \times (T/\text{K})$

P6.7 $K = 0.740$, $K = 5.71$, -103 kJ mol^{-1}

P6.9 +158 kJ mol^{-1}

P6.11 (a) 1.2×10^8 (b) 2.7×10^3

P6.13 (a) $CuSO_4$, $I = 4.0\times10^{-3}$, $ZnSO_4$, $I = 1.2\times10^{-2}$
(b) $\gamma_\pm(CuSO_4) = 0.74$, $\gamma_\pm(ZnSO_4) = 0.60$ (c) $Q = 5.9$
(d) $E_{cell}^\ominus = +1.102$ V (e) $E_{cell} = +1.079$ V

P6.15 2.0

P6.19 0.533

P6.21 $pK_a = 6.736$, $B = 1.997$, $k = -0.121$

P6.23 $\Delta_r G(T') = \Delta_r G(T) + (T - T')\Delta_r S(T) + \alpha(T', T) \times \Delta a + \beta(T', T) \times \Delta b + \gamma(T', T) \times \Delta c$, $\Delta_f G^\ominus(372\text{ K}) = -225.31$ kJ mol^{-1}

P6.27 (a) 41% (b) 75% (c) 55%

P6.29 (b) +0.206 V

P6.31 trihydrate

Chapter 7

P7.1 (a) 1.6×10^{-33} J m^{-3} (b) 2.5×10^{-4} J m^{-3}

P7.3 (a) $\nu = 223\bar{1}$ K, $\dfrac{C_{V,m}}{3R} = 0.0315$ (b) $\nu = 343$ K, $\dfrac{C_{V,m}}{3R} = 0.897$

P7.5 (a) 9.0×10^{-6} (b) 1.2×10^{-6}

P7.7 $x_{max} = a$

P7.11 $\dfrac{8\pi kT}{\lambda^4}$

P7.13 $\left(\dfrac{4}{c}\right)\sigma T^4$

P7.15 (a) $N = \left(\dfrac{2}{L}\right)^{1/2}$ (b) $N = \dfrac{1}{c(2L)^{1/2}}$ (c) $N = \dfrac{1}{(\pi a^3)^{1/2}}$ (d) $N = \dfrac{1}{(32\pi a^5)^{1/2}}$

P7.17 (a) yes (b) no (c) yes (d) no (e) no

P7.19 (a) no, no (b) no, $-k^2$ (c) no, no

P7.23 $\dfrac{(\hbar k)^2}{2m}$

P7.25 (a) $1.5a_0$, $4.5a_0^2$ (b) $5a_0$, $30a_0^2$

P7.31 5.35 pm

P7.33 (a) 811 K (b) 2.88 μm (c) 7.72×10^{-4} (d) 2.35×10^{-7}

Chapter 8

P8.1 $E_2 - E_1 = 1.24\times10^{-39}$ J, $n = 2.2\times10^9$, 1.8×10^{-30} J

P8.3 1.30×10^{-22} J, $\hbar$

P8.9 (a) $T = |A_3|^2 = A_3 \times A_3^* = \dfrac{4k_1^2k_2^2}{(a^2 + b^2)\sinh^2(k_2L) + b^2}$
where $a^2 + b^2 = (k_1^2 + k_2^2)(k_2^2 + k_3^2)$ and $b^2 = k_2^2(k_1 + k_3)^2$

P8.11 $g = \frac{1}{2}\left(\dfrac{mk}{\hbar^2}\right)^{1/2}$

P8.13 $0, \dfrac{3}{4}(2v^2 + 2v + 1)\alpha^4$

P8.15 (b) 0.0786

P8.17 $\alpha\left(\frac{v+1}{2}\right)^{1/2}, \alpha\left(\frac{v}{2}\right)^{1/2}$

P8.21 (a) $+\hbar$ (b) $-2\hbar$ (c) 0 (d) $\hbar\cos 2\chi, \frac{\hbar^2}{2I}, \frac{2\hbar^2}{I}, \frac{\hbar^2}{2I}, \frac{\hbar^2}{2I}$

P8.23 (a) 0, 0, 0 (b) $E=\frac{3\hbar^2}{I}, 6^{1/2}\hbar$

(c) $E=\frac{6\hbar^2}{I}, 12^{1/2}\hbar$

P8.25 $\theta=\arccos\frac{m_l}{\{l(l+1)\}^{1/2}}$, 54°44′, 0

P8.27 $\frac{\hbar}{i}\left(y\frac{\partial}{\partial z}-z\frac{\partial}{\partial y}\right), \frac{\hbar}{i}\left(z\frac{\partial}{\partial x}-x\frac{\partial}{\partial z}\right), \frac{\hbar}{i}\left(x\frac{\partial}{\partial y}-y\frac{\partial}{\partial x}\right), -\frac{\hbar}{i}\hat{l}_z.$

P8.31 (a) 3.30×10^{-19} J (b) 4.95×10^{-14} s^{-1} (c) lower, increases

P8.33 2.68×10^{14} s^{-1}

P8.35 (a) $E_{\pm5}=7.89\times10^{-19}$ J, $J_z=5.275\times10^{-34}$ J s (b) 5.2×10^{14} Hz

P8.39 5.8×10^{-11} N

Chapter 9

P9.1 $n_2\rightarrow 6$

P9.3 $R_{Li^{2+}}=987\ 663$ cm^{-1}, $\tilde{\nu}=137\ 175$ cm^{-1}, 185 187 cm^{-1}, $\tilde{\nu}=122.5$ eV

P9.5 $^2P_{1/2}$ and $^2P_{3/2}$, $^2D_{3/2}$ and $^2D_{5/2}$, $^2D_{3/2}$

P9.7 3.3429×10^{-27} kg, 1.000272

P9.9 (a) 0.9 cm^{-1} (b) small

P9.11 (b) $\tilde{\nu}=4.115\times10^5$ cm^{-1}, $\lambda=2.430\times10^{-6}$ cm, $\nu=1.234\times10^{16}$ s^{-1}

(c) $\frac{23}{2}a_0, \frac{3}{4}a_0, \frac{43}{4}a_0$

P9.13 ±106 pm

P9.15 (b) $\rho_{node}=3+\sqrt{3}$ and $\rho_{node}=3-\sqrt{3}, \rho_{node}=0$ and $\rho_{node}=4, \rho_{node}=0$ (c) $\langle r\rangle_{3S}=\frac{27a_0}{2}$

P9.19 (a) $\frac{Z}{a_0}$ (b) $\frac{Z}{4a_0}$ (c) $\frac{Z}{4a_0}$

P9.23 $\Delta l=\pm1$ and $\Delta m_l=0$ or ±1

P9.27 60 957.4 cm^{-1}, 60 954.7 cm^{-1}, 329 170 cm^{-1}, 329 155 cm^{-1}

P9.29 (a) receding, 1.128×10^{-3} c, 3.381×10^5 ms^{-1}

Chapter 10

P10.9 $E=\frac{-(\alpha_O+\alpha_N)\pm\{(\alpha_O+\alpha_N)^2-12\alpha_O\alpha_N\beta^2\}^{1/2}}{2}$,

$E_{deloc}=\{(\alpha_O-\alpha_N)^2+12\beta^2\}^{1/2}-\{(\alpha_O-\alpha_N)^2+4\beta^2\}^{1/2}$

P10.11 (b) 1.518β, 8.913 eV

P10.13 (b) $\Delta E/\text{eV}=3.3534+1.3791\times10^{-4}\ \tilde{\nu}/\text{cm}^{-1}$ (c) 30 937 cm^{-1}

P10.19 $\frac{2(k-Sj)}{1-S^2}$

P10.25 $j_0\{c_{Aa}^2(AB|AA)+c_{Aa}c_{Ba}(AB|AB)+c_{Ba}c_{Aa}(AB|BA)+c_{Ba}^2(AB|BB)\}$,
$j_0\{c_{Aa}^2(AA|AB)+c_{Aa}c_{Ba}(AA|BB)+c_{Ba}c_{Aa}^2(AB|AB)+c_{Ba}^2(AB|BB)\}$

P10.29 (a) linear relationship (b) -0.122 V (c) -0.174 V

Chapter 11

P11.1 (a) D_{3d} (b) D_{3d}, C_{2v} (c) D_{2h} (d) D_3 (e) D_{4d}

P11.3 $C_2\sigma_h=i$

P11.7 do not form a group

P11.9 (a) all five d orbitals (b) all except A_2 (d_{xy})

P11.11 (a) D_{2h} (b) C_{2h}, C_{2v}

P11.13 (a) $2A_1+A_2+2B_1+2B_2$ (b) A_1+3E (c) $A_1+T_1+T_2$
(d) $A_{2u}+T_{1u}+T_{2u}$

P11.15 $4A_1+2B_1+3B_2+A_2$

P11.17 (a) $7A_2+7B_1, \frac{1}{2}(a-a'), \frac{1}{2}(b-b'), \ldots, \frac{1}{2}(g-g'), \frac{1}{2}(a+a'), \frac{1}{2}(b+b'), \ldots, \frac{1}{2}(g+g')$

P11.21 $A_{1g}+B_{1g}+E_u$

P11.23 z-polarized transition is not allowed, x, y-polarized transitions are allowed

Chapter 12

P12.1 596 GHz, 19.9 cm^{-1}, 0.503 mm, 9.941 cm^{-1}

P12.3 $R_{CC}=139.6$ pm, $R_{CH}=108.\bar{5}$ pm

P12.5 $R(\text{HCl})=128.393$ pm, $R(^2\text{HCl})=128.13$ pm

P12.7 $R=116.28$ pm, $R'=155.97$ pm

P12.9 142.81 cm^{-1}, $D_0=3.36$ eV, 93.8 N m^{-1}

P12.11 (a) 2143.26 cm^{-1} (b) 12.8195 kJ mol^{-1} (c) 1.85563×10^3 N m^{-1}
(d) 1.91 cm^{-1} (e) 113 pm

P12.17 (a) 7 (b) C_{2h}, C_{2v}, C_2 (c) structure 2 is inconsistent with observation

P12.19 $1/\langle R\rangle^2=1/R_e^2$, $1/\langle R^2\rangle=\frac{1}{R_e^2}\left(\frac{1}{1+\langle x^2\rangle/R_e^2}\right)$, $\left\langle\frac{1}{R^2}\right\rangle=\frac{1}{R_e^2}\left(1+3\frac{\langle x^2\rangle}{R_e^2}\right)$

P12.21 $\tilde{D}=\frac{4\tilde{B}^3}{\tilde{\nu}^2}$

P12.27 230, 240, and 250 pm

P12.29 (a) $\Delta J=0$ is forbidden (c) 30 m

P12.31 $\tilde{B}=2.031$ cm^{-1}, $T=2.35$ K

Chapter 13

P13.1 49 36$\bar{4}$ cm^{-1}

P13.5 $\mathcal{A}=\frac{1}{2}\Delta\tilde{\nu}_{1/2}\varepsilon_{max}\sqrt{\pi/\ln(2)}$, $\mathcal{A}=1.3\bar{0}\times10^6$ dm^3 mol^{-1} cm^{-2}

P13.7 $D_0(B^3\Sigma_u^-)=6808.2$ cm^{-1} or 0.84411 eV, $D_0(X^3\Sigma_g^-)=5.08$ eV

P13.13 4×10^{-10} s or 0.4 ns

P13.25 6.37, 2.12

P13.27 $\Delta_fH^{\ominus}$(structure 2) $-\Delta_fH^{\ominus}$(structure 1) $=+28$ kJ mol^{-1}

Chapter 14

P14.1 $\mathcal{B}_0=10.3$ T, $\frac{\delta N}{N}\approx2.42\times10^{-5}$, β, ($m_I=-\frac{1}{2}$)

P14.3 300×10^6 Hz $\pm$ 10 Hz, 0.29 s

P14.5 $k=4\times10^2$ s^{-1}, $E_{II}-E_I=3.7$ kJ mol^{-1}, $E_a=16$ kJ mol^{-1}

P14.7 (b) $580-79\cos\phi+395\cos2\phi$

P14.9 158 pm

P14.11 6.9 mT, 2.1 mT

P14.13 0.10, 0.38 (a) 0.48 (b) 0.52, 3.8

P14.15 $\frac{e^2\mu_0Z}{12\pi m_ea_0}=1.78\times10^{-5}\,Z$

P14.19 $\frac{1}{2}\frac{A\tau}{1+(\omega_0-\omega)^2\tau^2}$

P14.21 $\omega_{1/2}=\frac{1}{\tau}$, $\omega_{1/2}=2(\ln 2)^{1/2}\left(\frac{1}{\tau}\right)$

P14.27 29 μT m^{-1}

Chapter 15

P15.5 7.41

P15.7 (a) (i) 5.00 (ii) 6.26 (b) 1.00, 0.80, 6.58×10^{-11}, 0.122

P15.9 (a) 0.641, 0.359 (b) 8.63×10^{-22} J, 0.520 kJ mol^{-1}

P15.11 (a) 1.049 (b) 1.548, $p_0=$ (a) 0.953 (b) 0.645, $p_1=$ (a) 0.044 (b) 0.230, $p_2=$ (a) 0.002 (b) 0.083

P15.15 {4,2,2,1,0,0,0,0,0,0}, $\mathcal{W}=3780$

P15.17 (a) $1+3\mathrm{e}^{-\varepsilon/kT}$ (b) $E_\mathrm{m}(T)=0.5245\,RT$, $\varepsilon=2.074$ J K^{-1} mol^{-1}, $S_\mathrm{m}=10.55$ J K^{-1} mol^{-1}

P15.19 $pV=nRT$

Chapter 16

P16.1 0.351, 0.079, 0.029

P16.3 $C_{V,\mathrm{m}}=4.2$ J K^{-1} mol^{-1}, $S_\mathrm{m}=15$ J K^{-1} mol^{-1}

P16.5 19.90

P16.7 199.4 J mol^{-1} K^{-1}

P16.13 $R\ln\left(\dfrac{2\pi\mathrm{e}^2 m\sigma_\mathrm{m}}{h^2N_\mathrm{A}\beta}\right)$, $\Delta S_\mathrm{m}=R\ln\left\{\left(\dfrac{\sigma_\mathrm{m}}{V_\mathrm{m}}\right)\times\left(\dfrac{h^2\beta}{2\pi m_\mathrm{e}}\right)^{1/2}\right\}$

P16.15 $U-U(0)=\dfrac{N\hbar\omega}{\mathrm{e}^x-1}$, $C_V=kN\left\{\dfrac{x^2\mathrm{e}^x}{(\mathrm{e}^x-1)^2}\right\}$, $H-H(0)=\dfrac{N\hbar\omega}{\mathrm{e}^x-1}$,

$S=Nk\left(\dfrac{x}{\mathrm{e}^x-1}-\ln(1-\mathrm{e}^{-x})\right)$, $A-A(0)=NkT\ln(1-\mathrm{e}^{-x})$

P16.17 $\mu_T=B(T)-T\dfrac{\mathrm{d}B(T)}{\mathrm{d}T}$.

P16.19 (a) $U-U(0)=nRT\left(\dfrac{\dot{q}}{q}\right)$, $C_V=nR\left\{\dfrac{\ddot{q}}{q}-\left(\dfrac{\dot{q}}{q}\right)^2\right\}$, $S=nR\left(\dfrac{\dot{q}}{q}+\ln\dfrac{eq}{N}\right)$

(b) 5.41 J K^{-1} mol^{-1}

P16.25 (a) $c_\mathrm{s}=\left(\dfrac{1.40RT}{M}\right)^{1/2}$ (b) $c_\mathrm{s}=\left(\dfrac{1.40RT}{M}\right)^{1/2}$ (c) $c_\mathrm{s}=\left(\dfrac{4RT}{3M}\right)^{1/2}$,

$c_\mathrm{s}=350$ m s^{-1}

P16.29 45.76 kJ mol^{-1}

Chapter 17

P17.1 (a) 1.1×10^8 V m^{-1} (b) 4×10^9 V m^{-1} (c) 4 kV m^{-1}

P17.5 $\alpha'=1.2\times10^{-23}$ cm^3, $\mu=0.86$ D

P17.7 $\alpha'=2.24\times10^{-24}$ cm^3, $\mu=1.58$ D, $P'_\mathrm{m}=5.66$ cm^3 mol^{-1}

P17.9 $\varepsilon=1.51\times10^{-23}$ J, $r_\mathrm{e}=265$ pm

P17.11 $P_\mathrm{m}=68.8$ cm^3 mol^{-1}, $\varepsilon_\mathrm{r}=4.40$, $n_\mathrm{r}=2.10$, $P_\mathrm{m}=8.14$ cm^3 mol^{-1}, $\varepsilon_\mathrm{r}=1.76$, $n_\mathrm{r}=1.33$

P17.13 (a) $\dfrac{6l^4q_1^2}{\pi\varepsilon_0 r^5}$ (b) $-\dfrac{9l^4q_1^2}{4\pi\varepsilon_0 r^5}$

P17.17 $r_\mathrm{e}=1.3598\,r_0$, $A=1.8531$

P17.21 -1.8×10^{-27} J $=-1.1\times10^{-3}$ J mol^{-1}

P17.25 (a) 3.5 (b) slope $=-1.49$, intercept $=-1.95$ (c) 1.12×10^{-2}

Chapter 18

P18.3 (a) $R_\mathrm{g}=\sqrt{\tfrac{2}{5}}\,a$, $R_\mathrm{g}/\mathrm{nm}=0.046460\times\{(\upsilon_\mathrm{s}/\mathrm{cm^3\,g^{-1}})\times(M/\mathrm{g\,mol^{-1}})\}^{1/3}$, $R_\mathrm{g}=1.96$ nm (b) $R_{\mathrm{g},\parallel}=\sqrt{\tfrac{1}{2}}\,a$, $R_{\mathrm{g},\perp}=\sqrt{\tfrac{1}{12}}\,l$, $R_{\mathrm{g},\parallel}=0.35$ nm, $R_{\mathrm{g},\perp}=46$ nm

P18.5 0.0716 dm^3 g^{-1}

P18.7 1.6×10^5 g mol^{-1}

P18.9 (a) $a=0.71$, $K=1.2\times10^{-2}$ cm^3 g^{-1}

P18.11 $\bar{M}_\mathrm{n}=155$ kg mol^{-1}, $B=13.7$ m^3 mol^{-1}

P18.15 (a) $R_\mathrm{rms}=N^{1/2}l$, $R_\mathrm{rms}=9.74$ nm (b) $R_\mathrm{mean}=\left(\dfrac{8N}{3\pi}\right)^{1/2}l$, $R_\mathrm{mean}=8.98$ nm

(c) $R^*=(\tfrac{2}{3}N)^{1/2}l$, $R^*=7.95$ nm

P18.17 $R_\mathrm{g}=\sqrt{\tfrac{N}{6}}\,l$

P18.21 (a) $B=\dfrac{16\pi}{3}N_\mathrm{A}(\gamma l)^3\left(\dfrac{N}{6}\right)^{3/2}$, 0.38 m^3 mol^{-1}

(b) $B=\dfrac{16\pi}{3}N_\mathrm{A}(\gamma l)^3\left(\dfrac{N}{3}\right)^{3/2}$, 1.1 m^3 mol^{-1}

P18.23 $-S\mathrm{d}T-l\mathrm{d}t$, $-S\mathrm{d}T+t\mathrm{d}l$, $-T\left(\dfrac{\partial t}{\partial T}\right)_l+t$

P18.29 65.6 kg mol^{-1}

P18.31 $S=5.40$ Sv, $M=63.2$ kg mol^{-1}

P18.37 (a) (1) $[\eta]=0.086$ dm^3 g^{-1}, $k'=0.37$ (2) $[\eta]=0.042$ dm^3 g^{-1}, $k'=0.35$ (b) (1) $\bar{M}_\mathrm{v}=2.4\times10^2$ kg mol^{-1} (2) $\bar{M}_\mathrm{v}=2.6\times10^2$ kg mol^{-1} (c) (1) $r_\mathrm{rms}=42$ nm (2) $r_\mathrm{rms}=33$ nm (d) (1) (2) $\langle n\rangle=2.3\times10^3$, $\langle n\rangle=2.5\times10^3$ (e) (1) $L_\mathrm{max}=5.8\times10^2$ nm, (2) $L_\mathrm{max}=6.2\times10^2$ nm

(f)

Solvent	$\langle n\rangle$	R_g/nm	$r_\mathrm{rms}^\mathrm{KR}$/nm	r_rms/nm
Toluene	$2.3\bar{0}\times10^3$	4.3	42	$10.\bar{4}$ or 7.4
Cyclohexane	$2.3\bar{0}\times10^3$	4.4	33	$10.\bar{8}$ or 7.6

Chapter 19

P19.1 118 pm

P19.3 face-centred cubic, $a=408.55$ pm, $\rho=10.507$ g cm^{-3}

P19.5 $\alpha_\mathrm{volume}=4.8\times10^{-5}$ K^{-1}, $\alpha_\mathrm{linear}=1.6\times10^{-5}$ K^{-1}

P19.7 $a=834$ pm, $b=606$ pm, $c=870$ pm

P19.9 $\rho=1.385$ g cm^{-3}, $\rho_\mathrm{Os}=1.578$ g cm^{-3}

P19.11 (a) tungsten has the bcc unit cell with $a=321$ pm, 139 pm (b) copper has the fcc unit cell with $a=362$ pm, 128 pm

P19.13 1.01 g cm^{-3}

P19.15 -146 kJ mol^{-1}

P19.17 0.254 cm^3 mol^{-1}

P19.19 $\dfrac{1}{d^2}=\left(\dfrac{h}{a}\right)^2+\left(\dfrac{k}{b}\right)^2+\left(\dfrac{l}{c}\right)^2$

P19.25 0

P19.29 (a) $\rho(E)=-\dfrac{(N+1)/2\pi\beta}{\left[1-\left(\dfrac{E-\alpha}{2\beta}\right)^2\right]^{1/2}}$

P19.35 $1-\left(\dfrac{1}{4(p/K)+1}\right)^{1/2}$, increases

P19.37 3.61×10^5 g mol^{-1}

Chapter 20

P20.3 (b) 1.8894 m (c) 1.8897 m

P20.5 7.3 mPa

P20.7 0.613 mS m^2 mol^{-1}

P20.9 13.82 μmol dm^{-3}

P20.11 $\dfrac{G_{H^+}}{\sum\limits_{i}^{\text{all ions}} G_i}=0.821, \dfrac{G_{H^+}}{\sum\limits_{i}^{\text{all ions}} G_i}=0.00279$

P20.13 (b) four, one to two

P20.15 (a) 0, (b) 0.0630 mol dm^{-3}

P20.17 $\Lambda_m^\circ(\text{NaI})=60.7\ \text{S cm}^2\ \text{mol}^{-1}$, $\Lambda_m^\circ(\text{KI})=58.9\ \text{S cm}^2\ \text{mol}^{-1}$, $\lambda(\text{Na}^+)-\lambda(\text{K}^+)=1.8\ \text{S cm}^2\ \text{mol}^{-1}$, $\Lambda_m^\circ(\text{NaI})=127\ \text{S cm}^2\ \text{mol}^{-1}$, $\Lambda_m^\circ(\text{KI})=150\ \text{S cm}^2\ \text{mol}^{-1}$, $\lambda(\text{Na}^+)-\lambda(\text{K}^+)=-23\ \text{S cm}^2\ \text{mol}^{-1}$

P20.19 0.83 nm

P20.21 $f(v)=\left(\dfrac{m}{kT}\right)ve^{-mv^2/2kT}, \left(\dfrac{\pi kT}{2m}\right)^{1/2}$ or $\left(\dfrac{\pi RT}{2M}\right)^{1/2}$

P20.23 $P=0.61$, (a) 39 per cent (b) 61 per cent (c) 53 per cent, 47 per cent

P20.25 $p=p_0e^{-t/\tau}$ where $\tau=\left(\dfrac{2\pi M}{RT}\right)^{1/2}\dfrac{V}{A}, t_{1/2}=\tau\ln(2)=\left(\dfrac{2\pi M}{RT}\right)^{1/2}\dfrac{V}{A}\ln(2)$

P20.27 $\langle x^4\rangle^{1/4}/\langle x^2\rangle^{1/2}=3^{1/4}$

P20.29 $\dfrac{1}{\Lambda_m}=\dfrac{1}{\Lambda_{m,\alpha=1}}+\dfrac{(1-\alpha)\Lambda_m}{(\alpha\Lambda_{m,\alpha=1})^2}, \Lambda_{m,\alpha=1}=\Lambda_m^\circ-\mathcal{K}(\alpha c)^{1/2}$

P20.31 $D=1.\bar{6}\times10^{16}\ \text{m}^2\ \text{s}^{-1}$, $\kappa=0.3\underline{4}\ \text{J K}^{-1}\ \text{m}^{-1}\ \text{s}^{-1}$

P20.33 (a) $5.4\times10^{-12}\ \text{cm}^3$ (b) 16 MK (c) 16 MK

P20.35 1.7×10^{-2} s

Chapter 21

P21.1 second-order, $k_r=0.059\bar{4}\ \text{dm}^3\ \text{mol}^{-1}\ \text{min}^{-1}$, 2.94 g

P21.3 $7.0\times10^{-5}\ \text{s}^{-1}$, $7.3\times10^{-5}\ \text{dm}^3\ \text{mol}^{-1}\ \text{s}^{-1}$

P21.5 55.4%

P21.7 first-order, $1.7\times10^{-2}\ \text{min}^{-1}$

P21.9 first-order kinetics, $7.2\times10^{-4}\ \text{s}^{-1}$

P21.11 $k_rK_1K_2[\text{HCl}]^3[\text{CH}_3\text{CH{=}CH}_2]$

P21.13 16.7 kJ mol^{-1}

P21.15 steady-state approximation

P21.19 $\dfrac{1}{I_f}=\dfrac{1}{I_{abs}}+\dfrac{k_Q[Q]}{k_fI_{abs}}$, $5.1\times10^6\ \text{dm}^3\ \text{mol}^{-1}\ \text{s}^{-1}$

P21.21 3.5 nm

P21.23 $[A]=\dfrac{k_r'([A]_0+[B]_0)+(k_r[A]_0-k_r'[B]_0)e^{-(k_r+k_r')t}}{k_r+k_r'}$,

$[A]_\infty=\left(\dfrac{k_r'}{k_r+k_r'}\right)\times([A]_0+[B]_0), [B]_\infty=\left(\dfrac{k_r}{k_r+k_r'}\right)\times([A]_0+[B]_0)$,

$\dfrac{[B]_\infty}{[A]_\infty}=\dfrac{k_r}{k_r'}$

P21.25 (a) $k_rt=\dfrac{2x(A_0-x)}{A_0^2(A_0-2x)^2}$ (b) $\left(\dfrac{2x}{A_0^2(A_0-2x)}\right)+\left(\dfrac{1}{A_0^2}\right)\ln\left(\dfrac{A_0-2x}{A_0-x}\right)$

P21.27 steady-state intermediate

P21.29 $\dfrac{k_1k_2k_3[A]_0}{k_1'k_2'+k_1'k_3+k_2k_3}$

P21.31 $(\langle M^2\rangle_N-\langle M\rangle_N^2)^{1/2}=\dfrac{p^{1/2}M_1}{1-p}, M_1\{kt[A]_0(1+kt[A]_0)\}^{1/2}$

P21.33 $k_r[\cdot M][I]^{-1/2}$

P21.35 $\left(\dfrac{I_a}{k_r}\right)^{1/2}$

P21.37 first-order, $k_r=0.00765\ \text{min}^{-1}=0.459\ \text{h}^{-1}$, $t_{1/2}=1.51\ \text{h}=91\ \text{min}$

P21.39 $v_0=6\times10^{-14}\ \text{mol dm}^{-3}\ \text{s}^{-1}$, $t_{1/2}=4.\bar{4}\times10^8\ \text{s}=14\ \text{yr}$

P21.41 $5.9\times10^{-13}\ \text{mol dm}^{-3}\ \text{s}^{-1}$

Chapter 22

P22.1 (a) $4.3\bar{5}\times10^{-20}\ \text{m}^2$ (b) 0.15

P22.3 $1.6\bar{4}\times10^{11}\ \text{mol}^{-1}\ \text{dm}^3\ \text{s}^{-1}$, 3.7 ns

P22.9 0.78, 0.38

P22.13 0.50, 0.150 A m^{-2}, 0.038 A m^{-2}

P22.17 $q_m^T/N_A=1.4=10^7$, $q_R(\text{nonlinear})\approx900$, $q_R(\text{linear})\approx200$, $q_R\approx q_E\approx1$ $A=6.3\times10^9\ \text{dm}^3\ \text{mol}^{-1}\ \text{s}^{-1}$, $A=3.3\times10^4\ \text{dm}^3\ \text{mol}^{-1}\ \text{s}^{-1}$ $P=2\times10^{-7}$

P22.19 $\dfrac{v^3}{(v^\ddagger)^2}e^{-\Delta E_0/RT}$, $2.7\times10^{-15}\ \text{m}^2\ \text{s}^{-1}$, $1.1\times10^{-14}\ \text{m}^2\ \text{s}^{-1}$

P22.21 (a) $1.37\times10^6\ \text{dm}^3\ \text{mol}^{-1}\ \text{s}^{-1}$ (b) $1.16\times10^6\ \text{dm}^3\ \text{mol}^{-1}\ \text{s}^{-1}$

P22.25 (a) $6.23\times10^9\ \text{dm}^3\ \text{mol}^{-1}\ \text{s}^{-1}$ (b) 4×10^{-10} m

P22.27 1.15 eV

P22.29 $9.5\times10^4\ \text{dm}^3\ \text{mol}^{-1}\ \text{s}^{-1}$

Chapter 23

P23.1 $v_{max}=2.31\ \mu\text{mol dm}^{-3}\ \text{s}^{-1}$, $k_b=115\ \text{s}^{-1}$, $k_{cat}=115\ \text{s}^{-1}$, $K_M=1.11\ \mu\text{mol dm}^{-3}$, $\eta=104\ \text{dm}^3\ \mu\text{mol}^{-1}\ \text{s}^{-1}$

P23.3 (b) pH = 7.0

P23.5 $V_{total}=-2.039\times10^{-18}$ J, $V_{total}=-2.20\times10^{-19}$ J, $V_{total}=-7.30\times10^{-19}$ J

P23.7 (a) $1.61\times10^{15}\ \text{cm}^{-2}$ (b) $1.14\times10^{15}\ \text{cm}^{-2}$ (c) $1.86\times10^{15}\ \text{cm}^{-2}$

P23.9 (a) $c=165$, $V_{mon}=13.1\ \text{cm}^3$ (b) $c=263$, $V_{mon}=12.5\ \text{cm}^3$

P23.11 $c_1=0.16$ g, $c_2=2.2$

P23.13 $K=0.138\ \text{mg g}^{-1}$, $n=0.58$

P23.15 $n_\infty=5.78\ \text{mol kg}^{-1}$, $K=7.02\ \text{MPa}^{-1}$

P23.17 (a) $\dfrac{[P]}{[P]_0}=(b+1)\dfrac{e^{at}}{1+be^{at}}$ (c) $-\dfrac{1}{a}\ln(b)$

(d) $A_0(A_0+P_0)kt=\left(\dfrac{y}{1-y}\right)+\left(\dfrac{1}{1-p}\right)\ln\left(\dfrac{p+y}{p(1-y)}\right)$,

$(A_0+P_0)^2kt_{max}=\tfrac{1}{2}-p-\ln 2p$

(e) $A_0(A_0+P_0)kt=\left(\dfrac{y}{p(p+y)}\right)+\left(\dfrac{1}{1+p}\right)\ln\left(\dfrac{p+y}{p(1-y)}\right)$,

$\dfrac{2-p}{2p}+(A_0+P_0)^2kt_{max}=\ln\dfrac{2}{p}$

P23.19 $v=\dfrac{v_{max}}{1+\dfrac{1}{K[S]_0}}$, $k_a'\gg k_b$

P23.23 it is described by the BET isotherm, 3.96, 75.8 cm^3

P23.25 (a) R values in the range 0.975 to 0.991 (b) $k_a=3.68\times10^{-3}$, $\Delta_{ad}H=-8.67\ \text{kJ mol}^{-1}$, $k_b=2.62\times10^{-5}\ \text{ppm}^{-1}$, $\Delta_bH=-15.7\ \text{kJ mol}^{-1}$

P23.27 (b) $K_F=0.164\ \text{mg g}_R^{-1}$, $n=1.14$ (c) $0.164c_{eq}^{-0.46}$

Index

(T) denotes a table in the Resource section.

A

B

M

N

O

P

Q

R

S